3rd International Conference on Mechanical, Electronic and Information Technology Engineering (ICMITE 2017)

MATEC Web of Conferences Volume 139 (2017)

Chengdu, China
16 - 17 December 2017

Part 1 of 2

Editors:

Bing Xu
Yinong Chen

ISBN: 978-1-5108-5350-8

Printed from e-media with permission by:

Curran Associates, Inc.
57 Morehouse Lane
Red Hook, NY 12571

Some format issues inherent in the e-media version may also appear in this print version.

Printed by Curran Associates, Inc. (2017)

For additional information, please contact EDP Sciences – Web of Conferences
at the address below.

EDP Sciences – Web of Conferences
17, Avenue du Hoggar
Parc d'Activité de Courtabœuf
BP 112
F-91944 Les Ulis Cedex A
France

Phone: +33 (0) 1 69 18 75 75
Fax: +33 (0) 1 69 28 84 91

contact-edps@webofconferences.org

Additional copies of this publication are available from:

Curran Associates, Inc.
57 Morehouse Lane
Red Hook, NY 12571 USA
Phone: 845-758-0400
Fax: 845-758-2633
Email: curran@proceedings.com
Web: www.proceedings.com

TABLE OF CONTENTS

PART 1

PART 2

Preface

It is our pleasure to welcome you to the 2017 3rd International Conference on Mechanical, Electronic and Information Technology Engineering (ICMITE 2017) in Chengdu, China. ICMITE 2017 is dedicated to issues related to Mechanical Engineering, Mechatronics Automation and Control, Intelligent Systems, Electronic and Embedded Systems, Applied Mathematics, Computational Methods and Algorithms, Data and Signal Processing, Information Technology and Computer Sciences. A major goal and feature of it is to bring academic scientists, engineers, industry researchers together to exchange and share their experiences and research results, and discuss the practical challenges encountered and the solutions adopted.

We hope you will have a technically rewarding experience, and use this occasion to meet old friends and make many new ones. Don't miss the opportunity to explore in Chengdu, China. And don't forget to take a sample of the many and diverse attractions in the rest of the China.

During the organization course, we have got help from different people, different departments, and different institutions. We would like to thank the organization staff, the members of the program committees and reviewers. They have worked very hard in reviewing papers and making valuable suggestions for the authors to improve their work. We also would like to express our gratitude to the external reviewers, for providing extra helps in the review process, and the authors for contributing their research result to the conference. Special thanks to EDP Publications for their kind and enthusiastic help and support for our conference.

I hope and believe that everybody here will have an academic enjoyment during this conference in Chengdu.

We look forward to seeing all of you next year at the conference.

Organizing Committee of ICMITE 2017

edp sciences

Web of Conferences

Statement of Peer review

In submitting conference proceedings to *Web of Conferences*, the editors of the proceedings certify to the Publisher that

1. They adhere to its **Policy on Publishing Integrity** in order to safeguard good scientific practice in publishing.
2. All articles have been subjected to peer review administered by the proceedings editors.
3. Reviews have been conducted by expert referees, who have been requested to provide unbiased and constructive comments aimed, whenever possible, at improving the work.
4. Proceedings editors have taken all reasonable steps to ensure the quality of the materials they publish and their decision to accept or reject a paper for publication has been based only on the merits of the work and the relevance to the journal.

Title, date and place of the conference

2017 3rd International Conference on Mechanical, Electronic and Information Technology Engineering（ICMITE 2017）

December 16-17, 2017, Chengdu, China

Proceedings editor(s):

Prof. Bing Xu
Prof. Yinong Chen

Date and editor's signature

November 16, 2017

Dynamic modeling and simulation of trailer with compliance steering system

Wang Qingyun[1], Wang Peng[2], Chen Ning [2*]

[1]*College of Automobile and Traffic Engineering, Nanjing Forestry University, Nanjing, 210037, China*
[2] *College of mechanical and electrical engineering, Nanjing Forestry University, Nanjing, 210037, China*

Abstract. Stability of a vehicle and tire wear is affected by the movement inference between the tractor and the trailer when the trailer is turning. Therefore, to solve the problem, Compliant Steering (CS) system is used. The system makes the rear-wheel of the trailer steer. After establishing a dynamic model of the trailer and the CS system, the simulation result shows that the system has a smaller lateral force. This smaller force at the hinge point indicates movement inference is reduced and motion response is far better. Furthermore, this paper contains the analysis of the influence of the key parameters of CS system and the speed on the moving characteristics of the trailer.

1 Introduction

Rear wheel CS technique is a passive steering technique not requiring steering mechanism that utilizes the lateral force and elasticity of the wheel and the rear suspension when the vehicle turns respectively [1]. That is when compared with the active steering technology, which is simple in structure and without energy consumption [2]. Currently, the research is abundant on CS technology used in large-scale vehicle [3].

A full trailer is made up of a tractor and a trailer. It is totally different from a car in terms of handling stability. On a small curvature road and turning in high speed, the full trailer with over-steer characteristics is easy to lead to folding or swing tail (trailer swing) [4]. Although low speed, problem of movement inference between the tractor and trailer during steering exists, thus, aggravating tire wear. In the past, due to small number of trailer, attention was not given. Currently, high rural mechanization and traffic, the number of trailer and accident is increasing gradually.

Recently, many scholars are researching and focusing on the dynamic analysis of trailers. After establishing a dynamic model of the trailer and the compliant steering system, simulation analysis of the trailer result shows the CS technology can reduce movement interference between the tractor and the trailer, improving the stability of the vehicle.

2 Dynamic Analysis of Full Trailer

2.1 Dynamic Equation of Articulated Full Trailer

The articulated full trailer model is shown in Fig. 1 [5, 6]. Here F_{yi} (i=1~4) represents tire lateral force, δ_1 is front wheel angle, $\Delta\varphi$ is the angle of deviation between the tractor and the trailer, F_{Tx}, F_{Ty} are the longitudinal and lateral forces of the hinge. Other annotations are shown in Tab 1.

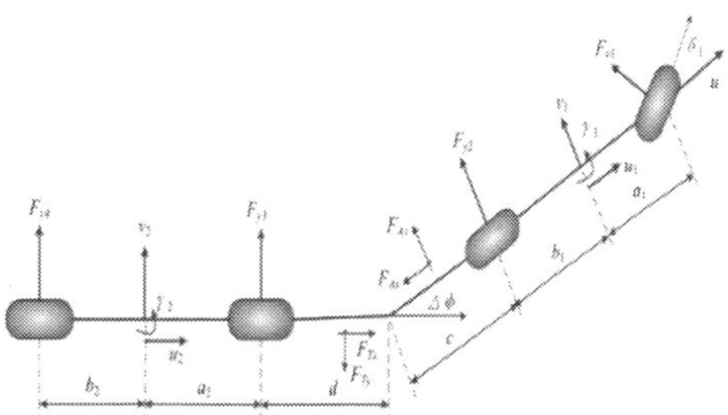

Fig.1. Mechanical analysis of the trailer when steering.

After applying a force-balance analysis [7] to the model in Fig.1, the dynamic equations can be derived. Then they are formulated in a standard state space form, thus

$$A\dot{X} = BX + Dq \tag{1}$$

where the state variable is taken as $X = [\beta_1 \quad \gamma_1 \quad \gamma_2 \quad \Delta\varphi]^T$, β_i is the vehicle side slip angle, γ_i is yaw rate (i = 1, 2 The tractor is labeled 1 and the trailer is 2), and q is the linear tire [8] lateral force vector

$$q = \left[F_{y1}(\alpha_1) \quad F_{y2}(\alpha_2) \quad F_{y3}(\alpha_3) \quad F_{y4}(\alpha_4)\right]^T$$

$$A = \begin{bmatrix} m_1 \cdot u \cdot (b_1 + c) & I_{Z1} & 0 & 0 \\ (m_1 + m_2) \cdot u & -m_2(b_1 + c) & -m_2(a_2 + d) & m_2 \cdot u \\ m_1 \cdot u \cdot (a_2 + d) & 0 & I_{Z2} & 0 \\ 0 & 0 & 0 & 1 \end{bmatrix},$$

* Corresponding author:chenning@njfu.com.cn

DOI: 10.1051/matecconf/201713900001

$$B=\begin{bmatrix}0 & -m_1\cdot u\cdot(b_1+c) & 0 & 0\\ 0 & -m_1\cdot u & -m_2\cdot u & 0\\ 0 & -m_1\cdot u\cdot(a_2+d) & 0 & 0\\ 0 & 1 & -1 & 0\end{bmatrix},$$

$$D=\begin{bmatrix}(b_1+c+a_1) & c & 0 & 0\\ 1 & 1 & 1 & 1\\ a_2+d & a_2+d & a_2 & -b_2\\ 0 & 0 & 0 & 0\end{bmatrix}.$$

2.2 Model of trailer with rear wheel compliant steering

From the dynamic analysis viewpoint, if the rear wheels of the trailer steer a small angle which is opposite to the turning direction [9], the stability of the trailer can be improved, and the lateral interference between the trailer and the tractor can be weakened [10, 11]. Therefore, the objective of the study is the full trailer with rear-wheel steering system. The CS system used here is shown in Fig.2, c_2, d_2 are its front and rear arm length, δ_4 is trailer rear wheel CS angle, regardless of the quality of the CS system and the effect of moment of inertia.

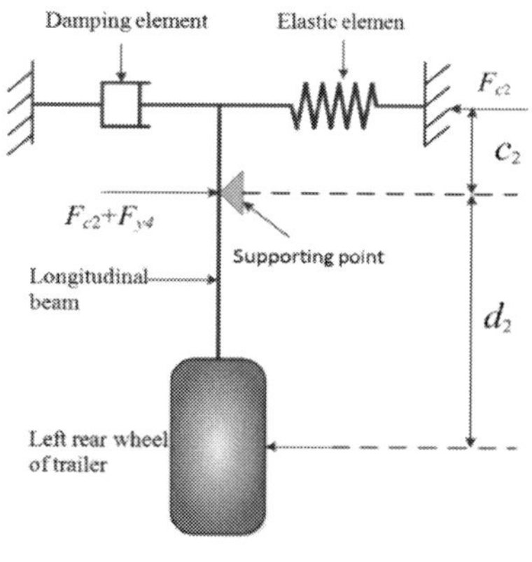

Fig. 2 Force diagrams of trailer CS system

F_{C2} is the elastic and damping force of the CS combined, thus

$$F_{c2}=-k_c\cdot c_2\cdot\delta_4-c_c\cdot c_2\cdot\dot{\delta}_4 \tag{2}$$

The CS angle meets the following equation:

$$\dot{\gamma}_2+\ddot{\delta}_4=-\frac{k_c d_2^2}{I_{cs2}}\cdot\delta_4-\frac{c_c d_2^2}{I_{cs2}}\cdot\dot{\delta}_4-\frac{c_2}{I_{cs2}}\cdot F_{y4} \tag{3}$$

Let $\delta_4=\phi,\dot{\delta}_4=\omega$, take the state vector as $\boldsymbol{X}=\begin{pmatrix}\beta_1 & \gamma_1 & \gamma_2 & \Delta\varphi & \phi & \omega\end{pmatrix}^T$. The state equation is

$$\boldsymbol{E}\dot{\boldsymbol{X}}=\boldsymbol{A}\boldsymbol{X}+\boldsymbol{N}\boldsymbol{p} \tag{4}$$

where

$$\boldsymbol{E}=\begin{bmatrix}1 & \frac{I_{Z1}}{(b_1+c)m_1u} & 0 & 0 & 0 & 0\\ 1 & -\frac{m_2(b_1+c)}{(m_1+m_2)u} & -\frac{m_2(a_2+d)}{(m_1+m_2)u} & \frac{m_2}{(m_1+m_2)} & 0 & 0\\ 1 & 0 & \frac{I_{Z2}}{(a_2+d)m_1u} & 0 & 0 & 0\\ 0 & 0 & 0 & 1 & 0 & 0\\ 0 & 0 & 0 & 0 & 1 & 0\\ 0 & 0 & 1 & 0 & 0 & 1\end{bmatrix},$$

$$\mathbf{A}=\begin{bmatrix}0 & -1 & 0 & 0 & 0 & 0\\ 0 & -\frac{m_1}{m_1+m_2} & -\frac{m_2}{m_1+m_2} & 0 & 0 & 0\\ 0 & -1 & 0 & 0 & \frac{k_c d_2^2}{(a_2+d)m_1u} & \frac{c_c d_2^2}{(a_2+d)m_1u}\\ 0 & 1 & -1 & 0 & 0 & 0\\ 0 & 0 & 0 & 0 & 0 & 1\\ 0 & 0 & 0 & 0 & -\frac{k_c d_2^2}{I_{cs2}} & -\frac{c_c d_2^2}{I_{cs2}}\end{bmatrix}$$

$$\mathbf{N}=\begin{bmatrix}\frac{a_1+b_1+c}{(b_1+c)m_1u} & \frac{c}{(b_1+c)m_1u} & 0 & 0\\ \frac{1}{(m_1+m_2)u} & \frac{1}{(m_1+m_2)u} & \frac{1}{(m_1+m_2)u} & \frac{1}{(m_1+m_2)u}\\ \frac{1}{m_1u} & \frac{1}{m_1u} & \frac{a_2}{(a_2+d)m_1u} & -\frac{b_2-c_2}{(a_2+d)m_1u}\\ 0 & 0 & 0 & 0\\ 0 & 0 & 0 & 0\\ 0 & 0 & 0 & -\frac{c_2}{I_{cs2}}\end{bmatrix}$$

3 Kinematics Simulation of Full Trailer

The main parameters affecting the handling stability of the trailer in the CS system are [12,13],

- The stiffness of the CS system k_s,
- the lengths of the front and rear arm c_2 and d_2, damping coefficient C_c or damping ratio ζ,
- the speed of the vehicle u.

Regardless of the specific effects of these parameters, comparing with the vehicle and without CS system, the following Tab.1 shows the specific parameter

Tab.1. Simulation parameters of a car

Parameter name	Values
Tractor, trailer weight m_1, m_2 (kg)	6804/5443
The longitudinal distance from the hinge point to the rear axle of the tractor c (m)	0.91
The longitudinal distance of the tractor center to the front axle and rear axle a_1, b_1 (m)	2.56/1.10
The longitudinal distance of the trailer center to the front axle and rear axle a2, b2 (m)	1.68/1.37

The longitudinal distance from the front axle of the trailer to the hinge point d (m)	2.44
Trailer rear wheel elastic components, rear wheel center of mass to support distance c2, d2 (m)	0.2/0.5
Trailer rear wheel CS stiffness k2s (n/rad)	45000
The damping ratio of CS system ζ	1
Tractor, trailer around the z-axis of inertia IZ1、IZ2 (kg.m)	21693/12541
Moment of inertia of the rear wheel CS system Ics2 (kg.m)	300

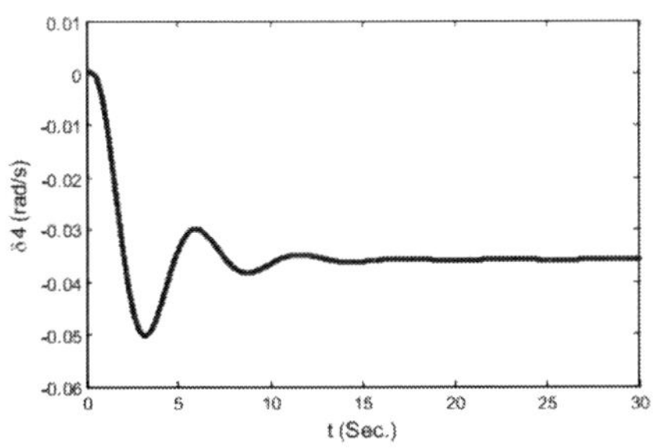

Fig.3. The curve of compliant rear-steering angle

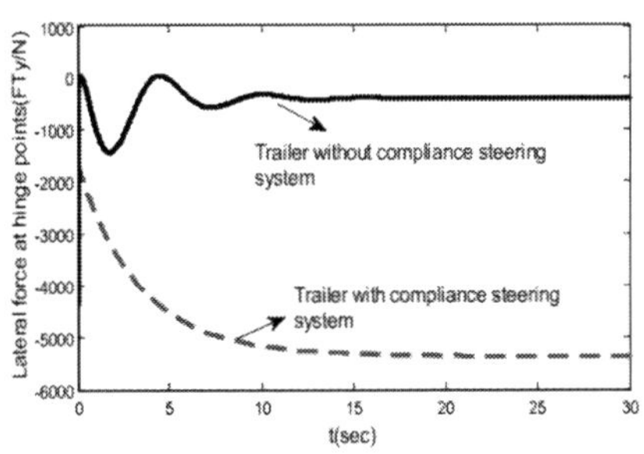

Fig.4. Lateral forces at hinge points

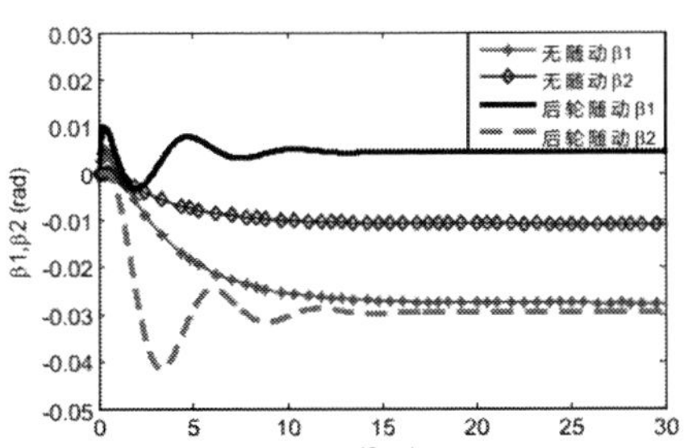

(a) The curves of side slip angle

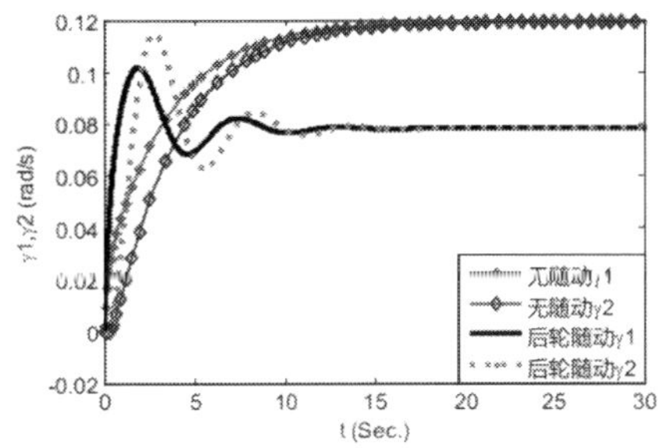

(b) The curves of yaw rate

Fig.5. Comparison of traditional trailer and trailer with CS system

When the vehicle with the CS turns, the rear wheels of trailer turn an angle opposite to the steering of the vehicle is shown in Fig.3. As shown in Fig.4, the generation of the compliant angle results in a significant decrease in the lateral forces at the hinge points. In Fig.5, it is visible that the yaw rate of the tractor and trailer and the side slip angle of the tractor are significantly improved.

The results verifies that when the rear wheel of trailer steers an angle, the lateral force between the tractor and the trailer is released, thereby reducing effectively the motion interference between the tractor and the trailer.

In order to analyze the specific characteristics of these parameters, packet simulation is required. Primarily, the effect of the sizes of c_2 and d_2 are discussed. Secondly, the influence of elastic element stiffness ks and damping ratio ζ is analyzed. Assuming that other parameters are the same, we only adjust the front and rear arm length c_2 and d_2. The values of the relevant parameters are: u=10 m/s, δ_1=0.02 rad, ζ=1, k_s=45000 n/m.

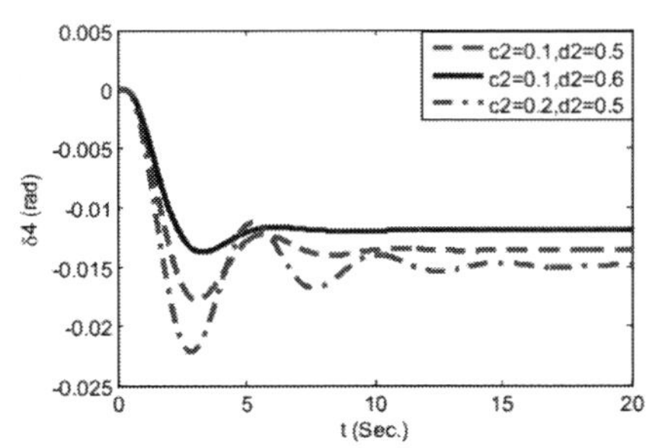

(a) The curves of compliant rear-steering angle

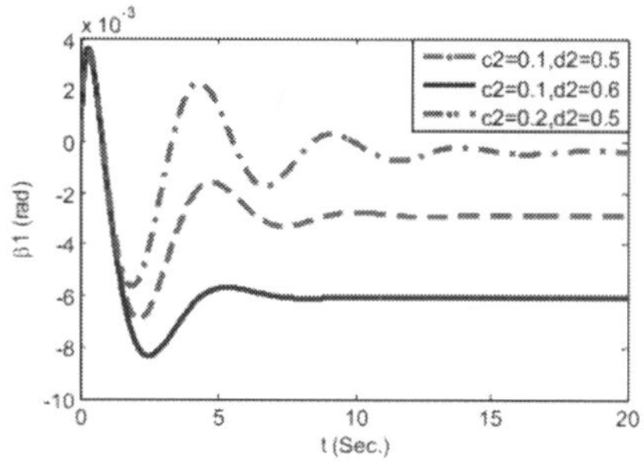

(b) The curves of side slip angle of tractor

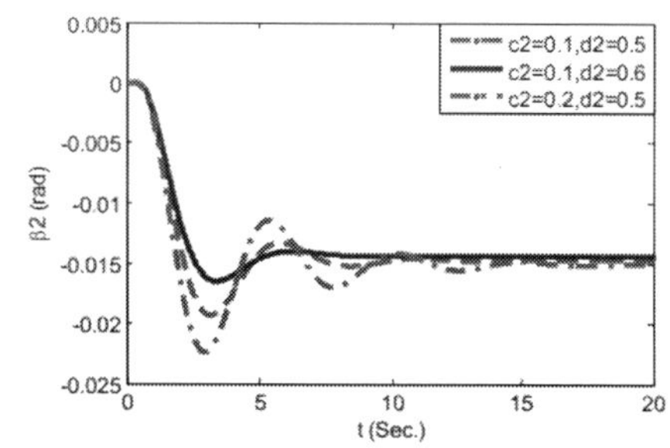

(c) The curves of side slip angle of trailer

Fig.7 The curves of compliant rear-steering

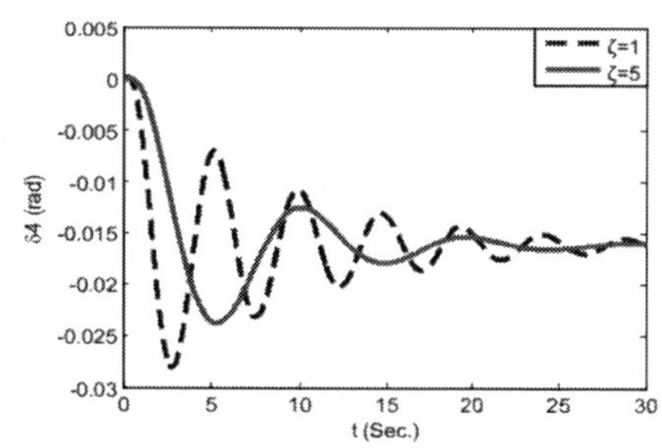

Fig.6. The curves when changing the support with different damping ratios point position angle

Theoretically, when the support point in the longitudinal beam on the relative position is declining, the same elastic force will produce a greater torque. As the relative position of the support point moves down (the value of c_2 increases and d_2 decreases relatively), the rear wheel angle of the trailer increases and the whole vehicle's vibration time and frequency are significantly increased. When the value of c_2 is greater than a certain value, the vehicle has been unstable. As shown in Fig 6, three sets of parameters are selected for comparison in the stable range. In Fig.6-a, when $c_2 = 0.1$ and $d_2 = 0.6$, it's compliant angle is relatively small, but its response is more stable. In Fig.6-b and Fig.6-c, the vehicle side slip angle is more stable, and the side angle of the tractor is relatively larger. When the parameter is chosen as $c_2 = 0.2$ and $d_2 = 0.5$, the system transient response is not good and the steady-state response is the best. Considering the effect of stiffness, we can refer to the dotted line in Fig.6-a and Fig.8-a as shown below. Through the design of specific parameters, these two CS angle response curves are obtained with the same parameters except the stiffness. With the increase of stiffness, the compliant angle response is more stable and the vehicle motion characteristics is improved. Fig.7 shows that when the damping ratio $\zeta = 1$, the system has a long-time oscillation by adjusting the parameters, and then keeping the other parameters unchanged, we only increase the damping ratio, the oscillation time and amplitude of the system are significantly reduced. Although the damping ratio changes, the steady-state response of the rear wheel compliant angle unchanged. The final stability value of the side slip angles of the tractor and the trailer are not changed with the change of the damping ratio. It is clear that the stiffness and damping ratio of the compliant angle system are variable parameters in the moving system structure, and adjusting the values of both are more effective in improving the motion characteristics of the trailer.

At a low speed (u=5 m/s), the CS angle of the trailer and the yaw rate of the tractor and the trailer is steady reached a fixed value. At this time tractor and trailer side angle is smaller. In the case of medium speed (u=10 m/s), the side angle of the tractor and trailer increases. Those shows that a group of parameters of CS system is correspond to a certain speed of the vehicle, so that the performances are better relatively, and the increase in speed will reduce the handling of the full trailer stability. When the speed reaches a certain value, the vehicle becomes instable. When a more stable performance at higher speeds is required, adjust the parameters of the CS system according to the analysis above.

4 Conclusion

(1) When the rear wheels of trailer steers, the lateral forces of the tires are released, so the lateral force at the hinge point is smaller and the movement interference between the tractor and the trailer is reduced.

(2) Based on the theory of CS technology to study the stability of the full trailer, the dynamic analysis of the

handling stability of the trailer with CS system is conducted. The specific impact of the various parameters and the speed on the performances of trailer are studied.

(3) The simulation results show that the angle of the rear wheel of the trailer can reduce the motion interference at the hinge point, and improve the stability of the vehicle, which is also consistent with the theoretical analysis. The key parameters of the CS system have a comprehensive effect on the vehicle, here the research methods and CS system parameters is provided for the study of the full trailer with compliant steering system.

Acknowledgements

This work was supported by the National Natural Science Foundation of China (Grant No. 11272159).

References

[1]Souharda R. Compliant Multi-link Vehicle Suspensions. South Carolina: Clemson University, 2008.

[2]Mu Xiangdong. Modeling and Test Verification of Citroen ZX Car Compliance Steering Characteristics. Journal of Highway and Transportation Research and Development, 2004, 21(9):127-130.

[3]Xu Xiaomei, Chen Ning, H. P. Lee. A Review on the Applied Research of Rear Wheel Compliance Steering. Automobile Technology, 2016, 7:1-6.

[4]Enomoto H, Akiyama K and Okuyama H. Advanced Safety Technologies for Large Trucks. In 14th Asia Pacific Automotive Conference, 2007.

[5]Yu Zhisheng. Automobile theory. Beijing: China Machine Press.

[6]Pacejka H B. Tire and Vehicle Dynamics (Second Edition). Burlington: Butterworth Heinemann, 2006.

[7]Wang Wei, Bei Shaoyi, Zhao Youqun, *et al.* Research on the development and future of vehicle handling and stability. Machine Design and Manufacturing Engineering, 2014, 43(10): 6-12.

[8]Wei Chaoyi, Chen Ke, Zhou Kongkang, *et al.* Stability Analysis of Operation for Passenger Car-trailer Based on Root-locus Method. Transaction of the Chinese Society for Agricultural Machinery, 2007,38(3): 15-18.

[9]Kharrazi S, Lidberg M, Roebuck R, *et al.* Implementation of Active Steering on Longer Combination Vehicle for Enhanced Lateral Performance. Vehicle System Dynamics. 2012, 50(12): 1949-1970.

[10] Kristopher J Seluga RMOI. Articulated Vehicle Yaw Stability During Braking — A parametric Study. In Rosemout, Illinois USA , 2004.

[11]Wu D H, Lin J H. Analysis of Dynamic Lateral Response for a Multi-Axle-Steering Tractor and Trailer. Heavy Vehicle Systems-International Journal of Vehicle Design. 2003, 10(4):281-294.

[12]Rangavajhula K, Tsao H. Effect of Multi-axle Steering on off-tracking and Dynamic Lateral Response of Articulated Tractor-trailer Combinations. International Journal of Heavy Vehicle Systems. 2007, 14(4): 376-401.

[13]Tica M, Dobre G, Mateescu V, et al. Influence of Compliance for an Elasto-kinematic Model of a Proposed Rear Suspension. International Journal of Automotive Technology, 2014, 15(6): 885-891.

[14]Yang Xiujian, Li Yaoping, Xiong jian. Analysis on the Lateral stability and Instability Mechanism of Tractor Semi-trailer Combination. Automobile Engineering, 2011, 33(06): 486-493.

[15]Bai Bin. Study on Lateral Control ability and Stability of Tractor Trailer. Jilin: Jilin University.

DOI: 10.1051/matecconf/201713900002

Development of a blocking logic checking apparatus for substation measurement and control devices

Li Chen, *Bingwei* Gao, *Feng* Qi, *Bifu* Qian, and *Lei* Zhang

Wenzhou Power Supply Company of SGCC, 325000 Zhejiang Wenzhou, China

Abstract. After transforming substation measurement and control devices, the traditional method of blocking logic check needs to go back and forth among the measurement and control devices, and repeatedly connect a large number of wires, which lead to tedious operation and low efficiency. This paper developed a logic verification device based on 4G wireless communication, which used the Modbus/TCP transport protocol to access Ethernet, to complete the real-time data communication with the mobile control terminal. And this paper made a request to the logic controller according to practicality and cost. And the terminal software interface was designed by the orthogonal experimental design idea. The development of this device provided a new and efficient method for logic verification.

1 Introduction

Maloperation prevention is an important item in substation operation management. In maloperation accident, it not only hurts operators severely, but also damages operating equipment seriously, even causes blackout [1].

When dispatching department and supervising system control the measurement and control device(MACD) remotely, MACD analyzes the maloperation prevention blocking(MPB) logic by acquired information and interlock information about other MACDs. If the condition is fulfilled, the contact will be closed, which means can be controlled remotely. Otherwise, it blocks operation.

Checking the correctness of bay level blocking logic is a considerable item for new MACD acceptance [2]. Traditional process of checking blocking logic demands that changing wires back and forth among interlocking MACDs for satisfying different requirements of tele-signaling input, as the blocking method is calculated by MACD with the state of contacts of field equipment [3]. It not only costs energies and time, but also likely causes "repeat checking" and "omit checking", and leads to potential risk. Therefore, this paper developed a blocking logic checking apparatus of substation MACD by using 4G communication technology. And it offered a new and efficient method for logic verification to engineers.

2 The principle of device

2.1 The interlocking principle of bays

The MPB of MACD consists of block in a bay and interlock among bays. The block in a bay is simple because information library of equipment is in one MACD. The interlock among bays is much more complex as data is derived from other bays besides local data [4]. The classic flow of blocking logic judgement is shown in Figure 1.

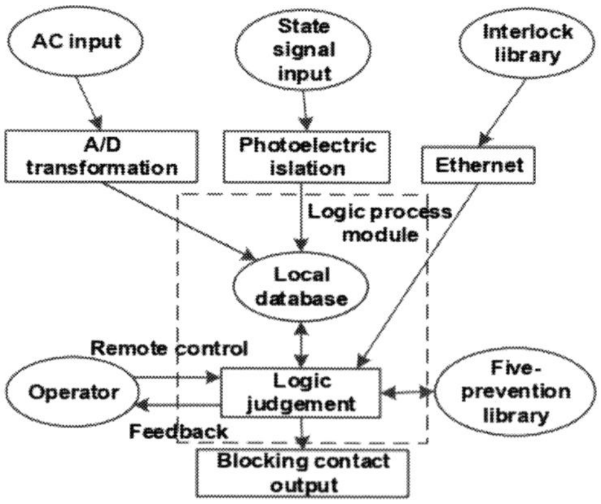

Fig. 1. The implementation process of blocking logic

2.2 The origin of the question

Traditional blocking logic verification consists of five segments according to the above principle. First, completing security precautions. Second, connecting and disconnecting circuits. Third, going back and forth among MACDs. Fourth, confirming input signal. Last, testing the tele-controlling.

Figure 2 is showing a part of primary electric wiring diagram in a 220kV substation. It contains three bays: Wenjiang 2391 Line, Wennan 2392 Line and bus coupler.

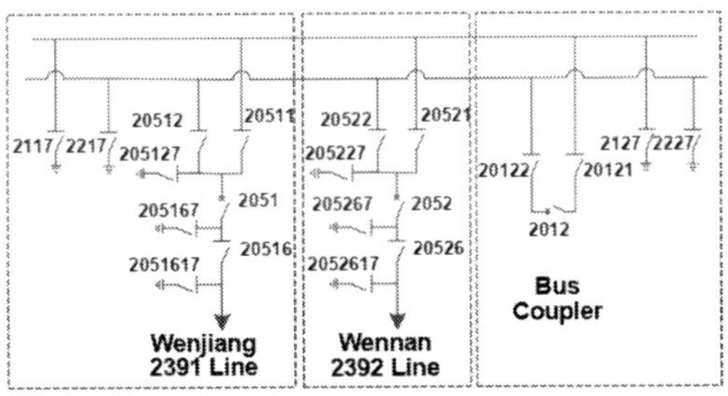

Fig. 2. The wiring diagram of multi-bay in a 220kV substation

For instance, according to requirement of MPB [5], the blocking logical expression of operating No.20512 disconnector of Wenjiang 2391 Line is showed below.

$$F_{CLs}(20512) = 2012(1) \,\&\, 20121(1) \,\&\, 20122(1) \,\&\, 20511(1) \,\&\, 2117(0) \,\&\, 2217(0) \,\&\, 2127(0) \,\&\, 2227(0) \text{ OR } 20511(0) \,\&\, 2051(0) \,\&\, 205127(0) \,\&\, 205167(0) \,\&\, 2217(0) \,\&\, 2227(0) \quad (1)$$

Where F_{CLs} means consequence of blocking logic, and (0) means open, and (1) means close.

According to above expression, when Wenjiang 2391 Line is switching bus, double bus should be in parallel operation, which means bus coupler(2012), I-bus disconnector(20122) of bus coupler and II-bus disconnector(20121) of bus coupler are close, besides the blocking of this line. Thus, to verify the blocking logic of No.20512 disconnector, engineers have to walk back and forth frequently among MACD of Wenjiang 2391 Line, MACD of bus bar and MACD of bus coupler for changing circuits in traditional method. Therefore, it will promote efficiency of verification greatly if there is a way that taking the place of changing circuits manually, and getting feedback of input in time.

2.3 Logic verification scheme based on 4G communication technology

4G technology has the advantages of no distance limitation and high quality transmission. Based on these advantages, this paper used 4G communication technology and developed a blocking logic controller with cross-platform control software, which achieved wireless verification of blocking logic of MACDs, and solved the low-efficiency problem caused by frequent changing circuits in the traditional method.

While logic verification of one MACD involved blocking information of other bays, this paper had to configure blocking logic controller in this MACD. Every controller had 16-channel contacts, and these contacts could be connected to tele-signaling circuit of MACD. The controller drove corresponding electric relays after that engineers issued tele-controlling orders. Then, relays changed the state of contacts for meeting the demand about blocking input.

3 The design of logic controller

3.1 The design of circuitry

Considering the utility of controller, economy and substation circumstances, the goals of the controller is shown below.

(1) Power-supply Module: For the convenience of engineering application, this paper chose Li-on battery as power supply because of small size and good heat dissipation. Its voltage output is 3V-24V, and can be used for ten days without charging.

(2) Electric Relay Module: Considering the electromagnet interference of substation, this paper selected AQW214S relay as low drive current, high voltage-endurance and low price.

(3) Communication Interface Module: Adopting M6310 4G communication module. It embeds entire TCP/IP protocol, and is antijamming, without distance limitation.

(4) Process Unit Module: This paper chose STM32F series MCU as the process unit. Its work frequency is 0~35MHz, and it could meet the requirement of the speed of transmitting TCP/IP data packets. Besides, it has low power consumption and low price.

(5) Other Modules: To avoid data loss while power cut, this paper selected EEPROM as memory chip.

The entire structure of the device is shown in Figure 3.

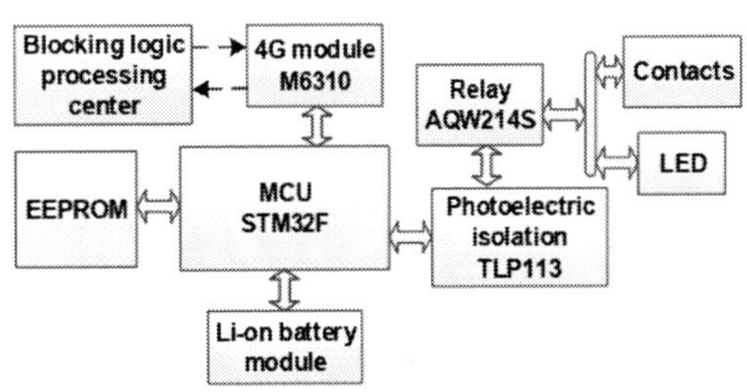

Fig. 3. The diagram of the hardware circuit

3.2 The design of communication interface

The logic controller based on standard Modbus/TCP protocol utilized Ethernet to achieve data interchange between logic controller and server by reliable peer-to-peer communication built on TCP protocol. Data frame received by M6310 module will be checked

 DOI: 10.1051/matecconf/201713900002

whether matching native IP address automatically. If the data packet was valid, it would be sent to buffer. And MCU would complete the analysis of tele-controlling order and the feedback of blocking contact. This process was crucial for communication between logic controller and system. The procedure of data interchange is shown in following diagram.

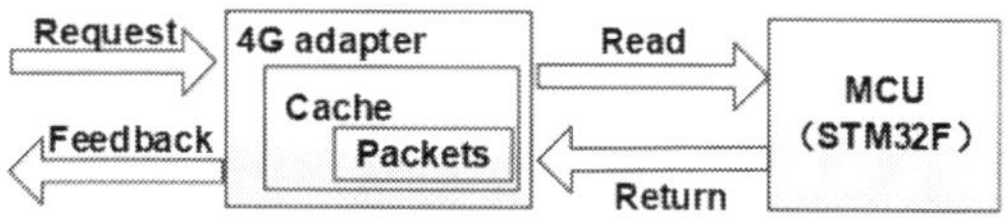

Fig. 4. The diagram of data interaction

4 The design of terminal application

4.1 The design of function interface

The development of this device aimed at simplifying the flow of logic verification, and promoting efficiency of checking. Therefore, this paper focused on utility and operability when designing function interface of the client. There were four major factors influencing operating efficiency, considering reality of logic verification.

(1) Association Mode of Channel Contacts
(2) Control Type of Contacts
(3) The Number of Local Control Units
(4) The Number of Channels

This paper chose an optimum combination through experiments. Parts of the function interface are shown in Figure 5.

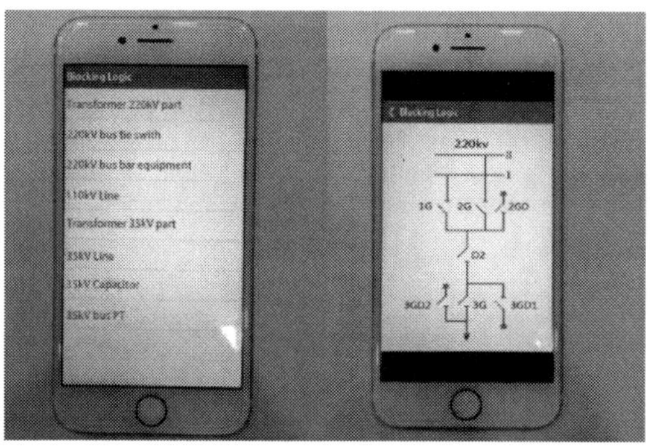

Fig. 5. The client function interface

4.2 Mechanism of mutually exclusive logic

The tele-signaling input usually indicates the state of on/off by dual position contact for most of MACDs. If a disconnector is close, the input of normal open contact is 1, and the input of normal close contact is 0. Similarly, when a disconnector is open, the signal of normally open contact and normal close contact is reverse. If one of contacts is adhesive or broken, the input will be either 1 or 0. This case is classified as uncertain state by MACD, and it leads to deadlock of MACD. Thus, this paper had to take correctness of contact output into consideration when simulating tele-signaling input by the logic controller. This application utilized the mechanism of mutually exclusive logic on principal and subordinate channels to simplify the operation and avoid deadlock situation. That was, principal and subordinate channels were related to one contact of logic controller respectively. And while engineers gave a tele-controlling order by the mobile terminal, values of principal and subordinate channels were reverse automatically.

For example, to close No.20516 disconnector of 220kV Wenjiang Line shown in Figure 3, the data frame of tele-controlling order sending to controller was 0x19 0x56. Converting it to binary number, and every two adjacent bits corresponded to control bit received from principal and subordinate channels(1 means close, 0 means open), except bit.14 and bit.15. The definition of control bits is shown in Table 1.

Table 1. The definition table of mutually exclusive logic

Control Bit(bit0~bit13)	No.	Definition
bit.0=0	20511	Close
bit.1=1		
bit.2=1	20512	Open
bit.3=0		
bit.4=1	205127	Open
bit.5=0		
bit.6=1	2051	Open
bit.7=0		
bit.8=1	205167	Open
bit.9=0		
bit.10=0	20516	Close
bit.11=1		
bit.12=1	2051617	Open
bit.13=0		

Storage of logic controller reserved current state information about every contact, and MCU compared received control bit with data of relevant channels. If the result of comparison is consistent, it will not change the state of contact; otherwise, it changed according to control bit.

5 Flow of verification

To ensure the integrality of checking MPB, and simplify the procedure, this paper compiled the standard operation procedure and logic verification scheme of this device. The specific flow is shown in Figure 6.

DOI: 10.1051/matecconf/201713900002

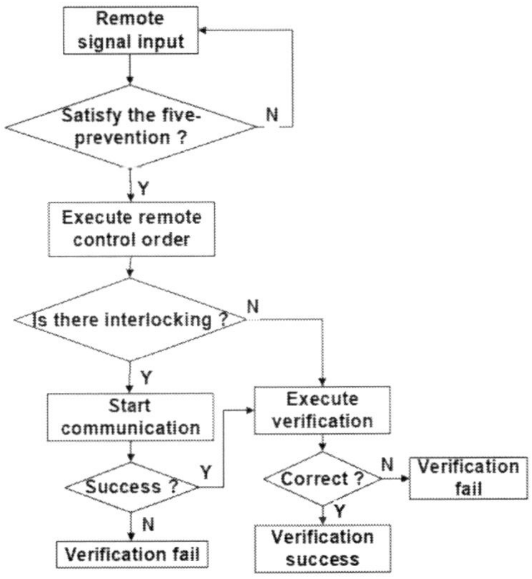

Fig. 6. Verification flow

(1) When checking a certain logic, this paper had to configure a logic controller in a relevant MACD first, and connected testing wires into tele-signaling input circuits.

(2) Controlling the state of contacts remotely through terminal application to meet the requirement of five-prevention. And this paper inspected it on supervising system or graphic interface of application.

(3) Giving a tele-controlling order by supervising system after completing the setting of contacts and meeting the demand of blocking logic. Once the MACD received instruction, it judged whether the equipment involved interlocking of other bays. If the interlock exited, it started interlocking communication first to acquire real time data; otherwise, it came to logic verification directly.

(4) The MACD computed MPB logic on the basis of local information and interlocking information of other bays, and sent the result of tele-controlling verification to the supervising system for testing the correctness of MPB logic.

(5) Judging the returned result to check the validity of MACD.

6 Test of achievement

In a retrofit project of MACD, this paper used traditional method and this device to verify the logic of the same bay, and wrote the time consumption down. The experimental data is shown in Table 2.

Comparing with the traditional method, as a result, wireless verification with this device promoted work efficiency greatly, and it assured the correctness and integrity of checking MPB.

Table 2. The test table of experiment

Project	220kV Liao 2Q94 Line	220kV 1# Transformer
Model	CSI-200E	NSD500
Number of Trials	5	3
Traditional Method / minute	146.6	131
New Method / minute	65.5	52.8
Logic Verification	Correct	Correct

7 Conclusions

The traditional method of checking MPB logic costs energies and time greatly. This paper provided a MPB logic verification device of MACD based on 4G communication. It's a new and efficient method for logic verification. Engineers could complete the verification through a mobile terminal. It promoted efficiency greatly. Meanwhile, it assured the safety of equipment. With the development of 4G and mobile devices, this method has favorable application prospect.

References

1. Y. Sun, L. Hou, P. Jie. Implementation of maloperation proof on platforms of relay and control units between bays. Automation of Electric Power Systems, **30**, 81-85 (2006)
2. L. Zhang, B. Qian, X. Liu. Some Problems Analysis of Transformation of CSI-200EA Measurement and Control Equipment. Diangong Dianqi, **04**, 35-37 (2015)
3. W. Zhou, L. Zhou. The Advantages and Disadvantages of Computer Monitoring and Control Device. Science & Technology Information, **10**, 123 (2012)
4. B. Duan, G. Chen, Y. Lin. On-Line Interlocking Implementation Based on Process Bus Communication of Digital Substation. Power System Technology, **19**, 37-43 (2009)
5. Y. Gu, W. Pi, C. Yang. Application analysis of anti-mistake defense system in substation. RELAY, **02**, 66-70 (2005)

MATEC Web of Conferences 139, 00003 (2017) DOI: 10.1051/matecconf/201713900003
ICMITE 2017

Kinematic Analysis on the Forehand Stroke of ATP Tennis Player Karen Khachanov

Chuanjia Du[1,*], *Jihe* Zhou[2], and *Shuai* Wang[1]

[1]Chengdu Sport Institute, 610041, Chengdu, Sichuan, China
[2]Chengdu Sport Institute, Sports Medicine Department, 610041, Chengdu, Sichuan, China

Abstract. By using the three-dimensional video analysis method to kinematic analyze the forehand stroke technique of the ATP professional tennis player, Karen Khachanov. Exploring the action mode and characteristics from him to provide a reference for all tennis learners to improve their forehand stroke. The results show that: (1) At the end of the preparation, Karen Khachanov's center of gravity was lower, and his body control was steady; (2) After the follow through, each joint of Karen Khachanov's racquet arm were stretched, so the swing radius is larger. His right leg was like a support point for the next stage---drive leg and rotate hip; (3) At the swing stage, the move of his driving leg and rotating hip was full, so the whole hitting action match the principle whipping technique movement. (4) The "wiper " style follow through was natural, coordinated, and smooth. After hitting the ball, his body of control was steady, the shoulder movement was obvious, and the hitting was fast and powerful. Conclusion: The forehand stroke of Karen Khachanov was a high quality hitting action, and it's worthy of tennis players to learn from.

1 Introduction

Tennis is a complex sport. The complexity is not only reflected in the game and the process of competitors' game, but more reflected in the technical action sequence structure of the correctness and stability. In all techniques, forehand stroke is the most frequently used technique with more than 70% usage[3]. Therefore, the study of forehand stroke is particularly important. This paper used three-dimensional video analysis method to analyze the forehand stroke of the ATP professional tennis player, Karen Khachanov. Collecting kinematic data, carry on the kinematics analysis, and explore the forehand movement mode, and the four major ways of the study to provide reference to learn forehand stroke theory.

2 Object and method

2.1 Object

Russian tennis star, Karen khachanov (refer to Table 1 for personal information). In July 2013, he won the under 18 European champion, and won the single match champion in the International ATP250 Chengdu Open 2016.

Table 1. Basic information of Karen Khachanov

Birth	Height (m)	Weigh (kg)	Dominate hand	Highest ranking
1996/5/21	1.98	88	Right	48

2.2Method

* Corresponding author: 549828421@qq.com

DOI: 10.1051/matecconf/201713900003

Three-dimensional video analysis: Two JVC-GC-PX10AC high-speed cameras (50fps) were used in the International ATP250 Chengdu Open 2016 to shoot Karen Khachanov's technique movements. The axis of the two cameras is about 50°(the position of cameras is showed in Fig. 1). The selected video was the last forehand stroke in the 2nd point of the first games first set. Inputting the selected video part in the 3-D Signal TEC V2.0C analysis system. Choosing the Dempster manikin (with 16 links, 21 joint parameters). In the meantime, adding the racket session for the study. Using the step by step analysis method and the low-pass filter to smoothing process to point data and get the results.

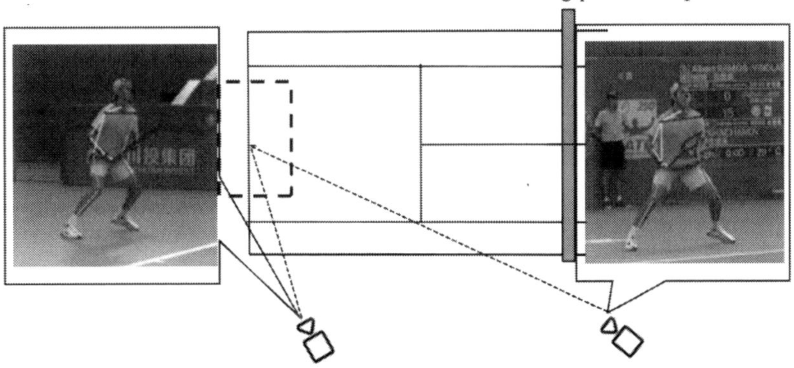

Fig.1. Diagram of cameras position

3 Results

3.1 Definition of moments, division of stages, selection and definition of joint

Definition of moments

T_1 The moment of the end of preparation: The moment that feet are split step and backswing.

T_2 The moment of backswing: The moment that the racket hand behinds the farthest.

T_3 The moment of touching the ball: The moment that the racket is going to touch the ball.

T_4 The moment of end of follow through: The moment that the angle is the maximum between shoulder and hip

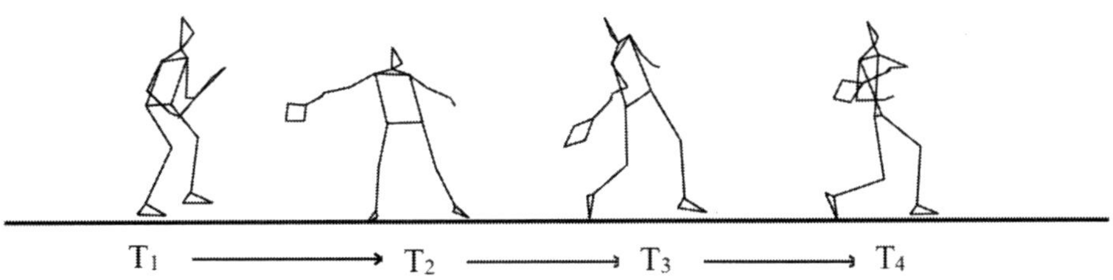

Fig. 1. division of stages

Note: T_1: The moment of the end of preparation, T_2: The moment of backswing, T_3: The moment of touching the ball, T_4: The moment of end of follow through. $T_{1\text{-}2}$: Backswing stage, $T_{2\text{-}3}$: Swing stage, $T_{3\text{-}4}$: Follow through stage.

Selection and definition of joints angle

* Corresponding author: 549828421@qq.com

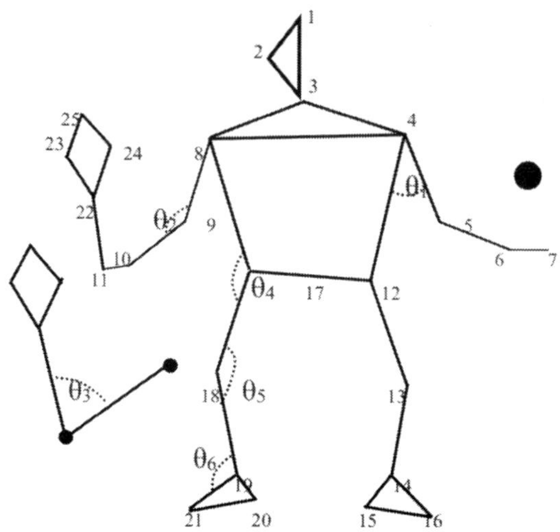

Fig. 2. Manikin and parts of parameters

Note: θ_1. Shoulder joint angle, θ_2. Elbow joint angle, θ_3. Racket-forearm angle, θ_4. Hip joint angle, θ_5.Knee joint angle, θ_6.Ankle joint angle
Shoulder angle: The angle between shoulder and upper arm.
Elbow joint angle: The angle between upper arm and forearm.
Racket-forearm angle: The angle between the forearm and racket.
Hip joint angle: The angle between hip and thigh.
Knee joint angle: The angle between thigh and crus.
Ankle joint angle: The angle between crus and ankle.

3.2 Analysis of the backswing stage

The backswing stage means the process that the player is prepared and backswing. Under the non-forced circumstance, the player would have more time to adjust the body positon, and to determine the hit line to improve the quality of return. Before adjusting the position, normally player have the split step movement to adjust the dance and backswing.

Table 2. Parameters about the end of preparation

Center of gravity (m)	Right shoulder (°)	Right elbow (°)	Racket forearm (°)	Hips (°)		Knees (°)		Ankles (°)	
				Left	Right	Left	Right	Left	Right
0.92	30.3	94.4	106.3	123.6	121.6	125.3	108.9	98.5	101.6

At the end of preparation, the height of Karen Khachanov's center of gravity was 0.92m, compared to his normal anatomy of the height, 1.20m, his height was lowered 0.28m. The bent of each lower limb joint showed the center of graving was down, at that time, the angles of Karen Khachanov's left and right hip joint angle were 123.6°, 121.6°. The left and right knee joint angles were 125.3°, 108.9°, and the left and right ankle joint angles were 98.5°, 101.6°. The hip, knees and angles of Karen Khachanov were bent, body control was steady, and compared to his left knee joint angle, the bent angle of his left knee joint was larger, according to the data. All these indicate that at the time of after preparation, Karen Khachanov already knew the coming ball from his competitor would point to his own forehand position. Therefore, his body adjusted in a shore time, and stayed in a relative lower center of gravity to finish all following movements better. His right shoulder joint angle and elbow joint angle were 30.3°, 124.4°, and the racket-forearm angle was 106.3°, which means Karen Khachanov's racket hand was naturally bent at that time.

After players forecast the position of the coming ball, they will move into the next stage, which is the backswing stage. Backswing stage is the reverse movement of forehand stroke. The purpose of backswing is to enlarge the radius and range of the movement. So the racket hand that is associated with muscle elongation and elasticity will help finish next stage movements. The kinematic parameters are important factors at the moment of the end of backswing stage that are used to determine the effect of the player's pre-elongated elasticity.

Table 3. Parameters about the end of backswing

Center of gravity (m)	Right shoulder (°)	Right elbow (°)	Racket forearm (°)	Hips (°)		Knees (°)		Ankles (°)	
				Left	Right	Left	Right	Left	Right
0.87	79.4	149.5	136.8	144.8	131.5	132.6	112.1	88.7	81.5

At the end of backswing, the right shoulder joint angle and right elbow joint angle of Karen Khachanov were 79.4°, 149.5°, and the racket-forearm angle was 136.8°. Related research pointed out[1] : The angle of right shoulder and elbow shows the player's hitting-arm stretch state at the end of backswing, compared to the trunk; At this moment, the right shoulder and the right elbow joint's angle of Andy·Murray were 46.5 °, 136.53 °. That means at this moment, Karen Khachanov's hitting-arm of the joints stretched fully and lead to the larger swing radius. At the end of backswing, Karen Khachanov chose the semi-open stance, and his left and right hip joint angles were 144.8°, 131.5°; the left and right knee joint angles were 132.6°, 112.1°, and the left and right ankle joint angles were 88.7°, 81.5°, the set of data shows that Karen Khachanov used his right leg as the main supporting function at this moment. The height of Karen Khachanov's center of gravity was 0.87m, compared with the after preparation time's center of gravity, which was 0.92m; the center of gravity down value was 0.05m. The main reason was Karen Khachanov deliberately to keep the center of gravity focus on the right leg. A literature shows [2] that the strength of forehand stroke comes from the ground, the right knee fast forceful extension to promote the formation of angular momentum, knee angular momentum deliver to the hip, shoulder, elbow, wrist and then get through racket head to form a complete power chain system.In the process, nevertheless, the premise of right knee fast forceful extension was that the center of gravity near to the right leg. The external performance of the center of gravity near to the right leg was that the right lower limb flexor than the left lower limb was more obvious. Therefore, through the kinematic parameters of the lower limbs show that Karen Khachanov's whole body center of gravity close to the right leg, which means established the advantageous condition for the next stage of hitting the ball.

3.3 Analysis of the swing stage

The swing stage refers to the process from the backswing to the moment to contact the ball. This is the functional action in the forehand stroke, and it is the core action of the forehand technique. In this stage, the players need to push off ground to get a larger angular velocity quickly and adequately, driving the hitting-arm to finish the stroke.

Table 4. Parameters about the moment of contact the ball

Center of gravity (m)	Right shoulder (°)	Right elbow (°)	Racket forearm (°)	Hips (°)		Knees (°)		Ankles (°)	
				Left	Right	Left	Right	Left	Right
0.94	62.3	107.1	105.4	157.8	166.8	165.0	154.6	124.2	115.0

At the moment of touching the ball, if the starting point is too close to the body, there will happen to "squeeze" phenomenon. The most intuitive performance shows that the shoulder angle is too small, and if the starting point away from the normal position, the shoulder and elbow angles will be too large, the backswing stage established by the larger waving radius advantage will be affected.

At the moment of touching the ball, the height of Karen Khachanov's center of gravity was 0.94m, the right shoulder joint, right elbow joint and racket-forearm angle were 62.3 °, 107.1 ° and 105.4 °. Through these data and combined with the video record we can see that: in this stroke, Karen Khachanov's control of the starting point was appropriate, which was an important prerequisite for this stroke to complete the winning points. At the same time, the data show that Karen Khachanov's the left and right hip angle were 157.8 °, 166.8 °; the left and right knee angle were 165.0 °, 154.6 °, the left and right ankle angle were 124.2 °, 115.0 °. Karen Khachanov's two knees are still in the buckling state, the height of the center of gravity was still lower than the standing position; observed from the Y-X plane, the projection point of Karen Khachanov's center of gravity in the left foot horizontal position, which shows that Karen Khachanov has been a significant kick his leg and then rotated his hip adequately, and the action means the center of gravity were rapid transfer.

Table 5. Parameters about the swing stage (m/s)

Center of gravity	Right knee	Right hip	Right shoulder	Right elbow	Right wrist	Racket head
0.86	1.32	1.47	2.68	5.46	11.21	22.61

A good swing action must be consistent with the principle of whipping action, that is, the speed of the proximal to the link of distal increasing in turn. There are some differences technical principles between the tennis forehand whipping and the whipping of throwing items. For example, in the process of throwing javelin movement, the javelin, as a whole part is a parallel movement. But the forehand in swing stage, the racket as a whole part around the player's frontal axis will do rotational movement, and then make the racket head get more acceleration. In the swing stage, the racket- arm angle used to be as an important indicator to judge the effect of whipping action and increasing the speed of the racket.

* Corresponding author: 549828421@qq.com

DOI: 10.1051/matecconf/201713900003

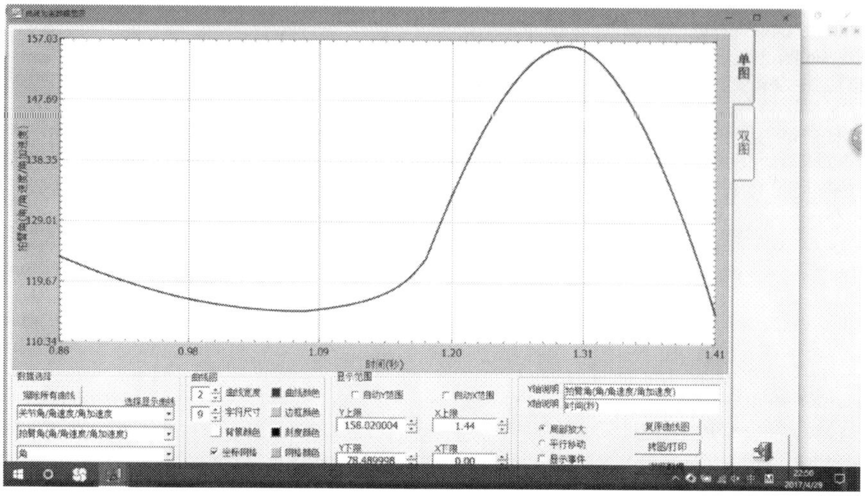

Fig. 2. Racket-forearm angle variation over time

We can see from Fig. 4, during the process of swing stage, the arm angle rapidly reduced to 106.7 °, at the moment of touching the ball, the arm angle instantaneously increased to 123.4 °. which means during the swing stage, Karen Khachanov's racket-forearm angle first gradually reduced and instantaneous increased in the moment of contacted the ball, so to achieve the effect of speeding the head. Data shows that at the moment to touch the ball, Karen Khachanov's right knee speed was 1.32m / s, the right hip speed was 1.47m/s, which also demonstrated that Karen Khachanov's kicked the hips and rotated his whole body significantly in the process of Swing stage. Furthermore, table 4 also showed: Karen Khachanov's right shoulder joint, right elbow joint, right wrist joint and racket head speed were 2.68m / s, 5.46m / s, 11.21m / s and 22.61m / s. From that, the proximal end of the link to the speed of the speed increases at the moment of touching the ball, the head speed was 22.61m / s, the technical action keep in line with the principle of whipping.

3.4 Analysis of the follow through stage

"Follow-through" means follow the previous action, make sure you finish all the movements and take the racket through the body to the sideways. The follow-through stage refers to the process from the moment of contact the ball to the end of the follow-through action. This action is also one of the essential stage in the forehand. The associated kinematic parameters of this stage are signs of the momentum transmission when swing the racket to contact the ball. In the follow-through stage, Karen Khachanov used to the "screenwiper" mode, that is, from the moment of the ball be contacted, quickly rotated his forearm in the direction of left sideway, with the purpose to increase the speed of the racket head in the frontal plane, and made the ball had strong topspin.

Table 5. Parameters about the end of the follow-through

Center of gravity (m)	Right shoulder (°)	Right elbow (°)	Shoulders -hips (°)	Hips (°)		Knees (°)		Ankles (°)	
				Left	Right	Left	Right	Left	Right
0.97	84.5	106.8	76.3	167.7	177.2	139.2	115.0	113.5	119.5

At the end of the follow through action, Karen Khachanov's right shoulder, right elbow angle were 84.5 °, 106.8 °, which means his shoulder and elbow were bent naturally. Shoulder-hip angle was 76.3 °, indicated that Karen Khachanov made full use of the rotational inertia that generated by twisted his torso, send the shoulder action obviously, and increased the role of this racket on the ball time, and then increased the power of distance. The left and right knee angles were 139.2 ° and 115.0 °, the left and right ankle angles were 113.5 ° and 119.5 °, and the center of gravity was 0.97m. By observing the original video, with the end of the follow through action, Karen Khachanov's lower limbs have completed the buffer action, then enter the next stroke to prepare the action.

* Corresponding author: 549828421@qq.com

4 Conclusions

According to kinematical analysis on the forehand stroke of Karen Khachanov and draw the following conclusions:

4.1 The backswing stage

At the end of the preparation, Karen Khachanov's hips, knees, ankles joints were bent and bent similar size, body control was more stable; compared to the height of normal anatomical body, his center of gravity height was lowered 0.28m, body center of gravity kept a low position, ready for the back position.

At the end of the backswing stage, Karen Khachanov's right shoulder, right elbow joint angle were 79.4 °, 149.5 °, racket-forearm angle was 136.8 °,that means the articulation of the joint arm was more fully and the waving radius was larger; From his hips, knees, ankles angle can be seen, Karen Khachanov's right lower limb flexion was more obviously, the right leg from the main support role for the next stage of the leg to establish the basis for rotated his hips.

4.2 The swing stage

At the moment of contact the ball, Karen Khachanov's center of gravity height was 0.94 m, the right shoulder joint, right elbow joint and the knee arm angle were 62.3 °, 107.1 ° and 105.4 °, respectively. through the data with the video recording, we can see that during the stroke process, Karen Khachanov controlled the starting point was more appropriate ; he rotated his hips adequately; the right shoulder, the right elbow, the right wrist joint and the head speed respectively, 2.68m / s, 5.46m / s, 11.21m / s and 22.61m / s, the proximal end to the distal end of the speed increases, in line with the principle of whipping.

4.3 The follow-through stage

At the end of the follow-through, Karen Khachanov's lower limb joints were still bent, the center of gravity height was 0.97m, the body control is smooth; the shoulder hip angle was 76.3 °, making full use of the torso twist generated by the moment of inertia. Sending the shoulder action is obviously, at the same time, it increased the role of this stage racket on the ball time, and then increased the work distance, the action is fast and powerful.

References

1. Tao Yang, Jihe Zhou. Kinematical analysis on the forehand stroke of Andy Murray [J].Sports vision, 2015，34（5）: 247-249.
2. Huifang Wang. Kinematical analysis on the forehand topspin stroke of ATP tennis players in chengdu[D]. Chengdu sport institute,2014.
3. Shengying Zhang, Wei Xu. Analysis on the technical characteristics of the World 's elite tennis players[J]. Sports Science and Technology in Hubei Province，2010,1,（29）： 73-74.
4. Ting Jiang, Bing Qi. Kinematic research on the forehand Stroke of different horizontal tennis players[J]. Journal of Beijing Sport University, 2014,9（37）: 128-132.
5. Hang Li. Kinematical analysis on the difference forehand stroke style: flat and topspin[D].Shenyang sport institute,2012.
6. Jian Guo. Biomechanical analysis and evaluation of tennis forehand stroke technique[J]. Athletics Forum,2009,12（17）:53-55.

* Corresponding author: 549828421@qq.com

MATEC Web of Conferences 139, 00004 (2017) DOI: 10.1051/matecconf/201713900004
ICMITE 2017

Optimization of vehicle compartment low frequency noise based on Radial Basis Function Neuro-Network Approximation Model

*Qi-guo HU[1] , Yu-qian WANG[1] *, Su-ping LI[2]*

[1]Department of Mechanical & Vehicle Engineering, Chongqing Jiaotong University, Chongqing, 400074, China
[2]Department of Traffic & Transportation, Chongqing Jiaotong University, Chongqing, 400074, China

Abstract. For reducing the vehicle compartment low frequency noise, the Optimal Latin hypercube sampling method was applied to perform experimental design for sampling in the factorial design space. The thickness parameters of the panels with larger acoustic contribution was considered as factors, as well as the vehicle mass, seventh rank modal frequency of body, peak sound pressure of test point and sound pressure root-mean-square value as responses. By using the RBF(radial basis function) neuro-network method, an approximation model of four responses about six factors was established. Further more, error analysis of established approximation model was performed in this paper. To optimize the panel's thickness parameter, the adaptive simulated annealing algorithm was im-plemented. Optimization results show that the peak sound pressure of driver's head was reduced by 4.45dB and 5.47dB at frequency 158HZ and 134Hz respec-tively .The test point pressure were significantly reduced at other frequency as well. The results indicate that through the optimization the vehicle interior cavity noise was reduced effectively, and the acoustical comfort of the vehicle was im-proved significantly.

1.INTRODUCTION

In addition to focus on the stability and safety of driving, customers pay more attention to the acoustic characteristics of vehicle compartment. One of the main reasons that affect the acoustic performance of the cab is the prominent low frequency noise which is difficult to be eliminated [1-3]. Previous investigation shows that the thin plates' structure radiation noise which is generated by external excitation in the frequency range of 0Hz~200Hz contributes significantly to the low frequency noise of the cab[1]. In order to reduce the noise in cab and improve vehicle acoustical comfort, the body structure that contributes tremendously to the acoustic should be optimized. The finite element calculation method for optimization is time consuming in view of the model size. Comparatively, the approximation model optimization method is of high efficiency[1-3]. In paper [3], the thickness of plate was designated as the design variable. The minimum pressure at the driver's right ear and minimum body mass was defined as the optimization objective. The approximation model was established by response surface method to optimize the acoustic performance. In paper [4], the second response surface approximation model was established for the optimization of acoustic and vibration. The method of using response surface method to establish the approximation model has the characteristics of simplicity expression and high computational efficiency. However, it fails to guarantee the accuracy of the model while the number of samples is limited.

This paper is organized as follows. In section 2, considering the low frequency noise of the cab, the optimal Latin Hypercube Sampling (LHS) method was used to collect sample data and the RBF neural network method was implemented to establish the approximation model. In this way the approximation model of high accuracy can be established with a small amount of sample data and the acoustic optimization efficiency can be improved. In section 3, based on the established approximation model, the adaptive simulated annealing algorithm was used to optimize the thickness parameters of the car body to improve the acoustic performance. In section 4, the simulation experiment results of vehicle compartment low frequency noise is conducted and more details are analyzed. In section 5, the conclusion is illustrated.

2.MODELING

2.1 Finite Element Model

The complex car body structure is composed of thin plates. The finite element model was simplified and meshed by shell element with the premise of assuring the main body structure's characteristics. Considering the calculation precision and the computation burden, the size of the grid unit is designated as 10mm. The closed chamber of the car body was meshed by the tetrahedral element. As the seats influenced on the acoustic characteristics of cavity sound field[3], the seat model was also established while establishing the finite element model. Based on the principle that the acoustic grid unit length contain at least six units in each wavelength, the grid unit length is designated as 50mm~140mm. The established finite element model of cavity sound field is shown in Figure 1, which contains 29467 tetrahedral elements and 6229 nodes, as the car body structure is composed of 444150 units and 464592 nodes.

* Corresponding author:haqiankaka@163.com

DOI: 10.1051/matecconf/201713900004

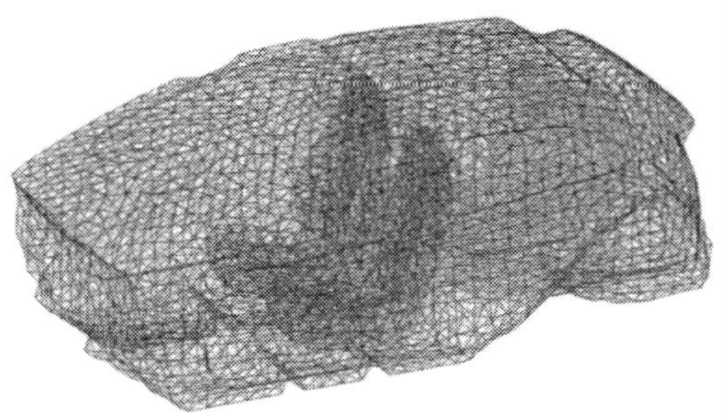

Fig. 1. Finite element model of cavity sound field

2.2 Approximation Model

The sound pressure response output point was defined at the driver's head D, deputy driver's head P, left and right rear passenger head RL, RR[6]. The engine force at 0Hz-300Hz and the suspension force caused by the random road incentive force was loaded at the same time as the external excitation. Through the finite element simulation, the A-level weighting acoustic pressure curve at test point is shown in Figure 2.

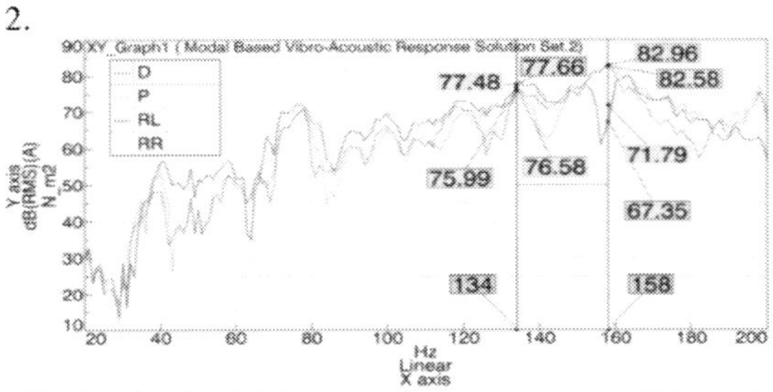

Fig. 2. A-level weighting acoustic pressure curves of four test point location

Based on figure 2, the response analysis of acoustic coupling based on the mode shows that at the low frequency of 0Hz-200Hz, the sound pressure of cab test points is high near 158Hz, 134Hz, 198Hz, 152Hz and 78Hz.

2.2.1 Collection of sample data

The analysis of the panel acoustic contribution[9] shows that where sound pressure frequency is high, the body floor, the ceiling around, the front door and the inner panel contribute mainly to acoustics performance of test point. In order to reduce the pressure in cab and avoid the excessive increase of body mass as well as the seventh rank modal frequency, the body mass, the seventh rank modal frequency, the sound pressure peak value and the RMS(root mean square) of sound pressure at the driver's head were taken as the response of test design. Each group's level value at test design matrix is designated as input parameter.

In the sampling experiment, the thickness parameter of cab floor(t_1),rear floor (t_2), the front panel (t_3), ceiling (t_4), front door inner plate (t_5) and rear door inner plate (t_6) were considered as a factor, whose design range is 0.5mm≤ti≤2mm. In order to establish the accurate approximation model by less sampling point, the optimal Latin Hypercube Sampling(LHS) method was used for test designing[10]. Taking 30 levels for each factor in the range, there are 30 level assemblies of design matrix through opt-LHS.

The optimal LHS improved the weakness of random LHS. It has the ability to allocate all test points well-distributed in design space. It has excellent filling property and proportionality. It can achieve a high precise between response and factor.

2.2.2 RBF(radial basis function) neural network approximation model

Based on the data achieved in part 1, six factors were designated as the input parameter and four responses were designated as output parameter. The approximation model was established by RBF neural network. Compared with other approximation model, the RBF neural network model has strong nonlinear approximation ability. It doesn't rely on mathematical assumptions. More over, it has the most optimal approximation ability for functions and can approximate any continuous function at any accuracy.

Taking the peak sound pressure D of the approximation model plate thickness t_1, t_2, t_3, t_4, t_5, t_6 as the example, the part of the three-dimensional surface is shown in figure 3.

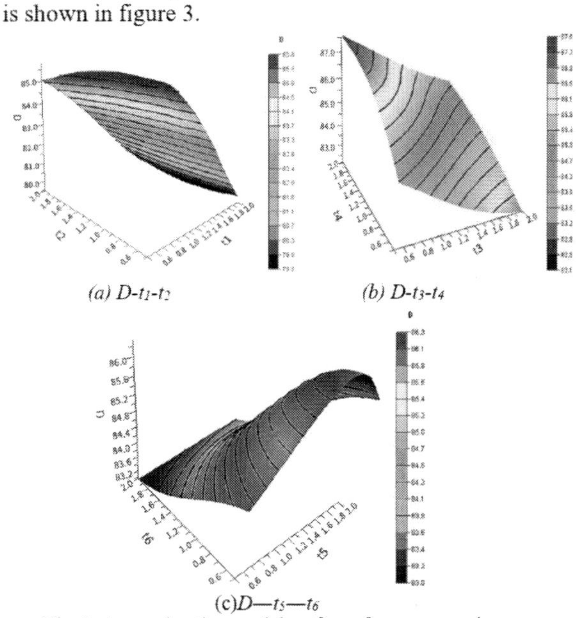

(a) D-t_1-t_2 *(b) D-t_3-t_4*

(c)D—t_5—t_6

Fig. 3. Approximation model surface of pressure peak

2.2.3*Error Analysis*

In order to estimate the accuracy and reliability of the RBF neural network approximation model, 20 non-interpolation points in the design variable space were used as error checking samples to analyze the error of the approximation model. The accuracy of the model is evaluated by four indexes: mean error, maximum error, RMS error and multiple correlation coefficients. The requested value range is listed in table 1.

Table 1.The requested value range

	Lower limit	Upper limit

 DOI: 10.1051/matecconf/201713900004

Mean error	0	0.2
Max error	0	0.3
RMS error	0	0.2
Mutli-correlation coefficient	0.9	1

The error analysis results of the RBF neural network model are shown in Table 2.

Table 2. Error analysis index

	mean error	maximum error	RMS error	multiple correlation coefficient
D	0.01153	0.03033	0.01423	0.9976
drms	0.01383	0.04563	0.01816	0.099608
m	0.01413	0.03206	0.01729	0.99531
sf	0.02195	0.06341	0.0259	0.9936

According to the error analysis index, mean error, maximum error and the RMS error of approximation model are far less than the standard error. The multiple correlation coefficient of the output response is very close to 1, which indicates that the model error is neglectable.

The corresponding input was designated as the input parameter of error check samples. All of the predicted value, such as the body mass, seventh rank modal frequency, peak pressure and acoustic pressure RMS at driver's head were calculated by approximation modal. The relationship between the true and predicted value of the four response points are shown in figure 4.

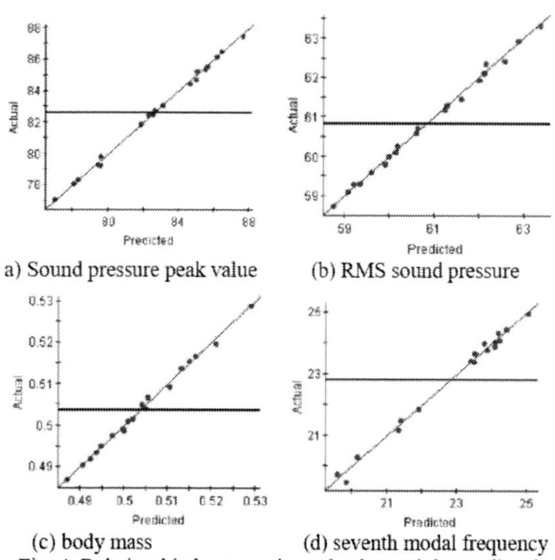

a) Sound pressure peak value (b) RMS sound pressure

(c) body mass (d) seventh modal frequency

Fig. 4. Relationship between the real value and the predicted value

As shown in Figure 4, there is tiny difference between the predicted value and the value calculated by the approximation model, which further proves that the model is of high accuracy.

Briefly, the sampling points were collected by opt-LHS test design. The approximation model was established by RBF neural network. The input parameter was designated as panel thickness. The output response was designated as the body weight, seventh rank modal frequency, the peak pressure and the acoustic pressure RMS at driver's head. Through the error analysis, the established approximation model has the characteristic of small error and high precision, which is able to meet the modeling requirements and replace the finite element model for optimization design.

3.OPTIMIZATION

3.1 Mathematical Optimization Model

Based on the established RBF neural network approximation model, the optimization objective function was created to minimum the peak value of sound pressure at driver's head. The design variables were defined as plate thickness parameters $t_i(1 \leqslant i \leqslant 6)$.

The RMS of sound pressure can reflect the overall level of cab pressure. In order to prevent the increase of cab sound pressure caused by the change of body structure, the RMS value of sound pressure at driver's head which is less than or equal to 60dB was defined as one of the constraints.

The change of the plate thickness would lead to the change of the weight of the car body. Considering the requirement of light-weighted design, the weight should be limited within a certain range. In the original design, the body structure mass is 0.4802t. In the optimization design, the body mass should be limited less than or equal to 0.5t.

As the change of body plate thickness, seventh rank modal usually locates at the front door while sampling the seventh rank body modal frequency response, whose main frequency is 24Hz. However, there are several individual factor level combination conditions at which seventh rank modal frequency is too low. If the modal frequency of car door is too low, the car door will be easily affected by external load excitation which causes a high cab sound pressure. So the limitation of seventh rank modal frequency is:

$$sf \geq 24\text{Hz}$$

sf denotes car body seventh rank modal frequency.

The mathematical function of the optimization problem can be expressed as:

$$\min\ D = f(t_1, t_2, t_3, t_4, t_5, t_6)$$

$$s.t. \begin{cases} drms \leq 60\text{dB} \\ m \leq 0.5\text{t} \\ sf \geq 24\text{Hz} \\ 0.5\text{mm} \leq t_i \leq 2\text{mm} \end{cases} \quad (1)$$

3.2 Optimization Theory

Based on the plate thickness, adaptive simulated annealing algorithm was used for optimization. Adaptive simulated annealing (ASA) is an efficient global optimization algorithm for solving nonlinear optimization problems with multiple peaks and non-smoothness [12].

The way of implementing ASA algorithm is:

(I).Generate an initial solution $\omega \in W$ randomly. The Initial temperature $T_{accept}(0)$ of acceptance probability function equals $J(\omega)$. The initial temperature of generated probability function is defined as $T_{i,gen}(0)$, $1 \le i \le n$. Both of them were set to 1. The control parameter c is given by the user. Annealing times $k_i (1 \le i \le n)$ and k_a are both set to 0.

(II).The new solution is generated on solution space by following equation.

$$\omega_i^{new} = \omega_i^{old} + q_i (U_i - L_i) \quad (2)$$

Where, $1 \le i \le n$, $\omega^{new} \notin W$.

$$q_i = \mathrm{sgn}(v_i - \frac{1}{2}) T_{i,gen}(k_i)[[1 + \frac{1}{T_{i,gen}(k_i)}]^{|2v_i - 1|} - 1]; \quad (3)$$

v_i denotes the uniformly random distributed variables from [0,1]. If the new generated solution $\omega^{new} \notin W$, abandon and regenerate a new solution till it satisfies $\omega^{new} \in W$.

The acceptance probability function of new solution ω^{new} is determined by the following equation.

$$P_{accept} = \frac{1}{1 + \exp((J(\omega^{new}) - J(\omega^{old}) / T_{accept}(k_a))} \approx \exp(-\Delta J / T_{accept}(k_a)) \quad (4)$$

From the above we obtain that $\Delta J = J(\omega^{new}) - J(\omega^{old})$. If $P_{accept} \ge P_{unif}$, the new solution is acceptable, otherwise abnegated. P_{unif} represents the uniformly distributed variable in [0,1].

(III).Every N_{accept}(th) times while the solution is accepted, tempering is processed. The following equation is implemented for calculating sensitivity of each variable, which describes the gradient variation.

$$s_i = \frac{J(\omega^{best} + e_i \delta) - J(\omega^{best})}{\delta}, 1 \le i \le n \quad (5)$$

Where ω^{new} is the optimal solution by now. δ is the step size. The i(th) element in n-dimension vector e_i is set to 1, else 0. Find $s_{max} = \max\{s_i;\ 1 \le i \le n\}$. The forming temperature $T_{i,gen}$ of each parameter is reset by factor s_{max}/s_i. The annealing time k_i is also reset.

$$\begin{cases} T_{i,gen}(k_i) = \frac{s_{max}}{s_i} T_{i,gen}(k_i) \\ k_i = [-\frac{1}{c} \log \left[\frac{T_{i,gen}(k_i)}{T_{i,gen}(0)}\right]]^n \end{cases} \quad (6)$$

At the same time, reset $T_{accept}(0)$ as final accepted cost function. Reset $T_{accept}(k_a)$ as $J(\omega_{best})$. The tempering time is reset by following equation.

$$k_a = [-\frac{1}{c} \log[\frac{T_{accept}(k_a)}{T_{accept}(0)}]]^n \quad (7)$$

(IV).The annealing process is proceeded at every N_{genera}(th) generating point.

$$\left.\begin{cases} k_i = k_i + 1 \\ T_{i,gen}(k_i) = T_{i,gen}(0) \exp(-c k_i^{\frac{1}{n}}) \end{cases}\right\} 1 \le i \le n$$

and

$$\begin{cases} k_a = k_a + 1 \\ T_{accept}(k_a) = T_{accept}(0) \exp(-c k_a^{\frac{1}{n}}) \end{cases} \quad (8)$$

The process ends if the parameters stay the same every continuously tempering and annealing time, or, the evaluation number of cost function reaches the minimum value, as well as the error be less than the given number. Otherwise returning to (II).

4.RESULT AND DISCUSSION

4.1 Optimization Results

The initial temperature is designated as 1°C. The convergence check interval is designated as 5. The decreasing relative ratio of temperature parameter is designated as 1. The temperature loss function drop ratio is designated as 1. The allowed calculation failure time is 5. After 10112 times of iteration, the optimal value of the plate thickness is obtained in the factor design space while the peak sound pressure at driver's head is the minimum.

The plate thickness of the finite element model was chosen by the determined value of the plate thickness parameter. The body mass, structural modal and sound pressure distribution of the cab were calculated. The response values calculated by the approximation model and the finite element model are compared in table 3.

Table 3. Comparison of response values

	Approximation model	Finite element model	Error ratio
D/dB	76.58609	78.43	-2.35%
drms/dB	58.70524	59.02441	-0.54%
m/t	0.499967	0.498	0.39%
sf/Hz	24.03601	24.1824	-0.61%

According to Table 4, the error rate of the response calculated by the approximation model and finite element model is very small. It shows that the approximation model established by RBF neural network is of high accuracy, and the optimization result is of high reliability.

4.2 Discussion

The before and after the optimization value including the thickness parameter of the body plate, body mass, seventh rank modal frequency of the vehicle body, peak value of sound pressure and RMS value of the sound pressure at the driver's head were compared in this part. The results are shown in table 4.

Table 4. Before and after optimization Comparison

Variable	Before optimization	After
t_1/mm	0.8	2
t_2/mm	0.7	0.77
t_3/mm	0.9	0.94
t_4/mm	0.8	0.5

DOI: 10.1051/matecconf/201713900004

t_5/mm	0.9	1.58
t_6/mm	0.9	0.58
D/dB	82.58	78.43
drms/dB	61.32021	59.02441
m/t	0.4802	0.498
sf/Hz	23.2706	24.1824

According to the comparison, the peak value of the maximum sound pressure at the driver's head D is decreased by 4.15dB, and the RMS sound pressure is decreased by 2.3dB. The weight of the car body is only increased by 17.8kg, which can meet the requirements of lightweight design. The seventh rank modal frequency of the body is 24.1824Hz, which can avoid the radiation noise caused by the presence of a lower local mode in the car door caused by external excitation. In order to analyze the change of sound pressure in cab cavity at 20Hz~200Hz, the before and after optimization A-level weighting acoustic pressure curve at the driver's head D were compared in figure 5.

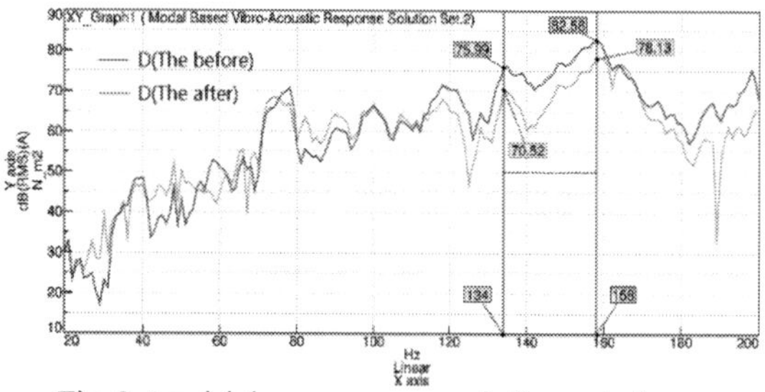

Fig. 5. A weighting pressure curves before and after optimization at driver's head

As shown in Figure 5, the sound pressure at the driver's head decreased significantly. The max sound pressure peak at 158Hz is decreased by 4.45dB. The sound pressure peak at 134Hz is decreased by 5.47dB. The measuring point sound pressure is decreased at different degree. It illustrates that the cab cavity noise is effectively decreased and the acoustic performance is improved through the optimization of body plate thickness.

5.CONCLUSION

(1) The optimal Latin hypercube sampling was used for acquiring sample data. The approximation model established by RBF neural network has the characteristics of small error and high precision.

(2) After optimization, the thickness parameter of the body plate, body mass, seventh rank modal frequency of the vehicle body, peak value of sound pressure and RMS value of the sound pressure were decreased at different degree. It illustrates that through the optimization of body plate thickness, the cab cavity noise is effectively decreased, and the acoustic performance is improved.

(3) Based on the established RBF neural network approximation model, the objective function was designated as the minimum of peak sound pressure at driver's head. Simulation results show that the adaptive annealing algorithm used for optimizing the plate thickness parameter to reduce the sound pressure of cab is a feasible and effective method.

ACKNOWLEDGEMENTS

This research is funded by Chongqing Research program of Basic Research and Frontier Technology (No. :cstc2015jycjBX0133) and National Natural Science Foundation of China(No. :51375519).

REFERENCE

1. Gonzalez A, Ferrer M, Diego M D, et al. Sound quality of low-frequency and car engine noises after active noise control. [J]. Journal of Sound & Vibration, 2003, 265(3):663-679.
2. Liu Yu, Yu Fan, Liu Jiang. Modal analysis of vehicle ride acoustic-structure interaction. [J] Journal of noise and vibration control, 2005, 67(5):38-40.
3. Bai Shengyong, Qi Xiaoxiong, Ding Yulan. Acoustic modal analysis of passenger car. [J] Journal of Tongji University, 2000, 28(1):206-209.
4. Zang Xianguo. Structural acoustic performance optimization based on modal shape and response surface method. [D]Changsha: Hunan University, 2011.
5. Zhu Caichao, Qin Datong. Coupling analysis and control of car body structure vibration and interior noise. [J] Journal of Mechanical Engineering. 2002, 38 (8): 54-58.
6. Sung S, Chao S, Lingala H, et al. Structural-Acoustic Analysis of Vehicle Body Panel Participation to Interior Acoustic Boom Noise[C]// SAE 2011 World Congress & Exhibition. 2011:650–657.
7. Qi Chang, Zhou Hong. Sound and vibration optimization of vehicle body panels based on the contribution of vehicle interior acoustic field. [J] Automotive Engineer, 2015, 37 (12):1438-1444,1432.
8. SHEN Yanyou, JIA Mingpin, ZHU Lin. Research on noise analysis and noise reduction optimization of centrifugal leaf suction machine based on Latin hypercube sampling. [J] Journal of Vibration and shock. 2016, 35(15):93-97.
9. LI supping, HU qiguo, HU haibo, LUO tianhong. Vehicle's Low Frequency Noise Prediction and Body Panel's Acoustic Contribution Analysis. [J] Journal of noise and vibration control. 2016, 36(4):103-107.
10. Lai Yuyang. Parameter optimization theory and examples based on Isight. [M] BeiJing: Beijing University of Aeronautics and Astronautics Press.2012.
11. Ingber L. Adaptive simulated annealing (ASA): Lessons learned. [J]. Control & Cybernetics, 2002, 25(1):33-54.
12. Liu P, Liu Y L, Fei-Hong Y U. Thin film optical parameters determination using adaptive simulated annealing algorithm [J]. Optical Instruments, 2008.

3D road surface reconstruction based on point clouds data assimilation algorithm

De-liang Chen[1, 2], and *Yan-yan* Lu[2,*]

[1]College of Geographic and Biologic Information, 210023 Nanjing University of Posts and Telecommunications, China
[2]College of Geographic and Oceanographic Sciences, 210023 Nanjing University, China

Abstract. Urban areas 3D model reconstruction is one of the major fields of application of 3D scanning technologies. In the future, vehicle-based laser scanning, here called mobile laser scanning system, should see considerable use for 3D road environment modelling in urban areas. In this context, one of the main limitations perceived by the mobile laser scanning system is the incompleteness of the sampling. Whenever we scan urban area road environment, the produced sampling usually presents a large number of missing regions. Many algorithmic solutions exist to close those gaps from specific hole filling algorithms to the drastic solution of using water-tight reconstruction methods. In this paper, a method for filling holes of road surface point clouds and generating 3D model of road surface from mobile laser scanning data is developed. The data is classified into road surface, on-road and off-road surface point clouds. Many large holes in the road surface point clouds are filled by using data assimilation algorithm. Then, the road surface is 3D modeled as a triangulated irregular network. It is shown that the whole road surface 3D model is integrated after data processing. The above mentioned steps are applied to a large set of mobile laser scanning data of urban area road environment, in order to obtain the whole urban road surface 3D model.

Keywords: 3D model reconstruction; Point cloud; mobile laser scanning system

1 INTRODUCTION

As scientists, engineers and planners require ever more detailed information about our urban areas the requirement for accurate 3D mapping of road and city structures is likely to increase considerably in the years to come. The technology available for providing 3D mapping may be applied from the ground using static instrumentation or applied from an air or space based platform. Advanced digital mapping tools and technologies are so-called 3D laser scanner are enablers for effective e-planning, consultation and communication of users' views during the planning, design, construction and lifecycle process of urban areas. The regeneration and transformation of cities from the industrial age to the knowledge age is essentially a 'whole life cycle' process consisting of: planning, development, operation, reuse and renewal. In order to enhance the implementation of urban areas solutions during the regeneration and transformation of cities, advanced digital applications can have a significant impact. However, the rapid development of laser scanning has shifted, since the middle of the first decade of the new millennium, from airborne laser scanning (ALS) systems and terrestrial laser scanning (TLS) systems towards the mobile laser scanning (MLS) systems. A MLS system integrate GPS/INS system and data acquisition sensors on a rigid moving platform for collecting laser scanning point clouds from the surroundings of the mapping system have been developed since the early 1990s. A good overview is presented in Barber's research article [1]. At present, the integrated GPS/INS/LS system is the key component in both MLS systems and ALS systems. MLS systems are different from ALS systems which assembled on an airplane or a helicopter, but fixed on platforms such as a car or a van. Compared to ALS systems, which typically collect point clouds with a resolution of 0.5-40 points/m^2 from an altitude of 100-3000 m, MLS systems provide point clouds with a resolution of hundreds or even thousands of points/m^2 from a distance of some dozens of meters. ALS and MLS data-sets complement each other in several ways as they have different viewing geometries. For example, ALS data-sets have become an important source for object extraction and reconstruction for various applications, such as urban analysis (the roofs of buildings) [2-4]; vegetation analysis [5]; landform mapping [6]; DTM generation [7, 8] and forest inventory [9-11]. However MLS data-sets are not only including the application of vegetation analysis [12-14], but best for detecting objects of urban areas, e.g., walls of building and collecting even more information from road surface [15], In the case of urban areas the detection and quantification of road surface is important for the implementation of urban areas solutions during the regeneration and transformation of cities. On the other hand urban road surface models are needed for accurate three-dimensional mapping of urban areas.

* Corresponding author: cdlyy52@163.com

The objective of this paper is to develop a method for automatically road surface point clouds detecting and road surface 3D modelling, and present a data assimilation algorithm for filling holes in road surface point clouds from mobile laser scanning data. The article is organized as follow: we will explain the proposed method which includes road surface extraction, holes filling in road surface point clouds and reconstruct 3D model of road surface in section 2. The experimental results are presented in section 3, and section 4 concludes the article.

2 METHOD

In this section, a developed method for automatic urban area road surface detection and 3D modelling is applied from the MLS point clouds. In our research, the aim is to detecting the road surface point clouds from the raw MLS data, and then modelling 3D road based on the road point clouds. The method is selected to deal with measurement noise (trees, building, car etc.), reorganizing point clouds (road, on-road and off-road) and missing data (holes in road point clouds). Our approach consisted of four processing steps:
(1) The raw MLS data are partitioned along road directions as sub regions called road parts, for which the road surface point clouds extraction is applied. The laser points are roughly classified into road, on-road and off-road. We eliminate the on- road point clouds and the off-road point clouds (e.g. buildings, cars, kerbstones, and parterre).
(2) The road segments are further examined if there is any missing point clouds data (holes). Especially the holes are caused by on-road point clouds. The holes filling method is based on a data assimilation algorithm.
(3) From the extracted road point clouds, we propose an algorithm to fill the scanning holes in the road surface point clouds.
(4) Finally we create a TIN and iteratively shaped the TIN into the smooth and intact road surface point clouds by filling the scanning holes.

2.1 Road surface extraction

The range scanner samples line profiles more than 40m on either side of the motion direction. Hence each profile has points pertaining to road surfaces and points pertaining to objects beyond the road such as trees, parking cars, or curbs within the scanning range. We segment the profiles to ignore points belonging to surfaces from those objects that are not of interest for the mapping. The on-road point clouds mean such objects as moving cars, people and trees which are on the road surface. The on-road point clouds differ from road surface point clouds in terms of their heights. The off-road point clouds mean outside both sides of road belong to such objects as buildings, curbs, parking cars, kerbstones, and parterre. There is a 'distance' between off-road point clouds (both sides) and road point clouds, and the off-road point clouds differ from road surface point clouds in terms of heights. The height information and distance information is then processed into segments that could be considered off-road point clouds. After that, we execute the following two-step algorithm for raw MLS point clouds to extract road point clouds:
(1) Make a distinction between road points and beyond road points. If there is a distance between some point clouds and road surface point clouds, and the height level of those point clouds has changed dramatically. We can make sure that those point clouds are off-surface points which should be assigned as point clouds of no interest;
(2) After step (1), we find out the top point of road surface point clouds, and the height (z) value of the top point is applied as a height threshold. We eliminate points which height value is higher than the threshold value, and the left point clouds are assigned as of interest. This procedure essentially filters out unnecessary data obtained from the laser scanners where the data are not of interest. This step is an important pre-processing step before we can execute surface reconstruction algorithms from 3D scattered point cloud data.

2.2 Algorithm for holes filling

Scan holes occur when parts of the scanned object are obscured or visually inaccessible. Well-planned scanning positions can minimize scan holes but complete avoidance is often, if not in most cases, unachievable. Urban area road environment components which are typically not visible to the scanner are detail surfaces of the decorative attachments, small cavities, ruined road surfaces, road surfaces behind trees and surfaces that are at a level higher or lower than the scanner. Incidentally, the same problem arises when acquiring urban area 3D model and photography for texturing the model. Filling scan holes, somewhat euphemistically referred to as surface augmentation fabricates information where there is none. This is highly questionable for urban road model construction and other applications where full surface detail is required. All these applications require reliable and faithful surface representations and do not tolerate fabricated data. Some hole filling occurs during the meshing process depending on the algorithm used, the choice of holes to be filled is not under the control of the software operator and therefore potentially detrimental. Holes filling can either be done automatically or manually. We present a new holes filling procedure to repairing missing part in the urban area road surface point clouds based on the MLS projection procedure.
We use a statistically-based algorithm that aims at refining the positional and radiometric accuracies of missing point clouds (holes) in the road surface point clouds. This algorithm utilizes the theory of data assimilation to enhance the 3D geo-referencing accuracy as well as fine-tuning the radiometric intensity by means of exploiting the correlation between two oppositely-collected datasets over the same study area.
Data assimilation is an optimization algorithm that is based on the principle of least squares analysis. The purpose of data assimilation is to combine two different datasets or models of the same phenomenon in order to

achieve the best estimate of the true state. This method was proved to be viable by Reichle who used it to integrate satellite remotely sensed snow data with ground field hydrology measurements [16]. Having found this method to be successful for the case of snow data which were collected by two different systems over different scales, the urban area MLS case will, presumably, has much higher probability of success as it was collected by the same laser sensor over the same area and scale under the same ambient conditions. Even more, the two MLS point clouds were collected over relatively same period of time unlike the case of snow data. Apparently, therefore, data assimilation will have better chances of recognizing the underlying correlation between the two MLS point clouds. Those factors, altogether, are the rationale behind choosing data assimilation to enhance the accuracy of road surface 3D modelling.

Suppose that we have two different point clouds V_i and V_j of a certain phenomenon. Let σ_i^2 and σ_j^2 be their respective variances. Now if V is the true state, then our goal to minimize the weighted sum of squared residuals:

$$J=w_i(V_i-V)^2+w_j(V_j-V)^2=(V_i-V)^2/\sigma_i^2+(V_j-V)^2/\sigma_j^2 \quad (1)$$

where

J is known as the cost function and w_i, w_j are the weights

The best estimate V' can be obtained by differentiating equation (1) and equating the derivative to zero, which yields:

$$V'=(\sigma_i^2+\sigma_j^2)^{-1}(\sigma_j^2V_i+\sigma_i^2V_j) \quad (2)$$

It might be of interest to demonstrate the robustness of this model by computing the variance (C) of V':

$$C(V')=(\sigma_i^2+\sigma_j^2)^{-1}C(\sigma_j^2V_i+\sigma_i^2V_j)$$

$$\sigma'^2<\sigma_i^2 \text{ or } \sigma_j^2 \quad (3)$$

which implies that the new model is closer to the true state than either point cloud.

In the event of having multivariable dataset, equation (2) can be generalized as follows (S is symbol of covariance):

$$V'=inv(S(V_i)+S(V_j))*(S(V_j)*V_i+S(V_i)*V_j) \quad (4)$$

As mentioned previously, the assimilation process uses the existing correlation between different observations of a particular phenomenon in order to obtain the best estimate of the true state.

The assimilation method is meant to integrate the attributes of two point clouds that are corresponding to the same geospatial positions where the two point clouds occupy the same 3D coordinates. However, in the case of MLS data, the geospatial positions themselves, in addition to the intensity, need to be assimilated. Since the 3D coordinates of the two point clouds do not coincide (they will coincide only if they are both true state), a new definition should be introduced to decide on which points are to be assimilated.

The MLS point cloud is irregularly spaced by nature, which will be inherited when the two point clouds are blended. The criterion of which point is to be assimilated should be based on the spatial neighborhood. When the two point clouds are merged, points from the first point cloud (point cloud M) might come closer to the points of the second point cloud (point cloud N), and then only the points that are sufficiently close will be assimilated. The status of sufficiently close will be declared for the points that are closer than the minimum spacing of each point cloud individually. If point A and point A' are two points from the forward and reversed data cloud then they will be assimilated if:

AA' < minimum distance of point cloud M AND minimum distance of point cloud N

The minimum distance of point cloud M (N) means the distance between one scan point to another scan point in point cloud M (N).

When that status occurs, the laser sensor is, most likely, targeting the same object (labeled as white), yet missing the exact position because of error measurements. The new point that results from assimilating A and A' will be higher accurate and closer to the true state as indicated by equation (3). However, there is a subtle downside for this approach as it suggests replacing two points with only one point which reduces the resolution of the MLS point cloud data when it is least expected to improve it. This situation can be avoided by changing the AND in the criterion to OR: AA' < minimum distance of point cloud M OR minimum distance of point cloud N.The following cases will be encountered during processing when A and A' are sufficiently close:

Case (I): The distance between A and A' is less than the minimum distance of M point cloud and lager than that of the N point cloud. In this case A will be replaced with the assimilated point while A' is retained. No point is lost.

Case (II): This is just the opposite of Case (I). Again, no point is lost.

Case (III): The distance between A and A' is less than the minimum distance of both the M point cloud and the N point cloud. In this case A' will be replaced with the assimilated point. To avoid reducing the resolution we can only change the height level of point A'.

Having done that, the total number of points will be exactly equal to the sum of the individual point cloud. Provided that the average distance of the MLS point clouds of this research is 5 cm and the minimum distance is about 1 cm, the positioning accuracy is expected to be within cm level.

2.3 Road surfaces reconstruction

A road surface model will be created by extracting all the road surface points in the point cloud. Then we create a triangulated irregular network (TIN) and iteratively shaped the TIN into a smooth surface by filling holes. The TIN model represents a surface as a set of contiguous, non-overlapping triangles. With each triangle the surface is represented by a plane. The TIN data structure is based on two basic elements: points

DOI: 10.1051/matecconf/201713900005

with x, y and z values and a series of edges joining these points to form triangles. In this paper, the triangles are made from point clouds. TIN's triangulation method satisfies the Delaunay criterion. Delaunay triangulation is a proximal method that satisfies the requirement that a circle drawn through the three vertices of a triangle will contain no other point. In preparation for triangulating the road surface, we find the outer bounds of the road based on the previously extracted road surface. A Delaunay triangulation is formed from the edge of road to the other side. From the initial road points, a Delaunay triangulation is created to represent the surface. For representing surface, the hole areas cannot create Delaunay triangulation should be found, and the hole should be filled based on the previously proposed algorithm. After filling all the holes failing to meet the requirement, we form a new TIN. This is done repeatedly for as long as the presenting road surface without holes.

3 EXPERIMENTAL RESULTS

3.1 Mobile Laser Scanner Datasets

The point cloud was collected with the FGI Roamer mobile laser scanning system developed at the Finnish Geodetic Institute. The Roamer consists of a Faro LS 880 laser scanner with a measurement frequency of 120 kHz and a NovAtel HG 1700 SPAN58 INS system. With slightly modified hardware for the standard FARO LS, it provides so-called tunnel mode, or profile measurements, synchronized with external positioning and data logging systems. This information is needed to derive the position and attitude information for each 3D point produced by the laser scanner. The mirror rotation frequency, or scan rate of the scanner on the Roamer can be set to 3-30 Hz, thus giving a vertical angular resolution of 0.0096-0.096 degrees (0.17-1.7 mrad), respectively. Corresponding point spacing at a typical scanning range of 15 meters in road mapping is thus 2.5-25mm in the scanning plane. The dataset consists of mobile laser scanning point clouds over a 1700 m closed loop at Espoonlahti neighbourhood in Espoo, about 15 km west of Helsinki. The data extent covers five road segments of Espoonlahdenkatu, Espoonlahdentie, Merenkulkijankatu, Kipparinkatu and Espoonlahdenranta streets in addition to Lippulaiva shopping mall (**Fig.** 1). The number of points in the raw MLS data is 22,559,451.

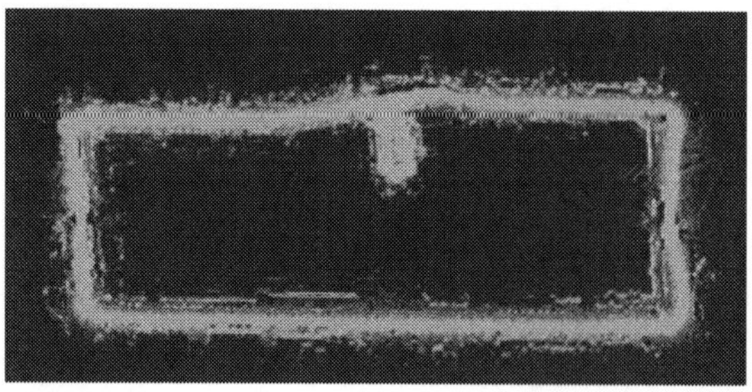

Fig. 1. Bird eye's view map of the raw MLS data extent covers five urban area road point clouds.

3.1 Mobile Laser Scanner Datasets

In this section the results of the testing of the presented method are showed. The resulting point clouds scan typically suffer from large amounts of missing data are focused. The proposed procedure by other researchers always ignores the surface reconstruction from incomplete point clouds. Based on visual inspection, the proposed data assimilation algorithm is able to fill hole using on-road point cloud, and the model reconstructed based on the hole filling point cloud is more completely than before (**Fig.** 2(a) and 7(b)).

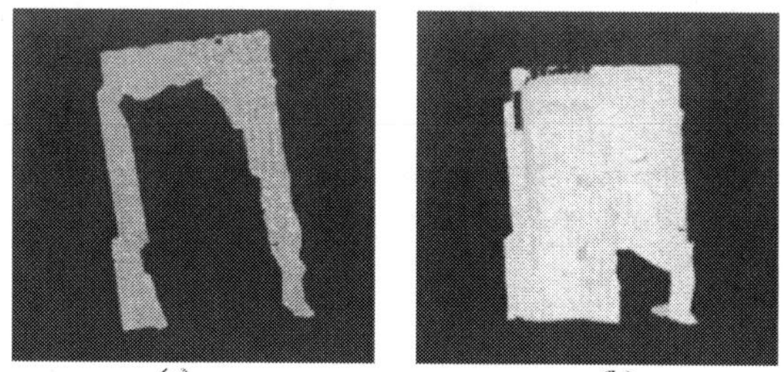

(a) (b)

Fig. 2. (a) The model is badly reconstructed based on the point cloud with a missing part. (b) The good model reconstructed after holes filling.

The road surface point clouds can be extracted after the on-road point clouds and off-road point clouds are eliminated. We check the road surface point clouds if there are holes areas or not. We fill the hole areas using data assimilation algorithm in order to obtain the road point cloud without incomplete part. The hole areas is a huge problem for creating TIN. After making sure there are no hole areas in the road surface point clouds, we create the complete road surface TIN. The final road surface 3D model bird's eye view map is shown in **Fig. 3** from the raw MLS data **Fig. 1**.

Fig. 3. Final road 3D model after road surface extraction, holes filling and surface reconstruction.

4 CONCLUSIONS

Accurate road surface information is needed in numerous applications, such as road maintenance, 3D city and noise modelling, location-based and driver assistance systems. MLS supplements conventional laser-based mapping techniques; it produces more accurate and denser data than ALS and is much more efficient than TLS when large areas are mapped. In this paper, a method for filling holes of road surface point clouds and generating model of road surface from MLS data are proposed. This is vital since such MLS system allows fast processing of long corridors, e.g. thousands of kilometers of road. Such information will be needed in future 3D road environment mapping, e.g. in automatic updating of the 2D and 3D navigation data for consumer and business applications. Visual experiments confirmed that the methods developed hole filling in road surface point clouds. Future studies will demonstrate the correctness of such methods with larger experimental sample. The methods will be further developed to be applicable for road crossings, road marking, building and other kinds of urban road environment. Also median islands and pedestrian refugees cause problems to the current methods.

Acknowledgments

This work was sponsored by National Natural Science Foundation of China (Grant No. 41601497), NUPTSF (Grant No. NY214195), Jiangsu Provincial Bureau of Surveying Mapping and Geoinformation Research Project (Grant No.JSCHKY201604), Jiangsu Graduate Student Research Innovative Projects (Grant No. KYZZ16_0046).

References

1. Barber D., Mills J. and Smith-Voysey S. Geometric validation of a ground- based mobile laser scanning system. ISPRS Journal of Photogrammetry and Remote Sensing **63**,1 (2008)
2. Rottensteiner F. Status and Further Prospects of Object Extraction from Image and Laser Data. Joint Workshop on Urban Remote Sensing: 2009 Joint Urban Remote Sensing Event. 20-22 May 2009, Shanghai, China
3. Huang X., Zhang L. P. and Gong W. Information fusion of aerial images and LIDAR data in urban areas: vector-stacking, re-classification and post-processing approaches. International Journal of Remote Sensing **32**, 1 (2011)
4. Kim J. and Muller J. P. Tree and building detection in dense urban environments using automated processing of IKONOS image and LiDAR data. International Journal of Remote Sensing **32**,8 (2011)
5. Reitberger J., Krzystek P. and Stilla U. Analysis of full waveform LIDAR data for the classification of deciduous and coniferous trees. International Journal of Remote Sensing **29**,5 (2008)
6. Clarke J., Gibson D. and Apps H. The use of LiDAR in applied interpretive landform mapping for natural resource management, Murray River alluvial plain, Australia. International Journal of Remote Sensing **31**,23 (2010)
7. Sithole G. and Vosselman G. Experimental comparison of filter algorithms for bare-Earth extraction from airborne laser scanning point clouds. ISPRS Journal of Photogrammetry and Remote Sensing **59**,1-2 (2004)
8. Liu X., Zhang Z., Peterson J. and Chandra S. Large area DEM generation using intensity airborne LiDAR data and quality control. The 8th International Symposium on Spatial Accuracy Assessment in Nature Resources and Environmental Sciences. 25-27 June 2008, Shanghai, China
9. Hyyppa J., Kelle O., Lehikoinen M. and Inkinen M. A segmentation-based method to retrieve stem volume estimates from 3-D tree height models produced by laser scanners. IEEE Transactions on Geoscience and Remote Sensing **39**,5 (2001)
10. Naesset E. Predicting forest stand characteristics with airborne scanning laser using a practical two-stage procedure and field data. Remote Sensing Environment **80**,1 (2002)
11. Naesset E. Estimating above-ground biomass in young forests with airborne laser scanning. International Journal of Remote Sensing **32**,2 (2011)
12. Jaakkola A., Hyyppa J., Kukko A., Yu X. W., Kaartinen H., Lehtomaki M. and Lin Y. A low-cost multi-sensoral mobile mapping system and its feasibility for tree measurements. ISPRS Journal of Photogrammetry and Remote Sensing **65**,6 (2010)
13. Puttonen E., Jaakkola A., Litkey P. and Hyyppa J. Tree Classification with Fused Mobile Laser Scanning and Hyperspectra Data. Sensors **11** (2011)
14. Rutzinger M., Pratihast A. K., Elberink S. J. O. and Vosselman G. Tree Modelling From Mobile Laser Scanning Data-Sets. Photogrammetric Record **26**,135 (2011)
15. Pu S., Rutzinger M. and Vosselman G. Recognizing basic structures from mobile laser scanning data for road inventory studies. ISPRS Journal of Photogrammetry and Remote Sensing **66**,6 (2011)
16. Reichle R. H. Data assimilation methods in the Earth sciences. Advances in Water Resources **31**,11 (2008)

Research and Application of Satellite Integration Design Based on MBD

Wu *Ruilan*, , *Yang Liuying*, *Xia Yongquan*, *Jin Xiaokang*, *Jiang Hao*
Institute of Telecommunication Satellite, China Academy of Space Technology, Beijing 100094, China

Abstract: As satellite integration design and implementation become more and more complex and difficult, a MBD model structure suit for satellite integration design is proposed in this paper, to ensure data consistency and the design work more efficiency. In order to help designers build normative MBD model efficient , a rapid MBD modeling tool has been developed and already been validated in several satellite development projects.

1 General Instructions

Satellite integration is the detailed design of mechanical aspects in satellite overall design. Integration implementation is the process that integration technologist and workers assemble the qualified onboard products into a complete satellite according to the integration design and technological requirements. [1]

With more tasks, shorter development cycle, while the mission is becoming more and more complex, satellite integration design and implementation is getting much more difficult than ever before. These challenges are meant to introduce a new way of satellite integration design.

Model-based definition (MBD) is an advanced digital definition methods that make all related design information structured inside native 3D models. A satellite integration design solution and implementation based on MBD was proposed in this paper, and a MBD modeling tool has been developed to help designers build MBD models of satellite products quickly and normatively.

2 Satellite Integration Design

Satellite integration design is the design process of planning and promoting demands on satellite AIT requirements, and finally writing them on documents and drawings, according to requirements from system, subsystem, and comprehensive test. [2] Following is the concrete design content of satellite integration:

- Mounting hole design of onboard product on deck, including the number, aperture size, depth, specification, and etc;
- Equipment assemble design;
- Cable route design and assemble design;
- Pipeline of propulsion system route design and assemble design;
- Grounding system design.
- Onboard subordinate parts design.

Satellite integration design and implementation, take almost half of the development cycle, is an important link in satellite development process. The content mentioned above is complex and complicated, traditional design method and expression can hardly meet more and more satellite program research needs:

- Integration design datum are scattered in different documents and drawings, which is not good for later maintenance.
- Easy to lead important information miss because integration technologists have to collect necessary information from several documents and drawings to compile process.
- Rely on documents and drawings too much that the presentation sometime is not accurate and easy to form misunderstanding.
- Low degree of automation that much data extraction are carried out manually which lead to low inefficient.

3 MBD Solution

3D models have been widely used in satellite integration design today, but 2D drawing is still used for detail description. Because of the existence of both 3D models and 2D drawings, designers can hardly ensure data from two sources consistency.

There are the advantages by adopting full 3D design and manufacturing mode without 2D drawing:

- Ensure data consistency;
- Reduce the possibility of mistakes by misunderstanding;
- Shorten product development cycle;
- Promote concurrent engineering and product lifecycle management.

Full 3D design and manufacture is not simply to add information expressed in 2D drawing into 3D models, but to explore a much more efficient design information representation that easy to understand by making full use of 3D model advantages.

MBD technology provides a way to make satellite full 3D digital integration design reality. With MBD, most of the data related to the product can be stored in the 3D CAD model. At present, there is not a unique approach for MBD scenarios, this paper focus on establishing an approach suitable for satellite integration design and implementation. There are two types of satellite product often used :

- Products only for satellite integration, such as the products delivered directly by the suppliers. During satellite integration design process, these products only need information of various interface like mechanical interface, thermal control interface, electrical interface, etc.
- Products for manufacture and satellite integration, such as onboard subordinate parts. Those products need both manufacturing definitions about explicit dimensions, functional tolerances, and manufacturing annotations for manufacture, and information of various interfaces for satellite integration.

From above description, not only manufacturing definition, but information of mass property, mechanical interface, thermal control interface, and electrical interface should be embedded in 3D models. A MBD structure based on satellite integration design was established, just as Figure 1 shows.

Following is the detailed data added in native 3D model:

- Basic information
  - product designation;
  - product code;
  - supplier;
  - Inheritance;
  - Maturity;
  - etc.
- MCI(mass, centroid, inertia) information
  - mass;
  - centroid location;
  - moment of inertia.
- Manufacturing oriented information
  - functional tolerancing and annotations (FT&A);
  - manufacturing requirements.
- Integration implementation oriented information
  - Mechanical interface information
  - Electrical interface information
  - Thermal control interface information
- Other information

DOI: 10.1051/matecconf/201713900006

Present CAD technologies make manufacturing oriented information easily added in the 3D model by FT&A module of 3D CAD software. However, integration implementation oriented information relies heavily on products' characteristics, production mode and process. There is no standard MBD model format and content for all the products. MBD model structure of satellite products should be customized based on their own characteristics.

4 MTERICS

In order to build normative MBD 3D models efficiently, METRICS(Mechanical Engineering Tri-dimensional Interface Control System) toolset was developed in CATIA environment. METRICS can help designers build the MBD model of an onboard product by writing all needed data into the native 3D model in document-driven form. The data written in is stored in geometric set form and exchanged while the model used. Figure 2 showed a typical MBD model structure built by METRICS.

Following is the data written in the MBD models by METRICS:

- Product basic information: such as product designation, boxcode, supplier,
- MCI(Mass, centroid, inertia) information,
- Mechanical mounting information;
- Mechanical grounding or insulation information;
- Basic thermal performance information;
- Electrical connector information;
- RF Ports definition and identification;
- Special announcement or description of the product.

Once satellite configuration and layout are completed, integration designer have to extract and summarize related information from product MBD models as soon as possible, then submit to structural designer or satellite integration technologist. Therefore, associated tools are developed:

- **METRICS_AIT** is used to extract mechanical mounting information from all models in structured tree.
- **METRICS_MCI** is used to extract MCI information from models in the local coordinate system, then convert to satellite mechanical coordinate system and do summary calculate.
- **METRICS_Drilling_Map** is used to extract mounting hole information from models in local coordinate system, then convert to satellite mechanical coordinate system and do summary calculate.

METRICS toolset has been validated in several satellite development projects and is constantly improved perfect according to the application. More data beside above mentioned can be involved in, to make MBD model more complete, not only used in satellite integration design, but satellite concurrent engineering and lifecycle management.

5 Conclusion

A MBD model structure of satellite products was proposed in this paper, which contained most onboard products' integration and implementation information. The MBD model used in satellite design made satellite lifecycle 3D digital design, manufacture and management reality.

A rapid MBD modeling toolset METRICS was developed to help integration designers build normative MBD 3D models efficiently and had been validated efficiently and accurately in several satellite programs.

In order to provide better service for satellite concurrent engineering and lifecycle management, more data beside this paper mentioned can be involved in the MBD model structure for the better application. It can be predicted that MBD is carrying more information for design, simulation and implementation to form an actual virtual prototype. Then TBD technology will be developed into MBSE technology.

Acknowledgements

This work has been supported by Project of Research and Application of Knowledge Management Technology in Aerospace Major Equipment Innovative Development Engineering under the Project Number 2015BAF18B01.

Reference

[1] M. Alemanni, F. Destefanis, E. Vezzetti. Model-based definition design in the product lifecycle management scenario[J]. International Journal of Advanced Manufacturing Technology, 2011(1/4): 1-14.

[2] Leilie Yin, Dunbing Tang, Qi Wang, Inayat Ulah, Haitao Zhang. Engineering Change Management of Product Design Using Model-Based Definition Technology[J].Journal of Computing and Information Science in Engineering, 2016,17(4): 041006-041006-19

[3] Wenhua Zhu, Matthieu Bricogne, Alexandre Durupt, Sebastien Remy, Baorui Li, Benoit Eyanard. Implementations of Model Based Definition and Product Lifecycle Management Technologies: a Case Study in Chinese Aeronautical Industry[J]. IFAC-Papers On Line,2016, 49(12):485-490

[4] Xu Fuxiang. Introduction to Satellite Engineering[M]. Beijing, China Astronautic Publishing House, 2003.

[5] Chen Xiangdong, Zhang Wangjun, Pan Yanhua. Study on Spacecraft Digital Integration Design[j]. Spacecraft Engineering, 2008, 17(6):64-67.

[6] Feng Tongneng, Wang Zhenyang, Mengjinhui. Application and Development of MBD in Digital Collaborate Manufacturing [J]. Journal of Nanjing University of Aeronautics & Astronautics, 2012,44:132-137.

[7] Chandrasegaran S.K., Ramani K., Sriram R.D., Horváth I., Bernard A., Harik R.F.. The Evolution, Challenges, and Future of Knowledge Representation in Product Design Systems[J]. Computer-Aided Design, 2013, 45(2): 204–228.

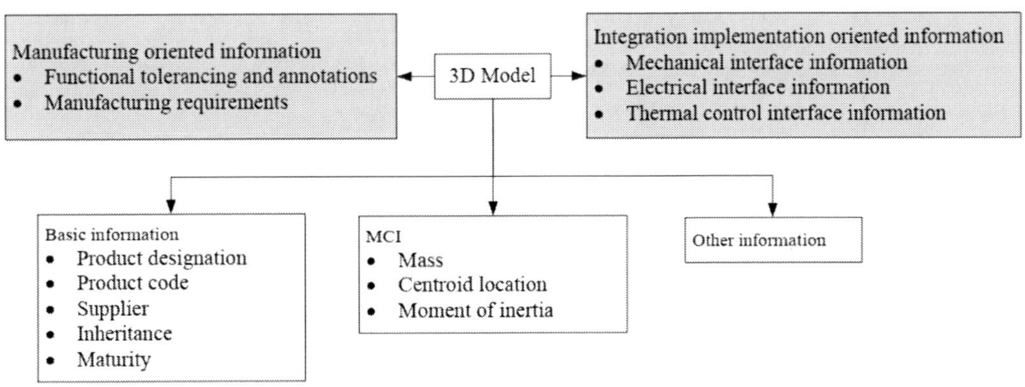

Fig.1 MBD Structure based on satellite integration design

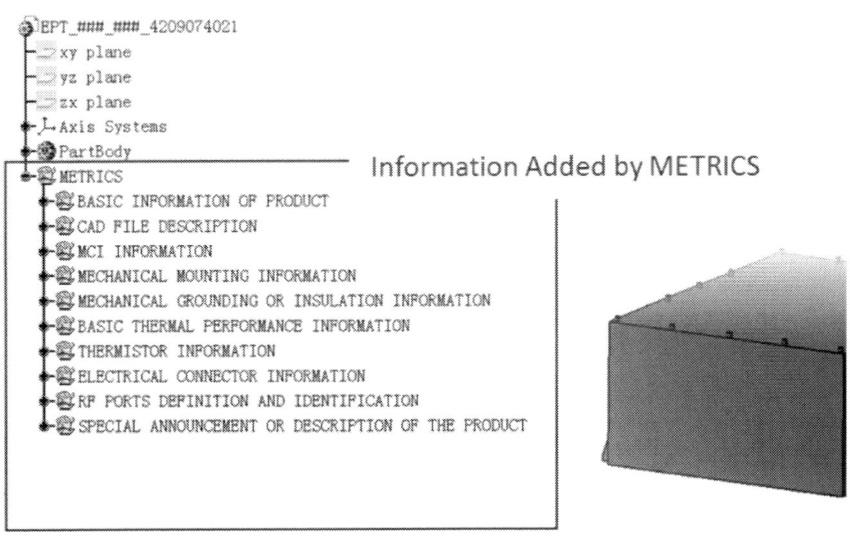

Fig.2 A typical MBD model structure built by METRICS

CNN-Based Vision Model for Obstacle Avoidance of Mobile Robot

Canglong Liu[1,2], Bin Zheng[2]*, Chunyang Wang[1]*, Yongting Zhao[2], Shun Fu[2], Haochen Li[2]

[1] School of Electronic and Information Engineering, Changchun University of Science and Technology, Changchun, 130022, China
[2] Chongqing Institute of Green and Intelligent Technology, Chinese Academy of Sciences, Chongqing, 400714, China

Abstract. Exploration in a known or unknown environment for a mobile robot is an essential application. In the paper, we study the mobile robot obstacle avoidance problem in an indoor environment. We present an end-to-end learning model based Convolutional Neural Network (CNN), which takes the raw image obtained from camera as only input. And the method converts directly the raw pixels to steering commands including turn left, turn right and go straight. Training data was collected by a human remotely controlled mobile robot which was manipulated to explore in a structure environment without colliding into obstacles. Our neural network was trained under caffe framework and specific instructions are executed by the Robot Operating System (ROS). We analysis the effect of the datasets from different environments with some marks on training process and several real-time detect experiments were designed. The final test result shows that the accuracy can be improved by increase the marks in a structured environment and our model can get high accuracy on obstacle avoidance for mobile robots.

Keywords. End-to-End Learning, CNN; Obstacle Avoidance, ROS

1 Introduction

With the continuous development of science and technology, various mobile robots have been widely used in different fields, such as life services, industrial production, education, entertainment and military area, etc. The mobile robot technology include control theory, mechanical design, computer technology. The ability of mobile robots to navigate and avoid obstacles is an important indicator of the robot's intelligence.

Autonomous navigation and obstacle avoidance for mobile robot need to equip with some range sensors and depend on complex algorithms[1]. Some typical range sensors such laser sensors, ultrasonic sensors, and visual sensors, while these sensors have their own limitations. For example, lasers are more expensive and traditional algorithms based vision are relatively complex.

In recent years, with the development of machine learning, especially deep learning[2], it has became a study hotspot that the robot avoid obstacles by self-learning[3]. Deep learning is an end-to-end learning approach, which is a mapping relationship from input to output through a deep learning network. That is to say, the system automatically learns the characteristics of the data by pouring a lot of data into the algorithm. Levine et al.[4] demonstrate that end-to-end learning is superior to the traditional approach that fixed vision layers . They presented the end-to-end application of CNN for motion planning of robot. They trained a end-to-end learning model based convolutional neural network. The image obtained from a camera as input and output the command for control the robot arm.

Gaya et al. showed the application of automatic obstacle avoidance based CNN for Autonomous Underwater Vehicles (AUVs) [1]. While they don't test the model in real-time and the network is not end-to-end since it consist of requirement of the intermediate procedures. Nicolai et al. use deep learning approach for odometry estimates based lidar data[5]. while their approach need to obtain laser data from expensive LIDAR. Giust et al study a problem of quadrotor trail a forest based CNN[6] . They acquire the dataset by a hiker who equipped with three cameras in head . And they train a CNN model to output control commands for a quadrotor trail the forest following a special path. Ross et al. [7] proposed a method that learn left/right controller for Micro Aerial Vehicles(MAV). They use monocular vision as its sensor, and extracted features from the raw image. The MAV can autonomously navigate through a forest environment. But the system only aimed to the left and right motion, it has to be operated by a person when need to forward motion command.

Compared to the traditional mobile robot obstacle avoidance method, the method of obstacle avoidance based end-to-end learning greatly simplifies the calculation. The system generate steering commands directly by the original pixels[8]. For the problem of robot explore environment, traditional algorithms steps are sense-plan-act. While our main work is try to use the end-to-end model. The main difference between traditional algorithms and end-to-end model is in the

* Corresponding author: zhengbin@cigit.ac.cn wangchunyang19@cust.edu.cn

DOI: 10.1051/matecconf/201713900007

path planning. Traditional algorithms are complicated, time-consuming and usually requires some sensors. By contrast, our method skips the traditional plan, that is sense-act[9].

In the paper, we present a mobile robot explore in an indoor environment, and the system can rapidly learn feature for avoid obstacles. We don't require high resolution camera, but only a cheap and common camera. The paper is organized as follows: Section 2 introduce the vision model of obstacle avoidance. Our experiment will present in Section 3, which include hardware platform, dataset, train model, a real-time test and other training details. Finally, conclusions and prospect are given in Section 4.

2 Vision model of obstacle avoidance

Deep convolution neural network has made great progress with the development of large-scale computation and GPU. More and more researchers take advantage of CNNs to solver some problems for robot in machine learning.

We control the robot by CNN, which take RGB images as its input. And it has three classes output, include go straight, turn left and turn right. The network architecture is shown in Fig.1, which consists of 5 convolutional layers and 3 fully connected layers. In our environment, extracting and learning feature from our dataset by using Caffe[10] , which is a deep learning open source frame. And we designed a CNN network based AlexNet , which was a winner in ILSVRC-2012 and r presented by Alex[11]. The last fully connected layer was adjusted to 3 nodes, which is consistent with the system included three turning commands.

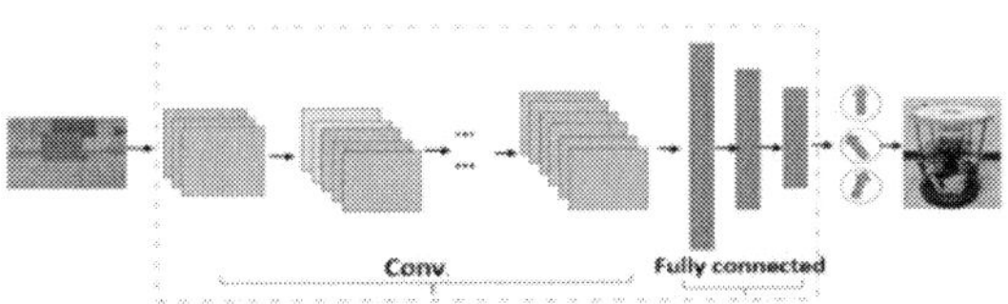

Figure 1. Structure of the CNN

The mobile robot system performed the instructions which was predicted with CNN based Robot Operating System (ROS). ROS is an open source framework for robot, which can provide a similar operating system for heterogeneous computer clusters[12].

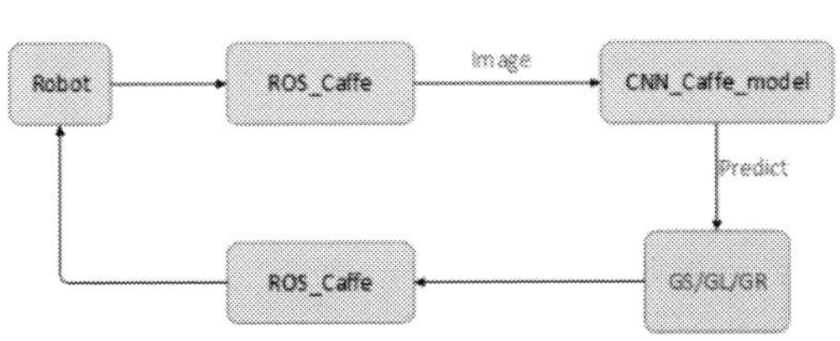

Figure 2. The flow chart of running

We use a package named ROS_caffe which provide a bridge between caffe and ROS. The flow chart is shown in Fig.2. The raw image was acquired by ROS from robot, a trained CNN model classifies the images to decide which direction will go next for the robot, and specific instructions are executed by the ROS.

3 Experiments and analysis

3.1 Hardware platform and environment

We built a light-weight and small mobile robot based iRobot Roomba and ROS, which is an open source frame. ROS provides a variety of software packages that can be applied to robots, and it played an important role in the design of mobile robot control systems. In addition, the system was equipped with a common camera and a Mini computer.

(a) (b)

Figure 3. Test environment

In order to demonstrate the problem more effectively, we constructed two types structured environment in indoor based KT foam board. The first type environment as shown in the Fig. 3(a), which was constructed using raw KT foam board. We put some black tape in another type environment, as shown in the Fig. 3(b). Besides, we placed a table in the center as a major obstacle.

3.2 Datasets

CNN need a massive dataset in order to train an effectively model. In these two environments, we operated the robot explore the environment and without colliding into obstacles. At the same time, we recorded the image and control commands by a tool of named rosbag in ROS.

Table 1. The number of training images

	TL	TR	GS	Total
Dataset 1	1892	1955	1888	5735
Dataset 2	1901	1999	2038	5938

We set the label for each frame by matching the clock of control commands and the clock of images, and the label include Turn left (TL), Turn right (TR), and Go straight (GS). The dataset 1 and dataset 2 was collected respectively in the first and second type environment. We sampled 5735 images from dataset 1, and 5938 images from dataset 2. The special information is shown in table 1. The dataset 1 was spilt in disjoint 5057 images for training and 678 images for testing. The experience use 5178 images for training and 760 images for testing in the dataset 2.

3.3 Training results

We start training based our model, as shown in Fig.1. The model was trained on a workstation equipped with an NVIDA GTX 1080 GPU and NVIDIA cuDNN. And the network can directly generate TL/TR/GS commands from images of camera after the completion of the training.

3.3.1 Training curves

The learning curves were plotted in the Fig. 4, which include test accuracy, train and test loss as the number of iterations. In the Fig. 4(a), the test accuracy eventually achieved 81.72%. And the model achieved a test accuracy 93.21% in the Fig. 4 (b).

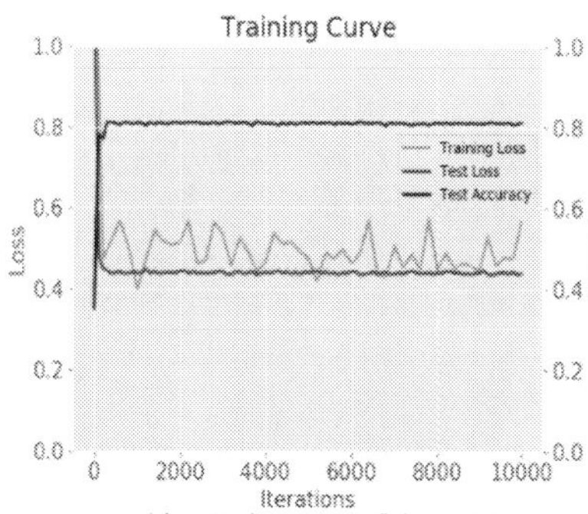

(a) Train curve of dataset 1

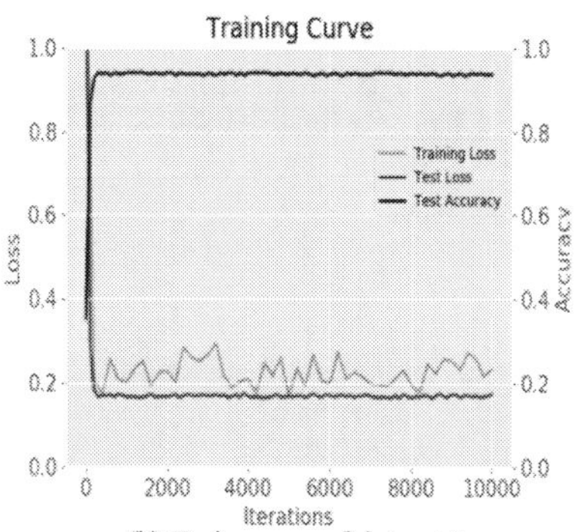

(b) Train curve of dataset 2

Figure 4. Training curves

We can see that the performance of dataset 2 outperform significantly dataset 1 from the training curves. Comparing two classes of the environment, we observe that the more easily the environment is identified, the higher the accuracy, which is similar to recognize the environment for human. The accuracy can be improve by change the environment mark in the constructed environment.

3.3.2 The confusion matrix

The confusion matrix of the dataset 1 classification result was shown in Fig.5. We can see that the overall accuracy of the dataset 1 is 81.72%. The accuracy rate of the GS class is the highest, which account for 96.08%. The classes TL and TR are low in accuracy. In addition, the error classification of the GR class is often misclassified into the GL class.

Output class

Target class	GL	GR	GS	
GL	20.65% 140	3.83% 26	2.21% 15	77.34%
GR	0.32% 70	24.93% 169	0.44% 3	69.83%
GS	1.47% 10	0.00% 0	36.14% 245	96.08%
				81.72%

Figure 5. The confusion matrix of the dataset 1 classification result

3.3.3 Test result for samples

We tested randomly some sample images with the trained model, shown in Fig. 6 .The average network prediction time of each frame is 4.03 ms. The raw input images are shown in the first line, and the second line are the corresponding predicted output which include turn left/right and go straight.

In the histograms, red mark represents the response probability values of three classes. We observe that CNN model is effectiveness, and it can extract valid information from raw input and predict the right commands.

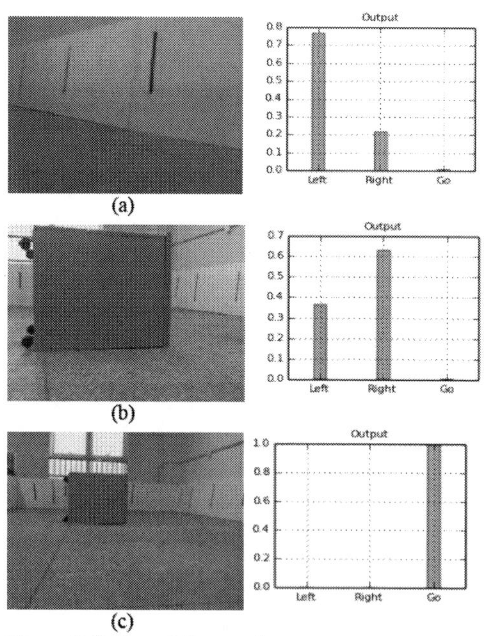

(a) (b) (c)

Figure 6. Test result for samples.

3.3.4 Comparison

We compare the result with some others in order to show the effectiveness of the method. As is shown in table 2, Lei Tai et al. tested deep-network solution towards obstacle avoidance in an indoor environment, their overall accuracy is 80.2%[13]. Comparing their method and result, our test was designed in the indoor environment with mark and without mark, respectively. The final test result shows that our model can get high accuracy on obstacle avoidance for mobile robots and the accuracy can be improved by increase the marks in the environment.

Table 2. Accuracy of obstacle avoidance

	Dataset 1	Dataset 2	[13]
Accuracy	81.72%	93.21%	80.2%

3.4 Real-time test

In our real-time-test experiment, we predicted the commands based the model that was trained with the dataset 2. The robot implements these outputs as constant rotational and/or translational velocities. We used a package named ROS_caffe which provide a bridge between caffe and ROS. The system performed the instructions which was predicted with CNN based ROS.

In the test, the robot predicted the instructions of the corresponding action by the trained model and perform the actions by ROS. Throughout the whole experiment, the robot successfully avoid square table in the center of the environment, and it didn't hit around the KT board. In real time testing, the network forecast outputs are shown in the Fig.7. The experiment achieve the goal of the mobile robot obstacle avoidance in explore the environment.

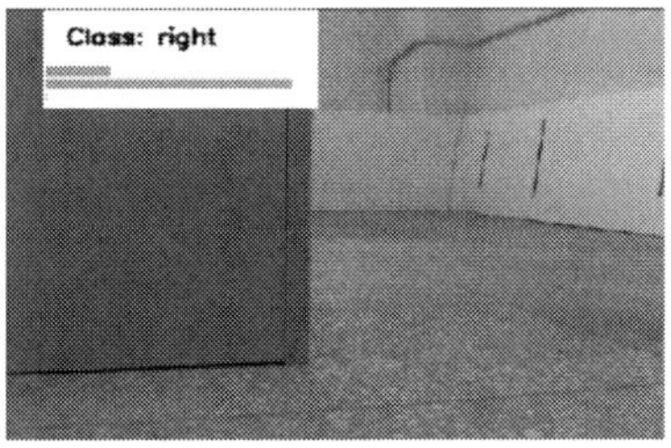

Figure 7. Real-time predict of the network

4 Conclusions and prospect

In the paper, we presented an approach to the mobile robot explore environment and obstacle avoidance using methods of end-to-end learning based CNN. A deep neural network was trained with the dataset and it converts directly the RGB images to steering commands. We also discussed how to improve the accuracy by change the environment mark in a constructed environment. The real-time test experiment our approach can get high accuracy on obstacle avoidance of mobile robots.

In the next work, we will attempt to do it in a more complex setting, include dynamic and non-structure environment. We will solver the tasks of robot navigation based CNN model and provide a significant contribution for the development of intelligence mobile robotic navigation.

References

1. Kovács L. Visual Monocular Obstacle Avoidance for Small Unmanned Vehicles[C]//Proceedings of the IEEE Conference on Computer Vision and Pattern Recognition Workshops. 2016: 59-66.
2. Deng L, Yu D. Deep learning: methods and applications[J]. Foundations and Trends® in Signal Processing, 2014, 7(3–4): 197-387.
3. Jia B, Feng W, Zhu M. Obstacle detection in single images with deep neural networks[J]. Signal, Image and Video Processing, 2016, 10(6): 1033-1040.
4. Levine S, Finn C, Darrell T, et al. End-to-end training of deep visuomotor policies[J]. arXiv preprint arXiv:1504.00702, 2015.
5. Nicolai A, Skeele R, Eriksen C, et al. Deep Learning for Laser Based Odometry Estimation[J].
6. Giusti A, Guzzi J, Cireşan D C, et al. A machine learning approach to visual perception of forest trails for mobile robots[J]. IEEE Robotics and Automation Letters, 2016, 1(2): 661-667.
7. Ross S, Melik-Barkhudarov N, Shankar K S, et al. Learning monocular reactive uav control in cluttered natural environments[C]//Robotics and Automation (ICRA), 2013 IEEE International Conference on. IEEE, 2013: 1765-1772.
8. LeCun Y, Muller U, Ben J, et al. Off-road obstacle avoidance through end-to-end learning[C]//NIPS. 2005: 739-746.
9. Pfeiffer M, Schaeuble M, Nieto J, et al. From Perception to Decision: A Data-driven Approach to End-to-end Motion Planning for Autonomous Ground Robots[J]. arXiv preprint arXiv:1609.07910, 2016.
10. Jia Y, Shelhamer E, Donahue J, et al. Caffe: Convolutional architecture for fast feature embedding[C]//Proceedings of the 22nd ACM international conference on Multimedia. ACM, 2014: 675-678.
11. Krizhevsky A, Sutskever I, Hinton G E. Imagenet classification with deep convolutional neural networks[C]//Advances in neural information processing systems. 2012: 1097-1105.
12. Quigley M, Conley K, Gerkey B, et al. ROS: an open-source robot operating system In: ICRA workshop on open source software[J]. IEEE, Kobe, Japan, 2009.
13. Tai L, Li S, Liu M. A deep-network solution towards model-less obstacle avoidance[C]//Intelligent Robots and Systems (IROS), 2016 IEEE/RSJ International Conference on. IEEE, 2016: 2759-2764.

Fault Detection and Diagnosis System in Process industry Based on Big Data and WeChat

Zengqiang Sun * and *Xinjian Lu*

Nanjing ChemCyber Technology Company LTD, nanjing 210036, PR china

Abstract. The fault detection and diagnosis information in process industry can be received, anytime and anywhere, based on bigdata and WeChat with mobile phone, which got rid of constraints that can only check Distributed Control System (DCS) in the central control room or look over in office. Then, fault detection, diagnosis information sharing can be provided, and what's more, fault detection alarm range, code and inform time can be personalized. The pressure of managers who worked on process industry can be release with the mobile information system.

1 Introduction

The fault detection and diagnosis system of the process industry can be analyzed from diagnosis technology and communication technology. There are mechanism model, empirical model and large data model in diagnostic technology, in traditional model, which is based on the mechanism of the process and the artificial experience of the accident report [1]. But, people could acquire different experience from the same accident, because of the choice of observation parameter. In another side, the utilization of on-line mechanism model is too low. So, the large data model is applied more and more.

On communication technology, the detection and diagnosis information could be received only through DCS in the control room many years ago, the operator and administrator can't receive production and fault information promptly and automatically after leaving the control room. Then, the short message service (SMS) was used to notice fault information to administrator of the factory promptly and automatically with the SMS link to DCS, but the shortcomings are also obviously, such as the information is incomplete, usually need to make a telephone call for fault confirmation after receiving the information. With the development of mobile phone network, E-mail notification on phone was used for fault information communication also, but the continuity of the information can't meet need in production. In general, the above means of communication could not offer the fault detection and diagnosis information anytime and anywhere, so the safety pressure in process industry can't reduce effectively. But, the ease of use and continuous information of WeChat will offer a new way.

In this paper, a fault detection and diagnosis model based on big data will be describes and the communication means of the system will be show. The structure is organized as follows: Section 2 presents the bigdata model, and describes the display of system. Section 3 describes the precondition, technical route, Algorithm and case of system implementation. Finally, Section 4 concludes the paper and points out further work.

2 The structure of system

2.1. The bigdata model of system

In each target operation unit or device which selected for bigdata modeling, historical data that including normal production and fault status are scan at first, then, key performance indicators are select, and the model is established after principal component analysis (PCA) [2], squared prediction error (SPE) and T^2 detection [3-5]. So, the fault data in process can be pick out with the control line of parameter after the online data directly into the model, and diagnose information can be obtain according to different parameter's contribution on the control line indicators.

The fault data can be identified from process data at a certain time in accident analysis, but, from the point of large data statistical analysis, there are two kinds of identified methods in the process of fault occurrence. one is the deviation of the parameter itself, in other word, the parameters that trigger the fault will deviate from the stable data set when the fault occurs. The other is relevance of the different parameter, which will change from stable conditions to fault conditions states.

Big data modeling is a process to find the regularity implied in numerous data initiative, which is quantitative.

* Corresponding author: zengqiang.sun@chemcyber.com

And then the model will be confirmed by expert, which part is qualitative. As the artificial confirmation is based on big data, so the model is more accurate. On the other hand, the small data modeling is a process to find the regularity from human experience, which is qualitative. And then the model will be confirmed by small data case, which part is quantitative. Therefore, large data modeling can minimize the impact of human subjective factors, making the model more comprehensive, detailed and reasonable in fault detection and diagnosis.

2.2 The display on Wechat of system

The fault detection and diagnosis information pushed by the system can meet the needs of the user anytime and anywhere, the content is more complete and continuous with the trend graph of the production data when the fault occurs.
The system will not meet the requirements of safe production if only one person to view the fault alarm information, the safety accident caused from someone negligence is not uncommon. But, WeChat will share the fault alarm information to different levels of management personnel. On the one hand, fault knowledge base or parameter model can be established early. On the other hand, fault Information can be diagnosed after expert consultation, and the system will update the diagnostic information to the knowledge base and parameter model, so the model can be improved continuously.
The manager who works on different positions pay different attention for different fault information, therefore, everyone should have the authority to adjust whether they are concerned about the fault information and the scope of the fault alarm. Such as the fault have more impact of your work should be adjust to a narrow alarm range, so that appropriate measures can be ready for the fault. In another way, the authority to adjust system personalized which keep operators and managers have a good rest.
As above characteristics, the comparison between SMS and WeChat communication system as follows:

Table 1. Comparison between SMS and Wechat

	SMS	Wechat
timeliness	Immediate arrival	Immediate arrival
Graphic display	No	Yes
Historical inquiry	No	Yes
Correlation analysis	No	Yes
personalization	Few	More
User number limit	Yes	No

3 The implementation of system

3.1 Precondition

The fault detection and diagnosis system based on large data and WeChat relies on real-time data at first, so that DCS, real-time database (RTD) or other control system are required. Then, real-time data should also be allowed to transport with Internet transmission or intranet transmission. Of course, the transmission process is security by the certificate and encrypted channel. In comparison, the cost, storage and computing of Internet transmission is lower and convenient.

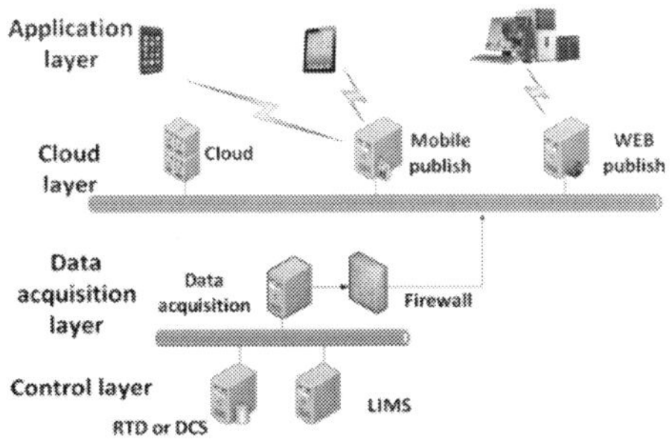

Fig. 1. Network topology diagram.

3.2 Technical route

Data acquisition safely is the first step for system implementation, so the acquisition device should be Connect to DCS or real-time database in a secure manner, such as adding a firewall or gatekeeper to ensure that the data does not leak or prevent the DCS from reversing control. In addition, the acquisition device could not affect the DCS or other information systems connected to the real-time database.
Data pretreatment can be used for increasing the safety in data transmission to the cloud servers, for example, data and corresponding physical explanation can be separate in transmission by pretreatment, so someone who gets the data through informal channels can do nothing without physical explanation. In another case, the data was deal by addition or multiplication before transmission to the cloud servers for safety, and would be restoration for modeling.
Real-time data could be stored to the cloud servers, which reduce the physical risk and cost of enterprises to buy physical storage servers. Then, the managers have no need to worry about power outages, host computer server crash and data disaster, because a professional team of cloud server suppliers to maintain. The last, real-time data of every enterprise in the cloud servers are stored separately, other enterprises have no right to access.
Model calculation, especially for high frequency calculation is a key step for real-time detection, the system achieves that through distributed and parallel cloud computing, in which computing tasks are distributed across resource pool of cloud servers that

constitute from a lot of computers, rather than local computers or remote servers. So, computational power, storage space, and information services of various application functions can to obtain. Moreover, the available resources are infinite expansion and can be readily available at any time.

In addition, the storage and calculation of data must follow a clear classification and unified data structure. So that users can easily obtain the required data. These classification and structure are also base for fault detection and diagnosis system.

The user can receive fault detection and diagnosis messages with pay close attention and register to system on WeChat, after the calculation result in cloud servers is published to the WeChat servers.

The roadmap of data flow as follows:

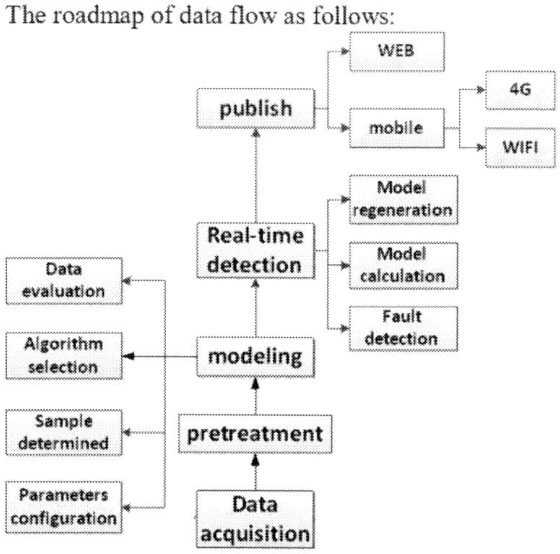

Fig. 2. Roadmap of data flow.

3.3 Algorithm

The basic algorithm used in the system, including principal component analysis and other statistics classical method, in addition, SPE and T^2 detection which used for fault detection and diagnosis. In this paper, these algorithms and some data pretreatment are combined to enhance the practicality of the results.

Steps are as follows:

(1) Collecting data from the real-time database to form a sample set for modeling;

(2) Extracting the key performance parameters which according to the physical model to form a new sample set for one specific operating unit;

(3) Normalizing the key performance parameters of the new sample set;

(4) Executing PCA of the normalized set of samples;

(5) Modeling with the lower limit or control line of SPE and T^2 when the confidence is 95%. The fault was identified when the value of real-time data about parameters are significantly higher than the two lower limits at the same time, such as 3 times of standard deviation of lower limit according to score matrix obtained by PCA.

(6) Comparing the contribution of different parameters to SPE and T^2, then the parameter which caused a malfunction could be diagnosis.

3.4 Case

Figure 3 happened at 10:43 on March 30, 2015, the DCS showed the flow value of non-purified wind which into the device of one enterprise in the normal range, but the instantaneous deviation was abnormal, as the point of blue data trend line breakthrough the red control line in figure 3. The operator didn't take anything action on DCS, but, the detection information from our system was sent to the technical staff, and the technical staff entered site for inspection, then, found a hidden danger. The parameter returned to normal after the hidden danger was elimination. If not, there would be a great impact on pipe network system, even device shutdown.

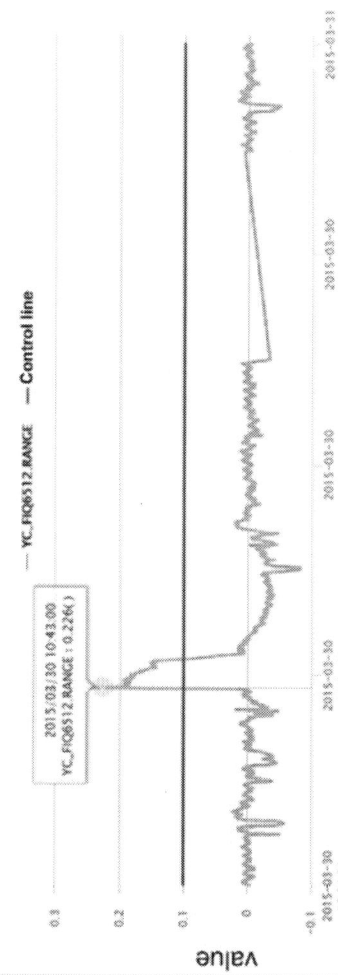

Fig. 3. Case 1 of data system

Figure 4 happened at 12:10 on January 12, 2015, the DCS showed a rapid decline of the external heaters material level, and then quickly restore the phenomenon, which didn't trigger alarm on DCS, so the operator

DOI: 10.1051/matecconf/201713900008

didn't take more care of that. But our system diagnosed there would be a fault of external heaters leak according to the comprehensive detection of relevant parameters, as the external heaters comprehensive detection parameters breakthrough the normal value obviously, although external heaters material level parameters was normal. Then sent the fault message to the technical staff, and the technical staff entered site for inspection of tube bundle in junction box of import and export on external heating device, then, found the tube 2# had leakage. The device returned to be normal after the tube 2# had be resection.

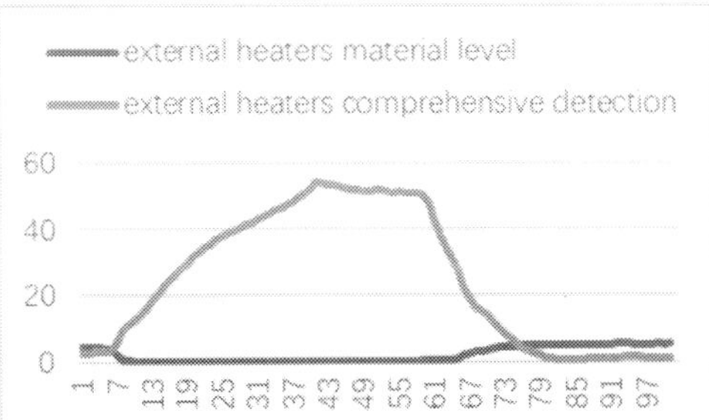

Fig. 4. Case 2 of data system

There would be a difficult for operator or technical staff to judge the fault manually, because of the not obvious of DCS data abnormal and then quickly restored. If not promptly dealt with, which could cause the regenerator pressure too high, affected the normal circulation of the catalyst, even lead to smoke machine vibration increasing if seriously, caused the smoke machine damage, and even caused the reaction feed was cut off, finally to the device emergency shutdown, which need 6 hours to recover and avoid the loss of 2.39 million yuan.

4 Conclusions

This paper introduces the fault detection and diagnosis system based on large data and WeChat developed by Nanjing ChemCyber Technology Co., Ltd., including technology route and algorithm realization, and provides the realization case. The system reduces the safety pressure of manager in process industry, so that safety management more convenient. And the continuous optimization of the device can be implementation with fault diagnosis information, so that the process industry can enjoy the changes and benefits by mobile information system.

References

1. L. Zhou, Z.H. Song, B.P Hou, Z.X. Fei. CIESC Journal, 68, (2017), 1109-1115.
2. M. Han, Z.K. Zhang. CIESC Journal, 66, (2015), 2139-2149,
3. K.X. Peng, L. Ma, K. Zhang, ACTA AUTOMATICA SINICA, 43, (2017), 349-365
4. B. Song, Y.X. Ma, Y.F. Fang, H.B Shi. CIESC Journal, 65, (2014), 620-627
5. A Carolos, M.A. Francisco, Q.C. Eduardo, G.M. Emilio, Journal of Loss Prevention in the process industries, 26, (2013), 660-665

Discussion on main points on designing of instructional software

Jiujiu Yu

College of Computer Engineering, Anhui SanLian University, 230601 Hefei, P. R. China

Abstract. Instructional software is a kind of software program or system, which reflects the teaching strategy according to the teaching objectives and teaching contents for the purpose on promoting teaching and learning by both computer technology and network technology. Firstly, the concept of instructional software, features and functions are introduced in the article. Secondly, making a detailed description on the current types of instructional software features for total eleven types, and discussing on main points on designing of instructional software for each type fully. Finally, an initial expectation on researches on directions of instructional software designing is given, which from three aspects on designing of teaching platform for SPOC(Small Private Online Course), designing of mobile learning resources, and the intelligence on designing of instructional software in the future.

1 Introduction

Instructional Software refers to the programs or software that is used in computer-assisted instruction, through the computer teaching content, structure, teachers teaching intent, guidance and control of teaching activities, to receive students' requirements and answers, storage of teaching materials for teaching and learning activities [1]. From the technical view, instructional software is a kind of computer software program that can display the teaching contents by displaying the text, sound, image, video, and other information media in an interactive way. From the perspective of content, the instructional software embodies the corresponding teaching strategy [1]. According to the teaching theory and the students' cognitive psychology, different instructional software has different characters and different ways of presentation. So, the main points on designing of different instructional software and its teaching purposes are also different.

2 Classification on the instructional software

2.1 Demonstration type

Instructional software of demonstration type refers to the teaching resources which introduce or show the content of teaching resources for the students. This kind of software is often designed by tools of PowerPoint, Authorware, Flash, and etc, with the text, pictures, animation and other forms of elements to display and explain the teaching content for students in multimedia classrooms of schools. The designing of instructional software for this type needs to follow some points:

2.1.1 Concise and Highlights

The present teaching of information on the page of software must be clear, and animation effect is not too much. Some additional effects which have nothing to do with the teaching contents, such as text, pictures, music and other information must be deleted.

2.1.2 Information appropriate

For the case of design of PPT, the words on each page cannot be too much, about 30 to 50 words are appropriate, and the quantities of concepts which transferred to students should not be more than five in one page.

2.1.3 Clear structure and friendly interface

The design on structure of the software should be provided with proper links between different pages, to display the content. Links for operation should be created on each page, and used to select the content of the presentation. Avoid redundant links and appearance of the page jump trek. The interface on software must be friendly, the contrast between background color and text on the pages should be strong and watched clearly for every student in classroom.

2.1.4 Enhance the interactivity

* Corresponding author: yjjyjL@163.com

The interactivity for software could be designed simply, the demo according to the preliminary design steps in the process of running are in turn. We suggest that appropriate increase interactivity can demonstrate problem sets in the link. For example, when students answer the questions correctly, the current page of software will be displayed the prompt information, the purpose of feedback on teaching information can be increased.

2.2 Individual counselling type

Instructional software of individual counselling type is based on method for individualized teaching, with the learning process of non-linear characteristics for learners. The learning content presented in the software does not divide the learning units into the chapters, but the content should be arranged from the learner's perspective. Learners can choose learning subject according to their own learning content. The designing of instructional software for this type needs to follow some points:

2.2.1 Independent learning unit

Each unit for contents of software should be have certain independence, and for different levels of students to learning, the sequence between each unit cannot be obviously.

2.2.2 Individual guidance on teaching strategy

The reasonable individual guidance on teaching strategy should be designed in software, and based on the knowledge points of context. The teaching strategy could be composed of a series of questions for testing. Through the test results of each student, it should be decided to let the students into the next phase of the learning process, or decided to the point for students to learning next.

2.2.3 Control on learning process

The control on learning process is the main body of students which is designed through the software for learning activities. For navigation buttons which are designed on pages of software to control page jumping. After a student finished a learning unit, the navigation can be used for summarizing the learning situation automatically. When the current unit is exit, at the same time, the following learning unit of content could be displayed for students automatically through the navigation by a decision strategy.

2.3 Practice evaluation type

Instructional software of practice evaluation type is focused on a particular course, and the main functions that designed are much simple. The main parts of software are composed of a lot of exercises for evaluation, and students could use the exercises to enhance their knowledge. The designing of instructional software for this type needs to follow some points:

2.3.1 Setting types for exercises

The type of exercises for testing are diversified, such as exercise types for choice, fills up the topic, true or false, and etc. At least two more types are included. The difficulty of the exercise should be reasonable, and the reference answer is needed to be provided also.

2.3.2 Arithmetic on exercises selecting

Random selection algorithm and adaptive selection algorithm are mainly included. Random selection algorithm is based on random selection on exercise, and suitable for evaluation for a large amount of students, but lack of pertinence. Adaptive selection algorithm is more targeted at one student, and suitable for the individual student's evaluation according to its own situation. The choice of exercise is more adaptability for a certain student through the software. Usually, the designer will select the corresponding algorithm according to users.

2.3.3 Feedback mechanism

Not only the function of evaluating for the software should be realized, but also the feedback mechanism needs to be designed. When the student finished exercises, the information for feedback should be displayed immediately, helping the student to know the result. The results of evaluating information include "right" or "wrong" for the answering. Some components of competition for participating could be designed in the software also, and the quiz show could be designed and realized between different students or between students and the computer.

2.4 Teaching simulation type

The main functions of instructional software of teaching simulation type is to using computer to simulate the phenomena on real world, using computer simulation technology or computer modeling techniques to show something of the structure and dynamics, to help students understand the changes in reality[2]. Students can observe phenomena on simulation by software, and to participate in the simulation process through playing a role in the process of simulation. Applications on this kind of software need to rely on professional virtual laboratory, the requirement on running for the software is not adaptable in ordinary classroom. The designing of instructional software for this type needs to follow some points:

2.4.1 Simulation import

Relatively, the instructional software for the type of teaching simulation is more complex about methods on using. Simulation import refers to displaying the background in the form of text or image to simulate for students before using the software, and the students could understand the transaction which is going to meet, the environment changes, for operation on equipments, etc.

2.4.2 Scenarios simulation

Scenario refers to the simulated objects at a particular time, by the time, place, characters and events. Usually, the scenario is characterized by a picture, or comes with 3D effect. Scenario simulation is refers to the software which is to simulated the above elements, and facilitate with students into the scenario as soon as possible, help them to understand the content for learning better.

2.4.3 Situation interaction

Situation interaction refers to the communication and information exchange between people and software interaction by the input/output devices of computer, such as request and response, etc. When a particular scene is presented by the software, students have to be responded in the scenario, and complete the corresponding operation. Operation type can be selected by projects or manipulation of the simulation objects, the reaction which is carried out on the events, etc. Of course, the content is decided by the function of the software which is provided.

2.4.4 Feedback

For instructional software, feedback refers to the system adjustment situation which is caused by the user action. That is to say, feedback is the correctness of the software to the user action degree of evaluation. For example, when the user's operation is error, the information such as "error, please ......" will be displayed to user automatically from the system with sound effects. Usually, feedback should be designed appropriate according to the teaching aim, teaching content and student's level from the software by designers.

2.4.5 Simulation exit

After a process of simulation finished, whether success or failure, the exit of the message must be designed to display for users. For example, displaying the message on encouraging user to continue or try again.

2.5 Subject resource for exploration

According to the specific teaching content and set up an opening website for teaching and learning. Usually, the instructional software of subject resource for exploration is in form of a website, and also be known as a specific website for learning. As a type of instructional software, the development direction of network learning is fully displayed to students. The main function of subject resource for exploration is connected with some special tasks for learning content, to inspire the students for exploring. Through active exploration on searching all kinds of the resources for learning and interaction with other students, the students could complete the task, to achieve the goal of learning [3]. The designing of instructional software for this type needs to follow some points:

2.5.1 Topic selection on exploration

The selection for topics on exploration should be extracted horizontal or vertical, according to the characters on teaching content which combined with the cognitive characteristics of students. The designing of topics can be based on chapter and section of textbook. And it also can be designed for appropriate extension of the content of the textbook, to form a topic which is comprehensive. Of course, the teacher can be allowed to design the topic interdisciplinary even, but the demands on capacities for students' learning are quite high.

2.5.2 Task on exploration

The task on exploration must have clear requirements. For the larger task, it should be broken into many subtasks. The goal of task design should be diversified, to have different goals and its evaluation criterion. In the process on the design of the task, the task is better to contained different profiles, and is advantageous for the division of cooperation, to achieve the goal of cooperative learning.

2.5.3 Design on learning tool

Exploration on learning is more depend on students' inquiry and collaborative communication, and it can't be departed from the support of learning tools. Students use the learning tools to complete the process on access to information resource, production and processing, to express their own thoughts, and communicate with others. So, the learning tools such as information searching tools (Baidu, Google), cooperative learning tools(BBS, WiKi, Message box), task statement tools(Word, Excel, PowerPoint) should be designed to integrate into the pages of software according to the purpose and content of the exploration on learning for the users.

2.5.4 Design on evaluation

During the process of the designing the exploration on subject resource, the way of the appropriate learning activity evaluation is need to designed according to the goal, task and the activities on learning, it play an important role to control quality for implementation on learning process. Learning process evaluation and result evaluation are included in evaluation, and the evaluation is constituted with self-evaluation, mutual evaluation between students and evaluation for teachers[1]. Teachers can set up the score ratio on each part in software according to the actual situation.

2.6 Teaching game

As a type of the instructional software, teaching game is in an independent existence for the development on the basis of the software of ordinary entertaining game. The essence of teaching games is computer software program yet, but the factors on interesting and competitive are added in the content for learning, to fulfilling the implied educational function for users. The teaching game is widely used in high schools and universities now. The designing of instructional software for this type needs to follow some points:

2.6.1The game object is duality

The design of teaching object for teaching game is duality. On one hand, the teaching object is the object for the game itself, on the game player's perspective, the task of the game should be designed for completing the game levels. On the other hand, the learning object should be designed through the passing game levels for students too. That is to say, from the perspective of learners, when the students finish one level which is set by the teaching game, the learning object is able to achieve for the students.

2.6.2 Scene space for the game

Game scene space is the activities of the virtual world for players, and several factors should be designed for game scene space. Firstly, the background of the game should be designed accord to the teaching content, such as the style type for modeling of the characters in the game space, landscape, buildings and props. Secondly, the corresponding with learning information need to be embedded in the game, it can help students to master the learning content better. Thirdly, the cognitive tools for the teaching game is to be designed also, such as map, mobile way, dialogue, speech, and etc, as for the guiding direction to learning and application on feedback on learners' achievements.

2.6.3 Rules of competition [4]

Rules of competition are also referred to the rules of the game, the decisions and actions that taken by game participants must be complied with rules. Rules on competition are the real core of game designing. Teaching goal for students, teaching regularity and the knowledge should be included in rules of teaching game[4].From the perspective on computer programming technology, competition of the rules is the realization of the images of various mathematical models.

2.6.4 Interaction

The interaction with learners and computer should be designed to promote the operation of the game. On the basis of rules on human-computer interaction for designing on normal computer game, the designer for teaching game should according to learning content, and design the appropriate interaction. For example, how to use a mouse or keyboard for learning behavior operation by students in the process of the game, how to make use of information to answer questions in teaching game in order to promote game process, how to quit or continue learning process according to own intend, etc.

2.6.5 Feedback and encourage

Positive feedback need to be designed in teaching game, to make evaluation on a certain behavior or a periodic task for the player, and the game scenario should be changed also. The importance on feedback makes the players know their behavior is right or wrong, and to encouraging players to continue to make efforts on finishing the next task for the game. By the way, the form of information on feedback should be diversification. For example, using numbers to display the current player's score, using pictures to identify the player's current position in the game, using music to display the result(success or failure) for game mission, etc.

2.7 Network course

Network course is a certain subject for teaching contents with the implementation of teaching activities by network. The teaching aim, teaching content and teaching environment for network supporting are included in network course[5]. Teaching environment refers to the teaching platform for teaching activities, and many different courses could be based on one teaching platform. Now, the mainly teaching platform such as Moodle, Blackboard, 4A are widely used in universities of many countries. Of course, many teaching platforms with simple functions have been designed in many colleges and universities too, according to actual teaching environment, and to be implicated on teaching by network. Network course is the development on education of information period. The designing of instructional software for this type needs to follow some points:

2.7.1 Design on learning object

Learning object is E-learning resources for helping learners to complete a learning process, which is independent and complete. Learning instruction, content, practice and evaluation at least four parts are included in one learning object to provide a complete learning process for students. The particle size for a learning object should be designed appropriate. Usually, a unit such as a chapter or a section of the course could be designed as a learning object. By the way, a learning object should not be too easy or too complicated, and it should be designed independent and reuse.

2.7.2 Design of learning resource

The design of learning resource is the important part of the network course, and is closely related to the course content. Electronic courseware, case materials, all kinds of auxiliary teaching resource, exercises and etc are mainly included in learning resource. Every learning object should be corresponded to learning resource for supporting. Learning resource should be designed to have openness, diversity, independence and reuse.

2.7.3 Strategies designing on self-learning

Learners are the main body of the designing on network course. According to the actual environment of teaching and the characters for students, varied activities on self-learning strategies should be designed in network course, in order to stimulate the students to explore knowledge actively and positively.

2.7.4 Design on evaluation

In network course, after each learning unit, appropriate study effect on evaluation should be designed for learners, such as online testing, practice, homework, questionnaire survey and etc. The network teaching platform is to be able to provided measures for evaluation supporting, data analysis and statistics are also be provided for teachers according to specific format.

2.8 Educational resource center

Generally, education resource center is developed for a specific course, and refers to the course of mass education resources collection. Education resource center is convenient to learners for online browsing, searching and accessing the learning resource, and achieve the goal of learning. Usually, education resource is in the form of course website on the internet, teachers can release or update learning resource which related to the course, and students can login the website and download learning resource. For some course website, it also has modules for online testing, discussion, BBS, and other functions. The designing of instructional software for this type needs to follow some points:

2.8.1 Privileges for four kinds of users

Different from the normal website, the four kinds of privileges for users to login should be designed on course website. First, system administrator(assignment and management for all kinds of user privileges, course resources to add, modify, delete, and so on). Second, reviewing expert (from the perspective on teaching, scientific and normalization to review all kinds of new learning resources which was uploaded by teachers).Third, teachers and students(terminal users for the system, they can login the website to search and download the learning resources. But the teachers could upload the learning resources to website, and students do not have privileges to upload resources).

2.8.2 Establish the service system as user-centered

Educational resource center is based on website for the course, B/S architecture is usually designed to establish the service system as user-centered. The front part of the system is the course website, and it is automatically generated according to the existing website templates. Page layout display technology is implicated on website, so the texts, pictures, tables and other elements in pages could be displayed beautiful. Educational resource center is the backstage of website in fact, and connected with the website by database system (SQL, DB2, Orcale). When Users (students or teachers) login the website, they can browse and search the learning resources conveniently, the operation is also simple for users. The information on update for the course will be automatically stored in the database, the system administrators can keep the course resources displaying on the website in time.

2.8.3 Design on visualization of website templates

The design process of the course site could be simplified by predefining the visual website template. Color templates and site layout templates are included in website templates, designers only need to provide course resources according to the actual teaching environment to apply the corresponding site template, and it is automatically generated website for the course, and simplifying the establishment of the website process.

2.9 Network teaching platform

The network teaching platform refers to the software system which is based on the Internet and provides the full support service for the network teaching. The network teaching platform is usually composed of the network teaching subsystem, the educational administration management subsystem, the network course collaborative development subsystem and the teaching resource management subsystem. At present, the development of functions on network teaching platform whatever domestic or abroad are relatively stable and mature. Various types of network teaching platform such as Blackboard, Moodle, Tsinghua Education Online, Nanjing Yi Xue sky and etcs are widely used in many domestic colleges and universities now. By contrast, it is much difficult to design of the network teaching platform, colleges and universities need to pay for installing and using. The designing of instructional software for this type needs to follow some points:

2.9.1 Design of mobile learning

Recently, mobile phone or other mobile device for online learning has become the mainstream of college students learning, and how to use the mobile device to finish the teaching process conveniently is also a popular topic of education research. Mobile devices must be able to access the network teaching platform successfully, and provide a variety of teaching services of network teaching platform for students. From the perspective on development on software, for the network teaching platform which based on B/S architecture, how to

 DOI: 10.1051/matecconf/201713900009

integrate Mobile Learning Management System (MLMS) into business logic layer on network teaching platform and the variety of service is able to converted to mobile devices completely is one of the Main Points on Designing of network teaching platform.

2.9.2 Collaborative development on network course

Collaborative development on network course is the development trend of network course construction. First, the unified courseware development tools need to be integrated in platform, course resource and resource management software from the platform should be provided to teachers. So, it can simplify the burden of the resource collection when designing the web course. Second, It is necessary to design a unified network course template for the same course, and network course prototype files can be generated automatically through the template. It will bring convenience on course structure design, knowledge presentation and teaching design for teachers.

2.10 Learning network community

Learning network community is network space for learning in Web2.0 period, which based on social software such as Blog, WiKi, BBS, QQ and etc. Openness, equality, dynamic, beyond the boundaries of time and space are main characters on learning network community. User-centered, attention to the interaction of knowledge, sharing and collaboration, learners from the traditional passive learning knowledge into active access to knowledge are the main functions of learning network community. The designing of instructional software for this type needs to follow some points:

2.10.1 Identity construction mechanism

The mutual trust between learners is an important guarantee for learning process and communication in learning network community. It is necessary to design a perfect identity construction mechanism for learners. For example, real-name system mechanism is designed for registered learners that can bring advantages to carry out the equality and trust on education, reduce the speech which is not conducive for learning, and be more conducive to the expansion of the community.

2.10.2Mechanism on motivating users to participate

It is important to design an effective mechanism on incentive users for participating the learning network community initiatively, and reward users to continue to participate in learning and discussion in community. For example, you can design score or point roll, for encouraging users to participate in posting, replying and other exchange discussion behaviors to get more scores, and increase the user's rank gradually to get more learning resources.

2.11 MOOC platform

MOOC(Massive Open Online Course) is a kind of online learning mode in the field of distance open education which is "learner-centered", and it is the product of educational idea fusion information technology in new situation. MOOC has emerged as an impact and opportunity of traditional universities [6]. As the mainly medium of open education, a strong support is provided by MOOC platform to carry out teaching activities for teachers. Since the three famous MOOC platforms (Coursera, Udacity and edX) which were opened from USA have been appeared by years of 2012, a strong development potent has been shown by MOOCs in the field of network teaching. More and more well-known domestic universities and educational management institutions are also active in developing the Chinese MOOC platform to promote work for distance education, such as Chinese Universities MOOCs, Cloud Class of Netease, Good University Online, and etc. It has the characters of large-scale learning (a large number of students) and open sharing (different regions, different backgrounds of registered learners can also share free teaching resources). The designing of instructional software for this type needs to follow some points:

2.11.1 The presentation of the MOOC video

Learning content and activities organization are the core elements of the MOOC, they are important factors which affect the learning process. Course video is the mainly way to present learning content in MOOC platform, and the learners can turn on different videos by navigating on the same page, reducing the distraction caused by populating multiple pages[7]. In addition, the completion of the states (Completed, Processing, Not start) should be marked during the process of video playing for different identification in order to improve the learners focus on learning and learning progress of self-control.

2.11.2 The presentation of the MOOC forum

Forum is the most frequent interactive tool for learners to achieve automatic posting subscription function in MOOC platform. From the view on ease of use, the topic list, discussion area and classification, sorting and other navigations should be designed in a page, which can reduce the jumping in different pages that caused by learning distraction[7]. In each of learning content and activity forum, the corresponding forum needs to be designed also. So, the learners can complete the targeted discussions, as well as a weekly discussion of the theme of the overall forum and public discussion area. In the end, a non-course content category of the forum which is classified by the MOOC platform should be set ahead atomically, without the manually setting by teachers.

2.11.3 Peer review mechanism

There are so many learners for online learning in MOOC platform, that the homework or exercise can not be assessed by a few teachers, the efficient peer evaluation mechanism need to be designed. A learner's homework is randomly assigned to another learner evaluation, and the process should be justice [8]. This requirement should be provided from MOOC platform, and to achieve thousands of learners to complete the evaluation function of peer review.

3 Future researches

With the development on educational information, on the current background of big data, mobile cloud and so on, the research on designing of instructional software needs to keep up with the pace on new technology in the future.

3.1 Design of teaching platform for SPOC

SPOC(Small Private Online Course) is the new development on web learning which is derivative of MOOC.SPOC is also the most important way for learning in Post-MOOC Era[9]. For small-scale learning crowd, focusing on campus localization, and to promote hybrid learning are the characters on SPOC. The teaching platform for SPOC can be a simplified design of the existing MOOC platforms, and is only applied on micro-curriculum construction. Colleges and universities can also design teaching platforms for hybrid teaching process which is based on SPOC concept according to the actual teaching environment.

3.2 Design of resources for mobile learning

With the popularity of 4G mobile networks, the learning by the using of mobile devices and wireless network has become the main learning methods. Mobile learning resources have the characters on simple interface, small and accurate of information content, and the design of interactivity, compatibility, applicability, and real-time on data transmission for various mobile client platforms (Android/Mac OS/Symbian) will be the trend on teaching software designing in the future [10].

3.3 Intelligence on designing of instruction software

With the development of teaching theory and artificial intelligence technology, especially the emergence of constructivism and Agent, learning activities have become a process of acquiring knowledge, including dialogue, situation, collaboration and meaning construction, and the direction of development on learning is to explore, cooperate and exchange on society. The design of Intelligent Tutoring System[11], Adaptive Hypermedia System(AHS), Interactive Learning Environment, Distributed Intelligent Learning System, and Educational Semantic Web will be the mainly directions for intelligent designing of instruction software in the future.

4 Acknowledgments

Author acknowledges financial support from the projects on teaching team of "Software Engineering course" of Anhui SanLian University (No. 15zlgc029) and "Massive Open Online Course on Software Engineering" of Anhui Province (No. 2015mooc104). Here, I would like to express my heartfelt gratitude to other colleagues from the both projects.

References

1. Yu Shengquan, The Handbook of Instructional Software Design, Beijing Tsinghua University Press, 2011.
2. Li Wenhao, Zheng Yan, Instructional Simulation: The New Approach in Integrating Technology in the Curriculum, Journal of modern educational technology, 2009, 16(4) :9-12.
3. Yin Rui, Action Research on the Web-based and Problem-based Cooperative Learning Activity Design, Journal of modern educational technology, 2005,15(4) :34-39.
4. Wang Guangxin, Dong Fei,Summary of Educational Games Basic Structure and Character, Journal of China Educational Technology, 2009,(7) :77-80.
5. Wu Fati, The Principle of Objectives-Driven E-Learning Course Design, Journal of China Educational Technology, 2006,(1) :17-20.
6. Cheng Zhu, LI Guilin, LIU Haitao,The Developments of Chinese MOOC Platforms,Journal of Higher Education Research, 2014,37(2) :15-16.
7. Wu Jinhui, Contrastive Analysis on Main MOOC Platforms in China, Journal of Library Work in Colleges and Universities,2015,35(1) :11-14.
8. Huang Ming, Liang Xu,Gu Xiaolin, Mass Open Online Course-MOOC Overview,Beijing:Publishing House of Electronics Industry, 2015.
9. He Bin, CaoYang, SPOC: MOOC-based Innovation of Teaching Processes, Journal of China Educational Technology, 2015,(3) :22-27.
10. Li Lei,Li Xiaoli, Research on M-learning Resource Design Based on IOS Handheld Device, Journal of China Educational Technology, 2014,(12) :93-97.
11. Xu Gaopan, Zeng Wenhua, Huang Cuilan, Research on intelligent tutoring system, Journal of Application Research of Computers, 2009,26(11) : 4019-4022.

Collaborative Filtering Recommendation Based on Trust Model with Fused Similar Factor

Li Ye[1], Chunming Wu[1,*], Min Li[1]

[1]College of Computer & Information Science, Southwest University, 400715, China

Abstract. Recommended system is beneficial to e-commerce sites, which provides customers with product information and recommendations; the recommendation system is currently widely used in many fields. In an era of information explosion, the key challenges of the recommender system is to obtain valid information from the tremendous amount of information and produce high quality recommendations. However, when facing the large mount of information, the traditional collaborative filtering algorithm usually obtains a high degree of sparseness, which ultimately lead to low accuracy recommendations. To tackle this issue, we propose a novel algorithm named Collaborative Filtering Recommendation Based on Trust Model with Fused Similar Factor, which is based on the trust model and is combined with the user similarity. The novel algorithm takes into account the degree of interest overlap between the two users and results in a superior performance to the recommendation based on Trust Model in criteria of Precision, Recall, Diversity and Coverage. Additionally, the proposed model can effectively improve the efficiency of collaborative filtering algorithm and achieve high performance.

1 Introduction

The tremendous growth in information and number of global Internet users pose many key issues for recommender system to solve. Take the following for example: extracting useful information from all the massive data, producing high quality recommendations for the user.

Collaborative Filtering (CF) is currently the most widely used, personalized recommendation technology. The traditional collaborative filtering algorithm attempts to exploit the similarity between users who have the same interests and use the similarity to predict users' interests. Thus, it works by building a dataset of users' similarity that is based on the users' past actions and interests, then uses the user similarity to determine a neighbor set. A new user who is matched against the neighbor set discovers neighbors, which is the algorithm used to make the recommendation to the user. However, this algorithm has its own problems -- cold start, sparse data and low scalability [6]. At the same time, the accuracy of similarity between users is the key factor affecting the recommended performance. To improve the accuracy of the similarity between users, scholars have done a number of corresponding studies. Wu et al. [1] proposed a ratio-based approach to calculate the similarity between users. Chen et al. [2] proposed an improved algorithm based on the optimization of user similarity was proposed by adding the equilibrium factor to the traditional Cosine Similarity algorithm. Bobadilla et al. [3] used the traditional similarity to calculate the degree of preference of two users for one item, and the greater the degree of preference implies a greater coherence, which is equivalent to an optimization similarity strategy. Huang et al. [4] studied the similarity between users by structural similarity in complex networks. BK Patra et al. [5] a similarity measure based on neighborhood CF was proposed. It used the rating made by neighboring users. Guo et al. [6-8] improved the similarity degree by adding the machine learning model on the basis of collaborative filtering algorithm, such as trust model, clustering model, feature model and association rule model. Although the methods mentioned above can improve the user similarity measure to a certain extent, they do not take into account the social trust in the similarity of users. To target this problem, this paper proposes a novel method: collaborative filtering recommendation based on Trust Model with Fused Similar Factor (TMFSF), which achieves higher precision and higher performance in making recommendations to users.

2 The user-based collaborative filtering

Collaborative filtering algorithm is divided into user-based collaborative filtering and item-based collaborative filtering. Among them, user-based collaborative filtering algorithm mainly includes the following three steps:

First, establish the user model, which namely is the users - items matrix. Suppose there are m users and n items, you can get a scale $m \times n$ matrix. As shown in Table.1, Where R_{ij} represents the ratings of user U_i for item I_j, and R_{ij} can be empty.

* Corresponding author: springsun@swu.edu.cn

Table 1. The users - items matrix.

	I_1	I_2	...	I_j	...	I_n
U_1		2				
U_2		4			3	
...	3					5
U_i		3		R_{ij}		
...				5		2
U_m		5				

Secondly, find the nearest neighbor of the target user. According to the user's rating on the project, find the similarity $Sim(U_i,U_j)$ between the target user and other users. Then, descend neighbors according to the similarity value, and get the nearest neighbor set by a given similarity threshold or "nearest" neighbors. The calculation method of user similarity mainly includes Cosine Similarity, Pearson Similarity and Modified Cosine Similarity. Among them, the Modified Cosine Similarity method is described in the formula (1):

$$Sim(U_i,U_j)=\frac{\sum_{c\in I_{i,j}}(R_{i,c}-\overline{R_i})(R_{j,c}-\overline{R_j})}{\sqrt{\sum_{c\in I_{i,j}}(R_{i,c}-\overline{R_i})^2}\sqrt{\sum_{c\in I_{i,j}}(R_{j,c}-\overline{R_j})^2}} \tag{1}$$

Where $Sim(U_i,U_j)$ is the similarity between user U_i and user U_j. $R_{i,c}$ and $R_{j,c}$ are the ratings of user U_i on item C. $\overline{R_i}$ and $\overline{R_j}$ are the average ratings of user U_i and user U_j respectively. $I_{i,j}$ indicates the item that user U_i and user U_j's collectively ratings.

Thirdly, predict the ratings and generate a recommended list. Evaluate the target user's rating on the unrated item according to the "nearest" neighbor's ratings, generating a item recommendation set. Assuming that the number of "nearest" neighbors of the target user U_i is n. The predictive method of ratings is shown in formula (2).

$$P_{i,c}=\overline{R_i}+\frac{\sum_{m=1}^{n}Sim(U_i,U_j)(R_{j,c}-\overline{R_j})}{\sum_{m=1}^{n}Sim(U_i,U_j)} \tag{2}$$

3 Trust Model with Fused Similar Factor

When facing a decision, we are more likely to accept viewpoints from credible sources. Hence, it is a valuable opportunity to improve recommendation quality by sufficiently and effectively utilizing trust information. The introduction of a trust model based on the trust relationship can adjust the weight of the similarity in the recommendation process [11], which in turn can eliminate the effectives caused by the sparseness original matrix, and improve the recommended quality to a certain extent.

In the traditional user-based collaborative filtering, the matrix of ratings is usually sparse and measurement of users' similarity always has low precision, the recommended quality is generally not satisfactory. However, the trust model does not have this limitation because it eliminates the precision error in measuring users' similarity

Basing on the above considerations, this section gives the "trust model" and "similar factor" definition firstly. And then, introduce the similar factor combining with the Modified Cosine Similarity. Finally, introduce the similarity of fusing similar factor combining with the trust model, which is used as the comprehensive weight to replace the similarity in traditional algorithm. According to the ideas in the paper [9], similarity expressed the degree of coincidence of interest. The similar factor of introduced by this paper is used to improve the accuracy of the similarity, and then combined with the trust model. The harmonic weight is calculated replace the similarity, so as to improve the recommendation quality.

Trust Model with Fused Similar Factor specific process shown in Fig.1.

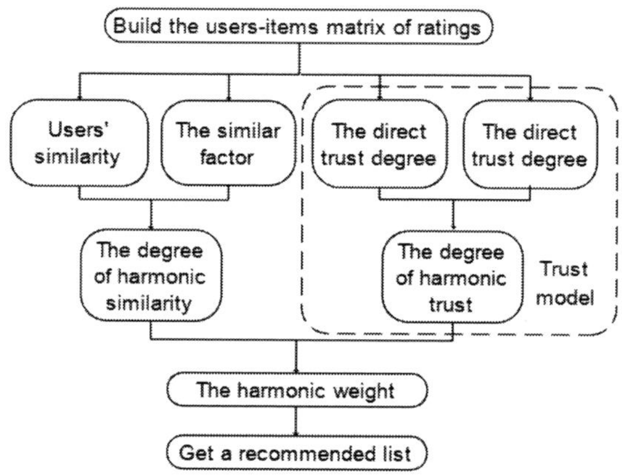

Fig. 1. Trust Model with Fused Similar Factor.

3.1 The Trust Model

The trust model is classified into direct trust degree and overall adjustment of trust degree [10]. The direct trust degree is the difference of two individual users' interests that is calculating directly by the users-items matrix. The overall adjustment of trust degree represents that user is trusted degree in the overall system. It is used to revise the direct trust. This paper combines of the direct trust degree and the overall adjustment of trust degree to get the degree of harmonic trust.

3.1.1 The Direct Trust Degree

In order to quantify the direct trust degree, this paper draws on the idea of the paper [10], using the formula (3) to calculate the direct trust degree $Trust1(U_i,U_j)$.

Definition 1: the direct trust degree [10].

$$Trust1(U_i,U_j)=\frac{Correct(U_i,U_j)}{C_{ij}} \tag{3}$$

where $Correct(U_i,U_j)$ is the correct number recommended by user U_j for user U_i, within the range of the number of common items rated by user U_i and user U_j. $C_{ij}=|C_i \cap C_j|$ represents the number of user U_i and user U_j common items of ratings.

The so-called "correct recommendation" means that the user U_j speculates that the speculative ratings of the user U_i generates a difference from the actual ratings of the user U_i. When the difference is less than the threshold e, it is considered to be the correct recommendation, see the formula (4).

$$\begin{cases} TR_{i,s} = R_i + \left(\mathrm{R}_j - \mathrm{R'}_j\right) * \mathrm{Sim'}\left(\mathrm{U}_i, \mathrm{U}_j\right) \\ |\, \mathrm{TR}_{i,s} - \mathrm{P}_i \,| < \mathrm{e} \end{cases} \tag{4}$$

where $TR_{i,s}$ denotes that user U_j estimates the speculative ratings of user U_i on item s. R_i represents the average ratings of user U_i. R_j represents the speculative rating of the item s by the user U_j. R'_j represents the actual rating of the item s by the user U_j. $Sim'(U_i,U_j)$ represents the harmonic similarity of user U_i and user U_j. P_i represents the actual rating of the item s by the user U_i. When e is a fixed value, when it is closer to 0, the correct recommendation is more accurate.

3.1.2 The Overall Adjustment of Trust Degree

In the recommendation system, there is a phenomenon that some users maybe become a lot of people "recent" neighbor, and some users maybe become very few people "recent" neighbors [10, 11]. It can be argued that the credibility of active users in the system is higher than that of inactive users. In calculating the trust degree of the target user to the neighbor user, in addition to considering the direct trust degree, it is also necessary to consider the performance of the neighbor user in all users [10].

Definition 2: The overall adjustment of trust degree. Set f_i be the total number of time s user U_i have evaluated in the system. q_i is the number of times a user has made a recommendation for other users. That is, the number of times the user become the " nearest" neighbor of other users.

$$Trust2(U_i) = \frac{2}{\dfrac{1}{1-1/\ln f_i} + \dfrac{1}{1-1/\ln(q_i+3)}} \tag{5}$$

3.1.3 The Degree of Harmonic Trust

The direct trust degree and the overall adjustment of trust degree combined to get the degree of harmonic trust, see the formula (6).

$$T\left(U_i,U_j\right) = \frac{2 * Trust1\left(U_i,U_j\right) * \mathrm{Trust}\,2\left(U_i\right)}{Trust1\left(U_i,U_j\right) + \mathrm{Trust}\,2\left(U_i\right)} \tag{6}$$

Where $T(U_i,U_j)$ represents the degree of harmonic trust between user U_i and user U_j, $Trust1(U_i,U_j)$ represents the direct trust degree of user U_j to user U_i, $Trust2(Ui)$ represents the overall adjustment of trust degree of user U_i to user U_j.

3.2 The Similar Factor

User's similarity indicates the degree of similarity between the user's preferences for the item. The accuracy of the similarity will directly affect the user's forecast ratings. Since the matrix of ratings is usually very sparse, there is insufficient information to deal with, which can lead to a significant reduction in the accuracy of similarity. If only calculating the similarity according to formula (1) will result in a large error [8]. Assume that two users rate items that are of interest to a group of items. It is obvious that if the number of common items rated by two people is more, the closer the interest preferences of the two users are. That is, the greater the similarity of users. The algorithm that introduces the trust model does not deal with the similarity between two users. Therefore, the similarity is introduced to preprocess the similarity.

Definition 3 : User ' s similar factors. Set $C_A=\{c_i|i \in [1, N_1]\}$ be the list of items evaluated by user U_A. Set $C_B=\{c_j|j \in [1, N_2]\}$ be the list of items evaluated by user U_B. $C_{AB}=|C_A \cap C_B|$ represents the number of user U_A and user U_B common items of ratings. S_{AB} represents the average of the user U_A and user U_B common items of ratings. Define the user similar factor, see formula (7).

$$\gamma_{ab} = \frac{c_{AB} + \alpha}{s_{AB} + \alpha} \tag{7}$$

Where α is a parameter of similar factor. Its value is to avoid the above formula in the denominator is 0 lead to the meaning of the formula, and it is also possible to prevent the value of γ_{ab} from 0 to cause the similarity to be 0. γ_{ab} represents a similarity factor between two users.

Based on the similar factor γ_{ab}, this article revises the user similarity, Based on the similar factor R, this paper revises the user similarity, named as the degree of harmonic similarity $Sim'(U_i,U_j)$, see formula 8.

$$Sim'\left(\mathrm{U}_i,\mathrm{U}_j\right) = \frac{2 * Sim\left(\mathrm{U}_i,\mathrm{U}_j\right) * \gamma_{ij}}{Sim\left(\mathrm{U}_i,\mathrm{U}_j\right) + \gamma_{ij}} \tag{8}$$

The degree of harmonic similarity $Sim'(U_i,U_j)$ replaces the similarity as the measure of the final users' similarity.

3.3 The Harmonic Weight

In the premise of revising the similarity of users, the introduction of trust model, the combination of the two can take full account of the user in the purchase of goods when the actual situation.

In this paper, combine the degree of harmonic similarity with the degree of harmonic trust, and get the harmonic weight instead of $Sim(U_i,U_j)$ in the formula (2), see the formula (9).

$$Weight(\mathrm{U}_i,\mathrm{U}_j) = \frac{2 * Sim'(\mathrm{U}_i,\mathrm{U}_j) * \mathrm{T}(\mathrm{U}_i,\mathrm{U}_j)}{Sim'(\mathrm{U}_i,\mathrm{U}_j) + \mathrm{T}(\mathrm{U}_i,\mathrm{U}_j)} \tag{9}$$

Where $Weight(U_i,U_j)$ represents harmonic weight. $Sim'(U_i,U_j)$ represents the degree of harmonic similarity with the similar factor , represents the degree of harmonic trust.

3.4 Generate Recommendations

The traditional user-based collaborative filtering algorithm in the similarity replaced by the harmonic weight. The predicted ratings are obtained according to the formula (10), and the *TopN* recommendation list is generated according to the predicted ratings.

$$P_{i,c} = \overline{R_i} + \frac{\sum_{m=1}^{n} Weight(U_i, U_j)(R_{j,c} - \overline{R_j})}{\sum_{m=1}^{n} Weight(U_i, U_j)} \quad (10)$$

4 Trust Model with Fused Similar Factor

By introducing the above ideas into the traditional user-based collaborative filtering algorithm, an improved weighted cooperative filtering algorithm based on trust model is proposed. The algorithm firstly improves the accuracy of similarity, and then combines with the trust model to get the harmonic weight.

4.1 Algorithm Description

The following gives the algorithmic description of the degree of harmonic trust, the degree of harmonic similarity and harmonic weights.

Algorithm 1 Implement the degree of harmonic similarity

Input: *Sim_matrix* (The matrix of similarity between users)
tot_matrix(The total matrix of ratings)
a (The parameter of similar factor)
Output: *Sim'*(U_i, U_j) (The degree of harmonic similarity)

(1) On the basis of the matrix of the user's ratings, the similarity $Sim(U_i, U_j)$ of the target user U_i and the other user U_j ($j \in [1, n]$, $U_i \neq U_j$) is obtained.
(2) The items of ratings for users U_i and U_j are obtained respectively, denoted as $itemI=\{k \mid item_k>0,\ k \in Z,\ k \in itemNumber\}$ and $itemJ=\{n \mid item_n>0,\ n \in Z,\ n \in itemNumber\}$.
(3) Use *Count* to record the number of common items rated by user U_i and user U_j, and if not, then *Count* is zero.
(4) Use *Average(Sum(itemI),Sum(itemJ))* to record the average of user U_i and user U_j common rating items.
(5) Calculate the similar factor $\gamma(U_i, U_j)$ according to formula (3).
(6) Calculate the degree of harmonic similarity $Sim'(U_i, U_j)$ according to formula (4).

Algorithm 2 Implement the degree of harmonic trust and harmonic weight

Input: *tot_matrix* (The total matrix of ratings)
bnc(The number of "nearest" neighbors per user)
correct(The correct number of recommendations)
Output: *Weight*(U_i, U_j) (Harmonic weight)

(1) Calculate the correct number of users U_i based on the matrix of the user's ratings. According to the formula (6) and its evaluation criteria, the correct evaluation number $Correct(U_i, U_j)$ $(j=1,2,...,i-1,i+1,...,n)$ is obtained.
(2) According to the formula (5) to obtain direct trust degree *Trust1*.
(3) Calculate user U_i to be the number of nearest neighbors for all other users, as $bnc=becomeNeighborCount(U_i)$.
(4) Get a collection of user U_i rated, as $itemI=\{k \mid item_k>0,\ k \in Z,\ k \in itemNumber\}$. The total evaluation number is calculated according to the item rating of user U_i, as *ReNumber=Sum(itemI)*.
(5) According to step (1), step (3) and formula (7) to get the user to the overall adjustment of trust degree *Trust2(bnc,ReNumber)*.
(6) According to the formula (8) to get the degree of harmonic trust *Trust(Trust1, Trust2)*.
(7) According to the formula (9) to get the harmonic weight $Weight(Sim'(U_i, U_j), Trust)$.

4.2 Metrics

In order to validate the algorithm, this paper uses the Precision[13], Recall[14], Diversity[16] and Coverage[17] to analyze the experimental process. And use MAE (Mean Absolute Error) [12] and RMSE (Root Mean Square Error) as the measure of similarity factor a.

The Precision is in relation to the predicted results, which refers to the percentage of the extracted correct information that accounts for all the recommended information. The Recall is in relation to the original sample, which refers to the percentage of the extracted correct information that accounts for all the recommended information in the sample. Diversity refers to the similarity between recommended lists. In other words, the overlap between the user's recommended lists defines the overall diversity. Coverage is the ratio of the number of items a system recommends to all users to the total number of items. Let *Ttest* represent all the recommended information in the sample. *Treco* represents all the information recommended by the recommender system for the user. That is ,the recommended list length. *TP* represents the user's favourite information. Precision(P), Recall(R), Diversity(D) and Coverage(C) are defined as follows:

$$P = \frac{TP}{Treco} \quad (11)$$

$$R = \frac{TP}{Ttest} \quad (12)$$

$$D = \frac{2}{Ttest(Ttest-1)} \sum_{i,j \in U, i \neq j} \frac{|Treco_i \cap Treco_j|}{|Treco_i|} \quad (13)$$

$$C = \frac{N_d(Treco)}{Ttest} \quad (14)$$

The MAE measures the accuracy of the forecast by the deviation between the user's predicted ratings and the user's actual ratings. The smaller the value of MAE, the smaller the difference between the predicted ratings and the actual ratings. This means that the forecast is more accurate and recommend higher quality. Set N to the number of items rated by the users in the test set. $P=\{i|P_i ,i=1,2,...,k\}$ represents the actual ratings of the user. $Q=\{i|Q_i ,i=1,2,...,k\}$ represents the user's predicted ratings, The MAE is defined as:

$$MAE = \frac{\sum_{i=1}^{N} | P_i - Q_i |}{N} \tag{15}$$

RMSE is a frequently used measure of the differences between values predicted by a model or an estimator and the values actually observed. RMSE can well reflect the precision of the measurement. Set N to the number of items rated by the users in the test set. $P=\{i|P_i ,i=1,2,...,k\}$ represents the actual ratings of the user. $Q=\{i|Q_i ,i=1,2,...,k\}$ represents the user's predicted ratings . The RMSE is defined as:

$$RMSE = \sqrt{\frac{\sum_{i=1}^{n} (P_i - Q_i)^2}{n}} \tag{16}$$

5 Simulation Experiment and Result Analysis

In order to verify the effectiveness of the proposed method, we conducted a series of experimental analysis using the *MovieLens* small-scale data set provided by the *Grouplens* Working Group. The dataset contains 100 thousand sets of ratings from 943 independent users on 1682 movies, and the rating is an integer from 1 to 5. In the experiment, establish the matrix of ratings for the data set firstly. Secondly, the 100 thousand ratings are divided into 80 thousand training set and 20 thousand test set. Then, the simulation experiment and contrast analysis are carried out based on the user-based collaborative filtering algorithm, collaborative filtering recommendation based on Trust Model and collaborative filtering recommendation based on Trust Model with Fused Similar Factor.

5.1 Complexity Analysis

The time complexity of computing the similar factor, the direct trust degree, the degree of harmonic similarity, the degree of harmonic trust and the harmonic weight are all $O(n^2)$. However, time complexity of the overall adjustment of trust degree is $O(n^3)$, because it has to calculate the number of times users have made recommendations for other users.

5.2 Determine the Parameter of the Similar Factor

In the experimental modeling process, we firstly need to determine the parameter α in the formula (3) to solve the similar factor. The paper [9] pointed out that MAE will appear the minimum when α ranges from 1.0 to 1.5. Therefore, in order to determine the specific value of α in [1.0, 1.5], and the following experimental setup was made: Set the number of neighbors in the model to 5, 10, 15, 20, 25, 30, 40, 60, 80, 100, and set the correct recommended threshold e = 0.5. According to the formula (8), we can calculate the degree of harmonic similarity, and the average values of MAE and RMSE corresponding to different neighbors are obtained at different values of α . The results are shown in Table.2.

Table 2. The average value of MAE and RMSE.

α	MAE	RMSE
1.0	0.8733	1.2128
1.1	0.8739	1.2147
1.2	0.8737	1.2140
1.3	0.8757	1.2165
1.4	0.8750	1.2156
1.5	0.8734	1.2132

Draw Fig.2 according to Table.2. Fig.2(a) and Fig.2(b) show the error curves for MAE and RMSE respectively. According to Fig.2 (a) and Fig.2 (b) can clearly see that when the α value of 1.0, MAE and RMSE are the minimum. Therefore, the value of α is set to 1.0.

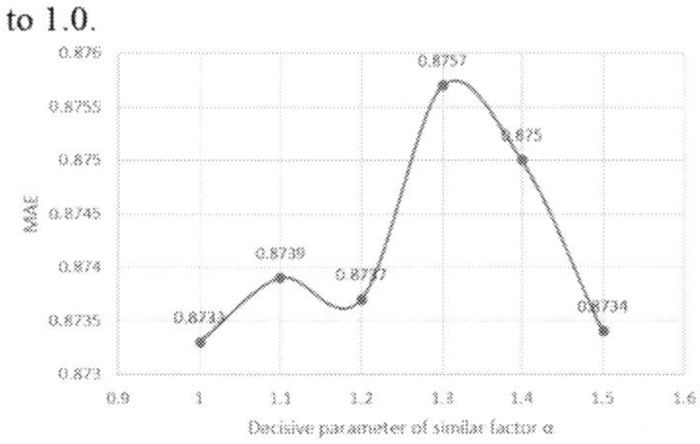

Fig. 2(a). The average value of MAE.

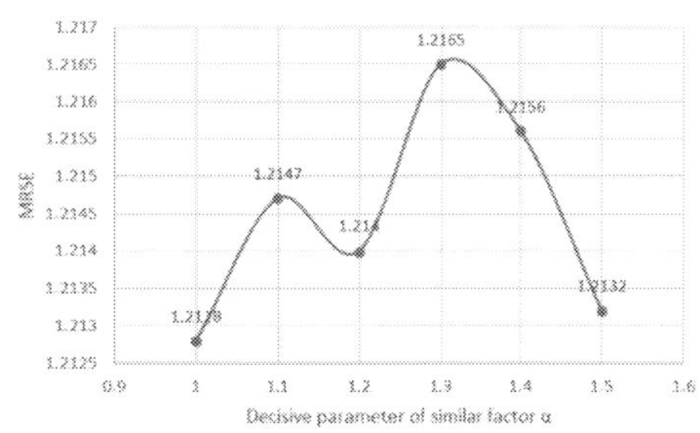

Fig. 2(b). The average value of RMSE.

5.3 Comparative Experiment

This paper compares collaborative filtering recommendation based on Trust Model with Fused Similar Factor (TMFSF) with collaborative filtering recommendation based on Trust Model (TM) and the User-Based collaborative filtering algorithm (UBCF). Precision, Recall, Diversity and Coverage are used in this article to measure the performance of the above algorithms. The experimental results are as follows:

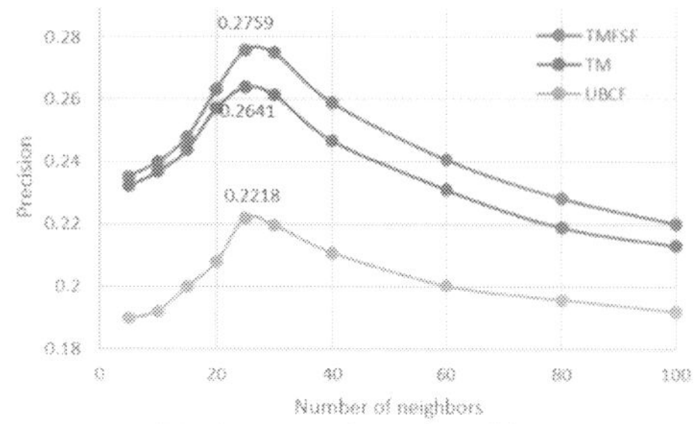

Fig. 3. Comparison of Precision.

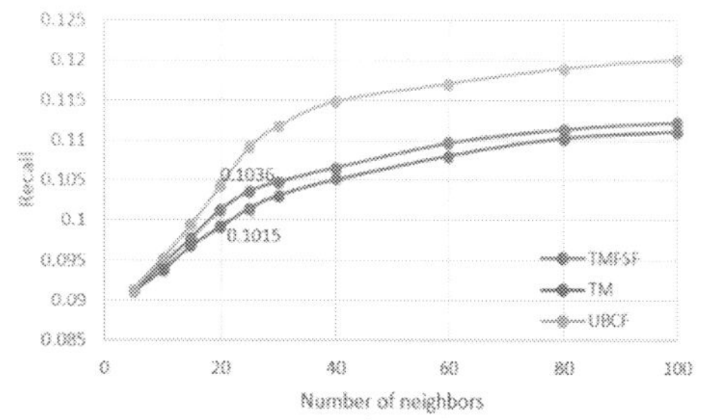

Fig. 4. Comparison of Recall.

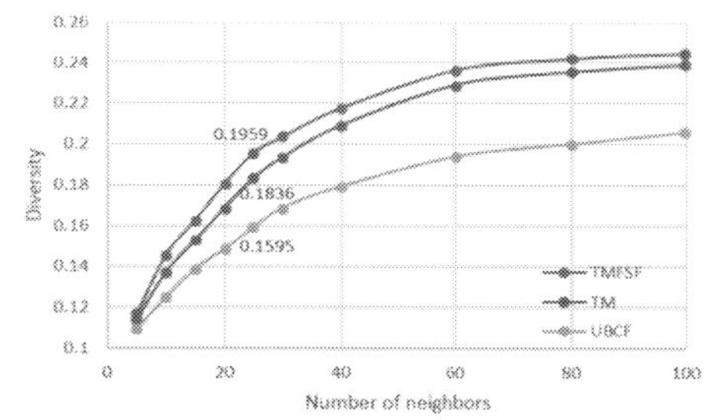

Fig. 5. Comparison of Diversity.

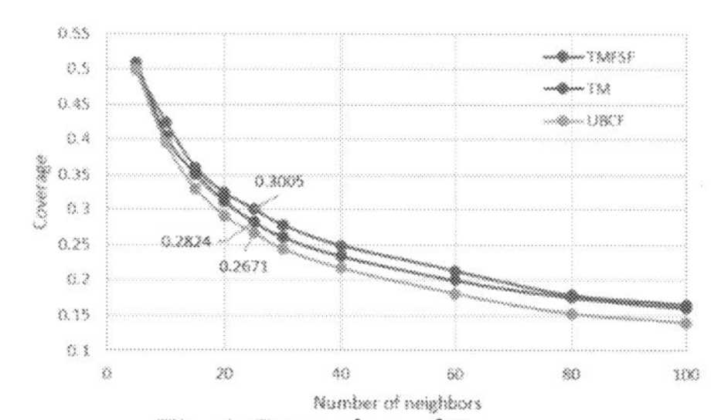

Fig. 6. Comparison of Coverage.

According to Fig.3, we can see, when the number of neighbors at 25, the Precision to reach the maximum. The TMFSF collaborative filtering algorithm is nearly 1.18% higher than the TM collaborative filtering algorithm. And it is 5.4% higher than the UBCF algorithm. The TMFSF collaborative filtering algorithm is superior to the UBCF algorithm and the TM algorithm.

As shown in Fig.4, the Recall of the TMFSF collaborative filtering algorithm is gradually increased as the number of neighbors increases. When the number of neighbors is 25, the TMFSF collaborative filtering algorithm is 0.21% higher than the TM collaborative filtering algorithm

Fig.5 is a comparison of the Diversity among the three groups. The TMFSF collaborative filtering algorithm is 1.23% higher than the TM collaborative filtering algorithm when the neighbor is 25. And it is 3.64% higher than the UBCF algorithm. This means that the TMFSF collaborative filtering algorithm can provide users with more choices.

Fig.6 is a comparison of the Coverage among the three groups. The TMFSF collaborative filtering algorithm is 1.81% higher than the TM collaborative filtering algorithm when the neighbor is 25. And it is 3.34% higher than the UBCF algorithm. Therefore, the TMFSF collaborative filtering algorithm can improve the user's satisfaction to some extent compared with the TM collaborative filtering algorithm.

The experimental results show that the TMFSF collaborative filtering algorithm in this paper has higher performance than the TM collaborative filtering algorithm, and its Precision, Recall, Diversity and Coverage have higher performance. Therefore, it can obtain High recommended quality.

6 Conclusions

In this paper, the concept of similar factor is introduced as an add-on to the collaborative filtering algorithm, which is based on the trust model. In calculating the user similarity, the impact of the number of common ratings between users is taken in to account. Similar factor is used to revise the similarity to make the similarity more accurate. The adjusted similarity is combined with the trust model to get the harmonic weights. Harmonic weights are combined with traditional collaborative filtering to produce higher quality recommendations. Consequently, we proved that the proposed algorithm Collaborative Filtering Recommendation Based on Trust Model with Fused Similar Factor is better than the traditional algorithm in Precision, Recall, Diversity and Coverage.

References

1. Wu, Xiaokun, B. Cheng, and J. Chen. "Collaborative Filtering Service Recommendation Based on a Novel Similarity Computation Method." IEEE Transactions on Services Computing **10**, (2017), 352-365
2. Chen. Hao, Z. Li, and W. Hu. An improved collaborative recommendation algorithm based on optimized user similarity. Journal of Supercomputing **72**, (2015), 2565-2578
3. Bobadilla, Jesús, F. Ortega, and A. Hernando. A collaborative filtering similarity measure based on singularities. Information Processing & Management **48**, (2012), 204-217
4. T Huang, R Huang. Improved Collaborative Filtering Recommendation Algorithm, Computer Science, **43**, (2016), 400-403
5. BK Patra, R Launonen, V Ollikainen, S Nandi.A new similarity measure using Bhattacharyya coefficient for collaborative filtering in sparse data, Knowledge-Based Systems,**82**, (2015), 163-177

6. G Guo,J Zhang,N Yorke-Smith.Leveraging multiviews of trust and similarity to enhance clustering-based recommender systems, Knowledge-Based Systems, **74**, (2015), 14-27
7. P Chen, C Chen. Incorporating Association Rules for Collaborative Filtering Recommendation Algorithm, Journal of Chinese Computer Systems, **37**, (2016), 287-292
8. Peng Zhenlian, Wang Jian, He Keqing, Tang Mingdong. A Requirements Elicitation Approach Based on Feature Model and Collaborative Filtering. Journal of Computer Research and Development, **53**, (2016), 2055-2066
9. GAO Quan-li, GAO Ling etc. Weighted Collaborative Filtering Algorithm with Fused Impact Facto [J]. Computer Engineering, **40**, (2014), 39-41
10. LU Kun, XIE Ling, LI Ming-chu. Research on Implied-trust Aware Collaborative Filtering Recommendation Algorithm [J]. Journal of Chinese Computer Systems, **37**, (2016), 241-243
11. GUO Yan-hong, DENG Gui-shi, LUO Chun-yu. Collaborative Filtering Recommendation Algorithm Based on Factor of Trust [J]. Computer Engineering, **34**, (2008), 1-3
12. WANG Jing, YIN Jian etc. Collaborative Filtering Recommendation Algorithm Based on Co-ratings and Similarity Weight [J]. Computer Science, **37**, (2010), 99-104
13. XU Xiang-yu, LIU Jian-ming. Collaborative Filtering Recommendation Algorithm Based on Multi-level Item Similarity [J]. Computer Science, **34**, (2016), 262-265
14. XIA Pingping , SHUAI Jianmei. Collaborative Filtering Algorithm Based on Similarity Extension and Interest Degree Scaling [J]. Computer Engineering, **42**, (2016), 199-202
15. WEN Jun-hao, SHU Shan. Improved Collaborative Filtering Recommendation Algorithm of Similarity Measure [J]. Computer Science, **41**, (2014), 68-71
16. C.Ziegler, S.M. McNee, J. A. Konstan et al. Improving recommendation lists through topic diversification. In: Proceedings of the 14th International Conference on World Wide Web. Chiba: ACM, (2005), 22-32
17. Ge M, Delgado-Battenfeld,Jannach D. Beyond accuracy: Evaluating recommender systems by coverage and serendipity. In:RecSys (2010): the 2010 ACM conference on Recommender systems. Barcelona:ACM,2010,257-260
18. Ge, Mouzhi, C. Delgado-Battenfeld, and D. Jannach. Beyond accuracy: evaluating recommender systems by coverage and serendipity. Recsys, (2010),257

Statistical Variation Analysis Using Pearson Distribution Family Based on Jacobian-Torsor Model

Siyi Ding[1,2], *Sun* Jin[1,2,*], *Zhimin* Li[1,2], *Fuyong* Yang[1], *Jia* Lin[2]

[1]State Key Laboratory of Mechanical System and Vibration, Shanghai Jiao Tong University, Shanghai, 200240, P.R. China
[2]Shanghai Key Laboratory of Digital Manufacture for Thin-walled Structures, Shanghai Jiao Tong University, Shanghai, 200240, P.R. China

Abstract. Assembly variations are unavoidable due to parts' geometrical errors. Statistical variation analysis is an effective method to quantitatively predict product quality in the original design stage. However, traditional methods can't handle the problem of abnormal distribution of the actual variation variables. Meanwhile, they are underdeveloped in regard to the complex geometrical errors in spatial 3D state. To overcome this problem, firstly, Jacobian-Torsor model is used to build the variation propagation, which is well suited to a complex assembly that contains large numbers of joints and geometric tolerances; secondly, Pearson distribution family is adopted to determine probability distribution pattern and build probability density function. By comparing results of the suggested method to the Monte Carlo method, it is observed that this novel method has the same accuracy, but much higher efficiency. The results also demonstrate that probability distribution types of the parts variations have a significant impact on the final assembling variation.

1 Introduction

Variation is inherent during any manufacturing and assembly process owing to the variability of the part, datum, positioner, fixture, etc. They cause minor variations in components from the nominal geometry [1]. Statistical variation analysis is an effective way to predict product quality quantitatively in the design stage [2]. Traditional methods are still based on engineering experience, extremum method [3, 4], or the mean square root method [5, 6]. The results obtained by extremum method are overly pessimistic which will cause high fabricating cost, while the mean square root method only gives results for the mean-square variation which is primarily used to solve the plane size-chain problem. These methods are underdeveloped in regard to the geometrical errors in spatial 3D state. Meanwhile, they widely adopt normality assumption which is convenient for statistical evaluation.

As is known to all, most variables of parts variations disobey normal distribution in the actual machining process. For instance, if tool wearing plays a leading role in the machining process, the actual variation variables will satisfy uniform distribution; if one manufacturing process is accomplished using two machine tools, the mixing variables of parts variations will follow bimodal distribution; when turning shaft journals or apertures, operators usually keep a certain allowance, and this will lead to skewed distribution. Monte Carlo (MC) simulation is suitable for the data of abnormal distributions. Currently, MC method has been developed into a core of some commercial softwares such as CETOL, VSA, 3DCS, etc. A big disadvantage to this approach, however, is that large amounts of data samples are required to achieve accurate predictions [7], which inevitably result in time-consuming computation. Thus there is a clear need to adopt a simple and rapid probabilistic approach for statistical variation analysis.

Yang et al. [8] studied the probability distribution of table-axis error in aero-engine assembly. Probability density functions (PDF) were derived based on Rayleigh distribution. Ghie [9] applied uncertain measurements into Jacobian-Torsor model. But the PDF deduced could not consider skewed distribution of variables. Ahsanullah and Hamedani [10] attempted to investigate various characterizations of certain univariate continuous distributions, providing a base of further research of multivariable continuous distributions. Kharoufehy and Chandraz [11] adopted a multivariate kernel density estimate to approximate the probability distribution. But it was difficult to determine the smoothing parameter (h). In this paper, we will introduce an innovation of Pearson distribution family [12] to model PDF that covers polymorphic probability distributions.

In addition, Jacobian-Torsor theory [13] will be used to model the variation propagation, which is well suited to a complex assembly that contains large numbers of joints and geometric tolerances. It combines the advantages of the torsor model which is suitable for tolerance representation and the Jacobian matrix which is suitable for tolerance propagation.

* Corresponding author: jinsun@sjtu.edu.cn

2 Related works

2.1 Jacobian–Torsor model

The unified Jacobian-Torsor model is composed of the Jacobian matrix and torsor model. Torsor model uses three translational vectors and three rotational vectors to express the position and orientation of an ideal surface. Supposing that S_0 is the nominal position and S_1 is an arbitrary variational surface relative to S_0 with tolerance t, the torsor model of S_1 which represents position and orientation can be represented as:

$$T_{1-0} = \begin{bmatrix} u & v & w & \alpha & \beta & \gamma \end{bmatrix}^T \tag{1}$$

where u, v, w are three specific vectors along the axes x, y and z in local reference system respectively; likewise, α, β and γ indicate the vectors around the axes x, y and z respectively.

These vectors are transferred to the objective controlled variable by virtue of the Jacobian matrix. Jacobian matrixes describe relative positions, as well as relative orientations between local reference frames and the global reference frame. The unified Jacobian-Torsor model can be expressed as:

$$\begin{bmatrix} (\underline{u},\overline{u}) \\ (\underline{v},\overline{v}) \\ (\underline{w},\overline{w}) \\ (\underline{\alpha},\overline{\alpha}) \\ (\underline{\beta},\overline{\beta}) \\ (\underline{\gamma},\overline{\gamma}) \end{bmatrix}_{FR} = \begin{bmatrix} [J]_{FE1} & \cdots & [J]_{FEn} \end{bmatrix} \times \begin{bmatrix} \begin{bmatrix} (\underline{u},\overline{u}) \\ (\underline{v},\overline{v}) \\ (\underline{w},\overline{w}) \\ (\underline{\alpha},\overline{\alpha}) \\ (\underline{\beta},\overline{\beta}) \\ (\underline{\gamma},\overline{\gamma}) \end{bmatrix}_{FE1} & \cdots & \begin{bmatrix} (\underline{u},\overline{u}) \\ (\underline{v},\overline{v}) \\ (\underline{w},\overline{w}) \\ (\underline{\alpha},\overline{\alpha}) \\ (\underline{\beta},\overline{\beta}) \\ (\underline{\gamma},\overline{\gamma}) \end{bmatrix}_{FEn} \end{bmatrix}^T \tag{2}$$

where α, β, γ, u, v and w are the same as Eq. (1); n is the number of functional elements (*FE*s); $(\underline{\alpha},\overline{\alpha})$ is tolerance interval where α should belong to; likewise, other torsors of components follow the similar way as α.

There are two types of *FE* pairs in assembly, i.e., internal *FE* (*IFE*) pairs which are composed of two *FE*s on the same part and contact *FE* (*CFE*) pairs which are made up of two *FE*s on different parts. Variations of each *FE* pair can be transferred to functional requirement (*FR*) with a Jacobian matrix, which of the ith *FE* can be expressed as:

$$[J]_{FEi} = \begin{bmatrix} [R_0^i]_{3\times3} \cdot [R_{Pti}]_{3\times3} & \vdots & [W_i^n]_{3\times3} \cdot ([R_0^i]_{3\times3} \cdot [R_{Pti}]_{3\times3}) \\ \cdots & \vdots & \cdots \\ [0]_{3\times3} & \vdots & [R_0^i]_{3\times3} \cdot [R_{Pti}]_{3\times3} \end{bmatrix} \tag{3}$$

where $[R_0^i]$ represents the local orientation of the ith frame with respect to the 0th frame which is global reference system; $[R_{Pti}]$ is a projection matrix designating the unit vectors along local axes respectively for tolerance zone tilted according to the direction of tolerance analysis; $[W_i^n]$ is a skew-symmetric matrix [13] allowing the representation of the vector among the ith and n^{th} frame (end point).

2.2 Pearson distribution family

Pearson system of distributions was firstly proposed by statistician K. Pearson in 1895, which included various types of distributions. Pearson system adopts four parameters to characterize each distribution. The probability density function (PDF) $f(x)$ depends on the following differential equation:

$$\frac{df(x)}{dx} = \frac{(a_0 + a_1 x) f(x)}{b_0 + b_1 x + b_2 x^2} \tag{4}$$

While a_1=1, the parameters a_0, b_0, b_1, and b_2 depend entirely on the first four order central moments of distributions μ_1, μ_2, μ_3 and μ_4. It can also be expressed as moment ratio, i.e. $\beta_1=\mu_3^2/\mu_2^3$ and $\beta_2=\mu_4/\mu_2^2$, as follows:

$$a_0 = -b_1 \tag{5}$$

$$b_0 = A\sigma^2(4\sigma^2\mu_4 - 3\mu_3^2) = B\sigma^2(4\beta_2 - 3\beta_1) \tag{6}$$

$$b_1 = A\mu_3(\mu_4 + 3\sigma^4) = B\sigma\sqrt{\beta_1}(\beta_2 + 3) \tag{7}$$

$$b_2 = A(2\sigma^2\mu_4 - 3\mu_3^2 - 6\sigma^6) = B(2\beta_2 - 3\beta_1 - 6) \tag{8}$$

where

$$A = -(10\sigma^2\mu_4 - 18\sigma^6 - 12\mu_3^2)^{-1} \tag{9}$$

$$B = -(10\beta_2 - 12\beta_1 - 18)^{-1} \tag{10}$$

The general solution of $f(x)$ are:

$$f(x) = C\exp\left\{-\int \frac{a_0 + a_1 x}{b_0 + b_1 x + b_2 x^2} dx\right\} \tag{11}$$

To be specific, Pearson system mainly contains seven distribution patterns, which are elaborated in [14]. These types can be distinguished according to the different distributions of roots of the quadratic equation $b_0+b_1x+b_2x^2=0$. Supposing $D=b_1^2-4b_0b_2$, $\lambda=b_1^2 \div (4b_0b_2)$, the Pearson distribution pattern can be judged by the value of D and λ, as provided in Table.1.

Table 1. Pearson distribution types.

a_1	D	λ	Distribution pattern	Remarks
0	0	0	Exponential distribution	
$\neq 0$	>0	$-\infty$	Pattern VII	$b_0=0$
		$(-\infty,0)$	Pattern I	
		0	Pattern II	
		$(1,\infty)$	Pattern VI	
		∞	Pattern III	
	$=0$	1	Pattern V	$b_1=b_2=0$
		Uncertain	Normal distribution	
	<0	(0,1)	Pattern IV	

3 Statistical analysis

According to the Pearson system's characteristics, distribution pattern can be determined and PDF can be derived while the first four central moments are known. As for arbitrary distributions, the first-order central moment μ_1 is identically equal zero; the second-order central moment μ_2 reflects data dispersion degree, which is equal to variance σ^2; the third-order central moment μ_3 is related to coefficient of skewness S, and $S=\mu_3/\sigma^3$; and the μ_4 is related to coefficient of kurtosis K, and $K=\mu_4/\sigma^4$. Due to the average value, variance, skewness and kurtosis of a probability distribution can visually reflect the shape of the distribution, we will solve these four parameters in this paper to indirectly gain the first four central moments.

Eq. (2) can be represented as $y_{FR}=f(u_i, vi, wi, \alpha i, \beta i, \gamma i,)$, $(i=1,2,\cdots,n)$, where y_{FR} is the *FR*'s variation, and (u_i, *vi, wi, αi, βi, γi,*) is the *FE*'s shifting variation or rotating variation in all directions. This representation is fit for deterministic variation analysis and beneficial to the error compensation. Supposing all of the *FE*'s variations are randomly distributed and independent of one another, we can solve the four parameters of the *FR* by means of the *FE*'s parameters. Then a_0, b_0, b_1, b_2 can be obtained with the known distribution parameters of *FR*. Using the suggested method (Table.1), distribution pattern can be distinguished and PDF can be solved, after which the distribution curve can be drawn and assembly qualified rate will be calculated. The analysis process is illustrated in Fig.1.

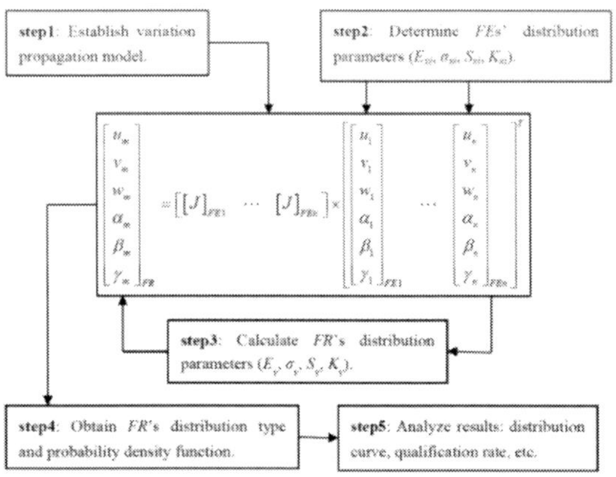

Figure. 1. Statistical variation analysis using Pearson distribution family.

4 Case study

To validate the practicability of the solution mentioned above, a realistic subassembly originating in an aero-engine is adopted for statistical variation analysis. The aero-engine subassembly is composed of four precise revolutionary parts and all the coordinate frames are specified in Fig.2 (a). Reference 0 is the global reference frame and the revolution axis is designated with a line passing through the base centre of the first component and is perpendicular to it.

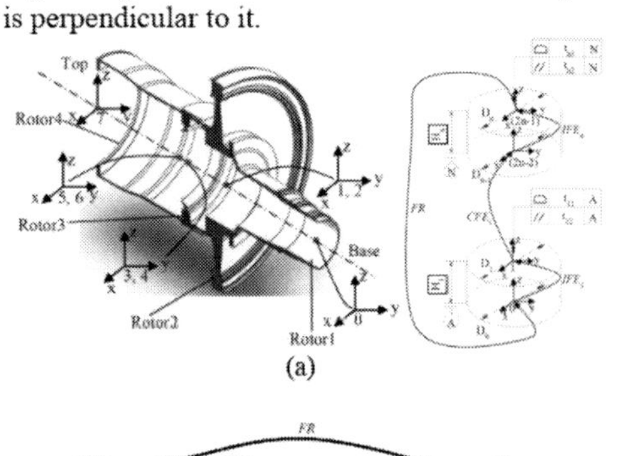

(a)

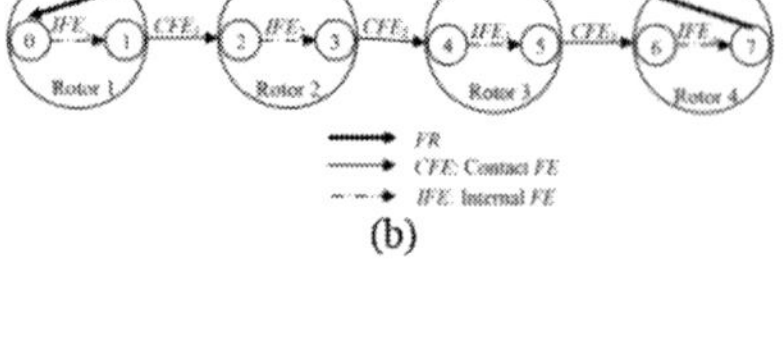

(b)

Figure. 2. Four-stage assembly: (a) parts drawing, (b) connection graph.

Each component's dimensions and composite tolerances which have been mentioned in Fig.2 (a) are listed in Table.2. The pose position and orientation of the final rotor which is accumulated by all variations of four components and three joints are very critical to concentricity control in aero-engine assembly technical requirement. Based on the 3D variation analysis method proposed in Section 3, the solving procedures are elaborated below:

Table 2. Dimensions and geometric tolerances of each component(Unit: mm).

Component	Base diameter	Top diameter	Height	Profile tolerance	Parallelism tolerance
1	107	405	315	0.020	0.015
2	405	410	135	0.020	0.015
3	410	400	165	0.020	0.015
4	400	440	365	0.020	0.015

Step 1: Build coordinate system and assembly graph. In this case, the global coordinate system (0) is set in the base centre of the rotor 1, and other local reference frames are built in the centre of related contact areas. The connection graph is shown in Fig.2 (b). *FR* is the pose of the top rotor in relation to the *0th* frame. As can be seen, there is a serial route of variation propagation for *FR*: IFE_1-CFE_1-IFE_2-CFE_2-IFE_3-CFE_3-IFE_4.

Step 2: Establish the variation propagation model based on the above connection graph. Using the unified Jacobian-Torsor theory, the Jacobian matrixes and corresponding tolerancing torsors, as well as their variation ranges can be obtained. Details of FE_1 are presented in Table.3, and other *FE*s have similar representations.

Table 3. Jacobian matrixes and tolerancing torsors.

FE_s	Jacobian matrixes	Tolerancing torsors	Variations
IFE_1	$\begin{bmatrix}1&0&0&0&315&0\\0&1&0&-315&0&0\\0&0&1&0&0&0\\0&0&0&1&0&0\\0&0&0&0&1&0\\0&0&0&0&0&1\end{bmatrix}$	$\begin{bmatrix}0\\0\\0\\0\\0\\0\end{bmatrix}$	0
CFE_1	$\begin{bmatrix}1&0&0&0&0.02&0\\0&1&0&-0.02&0&0\\0&0&1&0&0&0\\0&0&0&1&0&0\\0&0&0&0&1&0\\0&0&0&0&0&1\end{bmatrix}$	$\begin{bmatrix}0\\0\\w_1\\\alpha_1\\\beta_1\\0\end{bmatrix}$	$-0.020/2\le w_1\le 0.020/2$ $-0.015/405\le\alpha_1\le 0.015/405$ $-0.015/405\le\beta_1\le 0.015/405$

Next, on the basis of Eq. (2), the final results can be obtained by multiplying the Jacobian matrix and torsor values. In this example, we only focus on the eccentricity in a certain direction, for instance the *x* axis direction, which can be extracted from u_m and reformulated as Eq. (12).

$$y_{u_4}=0.02\beta_1+0.04\beta_2+0.06\beta_3+0.08\beta_4 \quad (12)$$

In addition, as listed in Table.3, all the variations of effective vectors in the torsor should be satisfied.

Step 3: Calculate distribution parameters of the *FE*s and then determine *FR*'s distribution type and its PDF. The distribution parameters of *FR* can be calculated as follows:

$$E_y = \sum_{i=1}^{n} a_i E_{x_i} \quad (13)$$

$$\sigma_y^2 = \sum_{i=1}^{n} a_i^2 \sigma_{x_i}^2 \quad (14)$$

$$S_y = \frac{E\left[y - E(y)\right]^3}{\sigma_y^3} = \frac{\sum_{i=1}^{n} a_i^3 S_{x_i} \sigma_{x_i}^3}{\sigma_y^3} \quad (15)$$

$$K_y = \frac{E\left[y - E(y)\right]^4}{\sigma_y^4} = (\sum_{i=1}^{n} a_i^4 K_{x_i} \sigma_{x_i}^4 + 6\sum_{i=1}^{n-1}\sum_{j=i+1}^{n} a_i^2 a_j^2 \sigma_{x_i}^2 \sigma_{x_j}^2)/\sigma_y^4 \quad (16)$$

where a_i is the sensitivity coefficient, E_{xi} is the *i*th *FE*'s mathematical expectation, and E_y is the *FR*'s mathematical expectation. σ_{xi} is the *i*th *FE*'s standard deviation, and σ_y is the *FR*'s standard deviation. S_i denotes the coefficient of skewness and K_i denotes the coefficient of kurtosis.

To verify the influence of probability distribution types on the final assembly variations, we suppose the distribution types of *FE*s follow normal distribution, skewed distribution and uniform distribution respectively, and four probability combinations will be used as shown in Table.4, where skewed distribution is simulated by virtue of Beta approximation.

Table 4. Distribution type and parameters of the *FE*s.

No.	Distribution pattern	Distribution parameter β_1	β_2	β_3	β_4
1	Normal	$\mu = 0$, $\sigma = \frac{0.005}{405}$	$\mu = 0$, $\sigma = \frac{0.005}{410}$	$\mu = 0$, $\sigma = \frac{0.005}{400}$	$\mu = 0$, $\sigma = \frac{0.005}{440}$
2	Beta(Right-skewed)	$\alpha = 2$, $\beta = 5$	$\alpha = 3$, $\beta = 4$	$\alpha = 2$, $\beta = 8$	$\alpha = 1.5$, $\beta = 3.2$
3	Beta(Left-skewed)	$\alpha = 5$, $\beta = 2$	$\alpha = 4$, $\beta = 3$	$\alpha = 8$, $\beta = 2$	$\alpha = 3.2$, $\beta = 1.5$
4	Mixing(Normal, Beta, Uniform)	$\mu = 0$, $\sigma = \frac{0.005}{405}$	$\alpha = 5$, $\beta = 2$	$\alpha = 2$, $\beta = 5$	$a = \frac{-0.015}{440}$, $b = \frac{0.015}{440}$

The results of variation distribution parameters of the *FR* can be worked out according to Eq. (13) to (16), as listed in Table.5. Based on Eq. (5) to (10), we can further obtain the values of a_0, b_0, b_1, b_2, and deduce the variation distribution pattern by reference to Table.1, which has been shown in the last column of Table.5.

Table 5. Distribution type and parameters of the *FR*.

Parameter	Combination 1	Combination 2	Combination 3	Combination 4
Mean	0	0.0604	0.1396	0.0457
Variance	0.0125	0.0189	0.0189	0.0115
Skewness	0	0.3734	-0.3734	0.2417
Kurtosis	3	2.8210	2.8210	2.9311
Pattern	IV	VI	VI	VI

Step 4: Solve PDF using Pearson distribution family formulas (pattern IV and VI). With the aid of Matlab function, we can easily get the probability denisty of the Pearsons distribution with mean 'mu', standard deviation 'sigma', skewness 'skew' and kurtosis 'kurt', evaluated at the values in X.

Calculate the qualification rate of four combinations, comparing with the MC results. In order to balance accuracy with efficiency, the simulation number of MC is set to 10000 in this case. With regard to the tolerance limits, *FR* should fall in the qualified range of 0.038mm in accordance with engineering requirements. Results are shown in Table.6, and Fig.3 illustrates the distribution curve as well as MC statistical histograms.

Table 6. Qualification rate of four combinations.

Method	Combination 1	Combination 2	Combination 3	Combination 4
Proposed method	99.88%	11.59%	0.00%	26.28%
MC method	99.84%	11.55%	0.00%	26.26%

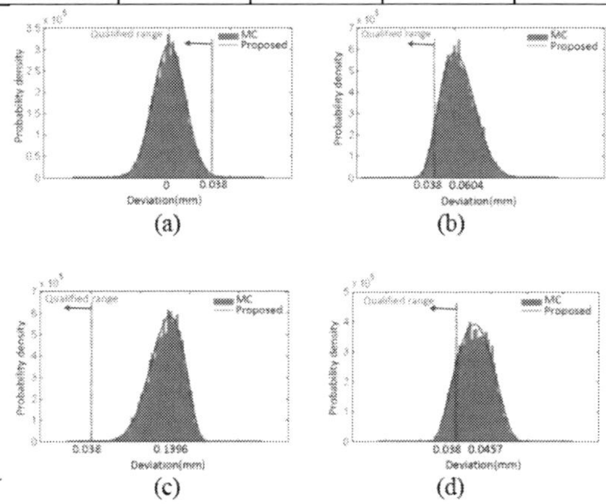

Figure. 3. Variation distribution curve and histogram: (a) combination 1, (b) combination 2, (c) combination 3, and (d) combination 4.

5 Comparison and discussion

By comparing results of the suggested method to the MC ones, it is observed that the probability density curves drawn by Pearson distribution family method agree with the generated histogram statistics using MC simulation fairly well, which have been illustrated in Fig.3 (a) to (d). Meanwhile in Table.6, the error between qualification rates produced by the two ways is less than 0.5 percent off. This demonstrates that the proposed method has the same accuracy of MC method. Additionally, the former will calculate only once in the entire simulation process while the latter will calculate ten thousand times repeatedly. Pearson distribution family method has vastly enhanced the work efficiency.

In addition, as can be seen from the diagram, when the distribution pattern of *FE*'s tolerance zone is changed, the shape of the variation distribution of the *FR* will bring about great changes, which appears to have happened at the assembly qualified rate as well. In the four groups of experiments, combination 1 (normal distribution) can guarantee the high pass percentage of 99.8% while combination 3 (left-skewed distribution) almost causes all products to be unqualified as a result of overlarge mean shift (=0.1396mm). All above results demonstrate that probability distribution types of the parts variations have a significant impact on the final assembling variation, and the traditional hypothesis of normal distribution cannot describe the true error conditions. Therefore, during the phase of product design, the practically processing technology and the level of processing must be taken into account to determine each component's real variation distribution. Only this can make results be more reasonable and reliable.

6 Conclusion

This work was attempting to propose a novel variation analysis method for complex assembly with multi joints. A calculating scheme was illustrated on a realistic case originating in aero-engine subassembly. Following conclusions could be drawn: first, for aero-engine rotors which contain lots of geometric tolerances, and planar and cylindrical features, Jacobian-Torsor theory can effectively solve the 3D variation modelling; second, compared to MC results with 10000 simulation times, Pearson distribution family method has the same accuracy, but much higher efficiency (runs only once); third, the result of *FR* depends on each *FE*'s probability distribution pattern significantly. In order to obtain more accurate predictions, the real distribution states of parts variations should be detected and used, which is beneficial to tolerance design and performance forecast.

Acknowledgments

This work was supported by the grants from the National Natural Science Foundation of China (Grant Nos. 51121063, 51175340) and the National Science & Technology Pillar Program during the 12th Five-year Plan Period (Grant No. 2012BAF06B03).

References

1. Mantripragada R, Whitney DE. Modeling and controlling variation in mechanical assemblies using state transition models. *IEEE International Conference on Robotics and Automation, 1998 Proceedings*2002, p. 219-26 vol.1.
2. Thimm G, Britton GA, Cheong FS. Controlling Tolerence Stacks for Efficient Manufacturing. *Int J Adv Manuf Technol*. 2001; 18: 44-8.
3. Forouraghi B. Worst-Case Tolerance Design and Quality Assurance via Genetic Algorithms. *Journal of Optimization Theory & Applications*. 2002; 113: 251-68.
4. Fei C, Bai G. Extremum selection method of random variable for nonlinear dynamic reliability analysis of turbine blade deformation. *Propulsion & Power Research*. 2012; 1: 58-63.
5. Kim C-B, Pulat PS, Foote BL, Lee DH. Least cost tolerance allocation and bicriteria extension. *International Journal of Computer Integrated Manufacturing*. 1999; 12: 418-26.
6. Lin EE, Zhang HC. Theoretical Tolerance Stackup Analysis Based on Tolerance Zone Analysis. *International Journal of Advanced Manufacturing Technology*. 2001; 17: 257-62.
7. Ata MY. A convergence criterion for the Monte Carlo estimates. *Simulation Modelling Practice & Theory*. 2007; 15: 237-46.
8. Yang Z, Mcwilliam S, Popov AA, Hussain T. A probabilistic approach to variation propagation control for straight build in mechanical assembly. *International Journal of Advanced Manufacturing Technology*. 2013; 64: 1029-47.
9. Ghie, Walid. Statistical analysis tolerance using jacobian torsor model based on uncertainty propagation method. *International Journal of Multiphysics*. 2016; 3: 11-30.
10. Ahsanullah, Hamedani, G. G. Characterizations of certain continuous univariate distributions based on the conditional distribution of generalized order statistics. *Pakistan Journal of Statistics*. 2012; 28: 253-8.
11. Kharoufeh JP, Chandra MJ. Statistical tolerance analysis for non-normal or correlated normal component characteristics. *International Journal of Production Research*. 2002; 40: 337-52.
12. Provost SB. A Brief Derivation of the Asymptotic Distribution of Pearson's Statistic and an Accurate Approximation to Its Exact Distribution. 2016.
13. Desrochers A, Ghie W, Laperrière L. Application of a Unified Jacobian—Torsor Model for Tolerance Analysis. *Journal of Computing & Information Science in Engineering*. 2003; 3: 2-14.
14. Karpov IG, Zyryanov YT. On modified pearson distributions and their identification. *Automatic Control & Computer Sciences*. 2015; 49: 366-72.

Knowledge push based on design flow and user capacity model

Hao Jiang[1,*], *Pu* Yin[1], *Lin* Guo[1], *Yibo* Wang[1]

[1] Institute of Telecommunication Satellite, China Academy of Space Technology, Beijing 100094, China

Abstract. The knowledge demand of the designer is determined by the knowledge structure of the knowledge user and knowledge usage scenario. Therefore, this paper proposes a knowledge push method based on design activities workflow model and user capability model. By constructing the workflow model of design activities, this paper determines the designer's knowledge demand scenario. The relationship between design knowledge and design activity is determined by calculating the similarity between the label of the design activity and the key words of design knowledge. By constructing user capability model, we describe the designer's knowledge structure and design ability .This paper determine the designer's personalized knowledge needs according to the user capability model. Finally, this paper presents the prototype system and proves the feasibility of this method with the example of the part design.

1 General Instructions

With the trend of economic globalization and increasing competition, knowledge has been a vital factor for business operations in an enterprise[1]. With the development of information technology, a growing number of knowledge-intensive corporations have raised more concern about Knowledge Management (KM) [2]. Knowledge management becomes the key factor for knowledge sharing in an organization. One of the recent trends in knowledge management is the knowledge push. Knowledge push provides R4 Service: the Right information at the Right time in the Right way to the Right person [3].

Many researchers pay close attention to knowledge management aiming at active personal knowledge push. Xu proposes a framework of an active knowledge service based on collaborative intent capture to assist product conceptual design[4]. Zhang proposes a framework of the knowledge push process including knowledge adapter and knowledge pusher based on context[5-6].Zhou presents the method of knowledge capture which based on ontology, and introduces the mechanism of knowledge pull and knowledge push[2].

Gu Xingjiang proposes a knowledge push method based on ontology and rough set combining with the designer's knowledge demand is proposed to solve the problem of large amount of knowledge in design activities, knowledge dispersion and low efficiency of knowledge push [7]. Wang modeled the design intention and realized the accurate push of the design knowledge by establishing the mapping relation between the design knowledge and the design intention [8]. According to the difference of knowledge demand of designers, Feng yixiong puts forward a knowledge push method based on semantic feature analysis [9]. Xu rongzheng uses frequent pattern mining to analyze the corresponding situation of knowledge and design tasks, and pushes the design knowledge to the designer through similarity calculation [10]. Mo rong thinks that the change of processing characteristics in process manufacturing process is the carrier of process knowledge and proposes the method of knowledge representation and push of machining process driven by geometric variation [11].

The current researchers have put forward a variety of knowledge matching methods to improve the accuracy of knowledge push. This paper argues that the key to improving the accuracy of knowledge push is to accurately grasp the knowledge demand of designers. The designer's knowledge needs are determined by two key factors: the current design task and the ability of the designer. The maximum set of knowledge requirements can be determined according to the current task. The set of knowledge that a designer needs to accomplish this task is a personalized subset of the maximum set. This paper proposed a knowledge push method based on design process and user capability model. First, we build the design activity workflow model and establish the mapping of design knowledge and design activities. Then, we build the user capability model to describe the designer's ability of design. Finally, the knowledge which is required by the designer is extracted from the knowledge set based on the user capability model.

2 Design process and user capability model

Accurate description of design task and user design ability is the basis of accurate design knowledge push。In order to realize the accurate push of design knowledge, we first need to model the design process and user ability.

2.1Design activities workflow model

A design process is a collection of ordered activities which is organized in a certain sequence。Design process is the main process of designer's design activities. The knowledge push determines the designer's knowledge demand according to the designer's current design activities, and provides corresponding design knowledge for designers. Figure 1 is a typical design process. In order to complete the design task of the parts, we need to complete a series of design activities such as demand analysis, weight optimization, operational analysis, process analysis, interference inspection, and mechanical analysis. Analytica the knowledge requirements of each design activity and establish the relevance of design knowledge and design activities will be elaborated in the next section.

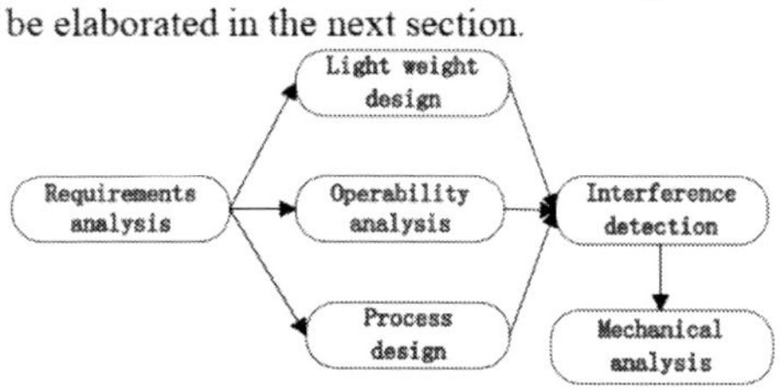

Fig. 1. Design activities workflow model.

2.2 User capability model

Even if the same design activity is completed, different designers have different design knowledge requirement due to their different knowledge structure and design ability. We need to provide personalized knowledge push service according to the designer's design ability. The designer's design ability is determined by his knowledge structure。This paper presents an extensible structured model of user capability to describe the designer's knowledge structure. Building user capability model is mainly for structured description the user's design knowledge structure, so its elements have a mapping relationship with the organization of knowledge. This paper constructs the basic framework of user capability model based on knowledge topic classification. The user capability model is shown in figure 2.

This model divides the designer's mastery of the knowledge of a design domain into three situations: proficient, qualified and beginner. "Proficient（P）" means that the designer is an expert in the field of design. "qualified（Q）" indicates that the designer is a qualified designer in this field and has the ability to complete the design task in dependently. "beginner（B）" indicates that the designer is a beginner in the field and has only a basic understanding of the field of design, which requires almost all design knowledge to complete the design task. This model allows you to extend it. As shown in figure 2, the parts design can be divided into mechanical analysis, weight optimization, process design and so on. The process design can be further refined into welding technology, machining technology, sheet metal process and casting process. Extensible user capability models allow designers to change and expand at any time. This feature allows users not to build a particularly accurate model at the beginning of their use. The designer can continuously improve the model which can improve the usability of the system.

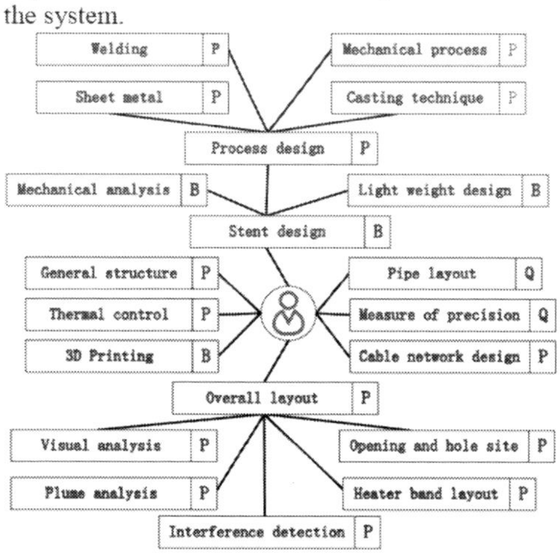

Fig. 2. User capability model.

3 Personalized knowledge push

After modeled the design process and user knowledge structure, it is possible to provide designers a set of personalized knowledge for their design tasks. In this paper, the knowledge push method based on the design process and user capability model firstly reorganizations the distributed knowledge in the field of knowledge. Then the knowledge requirement set is generated according to the knowledge requirement of the design activity, and the mapping of the knowledge requirement set and design activity is established. Finally, the personalized design knowledge requirement set is generated based on the user capability model, and it is pushed to the designer who is conducting the design activities. The process of knowledge push is shown in figure 3, mainly including knowledge reorganization, knowledge aggregation, knowledge linkage, knowledge matching and knowledge push.

3.1 Knowledge reorganization

At present, most enterprises construct various information systems, which lead to the design knowledge is stored in heterogeneous information systems. The design knowledge is stored in the enterprise's model base, document base, standard base, patent base, paper base, etc. When a designer needs to complete a design activity, it is necessary to find the design knowledge in multiple information systems, which greatly influences the design efficiency of the designer. In order to facilitate the retrieval and push of knowledge, this paper integrates distributed heterogeneous knowledge and reorganizes it according to the knowledge domain. A design domain contains all the knowledge of the field, including models and patents, standards, etc. The organization of design knowledge in the design field provides a good basis for the generation of design knowledge requirement set, because a design activity tends to correspond only to one design domain.

3.2 Knowledge aggregation and knowledge linkage

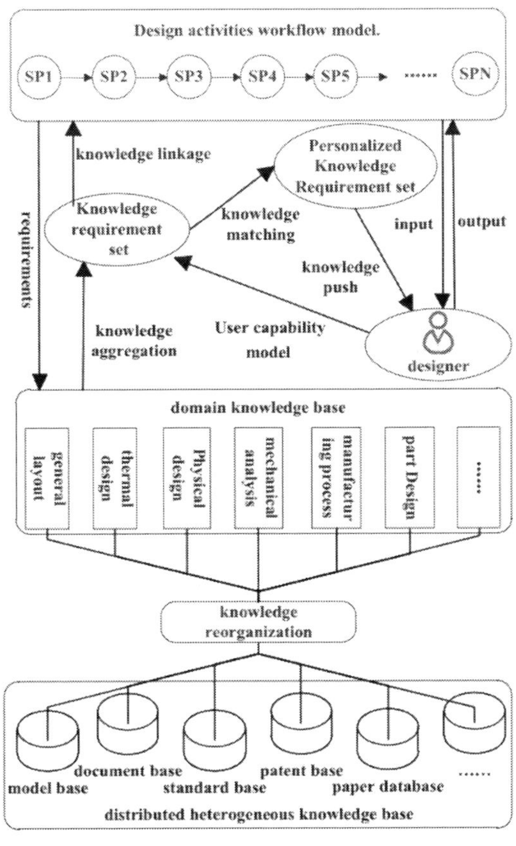

Fig. 3. Knowledge push control model.

Designers need different design knowledge to accomplish different design tasks. In order to realize the knowledge push for a certain design activity, it is necessary to retrieve specific series of design knowledge from the knowledge needs and characteristics of the design task. In this article, the designer has made a knowledge domain annotation for each design activity when building the design activities workflow model. We use multiple terms to describe the knowledge requirements for design activities. On the other hand, existing design knowledge also includes a set of keywords in its attributes. In this paper, the similarity of the design activity glossary and the design knowledge key word set is calculated by using the method of text similarity calculation. Then determine whether the design knowledge is required for this design activity based on the results of similarity calculation. The design activity glossary is $A = \{a_1, a_2, a_3, \ldots, a_n\}$. The design knowledge key word set is $B = \{b_1, b_2, b_3, \ldots, b_m\}$. The similarity calculation formula is as follows:

$$SIM(A, B) = \sum_{i=1}^{n} \max\{SIM(a_i, b_1), SIM(a_i, b_2), SIM(a_i, b_3), SIM(a_i, b_4), \ldots, SIM(a_i, b_m)\} \quad (1)$$

3.3 knowledge matching and knowledge push

The collection of knowledge requirements for specific design activities cannot be directly pushed to the designer. This collection is all the knowledge required for the design activity. Most designers don't need so much knowledge. We need to extract a set of personalized knowledge from this collection based on the user capability model of each designer. In the user capability model, designers are divided into three levels: "Proficient（P）"、"qualified（Q）"and "beginner（B）" in each design domain. This article determines whether the knowledge should be pushed to a specific designer based on the number of times the design knowledge is retrieved. This article pushes all design knowledge to the designer who is at the "beginner（B）" level. This article pushes the first 50% of the design knowledge to the designer who is at the "qualified（Q）" level. This article pushes the first 20% of the design knowledge to the designer who is at the "Proficient（P）" level. Finally, the generated personalized knowledge requirement set is pushed to the designer.

4 Application examples

We developed the prototype system based on the knowledge push method which is based on the design process and user capability mode. Take a part design as an example to show the result of knowledge push. After combing the parts design process, the design process model is constructed, as shown in figure 4. To complete the design task of the parts, we need to complete a series of design activities in an orderly manner. The knowledge requirements set of process design contains four principles: 1 Size marking should be marked with the design reference and the process reference; 2 The shape of the parts is as simple as possible; 3 considering the accessibility of machining tools; 4 Avoid deep hole. Figure 5 is the push result of designers who is at the "qualified（Q）" level. Figure 6 is the push result of designers who is at the "beginner（B）" level.

DOI: 10.1051/matecconf/201713900012

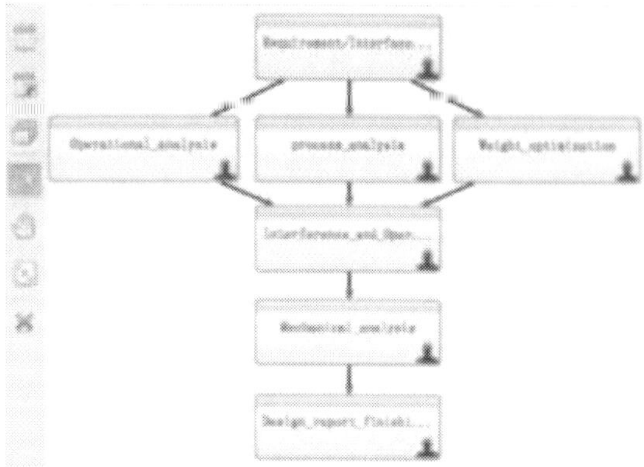

Fig. 4. Parts design process.

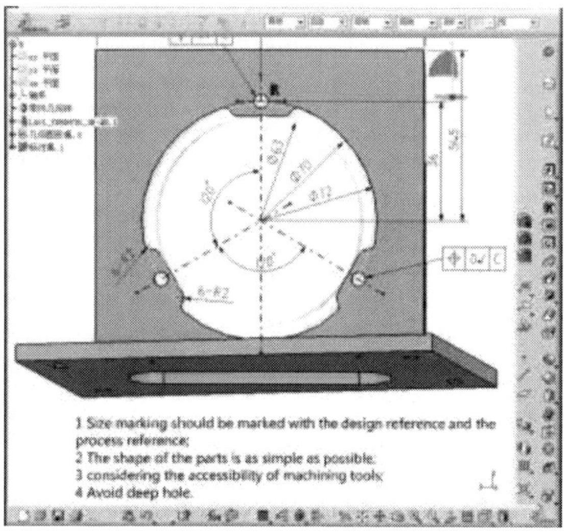

Fig. 5. Knowledge push for beginner.

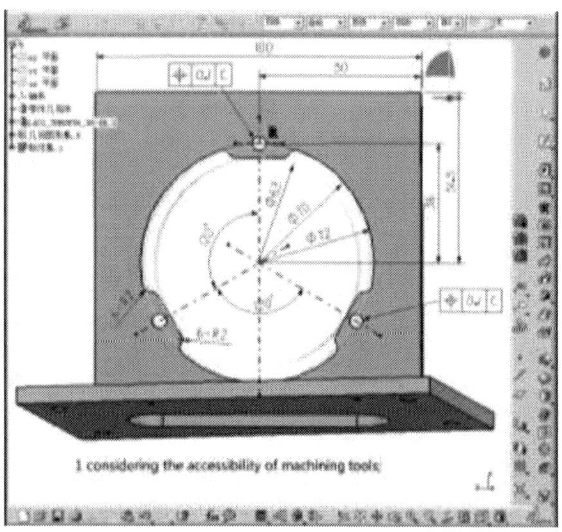

Fig. 6. Knowledge push for Proficient.

5 conclusion and adaptation prospect

This paper introduces the knowledge push method based on design process and user capability model. Constructing the design activities workflow model and user capability model are the important foundations for realizing the knowledge push. By constructing the design activities workflow model, the application scenario of design knowledge can be determined, which can help to determine the designer's knowledge demand for the design task. Extensible user capability model structured describe user knowledge structure and design capabilities. Its expandable features greatly improve the usability of the system. Finally, this paper presents the results of the knowledge push and proves the feasibility of this method with the example of the part design. How to build a more accurate correlation between design knowledge and design activities will be the next research focus. Automatic construction of user capability model will be another research focus.

ACKNOWLEDGMENT

National Key Technology Research and Development Program of the Ministry of Science and Technology of China （2015BAF18B01）

References

1. M.H.Shi, T.Wu, S.H.Jiang, Q.F.Wang, Y.D.Cheng, A Study of Knowledge Push System Based on Collaboration [A]. In: Proceeding of 2010 International Conference on Artificial Intelligence and Computational Intelligence [C],Nanjing,China,28-32,(2010)
2. M.J.Zhou, J.C.Tao, A Framework for Ontology-Based Knowledge Management [A]. In: Proceeding of 2011 Business Management and Electronic Information [C] ,Guangzhou,China,428-431,(2011)
3. Z.Y.Xin, J.Z.Zhao, C.H.Chi, J.G.Sun,. Information program generation for Webcasting [J].Data & Knowledge Engineering, 49(1),1-21, (2004)
4. Y.H.Xun, G.F.Yin, Y.Nie, Z.W.He, X.L.Dai, Research on an active knowledge push service based on collaborative intent capture [J]. Journal of Network and Computer Applications, 36, 1418-1430, (2013).
5. F.P.Zhang, L.Li, Research on Knowledge Push Method for Business Process Based on Multidimensional Hierarchical Context Model [A]. In: Proceeding of 2016 IEEE International Conference on Industrial Engineering and Engineering management [C], Bali, Indonesia, 656-660, (2016).
6. F.P.Zhang, L.Li, Research on Knowledge Push Method for Business Process Based on Multidimensional Hierarchical Context Model [J]. Journal of Computer-Aided Design & Computer Graphics,29(4),751-758,(2017)

7. X.Ji, X.J.GU, F.Dai, C.Y.Le, F.Y.Xu, Technology for product design knowledge push based on ontology and rough sets [J]. computer Integrated Manufacturing Systems, 19(1), 7-20, (2013)
8. Z.S.wang, L.Tian, W.R.Duan, Knowledge push technology based on design intent modeling [J]. computer Integrated Manufacturing Systems , 21(3), 606-617, (2015)
9. Y.X.Feng, S.Y.Zhang, Y.C.Gao, J.Cheng, J.R.Tan, Intelligent push method of CNC design knowledge based on feature semantic analysis [J]. computer Integrated Manufacturing Systems , 22(1), 189-201, (2016)
10. R.Z.Xu, Q.Gao, H.Wang, T.Xu, Product design knowledge recommendation based on sequential pattern mining [J]. computer Integrated Manufacturing Systems , 22(1), 189-201, (2016)
11. C.L.Li, R.Mo, Z.Y.Chang, H.C.Yang, D.L.Zhang, Y,Xiang, Knowledge representation and push of machining process driven by geometric variation [J]. computer Integrated Manufacturing Systems , 22(6), 1434-1446, (2016)

Modelling Method for Maintenance Decision-Making in Civil Aero Engines Based on Multiple State Parameters

Huiling Cao[1], *Zemin* Wu[2], and *Jiahe* Miao[1,*]

[1]School of Aeronautic Engineering, Civil Aviation University of China, Tianjin, 300300, P.R. China
[2]Uni-top Airlines Co., Ltd, Wuhan, 430302, P.R. China

Abstract. For airlines, a scientific and effective method for engine maintenance decision-making should be developed for the planning of aero engine maintenance and removal. The mathematical modelling method of maintenance decision-making for civil aero engines based on the currently widely used condition-based maintenance (CBM) strategy was mainly studied in this work, and the effects of multiple state parameters on the system operation were fully considered. Based on historical data for aero engine removal due to performance degradation, statistical regression modelling was used to establish a mathematical model of maintenance decision-making that can reflect a functional relationship between the engine state parameters and the time on wing. The model was based on the proportional hazards-proportional odds (PH-PO) model, combining two commonly used statistical regression models, the proportional hazards model (PHM) and the proportional odds model (POM), into a single new model form; as a result, the scope of application of the model was improved. Finally, the results of a case study of a specific example showed the high practical value of this method.

1 Introduction

With the increasing growth of airline fleet sizes, engine maintenance decision-making is becoming increasing complex. To reduce maintenance costs, flight delays and cancellation rates, it is important, among other engine maintenance activities, to draw engine removal plans, for which it is necessary to establish a model of maintenance decision-making.

The condition-based maintenance (CBM) [1] presents a new maintenance strategy that became more commonly used with the wide application of condition monitoring technology. In this strategy, the operation time and the current operational state of the engine were used to determine whether maintenance should be conducted, and as a result, maintenance decision-making based on the state of the engine was realized.

For civil aviation engines, performance degradation is an important cause of engine removal. Initially, the widely applied CBM decision-making method was used to conduct a simple trend analysis of performance degradation for a single performance parameter, such as the exhaust gas temperature margin (EGTM), and then, a pre-set monitoring threshold was used to determine whether the engine needed to be repaired. While this method is simple, it provides a relatively rough estimate and has some difficulty in reflecting the comprehensive effect of multiple influential factors.

Therefore, based on the CBM strategy, a statistical regression modelling method of reliability theory was used in this article. Based on research and analyses of historical maintenance data for an engine fleet, multiple state parameters were selected, and the time on wing of removal due to performance degradation was regarded as life data samples. Moreover, full consideration was simultaneously given to the effect of the necessary engine state information on the operation of the system. As a result, a comparatively accurate relationship was established among the system state parameters, the system reliability and the time on wing. Then, regarding the system reliability as a control objective, the maintenance decision-making model of the engine can be obtained and used to determine whether maintenance should be conducted based on the current state of the engine.

2 Mathematical form of the model

2.1. Typical statistical regression models

The PHM and the POM, as two main statistical regression models, are commonly applied in life management in the field of engineering maintenance.

Basic introductions of these two regression models are presented in Table 1.

However, when a severe deviation from the assumptions is present, it is difficult to obtain an accurate evaluation result. Later, based on the Aranda-Ordaz parameter family and its derivative models established by Huang, which extended the application of the model into the field of accelerated life testing, the proportional hazards-

* Corresponding author: miaojiahechn@gmail.com

proportional odds (PH-PO) [2] model was proposed. By using the transfer parameter, this model combined two kinds of commonly used statistical regression models, the PHM and the POM, into a single new model form, with these two kinds of models being two particular cases of the general model. As a result, the application scope and accuracy of assessment were improved [2]. Therefore, in this work, a model of maintenance decision-making was developed based on the PH-PO model.

Table 1. Basic introduction to PHM and POM.

	Proportional Hazards Model (PHM)	Proportional Odds Model (POM)
Reliability of the model definition function	$h(t;Z(t))=\frac{f(t;Z(t))}{R(t;Z(t))}$	$\theta(t;Z(t))=\frac{F(t;Z(t))}{R(t;Z(t))}$
Model equations	$h(t;Z(t))=h_0(t)\exp(\gamma\cdot Z(t))$	$\theta(t;Z(t))=\theta_0(t)\exp(\gamma\cdot Z(t))$
Baseline function equation	$h_0(t)=\frac{\beta}{\eta}(\frac{t}{\eta})^{\beta-1}$ (Weibull distribution)	$\theta_0(t)=(\lambda t)^p$ (log-logistic distribution)
Baseline equation and logarithmic form	$\ln h_0(t)=(\beta-1)\ln t+\ln(\frac{\beta}{\eta^\beta})$	$\ln\theta_0(t)=p\ln t+p\ln\lambda$
Basic assumptions of the model	$\frac{h(t;Z_1)}{h(t;Z_2)}=\frac{h_0(t)\exp(\gamma\cdot Z_1)}{h_0(t)\exp(\gamma\cdot Z_2)}=\exp[\gamma\cdot(Z_1-Z_2)]$	$\frac{\theta(t;Z_1)}{\theta(t;Z_2)}=\frac{\theta_0(t)\exp(\gamma\cdot Z_1)}{\theta_0(t)\exp(\gamma\cdot Z_2)}=\exp[\gamma\cdot(Z_1-Z_2)]$

2.2 PH-PO model

$g(t;Z(t))$ of the model is defined as follows [2]:

$$g(t;Z(t))=\frac{\frac{1}{R^c(t;Z(t))}-1}{c} \quad (1)$$

where c is the transformation parameter.

When $c\to 0$, this function reduces to the PH model as $g_{c\to 0}(t;Z(t))=\ln R(t;Z(t))$; and when $c=1$, the model is reduced to the PO model as $g_{c=1}(t;Z(t))=\frac{F(t;Z(t))}{R(t;Z(t))}$.

Hence, the PH-PO model can be written as

$$g(t;Z(t))=g_0(t)\exp(\gamma\cdot Z(t)) \quad (2)$$

where γ is an unknown parameter vector, $Z(t)$ is the covariate vector, and $g_0(t)$ is the baseline function of $g(t;Z(t))$ with an arbitrary form.

Inspection of Table 1 shows that the logarithmic forms of the baseline functions of the PHM and POM are similar. Therefore, we set the baseline function of the PH-PO model as follows :

$$\ln g_0(t)=\beta\ln t+\alpha \quad (3)$$

Rewriting Eq. (3), we obtain

$$g_0(t)=\exp(\alpha)t^\beta \quad (4)$$

where α and β are unknown parameters.

3 Establishment of the maintenance decision-making model

3.1 Modelling considerations

After confirmation of the model's mathematical form, the following issues must be considered in the establishment of the maintenance decision-making model.

(1) There are many reasons for engine removal, with the most common being performance degradation, lease period end, life limited parts (LLP), AD/SB requirements, foreign object damage (FOD) [3] and engine echeloned removal of fleet management. In this work, the establishment of the model was focused on engines removed due to performance degradation, so only removal caused by performance degradation was regarded as a real failure. However, considering that the complete data collected from current decision-making routes are still relatively deficient, in order to increase the sample size, multiclass data regarding removal can be considered for use in parameter estimation for the regression model. In cases of removal due to AD/SB, echeloned removal or LLP, the situation usually has not yet reached the level of performance degradation, so these cases can be regarded as censored data in the analysis of life data. Data for removal due to the end of the lease period are not considered in this article, because the time on wing of the engine is short in this case. FOD data are primarily caused by accidental factors and lack of regularity, so they are also ignored. Additionally, the model assumes that the faults occurring in the operation of the system can be remedied by maintenance, and then the system functionality after repair is as good as that of a new engine.

(2) For diverse types of engines, the structure and performance parameters are distinct; hence, the trends in monitoring parameters and the alert threshold are not the same. Therefore, in the model, statistical regression analysis must take the same type of engines as samples.

3.2 Covariate processing

There are many covariates that can be incorporated in the PH-PO model, but not all covariates have a significant correlation with life on wing; therefore, correlation analysis must first be performed between these covariates and life on wing, and then the most significant covariates are selected in the model. Additionally, the covariates are also required to be uncorrelated with each other in the model; in other words, the $Z(t)$ in the model equation must be independent. This requirement enables the use of principal component analysis (PCA) [4, 5]. This method can not only effectively eliminate correlations among covariates but can also achieve dimension reduction and can simplify the modelling.

For dimensionality reduction, the number of principal components can be obtained according to the cumulative contribution rate. Most often, when the cumulative contribution rate of the selected principal components exceeds 80%, the principal components include most of the information for the original variables.

3.3 Maximum likelihood estimation of parameters

The general form of the likelihood function is defined as

$$L(\alpha;\beta;\gamma)=\prod_{i=1}^{n} f(t_i;Z(t_i))^{\delta_i}\, S(t_i;Z(t_i))^{1-\delta_i} \qquad (8)$$

where n is the overall number of samples. δ_i is the indicator of the censored data, where $\delta_i=0$ indicates that the datum is a censored datum, and $\delta_i=1$ indicates that the datum is regarded as a real failure. $f(t;Z(t))$ is the density function of the failure probability, and $S(t;Z(t))$ is the survival function and is equivalent to the reliability function $R(t;Z(t))$.

From the formulation of the PH-PO model, the reliability function and probability density function are respectively obtained as

$$R(t;Z(t))=\exp\{\frac{1}{-c}\ln[c\exp(\alpha)t^{\beta}\exp(\gamma\cdot Z(t))+1]\} \qquad (9)$$

$$\begin{aligned} f(t;Z(t)) &= -\frac{dR}{dt}=\exp(\alpha)\beta t^{\beta-1}\exp(\gamma\cdot Z(t)) \\ &\cdot[c\exp(\alpha)t^{\beta}\exp(\gamma\cdot Z(t))+1]^{-\frac{1}{c}-1} \end{aligned} \qquad (10)$$

Substituting (9) and (10) into (8), the log-likelihood function can be written as

$$\begin{aligned} L(\alpha;\beta;\gamma) &= \prod_{i=1}^{q}\exp(\alpha)\beta t_i^{\beta-1}\exp(\gamma\cdot Z(t_i)) \\ &[c\exp(\alpha)t_i^{\beta}\exp(\gamma\cdot Z(t_i))+1]^{-1} \\ &\prod_{j=1}^{n}[c\exp(\alpha)t_j^{\beta}\exp(\gamma\cdot Z(t_j))+1]^{-\frac{1}{c}} \end{aligned} \qquad (11)$$

where q is the number of real failures.

From Eq. (11), the logarithmic form of the log-likelihood function can be obtained as

$$\begin{aligned} \ln L(\alpha;\beta;\gamma) &= q\alpha+q\ln(\beta)+(\beta-1)\sum_{i=1}^{q}\ln(t_i) \\ &+\sum_{i=1}^{q}\sum_{k=1}^{m}\hat{\gamma}_k Z_k(t_i)-\sum_{i=1}^{q}\ln\{c\exp(\alpha)t_i^{\beta}\exp[\sum_{k=1}^{m}\hat{\gamma}_k Z_k(t_i)]+1\} \\ &-\frac{1}{c}\sum_{j=1}^{n}\ln\{c\exp(\alpha)t_j^{\beta}\exp[\sum_{k=1}^{m}\hat{\gamma}_k Z_k(t_j)]+1\} \end{aligned} \qquad (12)$$

After the likelihood function is obtained, the unknown parameters can be determined by using a commonly used nonlinear optimization genetic algorithm (GA) approach [6, 7]. A GA is a probabilistic global optimization algorithm and shows a good global search ability. Hence, this algorithm can quickly search all solutions in the solution space and does not fall into the rapid decline locally optimal solution trap. Furthermore, this method takes full advantage of the GA's inherent parallelism, increasing the solution speed and obtaining the optimum solution. In addition, to increase the convergence efficiency and improve the calculation accuracy, the method in this paper is combined with the initial value estimates of the function, and then the search scope of the algorithm is restricted within a reasonable range close to the initial value. Finally, a favourable result was obtained in a case study, as described below.

3.4 Maintenance decision-making model

Based on the formulation of the PH-PO model, we have

$$R^{-c}(t;Z(t))-1=c\exp(\alpha)t^{\beta}\exp(\gamma\cdot Z(t)) \qquad (13)$$

Substituting the estimated parameters and rewriting Eq. (13), we obtain

$$\sum_{k=1}^{m}\hat{\gamma}_k Z_k=\ln[R^{-c}(t;Z(t))-1]-\ln(\hat{c})-\hat{\alpha}-\hat{\beta}\ln(t) \qquad (14)$$

Eq. (14) provides the function relating the system state parameters and the time on wing. Substituting a certain reliability value R^*, a logarithmic curve can be obtained with respect to time t as the X-axis and the comprehensive value of covariates $\sum_{k=1}^{p}\hat{\gamma}_k Z_k$ as the Y-axis. This step provides the system maintenance threshold under the reliability mentioned above. This threshold is different from the general linear threshold, providing a synthetical statement about the reliability and state of the system. We then multiply the current values of the covariates by the coefficient of covariates and plot the multiplication results in the figure using dots. A dot at the bottom of the curve indicates that the system can still operate on wing; however, if the dot is above the curve, the system's performance is seriously degraded and the engine must be repaired [8]. Obviously, this maintenance threshold corresponds to the control limit of maintenance decision-making with reliability as the control target.

4 Case study

Choosing historical data for the PW4077D engine fleet regarding removal from an airline during 1998~2013, a total of 70 data sets were obtained, as shown in Table 2. The data increases some state parameters at the moment of removal on the basis of historical records of the engine's life on wing. After the correlation analysis between covariates and life on wing, the selected parameters are *ΔΔEGT* (deviation of exhaust gas temperature), *EGTM* (exhaust gas temperature margin), *ΔΔFF* (deviation of fuel consumption), *ΔΔN1* (deviation of low pressure rotor speed), and *AVMH* (high pressure rotor vibration amplitude). Here, *TSO* refers to time on wing since overhaul, δ is an indicator of censored data, *SN* is the engine number, and "*ΔΔ*" represents the D-value between the removal moment deviation and the installation moment deviation.

Table 2. Partial samples from historical data for PW4077D aero engine removal.

SN	*TSO/h*	δ	*ΔΔEGT/C°*	*EGTM/C°*	*ΔΔFF /%*	*ΔΔN1/%*	*AVMH /ips*	**Removal reason**
222082	5959	0	23.8	49	3.2	0.7	0.1	Engine echeloned removal
222095	10879	1	56	27	7.1	0.3	0.2	Compressor blade damage
…	…	…	…	…	…	…	…	…
222083	8365	1	44.3	37	7.7	1.1	0	Turbine stator blade ablation

(1) Principal component analysis

Using MATLAB programming and calculation, we obtained

$$Z = \begin{pmatrix} z_1 \\ z_2 \\ z_3 \end{pmatrix} = \begin{pmatrix} -0.9290 & 0.6656 & -0.9287 & -0.7728 & 0.0004 \\ 0.0507 & -0.1413 & -0.0467 & -0.1261 & 0.9870 \\ -0.0268 & -0.6987 & -0.1358 & -0.4066 & -0.1570 \end{pmatrix} \begin{pmatrix} x_1 \\ \vdots \\ x_5 \end{pmatrix} \quad (15)$$

where $x_1,\ldots,x_5$ respectively represent *ΔΔEGT*、*EGTM*、*ΔΔFF*、*ΔΔN1*、and *AVMH*. *Z* is the extracted principal component matrix and has three compressed principal components that account for 55.32%, 20.29%, and 13.95%, respectively.

(2) Solution of unknown parameters

Using the single objective GA nonlinear optimization and MATLAB programming and calculation, the estimated values of the parameters are obtained as

$$\alpha = -24.9993; \beta = 3.0996; c = 0.2013;$$

$$\gamma_1 = 0.1743; \gamma_2 = 3.0751; \gamma_3 = -0.3033$$

Substituting these estimated values in the formulation of the PH-PO model, we obtain

$$g(t; Z(t)) = \exp(-24.9993)t^{3.0996} \cdot \exp(0.1743Z(t)_1 + 3.0751Z(t)_2 - 0.3033Z(t)_3) \quad (16)$$

$$R^{-0.2013}(t; Z(t)) - 1 = 0.2013\exp(-24.9993)t^{3.0996} \cdot \exp(0.1743Z(t)_1 + 3.0751Z(t)_2 - 0.3033Z(t)_3) \quad (17)$$

(3) Maintenance decision-making model

Substituting the historical engine removal data from Table 2 into Eq. (17), a series of reliability values can be calculated. For group incidents, in order to improve the accuracy of the reliability assessment, sample points that deviate strongly from the statistical regression area of the model should be eliminated. Using a conservative estimate, we set the R-value as 0.5 and 0.99 according to the maximum, minimum and distribution of the R-value. Then, on the basis of Eq. (14), the upper and lower control limits of the maintenance decision-making model can be obtained as follows:

The upper limit is

$$y_{\max} = \sum_{k=1}^{p} \hat{\gamma}_k Z_k = 39.3080 - 3.5101\ln(t) \quad (18)$$

The lower limit is

$$y_{\min} = \sum_{k=1}^{p} \hat{\gamma}_k Z_k = 26.4665 - 3.5101\ln(t) \quad (19)$$

According to the MATLAB simulation shown in Fig. 1, the following analysis results are obtained:

Decision-making 1: We multiply the current values of the covariates by the coefficient of covariates and then plot the multiplication results in the figure using dots. If a dot is at the bottom of the curve, the system can still operate on wing; however, if the dot is above the curve, the engine's performance is seriously degraded, and the engine must be replaced and corrective maintenance (CM) activities must be conducted.

Decision-making 2: When the dot is in the middle transition region, the engine should be subject to either intensive monitoring or preventive maintenance (PM) activities.

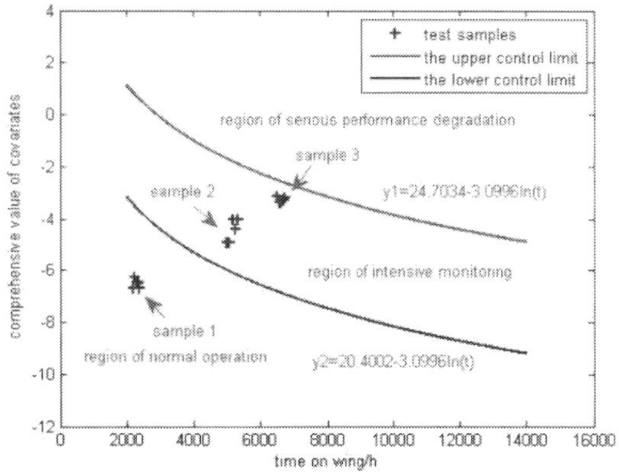

Fig. 1. Plot of maintenance decision-making model results.

(4) Model demonstration

Using the SN222159 engine as an example, the actual life on wing is 6753 hours. Three groups of sample points are selected during the period from the time of engine installation to the time of engine removal, as shown in Table 3. Among these samples, Sample 1 is taken from the initial stage of the engine's operation on wing, sample 2 is obtained from the mid-to-late stage, and sample 3 is extracted from the last stage prior to engine removal.

Table 3. Partial samples from historical data for SN22159 aero engine's operation on wing.

	TSO /h	*ΔΔEGT /C°*	*EGTM /C°*	*ΔΔF F/%*	*ΔΔN1 /%*	*AVM H /ips*
Sample 1	2192	17	61	2.4	0.5	0.2
	…	…	…	…	…	…
	2377	22	57	3.3	0.2	0.1
Sample 2	4978	31	45	3.6	0.4	0.3
	…	…	…	…	…	…
	5327	36	36	3.8	0.5	0.3
Sample 3	6517	46	27	4.6	0.9	0.4
	…	…	…	…	…	…
	6726	51	25	5.2	1.2	0.4

We multiply the covariate values of each line from Table 3 by the coefficient of covariates and use MATLAB to plot the multiplication results in a figure using dots. The results are shown in Fig. 1. To a certain extent, the results can reflect the actual operational situation.

4 Conclusion

A mathematical modelling method for maintenance decision-making regarding civil aero engines based on multiple state parameters was studied in this paper. Based on historical data for engine removal due to performance degradation, the PH-PO model was used to establish a mathematical model of maintenance decision-making that can reflect the relationship between the engine state parameters and the time on wing. Then, regarding the system's reliability as a control target, the maintenance decision-making model of the engine can be obtained and used to determine whether maintenance should be conducted based on the current state of the engine. Finally, in the case study of a specific example, the obtained results showed that the method has a high practical value and can provide a reliable basis for engine maintenance decision-making.

References

1. J. Moubray, *Reliability centred maintenance* (Butterworth-Heinemann, London, 1999)
2. T. Huang, E.A. Elsayed, T. Jiang, *The Proceedings of the 14th ISSAT International Conference on Reliability and Quality in Design*, 247-251 (2008)
3. G.R. Weckman, R.L. Shell, J.H. Marvel, Computers & Industrial Engineering, **40**, 51-63 (2001)
4. B. Schölkopf, A. Smola, K. Müller, Neural Computation, **10**, 1299-1319 (1998)
5. W.K. Härdle, L. Simar, *Applied Multivariate Statistical Analysis* (Springer-Verlag, Berlin, 2012)
6. Z. Michalewicz, *Genetic Algorithms + Data Structures = Evolution Programs* (Springer-Verlag, Berlin, 1996)
7. D.E.Goldberg, GENETIC *algoritms in search, optimization, and machine learning* (Addison-Wesley Publiching Company, New York, 1989)
8. H.F. Zuo, H.J. Zhang, X. Rong, Journal of Aerospace Power, **21**, 716-721 (2006)

Design Analysis Method for Multidisciplinary Complex Product using SysML

Jihong Liu[1,*], Shude Wang[1], and Chao Fu[1]

[1] School of Mechanical Engineering and Automation, Beihang University, 100191 Beijing, China

Abstract. In the design of multidisciplinary complex products, model-based systems engineering methods are widely used. However, the methodologies only contain only modeling order and simple analysis steps, and lack integrated design analysis methods supporting the whole process. In order to solve the problem, a conceptual design analysis method with integrating modern design methods has been proposed. First, based on the requirement analysis of the quantization matrix, the user's needs are quantitatively evaluated and translated to system requirements. Then, by the function decomposition of the function knowledge base, the total function is semi-automatically decomposed into the predefined atomic function. The function is matched into predefined structure through the behaviour layer using function-structure mapping based on the interface matching. Finally based on design structure matrix (DSM), the structure reorganization is completed. The process of analysis is implemented with SysML, and illustrated through an aircraft air conditioning system for the system validation.

1 Introduction

In the process of complex product design, system engineering is a kind of development methods which covers a wide range of applications across from system requirements analysis, function decomposition to physical construction, and so on. The purpose is to ensure that the user's needs are satisfied throughout the whole life cycle of the system and that the efficiency benefits are maximized [1]. With the increase of the complexity of the product, the document-based system engineering is difficult to meet the current R&D requirements, and the model-based system engineering (MBSE) becomes the best choice in the context of integrating the multidisciplinary and multidisciplinary background. The core idea is to model the development of all the information related to the system design from the beginning of the system research and development, and as a central system model throughout the product system life cycle.

However, the existing MBSE methodology only contains the modeling process of each stage, experience-based analysis process, and short of rigorous analysis theory and analysis methods [2]. The existing analysis theory is difficult to combine the characteristics of SysML model, and can't be good for SysML Expression. It is also difficult to support the whole process of model establishment. In this paper, by combining the conceptual design analysis theory and the expression characteristics of SysML model, a design analysis method is proposed to support the conceptual design stage of complex product in the multi-disciplinary field with the requirement analysis, function decomposition and function modeling, function structure mapping and evaluation, and structure reorganization. The whole process of the analysis method is aimed at providing designers with an effective analysis of ideas and theoretical support to reduce the bad design and improve design efficiency.

The paper is organized as follows: The second part of the article briefly introduces the relevant theoretical basis, the third part describes the specific implementation process of the method, and the forth part uses the aircraft air conditioning system as an example to verify the effectiveness of the method. The last part is the article summary.

2 Theoretical

2.1 MBSE methodology

The contribution of MBSE to short the development cycle and reduce the cost, not only rely on the ability to describe a strong system modeling language, but also need the support of scientific methodology. From a broad sense of conceptual perspective, MBSE methodology is a collection of processes, languages, and tools that support the full range of product development under "modeled" trends. At present, the mainstream methodologies include [3]: IBM Harmony-SE, INCOSE object-oriented system engineering method, using the "onion" model of the Vitech MBSE methodology, RUP SE methodology developed from the Software Development Unified Process Framework (RUP), Product system modeling methods based on the complex

* Corresponding author: ryukeiko@buaa.edu.cn

state analysis. Although the modeling order and the analysis process are different in different methodologies, they do not include specific analytical methods.

2.2 SysML

SysML has the characteristics of object-oriented, graphical, platform-independent, and can describe, analyse, design and verify the complex systems covering software, hardware and other information, and provide visual semantics expression for system requirements, behavior, structure and parameter modeling to ensure that model-based research and development can cover through the system design to the implementation of the various stages [4]. Combined with the characteristics of system engineering, SysML has been improved based on the UML, and model the whole conceptual design phase by expressing the four types of design information (requirement, behavior, structure and parameters) of the system through nine graphs. The requirement and parameter diagram is the new diagrams in SysML.

2.3 modern design methods

Modern design theory can provide the theoretical basis for MBSE analysis. There are many kinds of modern product design methods.

In the combination of modern design method and MBSE, Wang [2] studied the system engineering method and MBSE model system based on axiomatic design, and combined the axiomatic design process with MBSE modeling system. But the core theory of axiomatic design can't support the whole process of conceptual design.

Yuan and Liu [7] studied the function decomposition using SysML. And Giacomo [8] preposed a method to the development of mechatronic systems based on SysML. However, those methods are not able to support the the whole process of design.

3 Multidisciplinary complex product design analysis method

3.1 Requirement analysis based on quantization matrix

Requirements analysis is an important stage in the process of product design. Quality Function Deployment (QFD) which is a widely used design method is used in the conceptual design phase to ensure that the requirements from the customer or market is accurately transferred to the product life cycle by analysing the quality house each stage of the relevant technology and measures. In this paper, through the application of QFD thinking, the user needs into a standardized system requirements description, while the use of analytical matrix, analysis and evaluation, in order to get more in line with product system requirements. The specific process is shown in Fig.1.

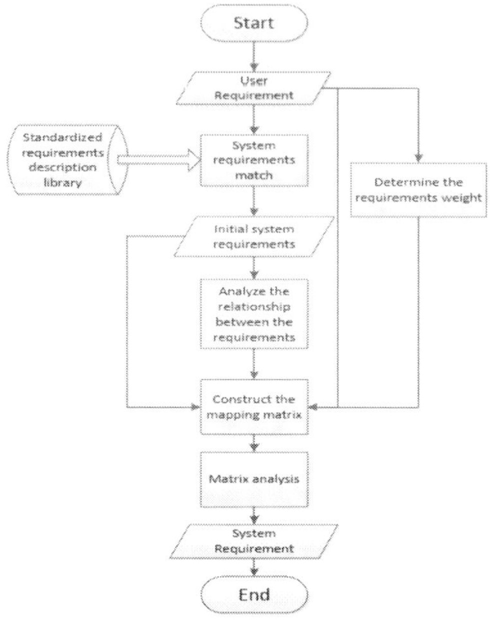

Fig.1. The process of requirement analysis

Step 1: according to the product properties, the technical staffs establish this type of product standardization requirements description library, such as air conditioning refrigeration, price and other content, to transport the non-standard requirement description provided by the customers, into the system requirements described with the standard technical terms in this domain.

Step 2: establishing the mapping matrix user needs to expand system requirements, the application of quality function of thought (QFD), the user needs to weigh mapped to the system requirements. At the same time, the description of the requirement relation is extended and the requirement relation of the system is analysed. The requirement relationship includes the inclusion, trace, derive, refine, satisfy, and verify, to quantify several relationships to determine the system requirements .The standard of quantifying is shown in the Table 1.

Table 1. The standard of quantifying

Relationship	Contain	Trace	Derive	Refine	Satisfy	Verify
Weight	1.0	0.9	0.7	0.5	0.3	0.1

Step 3, the resulting system requirements for attribute analysis, the property is defined as the functional requirements (describe the system function requirements type) and non-functional requirements (other types of needs, including price, size constraints), that is, system functions and Product constraints.

3.2 Function decomposition based on function knowledge base

The function modeling of the product is realized through the process of function decomposition. There are many ways to realize the function decomposition. According to

DOI: 10.1051/matecconf/201713900014

the characteristics of the SysML, this paper proposes the function decomposition method based on the function knowledge basis. In this method, the qualitative processing reasoning (QFR) [6] theory is applied. The change type is divided into four types: type change, numerical change, value space change and derivative change. The analysis of the change type is carried out to complete the function decomposition based on flow. The specific process is shown in Fig.2.

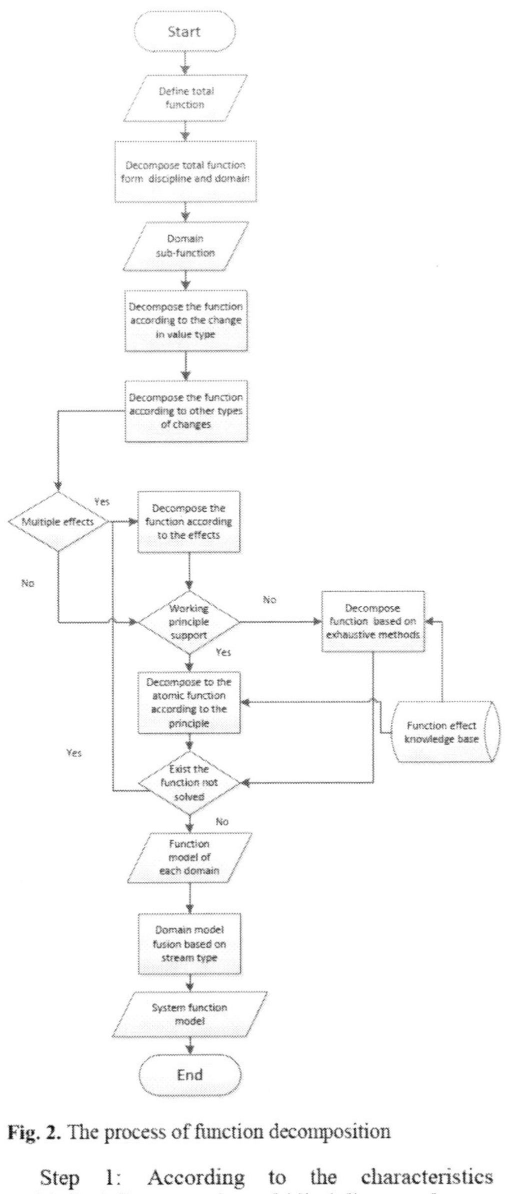

Fig. 2. The process of function decomposition

Step 1: According to the characteristics of multi-disciplinary and multidisciplinary, the overall function of the system is decomposed into the overall functions of the sub-systems according to the characteristics of the sub-systems, such as the control sub-system, the mechanical sub-system, and the total function of the subsystems defined by the input and output flow changes of the subsystems.

Step 2: Designers of different domains decompose the function of their sub-system according to the following principles: First, according to the change of the value of the stream, the change of the value type is defined as the result of the decomposition of the first layer. And then the decomposition of the sub-function on the next layer, according to the value of change, the value of space changes, derivative changes again function decomposition, the formation of the next level of function.

Step 3: Establish the function effect knowledge base, and define the atomic function (that is the function that isn't needed re-decomposition function), and the function based on the physical or chemical effects, such as the Joule effect, the effect of the input and output flow changes. The results of the previous analysis can be matched to support the effect of the change, and then decomposed into atomic functions. For the sub-functions which can't be matched the appropriate atomic function through the effect, all the atomic functions will be retrieved and matched according to the change type of input and output streams through the exhaustive method. For the function that isn't able to be deposed, expert opinion is needed to redefine the atomic function, and then need to be added to the knowledge base, so as to get a complete hierarchical function model.

3.3 Function-structure mapping based on interface matching

After the function modeling is completed, the function-structure mapping needs to be established. The result of the function model is mapped to the structure in order to obtain the overall structure model. The physical structure to realize the atomic function can be deduced through the function model [6], and the connections between the structure interfaces need to be established to achieve the complete system conceptual design scheme. The mapping process is divided into two categories. The first one is the function-structure direct mapping, mainly through the port matching relationship, which can provide an alternative design for the structure. The other makes use of the expected behavior as an intermediate transition layer, to realize the function-behavior-structure of the three layers model mapping. The former mainly focuses on the adaptive design process of the structure scheme with sufficient resources, and the latter can be carried on the innovative design through the principle selection process. The algorithm flow is shown in Fig.3.

In this paper, the establishment of function-structure knowledge base in the field will support the structure to the function of the mapping process. The knowledge base contains the following stuff:

1. The direct mapping relationship between the function and the structure, including the previous definition of the atomic function and the structure to support the function;

 DOI: 10.1051/matecconf/201713900014

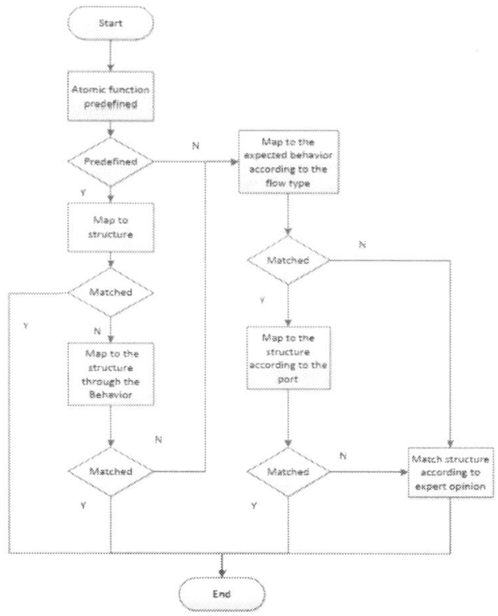

Fig. 3. The process of function-structure mapping

2. Function-behavior-structure mapping relationship, the complex atomic function, through the method of the behavior layer, then mapped to the structure.

The atomic functions decomposed in the previous step, which can be mapped to the predefined atomic functions in the original knowledge base, will be mapped to the specific structure according to the mapping relation in the function-structure knowledge base. For the absence of pre-defined atomic functions, it is necessary to analyse the behavior that supports the function, and then map to the specific structure. The mapping process is shown in the figure 3.

After the function-structure mapping is completed, the mapping results need to be evaluated [8]. The function-structure mapping matrix is established to analyze the relationship between functions and structures. The axioms of axiomatic design are applied to determine whether the matrix is coupled. If the coupling matrix, it is necessary to map the structure again to ensure the independence among functions.

3.4 Structure reorganization based on DSM

After determining the system infrastructure, it is necessary to carry out structure reorganization to ensure to build the system structure that has the high polymerization within the module, and low coupling between the modules. In this paper, the following two methods are proposed to determine the process to complete the system reorganization.

1, according to the function hierarchical structure of the reverse reorganization, structure modules are built by bottom up. The hierarchical structure of the functions decomposed by the independent function, so the functions are independent of each other and the coupling is low. The direct reverse reorganization can get a reasonable structure model

2, according to the design structure matrix (DSM). DSM is a powerful tool for product development, project planning, system engineering, organizational design, and so on. Among them, Product Architecture DSM (P-DSM) applies clustering algorithm to reduce the coupling between modules. In this paper, the resulting structures are clustered. A collection of substructures form a module to reduce the coupling between modules, improve the independence of the structure.

After the hierarchical module structure is formed, the system constraints need to be decomposed according to the system structure until the bottom layer. The design of the structure is accomplished.

4 Expression using SysML

Through the system requirements and the relationship between requirements that are obtained from the first step, the SysML requirements model is established. The relationship between the requirements is expressed by the requirement diagram. The use case diagram is used to define the stakeholders, Identify the system boundaries and determine the use case description of the total system function.

The function model will be expressed in the form of the activity diagram to determine the logical relationship (e.g. serial, parallel, input and output) between the different functions. At the same time, the hierarchy relationship between the activities will be expressed by the block definition diagram.

After the structure reorganization is completed, the hierarchical relationship between the structure modules is described by the package diagram. The relationship between the structure and the constraint is described by using the module definition chart and the internal module diagram. Finally, the structure is allocated to the activity diagram to build the white-box description of the functions.

4 Implementation and case study

The approach presented in this article is based on the release of No Magic's Cameo Systems Modeler 18.5 plug-in. By using the aircraft air conditioning system as an example, the analysis process is verified. The system interface is shown in Fig.4.

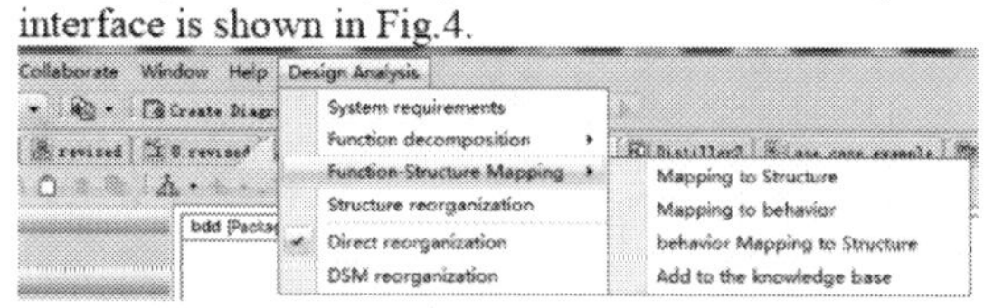

Fig. 4. Illustration of the plug-in menu

First of all, the requirements of air conditioning system are driven from the system requirements [11]. Therefore, we don't need to get those from the users. The requirements are shown in the Fig.5. After analysis, temperature control requirements are functional requirements, and the use case diagram is established.

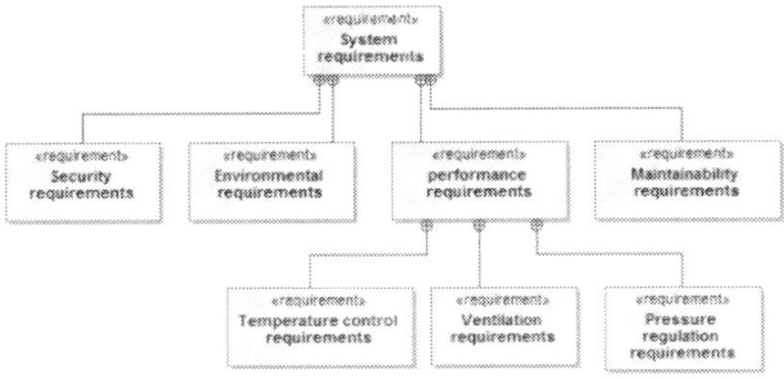

Fig. 5. The system requirement

The temperature control function is set to the total function, and decomposed. First, according to the flow type changes , it's divided into temperature changes, trim and control and warning. And then, based on the principle of decomposition, the cooling function is divided into heat exchange, heat energy into mechanical energy and so on. Therefore, it is decomposed into heat exchange and working outside. In the cooling process, the air vapor is condensed into liquid, so the moisture need to be discharged.

After the function decomposition is completed, the atomic function is mapped to the predefined structure. By matching with ports, it can be found that the drive turbine can be matched to the turbine structure. The heat exchange can be matched to the heat exchanger. Drain can be matched to the drain valve. Then the strunctures are allocated to the activities. The results need to be verified by independent axiom.

After the system structure reorganization, you can get a complete system hierarchy. The system is shown in Fig.6.

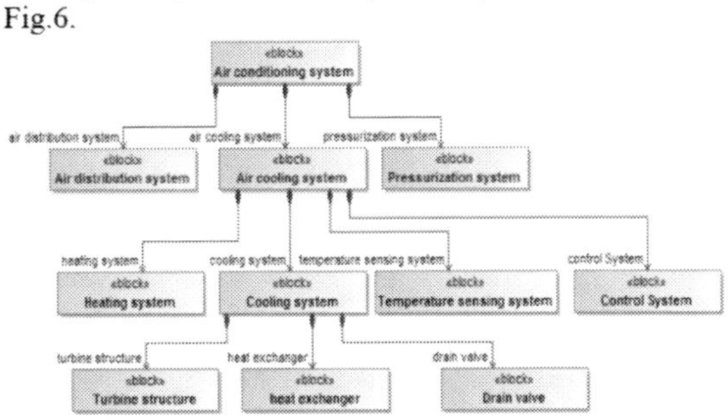

Fig. 6. The air conditioning System

5 Conclusion and future work

In this paper, the whole process of MBSE design analysis is established by integrating modern design method, which solves the problem that MBSE method lacks the support of scientific design theory, in order to provide designers with a complete analysis method and analysis process. Furthermore, this method makes the description of the analysis results with combining the characteristics of SysML, in order to establish a complete SysML model, and greatly improve the design efficiency and accuracy.

However, there are still limitations for this work. Analysis and mapping can only be done through a simple matching process, without the application of intelligent algorithms to provide support for the designer. The method relies on pre-defined atomic functions and structures, and lacks more creativity. The ability to support the model to automatically generate is poor. These factors will be considered and refined in the future work.

Acknowledgements

This work has been supported by Project of the National Key Technology Support Program, China (No.2015BAF17B01).

References

1. J. Holt, S. Perry, *SysML for Systems Engineering* (IET, 2008)
2. Haoqi W. Zhang X, System Engineering Method Based on Axiomatic Design and MBSE Modeling Framework, *ASME*, V01BT02A038 (2016)
3. J. A. Estefan. *Survey of Model-Based Systems Engineering (MBSE) Methodologies* (2008)
4. Object Management Group. *Systems Modeling Language specification*. (2009)
5. J. S. Gero. Design Prototypes: A Knowledge Representation Schema for Design. Ai Magazine, **11(4)**:26-361990 (1990)
6. Lin Y, Yusheng L, A hybrid approach for the automation of functional decomposition in conceptual design, J ENG DESIGN, **27:4-6**, 333-360 (2016)
7. B. Giacomo, F. Cesare, B. Roberto, A model-based design methodology for the development of mechatronic systems, MECHATRONICS, **24**, 833-843 (2014)
8. S Centre K. The Situated Function-Behaviour-Structure Framework, DESIGN STUD, **25(4)**, 373-391 (2001)
9. P. S. Nam., *Axiomatic Design – Advances and Applications* (China Machine Press, 2004)
10. S. D. Eppinger, T. R. Browning. *Design structure matrix methods and applications* (MIT press, 2012).
11. Xiaojin X, Shiquan L, Design Analysis of Civil Aircraft Air Conditioning System, CSTI, **16**, 188 (2014)
12. MagicDraw Open API User Guide 2017. No Magic, Inc.

A method for extracting design rationale knowledge based on Text Mining

Jihong Liu[1,*], *Jiaji* Wang[1], and *Kejian* Wang[1]

[1] School of Mechanical Engineering and Automation, Beihang University, 100191 Beijing, China

Abstract. Capture design rationale (DR) knowledge and presenting it to designers by good form, which have great significance for design reuse and design innovation. Since the 1970s design rationality began to develop, many teams have developed their own design rational system. However, the DR acquisition system is not intelligent enough, and it still requires designers to do a lot of operations. In addition, the existing design documents contain a large number of DR knowledge, but it has not been well excavated. Therefore, a method and system are needed to better extract DR knowledge in design documents. We have proposed a DRKH (design rationale knowledge hierarchy) model for DR representation. The DRKH model has three layers, respectively as design intent layer, design decision layer and design basis layer. In this paper, we use text mining method to extract DR from design documents and construct DR model. Finally, the welding robot design specification is taken as an example to demonstrate the system interface.

1 INTRODUCTION

DR knowledge is an explanation of why a product is so designed [1, 2]. DR knowledge includes explicit knowledge (such as design specifications, operating instructions, patent documents, etc.) and invisible knowledge. Explicit knowledge generated by the designer during the design of the workpiece. Invisible knowledge generated in the minds of the designer. Since the 1960s, DR began to develop, it has been recognized as providing important information and knowledge for design reuse, design reasoning and design evaluation [3]. For example, DRed (Design Rationale editor) developed by Bracewell (University of Cambridge Engineering Design and Research Center), by recording graphic design rational elements and content to establish a design rational model and support designers to browse the design information and access. DR is widely used in machinery, construction, software development and other fields.

In order to better manage and use the DR knowledge, many studies are devoted to the establishment of DR knowledge base. There are two sources of DR, one is to capture DR in the design process, and the other is to capture the DR from the design files that have been formed. Nearly 20 years, the DR acquisition method research mainly focuses on capturing the DR that generated during the design process. This method can record the explicit knowledge and tacit knowledge produced by engineering designers in detail, but need to spend a lot of time and energy. At present, there is relatively little research on DR captured in design documents. DR documents(such as design specifications, test report, patent document) because of its complex structure, complicated language, natural language is not easy to be computer recognition and other reasons have not been as an important source of DR. However, the design specifications, test reports, patent documents contain a lot of DR knowledge, for design reuse, design reasoning, design evaluation is very important [4]. Therefore, if there are some methods can accurately and efficiently extract DR from the design document, for the DR knowledge base, as well as the application of DR will have a great role in promoting. In view of the above mentioned problems, this paper proposes a method to extract the DR from the design document according to the Intent-Driven DR Knowledge model (IDDR) proposed by the team in the previous study. This method can improve design knowledge reuse efficiency.

Part one of this article describes the previous relevant research about DR. The second part puts forward the hierarchical model of design decision context. Part three presents an algorithm for obtaining DR from design documents and uses hierarchical model of design decision context to express. The fourth part uses the welding robot design manual as an example to demonstrate the system. The last part is the article summary.

2 DESIGN RATIONALE KNOWLEDGE HIERARCHY MODEL

A perfect DR model has a positive effect on the reuse of DR knowledge. At present, according to the different expression of DR, DR model is divided into two categories. The first approach is argumentation-based representation. In the argumentation-based DR

* Corresponding author: ryukeiko@buaa.edu.cn

representation model, the design process is taken as a problem solving process and it is composed of multiple subprocesses [5]. The argumentation-based representation approach is represented by IBIS (issue-based information system) which is the most mature expression in the field of DR. The IBIS model originated from Kunz and Rittel, and was later extended to different models such as PHI (Procedural Hierarchy of Issue), QOC (Question-Option-Criteria), DRL (Decision Representation Language) [6-8]. However, the argumentation-based DR model still has the shortcomings: first, the model divorce the product description, its expression is only for the design decision-making process, can not express the design itself ; second, the successful modeling process is based on the designer who have an accurate and full understanding of design issues, therefore it does not apply to all levels of the designer. But in the actual design process, the designer's understanding of the design problem is a process from scratch, the design problem in most cases is also a pathological problem, with an unclear initial state, unclear purpose and uncertain strategy. Therefore, the argumentation-based DR model can not fully reflect the typical characteristics of design activities such as design cognition and design iteration.

Intention-Driven Design Rationale knowledge modeling studies suggest that the designer's intentions interact with the environment, and co-evolution with it [9]. Intention-driven DR knowledge modeling can preserve design history by recording design intent evolution trajectories. Ganesha takes the design intent as a result of the continuous evaluation and refinement of the design process as a design rational knowledge [10]. However, the current intention-Driven design rational knowledge modeling research only concerned about the change process of the design intent, did not consider the design basis, reason and other design elements of design knowledge. Therefore the knowledge structure is simple, can not obtain comprehensive design rationale knowledge.

Our team based on the intention-driven ideas, put forward the design rationale knowledge hierarchy (DRKH) model. he DRKH model consists of three levels: design intention, design decision and design basis, as shown in Fig.1.

Design intent is the goal, plan and purpose of the designer to design thinking. It can explain the factors that affect the designer's problem solving, decision making and operation execution in the design process. Design intent is both a motivation for triggering designer thinking, and sometimes a direction for designers to think about development. Design intent can be decomposed into multiple levels of sub intent, when the realization of all sub intents, the parent's intent can achieve [11]. There is no need to re-decompose and can implement directly, called meta intent. Design decision refers to the designer referring to the relevant basis, according to the evaluation criteria, put forward, analyze and compare a number of design options, and get the final solution to achieve the design intent. The design basis is used to explain the reasons for decision making, including the basis, standards, and tradeoffs of the designer in terms of his own expertise, experience, preferences, and situational information.

DRKH model compared with other models, have no perplexing relationship, but can clearly and completely express a product design process from scratch. It can reflects cognitive behavior such as thinking, imagination, and decision making in the designer's mind throughout the design process Therefore, the method proposed in this paper is based on DRKH model.

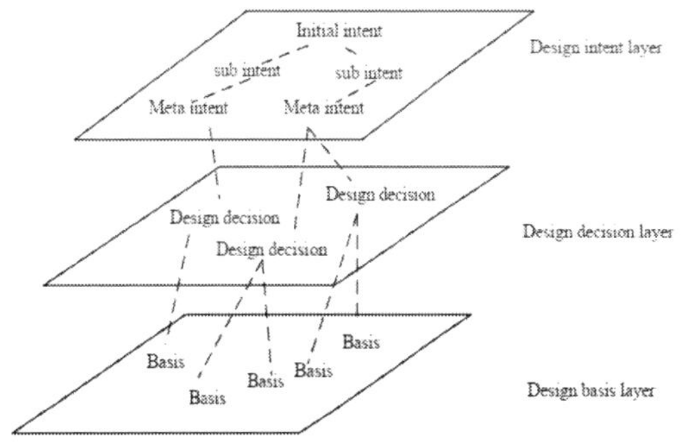

Fig. 1. Design rationale knowledge hierarchy model

3 DESIGN RATIONAL KNOWLEDGE EXTRACTION METHOD

3.1 Design intent layer generation

The task of design intent generation is to extract phrases that can express the designer's design motivation. After a lot of research and analysis, in the design document, design intent generally exist at different levels of the title, and the title of the higher level, the corresponding design intent in the design intention level at the higher level, so the extracted design document's title can be used for building design intent tree. The design document exists in the form of WORD, PDF format, so use the JAVA POI Library in Apache to identify and extract the title of the design document. The algorithm flow is shown in Fig.2.

Step 1: Design document preprocessing.

1) Input the design document W;

2) See if the design document W has a standard title format. If there is no standard title format, precede step 3, and if so, precede step 4;

3) Use the Apache POI to change the design document W to make its title standardized;

4) Use the Apache POI method to get all of the titles in the document and the inclusion relationships that exist between them;

5) All the titles obtained will be stored in the design intention candidate P according to their level;

Step 2: Design intent extraction.

The title stored in the design intent candidate set P is not all intended as a design intent, so it is necessary to delete the title that is not design intent.

1) Construct a word set D, in which words can not be used as design intents after a large number of experiments, and similar words are stored in dictionaries and stored in D;

2) Matches the title in the design intent candidate set P with the vocabulary in the word set D. If the title contains the words in the word set D, it is deleted and the design intent set Q is obtained;

Step3: Design intent layer generation.

After step 1, step 2, the design intent set is preserved in the Q, the title and its relationship between the header and their relations that can be used as the design intent layer are generated. Therefore the design intent layer is generated according to the Q.

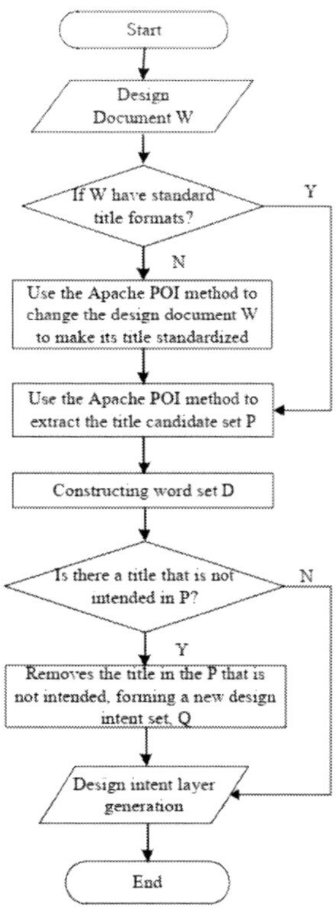

Fig. 2. Algorithm for design intent layer generation

3.2 Design decision layer generation

The design decision generation task is to extract design decision that complete a design intent. According to a large number research, features of the design decision statement is a verb plus noun form, namely verb object phrase, so that the design decision extraction is to extract relevant verb phrases related to design intent. In this paper, phrase structure based syntactic analysis is applied to extract verb object phrases from design decisions. The algorithm flow is shown in Fig.3.

Step 1: Design document preprocessing.

1) Input the design document W;

2) Divide design document W into n sentences $S = (x_1, x_2, \dots \dots x_n)$;

3) Segment each sentence in S into words and conduct part of speech tagging;

4) Using Apache POI method to determine the range of design decisions that will extraction, the design intention layer extraction algorithm mentioned each design intention corresponds to a title, therefore, the content between adjacent titles is the design decision and basis for realizing the design intention;

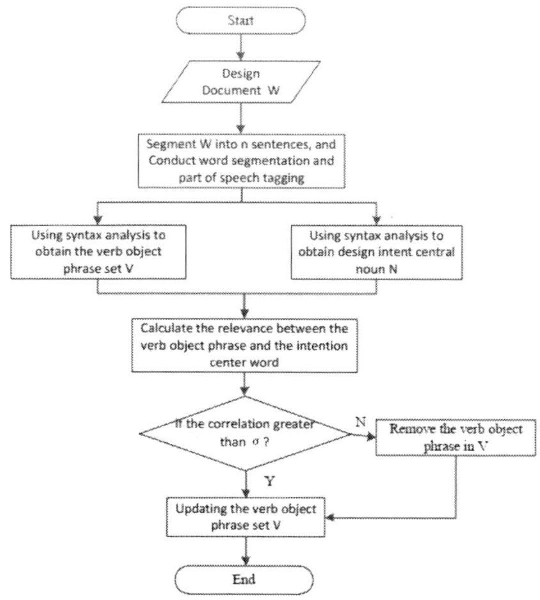

Fig. 3. Algorithm for design decision layer generation

Step 2: Design decision extraction;

1) Phrase structure based syntactic analysis is used to obtain the central word N of the design intent and all verb object phrases which are in the paragraph of a design intent. Put the verb object phrase into V, which is the candidate set of design decision;

2) Taking into account the design decision candidate set V obtained in 1), not all verb object phrases can be used as design decisions, so the verb object phrase that is not a design decision should be deleted from the V. In this paper, we adopt the method of correlation calculation, calculates the correlation between the words contained in adjacent titles and the relevance of the design intent central word N. Set the threshold σ, extract the relative degree is greater than the threshold σ word set C, match the verb object phrase V with the word set C. The verb object phrase that contains any of the words in the word set C is the design decision corresponding to the design intent. Calculation formula of correlation (1):

$$s(i,\ j) = \frac{tf(i,j)}{w(i,j)} \tag{1}$$

s (i, j) represents the degree of correlation between words i and j; tf (i, j) represents the number of times that the word i, j appears in a sentence at the same time; w(i,j) represents the relative distance of i, j, see equation (2):

$$w(i, j) = \frac{\text{The number of characters between } i \text{ and } j}{\text{The number of characters contained in the content}} \quad (2)$$

3) Delete the verb moving object phrases that are not design decisions in the design decision candidate set.

Step 3: design decision layer generation.

1) Update design decision candidate set V, that is the design decision set;

2) The verb object phrase of the design decision is connected with the corresponding design intention to form a design intention-design decision hierarchy.

3.3 Design basis layer generation

The design basis is used to explain the reasons for the decision, including the basis, the standards, and the tradeoffs that the designer makes on his own expertise, experience, preferences, and situational information. The design basis is explanatory sentences, so the design basis sentence usually contain "because......" "as a result of......" "on the basis of......" and other explanatory words and phrases In this process, we draw on the related algorithms of text mining. The algorithm flow is shown in Fig.4.

Step 1: Design document preprocessing.

1) Input the design document W;

2) Divide design document W into n sentences $\mathbf{S} = (x_1, x_2, \ldots \ldots x_n)$;

3) Establish a design basis vocabulary library O, then enter all the words in the dictionary that express, explain, and reason;

Step 2: design basis layer extraction.

1) According to vocabulary that stored in the design basis vocabulary O, extract the sentences containing any of the words between the adjacent titles as the design basis candidate set L.

2) The design decision between the corresponding adjacent titles obtained by the design decision extraction algorithm and the previous step acquired design basis candidate sentences L as the nodes, constructing semantic syntactic graph **G** (***B***, ***X***) [12]. The node ***B*** is defined as the vertex, and the connection weight is represented by the matrix X representing the semantic relevance of the node. $x(b_i, b_j)$ represents the correlation between the two sentences b_i and b_j, among which $x_{ii} = 0$. The similarity matrix X is treated with symmetric normalization, as follows formula (3):

$$M = Q^{-1/2} X Q^{-1/2} \quad (3)$$

Q is a diagonal matrix, $q_{ii} = \sum_{j=1,2,\ldots\ldots,n} w_{ij}$; sentence score vector $\mathbf{f} = [f_0, f_1 \ldots \ldots f_n]^T$, set initial value to $f_i = \frac{1}{n}$, define an n-dimensional vector $\mathbf{y} = [1,1 \ldots \ldots 1]^T$.

3) Calculate the weight between the design decision and the design basis. Cyclic calculation formula (4) until the calculation results converge. Where r represents the rth calculation, θ is the coefficient $(0 \le \theta \le 1)$.

$$f\ (r+1) = \theta M f(r) + (1-\theta) \quad (4)$$

Step 3: design basis layer generation

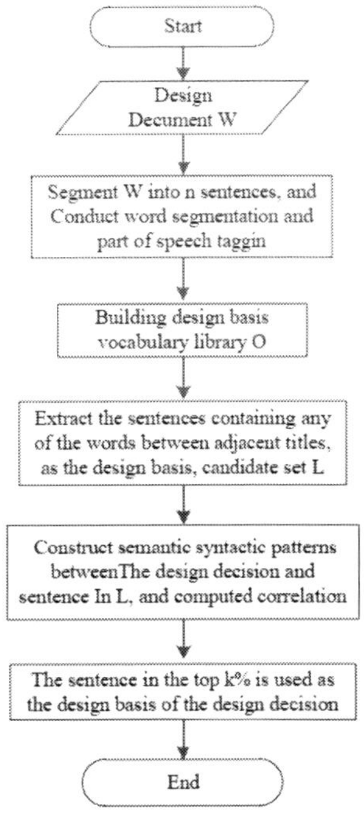

Fig. 4. Algorithm for design basis layer generation

Through step 2, the weight values between design decision and design basis are obtained. According to the weight value, the design basis which is most relevant to a design decision can be obtained. For each design decision take weight value in the former k% of the design basis, and build connection. Finally, the design intention-design decision-design basis hierarchy is formed.

4 Using design document as example

Fig. 5 shows the DR captured from a design document that focuses on welding robots. From the design intent layer, it indicates the motivations of a new design. These design intentions come from the title of the design document. In general, the design decision layer gives the knowledge of what to select to realize the design intent. The design basis layer shows the knowledge why to select the design decision to realize the design intent.

Fig. 6 shows our interface of DR knowledge extraction. The full name of this system is Cope DR Knowledge Model Editor (Cope-DRKM Editor). Cope-DRKM Editor is a system that extracts DR knowledge from design documents (word, PDF and other formats) and constructs a DR knowledge model. It has the functions of design intention extraction, design decision extraction, design basis extraction, DR knowledge model construction, and so on. The system supports the knowledge processing and the efficient reuse of DR knowledge.

The first step is to fill in the information related to project, the second step is to input the design document, the third step is to adjust the design document title format, the fourth step extraction design intent, the fifth step is extraction design decision, the sixth step is extraction design basis, the last step is to modify the model. Because the wizard system, both the new and old users can easily complete the extraction task.

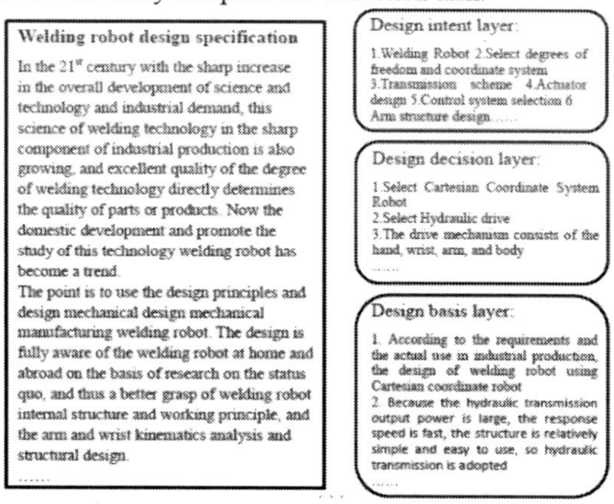

Fig. 5. An example of DR knowledge extracted

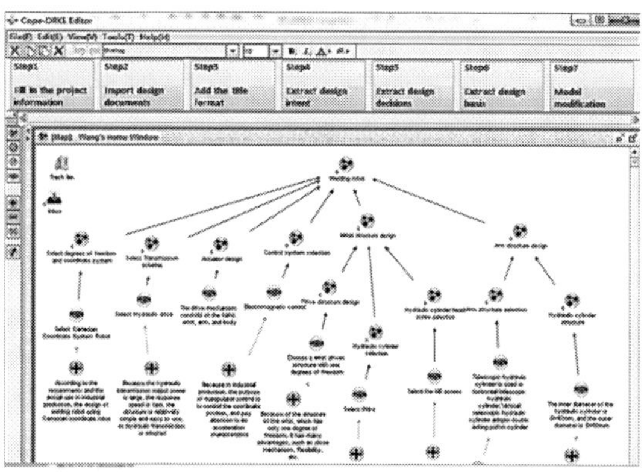

Fig. 6. Interface of DR knowledge extraction

5 Conclusion

This paper is based on the premise of not affecting the design process, extracting DR knowledge from a large number of historical design documents, and a structured and hierarchical DR knowledge model is used to express the DR knowledge. This method has promoted the acquisition efficiency and amount of DR, it lays a strong foundation for the DR knowledge base and the reuse of DR.

Because of computer recognition of natural language, resulting in model construction is not accurate enough, in the design decision layer and the design intent layer, there may be some lack of decision and intent, or there is a node that is not a decision and intent. Further experimental studies will focus on the improvement of the model, so that the system can achieve automatic extraction and improve the extraction accuracy.

Acknowledgements

This work has been supported by Project of National Science Foundation of China through approval No.51475027 and Project of National Key Technology R&D Program through approval No. 2015BAF18B01.

References

1. J. Lee, K. Lai, What's in design rationale. Human-Computer Interaction, **6** **(3)** :251—280 (2011)
2. J. Lee, Design rationale systems: understanding the issues. IEEE Expert, **12** **(3)** :78-85 (1997)
3. W.C. Regli, X. Hu, M. Atwood, W. Sun, A survey of design rationale systems: approaches, representation, capture and retrieval, Engineering with Computers, **16** **(3-4)** :209-235 (2000)
4. Y. Liu, Y. Liang, C.K. Kwong, W.B. Lee, A new design rationale representation model for rationale mining, J. Comput. Inform. Sci. Eng, **10** **(3)** :031009-1–031009-10 (2010)
5. Simon H. A, *The Sciences of the Artificial*, (M.I.T. Press, 1969)
6. McCall R. J., PHI: A Conceptual Foundation for Design Hypermedia, Design Studies, **12(1)**: 30-41 (1991)
7. MacLean A., Young R., Belloti V., et al., Questions, Options, and Criteria: Elements of Design Space Analysis, Human-Computer Interaction, **6(3-4)**: 201-250 (1991)
8. Lee J., Decision Representation Language (DRL) and Its Support Environment, Mit Artificial Intelligence Laboratory, **36** **(16)** :2307-2310 (1989)
9. Maher M., Tang H., Co-Evolution as a Computational and Cognitive Model of Design, Research in Engineering Design, **14(1)**: 47-64 (2003)
10. Ganeshan R., Garrett J., Finger S. A framework for representing design intent, Design Studies, **15(1)**: 59-84 (1994)
11. Liu J. H., Hu X. J. *Modeling the Evolving Design Rationale to Achieive a Shared Understanding*,(IEEE Press, 2012)
12. Y. Liu, Y. Liang, Learning the "Whys": Discovering design rationale using text mining - An algorithm perspective, Computer-Aided Design, **44** **(10)** :916-930 (2012)

A new model to express and capture the design rationale in the documents

Jihong Liu[1,*], *Jiaji* Wang[1], and *Kejian* Wang[1]

[1] School of Mechanical Engineering and Automation, Beihang University, 100191 Beijing, China

Abstract. As important design knowledge, design rationale（DR） knowledge plays an important role in design analysis, design reasoning and design innovation. There are generally two sources of DR knowledge, one is to capture DR in the design process, and the other is to extract DR from historical design documents, but the latter is always ignored. DR model is the foundation of DR. This paper aims at the deficiency of the previous DR model and requirements for DR knowledge acquisition from design documents. We propose design rationale knowledge hierarchy (DRKH) model. The model has three layers, design intent layer, design decision layer and design basis layer. The model also has three relationships, decomposed-into, achieved-by, refer-to. Using the DRKH model, we build the algorithm framework for extracting DR from design documents. Finally, we validate the feasibility of our model by extracting DR model from a welding robot design manual.

1 INTRODUCTION

Product design is the process of converting a person's purpose or need into a specific physical or tool. It is a process of expressing the plan, tentative idea, and problem solving method in an ideal form, through specific operations. DR knowledge is used to represent why a product or part of a product, is designed the way it is [1, 2]. Therefore, it is a good way to express the design process of product with DR knowledge. DR knowledge includes all relevant knowledge of product design, such as design intent decomposed for completing the product, design option made for completing the design intent, discussion of design decisions in the design process, reasons behind the design decisions, and so on [3].

In view of the importance of DR knowledge, many studies are devoted to establish DR knowledge base, i.e. express and store the acquired DR knowledge in a model and graphical language. The establishment of DR knowledge base consists of two processes: DR knowledge acquisition and DR knowledge representation. There are two sources of DR knowledge acquisition, one is stored the designer's thoughts about product design during the design process; the other is extracted DR knowledge from structured documents. At present, most of the studies are committed to capture DR knowledge during the design process, while ignoring the structured design document which as an important source of DR knowledge acquisition. This resulted in a great deal of waste of DR knowledge. If there is an access method that can effectively extract DR knowledge from design documents, it will play a significant role in promoting the establishment of DR knowledge base.

Since 1980s, many studies have made progress in the acquisition, expression and application of DR knowledge, and have made rapid progress in the DR system, however, there are still some problems in DR knowledge representation models. Taking into account the above DR knowledge acquisition source and the DR knowledge expression model problems, this paper proposes a new DR model for DR knowledge representation, which aims at extracting the DR knowledge in design documents. Section 2 introduces previous studies on DR and text processing. Section 3 presents our design rationale knowledge hierarchy (DRKH) model. In Section 4, the design specification of welding robot is taken as an example, demonstrate the process of extracting DR knowledge from a design document and expressing it with DRKH model. Section 5 concludes.

2 RELATED WORK

2.1 DR representation model

A clear and well-formed DR expression model is not only helpful for designers to understand and use the DR knowledge, more conducive to the effective reuse of design knowledge. DR models are generally divided into two categories, one is the argumentation-based DR representation model, and the other is the intention-Driven DR representation model. At present, most DR expression models belong to argumentation-based representation model. Argumentation-based DR representation model usually regards the design process as a problem solving process,

* Corresponding author: ryukeiko@buaa.edu.cn

and considers that the design process is composed of multiple sub processes: 1) identify design problems; 2) propose design alternative solutions to specific problems;3) form argument that supports or opposes certain design schemes;4) the decision to adopt or reject an alternative. Therefore, argumentation-based DR representation model research focuses on the process of design decision generation. It records the concrete formation process of design decision in detail with the three elements of problem, plan and argument.

The argumentation-based representation model is represented by IBIS (issue-based information system) which is the most mature expression in the field of DR . Many teams have developed graphical DR systems based on the IBIS model for the DR knowledge management. For example, graphical IBIS (gIBIS) is the first graphical IBIS tool, and Design Rationale editor (DRed) is a more mature IBIS tool at the moment [4]. McCall suggested the Procedural Hierarchy of Issues (PHI) model, which simplifies the connection form in IBIS and makes up for the lack of dependencies between design issues [5]. Liu et al. proposed an issue, solution and artifact layer (ISAL) representation model for DR capture from patent documents [7]. It extends the basic structure of IBIS and maps the expression of design thinking process to related product components, thus making up for the shortcomings of previous argumentation-based representation model.

The intention-Driven DR representation model believes that the designer's intention will interact with the environment and evolve with it, and hope to preserve the design history by recording the evolution path of design intention. A design intent modeling system developed by Arai et al. to describe the design thinking process through intention, operation, and design flow. Ganesha utilize the design intent continually evaluated and refined to record the design process [8]. In recent years, more and more researchers have realized that the design activity is essentially an evolving cognitive process, and the intention-Driven DR representation model can reflect the cognitive characteristics of the design process more accurately.

2.2 Document processing

Liu had an analysis of organized text retrieval technology [9]. They studied how to advance information admittance performance in document management. Li and Ramani proposed to use shallow natural language processing build an organized and semantics-based representation from design documents [10]. Romanowski and Nagi put forward a method to making generic bills of materials to backing varied design activities [11].

From the literature review above, we can see that there is still little research on capturing DR knowledge in structured design documents. Most of the existing DR system is used to capture the DR knowledge in the design process, or for the DR knowledge retrieval, however, there are few systems for DR knowledge capture in design documents. We consider that the captured document should not only have patent document, it should involve all structured design documents. Meanwhile, captured DR knowledge should be graphical representation, so that designers can quickly understand. Therefore, a new DR knowledge representation model is proposed to support the capture of DR knowledge in a variety of structured design documents, and express the DR knowledge in a graphical way, such a model is beneficial to the management and reuse of DR knowledge in subsequent design.

3 Design rationale knowledge hierarchy model

3.1 DRKH model for DR representation

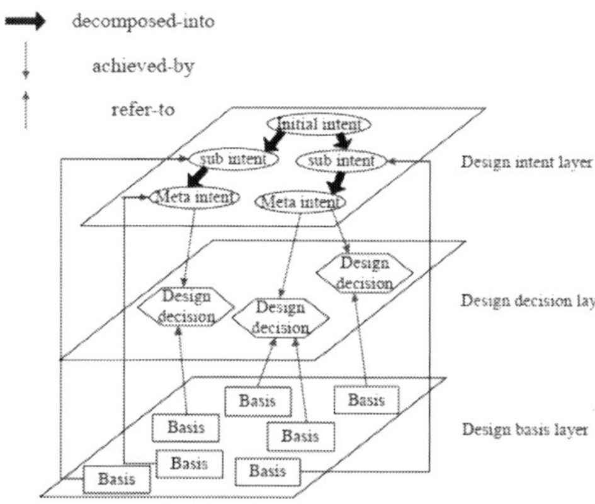

Fig. 1. Design rationale knowledge hierarchy model

The IBIS model consists of three types of elements, namely, issues, positions, arguments, and includes the relationships between the three elements. The issues are the discussion and reflection that needs to be done to complete a design goal, i.e., know how. Positions mean the method of solving the issues, that is, know how. Arguments stand for support or opposition to positions, that is, know why. They are connected by different rules to form an issue net. From the IBIS contains three elements can be seen IBIS model out of the product itself, it lacks the operation of the product. The net of issues formed by IBIS only solves an issue but there is no connection between them, and they are independent of each other. Although there are some problems in the IBIS model, it can still provide a basis for our model. Our DR knowledge model is mainly applied to capture DR knowledge in design documents, so we primarily consider three aspects in designing model:

1) The hierarchy of the model should conform to the DR structure in the design document. DR knowledge is generally composed of design issues, design options, design decisions made by the design option, design basis and design operations, but considering that the written of the design document is not record all the thoughts in the

designer's mind, so the design options can be removed from the model.

2) There should be a directional relationship between the layers of the model, and the meaning of each direction is formulated. The DR knowledge captured in product design has a huge structure, and must have clear direction and rules, in order to achieve efficient management and reuse of DR knowledge.

3) The model needs to be graphical. Compared to a large number of texts, graphic language allows designers to quickly understand the content expressed by Graphical DR models.

According to the problems discussed above, we propose a design rationale knowledge hierarchy (DRKH) model, which is shown in Fig.1. The model consists of three layers: design intent layer, design decision layer and design basis layer.

3.1.1 Design intent layer

Design intent is the goal and plan for the designer to design, it can explain the factors that affect the designer's problem solving, decision making and operation execution in the design process. Design intent is both a motivation for triggering designer thinking, and sometimes a direction for designers to think about development. Design intent can be decomposed into multiple levels of sub intent, when the realization of all sub intents, the parent's intent can achieve. The intents no need to re-decompose and can implement directly, called meta intent [12].

The design intent layer usually exists as a tree structure, as shown in Fig.2.The "Welding robot design" is initial intent; "Arm structure design", "Actuator design", "Control system selection", "Wrist structure design", "Select Transmission scheme", "Select degrees of freedom and coordinate system" are sub intent; "Arm structure selection", " Hydraulic cylinder structure"," Drive structure design"," Hydraulic cylinder head screw selection"," Hydraulic cylinder selection" are meta intent.

3.1.2 Design decision layer

Design decision refers to the designer referring to the relevant basis, according to the evaluation criteria, put forward, analyse and compare a number of design options, and get the final solution to achieve the design intent. Design decision refers to the designer referring to the relevant basis, according to the evaluation criteria, put forward, analyze and compare a number of design options, and get the final solution to achieve the design intent.

3.1.3 Design basis layer

The design basis is used to explain the reasons for decision making, including the basis, standards, and tradeoffs of the designer in terms of his own expertise, experience, preferences, and situational information. In the design document, feature of the sentence which as the design basis is very obvious, they usually have the words "because", "so", "therefore", "because", etc., which express the cause. The design basis layer is an important level to embody the value of DR knowledge, whether it is the basis of intent decomposition or the basis of choosing design decision, it is an important part of the designer's thinking about the design.

3.1.4 Connections between layers

In the model in Fig.1, we define three relationships respectively as, "decomposed-into"," achieved-by"," refer-to". Among the relationships, decomposed-into<design intent1, design intent2> represents intent2 is the sub intent of intent1; achieved-by<intent, decision> indicates that the decision can satisfy design intent; refer-to<basis, decision> indicates that the basis is the reason for choosing the decision, refer-to<basis, intent > represents the reason why the intent is decomposed.

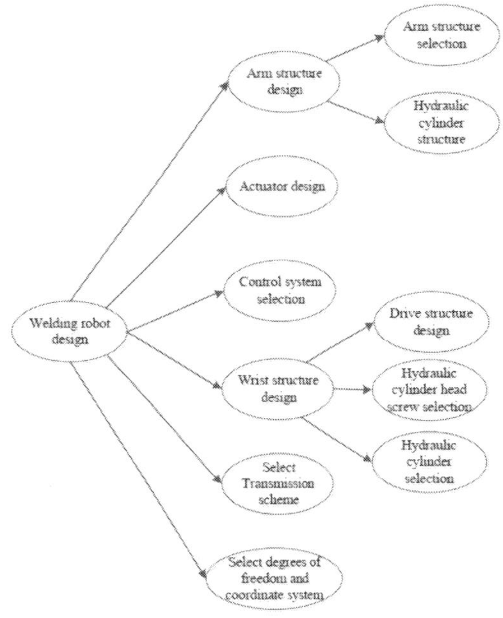

Fig. 2. Design intent tree structure

3.2 DRKH model for DR Extraction

3.2.1 Design intent extraction

Our model is applied in structured design documents to extract DR knowledge. The general form of design documentation is WORD and PDF, and structured design documents have header formats. Through extensive research, we find that the title of the design document is actually the source of the design intent. But not all titles can be used as design intent, so we will remove the title which is can not be the design intent.

The specific plan implementation framework is shown in Fig.3.

Step1.Preprocessing
Use the JAVA POI Library in Apache to identify and extract the title of the design document, and stored in the design intent candidate P;
Step2. Design intent extraction
Construct a word set D, in which words can not be used as design intents. Matches the title in the design intent candidate set P with the vocabulary in the word set D. If the title contains the words in the word set D, it is deleted and the design intent set Q is obtained.
Step3: Design intent layer generation
The design intent layer is generated according to the Q.

Fig. 3. The specific plan implementation framework of design intent extraction

3.2.2 Design decision extraction

The design decision knowledge in the design document has a characteristic, in order to realize the design intent, the design decision is usually a verb plus noun phrase. In this paper, phrase structure based syntactic analysis is applied to extract verb object phrases from design decisions. The specific plan implementation framework is shown in Fig.4.

Step1. Design document preprocessing
Divide design document W into n sentences. Segment each sentence into words and conduct part of speech tagging; Using Apache POI method to determine the range of design decisions that will extraction.
Step2. Design decision extraction
We adopt the method of correlation calculation, calculates the correlation between the words contained in adjacent titles and the relevance of the design intent central word N. Extracting the relative degree is greater than the threshold σ word set C. The verb object phrase that contains any of the words in the word set C is the design decision corresponding to the design intent.
Step3: Design decision layer generation
Update design decision candidate set , that is the design decision set;

Fig. 4. The specific plan implementation framework of design decision extraction

3.2.3 Design basis extraction

In the above analysis of the design basis, we refer to, design basis sentence usually contain "because......","as a result of......","on the basis of......" and other explanatory words and phrases. Therefore, we can establish a lexical library for expressing the reasons, and extract the words from the vocabulary library to build the design base layer. In this process, we draw on the related algorithms of text mining. The specific plan implementation framework is shown in Fig.5.

Step1. Design document preprocessing
Divide design document W into n sentences. Segment each sentence into words and conduct part of speech tagging. Establish a design basis vocabulary library O, then enter all the words in the dictionary that express explain and reason.
Step2. Design basis extraction
According to vocabulary that stored in the design basis vocabulary O extract the sentences as the design basis candidate set L. Constructing semantic syntactic graph G (B, X), The node B is defined as the vertex, and the connection weight is represented by the matrix X representing the semantic relevance of the node. The design decision and sentence in L is node B. Calculate the weight between the design decision and the design basis.
Step3: Design basis layer generation
For each design decision take weight value in the former k% of the design basis, and build connection.

Fig. 5. The specific plan implementation framework of design basis extraction

4 Example using design document

We use design document as our research data because design documents are important source of DR knowledge, and very easy to get. Fig.6 shows the DR knowledge model extracted from the welding robot design manual using the DRKH model.

In order to verify the applicability of the proposed model, we have done two experiments. The experiment is to let different designers to extract DR knowledge from the same design document with DRKH model. Before the designers use the model, we trained them in a unified way. We analysed the DR model of 30 designers, the 30 models are extracted from the same design document. The results are shown in Table 1.

Table 1. Model similarity data

Layer	Similarity
design intent layer	99%
design decision layer	85%
design basis layer	81%

From the data we can see that the designers are almost identical in their extraction of design intent, but the similarity that extraction design decisions and the design criteria is relatively low.

5 Conclusion

DR model is the foundation of DR. A good DR knowledge model is not only of great help to DR knowledge acquisition, moreover, it is of great significance for the efficient use of DR knowledge. Therefore, this paper aims at the deficiency of the previous DR model and requirements for DR knowledge acquisition from design documents. We propose design rationale knowledge hierarchy model. The model has three layers, design intent layer, design decision layer and design basis layer. The model also has three relationships, decomposed-into, achieved-by, refer-to.

From the data in Table 1, we can see that our model needs to be improved in terms of the design decision layer and the design base layer expression. In the following research, we will focus on the improvement of the model and achieve in the design document to extract the DR knowledge.

Acknowledgements

This work has been supported by Project of National Science Foundation of China through approval No.51475027 and Project of National Key Technology R&D Program through approval No. 2015BAF18B01.

References

1. J. Lee, K. Lai, What's in design rationale. Human-Computer Interaction, **6 (3)** :251—280 (2011)
2. J. Lee, Design rationale systems: understanding the issues. IEEE Expert, **12 (3)** :78-85 (1997)
3. J. Lee, The 1992 Workshop on Design Rationale Capture and Use, AI Mag., **14(2)** :24-26 (1993)
4. J. Conklin, M. Begeman, A Hypertext Tool for Exploratory Policy Discussion, ACM Transactions on Office Information Systems, **6(4)**: 301-331 (1988)
5. Fischer G., Lemke A., McCall R., et al., Making Argumentation Serve Design, Human-Computer Interaction, **6 (3)** :393-419 (1996)
6. Liu Y., Liang Y., Kwong C. K., et al., A New Design Rationale Representation Model for Rationale Mining, Journal of Computing and Information Science in Engineering, **10(3)**: 1-10 (2010)
7. Ganeshan R., Garrett J., Finger S. A framework for representing design intent, Design Studies, **15(1):** 59-84 (1994)
8. Feldman, R., Text Mining: *Theory and Practice*, (1988)
9. Liu, S., McMahon, C. A., and Culley, S. J., A Review of Structured Document Retrieval (SDR) Technology to Improve Information Access Performance in Engineering Document Management, Computers in Industry, **59 (1)** :3-16 (2008)
10. Li, Z., and Ramani, K., Ontology-Based Design Information Extraction and Retrieval, Artif. Intell. Eng. Des. Anal. Manuf., **21 (2)** :137-154 (2007)
11. Romanowski, C. J., and Nagi., A Data Mining Approach to Forming Generic Bills of Materials in Support of Variant Design Activities, ASME J.Comput. Inf. Sci. Eng., **4 (4)** :316-328 (2004)
12. Liu J. H., Zhan H. F., *A reconstruction method of the design rationale model based on design context*, (2013)

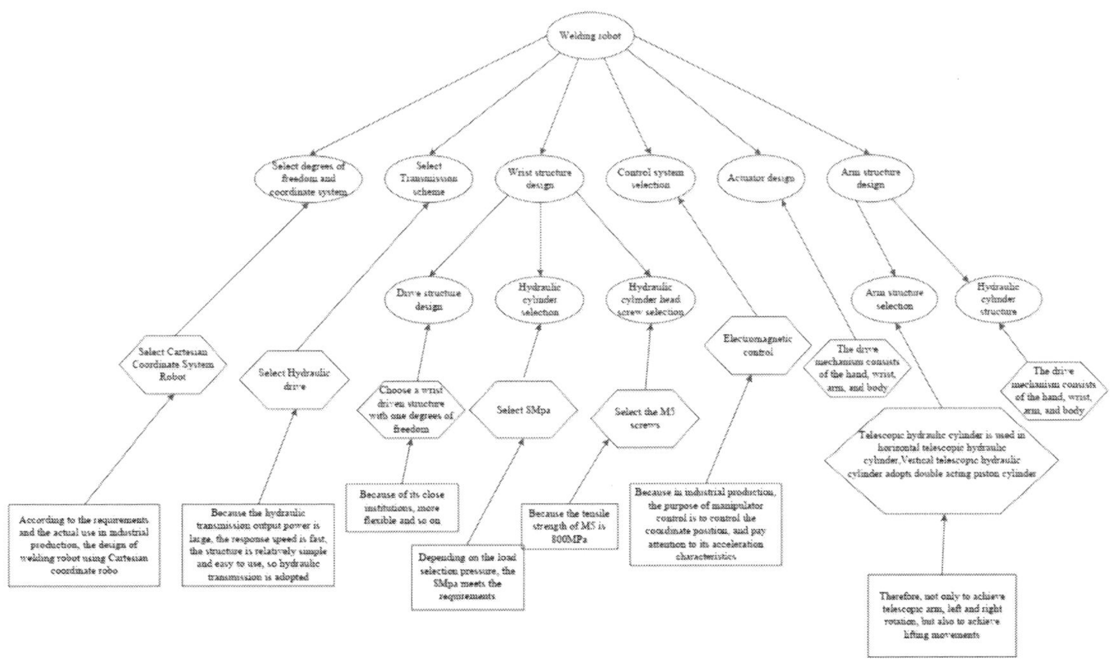

Fig. 6. The DR knowledge model extracted from the welding robot design manual

 DOI: 10.1051/matecconf/201713900017

MBD Attributes Template Method of Aeronautical Products

Xiumao Yang[1], *Zhengliang* Wu[1], *Yongzhu* Hou[2], and *Jihong* Liu[2,*]

[1]Hongdu Group Company of Aeronautical Industry, 330024 Nanchang, China
[2]School of Mechanical Engineering and Automation, Beihang Uiversity, 100191 Beijing, China

Abstract. Model-based Definition (MBD) expresses the technical information of traditional engineering drawings in a three-dimensional environment. The formed MBD dataset completely express the product information, what changed the product development model and improved product development efficiency. However, the integrity of the MBD dataset is difficult to guarantee and the efficiency to build it is low. In order to solve this problem, this paper proposes to use MBD attributes template method to solve the generation problem of attributes information in MBD dataset. Firstly, the concept and connotation of MBD attributes and MBD attributes template are described. Secondly, the type of aeronautical products and MBD attributes are combed to form MBD attributes template for different types of products. Thirdly, the process of MBD attributes template method is proposed. Finally, the validity of the proposed method is verified by the MBD attributes template construction and attributes values filling of a forging part. The results show that MBD attributes template method not only improves the efficiency of MBD attributes generation, but also improves the integrity and accuracy of MBD dataset.

1 Introduction

Three-dimensional model plays a vital role in engineering domain. It can drive product design, manufacturing, assembly and inspection stage of production development directly. The Model-based Definition (MBD) technology is proposed by the American Society of Mechanical Engineers based on its digital product definition specification (ASME Y14.41-2003) and combined with its digital manufacturing experiences [1]. MBD technology express technical information in the three-dimensional model, forming a completed product model drive product development activities, is a current research focus. For example: MBD design under the product lifecycle scenario [2]; through the MBD dataset to assist the construction of the MBD process model [3]; multi-level structured MBD model support the mechanical parts reuse [4].

MBD has been widely used in the whole aeronautical field. MBD-based 3D design has become the main development model of the current aeronautical enterprises. However, there are still has many management problems and technical problems need to be solved [5], including how to improve the efficiency of MBD dataset construction and ensure the integrity and accuracy of the MBD dataset.

In view of this problem, [6] studied the automatic annotation technology of 3D model based on GB/T 24734 to improve the efficiency of 3D model annotation. However, the MBD dataset contains MBD attributes besides annotation. MBD attributes contains both the management information needed for other product development phases and the technology information used to assist the MBD annotation to completely defining the product model. The attributes is of a wide variety and its account is large, thus: 1) there is incomplete problem when designers add MBD attributes, and is not conducive to management; 2) designers need to consult the relevant manual and standard to add MBD attributes what is time-consuming and laborious. In this paper, MBD attributes template method is proposed to solve the problem of MBD attributes generation. By combing the types of products and MBD attribute types, the MBD attributes template is formed according to different products types. And the MBD attributes values are itemized to support the construction of MBD attributes fast and accurate.

2 MBD attributes

MBD technology expresses the technical information in a three-dimensional environment, forming the MBD dataset that can fully describe the product. Technology information includes geometric dimensions and tolerances, roughness, datum, and other technical requirements once in the two-dimensional drawings. The physical form of MBD dataset is MBD model, which essence is adding the technical information to the three-dimensional model.

The ASME Y14.41-2003 standard specifies the contents of the MBD model firstly. As shown in Figure 1, the MBD dataset includes three parts: design model, annotation and attributes. The design model is the traditional meaning of the three-dimensional model which is consists of model geometry and supplemental

* Corresponding author: ryukeiko@buaa.edu.cn

geometry; Annotation is one type of technology information which is visible in the model of various dimensions and tolerances, roughness, datum and other content. They do not need to query or do other operations of the model. Attributes is other information which used to complete the definition of the product. The content of attributes is not visible in the model, but can be obtained through querying the product model.

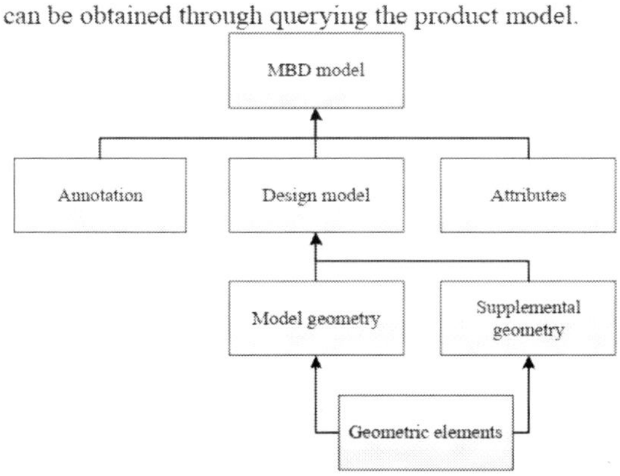

Fig. 1. Contents of MBD model.

In engineering practice, the implementation of MBD-based 3D design not only considering the integrity of the definition of product dataset, but also need to consider the collaborative of design and manufacture, model-based product lifecycle management and other factors. In this situation, it requires the attributes of MBD model both supplement the content of the annotation and include some basic information and a part of the management information of other development stage. In addition, engineering design is a process of repeated iterations and constantly validate, so the information that can be processed and obtained at different design stages is not the same. In the field of aviation, the design and manufacture of aircraft introduced the concept of maturity to describe the different stage of the parts design. After the previous stage is completed, the product design can enter a new stage of maturity. The maturity of product digital design usually divided into three stages: M0, M1 and M2.

In this paper, the MBD attributes is defined as the attributes of the MBD model which contains the technical information needed to design and manufacturing process and the management information in the product development process. This information needs to be defined in different maturity stages such as M0, M1 and M2.

In this paper, we use MBD attribute template to determine the MBD attributes content required by the different types of products. The MBD attributes content will be standardized, itemized and regularized in order to improve the MBD model design efficiency and the integrity and accuracy of MBD attributes.

MBD attributes template has two advantages:

(1) Build a unified MBD attributes template for different types of products. The designers no longer need to consider what attributes should to be added. They can just follow the template to fill in the contents according to the attributes template;

(2) For the attributes that can form a set of rules, the designer can directly select the corresponding attributes values rather than through their own search and entry.

3 Classification of products and MBD attributes

To achieve the MBD attributes template method we should firstly divided the type of products and MBD attributes. The classification of products is in order to distinguish the characteristics of different types of Aeronautical products, which is the basis of constructing different MBD attributes template. The classification of MBD attributes is to sort out the management information and technical information of aviation business needs and complete the MBD model content.

3.1 Classification of Products

The aeronautical products include eight types: part, assembly, piping, harness, deformer, virtual part, overall model and finished product. Each category can continue to be divided into multiple subtypes. The division of each category and its meaning are shown in Table 1.

Table 1. Classification of Products.

Type	Subtype	Meaning
Part	Sheet metal, Machine part, Forging part, Casting part, Welding part, Composite material part *.et al*	The design unit which constitute the assembly
Assembly	Riveting assembly, Clamp assembly, Welding assembly *.et al*	Composed of parts, usually as a system or subsystem
Piping	Tube, Conduit assembly	A component of one or more tubes
Harness	Geometric harness, Two - dimensional wire harness, Circuit diagram, *.et al*	The wiring parts used to connect the different parts
Deformer	Part deformation, Standard part deformation, *.et al*	Obtained through the deformation design by other types of products
Virtual Part	Fastener assembly, Annotation model, Catheter package model *.et al*	In the form of a geometric graph set, without a solid model

Overall Model	Overall skeleton model, Theoretical shape model, Aircraft benchmark model, *.et al*	Used to control the design range of system, sub-system
Finished Product	——	The final delivery of the design

3.2 Classification of MBD attributes

As with the division of products type, the MBD attributes can be divided into multiple types, each of which contains multiple subtypes. Table 2 lists the types and subtypes of the MBD attributes of the aeronautical products, and the maturity level that should be generated for each subtype attribute value. It should be noted that: 1) different subtypes of the same type may belong to different maturity stages, such as the material grade belongs to the M0 stage but the wool size belongs to the M1 stage, even though they both belongs material description type; 2) most of the technical information contained in MBD attributes exist in subtype general notes of the product notes, including manufacturing and inspection basis, heat treatment, surface treatment, non-destructive testing methods, sharp edge chamfering, inspection location, machining requirements, and so on.

Table 2. Classification of MBD attributes.

Type	Subtype	Maturity Stage
Security Description	Security Level Time Limit	M0 M0
Standard Annotation	Weight Unit Length Unit Copyright Notice	Automatic Generated
Product Description	Product ID Product Type Product Subtype Supply Characteristics Symmetrical or Not Symmetrical Description	M0 M0 M0 M0 M1 M1
Material Description	Material Grade Wool Size	M1 M2
Design Basis	Theoretical Shape Theoretical Circle Design Suggestions	M1 M1 M1
Product Version	Development Stage	M0
Product Characteristics	Single Element Type Long Period or Not Important Parts Amount Key Parts Amount	M1 M1 M2 M2
External Reference	——	M1
Auxiliary Geometry	——	M1
Product Notes	General Notes Structural Elements	M1 M1
	Wool Model	M2
Link Definition	——	M1
Weight	Theoretical Weight Model Weight	M2 M2

4 MBD attributes template method

As showed in Figure 2, the MBD attributes template method consists of six steps:

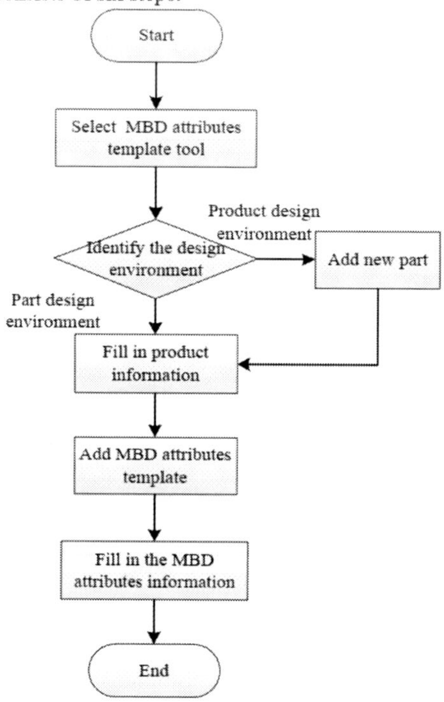

Fig. 2. MBD attributes template method.

Step 1: Select MBD attributes template tool: During the product design process, the MBD attributes template method is executed by the custom MBD attributes template tool;

Step 2: Identify the design environment: Automatically identify the current design environment, which mainly including product design environment and part design environment. Each type of design environment should have different MBD attributes template process. If the current design environment is product design environment, then the user needs to add a new part and build its MBD attribute template, go to step 3; if the current design environment is part design environment, then the user needs to directly build the corresponding MBD attributes template of the current part, go to step 4;

Step 3: Add new part: Create a new part in the product design environment automatically;

Step 4: Fill in product information: fill in the part type, subtype and part ID of the product will be created;

Step 5: Add MBD attributes template: Automatically find the corresponding MBD attributes template according to the filled part information and express the template in the product structure tree in a structured way;

Step 6: Fill in the MBD attributes information: Fill in the values of the various types of MBD attributes in the template based on the added MBD attribute template. The attributes values are filled in three ways: automatic creation of attribute values, selection of attribute value entries and manual entry.

5 Validation

In this paper, the MBD model of a forging part is used to illustrate the MBD attributes template method. This article use the VB language and VBScript language to achieve the development of MBD attributes template tool based on the CATIA platform, and through the CATIA macro function to provide custom MBD attributes template tool bar. As shown in Figure 3, MBD attributes template tool bar contains two functions: 1) create MBD attributes template; 2) fill in MBD attributes values.

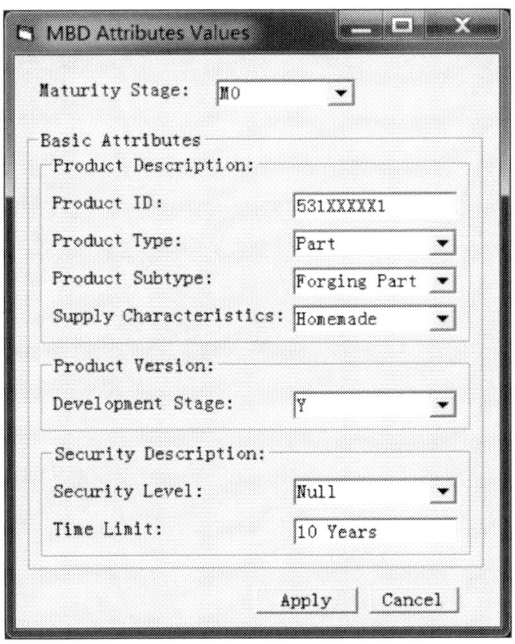

Fig. 4. Fill in MBD attributes values.

Firstly, enter the MBD attributes template creation interface through the button shown in Figure 3 marked ①. At the same time, the system automatically detects the current design environment is product design environment and requires users to fill in product-related information. When the user completes the information such as product ID, product type and product subtype, they can construct the relevant MBD attributes in the product structure tree automatically by clicking the "Apply" button.

After completing the creation of MBD attributes template, user can enter the MBD attributes value filling interface shown in Figure 4 by clicking the button shown in Figure 3 marked ②. The user needs to specify the stage of the current product design first (M0 stage in the Figure 4) before the MBD attributes value is filled in. After selecting the corresponding maturity stage, the system will loading the MBD attributes required for this phase. The user can check or fill in the relevant attributes value according to the given template. If all the attributes values are completed, the system updates the attributes values to the product structure tree.

When the product design completed the M0 stage, the user can continue to fill the MBD attributes values of other maturity stage such as M1 and M2 by the function of Filling MBD attributes values and ultimately complete the product design.

6 Conclusion

MBD technology brings both convenience and challenges to engineering design and manufacturing. How to construct MBD model accurately and rapidly is an essential engineering problem needed to be solve. This paper proposed the MBD attributes template method, and form different MBD attributes template for different product types by combing the aeronautical products and MBD attributes types. The MBD attributes template can constructed in a more completely and fast way. In addition, the article formed the MBD attributes values rules that can be formed. The designer can directly select the attributes values, thus reducing the workload of the designer, and improving the efficiency and accuracy of the MBD attributes. In fact, the MBD model also contains annotation information besides MBD attributes, which will be the focus of our future work.

Acknowledgements

This work has been supported by Project of National Key Technology R&D Program through approval No. 2015BAF17B01.

References

1. LIU J H., SUN L S, ZAHNG X, et al. Connotation and key problem of three-dimensional digital. Computer Integrated Manufacturing Systems, **20(3)**: 494-504 (2014)
2. Alemanni M, Destefanis F, Vezzetti E. Model-based definition design in the product lifecycle management scenario. International Journal of Advanced Manufacturing Technology, **52(1-4)**: 1-14(2011)
3. Wan N, Mo R, Liu L, et al. New methods of creating MBD process model: On the basis of

machining knowledge. Computers in Industry, **65(4)**: 537-549(2014)

4. Huang R, Zhang S, Bai X, et al. Multi-level structuralized model-based definition model based on machining features for manufacturing reuse of mechanical parts. International Journal of Advanced Manufacturing Technology, **75(5-8)**: 1035-1048(2014)
5. Quintana V, Rivest L, Pellerin R, et al. Will Model-based Definition replace engineering drawings throughout the product lifecycle? A global perspective from aerospace industry. Computers in Industry, **61(5)**: 497-508(2010)
6. An H, Yan G R, Lei Y. Automatic dimensioning of 3D model based on GB/T 24734. Journal of Beijing University of Aeronautics and Astronautics, **38(3)**: 416-421(2012)

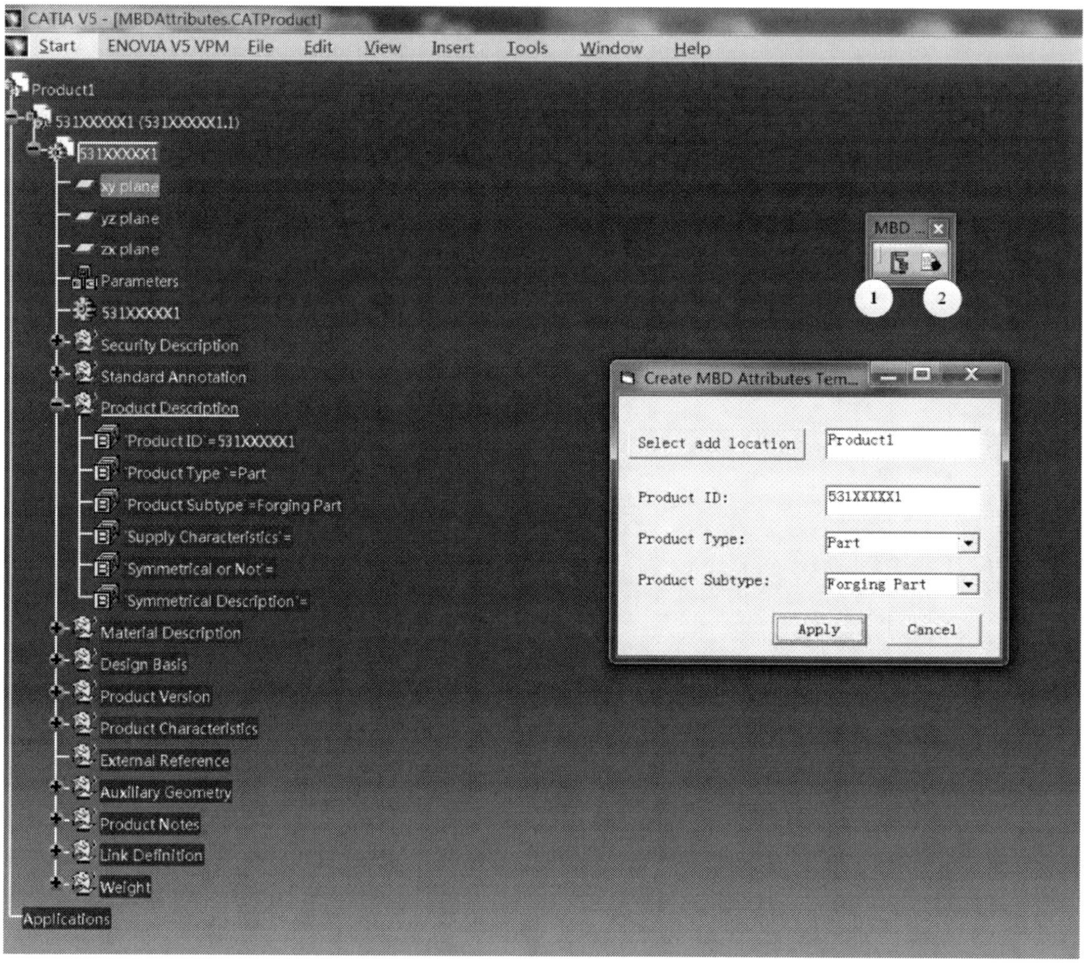

Fig. 3. Create MBD attributes template.

A 3D Model Search Method Based on Combination of Design Knowledge and Model Features

Jihong Liu[1,*], *Keyi Wang*[1], *Chao Fu*[1], *Yongzhu Hou*[1]

[1]School of Mechanical Engineering and Automation, Beihang University, 100191 Beijing, China

Abstract. 3D models play an important role in modern product design, and there has been a large number of models stored in enterprise database. Reuse of existing models takes a large proportion of new product design. Model search technology helps find existing models which is to be reused. At present there are three model search methods including text-based, content-based and semantic-based. Embedding knowledge into models filled models with abundant semantic information, which supports model search indexing. Embedding knowledge into models filled models with abundant semantic information, which supports model search indexing. A 3D model search method based on combination of design knowledge and model features is proposed. The technical connotation of the proposed method is introduced and an example is given to illustrate the effectiveness of the proposed method.

1 Introduction

Today, we are currently living in the era of knowledge-driven economy. The fierce competition in the modern global market requires that manufacturing companies be able to enhance their competitiveness in terms of time (T), quality (Q), cost (C), innovation (I) and service (S) [1]. The enhancement of competitiveness needs to be achieved through optimization of multi-domain knowledge applications. People have realized the importance of knowledge in the design and manufacture of products, so the enterprises need to effectively manage the existing knowledge, while the knowledge is quickly and accurately provided to the designers to achieve a comprehensive product optimization design [2]. With the development of computer technology in industrial, designers can build a 3D model to carry out more intuitive product design [3]. At the same time, the product of the 3D model in product simulation analysis and digital manufacturing is also in a very important position [4]. It is precisely rich expression of product information that the 3D model have, "full 3D" concept was raised [5]. As it is a representative technology in the full three-dimensional process, Boeing has introduced Model-Based Definition (MBD) technology to integrate the information of the product into a unified 3D model and has been applied in the B787 to achieve disappeared drawings of the design and manufacture.

The purpose of MBD is to annotate 3D models with engineering information on traditional 2D drawings. Those kinds of annotated information describe what the product is and how it should be manufactured. However, currently the design basis behind supportive information (design knowledge) has not been added into models. So integrating design process information into 3D models can assist designers to understand the product design and reuse them in new products. That is also an effective approach to preserve knowledge, which is precious for an enterprise, and make them more helpful.

The main purpose of the integration of the 3D model and the product design knowledge is to assist the design process through the reuse of the product model and the design process. The search is the most direct way to find the auxiliary resources. Whether the auxiliary design resources can be provided directly to the designer by analysing the search requirements is the main criterion for evaluating a method. At present, there are many researches on the search of 3D models. The existing model search technology is divided into three categories: text, content and semantics [6-9]. The model embedded with design knowledge [10, 11] makes the existing product model with richer semantic information, which also provides index support for model search. When the design knowledge is embedded in the 3D model, it provides the basis for the new search method.

2 The method associated with model and design knowledge

2.1 Classification and organization of knowledge

In order to give full play to the role of product model and engineering knowledge in product development, the literature in [12] describes classification and

* Corresponding author: ryukeiko@buaa.edu.cn

organization technology of the engineering knowledge, according to the structural characteristics of the design knowledge and its form of expression, The knowledge need to be marked to the product model and it should be classified and collated. At the same time, based on the knowledge classification method and knowledge level organization method of product structure and development process, the knowledge organization structure is achieved through XML, so as to realize the computer storage and expression of design knowledge. But it is also an important issue that how to use the marked knowledge to achieve engineering applications. For the three-dimensional model in which design knowledge has embedded, the most straightforward application is to retrieve the 3D model and the corresponding design knowledge related to the design through the 3D model of the embedded knowledge. This application pattern can be applied to multiple scenarios, such as collaborative design, design reuse, design learning and communication between design and manufacturing. It not only can improve the utilization of product design specifications, but also to promote the Information exchange between design process and manufacturing process. On this basis, this paper studies the 3D model search technology based on knowledge.

This paper associates explanatory object and the explanatory knowledge through the structure of a knowledge explanatory tree, as shown in Figure 1. Each explanatory object is explained by a knowledge explanatory tree. The knowledge explanatory tree has the following characteristics: (1) Each node is a knowledge packet, the knowledge package contains the corresponding knowledge item, each knowledge item includes multiple knowledge elements; (2) If a node knowledge package contains intermediate elements, its sub-knowledge package is used to explain its contents; (3) The leaf nodes in each tree shall not contain elements other than the design of the original information.

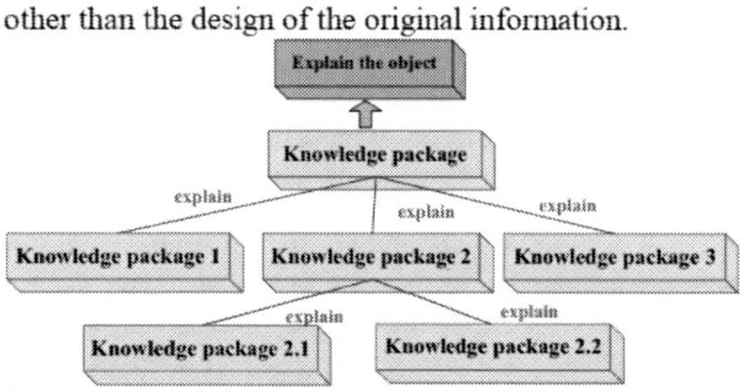

Figure 1. Knowledge explanatory tree

2.2 Knowledge embedded process

Based on the type of the design knowledge, the form of the knowledge representation, the embedding level of 3D model knowledge and organizational structure of 3D model design knowledge mentioned before, this section will elaborate the design knowledge embedding process of 3D model. The process will guide and regulate how the embedded knowledge is completed for an model which don't have embedded knowledge. The whole process is divided into five steps:

Step 1: Determine the model hierarchy. According to the specific model level that is required embedded knowledge, select the appropriate level. As mentioned above, here the model levels are divided into 5 levels, which are products, components (standard, non-standard), parts (standard, conventional, unconventional), features and surface.

Step 2: Select the explanatory object. Based on the level of the model determined in step 1, the explanatory object which is explained by the design knowledge are selected within the range of this hierarchy, such as the specific parts mentioned in the level of the parts, the specific size and tolerance mentioned in the level of the features and the specific surface in the level of the surface.

Step 3: Select the type of knowledge. After the specific explanatory object is determined, the corresponding knowledge package will be constructed by adding the knowledge item. The type of specific knowledge which is embedded is selected based on the knowledge types defined in the above. Each knowledge item belongs to a certain kind of knowledge.

Step 4: Select the form of knowledge representation. Each knowledge item is composed of one or more forms of knowledge, so it is necessary to select the form of knowledge before entering the specific design knowledge content. The forms of knowledge include text, formulas, charts and links.

Step 5: Input the content of knowledge. According to model level, explanatory object, knowledge type and knowledge form which are determined previously, input the corresponding and specific knowledge content, complete embedding of design knowledge.

A complete knowledge item can be formed from step 3 to Step 5. A complete knowledge packet can be formed from step 2 to Step 5, it. Through the iterations of the five steps, the embedding process of model knowledge can be achieved, The process will form multiple knowledge packets for different model levels and the explanations in which they are.

3 3D Model search method based on design intention and model features

In the 3D model search method, which combines the design intention and the model feature, the input content is divided into two parts, as shown in Figure 2. One part is the model which need to match, its essence is to describe the geometric requirements of the search through the model; the other part is the design intention description. Through the input of these two search requirements, the search results can be controlled by the two aspects, so that it can meet the more design requirements.

Figure 3 shows the basic flow of the search method. For the two input parts of the search needs are dealt respectively. The features sequence is extracted from the input model, and the natural language description of the design intention is used to construct the word sequence by using the word segmentation method.

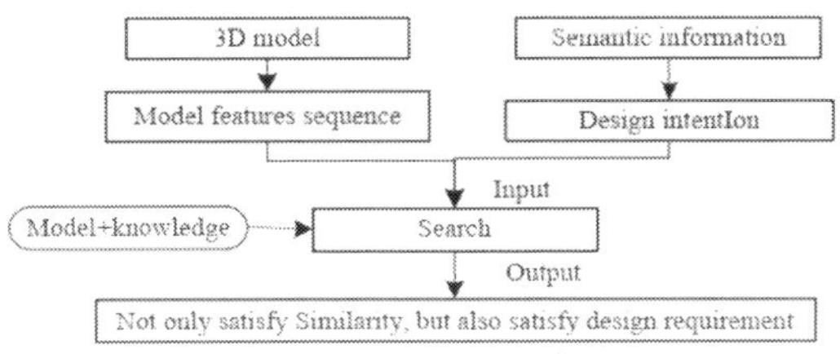

Figure 2. Ideas of Design intent and model characteristics combined with the model to retrieve

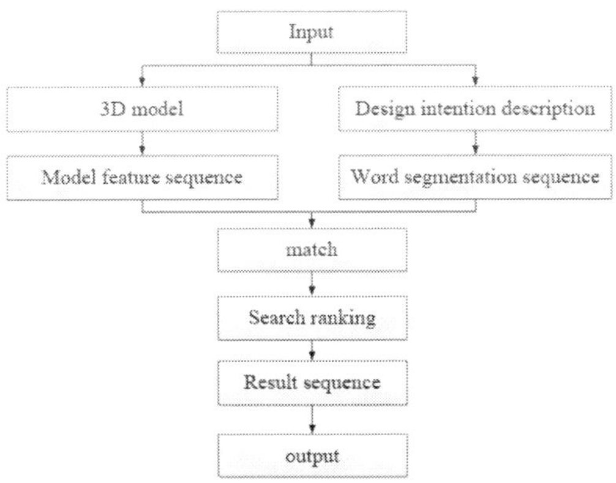

Figure 3. Flow of model search algorithm

Step 1: Input of Design Knowledge and 3D Model

The input content of the MBD model search is divided into two parts by using the method proposed by the paper. One is the models that need to be matched. The essence is to describe the geometric requirements of the search through the model, which is defined as 3D model and the other part is the design knowledge description, design knowledge is the interpretation for product geometry information and size function information, including the designer's product design requirements, design standards and so on.

Step 2: Model Matching Based on Design Knowledge Segmentation

The present invention adopts the method of keyword presence to match the model in the model library with the search request. Since the primary purpose of the design is to meet the design requirements and achieve the required functionality, the design intent is used for matching as the only control factor. On the other hand, for the characteristics of the model, the types of features are limited, and their combinations are relatively limited, so there is little effect on the matching results. Sequence of the segmentation $W=[w_1,w_2,\cdots,w_i,\cdots,w_n]$ and the knowledge focused index are searched, as long as there is any member of W, the model is added to the model set. The specific algorithm flow is shown in Figure 4.

Step 3: Geometric similarity calculation based on the longest common sub-sequence ratio

Step3.1: generation of Model feature sequence F

A set of feature sequences are arranged based on the modeling order. Here ft+1 is the characteristic that is generated after ft in the modeling process.

$$F=[f_1,f_2,\cdots,f_t,f_{t+1},\cdots,f_n] \quad (1)$$

The feature sequence is a set of feature sets arranged in time. The feature set itself can partially define the geometric information of the model, and the more accurate definition of the model geometry can be achieved by adding the time factor.

Step3.2: Calculate the longest common subsequence Fc

The longest common subsequence Fc of the characteristic sequence F and the input model i feature sequence F_i after matching, n_c is the length of Fc, Ni is the length of F_i, the ratio of the common subsequence of F to F_i is:

$$R_f=\frac{n_c}{N_i} \quad (2)$$

In particular, the number of sequences is n, and if a sequence T is a subsequence for each of the sequences, then T is referred to as a common subsequence of the sequence. If T is the longest of all the subsequences of the sequence, it is called the longest common subsequence.

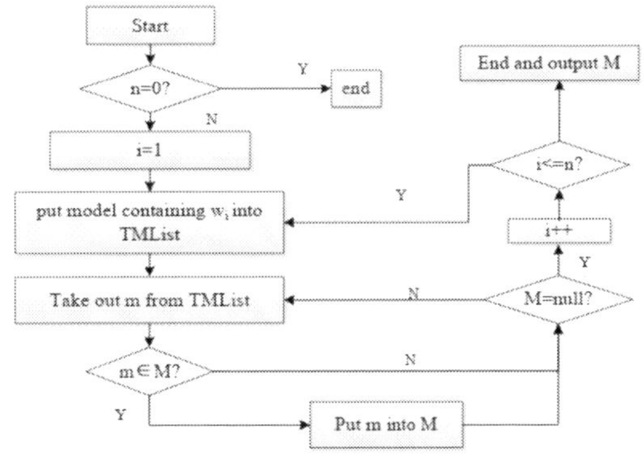

Figure 4. Flow of matching algorithm

Step 4: Calculating the Semantic Similarity of Basic Naive Bayesian

Step 4.1: Generate of the word segmentation sequence W of design knowledge

The word segmentation sequence is constructed by the order that each word appears in. Each word can be repeated, it is indicated that $W_i=W_j$ is allowed.

$$W=[w_1,w_2,\cdots,w_i,\cdots,w_j,\cdots,w_n] \quad (3)$$

The design knowledge segmentation sequence is an ordered set of vocabularies in the order, a set of vocabularies is an effective description of the input design information entered by the user.

Step 4.2: Naive Bayesian method transformation

The Naive Bayesian method is generally used in the text classification, and its classification process is roughly through the training of multiple types of text to get a certain amount of training parameters related to the frequency of words, and finally classified text will be used for classification of word segmentation and word frequency statistics, The Bayesian basic formula is used to calculate the probability that the document belongs to each type, and the type with the maximum probability value is selected as the final classification result. Each of these types of training documents generally includes a lot.

The method applies the Bayesian in the classification transformation. Regarding each model as a type, the semantic information included in each model is as a training text of a type, where there is only one training text of each type, The semantic information of demand is used as the sorted document, and the similarity of the semantic information of each model is calculated and the ranking of the results is achieved according to the similarity of semantic similarity. The modified Bayesian method is compared with the traditional Bayesian method used for classification as shown in Fig 5.

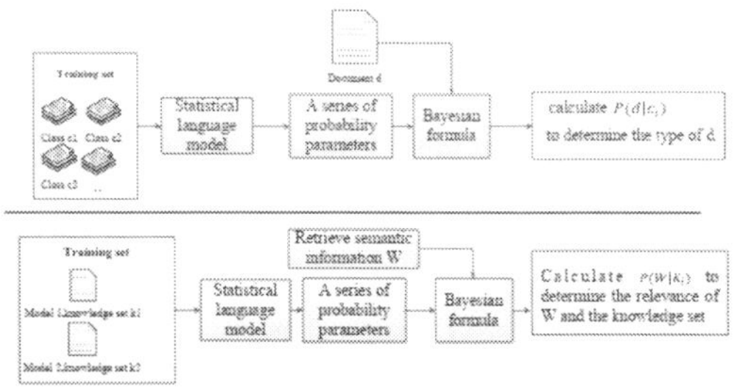

Figure 5. Naive Bayesian method for the comparison text classification and search

Step 4.3: Semantic similarity calculation based on Bayesian

The formula for calculating the semantic similarity using the Bayesian theorem is:

$$P(K_i \mid W)=\frac{P(W \mid K_i)P(K_i)}{P(W)} \tag{4}$$

Where K_i is the knowledge set corresponding to model i, W is the segmentation sequence after the design knowledge segmentation.

Since P(W) is the same as for any K_i, so $P(K_i)=\frac{T_i}{\sum T_i}$ is used to calculate $P(K_i)$. Where T_i is the total number of words in K_i and $\sum T_i$ is the sum of the summations included in all sorting models.

By simplifying the simple assumptions of class conditions,

By simple assumptions in which the class conditions are independent, it's known:

$$P(W \mid K_i)=P([w_1,w_2,\cdots,w_m] \mid K_i)=\prod_{j=1}^{m}P(w_j \mid K_i) \tag{5}$$

$$P(w_j \mid K_i)=\frac{t_j}{T_i}\approx\frac{t_j+1}{T_i+|v|} \tag{6}$$

Where t_j is the number of occurrences of w_j in K_i, T_i is the total number of words in K_i, and |V| is the total number of words in K. The semantic similarity R_w of the model can be calculated by formula (7) instead of P(W|K_i).

$$P(W \mid K_i)P(K_i)=\frac{t_j+1}{T_i+|v|}\bullet\frac{T_i}{\sum T_i} \tag{7}$$

Step 5: Relevance sequences based on design knowledge and model feature

Step 5.1: Normalization

Since the formulas R_f and R_w are used in two different calculation methods, the difference between the two calculated values will be large. In order to make the two values be closer, R_f and R_w are normalized to the range of [0,1]. Finally, the normalization of the two values is achieved by formulas.

$$R_f^Z=\frac{R_f-R_f^{\min}}{R_f^{\max}-R_f^{\min}} \tag{8}$$

$$R_w^Z=\frac{R_w-R_w^{\min}}{R_w^{\max}-R_w^{\min}} \tag{9}$$

In this paper, we eventually adopts the method of human participation in the value of α and β , which are mainly used to balance the semantic similarity R_w and the geometric similarity R_f. The relationship is:

$$\alpha+\beta=1 \tag{10}$$

Step 5.2: Selection of weight

The weight of the semantic similarity (R_f) is α and the weight of the model feature similarity (R_w) is β , semantic priority ($\alpha=0.2,\beta=0.8$), geometric priority ($\alpha=0.8,\beta=0.2$), the balance of semantic and geometric ($\alpha=0.5,\beta=0.5$), user-defined ($\alpha=x,\beta=1-x$).

Step 5.3: Calculate Comprehensive correlation

The linear correlation is used to calculate the correlation between the retrieved MBD model and the input

$$R=\alpha\times R_f+\beta\times R_w \tag{11}$$

Step 5.4: Output MBD model whose comprehensive correlation are front.

4 Instance verification

The following example illustrates this method and compares it with the other two methods. Select "connect two parts" as the retrieve intent. Through the word segmentation, we can get segmentation series ["connect", "two", "part"]. The part in figure 6 is selected as the input of search geometry information, and the characteristic sequence is ["cylinder", "cylinder"].

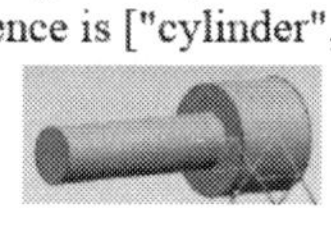

Figure 6 The part drawing

Retrieving the balance of semantic and geometric ($\alpha=0.5,\beta=0.5$) to get the results, as shown in Figure 7.

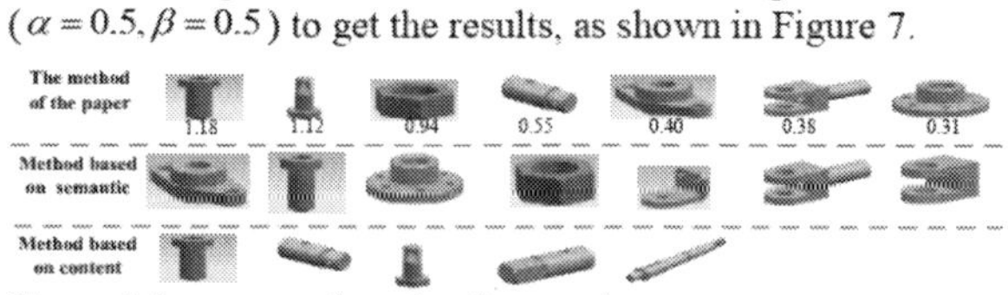

Figure 7 the compared results after search

The following values under the model are the final comprehensive similarity values. Table 1 also gives a comparison of the results obtained by simple semantic search and content-based method search. By contrast, we can find that the method of simple semantic search can find the model to satisfy the corresponding search semantic information, but there are some models in the result that the model is very different from the model geometry of the user. Therefore, these models are difficult to have practical meaning. The content-based approach can give a similarity of the search results model, but completely ignore the application of semantic information model, which makes a lot of models to meet the designer's geometric requirements, but it is difficult to directly meet the design requirements and can't be applied .

In comparison, the comprehensive search method based on design intention and model geometric information will be more satisfied with the user's functional requirements and geometric requirements, so as to be closer to the user's design needs. If it's applied in a larger model library, the advantages of the method will be more prominent.

5 Conlusion

This paper summarizes the shortcomings of the existing model search methods, and puts forward a new model search method for 3D models of embedded knowledge. The combination of design Knowledge and model Features can search the 3D model to meet the design functional requirements, also to meet the shape requirements of the model.

In the actual calculation, since the values of R_f and R_w vary greatly for different search requestment, the two values need to be normalized to control them between (0,1). The determination of the harmonic coefficients α and β needs further study.

Acknowledgements

This work has been supported by Project of National Key Technology R&D Program through approval No. 2015BAF17B01.

References

1 Lin Y J, Colello M, The International Journal of Advanced Manufacturing Technology, 23, (2004), 389-398.

2 Dias WPS, Knowledge-Based Systems, 20, (2007), 382-387.

3 Gómez-Pérez J.M., Erdmann M., Greaves M., et al, International Journal of Human-Computer Studies, 68, (2010), 641-668.

4 Quintana V., Rivest L., Pellerin R., et al, Computers in Industry, 61, (2010), 497-508.

5 Quintana V., Rivest L., Pellerin R., et al,Computers in Industry,63, (2012), 79-90.

6 YU W-D, HSU J-Y,Automation in Construction, 31, (2013), 65-74.

7 BARRA V, BIASOTTI S, Pattern Recognition, 46, (2013), 2985-2999.

8 BAI J, GAO S, TANG W, et al, Computer-Aided Design, 42, (2010), 1069-1084.

9 Bano, Muneera, and N. Ikram, Knowledge Management for Service Oriented Requirements Engineering, (2011):494-499.

10 F Xuan, R.J. Howlett, Frontiers in Artificial Intelligence and Applications, 149, (2006).

11 H Wang, Ceglarek D, Assembly Automation,29, (2009), 154-166.

12 Y Hou, J Liu, CAD&CG, 28, (2016), 1933-1939.

Table 1. similarity calculation of the model

List	Model	T_i	n_i	N_i	n_c	R_w^Z	R_f^Z
1		43	1	3	0	0.101448567	0
2		37	0	8	1	0.005492843	0.4375
3		59	1	5	1	0.065211437	0.7
4	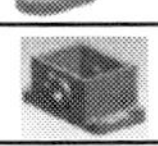	67	1	12	0	0.050219444	0
5	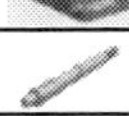	41	0	9	2	0.096390827	0.777777778
6		54	1	13	2	0.075563228	0.538461538

Aeronautics design and manufacture integrated enterprise model

Likang Song[1], *Xiumao* Yang[1], *Chao* Fu[2], and *Jihong* Liu[2,*]

[1]Hongdu Group Company of Aeronautical Industry, 330024 Nanchang, China
[2]School of Mechanical Engineering and Automation, Beihang Uiversity, 100191 Beijing, China

Abstract. In order to realize the integration and concurrent of design and manufacture, enhance the efficiency of aircraft development enterprises, a model centre based design and manufacturing integrated enterprise model is proposed. The model centre consists of central system models, product models and process model. The design and manufacture integrated development model based on the unified model centre is discussed. Finally, a maturity control method supporting design and manufacturing integration is proposed.

1 INTRODUCTION

Aviation product development is a comprehensive technology[1]. The design process of aircraft is a comprehensive trade-off and optimization of various systems and technologies. Practice has proved that the integrated design and manufacture of aircraft is an important technical to improve product quality and reduce product development cycle [2]. The core idea of design and manufacture integrated is the application of concurrent engineering, digital engineering and collaborative design technology etc. to realize seamless integration and parallel development of aircraft design and manufacture.

Aircraft development is complex system engineering. The enormous challenge for aeronautical manufacturing is to achieve the best in terms of performance, cost and planning cycle in aircraft development [3]. Modern aircraft design and manufacturing technology based on digitalization is changing the traditional way of aircraft development. Aircraft design and manufacturing technology from the digital three-dimensional design and manufacturing to the process as the centre of the integrated development model transition. A variety of individual digital technology, network technology, advanced manufacturing technology are gradually applied to the aircraft development process, and in the digital definition, product data management, information integration and other aspects of significant progress. However, the current aircraft development enterprises are faced with the following problems: cannot achieve the "demand - functional logic model - behaviour model - physical model" of a single data source management and association, trace management; cannot be established by the demand, functional logic model, behaviour model, physical model Constitute the BOM tree, cannot implement the model of the data baseline / version management and configuration technology state management; does not support the performance of the prototype (FDMU) in the ring design. So, a model centre based design and manufacturing integrated enterprise model is proposed. The model centre consists of central system models, product models and process model. The design and manufacture integrated development model based on the unified model centre is discussed. Finally, a maturity control method supporting design and manufacturing integration is proposed.

2 Model centre of designing and manufacturing integrated development enterprise

Model is the key and core of aircraft design and manufacture integrated enterprise. As shown in Figure 1, model centre that support the aircraft design and manufacture integration development include the central system model, product model, process model and knowledge model. Among them, the central system model is a special product lifecycle model. Knowledge model is divided into product knowledge model and process knowledge model, and it has no fixed form of expression. Integrated design and manufacture of aircraft is based on the system engineering framework, through the construction of enterprise products in the whole life cycle of product model, process model and knowledge model, and on this basis to carry out the corresponding engineering design, manufacture, and service.

2.1 Central system model

* Corresponding author: ryukeiko@buaa.edu.cn

DOI: 10.1051/matecconf/201713900019

The central system model is the primary work of system development. Unified central system model that defines system requirements and establishes traceability relationships between requirements and design at all levels. With the help of standard software environment and data exchange, the model of product model and process model in each stage of product lifecycle are integrated, which can achieve design, complex product manufacturing, services, scrap etc. under a single data source. The central system model first emphasizes the comprehensiveness, accuracy and consistency of the system description. By associating the related parameters of each model in the product life cycle and each stage, the specialized engineers of various disciplines and the subsequent technical and service personnel can work on the basis of the system architecture model. The data is taken from a common database to design, analyse and optimize using the discipline's model and software tools. That is to say, under the MBSE idea, the real single data source technology is based on the central system model.

2.2 Product model

Product is the main output of manufacturing enterprises. The product model mainly completes the functions stipulated in the application framework according to the requirements of the business framework, MBD model is the main form of product model. The purpose of MBD is to annotate 3D models with engineering information on traditional 2D drawings. Those kinds of annotated information descript what the product is and how it should be manufactured. However, currently the supportive information behind the design results (design knowledge) has not been added into models. This sort of supportive information for design explains how designers came up with the design results, while most of them are preserved in design instruction booklet and are rarely used. So to integrate design process information into 3D models can assist designers to understand the product design and reuse them in new products. That is also an effective approach to preserve knowledge, which is precious for an enterprise, and make them more helpful.

MBD data sets can be divided into geometric model, design information, simulation analysis information, and manufacturing information. Geometric model includes 3D entity and auxiliary geometry of the product; Design information includes product size, tolerances, assembly relations, material selection, technical requirements, and parts list; manufacturing information includes the process planning and assembly sequence of parts processing; Simulation analysis information includes finite element simulation, assembly process simulation and behaviour simulation.

2.3 Process model

The core value chain of modern enterprise is the business process. The process model is the description and coordinator of the relevant resources, information, tools and other elements in the enterprise to execute any process, support enterprises to change their products from concept to material and then to use. In general, the process model of enterprises can be divided into two categories: engineering process model and management process model.

The engineering process model is abstracted through engineering practice. Especially the development of complex products, such as aerospace, is an interdisciplinary complex system engineering oriented to the whole life cycle. A detailed analysis of the activities in various stages is required to describe the business process. While, the management process model aims to realize agile, refine and lean management. Through explicit, structured, and standardized business process, we can achieve the modelling, simulation and evaluation of enterprise management business processes and the implementation, monitoring, interaction and control of subsequent application processes. By modelling and understanding the behaviour of enterprise/system, through the simulation study and prediction of the future dynamics of the enterprise system, it provides a basis for the final construction of a new enterprise management method and mode.

3 The design and manufacture integrated development model based on the unified model centre

3.1 Model-based system engineering product development

Combining MBSE model based system engineering methodology, this paper aimed at the development characteristics of aircraft system development and the development of aircraft design and manufacturing, studied the model based system engineering product development system. Through the construction of products covering the whole life cycle of information technology system architecture, the business flow and data flow in the process of product development are realized, and the model-based demand engineering, model-based design engineering, model-based test (flight test) engineering, model based manufacturing engineering, and system life cycle management are developed for the product MBSE.

Through the model to support the requirements of the definition, design, manufacture, integration and validation. With the application of model - based system engineering, the model based on SysML is introduced in the conceptual stage of product development to modle the requirements definition, function analysis and architecture design. The correctness and completeness of requirements and functions are validated in advance by model execution. The model execution ensure collaboration between systems, and collaboration within each component of the system meets requirements. After the function logic of the system is modeled and verified, the function analysis loop is implemented. Through the establishment of a formal system engineering model to describe detailed definitions and system functions and control logic of the requirements of each stage of the aircraft development, this turn ways and experience of product development into structured knowledge; and with the system engineering model as the core, the digital expression model of different disciplines is integrated to achieve multidisciplinary collaboration; at the same time, using CAE technology to achieve simulation driven products, components and parts design, and from the point of view of system engineering, from demand management, functional analysis, simulation and virtual design simulation drive. On the other hand, virtual experiment is used to realize parts virtual test, component virtual structure and product simulation, and realize virtual testing, virtual synthesis and virtual verification of requirements, functions and design. And with the function (performance) virtual prototype bridge, the establishment of functional decomposition and virtual verification of the association to the implementation of the system engineering model to drive the model development plan and process execution, and ultimately the formation of double "V" system engineering development model. Through top-down, demand driven system development model, improve the process, methods and means to promote the efficiency of system development.

3.2 Design and manufacture integrated development based on Model Center

Design and manufacture integrated development enterprise based on model center is through the establishment of a single data source model center, multi-disciplinary, cross-sectorial product design, manufacturing and management, a fundamental reduction in product design, manufacturing and maintenance support time and cost. Its fundamental feature is based on the model, the core of knowledge as the driving force, for the whole life cycle of the system for business integration, data integration, parallel collaboration, agile and efficient product development with the guidance and support of the system engineering and knowledge engineering.

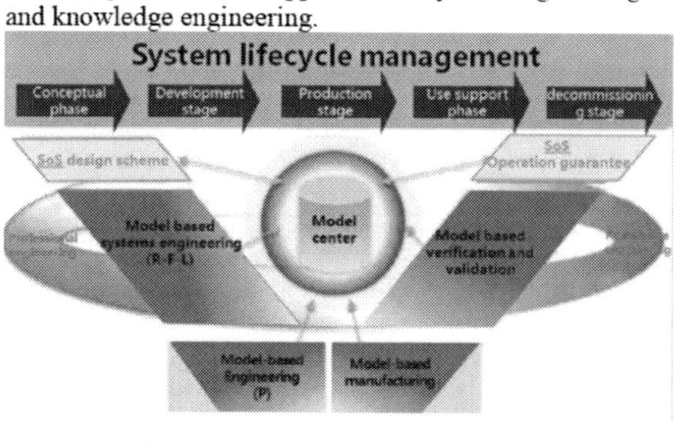

Figure 2. Enterprise development model based on unified model centre

As shown, in order to effectively use between various disciplines and the professional advantages, based on the model design and manufacturing integration in the centre of the research and development process is divided into different stages. And each stage needs to fully take advantage of the existing research, and use different disciplines software for the corresponding development activities.

■ Model based system engineering

In the conceptual phase, the requirements of the product are captured to define the operational parameters and performance that need to be satisfied during the product development process and form an entry document of requirements. In the process of functional analysis and architecture design, we can implement the analysis and refinement process from black box to white box, from product to subsystem and the formation of product design through the decomposition of requirements. And through the operation of the design scheme, the logical operation of the design scheme is completed. In this process, the SysML language can be used to analyse the product refinement process by the system engineering model method. Finally, the correctness of the logic in the design process is verified by the state diagram [4].

The results of the analysis are simulated by multi physics when using MBSE thought to analyse the design process of product. During the design process, the problems that may arise in the process of product development are understood and solved from the root. By establishing the product model and integrating it with the Modelica simulation software, the performance simulation of the product is realized, and the multi physical simulation in the design process is realized. As shown in Figure 3, through the data integration of DOORS, Rhapsody, SimulationX and VPM/CATIA, we can establish the BOM structure of R-F-L-P, which is composed of requirements, function logic model, behaviour model and physical model. This is used to support enhanced design capability by digital prototyping (DMU) to performance prototype (FDMU) design.

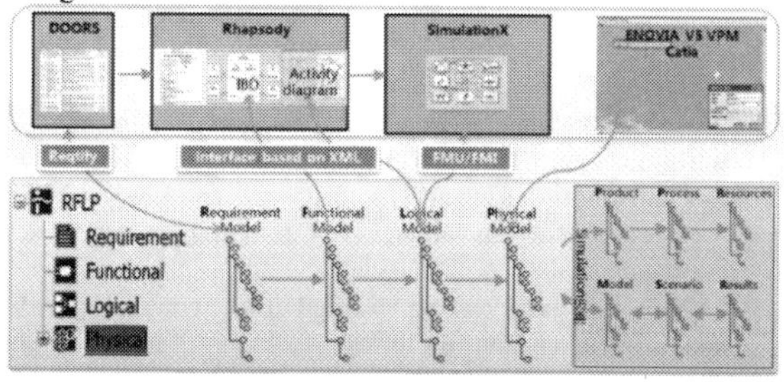

Figure 3 MBSE Integration model

■ Model-based engineering

Model-based engineering includes digital collaborative design and model-based analysis. Digital collaborative design is a digital optimization system with annotated 3D models at the core. Its purpose is by defining products via annotating models in the upstream of PLM to ensure the smooth progress of works in downstream. [5,6]. Model based analysis takes the PDM system as the platform; it integrates the 3D design tools and simulation analysis tools to build a proper design and analysis environment. This is in order to achieve unified management design model and analysis of data

DOI: 10.1051/matecconf/201713900019

and its version. Model based simulation uses the model to be divided into three categories: equipment model, cognitive model and message model. Model based simulation technology is based on the computer; take the principle of similarity, information technology and other related applications in the field of basic theory and technology as the foundation [7].

- Model - based manufacturing

Model based manufacturing includes model-based process design and model-based manufacturing execution [8]. Model based process planning is based on product design results, and the process planning of product manufacturing process is the direct basis of product manufacturing process. The process is based on CAPP software, including design and Simulation of 3D parts process, design and simulation of 3D assembly process, model based tooling design and management, model based process resource management and so on. Model based manufacturing is the development of modern manufacturing technology to meet the increasingly diversified product requirements, and the expression and application of product design and manufacturing data is the core issue of digital manufacturing.

- Model-based validation and delivery

Model based recognition and delivery is the digital model of existing design results that are presented to stakeholders in a visual form [9]. This avoids the delivery of the physical prototype, helping to shorten the entire life cycle of the product and accelerate the development of the product. This process uses the twin-model idea to digitize and virtualize the actual product and show the whole structure of the product.

4 Maturity control technology for design and manufacture integrated development

The model center solves the problem of uniform data source for collaborative processes of design and manufacture, and maturity control is the basic standard for concurrent of design and manufacture.

At present, there are many kinds of technology status control methods based on their own standards, such as stage, milestone, baseline and so on. Therefore, a model maturity management mechanism with 16 levels of maturity is proposed, which provides the necessary technical support for the concurrent collaborative design of design, fixture, and quality control.

For aircraft development projects and design and manufacturing departments who expects to realize concurrent and collaborative development of aircraft,Only according to the five stages of aircraft development to determine the organization, planning and management of the project must be too extensive. In order to shorten the development cycle of the aircraft and improve the development efficiency, the maturity model of product covering the whole life cycle of aircraft is proposed to explore collaborative work patterns and work content of the design and manufacturing department in different maturity level under the organizational model.

The process of the development of the aircraft project, from the start of the project to decommissioning, is divided into 16 Level maturity model based on major events and major assessment point of view. Fig 1 shows the maturity level of 16. ML0 start of the project, ML1 formal initiation of the project, ML2 Technical description of aircraft design, ML3 Determination Of preliminary overall technological scheme, ML4 Basic terminology, ML6 start of parts/fixture manufacturing, ML8 complete assembly of the fixture,ML9 01 plane into the final assembly, ML10 01 plane enter the test station, ML11 01 plane maiden flight, ML12 identification of flight test, ML15 halt production, ML16 decommissioning, also including ML5 appraisal of design scheme, ML7 detailed design examination, ML13 review of design finalization, ML14 review of production finalization and so on.

The product maturity model of the aircraft's full life cycle is established. In order to achieve the interdisciplinary concurrent collaborative development between the design, manufacturing and other professional departments, it is necessary to clarify the work content of each professional department under each maturity level. Based on the unified product life cycle model with maturity level, through the development of the corresponding standards, criterion and management system, it is necessary to combine with the actual situation of enterprises to achieve concurrent development in the final product design, process/fixture design and manufacturing. It becomes more efficient relying on PLM software platform.

5 Conlusion

As the main body of modern aircraft development, design and manufacturing integration is not only a technical problem, but also a social and technical problem involving many aspects, such as enterprise organization, business process, development strategy, planning and so on. This paper summarizes the model centers of R & D integrated manufacturing enterprises, including central system model, product model and process model. On this basis, the design and manufacturing integration development model using the unified model center is discussed, and the maturity control technology for concurrent design and manufacture is also introduced. Research on aircraft design and manufacturing integration enterprise model can guide the transformation of aircraft manufacturing enterprises and ensure the effectiveness of the development of design and manufacturing integration technology.

Acknowledgements

This work has been supported by Project of National Key Technology R&D Program through approval No. 2015BAF17B01.

References

1. Selvaraj, P., P. Radhakrishnan, and M. Adithan. An integrated approach to design for manufacturing and assembly based on reduction of product development time and cost, The International Journal of Advanced Manufacturing Technology, 42(1): 13-29 (2009).
2. Ulrich, Karl T., and David J. Ellison. Internalization and Integration of Design and Production. Production and Operations Management, 14(3): 315-330(2005).
3. Su, Daizhong, and Nariman Amin. A CGI-based approach for remotely executing a large program for integration of design and manufacture over the Internet. International Journal of Computer Integrated Manufacturing, 14(1): 55-65(2001).
4. Kaslow, David, et al. Integrated model-based systems engineering (MBSE) applied to the Simulation of a CubeSat mission. Aerospace Conference, 2014 IEEE.
5. Graignic, Pascal, et al. Complex system simulation: proposition of a MBSE framework for design-analysis integration. Procedia Computer Science,16(1): 59-68(2013).
6. Alemanni, M., F. Destefanis, and E. Vezzetti. Model-based definition design in the product lifecycle management scenario. The International Journal of Advanced Manufacturing Technology, 52(1): 1-14(2011).
7. Li C, Wong W H. Model-based analysis of oligonucleotide arrays: expression index computation and outlier detection. Proceedings of the National Academy of Sciences, 98(1): 31-36(2001).
8. Butterfield, J., et al. Optimization of aircraft fuselage assembly process using digital manufacturing. Journal of Computing and Information Science in Engineering, 7(3): 269-275(2007).
9. Estefan, Jeff A. Survey of model-based systems engineering (MBSE) methodologies. Incose MBSE Focus Group 25(8) (2007).

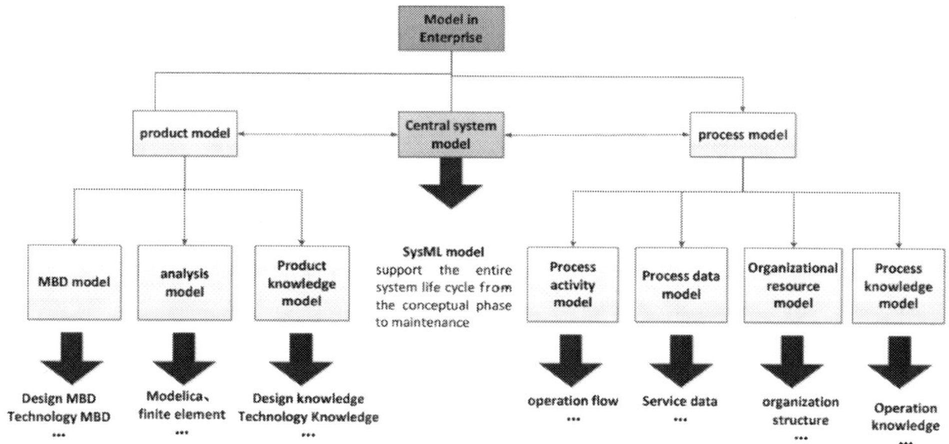

Figure 1. Model centre of designing and manufacturing integrated development enterprise

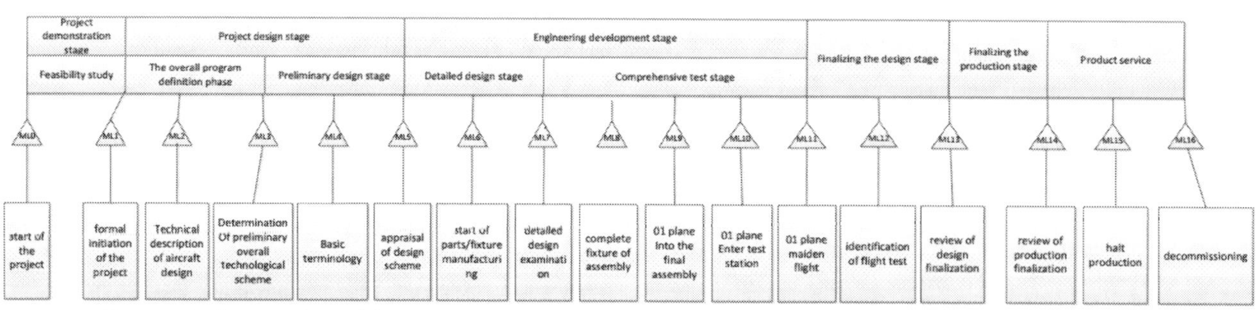

Figure 4. Product maturity model

Application of mind evolution Elman neural network model in gearbox fault diagnosis

Wang Hai jun,Jin Tao,Lin Zhi wei
Ordos Institute Of Technology, Inner Mongolia Ordos, 017000

Abstract. Gearbox as an important component of the power system of agricultural machinery, plays a vital role in the normal operation of agricultural equipment. At present, the diagnosis of gearbox fault is mainly based on expert experience, and the accuracy is not guaranteed. In view of the achievements of Elman neural network in fault diagnosis, Elman neural network is used as the basic model of gearbox fault diagnosis. Considering that the Elman neural network is sensitive to the initial weight threshold and easy to fall into the local minimum, the mind evolution algorithm is introduced into the parameter optimization of the Elman neural network fault diagnosis model. Experimental results show that this assumption is successful, and the new model not only reduces diagnostic errors but also improves stability.

1 Introduction

The gearbox as the main propulsion device of power system of all kinds of agricultural machinery and equipment to the tractor as the representative, is the core component of agricultural machinery power transmission gearbox, once the failure of mechanical system will inevitably lead to paralysis, lead to the agricultural machinery lost power, causing economic losses and even lead to accidents occurred. Studies have shown that the mechanical failure rate represented by a gearbox occupies all the 50%~70% except the engine. Therefore, the early diagnosis of agricultural machinery gearbox fault is of great significance to guarantee the power system of agricultural machinery. In the non disassembly condition, the traditional gearbox fault diagnosis often depends on the expert experience to judge. However, because the gearbox fault diagnosis model and feature vector is a very complex nonlinear relationship, the expert experience can not completely solve the gear box fault diagnosis problem [1].

The existing studies have shown that Elman neural network has very good application effect in fault diagnosis, Elman neural network is improved based on BP neural network with feedback, local structure, ability to handle dynamic information better. But because the Elman neural network also uses the BP neural network algorithm to correct the weight, there are some shortcomings, such as slow learning speed and easy to fall into local minimum[2,3]. Therefore, the emergence of a large number of Elman neural network based on swarm intelligent algorithm improved algorithm model, Huang Yuchun[4]using genetic algorithm to optimize Elman neural network to forecast the thermal error achieved good results, Chen Fafa[5]by optimization of Elman neural network in fault diagnosis of rotating machinery can significantly improve the accuracy of fault diagnosis of rolling bearing immune genetic algorithm. Can be seen from these studies that the swarm intelligence algorithm into the Elman neural network model can also improve operation efficiency.

In view of this, the Mind Evolutionary Algorithm (MEA) is introduced into the Elman model optimization, and the gearbox fault diagnosis model based on the thought evolution Elman neural network algorithm is constructed. The experimental results show that compared with the neural network model of Elman algorithm and BP algorithm based on mind evolutionary, the diagnosis model of mind evolutionary Elman neural network algorithm is superior to the former in the stability and accuracy of diagnosis based on operation.

2 Elman neural network

Elman neural network is proposed by J.L.Elman in 1990 which is a dynamic feedback network, compared with the traditional BP neural network, Elman neural network adds a layer of memory unit based on BP network, called undertake layer, through the internal storage of intermediate state, which can output to the hidden layer unit before the moment value of memory, so the network has a dynamic memory function. The working process of the Elman neural network is first initialized to the neural network the right threshold, the signal from the input layer to the hidden layer input, after weighting computation, the output results to the hidden layer and the output layer, the results are calculated, calculated results to undertake temporarily stored in the layer, and the calculated layer to undertake the output feedback back to the hidden layer re calculate the output output error, the final calculation of two times the output of the system, update the right threshold for next calculation, the calculation model is shown as the following formula [6].

$$x(k) = f(W^2 x_c(k) + W^1 u(k-1)) \quad (1)$$

$$x_c(k) = \alpha x_c(k-1) + x(k-1) \quad (2)$$

$$y(k) = g(W^3 x(k)) \quad (3)$$

Among them, $x(k)$ and $x_c(k)$ represent the output of the hidden layer and the receiving layer, respectively, $y(k)$ is the output of the output layer, u is the input vector, W^1 is the weight of the input layer to the hidden layer, W^2 is the weight of the hidden layer to the receiving layer, W^3 is the connection weight of the hidden layer to the output layer, $f(\cdot)$ and $g(\cdot)$ are the transfer functions of the hidden layer and the output layer, respectively, $0 \le \alpha \le 1$ is a feedback gainfactor for self connection.

* Corresponding author:wanghaijun11249@126.com

DOI: 10.1051/matecconf/201713900020

Let $y_r(k)$ indicate the actual output of step k of the system, Define the error objective function, as shown in formula (4)

$$E(k) = (\bar{y}(k) - y(k))^T(\bar{y}(k) - y(k))/2 \quad (4)$$

Then the dynamic learning algorithm of Elman neural network can be summarized as follows

$$\Delta w_{ij}^3 = \eta(\bar{y}_i(k) - y_i(k))g_i'(\cdot)x_j(k)$$
$$(i = 1,2,\cdots,m; j = 1,2,\cdots,n) \quad (5)$$

$$\Delta w_{jq}^2 = \eta\sum_{i=1}^{m}(\bar{y}_i(k) - y_i(k))g_i'(\cdot)w_{ij}^3 f_j'(\cdot)u_q(k-1)$$
$$(j = 1,2,\cdots,n; q = 1,2,\cdots,r) \quad (6)$$

$$\Delta w_{jl}^3 = \eta\sum_{i=1}^{m}(\bar{y}_i(k) - y_i(k))g_i'(\cdot)w_{ij}^3 \frac{\partial x_j(k)}{\partial w_{jl}^1}$$
$$(j = 1,2,\cdots,n; l = 1,2,\cdots,n) \quad (7)$$

$$\frac{\partial x_j(k)}{\partial w_{jl}^1} = f_j'(\cdot)x_l(k-1) + \alpha\frac{\partial x_j(k)}{\partial w_{jl}^1} \quad (8)$$

3 Mind evolution Elman neural network algorithm design

3.1 Algorithm core idea

The theoretical basis of the Elman neural network algorithm for mind evolution is the effective combination of the two algorithms through parameter transfer. The mind evolution algorithm takes the initial weight threshold parameter of the Elman algorithm as a solution group, which is called the initial population. In the process of evolution, the initial species group is divided into several sub groups, which are divided into two groups: the dominant group and the temporary group. Each sub group has high fitness value as the center, and generates M individuals according to certain rules. Within the sub population, individual competition becomes a winner, called a seed of a sub population. It is the basis for the next step of evolution. This process is called local convergence. The winners of each sub group compete with each other, according to the fitness value, the survival of the fittest, and find a new point of proportion in the solution space. This process is called "overall alienation". The convergence of operation to achieve the individual internal sub groups competing in the local domain, complete local search operation to achieve the global competition, alienation, guarantee group global search capability, reduce the probability of falling into local optimal, make the population toward the global optimum evolution[7,8]. After several generations of the convergence process of alienation eventually survived the initial solution, the optimal threshold of Elman algorithm is to find the right, then the threshold parameter to Elman algorithm model are two optimization, find the best solution. Figure 1 is the whole process of the algorithm combining the mind evolution algorithm with the Elman algorithm.

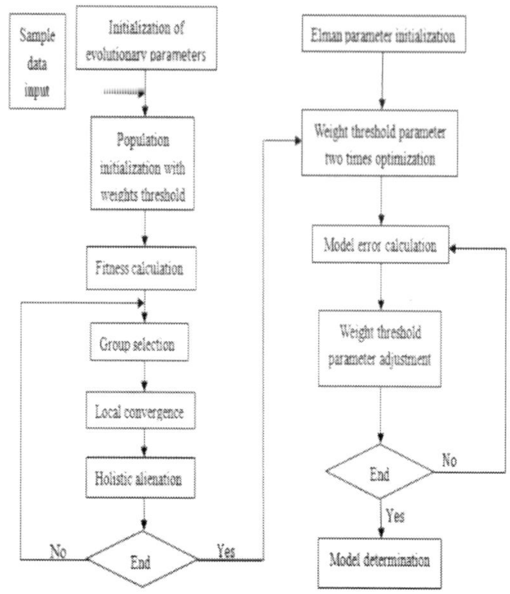

Figure 1 Mind evolution, Elman algorithm, flow chart

3.2 Algorithm execution step design

Step1 Population initialization——Using mind evolutionary algorithm to initialize population, let $W1$ be the input layer to the hidden layer weight、$W2$ is the weight from the hidden layer to the output layer、$W3$ is the weight of the hidden layer to the receiving layer、$B1$ is the threshold of the input layer to the hidden layer、$B2$ is the threshold from the hidden layer to the output layer.Because the Elman network is based on the BP gradient descent algorithm, therefore, only $W1$, $W2$, $B1$ and $B2$ are encoded and optimized here,$W3$ is randomly generated by the system. Figure 2 shows the basic body diagram of the Elman neural network used in this article, where each individual in the initial population is shown in the following formula.

$$\{W1, W2, B1, B2\} \quad (9)$$

In the group, the nodes in each individual are randomly generated in [-1,1] by the probability shown by formula (10).

$$P_{ij}=1/(S+1) \quad (10)$$

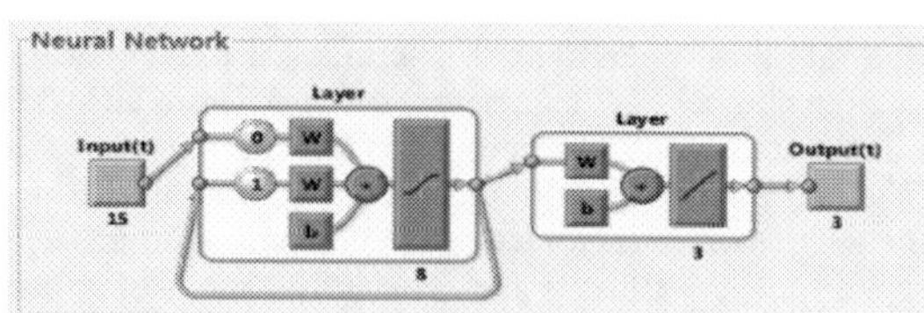

Figure 2 the basic structure of the Elman model

Step2 Construction of fitness function——The fitness value calculation process and external information interaction in the whole operation process of individual evolutionary algorithms, evolutionary function is the only standard evolutionary basis, therefore evolutionary function selection will directly affect the operation of the final result of the algorithm. The algorithm uses the reciprocal of the error squared between the actual value and the desired output of the gearbox fault diagnosis model and the reciprocal of the SE as the basis for the algorithm. This can effectively integrate the Elman algorithm with the evaluation criteria of the mind evolutionary algorithm. Evolutionary functions such as equation (11) are shown:

$$Fval = 1/SE = 1/\sum_{i=1}^{n}(y_i - t_i) \quad (11)$$

$Fval$ is the fitness value in the formula,y_i is the expected output of the Elman model, t_i is for actual output.

Step3 Group selection——Group selection is the key step for Population Division in the process of mind evolution, In this paper, the population size is set to *psize*=100,the winner group size *bsize* = 10,the provisional subgroup *tpsize* = 10. According to the formula (11), the individual fitness value of the population was calculated, and sort them from high to low, select *bsize+tsize* individuals as the winning group. Finally, the whole population is divided into bsize dominant sub groups and tsize temporary sub groups, each of which is *SG* in size.

$$SG=psize/(bsize+tpsize) \quad (12)$$

Step4 Individual convergence——After selecting the superior, a number of subgroups with normal size of $N(\mu,\Sigma)$ are generated with *SG* size, μ and Σ are central vectors and covariance matrices of normal distribution . μ is the individual location with the highest fitness value, that is the solution of the problem we ask. If Σ is diagonal matrix, this means that individuals in a subgroup are independent of each other. After i generation operation , $\{\sigma_{id}\}$ is a non-zero individual in Σ , among them, d=0,1,…,S represents the dimension of the solution space. Each one-dimensional variance of normal distribution is calculated as follows.

$$\sigma_{0d} = (H_d - L_d)/10 \quad (13)$$

In the formula H_d and L_d denote the upper and lower bounds of the d dimension respectively. In the iterative process, the highest score in the variance is the optimal solution. The formula for the variance of any two generations is as follows[9].

$$\sigma_{(i+1)d} = c_1\sigma_{id} + c_2\delta \quad (14)$$

Where c_1 and c_2 are inter generational adjustment coefficient. δ stands for the distance between the two generations. In this article, set c_1=1, c_2=0.5.set $\delta = rand(1,S)*2-1$ to increase the gap between the two generations and increase search speed.

Step5 Holistic alienation——After convergence, all groups changed, Therefore, the individuals in the swarm need to re calculate the order of fitness calculation in accordance with formula (11), individual fitness values of the better and temporary groups. The fitness value of the fitness group with the highest fitness value is T_{max},A set consisting of individuals with fitness values greater than T_{max} in the temporary group is denoted as $Temp=\{t_1,t_2,\cdots,t_p\}$,Find the fitness in the winner group, and the last p individuals are the set $Best=\{b_1,b_2,\cdots,b_p\}$. Replacing the *Best* with the *Temp* constitutes a new fitness group with the highest fitness value, and discards the *Best*.At the same time, a new set $Temp^*$ is generated randomly in the solution space, $Temp^*=\{t_1^*,t_2^*,\cdots,t_p^*\}$, and it was added to the temporary sub group to form a new temporary subgroup [10].

Step6 Iteration stop——The maximum iteration number is chosen as the convergence condition of the algorithm. If the convergence condition is not satisfied, the Step4 is returned to a new round of convergence and dissimilation operations until the iteration is stopped. The initial solution obtained by the optimization of the mind evolution algorithm is decoded, and then the Elman neural network loaded with the decoding value is trained and studied to form a diagnosis model.

4 algorithm simulation

4.1 sample selection of input and output

Most of the fault gear box are caused by gear fault, fault in the gear box parts, gear failure accounted for the largest proportion, accounting for about 60% of the time, bearing, so the emphasis of gearbox condition monitoring and fault diagnosis of gear fault diagnosis. This paper takes a tractor gearbox as the object of study, in the fault feature extraction, the amplitude $A_{i,j1}$, $A_{i,j2}$ and $A_{i,j3}$ are selected in the frequency domain characteristic signal at 2, 4 and 6 files in the frequency band family $f_s \pm nf_z$ of the 1, 2 and 3 axes, Where f_s represents the gear meshing frequency, and f_z is the frequency of the shaft, n=1,2,3, i=2,4,6 stands for stalls, and j=1,2,3 stands for serial numbers of axes, because in the 2 axis and 3 axis two of gear meshing, so with 1 and 2 respectively represent the two meshing frequency. In this way, the state of the gear forms a vector combination of 15 eigenvalues ($A_{1,2}$, $A_{1,4}$, $A_{1,6}$, $A_{2,21}$, $A_{2,22}$, $A_{2,41}$, $A_{2,42}$, $A_{2,61}$, $A_{2,62}$, $A_{3,21}$, $A_{3,22}$, $A_{3,41}$, $A_{3,42}$, $A_{3,61}$, $A_{3,62}$) which is used as the model input variable. Considering the Elman neural network model used in this paper to activate the function tansig (x) =2/ (1+exp (-2x)-1) and purelin (x) =x, comparison of the sensitivity of these two functions of the input data, the input data are normalized to table 1 for the input data after normalization. As for the output node, this study only studies the 3 states of the gear, so the corresponding eigenvalue of Table 2 is used as the output.

Table 1 input values for training samples

Sample / Enter	1	2	…	9
$A_{1,2}$	0.2286	0.2603	…	0.3171
$A_{1,4}$	0.1292	0.1715	…	0.1531

:	:	:	…	:
$A_{3,62}$	0.1828	0.0909	…	0.1125

Table 2sample output results

Gear state	Output result
Trouble free	(1,0,0)
Root crack	(0,1,0)
Broken teeth	(0,0,1)

Set the network training target of 0.0001, the learning coefficient is 0.1, Training times is 100.The hidden layer nodes are represented by formula (15) as the basic standard. After repeated experiments, they are determined as 15-8-3, table 3 is the network error corresponding to different hidden layer nodes.

$$mid = \sqrt{input + output} + \alpha \qquad \alpha = 1,2,\ldots,10 \quad (15)$$

Table 3 the corresponding error values of different hidden layer nodes

number	5	6	7	8	9
error	0.0375	0.0330	0.0259	0.0217	0.0248
number	10	11	12	13	14
error	0.0266	0.0314	0.0329	0.0278	0.0332

Since the Elman neural network is still essentially BP neural network, therefore, the corresponding training algorithm is consistent, a large number of studies have shown that the traingd algorithm based on gradient descent is slow in convergence and often fails to converge. Thus, a large number of improved algorithms appear, among which the Levenberg-Marquard (trainlm) algorithm is the most typical.Compared with other improved algorithms, faster convergence speed and higher efficiency with trainlm, figure 3 for the same problem, several BP algorithm performance comparison, can be seen from the figure, the trainlm algorithm is more efficient, therefore, trainlm algorithm is used as the Elman model training algorithm.

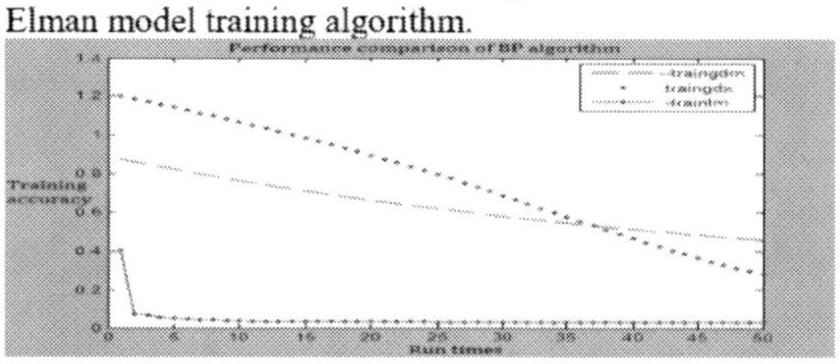

Figure 3 BP training result of improved algorithm

4.2 Result analysis

In view of the influence of initial parameters randomization of Elman model on the results, The Elman neural network model and the mind evolution BP model and the mind evolution Elman neural network model are used to diagnose the gear box faults respectively, and each runs 10 times. The final results are given in Table 4, and it can be seen from the table that the diagnostic error of the MEA-Elman neural network model is reduced by 0.0059 and 0.0112 compared to the MEA-BP model and the Elman model, Compared with the MEA-Elman neural network model, the MEA-BP neural network model does not reduce much error. But because the running stability of the MEA-Elman neural network model than the other two models have improved, so the fault diagnosis rate compared with the other two models were increased by 6.6% and 16.6%, we can see that the MEA-Elman neural network model in fault diagnosis accuracy is higher, the stability is better.

Table 4 Comparison of different model results based on different algorithms

Model type	MSE	Fault diagnosis rate
Elman	0.0935	66.7%
MEA-BP	0.0882	76.7%
MEA-Elman	0.0823	83.3%

5. Conclusion

The mind evolutionary algorithm to Elman neural network model parameter optimization, search ability to optimize the Elman neural network weight and threshold parameters by using its powerful global, and build on the gearbox mind evolution model of Elman neural network fault diagnosis model based on. The experimental results show that this assumption is successful, and the Elman neural network, which introduces the mind evolution algorithm, can well overcome the sensitivity of the Elman neural network to the initial weight threshold and is easy to fall into the local minimum. Therefore, the diagnosis model not only has higher prediction accuracy, but also improves the stability, and proposes a feasible research method for agricultural machinery fault diagnosis research.

Acknowledgements

This work has been supported by Scientific research project of Inner Mongolia higher education institutions NJZY16382.

Reference

1 Song Chong Zhi, Wu Yu Guo, Wang Lu.Research on fault diagnosis of engine gearbox based on modified Elman neural network[J]Chinese Journal of Scientific Instrument, 2008,29(7):1414-1417

2 Zheng Min,Ren Fang,Yang Zhao jian. Study on the Load Identification of the Rotor System With Self-Adaptive BP Neural Networks[J] Machinery Design & Manufacture, 2016,(6):85-89

3 Han Shu hua,Hou Xue yuan LI Wen qing. Study on Parameters Optimization of Injection Molding Based on Modified Particle Swarm Optimized Back Propagation Neural Network[J] Machinery Design & Manufacture, 2015,(10):98-102

4 Huang Yu chun,Tian Jian ping,Yang Hai li. Thermal Error Modeling for Machine Tool Based on Genetic Algorithm Optimization Elman Neural Network[J] Modular Machine Tool & Automatic Manufacturing Technique, 2015(4):74-77

5 ChenFa Fa,Tang Bao Ping,Huang Qing Qing.Rotating machinery fault diagnosis based on Elman neural network optimized by immune genetic algorithm[J]Journal of Chong qing University,2012, 35(5):7-13

6 Wang Xiao xia,Ma Liang yu,Wang Bing shu. Application of evolutionary Elman neural network in real-time data forecasting[J] Electric Power Automation Equipment, 2011,31(12):77-81

7 Liu J X, Wang F, Xie K M. Application of improved mind evolutio nary algorithm in wide band impendence transformer design[C]Procofthe 4th Int Confon Natural Com putation (ICNC. 08). Ji.nan, 2008: 428-432

8 WANG Fang,XIE Ke-ming,LIU Jian-xia. Swarm intelligence based MEA design[J]Control and Decision, 2010,25(1) : 145-148

9 Huang Jun,Feng Yong Jian.Improved Pressure Measurement Precision for Vacuum Insulation Panel with Mind Evolutionary Algorithm[J]Chinese Journal of Vacuum Science and Technology, 2015,35(5) : 528-532

10 Li Xiu Guang.Fault Diagnosis for Power Transformer Based on Mind Evolution Algorithm Optimized Neural Network[D]Taiyuan University of Technology,2010: 66-69

An Improved Equivalent Fixture Error Model for Machining Process

Fuyong Yang[1,2], *Sun* Jin[1,2,*], *Zhimin* Li[1,2], *Siyi* Ding[1,2], *Jia* Lin[1,2], *Shun* Liu[1,2]

[1]State Key Laboratory of Mechanical System and Vibration, Shanghai Jiao Tong University, Shanghai, 200240, P.R. China
[2] Shanghai Key Laboratory of Digital Manufacture for Thin-walled Structures, Shanghai Jiao Tong University, Shanghai, 200240, P.R. China

Abstract. Although the equivalent fixture error approach used in variation propagation can directly model the process physics regarding how datum-induced, fixture-induced and machine tool-induced errors generate the same error pattern on part features, the developed equivalent fixture error compensation technique did not consider machining-induced variations. Machining-induced variations are often caused by geometric-thermal effects, cutting force-induced variations, and/or cutting-tool wear, etc. Such machining-induced variations are an important factor that influences the part quality. Without considering machining-induced variations, the application of EFE model for error compensation will be limited. In order to overcome this limitation, this paper extends current equivalent fixture error to include machining-induced variations. This paper shows the benefits of the extended model through a case study.

1 Introduction

For a multistage machining process (MMPs), the part variation at certain operation is often due to two components: 1) the variation brought by current machine stage, such as fixture error and machine tool path error; and 2) the variation brought by datum feature error generated and propagated from previous stages [1]. The above errors account for the majority of machining errors in the machining process. Due to the existence of these main error sources, the features of the workpiece may deviate from their designed values to their actual values.

SPC is a very useful tool for monitoring process variation and reducing product variation [2-4]. Tsung et al. [5] gave a thorough review about the application of SPC techniques in MMPs, including MMPs and multistage service operations. However, SPC could not be directly used to model the relationship between the error sources and feature variations. It also cannot handle variation propagation problem in MMPs.

SoV is another useful technology for reducing product variation and can overcome the limitation of SPC. This technology can directly model how the variation in MMPs caused by datum, fixture, and machine path tool propagate and accumulate. The SoV methodology for MMPs can be described by employing the state space concept which has been extensively used in system or automatic control theories. The introduction of the state space modeling structure can successfully describe the relationship between the error sources and feature variations. A huge of literature can be found on SoV modeling. Jin et al. [6] introduced state space model to describe the dimensional variation accumulation and propagation for multistage body assembly processes. To obtain the SoV models for MMPs, lots of efforts have been done [1, 7-12]. Huang et al. [7] proposed an implicit non-linear SoV model to predict the variation accumulation and its propagation for MMPs. Djurdjanovic et al. [8] used Taylor series expansion to linearize Huang et al.'s non-linear SoV model. But the linearized model is yet implicit. Using differential motion vectors (DMVs), Zhou et al. [1] developed an explicit linear SoV model. Abellan-Nebot et al. [9] demonstrated that the absence of machining-induced variations in state space modeling of variation propagation for MMPs could be an important factor that influences the accurate in variation prediction. Using DMVs, Abellan-Nebot et al. [10] introduced such machining-induced variations including geometric-thermal effects, cutting force-induced variations, and/or cutting-tool wear, etc into the SoV model. Several review papers have introduced SoV modeling methods briefly [11-14].

To reduce this variation, Wang et al. [15, 16] proposed equivalent fixture error (EFE) concept. With this concept, datum error and machine tool error can be transformed to equivalent fixture locator errors at each operation and be compensated. However, only the datum errors and geometric errors of machine tool have been explicitly transformed to equivalent fixture locator variations at each operation in the reported works [15, 16], leaving the machining-induced variations unaddressed. Therefore, without considering machining-induced variations, the application of current EFE model for error compensation will be limited.

* Corresponding author: jinsun@sjtu.edu.cn

In order to overcome this limitation of current EFE compensation, this paper extends current EFE model to include machining-induced variations. By this extended EFE model, the machining-induced variations can be cancelled out by their corresponding EFE. Then, the machining quality can correspondingly improve. The limitation of current EFE model and the benefits of extending the EFE model through a case study will also be discussed in this paper.

2 Related works

2.1 EFE model

The concept of EFE introduced by [15, 16] for machining process can directly model the process physics regarding how datum error, and machine tool path error, and fixture error generate the same pattern on the machined features. Through the EFE model, datum error and machine tool path error can be transformed to equivalent fixture locator errors at each operation.

In Fig. 1(a), $\mathbf{X}_1^0$, $\mathbf{X}_2^0$, $\mathbf{X}_3^0$ and $\mathbf{X}_4^0$ (the solid line block with surfaces) are the nominal features in PCS. $\mathbf{X}_1^0$ and $\mathbf{X}_2^0$ are the designed primary and secondary datum features, respectively. When the datum features and the fixture are all at their designed positions, the part shown in Fig. 1(a) will be in its nominal setup position. However, when the datum feature $\mathbf{X}_1$ in Fig. 1(b) deviates from its nominal, the block will be transformed to its actual position. As shown in Fig. 1(c), the datum error can be equivalent to the fixture error Δd_1 and Δd_2.

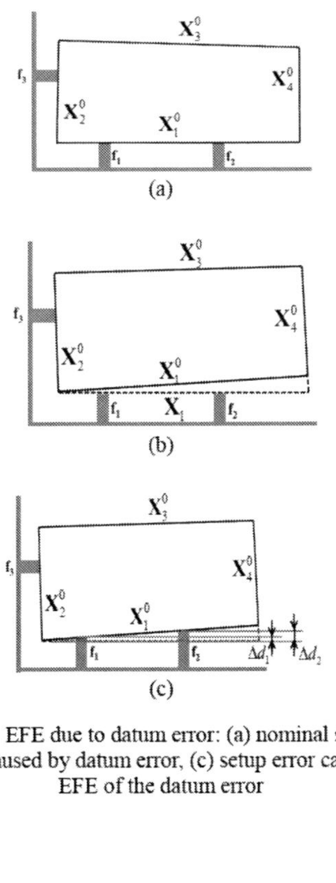

Figure. 1. EFE due to datum error: (a) nominal setup, (b) setup error caused by datum error, (c) setup error caused by the EFE of the datum error

Similarly, the machine tool path error shown in Fig. 2(a) can also be equivalent to its corresponding equivalent fixture locator errors shown in Fig. 2(b).

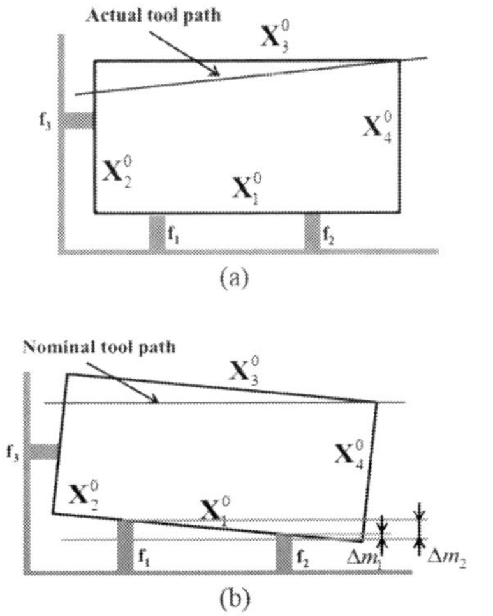

Figure. 2. EFE due to machine tool path error: (a) machining error caused by machine tool path error, (b) machining error caused by the EFE of the machine tool path error

The EFE model for datum error, fixture error and machine tool path error can be described as follows.

$$\mathbf{x}_i(k)= \mathbf{B}_i(k)[\Delta\mathbf{d}^{\mathrm{T}}(k)\ \ 0]^{\mathrm{T}} + \mathbf{B}_i(k)\mathbf{u}(k)+\xi_i(k) \tag{1}$$

where $\mathbf{B}_i(k)=\begin{bmatrix} -\mathbf{A}_{id}(k)\mathbf{J}^{-1}(k)\mathbf{\Phi}(k)\mathbf{E} & \mathbf{0}_{6\times1} \\ \mathbf{0}_{1\times6} & \gamma_i^0(k) \end{bmatrix}$ is the coefficient matrix linking the errors with the machined feature deviation. $rank(-\mathbf{A}_{id}(k)\mathbf{J}^{-1}(k)\mathbf{\Phi}(k)\mathbf{E}) \le 5$ and $\mathbf{u}(k)=[(\Delta\mathbf{m}(k)+\Delta\mathbf{f}(k))^{\mathrm{T}},\ \gamma_m(k)-1]^{\mathrm{T}}$. The expression of $\Delta\mathbf{d}(k)$ and $\Delta\mathbf{m}(k)$ are as follows.

$$\Delta\mathbf{m}(k)=\mathbf{K}\delta\mathbf{q}_m(k) \tag{2}$$

$$\begin{bmatrix}\Delta\mathbf{d}(k)\\ 1\end{bmatrix}=\begin{bmatrix}\mathbf{\Psi} & \mathbf{0}\\ \mathbf{0} & \mathbf{1}\end{bmatrix}_{7\times22}\mathbf{H}_{22\times22}\begin{bmatrix}\mathbf{G} & \mathbf{0}\\ \mathbf{0} & \mathbf{1}\end{bmatrix}_{22\times22}\begin{bmatrix}\mathbf{x}_1(k-1)\\ \mathbf{x}_2(k-1)\\ \mathbf{x}_3(k-1)\\ 1\end{bmatrix}_{22\times1} \tag{3}$$

where $\delta\mathbf{q}_m(k)$ is machine path tool error, $\mathbf{x}_1(k-1)$, $\mathbf{x}_2(k-1)$ and $\mathbf{x}_3(k-1)$ are the primary datum error, secondary datum error and tertiary datum error, respectively. The detailed expression of matrix $\mathbf{K}$, $\mathbf{\Psi}$, $\mathbf{H}$ and $\mathbf{G}$ can be found in [15].

2.2 Compensation strategy

To compensate the error sources, corresponding $\Delta\mathbf{d}$, $\Delta\mathbf{f}(k)$ and $\Delta\mathbf{m}$ should be calculated by EFE model. Denote the total EFE as $\mathbf{t}(k)=[\Delta\mathbf{d}(k)+\Delta\mathbf{f}(k)+\Delta\mathbf{m}(k)]$. In order to cancel out $\mathbf{t}(k)$, the amount of adjustment $\mathbf{c}(k)=-[\Delta\mathbf{d}(k)+\Delta\mathbf{f}(k)+\Delta\mathbf{m}(k)]$ should be applied to the corresponding fixture locators. The detailed compensation process by EFE model is shown in Fig. 3.

DOI: 10.1051/matecconf/201713900021

For example, to compensate the datum error shown in in Fig. 1(b), $\Delta\mathbf{d}$ should be calculated. When the amount of adjustment $-\Delta\mathbf{d}$ shown in Fig. 4 is applied to the corresponding fixture locators, $\Delta\mathbf{d}$ can be cancelled out. Then, the datum error can be compensated by this strategy. The same procedure can be performed to compensate the machine tool path error $\mathbf{u}_m$.

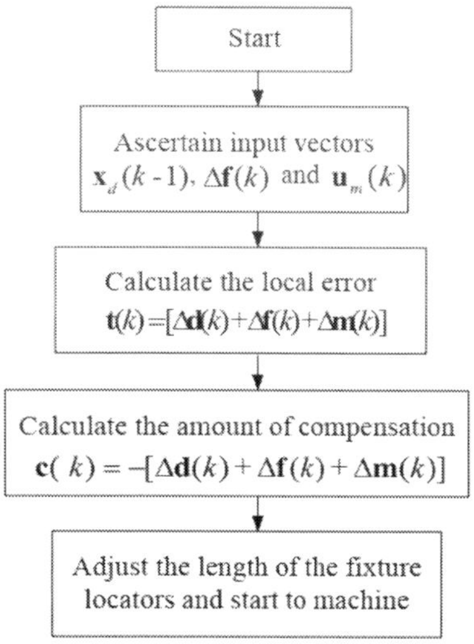

Figure. 3. The compensation process by EFE

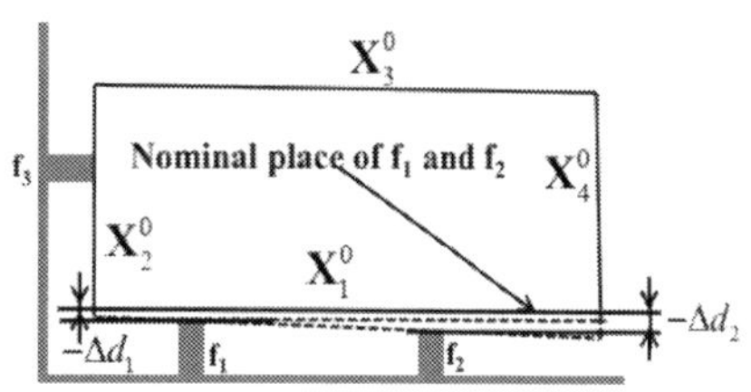

Figure. 4. The adjustment to compensate datum error

3 Improved Equivalent Fixture Error Model

In order to overcome the limitation of current EFE model, this paper extends current EFE model to include machining-induced variations. Abellan-Nebot et al. [10] presented a generic framework for machining-induced variation representation based on DMVs. The final machine tool path error caused by machining-induced variations is mathematically modeled as

$$\delta\mathbf{q}_m(k) = \mathbf{T}_{qi(k)}^{C_k}(\mathbf{T}_{C_k}^{S_k}(\mathbf{T}_{S_k}^{A_k^n}\mathbf{x}_{A_k^n}^{{}^oF_k} + \mathbf{x}_{S_k}^{A_k^n}) + \mathbf{x}_{C_k}^{S_k}) + \mathbf{x}_{qi(k)}^{C_k} \tag{4}$$

where $\mathbf{x}_{A_k^n}^{{}^oF_k}$ is the geometric-thermal variation of machine-tool axes; $\mathbf{x}_{S_k}^{A_k^n}$ represents the spindle-thermal variation; $\mathbf{x}_{C_k}^{S_k}$ refers to the cutting force-induced variation and $\mathbf{x}_{qi}^{C_k}$ is the cutting-tool wear induced variations; $\mathbf{T}_{qi(k)}^{C_k}$, $\mathbf{T}_{C_k}^{S_k}$ and $\mathbf{T}_{S_k}^{A_k^n}$ are the matrices to describe the transition of differential motion vectors can be found in [10].

It can be seen from Eq. 4 that when $\mathbf{x}_{A_k^n}^{{}^oF_k} = \mathbf{0}$, $\mathbf{x}_{S_k}^{A_k^n} = \mathbf{0}$ and $\mathbf{x}_{C_k}^{S_k} = \mathbf{0}$, the machining tool path error due to cutting-tool wear can be written as

$$\delta\mathbf{q}_m^{qi(k)}(k) = \mathbf{x}_{qi(k)}^{C_k} \tag{5}$$

Similarly, the machine tool path error due to cutting force-induced variation, spindle-thermal variation and geometric-thermal variation of machine-tool axes can be obtained as Eq. 6, Eq. 7 and Eq. 8, respectively.

$$\delta\mathbf{q}_m^{C_k}(k) = \mathbf{T}_{qi(k)}^{C_k}\mathbf{x}_{C_k}^{S_k} \tag{6}$$

$$\delta\mathbf{q}_m^{S_k}(k) = \mathbf{T}_{qi(k)}^{C_k}\mathbf{T}_{C_k}^{S_k}\mathbf{x}_{S_k}^{A_k^n} \tag{7}$$

$$\delta\mathbf{q}_m^{A_k^n}(k) = \mathbf{T}_{qi(k)}^{C_k}\mathbf{T}_{C_k}^{S_k}\mathbf{T}_{S_k}^{A_k^n}\mathbf{x}_{A_k^n}^{{}^oF_k} \tag{8}$$

The extended EFE model introduces the machining-induced variations into machine tool path error. Therefore, it is feasible to calculate the EFE for machining-induced variations. Then, the limitation of current EFE model can be overcome. The benefits of extending the EFE model and the limitation of current EFE model through a case study will also be discussed in the following section.

4 Case Study

A workpiece from [15] will be employed to compare current EFE model and improved EFE model. As show in Fig. 5, the origin of the PCS is O_P. In this case, FCS is taken the same as PCS for sake of simplicity. Only top surface EFGH is milled in this case. The workpiece is located by using the primary datum ABCD, secondary datum CDHG and tertiary datum BCGF to mill feature EFGH.

The nominal locations of the six locators expressed in ^{0}FCS are shown in Table. 1. The nominal location and orientation of the features in PCS are listed in Table. 2. In the machining, the input errors are intentionally added to the process. Using the above parameters listed in Table. 1 and Table. 2, corresponding EFE models can be established.

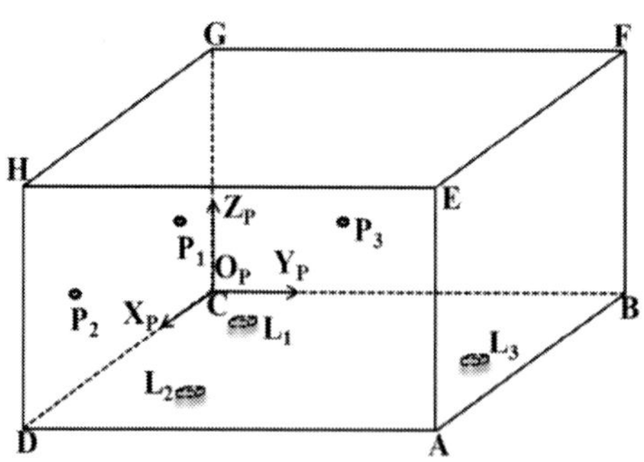

Figure. 5. Workpiece and locating scheme

Table 1. Nominal position of the locators in ^{0}FCS

Locator index	Datum feature index	Coordinates of locators w.r.t ^{0}FCS
1	primary datum ABCD	[38,19.1,0]
2	primary datum ABCD	[19,82.4,0]
3	primary datum ABCD	[57,82.4,0]
4	secondary datum CDHG	[19,0,8]
5	secondary datum CDHG	[57,0,8]
6	tertiary datum BCGF	[0,50.8,8]

Table 2. Nominal position and orientation of the features in ^{0}PCS

Feature index i	$^0\mathbf{w}_i$	$\mathbf{n}_i$	$^0\mathbf{t}_i$
1	[π ,0, 0]	[0,0,-1]	[38.1, 50.8, 0]
2	[π/2, 0, 0]	[0,-1,0]	[38.1, 0, 11.43]
3	[0, $-\pi$/2, 0]	[-1,0,0]	[0, 50.8, 11.43]
4	[0, 0, 0]	[0,0,1]	[0, 0, 20.32]

Two simple cases are used to discuss the limitation of current EFE model and the benefits of the improved EFE model. In the first case, the machining error without compensation, EFE and compensation value are predicted by current model. In the second case, the machining error without compensation, EFE and compensation value are predicted by the improved model.

In the first case, the datum error machine tool path error and fixture error are intentionally introduced. The datum errors are set as $\mathbf{x}_{d1}=[0\ \ 0\ \ 0\ \ 0\ \ 0\ \ -0.02]^{\mathrm{T}}$, $\mathbf{x}_{d2}=[0\ \ 0\ \ 0\ \ 0\ \ 0\ \ 0]^{\mathrm{T}}$ and $\mathbf{x}_{d3}=[0\ \ 0\ \ 0\ \ 0\ \ 0\ \ 0]^{\mathrm{T}}$. The input to the model that corresponds to the fixture locator error is $\Delta\mathbf{f}(k)=[0.01\ \ 0.01\ \ 0.01\ \ 0\ \ 0\ \ 0]^{\mathrm{T}}$. The machine path tool error in this case study includes only two parts: the cutting-tool wear induced error $\delta\mathbf{q}_m^{qi}=[0\ \ 0\ \ -0.015\ \ 0\ \ 0\ \ 0]^{\mathrm{T}}$, and the geometric-thermal variation of machine-tool axes induced error $\delta\mathbf{q}_m^{A''}=[0\ \ 0\ \ -0.015\ \ 0\ \ 0\ \ 0]^{\mathrm{T}}$, respectively. Since the machining-induced errors have not been explicitly calculated by current EFE model in [2], only datum error, machine tool path error due to geometric-thermal variation of machine-tool axes and fixture error can be predicted by current EFE model. The machined error of feature EFGH is $\mathbf{x}_{EFGH}^{P}=[0\ \ 0\ \ 0\ \ 0\ \ 0\ \ -0.045]^{\mathrm{T}}$. The EFE of machine tool path error calculated by current EFE model is $\Delta\mathbf{m}_c=[0.015\ \ 0.015\ \ 0.015\ \ 0\ \ 0\ \ 0]^{\mathrm{T}}$. The EFE of datum error calculated by Eq. 3 is $\Delta\mathbf{d}=[0.02\ \ 0.02\ \ 0.02\ \ 0\ \ 0\ \ 0]^{\mathrm{T}}$. By knowing $\Delta\mathbf{d}$, $\Delta\mathbf{m}$ and $\Delta\mathbf{f}$, the total EFE $\Delta\mathbf{t}$ can be calculated out as $\Delta\mathbf{t}=[0.045\ \ 0.045\ \ 0.045\ \ 0\ \ 0\ \ 0]^{\mathrm{T}}$. Then, the total compensation value for the above errors can be obtained as $\Delta\mathbf{c}=[-0.045\ \ -0.045\ \ -0.045\ \ 0\ \ 0\ \ 0]^{\mathrm{T}}$. After adjusting fixture locators under the compensation value $\Delta\mathbf{c}$, the feature EFGH will be milled. With error compensation by current EFE model, the total EFE be cancelled out is $\Delta\mathbf{t}=[0.045\ \ 0.045\ \ 0.045\ \ 0\ \ 0\ \ 0]^{\mathrm{T}}$, leaving the cutting-tool wear unaddressed.

In the second case, the inputs to the model with compensation that corresponds to the machine tool path error and datum error are set as the same as the previous case. Without error compensation, the machining error of feature EFGH predicted by the improved EFE model is $\mathbf{x}_{EFGH}^{P}=[0\ \ 0\ \ 0\ \ 0\ \ 0\ \ -0.06]^{\mathrm{T}}$. The EFE of machine tool path error calculated by Eq. 2 and Eq. 8 is $\Delta\mathbf{m}_i=[0.03\ \ 0.03\ \ 0.03\ \ 0\ \ 0\ \ 0]^{\mathrm{T}}$. The EFE of datum error $\Delta\mathbf{d}=[0.02\ \ 0.02\ \ 0.02\ \ 0\ \ 0\ \ 0]^{\mathrm{T}}$ can be calculated out by Eq. 3. The fixture error is $\Delta\mathbf{f}=[0.01\ \ 0.01\ \ 0.01\ \ 0\ \ 0\ \ 0]^{\mathrm{T}}$. By knowing $\Delta\mathbf{d}$, $\Delta\mathbf{m}$ and $\Delta\mathbf{f}$, the total EFE $\Delta\mathbf{t}$ can be calculated out as $\Delta\mathbf{t}=[0.06\ \ 0.06\ \ 0.06\ \ 0\ \ 0\ \ 0]^{\mathrm{T}}$. Then, the total compensation value $\Delta\mathbf{c}=[-0.06\ \ -0.06\ \ -0.06\ \ 0\ \ 0\ \ 0]^{\mathrm{T}}$ can be obtained. After adjusting fixture locators under the compensation value $\Delta\mathbf{c}$, feature EFGH will be milled. With error compensation by the improved EFE model, the total EFE be cancelled out will be $\Delta\mathbf{t}=[0.06\ \ 0.06\ \ 0.06\ \ 0\ \ 0\ \ 0]^{\mathrm{T}}$.

5 Comparison and discussion

By comparing the results of the current and improved method, it is observed that the improved model can capture more error sources. The machining error of feature EFGH predicted by current EFE model is $\mathbf{x}_{EFGH}^{P}=[0\ \ 0\ \ 0\ \ 0\ \ 0\ \ -0.045]^{\mathrm{T}}$. This machining error predicted through current model is only caused by datum error, fixture error and the geometric-thermal variation of machine-tool axes induced error, leaving the cutting-tool wear induced error unaddressed. Similarly, only datum error, fixture error and the geometric-thermal variation of machine-tool axes induced error can be equivalent to their corresponding EFEs and compensated by current model. Therefore, the total EFE for the above error sources is $\Delta\mathbf{t}=[0.045\ \ 0.045\ \ 0.045\ \ 0\ \ 0\ \ 0]^{\mathrm{T}}$.

However, the machining error predicted by improved EFE model is $\mathbf{x}_{EFGH}^{P}=[0\ \ 0\ \ 0\ \ 0\ \ 0\ \ -0.06]^{\mathrm{T}}$. The difference of the machining errors between current and improved model is understandable because current model can handle more error sources than current model. In addition, the improved EFE model can also cancel out more errors. The total EFE for the above error sources is $\Delta\mathbf{t}=[0.06\ \ 0.06\ \ 0.06\ \ 0\ \ 0\ \ 0]^{\mathrm{T}}$.

The total compensation values for current model is $[-0.045\ \ -0.045\ \ -0.045\ \ 0\ \ 0\ \ 0]^{\mathrm{T}}$ and for improved model is $[-0.06\ \ -0.06\ \ -0.06\ \ 0\ \ 0\ \ 0]^{\mathrm{T}}$. As also can be seen from the above two cases, after adjusting fixture locators under their corresponding compensation values for the current and improved models, the improved EFE model can cancel out more error sources than current EFE model. Therefore, after compensating errors by the improved model, machining quality can be improved obviously.

6 Conclusions

Current EFE compensation technique did not consider the machining-induced variations. Without compensating these machining-induced variations, the EFE based error compensation methodology will be limited. In order to overcome this limitation, this paper extends current EFE model that includes the machining-induced variations. A case study shows the benefits of the extended EFE model. The improved EFE model can compensate more errors than current EFE model. Compensating more errors by the improved model means that machining quality can be improved obviously.

Its open structure provides a quantitative framework for further expansion. In the future work, the EFE procedure based on DMVs can be applied to investigate error compensation methodology for a general 3-2-1 fixture setup rather than orthogonal 3-2-1 fixture layout. In addition, the EFE model can also expand to deformed workpiece. All of these potential works using EFE model based on DMVs will be pursued and reported in the future.

Acknowledgments

This work has been partially supported by the National Key Science and Technology Research Program of China (Grant No. 2014ZX04002011-3) and National Natural Science Foundation of China (Grant Nos. 51121063, 51175340, 51775346). The authors would like to thank these financial supports.

References

[1] Zhou S, Huang Q, Shi J. State space modeling of dimensional variation propagation in multistage machining process using differential motion vectors. *IEEE Transactions on Robotics and Automation*. 2003;19:296-309.

[2] Du S, Huang D, Lv J. Recognition of concurrent control chart patterns using wavelet transform decomposition and multiclass support vector machines. *Computers & Industrial Engineering*. 2013;66:683-695.

[3] Du S, Lv J. Minimal Euclidean distance chart based on support vector regression for monitoring mean shifts of auto-correlated processes. *International Journal of Production Economics*. 2013;141:377-387.

[4] Du S, Lv J, Xi L. On-line classifying process mean shifts in multivariate control charts based on multiclass support vector machines. *International Journal of Production Research*. 2012;50:6288-6310.

[5] Tsung F, Li Y, Jin M. Statistical process control for multistage manufacturing and ervice operations: A review and some extensions. *International Journal of Services perations and Informatics*. 2008; 3 (2), 191-204.

[6] Jin J, Shi J. State space modeling of sheet metal assembly for dimensional control. *Journal of Manufacturing Science and Engineering-Transactions of the ASME*. 1999;121:756-762.

[7] Huang Q, Shi J, Yuan J. Part dimensional error and its propagation modeling in multi-operational machining processes. *Journal of Manufacturing Science and Engineering-Transactions of the ASME*. 2003;125:255-262.

[8] Djurdjanovic D, Ni J. Dimensional errors of fixtures, locating and measurement datum features in the stream of variation modeling in machining. *Journal of Manufacturing Science and Engineering-Transactions of the ASME*. 2003; 125:716-30.

[9] Abellan-Nebot JV, Liu J, Romero F. Limitations of the current state space modeling approach in multistage machining processes due to operation variations. *3rd Manufacturing Engineering Society International Conference, Alcoy, Spain*. 2009.

[10] Abellan-Nebot JV, Liu J, Subiron FR, Shi J. State space modeling of variation propagation in multistation machining processes considering machining-induced variations. *Journal of Manufacturing Science and Engineering -Transactions of the ASME*. 2012;134: 1-13

[11] Liu J. Variation reduction for multistage manufacturing processes: A comparison survey of statistical-process-control vs stream-of-variation methodologies. *Quality and Reliability Engineering International*. 2010; 26: 645-661

[12] Shi J, Zhou S. Quality control and improvement for multistage systems: A survey. *IIE Transactions*. 2009; 41: 744-753

[13] Abellan-Nebot JV, Subirón FR, Mira JS. Manufacturing variation models in multi-station machining systems. *International Journal of Advanced Manufacturing Technology*. 2013; 64:63-83

[14] Yang F, Jin S, Li Z. A comprehensive study of linear variation propagation modeling methods for multistage machining processes. *International Journal of Advanced Manufacturing Technology*. 2017; 90(5): 2139-2151.

[15] Wang H, Huang Q, Katz R. Multi-operational machining processes modeling for sequential root cause identification and measurement reduction. *Journal of Manufacturing Science & Engineering*. 2005; 127:512-521

[16] Wang H, Huang Q. Error cancellation modeling and its application to machining process control. *IIE Transactions*. 2006; 38(4): 355-364

Study on Design and Test of Composite WIG Vehicle Considering on Impact Loading

Hyunbum Park[1,*]

[1]Department of Defense Science & Technology – Aeronautics, Howon University, 54058, 64 Howondae 3gil, Impi, Gunsan, Rep. of Korea

Abstract. This study proposed an example on preliminary structural design including trade-off study and design evaluation such as structural safety and stability analysis of a 20-seat composite WIG craft using the netting rule and the rule of mixture and a commercial FEM tool. The structural configuration of the wing adopted the skin-spar semi-monocoque type structure with foam sandwich, and the fuselage also used the semi-monocoque type structure with the skin to carry shear flow, the stringers to carry bending moments and the ring frames to carry local loads under consideration of various design local load cases. The used materials were Carbon/Epoxy (UD) for wing spar flanges and fuselage stringers and ring frames, Carbon/Epoxy (Fabric) for fuselage skins, the sandwich with Carbon/Epoxy face sheet (Fabric) and Al honeycomb core for the fuselage floor, and the sandwich with Carbon/Epoxy face sheet (UD) and Urethane foam core for wing skins and spar webs. Through the structural analysis, the structural safety and stability were confirmed.

1 Introduction

When a wing is closely flying to the ground or to the water surface within a couple of meters height, the lift force is greatly increased due to the ground-effect. Therefore, if a vehicle uses the wing having the ground effect, it is called a WIG (Wing-in-Ground Effect) craft. The WIG craft has a special feature, which has much wider wings than the traditional airplane wing and the hull type fuselage like a high-speed boat. The WIG craft borrows some merits from both airplane and ship, which can transport quickly many passengers or heavy payload. Since 1960, various types of WIG crafts have been vigorously developed by Russia for military or civil use [1, 2].

In Korea, recently studies on the WIG craft are lively progressing as a new generation maritime transportation system. For instance, KORDI (Korea Ocean Research & Development Institute) and some relating industries have developed several classes of WIG crafts such as the 4-seat, 6-seat and 20-seat small-scale WIG crafts and the 100-ton payload large-scale WIG craft.

This study carried out a preliminary structural design and analysis on wing, fuselage and control surfaces parts of the 20-seats small-scale WIG craft. Structural configuration adopted the skin-spar type structure with foam sandwich for the wing, and the semi-monocoque type structure with the skin to carry shear flow, the stringers to carry bending moments and the ring frames to carry local loads under consideration of various design local load cases for the fuselage. The used materials were Carbon/Epoxy (UD) for wing spar flanges and fuselage stringers and ring frames, Carbon/Epoxy (Fabric) for fuselage skins, the sandwich with Carbon/Epoxy face sheet (Fabric) and Al honeycomb core for the fuselage floor, and the sandwich with Carbon/Epoxy face sheet (UD) and Urethane foam core for wing skins and spar webs. The initial design was performed using the netting rule and the rule of mixture. Structural safety and stability evaluation on the design features was done by a commercial FEM code NASTRAN. Through several modifications, the final structural design features were obtained to meet the design requirements such as the system target weight, structural safety and stability.

2 Concept of Design and Analysis

The WIG craft to be treated in this study has a similar aerodynamic configuration to the fixed mid-wing airplane. For instances, upper and lower sections of the fuselage have an airplane configuration for reducing the air drag and a boat's hull configuration for reducing the water drag, and the wing is attached to the middle of the fuselage. The tail fins have a 'T' shape that the horizontal tail was vertically attached to the vertical fin. Because the WIG craft has an airplane like behaviour after taking-off, therefore load analysis must be firstly performed. For instance, the major design loads become shear forces and bending moments due to the aerodynamic lift and drag, the structural weight and the payload due to passengers and cargo under consideration of the maximum load factor in flight operation, and the impact load at landing. After load case analysis, the

* Corresponding author: swordship@daum.net

fuselage structure can be sized from the major design loads using selected composite materials [3].

After an initial design using the netting rule and the rule of mixture, structural analysis is firstly performed to confirm the structural safety and stability using a commercial FEM code PATRAN/NASTRAN. From the structural analysis on the first design configuration, some modifications can be drawn due to weak area on buckling and a bit heavier than the target weight. The final structural configuration is fixed through several repeated design modifications and analyses.

3 Impact Load and Structural Design

The splash down impact load can be defined as the quasi-static impact loads in nose, center (normal landing case) and tail direction, respectively. According to the previous study, it was found that the maximum load among the three cases of the splash down loads was the center body splash down landing case. However, at stress analysis step, all three splash down impact loads were considered. The quasi-static load splash down landing load can be expressed by the following equations:

$$A_{impact} = 9\times10^{-3} M_{\max}^{2/3} \quad (1)$$

$$P^e = \frac{F^e}{A_{impact}} \quad (2)$$

$$F^e = n^e M_{reduction} g \quad (3)$$

where P^e is pressure on impact area, A_{impact} – impact area, F^e – impact load, $n^e = 1{,}5$ – impact load factor, $M_{reduction}$ is reduction coefficient and g is gravitational acceleration[4].

Table 1 shows investigated impact load using the equations. The structural configuration was initially composed of 'I' type front spar and channel type rear spar including flange and web to avoid complexity of manufacturing of the selected Carbon/Epoxy composite laminate.

Table 2 show mechanical properties of carbon/epoxy UD prepreg and carbon/epoxy fabric prepreg. Preliminary structural design was initially performed by the netting rule and the rule of mixture [5, 6]. According to the netting rule, the principal load directional thickness of main spar flange and web can be sized by the crippling buckling strength.

In order to investigate the structural safety, the initial design feature of the fuselage was modelled as surfaces from the 3-D CATIA drawings after confirmation of detail configurations such as joint parts, spars, ring frames and so on. In the numerical structural analysis of the WIG fuselage, the static stress analysis, the structural stability analysis and the modal analysis were performed using a commercial FEM code PATRAN/NASTRAN. The applied failure theory for safety evaluation was the well-known Tsai-Wu failure criteria. Total number of elements for FEM mesh generation were 35546 including 19080 for the fuselage skin mesh. Fig. 1 shows FEM modelling result. Figures 2-4 and Table 3 show structural analysis result.

The required 20 years fatigue life of the final design feature was confirmed using the S-N diagram based on the reference carbon/epoxy materials [7, 8]. If WIG craft may operate with the following assumption such as 12 times 1-hour flight per a day, total number of flights during 20 years can be calculated as 87,600 times. If the safety factor of 3 may be considered, the total number of flights can be modified as 262800 flights. Therefore, the required mission fatigue cycles of 262800 for 20 years operation is less than the given material fatigue life at the same stress level, the designed structure may be safe for the required fatigue life.

Table 1. Landing impact loading of fuselage.

Load	Case 1	Case 2	Case 3
P^e	33 kN/m^2	43 kN/m^2	42 kN/m^2
F^e	124 kN		
A_{impact}	3,75 m^2	2,92 m^2	2,92 m^2

Table 2. Mechanical properties of carbon/epoxy composite

Property \ Material	Carbon/ epoxy UD	Carbon/ epoxy fabric
Longitudinal modulus (GPa)	145	63,4
Transverse modulus (GPa)	10	58
Axial shear modulus (GPa)	4,8	56,1
Poisson's ratio	0,25	0,17
Longitudinal tensile strength (MPa)	1240	635
Longitudinal compressive strength (MPa)	1240	572,9
Transverse tensile strength (MPa)	41	411,7
Transverse compressive strength (MPa)	170	304,4
In-plane shear strength (MPa)	80	114,5
Density (kgf/mm^3)	1,58E-6	1,58E-6
Thickness (1Ply) (mm)	0,14	0,2

Figure 1. Mesh generation for finite element analysis.

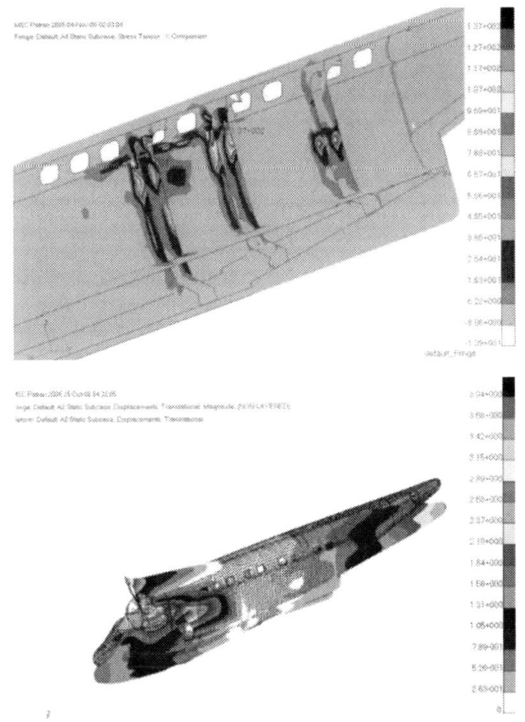

Figure 2. Stress contour and deformed shape of fuselage in load case 1.

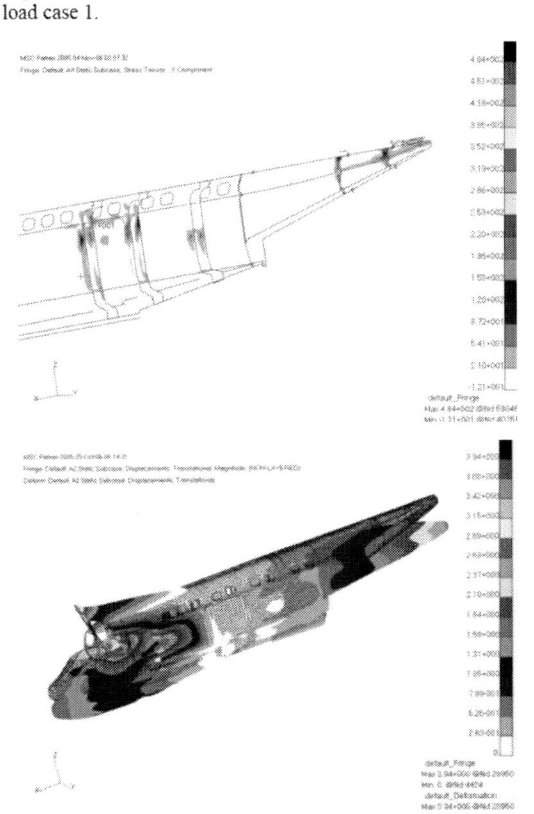

Figure 3. Stress contour and deformed shape of fuselage in load case 2.

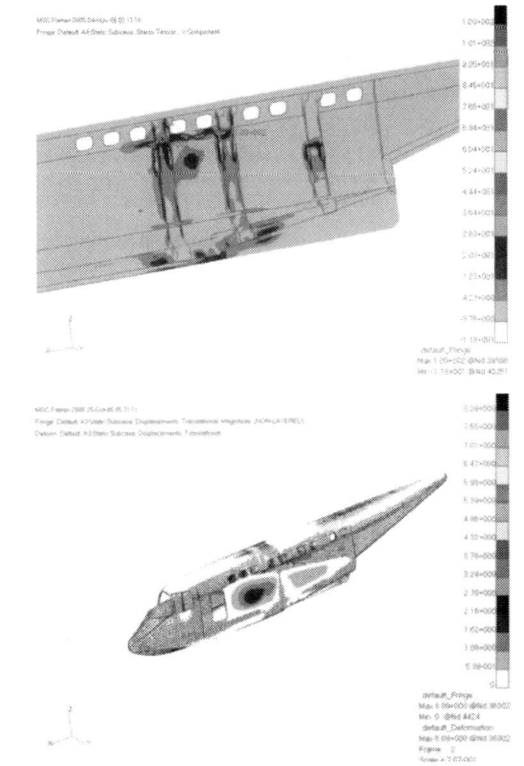

Figure 4. Stress contour and deformed shape of fuselage in load case 3.

Table 3. Structural Analysis Results.

Case of analysis Analysis result		Symmetric Load (Case 1)	Un-symmetric load (Case 2)	Splash Down (Case 3)
Max. stress, MPa	Ten.	137	484	109
	Com.	135	694	117
Max. disp., mm		3.84	3.94	8.9

4 Manufacturing and Test

A 6-seat WIG craft was developed as a subscale to minimize the development risk of the 20-seat WIG craft by KORDI-MOERI (Korea Ocean Research & Development Institute- Maritime and Ocean Engineering Research Institute), HANKUK FIBER CO. and others including authors. Table 4 shows the system specification of the 6-seat WIG craft and Figures 5-7 shows some views for manufacturing and flight test.

After an initial design using the netting rule and the rule of mixture, structural analysis is firstly performed to confirm the structural safety and stability using a commercial FEM code PATRAN/NASTRAN. From the structural analysis on the first design configuration, some modifications can be drawn due to weak area on buckling and a bit heavier than the target weight. The final structural configuration is fixed through several repeated design modifications and analyses. Before manufacturing the full scale the prototype, the structural test and flight test was performed by a sub-scale WIG craft for evaluation of the proposed structural design and analysis procedure. Through this comparison, even though there were some differences between them, it was confirm that the proposed design method is appropriate for the WIG's composite structure.

Table 4. System specification of 6-seat WIG Craft.

Length	12 m
Pay Load	6 passengers
Gross Weight	1200 kg
Maximum Speed	137 km/h
Engine Power	2×102 hp
Material	Carbon/Epoxy laminates + Al honeycomb core sandwich

Figure 5. Combination wing and fuselage of WIG craft.

Figure 6. Engine thrust test of WIG craft.

Figure 7. Flight test of 6-seat WIG craft.

5 Conclusion

In this study a composite WIG craft structure, which can satisfy the target weight as well as the structural safety and stability, was designed. Basic structural feature adopted the skin-spar type structure, and especially the foam sandwich composite was applied to upper and lower skins of the wing to improve buckling behavior and vibration absorption capability. The front spar adopted 'I' type beam and the rear spar adopted channel type beam to accommodate control surface structure. In order to improve the strength/weight ratio as well as the stiffness/weight ratio the carbon/epoxy composite material which is mostly used in aerospace vehicle design was selected.

The fuselage was designed by a semi-monocoque type structure with the skin to carry shear flow, the stringers to carry bending moments and the ring frames to carry local loads under consideration of various design load cases. The used materials were the carbon/epoxy laminates using the UD prepreg for stringers and ring frames and the fabric prepreg for the skin and the sandwich structure with Al alloy core and Carbon/Epoxy fabric face sheet for the floor.

Through the static stress analysis including the structural stability analysis, the fatigue life estimation and the modal analysis, the final design WIG craft structural feature with the required 20 years fatigue life and the proper vibration characteristics was confirmed.

Acknowledgements

This study was supported by research fund from KORDI (Korea Ocean Research & Development Institute), Hankuk Fiber and Howon University.

References

1. K. Rozhdestvensky, Wing-in-ground effect vehicles, Progress in aerospace sciences, **42** (2006)
2. N. Kornev, K. Matveev, Complex numerical modeling of dynamics and crashes of wing-in-ground-vechicles, AIAA **2003-600** (2003)
3. C. Kong, Structural investigation of composite wind turbine blade considering various cases and fatigue life, Energy **30** (2005)
4. B. Lee, C. Kim, M. Park, H. Jeong, ALE finite element analysis of the WIG craft under the water impact loads, Journal of the Korean society for aeronautical and space science **35** (2007)
5. S. Tsai: *Composites Design,* Think Composite, (1987)
6. M. Niu: *Airframe Structural Design*, Hong Kong Conmilit Press Ltd, (1988)
7. V. Vasiliev, E. Morozov, *Mechanics and analysis of composite materials,* Elsevier Science Ltd, (2001)
8. M. Hyer, Stress Analysis of Fiber-Reinforced Composite Materials, WCB/McGraw-Hill, (1998)

Research on the Growth of TFP of the Urban Commercial Banks Based on the Malmquist productivity index

Wen Hongmei[1] , *Sui* Xin[1,*] and *Lu* Shaopeng[1]

[1]School of Finance, Harbin University of Commerce, Harbin150000, China

Abstract. This paper measures the total factor productivity (TFP) growth of 40 urban commercial banks based on the Malmquist productivity index. On this basis, the first - order difference GMM is used to estimate the influencing factors of the TFP of urban commercial banks. This paper show that the technological progress during the sample period has promoted the overall upward trend of the TFP of China's commercial banks. And the rate of return on assets has a significant positive impact on the TFP, and the deposit and loan ratio has a significant negative impact on TFP. At the same time, we have also concluded that the factors affecting the TFP of the three major regions in East, West and West are different.

1 INTRODUCTION

When the first urban commercial bank of China has been listed since 1995, our urban commercial bank has shown a good momentum of development. Especially in 2004, with the promulgation of the outline of regulation and development of the city commercial bank, the urban commercial bank has entered a stage rapid of development. According to the statistics from China banking regulatory commission, by the end of 2015, the number of urban commercial bank institutional is 133, the total of assets is 22.68 trillion, it has increased nearly 4 times than 4.7 trillion in 2004, the assets accounted share for 11.38% of the whole banking financial institutions, and 1.4% of the non-performing loan ratio has been a lower level compared with the same commercial Banks. However, with the deepening reform of financial in our country, the development of the urban commercial bank turning, asset grow, non-performing loans increased, profit space narrowing and monomer differentiation acceleration and other issues make it into the downlink channel. In this context, the traditional extensive expansion mode has no longer adapts to the current development need, and the improvement of the efficiency of the bank is the key to the sustainable development of the city commercial bank. This paper aims to measure the total factor productivity to judge its core competitive power of city firms, and through the analysis of the influence factors to find out the effective ways to promote the TFP of Banks, thus to provide the realistic guidance for the future development of urban commercial bank .

2 Literature Review

The study of the efficiency of Banks can dates back to the 1950s, and the efficiency of Banks has been widely watched by scholars since then. The research on the efficiency of Banks in foreign countries has gone through three stages. The research focuses on the scale efficiency of the bank, the scope efficiency and the current frontier efficiency. The analysis of the dynamic efficiency of total factor productivity of Banks has become a hot topic in the current research. The method has been widely used in the research since the Malmquist production index method was constructed to calculate the total factor productivity in the study of foreign scholars Charnes and Cooper et al. (1978). In the case of empirical analysis, the methods for TFP of Banks in Spain, the United States and Kuwait were quantitatively calculated based on this method, such as those of the bank of Spain, the United States, and the Kuwaiti bank of Kuwait. Park (2006) research on the Korean banking TFP found that a sharp rise in its technical progress index led to a general increase in TFP in the banking sector. Domestic scholar Zhang Jianhua (2003) first used the Malmquist index to quantify the total factor productivity index of China's banking industry, and found that the TFP of China's banking industry showed an upward trend. Wang Fubiao (2006) also confirmed this conclusion based on empirical research on the same method. Different from the above research results, Zhu Chao (2006) and Cai Yuezhou, Guo Meijun (2009), which is based on different data samples using the same Malmquist index method, it is concluded that the average values of TFP in China's banking industry is on the decline.

From the grasp of the paper, most scholars on the research methods using TFP Malmquist productivity index of the bank, in the aspect of the research object is the TFP of listed commercial Banks more, there is little study on city business TFP. In this paper, based on the above research, and considering the features of China's

* Corresponding author: 513119931@qq.com
* International social science foundation(16BJL037)

DOI: 10.1051/matecconf/201713900023

three biggest region's history, selected the features three regions of 40 city business of large sample of panel data from 2011 to 2015, based on Malmqiust productivity index of city business in China and quantitatively analysis and decomposition of total factor productivity. At the same time, the influence factors of TFP were studied by using the first order difference GMM to determine the effective ways to improve the efficiency of the city. Through the study of this paper, the author tries to supplement and perfect the related research on total factor productivity of our city, and provide some suggestions for the sustainable development of the city.

3 Total Factor Productivity Measure and Decomposition of City Commercial Banks

3.1. index selection

1. Sample selection and data source

In order to measures the total factor productivity of urban commercial banks, the paper selects 40 urban commercial banks as samples., we chose the period from 2011 to 2015 during the sample survey period, the requirements for information disclosure of city businesses were higher, so the data's desirability and integrity were better. This paper selects 18 urban commercial banks in the eastern region, 13 in the central region, 9 in west western region, mainly from the annual reports of commercial banks, the annual report of the banking regulatory commission and related web data.

2. Selection of input and output indexes

Based on the production method and mediation method as the main basis of input and output index selection, choose the bank employees, total assets, bank branches and deposits as input indicators, with pre-tax profits and total loans as output indicators, descriptive statistics related index data can be found in the table 1.

Table 1 descriptive statistics of indicators selected

	Min	Max	Mean	Std.Dev.	Time
Bank branch	11	534	151	103.3852	200
The bank employee	694	14046	4189	2733.2514	200
The total assets of	84.39	18449.09	2595.5647	2875.2668	200
The total amount of the deposit	29.71	10223.00	1681.7286	1721.1046	200
Pre-tax profits	0.31	210.85	33.4947	34.0343	200
The total amount of the loan	11.54	13369.96	1123.1693	1487.3249	200

3.2. analysis of the calculation results

Table 2 shows the total factor productivity and decomposition results of the sample city business lines from 2011 to 2015. It can be concluded that, on the whole, the increase of technical level is greater than the technical efficiency, which leads to the growth of the TFP index of the sample cities in China. Second, from the point of time dimension, besides the TFP of value is less than 1 in 2014, the rest were greater than 1 year, so the city firm in the sample interval of TFP indexshowed up - down - the rising trend; Third, from the spatial dimension, the TFP of the three regions of eastern and western regions of the east and China are all showing an upward trend under the impetus of technological progress.

Table 2 TFP index and decomposition results obtained by time and region classification

	Simple size	TFP	TECH	EFF	PE	SE
simple	200	1.036	1.091	0.949	0.976	0.973
time						
2012	40	1.046	1.151	0.909	0.950	0.957
2013	40	1.075	1.289	0.834	0.940	0.887
2014	40	0.983	0.968	1.015	1.011	1.004
2015	40	1.041	0.988	1.054	1.005	1.049
Area						
East	72	1.044	1.108	0.978	0.981	0.973
Middle	52	1.255	1.253	0.982	0.977	0.967
weat	36	1.011	1.075	0.966	0.973	0.969

4 Study on the influence factors of TFP of urban commercial Banks

4.1. variable selection and model construction

a. Selection of variables

The study on the influence factors of TFP on urban commercial banks is mainly to clarify the effective way to improve TFP of city businesses by examining and analyzing the significance of influencing factors. Table 3 lists the definitions of each variable and the descriptive statistics.

Table 3 variable definition tables

Variable	mean	Mean	Std.Dev.	Min	Max
ln_tasset	Bank size	7.55	1.03	4.44	9.82
roa	profitability	1.11	0.25	0.05	1.83
crar	Security level	10.63	1.72	6.90	18.47
rld	Asset allocation	0.85	0.99	0.05	6.51
npl	Level of risk	1.00	0.59	0.00	5.59
district	Management ability	0.51	0.92	0.00	11.18

4.1. Model construction

In order to measures the influence factors of total factor productivity (TFP) growth of 40 urban commercial banks, TFP value calculated above city firms here is as explained variable, select the above six variables as explanatory variables. In addition, to investigate whether the above variables can have an impact on the TFP decomposition of city businesses, it is also considered that TECH and EFF are considered as explanatory variables. This paper builds a dynamic panel model based on the relevant data of sample commercial bank. Even the jade king (2010), for sample shorter dynamic panel data, using GMM estimation can obtain consistent estimator, so this article USES the first order difference GMM estimation to calculate the model, the model set is as follows:

$$y_{it} = \gamma y_{i,t-1} + x_{it}^{'}\beta + \alpha_i + \varepsilon_{it} \tag{1}$$

In this case, y_{it} is be explained variable, I is urban commercial banks, i=1,2,...,40, t means time; γ is coefficients of the first order delay $y_{i,t-1}$ of the explained variables; βi in the variable coefficient, $\beta=(\beta_1,\beta 2,\beta 3,\beta 4,\beta 5,\beta 6)'$; xit means interpreting variables; $x_{it}=(ln\_tasset,roa,crar,rld,npl,district)'$ εit random disturbance, $\varepsilon_{it} \sim i.i.d.N(0,\sigma_\varepsilon^2)$.

B. analysis of empirical results

a. Overall panel data regression

Dynamic panel based on the established model of regression, the estimated results as shown in the table 4, in view of this part mainly TFP is city firms and the influence factors of the decomposition of item analysis, so this list only a part of the measuring results. Table 4 reports three models that are explained separately with TFP, EFF, and TECH.

Table 4 TFP, TECH and EFF dynamic panel regression results

explained variable	ModelI Be explained variable: TFP			Model II Be explained variable: EFF			Model III Be explained variable: TECH		
	Coefficient	Standard error	Z value	coefficient	Standard error	Z value	coefficient	Standard error	Z value
ln_tasset	-0.276	0.187	-1.48	0.001	0.077	0.01	-0.328**	0.154	-2.13
roa	1.094**	0.477	2.29	0.453**	0.192	2.35	0.053	0.352	0.15
crar	0.022	0.079	0.28	-0.039	0.030	-1.29	-0.004	0.060	-0.07
rld	-2.801**	1.096	-2.55	-0.033	0.401	-0.08	-0.084	0.725	-0.12
npl	0.187	0.212	0.88	0.211**	0.087	2.44	-0.284**	0.155	-1.83
district	0.151	1.023	0.15	0.387	0.431	0.9	-0.319	0.804	-0.4
_cons	4.757***	1.858	2.56	0.368	0.688	0.53	4.782***	1.477	3.24
Wald chi2	16.92			15.37			14.33		
Prob>chi2	0.0179			0.0316			0.0456		

According to the empirical results can get the following conclusions: (1) about the influence factors of total factor productivity (TFP) study: return on assets (roa) and LDR (RLD) has a significant influence on city business of TFP, and total assets (ln_tasset), core tier 1 capital ratio (crar), non-performing loans (NPL) and fragmentation coefficient (district) failed to pass the test of significance. In particular, the yield of assets has a positive influence on TFP index of our city business bank, and the deposit and loan ratio has negative influence on it. (2) research on the impact factors of technical efficiency change (EFF) : asset and non-performing loan ratios have passed the significance test at a confidence level of 5%. (3) research on the impact factors of technical level change (TECH) : asset size and non-performing loan ratio have significant negative effects on technical level.

b. East and west panel data regression

At this stage, the economic and financial environment of the three regions of China's east and west are very different, and this paper makes a return to the sample cities and businesses in different regions based on this thesis. The parameters of the three-region dynamic panel are shown in table 5.

Table 5 regression results of TFP dynamic panel in eastern and western regions

explained variable	east		middle		west	
	coefficient value	Z value	coefficient value	Z value	coefficient value	Z value
ln_tasset	0.1171	1.04	-0.3810	-0.85	-0.0866	-0.80
roa	0.7560***	5.70	1.3127	1.08	0.9577***	2.91
crar	-0.0088	-0.33	0.0278	0.15	-0.0527	-1.28
rld	0.9480***	2.74	-5.0553**	-1.99	0.9484	1.13
npl	0.0745	1.49	0.1178	0.12	0.0667	0.50
district	0.1536	0.52	-0.8882	-0.24	0.1082	0.20
_cons	-0.7248	-0.68	6.8355	1.54	-0.1060	-0.13

Based on the estimated results, TFP in the eastern region is significantly affected by the ratio of asset yield and loan-to-deposit ratio. Among them, the coefficient of the variable of the asset return is positive, and its symbol direction is consistent with the overall regression result, so it is no longer detailed. LDR coefficient is positive shows high deposit-loan ratio can help improve the TFP of city business value, the conclusion can be answered from the perspective of bank profitability, because higher LDR is usually relative to the deposit of the loan proportion is larger, it spreads as the main source of income of city business profits will increase and therefore, to play a positive role in driving the TFP of city business. Secondly, the TFP in the central region of the city only has a significant test of deposit and loan ratio, and the coefficient of loan-to-deposit ratio is negative, indicating that the increase of the loan-to-deposit ratio is unfavorable to TFP of the city business. Finally, the coefficient of commercial bank in western regions is positive, which also indicates that the increase of asset yield has a positive effect on bank efficiency.

5 Conclusion

This paper measures the total factor productivity (TFP) growth of 40 urban commercial banks during 2011-2015 by the Malmquist productivity index. n this basis, the article first order difference GMM estimation for the influence factors of city business index of TFP and its decomposition has carried on the empirical research, the results showed that: (1) in the inspection period, 2011-2015 sample city firm's total factor productivity under the role of technological progress is on the rise as a whole. In terms of time dimension, except for the decline of TFP in the city in 2014, the other years showed a good upward trend. From the spatial dimension, TFP of the city in the central region is the fastest growing, the second in the east and the slowest in the west. (2) the return on assets to the firm's total factor productivity and technical efficiency has a significant positive effect, asset size has a significant negative influence on technical level, the significant negative impact on the deposit-loan ratio on total factor productivity non-performing loan ratio has a significant positive influence on technical efficiency, and has a significant negative influence on technical level. (3) return on assets and the deposit-loan ratio and total factor productivity to east city firm was significantly positively related to relationship between the LDR with central city business was significantly positively related to relationship between total factor productivity, return on assets and businesses in the western region city was significantly positively related to relationship between total factor productivity.

References

1. Charnes A , Cooper W W , Rhodes E. Measuring Efficiency of Decision Making Units[J]. European Journal of Operational Research, 1978 (3) : 429-444.
2. E.Grifell-Tatje, Lovell C.A.K. The Sources of Productivity Change in Spanish Banking[J]. European Journal of Operational Research , 1997 (5) : 364-380.
3. David C. Wheelock , Paul W. Wilson. Technical Progress , Inefficiency , and Productivity Change in US Banking, 1984-1993. Journal of Money , Credit , and Banking, 1999, 31 (2) : 212-234.
4. Darra A F C, Topuz T, Yousef. Assessing Cost and Technical Efficiency of Banks in Kuwait. Working Paper, 2002
5. Park K H, Weber L.A. Note on Efficiency and Productivity Growth in the Korean Banking Industry, 1992-2002. Journal of Banking and Finance, 2006, 30 (12) : 2371-2386.
6. Chung , Y. , R. Fare and S. Grosskopf. Productivity and Undesirable Outputs : A Directional Distance Function Approach[J]. Journal of Environment Management , 1997 (51) 229-240.
7. Fujji, H., S. Managi and R. Matousek. Indian Bank Efficiency and Productivity Changes with Undesirable Outputs : A DisaggregatedApproach[J]. Journal of Banking & Finance, 2014 (38) , 41-50.
8. Caves D, Christensen L, Diewert W. The Econonic Theory of Index Numbers and Measurement of Input , Output and Productiviy[J]. Journal of Econometrics, 1982 (6) : 1393-1414.

Quantitative Analysis of Range Image Patches by NEB Method

Wen Wang
School of Science,
Shandong Jianzhu University,
Jinan 250101 P. R. China.

Abstract. In this paper we analyze sampled high dimensional data with the NEB method from a range image database. Select a large random sample of log-valued, high contrast, normalized, 8×8 range image patches from the Brown database. We make a density estimator and we establish 1-dimensional cell complexes from the range image patch data. We find topological properties of 8×8 range image patches, prove that there exist two types of subsets of 8×8 range image patches modelled as a circle.

Keywords-**NEB; Range Image patch; Morse theory; cell complex; k-cells**

1 Introduction

Computational topology becomes an important method to analyze large sets of high-dimensional data for all branches of science [1], [2], [3], [4]. In order to study multi-dimensional data, we usually make a series of simplicial complexes from data to produce a presentation. The constructed simplicial complexes usually compose of lots of simplexes, often they are hard to compute. The authors utilize the NEB method to construct cell complexes by density functions [5], they built models for nonlinear data sets, and effectively detected the homology of the data.

In this paper, we find a circle model for 8×8 range image patches with the methods in [5], and give some topological characteristic of 8×8 range image patches. The range image patches used in this paper are from the Brown database.

2 BACKGROUND

We shortly introduce three topics: CW complexes, Morse theory, and the NEB method.

1.1 CW complexes.

First, A CW complex is a type of cell complex. For k a nonnegative integer, a k-cell is the closed ball $\{y \in R^k \,|\, \|y\| \leq 1\}$ of dimension k. So a 0-cell is a point, a 1-cell is a line segment, a 2-cell is a disk, etc. See Figure 1 for an example and [13] for further details.

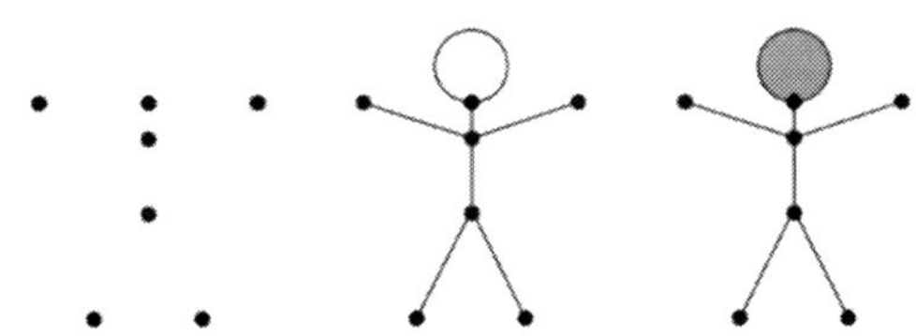

Figure 1 A stick figure represented as a CW complex containing eight 0-cells, eight 1-cells, and one 2-cell. The 0-skeleton is on the left, the 1-skeleton is in the center, and the full 2-skeleton is on the right.

1.2 *Morse theory.*

The following introduction to Morse theory is informal; see [13] for thorough treatment. Suppose M is a compact manifold of dimension d and Morse function f:M→R is smooth with non-degenerate critical points $m_1, \ldots, m_k \in M$ satisfying

$$a_0 < f(m_1) < a_1 < f(m_2) < \cdots < a_{k-1} < f(m_k) < a_k$$

the index λ_i of critical point mi is the number of linearly independent directions around mi in which f decreases. So a minimum has index 0, a maximum has index d, and a saddle point has index between 1 and d−1. Let $M^\alpha = f^{-1}((-\infty, \alpha])$ be the sublevel set corresponding to a∈R. Morse theory tells us that each M_{ai} is homotopy equivalent to a CW complex with one λ_i-cell for each critical point m_i. In particular, M_{ai} is homotopy equivalent to $M_{ai-1}$with a single λ_i-cell attached. For instance, M_{a1} is homotopy equivalent to a point and is obtained from M_{a0}, the empty set, by attaching a single 0-cell.

1.3 *NEB*

* Corresponding author: wen@sdjzu.edu.cn

To find saddle points in a high-dimensional space we use the nudged elastic band method (NEB) from computational chemistry. The nudged elastic band method is used to find minimum energy paths. An initial piecewise linear path (called a band) connecting two local minima of energy E is chosen, and forces move the band towards a minimum energy path [5].

An elastic band with N+1 images can be defined by $[U_0, U_1, ..., U_N]$, U_0 and U_N are initial and final states. The N−1 middle images are modified by an optimization algorithm [10]. The total force acting on each image is defined as following:

$$F_i = F_i^S|_{\parallel} - \nabla E(U_i)|\perp = (\|U_{i+1} - U_i\| - \|U_i - U_{i-1}\|)\tau_i - (\nabla E(U_i) - \nabla E(u_i)\tau_i) \quad (1)$$

the first part $F_i^S|_{\parallel}$ is called the spring force, the second part $\nabla E(U_i)|\perp$ is true force, and local tangent at image i. where E is the energy of the system.

The nudged elastic band method applies an optimization algorithm to shift the images depending to the force in (1) for finding the minimum energy path. For more details of NEB, please refer to papers [8], [9].

3 DATA SET OF RANGE IMAGE PATCHES

The Brown range image database by Lee and Huang is a set of 197 range images from indoor and outdoor scenes, mostly 444 × 1440 pixels. The database can be found at the following webpage:

http://www.dam.brown.edu/ptg/brid/index.html.

The operational range for the Brown scanner is typically 2–200 meters, and the distance values for the pixels are stored in units of 0.008 meters. From the Brown database we obtain a space of range image patches as in [5], see Figure2 for some examples of range image patches.

Let R be the resulting set of high-contrast, normalized, 8×8 range patches. Our data set is a random subset X⊂R of size 5,0000.

Figure 2 Sample range images from the Brown database

4 COMPUTING METHOD

In this part, the steps of calculation procedure are given. refer to the paper [5] for more details of the method. For $X \subset R^n$ from probability density function f: $R^n \to [0, \infty)$. We get superlevel sets $X^\alpha = f^{-1}([\alpha, \infty)) = \{x \in R^n | f(x) \geq \alpha\}$, important topological information is given by the high dense regions. We may make CW complex models Z^α to estimate X^α.

Only the 1-dimensional skeleton of the cell complex is given. We give a differentiable function to estimate the unknown density function. then we can get the partial maximum of the density estimate in order to make 0-cells, we make original bands randomly, then discover the convergent bands through NEB [10], so we can get the 1-cells.

A. Density estimator

For a data set $X \subset R^n$, let $\Phi_{x,\sigma}: R^n \to [0, \infty)$ be the probability density of a normal distribution centered at x ∈ X, we apply a differentiable density estimator $g(y) = |X|^{-1}\sum_{x\in X}\Phi_{x,\sigma}(y)$ to approach the unknown density.

B. 0-cells

To find 0-cells, we randomly select an initial point $y_0 \in X$, and iteratively define a sequence $\{y_0, y_1, ...\}$ with $y_{n+1} = m(y_n)$, where $m(y): R^n \to R^n$ is the mean shift function given by the formula:

$$m(y) = \frac{\sum_{x\in X}\Phi_{x,\sigma}(y)x}{\sum_{x\in X}\Phi_{x,\sigma}(y)}$$

The sequence $\{y_n\}$ converges to a local maximum of g [13]. we use single-linkage clustering to cluster the convergent points, and choose the densest member from each cluster as a 0-cell.

C. 1-cells

For two 0-cells, there is a 1-cell between them if we finda convergent band between them by using NEB. For an initial band $[U_0, U_1, ..., U_N]$, where U_0 and U_N are 0-cells. Using a similar formula as (1), the total force on each midterm node Ui is computed by

$$F_i = (\|U_{i+1} - U_i\| - \|U_i - U_{i-1}\|)\tau_i + c\nabla g(U_i)|\perp + F_{sm} \quad (2)$$

Where $c = \left(\sigma\sqrt{2\pi}\right)^n \sqrt{e}$ is the gradient constant. Let $h_{\theta,\phi}(x): [0, \pi] \to [0, 1]$ be a function defined by

$$h_{\theta,\phi}(x) = \begin{cases} 0 & x \leq \theta \\ \dfrac{\left(1 - \cos\left(\dfrac{x-\theta}{x-\phi}\right)\pi\right)}{2} & \theta < x < \phi \\ 1 & x \geq \phi \end{cases}$$

Let α_i be the angle between $(U_{i+1} - U_i)$ and $(U_i - U_{i-1})$, the smoothing force F_{sm} is computed by $h_{\theta,\phi}(\alpha_i)(U_{i+1} - 2U_i + U_{i-1})$, here $\theta = \pi/6$ and $\phi = \pi/2$.

DOI: 10.1051/matecconf/201713900024

5 EXPERIMENTAL RESULTS

Now we analyze subsets of X_8 with two types subsets of X_n: (1) $\overline{X_8}(50000)$ of X_n with 50000; (2) $\overline{X_n}(k,p)$ with k=200,300 and p=30. Table 1 is the detailed data of the subsets.

TABLE I. DATA SET INFORMATION

	$\overline{X_8}(50000)$	$\overline{X_8}(200,30)$	$\overline{X_8}(300,30)$
size of data set	50000	15000	15000
dimension n	64	64	64
standard deviation σ	0.36	0.32	0.34

For $\overline{X_8}(50000)$,with the deviation σ =0.36, we obtain four 0-cells with densities 97.15,97.83,105.7 and 150.8 respectively, and four 1-cells with densities 84.85,88.99,94.48 and 99.0, these cells make a loop. Therefore, for α = 84.85, the Z^{α} is a circle (Figure 3).

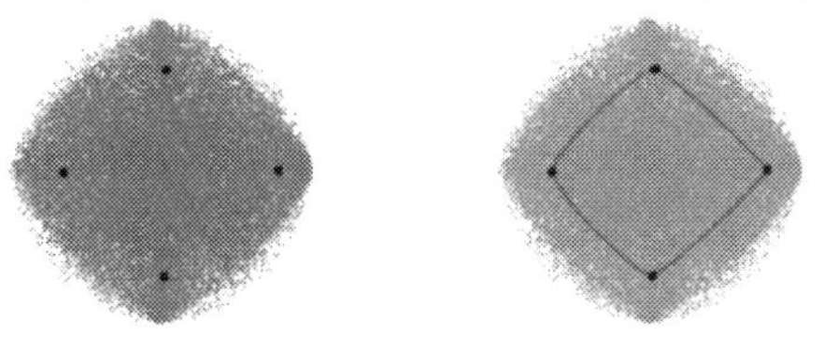

Figure 3 $\overline{X_8}(50000)$ and the circle $Z^{84.85}$, projected to a plane.

For $\overline{X_8}(200,30)$, we take the value of standard deviation σ = 0.32, we get four 0-cells having densities of 407.5,254.8,223.1and 192.8, and four 1-cells having densities of 198.7,182.9,146.0 126.8, all these cells make a circle too(Figure 4).

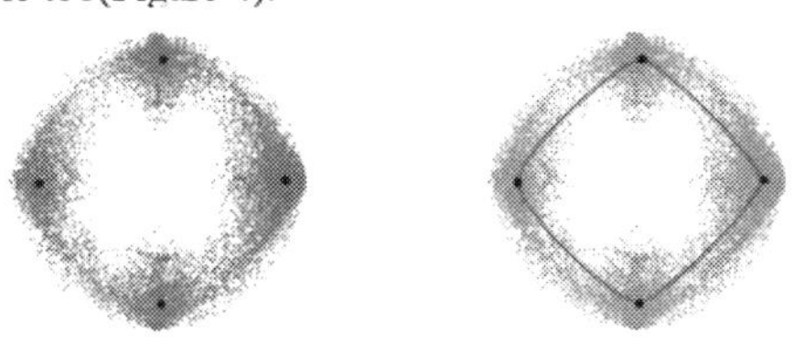

Figure 4 $\overline{X_8}(200,30)$ and the circle $Z^{126.8}$, projected to a plane.

For $\overline{X_8}(300,30)$, we take the value of standard deviation σ = 0.34, we get four 0-cells having densities of 194.9,225.4,256.8 and 412.3, and four 1-cells having densities of 128.0,146.7,185.6 and 201.1, all these cells make a circle too (Figure 5).

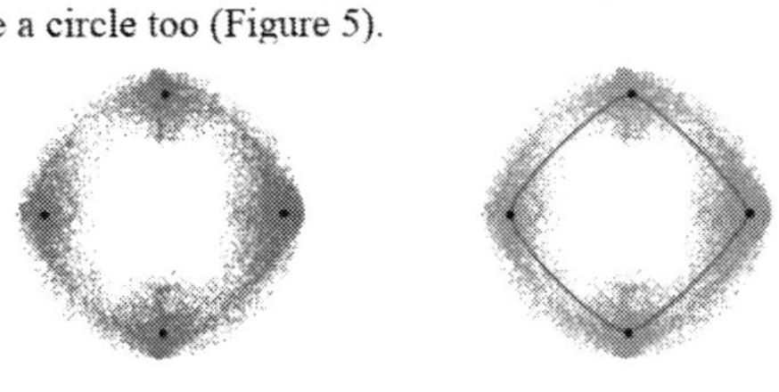

Figure 5 $\overline{X_8}(300,30)$ and the circle $Z^{128.0}$, projected to a plane.

6 CONCLUSIONS

We exploit the NEB technique to analyze 8×8 Range image patches, the results experimentally show that the spaces of 8×8 range image patches have different subsets modelled as a circle. We find four quarter-circular 1-cells forming a loop.

We test the approach on range image data sets and find compact complexes revealing important nonlinear patterns. But It's difficult to create higher dimensional cells with this method yet.

Acknowledgment
Project supported by the National Natural Science Foundation of China (Grant No.61471409).

References

1. H. Adams and G. Carlsson, On the nonlinear statistics of range image patches, SIAM J. Imag. Sci., 2 (2009), 110–117.
2. G. Carlsson, T. Ishkhanov, V. de Silva, A. Zomorodian, On the local behavior of spaces of natural images, Internat. J. Computer Vision, 76 (2008), pp. 1–12.
3. G. Carlsson, Topology and data, Bulletin (New Series) of the American Mathematical Society, 46 (2009), pp. 255–308.
4. A. Zomorodian, Topological data analysis, Advances in Applied and Computational Topology (Proceedings of Symposia in Applied Mathematics), A. Zomorodian, eds., American Mathematical Society, 70 (2012), pp. 1–39.
5. H. Adams, A. Atanasov, and G. Carlsson, Nudged elastic band in toplogical data analysis, Topological Methods in Nonlinear Analysis, 45 (2015), pp. 247–272.
6. A. B. Lee, K. S. Pedersen, and D. Mumford, The non-linear statistics of high-contrast patches in natural images, Internat. J. Computer Vision, 54 (2003), pp. 83–103.
7. H. Jegou, M. Douze, and C. Schmid, Hamming Embedding and Weak geometry consistency for large scale image search, Proceedings of the 10th European conference on Computer vision, October (2008), pp. 304–317.
8. D. Sheppard, R. Terrell, and G. Henkelman, Optimization methods for finding minimum energy paths, Journal Chemical Physics, 128 (2008), pp. 134106-1–10.
9. G. Henkelmana and H. Jonsson, ' Improved tangent estimate in the nudged elastic band method for finding minimum energy paths and saddle points, Journal Chemical Physics, 113 (2000), pp. 9978–9985.
10. G. Henkelman, B. Uberuaga, H. Jonsson, ' A climbing image nudged elastic band method for finding saddle points and minimum energy paths,

Journal Chemical Physics, 112 (2000), pp. 9901–9904.

11. S. Xia, A topological analysis of high-contrast patches in natural images, J. Nonlinear Sci. Appl., 9 (2016), pp. 126–138.
12. Y. Cheng, Mean shift, mode seeking, and clustering, IEEE Trans. Pattern Anal., 17 (1995), pp. 790–799.
13. A. Hatcher. Algebraic Topology. Cambridge University Press, Cambridge, 2002.
14. J. Milnor. Morse Theory. Princeton University Press, Princeton, 1965.

Analysis and Aesearch of White Spot on Creepage Discharge in oil-Pressboard Interface

YU Wen-bo[1] , ZHOU Xiao-xu[2], CHEN Fang-dong[1], CHEN Hua[1], SHANG Donghui[1], HAN Xu[1]

[1]State Grid Jibei Electricity Power Maintenance Company, Beijing
[2]BeiJing Power Equipment Group co.,LTD., Beijing

Abstract. Creepage discharges on oil impregnated pressboard is one of the most dangerous transformer insulation fault, which can cause the gas action even transformer housing cracking accident. "White spot" is creepage important phenomena in the process of insulation board along the surface, so it is necessary to study the insulation board along the surface of creepage "white spot" in the process of growth characteristics. This paper set up a insulating paper along the surface of creepage "white spot" in the process of growth characteristics of observation platform. Numerical example was calculated under different aging degree creepage cardboard along the surface of the "white spot" in the process of change, statistics and analysis of the "white spot" length, angle and size variation, which is very practical to the production practice.

I. Introduction

The oil paper insulation system is a commonly method used in the large transformer, many failure of high voltage transformer is due to the solid insulation oil (folding screen, stay, cardboard, etc.) is caused by the creeping discharge in china in recent years, so it is necessary to research the creeping discharge characteristics of aging under the different condition of the paper. Creeping discharge insulation board will form a "white spot", its essence is the discharge energy of oil gas inside the insulation board caused by decomposition and escape the cardboard, which belongs to a kind of insulation defects of recoverable.[1-6] "White spot" is creeping discharge electric breakdown of precursors, the direction and speed of growth is directly related to the remaining life of insulation board. This paper studied the growth of the "white spot" characteristics based on the different experimental conditions .[7-12]

II. White spot experiment method

It's has been more than 20 years of research work abroad in partial discharge degradation mechanism and development process of insulating materials, which is very deep especially in the research in the field of solid insulation bubble discharge, and a lot of achievements have been made. However, most studies focus on discharge types of pattern recognition, creeping discharge development process on the surface of the paper's research is less in creeping discharge in oil paper insulation structure. It is the reason that mainly oil paper insulation belong to solid liquid two phase insulation system, what's more, the phenomenon of partial discharge is far more complicated than the simple gas discharge and solid discharge, so it is very hard for microscopic observation. Border discharge insulation board caused by the second breakdown as shown in figure 1, the opposite has a similar situation.

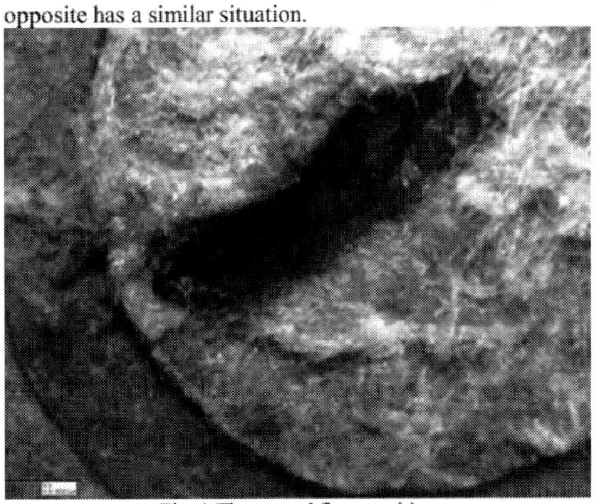

Fig. 1 The second floor positive

DOI: 10.1051/matecconf/201713900025

White spot is the produce of the oil discharge procedure in the board, then the bubble squeeze up cardboard surface. Therefore, we can observe the cardboard white spot of the bubble as a characterization of the location and scope of important indicators of growth status. Cardboard surface bubbles produced by creepage angle and creepage length can be as a characterization of the range and position of the white spot growth, as shown in figure 2.

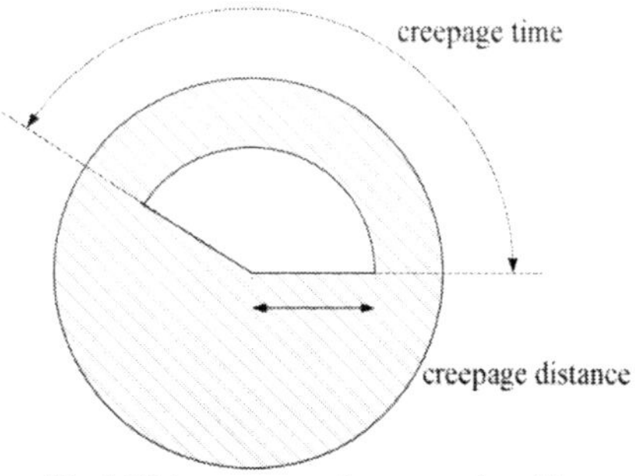

Fig. 2 White spot growth range and position

Where the creepage length calculation formula is R=L/D*80(mm), the most distant visible bubble cluster for the pick up point. Climbing angle just directly through measurement tools available for β. When there are multiple creepage points, white spots and a multiple white spots.

III. White spot analysis of the specific conditions

3.1 Change tendency of creepage area

Through real-time observation, the boards are divided into A, B, C three levels according to the degree of aging(A cardboard equal new transformer pressboard, B cardboard equal aged 10 years in transformer pressboard, B cardboard equal aged 20 years in transformer pressboard). According to the data of forensics can draw the trend chart cardboard creepage areas and surface creepage length trend chart.

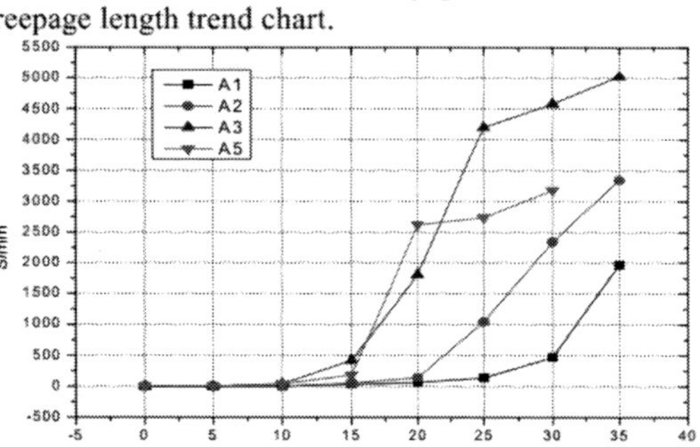

Fig 3. Group A creepage area S_A change tendency

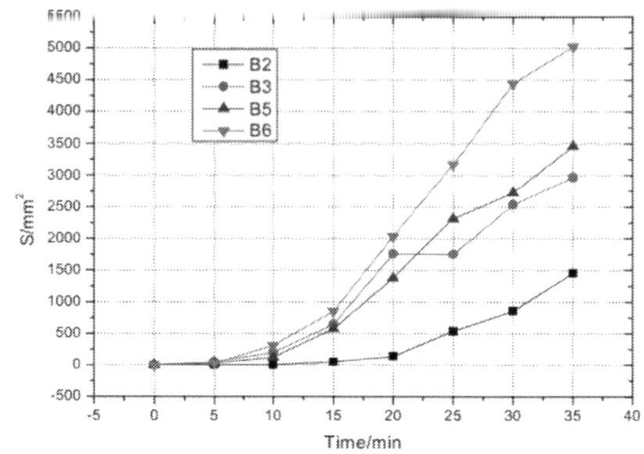

Fig 4. Group B creepage area S_B change tendency

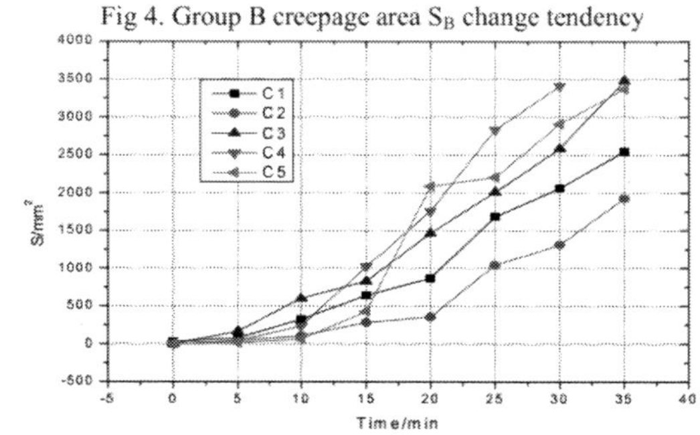

Fig 5. Group C creepage area S_C change tendency

We can find that creepage area in different aging board change trend varies greatly, cardboard in group A did not appear obvious 10 minutes before discharge, there is a sudden increase of process 15 minutes later, however, the old board in the discharge of 5 minutes will be in the cardboard surface showing clear bubble clusters. Aging board white spot is more smooth growth through the creepage area change tendency, what's more, there is a growing trend. Growth trends of the group C cardboard smoother than group B, the creepage area is larger than group B at the same time.

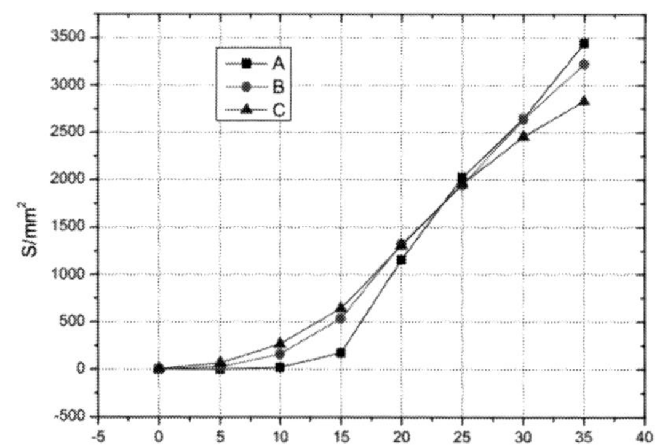

Fig 6. Groups A, B, C average cardboard creepage area S_{ave} change trend

The process continuous discharge for 20 minutes after beginning to discharge, the different aging conditions in paperboards showed different growth characteristics of white spots in white spots growth process. White spot of group C in turn grow faster than B, A cardboard. The trend can be seen from the the creepage area S_{ave} trends figure.

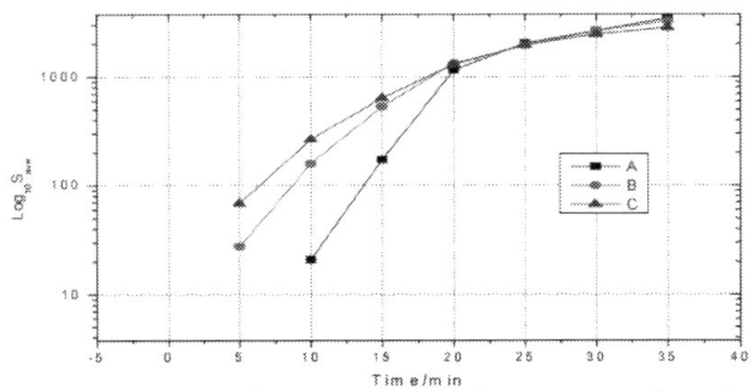

Fig 7. Groups A, B, C average cardboard creepage area $Log_{10}S_{ave}$ change trend

3.2 Trends of Creepage lengths

$Log_{10}S_{ave}$ trend figure is more intuitive description of the characteristics, but in late creepage, it continues 20 minutes until the breakdown of the process, the different aging conditions cardboard creepage area does not reflect the significant contrast changes. Cardboard along the creepage length R of the trend shown in Figure 8.

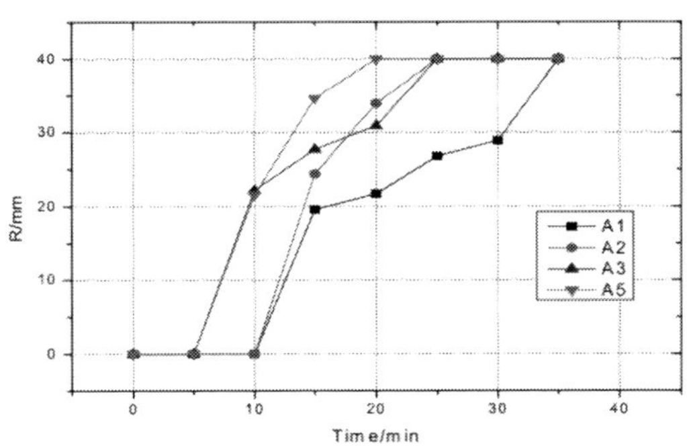

Fig 8. Group A creepage length R_A change trend

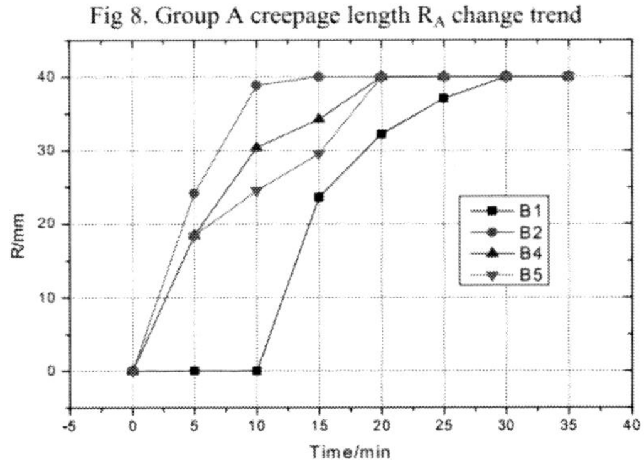

Fig 9. Group B creepage length R_B change trend

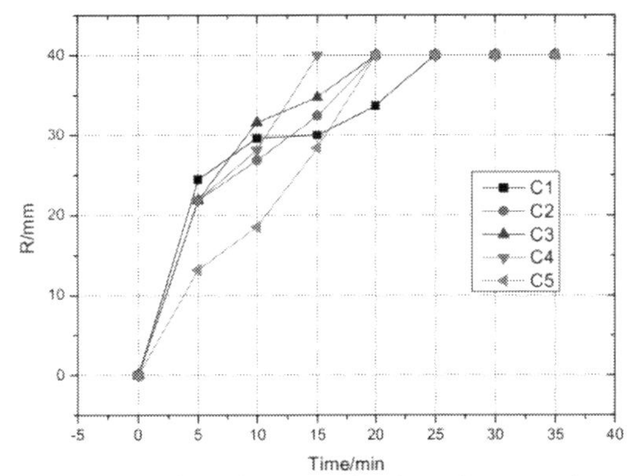

Fig 10. Group C creepage length R_C change trend

From the creepage length R in different aging board change trend, group A cardboard can be found in the discharge there was no obvious bubble cluster before 5 minute, creepage length is approximately equal to zero. It will gradually appear in cardboard surface of creepage and grow step by step after 5-10 minutes, however, aging board discharge appears in cardboard surface creepage sust as long as five minutes, we could calculate the creepage distance. From the point of creepage time, only A cardboard group stopped growing creepage distance after 20 minutes , but there are still a set of group B and C cardboard creepage distance is growing. Compared cardboard A with cardboard B and C, cardboard A creepage distance begin to grow later, creepage distance over the growth of the time is late, too. Compared with the aging cardboard, the non-aging one creepage needs more time.

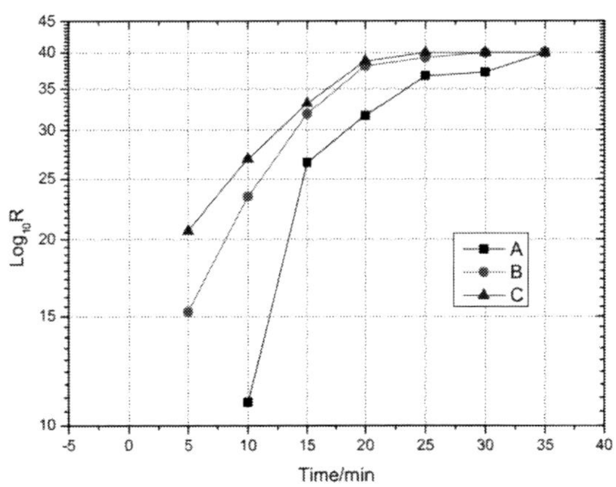

Fig 11. Trend of creepage length average R_{ave} of the group A, B, C Board $Log_{10}R_{ave}$

As we can see that A, B, C three sets of cardboard increase the creepage distance in turn before 30 minutes, and board B and C have longer creepage distance than board A . However, the difference is not very obvious although C cardboard longer than B cardboard between 15 minutes and 30 minutes.

IV. The law of development and comparative analysis

There is the detailed analysis and the final creepage length (breakdown) contrast by using comparison for three groups of cardboard surface creepage development law. The paper compared three groups of cardboard through specific example in 5 minutes, 25 minutes and 35 minutes of bubble cluster development respectively .

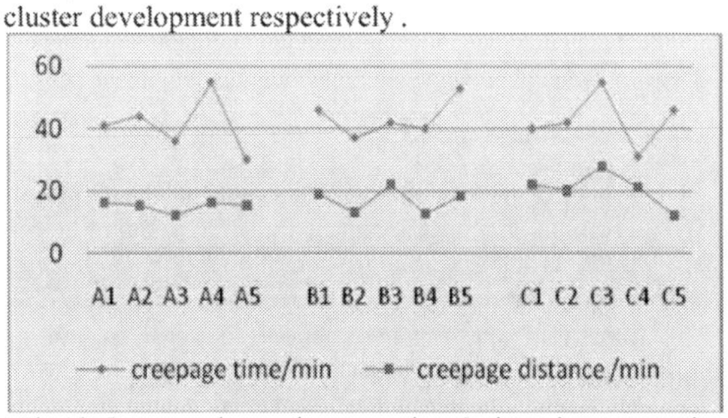

Fig 12. Creepage time and creepage length along the creepage in different cardboard

The intuitive characteristics is illustrated in figure 12 trend diagram. Three sets of cardboard creepage time averages about 43 minutes, but the final breakdown point of creepage length from A to C in turn increases at the same time.

V. The law of development and comparative analysis

Creeping discharge development process can be divided into early discharge in the preliminary formation of white spot, early electrical branch of creepage, cardboard internal branch to further expand and electrical breakdown of the branches along the longitudinal electric field from maintaining a constant voltage to test product breakdown.

In the process of preliminary white spot formation of early discharge, the oil within the board will gasify that is caused by high-energy flash discharge and then squeeze the cardboard. The obvious white spot and bubbles will appear in the cardboard surface near the high field strength, but the white spot area and the bubble cluster is relatively small at this point. The formation of white spot has a certain degree of damage on the board which encourages further discharget. The discharge will gradually develop from the surface of the slide flash into cardboard inside, it will form the initial electrical branch in strong magnetic field area of the cardboard lining, that is horizontal and vertical carbonization depth on the cardboard. The carbide prompted further extension of high field strength horizontally due to the formation of the channe, then lead to surface discharge further development. Electrical branch in the inner cardboard gradually extend, which Lead to discharge process of gas increase gradually. Bubbles on the surface of the cardboard cluster area expand step by step that discharge dense and intense. Longitudinal breakdown on board is in effect at the same time with the development of electrical branches of lateral extension to a certain degree. There will be breakdown at the forefront of electrical branches creepage points when longitudinal penetration reaches a certain critical worth time.

In the paper we can see that the aging of the cardboard with lower polymerization degree by comparring different aging time of creeping discharge process of cardboard, the gas overflow from inside the board easily when the discharge process produce a large number of bubbles.

REFERENCES

[1] A ATTEN P, SAKER A. Streamer propagation over a liquid/solid interface. IEEE Trans on Electrical In-sulation, 1993, 28(2),230-242.

[2] Barry H. Ward.A survey of new techniques in insulation monitoring of power transformers. IEEE Electrieal Insulation Magazine, 2001, 17(3): 16-23

[3] M. D. Judd, B. M. pryor, S. C. Kelly and B.F.Hampton. Transformer Monitoring Using The UHF Technique. IEE Conference on High Voltage Engineering. London. 1999:362-365

[4] R.Schifani, R. Candela. A New Algorithm for Mixed Weibull Analysis of partial Diseharge AmPlitude Distributions IEEE Trans On Dieleetrics and Eleetrical Insulation 1999 Vo1.6 No. 2 242-249

[5] A. S. Alghamdi and S. J. Dodd. The Influence of Absothed Moisture on PartialDischarge Pattems Measured During Tree Growth in an Epoxy Resin.2007Intemational Conference on Solid Dieleetries, Winchester, UK, Julys 8-13, 2007. P623-626

[6] J. V. Champion, S .J. Dodd, A. S. Vaughan, Y. zhao and S. .J.Sutton. The Effeet of Voltage, TemPerture And Morphology On Electrical Treeing In Polyethylene Blends .Dielectric Materials, Measurements and APPlications Conference Publieation No.473, IEEE 2000. P35-40

[7] DuY, Zahn M, Lesieutre B C,etal.Moisture equilibrium intransformer paper-oil systems[J]. Electrical Insulation Magazine,1999,15(1):11-20.

[8] Mitchinson P M, Lewin P L, Chen G, et al.. A new approach to the study of surface discharge on the oil-pressboard interface [C]//IEEE International Conference on Dielectric Liquids. Chasseneuil, France: IEEE. 2008.

[9] Dai J, Wang Z D, Jarman P. Creepage discharge on insulation barriers in aged power transformers[J]. IEEE Transactions Oil Dieletrics and Electrical Insulation, 2010, 17(4): 1327-1335.

[10]Pahlavanpour P, Eklund, Martins M A. Insulating paper ageing and furfural formation[C]. 2003 Electrical Insulation Conference and Electrical Manufacturing & Coil Winding Technology Conference Proceedings. Toronto, 2003: 283-28

[11]M.Caeeiari, A. Contin, G C. Montanari. Use of a Mixed-Weibull Distribution for the Identifieation of PD Phenomena IEEE Transaction on Dieleetrics and Eleetrieal Insulation. 1995 Vol. 2 No. 4614-627

[12]R.Sehifani, R. Candela. A New Algorithm for Mixed Weibull Analysis of partial Diseharge AmPlitude Distributions IEEE Trans On Dieleetrics and Eleetrical Insulation 1999Vol. 6 No. 2242-249

A Study on Structural Design and Test of Micro Scale Composite Material Blade for Wind Turbine

Hyunbum Park[1,*]

[1]Department of Defense Science & Technology – Aeronautics, Howon University, 54058, 64 Howondae 3gil, Impi, Gunsan, Rep. of Korea

Abstract. The purpose of the present study is to design a micro scale composite wind turbine blade. The blade airfoil of FFA-W3-211 was selected to meet Korean weather condition. The skin-spar-foam sandwich type structure was adopted for improving buckling and vibration damping characteristics. The design loads were determined at wind speed of 25m/s, and the structural analysis was performed to confirm safety and stability from strength, buckling and natural frequency using the finite element analysis method. The prototype was manufactured using the hand-lay up method and it was experimentally tested using the three point concentrated loading method. In order to evaluate the design results, it was compared with experimental results. According to comparison results, the estimated results such as compressible stress, max tip deflection, natural frequency and buckling load factor were well agreed with the experimental results.

1 Introduction

This study carried out the design and analysis on the aerodynamics and structures of a composite blade for a 500W-class micro scale horizontal-axis wind turbine with low wind speed starting characteristics and airfoils of excellent aerodynamic performance so that it could be applied to the geography of Korea.

It was designed to be advantageous to the area of low wind speed by determining low wind speed as rated wind speed as much as possible within a range that the blade's diameter does not excessively become larger. The FAA-W3-211 airfoil which characteristics such as Reynolds number, maximum lift coefficient, stall angle of attack, maximum lift-drag ratio and maximum thickness are relatively good was used, and according to the strip theory that uses the momentum theory and blade element theory together, the basic shape was designed to determine its airfoil's chord length and wing's torsion angle distribution. Even if there was a little inconvenience in manufacturing, it was selected a method to design as optimum angle of attack which could get higher efficiency [1].

The structural design was based on a sandwich basic structure of skin-spar-foam and made its thickness thinner towards the blade's tip to focus on weight lightening [2]. For the used materials, the glass/epoxy, which was produced by a Korean company and its material property was proven, was used. For structural analysis, the finite element method was used to analyze stress and strain, and a local buckling problem by the bending load was considered. In addition, the stability in operation was considered by checking its natural frequency.

2 Design and Analysis

The specification of the blade which aerodynamics was designed to be suitable for Korea's weather condition in this study is as Table 1, and its 3-dimensional shape is as Fig. 1.

The load mainly considered in terms of structural design is aerodynamic load, and the load condition was considered as Table 2.

Table 1. Aerodynamic design results of 500W class wind turbine blade.

Rated power	500W
Cut-in wind speed	3.0 m/s
Rated wind speed	10.0 m/s
Cut-out wind speed	25 m/s
Design tip speed ratio	6.6
Rated rotational speed	400 rpm
Number of blade	3
Rotor diameter	2.52 m
Aerodynamic profile	FFA-W3-211
Blade root chord	158 mm
Blade tip chord	44mm
Blade total twist angle	7.2°

* Corresponding author: swordship@daum.net

DOI: 10.1051/matecconf/201713900026

As a result of the load analysis, it was verified that the load case II caused the greatest load on the blade, and the structural design was carried out on the basis of this load [3].

Table 2. Load case for structural design.

Load Case	Case I	Case II	Case III
Reference wind speed	10.0 m/s	25.0 m/s	55.0 m/s
Gust condition (±20 m/s, ±40°)	Without gust	With gust	Storm condition
Rotational speed	400 rpm	450 rpm	Stop

3 Structural Design Concept

For the basic structure for structural design of composite blade, the precedent study results and literature were referred to select the sandwich structure as Fig. 1 [4]. This structure was laminated in the direction of ±45°, mainly in the longitudinal direction of the skin, which is responsible primarily for the shearing load, and the blade, composed of a spar and flange structure being responsible mainly for the bending load, and designed as a sandwich structure shape that internally applies polyurethane foam having an excellent vibration damping effect as well as significantly improving buckling strength. Table 3 shows the structural design results for blades.

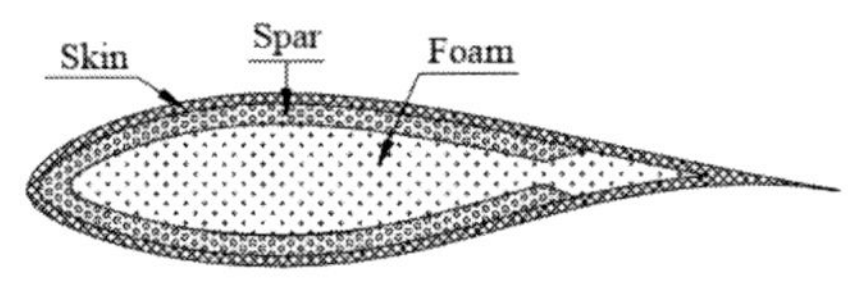

Figure 1. Skin-spar-form sandwich type structure.

Table 3. Structural design results of blade.

Station	Thickness (mm)	
	Upper surface	Lower surface
1	10t / Skin 1.25t / Spar 7.00t	
2	Skin 1.5t/Spar 6.25t	Skin 1.5t/Spar 6.25t
3	Skin 1.5t/Spar 6.25t	Skin 1.5t/Spar 6.25t
4	Skin 1.5t/Spar 6.25t	Skin 1.5t/Spar 6.25t
5	Skin 1.5t/Spar 5.00t	Skin 1.5t/Spar 5.00t
6	Skin 1.5t/Spar 5.00t	Skin 1.5t/Spar 5.00t
7	Skin 1.5t/Spar 5.00t	Skin 1.5t/Spar 5.00t
8	Skin 1.5t/Spar 2.75t	Skin 1.5t/Spar 2.75t
9	Skin 1.5t/Spar 2.75t	Skin 1.5t/Spar 2.75t
10	Skin 1.5t/Spar 2.75t	Skin 1.5t/Spar 2.75t

4 Structural Analysis

This study used the finite element analysis method to conduct the structural analysis [5, 6]. As a result of generating a mesh for analysing finite elements, it was composed of a total of 2,860 elements and 3,460 nodes. Table 4 is one that verified the result of the finite element linear static analysis according to the load case II and the structural stability for it, which was verified that it was structurally secure enough. Fig. 2 is the stress distribution for the design load (load case II) which is the maximum load.

Table 4. Structural analysis result.

Analysis result		Case II
Max. stress [Mpa]	Ten.	10.07
	Com.	10.06
Max. disp. [mm]		88.82
Max. stress failure criterion	Sxx/allow	0.054
	Syy/allow	0.013
	Sxy/allow	0.047

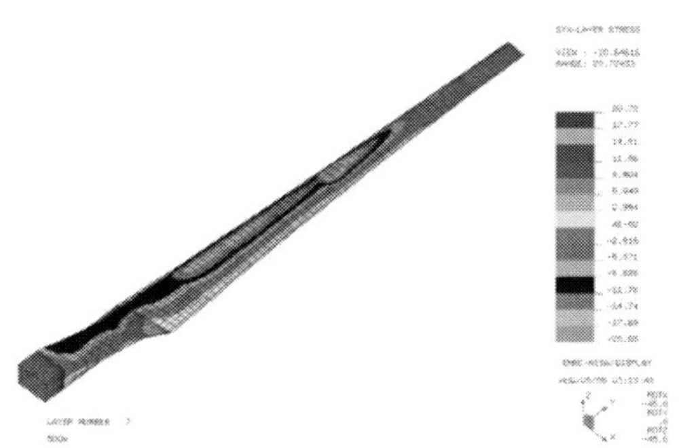

Figure 2. Stress analysis result of load case II.

For the natural frequency analysis, 1, 3 and 6 P.R.O are important because of the potential for resonance if it has three rotors [7]. Fig. 3 is one that draws the flap mode and shape. Examining the Campbell diagram of Fig. 4, it could be verified that the first flap mode's natural frequency did not cause resonance at 400 rpm which is the operational number of revolutions.

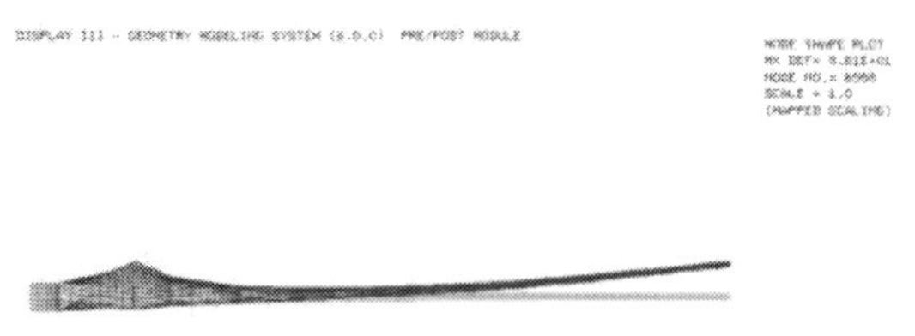

Figure 3. First flap mode shape and frequency.

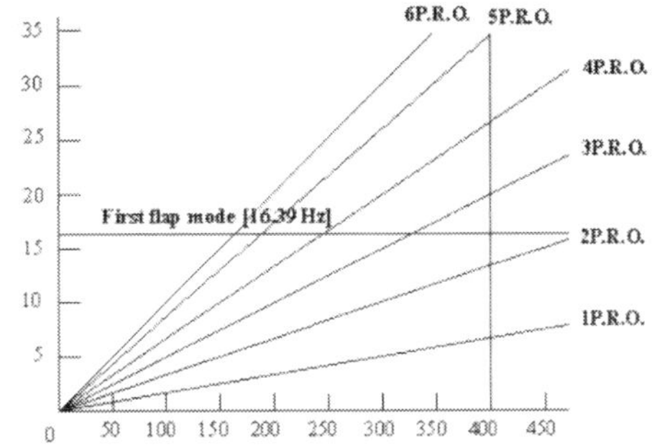

Figure 4. Campbell diagram.

Fig. 5 shows the result of buckling analysis for the load case II, which was verified that the minimum load factor was 15.36 and there was no buckling.

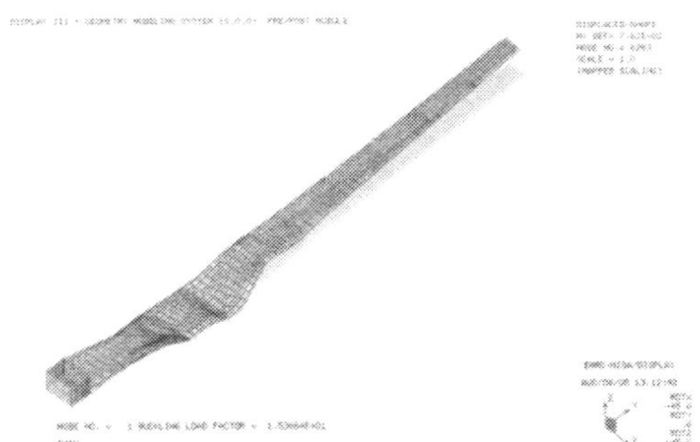

Figure 5. First buckling mode and load factor at load case II.

5 Prototype Manufacturing and Test

To make a prototype, a template for the exterior shape was manufactured, which was used to divide into the upper and lower sides to make a mold. After applying a mold release agent and spreading gel coat on the mold manufactured like this, glass fiber epoxy was laminated by the hand lay-up method. At that time, to keep the laminated fiber close to the mold, urethane foam was applied in advance to cut with heat rays, and then inserted to bond them. Fig. 6 is the manufactured blade prototype.

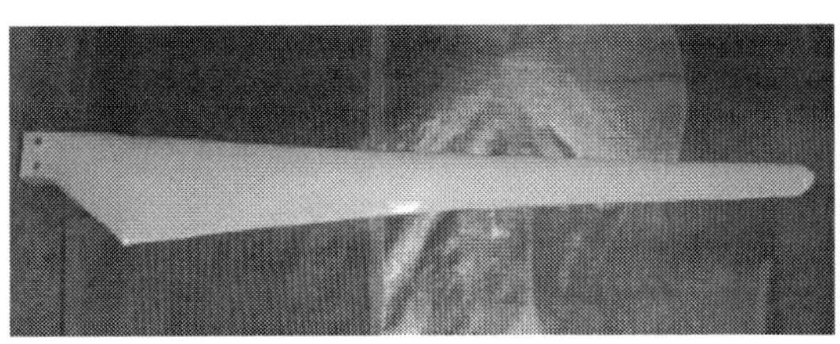

Figure 6. First blade prototype.

A structural test was carried out to evaluate the design result. First, to measure the natural frequency, a strain gauge was attached, and then an impact hammer was used to give impact. And the strain values varied at that time were collected through a data acquisition device (EDS400A, KYUWA). Fig. 7 is the collected data and the result of analysing them by FFT, the analysis and experiment results are 16.39Hz and 17.21Hz, respectively, and it could be known that they are almost the same.

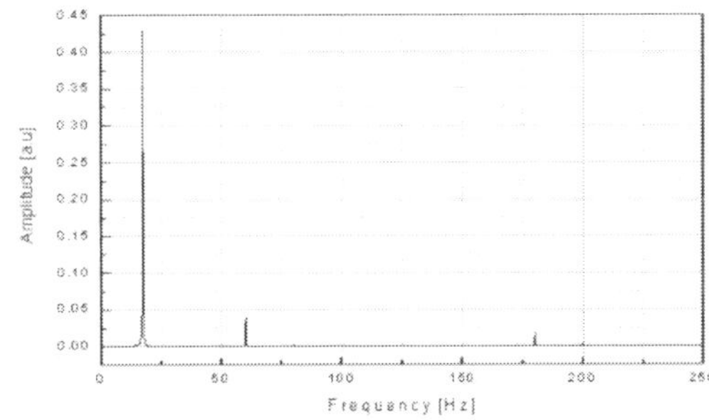

Figure 7. Eigenvalue test result.

For the static structure test, the design load was simulated as 4-point concentrated load as Fig. 8, and for the sandbag weights and load positions, 5, 11.5, 14 and 16 kg were used at 0.2, 0.5, 0.7 and 0.9 m, respectively. Table 5 is one that compares the structure test result with the finite element analysis result, which was verified that they are closely matched each other.

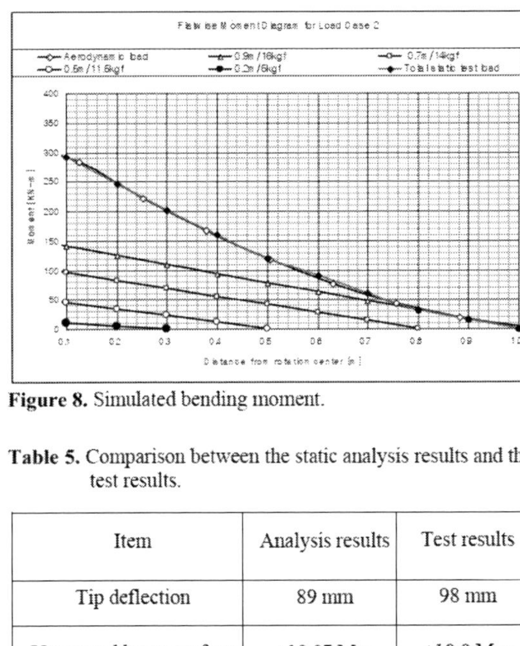

Figure 8. Simulated bending moment.

Table 5. Comparison between the static analysis results and the test results.

Item	Analysis results	Test results
Tip deflection	89 mm	98 mm
Upper and lower surface stresses at 0.2 r/R station	+10.07 Mpa -10.06 Mpa	+10.0 Mpa -10.4 Mpa

6 Conclusion

Through this study, the structure design and analysis of composite blades for a 500W-class micro scale wind power generation system applicable to the low wind

 DOI: 10.1051/matecconf/201713900026

speed geography like Korea was carried out, and the following conclusion was drawn.

As a result of the linear static analysis, it was verified that the blade, which structure design was finally confirmed, was a safe structure that secures 21.2 of safety coefficient even in the maximum load condition, and the blade's maximum displacement at that time was about 88.2 mm. As a result of drawing the Campbell diagram from the natural frequency analysis result to examine the potential for resonance, it was verified that there was no chance that resonance occurs during operation. As a result of the buckling analysis, the load factor for the primary buckling mode was 15.36, and it was verified that it was stable against the buckling. For the blade structure, it was selected the skin-spar foam sandwich structure type with excellent weight lightening, buckling and vibration damping characteristics. For the prototyping, it was selected the method that considers the ease of making to manufacture foam separately to bond. As a result of this study, it was examined that the blade design method performed in this study was reasonable.

Acknowledgements

This study was supported by research fund from Howon University.

References

1. C. Kong, Structural investigation of composite wind turbine blade considering various cases and fatigue life, Energy **30** (2005)
2. M. Niu, *Airframe Structural Design*, (Hong Kong Commilit Press Ltd, 1988)
3. Desire Le Gourieres, *Wind power plants theory and design*, (Pernamon Press, 1982)
4. V. Vasiliev, E. Morozov, *Mechanics and analysis of composite materials,* (Elsevier Science Ltd, 2001)
5. T. Y. Yang, *Finite element structural analysis*, (Prentice Hall, 1986)
6. Daryl L. Logan, *A first course in the finite element method*, (Brooks/Cole, 2002)
7. Leonard Meirovitch, *Analytical methods in vibrations*, (Macmillan Publishing Co., Inc., 1967)

Preliminary Coexistence Studies between IMT-2020 systems and inter-satellite service in 26 GHz

Haifeng Tan[1], Tan Wang[2], Jiajia Chen[2,a], Tianqi He[3], Zilong Deng[3]

[1] *Beijing University of Posts and Telecommunications, Beijing, China*
[2] *The State Radio Spectrum Management Center, Beijing, China*
[3] *North China Electric Power University, Beijing, China*

Abstract. This paper focuses on the sharing and compatibility studies of International Mobile Telecommunications 2020 (IMT-2020) and inter-satellite service (ISS) in 26 GHz frequency band, which was selected by International Telecommunications Union (ITU) as one of the most promising candidate frequency bands for 5G. Based on relevant ITU recommendations and reports, and current technical characteristics of both systems, a methodology is developed for the interference coexistence analysis between them. The spatial and temporal distribution analysis is proposed and the calculation of IMT-2020 aggregate interference takes the actual urban population into consideration. The results show that these two systems have considerable possibilities in coexistence in 26 GHz frequency band.

1 Introduction

Spectrum is one of the most basis factors of the development of IMT industry. As the demand for broadband wireless communication is growing at a high rate of speed, the deficits of spectrum requirement continue to increase. For the present, there is almost no more abundant frequency band can be used below 6 GHz, and then, finding several higher frequency bands for IMT-2020 is extremely necessary [1]. In this regard, International Telecommunications Union (ITU) has pointed out 11 candidate frequency bands (ranging 24.25-86 GHz) above 6 GHz during 2015 World Radiocommunication Conference (WRC-15) and asked for some feasibility verification on a worldwide scale [2]. ITU-R also set up Task Group 5/1 particularly and established Agenda Item (AI) 1.13 in WRC-19 to explore whether it is possible to share frequency bands between IMT and incumbent services.

24.25-27.5 GHz band (referred to as the 26 GHz band) is the lowest section in these candidate frequency bands of WRC-19 AI 1.13 and this band is already allocated to mobile service as a primary service in ITU Radio Regulations [3]. Thanks to its relatively low frequency and consecutively large bandwidth, 26 GHz band is attached great importance by the world and regional organizations, considering it to be utilized for the 5th generation mobile communication as soon as possible. However, there have been several incumbent services already. Before global commercial deployment, it is obligatory to ensure IMT system do not interfere with other incumbents.

Among the incumbents, inter-satellite service (ISS) operating in 25.25-27.5GHz gets much more concerns because of its complicacy and typicality [4]. It is necessary to conduct the sharing and compatibility studies to verify its coexistence possibility with IMT. This paper focuses on this issue. Based on ITU recommendations and reports, and current technical parameters of both systems, a methodology is developed for the interference coexistence analysis. The spatial and temporal distribution analysis is proposed and the calculation of IMT-2020 aggregate interference takes the actual urban population into consideration. The results show that these two systems have considerable possibilities in coexistence in 26 GHz frequency band. In the following, Section 2 briefly introduce the research background of ISS in 26 GHz band; Section 3 shows the technical characteristics of IMT and ISS in 26 GHz band; in Section 4, the methodology regarding IMT large-scale deployment is proposed; Section 5 gives the simulation results and summary is made in Section 6.

[a]Corresponding author: chenjiajia@srrc.org.cn

2 Allocation Information

According to the Radio Regulations of ITU, ISS is allocated in the frequency band 25.25-27.5 GHz as primary service listed in Table 1 as below.

Table 1. Frequency allcation in the 25.25-27.5 GHz frequency range

International Table		
Region 1	**Region 2**	**Region 3**
25.25-25.5 FIXED INTER-SATELLITE MOBILE Standard frequency and time signal-satellite (Earth-to-space)		
25.5-27 EARTH EXPLORATION-SATELLITE (space-to-Earth) FIXED INTER-SATELLITE MOBILE SPACE RESEARCH (space-to-Earth) Standard frequency and time signal-satellite (Earth-to-space)		

International Table		
Region 1	**Region 2**	**Region 3**
27-27.5 FIXED INTER-SATELLITE MOBILE	27-27.5 FIXED FIXED-SATELLITE (Earth-to-space) INTER-SATELLITE MOBILE	

ITU Radio Regulations advise that use of the 25.25-27.5 GHz band by the inter-satellite service is currently in accordance with space research and Earth exploration-satellite applications, and transmissions of data originating from industrial and medical activities in space.

According to [5], the inter-satellite service in 25.25-27.5 GHz works in uplink transmission, i.e., a satellite from geostationary satellite orbit (GSO) receives signal from a satellite or spacecraft from non-geostationary satellite orbit (non-GSO/NGSO). GSO is relatively static towards the Earth while its receiving antenna would keep tracking the location of non-GSO, which circles the Earth periodically. Because of the large height gap (GSO with 36,000 km above the ground and non-GSO with about 400-1000 km above the ground); the height of non-GSO is almost similar to that on the ground. And the interference occurs only when the GSO, which is tracking the non-GSO, is pointing also to the ground where there are considerable IMT network deployments (See Figure 1). To model the interfering scenario, it is critical to know how IMT system deploys and the GSO-NGSO tracking process.

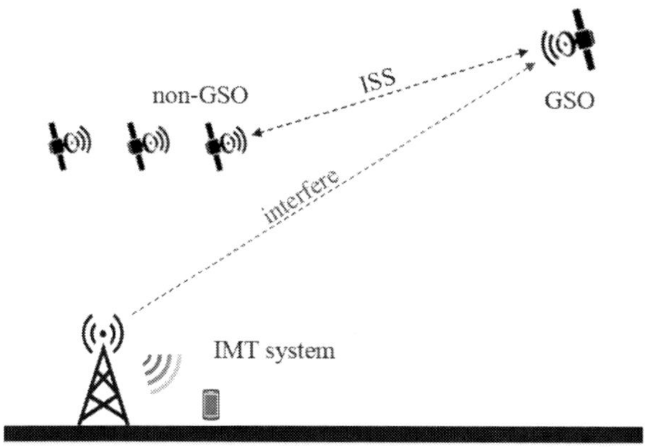

Figure 1. Typical scenario of interfered ISS

3 Technical Characteristics

This section provides the specific parameters used in the study. In particular, this paper mainly focuses on the interference from IMT to ISS in the same frequency band, which is the most important issue.

3.1 IMT system characteristics

Based upon many research cases [6], the interference from downlink is much more than that from uplink. Thus it is critical to study the interfering behavior of base station (BS) in IMT system. According to the study group of IMT in ITU [7][8], characteristics of BS for frequency sharing/interference analysis in the frequency range 24.25-33.4 GHz are listed in Table 2as below.

Table 2. Base station characteristics

Frequency	26 GHz
Duplex mode	TDD
Network topology density	30 BSs/km2
Network loading factor	20%
Antenna height	6 m (above ground level)
Sectorization	Single sector
Downtilt	10 degrees
Antenna pattern	Refer to Recommendation ITU-R M.2101
Element gain	5 dBi
Horizontal/vertical 3 dB beamwidth of single element	65° for both H/V
Horizontal/vertical front-to-back ratio	30 dB for both H/V
Antenna polarization	Linear ±45°
Antenna array configuration (Row × Column)	8x8 elements
Horizontal/vertical radiating element spacing	0.5 of wavelength for both H/V
Array Ohmic loss	3 dB
Conducted power (before Ohmic loss) per antenna element	10 dBm/200MHz
Base station maximum coverage angle in the horizontal plane	120 degrees

One obvious difference between 5G and 4G is that the massive MIMO is introduced in the sharing and compatibility studies. And in each time snapshot, the antenna main beam might point to different users, resulting in different interference level. When the electrical down-tilt steering and the electrical horizontal steering are both zero, a composite array pattern for 8x8 elements is shown in Figure 2. It is equivalent to that the served user equipment (UE) is located at a point in the normal direction of this antenna panel. The explicit formulas are referred in REC. M.2101 and the derivation procedures are shown in 3GPP TR 37.840 [9].

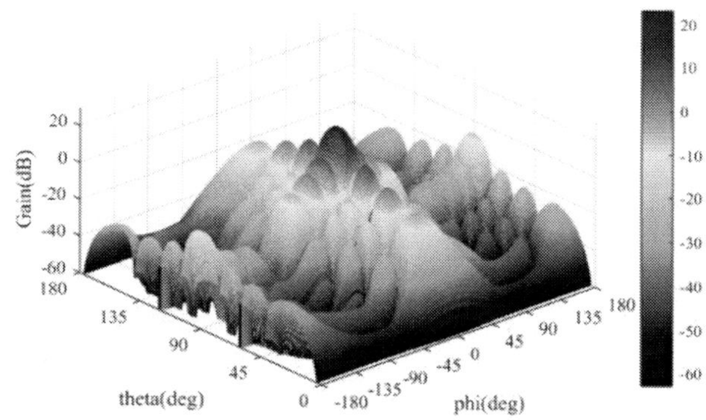

Figure 2.BS antenna gain pattern

20% network loading factor would normally represent a typical/average value for the loading of base stations across a large-scale network and therefore can be used for wide area analysis (province, national or larger satellite footprint, for example). It is necessary to introduce some modified parameters of the BS density to analyze large-scale deployment.

ITU has invented two parameters (Ra and Rb as shown in Table 3) to build a relationship between amounts of BS and land area where the IMT system would be. For instance, the land area of China is about 9.6 million km2 and it would be supposed that there are 1,008,000 (9,600,000km2*7%*5%*30BSs/km2) IMT-2020 base stations nationwide. Although this parameter represents an average value, not distinguished regarding the IMT demands from different countries, they are discussed and agreed by experts from the world in ITU.

Table 3. Large-scale deployment characteristics

Ra	7% for Urban
Rb	5%
Land area of China	9,600,000 km2

3.2 ISS system features

According to the ISS study group in ITU-R [10] and ITU-R REC. SA.1414 as Reference [5], the characteristics of ISS in the frequency range 25.25-27.5 GHz is shown in Table 4. This paper takes China's parameter as the case study, where the GSO is data-relay satellite (DRS) and non-GSO is referred to as spacecraft.

Table 4. Characteristics of existing DRS system

Transmitting spacecraft	
Network	China
Orbital locations	300-500 m (400 m for simulation)
Frequency range	25.25-27.50 GHz
Transmission rate	≤ 600 Mbit/s
Modulation	PSK
Polarization	Circular
Antenna size	≤ 0.8 m
Tx antenna gain	≤ 44.5 dBi
Tx antenna pattern	Rec. ITU-R S.672
Necessary bandwidth	≤ 600 MHz
Maximum e.i.r.p spectral density	-5.5 dBw/Hz
Receiving DRS	
Network	China
Orbital locations	Rec. ITU-R SA.1275 or Rec. ITU-R SA.1276
Antenna size	4.2 m
Rx antenna gain	57.5 dBi
Rx antenna pattern	Rec. ITU-R S.672
System noise temperature	1000 K
Interference criterion	I/N = -10 dB [11]

According to the satellite antenna radiation pattern from ITU-R REC. S.672 [12], Figure.3 represents the simulation result of receiving DRS/GSO antenna gain.

The propagation models for sharing and compatibility studies take the free space loss, atmospheric loss and clutter loss into account, based on the study group in ITU-R [13] and ITU-R REC. P.619 [14]. According to the elevation of each base station, the clutter loss is averaged under probabilities of different location.

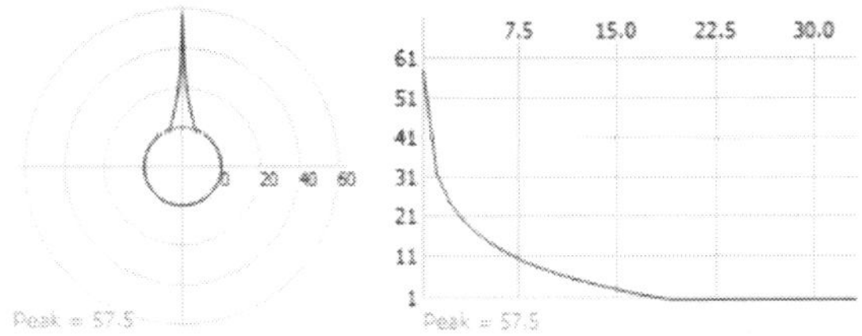

Figure 3. Receiving DRS antenna gain pattern

4 Analysis Process

To model the IMT aggregate interference towards the GSO receiver, it is difficult to simulating the BSs one by one, especially for a national scale. The approach embodied in this paper is developed referring ITU-R REC. F.1509 [15], where it is preferred to use a center base station (CBS) to replace a great area of IMT BSs. Each CBS's transmitting power is the sum of BS's power of all that coverage. In this paper, 36 cities are chosen as CBSs in mainland of China based on their special position (See Figure 4). Among them, 27 cities are provincial capitals, four are municipalities and five are specifically-designed cities in the state plan, as listed in Table 5.

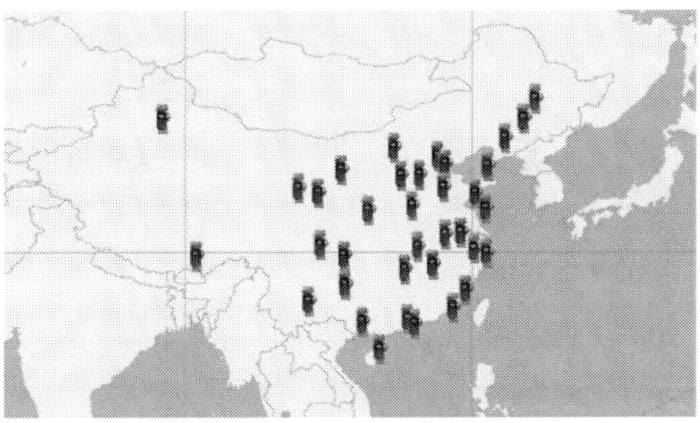

Figure 4. The locations of 36 center base stations

Table 5. Center base station information

City	Population (million)	Weight	Packing BSs numbers
Beijing	19.61	6.52%	65722
Shanghai	23.02	7.65%	77138
Guangzhou	12.70	4.22%	42561
Shenzhen	10.36	3.44%	34710
Chongqing	28.85	9.59%	96664
Nanjing	8.00	2.66%	26824
Xi'an	8.47	2.82%	28376
Wuhan	9.79	3.25%	32791
Zhengzhou	8.63	2.87%	28908
Tsingtao	8.72	2.90%	29205
Tianjin	12.94	4.30%	43356

City	Population (million)	Weight	Packing BSs numbers
Shenyang	8.11	2.69%	27164
Chengdu	14.05	4.67%	47074
Hangzhou	8.70	2.89%	29155
Xiamen	3.86	1.28%	12935
Ningbo	5.33	1.77%	17861
Dalian	6.69	2.22%	22418
Shijiazhuang	10.70	3.56%	35861
Harbin	10.64	3.54%	35642
Fuzhou	7.06	2.35%	23658
Jinan	7.06	2.35%	23658
Kunming	6.68	2.22%	22375
Lanzhou	4.02	1.33%	13456
Taiyuan	4.32	1.44%	14472
Changchun	7.73	2.57%	25900
Hefei	7.79	2.59%	26105
Nanchang	5.30	1.76%	17760
Changsha	7.31	2.43%	24501
Haikou	2.25	0.75%	7526
Guiyang	4.62	1.54%	15488
Xining	1.25	0.42%	4190
Hohhot	3.00	1.00%	10056
Nanning	6.99	2.32%	23411
Yinchuan	2.16	0.72%	7252
Lhasa	0.58	0.19%	1931
Urumqi	3.55	1.18%	11896
Total	300.80	100.00%	1008000

The packing BSs numbers and their weight are determined by population while the total number of BSs in mainland China is about 1,008,000 that have been mentioned in Section III. For example, the population of Beijing occupies 6.52% of the total population and its packing BSs number is 65,722 with the same percent of 1,008,000. The packing number would be turned into decibel and added to the transmitting power of a BS,hence creating one CBS's transmitting power.

Combined with the CBS deployment, simulation analysis can be divided into two different but complementary methods:

- Spatial distribution, where the aggregate interference to a DRS at a specified orbital location is computed as the high-gain receiving antenna of the DRS/GSO is scanning its whole visible area.
- Temporal distribution, where it is supposed to compute the temporal characteristics of the interference from IMT to DRS/GSO, when GSO is tracking non-GSO.

5 Simulation Result

The aggregate impact of BS is analyzed in the following steps. In both cases, the simulation takes into account:

1) Step 1: Choose a DRS/GSO at E59/113/169 degrees as the victim satellite in interfered ISS system.

2) Step 2: Distribute uniformly the horizontal direction of each CBS's array antenna panel.

3) Step 3: For each CBS, uniformly generate much enough users within 40 meters. And calculate the relative antenna beamforming gain towards the victim DRS/GSO satellite when its mean beam is pointing to one random selected user.

4) Step 4: Calculate the link budget from each CBS's transmitter to the victim DRS satellite receiver.

5) Step 5: Calculate the aggregate interference and use the ratio of interference and noise (I/N) in decibel as the interference result.

5.1 Spatial distribution analysis

It is assumed a low-orbiting satellite is moving over the land area of China while its DRS keeps tracing it and scanning the area of IMT system in a grid-based manner. The step size of the grid is 50 km throughout the scanned area, and every spot in this grid will be snapshotted for 500 times. Figure 5 to Figure 7 are the average interference-noise ratio results for the three victim DRS.

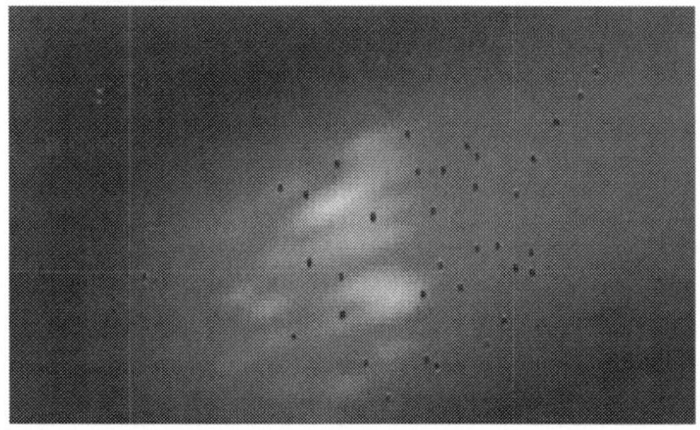

10dB 0dB -10dB -20dB -30dB -40dB

Figure 5. The spatial simulation result of the victim DRS at E59 degrees

10dB 0dB -10dB -20dB -30dB -40dB

Figure 6. The spatial simulation result of the victim DRS at E113 degrees

DOI: 10.1051/matecconf/201713900027

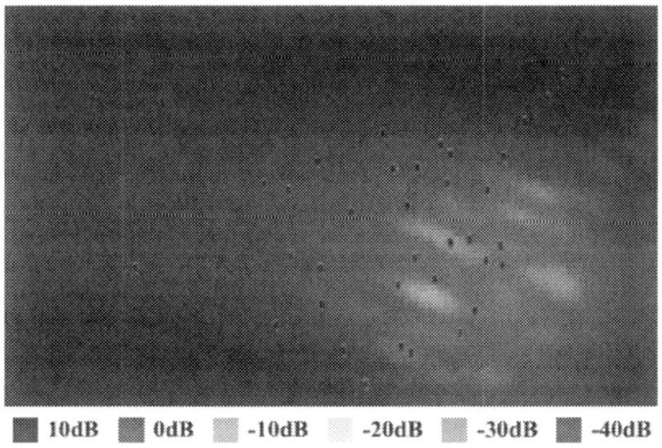

Figure 7. The spatial simulation result of the victim DRS at E169 degrees

The protection criterion of ISS is listed in Table 4, the spatial distribution results from these three interference scenarios are all under I/N = -10 dB. A few areas have peak value at about -20dB and the average of interference margin seems to be more than 10dB for these scenarios.

5.2 Temporal distribution analysis

In this case, a dynamic simulation is used. A DRS satellite with the receiving system is assumed to be located at a prescribed geostationary orbital location and to be tracking a moving low-orbiting satellite inclined by 57 degrees with respect to the equatorial plane (randomly selected). The simulation is time step based, with step size 1 second. At each step, the aggregate interference to the DRS receiving system from the 36 CBSs is calculated. The total simulation time is 365 days.

1) DRS & low-orbiting satellite: The DRS at E59/113/169 degrees is tracking a representative low-orbiting satellite respectively. The results inclueded the relationship of I/N and simulated time as well as its CDF scheme are shown in Figure 8 to Figure 10.

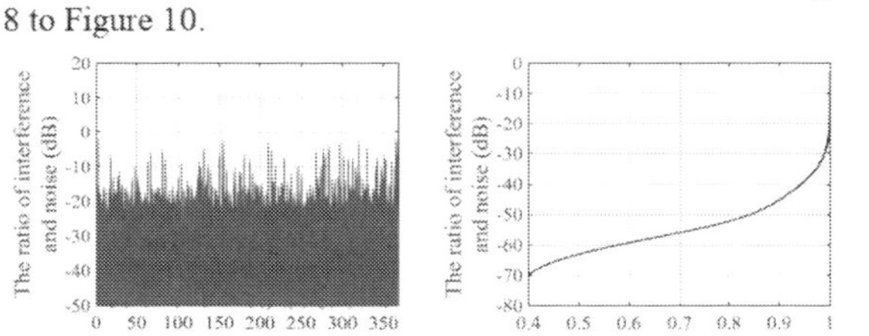

Figure 8. The temporal results of the DRS at E59 degree & a low-orbiting representative from ITU-R SA.1414.

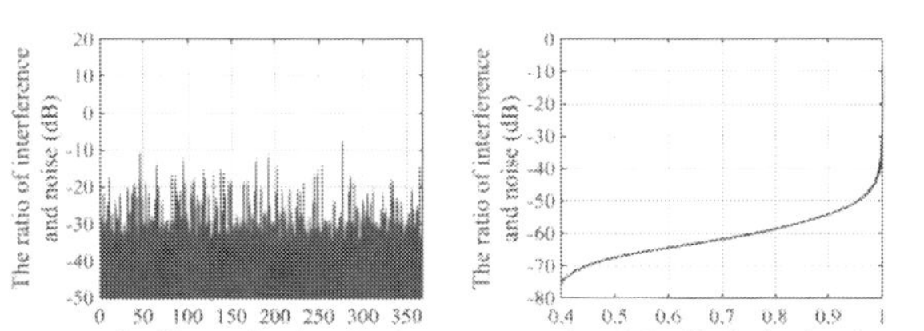

Figure 9. The temporal results of the DRS at E113 degree & a low-orbiting representative from ITU-R SA.1414.

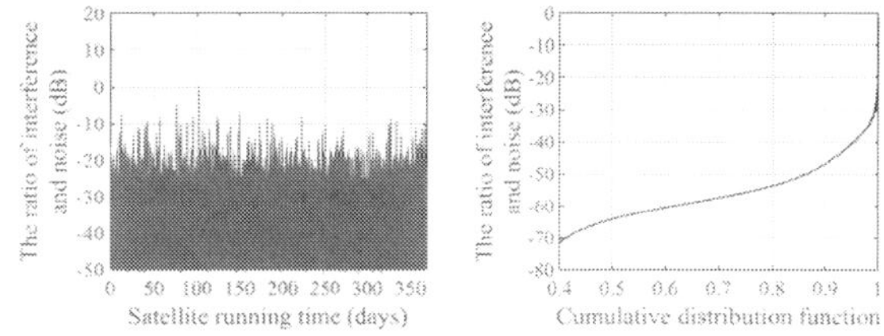

Figure 10. The temporal results of the DRS at E169 degree & a low-orbiting representative from ITU-R SA.1414.

2) DRS & low-orbiting satellite constellation: The DRS at E59/113/169 degrees is tracking a practical constellation, which constituted by 10 satellite units (See Figure11). The tracking strategy is to establish transmitting link with the nearest satellite unit in the in visible area range, as a scheduling method.

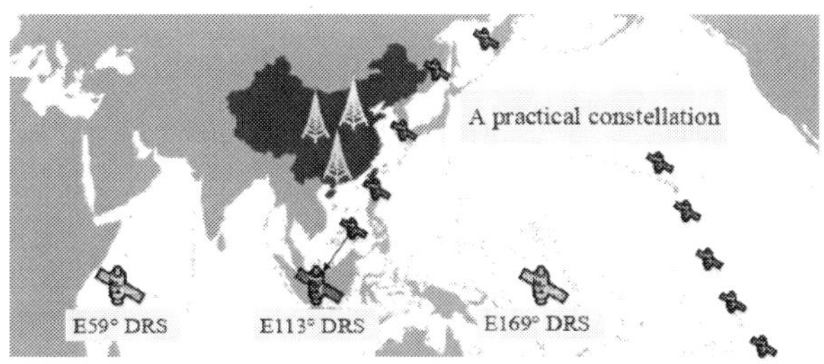

Figure 11. The sketch map of the DRS at E113 degrees tracking a ten-unit constellation

Figure 12 to Figure 14 present the simulation results of DRS at different orbit respectively.

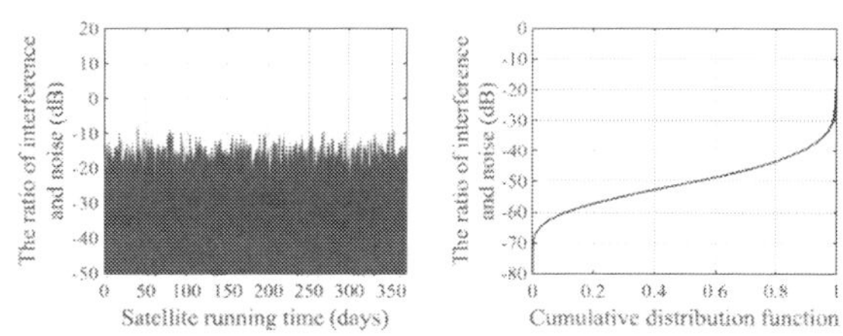

Figure 12. The temporal results of the DRS at E59 degree & practical constellation

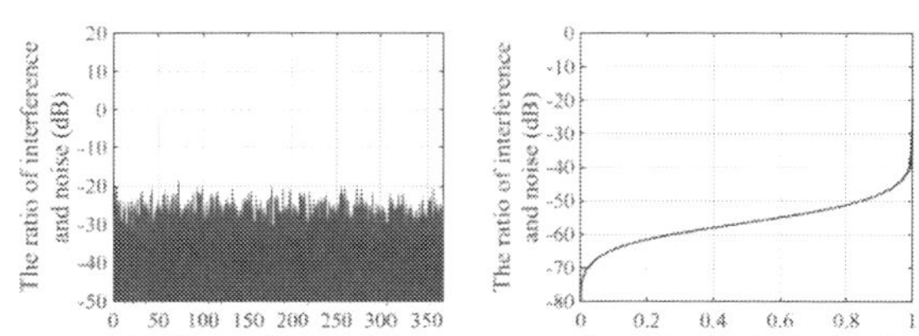

Figure 13. The temporal results of the DRS at E113 degree & practical constellation

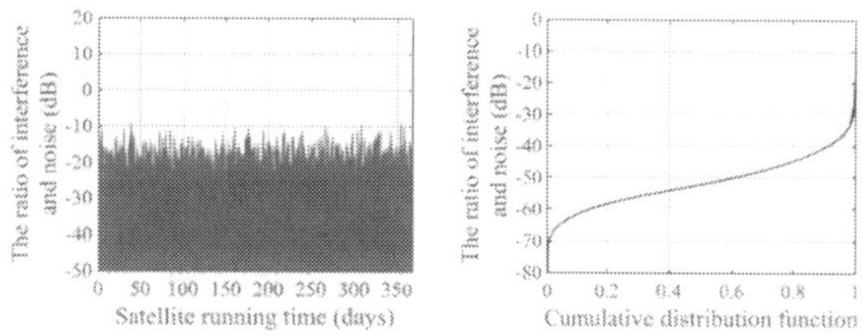

Figure 14. The temporal results of the DRS at E169 degree & practical constellation

To measure whether it exceeds the protection level or interference criterion of ISS, the percent of interfered time is computed in Table 6. Except for the criterion of interference value, there is also a ceiling margin (0.1%) of the percent of time with I/N > -10dB. As shown below, all results of the interference from IMT to the receiving DRS are far less than this percent margin.

Table 6. Interference result

Scenario	The percent of time with I/N>-10dB in satellite running time
GSO 59°, Custom non-GSO	0.00679%
GSO 113°, Custom non-GSO	0.00019%
GSO 169°, Custom non-GSO	0.00245%
GSO 59°, constellation	0.00057%
GSO 113°, constellation	0
GSO 169°, constellation	0.00038%

6 Conclusion

This paper introduces a methodology which aggregates large-area IMT base stations into a certain amount of center base stations properly, and takes China as an example to model IMT system interfering the DRS of inter-satellite service. For spatial distribution, the aggregate interference from downlink IMT system does not exceed the interference criterion of DRS receiver in ISS system. For temporal distribution, it is also under control in long-term operation. Based on these preliminary results, it is considered IMT will not cause harmful interference to ISS. Further studies may concentrate on the uplink aggregate interference from UEs, more complicated propagation model and larger area range with more CBSs.

Acknowledgment

This work was supported by the National Science and Technology Major Project under Grants No. 2015ZX03002008.

References

[1] Tan Wang, Gen Li, Jiaxin Ding, Qingyu Miao, Jingchun Li, and Ying Wang, "5G Spectrum: Is China Ready?", IEEE Communications Magazine, pp. 58-65, 2015.

[2] ITU-R. "Final Acts WRC-15", World Radiocommunication Conference, 2015, pp. 294-296.

[3] ITU-R. Radio Regulations of the International Telecommunication Union, 2015 ed., vol.1. Geneva: ITU, 2015, pp. 158-162.

[4] Qian, Zhaojun, Wang tan, Kang Long, and Li Bo. "Compatibility study between IMT and inter-satellite service systems in 25.25~27.5 GHz frequency band", Telecommunications Science, vol. 12, pp. 86-92, 2016.

[5] ITU-R. "Characteristics of data relay satellite systems", Recommendation ITU-R SA.1414-1, 2013.

[6] ITU-R. "Report on the sixth and final meeting of joint Task Group 4-5-6-7", Chairman Report, Document 4-5-6-7/715-E, 2014.

[7] ITU-R, "Pectrum needs and characteristics for the terrestrial component of IMT in the frequency range between 24.25 GHz and 86 GHz", Liaison statement from WP 5D to TG 5/1, 2017.

[8] ITU-R, "Modelling and simulation of IMT networks and systems for use in sharing and compatibility studies", Recommendation ITU-R M.2101-0, 2017.

[9] 3GPP, "Study of Radio Frequency (RF) and Electromagnetic Compatibility (EMC) requirements for Active Antenna Array System (AAS) base station", 3GPP TR 37.840 V12.1.0, 2013.

[10] ITU-R. "Reply liaison statement to Task Group 5/1 on parameters for WRC-19 agenda item 1.13", Liaison statement from WP 7B to TG 5/1, Document 5-1/28, 2016.

[11] ITU-R. "Protection criteria related to the operation of data relay satellite systems", Recommendation ITU-R SA.1155-1, 2013.

[12] ITU-R. "Satellite antenna radiation pattern for use as a design objective in the fixed-satellite service employing geostationary satellites", Recommendation ITU-R SA.672-4, 2013

[13] ITU-R. "Guidance on the use of ITU-R P-series recommendations for interference prediction and sharing studies", Liaison statement from WP 3J/3K/3M to TG 5/1, Document 5-1/38, 2017.

[14] ITU-R. "Propagation data required for the evaluation of interference between stations in space and those on the surface of the Earth", Recommendation ITU-R P.619-1, 1992.

[15] ITU-R. "Technical and operational requirements that facilitate sharing between point-to-multipoint systems in the fixed service and the inter-satellite service in the band 25.25-27.5 GHz", Recommendation ITU-R F.1509-3, 2015.

Sentence Similarity Research Based on Chinese FrameNet and Semantic Dependency Parsing

Cheng Li, *Yong Wu*, and *Binjun Wang*
College of information technology and cyber security,People's Public Security University of China,Beijing,China.

Abstract: According to the problems existing in the similarity comparison of Chinese sentences,this paper proposed a sentence similarity computing method which combined with advantages of Chinese frameNet method and semantic dependency parsing method.This method is based on the framework of semantic.And firstly,the method analyzed and calculated the similarity of two frameworks;Further,it analyzed semantic dependency relationship existing in the core frame elements from the two aspects of the meaning and the overall dependency relation;Finally,it put forward the fusion sentence similarity calculation formula.Experimental results show that compared with the method based on space vector model and based on HowNet and based on Frame Semantic Parsing, this method has higher accuracy in the similarity judgment of Chinese sentences.

Keywords: Chinese frameNet;semantic dependency parsing ;sentence similarity

1 Introduction

Sentence similarity research has shown great potential in the fields of machine translation[1],depth information retrieval[2],automatic question answering[3] and so on.

This paper will start from the Chinese FrameNet and identify target words of specific sentence firstly;next, calculate the similarity value of the two sentence framework name;and then calculate the semantic core frame elements in the dependency relation similarity;finally,measure the similarity of two sentences.

2 Related Work

At present,the research of Chinese sentence similarity is mainly divided into two aspects:One is the vector space model based on the word,and the other is the semantic sentence similarity method based on the word and syntax structure.Obviously,it is ideal to study text similarity based on syntax and semantics,But the research purely based semantic is difficult.Therefore,some experts have proposed semantic similarity research methods based on grammatical structure.At present,the more mature methods are the Chinese FrameNet method and the Semantic Dependency Parsing method.

Chinese FrameNet(CFN) is developed by Shanxi University,with Fillmore's frame semantics theory[4]and California University Berkeley FrameNet[5] as a reference method to construct a frame of semantic based on Chinese Corpus,and it includes frames,word elements,frame relationships,examples and chapters[6].

"The beggar pursues a dog.",as shown in figure 1.It is the result of using CFN semantic role automatic labeling tool.

Frame	Target word	Beggar	chased	a	dog.
following	chase	Thm	Tgt		Thm_c

Figure 1. Framework annotation example

The target word(Tgt) in the example is "chase" and the frame name is "following"."Beggar" is the Theme of this frame and it is usually a living entity.And the Cotheme represents another moving object.

Semantic Dependency Parsing(SDP)[7] is the analysis of semantic associations between linguistic components within a sentence and semantic associations are represented by dependency structures.Figure 2 shows the semantic dependency parsing of the example "he hears the explosion".

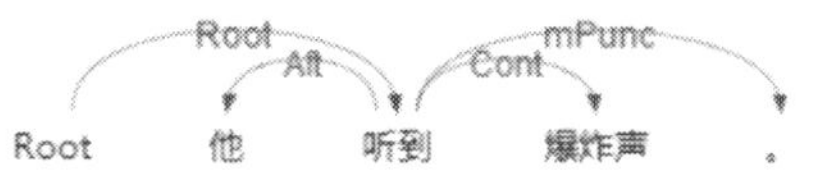

Figure 2. Semantic dependency analysis

The main dependency relation exist in above example is Affection(Aft) and Content(Cont),that is "hears→he" and "hears→explosion".

3 Sentence similarity computation based on Chinese FrameNet and Semantic Dependency Parsing

Through the above analysis,the Chinese FrameNet method is difficult to compare the similarities between different elements.For example,the example sentence shown in Figure 1 and "a dog chases a beggar.",their frames are "following",but the meaning of the sentences is completely different;The SDP method lacks a macro understanding of the whole sentence.For example,the example sentence shown in Figure 2 and "I smell dynamite.".the main semantic dependencies in the examples are Aft and Cont,but their frames are different,and the semantics of the two sentences are completely different.

Therefore,we propose a method of sentence similarity based on Chinese FrameNet and Semantic Dependency Parsing.

3.1 Similarity between frames

The CFN method first determines the target word of the sentence,and then determines the frame to which it is inspired based on the target word.In practical applications,we find that,the Chinese FrameNet and semantic computing research team of Shanxi University has many similar frames in the library of 323 frames accumulated over the years.If the target words of the two sentence can arouse similar frames,the two sentences may also have very high similarity.

Li Feng[8] proposed the semantic similarity algorithm of words,such as formula (1).

$$SimF(W_1, W_2)=\frac{a*\min(depth_{W_1},depth_{W_2})}{a*\min(depth_{W_1},depth_{W_2})+distance(W_1, W_2)} \quad (1)$$

The similarity between frames can be calculated in accordance with formula (1).Because CFN constructs the frame semantic resources and carries on the frame classification, and divides the same type of word element into a frame.So,in addition to some higher similarity frames,the similarity between most of the frame names are less than 0.4,and the similarity between more than half of frames is less than 0.1.

3.2 Similarity of sentence core frame elements based on Semantic dependency

If the two sentences have the same frames or a high degree of similarity is calculated in accordance with the 2.1,the semantic dependency relationships in the core frame elements within the sentence need to be further determined.

The similarity judgment of intra sentential semantic dependency is divided into two parts:single semantic dependency computation based on the meaning of words and the whole semantic dependency relation computation based on VSM.

3.2.1 Single semantic dependency computation based on the meaning of words

Semantic dependency parsing can be analyzed and determined by language technology platform of Harbin Institute of Technology[9].The results of the analysis are shown in figure 3,which are the example sentence shown in figure 1 and "a dog chases a beggar.".


```
<sent id="0" cont="乞丐追逐着一只狗。">
    <word id="0" cont="乞丐" pos="n" semparent="1" semrelate="Agt" />
    <word id="1" cont="追逐" pos="v" semparent="-1" semrelate="Root" />
    <word id="2" cont="着" pos="u" semparent="1" semrelate="mTime" />
    <word id="3" cont="一" pos="m" semparent="4" semrelate="Quan" />
    <word id="4" cont="只" pos="q" semparent="5" semrelate="Qp" />
    <word id="5" cont="狗" pos="n" semparent="1" semrelate="Pat" />
    <word id="6" cont="。" pos="wp" semparent="1" semrelate="mPunc" />
</sent>
<sent id="1" cont="狗追赶着乞丐。">
    <word id="0" cont="狗" pos="n" semparent="1" semrelate="Agt" />
    <word id="1" cont="追赶" pos="v" semparent="-1" semrelate="Root" />
    <word id="2" cont="着" pos="u" semparent="1" semrelate="mTime" />
    <word id="3" cont="乞丐" pos="n" semparent="1" semrelate="Datv" />
    <word id="4" cont="。" pos="wp" semparent="1" semrelate="mPunc" />
```


Figure 3. Semantic dependency analysis results of two example sentences

The above two sentences belong to the "following" frame,and the target words are "pursue/v" and "chase /v". The core frame elements are "beggars" and "dogs".According to the analysis of semantic dependency relation, two sentences' Agt are "beggars" and "dog";Pat in figure 1 is "dog" and Datv in "a dog chases a beggar."is "beggar".

In the two sentences,the similarity of the specific words in the same or similar semantic dependency relation can be expressed by formula (1) and represented by $Sim(R_{1i},R_{2i})$.Among them,"1" and "2" represent the first sentence and the second sentence respectively,and "i" represents the corresponding or similar semantic dependency in the two sentences.

3.2.2 The whole semantic dependency relation computation based on VSM

The number of times a semantic dependency "r" appears in the sentence is represented by the weight "W",then all the semantic dependencies in sentences S_1 and S_2 are represented by vectors such as R_1 and R_2,as shown in formula (2),and a semantic dependency represents a dimension in vector space model whose value is the weight of the dependency "W".

According to the theory and method of the vector space model cosine method[10],it can be similar to the similarity of the semantic dependencies of two sentences,such as formula (3).

$$R_1 = R_1(w_{11}, w_{12}, \ldots, w_{1n})$$
$$R_2 = R_2(w_{21}, w_{22}, \ldots, w_{2n}) \quad (2)$$

$$SimR(R_1, R_2) = \frac{R_1 \cdot R_2}{|R_1| * |R_2|} \quad (3)$$

3.3 Sentence similarity analysis

If taking into account the structure of the sentence, the time and cost will be enormous.so this paper only consider the semanti`c frame and core frame elements in the dependency relation between sentences,and semantic frame and core frame elements of the dependency relation is called the effective collocation of, it not only can eliminate fees for core frame elements interference calculation results,also can reduce the complexity of computing.Based on the analysis of 3.1 and 3.2,the similarity computation of Chinese sentence combining Chinese FrameNet network and Semantic Dependency Parsing is presented in this paper, such as formula (4).

$$SIM(S_1, S_2) = \alpha * SimF(F_1, F_2) + \beta * \frac{\sum_{i=1}^{n} SimF(R_{1i}, R_{2i})}{n}] + \gamma * SimR(R_1, R_2) \quad (4)$$

Among them,$SIM(S_1,S_2)$ represents the similarity of sentences S_1 and S_2;F_1 and F_2 represent the frame of two sentences,and $Sim(F_1,F_2)$ represents the similarity between the two sentence frame;"n" represents the amount of corresponding or similar semantic dependencies pairs existing in the two sentence;α,β and γ are empirical parameters,they represent the contribution of sentence frame,the single semantic dependency relationship based on word sense and the semantic dependency relationship based on VSM.Their values are determined by a large number of experiments,and $0 \le \alpha \le 1, 0 \le \beta \le 1, 0 \le \gamma \le 1, \alpha+\beta+\gamma=1$.

Considering the semantic dependency relationship between the frames and the core frame elements and the contribution of the semantic similarity relation to the sentence similarity,and after a large number of value comparison experiments,$\alpha=0.3,\beta=0.3,\gamma=0.4$.This not only shows the semantic expression of the frame to the whole sentence,but stresses the importance of semantic dependency in the sentence.

4 Experiment and result analysis

The final test corpus contains 790 Chinese sentences,of which 90 are experimental standard set,700 are experimental test sets,and the test set includes matching set of 270 sentences and noise set of 430 sentences.Each Chinese sentence in the standard set has 3 similar statements in the matching set.

This paper adopts the detection method proposed in document [11].Extracting a sentence from the standard set each time,and calculating its similarity to each sentence in the test set,then ranking the resulting similarity from large to small.If the 2~3 sentences with the highest similarity are standard sentences with similar sentences,then the result is correct.Correct rate calculation as shown in formula(5).

$$r = \frac{\text{The correct number of sentences for the test results}}{\text{Number of sentences to be measured}} \quad (5)$$

We Use VSM method,HowNet calculation method proposed in document [12],FrameNet semantic analysis method proposed in document [13] and the method

 DOI: 10.1051/matecconf/201713900028

proposed in this paper test Chinese sentences in the test corpus.The test example is shown in Table 1,and the test result is shown in table 2.

Table 1. Test case

Statement instance	Similarity value			
	1	2	3	This paper
爱夸张事实的孩子往往喜欢喜剧。	1.00	1.00	1.00	1.00
喜剧很受小学生欢迎。	0.38	0.73	0.49	0.56
孩子们爱看喜剧故事。	0.58	0.70	0.80	0.75
儿童喜欢喜剧。	0.39	0.68	0.50	1.00
孩子们有丰富的想象力。	0.43	0.35	0.05	0.19
母亲十分疼爱她的两个儿子。	0.46	0.20	0.21	0.53
丈夫总是不厌其烦地听着。	0.00	0.08	0.00	0.05
毛主席在杨家沟过生日。	0.00	0.05	0.01	0.01
凡是一般商店没有的，我们这里都有。	0.00	0.06	0.03	0.04
戏院的右边是书店。	0.05	0.01	0.02	0.03

(1: VSM;2:HowNet;3:FrameNet)

Table 2. Test results

Test Method	**The Number Of Test Sentences**	**Correct Sentences**	**Correct Rate**
VSM	90	37	41.11%
HowNet	90	66	73.33%
FrameNet	90	71	78.89%
This paper	90	76	84.44%

Can be seen from the experimental results,the method used in this paper is better than the first two methods,which analyses the dependency relation between the context and the framework inspired by the core frame elements, and don't analyze the word order and non core elements.By analyzing VSM method and HowNet method,we can indicate that the calculation result is not accurate while the sentence structure is complex and the verb is diverse.This makes it difficult for the algorithm to judge the keywords of sentences.After comparing the frame semantic dependency method and the other two methods in calculating correct sentences,we can indicate that when the sentence in the presence of Prod and Comp or two sentence words similarity is high,the method in this paper has higher accuracy.According to the basic understanding of Chinese,this method is more close to human understanding of Chinese.

5 Conclusion

This paper presents a calculation method of sentence similarity,which combines Chinese FrameNet and Semantic Dependency Parsing.This method across the surface grammatical structure and is more accurate to dig out the deep semantic relations of the sentences.This method only analyzes the core frame elements of sentences and ignores the analysis of non core framework elements when analyzing sentence components by Semantic dependency,and this can avoid interference from the unimportant components of the sentence on the results of the calculations.Next, we will further optimize the similarity calculation method,and propose an improved algorithm for the similarity computation of long difficult sentences,and further analyze how to deal with the sentences which contain formula symbols.

Reference

[1]Yang L. Shouxun Q L. Fuzzy Matching in Machine Translation Evaluation[J]. Journal of Chinese Information Processing, 2005.

[2]Mai Z Y. Jin P. Zeng S. The Computation of Chinese Word Similarity Based on Large Scale Corpus[J]. Journal of Zhongyuan University of Technology, 2010.

[3]YANG S C, CHEN J J.Research on Sentence Similarity Computing in Chinese Automatic Answering[J]. Journal of the China Society for Scientific & Technical Information, 2008, 27(1):35-41.

[4]FILLMORE C J. Frame Semantics[M]// Linguistics in the Morning Calm. 1982:111-138.

[5]BAKER C F, Fillmore C J, Lowe J B. The Berkeley FrameNet Project[C]// Meeting of the Association for Computational Linguistics and, International Conference on Computational Linguistics. Association for Computational Linguistics, 1998:86-90.

[6]LIU K Y. Research on Chinese FrameNet Construction and Application technologies[J]. Journal of Chinese Information Processing, 2011, 25(6):46-52.

[7]GUO J, CHE W X, LIU T. Chinese Semantic Dependency Parsing[J]. Intelligent Computer & Applications, 2011, 21(6):53-56.

[8]LI F.An New Approach Measuring Semantic Similarity in Hownet 2000[J]. Journal of Chinese Information Processing, 2007, 21(3):99-105.

[9]LIU T, CHE W,LI Z. Language Technology Platform[J]. Journal of Chinese Information Processing, 2011, 2(6):13-16.

[10]LI Y C,TIAN Z, YOU J. A new character feature vector similarity function[J]. Computer Engineering & Science, 2013, 35(5):93-99.

[11]LI B, LIU T,QING B, et al. Chinese Sentence Similarity Computing Based on Semantic Dependency Relationship Analysis[J].Application Research of Computers, 2003.

[12]CHENG C P,WU Z G. A Method of Sentence Similarity Computing Based on Hownet[J].Computer Engineering & Science,2012, 34(2):172-175.

[13]Ru L, Wang Z, Li S, et al. Chinese Sentence Similarity Computing Based on Frame Semantic Parsing[J]. Journal of Computer Research & Development, 2013, 50(8):1728-1736.

Research on the Intrusion Detection Method Based on Differentiated Cluster Center Offset Measure

Yu Chen[1,*], *Hao* Chun[2], and *Qing* Nian[2]

[1]School of Information Science and Engineering, Lanzhou University, Lanzhou, China
[2]Lanzhou Municipal Public Security Bureau, Lanzhou, China

Abstract. In this paper, a kind of local outlier mining method based on differentiated cluster center offset measure is proposed through which the outlier degree of sample can be calculated by use of the normal behavior model constructed by normal data sample and the preset anomaly threshold value, and whether the testing sample belong to intrusion behavior can thus be determined. Furthermore, KDD99 data set is also utilized to test the said method, and the experimental results show that the method proposed in this paper possesses higher detection rate and lower false alarm rate.

Key words: Intrusion detection; Anomaly detection; Outlier mining; Cluster center offset

1 Overview

Essentially, the intrusion detection is a classification problem, and the to-be-detected host audit records or network traffic data can be classified as normal behavior or intrusion behavior [1]. Depending on different detection methods, the intrusion detections can be divided into two categories, i.e. anomaly detection and misuse detection. The local outlier mining method studied in this paper belongs to one of such anomaly detection methods.

The hypothesis of anomaly detection is that the intruder activity is anomalous to the activity of normal subject [2]. The activity profile of normal activity of the subject can be established according to this concept, and comparisons can be made between the activity status of current subject and the activity profile. When the statistical law is violated, the activity may be considered as an intrusion. The difficult problem for anomaly detection is how to establish the activity profile of normal activity and how to design the statistical algorithm, so that normal operations are not considered as intrusion or the real intrusion behavior are not neglected.

Outlier mining technology is very suitable for the completion of anomaly-based intrusion detection. Through the analysis of network data characteristics, two facts can be drawn as below. Firstly, there are significant differences between normal behavior and anomaly behavior. Secondly, in practical applications, the number of anomaly behavior is much lower than that of the normal behavior. In respect of the entire network behavior, intrusion behavior belongs to anomaly data in small numbers, and it can be processed by treating it as outliers in the dataset, which can better reflect the nature of such invasion. At the same time, in respect of other intrusion detection methods, the anomaly detection method based on outlier mining can identify those new categories of attack samples which have not yet appeared [3], which represents the advantage which is not possessed by other detection methods. Therefore, the problem of intrusion detection can be transformed into the outlier mining problem in the network behavior data set.

2 Outlier Mining Method Based on Differentiated Cluster Center Offset Measure

Although many outlier mining methods have been proposed in recently-published literature, yet only a few of them are applied to the network anomaly detection. One of the key reasons leading up to this situation lies in the relatively high computational complexity in many popular outlier mining methods. For example, the well-known density-based local outlier coefficient method [4] and the angle-based outlier mining method [5] belong to the above situation. Furthermore, the directly distance-based mining method is only suitable for the global outlier problem. In the literature [8], a Network Anomaly Detection method based on Change of the Position of cluster centers (NADCP) is proposed. Firstly, the normal samples are clustered, and then the outlier degree corresponding to each of sample is measured by the

* Corresponding author: chenyu15@lzu.edu.cn

position offset of cluster center and with the cluster center taken as reference sample. In this way, the algorithm complexity and limitation can be better solved. However, the fact that the data distribution in reality is usually is usually no uniform, and the sample density of each cluster may be different after clustering, which determines comparatively large difference in the outlier degree in different intra-cluster samples. For this reason, there is no way to measure the outlier degree with a unified set of measurement parameters. Therefore, a kind of outlier mining method based on differentiated cluster center offset measure (IDDM) is proposed in this paper.

2.1 Selection of reference sample

For the anomaly detection method based on outliers, it is a very critical problem to determine and select reference samples. Reference samples can be selected directly from the data set, or obtained by making some calculations on the data set. Due to the fact that samples belonging to one category in a data set are likely to possess similar properties, the seeking of reference samples will be measured in category in this paper, that is, reference samples are selected or extracted from the sample set with normal flow. By analyzing, we consider selecting the cluster center as reference sample because it essentially serves as a linear combination of all samples corresponding to such category, and contains some of the properties that generally possessed by the samples of such category. In the literature [6], the class center is taken as the reference sample of such category in the data set, and the distance between each of sample and the class center to which category it belongs is also used to measure the outlier degree of such sample. Those samples which are further away from the class center are considered as outlier samples. This tactics can determine outliers in a data set with low computational complexity, but such algorithm can easily lead to the false negative of outlier data, and the detection rate of intrusion data is not high. In this paper, the acquisition means of reference sample can be described like this, kmeans clustering algorithm [7] is applied to divide the normal flow set into K unconnected clusters, i.e. C_1, C_2, C_3,..., C_k, and cluster centers of each cluster, i.e. c_1, c_2, c_3,..., c_k can be then extracted.

2.2 Defining the indicator of outlier degree

The influence caused by an addition of one outlier sample is greater than that caused by an addition of one normal sample. When this influence is measured by relative distance, it shows that if an outlier sample is newly added, the relative distance between the current "new" class center and the previous "old" class center is larger. If a normal sample is newly added, the relative distance will be smaller. Based on this idea, the position change of the class center can be observed by the way of "adding a sample" (i.e. "copying" an existing sample in one data set) and further determine whether such sample is an outlier. To be specific, when applied to the network anomaly detection, it is required to take the class center from the normal sample set as reference sample, and determine whether it is an anomaly sample by observing the position change of class center caused after the addition of samples to be detected.

In this paper, the Euclidean distance is used as a distance function to measure the similarity between samples. The formula is shown as follows:

$$D(x_i, c_j) = sqrt(\sum(x_i - c_j)^2) \tag{1}$$

After K cluster centers, i.e. c_1, c_2, c_3,..., c_k are extracted from the normal sample set and taken as the reference sample, the next step is to define and calculate the outlier degree score d of each sample relative to the reference sample. Assuming that x_i is a sample in X, it is required to firstly calculate the respective Euclidean distance between x_i and K cluster centers, i.e. c_1, c_2, c_3,..., c_k, extracted from X, find out the cluster center that has the nearest distance away from x_i (mark it as C_r and cluster center as c_r), and select it as the actual reference sample. Next, one target sample having the same characteristic value as x_i (also marked as x_i) is added, and the new cluster center of cluster C_r is marked as c_r' after the addition of such sample. In this case, c_r' can be calculated via the formula listed as below:

$$c_r' = \sum_{x_j \in (c_r \cup x_i)} x_j \tag{2}$$

After the new cluster center c_r' is calculated, the Euclidean distance can be used to measure the offset d between c_r and c_r' before and after the addition of x_i.

$$d = D(c_r, c_r') \tag{3}$$

In this paper, the offset d is used as outlier degree score which can measure the outlier degree of all samples in x_i relative to C_r having the nearest distance away from it. Meanwhile, the outlier degree of x_i relative to X can also be reflected. Therefore, the larger the d value is, the higher possibility that x_i is an outlier sample will be.

2.3 Establishment of normal behavior model

In order to establish the normal behavior profile, it is also necessary to define an exception threshold value m after the outlier degree score corresponding to each of normal sample has been calculated, so as to determine which what kinds of normal samples can be used to establish the normal behavior profile of the user. The specific means of judgment is to compare the outlier degree score corresponding to each of samples in X and the size of anomaly threshold m, and treat those normal training samples having the outlier degree score less than the anomaly threshold value as samples used to establish user normal behavior profile. At the same time, those samples having too high outlier degree scores will not be used to establish the normal behavior profile, as it may correspond to the noise data and affect the effect of model detection.

For a network anomaly detection method, different anomaly threshold values tend to have a very large impact on the detection performance. In general, if the anomaly threshold is set too high, the detection rate will be very low. On the other hand, if the anomaly threshold value is set too low, the false alarm rate will be very high. As we know, the data distribution in reality is usually not uniform, thus, the sample density of each intra-cluster is also different after the clustering, which determines the greater differences of outlier degree scores within different intra-cluster samples. For this reason, it is improper to use a unified anomaly threshold value to measure the outlier degree score of sample. The method adopted in this paper is to respectively set different anomaly threshold values m_1, m_2, m_3,..., m_k in K clusters based on the result of initial clustering. The concrete method is to respectively calculate the outlier degree scores corresponding to each sample in cluster C_1, C_2, C_3,..., C_k, and then sort in descending order for the outlier degree scores d set in each of clusters. Next, set up a global percentage parameter β, take the βth outlier degree score d in every outlier degree score set and separately assign values to anomaly threshold values m_1, m_2, m_3,..., m_k.

For a sample to be tested, i.e. x_i , the first step is to judge the cluster to which category it belongs according to the distance between it and the center of K clusters. Then, it is necessary to calculate the outlier degree score d of it relative to the cluster to which it belongs. Finally, comparison shall be made between such d and the anomaly threshold value m_i, to which cluster it belongs. If d is greater than m_i, such sample will be determined as anomaly data, and normal data conversely.

3 Implementation of Intrusion Detection Method Based on Differentiated Cluster Center Offset Measure

The outlier mining method based on cluster center offset measure is applied in the intrusion detection, and the specific intrusion detection process is shown in **Fig. 1**. To this end, the first step is to carry out preprocessing of testing samples, and realize data standardization. Next, the algorithm proposed in this paper is utilized to calculate the outlier degree of samples, and ultimately output the category labels of samples.

3.1 Data preprocessing

In this paper, KDD99 data set is adopted for experiment, and each entry of record of such data set comprises 41 feature attributes, including 38 numeric type features, and 3 nominal attribute type features, i.e. "protocol_type", "service", "flag", respectively. If the numerical computation can't be directly used in the nominal attributes, there is the need for mapping. The approach taken in this paper is to map the nominal attributes as binary attributes (0 and 1). For "service" attribute, as there are as many as 70 kinds of valuations, the characteristic dimension of data set will be greatly increased after the mapping, and the computational complexity may be increased and the classification effect may be affected. For this reason, it will be abandoned. For the two attributes of "protocol_type" and "flag", there are a total of 14 valuations respectively mapped as binary value attribute (0 and 1). Therefore, after the characteristic mapping, the original data set will be transformed from 41 feature attributes to 52 numeric type feature attributes.

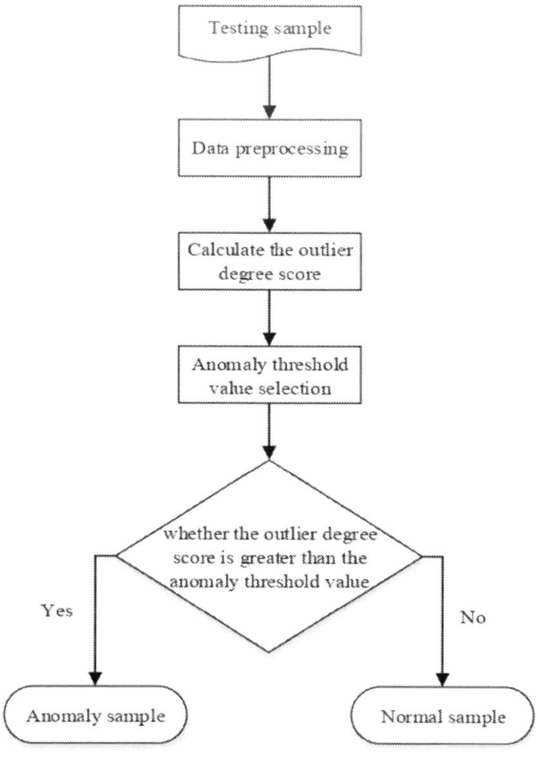

Fig. 1. intrusion detection process.

The data set after feature conversion is still not suitable for direct numerical calculation. As there are major differences in terms of the magnitude between every feature attribute, it is difficult to obtain good classification result. Thus, normalization adjustment of data is required. "Maximum-minimum" normalization method is adopted in this paper, enabling the end value to be mapped within the range of [0-1]. The transition function is listed as below:

$$x' = \frac{x - min}{max - min} \tag{1}$$

Among them, x stands for the original sample data, x' for the normalized sample data, max for the maximum value of sample data, and min for the minimum value of sample data.

3.2 Algorithm description

The outlier mining algorithm based on differentiated cluster center offset measure is described as follows:

Outlier mining algorithm based on differentiated cluster center offset measure

Input: X stands for the normal sample set used for training, K for the initial clustering number, m for anomaly threshold value, β for the percentage parameter, and x_t for the sample to be detected.

Output: Category label of the sample x_t to be detected.

1: Kmeans algorithm is applied to cluster the normal sample set X as $C_1, C_2, C_3, \ldots, C_k$, and output the cluster center of each cluster $c_1, c_2, c_3, \ldots, c_k$.
2: **for** each sample $x_i \in X$ **do**
3: Find its reference cluster center c_i from $c_1, c_2, c_3, \ldots, c_k$.
4: Calculate the outlier degree scores d_i corresponding to each x_i.
5: **end for**
6: The outlier degree scores d_i corresponding to each x_i shall be divided to $D_1, D_2, D_3, \ldots, D_k$ according to the cluster to which it belongs.
7: Sort the outlier degree scores d corresponding to $D_1, D_2, D_3, \ldots, D_k$ in descending order.
8: The percentage parameter β is applied to calculate anomaly threshold values of each cluster $m_1, m_2, m_3, \ldots, m_k$ corresponding to every outlier degree scores set D.
9: Find the reference cluster center c_t for x_t (the sample to be detected).
10: Calculate the outlier degree score d_t of x_t.
11: To compare between the outlier degree score d_t of x_t and the anomaly threshold value m_t of the cluster represented by the reference cluster center c_t.
12: **if** $d_t > m_t$ **do**
13: x_t will be determined as anomaly data.
14: **else**
15: x_t will be determined as normal data.
16: **end if**

4 Experiment and Result Analysis

KDD99 standard data set is selected in the experiment, and such data set comprises training set and testing set. All category labels are selected from "10% dataset" data as "normal" data sample, 97278 entries are used in training set for the construction of normal behavior model. There are more than 300000 lines of data samples in KDD99 standard test set ("corrected" file). Due to its overwhelming data size, only the front 50000 lines of data samples are selected as the testing set for this experiment. Different initial number of cluster K and anomaly threshold parameter β are applied to train these two parameters by use of experimental results. The testing performance of this experimental method for the testing set can be examined by use of three indicators, i.e. detection rate (probability of correct identification of invasion sample), false alarm rate (probability of normal sample identified as intrusion) and correct classification rate (probability of correct identification of sample).

In the training set, Kmeans algorithm is used to divide the dataset into clusters, the clustering number K is set as 10, and the anomaly threshold parameter β is respectively taken as 1%, 3%, 5%, 7% and 10%. The detection performance of the algorithm proposed in this paper is given in **Table 1.**

Table 1. Experimental results of detection performance at different β values (K = 10)

β (%)	detection rate (%)	false alarm rate (%)	correct classification rate (%)
1	94.45	0.55	95.09
3	94.57	1.88	95.02
5	94.62	3.49	94.86
7	94.67	4.73	94.74
10	94.69	7.13	94.46

As seen from **Table 1**, with the increase of β value the detection rate obtained by the test set continues to increase, but the false alarm rate also increases gradually. When β value is equal to 10%, the false alarm rate has obviously reached a very high level. The reason for this lie in the fact that along with the continuous decrease of anomaly threshold values, a portion of samples identified as normal samples when high anomaly threshold values are taken will be determined as anomaly threshold values when low anomaly threshold values are taken. However, these portions of samples contain not only a number of samples having aggressive real attributes, but also contain some normal samples having normal real attributes. Accordingly, both detection rate and false alarm rate have increased to some extent in this method. As revealed in the experimental results in Table 1, when β=1%, the detection performance is ideal. As revealed in further experiments under the condition of different clustering numbers K, when β value is less than 1%, the detection effect drops drastically.

Different initial clustering numbers K will also have a greater impact on the detection performance of test set. Specific experimental results will be given to different K values (4, 6, 8, 10, 14, 20), and the parameter of anomaly threshold value β will be taken as 1%.

Table 2. Experimental results of detection performance at different β values (β=0.01)

clustering number K	detection rate (%)	false alarm rate (%)	correct classification rate (%)
4	94.29	0.93	94.90
6	94.44	1.11	95.00
8	94.42	0.50	95.07
10	94.45	0.55	95.09
14	94.63	0.72	95.22
20	94.52	0.68	95.13

As revealed in the results of **Table 2**, when the values of clustering number *K* are taken within the range of 4-20, there are no significant fluctuations in the detection performance. It can be found by comparison that the detection performance can achieve a better effect if when the value of *K* is taken close to 8-10.

Table 3. Comparative results of detection performance obtained by other anomaly detection methods.

method	detection rate (%)	false alarm rate (%)	correct classification rate (%)
PSO-KM[9]	86	2.8	-
SOM[10]	91.5	14.5	-
CSI-KNN[11]	91.4	2.6	92.5
KSVM[12]	88.71	-	-
NADCP[8]	91.48	2.65	92.62
IDDM	94.45	0.55	95.09

Finally, the algorithm proposed in this paper is compared with several anomaly detection algorithms of the same category, and it is concluded that the algorithm proposed in this paper has obvious advantages in terms of detection performance.

5 Conclusion

In this paper, a kind of outlier mining method based on differentiated cluster center offset measure is proposed and also applied in the network intrusion detection. As revealed in the experimental results in KDD99 dataset, this method can help obtain higher detection rate and lower false alarm rate, and achieved better detection results when compared with similar methods. The next step is to combine the misuse detection means, further improve the detection effect, and apply this method to the network real-time intrusion detection.

References

1. A.A. Aburomman, M.B.I. Reaz. A survey of intrusion detection systems based on ensemble and hybrid classifiers. Computers & Security, **65**, (2017), 135-152.
2. M.H. Bhuyan, D.K. Bhattacharyya, J.K. Kalita. Network Anomaly Detection: Methods, Systems and Tools. IEEE Communications Surveys & Tutorials, (2013), 1-34.
3. R. Sonawane, T. Tajane, P. Chavan et al. Anomaly based intrusion detection network system. Software Engineering and Technology, **8(3)**, (2016), 66–69.
4. M.M. Breunig. H.P. Kriegel, R.T. Ng. LOF: identifying density-based local outliers. ACM, **29(2)**, (2000), 93-104.
5. H.P. Kriegel, M.S. Hubert, A. Zimek. Angle-based outlier detection in high-dimensional data. Acm Sigkdd International Conference on Knowledge Discovery & Data Mining, (2008), 444-452.
6. C.F. Tsai, K.C. Cheng. Simple instance selection for bankruptcy prediction. Knowledge-Based Systems, **27(3)**, (2012), 333-342.
7. T. Liang. Research on Intrusion Detection Technology Based on Clustering Analysis. Chongqing University, (2010).
8. C. Guo. Research on the Key Technology of Network Intrusion Detection Based on Data Mining [Ph.D. Dissertation]. Beijing: Beijing University of Posts and Telecommunications, (2014).
9. L. Xiao, Z. Shao, G. Liu. K-means Algorithm Based on Particle Swarm Optimization Algorithm for Anomaly Intrusion Detection. World Congress on Intelligent Control & Automation, **2**, (2006), 5854-5858.
10. H.G. Kayacik, A.N. Zincir-Heywood, M.I. Heywood. A hierarchical SOM-based intrusion detection system. Engineering Applications of Artificial Intelligence, **20(4)**, (2007), 439-451.
11. L. Kuang, M. Zulkernine. An anomaly intrusion detection method using the CSI-KNN algorithm. Acm Symposium on Applied Computing, (2008), 921-926.
12. U. Ravale, N. Marathe, P. Padiya. Feature Selection Based Hybrid Anomaly Intrusion Detection System Using K Means and RBF Kernel Function. Procedia Computer Science, **45(39)**, (2015), 428-435.

Model-plant Mismatch Detection of Nonlinear Processes Based on Multi-model LPV Model

Jiangyin Huang[1,*] and *Jing* Zhao[1]

[1] Department of Electrical Engineering and Automation, Xiamen University of Technology, 361024 Xiamen, Fujian, China

Abstract. The performance of model-based controller depends on the quality of the identified model. Accurate detection of the channel with model-plant mismatch can avoid re-identification of the entire multivariable system, thereby reducing the disturbance to normal production caused by identification test. A model-plant mismatch detection methodology for nonlinear systems based on LPV (Linear Parameter Varying) model was proposed in this work. The detection was performed only when the control performace becomes worse. Firstly, the LPV model based on multi-model interpolation was adopted to represent the nonlinear process. Then partial correlation coefficients between the model residuals and the inputs of the models at each of the operation points were analyzed to diagnose the model-plant mismatch of the local models. Finally, the LPV model was re-identified by updating the local mismatch models and re-estimating the model weighing parameters. The experimental results show that the partial correlation coefficient of the mismatch model is obviously larger than that of the exact model, which can point out the channel with model-plant mismatch correctly.The proposed method is suitable for the nonlinear processes which have relative steady states in their operating trajectorys.

1 Introduction

With the development of automatic control technology, the control algorithm based on process model, such as MPC (Model Predictive Control), has been widely used in industrial system control. The performance of such model based controller depends on the quality of the model which is identified and used in the controller. As the process model may change with the aging of components and the change of operating environment, the control performance will deteriorate as the model-plant mismatch increases. Although new models can be obtained by re-identification, however, for most complex industrial systems, identifying the entire large MIMO models is too costly, and may disturb the normal production because of the identification test signal. In order to solve this problem, in recent years, scholars have begun to study the methods which can be used to diagnose the exact channel with model-plant mismatch. Badwe et al. proposed a method which detects channels with significant mismatch using partial correlations between the model residuals and inputs [1]. It was followed by Loeff who improved this method by considering the influence of system delay [2]. Ji et al. use non-disturbing small sinusoidal test signals to obtain accurate estimated of process frequency response and then detect the model-plant mismatch by comparing the difference between the frequency response of the plant and the MPC model [3]. Two years later, a data driven methodology based on the statistical bands of Markov parameters are proposed [4]. In this method, the model mismatch is detected by evaluating the bias between the band of the normal case and that of the monitored case. Furthermore, the plant-model-ratio is adopted to identify the source of mismatch, namely gain, dynamics and delay mismatches from system closed-loop operation data [5]. A methodology based on a process model evaluation index is proposed in [6], the index is the ratio between the variance of the disturbance innovation and that of the model quality variable. All the detection methods mentioned above can only be used in linear system. However, most of the industrial processes are inherently nonlinear and linear models representing these processes are valid only over a limited range around the operating point. Because the process model will change as the operating point changes, model-plant mismatch detection should be performed according to the process running state, the control parameters are also need to be reset. There are only a few researchers who studied the model-plant mismatch detection of nonlinear systems. Chen et al. proposed a mismatch detection methodology based on mutual information [7]. The effectiveness of the proposed method is only verified on the system with nonlinear input signals. However, for the nonlinear system with changing models along its operation trajectory, the validity of the proposed method is not demonstrated.

In this work, a model-plant mismatch detection methodology for nonlinear systems is proposed. The

* Corresponding author: jiangyinhuang@xmut.edu.cn

LPV model based on multi-model interpolation is used to represent the nonlinear system first, and then the model-plant mismatch is diagnosed by analyzing the partial correlation coefficients between the model residuals and the system inputs of the local linear models at each of the operating points.

The paper is organized as follows: in Section 2, the problem of model-plant mismatch detection of nonlinear process is discussed and a closed-loop detection methodology based on partial correlation calculation for multi-model based LPV model is proposed. In Section 3, the validity of the proposed methodology is verified by simulating a LPV model. Section 4 gives concluding remarks and future directions.

2 Model-plant mismatch of LPV model

2.1. Problem definition

Consider a MIMO nonlinear system with m manipulate variables (MVs), n controlled variables (CVs) and l measureable disturbance variables (DVs). Suppose $\mathbf{G}_u$ and $\hat{\mathbf{G}}_u$ are the $n \times m$ plant and model between the MVs and CVs respectively, $\mathbf{G}_d$ and $\hat{\mathbf{G}}_d$ are the $n \times l$ plant and model between the DVs and CVs. Consider the closed-loop internal model control (IMC) structure of Fig.1, where $\mathbf{r}(k)$ is the vector of set points, $\mathbf{d}(k)$ is the vector of DVs, $\mathbf{y}(k)$ and $\hat{\mathbf{y}}(k)$ are the plant and model output vectors respectively, $\mathbf{e}(k)=\mathbf{y}(k)-\hat{\mathbf{y}}(k)$ is the vector of model residuals and $\mathbf{v}(k)$ is the vector of Gaussian disturbances acting on the process. $\mathbf{Q}$ is the multivariable controller and $\mathbf{u}(k)$ is the vector of MVs.

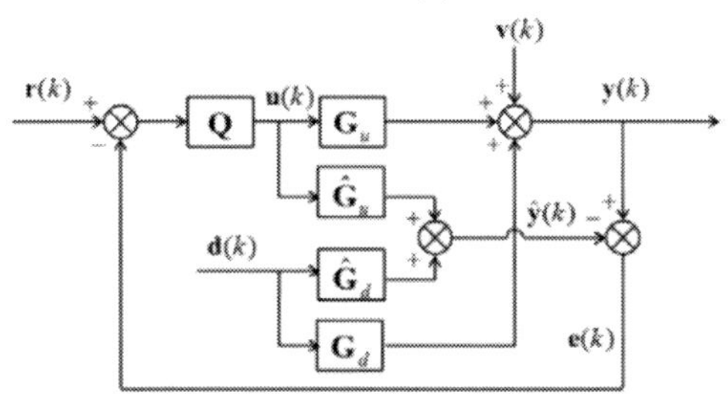

Fig. 1. Closed-loop system: IMC structure

Let $\mathbf{\Delta}_1=\mathbf{G}_u-\hat{\mathbf{G}}_u$ be the MV-CV model-plant mismatch, $\mathbf{\Delta}_2=\mathbf{G}_d-\hat{\mathbf{G}}_d$ be the DV-CV model-plant mismatch. The main objective of this work is to locate the channel in $\mathbf{\Delta}_1$ and $\mathbf{\Delta}_2$ which exists significant model-plant mismatch from the closed-loop data of the process. From Fig.1, the following expressions can be written:

$$\begin{aligned}\mathbf{e}(k)&=\mathbf{\Delta}_1\mathbf{u}(k)+\mathbf{\Delta}_2\mathbf{d}(k)+\mathbf{v}(k)\\&=[\mathbf{I}+\mathbf{\Delta}_1\mathbf{Q}]^{-1}\mathbf{\Delta}_1\mathbf{Q}\mathbf{r}(k)+[\mathbf{I}+\mathbf{\Delta}_1\mathbf{Q}]^{-1}\mathbf{\Delta}_2\mathbf{d}(k)+[\mathbf{I}+\mathbf{\Delta}_1\mathbf{Q}]^{-1}\mathbf{v}(k)\end{aligned} \tag{1}$$

$$\begin{aligned}\mathbf{u}(k)&=\mathbf{Q}[\mathbf{r}(k)-\mathbf{e}(k)]\\&=\mathbf{Q}[\mathbf{I}+\mathbf{\Delta}_1\mathbf{Q}]^{-1}\mathbf{r}(k)-\mathbf{Q}[\mathbf{I}+\mathbf{\Delta}_1\mathbf{Q}]^{-1}\mathbf{\Delta}_2\mathbf{d}(k)-\mathbf{Q}[\mathbf{I}+\mathbf{\Delta}_1\mathbf{Q}]^{-1}\mathbf{v}(k)\end{aligned} \tag{2}$$

According to the first line of Eq. (1), $\mathbf{\Delta}_1$ and $\mathbf{\Delta}_2$ can be detected by analyzing the correlations between $\mathbf{u}(k)$ and $\mathbf{e}(k)$ and the correlations between $\mathbf{d}(k)$ and $\mathbf{e}(k)$. However, there are three problems which should be solved in the analysis of such correlations.

(1)The disturbance vector $\mathbf{v}(k)$ is immeasurable in industrial process and is associated with model residuals. The correlations between $\mathbf{v}(k)$ and $\mathbf{e}(k)$ will influence the correlations between $\mathbf{e}(k)$ and the MVs/DVs. Therefore, these effects of $\mathbf{v}(k)$ on $\mathbf{e}(k)$ must be removed.

(2)There maybe correlations amongst the MVs and the DVs themselves because of the feedback loop and the controller. Such correlations will also confound the regular correlation analysis between $\mathbf{e}(k)$ and the MVs/DVs. Therefore, all of these correlations should also be decoupled.

(3)The plants $\mathbf{G}_u$ and $\mathbf{G}_d$ are both nonlinear, however, in order to reduce control difficulty, linear models are often used in internal model controller, namely, the models $\hat{\mathbf{G}}_u$ and $\hat{\mathbf{G}}_d$ are both linear. Therefore, the characteristics of $\mathbf{\Delta}_1$ and $\mathbf{\Delta}_2$ may be nonlinear and time-varying, which significantly increase the difficulty of analyzing the above correlations.

In order to solve the first two problems, an improved partial correlation analysis is adopted to remove the spurious correlations in this work. The traditional partial correlation analysis is often used to detect linear relations between some variables, especially when the variables are correlated amongst themselves. Abhijit et. al.[1] has adopted that method in the detection of model-plant mismatch for linear systems. However, if the plant is nonlinear and the model is linear, the relations between $\mathbf{e}(k)$ and $\mathbf{u}(k)$ and the relations between $\mathbf{e}(k)$ and $\mathbf{d}(k)$ are both nonlinear. In order to make the methodology of partial correlation analysis effective, LPV model based on multi model interpolation is used to represent the nonlinear plant.

2.2. Model-plant mismatch detection of LPV model

2.2.1 Multi-model based LPV model

Zhu and Xu [8] proposed a LPV model structure based on the interpolation of local linear models. An important assumption for such LPV model is that, although most industrial systems are nonlinear, their trajectories are composed of some steady states (operation points) and transitional stages. In the steady states, the plant is regarded as linear and is represented by a linear model, and then the models of the transitional stages are obtained by interpolating these local linear models. Consider the nonlinear MIMO process discussed in Section 2.1, without loss of generality, assume that there are 3 steady states on the operating trajectory of this process, $\hat{\mathbf{y}}^i(k)$ is the output vector of the identified linear model of the ith steady state:

$$\hat{\mathbf{y}}^i(k) = \hat{\mathbf{G}}_u^i\mathbf{u}(k) + \hat{\mathbf{G}}_d^i\mathbf{d}(k),\ (i = 1,2,3) \quad (3)$$

where $\hat{\mathbf{G}}_u^i$ and $\hat{\mathbf{G}}_d^i$ are the MV-CV and DV-CV linear model of the ith steady state. Then the output vector of LPV model $\hat{\mathbf{y}}(k)$ is obtained by interpolating the linear models in (3):

$$\hat{\mathbf{y}}(k) = \sum_{i=1}^{3}\alpha^i(w(k))\hat{\mathbf{y}}^i(k) \quad (4)$$

where $\alpha^i(w(k))$ are the model weights which are static functions of the scheduling variable $w(k)$ and satisfy $\sum_{i=1}^{3}\alpha^i(w(k)) = 1$. The scheduling variable is a measurable or a computable variable which can reflect the operating trajectory of the process. Namely, in the steady states, the value of the scheduling variable is constant. In extreme cases, when the process runs in a steady state, the weight of the corresponding linear model will reach its maximum value of 1 and the weights of all the other models are 0. It can be seen from Eq.(4) that the LPV model at each steady state is linear because the model weights are constant, therefore, the methodology of partial correlation analysis can be used to analyze the linear relations between the model residuals and MVs/DVs at the steady states.

2.2.2 model-plant mismatch detection using partial correlation analysis

There are two kinds of errors that could lead to model-plant mismatch of LPV model. One is the model-plant mismatch of the local linear models, the other one is the inaccurate parameters of model weights. In the identification of LPV model, the identification experiments are usually carried out in each steady state first to identify the local linear models. Then, the weighting parameters are estimated by minimizing the model output error. The accuracy of the local linear models not only affects the estimation of weighting parameters, but also affects the accuracy of the LPV model. Therefore, the detection of model-plant mismatch of LPV model should also involves two steps. First, the quality of the local models is detected; secondly, if there are mismatch local models, the weighting parameters should be re-estimated.

Abhijit et. al.[1]pointed out that in the absence of setpoint $\mathbf{r}(k)$ changes, it would not be possible to determine the extent of mismatch because of the direct causal relation between $\mathbf{e}(k)$ and $\mathbf{u}(k)$ .Therefore, when the mismatch diagnosis is performed, additional excitation signal should be superimposed on $\mathbf{r}(k)$,which will cause interference to the normal production. In order to minimize such disturbance the model mismatch diagnosis is started only when the control performance become worse. As we know, the performance of model-based control system depends a lot on the process model quality. Although the decrease in control quality does not always come from model-plant mismatch, e.g., it may also caused by inaccurate control parameters, defective valve or significant disturbance, the control error of the process can still be used as the criterion for the model mismatch detection. Considering the influence of the historical control error, define the integrated control error J of the process in (5):

$$J(k) = \sum_{i=1}^{n}\lambda_i J_i(k);$$
$$J_i(k) = \gamma e_{ci}^2(k) + \eta\sum_{j=1}^{L}\rho_j e_{ci}^2(k-j),\ (i = 1,2,\dots,n) \quad (5)$$

where $J_i(k)$ is the cumulative control error [9] of the ith output in finite time domain L . λ_i is the weight of $J_i(k)$ and satisfy $\sum_{i=1}^{n}\lambda_i = 1$. $e_{ci}(k)$ is the control error of the ith output at sampling time k , it depends on the difference between the setting value and the actual value of the corresponding output. γ and η are the weights of the current and historical control error respectively, ρ_j is the forgetting factor of the historical control error and satisfy $\sum_{j=1}^{L}\rho_j = 1$. All the weights $\lambda_i, \gamma, \eta, \rho_j$ can be set according to the prior knowledge of the process.

When $J(k)$ is greater than the specified threshold, it is reasonable to assume that the control performance is not good enough and model-plant mismatch of the LPV model need to be diagnosed.

According to the previous analysis, the mismatch of the local linear models of each of the steady states is diagnosed first. An improved partial correlation analysis is developed and the various steps in the proposed methodology are as follows:

Step1: Collect the process closed-loop data, including the setpoints, MVs, DVs, CVs and model residuals from each of the steady states. Make sure that the setpoints are sufficiently exciting, if not, continuous excitation signal, e.g. GBN (Generalized Binary Noise) signal need to be superimposed on the set point to ensure the estimation accuracy of the models which will be identified in Step2.

Step2: In order to reduce the influence of unmeasurable disturbance on the mismatch detection, the models between MVs and setpoints and the models between MVs and DVs are identified and then the disturbance free components of the MVs are found. Please note that the data of each steady state should be handled by the same way. The processing process of the data of each steady state is described as follows.

Let $\mathbf{Q}[\mathbf{I}+\boldsymbol{\Delta}_1\mathbf{Q}]^{-1} = \mathbf{S}_{ru}$ and $\mathbf{Q}[\mathbf{I}+\boldsymbol{\Delta}_1\mathbf{Q}]^{-1}\boldsymbol{\Delta}_2 = \mathbf{S}_{du}$, the expression in Eq.(2) can be written as

$$\begin{aligned}\mathbf{u}(k) &= \mathbf{S}_{ru}\mathbf{r}(k) - \mathbf{S}_{du}\mathbf{d}(k) - \mathbf{S}_{ru}\mathbf{v}(k)\\ &= \mathbf{u}^r(k) + \mathbf{u}^d(k) + \mathbf{u}^v(k)\end{aligned} \quad (6)$$

where $\mathbf{u}(k)$ is the MV data of the steady states, $\mathbf{u}^r(k)$, $\mathbf{u}^d(k)$, and $\mathbf{u}^v(k)$ are the components of $\mathbf{u}(k)$ which contain effects of the setpoints, measurable disturbances and unmeasurable disturbances respectively. Denote $\hat{\mathbf{S}}_{ru}$ and $\hat{\mathbf{S}}_{du}$ are the identified model matrix of $\mathbf{S}_{ru}$ and $\mathbf{S}_{du}$, then $\mathbf{u}^r(k)$, $\mathbf{u}^d(k)$ can be approximated from (7).

$$\hat{\mathbf{u}}^r(k) = \hat{\mathbf{S}}_{ru}\mathbf{r}(k);\quad \hat{\mathbf{u}}^d(k) = -\hat{\mathbf{S}}_{du}\mathbf{d}(k) \tag{7}$$

$\hat{\mathbf{u}}^r(k)$ and $\hat{\mathbf{u}}^d(k)$ are both free from $\mathbf{v}(k)$ because $\mathbf{r}, \mathbf{d}$ and $\mathbf{v}$ are uncorrelated to each other. According to the analysis in section 2.1, the model-plant mismatch $\mathbf{\Delta}_1$ and $\mathbf{\Delta}_2$ can be detected by analyzing the correlation between $\mathbf{e}(k)$ and $\hat{\mathbf{u}}^r(k)$ and the correlation between $\mathbf{e}(k)$ and $\hat{\mathbf{u}}^d(k)$. Because there are m MVs, n CVs and l DVs of the process, the components of $\mathbf{e}(k)$, $\hat{\mathbf{u}}^r(k)$ and $\hat{\mathbf{u}}^d(k)$ can be written as $\mathbf{e}(k) = [e_1(k)\, e_2(k) \cdots e_n(k)]$, $\hat{\mathbf{u}}^r(k) = [\hat{u}_1^r(k)\, \hat{u}_2^r(k) \cdots \hat{u}_m^r(k)]$, $\hat{\mathbf{u}}^d(k) = [\hat{u}_1^d(k)\, \hat{u}_2^d(k) \cdots \hat{u}_l^d(k)]$.

Step3: Diagnose the mismatch in $\mathbf{\Delta}_1$ of the local linear models at each of the steady states in turn by calculating the partial correlations between each item of $\mathbf{e}(k)$ and each item of $\hat{\mathbf{u}}^r(k)$. Take the channel $u_j - y_i$ as an example (u_j is the jth MV, y_i is the ith CV), the calculation process of the partial correlation is:

Firstly, find the component of $\hat{u}_j^r(k)$ that is uncorrelated with other MVs and the DVs. Assume the relations between $\hat{u}_j^r(k)$ and other MVs and the relations between $\hat{u}_j^r(k)$ and the DVs can be expressed as (8):

$$\hat{u}_j^r(k) = \tilde{\mathbf{u}}^r(k)\boldsymbol{\theta}_{u_j}^r + \hat{\mathbf{u}}^d(k)\boldsymbol{\theta}_{u_j}^d + \varepsilon_{u_j}(k) \tag{8}$$

where $\tilde{\mathbf{u}}^r(k)$ contains all the component of $\hat{\mathbf{u}}^r(k)$ except $\hat{u}_j^r(k)$, $\boldsymbol{\theta}_{u_j}^r$ and $\boldsymbol{\theta}_{u_j}^d$ are the parameter matrices which can be identified by Least-squares method from the experimental data. $\varepsilon_{u_j}(k)$ is the component of $\hat{u}_j^r$ that is free of effects from the other MVs as well as all DVs, the estimate of $\varepsilon_{u_j}(k)$ can be obtained as

$$\hat{\varepsilon}_{u_j}(k) = \hat{u}_j^r(k) - \tilde{\mathbf{u}}^r(k)\hat{\boldsymbol{\theta}}_{u_j}^r - \hat{\mathbf{u}}^d(k)\hat{\boldsymbol{\theta}}_{u_j}^d \tag{9}$$

Similarly, find the component of $e_i(k)$ that is uncorrelated with all the MVs expect $\hat{u}_j^r(k)$ and the DVs:

$$e_i(k) = \tilde{\mathbf{u}}^r(k)\boldsymbol{\theta}_{e_i}^r + \hat{\mathbf{u}}^d(k)\boldsymbol{\theta}_{e_i}^d + \varepsilon_{e_i}(k) \tag{10}$$

and obtain the estimate of $\varepsilon_{e_i}(k)$:

$$\hat{\varepsilon}_{e_i}(k) = e_i(k) - \tilde{\mathbf{u}}^r(k)\hat{\boldsymbol{\theta}}_{e_i}^r - \hat{\mathbf{u}}^d(k)\hat{\boldsymbol{\theta}}_{e_i}^d \tag{11}$$

The correlation between $\hat{\varepsilon}_{u_j}(k)$ and $\hat{\varepsilon}_{e_i}(k)$ indicates the extent of model-plant mismatch in the $u_j - y_i$ channel.

Then the model-plant mismatch in the $u_j - y_i$ channel is diagnosed by the cross correlation coefficient between $\hat{\boldsymbol{\varepsilon}}_{u_j}$ and $\hat{\boldsymbol{\varepsilon}}_{e_i}$. In order to make the diagnosis result more reliable, the cross correlation coefficient in (12) between the components of $\hat{\boldsymbol{\varepsilon}}_{u_j}$ and $\hat{\boldsymbol{\varepsilon}}_{e_i}$ with the interval of D is considered in turn:

$$R_{\varepsilon_{e_i}\text{-}\varepsilon_{u_j}}(k) = \frac{C_{\varepsilon_{e_i}\text{-}\varepsilon_{u_j}}(k)}{\sigma_{e_i}\sigma_{u_j}} \tag{12}$$

where $C_{\varepsilon_{e_i}\text{-}\varepsilon_{u_j}}(k)$ is the cross-covariance between $\varepsilon_{e_i}(k)$ and $\varepsilon_{u_j}(k)$ and satisfy

$$C_{\varepsilon_{e_i}\text{-}\varepsilon_{u_j}}(k) = \frac{1}{N-D}\sum_{t=1}^{N-D}[\hat{\varepsilon}_{e_i}(t+k) - \bar{\varepsilon}_{e_i}][\hat{\varepsilon}_{u_j}(t) - \bar{\varepsilon}_{u_j}] \tag{13}$$

σ_{e_i} and σ_{u_j} are the standard deviations of $\hat{\boldsymbol{\varepsilon}}_{u_j}$ and $\hat{\boldsymbol{\varepsilon}}_{e_i}$:

$$\sigma_{e_i} = \sqrt{\sum_{t=1}^{N-D}(\hat{\varepsilon}_{e_i}(t+k) - \bar{\varepsilon}_{e_i})^2};\quad \sigma_{u_j} = \sqrt{\sum_{t=1}^{N-D}(\hat{\varepsilon}_{u_j}(t) - \bar{\varepsilon}_{u_j})^2} \tag{14}$$

According to the concept of correlation coefficient, the range of $R_{\varepsilon_{e_i}\text{-}\varepsilon_{u_j}}(k)$ is $[-1, 1]$, if $R_{\varepsilon_{e_i}\text{-}\varepsilon_{u_j}}(k) \neq 0$, it can be considered that there exits model-plant mismatch in $u_j - y_i$ channel.

Step4: Diagnose the mismatch in $\mathbf{\Delta}_2$ of the local linear models at each of the steady states in turn by calculating the partial correlations between each item of $\mathbf{e}(k)$ and each item of $\hat{\mathbf{u}}^d(k)$. The calculation process is mostly the same as the process described in Sep3, so we won't go into much detail here.

All the mismatch local linear models should be re-identified after they are found. It is worth noting that the identification test signals are only superimposed on the inputs corresponding to the mismatch models. Since the whole MIMO process is not needed to be identified, the inference to the normal production of the process is greatly reduced. When all the mismatch local models are corrected, the new LPV model can be obtained by re-estimating the weighting parameters in Eq.(4) according to the weighting function it used. The detailed estimation method of LPV model with linear weight, polynomial weight and Gaussian weight is discussed in [10].

3 Simulation case studies

A MIMO nonlinear system with two MVs, two CVs and one DV is considered as our case study. For simulating the discrete process, a sampling period of 1min is used. The system is controlled by a multiple MPC controller module in MATLAB. The control parameters are adjusted automatically as the operation status changes. The plant is represented as (15):

$$\mathbf{y} = \begin{bmatrix} \dfrac{2w}{(w^2+20)s+1}e^{-5s} & \dfrac{2w^2}{(15w+12)s+1}e^{-4s} \\ \dfrac{1.6w+2}{(4w^2+10)s+1}e^{-6s} & \dfrac{w^3+2}{(5w^2+25)s+1}e^{-3s} \end{bmatrix}\mathbf{u} + \begin{bmatrix} \dfrac{0.5}{1.85s+1}e^{-s} \\ \dfrac{0.25}{2.65s+1}e^{-s} \end{bmatrix}d \tag{15}$$

where w is the scheduling variable which changes as the system's operation status changes. Assume there are three steady states and two transient processes along the operating trajectory of the plant. A closed-loop simulation test lasted for 5000 sampling points is performed and the trajectory of w is shown in Fig.2. The setting values of the CVs are GBN sequences. In order to simulate unmeasured process disturbance, filtered white noise sequences are added at the two CVs.

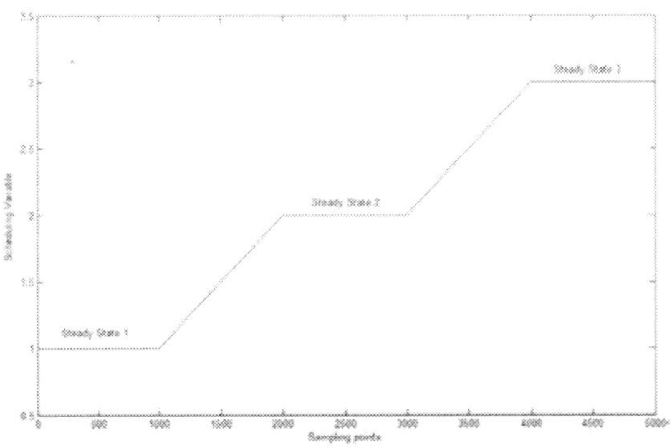

Fig. 2. The trajectory of the scheduling variable

Assume that there is no model-plant mismatch at the second steady state ($w = 2$) and some identified models which are used in MPC at the other steady states are not accurate. The differences between the plant and the model are listed in Table 1. The models which are not listed in the table are the same as the plant.

Table 1. Set of model-plant mismatch

Scheduling variable	Channel	Plant	Model
$w=1$	G_{12}	$\frac{2}{27s+1}e^{-4s}$	$\frac{4}{27s+1}e^{-4s}$
$w=3$	G_{21}	$\frac{6.8}{46s+1}e^{-6s}$	$\frac{8}{35s+1}e^{-5s}$
	G_{22}	$\frac{29}{70s+1}e^{-3s}$	$\frac{40}{78s+1}e^{-3s}$

Both the partial correlations between the model residuals and the MVs and between the model residuals and the DV at each of the steady states are calculated. The results are plotted in Figs.3, 4 and 5. It can be seen from these figures that most of the partial correlations of the channels without model-plant mismatch is within the range of [-0.15, 0.15]. The significant partial correlations between e_1 and MV2 in Fig.3 and between e_2 and MV1, MV2 in Fig.5 correctly point out the mismatch channel listed in Table 1. The BFTs (Best Fitting) of the mismatch LPV model and the re-identified LPV model are compared in Table 2. The comparison results show that the accuracy of the re-identified LPV model is much higher than that of the mismatch LPV model with respect to the best fittings of the two CVs.

$$R_{\mathrm{BFT}} = 100 * \max(1 - \frac{\|y(k) - \hat{y}(k)\|_2}{\|y(k) - \bar{y}\|_2}, 0) \tag{16}$$

where $y(k)$ is the process output, $\hat{y}(k)$ is the LPV model output and $\bar{y}$ is the average value of $y(k)$.

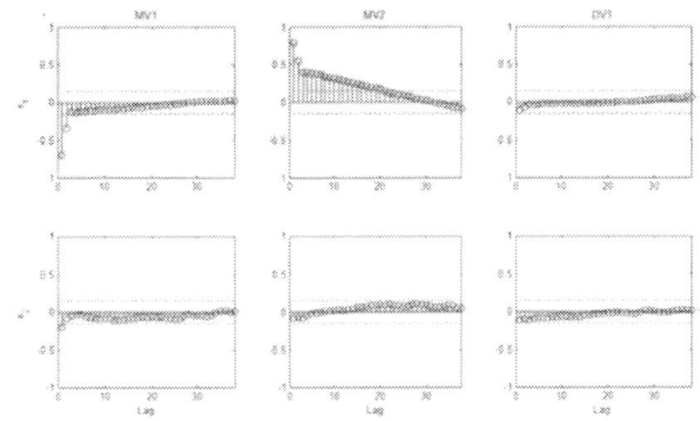

Fig. 3. Partial correlations plots at $w = 1$

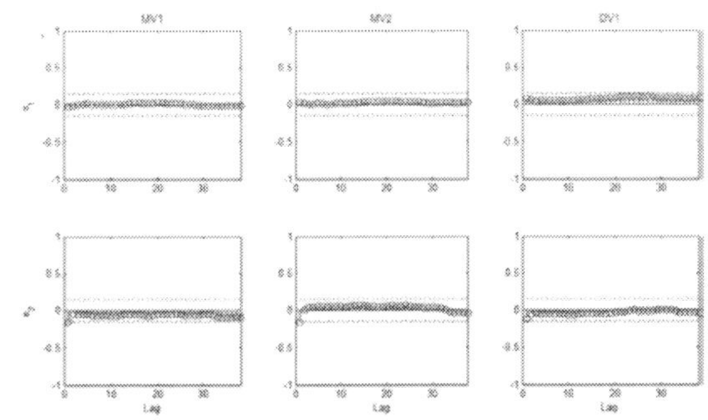

Fig. 4. Partial correlations plots at $w = 2$

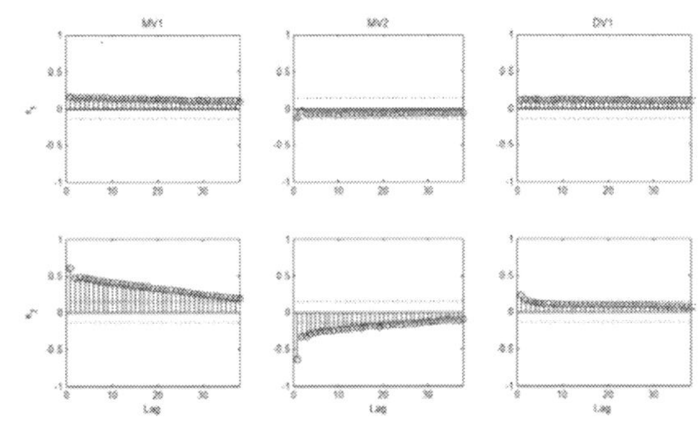

Fig. 5. Partial correlations plots at $w = 3$

Table 2. BFTs of the mismatch LPV models

	BFTs of the mismatch LPV model	BFTs of the re-identified LPV model
CV1	75.5%	92.4%
CV2	69.8%	91.6%

4 Conclusions

A methodology of model-plant mismatch detection for nonlinear processes based on LPV model is proposed in this work. The LPV model based on piecewise linear model interpolation is used to represent a nonlinear process, and then the model-plant mismatch of the local linear models at each of the operation points is diagnosed by partial correlation coefficient analysis. A new LPV model is obtained by estimating the weighting parameters after all the mismatch models are re-identified. The method proposed is suitable for most of the industrial processes which have relatively steady states in their operation trajectory. The main advantage of this method is that the mismatch of the local linear models at each operation point is diagnosed respectively and only the mismatch models need to be re-identified, which can reduce the cost of identifying the models of the entire process. The simulation results show that the partial correlation coefficient of the mismatch model is obviously larger than that of the accurate model, and the mismatch channel is correctly marked. Subsequent studies included the following aspects: (1) the simulation results show that the proposed method is sensitive to the mismatch of gain and time constant, but has a poor detection performance in time delay mismatch. The method should be improved so that the delay mismatch can be accurately detected. (2) The effectiveness of the proposed method in typical industrial processes should be verified. (3)The identification of the extent of model-plant mismatch and the impact of model-plant mismatch on controller performance should be studied. There is no need for re-identification when the mismatch is low or has little effect on the control performance.

Acknowledgements

This work is funded by the Fujian Natural Science Foundation of China(No. 2017J01509), the JK project of Education Department of Fujian(No. JK2015034) and the Opening Foundation of Advanced Process Control Laboratory for Light Industry of Jiangnan University (No. APCLI1601).

References

1. A.S. Badwe, R.D. Gudi, R.S. Patwardhan, S.L. Shah, S.C. Patwardhan, J. Process Control, **19**, 1305-1313 (2009)
2. M. V. Loeff, A. Kuramoto, and C. Garcia. *XX Congresso Brasileiro de Automática,*777-784,(2014)
3. G. Ji, K.K. Zhang, Y.C. Zhu, J. Process Control, **22**,635-642 (2012)
4. F. Yin, H. Wang, L. Xie, P. Wu, Z.H. Song,Comput. Elect. Eng, **40**,2178-2192 (2014)
5. S. Yerramilli, A.K. Tangirala, *IFAC-PapersOnLine,* **1,** 266-271(2016)
6. D. Ling, Y. Zheng, H.Zhang, W.D. Yang, B.Tao, J. Process Control, **57**,66-79 (2017)
7. G. Chen, L. Xie, J. Zeng, J. Chu, Y. Gu,Ind.Eng.Chem. **52**,1927-1938 (2013)
8. Y.C. Zhu, Z.H. Xu, *Proc. of the 17th IFAC World Congress, Seoul, Korea* (2008)
9. N. N. Nandola, S. Bhartiya, J. Process Control, **18**, 131-148(2008)
10. J.Y. Huang, G.L. Ji, Y.C. Zhu, P.V.D. Bosch, J. Process Control,**22**,1198-1208 (2012)

A Design of Long Jumping Device Based on Infrared Scanning and Data Collecting Technology

Zhigang Li[1,*], *Jie* Lv[2,3], *Xiaofei* Dong[3] and *Bin* Wang[3]

[1]Teaching Affairs Department, Suzhou Chien-Shiung Institute of Technology, 215411 Taicang, China
[2]School of Mecheanical Engineering, Jiangsu University, 212013 Zhenjiang, China
[3]School of Mecheanical Engineering, Yancheng Institute of Technology, 224051 Yancheng, China

Abstract. A standing long jump device based on infrared scanning technology was proposed for the aim of detecting and analysing the sports achievement of a group of people. The equipment is make of a standing long jump range finder, an infrared scanning range device, and a cloud platform for data management. The output of the measurement data is achieved through converting electrical signal produced by the instrument into a digital signal. The triangular formula and the minimum value algorithm were applied to deal with the distance data measured by the infrared light device. The function model of instantaneous data acquisition can effectively solve the measurement errors caused by the slight movement of tester's toes or heels. In addition, the device also has intelligent cross-line alarm function. Compared with the previous rangefinder equipment, this device can achieve a high level of measurement accuracy and upload real-time data to the cloud platform for analysis.

1 Introduction

In recent years, a variety of theoretical perspectives have been advanced to provide an understanding of the determinants of usage[1]. With the rapid development of Internet, Internet of Things[2], cloud computing[3] and other information technology in the global field, the rapid growth of data information has brought serious challenges and valuable opportunities to many industries. Therefore, today's information society has already entered the big data[4] era. The development of large data not only changed people's daily life and work, business mode of operation, but also led to the scientific model of the occurrence of a fundamental change[5]. Large data is a collection of data that can not be perceived, acquired, managed, processed, and serviced by conventional machines and hardware and software tools for a certain period of time. Network large data refers to the "human, machine, material" ternary world in the network space mutual interaction and integration generated by the Internet and access to large data, referred to as network data. By generating, collecting, storing and analyzing data on a large data basis, extracting useful values in different areas to provide advice or support decisions, resulting in different levels of potential value.

Intelligent system test system is a kind of automatic detection of a number of human body system, including long jump tester, finger pulse tester, vital capacity tester and many other subsystems. It is an important part of physical training as a test item for students' physical fitness test. Through the training of long jump, to promote the comprehensive development of young people has a very important role. "National Student Physical Health Standard" provides a long jump is junior high school, high school, college annual physical fitness test items. The traditional long jump test is painted on the ground jumper, next to the tape, the referee through the naked eye reading, and record the student's long jump results. This method of error is large, and data records, statistics, management are more trouble. In accordance with the Ministry of Education planning, legislation long jump and all test items should be universal access to intelligent instruments, and equipped with data management software to provide exercise prescription. The current use of conductive rubber and infrared on the tube ranging device is low, the installation of trouble, not equipped with data management system to provide exercise prescription. According to the existing devices have low accuracy, the start line is not set up alarm device, and the small movement of the foot after the jump caused by measurement errors or other issues. Some effective solutions has been given in this paper.

In this paper, an Internet-based student physical health intelligent test equipment has been designed, which involves infrared ranging[6-7] and large data technology. It intends to develop an intelligent terminal, through the intelligent development of young people's daily sports, not only enrich the extracurricular activities also collected young people's physical health data. The use of infrared sensors with fast response, long life advantages. Through the controller and processor control infrared sensor transmitter and receiver components to complete the data collection, processing, recording,

* Corresponding author: 234230348@qq.com

automatic completion of the data sort update and achieve the entire process of measurement automation. This means can effectively solve the current measurement means there are many inconvenience. Intelligent terminal through the collected data on the young athletes to give statistics, analysis, and uploaded to the cloud platform for comparing the different schools and regions of the physical quality of young people. It has an important reference role in the development of relevant national strategic development.

2 Device system structure

The program according to the principle of infrared light curtain, automatic measurement of long jump distance. The athlete is taking off at the takeoff. There is a separate infrared tube at the jumper line to determine whether to step on the line foul. When the athletes landed, the ground's infrared detection area was blocked. The system records all measured data. According to the algorithm to calculate the final long jump results. When the jump is finished, the athlete gets out of the detection area. The system waits for the next player to measure at the takeoff point. The measured data will be analyzed, counted and uploaded to the cloud platform.

Figure 1 is the legislation of the long jump range finder system structure diagram. The system consists of information reading device, data transmission device, function key group and data processing module and so on. The information reading device is composed of identity card identification, fingerprint identification, camera and the like, and is used for recognizing the identification information of the tester. The function of the data transmission device is to upload or download the measured data. Function key group is based on the different conditions of use of the tester for key operation, to achieve accurate measurement process. The data processing module realizes the acquisition, analysis and recording of the measurement data.

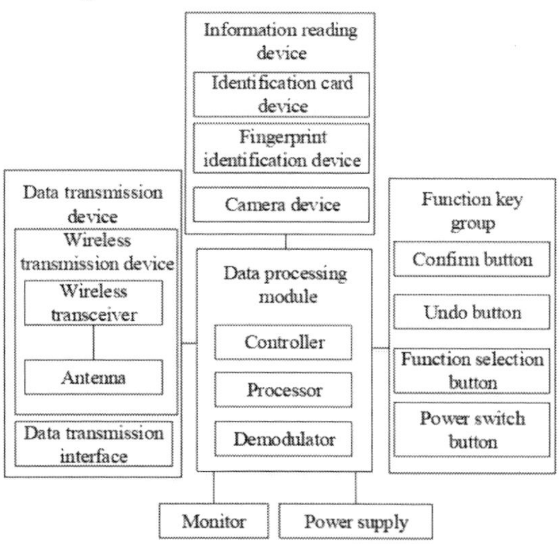

Fig. 1. System framework of the long jump range finder

Figure 2 shows a system for infrared scanning range devices. The circuit mainly includes infrared transmitting circuit, anti-over line module, modulation circuit, control circuit and so on. Infrared emission circuit for transmitting infrared beam sensing long jump distance. The anti-crossing module can effectively prevent the tester from crossing the start line when taking off. The modulation circuit is used to generate a data signal for transmission. The control circuit guarantees the overall function of the device.

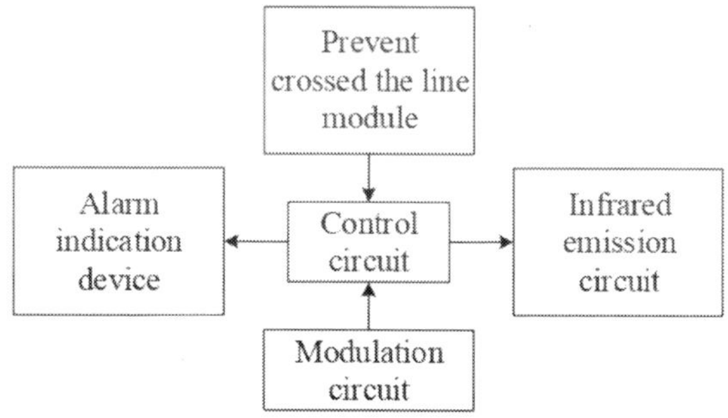

Fig. 2. System structure frame diagram of the infrared scanning range device

3 Installation of the device

The device (as shown in Fig. 3) consists mainly of three parts: (1) the long jump range finder (as shown in Fig. 4) for data read, store, upload; (2) the infrared scanning range device based on the infrared distance measurement principle (as shown in Fig. 5); (3) the data management cloud platform for large data statistics and analysis.

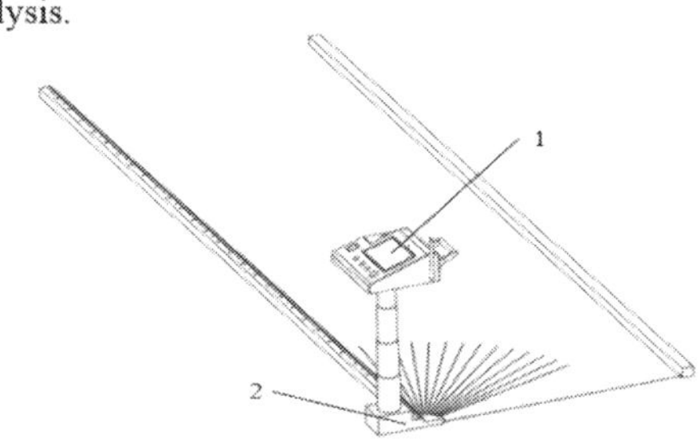

Fig.3. Simulation of distance measuring device

Where Device 1 is a Standing Long Jump Range Finder, and Device 2 is an Infrared Scanning Range Device, which radiates infrared lights to detect the distance data of human's feet. The long jump range finder (Fig. 4) remotely downloads the test list via the antenna. The tester performs personal information reading through the identification card or the fingerprint identification. Face Recognition Pinhole Camera Shoots the taster's face information. Test information is confirmed and tested. The results will be confirmed by the confirmation button to confirm the results, at the same time, the system will automatically store data and call out the next tester information. The display shows personal results. To retest, revoke the data with the Undo button. If someone is absent, press the Undo button to temporarily record the person's score as 0 and save it with the confirmation button. The function selection button allows you to select different modes. The saved data is uploaded to the data center via the antenna. The data can be stored in the range finder. The bracket is used to support the range finder, and the telescopic bracket can reduce the overall volume of the device, easy to install and store.

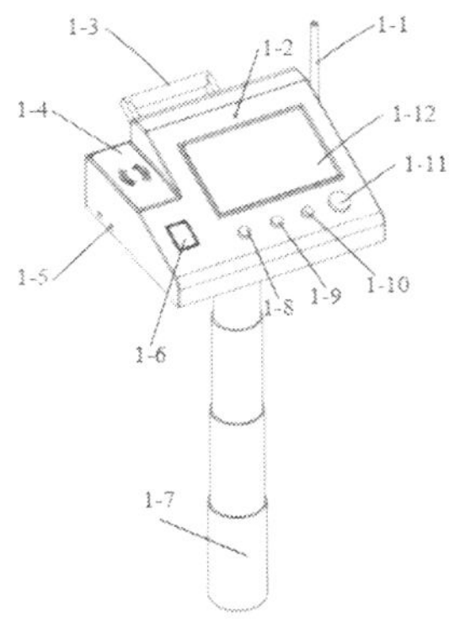

Fig. 4. Simulation of standing long distance rangefinder

The labelled parts in Fig. 4 are named as follows: 1-1. Antenna, 1-2. Pinhole Camera, 1-3. Handle, 1-4. Card Panel, 1-5. Data Transfer Port, 1-6. Fingerprint Identification, 1-7. Telescopic Stand, 1-8. Confirm button, 1-9. Undo button, 1-10. Function selection button, 1-11. Power switch, 1-12. Display.

Infrared scanning range device (Fig. 5) installed in the flat ground, by the rangefinder installation hole, infrared light screen range finder, infrared transmitter, reference scale and other components. The rangefinder mounting hole is used to install the range finder. Infrared light curtain rangefinder for fixed angle emission ranging infrared light curtain. Infrared emitters mounted close to the ground jumper position are used to prevent the tester's toes from crossing the start line. The reference scale is used for distance reference. When the tester's foot falls into the range of the rangefinder, the infrared scanning range device uses the infrared distance measurement principle to measure the distance between the tester's heel and the infrared light curtain. The device converts the resulting beam signal into a level signal and outputs it, and then generates a digital signal through the modulation circuit for data transfer. The processor calculates the triangular formula for the resulting measurement data and finally derives the actual jump distance of the tester.

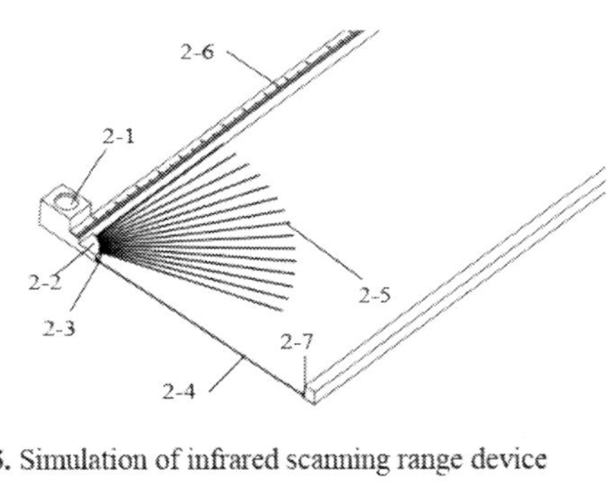

Fig. 5. Simulation of infrared scanning range device

The labelled parts in Fig. 5 are named as follows: 2-1. Mounting holes, 2-2. Infrared light curtain instrument, 2-3. Infrared emitter, 2-4. Infrared light beam, 2-5. Infrared light curtain, 2-6. Reference scale, 2-7. Infrared receiver.

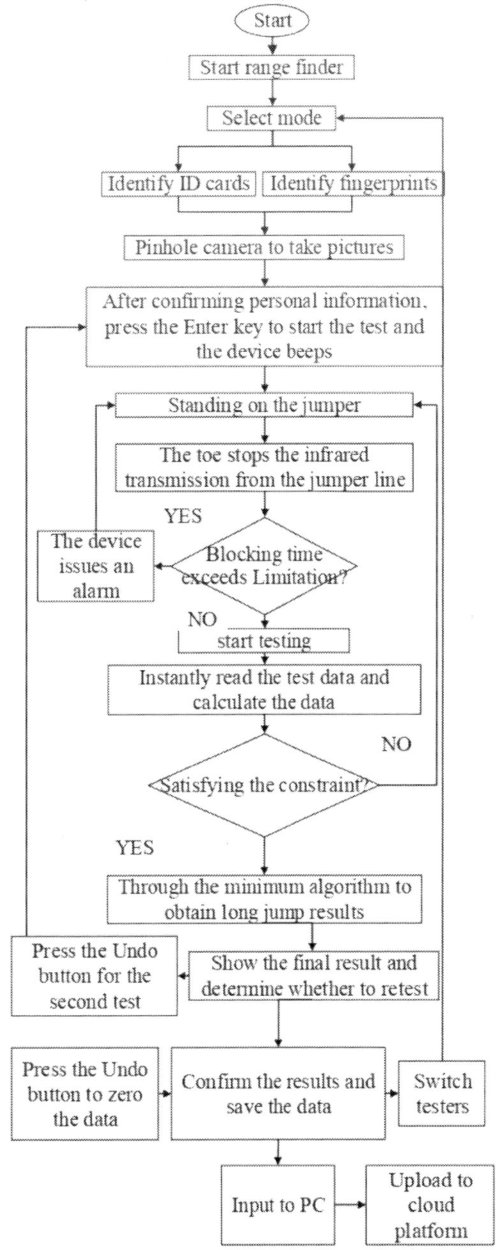

Fig. 6. Flow chart of long jump test

When all the tests are completed, the stored data can be uploaded to the PC by wireless or wired transmission for data statistics, analysis, and uploaded to the data management cloud platform. Cloud platform not only includes the test of the name, gender, age, performance and other details, but also can analyze the personal data and the overall data, showing the recent changes in their physical data trends and the difference with the average data. In addition, the system will give specific recommendations based on the individual's own situation. Through the improvement of diet or related exercise to help the testers to strengthen or improve physical fitness. In addition, parents can real-time view student physical data through the relevant mobile APP, so as soon as possible concerned about the healthy growth of students' physical fitness. Detailed test process is shown in Fig 5.

4 Principle of distance measurement

When the infrared photoelectric tube receives the pulse signal of the launch tube, the instantaneous change in the luminous flux causes the current to change. Infrared receiver through the op amp circuit differential, amplified signal and the threshold comparison circuit output pulse signal, and finally sent to the processor detection. When the luminous flux of the receiving tube changes, the reverse current will change, equivalent to the receiving tube resistance changes. As the receiving tube cathode through the bias resistor, so the voltage will change. The transmit modulation signal is a pulse with a very small duty cycle, and the input electrical signal of the receiving tube is also a pulse signal with very small pulse width. This signal is fed into the op amp differential circuit by capacitive coupling. Having a fast response, low noise, and output is not affected by the characteristics of the DC component of the photocurrent.

The use of infrared radiation is the principle of non-proliferation, because the infrared light through the other substances when the refractive index is very small, so long distance rangefinder will consider the infrared. Infrared transmission takes time. When the infrared ray is emitted from the range finder, the infrared ray is received by the distance meter when the reflector is reflected back. And then according to the infrared from the time of receipt to be received and the transmission speed of infrared can calculate the distance.

4.1 Innovation

4.1.1 Triangle formula-based distance measuring

Infrared light curtain into a fixed angle continuous launch ranging infrared light curtain. According to the formula:

$$Hn = Ln{\cdot}sin\alpha_n \tag{1}$$

where α_n $(n=1,2,3...)$ is regarded as the angle between the jumper and the measured infrared rays at the heel position, Ln is regarded as infrared beam length, *Hn* is regarded as jump distance (as shown in Fig 4). The formula draw the distance data (the result can be accurate to *0.1mm*). And then according to the minimum algorithm to compare the size of *H1* ~ *Hn*, the resulting minimum value for the measured heel and jumper between the distance, that is, long jump results. In addition, to avoid the measured data interference outside the working range, the constraint condition is added: $Ln\ cos\alpha_n < L$ (*L* is the starting line length).

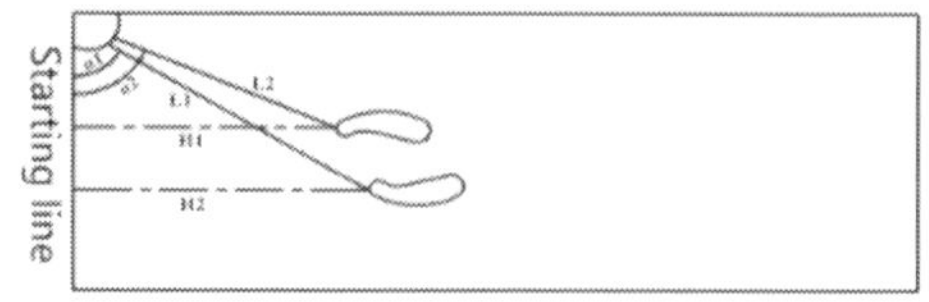

Fig. 7. Demonstration of data measurement

4.1.2 Minimum algorithm analysis

According to the C language circulatory thought, the obtained data are compared and the minimum result is obtained. The algorithm is as follows:

```
min=H[1];
    for(i=2;i<n;i++)
        if(H[i]<min)
            min= H[i];
printf(min);
```

where min is initial minimum parameter, *H[i]* *(i=1,2,3...n)* is the variable of each data obtained. Let *H[1]* be the initial minimum, and use the loop statement to compare the latter data with the previous one, and the minimum value is the result of the long jump.

4.1.3 Function of preventing crossing the line

In the actual long jump test process, because the tester intentionally or unintentionally lead to the toes beyond the jump line, causing foul behavior. The infrared emitter (shown at 2-3 in Fig 3) located at the ground jumper position is used to prevent the tester's toes from crossing the jumper. If the tester toes across the start line, that is, blocking the infrared beam, the device will issue an alarm tone. In order to avoid the tester in the take-off may pass the infrared beam lead to false alarm phenomenon, we add a timer program to the alarm system (very short time) according to the reasons for the faster acceleration and the shorter time when the tester takes off. The device will be alerted when it exceeds a certain amount of time.

4.1.4 Function of instantaneous reading

When the tester feet lands, to avoid the situation of tester's foot movement or other parts of the body into the scanning range due to the center of gravity instability. We have added a transient read measurement data program to the device. When the tester's feet fall within the scanning range, the infrared will instantly measure the distance between the heel and the instrument. The program can be collected in a very short period of time measurement data and save. The measured data will not affect the results originally obtained.

4.1.5 Data management cloud platform

Through the establishment of student physical health data management cloud platform, we can use large data technology for data analysis to get different regions, schools, age, gender, student physical growth. Different regions/provinces/national education departments and health departments through the cloud platform for the data of students' physical health statistics. It plays an important role in supporting national strategy.

4.2 Problems and solutions

4.2.1 Abnormal state processing

Long jump range finder in the actual use of the process, the equipment administrator or the tester may lead to a variety of abnormal state. If the infrared array in a long time in the useless scanning state, not only consume energy, but also affect the working life of the instrument. Previously classified according to various exceptions problems that may arise. When an exception occurs,. the finite state machine is used to switch the state to return to the normal state.

4.2.2 Multiple MCUs compete for power issues

This device requires a total of multiple single-chip synchronous work, synchronization signal's transmission and reception will be a single pin to complete the microcontroller. Due to the component parameters of the error and the power capacity of the original flash, there are individual microcontroller can not power or instantaneous power and power problems in the experimental process, resulting in synchronous high and low levels of the pin changes. As the single-chip receiver to receive the synchronization signal may be wrong to determine the synchronization signal to start work, so that the operation is not reliable. For this problem we can use dual power supply to solve.

4.2.3 Outdoor environmental impact

Outdoor test launch and receive spacing will be a greater degree of reduction, there are two main reasons impact. First, the sensitivity of the receiving head in the visible light environment will be reduced. Second, infrared wave attenuation in the outdoor environment is relatively large. So we can take the following solutions. First, the processing of cylindrical holes in front of the receiver and install the red filter can effectively reduce the impact of other visible light. The second is to increase the launch tube launch power. However, we need to adjust the duty cycle so that it does not exceed the rated power of the launch tube.

5 Conclusion

This field has designed an automatic distance measuring device based on infrared scanning. The use of infrared distance measurement principle for accurate long-distance jump. The triangular formula and the minimum algorithm are used to process the measured data to arrive at the final result. The function of preventing crossing the line and instantaneous reading ensure the effectiveness of the results. Data management cloud platform can achieve effective data sharing and application. After analysis, the device has the advantages of high efficiency, wide measurement range and stable performance.

On the basis of the existing research work, the circuit of the device function is designed by using other components such as single chip microcomputer. After the installation of the device is completed, cooperate with the local schools to carry out the testing work. The test results are analyzed to determine whether the device meets the intended design requirements. Optimizing the development of the cloud platforms and mobile devices APP. To make the program more convenient, humane and meet the needs of the masses of the use. On this basis as much as possible to reduce the cost of manufacturing equipment in order to better put into the market.

6 Acknowledgments

The authors gratefully acknowledge the support by Taicang Key Project of Research and Development Plan (Industry Perspective and Common Key Technologies) under Grant No. TC2016GY07 and Jiangsu Provincial Policy Guidance Program (Enterprise-University-Research Institute Cooperation)-Perspective Joint Research Program under Grant No. BY2016065-52.

References

1. T. Davenport, S. James. *The new industrial engineering: information technology and business process redesign* (1990)
2. H. Ma. *Journal of Computer Science and Technology*. **26**, 6 (2011)
3. X. Ye, X. Chen, H. Wan, X. Zeng. *Tsinghua Science & Technology*. **21**, 3 (2016)
4. G. Li, X. Cheng. *Bulletin of Chinese Academy of Sciences*. **27**, 6 (2012)
5. Y. Wang, X. Cheng. *Chinese Journal of Computers*. **36**, 6 (2013)
6. L. Wang, X. Zou, S. Liu. *Transactions of the Chinese Society for Agricultural Machinery*. **45**, 7 (2014)
7. A. Kumar, A. Makur. *2009 IEEE Region 10 Conference* (Singapore, 2009)
8. M. Gao, C. Jin, X. Wang, X. Tian, A. Zhou. *Chinese Journal of Computers*. **33**, 3 (2010)
9. T. Hey, S. Tansley, K. Tolle. *Microsoft Research* (2009)
10. D. Koller, N. Friedman. *Probabilistic Graphical Models: Prin-ciples and Techniques-Adaptive Com putation and M achineLearning* (Cam bridge, M A: The M IT Press, 2009)

Load capacity of a new rope-climbing robot

Hao Yin[1,2,*], *Peiqing* Ye[1], Tong Zhao[1] and *Hui* Zhang[1]

[1] School of Mechanical Engineering, Tsinghua University, Beijing 100084, China
[2] Qingdao Campus, Naval Aeronautical University, Qingdao 266041, China

Abstract. Theoretical analysis and experimental research are carried out on the load capacity of one new kind rope-climbing robot, this robot could lift load along the rope depends on the friction between rope and wheel of the robot. The classic Euler formula is based on the assumption of constant friction coefficient, and this paper establishes a new load capacity model based on non-constant friction coefficient model, presents a method to measure the distribution of friction coefficient, and builds the experimental platform to conduct experimental research on the friction coefficient and load capacity. Experimental results show that the friction coefficient decreases with the increase of wrap angle; the new model fits the results better than Euler formula.

1 Introduction

Friction transmission is widely used in civil and industrial fields, mine hoist, elevator and belt drive machine are some common typical forms of friction transmission[1-3], the rope-clibing robot to be studied in this paper is another application of friction transmission. The rope-climbing robot transports loads along the rope relied on the friction between its wheel and rope, and load capacity is a very important measure of the performance of the robot. For this kind rope-climbing robot, its load capacity equals to the maximum friction between rope and wheel, Euler formula is widely used to calculate the friction between wheel and rope or blet or other forms, but there is some deviation between the actual force and calculated values, especially for the rope-climbing robot. Most studies had applied Euler formula to different areas[1-5], some studies had modified Euler formulation for their special materials and structures[6,7], but it's assumed that the friction coefficient between rope and wheel keeps constant in these studies, the assumption is too simplified to correspond to reality. The rope-climbing robot as the research object, the load capacity of the robot and the distribution rules of friction coefficient between rope and wheel will be studied in this paper.

2 Theoretical analysis of load capacity

2.1 Theoretical analysis based on constant friction coefficient

Analyze the force of the rope wound on the wheel, mark the wrap angle φ, the radius of rope r and the wheel's R, wheel angular velocity ω, force at the end-A F_A and end-B F_B ($F_A > F_B$). Select a micro-segment of rope (radian $\mathrm{d}\alpha$) at point-D where arc-distance α from end-B, Fig.1 shows the forces acting on this micro-segment, these forces can be expressed by the following equation.

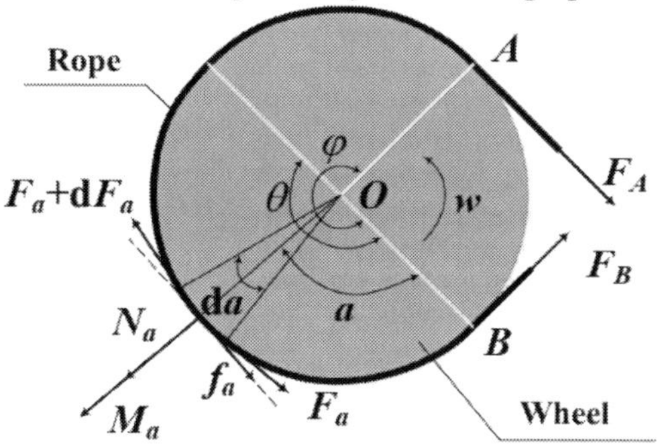

Figure 1. Working principle schematic diagram.

$$\left.\begin{aligned}&(F_\alpha + \mathrm{d}F_\alpha)\cos\frac{\mathrm{d}\alpha}{2} = F_\alpha\cos\frac{\mathrm{d}\alpha}{2} + f_\alpha\\&(2F_\alpha + \mathrm{d}F_\alpha)\sin\frac{\mathrm{d}\alpha}{2} = N_\alpha + M_\alpha\\&M_\alpha = \rho\cdot\omega^2\cdot r^2\cdot\mathrm{d}\alpha\\&f_\alpha = \mu_\alpha\cdot N_\alpha\end{aligned}\right\} \quad (1)$$

In the equation, f_α , M_α , N_α are the friction, centrifugal force and support force acting on the micro-segment rope, μ_α is the friction coefficient between the rope and wheel surface at this position and moment, ρ is the quality of the rope per unit length. Suppose that the friction coefficient is constant over the entire wrap angle[8],which could be described by formula $\mu_\alpha=\mu$, $\alpha\in[0,\ \varphi]$, owing to $\mathrm{d}\alpha\approx 0$, $\cos\frac{\mathrm{d}\alpha}{2}=1$, $\sin\frac{\mathrm{d}\alpha}{2}=\frac{1}{2}\mathrm{d}\alpha$, ignore the second order small quantity $\mathrm{d}F_\alpha\sin\frac{\mathrm{d}\alpha}{2}$,

* Corresponding author: yin-h14@mails.tsinghua.edu.cn

$$\frac{F_A\text{-}\rho\omega^2 r^2}{F_B\text{-}\rho\omega^2 r^2} = \mathrm{e}^{\mu\varphi} \tag{2}$$

Because ρ and ω are small, so M_a can be ignored,

$$\frac{F_A}{F_B} = \mathrm{e}^{\mu\varphi} \tag{3}$$

Eq.(3) is called Euler's formula because it was deduced by Euler firstly, as the classical formula for calculating the belt (or rope) - wheel friction, it is widely used in mine hoist, elevators, automotive belt transmission and so on. However, the mechanical properties of the rope used for the climbing robot is special. Coulomb friction hypothesis does not hold in the global context, so it is necessary to analyze the problem under the premise of non-constant friction coefficient .

2.2 Theoretical analysis based on non-constant friction coefficient

Although the friction coefficient is not constant in the global context, but focus on the micro-segment rope at the very moment, the local friciton coefficient could be seen as a constant and the relationship between the forces at both ends of the micro-segment could be described accurately by Euler Formula. Divide evenly the entire wrap angle to n copies, mark the friction coefficient in number i wrap angle $\mu_{i-1,i}$ and the average friction coefficient $\mu_{i,j}$ in wrap angle $\varphi_{i,j}$, they should satisfy the following equations:

$$\left.\begin{aligned}
F_1 &= F_0\mathrm{e}^{\mu_{0,1}\varphi_{0,1}} \\
&\vdots \\
F_i &= F_{i-1}\mathrm{e}^{\mu_{i-1,i}\varphi_{i-1,i}} = F_0\mathrm{e}^{\sum_{m=1}^{i}\mu_{m-1,m}\varphi_{m-1,m}} = F_0\mathrm{e}^{\mu_{0,i}\varphi_{0,i}} \\
&\vdots \\
F_k &= F_{k-1}\mathrm{e}^{\mu_{k-1,k}\varphi_{k-1,k}} = F_0\mathrm{e}^{\sum_{m=1}^{k}\mu_{m-1,m}\varphi_{m-1,m}} = F_0\mathrm{e}^{\mu_{0,k}\varphi_{0,k}} \\
&\vdots \\
F_j &= F_{j-1}\mathrm{e}^{\mu_{j-1,j}\varphi_{j-1,j}} = F_0\mathrm{e}^{\sum_{m=1}^{j}\mu_{m-1,m}\varphi_{m-1,m}} = F_0\mathrm{e}^{\mu_{0,j}\varphi_{0,j}} \\
&\vdots \\
F_n &= F_0\mathrm{e}^{\mu_{0,n}\varphi_{0,n}}
\end{aligned}\right\} \tag{4}$$

From Eq.(4), the friction coefficients satisfy the following equations:

$$\mu_{0,i}\varphi_{0,i} = \sum_{m=1}^{i}\mu_{m-1,m}\varphi_{m-1,m} \tag{5}$$

$$\mu_{0,i} = \frac{1}{i}\sum_{m=1}^{i}\mu_{m-1,m} \tag{6}$$

As Eq.(6) shows, the average friction coefficient of a wrap angle from beginning numerically equals to the equal weighted average of friction coefficients of all micro-segments rope, and it is applicable for the average friction coefficient of any wrap angle anywhere. Expand the equation F_j:

$$F_j = F_{j-1}\mathrm{e}^{\mu_{j-1,j}\varphi_{j-1,j}} = F_{j-2}\mathrm{e}^{\mu_{j-1,j}\varphi_{j-1,j}+\mu_{j-2,j-1}\varphi_{j-2,j-1}} = \cdots$$
$$= F_i\mathrm{e}^{\sum_{m=i+1}^{j}\mu_{m-1,m}\varphi_{m-1,m}} = F_i\mathrm{e}^{\mu_{i,j}\varphi_{i,j}} \tag{7}$$

$$\mu_{i,j}\varphi_{i,j} = \sum_{m=i+1}^{j}\mu_{m-1,m}\varphi_{m-1,m} = \ln F_j - \ln F_i \tag{8}$$

$$\mu_{i,j} = \frac{1}{\varphi_{i,j}}\sum_{m=i+1}^{j}\mu_{m-1,m}\varphi_{m-1,m} = \frac{1}{j-i}\sum_{m=i+1}^{j}\mu_{m-1,m} \tag{9}$$

Similarly:

$$F_j = F_k\mathrm{e}^{\mu_{k,j}\varphi_{k,j}} = F_k\mathrm{e}^{\sum_{m=k+1}^{j}\mu_{m-1,m}\varphi_{m-1,m}} \tag{10}$$

For any F_j, it can be seen obtained from initial force F_i after wrap angle $\varphi_{i,j}$ ($i<j$), or from initial force F_k after wrap angle $\varphi_{k,j}$ ($k<j$), or from any initial force after corresponding wrap angle, in other words, the effect of wrap angle and initial force are equivalent, initial force can be seen as a special case of wrap angle, thus discussion about the effect of wrap angle is sufficient. Denote the distribution function of friction coefficient along the wrap angle as $\mu(\varphi)$:

$$F_\varphi = F_0\mathrm{e}^{\int_0^\varphi \mu(\varphi)\mathrm{d}\varphi} \tag{11}$$

Or:

$$\int_0^\varphi \mu(\varphi)\mathrm{d}\varphi = \ln F_\varphi - \ln F_0 \tag{12}$$

Let F_0=1 N, so $\ln F_0 = 0$, thus Eq.(12) turns to:

$$\int_0^\varphi \mu(\varphi)\mathrm{d}\varphi = \ln F_\varphi \tag{13}$$

Or:

$$\frac{\mathrm{d}\ln F_\varphi}{\mathrm{d}\varphi} = \mu(\varphi) \tag{14}$$

Eq.(11) could be used to calculate the load capacity under the condition of non-constant friction coefficient, and Eq.(14) is the theoretical inference of the friction coefficient distribution.

3 Experimental Research

The theoretical derivation is given above; measure the static friction of different wrap angles in experiments, the function of friction coefficient along wrap angle could be obtained by computing, fitting and deriving the data. In actual use, the end tension (initial force) would be larger than 1 N, and its curve is part of the complete curve.

Fig.2 shows the experiment platform, in the experiments, end tension F_B would be applied by weights hung at one end of the rope, the signal of F_A is received by mechanical sensor fixed at the other end, and then transferred by a transmitter to PC.

DOI: 10.1051/matecconf/201713900032

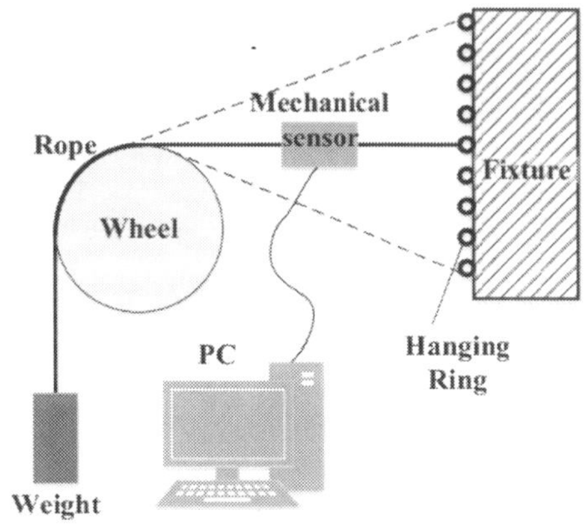

Figure 2. Experimental schematic diagram.

The experimental procedure is as follows:

1. Start the motor a few seconds after the weights hung stably;
2. Increase the motor's speed gradually until the rope slips globally in the wheel surface;
3. Shut down the motor.

Record the corresponding F_A signal of whole process above. Use a new rope segment for each test to avoid the effects of wear on the test results, repeat the test at least 15 times under the same conditions, the average F_A obtained from 15 values is seen as the load capacity under this condition. Change the wrap angle by adjusting the position of the sensor along the hanging ring rail on the fixture, and measure load capacity under new conditions.

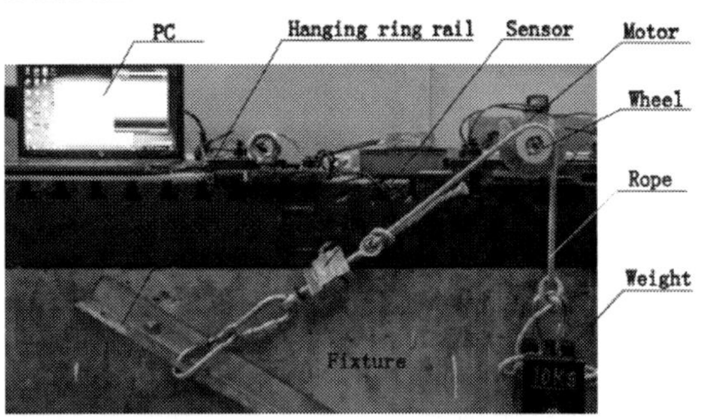

Figure 3. Experiment platform photo.

Draw the curve of load capacity changed with wrap angle, the results are shown in Fig.4. It is obvious that the load capacity and the wrap angle are approximately exponentially related, but the error will become larger as the wrap angle increases; the average coefficient of static friction decreases with the wrap angle increasing.

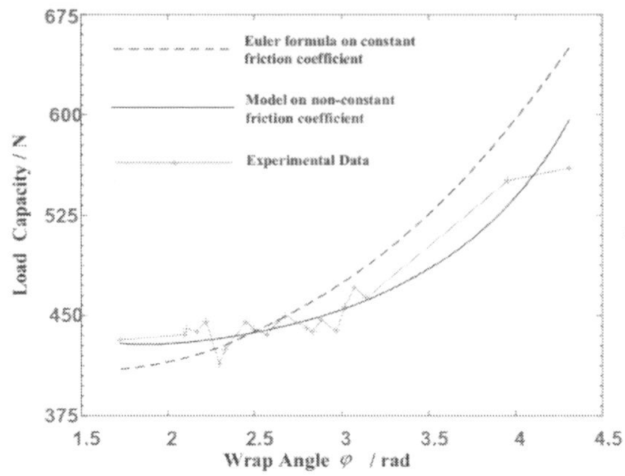

Figure 4. Load capacity trend with wrap angle.

It can be inferred that the function $\mu(\varphi)$ decreases monotonically with the wrap angle and tends to a constant value, that is, its first derivative should be negative and gradually approach 0; and its integral function should increase monotonically according to its physical meaning, construct a $\mu(\varphi)$ function of the following form:

$$\mu(\varphi)=ae^{b\varphi}+c \tag{15}$$

In the above formula, there is $a>0$, $b<0$, $c>0$. Select the appropriate parameter, the error between model predictive value and actual value is very small, it indicates that when the wrap angle increases to a certain extent, the effect of wrap angle on friction is very slight, $\ln F_A$ increases linearly with the wrap angle.

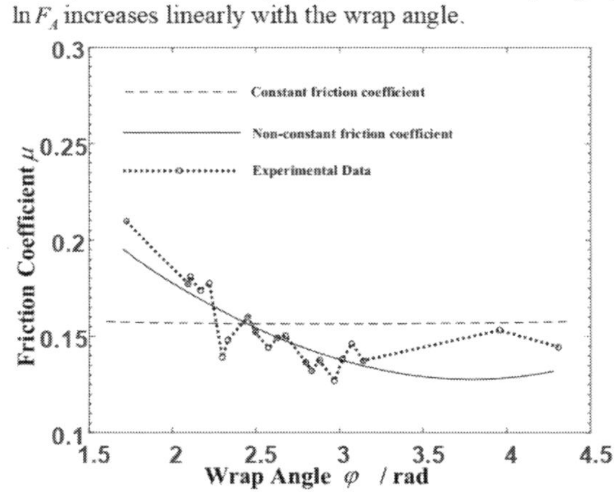

Figure 5. Friction coefficient trend with wrap angle.

3 New model of load capacity

According to Eq.(13) and Eq.(14), the curve in Fig.5 is just a part of the complete curve starts from point (0,1). On the complete curve, the load force corresponds to the wrap angle, as shown in Fig.6, a force F_B corresponds to angle φ_B, and the warp angle is φ, the start point of the integral is $B(\varphi_B,\mu_B)$, and the end point is $A(\varphi_A,\mu_A)$, where φ_A and φ_B satisfy $\varphi_A=\varphi_B+\varphi$.

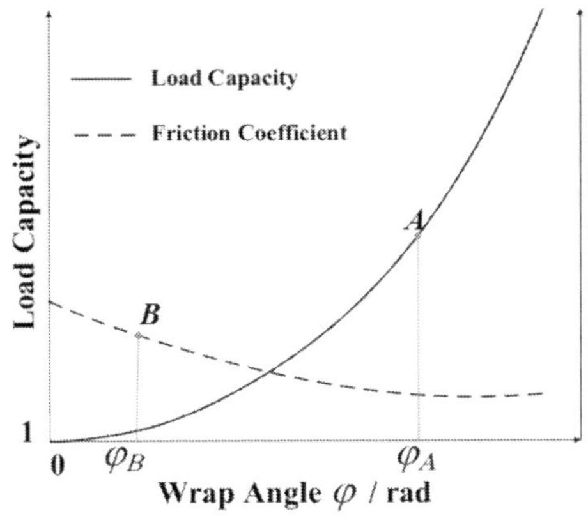

Figure 6. The complete curve.

Eq.(15) show the new model of the distribution of friction coefficient along the wrap angle, and the force can be calculated by Eq.(16), the two formula together form the load capacity mathematical model.

$$F_A = F_B e^{\int_{\varphi_B}^{\varphi_A} \mu(\varphi)\mathrm{d}\varphi} \tag{16}$$

4 Conclusion

Theoretical derivation and experimental analysis are carried out on the load capacity of a new kind of rope-climbing robot, mathematic model of the load capacity is obtained considering non-constant friction coefficient, and a measuring method of the friction coefficient distribution is proposed. The experimental results show the friction coefficient decreases with the increase of wrap angle; and the new model has higher precision than Euler formula.

5 Acknowledgments

This work was financially supported by the Foundation of The State Key Laboratory of Tribology, Tsinghua University. Project No. SKLT09B08.

References

1. Berna Balta, Fazil O. Sonmez, Abdulkadir Cengiz. Speed losses in V-ribbed belt drives[J]. Mechanism and Machine Theory, 2015, 86: 1-14.
2. Nilabh Srivastava, Imtiaz Haque, Transient dynamics of metal V-belt CVT: Effects of band pack slip and friction characteristic[J]. Mechanism and Machine Theory, 2008, 43(4): 459-479.
3. Gang Sheng, Jonah H. Lee, Vikram Narravula, et al. Experimental characterization and analysis of wet belt friction and the vibro-acoustic behavior[J]. Tribology International, 2011,44(3): 258-265.
4. Davies P, François M, Lacotte N, et al. An empirical model to predict the lifetime of braided HMPE handling ropes under cyclic bend over sheave (CBOS) loading[J]. Ocean Engineering, 2015, 97: 74-81.
5. Kastratović G, Vidanović N, Bakić V, et al. On finite element analysis of sling wire rope subjected to axial loading[J]. Ocean Engineering, 2014, 88: 480-487.
6. Weller S D, Davies P, Vickers A W, et al. Synthetic rope responses in the context of load history: Operational performance[J]. Ocean Engineering, 2014, 83: 111-124.
7. Leech C M. The modelling of friction in polymer fibre ropes[J]. International Journal of Mechanical Sciences 44 (2002) 621–643.
8. Valentin L. Popov. Contact Mechanics and Friction Physical Principles and Applications[M]. Tsinghua University, Firsted. Beijing, 2011.

Porcellio scaber algorithm (PSA) for solving constrained optimization problems

Yinyan Zhang[1,*], *Shuai* Li[1,**], and *Hongliang* Guo[2,***]

[1] Department of Computing, The Hong Kong Polytechnic University, Hong Kong, China
[2] Center for Robotics, School of Automation Engineering, University of Electronic Science and Technology of China, Chengdu, China

Abstract. In this paper, we extend a bio-inspired algorithm called the porcellio scaber algorithm (PSA) to solve constrained optimization problems, including a constrained mixed discrete-continuous nonlinear optimization problem. Our extensive experiment results based on benchmark optimization problems show that the PSA has a better performance than many existing methods or algorithms. The results indicate that the PSA is a promising algorithm for constrained optimization.

1 Introduction

Modern optimization algorithms may be roughly classified into deterministic optimization algorithms [1–5] and stochastic ones. The former is theoretically sound for well-posed problems but not efficient for complicated problems. For example, when it comes to nonconvex or large-scale optimization problems, deterministic algorithms may not be a good tool to obtain a globally optimal solution within a reasonable time due to the high complexity of the problem. Meanwhile, while stochastic ones may not have a strong theoretical basis, they are efficient in engineering applications and have become popular in recent years due to their capability of efficiently solving complex optimization problems, including NP-hard problems such as the travelling salesman problem. Bio-inspired algorithms take an important role in stochastic algorithms for optimization. These algorithms are designed based on the observations of animal behaviors. For example, one of the well known bio-inspired algorithm called particle swarm optimization initially proposed by Kennedy and Eberhart [6] is inspired by the social foraging behavior of some animals such as the flocking behavior of birds.

There are some widely used benchmark problems in the field of stochastic optimization. The pressure vessel design optimization problem is an important benchmark problem in structural engineering optimization [7]. The problem is a constrained mixed discrete-continuous nonlinear optimization problem. In recent years, many bio-inspired algorithms have been proposed to solve the problem [8–11]. The widely used benchmark problems also include a nonlinear optimization problem proposed by Himmelblau [12].

*e-mail: yinyan.zhang@connect.polyu.hk
**e-mail: shuaili@polyu.edu.hk
***e-mail: guohl1983@uestc.edu.cn

Recently, a novel bio-inspired algorithm called the porcellio scaber algorithm (PSA) has been proposed by Zhang and Li [13], which is inspired by two behaviors of porcellio scaber. In this paper, we extend the result in [13] to solve constrained optimization problems. As the original algorithm proposed in [13] deals with the case without constraints, we provide some improvements for the original PSA so as to make it capable of solving constrained optimization problems. Then, we compare the corresponding experiment results with reported ones for the aforementioned benchmark problems as case studies. Our extensive experiment results show that the PSA has much better performance in solving optimization problems than many existing algorithms. Before ending this introductory section, the main contributions of this paper are listed as follows:

1) We extend the PSA to solve constrained optimization problems, including the constrained mixed discrete-continuous nonlinear optimization problem.
2) We show that the PSA is better than many other existing algorithms in solving constrained optimization problems by extensive numerical experiments.

2 Problem Formulation

The constrained optimization problem (COP) considered in this paper is presented as follows:

$$\begin{aligned} &\text{minimize } \breve{f}(\mathbf{x}), \\ &\text{subject to } g_j(\mathbf{x}) \leq 0, \\ &\qquad l_i \leq x_i \leq u_i, \end{aligned} \tag{1}$$

with $i = 1, 2, \cdots, d$ and $j = 1, 2, \cdots, m$, where $\mathbf{x} = [x_1, x_2, \cdots, x_d]^{\mathrm{T}}$ is a d-dimension decision vector; l_i and u_i are the corresponding lower bound and upper bound of the ith decision variable; $\breve{f}(\mathbf{x}) : \mathbb{R}^d \to \mathbb{R}$ is the cost function

Algorithm 1 Original PSA

Cost function $f(\mathbf{x})$, $\mathbf{x} = [x_1, x_2, \cdots, x_d]^{\mathrm{T}}$
Generate initial position of porcellio scaber $\mathbf{x}_i^0$ $(i = 1, 2, \cdots, N)$
Environment condition $E_{\mathbf{x}}$ *at position* $\mathbf{x}$ *is determined by* $f(\mathbf{x})$
Set weighted parameter λ *for decision based on aggregation and the propensity to explore novel environments*
Initialize f_* *to an extremely large value*
Initialize each element of vector $\mathbf{x}_* \in \mathbb{R}^d$ *to an arbitrary value*
while $k < MaxStep$ **do**
  Get the position with the best environment condition, i.e., $\mathbf{x}_b = \arg\min_{\mathbf{x}_j^k}\{f(\mathbf{x}_j^k)\}$ *at the current time among the group of porcellio scaber*
  if $\min_{\mathbf{x}_j^k}\{f(\mathbf{x}_j^k)\} < f_*$ **then**
    $\mathbf{x}_* = \mathbf{x}_b$
    $f_* = \min_{\mathbf{x}_j^k}\{f(\mathbf{x}_j^k)\}$
  end if
  Randomly chose a direction $\tau = [\tau_1, \tau_2, \cdots, \tau_d]^T$ *to detect*
  Detect the best environment condition $\min\{E_{\mathbf{x}}\}$ *and worst environment condition* $\max\{E_{\mathbf{x}}\}$ *at position* $\mathbf{x}_i^k + \tau$ *for* $i = 1 : N$ *all N porcellio scaber*
  for $i = 1 : N$ *all N porcellio scaber* **do**
    Determine the difference with respect to the position to aggregate i.e., $\mathbf{x}_i^k - \arg\min_{\mathbf{x}_j^k}\{f(\mathbf{x}_j^k)\})$
    Determine where to explore, i.e., $p\tau$
    Move to a new position according to (2)
  end for
end while
Output $\mathbf{x}_*$ *and the corresponding function value* f_*
Visualization

to be minimized. For the case that the problem is convex, there are many standard algorithms to solve the problem. However, for the case that the problem is not convex, the problem is difficult to solve.

3 Algorithm Design

In this section, we modify the original PSA [13] and provide an improved PSA for solving COPs.

3.1 Original PSA

For the sake of understanding, the original PSA is given in algorithm 1 [13], which aims at solving unconstrained optimization problems of the following form:

$$\text{minimize } f(\mathbf{x}),$$

where $\mathbf{x}$ is the decision vector and f is the cost function to be minimized. The main formula of the original PSA is given as follows [13]:

$$\mathbf{x}_i^{k+1} = \mathbf{x}_i^k - (1-\lambda)(\mathbf{x}_i^k - \arg\min_{\mathbf{x}_j^k}\{f(\mathbf{x}_j^k)\}) + \lambda p\tau, \quad (2)$$

where $\lambda \in (0, 1)$, τ is a vector with each element being a random number, and p is defined as follows:

$$p = \frac{f(\mathbf{x}_i^k + \tau) - \min\{f(\mathbf{x}_i^k + \tau)\}}{\max\{f(\mathbf{x}_i^k + \tau)\} - \min\{f(\mathbf{x}_i^k + \tau)\}}.$$

Evidently, the original PSA does not take constraints into consideration. Thus, it cannot be directly used to solve COPs.

3.2 Inequality constraint conversion

In this subsection, we provide some improvements for the original PSA and make it capable of solving COPs. As the original PSA focuses on solving unconstrained problem, we first incorporate the inequality constraints $g_j(\mathbf{x}) \leq 0$ $(j = 1, 2, \cdots, m)$ into the cost function. To this end, the penalty method is used, and a new cost function is obtained as follows:

$$\check{f}(\mathbf{x}) = f(\mathbf{x}) + \gamma \sum_{i=1}^{m} g_i^2(\mathbf{x}) h(g_i(\mathbf{x})), \quad (3)$$

where $h(g_i(\mathbf{x}))$ is defined as

$$h(g_i(\mathbf{x})) = \begin{cases} 1, & \text{if } g_i(\mathbf{x}) > 0, \\ 0, & \text{if } g_i(\mathbf{x}) \leq 0, \end{cases}$$

and $\gamma \gg 1$ is the penalty parameter. By using a large enough value of γ (e.g., 10^{12}), unless all the inequality constraints $g_i(\mathbf{x}) \leq 0$ $(i = 1, 2, \cdots, m)$ are satisfied, the term $\gamma \sum_{i=1}^{N} g_i^2(\mathbf{x}) h(g_i(\mathbf{x}))$ takes a dominant role in the cost function. On the other hand, when all the inequality constraints $g_i(\mathbf{x}) \leq 0$ $(i = 1, 2, \cdots, m)$ are satisfied, $h(g_i(\mathbf{x})) = 0$, $\forall i$, and thus $\check{f}(\mathbf{x}) = f(\mathbf{x})$.

3.3 Addressing simple bounds

In terms of the simple bounds $l_j \leq x_j \leq u_j$ with $j = 1, 2, \cdots, d$, they are handled via two methods. Firstly, to satisfy the simple bounds, the initial position of each porcellio scaber is set via the following formula:

$$x_{i,j}^0 = l_j + (u_j - l_j) \times rand(0, 1) \quad (4)$$

where $x_{i,j}^0$ denotes the initial value of the jth variable of the position vector of the ith (with $i = 1, 2, \cdots, N$) porcellio scaber; $rand(0, 1)$ denotes a random number in the region (0, 1), which can be realized by using the *rand* function in Matlab. The formula (4) guarantees that the initial positions of all the porcellio scaber satisfy the the simple bounds $l_j \leq x_j \leq u_j$ with $j = 1, 2, \cdots, d$.

Secondly, if the positions of all the porcellio scaber are updated according to (2) by replacing $f(\mathbf{x})$ with $\check{f}(\mathbf{x})$ defined in (3) for the constrained optimization problem (1), then the updated values of the position vector $\mathbf{x}_i^k$ may violate the simple bound constraints. To handle this issue, based on (2), a modified evolution rule is proposed as follows:

$$\mathbf{x}_i^{k+1} = P_{\Omega}(\mathbf{x}_i^k - (1-\lambda)(\mathbf{x}_i^k - \arg\min_{\mathbf{x}_j^k}\{\check{f}(\mathbf{x}_j^k)\}) - \lambda p\tau), \quad (5)$$

Algorithm 2 Algorithm for the evaluation of $P_\Omega(\mathbf{x})$ with $\mathbf{x} = [x_1, x_2, \cdots, x_d]^\mathrm{T}$

---

for $i = 1 : d$ **do**
  if $x_1 < l_i$ **then**
    $x_i = l_i$
  end if
  if $x_i > u_i$ **then**
    $x_i = u_i$
  end if
end for
return $\mathbf{y} = [x_1, x_2, \cdots, x_d]^\mathrm{T}$

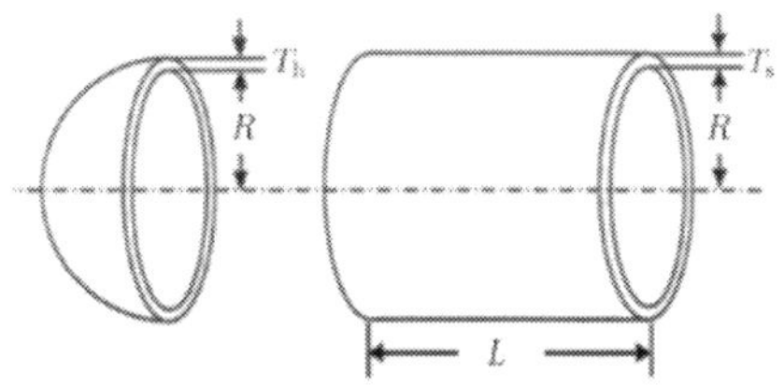

Figure 1. A diagram showing the design parameters of a pressure vessel [14].

where $\lambda \in (0, 1)$, τ is a vector with each element being a random number, and

$$p = \frac{\check{f}(\mathbf{x}_i^k + \tau) - \min\{\check{f}(\mathbf{x}_i^k + \tau)\}}{\max\{\check{f}(\mathbf{x}_i^k + \tau)\} - \min\{\check{f}(\mathbf{x}_i^k + \tau)\}}.$$

Besides, P_Ω is a projection function and make the updated position satisfy the simple bound constraints, where $\Omega = \{\mathbf{x} \in \mathbb{R}^d | l_i \le x_i \le u_i, i = 1, 2 \cdots, d\}$. The mathematical definition of $P_\Omega(\mathbf{x})$ is $P_\Omega(\mathbf{x}) = \arg\min_{\mathbf{y}\in\Omega} \|\mathbf{y} - \mathbf{x}\|_2$ with $\|\cdot\|_2$ denoting the Euclidean norm. The algorithm for the evaluation of $P_\Omega(\mathbf{x})$ is given in Algorithm 2.

3.4 PSA for COPs

Based on the above modifications, the resultant PSA for solving COPs is given in Algorithm 3. In the following section, we will use some benchmark problems to test the performance of the PSA in solving COPs.

4 Case Studies

In this section, we present experiment results regarding using the PSA for solving COPs.

4.1 Case I: Pressure vessel problem

In this subsection, the pressure vessel problem is considered. The pressure vessel problem is to find a set of four design parameters, which are demonstrated in Fig. 1, to minimize the total cost of a pressure vessel considering the cost of material, forming and welding [6]. The four design parameters are the inner radius R, and the length L of the cylindrical section, the thickness T_h of the head, the thickness T_s of the body. Note that, T_s and T_h are integer multiples of 0.0625 in., and R and L are continuous variables.

Algorithm 3 PSA for COPs

Cost function $\check{f}(\mathbf{x})$ as defined in (3), $\mathbf{x} = [x_1, x_2, \cdots, x_d]^\mathrm{T}$
Generate initial position of porcellio scaber $\mathbf{x}_i^0$ ($i = 1, 2, \cdots, N$) *according to (4)*
Environment condition $E_\mathbf{x}$ *at position* $\mathbf{x}$ *is determined by* $\check{f}(\mathbf{x})$
Set weighted parameter λ *for decision based on aggregation and the propensity to explore novel environments*
Set penalty parameter γ *in* $\check{f}(\mathbf{x})$ *to a large enough value*
Initialize f_* *to an extremely large value*
Initialize each element of vector $\mathbf{x}_* \in \mathbb{R}^d$ *to an arbitrary value*
while $k < MaxStep$ **do**
  Get the position with the best environment condition, i.e., $\mathbf{x}_b = \arg\min_{\mathbf{x}_j^k}\{f(\mathbf{x}_j^k)\}$ *at the current time among the group of porcellio scaber*
  if $\min_{\mathbf{x}_j^k}\{f(\mathbf{x}_j^k)\} < f_*$ **then**
    $\mathbf{x}_* = \mathbf{x}_b$
    $f_* = \min_{\mathbf{x}_j^k}\{f(\mathbf{x}_j^k)\}$
  end if
  Randomly chose a direction $\tau = [\tau_1, \tau_2, \cdots, \tau_d]^T$ *to detect*
  Detect the best environment condition $\min\{E_\mathbf{x}\}$ *and worst environment condition* $\max\{E_\mathbf{x}\}$ *at position* $\mathbf{x}_i^k + \tau$ *for* $i = 1 : N$ *all N porcellio scaber*
  for $i = 1 : N$ *all N porcellio scaber* **do**
    Determine the difference with respect to the position to aggregate i.e., $\mathbf{x}_i^k - \arg\min_{\mathbf{x}_j^k}\{f(\mathbf{x}_j^k)\}$)
    Determine where to explore, i.e., $p\tau$
    Move to a new position according to (5) where $P_\Omega(\mathbf{x})$ *is evaluated via Algorithm 2*
  end for
end while
Output $\mathbf{x}_*$ *and the corresponding function value* f_*
Visualization

Let $\mathbf{x} = [x_1, x_2, x_3, x_4]^\mathrm{T} = [T_\mathrm{s}, T_\mathrm{h}, R, L]^\mathrm{T}$. The pressure vessel problem can be formulated as follows [14]:

$$\begin{aligned}
\text{minimize } f(\mathbf{x}) &= 0.6224x_1x_3x_4 + 1.7781x_2x_3^2 \\
&\quad + 3.1661x_1^2x_4 + 19.84x_1^2x_3, \\
\text{subject to } g_1(\mathbf{x}) &= -x_1 + 0.0193x_3 \le 0, \\
g_2(\mathbf{x}) &= -x_2 + 0.00954x_3 \le 0, \\
g_3(\mathbf{x}) &= -\pi x_3^2x_4 - \frac{4}{3}\pi x_3^3 + 1296000 \le 0, \\
g_4(\mathbf{x}) &= x_4 - 240 \le 0, \\
x_1 &\in \{1, 2, 3, \cdots, 99\} \times 0.0625, \\
x_2 &\in \{1, 2, 3, \cdots, 99\} \times 0.0625, \\
x_3 &\in [10, 200], \\
x_4 &\in [10, 200].
\end{aligned}$$

Table 1. Comparisons of best results for the pressure vessel problem

	x_1	x_2	x_3	x_4	$g_1(\mathbf{x})$	$g_2(\mathbf{x})$	$g_3(\mathbf{x})$	$g_4(\mathbf{x})$	$f(\mathbf{x})$
[9]	0.8125	0.4375	42.0984	176.6366	8.00e-11†	-0.0359	-2.724e-4	-63.3634	6059.7143
[14]	0.7782	0.3846	40.3196	200.000	-3.172e-5	4.8984e-5†	1.3312†	-40	5885.33
[8]	0.8125	0.4375	42.0984	176.6366	8.00e-11†	-0.0359	-2.724e-4	-63.3634	6059.7143
[15]	1.125	0.625	58.2789	43.7549	-0.0002	-0.06902	-3.71629	-196.245	7198.433
[16]	1.125	0.625	48.97	106.72	-0.1799	-0.1578	97.760	-132.28	7980.894
[17]	1.125	0.625	58.1978	44.2930	-0.00178	-0.06979	-974.3	-195.707	7207.494
[18]	0.8125	0.4375	40.3239	200.0000	-0.034324	-0.05285	-27.10585	-40.0000	6288.7445
[19]	0.9375	0.5000	48.3290	112.6790	-0.0048	-0.0389	-3652.877	-127.3210	6410.3811
[20]	1.125	0.625	58.291	43.690	0.000016	-0.0689	-21.2201	-196.3100	7198.0428
[21]	0.8125	0.4375	41.9768	182.2845	-0.0023	-0.0370	-22888.07	-57.7155	6171.000
[22]	1.000	0.625	51.000	91.000	-0.0157	-0.1385	-3233.916	-149	7079.037
[23]	0.8125	0.4375	42.0870	176.7791	-2.210e-4	-0.03599	-3.51084	-63.2208	6061.1229
[24]	1	0.625	51.2519	90.9913	-1.011	-0.136	-18759.75	-149.009	7172.300
[25]	0.8125	0.4375	42.0984	176.6378	-8.8000e-7	-0.0359	-3.5586	-63.3622	6059.7258
PSA	0.8125	0.4375	42.0952	176.8095	-6.2625e-5	-0.0359	-738.7348	-63.1905	6063.2118

† means that the corresponding constraint is violated.

Algorithm 4 Algorithm for $P_\Omega(\mathbf{x} = [x_1, x_2, x_3, x_4]^{\mathrm{T}})$ in the pressure vessel problem

```
y1 = round(x1/0.0625) × 0.0625
if y1 < 0.0625 then
    y1 = 0.0625
end if
if y1 > 99 × 0.0625 then
    y1 = 99 × 0.0625
end if
y2 = round(x2/0.0625) × 0.0625
if y2 < 0.0625 then
    y2 = 0.0625
end if
if y2 > 99 × 0.0625 then
    y2 = 99 × 0.0625
end if
if x3 < 10 then
    y3 = 10
end if
if x3 > 200 then
    y3 = 200
end if
if x4 < 10 then
    y4 = 10
end if
if x4 > 200 then
    y4 = 200
end if
return y = [y1, y2, y3, y4]^T
```

Evidently, this problem has a nonlinear cost function, three linear and one nonlinear inequality constraints. Besides, there are two discrete and two continuous design variables. Thus, the problem is relatively complicated. As this problem is a mixed discrete-continuous optimization, the projection function $P_\Omega(\mathbf{x})$ is slightly modified and presented in Algorithm 4. Besides, the initialization of the initial positions of porcellio scaber is modified as follows:

$$x^0_{i,1} = 0.0625 + floor((99-1) \times rand) \times 0.0625,$$
$$x^0_{i,2} = 0.0625 + floor((99-1) \times rand) \times 0.0625,$$
$$x^0_{i,3} = 10 + floor(200-10) \times rand,$$
$$x^0_{i,4} = 10 + floor(200-10) \times rand,$$

where $x^0_{i,j}$ denotes the jth variable of the position vector of the ith porcellio scaber; $floor(y) = \mathrm{argmin}_{x\in\{0,1,2,\cdots\}}\{x + 1 > y\}$, i.e., the $floor$ function obtains the integer part of a real number; $rand$ denotes a random number in the region $(0, 1)$. The functions $floor$ and $rand$ are available at Matlab.

The best result we obtained using the PSA in 1000 instances of executions and those by using various existing algorithms or methods for solving this problem are listed in Table 1. Note that, in the experiments, 40 porcellio scaber are used, the parameter λ is set to 0.6, and the $MaxStep$ is set to 100000 with τ being a zero-mean random number with the standard deviation being 0.1. As seen from Table 1, the best result obtained by using the PSA is better than most of the existing results. Besides, the difference between the best function value among all the ones in the table and the best function value obtained via using the PSA is quite small.

4.2 Case II: Himmelblau's nonlinear optimization problem

In this subsection, we consider a nonlinear optimization problem proposed by Himmelblau [12]. This problem is also one of the well known benchmark problems for bio-inspired algorithms. The problem is formally described as

Table 2. Comparisons of best results for Himmelblau's nonlinear optimization problem

	x_1	x_2	x_3	x_4	x_5	$g_1(\mathbf{x})$	$g_2(\mathbf{x})$	$g_3(\mathbf{x})$	$f(\mathbf{x})$
[26]	78.0	33.0	27.07997	45.0	44.9692	92.0000	100.4048	20.0000	-31025.5602
[27]	78.00	33.00	29.995	45.00	36.776	90.7147	98.8405	19.9999†	-30665.6088
[28]	81.4900	34.0900	31.2400	42.2000	34.3700	90.5225	99.3188	20.0604	-30183.576
[12]	78.6200	33.4400	31.0700	44.1800	35.2200	90.5208	98.8929	20.1316	-30373.949
[29]	78.00	33.00	29.995256	45.00	36.775813	92	98.8405	20	-30665.54
PSA	79.9377	33.8881	28.5029	41.3052	41.7704	91.6157	100.4943	20.0055	-30667.8113

† means that the corresponding constraint is violated.

follows [12]:

$$\begin{aligned}
\text{minimize } f(\mathbf{x}) =& 5.3578547x_3^2 + 0.8356891x_1x_5 \\
&+ 37.29329x_1 - 40792.141, \\
\text{subject to } g_1(\mathbf{x}) =& 85.334407 + 0.0056858x_2x_5 \\
&+ 0.00026x_1x_4 - 0.0022053x_3x_5, \\
g_2(\mathbf{x}) =& 80.51249 + 0.0071317x_2x_5 \\
&+ 0.0029955x_1x_2 + 0.0021813x_3^2, \\
g_3(\mathbf{x}) =& 9.300961 + 0.0047026x_3x_5 \\
&+ 0.0012547x_1x_3 + 0.0019085x_3x_4, \\
&0 \le g_1(\mathbf{x}) \le 92, \\
&90 \le g_2(\mathbf{x}) \le 110, \\
&20 \le g_3(\mathbf{x}) \le 25, \\
&78 \le x_1 \le 102, \\
&33 \le x_2 \le 45, \\
&27 \le x_3 \le 45, \\
&27 \le x_4 \le 45, \\
&27 \le x_5 \le 45,
\end{aligned}$$

with $\mathbf{x} = [x_1, x_2, x_3, x_4, x_5]^{\mathrm{T}}$ being the decision vector. In this problem, each double-side nonlinear inequality can be represented by two single-side nonlinear inequality constraints. For example, the constraint $90 \le g_2(\mathbf{x}) \le 110$ can be replaced by the following two constraints:

$$\begin{aligned}
-g_2(\mathbf{x}) &\le -90, \\
g_2(\mathbf{x}) &\le 110.
\end{aligned}$$

Thus, this problem can also be solved by the PSA proposed in this paper.

The best result we obtained via using the PSA in 1000 instances of executions, together with the result obtained by other algorithms or methods, is listed in Table 2. In the experiments, 40 porcellio scaber are used, the parameter λ is set to 0.6, and the $MaxStep$ is set to 100000 with τ being a zero-mean random number with the standard deviation being 0.1. Evidently, the best result generated by the PSA is ranked No. 2 among all the results in Table 2.

By the above results, we conclude that the PSA is a relatively promising algorithm for solving constrained optimization problems. The quite smalle performance difference between the PSA and the best one may be the result of the usage of the penalty method with a constant penalty parameter.

5 Conclusions

In this paper, the bio-inspired algorithm PSA has been extended to solve nonlinear constrained optimization problems by using the penalty method. Case studies have validated the efficacy and superiority of the resultant PSA. The results have indicated that the PSA is a promising algorithm for solving constraint optimization problems. There are several issues that requires further investigation, e.g., how to select a best penalty parameter that not only guarantees the compliance with constraints but also the optimality of the obtained solution. Besides, how to enhance the efficiency of the PSA is also worth investigating.

Acknowledgement

This work is supported by the National Natural Science Foundation of China (with numbers 91646114, 61370150, and 61401385), by Hong Kong Research Grants Council Early Career Scheme (with number 25214015), and also by Departmental General Research Fund of Hong Kong Polytechnic University (with number G.61.37.UA7L).

References

[1] L. Jin, S. Li, B. Liao, and Z. Zhang, "Zeroing neural networks: A survey," *Neurocomputing*, vol. 267, pp. 597–604, 2017.

[2] L. Jin and Y. Zhang, "Continuous and discrete Zhang dynamics for real-time varying nonlinear optimization," *Numerical Algorithms*, vol. 73, no. 1, pp. 115–140, 2016.

[3] B. Liao, Y. Zhang, and L. Jin, "Taylor $O(h^3)$ discretization of ZNN models for dynamic equality-constrained quadratic programming with application to manipulators," *IEEE Trans. Neural Netw. Learn. Syst.*, vol. 27, no. 2, pp. 225–237, 2016.

[4] L. Jin and Y. Zhang, "Discrete-time Zhang neural network for online time-varying nonlinear optimization with application to manipulator motion generation," *IEEE Trans. Neural Netw. Learn. Syst.*, vol. 26, no. 7, pp. 1525–1531, 2016.

[5] L. Jin, Y. Zhang, S. Li, and Y. Zhang, "Modified ZNN for time-varying quadratic programming with inherent tolerance to noises and its application to kinematic redundancy resolution of robot manipulators," *IEEE Trans. Ind. Electron.*, vol. 63, no. 11, pp. 6978–6988, 2016

[6] J. Kennedy and R. Eberhart, "Particle swarm optimization," in *Proc. IEEE Int. Conf. Neural Netw.*, 1995, 1942–1948.

[7] A. H. Gandomi, X. Y. Yang, "Benchmark problems in structural optimization," in *Computational Optimization Methods And Algorithms*, Berlin: Springer-Verlag, 2011, pp. 259–281.

[8] S. He , E. Prempain, and Q. H. Wu, "An improved particle swarm optimizer for mechanical design optimization problems," *Eng. Opt.*, vol. 36, no. 5, pp. 585–605, 2004.

[9] A. H. Gandomi, X. S. Yang, and A. H. Alavi, "Cuckoo search algorithm: A metaheuristic approach to solve structural optimization problems," *Eng. Comput.*, vol. 29, pp. 17–35, 2013.

[10] O. Pauline, H. C. Sin, D. D. C. V. Sheng, S. C. Kiong, and O. K. M., "Design optimization of structural engineering problems using adaptive cuckoo search algorithm," in *Proc. 3rd Int. Conf. Control Autom. Robot.*, 2017, pp. 745–748.

[11] V. Garg and K. Deep, "Effectiveness of constrained laplacian biogeography based optimization for solving structural engineering design problems," in *Proc. 6th Int. Conf. Soft. Comput. Problem Solv.*, 2017, pp. 206–219.

[12] D. Himmelblau, *Applied Nonlinear Programming*, New York: McGraw-Hill, 1972.

[13] Y. Zhang and S. Li, "PSA: A novel optimization algorithm based on survival rules of porcellio scaber," Available at https://arxiv.org/abs/1709.09840.

[14] X. S. Yang, "Firefly algorithm, stochastic test functions and design optimisation," *Int. J. Bio-Inspired Comp.*, vol. 2, no. 2, pp. 2, pp. 78–84, 2010.

[15] K. S. Lee and Z. W. Geem, "A new meta-heuristic algorithm for continuous engineering optimization: Harmony search theory and practice," *Comput. Methods Appl. Mech. Engrg.*, vol. 194, pp. 3902–3933, 2005.

[16] E. Sandgren, "Nonlinear integer and discrete programming in mechanical design optimization," *J. Mech. Des. ASME*, vol. 112, 223–229, 1990.

[17] S. J. Wu and P. T. Chow, "Genetic algorithms for nonlinear mixed discrete-integer optimization problems via meta-genetic parameter optimization," *Engrg. Optim.*, vol. 24, 137–159, 1995.

[18] C. A. C. Coello, "Use of a self-adaptive penalty approach for engineering optimization problems," *Comput. Ind.*, vol. 41, no. 2, pp. 113–127, 2000.

[19] K. Deb, "GeneAS: a robust optimal design technique for mechanical component design", in: D. Dasgupta, Z. Michalewicz Eds., *Evolutionary Algorithms in Engineering Applications*, Springer-Verlag, Berlin, 1997, pp. 497–514.

[20] B. K. Kannan and S. N. Kramer, "An augmented Lagrange multiplier based method for mixed integer discrete continuous optimization and its applications to mechanical design," *J. Mecha. Design, Trans. ASME*, vol. 116, pp. 318–320, 1994.

[21] S. Akhtar, K. Tai, T. Ray, "A socio-behavioural simulation model for engineering design optimization," *Eng. Optmiz.*, vol. 34, no. 4, pp. 341–354, 2002.

[22] J. F. Tsai, H. L. Li, and N. Z. Hu, "Global optimization for signomial discrete programming problems in engineering design," *Eng. Optmiz.* no. 34, no. 6, pp. 613–622, 2002.

[23] C. A. C. Coello and N. C. Corteś, "Hybridizing a genetic algorithm with an artificial immune system for global optimization," *Eng. Optmiz.*, vol. 36, no. 5, pp. 607–634, 2004.

[24] H. L. Li and C. T. Chou, "A global approach for nonlinear mixed discrete programming in design optimization," *Eng. Optmiz.* vol. 22, pp. 109–122 1994.

[25] A. Kaveh and S. Talatahari, "An improved ant colony optimization for constrained engineering design problems," *Eng. Comput.* vol. 27, no. 1, pp. 155–182, 2010.

[26] G. H. M. Omran and A. Salman, "Constrained optimization using CODEQ," *Chaos Soliton. Fract.*, vol. 42, pp. 662–668, 2009.

[27] A. Homaifar, S. Lai, X. Qi, "Constrained optimization via genetic algorithms," *Simulation*, vol. 62, no. 4, pp. 242–253, 1994.

[28] M. Gen and R. Cheng, *Genetic Algorithms and Engineering Design*, Wiley, New York, 1997.

[29] G. G. Dimopoulos, "Mixed-variable engineering optimization based on evolutionary and social metaphors," *Comput. Method. Appl. Mech. Eng.*, vol. 196, pp. 803–817, 2007.

Design of an instantaneous optimal energy management strategy for a dual-axis-parallel plug-in hybrid electric vehicle

YAN Zhengjun[1,2], WANG Weida[1,*], LI Hongcai[1], HAN Lijin[1], LV Yong[2], ZHANG Sheng[2], and LIN Yi[2]

[1]School of Mechanical Engineering, Beijing Institute of Technology, Beijing 100081, China
[2]PLA Army Engineering University, Wuhan 430070, China

Abstract. The research took a dual-axis-parallel plug-in hybrid electric vehicle (PHEV) as its research object: the engine and electric motor were connected with an automated mechanical transmission (AMT) through two different axles. As the torques of the engine and the electric motor converge through a variety of gear combinations selected within the AMT, the gearshift schedule has a high degree of coupling with the energy management strategy and therefore they jointly determine various aspects of the vehicle performance. Under the coupling constraints of torque and speed of the engine and electric motor, an instantaneous optimal energy management strategy was proposed by considering the influences of torque distribution and gear-matching on the performance of the whole vehicle. By adjusting the torques of the engine and the electric motor, as well as the gear of the AMT independently, the strategy allowed both the engine and the electric motor to operate at a higher efficiency. In this way, the engine, the electric motor, and the AMT can be comprehensively controlled and optimised. The feasibility and effectiveness of the proposed integrated optimisation strategy was proved by simulation on the AVL Cruise and MATLAB™/Simulink platform.

1 Introduction

Owing to its being equipped with a high battery capacity and a high-power drive motor, the plug in hybrid electric vehicle (PHEV) can not only travel far with the only support from electrical power, but it can also reach an optimal fuel consumption pattern by adjusting the parameter range of the engine through its electric motor. Thus, a PHEV is an ideal product in the transition period from traditional vehicles to purely electric vehicles [1].
As one of the core technologies of PHEV, the energy management strategy learns from that of an ordinary HEV while it also differs from the latter. At present, there are mainly rule-based and optimisation-based strategies available in which the rule-based energy strategy uses the "CD-CS" mode [2]. This means that it is better to choose the motor drive to acquire a low operating cost when the state of charge (SOC) of the power battery is good, and apply the same strategy as used in an ordinary HEV to improve the working condition of the engine and maintain the power balance when the SOC is poor [3]. Such a strategy shows both the advantages involving simplicity, effectiveness, and small computational burden and one disadvantage: a failure to reach optimal effect. The optimisation-based energy management strategy [4] is divided into global optimisation [5] represented by the dynamic programming algorithm (DP) and instantaneous optimisation [6-7] represented by the equivalent consumption minimisation strategy (ECMS). As the former requires an a priori understanding of the working conditions, although the globally optimal solution can be acquired, it offers poor real-time performance and requires many calculations. The latter weighs the fuel consumption of engine and electric consumption of the electric motor and finds the optimal mode for energy distribution, in real-time, by exploring the operating points with the lowest cost function in a feasible domain. Although it can only obtain a locally optimal solution, the strategy offers strong real-time performance.
The traditional ECMS is proposed based on an HEV. As electrical energy from the HEV power battery is all derived from fuel [8], the electrical consumption of battery power is transformed to fuel consumption in the engine through use of an equivalent factor used to find the optimal operating points of both the engine and electric motor aimed at minimising instantaneous fuel consumption. For a PHEV, the electrical power from the battery is divided into two streams: that acquired from the power grid and that transformed by fuel [9]. Thus, it is necessary to modify the traditional ECMS by combining the characteristics of the PHEV while ECMS is applied therein: considering this, an instantaneous optimal energy management strategy based on the driving cost minimisation strategy (DCMS) of the whole vehicle is proposed by considering costs of fuel and electricity consumption as well as the effects of gear selection in the automated mechanical transmission (AMT) on the performance. In this way, the torques of the engine and electric motor, as well as gears, are

* Corresponding author: wangwd0430@bit.edu.cn

 DOI: 10.1051/matecconf/201713900034

dynamically distributed in real-time to realise driving cost minimisation for the whole vehicle. On the basis of a dual-axis-parallel PHEV, the study verified the rationality of the proposed strategy by using a simulation on the AVL Cruise and MATLAB™/Simulink platform.

2 Structure and basic parameters of the dual-axis-parallel PHEV

The structural mechanism of the dual-axis-parallel PHEV is shown in Figure 1. The structure is composed of an engine, an electric motor, a clutch, and an AMT with a dual-axis input. It can work under four modes including pure electric, hybrid drive, driving charge, and regenerative braking.

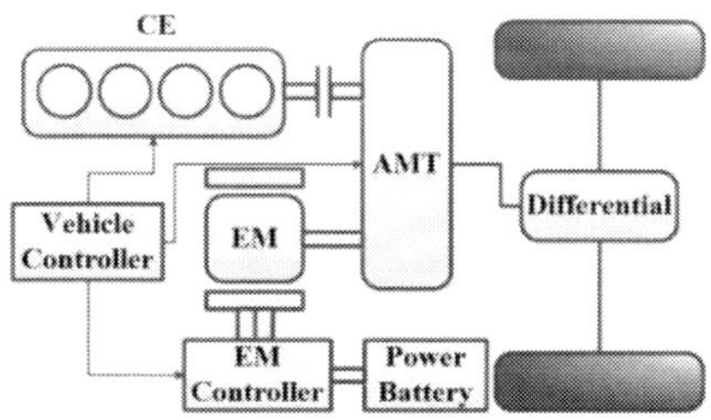

Figure 1 The dual-axis-parallel PHEV

While changing speed and torque, the torques of the engine and electric motor are coupled inside the AMT, while mechanical and electrical gears upshift alternatively to complete the gear-shift without power failure [10]. The transmission of the AMT is shown in Figure 2 and the transmission ratios of various gears are listed in Table 1.

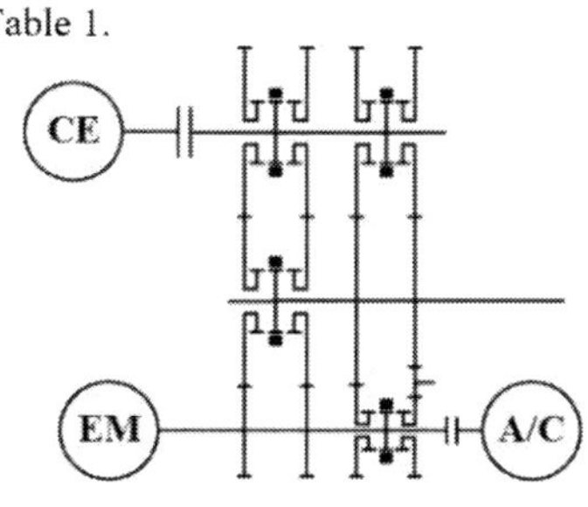

Figure 2 The AMT system

Table 1 Transmission ratios of various AMT gears

Gear/-	i_{ge}/-	i_{gm}/-	i_0/-
1	3.369	2.333	
2	1.815	2.333	
3	1.815	0.867	
4	1.252	0.867	
5	0.978	0.867	4.053
6	0.784	0.867	
7	0.784	0.441	
8	0.636	0.441	
9	0.497	0.441	

3 Optimisation objective of DCMS

The price cost is obtained by summation of the costs of fuel and electricity consumption in the proposed DCMS. By considering maintaining the electric energy of battery, the driving cost function of the whole vehicle is established by using the weighting method. In this way, the aforementioned multi-objective problem is transformed into a single-objective optimisation problem, which is regarded as the control objective of the instantaneous optimal control strategy.

3.1 Definition of price cost

After investigating market fuel and electricity prices, the price cost of the whole vehicle, during operation, is defined in Equation 1:

$$F_{cost}(t)=\begin{cases}\left(\dfrac{j_f Q_f(t)}{\rho_f}+\dfrac{j_e P_b(t)}{\eta_g}\right)\Delta t, P_b(t)\geq 0 \cap SOC(t)\geq SOC_{ref} \\ \left(\dfrac{Q_f(t)}{\rho_f}+P_b(t)Q\right)j_f\Delta t, P_b(t)\geq 0 \cap SOC(t)< SOC_{ref} \\ \dfrac{Q_f(t)\left(1+\dfrac{P_b(t)}{P_e(t)}\right)j_f\Delta t}{\rho_f}, P_b(t)<0\end{cases} \quad (1)$$

Where, j_f, je, Qf(t), P_b(t), and ρ_f refer to the price of fuel per litre (yuan/L), the price of electricity per kilowatt hour (yuan/kW•h), rate of consumption of fuel in the engine at time t (g/s), the power of the battery at time t (kW), and the mass density of gasoline (g/L), respectively. Moreover, η_g, Q, Δt, and SOC_{ref} represent the charge efficiency of the power grid and the average fuel consumption of the engine for generating one kilowatt hour of electricity (L), the controlled step-length (s), and the objective value of power in the current battery SOC, respectively. It is worth noting that when $SOC < SOC_{ref}$, the cost should be calculated according to the fuel price because the electrical energy consumed by the battery is transformed from fuel: additionally, the battery is discharged or charged when the power P_b of the battery is greater, or less, than zero, respectively.

$$P_b(t)=\begin{cases}\dfrac{T_m(t)n_m(t)}{9550\eta_{dis}\eta_{bat}}, T_m(t)\geq 0 \\ \dfrac{T_m(t)n_m(t)\eta_{chg}\eta_{bat}}{9550}, T_m(t)<0\end{cases} \quad (2)$$

Where, $T_m(t)$ refers to the torque (Nm) of the electric motor at time t, which is positive or negative at electric, or electric generating, modes, respectively. Moreover, $n_m(t)$, η_{dis}, η_{chg}, and η_{bat} represent the rotational speed (rpm) of the electric motor at time t, the motor efficiency of the electric motor, the generating efficiency of the electric motor, and the average charge and discharge efficiency of the battery, respectively.

3.2 Driving cost function

After the battery discharges to a certain extent, the PHEV has to maintain the amount of electrical energy in its battery and allow this to fluctuate only within a certain range. By doing so, the service safety of the battery is guaranteed so as to prolong service life and also the economy of the whole vehicle can be improved. Therefore, the degree of SOC deviation from the objective value as the penalty term is incorporated into the driving cost of the whole vehicle, which is defined as follows:

$$f_{SOC}(t)=\alpha sign\left[P_b(t)\right]\left[SOC(t)-SOC_{ref}\right] \quad (3)$$

Where, α refers to the penalty-coefficient of SOC change, and is a negative value; *sign* represents a symbol function, which takes values of 1, -1, or 0 when P_b is greater than, less than, or equal to zero, respectively.
The objective function of the DCMS control strategy after considering the SOC state constraint is expressed as follows:

$$J=F_{cost}+f_{SOC} \quad (4)$$

3.3 Instantaneous optimal algorithm

The instantaneous optimal algorithm based on driving cost minimisation of the whole vehicle aims to calculate the controlled quantity u which minimises the operation cost J of the whole vehicle in each control period.

$$u^{opt}=\arg\min_{u}(J) \quad (5)$$

The critical components need to satisfy the following constraints:

$$\begin{cases} T_{e\_min} \le T_e(t) \le T_{e\_max} \\ n_{e\_min} \le n_e(t) \le n_{e\_max} \\ T_{m\_min} \le T_m(t) \le T_{m\_max} \\ n_{m\_min} \le n_m(t) \le n_{m\_max} \\ gear \in \{1,2,3,4,5,6,7,8,9\} \end{cases} \quad (6)$$

The solution process for DCMS instantaneous optimisation is shown in Figure 5.

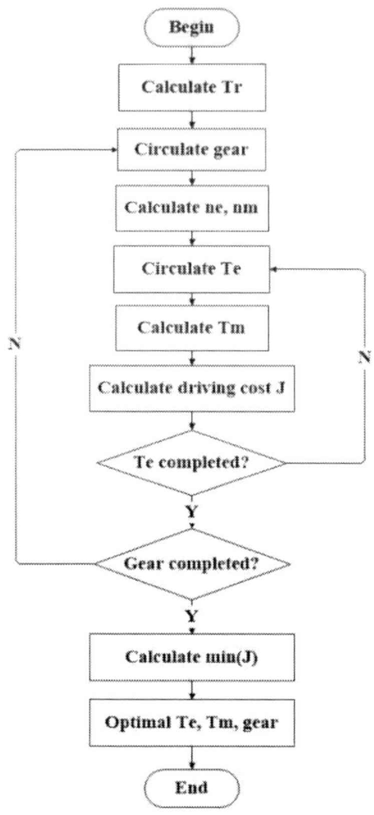

Figure 5 DCMS: instantaneous optimisation

4 Simulation results and analysis

To verify the effectiveness of the proposed instantaneous optimal strategy, the kinetic model and vehicle controller model are separately established for the dual-axis-parallel PHEV based on the AVL Cruise and MATLAB™/Simulink platform to construct the combined simulation model (Figure 6). On the condition of having identical parameters, the simulation comparison is conducted by using the proposed instantaneous optimal, and rule-based, energy management strategies, respectively.

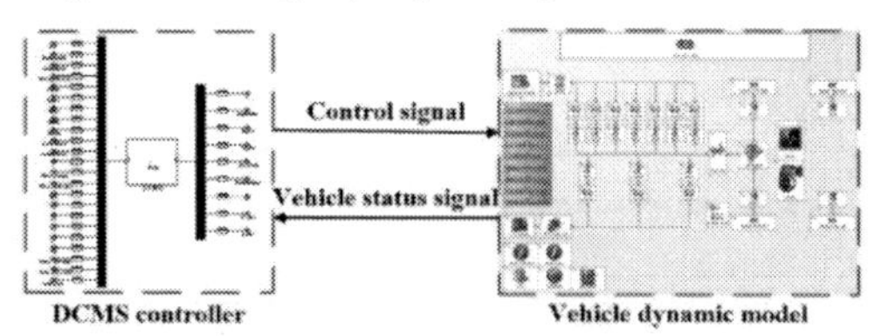

Figure 6 Control flowchart: combined simulation

4.1 Results of combined simulation

The economical simulation is conducted mainly on the basis of the cyclic working conditions of the NEDC. In

 DOI: 10.1051/matecconf/201713900034

terms of a PHEV, it is necessary to satisfy the requirements of travelling over the whole driving range (designed to be 55 km) using only the pure electrical motor while determining the SOC_{ref}. The study conducted the simulation based on pure electrical working conditions and assumed a value of SOC_{ref} of 25%. Considering the consumption and maintenance processes of electrical energy, and also to shorten the simulation time, the original SOC was set to a value of 30%. The result of this simulation of the rule-based energy management strategy is shown in Figure 7.

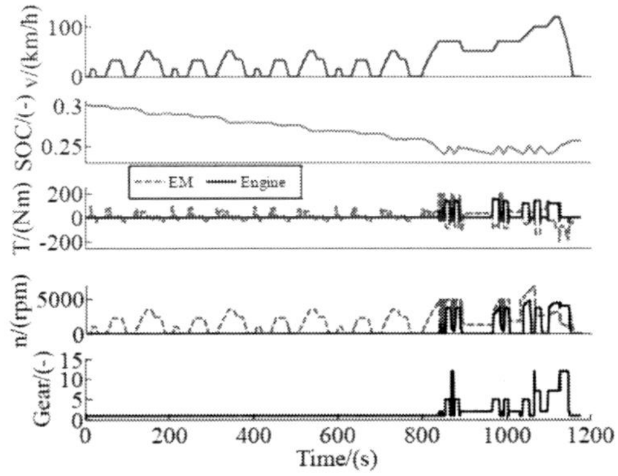

(a) Simulation results: key parameters

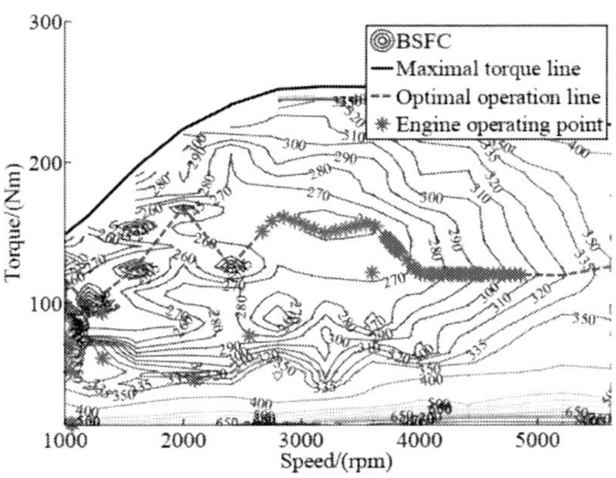

(b) Distributions of engine operating points

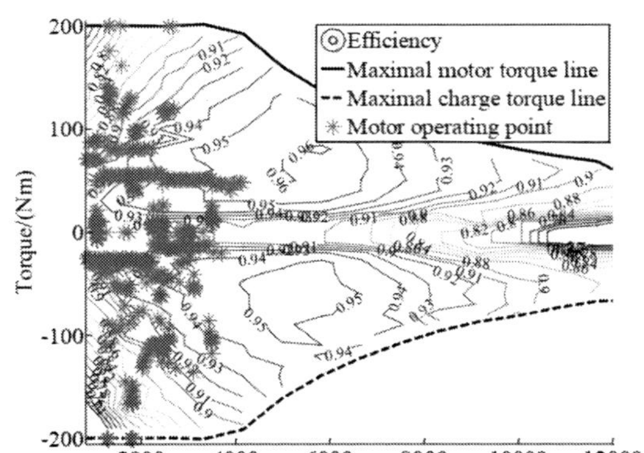

(c) Distribution of electric motor operating points

Figure 7 Simulation of rule-based strategy

The result of simulation of the DCMS-based energy management strategy is shown in Figure 8.

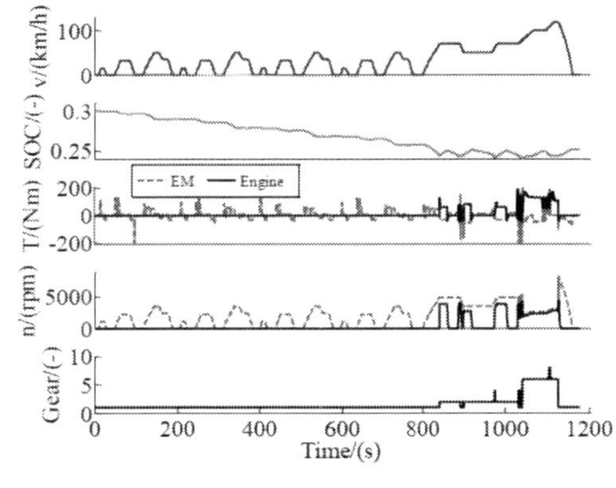

(a) Simulation results: key parameters

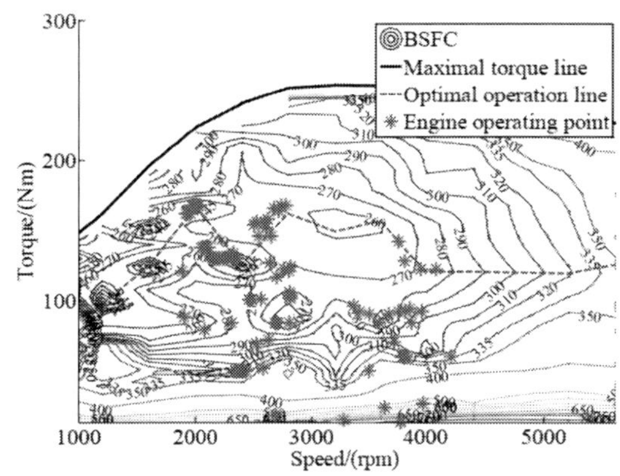

(b) Distribution of engine operating points

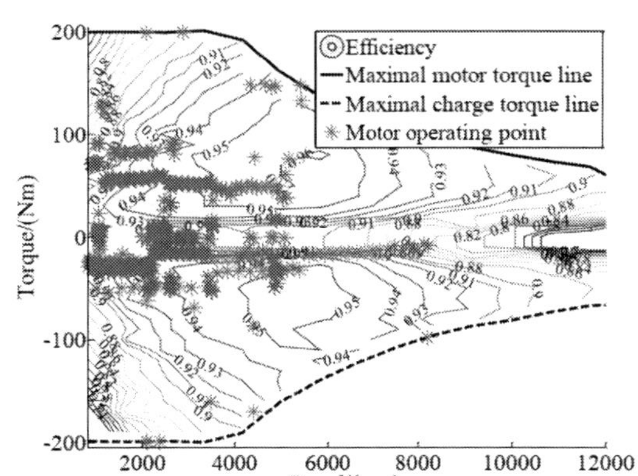

(c) Distribution of electric motor operating points

Figure 8 Simulation of DCMS-based strategy

4.2 Comparative analysis of simulation results

As shown in Figures 7(a) and 8(a), the decrease-maintenance of electrical energy in the battery can be realised by using the two selected strategies. After comparing Figures 7(b) and 8(b), the operating points of the engine are mainly distributed along the optimal torque curve when employing a rule-based energy management strategy. In contrast, the DCMS-based strategy allows the operating points of engine to be clustered in a high-efficiency, low-power area due to consideration of the fuel consumption cost. Therefore, there is an optimal fuel economy point. By comparing Figures 7(c) and 8(c), it can be seen that, owing to the economy (performance) of the electric motor being taken into consideration, the operating points of the electric motor under a DCMS-based strategy are optimised and therefore were mostly distributed in a high-efficiency area. Therefore, the electrical, and generating, losses of electric motor are reduced to further decrease the driving cost of the whole vehicle by about 21.7% (Table 2).

Table 2 Simulation of driving costs

Control strategy	Fuel consumption/ (L)	Electricity consumption/ (kWh)	Cost/ (yuan)
Rule-based	0.62	0.50	4.10
DCMS-based	0.47	0.57	3.21

5 Conclusions

(1)An instantaneous optimal energy management strategy based on the driving cost of the whole vehicle is proposed by considering the influence of the torque distribution of both the engine and electric motor and gear-matching in the AMT, on performance of the whole vehicle.
(2)The change in electrical energy in the battery is incorporated into the instantaneous optimal objective function as a penalty factor to make the battery SOC balance.
(3)The DCMS-based and rule-based energy management strategies are compared by establishing a combined simulation platform based on AVL Cruise and MATLAB™/Simulink. The result shows that the proposed DCMS-based energy management strategy has an optimal effect on the operating points of both engine and electric motor of a dual-axis-parallel PHEV and therefore greatly improves the performance of the whole vehicle.

This work was supported by the National Natural Science Foundation of China (Grant nos 51575043 and U1564210).

References

1. Cheng Qingquan, Sun Fengchun, Zhu Jiaguang. Modern electric vehicle technology. Beijing: Beijing Institute of Technology Press, (2012)
2. Jiankun Peng, Hongwen He, Rui Xiong. Rule based energy management strategy for a series-parallel plug-in hybrid electric bus optimized by dynamic programming. Applied Energy, **185**, 1633-1643 (2017)
3. Wang Weida, Xiang Changle, Han Lijin. Integrated Control Strategy of Electro-mechanical Transmission System. Journal of Mechanical Engineering, **47**, 20:152-158 (2011)
4. WIRASINHA S G, EMADI A. Classification and review of control strategies for plug-in hybrid electric vehicles. IEEE Trans. on Vehicular Technology, **60**, 1:111-122 (2011)
5. Yalian Yang, Xiaosong Hua, Huanxin Pei. Comparison of power-split and parallel hybrid powertrain architectures with a single electric machine: Dynamic programming approach. Applied Energy, **168**, 683-690 (2016)
6. Huang Shuo, Li Liang, Yang Chao. Rule Correction-based Instantaneous Optimal Energy Management Strategy for Single-shaft Pa0rallel Hybrid Electric Bus. IEEE Trans. on Vehicular Technology, **50**, 20:113-121 (2014)
7. QIN Datong, WEI Hanbing, DUAN Zhihui, et al. Amulti-objective real-time optimal control strategy for Fuel consumption and emissions of full hybrid electric vehicle. Journal of Mechanical Engineering, **48**, 6:83-89 (2012)
8. Paganelli G, Delprat S, Guerra T M, et al. Equivalent consumption minimization strategy for parallel hybrid powertrains. Vehicular Technology Conference, 2002. VTC Spring 2002. IEEE 55th. IEEE, **4**, 2076-2081 (2002)
9. Qin Datong, Yang Guanlong, Liu Yonggang. Energy Management Optimization Control Strategy of Plug-in Parallel Single-motor Hybrid Electric Vehicle. China Journal of Highway and Transport, **26**, 5:170-176. (2013)
10. Wu Haikang, Zhong Zaimin, Yu Zhuoping. Power Shift Feature Research of a Novel Active Transmission for Parallel Hybrid Vehicle Use. Journal of Tongji University, **42**, 12 (2014)

Insulator recognition based on convolution neural network

Yanli Yang* and *Lijuan* Wang

Tianjin Key Laboratory of Optoelectronic Detection Technology and Systems, Tianjin Polytechnic University, 300387, Tianjin, China

Abstract. Insulator fault detection plays an important role in maintaining the safety of transmission lines. Insulator recognition is a prerequisite for its fault detection. An insulator recognition algorithm based on convolution neural network (CNN) is proposed. A dataset is established to train the constructed CNN. The correct rate is 98.52% for 1220 training samples and the accuracy rate of testing is 89.04% on 1305 samples. The classification result of the CNN is further used to segment the insulator image. The test results show that the proposed method can realize the effective segmentation of insulators.

1 Introduction

As a special insulation control, insulators play an important role in transmission lines. Once the insulator breaks down, it will damage the use and the running life of lines. Therefore, insulator fault detection is very important in maintaining the safety of the whole line [1]. Manned helicopter inspection was first used in some developed countries [2]. However, its cost is high, and there are safety problems [3]. Unmanned aerial vehicle (UAV) inspection has more advantages, such as convenient carrying, high efficiency, simple, reliable, quick response, low cost, rich load, using broadly, low environmental request to take-off and land, autonomous flight and so on [4]. Thus it receives widespread attention. With the development of computer vision technology, it has become the mainstream of inspecting transmission line by using UAV to generate report of various faults automatically [5].

In order to identify and detect faults of aerial insulator, some methods are already designed. It is shown in [6] that colour features and gradient characteristics are used to segment the insulator image, and then adaptive morphology is applied to identify fault for segmented insulator. The insulator image is segmented by using the maximum between-class variance method (OTSU) and faults are located by calculating the distance between adjacent insulator pairs [7]. It is reported in [8] that insulators are segmented by using the improved edge direction histogram, and faults are judged by using the normalized histogram distance. The insulator contour is detected by Hough transform, and then fault detection is judged with the block texture feature sequence [9]. In addition, the image is segmented by the difference of gray value, and the insulator status is judged by the difference between normal insulator and fault insulator [10]. The characteristic of these detection methods is that insulator images are segmented firstly and then faults are identified. Thus it can be found that the basis of fault detection is insulator image segmentation [11]. And there are many methods for image segmentation, such as threshold-based [12, 13], texture-based [14], edge-based, morphological-based, local mean based [15], neural network based [16], automatic cooperative segmentation with Hough detection and CV model [17] and so on. At present, there is not a generally and effective image segmentation method for the insulator images with complex and changeable background. So it is undoubtedly a difficult problem to design an image segmentation method with adaptive.

Insulator segmentation is usually designed for images that contain insulators. However, the inspection objects of UAV include not only insulators, but also wires, wire clamps, leading wires, tower poles, strained tube and line corridors etc [18]. The efficiency of relying on workers to observe and judge insulator images is very low because the images collected are varied [19]. In order to realize the automatic detection of insulator faults, it is necessary to select out the insulator image first.

Convolution neural network (CNN) has gained great success in the field of image processing, and it receives more and more attentions [20]. In this paper, an algorithm for insulator recognition based on CNN is proposed, which can automatically extract the feature of insulators and distinguish the insulator and background.

2 Insulator image recognition algorithm

The flowchart of insulator image recognition algorithm based on CNN is shown in Figure 1. Firstly, the image is cut to make a training dataset. Then a CNN is build and it is trained by using the dataset. The trained CNN is

* Corresponding author: yyl070805@163.com

tested and the misjudged sample is found out to update the dataset. Finally, the CNN is trained again on the updated dataset.

CNN is mainly composed of convolution layers and pooling layers. A pooling layer usually follows a convolution layer. Convolution layer uses convolution kernels to extract the feature of previous layer through linear convolution. Pooling layer extracts the correlation feature in neighboring pixels by the maximum pooling or the mean pooling. It can effectively reduce the computation load while remaining unchanged for the extracted features.

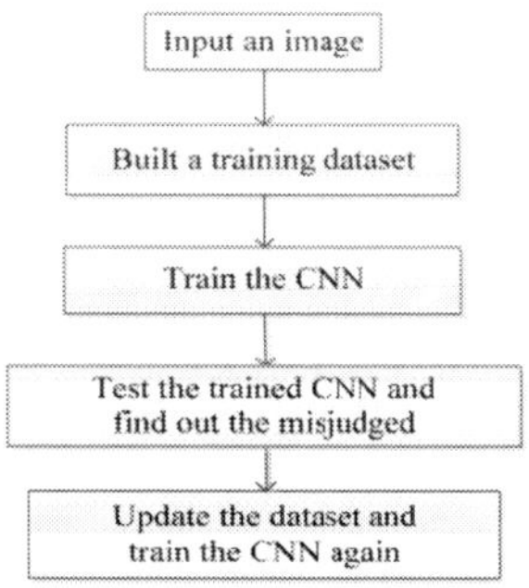

Figure 1. Flowchart of insulator image recognition algorithm based on CNN.

The CNN is constructed in this paper as shown in Figure 2, which consists of 7 layers, including one input layer, two convolution layers, two pooling layers, one fully connected layer and one output layer. Input layer is the gray image of 48*48. Six convolution kernels of 5*5 are made convolution with input layer to obtain six feature maps of 44*44. Then, six pooling maps of 22*22 are generated by mean pooling for the features extracted by the first convolution layer. Repeat the previous steps, the second convolution layer has 12 feature maps of 18*18, and the followed pooling layer has 12 pooling maps of 9*9. These features are input into a common neural network for classification. Fully connected layer has 972 neurons, and output layer has 2 neurons corresponding to background and insulator, respectively.

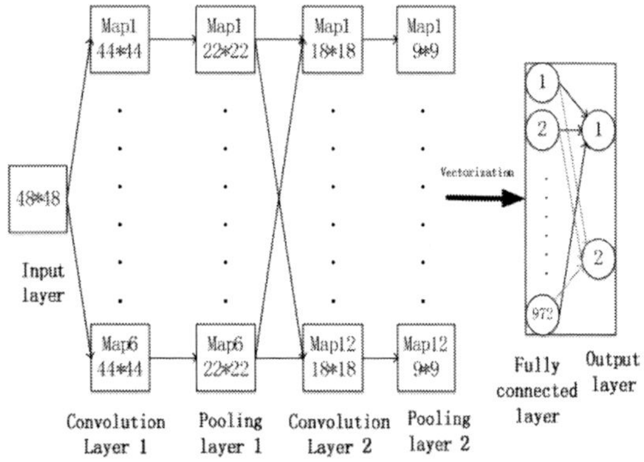

Figure 2. The constructed CNN

3 Experiment analysis

The constructed CNN is trained and tested with aerial images, some images as shown in Figure 3. The first column is the images without insulators (negative sample), and the second column is the images containing insulators (positive sample). A common feature of these images is that the background is complex and changeable.

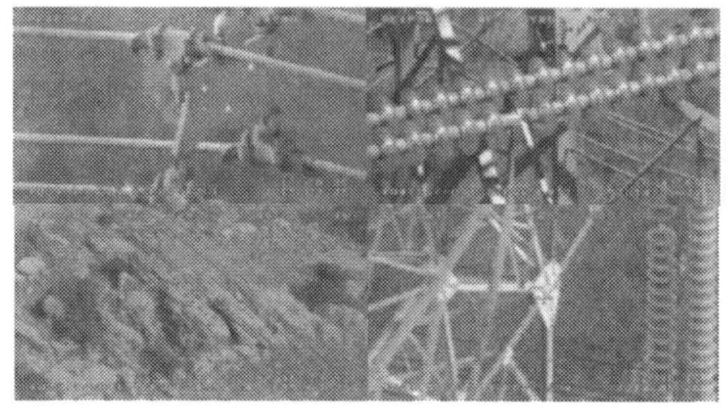

Figure 3. Partial images for experimental analysis

At present, there is not a common database for insulator detection. In this paper, a database containing 151 aerial images (1280*720) is constructed, including 133 images containing insulator and 18 images without insulator. The raw images are too big to deal with by the CNN. So, we cut them into sub-images by using a slide window. We pick out 64 aerial images from the database to establish the training dataset. These raw images are cut through a sliding window, and then the sub-images are compressed and converted to gray images. The training dataset I including 960 sub-images which size is 48*48. The training dataset includes two kinds of images, one is background and the other is insulator. Some training samples are shown in Figure 4.

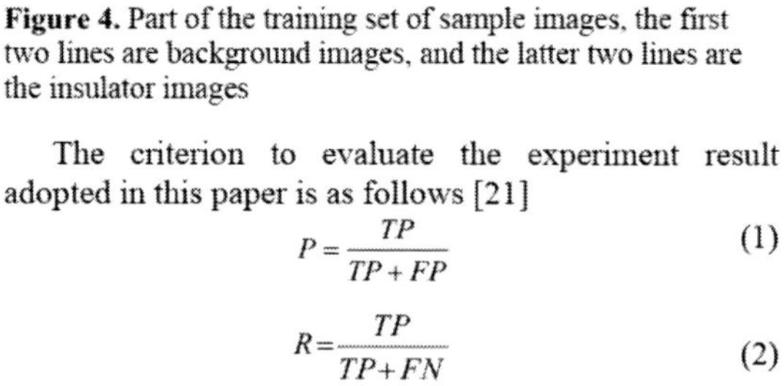

Figure 4. Part of the training set of sample images, the first two lines are background images, and the latter two lines are the insulator images

The criterion to evaluate the experiment result adopted in this paper is as follows [21]

$$P = \frac{TP}{TP + FP} \tag{1}$$

$$R = \frac{TP}{TP + FN} \tag{2}$$

$$AC = \frac{TP + TN}{TP + TN + FP + FN} \tag{3}$$

where P is the precision rate, R is the recall rate, and AC is the accuracy rate. TP represents the number of positive samples that are correctly identified. TN represents the number of negative samples that are correctly identified.

FP represents the number of negative samples that are judged as positive samples. *FN* represents the number of positive samples that are judged as negative samples.

The CNN is trained by using the training set I. The convergence curve for training set I is shown in Figure 5, and the training accuracy is 98.96%. The error converges at 361 epoch and tends to stabilize.

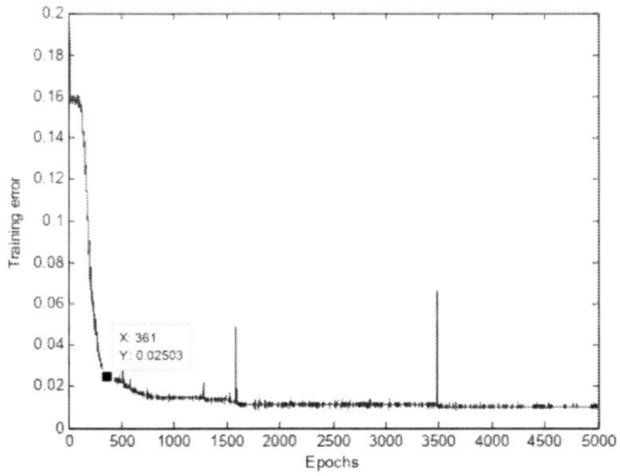

Figure 5. Training error of training dataset I

The trained CNN is tested by using the rest 87 raw images from the database. These raw images are cut into 1305 new sub-images, including 307 insulator images and 998 background images. The test result is shown in Table 1, which shows 99 insulator images and 949 background images are identified correctly. The accuracy rate, the recall rate, and the precision rate are 80.31%, 32.25% and 66.89%, respectively. It costs 21.98s for dealing with all the 151 raw images. The average processing time of each image is 0.15s.

Table 1. Test results of training models for training set I

dataset	307positive samples	998negative samples	Accuracy rate	Recall rate	Precision rate
I	99	949	80.31%	32.25%	66.89%

We pick out the 257 misjudged samples and add them to the training dataset I to form training dataset II. For the convenience of setting the parameters, we add 3 background images to training dataset, so the number of training dataset II is 1220. The CNN is trained again with training dataset II. The training error of the training dataset II is shown in Figure 6. The training accuracy rate is 98.52%, and the error converges from the 433 epoch.

Similarly, the CNN trained on the training dataset II is tested by the previous 1305 samples. Its result is shown in Table 2. We can see from this Table that 274 insulator and 888 background images are correctly identified, respectively. Compared to the test result on dataset I, we can get that the accuracy rate, the recall rate and the precision increase to 89.04%, 89.25%, and 71.35%, respectively. It costs 26.1s with all the 151 raw images. The average processing time of each image is 0.17s.

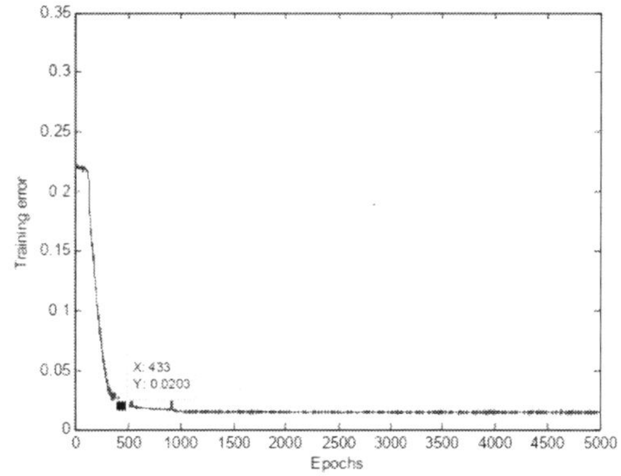

Figure 6. Training error of training dataset II

Comparing Table 1 with Table 2, we can get that the test accuracy rate increases by 8.73% and the recall rate increases by 57%, and the precision rate increases by 4.46%, when the number of training samples increases from 960 to 1220. The average time of processing an image is about 0.15s~0.17s. Thus, the method proposed in this paper can effectively identify the insulator and background in aerial images. It can provide convenience for insulator segmentation and faults detection by selecting the image containing insulators.

Table 2. Test results of training models for training set II

dataset	307positive samples	998negative samples	Accuracy rate	Recall rate	Precision rate
II	274	888	89.04%	89.25%	71.35%

4 Insulator images segmentation

In view of the method proposed in this paper can quickly identify insulators, we use it to segment the insulator image. The image segmentation method based on CNN and OTSU is shown in Figure 7. An insulator image is first cut into small sub-images, and then put into the trained CNN for classification. If the sub-image is judged to be background, the pixel value of the sub-image is set to 0, otherwise it is segmented by OTSU.

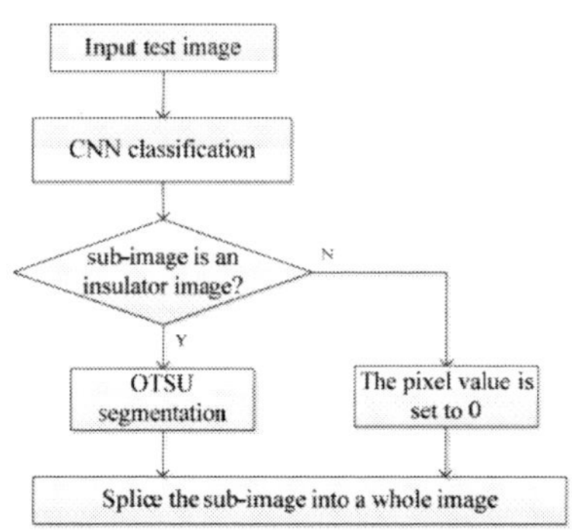

Figure 7. Flowchart of insulator images segmentation

The algorithm is tested with a few samples which are shown in Figure 8. The first line of Figure 8 is the original image, the second line is the segmentation result by OTSU, and the third line is the segmentation result by our method. From this Figure, we can see that the segmentation result of our method only has few backgrounds in the binary images comparing with the results of OTSU. Especially, the segmentation result is more obvious when the insulator and the tower pole are non-overlapping. Hence, we can conclude that the method proposed in this paper can effectively remove the residual towers, grassland and other background, and get a better segmentation result.

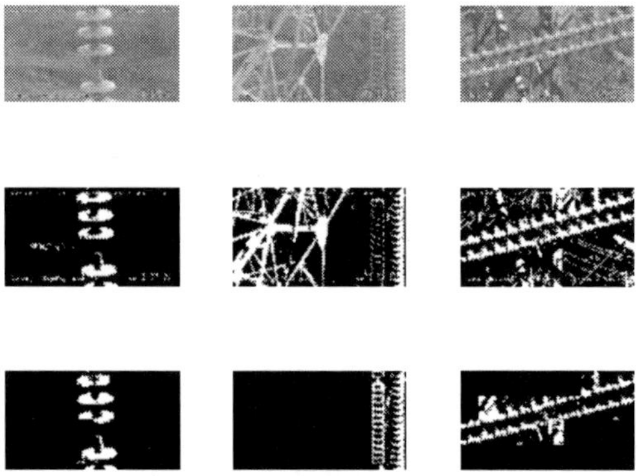

Figure 8. Example of insulator images segmentation. The first, the second and the third lines are, respectively, the original image, the OTSU result, and our method result.

Nevertheless, one point must be pointed out that the size of sub-images can affect the result of segmentation. The removal of the background in some areas is not complete if the sub-image is too large. As shown in the third column of Figure 8, it has both insulator and a lot of background in some areas. The sub-image is not too small either, otherwise the insulator cannot be identified. Therefore, in practical engineering applications, the segmentation can be improved by increasing the number of training samples and improving the classification ability of CNN.

5 Conclusions

In this paper, an insulator image recognition method based on CNN is proposed to detect images containing insulator. The experiments results show that the recognition rate of training samples can reach 98.52%, and the accuracy rate of testing increase from 80.31% to 89.04% with increasing the number of samples in the training dataset. The method can effectively distinguish the image with and without insulators. By using the classification result of the CNN, an image segmentation algorithm is designed. The test results show that the proposed algorithm can improve the image segmentation result. The proposed insulator image recognition method helps to insulator intelligent fault detection, so it has high value in engineering application.

References

[1] J.Y. Liu, North China Electric Power University (Hebei), (2010)

[2] K. Jaka, P. Franio, L. Bostjan, IEEE Transactions on Power Delivery, 25, **1**, 485-493, (2010)

[3] H. Lin, Z.H. Lin, M.W. Tang, L.H. Dai, East China Electric power, 39, **10**, 1657-1660, (2011)

[4] W.L. Wei, North China Electric Power University, (2014)

[5] T. Fang, C. Dong, X.L. Hu, Y. Wang, Journal of Shanghai Jiao Tong University, 47, **12**, 1818-1822, (2013)

[6] Y.J. Zhai, D. Wang, M.L. Zhang, J.R. Wang, F. Guo, Multimedia Tools & Applications, 76, **9**, 12051-12064, (2017)

[7] S.P. Zhang, Z. Yang, X.N. Huang, H.Q. Wu, G.Y. Zheng, Journal of Terahertz Science and Electronic Information Technology, **4**, 609-613, (2013)

[8] L. Shi, North China Electric Power University (Beijing), (2013)

[9] X.Y. Zhang, J. B. An, F. M. Chen, Wuhan, 2010 Second WRI Global Congress on Intelligent Systems, 200-203, (2010)

[10] W. Yang, North China Electric Power University (Beijing), (2016)

[11] C. Zhong, Dalian Maritime University, (2014)

[12] Z.B. Zhao, L. Wang, Chinese Journal of Scientific Instrument, 35, **3**, 558-565, (2014)

[13] Z.B. Zhao, S.X. Jin, Y. C. Liu, Chinese Journal of Scientific Instrument, 33, **9**, 2045-2052, (2012)

[14] C.R. Yang, Electric Technology, **7**, 46-48, (2010)

[15] J.L. Xu, J. Cao, K.B. Yang, Computer Engineering, 42, **9**, 262-267, (2016)

[16] T. Zheng, Dalian Maritime University, (2011)

[17] Z.B. Zhao, L. Xu, Y.C. Qi, Y.P. Cai, Chinese Journal of Scientific Instrument, 37, **2**, 395-403. (2016)

[18] X.G. Chen, G. Dong, B.H. Wang, B. Tian, Y.H. Qiao, Shandong Electric Power, **5**, 1-5, (2011)

[19] Z.H. Feng, L. X. Wang, W. Y. Liang, J. Tu, J.W. Wan, Insulators and Surge Arresters, **4**, 22-26. (2016)

[20] K.Yu, L. Jia, Y.Q. Chen, W. Xu, Journal of Computer Research and Development, 50, **9**, 1799-1804, (2013)

[21] S.L. Liao, J.B. An, IEEE Geoscience and Remote Sensing Letters, 12, **5**, 963-967, (2015)

MATEC Web of Conferences 139, 00036 (2017) DOI: 10.1051/matecconf/201713900036
ICMITE 2017

System design and application of sewage treatment and the use of recycled water in Dianchi Lake Basin based on WebGIS

Qiaoling Tang[1] , Hong Liang[1] , Yumei Guo[2] , Ruitao Cun[2]

[1]College of Information Science & Engineering, Yunnan University, Kunming, China
[2]Kunming Drainage Facilities Management Co., Ltd , Kunming, China

Abstract. In this paper, due to the current problem of sewage treatment in Dianchi Lake Basin in Kunming and the practical demand for the integrated solution of the sewage treatment and reclaimed water use, our team has developed the online management system based on the WebGIS. The online management system is designed based on geographic spatial information mapping technology, WebGIS technology and spatial database technology and finally displayed in the form of thematic map called "Visual information platform for Dianchi Lake Basin sewage treatment and recycling water use". The system has achieved scientific and informative management for Dianchi Lake Basin sewage treatment and recycling of water use. It also provides comprehensive and accurate basic data and decision support for the improvement of urban water environment and pollution control in Dianchi Lake in Kunming. The online system improves the management efficiency, management level and decision-making capacity of sewage treatment and recycling industries in Kunming, and provide a good demonstration and reference in infrastructure of the city.

1 Introduction

Water is the basic and strategic resource of social development, and it is also an important ecological environment for human beings to survive. The cities, as the major human settlements, are showing more and more dependency for the water environment and water resources. In recent years, with the rapid development of urban economic construction, the city expanding, and extensive increasing urban population, the urban water environment has deteriorated. Modern cities are facing the dual pressures of water pollution and water shortage. As a strategic hub and axis of China's southwest open bridgehead, Kunming, the water environment pollution and water shortage is particularly prominent. With the rapid development of economy in Kunming and the rapid expansion of the scale of the city, a large number of domestic sewage and industrial wastewater had been released into Dianchi Lake, resulting in water pollution which did great damage to water function and water quality.

The system in this paper is developed based on ArcGIS produced by ESRI. It is a large GIS software platform with comprehensive, complete and scalable function. In addition to the powerful processing and analysis of spatial data, the platform also provides a rich development interface including Python, Javascript, .NET and so on, which provides strong support for the research and development of this paper. At the desktop, server, browser, mobile and cloud, ArcGIS has the corresponding services and products to provide users with the complete open GIS user experience which make0 it has become the most popular geographic information system platform software.

The two main applications of this software are desktop products ArcMap and server-side products ArcGIS Server. ArcMap is mainly for implementation of data processing and other functions, ArcGIS Server is responsible for publishing good map data, a map service for later development and use. In this paper, ESRI's ArcGIS Server is used to implement Web-based map publishing technology. It uses C / S and B / S hybrid network architecture to implement programming and integration with Javascript as programming language, and realize the urban sewage treatment and resource recycling Information management platform in the research and application [1].

This system can realize the remote and dynamic visual management and analysis of sewage treatment on-line monitoring, sewage treatment water quality analysis data management, reclaimed water use management, and sludge disposal, and provide comprehensive and accurate sewage treatment and resource recycling information for management and decision support. (Including sewage treatment plants, sewage treatment facilities, sludge disposal plants, reclaimed water treatment stations, recycled water users, recycled water users, reclaimed water intake points, etc.) for the relevant management of Dianchi Lake, which is related to the management of Dianchi Lake sewage treatment and resource recycling.) Efficient storage and management of data resources, as well as data modification, editing, updating and maintenance, to ensure the real-time data, accuracy and integrity, to achieve data sharing.

2 The establishment of database

2.1 Spatial data model

In this paper, a new generation of object-oriented spatial data mode based on GeoDatabase as Dianchi Lake sewage treatment and resource recycling network information management platform for spatial data model. The model uses standard relational database technology to express the data model of geographic information, and integrates spatial data and attribute data into the standard relational database in the background to realize the seamless integration of spatial data and attribute data. It is based on RDBMS On the intelligent spatial data model.

GeoDatabase uses object-oriented technology to abstract the real world into feature classes (points, lines, faces) that contain spatial information and object classes (feature attributes) that do not contain spatial information. Each feature class and object class has attributes, behaviors, and rules that are associated with objects such as relational classes, geometric networks, etc.

2.2 Design and Implementation of Spatial Database

Dianchi Lake sewage treatment and resource recycling spatial database design and construction is the core of the project implementation of the basic work. Spatial database design, it is necessary to fully consider the relational database of standardized theory, the establishment of a correct reflection of the objective reality of the relationship between the reduction of data redundancy and duplication, but also taking into account the Dianchi Lake sewage treatment and resource recycling characteristics and drainage management related management Business application requirements.

As for the new generation object - oriented spatial database model - Geodatabase, the platform designs and builds a reasonable spatial data element class, object class and relation class, and designs an efficient, stable, safe and practical spatial database of urban drainage system.

Dianchi sewage treatment and resource recycling Geodatabase is in accordance with certain models and rules combined geographical elements of data sets, object classes and relational class collection to manage and use the geographic elements of data. That is, Dianchi Lake sewage treatment and resource recycling spatial database use hierarchical data objects to organize geographic data, these data objects include feature data sets (Feature Dataset), feature class (Feature Class) and object class (Object Class) [2]

In order to make the database structure clear, easy to design and development and maintenance management, Dianchi Lake sewage treatment and resource recycling spatial database use hierarchical data objects to organize geographic data, which means, the entire geographic database is divided into the following layers: basic data, sewage treatment plant (point elements), sewage treatment facilities(point elements), pollution(point elements), reclaimed water (point elements), recycled water users (point elements), reclaimed water delivery pipe network (line elements), reclaimed water intake points (point elements), as shown in Figure 1. Among them, the basic data layer for the Kunming City is the basic electronic map, including Kunming roads, neighborhoods, water systems, rivers, units, green and so on.

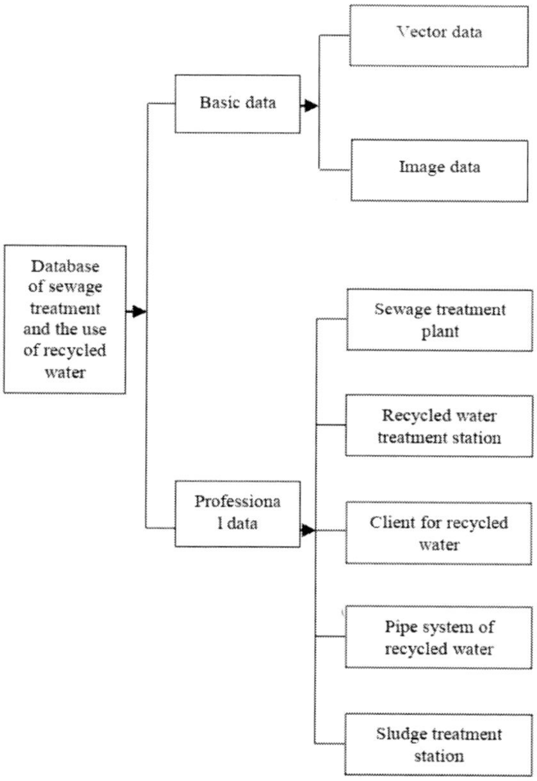

Figure 1. Spatial database layer structure

In the Dianchi Lake sewage treatment and resource recycling spatial database, the feature data set is a collection of the feature class of the shared space reference system. In the spatial database, all feature classes in the same geographic feature dataset must have the same spatial reference, which means its coordinate system, spatial domain, and the precision are the same. Spatial reference is an important part of database design. Dianchi Lake sewage treatment and resource recycling space database built in Microsoft SQL Server 2008 SDE database, achieving the creation of spatial database process through ArcCatalog. Dianchi Lake sewage treatment and resource recycling spatial database layer structure is shown in Figure 1.

The designing for the database of the sewage treatment and recycled water is shown in Figure 2:

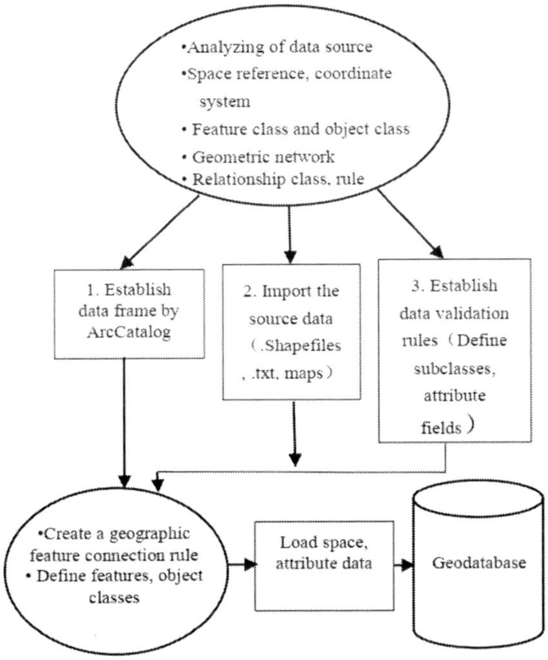

Figure 2. Designing for the database of the sewage treatment and recycled water

3 WebGIS

WebGIS (Web Geographic Information System) refers to the Internet-based platform using geographic information systems on the Internet, whose client application software using network protocols. WebGIS achieve spatial data sharing and interoperability through the Internet to publish and apply geospatial data, such as GIS information online query and business processing. WebGIS is a new technology to expand and perfect GIS by using Internet technology. Its core is to embed the application standard of HTTP standard in GIS, and realize spatial information management and distribution in Internet environment. WebGIS can be distributed with multiple hosts, multi-database, through the Internet / Intranet to achieve interconnection, with a browser / server (B / S) structure. The server provides information and services to the client, the browser (client) has a variety of spatial information and application functions.

WebGIS is the product of Internet technology applied to GIS development. From any node on the Web, Internet users can browse the spatial data in the WebGIS site, create thematic maps, and perform various spatial retrieval and spatial analysis.

3.1 The structural model of WebGIS

WebGIS has the three-tier structure with client, application server and database composition, as shown in Figure 3, the client request data services through the HTTP protocol item web server, and server return data results in web.

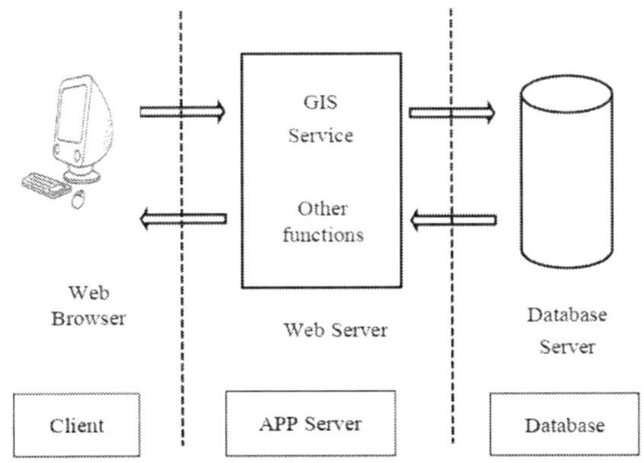

Figure 3. The structural model of WebGIS

3.2 Design and Implementation of WebGIS

3.2.1 Server policy

In this WebGIS architecture, the server finishes all the work of spatial data processing, the client is only used to issue the final result of the service request and display server, the performance requirements of the server are much higher than other situation [3].

The advantage of server policy is that the requirement for client PC hardware is much lower when running complex GIS operations and also can take full advantage of the existing GIS resources.

The disadvantage is that each step of the client needs sending request to the server through the network and the GIS server will also return processed results to the user in the form of static images, resulting in heavy network transmission burden. Meanwhile, the burden on the server is also heavy, and the client's operability is poor.

The architecture of the server policy is shown in Figure 4:

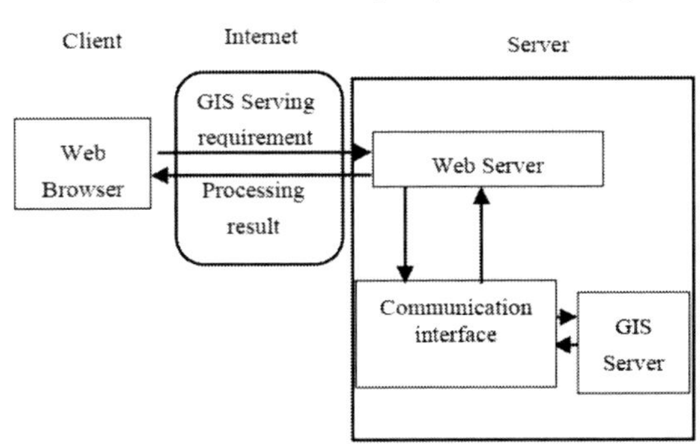

Figure 4. Architecture of the server policy

3.2.2 Client policy

Client software can be built in three different technologies, namely Plug-in, ActiveX controls and Java applets [3].

The advantage for this policy is that GIS is running fast. Because all GIS operations, such as map scaling, attribute queries, etc., are done locally by the client software, so the system responds quickly. And because the server only provides GIS data services, while the server and the client do a data transmission only once, it can greatly reduce the burden of server and network transmission.

The disadvantage is that this policy use WebGIS developed based on ActiveX technology which can only be carried out under the IE browser of Microsoft`s Windows; and client need to install the client plug-in before use if the WebGIS is developed based on Plug-in technology.

The architecture of the client policy is shown in Figure 5:

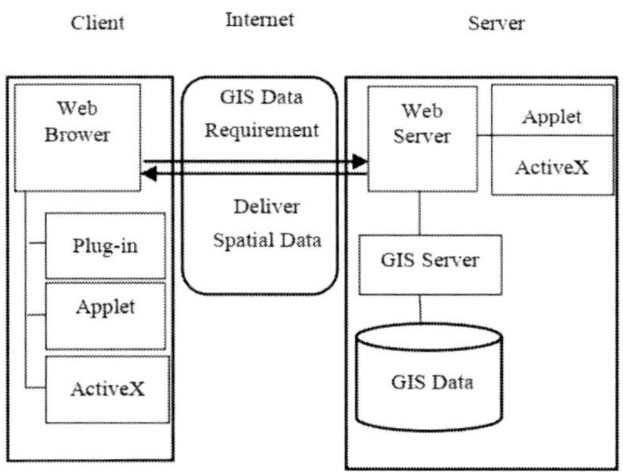

Figure 5. Architecture of client policy

3.2.3 Mixed policy

From the above analysis, we can see that, a simple server policy and client policy always has some limitations. When server-side policies involve frequent data transfers, their efficiency is severely affected by network bandwidth and network load. In the client policy, when the processing and processing capacity is inconsistent, by the computing power of the impact of certain tasks may run slower, or even impossible to complete [3].

So the mixed policy came into being, when it involves a lot of data operations and complex analysis tasks, you can make a strong computing power to deal with the server, when it comes to the user interaction more tasks, you can use the client to deal with, in this case, servers and clients can share their performance and data processing capabilities, thereby rationally allocating data and data processing procedures to make the overall performance of the system better.

4 Thematic map

Thematic maps, also known as special maps, are focus on one or several natural elements or socio-economic phenomenon of the map. The map area of the thematic map is broad, and information data with spatial attributes can be used to express it. The content of the thematic map consists of two parts:

1) Topic content

The natural or socioeconomic phenomenon highlighted in the figure and its related characteristics.

2) Geographic basis

To mark the thematic elements of the spatial location and geographical background of the general map content, mainly coordinate system, water, realm, residential and so on.

The use of thematic maps can study the distribution characteristics and laws of various thematic phenomena, can also study the interrelationships and dynamic changes of various phenomena., predict the coming trends and changes, do the comprehensive evaluation, zoning and planning.

4.1 Method of Realizing Feature Map Function

WebGIS implementation of the thematic map function can be achieved using the following methods: XSLT technology, XSLT used to convert an XML to other structures of the XML, the XSLT processor use XSLT style form of the application to the XML document to output XHTML, SVG or other documents in XML format. Since the SVG file conforms to the XML standard, you can write a good XSL style sheet file according to the different requirements, and then apply the style sheet to the original map SVG file to generate a new SVG file containing the thematic map to the user. Another way is that, because the SVG file is also an XML file, according to the user's thematic map needs, directly to the file analysis and modification, and then modify and save the map SVG file back to the user. The third method, you can write the Javascript code for the SVG file document object model that DOM operation to achieve the thematic map function, this article uses this method to achieve the thematic map function [4].

4.2 The realization of thematic map function

The thematic map is an important function of GIS, which can intuitively express one or more data related to the topic, making the data more convincing. The creation of thematic maps is generally divided into two ways: one way is to change the existing layer, for example, change the color of each region in a layer, with different colors on behalf of the size of different data; another way is based on the existing layer to add some new layers, these layers include some dots, circles and other graphic symbols, with these graphical symbols to the image of the amount of data[5]. This article uses the second approach. Figure 6 is a thematic map showing all the pump stations in circles.

Figure 6. Thematic map

5 Conclusions

The WebGIS system allows people to easily browse or retrieve a variety of distributed geospatial data on the web from any node on the internet and conduct a variety of on-line geospatial analysis. The emergence of the WebGIS system enables GIS to truly move towards society and serve social groups. The design and application of WebGIS in Dianchi Lake Basin has realized the scientific and informative management of wastewater treatment and reclaimed water in Dianchi Lake Basin, and provided comprehensive and accurate basic data and scientific decision for Kunming water environment improvement and Dianchi Lake pollution control at the same time.

References

1. Huang, Shao Hua, and R. T. Huang. "Analysis and Design of Wuhan Tourism Information System Based on WebGIS." *Bulletin of Surveying & Mapping* (2003).
2. Choi, Jinmu, W. Cooke, and M. Stevens. "Designing a Geodatabase for Fire Risk Assessment.".
3. Zhou, Kun, et al. "RESEARCH AND APPLICATION OF WEBGIS BASED ON SVG TECHNOLOGY." *Contributions to Geology & Mineral Resources Research* 21.1(2006):49-53.
4. Zhang, Yihan. "DESIGN AND DEVELOPMENT OF ENVIRONMENTAL MONITORING DATA MANAGEMENT SYSTEM BASED ON WEBGIS."*Computer Applications & Software* (2007).
5. Lin, Shouyi, et al. "Design and Implementation of Lujiazui Land Management Information System Based on WebGIS." *International Conference on E-Business and E-Government* IEEE Computer Society, 2010:667-670.

An Efficient Heterogeneous Signcryption Scheme from Certificateless to Identity-based Cryptosystem

Shufen Niu[1], *Zhenbin Li*[1,*], Miao Tian[1] , Caifen Wang[1] and Xiangdong Jia[1]

[1] College of Northwest Normal University of Computer Science and Engineering , Lanzhou, 730070, PR China

Abstract. The heterogeneous signcryption can not only realize the confidentiality and un-forgeability of the information transmission between different public key cryptography environments, but also reduce the communication cost. The proposed scheme uses bilin-ear pairings between certificateless cryptography and identity-based cryptography to construct a heterogeneous signcryption scheme. In addition, two cryptography systems use different secret master keys. Furthermore, the proposed scheme has the indistin-guishability against adaptive chosen ciphertext attacks and existential unforgeability against adaptive chosen-message attacks in the random oracle model. Through the data analysis, this scheme is more effective than other similar types of scheme.

Keywords. Heterogeneous signcryption, Certificateless cryptography, Identity-based cryptography, Different secret master, Secure

1 Introduction

Data confidentiality and authentication should be ensured when transmitting data in public networks. The traditional approach, signature then encryption involves numerous calculations, costly communication, and low computational efficiency. To improve the computational efficiency, Zheng[1] proposed the notion of signcryption in 1997. Signcryption can simultaneously perform signature and encryption, which enhances the efficiency of communication systems.

In the future, heterogeneous network technology is crucial to secure communication. Generally, different cryptosystems should be used in heterogeneous networks. Heterogeneous Signcryption (HSC) should be explored to ensure the confidentiality and authentication of messages in a heterogeneous network.

In 2010, Sun[2]. proposed the HSC scheme that communicates from Traditional Public Key Instruction (TPKC) to Identity-based Public Key Cryptography (IDPKC)[3]. However, the scheme only addresses external security threats[4]. In 2011, Huang et al.[5] proposed a HSC scheme with key privacy, which satisfies internal security[4]. In 2013, Fu et al.[6] proposed an IDPKC-to-TPKC construction of a multi-receiver signcryption scheme. In the same year, Li et al. [7] proposed a two-way HSC scheme. However, these HSC schemes only consider the signcryption problem between IDPKC and TPKC. Although these schemes use different public key cryptosystems, the sender and the receiver use the same system parameters. In 2016, Zhang et al.[8] proposed a HSC scheme that communicates from Certificateless Public Key Cryptography (CLPKC) [9, 10] to TPKC. To the best of our knowledge, only four HSC schemes have been developed, and the signcryption problem between IDPKC and CLPKC has not been addressed in the literature.

Existing heterogeneous signcryption schemes use similar master secret keys. In 2016, Li et al.[11] proposed a multi-receiver signcryption scheme for heterogeneous systems. Their work provided inspiration for the current study.

Based on Li[12], the proposed scheme develops a scheme from CLPKC[13] to IDPKC. The present paper introduces a formal security model called Efficient CLPKC-to-IDPKC Heterogeneous Signcryption Scheme and proves the model's semantic security and existential unforgeability in the random oracle model [14]. Our scheme is more secure and efficient than other existing schemes. The proposed scheme has the following features.

• This scheme uses different master secret keys in different cryptosystem systems, and improve the security of the system.

• Compared with existing heterogeneous schemes, this scheme is reduced system uptime through reducing the number of bilinear pairs. This scheme is more secure and efficient.

• In the random oracle model, this scheme ensures the confidentiality and un forge ability of data in the context of the Variants Decisional Bilinear Dif-fie-Hellman Problem (VDBDHP) ,Variants Computational Bilinear Dif-fie-Hellman Problem (VCBDHP) and Discrete Logarithm Problem (DLP).

The rest of this paper is organized as follows. Related knowledge are reviewed in Section 2.The formal

* Corresponding author: 775627945@qq.com

definition and security model of the proposed scheme are described in Section 3. The heterogeneous signcryption scheme is presented in Section 4. The security analysis of the scheme is discussed in Section 5. The performance analysis is discussed in Section 6. Finally, Section 7 concludes the paper.

2 Related knowledge

This section briefly describe bilinear maps and hard problems.

Let G_1 be a cyclic additive group generated by P, with a prime order q, and G_2 be a multiplicative group of the same order. The bilinear pairing is a map $e: G_1 \times G_1 \rightarrow G_2$ with the following properties.

•Bilinearity: $e(aP,bQ)=e(P,Q)^{ab}$ for all $P,Q \in G_1$ and $a,b \in Z_q^*$.

• Non-degeneracy: $P,Q \in G_1$ such that $e(P,Q) \neq 1$.

• Computability: An efficient algorithm is available for computing $e(P,Q)$ for any $P,Q \in G_1$.

The hard problems involved in the security of the proposed scheme are described as follows.

2.1. Hard Problems

• Variants Decisional Bilinear Diffie-Hellman (VDBDH) problem

Given $(P,aP,bP,cP,dP,c^{-1}P,d^{-1}P)$ unknown $a,b,c,d \in Z_q^*$, compute for $e(P,P)^{abc^{-1}}$.

•Variants Computational Bilinear Diffie-Hellman (VCBDH) problem

Given $(P,aP,bP,cP,dP,c^{-1}P,d^{-1}P)$ with unknown $a,b,c,d \in Z_q^*$ compute for $e(P,P)^{abd^{-1}}$.

• Discrete Logarithm Problem (DLP)

Given a cyclic additive group G and generator P, the DLP is defined that given (P,aP), compute $a \in Z_q^*$.

According to the existing schemes [15,16], it is easy to known that both VDBDHP and VCBDHP are hard. Obviously, Discrete Logarithm Problem (DLP) is hard.

3 Semantic security and security model

3.1. Semantic security

A CLC$\rightarrow$IBC heterogeneous signcryption scheme generally includes the following algorithm:

• **Setup.** Choose a security parameter l to input. The PKG and KGC generate their own master key, and outputs the system's parameters *params*.

• **CLPKC-KG.** The algorithm runs by the KGC of the CLPKC system. Submit the identity ID_U and a secret value x_U, the corresponding sk_s is generated as the private key of the CLPKC user.

• **IDPKC-KG.** The algorithm runs by the PKG of the IDPKC system. By inputting a user's identity ID_U, the corresponding private key S_U is generated.

• **Signcryption.** Given the sender's identity ID_s, the receiver's identity ID_r, the private key S_U and the message m, the sender runs the signcryption algorithm to generate a ciphertext σ.

• **Unsigncryption.** Given the sender's identity ID_s, the private key S_U of the receiver and a ciphertext σ, the receiver computes and outputs the plaintext m, or the symbol $\perp$.

3.2. Security Models

3.2.1 Confidentiality

A heterogeneous signcryption scheme has the IND-CCA2 property if no probabilistic polynomial time adversary A has a non-negligible advantage in the following game:

Setup. Challenger C runs the setup algorithm with a security parameter l and send system parameters and public keys to the A, while the master key is kept secret.

Phase 1. A can ask several kinds of queries to the following random oracles.

• **Key generation query.** A submits an identity ID_U, C runs the IDPKC-KG algorithm to generate the private key S_U and send it to A.

• **Unsigncryption query.** A submits a signcrypted message σ and an identity ID_s. C computes the private key S_U. If σ is a valid ciphertext, C returns a message $m = Unsigncrypt(\sigma, params, S_U)$; otherwise, an error symbol $\perp$ is returned.

Challenge. A decides when Phase 1 ends. A selects two plaintexts m_0 and m_1 of the same length, a sender's identity ID_A, and a receiver's identity ID_B^*, which it wants to challenge. A is not allowed to ask for the private key to ID_B in the first stage. C then selects $b \in \{0,1\}$ and runs the corresponding algorithms to obtain the ciphertext σ . σ is transmitted to A.

Phase 2. A can perform the queries similar to those in Phase 1. A cannot query the key extraction for the target identities. A should not query the Unsigncryption for σ.

Guess. Finally, A produces a bit b', A wins the game if $b' = b$.

The advantage of A is defined as $Adv(A)$=|Pr[b'=b]-1| ,where Pr[b'=b] denotes the probability that $b' = b$.

3.2.2 Unforgeability I

A CLPKC$\rightarrow$IDPKC heterogeneous signcryption scheme is referred to as EUF-CMA-I. The game is defined between a challenger C and an adversary F_I. If no probabilistic polynomial time adversary F_I has a non-negligible advantage against a challenger in the following game:

Setup. The challenger C runs the setup algorithm. It sends public parameters $params$ to the adversary F_I, while the master keys are kept secret. When F_I receives $params$, F_I outputs a target identity ID^*.

Attack. F_I issues several kind of queries.

• **Extract partial private key query.** F_I selects an identity ID_i, C computes the partial private key S_i, and transmits it to F_I.

• **Extract secret key query.** F_I produces an identity ID_i, C computes the se cret key x_i, and transmits it to F_I.

• **Request public key query.** When F_I receives a public key extraction query for an identity ID_i, C runs the CLPKC-KG algorithm to the public key pk_i, and transmits it to F_I.

• **Replace public key query.** For any identity ID_i and a valid public key pk_i', C replaces pk_i with a value pk_i'.

• **Signcrypt query.** F_I produces a message m, a sender's identity ID_s, and a re ceiver's identity ID_r, C first finds the sender's private key sk_s from the CLPKC-KG algorithm. C then runs signcryption algorithm and sends σ to F_I.

Forgery. F_I produces a tuple $(ID_s^*, \sigma^*, ID_r^*)$. ID_s^* cannot be queried a partial private key extraction. F_I wins if the unsigncryption returns m.

The advantage of F_I is defined as the probability that F_I wins the EUF-CMA-I game.

3.2.3 Unforgeability II

A CLPKC→IDPKC heterogeneous signcryption scheme is referred to as EUF-CMA-II , The game is defined between a challenger C and an adversary F_{II}. If no probabilistic polynomial time adversary F_{II} has a non-negligible advantage against a challenger in the following game:

Setup. The challenger C runs the setup algorithm. It generates master secret keys s_{KGC}, s_{PKG} and system's public parameters $params$ to the adversary F_{II} while s_{PKG} is kept secret. When C receives the public parameters, F_{II} outputs a challenge identity ID^*.

• **Extract secret key query.** F_{II} produces an identity ID_i, C computes the secret key x_i from the CLPKC-KG algorithm, and transmits it to F_{II}.

• **Request public key query.** When F_{II} receives a public key extraction query for an identity ID_i, C runs the CLPKC-KG algorithm to the public key pk_i, and transmits it to F_{II}.

• **Signcrypt query.** F_{II} produces a message m, a sender's identity ID_s and a re ceiver's identity ID_r. C first finds the sender's private key sk_s from the CLPKC-KG algorithm. C then runs the signcryption algorithm and sends σ to F_{II}.

Forgery. F_{II} produces a tuple $(ID_s^*, \sigma^*, ID_r^*)$. ID_s^* cannot be queried a partial private key extraction. F_{II} wins if the unsigncryption returns m. σ was not produced by any signcrypt query.

The advantage of F_{II} is defined as the probability that F_{II} wins the EUF-CMA-II game.

4 Describe the scheme

The heterogeneous signcryption scheme is described in this section. It can be allowed as the sender in the CLPKC system and as the receiver in the IDPKC system. The proposed scheme is described as follows.

Setup. In the heterogeneous system, let G_1 be a cyclic additive group of prime order q and G_2 be a cyclic multiplicative group of the same order. Let P be the generator of G_1. A bilinear map is defined as $e: G_1 \times G_1 \rightarrow G_2$. Three cryptographic hash functions are chosen,namely, $H_1: \{0,1\}^* \rightarrow G_1$, $H_2: G_2 \rightarrow \{0,1\}^{l_m}$ and $H_3: \{0,1\}^* \rightarrow Z_q^*$. where l_m is the length of a message. The KGC selects $s_1 \in Z_q^*$ as the master secret key and sets $P_1 = s_1 P$ as the public key of the CLPKC system . The PKG selects $s_2 \in Z_q^*$ as the master secret key and sets $P_2 = s_2 P$ as the public key of the IDPKC system . The global public parameters are $param = \langle G_1, G_2, e, P, P_1, P_2, H_1, H_2, H_3 \rangle$ and the master secret key are s_1 and s_2.

CLPKC-KG. As a sender in the CLPKC system, it uses an identity ID_A and com putes the private key using the following algorithm.

• Compute $Q_A = H_1(ID_A)$ and set the partial private key $S_A = s_1^{-1} Q_A$.

• Choose a random secret value $x_A \in Z_q^*$.

• Compute the public key $pk_A = x_A P$ and the private key $D_A = x_A S_A$.

IDPKC-KG. The PKG computes the private key of an identity ID_B as $S_B = s_2^{-1} Q_B$ where $Q_B = H_1(ID_B)$.

Signcryption. The system parameters, namely, the message m of length l_m, the receiver's identities ID_B are given, Then the sender uses the following algorithm:

• Choose a random number $r \in Z_q^*$ and compute $U_1 = rP_2$, $U_2 = rQ_A$.

• Compute $R = e(P, Q_B)^r$ and $k = H_2(R)$.

• Compute $c = m \oplus R, h = H_3(m, k, U_1, U_2), V = (r+h)D_A$.

• Output ciphertext as $\sigma = (c, U_1, U_2, V)$.

Unsigncryption. When the system receives a ciphertext $\sigma = (c, U_1, U_2, V)$ and the system parameters, the receiver with an identity ID_B unsigncrypts the ciphertext σ as follows:

• Compute $R = e(U_1, S_B)$ and $k = H_2(R)$.

• Recover $m = c \oplus R$.

• Compute $h = H_3(m, k, U_1, U_2)$.

•Accept the message if and only if the equation holds: $e(P_1, V) = e(pk_A, U_2 + hQ_A)$.

Correctness. The following equations show the correctness of the scheme.

When receiving the ciphertext, the receiver can decrypt the ciphertext as follows:

$$R = e(U_1, S_B) = e(rP_2, s_2^{-1}Q_B) = e(P, Q_B)^r$$

Simultaneously, the receiver can verify the following equalities to show:

$$e(P_1, V) = e(s_1P, (r+h)D_A) = e(s_1P, (r+h)x_AS_A) = e(x_AP, (r+h)s_1S_A) = e(pk_A, (r+h)Q_A) = e(pk_A, rQ_A + hQ_A) = e(pk_A, U_2 + hQ_A)$$

The scheme can be modified into environments with different systems parameters. For example, the PKG has system parameters $\langle G_1^*, G_2^*, P^*, H_1, H_2, H_3 \rangle$ and the KGC has the system parameters $\langle G_1, G_2, P, H_1, H_2, H_3 \rangle$.The scheme will remain secure.

5 Security Analysis

This section will show the scheme's security analysis, including confidentiality and unforgeability.

5.1. Confidentiality

Theorem 1. In the random oracle, if an IND-CCA2 adversary C can attack our proposed scheme, then there exists an algorithm C that can solves the VDBDH problem.

Proof. The distinguisher C inputs $(P, aP, bP, cP, dP, c^{-1}P, d^{-1}P, \theta)$ and attempts to decide whether $\theta = e(P, P)^{abc^{-1}}$ or not. C will run A as a subroutine and act as $A's$ challenger in the IND-CCA2 game. We model H_i ($i = 1,2,3$) as random oracles. C keeps three lists L_1, L_2, L_3 to store the answers.

Setup. At the beginning of the game, C runs the setup algorithm and set $P_b = cP$ and provides the system parameters to attacker A.

Phase 1. A can request several queries. To respond to these queries, H_1, H_2, H_3 are random oracles controlled by C as follows.

• H_1 **-query**: C selects a random number $j \in \{1, 2, \cdots, qH_1\}$. A asks for a polynomial bounded number of H_1 queries on identities of his choice. At the $j-th$ H_1 query, C answers by $H_1(ID_j) = bP$. For queries $H_1(ID_i)$ with $i \neq j$, C selects $e_i \in Z_q^*$, puts the pair (ID_i, e_i) in list L_1 and answers $H_1(ID_i) = e_iP$.

• H_2, H_3 **-queries**. When A asks queries on these hash values, C checks the list. If an entry for the query is found, it will be return to A. Otherwise, a random value is chosen to send to A, and the answer will then be stored in the list.

• **Key extraction query.** When A asks a key extraction query in identity ID_i, if $ID_i = ID_j$, then C fails and stops. If $ID_i \neq ID_j$, C finds (ID_i, e_i) from H_1 (which indicates that C previously answered $H_1(ID_i) = e_iP$ on an H_1 query on ID_i). C updates the list, and then computes the corresponding private key $e_ic^{-1}P$. C returns it to A.

•**Unsigncryption queries.** For an unsigncryption query in a ciphertext σ' for iden tities ID_A and ID_B. C follows the steps below:

Case 1: $ID_B = ID_j$. C always answers A that σ' is invalid. If σ' is valid from $A's$ viewpoint, C will fail.

Case 2: $ID_B \neq ID_j$. C computes $R' = e(U_1', S_B)$ (C could obtain S_B from the key extraction algorithm because $ID_B \neq ID_j$). C then runs the H_2 simulation algorithm to obtain $k' = H_2(R')$ and computes $m' = c \oplus R'$.

Finally, C runs the H_3 simulation algorithm to generate $h' = H_3(m', k', U_1', U_2')$ and checks if $e(P_1, V') = e(pk_A, U_2' + h'Q_A)$ holds. If the above equation does not hold, C returns $\perp$. Otherwise, C returns m'.

Challenge. A outputs two plaintexts m_0, m_1 to C. C randomly chooses $b \in \{0,1\}$ and computes the signcryption of message m_b as follows.

• It sets $U_1^* = aP$, obtains $k^* = H_2(\theta)$ (where θ is C candidate for the VDBDHP).

• Compute $c_b = m_b \oplus R$.

• Randomly choose $x^*, h^* \in Z_q^*$.

• Compute $U_2^* = -h^*Q_A + x^*pk_A$ and $V^* = x^*P$.

• Adds (c^*, U_1^*, U_2^*, V^*) to list L_3.

Finally, C produces the challenged ciphertext $\sigma^* = (c_b, U_1^*, U_2^*, V^*)$ and send it to A.

Phase 2. A performs new queries that are treated the same way as in Phase 1.

Guess. A produces a bit b' as its guess. At the moment, if $b = b'$, C outputs $\theta = e(U_1^*, S_{ID_j}) = e(aP, c^{-1}bP) = e(P, P)^{abc^{-1}}$ as a solution of the VDBDHP. Otherwise, C stops and outputs "failure".

5.2 Unforgeability I

Theorem 2. In the random oracle model, if an EUF-CMA-I adversary F_I can attack our proposed scheme, then there exists an algorithm C that can solve the VCBDH problem.

Proof. If an EUF-CMA-I adversary F_I exists, then we construct a simulator C that uses F_I to compute $e(P,P)^{abd^{-1}}$ from an instance $(P,aP,bP,cP,dP,c^{-1}P,d^{-1}P)$ of the VCBDHP.

Setup. The challenger C runs the setup algorithm. It sets $P_1 = dP$, and sends the system parameters $params = \{G_1,G_2,e,P_1,P_2,H_1,H_2,H_3\}$ to F_I , then F_I outputs a challenge identity ID^*.

• H_1 -query. When C receives a query $(ID_j,\alpha_j,H_1(ID_j),D_j)$. If $(ID_j,\alpha_j,H_1(ID_j),D_j)$ exists in L_1 , then h_j is returned to F_I. Otherwise, C does as follows:

(1) If $ID_j \neq ID^*$, select $\alpha_j \in Z_q^*$ at random, then compute $H_1(ID_j) = \alpha_j P$, $D_j = \alpha_j P_1$;otherwise, set $\alpha_j = D_j = \perp$, $H_1(ID_j) = bP$.

(2) Put $(ID_j,\alpha_j,H_1(ID_j),D_j)$ into L_1, return $H_1(ID_j)$.

• $H_i(i = 2,3)$ **-query.** When C receives a query, if the corresponding query exists in $L_i(i = 2,3)$, C returns it to F_I. Otherwise, C randomly chooses an number as the query answer and return to F_I. Meanwhile, C puts the query result into L_i.

• **Public key query.** When C receives a query (ID_j,x_j,pk_j) . If (ID_j,x_j,pk_j) exists in the public key-list, then pk_j is returned to F_I, Otherwise, C chooses $x_j \in Z_q^*$ at random, computes $pk_j = x_j P$, then puts (ID_j,x_j,pk_j) into the public key-list and return pk_j as answer.

• **Extract secret key query.** When C receives a query ID_j. If C replaces the public key of ID_j, then it returns $\perp$. Otherwise, (ID_j,x_j,pk_j) exists in the key-list, and C returns x_j as answer.

• Extract partial private key query. When C receives a query (ID_j,x_j,D_j), C does as follows:

(1) If $ID_j = ID^*$,then C aborts; if $(ID_j,\alpha_j,H_1(ID_j),D_j)$ exists in L_1, D_j is returned to F_I; otherwise, C first makes a query to H_1.

(2) Put (ID_j,x_j,D_j) into L_1; return D_j.

• **Replace public key query.** When C receives a replace public key query (ID_j,pk_j') , C first finds (ID_j,x_j,pk_j) on the public key-list, then C updates the public key-list with tuple $(ID_j,\perp,pk_j')$, and sets $x_j = \perp$, $pk_j = pk_j'$.

• **Signcryption query.** When C receives a signcryption query, let ID_A, ID_B be the identies of the sender and receiver, respectively, m be the plaintext, C will do as follows:

(1) If $ID_A \neq ID_j$. C computes the private key S_{ID_A} corresponding to ID_A by running the key extraction query algorithm. Then C answers the query.

(2) If $ID_A = ID_j$, C simulates the signcryption algorithm to create a signcryption in the following steps. C chooses $x,h \in Z_q^*$ randomly, and computes $U_1 = xP_b$, $R = e(U_1,S_B)$ (C could obtain S_B from the key extraction algorithm because $ID_B \neq ID_j$). C runs the H_2 simulation algorithm to find $k = H_2(R)$ and computes $c = m \oplus R$. Then C computes $U_2 = -hQ_A + xpk_A$ and $V = xP$, and adds (c,U_1,U_2,h) to list L_3. If a collision happens, C outputs fail and exits. Otherwise, C gives the ciphertext σ to F_I.

Attack. F_I adaptively performs a polynomially bounded number of queries to the various oracles in this phase.

Forge. Eventually, F_I outputs a tuple $(m^*,\sigma^*,ID_s^*,pk_s^*)$. If $ID_s^* \neq ID^*$, C aborts. Otherwise, by forking lemma, C chooses different h,h^* and interacts with F_I the same random tape, then the adversary can give a different forger σ^*. σ^* and σ should verified by the equation $e(P_1,V) = e(pk_A,U_2 + hQ_A)$ and $e(P_1,V^*) = e(pk_A,U_2 + h^*Q_A)$. Where if $P_1 = dP$ and $pk_A = \beta P$, C can compute $bd^{-1}P = \frac{V - V^*}{\beta(h - h^*)}$, Simultaneously, C delivers the value of $e(P,P)^{abd^{-1}}$ as $e(aP,bd^{-1}P)$. Hence, C successfully solves the VCBDHP.

5.3 Unforgeability II

Theorem 3. In the random oracle model, if an EUF-CMA-II adversary F_{II} can attack our proposed, then there exists an algorithm C that can solve the Discrete Logarithm Problem (DLP).

Proof. If an EUF-CMA-II adversary F_{II} exists, then we construct a simulator C that uses F_{II} to compute $a \in Z_q^*$ from an instance $\langle P,aP \rangle$ of the DLP.

Setup. C first runs the setup algorithm to generate s_1,s_2 as the master keys, and sets $P_2 = s_2 P$, then gives s_1 , $params$ to the attacker F_{II}, s_2 is kept secret. F_{II} outputs a challenge identity ID^*.

• $H_i(1,2,3)$ **-query.** When C receives a query, if the corresponding query exists in H_i-list, C returns it to F_{II}. Otherwise, C randomly chooses an integer as the query result and returns it to F_{II}. Meanwhile, C puts the query result into the H_i-list.

• **Public Key query.** When C receives a query (ID_j, x_j, pk_j). If (ID_j, x_j, pk_j) exists in the public key-list, then C returns pk_j. Otherwise, C chooses $x_j \in Z_q^*$ at random, computes $pk_j = x_j P$, then puts (ID_j, x_j, pk_j) into the public key-list and returns pk_j as answer.

• **Signcryption query.** The same as Unforgeability I's Signcryption query.

Attack. F_{II} adaptively performs a polynomially bounded number of queries to the various oracles in this phase.

Forge. Eventually, F_{II} outputs a tuple $(m^*, \sigma^*, ID_s^*, pk_s^*)$. If $ID_s^* \neq ID^*$, C aborts; otherwise, C randomly chooses $h \in Z_q^*$, then the adversary can give a forger σ. We know that σ should verified by the equation $e(P_1, V^*) = e(pk_A^*, U_2^* + hQ_A^*)$. And C set $pk_A^* = aP$ and compute the equation, then C obtain the result $a = \frac{s_1 V^*}{U_2^* + hQ_A^*}$ as the solution to given DLP. Hence, C successfully solves the DLP.

6 Performance analysis

We compare the proposed scheme with other existing signcryption schemes [2],[7],[18]. The Table 1 shows the efficiency and distinction of the schemes. We denote Pa as the total number of pairing computations required, mark the Mu as the total number of point multiplications required and denote the Ex as the total number of exponentiations required. We compare our scheme with scheme [2],[7],[18] even though these schemes are all in the same cryptosystems. The results are shown in Table 1 and Table 2. As shown in Table 1 and Table 2, our scheme has a lower computation cost than other schemes.

Table 1. Comparison with existing scheme for signcryption.

Scheme	Pa	Mu	Ex	Master key
Sun[2]	1	3	1	Same
Li[7]	2	3	2	Same
F Li[18]	1	3	1	Same
Our scheme	1	2	1	Different

Table 2. Comparison with existing scheme for unsigncryption.

Scheme	Pa	Mu	Ex	Master key
Sun[2]	4	1	0	Same
Li[7]	3	2	2	Same
F Li[18]	5	0	0	Same
Our scheme	3	1	0	Different

In the simulation, we show the time by measuring the performance of the Signcryption and Unsigncryption. Our program is written in with the PBC library [17], and we run on a personal computer with 3.10GHz CPU and 2GB of RAM, using Linux operating system. In order to the effect of comparison, we will adopt Y Li et al [11]. Both schemes are set up in heterogeneous system, and both the schemes use different system parameters. We can get the result as Fig 1, which SC means signcryption operation and USC means unsigncryption operation. In the performance of the Signcryption and Unsigncryption, our scheme has been improved in terms of efficiency.

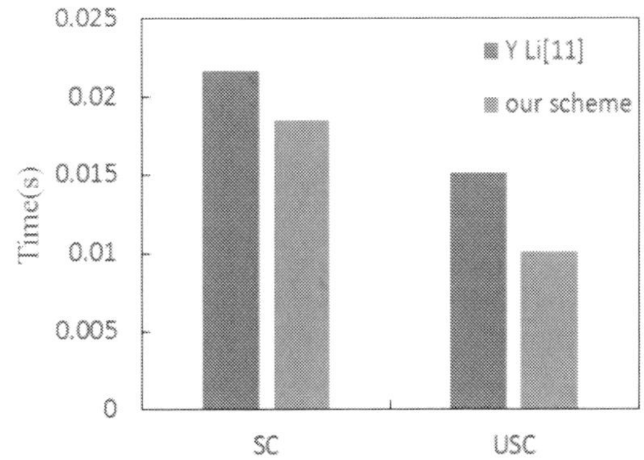

Fig. 1. Performance comparison between Li's and our schemes.

7 Conclusion

We have developed an efficient signcryption scheme for the CLPKC→IDPKC heterogeneous system. Compared with the existing signcryption schemes, the proposed scheme chooses different master secret keys in different systems. The security models are provided, and the proposed scheme ensures the confidentiality and unforgeability of data in the random oracle model. The proposed is more secure because it uses different master keys. Furthermore, the proposed scheme can be used in environments with different system parameters.

Acknowledgements

The work was supported by the National Natural Science Foundation of China under grant 61562077, 61462077, 61662071, 61662069, and Natural Science Foundation of Gansu Province for Distinguished Young Scholars 1308RJDA007.

References

1. Zheng Y. Digital signcryption or how to achieve cost in AdvancesinCryptology'97 on Springer, 1997. pp.165-179.
2. Sun, Li H.Efficient signcryption between TPKC and IDPKC and its multi-receiver construction, Science China Information Sciences, Vol 53, 2010, pp.557-566.
3. Shamir. Identity-Based cryptosystems and signature schemes, Springer, Heidelberg, 1985, pp.47-53.
4. An J H, Dodis Y, and Rabin T. On the Security of Joint Signature and Encryption, Springer, Berlin, 2002, pp.83-107.
5. Huang Q, Wong D S, and Yang G. Heterogeneous signcryption with key Privacy, Computer Journal, Vol 54, 2011, pp.525-536.
6. Fu, Li, and Liu W. IDPKC-to-TPKC construction of multi-receiver signcryption, Proceedings of the INCoS (5) on IEEE, 2013, pp.335- 339.
7. Li, Zhang H, and Takagi T. Efficient signcryption for heterogeneous systems, IEEE Systems Journal, Vol 7, 2013, pp.420-429.
8. Zhang et al, CLPKC-to-TPKI Heterogeneous Signcryption Scheme with Anonymity, Acta Electronica Sinica, Vol 44, 2016, pp.2432-2439.
9. Al-Riyami S S, and Paterson K G. Certificateless public key cryptography, Advances in Cryptology-Asiacrypt on Springer, 2003, pp.452-473.
10. Zhou Y W, Yang B, and Zhang W Z. Provably secure and efficient certificateless generalized signcryption, Chinese Journal of Computers, Vol 39, 2016, pp.543-551.
11. Li Y, Wang C, Zhang Y, and Niu S. Privacy preserving multi - receiver signcryption scheme for heterogeneous systems, Security & Communication Networks, Vol 9, 2017, pp.4574-4584.
12. Li F, Shirase M, and Takagi T. Efficient Multi-PKG ID-Based Signcryption for Ad Hoc Networks, Information Security and Cryptology on Springer, 2009, pp.289-304.
13. Al-Riyami, S. S, Paterson, and K. G. Certificateless public key cryptography. International Conference on the Theory and Application of Cryptology, Springer, Berlin, 2003, pp.452-473.
14. Bellare, M. and Rogaway P. Random Oracles are Practical, 1993, pp.366–368.
15. Bao F, Deng R H, and Zhu H F. Variations of Diffie-Hellman Problem, in Information and Communications Security, Huhehaote, China, 2003, pp.301-312.
16. Chow S S M, Yiu S M, and Hui L C K. Efficient Forward and Provably Secure ID-Based Signcryption Scheme with Public Verifiability and Public Ciphertext Authenticity, Vol 2971, 2004, pp.352-369.
17. PBC Library. http://crypto.stanford.edu/pbc/.
18. Li, Fagen, Y. Han, and C. Jin. Practical access control for sensor networks in the context of the Internet of Things , Computer Communications, Vol 89,2016, pp.154-164.

MATEC Web of Conferences **139**, 00038 (2017) DOI: 10.1051/matecconf/201713900038
ICMITE 2017

The development of stereoscopic warehouse stacker control system based on motion controller

Xuan Sun[*], *Zhedong Ma*, *Zhiyong Wang* and *Changsheng Ai*

School of Mechanical Engineering, University of Jinan, Jinan, China

Abstract. In order to meet the requirement of high efficiency, high accuracy, high reliability and high expansibility of stacker. The stacker control system based on motion controller is developed,which uses digital control technology of NC machine tools for reference and adopts the half closed loop control system of motion controller and servo drive system to realize high efficiency, high precision and high reliable motion control of stacker.Using the network interface of controller interact with the logistics system server, storage management, computer information management system of ERP, AGV car, palletizing robot and other equipment.Make it becomes the organic part of the logistics system,which provides a good solution to improve the level of enterprise logistics automation.

1 Introduction

Stereoscopic warehouse is an important component of the intelligent logistics system. It is the key link of the warehousing and transportation system. It is the result of the development of logistics and storage system and the progress of information technology [1].Now the warehouse has been developed integrated software product that consists of the mechanical and electrical combination, combination of strong current and weak current control, automatic feeding system, automatic sorting system, automatic control system, AGV transport system, automatic stacking stack system and host computer control software [2].Compared with the traditional platform warehouse and floor warehouse, it has incomparable advantages.It is the key node in the intelligent factory and Internet of things system.

2 Requirement analysis of mechanical structure and motion control of stacker

The mechanical of the stacker is composed of operating mechanism, lifting mechanism, extension fork mechanism, loading platform, frame and various safety protection devices [3].The structure is shown in Fig.1.

In order to complete the incoming and outbound operations of the goods, the stacker needs to meet the design standards :Need the direction of three degrees of freedom, which is moving in horizontal direction, realize the column positioning of the goods listed in warehouse;The vertical movement of the cargo platform to realize the goods layer positioning in warehouse;Bidirectional telescopic movement of goods fork achieves the access to the goods.It has the safety protection device for ensuring the safety of the stacker body and the operation person.There are chain breaking protection mechanism, overload protection device and limit protection device for each movement direction.The structural and mechanical properties of stacker meet the principles of strength, stiffness and reliability principle.The weight of each part should be reduced to minimize the impact of inertial impact on the acceleration, braking and operation of the stacker.At the same time ,manufacturing costs can also be reduced [4].

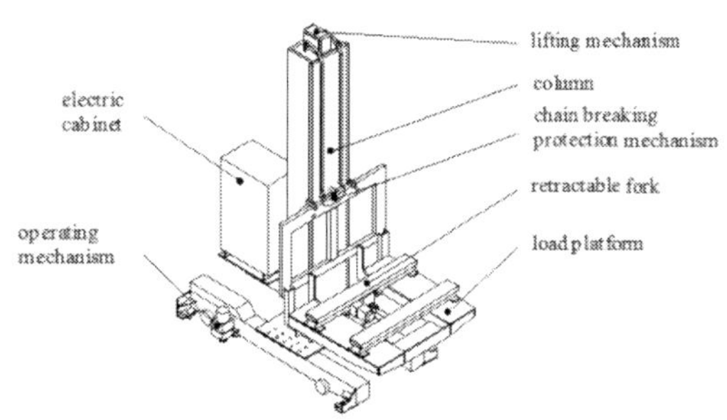

Fig. 1. Stacker structure

2.1 Operating mechanism

The major function is servo driver inside the electrical cabinet drive the servo motor that is mounted on the lower cross member of stacker.The gear on the motor output shaft meshes with the rack on the floor rail so that the stacker can move horizontally in order to realize the column positioning of the goods in storage.A walking wheel is arranged on one side of the lower cross beam, and a guide wheel is arranged on the other side, which plays a good guiding role in the smooth operation of the stacker.When the overload and stall conditions, servo driver will execute the overload protection, the motor will perform brake.

[*] Corresponding author: me_sunx@ujn.edu.cn

2.2 Lifting mechanism

The lifting mechanism is the mechanism that realizes the vertical lifting and power transmission.Because of the heavy load of the stacker, the counterweight mechanism which consists of reduction box and the counterweight block is designed to balance the load of the loading platform.Lifting mechanism adopts chain drive has the advantages of stable transmission, accurate average drive ratio, large transmission power and high overload capacity.The servo motor is mounted on the top of the column to drive the gear reducer to rotate. It is connected with the counterweight block in the column by the chain transmission, so as to pull the truck in the vertical direction.

2.3 Fork part

In order to improve the ground utilization ratio of stereoscopic warehouse and the utilization ratio of stacker ,so as to reduce the number of stacker used in the same area, it is decided to adopt double extension fork.The mechanism is shown in Fig.2.The double extension fork adopts three level linear expansion mechanism, which is composed of drive motor, reducer, coupling, upper fork, middle fork, lower fork, chain wheel and transmission device [5].

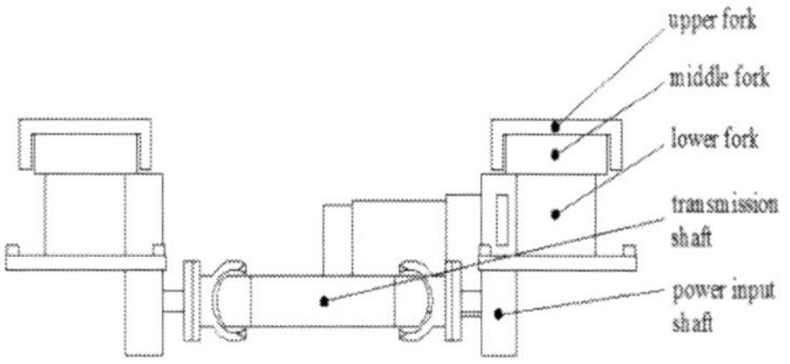

Fig. 2. Fork structure

2.4 Cargo platform

The cargo platform is an important goods bearing unit is mainly composed of a guide wheel, wheel frame, truck body and a guide seat , which is provided with a double extension fork and anti chain breaking device mechanism [6].Under the condition of meeting the strength and stiffness, on both sides of the ribs are opening with circular holes as far as possible to reduce its quality and effectively reduce the vertical motor dragging weight.The guiding mechanism can ensure the stable movement of the cargo platform in the vertical direction of the column, reduce the friction resistance and improve the positioning accuracy of the cargo table movement.

3 Analysis of reliability and safety requirements for stacker crane

Any hidden danger of insecurity is likely to cause incalculable losses to enterprises and users,so take the necessary protective measures and safety is very important.

3.1 Limit position protection

In the horizontal direction of the stacking machine, not only the zero limit switch is set, but also the limit position is set at both ends of the guide rail.The zero limit switch is arranged in the vertical direction, and the limit switch is also arranged to prevent the loading platform from moving beyond the safe stroke, and the falling prevention mechanism is arranged to prevent the accident of the motor from being broken and the chain breaking [7].

3.2 Software protection

In the stacker control software, the fork can perform the soft protection of the extended and retracted movements after the stacker is moving in a horizontal and vertical direction to a safe position.At the same time, the stacker can move horizontally and vertically when the fork moves to a safe position, which effectively prevents the collision between the stacker and the warehouse shelf.

3.3 Power-off protection

In case of accidental power failure, each axis servo motor has brake function effectively prevent the loss of power to each shaft accidents.This function makes each axis keep its original state to prevent the stacker goods falling and protect the personal safety of the operator.

4 Stacker control system design

4.1 Stacker work flow analysis

In stereoscopic warehouse ,Stacker as an important out / storage of goods equipment that equivalent to the arms in a stereoscopic warehouse.It needs to complete the operation of warehousing and outbound operations.

4.1.1 Warehousing operation

After the stacker detects the arrival of the AGV, the stacker performs the action to read the tag information on the tray, reads the information to the stacker controller and displays it on the display screen.According to the rules of the nearest and uniform, the stereoscopic warehouse management system automatically assigns the pallet goods a target position.When the stacker performs the insertion and removal of the goods, it sends the completion notification to the host computer. The host computer tells AGV to leave.The pallet load is placed in the warehouse.If you read the electronic tag on the tray error, the system will automatically alert the alarm information.Waiting for the completion of manual processing, and then perform the appropriate action.

4.1.2 Outbound operation

Warehouse has two kinds of outbound form,which is delivery of cargo from storage by position or type.The specific process is to enter the type of goods needed to be shipped out in the upper computer, and then inquire about the quantity or type of the goods stored in the warehouse,enter the quantity or type needed to be shipped out of the warehouse.The upper computer will send out the warehouse command to the stacker according to the FIFO rule, and the stacker will perform the corresponding action.

4.2 Design of control hardware

The warehouse control system hardware consists of six axis motion controller GUS, demonstrator, servo motor, servo driver, limit switch, origin switch, RFID electronic label reader, barcode scanner, outbound PC etc [8].Hardware diagram of stacker control system,as shown in Fig.3.

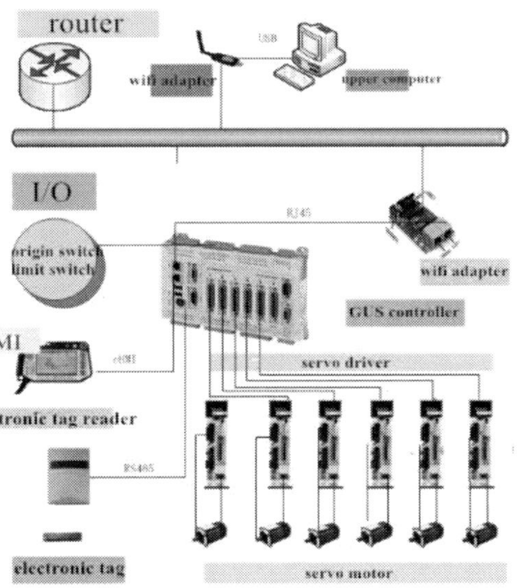

Fig. 2. Hardware composition of control system

The system uses the integrated motion controller as the core of the control system of the stereoscopic warehouse.It is connected with each axis servo control system through its axis interface.Receive user commands from the keyboard and mouse through the PS/2 interface.Various sensor signals and alarm signals are received through the input/output interfaces.The graphical interface program is displayed by the demonstrator and interacts with the user to return the result of the operation [6].

4.3 Design of control software

The automatic warehouse control system software is divided into eight functional modules:

- Return to zero: When the system is switched on, the stacker performs a zero return operation to establish a warehouse coordinate system to determine the distance of movement to each position;
- Back to standby point;
- The motion control of loading and unloading dispatch of the warehouse: automatically /manually loading and unloading dispatch of the warehouse;
- Back to standby point;
- Operation information management;
- Inventory information management:Users can view the current usage of warehouse, inventory model, storage time, storage amount in the warehouse browsing function in real time.The user can modify the stock information of the position in the software, and transmit the information to the stereoscopic warehouse stacker, which is convenient for users to view the current information in real time.
- Historical data:Users can view the software login, information status, login time.You can check the number of outbound and outgoing time of a certain model according to the time period.You can also check the delivery time and quantity of all the models in a period of time;
- System debugging: The parameters such as the zero return speed, the running speed, the position of the standby point and the extending distance of the fork are set up for each axle of the stacker When the whole system is debugged on-line. The user can operate the axes manually;
- System operation parameter setting:Set the stacker's zero return speed and acceleration, manual and automatic warehousing speed and acceleration, fork speed, acceleration and other parameters in order to achieve the smooth running of each axis of the whole system;
- Fault alarm and diagnosis: In the course of stacker operation, the corresponding warning information will be displayed on the handheld display device when the axle overload and servo alarm are failed.The occurrence of major errors also need to stop stacker. This method facilitates the operators to find the cause of the trouble and to eliminate the trouble in time.These alarms mainly including servo driver alarm, communication failure alarm, abnormal shutdown alarm, command execution failed alarm.

According to the functional requirements of the management system, the function diagram of the software of the stereoscopic warehouse management system is designed, as shown in Fig.4.

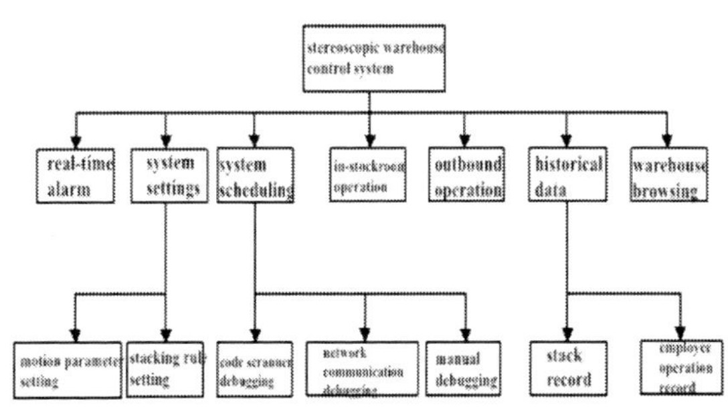

Fig. 4. Management system software structure

 DOI: 10.1051/matecconf/201713900038

4.4 Control system speed and position control

The motion of each axis of the control system is driven by a servo driver, servo motor drives the corresponding shaft movement, and the speed control sets corresponding parameters through the display screen.The movement of each axis is controlled by acceleration and deceleration. The advantage of this method is that the system runs smoothly, the impact load is smaller, the torque is large, the position control is accurate, and the mechanical characteristics are hard.

In the direction of movement of the X axis:At first, the acceleration was accelerated at 100mm/s2, uniform motion after speed up to 21m/min , in the target position still has one position distances, with 100mm/s^2 acceleration deceleration to the target position.In the direction of Z axis movement, there are two cases of loading and no-load in the loading platform.Two speed parameters are set for the two cases respectively.

At no-load, the Z axis at first to 120mm/s^2 acceleration operation, uniform operation after accelerated to 6m/min .At a relatively short distance from the target position, slow down at 80mm/s^2 acceleration to target position.At load,the Z axis at first to 80mm/s^2 acceleration operation, uniform operation after accelerated to 5.5m/min.At a relatively short distance from the target position, slow down at 80mm/s^2 acceleration to target position.The stacker speed curve is shown in Fig.5.

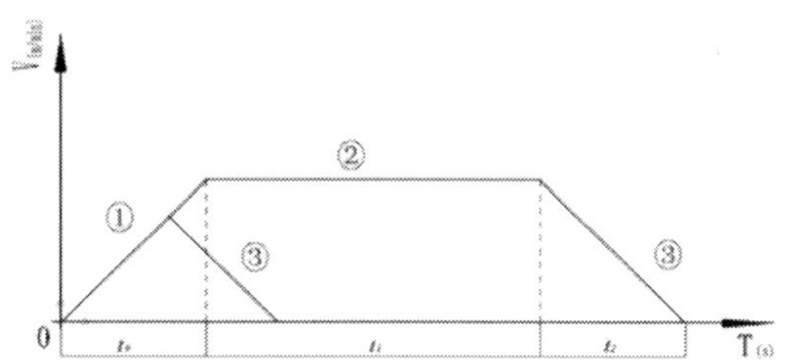

Fig. 5. Stacker operating speed curve

The warehouse uses a layout of the unit goods format, which requires an accurate location for each cell.There are three common positioning methods in stereoscopic warehouse, they are location of target address, location of laser range and location of pulse count [8].The position setting value of the actuator can be adjusted by man-machine interface in the position control process of the axle positioning, and then modified in the program.

The servo motor drives the stacker movement and converts the circumferential motion into linear motion through the rack and pinion mechanism.The position information of the actuator is fed back to the GUS motion controller via an encoder of the motor.The motion controller calculates the motor speed periodically according to the setting value and the actual value.The value is output to the servo drive via the pulse output port to form a half closed loop control that improves positioning accuracy [5].The stereo storage system uses servo motors with encoders. After practical testing, the positioning accuracy is 1mm ,which can achieve the positioning function well and meet the requirements of the positioning accuracy of the system.

5 Stacker software function module

5.1 Return reference point function module

The incremental encoder is arranged on the servo motor of the stacker crane of the stereoscopic warehouse. This device has the advantages of high precision and good thermal stability, but the disadvantage that the zero point and the absolute position of the reference point can not be recorded after the power failure is taken. This requires each time you start the machine, you need to go back to the reference point and establish a position coordinate system to ensure the accuracy of the position when the goods are loading and unloading dispatch of the warehouse.

According to the system accuracy requirements (1mm) and the actual situation of the system hardware, the system uses the Home switch to search the reference point.The origin search principle is shown in Fig.6.

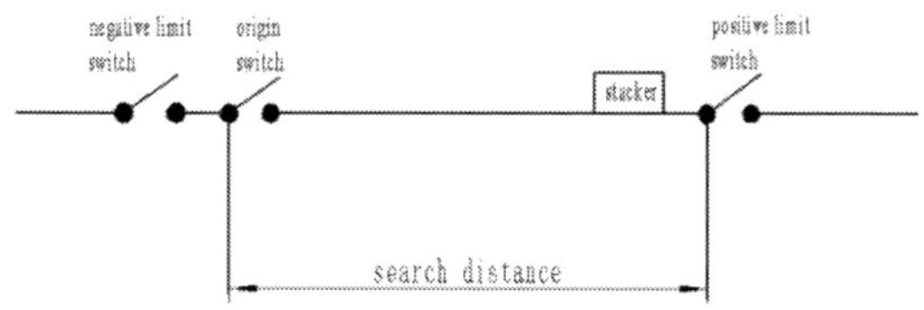

Fig. 6. Origin search principle

5.2 Online automatic storage module

When the AGV of the welding machine reaches the storage turntable and stops smoothly, the AGV sends in place signals to the storage management server through the router.The server reads the current state of the stacker after receiving the information.If the current stacker is idle, the management computer sends the storage operation instructions to the stacker controller, and then the stacker starts the automatic storage process.

5.3 Manual storage function module

Companies need to put some of the less commonly used or backlog of goods into the designated positions, or because of other reasons need to specify warehousing positions, so the need to develop manual warehousing functions to put the small volume goods in warehouse.In the manual storage case, the user need to manually enter the amount,type and target position. The controller checks the parameters entered by the user.If the parameter is valid, the stacker performs warehousing operations and puts the goods in the target position.After completion the storage, the stacker will return to standby and wait for the next order.

6 Conclusions

In order to adapt to the rapid development of intelligent logistics industry, the stacker control system to carry out the study play an important role in the modern logistics management system, which can meet the automatic stereo warehouse to high speed, heavy load, high precision, high reliability, complex system development urgent needs.The system uses GUS embedded motion controller and servo system, the controller software development platform of OTOSTUDIO has rich library functions and powerful instruction systems.Developed a set of automated warehouse control system with high positioning accuracy, high efficiency of storage and delivery, stable and reliable operation. Combine it with computer software, which promotes the improvement on the control system of the logistics industry.After the actual operation test, the whole system has good coordination, reliable operation, accurate movement and high efficiency. It meets the requirements of each function and performance index of the stacker.

References

1. Riccardo Manzini, Mauro Gamberi. Design and control of an AS/RS. Int J Adv Manuf Technol. 2006, 11(28): 766-774.
2. Yasukawa, Machiko, Takaki, Masayoshi, et al. A study on slip and fall of stacked cargoes in automated warehouse.Structural and Construction Engineering, 2017, 82(732): 183-192.
3. Colla V, Nastasi G, Matarese N, et al. Ga-based solutions comparison for storage strategies optimization for an automated warehouse.Intelligent Systems Design and Applications, 2009. ISDA'09. Ninth International Conference on. IEEE, 2009: 976-981.
4. Eldemir F, Graves R J, Malmborg C J. New cycle time and space estimation models for automated storage and retrieval system conceptualization. International Journal of Production Research, 2004, 42(22):4767-4783.
5. Salah B, Janeh O, Noche B, et al. Design and simulation based validation of the control architecture of a stacker crane based on an innovative wire-driven robot. Robotics and Computer-Integrated Manufacturing, 2017, 44: 117-128.
6. Kolla S, Michaloski J, Rippey W. Evaluation of component-based reconfigurable machine controllers.Automation Congress, 2002 Proceedings of the 5th Biannual World. IEEE, 2002, 14: 625-630.
7. Eldemir F, Graves R J, Malmborg C J. A comparison of alternative conceptualizing tools for automated storage and retrieval systems. International journal of production research, 2003, 41(18): 4517-4539.
8. Piasecki D. Warehouse Management Systems: Is It Worth the Trouble. World Trade, 2004, 17: 60-69.

Improved Greedy Algorithm Based Matrix Satellite Cable Connection Optimal Design

Yiyi Wang[1], *Zhifu Wang*[1], *Ke Zong*[1], *Mian Ye*[1], *Xi Zhou*[1] *and Zhijie Wang*[1]

[1] Institute of Telecommunication Satellite, China Academy of Space Technology, Beijing 10094,China

Abstract. An improved greedy algorithm based matrix satellite cable optimal design method is presented in this paper aim to overcome the difficulties including designing complexity, designing cycle and iterative designing process. An optimal algorithm model is built to help designer not only to accomplish design correctly and efficiently, but also to obtain better designing outcomes. Firstly, matrix cable designing model is pre-translated by turning electrical interface to equivalent digital model. Then the digital model is turned into computer language. At last, optimal design rules and method are extracted according to improved greedy algorithm. Real project data are used to prove the feasibility and validity of the proposed method.

1 Introduction

Matrix circuit including matrix telemetry and command is a core technology of on-board integrated electronic technology [1] to reduce the amount of electronic components and cables. This technology is widely used in all kinds of satellites to observably lose the weight of satellite platform. The devices with matrix telemetry and command can be connected with integrated electronic subsystem by matrix cable to send commands and receive telemetries. Matrix cable design has great possibility to be improved by reducing the usage of cables because matrix circuit usually has a large number of brunches which may cause the redundancy of design.

Consequently, an improved greedy algorithm based matrix satellite cable optimal design method in presented in this paper aim to overcome the difficulties listed above. Computer based design can acquire correct and preferred outcome, meanwhile, improve the work efficiency.

2 Matrix cable contact design principle

Matrix command contact circuit comprises command drive device and command load device. Command drive device can control m×n pieces of load device by control m row cables and n column cables [2]. It takes 2(m+n) cables to control m×n pieces of load device if backup is included. Matrix telemetry contact circuit is using the same principle. Matrix cable contact design principle is to build a point-to-point mapping relation between drive devices and load devices. By building a physical cable connection, drive devices can control load devices to executive commands and receive telemetries.

3 Improved greedy algorithm based satellite matrix cable connection optimal design

1.1 Design model pre-processing

Circuit model is tuned into structural digital model to achieve optimal design. Assuming that a command load device has 9 matrix commands controlled by 3 row cables and 3 column cables, it can be shown as $M_{RC}=\begin{bmatrix} M_{11} & M_{12} & M_{13} \\ M_{21} & M_{22} & M_{23} \\ M_{31} & M_{32} & M_{33} \end{bmatrix}$. Similarly, drive device can be shown as $N_{RC}=\begin{bmatrix} N & \cdots & N \\ \vdots & \ddots & \vdots \\ N & \cdots & N \end{bmatrix}_{R\times C}$.

From the knowledge mentioned above, row 1 and column 1 take control of M11. According to the principle, circuit connection can be translate into mapping relation between drive device model and load device model. If NRC= $\begin{bmatrix} N & \cdots & N \\ \vdots & \ddots & \vdots \\ N & \cdots & N \end{bmatrix}_{5\times 5}$,the relation can be shown as $\begin{bmatrix} N & M & M & M & N \\ N & M & M & M & N \\ N & M & M & M & N \\ N & N & N & N & N \\ N & N & N & N & N \end{bmatrix}$.

1.2 Formula expression of design model

* Corresponding author: wyywangyiyi@126.com

Formula expression of design model is { Rx, Cy,k}.

Rx represents for row cable, Cy represents for column cable, k represents for the number of matrix model on one device.

1.3 Improved greedy algorithm based matrix cable connection optimal design method

Optimal design is achieved by algorithm based programming on the basis of digital model and formula expression input.

A greedy algorithm does not consider the overall result, but aim to find a partial optimum. In most cases, a greedy strategy follows an optimal principle and takes partial optimum step by step. The greedy algorithm structure is as follows.

Beginning with an original solution for a problem,

While (making one step forward to final goal)

{

Use available strategies to find a solution

}

Get a final solution by combining all the solutions

VLSI routing problem [3] and 2D packing problem [4] chooses different row cable and column cable will get different matrix connections. Travelling salesmen problem [5] has capacity restriction and mutual restriction. Unlike VLSI or TSP problem, this case is a 2D vehicle routing problem [6] with confliction and capacity restriction. To solve this problem, this article raised an improved greedy algorithm with complexity of O (M*n). M represents the maximum of row or column, N represents the amount of matrix.

1.3.1 Optimal Design rules summary

Step 1 List all Matrix signals. {T} is the set of telemetries; {C} is the set of commands. {T_M} is the set of matrix telemetries; {C_M} is the set of matrix commands. $T_M \in T$, $C_M \in C$. {T_M} and {C_M} should be managed separately.

Step 2 Separate all devices into different secions. Put devices into correspoding section lists by recognize their coordinates.

$$T_M\begin{cases}\text{South Part}\begin{cases}+X\\-X\end{cases}\\ \text{North Part}\begin{cases}+X\\-X\end{cases}\end{cases}\qquad C_M\begin{cases}\text{South Part}\begin{cases}+X\\-X\end{cases}\\ \text{North Part}\begin{cases}+X\\-X\end{cases}\end{cases}$$

Definition 1 T_{Max} is the matrix telemetry with maximum of row or column in {T_M}.

Definition 2 C_{Max} is the matrix command with maximum of row or column in {C_M}.

Definition 3 T_{Sum} is the matrix telemetry with maximum sum of row and column in a specific set of telemetries.

Definition 4 C_{Sum} is the matrix command with maximum sum of row and column in a specific set of commands.

Step 3 Find TMax and C_{Max}. If there are more than one T_{Max} or C_{Max}, we get {T_{Max}} or{ C_{Max} }.Then we find T_{Sum} or C_{Sum}. If there are more than one T_{Sum} or C_{Sum}, the priority is random.

Rule 1 Using row cable is prior to column cable if the total increment and partial increment are fixed.

Definition 5 If one device has been appointed to be D_{Aim}, algorithm cannot move on to other device until all the matrix signals are allocated.

Definition 6 The set of all matrix telemetries of D_{Aim} except T_{Max} is T_{Next}. The set of all matrix commands of D_{Aim} except C_{Max} is C_{Next}.

Step 4 When T_{Max} or C_{Max} is chosen, the device T_{Max} or C_{Max} becomes D_{Aim}. T_{Next} of C_{Next} must be assembled with T_{Max} or C_{Max} to get the minimum sum of rows and columns in order to use cables as less as possible.

Definition 7 The set of devices on the same branch of D_{Aim} except D_{Aim} is {D_{Next}}.

Definition 8 T_{Aim} or C_{Aim} is the matrix signal which is going to be allocated then.

Step 5 When finished D_{Aim}, repeat step 3 and 4 for {D_{Next}}.

Rule 2 The second junction point is the flag of where a branch starts.

Rule 3 If there are more than one position with minimum cost increment, length increment is the solution to find the best choice.

Step 6 When {D_{Next}} is finished, repeat step 1 to 5 for the rest.

Rule 4 Choose matrixes which have the same structure as T_{Aim} or C_{Aim} to assemble if priority cannot be recognized so that unnecessary blank can be avoided as far as possible. This rule is also made to reduce the usage of cable and shorten cable length.

Step 7 After all matrix signals were allocated, do a final check to see if any adjustment can be done to get a better result.

1.3.2 Improved greedy algorithm based matrix cable connection optimal design

Definition 9 $S_{R\times C}$ is the capacity of matrix cable allocation.

Definition 10 $\Delta Cost$ is the cost increment when choosing position.

Definition 11 $\Delta Length$ is the length increment.

Definition 12 P is insertion point. P is on the first line and first row of T_{Aim} or C_{Aim}.

If $S_{R\times C}=\begin{bmatrix} C & \cdots & C \\ \vdots & \ddots & \vdots \\ C & \cdots & C \end{bmatrix}_{16\times 24}$, 16 row lines and 24 column lines can be shown in a sketch map as follows. In figure 1, judgment flags as TRUE or FALSE are set at junction point, such as M, N, P.

According to figure 1, $C_{Max}=\begin{bmatrix} C & C & C & C \\ C & C & C & C \\ C & C & C & C \\ C & C & C & C \end{bmatrix}_{4\times 4}$, $\{C_{Next}\}$ has $[C_{N1} \quad C_{N1}]_{1\times 2}$ and $[C_{N2} \quad C_{N2}]_{1\times 2}$. We can get $\begin{bmatrix} C & C & C & C \\ C & C & C & C \\ C & C & C & C \\ C & C & C & C \\ C & C & C & C \end{bmatrix}$ to make $\Delta Cost$ least. Now, $C_{Max}=\begin{bmatrix} C & C & C \\ C & C & C \\ C & C & C \\ C & C & C \end{bmatrix}$. P has 10 probable positions as $\begin{bmatrix} C & C & C & C & P \\ C & C & C & C & P \\ C & C & C & C & P \\ C & C & C & C & P \\ C & C & C & C & P \\ P & P & P & P & P \end{bmatrix}$. To make both $\Delta Cost$ and $\Delta Length$ least, the final position is $\begin{bmatrix} C_{Max} & C_{Max} & C_{Max} & C_{Max} & C_{Aim} & C_{Aim} & C_{Aim} \\ C_{Max} & C_{Max} & C_{Max} & C_{Max} & C_{Aim} & C_{Aim} & C_{Aim} \\ C_{Max} & C_{Max} & C_{Max} & C_{Max} & C_{Aim} & C_{Aim} & C_{Aim} \\ C_{Max} & C_{Max} & C_{Max} & C_{Max} & C_{Aim} & C_{Aim} & C_{Aim} \\ C_{N1} & C_{N1} & C_{N2} & C_{N2} & & & \end{bmatrix}$.

Total cable length that C_{Max} used: $8\times QA=8\times 8=64$

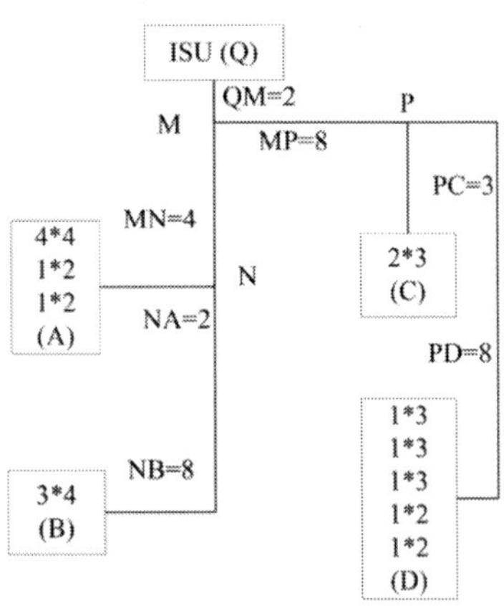

Figure 1 Cable layout sketch map

C_{N1} and C_{Max} belong to the same device, they share same column lines. C_{N1} brings a new row line.So $\Delta Cost=1$, $\Delta Length=QA=8$. Then C_{N2} and C_{Max} belong to the same device, C_{N2} share the same column lines with C_{Max}, meanwhile, share the same row line with C_{N1}.So $\Delta Cost=0$, $\Delta Length=0$。Total length of D_{Aim} is 64+8=72. In $\{D_{Next}\}$, $C_{Max}=\begin{bmatrix} C & C & C & C \\ C & C & C & C \\ C & C & C & C \\ C & C & C & C \end{bmatrix}$. $\Delta Cost=0+3=3$, $\Delta Length=4\times NB+3\times BQ=4\times 8+3\times 14=32+52=84$。

After D_{Aim} is allocated, the judgment flags of M and N turned to TRUE as shown in figure 2 because these junction points have been used once. When moving to next step, algorithm searches the first TRUE point to the direction back to the driver. At last, we got

$$\begin{matrix} C & C & C & C & C & C & C \\ C & C & C & C & C & C & C \\ C & C & C & C & C & C & C \\ C & C & C & C & C & C & C \\ C & C & C & C & C_{Max} & C_{Max} & C_{Max} \\ & & & & C_{Max} & C_{Max} & C_{Max}. \end{matrix}$$

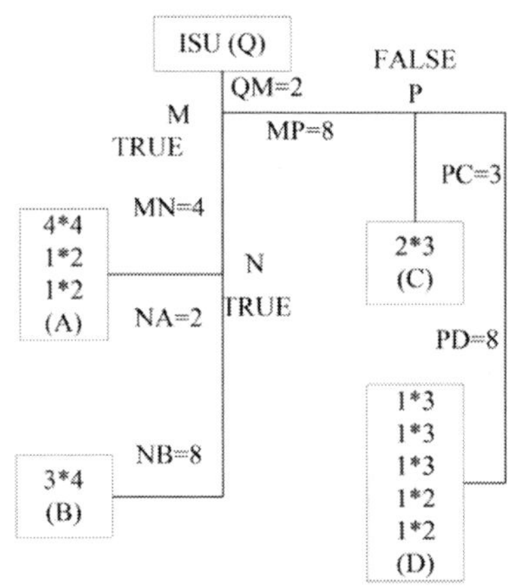

Figure 2 branch judgment flag

4 Design example

4.1 Programming

Interface 1 CountWidth
Interface 2 HeightComparator
Interface 3 Cell
Interface 4 SortCellUtil
The programming sturcture for interfaces is showed in figure 3.

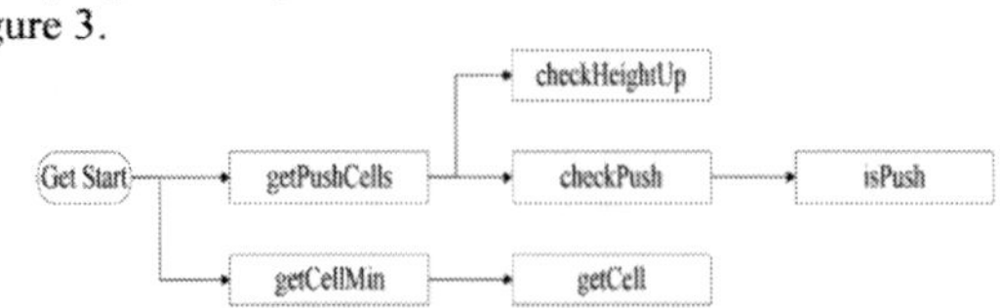

Figure 3 structure sketch

 DOI: 10.1051/matecconf/201713900039

The programming logic is showed in figure 4.

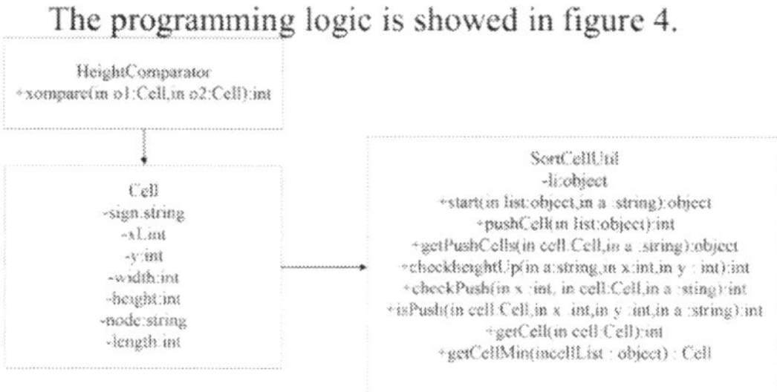

Figure 4 Programming logic

4.2 Real project data verification

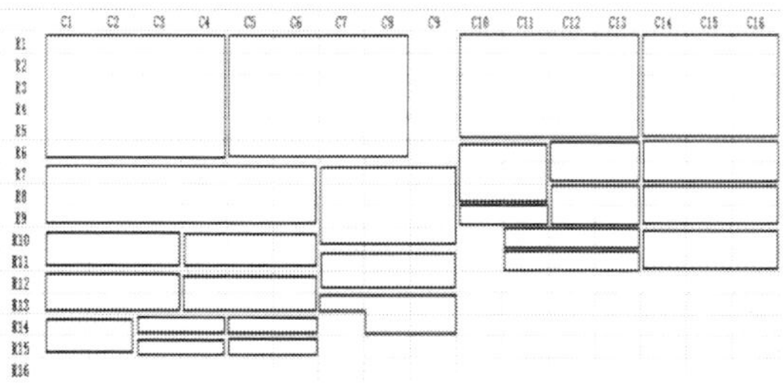

Figure 5 Optimal design result

Comparing with original design result, the optimal design result reduces total cable weight by 8% which is 0.416 kg.

5 Conclusion

An improved greedy algorithm based matrix cable connection optimal design method is proposed in this article and the method is tested by real data project to prove its reasonableness and effectiveness. The result shows that optimal design result reduced weight by 8% comparing with original design result. Some complicated situation may occur in real projects such as branch terminals and layout adjustments. Later we will make further improvement of this optimal design method.

Acknowledgments

This work has been supported by Project of the Program of National Natural Science Foundation of China under the Grant Number 61471360.

This work has been supported by the 13th Five-Year Plan Pre-research special technology project from Space System Division.

References

1. Mingfeng Li, Yanyan Hao, Xiao Chen. New telecommunication satellite integrated electronic system research [J]. Space International, 6, 46-50,(2013)
2. Ke Zong, Zhifu Wang, Hangyan Xu. Optimal Mechatronic Design Research for Matrix Cables Used on Satellite [J]. Spacecraft Engineering,8(26),15-21 (2017)
3. Xiaoping Tang, Ruiqi Tian, and Martin D.F.Wong. Minimizing wire length in floorplanning[J]. IEEE Transaction on Computer-Aided Design of Integrated Circuits and Systems, 25(9), 1744-1753,(2006)
4. Gerhard Wäscher, Heike Haubner, Holger Schumann. An improved typology of cutting and packing problems [J]. European Journal of Operational Research, 183(3), 1109-1130,(2007)
5. Tolga Bektas, The multiple traveling salesman problem: An overviewof formulations and solution procedures [J]. Omega, 34(3),209-219,(2006)
6. Kris Braekers, Katrien Ramaekers, Inneke Van Nieuwenhuyse. The vehicle routing problem: State of the art classification and review [J]. Computers & Industrial Engineering, 99,300–313,(2016)

Poles Extraction of Radar Target Based on Narrowband Frequency Response Approximation

Xuran Xing[1,*], *Hongzhong* Zhao[2], and *Xin* Jia[3]

[1]Sapace Engineering University, Beijing, 101416, China
[2]Sapace Engineering University, Beijing, 101416, China
[3]Sapace Engineering University, Beijing, China

Abstract. Pole is the most important feature of radar target in resonance region, and it is attitude insensitive. The poles extraction requires UWB (ultra-wideband) late-time response signal or frequency response data, which is difficult to achieve through general equipment. This paper proposed a method for extracting poles from narrowband data. At first, the frequency response equation of the resonance region was derived. Then, the equation parameters were solved by the interior point method through the narrowband data. At last, the zeros of the transfer function denominator were calculated. FEKO simulation data test verifies that the method has extrapolation ability.

1 Introduction

Radar target recognition technology determines the target properties and categories based on the target electromagnetic scattering properties. When the target size and the wavelength size are at the same order of magnitude, the target is in the resonant zone and the main electromagnetic scattering property is the pole. The pole is insensitive of the observation angle and the polarization of the radar. It is only related to the size, shape and material of the target itself, so it is a robust electromagnetic feature for target recognition.

One of the conditions for poles extraction is UWB signals, but most radar devices do not have the capability of ultra-wideband. Existing poles extraction algorithms, such as prony method[1], matrix pencil method[2]and iterative method[3], do not have the ability to extract poles from narrowband data. This paper proposed a poles extraction method from narrowband frequency response data. The transfer function equations of the resonance region was derived according to the singularity expansion method (SEM) theory. Then, the primal-dual interior point method solved the ill-conditioned equations and obtained the transfer function coefficients. The zeros of the transfer function denominator are the poles. FEKO calculated the far-field data of thin-cylinder, rectangle, and a complex target, and validated the algorithm.

2 Electromagnetic characteristics of radar target in resonance region

The singularity expansion method (SEM) was proposed when Baum studied the back response of radar echoes in 1971[4]. SEM describes the characteristics of the echo and transfer function in the resonance region systematically. The ultra-wideband pulse signal is usually used as the excitation when studying the echo signals in resonance region, so the response signal is very short in time domain and it is divided into two parts: the early-time response and the late-time response. The early-time response describes the scattering center characteristics of the target, varying with the observation angle; the late-time response describes the poles characteristics of the target, and the position of the poles does not change with the observation angle. The late-time response can be represented as follows:

$$y(t)=\sum_{m}^{M} r_m e^{s_m t} \tag{1}$$

where r_m are the residues, $s_m=\sigma_m+\omega_m j$ are the poles. The real part σ_m is the attenuation factor and the imaginary part ω_m is the angular frequency. The residues change when changing the observation angle, but the poles remain stable. The transfer function is modelled as the sum of a residue series plus and an entire function:

$$H(s)=\sum_{i=1}^{M}\frac{r_i}{s-s_i}+C(s) \tag{2}$$

In time domain, the residue series plus corresponds to late-time response, and the entire functions corresponds to early-time response.

* Corresponding author: xxr01839@mail.ustc.edu.cn

There are two ways to acquire the late-time response signal. One is exciting target by UWB pulse signal in time domain, this method requires wide bandwidth and high sampling rate of equipment. The other way is transmitting a series of simple-frequency signal and calculating the frequency response value at each frequency point. Then, the late-time signal was reconstructed through inverse Fourier transform to the frequency response. The second methods are easier to implement and have greater application value. Existing poles extraction algorithms rely on the late-time signals, so UWB frequency response data is necessary. When the UWB condition is not satisfied, even if the data near the poles are acquired, the poles can not be accurately estimated. This constraint limits the application of poles as a target identification method.

3 Poles extraction of radar target under narrow-band conditions

3.1 Transfer function equation

The entire functions is complex and its energy is small, so it is usually neglected when we concern about the poles. Formula 2 is simplified as:

$$H(s)=\sum_{i=1}^{M}\frac{r_i}{s-s_i}=\frac{\sum_{i=0}^{M-1}b_i s^i}{1+\sum_{i=1}^{M}a_i s^i} \tag{3}$$

Multiplying both sides by the denominator and transposition

$$H(s)\times\sum_{i=1}^{M}a_i s^i-\sum_{i=0}^{M-1}b_i s^i=-H(s) \tag{4}$$

Discretization of formula 5 is expressed as below:

$$A\cdot x=b \tag{5}$$

Where

$$A=\begin{bmatrix} H_1 s_1 & H_1 s_1^2 & \cdots & H_1 s_1^M & -s_1^0 & -s_1^1 & \cdots & -s_1^{M-1} \\ H_2 s_2 & H_2 s_2^2 & \cdots & H_2 s_2^M & -s_2^0 & -s_2^1 & \cdots & -s_2^{M-1} \\ \vdots & \vdots & \ddots & \vdots & \vdots & \vdots & \ddots & \vdots \\ H_N s_N & H_N s_N^2 & \cdots & H_N s_N^M & -s_N^0 & -s_N^1 & \cdots & -s_N^{M-1} \end{bmatrix},$$

$x=[a_1 \;\; a_2 \;\; \cdots \;\; a_M \;\; b_0 \;\; b_1 \;\; \cdots \;\; b_{M-1}]^T$ is the parameter vector and $b=-H=-[H(jw_1)\;\; H(jw_2)\cdots H(jw_N)\;]^T$ is the response data collected in frequency domain. The parameterized transfer function is realized by solving the parameter vector x.

3.2 Solving the transfer function equation

For the overdetermined equation of Formula 5, the usual solving method is calculating the least square solution $x=(A^T A)^{-1}A^T b$ .But this method is not applicable to ill conditioned equations. The problem of poles estimation is a ill conditioned problem[5].Ill conditioned means that the matrix A approach to singularity, at this point, the solution of the equation is unstable.

In order to obtain accurate parameters, the problem of parameter solving is considered as a nonlinear programming problem.

$$\begin{aligned} &\min \|Ax-b\|_2 \\ &s.t. \quad Ax=b \end{aligned} \tag{6}$$

It is also a quadratic programming problem as follows

$$\begin{aligned} &\min \; x^T Qx - 2c^T x \\ &s.t. \quad Ax=b \end{aligned} \tag{7}$$

Where $Q=A^T A$ and $c_i=b_i\sum_{j=1}^{2M+1}a_{ij}$.

Define $X=x\cdot x^T$, the quadratic programming problem has the following relaxation model [6]

$$\begin{aligned} &\min \; Q\cdot X \\ &s.t. \quad Ax=b \\ &X\geq 0 \end{aligned} \tag{8}$$

At this point, the problem has been transformed into a linear semidefinite programming problem. Semidefinite programming is an extension of programming problems. It is a new branch of mathematical programming developed in recent 20 years, with good mathematical properties and extensive engineering applications[7]. When the ill conditioned equations are transformed into linear semidefinite programming problems, some sophisticated optimization algorithms can be used to guarantee the accuracy and efficiency. This paper called the SDPT3, and used the primal-dual interior point method to solve above problem[8].

Interior point method constructs the objective function with penalty function, then the constrained optimization problem is transformed into an unconstrained optimization problem. The penalty function is small within the feasible region, and it is large at the boundary of the feasible region, so that the parameters in each iteration do not go beyond the feasible domain. The procedure of the primal-dual interior point method is described below.

For standard linear semidefinite programming problem as follows

$$\begin{aligned} &\min \; C\bullet X \\ &A\bullet X=b_k, k=1,\ldots,m \\ &X\geq 0 \end{aligned} \tag{9}$$

and its dual problem

$$\begin{aligned} &\max \; b^T y \\ &\sum_{k=1}^{m} ykA_k+Z=C \\ &Z\geq 0 \end{aligned} \tag{10}$$

DOI: 10.1051/matecconf/201713900040

The primal-dual interior point method can be summarized as following steps[8]:

1.Compute the duality gap $\mu = \boldsymbol{X} \bullet \boldsymbol{Z} / n$ and stop the iteration if it is smaller than the threshold;

2.Compute the search direction $(\Delta \boldsymbol{X}, \Delta \boldsymbol{y}, \Delta \boldsymbol{Z})$;

3.Update $(\boldsymbol{X}, \boldsymbol{y}, \boldsymbol{Z})$ to $(\boldsymbol{X} + \alpha \Delta \boldsymbol{X}, \boldsymbol{y} + \beta \Delta \boldsymbol{y}, \boldsymbol{Z} + \beta \Delta \boldsymbol{Z})$, where $\alpha = \min(1, -\frac{\gamma}{\lambda_{\min}(\boldsymbol{X}^{-1} \Delta \boldsymbol{X})})$ and $\beta = \min(1, -\frac{\gamma}{\lambda_{\min}(\boldsymbol{Z}^{-1} \Delta \boldsymbol{Z})})$;

4.Set step-length parameter $\gamma = 0.9 + 0.09 \min(\alpha, \beta)$ and return to step 1.

3.3 Model order determination

The order of the transfer function is the number of poles, and it is unknown for the unknown target. In the case of narrowband data, the algorithm does not care about the number of all poles, because the purpose is to extract the poles near the narrow band. However, we are concerned with the effect of order on the fitting degree. If the order is too small, the transfer function model can not fit the collected narrowband data, and the poles can not be accurately extracted; if the order is too large, the ill conditioned is too serious, result in arithmetic overflow. In order to determine the proper order, the different order equations are constructed, and then the frequency response $H_i(jw)$ is reconstructed by the parameters respectively. Calculate the errors of the reconstructed responses between the original data as follows:

$$error(i) = \left\| H_i(jw) - H(jw) \right\|_2 \qquad (11)$$

The order with the least error is considered as the order of the transfer function.

3.4 Poles estimation

The poles can be computed after obtain the parameterized transfer function. Calculate the zeros of the transfer function denominator as below

$$a_{M+1} s^{M+1} + a_M s^M + \cdots + a_1 s^1 + 1 = 0 \qquad (12)$$

Since the equations are constructed by narrowband data, the solutions near the narrow band are accurate, while the solutions of other frequency bands are wrong, so the solutions must be cleaned up. Solutions both near the narrow band and in the second quadrant of the Laplace domain are preserved as the poles. To sum up, the algorithm proposed in this paper is summarized as follows:

(1) Set the model order $i = i_0$;

(2) Construct and solve the equation as formula (5);

(3) Reconstruct the frequency response by the parameters of step2, then calculate the error by formula (11);

(4) If the equation is overflow, proceed to next step, and if not, set $i = i + 1$ and return to step 2;

(5) Compare the errors, the solution with the minimum error is chosen as the final parameters of the transfer function;

(6) Clear the result and preserve the poles;

4. Simulation

4.1 Thin-cylinder

The frequency response data of thin-cylinder with long 1m and radius 0.005m are calculated by FEKO. The material is ideal conductor, the incident angle is 60 degrees, and the simulation band is 0~1.5GHz. 31 frequency points in the 0.85~1GHz are regarded as narrowband data.

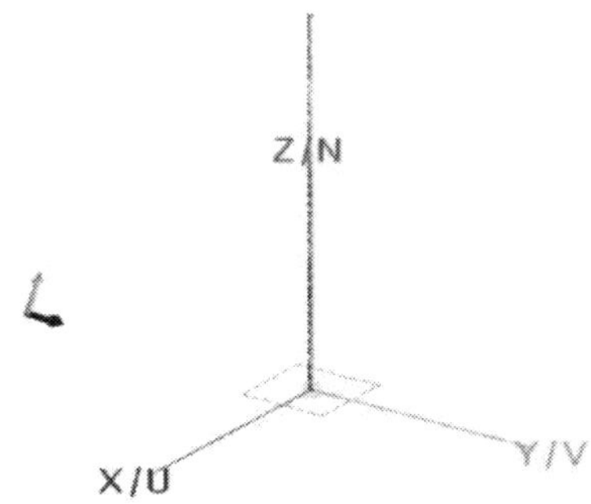

Fig.1. The mesh of thin-cylinder

The error of reconstructed frequency response in different order is shown in the following figure to obtain the appropriate model order.

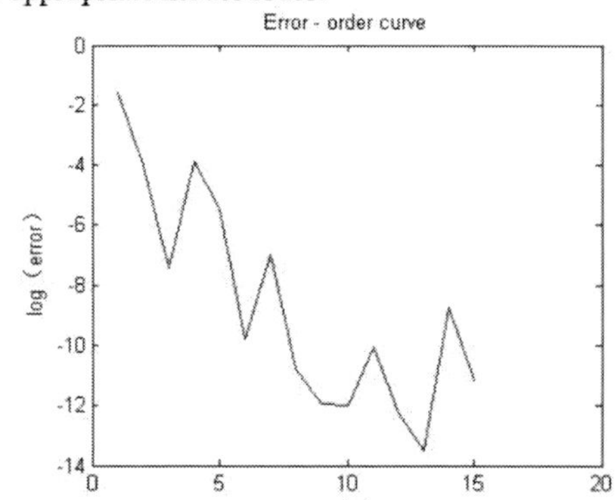

Fig.2. The Error-order curve

As shown in the Fig.2, the chosen order is 13. The reconstructed frequency response data is compared with the actual data. The reconstructed data is basically coincident with the actual data near the sampling narrow band. In other band, the error is very large, but it do not affect the accuracy of poles estimation near the narrowband.

DOI: 10.1051/matecconf/201713900040

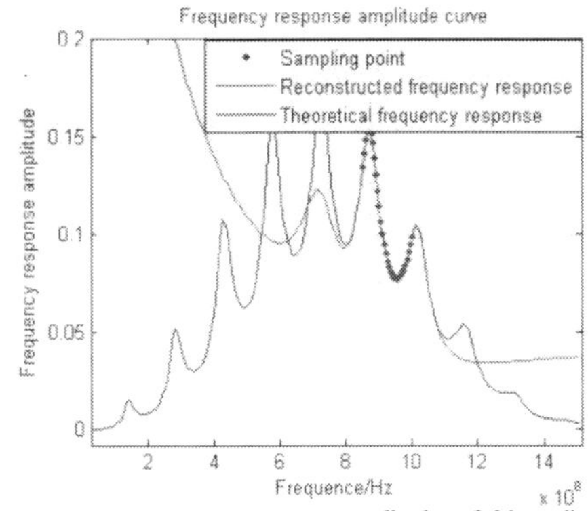

Fig.3. Frequency response amplitude of thin-cylinder

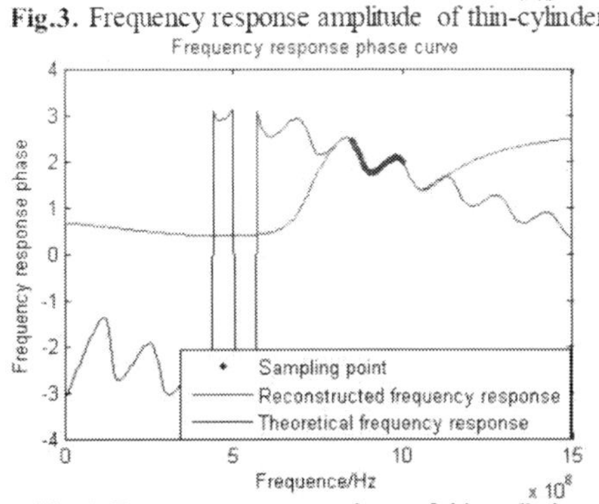

Fig.4. Frequency response phase of thin-cylinder

The poles were computed directly by formula 12, and the results are as follows:

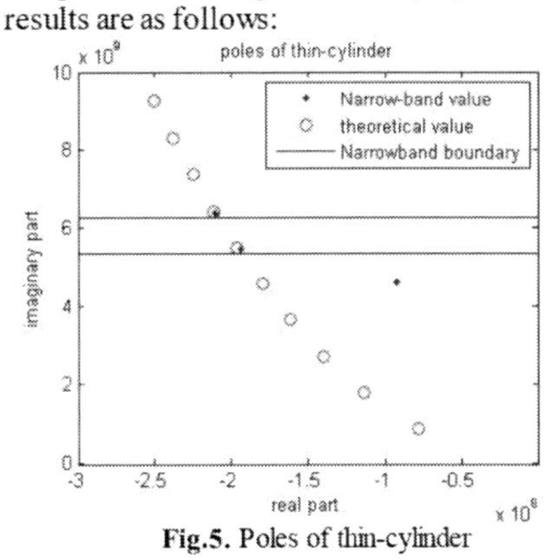

Fig.5. Poles of thin-cylinder

It can be seen that the algorithm extracts both the poles in the sampling narrow band and the poles near the sampling narrow band.

4.2 Cuboid

Because the thin-cylinder has only one characteristic length and regular frequency response, the simulation of cuboid is given in this paper. The frequency response data of cuboid with long 0.8m, width 0.5m and high 0.35m are calculated by FEKO. The material is ideal conductor and the simulation band is 0~1GHz. 21 frequency points in the 125~225MHz are selected as narrowband data.

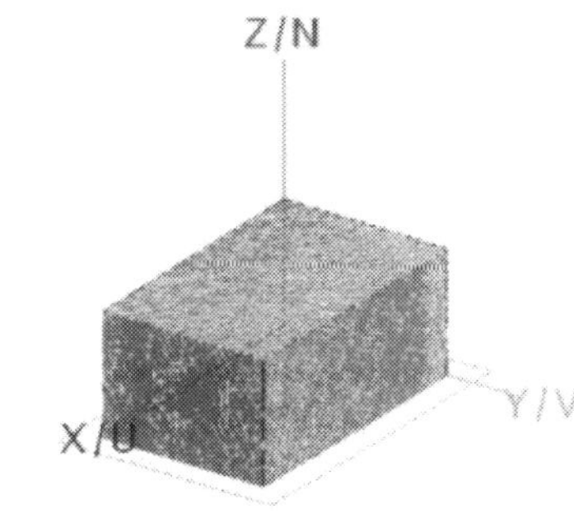

Fig.6. The mesh of cubiod

The poles of the cuboid as Fig.7, and the results are close to the theoretical values.

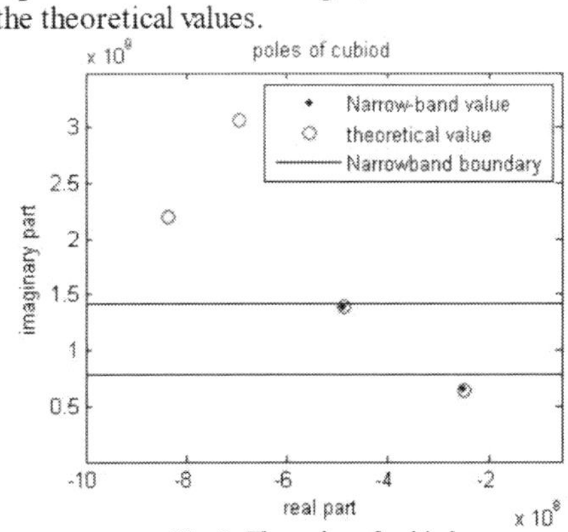

Fig.7. The poles of cubiod

4.3 Complex target

A small satellite is taken as an example to verify the effectiveness of the algorithm on complex targets. The observation angle and the mesh is shown in Fig.8. The simulation frequency band is 0~1GHz.Select two narrowband in 100MHz~300MHz and 500MHz~700MHz.

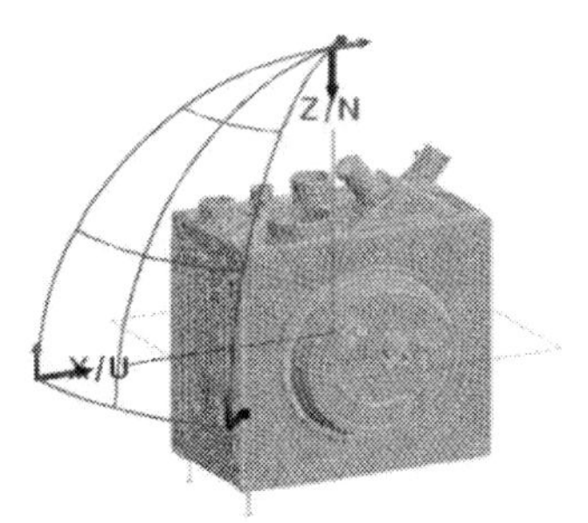

Fig.8. The mesh and observation angles of the satellite

According to the observation angle insensitivity characteristic of poles, the results calculated at different angles should be consistent. The results of this algorithm under three incident angles and the results of wideband data are presented in Fig.7, and the results are consistent.

The wideband results was calculated by the algorithm of Ref. [9].

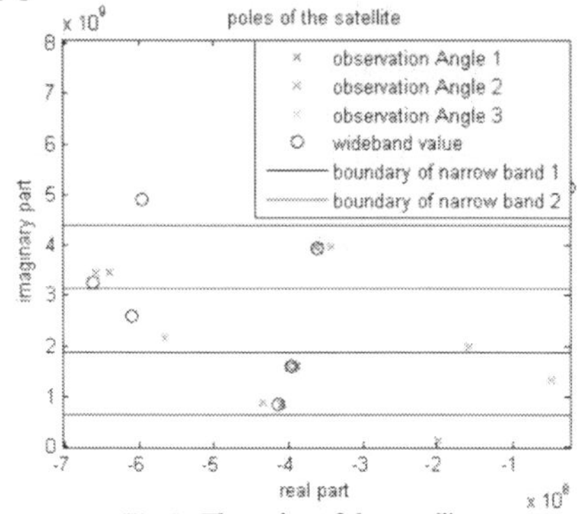

Fig.9. The poles of the satellite

5 Conclusion

In this paper, we proposed a poles extraction algorithm which can extract poles from narrow band data. The results show that the method can estimate the poles both in the sampling frequency band and near the sampling frequency band, which means the algorithm has extrapolation ability. It is an effective target recognition method when ultra-wideband data is not available.

Reference

1. Van Blaricum M, Mittra R. *A technique for extracting the poles and residues of a system directly from its transient response*,IEEE Transactions on Antennas and Propagation, **23(6)**.1975, 777-781.
2. Man, Liang, et al. *Poles Extracting and Analyzing of Complex Stealth Target Based on Matrix Pencil Method*.IEEE International Conference on Computer and Information Technology IEEE Computer Society, (2014)894-898.
3. Choi I S, Lee J H, Kim H T, et al. *Natural frequency extraction using late-time evolutionary programming-based CLEAN*. IEEE Transactions on Antennas & Propagation, **51(12)**, (2003),3285-3292.
4. Baum C E. *On the Singularity Expansion Method for the Solution of Electromagnetic Interaction Problems*. (1971).
5. Yang Songyan. *Research on High Frequency Band Radar Target Features Extraction and Identification*. Harbin Institute of Technology,(2016).
6. Jiang Yaowei. *Research of semidefinite Programming and its Application*. Xidian University,(2009).
7. Xiu Naihua,Luo Ziyan. *Semidefinite Programming*. Beijing Jiaotong University Press,(2014).
8. K. C. Toh, M. J. Todd, R. H. Tütüncü. *SDPT3 — A Matlab software package for semidefinite programming, Version 1.3*. Optimization Methods & Software, **11(1-4)**,(1999),545-581.
9. Sarkar, T. K., et al. *Application of the matrix pencil method for estimating the SEM (singularity expansion method) poles of source-free transient responses from multiple look directions*. IEEE Transactions on Antennas & Propagation **48.4**,(2002),612-618.

An efficient 3D point cloud data denoising algorithm for ship block visual measurement

Liu Tao [1,*], Du Hongwang[1]

[1]College of Automation, Harbin Engineering University, Harbin 150001, China

Keywords: ship block measurement; point cloud data; denoising; least square fitting

Abstract. For the 3D point cloud data aquired by laser scanning measurement of ship block, an efficient denoising algorithm based on image and normal vector threshold judgement is proposed. Firstly, large scale noise points are eliminated using global threshold judgement based image, then Kuwahara filter algorithm is used for data smoothing and a denoising algorithm based on normal vector threshold judgement is proposed to eliminate noises point excluding ship manufacture sections. The experiment result demonstrates that not only the proposed denoising algorithm keeps key data points but also avoids bluring point cloud boundary and eliminates noise points effectively.

1 Introduction

As manufacturing sections of ship block are composed of a lot of steel plates, 3D point cloud data obtained by structured light vision measurement sensor is not a geometry distribution to some extent, which is scattered and disordered. A few noise point cloud data excluding the ship segmentation during vision measurement procedure will be obtained simultaneously. So point cloud data denoising is an essential alogorithm in reverse engineering; point cloud denoising processing will directly affect the quality of subsequent 3D model reconstruction, which significantly affects the analysis and manufacture of results using the 3D model.

Common method of data smoothing and denoising are filtering technology in digital image processing, which are divided into space domain and frequency domain filtering. In recent years, many scholars have put forward various point cloud denoising algorithms[1-4]. These methods are generally facing the challenge both to reduce noise and to ensure the point cloud important feature information unaffected. According to the characteristics of different 3D objects, different point cloud denoising algorithms need to be putted forward. During the point cloud data processing of ship sections, some useless outlier points are needed to removed, and the background points need processing in order to the point cloud feature extraction subsequently. According to the point cloud data generated of ship sections visual measurement, a point cloud denoising algorithm is putted forward. Firstly based on image the large areas background noise points are removed, then using Kuwahara[5] filtering algorithm for data smoothing, finally a denoising algorithm based on normal vector threshold is proposed to eliminate noise data outside of manufacturing section. The algorithm firstly calculate K neighborhood point, then calculate plane tangent of the least square fitting for K neighborhood point, finaly use the angle of measuring point with tangent plane normal vector and the setting plane vector direction to distinguish whether to remove the measuring points.

2 Point cloud data denoising based on image

The feature point cloud of ship block manufacturing section obtained by digital measurement is basically in a section, so data depth has little difference, and the target data depth value basically than the noise data is smaller. Using the above information, based on the concept of digital image processing, the obtained data points are taking as image data, the measurement data points depth value as the pixel gray values in the image. Using the depth value distribution of the statistical point cloud data in histogram image processing, target data depth range ranges were counted, in order to remove large area noise point.The global thresholding method is as follow:

Step1 Choose an initial estimate T_0 for T in the middle of the maximum and minimum value.

Step2 Uses T to segmentation point cloud data. Achieve two sets of point cloud data: G_1 composed of all $z \geq T$ pixel, G_2 composed of all $z < T$ pixel.

Step3 Calculates the average z with point cloud data μ_1 and μ_2 within the range of values G_1 and G_2.

Step4 Calculates a new threshold: $T = \frac{1}{2}(\mu_1 + \mu_2)$

Step5 Repeat steps 2 through 4 successive iteration, until the T value is smaller than the pre specified parameters T_0.

* Corresponding author: hrbeu411jys@163.com

For point cloud data obtained by laser scanning measurement, common filtering algorithms cannot remove noise and preserve edge also has good effect. In order to obtain the boundary characteristics accurately, which can eliminate the noise but also to ensure the edge, corner feature is not smooth, using the Kuwahara filtering algorithm [5]. Kuwahara filtering algorithm is based on the depth image which is the famous boundary preserving smoothing method.

2 Denoising method of point cloud data based on normal vector threshold judgement

2.1 The K neighbourhood calculation

The three-dimensional point cloud of block visual measurement, have data redundancy, the existence of incomplete data and point cloud data with unknown topology. The establishment of geometric topology relationship between measuring point cloud, is beneficial to reduce the search range data and improve the dense scattered point cloud modeling speed. The common establishment method of data point cloud topological relationship is space grid method, such as proposed 3D scattered data nearest neighbor search algorithm[6]. The ship section measuring point cloud data processing used a fast searching large data sets K nearest neighbor spatial sphere algorithm [7], which is easy to realize and can quickly detect the point nearest neighbor.

2.2 Calculation of tangent plane based on least square fitting

Normal vector and curvature are the basic characteristics of surface, and are important bases of surface feature recognition. The normal vector can be determined by the tangent plane of the point[8]. Manufacturing section model of ship sections were positioned substantially in the plane, so the plane can be obtained by using least squares fitting.

For any point P in space, find its K neighbor and include the point denoted $Nbhd(P)$, ensure the minimum square of the distance from the K neighborhood of the point to the tangent plane, the normal vector can be calculated, and calculate the point approximation tangent plane by plane equation.

Set $Nbhd(P)$ as approximation of planar equation for a:

$$Ax + By + Cz + D = 0 \tag{1}$$

The unit normal vector plane is $N(A,B,C)$, then the distance formula of $Nbhd(P)$ point and plane $P_i(x_i,y_i,z_i)$ can be expressed as follows:

$$D_i = | Ax_i + By_i + Cz_i + D | \tag{2}$$

K neighborhood point set is $P_i(i = 1,2,3,\cdots,k)$,by the distance minimum square of these points and the least square plane ,the objective function is established:

$$F(A,B,C,D) = \sum_{i=1}^{k}(Ax_i + By_i + Cz_i + D)^2 \tag{3}$$

To make F minimum, dealing with the F derivative, which is equal to zero, get:

$$D = -(Ax_c + By_c + Cz_c) \tag{4}$$

Among: $x_c = \frac{1}{k}\sum_{i=1}^{k} x_i$, $y_c = \frac{1}{k}\sum_{i=1}^{k} y_i$, $z_c = \frac{1}{k}\sum_{i=1}^{k} z_i$ 。

(x_c, y_c, z_c) is the center coordinates of point set P_i .

Eq (3) has:

$$F(A,B,C,D) = \sum_{i=1}^{k}[A(x_i - x_c) + B(y_i - y_c) + C(z_i - z_c)]^2 \tag{5}$$

The Eq (5) derivative is:

$$\begin{bmatrix} a_{00} & a_{01} & a_{02} \\ a_{10} & a_{11} & a_{12} \\ a_{20} & a_{21} & a_{22} \end{bmatrix} \begin{bmatrix} A \\ B \\ C \end{bmatrix} = 0 \tag{6}$$

Introduce standard constraint condition in the plane normal vector $A^2 + B^2 + C^2 = 1$ The problem is under the constraint condition of extremum. The constraint conditions are:

$$G(A,B,C) = A^2 + B^2 + C^2 - 1 \tag{7}$$

By Lagrange multiplication:

$$\nabla F(A,B,C) = \lambda \nabla G(A,B,C) \tag{8}$$

$$\begin{cases} \begin{bmatrix} a_{00} & a_{01} & a_{02} \\ a_{10} & a_{11} & a_{12} \\ a_{20} & a_{21} & a_{22} \end{bmatrix} \begin{bmatrix} A \\ B \\ C \end{bmatrix} = \lambda \begin{bmatrix} A \\ B \\ C \end{bmatrix} \\ A^2 + B^2 + C^2 = 1 \end{cases} \tag{9}$$

The expansion is:

$$\begin{cases} \sum A(x_i - x_c)[A(x_i - x_c) + B(y_i - y_c) + C(z_i - z_c)] = \lambda A^2 \\ \sum B(y_i - y_c)[A(x_i - x_c) + B(y_i - y_c) + C(z_i - z_c)] = \lambda B^2 \\ \sum C(z_i - z_c)[A(x_i - x_c) + B(y_i - y_c) + C(z_i - z_c)] = \lambda C^2 \end{cases} \tag{10}$$

The sum of all kinds:

$$\sum[A(x_i - x_c) + B(y_i - y_c) + C(z_i - z_c)] = \lambda \tag{11}$$

According to the center coordinates and the above method the tangent plane equation can be determined.

2.3 Data processing based on normal vector threshold judgement

Setting normal vector threshold on feature regions, greater than the threshold point cloud grids are delete, so we can get a single value plane (space plane projection direction does not coincide). Set reference plane normal vector n of hip manufacturing section, each measuring point x_i nearby is calculated and the close points are fitted as the constraint tangent plane $P(x_i)$ of the point nearby. Keeping minimum distance square sum of the K neighborhood to tangent plane, calculate the normal vector n_i of the tangent plane, use α_i to show normal s vector angle of the measuring threshold. If $v_1 > \alpha_i > v_2$ is present, measuring point is removed.

By the tangent plane projection, 2D projection point set can beobtained. The coordinates of the point are (x', y', z'). The projection Coordinates of points in the plane are (x, y, z). The line is through a point (x', y', z') and direction vector $N(A, B, C)$ .The Eq(18) is obtained and By the Eq (1) and Eq(7) the simultaneous solution of the projection point coordinates can be obtained

$$\frac{x-x'}{A}=\frac{y-y'}{B}=\frac{z-z'}{C}=k \qquad (12)$$

After finding the tangent plane, through coordinate transformation to the coordinates of the points into the two-dimensional coordinates, you can consider only the two coordinate value without considering the third direction coordinate values, to simplify the problem. Of course, change the coordinates of the projection plane do similarly applied to all spatial data sets, to maintain data consistency and projection plane.

3 Experiments

The obtained point cloud data of ship block visual measured in different parts of the model is denoising, and the experimental results are analyzed. Figure 1(a) and (c) show point cloud obtained of ship model frame scan. The scanned point cloud includes the local data points and large areas noise point block which can be seen, so the denoising is very necessary for data processing. Figure 1 (b) and (d) show the experimental results for denoising the data point cloud by using this proposed algorithm. Figure 2 (a) shows the plane graph of point cloud data denosing of ship block model, Fig. 2 (b) is shown amplification maps in Figure 2 (a) arrow site. Through the denoising point cloud data the large background noise are effectively eliminated.

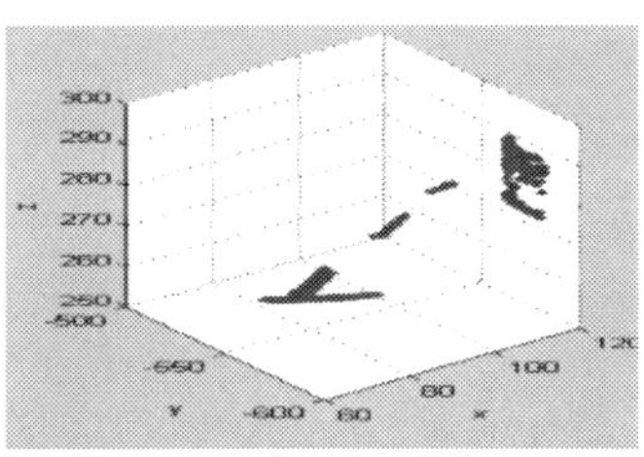

(a) the rib cage point cloud data measured

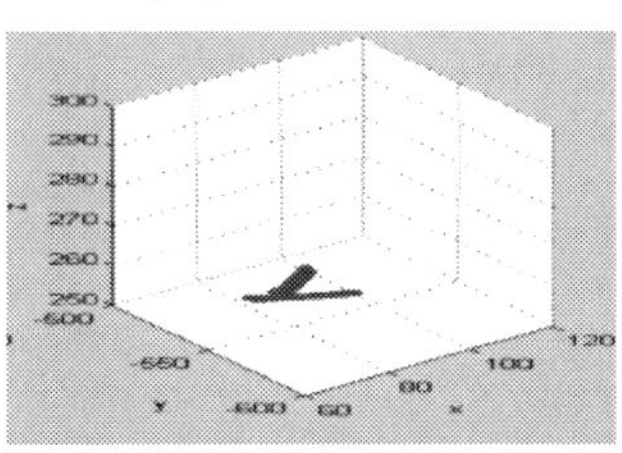

(b) the rib cage point cloud data after denoising

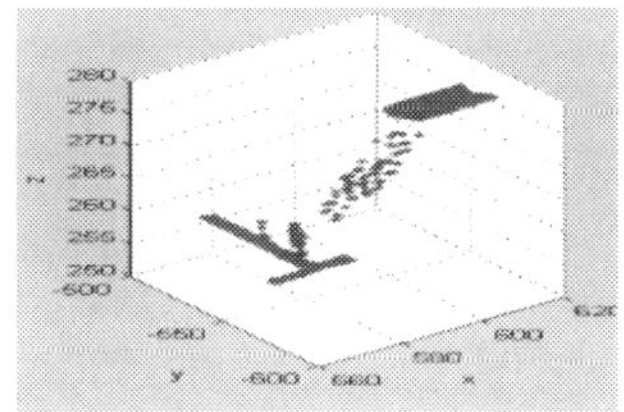

(c) T type of rib point cloud data measured

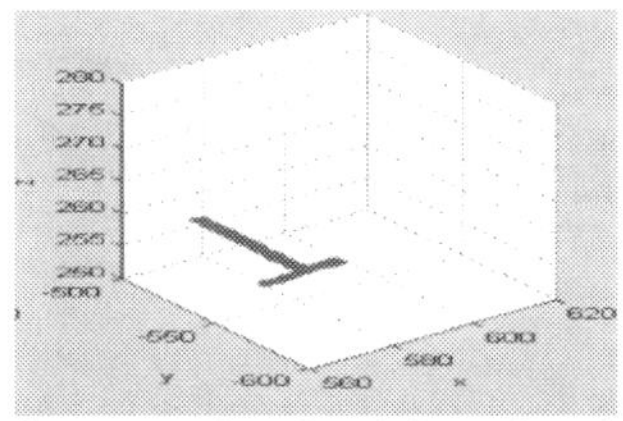

(d) the rib cage point cloud data after denoising

Fig.1 Schematic diagram of denoising for the point cloud of ship block measurement

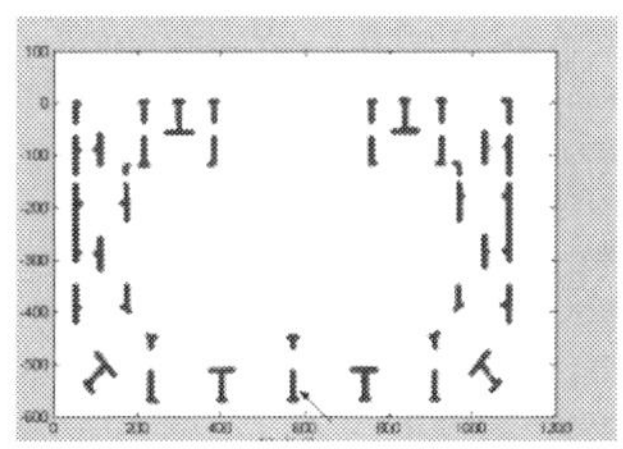

(a) the plane graph of point cloud data denosing

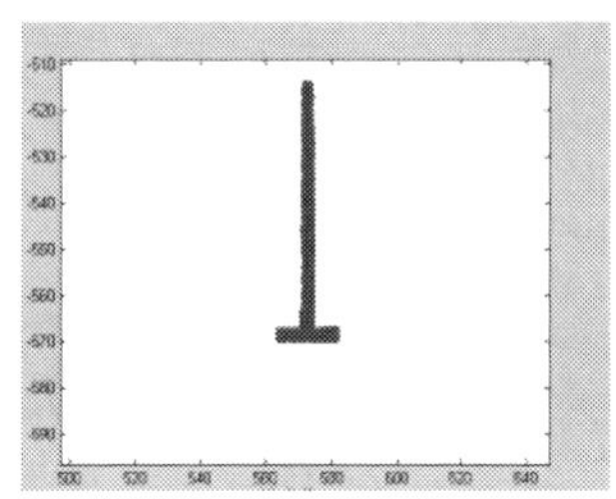

(b) the point cloud data locally amplified of ship block model

Fig. 2 Schematic diagram of de-noising for point cloud of ship block model

4 Conclusions

For 3D point cloud data of the ship block model vision measurement, the efficient point cloud data denoising algorithm has been researched deeply. The algorithm firstly calculate K neighborhood point, then calculate tangent plane tangent of the least square fitting for K neighborhood point, finaly use the angle of measuring point with tangent plane normal vector and the tangent plane vector direction to distinguish whether to remove the measuring points. The experimental results validate the effectiveness of the proposed algorithms.

5 Acknowledgements

This project was financially supported by the Heilongjiang Province Natural Science Foundation (F201342), and National Natural Science Foundation of China (51409059), Heilongjiang Province Postdoctoral Research Foundation(LBH-Q16066) and China Scholarship Council .

References

[1] Kai W, Dieter R, Dirk S. An approach to computer-aided quality control based on 3D coordinate metrology. Journal of Materials Processing Technology. 2000,107: 96-110P

[2] Li jiangxiong, Ke yinglin . Reverse engineering of complex surface based on feature. CHINESE JOURNAL OF MECHANICAL ENGINEERING. 2000,36(5):18-21P

[3] ZHANG Xue-chang etc. Study on the Choice of Disciplinal Function During Point Cloud Smoothing Based on Anisotropic Heat Diffusion. JOURNAL OF SHANGHAI JIAOTONG UNIVERSITY. 2008,42(10):1687-1691

[4] Liu dafeng etc. Scattered point cloud to research and implementation of noise algorithm. Journal of Southeast University(Natural Science Edition). 2007,37(6):1108-1112

[5] M.Kuwahara, K.Hachimura, S.Eiho, M.Kinoshita. Digital Processing of Bionedical Images. 1976

[6] Xiong bangshu etc. Three dimensional scattered data K nearest neighbor search algorithm. Journal of Computer Aided Design & Computer Graphics. 2004,16(7):909-912

[7] Wei wei etc. A quick search of the massive data sets K nearest neighbor spatial sphere algorithm. Acta Aeronautica Et Astronautica Sinica. 2006,27(5):944-948

[8] F.P Ferrie, J.Lagarde, P.Whaite. Darboux frames, snakes, and super- quadrics: geometry from the bottom up. IEEE Transaction s on Pattern Analysis and Machine Intelligence. 1993,15(8):771- 783

Study on Autonomous Obstacle Avoidance Based On Guidance

Zhizhang Hu[1,*], *Guangdi* Hu[1], *Xiaoyuan* Zhu[2], *Zihao* Lei[1], *and Haipeng* Wan[2]

[1]School of Mechanical Engineering, Southwest Jiaotong University, No.999 Xi'an Road, Chengdu, China
[2]School of Information Science and Technology, Southwest Jiaotong University, No.999 Xi'an Road, Chengdu, China

Abstract. DJI Guidance is an entirely new visual sensor navigation system, which has a function of real-time transmission of the depth of field images and other data. It is a robust optical sensing platform which significantly promotes the development of unmanned aerial vehicle (hereinafter referred to as UAV). On this basis, we have further studied and developed Guidance in application to the field of UAVs to also cut a striking figure, by applying the threshold and binaryzation to the single frame of a static image generated from Guidance. Utilizing OpenCV's contour detection algorithm, we find all the closed contour information and re-plan the feasible region. Also, we specially re-project the depth information to achieve three-dimensional reconstruction of the image. Then combining the re-planed feasible domain with three-dimensional reconstruction, we can perform real-time obstacle avoidance and mapping. The results demonstrate that our improvement can efficiently complete the obstacle avoidance and real-time mapping in the specific circumstances.

INTRODUCTION

In recent years, more and more attention has been paid to UAVs as advanced technology, and the research of UAV autonomous obstacle avoidance has become one of the focuses of the study. For an obstacle avoidance system, the detection and location of obstacles is the first step and the most critical step to obtain the surrounding environment. Currently, the widely used methods of detection and location can be classified into two categories: one is based on UAV's simultaneously localization and mapping (hereinafter referred to as SLAM) [1] to surrounding environment. Thus the accurate location information of UAV and stereo relationship between UAV and obstacles is obtained. The other is based on precise location information provided by external facilities, such as global positioning system (hereinafter referred to as GPS) and motion detection system. Apparently, only the dynamic obstacle avoidance system can solve the UAV collision problem in the real world. In the realm of autonomous obstacle avoidance system, considering both of the overall cost and effectiveness, the visual sensor is a better choice. Guidance, introduced by DJI, is an innovative visual-sensing platform but there is few related research at present. In this paper, the obstacle avoidance of UAV is explored by use of DJI Guidance.

1. Feasible Region Decision

Utilizing binocular vision system of Guidance and image processing, location and distance information of obstacles is obtained. In this way, UVA can decide the feasible region.

1.1 Binocular vision stereo correction

In view of the binocular vision technology, it is generally using two cameras to obtain two images of the same scene in different spatial locations, then matching feature points and obtaining internal parameters of cameras by calibration. In this way, the coordinated information from disparity map could be computed. When the two image plane is completely line aligned, the calculation of disparity is the most simple. However, due to the multiple unavoidable errors, it is impossible for the two cameras to complete line alignment accurately. Therefore, the stereo correction has become indispensable. By stereo correction, the image planes of two cameras can be projected back to the same plane. Since the front parallel planes are finite, there are bound to be constrained by selection, including maximum view overlapping and distortion minimization. There are many algorithms to help implement selection. In this paper, we choose OpenCV machine vision language as the development platform. There are two main algorithms: one is the Hartley algorithm, and the other is the Bouguet algorithm. In light of the calibration mode that Guidance can use, we choose the Bouguet algorithm that generates fewer distorted images generation[2] and maximizes image overlap. [3]

First, the rotation matrix and the translation vector between the two cameras of the binocular camera can be represented as follows:

* Corresponding author: kevinhoo@my.swjtu.edu.cn

$$R = R_r(R_l)T$$
$$T = T_r - R(T_l) \quad (1)$$

Where: R denotes the rotation matrix to be calculated, and T denotes the translation vector to be computed. The subscript l and r refers to the parameters of the left camera and the parameters of right camera, respectively.

The spatial position relation is shown in Figure 1:

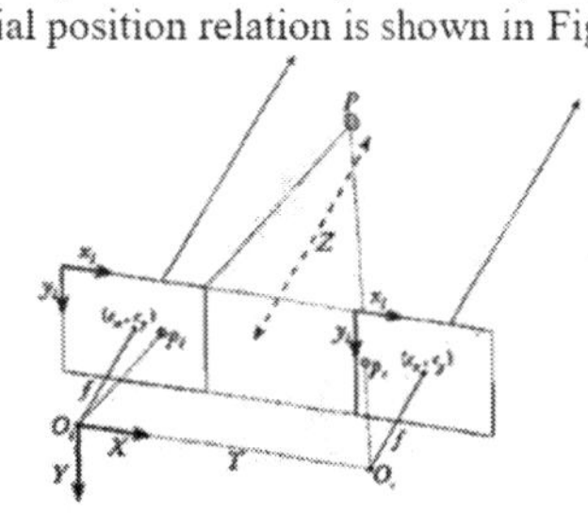

Fig. 1. Spatial Position Relationship.

We set the R matrix to rotate the right camera image plane to the left camera image plane. In order to minimize the distortion, the R matrix needs to be divided into two parts, called synthetic rotation matrix r_1 and r_r. Rotating each camera in half so that the camera can be coplanar, but line alignment is still not available, therefore a new matrix R_{rect} is needed, which transforms the left camera pole to infinity and aligns the pole line horizontally. In order to calculate R_{rect}, it is necessary to establish a rotation matrix starting with the pole e_1. The main point (c_x, c_y) is the origin of the left image, so the direction of the pole can be:

$$e_1 = \frac{T}{\|T\|} \quad (2)$$

Without any other restrictions, the next vector e_2 must be orthogonal to e_1. So one can choose the orthogonal direction of the main light, that is:

$$e_2 = \frac{\left[-T_y T_x 0\right]^T}{\sqrt{T_x^2 + T_y^2}} \quad (3)$$

The third vector is the cross product of the first two vectors. In this way, the third vector is:

$$R_{rect}\begin{bmatrix}(e_1)^T \\ (e_2)^T \\ (e_3)^T\end{bmatrix} \quad (4)$$

From this, line alignment can be achieved by setting: $R_l = R_{rect} r_1$, $R_r = R_{rect} r_r$.

Eventually, one can calculate the projection matrix, which transforms the three-dimensional points under the coordinates into the two-dimensional points under two-dimensional homogeneous coordinate system:

$$P\begin{bmatrix}X \\ Y \\ Z \\ 1\end{bmatrix} = \begin{bmatrix}x \\ y \\ w\end{bmatrix} \quad (5)$$

1.2 Feature Point Matching

In general, one feature in an image may have many candidate matching pairs in another image, and there is only one structure with the same name, so ambiguous matching occurs. [4] In this case, it is necessary to complete the matching according to a specified matching algorithm and experience. Simultaneously, the selection of feature points is critical, since it plays a critical role in the accuracy and rapidity of binocular distance measurement. If the feature points are too few, the judgment of contours of the objects would be inaccurate, and too many feature points will cause the operation speed to decrease and increase the probability of matching errors. Based on the OpenCV language, we tried two algorithms: 1. Block stereo matching algorithm that is, using the sum of absolute differences (hereinafter referred to as SAD) to find the matching points between the two images that have been corrected. [5] 2 SIFT feature extraction algorithm. SIFT features have scale invariance, that is, the matching effect is less affected by the angle of the shot and the change of light. [6] In order to compare these two algorithms, we use VisualStudio in a desktop computer with Intel Core i7-4790CPU (3.60GHz) to build a simulation test environment based on OpenCV. By comparing the samples of 10 pixels to 320*240, we found that the speed and effect of the SAD algorithm in the texture-rich region were better than that of SIFT algorithm. However, in the low texture region, the matching error rate of the SAD algorithm is high, the need for late interpolation correction kept enhancing, thus reducing the speed of operation. In contrast, the matching output of the SIFT algorithm are all within the acceptable range, and the average processing speed is 17% higher than the SAD algorithm. Therefore, we select SIFT algorithm to match feature points.

1.3 Image Processing

With image processing, the feasible region can be obtained, and the related process is shown in table one:

Table 1. Processing Process

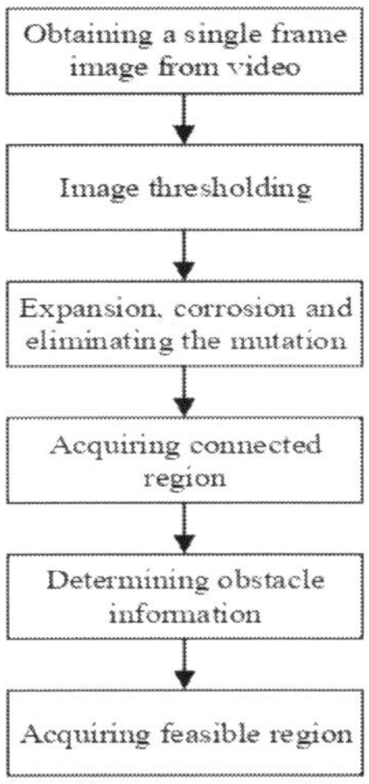

For a scene, binocular cameras of Guidance obtains two images with a tiny difference of shooting angle simultaneously, with which the depth image could be computed. Depth image contains the distance information of each pixel, in which the more extensive the gray scale is, the closer the distance between UVA and obstacle is. For the video information returned by Guidance, it is processed through each single frame static image in this paper. The video captured by Guidance can obtain a single frame static image in real time and can be stored directly in the Mat structure of OpenCV so that it is convenient for further processing and calculation. In the laboratory, we used the Guidance binocular camera to capture two static images at the same time as shown in Figure 2 and Figure 3:

Fig.2. Left Camera Image

Fig.3. Left Camera Image

After the selection and matching of feature points in two pictures, the disparity image is obtained. Then the depth image can be computed from the disparity image and internal parameters of the binocular camera obtained from calibration. The derivation of the relation between depth image and disparity image is shown in Figure 4 and related formula:

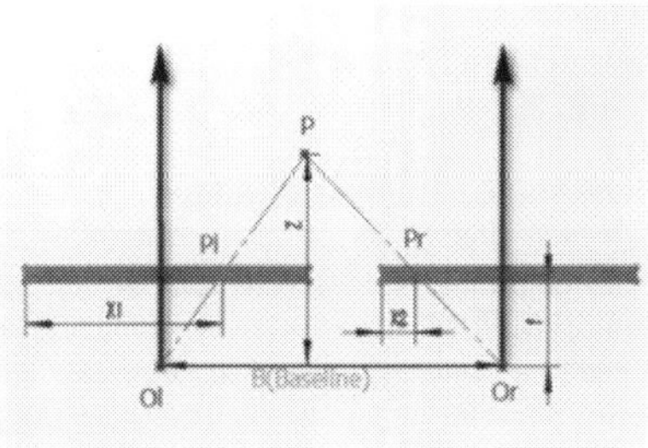

Fig.4. Relationship Derivation

In Figure 4, Ol and Or respectively represent the position of two camera optical center, the obstacle P's position in the two camera image plane are apiece 1 and 2, and disparity:

$$\Delta x = x_l - x_r \tag{6}$$

The mathematical expression of depth z can be obtained from similar triangle relations:

$$\begin{cases} \dfrac{x_l}{f} = \dfrac{x}{z} \\ \dfrac{x_r}{f} = \dfrac{x-b}{z} \end{cases} \tag{7}$$

Moreover, then depth z can be calculated:

$$z = \frac{bf}{x_l - x_r} \tag{8}$$

From (8) we can see that depth z is inversely proportional to the disparity Δx. Therefore, the distance information of the obstacle can be calculated by matching the feature points measured in the binocular vision. In the experiment, disparity image (Figure 5) and depth image (Figure 6) is obtained.

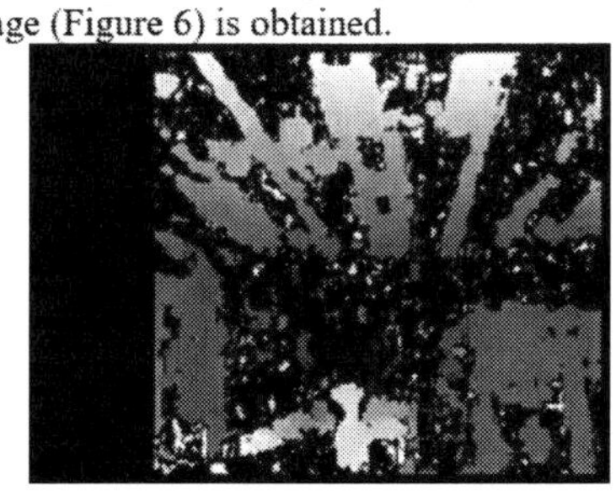

Fig.5. Disparity Image

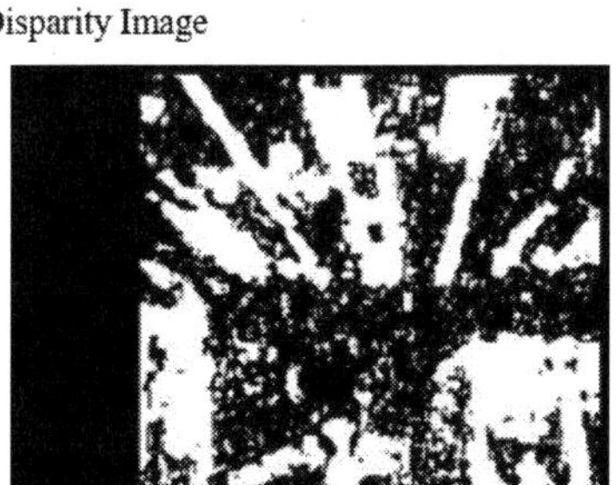

Fig.6. Depth Image

In this paper, the obtained depth images are processed by thresholding. Since the grayscale information in the image represents the fixed distance quantity, we selected 125 (corresponding distance is 4.5m) and 60 (corresponding distance is 2.0 meters) as the threshold respectively for first and second thresholding processing to the same image after experiments. [7] Taking the instance of selecting 125 as the threshold to illustrate: During the processing, a pixel with a value higher than 125 is considered an obstacle-free object, i.e., at this point, the region is a feasible region and marked as 255. On the contrary, a pixel with the value that below 125 is considered an obstacle at that moment, labeled as 0 and as an infeasible region.

$$\begin{cases} f_s(i,j)=255 & f_s(i,j)\geq 125 \\ f_s(i,j)=0 & f_s(i,j)<125 \end{cases} \tag{9}$$

By processing the image in real time, Guidance is able to obtain the feasible region without obstacle in the distance between 2m and 4.5m simultaneously, to realize the real-time monitoring and avoidance of the obstacle object. The purpose of threshold processing is to reduce the amount of data and improve the speed of real-time processing. In the experiment, processing speed has been enhanced by 7% compared to full gray scale processing.

Through the above threshold processing, we can get a binary image, as shown in Figure 7:

Fig.7. Binary Image

In Figure 7, one can see that there is some noise with erroneous distance information in the image, which can affect the judgment of the position of the obstacle information. Therefore, it is necessary to process the obtained binary image. In this paper, the mean filtering process is utilized. The mean filter can efficiently eliminate the salt and pepper noise and reduce the error distance information. Meanwhile, the filtering speed is faster and has better real-time performance. [8] By mean filtering, some of the noise in the image is effectively suppressed, as shown in Figure 8:

Fig.8. Filtered Image

In the image after the mean filtering, there is still a small area of unreasonable jump value in some regions, forming a slight mutation connected domain. The reason for this is the distance measurement error caused by light, shadow, texture and misalignment, which could be avoided by eliminating the smaller connection domain. [9] The size of the domain to be eliminated is related to the actual area corresponding to it. For example, in this paper, 125 is the threshold value, and the obstacle distance is 4.5m. In this way, 300 is selected as the size of the eliminating domain. After eliminating, the effect is shown in Figure 9.

Fig.9. Image after Eliminating

For the optimized binarized image, all the obstacles can be marked out by the contour detection algorithm, and the size and position information of the obstacle can be calculated. The results are shown in Figure 10, with a rectangular mark out the obstacles:

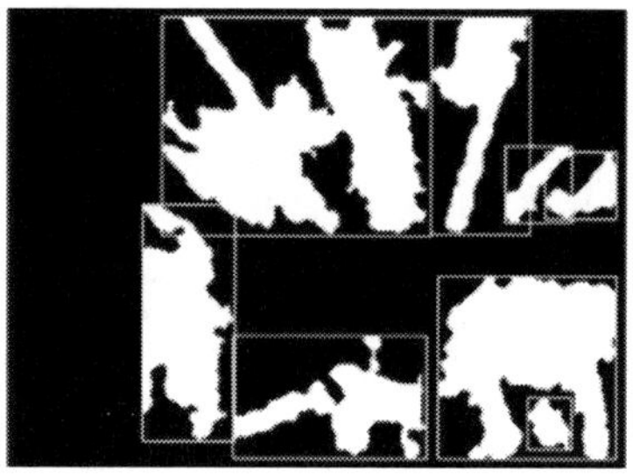

Fig.10. Contour Detection

In this image, all the obstacles of the location information can be obtained, and feasible region according to the size of the UAV in the distance that is within 4.5m can be obtained. Similarly, selecting 60 as the threshold of the image for the same treatment, feasible region within 2m is accessible. In this way, the real-time obstacle avoidance can be realized by simultaneously processing of continuous images in the real-time feedback of the UAV.

2 Three-dimensional Reconstruction

Due to the lack of environmental holism in the approach proposed earlier, implementing three-dimensional reconstruction (hereinafter referred to as 3D reconstruction) with feasible region detection to improve environmental wholeness is introduced.

2.1. 3D Reconstruction Technology

If the camera's internal parameter matrix is known, the reprojection matrix Q is available.

$$Q=\begin{bmatrix} 1 & 0 & 0 & -C_x \\ 0 & 1 & 0 & -C_y \\ 0 & 0 & 0 & f \\ 0 & 0 & \frac{-1}{T_x} & (C_x - C_x')T_x \end{bmatrix} \quad (10)$$

Where : C_x' denotes the x coordinates of the main point in the right image, and the other parameters all come from the left image. If the main light intersects at infinity, then $C_x = C_x'$, and the lower right corner of the item is 0. [2] Present, given a two-dimensional homogeneous point and the disparity d associated with it, one can project this point into three dimensions:

$$Q\begin{bmatrix} x \\ y \\ d \\ 1 \end{bmatrix} = \begin{bmatrix} X \\ Y \\ Z \\ W \end{bmatrix} \quad (11)$$

The 3D coordinates are:

$$\left(\frac{X}{W}, \frac{Y}{W}, \frac{Z}{W}\right) \quad (12)$$

Based on this, 3D reconstruction can be achieved by spatial projection of the acquired depth information. In accordance with the processing order, the experimental results are as follows.

Fig.11. Left Camera Image

Fig.12. Right Camera Image

Fig.13. Disparity Image

Fig.14. Depth Image

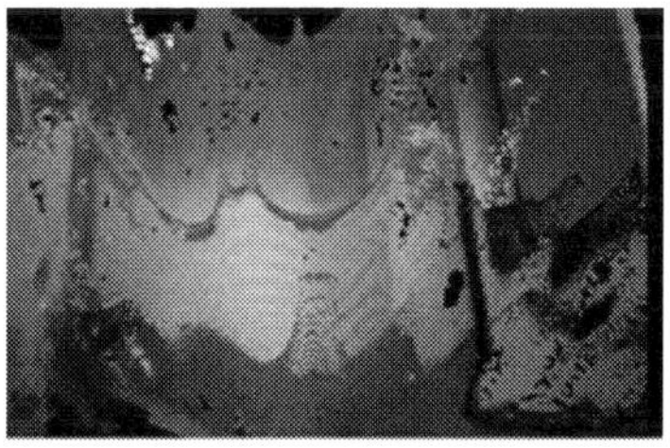

Fig.15. Result of 3-D Reconstruction

2.2. Accuracy Correction

Since the limited computing power of the Guidance itself can be very limited in response to the need for high precision reconstruction, according to the actual demand of flight and Guidance hardware conditions, after the experiment, we found the reconstruction accuracy can be reduced to a continuous reconstruction needs. The reason is that in the UAV operations, Guidance only needs to let the UAV know what in the flight environment is obstacles, as well as the obstacle and the UAV itself position relationship. There is no need to distinguish what obstacle is. At the same time, when each pixel of the disparity map is projected to the world coordinate system, knowing its precise location is unnecessary. What only needs to know is the region of location, into the treatment of some areas. In this study, the area length and width dimensions were 30 cm, and the height of the area was determined from the height histogram of the pixels falling in the area, and the risk of the region was determined according to the region and its coordinate distance. The implement of this method may result in the expansion of obstacle in the depth and the width maximum to 60cm so that the maximum diameter of the feasible region can be reduced by 60cm. Although the diameter of the feasible region can be reduced, in determining the safety zone, the feasible region diameter must have been more than the fuselage diameter of a certain margin. This method not only simplifies the data processing but also improves the real-time performance without affecting the performance of obstacle avoidance of the UAV. After measurement, each 3D reconstruction period is 0.6 seconds, which can meet the security requirements of reconstruction.

2.3. 3D Global Map Building

The perceived information obtained during the UAV flight is relative to the body freezing coordinates, so the three-dimensional information generated by the current reconstruction is the three-dimensional data under the local coordinates generated at the time of reconstruction. Therefore, the global information of each time can be integrated into a global map by providing the global coordinates of the global position when the UAV is reconstructed. [10] Considering that the reconstruction cycle is short, the map missing parts can be directly connected between each reconstruction cycle, and the constructed map primarily conforms to the real three-dimensional scene in the space. In this way, SLAM is available.

2.4. Conjunction of Obstacle Detection and 3D Reconstruction

Implementing obstacle detection in conjunction with 3D reconstruction could realize high-performance path planning. Path planning is to find a safe and efficient collision avoidance path for the autonomous vehicle between the given starting point and the target point in the concurrent scene. [11] Due to the limited computing power of Guidance equipment and the specific requirements of the actual UAV flight operation, the path planning strategy of Guidance requires security first, followed by low computation. We use the feasible domain detection combined with 3D reconstruction for path planning. In each local 3D reconstruction cycle (0.6 seconds), matching the connected feasible domain with the 3D modeling completed in the same cycle. In this way, Guidance could mark the feasible region obtained from the binary image in the 3D environment, then calibrate the feasible region with 3D information. If the feasible region is judged as safe in its corresponding 3D environment, it can be regarded as a safe flight airspace.

3 Conclusions

In this paper, we propose our improvement and research based on Guidance multi-information sensor. First, we focus on the processing of depth image and binary image, then re-plan the feasible region. Additionally, we imitate the principle of human eye imaging with the utilization of Guidance's binocular camera system to complete the three-dimensional reconstruction of the front space, so as to be more effective for obstacle avoidance. Finally, combining three-dimensional reconstruction with feasible region planning, we propose a new idea for obstacle avoidance. Because of the low precision matching of the feasible region obtained from binary image and three-dimensional reconstruction, the actual avoidance effect will be limited, which is our deficiency at this stage. How to match accurately is the breakthrough point of our future work. Simultaneously, combining neural network algorithm and genetic algorithm to optimize path planning is also the focus of prospective research.

Acknowledgements

This work has been supported by 2016 National-Funded Student Research Training Program under the Grant Numbers 201610613031.

References

1. Artieda, J., Campoy, P., Correa, J. F., & Olivares, M.. Visual 3-d slam from UAVs. Journal of Intelligent & Robotic Systems, **55**(4-5), 299-321. (2009)
2. Bradski, G. R., & Kaehler, A. *Learning OpenCV - computer vision with the OpenCV library: software that sees.* (DBLP. 2008)
3. Xin, X.. *Embedded System Design Based on FPGA.* (CMP. 2005)
4. Xiaohua, W.. Research on 3D Reconstruction Technology Based on Computer Vision. (Doctoral dissertation, SDUST, 2004)
5. Yang, Q., Rong, X., Yu, X., & Guan, C. Intensity Weighted Stereo Matching Algorithm. Affective Computing and Intelligent Interaction, **137**,93-100.(2012)
6. Bingxiong L. An improvement of matching criteria and method of SIFT feature and its application. (Doctoral dissertation, DLUT, 2010)
7. Tony Lindeberg. Scale-space theory: a basic tool for analyzing structures at different scales. Journal of Applied Statistics, **21**(2), 224-270, (1994)
8. Kejun, W., Xinyan, X., Zhen, R.. Highly efficient mean filtering algorithm. Application Research of Computers, **27**(2), 434-438, (2010)
9. Zhizao, L., Yingwu, Z., & Zhongkai, Zhen. Optimizing Algorithm of Labeling Connected Components in Binary Images. Journal of Anqing Normal University(Natural Science Edition), **16**(4), 34-39. (2010)
10. Huahua, C. Research on the Key Technology of Vision-based Navigation: Stereo Vision and Path Planning. (Doctoral dissertation, ZJU, 2005)
11. Toogood, R., Hong, H., & Chi, W.. Robot path planning using genetic algorithms. IEEE International Conference on Systems. **1.1**,489-494. (1995)

Heliostat attitude angle detection method based on BP neural network

LIU Guangyu[1,*],CAI Zhongkun[1]

[1]School of Automation Engineering, Hangzhou Dianzi University, Hangzhou 310018, China

Abstract: Traditional fossil fuels have dried up, global warming and sustained economic development have led to the rapid growth of clean energy resources. Tower thermal power generation has attracted much attention due to its ability to generate electricity during the night. The traditional tower thermal power generation adopts open-loop control which requires very high mechanical accuracy. In the operation of power station and there may be a settlement, wind load or other factors make the heliostat skew phenomenon. It will eventually lead to a decline in power generation efficiency. Thus, we propose a closed-loop feedback control method based on machine vision and optical reflection principle based on the method of using the correction of heliostat spot acquisition board. To identify the spot and the ellipse fitting method for spot feature extraction using image processing technology, we propose a heliostat to determine the characteristics of the corresponding spot mapping the attitude angle method based on BP neural network. Thus we can provide direct feedback control of heliostat errors. The new method can effectively increase the heliostat tower power generation efficiency and also can make the tower heliostat thermal power generation cost reduced with the popularization and application of significance.

1 Introduction

As a clean and renewable energy, solar energy attracts more and more attention. Solar power is divided into two categories: photo-voltaic and photo-thermal power generation. And the technology faces enormous pressure to reduce the cost of electricity as soon as possible[1]. Because of the thermal power generation tower has special advantages, has attracted much attention, but because the control precision, heliostat operation stability, safety and reliability and the construction cost is limited. In order to reduce the cost of tower thermal power generation[2,8,9,12].

Tower power by elevation and azimuth tracking control of heliostat. The heliostat tracking control method for main program control and sensor control and sensor program, hybrid control in three ways[5]; in the practical application, mainly adopts open-loop control based on program control. The procedure control and the use of sensor data is gradually become the focus of the heliostat [3,4,13].

Heliostat traditional open loop control directly using the clock to control the rotation angle of the heliostat, very high requirements for heliostat support mechanical precision, resulting in high construction costs . The literature[5] used the camera as the day to achieve accurate tracking sun trajectory tracking, but because of the weather, the application will have certain limitations and difficulties in practical application; Literature[6] proposes the use of Whiteboard Based correction methods. But because of the way based on spot correction, not directly to optimize the layout of the heliostat field to provide effective data.

The above mentioned in the literature are not involved with heliostat attitude, but in practical application, heliostat attitude angle correction for heliostat reflection is very important[18]. In the tower heliostat thermal power generation, because of the heliostat tower distance target center is far away, it will cause deviation of the heliostat spot cannot be accurately reflected to the target position, if the deviation is too large, it will even cause the spot projected on the target position.

This paper presents a method of heliostat attitude angle from the spot of direct mapping. In many applications of new energy systems, such as fault diagnosis, neural network method can be used as a powerful tool for processing, and has a very good effect[14,19,20].The complex neural network structure can approximate any nonlinear function, and can solve

* Corresponding author: g.liu@hdu.edu.cn

the complex nonlinear mapping problem very well. This paper uses BP neural network mining implicit mapping relationship between spot feature and heliostat attitude angle. Based on the small biaxial heliostat tracking platform for data acquisition and verification experiment, the method can accurately identify the heliostat spot with the characteristics of the corresponding attitude angle.

2 Problem description

In the tower heliostat thermal power generation, we need the heliostat reflection light reflected to the specified position of the target tower to achieve maximum heat collecting efficiency and Maximize output [6]. In engineering applications, due to the installation error, mechanical wear and other irresistible factors will make the heliostat cannot be accurately reflected to the target area. Therefore, in practice, is the attitude angle of the heliostat need correction, this paper proposes a correction method for high precision heliostat attitude angle based on BP neural network.

According to the physical model of heliostat reflection [6,7]. At the specified coordinates by the relevant input parameters, we can get the heliostat azimuth and elevation angle. The reflected light spot is projected to the target area, and then the original image data is collected. In order to use the data effectively and get the best spot, we need to preprocess the original data [10,15,16,17].After the image processing, the smaller size of the reflected spot image is obtained. Based on the least square method, ellipse fitting is used to obtain the general characteristic equation of fitting ellipse, and the algebraic form of conic equation is expressed as formula (1)

$$Ax^2+Bxy+Cy^2+Dx+Ey+F=0 \quad (1)$$

The five eigenvalue representations of the ellipse can be obtained from the characteristic equation of the ellipse, that is, the coordinates of the central point of the ellipse, the length of the long axis, the axis of the ellipse and the inclination of the ellipse. Its algebraic expressions, such as formula (2) to (6)

$$x_c=\frac{BE-2CD}{4AC-B^2} \quad (2)$$

$$y_c=\frac{BD-2AE}{4AC-B^2} \quad (3)$$

$$a=2\sqrt{\frac{-2F}{A+C-\sqrt{B^2+(\frac{A-C}{F})^2}}} \quad (4)$$

$$b=2\sqrt{\frac{-2F}{A+C+\sqrt{B^2+(\frac{A-C}{F})^2}}} \quad (5)$$

$$\theta=\frac{1}{2}\arctan\frac{B}{A-C} \quad (6)$$

The characteristic value of the ellipse is equivalent to the feature of the reflected facula. Thus, the feature extraction of the spot is completed. Based on the data source for the identification of heliostat attitude angle.

Since the establishment of the model directly from the heliostat reflection spot a heliostat precise attitude angle is difficult. Considering only the heliostat elevation angle and azimuth axis moving, so under the given conditions there will be corresponding with the only spot[11]. There is a map that reflects the pattern characteristics of heliostat attitude angle.

In view of the above situation, this paper proposes a data driven model identification method based on BP neural network. To solve this problem, this paper use the method of image processing based on the spot processing target spot and find the corresponding elliptic equation by the method of ellipse fitting. five features: Based on ellipse long axis a, short axis b, inclination of the long axis θ, the coordinates of the center of x_c and y_c. In the specific time and location and coordinates, this paper presents a mathematical mapping for two spot attitude characteristics and heliostat angle between δ_a and δ_z. The following mapping:

$$f_a(t):(a(t),b(t),x_c(t),y_c(t),\theta(t))\rightarrow\delta_a(t) \quad (7)$$

$$f_z(t):(a(t),b(t),x_c(t),y_c(t),\theta(t))\rightarrow\delta_z(t) \quad (8)$$

3 Heliostat angle detection method based on BP neural network

3.1 Method design

In order to solve the above problems, a flow chart is designed using the method shown in the diagram. Among them, the BP neural network model is the main problem solving tool used in this paper.

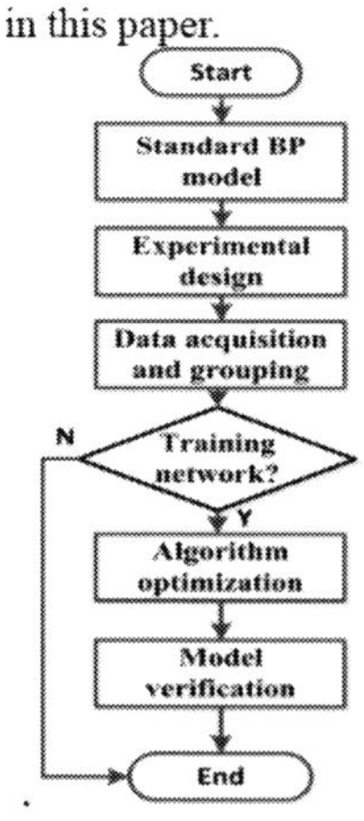

Figure. 1 flow chart of method design

It can be seen from the picture, the mapping between the unknown heliostat reflection spot feature and heliostat attitude angle. The model identification process mainly includes the following five parts:

Step one: model identification using standard BP neural network, aiming at the problem of the analysis results, application of this issue to be resolved for the mapping relationship between heliostat spot feature and heliostat attitude angle, the model identification using standard BP neural network.

Step two: experimental design, in order to obtain experimental data for training neural networks, we use a small biaxial experimental device console, which can real-time sunlight reflected to the specified target position. On the whiteboard to take pictures to get the image with the spot using industrial camera, and processing technology to obtain the characteristic spot value of the image by image processing, according to the different attitude angle to obtain the heliostat spot feature data

Step three: Data acquisition. Different target points are input to the biaxial experimental console to simulate the position of the light spot in different targets. At the same time, with the change of time, spot feature data recording attitude angle of heliostat and image.

Step four: Determine the training set and validation set. The data were randomly divided into two groups, and a set of data was used to train the neural network structure. For the trained neural network, another set of data can be used to verify the accuracy of the model.

Step five: Algorithm optimization. The obtained training data is applied to the given neural network structure, and the optimization algorithm is used to obtain the weight matrix and threshold matrix of the network structure.

Step six: Model verification. In view of the well trained network model, we use another set of data to verify. Then the trained neural network model can accurately spot according to the characteristic of image acquisition attitude angle corresponding to the heliostat

3.2 LM-BP neural network method

The theory of BP algorithm has the advantages of reliable basis, rigorous derivation process, high precision and good universality. But the standard BP algorithm has the following disadvantages: slow convergence speed, easy to fall into local minimum and it is difficult to determine the number of hidden layer and hidden layer nodes. In practical applications, the BP algorithm is difficult to do. Therefore, there are many improved algorithms. In this paper, the BP neural network (BPNN) algorithm based on L-M (Levenberg-Marquardt) algorithm is used.

The algorithm includes two aspects: the input beam characteristic value of transmission signal propagation and heliostat attitude angle error of direction. That is, the actual output is calculated from the input to the output direction, and the weights and thresholds are corrected from the output to the input direction. Table 1 lists the mathematical symbols used in the model and their implications.

Table 1 ANN Symbol specification

Symbol	Implication
η	Learning rate
p	Sample p
w_{ij}	The connection weight between the input layer and the hidden layer neuron i through the meta j
w_{ki}	The connection weights between the hidden layer meta i and the output layer neuron k
o_j^p	The j output of the first sample input layer node p
θ_i	Threshold of hidden layer neuron i
θ_k	The threshold of the output layer neuron k
o_i^p	The output of the hidden layer node i at the action of the sample p
o_k^p	Output layer node k output at the action of sample p
δ_k^p	The gradient direction of the output layer neurons
δ_i^p	The gradient direction of hidden layer neurons
N	All training samples of the neural network

L-M algorithm optimization BP neural network implementation steps are as follows:

Step one: initialize network parameters. Initialize weights ω_{ij}, ω_{ki} and Threshold θ_i, θ_k.

Step two: calculate the input and output of the hidden layer and the output layer. The formulas are as follows (9) to (12).The input net_i^p and the output o_i^p of the i node of the hidden layer are respectively:

$$net_i^p = \sum_{j=1}^{M} \omega_{ij} o_j^p - \theta_i \left(i = 1, 2, \ldots q \right) \tag{9}$$

$$o_i^p = g\left(net_i^p\right) \ \left(i = 1, 2, \ldots, q\right) \tag{10}$$

The input net_k^p and output o_k^p of the output node k of the output layer are, respectively:

$$net_k^p = \sum_{i=1}^{q} \omega_{ki} o_i^p - \theta_k \left(k = 1, 2, \ldots L \right) \tag{11}$$

$$o_k^p = f\left(net_k^p\right) \ \left(k = 1, 2, \ldots, L\right) \tag{12}$$

Step three: calculate the output error. The mathematical expression for the error of the output layer is $t_k^p - o_k^p$, and the error function is set as the two order objective function.

Step four: use the iterative formula based on the L-M algorithm to correct the weights and thresholds. The formulas are as follows (13) to (16). The weight and threshold correction formula of the first node of the output layer is:

$$\omega_{ki}(k+1) = \omega_{ki}(k) + [J^T(w_{ki})J(w_{ki}) + \mu I]^{-1} J^T(w_{ki}) e(w_{ki}) \tag{13}$$

$$\theta_{ki}(k+1)=\theta_{ki}(k)+[J^T(\theta_{ki})J(\theta_{ki})+\mu I]^{-1}J^T(\theta_{ki})e(\theta_{ki}) \quad (14)$$

The weight and the threshold correction formula of the k node in the hidden layer are:

$$\omega_{ij}(k+1)=\omega_{ij}(k)+[J^T(\omega_{ij})J(\omega_{ij})+\mu I]^{-1}J^T(\omega_{ij})e(\omega_{ij}) \quad (15)$$

$$\theta_{ij}(k+1)=\theta_{ij}(k)+[J^T(\theta_{ij})J(\theta_{ij})+\mu I]^{-1}J^T(\theta_{ij})e(\theta_{ij}) \quad (16)$$

Step five: determine the termination condition. If the termination condition is reached, the algorithm is finished and the network training is completed. Otherwise, go back to step two.

4 Experiment and analysis

4.1Experimental device design

In order to obtain the experimental data of the training neural network, we use a small double axis image processing technology and other related technology to process the image to obtain the characteristic value of the spot. According to the different attitude angle to obtain the heliostat spot feature data. These data and the heliostat attitude angle data stored in the computer, as the original experimental data.

The experimental device has a programmable microcontroller, the use of heliostat algorithm can be based on local GPS input information and target coordinate information in real time the sun spot projected to the center of the target area. To realize the sun tracking function, and can simulate the solar facula projected onto the target near the center in order to achieve non precise target center projection. The use of four horns with rectangular black box as target, using projection whiteboard industrial camera photographs obtained using image processing technology for image processing features obtained spot value image, with a spot on the whiteboard. According to the different attitude angle to obtain the heliostat spot feature data

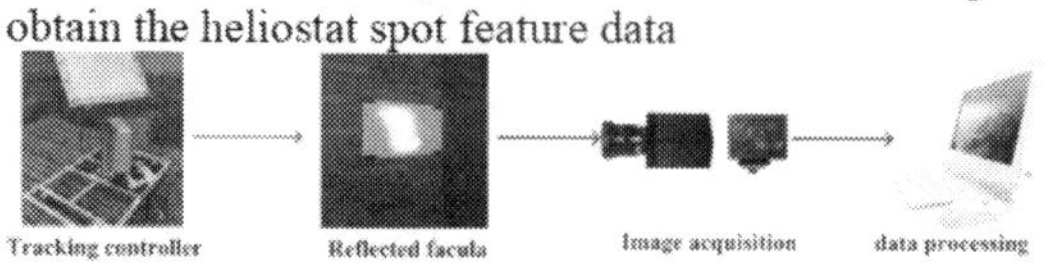

Figure. 2 Experimental test（Hangzhou Zhejiang）

4.2 Experimental design and data acquisition

In order to calculate the weights and thresholds between neurons, we need a set of data to train the designed BP neural network. The relationship between the experimental data and can reflect the characteristics and the heliostat mirror facula attitude angle Tingri effectively. In this experiment the altitude angle and the azimuth angle of the heliostat were collected 200 groups of input and output value, because of the large amount of data, it is inconvenient to list all of them. A typical example of the data is collected from 9:51 a.m. to 11:26 April 23, 2017, and data is collected every 5 minutes. The target spot characteristic data is obtained from the Descartes coordinate system based on the whiteboard plane. Using the SPA algorithm [3] to obtain the solar altitude angle and azimuth angle, the algorithm calculates the sun position error is less than 0.0003 degrees, so its reliability is high

Table 2 Heliostat controller input parameters

Time zone	Longitude(◦)	Latitude(◦)	Time	Heliostat center coordinates(m)	Target center coordinate(m)
+8	120.20	30.27	2017.4.23	(-100,0,0)	(0,0,26)

Table 3 Partial data acquisition

Time t	Coordinate X	Coordinate Y	Long axis length a(mm)	Short axis length b(mm)	Inclination of long axis θ (°)	Altitude angle δ_a (°)	Azimuth angle δ_z (°)
9:51	480.861	444.169	1081.28	518.242	161.28	14.63	-13.32
9:56	489.3913	450.3896	1084.79	523.3745	165.2981	15.38967	-10.9208
10:01	482.0942	446.0081	1083.68	522.4147	161.7765	23.65716	-3.87213
10:06	485.7696	449.0615	1084.657	527.2425	164.9725	15.74203	-5.51748
10:11	484.7584	446.5859	1085.319	519.2065	162.5997	24.05051	-3.75865
10:16	486.6131	444.7668	1083.628	521.7736	169.4919	14.78403	-12.8898
10:21	482.5509	450.6602	1088.597	524.7195	165.7892	20.10009	-10.3568
10:26	488.3079	446.0586	1088.148	520.0771	164.9648	20.88619	-5.51773
10:31	481.6723	453.4629	1089.037	523.1099	165.6386	19.09784	-10.2565
10:36	485.9461	449.2767	1089.456	526.1903	167.7232	18.41609	-5.2042
10:41	486.1893	447.6763	1090.67	527.0014	166.7816	20.85475	-7.44955
10:46	482.9384	447.1815	1085.989	520.5469	169.7231	16.57764	-11.0608
10:51	482.5681	446.4456	1085.637	521.353	170.5138	18.93207	-11.4718
10:56	489.9098	453.9665	1085.669	519.3532	163.8606	18.7172	-7.37104
11:01	483.4831	450.1974	1088.392	520.4595	162.4542	17.59676	-10.1322
11:06	485.1027	449.2476	1082.135	520.8668	169.2901	14.9222	-4.03146

11:11	488.1643	449.0551	1087.065	520.6148	165.8685	24.26089	-7.85194
11:16	486.0724	446.4849	1086.169	524.4826	168.0714	18.58515	-9.64563
11:21	490.7408	444.5464	1090.132	527.3749	169.2418	15.61712	-10.7013
11:26	484.2146	450.9663	1082.646	525.4543	162.3476	21.16757	-8.37826

4.3 Application of BP neural network

The network sample set consists of training samples and test samples. The training sample set is used to train the network and the test sample set is used to detect the effect of network training and the generalization ability. The training sample set and the test set are composed of the input data vector and the target output vector。

Since BP network is trained by teachers, its sample set is composed of input data and target data. Here, the long axis a, the short axis b, the long axis tilt θ, the central coordinate x_c and the y_c are used as the neural network to input 5 components of the data, and the angle value is used as the one component of the network target data. The experimental data were altitude angle and the azimuth angle of the heliostat were randomly divided into two groups, the implementation of network training experiment.

BP neural network is established to determine the structure of the network and the transfer function of each layer of neurons. The structure of BP network mainly includes the number of hidden layers and the number of nodes in the hidden layer. Because the input vector dimension is not too much, the number of hidden layers arranged in one layer; after above the experimental data collection and analysis, the number of the input neurons of the neural network structure is five, the output layer neuron number is one, the number of hidden nodes using the empirical formula method is set to eight.

The activation functions of the hidden layer and the output layer are usually nonlinear functions to enable the nonlinear mapping relation between input and output. Generally speaking, for BP neural networks, there are several commonly used nonlinear functions. For the hidden layer, we use the tan-sigmoid function to complete the data transfer from the input layer to the hidden layer, and its function is in the form of formula (17):

$$g(\mathrm{x}) = \tan\left(\frac{1}{1+e^{-x}}\right) \tag{17}$$

The activation function of the output layer uses the log-sigmoid function, and its function is in the form of formula (18):

$$f(\mathrm{x}) = \log\left(\frac{1}{1+e^{-x}}\right) \tag{18}$$

4.4 Training BP neural network

Before the network training, the training parameters of the network need to be set up. The main parameter settings include learning rate, training time, training target and the maximum number of iterations. The specific parameters set in this article are shown in table 4.The training error curve of the network is shown in Figurer 3, in which the abscissa represents the training times of the network, and the ordinate is expressed as the error precision of the network. The transverse line in the graph is expressed as the expected error of the neural network. As you can see from Figure 3, the network has achieved the desired performance after 9 training sessions. The actual output and the expected output between regression coefficient was 0.97978, indicating that the network can be very good for prediction of output to input data, so that the network has completed training, can be used as a predictive model of heliostat attitude angle.

Table 4 Network training parameters

parameters	values
learning rate	0.01
training time	inf
Training target	0.01
Maximum iterations	1000

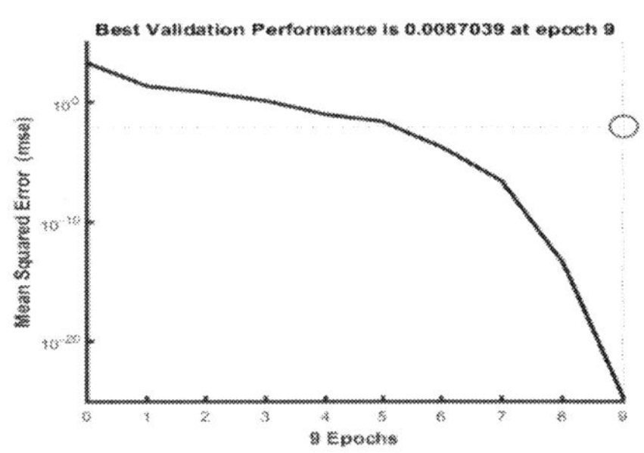

Figure 3 Network error training curve

For the network performance function, the mean square error performance function (MSE) is used in this paper. Through the above analysis, we can complete the establishment of BP neural network. For the establishment of a good model, the network weights and thresholds are in the initial state, the network also can not work normally, so it is necessary to adopt the algorithm for training. Training is completed and achieved the desired goals of the network can be very good according to the given input to calculate the desired output value.

5 Experimental result analysis

For the BP neural network model with training performance and performance target, we can obtain the connection weights and threshold matrix of each neuron in the network, as shown in Figurer 4 and Figurer 5. At this time, we think that the neural network model can be used as a mathematical model of the heliostat attitude angle recognition.

$$\omega_{ij} = \begin{bmatrix} 1.1188 & -1.0502 & 0.3531 & -1.2760 & 2.5572 \\ -5.5290 & 8.8678 & -3.5834 & -3.0928 & 5.4880 \\ -3.9073 & 0.4600 & 1.1089 & -2.5333 & -2.2704 \\ -5.0056 & -4.9340 & -4.1532 & 7.1648 & 4.3677 \\ 3.0923 & -4.6064 & -4.1051 & 0.1683 & -2.4861 \\ 3.9458 & -7.7439 & 5.4709 & -10.4115 & 3.7717 \\ 1.2974 & -8.0590 & -3.2221 & -4.6303 & 3.9264 \\ 1.3851 & -8.7409 & 5.2735 & -9.8288 & 0.5519 \end{bmatrix} \qquad b_i = \begin{bmatrix} -0.4258 \\ 6.8467 \\ 0.6817 \\ -2.1703 \\ 4.9563 \\ 7.7854 \\ -3.8451 \\ -1.1217 \end{bmatrix}$$

$$\omega_{ki} = \begin{bmatrix} 0.2206 \\ -0.3667 \\ 0.0467 \\ -0.0804 \\ -0.2663 \\ -0.1971 \\ -0.4109 \\ 0.2340 \end{bmatrix}^T \qquad b_k = [0.3962]$$

Figure. 4 heliostat elevation angle weights and threshold matrix

$$\omega_{ij} = \begin{bmatrix} 1.1188 & -1.0502 & 0.3531 & -1.2760 & 2.5572 \\ -5.5290 & 8.8678 & -3.5834 & -3.0928 & 5.4880 \\ -3.9073 & 0.4600 & 1.1089 & -2.5333 & -2.2704 \\ -5.0056 & -4.9340 & -4.1532 & 7.1648 & 4.3677 \\ 3.0923 & -4.6064 & -4.1051 & 0.1683 & -2.4861 \\ 3.9458 & -7.7439 & 5.4709 & -10.4115 & 3.7717 \\ 1.2974 & -8.0590 & -3.2221 & -4.6303 & 3.9264 \\ 1.3851 & -8.7409 & 5.2735 & -9.8288 & 0.5519 \end{bmatrix} \qquad b_i = \begin{bmatrix} -0.4258 \\ 6.8467 \\ 0.6817 \\ -2.1703 \\ 4.9563 \\ 7.7854 \\ -3.8451 \\ -1.1217 \end{bmatrix}$$

$$\omega_{ki} = \begin{bmatrix} 0.2206 \\ -0.3667 \\ 0.0467 \\ -0.0804 \\ -0.2663 \\ -0.1971 \\ -0.4109 \\ 0.2340 \end{bmatrix}^T \qquad b_k = [0.3962]$$

Figure. 5 heliostat azimuth weights and threshold matrix

The trained neural network, we use the test data of heliostat altitude angle and the azimuth angle is verified. The verification results are shown in Figurers 6 and 7, in which Figure. 6 is the verification result of height angle, and Figure. 7 is the verification result of azimuth. From the chart we can clearly see that the gap between the output of the neural network and the actual output is very small. It shows that, for the specific coordinates in the heliostat spot are given, we can determine the spot fitting ellipse long axis a, short axis b, axis tilt the special function of the relationship between θ and the coordinates of the center of x_c and y_c and the heliostat attitude angle δ_a or height angle δ_z. We use the neural network method based on data driven can effectively obtain the heliostat attitude angle recognition equation. It is difficult for the mathematical modeling of heliostat attitude angle to the relationship between the spot features, we can use the neural network modeling method based on data driven.

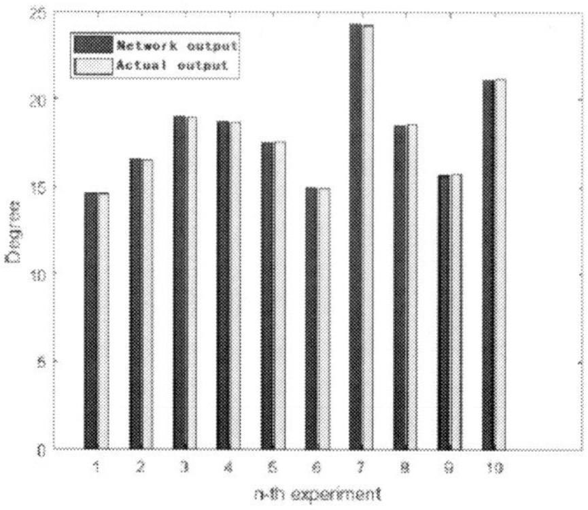

Figure. 6 Comparison of height angle network output and actual output

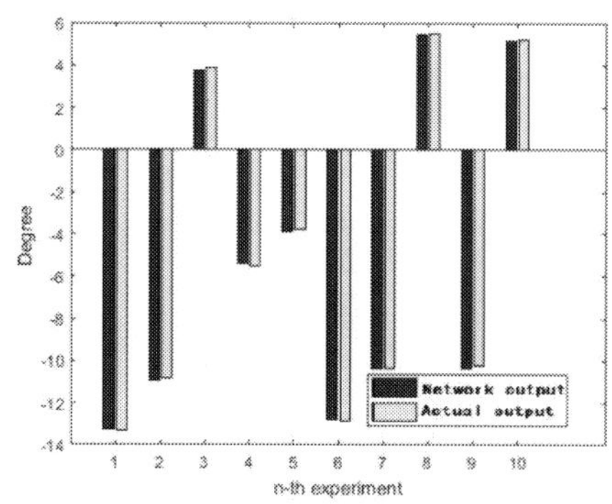

Figure. 7 Comparison of azimuth network output and actual output

6 Conclusion

In view of the tower thermal power heliostat attitude angle corrections to the problem, this paper proposes a detection method of heliostat attitude angle based on BP neural network. This method avoids the complex modeling of heliostat reflection system and accurate modeling of the spot. Based on the method of image processing and BP neural network, the function relation between the reflected spot of heliostat mirror and the attitude angle of heliostat mirror is solved. The experimental results show that the method can obtain the attitude angle of the heliostats according to the characteristics of the reflected light spot. Based on the data driven to achieve the heliostat attitude angle recognition, and have a high accuracy.

Acknowledgements

This work is supported by Natural Science Foundation of Zhejiang, China (Grant No. LR14F030001) for distinguished young scientists and Natural Science Foundation of China (Grant No. 61174074 and Grant No. 61427808).

Reference

[1] Janet L., Sawin, Eric Martinot. Renewables Bounced Back in 2010, Finds REN21 Global Report [R]. Renewable Energy World(2011)

[2] Solar Energy Development Programmatic EIS. Concentrating Solar Power Technologies [DB].http://solareis.anl.gov/guide/solar/csp/index.cfm.

[3] National Renewable Energy Laboratory (NREL). Concentrating Solar Power Projects [EB]

[4] http://www.nrel.gov/csp/solarpaces/project—detail.cfm/project ID=38, 2012.8.24/2012.9.14

[5] Liu Yu.Study on the Visual Measurement Technology of Mirror Reflective Spot Simulation and Settlement Error [D] .Department of Computer Science, Beijing University of Technology, Beijing 100083, China.

[6] McDonnell Douglas Corp. Automatic Heliostat Track Alignment Method[P].US4564275A,1986.1.14.

[7] Reznik Dan S.,Azarchs Adam D.,Csaszar Ambrus. Calibration and Tracking Control of Heliostats in A Central Tower Receiver SolarPowerPlant[P],US2009107485A1,2009.4.30.

[8] M.Berenguel,F.R.Rubio. An Artificial Vision-based Control System or Automatic Heliostat Positioning Offset Correction in A Central Receiver Solar Power Plant [J]. Solar Energy, 2004,76(5): 563-575

[9] Marc Roger, Chirstof Prahl, Steffen Ulmer. Fast Determination of Heliostat Shape and

[10] Orientation by Edge Detection and Photogrammetry [J]. Proceedings of the 14th BiennialCSP SolarPACES Symposium(2008)

[11] Steffen Ulmer. Automated High Resolution Measurement of Heliostat Slope Errors [J].Solar Energy, 2011, 85(4): 681-687

[12] Advances in tower solar thermal power generation technology and its application prospect in China m. Tai Neng, 2008,11: 33-37. ZHANG Yu-fang, ZHU Tian-yu, WANG Xi-chen

[13] China Huadian Engineering (Group) Co., Ltd. to improve the fixed mirror mirror automatically track the accuracy of the sun trajectory [P]. (2010)

[14] Jain AK., Mao JC.,DuinRPW.Statistical Pattern Recognition: A Review [J]. IEEE Transactions on Pattern Analysis and Machine Intelligence, 2000, 22(1): 4-37

[15] Ehrgardt R. Morphology Filter Design with Genetical Algorithm [J]. SPIE, 1994, 23: 2-12.

[16] Gasteratos A., Tsalides S. Fussy Soft Mathematical Morphology [J]. Image SignalProcessing, 1998, 145(1): 41-49

[17] Otsu N. A. Threshold Selection Method from Gray-level Histograms [J]. IEEE Transactionson System Man and Cybernetic, 1979, 9(1): 62-66

[18] Yuan Li. Measurement of Camera Field of View [J]. Science and Technology Information, 2011,32: 38.

[19] F. Yu, X. Xu, A short-term load forecasting model of natural gas based on optimized genetic algorithm and improved BP neural network, Appl. Energy 134 (2014)102–113

[20] Widrow, Bernard, Rumelhart, David E., Lehr, Michael A. Neural networks: applications in industry, business and science. Communications of the ACM, 1994, 37 (3):93-105

DOI: 10.1051/matecconf/201713900044

Research on a Compound Control Strategy of Three-Phase Inverter for Unbalanced Loads

Gaili Yue, Wen Li[a], Kena Li

College of Electrical and Control Engineering, Xi'an University of Science and Technology, Xi'an 710043, China

Abstract: To solve the problem of unbalanced output voltage of inverter under three-phase unbalanced load. In this paper, the main circuit topology using three-phase three-wire inverter topology. Established the mathematical model in several different coordinate system of the three-phase inverter. A compound control method is proposed by combining voltage current dual loop control method and repetitive control method, The control method is realized in three - phase 400 Hz inverter, and can meet the demand of unbalanced load to inverter, the reliability of the inverter circuit is improved. At the end of this paper, Matlab simulation analysis and experimental results are given, it is proved that the composite control method designed in this paper has a strong ability of unbalanced load.

1 Introduction

With today's power electronics technology continues to advance and rapid development in the petroleum, ship, communications switches and other fields, inverter power supply has a very wide range of applications[1]. More and more applications require three-phase inverters with very strong unbalanced load supply capability. Normally under unbalanced load conditions, three phase inverter power supply can produce unbalanced output voltage[2-3]. Thus, the three-phase inverter power supply cannot be very good normal work. Therefore, it is of great significance to improve the analysis and research of unbalanced load capacity of three-phase inverter power supply.

At present, measures for correcting output voltage asymmetry of three phase inverter under unbalanced load, the main circuit mainly includes the following topological forms: three-phase split-capacitor inversion topology[4], three-phase four-leg inverter topology[5], combined three-phase inverter topology, insertion Δ / Y transformer topology[6-7]. But the four topologies have their flaws, The three phase splitting capacitance inverter topology requires a large number of DC filter capacitance, and there is a problem of the capacitor pressure, and the utilization of DC voltage will be very low[8]. The three-phase four-bridge arm contravariant topology, under the condition of adding a bridge arm, to some extent to achieve the effect of suppressing the output voltage imbalance, but also makes the control more complex, difficult to design parameters. The combined three-phase inverter topology is mainly composed of three single-phase inverters which are independent of each other. So the number of switches will be more use, increasing the difficulty of control[9]. Insert Δ/Y transformer topology structure and control methods are relatively simple. However, the weight, volume and cost of the inverter will be increased, and it is not suitable for use in high power applications [10].

The symmetrical output voltage of three phase inverter under unbalanced load is discussed in this paper, established and analyzed the mathematical model of three-phase three wire inverter. A composite control

[a] Corresponding author:710737995@qq.com

 DOI: 10.1051/matecconf/201713900044

method is proposed, mainly combined with the repeated control and double-loop control of these two control methods. The improvement of the control method is realized, and the unbalanced load capacity of the inverter is obviously improved [11]. Simulation studies show that, the inverter operates normally under various loads and can output very symmetrical voltages.

2 Mathematical model of three-phase inverter

As shown in Figure 1, the three phase three-wire inverter topology is composed of capacitance C, three-phase filter inductance L and three-phase inverter bridge, r is equivalent impedance and loss. The output voltage of the inverter is U_{AB}, U_{BC}, U_{CA}, the capacitance voltage is u_{ab}, u_{bc}, u_{ca}, the inductor current isi$_A$, i_B, i_C, the load current is i'_a, i'_b, i'_c.

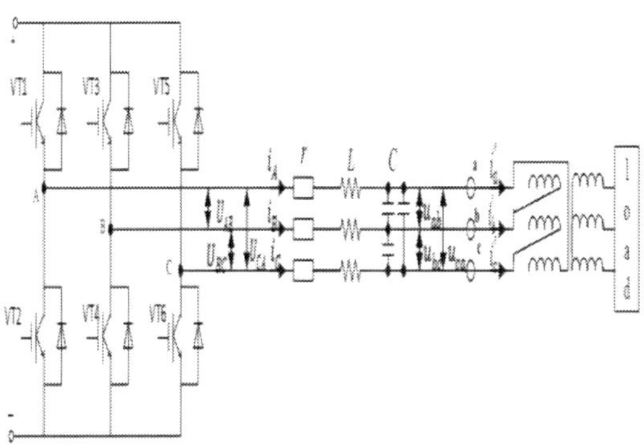

Fig. 1. Three - phase inverter main circuit topology diagram

The inductance current and capacitance voltage are treated as state quantity, can get the following equation of state:

$$\begin{cases} 3L\dfrac{di_A}{dt} = -u_{ab} + u_{ca} - 3ri_A + 2U_A - U_B - U_C \\ 3L\dfrac{di_B}{dt} = -u_{bc} + u_{ab} - 3ri_B + 2U_B - U_A - U_C \\ 3L\dfrac{di_C}{dt} = -u_{ca} + u_{bc} - 3ri_C + 2U_C - U_A - U_B \end{cases} \quad (1)$$

$$\begin{cases} 3C\frac{du_a}{dt} = i_A - i'_a \\ 3C\frac{du_b}{dt} = i_B - i'_b \\ 3C\frac{du_c}{dt} = i_C - i'_c \end{cases} \quad (2)$$

Formula 1 and Formula 2 are transformed by CLARK as follows:

$$\begin{cases} C\dfrac{du_\alpha}{dt} = i_\alpha - I_\alpha \\ C\dfrac{du_\beta}{dt} = i_\beta - I_\beta \end{cases} \quad (3)$$

$$\begin{cases} L\dfrac{di_\alpha}{dt} = U_\alpha - u_\alpha - ri_\alpha \\ L\dfrac{di_\beta}{dt} = U_\beta - u_\beta - ri_\beta \end{cases} \quad (4)$$

The CLARK transform can be used to derive the mathematical model of the three-phase inverter in the stationary coordinate system, then type 3 and 4 through the Park transformation, available:

$$\begin{cases} C\frac{du_d}{dt} = \frac{1}{3}i_d - \frac{1}{3}I_d + \omega Cu_q \\ C\frac{du_q}{dt} = \frac{1}{3}i_q - \frac{1}{3}I_q - \omega Cu_d \end{cases} \quad (5)$$

$$\begin{cases} L\frac{di_d}{dt} = U_d - u_d - ri_d + \omega L_q \\ L\frac{di_q}{dt} = U_q - u_q - ri_q - \omega L_d \end{cases} \quad (6)$$

The mathematical model of the three-phase inverter in the dq rotating coordinate system is obtained. The following vector diagram can be obtained by converting the three-phase still abc coordinate system into two phase-rotating dq coordinate systems.

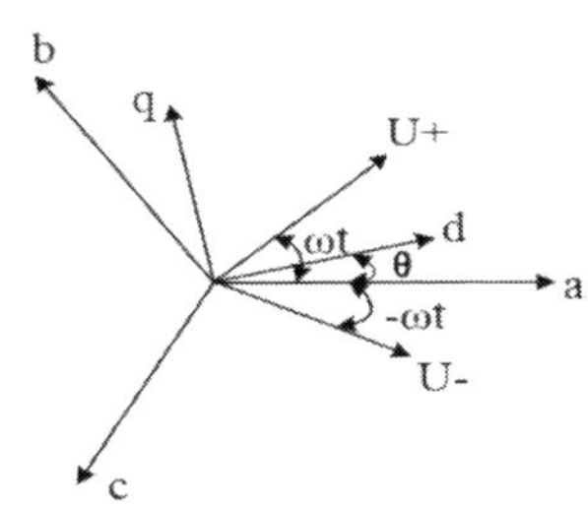

Fig. 2. abc coordinate system to dq coordinate system transformation

If the output voltage is set as follows:

$$\begin{cases} u_a = U_m \cos \omega t \\ u_b = U_m \cos(\omega t + \frac{2}{3}\pi) \\ u_c = U_m \cos(\omega t - \frac{2}{3}\pi) \end{cases} \quad (7)$$

The inverter's three-phase output voltage is converted to two-phase rotation dq coordinate system:

$$\begin{bmatrix} u_b \\ u_q \end{bmatrix} = C_{PARK} \cdot C_{CLARK} \cdot \begin{bmatrix} u_a \\ u_b \\ u_c \end{bmatrix} = \begin{bmatrix} U_m \\ 0 \end{bmatrix} \quad (8)$$

In formula 8, C_{CLARK} and C_{PARK} are CLARK transform matrices and PARK transform matrices

$$C_{CLARK} = \frac{2}{3}\begin{bmatrix} 1 & -1/2 & -1/2 \\ 0 & \sqrt{3}/2 & -\sqrt{3}/2 \end{bmatrix} \quad (9)$$

$$C_{PARK} = \begin{bmatrix} \cos\theta & \sin\theta \\ -\sin\theta & \cos\theta \end{bmatrix} \quad (10)$$

The mathematical model of the three-phase inverter can be seen from each of the above coordinate systems, in the three-phase stationary coordinate system and the dq rotating coordinate system, there is a coupling phenomenon between the inverters, but there is no coupling relation in alpha beta coordinate system. The inverter outputs three-phase voltage to the rotating coordinate system, the straight axis component is the phase voltage peak， the component of the cross axis is zero, thus there is only the DC component.

3 Inverter control system design

When the three-phase inverter is operating in a non-linear load and unbalanced load， the output voltage has periodic disturbances, and use of repetitive control methods to suppress this disturbance. And in order to improve the dynamic response of the system, a compound control algorithm with repetitive control and voltage and current double loop control is adopted. Figure 3 is a system block diagram of composite control.

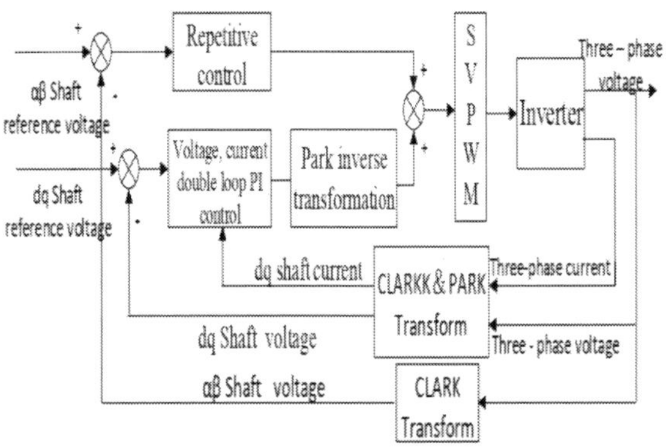

Fig. 3. Composite control system diagram

3.1 Design of Closed - loop Controller for Voltage and Current

The double closed-loop controller consists mainly of two parts， one part is the current inner loop controller, and the other part is the voltage outer ring controller. Among them, the current loop controller can make the inverter fast tracking the actual output current, thus， a quick response is achieved. The voltage outer ring controller can maintain stable output voltage. For the coupling of inverter, a feedforward decoupling control method is adopted to decouple voltage current. The PI regulator is used to control the current inner loop.

The current control equation is:

$$\begin{aligned} u_d^* &= \left(K_{Pi} + \frac{K_{Ii}}{s}\right)(i_d^* - i_d) + u_d - \omega L i_q \\ u_q^* &= \left(K_{Pi} + \frac{K_{Ii}}{s}\right)(i_q^* - i_q) + u_q - \omega L i_d \end{aligned} \quad (11)$$

The voltage control equation is:

$$\begin{aligned} i_d^* &= 3\left(K_{Pv} + \frac{K_{Iv}}{s}\right)(u_d^* - u_d) + i_d' - 3\omega C u_q \\ i_q^* &= 3\left(K_{Pv} + \frac{K_{Iv}}{s}\right)(u_q^* - u_q) + i_q' - 3\omega C u_q \end{aligned} \quad (12)$$

In the formula, the K_{Pi} is the PI regulator and the K_{Pv} is the integral factor of the PI regulator. The i_d^* and i_q^* are reference currents for the inner loop of the current.

Controlled by current control equation 11 and voltage control equation 12, available:

$$\begin{cases} u_d = \frac{1}{Cs}[\left(K_{Pv} + \frac{K_{Iv}}{s}\right)(u_d^* - u_d)] \\ u_q = \frac{1}{Cs}[\left(K_{Pv} + \frac{K_{Iv}}{s}\right)(u_q^* - u_q)] \\ i_d = \frac{1}{Ls}[\left(K_{Pi} + \frac{K_{Ii}}{s}\right)(i_d^* - i_d) - r i_d] \\ i_q = \frac{1}{Ls}[\left(K_{Pi} + \frac{K_{Ii}}{s}\right)(i_q^* - i_q) - r i_q] \end{cases} \quad (13)$$

It can be seen by equation 13 that after decoupling control, the decoupling of voltage and current is completed. Figure 4 is the block diagram of d axis current closed loop control.

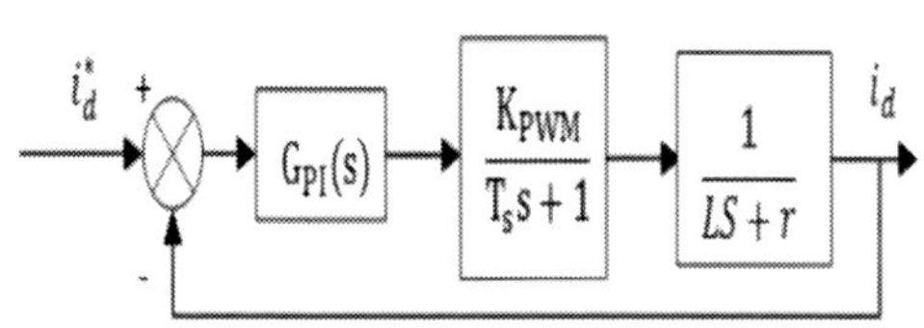

Fig. 4. D-axis current closed-loop control block diagram

In figure 4, the Ts is the current sampling period (switching cycle of the inverter). The L is the filter inductance, the K_{PWM} is the modulation ratio of the inverter, and the r is the equivalent impedance, the G_{Pi} (s) is the transfer function.

The inverter is under the control of the current loop, in steady-state operation, i_q is equal to zero, and the change of i_d in dynamic process is very small. And before the big change in the DC voltage, i_q has completed its transient process to zero, therefore. Transfer function of loop closed loop:

$$C(s) = \frac{1}{T_{is}+1} \quad (14)$$

Formula of T_{is}=L/$K_{pi}K_{PWM}$, the voltage loop control block diagram can be drawn from the above analysis, as shown in figure 5

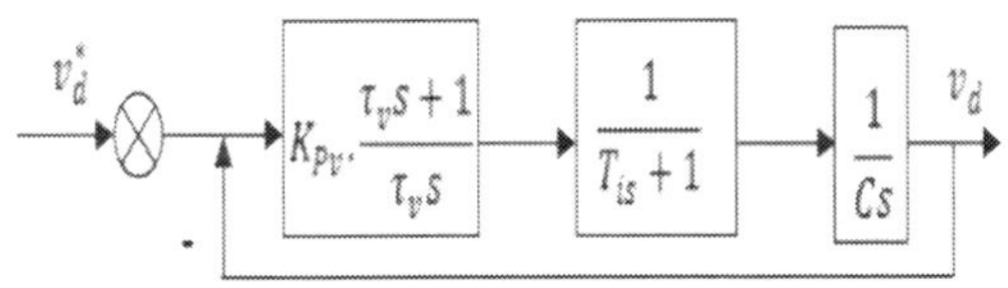

Fig. 5. Voltage loop control block diagram

It can be seen from the formula 14 that there is no coupling between the voltage and current on the dq axis, so the control block diagram of the decoupling system is as follows.

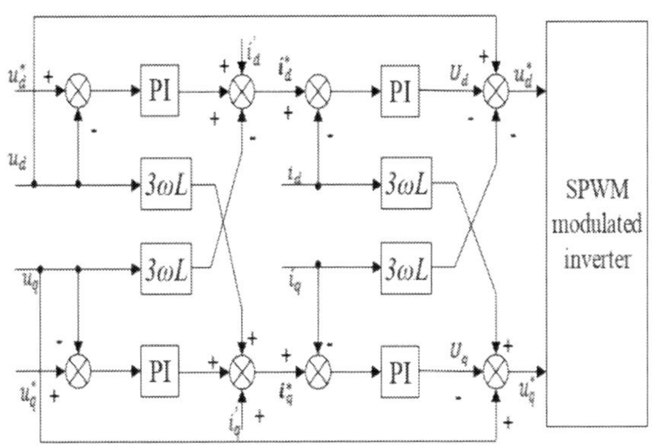

Fig. 6. Control diagram of the decoupling system

3.2 Repetitive controller design

Repetitive control in inverters has a wide range of applications. The main control method of the system is based on the principle of system internal mode, it is to add an "internal model" that can better describe the characteristics of the system under the control system. Therefore, the control accuracy can be improved by solving the tracking of the controlled signal.

So the core of the repetitive control system is the internal model, the performance of repetitive control lies in the establishment of internal model. Figure 7(a) is a block diagram of an ideal structure, which is equivalent to an integral of a step size with certain cycles, so that a given signal is tracked. If the repeat control system is stable, the constraint is $\| 1 - \mathrm{P} \| \infty < 1$. However, it is difficult for the system to satisfy this constraint during the operation of the ideal internal mode. Therefore, internal model structure should be improved. As shown in Figure 7(b), in the feedback loop, the auxiliary compensator $1 - z^{-N}$ is replaced by Q(z) to ensure the stability of the system.

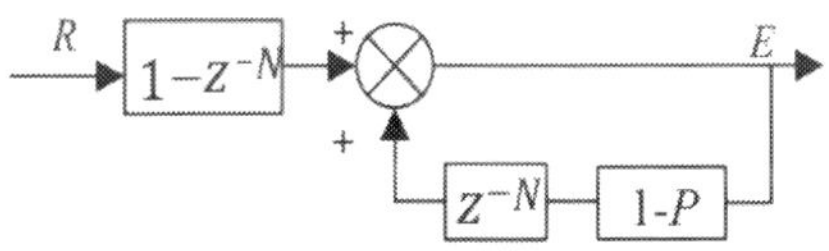

Fig. 7 (a). the ideal structure block diagram

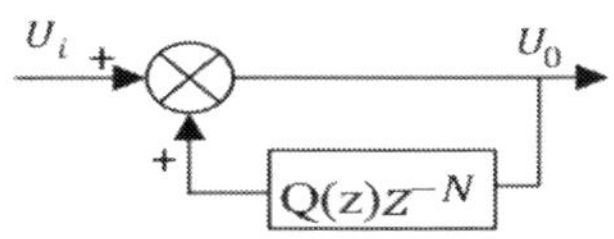

Fig. 7 (b). Improved internal model block diagram

$$G(z) = \frac{U_0(z)}{E(z)} = \frac{1}{1 - Q(z)z^{-N}} \quad (15)$$

The N is the fundamental period sampling number, and the Q(z) is a low-pass filter constant, or a function, to enhance system stability, this paper takes 0.95 and z^{-N} as the periodic delay link, the E(z) is the input and output error, the U_i(z) is the internal model output.

In the improved internal model structure, the auxiliary compensator S(z) and the periodic delay link Z^{-N}. The complete repeat controller is constructed, as shown in figure 8.

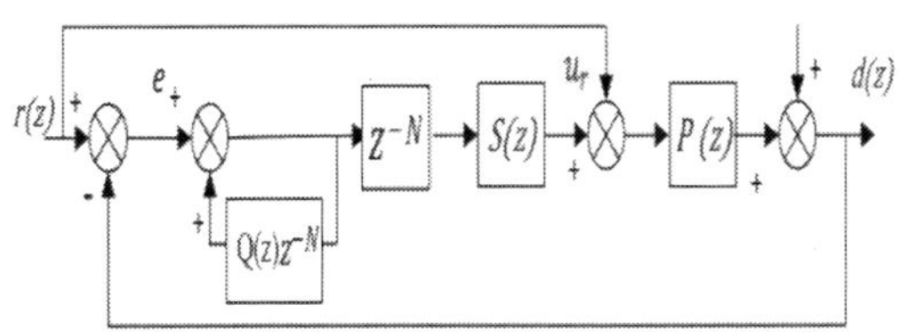

Fig. 8. Repetitive control system block diagram

The P(z) is the control object, r is the reference sinusoidal input signal, and Z^{-N} is the periodic delay link, to achieve the delay control, d is the interference signal. But If the object P(z) has a zero outside the unit circle, the compensator is $S=P^{-1}$ there will be a pole outside the unit circle. Therefore, the compensator is unstable at this time, resulting in instability of the repetitive control system. For this problem, the form of compensator is improved, the modified form is shown in equation 16, in advance, compensating for the phase of internal mode output signal, and then in series with a low-pass filter, this makes up a highly stable controller.

$$S(z) = K_r . z^k . S'(z) \tag{16}$$

In the formula, Kr is the compensator gain to adjust the output amplitude, generally $0<Kr<1$, the main function is to compensate filter phase, S'(z) for low pass filter, To eliminate the high resonance peak of P(z).

4 Simulation analysis

The three-phase inverter power supply model built in this paper is simulated on the MATLAB/Simulink simulation software. Main parameters of the system: DC input voltage U_i=200V, filter capacitorC_f=4.7μF, R_f=0.07Ω, filter inductance L=11.25mH, filter inductance internal resistance R=2mΩ, rated output frequency f_0= 400HZ, switching frequency is 10KHz.

Figure 9(a) is a three-phase inverter under the compound control mode, A, B, C three-phase load in which two phases for the rated load, and the output voltage simulation waveform of one phase is empty. In order to contrast, in this paper, the simulation waveforms of one phase no-load voltage in the traditional double loop control mode are shown in Figure 9(b). It is obvious that the compound control in one phase no-load of the situation does not have a big impact on the output, the balance of output voltage of inverter is obviously better and distortion is small.

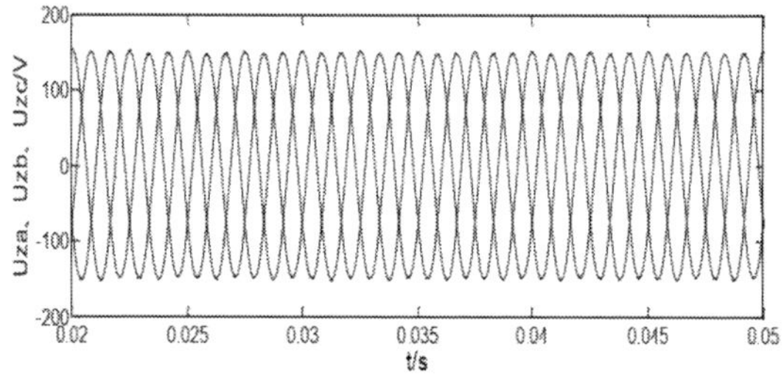

Fig. 9 (a). Voltage simulation waveforms of compound control in one phase no-load operation

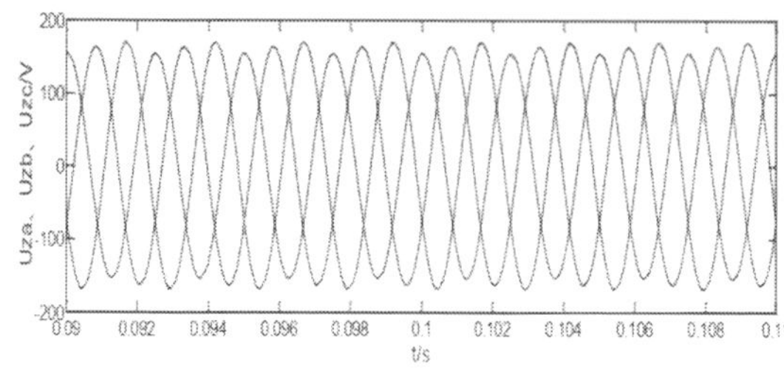

Fig. 9 (b). Double loop control voltage simulation waveform at one phase no-load operation

Figure 10 shows the case of load mutations, in the process of system two phase load operation, In 0.1s when the sudden increase in one load when the composite control system and dual-loop control system voltage and current output of the simulation waveform. It is possible to see that the voltage waveform of the composite control system has not changed much before and after 0.1s. That is, the composite control system in the absence of one phase load still maintain the output voltage balance. The voltage output waveform under the double loop control system is shown in figure (b), there was a severe voltage imbalance before 0.1 s, and the balance was restored after 0.1 s. It can be seen that the output waveform of the composite control system is better.

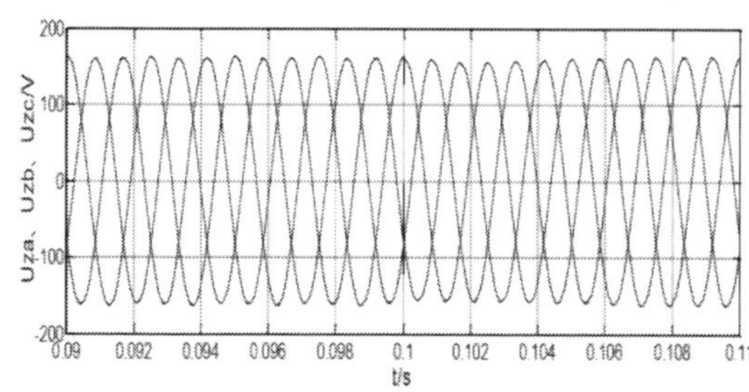

Fig. 10 (a). Composite control sudden plus one phase load voltage waveform

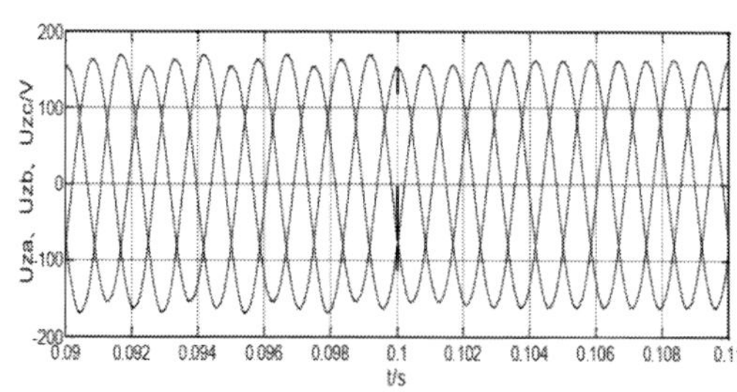

Fig. 10 (b). Double loop control sudden plus one phase load voltage waveform

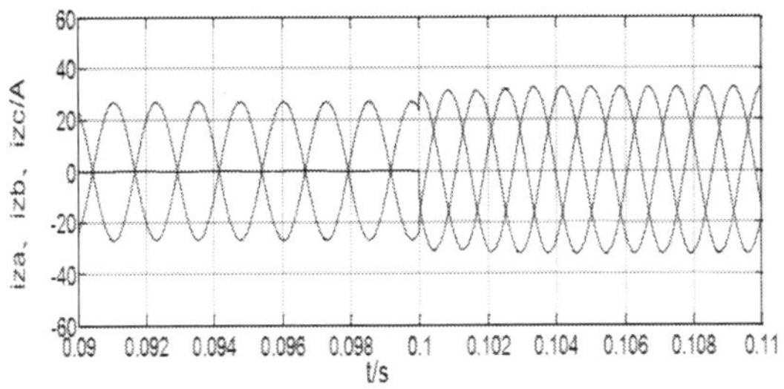

Fig. 10. (c) Compound controls sudden plus one phase load current waveform

5 Conclusion

In this paper, the mathematical model of three-phase three wire inverter is constructed, after analyzing and designing two kinds of inverter control methods, repetitive control and double-loop control, a composite control method combining them is proposed. And through MATLAB/Simulink simulation test verification, comparison of several simulation waveforms, it can be concluded that the dual loop control is much better than the compound control in suppressing the output voltage distortion due to unbalanced loads. The three-phase 400Hz inverter can maintain the voltage balance better under the compound control. And the system has fast response speed, easy to control and fast response, and the structure of the control method is relatively simple.

Reference

1. Li Cui. Control system analysis of single phase 400Hz inverter power supply [D]. Wuhan: Wuhan University of Science and Technology, 2015.
2. SUN Chi, MA Weiming, LU Junyong. Analysis of the unsymmet-rical output voltages distortion mechanism of three-phase inverterand its correction [J] Proceedings of the CSEE, 2006, **26** (21) :57 — 64.
3. Cao Taiqiang, Qi Qiang, Wang Jun. Study on unbalanced load of three-phase inverter[J]. Electric Machines and Control, 2012, **16**(4): 50-55(in Chinese).
4. Jeong C Y, Cho J G, Kang Y, et al. A 100k VA power conditioner for three-phase four-wire emergency generators[C]. IEEE PESC'98, Fukuoka, Japan, 1998.
5. Sun Chi, Lu Junyong, Ma Weiming. A novel control method forthree-phase Four-leg Invert [J]. Transaction of China Electrotechnical So-ciety, 2007 **22**(2) : 57-63.
6. Uffe Borup Jensen, Prasad N Enjeti, Frede Blaabjerg. A new space vector based control method for ups systems powering nonlinear and unbalanced loads[J]. IEEE Trans on IA. 2001, **37**(6): 1864-1870.
7. Kong Xuejuan, Wang Jingjiang. Waveform Control Technology of Three - phase Voltage Source Inverter Based on Internal Model Principle [J] Chinese Journal of Electrical Engineering, 2003, **23**(7):67-70.
8. Kukrer O, Komurcugil H, Doganalp A. A three-level hysteresis function approach to the sliding-mode control of single-phase UPS inverters[J]. IEEE Trans on Industrial Electronics, 2009, **56**(9): 3477-3486. .
9. Chen Daolian, Li Xun, Zhang Rong. Research on Combined Three - phase High Frequency Pulse DC Link Inverter [J]. Chinese Journal of Electrical Engineering, 2005, **25**(8): 75-79.
10. Ilavarasi V, Rajan C C A. quality improvement in grid connected system using four leg VSI[C]//Advances in Engineering, Science and Management. Tamil Nadu, India: ICAESM, 2012: 540-546.
11. Zheng Zhiyao, Sun Qinpei , Wang Jiangbo. Synchronous control method of combined three phase inverter [J] . Journal of Agricultural Engineering, 2015, **31** (5): 146-151.

DOI: 10.1051/matecconf/201713900045

A Novel Boost Conversion with High Voltage Ratio

Yu Chai, Xiaonan Li

College of Electrical and Control Engineering,Xi'an University of Science and Technology,Xi'an 710054 China

Abstract. When the traditional Boost converter duty cycle exceeds a certain value, the output voltage is not only increased, but will be reduced. Therefore, a new Boost converter with intermediate tap is proposed. By replacing the ordinary inductance of the traditional Boost converter with the inductance with the intermediate tap, the theoretical comparison is made with respect to the traditional boost converter. Under the same duty cycle, The Boost converter has a significant high gain ratio. Finally, the simulation and experimental analysis of the proposed new Boost converter are carried out by simulation software Pspice. The experimental results show that the new Boost converter has a significant high gain ratio compared with the traditional boost converter.

1 Introduction

In the development and utilization of new energy, solar energy and power generation, the need to pass the energy of solar modules to the grid, however, the solar panels generally produce lower voltage, generally below 60V [1], and for the output of 220V AC grid voltage grid-connected inverter, the single-phase full-bridge topology required for the bus voltage of 380V or even Higher [2-3]. There is therefore a need for a DC power converter with a high boost function to boost the voltage of the solar panel to match the voltage of the grid bus. If the traditional boost circuit to achieve this high voltage conversion, you need to work in a large duty cycle state, which will lead to the circuit switch and diode loss increases, affecting the conversion efficiency of the converter [4]. Based on this, a great deal of research has been done on the high-voltage converter at home and abroad, and a variety of converters are proposed for high boost.

The literature [5-8] proposed a built-in transformer unit circuit (Built-In Transformer Cell, BITC) high gain converter, but in order to eliminate the voltage spikes of the switch, the need to increase the absorption circuit, increasing the complexity of the circuit. The literature [9-11] proposed a cascade high gain converter, in the post-converter using coupling inductance technology, can effectively achieve the converter's high gain characteristics [12-13], but the converter cascade will lead to low efficiency.

Based on this, a high gain converter based on a conventional boost converter is proposed. The conventional inductor of the conventional Boost converter is replaced with an inductor with an intermediate tap [13-14], and the voltage gain is effective by selecting the appropriate turn ratio. To increases the voltage gain ratio. And this topology is simple and easy to implement [15]. Therefore, in the new energy power generation system has a good application value. The working principle and steady state characteristics of the proposed converter are analyzed theoretically. Finally, the circuit is built by the simulation software, and the correctness of the theoretical analysis is verified.

Corresponding author: larrynance@126.com

2 New high boost converter

2.1 Traditional boost converter

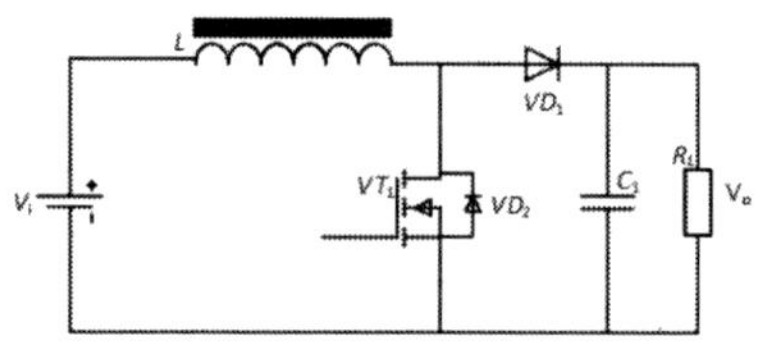

Figure 1. Traditional Boost converter circuit

The traditional Boost circuit shown in Figure 1, when VT1 conduction, the diode VD withstand back pressure and off, the equivalent circuit shown in Figure 2a, the current i_L flow through the inductor coil L, the current linear increase in energy to magnetic energy The form is stored in the inductance coil L. At this time, the capacitor C discharge, only by the capacitor to the load R_L power supply, load current I_o.

When VD is turned off, the equivalent circuit shown in Figure 2b, because the current i_L in the inductor L can not be mutated, forced diode VD conduction flow, so that the inductor L and power V_i at the same time to the

DOI: 10.1051/matecconf/201713900045

capacitor C, load R_L power supply. When the inductor current is higher than I_o, the capacitor is charged and at the same time to provide power to the load; when the inductor current is less than I_o, the inductor and capacitor discharge to the load R_L at the same time, to maintain V_o unchanged.

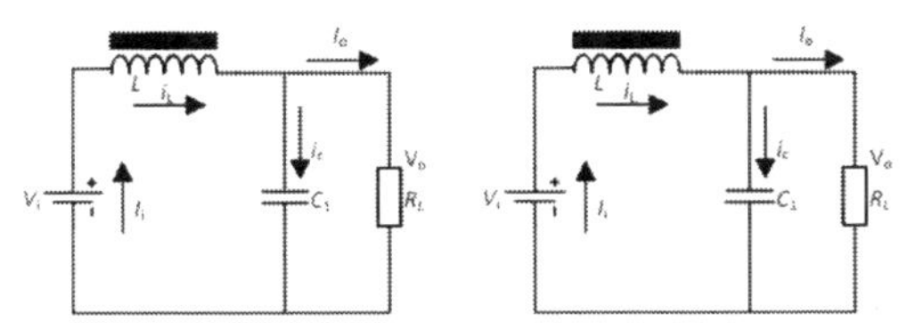

a) Switch VT1 turns on b) Switch VT1 is off

Figure 2. Equivalent circuit for conventional boost converter operation

Define the duty cycle D as:

$$D = \frac{T_{on}}{T_{on} + T_{off}} \quad (1)$$

Where: T_{on} is the time of VT1 conduction, and T_{off} is the time of VT1 turn-off.

By analyzing the VT1 turn-on and turn-off of a period T, apply the volt-second balance to the inductor L:

$$V_i DT = (V_o - V_i)(1 - D)T \quad (2)$$

From (2) the solution of the voltage gain ratio M:

$$M = \frac{V_o}{V_i} = \frac{1}{1 - D} \quad (3)$$

Type (3) is the traditional Boost converter voltage gain ratio formula, we can see that when the duty cycle is 0.5, the output voltage V_O is only 2 times the input voltage V_i, and the duty cycle can not be too large, generally less than 0.8.

2.2 A boost circuit with a snubbed circuit for coupling inductors

Application of coupling inductance with a buffer circuit boost circuit shown in Figure 3, inductance L_1 and inductance L_2 set the coupling coefficient of 1 can be considered an ideal transformer. Where L_1 turns are N_1, L_2 turns are N_2, and turns ratio N = N_2 / N_1. Capacitor C_1 and Diode D_1 clamp the capacitor and clamp diodes, respectively, and D_2 is the rectifier diode in the snubbed circuit.

The basic working principle is: When the switch VT_1 conduction, the power supply to the L_1 charge, which inductor L_1 current linear growth, L_2 on the induced voltage through the clamping capacitors C_1 and D_2 and switch VT_1 charge circuit to charge the capacitor C_2 charge , As shown in Figure 4a.

When the switch VT_1 shut off, the output diode D_o conduction, D_2 is off. The supply voltage V_i, the voltages on the inductor voltages V_{L1} and V_{L2}, and the voltage on the capacitor C_2 form the charge loop to supply the load R_L to charge the capacitor C_o, as shown in Figure 4b.

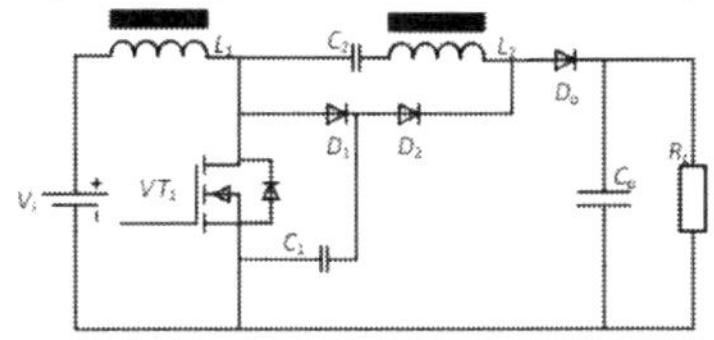

Figure 3. A boost circuit with a snubbed circuit for coupling inductors

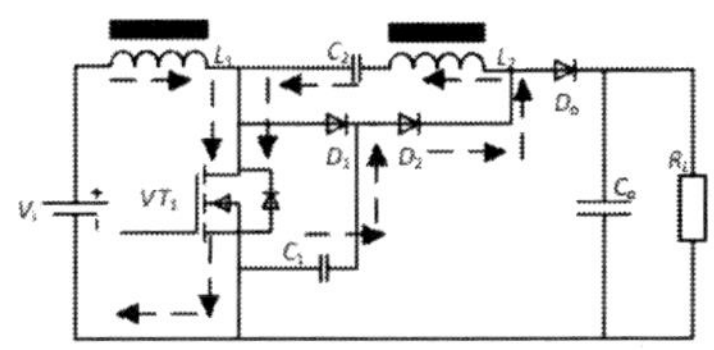

a) Switch VT1 turns on

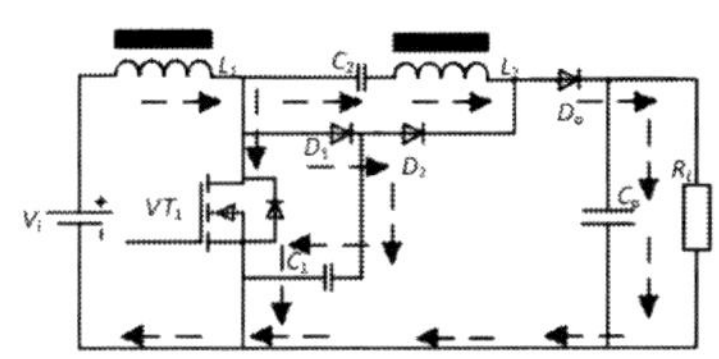

b) Switch VT1 is off

Figure 4. Working state diagram

By calculating the boost ratio M of the circuit is:

$$M = \frac{V_O}{V_i} = \frac{2 + NK}{1 - D} + \frac{D(1 - K)(N - 1)}{1 - D} \quad (4)$$

In equation (4): D is the duty cycle, K is the coupling coefficient, and N is the turn ratio.

Therefore, when the coupling coefficient K is 1, M is:

$$M = \frac{V_o}{V_i} = \frac{2 + N}{1 - D} \quad (5)$$

From the formula (5) can be obtained when the duty cycle is 0.5, the output voltage for the input voltage of (4 +2N) times, we can see that the circuit has a greater boost, but the circuit is relatively complex.

2.3 New boost converter with intermediate tap inductor

As shown in Figure 5, the boost converter with an inward tap inductor converts the inductance of a conventional boost converter with an inductor with an intermediate tap, and L_1 and L_2 are coiled on a core to form a coupling inductor with an intermediate tap. The L_1 as the primary, the number of turns n_1, L_1 as the secondary, the number of turns n_2, in a switching cycle, the circuit works as follows:

When the switch VT_1 conduction, the diode D by the back pressure off, the power V_i to L_1 charge, as shown in Figure 6a. When the switch is off, the diode turns on; the inductors L_1 and L_2 are connected to the supply V_i circuit to supply the load R_L, as shown in Figure 6b.

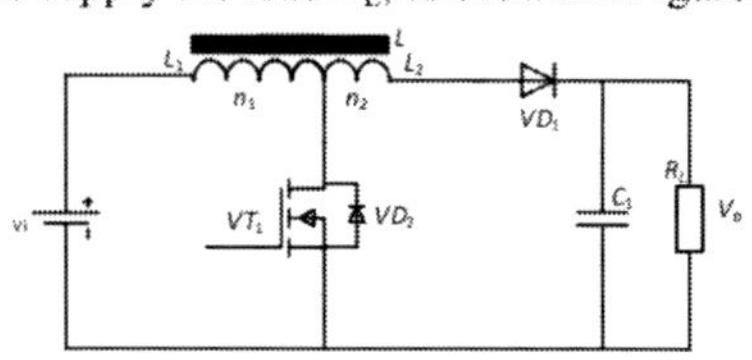

Figure 5. Boost converter circuit with intermediate tap inductor

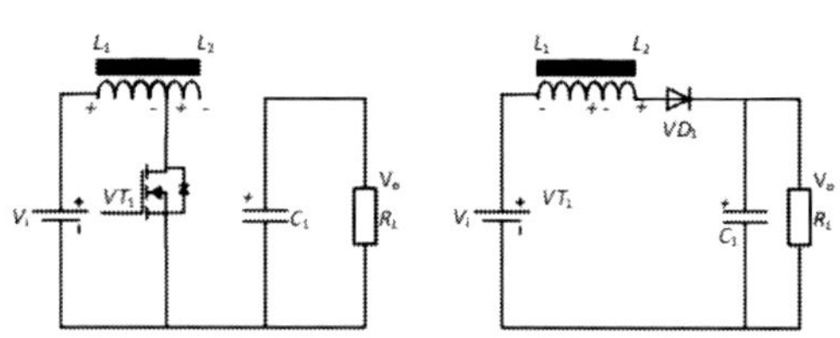

a) Switch VT1 turns on b) Switch VT1 is off

Figure 6. Working state diagram

The boost ratio of the boost converter with intermediate tap inductance can be calculated by changing the flux of the magnetic flux in the tap inductor L for one cycle of turning on the switch VT1:

$$M = \frac{V_o}{V_i} = \frac{1 + ND}{1 - D} \quad (6)$$

From the formula (6) can be obtained when the duty cycle is 0.5, the output voltage for the input voltage of (2 + N) times, we can see that the circuit relative to the traditional boost converter has a greater boost ratio, the relative band coupling inductance has Buffer circuit of the boost converter, the circuit structure is simple, easy to control, so it has a wide range of applications.

3 Simulation and verification

In order to verify the correctness of the theoretical analysis, Pspice software is used to simulate the traditional boost converter, the boost circuit with the buffer circuit and the new boost converter circuit with the middle tap inductance. Figure 7, Figure 8, and Figure 9 shows the three circuits in the input voltage 100V, N = 2, D = 0.5, when the output voltage waveform. Through the simulation of the waveform, we can see that the traditional boost converter boost ratio is probably, the output DC voltage of 186V or so, the theoretical value of 200V, so the circuit can improve the boost ratio, but the boost capacity is limited; Application coupling inductance with a buffer circuit output DC voltage of 690V or so, the theoretical value of 800V, and the theoretical value of some gaps, which is related to the value of the capacitor inductance, but the circuit can significantly achieve the purpose of boost; The middle tap inductance of the new boost converter output voltage of 386V or so, the theoretical value of 400V, and the theoretical value of the basic line.

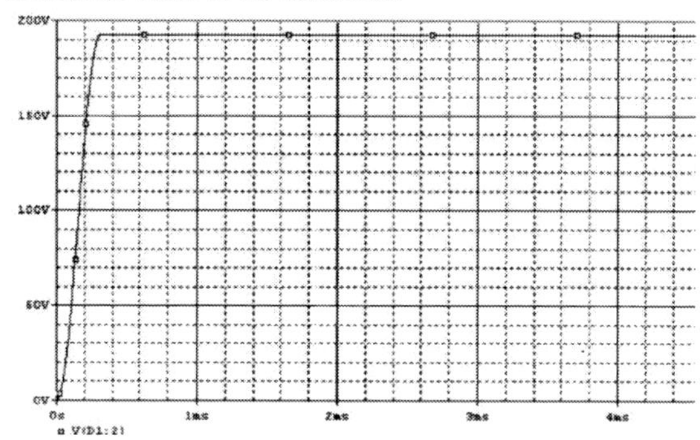

Figure 7. Traditional boost converter output voltage simulation waveform

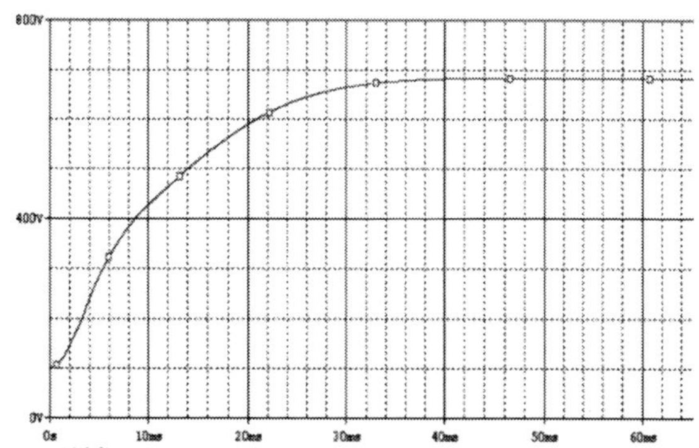

Figure 8. Application of coupling inductor with buffer circuit output voltage simulation waveform

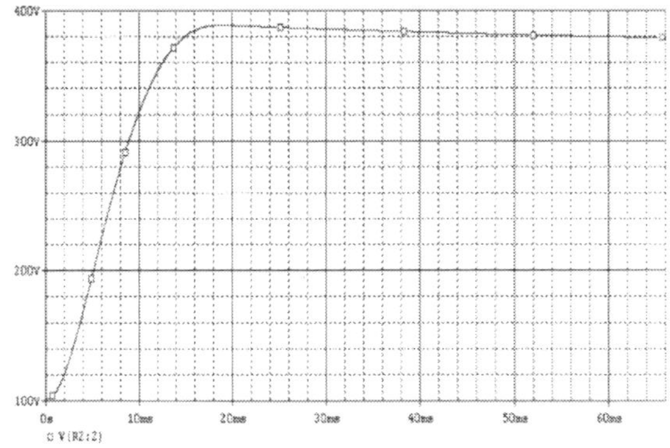

Figure 8. New boost converter output voltage simulation waveform with intermediate tap inductor

At the same time, the three circuit structures mentioned above and the devices used are compared, as shown in Table 1.

Table 1. Three kinds of boost circuit structure complexity comparison

拓扑类型	电路结构复杂度	升压比
传统 Boost 变换器	简单	$M = \frac{1}{1-D}$
应用耦合电感的具有缓冲电路的变换器	复杂	$M = \frac{1+ND}{1-D}$
带中间抽头电感的 Boost 变换器	简单	$M = \frac{2+N}{1-D}$

From the above table can be seen, through the cost of the circuit complexity, boost size than the size of the proposed new boost circuit, you can take into account the circuit cost and boost the advantages of relatively large, so this circuit has a larger research value.

4. Conclusions

Based on the theoretical analysis and simulation of the three booster circuits, it can be concluded that the proposed Boost converter with intermediate tap inductance has a higher boost ratio than other booster circuits, and the circuit structure is also relatively Simple and easy to control. So it has a high practical value in the application of distributed photovoltaic power generation system, uninterrupted power supply system and so on.

References

1. Yang B, Li W, Zhao Y, et al. Design and Analysis of a Grid-Connected Photovoltaic Power System[J]. IEEE Transactions on Power Electronics, 2010, 25(4):992-1000.
2. Zhao Q, Lee F C. High-efficiency, high step-up DC-DC converters [J]. IEEE Transactions on Power Electronics, 2003, 18(1):65-73.
3. Yang L S, Liang T J, Chen J F. Transformerless DC–DC Converters With High Step-Up Voltage Gain[J]. IEEE Transactions on Industrial Electronics, 2009, 56(8):3144-3152.
4. Zhang Zhansong. Switching power supply principle and design [M]. Electronic Industry Press, 2004.
5. Li W, Wu J, Xie R, et al. A non-isolated interleaved ZVT boost converter with high step-up conversion derived from its isolated counterpart[C]// European Conference on Power Electronics and Applications. IEEE, 2008:1-8.
6. Li W, Li W, He X. Zero-voltage transition interleaved high step-up converter with built-in transformer[J]. Iet Power Electronics, 2011, 4(5):523-531.
7. Li W, Xiang X, Li C, et al. Interleaved High Step-Up ZVT Converter With Built-In Transformer Voltage Doubler Cell for Distributed PV Generation System[J]. IEEE Transactions on Power Electronics, 2012, 28(1):300-313.
8. Chen Zhangyong, Xu Jianping, Wu Jianxue, et al.High-boost gain Boost converter of coupled inductance voltage doubling unit based on LC absorption circuit [J] .Journal of Electrical Engineering, 2016, 31 (2): 78-85.
9. Zhang Shiyu, XU Jianping, Yang Ping.Study on a Novel Single-switch High-gain Boost Converter [J]. New Technology of Electric Power, 2013, 32 (3): 000012-15.
10. Chen S M, Liang T J, Yang L S, et al. A Cascaded High Step-Up DC–DC Converter With Single Switch for Microsource Applications[J]. IEEE Transactions on Power Electronics, 2011, 26(4):1146-1153.
11. Hu X, Gong C. A High Voltage Gain DC–DC Converter Integrating Coupled-Inductor and Diode–Capacitor Techniques[J]. IEEE Transactions on Power Electronics, 2013, 29(2):789-800.
12. ZHOU Yu-fei, HAN Wen-xin. Coupled Inductance Single-stage Boost Inverter [J]. Proceeding of the CSEE, 2011, 31 (33): 61-67.
13. Cao Taiqiang, Liu Wei, Guo Xiaoying, et al. Dual Coupled Inductance Quadrature High Boost Gain DC-DC Converter [J] .Journal of Electrical Engineering, 2015, 30 (8): 104-112.
14. Hua Wei.Study on correction factor of chopper boost power factor with center tap inductor [J]. Electrical Drive Automation, 1999 (1): 32-35.
15. WANG Hao.Study on Topology of High Gain DC-DC Converters for Building Integrated PV System [J]. Inverter World, 2012 (4): 56-61.

A Recognition Method for MIMO DUSTC Signals

Bo Qian[1], *Jinhan* Huang[1], *Siyao* Xie[1], and *Yongxin* Feng[2,*]

[1]School of Information Science and Engineering, Shenyang Ligong University, 110159 Shenyang, China
[2]School of Graduate Studies, Shenyang Ligong University, 110159 Shenyang, China

Abstract. For recognizing the MIMO DUSTC signals without a priori knowledge, a blind recognition method is presented. By estimating the number of transmitting antennas, separating the mixed signals, and recognizing modulation, the unitary matrices $\boldsymbol{S}$ is estimated by statistics and DUSTC signals are recognized. The simulation results showed that the method was effectively to recognize DUSTC signals with the SNR no less than 4dB.

1 Introduction

The blind recognition of communication signals is an important research topic in the communication countermeasure fields. In the last decade, the Multiple-Input Multiple-Output (MIMO) technology is widely used in the field of civilian and military communication [1-4]. The capacity and performance of system are enhanced by Space-Time Codes (STC). The composite signals are received by each receiving antenna because the MIMO signals are transmitted by multi-antenna array, so it increases the complexity and difficulty of recognition MIMO signals. So the blind recognition methods for MIMO signals are a key research issue in communication countermeasure fields.

At present, there are many algorithms for recognizing STC. The methods of recognizing STBC are proposed using the second-order and higher-order statistics in [5,6]. In [7], a blind recognition algorithm of orthogonal space-time block codes (OSTBC) is proposed based on a cross-correlation analysis over linear combinations of the transmitted signals. The most existing methods focused on recognition STBC. To our knowledge, no method exists in literature for the blind recognition of DUSTC.

In this paper, we propose a method to recognition differential unitary space-time code but without a priori knowledge about the number of antenna array, the modulation and so on. Section II of this paper presents the signal model and characterization of the DUSTC. In Section III, the blind recognition method for DUSTC is presented. Finally, the performances of our proposed method are simulated and analyzed in section IV.

2 Signal Model

DUSTC is an approach to differential modulation for multi-antennas, based on space-time block codes with a group structure [8-10]. Consider the MIMO system with t transmit antennas and constellation C. let $\boldsymbol{S}$ be any group of $n\times n$ unitary matrices, where $SS^H=S^HS=I$ for all $S\in\boldsymbol{S}$. Let D be a $t\times n$ matrix such that $DS\in C^{t\times n}$ for all $S\in\boldsymbol{S}$. A matrix collection M is defined as

$$M=\{DS:S\in\boldsymbol{S}\} \tag{1}$$

where D is a fixed matrix that satisfies $DD^H=tI$. M is a group code of length n over the constellation C.

At time k=0, the block M_0=D is sent to initialize the transmission. In successive blocks, the matrix block M_k is differentially encoded which is given by

$$M_k=\begin{cases} D\ , & k=0 \\ M_{k-1}S_k, & k=1,2,3,\cdots \end{cases} \tag{2}$$

where S_k is the $n\times n$ unitary matrices which is determined by transmitted data.

At the receiver, the signal $\boldsymbol{Y}$ which is received by antenna is given by

$$\boldsymbol{Y}=\boldsymbol{HM}+\boldsymbol{N} \tag{3}$$

where $\boldsymbol{H}$ is the channel transfer matrix, $\boldsymbol{M}$ is the matrix block collection $\{M_0, M_1, M_2, \ldots\}$ and $\boldsymbol{N}$ is zero-mean Gaussian noise.

It can be seen from the formula (3) that each receiving antenna received the multi composite signal which transmitted by multi-antennas. The multi-component signal can lead to signal that's difficult to distinguish, so it increase the difficulty to recognize DUSTC signal.

3 DUSTC Recognizing Mechanism

3.1 Signals Separation

Table 1. The transmitted M_k

k	0	1	2	…	K
M_k	D	DS_1	DS_1S_2	…	$DS_1S_2\cdots S_K$

* Corresponding author: fengyongxin@263.net

According to the formula (2), the transmitted block M_k is showed as in table 1 with k increase from 0 to K.

At the transmitter side, when we multiply the $M_{k\text{-}1}$ by S_1, we get the matrix block M_k. The reason that the receiving antenna received multi composite signal is because the multi signals are transmitted in wireless channel. So if we can get the matrix block M_k, we can estimate the S_k as

$$S_k = M_{k-1}^{-1} M_k = (DS_1S_2 \cdots S_{k-1})^{-1} DS_1S_2 \cdots S_{k-1}S_k \quad (4)$$

In order to get M_k, we use the blind source separation (BSS) [11,12]. BSS is the approach to estimate original source signals without any prior information. There are many algorithms to separate the multi composite signal, such as FAST ICA, JADE and so on. The process of blind source separation is showed as Fig.1.

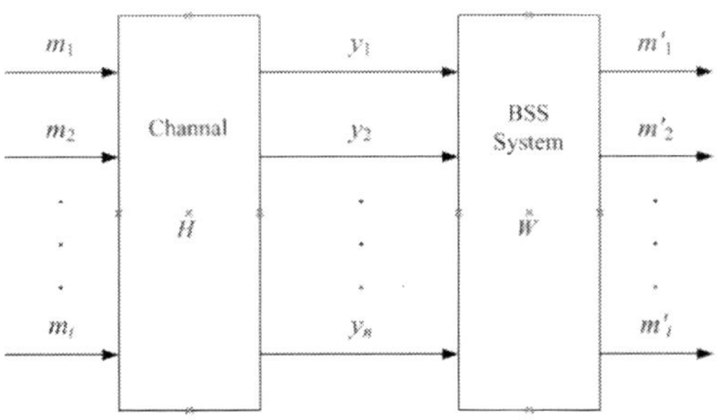

Fig. 1. The process of blind source separation

In Fig.1, m_t is the transmitted signal, y_n is the received signal, and m'_t is the signals after separating. The BSS algorithms need to meet that the number of receiver antennas is no lower than that of transmitter, and the number of signal sources is known. We can increase the number of receiver antennas to meet the first requirement. Based on Minimum Description Length (MDL) principle [13], the number of signal sources can be estimated.

The objective function of MDL principle is given by

$$MDL(k) = -\ln[(\prod_{i=k+1}^{M} \lambda_i^{\frac{1}{M-k}}) \Big/ (\frac{1}{M-k}\sum_{i=k+1}^{M}\lambda_i)]^{(M-k)N} + \frac{k}{2}(2M-k)\ln N \quad (5)$$

where M is the number of receiver antennas for separating signals, λ_i is the eigenvalue of covariance matrix of received signal Y, and N is the number of data. As k increase from 0 to M-1, we find the minimum value of $MDL(k)$ when k is equal to k', and then k' is the number of source signals. It is given by

$$k' = \min\{MDL(k),\quad k = 0,1,2,\cdots,M-1\} \quad (6)$$

By using MDL principle and BSS, we can separate the multi composite signal, and get the signal that send by each transmitted antenna.

3.2 Separated Signal Demodulation

Because the DUSTC signal is a digital modulation wave, so we need to demodulate the signal to get the transmitted constellation. Without a priori knowledge, the modulation and carrier frequency is unknown. For demodulating the separated signal, we need to estimate the carrier frequency and recognize the modulation.

For estimating the carrier frequency, we can use the Fast Fourier Transmit, power spectrum estimation, and so on [14,15]. For recognizing the modulation, these methods are divided into two categories [16-18]. One is the decision theoretic approaches, and the other is pattern recognition. One classical method for modulation recognizing is proposed in [16]. There are five parameters such as the maximum value of the spectral power density of the normalized-centered instantaneous amplitude r_{max}, the standard deviation of the absolute value of the non-linear component of the instantaneous phase σ_{ap}, the standard deviation of the nonlinear component of the direct (not absolute) instantaneous phase σ_{dp}, the standard deviation of the absolute value of the normalized-centered instantaneous amplitude σ_{aa}, and the standard deviation of the absolute value of the normalized centered instantaneous frequency σ_{fa}. The 2FSK, 4FSK, 2ASK, 4ASK, BPSK and QPSK signals can be recognized by five parameters. The detailed pictorial representation for digital modulation identification is shown in Fig.2.

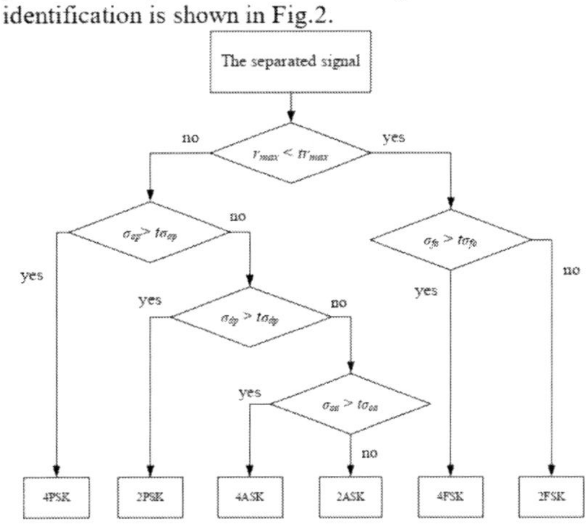

Fig. 2. Digital modulation identification

The separated signal is a pure signal by blind source separation, and the interference in identical, adjacent and multiple frequencies are eliminated. So it is suit for estimating the carrier frequency by power spectrum estimation and recognizing the modulation by using the method in [16].

3.3 DUSTC Signals Recognition

After demodulating each separated signal, we can get the constellation M_k which sent by each transmitting antenna. According to the principle of DUSTC, the successive constellations M_k is equal to $M_{k\text{-}1}$ multiply by S_k. If the separated signals are DUSTC signal, we will get S_k by formula (4). If we get enough separate signals, we can estimate $\boldsymbol{S}$. If the separated signals are not DUSTC signal, there isn't S_k but random matrix. According to whether there is DUSTC signal, it is given by

$$H_1 : S_k = M_{k-1}^{-1} M_k, \quad S_k \in \mathbf{S}$$
$$H_0 : random\,matrix = M_{k-1}^{-1} M_k \tag{7}$$

The unitary matrices $\boldsymbol{S}$ can be estimated by statistical symbols, and the DUSTC is recognized. The process of recognition DUSTC is as shown in figure 3.

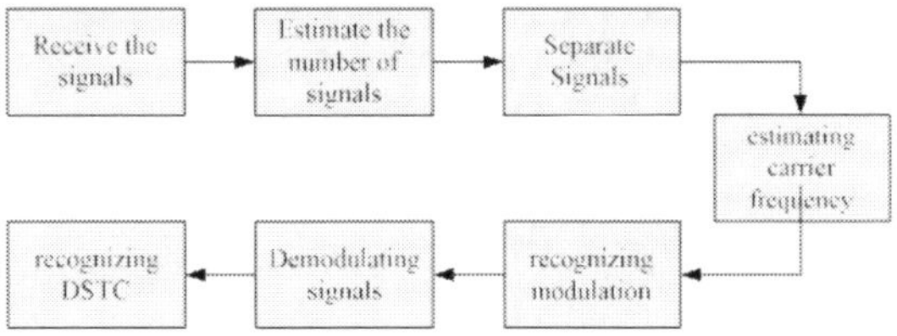

Fig. 3. The process of recognition DUSTC signals

4 Performance analysis

In order to verify that the method for recognizing DUSTC signals is effective, it is simulated in this paper. The number of transmitting antennas is 2, and the number of receiving antennas for recognizing is 8. The modulation is QPSK, and the number of sampling data is 120000.

Table 2 shows the results of the objective function of MDL principle with CNR=15dB. It is evident that $MDL(k)$ is the smallest when k is 3. Then, the estimated number of transmitting antennas is 2.

Table 2. The objective function of MDL principle

k	0	1	2	3
MDL(k)	1397.44	799.17	100.36	139.62
k	4	5	6	7
MDL(k)	171.77	196.79	214.65	225.37

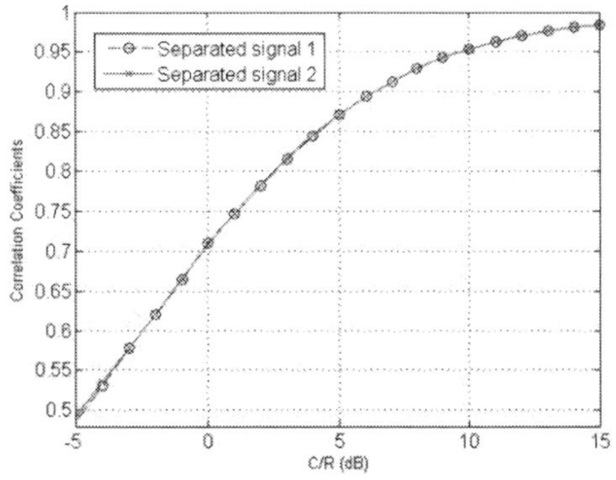

Fig.4. The correlation coefficients between original signals and separated signals

Figure 4 shows the correlation coefficients between original DUSTC signals and separated signals. From Fig. 6, we can see that the separated signals are similar to the original signals with a higher CNR. The correlation coefficients decrease as the CNR gradually decreases.

Figure 5 shows the results of r_{max} and σ_{ap} with different CNR. In this paper, because the modulation of DUSTC signals is QPSK, so it can be recognized by r_{max} and σ_{ap} parameters. The thresholds are 39 and 0.5, for R_{max} and σ_{ap} respectively. From Fig.5 it is clear that for CNR > 6dB it is distinguished effectively FSK from other demodulations by R_{max}, and for CNR>0dB it is distinguished effectively QPSK from BPSK and ASK by σ_{ap}.

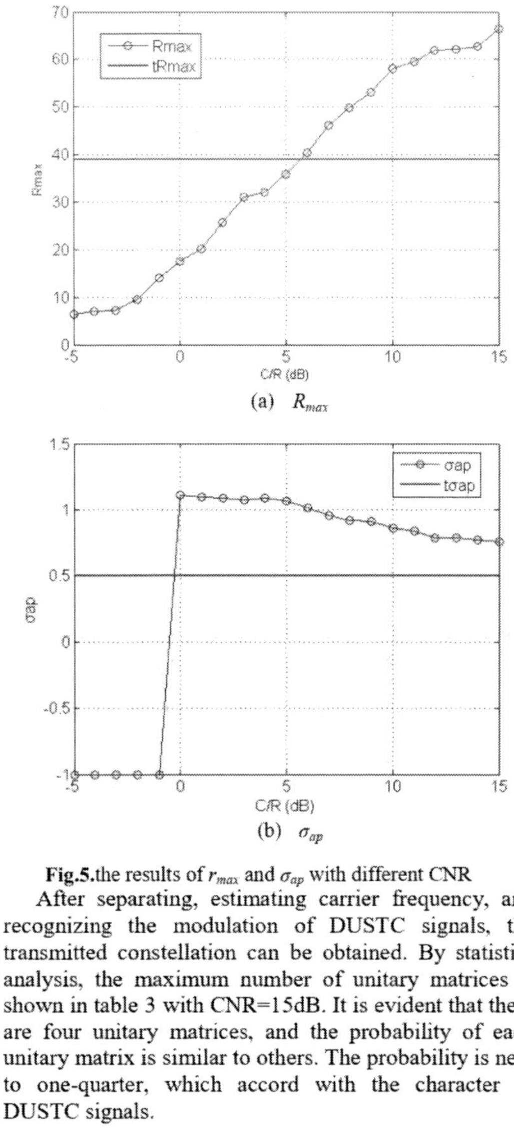

Fig.5.the results of r_{max} and σ_{ap} with different CNR

After separating, estimating carrier frequency, and recognizing the modulation of DUSTC signals, the transmitted constellation can be obtained. By statistics analysis, the maximum number of unitary matrices is shown in table 3 with CNR=15dB. It is evident that there are four unitary matrices, and the probability of each unitary matrix is similar to others. The probability is near to one-quarter, which accord with the character of DUSTC signals.

Table 3. The probability and constellation of unitary matrices

NO.	1	2	3	4
Constellation	1 0 0 1	-1 0 0 -1	j 0 0 j	-j 0 0 -j
Probability	0.2367	0.2317	0.2683	0.2617

Figure 6 shows the recognition probability of DUSTC with different CNR. From Fig.6, we can see that the DUSTC signals can be recognition effectively with CNR > 6dB. For CNR<4dB, the recognition probability decrease rapidly.

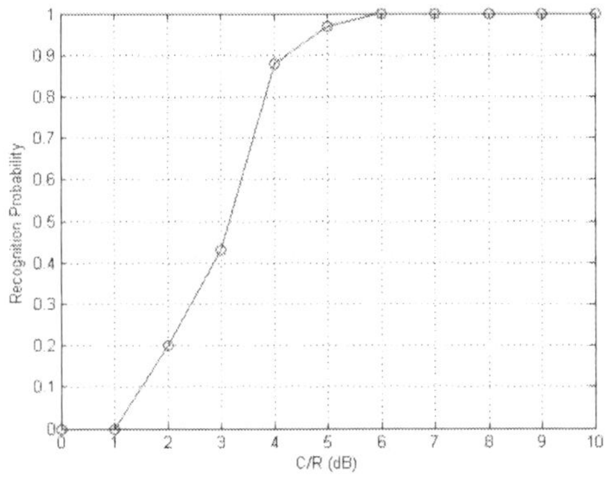

Fig.6. The recognition probability of DUSTC signals

Conclusion

The simulation results showed that the method of recognition DUSTC signals was effectively in this paper. However the recognition method is carried out with 2×2 antennas array. The further study will be needed to recognize DUSTC signals with more antennas.

This work is supported by The Science and Technology Foundation from Liaoning Education Department (No.L2015462, No.LYB201620), Shenyang Ligong University Liaoning Province Information Network and Countermeasure Technology Key Laboratory Open Foundation (2015), and Shenyang Ligong University Postdoctoral Science Foundation.

References

1. Lamare R C D. Massive MIMO systems: Signal processing challenges and future trends. *Ursi Radio Science Bulletin*, **86(4)**, 8-20(2017)
2. Jindal N. MIMO Broadcast Channels With Finite-Rate Feedback. *IEEE Transactions on Information Theory*, **52(11)**,5045-5060(2006)
3. Gao X, Edfors O, Rusek F, et al. Massive MIMO Performance Evaluation Based on Measured Propagation Data. *IEEE Transactions on Wireless Communications*, **14(7)**,3899-3911(2015)
4. Björnson E, Hoydis J, Kountouris M, et al. Massive MIMO Systems With Non-Ideal Hardware: Energy Efficiency, Estimation, and Capacity Limits. *IEEE Transactions on Information Theory*, **60(11)**, 7112-7139(2015)
5. G Qian, L Li, M Luo, et al. Blind recognition of space-time block code in MISO system. *Eurasip Journal on Wireless Communications & Networking*, **1**, 164(2013)
6. W Yan. Blind Recognition of Orthogonal Space-time Block Coding (OSTBC) Using Higher-order Cumulants. *Journal of Information & Computational Science*, **12(15)**, 5807-5819(2015)
7. A Mazeika, K Yao, L Collin, et al. Blind Recognition of Orthogonal Space-Time Block Codes Based on the Kurtosis of Cross-Correlations Lags. *Military Technical Academy Review*, 109-134 (2011)
8. P K M Pun, P K M Ho. Fano Multiple-Symbol Differential Detectors for Differential Unitary Space–Time Modulation. *IEEE Transactions on Communications*, **55(3)**, 540-550(2007)
9. Bertrand M. Hochwald, Wim Sweldens. Differential Unitary Space Time Modulation. *IEEE Transactions on Communications*, 48(12), 2041-2052(2000)
10. Yuen C, Guan Y L, Tjhung T T. Unitary Differential Space–Time Modulation With Joint Modulation. *IEEE Transactions on Vehicular Technology*, **56(6)**, 3937-3944(2007)
11. Congedo M., Gouy-Pailler C., Jutten, C.. On the blind source separation of human electroencephalogram by approximate joint diagonalization of second order statistics. *Clinical Neurophysiology*, **119(12)**, 2677 (2008)
12. Mitsui Y, Kitamura D, Takamichi S, et al. Blind source separation based on independent low-rank matrix analysis with sparse regularization for time-series activity. *IEEE International Conference on Acoustics, Speech and Signal Processing*, 21-25(2017)
13. Wajid B, Serpedin E, Nounou M, et al. Optimal reference sequence selection for genome assembly using minimum description length principle. *Eurasip Journal on Bioinformatics & Systems Biology*, **2012(1)**, 18(2012)
14. L Wang, W Zhang, Z P Zhang, et al. Spinning frequency estimation algorithm of MEMS gyro's output signal based on FFT coefficient. *Microsystem Technologies*, 1-5(2017)
15. W T Sun, L V Jun, L I Tong. An Algorithm of Frequency Estimation for ESPRIT Frequency-Hopping Signal Based on FFT Rotational Invariance. *Measurement & Control Technology*, **23(6)**,347 (2014)
16. Azzouz E E, Nandi A K. Procedure for automatic recognition of analogue and digital modulations. *IEEE Proceedings Communications*, **143(5)**, 259-266(1996)
17. Zhao X, Guo C, Jingchun L. Mixed Recognition Algorithm for Signal Modulation Schemes by High-order Cumulants and Cyclic Spectrum. *Journal of Electronics & Information Technology*, **38(3)**, 674-680(2016)
18. Zhu X, Lin Y, Dou Z. Automatic recognition of communication signal modulation based on neural network. *IEEE International Conference on Electronic Information and Communication Technology*.223-226(2017)

Optimal Design of Grooved Cam Profile Using Non-uniform Rational B-splines

Guantao Xuan[1,2], *Yuanyuan* Shao[1,*], and *Lei* Liu[1]

[1]College of Mechanical and Electronic Engineering, Shandong Agricultural University, 271018 Tai'an, China
[2]Shandong Province Key Laboratory of Horticultural Machinery and Equipment, 271018 Tai'an, China

Abstract. In order to reduce the fatigue damage in grooved cam mechanisms, grooved cam profile was reconstructed with non-uniform rational B-splines (NURBS). Considering joint friction, dynamic model of grooved cam mechanisms was established and the contact stress between grooved cam and follower was calculated using Hertz contact theory. Taking the minimum contact stress and the minimum acceleration as optimal objectives, integrated design model for respective kinematic and dynamic design approaches was set up. The integrated design mode was optimized to search Pareto-optimal solution by an improved artificial fish swarm algorithm, and optimized NURBS profile for grooved cam was acquired. The results show NURBS profile has better kinematic and dynamic performances. The impacts on grooved cam mechanism are reduced and wear characteristics are improved.

1 Introduction

As a kind of cam mechanisms, grooved cam mechanisms are widely used in various automatic machines, such as assembly manipulator, packing and agricultural machinery, etc. Mechanical properties of grooved cam mechanisms are concerned in industrial practices and hence frequently studied in literature [1-3]. The contact stress at the cam-follower interface is larger due to the line-contact higher pair, which leads to contact fatigue wear, influence the working performance, even failure. The lager acceleration also exaggerates the shaking forces and moments, which results in vibrations, poor motion precision, noises and fatigue damage. Recently, there are a series of studies to improve the mechanical properties of cam mechanisms by optimizing the structure parameters and using excellent motion curves such as polynomial curve [4], spline curve [5-6], non-uniform rational B-splines [7-8] and other curves, but all these researches are rarely related to grooved cam mechanisms.

In this paper, non-uniform rational B-splines was adopted to reconstruct the grooved cam profile and dynamic model of grooved cam mechanisms was established. Considering the kinematic and dynamic integration method, multi-objective optimization model of grooved cam mechanisms was set up for reducing the peak value of the contact stress and acceleration. An optimization simulation example was given to compare NURBS profile with traditional cam profile in the end.

2 Grooved cam mechanisms dynamic analysis

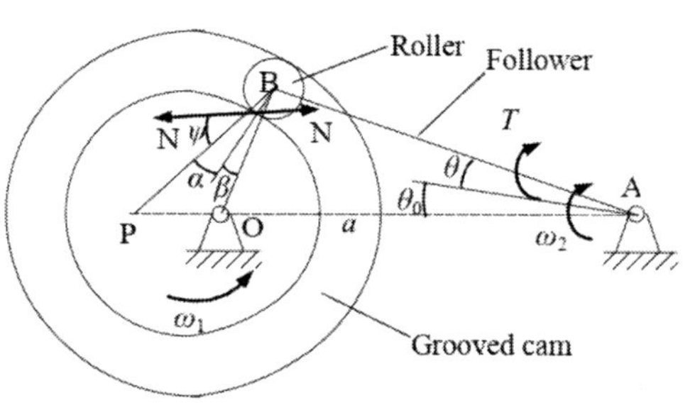

(a) Same direction of rotation (m=-1)

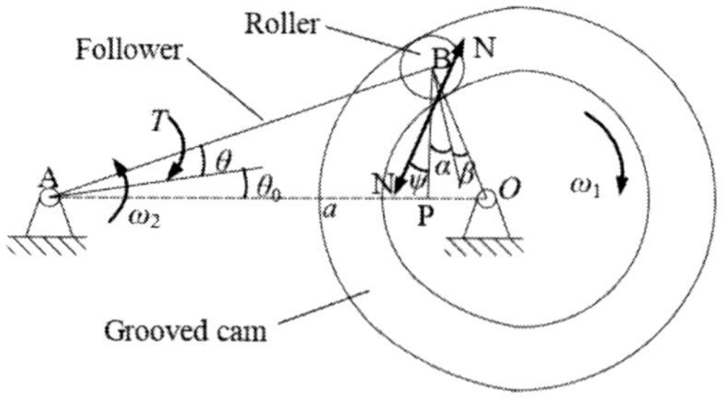

(b) Different direction of rotation (m=1)

Fig. 1. Mechanical model of grooved cam mechanism

* Corresponding author: xoyo111@163.com

Grooved cam mechanisms can be regarded as mechanical model shown in Fig. 1. Where, O, A is the centre of grooved cam and the oscillation centre of the follower respectively, the length of OA is a , the center of the roller is B, length of the follower AB is l , radius of the roller is r_0 , θ and φ is the angular displacement of the oscillating follower and the grooved cam respectively. Groove cam rotational speed ω_1 can be defined as $\omega_1 = d\varphi / dt$, the oscillating follower angular velocity ω_2 be defined as $\omega_2 = d\theta/dt$, P is the instantaneous center of the oscillating follower and the groove cam. Applying Kennedy Aronhold the orem to grooved cam mechanisms, Equation (1) and (2) are deduced as follows.

$$\overline{OP} = \frac{a\omega_2}{\omega_1 + m\omega_2} = \frac{a\dfrac{d\theta}{d\varphi}}{1+m\dfrac{d\theta}{d\varphi}} \tag{1}$$

$$\overline{AP} = \frac{a\omega_1}{\omega_1 + m\omega_2} = \frac{a}{1+m\dfrac{d\theta}{d\varphi}} \tag{2}$$

In Equation (1) and (2), m=-1 when the cam rotates in the same direction with the oscillating follower as shown in Fig. 1 (a); else, m=1 as shown in Fig. 1 (b).

Refer to Fig. 1, let r_b denote the radius of base circle, and $\theta_0 = \arccos\left(\dfrac{a^2+l^2-r_b^{\,2}}{2al}\right)$ is the initial angle. The pressure angle α at the contact point can be derived from Equation (1) and (2).

$$\alpha = \arctan\left(m\frac{a\cos(\theta_0+\theta) - l\left(1+m\dfrac{d\theta}{d\varphi}\right)}{a\sin(\theta_0+\theta)} \right) \tag{3}$$

In Fig. 1, considering frication of rotary pair and motion pair in grooved cam mechanisms, resultant force N can be derived by applying torque balance principle.

$$N = \frac{T}{l\cos(\alpha+\psi) + mr_o\sin\psi + m(\rho_1 - \rho_2)} \tag{4}$$

Where $\psi = \arctan f$, which is angle of friction between the grooved cam and the oscillating follower, T is external torque exerted on the follower, f is the friction coefficient, ρ_1 , ρ_2 represent the radius of friction circle of the roller and the oscillating follower rotating center A respectively.

For grooved cam mechanisms with oscillating follower, there is line contact between the grooved cam and the roller in the theory. But contact surface is actually a tiny rectangular contact region due to the elastic deformation of the material when load is imposed on grooved cam mechanisms. According to Hertz contact theory, the peak value of the contact stress occurs in the mid-line of contact region, contact stress σ_H is defined as follows.

$$\sigma_H = \sqrt{\frac{F_N}{\pi b}\cdot\frac{\dfrac{1}{R_c}+\dfrac{1}{R_f}}{\dfrac{1-\mu_c^{\,2}}{E_c}+\dfrac{1-\mu_f^{\,2}}{E_f}}} \tag{5}$$

In Equation (5), $F_N = N\arccos\psi$ is normal pressure of the grooved cam pair, b is the length of contact line between grooved and roller, μ_c , μ_f are Poisson ratio of material for grooved cam and roller, and E_c , E_f are elastic modulus of material for grooved cam and roller respectively. For common steel grooved cam pair, set $\mu_c = \mu_f = 0.3$, $E_c = E_f = 206\text{GPa}$.

Refer to Fig. 1, radius of curvature of the theoretical profile for grooved cam can be written as follows.

$$R_B = \sqrt{a^2 + l^2 - 2al\cos(\theta_0+\theta)} \tag{6}$$

3 Grooved cam profile reconstruction using non-uniform rational B-splines

Excessive contact stress between grooved cam and oscillating follower can cause unnecessary abrasion of the contact surfaces and distortion of the components. To improve mechanical properties of grooved cam mechanisms, non-uniform rational B-Splines (NURBS) was used to fit motion curves of the oscillating follower.

The NURBS profile of grooved cam is defined as follows.

$$\theta(\varphi) = \sum_{i=0}^{n} d_i R_{i,k}(\varphi) \tag{7}$$

$$R_{i,k}(\varphi) = \frac{w_i B_{i,k}(\varphi)}{\sum_{i=0}^{n} w_i B_{i,k}(\varphi)} \tag{8}$$

Where, d_i is the control point, w_i is the factor, $R_{i,k}(u)$ is k rational B-splines, and $B_{i,k}(u)$ is B-splines rational basis function depending on $\boldsymbol{U} = [u_0, u_1, \ldots, u_{n+k+1}]$.

Refer to Equation (7) and (8), by facilitating sampling points of the profile, control points and weight factors are calculated and adjusted to optimize NURBS profile. Angular velocity curve ($\theta'(\varphi)$), Angular acceleration curve ($\theta''(\varphi)$) and Angular jerk curve ($\theta'''(\varphi)$) are formulated as follows.

$$\theta'(\varphi) = \sum_{i=0}^{n} d_i R'_{i,k}(\varphi) \tag{9}$$

$$\theta''(\varphi) = \sum_{i=0}^{n} d_i R''_{i,k}(\varphi) \tag{10}$$

$$\theta'''(\varphi) = \sum_{i=0}^{n} d_i R'''_{i,k}(\varphi) \tag{11}$$

Where, $\theta'(\varphi)$, $\theta''(\varphi)$, $\theta'''(\varphi)$ represent the 1st-order, the 2nd-order and 3rd-order derivative of rational B-splines respectively. If $3 \le k \le 5$, the acceleration curve is continuous, and rigid and flexible impact does not exist in the theory. Furthermore, maximum angular velocity and maximum angular acceleration are reduced by optimizing cam profile with NURBS curve.

4 Integrated kinematic and dynamic optimal design

The peak value of contact stress $\delta_{H\max}$ and inertial forces caused by the peak value of angular acceleration $\theta''(\varphi)$ have great influence upon mechanical properties of grooved cam mechanisms. Thinking d_i and w_i as design variables, integrated kinematic and dynamic optimization mathematical model of grooved cam NURBS profile is expressed as Equation (12).

$$\min F(X)=[\delta_{H\max}, \theta''_{\max}]^T \quad (12)$$

Grooved cam mechanisms do not lock in operation to meet the following constraints.

$$\alpha_{s\max} < [\alpha]_s = 35° \quad (13)$$

$$\alpha_{h\max} < [\alpha]_h = 70° \quad (14)$$

Where, $[\alpha]_s$ 、 $[\alpha]_h$ are the permissive pressure angle of rise and fall periods respectively, $\alpha_{s\max}$ 、 $\alpha_{h\max}$ are also the maximum pressure angle of rise and fall periods respectively.

In order to avoid interference, the minimum radius of curvature $\rho_{\min}$ is limited to Equation (15).

$$\rho_{\min} - r_0 \geq 3 \quad (15)$$

Moreover, Grooved cam mechanisms are required to meet the boundary conditions shown in Table 1.

Table 1. Dwell-Rise-Dwell-Fall kinematics boundary conditions

Cam rotation angle $\varphi/(°)$	0	150
Follower angular displacement $\theta/(°)$	0	20
Follower angular velocity θ'	0	0
Follower angular acceleration θ''	0	≠0
Follower Angular jerk θ'''	≠0	0

5 Improved artificial fish swarm algorithm

Since artificial fish swarm algorithm [9-10] easily fell into local solutions, improved artificial fish swarm algorithm （IAFSA） was proposed by combining the taboo search strategy and variable parameters optimization effectively as shown in Fig. 2.

As shown in Fig. 3, in order to speed up searching, the first IAFSA optimization was performed in early algorithm by setting parameters P_1: artificial fish perception range *visual*、 the maximum moving distance *step*, congestion factor δ and number of fish N for greater values, feeding frequency *Try_number* for smaller value. Pareto-optimal solution was locked using taboo criteria, and no Pareto-optimal solution domain was tabooed to avoid the second optimization blindness. In the algorithm later, in order to improve the searching precision, the second IAFSA optimization was carried out by setting parameters P_2: *visual, step*, δ and N for smaller values, *Try_number* for larger value. Finally the Pareto-optimal solution was placed in the bulletin board and outputted.

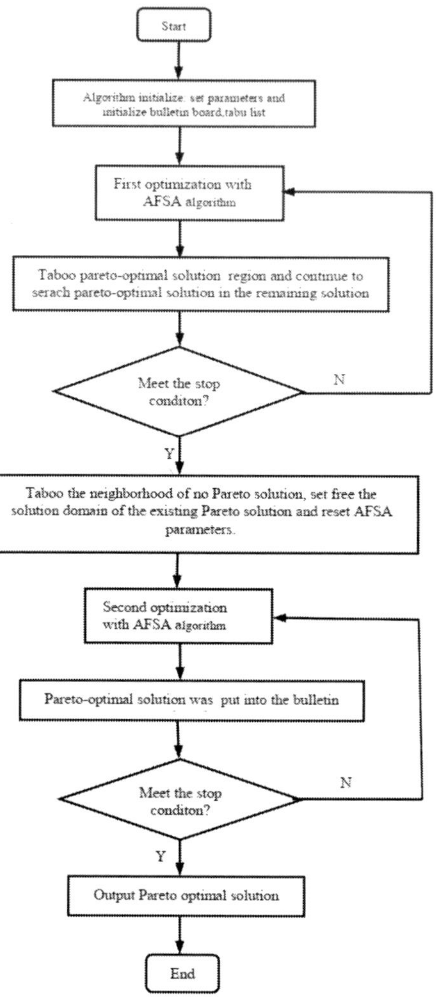

Fig. 2. Flow diagram of Improved Artificial Fish Swarm Algorithm

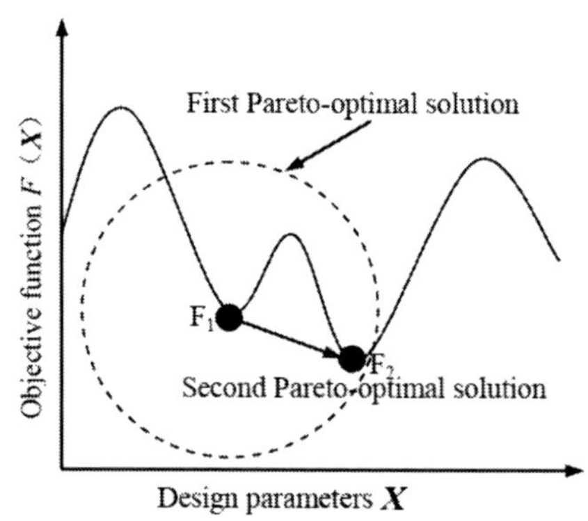

Fig. 3. Optimization twice with variable parameters

6 Design example

The related specifications of the grooved cam mechanism were T =27N•m, r_0 =54mm, l =200mm, b =20mm, a =300mm, r_b =135, f =0.05, ρ_1 =0.5, and ρ_2 =1. Of the follower motion curves, modified sine (MS) curve has low velocity and acceleration, which is suitable for improving mechanical properties of grooved cam mechanism. MS profile was reconstructed using 5-order NURBS curve, $n=20$. IAFSA optimization algorithm parameters were set as follows.

Each artificial fish state $X=(d_0, w_0, d_1, w_1, \ldots d_n, w_n)$, current food concentration $\min F(X)=[\delta_{H\max}, \theta''_{\max}]^T$, taboo table length L=50, N=150, maximum number of iterations Gm=500, P_1 : $visual$=5, $step$=2, δ =3, and Try_number=1, P_2: $visual$=1, $step$=0.5, δ=1 and Try_number=5. One of the Pareto-optimal solutions was obtained as shown in Table 2, and mechanical properties comparison between MS curve and NURBS curve have been shown in Fig. 4-Fig. 8.

Table 2. Mechanical properties of MS and NURBS curve

Mechanical properties	$\theta'_{\max}$	$\theta''_{\max}$	$\theta'''_{\max}$	$\delta_{H\max}$
Single stop MS	0.221	0.0047	0.00039	98.8637
NURBS	0.223	0.0045	0.00021	93.7658

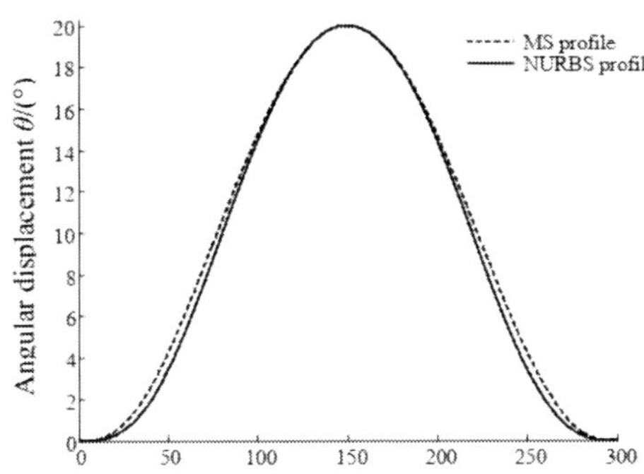

Fig. 4. Angular displacement curve

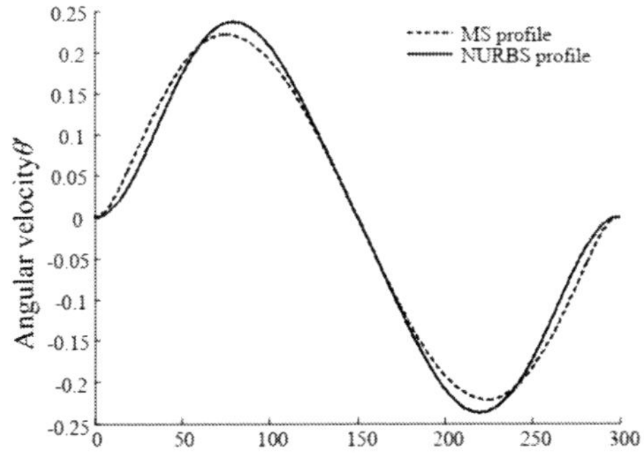

Fig. 5. Angular velocity curve

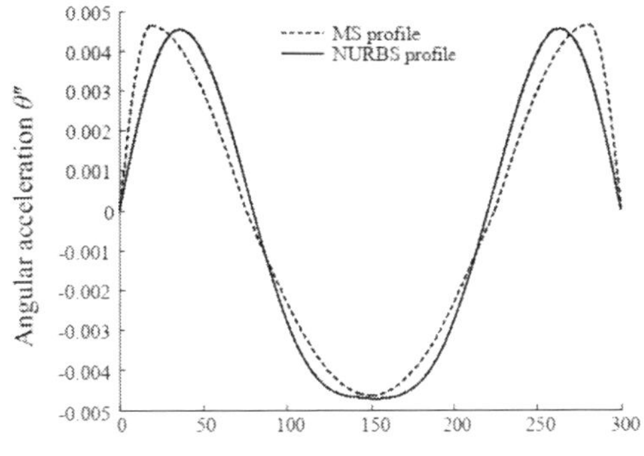

Fig. 6. Angular acceleration curve

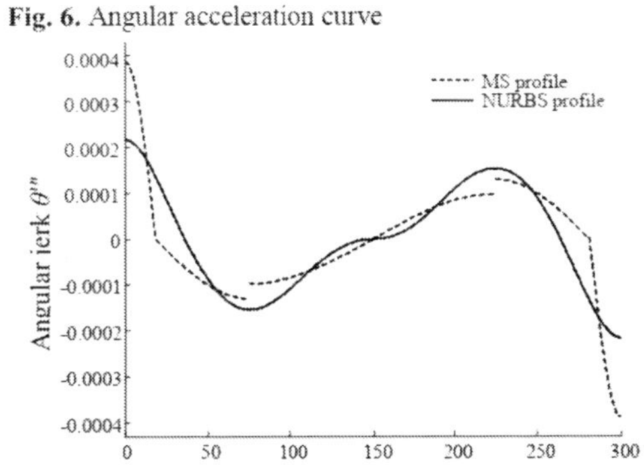

Fig. 7. Angular jerk curve

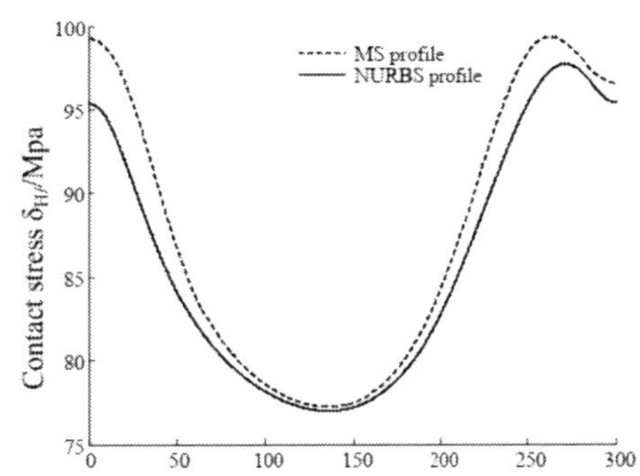

Fig. 8. Contact stress curve

It shows the motion curve of the follower is smoother by applying NURBS curve to optimize MS profile. Especially angular jerk curve is continuous, and the peak value of angular jerk is smaller in comparison with MS profiles. Fig. 6 also illustrates, the peak value of angular acceleration curve is decreased from 0.0047 to 0.0045 by 4.3%. Meanwhile, the contact stress between grooved cam and follower is decreased from 98.8637Mpa to 93.7658Mpa by 5.16% in Fig. 8. For grooved cam mechanisms, kinematic and dynamic characteristics of NURBS profile have been greatly improved compared with MS curve. With NURBS profile applied, shock and wear in grooved cam mechanisms are reduced, which improves the working reliability of grooved cam mechanisms, extends the service life.

7 Conclusions

Dynamic analysis was first carried out to provide the expression for the contact stress in grooved cam mechanisms. By adjusting the node vector, control points and power factors, grooved cam profile was reconstructed with NURBS curve. Integrated optimum design model which decreased the maximum contact stress and angular acceleration was developed sequentially. A design example was demonstrated in the end. The design results indicate that the peak value of angular acceleration and contact stress can be effectively reduced, which improve mechanical properties of grooved cam mechanism to some extent.

Acknowledgements

We would like to thank "National Natural Science Foundation of China (Nos.31671632, 31701325)", "Natural Science Foundation of Shandong Province (BS2015NY012)" and "Key Research and Development Program of Shandong Province (2016GNC112007)" for financial supports.

References

1. Demeulenaere, B. and De Schutter, J., Input torque balancing using an inverted Cam mechanism, ASME transactions, Journal of Mechanical Design, **127** (2003).
2. Demeulenaere, B., Spaepen, P., and De Schutter, J., Input torque balancing using a Cam-based Centrifugal Pendulum:Design procedure and Example, Journal of Sound and Vibration **283**, 1-2 (2005).
3. Nayak Nagaraj, Lakshminarayanan P.A., Babu M.K. Predictions of cam follower wear in diesel engines, Wear, 2006.
4. Chen F. Y., Mechanics and Design of Cam Mechanisms, Pergamon Press, New York, 1982.
5. Tsay D. M., Huey C. O., Cam motion synthesis using spline function", Trans. ASME, J. of Mech., Trans., Auto. in Des. **llO**, 2 (1988).
6. Yoon K., Rao S. S., Cam motion synthesis using cubie spline, Trans. ASME, J. of Mech. Des. **115**(3), 3 (1993).
7. Guantao Xuan, Yuanyuan Shao, Zhaoqin Lü, Profile Reconstruction of Spatial Indexing Cam and Simulation Based on NURBS, Transactions of the Chinese Society for Agricultural Machinery **43**, 6 (2012).
8. Rongyu Ge, Shiwei Li, Li Liu, Flexible cam curve expressed by NURBS and mul-tiobjective Optimization based on genetic algorithm, Transactions of the Chinese Society for Agricultural Machinery **39**, 2 (2008).
9. Na Fang, Jianzhong Zhou, Rui Zhang, A hybrid of real coded genetic algorithm and artificial fish swarm algorithm for short-term optimal hydrothermal scheduling, International Journal of Electrical Power & Energy Systems, **62**(2014).
10. X. Li, Z. Shao, J. Qian, An optimizing method based on autonomous animats: fish-swarm algorithm, Syst Eng Theory Pract **22**, 11(2002).

Study on Small Component Thermal Ignition Compliance meeting IEC Standard

Qin Xue

Xi'an Research Institute of China Coal Technology Engineering Group Corp, Xi'an, China

Abstract. With the supplement and revision of IEC standard, the thermal ignition test of small components has been stipulated. Small component ignition test apparatus is testing equipment of intrinsically safe electrical products about thermal ignition. Its research and application is an important link of energy storage components for energy control and temperature control. This paper starts from the analysis of restraining measures of capacitor and inductance energy storage, its system composition, main technical requirements and working principle of small component ignition test apparatus are systematically introduced. Then, the influence of the surface area of heat source in the calibration circuit on the minimum ignition temperature of the mixture of methane and air is demonstrated. From the running situation and the experimental data, the apparatus has the ability to carry out various small components ignition inspection.

1 General Instructions

With continuous improvement of the level of industrial automation, explosive areas of intrinsic safety electrical equipment is widely used in coal and petroleum chemical industry. The method of intrinsically safe equipment design and inspection is to limit the electric spark and heat energy of the two light source. In order to evaluate intrinsically safe performance of the electrical circuit, the intrinsically safe equipment must be carried out ignited evaluation and test validation in an inspection institution with relevant qualifications. After passing the inspection, it can be used in a potentially explosive environment [1]. So far the evaluation method and apparatus of spark ignition have become mature, however, there is little mention of the thermal effect ignition evaluation methods and test apparatuss.

In recent years, with the supplement and revision of IEC standard, the ignition test of small components has been stipulated [2]. The thermal effect is controlled by surface temperature of the component, it can exceed the standard surface temperature for small components. But it is necessary to test the resistance, potentiometer and other components of the electrical equipment in the explosive gas environment to ignite the flammable gas mixture by the small component ignition test apparatus.

Due to the specific description of IEC standard only provides the technical requirements without testing apparatus, the development of small component ignition test apparatus can fill detecting ability of detection institutions.

2 Thermal effect control of small component

The design of equipment protection by intrinsic safety "*i*" must start from the limit of energy, through reliably limit voltage and current in circuit within the permissible range. In order to ensure that the equipment in normal operation or the damage of small components and other circumstances, the thermal effect cannot cause the dangerous gas explosion around it. Therefore, in order to control the maximum surface temperature, the following measures can be adopted to reduce the thermal effect of small components and wires, so as to ensure that the explosive gas can avoid explosion caused by the exposure to the thermal surface higher than its natural temperature.

Selection of small components. In the design of intrinsically safe circuit, maximum power of components must be calculated in the most unfavourable working conditions or fault condition. At the same time, the value is not greater than the 2/3 of the selected component rating [3]. The surface temperature of the component should be measured to determine whether allowable temperature requirement is met.

Encapsulation of small components. If some small components temperature is too high to meet the

* Corresponding author: xueqin@cctegxian.com

temperature requirement of their group, the heat dissipation measures can be used to meet the temperature requirements. For example, the method of enlarging the encapsulation area and metal heat sink can be adopted. The volume and thickness of sealant shall reduce the maximum surface temperature of sealed component to the prescribed temperature. At the same time, the rated continuous operating temperature of sealant shall not be less than maximum temperature of sealed component.

3 Thermal ignition assessment of small component

Methods of circuit analysis and theoretical calculation are usually used in thermal ignition assessment. According to IEC 60079-0 standard requirements, the calculation method of thermal resistance constant and linear interpolation method are presented in this paper. It determines temperature group and power dissipation of components by means of total surface area greater than 20mm^2. Otherwise, the applicable assessment method can be judged according to whether thermal ignition is limited by encapsulation.

3.1 Thermal resistance constant method

According to definition of thermal resistance constant, hindrance ability of heat conduction is proportional to the length of conduction path, and inversely proportional to the thermal conductivity of material. Therefore, it is possible to derive the formula for calculating component surface temperature under the fault condition by using the thermal resistance constant.

The surface temperature (T) is equal to product of component dissipation power under normal operation and prescribed fault condition (P) and thermal resistance constant (R), additionally the maximum environmental temperature (T_a) of the whole machine is added [5].

Some small components are selected to organize test according to the derived formula, The experimental conditions is controlled (ambient temperature is 24℃). By using thermal resistance constant, maximum surface temperature of small component in fault condition is obtained, as shown in Table 1.

Table 1. Thermal resistance constant method record table.

Component Name	Test power /W	Thermal resistance constant /℃W^{-1}	surface temperature /℃
Resistance	0.3513	39.98	54.1
Light-emitting diode	0.2119	89.24	58.9
diode	0.2148	56.13	52.1
Field-effect transistor	0.3711	16.62	46.2

3.2 Linear interpolation method

The resistance constant method is not applicable to judge surface temperature of embedded small component. In design process, it considering the number of electronic components including wire wound resistor, the surface temperature rise and power dissipation can be equivalent to a linear relationship [6], so you can find a way to take value without knowing thermal resistance constant, only judging component temperature rise by experiment.

Suppose that Test power P_1 and P_2 are component power in two working states, and component surface temperature can be measured by thermocouple, T_1 and T_2 are component surface temperature corresponding to above power.

The linear interpolation method can be used to obtain component surface temperature under fault condition, and encapsulation compound surface temperature can also be obtained directly by this method. The maximum surface temperature of small component is obtained by linear interpolation method in fault condition, as shown in Table 2.

Table 2. linear interpolation method record table.

Component Name	Test power (P_1) /W	Test power (P_2) /W	Temperature (T_1) /℃	Temperature (T_2) /℃	surface temperature /℃
Resistance	0.1248	0.1006	29	26	54.5
Field-effect transistor	0.3711	0.2504	29	26	45.8
diode	0.1515	0.0939	32	30	52.1
Three terminal regulator	0.5012	0.3502	35	29	48.1

From the data of Table 1 and table 2, it can be seen that obtaining the surface temperatures values by two methods are in agreement. Therefore, it can determine surface temperature of small components in the fault condition by using thermal resistance constant calculation method and linear interpolation test method.

4 Thermal ignition test apparatus

For intrinsically safe equipment, the inspection process is authenticated by first evaluation and then validated. Small component is evaluated by thermal ignition, and when the surface temperature exceeds the corresponding temperature group, the small component ignition test apparatus is required to be tested and verified, as long as no ignition is considered.

4.1 Apparatus components

Thermal ignition test apparatus of small component is mainly composed of four parts, including ignition test flume is placed in constant temperature test chamber, gas mixing system with gas distribution facilities and display panel, control circuit unit and calibration system unit with operation and display monitoring, as shown in figure 1.

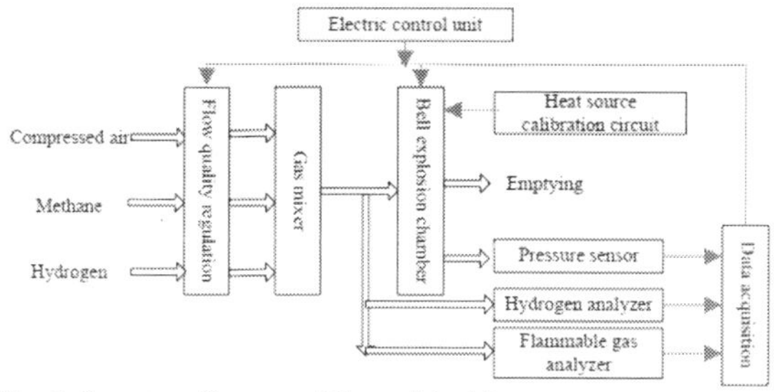

Fig. 1. Structure diagram of thermal ignition test apparatus.

Explosion chamber. It having a volume of at least 250 cm^3, it can withstand blast pressure of 1.5MPa. For group I apparatus, the test gas should be a mixture of methane and air with a methane concentration of 6.2% - 6.8%. There are 1 pairs of small element ignition test electrodes, 1 pairs of heating calibration electrodes and the electrode calibration temperature of 1 pairs of thermal resistance sensing electrodes in the chamber. The 3 pairs of electrodes are passed through the outer terminals of explosion-proof connectors to their respective circuits. In the test, the test component is connected to the small component ignition test electrode in the test chamber.

Gas mixing system with gas distribution apparatus and display panel. A gas distribution unit is unit gas test filter pressure after entering synchronous voltage stabilizing system, we need to control various gas flow by adjusting flow regulating valve, mix gas through flow control into mixing chamber and achieve desired concentration of mixed gas configuration [4].

Control circuit with operation and display monitoring. The control part is provided by the microcomputer to assemble circuit into traditional computer, and interface circuit establishes the interaction between control unit and electromagnetic valve, switching relay, photoelectric sensor and calibration circuit of heat source. The concentration detection of mixed gas is carried out at the outlet of gas mixture of apparatus, and the gas analyser is connected to the reserved gas pipeline to test whether the concentration ratio meets the test requirements.

Heat source calibration system. The apparatus uses Ni-Cr alloy resistance as heating element to produce high temperature in the self-correcting circuit. After the temperature reaches a certain value, the gas is ignited. Before and after of each series test, sensitivity of test apparatus are to check is qualified under the specified conditions in calibration circuit.

4.2 Working principle of apparatus

The basic working principle of apparatus is to vacuum and re-inject explosive gas mixture, then the energized test of test component. Components should be installed in electrical equipment according to the specified conditions in the test and take effective measures to ensure temperature of test mixture and flow of mixture around the components.

When the test component does not ignite or detonate in the specified test time, the calibration timing test is immediately transferred to the calibration component. It is proved that the test of small components is qualified after ignition or detonation occurs within the prescribed calibration time. Otherwise it is proved to be unqualified. When the test failed, PLC immediately control the corresponding relay to cut off two electrodes in the test circuit and open exhaust valve, and timely cleaning test chamber, clearly and completely display test conditions and test results on the display. The specific block diagram of the small component ignition test apparatus is shown in figure 2.

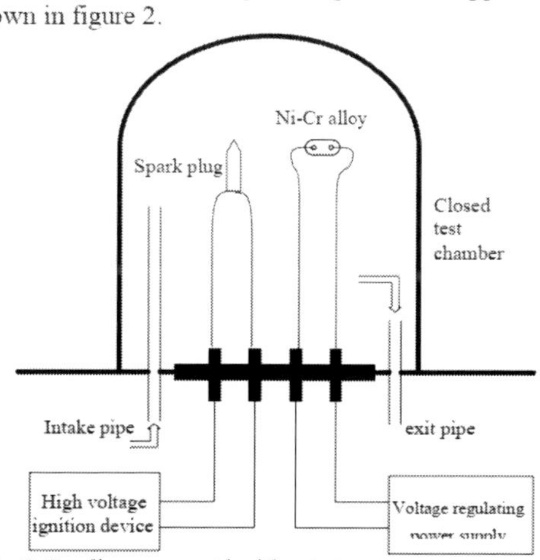

Fig. 2. Small component ignition test apparatus.

4.3 Test verification

The IEC standard stipulates that test should be carried out on small components when the surface temperature reaches the maximum value under the specified fault condition in special explosion-proof type. Safety margin should be obtained by improving the test environment temperature. Group I apparatus work in coal mine under the environment temperature of -20 to +40℃, the safety margin is 25℃. According to requirements, the highest surface temperature should be 150 degrees, and the small components will be placed in the thermostat in the explosion container during testing, so that the small components work at the highest temperature.

The sensitivity of the apparatus is the main judgment basis for the test verification. The sensitivity of the heat source calibration circuit is related to heating-up time of Ni-Cr alloy and ignition delay. Taking into account the criticality of ignition temperature and combustion, the self-calibration time is five minutes. The thermocouple end is placed in the small component test environment, and the other end is connected to the temperature recorder. The explosive mixture was ignited, and the sensitivity of the device was qualified. The verification record of the heat source calibration circuit is shown in table 3.

Table 3. Small components test verification record.

Component	Gas mixture /%	Ignition delay /s	Ignition temperature /℃
Resistance	6.3%	32	690
diode	6.7%	31	691
Field-effect transistor	6.5%	28	697

5 Conclusion

Through thermal effect control and evaluation method to determine small component temperature group, when the temperature group is not suitable to determine its safety, the small group ignition test apparatus is used for the test verification. In order to determine test results accurately, it uses the naked eye and thermocouple to determine the method together. Through a large number of small components testing, it shows that the apparatus meets the requirements of the stipulated inspection items by IEC standard. The main technical indicators have achieved satisfactory test results.

References

1. Gang Zhang, Jianping Xu, Shuling Fu, et al. *GB 3836.4-2010 explosive gas environment*. Beijing: China Standard Press. 2011.
2. International Electrotechnical Commission. *IEC60079-0 Explosive atmospheres-Part 0: Equipment – General requirements* [M]. International Standard, Edition 6.0. (2011)
3. International Electrotechnical Commission. *IEC60079-11 Explosive atmospheres-Part11: Equipment protection by intrinsic safety "i"* [M]. International Standard, Edition 6.0. (2011)
4. Chaolong Yu, Qing Liu. *Development on Tri-gas dis-tribution device for intrinsically safe apparatus*. Electric Explosion Protection (3): **40-43**(2008)
5. Yu Zheng, Zhengyong Zhang. *Application of constant temperature detection method in detection of mixed gas*[J]. Industrial automation (7) : **57-61** (2010)
6. Xiaonian Zhang. *Analysis of the relationship between surface temperature rise and power of wirewound resistors*[J]. China electronic business (3) :**70-72**.2012

A time-encoded approach to detect cooperative targets for UAV landing

Wang Zi[1]⋆, *Zhang* HongLiang[1], and *Shang* Yang[1]

[1] *College of Aerospace Science and Engineering, National University of Defense Technology, Changsha 410082, PR China*

Abstract. To detect visible light cooperative targets during UAV landing, we use a time relay to encode the intensity of lamps. The change pattern of the lamps intensity is presented in this paper. The frames with strong and weak lamps light are called strong frames weak frames respectively. In continuous sequence images, we take weak frames as background and subtract background using strong fames. The geometric transformations are considered when subtracting background and a float threshold value are used to threshold the difference images. We also design several geometric constraints to eliminate interference points. The approaches to find the first weak frames and to track lamps in a sequence images are also presented. The experiments show that our method is robust to scale change, interference points and moving object in background.

Keywords. UAV landing, cooperative targets, detection.

1 Introduction

The Unmanned Aerial Vehicle (UAV) takes on many military and civilian missions, such as early warning, reconnaissance, surveillance, communication, relay, target indication and damage assessment, and makes it an important tool in modern naval warfare [1]. Because of the complex landing environment and the limited space, UAV landing is one of the most dangerous phases in a flight mission. To solve this problem, several schemes has been proposed, such as net-catch landing, parachute landing, GPS-based landing, IMU-based landing and manual remote control landing [3–5]. Among these proposals, GPS navigation are likely to be disabled in wartime, and it is difficult for other proposals to guarantee security, accuracy, speed and autonomy.

For the last decades, computer vision has been widely used in UAV navigation and landing, due to the development of videometrics and image processing. Vision navigation gets the image of the ideal landing site through airborne camera, detect and recognize the cooperative targets, and obtain the relative position and attitude between UAV and the World Coordinate through coordinate transformation. The accuracy and precision of detection of the cooperative targets are very important in vision-based UAV landing. Although a number of cooperative target schemes have been proposed, infrared lamps are the most popular cooperative targets for fixed wing UAV landing. Gui et al.[2] presented a UAV landing system which consisted of four infrared lamps and a airborne camera with a 940 nm optical filter, and detected the centers of infrared lamps using NLOG operator. In this method, The measurement range of distance between the UAV and the ideal landing point was about 450 m – 5 m. Hao et al [6] proposed a similar scheme, which consisted a five infrared lamps and a 850 nm optical filter. In his method, the lamp can be detected as far as 80 m. Using infrared lamps and optical filter, cooperative targets detection is robust to illusion, viewpoint, distance and background changing. Actually, an IR filter have to be installed in front of the airborne camera, resulting that the camera is only sensitive to the rays with special wavelength. For most UAVs, it is very hard to add a new camera. So, how to use visible light lamp as cooperate target and how to detect it are valuable engineering problems.

In airborne vision-based landing, visible light lamp detection is different from moving object detection and salient object detection. In moving object detection [7], the background is salient still or dynamic while the object has special motion mode; in salient object detection [8], the object is non-cooperative. Compared with the above two tasks, the cooperative visible lamp detection has three characteristics:

- The lamps are stationary relative to the background.
- The camera is moving, and the scale becomes larger as the UAV approaching the ideal landing site.
- The spatial position of the lamps is fixed, but the temporal intensity can be adjusted.

In this paper, we present a time-encoded visible lamp detection approach in UAV landing. In our method, the intensity of all lamps in continuous video frames changes at a fixed frequency. The video frames with higher intensity of lamp are called the strong frame, while those with lower intensity of lamp are called the weak frame. Because the frequencies of lamps intensity and camera video are fixed, the strong frames and the weak frames

⋆e-mail: wangzitju@163.com

appear at fixed frequency together. We get background from the weak frames, and remove background in strong frames using difference image. Consider the disturbance points and scale changing in the video during UAV landing, we design several geometric constraints and compensation method. Finally, the effectiveness of the approach is verified by experiments.

2 Structure of Proposed Approach

The equipments used in our approach mainly consist of a camera, four visible light lamps and a digital time relay. The parameters of lamps are shown in Table 1. It should be noted that the visible distance is determined by camera's focal length and aperture. The real lamp is shown in Figure 1. The maximum frequency of time relay is 5 Hz, and the minimum on-off step is 0.1s. The camera we used is Sony FCB-EX2700P, its main parameters are shown in Table 2. The four lamps are installed on an iron shelf, which is shown in Figure 2. The shelf has three arms, which are not in the same plane. The length of the arm towards up is 2.1 m, and the other two are 2.1 m and 3.1 m respectively.

Table 1. The parameters of lamp

power	diameter	emitting angle	visible distance
50 watts	20 cm	30°	300 m

Table 2. The parameters of camera

focal length	aperture	frame rate	video size
3.06 – 122.4 mm	1.6 – 4.6	25 Hz	288 × 352

We use a time relay to blink four lamps synchronously at 5 Hz. The frame rate of the video is 25 Hz. Thus the weak frame should appear every 5 frames. We note **F** as the set of a period of continuous video frame, I_i $(1 \le i \le M)$ as the i_{th} frame, and I_{i_0} as the first weak frame in **F**. It should be note that $1 \le i_0 \le 5$ and all $I_{i_0+n\times 5}$ $(n \in N)$ are weak frames.

Figure 1. One visible light lamp.

Figure 2. Four visible light lamps with the iron shelf.

The basic idea in this paper is getting background from the weak frames and subtracting the background in the following strong frames to get the position of the lamps. So the first assignment is to find i_0, in other words, the first weak frame. We assume that the background in adjacent frames are similar. The assumption is tenable, because the camera image the same scene during UAV landing. But the camera is moving, it is impossible that adjacent frames have the same background. We further clarify the assumption that there are translate and scale transformation of background between adjacent frames. The reasons why we not use affine and projective transformation are

- affine and projective transformation have too many parameters;
- it is difficult to get a lot of stable feature points during UAV landing, because of the spurious feature points of ideal landing site;
- the dissimilarity between adjacent frames is relative small as long as UAV's attitude doesn't change drastically.

The second assignment is the estimation of the transformation from weak frame I_{i_0+5n} to the following four strong frame I_{i_0+5n+j} $(j = 1, 2, 3, 4)$ respectively. And the transformation matrix is noted as $\mathbf{S}_n^j$, where $n = 0, 1, 2, 3\ldots$ and $j = 1, 2, 3, 4$. Then the transformed weak frames I'_{i_0+5n} can be obtained using geometric transformation:

$$I'_{i_0+5n} = \mathbf{T}(I_{i_0+5n}, \mathbf{S}_n^j) \tag{1}$$

where $\mathbf{T}(\cdot)$ is geometric transformation. And the background can be removed from the strong frames using difference image:

$$D_{i_0+5n+j} = I_{i_0+5n+j} - I'_{i_0+5n} \quad (j = 1, 2, 3, 4) \tag{2}$$

where D_i is the difference of I_i and the I'_{i-1}. Then we threshold the difference image, extract 8-connect regions and calculate their centers. The binary image of D_i is noted as B_i. We take all the centers as the candidates of the position of center. Finally, several geometric constraints of the relative position relation between lamps are presented, to

DOI: 10.1051/matecconf/201713900049

determine lamp position from the candidates. We adopt a floating threshold when thresholding the difference image:

- if there are too many candidates, we increase the increase threshold value.
- if there is no combination of four centers satisfying the constraints, we decrease the threshold value.

The work flow of the proposed method is shown in Figure 3.

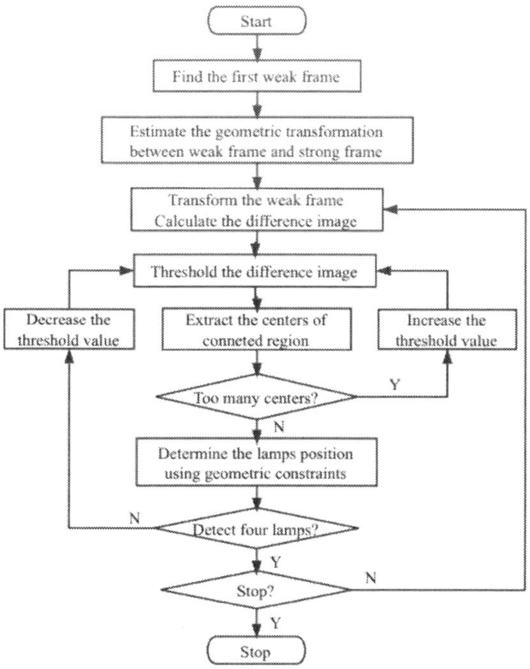

Figure 3. The work flow chart of the visible light lamp detection.

3 Principle and technical implementation

Before mentioned, several assignments have to be tackled to implement the proposed method. In this chapter, we present the approaches for these problems.

3.1 Estimation of transformation

We take each weak frame as the background of the next four strong frames. Because of the motion of the camera, the transformation relationships between weak and strong frames need to be estimated. During UAV landing, only the image of the scene around the ideal landing site or the lamps are stable, due to dynamic background. Naturally, we should only use the image of the ideal landing site or scene around the lamps to estimate the transformation.

For a long time, SIFT+BFMacth+RANSAC has been the most used method for estimation of geometric transformation between image pairs, which is a general approach definitely. Recently, a new method for feature matching in video application, Grid-based Motion Statistics(GMS)[9], has been proposed. GMS uses ORB features and provides a real-time, ultra-robust correspondence system. We use ORB+BFMatch+GMS to detect and match the feature points in weak and strong frames. We note the set of matched feature points in weak and strong frames as $\{p_i = (x_i, y_i, 1)^T\}$ and $\{p'_i = (x'_i, y'_i, 1)^T\}$ respectively. The transform matrix from weak frame to strong frame is given as:

$$\mathbf{S} = \begin{bmatrix} s & 0 & d_x \\ 0 & s & d_y \end{bmatrix} \tag{3}$$

where s is scale factor while d_x and d_y are translation factor. So, we get the equation:

$$p'_i = \mathbf{S}p_i \tag{4}$$

for N pairs matched points, we have $2N$ equations actually:

$$\begin{bmatrix} x'_1 \\ y'_1 \\ x'_2 \\ y'_2 \\ \vdots \\ x'_N \\ y'_N \end{bmatrix} = \begin{bmatrix} x_1 & 1 & 0 \\ y_1 & 0 & 1 \\ x_2 & 1 & 0 \\ y_2 & 0 & 1 \\ \vdots & \vdots & \vdots \\ x_N & 1 & 0 \\ y_N & 0 & 1 \end{bmatrix} \begin{bmatrix} s \\ d_x \\ d_y \end{bmatrix} \tag{5}$$

the equation (5) can be easily solved by Least Square method.

3.2 Geometric constraints

If we find N connected region centers and $N \geq 4$, C_N^4 candidates of the position of four lamps need to be examined. We design several geometric constraints according to the relative position of lamps. Figure 4 is the binary image of the difference image.

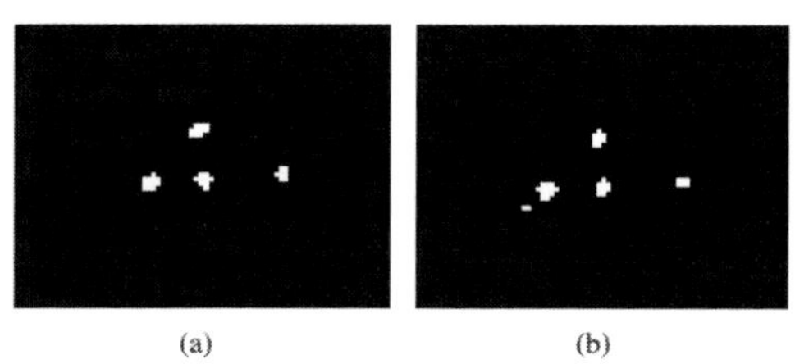

Figure 4. The binary image of difference image. (a): without interference point; (b): with interference point.

To examine four points in a candidates, the first step is numbering them as Table 3. To illustrate our approach effectively, we note the position of the coordinates of the numbering centers as (x_U, y_U), (x_L, y_L), (x_M, y_M) and (x_R, y_R) respectively. The second step is designing several relative position constraints:

1. $x_L \leq x_U \leq x_R$
2. $1/th_1 \leq Dis(U, M)/Dis(L, M) \leq th_1$

3. $\left|(U-M)^T(L-M)/\left(Dis(U,M)\cdot Dis(L,M)\right)\right| \le th_2$

4. $(L-M)^T(R-M)/\left(Dis(L,M)\cdot Dis(R,M)\right) \le th_3$

5. $th_4 \le Dis(R,M)/Dis(L,M) \le th_5$

The $Dis(\cdot)$ above is L_2 distance. The constraint 1 means that x coordinate the top lamp should be smaller than the right one's and be bigger than the left one's; the constraint 2 and 3 mean that the three centers on the left are similar to an equilateral triangle; the constraints 4 means that the three centers on the bottom are in a line; the constraint 5 means that the ratio the distances from the left one and the right one to the middle one are approximately equal to 1.

Table 3. The approach for numbering centers

number	position
U	the center with the smallest y
L	the center with the smallest x expect U
R	the center with the biggest x expect U and L
M	the remaining one

3.3 Track centers of lamps in sequence images

If four centers coordinates satisfy the constraints above, we use a similar method in [2] to track lamps in sequence images. But the different thing is we only detect lamps in strong frames. As four centers in a strong frame satisfy the constraints above, the maximum distance D_{xmax} and D_{ymax} in x and y axis among four centers and their average coordinates is computed. The average coordinates are computed as:

$$\begin{cases} \bar{x} = (x_U + x_L + x_M + x_R)/4 \\ \bar{y} = (y_U + y_L + y_M + y_R)/4 \end{cases} \tag{6}$$

In the next strong frame, the searching is done in the region which takes $(\bar{x},\bar{y})$ as the center and $3D_{xmax}$ and $3D_{ymax}$ as width and height respectively. However, it is difficult to detect and match feature points in Section3.1 when the searching area is very small. So we choose a threshold for the width and height of searching region. We donate the width and height as W and H respectively, then they are given as:

$$W = \begin{cases} 3D_{xmax}, & \text{if } 3D_{xmax} \ge 80 \\ 80, & \text{otherwise.} \end{cases} \tag{7}$$

$$H = \begin{cases} 3D_{ymax}, & \text{if } 3D_{ymax} \ge 60 \\ 60, & \text{otherwise.} \end{cases} \tag{8}$$

Because of the shake of the UAV, the scene of the lamps is likely to be out of the camera during UAV landing. In this case, if no centers satisfy the constraints in continuous four strong frames, the searching region are set to the whole image otherwise the searching region is not updated.

3.4 Further details

We stated earlier that the first assignment is find the first weak frame. Because two adjacent of a weak frame are strong frames, there should be four centers satisfy the constraints in the binary images of the difference images of a weak frame and the previous frame and the next frame. In other words, if I_i is a weak frame, I_{i-1} and I_{i+1} are strong frames. Therefore, there must four centers satisfy the constraints above in binary image B_i and B_{i+1}. So our approach to detect the first weak is find i_0 so that B_{i_0} and B_{i_0+1} contain four centers satisfy the constraints and the searching region is the whole image.

After thresholding the difference image, some morphological operations are taken to improve the robustness of our approach. The first operation is eliminating connected region whose area is smaller than 4. Then morphological dilation is taken to connect disconnected regions caused by subtracting images. Finally, the connected regions are counted and each center is computed as:

$$p_{center} = \frac{1}{Area}\sum_{p_i \in region} p_i \tag{9}$$

Once i_0 is obtained, the positions of four lamps are also obtained and the searching region for all the following strong frames can be reduced as described in Section 3.3.

4 Experimental results and discussion

To prove the validity of the method, experiments were carried out. The videos were taken by the equipments mentioned in Section 2. We did not optimize any component of our final method. The program was written by OpenCV 2.4.13 using C++ and run on Intel 7700k at 4.2 Hz with 16 GB RAM. The distance from lamps to the camera is about 200 m, measured by a total station.The parameters used in Section 3.2 are shown in Table 4, which is determined by experiments.

Table 4. The parameters of geometric constraints

th_1	th_2	th_3	th_4	th_5
1.5	0.2	-0.95	1.0	2.0

The result of our approach is shown in Figure 5. The original image size is 288 × 352 and the result images in Figure 5 and 6 are cropped in order to demonstrate our method effectively. The intensity of the lamp changes with a fixed mode, which can be seen visually. The 217th and 222th frame are weak frames while the other four among them are strong frames. The scale change can be seen visually, caused by the change of focal length, as the person and the shrub become bigger. To illustrate the robust of our method, the difference images and binary images are also shown. Unfortunately, the detection of the right lamp was failure in 219th frame, because the interference point and the other three lamps in the the binary image satisfy the constraints in Section 3.2. However, this failure only effect the next searching region's position and

 DOI: 10.1051/matecconf/201713900049

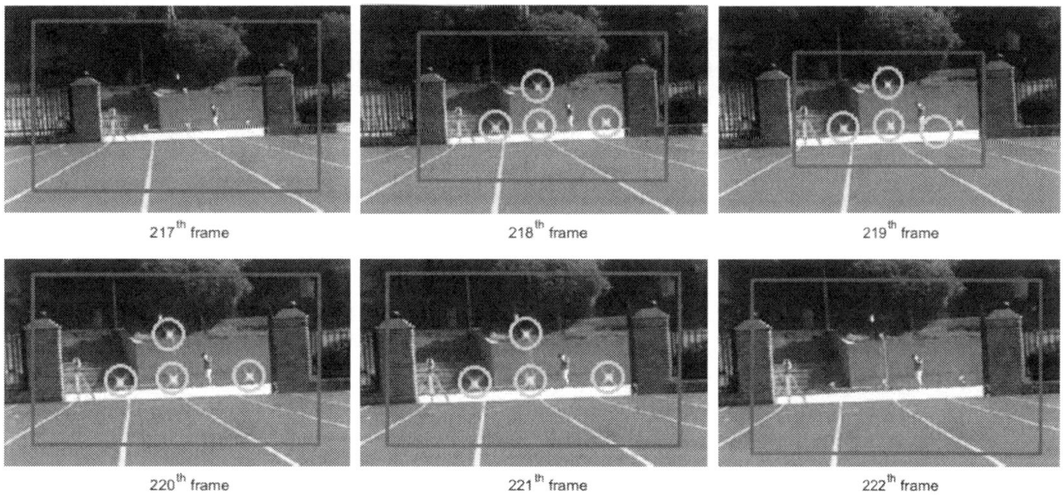

(a) the detection result(the images are cropped)

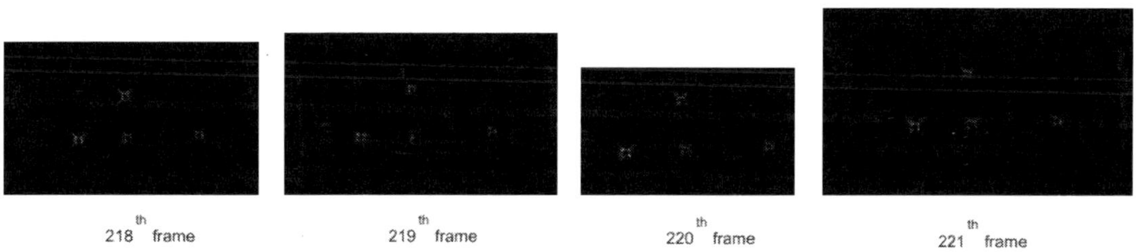

(b) the difference image between strong frames and the weak frame

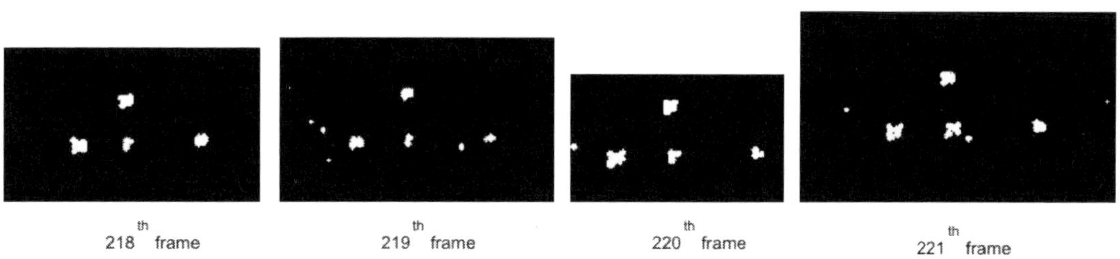

(c) the binary images of difference images

Figure 5. The detection of lamps in scale changed sequence images.

Figure 6. The detection of lamps in sequence images with a white car going through searching regions(the images are cropped).

size. And the success of 220th and 221th frame show that our method is robust to such failure. In 220th and 221th frame, the binary images also show the effectiveness of the geometric constraints in Section 3.2. In Figure 6, a white car went through four searching regions, but the lamps were detected successfully, which show the method is robust to such interference.The computation time of one frame ranges from 40 ms to 100 ms, determined by the size of searching region. The average computation time is about 70ms for one frame, which satisfy the processing frequency of the airborne control system proposed in [2].

5 Conclusion

To utilize visible light lamps as the cooperative targets for UAV landing, we use a time delay to change the intensity of lamps temporally and analysis the pattern of change. In our experiments, there is an image in which the intensity of lamps is weak among five continuous sequence images, while the intensity of lamps in the other four image is strong. The former type frames are called weak frames and the latter ones are called strong frames. We take weak frames as background and subtract background from strong frames. The geometric transformations between weak and strong frames are considered and a float threshold value strategy is used. In order to find lamps, we design a approach to eliminate interference points. The method for tracking lamps and finding the first weak frame in sequence images are presented in this paper.

The experiments show that the proposed method is robust to moving object in background, scale change and interference points. And the computation time is acceptable. In conclusion, our method is suitable for detecting visible light cooperative targets during UAV landing. The further work can be done on the location accuracy of lamps and the constraints to eliminate interference points.

Acknowledgement: This work was supported by Scientific Research Program of National University of Defense Technology(No.ZK16-03-27).

References

1. Kumar V, Yong H, Min D, et al. Auto landing control for small scale unmanned helicopter with flight gear and HILS. International Conference on Computer Sciences and Convergence Information Technology. IEEE, 2011:676-681.
2. Gui Y, Guo P, Zhang H, et al. Airborne Vision-Based Navigation Method for UAV Accuracy Landing Using Infrared Lamps. Journal of Intelligent & Robotic Systems, 2013, 72(2):197-218.
3. Coutard L, Chaumette F. Visual detection and 3D model-based tracking for landing on an aircraft carrier. IEEE International Conference on Robotics and Automation. IEEE, 2011:1746-1751.
4. Zhang Y, Zhao S S, Wang Y S, et al. Study on Guidance Law of UAV Net Recovery System. Journal of China Ordnance, 2007, 3(1):64-67.
5. Guo L, Zhang H Y, Tong M B. Dynamic analysis on parachute recovery of aerial vehicle. Journal of Najing University of Aeronautics & Astrounautics, 2012, 44(1):14-19.
6. Hao S, Cheng Y, et al. Robust detection and matching algorithm oriented to near infrared cooperative objects. System Engineering and Electronics, 2014, 36(9):1854-1859.
7. Shantaiya S, Verma K, Mehta K. A Survey on Approaches of Object Detection.pdf. IEEE Electron Device Letters, 2013, 28(5):419-421.
8. Borji A, Cheng M M, Jiang H, et al. Salient Object Detection: A Survey. Eprint Arxiv, 2014, 16(7):3118.
9. Bian J W, Lin W Y, Matsushita Y, et al. Gms: Grid-based motion statistics for fast, ultra-robust feature correspondence. IEEE Conference on Computer Vision and Pattern Recognition. 2017.

Identification of Blueberry Beverage Using Vis/NIR Spectroscopy

Dongze Li, *Guifang* Wu*, *Hai* Ma, *Zhao* Liu, *Guiquan* Liu, and *Junhua* Hou

College of Mechanical and Electrical Engineering, Inner Mongolia Agricultural University, Hohhot, China

Abstract. Four kinds of blueberry beverage from different varieties, a total of 140 samples were acquired and analyzed by applying of spectrum technology. Using Savitzky-Golay spectral smoothing and multiplicative scatter correction (MSC) on the sample data pretreatment, four varieties of blueberry beverage were cluster analyzed by using principal component analysis method (PCA),a three-dimensional score view was achieved by the first 3 principal components of all samples (PC1, PC2 and PC3), which shows an obvious classification effect on the blueberry beverage. The first three principal components of the load diagram analysis, the characteristic bands related with the blueberry beverage varieties were 420-430nm, 490-500nm, 570-580nm and 1350-1365nm. According to the cumulative contribution rate (99.20%) of the first 6 principal components, the first 6 principal components was choosed as the input of multilayer perceptron (MLP) neural network, 100 samples in all the blueberry beverage samples were selected as a training set, and the remaining 40 samples were used as the prediction set. Training set were trained and prediction set were predicted by applying the multilayer perceptron neural network, and the correct rate of prediction were 100%.Research shows, using principal component analysis combined with multilayer perceptron neural network to identify the varieties of blueberry beverage is feasible.

Key words: Blueberry beverage, Vis/NIR spectroscopy, PCA, MLP neural network

1 Introduction

Blueberry, also known as bilberry, is called the King of Berries because of their special texture, sweet-tart taste, and a nutrious abundance in VC[1-2].The earliest cultivation of blueberries took place in North America, but that happened less than a hundred years ago[3]. In China, the cultivation area of blueberries is mainly in Greater Khingan Mountains, Lesser Khingan Mountains, which are in northeast of China, and Shandong Peninsula. In Zhejiang, Hubei and Sichuan Basin and other areas also have a small amount of cultivation. With an abundance in vitamin, Superoxide Dismutase (SOD), arbutin and flavonoids compounds, blueberries are antioxidant food. It not only works in erythropsine synthesis, vision improvement, immune enhancement and cardiac strengthen[4], but also helpful in medical efforts against aging, ulceration, inflammation and cardiovascular diseases[5].Wild blueberries contain many great substances such as folic acid and ursolic acid that are good for hypertension treatment, and anthocyanin that is capable of preventing and curing inflammation[6-7]. Blueberries are also natural anti-cancer fruit because of the existence of many active constituents that inhibit the activity of cancer cell growth and even accelerate their apoptosis[8]. For Healthy adults, a moderate drinking of blueberry beverage everyday could not only reduce the memory of the recession, but also can improve the body's ability to resist oxidative stress, while reducing the loss of lymphocyte DNA[9].

Blueberry beverage is mainly made from blueberries, which gains much popularity among people. However, blueberry beverage in market is various and the identifications of blueberry beverage varieties are also complex and diverse[10]. To better develop blueberry beverage, it's vital and necessary to find a fast and effective way among these complex and diverse identifications

Visible/near infrared spectroscopy is a spectroscopic method that bases on different results of the absorption of electromagnetic waves against different substances, which is typically applied in the field of chemistry, medical, agriculture and detection of agricultural products[11-12]. And if NIR Spectroscopy is applied in the identification of blueberry beverage, there will be many advantages such as fast speed, high efficiency, no damage, high stability and accuracy[13-14].

Principal Component Analysis (PCA), can effectively find the most important elements and structure among a large number of data. It can also remove the noise and redundancy to deal with the original complex data for dimension reduction processing so that the simple structure inside the complex data can be extracted. That is, without affecting the main spectral information, many original variables can be replaced with few variables. As one of neural network[15-16]. Multilayer Perceptron (MLP) neural network, which is widely used in pattern recognition

* Corresponding author: wgfsara@126.com

with good distribution ability, can solve the complex classification problem of pattern distribution[17-18].

In this study visible/ near infrared spectroscopy (NIRS) and principal Component Analysis (PCA) with Multilayer Perceptron (MLP), as a new way for blueberry beverage identification, are used to analyze different kinds of blueberry beverage.

2 Materials and Methods Network

2.1 Instrument and Equipment

American ASD (Analytical Spectral Device) company Quality Spec spectrometer, its probe field of view angle is 25 degrees, the resolution is 3.5nm, spectral range of 350 ~ 1830nm, which sampling interval are 1.4nm @ 350 ~ 1000nm, 2nm @ 1000 ~ 1830nm, spectral resolution are 3nm @ 700nm, 10nm @ 1400nm. Analysis software is ASD View Spec Pro 6.0, Unscrambler 9.7, SPSS19.0.

2.2 Sample Source and Spectral Scan

A total of 35 blueberry beverages of four varieties of Genhe blueberry juice, Ye Laoda, Ye Shanpo and Wild blueberry were collected randomly, samples were collected before the blueberry beverages were shaken well and 140 samples were prepared. The spectral scanning adopts transmission method with an optical path of 2 mm, each sample was scanned 30 times and three spectral curves were preserved, the average spectrum of the three spectral curves was used as the final transmission spectrum.

2.3 Preprocessing of spectral data

In order to reduce or eliminate the effects of noise, baseline drift and sample nonuniformity during spectral scanning, the original spectrum needs to be mathematically transformed to improve the accuracy and stability of the estimation model.

2.3.1 Smooth method

Savitzky-Golay convolution smoothing is an improvement on moving smoothing by multiplying the measured value by the smoothing factor to reduce the effect of smoothing on useful information and then fitting of the least squares method. The concrete formula is:

$$x_{k,s} = x_k = \frac{1}{H}\sum_{i=-w}^{+w} x_{k+i}h_i \tag{1}$$

In formula (1), h_i represent the smoothing factor, H represent the normalization factor.

2.3.2 Multiplicative scatter correction

When calculating the multiplicative scatter correction (MSC) for the acquired spectrum, the average spectrum is used instead of the "ideal" spectrum to the least squares method fit of each spectrum and the average spectrum so that the average spectrum can as much as possible becomes as linear relationship. The concrete formula is:

2.3.2.1 Calculate the average (ideal) spectrum of the sample

$$\overline{X} = \frac{\sum_{i=1}^{n} X_i}{n} \tag{2}$$

2.3.2.2 Perform linear regression and the regression coefficients a_i and regression constants b_i were obtained by least squares method

$$X_i = a_i \overline{X} + b_i \tag{3}$$

2.3.2.3 Correct each spectrum

$$X_{i,M} = \frac{X_i - b_i}{a_i} \tag{4}$$

In formula(2)(3)(4), X_i represent the sample spectrum, n represent the number of samples, $\overline{X}$ represent the average spectrum, a_i represent the regression coefficients, b_i represent the regression constants, $X_{i,M}$ represent the corrected spectrum.

2.3.3 Preprocessed spectra

In the Unscrambler 9.7 software, Savitzky-Golay smoothing is used, the smoothing point is 9, and the multiple scattering correction (MSC) is used to preprocess the sample. The processing effect is the best. In order to remove the noise effects of the first and the end of the spectral curve, use the spectrum of the band 400 ~ 1800 nm[19-21]. The spectral curves of the four blueberry beverages were obtained after pretreatment. As shown in Fig. 1, the spectral lines of one sample were arbitrarily selected in each blueberry beverage sample.Taking the abscissa as the wavelength of the spectrum and the ordinate as the absorbance of each sample.

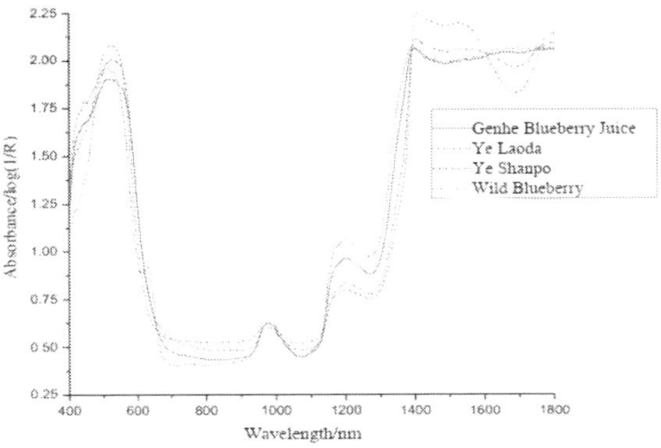

Fig. 1. Spectral curve four varieties of blueberry beverage.

2.4 Principal component analysis

A total of 140 samples were cluster analyzed by using principal component analysis method of four blueberry beverages, blueberry juice, Ye Laoda, Ye Shanpo and Wild blueberry. As shown in Fig.2, in which the X, Y and Z axes respectively represent the scores of the first principal component (PC1), the second principal component (PC2) and the third principal component (PC3).The four kinds of blueberry beverages in Fig.2 were divided into four categories, indicating that PC1, PC2 and PC3 have very good clustering effect on four kinds of blueberry beverages also can qualitatively characterize the characteristics of the four blueberry beverages. But the edge sample distinction of Genhe blueberry juice, Ye Laoda and Ye Shanpo are not obvious.In order to improve the prediction accuracy, using principal component analysis combined with multilayer perceptron neural network to establish detection and analysis model of four kinds of blueberry beverage.

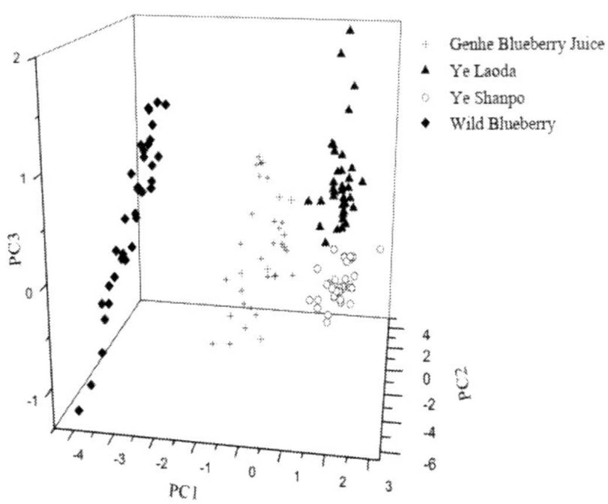

Fig. 2. Blueberry beverage principal component cluster diagram.

3 Test Results and Analysis

3.1 Characteristic bands selected

According to the principal component 1, 2, 3 in the entire wavelength range of the load value shown in Fig. 3.Taking abscissa as the wavelength of 400 ~ 1800nm, the ordinate as the load of different principal components in each wavelength variable values. The correlation between the principal component and the absorbance of the blueberry beverage at each wavelength. From Figure 3 can be obtained that range in 420 ~ 430nm, 490 ~ 500nm, 570 ~ 580nm and 1350 ~ 1365nm have maximum correlation with principal component. The main component 4-20 has maximum correlation with the absorbance of these four blueberry beverage's bands. Therefore, the four wavelength bands were selected as the characteristic bands of near infrared spectrum of the blueberry beverage over the entire wavelength range.

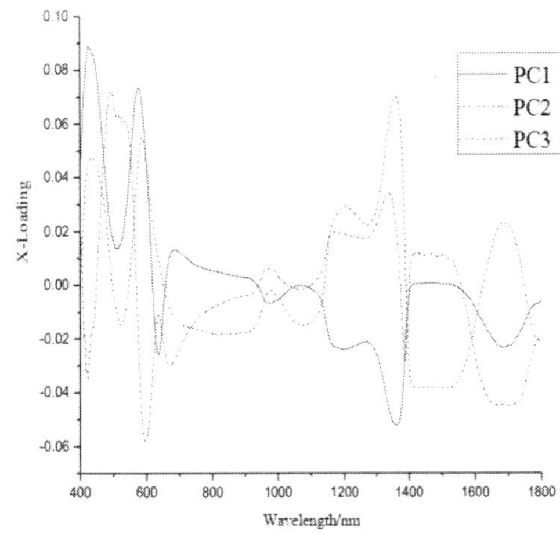

Fig. 3. The load of the principal component values.

3.2 Principal component analysis results

The cumulative contribution rate of the first six principal components has reached 99.204% by the principal component analysis method. As shown in Table 1,the spectral data of 140 samples were replaced by the first six principal components. All the samples were analyzed by the first six principal components.

Table 1. The cumulative contribution rate of the principal component.

Principal Component	PC1	PC2	PC3	PC4	PC5	PC6
Cumulative Contribution Rate	53.65 %	91.16 %	95.67 %	97.83 %	98.64 %	99.20 %

3.3 Identification of blueberry beverage based on multilayer perceptron neural network

The first six principal components obtained by principal component analysis are used as inputs to the multilayer perceptron (MLP) neural network. The three-layer network structure is chosen as the best training structure, and its structure is 6 nodes, the hidden layer is 5 nodes and the output layer is 4 nodes, with hyperbolic tangent as the activation function of the hidden layer, softmax as the activation function of the output layer. A total of 100 blueberry beverage samples were used as training set for training. The remaining 40 blueberry beverage samples were used as test set. The results show that, as shown in Table 2 ("1" is Genhe blueberry juice, "2" is Ye Laoda, "3" is Ye Shanpo, "4" is Wild blueberry) accuracy of four kinds of blueberry beverage classification has reached 100%.

Table 2. Blueberry beverage classification prediction accuracy.

Blueberry Beverage Varieties	1	2	3	4	Accuracy Rate
1	10	0	0	0	100%
2	0	10	0	0	100%
3	0	0	10	0	100%
4	0	0	0	10	100%

4 Conclusion

Based on the combination of principal component analysis and multilayer perceptron(MLP) neural network, the model of blueberry beverage brand identification was established, and the accuracy of classification prediction was 100%.This fully illustrates the use of near infrared spectroscopy technology can quickly and accurately identify blueberry beverage varieties. This method of principal component analysis combined with multilayer perceptron (MLP) neural network is use the principal component as the input of multilayer perceptron(MLP) neural network, which can reduce the calculation of neural network, accelerate the training of the sample and eliminate the spectral signal interference, greatly improving the accuracy of the forecast. At the same time, the characteristic bands of the blueberry beverage brand are 420 ~ 430nm, 490 ~ 500nm, 570 ~ 580nm and 1350 ~ 1365nm., which are closely related to the beverage brands of Blueberry. Therefore, it is feasible to use the principal component analysis and the multi-layer perceptron neural network to identify the varieties of blueberry beverages. It also opens up a good prospect for the future development of blueberry beverage testing equipments.

Acknowledgment

This work was supported by National Natural Science Foundation of China (6126511); Higher School Doctoral Discipline Special Research Fund (20111515120004); Inner Mongolia Natural Science Foundation of China (2102MS0915); Inner Mongolia Agricultural University doctoral research start fund (BJ09-18).

References

1. A. Han Zhang. *Wild blueberry health care value*. Heilongjiang STI. **13**, 87 (2013)
2. K. Kim Mina, Han Sub Kwak. *Influence of functional information on consumer liking and consumer perception for blueberry functional beverages*. IJFST. **50**, 70-76 (2013)
3. Shiow Y. Wang, Hangjun Chen, J CAMP MARY, et al. *Genotype and heavy season influence blueberry antioxidant capacity and other quality attributes*. IJFST. **47**, 1540- 1549 (2012)
4. J. Barba Francisco, J. Esteve Maria, Frigola Ana. *Physicochemica and nutritional characteristics of blueberry juice after high pressure processing*. FRI. **50**, 545-549 (2013)
5. Ying Wang, Shu Wang, Chunguang Ren, etc. *Blueberry wine fermentation process and quality of the relationship between the study*. Guizhou Science. **31**, 75-78 (2013)
6. Pengxiang Han, Bei Zhang, Xuqiao Feng, etc. *Blueberry nutrition and health functions and its development and utilization*. FIT. **36**, 370-375,379 (2015)
7. Yangcheng Song, Yujuan Chen, Hao Li, etc. *Blueberries in the rapid detection of anthocyanins*. Food Science. **31**, 334-336 (2010)
8. Lihua Pan, Jianfei Wang, Xinggan Ye, etc. *Blueberry anthocyanin extraction process and its immunomodulatory activity*. Food Science. **35**, 81-86 (2014)
9. Haiyan Gao, Long Xu, Hangjun Chen, etc. *Blueberry postharvest quality control and anti-oxidation research progress*. Chinese JFS. **6**, 1-8 (2013)
10. Haige Li, Yanjun Wu, Yange Yang, etc. *Advances in technology research to identify the authenticity of berry juice*. Food Science. **37**, 243-250 (2016)
11. Guifang Wu, Hai Ma, Xin Pan. *Identification of varieties of natural textile fiber based on vis / NIR infrared spectroscopy*. IAEAC.Institute of Electrical and Electronics Engineers Inc, 585-589 (2015)
12. R. BEGHI, V. GIOVENZANA, A.SPINARDI, etc. *Derivation of a blueberry ripeness index with a view to a low-cost, handheld optical sensing device for supporting harvest decisions*. Transaction of the ASABE. **56**, 1551 -1559 (2013)
13. Guifang Wu, Wei Jiang, Chunguang Wang, etc. *Quality inspection of tomato shelf life based on TPA and Vis / NIR*. MFST. **31**, 290-294 (2015)
14. Guifang Wu, Chunguang Wang. *Investigating the effects of simulated transport vibration on tomato tissue damage based on Vis / NIR spectroscopy*. PBT. **98**, 41-47 (2014)
15. Shirong Ai, Ruimei Wu, Yan Wu. *Study on the Identification of Tea Beverage by Near Infrared Spectroscopy Based on BP Neural Network*. Anhui JAS. **14**, 7658-7659, 7662 (2010)

16. Daniel Cozzolino, Wies Cynkar, Nevil Shah, etc. *Varietal differentiation of grape juice based on the analysis of near- and mid-infrared spectral data.* Analyticical Methods. **5**, 381- 387 (2012)
17. Royston Goodacre, David Hammond, B. Kell Douglas. *Quantitative analysis of the adulteration of orange juice with sucrose using pyrolysis mass spectrometry and chemometrics*. JAAP. **40-41**, 135-158 (1997)
18. I. Suarez Gracirla, A.Ortiz Oscar, M. Aballay Pablo, etc. *Adaptive neural model predictive control for the grape juice concentration process.* // International Conference on Industrial Technology. Universidad Tecnica Federico Santa Maria. Institute of Electrical and Electronics Engineers Inc. 57-62 (2010)
19. Guifang Wu, Yixiang Jiang, Yanyan Wang, etc. *Identification of dry red wine varieties based on independent principal component and BP neural network*. SSA. **29**, 1268-1271 (2009)
20. Jianhu Wu, Juntao Lei, Qi Yang. *Identification of Dry Date by Using Visible / Near Infrared Spectroscopy*. JFSQI. **7**, 1870-1875 (2016)
21. Yulu Wang, Hui Guo, Changjie Han. *Study on identification method of different mutton based on visible / near infrared spectroscopy*. FST. **40**, 298-302 (2015)

MATEC Web of Conferences **139**, 00051 (2017) DOI: 10.1051/matecconf/201713900051
ICMITE 2017

Carrier tracking algorithm based on joint acquisition of frequency locked loop and phase locked loop

Kang Gao[1,*], Qiang Zhao [1], BaoTong Peng[1]

[1] *Department of Optical and Electronic Equipment, Aerospace Engineering University, Beijing, 101416, PR China*

Abstract. Aiming at the problem of frequency step in the frequency lock loop (FLL) - phase lock loop (PLL) carrier tracking algorithm's conversion state, presenting an improved algorithm: PLL and FLL joint acquisition to replace the single FLL acquires frequency, and deduce the loop state transition threshold. The simulation results show that the improved algorithm is more stable in the conversion process, and the loop performance is optimized. When the SNR is -10dB, and has the acceleration rate, the tracking loop does not have a frequency step at the time of conversion, achieving the design purpose.

1 Introduction

In the space measurement and control system, due to the long distance transmission loss and the high-speed movement of the target carrier, the received signal has the characteristics of low SNR and high dynamic. At present, the receivers in the measurement and control system are mostly coherent receivers, the principle of which is to obtain the coherent carrier with the same frequency and phase to the received signal through the carrier tracking. In the common carrier tracking algorithms, FLL-PLL carrier tracking algorithm has the advantages, such as simple implementation, good dynamic performance and high tracking accuracy. FLL-PLL carrier tracking algorithm is one of the tracking algorithms which are widely used.
However, the traditional carrier tracking algorithm of FLL-PLL has the problem of frequency step when converting FLL to PLL. In [1] and [2] which used to adjust the conversion threshold to solve that problem, although this method can improved the frequency step problem, but the effect is not ideal.

This paper aims at the problem of discontinuous transition state, proposed the improved algorithm: FPLL-PLL carrier tracking algorithm, the algorithm uses FLL and PLL joint acquisition during the acquisition stage, when converting smoothly into PLL tracking stage. At the same time, the conversion threshold of the algorithm is analysed, which provides a theoretical basis for the design of conversion threshold.

2 FLL-PLL carrier tracking algorithm

Figure 1 shows the principle of FLL-PLL tracking loop, which consisted of the discriminator, loop filter (LF) and numerical control oscillator (NCO). From the Figure 1 when the input signal enter the loop, firstly, input signal mixes with the local signal which generated by the NCO, then integral cleaning of the mixed signal, the discriminator can get the phase error or the frequency error from mixed signal, finally, putting the phase error or the frequency error into the loop filter can obtained the NCO frequency control, which makes the output signal of NCO has the same frequency and phase with the received signal, completing carrier tracking.

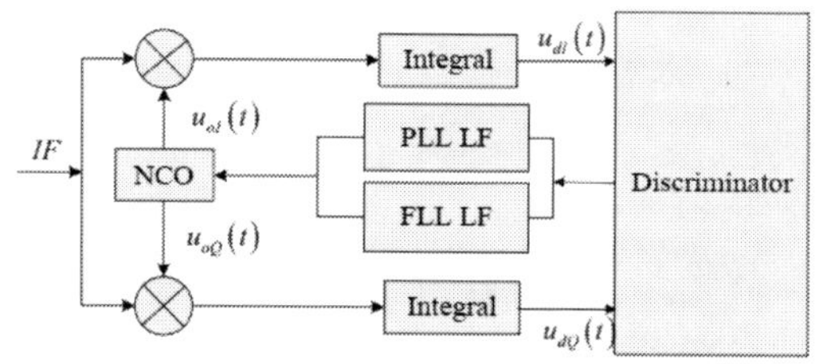

Figure 1. FLL-PLL tracking loop schematic.

2.1. The design of the discriminator

FLL-PLL tracking loop discriminator including frequency discriminator and phase discriminator, there are several identification algorithms compared in table 1 and table 2. From table 1 and table 2 can get the form of loop discriminator[3].

Table 1. Commonly used phase recognition algorithm.

Phase algorithm	Output	Character
$u_{dQ} \times u_{dI}$	$\sin(2\theta_e)$	In the low SNR close to the best, moderate amount of calculation.
$u_{dQ} \times sign(u_{dI})$	$\sin(\theta_e)$	In the high SNR close to the best, low computational requirements.
u_{dQ}/u_{dI}	$\tan(\theta_e)$	In the low SNR is the best, and high SNR is good, high computational requirements.
$ATAN(u_{dQ}/u_{dI})$	θ_e	Two quadrant arc tangent, in the high and low SNR are the best, high computational requirements.

Table 2. Commonly used frequency recognition algorithm

* Corresponding author: wouwouwou@126.com

frequency algorithm	Output	Character
$\frac{P_{cross}}{t_2 - t_1}$	$\frac{\sin[(\theta_2 - \theta_1)]}{t_2 - t_1}$	In the low SNR close to the best, moderate amount of calculation.
$\frac{P_{cross} \times P_{dot}}{t_2 - t_1}$	$\frac{\sin[2(\theta_2 - \theta_1)]}{t_2 - t_1}$	In the high SNR close to the best, low computational requirements.
$\frac{ATAN2(P_{cross}, P_{dot})}{t_2 - t_1}$	$\frac{\theta_2 - \theta_1}{t_2 - t_1}$	Four quadrant arc tangent, in the high and low SNR are the best, high computational requirements.

As can be seen from table 1 and table 2 that the discrimination performance of the tangent detection algorithm and the cross product frequency discrimination algorithm are optimal. The structure diagram of discriminator is shown in Figure 2.

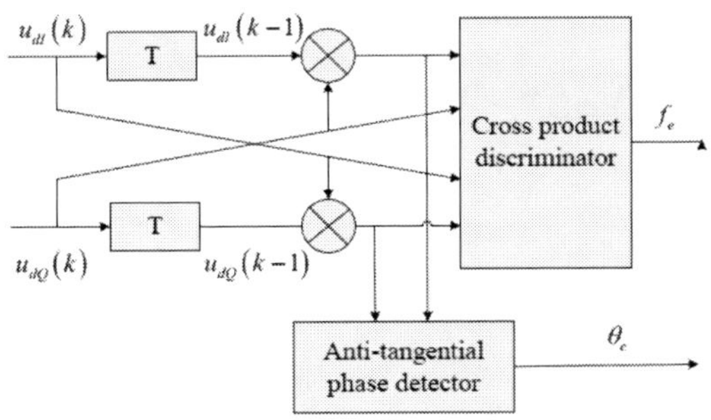

Fig. 1. Discriminator structure.

The received signal is assumed to be of the form

$$u_i(t) = A\cos(\omega_0 t + \varphi_0) \quad (1)$$

The NCO output is assumed to be of the form

$$u_{oI}(t) = \cos(\omega_1 t + \varphi_1) \quad (2)$$

$$u_{oQ}(t) = \sin(\omega_1 t + \varphi_1) \quad (3)$$

After the mixing and integral cleaner[4]

$$u_{dI}(t) = \frac{1}{T_S}\int_{(k-1)NT_S}^{kNT_S} \frac{A}{2}\cos(\Delta\omega t + \Delta\varphi)dt \quad (4)$$

$$u_{dQ}(t) = \frac{1}{T_S}\int_{(k-1)NT_S}^{kNT_S} \frac{A}{2}\sin(\Delta\omega t + \Delta\varphi)dt \quad (5)$$

where: $t_k = kNT_S$, $t_{k-1} = (k-1)NT_S$, $T = t_k - t_{k-1}$

$$\begin{aligned} u_{dI}(t) &= \frac{1}{T_S}\int_{(k-1)NT_S}^{kNT_S} \frac{A}{2}\cos(\Delta\omega t + \Delta\varphi)dt \\ &= \frac{A}{2T_S\Delta\omega}\sin(\Delta\omega t + \Delta\varphi)\Big|_{(k-1)NT_S}^{kNT_S} \\ &= \frac{AN}{2}\operatorname{sinc}\frac{\Delta\omega T}{2}\cos\left(\Delta\omega t_{k-1} + \frac{\Delta\omega T}{2} + \Delta\varphi\right) \end{aligned} \quad (6)$$

and

$$u_{dQ}(t) = \frac{AN}{2}\operatorname{sinc}\frac{\Delta\omega T}{2}\sin\left(\Delta\omega t_{k-1} + \frac{\Delta\omega T}{2} + \Delta\varphi\right) \quad (7)$$

P_{cross} and P_{dot} are given blow

$$\begin{aligned} P_{cross} &= u_{dI}(t_{k-1})u_{dQ}(t_k) - u_{dI}(t_k)u_{dQ}(t_{k-1}) \\ &= \frac{(AN)^2}{4}\operatorname{sinc}^2\frac{\Delta\omega T}{2}\sin\Delta\omega T \end{aligned} \quad (8)$$

$$\begin{aligned} P_{dot} &= u_{dI}(t_{k-1})u_{dI}(t_k) + u_{dQ}(t_k)u_{dQ}(t_{k-1}) \\ &= \frac{(AN)^2}{4}\operatorname{sinc}^2\frac{\Delta\omega T}{2}\cos\Delta\omega T \end{aligned} \quad (9)$$

The two quadrant tangent phase detection algorithm is used to obtain the phase discrimination error

$$\theta_e = \arctan(P_{cross}/P_{dot}) = \Delta\omega T \quad (10)$$

The four quadrant tangent frequency detection algorithm is used to obtain the frequency discrimination error

$$f_e = \frac{\arctan 2(P_{cross}/P_{dot})}{(t_k - t_{k-1})2\pi} = \frac{\theta_e}{2\pi T} = \frac{\Delta\omega}{2\pi} \quad (11)$$

The phase discrimination error and the frequency discrimination error are optimal under the condition of high and low SNR, and the slope of the identification curve is independent of the signal amplitude.

2.2 Design of Loop Filter

When design a carrier tracking loop, the order of the loop filter determines the order of the carrier tracking loop. According to the high dynamic characteristics of the signal in the application background, the FLL-PLL tracking loop needs to overcome the influence of the acceleration and jerk. So it is necessary to design the order of the loop.

Take the second order FLL as an example. Since the FLL tracking loop contains an integrator more than the same order PLL, therefore the error transfer function of the second order FLL with active proportional integral filter is

$$H_e(s) = \frac{Ks^3}{Ks^3 + 2\xi\omega_n s + \omega_n^2} \quad (12)$$

where: K is loop gain, ω_n is angular frequency of natural oscillation, ξ is damping coefficient

Under the actual conditions, the mathematical model of the acceleration is the frequency ramp signal, the mathematical model of the jerk is the frequency acceleration signal, the steady-state error of the second order FLL is calculated according to the terminal value theorem in the case of frequency ramp and frequency acceleration:

When the input signal is a frequency ramp signal

$$\begin{aligned} f_e(\infty) &= \lim_{t\to\infty} f_e(t) = \lim_{s\to 0} s \cdot f_e(s) \\ &= \lim_{s\to 0} s \cdot \frac{Ks^3}{Ks^3 + 2\xi\omega_n s + \omega_n^2} \cdot \frac{\dot{f}_d}{s^3} \\ &= 0 \end{aligned} \quad (13)$$

When the input signal is a frequency acceleration signal

$$f_e(\infty)=\lim_{t\to\infty} f_e(t)=\lim_{s\to 0} s\bullet f_e(s)$$
$$=\lim_{s\to 0} s\bullet\frac{Ks^3}{Ks^3+2\xi\omega_n s+\omega_n^2}\bullet\frac{\ddot{f}_d}{s^4} \quad (14)$$
$$=\frac{K\ddot{f}_d}{\omega_n^2}$$

From (13) and (14), we can see that the second order FLL can track the frequency ramp signal without error, but the steady frequency error occurs when tracking the frequency acceleration signal.

Take the third order PLL as an example. The error transfer function of the third order PLL with three parameter filter[5] is

$$H_e(s)=\frac{Ks^3}{Ks^3+2\xi\omega_n s+\omega_n^2} \quad (15)$$

where:

K is loop gain,

$a=(m+2)\xi\omega_n/K$,

$b=(1+2m\xi)\omega_n^2/K$,

$c=m\xi\omega_n^3/K$

When the input signal is a frequency ramp signal

$$\theta_e(\infty)=\lim_{t\to\infty}\theta_e(t)=\lim_{s\to 0} s\bullet\theta_e(s)$$
$$=\lim_{s\to 0} s\bullet H_e(s)\bullet\frac{\dot{f}_d}{s^3}$$
$$=\lim_{s\to 0} s\bullet\frac{s^3}{s^3+s^2Ka+sKb+Kc}\bullet\frac{\dot{f}_d}{s^3} \quad (16)$$
$$=0$$

When the input signal is a frequency acceleration signal

$$\theta_e(\infty)=\lim_{t\to\infty}\theta_e(t)=\lim_{s\to 0} s\bullet\theta_e(s)$$
$$=\lim_{s\to 0} s\bullet H_e(s)\bullet\frac{\ddot{f}_d}{s^4}$$
$$=\lim_{s\to 0} s\bullet\frac{s^3}{s^3+s^2Ka+sKb+Kc}\bullet\frac{\ddot{f}_d}{s^4} \quad (17)$$
$$=\frac{\ddot{f}_d}{Kc}$$

The computed results show that in the traditional FLL-PLL carrier tracking algorithm, firstly, the FLL to capture, from (14) we can see that the second order FLL will have a stable frequency error, at this time FLL compensates for acceleration and jerk, for the third-order PLL there is only a frequency step to track, so in the third-order PLL separate work stage, PLL can be accurate tracking without error. Therefore, FLL and PLL loop filter design process, as discussed above, can determine the second-order FLL and third-order PLL combination of the form.

3 FPLL-PLL carrier tracking algorithm

On the basis of guaranteeing the locking speed and tracking accuracy, the FPLL-PLL carrier tracking algorithm can make the conversion state more continuous and the loop is more stable. In the FLL-PLL carrier tracking algorithm, FLL captures Doppler frequency at first, when being stable to only a steady-state frequency error, then the loop is converted to PLL for accurate tracking. Because of the conversion mode for the forced convert, it is clear that the transition of the loop state will not continuous, thus affecting the tracking performance of the loop, or even cause the phenomenon of loss of lock.

The FPLL-PLL carrier tracking algorithm uses the FLL and PLL joint acquisition. At this stage, the PLL assists FLL to acquire Doppler frequency, in this way can avoid the problem of frequency step due to forced conversion and then the loop converts to eh exact tracking mode for the PLL, until the loop error is too large, resulting in loss of lock.

3.1. Dynamic performance of FPLL

In order to ensure the locking speed of FPLL-PLL carrier tracking algorithm, it is necessary to analyse the jerk stress threshold that FPLL and compare with FLL. The jerk stress threshold under different carrier-to-noise ratios can be deduced from the total tracking loop measurement error. The jerk stress threshold of second-order FLL and third-order PLL are given blow.

The dominant source of the FLL frequency error are frequency jitter due to thermal noise and dynamic stress errors. The rule of thumb for tracking threshold is not exceed one-fourth of the frequency pull-in range of the FLL discriminator[6]. Therefore, the general rule of thumb for FLL tracking threshold is:

$$\sigma_{FLL}=\sigma_{tFLL}+f_e/3\le 1/12T \quad (18)$$

where:

σ_{tFLL} = 1-sigma thermal noise frequency jitter

f_e = dynamic stress error in the FLL tracking loop

The FLL tracking loop jitter due to thermal noise is:

$$\sigma_{tFLL}=\frac{1}{2\pi T}\sqrt{\frac{4FB_L}{C/N_0}\left[1+\frac{1}{TC/N_0}\right]} \quad (19)$$

where:

$F=1$ at high C/N_0

=2 near threshold

The dynamic stress error is:

$$f_e=0.2809\frac{\ddot{f}_d}{B_L^2} \quad (20)$$

From (18), (19), (20) the jerk stress threshold of FLL is defined as follows:

$$\ddot{f}_d\le\frac{3}{0.2809}\left(\frac{B_L^2}{12T}-\frac{B_L^2}{2\pi T}\sqrt{\frac{4FB_L}{C/N_0}\frac{TC/N_0+1}{TC/N_0}}\right) \quad (21)$$

In the case of a dataless PLL four-quadrant arctangent discriminator whose phase pull-in range is 360°, the 3-sigma rule threshold is therefore 90° [6]. Therefore, the general rule of thumb for FLL tracking threshold is:

$$\sigma_{PLL} = \sigma_{tPLL} + \theta_e / 3 \leq 30 \quad (22)$$

where:

σ_{tPLL} = 1-sigma thermal noise phase jitter

θ_e = dynamic stress error in the PLL tracking loop

The FLL tracking loop jitter due to thermal noise is:

$$\sigma_{tPLL} = \frac{360}{2\pi}\sqrt{\frac{B_L}{C/N_0}\left[1+\frac{1}{2T\,C/N_0}\right]} \quad (23)$$

The dynamic stress error is:

$$\theta_e = 0.4828\frac{\ddot{f}_d}{B_L^3} \quad (24)$$

From (21), (22), (23) the jerk stress threshold of PLL is defined as follows:

$$\ddot{f}_d \leq \frac{3}{0.4828}\left(30B_L^3 - \frac{360B_L^3}{2\pi}\sqrt{\frac{B_L}{C/N_0}\left[1+\frac{1}{2T\,C/N_0}\right]}\right) \quad (25)$$

According to the Monte Carlo simulation experiment and the experience value of the engineering experiment, we can get the relationship between the jerk stress threshold and the loop bandwidth of FLL and FPLL under SNR = -10dB[7], as shown in the following figure

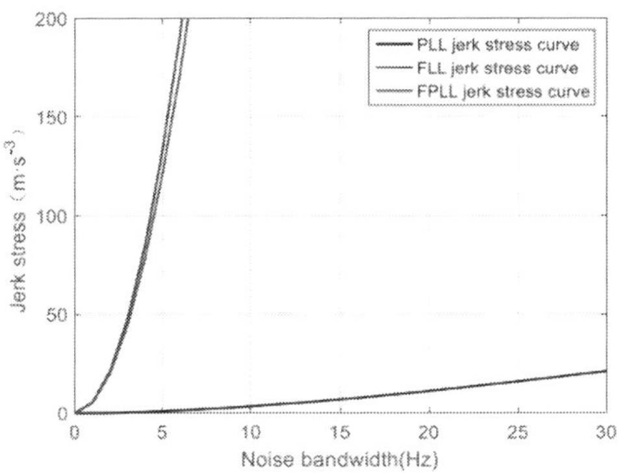

Fig 3. Jerk stress threshold of FLL and FPLL.

In summary, compared the jerk stress threshold of FPLL and FLL, FPLL's jerk stress threshold meets the actual conditions and enable guarantee loop locking.

3.2. Transition threshold

Whatever the traditional FLL-PLL carrier tracking algorithm or the FPLL-PLL carrier tracking algorithm, the choice of conversion timing is a key factor affecting loop tracking performance. Premature or late switch to the PLL accurate tracking state will affect the performance of the loop, and even cause the loop to lose the lock.

The threshold for converting the FPLL state to the PLL exact tracking state can be set to fast capture of the PLL. Because in the fast catch band, the PLL can't be jumped to reach the lock.

The high frequency component of the three-parameter filter transfer function can be shown to be

$$\lim_{s\to\infty} F(s) = \lim_{s\to\infty}\frac{as^2+bs+c}{s^2} = a \quad (26)$$

where: $m = 5$, the gain of the high frequency part of the loop is can be shown to be

$$K_H = Ka = (m+2)\zeta\omega_n \quad (27)$$

(27) is the maximum direct NCO frequency control range that is fast capture, can be shown to be

$$\Delta\omega_L = Ka = (m+2)\zeta\omega_n \quad (28)$$

In summary, when the FPLL's steady-state error reaches the threshold, the FLL in the FPLL stop working, the loop is converted to PLL accurate tracking stage.

4 Simulation and analysis

The analytical simulation results for verification. For example, a sinusoidal signal with a frequency of 70MHz, and the sampling rate is 40MHz, the initial Doppler frequency difference is 200Hz, the Doppler first order change rate is 50Hz/s, the Doppler second order change rate is $5\,\mathrm{Hz/s^2}$, under the condition of SNR is -10dB, using FLL-PLL and FPLL-PLL tracking the above signal. The simulation results are shown in the following figure

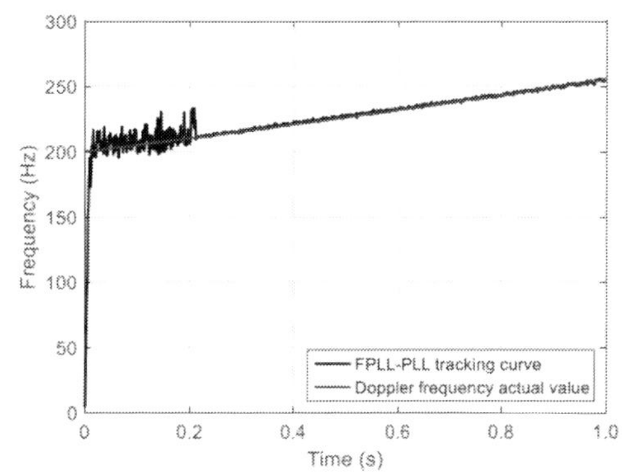

Fig 4. Output of FPLL-PLL loop filter.

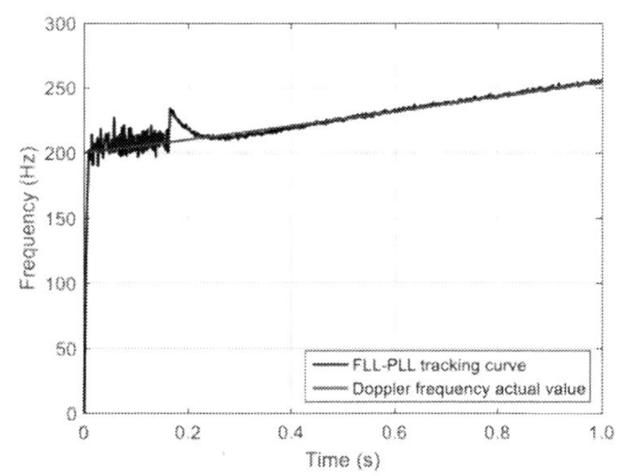

Fig 5. Output of FLL-PLL loop filter.

It can be seen from Figure 4 and Figure 5 that the conversion state of the FPLL-PLL carrier tracking algorithm is more continuous and the loop is more stable

DOI: 10.1051/matecconf/201713900051

under the low SNR. The traditional FLL-PLL carrier tracking algorithm takes a frequency step in the conversion state, the loop stability is relatively poor.

5 Conclusions

The FPLL-PLL carrier tracking algorithm proposed in this paper uses the method of PLL and FLL joint acquisition , improving the loop conversion state, compared with the traditional FLL-PLL carrier tracking algorithm, the loop conversion is more continuous and the loop stability is higher.

However, this paper only has improved the frequency step problem of carrier tracking loop conversion, and it has not been studied for the variable bandwidth problem in the precise tracking stage. This will be the next step to be improved.

References

1. Z. P. Xu, C. Cui, J . She, Telecommunication Engineering, **52**, (2012), 558-561.
2. H. M. Wu, Y. Yong. Su, A. J. Liu, Science Technology and Engineering, **10**, (2010), 4645-4650.
3. H. B. Lu, W. W. Wang, Q. C. Gu, Information and Control, **39**, (2010), 451-454.
4. F. D. Natali, AFC tracking algorithms[J]. IEEE Transactions on Communications, **32**, (1984), 935-947.
5. Y. Z. Wang, Telecommunication Engineering, **48**, (2008), 51-55.
6. E. D. Kaplan, Journal of Atmospheric and Solar-Terrestrial Physics, **59**, (2006):598-599.
7. X. D. Deng, S. Wu, Modern Defence Technology, **38**, (2010), 137-141.

An Improved Particle Swarm Optimization(PSO)-Based MPPT Strategy for PV System

Tianmeng Wei[1], *Dongliang Liu*[1,2,a], and *Chuanfeng Zhang*[1]

[1]School of Automation Engineering, Hangzhou Dianzi University, Hangzhou, China
[2]Wolong ELectric Group Co.,Ltd, Shangyu, China

Abstract. Under partially shaded conditions, the P-U curve of PV array contains multiple extreme points. General MPPT methods may misjudge the MPP and trap in the local extreme point, which will cause low working efficiency. Although the traditional PSO algorithm can accurately track the maximum power point under this condition, the optimizing process fluctuates obviously and the tracking speed can be improved. In order to solve these problems, an improved PSO algorithm is proposed. The initial positions of the particles are located by analysing the relationship of the I-U and P-U characteristic curves. It is more closed to the maximum power point. So the efficiency of PSO algorithm is improved. To evaluate the effectiveness of this method, the simulation model is established in MATLAB/Simulink. Under partially shaded conditions the algorithm can track the maximum power point quickly and accurately.

Keywords*—Maximum power point Tracking(MPPT), partial shading, particle swarm optimization(PSO), photovoltaic(PV) system.*

1 Introduction

With the rapid increase in the environmental costs of fossil fuels, many countries are seeking to meet their energy needs through renewable energy sources, particularly solar energy. However, in the developing process of solar energy, a number of problems are eager to be solved. One of the problems, experts in the whole word focusing on, is that the efficiency of the optimization process of maximum power point tracking. In recent years, numerous MPPT methods are presented to track MPPs correctly and to improve the dynamic and steady-state tracking performance. Perturb and Observe (P&O) is the most popular among these strategies. However, in the actual cases, due to the influence of partial shading, mismatch between components and ageing of solar panels, P-U characteristic of PV arrays appears multi extreme points. In this case, traditional method will trap in local maximum points and cause fail of optimization. To solve this problem, experts did a lot of researches. Reference [4] proposed an improved global scanning method. This method can track MPPT more accurately, but the optimization process required for a long time, which is easy to cause power loss. Reference [5] using Fibonacci MPP optimization method. Although this method can adapt for the changing condition, it convergence for a long time and lack of practicality. Reference [6] using traditional particle swarm optimization algorithm, it chooses the randomness position as initial value of particles. The optimization process is slow and easy to lose MPP.

As discussed before, a new and feasible MPPT method need to be developed. In this paper, an improved particle swarm optimization algorithm is proposed. Based on traditional particle swarm optimization algorithm, this method initializes the initial value of particles by analysing the derivative of PV arrays' I-U curve. It can make sure that the improved method will not trap in the local maximum power point and find the global maximum power point faster than before. That means the efficiency of PSO algorithm has been improved. In addition, different from the general particle swarm algorithm the fitness value of this improved algorithm is the real-time sampling of PV array's output power. It leaves out the trouble of repeated iteration and improves the operation efficiency; As same as the general PSO algorithm, the termination and restart conditions are set to avoid the vibration tending to the maximum power point. To evaluate the effectiveness of the proposed method, MATLAB simulations are carried out under the multi extreme point conditions. It is verified that the proposed method can quickly complete the MPP optimization process, improve the efficiency of the PV system and enhance the capacity of the system.

2 Modeling of PV array

In this paper, a {3×3} PV array as shown in Fig.1 is the PV array structure of simulation. To avoid the hot spot effect of PV products, each component is paralleling a bypass diode. At the same time for each series branch has a blocking diode in series, which can protect the operation of the PV array [11-12]. In the simulation, the component

[a] Corresponding author: liudl@hdu.edu.cn

parameters are set following the parameters of solar panel JLS60Pj(250). Under standard condition (Temperature is 25°C, Sunlight is 1000W/m2) solar panels' open circuit voltage, short circuit current of PV panels, MPP's current value and the voltage value of the MPP are 36.5V, 30.4V, 9.1A and 8.2A respectively.

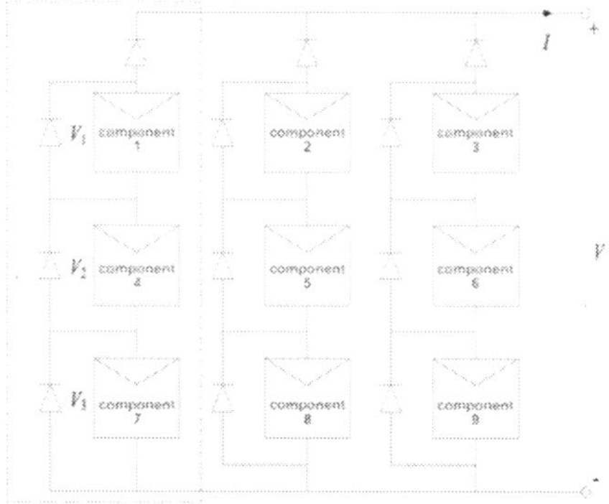

Fig.1 {3×3} Configuration of PV array

To facilitate the analysis and understanding, we choose the double peak situation in this paper. The first peak is local maximum power point, and the second peak is the global maximum power point. We set the parameters of components 1 to 6 as standard condition and the parameters of components 7 to 9 as T=25°C, S=700W/m2. In order to get the PV array's output characteristic equation, it is necessary to analyse the internal physical structure of PV array. Because the three branches of PV array have the same characteristic, we choose the components 1, 4, and 7 (as shown in the dotted box in Fig.1) as the physical model to get the mathematical model of PV array. Fig.2 shows the structure of circuit model.

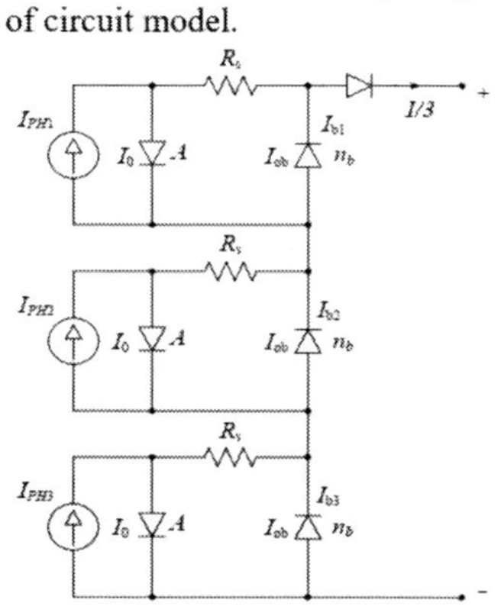

Fig.2 Circuit model of serial module with bypass diode

In this model, three pieces of photovoltaic cells' photo-generated current are I_{PH1}、I_{PH2}、I_{PH3} respectively. The current flowing through this branch is $I/3$. R_s is the resistance of the solar cell. A、n_b is the diode's impact factor. I_0、I_{ob} is the diode's reverse saturation current. T is absolute temperature K and q is electronic charge constant. Because component 1 and 4 are in the same temperature and radiation conditions, I_{PH1} is equates to I_{PH2}. According to the previous settings, the radiation condition of component 7 is less than component 1 and 4, so $I_{PH1}>I_{PH3}$. Commonly used simplified current equation of photovoltaic cell can be expressed as follow:

$$I = I_{PH} - I_0[\exp(\frac{U+IR_s}{AkT})-1] \tag{1}$$

Because I_{PH1} is greater than I_{PH2}, when the output current is very big the bypass diode of component 7 is in a conducting state. Then the excess current of component 1 and 4 will flow through this diode to the external. At this stage, only component 1 and 4 output power. As for component 7, it is absorbing part of the power from component 1 and 4; When the output current of component 1 and 4 is equal to component 7's, the bypass diode of component 7 will in blocking state. Then all the three components output power. Based on the above analysis, under shading condition the solar cell's output characteristic equation is shown as follow[13]:

$$U=\begin{cases} 2\frac{AkT}{q}\ln(\frac{I_{PH1}-I/3}{I_0}+1)-2IR_S/3- \\ \frac{n_b kT_b}{q}\ln(\frac{I/3-I_{PH3}}{I_{ob}}+1), I_{PH3}<I/3<I_{PH1} \\ 2\frac{AKT}{q}\ln(\frac{I_{PH1}-I/3}{I_0}+1)-IR_s+ \\ \frac{AkT}{q}\ln(\frac{I_{PH3}-I/3}{I_0}+1), 0<I/3<I_{PH3} \end{cases} \tag{2}$$

According to the circuit equation P=UI and equation (2), the power output function is able to get, as shown in equation (3).

$$P=\begin{cases} 2\frac{AkTI}{q}\ln(\frac{I_{PH1}-I/3}{I_0}+1)-2I^2R_S/3- \\ \frac{n_b kT_b I}{q}\ln(\frac{I/3-I_{PH3}}{I_{ob}}+1), I_{PH3}<I/3<I_{PH1} \\ 2\frac{AKTI}{q}\ln(\frac{I_{PH1}-I/3}{I_0}+1)-I^2R_s+ \\ \frac{AkTI}{q}\ln(\frac{I_{PH3}-I/3}{I_0}+1), 0<I/3<I_{PH3} \end{cases} \tag{3}$$

The corresponding I-U and P-U output characteristic curve can be draw according to the equation (2) and (3), as shown in Fig.3 and Fig.4 respectively. By observing Fig.3, the variation tendency of the curve shows that the current value is almost unchanged. And when the voltage increases to a certain value, the current begins to sharply reduce. It means the slope is zero at the very beginning and as the increasing of the voltage, the slope rapidly decreases.

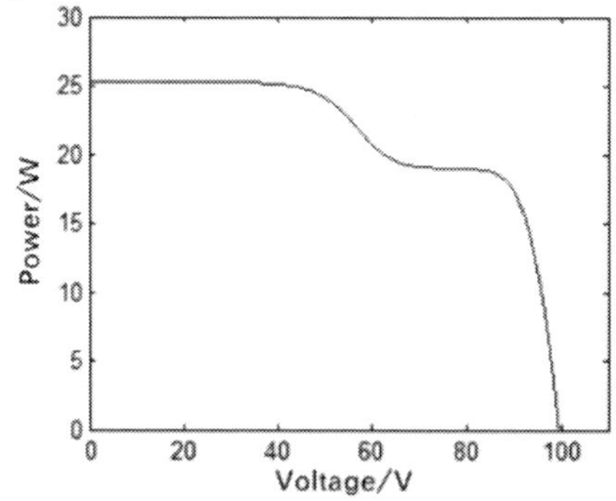

Fig.3 I-U characteristics of shaded array

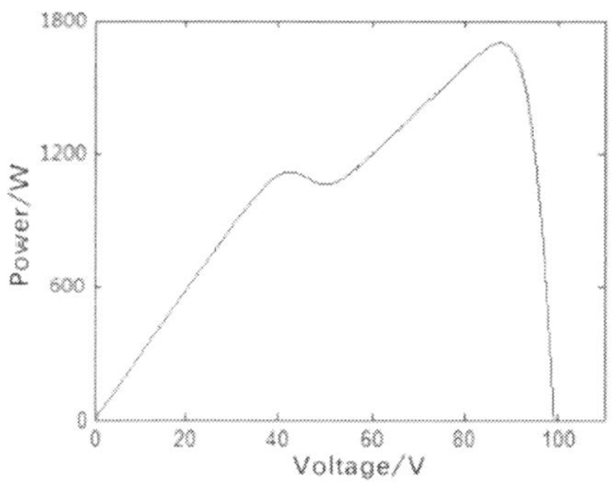

Fig.4 P-U characteristics of shaded array

By observing these characteristics, the maximum power point shows up only after the slope start to quickly decline. Then it is reasonable to initialize the particles of the PSO algorithm at a voltage value where the slope starts to quickly change. Here we choose the voltage at the point 'slope=-3' as the initial value of particles. Then instead of searching the whole voltage, the PSO algorithm just need to search the voltage from this initial value to the end. By doing so, it can helps narrow the searching area and track MPP faster.

3 PSO-Based MPPT

3.1 General Overview of PSO

Particle Swarm Optimization technique is an intelligent optimization technique developed by Dr. Eberhart and Dr. Kennedy[14], inspired by social behavior of bird foraging. The basic idea of Particle Swarm Optimization algorithm is through the cooperation and competition between particles to find the optimal solution.

PSO is initialized with a group of random particles and searches the best solution by updating generations. In each iteration, every particle is updated by private best value and global best value. Private best value is called P_{best}, which is the best solution (fitness) this particle has achieved so far. The global best value (G_{best}) is the best solution (fitness) of the whole group. In each iteration, the particle knows where to go next by compare P_{best} and G_{best} and updates the velocity and the position by the following equations [15].

$$v_i^{k+1} = wv_i^k + c_1r_1(P_{best_i} - s_i^k) + c_2r_2(G_{best} - s_i^k) \quad (4)$$

$$s_i^{k+1} = s_i^k + v_i^{k+1} \quad (5)$$

In the above equations, i is the number of particle, k is the itration time. v_i^k is the velocity of ith particle within k iterations. s_i^k is the position of ith particle within k iterations. The ω is used to keep the balance between global and private search. c_1 and c_2 are learning factors. r_1 and r_2 are two random numbers in [-1,1].

3.2 Improved PSO algorithm in MPPT

As discussed before, due to the influence of partial shading and the mismatching of components, the output characteristics always tend to appear multi extreme situation. In this paper, we are focus on this situation and the others are similar. MPPT method control the output power of PV array by controlling the output voltage. The aim of MPPT method is to make sure the PV system works at a most efficient state.

PSO algorithm is one of the EA algorithms. It is widely used because of it's simple process, ease of implementation, and speed of calculation. Some researchers find that because of particle swarm optimization is based on search optimization, it must be capable of finding the maximum power point whatever the environment changes. The process of PSO algorithm, essentially, is an iterative process. In each iteration, follow the better fitness value, the particle moves toward to G_{best}. After several iterations, the velocity becomes very small or equals to zero, then the maximum power point is fund. In this paper, the objective function is defined as the measurement of PV array's output power and the position of the particle represents PV array's output voltage.

3.2.1 Parameter setting and adjustment

ω is inertia weight, which reflects the degree of inheritance of particle velocity. Smaller inertia weight makes the algorithm trapped in local extreme point and the larger inertia weight facilitate global search. But actually it is not easy to get the exact solution, therefore, to improve the efficiency of optimization, inertia weight is usually set from large to small. Generally c_1 is set from large to small and c_2 is set from small to large. So it is obviously that the adjustment of parameters plays a vital role in the optimization of PSO algorithm. In this paper, we set parameters: ω=1, c_1=c_1=1.49445.

3.2.2 Particle initialization

General the choice of particle's initial value is arbitrary or according to the experience. The PSO methods in reference [8-10] select a number of points, which will reduce the optimization efficiency; The initializing of the particles in these methods is random, which is easy to cause reducing of optimization efficiency. As discussed above, to narrow the searching area, we choose the voltage at the point 'slope=-3' as the initial value of particles. By comparing to the traditional PSO algorithm, this method can track MPP faster.

3.2.3 Terminate and restart conditions of PSO

Similar to the traditional PSO algorithm, the proposed method also need to set up the terminate and restart conditions. The terminate condition is applied to avoid the oscillation near MPP. And the restart condition is used to restart PSO algorithm when the environment changed.

$$\frac{|P(s_{i+1}) - P(s_i)|}{P(s_i)} > \Delta P \quad (6)$$

In equation (6), $P(s_i)$ is the current maximum power. $P(s_{i+1})$ is the output power of the next moment. ΔP is power variation. And we set ΔP=0.15, which means PSO algorithm will be restarted when the power variation is greater than 0.15.

4 Simulation and Results

4.1 MATLAB simulation model Positioning

MATLAB simulation model, as shown in Fig.5, is established in accordance with the contents mentioned above. To facilitate the analysis, simulation model of MPPT is established based on the BOOST circuit. The improved PSO algorithm is implemented by S-Function of MATLAB. In this study, we set DC bus voltage capacitanc C_2=1000μF. The inductance and capacitance of BOOST circuit are L=0.1mH and C_1=1000mH, The load resistance is set to R=8.2Ω.

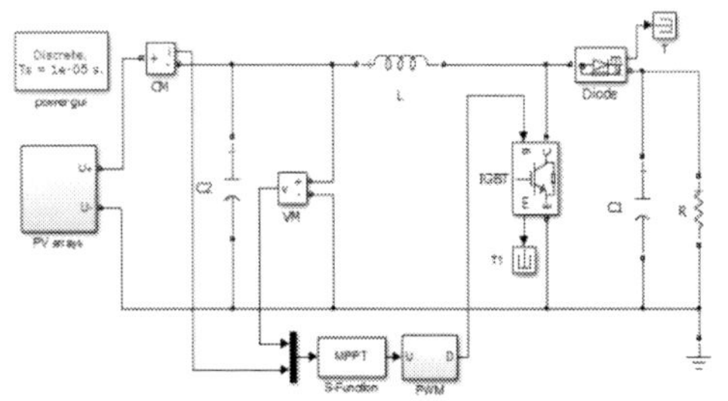

Fig.5 Simulation model of MPPT based on improved PSO

In this study, we set the output voltage of PV array to U and the duty cycle of BOOST circuit to D. Load the booster of voltage U0 charged with constant value = 120 v, the calculation formula of duty cycle is as follow:

$$D = 1 - U / U_0 \qquad (7)$$

Before running the PSO algorithm, the BOOST circuit's duty cycle is set to 0 to get the open-circuit voltage, which is used to determine the bound of particle's position. In the optimization process, the positions of particles are updated by equation (4) and (5). Then the equation (7) is used to update duty cycle. Fitness values are updated by the output power of PV array. Finally, the method can make sure the PV system works at a most efficient state by optimizing the fitness value of each particles.

Under shading condition, the PV array's output characteristic are shown in Fig.3 and Fig.4. By observing the I-U characteristic, the PV system has two extreme value points. These two points are suitable as the initial value of the particles. From the I-P characteristic we can find that under this condition system does appear two extreme value point. One is local extreme value point $P_1 = 1098.4W$ and another one is global extreme value point $P_2 = 1741.5W$. Therefore, to improve the tracking efficiency of the PSO algorithm, it is reasonable to initialize the particles' position according to the I-U curve.

4.2 Results

Two versions of PSO are implemented in this study. The results are shown as below.

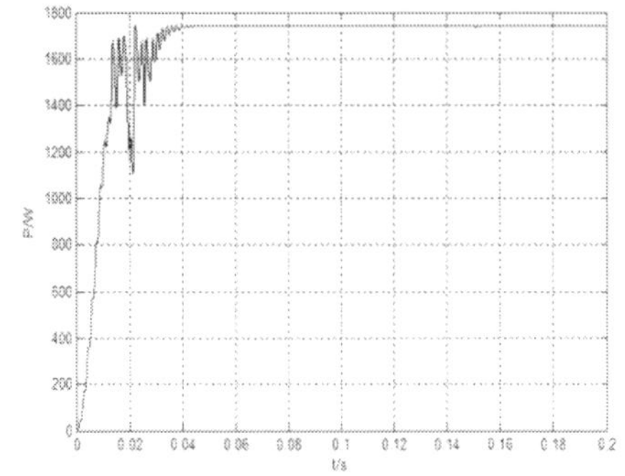

Fig.6 The tracking result of PSO under shading

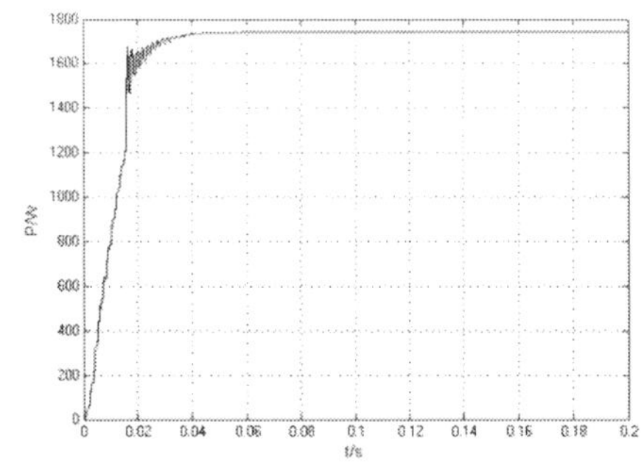

Fig.7 The tracking result of improved PSO under shading

Compared the basic PSO algorithm and improved PSO algorithm, we can find that under the condition of the partial shading both can accurate tracking to the MPP. Fig. 6 shows that when employing the basic PSO algorithm, we get large fluctuations in the optimization process as a result, which means it may cause misjudgment of MPP, and it took 0.044 seconds to find the MPP. The improved PSO algorithm shown in Fig.7 shows a visibly smoother curve and took only 0.037 seconds to find the MPP, which is an improvement of 0.007 seconds. Therefore, this is definitive proof that the optimization makes the process more stable, and thus we can conclude that the improved PSO algorithm is more advantageous to the overall performance of the system.

5 Conclusion

In this paper, an improved PSO algorithm is used to track the MPP of a PV system. In order to improve the tracking speed, an initial value, determined by the analysing of the I-U and P-U characteristics curve, is given to the particles of PSO. In this proposed method, the fitness value is no longer determined by iteration and it is determined by the real-time sampling of output power. It is shown that the proposed method can track MPP faster and has less fluctuation. The efficiency and practicability of the algorithm has been improved.

Acknowledgement

The authors would like to thank the Natural Science Foundation of Zhejiang, China (Grant No. LYE0700006).

Reference

1. A Chen, L Feng, C S Du, C H Zhang, Modeling of Photovoltaic Array Based on Support Vector Machines Under Partial Shaded Conditions, Transactions of China Electrotechnical Society, Vol.26, No.3 (2011)
2. Y H Yang, K L Zhou, Photovoltaic Cell Modeling and MPPT Control Strategies, Transactions of China Electrotechnical Society, Vol.26, Sup.1 (2011)
3. M Dong, J Yang, K Peng, An Luo, Zero Average Incremental Conductance Maximum Power Point Tracking Control for Photovoltaic System, Proceedings of the CSEE, Vol.30, No.21 (2010)
4. Parlak K S, Can H. A new MPPT method for PV array system under partially shaded conditions[C]. IEEE International Symposium on Power Electronics for Distributed Generation Systems (PEDG), 437-441 (2012)
5. X Zhang, R X Cao, Solar Photovoltaic(PV) Grid Generation and Its Inverter Control[M], Mechanical Industry Press (2011)
6. Ramaprabha R, Mathur B, Ravi A, et al. Modified Fibonacci search based MPPT scheme for SPVA under partial shaded conditions[C]//3rd International Conference on Emerging Trends in Engineering and Technology. Goa, India: IEEE, 379-384 (2010)
7. Z Cheng, H Zhou, H Z Yang. Research on MPPT control of PV system based on PSO algorithm[J]. Control and Decision Conference (CCDC), Chinese, 887-892 (2010)
8. Y W Zhu, X C Shi, Y Q Dan, P Li, W Y Liu, D B Wei, C Fu, Application of PSO Algorithm in Global MPPT for PV Array, Proceedings of the CSEE, Vol.32, No.4 (2012)
9. Miyatake M, Veerachary M, Toriumi F. Maximum power point tracking of multiple photovoltaic arrays : a PSO approach[J], IEEE Transactions on Aerospace and Electronic Systems, 47(01) :367-380 (2011)
10. Phinmasone V, Kondo Y, Kamejima T, et al. Evaluation of Extracted Energy from PV with PSO-based MPPT against Various Types of Solar Irradiation Changes[C]. International Conference on Electrical Machines and Systems (ICEMS). IEEE, 487-492 (2010)
11. H H Bian, Q S Xu, S Gao, Kazuto Yukita, Katsuhiro Ichiyanagi, Operation Mismatches of Photovoltaic Array Considering Random Shadows, Transactions of China Electrotechnical Society, Vol.25, No.6 (2010)
12. L Tang, C B Zeng, H Miao, W Xu, Y H Zhang, Y Y Liu, A Novel Maximum Power Point Tracking Scheme for PV Systems Under Partially shaded Conditions Based on Monte Carlo Algorithm, Transactions of China Electrotechnical Society, Vol.30, No.1 (2015)
13. Z T Zhai, X F Cheng, J L Ding, J Zha, Characteristics of partially shadowed serial PV module ouput, Journal of Unversity of Science and Technology of China, Vol.39, No.4 (2009)
14. Khanesar M A, Teshnehlab M, Shoorehdeli M A. A novel binary particle swarm optimization[C], Proceedings of IEEE Conference on Control&Automation, Mediterranean IEEE Control Systems Society, 1-6 (2007)
15. Phinmasone V, Kondo Y,Shiota N, et al. The Demonstration Experiments to Verify the Effectiven-ess of the Improved PSO-based MPPT Controlling Multiple Photovoltaic Arrays[C].ECCE Asia Downu-nder(ECCE Asia).IEEE, 86-92 (2013)

An approved frequency tracking method in connected element interferometry

Qiang Zhao*, Kang Gao, and Xiaomin Hou

Department of Optical and Electronic Equipment, Aerospace Engineering University, Beijing, 101416, PR China

Abstract. Carrier tracking loop is a basic method of frequency tracking in CEI (connected element interferometry), but the tracking error is very large or even worse at high dynamic. To solve this problem, a frequency tracking method assisted by open loop frequency estimation is proposed in this paper. First, an open-loop frequency estimation algorithm is used to estimate the frequency of signals at different moments, then, the least squares fitting method is used to estimate the corresponding dynamic parameters of different orders. Next, a local signal is generated to compensate the received signal. After eliminating most of the dynamic characteristics, the signal is fed into PLL (Frequency locked loop) and high-precision frequency tracking results can be obtained. Simulation results show that, in the condition of low SNR, the new method has higher accuracy than the method assisted by FLL (Frequency locked loop), and the processing time is shorter. When there is a two order Doppler frequency change, the open-loop assisted method can overcome the steady phase error of the PLL, and is of great significance in the applications of connected elements interferometry.

1 Introduction

CEI is a method of target azimuth measurement, which has the advantages of high accuracy, short baseline and flexible station arrangement [1]. However, it has a very strict requirement for the measurement accuracy of signal frequency and phase [2]. Frequency tracking loop is a common method, but its accuracy is difficult to meet the requirements under high dynamic condition [3-5]. How to balance the contradiction between dynamic stress and tracking accuracy is a difficult and hot issue.

At present, the carrier tracking method under high dynamic condition including the method based on kalman filter, the method assisted by FLL, and the method assisted by maximum likelihood estimation. Among them, kalman filter is used to estimate carrier phase difference, Doppler frequency and its change rate, and can adaptively adjust the loop parameters, but the uncertainty of the environment factors on the stability of the system greatly, and the computation is high, not easy to be implemented [6,7]. FLL has good resistance to dynamic stress, is responsible for eliminating most of the dynamic characteristics, but its tracking accuracy is low. The principle of this method is simple, but because of the large bandwidth of FLL loop, more noise is introduced into the tracking loop [8,9]. Maximum likelihood estimation is a kind of asymptotically unbiased estimation. The variance of parameter estimation can be close to the Cramer-rao lower bound (CRLB), but the storage and computation are relatively large, and the real-time application is greatly restricted [10, 11]. This paper introduces the least square fitting method [12, 13], to predict continuous change of signal frequency by estimating signal frequency of partial period. And then the dynamic characteristic parameters of Doppler can be calculated, on this basis, generating a local signal to compensate Doppler frequency and eliminate most of the dynamic characteristics of the received signal. Finally accurate synchronization is finished by PLL.

2 Traditional frequency tracking loop in CEI

PLL is the common frequency tracking loop in CEI. The tracking error of PLL loop caused by thermal noise is [14]:

$$\sigma_T = \frac{1}{2\sqrt{2}\pi\tau}\sqrt{\frac{N_0}{P_s}B_L} \tag{1}$$

In formula (1), τ is integration time, N_0 is noise power spectral density, P_s is signal power and B_L is bandwidth of loop filter.

It can be seen from the formula (1) that the loop bandwidth determines the tracking accuracy when integration time is fixed. In high dynamic conditions, in order to ensure that the loop does not mismatch, a larger loop bandwidth is generally set, which to some extent sacrifice the tracking accuracy. On the other hand, when there is a two order Doppler frequency change rate, the three order PLL is usually used. At this point, there will be a steady phase error. Assuming the two order Doppler

* Corresponding author: zqaniu@163.com

frequency jerk is R_2, acceleration is R_1 and initial value is f_{d0}, the instantaneous frequency of the signal can be expressed as a formula (2):

$$f_d = 1/3 R_1 t^2 + 1/2 R_2 t + f_{d0} \qquad (2)$$

After the derivation, when the frequency has a parabolic variation, the steady phase error of the three order PLL is:

$$\begin{aligned}\theta_e(\infty) &= \lim_{s\to 0} s\cdot\theta_e(s) = \lim_{s\to 0} s\cdot H_e(s)\cdot\theta_1(s) \\ &= \lim_{s\to 0} s\cdot\frac{s^3}{s^3+s^2Ka+sKb+Kc}\cdot\frac{2R_2}{s^4} \\ &= \frac{2R_2}{Kc}\end{aligned} \qquad (3)$$

In above formula, $H_e(s)$ is error transfer function, $\theta_1(s)$ is Laplace transform of input signal, K is gain of loop, a, b, c represent the coefficients of the three order loop filter respectively. $c = m\cdot\xi$,generally, $m = 5$, $\xi = 0.707$ (damping coefficient).From formula (4), when the filter coefficient and loop gain are fixed, the larger the jerk, the larger the steady-state phase error.In CEI, it's an important error term and must be elimanated.

In order to improve the frequency tracking precision under high dynamic conditions, the PLL tracking method assisted by FLL is more widely used in space TT & C, and its principle is shown in figure 1.

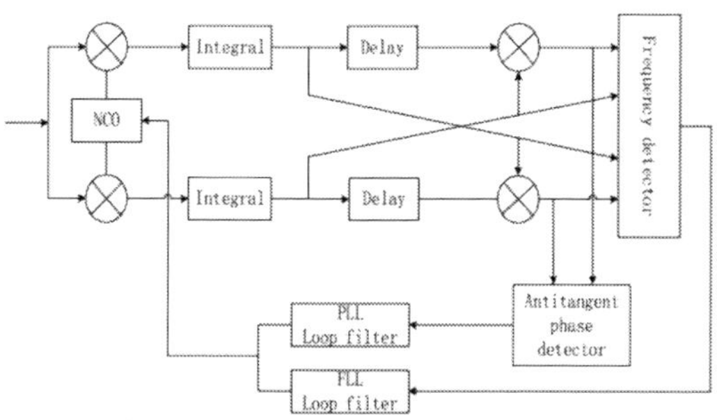

Figure.1. Principle of FLL-PLL frequency tracking loop

FLL is used to eliminate most of the dynamic structure the PLL determines the final tracking accuracy in Figure 1. The structure of this method is relatively simple, but because the FLL loop bandwidth is larger, PLL will also introduce more noise power. When SNR is low the tracking accuracy is low or even lock will be lost.

3 High dynamic frequency tracking algorithm assisted by open loop frequency estimation

In this paper, a high dynamic frequency tracking algorithm based on Rife open loop frequency estimation algorithm and least squares fitting is designed.

3.1 Rife algorithm

Rife algorithm is a classical open-loop frequency estimation algorithm for sinusoidal signals [15, 16], which effectively overcomes the fence effect of the traditional DFT and improves the estimation accuracy greatly. Its expressions is as follows:

$$\hat{f}_c = \Delta f\cdot\left(k_0 - 1 + \alpha\cdot\frac{|G_{k_0+\alpha}|}{|G_{k_0+\alpha}| + |G_{k_0}|}\right) \qquad (4)$$

In formula (4), Δf is frequency resolution of FFT, k_0 is the corresponding position of maximum spectrum, G_{k_0} is the maximum value, $G_{k_0+\alpha}$ is sub large value of spectrum, $\alpha = 1$ or -1 .

3.2 Least squares fitting

The idea of least square fitting is expressed as formula (5):

$$\min_{\varphi} \sum_{i=1}^{m} \delta_i^2 = \sum_{i=1}^{m} (\varphi(x_i) - y)^2 \qquad (5)$$

Assuming that the fitting polynomial is

$$y = a_0 + a_1 x + \cdots + a_k x^k \qquad (6)$$

Then

$$\begin{bmatrix} 1 & x_1 & \cdots & x_1^k \\ 1 & x_2 & \cdots & x_2^k \\ 1 & \vdots & \ddots & x_3^k \\ 1 & x_n & \cdots & x_4^k \end{bmatrix} \begin{bmatrix} a_0 \\ a_1 \\ \vdots \\ a_k \end{bmatrix} = \begin{bmatrix} y_1 \\ y_2 \\ \vdots \\ y_n \end{bmatrix} \qquad (7)$$

It can be written as $X * A = Y$, $A = (X' * X) - 1 * X' * \text{Y}$, thus, the coefficient matrix A can be obtained. Furthermore, the fitting curve can be obtained.

The principle of high dynamic frequency tracking algorithm based on Rife algorithm and least square fitting designed in this paper as shown in figure 2.

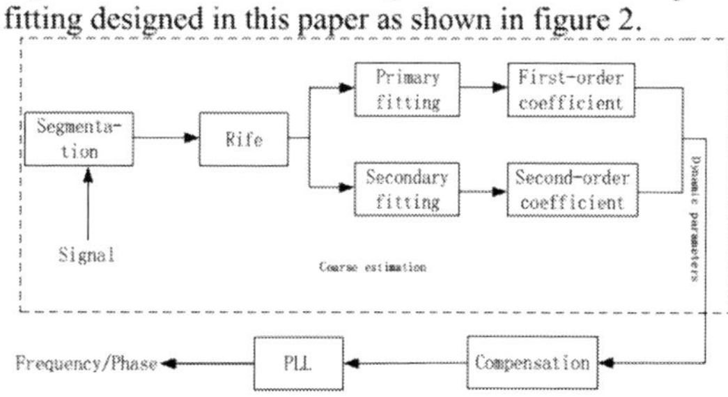

Figure.2. Principle of improved frequency tracking loop

The basic idea is to estimate the instantaneous frequency of different sampling time by using the high precision open loop frequency estimation algorithm, then, the curve of Doppler frequency with time is obtained by fitting the frequency at different time points by least square fitting method. At this time, the Doppler frequency dynamic parameters of different orders can be estimated. These coefficients are used to construct the local signal to compensate the received signal, eliminate most of the dynamic. At last, PLL is used to get a high accuracy frequency tracking result.

4 Simulation experiment

4.1 High dynamic model and simulation parameters

The high dynamic model of JPL is adopted in this paper. JPL proposed that the high dynamics is a relative concept, and considered that when the speed and acceleration value of the target is relatively large in reality, it is the high dynamic situation [4], and the high dynamic model given by it is shown in figure 3.

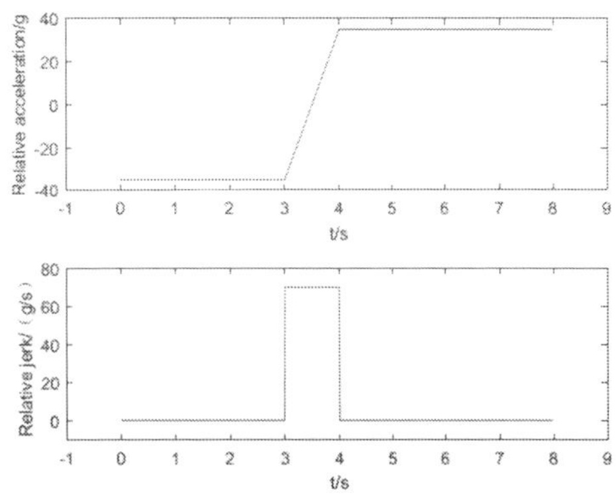

Figure.3. High dynamic acceleration model of JPL

In Figure 3, the acceleration corresponds to the first order change rate of the Doppler frequency, and the jerk rate corresponds to the two order change rate of the Doppler frequency. In practice, although there may be three or even higher order information, but the value is generally very small, therefore, only two order change rate is considered.

According to the JPL model, the Doppler frequency is set as follows:

$$f_d(t)=\begin{cases}-7500t+f_{d1} & 0\le t<3\\ 15000t^2-7500t+f_{d2} & 3\le t<4\\ 7500t+f_{d3} & 4\le t<8\end{cases} \quad (8)$$

f_{d1}, f_{d2} and f_{d3} represent the initial values of Doppler frequency at the beginning of three periods respectively. The received signal is a standard sinusoidal signal, which is expressed as follows:

$$s(t)=A\cos\left(\left(f_c+f_d(t)\right)t+\Phi\right)+N(t) \quad (9)$$

In formula (9), A is signal amplitude, f_c is carrier frequency, Φ is initial phase, $N(t)$ is additive Gauss white noise.

The simulation is divided into three stages according to the Doppler frequency change and parameter settings are shown in Table 1.

Table 1. Simulation parameter settings.

Parameter	Value	Parameter	Value
f_{d1}	500Hz	f_c	14MHz
f_{d2}	500Hz	f_s	56MHz
Φ	0	Segment time of Rife	0.01s
C/N_0	50	Integral time of loop	0.001s

4.2 Simulation results

4.2.1 Stage one

When PLL is used for tracking directly, $B_L=55\text{Hz}$ .The tracking results are shown in Figure 4.

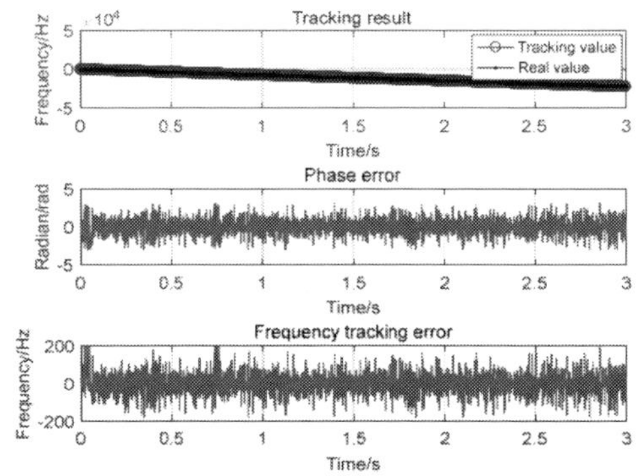

Figure.4. PLL direct tracking results

After fitting the tracking results, the Doppler frequency acceleration is -7.486 KHz/s, and the Doppler frequency initial value is 466.9 Hz.

The fitting results of estimated frequency from Rife of 300 segment signals are shown in Figure 5.

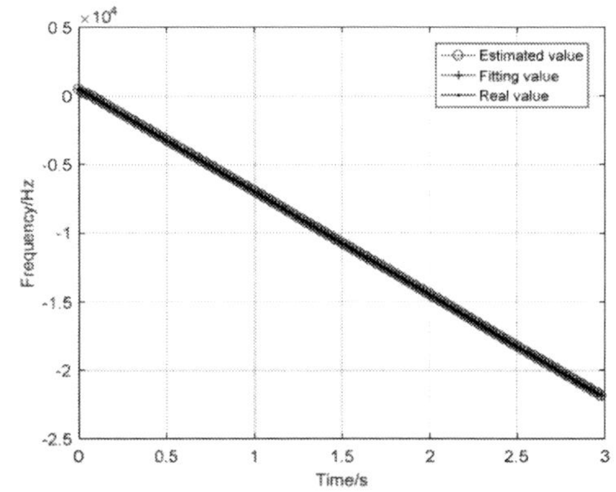

Figure.5. Frequency estimation results of Rife

The Doppler frequency acceleration estimation value is -7.469 KHz/s, and the Doppler frequency initial estimation value is 503.9 Hz. After the signal is compensated, the PLL tracking results are shown in Figure 6. At this time, $B_L=5\text{Hz}$.

After fitting the tracking results, the Doppler frequency acceleration is -30.9 Hz/s, and the Doppler frequency initial value is -3.88 Hz.

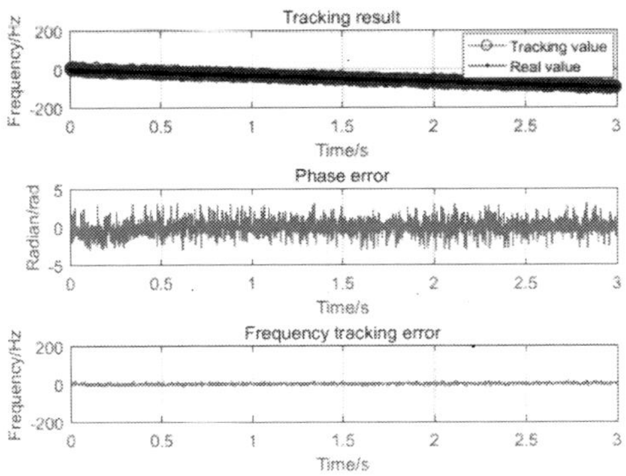

Figure.6. PLL tracking results after compensating

4.2.2 Stage two

When PLL is used for tracking directly, $B_L = 60\text{Hz}$ .The tracking results are shown in Figure 7.

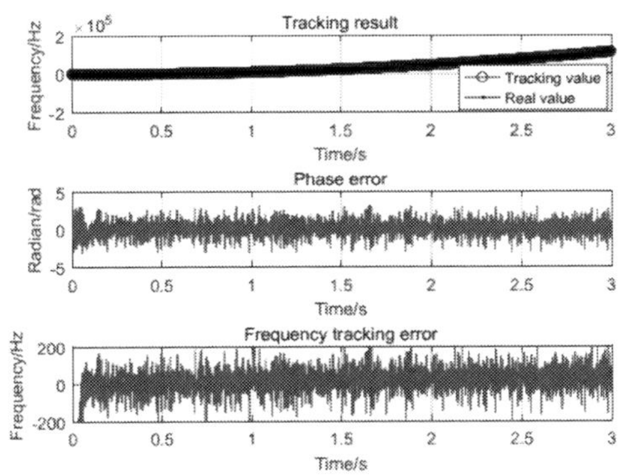

Figure.7. PLL direct tracking results

After second order polynomial fitting the tracking results, the Doppler frequency jerk is 14.987KHz/s^2, the Doppler frequency acceleration is -7.437 KHz/s, and the Doppler frequency initial value is 459.7 Hz.

The fitting results of estimated frequency from Rife of 100 segment signals are shown in Figure 8.

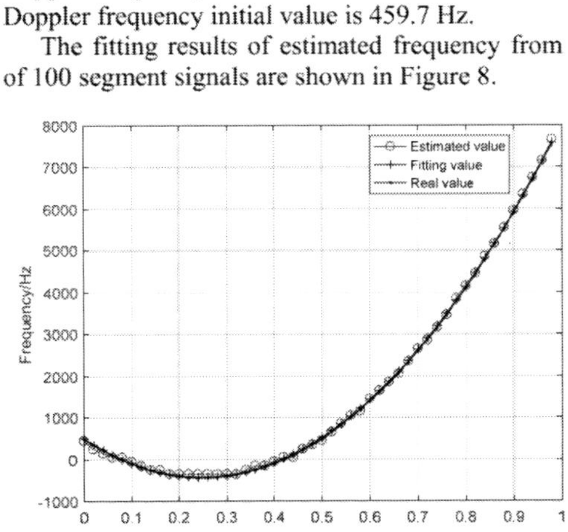

Figure.8. Frequency estimation results of Rife

The Doppler frequency jerk estimation value is 14. 966 KHz/s^2, acceleration estimation value is -7.407 KHz/s, and the initial estimation value is 499.7 Hz. After the signal is compensated, the PLL tracking results are shown in Figure 9. At this time, $B_L = 30\text{Hz}$.

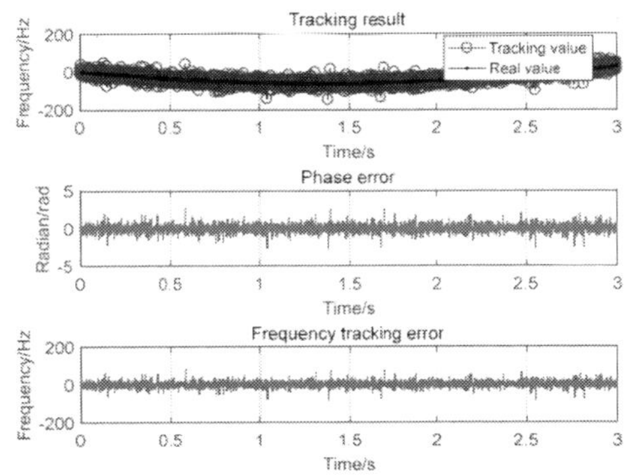

Figure.9. PLL tracking results after compensating

After fitting the tracking results, the Doppler frequency jerk is 33.91 Hz/s^2, acceleration is -92.7 Hz/s, and initial value is 0.18 Hz.

4.2.3 Stage three

The processing method of stage three is the same as stage one.

It can be seen from the above simulation results that the tracking accuracy is improved obviously and the steady phase error is smaller when there exists a jerk.

4.3 Comparison of estimation accuracy of FLL and Rife for dynamic characteristic parameters

Compared with the FLL-PLL frequency tracking method, the difference of the method proposed in this paper is that the open-loop frequency estimation (Rife algorithm) and the fitting method are used to eliminate most of the dynamic characteristics of the signal instead of FLL. Therefore, the final frequency tracking precision determined by FLL and Rife algorithms. The Doppler dynamic parameter estimation errors of FLL and Rife algorithms under different SNR are shown in Table 2 (1000 Monte Carlo simulations):

It can be seen from table 2 that when the SNR is low, the estimation accuracy of Rife algorithm is much higher than that of FLL. With the increase of SNR, the estimation accuracy of the two algorithms is improved significantly and tends to be stable. When the SNR is high, the dynamic parameter estimation performance of Rife algorithm has little difference with that of FLL.

Table 2. Doppler dynamic parameter estimation errors of FLL and Rife (B_L=10Hz)

C/N_0 (dB/Hz)	Algorithm	Initial Value(Hz)	Acceleration (Hz/s)	Jerk (Hz/s^2)

50	FLL	93	311	244
	Rife	5	83	65
55	FLL	41	140	104
	Rife	4	83	61
60	FLL	24	71	46
	Rife	4	83	61
65	FLL	24	69	46
	Rife	4	83	61
70	FLL	23	66	45
	Rife	4	83	61

5 Conclusion

This paper proposes an approved frequency tracking method in connected element interferometry. The estimation and compensation of Doppler dynamic parameters overcome the limitation of the dynamic stress on the traditional PLL tracking capability. Simulation results show that the accuracy of the new method is much higher than that of the traditional PLL method. At the same time, the dynamic parameter estimation performance of Rife algorithm is better than that of FLL under low SNR.

References

1. Edwards C. Goldstone intracomplex connected element interferometry. TDA Progress Report, 1990 ,42-101.
2. L. Huang, H.T. Li, W.H. Hao. Journal of Spacecraft TT&C Technology, **33**(2014),371-376.
3. P. Lian. Optical Engineering,**43**(2004),1552-1560.
4. Kamel M, Borio D, Nielsen J, et al. The Institute of Navigation GNSS, San Diego, 2011, 374-383.
5. X. Zhang,C. Guo, J. Cao. International Conference on Communications, Circuits and Systems, 2014,243-246.
6. W. Li,S. Liu,C. Zhou,et al. High Dynamic Carrier Tracking Using Kalman Filter Aided Phase-Lock Loop,2007,673-676.
7. F. Shen, R. He, D.Z. Lv, et al. Journal of Astronautics, **33**(2012),1041-1047.
8. X.D. Deng, W. Sun. Modern Defence Technology, **38**(2010), 137-141.
9. W.Q. Mou, R.K. Liu, Q.P. Song, et al. Journal of Beijing University of Aeronautics and Astronautics, **41**(2015),1918-1924.
10. H. Xun, X.R.. Tao, R. Tao, et al. Journal of Electronics & Information Technology, **36**(2014), 577-582.
11. J.R. Cheng, G.B. Liu, Q. Zhang, et al. Journal of Astronautics, **36**(2015),103-108.
12. Rife D C, Boorstyn R R. IEEE Trans Inform Theory, **20**(1974), 591-598.
13. X.D. Wang, Y. Liu, Z.M. Liu. Systems Engineering and Electronics, **20**(2008),621-624.
14. J.X. Liu. Spacecraft TT & C and information transmission technology, National Defense Industry Press, 2011, 74-75.
15. L.D. Li.. Journal of Liaoning Technical University(Natural Science), **30** (2011),202-204.
16. C. Wang, H.P. Wang, W. Xiong, et al. Acta Aeronautica et Astronautica Sinica, **37**(2016), 1603-1613.

Research on variant design based on topological entity compression algorithm

Yulin Shuai [a], *QiuJiao* Wang, and *Juan* Chen
School of Mathematics, Southwest Jiaotong University, Chengdu 611756, China

Abstract. In order to improve the efficiency of product design and shorten the design cycle of product, the research progress in recent years is introduced. The basics of variant design such as the entities of topology, the equation's structure and the compression/decompression are analyzed. An algorithm, which can compress redundant parts, based on the conditions of algebraic equations completely is put forward. A variant design system of volumetric heat exchanger is designed by using VB and SolidWorks API (Application Programing Interface). The results show that the variant design system based on algebraic equations can be quickly and accurately modeled according to different process parameters, which can improve the efficiency of product design and shorten the design period.

1 Introduction

In 3D graphics, entity refers to all objects composed of visual graphics, such as point, line and surface elements, stretching, rotation and other elements, product components, with the topology relationship between various entities. The current mainstream design software of CAD (Computer Aided Design) to the present and the design of visual graphic, such as CATIA SolidWorks and Pro/e are using the way of topological entities. The variant design refers to the basic principle of the product does not change, only through specific parameters changing entities or different entities topological relation, and the rapid design of new products [1-2].

Qin chao ruan[3],Li fen liang [4] have studied the variant design based on parameters configuration, by giving some parameters multiple discrete values, establishing different parameters, Form different configurations of an entity. When the technological parameters are changed, selecting the corresponding configuration of physical parameters to change the parameters. However, the method can only assign discrete values to the entity parameters and parameter values can't be changed continuously. Chen qiang luo [5] of the variant design based on the two development of the way, mainly through the VB, C++ and other high-level programming language called 3D software driver software of API drawing, can achieve larger. But the method involves variant design function of API, many built-in the complex structure and long development cycle.

The variant design based on algebraic equation can establish algebraic equation in different parameters, which can avoid the complicated programming, and can make the parameters continuous change, can avoid the deficiency of the 2 kinds of variant design., Gao Jie[6] ,Song Long [7] of variant design based on algebraic equations, greatly improves the design efficiency of graphics. But only in the entity hierarchy which topological entities are not involved in assembly level compression.

In this paper, based on the document [6-7], under the SolidWorks software environment, a topology entity compression algorithm is proposed, and the variant design based on topology entity compression is studied at the assembly level.

2 Solidworks orks variant design basis

2.1 Entity topology structure

The SolidWorks' entities topology structure in general as shown in Figure 1, the S assembly contains a sub assembly S1 ,S2 and nesting child assembly S3, and the component P1, P2. The cooperation parts of components belongs to the two common parent file [8], such as with the relationship between P1 and P2 which belong to S3, if P1 move up one level, and tied with S3,it will move to S1; if the P1 moved to S2, with the relationship will move to the top assembly S.

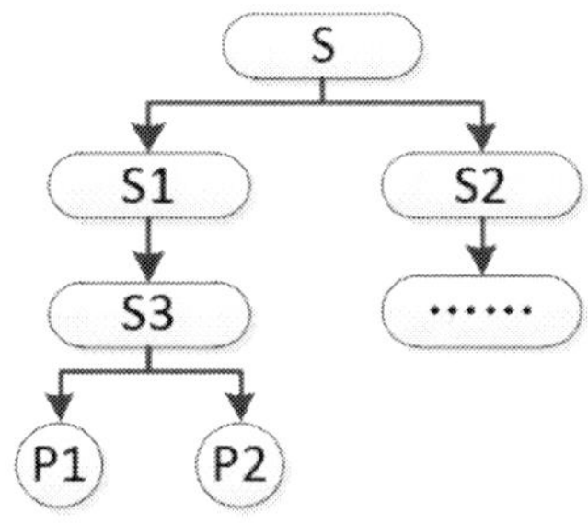

Figure. 1

2.2 Assembly equation structure

In SolidWorks assembly, the commonly used equation forms are mainly 3 kinds of "global variables", "characteristic" and "equation - Parts", as shown in Figure2.

"Global variables" refers to change the underlying active parameters of the parameters, the representation is A=P; "features" for the compression or decompression parts,the representation is B=S;" equation - Parts" refers to the size parameters of parts, the representation is C@D@B-i=P

Among

A -- Global variable name

B -- The name of the component

C -- Parameter names for parts sketches or features, such as length, width, and height

D -- Part sketch or feature name

P -- The numerical of a variable or algebraic equation

S -- Parts status. There are 2 states of compression and decompression

I -- Identification code of parts .In SolidWorks assembly will be used in multiple instances of a component, which is to add multiple parts to the same assembly, parts identification code for each instance to distinguish between parts, according to the order assignment is added to the assembly of 1, 2, 3...... .

Examples of the 3 representations are as follows:

" Diameter"=40

"Part1-1. part"=suppressed

"H@Sketch1@Part-1. Part"=unsuppressed

2.3 Compression and decompression

Compression of topological entities (suppressed) refers to the parts removed from the model in computer memory, and hidden from the assembly, but does not remove it, before relieving compression (unsuppressed), it makes temporarily unavailable. Compression can reduce the storage and memory, compression plays an important role in the graphic image file storage, transmission and processing.

There are 2 general ways of compression or decompression of an entity, unconditional execution command, such as the above formula gives the suppressed or unsuppressed command directly on the right end; second is the condition of compression, that meet compression or compression of entities will be existed if certain conditions satisfied.

"Part2-1. Part"=IF("Diameter"
<50, suppressed, unsuppressed)

In this paper, an algorithm for condition compression of topological entities is proposed. The components used in the previous stage are added to the assembly, and the constraints are set. The input condition is used to determine whether to compress or decompress of the entity.

3 Heat exchanger structure

DOI: 10.1051/matecconf/201713900054

3.1 Composition of heat exchanger

Volume type heat exchanger is a key product of the centralized water supply system, widely used [10] with the water storage capacity is large. Its shape as shown in Figure 3, which is mainly composed of a cylinder body 1, cylinder head 2, the heat exchanger core component 3 (including tube box, flange head, etc.), hot import and export component 4, the refrigerant components import and export group5, tube box and outlet6, lug 7, pressure gauge8, instrument tube 9, emission components 10, base 11 and other parts. There are a large number of driven parameters, whether early design or late manufacturing engineering drawings, all need to spend a lot of time, variant design of its products is necessary.

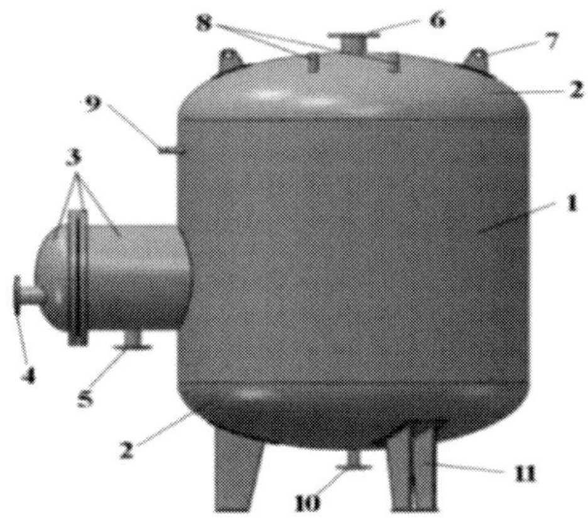

Figure. 3

3.2 Ellipsoidal head

The head of heat exchanger is usually ellipsoidal head, whose plane projection is elliptical, and the semi long axis a and half short axis b meet a=2b

Elliptical semi long axis a and barrel diameter D meet a =D/2

As shown in Figure 4,the OA is X axis, the OB is Y axis, establishment of coordinate system, and safety valve pipe length is L_0, coordinates of M with the highest point branch need to obtain, and then taking M as the starting point down the stretch entity to the exchanger head.

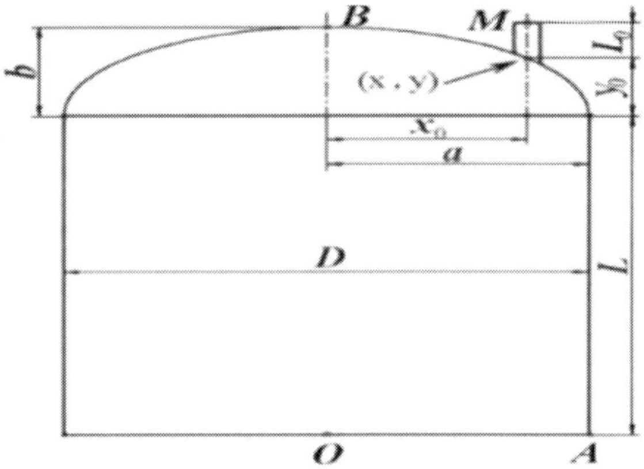

Figure. 4

The distance between one point of the ellipse (x, y) and the long axis and the short axis is y_0 and x_0 respectively, and the two relations satisfy the relation

$$y_0 = \frac{b}{a}\sqrt{\left(a^2 - x_0^{\ 2}\right)}$$

The longitudinal coordinate y_M of the branch starting point M is satisfied

$$y_M = L + y_0 + L_0$$

The abscissa x_M is satisfied with the diameter of the cylinder D:

$$x_M = kD$$

Among them, k is the constant factor .

By combining the above formula with figure 5, we can calculate the relation between the ordinate y_M and the global variable D.

$$y_M = L + \frac{D}{4}\sqrt{\left(1 - 4k^2\right)} + L_0$$

The relation is expressed as follows in the SolidWorks algebraic equation:

$$"Ordinate@Branch@Barral-1.Part" =$$

$$"Length"+"Diameter"/4*sqr(1-4*k*k)+L_0$$

The designer can evaluation L_0 and k according to

 DOI: 10.1051/matecconf/201713900054

the specific process requirements.

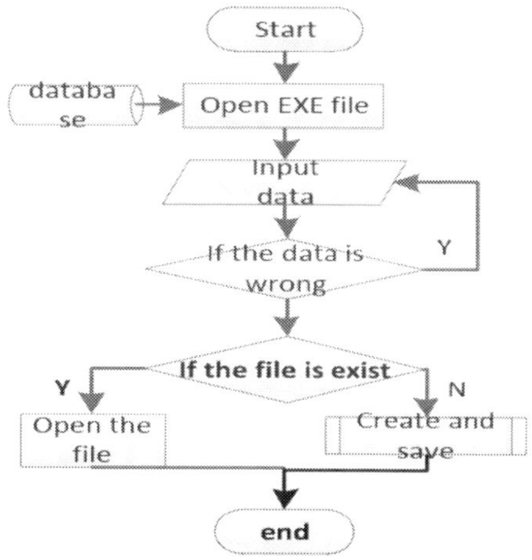

(a)

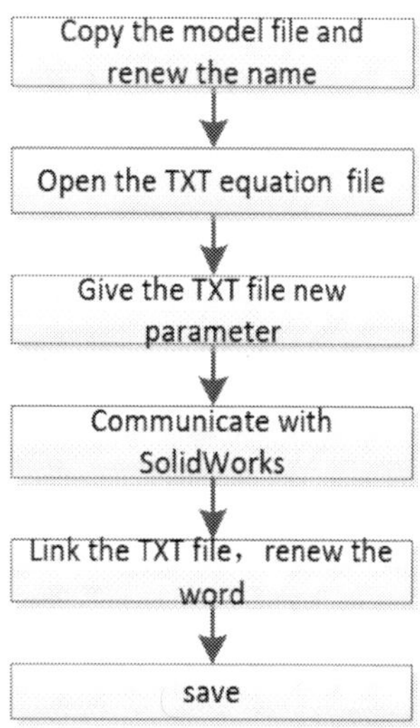

(b)

Figure.5

4 Conditional compression algorithm for topological entities

4.1 Algorithm flow

Although there are quite a number of the heat exchanger parameters, most of the parameters are driven parameters. Using the variant design based on algebraic equations, the parameter setting can be active in the global variable, then driven parameters and establish algebraic equations. When the parameters change ,it can only modify the active parameters, all driven parameters associated with changes so that form the new products.

The algebraic equation can be used with external TXT text file format for data transmission. The programming language generated EXE file, modify the TXT file in the form of expression data to modify global variables and equations. Using API in swEqnMgr:: FilePath links graphics files and TXT files for external.

The algorithm flow chart in figure 5- (a), which including the new sub process and save the document in figure 5- (b). If the model file of the required parameters is already exists, then open the file directly. If the file does not exist, the TXT data file is the first copy the model files and links, and named the model file as an input parameter,it is easy to search; and then open the TXT file, saving the TXT document after modify the parameters corresponding to the value of the global variable; finally open model file reconstruction (refresh) model and save. Modify global variables, all the equations solved automatically.

4.2 Main code of algorithm

To achieve the above algorithm Visual Basic program, the core code is as follows:

```
......
DocName = "RH"
N = combo.UBound
For i = 1 To N
DocName = DocName & "-" & combo (i)   ' Use the file's own parameters to name the file
Next i
DocPath = App.Path & DocName & ".SLDASM"   ' File (assembly) path
If Dir(DocPath) <> "" Then   ' Determine if the file already exists
Msgbox "The file already exists!"
......
Else   ' If it does not exist, new
Open App.Path & "\ RH.txt" For Input As #1   ' Open file
......  'Change the original parameter to the new parameter and write to the TXT file
Set SWAPP = Create Object ("SldWorks. Application") '
Communicate with SolidWorks
Set Model = SWAPP.NewDocument (App. Path & "\" & DocName & ".SLDASM", 0, 0, 0)   ' Open file
```

 DOI: 10.1051/matecconf/201713900054

```
Set swEqnMgr = Model.GetEquationMgr
swEqnMgr.FilePath = App.Path & "\" & DocName &
".txt"   ' Link external TXT files
boolstatus = Model.EditRebuild3()   ' Reconstruction
model
End if
......
```

4.3 Examples of conditional contraction of topological entities

The nominal diameter of the conventional volumetric heat exchanger is generally 800mm~1600mm, and the nominal diameter of the outlet of the heat exchanger with different parameters is selected in the set $Q = \{32,40,50,65,80\}$, and the nominal diameter of the flange is also one of the 5 parameters mentioned above

At the beginning of the design template, the flanges of the above 5 parameters are added to the assembly and are exactly the same as the drain pipe. Then the following equations are set

"*Flange name*-identification code. Part"
=IF("Outlet diameter"=Specificparameters , unsuppressed, suppressed)

For example, the flange is named "FL- nominal diameter", and the identification code for the discharge flange is 2. The algebraic equation for the flange with nominal diameter 32mm is shown as follows

"FL-32-2. Part"=IF("Outlet diameter" =32, unsuppressed, suppressed)

When the nominal diameter of the discharge pipe is equal to 32mm, the flange is decompress. Otherwise it has been in a compression state. Thus, the flange assembled with the discharge pipe will change accordingly with the change of the parameters of the sewage pipe.

4.4 Multiple conditional judgment

For parameters that vary in gradient according to global variables, the IIF function can be used to establish equations for handling multiple conditions, such as the height of the bearing and the nominal diameter, which correspond to the relation shown in Table 1.

Table 1

Nominal diameter of heat exchanger	Bearing height
800 ~ 1000	350
1100 ~ 1400	420
1500 ~ 1800	460

For the height of the support, write the algebraic equations as shown below so that they vary automatically according to the input parameters.

$$"h@H@Part-1.Part" = IIF("Diam"\langle=1000,350, IIF("Diam"\langle=1400,420,460))$$

4.5 Algorithm analysis

The variant design provided the mainstream CAD the design software, variant design based on the parameter configuration only give to the entity parameters discrete values ,parameter values can't be changed continuously, the latter use is restricted, the life cycle of variant design system is short, the preliminary design is mainly applied to small variant entities; variant design based on API develop of two times need to use advanced programming language called software's API, and the API number is large, complex structure, the long development cycle.

The proposed algorithm based on topological entity compression between the parameters of different parts established algebraic equation, the parameter changes continuously, no parameter group can significantly extend the discrete additions, prolong the system life cycle; only by programming in terms of parameter transfer and interactive interface, avoids the complex API programming, greatly shorten the system design cycle. In addition, the algorithm of top-down, variant design with global in the assembly level, is suitable for the sketch, parts and assembly, also avoids the shortcomings of the algorithm only applies to parts of entities in the literature [6-7].

5 Conclusion

DOI: 10.1051/matecconf/201713900054

A algorithm based on redundant topology entity conditions compression, the study of variant design in assembly hierarchical top-down, parameters in different parts were correlated by algebraic equations. Using VB language to implement the algorithm, and design the heat exchanger variant design system. The variant design system base on topological entities compressed can quickly and accurately process variant design according to different parameters, which can avoid the complex programming and a large number of parameters calculate, and can make the parameters change continuously, improve the efficiency of product design, shorten the design cycle of the system.

This work was partially supported by the Basic Research Fund of Central University (No.2682014ZT28).

E-mail addresses:1021734635@qq.com(Y.Shuai)

Reference

[1]T. T. qian,Study on variant design method and application of hydraulic support [D]. Suzhou: Suzhou University, (2015).

[2]C.G. Yao, Y. H. Sun, B.Kuang, et al. Variant design of assembly parts based on parametric technology [J]. Computer system applications,**21,**(2012), 160-163.

[3]Q.C.Ruan, W. P.Dong. Parametric modeling of SolidWorks based wedge mechanism [J]. Journal of Shanghai University of Engineering Science, **30**(2016), 149-154.

[4]L.F.Liang, Z.Y.Wang,S.F.Wu, et al. Parametric design of reducer case structure based on SolidWorks configuration technology [J]. Mechanical transmission,**39** (2015), 80-83.

[5]C.Q.Luo,L.W.Zhong,J.Zhu,Rapid design system of crane lifting mechanism based on configuration technology [J]. Computer system applications, **21** (2012) ,18-21.

[6]J.Gao, Research and development of construction hoist CAD system based on [D]. Solidworks Xi'an: Chang'an University, (2013).

[7]L.Song,L.Ma,Z.X,Jia. Research on parametric technology of drive mechanism of crank press based on [J].. SolidWorks forging technology, **40** (2015) ,107-111.

[8]SolidWorks® company SolidWorks part and assembly tutorial [M]. Beijing: Mechanical Industry Press,(2014): 328-394.

[9]C. C.Wang,L.J.Wang,Y.Z.Zhao, et al. Study on automatic assembly technology of Modular Parallel Robot Based on [J].SolidWorks mechanical design and manufacture, (2009), 166-168.

[10]C.Lu,D. Z.Zhu. The experimental study on water supply by coil and volumetric heat storage water tank [J]. Journal of solar energy,**29**(2008), 569-573

Name	Value / Equation	Evaluates to
⊟ Global Variables		
"Diameter"	= 40	40
Add global variable		
⊟ Equations - Components		
"D2@Sketch5@DZ-1100-10-1.Part"	= "Diameter" / 2	20deg
Add equation		
⊟ Dimensions - Top Level		
D3@LocalCirPattern1	180deg	180deg
D1@LocalCirPattern1	2	2
D1@Distance1	200mm	200mm

Figure.2

Heterogeneous modelling and finite element analysis of the femur

*Binkai Zhang, Xiang Wang**, *Yangyang Yu, Jinjin Zheng.*

Department of Precision Machinery and Precision Instrumentation, University of Science and Technology of China, Hefei, Anhui 230026, China

Abstract. As the largest and longest bone in the human body, the femur has important research value and application prospects. This paper introduces a fast reconstruction method with Mimics and ANSYS software to realize the heterogeneous modelling of the femur according to Hu distribution of the CT series, and simulates it in various situations by finite element analysis to study the mechanical characteristics of the femur. The femoral heterogeneous model shows the distribution of bone mineral density and material properties, which can be used to assess the diagnosis and treatment of bone diseases. The stress concentration position of the femur under different conditions can be calculated by the simulation, which can provide reference for the design and material selection of prosthesis.

1 Introduction

Hip joint and knee joint are important load-bearing parts of the human body, and also vulnerable to injury and inflammation. Joint replacement is a prevalent approach to achieve function restoration of degenerative joint diseases [1]. Therefore, the analysis of the mechanical properties of the femur is of great significance in the design and manufacture of the implantable prosthesis [2].

With the development of computer technology, finite element analysis is more and more applied to the research of bone biomechanics [3]. To obtain better analysis results, the skeleton finite elements are required to describe the geometry and material properties in detail. Compute tomography (CT) has a high contrast between bone and surrounding soft tissue and can accurately describe bone geometry. Moreover, the CT value (Hounsfield unit, Hu) has an approximate linear relationship with bone density and intensity. Therefore, CT data are widely used in bone finite element modelling [4].

In this paper, in consideration of the heterogeneity of the femoral material, the Mimics and ANSYS software were used to determine the distribution of bone mineral density and material properties through CT image information. This femoral heterogeneous model can be rapidly reconstructed and provide data support and theoretical guidance for the diagnosis and treatment of bone diseases. By the finite element analysis in different simulations, we can find out the mechanical properties and weak points of the femur, which are is helpful for design and material selection of prosthesis.

2 Surface reconstruction

Mimics (Materializes Interactive Medical Image Control System) is a medical image processing software for 3D design and modelling, developed by Materialise NV. Mimics is used to create 3D surface models from stacks of 2D image data for a variety of engineering applications.

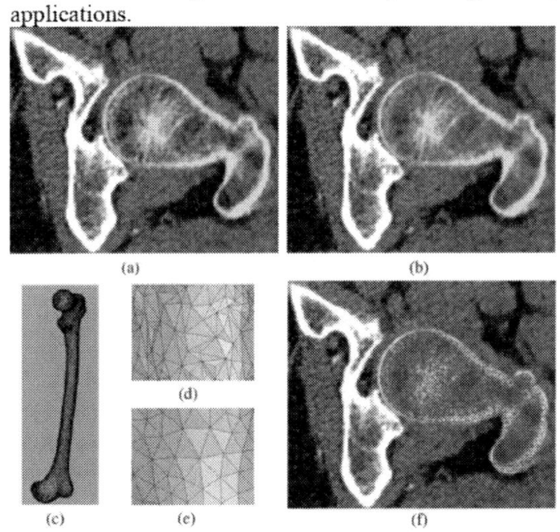

Fig. 1. 3D reconstruction of the femur by Mimics. (a)The segmentation mask after "Thresholding" and "Region Growing"; (b) The segmentation mask covered the whole femur after editing; (c)The final surface triangular mesh of the femur; (d) The partial enlargement of the surface before mesh optimization; (e) The partial enlargement of the surface after mesh optimization; (f) The cross section of the tetrahedral mesh on the image plane.

* Corresponding author: wxyf@ustc.edu.cn

Here, we used Mimics software to process the CT series scanned from a female patient aged 63 years. After importing the CT data, the mask was developed to describe the femur by "Thresholding" and "Region Growing" (as shown in Fig. 1(a)). As the default threshold was set for compact bone, the mask needed to be edited to cover the whole femur (as shown in Fig. 1(b)).

Then the femoral surface triangular mesh could be reconstructed based on the mask by "3D Calculate". However, the regularisation of this initial mesh was not high as it is shown in Fig. 1(d). Therefore, the "Remesh" function was then executed to improve the regularisation by reducing triangles and smoothing the mesh, and the detail can be seen in Fig. 1(e). In this way, the femoral surface mesh was extracted from the CT series (shown in Fig. 1(c)).

3 Heterogeneous modelling

The accurate femoral model could be obtained by surface reconstruction using Mimics. However, in the actual finite element analysis, the distribution of the bone mineral density and mechanical properties is heterogeneous. Therefore, we reconstructed the heterogeneous model of the femur with ANSYS software.

To achieve the heterogenous modelling, the triangular mesh should be transformed to tetrahedral mesh to depict the distribution of internal materials. In ANSYS, we imported the surface triangular mesh and then performed creating tetrahedral mesh from area elements and defined the element type as "SOLID92". Then the model was exported as a tetrahedral mesh model to Mimics and the internal nodes distribution could be observed in Fig. 1(f).

After that the material property assignation should be determined according to the CT information. Through the interaction with Mimics, the internal nodes of the tetrahedral mesh can correspond to the CT image information. To achieve heterogeneous modelling of the femur, the material property of the elastic modulus, density and Poisson's ratio of model nodes could be assigned on the basis of the CT value at the corresponding location.

According to the Hu distribution histogram shown in Fig. 2(a), the internal material of the femur was divided into 10 types with various colours (Fig. 2(b)). The bone mineral density and elastic modulus of the femur are approximately linear with CT value [5,6], and can be calculated by the following empirical formula:

$$Density = 1041395 + 1017Hu, \quad (1)$$

$$E\text{-}Modulus = -388.8 + 5925Density, \quad (2)$$

where *Hu* indicates Hounsfield unit, and *Density* indicates the bone mineral density with the unit of g/m^3, and *E-Modulus* indicates the elastic modulus with the unit of Pa.

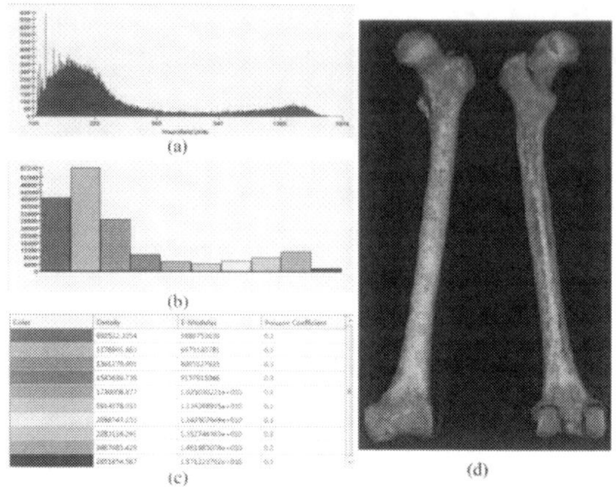

Fig. 2. The process of the femoral heterogeneous modelling. (a) Hu distribution histogram of the tetrahedral mesh; (b)The internal materials divided into 10 types indicated and distinguished by colour; (c)The assignment of mechanical properties of the materials; (d)Distribution of the femoral material properties from different perspectives.

As shown in Fig. 2(c), the femoral material properties such as bone mineral density, elastic modulus and Poisson's ratio were assigned in the tetrahedral mesh on the basis of Hu distribution, and the reconstruction of femoral heterogeneous model was completed (shown in Fig. 2(d)).

By calculating the volume of different materials, the average bone mineral density of the femur was 1.49 g/cm^3, which belongs to a low level. From the Fig. 2(b) and Fig. 2(d), it can be seen that the proportion of cancellous bone in the femur is the largest, which is consistent with the fact of the subject aged 63 years. Considering that the bone mineral density is an important indicator for the diagnosis of osteoporosis [7], the femoral heterogeneous model can provide data support for the diagnosis and treatment of bone diseases.

4 Finite element analysis for the femur

On the femoral heterogeneous model, finite element mechanical analysis can be carried out to simulate the stress distribution of the femur under different conditions. In ANSYS, we simulated the stress characteristics of the femur under the effect of gravity and horizontal impact force by defining different loads.

To analyse the effect of gravity, a vertical downward force of 500N was applied to the femoral head with the displacement of the lower extremity fixed (as shown in Fig. 3(a)). On this condition, the simulation result shows that the maximum of von Miss stress is 11.8MPa and the maximal deformation is 2.23mm. From the distribution of von Miss stress (shown in Fig. 3(b)) and deformation (shown in Fig. 3(c)), it can be seen that the large stress is concentrated on the upside of the femoral shaft.

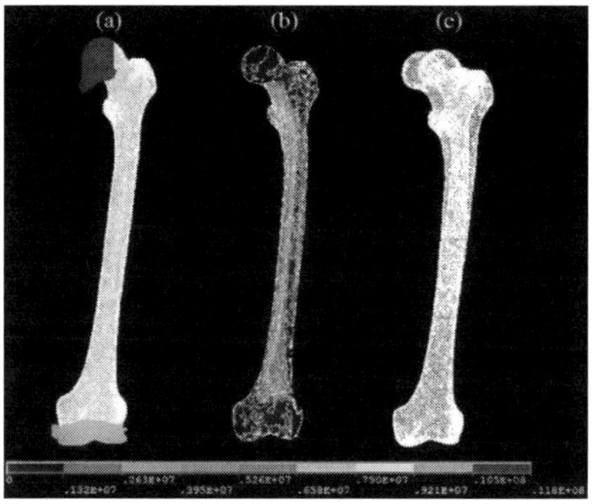

Fig. 3. Simulation when the vertical downward force applied to the femoral head with the displacement of the lower extremity fixed. (a)Applied loads; (b) Distribution of von Miss stress; (c) Deformed shape.

To simulate the effect of horizontal impact, we applied the forces in the horizontal direction of 500N to three positions of the femur (shown in Fig. 4(a), Fig. 5(a) and Fig. 6(a)). The simulation results show that the maximums of von Miss stress are 15.7MPa, 12.7MPa and 6.3MPa (shown in Fig. 4(b), Fig. 5(b) and Fig. 6(b)), and the maximums of the deformation are 0.156mm, 0.283mm and 0.096mm (shown in Fig. 4(c), Fig. 5(c) and Fig. 6(c)), respectively. It can be seen that the maximal stress is greater on an upper position of the femur, while the maximal deformation is greater at the middle position.

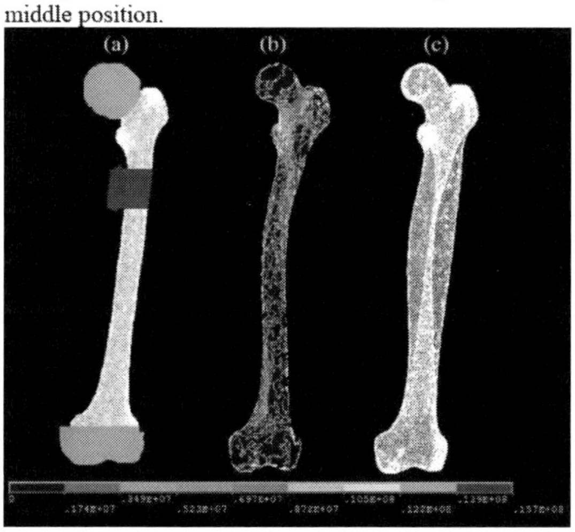

Fig. 4. Simulation when the horizontal force applied to the upper thirds of the femur with the displacement of the femoral head and the lower extremity fixed. (a)Applied loads; (b) Distribution of von Miss stress; (c) Deformed shape.

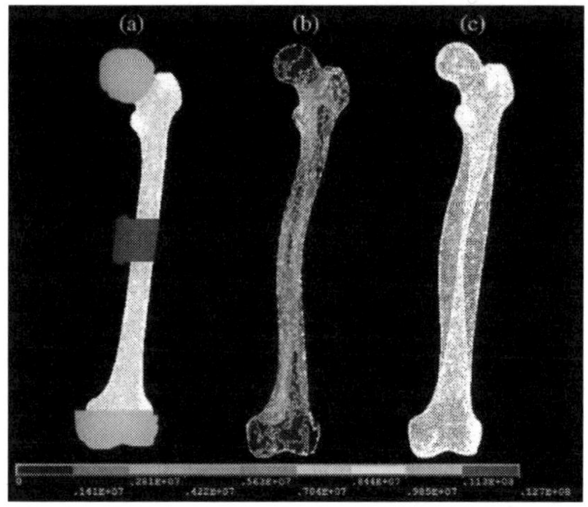

Fig. 5. Simulation when the horizontal force applied to the middle of the femur with the displacement of the femoral head and the lower extremity fixed. (a)Applied loads; (b) Distribution of von Miss stress; (c) Deformed shape.

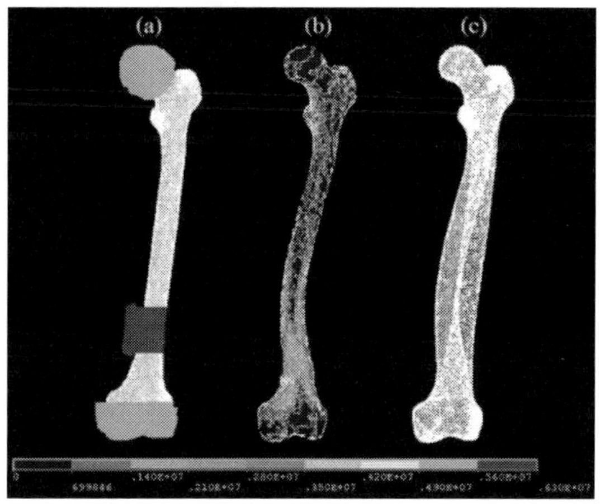

Fig. 6. Simulation when the horizontal force applied to the lower thirds of the femur with the displacement of the femoral head and the lower extremity fixed. (a)Applied loads; (b) Distribution of von Miss stress; (c) Deformed shape.

By comparing the stress distribution, it can be found that the stress concentration appears at the upper and lower sides of the femoral shaft on all the three conditions. These positions are both weak points of the femur where the fractures are prone to occur. In the design of the prosthesis, the structural strength of the stress concentration positions should be taken into consideration to avoid the periprosthetic fractures [8].

5 Conclusion

In this study, we proposed a method to reconstruct the femoral heterogeneous model on the basis of Hu distribution of CT images with Mimics and ANSYS software. Because the heterogeneous model of the femur indicates the distribution of bone mineral density and material properties, it can provide data support for the diagnosis and prediction for osteoporosis. In the finite element analysis, we analysed the effect of gravity and impact force by applying various loads. The stress concentration position of the femur under different conditions was calculated in the simulation, which can provide reference for the design and material selection of prosthesis.

Acknowledgements

The authors are grateful to NSFC-CAS Joint Fund [grant number U1332130], 111 Projects [grant number B07033], 973 Project [grant number 2014CB931804], and the Key Research and Development Program of Anhui [grant number 1704a0902051] for their financial support.

References

1. Peng M J Q, Chen H Y, Hu Y, et al. Finite Element Analysis of porously punched prosthetic short stem virtually designed for simulative uncemented Hip Arthroplasty. BMC Musculoskeletal Disorders, 2017, 18(1): 295.
2. Yerasimides J G. Use of the Fitmore® hip stem bone-preserving system for the minimally invasive anterior-supine approach in hip replacement. American journal of orthopedics (Belle Mead, NJ), 2010, 39(10 Suppl): 13-16.
3. Janssen D, Mann K A, Verdonschot N. Finite element simulation of cement - bone interface micromechanics: A comparison to experimental results. Journal of Orthopaedic Research, 2009, 27(10): 1312-1318.
4. Bougherara H, Zdero R, Miric M, et al. The biomechanics of the T2 femoral nailing system: a comparison of synthetic femurs with finite element analysis. Proceedings of the Institution of Mechanical Engineers, Part H: Journal of Engineering in Medicine, 2009, 223(3): 303-314.
5. Rho J Y, Hobatho M C, Ashman R B. Relations of mechanical properties to density and CT numbers in human bone. Medical engineering & physics, 1995, 17(5): 347-355.
6. Ciarelli M J, Goldstein S A, Kuhn J L, et al. Evaluation of orthogonal mechanical properties and density of human trabecular bone from the major metaphyseal regions with materials testing and computed tomography. Journal of Orthopaedic Research, 1991, 9(5): 674-682.
7. Oden Z M, Selvitelli D M, Bouxsein M L. Effect of local density changes on the failure load of the proximal femur. Journal of Orthopaedic Research, 1999, 17(5): 661-667.
8. Beals R K, Tower S S. Periprosthetic fractures of the femur: an analysis of 93 fractures. Clinical orthopaedics and related research, 1996, 327: 238-246.

Design and Simulation for a High-efficiency Class-B Power Amplifier

Wei Jiang [1], *Qian Lin* *[1], *Haifeng Wu* [2]

[1] College of Physics and Electronic Information Engineer, Qinghai University for Nationalities, Xining, 810007, China;
[2] Chengdu Ganide Technology Co Ltd, Chengdu, 610000,China;

Abstract. A class-B power amplifier with high output power and high power added efficiency (PAE) is designed in this paper. Based on the ADS tools of load-pull and source-pull, the optimal load impedance and source impedance can be obtained, respectively. Then, the matching circuit have been achieved by combining the impedance matching technique. Simulation results show that the class-B power amplifier has the PAE of 69.386% and output power of 45.32dBm with the working frequency of 960MHz. Therefore, It is a high-efficiency class-B power amplifier.

1 Introduction

Nowadays, all the major wireless standards-GSM, DMA,WCDMA, 802.11 a/b/g/n, WiBro and WiMax- are being integrated into a single package and widely adopted in the wireless communication system. A radio that integrates all of this functionality onto a single chip is urgently needed. One of the major impediments to addressing all these needs is the RF front end, specifically, the power amplifier (PA). That is to say, the compact and fully integrated radio frequency (RF) and microwave front-end products are the indeed demand for the rapid development of the wireless communication system. As a result, achieving simultaneously high efficiency and high linearity remains one of the major challenges in microwave PAs.

To our knowledge, PA is the core component of the wireless transmitter, its main function is to increase the RF power enough, which is effectively transmitted to a receiver through air medium. As the key equipment for the modern wireless communication, PAs have been widely applied in satellite communication, radar, wireless communications, navigation, electronic counter measures etc [1].

As mentioned above, in order to obtain the largest output power within a certain frequency range, the power transistor often works near the saturated state. At this point, its S parameters often change with the variation of the input signal, and the power gain will become smaller. In other words, the output/input of conjugate match gradually will become mismatching due to the small signal input. Then, it can not be able to obtain the maximum output power [2]. Therefore, the key point for PA design mainly lies in the design of matching network.

The technology of load-pull is the major method to find out the optimum load impedance corresponding to the largest output power. Meanwhile, the optimal source impedance can be obtained by the technology of source-pull. Finally, the output and the input matching circuit can be achieved, respectively.

Furthermore, comparing with the other PAs, Class-B PA has better linearity and higher power efficiency, for which its PAE can reach about 78.5% theoretically [3]. The DC working point for the power transistor of the class-B PA is often biased at the breakpoint [4], then the DC output current is zero. When the AC signal is used as the input, the power transistor switches on in a half cycle and switches off in the negative half cycle of AC signal. This operation mode of half cycle for class-B PA has remarkable advantages over Class-A PA. Moreover, the microwave field effect transistor (FET) of class-B PA can cause the increase of power efficiency and substantial reduction of both DC power and heat dissipation. Therefore, how to keep high efficiency and high power is the major challenge for the class-B PA.

The design strategies and implementation of a high efficiency class–B power amplifier based on the Freescale FET of MRFE6S9046 are presented in this paper. The simulation results show that the power additional efficiency (PAE) is 69.39% with output power of 45.32dBm when the input power is 26.5dBm in the working frequency of 960 MHZ. Thus, the demands of high performance and high efficiency can be achieved by comparing with the class-B PAs reported in exiting literatures.

This paper is organized as follows. The design of a high efficiency class-B PA is introduced in detail in Section II. Then its matching circuits of the source input impedances and the output impedances have been presented in Section III. In Section IV, the measured results of this PA are discussed as well as the conclusions.

* Corresponding author: linqian@tju.edu.cn

2 Design of the class-B power amplifier

The circuit schematic of the class-B PA based on the FET of Freescale MRFE6S9046 is given in Fig.1. It can be observed that it is consisted with the signal source, input matching circuit, power transistor, DC bias circuit, the output matching circuit and the load. Furthermore, based on the datasheet of this transistor, it can be known that when the PA works in the typical GSM mode. In this way, the DC bias circuit is used to set the appropriate working point for the power transistor. After simulation in ADS, The drain and gate of transistors were biased at 28V and 2.3V, respectively.

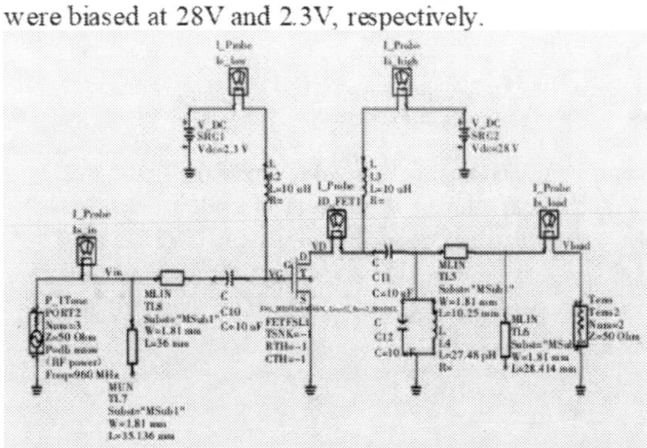

Fig.1 Circuit schematic of the class-B power amplifier

The PORT2 and Term2 in Fig.1 is the signal source and load, respectively. The DC bias circuit of the gate is consisted with SRC1, L2 and C10, where the voltage of SRC1 is the clip breakpoint voltage at the gate. L2 is used to switch on the DC signal and switch off the AC signal, and C10 is used to block the DC signal and conduct the AC signal. Similarly, the DC bias circuit of the drain is consisted with SRC2, L3 and C11. Where the voltage of SRC2 is 28V, L3 is used to conduct the DC signal and cut off the AC signal, and C11 is used to block the DC signal and conduct the AC signal.

Because the input matching circuit aims to match the optimal source impedance to 50 ohms, and the output matching circuit is used to match the 50 ohm to the optimal load impedance. In this way, TL5 and TL6 in Fig. 1 is used to match the optimal load impedance of 50 ohm based on the structure of micro-strip line, while TL7 and TL8 is used to achieve the matching for the optimal impedance to 50 ohm. The parallel resonance network which is consisted with L4 and C12 is used to determine the work frequency of the PA, which is 960 MHZ. And the parallel network is equivalent to open circuit at the working frequency, but it is equivalent to the short circuit for high order harmonic[5-6]. As a result, the PA has better function of filter.

After simulation in ADS, the optimal load impedance and source impedance corresponding to the maximum output power can be obtained using the load-pull and source-pull template. Here, the optimal load impedance and source impedance are 9.976-j*1.818 and 3.21-j*5.65, respectively.

3 Simulation optimization and analysis

To evaluate the S-parameters and validate the matching degree of the input and output matching circuit, this PA has been measured with small-signal and large-signal in ADS. Then the simulation curves of S11 and S21 at the room temperature of 25°C for input match circuit and output match circuit can be obtained as shown in Fig.2 and Fig.3, respectively.

It can be seen that the PA achieves a gain approximately 18.2dB with S_{21} of less than −0.091dB and S_{11} of -39.55dB in the input matching circuit, as shown in Fig.2. Therefore, it can be concluded that the input matching network can realize the matching of optimal source impedance to 50 ohm.

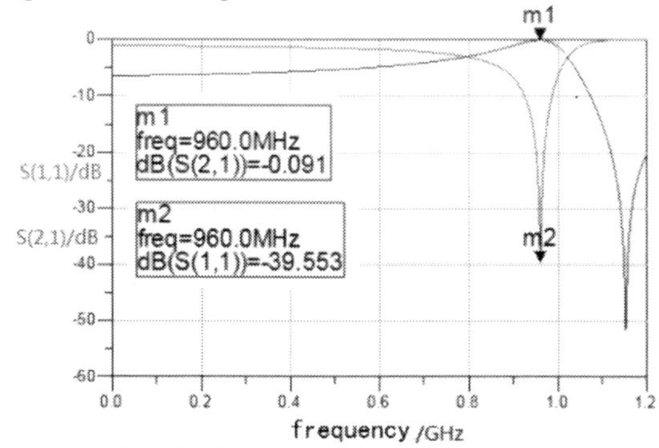

Fig.2 Performance of input match circuit.

Similarly, it can be found that the S21 is -0.026 and S11 is -38.86 at the frequency of 960 MHz in the output matching circuit from the Fig. 3. Therefore, the output matching network can realize the matching of 50 ohm to the optimal load impedance.

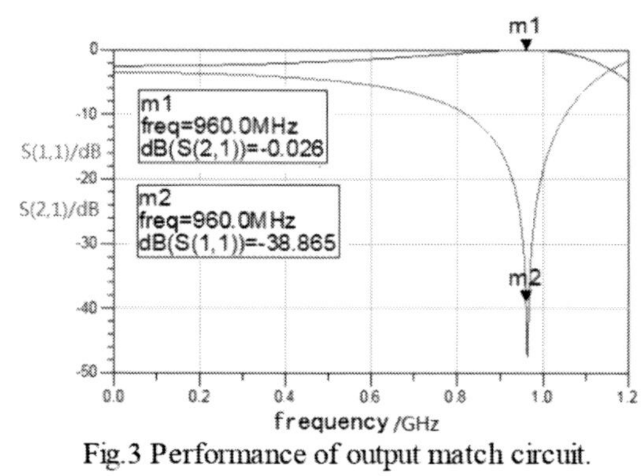

Fig.3 Performance of output match circuit.

Except for S parameter, PAE and output power are two important performance indexes for PA. Where PAE is used to measure the ability of converting the DC power into AC power for PA. The higher PAE means the better conversion efficiency [7]. Output power is the load power for the fundamental wave frequency signal. The higher output power, the greater power for the load and there is enough energy to emit the fundamental wave signal. Therefore, the large-signal RF power has been measured under different frequencies of driving continuous wave (CW) signal to evaluate the RF performance of this PA. Thus, the relationship of PAE and output power against input power for this PA can be obtained as shown in Fig.4 and Fig.5, respectively. From the Fig.4, it can be seen that PAE increases rapidly with the rise of input power, and then reaches to the maximum, finally it decreases with the continuous rise of the input power. Especially, when the input power is 26.5dBm, PAE is 69.386%. This is rather good for class-B PA.

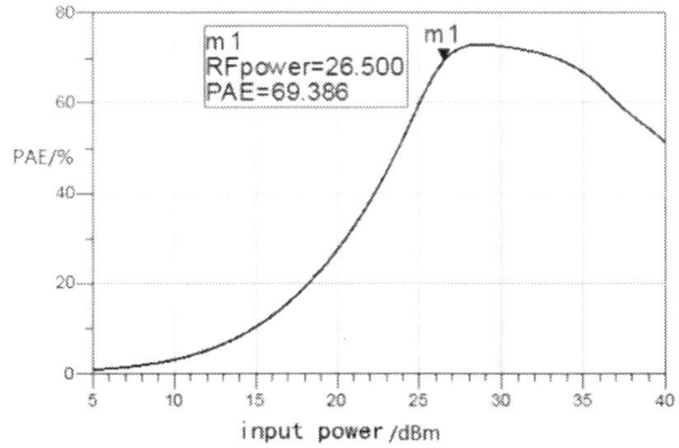

Fig.4 PAE against input power

At the same time, it can be observed from the Fig.5 that the output power increases linearly with the input power at the beginning. When input power reaches to 26.0dBm, the output power is almost invariant and keeps at the 47dBm. Especially, when input power is 26.5dBm, the output power is 45.32dBm. It is the most considerable index for class-B PA.

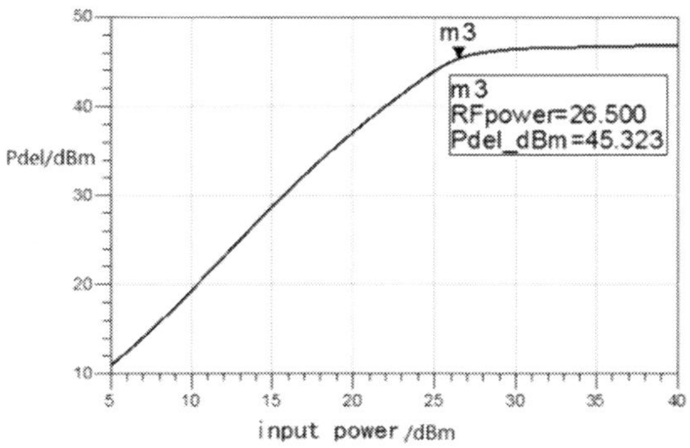

Fig.5 Output power against input power

Meanwhile, by scanning the input power, the waveform of U_{DS} and I_{DS} can be obtained as shown in Fig.6. The red sine curve clusters near the DC voltage of 28V are the waveforms of U_{DS}, While the blue curve clusters are the waveforms of I_{DS}, it is similar to the half sine wave in the vicinity of 0A. Moreover, because the transfer characteristic of the power transistor is not a straight line strictly, it is not the ideal half sine wave. At the same time, it can be seen that there is a interval of half cycle between the waveform of U_{DS} and I_{DS}. That is to say, when U_{DS} is big I_{DS} will become small and vice versa. In this way, the power of this FET would become smaller, then the DC power will be inverted into more useful load power. As a result, the power efficiency of this class-B PA can be significantly improved [8].

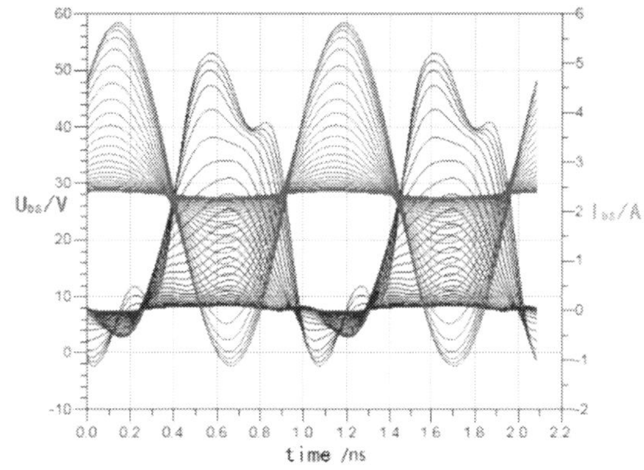

Fig.6 Waveforms of U_{DS} and I_{DS}

4 Results and discussions

In order to verify the truth that this class-B PA is high performance, the performace comparsion for all the designed class-B PA in the reported literatures has been summaried as shown in Table 1.

Table 1 Performance comparison of the class-B PA

References	Frequency (Hz)	PAE(%)	Output Power (dBm)	Gain (dB)
[9]	1G	45%	30.8	5.8
[11]	1.65G	40%	27.4	7.3
[10]	1.95G	27.4%	29.2	14.2
[12]	4G	44%	20	14.5
[13]	2.1G	72.8%	30.5	13.1
This paper	960M	69.39%	45.32	18.82

It can be observed that there are few research about the class-B PA and the operating frequency can reach several GHz with the increase of gain. The gain of this PA is the highest and the PAE is nearly 70%. It has achieved the high performance, but the frequency still needs to be improved.

5 Conclusions

A high-efficiency class-B PA based on the Freescale RF power FET of MRFE6S9046 is designed in this paper. After simulating and optimization in the ADS, it can work at 960 MHZ with output power of 45.32dBm and PAE of 69.386% when input power is 26.5dBm. Therefore, it can realize the high power efficiency and large output power and can be called the high-efficiency class-B PA.

Acknowledgements

This work is supported by Chun Hui Project of Education Ministry of China under the Grant Number of (Z2015033) and (Z2016071), the Apply Base Research Plan of Qinghai the Grant Numbers of (2017-ZJ-753), and the Undergraduate Teaching Research Project of Qinghai National University (2016-BKJXYB-09).

References

1. Gilmore R, Besser L. Practical RF circuit design for modern wireless systems[M]. Artech House, (2003).
2. Paidi V, Xie S, Coffie R, et al. High linearity and high efficiency of class-B power amplifiers in GaN HEMT technology[J]. Microwave Theory and Techniques, IEEE Transactions on, (2003), 51(2): 643-652.
3. Medrel P, Ramadan A, Nebus J M, et al. High efficiency class B GaN power amplifier with dynamic gate biasing for improved linearity[J]. Electronics letters, (2012),48(18): 1136-1137.
4. Khan M A, Kalim D, Negra R. Analysis and design of an unconditionally stable common-drain class-B RF power amplifier in 90 nm CMOS technology[C] //Integrated Nonlinear Microwave and Millimetre-Wave Circuits (INMMIC), Workshop on. IEEE, 1-4,(2011).
5. Birafane A and Kouki A B. On the linearity and efficiency of out phasing microwave amplifiers[J]. IEEE Transactions on Microwave Theory and Techniques, (2004), 52(7):1702-1708.
6. Cohn M, Degenford J E, Freitag R G. Class B operation of microwave FETs for array module applications[C]// MTT-S International Microwave Symposium Digest. Dallas Texas: IEEE, (1982).
7. Shervin M, Abidi A A. The Outphasing RF Power Amplifier: A Comprehensive Analysis and a class-B CMOS Realization[J]. IEEE Journal of Solid-State Circuits, (2013), 48(6):1357-1369.
8. Jiang Xin, Martin A L, Mortazawi A. A class-B push-pull power amplifier based on an extended resonance technique[J]. IEEE Microwave and Wireless Components Letters, (2003), 13(12):550-552.
9. Lane J R, Freitag R G, Degenford J E, Cohn M. High Efficiency One, Two, and Four Watt Class B FET PowerAmplifiers[C]// MTT-S International Microwave Symposium Digest. Baltimore, USA: IEEE, (1986).
10. Asbeck P, Fallesen C. A 29 dBm 1.9 GHz class B power amplifier in a digital CMOS process[C]// ICECS 7th International Conference on Electronics, Circuits and Systems. Jounieh, Lebanon: IEEE (2000).
11. Eccleston K W, Kyaw O. Analysis and design of class-B dual fed distributed power amplifiers[J]. IEEE Proceedings - Microwaves, Antennas and Propagation, (2004),151(2): 104-108.
12. Tyson S W, Lawrence E L. A 1 Watt 1-5 GHz Class B push-pull Si/SiGe HBT power amplifier[C]// Custom Integrated Circuits Conference. San Jose, California.: IEEE,(2009).
13. Mimis K, Bensmida S, Morris K A, McGeehan J P. Continuous harmonically tuned class-B power amplifier: Aclosed form equation design approach[C]// MTT-S International Microwave Symposium Digest, Montreal, Quebec, Canada: IEEE, (2012).

Tracking Algorithm with Adaptive Bandwidth of Kernel Function for UAV Autonomous Aerial Refueling

ZhuoYa Wang[1], Yaohong Qu[1,*], Jiaju Wu[2], Bing Xiao[1] and Jianguo Yan[1]

[1]Northwestern Polytechnical University, Xi`an, shaanxi,China,710129
[2]AVIC The First Aircraft Institute, Xi`an, shaanxi, China, 710089

Abstract. The use of machine vision to track the receiver aircraft's receptacle is one of the key steps in automated aerial refueling. Traditional Meanshift tracking algorithm's kernel bandwidth is fixed, when the target size changes, it will produce with the bias even with the loss of the situation. Aiming at this problem, this paper proposes a kernel bandwidth adaptive algorithm based on corner matching. First, use backward tracking method to amend the core window center. Then according to the Harris algorithm to detect the corner, use small-scale approach for the two points between frames to match and regression analysis, namely determine the update parameters of nuclear window width. Experiments show that the algorithm can track the target accurately when the size of tracking target change, and the kernel window width can be better suited to the target size.

Keywords. Aerial refueling, Meanshift, Kernel bandwidth adaptive algorithm, Corner matching

1 Introduction

UAV (unmanned aerial vehicles) play an increasingly important role in the field of civilian and military fields because of its small size, low cost, easy to use, etc. However, weak load capacity, short life time and other issues limit the use of UAV. UAV with independent air refueling capabilities has a large flight distance and a long blank time, expanding the use of UAV.

One of the key technologies of autonomous air refueling is real-time access to the location of the receiver aircraft's receptacle. Traditional UAV navigation system is generally based on GPS, INS and integrated navigation system, which does not meet the location accuracy requirements. In recent years, the development of machine vision provides an effective support for the UAV autonomous air refueling. Introduce the machine vision into the autonomous air refueling process can get more accurate location information. In many machine vision tracking methods, many scholars have studied Meanshift because it is an efficient pattern matching algorithm with the advantages of high robustness, real-time and easy realization. Document [3] use RGB linear fusion and add Kalman Filter algorithm to estimate the location of moving conditions, which can track the target accurately when the size of target is constant. When the tanker closes to receiver aircraft's receptacle, target will gradually become larger in the tracking video. Then the tracking is not accurate because the method kernel function bandwidth is fixed. Document [4] uses the method of comparing Bhattacharyya coefficients to change the bandwidth of kernel function. However, this method needs to do multiple times of Meanshift algorithm with different kernel bandwidth. The calculation is large, and the search box is proportional change, which cannot adapt to the target changes. Document [5] uses positive and negative 10% increments to change the kernel function bandwidth. It needs three times of Meanshift tracking based on three kernel bandwidth in the tracking of each frame. It will get a good tracking effect when the target is smaller, not suitable for autonomous air refueling process by the receiver aircraft's receptacle tracking. Document [6] changes the kernel bandwidth by the backtracking method based on the corner feature, but the tracking is not accurate enough only when the target feature is described.

Aiming at this problem, this paper proposes a kernel bandwidth adaptive algorithm based on corner matching. Use backward tracking method to amend the core window center and make the tracking window aim at the target heart. According to the Harris algorithm to detect the corner, use small-scale approach for the two points between frames to match, which reduce the complexity of corner matching. Obtain the affine parameters based on the result of corner matching and modify the kernel function. In addition, when using the Meanshift tracking algorithm, add spatial information and color information

* Corresponding author: qyh0809@126.com

to describe the target model comprehensively, and ensure the tracking effect is more accurate.

2 Meanshift Algorithm Theory

Figures and tables, as originals of good quality and well contrasted, are to be in their final form, ready for reproduction, pasted in the appropriate place in the text. Try to ensure that the size of the text in your figures is approximately the same size as the main text (10 point). Try to ensure that lines are no thinner than 0.25 point.

Given n number sample points x_i in a d-dimensional space R^d. Meanshift vector at the point of x is defined as:

$$M_h(x)=\frac{1}{k}\sum_{x_i\in S_h}(x_i-x) \quad (1)$$

Where, k represents k number points in n number sample points fall in the selection area. x_i is the pixel, x is the initial position, S_h is a high-dimensional ball with the radius h, which is the set point of y satisfying the relationship.

$$S_h=\left\{y:(y-x)^T(y-x)\le h^2\right\} \quad (2)$$

It can be seen (x_i-x) is the shift vector of sample x respect to the initial position. Figure 1 is Meanshift iteration process. The average offset vector $M_h(x)$ will point to most sample distribution area, that is, the gradient direction of the probability density function. Meanshift algorithm will eventually converge to the actual location of the target through constant iteration.

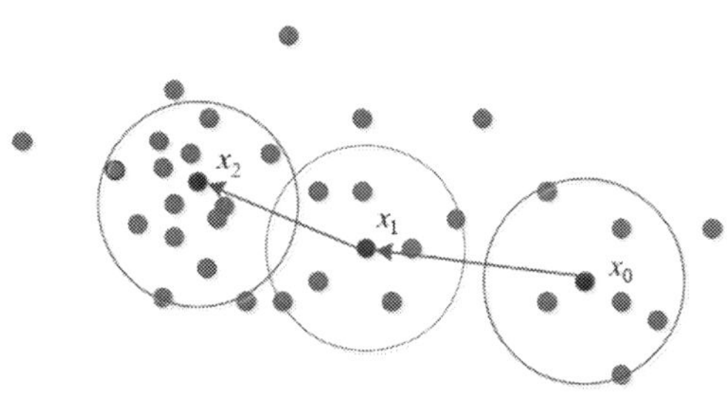

Fig. 1. Meanshift iteration process

3 Meanshift Tracking Algorithm with Adaptive Bandwidth

3.1. Target model and similarity measure

Add the spatial location information of the pixel, and synthesize the color feature and the spatial feature to describe the target template. According to the distance between each pixel and the reference point, the kernel function is used to assign different weights to different pixels, and the weighted color histogram is constructed. The weight of each pixel is calculated by the following kernel function:

$$K(x)=1-\frac{(x_i-x_{mid})^2+(y_i-y_{mid})^2}{(2*x_{mid})^2+(2*y_{mid})^2} \quad (3)$$

Where, x_{mid} 、 y_{mid} are the x-axis and y-axis center coordinate of the selected graphic template. x_i, y_i , respectively, are horizontal and vertical coordinates of point x . $(2*x_{mid})^2+(2*y_{mid})^2$ is the square of the radius of the kernel function. When point x is at the center of the rectangular template, $K(x)$ is 1. When point x is at the border of the rectangular template, $K(x)$ is 3/4. The more the distance from center pixel template is, the smaller the value is.

The probability density of the target template is estimated as:

$$q_u=C\sum_{i=1}^{M}K\left(\left\|\frac{x-x_i}{h}\right\|^2\right)\delta\left[b(x_i)-u\right] \quad (4)$$

Where, q_u is the probability density of the target template histogram, h is kernel function bandwidth, $\delta\left[b(x_i)-u\right]$ is histogram index function, M is the number of pixels in a given space, x_i is any pixel in the template, C is normalization coefficient.

Similarly, the probability density of the sample template with center point x^j is estimated as:

$$p_u(x^j)=C\sum_{i=1}^{M}K\left(\left\|\frac{x^j-x_i}{h}\right\|^2\right)\delta\left[b(x_i)-u\right] \quad (5)$$

Introduce BH coefficients to describe the similarity between the target template and the sample template:

$$BH\left[p^j,q\right]=\sum_{u=1}^{m}\sqrt{p^j_{u.}(x^j)q_{u.}} \quad (6)$$

Where q_u is the probability density histogram of the target template, x^j is the j th particle of the particle set, $p^j_u(x^j)$ is the probability density histogram of the sample template centered on x^j, and m is the number of color histograms u. The higher the value of BH, the higher the similarity between the target template and the sample template is, and vice versa.

3.2 Kernel function bandwidth correction

3.2.1 Affine model

Affine transformation is a basic transformation in affine geometry, and the affine transformation of two-dimensional continental space can be expressed as:

$$\begin{bmatrix} x_{i+1} \\ y_{i+1} \end{bmatrix} = \begin{bmatrix} a_x & a_{xy} \\ a_{yx} & a_y \end{bmatrix} \begin{bmatrix} x_i \\ y_i \end{bmatrix} + \begin{bmatrix} l_x \\ l_y \end{bmatrix} \quad (7)$$

Where $(x_i, y_i)^T$ and $(x_{i+1}, y_{i+1})^T$ are the position coordinates of any feature point in the i frame and the $i+1$ frame, respectively. $(l_x, l_y)^T$ is the translation vector, $\begin{bmatrix} a_x & a_{xy} \\ a_{yx} & a_y \end{bmatrix}$ is a synthetic matrix for rotation. Vector $a = [a_x, a_{xy}, a_{yx}, a_y, l_x, l_y]$ determines the coordinate transformation between two feature points. Here only consider the most common transformations in the air refueling: translation and expansion. Make $a_{xy} = a_{yx} = 0$, the affine model is defined as:

$$\begin{bmatrix} x_{i+1} \\ y_{i+1} \end{bmatrix} = \begin{bmatrix} a_x & 0 \\ 0 & a_y \end{bmatrix} \begin{bmatrix} x_i \\ y_i \end{bmatrix} + \begin{bmatrix} l_x \\ l_y \end{bmatrix} \quad (8)$$

Then the kernel function bandwidth can be modified according to Eq. (9):

$$h_{i+1} = h_i \cdot \sqrt{a_x^{\ 2} + a_y^{\ 2}} \quad (9)$$

Where, h_i and h_{i+1} are the kernel function bandwidth of the i frame and $i+1$ frame respectively.

3.2.2 Kernel window centre registration

Set the frame i of the tracking window (namely kernel window) as T_i, the center of the window is C_i, the target heart is O_i. It is known from the literature [6] that when the target becomes smaller, can obtain the best match window through the Meanshift iteration, as long as the expansion and contraction of target in the frame $i+1$ are in the T_i range. That is, the center C_{i+1} of the tracking window will always lock the target's centroid. In most cases, Meanshift tracking algorithm can lock the target centroid as long as most of the region of the target is within the range of T_i. When the target gradually increases, it will produce positioning bias, and then need to register the center of the window registration.

When the target is gradually increased, if the $i+1$ frame target is larger than the tracking window range, the tracking window T_{i+1} contains only part of the target, the current total target centroid O_{i+1} and the tracking window center C'_{i+1} that has not been registered at this time will have a certain offset.

$$d = C'_{i+1} - O_{i+1} \quad (10)$$

At this point the tracking window T_{i+1} contains only part of the target; the corresponding tracking window center C'_{i+1} is actually the center O'_{i+1} of the target. In order to register the center of the window to find the current target heart, need to compensate for the offset d .As mentioned above, when the target is gradually smaller, there is no deviation between the tracking window and the centroid. Therefore, use the backward tracking, and the part of the tracking window of the $i+1$ frame is used as the new target for Meanshift tracking. Get the tracking window T_i', center C_i' of T_i' is the part of the target in the i frame of the centroid, and then the centroid deviation of the target and the total target in the i frame part is:

$$d' = C_i - C_i' \quad (11)$$

Since the interval between two adjacent frames is very short, so use d' to make up the offset d, the $i+1$ frame track window after the registration is estimated as follows:

$$C_{i+1} = d' + C'_{i+1} \quad (12)$$

3.2.3 Corner detection and matching

Corner detection of the target template, match the corner points between the two frames by a small range of fringes, and the affine model parameters are estimated according to the corner matching results.

Use Harris algorithm to detect corner [8-9], judge whether a point is a corner point by analyzing the autocorrelation function characteristic matrix of the surrounding area of the pixel, the autocorrelation function is:

$$E(u,v) = \sum_{x,y} w(x,y)\left[I(x,y) - I(x+u, y+u)\right]^2 \quad (13)$$

Where u、v are small offset in the x、y direction, $I(x,y)$ is the gray value at (x,y) pixel, $w(x,y)$ is filter, the autocorrelation function for Taylor expansion can be obtained from the correlation matrix:

$$M = \sum_{x,y} w(x,y) \otimes \begin{bmatrix} I_x^2 & I_x I_y \\ I_y I_x & I_y^2 \end{bmatrix} \quad (14)$$

Corner response function is:

$$R = |M| - k \cdot tr^2(M) \quad (15)$$

Among them, the value of scale factor k is generally 0.04-0.06, and then can use the value of R to determine whether the pixel point is corner. When absolute value of R is smaller, it represents a flat area. R is negative when it represents boundary and positive when it is corner. Roughly suppress R to retain more than the threshold T and the local maximum point as a corner point.

Use corner match after corner point detection. Set the tracking window T_i of i frame has N corners; the tracking window T_{i+1} in $i+1$ frame after center matching has N' corners. T_i is same as T_{i+1} in size. Center of the tracking window is consistent with the target centroid, with using small-scale approach for corner matching. For any corner P_i^j in T_i, find the corresponding corner in the $i+1$ frame. First give a 5 * 5 small window in the $i+1$ frame with the center of corner P_i^j, find the brightness corner close to P_i^j. Then the point is the corresponding corner in $i+1$ frame, regression analysis the corner coordinates to obtain a_x and a_y, complete the kernel function of the bandwidth correction.

4 Tracking Algorithm

Use backward tracking method to amend the core window center. According to the Harris algorithm to detect the corner, use small-scale approach for the two points between frames to match. The affine parameters are obtained based on the result of corner matching and the kernel function is modified. The algorithm is as follows:

Step 1: Initialization: manually select the target area and determine the target template histogram.

Step 2: Based on the backward tracking method, according to the formula (11) to calculate the error compensation value d', according to formula (12) to calculate the window centre C_{i+1} of modified $i+1$ frame.

Step 3: According to formula (15) to calculate the corner points in the tracking window of the i frame and the $i+1$ frame respectively, and match the corresponding corner points by the small range adjacent method.

Step 4: regression analysis of the matching corners to determine the affine parameters a_x、a_y, amend the kernel bandwidth.

Step 5: track and correct the next frame, go to step (2).

5 Experiment Results

The effectiveness of the algorithm is verified by the semi-physical simulation platform. We use F-16 model as tracking target. In the x, y-axis direction, it is headed by a stepper motor driving a ball screw. The movement holder is shown in Fig 2.

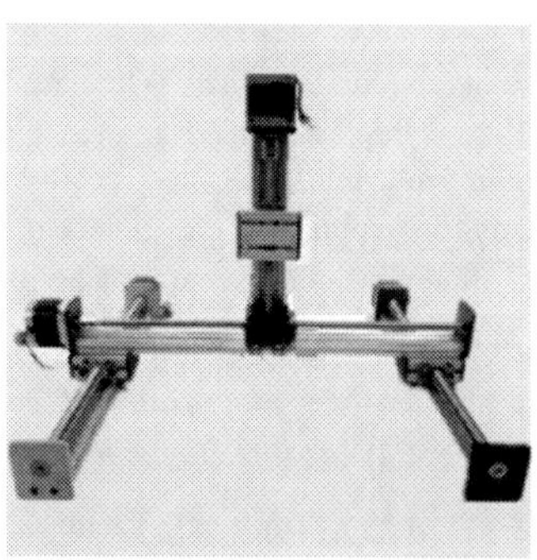

Fig. 2 Target motion trestle

Tracking results of receiver aircraft's receptacle which use traditional Meanshift algorithm are shown in Fig. 3:

(a) Frame 10 (b) Frame 80

(c) Frame 150 (d) Frame 220

Fig. 3 Tracking results of traditional MS algorithm

As shown in Fig. 3 (a) (b), when the target is slightly larger, the tracking window (kernel window) has a small deviation, the tracking result is still possible, in Fig. 3 (c) target continues to become larger, tracking error becomes larger, and the tracking window is slightly smaller with the target. As shown in Fig. 3 (d), as the target becomes larger, the tracking window is significantly smaller than the target and has a significant deviation.

Kernel bandwidth adaptive algorithm based on corner matching tracking results is shown in Figure 4:

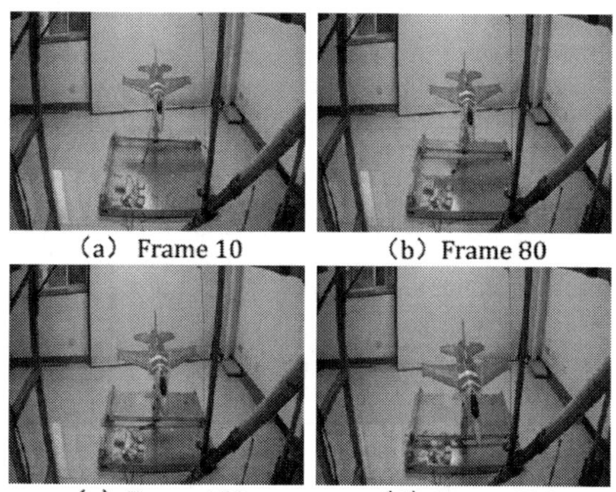

(a) Frame 10 (b) Frame 80

(c) Frame 150 (d) Frame 220

Fig. 4 Tracking results of kernel bandwidth adaptive algorithm

As shown in Fig.4, with the change of the target size, the algorithm mentioned in this paper can accurately track the target, the tracking window can also adapt to the target size changes.

6 Conclusions

This paper proposes a kernel bandwidth adaptive algorithm based on corner matching. Use backward tracking method to amend the core window centre and make the tracking window aim at the target heart. According to the Harris algorithm to detect the corner, use small-scale approach for the two points between frames to match. The affine parameters are obtained based on the result of corner matching and the kernel function is modified. In addition, when using the Meanshift tracking algorithm, add spatial information and colour information to describe the target model comprehensively, and ensure the tracking effect is more accurate. Experiments show that the algorithm proposed in this paper can adapt the core window centre to the target centroid according to the change of the target size, and correct the kernel bandwidth according to the size of the target.

Acknowledgements

This work has been supported by Project of the Program of National Natural Science Foundation of China under the Grant Numbers 61473229.

References

1. SUN ZH S, SUN J X, SONG J ZH, et al.. Antiocclusion arithmetic for moving object tracking[J].Opt. Precision Eng.,2007,15(2):267-271.
2. ZHOU Shang bo, LI Kun. Algorithm of vedio object tracking based on mean shift with adaptive bandwidth of kernel function [J]. Computer Engineering and Design, 2012, 33(07):2776-2781.
3. Wu J, Yan J, Wang Z, et al. Object tracking algorithm for UAV autonomous Aerial Refueling[C]// Advanced Information Management, Communicates, Electronic and Automation Control Conference. IEEE, 2017:1665-1669.
4. Bai X, Li A, Li X, et al. Mean-Shift tracking algorithm based on adaptive band width [J]. Journal of Computer Applications, 2011, 31(01):254-257.
5. Comaniciu D, Ramesh V, Meer P. Kernel-Based object tracking. IEEE Trans. on Pattern Analysis and Machine Intelligence, 2003, 25(5):564−575.
6. Peng NS, Yang J, Liu Z, Zhang FC. Automatic selection of kernel-bandwidth for Mean-Shift object tracking [J]. Journal of Software, 2005,16(9):1542−1550.
7. Yang Xiaofeng, Zhang Guilin. A Target Tracking Algorithm Based On Affine Transform Model [J]. Computer and Digital Engineering,2005,(12):30-34.
8. Liu Chuang, Gong Shengrong. Tracking Multiple Objects Method on Harris corner Sampling[J]. Journal of Image and Graphics,2008,13(10):1873-1877.
9. LIU Yu-ran, ZHANG Yu-dong, RAO Xue-jun, HOU Ming-lian. Method of Corner Matching Used in Image Registration [J]. Opto-Electronic Engineering,2008,(01):100-104.
10. XU Wei, WANG Wei. Corner Matching-based Approach of View Synthesis [J]. Science Technology and Engineering, 2007, (14):3263-3265+3280.

Aerodynamic analysis and elastic suppression of refueling boom

Yaohong Qu*, *Xue* Mu, *Zhuoya* Wang, and *Bing* Xiao

Northwestern Polytechnical University, Xi'an, Shaanxi, China

Abstract. This paper studies on the refueling boom aerodynamic characteristics and proposes an elastic suppression method for the refueling boom with the property of large length-diameter ratio. In order to obtain the accurate frequency response, modal analysis for the refueling boom is conducted based on ANSYS MODAL. Then the dynamics and kinematic equations of the elastic refueling boom are deduced, which contain the frequency parameter. To complete the equations, the aerodynamic performance is solved in ANSYS FLUENT. In addition, based on the mixed H_2/H_∞ theory, an optimal PID controller is proposed to suppress the elastic deformation. Furthermore, the parameters of the PID controller are optimized with the differential evolution algorithm. Simulation shows that the different elastic mode has different effect on the control of the refueling boom. The elastic deformation is suppressed obviously for each mode.

Keywords: refueling boom; aerodynamic analysis; modal analysis; elastic suppression; mixed optimal PID controller

1 Introduction

In recent years, modeling and controlling of the refueling boom have been widely attention and studied. In [1], a preliminary aerodynamic model of the refueling boom under the background of EU FP7 RECREATE is putting forward. In order to improve the model, the hinge joint is taken into account in [2-3], but the estimation of aerodynamic parameters is rough. In [4], the finite element analysis is used to obtain the model of the structural force of the refueling boom. Furthermore, several dynamic models of the refueling boom are considered, which are closer to the actual conditions. One of them is a mathematic refueling boom model with changing mass caused by the fuel transfer [5-6]. The other introduces the boom extension of the dynamic model, where the static model and moving model are established [7].

It can be seen from the researches that the elastic deformation of the refueling boom is not considered. Nonetheless, extensive studies on the elastic aircraft have been made. Waszak and Schmidt propose the nonlinear equations of motion for an elastic aircraft based on the Lagrange equation and the principle of virtual work [8]. In [9], Schmidt and Raney extend this work and demonstrated the importance of coupled flight dynamics and structural dynamics [9]. On basis of it, Shearer develops an integrated flight dynamic model for a representative high altitude long endurance vehicle [10]. In order to simplify the model without changing performance, a reduced-order aeroelastic model is proposed [9].

In general, the researches to the elastic problem, especially the elastic refueling boom, are not mature enough. In this paper, the modeling and control of elastic refueling boom are investigated. For this purpose, we analyse the modes of elastic refueling boom based on ANSYS MODAL. Then, based on the simulation of the aerodynamic parameters in FLUENT, the model of the refueling boom is improved. Further, using the parameters, we deduced the dynamics and kinematic equations of the elastic refueling boom. On basis of it, the optimal PID controller to control the attitude and suppress the elastic deformation of the refueling boom is proposed. Simulation results verify the effectiveness of the proposed method.

2 Modal analysis of the refueling boom

2.1. Theoretical analyses of the mode

As the vibration characteristic of the object structure itself, the mode possesses the following parameters, inherent frequency, mode shapes and damping ratio. These parameters are mainly obtained by two approaches, mathematical method and experiment. This paper takes the method of mathematic to deduce the parameter characteristics of the mode.

Suppose there are n freedoms to the refueling boom, whose vibration differential equation is

$$M\ddot{X}+C\dot{X}+KX=F \tag{1}$$

where M, C, K and F is the mass matrix, damping matrix, stiffness matrix and the external force matrix of

* Corresponding author: qyh0809@126.com

the elastic refueling boom respectively. $\ddot{X}$, $\dot{X}$ and X is respectively the acceleration vector matrix, the velocity vector matrix and the displacement vector matrix of the elastic refueling boom.

Assume the external force is zero. According to Fourier transform

$$\left(K-\omega^2 M\right)X(\omega)=0 \tag{2}$$

Define a set of generalized coordinates

$$\{q_1(\omega) \quad q_2(\omega) \quad \dots \quad q_n(\omega)\} \tag{3}$$

Then $X(\omega)$ can be represented as follow.

$$X(\omega)=\begin{bmatrix}\phi_1 & \phi_2 & \dots & \phi_r & \dots & \phi_n\end{bmatrix}\begin{bmatrix}q_1(\omega)\\ q_2(\omega)\\ \vdots \\ q_n(\omega)\end{bmatrix}=\Phi Q \tag{4}$$

where $\phi_r=\begin{bmatrix}\varphi_{1r} & \varphi_{2r} & \dots & \varphi_{nr}\end{bmatrix}^T$ is the r th vibration mode of the elastic refueling boom, Φ is the modal matrix of the elastic refueling boom.

$$\left(K-\omega^2 M\right)\Phi=0 \tag{5}$$

Based on the solvable conditions of equation, characteristic equation Eq. (8) is deduced.

$$\det\left|K-\omega^2 M\right|=0 \tag{6}$$

Suppose the characteristic equation has no multiple root. We arrange n positive different roots ω_r , which solved by Eq. (8), in ascending order.

$$0<\omega_1<\omega_2<\dots<\omega_r<\dots<\omega_n \tag{7}$$

where ω_r is the r th inherent frequency of the refueling boom. For each inherent frequency, the corresponding vibration mode can be solved.

In theory, infinite number of eigenvalues and corresponding eigenvectors can be obtained in the solving processes. Nevertheless, the eigenvalues and eigenvectors arranged in the first few modes play the major role to the actual vibration system.

2.2. Modal analysis base on the software

In term of the conclusion in previous section, there are infinite number of modes for the elastic refueling boom. With that, it's impossible to realize manual computation. Therefore, this paper utilizes several software to solve the elastic modal. The airfoil data of the winglet which installed on the refueling boom are generated in PROFILI. Then the winglet is generated by importing the airfoil data into CATIA and is assembled on the created three-dimensional model of refueling boom. The analysis of the elastic mode is carried out in ANSYS MODAL at last.

Based on the results of modal analysis in ANSYS, inherent frequencies of six lowest orders that corresponding to the first six order modes of refueling boom are obtained. Table 1 shows inherent frequencies of refueling boom.

Fig. 3 implies the deformation diagrams of the six lowest order vibration modes by ANSYS software. It shows that, for every mode, the vibration of refueling boom is violent. However, there are no vibration modes like this in real world. The reason is that different mode has different weight and the participation factor of high order mode is extremely small [10].

Figure. 1. The airfoil of the winglet generated in PROFILI.

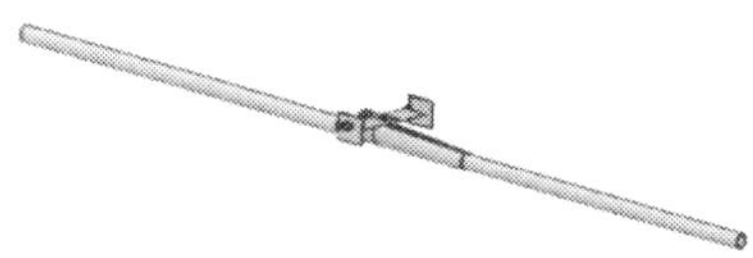

Figure. 2. Model of refueling boom created by CATIA.

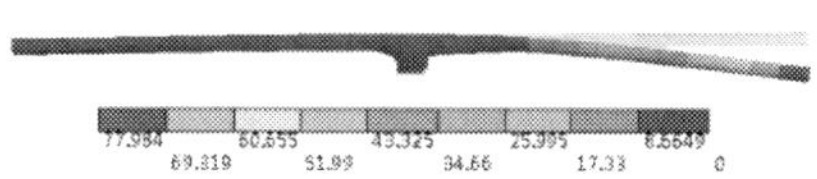

(a) The 1st order vibration mode.

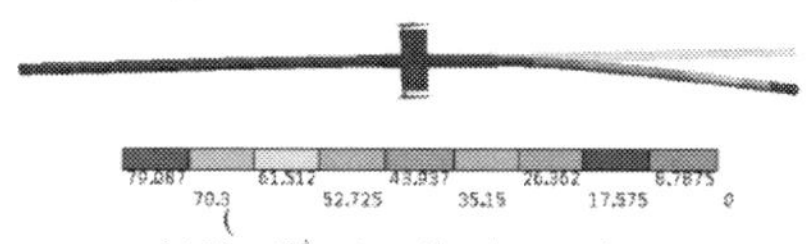

b) The 2nd order vibration mode.

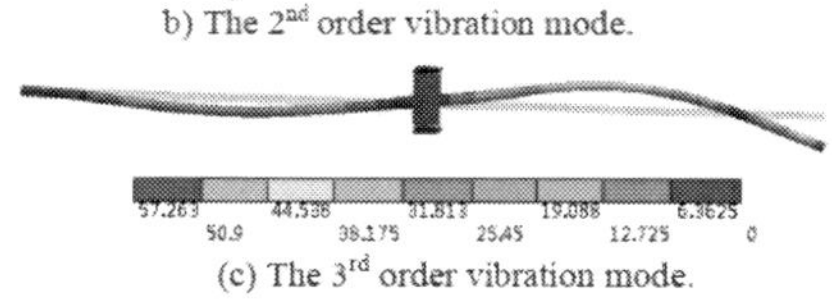

(c) The 3rd order vibration mode.

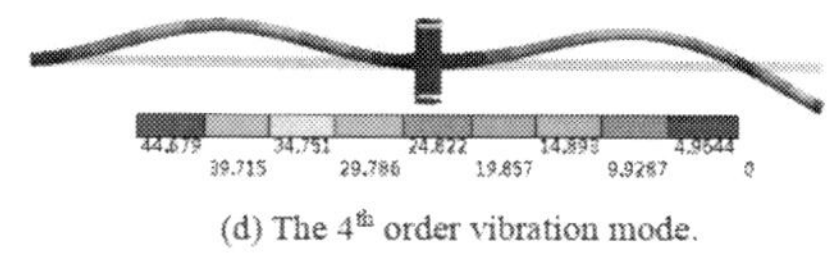

(d) The 4th order vibration mode.

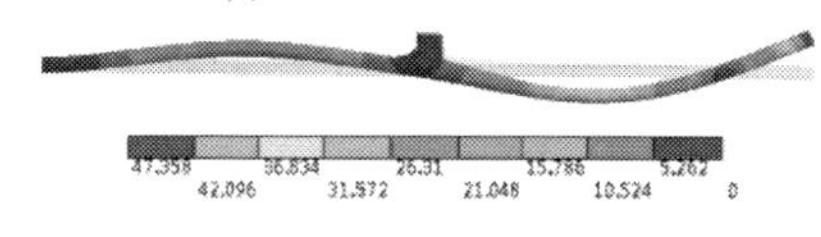

(e) The 5th order vibration mode.

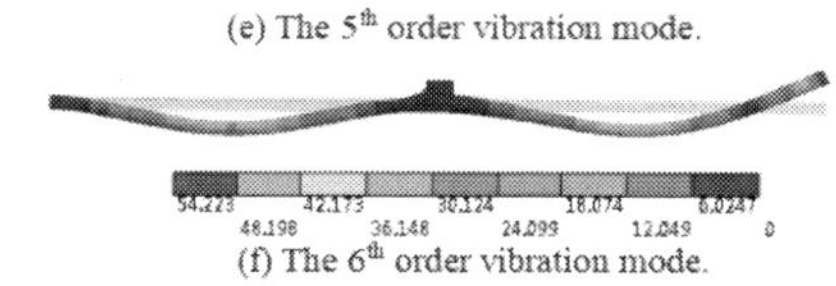

(f) The 6th order vibration mode.

Figure. 3. The vibration modes of six lowest orders.

2.3. Simulation of aerodynamic parameters

Take the first mode of the elastic refueling boom as an example. The first order deformation diagram generated by MODAL is exported as a file with stp format. Then import the file to the ANSYS ICEM. After meshing the refueling boom and the fluid area of the atmospheric environment based on ICEM, the simulation of aerodynamic parameters is proceeded.

To acquire the aerodynamic force generated by deformed refueling boom, k-epsilon turbulence model is chosen as the flow field model of the atmospheric environment. Suppose the atmospheric environment surrounding the refueling boom is compressible ideal gas, the velocity of the airflow is 0.75 *Mach* and the angle of attack α changes from 0° to 16° in 2° increments. Then, the aerodynamic parameters are obtained. Table 2 to Table 4 show the coefficients of fitted curves to each mode. In Table 2 to Table 4, the general formulation is $y_i = ae^b$, where i and y denote the order of the mode and the fitted curve, respectively. In each aerodynamic parameter calculation, the iterative convergence condition of the value must satisfy the requirement. Fig. 4 is the residual convergence diagram of aerodynamic parameters in once iterative process.

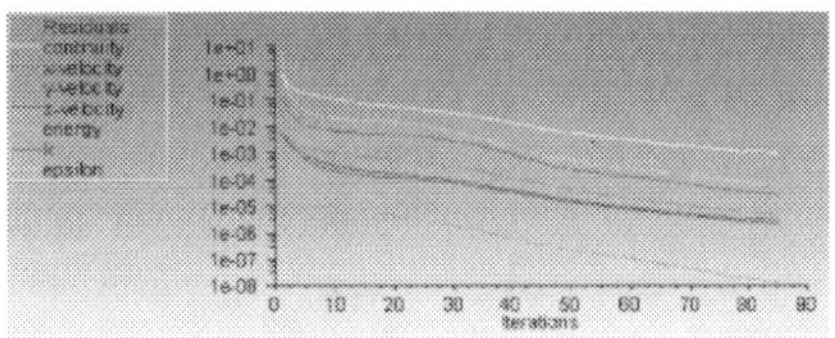

Figure. 4. The residual convergence diagram of aerodynamic parameters in once iterative process.

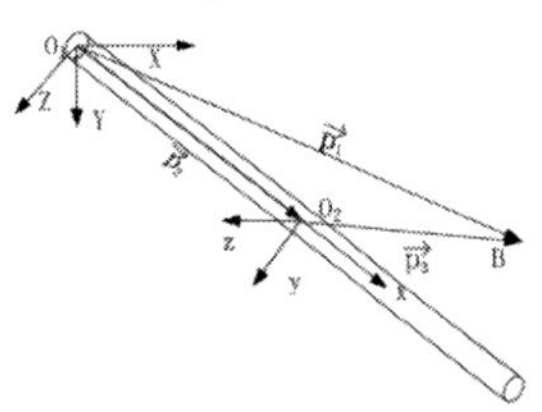

Figure. 5. Schematic diagram of refueling boom.

Table 1. Six lowest inherent frequencies of refueling boom

Mode	1	2	3	4	5	6
Inherent frequency	0.8878	1.0264	3.5738	5.6327	8.4683	10.2681

Table 2. Coefficients to each mode about lift moment

Order	Lift Coefficients	
	a	b
1	0.1134	0.0985 α
2	0.1236	0.0985 α
3	0.1371	0.0985 α
4	0.1449	0.0985 α
5	0.1501	0.0985 α
6	0.1601	0.0985 α

Table 3. Coefficients to each mode about resistance moment

Order	Resistance Coefficients	
	a	b
1	0.5910	$-[(\alpha-15.16)/7.724]^2$
2	0.5330	$-[(\alpha-15)/7.724]^2$
3	0.4481	$-[(\alpha-14.78)/7.724]^2$
4	0.3999	$-[(\alpha-16)/7.724]^2$
5	0.2223	$-[(\alpha-15.99)/7.724]^2$
6	0.1123	$-[(\alpha-13.56)/7.724]^2$

Table 4. Coefficients to each mode about pitch moment

Order	Pitch Coefficients	
	a	b
1	0.02666	$-[(\alpha-0.2991)/7.337]^2$
2	0.02899	$-[(\alpha-0.3695)/7.595]^2$
3	0.03536	$-[(\alpha-0.4991)/7.837]^2$
4	0.03676	$-[(\alpha-0.5991)/8.137]^2$
5	0.03996	$-[(\alpha-0.6991)/8.537]^2$
6	0.04529	$-[(\alpha-0.7991)/8.637]^2$

3 Problem formulation

3.1. The refueling boom with elastic deformation

The body coordinate frame (O_1XYZ) of the tanker and the coordinate frame of refueling boom (O_2xyz) are built in Fig. 5, where O_1 is at the universal joint which between tanker and refueling boom, O_2 is the center of mass for refueling boom, where $\vec{p}_1 = \vec{p}_2 + \vec{p}_3$.

There is angular velocity $\vec{\omega}$ between two coordinate frames O_1XYZ and O_2xyz. The kinetic energy of the elastic refueling boom can be represented

$$E_k = \frac{1}{2}\int_V \left\{ \frac{d\vec{p}_2}{dt}\frac{d\vec{p}_2}{dt} + \frac{\delta\vec{p}_3}{\delta t}\frac{\delta\vec{p}_3}{\delta t} + (\vec{\omega}\times\vec{p}_3)(\vec{\omega}\times\vec{p}_3) + 2\frac{d\vec{p}_2}{dt}\frac{\delta\vec{p}_3}{\delta t} + 2\frac{d\vec{p}_2}{dt}(\vec{\omega}\times\vec{p}_3) + 2\frac{\delta\vec{p}_3}{\delta t}(\vec{\omega}\times\vec{p}_3) \right\}\rho dV \quad (8)$$

where ρ, V, ρdV mean the density of refueling boom, the volume of refueling boom, and the elementary mass. $\delta/\delta(\cdot)$ represents the time derivative of the coordinate frame for refueling boom.

According to the theory of mean axis, the kinetic energy and the gravitational potential energy of the elastic refueling boom can be rewritten as

$$E_k = \frac{1}{2}M\frac{d\vec{p}_2}{dt}\frac{d\vec{p}_2}{dt} + \frac{1}{2}\int_V \frac{\delta\vec{p}_3}{\delta t}\frac{\delta\vec{p}_3}{\delta t}\rho dV + \frac{1}{2}\omega^T[I]\omega \quad (9)$$

$$E_{pg} = -\vec{p}_2 Mg \quad (10)$$

where M and I are respectively the mass of the refueling boom and the inertia tensor matrix.

$$E_{pe} = \frac{1}{2}\sum_{i=1}^{\infty}\omega_i^2 q_i^2 M_i \quad (11)$$

$$E_k = \frac{1}{2}M\frac{d\vec{p}_2}{dt}\frac{d\vec{p}_2}{dt} + \frac{1}{2}\sum_{i=1}^{\infty}M_i\dot{q}_i^2 + \frac{1}{2}\omega^T[I]\omega\ ,\ M_i = \int_V \vec{\phi}_i\vec{\phi}_i\rho dV \quad (12)$$

Based on the facts that the deformation quantity is extremely small essentially and the modes are orthogonal, the final forms of the elastic potential energy and the kinetic energy are presented as Eq. (15), (16), where q_i is the generalized coordinate, and ω means the modal frequency.

By the principle of dynamics, $\vec{p}_2$, $d\vec{p}_2/dt$ and $\vec{\omega}$ can be defined as follows

$$\vec{p}_2 = x\vec{i} + y\vec{j} + z\vec{k} \quad (13)$$

$$d\vec{p}_2/dt = u\vec{i} + v\vec{j} + w\vec{k} \quad (14)$$

$$\vec{\omega} = p\vec{i} + q\vec{j} + r\vec{k} \quad (15)$$

where $p=\dot{\phi}$, $q=\dot{\theta}\cos\phi$, $r=-\dot{\theta}\sin\phi$, $x=l\cos\theta$, $y=l\sin\theta$, $z=0$. θ represents the roll angle and ϕ denotes the pitch angle. We defined l as the distance from the universal joint to the center of mass for refueling boom.

By the formulations of E_k and E_{pg} , it can be deduced as follows

$$E_k=\frac{1}{2}M(u^2+v^2+w^2)+\frac{1}{2}\sum_{i=1}^{\infty}M_i\dot{q}_i^2+\frac{1}{2}[p \quad q \quad r][I][p \quad q \quad r]^T \tag{16}$$

$$E_{pg}=-Mg(-x\sin\theta+y\sin\phi\cos\theta+z\cos\phi\cos\theta) \tag{17}$$

The following equations (18-25) can be deduced by combining Lagrange equations with the formulations of virtual work

$$M(\dot{u}-vr+wq+g\sin\theta)=X \tag{18}$$

$$M(\dot{v}-wp+ur-g\sin\phi\cos\theta)=Y \tag{19}$$

$$M(\dot{w}-uq+vp-g\cos\phi\cos\theta)=Z \tag{20}$$

$$\dot{p}I_x-\dot{r}I_{xz}+qr(I_z-I_y)-pqI_{xz}=M_x \tag{21}$$

$$\begin{aligned}&\dot{q}(\cos\phi-p\sin\phi)I_y+(pr\cos\phi-\dot{r}\sin\phi)I_z+(\dot{p}\sin\phi+p^2\cos\phi)I_{xz}\\&+Mg\sin\phi(x\sin\phi\cos\theta+y\sin\theta)+M\cos\phi(vpx-upx+wqz+vrz\\&-wqz-wry)+M\sin\phi(\dot{u}y-vry+wpy-\dot{v}x-urx+vqz)=\cos\phi M_y\end{aligned} \tag{22}$$

$$M_i\ddot{q}_i+M_i\omega_i^2q_i=Q_{qi} \tag{23}$$

$$M_x=l_1Qs_1C_{l\delta_r}\delta_r+lQs\sum_{i=1}^{\infty}C_{lq_i}q_i \tag{24}$$

$$M_y=l_1Qs_2C_{m\delta_e}\delta_e+lQs\sum_{i=1}^{\infty}C_{mq_i}q_i \tag{25}$$

where M_x and M_y are the pitch moment and the roll moment of refueling boom, respectively. $C_{m\delta_r}$, C_{lq_i} , $C_{l\delta_e}$ and C_{mq_i} are relative aerodynamic coefficients of refueling boom, where $C_{l\delta_e}$ and $C_{m\delta_r}$ are effected by the deflection of the control surface δ_e and δ_r , respectively. C_{lq_i} , C_{mq_i} represent the aerodynamic parameters of the boom under the i th mode. We acquire these parameters by fitted curves. s , s_1 , s_2 are respectively the reference area of the refueling boom, the elevator area, the rudder area. l_1 represents the distance from the top of the refueling boom to the winglet. Q_{qi} can be found in [8].

4 Controller design

4.1. The mixed H_2/H_∞ optimal PID controller with the differential evolution algorithm

In this section, on the basis of elastic refueling boom model after linearization, a controller will be developed for the elastic suppression.

There are 16 state variables in the proposed model. For convenience of simulation and calculation, model reduction is implemented using balanced realization method [11] and Hankel singular value method [12]. The reduction model is Eq. (26).

$$\begin{aligned}\dot{\bar{\mathbf{x}}}_1&=\bar{A}_{11}\bar{\mathbf{x}}_1+\bar{B}_1\mathbf{u}\\ \mathbf{y}&=\bar{C}_1\bar{\mathbf{x}}_1\end{aligned} \tag{26}$$

where $\bar{\mathbf{x}}_1=[\theta \quad \dot{\theta} \quad \phi \quad \dot{\phi} \quad q_i \quad \dot{q}_i]^T (i=1,2,3,4,5,6)$ is state variable. $\mathbf{u}=[\delta_e \quad \delta_r]^T$ is input, and $\mathbf{y}=[\theta \quad \phi]^T$ is output.

The optimal PID controller for the mixed H_2/H_∞ with the differential evolution algorithm is a method to improve the robustness of PID controller.

$$K(s)=k_p+\frac{k_i}{s}+k_ds \tag{27}$$

$$S(s)=(1+K(s)F(s))^{-1} \tag{28}$$

$$T(s)=K(s)F(s)(1+K(s)F(s))^{-1} \tag{29}$$

$W_1(s)$, $W_2(s)$ should meet the conditions

$$\bar{\sigma}[S(j\omega)]\le\bar{\sigma}[W_S^{-1}(j\omega)] \tag{30}$$

$$\bar{\sigma}[T(j\omega)]<\bar{\sigma}[W_T^{-1}(j\omega)] \tag{31}$$

where $F(s)$, $K(s)$ are respectively the controlled object and the PID controller. $W_1(s)$, $W_2(s)$ are the weight functions of sensitivity $S(s)$ and complementary sensitivity $T(s)$ respectively.

The singular values of $S(s)$, $T(s)$ are expected to reach the minimum. For $S(s)$ and $T(s)$, there are two optimization indicators.

$$J_S=\|W_s(s)S(s)\|_\infty \tag{32}$$

$$J_T=\|W_T(s)T(s)\|_\infty \tag{33}$$

The optimization indicator is obtained further

$$J_\infty=(J_T^2+J_S^2)^{1/2} \tag{34}$$

$$J=J_\infty+J_2 \tag{35}$$

where J is the comprehensive optimization indicator.

The differential evolution theory is used to minimize the optimization indicator to optimize the three parameters of the optimal PID controller. Furthermore, we obtain the mixed H_2/H_∞ optimal PID controller. Suppose the individual $\mathbf{x}=[k_p \quad k_i \quad k_d]$, whose initialization is $\mathbf{x}_i=(x_{i1},x_{i2},x_{i3})$, where $i\in[1,50]$, $x_i\in[0,20]$.

The iteration begins:

Step 1: Get variation individuals

$$v_i=x_i+F(x_{r1}-x_{r2}) \tag{36}$$

where $F=0.5$.

Step 2: Obtain hybrid individuals

$$u_{ij}=\begin{cases}v_{ij} & rand[0,1]\le P_r,\quad or\ j=j_{rand}\\ x_{ij} & else\end{cases} \tag{37}$$

where $P_r=0.3$.

Step 3: Select the next generation of individuals

$$x_i^{g+1}=\begin{cases}u_i & f(u_i)<f(x_i^g)\\ x_i^g & f(u_i)\ge f(x_i^g)\end{cases} \tag{38}$$

where $f(\cdot)$ is the optimizing index for the optimal PID controller.

4.2. Simulink the optimal controller designed

On the basis of the previous section, select the weight function of sensitivity and complementary sensitivity according to the principle of weight function selection.

$$W_1(s) = \frac{200}{175s+1} \tag{39}$$

$$W_2(s) = \frac{0.2146s+0.0011}{0.45s+1} \tag{40}$$

After 100 iterations of the differential evolution algorithm, the optimal PID controller parameters with mixed H_2 / H_∞ are k_p=20.356 , k_i=10.928 , k_d=1.063 .

Assume the flight velocity of the tanker is 200 m/s . Then, we simulate the proposed optimal PID controller. The simulation results are shown in Fig. 6.

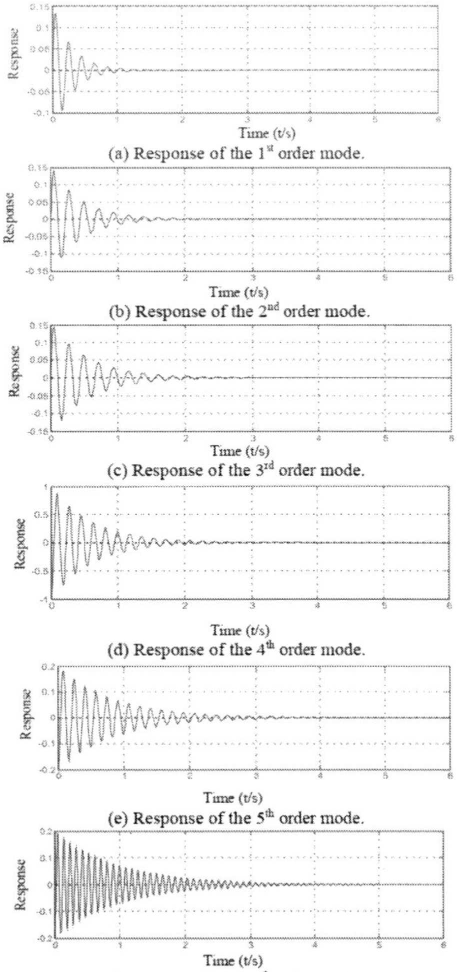

(a) Response of the 1st order mode.

(b) Response of the 2nd order mode.

(c) Response of the 3rd order mode.

(d) Response of the 4th order mode.

(e) Response of the 5th order mode.

(f) Response of the 6th order mode.

Figure. 6. Response curves at the objective $(30^\circ, 0^\circ)$ to $(32^\circ, 0^\circ)$.

5 Conclusion

In this paper, the modal analysis and aerodynamic analysis of elastic refueling boom have been developed by ANSYS. Through the analysis of the elastic refueling boom, the accurate frequency response and aerodynamic characteristics are obtained. Further, a proposed model is deduced and with that an optimal PID controller to suppress the elastic deformation is proposed. The validity of the proposed controller is verified by simulation. That is, the elastic deformation of each mode is also obviously suppressed.

Acknowledgements

This work is supported by the National Natural Science Foundation of China (No. 61473229).

References

1. Manfriani L, Righi M. Aerodynamic modeling of a refueling boom. 29th Congress of the International Council of the Aeronautical Sciences, ICAS 2014.
2. Chang L, Jia YM. Analog strategy of a receiver aircraft tested for boom and receptacle refueling. Chinese Control Conference. 2015,5857-5862.
3. Yuan DL, Zhao CX, Xiao ZB. Research on path planning of manipulator for aerial boom refueling test system. Applied Mechanics and Materials, 2014, 538:159-162.
4. Yang Q, Li C, Ji Q. Application of non-cancellation decoupling in boom refueling control. International Conference on Mechatronics and Automation, 2009:4203-4207.
5. David I.L, Bernard F. Visual comparisons of selected shores affected by the Prestige oil spill, Northern Spain 2002-2004. 2005 International Oil Spill Conference, 2005:10514-10517.
6. Zhu Z.H, Meguid S.A. Elastodynamic analysis of aerial refueling hose using curved beam element. AIAA Journal, 44(6): 1317-1324.
7. Yang C, Yang Y, Lu Y. Modeling and simulation for refueling boom and receiver in coupled mode. Transactions of Nanjing University of Aeronautics and Astronautics, 2017, 34(2):143-151.
8. Waszak M.R, Schmidt D.K. Flight dynamics of aeroelastic vehicles. Journal of Aircraft, 1988, 25(6): 563–571.
9. Baghdadi N, Lowenberg M.H, Isikveren A.T. Analysis of flexible aircraft dynamics using bifurcation methods. Journal of Guidance, Control, and Dynamics, 2011, 34: 795-809.
10. Skogestad S, Postlethwaite I. Multivariable feedback control. New York: Wiley, 2005.
11. Wortelboer P, Odstveen H.V. Model reduction guided by Hankel singular value intervals. Identification, Model and Control, 1990, 1:17-26.
12. Ho SJ, Ho SY, Hung MH, Shu LS, Hung HL. Designing structure-specified mixed optimal H_2 / H_∞ controllers using an intelligent genetic algorithm IGA. IEEE Trans on Control Systems Technology, 2005, 13(6): 1119- 1124.

Research on the image of sweeping robot based on the Artificial Neural Network

Chang Song[1,a*], *Yoshito* Ogata[2]

[1]Kyushu University, Graduate School of Design, Japan,and Sichuan University of Media and Communications,China
[2]Kyushu University, Faculty of Design, Japan

Abstract. Based on the theory of Artificial Neural Network and Kansei Engineering, the image of sweeping robots are formed using the content analysis method, and propose four kinds of sweeping robot as the experimental samples, which have a strong influence on the market. The image questionnaires are compiled by the semantic differences methods. 200 office workers, half men and half women, are chose as the survey respondents. And use SPSS statistical software for data analysis. Afterwards, the BP Artificial Neural Network model is established by Matlab based on the questionnaire results, and the optimized design scheme with image feature combination for sweeping robot products is generated on the basis of BP Artificial Neural Network model. This study construct the emotional demands on the image level, and carry out experiments and statistical analysis, which lays a solid foundation for the study of product image in theory and approach.

Keywords. Artificial Neural Network, Image, Sweeping Robot, Kansei

1 Introduction

With the development of human society, most parts of life and time are occupied by heavy work. In order to lighten the burden on people' life, the sweeping robots are gradually used to help people do housework. Many kinds of sweeping robot brands appearing on the market, only Jingdong mall has more than 60 brands on sale. Moreover, the more and more diversified and personalized trends are formed to meet the needs of people. And the product image is an effective way to express the people's needs. For a long time, designers are relying on their own experience, intuition and inspiration to solve the problem of product image without the effective theoretical and experimental support. it is difficult to accurately and efficiently meet the emotional needs of consumers.

The image questionnaire was compiled by the content analysis and emantic differences methods based on the theory of Artificial Neural Networks and Kansei Engineering in this study. And use SPPSS statistical software for data analysis. Afterwards, the BP Artificial Neural Network model is established by Matlab, and he optimized image design scheme for sweeping robots are generated as references for the products image of sweeping robots, and lay a sound foundation for the product image.

2 Theory discussion

2.1 Discussion on image

First of all, it must make a connection between "image" and the "Kansei Engineering". In the aspect of significance, (In a sense), the "sensibility" is an important part of the study of Kansei Engineering ", which is exactly matched with the" image ". The former belongs to philosophy, and the process is the focus of the research.

While the latter is part of psychological aspects, which is used for referring to psychological characteristics. Secondly, the clear boundaries are not observed between the" image "and" Kansei Engineering ", both have the ability to present the audience' emotion to the products.

The image is a mental image, and it is the merged crystallization between subjective mind image and environmental factors. Sometime, it is the results of the combination between instantaneous feeling and past experiences. The investigated image involves feeling, perception, style and other areas. Figure 1 shows the image formation process.

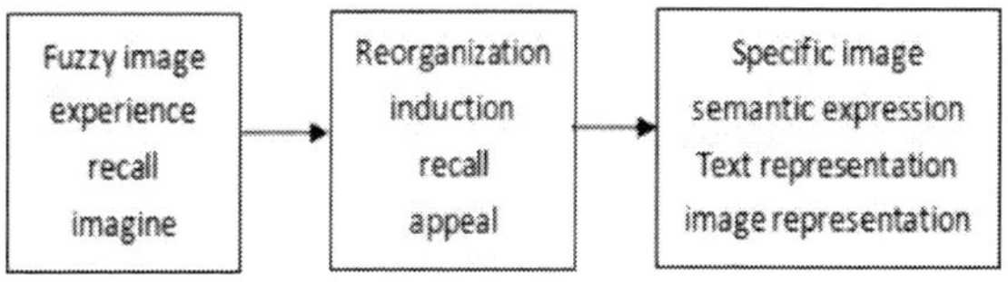

Fig. 1. Process of image formation

* Corresponding author: schang025@163.com (Chang Song)

In conclusion, the image can be interpreted as the formed mental imagery based on past experience or feel in their situation. It may be an intuitive response or the memory of past experience. In the design process, the purpose of design image is giving the product image according to the users' demand for products, feeling and thoughts. Product image is formed gradually through the users' perceptual cognition. It can be divided into function, structure, shape, color, material and other aspects. Generally, the product images need to be conveyed based on the needs and feelings of people. The studies on the preference of sweeping robots' image is an important communicating bridge between designers and consumers, which will greatly give the designers much assistance and hints.

2.2 Artificial Neural Networks

Artificial Neural Networks, abbreviated as neural networks or artificial neural network, is an information processing system for imitating the biological function of the nervous system. The network consists of input layer, output layer and the hidden layer, as shown in the figure 2.

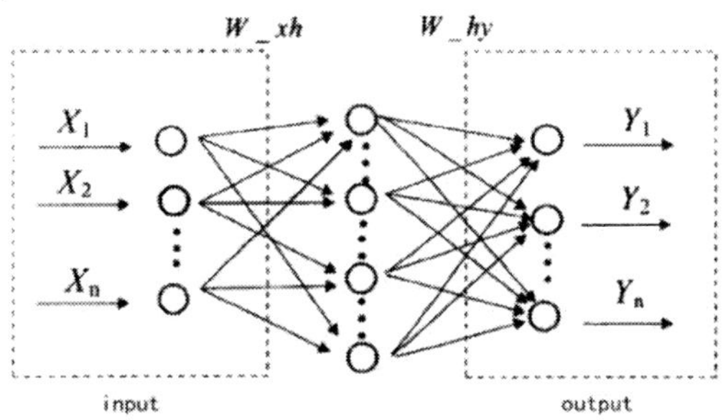

Fig. 2. Artificial Neural Networks of systems

The BP artificial neural network is supervised. Among the study of artificial neural network, learning and memory is the two main parts its basic framework is multilayer network, and multilayer architecture is to increase the nonlinearity, hence it can handle nonlinear problems which cannot be rendered by regression analysis. The BP artificial neural network can be realized with MATLAB programming for forecasting and fitting, and the function is very powerful. In recent years, in terms of product design, design and research personnel will transfer the application into perceptual design through the computer simulation, and the preferences, product function, form and color for users are investigated systematically. The product image is presented in a structured way.

3 Research method

This paper determines the dimensions of the sweeping robot image and extraction of experimental samples through content analysis methods, the drafting group measured and the image questionnaire by the semantic differential method; statistical analysis proceeds using SPSS; finally, Matlab is used to establish BP Artificial Neural Network model in order to accomplish the image design.

3.1The sample and the tested population

According to the content analysis method, the image of sweeping robot is divided into function, structure, shape, color and material image space. Meanwhile, the highest acceptability price (1688 - 2340 RMB) is selected from the Jingdong mall, and the four types of sweeping robot are chose from four brands based on the sales volumes, as presented in table 1.

Table 1. The sweeping robot.

iRobot	MIJIA
	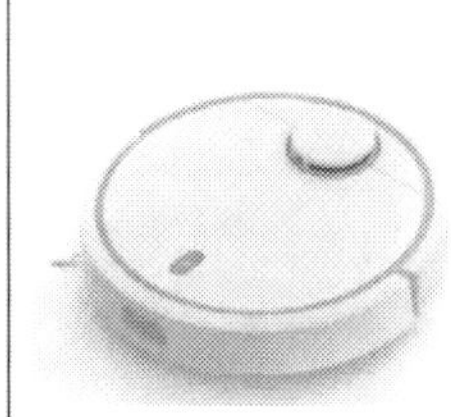
Proscenic	TOMEFON
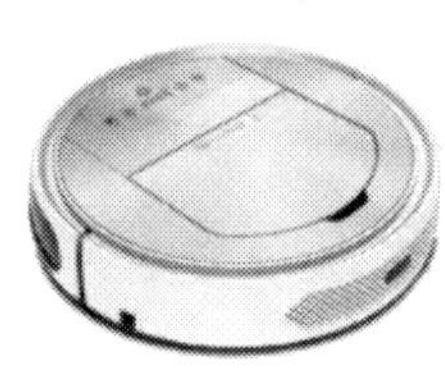	

The young people who are usually busy working make up the majority of the sweeping robot users. Therefore, 200 office workers, half men and half women, are chose as the survey respondents to explore the sweeping robot image.

3.2 Image Questionnaire

This study is based on Kansei Engineering and image in product design, and the adjective phrases of each image space are obtained by the semantic differential method, as shown in Table 2. Likert 5 point scale is applied to assess the opposite words, and interval scale scores were expressed in 2,1,0,1 and 2, respectively.

Table 2. Adjective with image .

Functional image	**Structural image**
Novel & obsolete	Stable & adjustable
Versatile & single	Firm & loose
Easy to use & difficult to use	Independent & systematic

Intelligent & rigid	Complex & concise
Suitable & absurd	Dangerous& Safe
Morphological image	**Color image**
Beautiful & ugly	Passionate & callous
Science fiction & reality	Dull & bright
Round & sharp	Simple & fancy
Unique & common	Gorgeous & artless
Complex & concise	Warm & cold
Material image	
Good & poor	
Hard & soft	
Durable & easy to break	
Smooth & rough	
Elegant & vulgar	

3.3 Statistical analysis

After the questionnaires are collected, the scores are coded and the 2,1,0,1,2 are changed into 1,2,3,4,5 in sequence. The One-sample T T test is used to verify five images of the data through statistical software SPSS to obtain the significance and the bias analysis of image phrases.

3.4 BP Artificial Neural Network model

The BP Artificial Neural Network model is established by Matlab based on the results of the image questionnaires. Learning and recalling, analyzing and establishing the image data model, provide a data validation model for the design of new products.

4 Results and Discussion

4.1 One-sample T test

200 young workers, half men and half women are questioned to obtained the data. The 200 valid questionnaires are analyzed by statistical software SPSS. One-sample T test is carried out on the five space images of the sweeping robots .

According to the results of One-sample T test of the sweeping robot image spaces (Table 3), we can know that the adjective vocabulary is significant, and the function of image is novel, versatile, easy to use, intelligent, suitable image (the versatile aspect is the most significant, M=1.710).

The image of structure tends to be stable, firm, systematic, concise, safe (the firm aspect is the most significant, M=1.530).

The image of morphological tends to be beautiful, science fiction, round, common, concise (the aspect of science fiction is the most significant, M= 1.600).

The image of color space tends to be passionate, bright, simple, gorgeous, warm (the passionate aspect is the most significant, M=1.985).

Finally, the image of material tends to be superior, hard, durable, smooth, elegant (the durable aspect is more significant, M=1.420).

Table 3. One-Sample Test

Image	Adjective with image	M	Sig.
Functional	Novel & obsolete	2.205	0.000***
	Versatile & single	1.710	0.000***
	Easy to use & difficult to use	2.020	0.000***
	Intelligent & rigid	2.235	0.000***
	Suitable & absurd	1.740	0.000***
Structural	Stable & adjustable	1.815	0.000***
	Firm & loose	1.530	0.000***
	Independent & systematic	4.230	0.000***
	Complex & concise	4.000	0.000***
	Dangerous& Safe	4.330	0.000***
Morphological	Beautiful & ugly	1.760	0.000***
	Science fiction & reality	1.600	0.000***
	Round & sharp	2.360	0.000***
	Unique & common	4.000	0.000***
	Complex & concise	4.305	0.000***
Color	Passionate & callous	1.985	0.000***
	Dull & bright	3.615	0.000***
	Simple & fancy	2.375	0.000***
	Gorgeous & artless	2.365	0.000***
	Warm & cold	2.060	0.000***
Material	Good & poor	1.825	0.000***
	Hard & soft	2.525	0.000***
	Durable & easy to break	1.420	0.000***
	Smooth & rough	1.895	0.000***
	Elegant & vulgar	1.950	0.000***

*Sig.<0.05; **Sig.<0.01; ***Sig.<0.001

DOI: 10.1051/matecconf/201713900059

4.2 Artificial Neural Network

Transfer the average value of M obtained by the statistical software SPSS into the interval (0,1). Afterwards, using Matlab as a BP artificial neural network training software, the image M value is trained and input into program until the BP artificial neural network convergence is obtained. Then the sweeping machine image data model can be established and the image database is constructed completely, which provide the data model for the new products verification.

Firstly, the Excel data table is set up with the average value of the single sample T test, and P1, P2, P3 and P4 are used to represent the four sweeping robot samples. Then the data is input into the Matlab program again, and the initial eigenvalue is obtained to prepare for normalization.

Then, the feature matrix is normalized, and the normalized command program is input into the Matlab program command box:

```
for i=1:1:25;
for j=1:1:5;
a1(i,j)=(a(i,j)-min(a(i,:)))/(max(a(i,:))-min(a(i,:)));
end
End
```

Thus, the normalized values of 25 image eigenvalues are obtained, And then the BP artificial neural network is established. The detailed program coding was as follows:

```
P=[a1(:,1) a1(:,2) a1(:,3) a1(:,4)];

T=[1 0 0 0;0 1 0 0;0 0 1 0;0 0 0 1];
x=ones(size(P,1),2);
x(:,1)=0;
net=newff(x,[5,4],{'tansig','logsig'},'traingdx');
net.trainParam.show=50;
net.trainParam.epochs=5000;
net.trainParam.goal=0.005;
net=train(net,P,T);
```

The main process of Matlab Net is the new BP artificial neural network. The training process report is shown in Figure 3, and it presents that the BP artificial neural network convergence can be observed clearly. Figure 4 shows the training convergence curve of BP artificial neural network is continuous, and finally tend to a point, the image data model of the BP artificial neural network is established completely.

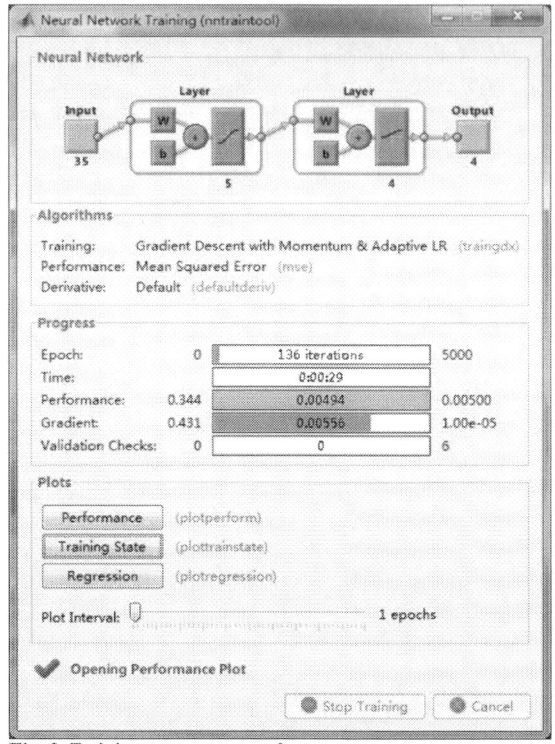

Fig. 3. Training process report chart.

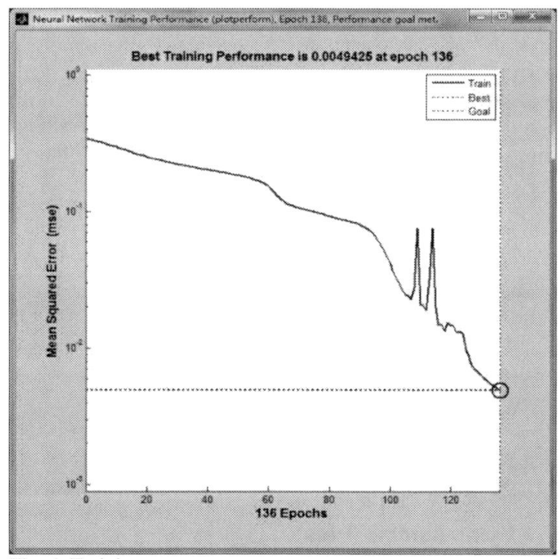

Fig. 4. Training convergence.

5 Conclusion

In summary, the four kinds of sweeping robots are selected as the experimental samples. And the One-sample T test is conducted by SPSS based on the theory of Kansei Engineering, and then Matlab are used to establish BP artificial neural network image model, which will provide a reference for the study of sweeping robot product image.

Acknowledgements

Supported by China Scholarship Council (CSC).

References

1. Hung-Yuan Chen , Yu-Ming Chang, Extraction of product form features critical to determining consumers' perceptions of product image using a numerical definition-based systematic approach, International Journal of Industrial Ergonomics 39 ,133–145(2009)
2. Nagamachi, M., Kansei engineering and comfort. International Journal of Industrial Ergonomics 19 (2), 79–80(1997)
3. Nagamachi, M., Kansei engineering as a powerful consumer-oriented technology for product development. Applied Ergonomics 33 (3) ,298–294 (2002)
4. Crilly, N., Moultrie, J., Clarkson, P.J., Seeing things: consumer response to the visual domain in product design. Design Studies 25 (6), 547–577 (2004)
5. S. Bornholdt and D. Graudenz. General asymmetric neural networks and structure design by genetic algorithms. Neural Networks, 5:327–334,(1992)
6. L Liu ,M Deng,An Evolutionary Artificial Neural Network Approach for Breast Cancer Diagnosis,International Conference on Knowledge Discovery& Data Mining , 25 (3) :593-596(2010)
7. L Liu ,M Deng,An Evolutionary Artificial Neural Network Approach for Breast Cancer Diagnosis,International Conference on Knowledge Discovery& Data Mining , 25 (3) :593-596(2010)
8. Chang Song ,Chen Tien-li,Image Study of Long-standing universities based on international students — In National Taipei University of Technology's case,2011 IEEE the 12th International Conference on Computer-aided Industrial Design & Conceptual Design, 211-215(2011)

Robust Robot Grasp Detection in Multimodal Fusion

Qiang Zhang[1,2,3,*], Daokui Qu[1,2,3], Fang Xu[1,2,3], and Fengshan Zou [3]

[1] *State Key Laboratory of Robotics, Shenyang Institute of Automation, Chinese Academy of Sciences, Shenyang, No.114 Nanta Street, Shenhe District, Shenyang 110016, P.R. China*
[2] *University of Chinese Academy of Sciences, No.19A Yuquan Road, Beijing 100049, P.R. China*
[3] *SIASUN Robot & Automation Co., LTD., NO.16 Jinhui Street, Hunnan District, Shenyang 110168, P.R. China*

Abstract. Accurate robot grasp detection for model free objects plays an important role in robotics. With the development of RGB-D sensors, object perception technology has made great progress. Reach feature expression by the colour and the depth data is a critical problem that needs to be addressed in order to accomplish the grasping task. To solve the problem of data fusion, this paper proposes a convolutional neural networks (CNN) based approach combined with regression and classification. In the CNN model, the colour and the depth modal data are deeply fused together to achieve accurate feature expression. Additionally, Welsch function is introduced into the approach to enhance robustness of the training process. Experiment results demonstrates the superiority of the proposed method.

1 Introduction

With the development of visual sensors, robot vision perception technology has made great progress. Especially in recent years, robots can perceive the colour and the distance properties of the environment due to the development of RGB-D sensors like Kinect and Xtion. In the field of home service robot applications, vision based robot grasp detection has been an important research direction in robot technology for the reason that it can improve the level of human-computer interaction.

For the service robot, the objects to be operated are divided into two categories, the known and the unknown objects. The known objects mean that robots have stored the model of the object. Under this circumstance, robot grasp detection procedure can be separated into two processes. The first process is to detect the specific object and the second one is to estimate the object's pose and find out the proper grasp point. However, on the other hand, the robot cannot always store models for all objects. For example, when the service robot enters a strange environment, objects in this exception would be unknown for the robot. Then, grasping the unknown object or the model free object can be a difficult task for the robot.

During last few years, several significant approaches [1-9] have been proposed to solve the model free grasp detection problem. In 2006, 2D grasping point representation method was proposed by A. Saxena [1]. In 2010, Q. V. Le [2] introduced a method by using multiple contact points to represent grasp locations. In 2011, Y. Jiang [3] presented a new representaiton mehtod which describe the robot grasp as a 5-dimensiaonal oriented rectangle. In order to learn the representation, the SVM ranking algorithm was introduced in the learning process. With the development of deep learning, convolutional neural networks has been introduced as powerfull visual models [4-5]. In 2013, Lenz [6] proposed a convolutional networks based method for robot grasp detection. The method using sliding windows to generate multiple candidate grasps firstly and true grasps could be retained by the classifier. In 2016, J. Wei [7] proposed a multimodal fusion based deep extreme learining machine for robot grasp recognition. L. Trottier presented a dictionary learning method in the same year. However, these three methods have disadvantages of high complexity. In 2015, J. Redmon [9] introduced a real-time detection method by using single-stage regression. The end-to-end solution reduced the training difficulty. However, the coarse data fusion leads to low detection accuracy. Meanwhile, the training process has slow convergent speed because of outliers.

In order to improve robot grasp detection accuracy, we adopted an improved approach. The contribusions are the following. First, we proposed a robust loss function by taking Welsch function into consideration. Second, we introduced Atrous convolution algorithm to our architecture to improve the local expression ability of features. Both these two contributions can improve robot grasp detection accuracy.

The rest of the paper is orgnised as follows. Section 2 describes the deep regression model for robot grasp detection. Details of the proposed method are discussed

* Corresponding author: zhangqiang@sia.cn

in Section 3. Section 4 evaluates performance of the proposed method. Finally, Section 5 gives the summary and conclusion.

2 Related works

Generally, object models should be constructed before robotic grasping. For instance, the specific object can be expressed by invariant local features [10], sparse 3D point cloud model [11] and dense 3D point cloud [12]. However, building object model is difficult and time consuming. This limits the ability of the robot to adapt to the environment. Recently, category based object detection [13-15] has been applied into robot applications. It is still difficult to determine object's pose.

Y. Jiang [3] proposed a rectangle representation method for robot grasp, which skips object detection and pose estimation process. Each grasp is described by a rectangle with its central coordinates, size, and orientation. This expression method greatly simplifies the complexity of the model. With the development of deep learning, convolutional neural networks based grasp detection method was proposed by I. Lenz [6]. However, the sliding window approach decreases the detection efficiency. L. Trottier [8] proposed a dictionary learning method for robot grasp detection. Although it takes the advantage in high accuracy, proposal procedure makes it inefficient. J. Redmon [9] presented an end-to-end solution for real time grasp detection. Nevertheless, the shallow data fusion architecture can affect the accuracy of the detection. At the same time, full-connected layers reduced the local feature representation ability. D. Guo [16] introduced a pipeline using reference rectangles on the feature map. For that only colour information was employed, the detection accuracy is still not high enough.

3 Robust robot grasp detection

3.1. Architecture

In our work, the robot grasp is represented by a grasping rectangle which is a 5-dimontional vector $\{x, y, w, h, \theta\}$ the same as in [5]. In this vector, x and y denotes the central coordinates, w and h denotes the rectangle size, θ represents the angle between the rectangle and the horizontal direction. Inspired by [9], robot grasp detected in the regression way. The proposed architecture is shown in Figure 1.

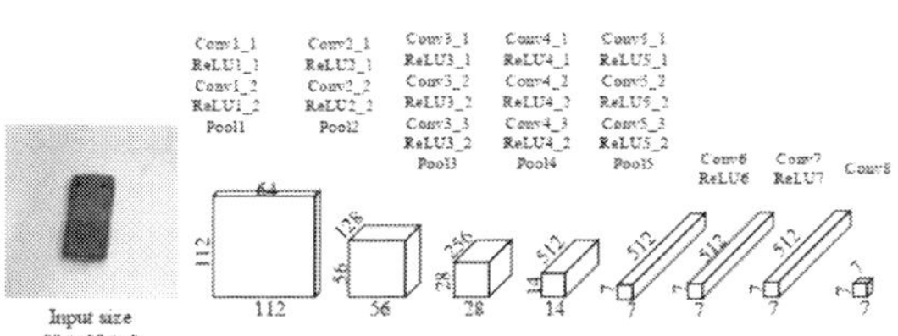

Fig. 1. The robot grasp detection architecture

The proposed architecture is a modification of the VGG-16 networks [17]. The full connection layers of VGG-16 were replaced by three convolutional networks aiming at improving the expression ability of local information according to hole filling algorithm [18]. In the architecture, all convolutional layers have the same kernel size 3×3 and the same stride 1. After each convolutional layer, batch normalization was employed to improve detection accuracy. All pooling layers are maxpooling layers and the same stride is 2. In the architecture, the input colour image size is 224×224×3 and the output matrix size is 7×7×7.

3.2. Detection methodology

The detection architecture is designed aiming at combining classification with regression. For each input image, we detect 49 (7×7) results, and each result is a 7 dimensional vector. As can be seen in Figure 2, the vector indicates the graspable probability of the detection result. Meanwhile, it represents the location of the graspable rectangle. The detection result is illustrated in Figure 2.

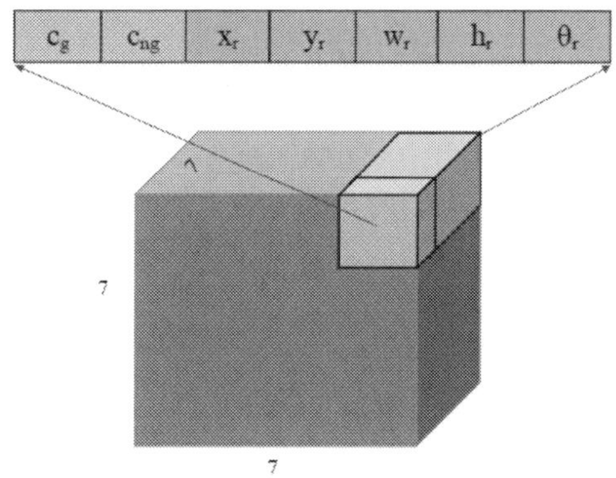

Fig. 2. Detection result in each vector

The loss function of the proposed architecture is as follows.

$$L(c,r)=\sum_{i=1}^{N}\left[L_c(c_i)+\lambda L_r^g\left(x_i,y_i,w_i,h_i,\theta_i\right)\right] \quad (1)$$

In equation(1), the loss function is composed by two parts: the classification part and the regression part. L_c is the Softmax Loss between the classification result and the truth label. The truth label is 1.0 when the graspable rectangle centre is located in the block, otherwise the value is 0. L_r^g indicates differences between the regression result and the true rectangle locations. The weight term λ is set to 1.0 according to validation result. In the training process, truth values of $\{x, y, w, h\}$ are all normalized as:

$$\begin{cases} x_n = \mathrm{mod}(x,32)/32 \\ y_n = \mathrm{mod}(y,32)/32 \end{cases} \quad (2)$$

$$\begin{cases} w_n = w / w_{max} \\ h_n = h / h_{max} \end{cases} \quad (3)$$

In equation (2), the value 32 indicates the sizes of each separate block. In equation (3), w_{max} and h_{max} are the largest width and height of the graspable rectangles.

During training process, back propagation is to adjust parameters to meet detection requirements. Outliers can lead to slow convergence rates. Meanwhile, small residuals contribute less to the back propagation process. Inspired by [19], we introduced a more robust loss function named Welsch to retain relative large residuals. The function is designed as follows:

$$f(x) = \frac{\alpha^2}{2}\left[1 - e^{-(x/\alpha)^2}\right] \quad (4)$$

In equation (4), α=2.9846. The function and its derivatives are shown in Figure 3.

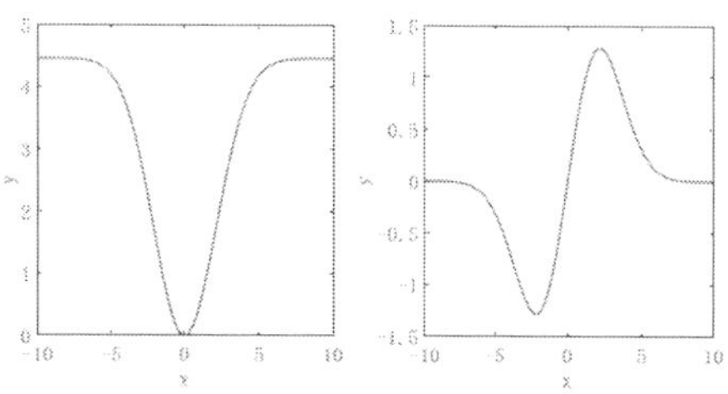

Fig. 3. Welsch function and its derivative

Then, the full loss function L_r^g is designed as follows.

$$\begin{aligned} L_r^g = \sum_{reg \in pos} \{ & f(x_n - x_r) + f(y_n - y_r) \\ & + f[\log(w_n) - \log(w_r)] + f[\log(h_n) - \log(h_r)] \\ & + f[\sin(\theta_n - \theta_r)] \} \end{aligned} \quad (5)$$

3.3. Multimodality fusion

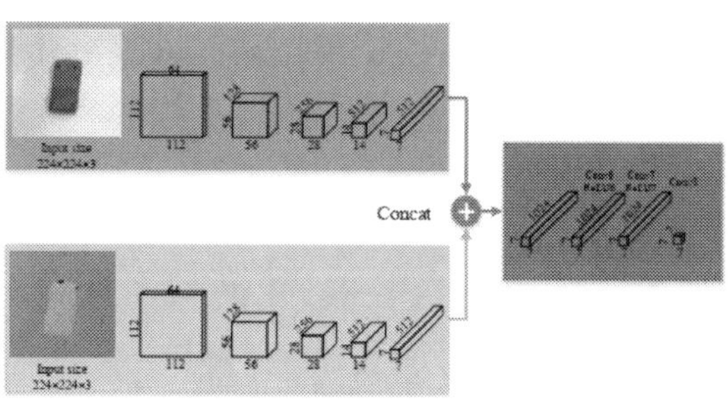

Fig. 4. Feature fusion principle in the regression architecture

Nowadays, RGB-D sensors make it possible for robots to get distance information of the scene. For robot grasp, colour and depth data fusion has been a research hotspot. In our work, we proposed a data fusion method based on feature fusion. We first trained the RGB based neural networks. The same as RGB image based networks training, we feed depth image to the similar networks. In order to simplify the training complexity, each depth image was expanded to 3D. After all, we concatenate the features from the RGB image and the depth image, and retrain layers Conv6, Conv7 and Conv8. Feature fusion methodology is illustrated in Figure 4.

4 Experiments

4.1. Experimental methodology

We adopted the Cornell Grasping Dataset to evaluate the proposed method. The dataset contains 240 different objects. The total number of the samples is 885. For each sample, RGB image and 3D dense cloud image is provided. The 3D dense cloud images were first transformed to depth images. For the reason that each image contains multiple graspable and non-graspable locations, the proposed method can be applied to these multiple label detection task. In order to improve the robustness of the networks, we performed an extension for the original dataset in image cropping, rotation. The number of the training examples was extended to 3540 and the number of the graspable rectangles was over 20,440. The dataset was separated into two parts, 80% was the training set and the remaining 20% was test set. Moreover, all images were cropped into the size of 224×224.

In order to evaluate the performance of the proposed method and the conventional methods, the grasp intersection over union (IoU) metric was employed in our work. The IoU is defined as:

$$IoU = \frac{G_{detection} \cap G_{truth}}{G_{detection} \cup G_{truth}} \quad (6)$$

In the upper equation, the numerator and the denominator of the right part denotes the overlap area and the union area respectively. The grasp was seen as a right one only when the IoU was above 0.25 and the angle difference between the detected grasp and the truth was less than 30°. Two experiments were conducted in order to evaluate the image-wise and object-wise detection accuracy.

4.2. Training details

As mentioned previously, our proposed architecture is a modification of VGG-16 networks. Parameters of the last three layers were generated randomly, other parameters were initialized by the pre-trained VGG-16. Then we converted the model to perform robot grasp detection on the RGB and depth image dataset respectively. Features corresponding to the RGB and depth images were extracted and concatenated to perform multiple modalities fusion. The last three convolutional neural network layers were re-trained accordingly.

During training and validation process, Caffe deep learning library was employed. The experiment was performed based on two NVIDIA Tesla K80 GPUs with 24GB memory. In order to speed up the training process, the NVIDIA CuDNN library was taken into our work. In the training procedure, we took stochastic gradient descent method to optimize the model.

4.3. Results and analysis

We compared our proposed method with algorithms proposed by Y. Jiang [3], I. Lenz [6] and J. Redmon [9]. Experimental results can be seen in Table 1. In Table 1, the proposed method has the better detection accuracy of 88.90% in image-wise split and 88.20% in object-wise split. The detection speed of the proposed method is 117ms per image. Some detection results can be seen in Figure 5.

Table 1. Detection results of different algorithms.

Algorithm	Detection accuracy		speed
	Image wise split	Object wise split	
Jiang	60.50%	58.30%	-
Lenz	73.90%	75.60%	13,500ms
Redmon	88.00%	87.10%	76ms
Ours	88.90%	88.20%	117ms

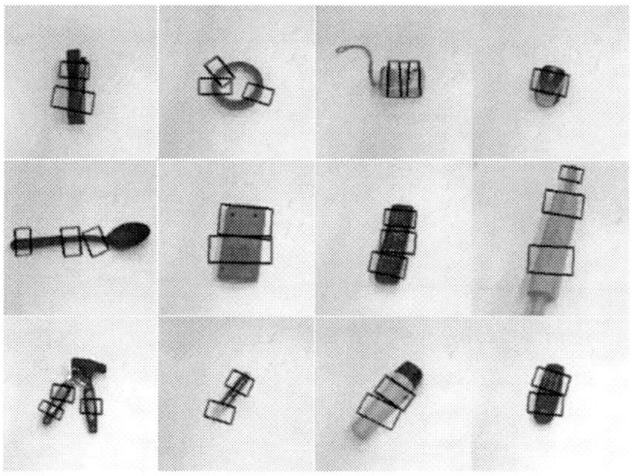

Fig.5 Detection results of the proposed method

According to the experimental results, the detection accuracy is higher than that of the traditional convolutional neural networks. The accuracy of the proposed method is increased for the reason of the reasonable loss function and the modified neural networks. The speed of our proposed method is slightly slower than Redmon's because that the proposed architecture is much more complex. However, the speed is still acceptable. Moreover, the end-to-end convolutional neural network structure takes the advantage of high speed compared with auto-encoder based learning method.

We also applied the proposed method to our application. The detection accuracy reached to 82%. Some detection examples are shown in Figure 6.

5 Conclusions

Robot grasp improves the ability of human robot communion. This paper proposes a robust method to detect robot grasps. The main contributions of the paper are: (1)We propose a new robust loss function for robot grasp detection using deep neural networks. By taking Welsch function into consideration, back propagation by interior points with small contributions and outliers can be reduced. This can lead to robust training progress. (2)Hole filling algorithm is introduced to improve the expression ability of local features. Experimental results show that the proposed method achieves the superior performance in robot grasp detection.

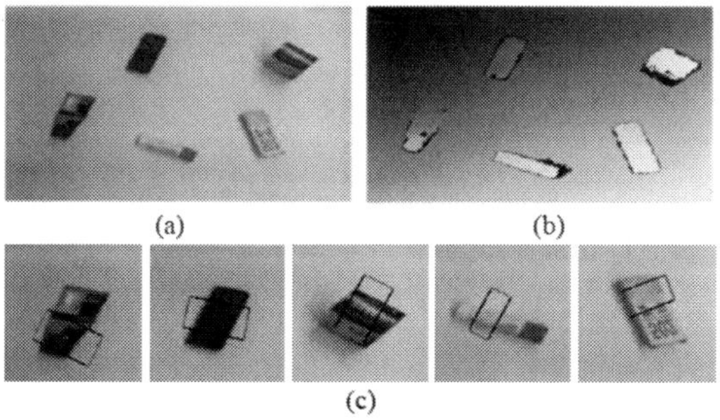

Fig.6 Detection results in our application: (a) The RGB image; (b)The depth image; (c)Detection results

Acknowledgements

We give our thanks to Shenyang SIASUN Robot Automation Co., LTD for funding the research program. This work was supported by the National Key Research and Development Program of China under Grant No. 2016YFF0202701.

References

[1]Saxena A, Driemeyer J, Kearns J, et al. Robotic grasping of novel objects[C]//Advances in neural information processing systems. 2007: 1209-1216.

[2]Le Q V, Kamm D, Kara A F, et al. Learning to grasp objects with multiple contact points[C]//Robotics and Automation (ICRA), 2010 IEEE International Conference on. IEEE, 2010: 5062-5069.

[3]Jiang Y, Moseson S, Saxena A. Efficient grasping from rgbd images: Learning using a new rectangle representation[C]//Robotics and Automation (ICRA), 2011 IEEE International Conference on. IEEE, 2011: 3304-3311.

[4]Krizhevsky A, Sutskever I, Hinton G E. Imagenet classification with deep convolutional neural networks[C]//Advances in neural information processing systems. 2012: 1097-1105.

[5]Girshick R, Donahue J, Darrell T, et al. Rich feature hierarchies for accurate object detection and semantic segmentation[C]//Proceedings of the IEEE conference on computer vision and pattern recognition. 2014: 580-587.

[6]Lenz I, Lee H, Saxena A. Deep learning for detecting robotic grasps[J]. The International Journal of Robotics

Research, 2015, 34(4-5): 705-724.

[7]Wei J, Liu H, Yan G, et al. Robotic grasping recognition using multi-modal deep extreme learning machine[J]. Multidimensional Systems and Signal Processing, 2017, 28(3): 817-833.

[8]Trottier L, Giguère P, Chaib-draa B. Dictionary Learning for Robotic Grasp Recognition and Detection[J]. arXiv preprint arXiv:1606.00538, 2016.

[9]Redmon J, Angelova A. Real-time grasp detection using convolutional neural networks[C]//Robotics and Automation (ICRA), 2015 IEEE International Conference on. IEEE, 2015: 1316-1322.

[10]Kurzejamski G, Zawistowski J, Sarwas G. Robust method of vote aggregation and proposition verification for invariant local features[J]. arXiv preprint arXiv:1601.00781, 2016.

[11]Hsiao E, Collet A, Hebert M. Making specific features less discriminative to improve point-based 3D object recognition[C]//Computer Vision and Pattern Recognition (CVPR), 2010 IEEE Conference on. IEEE, 2010: 2653-2660.

[12]Yi Z, Li Y, Gong M. An Efficient Algorithm for Feature-Based 3D Point Cloud Correspondence Search[C]//International Symposium on Visual Computing. Springer International Publishing, 2016: 485-496.

[13]Tychsen-Smith L, Petersson L. DeNet: Scalable Real-time Object Detection with Directed Sparse Sampling[J]. arXiv preprint arXiv:1703.10295, 2017.

[14]Song S, Xiao J. Deep sliding shapes for amodal 3D object detection in RGB-D images[C]//Proceedings of the IEEE Conference on Computer Vision and Pattern Recognition. 2016: 808-816.

[15]Kehl W, Tombari F, Navab N, et al. Hashmod: a hashing method for scalable 3D object detection[J]. arXiv preprint arXiv:1607.06062, 2016.

[16]Guo D, Sun F, Kong T, et al. Deep vision networks for real-time robotic grasp detection[J]. International Journal of Advanced Robotic Systems, 2016, 14(1): 1729881416682706.

[17]Simonyan K, Zisserman A. Very deep convolutional networks for large-scale image recognition[J]. arXiv preprint arXiv:1409.1556, 2014.

[18]Chen L C, Papandreou G, Kokkinos I, et al. Deeplab: Semantic image segmentation with deep convolutional nets, atrous convolution, and fully connected crfs[J]. arXiv preprint arXiv:1606.00915, 2016.

[19]Lee C C, Chiang Y C, Shih C Y, et al. Noisy time series prediction using M-estimator based robust radial basis function neural networks with growing and pruning techniques[J]. Expert Systems with Applications, 2009, 36(3): 4717-4724.

MATEC Web of Conferences 139, 00061 (2017) DOI: 10.1051/matecconf/201713900061
ICMITE 2017

Study on key algorithm for Thermal Imaging System Simulation

Huan Zhang [1,*], *Gao-feng* Cheng [1], *Jun-qiang* Gao [1]

1.Department of Control Engineering, Academy of Armored Forces Engineering, Beijing, China

Abstract: With the fast development of the armored equipment, the new thermal imaging devices are more and more complex, but the army guaranteeing personnel is lack of the basic training means, which can not provide the technical guarantee for the new thermal imaging equipment effectively and timely. To solve the difficult problem on maintenance training on the thermal image system, semi-physical simulation system for the thermal system is used in maintenance training instead of the real equipment to improve the support capability to the thermal equipment. The fast algorithm about the main functions of the thermal image system is researched in this paper, which established the foundation of the simulation of the thermal image system. The principle of focusing、the view switch and image adjustment function in thermal imaging system are researched in this paper, which is the foundation of how to simulate the thermal imaging system.

1 Introduction

As one of the most important part of the observing and sighting system, the thermal image system provides an important means for the crew of the armored vehicle to observe, search, aim and fire to the target at night and rain, snow, fog, haze and other bad weather conditions by translating the infrared radiation spatial distribution due to the different temperature and emissivity into a thermal image. At present, the troops of thermal equipment maintenance training is widely used the mode of theory teaching and the practical training, and the practical training is effective which can help the user master the method of application quickly, but having the disadvantages of high training cost limited training time and place and hard to set fault to the equipment [1].Using the semi-physical simulation technology in the thermal equipment maintenance training is one of the most effective method to form fighting capacity and support capability. Based on the principle of the thermal imaging system, a series of algorithms to focusing, view and image adjustment of the thermal image system are established, which is the foundation to the simulation for the thermal image system.

2 The function of thermal imaging system

The thermal imaging system collects and detects the target radiation within the waveband, and forms the thermal image corresponding to the radiation distribution of the target scene by rearranging the signals from the detector, and this thermal image recreates the radiation changes of each part of the target scene, so it can show the characteristics of the target scene. The thermal imaging system is a device that converts invisible infrared radiation into visible images, which extends the visual range of human eyes to the far infrared region.

The main function of the thermal image system included view switch, image adjustment and focusing[2]. In dual field infrared telescope, the wide view($6°\times6°$) is used to detecting and searching targets and the narrow view ($3°\times3°$) is used to identifying and aiming at targets. Image adjustment includes brightness adjustment, contrast adjustment and the polarity switch, changing the brightness is to change the temperature of the center of the target object, changing the contrast is to change the width of the temperature window(temperature difference between the highest and the lowest temperature) , polarity switch is to choice displaying in incandesce or black favor according to the observer's visual sensitivity, the environment of target and the temperature of the targets. Focusing is used to obtain the clear image of the target by changing the adjustment amount when the infrared telescope is switched to narrow-field-view.

3 Key Algorithm Design for the Major Function of Thermal Imaging System

3.1 The simulation algorithm of focusing

The method of focusing simulation for the thermal image system is realized by image processing technology with the defocusing parameter.

3.1.1 The reason of the phenomenon of defocus fuzzy

The ideal optical system is the imaging of a pair of conjugate planes, that is, in the object space, only the point on the object plane conjugate with the image plane can form a point image. In **Fig. 1.**, F is the focal point, H is the principal point, point A' is the image point of point A, the plane where point A' located is the image plane, and the conjugate plane is the object plane where the point A located.

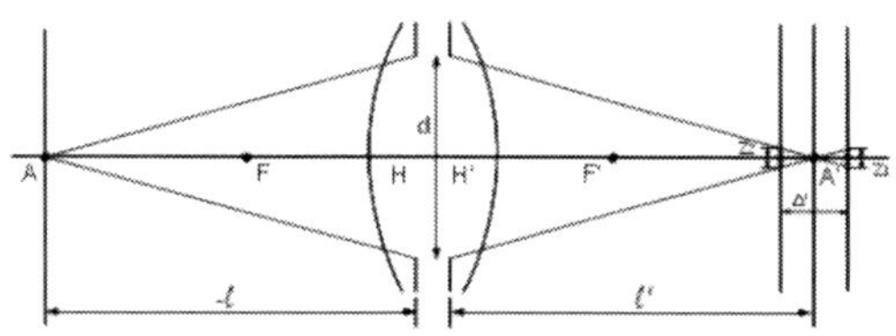

Fig. 1. The principle of focusing in the optical system

The limit resolution of infrared detector in thermal image system, so to a point A in the object plane, the diameter of the diffuse spot z_1、z_2 are less than z' (limit resolution), that means it also can get a clear image of point A when the image plane is moving in the range Δ'. The process of adjusting the position of the image plane according to the position of the optical object is called focusing, which can get a clear image, and the unfocused fuzzy is caused by inaccurate focusing.

As an optical system, the result of the unfocused fuzzy can be described by information transform, and supposed the optical system is liner. According to convolution imaging principle of optical system, image function $f'(y,z)$ is the convolution by object function $f(y,z)$ and the point diffusion function $h(y,z)$, an shown in Eq.1,

$$\begin{aligned} f'(y,z) &= T\{f(y,z)\} \\ &= f(y,z)*h(y,z) \end{aligned} \quad (1)$$

Specially, the transform in Eq.1, when the object plane is conjugating with the image plane, here is $h(y,z)=\delta(y,z)$, that is point resource makes point image; when it is defocus, $h(y,z)$ is a diffuse spot with special shape, here is $h(y,z)\neq\delta(y,z)$, there is defocus fuzzy[3]。There are two templates for the point diffusion function which are circle type and Gaussian type in the research on defocus fuzzy now .

3.1.2 The algorithm of defocus fuzzy

In order to get the template of point diffusion function, circle type point diffusion function is approximated to a rectangle type point diffusion function. Rectangle and Gaussian type point diffusion function can be decomposed into one-dimensional component in horizontal and vertical, which are shown in **Fig. 2.**

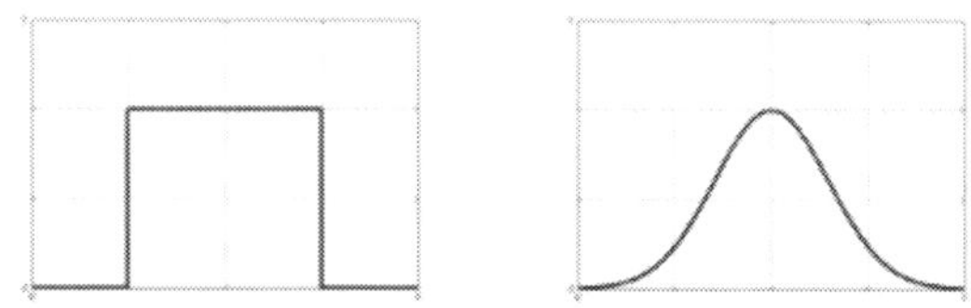

(a) Rectangle type (b) Gaussian type

Fig. 2. The one dimensional component of point diffusion function

1	1	1	1	1

(a) Rectangle type

1	4	6	4	1

(b) Gaussian type

1	1	1	1	1
1	1	1	1	1
1	1	1	1	1
1	1	1	1	1
1	1	1	1	1

(c) Rectangle type

1	4	6	4	1
4	16	24	16	4
6	24	36	24	6
4	16	24	16	4
1	4	6	4	1

(d) Gaussian type

Fig. 3. The template of point diffusion function

Taking a 5×5 pixels template of the point diffusion function as an example, the coefficients in the convolution template are the product of the corresponding positions of the horizontal and vertical one-dimensional component of the templates, the one-dimensional component of the rectangle type point diffusion function is in **Fig. 3.** (a) and the corresponding convolution template is in **Fig. 3.** (c); the one-dimensional component of the Gaussian type point diffusion function is 1-D Gaussian function, the simplest approximation of Gauss's function can be achieved by YANG Hui Triangle, so the one-dimensional component of the Gaussian type point diffusion function is in **Fig. 3.** (b) and the corresponding convolution template is in **Fig. 3.** (d) .In the convolution operation, the template rotation can be ignored because of the templates of rectangular and Gaussian types point diffusion functions are horizontal and vertical symmetry。

DOI: 10.1051/matecconf/201713900061

The MATLAB simulation results for focusing are shown in **Fig. 4.**, which the original image and the results of convolution using 5×5 template、9×9 template and 13×13 template are from left to right。

(a) The point diffusion function using rectangle template

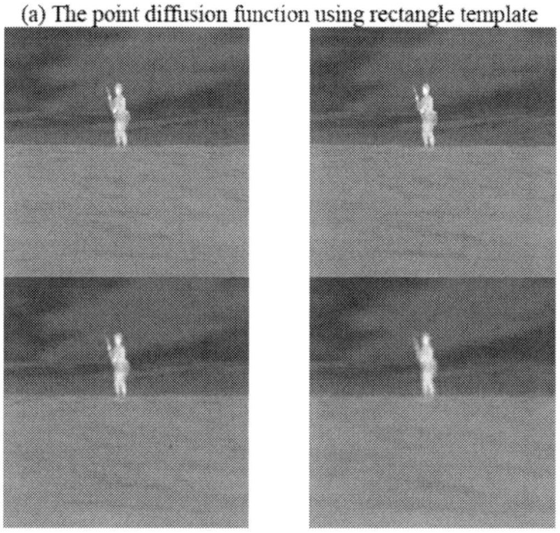

(b) The point diffusion function using Gaussian template

Fig. 4. The MATLAB simulation results for focusing

The simulation results show that the decomposable template convolution algorithm can simulate the phenomenon of defocus fuzzy well, and the phenomenon of fuzzy is more clear by using rectangle template.

3.2 The simulation algorithm of image adjustment function

Image adjustment includes brightness adjustment、contrast adjustment and polarity switch.

3.3.1 The principle of image adjustment

(1) The principle of brightness adjustment

The brightness of the infrared features image of the targets refers to the center temperature of the object, and the contrast refers to the width of the temperature window, the adjustment of the brightness and the contrast can be realized by the brightness and the contrast transform function.

The brightness transform function is shown in Eq.2, Y_i is the brightness value of a point in the input image, and Y_o is the brightness value of the corresponding point in the output image, T refers to the transform function, which adjusts the brightness of the image by add the regulated value ΔY on the original brightness value.

$$Y_o = T(Y_i) = Y_i + \Delta Y \qquad (2)$$

(2) The principle of contrast adjustment

The contrast transform function in Eq.3 shows that the part of the input image which the brightness value is less than m will be compressed to a narrower and darker range, by contrast, the part of the input image which the brightness value is more than m will be compressed to a wider and brighter range. Finally, we can get a high contrast output image which the bright part of the image became more brighter and the dark part became more darker:

$$Y_o = T(Y_i) = \frac{1}{1+(m/Y_i)^E} \qquad (3)$$

According to the definition of the contrast of the infrared features image of the targets, the transform function of the contrast shown in Eq.3 can be approximated to piecewise function shown in Eq.4, and the value of the contrast is unchanged when C=0, increased when C>0 and reduced when C<0。

$$Y_o = T(Y_i) = CY_i \qquad (4)$$

So, the transform function of the brightness and the contrast of the image can be shown in Eq.5 finally, and change the value of ΔY and C to adjust the brightness and contrast of the infrared features image.

$$Y_o = T(Y_i) = CY_i + \Delta Y \qquad (5)$$

(3) The principle of polarity switch

The polarity of the infrared features image of the targets can be candescence image and black thermal image, which the thermal target is white (high grayscale value) in candescence and the thermal target is black (low grayscale value) in black thermal image. The simulation of polarity switch can be realized by image reversal when the value of greyscale of an image between [0, L-1], as the Eq.6, in which r and s are the value of pixel before processing and after processing.。

$$s = T(r) = L-1-r \qquad (6)$$

3.2.2 The algorithm of image adjustment

The simulation of image adjustment is processing every pixel one after another from the first pixel of the original image, we use a 8 bits grayscale image which is 256×256 pixels to verify the correctness of the algorithm of image adjustment, and the MATLAB simulation results are shown in **Fig. 5.**、 **Fig. 6.** and **Fig. 7.** respectively.

(a) ΔY=-50 (b) ΔY=-25 (c) ΔY=0

(d) ΔY=+25 (e) ΔY=+50

Fig. 5. The MATLAB simulation results for brightness adjustment

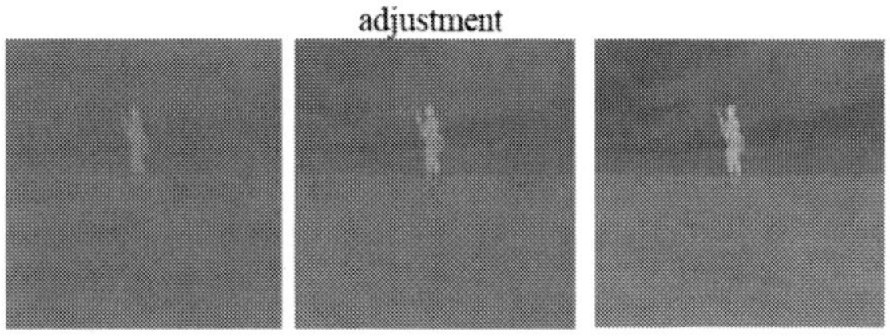

(a) C=0.5 (b) C=0.75 (c) C=1

(d) C=1.25 (e) C=1.5

Fig. 6. The MATLAB simulation results for contrast adjustment

(a) candescence image (b) black thermal image

Fig.7. The MATLAB simulation results for polarity switch

3.3 The simulation algorithm of view switch

The method of view switch simulation is realized by processing the infrared feature image of the targets from wide-field-of-view to narrow-field-of-view when receiving the command of view switch.

3.3.1 The principle of view switch

The function of view switch is realized by switching the objective lens of the dual field infrared telescope, the wide-filed-view and narrow-filed-view are 6°×6° and 3°×3° respectively. The relationship between visual magnification and the view angle is shown in Eq.7, which the visual magnification is the ratio of the tangent function of the view angle from telescope and direct observation, $f_{物}'$ and $f_{目}'$ are the focal length of the objective lens and eyepiece respectively.

$$\Gamma = \frac{\tan\omega'}{\tan\omega} = -\frac{f_{物}'}{f_{目}'} \qquad (7)$$

The focal length of the eyepiece is unchanged when the view switched, and the reactions between the focal length of the objective lens and the view angle is shown in Eq.8, we can get the conclusion that the visual magnification of the narrow-filed-view is 4 times to the wide-filed-view by putting $2\omega_{窄} = 3^\circ$、$2\omega_{宽} = 3^\circ$ into Eq.8, that means the optical effect of view switch can be realized by magnifying the center of the infrared feature image of the targets 2×2=4 times and outputting with the original resolution.

$$\begin{cases} \Gamma_{窄} = \dfrac{\tan\omega'}{\tan\omega_{窄}} = -\dfrac{f_{窄}'}{f_{目}} \\ \Gamma_{宽} = \dfrac{\tan\omega'}{\tan\omega_{宽}} = -\dfrac{f_{宽}'}{f_{目}} \\ M = \dfrac{\Gamma_{窄}}{\Gamma_{宽}} = \dfrac{\tan\omega_{宽}}{\tan\omega_{窄}} = \dfrac{\omega_{宽}}{\omega_{窄}} \end{cases} \qquad (8)$$

3.3.2 The algorithm of view switch

The typical image magnification algorithm includes nearest neighbor interpolation、bilinear interpolation and trilinear interpolation [4]。Considering the amplification effect、 the complexity of the algorithm、the speed of calculation and so on, the method of bilinear interpolation is used to simulate the phenomenon of the view switch.

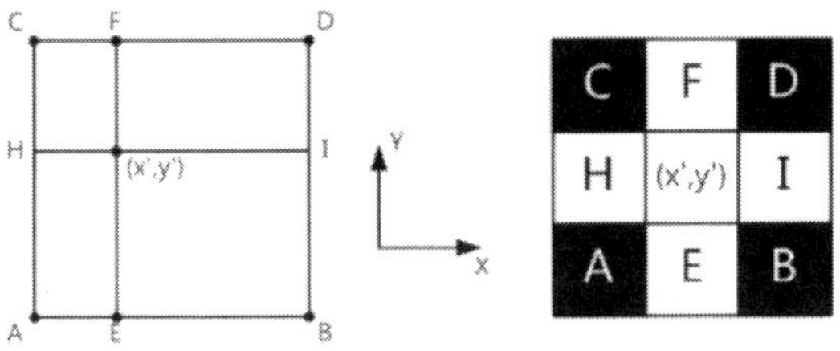

(a) The principle of bilinear interpolation

(b) Magnify to 4 times by hardware

Fig. 8. The method of bilinear interpolation

The principle of bilinear interpolation is shown in **Fig. 8.** (a), the grayscale value $g(x',y')$ of the interpolation point (x',y') is weighted calculated by the grayscale value of the 4 points around it. Known from the front section, it need to magnify 2 times in the horizontal and vertical of the thermal feature image of the targets, so the interpolation point is just at the center of the 4 points, the hardware bilinear interpolation algorithm became **Fig. 8.** (b), which is getting the greyscale value of the point E、F、H、I and (x',y') by interpolation（Eq.9）.

$$\begin{cases} g(E)=\frac{1}{2}[g(A)+g(B)] \\ g(F)=\frac{1}{2}[g(C)+g(D)] \\ g(H)=\frac{1}{2}[g(A)+g(C)] \\ g(I)=\frac{1}{2}[g(B)+g(D)] \\ g(x',y')=\frac{1}{2}[g(E)+g(F)] \end{cases} \quad (9)$$

The method of bilinear interpolation needs lots of data in three lines of the image and takes a lot of storage space, so it is so complexed to achieve the instantaneity and continuity, so an improvement is made to the method of bilinear interpolation which is realized the view switch by 3 steps, the results of the simulation by a 8×8 image are shown in **Fig. 9.**.

（1）To reset the position of all of the pixels that in the area want to magnify of the original image in order to get the first image;

（2）Do Interpolation operation with every line of the first image in order to get the second image;

（3）Do Interpolation operation with every column of the second image in order to get the final magnified image.

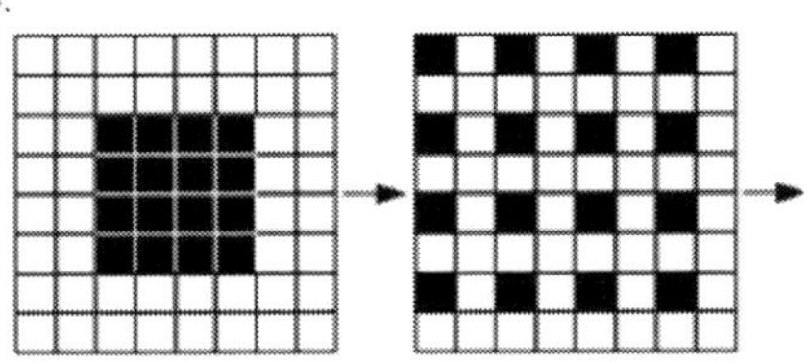

(a) The original image (b)The first image

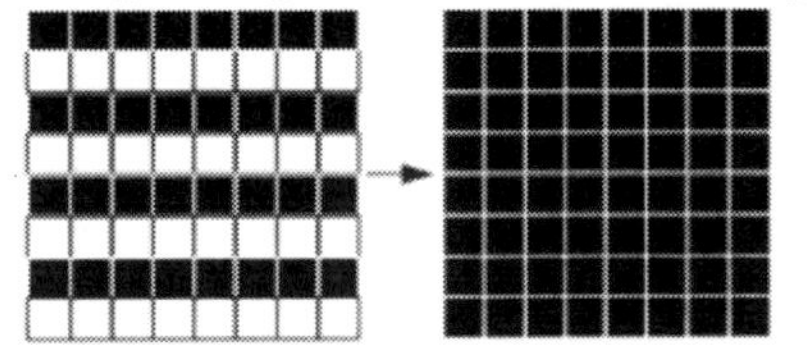

(c)The second image (d) The final image

Fig. 9. The algorithm of the view switch by hardware

In order to verify the correctness of the algorithm of the view switch function, we do the interpolation operation following the above steps using a 256×256 pixels gray image, and the **Fig. 10.** showed that the algorithm can simulate the optical effect of the view switch well.

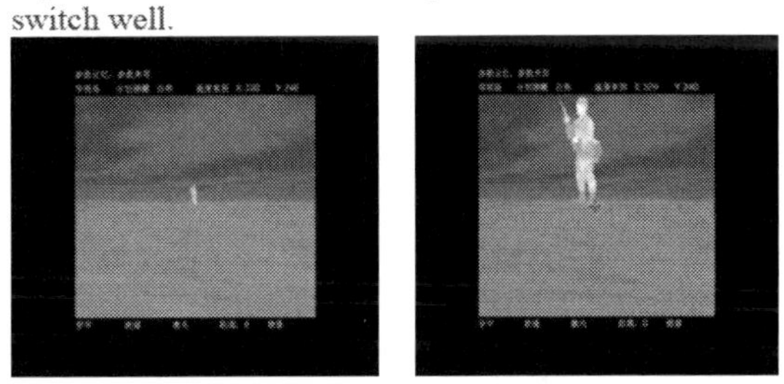

(a) Wide-field-of-view (b) Narrow-field-of-view

Fig. 10. The MATLAB simulation results for view switch

4 Conclusion

The application of virtual simulation technology gives a convenient、efficient and low cost method in maintenance teaching and training for the thermal equipment, which is out of the dependent of the complex equipment mounting. This paper gave the simulation algorithms of the main function of the thermal system by MATLAB, including focusing、image adjustment and polarity switch function. The research in this paper established the foundation using the scientific training methods in the maintenance teaching of the thermal image system in school and troops.

References

[1]Li Lin，Wu Wei-ling, Huang Quan-xin. Survey of Virtual Maintenance[J].Technology Information，2009(33): 847-848.
[2]Han Zhao-fu. The principle and Operation of 96 Tank Watching-aiming System [M]. Beijing: Department of Control Engineering, Academy of Armored Forces Engineering，2003.
[3]Mao Wen-wei. Fundamentals of Optical Engineering[M].Beijing: Tsinghua University Press，2006
[4]Zhao Hai-feng, Zhou Yong-fei, Comparison of Image Amplifying Method[J].Technology of Computer Application，2010（24）: 33-36

The optimal financing mode in a three-stage supply chain under capital constraint of retailers

Yuanyuan Zhang[1], *Lulu Ren*[2]

[1] School of Managemant, Shanghai University, Shanghai, 200444, P.R. China
[2] School of Managemant, Shanghai University, Shanghai, 200444, P.R. China

Abstract. In real life, there is a problem of capital fracture in some enterprises especially small and medium enterprises in the upstream and downstream of the supply chain. In order to research how retailers choose the optimal financing mode, this paper analyzes the double channel and three- stage supply chain under capital constraint of retailers, uses multi-objective nonlinear programming method, constructs the delayed payment financing model and the loan financing model respectively and gives the optimal decentralized decisions of suppliers, manufacturers and retailers under the two modes. The research shows that under the coexistence of the delayed payment financing model and the loan financing model, when the delayed payment rate is equal to the lending rate, if the retailers choose the delayed payment model, then it can not only increase the profits but also improve the market competitiveness and expand the market. This provides certain theory and numerical reference basis for retailers to choose a financing model.

1 Introduction

Along with the change of the global competition environment and the development of information technology, the development of the industrial economy has shifted from the competition between enterprises to the competition among the supply chains. In the practice of supply chain management, core competitive advantage is not only reflected in the core enterprise or large enterprises but also in the small and medium enterprises. Small and medium enterprises have the small scale, low credit and less financing channels. So financing is very difficult and the cost is relatively high. This may result in capital strand breaking, which affects the development of the entire supply chain, the whole industry and the whole national economy. Therefore, it is of high value to study the financing and operation decisions of the supply chain. Especially it is more practical to study on the financing decisions of the small and medium enterprises.

In recent years, many scholars begin to research on operation and financing decisions of participants in the supply chain under financial constraints. Existing literature research is divided into two pieces. One is the supplier (manufacturer) under capital constraint and the other is the retailer under capital constraint. Among them, the study for operation and financing decisions of suppliers in the supply chain under funding constraints is much more. Tu et al. (2011) study the problem of financing and production decision in the supply chain under capital constraint. The manufacturer can supply the market directly through bank loan financing [1]. Zhao and Li (2012) consider two level supply chain suppliers face financial constraints and research financing and pricing decisions in the noncompetitive market when suppliers have different initial funding [2].

In the existing literature, supply chain financing decisions for retailers under the capital constraint is relatively few. Chen et al (2016) study supply chain financing and operation decisions, which is composed of a single supplier and a single capital constrained retailer. When retailers are under the capital constraint, they get the loan from the capital market to obtain the financing service. [3] Yan and Sun (2013) study the Stackelberg game model of retailers under the capital constraint [4]. In order to make the model more realistic, Li and Chen (2015) study a multi-channel supply chain model consisting of a manufacturer and multiple retailers and analyze the optimal decisions in borrow financing and deferred financing mode for the manufacturers and retailers under the capital constraint of the retailer [5].

In summary, from previous literature research, about financing and operational decision making in the supply chain under the financial constraints of retailers, it is found that literatures are based mostly on a two-stage supply chain consisting of a single supplier and a single retailer or a single manufacturer and a single retailer and rarely based on a three-level supply chain, ignoring the multilevel and complexity of supply chain in real management practice. Therefore, it is worth exploring the practicality of theoretical and model results in the multi-stage supply chain. In this paper, combining the management practice in real life, based on the research of scholars at home and abroad, the research on financing decision of capital constraint is extended to the three-stage supply chain. The structure is organized as follows: Considering the three-stage supply chain consisting of single supplier, single manufacturer and single retailer and dual channels sale as the research object, section 2 uses the Stackelberg game theory to model and analyze, and explores the operation mode of the three-stage supply chain, the optimal decision-making problem of supply chain members and the impact of the supply chain members' benefits under the operation mode of the loan financing and deferred payment. Section 3 performs case study and simulates the model quantitatively through the EXCEL data analysis software. Finally, Section 4 concludes the paper, providing a theoretical and numerical basis for the actual decision and points out further work.

2 Fundamental model

2.1 Variable definition and relevant assumptions

Assuming in a three-stage supply chain consisting of a single supplier, a single manufacturer and a single retailer, manufacturers process products providing by suppliers. The entire supply chain exists two sales channels of manufacturers direct sales and retail outlets. It only manufactures and sales one product and there is no shortage phenomenon. Assuming that the retailer's own initial capital is B, there is no shortage of funds for suppliers and manufacturers. At the beginning of the sales cycle, retailers can apply for deferred payment from the manufacturer or apply for loans from the bank if there are insufficient funds, regardless of the repurchase. Specific variables are defined and described as follows:

- p_m: Manufacturer direct selling price;
- p_r: Retail unit price;
- d_m: Consumers demand for manufacturer selling directly to;
- d_r: Consumer demand for retailer sales;
- w_s: Wholesale prices of suppliers to manufacturers;
- w_m: Wholesale price of manufacturer to retailer;
- c: Unit cost of production;
- c_m: Total production cost per unit of manufacturer;
- B : Initial free funds for retailers;
- δ : Interest rates for manufacturers postponing payment;
- r_b: Bank lending rate;
- π_s: Supplier's sales profit;
- π_m: Manufacturer's sales profit;
- π_r: Retailer's sales profit.

For the sake of simplicity, make the following assumptions:

Hypothesis 1: $d_m = (1-t)a - b_1 p_m + b_2 p_r$, $d_r = ta - b_1 p_r + b_2 p_m$, where a is the total potential consumer demand in the market, t is the proportion of retail market in the total market, b_1 is price elasticity of demand on two sales channels of supply. b_2 is the degree of consumer transfer due to the different prices of the two channels, $b_1 > b_2$;

Hypothesis 2: $c_m = w_s + m$, where m is the unit processing cost of manufacturers making the raw material into the finished product;

Hypothesis 3: Suppliers, manufacturers, retailers and banks all have information that is symmetrical, and the risk-free interest rate of the company's funds is zero.

2.2 Decentralized decision model of supply chain with sufficient funds for retailers

${w_s}^0$, ${w_m}^0$, ${p_m}^0$, ${p_r}^0$ represent the optimal decisions of the supplier, manufacturer and retailer respectively when the retailer's initial free capital is adequate. The profit function of the three is expressed as π_s^0, π_m^0, π_r^0 and the specific functions are as follows:

$$\pi_s^0 = (w_s - c)(d_m + d_r), \tag{1}$$

$$\pi_m^0 = (p_m - c_m)d_m + (w_m - c_m)d_r, \tag{2}$$

$$\pi_r^0 = (p_r - w_m)d_r. \tag{3}$$

In decentralized decision-making, suppliers first determine the supplier wholesale price w_s . The manufacturer's wholesale price w_m and the manufacturer's direct selling price p_m are jointly determined by the manufacturer according to the supplier's wholesale price decision and direct selling demand function. Finally, the retail price p_r is determined by the retailer based on the manufacturer's decision and demand function, in order to achieve the optimal profit in the initial capital adequacy. The optimal decentralized decision in the case of capital adequacy is introduced by using the Stackelberg game theory and reverse order decision research.

Retailers determine the optimal retail price based on the manufacturer's direct selling price of p_m and the manufacturer's wholesale price w_m (and also determines its order quantity according to the demand function). Plugging $d_r = ta - b_1 p_r + b_2 p_m$ into the retailer's profit function type (3), we can obtain the following new retailer's profit function:

$$\pi_r^0(p_r) = -b_1 {p_r}^2 + (ta + b_2 p_m + b_1 w_m)p_r \\ -(ta + b_2 p_m)w_m.$$

Take derivative of the target function $\pi_r^0(p_r)$ for p_r of first-order derivative and second-order derivative. It is easy to prove that π_r^0 is a concave function for p_r . The process is as follows:

$$\pi_r^{0'} = -2b_1\, p_r + ta + b_2 p_m + b_1 w_m,$$

$$\pi_r^{0''} = -2b_1 < 0.$$

Because the two order derivative of the objective function $\pi_r^{0''}$ is less than zero. That is a concave function π_r^0 and it has a maximum value. Make $\pi_r^{0'} = 0$, optimal value of p_r can be obtained with ${p_r}^0$ indicating the following:

$${p_r}^0 = \frac{ta + b_2 p_m + w_m b_1}{2b_1}.$$

$d_m = (1-t)a - b_1 p_m + b_2 p_r$, $d_r = ta - b_1 p_r + b_2 p_m$ substitute profit function of manufacturers (2). At the same time, put algebraic expression ${p_r}^0$ about w_m, p_m into manufacturer's profit function (2), and new profit function of manufacturers is obtained as follows:

$$\pi_m^0(w_m,p_m) = (p_m-c_m)\left[(1-t)a-b_1p_m + b_2\frac{ta+b_2p_m+w_mb_1}{2b_1}\right] + (w_m-c_m)\left[ta-\frac{ta+b_2p_m+w_mb_1}{2} + b_2p_m\right].$$

Take derivative of the target function $\pi_m^0(w_m,p_m)$ for manufacturers wholesale prices w_m and manufacturers direct price p_m of first-order derivative and second-order derivative. Then the results can be obtained as follows:

$$\frac{\partial_{\pi_m^0}}{\partial_{p_m}} = (1-t)a - b_1p_m + \frac{b_2}{2b_1}(ta+b_2p_m+b_1w_m) + \left(\frac{b_2^2}{2b_1}-b_1\right)(p_m-c_m) + \frac{b_2}{2}(w_m-c_m),$$

$$\frac{\partial_{\pi_m^0}}{\partial_{w_m}} = \frac{b_2}{2}(p_m-c_m) + ta - \frac{1}{2}(ta+b_2p_m+b_1w_m) + b_2p_m - \frac{1}{2}b_1(w_m-c_m),$$

$$\frac{\partial^2{\pi_m^0}}{\partial_{p_m}{}^2} = \frac{b_2^2-2b_1^2}{b_1}, \qquad \frac{\partial^2{\pi_m^0}}{\partial_{p_m}\partial_{w_m}} = b_2,$$

$$\frac{\partial^2{\pi_m^0}}{\partial_{w_m}{}^2} = -b_1, \qquad \frac{\partial^2{\pi_m^0}}{\partial_{w_m}\partial_{p_m}} = b_2.$$

From the results above, it is can be proved that π_m^0 is the joint concave function *about* w_m、p_m and the function π_m^0 can get the optimal solution. Make $\frac{\partial_{\pi_m^0}}{\partial_{p_m}} = 0$ and $\frac{\partial_{\pi_m^0}}{\partial_{w_m}} = 0$ get solution together and put the solution into the algebraic expression $p_r{}^0$. The optimal decisions of manufacturers and retailers $w_m{}^0$, $p_m{}^0$, $p_r{}^0$ denoted by c_m can be obtained. They are as follows:

$$w_m{}^0 = \frac{1}{2(b_1{}^2-b_2{}^2)}\left[(1-t)ab_2 + tab_1 + (b_1{}^2-b_2{}^2)c_m\right],$$

$$p_m{}^0 = \frac{1}{2(b_1{}^2-b_2{}^2)}\left[(1-t)ab_1 + tab_2 + (b_1{}^2-b_2{}^2)c_m\right],$$

$$p_r{}^0 = \frac{1}{4b_1(b_1{}^2-b_2{}^2)}\left[2(1-t)ab_1b_2 + (3b_1{}^2-b_2{}^2)ta + (b_1{}^3-b_1b_2{}^2+b_1{}^2b_2-b_2{}^3)c_m\right].$$

From $c_m = w_s + m$, optimal decision of manufacturers and retailers $w_m{}^0$, $p_m{}^0$, $p_r{}^0$ can be obtained as the algebraic expression w_s. $d_m = (1-t)a - b_1p_m + b_2p_r$, $d_r = ta - b_1p_r + b_2p_m$ are substituted into the supplier's profit function type (1). At the same time, $p_m{}^0$, $p_r{}^0$ about w_s are substituted into profit function of suppliers (1). Then new profit function of suppliers $\pi_s^0(w_s)$ is obtained. Take derivative of it for the wholesale price of suppliers w_s, and similarly we can get the optimal wholesale price of the suppliers $w_s{}^0$. Then put $w_s{}^0$ into $w_m{}^0$, $p_m{}^0$, $p_r{}^0$ and we can get the following conclusions:

Proposition 1. In decentralized decision making, for the dual channel supply chain of suppliers, manufacturers and retailers in initial capital adequacy, the optimal decision-making suppliers, manufacturers and retailers $w_s{}^0, w_m{}^0$, $p_m{}^0$, $p_r{}^0$ are expressed as follows respectively:

$$w_s{}^0 = \frac{1}{2(3b_1+b_2)(b_1-b_2)}[2ab_1 - ta(b_1-b_2)] + \frac{1}{2}(c-m),$$

$$p_m{}^0 = \frac{1}{2(b_1{}^2-b_2{}^2)}[ab_1(1-t)+tab_2] + \frac{1}{4}(c+m) + \frac{1}{4(3b_1+b_2)(b_1-b_2)}[2ab_1 - ta(b_1-b_2)],$$

$$w_m{}^0 = \frac{1}{2(b_1{}^2-b_2{}^2)}[ab_2(1-t)+tab_1] + \frac{1}{4}(c+m) + \frac{1}{4(3b_1+b_2)(b_1-b_2)}[2ab_1 - ta(b_1-b_2)],$$

$$p_r{}^0 = \frac{1}{4b_1(b_1{}^2-b_2{}^2)}\left[2ab_1b_2(1-t) + ta(3b_1{}^2-b_2{}^2)\right] + \frac{b_1+b_2}{8b_1}(c+m) + \frac{b_1+b_2}{8b_1(3b_1+b_2)(b_1-b_2)}[2ab_1 - ta(b_1-b_2)].$$

2.3 Decentralized decision model of supply chain under the mode of retailer delayed payment

The postponement payment mode refers to in order period, retailers of insufficient funds firstly pay to manufacturers with the existing initial free money, and after they sell the products and receive cash, then pay to manufacturer of the remaining balance of payment at a certain interest rate. On the assumption that the risk-free interest rate is zero, namely, capital operation of supply chain participants does not consider the time cost. Further studies can consider the issue of funds discount. The discount of the balance of order, retailer and manufacturer's sales and the impact of risk-free interest rate on financing decision making can be discussed.

Capital deficiency of the retailers refers to on ordering stage the initial funding amount $B \le w_m{}^0 d_r$, and then retailers choose to defer payment in order to obtain more revenue. Delayed payment of retailers refers to in the ordering phase, retailers of insufficient funds first pay a partial order payment with initial capital of free. Goods of $\frac{B}{w_m}$ units are obtained at the wholesale price w_m and the payment $w_m d_r - B$ is required for the remaining $d_r - \frac{B}{w_m}$ unit commodity. The manufacturers allow retailers to defer payment at per unit price $(1+\delta)w_m$,

$(\delta \geq 0)$ after the implementation of the requirements. Delayed payment financing model can be regarded as a model of trade credit financing.

When the retailer's initial free funds are in short supply and $w_s{}^1$, $w_m{}^1$, $p_m{}^1$, $p_r{}^1$ is denoted optimal decision of the suppliers, manufacturers and retailers in delayed payment financing mode respectively. The profit function is $\pi_s{}^1$, $\pi_m{}^1$, $\pi_r{}^1$ respectively, and they are expressed as follows:

$$\pi_s{}^1 = (w_s - c)(d_m + d_r), \quad (4)$$

$$\pi_m{}^1 = (p_m - c_m)d_m + (w_m - c_m)\frac{B}{w_m} + [(1+\delta)w_m - c_m]\left(d_r - \frac{B}{w_m}\right), \quad (5)$$

$$\pi_r{}^1 = \frac{B}{w_m}(p_r - w_m) + [p_r - (1+\delta)w_m]\left(d_r - \frac{B}{w_m}\right). \quad (6)$$

When the initial free funds of retailers are in short supply, in the decentralized decision-making, the analysis process of the sequence of decisions for suppliers, manufacturers, retailers is similar to the situation when the capital is abundant above. The optimal decentralized decision making of the suppliers, manufacturers and the retailers is obtained with the research method of reverse order decision making when the retailers are of capital shortage. $d_r = ta - b_1 p_r + b_2 p_m$ is substituted into the retailer's profit function (6), then we can obtain the following new profit function of retailers:

$$\pi_r{}^1(p_r) = \frac{B}{w_m}(p_r - w_m) + [p_r - (1+\delta)w_m]\left(ta - b_1 p_r + b_2 p_m - \frac{B}{w_m}\right).$$

Take derivative of the target function $\pi_r{}^1(p_r)$ for p_r of first-order derivative and second-order derivative and the results are as follows:

$$\pi_r{}^{1'} = (1+\delta)b_1 w_m + b_2 p_m + ta - 2b_1 p_r,$$

$$\pi_r{}^{1''} = -2b_1.$$

Due to $\pi_r{}^{1''} < 0$, it can be proved that $\pi_r{}^1$ is a concave function of p_r. So make $\pi_r{}^{1'} = 0$, the best decision $p_r{}^1$ can be got as follows:

$$p_r{}^1 = \frac{(1+\delta)b_1 w_m + ta + b_2 p_m}{2b_1}.$$

$d_m = (1-t)a - b_1 p_m + b_2 p_r$, $d_r = ta - b_1 p_r + b_2 p_m$ are substituted into manufacturer's profit function (5), and at the same time algebraic $p_r{}^1$ about w_m, p_m is substituted into manufacturer's profit function (5). Then the new manufacturers profit function can be obtained as follows:

$$\pi_m{}^1(w_m, p_m) = (p_m - c_m)[(1-t)a - b_1 p_m + b_2 p_r] + (1+\delta - c_m)(a - b_1 p_r + b_2 p_m) - \delta B.$$

Take derivative of the target function $\pi_m{}^1(w_m, p_m)$ for manufacturers wholesale prices w_m and manufacturers direct price p_m for the first-order derivative and second-order derivative respectively and the results are as follows:

$$\frac{\partial_{\pi_m{}^1}}{\partial_{p_m}} = (1-t)a + \left(\frac{b_2{}^2}{b_1} - 2b_1\right)p_m + (1+\delta)b_2 w_m + \frac{2b_1{}^2 - b_2{}^2 - b_1 b_2}{2b_1}c_m + \frac{tab_2}{2b_1}$$

$$\frac{\partial_{\pi_m{}^1}}{\partial_{w_m}} = (1+\delta)b_2 p_m - (1+\delta)^2 b_1 w_m + \frac{1}{2}(1+\delta)(ta + b_1 c_m - b_2 c_m)$$

$$\frac{\partial^2{}_{\pi_m{}^1}}{\partial_{p_m{}^2}} = \frac{b_2{}^2}{b_1} - 2b_1; \quad \frac{\partial^2{}_{\pi_m{}^1}}{\partial_{w_m{}^2}} = -(1+\delta)^2 b_1$$

$$\frac{\partial^2{}_{\pi_m{}^1}}{\partial_{p_m}\partial_{w_m}} = (1+\delta)b_2; \quad \frac{\partial^2{}_{\pi_m{}^1}}{\partial_{w_m}\partial_{p_m}} = (1+\delta)b_2$$

From the above results, we can prove that $\pi_m{}^1$ is a joint concave function about w_m、p_m and the function $\pi_m{}^1$ can get the optimal solution. Make $\partial_{\pi_m{}^1}/\partial_{p_m} = 0$ and $\partial_{\pi_m{}^1}/\partial_{w_m} = 0$ joint solving, and put the solution $w_m{}^1$, $p_m{}^1$ into algebraic $p_r{}^1$. The optimal decision of manufacturers and retailers $w_m{}^1$, $p_m{}^1$, $p_r{}^1$ represented by c_m can be obtained as follows:

$$w_m{}^1 = \frac{1}{2(1+\delta)(b_1{}^2 - b_2{}^2)}\left[(1-t)ab_2 + tab_1 + (b_1{}^2 - b_2{}^2)c_m\right],$$

$$p_m{}^1 = \frac{1}{2(b_1{}^2 - b_2{}^2)}\left[(1-t)ab_1 + tab_2 + (b_1{}^2 - b_2{}^2)c_m\right],$$

$$p_r{}^1 = \frac{1}{4b_1(b_1{}^2 - b_2{}^2)}\left[2(1-t)ab_1 b_2 + (3b_1{}^2 - b_2{}^2)ta + (b_1{}^3 - b_1 b_2{}^2 + b_1{}^2 b_2 - b_2{}^3)c_m\right].$$

Because $c_m = w_s + m$, the optimal strategies of the manufacturers and the retailers $w_m{}^1$, $p_m{}^1$, $p_r{}^1$ about w_s can be obtained. $d_m = (1-t)a - b_1 p_m + b_2 p_r$, $d_r = ta - b_1 p_r + b_2 p_m$ are substituted into supplier profit function (4), and $p_m{}^1$, $p_r{}^1$ about w_s are substituted into supplier profit function (4). Then a new supplier's profit function $\pi_s{}^1(w_s)$ can be obtained. Take derivative of the supplier's wholesale price w_s, and the wholesale price of the supplier optimal decision $w_s{}^1$, can be obtained. Then $w_s{}^1$ are substituted into $w_m{}^1, p_m{}^1$, $p_r{}^1$ and the following conclusions can be got.

Proposition 2. In decentralized decision making, for the two channel supply chain of the retailer's initial free funds shortage, when the interest rate δ is taken to postpone the payment method, the optimal decision of the suppliers, the manufacturers and the retailers $w_s{}^1, w_m{}^1, p_m{}^1, p_r{}^1$ is obtained respectively:

$$w_s^{\,1} = \frac{1}{2(3b_1 + b_2)(b_1 - b_2)} [2ab_1 - ta(b_1 - b_2)] + \frac{1}{2}(c - m),$$

$$p_m^{\,1} = \frac{1}{2(b_1^{\,2} - b_2^{\,2})}[ab_1(1-t) + tab_2] + \frac{1}{4}(c + m) + \frac{1}{4(3b_1 + b_2)(b_1 - b_2)}[2ab_1 - ta(b_1 - b_2)],$$

$$w_m^{\,1} = \frac{1}{2(b_1^{\,2} - b_2^{\,2})(1+\delta)} [ab_2(1-t) + tab_1] + \frac{1}{4(1+\delta)}(c+m) + \frac{1}{4(3b_1 + b_2)(b_1 - b_2)(1+\delta)} [2ab_1 - ta(b_1 - b_2)],$$

$$p_r^{\,1} = \frac{1}{4b_1(b_1^{\,2} - b_2^{\,2})}[2ab_1b_2(1-t) + ta(3b_1^{\,2} - b_2^{\,2})] + \frac{b_1 + b_2}{8b_1}(c+m) + \frac{b_1 + b_2}{8b_1(3b_1 + b_2)(b_1 - b_2)} [2ab_1 - ta(b_1 - b_2)]$$

From the comparison of $w_s^{\,1}$, $w_m^{\,1}$, $p_m^{\,1}$, $p_r^{\,1}$ and $w_s^{\,0}$, $w_m^{\,0}$, $p_m^{\,0}$, $p_r^{\,0}$, it can be found that dual channel supply chain decentralized decisions of retailers capital constraints on defer payment financing mode, manufacturers provide lower the manufacturer's wholesale price $(w_m^{\,0} = (1+\delta)w_m^{\,1},\ \delta > 0)$ to attract retailers to increase the order quantity. It can be seen by proposition manufacturers will determine the wholesale price basing on the deferred interest rate δ, and it's $1/(1+\delta)$ times the wholesale price when the retailer is well-funded .That is the higher the interest rate, manufacturer's wholesale price will be lower.

At the same time, the manufacturers wholesale prices, the optimal retail price, manufacturers selling price and supplier's wholesale prices will not be affected by the influence of retailer capital constraints, and the optimal decisions are the same with the retailer capital adequacy $(w_s^{\,1} = w_s^{\,0}$, $p_m^{\,1} = p_m^{\,0}$, $p_r^{\,1} = p_r^{\,0})$. The consumption demand function shows that the total demand of the dual channel supply chain will not change because of lack of funds for retailers. Due to the constant demand of manufacturers and retailers $(p_m^{\,1} = p_m^{\,0},\ p_r^{\,1} = p_r^{\,0})$, so the product demand of manufacturer to supplier's will be unchanged, and supplier's wholesale price will also be unchanged $(w_s^{\,1} = w_s^{\,0})$. Therefore the ultimate benefit for suppliers will not be affected.

2.4 Decentralized decision model of supply chain under the mode of retailer credit payment

Retailers loan payment model refers to that underfunded retailers are required to loan D from the bank at interest rate r_b before ordering in order to pay the ordering cost $w_m d_r - B$ and return principal and interest $D(1 + r_b)$ after sales. In this mode, the suppliers, manufacturers and retailers make decisions respectively. Because the funds risk-free interest rate is generally lower than the bank loan interest rate (in this paper, it is assumed that the risk-free interest rate is 0) and there is a hypothesis that the demand function is predictable (demand is a function of potential market demand and sales price), information is symmetric and bank loan don't need to consider the risks of uncertain demand.

When the retailer's initial free funds is short, on this financing mode, $w_s^{\,*}$, $w_m^{\,*}$, $p_m^{\,*}$, $p_r^{\,*}$ indicates the optimal decision of suppliers, manufacturers and retailers respectively. The profit function is $\pi_s^{\,*}, \pi_m^{\,*}, \pi_r^{\,*}$ respectively, and specific functions are as follows:

$$\pi_s^{\,*} = (w_s - c)(d_m + d_r), \tag{7}$$

$$\pi_m^{\,*} = (p_m - c_m)d_m + (w_m - c_m)d_r, \tag{8}$$

$$\pi_r^{\,*} = (p_r - w_m)d_r - (w_m d_r - B)r_b. \tag{9}$$

When the initial free funds of retailers is in short supply, in the decentralized decision-making, the analysis process of the sequence of decisions for suppliers, manufacturers, retailers is similar to the situation when the capital is abundant above. The optimal decentralized decision making of the suppliers, manufacturers and the retailers is obtained with the research method of reverse order decision making when the retailers are of capital shortage. $d_r = ta - b_1 p_r + b_2 p_m$ is substituted into the retailer's profit function (9), then we can obtain the following new profit function of retailers:

$$\pi_r^{\,*}(p_r) = (p_r - w_m)(ta - b_1 p_r + b_2 p_m) - [w_m(ta - b_1 p_r + b_2 p_m) - B]r_b.$$

Take derivative of the target function $\pi_r^{\,*}(p_r)$ for p_r of first-order derivative and second-order derivative and the results are as follows:

$$\pi_r^{\,*\prime} = (1 + r_b)b_1 w_m + b_2 p_m + ta - 2b_1 p_r,$$

$$\pi_r^{\,*\prime\prime} = -2b_1.$$

Due to $\pi_r^{\,*\prime\prime} < 0$, it can be proved that $\pi_r^{\,*}$ is a concave function of p_r. So make $\pi_r^{\,*\prime} = 0$, the best decision $p_r^{\,*}$ can be got as follows:

$$p_r^{\,*} = \frac{(1 + r_b)b_1 w_m + ta + b_2 p_m}{2b_1}.$$

$d_m = (1-t)a - b_1 p_m + b_2 p_r$, $d_r = ta - b_1 p_r + b_2 p_m$ are substituted into manufacturer's profit function (8), and at the same time algebraic $p_r^{\,*}$ about w_m, p_m is substituted into manufacturer's profit function (8). Then the new manufacturers profit function can be obtained as follows:

$$\pi_m^{\,*}(w_m, p_m) = (p_m - c_m)\left[(1-t)a - b_1 p_m + b_2 \frac{(1 + r_b)b_1 w_m + ta + b_2 p_m}{2b_1}\right] + (w_m - c_m)\left[ta - \frac{(1 + r_b)b_1 w_m + ta + b_2 p_m}{2} + b_2 p_m\right].$$

Take derivative of the target function $\pi_m{}^*(w_m, p_m)$ for manufacturers wholesale prices w_m and manufacturers direct price p_m for the first-order derivative and second-order derivative respectively and the results are as follows:

$$\frac{\partial_{\pi_m^*}}{\partial_{p_m}} = (1-t)a + \left(\frac{b_2^{\ 2}}{b_1} - 2b_1\right)p_m + \frac{(2+r_b)}{2}b_2 w_m + \frac{2b_1^{\ 2} - b_2^{\ 2} - b_1 b_2}{2b_1}c_m + \frac{tab_2}{2b_1}$$

$$\frac{\partial_{\pi_m^*}}{\partial_{w_m}} = \frac{(2+r_b)}{2}b_2 p_m - (1+r_b)b_1 w_m + \frac{(1+r_b)}{2}(b_1 - b_2)c_m + \frac{ta}{2}$$

$$\frac{\partial^2_{\pi_m^*}}{\partial_{p_m^2}} = \frac{b_2^{\ 2}}{b_1} - 2b_1; \qquad \frac{\partial^2_{\pi_m^*}}{\partial_{w_m^2}} = -(1+r_b)b_1$$

$$\frac{\partial^2_{\pi_m^*}}{\partial_{p_m}\partial_{w_m}} = \frac{(2+r_b)}{2}b_2; \qquad \frac{\partial^2_{\pi_m^*}}{\partial_{w_m}\partial_{p_m}} = \frac{(2+r_b)}{2}b_2.$$

From the above results, we can prove that $\pi_m{}^*$ is a joint concave function about w_m、p_m and the function $\pi_m{}^*$ can get the optimal solution. Make $\partial_{\pi_m^*}/\partial_{p_m} = 0$ and $\partial_{\pi_m^*}/\partial_{w_m} = 0$ joint solving, and put the solution $w_m{}^*$, $p_m{}^*$ into algebraic $p_r{}^*$. The optimal decision of manufacturers and retailers $w_m{}^*$, $p_m{}^*$, $p_r{}^*$ represented by c_m can be obtained as follows:

$$w_m{}^* = \frac{1}{b_1 M}\left[2b_1 b_2(2+r_b)(1-t)a + \left(4b_1^{\ 2} + b_2^{\ 2}r_b\right)ta + (b_1 - b_2)\left(4b_1^{\ 2} + 4b_1 b_2 - b_2^{\ 2}r_b + 2b_1 b_2 r_b + 4b_1^{\ 2}r_b\right)c_m\right],$$

$$p_m{}^* = \frac{1}{M}[4(1-t)(1+r_b)ab_1 + (4+3r_b)tab_2 + (b_1 - b_2)(4b_1 + 4b_2 + r_b b_2)(1 + r_b)c_m],$$

$$p_r{}^* = \frac{(1+r_b)}{b_1 M}[b_1 b_2(4+r_b)(1-t)a + 2\left(3b_1^{\ 2} - b_2^{\ 2}\right)ta + (b_1 - b_2)[2(b_1+b_2)^2 + b_1 r_b(2b_1 + b_2)]c_m],$$

where $M = 8\left(b_1^{\ 2} - b_2^{\ 2}\right)(1+r_b) - r_b^{\ 2}b_2^{\ 2}$.

Because $c_m = w_s + m$, the optimal strategies of the manufacturers and the retailers $w_m{}^*$, $p_m{}^*$, $p_r{}^*$ about w_s can be obtained. $d_m = (1-t)a - b_1 p_m + b_2 p_r$, $d_r = ta - b_1 p_r + b_2 p_m$ are substituted into supplier profit function (7), and $p_m{}^*$, $p_r{}^*$ about w_s are substituted into supplier profit function (7). Then a new supplier's profit function $\pi_s{}^*(w_s)$ can be obtained. Take derivative of the wholesale price of suppliers w_s of first-order and second-order derivative. The results are as follows:

$$\pi_s{}^*(w_s)' = \frac{1}{M}(b_2 - b_1)(1-t)(1+r_b)a[4(b_1+b_2) + r_b b_2] + \frac{1}{M}(b_2 - b_1)ta\left[b_2(4+3r_b) + \frac{2}{b_1}(1+r_b)\left(3b_1^{\ 2} - b_2^{\ 2}\right)\right] + (b_2 - b_1)(A+D)(2w_s + m - c) + a,$$

$$\pi_s{}^*(w_s)'' = 2(b_2 - b_1)(A+D) < 0.$$

Among them, $A = \frac{1}{M}(b_1 - b_2)(1+r_b)(4b_1 + 4b_2 + r_b b_2)$, $D = \frac{1}{b_1 M}(b_1 - b_2)(1+r_b)[2(b_1+b_2)^2 + b_1 r_b(2b_1 + b_2)]$, $M = 8\left(b_1^{\ 2} - b_2^{\ 2}\right)(1+r_b) - r_b^{\ 2}b_2^{\ 2}$.

From $\pi_s{}^*(w_s)'' < 0$, it can be obtained that $\pi_s{}^*(w_s)$ is a concave function on w_s. So make $\pi_s{}^*(w_s)' = 0$, we can get the optimal decision, namely, you can get the optimal decision $w_s{}^*$ of suppliers wholesale price. Then it is substituted into $w_m{}^*$, $p_m{}^*$, $p_r{}^*$ and the following conclusions can be obtained:

Proposition 3. In decentralized decision making, for the two channel supply chain of the retailer's initial free funds shortage, when the interest rate r_b is taken to the loan payment method, the optimal decision of the suppliers, the manufacturers and the retailers $w_s{}^*$, $w_m{}^*$, $p_m{}^*$, $p_r{}^*$ is obtained respectively:

$$w_s{}^* = \frac{a}{2(A+D)(b_1 - b_2)} - \frac{a}{2M(A+D)}\left[(1-t)(1+r_b)[4(b_1+b_2) + r_b b_2] + t\left[b_2(4+3r_b) + \frac{2}{b_1}(1+r_b)\left(3b_1^{\ 2} - b_2^{\ 2}\right)\right]\right] + \frac{1}{2}(c-m),$$

$$p_m{}^* = \frac{aA}{2(A+D)(b_1 - b_2)} + \frac{a}{M}[4(1-t)(1+r_b)b_1 + (4+3r_b)tab_2] - \frac{aA}{2M(A+D)}\left[(1-t)(1+r_b)[4(b_1+b_2) + r_b b_2] + t\left[b_2(4+3r_b) + \frac{2}{b_1}(1+r_b)\left(3b_1^{\ 2} - b_2^{\ 2}\right)\right]\right] + \frac{1}{2}A(c+m),$$

$$w_m^* = \frac{1}{b_1 M}\Bigg[2b_1b_2(2+r_b)(1-t)a + (4b_1^2 + b_2^2 r_b)ta + (b_1-b_2)(4b_1^2+4b_1b_2-b_2^2 r_b + 2b_1b_2r_b + 4b_1^2 r_b)\Bigg[\frac{a}{2(A+D)(b_1-b_2)} - \frac{a}{2M(A+D)}\Bigg[(1-t)(1+r_b)[4(b_1+b_2)+r_bb_2] + t\Bigg[b_2(4+3r_b) + \frac{2}{b_1}(1+r_b)(3b_1^2-b_2^2)\Bigg]\Bigg] + \frac{1}{2}(c+m)\Bigg]\Bigg],$$

$$p_r^* = \frac{aD}{2(A+D)(b_1-b_2)} + \frac{a(1+r_b)}{b_1 M}[b_1b_2(4+r_b)(1-t) + 2(3b_1^2-b_2^2)t] - \frac{aD}{2M(A+D)}\Bigg[(1-t)(1+r_b)[4(b_1+b_2)+r_bb_2] + t\Bigg[b_2(4+3r_b) + \frac{2}{b_1}(1+r_b)(3b_1^2-b_2^2)\Bigg]\Bigg] + \frac{1}{2}D(c+m).$$

Among them, $A=\frac{1}{M}(b_1-b_2)(1+r_b)(4b_1+4b_2+r_bb_2)$, $D=\frac{1}{b_1M}(b_1-b_2)(1+r_b)[2(b_1+b_2)^2+b_1r_b(2b_1+b_2)]$, $M=8(b_1^2-b_2^2)(1+r_b)-r_b^2b_2^2$.

By comparing the optimal pricing decision of the loan payment financing model with the deferred payment financing model, the following conclusions can be obtained:

Proposition 4. When retailers of dual channel three-tier supply chain are lack of funds, if the deferred payment interest rates are equal to the lending rates, that is $\delta = r_b$, there are $p_r^1 - p_r^* \le 0$, , $w_m^1 - w_m^* \le 0$, $d_r^1 - d_r^* \ge 0$.

From the conclusion above, when deferring payment interest rate is equal to lending rate, comparing with the lending mode, if the retailers choose to defer the payment, retailers can obtain goods at lower wholesale prices and consumers of retail channels can also benefit from it (at a lower price to obtain the necessary goods). Because lower retail price attracts more demand, the demand for retail sales channels is increased, the market for retailers is expanded and market share is improved. Thus generally, there are:

Corollary 1. If $\delta = r_b$, in the decentralized decision-making of the dual channel supply chain under the retailer funds constraints, the underfunded retailers choose to postpone payment.

Proof. If we want to prove that retailers will choose to defer the payment mode, we only need to prove that under the optimal decision of the two models, more profit can be obtained with retailers delayed payment model. That is $\pi_r(\delta)^* \ge \pi_r(r_b)^*$. When the delayed payment interest rate is equal to the lending rate$\delta = r_b$, the profit function of retailers decentralized decision can be expressed as follows:

$$\pi_r(\delta)^* = \frac{B}{w_m(\delta)^*}(p_r(\delta)^* - w_m(\delta)^*) + [p_r(\delta)^* - (1+\delta)w_m(\delta)^*]\left(d_r(\delta)^* - \frac{B}{w_m(\delta)^*}\right) = (p_r(\delta)^* - (1+\delta)w_m(\delta)^*)d_r(\delta)^* + \delta B,$$

$$\pi_r(r_b)^* = (p_r(r_b)^* - w_m(r_b)^*)d_r(r_b)^* - (w_m(r_b)^* d_r(r_b)^* - B)r_b = (p_r(r_b)^* - (1+r_b)w_m(r_b)^*)d_r(r_b)^* + r_b B.$$

Proposition 4. It shows that $d_r(\delta)^* - d_r(r_b)^* \ge 0$, so when $\delta = r_b$, we need to prove that in delayed payment mode, insufficient funds of retailers will have higher marginal revenue. That is $p_r(\delta)^* - (1+\delta)w_m(\delta)^* \ge p_r(r_b)^* - (1+r_b)w_m(r_b)^*$, and we can just substitute the specific decision variables into it.

In the decentralized decisions of the dual channel supply chain under capital constraint of retailers, retailers have the right to choose the financing mode of solving the problem of lack of funds. According to proposition 4 and corollary 1, it can be obtained that when delay payment interest rate is equal to lending rate, retailers under delay payment financing mode will get more profits, so retailers will give preference to delay payment financing mode. In addition, in order not to affect the sales demand, compared with capital adequacy, in the delayed payment financing mode, manufacturers will reduce wholesale price at a rate and increase demand for retailers. Sales price and demand of retail channel and direct selling channel and suppliers' wholesale price will not be affected by the retailers' capital constraints. Therefore, the total revenue of suppliers will not be affected. However, whether the retailer can choose to defer payment also need to consider the manufacturer's decisions and whether the manufacturer is willing to provide the opportunity to delay payment, we need to discuss the manufacturers' earnings under the deferred payment and lending model respectively.

3 Example analysis

In the last Chapter, we first construct the pricing decision models of suppliers, producers and retailers in supply chain under sufficient capital. Secondly, we discuss the pricing decisions of suppliers, manufacturers and retailers through theoretical analysis under the insufficient initial free funds of retailers B, in the delayed payment and bank loan financing mode. This section further explores the relationship between the decision variables and the optimal profit through numerical simulations. Make model parameter values for $a = 1000$, $t = 0.6$, $c = 3$, $b_1 = 8$, $b_2 = 2$, $m = 1$.

3.1 The comparison of capital adequacy pricing decisions and deferred payment pricing decisions

First compare in the decentralized decision making well-funded pricing decisions of the supply chain with the pricing decisions under the delayed payment financing mode. And through numerical substitution and calculation, it can be obtained that the optimal selling price of manufacturers and retailers under capital constrained is equal to the optimal decision of well-funded retailers, as follows:

$$p_m{}^0 = p_m{}^1 = 57.54, \quad p_r{}^0 = p_r{}^1 = 78.46.$$

And in the delayed payment the optimal wholesale price of the manufacturers $w_m{}^1$ and the optimal decision-making price $w_m{}^0$ under adequate capital meet the following equation:

$$w_m{}^1 = \frac{w_m{}^0}{1+\delta} = \frac{67.54}{1+\delta}.$$

Because information is symmetrical among enterprises in dual channel supply chain, if retailers choose to defer the payment, direct selling channel price and retail price are not affected by retailers' capital constraints, and retail and direct sales demand will also not affected by retailers' capital constraints. So there are $1 = d_m{}^0 = 96.62$, $d_r{}^1 = d_r{}^0 = 87.38$. Therefore, direct sales revenue of manufacturers and sales revenue of retailers are unchanged.

However, the financial constraints of retailers will reduce manufacturers' wholesale price for the retailers. The increase of delayed payment interest rate will reduce the marginal profit of retailers selling goods, which will reduce the retailers ordering enthusiasm and affect the manufacturer's wholesale revenue. Therefore, manufacturers will correspondingly lower wholesale prices for retailers and encourage retailers to improve commodity ordering quantity.

Through the calculation it can be known that in the deferred payment financing model, the supplier's revenue is $\pi_s{}^1 = \pi_s{}^0 = 6944.90$, the manufacturer's revenue is $\pi_m{}^1 = \pi_r{}^0 - \delta B = 3779.90 - \delta B$, and the retailer's revenue is $\pi_r{}^1 = \pi_r{}^0 + \delta B = 954.54 + \delta B$.

3.2 Table analysis of delay payment financing pricing decision and loan financing pricing decision

According to optimal decision expression of suppliers, manufacturers and retailers under defer payment model and loan payment model (proposition 2.2, 2.3), it can be obtained that through numerical substitution, the relationship between the optimal decision variable and the rate of payment (including the delay payment interest rate and lending rate) for suppliers, manufacturers and retailers in decentralized decisions making. When $\delta = r_b$ takes 0.1, 0.2, 0.3, 0.4, it can be obtained that the relationship between payment rate and decision variable and the optimal profit of the retailers, manufacturers and suppliers in the delay payment model and the loan payment mode, respectively indicating as Table 1.

The relationship between the optimal decision variables and their profit in $\delta = r_b$ under the deferred payment model and the credit payment model is shown in Table 2. Among them, $\Delta w_s{}^* = w_s{}^1 - w_s{}^*$, $\Delta w_m{}^* = w_m{}^1 - w_m{}^*$, $\Delta p_m{}^* = p_m{}^1 - p_m{}^*$, $\Delta p_r{}^* = p_r{}^1 - p_r{}^*$, $\Delta d_r{}^* = d_r{}^1 - d_r{}^*$, $\Delta d_m{}^* = d_m{}^1 - d_m{}^*$, $\Delta \pi_r{}^* = \pi_r{}^1 - \pi_r{}^*$, $\Delta \pi_m{}^* = \pi_m{}^1 - \pi_m{}^*$, $\Delta \pi_s{}^* = \pi_s{}^1 - \pi_s{}^*$.

From table 3, when the delayed payment interest rate is equal to the lending rate, the optimal profit of suppliers and retailers in the delayed payment financing model is more than that in the loan financing mode. And with the increase of the interest rate, each can get more profits. However, manufacturer's optimal profit depends on the number of initial free money of the retailers. When initial free funds of retailers are fixed value, there is a maximal value of interest rate. If the delayed payment interest rates and lending rates are higher than the maximal interest rate at the same time, manufacturers can make more profits in the loan financing mode of the retailers. Therefore, in this situation, the manufacturers will not provide the delayed payment financing mode.

3.3 Graphical analysis of the pricing decision making in the deferred payment financing mode and the loan financing mode

In order to further clearly describe how effectively the financing and pricing decisions are making for the suppliers, manufacturers and retailers in different payment rates. Next, we will extend the payment rate δ, r_b to the interval (0, 0.8), and take the payment rate as the horizontal coordinate, and the price, the demand or its profit as the ordinate. And we can get the price decision, demand and profit scatter plot with smooth line of suppliers, manufacturers and retailers in the defer payment and loan financing mode using excel statistical tool.

Figs. 1 ~ 4 show the curve of the relationship between the pricing decisions of suppliers, manufacturers and retailers and different interest rates in delayed payment and loan financing mode. Figs 5 and 6 give the curve of the relationship between the sales demand of manufacturers and retailers and different interest rates in delayed payment and loan financing mode. Figs. 7 and 8 also give the curve of the relationship between the profits of suppliers and retailers and different interest rates in delayed payment and loan financing model. Among them the solid line represents the deferred payment financing model, and dashed line represents the loan financing model.

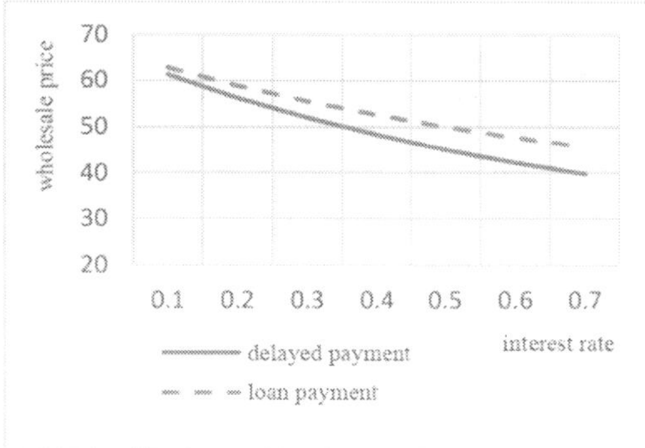

Fig.1. Wholesale price of suppliers

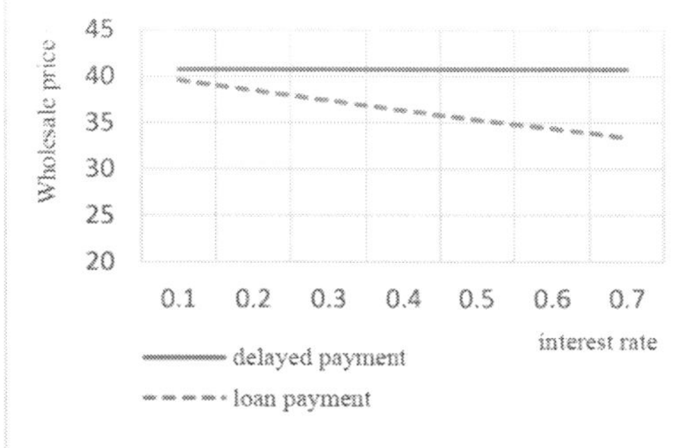

Fig.2. Wholesale price of manufacturers

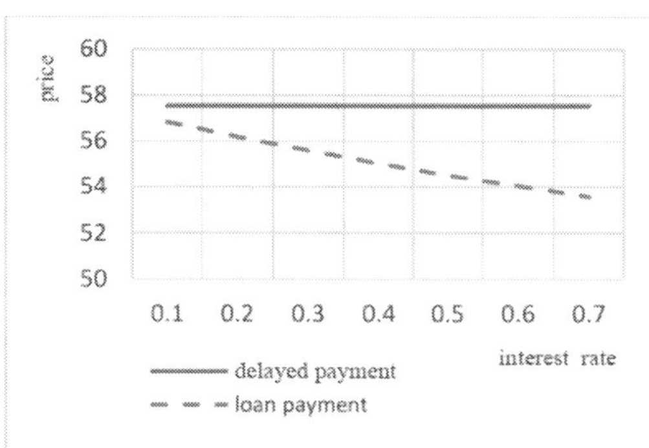

Fig.3. Direct selling price of manufacturers

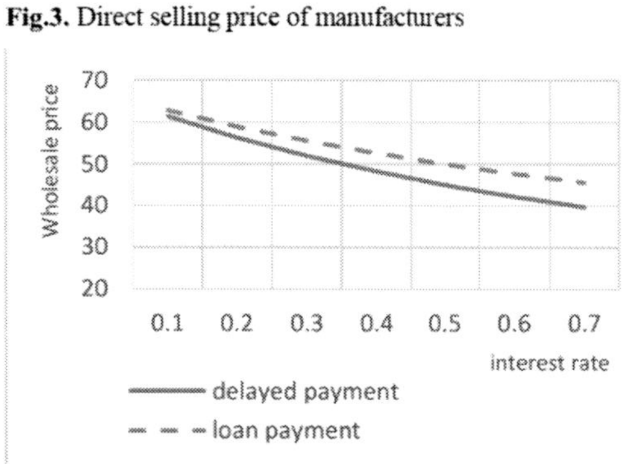

Fig.4. Retail price

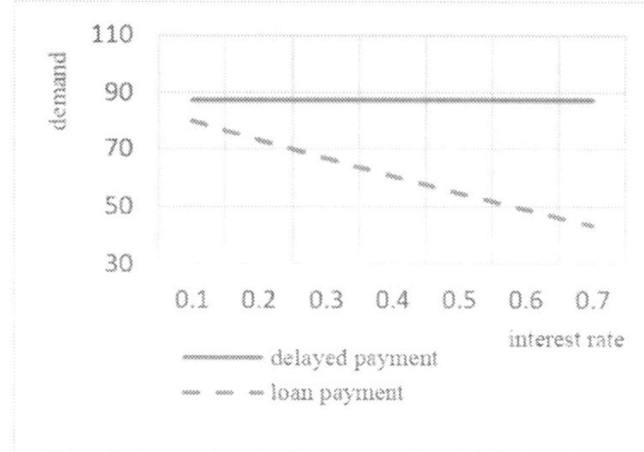

Fig.5. Demand of retailers

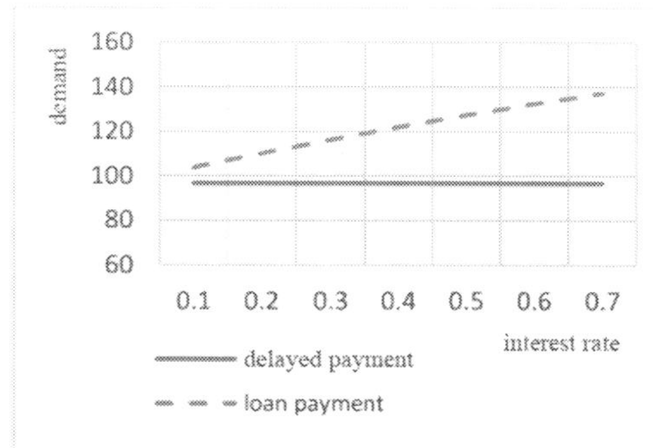

Fig.6. Demand of manufacturers

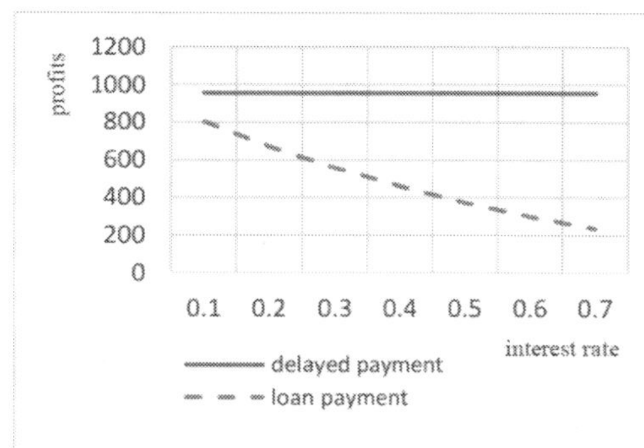

Fig.7. Profits of retailers

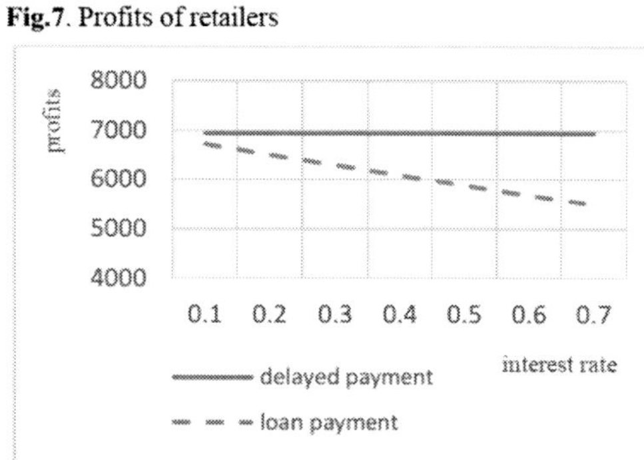

Fig. 8. Profits of suppliers

From table 2 and figure (1) ~ (4), figure (5) and (6), figure (7) and (8), in general the following conclusions can be obtained: On the premise that the delayed payment interest rate δ is equal to the loan interest rate r_b, retailers prefer to the defer payment financing mode. At the same time the supplier also hope that retailers choose to defer payment financing mode. As for the manufacturers, they will make decisions according to the initial free funds of retailers. The specific analyses are as follows:

For the retailers, on the one hand, in the delayed payment financing mode, retailers of insufficient funds obtain goods at relatively lower wholesale prices of manufacturers, and sell the goods at a discount of the retail price so that consumers of the retail channel can get benefits, more customers can be attracted, the demand of retail channel will be expanded and market share and influence of the retailers will be improved; on the other hand, under the deferred payment financing model, the profit benefit is greater than that of the loan financing model. So retailers will choose the financing model which are conducive to maximize revenue. Therefore, in the decentralized decision-making, Capital-constrained retailers choose to work with manufacturers to accept the deferred payment financing mode.

For the suppliers, if retailers accept the deferred payment financing model, suppliers can sell goods at higher wholesale prices to manufacturers and the total sale of suppliers is unchanged. Through comparison, we can see an increase in supplier's optimal profits. Therefore, at the same rate, suppliers will also wish that the retailers choose the delayed payment financing mode.

For manufacturers, at a certain rate of payment in the delayed payment mode, optimal profit of the manufacturers is inversely proportional to the initial free capital of the retailers; by comparing the optimal profits in the delayed payment financing mode and loan financing mode, whether manufacturers cooperate with retailers to offer the deferred payment financing mode depends on the number of the retailer's initial free funds. When the interest rate is certain, if the initial free capital of the retailers is not greater than a certain value, the manufacturers can make more profits in the deferred payment financing mode. And the smaller the initial free capital of the retailers is, the greater profits the manufacturers will gain. It also shows that manufacturers are more willing to work with retailers to provide the delayed payment financing mode when the retailers are severely underfunded.

4 Conclusions and future work

This paper chooses the three-stage supply chain consisting of a single supplier, a single manufacturer and a single retailer as the research object, and makes theoretical research and numerical analysis on the financing decisions and pricing in the supply chain of retailers under the capital constraints. Firstly, the profit model for each member of the three-tier supply chain of dual channel sales is constructed under capital adequacy. Then, the profit model of each member of the supply chain under the deferred payment financing mode and loan financing mode is constructed respectively to compare which is more advantageous between the two financing modes. It can be found through comparison analysis of theoretical model and the method of numerical simulation as follows:

In the premise that the delayed payment interest rate δ is equal to the loan interest rate r_b, the retailers will prefer to choose the delay payment financing mode. On the one hand, the retailers will gain more profits under the deferred payment model than the loan financing model; On the other hand, in the delayed payment financing mode, retailers of insufficient funds obtain goods at relatively lower wholesale prices of manufacturers, and sell the goods at a discount of the retail price so that consumers of the retail channel can get benefits, more customers can be attracted, the demand of retail channel will be expanded and market share and influence of the retailers will be improved. At the same time the supplier also hope that retailers choose to defer payment financing mode. As for the manufacturers, they will make decisions according to the initial free funds of retailers. But manufacturers are more willing to work with retailers to provide the delayed payment financing mode when the retailers are severely underfunded. And then, the optimal choice of the manufacturers is the deferred payment contract without compensation (deferred payment interest rate is 0) to make the profits optimal.

Although this paper comes to some significant conclusions through model studies, the research still has its limitations. First, this paper studies the decentralized decision-making of the supply chain members, and is not related to the centralized decision-making problems of supply chain members. When analyzing the capital constraints problems of retailers, it can be obtained that the retailers can get more profits in the deferred payment financing mode than the loan financing mode. Under certain conditions of decentralized decision-making, the manufacturers are willing to provide the deferred payment financing mode. However, we also need to consider the decisions of the suppliers from the centralized decision-making to study if they are still willing to provide the delayed payment financing mode. This is what can continue to be studied in the future. Secondly, this paper focuses on the supply chain of a single supplier, a single manufacturer and a single retailer, which is still simplified compared with the supply chain in the real environment. Therefore, in the later work, the research object can be extended to multiple suppliers, multiple manufacturers and multiple retailers, and initial funding constraints can also be extended to the manufacturers and retailers at the same time. In addition, this paper shows simple ways of finance services the delayed payment financing mode and the loan financing mode. With the continuous innovation and development of supply chain finance, the financing modes will become diversified. Therefore, in the later work, the issue of pricing decision and operation research with capital constraints under other new financing modes can also be considered.

References

1. H.Y. Tu, G.W. Hua, S.Y. Wang. Research on financing and production decisions in the supply chain of capital constraints. Systems Science and Mathematics, 2011, 31(11): 1414-1422.
2. A.M. Zhao , M. Li. Research on the financing and pricing of the supply chain under the condition of noncompetition. Management Science in China, 2012, 20: 521-527.
3. X.F. Chen, D.L. Zhu, W.J. Ying. Research on capital constraint and financing and operation comprehensive decision-making in the supply chain. Journal of Management Science, 2016, 11(3): 70-77.
4. N. Yan, B. Sun. Coordinating loan strategies for supply chain financing with limited credit. OR Spectrum, 2013, 35: 1039-1058.
5. K. Li, W.H. Chen. Pricing policy of multi-channel supply chain for funds constraints of retailers. Operations Research and Management, 2015, 36 (10): 1511-1515.
6. Y.Q. Song, W.L. Wang. Financing model and risk prevention of suppliers under capital constraints. Commercial Economy, 2017, (12): 46-48.
7. Y.G. Zhang, X.W. Tang. Study on the wholesale price contract of short life cycle products under the deferred payment. Management Science, 2011, 19(3): 63-69.
8. X. Zhang, Q. Lin. Research on the optimal decision making in the three-stage supply chain with the delayed payment and order financing. Journal of Systems Engineering, 2015, 39(02): 405-409.
9. Y.D. Gong, B.Y. Li, W.X. Li. Research on the three-stage supply chain coordination and profit distribution mechanism. Statistics and Decision, 2013, 23(1): 185-186.
10. H. Wang, J. Zhou. Research on decision making of dual channel supply chain under different price modes. China Management Science, 2009, 17(6): 84-90.
11. V. Babich, G. Aydin, Y. Brunet. Risk, financing and the optimal number of suppliers. Supply Chain Disruptions, 2012, (32): 195-240.
12. W.L. Wang, J.W. Luo. Research on the prepayment financing strategy of supply chain based on price discount. Journal of Management Science, 2014, 17(11): 20-32.
13. J.Z. Zhan, X.Z. Zhou, Y.D. Gong. The optimal financing and production decisions of the supply chain of production capital constraints. Journal of Systems Engineering, 2015, 30(2): 190-201.
14. S.Z. Bai, N. Xu, Z.H. Yan. Research on supply chain coordination based on inventory pledge financing mode. Operations Research and Management, 2013, 22(3): 185-193.
15. Y.H. Fu, X.F. Wang, G.H. Chen. Financing options in the supply chain for inventory financing and credit loan under the buyback coordination. System Engineering, 2014, 32(11): 44-49.

Table 1. The decision variables and profits in the deferred payment model.

δ	w_s^1	w_m^1	p_m^1	p_r^1	d_r^1	d_m^1	π_r^1	π_m^1	π_s^1
0.1	40.74	61.40	57.54	78.46	87.38	96.62	954.54 + 0.1B	3,779.90 - 0.1B	6944.90
0.2	40.74	56.28	57.54	78.46	87.38	96.62	954.54 + 0.2B	3,779.90 - 0.2B	6944.90
0.3	40.74	51.95	57.54	78.46	87.38	96.62	954.54 + 0.3B	3,779.90 - 0.3B	6944.90
0.4	40.74	48.24	57.54	78.46	87.38	96.62	954.54 + 0.4B	3,779.90 - 0.4B	6944.90

Table 2. The decision variables and profits in the loan financing model.

r_b	w_s^*	w_m^*	p_m^*	p_r^*	d_r^*	d_m^*	π_r^*	π_m^*	π_s^*
0.1	39.59	62.87	56.84	79.19	80.19	103.67	803.83 + 0.1B	3472.15	6726.60
0.2	38.46	58.93	56.19	79.88	73.35	110.21	672.54 + 0.2B	3271.83	6509.31
0.3	37.37	55.53	55.60	80.55	66.83	116.32	558.30 + 0.3B	3149.94	6295.55
0.4	36.33	52.58	55.04	81.19	60.60	122.04	459.11 + 0.4B	3086.07	6086.91

Table 3. The relationship between the optimal variable value of the deferred payment model and the loan payment model in $\delta = r_b$.

$\delta = r_b$	Δw_s^*	Δw_m^*	Δp_m^*	Δp_r^*	Δd_r^*	Δd_m^*	$\Delta \pi_r^*$	$\Delta \pi_m^*$	$\Delta \pi_s^*$
0.1	1.16	-1.48	0.70	-0.72	7.19	-7.05	150.71	307.75 -0.1B	218.30
0.2	2.28	-2.64	1.35	-1.42	14.03	- 13.59	282.00	508.06 -0.2B	435.59
0.3	3.37	-3.58	1.94	-2.08	20.55	- 19.70	396.24	629.95 -0.3B	649.34
0.4	4.42	-4.34	2.50	-2.72	26.78	- 25.42	495.43	693.82 -0.4B	857.99

Face recognition based on improved BP neural network

Gaili Yue, Lei Lu[a],

College of Electrical and Control Engineering, Xi'an University of Science and Technology, Xi'an 710043, China

Abstract. In order to improve the recognition rate of face recognition, face recognition algorithm based on histogram equalization, PCA and BP neural network is proposed. First, the face image is preprocessed by histogram equalization. Then, the classical PCA algorithm is used to extract the features of the histogram equalization image, and extract the principal component of the image. And then train the BP neural network using the trained training samples. This improved BP neural network weight adjustment method is used to train the network because the conventional BP algorithm has the disadvantages of slow convergence, easy to fall into local minima and training process. Finally, the BP neural network with the test sample input is trained to classify and identify the face images, and the recognition rate is obtained. Through the use of ORL database face image simulation experiment, the analysis results show that the improved BP neural network face recognition method can effectively improve the recognition rate of face recognition.

1 Introduction

Face recognition method is one of the popular research topics in the field of pattern recognition and computer vision. Face recognition technology is widely used in identity authentication, finance, file management, human-computer interaction and other fields, has a very large practical value[1-2]. The face recognition process consists of four main parts: face extraction, face image preprocessing, feature extraction, classification recognition. Improving the recognition rate is the main research content of face recognition. After decades of development, face recognition has many classical recognition algorithms, such as principal component analysis method for face recognition, geometric feature matching method based on board, PCA, eigenface, wavelet transform[3], Hidden Markov model. In recent years, with the deepening of artificial neural network research, artificial neural network is widely used in pattern recognition, system identification and so on. In the artificial neural network, the most widely used network structure is BP neural network. BP neural network using the classical error back propagation algorithm for training, can deal with linear indivisibility.

Combined with the characteristics of BP neural network, an algorithm based on histogram equalization[4], PCA and BP neural network fusion is proposed. Used to improve the recognition rate of face images.

2 Histogram equalization

Due to the collection of face images will be subject to other factors such as lighting causes the image quality is not uniform, the image contrast is low. Histogram equalization is an image enhancement method that adjusts the contrast of an image by an image histogram. After processing the histogram of the image can be presented evenly distributed features, the image contrast is increased.

The process of histogram equalization: First, we obtain the probability distribution of all gray levels in grayscale image in turn. The f_i represents the gray level, n_i represents the number of occurrences of the ith gray level, N represents the total number of pixel pixels, P (f_i) represents the ith gray level of the frequency, then:

$$P(f_i) = \frac{n_i}{N}, i \in (0, 1, \dots, 255) \quad (1)$$

Then calculate the cumulative distribution function:

$$S_k = \sum_{i=0}^{k} P(f_i) \quad (2)$$

Remember g_j for the histogram equalization after the gray level, then:

$$g_j = INT[(g_{max} - g_{min})S_k + g_{min} + 0.5] \quad (3)$$

Where INT is the rounding operation. Histogram equalization is a mapping of g_j and fi. The gray scale of the image after the equalization is evenly distributed, thus increasing the contrast of the image and facilitating the processing of the later algorithm of face recognition.

3 Principal component analysis

Principal Component Analysis (PCA)[5-6] is a very effective way to reduce the dimensionality of high-dimensional data. The basic principle of principal component analysis is to extract the main components of the data by K-L transformation with the minimum mean square error as the criterion, and constitute the feature space. Turk and Pentland A were the first researchers to use the PCA method for face recognition. Because of the large number of face samples and the large amount of data in face recognition, the reduction and compression of the sample data by PCA method can significantly reduce the data processing time and recognition time.

The training face grayscale images have N, each picture matrix is processed into a row vector, N row vector composed of N*1 dimensional matrix $=(x_1, x_2, \dots, x_N)^T$.

Find the mean of matrix x:

$$\psi = \frac{1}{N}\sum_{i=1}^{N} x_i \quad (4)$$

The difference between each row vector and the mean of the x matrix is:

$$\varphi_i = x_i - \psi \quad (5)$$

Calculate the covariance matrix for the x matrix:

$$S = \frac{1}{N}\sum_{i=1}^{N}(x_i - \psi)(x_i - \psi)^T \quad (6)$$

The covariance S is a matrix of N * N. In general, it is very difficult and time-consuming to obtain the eigenvalues and eigenvectors of the covariance matrix S directly because of the large number of samples to be trained (N is large). Therefore, the SVD theorem is used to obtain the eigenvalues of matrix S and its corresponding eigenvectors.

Let the matrix$H_{m*n} = U\sqrt[2]{F}V^T$, where F is the diagonal matrix, U, V is the orthogonal matrix, its form is as follows.

$$U = (u_1, u_2, \dots, u_m), U^TU = 1 \quad (7)$$

$$V = (v_1, v_2, \dots, v_m), V^TV = 1 \quad (8)$$

$$F = \mathrm{diag}(\lambda_1, \lambda_2, \dots, \lambda_m), \lambda_1 \geq \lambda_2 \geq \dots \geq \lambda_r \quad (9)$$

Where, r=min(m, n), i=(1,2,…,r), denotes the non-zero eigenvalues of the matrix HH^T and H^TH.

U_i and V_i represent the eigenvectors corresponding to the nonzero eigenvalues of the matrix HH^T and H^TH, respectively. Use $\sqrt{\lambda_i}$ to represent singular values.

According to the SVD theorem, when we know that the r nonzero eigenvalues of A^TA and the corresponding eigenvectors are denoted by λ_i (i=1,2, ..., r) and v_i respectively, we can compute the eigenvectors of A^TA u_i is:

$$u_i = \frac{1}{\sqrt{\lambda_i}} A^T v_i, \quad i = 1, 2, \ldots, r \tag{10}$$

The eigenvector of the image can be calculated, and the eigenvector corresponding to the larger eigenvalue is called the principal component of the image matrix. The reconstructed image of the principal component of the face image matrix becomes the eigenface.

4 BP neural network

BP neural network is a network containing multiple hidden layers, with the ability to deal with linear indivisibility problems. BP neural network is the core of the forward neural network, which is the essence of the whole artificial neural network. BP neural network is used by error back propagation algorithm, the algorithm to solve the multi-layer neural network learning problems[7]. The actual use of BP neural network structure is very common, accounting for about 80%, which is widely used in classification and identification, system identification, compression and other fields. In this paper, BP neural network is used to classify and identify face images. Train the BP neural network to make it a classifier for face images.

4.1 Regular BP neural network

The error back propagation algorithm used in the learning process of BP neural network. The training process of neural network is the process of correcting the weights of each layer. When the error back propagation algorithm is used for network training, the data is propagated from the input layer to the output layer. The modification of the weights is modified from the output layer to the direction of the error reduction by the hidden layer.

The structure of BP neural network includes input layer, hidden layer and output layer. The hidden layer can be one or more layers. The structure is shown in Fig. 1.

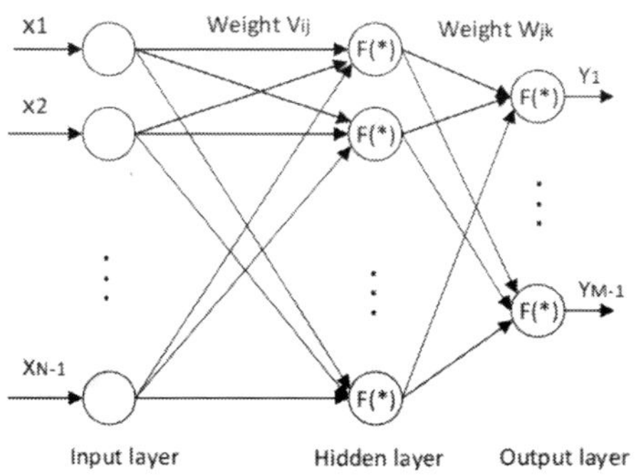

Fig.1. three-tier BP neural network structure

4.2 Improved weight adjustment method

The conventional BP neural network uses the gradient descent method to adjust the weight, but there will be a problem that the convergence rate of the network is slow and easy to fall into the local minimum point. For the problem in the conventional BP network, a method of weight adjustment based on the combination of the additional momentum[8] and the elastic gradient is adopted. This improved weighting method is used to improve the convergence speed and learning accuracy of the network.

The weight adjustment method of the improved weight adjustment method at t + 1 is:

$$W(t+1) = \begin{cases} W(t) - sign\left(\frac{\partial E(t)}{\partial w}\right)\Delta t & \frac{\partial E(t)}{\partial w} \neq 0 \\ 0 & \frac{\partial E(t)}{\partial w} = 0 \end{cases} \tag{11}$$

When the iteration gradient is the same for two consecutive times, the update value of the weight increases and the adjustment direction of the weight remains unchanged. Update the value of Δt to adjust the formula:

$$\alpha \times (1 - mc) \times \Delta(t-1) + mc \times \Delta(t-1) \quad (12)$$

Where mc is the momentum coefficient, mc $\in(0,1)$.

When the successive two iterative gradient directions are opposite, the updated value of the weight decreases and the weight remains constant with a certain probability P. The probability P is defined as the number of changes over time $P = \beta/((1+\beta)(e^{0.2t}+1))$, $P \in (0, 1)$. The adjustment direction of the weights is constant with a certain probability P, which can improve the probability that the network will jump out of the local optimal solution. The adjustment formula for updating the value Δt at this time is:

$$\beta \times (1 - mc) \times \Delta(t-1) + mc \times \Delta(t-1) \quad (13)$$

Where $mc \in (0,1)$.

In other cases, the adjustment formula remains unchanged. Through this method, the adverse effect of the size of the gradient on the convergence rate of the network can be solved, and the convergence rate can be improved. When the network error is a big change, it can effectively reduce the network into the local optimal solution, thus improving the network learning accuracy.

5 Simulation and result analysis

5.1 simulation

The experiment is programmed on software Matlab 2012a. In this paper, we selected 400 face images (40 people, 10 per person) of the ORL face database with 112 * 92 as the training and test sample set. Each person's 10 images, facial expressions and gestures have varying degrees of change, the scale of the image changes in about 10%. In the course of the experiment, each of the 10 images are divided into two groups, each group of five. The training samples were the first five images of each set of image samples, and the last five images of each sample of the sample were tested.

5.2 results analysis

(1) The face image feature has been enhanced to enlarge the dynamic range of image gray scale, after histogram equalization. The original image and the image histogram equalization contrast, as show in fig.2.

Fig.2. Original (left) histogram equalization (right)

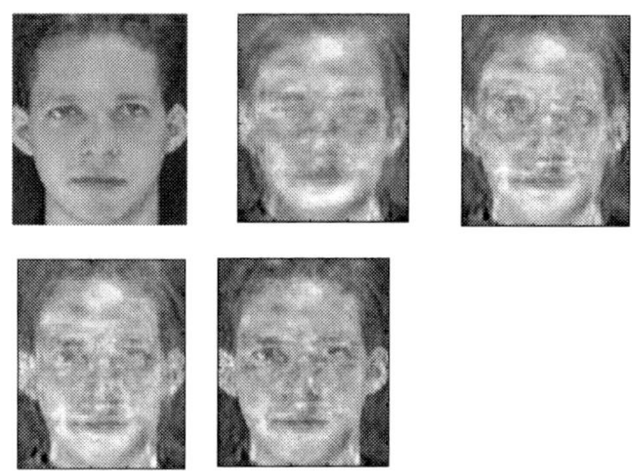

Fig.3. Face images corresponding to different principal components

(2) PCA algorithm can effectively achieve the image matrix dimensionality reduction. After selecting the dimension of the eigenvector, the eigenvector of the original image of the face image drops from 10304 dimension to the dimension of the selected eigenvector. The face map for the different principal components of a single face is shown in Fig.3. The face map for the different principal components of a single face is shown in Fig.3. From left to right are the original image, the principal component 20, the principal component 40, the principal component 60, and the principal component

80.It can be seen from the figure, the larger the main component of the details of the more comprehensive part of the image.

(3) The parameters of the BP neural network are as follows: the number of hidden layer nodes is 150, the output node is 40, the momentum coefficient mc is 0.6, the learning rate is 0.03, a is the random number between 0.5 and 1, the number of training times is 1000 , The coefficient α = 1.2, the coefficient β = 0.5. Use the ORL face database to simulate 400 images. In this paper, the recognition results of PCA + BP neural network in different feature vectors are shown in Fig.4.

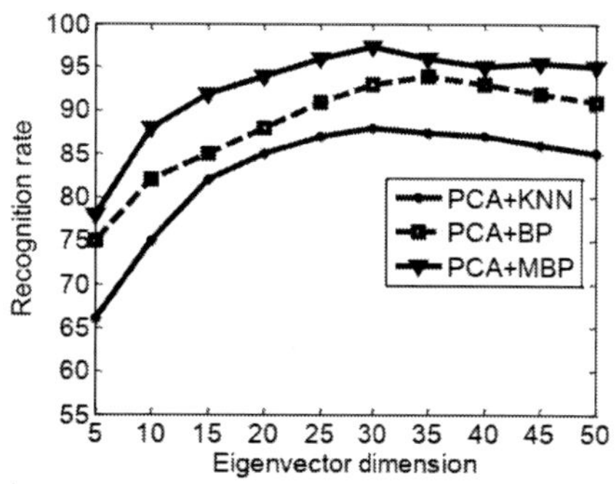

Fig.4. Comparison of algorithm (PCA + MBP) with traditional recognition method of face recognition algorithm.

It can be seen from Fig. 4 that the algorithm used in this paper is superior to the traditional PCA + BP neural network and PCA + KNN algorithm. In this paper, the algorithm achieves a recognition rate of 97.4% when the feature vector dimension is 30.When the feature vector dimension is greater than 30, the recognition rate fluctuates in a small range.

6 Conclusion

The main feature data of the image is obtained by PCA algorithm, and then the parameters of BP neural network are selected according to the extracted feature data. The improved weight adjustment method is used to train the network. Finally, the training of BP neural network is used to classify and identify the face images. Using ORL face database of face images in MATLAB 2012a software simulation experiments. According to the simulation results, the improved BP neural network face recognition method can effectively improve the recognition rate of face recognition.

References

1. Haibo Wang. Research on face recognition method based on deep learning[D]. HeFei University of Technology,2014.
2. Yanyan Huo. Face recognition based on improved algorithm of PCA and LBP [D]. Harbin University Of Science And Technology, 2015.
3. Yanxuan Li, Rong Wang. PCA face recognition method based on Gabor wavelet feature extraction[J]. Computer Knowledge and Technology, 2015,**11**(32):138-141.
4. Kenneth RCastleman. Digital image processing [M]. Beijing: Publishing House of Electronics Industry,1998.
5. Turk M, Pentland A. Eigenfaces for recognition [J]. Journal of Cognitive Neuroscience, 1991, **3**(1): 71-86.
6. Xin Zhao, Weijia Wang, Yayun Pu, Caiwei Xiong, Yanjia Ren. A new face recognition algorithm based on improved modular PCA [J].Computer Engineering and Applications, 2015,**51**(2): 54-56.
7. Liyan Xu. Research on face recognition technology[D].Tianjin University,2010.
8. Kangshun Li, Kai Li, Wensheng Zhang. PCA face recognition algorithm based on improved BP neural network[j].Computer Applications and software, 2014, 1 (31) 158-161.

A Knowledge Context Fuzzy Clustering Method Based on Genetic Algorithm

Faping Zhang[1], *Li* Li[1,*], and *Cuixiang* Zhou[2]

[1]School of Mechanical Engineering, Beijing Institute of Technology, Beijing, China
[2]No.208 Research Institute of China Ordnance Industries, Beijing, China

Abstract. A fuzzy clustering method based on genetic algorithm is proposed aiming at the problem of automatic clustering of knowledge context. Firstly, the knowledge context model is constructed to determine the similarity measure of knowledge context. Then the initial clustering centers are obtained based on the density peak method. Then the fuzzy C mean clustering result is solved by genetic algorithm, and the clustering of knowledge context is realized. Finally, the knowledge context clustering of an aircraft part design process is taken as an example to illustrate the effectiveness of the algorithm.

1 Introduction

Knowledge plays an increasingly important role as an intangible resource in improving the production efficiency, reducing product costs, especially in knowledge-intensive enterprises.

Knowledge context is the relevant background environment on which knowledge acquisition, understanding, application, evaluation, sharing depend on.

Enterprise knowledge management system, records the user's knowledge behaviour in different knowledge context, and gets the internal relationship between the knowledge context and knowledge through data mining and other ways.

When the business people face a new business context, the system calculates the similarity between the current context and all the knowledge contexts, finds out a set of the most similar knowledge contexts, and then pushes the knowledge associated with the set of knowledge contexts to the users. However, with the accumulation of the number and types of knowledge contexts, the computational similarity of traversal undoubtedly greatly affects the search efficiency. In order to solve this problem, this paper proposes clustering similar knowledge contexts, which only need to be calculated in the class of the current context to get the similarity, so as to improve the computational efficiency and knowledge reuse efficiency.

The main clustering methods include hierarchical clustering [1], density- based [2], grid-based [3] and model-based methods [4], and have been widely used in many fields such as pattern recognition [5], data analysis [6], image processing [7], market research [8] and so on.

Aiming at the clustering problem of knowledge context, this paper proposes a knowledge context fuzzy clustering method based on genetic algorithm. Firstly, the knowledge context model is constructed and the similarity measure method is put forward. Then, the cluster is clustered by the density peak method as the initial solution. Then the genetic fuzzy C-means clustering algorithm is used to realize the clustering of knowledge context. Finally, an example is given to illustrate the effectiveness of the algorithm.

2 Knowledge context model and the similarity measure

2.1. Knowledge context model construction

In the enterprise knowledge management process, the knowledge context not only includes the corresponding business process and task characteristics, but also includes the personality characteristics of users. The knowledge context model is used to describe the context in the form of extensible framework. The enterprise environment is complicated, so it is necessary to consider the characteristics of the task itself, the relationship with the business process, users' cognitive level and knowledge preference.

This section constructs a multi-dimensional multi-attribute context model for the business process of the enterprise in order to describe the context elements of different dimensions more accurately and make the context model more manageable.

The context model is characterized by four main dimensions: person, task, object, and software.

Context={Person, Task, Object, Software}

There are specific portrayals under the four main dimensions:

The person, including personnel roles, personnel skills level, the field described, is

Person = {P-position, P-skills, P-domain}.

* Corresponding author: bitlili@163.com

The task, including the task name, task objectives, tasks required resources, the relationship between tasks and the process, is

Task = {T-name, T-goal, T-resource, T-relation}.

The object, including the object name, object material, and object technical characteristics, is

Object = {O-name, O-material, O-TecFeatures}.

The software, including the software name, software version, software use, is

Software = {S-name, S-version, S-usage}.

After the refinement, the context model is

Context= {P-position, P-skills, P-domain, T-name, T-goal, T-resource, T-relation, O-name, O-material, O-TecFeatures, S-name, S-version, S-usage}.

It should be noted that the enterprise can be constructed according to the actual environment, including but not limited to the above attributes.

2.2 Knowledge context similarity

Clustering is a process that divides a data set into different categories or clusters and makes the properties of objects in the same category the same or similarities, and the properties of the objects in different categories are significantly different. The distance between the knowledge context is measured by the knowledge context similarity, the higher the degree of similarity, the closer the distance is. This section describes the similarity of knowledge context.

Knowledge contexts are portrayed from multiple dimensions, and the attributes of different dimensions may not be the same. The attribute type is divided into four types: numeric attribute, vector type attribute, collection type attribute, text type attribute. Numerical types and vector types can be calculated directly, and the collection type and text type attributes need to be vectorised first.

The vectorization process is:

Step1: the construction of the domain vocabulary. For example, the total number of a set of possible vocabulary is m, then the m words in order to form a vocabulary.

Step2: vectorization.

This type is described by a m-dimensional vector. For a collection type, the {0,1} value indicates whether a vocabulary is present in the set. For text attributes, the TF-IDF value for each word in the text as the value is the vector component.

After the pre-treatment, the n knowledge context models can be expressed by the knowledge context model vector matrix:

$$\Sigma = \begin{bmatrix} \overset{\text{the first attribute}}{x_{11}} & \overbrace{x_{12} \quad \cdots \quad x_{1k}}^{\substack{\text{the second attribute} \\ (k\text{-1 dimension vector})}} & \cdots & x_{1h} \\ x_{21} & x_{22} \quad \cdots \quad x_{2k} & \cdots & x_{2h} \\ \vdots & \vdots \qquad\quad \vdots & & \vdots \\ x_{n1} & x_{n2} \quad \cdots \quad x_{nk} & \cdots & x_{nh} \end{bmatrix}$$

For vector type attributes and numeric type attributes, different methods are used for the calculation.

a vector attribute calculation

$$sim_v(v,v') = \frac{\sum_{i=1}^{m}(q_i-\bar{q})(p_i-\bar{p})}{\sqrt{\sum_{i=1}^{m}(q_i-\bar{q})^2}\sqrt{\sum_{i=1}^{m}(p_i-\bar{p})^2}}$$

where v and v 'denote the m-dimensional vector type attributes, the values are (q_1, q_2, q_3, ..., q_m) and (p_1, p_2, p_3, ..., p_m), The similarity of the vector type attribute is in the range of [-1,1]. The larger the value, the higher the similarity.

b numerical attribute calculation

$$sim_n(v,v') = \frac{\min\{n_1,n_2\}}{\max\{n_1,n_2\}}$$

Where v, v 'denote the numeric type attribute, the values are n_1、n_2, n_1、n_2> 0. The value range of the value type attribute is (0,1). The larger the value, the higher the similarity.

The context similarity is the weighted average of the similarity of each attribute:

$$sim(V,V') = \frac{\sum_{i=1}^{s} w_i \cdot sim(v_i, v_i')}{\sum_{i=1}^{s} w_i}$$

Where V, V'denote two contexts, there are s attributes, v_i, v_i' denote the i-th attribute, and w_i the weight of the i-th attribute.

3 Improved fuzzy C - means clustering algorithm for initial value selection

3.1 Fuzzy C-means clustering algorithm

Ruspini[9] introduced the Zadeh fuzzy mathematical theory[10] into clustering analysis, and proposed the concept of fuzzy partitioning. The membership degree $u_{ij} \in$ {0,1} in traditional hard clustering is improved to $u_{ij} \in$ [0,1] continuous interval. Later Dunn and Bezdek promoted it to form the widely used fuzzy c-means algorithm (FCM).

The FCM algorithm determines the membership of the sample by minimizing the objective function to further determine the class of the sample.

Assume Knowledge Context Data Set

$$X = \{x_1, x_2, \ldots, x_k, \ldots, x_n\} \subset R^h$$

Where n is the number of elements in the data set X, the sample has s attribute value, h attribute dimension (h ≥ s).

The matrix V of the c clusters is:

$$V = \begin{bmatrix} v_{11} & v_{12} & \cdots & v_{1h} \\ v_{21} & v_{22} & \cdots & v_{2h} \\ \vdots & \vdots & & \vdots \\ v_{c1} & v_{c2} & \cdots & v_{ch} \end{bmatrix}$$

Where c is the number of clusters, $c \in [2, n)$. v_{ij} is the value of the j-th attribute dimension of the i-th cluster center.

The objective function is expressed as:

$$V = \begin{bmatrix} v_{11} & v_{12} & \cdots & v_{1h} \\ v_{21} & v_{22} & \cdots & v_{2h} \\ \vdots & \vdots & & \vdots \\ v_{c1} & v_{c2} & \cdots & v_{ch} \end{bmatrix}$$

Where c is the number of clusters, $c \in [2,n)$. the v_{ij} is the value of the j-th attribute dimension of the i-th cluster center.

The objective function is expressed as:

$$J_m(U,V) = \sum_{i=1}^{c} \sum_{j=1}^{n} u_{ij}^m . d_{ij}^2 \tag{1}$$

Constraints:

$$\begin{cases} \sum_{j=1}^{c} u_{ij} = 1 & 1 \le j \le n \\ 0 \le u_{ij} \le 1 & 1 \le i \le n, 1 \le j \le c \\ 0 \le \sum_{i=1}^{n} u_{ij} \le n & 1 \le i \le c \end{cases}$$

Among them, uij is the membership degree of the i-th sample belongs to the j-th cluster center, and m is the fuzzy index. the larger the m, m≥1, the lower the degree of fuzzy clustering. $U=[u_{ij}]_{n\times c}$ is the membership matrix.

The d_{ij} represents the distance from the i-th sample to the j-th cluster center, where the distance is characterized by similarity (section 2.2).

The Lagrangian multiplier method is used to solve the membership degree. The formula (2) is used to calculate membership, and the clustering center can be updated by the formula (3):

$$u_{ij} = \frac{1}{\sum_{k=1}^{c} \left(\frac{d_{ij}}{d_{ik}} \right)^{\frac{2}{m-1}}} \tag{2}$$

$$v_j = \frac{\sum_{k=1}^{n} \left(u_{kj}\right)^m x_k}{\sum_{k=1}^{n} \left(u_{kj}\right)^m} \tag{3}$$

In the formula (2) and (3), i = 1, ..., n, j = 1, ..., c.

3.2 Algorithm to improve the initial value of the selected

The FCM algorithm is sensitive to the selection of the initial clustering center. The choice of the initial solution is closely related to the iterative time and the iterative result. If the initial solution is improperly chosen, the iteration time is longer and it is easy to fall into the local optimum. In general, the initial clustering center, that is, the initial solution for the objective function formula (1), is randomly generated within a range of values, which greatly affects the clustering effect.

In this paper, the density peaks clustering proposed by Rodriguez is used in the selection of the initial solution. This method is based on the analysis of the characteristics of the clustering center, and characterizes the data points by local density ρ (x_i) and distance δ (x_i). In general, the clustering center of these two parameters are higher.

Define the local density as:

$$\rho(x_i) = \sum_j \chi(d_{ij} - d_c), \qquad \chi(\Delta d) = \begin{cases} 1 & \Delta d < 0 \\ 0 & \Delta d \ge 0 \end{cases} \tag{4}$$

Where $\rho(x_i)$ is the number of data points which the distance with xi is less than d_c, and d_c is the truncated distance, given by the user. The d_{ij} is the distance between the data points x_i and x_j, expressed by similarity (section 2.2).

The data point distance δ (x_i) is the distance from the nearest point of xi whose local density is greater than xi, which is defined as follows:

$$\delta(x_i) = \begin{cases} \min\limits_{j:\rho(x_j)>\rho(x_i)} (d_{ij}), & \rho(x_i) \ne \rho_{\max}; \\ \max\limits_{j}(d_{ij}), & \rho(x_i) = \rho_{\max} \end{cases} \tag{5}$$

After the local density and distance of the data points are obtained, the relationship between the two parameters is shown graphically, called the decision graph. It is convenient and quick to determine points with high density and distance with the help of decision graphs. The clustering center determined by this method is used as the initial clustering center of FCM algorithm.

4 Genetic optimization FCM clustering algorithm

Genetic algorithm is a random parallel search algorithm based on natural selection and genetics. It is an efficient method to find the global optimal solution without any initial information. The solution of the problem is treated as a population, and the quality of the solution is getting better and better through selection, crossover, mutation and other genetic operations.

The process of genetic algorithm optimization as shown below.

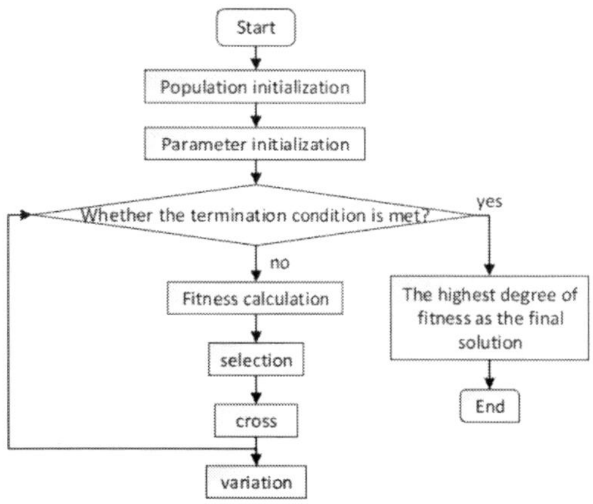

Fig. 1. Genetic algorithm flow chart

The coding, fitness function, population initialization, termination condition, crossover operator need to be determined when the clustering algorithm is optimized by genetic algorithm. The following is a detailed description of the fuzzy clustering problem which is used to identify the knowledge context.

4.1 Encoding

The solution for the fuzzy clustering problem of knowledge context is the matrix of clusters, which is expressed with V. The original data points are numbered in order to facilitate the calculation, and the representation of the data points is numbered instead of vector. If an individual is represented as (3, 6, 8, 10, 18), the meaning is that the cluster center is made up of dots numbered 3, 6, 8, 10, 18. The decimal number is converted to a binary number in actual calculations.

4.2 Population initialization

The density peak method is used to determine the initial population. The density peak method determines the clustering center with a certain subjectivity, and the same decision map may yield different results. Therefore, we can determine the m solution according to the decision graph. As the initial solution set, the number of clusters in each solution should be consistent.

4.3 Fitness calculation

The optimal result of the clustering algorithm corresponds to the minimum value of the objective function, that is, the better the clustering result, the smaller the objective function. In the genetic algorithm, the fitness of the global optimal solution should be the largest. So, the fitness function is defined as:

$$f(v_i) = \frac{1}{J_m(U,V) + \varepsilon_0} \tag{6}$$

ε_0 is a sufficiently small positive number, which avoids the denominator to zero. When the fitness calculation is carried out, the data set X and a set of clustering centers are known. Then, the degree of membership of the data set point to the cluster center is calculated according to formula (2), and then calculated $J_m(U,V)$ according to formula (1). Finally, the degree of fitness is obtained by formula (6).

4.4 Selection

The selection operation makes it possible to select the better individuals in the current population. First, select the individuals with the best fitness in the current population and do not enter the next generation directly. Then, the remaining individuals are selected according to the "proportional selection method". This method is to calculate the individual fitness, the probability that each individual is selected is:

$$P(V_i) = \frac{f(V_i)}{\sum_{j=1}^{N} f(V_j)}$$

i = 1, ..., n. the n is the population size.

4.5 Cross

Cross-operation is the main method of generating new individuals in genetic algorithm. The design of the crossover operator includes how to determine the position of the cross and how to cross it.

Common cross-cutting methods include single-point crossover, two-point crossover, multi-point crossover, uniform crossover and so on. The single-point crossover is used in this case.

The cross formula is as follows:

$$v_A^{t+1} = \alpha v_B^t + (1-\alpha) v_A^t$$

$$v_B^{t+1} = \alpha v_A^t + (1-\alpha) v_B^t$$

The v^t is the value before crossover, and the v^{t+1} is the value after crossover.

4.6 Variation

Variation can maintain group diversity, and it is an auxiliary method that generate a new generation of individual. Set a variation probability P_m, randomly generate a random number r_m, if $r_m < P_m$, for mutation. If the individual adaptation value after mutation is greater than the parent, the individual substitute parent, and become the offspring individual.

4.7 Termination conditions

In this case, the genetic termination condition is set to: the iteration is stopped when the solution of the best fitness is no longer in the solution set, or the iteration number reaches the maximum iteration times L.

5 Example

The knowledge contexts of the development of an aircraft parts are clustered and the number of extraction contexts is 63. This problem is solved by MATLAB iterations. The density peak method is used to calculate the density and distance of each data point to determine the initial solution. The decision chart is shown below:

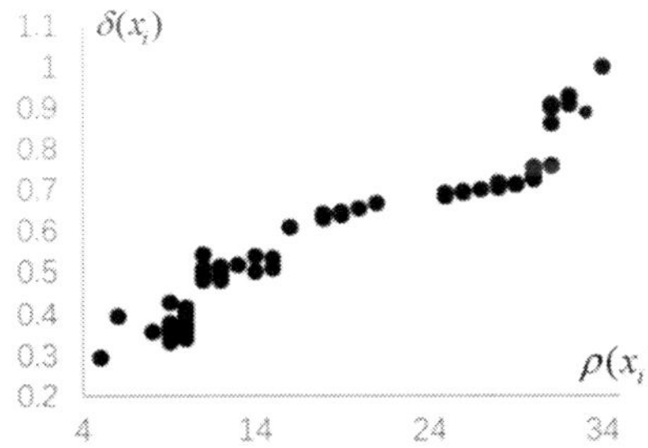

Fig. 2. "$\rho(x_i)$-$\delta(x_i)$"decision chart.

The number of population is 20. We selected 10 groups as the initial values in the upper right area of the decision graph. The other 10 groups are randomly generated to form the initial solution.

The probability of crossing is 0.6, the probability of mutation is 0.01, and the number of iterations is 100. The results of the iteration are as follows:

Table 1. Iterative results.

	value
optimal fitness	0.2932
the algebra of the optimal result	8

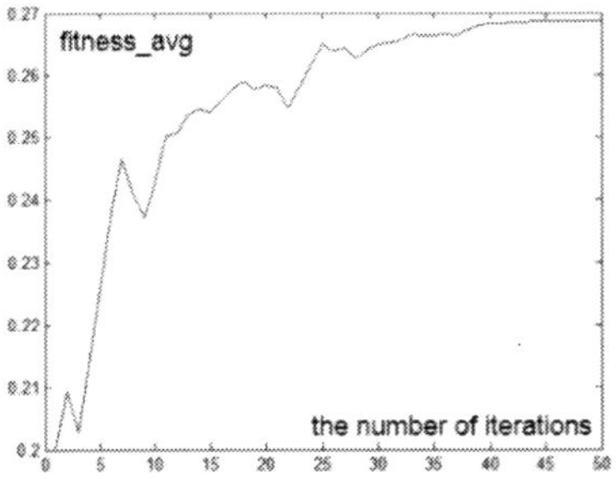

Fig. 3. The process of algorithm convergence.

6 Conclusion

The knowledge context clustering method proposed in this paper is helpful to the construction and reorganization of the knowledge base, which provides a strong support for the knowledge service. In this paper, the density peak method is used to determine the initial solution, then the genetic FCM method is used to solve the problem. In this way, the satisfactory solution can be obtained in a short time.

However, there are still many shortcomings in this paper. The weight of each dimension in knowledge context is not measured by objective data, and the number of clustering centers is determined by subjectivity. In the future, we will use the clustering results of knowledge context to solve the specific problems, and extract the knowledge application cases related to the business process to guide the business practice.

References

1. Murtagh F, Contreras P. Algorithms for hierarchical clustering: an overview[J]. Wiley Interdisciplinary Reviews Data Mining & Know,2012,2(1): 86-97.
2. Mansoori E G. GACH: a grid-based algorithm for hierarchical Clustering of high-dimensional data [J]. Soft Computing,2013, 18 (5): 905-922.
3. Kriegel H P, Kroger P, Sander J, et al. Density based Clustering [J]. Wiley Interdisciplinary Reviews Data Mining & Knowledge Discovery, 2011,1 (3): 231-240.
4. Bouveyron C, Brunet-SA C. Model-based clustering of high-dimensional data: a review [J]. Computational Statistics& Data Analysis, 2013,71 (1): 1-27.
5. Young J A, Fivelman Q L, BLAIR P L, et al. The plasmodium Falciparum sexual development transcriptome: a microarray Analysis using ontology-based pattern identification[J]. Molecular & Biochemical Parasitology, 2005, 143 (1):67-79.
6. Ren T, Zeng W, Wang N, et al. A novel approach for fMRI data analysis based on the combination of sparse approximation and affinity propagation clustering [J]. Magnetic Resonance Imaging, 2014, 32 (6): 736-746.
7. Moulick H N, Ghosh M. Image compression using k means Clustering and nuclear medicine image processing [J]. International Journal of Innovative Research in Computer & Communication Engineering, 2013, 1 (4): 869-877.
8. Grekousis G, Hatzichristos T. Fuzzy clustering analysis in geomarketing research [J]. Environment & Planning B Planning & Design, 2013, 40 (1): 95-116
9. Ruspini E H. A new approach to clustering[J], Information and control, 1969,15(1): 22-32.
10. Zadeh L. Fuzzy sets[J]. Information and Control, 1965, 8(3): 338-353.
11. Rodriguez A, Laio A. Clustering by fast search and find of density peaks [J]. Science, 2014, 344 (6191): 1492-1496.

DOI: 10.1051/matecconf/201713900065

An application of shortest path algorithm: optimal assortments for network goods with vertical differentiation

Yinbo Feng[1,*] and *Ping* Wu[1]

[1]School of Management, Fudan University, Shanghai 200433, China

Abstract. This paper presents an efficient algorithm based on the *shortest path algorithm* for a directed weighted graph where all arc cots depend on which node is thought of as the source node. This approach can be applied well to the optimal assortment problem for a network good industry with vertical differentiation. The assortment problem was introduced for decision-makers to select what products to be produced or stocked from a large product category when it is not desirable to stock all of them. A network good is a good with *network effect*—the effect that the consumption of a good by one has an impact on the value of that product to others. Vertical differentiation occurs when the several goods can be ordered according to their objective quality in the sense that a high-quality good is "better" than the low-quality one. It is shown that an optimal assortment plan can be identified in a polynomial time by our algorithms. The theoretical results also provide qualitative managerial insights into the optimal assortment plan for decision-makers.

1 Introduction

The shortest path problem, in graph theory, involves finding a path between two vertices (or nodes) in a graph such that the sum of the weights of its constituent arcs is minimized. The most important algorithms for solving this problem are summarized in [1]. Shortest path algorithms are widely applied in practices, such as finding driving directions on web mapping websites like Google Maps, finding the minimal delay path in telecommunication networks, etc.

This paper applies a shortest path algorithm to the optimal assortment problem for a network good industry with vertical differentiation. A network good is a good with *network effect*—the effect that the consumption of a good by one has an impact on the value of that product to others. Vertical differentiation occurs when the several goods can be ordered according to their objective quality. All else equal, customers prefer a product with a higher quality level to a product with a lower quality level, but they hold inherently heterogeneous valuations for product quality. When facing multiple choices, different customers would purchase a product with a different quality to maximize their own utilities. Note that, with network effect, consumers benefit from not only the product they purchased but also the user base induced by all products offered. Consider a product category with hundreds, possibly thousands, of products with network effect. A seller (or a decision maker) has to choose what products to carry from the product category to maximize his total profit, which is called the *assortment problem*. This kind of problem was first studied by [2]. Different from their work, this paper incorporates the network effect into analysis. For the ease of notation, most of the mathematical symbols are the same as in [2].

This assortment problem can be modeled as a shortest path problem in a graph. Unlike classic shortest path problem in which all arc costs are priorly given, in the graph modeled by this paper, all arc costs depend on which node is thought of as the source node. Therefore, a modified approach based on the shortest path algorithm has been developed to solve this problem and to further investigate how network effects influence the assortment decision.

2 Model formulation

Consider a single-period product market operated by a monopolist (seller). The seller has a set of feasible products with vertically differentiation, labeled by $\{1, 2, \dots, n\}$. Let q_i, p_i, and c_i respectively denotes the quality, the price, and the production cost of product i, $1 \le i \le n$. Without loss of generality, it is assumed that $0 < q_1 < q_2 < \cdots < q_n$ and $0 < c_1 < c_2 < \cdots < c_n$, that is, these products can be ordered according to their objective quality from the lowest to the highest and the production cost for a low quality product is low. The seller has to choose what products to carry from this product category.

The utility for a customer who purchases the product i is assumed to be

$$u_i(\theta, Q) = \theta q_i - p_i + s(Q),$$

where θ measures the customer's willingness to pay for one unit of quality, and Q is the size of the product i's network. The first two terms represents the *stand-alone valuation*, which is the customer's intrinsic valuation of the

*e-mail: yinbofeng13@fudan.edu.cn

product. The third term $s(\cdot)$, an concavely increasing function with $s(0) = 0$, represents the *network benefit*. Assume all products offered share the same network and hence Q is simply the seller's total unit sales. This paper adopts an additive form of the two kind of utilities. One can check that the algorithms and results in this paper can also carry over to the case with a multiplicative form. Given an assortment $S = \{j_1, \ldots, j_m\}$, a subset of $\{1, \ldots, n\}$, each customer who is interested in buying at most one product will select the product with the highest utility or select nothing with zero utility. In this paper, it is assumed that the seller cannot identify the specific θ value for each customer but knows the distribution of θ. Let $f_Q(\theta)$ and $F_Q(\theta)$ with support $[\underline{\theta}, \overline{\theta}]$ denote the probability density function and cumulative density function of θ, where $0 \leq \underline{\theta} \leq \overline{\theta} \leq \infty$. The subscript Q in $f_Q(\cdot)$ and $F_Q(\cdot)$ means they may depend on the network size Q.

Given an assortment S and customers' belief Q, the probability that a random customer purchases the product $j \in S$ can be characterized as

$$P_j(S, Q) = \mathbb{P}\{u_j(\theta, Q) \geq u_i(\theta, Q), \forall i \in S \text{ and } u_j(\theta, Q) \geq 0\}.$$

This paper adopts the Fulfilled Expectations Equilibrium [3] as our solution concept. That is, facing a set of products offered, rational consumers form an identical expectation regarding the ultimate network size, Q, based on which customers make purchase decisions. A fulfilled expectations equilibrium requires that the realized network size is the same as consumers' prior expectation. The total market size is normalized to be 1 and hence we must have

$$Q = \sum_{j \in S} P_j(S, Q), \; 0 \leq Q \leq 1. \qquad (1)$$

Let $Q^*(S)$ denote the equilibrium, which is definitely a function of S, or the largest equilibrium if it has multiple equilibria. The largest equilibria is known as a Pareto Equilibrium [4]. Thus the ultimate purchase probability for product $j \in S$ can be written as

$$P_j(S) = P_j(S, Q^*(S)).$$

The expected profit for the seller is as follows:

$$\Pi(S) = \sum_{j \in S} P_j(S)(p_j - c_j) - K \cdot |S|$$

where $|S|$ denotes the number of products in S, and $K > 0$ denotes the fixed cost for each product offered, including the costs of producing, stocking, and advertising a product. If prices are priorly given, the seller's problem is find an optimal assortment S^* to maximize

$$\Pi^* = \max_{S \subseteq \{1, \ldots, n\}} \Pi(S).$$

If prices are not priorly given, the seller's problem is to find S^* and corresponding prices $p^*(S^*)$ to maximize

$$\Pi^* = \max_{\substack{S \subseteq \{1, \ldots, n\} \\ p(S)}} \Pi(S).$$

3 Analysis and algorithms

This section developed two algorithms to efficiently solve the problems defined in Section 2. One algorithm is designed to solve the optimal assortment problem with exogenous prices which are determined by the market environment and the seller cannot alter them by himself; and the other aims at solving the optimal assortment problem with endogenous prices which can be decided by the seller.

3.1 Exogenous pricing

This subsection studies the optimal assortment problem with exogenous prices. One can enumerate all possible assortments then identify the optimal one. However, the complexity of this naive enumeration method is $O(2^n)$. Consequently, this subsection develops an efficient algorithm to obtain the optimal solution. The following lemma shows some important properties of the optimal assortment which are useful to design the algorithm. In what follows, we write Q^* for $Q^*(S^*)$, $f(\cdot)$ for $f_{Q^*(S^*)}(\cdot)$, and $F(\cdot)$ for $F_{Q^*(S^*)}(\cdot)$.

Lemma 1 *In an optimal assortment $S^* = \{j_1, \ldots, j_m\}$ such that $j_1 < j_2 < \cdots < j_m$, the following properties must hold:*

$$Q^* = \sup\{0 \leq Q \leq 1 : F_Q^{-1}(1-Q)q_{j_1} - p_{j_1} + s(Q) \geq 0\}, \quad (2)$$

$$\max\left\{\underline{\theta}, \frac{p_{j_1} - s(Q^*)}{q_{j_1}}\right\} < \frac{p_{j_2} - p_{j_1}}{q_{j_2} - q_{j_1}} < \cdots < \frac{p_{j_m} - p_{j_{m-1}}}{q_{j_m} - q_{j_{m-1}}} \leq \overline{\theta}, \quad (3)$$

$$p_{j_1} < p_{j_2} < \cdots < p_{j_m}, \quad (4)$$

$$\frac{p_{j_1} - s(Q^*)}{q_{j_1}} < \frac{p_{j_2} - s(Q^*)}{q_{j_2}} < \cdots < \frac{p_{j_m} - s(Q^*)}{q_{j_m}}, \quad (5)$$

$$p_{j_1} - c_{j_1} < p_{j_2} - c_{j_2} < \cdots < p_{j_m} - c_{j_m}, \quad (6)$$

$$P_{j_i}(S^*) = F\left(\frac{p_{j_{i+1}} - p_{j_i}}{q_{j_{i+1}} - q_{j_i}}\right) - F\left(\frac{p_{j_i} - p_{j_{i-1}}}{q_{j_i} - q_{j_{i-1}}}\right), \; i = 1, \ldots, m, \quad (7)$$

where p_{j_0}, q_{j_0}, $p_{j_{m+1}}$, and $q_{j_{m+1}}$ are respectively defined as $s(Q^*)$, 0, $+\infty$, and $q_m + 1$.

All proofs are relegated to the appendix. We can see from Lemma 1 that, in the optimal assortment, the prices, price minus network benefit–quality ratios are strictly increasing in the quality level. The condition (3) guarantees that all products in S^* can get a positive purchase probability. By (2) (3) and (7), the network effect helps to increase the total sales volume because a high-attraction product will be chosen to be the first product in the assortment in order to lower the barrier for customers to entry. Equation (6) indicates that the profit margins are increasing in the quality levels. The intuition behind is that offering a high-quality product leads to a higher profit margin than a low-quality product; if not, the seller can increase his profit by removing the high-quality product and inducing those customers who are willing to buy the high-quality product to buy the low-quality one with higher profit margin.

By Equation (7), the seller's problem with exogenous prices can be rewritten as

$$\Pi^* = \max_{\substack{j_1<\cdots<j_m\in\{1,\ldots,n\}\\ 1\le m\le n}} \sum_{i=1}^{m} \left\{ \left[F\left(\frac{p_{j_{i+1}}-p_{j_i}}{q_{j_{i+1}}-q_{j_i}}\right) - F\left(\frac{p_{j_i}-p_{j_{i-1}}}{q_{j_i}-q_{j_{i-1}}}\right)\right] \cdot (p_{j_i}-c_{j_i}) - K \right\} \text{ with constraints (3)}$$

$$= \max_{\substack{j_1<\cdots<j_m\in\{1,\ldots,n\}\\ 1\le m\le n}} \sum_{i=2}^{m} \left\{ \bar{F}\left(\frac{p_{j_i}-p_{j_{i-1}}}{q_{j_i}-q_{j_{i-1}}}\right)(p_{j_i}-c_{j_i}-p_{j_{i-1}}+c_{j_{i-1}}) - K \right\} + Q^*(p_{j_1}-c_{j_1}) - K \text{ with constraints (3)}, \quad (8)$$

where $\bar{F}(\cdot) = 1 - F(\cdot)$, and constraints (3) are equivalent to constraints (4) and (5). It is noteworthy that the contribution to seller's objective by product j_i in an optimal assortment depends on products j_1, j_{i-1}, j_{i+1} and itself. Notice that product j_1 determines $Q^*(S)$ for an assortment S, and the contribution by any other product in S depends on $Q^*(S)$. As a consequence, we are able to model this problem as a shortest path problem if the product with smallest label (the first product) in S^* is known.

We can construct a graph $G = (V, A)$, where V is the set of nodes and A is the arc set of G. Node set V consists of pairs of products (i, j) such that product j has a higher price, quality, price minus $s(Q^*)$–quality ratio, and profit margin than product i. A node (i, j) means product j could be offered if product i has been offered. To facilitate the elaboration, some fictitious nodes, $(0, k)$ and $(k, n+1)$ for $i = 1, \ldots, n+1$, are introduced. For product 0 and $n+1$, p_0, q_0, p_{n+1}, and q_{n+1} are respectively defined as $s(Q^*)$,0,$+\infty$, and $q_n + 1$. The node $(n+1, n+1)$ is the unique terminal node. Arc set A consists of those arcs from $(i, j) \in V$ to $(j, k) \in V$. We cannot directly find the optimal assortment by finding a shortest path in the graph G, because all arc costs depends which node is considered to be the source node. Fortunately, the source node has only $n+1$ possibilities, i.e., $(0, k)$, $k = 1, 2, \ldots, n+1$, where if product $n+1$ is chosen as the source node then the seller sells nothing to the market. Hence we can find a shortest path from each possible source node to the terminal node then we will obtain the optimal assortment by comparing the costs of the $n+1$ shortest paths. Algorithm 1 represents the procedure of finding an optimal assortment.

Theorem 1 *Algorithm 1 gives an optimal assortment for the exogenous pricing case. Suppose the complexity of the problem (2) is X, then the complexity of Algorithm 1 is* $O(nX + n^4)$.

3.2 Endogenous pricing

This section presents the model with endogenized prices. The seller needs to choose what products as well as their prices before customers arrive. In this subsection, $\frac{f_Q(\theta)}{\bar{F}_Q(\theta)}$ (the failure rate) is increasing in θ, for any $Q \in [0, 1]$. This assumption is satisfied by most common distributions, for example, uniform, normal, exponential, etc, and is adopted by many studies in the field of operations management [5].

Algorithm 1 Exogenous pricing case

---

step 1. Construct the node set V. $V = \{(0, k), (k, n+1), k = 1, \ldots, n+1\} \cup \{(i, j) : p_i < p_j, q_i < q_j, p_i - c_i < p_j - c_j, \frac{p_i - s(Q^*)}{q_i} < \frac{p_j - s(Q^*)}{q_j}\}$.

step 2. Construct the arc set A. $A = \{$arc from $(k, n+1)$ to $(n+1, n+1), k = 0, 1, \ldots, n\} \cup \{$arc from (i, j) to $(j, k) : 0 \le i < j < k \le n+1\}$.

step 3. For $h = 1, 2, \ldots, n$, consider $(0, h)$ as the source node and compute the corresponding network size

$$Q^*(h) = \sup\{0 \le Q \le 1 : F_Q^{-1}(1-Q)q_h - p_h + s(Q) \ge 0\}.$$

Then define arc costs:

$$C^h_{(i,j),(j,k)} = \begin{cases} K - \bar{F}\left(\frac{p_k-p_j}{q_k-q_j}\right)(p_k - c_k - p_j + c_j), \\ \quad \text{if } i = 0, j = h, j < k \le n \\ \quad \text{or } h \le i < j < k \le n, \\ \quad \text{and } \frac{p_k-p_j}{q_k-q_j} > \frac{p_j-p_i}{q_j-q_i}, \\ 0, \quad \text{if } k = n+1, \\ +\infty, \quad \text{otherwise.} \end{cases}$$

Then solve the shortest path problem from $(0, h)$ to $(n+1, n+1)$. Denote the cost of a shortest path by v^h.

step 4. Minimize the following problem

$$\min_{1\le h\le n} \{v^h + K - Q^*(h)(p_h - c_h)\}.$$

If the minimum is negative, then h^* minimizing the problem is the product with smallest label in the assortment. Other products in the assortment can be found according to nodes on the shortest path from $(0, h^*)$ to $(n+1, n+1)$. If the minimal value of the above problem is nonnegative then the assortment set is empty.

Suppose $S = \{j_1, \ldots, j_m\}$ is a assortment with endogenized prices $\{p_{j_1}, \ldots, p_{j_m}\}$. Define $\theta_{j_1} = \frac{p_{j_1} - s(Q^*)}{q_{j_1}}$ and $\theta_{j_i} = \frac{p_{j_i} - p_{j_{i-1}}}{q_{j_i} - q_{j_{i-1}}}$ for $i = 2, \ldots, m$, where Q^* is the network size for this assortment. A customer with valuation θ_{j_i} gets the same utility from buying product j_i and product j_{i-1}, $i = 1, \ldots, m$, where the product j_0 can be thought of as the no purchase option. Since $\{p_{j_1}, \ldots, p_{j_m}\}$ and $\{\theta_{j_1}, \ldots, \theta_{j_m}\}$ can be uniquely determined by each other, in what follows, $\{\theta_{j_1}, \ldots, \theta_{j_m}\}$ can be thought of as the seller's decision variables, given an assortment S. The following lemma presents some properties for the endogenous pricing case when an optimal assortment is given.

Lemma 2 *Given an assortment* $S^* = \{j_1, \ldots, j_m\}$, *the optimal prices for products in* S^* *must satisfy the following properties:*

$$Q^* = \operatorname*{argmax}_{0\le Q\le 1} Q(F_Q^{-1}(1-Q)q_{j_1} - c_{j_1} + s(Q)), \quad (9)$$

$$\theta^*_{j_1} = 1 - Q^*, \tag{10}$$

$$\theta^*_{j_i} = \frac{\bar{F}(\theta^*_{j_i})}{f(\theta^*_{j_i})} + \frac{c_{j_i} - c_{j_{i-1}}}{q_{j_i} - q_{j_{i-1}}},\ i = 2, \ldots, m, \tag{11}$$

$$\theta^*_{j_1} < \theta^*_{j_2} < \cdots < \theta^*_{j_m} \leq \bar{\theta}.$$

In Lemma 2, (9) and (10) illustrate the role of the network effect in pricing the products when an assortment is given. With the network effect, the first product in an optimal assortment is tends to be with a low price to attract more customers with low valuations to join the network and increases the network value of the high-end product, allowing the seller to extract greater surplus. Analogous to Lemma 1, the following lemma provides some properties of the optimal assortment, which are important to develop the algorithm for endogenous pricing case.

Lemma 3 *In an optimal assortment $S^* = \{j_1, \ldots, j_m\}$ such that $j_1 < j_2 < \cdots < j_m$, Q^* can de characterized by (9) and the following conditions must hold:*

$$\frac{c_{j_1} - s(Q^*) - s'(Q^*)Q^*}{q_{j_1}} < \frac{c_{j_2} - c_{j_1}}{q_{j_2} - q_{j_1}} < \cdots < \frac{c_{j_m} - c_{j_{m-1}}}{q_{j_m} - q_{j_{m-1}}}, \tag{12}$$

$$c_{j_1} < c_{j_1} < \cdots < c_{j_m} \tag{13}$$

$$\frac{c_{j_1} - s(Q^*) - s'(Q^*)Q^*}{q_{j_1}} < \cdots < \frac{p_{j_m} - s(Q^*) - s'(Q^*)Q^*}{q_{j_m}}. \tag{14}$$

Lemma 3 shows that, in the optimal assortment, the production costs, production costs minus $s(Q^*) + s'(Q^*)Q^*$–quality ratios are strictly increasing in the quality level. The condition (12) guarantees that all products in S^* can get a positive purchase probability. Given an optimal assortment S^* and corresponding prices, we can find that the prices and profit margins for products in S^* must be increasing in the quality due to Lemma 1. By the definition of θ_{j_i}s, the seller's problem with endogenous pricing can be rewritten as

$$\Pi^* = \max_{\substack{j_1 < \cdots < j_m \in \{1, \ldots, n\} \\ 1 \leq m \leq n}} \sum_{i=2}^{m} \Big\{ \bar{F}(\theta_{j_i})[\theta_{j_i}(q_{j_i} - q_{j_{i-1}}) - (c_{j_i} - c_{j_{i-1}})] - K \Big\} + Q^*(\theta_{j_1} q_{j_1} - c_{j_1} + s(Q^*)) - K \text{ with constraints (12).} \tag{15}$$

As analogous as in Section 3.1, we are able to model this problem as a shortest path problem if the product with smallest label (the first product) in S^* is determined. The construction of the graph $G = (V, A)$ is similar with that in Section 3.1. Algorithm 2 represents the procedure of finding an optimal assortment. In this algorithm, we set $c_0 = s(Q^*) + s'Q^*$ and $q_0 = 0$, where Q^* can be computed by (9) if the first product in an assortment is known.

Theorem 2 *Algorithm 2 gives an optimal assortment for the endogenous pricing case. Suppose the complexity of the problem (9) is X, then the complexity of Algorithm 2 is $O(nX + n^4)$.*

Algorithm 2 Endogenous pricing case

---

step 1. Construct the node set V. $V = \{(0, k), (k, n+1), k = 1, \ldots, n+1\} \cup \{(i, j) : c_i < c_j, \frac{c_i - s(Q^*) - s(Q^*)Q^*}{q_i} < \frac{c_j - s(Q^*) - s(Q^*)Q^*}{q_j}\}$.

step 2. Construct the arc set A. $A = \{$arc from $(k, n+1)$ to $(n+1, n+1), k = 0, 1, \ldots, n\} \cup \{$arc from (i, j) to $(j, k) : 0 \leq i < j < k \leq n+1\}$.

step 3. For $h = 1, 2, \ldots, n$, consider $(0, h)$ as the source node and compute the corresponding network size

$$Q^*(h) = \operatorname*{argmax}_{0 \leq Q \leq 1} Q(F_Q^{-1}(1 - Q)q_h - c_h + s(Q)).$$

Then define arc costs:

$$C^h_{(i,j),(j,k)} = \begin{cases} K - \bar{F}(\theta_k)[\theta_k(q_k - q_j) - (c_k - c_j)], \\ \quad \text{if } i = 0, j = h, j < k \leq n \\ \quad \text{or } h \leq i < j < k \leq n, \\ \quad \text{and } \frac{c_k - c_j}{q_k - q_j} > \frac{c_j - c_i}{q_j - q_i}, \\ 0, \quad \text{if } k = n + 1, \\ +\infty, \quad \text{otherwise,} \end{cases}$$

where θ_k is defined by Equation (11). Then solve the shortest path problem from $(0, h)$ to $(n+1, n+1)$. Denote the cost of a shortest path by v^h.

step 4. Minimize the following problem

$$\min_{1 \leq h \leq n} \{v^h + K - Q^*(h)(\theta_h q_h - c_h + s(Q^*))\}.$$

If the value is negative, then h^* minimizing the problem is the product with smallest label in the assortment. Other products in the assortment can be found according to nodes on the shortest path from $(0, h^*)$ to $(n+1, n+1)$. If the value of the above problem is nonnegative then the assortment set is empty.

---

4 Conclusion

This paper presents a modified shortest path algorithm for a directed weighted graph where all arc cots depend on which node is thought of as the source node, and applies this approach to an optimal assortment problem for a network good industry with vertical differentiation.

In our model, we examine two cases: (i) the exogenous pricing case in which the prices for all products are determined by the market environment and the seller cannot alter them by himself; and (ii) the endogenous pricing case in which the seller need not only to choose what products to offer from a given product category but also to make price decisions for the products offered. For each case, we develop a efficient algorithm based on the shortest path method to obtain an optimal assortment. It is shown that each algorithm can get an optimal solution in a polynomial time.

The theoretical results, e.g., Lemma 1 and Lemma 3, show (i) that, in an optimal assortment with exogenous prices, the product prices, profit margins, and price minus network benefit-quality ratios are increasing in the quality; and (ii) that in an optimal assortment with endogenous prices, the prices, production costs, production cost minus network benefit-quality ratios, and profit margins are increasing in the quality; and (iii) that, with network effect, the seller always has incentive to choose a high attractable product as the low-end product in order to expand the market coverage then to increase the value of the high-end product for customers. These observations provide the qualitative insights into the optimal assortment plan for decision-makers.

Appendix. Proofs

Proof of Lemma 1

Q^* is the largest fixed point of Equation (1), given S^*. To solve this largest fixed point is equivalent to solve the problem (2).

Since $K > 0$, every product in an optimal assortment S^* must have a positive purchase probability because, if not, removing the product with zero demand does not affect the total demand but decreases the total fixed cost. From the definition of $P_{j_i}(S)$, we know that a customer with valuation $\frac{p_{j_{i+1}}-p_{j_i}}{q_{j_{i+1}}-q_{j_i}}$ gets the same utility from buying j_i and j_{i+1}, for $i = 1, \ldots, n-1$; and a customer with valuation $\frac{p_{j_1}-s(Q^*)}{q_{j_1}}$ is indifferent from buying 1 or nothing. Hence, to guarantee $P_{j_i} > 0$ for any $j_i \in S$, Equation (3) must hold. Equations (4) and (5) and (7) follow directly from Equation (3). Finally, Equation (6) can be seen from the seller's problem (15).

Proof of Theorem 1

According to Lemma 1 and the construction of Algorithm 1, we can see that any path with nonpositive cost at Step 3 uniquely corresponds to an assortment satisfying conditions in Lemma 1 and vice versa. Hence, the assortment given by Algorithm 1 must be one of the optimal assortments by enumerating all possibilities satisfying conditions in Lemma 1. That is, Algorithm 1 gives an optimal assortment.

By Corollary 1 in [2], the complexity of the shortest path problem at Step 3 in algorithm 1 is $O(n^3)$ if the source node is given. This kind of shortest path problem would be run n times. Therefore, the complexity of this algorithm is $O(nX + n^4)$.

Proof of Lemma 2

The seller's problem can written as

$$\Pi^* = \max_{j_1,\ldots,j_m} \sum_{i=2}^{m} \Big\{\bar{F}(\theta_{j_i})[\theta_{j_i}(q_{j_i} - q_{j_{i-1}}) - (c_{j_i} - c_{j_{i-1}})]$$

$$- K\Big\} + Q(\theta_{j_1} q_{j_1} - c_{j_1} + s(Q)) - K \text{ with constraints (12).}$$

We have known that, when an assortment is given, choosing optimal prices for these products is equivalent to choosing $\{\theta_{j_1}, \ldots, \theta_{j_m}\}$ to maximize the seller's profit. Since $Q = \bar{F}(\theta_{j_1})$ in equilibrium, taking derivatives of the objective function with respect to $Q, \theta_{j_2}, \ldots, \theta_{j_m}$, we obtain Equations (9), (10) and ((11)).

The condition $\theta^*_{j_1} < \theta^*_{j_2} < \cdots < \theta^*_{j_m} \le \bar{\theta}$ is necessary to guarantee each product in the optimal assortment can have a positive purchase probability. Therefore, this lemma is proved.

Proof of Lemma 3

Given $S^* = \{j_1, \ldots, j_m\}$, Equation (11) gives the definition of θ_{j_i} for $i = 2, \ldots, m$. For $i = 1$, taking the derivative of the objective function with respect to θ_{j_1}, we have that

$$\theta_{j_1} = \frac{\bar{F}(\theta_{j_1})}{f(\theta_{j_1})} + \frac{c_{j_1} - s(Q) - s'(Q)}{q_{j_1}},$$

where $Q = \bar{F}(\theta_{j_1})$. Then by the last two conditions in Lemma 2 we can know that Equation (12) holds to guarantee each product in the optimal assortment can have a positive purchase probability. Equations (13) and (14) directly follow Equation (12).

Proof of Theorem 2

The proof is analogous to the proof of Theorem 1.

References

[1] Cherkassky, B. V., Goldberg, A. V., Radzik, T. Shortest paths algorithms: theory and experimental evaluation (eng). Mathematical Programming, **73**(2): 129–174 (1996).

[2] Pan, X. A., Honhon, D. Assortment planning for vertically differentiated products. Production & Operations Management, **21**(2): 253–275 (2012).

[3] Jing B. Network externalities and market segmentation in a monopoly. Economics Letters, **95**(1):7-13 (2007).

[4] Cooper R. Equilibrium Selection in Imperfectly Competitive Economies with Multiple Equilibria. Economic Journal, **104**(426):1106-1122 (1994).

[5] Lariviere M A, Porteus E L. Selling to the Newsvendor: An Analysis of Price-Only Contracts. Manufacturing & Service Operations Management, **3**(4):293-305 (2001).

Fabrication of depletion type micro-ring modulator with high extinction ratio and high coupling quality factor

Yun Shen[1], *Xiaodong Wang*[1], *Wei Zhang*[1], *Ciyuan Qiu*[2], *Xiulan Cheng*[1,*]

[1]Center for Advanced Electronic Material and Devices, Shanghai Jiao Tong University, Shanghai 200240, China
[2]State Key Lab of Advanced Optical Communication Systems and Networks, Department of Electronic Engineering, Shanghai Jiao Tong University, Shanghai 200240, China

Abstract. We experimentally demonstrate an optical micro-ring modulator based on a reverse-biased PN junction which is integrated into a silicon-on-insulator (SOI) rib waveguide. The performance of micro-ring resonator with different sizes is simulated with Lumerical MODE Solutions. The fabricated device provides a high extinction ratio (ER) of 20 dB and coupling quality factor (Q) of $2*10^4$. Modulation rate is higher than 11 Gbit/s with only 2 V peak-to-peak (Vpp) driving voltage. The theoretical modulation rate can be higher than 25 Gbit/s by the optimizing the matching resistor and the test system.

1 Introduction

High speed electro-optic modulator is a key component in photonics systems for datacom applications. Various mechanisms for realizing high-speed modulation in silicon material have been investigated throughout the years. The main method is altering the refractive index of Silicon, which is based on thermo-optic effect or plasma dispersion effect [4]. The thermo-optic effect is too slow to realize modulation frequencies higher than 1-MHz [8]. Thus, the plasma dispersion effect is considered as the most effective way of refraction index modulation in silicon [4]. Few structures of electro-optic modulator have been reported, such as P-I-N diode in injection type [1-3, 5-6], PN diode in depletion type [7-12, 15-17, 21], MOS capacitor [13] and so on. The mechanism of carrier-injection-based P-I-N modulator is to change the refraction index of silicon material with the free-carrier dispersion. Its modulation rate is limited due to the slower diffusion rate of minority carriers [2]. However, the carrier-depletion silicon ring modulators employing reverse-biased PN diode have been demonstrated to have desirable outstanding performances including high speed higher than 10 Gb/s, high extinction ratio (ER) more than 5 dB, low peak-to-peak driving voltage (V_{pp}) less than 2 V, and insensitive to carrier lifetime [9], etc.

Most of silicon-based modulators are incorporated into Mach-Zehnder Interferometer (MZI) structures in order to obtain a wider optical bandwidth, high thermal stability, lower chirp, more reasonable tolerance to fabrication variations and no photon cavity lifetime limitation [12]. Compared to MZI structure, the micro-ring resonator structure has some obvious advantages like small footprint, flexible structure, low power consumption, simple fabrication and easier integration.

The modulation rates of depletion type micro-ring modulators have been demonstrated up to 40 Gbit/s [15, 16]. However, most of the domestic researches on this field are no experiment demonstration due to the limited fabrication condition. In this work, relying on the advanced processing condition of Center for Advanced Electronic Material and Devices (AEMD) in Shanghai Jiao Tong University, we designed and fabricated a depletion type micro-ring modulator integrated with high-index-contrast SOI rib waveguides. The ring resonator exhibits a high ER more than 20 dB and a high quality factor (Q) more than $2*10^4$. Modulation rate is higher than 11 Gbit/s with $2V_{pp}$ from -1.6 V to 0.4 V.

2 Device modelling

2.1. Structure design

The schematic of the electro-optic modulator is shown Figure 1. The main components in our device include a ring resonator coupled to a single rib waveguide. The input light is injected from a tunable laser into a surface TE mode supported grating coupler which was etched into the waveguide. At the other end of the device, the output light is collected by another grating coupler. The transmission of the waveguide is highly sensitive to the signal wavelength and is greatly reduced at wavelengths in which the ring circumference corresponds to an integer number of guided wavelengths [1]. The resonance wavelength is modified by tuning the effective refractive index of the ring waveguide, inducing a strong modulation of transmitted signal. The effective refractive index of ring is altered by applying different reverse-bias voltages to PN diode, which changes the width of depletion region of PN junction.

* Corresponding author: : xlcheng@sjtu.edu.cn

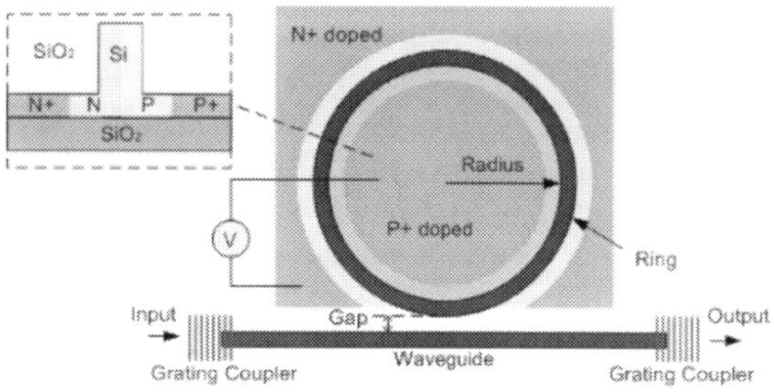

Fig. 1. Schematic layout of the micro-ring modulator.

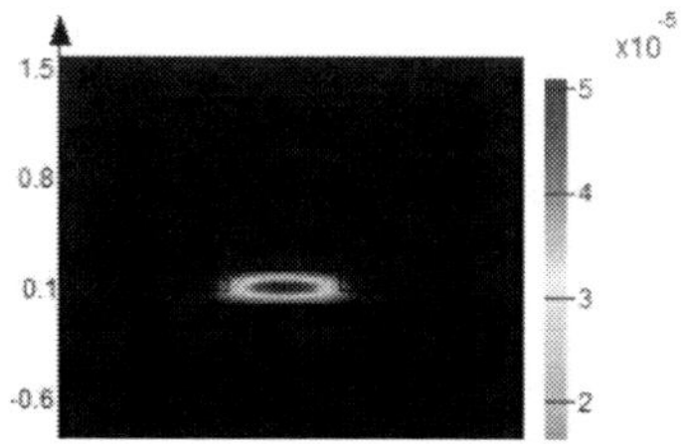

Fig. 2. Simulation about light field distribution(TE0 mode) of rib waveguide.

As shown in Figure 1, Ring resonator has two key parameters, ring radius as well as the gap between ring and straight waveguide. To obtain the best performance of ring resonator, we have simulated several ring resonators with different parameters by Lumerical Mode Solutions. It turned out that ring resonator with 10 μm radius and 100 nm gap performed kind of best around 1.55 μm wavelength. However, in actual process, the layout should cover different parameters to ensure that we can get the best one among all the final devices.

The waveguide structure can affect the overall performance of the device. The device uses a rib waveguide with 220 nm height, 500 nm width and 90 nm slab thickness based on SOI substrate, in which only quasi-TE mode was supported [19]. Figure 2 is the simulation about light distribution of ridge waveguide, most of which is TE0 mode (quasi-TE mode). A rib structure leads to vertical confinement of the optical mode as a result of the high refractive index difference between the air/silicon/SiO_2 and horizontal confinement results from the etched rib [18].

2.2 Process simulation

To achieve strong efficiency modulation, the target ion concentration of PN junction was 1E18 cm^{-3}. The P region was doped twice to provide a uniform doping, 5E12 cm^{-2} of boron with doping energy of 16 KeV and 40 KeV respectively. Similarly, the N region was doped twice (3E12 cm^{-2} of phosphorus with doping energy of 45 KeV and 110 KeV respectively). The ohmic contacts were formed by highly doped P+/N+ regions (4E15 cm^{-2}) of boron and phosphorus, suited in the slab on both sides of the waveguide, 0.8 μm away from the center of the waveguide. SILVACO has been employed to optimize the overlap between the injected free carriers [18]. Figure 3 is the concentration distribution of ridge waveguide cross section. The color depth indicates the level of ion concentration. It shows that the ion concentration of P+ and N+ region have reached the order of magnitude of 1E20 cm^{-3}.

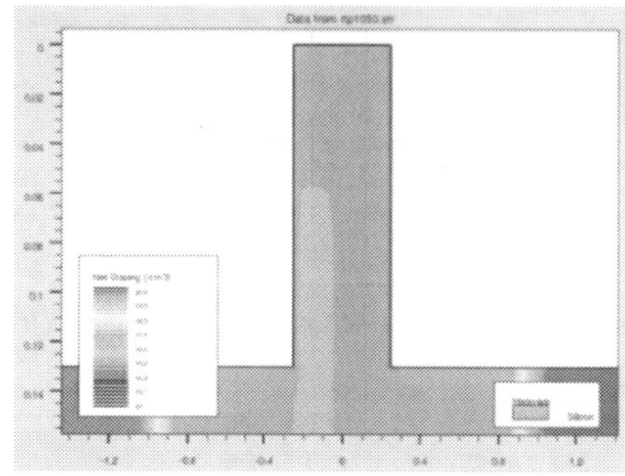

Fig. 3. Concentration distribution in ridge waveguide.

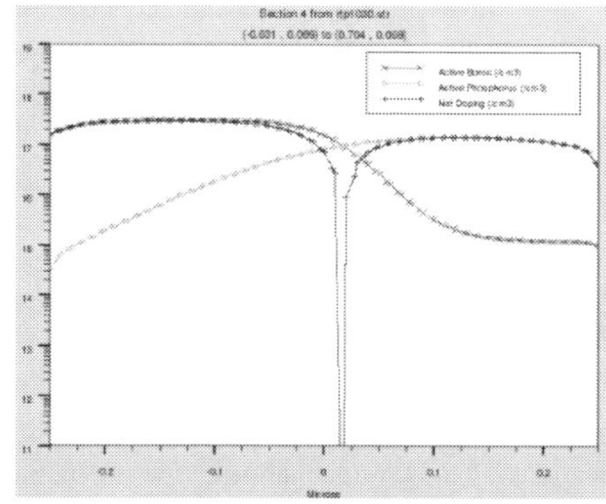

Fig. 4. Concentration curve in waveguide cross section.

Figure 4 shows the concentration curve in waveguide cross section. We can clearly see the depletion region in the middle of waveguide and the concentration distribution is uniform. The depletion region coincides with the main area of the optical mode field, which meets our design requirement.

3 Device fabrication

The fabrication process of electro-optical modulator is demonstrated in Figure 5. The process started with a SOI wafer, containing a 3 μm thick buried oxide layer and a 220 nm thick top silicon layer. Firstly, define the alignment markings by deep-UV lithography process, which was then etched into the silicon layer by Inductively Coupled Plasma (ICP) [20]. Except for the alignment marking process, all the other resist windows were defined by Electron-beam lithography system (EBL). Then define the grating coupler by EBL and etch 70 nm thickness top silicon. Next, etch grooves along the outside of the device to define the straight waveguide and micro-ring resonator region. The grooves are 3 μm width and 220 nm depth (top silicon all etched). Following the etching step, define the slab region by EBL and etch 150 nm thickness top silicon. To this step, the rib waveguide has been fabricated, as shown in Figure 5(b).

Then define the P+PNN+ doping region by four times of EBL and ion implantation with different implantation doses and energies. Firstly, define the resist window of N+ doping region outside the ring and implant phosphorus (4E15 cm^{-2}, 30 KeV) into the N+ slab region. The P+ doping region (boron, 4E15 cm^2, 4 KeV) was formed in the same way. To minimise interaction with the optical mode and absorption losses, the gap between the waveguide and the heavy doping region was set to be 0.8 μm. Then in the same way the other two light doping regions were doped symmetrically to form the PN junction. During doping process, the thickness of photoresist should be at least 2 μm. Following the implantation process, rapid thermal annealing (RTP) at 1030 °C for 8 seconds is needed to activate the doping regions, repair lattice damage and form a uniform P/N type background doping. RTP is better than other annealing processes since it results in a high level of electrical activation whilst minimising dopant diffusion [2]. As illustrated in Figure 5(c), the main structure of ring resonator was completed.

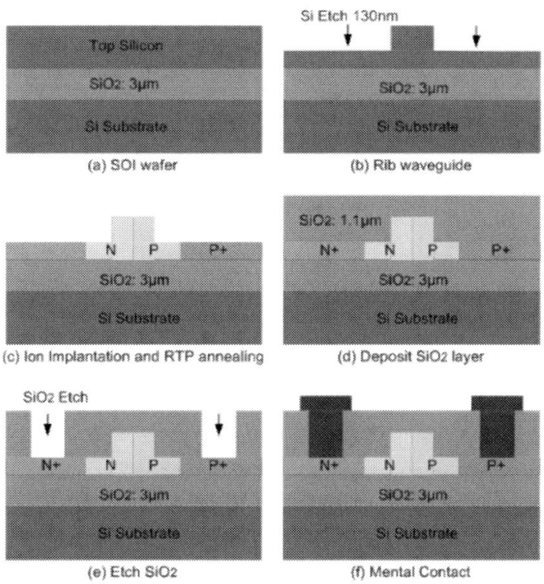

Fig. 5. Fabrication process of modulator.

The final process is to fabricate the contact electrodes. Its position is critical to the performance of modulator since the frequency response is limited by the RC cut off frequency resulting from capacitive effects within the junction and the resistance of the doped regions and metal contacts [21]. Then the sample was covered by a 1.1 μm thick SiO_2 layer using plasma-enhanced chemical vapor deposition (PECVD), see Figure 5(d). As shown in Figure 5(e), contact windows were opened in the SiO_2 layer above the P+ and N+ doping region by dry etching for 1.0 μm and wet etching (HF solution) for 0.1 μm. Wet etching has advantages such as slower etching rate, better selectivity, which can effectively avoid over etching. The electrodes were formed by lift-off process. The photoresist we used was PMMA (A9), which was spin-coated with a slow rotate speed to provide a thick photoresist layer, enabling the lift-off process easier. After electrode areas defined by EBL, dip it in HF solution for a few seconds to remove the native oxide layer, then a 40 nm thick titanium layer and a 100 nm thick TiN film were sputtered by Multi-target Magnetic Control Sputtering System. In the same way 600 nm thickness of aluminum was sputtered on the TiN film. Dip the sample in acetone solution to remove the photoresist along with the metal layers above. Because the photoresist on electrode areas was developed, the electrodes were not removed while dipping. Finally, these multilayer metallizations were annealed in nitrogen at 450 ℃ for 30 minutes, enabling the thin Ti layer to transfer to $TiSi_2$ at the Si/Ti interface, resulting in a lower electric resistance. As for the TiN film, it is used as a diffusion barrier in case of interdiffusion of silicon and aluminum during annealing [22]. Figure 5(f) shows the fabricated device cross section.

4 Experimental results

A top-view microscope image of the fabricated device is shown in Figure 6. It turned out that a micro-ring resonator with 10 μm ring radius and 300 nm gap space performed best among all the devices. The tunable laser injected light with wavelength from 1530 to 1570 nm into the waveguide. As demonstrated in Figure 7, this resonator exhibits a free spectra range (FSR) of 9.44 nm, 3 dB bandwidth of 0.074 nm, ER higher than 20 dB and high Q better than $2*10^4$.

Fig. 6. Microscope top-view image of the fabricated modulator.

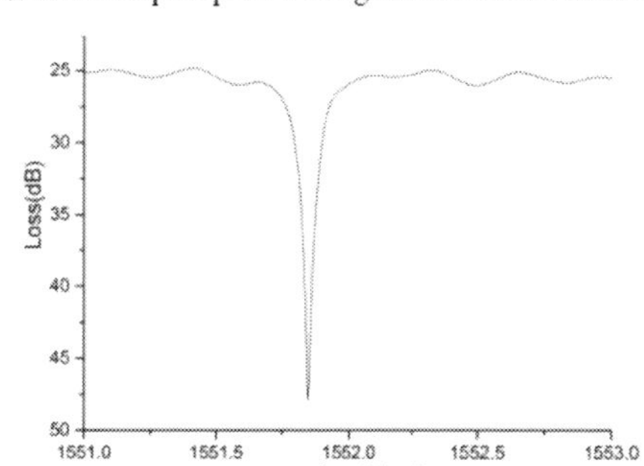

Fig. 7. Transmission spectra of the ring resonator.

The first step was to test the static characteristics of the modulator. Figure 8 shows the measured resonance spectra shift at different DC bias voltages, from 0.7 V to

-3 V. With different reverse-bias voltages applied to the PN diode, the effective index of rib waveguide was changed due to the drift motion of majority carriers, which made the resonance wavelength shift. Assuming voltage swing from 0 V to -1 V (1V_{pp}), we can see in Figure 8 that the resonance shift is about 10 pm. The ring modulator demonstrates extinction ratio of 19.5 dB and quality factor of 11820 at -2 V reverse bias voltage. The operation wavelength can be set at around 1549.435 nm.

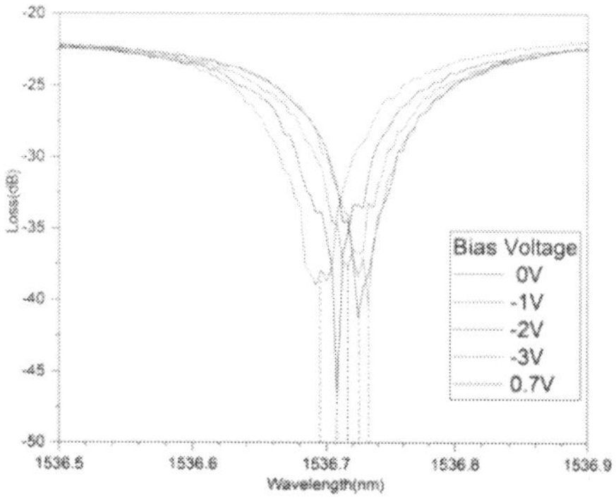

Fig. 8. Transmission spectra with different voltages applied to the PN diode.

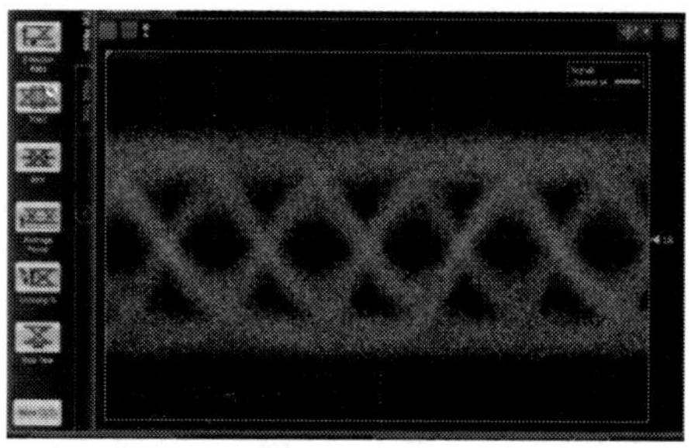

Fig. 9. Eye diagrams obtained by the modulator using 11 GHz OOK signal with 2 V_{pp}.

Then we tested the dynamic characteristics of the modulator. 2V_{pp} externally driven voltage (from -1.6 V to 0.4 V) was applied to the electrodes via the high-speed prober from signal generator. The light was modulated and transferred into electrical signal by photodiode. The electrical signal had the same frequency as driven voltage and its waveform was displayed on the oscilloscope. In the meantime, the eye diagram was obtained. As shown in Figure 9, an open eye can be observed at 11 Gbit/s with 2V_{pp} swinging from -1.6 V to 0.4 V. From the eye diagram, the width of eyeline is slightly wide. It may be caused by the noises since the Erbium Doped Fiber Amplifier (EDFA) was not added into the test system.

4 Conclusions

In this work a micro-ring modulator based on the depletion of a PN junction integrated into a SOI rib waveguide at 1.55 μm operation wavelength was demonstrated. The detailed fabrication process was illustrated and key procedures were simulated in SIVALCO TCAD. The device exhibits an extinction ratio of more than 15 dB, quality factor more than 10^4 and modulation rate is no less than 11 Gbit/s. Limited by the condition of test system, the modulation rate has not reached its limitation. This is the first attempt to fabricate the silicon micro-ring electro-optic modulator based on a domestic platform. By the optimization of matching resistor, device structure and process parameters, the modulation rate can be higher than 25 Gbit/s.

References

1. Xu Q. F., Schmidt B, Pradhan S, et al. Nature, **435**(2005): 325-327
2. Xu Q. F., Manipatruni S., Schmidt B., et al. Opt. Express, **15**(2007): 430-436
3. Manipatruni S, Xu Q. F., Schmidt B., et al. Opt. Express, (2007): 537-538
4. Soref R., Bennett B. Quantum Electron, **23**(1987): 123-129
5. F. Gan, F. X. Kartner. *Conference on Lasers and ElectroOptics,* (2005).
6. L. Zhou, A. W. Poon. Opt. Express, **14** (2006) :6851-6857.
7. A. Liu, L. Liao, D. Rubin, H. Nguyen, B. Ciftcioglu, Y. Chetrit, et al. Opt. Express, **15**(2007) : 660-668.
8. F. Y. Gardes, G. T. Reed, N.G. Emerson. Opt. Express, **13**(2005) : 8845-8854.
9. Guoliang Li, Xuezhe Zheng, Jin Y, Hiren Thacker, et al. Opt. Express, **19**(2011) :20435-20443.
10. D. J. Thomson, F. Y. Gardes, et al. Opt. Express, **20**(2012) : 12014-12020.
11. D. J. Thomson, F. Y. Gardes, et al. Opt. Express, **19**(2011) : 11804-11814.
12. D. J. Thomson, H. Porte, B. Goll, et al. Laser & Photonics, **8**(2014) : 180-187.
13. Liu A S, et al. Nature, **427**(2004) : 615-618.
14. G. T.Reed, G. Z. Machanovich, F Y. Gardes, et al. Nanophotonics, **3**(2014).
15. Guoliang Li, Xuezhe Zheng, et al . *2012 IEEE.*
16. D. Marris-Morini et al. Opt. Express. **21**(2013) :22471-22475.
17. M. Pantouvaki, P. Verheyen, J. De Coster, G. Lepage, et al. *Optical Communication*, (2015).
18. P. D. Hewitt, G. T. Reed. Lightwave Technology, **18**(2000) :443-450.
19. Qiu C. Y., Ye X., Soref R., Yang L., et al. Opt. Letter, **37**(2012) : 3942-3944.
20. Qiu C. Y., Gao W. L., R. Vajtai, et al. NanoLett, **14**(2014) : 6811-6815.
21. F.Y. Gardes1, A. Brimont, P. Sanchis2, G. Rasigade, et al. Opt. Express, **17**(2009) :21986-21991.

Dynamic Modeling of a Cubical Robot Balancing on Its Corner

Zhigang Chen[1,2,*], *Xiaogang Ruan*[1,2] *and Yuan Li*[1,2]

[1] Faculty of Information Technology, Beijing University of Technology, Beijing 100124, China
[2] Beijing Key Laboratory of Computational Intelligence and Intelligent System, Beijing, 100124, China

Abstract. Aiming at the problem of self-balancing control of a cubical robot, this paper makes a research on the dynamic modeling of the cubical robot balancing on its corner. Using the prototype of cubical robot we built as the research object, the dynamic model is derived with Lagrangian method on foundation of analysis of coordinate transformation relation of cubical robot. The correctness of the model is verified in theory by numerical simulation. The controller, designed based on reaction torque characteristic of inertia wheels, is used in balance control of the cubical robot. The effectiveness of the controller is verified again with the obtained expect effect. The dynamic model developed and controller can provide a base for further study of balance control of a cubical robot.

1 INTRODUCTION

The cubical robot is a robot device that shaped like a cube and driven by internal moment generator. The cubical robot could balancing on its edge or corner, and is a typical nonlinear, unstable and multi freedom spacial inverted pendulum, which can be used as an ideal verification platform for controlling theoretical verification [1-4]. The cubical robot can jump up by the inner torque producer device. And multi cubical robot can be used for space exploration and other fields as group robot could realize self assembly.

Scholars pay much attention to the cubical robot for its novel features. In [5] researched on the cubical robot system that has an internal active pendulum installed, and a balancing controller was designed by means of fuzzy control method in 2006. Based on this system, a nonlinear reduced order state observer was designed in [6]. In 2012, a cubical robot that actuated by reaction wheels mounted on three faces of the cube was proposed by Gajamohan et al7-8. In [7], [8] the dynamic model of the cubical robot both balancing on its edge and balancing on its corner was derived respectively, and the parameters of nonlinear systems were identified using frequency domain based approach while the cubical robot balancing on its edge with a nominal controller, the corner balancing using a linear feedback controller was presented. In [9], [10], dynamic model of the cubical robot balancing on its edge and balancing on its corner was derived respectively using concept of generalized momenta, and a backstepping based controller was derived for balancing.

An accurate dynamic model is the important foundation for balancing control of the cubical robot. In the above work, Gajamohan, Muehlebach, et al. derived the dynamic model of the cubical robot balancing on the corner, but the specific dynamic model hasn't been given[7-10].

Aiming at the problem of self-balancing control of a cubical robot balancing on its corner. Using the prototype of the cubical robot we built as the research object, In first place, this paper deduce the dynamic model of the cubical robot with Lagrange method, and give the specific dynamic equation. Then, the correctness of the model is verified by numerical simulation. A balancing controller is proposed based on reaction torque characteristics of reaction wheels, and is used in balancing control of cubical robot and obtained expect effect. At last, the research results are given.

2 CUBICAL ROBOT

The cubical robot designed in this paper as show in Fig.1., that composed by a cube and reaction wheels mountain on three faces of the cube. Reaction torques are applied to make the cubical robot balancing on its corner by control the speed of reaction wheels.

Figure 1. Cubical robot prototype.

From the modeling point of view, the cubical robot consists of four rigid bodies with cube and reaction wheels on three faces of cube as show in Fig.2., where o-xyz denotes the Newtonian frame fixed in space, o-$x_b y_b z_b$ the body frame whose origin is located at corner

* Corresponding author: chenzgbs@126.com

of the cube, c-$x_f y_f z_f$ the reaction wheels coordinate frame attaches at the center of the cube, and whose axes coincide with the rotation axis of reaction wheels. Next, let m and l describe the mass and the distance from mass center point c to corner point o of the cubical robot, I_x , I_y and I_z denote the system's total moment of inertia around the corner point in the body fixed coordinate frame respectively, I_w denotes the reaction wheel moment of inertia around the rotation axis in reaction wheels coordinate frame. Let α , β and γ denote the deflection angle from body coordinate frame to space coordinate frame in axes respectively, $\dot{\theta}_A$, $\dot{\theta}_B$ and $\dot{\theta}_C$ denote the angle velocities of reaction wheels respectively.

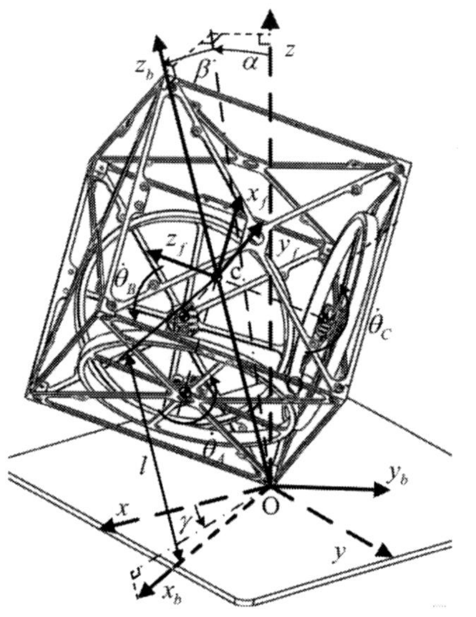

Figure 2. Coordinate system definition of cubical robot.

3 DYNAMICS MODEL OF CUBICAL ROBOT

3.1 Coordinate transformation

In system of the cubical robot, space coordinate frame o-xyz , body coordinate frame o-$x_b y_b z_b$ and reaction wheels coordinate frame c-$x_f y_f z_f$ are defined. The body coordinate frame coincide with the space coordinate frame, while the cubical robot turned α , β and γ degree relative to the axes of it respectively. Than we have the transformation matrix T_{ob} from body coordinate frame to space coordinate frame as

$$T_{ob} = R_z(\gamma)\cdot R_x(\alpha)\cdot R_y(\beta) \tag{1}$$

The body coordinate frame coincide with reaction wheels coordinate frame while the frame turn 45°relative to axis x, 35.3°relative to axis y. Then we have the transformation matrix T_{bf} from body coordinate frame to reaction wheels coordinate frame as

$$T_{bf} = R_y(-35.3°)\cdot R_x(45°)\,. \tag{2}$$

with

$$R_x(\theta)=\begin{bmatrix}1 & 0 & 0\\ 0 & c\theta & -s\theta\\ 0 & s\theta & c\theta\end{bmatrix},\ R_y(\theta)=\begin{bmatrix}c\theta & 0 & s\theta\\ 0 & 1 & 0\\ -s\theta & 0 & c\theta\end{bmatrix},$$

$$R_z(\theta)=\begin{bmatrix}c\theta & -s\theta & 0\\ s\theta & c\theta & 0\\ 0 & 0 & 1\end{bmatrix},\ c\theta=\cos\theta\,,\quad s\theta=\sin\theta\,.$$

3.2 Lagrange method

3.2.1 Energy Calculation

In applying the Lagrangian modeling approach, we need to exprss the kinetic energy and potential energy of the rigid bodies as the functions of generalized coordinates[11,12]. First, the kinetic energy of the cubical robot can be written as

$$T_b = \frac{1}{2}(I_y\dot{\alpha}^2 + I_x\dot{\beta}^2 + I_z\dot{\gamma}^2) \tag{3}$$

By using the mass moment of inertia of the reaction wheel, the rotational kinetic energy of the three ones is given by

$$T_f = \frac{1}{2}I_w(\dot{\theta}_A^{\,2} + \dot{\theta}_B^{\,2} + \dot{\theta}_C^{\,2}) \tag{4}$$

When the cubical robot standing up on its corner, the vertical motion happens both by the angle α and β . Thus, the potential energy of the cubical robot is

$$V = mgl\cos\alpha\cos\beta \tag{5}$$

Now, the Lagrangian is defined by

$$L = T_b + T_f - V \tag{6}$$

3.2.2 Generalized Force Calculation

When the cubical robot is working, the reaction torques produced by the acceleration of three reaction wheels impact on the cube, and the reaction wheels are impacted by Coriolis inertia force simultaneously. The angle acceleration of cubical robot are written as $\ddot{\alpha}$, $\ddot{\beta}$ and $\ddot{\gamma}$, the angle acceleration of reaction wheels can be written as $\ddot{\theta}_A$, $\ddot{\theta}_B$ and $\ddot{\theta}_C$. Thus, in the body coordinate frame, the components along the axes of reaction torque impacted on cube are denoted by τ_α , τ_β and τ_γ , that is

$$\begin{bmatrix}\tau_\alpha\\ \tau_\beta\\ \tau_\gamma\end{bmatrix} = -I_w\cdot T_{bf}\cdot\begin{bmatrix}\ddot{\theta}_A\\ \ddot{\theta}_B\\ \ddot{\theta}_C\end{bmatrix} \tag{7}$$

In reaction wheels coordinate frame, the components along the axes of Coriolis inertia force impacted on reaction wheels are denoted by τ_{θ_A} , τ_{θ_B} and τ_{θ_C} , that is

$$\begin{bmatrix}\tau_{\theta_A}\\ \tau_{\theta_B}\\ \tau_{\theta_C}\end{bmatrix} = \begin{bmatrix}\tau_A\\ \tau_B\\ \tau_C\end{bmatrix} - I_w\cdot T_{ob}\cdot\begin{bmatrix}\ddot{\alpha}\\ \ddot{\beta}\\ \ddot{\gamma}\end{bmatrix} \tag{8}$$

With

τ_j - Motor torque of reaction wheels, $j = A,B,C$.

Due to control of reaction wheel motors by voltage control method base on PWM, denote the moment constant of motor as K_t, the back electromotive force constant as K_b, the armature resistance as R_m, the armature inductance as L_m, armature voltage as v_j. The input torque $\tau_j = K_t i$ generated by the electric motors, where i is the armature current, is obtained from the differential equation of the armature electric circuit:

$$v_j = R_m i + L_m \frac{di}{dt} + K_b \dot{\theta}_j \tag{9}$$

As some parameters are negligible, in particular the armature inductance is neglected(L_m=0), the input torques are derived as

$$\tau_j = \frac{K_t}{R_m} v_j - \frac{K_t^2}{R_m} \dot{\theta}_j \tag{10}$$

3.2.3 Lagrangian equation of motion

Denote six generalized coordinates of the system as

$$q_1=\alpha,\ q_2=\beta,\ q_3=\gamma,\ q_4=\theta_A,\ q_5=\theta_B,\ q_6=\theta_C \tag{11}$$

The above Lagrangian becomes the function of the generalized coordinates. The Lagrangian equation of motion is given by

$$\frac{d}{dt}\left(\frac{\partial L}{\partial \dot{q}_i}\right) - \frac{\partial L}{\partial q_i} = \tau_i, \quad (i = 1 \sim 6) \tag{12}$$

With

$\tau_1 = \tau_\alpha$, $\tau_2 = \tau_\beta$, $\tau_3 = \tau_\gamma$, $\tau_4 = \tau_{\theta_A}$, $\tau_5 = \tau_{\theta_B}$, $\tau_6 = \tau_{\theta_C}$ 。

Now, the dynamic model of system can be derived by substituting (1)-(11) into (12) and solving $\ddot{q}_i$. That is

$$\begin{cases} \ddot{\alpha} = mglR_m \sin\alpha\cos\beta / [(I_w - I_y)R_m] \\ \quad + \dfrac{\sqrt{6}K_t[v_b + v_c - 2v_a + M(2\dot{\theta}_A - \dot{\theta}_B - \dot{\theta}_C)]}{6(I_w - I_y)R_m} \\ \ddot{\beta} = \dfrac{-2mglR_m \cos\alpha\sin\beta + \sqrt{2}K_t[(v_b - v_c) - M(\dot{\theta}_B - \dot{\theta}_C)]}{2(I_w - I_x)R_m} \\ \ddot{\gamma} = \dfrac{K_t[(v_a + v_b + v_c) - K_t(\dot{\theta}_A + \dot{\theta}_B + \dot{\theta}_C)]}{\sqrt{3}(I_w - I_z)R_m} \\ \ddot{\theta}_A = \sqrt{6}mgl \sin\alpha\cos\beta / [3(I_w - I_y)(I_w - I_z)] \\ \quad + \dfrac{K_t\{3I_yI_zV_a - I_w[I_y(2v_a - v_b - v_c) + I_z(v_a + v_b + v_c)]\}}{3I_w(I_w - I_y)(I_w - I_z)R_m} \\ \quad + \dfrac{K_t^2[(I_w(2I_y + I_z) - 3I_yI_z)\dot{\theta}_A - I_w(I_y - I_z)(\dot{\theta}_B + \dot{\theta}_C)]}{3I_w(I_w - I_y)(I_w - I_z)R_m} \\ \ddot{\theta}_B = K_1[K_2 + K_3 + K_4] \\ \ddot{\theta}_C = K_1\{K_2 + K_3 - K_4 \\ \quad + 6I_x(I_w - I_y)(I_w - I_z)K_t[(v_b - v_c - K_t(\dot{\theta}_B - \dot{\theta}_C)]\} \end{cases} \tag{13}$$

With

$$K_1 = 1/(6I_w(I_w - I_x)(I_w - I_y)(I_w - I_z)R_m),$$

$$K_2 = 2K_tI_w(I_w - I_x)(I_y - I_z)v_a - K_5K_tv_b + I_w[I_w(3I_x - I_y - 2I_z) + 3I_yI_z - I_x(2I_y + I_z)]K_tv_c,$$

$$K_3 = K_t^2\{-2I_w(I_w - I_x)(I_y - I_z)\dot{\theta}_A + K_5\dot{\theta}_B - I_w[I_w(3I_x - I_y - 2I_z) + 3I_yI_z - I_x(2I_y + I_z)]\dot{\theta}_C\},$$

$$K_4 = \sqrt{6}mglI_w(I_w - I_z)R_m[\sqrt{3}(I_w - I_y)\cos\alpha\sin\beta - (I_w - I_x)\sin\alpha\cos\beta],$$

$$K_5 = 6I_xI_yI_z + I_w^2(3I_x + I_y + 2I_z) - I_w(4I_xI_y + 5I_xI_z + 3I_yI_z).$$

3 Simulation and Experimental

Simulation the response of system in zero input and response in situation of zero state respectively to validate the correctness of the dynamic model of system, based on the above dynamic model.

3.1 Response of Zero Input

Observe the system response of zero input, x denote the state prameres, u denote the input pramenters.for specific: First, let the cubical robot stand up on its corner, and start motion from state that x = $[\alpha_0, 0, \beta_0, 0,0,0,0,0,0]$,where there are angles α_0, β_0 from the body coordinate frame to the space coordinate frame, and with no input torque. According to the physical system, without the restriction of ground, the cubical robot will rotate towards direction of the deflection angle enlargement, than, the cubical robot do damping motion about corner point o along with the energy loss of system, at last, the cubical robot stabilized in vertical downward. Simultaneously, the corresponding reaction wheels do damping motion toward inverse direction with cube under the common action of electromagnetic force and Coriolis force. Than, let the cubical robot stand up on its corner without input torque, and start motion from station where there is angle γ_0 along with z axis from the body coordinate frame to the space coordinate frame, both the cube and reaction wheels will keep stable.lower case.

Simulate the above case, and the main parameters of the cubical robot prototype show as table 1. In experiment, the initial angle of the cubical robot along with axis x is $\alpha_0 = 20°$, the input voltage of motors are zero, and the simulation result show as figure 3. It can be see from Fig.3. that the cube start motion with angle $\alpha = 20°$, then, do damping motion about position of $\alpha = 180°$, and stop toward vertical downward at last, and the detection angle β, γ keep zero all the time. Simultaneously, along with the rotation of cube, the reaction wheel A and B, C rotate toward inverse direction, then, the angle velocities of the reaction wheels damped oscillation about zero position, and reduce to zero at last. And the angle velocity of reaction wheel A $\dot{\theta}_A$ is two time large as the angle velocity of reaction wheel B and C $\dot{\theta}_B$, $\dot{\theta}_C$. The experimental results are consistent with the physical phenomenon.

Table 1. Main parameters of the cubical robot prototype.

Symble	Value	Symble	Value
m	1.4 kg	I_w	0.00155 $kg \cdot m^2$
l	0.135m	K_t	0.0251 N·m/A
I_x	0.0347 $kg \cdot m^2$	R_m	0.464 Ω
I_y	0.0347 $kg \cdot m^2$	g	9.81 m/s^2
I_z	0.0116 $kg \cdot m^2$	-	-

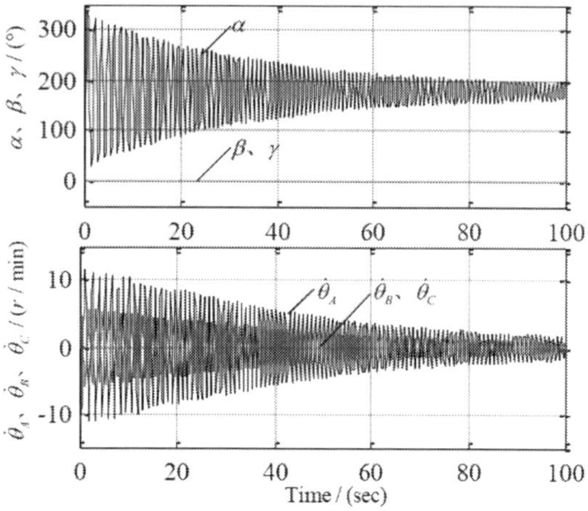

Figure 3. Response of zero input at $\alpha_0 = 20°$.

In experimental, the initial detection angle of the cubical robot along with axis y is $\beta_0 = 20°$, and the simulation result show as figure 4. It can be see from the figure that the detection angle β start at 20° , and damped oscillation about position of $\beta = 180°$, keeping at position of $\beta = 180°$ at last, the detection angle α , γ keep zero all the time. Simultaneously, the angle velocities of reaction wheels B and C $\dot{\theta}_B$, $\dot{\theta}_C$ are always the same as the other but in the opposite direction, and damped oscillation about zero. And the angle velocity of reaction wheel A $\dot{\theta}_A$ keeps zero all the time. The experimental results are consistent with the physical phenomenon.

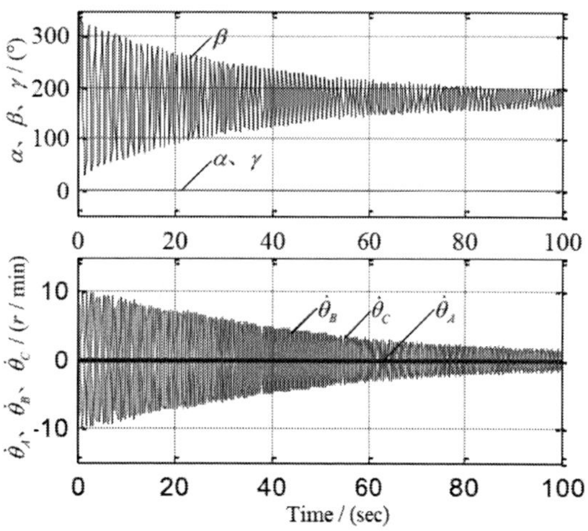

Figure 4. Response of zero input at $\beta_0 = 20°$.

4.2 Response of Zero State

Observer the system response of zero state, for specific: let cubical robot stand up on its corner o and the cubical robot is in balancing condition at initial state, $x = [0,0,0,0,0,0,0,0,0]$, then, give step input voltage to motors. According to the physical system and neglecting the restriction of ground and impact of dead zone of motors. The physical phenomenon should be that along with the acceleration rotation of the reaction wheel motors, the cube rotate around the corner under reaction torque. As the motors reaching matching speed with input voltage, angle acceleration drop to zero, and the motor into the uniform rotation, and cube do damped oscillation around the equilibrium position.

The correspondence between the angle velocities of reaction wheels and the control torque of motors is derived by substituting (2) into (7), that

$$\begin{bmatrix} \tau_\alpha \\ \tau_\beta \\ \tau_\gamma \end{bmatrix} = -I_w \cdot \begin{bmatrix} \frac{\sqrt{6}}{6} & \frac{\sqrt{2}}{2} & \frac{\sqrt{3}}{3} \end{bmatrix} \begin{bmatrix} -2\ddot{\theta}_A + \ddot{\theta}_B + \ddot{\theta}_C \\ \ddot{\theta}_B - \ddot{\theta}_C \\ \ddot{\theta}_A + \ddot{\theta}_B + \ddot{\theta}_C \end{bmatrix} \quad (13)$$

It can be see from (13) that, while the angle velocity of reaction wheel A is two time of the angle velocities of reaction wheels B, C but in opposite direction, the direction of the resultant reactionary moment impact on cube is around the axis x; when the angle velocity of reaction wheel A is zero and angle velocities of reaction wheels B, C are same as each other but in opposite direction, the direction of the resultant reactionary moment impact on cube is around the axis y; when the angle velocities of reaction wheels are same in size and direction, the direction of the resultant reactionary moment impact on cube is around the axis z.

Simulation the above case, that in experiment, the cubical robot initial with balance state, and give step input of voltage $u = [0.5,-0.25,-0.25]$(V) to the motors at 1 second, response of zero state show as figure 5.

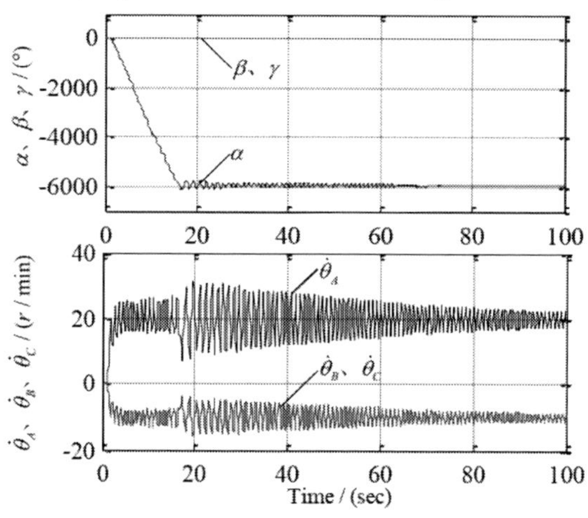

Fig. 5. Response of zero state at $u = [0.5,-0.25,-0.25]$(V)

It can be see that the angle velocity of the reaction wheel A accelerates to 20r/min, and the angle velocities of reaction wheel B accelerate to -10r/min after 1 second. And then the angle velocities of reaction wheels damping oscillate. Simultaneously, the cube turn 5940 degree around axis x under the action of resultant reactionary moment; and then, damped around the vertical downward. The experimental results are consistent with the physical phenomenon.

Let the cubical robot initial from balancing state, and give step input of voltage $u=[0,0.5,-0.5](V)$to the motors at 1 second, response of zero state show as figure 6. It can be see that the angle velocity of the reaction wheels B and C accelerate to 20r/min, but in opposite direction after 1 second, and the angle velocities of reaction wheel A keep be zero. And then, the angle velocities of reaction wheels damping oscillate. Simultaneously, the cube turn 2340 degree around axis y under the action of resultant reactionary moment; and then, damped around the vertical downward. The experimental results are consistent with the physical phenomenon.

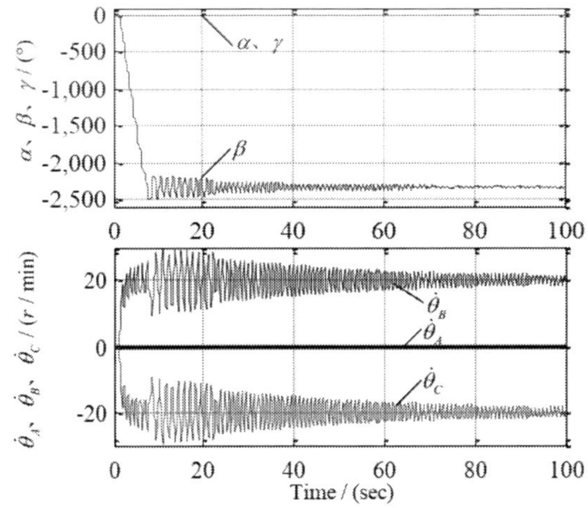

Figure 6. Response of zero state at $u=[0,0.5,-0.5](V)$.

Let the cubical robot initial from balancing state, and give step input of voltage $u=[0.25,0.25,0.25](V)$ to the motors at 1 second, response of zero state show as figure 7.

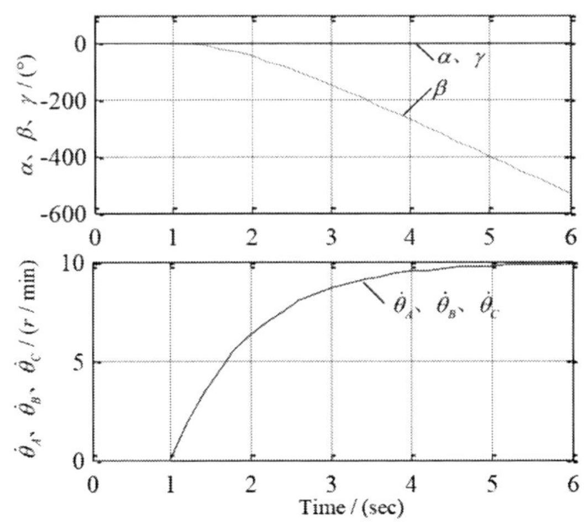

Figure 7. Response of zero state at $u=[0.25,0.25,0.25](V)$.

It can be see that the angle velocity of the reaction wheels A, B and C accelerate to 10r/min after 1 second, and keep stable. Simultaneously, the cube accelerated turn -25.4 degree around axis z under the action of resultant reactionary moment during 1s to 1.7s; and then, keep the uniform angle velocity around axis z. The experimental results are consistent with the physical phenomenon.

As results of the simulation of cubical robot both zero input response and zero state response. It show the consistency of experimental results and physical phenomenon and the correctness of the dynamic model.

5 CONCLUSION

The dynamic model of cubical robot balancing on its corner by Lagrangian method based on solve the energy function. Observer the simulation of zero state input response and zero state simulation based on the above dynamic model, the simulation results are consistent with physical phenomenon, and the correctness of the dynamic model is verified. The dynamic model developed can provide a base for further study of balance control of a cubical robot.

We acknowledge support from Beijing Natural Science Foundation（4174083）; National Natural Science Foundation of China (No. 61375086); Key Project (No. KZ201610005010) of S&T Plan of Beijing Municipal Commission of Education.

References

1. J.J. Wang, D. Liu, B. Wang, Acta Automatica Sinica **39**, (2013), 92-96.
2. Y.L. Zhang, Ph.D. thesis, Dalian University of Technology, (2011).
3. H.Q. Wang, M, Applied Mechanics and Materials, 195, (2012), 17-22.
4. T.C. Tang, D.Y. Zhou, J. Wen, PLOS ONE, 11, (2016), e0160416.
5. L. Tian, Ph.D. thesis, Nanjing University of Science and Technology (2006).
6. Z.B. Qiu, Ph.D. thesis, Nanjing University of Science and Technology (2009).
7. M. Gajamohan, M. Merz, I. Thommen, R. Dandrea, (2012), 3722-3727.
8. M. Gajamohan, M. Muehlebach, T. Widmer, R. D'Andrea, The Cubli: A reaction wheel based 3D inverted pendulum, in Control Conference (2013), 268-274.
9. M. Muehlebach, G. Mohanarajah, R. Dandrea, (2013), 1283-1288.
10. M. Muehlebach, R. Dandrea, IEEE Transactions on Control Systems and Technology 25, (2017), 235-246.
11. S.t. Kim, S.J. Kwon, International Journal of Control, Automation and Systems, 13, (2015), 926-933.
12. M. Olivares, P. Albertos, Isa Transactions, 53, (2014), 1-8.

Monocular Visual-Inertial State Estimation for Micro Aerial Vehicles

Yao Xiao[1,*], *Xiaogang* Ruan[1], *Xiaoping* Zhang[1], and *Pengfei* Dong[1]

[1] *Faculty of Information Technology, Beijing University of Technology*

Abstract. Autonomous micro aerial vehicles (MAVs) equipped with onboard sensors, are idea platforms for missions in complex and confined environments for its low cost, small size and agile maneuver. Due to the size, power, weight and computation constraints inherent in the filed of MAVs, monocular visual-inertial system that consist of one camera and an inertial measurement (IMU) are the most suitable sensor suit for MAVs. In this paper, we proposed a monocular visual-inertial algorithm for estimating the state of a MAV. Firstly, the Semi-Direct Visual Odometry (SVO) algorithm used as the vision front-end of our framework was modified so that it can be used for forward-looking camera case. Second, an Error-state Kalman Filter was designed so that it can fuse the output of the SVO and IMU data to estimate the full state of the MAVs. We evaluated the proposed method with EuRoc Dataset and compare the results to the state-of-the-art visual-inertial algorithm, VINS-Mono. Experiments show that our estimator can achieve comparable accurate results.

1 Introduction

MAVs are ideal platforms for missions such as exploration, inspecting, search and rescue in complex and confined environments due to its low cost, small size and agile maneuver. To achieve these goals, it is essential that the MAVs is capable of autonomous navigation in unknown environments, which include reliable state estimation, control, environment mapping, planning and obstacles avoidance. Among these, state estimation is the first and most critical components for autonomous flight. Despite GPS-based autonomous flight have been successful apply in commercial MAVs, it is unavailable when operate in indoor or tunnel-link environments due to effects like shadowing.

At present, the most common used sensors for state estimation of MAVs are monocular [1–4], stereo [5–7] and RGB-D [8, 9] cameras as well as laser scanners [10]. However, MAVs usually come with tight size, weight and power constraints, limiting their ability to carry active but heavy sensors such as radars or LiDARs. The dimensions of a small platform also limit its ability to carry stereo or multicamera systems due to insufficient baseline length. As the platform becomes smaller, a monocular visual-inertial navigation system (VINS), consisting of only a low-cost inertial measurement unit (IMU) and a camera, becomes the only viable sensor suite allowing autonomous flights with sufficient environmental awareness [4].

Recent research on VINS for MAVs have yielded a number of significant results[1–4, 11]. Most of these existing approach can be classified into filter-based and optimization-based systems. In [1], a parallel tracking and mapping (PTAM) algorithm from the Simultaneous Localization and Mapping (SLAM) community is used as front-end of the VINS system which estimate a 6DOF with arbitrarily scaled and drifting camera pose measurement, and an Extended Kalman Filter is used as the back-end which can fuse the output of the PTAM and the IMU information to get a full state metric estimation of the MAVs. In [11], Faessler et al. present a system that enables a monocular-vision–based quadrotor to automatically recover from any unknown, initial attitude with significant velocity, such as after loss of visual tracking due to an aggressive maneuver. The images from the camera are processed by means SVO pipeline [12]. The visual-odometry pipeline outputs an un-scaled pose that is then fused with the IMU readings in an Extended Kalman Filter framework (Multi Sensor Fusion, MSF [13]) to compute a metric state estimate. This work have been extended in [3]. However, SVO is a direct-based method and designed for MAVs with downward-looking camera, which made the system cannot used for forward-looking case. In [2], Shen proposed a optimization-based VINS framework for rotorcraft MAVs. This estimator tightly couples the vision and IMU measurements and allows the MAV to execute trajectories at 2 m/s with roll and pitch angles up to 30 degrees. To the best our knowledge, this is the first tightly-coupled nonlinear optimization-based monocular VINS estimator successful apply for autonomous flight. This work have been extended in their work in [4]. However, tightly-coupled nonlinear optimization-based method is more computation-intensive.

In this paper, we proposed a monocular visual-inertial algorithm for estimating the state of a micro aerial vehicles (MAV). The SVO algorithm was modified so that it can be used for forward-looking camera case. And an Error-state Kalman filter was designed which can fuse the output of

*e-mail: xiaoyao1103@mails.bjut.edu.cn

 DOI: 10.1051/matecconf/201713900068

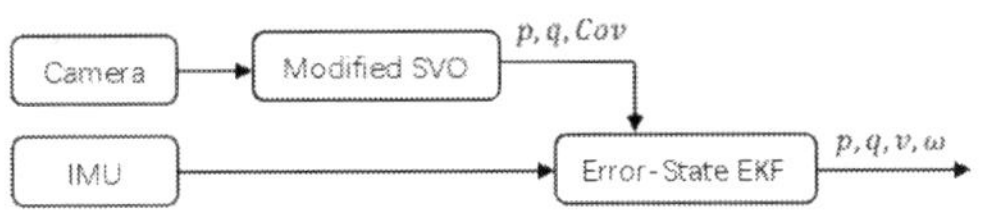

Figure 1. Sofware arichiteture of our algorithm

the SVO and IMU data to estimate the full state of the MAVs.

One contribution of this paper is that we modified the original SVO algorithm such that it can be used in the forward-looking camera case. As a second contribution, we formulate a fusion framework based on error-state Kalman filter which can fuse the output of the SVO and the IMU data to get the full state estimate of the MAVs. Compare to the work of [1, 14, 15], we use the SVO as the front end of the frame work which would less computational demand. And compare to the work of [3], our estimator can be run in the case of forward-looking camera. As a third contribution, we compare our propose algorithm against the state-of-the-art visual-inertial algorithm (Mono-VINVS) and demonstrate its comparable performance.

The rest of this paper is organized as follows: Section 2 presents a general overview of the software architecture. Section 3 describes what we have improve from the original SVO; Section 4 describes the details of the design the error-state Kalman filter. Section 4 consists the experiments to evaluate our proposed approach. Finally, Section 6 concludes the work.

2 Software Architecture

The Software Architecture of our algorithm is shown in Figure 1. The p, q, v, ω represent the position, attitude, translation velocity, angular velocity of the MAVs respectively. And Cov represents the covariance of p and q. The software consists two components. The first one is the modified SVO, which input is the image scream capture from the camera. The Error-State EKF component fuses the output of the SVO and the IMU data to estimate the full state of the MAVs.

3 Modified SVO

3.1 Keyframe Selection

The original version of SVO algorithm was designed for downward-looking case. And a keyframe is selected if the Euclidean distance of the new frame relative to all keyframes exceeds 12% of the average scene depth [12]. However, if the camera is facing forwards, the scene depth is very large and no new keyframes are selected.

Generally, a keyframe should be required in there cases:

(1) The field of view changes with a threshold.

(2) The camera translate exceeds a threshold.

(3) The environment light condition changes or the camera exposure time changes.

Compare to the feature-based method, SVO is more sensitive to the light condition change or exposure time change as it is a direct-based method. However, determining thresholds of the three cases is a problem in reality. As a second analyze of the cases above, we could find that all of the cases would cause tracking features to be lost. According to this idea, in this paper we set the keyframe selection strategy as a simple criteria, that is, a keyframe is selected if the percent of the lost tracking feature respect to the last keyframe exceeds 15%.

3.2 Tracking Failure Process

In the original version of SVO tracking pipline, the pose of a new coming frame is initialize through sparse model-based image alignment. The camera pose relative to the last frame is found through minimizing the photometric error between pixels corresponding to the projected location of the same 3D points [12]. And if one frame tracking failure, the system will try to do image alignment with the last keyframe. This would be fine if the frame rate is high enough (>50Hz). Experiments showed that one or two continue frame may be very easily failed for tracking due to focus losing or motion blur. In this case, if the SVO are run on low frame rate (~ 20 Hz), it would be very difficult to re-localzie the new frame and continue the tracking again. This maybe for several reasons:

(1) The image align process dose not wrap the patches used to align for computing speed reasons.

(2) Compare to two consecutive frames, the change of field of view or translation between the last keyframe is more larger.

(3) If more than two consecutive are failed for tracking, the change of field of view or translation between the last keyframe become larger and larger. As a result, it is more and more difficult to re-localize the new frame.

In our modified version of SVO, the strategy we use is that if one frame is failed for tracking, we would drop out the current frame and keep the last frame unchanged, and after the new frame arrive, we would align with the last frame. As we discuss above, the new coming frame would be more easier to re-localize with this strategy.

4 Error-State Kalman Filter

The unit quaternion has the lowest dimension of globally nonsigular attitude parameterization and represents the attitude matrix as a homogeneous quadratic function. Due to the quaternion norm constraint, the 4×4 quaternion covariance is assumed to have rank three [16]. As a result, we cannot simply apply the standard Exetend Kalman filter for the system. In this paper, we use the Error-State Kalman Filter to fuse the IMU data and the output of SVO.

4.1 Notations

There are several quaternion conventions. In this paper, the Hamilton convention is used, which is defined as:

$q = q_w + q_x i + q_y j + q_z k$, and $i^2 = j^2 = k^2 = ijk = -1$. The quaternion q_{ab} describes a ratation from frame a to the frame b, and $q_{ac} = q_{ab} \otimes q_{bc}$. R_{ab} is the corresponding rotation matrix of q_{ab}.

The subscript w, i, v of the position p, quaternion q and velocity v denote the world frame, IMU frame, vision frame, camera frame respectively, such as p_{wi} means the IMU position of the MAV express in the world frame. We assume the IMU frame is coincide with the body frame, which is the general case in reality. The skew symmetric matrix of w is represented as $[w]_\times$. The hat of a variable x means its estimate value $\hat{x}$.

4.2 System Kinematics

We assumed the IMU measurements contain a slowly varying bias b and white Gaussian noise n:

$$\omega = \omega_m - b_\omega - n_\omega \quad a = a_m - b_a - n_a \tag{1}$$

$$\dot{b}_\omega = n_{b_\omega} \quad \dot{b}_a = n_{b_a} \tag{2}$$

Where, a, ω is the real acceleration and angular velocity respectively. The a_m, ω_m denotes the measurements value.

Define the system state as

$$X = \left\{ p_{\text{wi}}^T \quad q_{\text{wi}}^T \quad v_{\text{wi}}^T \quad b_w^T \quad b_a^T \quad \lambda \right\} \tag{3}$$

The system kinematics is represented as following differential equations:

$$\dot{p}_{\text{wi}} = v_{\text{wi}} \tag{4}$$

$$\dot{q}_{\text{wi}} = \frac{1}{2}\Omega(\omega_m - b_\omega - n_\omega) q_{\text{wi}} \tag{5}$$

$$\dot{v}_{\text{wi}} = R_{\text{wi}}(a_m - b_a - n_a) - g \tag{6}$$

$$\dot{b}_w = n_{b_w} \quad \dot{b}_a = n_{b_a} \quad \dot{\lambda} = 0 \tag{7}$$

The nominal-state kinematics is defined as:

$$\dot{\hat{p}}_{\text{wi}} = \hat{v}_{\text{wi}} \tag{8}$$

$$\dot{\hat{q}}_{\text{wi}} = \frac{1}{2}\Omega\left(\omega_m - \hat{b}_\omega\right) \hat{q}_{\text{wi}} \tag{9}$$

$$\dot{\hat{v}}_{\text{wi}} = R_{\text{wi}}\left(a_m - \hat{b}_a\right) - g \tag{10}$$

$$\dot{\hat{b}}_w = 0 \quad \dot{\hat{b}}_a = 0 \quad \dot{\hat{\lambda}} = 0 \tag{11}$$

4.3 Error State Kinematics

Define $q_{\text{wi}} = \hat{q}_{\text{wi}} \otimes \delta q_{\text{wi}}$, which means

$$\delta q_{\text{wi}} = q_{\tilde{i}i} = \hat{q}_{\text{wi}}^{-1} \otimes q_{\text{wi}} \approx [1 \ \frac{1}{2}\delta\theta_{\text{wi}}]^T \tag{12}$$

Define error state of the system as

$$\delta x = \left\{ \delta p_{\text{wi}}^T \quad \delta\theta_{\text{wi}}^T \quad \delta v_{\text{wi}}^T \quad \delta b_w^T \quad \delta b_a^T \quad \delta\lambda \right\} \tag{13}$$

The error state kinematics is represented as following differential equations:

$$\delta\dot{p}_{\text{wi}} = \delta\hat{v}_{\text{wi}} \tag{14}$$

$$\delta\dot{\theta}_{\text{wi}} = -[\hat{\omega}]_\times \delta\theta_{\text{wi}} - \delta b_\omega - n_\omega \tag{15}$$

$$\delta\dot{v}_{\text{wi}} = -R_{\text{wi}}^T [\hat{a}]_\times \delta\theta_{\text{wi}} - R_{\text{wi}}^T \delta b_a - R_{\text{wi}}^T n_a \tag{16}$$

$$\delta\dot{b}_w = n_{b_w} \quad \delta\dot{b}_a = n_{b_a} \quad \delta\dot{\lambda} = 0 \tag{17}$$

This can be summarized to the linearized continous-time error state equation

$$\delta\dot{x} = F_c \delta x + B_c u \tag{18}$$

with u being the noise vector. Discretize the error state kinematics we can get F_d:

$$F_d = \exp(F_c \Delta t) = I_d + F_c \Delta t + \frac{1}{2!} F_c^2 \Delta t + \ldots \tag{19}$$

$$F_d = \begin{bmatrix} I_3 & A & \delta t & B & -C\frac{\Delta t^2}{2} \\ 0 & C & 0 & D & 0 \\ 0 & E & I_3 & F & -E\Delta dt \\ 0 & 0 & 0 & 1 & 0 \\ 0 & 0 & 0 & 0 & 1 \end{bmatrix} \tag{20}$$

Where:

$$A = R_{\text{wi}}^T [\hat{a}]_\times \left(-\frac{\Delta t^2}{2} + \frac{\Delta t^3}{3!} [\omega]_\times - \frac{\Delta t^4}{4!} [\omega]_\times^2 \right) \tag{21}$$

$$B = R_{\text{wi}}^T [\hat{a}]_\times \left(\frac{\Delta t^3}{3!} - \frac{\Delta t^4}{4!} [\omega]_\times + \frac{\Delta t^5}{5!} [\omega]_\times^2 \right) \tag{22}$$

$$C = I_3 - \Delta t [\omega]_\times + \frac{\Delta t^2}{2} [\omega]_\times^2 \tag{23}$$

$$D = -\Delta t + \frac{\Delta t^2}{2} [\omega]_\times + \frac{\Delta t^3}{3!} [\omega]_\times^2 \tag{24}$$

$$E = R_{\text{wi}}^T C \tag{25}$$

$$F = -A \tag{26}$$

4.4 Update

Measurement model of the system is described as

$$z_p = p_{\text{vc}} = R_{\text{vw}} (p_{\text{wi}} + R_{\text{wi}} p_{\text{ic}}) \lambda + n_p \tag{27}$$

The measurement error can be define as:

$$\begin{aligned} \tilde{z}_q &= z_q - \hat{z}_q \\ &= R_{\text{vw}} (p_{\text{wi}} + R_{\text{wi}} p_{\text{ic}}) \lambda + n_p \\ &\quad - R_{\text{vw}} \left(\hat{p}_{\text{wi}} + \hat{R}_{\text{wi}} p_{\text{ic}} \right) \hat{\lambda} \end{aligned} \tag{28}$$

Which can be linearized to

$$\tilde{z}_q = H_p \delta x \tag{29}$$

with

$$H_p^T = \begin{bmatrix} R_{\text{vw}} \hat{\lambda} \\ -R_{\text{vw}} \hat{R}_{\text{wi}} [p_{\text{ic}}]_\times \hat{\lambda} \\ 0_{3\times3} \\ 0_{3\times3} \\ 0_{3\times3} \\ \left(R_{\text{vw}} \hat{p}_{\text{wi}} + R_{\text{vw}} \hat{R}_{\text{wi}} \hat{p}_{\text{wi}} \right) \end{bmatrix}^T \tag{30}$$

Measurement model of rotation:

$$z_q = q_{\text{vc}} = q_{\text{vw}} \otimes q_{\text{wi}} \otimes q_{\text{ic}} \tag{31}$$

Which yields for the error measurement:

$$\begin{aligned} \tilde{z}_q &= (q_{\text{vw}} \otimes \hat{q}_{\text{wi}} \otimes q_{\text{ic}})^{-1} \otimes (q_{\text{vw}} \otimes q_{\text{wi}} \otimes q_{\text{ic}}) \\ &= \begin{bmatrix} 1 & 0 \\ 0 & R_{\text{ic}}^T \end{bmatrix} \begin{bmatrix} 1 \\ \delta\theta \end{bmatrix} \end{aligned} \tag{32}$$

4.5 Filter Pipeline

(1) Propagate nominal-stated with the IMU measurements according to equations (8)-(11).

(2) Compute the progated covariance matrix according to:

$$P_{k+1|k} = F_d P_{k|k} F_d + Q_d \tag{33}$$

(3) Compute the residual according to equation (28)

(4) Compute the innovation:

$$S = HPH^T + R \tag{34}$$

(5) Compute the Kalman gain:

$$K = PH^T S^{-1} \tag{35}$$

(6) Compute the correction:

$$\hat{\tilde{x}} = K\tilde{z} \tag{36}$$

(7) Correct the propagate nominal-stated, where the attitude is corrected by:

$$\hat{q}_{wi} = \hat{q}_{wi} \otimes \hat{\tilde{x}}_{(q_{wi})} \tag{37}$$

5 Experiment Results

We evaluated the proposed algorithm with the Machine Hall 01 of EuRoc Visual-Inertial Dataset. The dataset are collected on-board a Micro Aerial Vehicle (MAV), which contain stereo images(Aptina MT9V034 global shutter, WVGA monochrome, 2 × 20 FPS), synchronized IMU measurements(ADIS16448, angular rate and acceleration, 200 Hz), and accurate ground-truth states. We only use the left camera of the stereo images set.

The results do not contain the comparison of the performance difference between the original version of SVO and the our modified SVO as the the original version of SVO cannot run on this dataset. We compare the performance of the proposed algorithm with ground-truth and the VINS-Mono framework [4].

The whole trajectories are shown in Figure 2. Compare to the ground-truth, both of the proposed method and VINS-Mono exist the drift in the trajectories as the time goes on. It is hard to say which one drift more. However, benefit from the SVO, the trajectory of our propose method nearly return to the starting point at the end of flying. In contrast, the result of VINS-Mono show more drift at the end of the trajectory.

Figure 3 shows the translation error of the trajectories in x, y, z axises respectively. It shows that the VINS-Mono achieve more accurate in y axis. However, proposed algorithm shows less drift on z axis.

The attitude errors expressed in yaw, roll, pitch respectively are showed in Figure 4. It is obvious that the proposed method achieve more accuracy with the attitude estimation compare to the results ofvalidity VINS-Mono.

Figure 5 shows the velocity error. It can be seen that Both the method achieve accurate velocity estimating. It is obvious that there is a abrupt jump in the measurements of our propose methods, this should be caused by tracking failure. However, the vision estimator have successful relocalize the subsequent frames, which shows the effectiveness our modified tracking failure process.

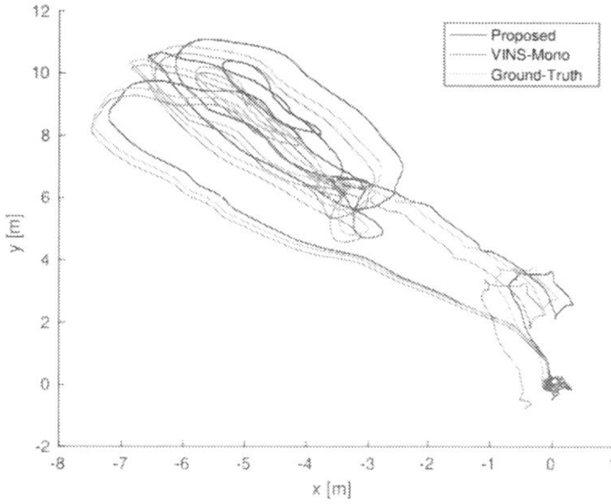

Figure 2. Trajectory.

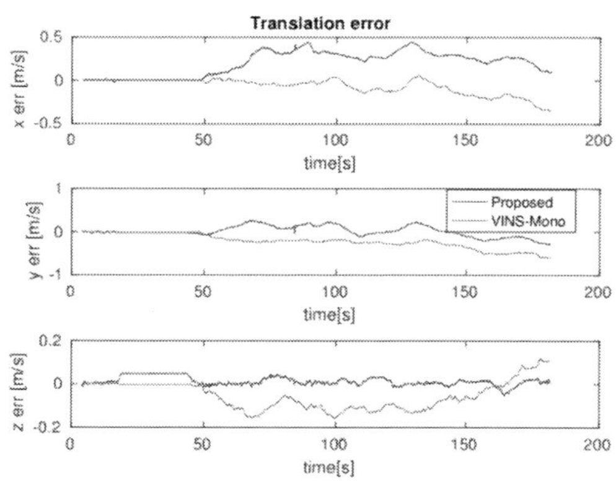

Figure 3. Translation error.

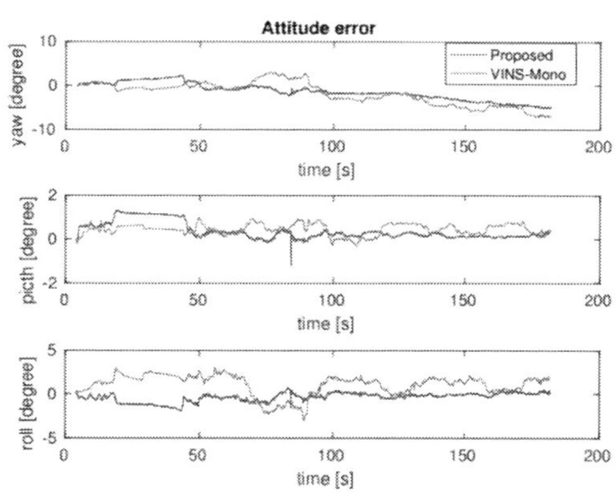

Figure 4. Attitude error.

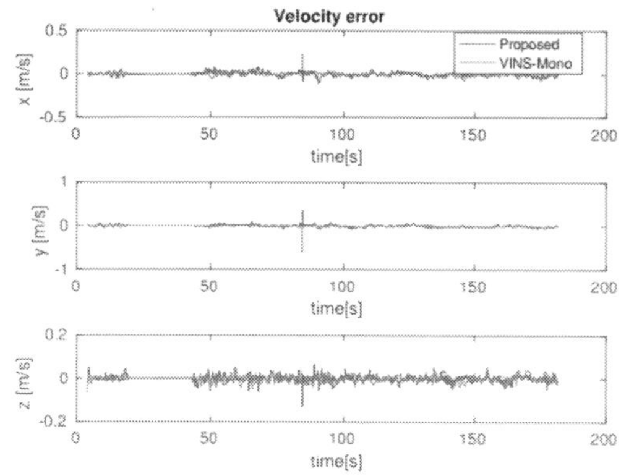

Figure 5. Velocity error.

6 Conclusion

In this paper, we presented a monocular visual-inertial state estimation algorithm based on the modified SVO and error-state Kalman Filter. The proposed method can be used for state estimation of MAVs, the result of which can be used in the higher lever task such as controlling, mapping, planing and trajectory following. The performance of the proposed method was validate on the EuRoc dataset and the results was compared to the ground-truth and the state-of-the-art algorithm, VINS-Mono. However, two component of our framework are work independent at present, which means the estimate result do not feedback to the vision-pipeline for failure detecting or dirft suppression. Feature will investigate to tightly integrate the filter component in the vision-pipeline to get more accurate results.

Acknowledgements

This work has been supported by National Natural Science Foundation of China (No. 61375086), Key Project (No. KZ201610005010) of S&T Plan of Beijing Municipal Commission of Education, Beijing Natural Science Foundation (4174083)

References

[1] S. Weiss, M.W. Achtelik, S. Lynen, M.C. Achtelik, L. Kneip, M. Chli, R. Siegwart, Journal of Field Robotics **30**, 803 (2013)

[2] S. Shen, N. Michael, V. Kumar, *Tightly-coupled monocular visual-inertial fusion for autonomous flight of rotorcraft MAVs*, in *2015 IEEE International Conference on Robotics and Automation (ICRA)* (IEEE, 2015), pp. 5303–5310

[3] M. Faessler, F. Fontana, C. Forster, E. Mueggler, M. Pizzoli, D. Scaramuzza, Journal of Field Robotics **33**, 431 (2016)

[4] Y. Lin, F. Gao, T. Qin, W. Gao, T. Liu, W. Wu, Z. Yang, S. Shen, Journal of Field Robotics pp. n/a–n/a (2017)

[5] L.R. García Carrillo, A.E. Dzul López, R. Lozano, C. Pégard, Journal of Intelligent & Robotic Systems **65**, 373 (2012)

[6] L. Heng, D. Honegger, G.H. Lee, L. Meier, P. Tanskanen, F. Fraundorfer, M. Pollefeys, Journal of Field Robotics **31**, 654 (2014)

[7] K. Schmid, P. Lutz, T. Tomić, E. Mair, H. Hirschmüller, Journal of Field Robotics **31**, 537 (2014)

[8] A. Huang, A. Bachrach, P. Henry, M. Krainin, D. Maturana, D. Fox, N. Roy, *Visual odometry and mapping for autonomous flight using an RGB-D camera*, in *Int. Symposium* (2011), pp. 1–16

[9] S. Shen, N. Michael, V. Kumar, *Autonomous indoor 3D exploration with a micro-aerial vehicle*, in *2012 IEEE International Conference on Robotics and Automation* (IEEE, 2012), pp. 9–15

[10] M.W. Achtelik, A. Bachrach, R. He, S. Prentice, N. Roy, *Stereo Vision and Laser Odometry for Autonomous Helicopters in GPS-denied Indoor Environments*, in *SPIE Conference on Unmanned Systems Technology* (2009), pp. 19–29

[11] M. Faessler, F. Fontana, C. Forster, D. Scaramuzza, *Automatic re-initialization and failure recovery for aggressive flight with a monocular vision-based quadrotor*, in *2015 IEEE International Conference on Robotics and Automation (ICRA)* (IEEE, 2015), June, pp. 1722–1729

[12] C. Forster, M. Pizzoli, D. Scaramuzza, Proceedings - IEEE International Conference on Robotics and Automation pp. 15–22 (2014)

[13] S. Lynen, M.W. Achtelik, S. Weiss, M. Chli, R. Siegwart, *A robust and modular multi-sensor fusion approach applied to MAV navigation*, in *IEEE International Conference on Intelligent Robots and Systems* (2013), pp. 3923–3929

[14] S. Weiss, R. Siegwart, *Real-time metric state estimation for modular vision-inertial systems*, in *Proceedings - IEEE International Conference on Robotics and Automation* (2011), Vol. 231855, pp. 4531–4537

[15] D. Scaramuzza, M.C. Achtelik, L. Doitsidis, F. Friedrich, E. Kosmatopoulos, A. Martinelli, M.W. Achtelik, M. Chli, S. Chatzichristofis, L. Kneip et al., IEEE Robotics & Automation Magazine **21**, 26 (2014)

[16] F.L. Markley, Journal of Guidance Control and Dynamics **26**, 311 (2003)

[17] Z. Yang, S. Shen, IEEE Transactions on Automation Science and Engineering **14**, 39 (2017)

[18] Z. Yang, S. Shen, *Monocular visual-inertial fusion with online initialization and camera-IMU calibration*, in *2015 IEEE International Symposium on Safety, Security, and Rescue Robotics (SSRR)* (IEEE, 2015), pp. 1–8

[19] R. Mur-Artal, J.D. Tardos, IEEE Robotics and Automation Letters **2**, 796 (2017)

A Smart Home Center Platform Solution Based on Smart Mirror

Xibo DENG [1], Zhiran PENG [2], and Wenquan WU [1]

[1] *School of Electronic Engineering, Navy University of Engineering, Wuhan, 430000, P.R. China*
[2] *School of Electrical Engineering, Navy University of Engineering, Wuhan, 430000, P.R. China*

Abstract. With the popularization of the concept of smart home, people have raised requirements on the experience of smart living. A smart home platform center solution is put forward in order to solve the intelligent interoperability and information integration of smart home, which enable people to have a more intelligent and convenient life experience. This platform center is achieved through the Smart Mirror. The Smart Mirror refers to a smart furniture, on the basis of the traditional concept of mirror, combining Raspberry Pi, the application of one-way mirror imaging principle, the touch—enabled design, voice and video interaction. Smart Mirror can provide a series of intelligent experience for the residents, such as controlling all the intelligent furniture through Smart Mirror; accessing and displaying the weather, time, news and other life information; monitoring the home environment; remote interconnection operation.

1 The overall framework and functionality of the solution

Smart home is intelligent home. In the current smart home solutions, the vast majority of them are through the embedded gateway or independent central processor to communicate with the outside, relying on the Internet of things(IoT), all kinds of household things linked together. Smart home does not have a complete system of technical standards and solutions in the international and domestic [1]. But with the development of Smart home, more and more users need an integrated system platform, which can provide a convenient control operation and human-computer interaction for people. At the same time, it also provides people with various social life information. This article is based on the mirror used in home life, designed and implemented a Smart Mirror, which based on the home center intelligent platform solutions. This center platform includes showing real-time weather, news, monitoring of family health, monitoring of home environments, burglar alarm and so on. Not only that, all the home appliances and related intelligent devices access to the Internet, real-time feedback home status parameters.

Nowadays, all household appliances have corresponding intelligent products in connection with LAN or Internet, such as small alarm clock, dust collector, television, washing machine, with different grades at different prices. The mirror is used as the solution to satisfy people's demands for intelligent home. Main reasons are shown as follows:

(1) As the mirror usually hangs in contrast with other electrical appliances, it can spread throughout each independent living space.

(2) As the natural imaging curtain, the mirror occupies a small space, which is convenient for imaging and saves the living space at home.

(3) Different from other large household appliances, the mirror is not limited by space, and can meet different requirements in different environments. It is advantageous in integrating different functions in different laying space. Three typical applications are taken as an example to elaborate overall framework of function and solution as Fig. 1.

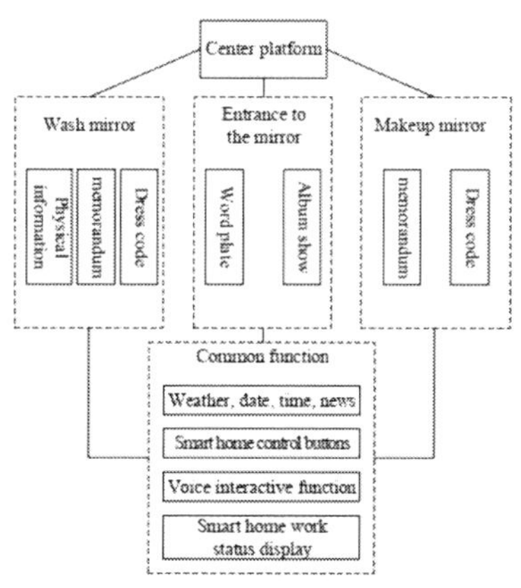

Figure 1. Intelligent mirror center platform overall framework

* Corresponding author: 347718046@qq.com

2 Hardware design

2.1. Raspberry Pi

Raspberry Pi was developed by British charitable organization Raspberry Pi foundation. Eben Upton of University of Cambridge officially launched computer version of Raspberry Pi on the basis of Linux in March 2012 [2]. Raspberry Pi was originally designed for the students in computer programming education, but admired by numerous computer fanciers after coming out. It is a micro computer motherboard based on ARM, with SD/ Micro SD as storage hard disk, in connection with keyboard, mouse and cable, furnished with TV output interface with video analog signal and HDMI HD video output interface. These components are integrated in the motherboard slightly larger than credit card, with all the basic functions of PC [3]. There are A/A+/B/B+/second B type /third B type, and three kinds of internal storage, including 256MB, 512MB and 1GB. Raspberry Pi takes Python as the main programming language. The paper is based on "Raspberry Pi B+" released in July 2014, which was equipped with internal storage of 512MB, 40 pins of GPIO and 4 USB interfaces. In contrast with the previous generation, the new one renovates hot plug and over-current protection, replaces old SD socket with Micro SD, reduces power consumption to 0.5 ~ 1W, and adopts low-noise power supply in audio circuit [4]. Contrasting with taking micro-controller [5], no much extra money is spent on developing Raspberry Pi, and peripheral circuit design is saved, which means that it is a good choice to create excellent multimedia center.

2.2 Imaging control system

The imaging control system lays foundation for display of Smart Mirror and human-computer interaction. In addition to retention of imaging characteristics of the mirror, daily information, such as weather, news and time, is obtained directly by the Internet. At the same time, synchronously memo, changing of teleprompter board and monitoring the mirror image can be achieved through the user's mobile phone. Users can even directly control the intelligent home, selfie in front of the mirror and look through built-in family photos through directly touching. The imaging control system is mainly based on display technology, by which expected information and control unit are obtained in home mirror. The imaging control system of Smart Mirror is shown by Fig. 2, mainly consisting of following parts: unidirectional mirror, driving plate, display screen and infrared frame.

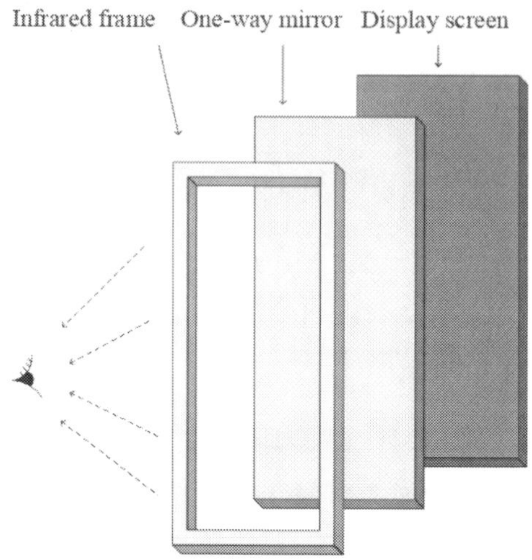

Figure 2. Imaging system structure

2.2.1 One-way mirror

A one-way mirror is also called single lens, atomic mirror. According to the principle of human eye capturing high light objects in visual range, black room environment is constructed on one side of mirror, supplemented by high light on the other side. In this case, people on high light side take glass as a mirror, but look objects though glass in the dark environment. The specific observation environment is demonstrated in Fig. 3. Currently, the one-way mirror adopts magnetron sputtering coating technology, with thin semi reflective layer in the middle of two pieces of glass attached by PVB glue, as shown in Fig. 4.

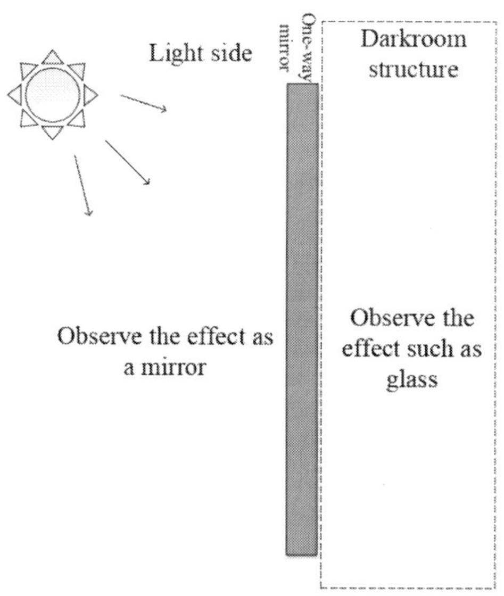

Figure 3. Imaging system structure

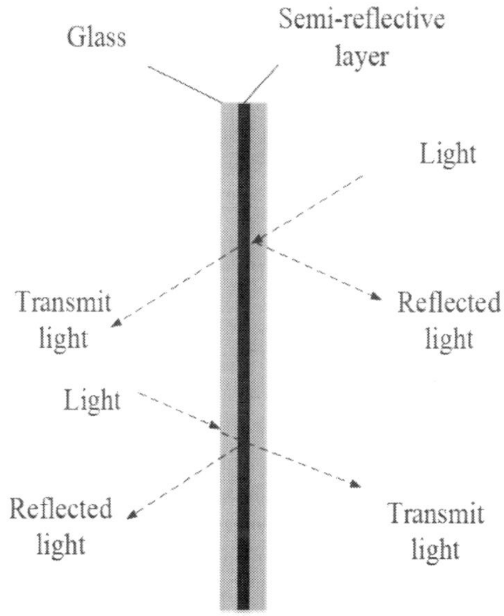

Figure 4. One-way mirror structure

2.2.2 Display screen

In the term of selection of screen, this paper chooses TFT type liquid crystal display after taking both contrast and color saturation into account [6]. This kind of monitor includes liquid crystal materials, thin film transistor, polarizing plate, light guide plate, which can display graphics in full color, and ensure 70% of all black display at high utilization rate through interface design. This design is conducive to the chamber structure behind the mirror, realizing mirror imaging effect [7].

2.2.3 General drive board

The generic driver board is also known as the LCD driver board. There are a lot of brands of common driver plate in market, such as Leroy, DingKe, Triumph and Yuekang. Although components of general driver board are inconsistent with those of original driver board, as long as by drivers are loaded into general drive board by LCD programmer and lines are modified, different LCD screens can be driven, which reduces costs very low cost, is easy to develop and accepted by the customers.

2.2.4 Infrared frame

In order to solve the basic problem of touch interaction, infrared frame is adopted, which can connect with Raspberry Pi through the USB port. Infrared network is constructed inside the infrared range, and determines the touch position through touching objects, so as to achieve the touch interaction purpose.

3 Implementation plan

3.1. Interactive function implementation

3.1.1 Touch interaction

The touch control mode of Smart Mirror is realized by the infrared frame. The interaction solution is simple, which only needs to carry on the related position debugging after being connected with the infrared frame. Besides, the design is less demanded on the mirror itself, the mirror can be developed on the basis of the existing without using those expensive capacitive touch screens or resistance touch screens. In the meanwhile, the infrared frame mode requires less on the user's touch side compared with the capacitive touch screens or resistance touch screens, wet fingers, non-conductor touch bar, and any other thin rod shape objects are available with the Smart Mirror.

3.1.2 Voice interaction

At present, the development of cloud computing enables many terminal devices to interact with each other by cloud computing efficiently and accurately [8]. In the voice interaction part, this solution is based on the open source development platform of Baidu Voice Assistant, setting the voice recognition part in the cloud, and the voice recognition part is accomplished off line. The development interface is shown in Fig. 5. The benefits are mainly embodied in three following aspects: 1. The response mode is fixed, family members can set on their own mobile phones or PCs; 2. Processing speed is faster than traditional speech recognition, comparison test results are shown in Fig. 6; 3. Making the central platform safer, closed-source voice response processing makes the Smart Mirror emerging in the network security protection.

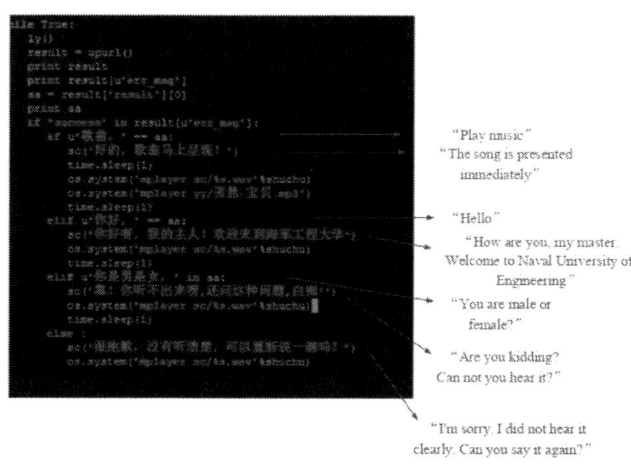

Figure 5. Voice processing

DOI: 10.1051/matecconf/201713900069

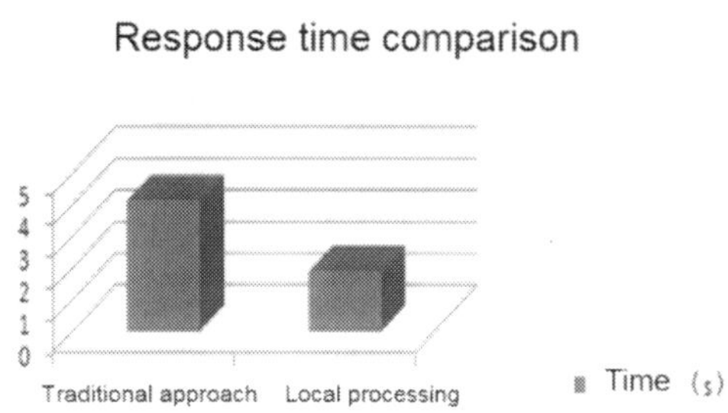

Figure 6. Response time comparison

3.2. Infrared wake up

The infrared wake-up function is effective in two aspects, user experience and energy saving. Raspberry Pi with a dedicated external infrared sensor module can make the infrared sensor through the Raspberry Pi by pin. The infrared control terminal is opened by the Raspberry Pi program, allowing the Smart Mirror wake up when people move or moving subjects are in front of the mirror, when the mirror is unmanned, the mirror acts as a common one.

3.3. Anti - theft monitoring and image acquisition

The Raspberry Pi used as the central processing unit has 4 USB communication ports, one of which is assigned to the camera so that the mirror has the function of capturing the scene in front of the mirror in real time. Based on the open source public number development platform of WeChat public client, the home environment can be real-time monitored by mobile digital terminals such as mobile phone, and users can decide the face recognition mode positive or not to control the opening of the intelligent function, which can increase the safety of home furnishing equipment.

3.4. The design of portable terminal

The solution proposed in this paper is not only valid in the development and function integration of intelligent home furnishing, but also provides a more convenient and fresher way for user's control. The main method is based on the development of mobile digital terminals and the intelligent home furnishing network. Nowadays, Nowadays, the electronic device that most people cannot do without in daily life is the mobile phone, laptop, tablet computer and vehicle receiving terminals are also included, with the help of them, people can synchronize daily reminders, refresh WeChat public service number information, control others intelligent home furnishings and so on. The interfaces are shown as Fig.7 and Fig.8.

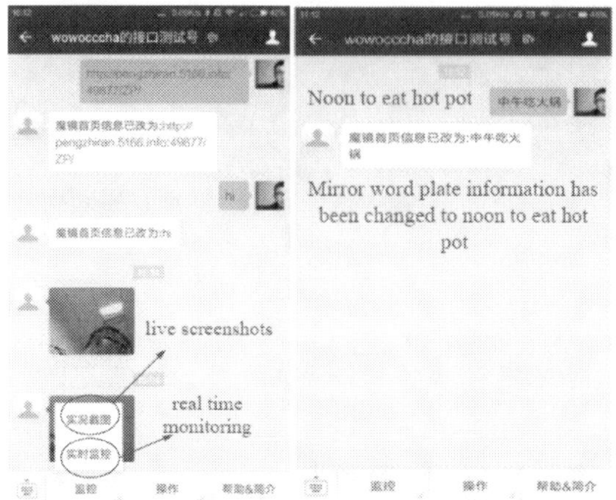

Figure 7. WeChat platform development

Figure 8. Memo sync

4 Software design

To carry out the relevant preparatory work in the development platform before the software design. The specific tasks are as follows: 1.The solution uses a 2G SD card, preferably a high-speed card that recommended more than Class4, due to the card directly affects the speed of Raspberry Pi faction running. 2. Download the Raspberry Pi system imaging file. 3. Configure the SD card, connect the keyboard and keyboard, power cord, HDMI cable and other hardware devices.

As a result of daily household life, intelligent home system of the various parts of the work is relatively independent [9], so in this solution the application is modular design. According to the smart home center platform of the various requirements, the system workflow of Smart Mirror is shown in Fig. 9.

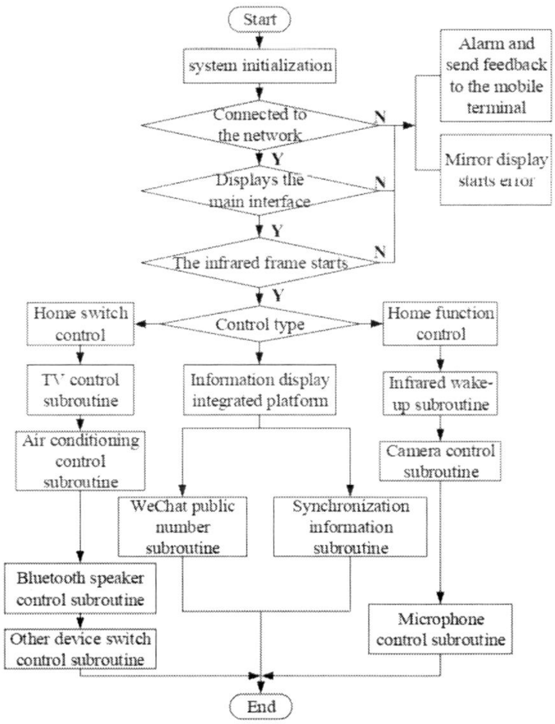

Figure. 9. System software design process

5 summary

Smart Mirror, as a solution of smart home center platform, can provide a physical terminal to the user in the operation and experience. At the same time without additional furniture, under the premise of efficient and intelligent it can provide users with high quality of life. Smart Mirror as a home center platform has many incomparable advantages which other devices do not have, such as relatively low cost, do not need additional development of the display screen. In addition, it does not contain the electromechanical system, so the external control system development is simple and so on. Overall, the Smart Mirror can provide a central platform for smart home in the development, application and so on. Based on this concept and design principles we designed the" Smart Magic Mirror" works and won the" first prize" in the national final teams of the eleventh "Huawei Cup" The China Graduate Electronics Design Contest; personal "second prize"; "first prize" in the national finals Smart Home Group of 2017 National College Students Connected Smarter System Innovation Competition. And the project is currently pending patent applications.

References

1. Zhang Weihua, Huang Xiaolin, Research on Intelligent Home Technology and Standardization of Internet of Things [J]. Information Technology & Standardization, 2012, 7: 35-38.
2. Liu Jiyuan, Application of Internet of Things Based on Raspberry [J]. Electronic World, 2016 (8): 24-25.
3. DAI Yupei, ZHANG Yu, LIU Yunfei, Intelligent home and building real-time monitoring system design [J]. MCU and embedded systems applications, 2017, 17 (2): 49-52.
4. Hao Linwei, Liang Ying, Research and Practical Application of Network Monitoring and Home Control System Based on Raspberry + Cloud Server [J]. Internet of Things Technology, 2016, 6 (9): 45-47
5. Zeng Guijing, Song Yue, He Zhihui, Hardware Design of a Intelligent Home Remote Control System [J]. Electronic Technology Applications, 2011, 37 (4): 81-84.
6. Huang ximin, TFT LCD - Future mainstream display technology [J]. World Electronic Components, 2001 (11): 9-10.
7. Li Lianyi, TFT-LCD liquid crystal display technology and application [J]. Journal of Shaoguan University, 2012, 33 (4): 46-49.
8. Chen Wei, Qin Huibin, Cao Shuguang, etc. Based on the Android platform intelligent home system design [J]. Electronic Technology Applications, 2015, 41 (10): 158-160.
9. Xin Hai-liang, Zhong Peisi, Zhu Shaoyi, et al. Intelligent home control system based on ZigBee [J]. Electronic Technology Applications, 2013, 39 (12): 79-81

A motion-planning method for dexterous hand operating a tool based on bionic analysis

Bo Wei, Congying Deng[*] *, Jie Zhao and Yi Zhang*

Chongqing University of Posts and Telecommunications, Chongqing, China

Abstract. In order to meet the needs of robot's operating tools for different types and sizes, the dexterous hand is studied by many scientific research institutions. However, the large number of joints in a dexterous hand leads to the difficulty of motion planning. Aiming at this problem, this paper proposes a planning method abased on BPNN inspired by human hands. Firstly, this paper analyses the structure and function of the human hand and summarizes its typical strategy of operation. Secondly, based on the manual operation strategy, the tools are classified according to the shape and the operation mode of the dexterous hand is presented. Thirdly, the BPNN is used to train the humanoid operation, and then output the operation plan. Finally, the simulating experiments of grasping simple tools and operating complex tools are made by MATLAB and ADAMS. The simulation verifies the effectiveness of this method.

1 Introduction

As one of the greatest human inventions in the 20th century, robots are widely used in many fields to replace or assist the human on work. To operate the complex labour tools, robots need a dexterous hand to accomplish so many kinds of tasks. The traditional end-effector of robot is too simple to operate kinds of tools. However, the dexterous hand with multiple joints has to face the problems of complex Motion planning. Therefore, this paper presents a planning strategy based on BP Neural Network which enables the dexterous hand to use a wide range of tools like human and complete the operation task.

In recent decades, a large number of dexterous hands are invited and some of them have been applied in the fields of military and outer space[1,2,3]. In 2010, Aaron m. Dollar and Robert d. Howe proposed a very simple manipulator design concept[4]. Although the manipulator can't complete the delicate operation, this design concept showed a manipulator ' s high adaptability to operate[5].Engineers from the German Space Center have developed the DLR Hand Arm System based on the full analysis of the human body's physiological structure. This hand is very flexible, but it also brings 38 motors and complex kinematic[6]. The robot astronaut Robonaut 2, developed by NASA, has a dexterous hand with 12 Dofs[7,8]. Shadow Company began to develop a five-finger hand with 24 joints in 2004[9,10].

In this paper, the motion planning of multi-degree-of-freedom manipulative dexterous hand is studied. First of all, the bionic system of the human hand is analyzed, including structure and function. Then, six classic grasping methods are summarized. Secondly, the operation mode of human hand is classified, and it is classified according to the shape of different tools. Thirdly, based on classification, BP neural network is used to plan the motion of dexterous hand. Finally, the simulation experiment proves the correctness of the operation plan.

2 BIONICS OF HUMAN HANDS

Nowadays, in the multi-fingered hand research, there is a contradiction between the mechanical configurations and the operating function. The contradiction can be described as if the dexterous hand is simple, its reliability is high relatively. But the hand will not complete accurate operation mission - and vice versa. After thousands of years of human evolution, human hands have found the balance between mechanical configurations and operating function.

Based on this, we analysed both the topological structure and function of a man's hand, and extract the characteristic of bionics significance, such as the functional analysis of a man's hand, coupling finger gestures with hands and finger strength and flexibility analysis. Through the analysis of hand function, this paper uses BPNN(back propagation neural network) to make the motion planning for dextrous hand. This method could improve the operation ability, and at the same time, the difficulty of control is reduced.

2.1. Structure of human hand

The structure of the hand is very complex. With the development of anatomy, modern medical research on human hand's skeleton structure has been perfected.

[*] Corresponding author: bowei@cqupt.edu.cn

DOI: 10.1051/matecconf/201713900070

A human hand consists of the thumb, forefinger, middle finger, ring finger and little finger, as well as a palm. That consists of 44 tendons and 27 bones. The skeletal structure is shown in Fig.1.

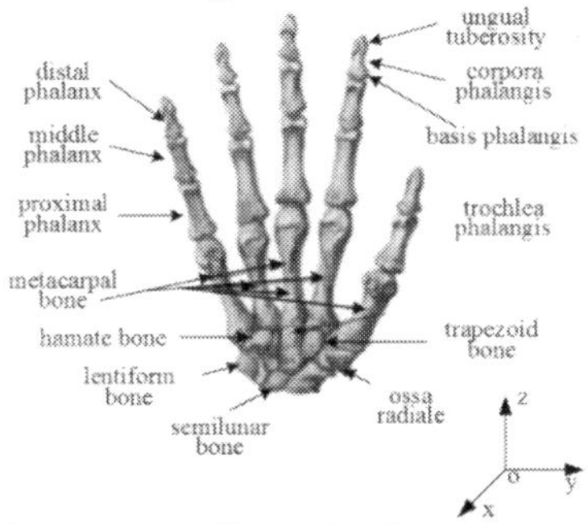

Fig.1. Skeletal structure of human hand

As is seen in Fig.1, three orthogonal axes are defined to describe the motion trajectory. The x axis is perpendicular to the ground and the direction is from top to bottom. The Y-axis is parallel to the ground, and the direction is horizontal extension. The z axis is parallel to the ground and the direction is from front to back. Motions of the hand are as below.

a) Internal Rotation and Outward Rotation. That joint rotats long the X-axis. If the joint roates towards the inner circle, it's called the inner rotation. If the joint roates towards the outer circle, it's called the outer rotation.

b) Flexion and Extension. That joint rotats long the Y-axis. If the angle between the two joints becomes smaller, it's called flexion. If the angle between the two joints becomes larger, it's called extension.

c) Inward and Outward Extending. that joint rotats long the Z-axis. If the joint rotats toward the origin point, it's called inward extending. The opposite is called outward extending.

d) Circular motion. The joint is rotated along the axial direction.

The base joint of human hand's finger has two degrees of freedom in two directions, which makes the griping more stable and the operation more dexterous.

When the operating target is so large that it cannot be fully enveloped by palm(Fig.2(a)), each finger moves sideways and the palm opens to increase the covering area of the hand to achieve a more reliable grip. When the operation is required to be precise, such as playing a guitar(Fig.2(b)) , the four fingers can make sure the hand to complete the precise operation.

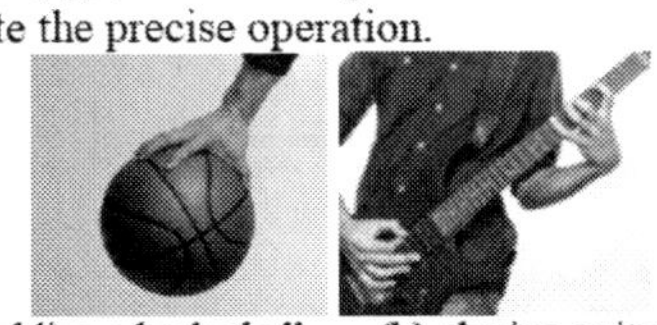

(a) holding a basketball (b) playing guitar

Fig.2. Motion ability of human fingers

2.2 Function of Human Hand

Even though the sizes and shapes of different human hands are different, the hands have similar functions. This is mainly due to the very reasonable structure and proportion of the hands. As can be seen in Fig.3, the shapes of two hands are different, but they have the same degree of freedom configuration and the proportions of each part are roughly the same. As a result, for the same object, the two hands operate roughly the same way[12].

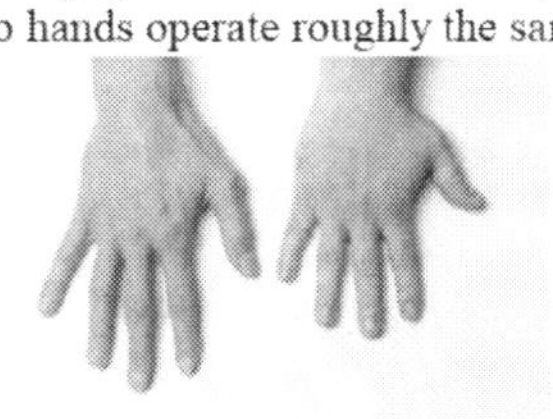

Fig.3. The sizes of different don's influence the operating function

In the study of human hand anatomy, there are many kinds of opration mode classification. The hand opration modes can be divided into six types, as shown in Fig.4.

a) Strong grasp. The thumb is relative to the other four fingers and is completely bent, such as holding a tool.

b) Pinching. The thumb and forefinger touch each other together, such as pinching a needle.

c) Lift. The four fingers, beside the thumb, are used to lift a target, such as a bag.

d) Tripod pinc. The index finger, thumb and middle finger are combined to opearte somthing, such as pen.

e) Sphere grasp. All five fingers bend and grasp, such as grasp a ball.

f) Lateral pinch. A side of the index finger is opposite to the inside of the thumb, such as pinching a key.

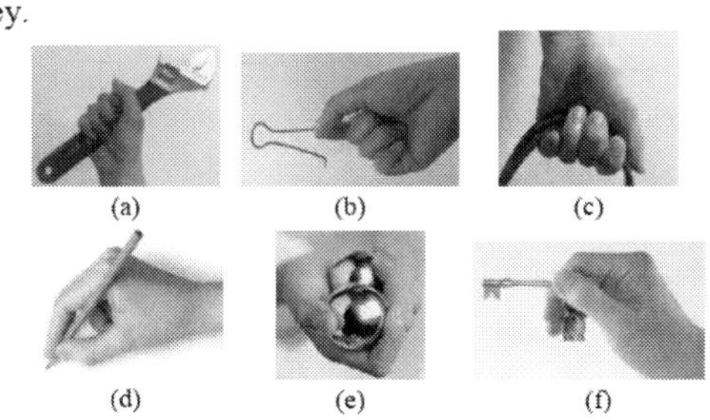

Fig.4. Six operating modes of hunman hand

3 Operation strategy of dextrous hand

For different tools, the way dextrous hand operates is different. And for the same kind of tool, because of the tool size and its open-close state, the operating position and the fingers' degrees of motion are different. In the previous section, the operation modes of human hand has been analysed. In this section, the operation mode of dextrous hand is judged by BP neural network and the joint angles of each joint are also given by it.

3.1. Operation mode analysis of Dexterous hand

After analyzing the function of human hand, the operation methods vary from tool to tool. Therefore, in order to satisfy the operation demand of different kinds

DOI: 10.1051/matecconf/201713900070

of tools, the operating methods of tools in different shapes are classified in Fig.5.

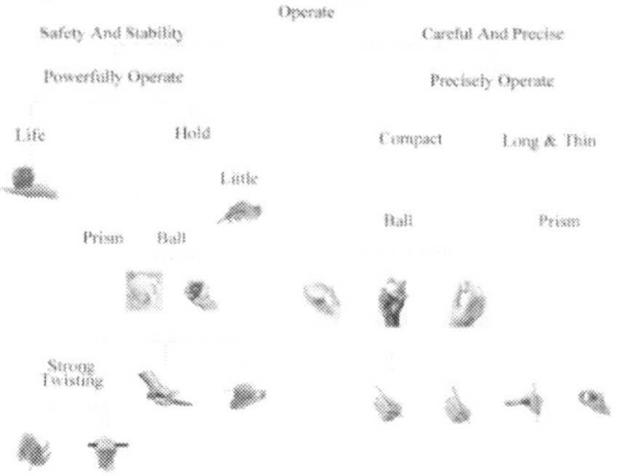

Fig.5. Operating methods of tools in different shapes

3.2 Motion planning based on BP neural network

The way how human hands operate a tool mainly depends on its shape, size and current state. Mostly, the operation mode is based on knowledge base. However, it takes a long process to build a knowledge base and it is difficult to guarantee the integrity of the knowledge base. The artificial network has the ability to learn rapidly, complete the complex nonlinear mapping and deduce the very approximate conclusions from incomplete input. Therefore, inspired by the way human hands operation, the artificial neural network is used to make the motion plan of dexterous hand. And because BP neural network has good self-learning ability, generalization ability and fault tolerance, the BP neural network(BPNN) is selected to complete the motion planning.

A pattern recognition system mainly includes training process and identification process. The training process is mainly based on a large number of samples and a designed classifier. The identification is the process of matching unknown patterns with trained classifiers. The core of BPNN the correction of the network weight value and the threshold value, so that the error function decreases along the gradient direction.

In this paper, we selected a three-layer BPNN for identification. As can be seen in Fig.6, the tools are divided into three categories: gun style tool, clamp style tool and stick style tool. The transformation function of implicit layer is generally nonlinear function, such as s-type function or hyperbolic tangent function. The transformation function of the output layer can be non-linear or linear, which is determined by the need for input and output mapping relationships.

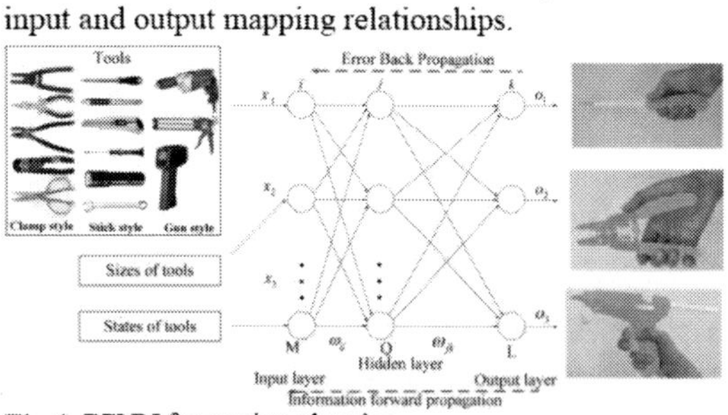

Fig.6. BPNN for motion planning

3.3 BPNN for dextrous hand operation

According to the task requirements of the dexterous hand operation tool, the neural network structure is set as follows. The input layer has M nodes. The hidden layer has Q nodes, which are the connection weights between the input layer and hidden layer. The output layer has L nodes, which are the connection weights between the hidden layer and output layer. The input values of the hidden layer and output layer are the weighted sum of the former layer. The degree of excitation of each node is determined by its excitation function. The Constant distance characteristics are put in the network and the sampling training is carried, then, sample full training is carried.

If $E<0.01$ for 5 consecutive times, the training is finished. If the training times are over 5000 times, the training is failure. In order to increase the robustness of the system, the output of the BPNN is the grade of membership of a tool. If the maximum output is smaller than 0.6, that means the target is not belonging any style. If there are several maximum outputs, that means the judge is wrong.

4 Eexperiments and Results

4.1 Parameters of BPNN

4.1.1 Input parameters

The tools are classified as 3 types. The type is defined as t(t=1,2,3). t=1 means clamps style, t=2 means stick style, t=3 means gun style.
Classification of tools, tool type t(t=1,2,3) as an input to BP neural network.

Even though the tools are the same type, they may have different sizes. Therefore, the tool size is also an input to the neural network. The tool size is set as the minimum envelope radius R in the plane of the tool.

Besides, the open and closed state of the clamping tool also affects the angles of finger joints. So, the open and closed state is also the input of the neural network.

In addition, the open and closed state of the clamping tool also affects the Angle of finger grasping, so it is also the input of the neural network.

4.1.2 Output parameters

The output of the neural network is defined as the position of the fingertip relatively to the wrist. Then the angles of the finger joints are calculated through inverse kinematics.

4.1.3 Setting of network structure

The input layer includes the tool type, open-close state and the minimum envelope circle radius. So the node number of input layer m is 3.

The node number of the output layer is related to the structure of the dexterous hand. In order to simplify the neural network, the output is set depending on the operation gesture. There are 15 kinds of gesture types in Fig.5. Adding the open-close state, the node number of output layer n is 16

The hidden number of the output layer is determined by empirical function $s=\sqrt{m+n}+a$ or $s=\sqrt{0.43mn+0.12n^2+2.54m+0.77n+0.35}+0.51$, where a is an integer between 1 and 10. According to experience, the number of hidden layer nodes is 10.

The transfer function between the input layer and the hidden layer is a kind of log-sigmoid function. The transfer function between the hidden layer and the output layer is a linear function.

4.1.4 Network value adjustment

In the iterative process, the change of the error function is used to adjust η .

$$\eta(t)=\begin{cases}1.2\eta(t-1)\cdots E(n)<E(n-1)\\0.8\eta(t-1)\cdots E(n)<1.05E(n-1)\\1.2\eta(t-1)\cdots E(n-1)<E(n)1.05E(n-1)\end{cases}\tag{1}$$

In order to ensure that each learning can be effectively corrected and accelerate the convergence process, the output of log-sigmoid is limited.

$$f(x)=\begin{cases}0.01\cdots f(x)<0.01\\0.99\cdots f(x)>0.99\\f(x)\cdots 0.01<f(x)<0.99\end{cases}\tag{2}$$

The parameters are initialized as:

$$\eta=\frac{1}{\sqrt{s}}\tag{3}$$

$$\alpha=0.8\tag{4}$$

4.2. Simulation results

MATLAB and ADAMS are used to simulate dexterous hand to operate tools. As can be seen in Fig.6, firstly, tools and dexterous hand are established in the ADAMS. Secondly, the shape, size and the open-close state are input to MATLAB. Thirdly, the positions between the fingertips and the wrist are deduced from BPNN. Fourthly, the angles of each finger's joints are calculated by inverse kinematics. Finally, the angles are input to ADAMS and the hand is motioned.

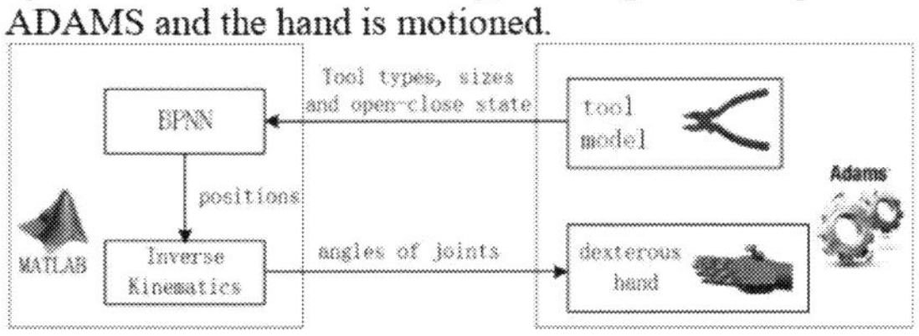

Fig.7. Dexterous hand operation simulation platform

4.2.1 Tools in simple shape

Three types of tools (long-cylinder, square, plate) are selected for the simulation experiments.

A. Cylinder

Select the cylindrical object to carry out the simulation experiment. The flashlight and the handle are tools in cylindrical shape.

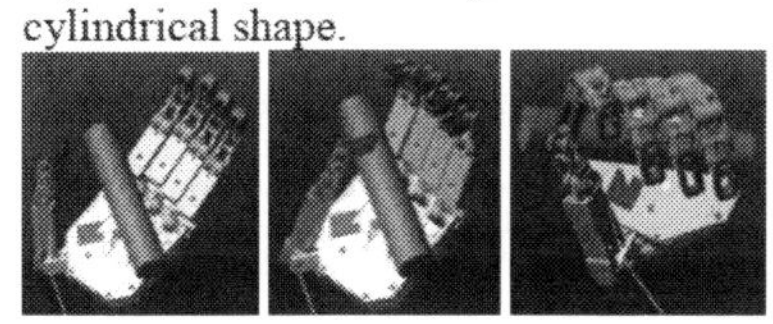

Fig.8. The grasping process of a cylindrical object

It can be seen from Fig.8 that when the dexterous hand is grasping the cylindrical object, the finger joints can be bent in order to reach the surface of the cylindrical object.

B. Square

Select the square object to carry out the simulation experiment. Square objects are common in experimental tools, such as boxes.

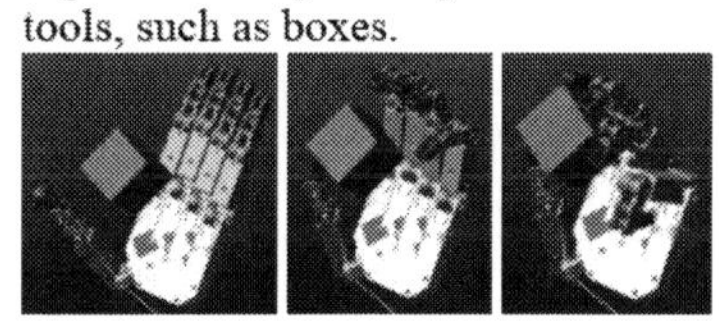

Fig.9. The grasping process of a square object

As can be seen from Fig.9, when grasping the target is as a square object, the fingers cannot make the knuckle close to the object as it can grasp a cylindrical object. Therefore, according to the gesture got from the BP neural network, the fingers bend to make sure the target is enveloped as much as possible.

C. Plate objects

Select the Board object to carry out the simulation experiment. In commonly used tools, the plate objects are often used as operating objects, such as planks, bricks, etc.

Fig.10. The grasping process of a plate object

As can be seen from Fig.10, the fingers cannot completely envelop the target in plate shape. In order to make the grasp more stable, the end knuckle should be as close as possible to the target.

4.2.2 Tools in Complex shape

Set up clamp in ADAMS, and select two different open and closed states.

Table 1 shows the size and the closed state of the tool. Table 2 shows the angles of finger joints after MATLAB calculation. Fig.11 shows the operation motion of the dexterous hand.

 DOI: 10.1051/matecconf/201713900070

Table 1. Size and closed state of the clamp

size	X	8.3mm
	Y	14.7mm
open angle	0.9°	

Table 2. Angles of finger joints

Finger	joint 1(°)	joint2(°)	joint 3(°)
1	103.3	98.4	88.6
2	25.1	68.9	100.2
3	29.6	70.4	90.4
4	15.1	69.8	80.2
5	10.3	40.2	90.7

Fig.11. The dexterous hand operate a clamp in closed state

Table 3 shows the size and the open state of the tool. Table 4 shows the angles of finger joints after MATLAB calculation. Fig.12 shows the operation motion of the dexterous hand.

Tab.3. Size and closed state of the clamp

size	X	14mm
	Y	14mm
open angle	21°	

Tab.4 Angles of finger joints

Finger	joint 1(°)	joint2(°)	joint 3(°)
1	102.9	98.2	89.1
2	-5.2	29.8	90.4
3	0.3	27.6	89.7
4	5.4	0.1	88.9
5	5.1	0.3	10.2

Fig.12. The dexterous hand operate a clamp in open state

It can be seen from the simulation results that motion planning method base on BPNN method is able to perform the motion task of planning finger joint angles and finally operate the tool.

5. Conclusion

The motion planning is an important issue in the robotic system. In this paper, BPNN was used to plan the rotation angles of each joint when the dexterous hand operate a tool combining with inverse kinematics of fingers. Finally, the simulation experiments verifies the effectiveness of this method. In the future, we tend to make a prototype of the dexterous hand and experiment this method on the hand of the machine.

Acknowledgments

This work was supported in part by the National Natural Science Foundation of China under Grant 61703067 and 51705058, Scientific and Technological Research Program of Chongqing Municipal Education Commission under Grant KJ1704072, Chongqing Research Program of Basic Research and Frontier Technology under Grant cstc2017jcyjAX0212.

References

1. M. T. Mason, J. K. Salisbury. Robot Hands and the Mechanics of Manipulation. MIT Press, Cambridge, USA, 1985: 3~93.
2. Lachnit H, Piepar W. Speed and accumcy effects of fingers and dexterity in 5-choice reaction tasks[J] Ergonomics, 1990, 33(12): 1443—1454.
3. Martin BJ, Armstrong TJ, Foulke JA, et d. Key — board reactical force and finger flexor electromyograms during OoInputer keyborad work[Jj. Human Factors, 1996, 38(4): 654—664.
4. Dollar A M. The Highly Adaptive SDM Hand: Design and Performance Evaluation[J]. International Journal of Robotics Research, 2010, 29:585-597.
5. Cutkosky M R, Kao I. Computing and controlling compliance of a robotic hand[J]. Robotics & Automation IEEE Transactions on, 1989, 5(2):151-165.
6. Grebenstein M, Chalon M, Friedl W, et al. The hand of the DLR Hand Arm System: Designed for interaction[J]. The International Journal of Robotics Research, 2012, 31:1531-1555.
7. Lovchik C, Diftler M. The robonaut hand: a dexterous robot hand for space[C]//Proceedings of the IEEE International Conference on Automation and Robotics. Detroit, Michigan, 1999:907-912.
8. Bridgwater L B, Ihrke C A, Diftler M A 等. The Robonaut 2 hand - designed to do work with tools[C]. //Robotics and Automation, IEEE International Conference on. IEEE, 2012:3425 – 3430.
9. Rothling F, Haschke R, Steil J J 等. Platform portable anthropomorphic grasping with the bielefeld 20-DOF shadow and 9-DOF TUM hand[C]. //Intelligent Robots and Systems, 2007. IROS 2007. IEEE/RSJ International Conference on. IEEE, 2007:2951 - 2956.
10. Shadow Robot Company. Shadow Dextrous Hand Technical Specification [Z]. 2005,2.

Strategic project selection based on evidential reasoning approach for high-end equipment manufacturing industry

Guangyan Lu[1,*], *Min* Xue[2,3], *Yin* Liu[2,3], and *Wenjun* Chang[2,3]

[1]Department of Operation Management, CRRC Qishuyan Institute Co., Ltd, Changzhou 213011, Jiangsu, P.R.China
[2]School of Management, Hefei University of Technology, Hefei, Box 270, Hefei 230009, Anhui, P.R. China
[3]Key Laboratory of Process Optimization and Intelligent Decision-making, Ministry of Education, Hefei, Box 270, Hefei 230009, Anhui, P.R. China

Abstract. With the rapid development of science and technology, emerging information technologies have significantly changed the daily life of people. In such context, strategic project selection for high-end equipment manufacturing industries faces more and more complexities and uncertainties with the consideration of several complex criteria. For example, a group of experts rather than a single expert should be invited to select strategic project for high-end equipment manufacturing industries and the experts may feel difficulty to express their preferences towards different strategic projects due to their limited cognitive capabilities. In order to handle these complexities and uncertainties, the criteria framework of strategic project selection is firstly constructed based on the characteristics of high-end equipment manufacturing industries and then evidential reasoning (ER) approach is introduced in this paper to help experts express their uncertain preferences and aggregate these preferences to generate an appropriate strategic project. A real case of strategic project selection in a high-speed train manufacturing enterprise is investigated to demonstrate the validity of the ER approach in solving strategic project selection problem.
Keywords. Strategic project selection, high-end equipment manufacturing, evidential reasoning approach, multiple criteria group decision making

1 Introduction

High-end equipment manufacturing industries mainly including aviation industry, satellite industry, rail transportation equipment industry, ocean engineering equipment and intelligent manufacturing equipment industry, are considered as an important symbol for the level of science and technology in a nation [1-2]. Emerging information technologies which have driven a new round of revolution in almost all aspects of high-end equipment manufacturing industries, not only create a large amount of opportunities for the development of high-end equipment manufacturing but also bring severe challenges to them. On the one hand, the development of high-end equipment manufacturing enterprises has gradually stepped into a stable stage after experienced a rapid growing stage. On the other hand, the profits and space of the existing market have been constantly compressed and meanwhile the marker shares tend to stabilization.

In order to copy with such challenges to achieve sustainable profits and development, a good choice for high-end equipment manufacturing enterprises is to exploit a new strategic industry or business by selecting an appropriate strategic project [3]. An appropriate strategic project will help high-end equipment manufacturing enterprises expand market shares, increase market profits, improve market reputation to achieve long term and stable competitiveness. However, the problem of selecting a strategic project is not easy in practice [4-5]. Many complex factors should be considered in the process of solving this problem such as the ability to conform the policies of the nation and the industry, market potential, market profitability, the maturity of design and manufacturing technology, human resources and capital resource (See Section 2). Under such condition, it is difficult for a decision maker alone to directly select an appropriate strategic project and a group of experts rather than a single expert alone should be invited. Meanwhile, due to the limited knowledge, experience and cognitive capability of the experts, they may feel uncertain to provide exact preferences about strategic projects. From the above analysis, the strategic project selection can be modeled as a multiple criteria group decision making (MCGDM) problem with uncertainties [6]. An appropriate method is needed to copy with this problem.

With the aim of handling this problem, in this paper, group evidential reasoning (ER) approach, as a useful uncertain MCGDM method, is introduced to help the decision makers select an appropriate strategic project. ER approach [7] is developed based on the decision theory and Dempster-Shafer theory of evidence [8] and allows the decision makers provide their assessments in terms of their limited knowledge, experience and cognitive capability by using a distributed belief framework. That is, the decision makers can express their preferences about the degree to which a strategic project is poor, good and average simultaneously. More importantly, the kernel of the ER approach is an ER algorithm to combine assessments provided by each decision maker on each attribute with corresponding weights objectively in order to obtain the collective assessment of each strategic project. Utility is introduced in ER approach to portray the incomplete cognition of the decision maker towards to decision problem. In this situation, utility interval will be generated to generate an appropriate strategic project with the help of minimal satisfaction rule. Finally, a real strategic project selection problem of a high-speed train manufacturing enterprise located in Changzhou, a city in Jiangsu Province of China, is solved by the ER approach.

The rest of this paper is organized as follows. Section 2 constructs the criterion framework for selecting strategic project. Section 3 introduces the ER approach. Section 4 demonstrates a real case of the strategic project selection problem. The paper is concluded in Section 5.

2 Criteria for strategic project selection

Strategic project selection plays a key role in the long-term development of high-end equipment manufacturing enterprises. More importantly, how to select an appropriate strategic project is considered as a complex MCGDM problem as mentioned in Introduction. With the aim of solving this problem, criteria framework is firstly constructed by reviewing the existing studies and investigating some enterprises, which includes sixteen criteria belonging to the four perspectives such as business, market, technique and resource as shown in Table 1.

Table 1. Criteria framework for strategic project selection

Perspective	Criteria
Business	Ability to conform the policies of the nation (PN)
	Ability to conform the policies of the industry (PI)
	Capability of being in line with the development strategy of the enterprise (DSE)
Market	Potential capacity of market (CM1)
	Potential growth rate of market (GRM)
	Competition of market (CM2)
	Market Profitability (PM)
Technical maturity	Maturity of design technology (DT)
	Maturity of material technology (MT1)
	Maturity of experimental testing technology (ETT)
	Maturity of manufacturing technology (MT2)
Resource availability	Capital resource (FR)
	Human resource (HR)
	Equipment resource (ER)
	Resource from supply chain (SCR)
	Outside Technical support (TS)

Table 1 shows that sixteen criteria are included in the framework for the strategic project selection. Several criteria belonging to different perspectives are selected as typical examples to explain their meanings. PN and DSE belonging to the perspective of business mean that the extent to which a project conforms nation policy and development strategy of the enterprise, respectively. PDM and PM belonging to the perspective of market mean the potential size of a market for products and the maximum possible return on an investment, respectively. DT belonging to the perspective of technique means how advanced the proposed design technology is compared with existing design technology. FR and HR belonging to the perspective of resource mean that how much capital and manpower can be obtained, respectively.

3 Strategic project selection based on ER approach

In order to copy with strategic project selection, ER approach as a useful MCGDM method is introduced in this section.

3.1. Modeling of strategic project selection problem

Suppose that a strategic project selection problem includes T experts t_j (j = 1, …, T). The relative weights of the T experts on criterion e_i for project a_l are denoted by $\lambda(e_i) = (\lambda^1(e_i), \lambda^2(e_i), \ldots, \lambda^T(e_i))$ such that

$$0 \leq \lambda^j(e_i) \leq 1 \text{ and } \sum_{j=1}^{T} \lambda^j(e_i) = 1. \quad (1)$$

All experts are invited to handle a common strategic project selection problem which has M projects a_l (l = 1, …, M) and L criteria e_i (i = 1, …, L). The relative weights of the L criteria are denoted by $w = (w_1, w_2, \ldots, w_L)$ such that

$0 \le w_i \le 1$ and $\sum_{i=1}^{L} w_i = 1$. (2)

In general, it may be difficult for the experts to provide completely exact assessments of strategic projects due to their limited knowledge, experience, and cognitive capability. Under such condition, belief distribution in ER approach is a useful tool to help express uncertain assessments.

In ER approach, assume $\Omega = \{H_1, H_2, \ldots, H_N\}$ symbolizes a set of grades which is increasingly ordered from worst to best. That is, the utilities of grades $u(H_n)$ ($n = 1, \ldots, N$) satisfy the constraint $0 = u(H_1) < u(H_2) < \ldots < u(H_N) = 1$. The M projects are assessed at the L criteria using H_n ($n = 1, \ldots, N$). Let expert t_j evaluate project a_l on criterion e_i to grade H_n with a belief degree of $\beta^j_{n,i}(a_l)$, then the assessment can be profiled by $B^j(e_i(a_l)) = \{(H_n, \beta^j_{n,i}(a_l)), n = 1, \ldots, N; (\Omega, \beta^j_{\Omega,i}(a_l))\}$, where $\beta^j_{n,i}(a_l) \ge 0$, $\sum_{n=1}^{N}\beta^j_{n,i}(a_l) \le 1$, and $\beta^j_{\Omega,i}(a_l) = 1 - \sum_{n=1}^{N}\beta^j_{n,i}(a_l)$ represents the degree of global ignorance. If $\beta^j_{\Omega,i}(a_l) = 0$, the assessment is complete; otherwise, it is incomplete.

After all projects are assessed by each expert on L criteria, the belief decision matrices will be generated as

$$D^j = B^j(e_i(a_l))_{L\times M}. \quad (3)$$

3.2. Aggregation of assessments based on ER approach

When decision matrices are obtained, the next step is to aggregate assessments of each project provided by T experts on L criteria to generate the collective assessment of each project. In other words, the generation of the collective assessment of project a_l includes two steps:

(1) $B^j(e_i(a_l))$ is combined using the ER approach with $\lambda^j(e_i)$ to generate the aggregated group assessment of project a_l on criterion e_i denoted by $B(e_i(a_l)) = \{(H_n, \beta_{n,i}(a_l)), n = 1, \ldots, N; (\Omega, \beta_{\Omega,i}(a_l))\}$.

(2) $B(e_i(a_l))$ is combined using the ER approach again with w_i to generate the collective assessment of project a_l denoted by $B(a_l) = \{(H_n, \beta_n(a_l)), n = 1, \ldots, N; (\Omega, \beta_\Omega(a_l))\}$.

In order to complete above aggregation, the ER approach is demonstrated as follows.

Given $D^j = B^j(e_i(a_l))_{L\times M}$ where $B^j(e_i(a_l)) = \{(H_n, \beta^j_{n,i}(a_l)), n = 1, \ldots, N; (\Omega, \beta^j_{\Omega,i}(a_l))\}$ as modeled in Section 3.1, some associated concepts can be firstly defined for project a_l on criteria e_i.

$$m_{n,j} = \lambda^j(e_i)\cdot \beta^j_{n,i}(a_l),\ n = 1, \ldots, N, j = 1, \ldots, T, \quad (4)$$

$$m_{\Omega,j} = 1 - \sum_{n=1}^{N} m_{n,j} = 1 - \lambda^j(e_i)\cdot \sum_{n=1}^{N}\beta^j_{n,i}(a_l),\ j = 1, \ldots, T, \quad (5)$$

$$\bar{m}_{\Omega,j} = 1 - \lambda^j(e_i), j = 1, \ldots, T, \quad (6)$$

$$\hat{m}_{\Omega,j} = \lambda^j(e_i)\cdot(1 - \sum_{n=1}^{N}\beta^j_{n,i}(a_l)), j = 1, \ldots, T, \quad (7)$$

$$\text{with } m_{\Omega,j} = \bar{m}_{\Omega,j} + \hat{m}_{\Omega,j}, \quad (8)$$

where, $m_{n,j}$ means a basic probability mass profiling the degree to which the jth expert supports the hypothesis that project a_l on criteria e_i is assessed to the nth grade H_n and $m_{\Omega,j}$ denotes the remaining probability mass unassigned to any individual grade after all N probability masses $m_{n,j}$ ($n = 1, \ldots, N$) have been assessed.

Based on Eqs.(4)-(8), $B(e_i(a_l))$ can be generated by using the ER approach [7, 9] as follows. Given any two assessments denoted by $B^1(e_i(a_l))$ and $B^2(e_i(a_l))$, their aggregated assessment can be represented by $B(2) = \{(H_n, \beta_{n,B(2)}), n = 1, \ldots, N; (\Omega, \beta_{\Omega,B(2)})\}$ with

$$\beta_{n,B(2)} = \begin{cases} 0 & n = \varnothing \\ \dfrac{\hat{m}_{n,B(2)}}{\sum_{D\subseteq\Omega}\hat{m}_{D,B(2)}} & n \subseteq \Omega, n \ne \varnothing \end{cases} \quad (9)$$

$$\hat{m}_{n,B(2)} = [(1 - w_2)\cdot m_{n,1} + (1 - w_1)\cdot m_{n,2}] + \sum_{E\cap F=n} m_{E,1}\cdot m_{F,1},\ \forall n \subseteq \Omega \quad (10)$$

The residual support of $B(2)$ symbolized by $\hat{m}_{P(\Omega),B(2)} = (1 - w_2)\cdot(1 - w_1)$ is reassigned back to all elements of Ω and thus not involved in $B(2)$, as presented in Eq. (9). When the first j assessments are combined, the recursive ER approach is described as follows.

$$\beta_{n,B(j)} = \begin{cases} 0 & n = \varnothing \\ \dfrac{\hat{m}_{n,B(j)}}{\sum_{D\subseteq\Omega}\hat{m}_{D,B(j)}} & n \subseteq \Omega, n \ne \varnothing \end{cases}, \quad (11)$$

$$m_{n,B(j)} = \begin{cases} 0 & n = \varnothing \\ \dfrac{\hat{m}_{n,B(j)}}{\sum_{D\subseteq\Omega}\hat{m}_{D,B(j)} + \hat{m}_{P(\Omega),B(j)}} & n \subseteq \Omega, n \ne \varnothing \end{cases}, \quad (12)$$

$$m_{P(\Omega),B(j)} = \frac{\hat{m}_{P(\Omega),B(j)}}{\sum_{D\subseteq\Omega}\hat{m}_{D,B(j)} + \hat{m}_{P(\Omega),B(j)}}, \quad (13)$$

$$\hat{m}_{n,B(j)} = [(1 - w_i)\cdot m_{n,B(j-1)} + m_{P(\Omega),B(j-1)}\cdot m_{n,j}] + \sum_{E\cap F=n} m_{E,B(j-1)}\cdot m_{F,j},\ \forall n \subseteq \Omega, \text{ and} \quad (14)$$

$$\hat{m}_{P(\Omega),B(j)} = (1 - w_i)\cdot m_{P(\Omega)B(j-1)}, \quad (15)$$

where $0 \le p_{n,B(j)}$, $m_{n,B(j)} \le 1$, $\forall n \subseteq \Omega$, $0 \le m_{P(\Omega),B(j)} \le 1$, and $\sum_{n\subseteq\Omega} m_{n,B(j)} + m_{P(\Omega),B(j)} = 1$ for $j = 2, \ldots, T$ recursively.

By using Eqs. (4)-(15), $B(e_i(a_l))$ can be calculated. After that, $B(a_l) = \{(H_n, \beta_n(a_l)), n = 1, \ldots, N; (\Omega, \beta_\Omega(a_l))\}$ will be generated by using the recursive ER approach in Eqs. (4)-(15) again.

3.3. Generation of an appropriate strategic project

With the aim of generating an appropriate strategic project, the minimum and maximum expected utilities of project a_l (Chin and Fu 2014) are calculated as follows.

$$u_{min}(a_l) = \sum_{n=2}^{N} \beta_n(a_l)u(H_n) + (\beta_1(a_l) + \beta_\Omega(a_l))u(H_1) \quad (16)$$

$$u_{max}(a_l) = \sum_{n=1}^{N-1} \beta_n(a_l)u(H_n) + (\beta_N(a_l) + \beta_\Omega(a_l))u(H_N) \quad (17)$$

With the aid of $u_{min}(a_l)$ and $u_{max}(a_l)$, the minimal satisfaction of project a_l can be obtained as

$$V(a_l) = u_{\min}(a_l) - \max_{m \neq l}\{u_{\max}(a_m)\} \; (l = 1, \ldots, M) \quad (18)$$

where $V(a_l) \in [-1, 1]$ due to $0 \le u_{\min}(a_l) \le u_{\max}(a_l) \le 1$ (l = 1, …, M). $V(a_l)$ is used to measure the gain from selecting the project a_l under the worst case scenario when there is unknown in the performances of any projects. A project with larger minimal satisfaction is more preferred to other project.

4 Case study

In this section, a real strategic project selection problem is investigated by the above ER approach.

4.1. Description of the strategic project selection problem

In 2016, "the Thirteenth Five-Year Plan for National Economic and Social Development of the People's Republic of China" is enacted to promote the rapid development of the economy and society and improve the level of living in China. The plan puts forward some new requirements for the development of manufacturing industries where an important one is to exploit new industries for the purpose of helping manufacturing industries achieve sustainable competitiveness. Under such condition, a high-end equipment manufacturing enterprise in Changzhou, a city in Jiangsu Province of China, aims to select a strategic project to develop a strategic emerging industry, which is crucial to the long-term development of the enterprise. Following the goal, the enterprise has analyzed its industrial foundation and advantage and further identified four projects as the candidates of strategic projects. They are the driving system of new energy automobile (DSNEA), the RV reducer of industrial robot (RVRIR), the reducer of shield-driven machine (RSDM), and the cleaning machine with high pressure water jet (CMHPWJ).

In order to solve this selection problem, five experts are firstly invited to be the decision makers from three departments which are the department of operation and management, the department of integrated technology and equipment and the department of technology planning and development. Suppose that the five decision makers are denoted by t_j (j = 1, …, 5), the four projects are denoted by a_l (l = 1, …, 4), and the sixteen criteria are denoted by e_i (i = 1, …, 16). After studying the documents concerning the criteria, the decision makers specify the criterion weights denoted by **w** = (0.1, 0.03, 0.045, 0.1, 0.08, 0.08, 0.06, 0.06, 0.05, 0.04, 0.05, 0.0625, 0.0625, 0.03, 0.1, 0.05). The four projects are assessed by each decision maker on all criteria listed in Table 1 by using a set of assessment grades $\Omega = \{H_n, n = 1, \ldots, 5\}$ = {*Very Poor*, *Poor*, *Average*, *Good*, *Very Good*} = {*VP*, *P*, *A*, *G*, *VG*}. Due to the limited length of this paper, assessments of a_l (l = 1, …, 4) on e_1 and e_2 provided by t_j (j = 1, …, 5) are demonstrated as an example shown in Table 2. Other assessments are omitted.

Table 2. Assessments on a_l (l = 1, …, 4) on e_i (i = 1, 2) provided by t_j (j = 1, …, 5)

Criterion	a_1	a_2
e_1	t_1: {*VG*, 0.8; *G*, 0.2} t_2: {*VG*, 0.9; *A*, 0.1} t_3: {*VG*, 1} t_4: {*VG*, 0.9; *A*, 0.05; *P*, 0.05} t_5: {*VG*, 1}	t_1: {*VG*, 0.9; *G*, 0.1} t_2: {*VG*, 0.8; *A*, 0.2} t_3: {*VG*, 1} t_4: {*VG*, 1} t_5: {*VG*, 1}
e_2	t_1: {*VG*, 0.9; *G*, 0.1} t_2: {*VG*, 0.8; *A*, 0.2} t_3: {*VG*, 0.2; *G*, 0.8} t_4: {*VG*, 0.8; *G*, 0.1; *VP*, 0.1} t_5: {*G*, 0.8; *VP*, 0.2}	t_1: {*VG*, 0.9; *G*, 0.1} t_2: {*VG*, 0.8; *A*, 0.2} t_3: {*VG*, 0.2; *G*, 0.8} t_4: {*VG*, 0.95; *A*, 0.05} t_5: {*VG*, 1}

Table 2. (continued) Assessments on a_l (l = 1, …, 4) on e_i (i = 1, 2) provided by t_j (j = 1, …, 5)

Criterion	a_3	a_4
e_1	t_1: {*G*, 0.6; *A*, 0.3; *Ω*, 0.1} t_2: {*G*, 0.6; *A*, 0.4} t_3: {*VG*, 0.5; *G*, 0.5} t_4: {*G*, 0.5; *A*, 0.5} t_5: {*VG*, 0.8; *G*, 0.2}	t_1: {*G*, 0.5; *A*, 0.5} t_2: {*VG*, 0.8; *A*, 0.2} t_3: {*VG*, 0.6; *G*, 0.4} t_4: {*VG*, 0.9; *G*, 0.1} t_5: {*A*, 0.9; *Ω*, 0.1}
e_2	t_1: {*G*, 0.6; *A*, 0.4} t_2: { *G*, 0.6; *A*, 0.4} t_3: { *VG*, 0.3; *G*, 0.7} t_4: { *G*, 0.5; *A*, 0.5} t_5: { *G*, 0.7; *Ω*, 0.3}	t_1: {*A*, 0.3; *P*, 0.2; *Ω*, 0.3} t_2: { *VG*, 0.7; *A*, 0.3} t_3: { *VG*, 0.3; *G*, 0.7} t_4: { *VG*, 0.9; *G*, 0.1} t_5: { *A*, 0.9; *Ω*, 0.1}

4.2. Generation of a solution to strategic projects selection problem

To find the solution to the project selection problem, assessments provided by each decision maker on all criteria are aggregated to form collective assessment $B(a_l)$ (l = 1, …, 4) by using Eqs. (4)-(15) in Table 3.

Table 3. Collective assessments of the four projects

l	$B(a_l)$ (l = 1, …, 4)

1	{(VP, 0.0198), (P_2, 0.052), (A_3, 0.2873), (G_4, 0.3226), (VG_5, 0.299), (Ω, 0.0194)}
2	{(VP, 0.0261), (P_2, 0.1136), (A_3, 0.1767), (G_4, 0.2044), (VG_5, 0.4494), (Ω, 0.0298)}
3	{(VP, 0.0098), (P_2, 0.1196), (A_3, 0.3315), (G_4, 0.3365), (VG_5, 0.1676), (Ω, 0.035)}
4	{(VP, 0.0069), (P_2, 0.0672), (A_3, 0.2485), (G_4, 0.3197), (VG_5, 0.3014), (Ω, 0.0565)}

The utilities of the grades are provided by the decision maker as follows: $u(VP) = 0$, $u(P) = 0.25$, $u(A) = 0.5$, $u(G) = 0.75$, $u(VG) = 1$. By using Eqs. (16) – (18), the minimal satisfaction of each project are calculated in Table 4.

Table 4. The rank-order of the four projects in the strategic project selection problem

Projects	**Minimal satisfaction**	**Ranking**
DSNEA	-0.0516	2
RVRIR	-0.0192	1
RSDM	-0.1336	4
CMHPWJ	-0.0671	3

The ranking order is generated as RVRIR $\succ$ DSNEA $\succ$ CMHPWJ $\succ$ RSDM where RVRIR is the most appropriate strategic project in this enterprise. That is, RVRIR is the most appropriate strategic project for this high-speed train manufacturing enterprise.

4.3. Sensitivity analysis

In the above problem, the weights of criteria are determined by the group of all experts. Different weights may influence the final values of minimal satisfaction or even final result of the strategic project selection. Thus, two controversial criteria will be further analysed which are capability of being in line with the development strategy of the enterprise (w_3) and market profitability (w_7).

Here, w_3 is changed from 0.045 to 0.32 with a step of 0.055. With the variation of w_3, minimal satisfaction of four projects derived from Eqs. (16) – (18) are changed as plotted in Figure 1. Although the ranking order in Figure 1 is not changed, the values of minimal satisfaction of four projects are significantly changed.

Then, w_7 is changed from 0.06 to 0.26 with a step of 0.04. With the variation of w_7, minimal satisfaction of four projects also derived from Eqs. (16) – (18) are changed as plotted in Figure 2. Figure 2 shows that a_4 replaces a_2 to be the most appropriate strategic project when the value of w_7 increases to about 0.16. That means the minimal satisfaction is sensitive to the variation of w_7. In other words, the different importance of market profitability will influence the final solution to the problem of selecting strategic projects.

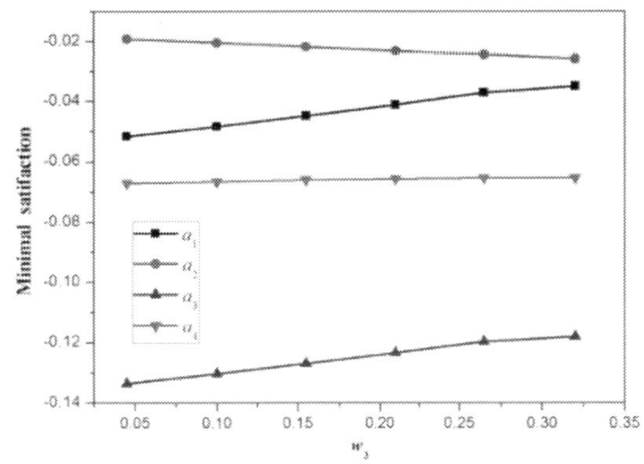

Figure 1. Movement of the minimal satisfaction of the four strategic projects with variation in the value of w_3

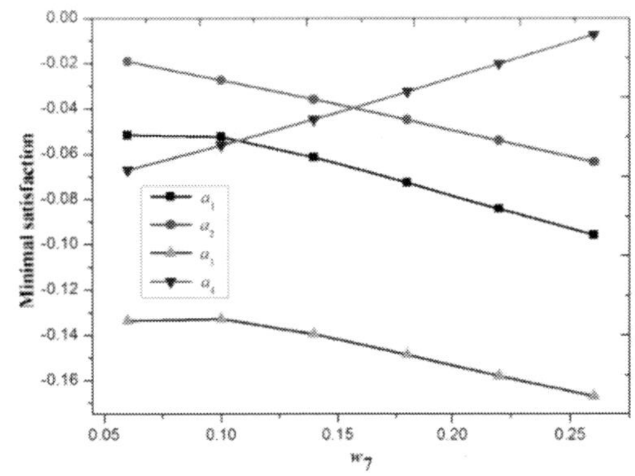

Figure 2. Movement of the minimal satisfaction of the four strategic projects with variation in the value of w_7

From the above analyses, weights of criteria are very important for selecting an appropriate strategic project. In particular, when the experts consider the weight of market profitability (w_7) should be more than 0.16, RSDM (a_4) will be the the most appropriate strategic project. The determination of weight of market profitability (w_7) should be more careful.

5 Conclusion

In this paper, the strategic project selection problem of high-end equipment manufacturing enterprises is analyzed and investigated based on ER approach. The contributions of this paper are summarized as follows. (1) The criteria framework of selecting strategic project is constructed by analyzing the existing studies. (2) Belief distribution is introduced to help the decision makers express their preferences in strategic project selection flexibility. (3) ER algorithm is demonstrated to combine assessments with weights to generate the best project with the help of utility theory. (4) A real problem of selecting strategic project in a high-speed train manufacturing enterprise is solved by using the ER approach.

In the future, the selected strategic project will be continued analyzed to realize the detailed implement of the project. Moreover, the problem of strategic project selection in more high-end equipment manufacturing enterprises will be investigated to further verify the validity of the ER approach. In addition, more complex factors will be considered in the future work of strategic project selection for high-end equipment manufacturing enterprises such as the group consensus requirements and reliability of the experts.

Acknowledgements

This research is supported by the National Natural Science Foundation of China (Grant Nos. 71571060).

References

1. T.M. Froese, The impact of emerging information technology on project management for construction, Automation in Construction, **19(5)**, 531-538, (2010).
2. K. Li, C.S. Shi, Research on classification of suppliers of high-end equipment manufacturing enterprises based on the degree of interdependence. 2013 International Conference on Management Science & Engineering, 17-19, (2013).
3. K. Khalili-Damaghani, S. Sadi-Nezhad, M. Tavana, Solving multi-period project selection problems with fuzzy goal programming based on TOPSIS and a fuzzy preference relation, Information Sciences, **252**, 42-61, (2013).
4. D.J.F Feng, K.H. Huang, Strategic project portfolio selection for national research institutes, Journal of Business Research, **68**, 2305-2311, (2015).
5. M.N.M Arratia, I.F. Lopez, S.E. Schaeffer, L. Cruz-Reyes, Static R&D project portfolio selection in public organizations, Decision Support Systems, **84**, 53-63, (2016).
6. M. Tavana, K. Khalili-Damaghani, S. Sadi-Nezhad, A fuzzy group data envelopment analysis model for high-technology project selection: A case study in NASA, Computers & Industrial Engineering, **66**, 10-23, (2013).
7. Yang J B, Xu D L. On the evidential reasoning algorithm for multiple attribute decision analysis under uncertainty, IEEE Transactions on Systems, Man and Cybernetics-Part A: Systems and Humans, **32(3)**, 289-304, (2002).
8. Fu C, Yang J B, Yang S L. A group evidential reasoning approach based on expert reliability. European Journal of Operational Research, **246**, 886-893, (2015).
9. Yang J B, Xu D L. Evidential reasoning rule for evidence combination. Artificial Intelligence, **205**, 1-29, (2013).

Analysis of identification of unsafe behavior in mechanical operation of first line workers

Jian Shen*, *Hong-xia* Li

Xi'an University of Science and Technology, School of Management, 710054 Xi'an Shaanxi, China

Abstract. The occurrence of mechanical accidents in petroleum enterprises is closely related to the operatio n of frontline workers. Based on the behavior observation method, the paper takes a petroleum enterprise as objective to identify the unsafe behavior of the operation workers and its contributing factors. The influence s of the different factor on the mechanical operation are investigated by using the analytic hierarchy process. The results show that there are eight kinds of unsafe behaviors of the mechanical operation in the petroleum enterprise. The factors that induce the unsafe behaviors include safety knowledge, safety consciousness, em otion, physiological conditions, safety attitude, manager behavior, management system, production task and human environment. Among them, safety attitude, safety consciousness, manager behavior and production task are the most important factors. Therefore, the safety production capacity of the frontline workers in petr oleum enterprises should be improved through safety education training, manager lead demonstration, enter prise safety culture construction.

1 Introduction

The characteristics of oil industries is flammable, explosive, toxic, harmful, high temperature, high pressure and continuous production, which leads to frequent production accidents[1]. Safety management has become the focus of sustainable development of petroleum enterprises.

Compared with other high-risk industries, there are many machineries and equipments, and the high level of automation in the petroleum industry. Once the operation fails, it will lead to shutdown, fire, explosion, environmental pollution, personal injury and other accidents. According to the accident statistics, it indicates that most of the mechanical accidents are directly related to the human operation[2]. The reliability of the human operation has become a decisional factor that affect the normal operation of the equipment and the prevention of accidents. In order to realize the safety production and reduce the accident, it is necessary to prevent and reduce the unsafe behavior of the workers in the operation of machinery and equipment. Therefore, taking a petroleum enterprise as the example, the paper identifies the unsafe behaviors in mechanical operation and its influencing factors by using the observation method. The influence degree of the influencing factors are analyzed by using Analytic hierarchy process method. The effective measures are taken to reduce the unsafe behaviors to improve the safe production capacity of petrochemical industry enterprise workers.

2 Identification of unsafe operation of machinery and equipment for petroleum workers

2.1 Identification methods and data sources

Because the operation behavior of mechanical equipment usually occurs on the job site, the behavior safety observation method is used to identify the unsafe behavior of employees and its contributing factors. Behavior safety observation method is that the observers directly observe the behaviors and reactions of the objects of study by using their senses and auxiliary tools in the field based on the research purpose[3], the research outline or behavior safety observation table, in order to obtain data. The method is widely used in human unsafe behavior management in coal mines and other high-risk enterprises. It can effectively detect the hidden danger and danger in safety production staff.

The implementation of the behavior safety observation can be divided into three parts: analysis of tasks, implementation of safety behavior observation and recording of observation results. Among them, the implementation of behavior safety observation is the core, including the determination of observation objectives, the determination of the scope of behavior observation, field observation, and communication. In order to deeply analyze the causal factors of unsafe behavior in mechanical operation, the depth interviews with key personnel should be conducted during communication.

* Corresponding author: 765826364@qq.com

2.2 Data sources

All work stations of oil production plant affiliated to PetroChina Co Ltd are taken as research object. The company is mainly engaged in oil and gas exploration and production, transportation and storage, the exploration and drilling, and planning research. The annual production capacity is more than 400 Million tons of crude oil, and the asset of oil and gas is of 200Billion Yuan RMB. It has a strong representation in the scale of production and management.

In order to ensure the comprehension and extension of the investigation, five production areas, six auxiliary production units are selected in accordance with the production and management of oil production plant, including the oil production area, operation area, oil tubing maintenance station, operation team, engineering, Engineering Battalion and brigade electric pump center.

The whole data collection process took 45 days. Prior to the survey, the contents of the implementation and the interview of the behavior safety observation are determined by combining with the existing literature before survey, and then the preliminary research program is formed. The final draft of the interview outline is formed by consulting five experts in security management field. During survey, the unsafe behaviors in the operation field are observed, identified and recorded by the behavior observation method. According to the observation results, the stuffs observed with unsafe behavior are chosen to interview face-to-face. The interviews with the different levels, different emphases are conducted, which are recorded by the corresponding text and recording.

In order to ensure the authenticity of the collection data, the data alternate collection and analysis. Each data collection makes new adjustments according to the analysis of the existing data. During the analysis process, the data is also checked to continue to enrich the information and fill the void data. It is ensured that information is true and accurate to reflect the business conditions and realize that the data can reflect the research problem.

2.3 Identification results

2.3.1 Concrete manifestation of unsafe behavior in mechanical operation of first line workers

(1) The concrete manifestation of unsafe behavior.

On the basis of the existing research results, and combining with the field observation results, the unsafe behavior of mechanical operation of the front-line workers in the oil production plant is determined. There are eight main forms of mechanical operation unsafe behavior:

① Operation under the condition of no danger elimination, lack of protective measures;

②The operation under unsafe speed;

③Operating unsafe equipment;

④Unsafe operation of equipment;

⑤In unsafe position;

⑥Using unsafe operation posture;

⑦Working on dangerous equipment in operation;

⑧Not operate according to the rules of procedure.

(2) Frequency of unsafe behavior

According to statistics, the unsafe behavior is mainly due to no operating according to the rules of procedure, and followed by unsafe operation of equipment, In unsafe position, working on dangerous equipment in operation, operation under the condition of no danger elimination, lack of protective measures, operating unsafe equipment, using unsafe operation posture, and operation under unsafe speed. As shown in Fig. 1.

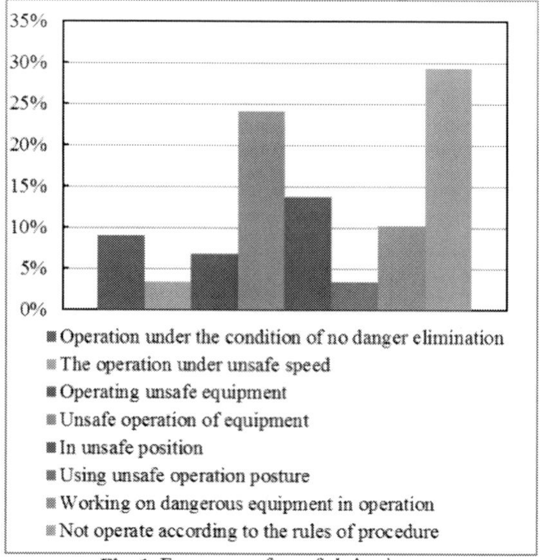

Fig. 1. Frequency of unsafe behavior

2.3.2 Contributing factors of unsafe behavior due to mechanical operation of first line workers

According to the results of behavior observation, and through interviews with key personnel, it is determined that the factors causing unsafe behavior of workers' mechanical operation mainly include two factors, the worker's own factor and the external factor.

(1)Worker own factors

①Knowledge and skills

The survey indicates that most of worker having unsafe behavior have the following features: the low level of culture, lack of oil production safety knowledge, poor skill, misunderstand relevant provisions, weak judgment ability to danger. When the environmental conditions change or new methods and new processes are in use, the occurrence of unsafe behavior is due to the misunderstood and unaccepted by these workers.

②Safety attitude

Some employees do not pay attention to safety in mind, and even think that some security systems, regulations, measures will have an impact on their work.

They know the existing danger, but they do not take safety measures due to feeling trouble.

③Emotional factors

The emotion also has a great influence on employees' behaviour[4].Because of family, work pressure, interpersonal relations and other reasons, some of employee have a lot of emotional problems. The fact results in job irritability, mental concentration and other phenomena, and result problems in the work.

④Physiological factors

Because of the physiological factors, some workers cannot discover some accident sign in production process and production environment. especially for some employees with physical discomfort, they cannot recognize and adapt to the environment, and cannot discover the omens of risk factors and the environment and equipment accidents, which results in unsafe behavior.

⑤Safety consciousness

The Interviews found that some of workers have weak safety awareness, and safety concept is transferred from the passive stage to the active stage. They blindly believe that the existing measures and technology have been more perfect and slightly unsafe behavior cannot cause the accident, even if the accident could not happen in their own on the body. Some others think that those enterprises safety regulations and safety measures are used to lead management and remediation of employees, which lead them relax their vigilance in the work process, and the violations seriously.

(2)External factors

①Management rules and regulations

The survey indicates that the enterprise has many rules and regulations related to safety operation, but the implement shows poor execution in staffs. According to interviews with workers, it is found that there are too many rules, and the rules are often changed, and it results that some rules is not fully understood and replaced. on the other hand, the lack of effective training and explanation to the issued rulers lead to the fact that the specific requirements in rulers cannot be fully understood. The deviations in the implementation process happen and affect the policy making effect.

②Manager behavior

Managers' exemplary role plays an extremely important role in the behavior of employees, and is the criterion and reference of workers. If the managers themselves attach great importance to safety in production, strictly follow the norms in the job, and in the safety supervision in accordance with the requirements of strict management, the staff will pay attention to their behavior in the work.

But the survey found that some managers do not pay enough attention to the safety, and lack the safety awareness and habit. They do not care about security, and operate in accordance to work experience and habits. They do not know their security responsibilities, and assume that security should only belong to department of safety and safety supervision personnel matters. They have passive attitudes to the safety requirements and cannot strictly control the unsafe behavior of employees. These factors affect the correct execution of employee behavior.

(3) Production tasks and benefit factors

The extensive and long duration of management results in the dispersed operation and the lack of operators. The enterprise's own production task is also heavier, and the fact lead to the managers ignore the safety to complete the production tasks. On the other hand, in order to complete the production tasks, the employees work continuously in a state of tension. Especially, when high yield wells are found, the employees may violate the operation rules in order to catch the progress schedule and improve the volume of products, which produce some hidden dangers for the enterprise production safety accidents.

(4)Influence of environmental factors

Behavior is the result of interaction between human and environment, and changes with the change of human and environment. Therefore, the production environment has a direct impact on the emergence of unsafe behavior. The observation indicates that the enterprise has the clean and orderly working environment, advanced production equipment and other advanced hardware conditions. The environmental factor which affects the behavior of employees is the human environment, i.e. the relationship and the behavior among the workers. The disharmony behaviors between workmates are the main causes of unsafe behaviors.

3 Impact of unsafe behavior of mechanical behaviors

3.1 Research method

Analytic hierarchy process (AHP) was used to analyze the influence of above internal and external factors on unsafe behavior of mechanical operation of first line workers. AHP is a system analysis method proposed by Professor T.L.Saaty. Compared to other methods, it can measure qualitative factors to compare different factors and determine the influence degree of the different factors, which can ensure the accuracy of analysis results[4].

3.2 Establishment of a hierarchical structure model

An analytic hierarchy process model is generally divided into target layer, standard layer and decision layer. Combined with the previous analysis of constraint factors, a hierarchical model is established to consider the social responsibility information disclosure, as shown inTable 1.

3.3 Construction of judgment matrix and consistency test

The comparison is conducted by using the matrix scaling method (see Table 2) for comparison. The relative importance of factors and their scores are determined by using expert scoring method. All judgment matrixes in various levels are given. The weight vectors are calculated and the consistency is tested.

Table 1 The hierarchical model of information disclosure

Target layer	**Standard layer**	**Decision layer**
Unsafe behavior of mechanical behaviors	External factor G_1	Production tasks and benefit factors G_{11}
		Manager's behavior G_{12}
		Humanistic environment factors G_{13}
		Management rules and regulations G_{14}
	Internal factor G_2	Knowledge and skills G_{21}
		Safety attitude G_{22}
		Emotional factors G_{23}
		Physiological factors G_{24}
		safety consciousness G_{25}

Table 2 Scale rules of judgment matrix

Scaling a_{ij}	**Mean**
1	Factors c_i and c_j have same effect
3	Effect of factors c_i is slightly greater than that of c_j
5	Effect of factors c_i is greater than that of c_j
7	Effect of factors c_i is strongly greater than that of c_j
9	Effect of factors c_i is extremely greater than that of c_j
2、4、6、8	Take the intermediate value between two adjacent numbers as c_i and c_j is near.
Reciprocal	When Effect of factors c_j is slightly greater than that ofc_i, the reciprocal values are taken

The consistency test of the judgment matrix is as follows:

(1) The judgment matrix of the influence degree of internal factors and external factors on social responsibility information disclosure is as follows.

G	G_1	G_2
G_1	1	1/3
G_2	3	1

The judgment matrix $A=\begin{bmatrix}1 & 1/3\\ 3 & 1\end{bmatrix}$, the eigenvalue value is calculate by using MATLAB: Wa= {0.25,0.75}, $\lambda max=\sum_{i=1}^{n}\frac{(AW)_i}{nW}=2$, the judgment matrix is completely consistent. The weight in the matrix Wa is the weight value of the internal and external factors.

(2) According to the relationship between the external factors, the judgment matrix is as follows:

G_1	G_{11}	G_{12}	G_{13}	G_{14}
G_{11}	1	1	5	7
G_{12}	1	1	5	7
G_{13}	1/5	1/5	1	3
G_{14}	1/7	1/7	1/3	1

The eigenvalues of the matrix are W={W1, W2, W3, W4}={0.41, 0.41, 0.13, 0.05}, and λmax=4.129, $CI=\frac{(\lambda max-n)}{n-1}=\frac{4.129-4}{3}=0.043$, RI=0.90, $CR=\frac{CI}{RI}=\frac{0.043}{0.9}=0.048<0.1$. The consistency can be accepted.

(3) According to the relationship between the internal factors, the judgment matrix is as follows:

G_2	G_{21}	B_{22}	G_{23}	G_{24}	G_{25}
G_{21}	1	1/5	1	5	1/5
G_{22}	5	1	4	7	1
G_{23}	1	1/4	1	3	1/3
G_{24}	1/5	1/7	1/3	1	1/7
G_{25}	5	1	3	7	1

The eigenvalues of the matrix are W={0.152, 0.369, 0.094, 0.037, 0.348}, the maximum eigenvalue λmax=5.351, and the consistency index is $CR=\frac{0.088}{1.12}=0.078<0.1$. the consistency is acceptable.

All single sorted CR<0.1, therefore, the consistency of each judgment matrix is acceptable.

3.4 Level total ordering and consistency test

According to the above analysis results, the factors causing the unsafe behavior of the first line workers are sorted out as shown in Table 3.

Table 3 total factor ranking of the unsafe behavior of mechanical behavior

Target layer	**Standard layer**		**Decision layer**		**Total weight**	**Weight sorting**
Unsafe mechani-cal behavior	G_1	0.25	G_{11}	0.41	0.1025	3
			G_{12}	0.41	0.1025	3
			G_{13}	0.13	0.0325	6
			G_{14}	0.05	0.0125	8
	G_2	0.75	G_{21}	0.152	0.114	4
			G_{22}	0.369	0.2768	1
			G_{23}	0.094	0.0705	5
			G_{24}	0.037	0.0277	7
			G_{25}	0.348	0.261	2

3.5 result analysis

According to the above results, the factors that cause the unsafe behavior of the front-line employees in oil enterprises are safety attitude, safety awareness, manager behavior and production. The main reason is the safety attitude, which is the evaluation of safety and tendency of behavior that people make in their own morality and values. The safe attitude has directly, instructive and dynamic effect on the behavior of people. Safety awareness is the safety concept set up in people's minds. The weak safety awareness will produce an incorrect cognition and lead to unsafe behavior. The behaviors of managers and production tasks have an indirect effect on the behavior of staff by a mean of model effect and external pressure. , if the managers obey safety regulations and the production task is light, the behavior of employees will be relatively safe.

The above analysis indicates that in order to control the unsafe behavior of mechanical operation, the following aspects should be done: (1)strengthening the safety education and training, (2)managers play a leading role model, (3)reasonable arrangement of work tasks and establishment of the concept of effective security, and (4)promote the construction of enterprise safety culture.

4 Conclusions

(1) The eight unsafe behavior is observed on site as follows: not operate according to the rules of procedure, unsafe operation of equipment, in unsafe position, working on dangerous equipment in operation, operation under the condition of no danger elimination, lack of protective measures, operating unsafe equipment, using unsafe operation posture, and operation under unsafe speed.

(2) Self factors induce the unsafe mechanical operation behaviors of the front-line workers has its internal factors and external factors. The internal factors include safety knowledge, safety awareness, emotion, physiological and safety attitude. The external factors include management behavior, management system, production task and humanistic environment.

(3) The influence degree is analyzed by using AHP. The result indicated that the main influence factors are safety attitude, safety awareness, management behavior and production tasks. Other factors are safety knowledge and skills, humanistic environment, emotion, and management rules and regulations.

Acknowledgements

The work is supported by Shaanxi Soft Science Research Program(2017KRM068) and Educational Commissiom of Shaanxi Province of China(15JZ036)

References

1.Li Zhi-hong.Cause analysis on the safety accidents in petro chemical enterprises and research on the counter measures. Petro chemical safety and environ mental protection technology, 6,29 (2013)

2.Erik H.Cognitive Reliability and error Analysis Method. Oxford：Alden Group,(1998)

3.Mark F， Ronny L. Promoting best practice in behavior-based safety. Symposium series：IchemE，(2001)

4.Li huan,Tian Shui-cheng, Experimental Study between the Emotion and Coal Unsafe Behavior miners. Shaanxi: Xi'an University of Science and Technology(2015)

5.Guo Jin-yu,Zhang Biao-zhong. Study and Applications of Analytic Hierarchy process. China Safety Science Journal.5,18(2008)

 DOI: 10.1051/matecconf/201713900073

Link prediction in author collaboration network based on BP neural network

Chaoqun Chen [1,2] ,*Sanhong* Deng [1,2] ,*Jing* Lu[3]

[1]School of Information Management, Nanjing University, Nanjing, 210023, China
[2]Jiangsu Key Lab of Data Engineering and Knowledge Service, Nanjing, 210023, China
[3]Jiangsu Institute of Geological Survey, Nanjing, 210018, China

Abstract. Recently, more and more authors have been encouraged for collaboration because it often produces good results. However, the author collaboration network contains experts in various research directions within various fields, and it is difficult for individual authors to decide which authors are best suited to their expertise. This paper uses the relationships among authors to predict new relationships that may arise, recommending each author with the collaborators they may be interested in. The data source comes from 4-year data in DBLP from 2001 to 2004. After data cleaning, the training set and test set are constructed and then used BP neural network to build model. At the same time, this article compares the performance with Logistic Regression, SVM and Random Forest. The experiment shows that the BP neural network can get better result, and it is feasible to predict links in the author collaboration network.

1 INTRODUCTION

Link prediction is a very important research direction in social networks, such as Facebook, Twitter and Flicker and other social platforms, which recommend the friends to their users through link prediction. A link may mean a friend relationship in a social network, or it may mean a collaboration relationship in the collaboration network, or it may mean interaction in a protein network. Any kind of relationship between two nodes in a network can be considered a link. A definition of link prediction: Given a snapshot map of a social network at a time G = <V, E> and node V_i, V_j , predict the probability of a link between nodes V_i, and V_j [1]. It can be seen that the task of link prediction is divided into two categories: one is to predict the link that will appear in the future time, the second one is to predict the hidden unknown link in the space, this article discusses the former.

Link prediction has two main approaches: a score-based approach and a machine learning approach. The score-based approach is to consider link predictions as a regression problem by calculating the similarity scores for each pair of nodes and then sorting them, the order determines the likelihood of forming future links. The score-based approach is a simple and effective method, but this approach is sensitive to different features' weights. The machine learning approach is effective to use a variety of attributes to predict the formation of links, and does not need to give the weight of each feature manually, and it is easy to expand. Therefore, the method of machine learning has been widely used in link prediction. The machine learning approach regards the link prediction problem as a

* Corresponding author: sanhong@nju.edu.cn

classification problem, and then uses the SVM, decision tree, Boosting, Logistic regression and other methods [2].

Neural network (NN) is a powerful machine learning approach for nonlinear prediction, composed of independent units of neurons. Each processing unit sums the weighted inputs, and then applies the result to linear or non-linear functions to determine the outputs. In the neural network algorithm, BP neural network (Backpropagation network) has been very concerned about, with a strong non-linear mapping capabilities. BP neural networks have been widely used in various fields in recent years, including the prediction of the quality or robustness of plastic parts based on key process variables and material grade changes [3]. Besides, the use of BP neural networks for accurate short-term load forecasting (STLF) has played an important role in the national and regional power system management [4]. However, in the field of link prediction, few scholars use the neural network algorithm, so the author will apply BP neural network to predict new links in author collaboration network, then explore whether it can achieve good results in this field.

2 THE BASIC KNOWLEDGE ON BP NEURAL NETWORK

This article uses a typical three-layer BP neural network, including input layer, hidden layer and output layer. The layers are full connected with edges and each edge corresponds to a weight w. At the same time, in addition to the neuron in input layer, neurons in other layers also include a bias b, those neurons will have an input value z which received from weighted summation and an output value which received from nonlinearity transform z through the activation function. The activation function used in this article is the Sigmoid function($f(x) = \frac{1}{1+e^{-x}}$). The input value of the neuron j in the first-layer is expressed as: $z_j^l = \sum_{i=1,2,...,n} w_{ij} \cdot a_{ij}^{l-1} - b_j^l$ .The variable i and the variable j represent neurons. ij represents the connection from the neuron i to neuron j, w_{ij} means the weight of the connection. a_{ij}^{l-1} represents the output of the neuron i in the l-1 layer, that is, the i-th input of the neuron j in the l layer. b_j^l represents the bias of the neuron j in the l layer. Then the output of the neuron j in the l layer is expressed as :$a_j^l = f(z_j^l) = \frac{1}{1+e^{-z_j^l}}$.

In practice, the BP neural network needs artificially specify the number of layers and the number of neurons in each layer. At the same time, the BP neural network needs to initialize the weight and bias of each edge in the initial stage and then the BP neural network algorithm will perform the forward transmission to get the predictive value for each sample of the training set. Then, according to the error value between the true value and the predicted value, the weight value of each connecting edge and the bias value of each layer in the neural network will be updated by backpropagation. The stopping conditions can be one of the following: a. The iterations times reach the given value; b. The error value is less than a given threshold. In this article, the stopping condition is to reach the given value of iterations times, that is, the number of times the parameter was updated reaches the given value.

Assuming that a given training set contains m training samples, we can use the gradient descent algorithm to solve the BP neural network. For example, for a training sample (x, y), define the cost function (the difference between the predicted value and the true value) is: $J(W,b;x,y) = \frac{1}{2}\left\|h_{W,b}(x) - y\right\|^2$, the cost function on the training set is: $J(W,b) = \left[\frac{1}{m}J(W,b;x^{(i)},y^{(i)})\right] + \frac{\lambda}{2}\sum_{l=1}^{n_l-1}\sum_{i=1}^{s_l}\sum_{j=1}^{s_l+1}\left(W_{ji}^{(l)}\right)^2$. It consists two parts: The first part is the sum of error mean square and the latter part is the regularization, which mainly to prevent the overfitting. In this article, the link prediction problem is transformed into a classification problem. The y value represents 0 or 1, and $h_{W,b}(x)$ is the range of [0,1] of the sigmoid function. The optimization problem of BP neural network is to obtain the W, b that minimizes the cost function. In general, random values near 0 are used to initialize weights W and bias b, and then the value of W and b is updated by optimization algorithm such as gradient descent algorithm: $W_{ij}^{(l)} = W_{ij}^{(l)} - \alpha \frac{\partial J(W,b)}{\partial w_{ij}^{(l)}}, b_i^{(l)} = b_i^{(l)} - \alpha \frac{\partial J(W,b)}{\partial b_i^{(l)}}$. The new cost function value can be obtained with the new W and b values and the cost function value will keep close to its minimum value gradually. In the above formula, α indicates the learning rate which can be understood as the pace of each gradient, that is, the decreasing rate of cost function. Its value is very important. If the value is oversized, the decreasing rate may be too fast and missed the minimum value of cost function. If the value is too small, the cost function may be converged very slowly. The general α value is between 0.001-0.3. At the same time, it can be seen that when the sample size is relatively large, if updated value W and b by all the data set each time, the efficiency will be relatively low. Therefore, this article uses Stochastic Gradient Descent (SGD), and each iteration process just selects a part of dataset randomly.

In order to evaluate the performance of BP neural network in the author's collaboration network linkage prediction, this paper compares the results of the three algorithms: Logistic Regression, SVM and Random Forest.

3 EXPERIMENTAL RESULTS AND ANALYSIS

3.1 Dataset

The dataset used in this article is an open social network dataset DBLP (Digital Bibliography and Library Project) which is an integrated database system of English literature taking the author as the core on research results in the computer science field. The author's research achievements, including published papers in international journals and conferences, are listed according to the year. The data set of DBLP is very large. Considering the time cost, the author only experimented with data from 2001 to 2004, and removed the data which only has one author. Divided the four-year data into two periods, 2001-2002 and 2003-2004, this article constructed the features by 2001-2002 data and the labels by 2003-2004 data. If the authors have no collaboration relationship in 2001-2002 while they have the partnership in 2003-2004, the label is y=1, which indicating that a new link appears between the author's pair. Otherwise the label is y=0, which indicating that no new links appear between the authors. It is noteworthy that, during the feature construction, the selected author's pair should have no collaboration relationship in 2001-2002, and then to predict whether they would collaborate in 2003 or 2004. At the same time, the author notes that some authors are active only in 2001-2002 and no longer appear in 2003-2004; Conversely, some authors are active only in 2003-2004 while no longer appear in 2001-2002. In the author collaboration network, it will result in the mismatched between features and labels. To avoid this situation, this article considers only the authors that are included both in the two time periods. At the same time, due to the fact that the number of y=0 is far greater than y=1, namely the imbalanced learning problem. Therefore, in the experiment, the author randomly selected minority examples in the negative samples (the samples which y=0), this article obtains the negative samples which has the same quantity with positive samples (the samples which y=1), both are 5,504 samples. As a result, the social network constructed in this article is an undirected graph without weight with 8,363 nodes in the network. In the process of splitting training set and test set, 70% samples are randomly divided as training sets and the remaining 30% are test sets, in which there are 7,707 samples in training set, and 3,302 samples of test sets.

For the feature vector, the author refers to the researches of other scholars [5~7] and selects the following features as shown in Figure 1, these features are derived from the author collaboration network where each node represents an author:

3.2 Features

Feature name	Formula
Total neighbors	$\lvert \Gamma(v_i) \cup \Gamma(v_j) \rvert$
Common neighbors	$\lvert \Gamma(v_i) \cap \Gamma(v_j) \rvert$
Shortest path	$\min\{s \mid path_{ij}^{s} > 0\}$
Jaccard's coefficient	$\frac{\lvert \Gamma(v_i) \cap \Gamma(v_j) \rvert}{\lvert \Gamma(v_i) \cup \Gamma(v_j) \rvert}$
Adamic/Adar	$\sum_{v_k \in \lvert \Gamma(v_i) \cap \Gamma(v_j) \rvert} \frac{1}{log \lvert \lvert \Gamma(v_k) \rvert \rvert}$
Preferential attachment	$\lvert \Gamma(v_i) \rvert \cdot \lvert \Gamma(v_j) \rvert$
$\mathrm{PageRank}_d^{min}$	$\min\{\mathrm{PageRank}_d(v_i), \mathrm{PageRank}_d(v_j)\}$
$\mathrm{PageRank}_d^{max}$	$\max\{\mathrm{PageRank}_d(v_i), \mathrm{PageRank}_d(v_j)\}$
communities	1 if v_i and v_j belong to the same community,0 else
communities member number	number of communities member

Fig1. Features computed for each author pair(V_i,V_j)

The Total neighbors of the two nodes v_i and v_j is defined as the total numbers of their neighbors, and the common neighbor calculates only once. Common neighbors calculate the number of common neighbors that two nodes have. For example, for two authors of a collaboration network, this is the number of other authors they have worked together. Shortest path is the minimum number of edges connected with two nodes. If there is no such connection path, this feature value is usually considered to be infinity. Jaccard's coefficient [8] is the normalized metric for Common neighbors, which calculates the proportion of the two authors common neighbors in their all neighborhoods. This feature sometimes has better performance than Common neighbors, especially when the number of one node's neighbors is much greater than the other node's. Adamic/Adar [9] measures the similarity of two nodes by assigning weights to common neighbors. The basic principle is that the smaller the neighbors number of the common neighbor, the greater the contribution to the similarity, that is to say, the more likely the two nodes are to be linked. Preferential attachment [10] indicates that a node with a higher centrality degree is more likely to produce new links. PageRank is the core algorithm used by Google to rank search results [11]. The PageRank value of webpage v_i indicated by formula: $PageRank_d(v_i) = (1 -$

$d) + d\sum_{v_m \in \Gamma(v_i)} \frac{PageRank_d(v_m)}{\Gamma(v_m)}$,where $\Gamma(v_i)$ refers to the webpages link to webpagev_i. In fact, in the graph structure, the PageRank value of a node is a probability used to represent the likelihood that a person randomly walking on edges will arrive at any particular node. d is a damping factor which mainly considering that there are some nodes in the directed graph do not point to other nodes, which may result in a dead loop when accessing this node. Therefore, it is necessary to modify the PageRank formula, that is, to add the damping factor. The value is 0.85 in general. In this article, PageRank is used for the undirected graph. $\Gamma(v_i)$ refers to all the neighbor nodes of the node. Because the authors pair is considered, the larger PageRank value between two authors is expressed as $pagerank_d^{max}$ while the smaller PageRank value is $pagerank_d^{min}$. Many complex networks can be divided into communities, and the nodes in the same community are more closely linked. For authors collaboration networks, authors in the same community are more likely to have partnerships. This feature of the Communities is used to indicate whether the two authors are belonging to the same community. This article uses the clique percolation method in Python to discover community [12]. For authors in the same community, the communities' member number represents the number of authors in the community. Generally speaking, the smaller the number of authors in a community, the more likely that two authors in this community will collaborate.

3.3 Performance metrics

The performance metrics of this article adopts to evaluate and compare algorithms are precision, recall and F1 rates. Several terms must be identified before introducing specific indicators: TP represents true-positive which means a positive sample is predicted as positive by the model; FN represents false-negative when a positive sample is predicted as negative wrongly by the model; In the negative case, if a sample is predicted correctly,the prediction is said to be true-negative (TN);otherwise it is false-positive(FP). P represents the positive sample, namely there is a collaboration relationship between authors. N represents a negative sample, namely there is no collaboration relationship between authors. Precision is expressed by formula:$\frac{|TP|}{|TP|+|FP|}$; Recall is $\frac{|TP|}{|TP|+|FN|}$and the F rate is $\frac{1}{\partial\frac{1}{precision}+(1-\partial)\frac{1}{recall}}$. The ∂ is between 0 and 1. When $\partial = 0.5$, it indicates that the precision and recall rate are equally important; when $\partial > 0.5$, it indicates that precision is more important; in this paper, the precision and recall rate are treated as the same important, that is, $\partial = 0.5$, and the F rate is called F1 rate at this time.

3.4 Experient result

In the experiment, the number of neurons in the input layer, hidden layer and output layer of BP neural network were 10, 50 and 2. In practice, the iterations times epoch and the learning rate eta have an effect on the performance of BP neural network. This article has carried on several experiments. Under the condition of epoch=200, this article respectively chooses the learning rate including 0.001, 0.003, 0.01, 0.03, 0.1 to carry on the experiment. It can be found that the precision rate of the model on the test set is increasing, while the recall rate is decreasing because the model predicts part of negative samples as positive with the increase of learning rate (Fig 2). And in terms of F1, under the learning rate of 0.003 BP neural network can perform better. Therefore, the author choses the learning rate of 0.003 and does experience in the iterative times of 100, 200, 300, 400 and 500 to obtain BP neural network performance metrics under different iterations times. It can be seen that the recall rate of the model is at the maximum in the test set when the iteration number is 100 and the F1 rate at this time also gets the maximum value (Fig. 3). Therefore, we chose 100 as iteration times and 0.003 as the learning rate finally. At the same time, the BP neural network is compared with the Logical Regression, SVM and Random Forest in the same test set, and the result is shown in Figure 4. The recall rate of BP neural network is the highest, and the F1 rate is higher than Logistic Regression, which is approximately the same as the Random Forest.

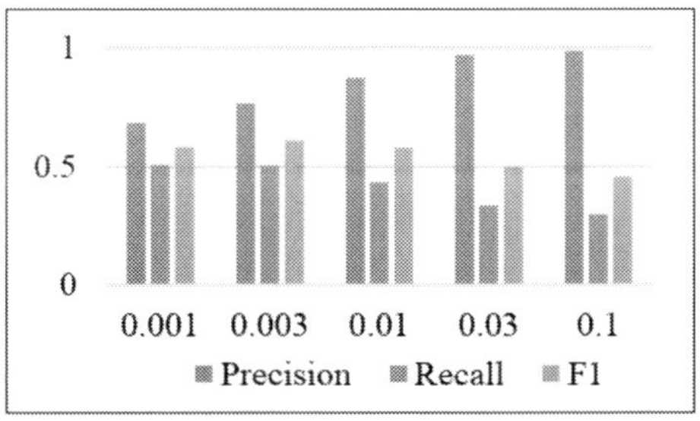

Fig. 2. BP neural network classification results on different learning rate (epoch=200)

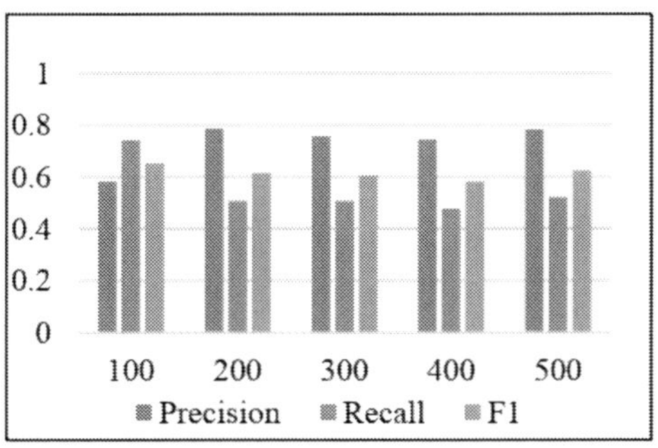

Fig. 3. BP neural network classification results on different iteration times(eta=0.003)

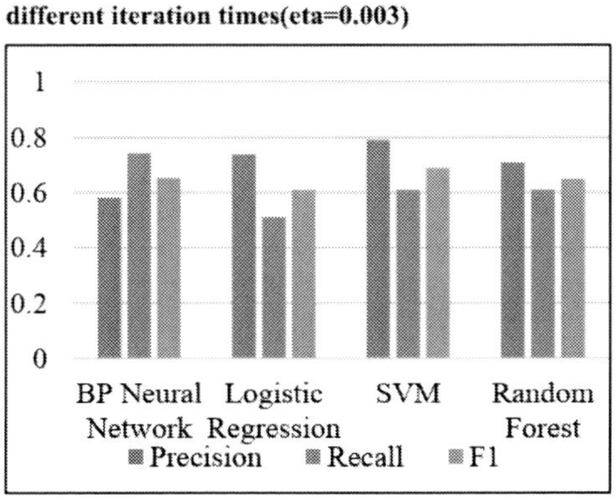

Fig. 4. Classification results for different algorithms

4 Conclusion

This article takes the data from 2001 to 2004 in DBLP as dataset and constructs 10 attributes as feature vector such as total number of neighbors, common neighbors and Adamic/Adar. Next, this article uses BP neural network to predict links in the author collaboration network and compares the performance metrics with Logistic Regression, SVM and Random Forest. Finally, it is found that the model trained by BP neural network outperforms other algorithms on recall rates, and the F1 rate is also good, namely, there is valuable information to be learned from the structure of the author collaboration network. Therefore, it is reasonable to apply BP neural network to link prediction in social network. Of course, there are deficiencies in this article. The first is that the features used to describe the structure of a collaboration network in the paper is not comprehensive, and some features those may affect authors' collaboration relationship may not take into account. The second is that the F1 rate is not particularly good and need to improve. In future, we can increase the accuracy of the predictor by adding the features, the number of the hidden layers and the number of neurons in BP neural network, and optimizing the parameter initialization.

This work has been supported by the Youth Project of the National Natural Science Foundation of China under the Grant Numbers 71503121.

References

1. H. YIN. Link Prediction in Social Network". Master's Thesis, Jilin University, Jilin, (2012).

2. C. Ahmed, A. Elkorany, R. Bahgat. A supervised learning approach to link prediction in Twitter. Social Network Analysis & Mining, **6**, (2016), 1-11.
3. B.H.M. Sadeghi. A BP-neural network predictor model for plastic injection molding process. Journal of Materials Processing Technology, **103**, (2000), 411-416.
4. Z. Xiao, S.J. Ye, B. Zhong, et al. BP neural network with rough set for short term load forecasting. Expert Systems with Applications, **39**, (2009), 273-279.
5. D. Liben-Nowell, J. Kleinberg. The link-prediction problem for social networks[J]. Journal of the Association for Information Science & Technology, **58**, (2007), 1019–1031.
6. Hasan, A. Mohammad. Link Prediction using Supervised Learning[J]. Proc of Sdm Workshop on Link Analysis Counterterrorism & Security, **30**, (2006),798-805.
7. M. Pavlov, R. Ichise. Finding experts by link prediction in co-authorship networks. International Conference on Finding Experts on the Web with Semantics. CEUR-WS.org, (2007), 42-55.
8. G. Salton, M.J. McGill. Introduction to Modern Information Retrieval. Program, **55**, (2004), 239-240, 2004.
9. L.A. Adamic, E. Adar. Friends and neighbors on the Web. Social Networks, **25**, (2003), 211-230.
10. M.E.J. Newman. Clustering and preferential attachment in growing networks. Physical Review E Statistical Nonlinear & Soft Matter Physics, **64**, (2001), 025102.
11. S. Brin, L. Page. Reprint of: The anatomy of a large-scale hypertextual web search engine. Computer Networks, **56**, (2012), 3825-3833.
12. G. Palla, I. Derényi, I. Farkas, et al. Uncovering the overlapping community structure of complex networks in nature and society. Nature, **435**, (2005), 814.

Reliability modelling for the GIL subject to competing failure modes

Songhua Hao[1], *Jun Yang*[1,*], *Bo Cheng*[2], *Yun Chen*[3] and *Boyuan Cui*[3]

[1]School of Reliability and Systems Engineering, Beihang University, Beijing, 100191, China
[2]Nuclear Power Institute of China, Chengdu, 610213, China
[3]China Electric Power Research Institute, Beijing, 100085, China

Abstract. Due to necessity of enforcing the electrical transmission network, a long distance gas insulated transmission line (GIL) has been developed as a new alternative underground technology, and has shown an ever-increasing usage worldwide. The main failure modes of the GIL are performance degradation due to air leakage, and random shocks caused by partial discharge. The failure of the GIL subject to both soft failure and hard failure is the result of their competition. Considering that the degradation occurs only after a certain delay, i.e. there is a random degradation initiation time before the air in the GIL begins to leak, this paper presents a new reliability model for the GIL subject to competing failure modes. Its reliability is assessed through analytical derivation and numerical calculation methods. And a numerical study is conducted to illustrate the implementation of the proposed model.

Key words. the GIL, partial discharge, air leakage, degradation initiation time, reliability.

1 Introduction

To meet the necessity of enforcing the electrical transmission network, and considering that erecting new overhead lines may encounter many insurmountable difficulties for some specific environment and terrain, a new alternative underground technology is developed as a long distance GIL [1]. Based on the successful SF6 tubular conductor technology, the first GIL was commissioned in 1975 at Wehr pumped storage station in the Black Forest in Southern Germany [2]. Since then, the GILs at rated voltages of 110, 225, 420, 550 and up to 1100kv have shown an ever-increasing usage worldwide, and the GIL technique has been proven to be in high reliability in terms of mechanical and electrical failures [3].

Similar to the gas insulated transmission switchgear (GIS), the failure of the GIL is mainly due to partial discharge and air leakage. Due to the existence of defects like abnormal conductor surface roughness and metallic particles, partial discharge occurs and finally leads to a break down [4]. Therefore, guidance for the design, installation, testing, operation and other technical aspects of the GIL is provided in [5], including the partial discharge tests on the insulators. And newly developed measuring system are studied in many literature such as [6]. Besides hard failure caused by partial discharge, soft failure caused by air leakage can also be found in many applications, e.g., Jinping-I Hydropower Station in China [7] and Hainan nuclear power unit 1 in China [8]. And the main reason of SF6 leakage of the GIL is the degradation of the sealing ring [9].

Based on practical failure analysis in [7], it can be indicated that the defect of the sealing ring will not cause air leakage as soon as the GIL is put into service. In practice, sometimes the degradation occurs only after a certain delay, called the onset time or the initiation time. In [10], Peng and Feng developed a reliability model for Ultrathin Gate Oxides subject to Verhulst logistic growth model with random onset time. Based on a Wiener degradation process and exponentially distributed random initiation time, Paroissin studied the corresponding parameter estimation methods and some asymptotic results [11]. To characterize the occurrence of failure caused by partial discharge, Wang et.al presented a compound Poisson process and planned the maintenance strategy under comprehensive consideration of optimal reliability and economic efficiency [12].

Considering the competing failure modes of the GIL, i.e., hard failure due to partial discharge and soft failure caused by air leakage, this paper proposes a new reliability model with random degradation initiation time before the air in GIL begins to leak. A nonlinear general path model is developed to model the aging of the GIL sealing rings, which is assumed to be the primary cause of air leakage. Besides, a Poisson process is utilized to model the occurrence of partial discharge, which will

* Corresponding author: tomyj2001@buaa.edu.cn

cause the GIL failure according to the extreme shock model [13], i.e., the voltage exceeds the partial discharge inception voltage (PDIV) of the GIL. Finally, a competing failure model is proposed to assess the reliability of GIL system.

The rest of this paper is organized as follows. Section 2 describes the competing failure modes of the GIL. System reliability model for the competing failure modes is proposed in Section 3. A numerical example is developed in Section 4 to demonstrate the implementation and effectiveness of the proposed model. And some conclusions are finally given in Section 5.

2 Competing failure modes of the GIL

As shown in Figure 1, the failure of the GIL is the competing result of soft and hard failures. A degradation based soft failure occurs when the air leakage, which begins after a random initiation time, exceeds a critical threshold level H. In addition, a shock based hard failure occurs once a partial discharge with large voltage arrives. The whole description of the system and relevant denotations are expounded below.

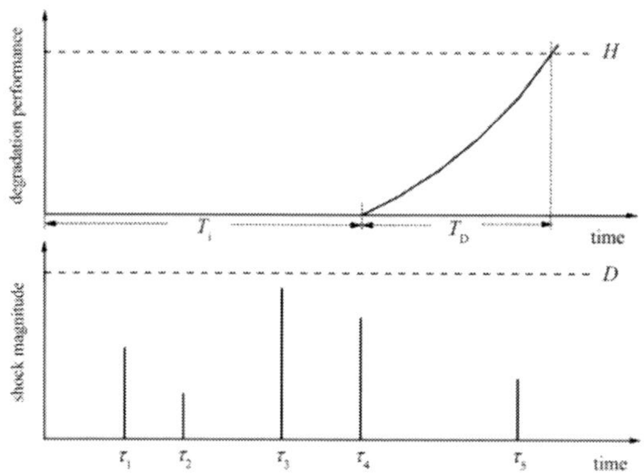

Figure 1. Competing failure modes of the GIL.

1) The degradation process $X(t)$ is modeled by a nonlinear general path model [14] with random initiation time T_i. After the degradation initiation time, i.e., in the degradation propagation stage, the degradation performance can be defined as

$$X(t) = \lambda\Lambda(t) + \varepsilon \quad (1)$$

where $\Lambda(t)$ is the time-transformed function of degradation path. $\lambda \sim N(\mu_\lambda, \sigma_\lambda^2)$ is the random effect parameter representing the unobservable endogenous factors of samples. $\varepsilon \sim N(0, \sigma_\varepsilon^2)$ is the model error term.

2) Similar to the assumption in literature [11] and [15], T_i is assumed to obey an exponential distribution in this article, i.e., $T_i \sim E(\lambda_i)$. Considering that the air leakage of the GIL is mainly due to the aging of sealing rings, an exponential form is utilized to characterize it, i.e., $\Lambda(t) = e^{Kt} - 1$, which is common in the degradation modelling for rubber sealing rings such as [16] and [17]. Therefore the nonlinear general path model (1) can be further expressed as

$$X(t) = \lambda\left(e^{Kt} - 1\right) + \varepsilon \quad (2)$$

3) Shocks arrive at a homogenous Poisson process with a constant rate λ_0. The number of arrived shocks before time t is denoted by $N(t)$, and $P\{N(t) = k\} = \frac{(\lambda_0 t)^k e^{-\lambda_0 t}}{k!}, k = 0, 1, 2, \cdots$. The magnitude W of shocks is independent and normally distributed variable $W \sim N(\mu_W, \sigma_W^2)$, and its cumulative distribution function (cdf) is denoted by $F_W(x)$;

4) The threshold levels for soft and hard failures are constants H and D, respectively. Once the overall degradation performance exceeds H, or the magnitude of a shock is greater than D, the GIL fails.

3 Reliability modeling for the GIL

The failure of the GIL subject to both soft failure and hard failure is the result of their competition. Considering these two failure modes are independent with each other, this Section respectively derives the reliability functions for soft and hard failures, and the system reliability is finally obtained as the product of them.

$$R(t) = R_{SF}(t) \cdot R_{HF}(t) \quad (3)$$

Let $F_i(t)$ and $f_i(t)$ be the cdf and pdf of exponentially distributed degradation initiation time T_i, respectively. Then it can be derived that:

$$f_i(t) = \lambda_i e^{-\lambda_i t}, t > 0 \quad (4)$$

Based on the degradation model (2), and considering that λ, ε are independent normally distributed variables, it can be derived that $X(t)$ also follows a normal distribution with mean $\mu_\lambda\left(e^{Kt} - 1\right)$ and standard deviation $\sqrt{\sigma_\lambda^2\left(e^{Kt} - 1\right)^2 + \sigma_\varepsilon^2}$. Denote the degradation propagation time before failure by T_D, then its cdf $F_D(t)$ can be derived as

$$\begin{aligned} F_D(t) &= P\{T_D < t\} \\ &= P\{\lambda(e^{Kt}-1)+\varepsilon < H\} \\ &= \Phi\left\{\frac{H-\mu_\lambda(e^{Kt}-1)}{\sqrt{\sigma_\lambda^2(e^{Kt}-1)^2+\sigma_\varepsilon^2}}\right\} \end{aligned} \tag{5}$$

As shown above in Figure 1, the soft failure life T is determined by the overall degradation process, which consists of degradation initiation time T_i and degradation propagation time T_D. Therefore, the pdf of T can be obtained by using a convolution integral of T_i, T_D.

$$\begin{aligned} R_{SF}(t) &= P\{T > t\} = P\{T_i + T_D > t\} \\ &= \int_0^{+\infty} P\{v+T_D > t\} f_i(v)dv \\ &= \int_0^{+\infty} P\{T_D > t-v\} f_i(v)dv \\ &= \int_0^{+\infty} F_D(t-v) f_i(v)dv \\ &= \int_0^{+\infty} \Phi\left\{\frac{H-\mu_\lambda\left[e^{K(t-v)}-1\right]}{\sqrt{\sigma_\lambda^2\left[e^{K(t-v)}-1\right]^2+\sigma_\varepsilon^2}}\right\} d\left[F_i(v)\right] \end{aligned} \tag{6}$$

The complexity of the integration function forms a challenge on the computation of Equation (6). Considering the fact that $v \in (-\infty, +\infty)$ corresponds to $F_i(v) \in [0,1]$, a numerical integration method can be adopted to calculate Equation (6) as follows:

$$\begin{aligned} R_{SF}(t) &= \int_0^{+\infty} \Phi\left\{\frac{H-\mu_\lambda\left[e^{K(t-v)}-1\right]}{\sqrt{\sigma_\lambda^2\left[e^{K(t-v)}-1\right]^2+\sigma_\varepsilon^2}}\right\} d\left[F_i(v)\right] \\ &= \int_0^1 \Phi\left\{\frac{H-\mu_\lambda\left[e^{K\left(t-F_i^{-1}(u)\right)}-1\right]}{\sqrt{\sigma_\lambda^2\left[e^{K\left(t-F_i^{-1}(u)\right)}-1\right]^2+\sigma_\varepsilon^2}}\right\} du \\ &\cong \frac{1}{N_0}\sum_{j=1}^{N_0} \Phi\left\{\frac{H-\mu_\lambda\left[e^{K\left(t-F_i^{-1}\left(\frac{j}{N_0}\right)\right)}-1\right]}{\sqrt{\sigma_\lambda^2\left[e^{K\left(t-F_i^{-1}\left(\frac{j}{N_0}\right)\right)}-1\right]^2+\sigma_\varepsilon^2}}\right\} \end{aligned} \tag{7}$$

Considering that the reliability for hard failure varies when different numbers of shocks arrive, the probability of surviving from soft failure can be derived on condition of a fixed $N(t)$.

$$\begin{aligned} R_{HF}(t) &= \sum_{k=0}^{+\infty} R_{HF}(t \mid N(t)=k) \cdot P(N(t)=k) \\ &= \sum_{k=0}^{+\infty} \left[F_W(D)\right]^k \cdot \frac{(\lambda_0 t)^k e^{-\lambda_0 t}}{k!} \\ &= \sum_{k=0}^{+\infty} \left[\Phi\left(\frac{D-\mu_W}{\sigma_W}\right)\right]^k \cdot \frac{(\lambda_0 t)^k e^{-\lambda_0 t}}{k!} \end{aligned} \tag{8}$$

Therefore, the system reliability function can be derived as.

$$\begin{aligned} R(t) &= R_{HF}(t) \cdot R_{SF}(t) \\ &\cong \frac{1}{N_0}\sum_{j=1}^{N_0} \Phi\left\{\frac{H-\mu_\lambda\left[e^{K\left(t-F_i^{-1}\left(\frac{j}{N_0}\right)\right)}-1\right]}{\sqrt{\sigma_\lambda^2\left[e^{K\left(t-F_i^{-1}\left(\frac{j}{N_0}\right)\right)}-1\right]^2+\sigma_\varepsilon^2}}\right\} \\ &\times \sum_{k=0}^{+\infty} \left[\Phi\left(\frac{D-\mu_W}{\sigma_W}\right)\right]^k \cdot \frac{(\lambda_0 t)^k e^{-\lambda_0 t}}{k!} \end{aligned} \tag{9}$$

4 Numerical example

To illustrate the implementation of the proposed reliability model, a realistic example is studied in this Section. Based on some reasonable assumptions for illustration, Table 1 lists the relevant model parameter values. And according to Equation (9), the reliability assessment results of the GIL, from its installation time up to 20 years, are calculated and presented in Figure 2. It is shown that the reliability of the GIL remains at a relatively high level at first for about 5 years, which is mainly due to the degradation initiation time, whose mean is identical to 5 too. After that, the reliability curve presents a sharp decline until the 20th year, when the sealing rings of the GIL degrade greatly and the SF6 in the GIL are insufficient to maintain a good arc-quenching property.

Table 1. Model parameter values.

Parameter	Value	Parameter	Value
μ_λ	2	H	2.5
σ_λ	0.3	λ_0	3
K	0.2	μ_W	1.2
σ_ε	0.2	σ_W	0.1
λ_i	5	D	1.5

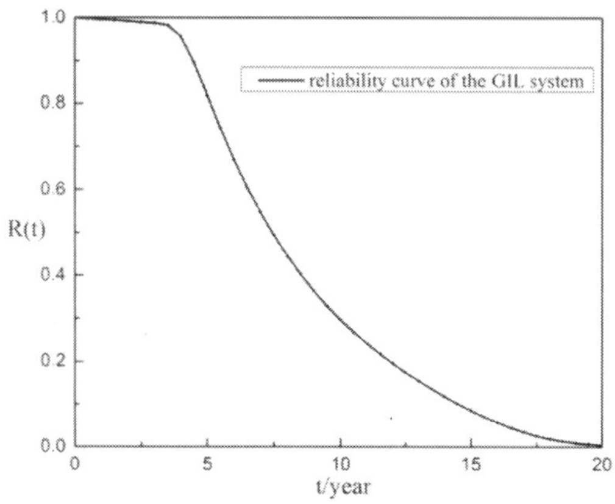

Figure 2. Reliability curve of the GIL.

Furthermore, considering λ_i and λ_0 are two critical parameters involving the design and manufacturing process, sensitivity analysis is performed to measure the effect of them on the reliability, and the analysis results are plotted in Figures 3 and 4. It can be indicated that the reliability assessment results show remarkable sensitivity on λ_i, and the effect of λ_0 cannot be negligible, either. A larger λ_i leads to a longer degradation initiation time and thus resulting in higher system reliability. And as λ_0 increases, the frequency of random shocks enlarges and the probability of hard failure increases, which makes the reliability of the GIL decreases.

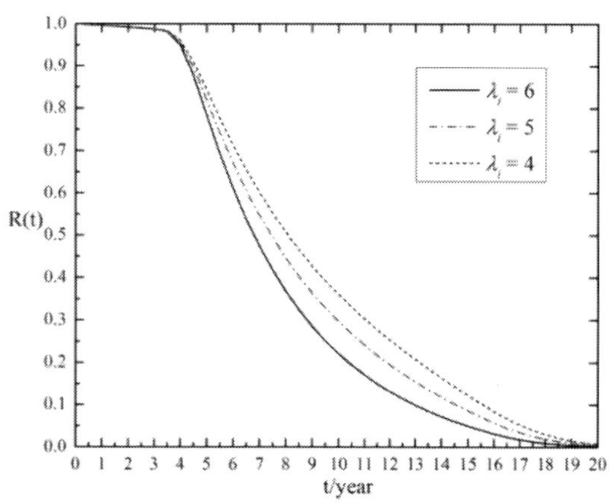

Figure 3. Sensitivity analysis of λ_i.

Guided by the sensitivity analysis results, in order to obtain larger system reliability, efforts should be made to increase the value of λ_i and decrease the value of λ_0. Some suggestions can be: 1) strengthening the quality control and management of the manufacturing process, which can reduce the quantity of potential defects in the GIL and prolong the degradation initiation time; 2) improving system design and take steps to reduce the occurrence probability of partial discharge, e.g., setting a particle catcher to collect charged particles in the GIL.

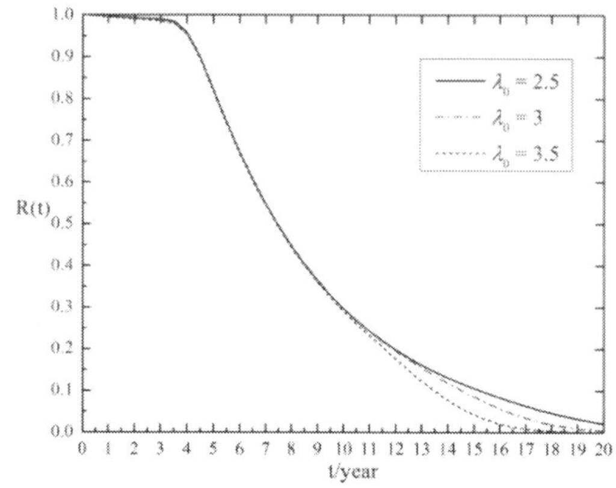

Figure 4. Sensitivity analysis of λ_0.

5 Conclusions

The primary competing failure modes of the GIL are soft failure due to air leakage, and hard failure caused by partial discharge. Taking into account random degradation initiation time before the air in the GIL begins to leak, this paper proposes a new competing failure model, and derives the reliability assessment results through analytical derivation and numerical calculation methods. Furthermore, a realistic example is studied to illustrate the implementation of the proposed model. Sensitivity analysis is conducted to measure the effect of λ_i and λ_0 on the reliability assessment results. It is indicated that a larger λ_i or a smaller λ_0 will lead to higher system reliability. For obtaining larger reliability of the GIL, it is suggested to make effort to increase the value of λ_i and decrease the value of λ_0, e.g., strengthening the quality management of the manufacturing process and reducing the quantity of potential defects, or improving the system design by setting a particle catcher to collect charged particles.

Acknowledgement

This work was supported partially by National Natural Science Foundation of China (71672006), Science and Technology Project of State Grid Corporation of China (SGTYHT/15-JS-191) with title "Reliability analysis of long-distance ultra-high voltage GIL equipment".

References

1. R. Benato and D. Napolitano, 'Reliability Assessment of EHV Gas-Insulated Transmission

Lines: Effect of Redundancies', IEEE Trans. Power Deliv., vol. 23, no. 4, pp. 2174–2181, Oct. 2008.

2. G. P. Baer, A. Diessner, and G. F. Luxa, '420 kV SF6-insulated tubular bus for the Wehr pumped-storage plant electric tests', IEEE Trans. Power Appar. Syst., vol. 95, no. 2, pp. 469–477, Mar. 1976.
3. H. Koch and A. Schuette, 'Gas insulated transmission lines for high power transmission over long distances', Electr. Power Syst. Res., vol. 44, no. 1, pp. 69–74, Jan. 1998.
4. H. Okubo, M. Yoshida, T. Takahashi, T. Hoshino, M. Hikita, and A. Miyazaki, 'Partial discharge measurement in a long distance SF6 gas insulated transmission line (GIL)', IEEE Trans. Power Deliv., vol. 13, no. 3, pp. 683–690, Jul. 1998.
5. 'IEEE Guide for Application and User Guide for Gas-Insulated Transmission Lines, Rated 72.5 kV and Above', IEEE Std C371224-2016, pp. 1–49, May 2017.
6. A. Miyazaki et al., 'Results of Partial Discharge Measurements in a Long-Distance 275 kV GIL', IEEE Power Eng. Rev., vol. 22, no. 10, pp. 58–58, Oct. 2002.
7. Y. Li et.al, 'Two cases of failure reason and treatment analysis for gas insulated transmission line (GIL)', Electronic age, vol. 0, no. 11, pp. 66 - 67, 2016. (In Chinese)
8. C. Pan, 'The analysis on the reason of the GIL air leakage in 220kV system', Guizhou Electronic Power technology, vol. 17, no. 6, pp. 71 - 73, 2014. (In Chinese)
9. R. Huang et al., 'Study on aging of material for GIS sealing ring', in TENCON 2015 - 2015 IEEE Region 10 Conference, 2015, pp. 1–4.
10. H. Peng and Q. Feng, 'Reliability Modeling for Ultrathin Gate Oxides Subject to Logistic Degradation Processes with Random Onset Time: Reliability Modeling for Ultrathin Gate Oxides', Qual. Reliab. Eng. Int., vol. 29, no. 5, pp. 709–718, Jul. 2013.
11. C. Paroissin, 'Inference for the Wiener Process With Random Initiation Time', IEEE Trans. Reliab., vol. 65, no. 1, pp. 147–157, Mar. 2016.
12. Q. Wang, Z. He, S. Lin, and Z. Li, 'Failure Modeling and Maintenance Decision for GIS Equipment Subject to Degradation and Shocks', IEEE Trans. Power Deliv., pp. 1–1, 2016.
13. P. Cirillo and J. Hüsler, 'Extreme shock models: An alternative perspective', Stat. Probab. Lett., vol. 81, no. 1, pp. 25–30, Jan. 2011.
14. C. J. Lu and W. O. Meeker, 'Using Degradation Measures to Estimate a Time-to-Failure Distribution', Technometrics, vol. 35, no. 2, pp. 161–174, May 1993.
15. Y. Zhang, H. Liao, and Y. Hong, 'Planning accelerated destructive degradation tests with initiation time', in 2015 Annual Reliability and Maintainability Symposium (RAMS), 2015, pp. 1–6.
16. J. Pan, S. W. Xu, W. H. Chen, X. Y. Wang, P. Qian, and Q. W. Hu, 'Accelerated Aging Test and Study of Storage Life Prediction of NBR O-Ring', Adv. Mater. Res., vol. 415–417, pp. 184–190, Dec. 2011.
17. C. Peng, X. H. Gu, and K. Xiao, 'Prediction for O-Ring Type Sealing Ring's Storage Life Based on Accelerated Ageing Data', Appl. Mech. Mater., vol. 433–435, pp. 2096–2100, Oct. 2013.

Random demodulation for structural health monitoring excited by the five-cycle sine burst

Xing Li, Dong Liang, Yishou Wang, Xinlin Qing, Hu Sun*

School of Aerospace Engineering, Xiamen University, Xiamen 361005, China

2458202237@qq.com

Abstract. Nowadays, the Structural Health Monitoring (SHM) has been paid more and more attention. The five-cycle sine burst is widely used as the exciting signal in SHM and the sensors' responded signals are analyzed to research the damage. In the sensor network, there will be many sensors which mean many responded signals will be sampled, restored and sometimes transferred. In the traditional way which is known as Nyquist sampling theorem, the sampling rate must be more than twice the highest rate of the original signal. In this way, the amount of data will be huge. As the result, the costs will be very expensive and the equipment may be huge and heavy, which is especially unaccepted in the aircraft. It is necessary to do some research to compress the signal. The Compressing Sensing (CS) theory provides new methods to compress the signals. The Random Demodulation (RD) is a specific method which can accomplish the physical implementation of CS theory. In this paper, according to the structure of RD, we chose some chips to build a RD system. And we did some experiments to verify the method through the system. We chose the Orthogonal Matching Pursuit (OMP) as the construct algorithm to recover the signal.

1. Introduction

SHM offers an online and real-time monitoring. It can discover the structural damage or fatigue timely and avoid the risk. However, in practical applications of large structural health monitoring, a large number of distributed sensors are usually adopted to monitor the big dimension structures. Hence, how to obtain a fast and accurate impact signal from the big data is an important problem for the damage assessments. According to the different functions of the system, there are two types of monitoring. One is active monitoring, the other is passive monitoring [1-2]. In this paper, the research is on the active monitoring.

Recently, the CS theory has been researched by many scholars. It has a lot of advantages on signal compressing, like independent acquisition and reconstruction, simple computation and good compression. So the CS has a great application prospects on SHM [3-5]. In this paper, we used the CS in the active monitoring, aiming to achieve a wide range of high-resolution data acquisition in large-scale structure quickly and efficiently. The research is about the five-cycle sine burst compressed by the RD system.

The paper has 7 parts. Next, Section 2 introduces the background knowledge of RD system. In Section 3, the signal used in SHM is introduced. Section 4 gives simulations. Section 5 is the experimental part. And section 6 analyses the experimental results. Finally, section 7 makes a conclusion for the paper.

*Corresponding author: 476984750@qq.com

2. Random demodulation (RD)

According to the CS theory, if a signal can be expressed as a product of a vector and a martrix

$$x = \Psi\alpha. \quad (1)$$

And the (1) can also be written as

$$x(t) = \sum_{i=1}^{N} \alpha_i \Psi_i(t) \quad (2)$$

If there is a few big coefficients in α, then the $x(t)$ is sparse in the basis of Ψ .And the $x(t)$ can be compressed and sampled at the same time by the method of global observation.

RD is a physical implementation of the CS. The structure of the RD is shown in Fig. 1.

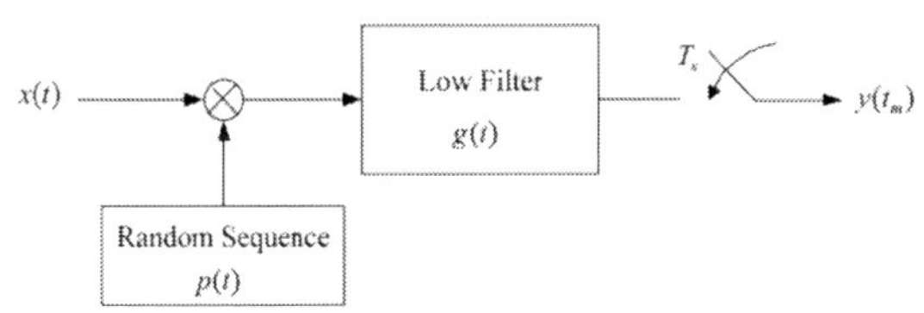

Fig. 1. The structure of RD

Its structure is simple and includes three main parts: mixing, filtering and sampling. The process expressed mathematically as

$$y(t_m) = \int_{-\infty}^{+\infty} x(\tau)p_c(\tau)h(t-\tau)\mathrm{d}\tau \Big|_{t=t_m=mT_s} \quad (3)$$

Substitute (2) into (3), the equation (3) is written as

$$y(t_m) = \sum_{n=1}^{N} \alpha_n \int_{-\infty}^{+\infty} \Psi_n(\tau)p_c(\tau)h(mT_s-\tau)\mathrm{d}\tau = \sum_{n=1}^{N} \alpha_n v_{m,n} \quad (4)$$

where $v_{m,n} = \int_{-\infty}^{+\infty} \Psi_n(\tau)p_c(\tau)h(mT_s-\tau)\mathrm{d}\tau$, where $h(t)$ is the unit impulse response of the filter. For the discrete signal, the output vector y is equivalent to multiply matrix V by the input vector α for the output vector $y = V\alpha$,

where $V = \begin{bmatrix} v_1(t_1) & v_2(t_1) & \dots & v_N(t_1) \\ v_1(t_2) & v_2(t_2) & \dots & v_N(t_2) \\ \dots & \dots & \dots & \dots \\ v_1(t_m) & v_2(t_m) & \dots & v_N(t_m) \end{bmatrix}$. In this paper, the Ψ is the Inverse Discrete Fourier Transform base.

$$\Psi = \begin{bmatrix} 1 & 1 & 1 & \cdots & 1 \\ 1 & \overline{w} & \overline{w}^{2} & \cdots & \overline{w}^{N-1} \\ \vdots & \vdots & \vdots & \cdots & \vdots \\ 1 & \overline{w}^{N-1} & \overline{w}^{2(N-1)} & \cdots & \overline{w}^{(N-1)(N-1)} \end{bmatrix}_{N\times N},$$

where $\overline{w} = e^{\frac{2\pi i}{N}}$.

There is a few big coefficients in α, so the solution of the underdetermined Equations $y = V\alpha$ problem can be converted into the minimum 0-norm problem:

$$\hat{\alpha} = \arg\min \|\alpha\|_0 \text{, s.t. } y = V\alpha. \quad (5)$$

To solve the equation (5), the orthogonal matching tracking algorithm (OMP) is chosen, because the speed of the reconstruction is fast[6-13].

3. The narrowband waves used in SHM

Recently, the five-cycle sine burst is widely used as the exciting signal in SHM. It is a narrow-band signal which can be expressed as

$$f_a(t) = A\sin(2\pi f_c t)(1-\cos(\frac{2\pi f_c t}{5})) \quad (6)$$

where A denotes the amplitude and f_c represents the central frequency of the signal[14]. The exciting signal is the first cycle from the beginning of the x-axis of the $f_a(t)$. The rest is 0. The Fig. 2 shows the exciting signal.

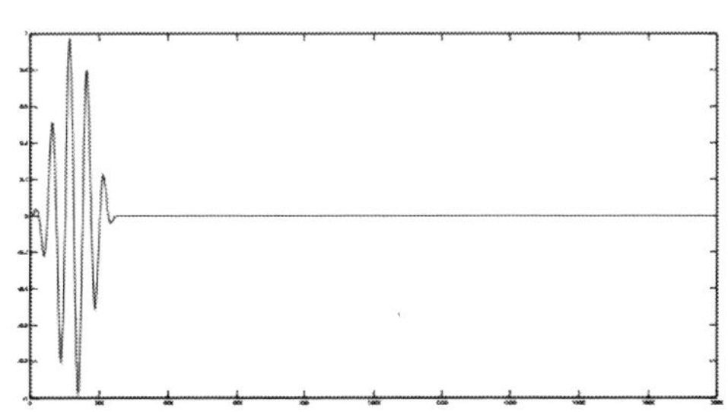

Fig 2. The five-cycle sine burst

4. Simulation

4.1 The simulation instructions

The simulation is done by the MATLAB. The random sequence produced by function randsrc(N,1) which can produce a sequence that is N-row and 1 column with value of ±1. The Butterworth is selected as the type of the filter using three functions, buttord, butter and buttap.

In the simulation, we use the Signal-Noise Ratio (SNR) to evaluate the effect of reconstruction. The SNR expressed as

$$SNR = 20\log_{10}\frac{\|x\|_2}{\|x-\hat{x}\|_2}, \quad (7)$$

where $\|\bullet\|_2$ is defined as $\|x\|_2 = \sqrt{(|x_1|^2 + |x_2|^2 + \cdots + |x_n|^2)}$, $\hat{x}$ is the reconstructed signal while x is the original signal.

The simulations are about sin waves and the five-cycle sine burst. In the simulation, the sampling rate of generating signal is 100K, so the sampling rate of the digital filter is 100K too. We stipulate the reconstruction is successful when the SNR is higher than or equal to 15dB. The probability of successful reconstruction is another factor to evaluate the result and defines as the ratio of successful times and total times. And the compression ratio is 10.

4.2 The results

The results of sin waves are very good and the SNR is very high. And the frequency of the signal reconstructed well can be very high, nearly the half of the sampling rate. The results are shown in Fig. 3.

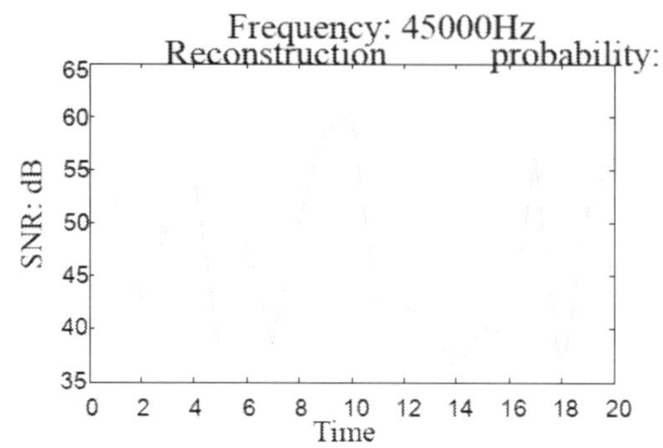

Fig. 3 The reconstructed result of sine waves

However, the results are not so good for the five-cycle sine burst. As the Fig. 4 shows, as the frequency becomes increased, the SNR and the reconstruction probability are decreased. When the frequency is 2000Hz, the reconstruction probability is even 0. The relation between the frequency and the reconstruction probability is shown in Table. 1.

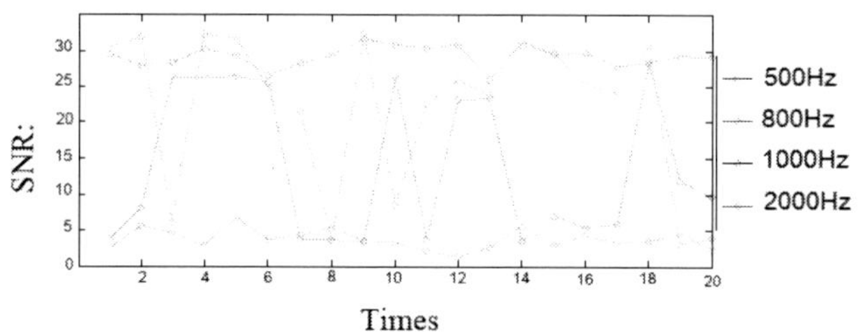

Fig. 4 The frequency and the SNR when the sampling rate is 100K Hz

Table 1. The frequency and the reconstruction probability when the sampling rate is 100K

frequency (Hz)	500	800	1000	2000
reconstruction probability (%)	100	75	40	0

The results show that the five-cycle sine burst has frequency limits on using in the RD compared to the sine waves. The reason may be the sparsity of signals in the Inverse Discrete Fourier Transform base.

There is only one nonzero factor in the spectrum of the sine waves. But in the spectrum of the five-cycle sine burst, it's a limited band. As shown in Fig. 5.

DOI: 10.1051/matecconf/201713900075

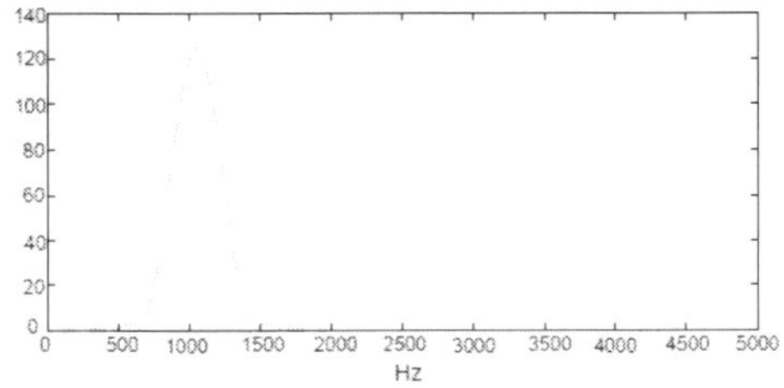

Fig. 5 The spectrum of the five-cycle sine burst

However, the low frequency can be reconstructed well, like 500Hz and 800Hz. The frequency is higher, the results are worse. The reason is that as the frequency increased, the nonzero width of spectrum is wider, as the Fig. 6 shows.

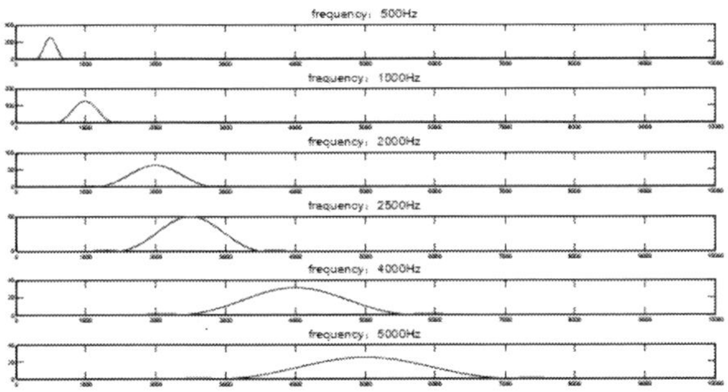

Fig. 6 The spectrum of different frequency of the five-cycle sine burst

According to the CS theory, obviously, there are less nonzero factors in the spectrum, the reconstructed results will be better. However, the sparsity is relative, but it is not to total points of signal, according to the simulation, it's to the sampling rate.

If we change the sampling rate of generating signal to 1000 KHz, the results' frequency increases 10 times too. The results are shown in the Fig. 7 and Table 2.

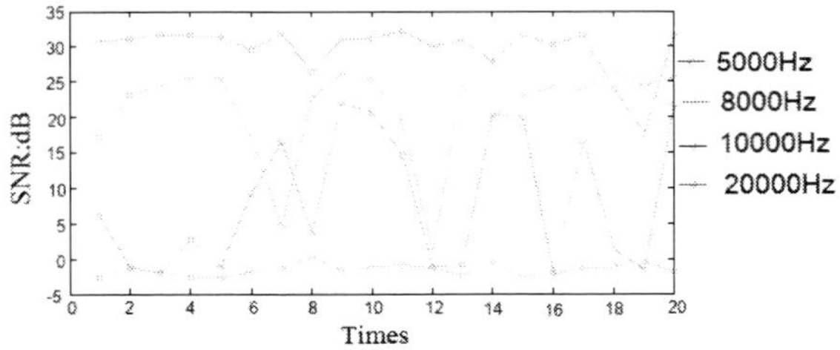

Fig. 7 The frequency and the SNR when the sampling rate is 1000K

Table 2. The frequency and the reconstruction probability when the sampling rate is 1000K

Frequency（Hz）	5000	8000	10000	20000
Reconstruction probability（%）	100	90	35	0

It seems that the ratio of frequency and the sampling rate of generating signal must be in a specific range, the results will be good.

What's more, the cutoff frequency of the filter can influence the results. When the signal's frequency increases, increasing the cutoff frequency from the low frequency's cutoff frequency is helpful. But it's not positive correlative. The cutoff frequency must be suitable.

5. Experiments

5.1 Experiment system

We set up a system to put the RD in the active monitoring system. The structure is shown in Fig. 8.

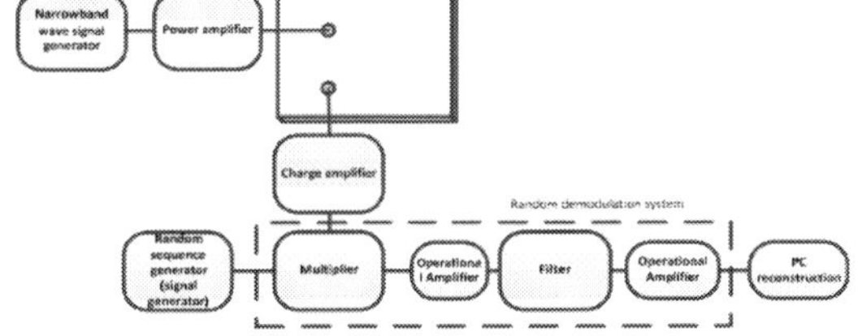

Fig. 8 The RD in the active monitoring system

As the Fig shows, the five-cycle sine burst is amplified by the power amplifier, then to one PZT sensor. The other sensor receives the reflected signal and is through the charge amplifier. Then the signal enters into the RD system, compressed, sampled and reconstructed.

5.2 The method of experiment

The reponse signal of sensor has many reflected waves, however, upon most occasions, only the

DOI: 10.1051/matecconf/201713900075

first main response five-cycle sine burst wave is used in SHM. As shown in Fig. 9.

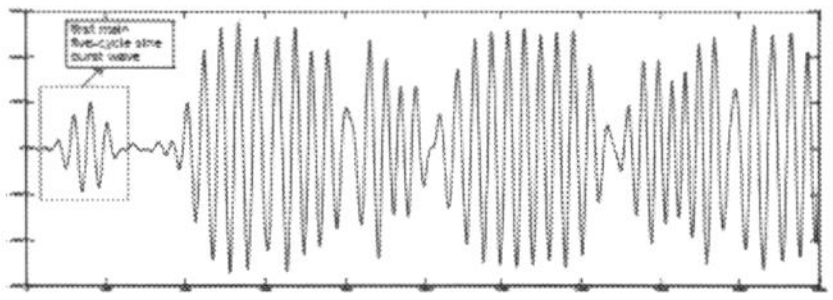

Fig. 9 The reponse signal of sensor

And the first main wave is the same as the exciting wave. For simplicity, firstly, we used the five-cycle sine burst produced by the signal generator instead of the response signal of sensor as the original signal to verify the method, as shown in Fig. 10 .

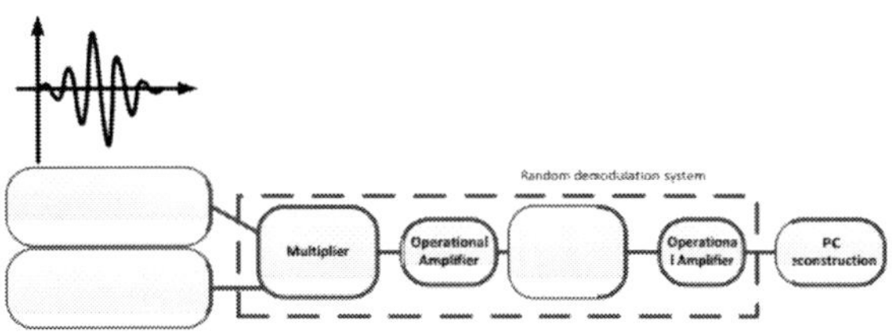

Fig. 10 The experimental system

The five-cycle sine burst is generated by NI PXIe-5442 wave generator; the random sequence is generated by NI PXIe-6115 and the sampling is done by NI PXIe-5122. The three cards are all in a chassis and transferred by the LabVIEW program. And the OMP algorithm is done by MATLAB and embedded in the LabVIEW program.

5.3 The design of RD system

The AD633 is used as the multiplier and the MAX275 is used as the filter. The operational amplifier after multiplier is OPA627, after filter is MAX44241.

The **m** sequence is chosen as the random sequence. **m** sequence is referred to as the longest linear feedback shift register sequence. It is a pseudo-random binary sequence and its magnitude approximately randomly jumps to 1 or -1, which make its frequency spectrum looks similar to the broadband noise spectrum and cover almost the entire frequency axis. The **m** sequence is easily produced, regularity and has many excellent performance.

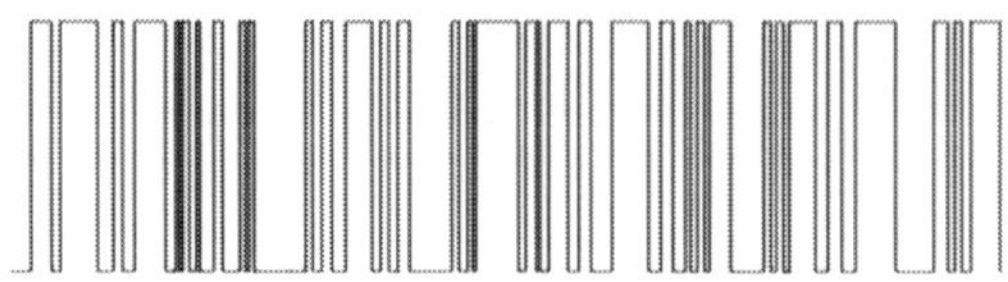

Fig. 11 m sequence

6. Results of experiments and analysis

6.1 results

The results of experiments' SNR are worse than the simulation. However, the reconstruction probability is similar to the simulations. Some results are shown in Fig. 12.

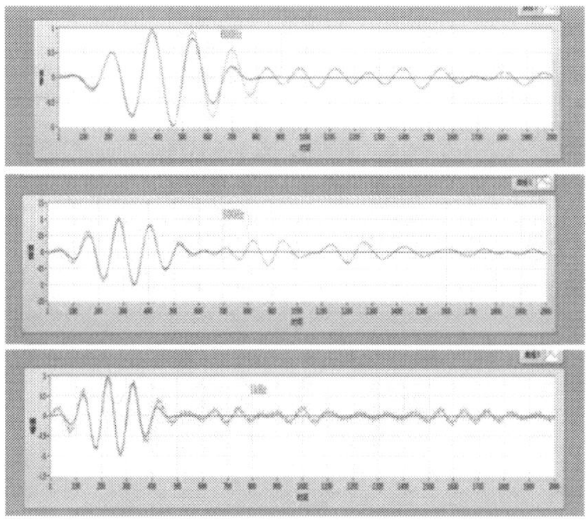

Fig. 12 The results of the experiments: the red is reconstructed signal, the blue is original one

As the results shown, the reconstructed signal has the roughly outline of the original signal. But there are still some differences in the details. Those differences in detail result in the lower SNR compared to the simulations. One important reason is the unit impulse response of the RD system.

6.2 The unit impulse response

In the experiments, we used the unit impulse response of the RD system rather than the filter in the simulation. And the unit impulse response has a great influence on the reconstructed results [15].

In this paper, we use the **m**-sequences for impulse response measurement. There are many noises in the impulse in one measurement. So we measure and calculate several times to get its average. There is less noise in the impulse response, the reconstructed results will be better.

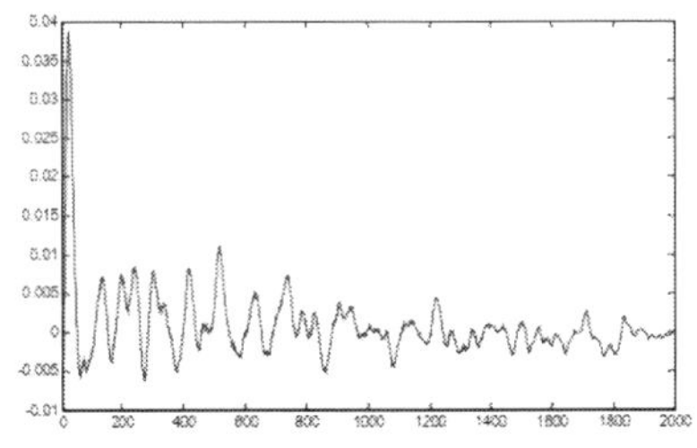

(a)

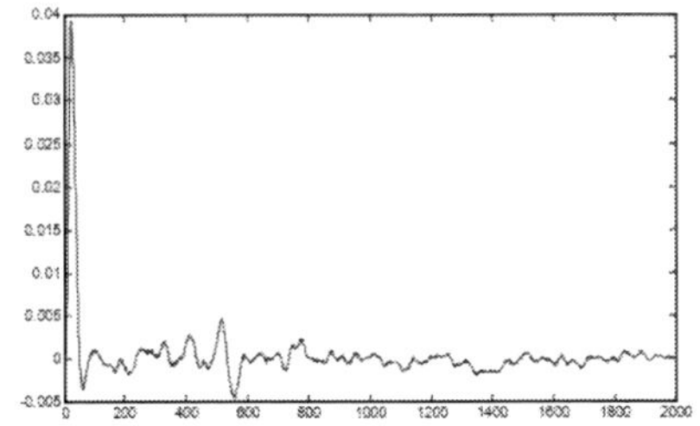

(b)

Fig. 13 The unit impulse response: the (a) is measured by once, the (b) is the average measured by ten times.

6.3 The other reasons

The noise in the unit impulse response is influenced by the chips, generator or sampling. The chips may produce the noise when they work. And there may be some mistakes in generating **m** sequence and sampling the output signals of the system. Those noise and mistakes result in the noise in the unit impulse response. What's more, the system may have some offsets. All those nonideal factors will lead the mistakes in the reconstructed results.

7. Conclusion

In this paper, we have done some research on using RD system to compress and sample the five-cycle sine burst. For now, the frequency of the five-cycle sine burst that can be well reconstructed is relatively low. Then we analyzed the reason of the problem. The main reason is that the five-cycle sine burst's spectrum is a limited band and as the frequency increases, the spectrum is wider. In the experiments, we found some factors that influence the practical application of the RD. The unit impulse response matters a lot and the accuracy of instruments and chips affects the results too.

Acknowledgements

This work is supported by the National Natural Science Foundation of China (Grant no. 51405409) and the Fundamental Research Funds for the Central Universities.

References

[1] K. Diamanti, C. Soutis, Structural health monitoring techniques for aircraft composite structures, *Progress in Aerospace Sciences* 46 (8) (2010) 342-352.

[2] Qing X, Beard S, Ikegami R, Chang F-K, Boller C. Aerospace Applications of SMART Layer Technology[M]. *Encyclopedia of Structural Health Monitoring*, Wiley, 2009.

[3] Alessandro Perelli, Sevan Harput , Luca De Marchi and Steven Freear. Frequency Warping Compressive Sensing for Structural Monitoring of Aircraft Wing[J], *IEEE Transactions on Ultrasonics, Ferroelectrics, and Frequency Control,* Vol: 60, pp.2089-2097, October. 2013.

[4] Yuan Mei, Wang Shujuan, Dong Shaopeng and Pang Zhuo. "Reconstruction of Undersampled Damage Monitoring Signal Based on Compressed Sensing". Guidance, Navigation and Control Conference (CGNCC), 2014 IEEE Chinese, Yantai, China, 2014, pp.2243-2248.

[5] Ju-min Zhao, Ding Feng, Deng-ao Li, and Bao-feng Zhao, " Power-free Structural Health Monitoring via compressive sensing", Computing and Communications Conference (IPCCC), 2015 IEEE 34th International Performance, Nanjing, China, 2015.

[6] D. Karampoulas , S.M.Kouadri , L.S.Dooley, "A

Novel Precolouring-Random Demodulator Architecture for Compressive Spectrum Estimation", Intelligent Signal Processing Conference 2013 (ISP 2013), IET, London, UK, 2013.

[7] Tamer Ragheb, Jason N. Laska, Hamid Nejati and Sami Kirolos, "A Prototype Hardware for Random Demodulation Based Compressive Analog-to-Digital Conversion", 51st Midwest Symposium on Circuits and Systems, Knoxville, TN, USA, 2008.

[8] Ning Fu, Pingfan Song, Peizhuo Liu, and Jingchao Zhang, "Boost the Efficiency of Spectrum Sensing Using Synchronized Random Demodulation", 19th International Conference on Digital Signal Processing, Hong Kong, China, 2014, pp.525-530.

[9] Ning Fu, Tingting Yao, Hongwei Xu, "COMPRESSIVE BLIND SOURCE RECOVERY WITH RANDOM DEMODULATION", 22nd European Signal Processing Conference (EUSIPCO), Lisbon, Portugal, 2014, pp.746-750.

[10] E. Astaiza, H. F. Bermúdez, W. Y. Campo, "Efficient Wideband Spectrum Sensing Based on Compressive Sensing and Multiband Signal Covariance", IEEE Latin America Transactions, vol:15, pp.393-399, 2017

[11] Smaili S, Massoud Y, "Accurate and efficient modeling of random demodulation based compressive sensing systems with a general filter", Circuits and Systems (ISCAS), 2014 IEEE International Symposium on. IEEE, 2014, pp. 2519-2522.

[12] Kirolos S, Laska J, Wakin M, et al, "Analog-to-information conversion via random demodulation", Design, Applications, Integration and Software, 2006 IEEE Dallas/CAS Workshop on. IEEE, 2006, pp.71-74.

[13] Ning Fu, Pingfan Song, Jingchao Zhang and Ying Liu, "A Random Demodulation Hardware System with Automatic Synchronization Function", IEEE International Instrumentation and Measurement Technology, Minneapolis, MN, USA, 2013, pp.1554-1558.

[14] Shan S, Qiu J, Zhang C, et al. Multi-damage localization on large complex structures through an extended delay-and-sum based method[J]. Structural Health Monitoring, 2016, 15(1): 50-64.

[15] Pawel J. Pankiewicz, Thomas Arildsen, Torben Larsen, "SENSITIVITY OF THE RANDOM DEMODULATION FRAMEWORK TO FILTER TOLERANCES", Signal Processing Conference, European, 2011, pp.534-538.

The research on the influence of kinetic parameters on the vibration of blood pump in viscoelastic implantation environment

Zhong Yun, Chuang Xiang* and Liang Wang

School of Mechanical and Electrical Engineering, Central South University, Changsha 410083, R.P. China

Abstract. The implantable axial flow blood pump would be vibrated when in actual use because of the viscoelastic implantation environment, unbalanced driven force, high speed rotor and so on. The blood pump researched in the paper was designed to be implanted in ascending aortic and driven by external magnetic force. The kinetic equation of blood pump worked in viscoelastic implantation environment was built, and then the blood pump's vibration influenced by viscoelasticity of implantation environment, running speed of blood pump, weight of blood pump were studied. Appropriate damping coefficient of implantation environment, running speed of blood pump and weight of blood pump for blood pump's more stable working were found in the paper. The paper could provide theoretical basis and research method for blood pump's structure parameters and work performance improving.

1 Introduction

Blood pump can take place the heart of heart patient temporary or long-term [1-4]. In recent years, most of the researches on blood pump focused on the impeller structure, flow channel, drive system, hemolysis, suspension mechanism and so on [5-9]. But the research on the kinetic system of blood pump in implantation environment was few. The implantable blood pump was often implanted in the chest, aortic valve and ascending aortic [10, 11]. There were blood vessel, myocardium, connective tissue and other tissues around the blood pump. These tissues were all viscoelastic material [12, 13], would be deformed when being compressed or stretched. In order to be implanted easily, the smaller was the blood pump the better, so the speed of blood pump would be very high (about 10000r/min) [14,15] for supplying enough blood flow. The driven system of blood pump often was brushless direct-current motor [16]. The blood pump system of this paper was large gap external magnetic driven system. The driven force was changed periodically for providing periodical driven moment. The changed driven force and viscoelastic implantation environment would cause the blood pump vibration when working.

The blood pump researched in this paper was miniature implantable axial blood pump and the position of being implanted was ascending aortic. The paper modeled the implantation environment of blood pump, and then built the kinetic equation. Through solving the kinetic equation in SIMULINK, the vibration acceleration, velocity and displacement of blood pump influenced by implantation environment, rotor speed and weight of blood pump were researched. The research in this paper could provide theoretical evidence for the optimization of blood pump.

2 Modelling and method

The blood pump researched in this paper was implanted in ascending aortic. There were vessel, connective tissue and other tissues wrapped the blood pump. As shown in Fig.1, the implantation environment was modeled to be two coaxial cylinders, one represented the vessel wall and the other represented the myocardial tissue, connective tissue and other tissues, for simplifying the calculation. The vessel's inner diameter is D_1=20 mm (blood pump's outer diameter), the thickness of vessel wall [13] is about 1 mm (vessel's outer diameter D_2=22 mm), the tube's outer diameter is D=40 mm, the axial length is L=80 mm (the length of the blood pump), the outer wall is fixed. All the material was regarded as isotropic [13]. Because the external driven force was perpendicular to the axis of blood pump, so the axial length of cylinder which was equal to the length of blood pump was considered in the paper.

* Corresponding author: 08105077@163.com

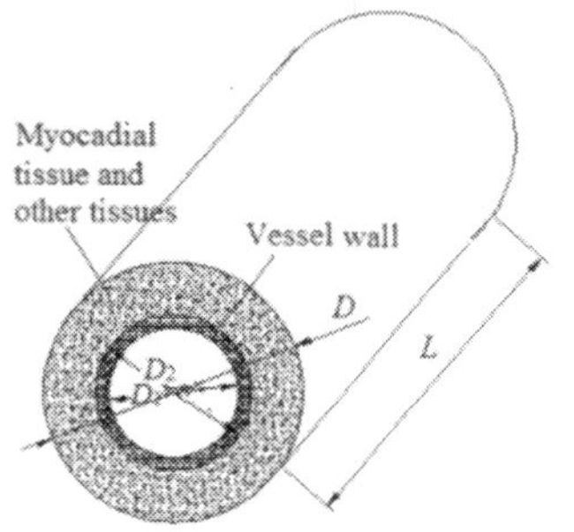

Fig. 1. The model of blood pump's implantation environment. D was the outside diameter of cylinder which represented the myocardial tissue and other tissues. D_1 was the inside diameter of vessel wall and D_2 was the outside diameter. L was the length of the cylinders.

In order to build the kinetic model of blood pump in implanted environment, the spring-damping model [13] was used to simulate the vessel and the tissues around vessel. According to the positional relationship between the ascending aortic vessel wall and tissue, the viscoelastic model of the implanted environment was established which was shown in Fig. 2.

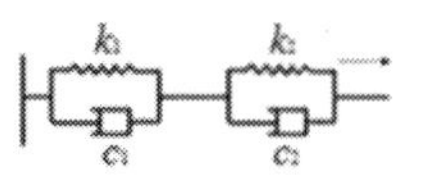

Fig. 2. Spring-damping model of blood pump's implantation environment. $c_1$and c_2 were damping coefficient. k_1 and k_2 were elasticity coefficient.

The elasticity coefficient and damping coefficient of the implanted environment's viscoelastic model can be calculated by the formulas:

$$k = 2\frac{k_1 k_2}{k_1 + k_2} \tag{1}$$

$$c = 2\frac{c_1 c_2}{c_1 + c_2} \tag{2}$$

Where, k, k_1, k_2 were elasticity coefficient of implantation environment, myocardial tissue and other tissues, blood vessel wall respectively. c, c_1, c_2 were the damping coefficient of implantation environment, myocardial tissue and other tissues, blood vessel wall respectively.

Substituting numerical value [17] into the Eq. (1) and Eq. (2), the elasticity coefficient of blood pump's implanted environment is k=981.7 N/m, the damping coefficient of blood pump's implanted environment is c=21.4 N•s/m.

Considered the driven force was perpendicular to the axis of blood pump, the kinetic model of blood pump working in viscoelastic implantation environment can be simplified to be two force components, two spring-damping system and a mass point which were shown in Fig.3. The mass point represented the blood pump. Two force components represented the driven force divided into horizontal and vertical directions. Two spring-damping systems represented the implantation environment's resistance to the blood pump's movement which was divided into horizontal and vertical directions. Because the material is isotropic, the parameters of two spring-damping systems were same, $c_1'=c_2'=c$, $k_1'=k_2'=k$.

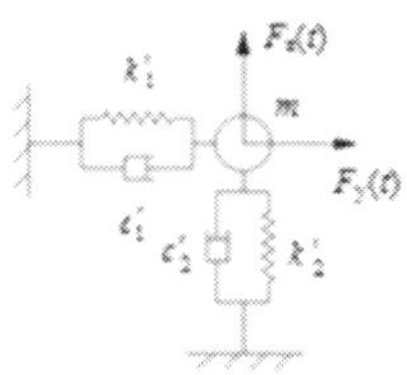

Fig. 3. The kinetic system of blood pump in implantation environment.

Based on the Newton's second law, the kinetic equation of blood pump working in viscoelastic implantation environment was:

$$\begin{bmatrix} m & 0 \\ 0 & m \end{bmatrix}\{\ddot{s}\} + \begin{bmatrix} c_1' & 0 \\ 0 & c_2' \end{bmatrix}\{\dot{s}\} + \begin{bmatrix} k_1' & 0 \\ 0 & k_2' \end{bmatrix}\{s\} = \begin{Bmatrix} F_y(t) \\ F_z(t) \end{Bmatrix} \tag{3}$$

m was the weight of blood pump. $s=[y, z]^T$, represented the displacement of blood pump, could be divided into horizontal direction y and vertical direction z. The driven force was divided into horizontal direction $F_y(t)$ and vertical direction $F_z(t)$ which were shown in Fig.4 [18].

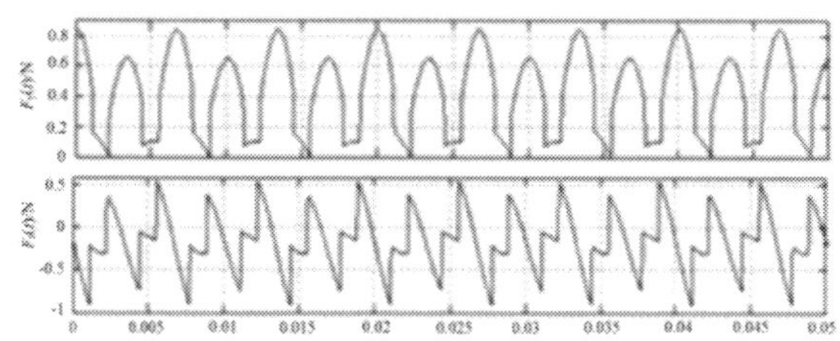

Fig. 4. The driven force which was divided into horizontal direction $F_y(t)$ and vertical direction $F_z(t)$.

The natural frequency of the damping system ω_d can be calculated by the formula:

$$\omega_d = \sqrt{(1-\xi^2)\frac{k}{m}} \tag{4}$$

In the Eq.(4), $\xi=c/(2(mk)^{1/2})$. The kinetic system's natural frequency was calculated to be $fd1=fd2=\omega_d/(2\pi)$ $\approx$9.2 Hz.

3 Results and discussion

3.1. Vibration response

Modules in SIMULINK were used to solve the kinetic equation. Fig.5 showed the vibration response in the direction of Y and Z. The blood pump's displacement increased rapidly in the initial stage. Then blood pump achieved the maximum displacement at about 0.06s, both in the direction of Y and Z. The displacement fluctuation decreased gradually later. At last he blood pump reached a dynamic balance and vibrated periodically with tiny amplitude relative to the equilibrium position. The cycle of the balance vibration and exciting force are the same, while the equilibrium position in the direction of Y and Z were respectively 4.5×10^{-3} m and -1.6×10^{-3} m.

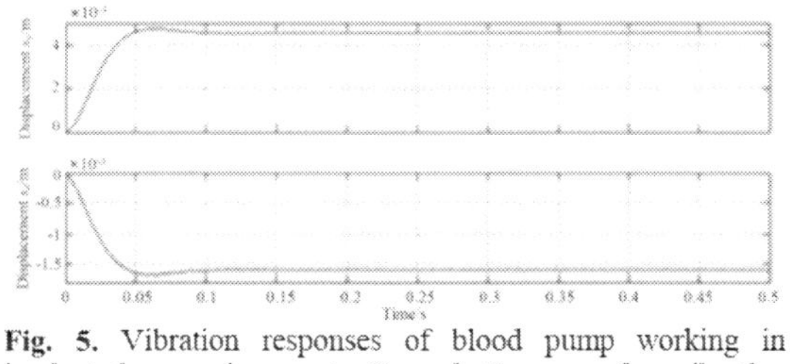

Fig. 5. Vibration responses of blood pump working in implantation environment. S_y and S_z were the vibration displacement in the direction of Y and Z respectively.

3.2 Vibration influenced by implantation environment

The parameters of implantation environment vary from person to person, and are different when the person in different ages. According solving the Eq.(3) when the damping coefficient varied from 1 to 30 N•s/m, the paper researched the blood pump's vibration acceleration A_a, velocity A_v and displacement A_s affected by damping coefficient which was shown in Fig.6. As shown in Fig.6, the A_a, A_v and A_S were all divided along horizontal and vertical directions into A_{ay}, A_{az}, A_{vy} A_{vz}, A_{sy} A_{sz} respectively. As the damping coefficient increasing, the vibration acceleration first decreased, then began to rise at about 8 N•s/m; The vibration velocity decreased all the time. When the damping coefficient was small, the amplitude was larger, and the change in the amplitude was also larger. But when the damping coefficient was bigger than 4 N•s/m, the change of damping coefficient had little effect to the vibration displacement.

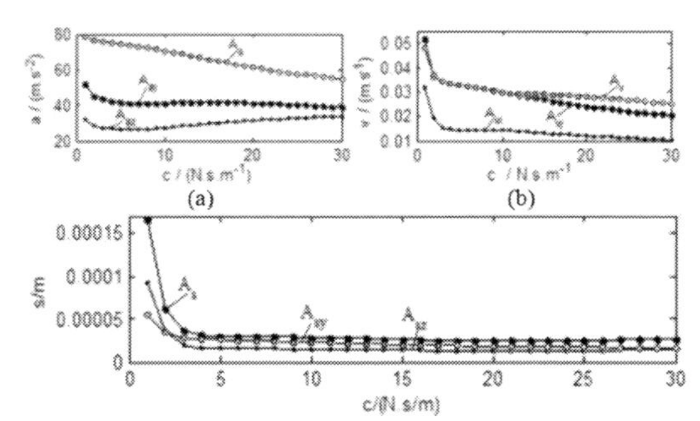

Fig. 6. The vibration acceleration A_a (a), velocity A_v (b) and displacement A_S (c) affected by damping coefficient. A_{ay} and A_{az} were components of A_a. A_{vy} and A_{vz} were components of A_v. A_{sy} and A_{sz} were components of A_s.

3.3 Vibration influenced by rotor speed

The speed of blood rotor would be change in order to supply different blood flow for person who was implanted the blood pump in different movement state. The rotor speed had direct relationship with the frequency of exciting force. So, the rotor speed was an important parameter for kinetic system of implantation environment. The vibration acceleration, velocity and displacement of blood pump influenced by rotor speed which varied from 1000 to 12000r/min were researched in the paper. Fig.7 showed the relationship between rotor speeds with vibration acceleration. When the rotor speed increased, the blood pump's vibration acceleration, velocity and displacement all increased at low speed, and then decreased. The affect of rotor speed to the blood pump's vibration acceleration, velocity and displacement was obvious at low speed (<4000 r/min), small at middle speed (4000~8000 r/min), and not obvious at high speed (>8000 r/min). When the rotor speed at 1000-2000 r/min, the vibration amplitude of blood pump was large, the blood pump should reduce work time in this speed range. The blood pump had a good work performance when rotor speed was higher than 8000 r/min.

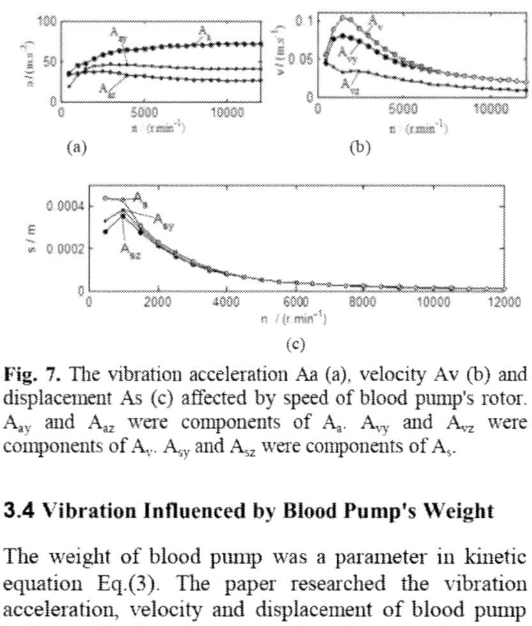

Fig. 7. The vibration acceleration Aa (a), velocity Av (b) and displacement As (c) affected by speed of blood pump's rotor. A_{ay} and A_{az} were components of A_a. A_{vy} and A_{vz} were components of A_v. A_{sy} and A_{sz} were components of A_s.

3.4 Vibration Influenced by Blood Pump's Weight

The weight of blood pump was a parameter in kinetic equation Eq.(3). The paper researched the vibration acceleration, velocity and displacement of blood pump influenced by blood pump's weight changing from 0.01 Kg to 0.04 Kg. The results were shown in Fig. 8. As the blood pump's weight increasing, the vibration acceleration, velocity and displacement of blood pump decreased, and the degree of decrease was smaller and smaller. It seemed that the heavier of the blood pump, the kinetic system of blood pump would be more stable. But the increase of blood pump's weight would lead more difficult implantation and worse comfortable. Form Fig. , we knew the weight of blood pump affected vibration displacement not obviously when the weight of blood pump was heavier than 0.03 Kg. So, it's better for the weight of blood pump be about 0.03 Kg.

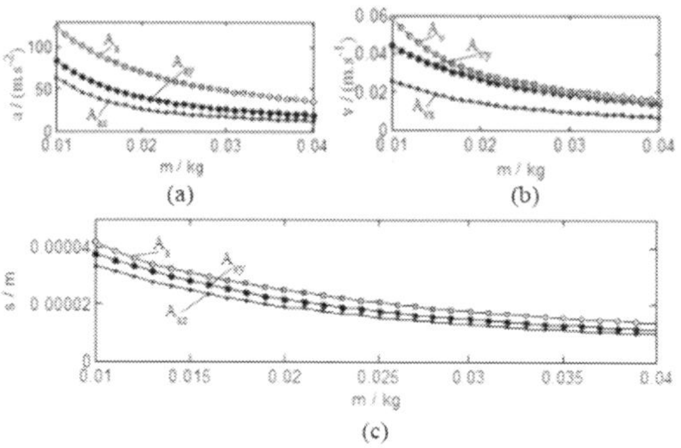

Fig. 8. The vibration acceleration A_a (a), velocity A_v (b) and displacement A_s (c) influenced by weight of blood pump. A_{ay} and A_{az} were components of A_a. A_{vy} and A_{vz} were components of A_v. A_{sy} and A_{sz} were components of A_s.

4 Conclusions

The spring-damping model was used in the paper to simulate the viscoelasticity implantation environment of blood pump. By solving the kinetic equation of blood pump in implantation environment through SIMULINK, the vibration acceleration, velocity and displacement of blood pump influenced by implantation environment, rotor speed and weight of blood pump were researched. The research found the kinetic system of blood pump in viscoelasticity implantation environment would be more stable when the damping coefficient of implantation environment was larger than 4 N•s/m, the rotor speed of blood pump was higher than 8000 r/min, and the weight of blood pump was about 0.03 Kg. In the research, the implantation environment was simplified and the driven force only considered. There were other complicated tissues around blood pump, fluid force, motion excitation and other factors should be considered in kinetic system of blood in implantation environment in future work.

This work was supported by the National Natural Science Foundation of China (Grant No. 31670999, 31271057, 51475477).

References

1. M. Abshire, C. R. Dennison, S D Russell, J. Card. Fail., **20**, 12(2014), 973-983.
2. N. Moazami, K. Fukamachi, M. Kobayashi, N. G. Smedira, K. J. Hoercher, J. Heart Lung Transpl., **32**, 1(2013), 1-11.
3. L. Bellumkonda, P. Bonde, International Anesthesiology Clinics, **50**, 3(2012), 123-145.
4. S. P. Pinney, J. Card. Fail., **21**, 10(2015), 825-834.
5. N. Moazami, R. J. Steffen, Y. Naka, U. Jorde, S. Bailey, S. Murali, Ann. Thorac. Surg., **98**, 2(2014), 541–547.
6. H. Horiguchi, T. Tsukiya, T. Takemika, T. Nomoto, Y. Tsujimoto, International Journal of Fluid Machinery and Systems, **8**, 1(2015), 1-12.
7. T. Murashige, R. Kosaka, M. Nishida, O. Maruyama, T. Yamane, K. Kuwana, Y. Kawaguchi, 35th Annual International Conference of the IEEE EMBS, (2013),2732-2735.
8. J. C. Wu, J. F. Antaki, J. Verkaik, S. snyder and M. Ricci, J. Fluid. Eng., **134**, 4(2012),379-389.
9. S. Saito, K. Yamazaki, T. Nishinaka, Y. Ichihara, M. Ono, S. Kyo, J. Heart Lung Transpl., **33**, 6(2014), 599-608.
10. E. J. Molina, S. W. Boyce, Semin. Thorac. Cardiovasc. Surg., **25**, 1(2016), 56-63.
11. E. Okamoto, Y. Ishida, T. Yano, Y. Mitamura, J. Artif. Organs, **18**, 2(2015), 181-184.
12. Q. Yu, Am. J. physiol., **265**, (1993), 1052-1060.
13. Z. Tao, *Biofluid Mechanics* (Beijing, Science Press, 1984).
14. M. S. Slaughter, I. I. Michael, D. Tamez, T. Horrell, J. Graham, S. P. Patroklos, Tex. Heart I. J., **36**, 1(2009), 12.
15. G. R. Li, X. D. Zhu, B. S. Tian, Z. C. Hao, H. F. Cheng, China Medical Device Information, **6**, (2013), 20-24.
16. B. Pratondo, I. Andi, R. A. Akbar and F Ario, Procedia Computer Science, **72**, (2015), 217-224.
17. Y. Lin, *Studies on the Characteristics of the Implantable Axial Blood Pump's Vibration Caused by Magnetic Force*,(Changsha, Central South University, 2015).
18. Z. Yun, Y. Lin, C. Xiang, J. Huazhong Univ. of Sci. & Tech. (Natural Science Edition), **41**, 9(2013), 97-101.

MATEC Web of Conferences **139**, 00077 (2017)
ICMITE 2017
DOI: 10.1051/matecconf/201713900077

Temperature characteristics modeling of Preisach theory

Hao Chen[1], *Qifeng* Xu[1,*], *Yukai* Xiang[1] and *Yifan* Huang[1]

[1]College of Electrical Engineering and Automation, Fuzhou University, Fuzhou, China

Abstract. This paper proposes a modeling method of the temperature characteristics of Preisach theory. On the basis of the classical Preisach hysteresis model, the Curie temperature, the critical exponent and the ambient temperature are introduced after which the effect of temperature on the magnetic properties of ferromagnetic materials can be accurately reflected. A simulation analysis and a temperature characteristic experiment with silicon steel was carried out. The results are basically the same which proves the validity and the accuracy of the method.

1 Introduction

It is important to accurately grasp the magnetic properties of ferromagnetic materials for modeling and simulation of power transformers, voltage transformers and current transformers. However, as the hysteresis loop will be affected by temperature (in actual applications there is widespread temperature fluctuation in these transformers), it is necessary to analyze the influence of temperature on the hysteresis loop. The native Preisach operator is static and does not incorporate time or temperature dependent effects. Therefore, a modeling method of the temperature characteristics of Preisach theory is proposed in this paper in which the Curie temperature, the critical exponent, and the ambient temperature of magnetic material are introduced into the classical Preisach hysteresis model. This method can be used to simulate the magnetic properties of ferromagnetic materials at different temperatures to make up the defect that the classical Preisach theory can not simulate the temperature characteristics of ferromagnetic materials.

2 Preisach Hysteresis Model

According to the Preisach theory, magnetic material is composed of a large number of magnetic dipoles with the positive reversal threshold value α and the negative reversal threshold value β of dipoles being statistically distributed [1-3]. The distribution density of the magnetic dipole is expressed by the non-negative two-variables function $\mu(\alpha,\beta)$, which is called Preisach function as shown below:

$$\begin{cases} \mu(\alpha,\beta)=0(\alpha>H_{sat}\ \ or\ \ \beta<-H_{sat}\ \ or\ \ \alpha<\beta) \\ \mu(\alpha,\beta)=\mu(-\beta,-\alpha) \end{cases} \quad (1)$$

The dipole magnetization characteristics described by the Preisach theory is shown in Fig.1. The region S+ indicates that the dipole is in the +B state, and the region S- indicates that the dipole is in the -B state.

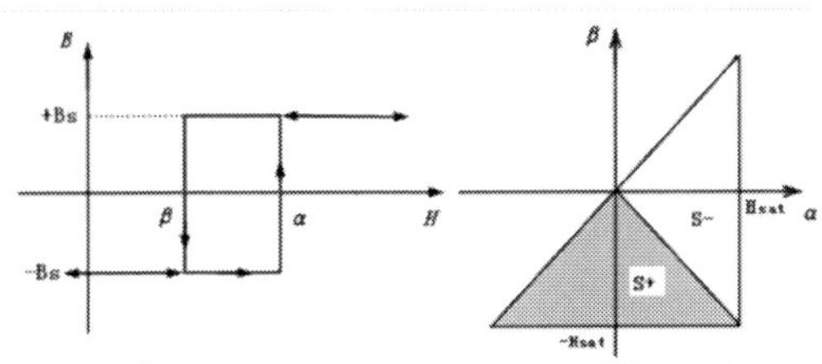

Figure 1. Magnetization characteristics of the dipoles of Preisach Model

In the case where the magnetic material is not saturated, it is required to use the equation (2) to double integral calculation. Assuming the Preisach function variables can be separated [4], so:

$$B_S\mu(\alpha,\beta)=\mu_\alpha(\alpha)\mu_\beta(\beta) \quad (3)$$

The hysteresis loop equation can be derived under different conditions [5, 6]. If the ferromagnetic material is magnetized by the complete demagnetization state, it will run in the initial magnetization curve:

$$B_i(H)=\left[\frac{B_d(H)+B_u(H)}{2}\right]^2\left[B_d(H)\right]^{-1} \quad (4)$$

Making

$$F(H)=\int_H^{H_{sat}}\mu(\alpha)d\alpha \quad (5)$$

Then

$$\int_{-H_{sat}}^{H}\mu(-\beta)d\beta=-F(H) \quad (6)$$

If (H_0,B_0) is the extreme point of the magnetization locus, the ascending and descending branches starting from the extreme point are:

$$B_{H_0,B_0}^{\mu}(H)=-B_d(-H)-B_d(H_0)+B_0+2F(H_0)F(H) \qquad H>H_0 \quad (7)$$

$$B_{H_0,B_0}^{\mu}(H)=B_d(H)+B_d(-H_0)+B_0-2F(-H_0)F(-H) \qquad H<H_0 \quad (8)$$

* Corresponding author: Ranger123098@163.com

Where:

$$B_\mu(H) = -B_d(-H) \tag{9}$$

$$F(H) = \frac{B_d(H) - B_\mu(H)}{2\sqrt{B_d(H)}} \qquad H \geq 0 \tag{10}$$

$$F(H) = \sqrt{B_d(-H)} \qquad H < 0 \tag{11}$$

Each branch is related to the function *F(H)*, and *F(H)* is related to the function of descending branch $B_d(H)$. Therefore, as long as the data are obtained, the hysteresis loops under various circumstances can be described by the equations (4)-(11).

3 Temperature Correction Method for Preisach Model

3.1 Analysis of temperature characteristics of ferromagnetic materials

The M_S-T curves of saturation magnetization of ferromagnetic materials varies with temperature as shown in Fig.2. It can be found that the saturation magnetization M_S decreases continuously with the increase of temperature. When the temperature increases to the Curie temperature the saturation magnetization M_S decreases to zero and the ferromagnetic material changes from ferromagnetic to paramagnetic [7].

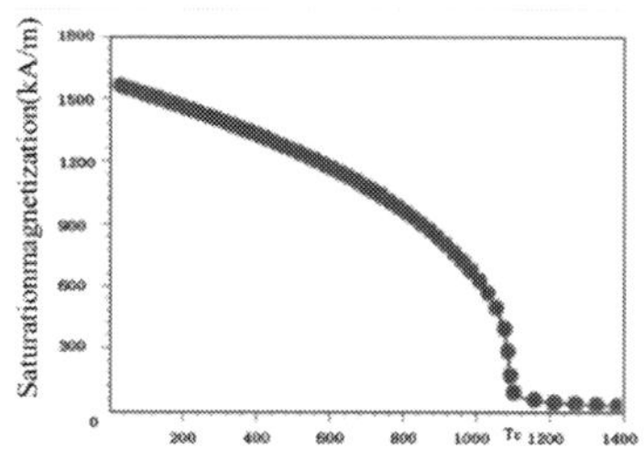

Figure 2. Variation of spontaneous magnetization with temperature

Generally, the Curie temperature of ferromagnetic material is very high. For example, the temperature of silicon steel can reach 700℃. However, under the general operating temperature of ferromagnetic material it is difficult to reach the Curie temperature, Therefore, it is usually only considered that operating temperature of ferromagnetic material is lower than the Curie temperature. In this case, the ferromagnetic material is in a ferromagnetic state.

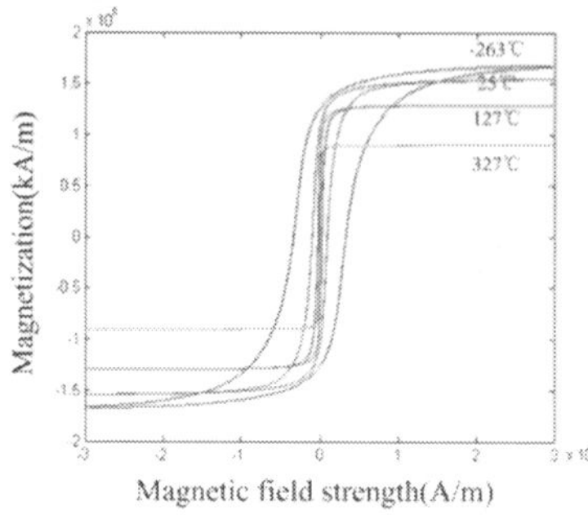

Figure 3. Variation of hysteresis loop with temperature

The temperature variation of M-H curve is shown in Fig.3. It can be seen that under the same magnetic field strength *H*, the magnetization *M* decreases continuously with the increase of temperature *T*. This is consistent with the variation law of Fig.2. In the normal operating temperature range, the hysteresis loops are absolutely different. For example, in the Fig.3 it is clear that the hysteresis loop is different at 25℃ compared to 127℃. Therefore, it is necessary to build a temperature characteristic model to analyze the hysteresis loop of ferromagnetic materials.

3.2 Temperature characteristics modeling of Preisach theory

The temperature characteristics of ferromagnetic materials are studied according to the Weiss theory and the mean field theory [8, 9]. Below using the Curie temperature, the critical exponent γ and the Curie temperature T_C are introduced to obtain the approximate expression of Ms-T curve:

$$M_S(T) = M_S(0)\left(\frac{T_C - T}{T_C}\right)^\gamma \qquad T < T_C \tag{12}$$

Where *Ms(0)* expresses the saturation magnetization of the magnetic material at absolute zero temperature (K). The temperature T_0 is taken as a reference temperature ($T=T_0$=298K) and it is introduced into the equation (12):

$$M_S(T_0) = M_S(0)\left(\frac{T_C - T_0}{T_C}\right)^\gamma \qquad T < T_C \tag{13}$$

Simultaneously using equation (12) and (13), then eliminate *Ms(0)*:

$$M_S(T) = M_S(T_0)\left(\frac{T_C - T}{T_C - T_0}\right)^\gamma \qquad T < T_C \tag{14}$$

The relation between saturation magnetization M_S and magnetic induction intensity $B_S(T)=\mu_0[H_m+M_S(T)]$ is plugged into equation (2):

$$B(H,T) = \iint_{S_+} B_S(T)\mu(\alpha,\beta)d\alpha d\beta - \iint_{S_-} B_S(T)\mu(\alpha,\beta)d\alpha d\beta \tag{15}$$

For the separated Preisach function only its descending branch is required, so it is easy to get:

$$B_d(H,T) = B_d(H,T_0)\frac{M(T)+H}{M(T_0)+H} \tag{16}$$

The modified Preisach model with the temperature correction can be obtained through plugging equation (14) and (16) into equation (4)-(11). By measuring the hysteresis loop at normal atmospheric temperature, the hysteresis loop can be obtained at any temperature below the Curie temperature.

3.3 Temperature parameters acquisition method

It is found that the two temperature parameters of the critical exponent γ and the Curie temperature T_C are key to this Preisach temperature model. There are various methods to obtain the Curie temperature, such as the M_S-T curve method, the induction method, the initial permeability curve method, the magneto resistance effect method and so on [7, 10]. At present, the M_S-T curve method using physical property measurement system (PPMS) to measure T_C is most accurate. This paper uses the M_S-T curve method with the curve shown in Fig.2. When the M_S in the M_S-T curve is close to zero, the corresponding temperature is the Curie temperature T_C [11, 12].

As for the critical exponent γ, two M-H curves are measured respectively at the normal atmospheric temperature T_0 and another temperature T experimentally, then the saturation magnetization $M_S(T_0)$ and $M_S(T)$ can be obtained. Finally, the initial critical exponent γ_0 can be obtained by importing $M_S(T_0)$ and $M_S(T)$ into the equation (14). There will be an error in this initial critical exponent, so the initial critical exponent should be further optimized through a successive approximation method [13] as shown in Fig.4.

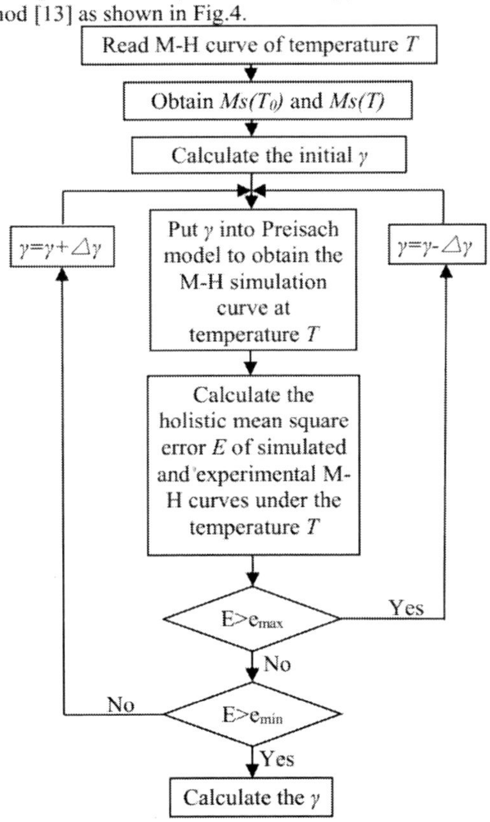

Figure 4. The successive approximation method

The software flow chart of successive approximation method to get the critical exponent is shown in Fig.4.

(a) Take the initial critical exponent γ_0 into the Preisach temperature model and the M-H simulation curve can be obtained at temperature T.

(b) Calculate the mean square error E of the simulated M-T curve and the M-T curve obtained by experiment at temperature T.

(c) Determine whether the mean square error is in the required error range emin-emax. If the error is too large, $\gamma_1=\gamma_0-\triangle\gamma$ will be substituted into step (1) for two iterations. If the error is too large, $\gamma_1=\gamma_0+\triangle\gamma$ will be substituted into the step (a) for the next iteration. Only when the mean square error is within error range, the critical exponent γ which is suitable for the whole M-H curve will be output.

4 Simulation and Experimental Verification

4.1 Experimental principle and parameters

Figure 5. Iron core coil

The experiments to obtain the hysteresis loop using the same coil at different temperatures has been carried out with Z110 (cold-rolled silicon steel sheet) manufactured by Nippon Steel. The thickness of the silicon steel sheet is 0.23mm, the inner radius of iron core coil is 30mm, the outer radius of iron core coil is 70mm and the width of iron core coil is 40mm, as shown in Fig.5.

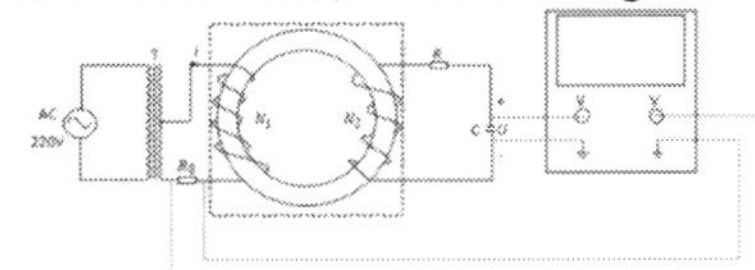

Figure 6. Experimental schematic

The B-H curve is measured by oscilloscope. The experimental principle is shown in Fig.6, and the related parameters are shown in table 1.

Table 1. Experimental circuit data

Parameters	Value
Sample resistance (Ω)	12
Resistance (kΩ)	10
Capacitance (μΩ)	10
N1(Turns)	300
N(Turns)	30

By adjusting the transformer T to generate an input current to provide a magnetic field strength of H the primary current I can be measured by a current probe of oscilloscope (Tektronix A622), where the voltage U_{Rs} on the resistor R_S and the magnetic field strength H have a positive proportional relation $H=N_1*U_{Rs}/LR_S$. The voltage probe of the oscilloscope (Tektronix TPP0100) is connected to two sides of the capacitor C on the secondary side to measure the voltage U_C. U_C and the magnetic induction intensity B have a positive proportional relation of $B=(RCU_C)/(N_2A)$. In such a way the B-H curve can be obtained by oscilloscope [14, 15]. The temperature experiment only demands the iron core coil to be put into the temperature test box.

The temperature metrical method is pointed out in section 4.1 in which the M_S-T curve of the silicon steel sheet is obtained by using PPMS-9T, and the Curie temperature T_C is also determined. The initial critical exponent γ_0 is obtained by using the MS-T curve and the equation (14), then γ_0 would be further optimized through the method proposed in section 3.3 to get γ. The specific values of the relevant parameters identified by the material are shown in table 2.

Table 2. Parameters for Fe-Si Magnetic Sheet

Parameters	Value
Reference temperature	298
Curie temperature (K)	1104
Critical exponent	0.3536
Mass density(kg/m3)	7650
Silicon content(%)	3.2
Saturation magnetization at room temperature(kA/m)	1365

4.2 Experiment and simulation results

The working temperature of power equipment is generally much lower than the Curie temperature. Taking the normal atmospheric temperature of 25℃ as a reference to the actual operating temperature of general power system equipment (The general operating temperature range of transformers is -15℃ to 85℃), the simulation of the limit hysteresis loop is obtained by selecting the temperature points of -15℃, 0℃ and 85 ℃, as shown in the Fig.7. In practical applications, the hysteresis loops in a limited temperature range (-15℃ to 85℃) is considered to be approximately reversible when the temperature is close to the normal atmospheric one. The experimental and simulation studies are carried out on the basis of those conditions [16-18].

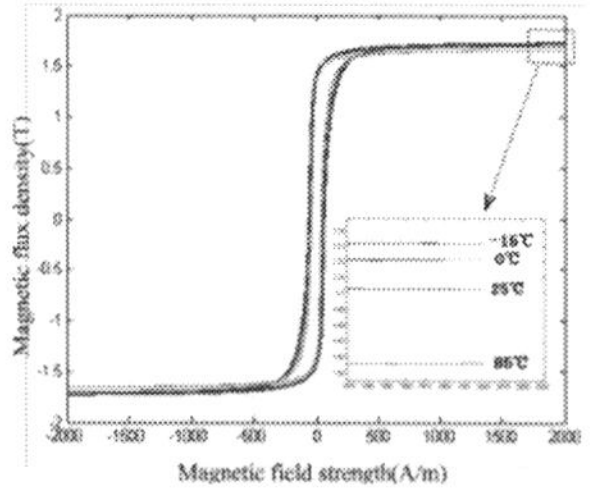

Figure 7. Hysteresis loop with different temperature

The simulation results are compared with the experimental ones as shown in Fig.8. Fig.8(a) shows the basic magnetization curve at normal atmospheric temperature. Fig.8(b) shows a comparison of experimental results with simulation results in the descending part of -15℃ and 85℃. Fig.8(c) shows a comparison between the experiment and the simulation at 0℃. Fig.8(d) shows the magnetization curves of saturated section at -15℃, 0℃, 25℃ and 85℃ respectively.

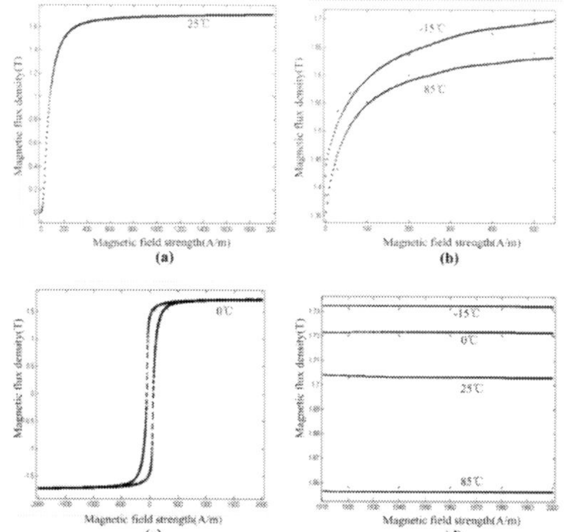

Figure 8. Experimental results

Magnetization decreases with the increase of temperature under the same magnetic field strength. The changing regularity in the experimental curve is consistent with simulation results and the experimental data fit well with simulation results.

4.3 Error analysis

The standard errors θ of simulation and experimental data at -15℃ and 85℃ are calculated respectively.

$$\theta = \sqrt{\frac{\sum_{i=1}^{n} (X_i^* - X_i)^2}{n}} \quad (17)$$

The error in the experimental results with reference to the simulation data is 3.214% at -15℃ and 3.269% at 85℃. This means that the model proposed in the paper can reflect the influence of temperature on the hysteresis loop of ferromagnetic material very accurately. The causes of error is that the Preisach temperature model is only an approximate fitting one which can not with strict accuracy express the effect of temperature on the hysteresis loop.

5 Conclusions

In this paper the modeling method of Preisach theory with temperature characteristics is proposed, and the hysteresis loop model with temperature characteristics is built up. The Preisach model can be employed to simulate the hysteresis loop at different temperatures by introducing the critical exponent, the Curie temperature and the ambient temperature. The comparison between experimental data and simulation results shows that the model is effective and accurate. The model is suitable for the coupling simulation of the magnetic properties and the temperature characteristics of electromagnetic equipment such as electric transformers.

References

1. Li Fuhua, Liu Dezhi, Chen Junquan, Wang Dong and Guo Yun. Identification of a Preisach Hysteresis Model With Lorentzian Function and Its Verification. Transactions of China Electrotechnical Society, **26**, (2011), 1-7.
2. Zhao Guosheng and Li Langru. A Nonlinear Vector Preisach Model Considering Reversibility of the Hysteresis. Proceedings of the CSEE, **20**, (2000), 4-10.
3. Li Zhen, Li Qingmin, Li Changyun, Sun Qiuqin and Lou Jie. Queries on the J-A Modeling Theory of the Magnetization Process in Ferromagnets and Proposed Correction Method. Proceedings of the CSEE, **31**, (2011), 124-131.
4. S.R. Naidu. Simulation of the hysteresis phenomenon using preisach's theory. IEE Proc-C, **137**, (1990),73-79.
5. Zhang Xingang and Wang Zezhong. Research on CT Modeling Based on the Preisach Theory. Proceedings of the CSEE, **25**, (2005), 68-72.
6. Zhao Yongfu. Research on Modeling of High-voltage Current Transformer and Effect of Its Saturation on Differential Protection. Master dissertation, North China Electric Power University, Baoding, (2009).
7. Xie Guangbin, Jin Xi, Lang Tong, Jiang Lei, Chen Xuemin and Cai Yongbin. The Research of Contact Feeling Temperature Indicator Based on Curie Temperature for Transmission Line. Electrical Measurement and Instrumentation, **49**, (2012), 59-63.
8. P.Andrei, A. Stancu, H. Hauser and P. Fulmek. Temperature, Stress and Rate Dependent Numerical Implementation of Magnetization Processes in Phenomenological Models. Journal of optoelectronics and advanced materials, **9**, (2007), 1137-1139.
9. A. Arrott and J. Noakes. Approximate Equation of State for Nickel Near Its Critical Temperature. Phys. Rev. Lett., **19**, (1967), 786-789.
10. Ji Song, Qian Kunming, Tan Suokui and Zhang Yansong. Development of Computer Aided Curie Temperature Metrical Instrument for Soft Magnetic Materials Applying Hopkinson Effect. Journal of Magnetic Materials and Devices, **36**, (2005), 34-37.
11. B. R. Knight, J. A. Bain and T. E. Schlesinger. Hamr Adjacent Track Stability in the Presence of a Medium Curie Temperature Distribution. IEEE Transactions on Magnetics, **46**, (2010), 2462-2465.
12. Wang Yiming and Zhu Jiangang. SNR Impact of Media Anisotropy Near Curie Temperature in Heat-Assisted Magnetic Recording. IEEE Transactions on Magnetics, **47**, (2011), 2368-2370.
13. Li Xiaoping, Peng Qingshun, Li Jinbao, Wen Xishan and Ru Hailiang. Parameter Identification of Hysteresis Loop Model for Transformer Core. Power System Technology, **36**, (2012), 200-205.
14. Ma Yuli and Dai Xinrui. Optimized design of apping method for dynamic hysteresis loop of ferromagnetic material. Physics and Engineering, **22**, (2012), 32-34.
15. Zhao Xiuke. Switching Power Supply of Magnetic Components, Nanjing University of Aeronautics and Astronautics Institute of Automation: Nanjing, (2004), 48-49.
16. Aina He, Anding Wang, Shiqiang Yue, Chengliang Zhao and Chuntao Chang et al. Dynamic magnetic characteristics of Fe78Si13B9 amorphous alloy subjected to operating temperature. Journal of Magnetism and Magnetic Materials, **408**, (2016), 159-163.
17. Yongfeng Liang, Junpin Lin, Feng Ye, Yanli Wang and Guoliang Chen. Effect of Heat Treatment on Microstructure and Properties of Heavily Cold Rolled Fe-6.5 wt %Si Alloy Sheet. Metallic Functional Materials, **17**, (2010), 43-47.
18. Yufeng Du. Research on Coupling Simulation of Magnetic Field and Temperature Field Based on Temperature Characteristics Test of Ferromagnetic Materials, M.S. dissertation, Shenyang University of Technology, (2016).
19. He Yuan, Li Xin and Luo Jian. Distinguishing Magnetizing Inductance of Power Transformer Based on Hysteresis Loop. Power System Protection and Contro, **41**, (2013), 19-24.
20. P. R. Wilson, J. N. Ross, and A. D. Brown. Simulation of Magnetic Component Models in Electric Circuits Including Dynamic Thermal Effects. IEEE Transactions on Power Electronics, **17**, (2002), 55-65.
21. Vittorio Basso, Cinzia Beatrice, Martino LoBue, Paola Tiberto, and Giorgio Bertotti. Connection between hysteresis and thermal relaxation in magnetic materials. PHYSICAL REVIEW B, 61, (2000), 1278-1285.
22. Alexander Sutor, Stefan J. Rupitsch, Shasha Bi, and Reinhard Lerch. A modified Preisach hysteresis operator for the modeling of temperature dependent magnetic material behavior. JOURNAL OF APPLIED PHYSICS, **109**, (2011), 07D338-07D340.

Development of remote control software for multiformat test signal generator

Yang Gao[1],*, Jiefeng Liu[1]

[1]Commuication University of China, School of Science and Technology, Beijing, China

Abstract. The multi format test signal generator mentioned in this paper is the video signal generator named TG8000 produced by Tektronix Company. I will introduce the function about remote control for signal generator, how to connect the computer to the instrument, and how to remote control. My topic uses my computer to connect the instrument through the 10/100/1000 BASE-T port on the rear panel of TG8000. Then I write program to transmit SCPI (Standard Commands for Programmable Instruments) to control TG8000. The application is running on the Windows operating system, the programming language is C#, development environment is Microsoft Visual Studio 2010, using the TCP/IP protocol based on Socket. And the method of remote control refers to the application called TGSetup which is developed by Tektronix Company. This paper includes a brief summary of the basic principle, and introduce for details about the process of remote control software development, and how to use my software. In the end, I will talk about the advantages of my software compared with another one.
Keywords: TG8000, C#, Socket, SCPI

1 Introduction

Test signal generator is mainly applied in generating synchronization signal or test signal for final-stage production of TV program sand broadcasting television system. TG8000 is a major player on test signal generator market especially for TV station and university. Due to the high cost of signal generator, is not easy to replace, some of the button function has been damaged. In order to solve the problem of the loss of function caused by the damage of the button, it is necessary to develop one way to remote control by computers. Tektronix developed TGSetup software for remote control. However, this software is not open source, the operating system version of the system has a lot of limitations, has a long latency time and so on shortcomings. Worst of all, operating system often crashes when TGSetup running. This paper designs an improved remote control method, based on the Visual Studio environment, almost no delay. It is worth mentioning that the main function of the software is to remotely control the signal generator to generate the required signal.

2 SoftwareDevelopment

2.1 The Way to Solve the Communication Problem Between PC and TG8000

With the rapid development of modern science and technology, information and technology updates faster and faster, to meet the people of higher quality, more content, more information and higher level visual auditory experience requirements, as carrying the arduous task of dissemination of information of the radio and television industry is also undergoing a change rapidly changes. Video signal generator is an important tool in the radio and television industry.

TG8000 is a sophisticated multi format analog and digital signal generator platform, it provides synchronous pulse and time code signal for radio and TV applications, and can provide reference for the application of video test signal testing equipment. TG8000 is mainly used to generate the pulse signal and test signal of the post production and broadcast television broadcasting system, to study and develop the test signal, and to design and maintain the equipment.

Because of the lack of information on how to connect PC with TG8000 to send commands, I must refer to TGSetup. First of all, I need to connect my PC to the TG8000 through LAN port with a cable [1]. There is a 10/100/1000 BASE-T port on the rear panel of TG8000 which is designed for remote control [2]. Then I get the IP address of TG8000 through its front panel and set the IP address of my computer to the same segment. Open TGSetup, enter the IP address of TG8000 and click the connect button in the dialog box that appears, about ten seconds after the LCD display of TG8000 turn from light to dark, and the TGSetup access control interface. This phenomenon represents a state of remote control.

From these steps, it is no difficult to find that after the establishment of a physical connection, the software send a series of commands packed into IP packets to TG8000 through the network. So I analyzed the commands sent by the software by using Wireshark software. This software is used to follow TCP stream. There two colour letters in the screen. The red letters matched the data sent from the PC to the TG8000 while the blue letters matched the feedback of the TG8000. The software sent packets to control the instrument, so it is no need to be focus on the blue letter. The TCP stream shows that SCPI is encapsulated in TCP data packet and sent to the TG8000 to make it response. SCPI (Standard Commands for Programmable Instruments) is a software standard independent of hardware, so many of these commands can be consulted. SCPI is a hardware independent software standard, the string can be transmitted through any instrument interface. It defines the content and structure of the controller to the instrument and the instrument to exchange information between the controller layer messages, which is a standard way to describe the various functions of the instrument; its purpose is to reduce the automatic test equipment program development time. The SCPI command consists of three parts:

•Grammar and style -a description of the basic principles of the application and development of real language,

•Command structure -describes the actual language structure in the instrument,

•Data interchange format -a standard representation of a data set that can be used between instruments and applications, between applications and applications or between instruments and devices.

SCPI can be divided into two commands: the general command and the instrument specified command. For example, "*IDN?" is a general command. It asks the instrument to return the version information and if the TG8000 receives this command successfully, it will reply" TEKTRONIX, TG8000, 0, 0". Besides these request commands, there are also some control commands. They end with a period, while those request commands end with a question mark [3]. Obviously, we can use these control commands to achieve remote control. The most important work of our software is to control the TG8000 to send signal that we need. In order to accomplish this, we will use following commands:

"*OPC?" It is a general command, and its main function is to stop receiving any commands until all the commands before had been completed. This command ensures the complete execution of the key commands.

"MMEMory:LOAD:SIGNal<module_name>,<signal _name>",This is a control command without a return value. It's no difficult to see that this command is used for loading signal. When analyzing this command, I found that the type and the format of a signal appear as a whole.Forexample,MMEM:LOAD:SIGN3,"/NTSC/OTHER/NTC7^Combination.cmp.

This provided a clear idea for my software. The signal format of AVG module includes NTSC，PAL-1，PAL-2，NTSC NO SETUP，625 YPbPr，525 GBR，625 GBR，525 YPbPr，525 Beta [4]. These formats are corresponding to different specific signals. And the signals (expressed in the form of a SCPI parameter) are:

OTHER/NTC7^Combination.cmp，
OTHER/FCC^Multiburst.cmp，
OTHER/NTC7^Composite.cmp,
OTHER/Matrix 1.cmp,
OTHER/FCC^Composite.cmp,
OTHER/SNG^ColorBars.cmp,
SWEEP/100%^Sweep.cmp,
SWEEP/Chroma^Frequency^Response.cmp,
SWEEP/60%^Sweep.cmp,
COLOR BAR/100%^ColorBars.cmp,
COLOR BAR/75%^ColorBars.cmp,
COLOR BAR/SMPTE^ColorBars.cmp,
FLAT FIELD/10%^Flat Field.cmp,
FLAT FIELD/Field^SquareWave.cmp,
FLAT FIELD/100%^Flat Field.cmp,
FLAT FIELD/Black Burst^with^FieldRef.cmp,
FLAT FIELD/40%^Flat Field.cmp,
FLAT FIELD/0%^Flat Field.cmp,
FLAT FIELD/50%^Flat Field.cmp,
FLAT FIELD/Black Burst.cmp,
MULTIBURST/60%^Multiburst.cmp,
MULTIBURST/Multipulse.cmp,
MULTIBURST/100%^Multiburst.cmp,
MONITOR/White^Window.cmp,
MONITOR/2 Level^Pedestal^&Pluge.cmp,
MONITOR/Monitor^Setup.cmp,
MONITOR/Convergence^Pattern.cmp,
MONITOR/Safe^Area.cmp,
MONITOR/75% Red^Field.cmp,
MONITOR/Gray^Window.cmp,
MONITOR/4 Level^Pedestal^&Pluge.cmp,
MONITOR/100% Red^Field.cmp,
LINEARITY/5 Step.cmp,
LINEARITY/Modulated^5 Step.cmp,
LINEARITY/Ramp.cmp,
LINEARITY/Modulated^Pedestal.cmp,
LINEARITY/10 Step.cmp,
LINEARITY/Shallow^Ramp.cmp,
LINEARITY/Modulated^Ramp.cmp,
PULSE BAR/Window 2T^Pulse &Bar.cmp,
PULSE BAR/SinX`sX.cmp.

These are all the signals that could be used for software. I can take advantage of these combinations to create the signals that we need.

2.1.1 The Design Idea of the Software

DOI: 10.1051/matecconf/201713900078

After Part 1 had understood the working principle of TGSetup, Part 1 focused on the improvement method. TGSetup can only be used on Windows7 or Windows XP, so I write my program on Visual Studio so that it could be used on most operating systems. The biggest shortcoming of TGSetup is the long delay. It takes at least ten minutes to connect, and worst of all, it costs more than thirty minutes to set a signal. So I analysed data by Wireshark.

My conclusion is that two-way communication takes too long, so I tried to communicate one way most time. Two-way communication is to ensure each command is received correctly, so the software each sends a command an additional request, but this greatly reduces efficiency. My software retains the request that follows the control command, if the return value is an error, it will inform the user control error. If any other kind of error makes the computer fail to receive the return value, it will automatically report the error and terminate the program after a period of time. On the basis of the error reporting function, this method greatly speeds up the execution of the program.

After solving these problems, my design idea is completed. First, I will build a virtual panel in C#. The virtual panel is similar with the real panel but more simple and intuitive. Then I will connect a computer to the TG8000 by using the Socket programming. The last step is to write functions to send SCPI. Once TG8000 receives the message, it will make response.

2.1.2 The Process of Writing of the Software

The first step is to build a virtual panel. Since there are three parts in the command, my software is made up by three windows as well, as shown in Figure 1, Figure 2, and Figure 4.

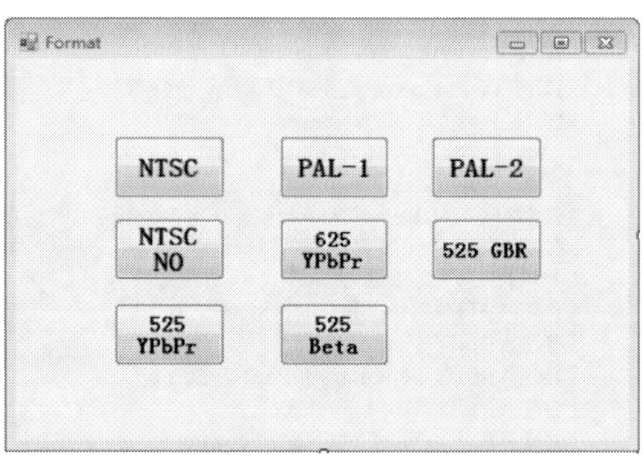

Figure.1. Format interface

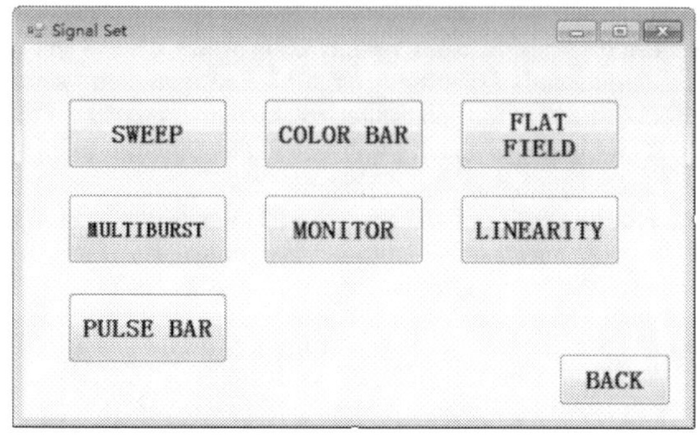

Figure.2. Signal Set interface

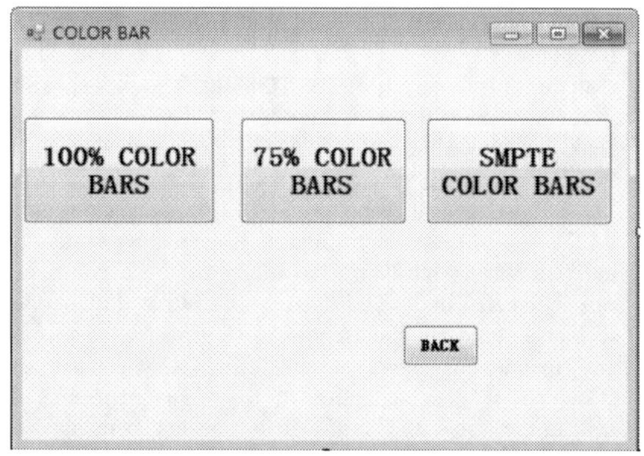

Figure.3. Specific signal interface

These three parts can give command "MMEMory:LOAD:SIGNal<module_name>,<signal_name>" a clear parameter. The only thing left to do is writing functions for the buttons. The software need to send commands through network, so I chose to use Socket to achieve this process.

Socket uses the client / server mode to solve the problem of establishing the communication. The connection process between the sockets can be divided into three steps: server monitoring, client request, connection confirmation.

•Server monitoring: the server socket does not locate a specific client socket, but is in the state of waiting for the connection,

•Client request: refers to the client socket connection request, the target is to connect to the server socket. To do this, the client socket must first describe the socket of the server to which it is connected, point out the address and port number of the server socket, and then make a connection request to the server socket,

•The connection confirmation: refers to the server socket to listen to or received by the client socket connection request, it will respond to the client request socket, create a new thread, the server socket description to the client. Once the client confirms this description, connection is established. While the server socket continues to be in the listening state, continue to receive other client socket connection request [5].

For my software, I know the IP address and the port number of the TG8000 and my computer, so I can easily build their connection. Besides, there are two Socket functions playing a key role. After setting the parameters, function "clientSocket.connect( )" make my computer connect to the TG8000. Then I write the SCPI into string, and use function "clientSocket.send( )" to send SCPI to the TG8000 [6].

2.1.3 Software Demonstration

After Part 1have finished all the code, the last thing I need to do is to test the software. For example, if I want to set 100%SWEEP in PAL-1 format, I should run the program and chose PAL-1, as shown in Figure 4.

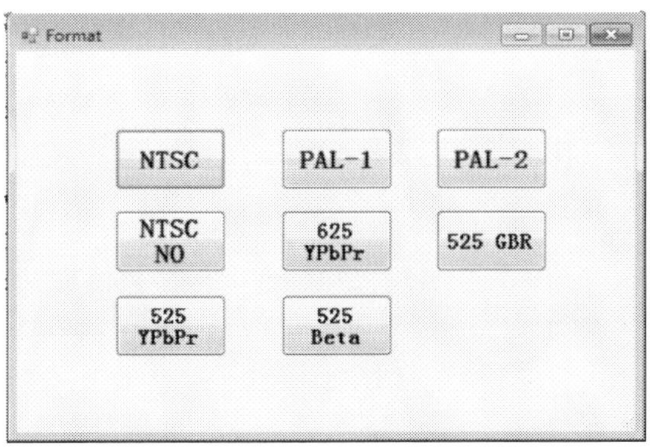

Figure.4. Final test 1

Then I will enter next screen and chose SWEEP, as shown in Figure 5.

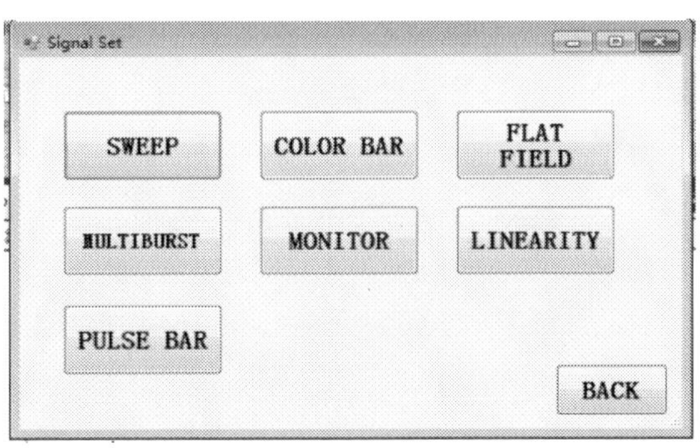

Figure.5Final test 2

I chose 100%SWEEP, as shown in Figure 6.

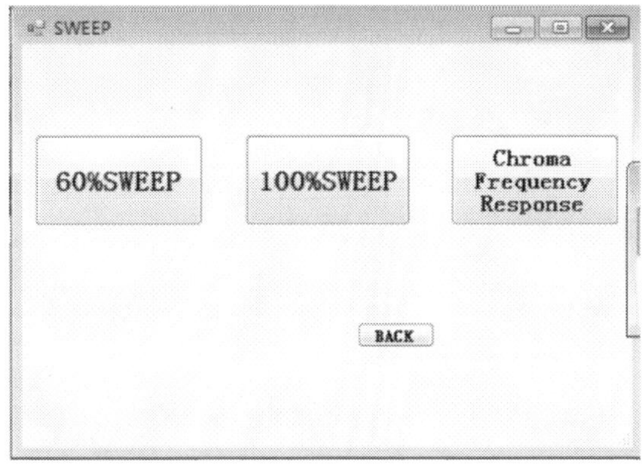

Figure.6 Final test 3

There is one way to verify this result. We can connect the TG8000 to an Oscilloscope, like VM700, as shown in Figure 7

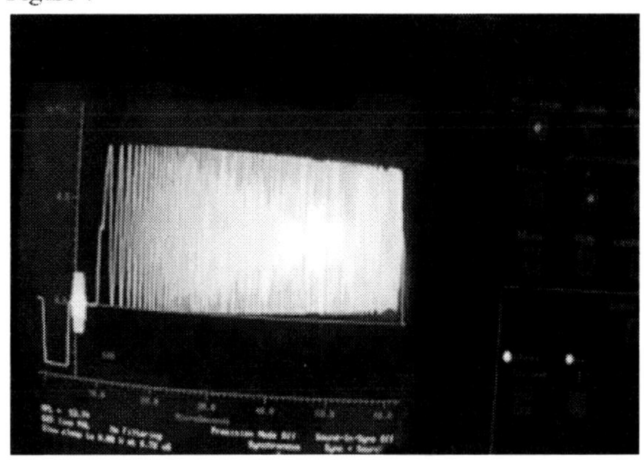

Figure.7 Result in VM700

3 CONCLUSION

Table 1. Delay (ten times average)

	TGSetup	**My software**
Connect(s)	9.2	<1
Change signal(s)	32.8	3.6

We can see from the comparison that there is almost no delay for this remote control software. It is worth mentioning that there is no running error or system breakdown in my test.

Test signal generator is now widely used and a lot of problems followed. Now this software has been put into use in our university. Some instruments with not working button could be reused because of the remote control software. It is also used in experimental lessons when teachers start to demonstrate operation so that the students can watch the operating process clearly. So I can say the remote control software expand the use of signal generator and improve the efficiency of users.

REFERENCES

1. Tektronix.TG8000 Multiformat Test Signal GeneratorPC Tools Technical Reference (2014)
2. Tektronix.TG8000-Multiformat-Test-Signal-Generator-Programmer-Manual (2014)
3. Liu Zhen. Design and implementation of SCPI parsing platform [J] (2012)
4. Tektronix.TG8000-Multiformat-Test-Signal-Generator-User-Manual_0 (2014)
5. Wang Wei Guo, Yang Guang Yu, Zhou HaiFeng. Application driven development of programmable instrument [J] (2010)
6. Carter. Socket principle and programming foundation [J] (2011)

Research and Implementation of Distributed Database HBase Monitoring System

Lisi Guo[1,*], Tao Chen[2], Yanming Chen[2], and Xinlong Luo[1]

[1]Beijing University of Posts and Telecommunications, Beijing, China
[2]China Mobile Group Design Institute Co.,Ltd, Beijing, China

Abstract. With the arrival of large data age, distributed database HBase becomes an important tool for storing data in massive data age. The normal operation of HBase database is an important guarantee to ensure the security of data storage. Therefore designing a reasonable HBase monitoring system is of great significance in practice. In this article, we introduce the solution, which contains the performance monitoring and fault alarm function module, to meet a certain operator's demand of HBase monitoring database in their actual production projects. We designed a monitoring system which consists of a flexible and extensible monitoring agent, a monitoring server based on SSM architecture, and a concise monitoring display layer. Moreover, in order to deal with the problem that pages renders too slow in the actual operation process, we present a solution: reducing the SQL query. It has been proved that reducing SQL query can effectively improve system performance and user experience. The system work well in monitoring the status of HBase database, flexibly extending the monitoring index, and issuing a warning when a fault occurs, so that it is able to improve the working efficiency of the administrator, and ensure the smooth operation of the project.

1 Introduction

With the development of technology, nowadays the companies are producing rich types of data at an unprecedented speed and scale. Data storage increase dramatically.To meet the demand for the huge data, the NoSQL database came into being. The NoSQL database has good reading and writing abilities, and flexible database data model.In the existing NoSQL database, HBase is a highly reliable, column oriented, and scalable distributed database with high performance,gaining the superiority in offline data batch processing. As a distributed database, HBase , which disperses the data originally stored in a centralized database to multiple data storage node, have faster data access,greater scalability, and higher concurrent traffic. Thus, it can better meet current needs for data storage.

A well-running database play an important role in the era of big data, so database monitoring becomes a very important part of database usage. There are two general ways of database monitoring:artificial monitoring and software monitoring.

Artificial monitoring tends to be delayed in detection of malfunction.It's susceptible to data loss, abnormal operation of project and other problems.Database error or stop working even for a short period of time may cause huge losses.In addition, this approach requires a lot of labor,which is inability to cope with the increased management difficulty and workload of the database.The other method is software monitoring.The main open source monitoring software for large data analysis system is Ganglia, Nagios/Centreon, etc. Existing monitoring systems for large open source components is coarse-grained. Acquisition ability of monitoring index can't meet all the needs of big data platform system. They can not concisely show indicators the enterprise most want to see and meet the new monitoring needs constantly emerging in practical applications.

In summary, the development of an on-demand HBase database monitoring system has a very important practical significance. This article will provide relevant researchers with our design and implementation ideas of an on-demand extension of the HBase database monitoring system. On one hand, by means of modulating monitoring indicators of different type, this system can flexibly adding monitoring indicators by configuration file based on the actual needs of the project.On the other hand, It can detect alarm situation regularly and inform administrators immediately when problems occur, which can significantly improve the efficiency of management staff.Besides,Through the charts in front-end which show the performance of the latest data and historical data, managers can fully aware of the changes of database performance.

2 Research Objective and The logic Principle of Design

2.1 General Idea

The end-to-end monitoring to each node of the distributed database HBase is the premise to realize automatic opera-

*Corresponding author:greece_gls@163.com

tions and maintenance. Its implementation is based on collecting key data from the HBase sub-nodes and the server host where they reside.

There are three dimensions of key data information can be acquired: the database performance index (read and write traffic, downtime issues, etc.), the host performance metrics (CPU, memory, hard disk, etc.) and network connectivity (ping).

The monitoring of HBase database cluster operation can be achieved through managing, analyzing and presenting the collected data, thus guiding the storage of the new coming data according to the results. Therefore, the situation where the downtime and other serious consequences occur can be avoided, since the overloaded database or host will not continue to store data.

By setting the alarm rules and analyzing the collected data on the back-end, the risk management of the database can be realized. Once a server trigger a warning, the administrator would be notified by email to take reasonable steps to protect the normal operation of the database, ensuring the normal operation of the actual production system.

In all parts, data acquisition is the premise, centralized management and logical analysis of data is the core,and data application is the purpose. The implementation architecture of HBase monitoring system is shown in Figure 1.

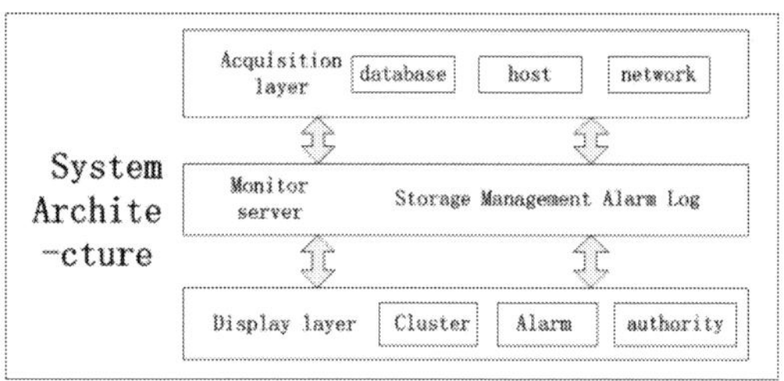

Figure 1. the architecture of HBase monitoring system

Based on the goal of building a distributed database HBase monitoring system and the actual operation and performance requirements of the project, the system is supposed to possess the following two capacity.

First, the ability to monitor the normal indexes of each sub-node of HBase database.

Second, the capabilities to manage and analyze data on server side.

2.2 The Monitoring Ability

The normal index of HBase database is the basis of monitoring system. The normal indexes of collected include physical hardware index, database performance index and network connectivity index.The normal indicators are shown in Figure 2.

2.2.1 physical hardware monitoring

CPU utilization, disk usage, memory usage and other hardware information on the server are monitored in the physical hardware monitoring. The server is the foundation of all business operation. Once the physical hardware fails, the server can't provide services normally, and it even causes very serious consequences, such as server downtime.

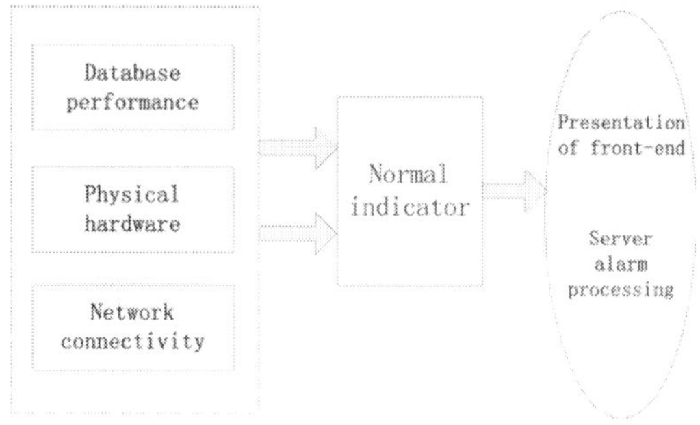

Figure 2. the normal indicators

2.2.2 database performance monitoring

These indicators include HRegionServer's operating status(lived or not), reading and writing visits and so on. These information presents the running state of the database, and provide services for database optimization.

2.2.3 network connectivity monitoring

The normal network is a prerequisite for the operation of a distributed database. Therefore, smooth network status is essential to the normal operation of the database.

2.3 The data management and analysis capabilities

The server mainly undertakes four tasks: data processing, fault alarm and logging.

2.3.1 data processing

After receiving the reported data, the server would store the data into the database respectively according to their different types. The front-end acquires the data from the server through the corresponding interface. The server would make a summary of the data per hour, obtaining the total average of that hour. In addition, to prevent database overloading, the system would remove data two days ago.

2.3.2 fault alarm

The server would initiate the alarm check task according to the specified frequency or time node. If the alarm threshold is reached, the system will automatically send a mail to notify administrator for subsequent processing.

2.3.3 Logging

The server would regularly record the daily operation of the database every day so that the administrator can check its running history. The server would record the time and details as long as an alarm occurs, otherwise record that the Hbase is running well today would be written.

3 Key Technologies and Key Issues

3.1 proxy implementation

The client-side deploys the proxy to collect data on a recurrent acquisition and initiative report basis.

The proxy is an extensible data acquisition module under modularized management, which can be used to expand the new functions according to the actual needs and configure the start-stop of various types of data in the configuration file. It adds types of indicator data to the task queue by different jobs, regularly initiating the data collection tasks and reporting to the server in accordance with the specified frequency or time node. The physical hardware information acquisition modules need to be deployed to the corresponding server. The specific process is shown in the Figure 3.

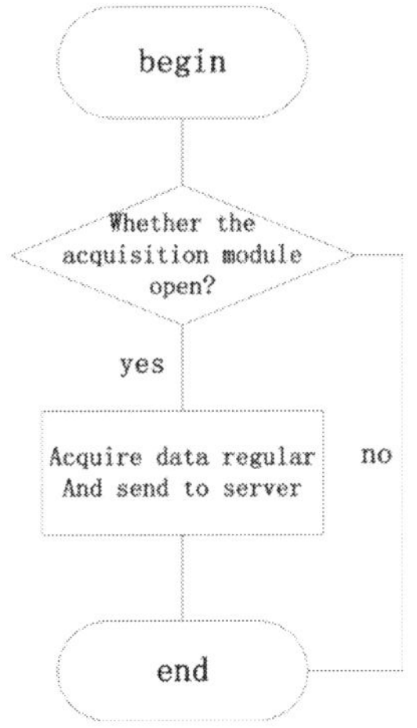

Figure 3. Data collection flowchart of the proxy

Among the normal indicators of the system, the physical hardware indicators are collected by Sigar API, database performance indicators are acquired through the HBaseAdmin interface, and the network connectivity is obtained through the ping connectivity detection.

3.2 the implementation of server

The server side adopts SSM architecture to develop. As a lightweight framework with simplicity, testability and loose coupling, the framework Spring has the features of Inversion of Control(IOC) and Aspect Oriented Programming(AOP). SpringMVC is more customizable because it separates the role of controller, model object, dispatcher and handler object. MyBatis provides the persistence layer framework which include SQL Maps and Data Access Object (DAO),almost eliminating the JDBC code and manual configuration parameters.It uses XML for configuration and original mappings, mapping the interface and common Java objects to records within the database, thereby unifying the database query management. Besides, the database adopt MYSQL in this system. The timing tasks use quartz for periodical summary, alarm detection and logging. Mail system uses the Postfix server, which can send the mails after finishing the configuration on corresponding server.

3.2.1 database table design

The database includes the following four tables:Monitordata, Monitordatahour, Cluster and Alarmdata.The table Monitordata storages the all monitor data.The table Monitordatahour makes summary for the monitor data per hour.The table Cluster storages the information of cluster.And table Alarmdata records the information of alarm data.Details are as shown from Table1 to Table4:

Table 1. the monitordata

column name	column discription	example
eventid	id for event	5365
objetid	type of monitoring data	hbase
timestamp	monitoring timestamp	2017-07-5 11:23:30
location	IP of monitoring location	10.254.201.204
content	monitoring value	{"serverName":"NBG-204", "serverStatusSign":"on", "memoryMap": {"total":23414,"used":6666}, "visits":0.0, "ReadRequestCount":43457384, "WriteRequestCount":357045, "hbaseTest":"true"}

Table 2. the monitordatahour

column name	column discription	example
eventid	id for event	5365
objetid	type of monitoring data	hbase
timestamp	monitoring timestamp (that hour)	2017-07-5 11:00:00
location	IP of monitoring location	10.254.201.204
content	monitoring value	{"serverName":"NBG-204", "serverStatusSign":"on", "memoryMap": {"total":23414,"used":6642}, "visits":0.0, "ReadRequestCount":43275952, "WriteRequestCount":357045, "hbaseTest":"true"}

Table 3. the cluster

column name	column discription	example
id	id for event	1
hostip	IP of server	10.254.201.201
role	the role that server contains	hbase
hostname	name of the server	"NBG-201"

Table 4. the alarmdata

column name	column discription	example
eventid	id for event	5333
objetid	objective of alarm data	cpu
timestamp	alarm timestamp	2017-07-5 11:05:30
location	host IP of alarm occured	10.254.201.204
alarmname	alarm name	high cpu usage
alarmdetail	alarm details	cpu usage over 90%
level	alarm level	high
status	alarm processing status	send email

3.2.2 The key technologies of the SSM architecture

The control layer (Controller), which is responsible for data management, is the core of server-side. The main tasks of the control layer are: Parsing and storing received data into the database, and directly using corresponding Java interface of the mapper layer to process data in the database; Providing data interface to the front-end. The control layer passes data to the front-end by post. It sends parameterless request when the front-end requests for loading the first page data, while it sends request with parameters when the front-end requests for certain types of data.

The Service is a processing layer for business logic, which is mainly responsible for alarm checking and logging. Alarm detection and log records will be triggered regularly. Basic warning rules are listed below. 1. Hardware: the CPU,memory or disk is too high. The threshold can be set by the administration.2. HBase: Once down, the alarm will be sent. 3. Network connectivity: A warning will be sent when there is a host fail to ping. Warning rules can also be increased on the front-end according to actual demands. Log records are carried out daily. The time and details would be recorded when alarms occur. Otherwise it records the good condition that day.

The Mapper layer provides the mapping between Java interfaces and SQL, and the original SQL queries are all defined in the Mapper's XML file.

3.3 The resolution to the key issues

The rendering speed of the web front end has an extremely important impact on user experience. In the actual development of a project, excessive SQL queries will be time-consuming, resulting in slow page rendering and poor user experience. Therefore, it is of great importance to reduce SQL query statements to improve system performance and user experience.

In this system, because the monitored database is a distributed database, three different sets of data exist if there are three server hosts. Each group of data is divided into multiple types of data, such as host data, HBase data and ping data. Supposing that different types of data were acquired by SQL queries respectively, getting all the information from the latest database would go through 3 * 3 times SQL queries. If the cost of one query is 0.25s, it will cost 0.25 * 9s when the front end send requests to the server asking for data.Excessive SQL queries bring too much time to consume. Therefore, by using one SQL query to get all types of data from one host and distinguishing various types on the server side, we can reduce the time overhead. The time overhead this method requires is 0.25 * 3s, which is one third of the previous one. The front-end takes less than one second to render, while it costs more than two seconds before, thus the user experience is greatly improved.

4 Results

Among three months of actual operation, the system worked well to detect the latest state of the host, the database, and the network. It can also longitudinally perceive the changes of the database operation by checking the historical data. Besides, the system have done well in alarm warning, which effectively improved the working efficiency of the administrator. Part of the front-end interface is shown in the Figure 4,Figure 5 and Figure 6.

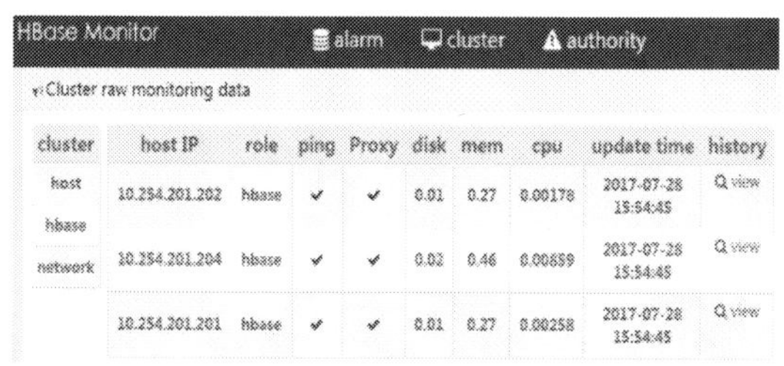

Figure 4. host information

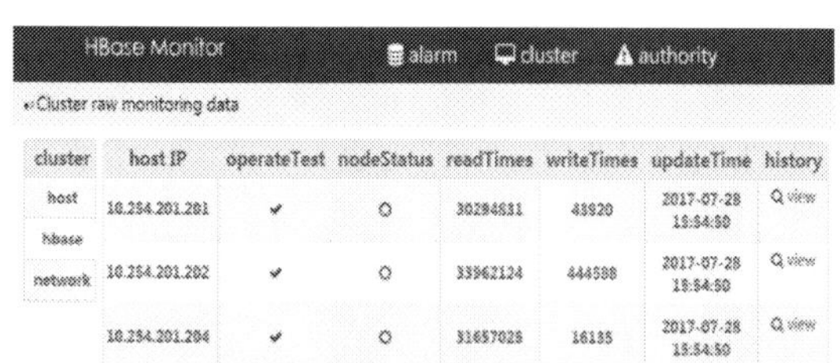

Figure 5. HBase information

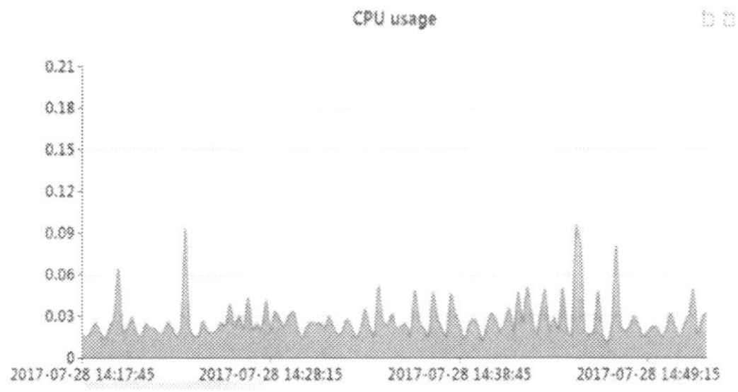

Figure 6. the historical cpu usage

5 Conclusion

In recent years, with rapid growth of data quantity, the distributed database have begun to play an important role. Monitoring database becomes important means to assure the safety of data storage and smooth operation of project. This paper dealt with a certain operator's actual demand of the distributed database HBase in its projects. We have designed and realized an on-demand HBase monitoring system which can extend monitoring indexes flexibly. In addition, given the low loading speed in the process of practical operation, we work out a solution to reduce the SQL query, providing a reference for related researchers. Practice shows that reducing the SQL queries can effectively improve system performance and user experience. The system works well in monitoring the HBase database's running state and sending alarms in time, which improve the efficiency of the administrators and ensure the smooth running of the projects. Nevertheless, this paper only involves data collection, management and presentation. The application of data remains further research. Studying the historical monitoring data and guide the fault prediction would be a new direction which is worth exploring.

References

[1] Bartosz Balis,Marian Bubak,Daniel Harezlak,Piotr Nowakowski,Maciej Pawlik,Bartosz Wilk. "Towards an operational database for real-time environmental monitoring and early warning systems[J]," Procedia Computer Science,2017,108.

[2] Jun Wu Xu,Jun Ling Liang. "Research on a Distributed Storage Application with HBase[J]," Advanced Materials Research,2013,2200(631):

[3] Mavromichalaki, H.,Papaioannou, A.,Plainaki, C. et al."Applications and usage of the real-time Neutron Monitor Database[M]," Advances in Space Research: The Official Journal of the Committee on Space Research(COSPAR),2011,47(12):2210-2222.

[4] Seung Kim,Nam Wook Cho,Young Joo Lee,Suk-Ho Kang,Taewan Kim,Hyeseon Hwang,Dongseop Mun. "Application of density-based outlier detection to database activity monitoring[J]," Information Systems Frontiers,2013,151:.

[5] Bartosz Balis, Robert Brzoza-Woch, Marian Bubak, Marek Kasztelnik, Bartosz Kwolek, Piotr Nawrocki, Piotr Nowakowski, Tomasz Szydlo, Krzysztof Zielinski. "Holistic approach to management of IT infrastructure for environmental monitoring and decision support systems with urgent computing capabilities", Future Generation Computer Systems (2016) In press.

[6] Anonymous. Noble Gases; Application Security, Inc." Announces Database Activity Monitoring for DB2 on IBM z/OS[J]," Computers, Networks & Communications,2011.

[7] GaoYong-ping,Guan Fen-fen."Based on MVC to Monitor the Performance of Oracle Database[C]," Proceedings of 2012 international conference on electronics information and electrical engineering.2012:325-327

[8] Chun-ying WU,Li-yan ZHANG,LieCAO et al."Distributed Spinning Frame Monitor System Based on DSP and PROFIBUS[C],"//2010 2nd International Conference on Computer Engineering and Technology (ICCET 2010). v.3.2010:1872-1875.

Assessment Method of Wind Farm Harmonic Emission Value Based on Improved Complex Linear Regression Model

Tiantian Chen[1,*], *Peng Zhang*[1], *Ling Pan*[1], *Peng Tan*[2] and *Honggen Yang*[2]

[1]State Grid Shanghai Municipal Electric Power Company, Shanghai, 200437, China
[2]School of Electrical and Information, Sichuan University, Chengdu, 610065, China

Abstract. Wind turbine filter and reactive power compensation devices lead to the harmonic impedance of wind farm is not much larger than that of the utility, so the influence by the wind farm harmonic impedance can not be neglected while assessing the harmonic emission value of wind farm A method based on improved complex linear regression is proposed in this paper for assessing the harmonic emission value of wind farm. The linear regression model is established by using the harmonic current at PCC point as the explanatory variable and the harmonic voltage of the wind farm as the explanatory variable. The utility harmonic impedance is calculated by complex least squares method. For various of the topology of wind farm feeder network, an equivalent method of feeder network is proposed to calculate the wind farm harmonic impedance. Errors are analyzed by using the error marginal effect of the dispersion parameter. Simulation and measured data verify the effectiveness of the proposed method

1 Introduction

LCL type grid-connected inverter of wind turbine will produce harmonics, witch lead to power quality and stability problems [1].The harmonics of the point of common coupling of wind farm is the result of the internal harmonic source of wind farm and the harmonic source of external power network. Therefore, it is necessary to reasonably classify the harmonic pollution responsibility of PCC point, the premise is to reasonably evaluate the harmonic emission level on both sides of PCC point [2].

A large amount of research on harmonic emission level evaluation methods was carried out. The customer and the utility reference impedance method was proposed in [3]. However, the operating conditions of generator of utility and the load of customer are changing. So, it is hard to calculate the reference impedance accurately according to the utility and customer parameters. Therefore, the non-invasive harmonic emission level assessment method based on the measurement data was proposed, including fluctuation method [4], linear regression method [5], independent random vector method [6], maximum likelihood estimation method [7] and independent component analysis method [8].The complex linear least square method proposed in [9] is the real solution in the complex domain calculation of least squares regression model, which maintains the strict linear relationship of the regression model and improves the weakness of least squares solution of regression model in the real domain. However, the accuracy of regression results are affected by the collinearity between the PCC harmonic current and utility harmonics. All the methods above assume that the customer harmonic impedance is much larger than the utility harmonic impedance, so that the customer harmonic impedance is neglected when calculating the harmonic emission level of customer. For typical industrial nonlinear customers (Industrial rectifier loads, arc furnaces, etc.), the above hypothesis holds. However, the wind turbine inverter is equipped with LCL-filter, and the collector point is equipped with reactive power compensation devices, which may cause the harmonic impedance of the wind farm is not much larger than that of the utility, so the influence by the wind farm harmonic impedance can not be neglected.

The harmonic impedance of the wind farm is mainly determined by the wind turbine LCL-filter, transformer, feeder line and reactive compensation capacitor, all of which have definite harmonic impedance model. The wind turbine of same type and capacity is equipped with LCL filters whose parameters are basically consistent. Only the length of line between the wind turbine and collecting point is dispersive, so the structure of the feeder network can be simplified and the harmonic impedance of the wind farm side can be calculated.

In this paper, an improved linear regression model is proposed by using the PCC point harmonic current as the explanatory variable and the harmonic voltage of the wind farm as the explanatory variable. The utility harmonic impedance is calculated by using the complex least square method. Aiming at the variability of topology structure of wind farm feeder network, an equivalent method of collector network is proposed to calculate the wind farm harmonic impedance. The error

* Corresponding author: author@e-mail.org

DOI: 10.1051/matecconf/201713900080

margin effect of the dispersion parameter is used to analyze the error. Simulation and field data verify the effectiveness of the proposed method.

2 Theory of improved complex linear regression model

In this paper, Norton equivalent circuit is commonly used to perform harmonic analysis as shown in Fig.1,where I_U and Z_U are the utility-side equivalent harmonic current and impedance, I_C and Z_C are the wind farm-side equivalent harmonic current and impedance, and U_{PCC} and I_{PCC} are the harmonic voltage and current measured at the PCC.

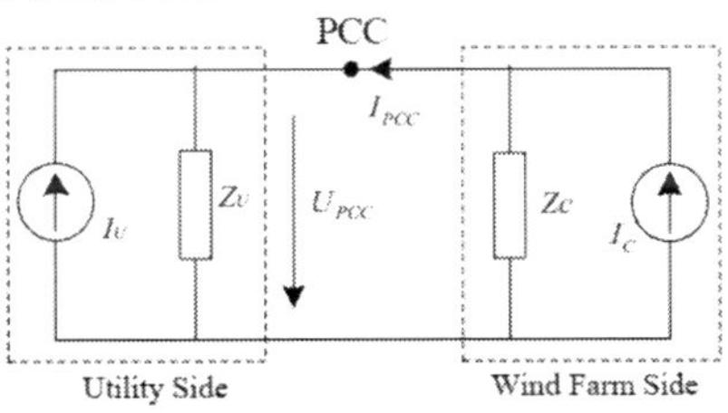

Figure 1 Norton equivalent circuit of utility and wind farm

Referring to Fig.1, the I_{PCC} can be expressed as

$$I_{PCC}=\frac{Z_C}{Z_U+Z_C}I_C-\frac{Z_U}{Z_U+Z_C}I_U \tag{1}$$

According to Thevenin and Norton transformation, I_{PCC} can be expressed as (2), where V_C and V_U are the equivalent utility-side and wind farm-side harmonic voltage.

$$I_{PCC}=\frac{V_C}{Z_U+Z_C}-\frac{V_U}{Z_U+Z_C} \tag{2}$$

According to Kirchhoff's current law, the wind farm-side equivalent voltage V_C can be calculated as

$$V_C=U_{PCC}+Z_C I_{PCC} \tag{3}$$

The wind farm-side equivalent harmonic impedance can be calculated by parameters of wind farm.

Due to the error of the measured data, the following linear relationship exists between I_{PCC} and V_C can be expressed as (4), where ε is the measurement error.

$$I_{PCC}=\frac{1}{Z_U+Z_C}V_C+\frac{-V_U}{Z_U+Z_C}+\varepsilon \tag{4}$$

A linear regression equation between I_{PCC} and V_C can be established based on (4) as

$$I_{PCC}=\beta_1 V_C+\beta_0+\varepsilon \tag{5}$$

where β_1 and β_0 are partial regression coefficient. $\beta_1=1/(Z_U+Z_C)$, $\beta_0=-V_U/(Z_U+Z_C)$.

During a measurement period, a n group data of PCC point harmonic currents and wind farm side harmonic voltages can be expressed as

$$\begin{cases} I_{PCC}(1)=\beta_1 V_C(1)+\beta_0+\varepsilon_1 \\ I_{PCC}(2)=\beta_1 V_C(2)+\beta_0+\varepsilon_2 \\ \cdots \\ I_{PCC}(n)=\beta_1 V_C(n)+\beta_0+\varepsilon_n \end{cases} \tag{6}$$

(6) can be represented as a matrix equation

$$y=Ax+\varepsilon \tag{7}$$

Where $y=[I_{PCC}(1)\ I_{PCC}(2)\ \ldots\ I_{PCC}(n)]^T$,

$$A=\begin{bmatrix} 1 & \dot{I}_C(1) \\ 1 & \dot{I}_C(2) \\ \vdots & \vdots \\ 1 & \dot{I}_C(n) \end{bmatrix}, x=\begin{bmatrix} \beta_0 \\ \beta_1 \end{bmatrix}, \varepsilon=\begin{bmatrix} \varepsilon_1 \\ \varepsilon_2 \\ \vdots \\ \varepsilon_n \end{bmatrix}.$$

In order to minimize the sum of squares of ε, the complex least squares method in [9] is used by

$$x=(\bar{A}^T A)^{-1}\bar{A}^T y \tag{8}$$

where "$^{-}$" stands for conjugate vector.

Compared with the method in [9], there is no correlation between the harmonic current of the wind farm-side and the utility-side harmonic voltage. For the regression results of the equation (7) are not interfered with by the collinearity factor, the regression result is more accurate.

Based on the regression result of β_1, the utility harmonic impedance can be calculated by

$$Z_U=\frac{1}{\beta_1}-Z_C \tag{9}$$

By the equivalent circuit, the harmonic voltage of PCC point can be expressed as

$$U_{PCC}=\frac{Z_C Z_U}{Z_U+Z_C}I_C+\frac{Z_C Z_U}{Z_U+Z_C}I_U \tag{10}$$

According to the principle of superposition, the harmonic voltage emission value of the wind farm side is

$$U_{PCC\text{-}C}=\frac{Z_U Z_C}{Z_U+Z_C}\left(\frac{U_{PCC}}{Z_C}+I_{PCC}\right) \tag{11}$$

And the harmonic voltage emission value of the utility side is

$$\dot{U}_{PCC\text{-}S}=\dot{U}_{PCC}-\dot{U}_{PCC\text{-}C} \tag{12}$$

3 Calculation of the wind farm side harmonic impedance

2.1 Harmonic model of wind farm

Taking an actual wind farm as an example, the wind farm consists of two parts, one part is the feeder network, which is composed of the wind turbine, transformer and feeder line, the other part is the reactive power compensation device, as shown in Fig.2

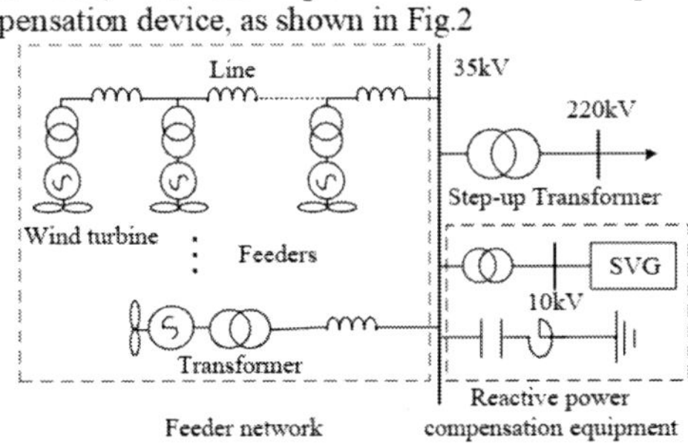

Figure 2 Topology of wind farm power station

2.2 Harmonic Impedance Model of PMSG

The permanent magnet synchronous generator (PMSG) consists of wind turbine, permanent magnet synchronous generator, generator side converter, grid side converter and LCL filter. Since the generator side converter and the grid side converter are separated by DC capacitor, the harmonic characteristics of wind turbine are determined only by the harmonic characteristics of the LCL inverter on the grid side.

The harmonic voltage of grid side inverter consists of two parts, one part U_{inv}, is generated by the control system of inverter the other part U_{pwm}, is generated by the dead time and modulation process of inverter. Using the inverter harmonic voltage source and LCL filter to represent the wind turbine harmonic impedance model [10] as shown in Fig.4.

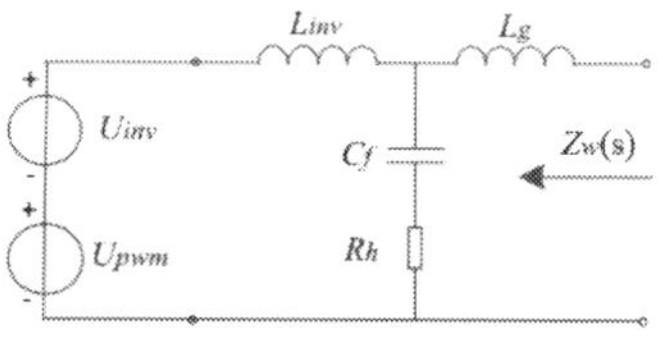

Figure 4 Harmonic model of PMSG-based wind turbine

Where, L_{inv} and L_g are the inverter side and grid side inductance, C_f is the filter capacitor, R_h is the damping resistance.

The harmonic impedance of wind turbine can be obtained by the LCL filter parameter as follows:

$$Z_W(w) = jwL_g + \frac{-w^2R_hC_fL_{inv} + jwL_{inv}}{-w^2L_{inv}C_f + jwR_hC_f + 1} \quad (12)$$

Where, w stands for the frequency.

2.3 Harmonic Impedance Model of SVG

The SVG outputs the reactive current through the inverter,in order to suppress the harmonics produced by SVG inverters, L filters are usually connected between the grid and the inverter. In practical engineering, the SVG connection transformer is commonly used as the filter inductance. Accordingly, the harmonic impedance model of SVG is a series harmonic model consisting of the equivalent harmonic voltage source of the inverter and the short-circuit impedance of the transformer connected with the SVG. It can be de defined as follows:

$$Z_{SVG}(w) = jwL_T + R_T \quad (13)$$

Where L_T and R_T are the equivalent inductance and resistance of transformer.

2.4 Harmonic Impedance of wind farm side

The impedance of the feeder network can be accurately calculated according to the parameters of the LCL filter, the line impedance and the detailed topology of the feeder network. The harmonic impedance of reactive power compensation device can be calculated, according to the detail parameters. When the measurement point is located on the collector point, the current measured is the total current of the feeder network and the reactive compensation device, the harmonic impedance of the wind farm-side is the parallel total impedance of the feeder network and reactive power compensation devices.

4 Approximate calculation of the wind farm-side harmonic impedance

In the actual project, it is difficult to obtain the detailed topology of the feeder network, which needs to approximate the harmonic impedance of the feeder network.

4.1 Aggregation equivalence method of feeder network

The topology of the feeder network is changeable, but the parameters of the LCL filter are consistent, only the length of line between the wind turbine and collecting point is dispersive. The equivalent circuit of a single feeder is shown in Fig 5.

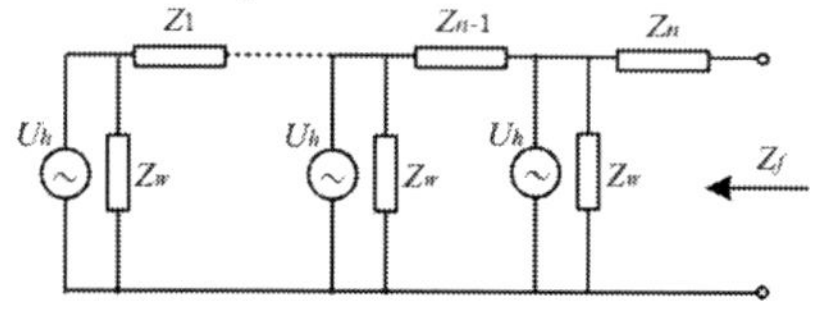

Figure 5 Equivalent circuit diagram of single feeder

Where n is the number of the wind turbines, Z_W is the impedance of a single wind turbine, U_h is the harmonic voltage from a single wind turbine. Z_n (n = 1, 2, 3 ... n) is the impedance of the feeder line between adjacent wind turbines, whose sum is the impedance of the feeder line Z_L. Z_f is the impedance of a single feeder.

The impedance of the feeder line is mixed by the wind turbine impedance and the line impedance. Considering that Z_W is much larger than Z_L, the impedance of single feeder can be simplified as follows:

$$Z_f = \left(Z_W + Z_{eq}\right)/n \quad (14)$$

Where, Z_{eq} is the equivalent line impedance of a single feeder, it can be obtained by

$$Z_{eq} \approx \sum_{i=1}^{n} iZ_i = nZ_L - \sum_{i=1}^{n}(n-i)Z_i \quad (15)$$

By (15), when $Z_i = 0$ (i = 1,2 ... n-1), $Z_n = Z_L$, that is, the wind turbines are arranged at the end of the feeder, Z_{eq} obtains the maximum value as nZ_L. when $Z_i = 0$ (i = 2,3 ... n), $Z_1 = Z_L$, that is, only one wind turbine is arranged at the end of the feeeder, and the other wind turbines are arranged at the head of the feeder, Z_{eq} obtains the minimum value as Z_L. The boundary value of Z_{eq} can be expressed as follows:

$$Z_L \le Z_{eq} \le nZ_L \quad (16)$$

In order to minimize the impedance error of the feeder network between the approximation and the true values, the impedance approximation of feeder network Z_{FJS} is defined as the mean value of the maximum and minimum by

$$Z_{FJS} = \left(Z_W + \frac{Z_{L\min} + n_{\max}Z_{L\max}}{2}\right)/N \quad (21)$$

Where, Z_{Lmin} and Z_{Lmax} are the impedance of the shortest and the longest feeder lines. n_{max} is the maximum number of wind turbines connected to a single feeder.

4.2 Approximate Error of Harmonic Impedance in feeder Network

Take an actual wind farm as an example,there is 5 feeders in the wind farm, each of which is equipped with 10 wind turbines. The length of feeder line is 4.7km, 8.0km, 11.2km, 13.5km, 15.5km, respectively. Parameters of wind turbine and LCL filters is shown in Table1. The short circuit inductance of step-up transformer is used as the grid side inductance of LCL filters, whose value is 0.041mH.

Table 1. Parameters of wind turbine and filter

Parameters	Value
Rated Power P/MW	2.0
Rated Voltage U_g/kV	0.69
Rated Capacity S_T/MVA	2.2
Rated Voltage of transformer U_T/kV	0.69/37
Short-circuit impedance of transformer %	6%
Inverter side inductance L_{inv}/mH	0.18
Grid side inductance L_g/mH	0.04
Filter capacitor C_f/uF	700
Filter resistor R_h/Ω	0.1

Consider the extreme conditions that may exist in the topology of feeder network: the wind turbines adopts an end or head arrangement, and the length of the feeder 1 is the shortest or the longest. The error of primary harmonic approximate impedance of the feeder network is shown in table2.

Table 2. Errors of harmonic impedance approximation of feeder network

Errors of Harmonic Impedance Approximation of Feeder Network (%)					
Harmonic order		*5th*	*7th*	*11th*	*13th*
End arrangement	*Longest line*	-0.5	-0.35	0.5	1.1
	Shortest line	10.0	6.2	-6.9	-16.1
Head arrangement	*Longest line*	-5.1	-.32	3.8	10.2
	Shortest line	-4.1	-2.6	3.1	7.8
Accurate value		-1.1	-0.7	0.7	2.0

Taking the 5th harmonic as an example, in order to show the relationship between the feeder network impedance and the number of wind turbines, the harmonic admittance of the feeder network is shown in Fig.6.

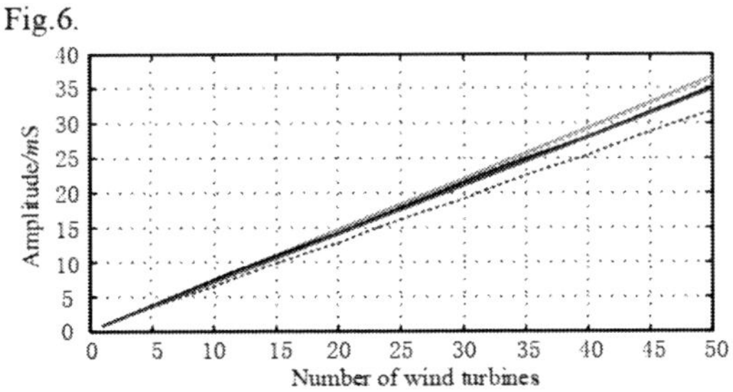

Figure 6 Amplitude of feeder network of 5th harmonic admittance

Where, the red curve represents the approximate value of admittance, and the black curve represents the exact value. The green and blue curves represent the end and head arrangement of wind turbines, respectively. The corresponding solid lines and dashed lines represent the conditions of shortest and longest lines.

Obviously, with the increase of the number of the wind turbine, the accurate harmonic admittance of the feeder network almost increases linearly, and the maximum error of the approximate harmonic impedance value is only -16.1%.

4.3 Approximation of wind farm-side harmonic impedance

Other parameters of the wind farm are shown in table3.

Table 3. Parameters of wind farm

Parameters	Parameter Description
Short-circuit Capacity	35kV:766MVA 220kV:11090MVA
Step-up Transformer	Rated Voltage 230±8×1.25%/37kV, Rated Capacity 100MVA, Connection Sets: Ynyn0, Short-circuit Impedance of Transformer: 6%
Fixed Capacitor	Rated Capacity 12Mvar, Capacitor Series Reactance Rate: 6%, Quality Factor: 100.
SVG	Rated Capacity 9Mvar, Rated Voltage: 10kV
SVG Transformer	Rated Voltage 37±2×2.5%/10kV, Rated Capacity 10MVA, Connection Sets: Ynyn0, Short-circuit Impedance of Transformer: 10%

When the measurement point is located in the 35kV bus (collecting point), the minimum, maximum, accurate and approximate values of the primary harmonic impedances of wind farms side are shown in table 4

Table 4. Harmonic impedance boundary value of wind farm

Harmonic impedance boundary value of wind farm(Ω)				
Harmonic order	*Minimum value*	*maximum value*	*Accurate value*	*Approximate value*
5th	7.8∠88°	8.1∠88°	7.9∠88°	7.9∠88°
7th	20.1∠86°	20.8∠86°	20.3∠86°	20.4∠86°
11th	64.4∠17°	79.6∠19°	74.9∠18°	74.2∠18°
13th	26.3∠-14°	40.9∠-30°	35.7∠-25°	34.9∠-25°

From table 4, the maximum amplitude error of the 5th, 7th, 11th, 13th harmonic impedance approximation is 2.5%, 1.9%, 13.2% and 24.6%, respectively.

The utility harmonic impedances can be roughly calculated by the short-circuit capacity of measuring point, which are 9, 13, 20, 23Ω for 5th, 7th, 11th, 13th harmonic respectively. So, the amplitude of wind farm harmonic impedance is not much larger than the utility side.

5 Harmonic impedance boundary value of wind farm with parameter error

There is an error between the actual parameters and the rated parameters of the component, resulting in a change in the impedance of the wind turbine within a certain range. There is a error in the capacity of filter inductance and filter capacitor, which is generally believed to 5% as maximum [11]. And there is a error in short-circuit impedance of transformer, which is generally believed to 10% as maximum.

The inductance of the filter, the capacity of the capacitor, and the short-circuit impedance of transformer may increase or decrease at the same time. Based on analysis and comparison, the 5th, 7th harmonic impedance increase and the 11th, 13th harmonic impedance decrease, when L_{inv}, C_f and L_g increase, vice versa.

When the filter parameter error is the maximum value, the boundary value of single wind turbine is shown in Fig7.

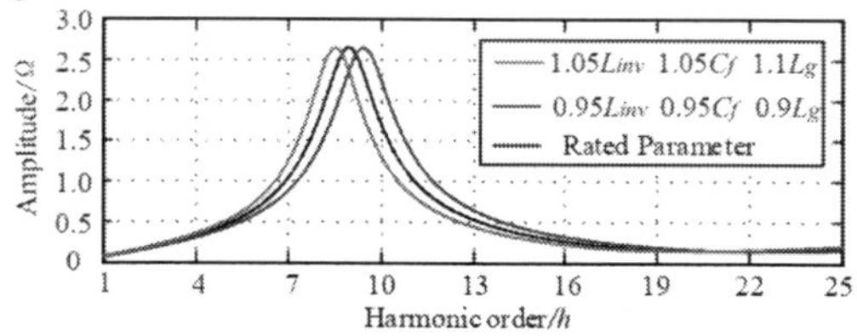

Figure 7 Amplitude of wind turbine harmonic impedance

In order to avoid resonance between filter and power grid, the tuning point is designed on the frequency of non integer harmonic according to the principle of partial resonance.

As shown in Fig8, the actual resonance point of the filter is designed between 430Hz-471Hz with parameter errors of LCL filter, witch is from the primary integer harmonics.

Therefore, the error of filter parameters will not lead to changes in impedance properties of wind turbine. So, there is no interference between the inductive and capacitive impedances of wind turbines.

The boundary value of the impedance of the feeder network can be obtained by the boundary value of the impedance of each wind turbine.

Considering the parameter error of component and the approximation error of the feeder network, the boundary error of the harmonic impedance of the feeder network and the corresponding boundary value of harmonic impedance of the wind farm can be calculated as shown in table 5 and table 6, respectively.

Table 5. Harmonic impedance boundary error of feeder network

Harmonic impedance boundary error of feeder network(%)				
Harmonic order	*5th*	*7th*	*11th*	*13th*
Minimum value	-13.6	-17.9	-28	-32
maximum value	19.2	25.1	34.2	33.9

Table 6. Harmonic impedance boundary value of wind farm with parameter error

Harmonic impedance boundary value of wind farm(Ω)				
Harmonic order	*Minimum value*	*maximum value*	*Accurate value*	*Approximate value*
5th	7.6∠88°	8.3∠88°	7.9∠88°	7.9∠88°
7th	19.0∠86°	21.8∠85°	20.3∠86°	20.4∠86°
11th	52.7∠3°	78.9∠37°	74.9∠18°	74.2∠18°
13th	20.2∠-10°	52.9∠-24°	35.7∠-25°	34.9∠-25°

From table 6, the maximum error of approximate value of wind farm impedance is 5%, 7%, 30% and 52% for 5th, 7th, 11th, 13th harmonic respectively.

6 Simulation and field data analysis

6.1. Simulation analysis

A computer simulation study was performed by Matlab7.0, which was based on the Norton equivalent circuit shown in Fig.1, including 1440 sample points at PCC.

The parameter was set as follows (the percentage of disturbance indicates the maximum of uniform disturbance comparing to initial value):

1) The amplitude of I_C is 150A, and that of I_U is 0.5 times smaller. Phase angles of I_C and I_U are respectively and 30° and 50°. 10% sine fluctuation and 10% random disturbance are added to the phase angles of I_C.5% sine fluctuation and 20% random disturbance are added to the amplitude of I_C. 10% sine fluctuation and 10% random disturbance are added to the phase angles of I_U.5% sine fluctuation and 10% random disturbance are added to the amplitude of I_U

2) The amplitude of Z_U is set as the rough values of utility harmonic impedance in the foregoing paragraph. Phase angles of Z_U is 70°. 10% sine fluctuation and 20% sine fluctuation are added to amplitude and angles of Z_U respectively. Z_C is set as the values in tableV and tableVI.

With 100 sample points for each subinterval, sliding analysis is performed. The results errors of method in [9] (method a) and method proposed in this paper (method b) was compared.With the approximate error of feeder network is taken into account, the errors of utility impedance and harmonic voltage emission value of wind farm are shown in table 7 and table 8.

Table 7. Errors of utility harmonic impedance amplitude

Root mean square value of the relative error (%)						
Zc	*Minimum value*		*maximum value*		*Accurate value*	
Method	*a*	*b*	*a*	*b*	*a*	*b*
5th	10.2	5.1	10.5	5.3	9.0	4.1
7th	8.9	5.4	9.2	5.3	8.7	3.5
11th	7.8	4.6	5.6	3.6	7.2	3.2
13th	8.6	4.4	9.0	4.5	7.5	3.8

Table 8. Errors of harmonic voltage emission value of wind farm

Root mean square value of the relative error (%)						
Zc	*Minimum value*		*maximum value*		*Accurate value*	
Method	*a*	*b*	*a*	*b*	*a*	*b*
5th	52.2	3.1	49.8	3.5	50.9	3.4
7th	30.8	3.2	28.1	2.8	27.8	3.0
11th	5.6	3.6	5.2	3.9	4.8	3.2
13th	28.2	10.8	29.1	5.4	30.4	4.6

With the approximate error of feeder network and the parameter errors of LCL filters are taken into account, the errors of utility impedance and harmonic voltage emission value of wind farm are shown in table 9 and table 10.

Table 9. Errors of utility harmonic impedance amplitude

Root mean square value of the relative error (%)						
Zc	*Minimum value*		*maximum value*		*Accurate value*	
Method	*a*	*b*	*a*	*b*	*a*	*b*
5th	8.9	4.8	11.3	5.9	9.0	4.1
7th	8.0	5.3	8.9	6.1	8.7	3.5
11th	7.5	4.3	10.3	6.0	7.2	3.2
13th	8.0	5.0	6.8	4.3	7.5	3.8

Table 10. Errors of harmonic voltage emission value of wind farm

Root mean square value of the relative error (%)						
Zc	*Minimum value*		*maximum value*		*Accurate value*	
Method	*a*	*b*	*a*	*b*	*a*	*b*
5th	52.1	3.6	48.2	3.5	50.5	3.0
7th	30.6	2.2	25.0	5.7	29.1	6.7
11th	5.3	3.2	7.2	5.8	4.6	3.3
13th	28.2	17.4	26.6	4.7	29.8	4.3

It can be easily learned from table VII-X that method b has the smaller errors than method a, no matter the harmonic impedance of wind farm is the accurate value or the boundary value.

The actual impedance of the wind farm is within the range of the maximum and minimum boundary values. So, the method proposed in this paper has smaller errors of the utility harmonic impedance and the harmonic voltage emission value.

6.2. Field data analysis

To further verify effectiveness of the proposed method, field data are used to perform calculation. The measured data are drawn from the collecting point of the actual wind farm described above. The harmonic current of the feeder network and the harmonic voltage of PCC point is sampled. Harmonic data are acquired by using FFT algorithm to analyze sample data per minute. The 5th harmonic current and voltage is shown in Fig8.

Method proposed in this paper and method in [9] are used to calculate the utility impedance and the harmonic voltage emission value of wind farm as shown in Fig9.

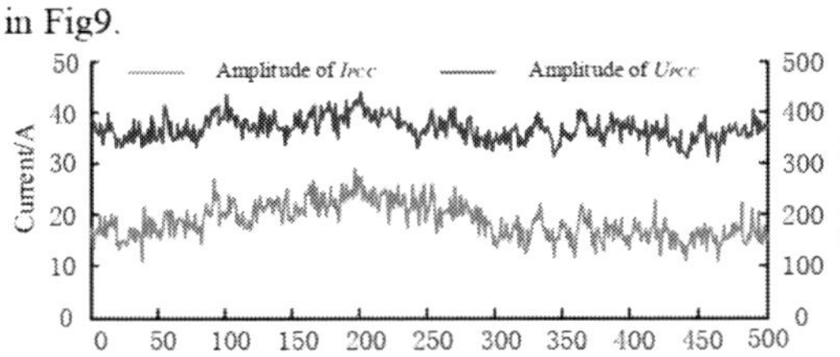

Figure 8 5th harmonic voltage and current of PCC

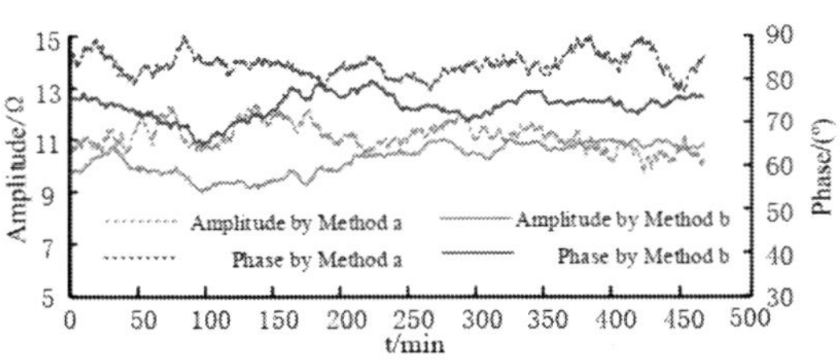

Figure 9 Estimates of the 5th utility harmonic impendance

The utility and wind farm side harmonic voltage emission value (95% probability value) obtained by the method in [9] are respectively 231V and 174V. However the utility and wind farm side harmonic voltage emission value (95% probability value) obtained by the method proposed by this paper are respectively 308V and 79V, which is quite different. Based on the analysis result of the simulations, the method proposed is more accurate.

6 Conclusion

In this paper, an assessment method of wind farm harmonic emission value based on improved complex linear regression model is proposed. Simulation and field data analysis verify the effectiveness and superiority of the proposed method.How to calculate the impedance of wind farm easily in engineering survey, through the component parameters acquired by the wind farm is the further research direction

References

1. Shuai Z, Liu D, Shen J, et al. Series and Parallel Resonance Problem of Wideband Frequency Harmonic and Its Elimination Strategy[J]. IEEE Transactions on Power Electronics, 2013, 29(4): 1941-1952.
2. Yang Honggeng, Wang Lei. Assessing harmonic emission levels from non-loads using Laguerre polynomials [J]. Proceedings of the CSEE, 2005, 25(7): 81-85.
3. Xu W, Liu Y. A method for determining customer and utility harmonic contributions at the point of common coupling [J]. IEEE Transactions on Power Delivery, 2000, 15(2):804-811.
4. Hui Jin, Yang Honggeng, Ye Maoqing. Assessing harmonic emission level based on the impedance gathering trend discrimination [J] Proceedings of the CSEE, 2011, 31(10): 73-80.
5. Zhang Wei, Yang Honggeng. A method for assessing harmonic emission level based on binary linear regression [J]. Proceedings of the CSEE, 2004, 4(6): 50-53.
6. Hui J, Yang H G, Lin S F, et al. Assessing utility harmonic impedance based on the covariance characteristic of random vectors [J]. IEEE Transactions on Power Delivery, 2010, 25(3): 1778-1786.
7. Hua Huichun, JiaXiufang, Cao Dongsheng, et al. A maximum likelihood method for harmonic impedance estimation [J]. Proceedings of the CSEE, 2014, 34(10): 1692-1699.
8. Zhao X, Yang H. A New Method to Calculate the Utility Harmonic Impedance Based on FastICA [J]. IEEE Transactions on Power Delivery, 2016, 31(1): 381-388.
9. Jia Xiufang, Hua Huichun, Cao Dongsheng, et al. Deterrmining harmonic contributions based on complex least squares method [J]. Proceedings of the CSEE, 2013, 33(4): 149-155.
10. Liu Z, Rong J, Zhao G, et al. Harmonic Assessment for Wind Parks Based on Sensitivity Analysis [J]. IEEE Transactions on Sustainable Energy, 2017, PP(99):1-1.]
11. Xu Mingjin, Ji Lin, Ge Xiaowei, et al. LCL-filter optimization design with consideration of inverter-side current feedback control impacts [J]. Proceedings of the CSEE, 2016, 36(17): 4656-4664

Self-adjusted multi-sensor information fusion electric energy measuring based on neural networks

ZhuFeng Li[1],Yang Li[2]

[1]College of Application Physics and material ,WuYi Univ. 529020 jiangmeng Guangdong China
[2] Student , College of Biological Sciences , State university of Minnesota . 55401 st.paul Minnesota United States

Abstract : In this article ,self-adjusted Multi-sensor Information Fusion measuring method of electric energy based on neural networks has been thoroughly given . This paper studies the method of automatic error correction of electric power measurement also . The effective learning algorithm of the neural network based on gradient algorithm and Newton algorithm is combined with the LEA discriminant method.The results show that the method can improve the learning efficiency. The hardware model of adaptive real-time fast power measurement is constructed by using DSP device. The experimental results show that the adaptive power measurement model is better than the traditional power meter.

KeyWords : neural networks、power energy measuring、self-adjusted 、DSP

1 Introduction

High precision measurement of electric energy in complex field environment is an urgent problem that has not been solved. Such as temperature - 30 - +40 ^{0}C, humidity of 40% - 80%, frequency, voltage, current, power factor, external magnetic field, waveform distortion, voltage line loss and so on. These factors seriously affect the accuracy of electrical energy measurement.The economic loss caused by inaccurate measurement of electric energy can only be artificially estimated, which is unscientific and unfair. The failure of the metering device can not be found in time and non manual. Fault can only be detected when waiting for cycle check, Resulting in time delays and economic losses. How to improve the measurement error of the electric power measurement device in a complex and changeable environment,It is of great significance to realize the real-time intelligent correction, to improve the accuracy of the electric energy meter in the complex and changeable environment, and to carry out the intelligent diagnosis of the fault of the metering device.The multi sensor information fusion based on the optimized neural network can effectively correct the influence of the environment on the electric energy measurement, and can realize the real-time, dynamic and high-precision intelligent electric energy measurement[1].Optimization of neural network is to determine a reasonable neural network structure, the optimal neural network weight learning algorithm.According to the actual situation, the establishment of a reasonable data sample library of environmental factors and power relations, measuring device status information and fault classification samples，It is of great academic and practical value to realize the accurate measurement and fault diagnosis of electric field measurement device[2,3,4].

2 In the complex scene,basic principle of electric energy measurement and fault diagnosis by the optimization of multi sensor information fusion based on neural network method

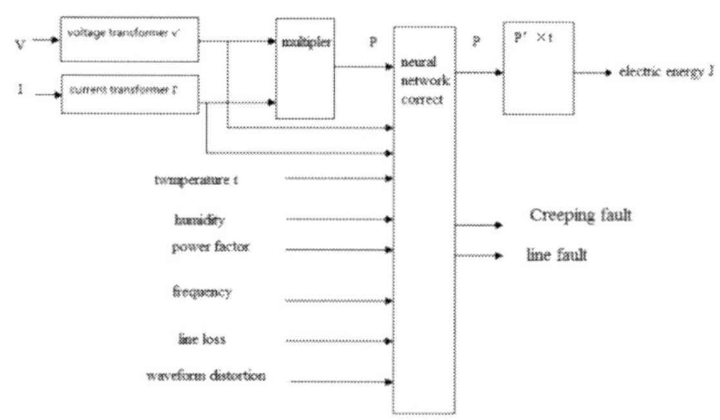

Fig.1 *Electric energy measurement and fault diagnosis system scheme*

Based on the optimized neural network and multi sensor information fusion ,electric energy measurement and fault diagnosis system scheme is shown in figure 1.The voltage transformer transforms the high voltage into a smaller voltage V'. The current transformer transforms the high current into a small current I'.The electrical power P can be obtained by multiplying the voltage V' with the current I'. According to the measurement results of multi sensors for field environment (temperature, humidity, voltage, frequency, current, power factor, magnetic field, waveform distortion, voltage line loss, etc.), real time and error corrected power P' can be obtained by neural network computing. P 'multiplied by time t is the result of electrical energy E.

The neural network needs to be optimized and trained before use.The sample database that maps the actual data relationship needs to be collected to built. Firstly, the fitness function of the network structure is established. Then, based on the genetic algorithm to optimize the weights of neural network, the neural network structure is optimized , To search for a reasonable neural network structure for the application of

*Corresponding author: zhufenglee@126.com

multi sensor information fusion in complex field environment. According to the state information (power, current, voltage, power factor, line loss, etc.),real time fault diagnosis of electric energy measurement device can be completed by neural network.Creeping , large line loss and disconnection are part of the fault state information .Neural network needs to complete information fusion learning,clustering learning,finally realize the diagnosis function[5,6].

3 Optimization methods of neural network structure

According to the complexity of the specific problems, the appropriate neural network structure can be selected, that is, to ensure the network of data induction, but also save resources and time.The number of nodes in the input layer is determined by the physical quantity of the detection.The number of nodes in the output layer is determined according to the number of outputs.It is difficult to determine the number of hidden nodes.Considering the network capacity and function approximation , the number of hidden units should be chosen more.However, considering the generalization ability of the network, the number of hidden units should not be too much.So the number of hidden units should have the best value.The initial number of hidden units should be given first, and then the correlation pruning algorithm is used to reduce the number of hidden nodes until the number of hidden nodes is determined.Finally, the number of nodes of hidden units with less and reasonable number is determined.The initial number of hidden units is determined by subtractive clustering method[7].

The method of determining the initial number of hidden units (subtractive clustering method).

Set m dimensional space n data points (X_1, X_2,... , X_n), subtractive clustering process is as below. first, for each point X_i of the data set , the density index is calculated.The highest density data X_{ci} is the first cluster center.

$$D_i = \sum_{j=1}^{n} \exp\left[-\|x_i - x_j\|^2 / (0.5r_a)^2\right] \text{........(1)}$$

X_{ck} is selected as the cluster center of K, and the corresponding density index is D_{ck}, and the density function of each data point is modified by the following formula

$$D'_i = D_i - D_{ck} \exp\left[-\|x_i - x_{ck}\|^2 / (0.5r_b)^2\right] \text{........(2)}$$

The $X_{c(k+1)}$, which has the highest density index, is a new clustering center.

If the condition $D_{c(k+1)} / D_1 < \delta$ is not established, the application formula (2) can be used to select the next cluster center.If it is set up, the clustering results and the number of clusters (i.e. the initial number of hidden layers) are given.the parameters $\delta<1$ is based on empirical parameters.This parameter determines the final number of initialization cluster centers, the smaller the δ, the more the number of clusters.

$$r_a = r_b = \frac{1}{2} \min_k \left\{ \max_i \{\|x_i - x_k\|\} \right\} \text{.........(3)}$$

The neural network consists of three layers: input layer, hidden layer and output layer.Neural network has a network structure with parallel and distributed information processing , With learning, induction, extraction and other functions.The input layer receives the input letter from the system.According to the input information, the hidden layer can be used for internal learning and processing, which is connected with the input layer and the output layer through the right.The output layer obtains the desired result according to the input and adjustment weight coefficients.Input layer input information is power, temperature, humidity, frequency, voltage, current, magnetic field strength etc.The output layer uses the actual power output value Y1, and the other outputs are used for function development.The actual power output value Y1 is multiplied with time and accumulated into the measured power energy.Power expectation B is provided by standard power meter. The accuracy of the standard power meter is relatively small, and it is only used in the neural network learning stage to determine the connection weights of the neural network.The standard power meter is not used for the actual measurement of electric energy.The actual power output Y1 should be close to the corresponding power expectation value B under various conditions calculated by neural network .According to the input value of power measurement X1 and the change of the environment quantity, by the neural network calculating, and the correction value adjusted self-adaption,high precision power can be obtained.As a result, it can realize the adaptive adjustment of the error in the power measurement. finally the adaptive error correction of the electric energy measurement is realized.[7,8]

4 learning methods

BP neural network model shown in figure 2.The learning process of the network algorithm consists of two parts: forward and backward propagation. In the process of forward propagation, the input information is transmitted from the input layer to the output layer by layer by layer. The state of each neuron affects only the state of the next neuron. If the desired output is not obtained in the output layer, then it is transmitted back, and the error signal is returned along the original path, and the error signal is minimized by modifying the weights of the neurons in each layer.[9,10,11]

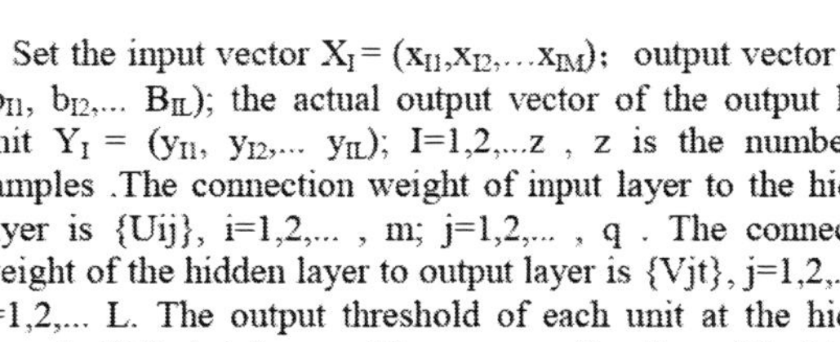

Fig . 2 *Bp neural network model*

Set the input vector $X_I = (x_{I1}, x_{I2}, \ldots x_{IM})$; output vector $B_I = (b_{I1}, b_{I2}, \ldots B_{IL})$; the actual output vector of the output layer unit $Y_I = (y_{I1}, y_{I2}, \ldots y_{IL})$; I=1,2,...z , z is the number of samples .The connection weight of input layer to the hidden layer is {Uij}, i=1,2,... , m; j=1,2,... , q . The connection weight of the hidden layer to output layer is {Vjt}, j=1,2,... , q; t=1,2,... L. The output threshold of each unit at the hidden layer is {Oj}, j=1,2,... , q. The response function of the hidden layer nodes is

DOI: 10.1051/matecconf/201713900081

$f(x)=(1+e^{-x})^{-1}$ .The response function of the output notes is liner.

Formula (4) is the error energy function. That is, the output vector B_I = (b_{I1}, b_{I2},... B_{IL}) subtract the actual output vector of the output layer unit Y_I= (y_{I1}, y_{I2},... y_{IL}). Then ,the error squared is added.

$$E=\frac{1}{2}\sum_{I=1}^{z}\sum_{t=1}^{L}\left[\sum_{j=1}^{q} v_{jt} f\left(\sum_{i=1}^{M} u_{ij} x_{Ii}+o_j\right)-b_{It}\right]^2+\alpha e^{-\frac{1}{q+fm}}\text{.......(4)}$$

By adjusting the weights {u_{ij}}, {v_{jt}} and the number of hidden layer nodes Q, n number of network connection weights, the error energy function E ($_W$) can reach the minimum, to realize the optimization of the network selection.Then E is a function of w, *E=E (w)*.[12,13,4]

4.1 Steepest gradient descent learning algorithm

W is a vector, E is a multivariate function.Through the network learning, w value can be determined, at this time the function E to achieve the minimum value.Gradient G (w) is vector. If E (w) produces the greatest change, the corresponding direction is G (w) direction.The mode is the maximum change rate of E (w) at M point.Vector G (w) is the gradient of the number field E (w) at the w point.grad E $|_M$=G(W)=▽E.

$$gradE=\frac{\partial E}{\partial x}\vec{i}+\frac{\partial E}{\partial y}\vec{j}+\frac{\partial E}{\partial z}\vec{k}+\ldots \qquad \nabla=\vec{i}\frac{\partial}{\partial x}+\vec{j}\frac{\partial}{\partial y}+\vec{k}\frac{\partial}{\partial z}+\ldots$$

. ▽ is Hamiiton operator.

The gradient descent learning algorithm is shown as follows.The initial position W (k) is chosen randomly, and the moving direction and step size of W (k+1) are determined according to the gradient of W (k) point, so that the error function E (W (K+1)) is decreased rapidly than E (W) (K).Then W (k+1) is used as the starting point and the steepest descent method is used repeat to determine the W (k+2). Finally, the point W (k+n), which former the smallest E (W) value， is found as the solution of the neural network weights.The iterative formula of steepest descent learning algorithm is formula (5).A (k) is a positive number, also known as the step size. At each iteration, a (k) is determined by the one dimensional linear search.The steepest gradient descent learning method decreased rapidly at the early stage of learning, and learning in the later there will be a severe vibration phenomenon that the learning speed is very slow, because two times the search direction of the adjacent gradient algorithm is orthogonal, the search process will appear "seesaw" phenomenon. Therefore, in the later stage of learning the Newton method is used.

$$W(k+1)=W(k)-a(k)\nabla E(W(k))\text{..........(5)}$$

Newton iteration method

According to the BP neural network in Figure 2, the weights of BP neural network can satisfy the solution of nonlinear equations.The nonlinear equations are formula (6).

$$\sum_{j=1}^{q} v_{jt} f\left(\sum_{i=1}^{M} u_{ij} x_{Ij}+o_j\right)-b_{It}=0=F(W)\text{........(6)}$$

（t=1,2,..L），（I=1，2，…，Z）,.

$$W=(v_{11},v_{12},\cdots,v_{qL},u_{11},u_{12},\cdots,u_{mq},o_1,o_2,\cdots,o_q)$$

W is a vector of unknown weights.F=（f_{11}，f_{12}，…f_{ZL}）^{T}.F is a nonlinear function.The nonlinear equations (6) have q (M+L+1) unknown variables and ZL nonlinear equations.The nonlinear equations (6) is the objective function of the Newton iterative learning algorithm for solving the weights of neural networks.

$$F'(W(k))=\left[\frac{\partial f_{It}}{\partial w_{ij}}\right]_{ZL\cdot q(M+L+1)}$$

F '(W (k)) is vector.The partial derivative of the function F is calculated separately for each variable.A matrix of partial derivatives is called the Jacobi matrix. Newton iterative method to calculate the weights of neural networks is formula (7).

$$W(k+1)=W(k)-\left[F'(W(k))\right]^{-1}F(W(K))\text{........(7)}$$

$[F'(W(k))]^{-1}$ is generalized inverse matrix of F'(W(k)).In order to avoid the partial derivative calculation, the variation value can be used to replace the partial derivative in the Jacobi matrix， that is, using the approximate Newton method to complete the calculation.The applicable conditions of Newton method are as follows.The applicable conditions of Newton method are as follows. The choice of the initial point is sufficiently close to the extreme point (w*), which is generated by the iterative computation to produce the sequence {W (k)}, which is the solution of the nonlinear equations.If the initial point is far away from the pole, the Newton iteration method is likely to fail.As a result, the sequence obtained by iterative computation does not converge.Newton iterative method has the advantages of fast convergence speed and two order convergence rate.The gradient iteration method has only one order convergence rate.[15,16,17]

4.2 The method of calculating the ratio of dislocation accumulation (CRDA method).

When the gradient method fails, Newton method is used. CRDA method is a method to determine how to choose different methods to realize the combination of gradient method and Newton method.The CRDA method is proposed by the author of this thesis.For the gradient method, the error function in the initial stage of the neural network is decreased rapidly, and the selection of the initial point is not required.In the later period of study, there will be a violent vibration, which makes the learning speed slow. Newton iterative method has the advantage of fast convergence speed and two order convergence rate, but it is necessary to calculate the initial point near the poles.First, the gradient method is used to calculate the sequence point until it falls into the convergence domain of Newton method. Then, Newton method is used to realize the combination of gradient method and Newton method.An effective method is to give full play to the advantages of the two algorithms and make up for their deficiencies.The key question is when the gradient method is converted to Newton's method, which has a great influence on the learning effect.

In the case of the K iteration calculation, the sum of squared error function E (W (k-i)) is summed up. Only the fixed number of times (n) is calculated.A (k) represents the calculated results.R (K) is the ratio of the function E (W (k-1-i)) with the function E (W (k-i)) .That is, the location of the dislocation, to accumulate, the ratio, known as the dislocation plus CRDA method.

$$A(k)=\sum_{i=0}^{n}E(W(k-i))$$

$$R(k)=\frac{A(k-1)}{A(k)}=\frac{\sum_{i=0}^{n}E(W(K-1-i))}{\sum_{i=0}^{n}E(W(k-i))}$$

When $R\ (k)\rightarrow 1$, the gradient iteration method can be considered as a "seesaw" state, and the gradient algorithm is replaced by the Newton iteration method.

Formula(8) is the combination of gradient method and Newton method.When R (k) > 1+ ε, ε is a small positive number greater than 0, $\lambda_k=1$ is used and gradient iteration method is adopted.When R (k) = 1+ ε, $\lambda_k=0$ is selected, using Newton iterative method[29,30,31,32].The selection of epsilon will affect the iterative steps of the gradient iteration method. The smaller the epsilon is, the more the gradient method is used to iterate, and the higher the accuracy of judging the gradient iteration failure, the longer the computation time of the machine.Thus, by using the CRDA method, the point at the end of the gradient iteration method falls into the convergent domain of the Newton iterative method to be used.The effective combination of gradient iteration method and Newton iteration method is achieved.

$$W(k+1)=W(k)-\left[\lambda_k a_k \nabla E(w(k))+(1-\lambda_k)\left[F'(W(k))\right]^{-1}F(W(k))\right] \quad ....(8)$$

5 DSP technology is adopted to realize the real-time fast measurement of electric energy

Digital signal processor DSP (Digital Signal Processing) is particularly suitable for digital signal processing operations, the main feature is that real-time and rapid implementation of a variety of digital signal processing algorithms. DSP devices work well in common convolution, filtering, FFT and matrix operations.The DSP device produced by the American Simulated company ADSP-21060 is designed for high performance multiprocessing applications such as medical imaging, 3D graphics acceleration, radar, electronic signal processing, etc.It represents the best performing 32 bit DSP chip at present.ADSP-21060 is a high performance floating-point DSP chip, the super Harvard architecture, with four independent bus (two data bus, a program bus and a I/O bus); 4M memory, with DMA controller and I/O processor, allowing flexible and high speed data transmission without overhead, the rate is 240Mbytes/s can provide; and 16 32 bit processor interface; multi channel serial port with 40Mbit/s.In this paper, the DSP devices are used to receive the data of various environmental sensors and perform neural network calculation. The corrected power value is calculated, and then the electric energy is calculated by the integral calculation of power and time. All the control and operation are completed by DSP, and the adaptive measurement of electric energy is realized. The neural network weights are obtained by using gradient Newton combined learning method.

6 Experimental results

Table 1 *Comparison of experimental results based on CRDA algorithm with other algorithms*

Error precision	Maximum number of learning steps	The steepest descent gradient algorithm		The ordinary gradient Newton method				CRDA method			
		Number of learning steps	Convergence ratio	Total learning steps	Convergence ratio	The number of step gradient	The number of step Newton	Total learning steps	Convergence ratio	The number of step gradient	The number of step Newton
0.1	2000	177	80%	99	100%	94	5	96	100%	88	8
0.01	4000	499	90%	297	85%	291	6	102	92%	90	12
0.001	8000	1221	86%	460	95%	448	12	119	98%	91	28

Table . 2 *Environmental impact experiment of power measurement*

error \ environmental / Meter	Temperature change 40℃	Moisture change 40%	Voltage change 80%	Current change 100%	External magnetic field change 30%
Adaptive power meter error change	0.02	0.02	0.01	0.02	0.02
General energy meter error change	0.18	0.15	0.12	0.17	0.11

The experimental results of different algorithms are shown in table 1.

Repeated studies by Random selection of initial values are completed 100 times, and average results are shown in table 1.We can see from the experimental results of table 1. The CRDA method to judge the tonsure Newton algorithm significantly reduced the number of learning steps, improve the rate of convergence. The effectiveness of this method is demonstrated.The environmental impact experiment of electric power adaptive measuring instrument based on neural network is carried out. The error of adaptive watt hour meter and general electric energy meter is measured by 0.02 level standard watt hour meter, and the change of the error of two kinds of watt hour meter is compared with the change of environment quantity. The result is shown in table 2. From the experimental results of Table 2, it can be seen that the error of electric energy measurement in general electric energy meter will cause great additional error when the environment changes.However, the added error of the adaptive energy meter when the environment changes is very small. It basically matches the standard electric energy meter to measure the electric energy. The validity of neural network based adaptive measurement method for electric power is verified.

7 Conclusion

The adaptive measurement method of power and energy based on neural network can effectively modify its own measurement errors adaptively and in real-time in complex environment. Real time intelligent correction is realized to improve the accuracy of the meter in complex environment. In the neural network learning, the CRDA method is used to reduce the point at the end of the gradient iteration method into the convergent domain of the Newton iterative method to be used, and the gradient iteration method and the Newton iteration method are combined effectively. The adaptive electric energy measurement method will effectively change the traditional principle and calibration method of electric energy measurement, and improve the accuracy, range of application and environmental impact resistance of the meter. This method is of great significance.

8 Reference

1. Sulan .Xia , Li.Zhao , Research of multi-sensor information fusion based on BP neural networks [J]computer measurement & control . 23(5),1823-1826(2015).
2. Xian z .W , Application of fuzzy adaptive back-propagation neural networks in thermal conductivity gas analyzer [J]Neural computing.7(3),579-583(2010).
3. Zhangyang. L , Research on multi-sensor date fusion based on improved bp algorithim.[J]Journal of Southeast University(Natural Science Edition) .38(1),258-261(2008).
4. Napolitano. M R, Windon. D A, Casanova J L, et al. Kalman filters and neural-network schemes for sensor validation in flight control systems[J]. IEEE Transactions on Control Systems Technology, 6(5),596-611(1998).
5. Chen.C.L.,Liu'.J.Wen.G.X. Fuzzy neural network based adaptive control For a class of uncertain nolinear stochastic system [J].IEEE Transactions on Cybernetics,2014,44(5):583-593.
6. Fu Yue,Chai Tianyou, Nonlinear multivariable adaptive control using multiple models and neural networks[J],.Automatica, 2007(43):101-111.
7. Narendira K.S.,Parthasarathy K.,Identification and control of dynamical systems using neural networks, [J].Neural Networks 、 IEEE Transactions on,1(1),1990,4-27.
8. Rovithakis G.A.,Christodoulou M.A.,Direct adaptive regulation of unknown nonlinear dynamical systems via dynamic neural networks [J].IEEE Transactions on Systems,Man and Cybernetics,25(^n),1995,1578-1594.
9. A.S.Poznyak and E.N.Sanchez,Nonlinear system approximation by neural networks:error stability analysis[J].International Journal of Intelligent Intelligent Automation and Soft Computing,1995,247-258.
10. A.S.Poznyak,Learning for Dynamic Neural Networks [C].In 10th Yale Workshop.Adaptive Learning Systems,1998,38-47.
11. A.S.Poznyak,W.Yu,E.N.Sanchez,and J.R Perez,Stability analysis of dynamic neural control[J].Expert System Application, 14(1),1998,227-236.
12. A.RovMiakis and M.A.Christodoulou,Adaptive control of unknown plants using dynamical neural networks[J].IEEE Transaction on Systems,Man and Cybernetics,1994,(24):400-412.
13. Rovithakis G.A.,Christodoulou M.A.A robust direct adaptive regulation Architecture using dynamic neural network models[C]IEEE International Conference on Systems,Man,and Cybernetics,1994.Humans,InformationandTechnology.1994,(2):1110-115.
14. LiangN.Y.,HuanG.B.,Saratchandran P,et al.A fast and accurate online sequential learning algorithm for feedforward networks[J].Neural Networks,IEEE Transactions on,2006,17(6):1411-1423.
15. Huang G.B.,Wang D.H.,Lan Y.Extreme leaming machines:a survey.International Journal of Machine Learning and Cybemetics[J],2011,2(2):107-122.
16. Huang G.B.,Zhu Q.Y.,Siew C.K.Extreme learning machine:a new Learning scheme of feedforward neural networks[C].Neural etworks,IEEE Intenational Joint Conference on,2004,(2):985-990.
17. Liu G.R,Kadirkamanathan V.Predicti vecontrol for non-linear systems useing neural networks[J].Intemational Journal of control,1998,71(6):1119-1132.

Acknowledgments: This work was supported by the found of wuyi university (grant 2013-7-112).

Zhufeng Li . 1965. men . Senior laboratory technician. Master. Mainly engaging in physical experiment , electronic measuring techonlogy and neural net research . Undertake 2 scientific and technological projects in Jiangmen city, 4 research projects funded by Wuyi University. More than 30 papers were published.
Communicate address: 207# College of Application Physics and material,WuYi Univ. jiangmeng Guangdong China . Zip code: 529020,Telephone: 0750-3299401(o)
Email: zhufenglee@126.com

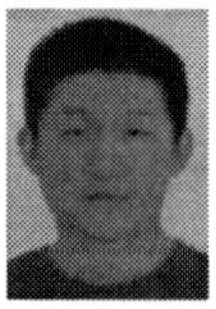

Yang Li. 1991. men .born in harbin china . In 2016, the Bachelor's degree has been awarded from college of life science , state university of Minnesota United States . Studying and internship in state university of Minnesota United States now . Mainly engaging in life science , Genetic engineering , physics , chemical and light application research .In 2017, 1 invention patents were awarded from People's Republic of China, ranking fourth,The patent number is "ZL 201310695565.1". 3 papers were published.

Analysis and Compensation of the Inner-arm Error in the North-finder Syetem of SINS

Lei Wang[1], Zhou Wang [2], Yuan Ren[2]

[1]*Graduate School, Aerospace Engineering University, Beijing 101416, China;*

[2]*Department of Space Equipment, Aerospace Engineering University, Beijing 101416, China*

Abstract: For the high precision strapdown inertial navigation system in north, the rotary modulation will not get rid of the inner-arm error.In the north-seeking process, because of the rotating device,it will introduce the error into the north navigation calculation. In this paper, first, the error of the within arm is modeled and analyzed, and the mathematical expression of the error effect is deduced. And two factors affecting the error of the within arm are analyzed and determined, then error compensation method of the within arm is put forward. Finally the simulation results are verified.

1 Introduction

The IMU rotation is used to drive the IMU spin and the navigation solution is used in the conventional strap-homing system. According to the traditional error compensation method, the common error can be removed by rotation modulation[1]. In the course of the study, the emphasis should be placed on the error which can not be demodulated with the rotation modulation, or the research and analysis of some new errors introduced after the rotation device is added.
Because of the mechanical structure and the reason of the system, the three accelerometers in the IMU are not superimposed. The three accelerometers measure the ratio of each acceleration meter to the direction of its input axis. In the process of angular motion, the error of the accelerometer is bound to result in the error of finding the north navigation solution, which is commonly referred to as the inner-arm error (also known as the size effect)[2]. In the process of this study, using the gyroscope for high precision laser gyro, due to the physical structure of the optical does not exist in the mechanical rotor, is not sensitive to the change of the acceleration, therefore optical gyro output is not affected by the influence of size effect, there is no error of the lever arm. However, in the navigation solution calculation, the measurement error caused by the error effect of the inside lever arm of the accelerometer can not be ignored, which can affect the north navigation results.

2 The analysis and modeling of inner-arm error

If you look at the IMU as a rigid body, according to the rigid body theory, in the IMU motion, the angular velocity of each point on it is the same. So for laser gyro in the company, the installation position and orientation of three gyro can, as long as there is no parallel between three axis gyro sensitive relationship can, after calibration can achieve three-dimensional angular velocity measurement. But for accelerometer, due to the three accelerometer physical size and the limitation of the actual installation position, makes them measurement of acceleration is the direction of the IMU differences, if do not take this factor into consideration, the three accelerometer to measure the acceleration as the output of the "point measuring component" for north and navigation, so will lead to navigation errors. If there is a large internal lever error, the angular acceleration measured by the accelerometer will cause a certain error to the north seeking result, and the result will be affected. In inertial navigation system on the filter, if use speed as observed quantity, if the company is larger in the lever arm, in the case of the rotating modulation, plus the indexing mechanism of IMU continuously rotates, it will bring bigger error on observation, affect the alignment accuracy.
In the inertial measurement unit, the three accelerometers are installed in three direction positions on the northeast sky axis respectively, and the relative position remains constant. The extension cord of the three accelerometer's sensitive axis intersects at the origin O point of the coordinate system. The specific location installation is shown in figure 1:

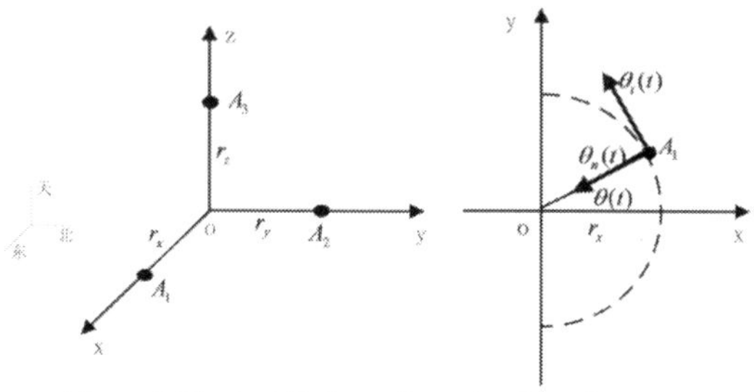

Figure 1. The installation schematic of accelerometer

Because the two vertical installations of the three accelerometers. In the process of analysis, an example is given in the X-axis direction, and the partial error is used in the process of seeking north. Assuming that the IMU rotates around the Z axis, in the xy axis that forms the horizontal plane, assuming that the Angle between the accelerometer A and Ox is the sinusoidal Angle motion:

$$\phi(t) = \phi_0 \sin(\omega t) \quad (1)$$

In the formula: the ϕ_0 is to define the amplitude of vibration; ω is the rotation angular velocity; the oscillation period is $T = 2\pi/\omega$.

Taking the derivative of the upper formula, the angular velocity of rotation can be obtained:

$$\dot{\phi}(t) = \phi_0 \omega \cos(\omega t) \quad (2)$$

According to formula (1) and formula (2) ,we can have the normal velocity, the centripetal acceleration, the tangent velocity and tangential acceleration of accelerometer A:

$$\begin{aligned} & v_n(t) = 0 \\ & a_n(t) = r_A(\dot{\phi}(t))^2 = r_A \phi^2{}_0 \omega^2 \cos^2(\omega t) \\ & v_t(t) = r_A \dot{\phi}(t) = r_A \phi_0 \omega \cos(\omega t) \\ & a_t(t) = (v_t(t))' = -r_A \phi_0 \omega^2 \cos(\omega t) \end{aligned} \quad (3)$$

In the formula (3), the velocity component can be obtained by decomposing the projection of the normal velocity and tangent velocity along the axis xyz :

$$\begin{aligned} v_x(t) &= v_{nx}(t) + v_{tx}(t) = 0 - v_t(t)\sin[\phi(t)] \\ &= -r_A \phi_0 \omega \cos(\omega t) \sin[\phi_0 \cos(\omega t)] \\ v_y(t) &= v_{ny}(t) + v_{ty}(t) = 0 + v_t(t)\cos[\phi(t)] \\ &= r_A \phi_0 \omega \cos(\omega t)\cos[\phi_0 \cos(\omega t)] \end{aligned} \quad (4)$$

In formula (3), the acceleration projection can be obtained by decomposing the tangent acceleration and the normal acceleration along the axis:

According to the above formula (4) , easy to know:

$$v_x(0) = v_x(T) = o; v_y(0) = v_y(T) = r\phi_0 \omega$$

The following relation can be obtained by taking the derivative of formula (4) :

$$\begin{aligned} v_x(t) &= v_{tx}(t) + v_{nx}(t) \\ v_y(t) &= v_{ty}(t) + v_{ny}(t) \end{aligned} \quad (5)$$

$$\begin{aligned} a_{nx}(t) &= -a_n(t)\cos[\phi(t)] \\ &= -r_A \phi^2{}_0 \omega^2 \cos^2(\omega t)\cos[\phi_0 \sin(\omega t)] \\ a_{ny}(t) &= -a_t(t)\sin[\phi(t)] \\ &= -r_A \phi^2{}_0 \omega^2 \cos^2(\omega t)\sin[\phi_0 \sin(\omega t)] \\ a_{tx}(t) &= -a_t(t)\sin[\phi(t)] \\ &= r_A \phi_0 \omega^2 \cos(\omega t)\sin[\phi_0 \sin(\omega t)] \\ a_{ty}(t) &= a_n(t)\cos[\phi(t)] \\ &= -r_A \phi_0 \omega^2 \cos(\omega t)\cos[\phi_0 \sin(\omega t)] \end{aligned} \quad (6)$$

Defining function:

$$\begin{aligned} f_1(x,t_1,t_2) &= \int_{t_1}^{t_2} x \sin t \sin(x \sin t)\,dt \\ f_2(x,t_1,t_2) &= \int_{t_1}^{t_2} x \sin t \cos(x \sin t)\,dt \end{aligned} \quad (7)$$

According to the symmetry and periodicity of the integrand, it can be obtained:

$$\begin{aligned} f_1(x,o,\pi/2) &= f_1(x,\pi/2,\pi) = f_1(x,\pi,3\pi/2) \\ &= f_1(x,3\pi/2,2\pi) \\ f_2(x,o,2\pi) &= f_2(x,\pi/2,3\pi/2) = 0 \end{aligned} \quad (8)$$

In one cycle, the acceleration integral along the coordinate axis can be obtained by the increment of velocity:

$$\begin{cases} \Delta v_{tx}(T) = \int_o^T a_{tx}(t)dt = 4r_A \omega f_1(\phi_0, 0, \pi/2) \\ \Delta v_{ty}(T) = \int_o^T a_{ty}(t)dt = -f_2(\phi_0, 0, 2\pi) = 0 \\ \Delta v_{nx}(T) = \int_o^T a_{nx}(t)dt = -4r_A \omega f_1(\phi_0, 0, \pi/2) \\ \Delta v_{ny}(T) = \int_o^T a_{ny}(t)dt = 0 \end{cases} \quad (9)$$

3 Error analysis of the inner-arm effect

In the rotation process, the three accelerometers are assumed to be perpendicular to each other and the sensitive axis intersects at point o_b . Assume that the rotation axis and the rotation axis are not at the same point, so the axis of rotation is perpendicular to $o_b z_b$, the rotation angular velocity is ω , and the amplitude of vibration is ϕ_0 .

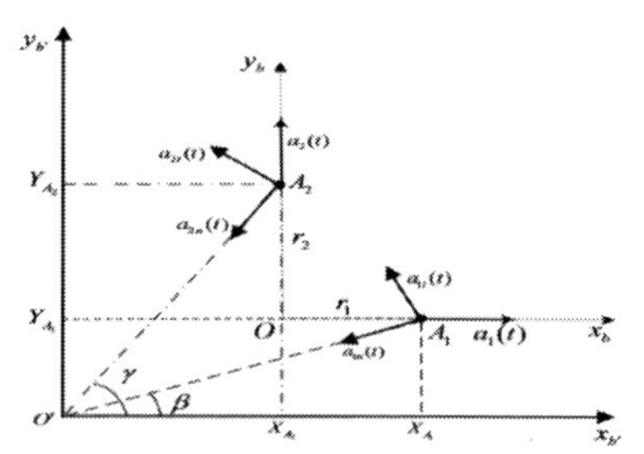

Figure 2. Schematic diagram of installation in the xy plane

The measured value of the accelerometer is the sum of the projection of the centripetal acceleration and tangent acceleration at a point in the direction of the accelerometer.

The centripetal acceleration andtangential acceleration at the accelerometer of A_1 can be expressed as follows:

$$\begin{aligned} a_{1n}(t) &= R\phi_0{}^2\omega^2\left[\cos(\omega t)\right]^2 \\ a_{1t}(t) &= -R\phi_0\omega^2\sin(\omega t) \end{aligned} \quad (10)$$

The measurement value of accelerometer A_1:

$$\begin{aligned} a_1(t) &= -a_{1n}(t)\cos\beta - a_{1t}(t)\sin\beta \\ &= -\left|O'X_{A_1}\right|\phi_0{}^2\omega^2\cos^2(\omega t) \\ &+\left|O'Y_{A_1}\right|\phi_0\omega^2\sin(\omega t) \end{aligned} \quad (11)$$

The measurement value of accelerometer A_2:

$$\begin{aligned} a_2(t) &= -a_{2n}(t)\sin\gamma + a_{2t}(t)\cos\gamma \\ &= -\left|O'Y_{A_2}\right|\phi_0{}^2\omega^2\left[\cos(\omega t)\right]^2 - \left|OX_{A_2}\right|\phi_0\omega^2\sin(\omega t) \end{aligned} \quad (12)$$

In the solution of the above equation, the measurement values of the accelerometer of A_1 and A_2 can be decomposed into the x and y axis of the coordinate system $o_b'x_b'y_b'$, which is equivalent to the sum of the virtual tangential acceleration and the normal acceleration of the coordinate axes. The measurement value of A_1 is the synthesis of the virtual accelerometer in X_{A_1} and X_{A_2}, and the measurement value of A_2 is synthesized from the virtual accelerometer in Y_{A_1} and Y_{A_2}.

Y_{A_1} is the virtual tangential acceleration:

$$a_{A_1t}(t) = \left|O'Y_{A_1}\right|\phi_0\omega^2\sin(\omega t)$$

X_{A_1} is the acceleration of the virtual method:

$$a_{A_1n}(t) = -\left|O'X_{A_1}\right|\phi_0^2\omega^2\cos^2(\omega t)$$

Y_{A_2} is the virtual tangential acceleration:

$$a_{A_2t}(t) = -\left|O'X_{A_2}\right|\phi_0\omega^2\sin(\omega t)$$

X_{A_2} is the acceleration of the virtual method:

$$a_{A_2n}(t) = -\left|O'Y_{A_2}\right|\phi_0^2\omega^2\cos^2(\omega t)$$

Refer to the above analysis, the derivation of the virtual tangential acceleration and normal acceleration along the fixed coordinate system of each virtual axial projection, and a point in a cycle of integral sum, we can get the navigation axial velocity increment :

$$\begin{aligned} \Delta v_x(T) &= \int_0^T a_x(t)dt \\ &= \int_0^T\left[a_{A1nx}(t) + a_{A1tx}(t) + a_{A2nx}(t) + a_{A2tx}(t)\right]dt \\ &= -4\left|X_{A_1}X_{A_2}\right|\omega f_1(\phi_0, 0, \pi/2) \\ \Delta v_y(T) &= \int_0^T a_y(t)dt \\ &= \int_0^T\left[a_{A1ny}(t) + a_{A1ty}(t) + a_{A2ny}(t) + a_{A2ty}(t)\right] \\ &= -4\left|Y_{A_1}Y_{A_2}\right|\omega f_1(\phi_0, 0, \pi/2) \end{aligned} \quad (13)$$

From the above analysis, the navigation speed increment is independent of the rotation center in one period, which is independent of the origin. When the amplitude is fixed, the acceleration of the navigation speed is proportional to the distance from the origin and the distance between the axial velocity and the rotation angular velocity [5]. In practice, if the accelerometer is used as A point measurement component, the navigation error will be generated from the principle, and the calculation error will be formed. We will call that $r_1 = \left|X_{A_1}X_{A_2}\right|$ 、 $r_1 = \left|Y_{A_1}Y_{A_2}\right|$ is the inner arm length.

Using vector analysis method: if there is no internal lever error, then accelerometer of P point:

$$a_P = a_0 \quad (14)$$

In fact, if there is an internal lever error, the acceleration at P point can be obtained as:

$$a_P = a_0 + \dot{\omega}\times r_P + \omega\times(\omega\times r_P) \quad (15)$$

$\omega\times r_p$ and $\omega\times(\omega\times r_P)$ respectively represent the tangent acceleration and normal acceleration of P points relative to O_b points.

The inner rod effect refers to the centripetal force and tangential force that the accelerometer detected during the movement of the carrier due to the relative ideal location[6].

As a result of the three accelerometer sensitive shaft extension cord to the intersection point, so the lever arm within error of accelerometer is introduced by $\omega\times(\omega\times r_P)$. Assume that the accelerometer's sensitive axes coincide with the vector coordinate system, due to the sensitive sensitive accelerometer only along the axis direction of the acceleration, the projection on the axis of accelerometer lever arm in the corresponding error can be represented as:

$$a_e = \begin{bmatrix} a_{ex} & a_{ey} & a_{ez} \end{bmatrix}^T \quad (16)$$

$$\begin{cases} a_{ex} = -\left(\omega_y^2 + \omega_z^2\right)r_x \\ a_{ey} = -\left(\omega_x^2 + \omega_z^2\right)r_y \\ a_{ez} = -\left(\omega_y^2 + \omega_x^2\right)r_z \end{cases} \quad (17)$$

In particular, when there are oscillatory motion, the oscillatory motion passes through the rectifier, it will

DOI: 10.1051/matecconf/201713900082

produce a constant acceleration error.
when $\omega_y = \omega\theta_y \sin\omega t$, $\omega_z = \omega\theta_z \sin(\omega t + \varphi)$:

$$\begin{aligned} a_{ex} &= -\left\{\omega^2\theta_y^2 \sin^2 \omega t + \omega^2\theta_z^2 \sin^2(\omega t + \varphi)\right\} r_x \\ &= -\tfrac{1}{2}\omega^2(\theta_y^2 + \theta_z^2) r_x \\ &\quad + \tfrac{1}{2}\omega^2\left\{\theta_y^2 \cos 2\omega t + \theta_z^2 \cos(2\omega t + 2\varphi)\right\} r_x \end{aligned} \tag{18}$$

The upper formula indicates that the size effect causes a constant acceleration zero deviation, which is $\omega^2(\theta_y^2 + \theta_z^2)r/2$. And when the amplitude of the periodic motion is 0.1, and the frequency is 50HZ. The position deviation of 10cm will produce about 1.5mg of the acceleration zero deviation. Therefore, the error of the accelerometer size effect will not be avoided in the process of finding the north.
The measurement output of the accelerometer can be expressed as:

$$a_P = a_0 - a_e = \begin{bmatrix} a0x \\ a0y \\ a0z \end{bmatrix} - \begin{bmatrix} (\omega_y^2 + \omega_z^2) r_x \\ (\omega_x^2 + \omega_z^2) r_y \\ (\omega_y^2 + \omega_x^2) r_z \end{bmatrix} \tag{19}$$

The form of vector notation is:

$$a_0 = a_P - \omega \times (\omega \times r_P) \tag{20}$$

The upper equation is the compensation form for the error of the inner rod arm. If the O point is viewed as the output point of the accelerometer, the inner rod arm error will not exist.And the measurement result of the addition will not be affected by the principle.
When the three accelerometer sensitive shaft extension cord are not intersect in one point, the size effect error will inevitably related to the value of the derivative of angular velocity, which can cause the tangential acceleration. This article does not consider this situation. This requirement can be achieved in the current assembly process. Therefore, this study is not studied in this paper.

4 The error of inner rod is affected by single axis rotation

Based on the single-axis IMU turntable in the laboratory, the accuracy of the error modeling and the validity of the error compensation are verified in this paper. The effect of the inner rod effect on the north precision of the system is analyzed.

Figure 3. Experimental setup diagram

The test process is as follows: (1) Initial alignment of work mode: divided into coarse alignment and precise alignment process. Using the method of document 5, we have added the compensation scheme for the error of the inner rod arm in the scheme. (2) The autonomous inertial navigation mode: collect the output data of IMU for navigation in the navigation computer, analyzing the validity of error compensation scheme after modulation results. The experiment principle as shown in figure 4.

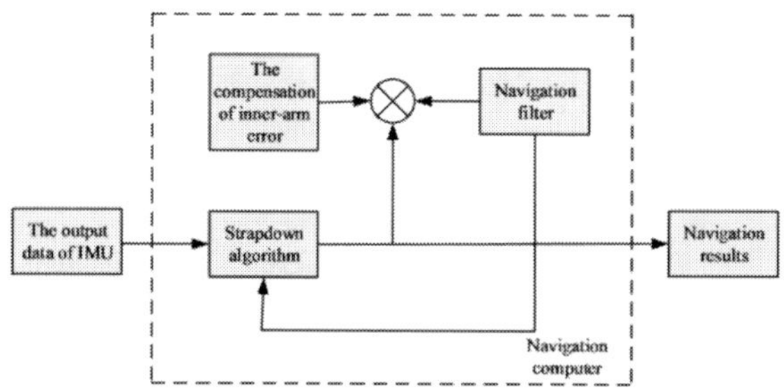

Figure 4. The principle block diagram of the SINS navigation system

During the experiment, the rotary table is guaranteed to be in constant position, position and orientation, and conduct multiple navigation experiments. By error analysis and data processing of the results of the navigation settlement, the error suppression is carried out by using kalman filter algorithm, and the results are shown as:

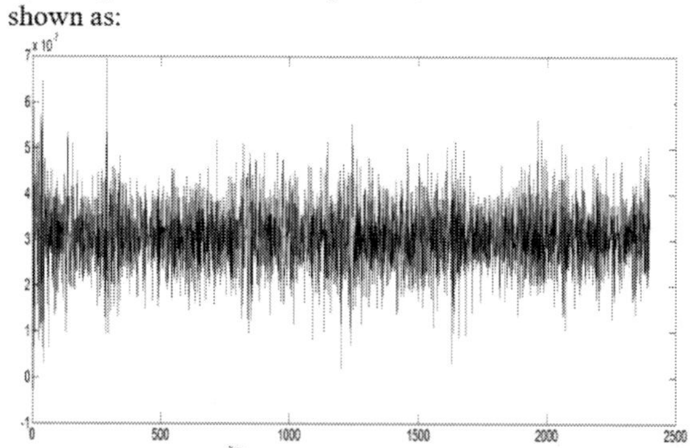

Figure 5. Filter simulation results
(Blue-the original data; Red-the post-filtered data)

After the rod arm error compensation, it can be seen

from fig.5 that the error of output data is significantly reduced. In the process of uniaxial rotating find north, the existence of the rotating device will increase the size effect of error, this is equivalent to the introduction of the new constant drift, cause acceleration got an additional acceleration signal measuring. During the long rotation process, the existence of the rotating device can realize the offset of the error, which will not lead to the divergence of navigation positioning error. The theoretical analysis and simulation results show that the size error of the accelerometer will not lead to the divergence of positioning error, but it will increase the oscillation amplitude of the undamped system speed error.

5 Conclusion

In this paper, under the condition of the accelerometer perpendicular installation, detailed in the principle of lever arm effect error formula was deduced and analyzed the causes, the analysis results: navigation speed increment within lever-arm effect has nothing to do with the center of rotation, and is proportional to the inside lever arm length, and is proportional to the IMU rotation angular velocity. The influence of the error effect of the inner rod arm in the process of the homing system is analyzed. This article only for vertical installation situation carries on the analysis and discussion, if three accelerometer is not vertical or sensitive shaft extension cord is not intersect in one point, in this case the error analysis and compensation discussion is more complex, needed further in-depth analysis.

References:

1. Yuan Bao-lun, Rao Gu-yin.On the theory of Optical Gyro Rotating Inertial navigation System[J].Journal of National University of Defense Technology, 2007,28(6):76-80.
2. David H Titterton , John L Weston.Strapdown Inertial Navigation Technology[M]. Zhang Tian-guang, Wang Xiu-ping,et al.BeiJing: National Defense in Dustry Press, 2007.
3. Yan Gong-min, Yan Wei-sheng, Xu De-min. Analysis and Compensation on Inner Lever Arm Effect of Strapdown Inertial Measurement Unit[J]. Journal of Chinese Inertial Technology, 2008, 16(2):148-153.
4. Chen Yong-bing, Zhong Bin.Inertial Navigation Principle[M].Bei Jing:National Defense in Dustry Press,2007.
5. Yu Xu-dong,Wang Yu,Zhang Peng-fei,et.Influence of Single-axial Rotation on INS Error Characteristics[J].Journal of Chinese Inertial Technology, 2008,16(6):643-648.
6. Huang Xi-cheng,Research of the High Precision North Finding Technology Using Ring Laser Gyroscope[D] :Harbin Institute of Technology, 2013,09.

Research of Smart Payment System of Power Grid Using Strongly Sub-feasible SQP Algorithm

Fang Yang[1], Lei Cao[1], Qi Jiang[2]

[1] *Hanzhong Vocational and Technical College, Hanzhong 723002,China*
[2] *State Grid Fujian Electric Power Company, Fuzhou 350003, China*

Abstract. With the continuous development and perfection of "Internet + Electricity", the regional grid operation has gradually realized the Internet-based automation. In order to improve the smart level of regional grid operation, this paper analyzes the status quo of power grid terminal in Fujian local power (group) company, and introduces the strongly sub-feasible sequence quadratic programming (SQP). The smart payment system based on strongly sub-feasible SQP algorithm is described by its structure, function and implementation process. Through the information technology to improve the efficiency of the service, so that payment staff and smart terminal of self-service payment system has been information between the interactive mode, the actual operation effe ct is good.

1 Introduction

In the increasingly developed modern information technology, It is the inevitable choice of power supply enterprises to deepen and enhance the level of smart management of power grids for electricity customers to provide better service by information technology means.In china there are six types of electric payment terminal: lobby-style self-service terminal; through the wall-type self-service terminal; through the wall self-service terminal; POS terminal; financial institutions self-service terminal; third party terminal. In the above-mentioned various power payment terminals, the wall-mounted self-service terminal, the POS terminal, the financial institution self-service terminal, the third party terminal can only complete the payment function, the lobby type self-service terminal and the wall-type self-service terminal can complete the payment function To provide historical power tariff detailed inquiries.If 24-hour self-service business hall installed through the wall-type self-service terminal expands its function, it will also display the message of power outage, install the voice and video system can provide remote assistance services. There are different needs on the location choice for the power companies, the financial institutions and the third-party agents. Power companies want to provide customers with more options to facilitate the nearest customer payment, to minimize the customer non-malicious arrears caused by power outages, and to speed up the recovery of electricity fee .

The main problems of current payment terminal: the table after the opening of the network layout issues; own network terminal deployment planning issues; power payment terminal installation and commissioning work; power self-service payment terminal business status monitoring issues; power self-service payment terminal Equipment operation monitoring problem; some payment terminal customers can not get invoice and payment success message. In this paper, we propose a grid payment system based on strong and feasible SQP algorithm, which is based on Qt, Oracle software, and so on. In this paper, we propose a scheme of power payment system based on strong and feasible SQP algorithm. With the interactive functions, the developed system has a very good versatility and achieve good results in practical operation.

2 Mathematical Model of Smart Payment System for Power Grid

The payment terminal is the core component of the smart grid operation system. The user can judge whether the payment is made by checking the electricity bill and the personal account balance. At the same time, the payment method is provided, which includes the artificial electricity fee, the online payment and the bank direct deductions and other ways, and the system will be based on the user to set the reminder function for self-service payment management. For the optimal solution of the multiobjective nonlinear problem, the self-service payment system model is as follows.

Assuming that λ is the user's monthly electricity bill, θ is the electricity owed by the user, μ is the account balance of the user, ξ is the upper limit of the electricity charge for the user to forcibly.

$$\xi=\Sigma M_{i=1}\varphi_i\lambda_i,\ \varphi_i=\begin{cases}1/\theta_i & \theta_i\neq 0\\ 0 & \theta_i=0\end{cases} \quad (1)$$

On behalf of the calculation of the weight, its value depends on the amount of electricity owed by the user, M refers the number of months, generally 12 months. According to the system requirements, It is necessary to meet the following constraints in order to use the self-service payment function:

$$\lambda+\theta\geq\xi \quad (2)$$

$$\theta+\varphi\leq\mu \quad (3)$$

* Corresponding author: 1770320208@qq.com

For manual payment, you need to manually control the user according to individual needs and also need to use remote data access technology for data update operation.

3 The Principle of SQP Algorithm

The SQP (sequence quadratic programming) algorithm is one of the most effective algorithms to solve the nonlinear constrained optimization. The feasible SQP algorithm is one of the most important algorithms. The SQP (sequence quadratic programming) algorithm is one of the most effective algorithms for solving nonlinear constrained optimization. Because it can produce feasible iterative point sequence, It has a very important practical application in engineering design and real-time control.

Strongly feasible SQP algorithm principle General inequality constraint optimization problem can be expressed as:

$$\begin{cases} Min f(x) & (1) \\ s.t. g_j(x) \le 0, j \in I = 1,2,\ldots,n \end{cases}$$

$f(x)$ is the objective function; x is the design variable; I is the subscript set; g_j (x) is the constraint function; j is the number of the constraint function. The kth iteration point $x^k \in Rn$ is defined as follows.

$$\begin{cases} \Gamma(x^k) = \{j \in I: g_j(x) \le 0\} & (2) \\ I^+(x^k) = \{j \in I: g_j(x) > 0\} \\ \varphi(x) = max \ \{0, g_j(x), j \in I\} & (3) \end{cases}$$

The inequality constraint optimization problem (1) can be solved by solving the quadratic programming problem of the following formula (4), and finally the main search direction d_0^k is obtained.

$$\begin{cases} Min \ \nabla f(x^k)^T d + \frac{1}{2} d^T B^k d \\ s.t. \ g_j(x^k) + \nabla g_j(x^k)^T d \le 0, j \in (I^k)^- & (4) \\ s.t. \ g_j(x^k) + \nabla g_j(x^k)^T d \le \varphi_k, j \in (I^k)^+ \end{cases}$$

Where $B^k \in Rn \times n$ is the approximation of the Hessen matrix of the Lagrangian function at x^k. Since the formula (4) has a feasible solution d = 0, d_0^k is its optimal solution, but not necessarily a feasible solution direction, the algorithm uses the generalized projection technique to obtain the new improved display correction direction d^k, and give the new Of the high-order display correction direction ~ d^k, the specific formula can be seen in details in the reference.

The iterative formula for the design variables is:

$$x^{k+1} = x^k + \lambda^k d^k + (\lambda^k)^2 \tilde{d}^k \qquad (5)$$

The step length λ^k is found by constructing a new Armijio curve. When $(d_0^k, \varphi_k) = (0,0)$, x^k is the optimal solution of equation (1).

The strong and feasible SQP algorithm can first ensure that the algorithm has been subjected to finite iteration, the iterative point falls into the feasible domain, and then automatically becomes a feasible SQP algorithm, and thus overcome a strong assumption (that is, the maximum constraint violation function is the higher order of QP subproblems.) Second, the strict complementary conditions are removed, which greatly reduces the amount of computation. Finally, the algorithm has global and superlinear convergence. The preliminary numerical experiments show that the algorithm is stable and effective.

4 Design of Smart Payment System for Power Grid and Realization of Main Function

Power grid intelligent payment system in the overall design of the modular structure of the system, divided into six modules: manual payment module, bank card payment module, file and electricity tariff query module, electricity invoice and electricity bill printing module.

4.1 Manual Payment Module

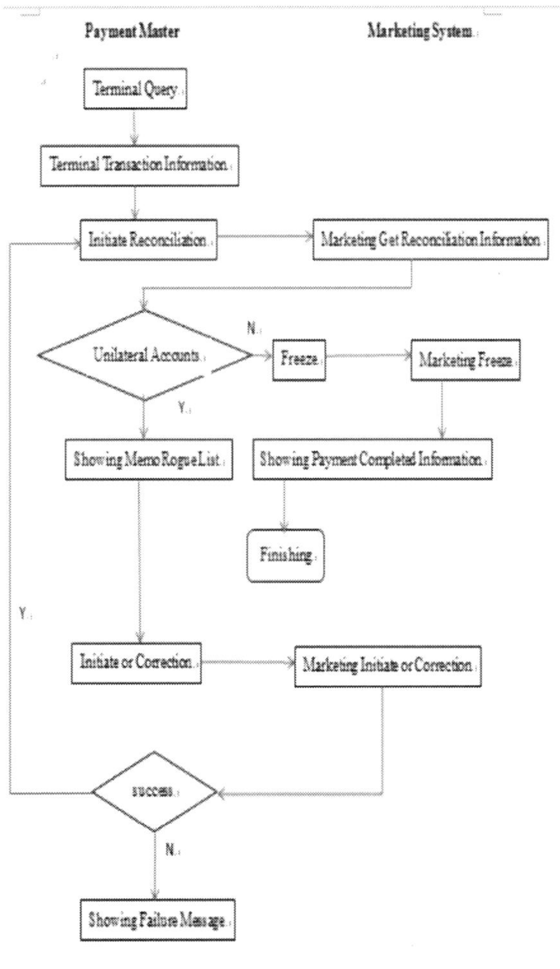

Fig 1 Artificial solution flow chart

 DOI: 10.1051/matecconf/201713900083

Manual payment is to provide the administrator to use the function when payment cannot be solved because of the network or machine failure at the terminal .As the process shown in Figure 1, the module consists of pay the main station and marketing system 2 parts. This function is only open to the administrator, the administrator can enter the wenbrowser interface in the main station and input the needed terminal number of payment, SG186 account, find out the current records of the unpayment at the corresponding terminal, and show detailed transaction records and the total amount ,the total number of pen.

As shown in Figure 2 , in order to solve the problem at payment terminal which is because of network barrier at the time of the boot ,but the consumers due to the terminal fault, the boot module function at payment terminal is added.

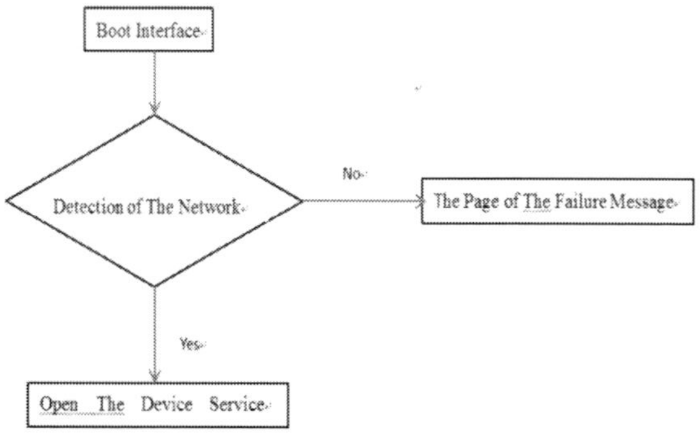

Fig 2 Payment terminal flow chart

Terminal start when the first start a QT prepared by the client program, the interface shows "is initializing the payment process", this procedure is to pay the main station system network check, if the network is smooth, then start the device service program by the device service program to start Firefox. If the network does not work, the page prompts "power network interruption, please contact the network administrator to view the network connection status".

This boot interface displays the "restart", "shutdown" button, after the firefox browser is opened, the boot interface is set at the end, while adding a button "restart the payment process", click this button to restart the device service process, After the firefox runs into fault,the user will see the boot interface, and click "restart the payment process" to restart the program.

4.2 Bank Card Payment Modul

The bank card payment module mainly implements the bank card payment function, and the terminal directly communicates with the CUP system to complete the data interaction and the interactive data is 8583 packets. Bank card payment authority is for electricity customers. The basic processes : Enter the number -> Show under the charge information -> Enter the amount of payment -> Insert the bank card -> enter the bank card password -> payment success.

4.3 File and Electricity Tariff Query Module

(1) file query module, enter the user account number, you can query the file information. The basic processes : Enter the number and password -> Show file information. The file information mainly includes the customer's basic information (customer number, account name, contact person, contact telephone number, document name, document number, payment method, withholding bank, bank account number), customer electricity information (electrical address, copy number , Power supply units, industry classification, measurement methods, voltage levels, electricity categories, user classification, contract capacity, production frequency, into the line) data.

(2) power tariff query module mainly through the user input power account number, customers can query their own nearly a year of electricity consumption information, this authority is for electricity customers, The basic processes : enter the number and password -> show the month Electricity situation -> click on the specific month to check the power situation. File query module constraints for the electricity situation A page shows 6 months, you can click on the previous page or next page to see the last 6 months or the next 6 months. The electricity situation includes: meter reading detail (meter date, the type of display, magnification, last month only, this month only), electricity price details (electricity price unit price, power billing power, electricity tariff) Rate details (to adjust the amount of electricity, adjust the coefficient, the power factor to adjust the electricity), if the user has the corresponding details , it will show, Otherwise no show.

4.4 Electricity Bill and Electricity Bill Print Module

(1) electricity invoice module provides the effective fee vouchers for the electricity customers. After the customer payment is successful, the self-service payment window prompts to print the invoice.

(2) The tariff list print module is to enter the user's account number, select the year to be printed, the list printing, the basic process input number and password -> select the year to print -> to print. The tariff list print module constraint allows customers to print only once for a month (such as 201701) during the user's login session, but also allows the user to print the list for the month after the next login.

5 System Operation Example Analysis

In the current Jinjiang power payment system, for example, when the power grid intelligent payment system is running, the operator firstly judges whether the network terminal is operating normally. It is necessary to add a judge, when the user input SG186 account, you need to join the terminal number to make judgments.if the customers input SG186 account which does not correspond to the machine's terminal number, it will show: the SG186 account and the terminal does not match. The administrator can initiate the reconciliation function manually and can initiate a freeze operation if there is no unilateral account. As shown in Figure 3, if there is a one-sided account can be judged by the system program out of the single-sided account to fill or red positive action. After the manual payment is completed, the administrator also needs to contact the operating room counter in time to solve the situation, and inform the amount of money and the number of pen to facilitate the opening of the money box to check the amount.

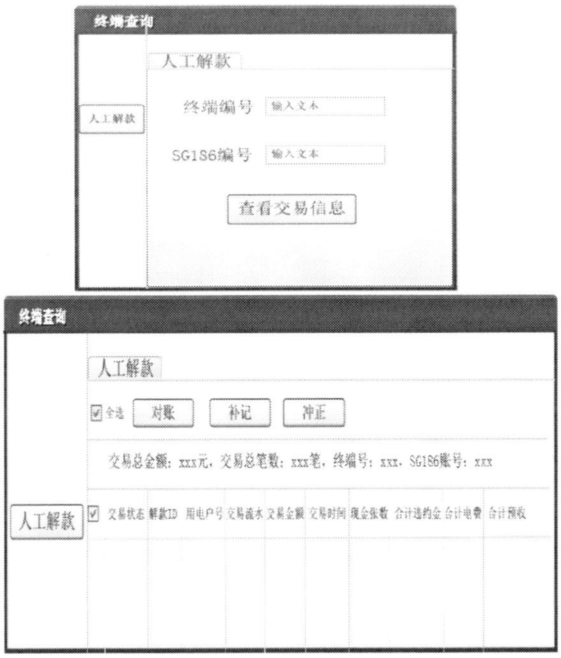

Fig 3 Terminal query and terminal list

After successfully registering the payment terminal, it displays the main interface function area of the system. As shown in the figure, there are six parts: cash payment fee, bank card payment fee, electricity bill invoice printing, electricity bill printing, file inquiry and electricity tariff query.

Fig 4 The main interface of grid smart payment system

Power customers choose electricity tariff query function, enter the electricity account number, the system shows the arrears of information, and prompted the amount required to pay. The system prompts the customer to pay the bank card, after entering the bank card password, the terminal automatically networked bank deductions.

As shown in Figure 5, payment system prompts that customers pay successfully, invoices can be printed. If the power customers forget the user number, as shown in Figure 6 , log on the payment system by swiping the account holder's ID card. In this case, the data obtained from the intelligent payment system of the power grid is consistent with the actual on-site payment data, and the reliability of each module is verified.

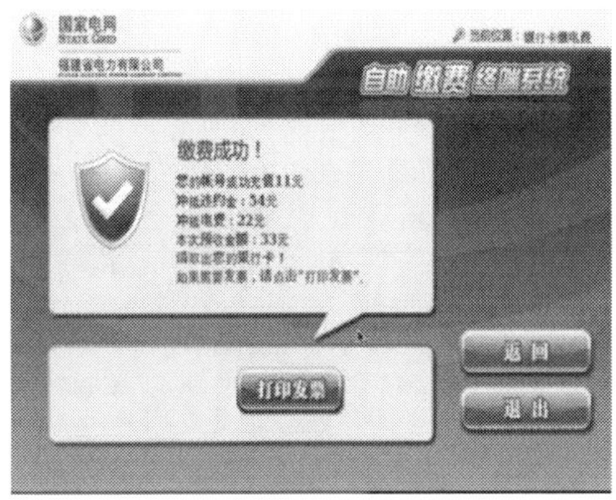

Fig5 Payment Successful to print Invoice

Fig 6 Login mode of ID card

6 Conclusions

The intelligent payment system of power grid mainly focuses on the existing problems of self-service terminal of power payment. It is more convenient and quick to pay electricity and improve the application level of power payment terminal. It adopts the power grid intelligent payment system architecture based on strong and feasible SQP algorithm, Armi jio curve search and two new display correction directions with the same inverse matrix structure, which reduces the computational complexity and ensures the validity of the data. Qt, Oracle software's powerful interactive function has developed a complete application system.

Power grid intelligent payment system is one of the key technologies of smart grid development, and is the cornerstone of achieving the smart grid "informatization, digitization, automatization, interactivity". The system has a good versatility and good results in operation. It has a wide range of development prospects in practical engineering applications. The upgrading and updating of the system equipment are to be in further and deep discussion.

Acknowledgment

This work was supported by the Key Project of Demonstration School of Hanzhong Vocational and Technical College under Grant No. HZY2017B7.

References

1.Journal of Mathematical Physics, 2001,21 (2): 268-277 (in Chinese with English abstract) [J] .Journal of Mathematical Physics, 2001,21 (2): 268-277. JIN Jin-bao. Strongly feasible SQP for hyperlinearity and quadratic convergence of inequality constraints optimization.

2.Tang Chunming.Study on Quadratic Programming Sequence Quadratic Constraints [D]. Shanghai: Shanghai University, 2008.

3.Wang Junyan. Electric power payment terminal application analysis [J]. China Electric Power Enterprise Management, 2011,3: 110-112.

4.Shi Jin, Tu Guangyu, Luo Yi. Complex network characteristics analysis and model improvement of power system [J]. Proceeding of the CSEE, 2008,28 (25): 93-98.

5.Chen Guozhang, Xiang Bing, Jiang Weichao.Evaluation model and optimization of power system based on scale-free network [J]. Electrical Measurement and Measurement, 2010,47 (s1): 39-42.

6.Wang Guangzeng, Cao Yijia, Bao Zhejing. A new local evolution model of power network [J]. Acta Physica Sinica, 2009,58 (6): 3597-3602.

7.Ding Lijie, Cao Yijia, Liu Meijun. Dynamic model and analysis of chain fault in complex power network [J]. Journal of Zhejiang University (Engineering Science), 2008,42 (4): 641-646.

8.Zhang Yunzhou, Cheng Lu. China Electric Power "thirteen" and the long-term development of major issues [J]. China Power, 2015,48 (1): 1-5.

9.Li Yujia, Dong Chuan. Low-carbon active distribution network planning based on two kinds of bionic algorithms [J]. China Electric Power, 2017,4 (50): 66-70.

The Preventive Maintenance of Highway Based on Data Mining

Shunzhou Xiao[1,2],and *Mingxin* Nie[1,2]

[1]School of Information Engineering, Wuhan University of Technology, Wuhan, 430070, China

[2]Key Laboratory of Fiber Optic Sensing Technology and Information Processing, Ministry of Education, Wuhan University of Techn ology, Wuhan, 430070, China

Abstract: Judging from the current situation of Chinese highway maintenance,only after there is highway distress would the staff have a repair, that results in the poor efficiency of highway maintenance. In order to improve the efficiency of highway maintenance,this paper will use data mining technology to predict the pavement performance of highway and analyze the main factors of pavement performance attenuation,so that the preventive maintenance can be carried out.We will provide data support to the preventive maintenance of highway by using the isolation Forest anomaly detection algorithm to have a data pretreatment, the regression model and time series GM (1,1) model to predict the pavement performance and the association rule analysis and isolation Forest to analyze the main factors of pavement performance attenuation.

1.Introduction

In recent years, with the rapid development of information technology, the highway maintenance management has realized informatization.Technology status assessment of highway, traffic flow and other information have accumulated a large amount of data with the highway operation.On the other hand, our road maintenance management is still using traditional maintenance methods--passive maintenance, only if the pavement performance dropped to a level of fairly low will a series of repair measures be carried out. This kind of maintenance dose not make full use of our existing data information and it can just improve the pavement performance a little, what's more, the rate of performance decaying again is also very fast. That results in the extremely low efficiency of highway maintenance and the waste of maintenance funds. According to these problems above, this paper will apply data mining technology to highway maintenance work to fully explore the potential value of technology data that already existed for the purpose of improving the efficiency of highway maintenance[1].

In this paper, the iForest anomaly detection algorithm is used to preprocess the existed data of highway pavement performance. Then, we predict the pavement performance like as pavement damage index (PCI), riding quality index (RQI), road rutting depth index (RDI) and pavement comprehensive quality index (PQI) by using the regression model and the time series model GM (1,1)[2]. Finally, it uses the association rule algorithm and iForest algorithm to analyze the main factors of pavement performance decay and have practical application with existed pavement performance data of

Hubei Province.The pavement performance prediction results and the mining of main factors of highway performance decay can allow us to locate the target that needing maintenance and carry out preventive maintenance.Further,preventive maintenance can prevent or delay the rapid deterioration of highway diseases, effectively extend the road life and save a lot of road maintenance funds.

2.Key technology of preventive maintenance

2.1IForest anomaly detection model

The iForest algorithm consists of a large number of binary trees, called isolation tree and iTree for short, which are the basis of the iForest algorithm and are constructed as follows.

Assuming that the data set has N records,randomly selecting ψ records from the data set as the training samples to build an iTree,usually sampling without replacement.

In the samples, an attribute q and a split value p is randomly selected within the range of all values of the attribute(between the minimum and maximum). Then the samples is divided into two parts by attribute q and split value p. The instance whose q is smaller than p will be divided into left subtree and the instance whose q is greater than or equal to p will be divided into right subtree.That results in a splitting condition and a data set on the left and right subtree. We recursively divide subtree according to the method above,until either:(i)the tree reaches a height limit,(ii)the subtree has only one node,(iii)all data in subtree have the same values.

After the amount of iTree reach to t, iForest is ready to work. We can evaluate the degree of anomaly of test instance x by using the generated isolation forest above, the process of testing as follows: (i) instance x traverses an iTree from the root node until the traversal is terminated at an external node,(ii)make the test data x in each iTree along the corresponding branch conditions to go down to the leaf node and calculate the height $h(x)$ of the end node in the iTree,(iii)calculate the anomaly score of instance x using the following Equation(1).

$$s(x,n) = 2^{-\frac{E(h(x))}{c(n)}} \quad (1)$$

where $E(h(x))$ is the average of $h(x)$ from a collection of isolation trees.$c(n)$ is the average height of the unsuccessful search in binary search tree, and can be calculate by the equation as follows:

$$c(n) = 2H(n-1) - (2(n-1)/n) \quad (2)$$

where $H(i)$ is the harmonic number and it can be estimated by $ln(i) + 0.5772156649$.We can calculate the anomaly score by Equation(1).Then we are able to make the following assessment:

(a) if instances return s very close to 1, then they are definitely anomalies[3],

(b) if instances have s much smaller than 0.5, then they are quite safe to be regarded as normal instances[3],

(c) if all the instances return $s \approx 0.5$, then the entire sample does not really have any distinct anomaly[3].

2.2The model of regression prediction

Through research and practical observations, we find that the attenuation of pavement performance is not linear, and it is very slow in the beginning period of highway operation, but it will begin to drop sharply when the performance drops to a certain value. According to this characteristic, we use the nonlinear regression equation shown in Equation(3) as the model[4].

$$PPI = PPI_0\{1 - exp[-(\tfrac{\alpha}{t})^{\beta}]\} \quad (3)$$

where PPI_0 represents the initial value of the pavement performance (usually 100), α is the parameter which controls the time that the pavement performance decays to 63.2% of the initial value, and β is the parameter which controls the decay nature of the curve, t is the age of highway.Then Equation(3) can be transformed to the format as Equation (4):

$$\beta \ln t + \beta \ln\alpha^{-1} = \ln[\ln(1 - \frac{PPI}{PPI_0})^{-1}]^{-1} \quad (4)$$

Then we can do variable substitution on Equation(4), we replace lnt with variable x and replace β with variable

a,replace $\beta ln\alpha^{-1}$ with b, replace $\ln[\ln(1-\frac{PPI}{PPI_0})^{-1}]^{-1}$ with y, then the Equation(4) can express as $y = ax + b$. According to our historical pavement performance data PPI and t, we can get the corresponding(x, y), so we can estimate the parameters a and b through the least squares method as follows Equations[5] :

$$a = \frac{n\sum_{i=1}^{n} x_i y_i - \sum_{i=1}^{n} x_i \sum_{i=1}^{n} y_i}{n\sum_{i=1}^{n} x_i^2 - (\sum_{i=1}^{n} x_i)^2} \quad (5)$$

$$b = \frac{\sum_{i=1}^{n} x_i^2 \sum_{i=1}^{n} y_i - \sum_{i=1}^{n} x_i \sum_{i=1}^{n} x_i y_i}{n\sum_{i=1}^{n} x_i^2 - (\sum_{i=1}^{n} x_i)^2} \quad (6)$$

After we get the values of a and b, we can estimate the values of α and β according to the equation as follow.

$$\beta = a \quad (7)$$

$$\beta \ln \alpha^{-1} = b \quad (8)$$

then we can use the Equation(3) to estimate the future trend of highway pavement performance.

2.3The model of time series prediction

The factors that influence pavement performance are too many and uncertain[6], so grey theory can be used to predict and analyze the pavement performance. And GM (1,1) is a more commonly used prediction model, with the modeling process as follows[7].

Given an original discrete sequence X_0={x0(1),x0(2),x0(3),...,x0(n)},making a cumulative calculation to generate a new series X_1={x1(1),x1(2),x1(3),...,x1(n)} through the Equation(9),

$$x_1(k)=\sum_{i=1}^{k} x_0(i) \quad (9)$$

then calculating generation sequence of consecutive neighbors of X_1,as Z_1={z1(1),z1(2),...,z1(n)},by the Equation(10) as follow.

$$z_1=0.5x_1(k)+0.5x_1(k-1) \quad (10)$$

Then basic equation of the GM (1,1) model and the albinism differential equation are shown as follows:

$$x_0(k) + az_1(k) = b \quad (11)$$

$$\frac{dX_1}{dt} + aX_1 = b \quad (12)$$

where a, b are the parameters need to be estimated.We can use the least squares method to estimate the value of a and b by Equation(13)(14) as follows.

$$\bar{a}=\frac{\sum_{i=2}^{n} z_1(k)\sum_{i=2}^{n} x_0(k)-(n-1)\sum_{k=2}^{n} z_1(k)x_0(k)}{(n-1)\sum_{i=2}^{n} z_1^2(k)-(\sum_{i=2}^{n} z_1(k))^2} \quad (13)$$

$$\bar{b}=\frac{\sum_{i=2}^{n} z_1^2(k)\sum_{i=2}^{n} x_0(k)-\sum_{k=2}^{n} z_1(k)\sum_{k=2}^{n} z_1(k)x_0(k)}{(n-1)\sum_{i=2}^{n} z_1(k)^2-(\sum_{i=2}^{n} z_1(k))^2} \quad (14)$$

Then solve the differential Equation(12), and take $x_1(0) = x_0(1)$, we can obtain the prediction model of X_1:

$$\overline{x_1}(k+1) = (x_0(1) - \tfrac{b}{a})e^{-ak} + \tfrac{b}{a} \quad (15)$$

According to the relationship between X_1 and X_0,we can estimate the value of $\overline{x_0}(k+1)$ through $\overline{x_1}(k+1)$ using the formula as follow:

$$\overline{x_0}(k+1) = \overline{x_1}(k+1) - \overline{x_1}(k) \quad (16)$$

3.Practical application of data mining in preventive maintenance

Taking the data set from a common trunk highway section of Hubei Province as an example, the regression model and the GM (1,1) model are used to forecast the performance of the pavement. The pavement performance index from 2011 to 2016 as follows.

Table 1. The pavement performance index of one section

index	2011	2012	2013	2014	2015	2016
PQI	95.11	94.93	93.62	89.73	85.54	82.27
PCI	94.30	91.22	88.97	86.63	84.12	80.93
RQI	96.38	95.29	93.55	90.61	86.86	84.17
RDI	95.03	94.11	93.01	90.45	85.86	82.98

3.1 Pavement performance prediction based on regression model

According to the existed data of pavement performance index of highway,we apply the regression model described in subsection 2.2. Because the application of the four index are similar, we only select the pavement condition index (PCI) as an example to predicting.

Using the PCI value of 2011 as the initial value PPI_0,according to the principle of variable substitution about Equation(4),we can get value pairs(x_i,y_i) as follows:{(0,-1.509),(0.693,-1.207),(1.098,-1.301),(1.386, -0.889),(1.609,-0.742)};then calculate the estimates of a and b according to Equation(5) and Equation(6) as $a\approx$ 0.4668, $b\approx$-1.5227;next,we can estimate the values of α and β according to the relation between α, β and a,b.as a result that $\alpha\approx$26.0985,$\beta\approx$0.4668;finally,the predictive values of PCI index can be calculated according to Equation(3).The prediction results of PCI from 2012 to 2016 are shown in Table 2.

Table 2. Predictive values of PCI based on regression model

year	actual values of PCI	predictive values of PCI	differrences
2012	91.22	91.28	-0.06
2013	88.97	88.87	0.10
2014	86.63	86.30	0.33
2015	84.12	83.86	0.26
2016	80.93	81.62	-0.67

The comparison between the predicted value and the actual value of PCI and variation trend based on regression model are shown in Fig 1.

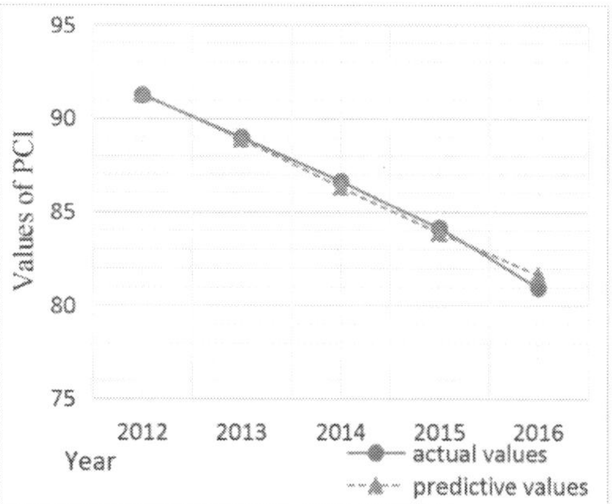

Fig 1. Comparison of regression prediction and actual values

Table 2 and Figure 1 show that the predictive values from regression model is basically consistent with the actual values of pavement condition index, and the max differences do not exceed 1%,futher, the PCI index is declining year by year.According to the current trend， it will be reduced to under 80 in the next year, so we have to take measures to have a preventive maintenance in advance, for the purpose of preventing further deterioration of road performance[8].

3.2Pavement performance prediction based on time series model

In order to ensure the credibility of the prediction results, the time series GM (1,1) model is also used to predict the pavement performance indicators, so that the two predictions can be confirmed mutually. We still used the PCI values in Table 1 as an example to forecast PCI's future tend[9].

According to Table 1, there are original discrete sequences X_0 = {94.30,91.22,88.97,86.63,84.12,80.93}; next,calculating the accumulated sequences of X_0 as a result X_1={94.30,185.52,274.49,361.12,445.24,526.17}; then we can obtain generation sequence of consecutive neighbors of $X1$ as $Z1$={47.15,139.91,230.005,317.805, 403.18,485.705};futher,we can estimate the parameteres of model through Equation(12)(14) as a result a=0.029,b=95.633;finallly,we estimate the predictive values of PCI according to Equation(15)(16) and shown in Table 3.

Table 3. Predictive values and differences based on GM(1,1)

year	actual values of PCI	predictive values of PCI	differrences
2012	91.22	91.51	-0.28
2013	88.97	88.87	0.10
2014	86.63	86.29	0.34
2015	84.12	83.8	0.32
2016	80.93	81.37	-0.44

The comparison between the predicted value and the actual value of PCI and variation trend based on GM(1,1) model are shown in Fig 2.

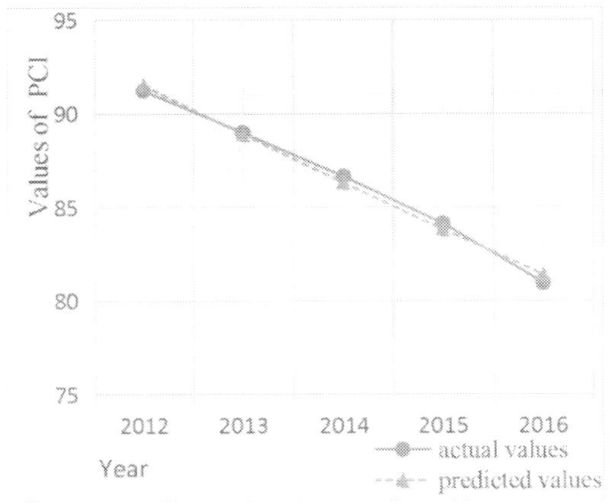

Fig 2. Comparison of GM(1,1) prediction and actual values

It can be seen from Table 2 and Figure 2 that the predictive values from GM(1,1) model is basically consistent with the predictive values from regression model and it also quite similar with the actual value, the max differences do not exceed 1%,futher, the PCI index is declining year by year.According to the current trend, it will be reduced to under 80 in the next year, so we have to take measures to have a preventive maintenance in advance, for the purpose of preventing further deterioration of road performance.we get the same conclusion with regression model.

4. Conclusions

The combination of data mining technology and highway maintenance will bring great surprises to China's highway maintenance and management system, which will change our traditional passive maintenance mode into the period of active preventive maintenance. As a result,we predict the future trend of the highway and the main factors of influencing pavement performance to find out the potential or possible highway distress in advance,so that we can locate the particular disease of highway and be well prepared before the highway pavement perormance reach to a bad status.

The research in this paper shows that the pavement performance predictive values based on the regression model and the time series model is consistent with the actual value and the error is very small. To a certain degree,it reflects the fact that the data mining technology can accurately predict the variation tendency of pavement performance and provide reliable data support for the highway preventive maintenance[10]. In short, Preventive maintenance of highway can greatly improve the efficiency of highway maintenance and save a lot of funds used in highway maintenance.

Acknowledgements

This work has been supported by Project of Hubei Provincial Highway Bureau under Grant Numbers 20141h0288.

References

1. S.L. Hong,Y.H. Zhuang,K. Li,*Data mining technology and engineering practice*. Beijing: China Machine Press,105-136,(2014).
2. *Ministry of Transport of the People's Republic of China*,technical specifications of Highway maintenance (JTG H10-2009).
3. F.T. Liu,K.M. Ting,Z.H. Zhou.*Isolation Forest*.Eighth IEEE. ICDM.(2008).
4. L.J. Sun,*Asphalt pavement structure behavior theory*,People's Publishing Press,380-472,(2005).
5. X.J. Liu,Y.J. Zheng.*Study on multi-index forecasting of asphalt pavement performance based on gray theory*.HighWay,**04**: 233-237, (2012).
6. G. Wang.*The applications of data mining in urban highway traffic*. Chongqing: Chongqing University, (2016).
7. T. Cheng,*The application of data mining in traffic accident*. Harbin Institute of Technology ,(2009).
8. M.G. Tan,*research and application of data mining in Urban highway traffic*. Fudan University, (2010).
9. Q. Ma,*the research and application on Association rules algorithm*, Taiyuan University of Technology,(2007).
10. B.B. Meng,*The application of data mining technology in intelligent traffic detection system*,Wuhan University of Technology ,(2008).

A Method to Detect Hazards in Pipeline Processor

Yihui He[1], *Han* Wan[1,*], *Bo* Jiang[1] and *Xiaopeng* Gao[1]

[1]School of Computer Science and Engineering, Beihang University, Beijing, China
{zy1506229, wanhan, jiangbo, gxp}@buaa.edu.cn

Abstract. In order to improve the throughput of the processors, pipeline technique is widely used to implement the instruction-level parallelism. However, this technique also leads to data hazards which has a great influence on the performance. This paper proposed a method called supply-matching to detect and solve data hazards efficiently. The logic of bypassing and stalling can be easily realized through this method. Furthermore, an RTL description of instructions was also introduced in this paper to reduce resource utilization. The case study was conducted through a five-stage microprocessor based on the PowerPC architecture with different approaches. Experiment results show our method requires less resources and achieves better performance.

1 Introduction

Pipeline technique is widely used in microprocessor design to improve performance. A pipeline processor can execute multiple instructions within a clock cycle. However, hazards arise if pipeline architecture is used, including data hazards, control hazards and structural hazards. Moreover, the problem becomes more complex when pipeline depth increases. Forwarding and stalling are effective solutions to resolve hazards for processors in embedded system. However, the complexity of detecting hazards increases rapidly as the number of instructions increases. In this case, many combination of instructions may lead to hazards in unanticipated way. Performing a highly effective method to resolve huge plenty of hazards oriented from deep pipeline and large instruction set is necessary.

Many studies have proposed some methods in pipeline processor design motivated by these situations mentioned above. Amit Pandey and Yu Qiaoyan used class-based method to detect data hazard [1, 2]. This method divides the whole instruction set into several parts so big problem are broken up into smaller ones. P. Bernardi and D. Boyang proposed a SBST algorithm [3]. Jiajing Lu designed a dynamic scheduling algorithm to improve the pipeline efficiency, which only increases one single-instruction buffer and some combination logic [4]. Also, Schönherr J, Schreiber I, and Fordran E proposed a method using symbolic model checking to detect hazards in pipelined processor [5].

In this work, we aim to find solutions to resolve huge plenty of hazards oriented from deep pipeline and large instruction set. Firstly, we briefly introduce the architecture, controllers and discuss our method of generating datapath according to the RTL description in Section 2. Next, a method called supply-matching is introduced to detect and resolve data hazards completely and efficiently in Section 3. Then we introduce control hazards and structural hazards in Section 4. Finally, we do an experiment to verify our method in Section 5.

2 Preliminary

2.1 Architecture

In this paper, processor adopts the architecture with five pipeline stages, as shown in Fig.1 (The italic words mean the name of the core units. Other words mean the name of interface and data).

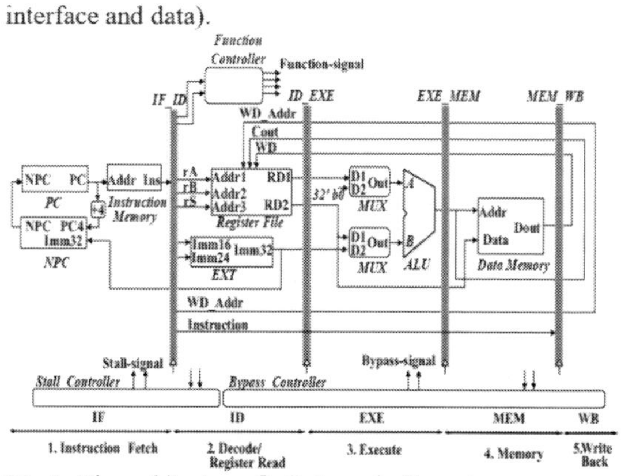

Fig. 1. The architecture of a 5-stage-pipeline microprocessor

IF stage means fetching instructions from program memory. PC, Next PC (NPC) and Instruction Memory (IM) are assigned in IF stage. ID stage is assumed to decode the instruction and read or write register file. Also, the operation of expanding immediate is done in ID stage. EXE stage is used to execute arithmetic operations and logic operations. MEM stage is supposed to read or write memory. And WB stage means the execution result of the

* Corresponding author: wanhan@buaa.edu.cn

instruction will be wrote back to register file. Four pipeline registers are distributed in the pipeline architecture to store data tentatively.

2.2 Notations and datapath

An RTL description of instructions was also introduced in this paper to reduce resource utilization. The rule is defined as follow: (1) A.B means port B of core unit A. (2) X_Y means pipeline register. For example, IF_ID means pipeline register between IF stage and ID stage. (3) B@X_Y means data B locked in pipeline register X_Y. (4) Z[z] means field z of unit Z. For example, Ins@IF_ID[Imm16] means 16-bits immediate operand of instruction stored in IF_ID. (5) A.B → C.D means data transfer from port B of unit A to port D of unit C.

It is more easily to structure datapath by using the RTL description. A datapath is a collection of functional units (such as ALU or multipliers), registers and buses. Follow the RTL rule, the data flow of each instruction is clear and all units have been linked. Then merging the data flow in the vertical direction to remove the repeated data flow of the whole instruction set. Adding MUX unit if the core unit has multiple inputs. The MUX control signal is generated by the controller which described in Section 2.3. Fig.2 shows the result of the method which mentioned above that aimed at ADD, SUBF, STW, LWZ, B instructions of PowerPC instruction set.

IF	ID	EXE
PC.PC → IM.Addr	Ins@IF_ID[11:15] → RF.Addr1	RD1@ID_EXE → ALU.A
PC.PC + 4 → NPC.PC4	RF.RD1 → ID_EXE	RD2@ID_EXE → ALU.B
IM.Ins → IF_ID	Ins@IF_ID[16:20] → RF.Addr2	ALU.Cout → EXE_MEM
NPC.NPC → PC.NPC	RF.RD2 → ID_EXE	Ins@ID_EXE → EXE_MEM
	Ins@IF_ID → ID_EXE	(Ins@ID_EXE[11:15]==0)?32'd0:RD1@ID_EXE → ALU.A
	Ins@IF_ID[6:10] → RF.Addr3	Imm32@ID_EXE → ALU.B
	RF.RS → ID_EXE	RS@ID_EXE → EXE_MEM
	Ins@IF_ID[16:31] → EXT.Imm16	
EXT.Imm32 → NPC.Imm32	Ins@IF_ID[6:29] → EXT.Imm24	

MEM	WB
Cout@EXE_MEM →MEM_WB	Cout@MEM_WB → RF.WD
Ins@EXE_MEM → MEM_WB	Ins@MEM_WB[6:10] → RF.WD_Addr
Cout@EXE_MEM → DM.Addr	
RS@EXE_MEM → DM.Data	
DM.Dout → MEM_WB	Dout@MEM_WB → RF.WD

Fig. 2. RTL description of five instructions

2.3 Controller design

A three-controller architecture which includes function-controller, bypass-controller and stall-controller is used in this paper, as shown in Fig.1. Function-controller is responsible for decoding the instruction and creating the function-signals to indicate what the core units should do. For example, if the instruction is ADD, the function-controller would create signal like DM_Wr to denote whether Data Memory is wrote or not. Also, function-controller determines the data source of core units by creating selecting signal of function-multiplexers. This function-signals are certain since they are directly appeared in RTL description for every instruction. So the final logic of the signals can be created by integrating all instructions' RTL directly.

Bypass-controller is mainly in charge of bypass-signals which choose the right data source as input to bypass-multiplexers. Bypass-controller design is the key to realize bypassing technique. The logic that generates the bypass-signal is more complicated than the logic of function-signals as hazard detection becomes more difficult when the sum of instruction or the pipeline depth increases. This paper will introduce an effective method to detect hazards in the next chapter, so the logic of bypass-signals can be much easier to get. Stall-controller is responsible for pipeline stalling as some hazards situation cannot be resolved by using bypass technique. In this situations, stall-controller generates stall-signals to stall IF stage and ID stage. For pipeline registers, ID_EXE clears all data and IF_ID remains unchanged. Also, PC should remain the value of PC+4 to ensure the correctness of the order in which instructions are executed.

3 Data hazard

3.1 Problem definition

Data hazards are the hazards which are most frequently occurring in pipeline processor. Forwarding and stalling are effective solutions to resolve this problems. However, the complexity of detecting hazards increases rapidly as the number of instructions increases. In this situation, many combination of instructions may lead to hazards in unanticipated way. It is imperative to take completeness detection of data hazards to ensure that all hazards combinations are considered.

3.2 Solution

This paper proposed a method called supply-matching to detect and solve data hazards completely and efficiently. The method can be divided into two steps.

- Build T_{use}-T_{new} matrix of all instructions for specified instruction set. Then all value of T_{use} of registers and all value of T_{new} for the processor can be got by synthesizing all records of T_{use}-T_{new} matrix.
- Build register strategy matrix according to T_{use}-T_{new} matrix. A T_{use}-T_{new} record can be uniquely determined for specific register. According to the supply-matching model which is described in section 3.2.1, any data hazards can be detected and resolved by using formula (1).

3.2.1 Supply-matching model

Data hazards occur when instructions that exhibit data dependence modify data in different stages of a pipeline. Therefore, data hazards detection can be transformed into the detection of relationship between data demand and data supply. In this case, provider is the pipeline register which saved the execution result of last instruction. For example, EXE_MEM pipeline register and MEM_WB pipeline register are the provider for all operation instructions. Demander is the components which need the most up-to-date value saved in provider at present. For example, ALU is the demander for all operation instructions.

Two basic principles are defined in this method.

DOI: 10.1051/matecconf/201713900085

- The stage of the instruction decoding must not be earlier than ID stage whether the centralized decoding mode or distributed decoding mode is adopted.
- Forwarding technology has higher priority when both forwarding and stalling technology can be used to resolve data hazards.

Besides, two parameters called T_{use} and T_{new} are defined.

- T_{use} means the number of clock cycles that a certain functional unit will use the value saved in register after the instruction enters ID stage. T_{use} is a static value and an instruction can have multiple T_{use} according to the number of operands of the instruction. Meanwhile, $T_{use} \geqslant 0$.
- T_{new} means the minimum number of clock cycles that the instructions which at stages after ID stage will produce the result that will be wrote back to registers. T_{new} is a dynamic value. The value reduces by 1 as instruction flows through the pipeline stage and the value will no longer change once the value is 0. So an instruction has different T_{new} at different stage. Meanwhile, $T_{new} \geqslant 0$. The management of T_{new} and T_{use} in pipeline processor is shown in Fig.3.

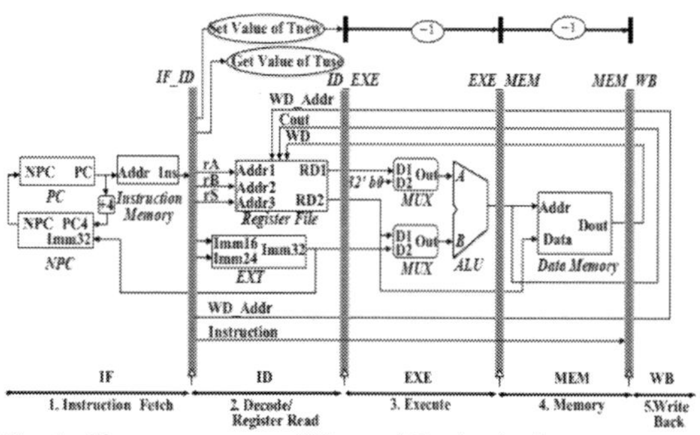

Fig. 3. The management of T_{new} and T_{use} in pipeline processor

Based on the above, solution of data hazards becomes digital. According to the definition of T_{use} and T_{new}, each instruction has a set of digits which represent T_{use} and T_{new}. For a specific instruction set, the comparison between T_{use} and T_{new} of different instructions reflects data dependence. If the $T_{new} > T_{use}$, data hazards can only be solved by stalling as the result writes back too late. If the $T_{new} \leqslant T_{use}$, data hazards can be solved by forwarding. The formula is as followed.

$$\text{Operation} = \begin{cases} \text{Stall} & Tuse < Tnew \\ \text{Forward} & Tuse \geq Tnew \end{cases} \tag{1}$$

3.2.2 T_{use}-T_{new} matrix

The T_{use}-T_{new} matrix is used to find all the value of T_{use} and T_{new} of all instructions. The T_{use}-T_{new} record of an instruction is certain depending on the definition of T_{use} and T_{new} if the pipeline architecture stays the same. Therefore, a T_{use}-T_{new} matrix is certain too if all instructions of specific instruction set are taken into account. In this situation, all the work is focusing on the execution semantics of each instruction and completing the matrix line by line, rather than concentrating on the specific classification of the instructions. This method greatly reduces the complexity of hazards detection even if the number of instructions increases. For example, supposing the instruction set includes instructions of ADD, SUB, ANDI, ORI, LW, SW and BEQ. Concerning the five-stage pipeline, the T_{use}-T_{new} matrix is shown in Table 1.

Table 1. T_{use}-T_{new} matrix

Target / Ins	T_{use}_1	T_{use}_2	T_{new}		
	RS	RT	EXE	MEM	WB
ADD	1	1	1	0	0
SUB	1	1	1	0	0
ANDI	1	Null	1	0	0
ORI	1	Null	1	0	0
LW	1	Null	2	1	0
SW	1	2	Null	Null	Null
BEQ	0	0	Null	Null	Null
	{0,1}	{0,1,2}	{1,2}	{0,1}	{0}

Using the supply-matching model can easily build T_{use}-T_{new} matrix. The procedure also applies to processor which has deeper pipeline stage or larger instruction set.

3.2.3 Register strategy matrix

Register strategy matrix provides the resolution strategy of data hazards for specific register. Based on the T_{use}-T_{new} matrix, a complete strategy matrix can be built for specific register as the T_{use}-T_{new} matrix considers all instructions and all pipeline stages except the stage before ID stage (Basic principle 2 makes the rule. If the number of stages which before ID stage more than one, some additional work should be done to detect and resolve the data hazards for this part). Formula (1) is used to structure strategy matrix. Taking the RT register as example, the result is shown in Table 2.

Table 2. Strategy matrix for RT register

T_{new} / T_{use}	EXE		MEM		WB
	1	2	0	1	0
0	Stall	Stall	Bypass	Stall	Bypass
1	Bypass	Stall	Bypass	Bypass	Bypass
2	Bypass	Bypass	Bypass	Bypass	Bypass

This strategy matrix is correct and can be proofed. We can prove it from follow aspects:

1. According to the definition of T_{use} and T_{new}, $T_{use} \geqslant 0$, $T_{new} \geqslant 0$. So there just three relationship between T_{use} and T_{new}, which are $T_{use} > T_{new}$, $T_{use} < T_{new}$ and $T_{use} = T_{new}$.
2. If $T_{use} > T_{new}$, it indicates that the instruction before has already finished computing the write-back data which the current instruction needs. So bypassing can be established correctly.
3. If $T_{use} < T_{new}$, it means that current instruction cannot get the dependency data from the subsequent stages. Thus stalling is used to ensure processor performs correctly.

4. If $T_{use} = T_{new}$, it shows that current instruction can get related data immediately as the instruction before finish computing at the same time. Forwarding technology can be used here.

A complete stalling control signal can be created by using logic operation OR to integrate all the stalling conditions. In other cases, data hazards can be resolved by forwarding. However, there may have many forwarding sources represented data from multiple pipeline stages when using forwarding technology. In this situation, the priority of forwarding source should be set. This paper adopts a forwarding strategy based on the pipeline priority. The strategy sets the stage which T_{use} stages behind ID stage has the highest priority among the whole pipeline stages and the priority of other stages behind it are decreasing in turn. Based above, just selecting forwarding source which has the highest priority when there are multiple forwarding sources.

4 Control hazard and structural hazard

Control hazards (branch hazards) cause by branch instructions. There are two main techniques to resolve branch hazards, including branch prediction and branch delay slot. Other studies have already done this part efficiently. Please refer to [6, 7, 8] for details.

Structural hazards occur when a part of the processor's hardware is needed by two or more instructions at the same time. The strategy for resolving this hazards is simple. Just stalling the pipeline or copying the basic unit.

5 Experiment

5.1 Framework

The methods mentioned above were adopted to implement a five-stage PowerPC microprocessor which supports 72 instructions. It is significant to realize the whole microprocessor rather than the logic of hazards detection alone because microprocessor cannot work normally if the system only supports the logic of hazards detection.

There are two decoding modes which are centralized decoding and distributed decoding in processor design. In centralized decoding mode, controllers are assigned at ID stage. Then control signals created by controllers transfer through the pipeline. But in distributed decoding mode, controllers are distributed in multiple stages and controllers only create the control signal which related to the core units in the same stage [9, 10]. In this paper, the supply-matching method uses a hybrid decoding mode which means T_{use} uses centralized decoding mode and T_{new} uses distributed decoding mode. However, in order to set up a control experiment, we also implemented the supply-matching method only using centralized decoding mode.

The framework for the experiment as shown in Fig.4. Firstly, instruction set, core units and pipeline architecture were determined according to architecture specification. After that, core unit modules, controllers, datapath, and multiplexers were implemented with Verilog HDL. Meanwhile, hazards in pipeline are resolved based on our method. Finally, binary codes were got from compiling the programs written in C or assembly by GCC. The result of registers after executing every instruction is obtained from QEMU in single-step mode. The correctness of the processor can be determined by comparing the value of registers with the result from QEMU during the simulation. Spartan6-6SLX150FGG484 made by Xilinx is the FPGA used to synthesize and implement with ISE in our experiment.

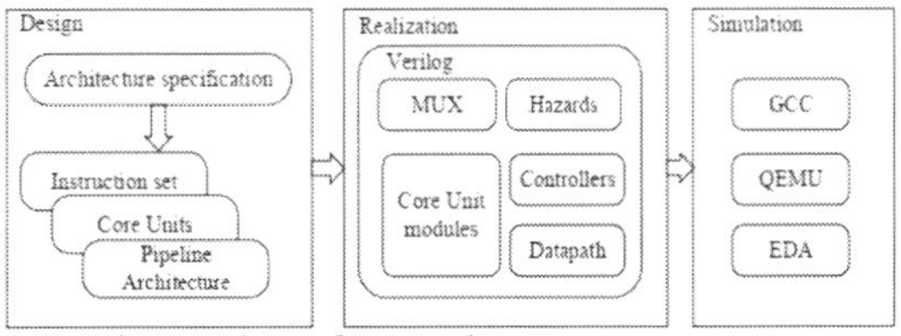

Fig. 4. The experiment framework

5.2 Result and analysis

By comparing the value of registers with the result from QEMU, the correctness of the processor has been proved. This also indicates the correctness of the supply-matching method. Meanwhile, we have compared the synthesize reports of two microprocessors which implemented the supply-matching method using different decoding modes in three aspects: clock frequency (MHz), the number of Flip-Flops (FF), the number of BELs (which includes all basic logic primitives like LUT, MUXCY, etc). The result is shown in Table 3.

Table 3. Comparisons of method implementation using different decoding mode

Parameter / Mode	Clock	FF	BEL
Using centralized decoding mode	8.46	3525	12844
Using hybrid decoding mode	59.46	3481	10380

It is obvious to conclude that our method which using hybrid decoding mode can gain faster clock frequency with less resources.

6 Conclusions

In this paper, an efficient method called supply-matching to completely detect and resolve data hazards has been proposed. The logic of bypassing and stalling can be easily integrated and implemented due to this method. Finally, we conducted extensive experiments based on a state-of-art microprocessor, PowerPC architecture with five stage pipeline. Experiment results proof that our method can achieve faster clock frequency with less resource than the well-known method called stage-decoding.

Acknowledgment

This work has been supported by project of the Professional Group Construction of Beijing Educational Committee.

References

1. Pandey A. Study of data hazard and control hazard resolution techniques in a simulated five stage pipelined RISC processor. *International Conference on Inventive Computation Technologies. IEEE*, 1-4 (2017).
2. Qiao-Yan Y, Peng L, Qing-Dong Y. A data hazard detection method for DSP with heavily compressed instruction set. (2004).
3. Bernardi P, Boyang D, Ciganda L, et al. A functional test algorithm for the register forwarding and pipeline interlocking unit in pipelined microprocessors. *Design and Test Symposium. IEEE*, 1-6 (2014).
4. Lu J, Zhou X, Wang J. A novel dynamic scheduling algorithm of data hazard for embedded processor. *International Conference on Asic. IEEE*, 28-31 (2007).
5. Schönherr J, Schreiber I, Fordran E, et al. Hazard Checking in Pipelined Processor Designs Using Symbolic Model Checking. *Euromicro Conference, 1999. Proceedings. IEEE*, **1**, 75-78 (1999).
6. E. Nurvitadhi, J. C. Hoe, T. Kam, and S. Lu. Automatic pipelining from transactional datapath specifications. *In Design, Automation and Test in Europe, DATE 2010, Dresden, Germany, March 8-12, 2010*, 1001-1004 (2010).
7. E. Nurvitadhi. Automatic pipeline synthesis and formal verification from transactional datapath specifications. *Terapevticheski Arkhiv*, **80(4)**, 73-6 (2008).
8. E. Nurvitadhi, J. C. Hoe, T. Kam, and S. Lu. Automatic pipelining from transactional datapath specifications. *IEEE Trans. on CAD of Integrated Circuits and Systems*, **30(3)**, 441-454 (2011).
9. P. Yiannacouras, J. G. Steffan, and J. Rose. Exploration and customization of fpga-based soft processors. *IEEE Trans. on CAD of Integrated Circuits and Systems*, 266–277 (2007).
10. Yiannacouras, Peter, Jonathan Rose, and J. Gregory Steffan. The microarchitecture of FPGA-based soft processors. *Proceedings of the 2005 international conference on Compilers, architectures and synthesis for embedded systems*, 202-212 (2005).

A High Efficiency RF Power Amplifier Using Linearity-Enhanced Method in 40nm Standard CMOS Process

Wang Cheng [1] ,Ting-Ting Mo*

1 Center for Analog/RF Integrated Circuits (CARFIC)
School of Electronic Information and Electrical Engineering,
Shanghai Jiao Tong University, Shanghai 200240, China
* Email: wangcheng1992@sjtu.edu.cn

Abstract. A two-stage RF CMOS power amplifier with high linearity for WLAN is presented in this paper.The proposed PA consists of a programmable gain amplifier and a high power stage which is composed of a main amplifier with class AB bias and an auxiliary amplifier with class C bias. To improve the linearity ,an integrated diode linearition circuit provides a compensation mechanism for the input capacitance variation of the active devices,improving the linearity from the gain compressing. Moreover ,based on the un-even bias scheme,the power stage can improve linearity and reduce current consumption in the low power region. In order to demonstrate the feasibility of the technique,two types of PAs have been designed.The improved PA at 3.3V supply voltage,has a 37dB of power gain, 1.1 dBm increase of P,8.3% increase of PAE@Pand dB increase of ACPR for 802.11g WLAN,respectively,as compared with the traditional PA.

1.Introduction

Due to low cost and ease of integration complimentary metal oxide semiconductor(COMS) is becoming the technology of choice for most wireless application. The past few years have seen a lot of advancement in CMOS circuit design as well as device technology,which have made it possible to implement RF circuit blocks in CMOS with performance just as good as any other technology. RF blocks such as low noise amplifiers(LNA),mixers and voltage-controlled oscillators(VCO) are readily implemented in CMOS.However, RF power amplifiers (PAs) are a major roadblock in these efforts to create highly integrated system-on-chip,due to both the low breakdown voltage of the transistor and lossy substrate associated CMOS technology.Therefore, the PAs for high data rate wireless communication systems have been required high efficiency and linearity at the back-off power levels to efficiently amplify a multiplexing signal.

Thus,aware of the above problems,a cascode configuration and thick gate-oxide transistors have beenused to eliminate the effects of oxide breakdown voltage and carrier degradation,allowing the use of a larger supply voltage. Moreover, the output-stage is composed of a main amplifier with class AB bias and an auxiliary amplifier with class C bias,which can improve linearity and reduce current consumption in the low power region.In particular,an integrated diode linearization technology provides the compensation mechanism for the input capacitance variation of the active devices to enhance the linearity of the CMOS PAs[1]. A second harmonic tank and optimum gate biasing point are applied for the cancellation of the nonlinear harmonic generated by [2].

Thus, this paper describes some improved PA design techniques to realize high power and high efficiency in a small area using the standard 40nm CMOS technology. In order to demonstrate the improved methods, a conventional PA and an improved one are compared for the P ,PAE.

The paper is organized as follows. section 2 describes the configuration and the operation principle of the proposed un-even biased HPA . The circuit design methodology will be discussed in Section 3, The simulation results are presented in section 4. Finally, section 5 provides a conclusion of this paper.

2.Design and Implementation

Fig.1 shows the simplified block diagram of the proposed PA.It is a two-stage design of a programmable gain amplifier(PGA) for output power adjustment and a high power amplifier(HPA) for high linear operation.In addition, an balun is designed for impedence transforming and for combining the differential signals to a single load.

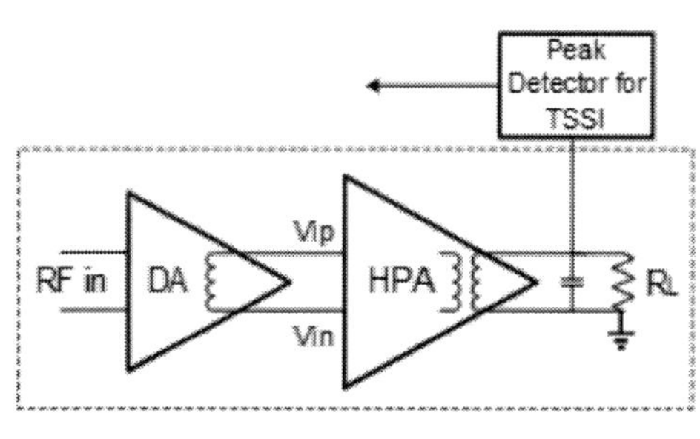

Figure. 1. Simplified block diagram of the two-stage PA design

* Corresponding author:wangcheng1992@sjtu.edu.cn

DOI: 10.1051/matecconf/201713900086

The HPA stage and PGA stage employ a pseudo differential double cascode structure .The differential configuration is less sensitive to the ground inductance and has better stability compared with a single-ended version.The transformer is designed using the HFSS to provide the optimal differential load impedance from a 50 Ohm antenna at the 2.4GHz center frequency based on load-pull technology.

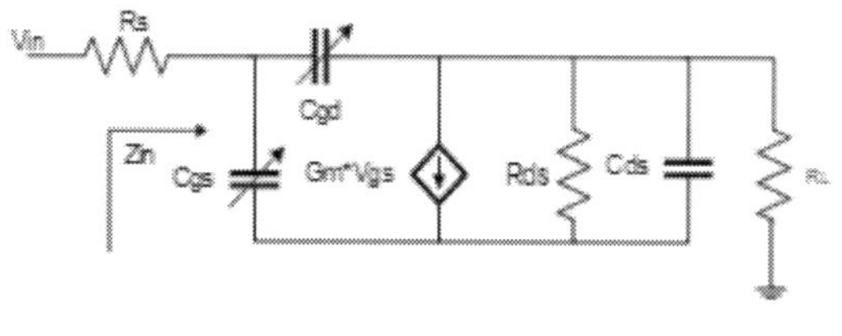

Figure.2. Nonlinearity model of NMOS power amplifier

As shown in Fig2.the two main sources of nonlinearity are Cgs and gm. The input impedance Zin is approximately as follows(1):

$$Z_{in}=\frac{1}{j\omega\omega_{in}}=\frac{1}{j\omega[C_{gs}+(1+Av)C_{gd}]} \quad (1)$$

where Av is the voltage gain of the power amplifier. The relationship of Vgs and Vin is given by the following equation:

$$Vgs=\frac{Z_{in}}{Z_{in}+Z_S}V_{in} \quad (2)$$

As shown in equation(2), Vgs does not linearly increase with Vin and produce significant nonlinearity because Zin is not a constant parameter.

2.1 Operation Principle of Proposed un-even biased PA

Fig. 3 shows the principle of an un-even biased HPA. The input transistors of the output stage are separated into auxiliary and main transistor, and the gates of main and auxiliary transistors are biased at class AB and class C, respectively. The un-even biased method of combining two Gm stages (M1 and M2) with different gate bias is adopted for better linearizition.M1and M2 transistors have different nonlinear characteristic and generate opposite polar third harmonics. According to [5],[6], the third harmonic nonlinearities are cancelled out by opposite phase combination, as shown in Figure.3. Moreover, when the input power is at the low power region, M2 does not consume the current . thus, the average efficiency can be improved in the overall power range.

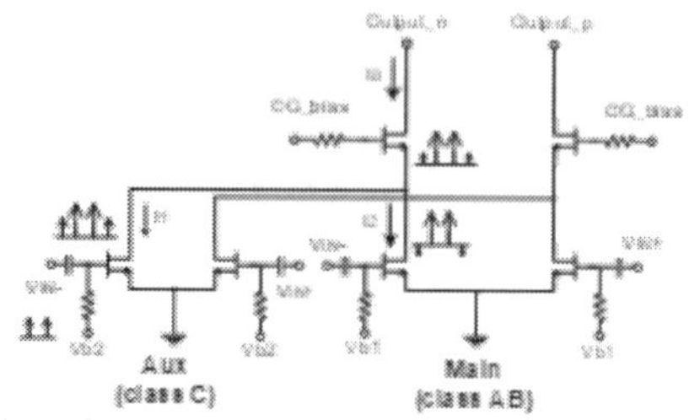

Figure.3 Simplified schematic of HPA and principle of gm3 cancellation

To abtain the optimized bias voltage ,the nonlinearity current(i_{ds}) can be expressed by the third-order Taylor expansion as follows:

$$i_{ds}=g_{m1}\cdot v_{gs}+\frac{g_{m2}}{2!}\cdot v_{gs}^2+\frac{g_{m3}}{3!}\cdot v_{gs}^3+\wedge \quad (3)$$

From the equation(3),the second harmonic and third-order inter-modulation can be expressed by the gm2 and gm3 as follows:

$$i_{ds,2\omega_2-\omega_1}=\frac{3}{4}g_{m3}v_{gs,\omega_2}^2 v_{gs,\omega_1}^* \quad (4)$$

According to [4], the third-order inter-modulation is generated from gm3. Thus, the strong nonlinear behavior can be reduced by setting the gate biasing of main and aux transistors close to A point and B point as indicated in Fig.4.

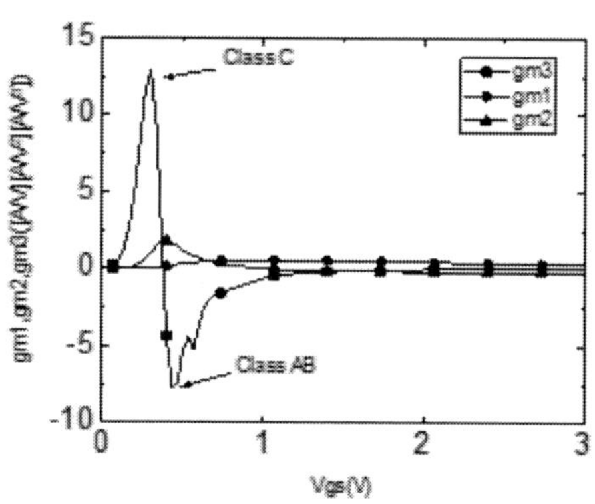

Figure.4. Calculated power expansion cofficients of gm at Vds=3.3V

However, due to the bias voltage of M2 is smaller than that of M1, the size of M1 and M2 should be optimized to obtain the proper linearization effect. though simulating drain current and power gain of the designed PA, respectively. In particularly, the proposed un-even biased technique can improve the average efficiency and linearity simultaneously by using M1 and M2 transistors with a 1:2 size ratio.

2.2The input-capacitance compensation technique for improving linearity

The integrated diode linearization technique is proposed. The advantages of this technique are small size and no extra dc current consumption.
To solve this problem, an integrated diode, linearization circuit is added between the bias and the gate of NMOS, as shown in Fig.5. Since the Vin gradually rises, the Cgs of input NMOS rises while the Cgs of the added transistor decreases. Thus the added transistor provides a cancellation mechanism for the distortion due to the nonlinearity of the input capacitance. As a result, the linearity of the HPA is improved

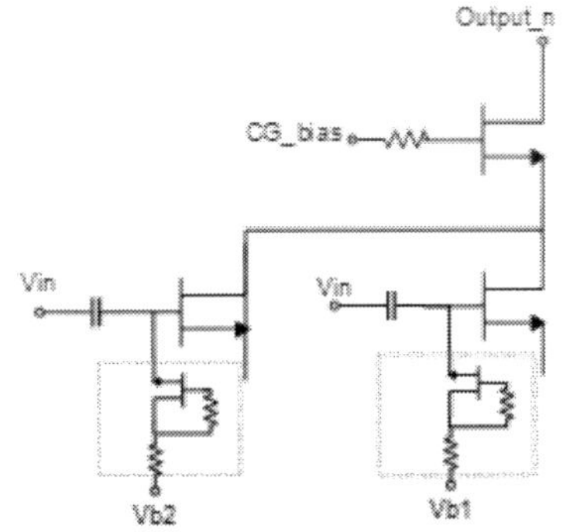

Figure.5. Simplified schematic of input-capacitance compensation technique

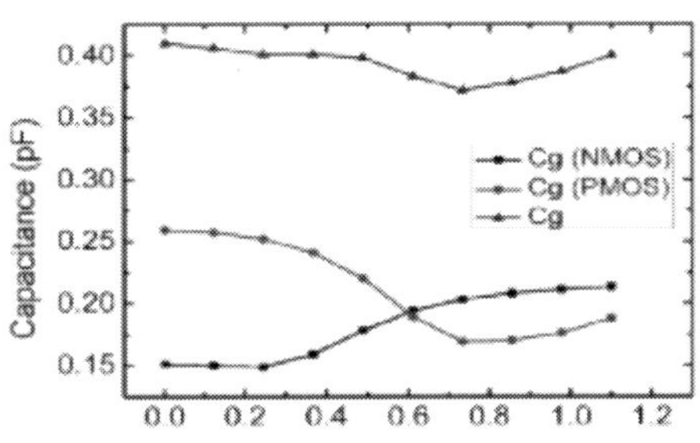

Figure.6. The capacitance of NMOS and PMOS

3. The simulation result

Fig.7 shows an overall schematic of the designed CMOS drive amplifier(DA) and high power amplifier (HPA).

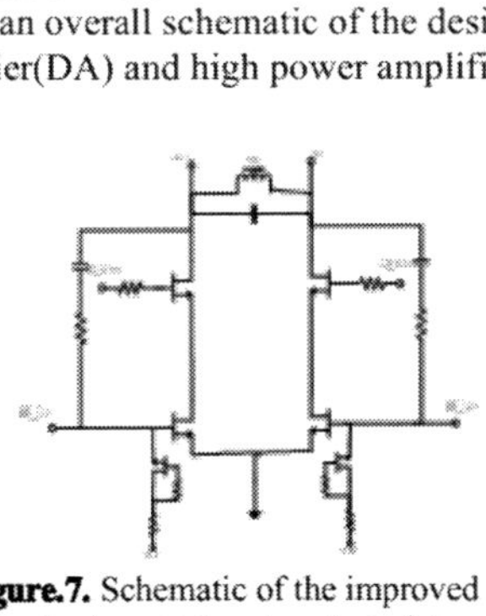

Figure.7. Schematic of the improved DA

This proposed circuit is simulated by the cadence software. To confirm the function of these improved techniques, a comparison of improved and conventional cuircuit in terms of 1dB compressing point and PAE is

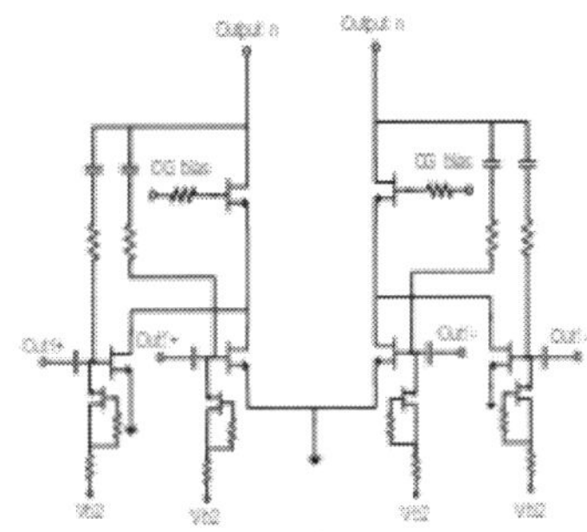

Figure.8. Schematic of the improved PA

shown in Fig 8 and Fig 9.The measurement was performed with a continuous wave single tone at a center frequency of 2.445 GHz. It reveals output P1db of 28.2dBm in the improved PA compared to the conventional PA, in which P1db of 27.1 dBm. This result shows the improved PA has 1.1dB higher linear range. As described in Figure 8, the PAE@P1db of improved PA and conventional PA is 32.4% and 26.1%,respectively.

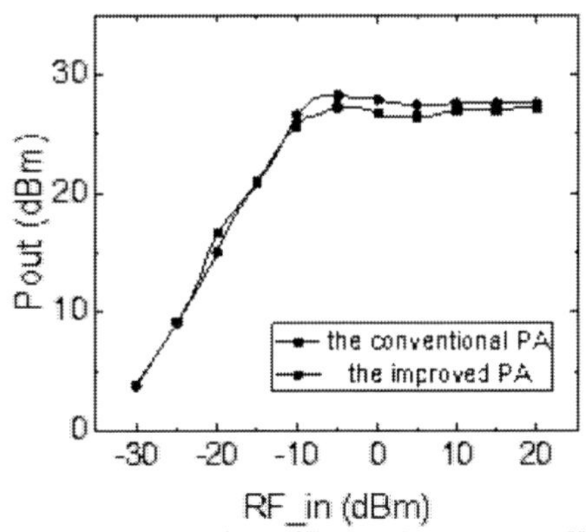

Figue.9. Comparison of Pout of the two types of PA

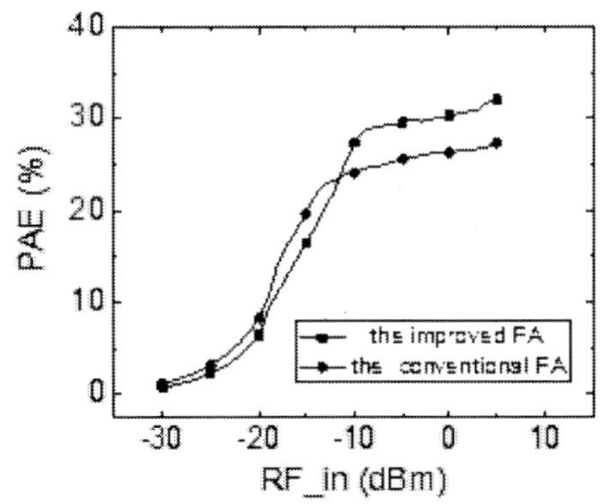

Figure.10. Comparison of PAE of the two types of PA

Finally, to compared the characteristic of the improved and conventional PA, the OFDM signal with a 20MHz channel bandwidth of 802.11g WLAN was applied. Fig.10 shows the measured ACPR at 25 dBm output power, the ACPR of the improved PA is 12.9dB better than the conventional ,which confirms the two improved technique improving the PA linear characteristic. The performance of the improved PA is compared to the conventional PA in Table I. By using the proposed scheme in this work, the PA was achieved better performance.

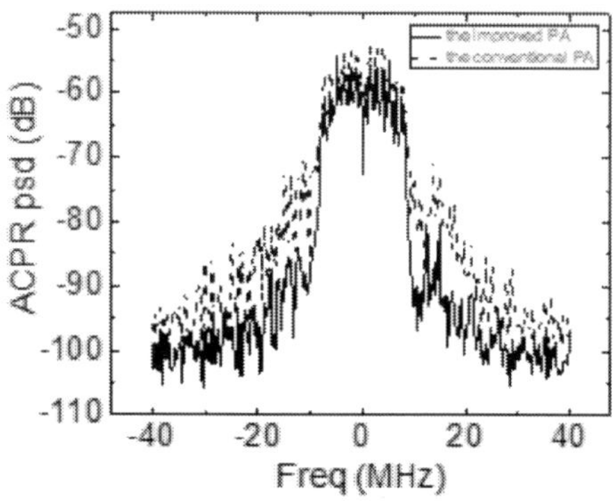

Figure.11. Comparison of ACPR of the two types of PA

Table 1. Comparison of PA

	Conventional PA	Improved PA
Supply voltage	3.3V	3.3V
(dBm)	27.1	28.2
PAE@P1dB	26.1%	32.4%
Power Gain@P1dB	37.5dB	36.4dB
Technology(CMOS)	40nm	40nm

Acknowledgements

This research was supported by Center for Analog and Radio Frequency Integrated Circuit (CARFIC) in Shanghai Jiao Tong University(SIP).A Huawei Company, and National Natural Science Foundation of China(2014zx03001016)

References

[1] S.H.-L.Tu,S.C.-H.Chen,IEEE Conference, p.421-424 (2007)
[2] Jongchan Kang, Jehyung Yoon, Kyoungjoon Min, Daekyu Yu, Joongjin Nam, Youngoo Yang, and Bumman Kim, IEEE Journal of Solid-State Circuits, **vol.41, no.9,** p.1314-1321(2006)
[3] T.Joo, B.Koo, and S.Hong, IEEE, Trans. Microw. Theory Tech, **vol.61, no. 3,** p.1272-1279 (2013)
[4] A.Afsahi and L. E.Larson, IEEE Custom Integr. Circuits Conf. (CICC), p. 19–22 (2010)
[5] Chengzhou Wang, Lawrence E. Larson, and Peter M. Asbeck, IEEE Radio Frequency Integrated Circuits Sumposium.,p. 39-42(2001)

Study of "acoustic tweezer" chip for particles driving

Zhao Weibing Zhu Xinhua Su Yan Wan Ying

Nanjing University of Science and Technology, Nanjing 210094

Abstract: Surface Acoustic Wave (SAW) is one of the sound waves that travel parallel to the surface of an elastic material, and is capable of manipulating microparticles. This SAW device was named acoustic tweezers. Acoustic tweezers possess advantages of energy-efficiency, small size and easy-to-produce. Herein, we developed an acoustic tweezer chip to sort microparticles of different sizes. The chip consists of a pair of interdigital transducer (IDT) and a PDMS microfluidic channel. The IDTs generate the surface acoustic wave which acts on the particles in the microchannel. Finally the sorting of particles can be seen.

Key words: manipulating microparticles, acoustic tweezer, low energy consumption, interdigital transducer

1 Introduction

Acoustic tweezers, the system which use sound as tweezers, are so small and can be placed on a chip to manipulating micron or nanometer sized particles [1, 2]. The energy consumption of acoustic tweezers is less than 500 thousand times [3], which is much less than that of optical tweezers. Due to its small size, acoustic tweezers can be made by standard chip processing technology, so it has the possibility to be widely used.

Acoustic tweezers work through establishing a continuous surface acoustic wave. If the two sound sources are interrelated and each sound source produces the sound of same wavelength, then there will be a point which makes the related sounds cancel each other out, this point can be seen as the trough. Because sound waves have pressure, which can drive very small objects, so micro or nano particles move with the sound waves until the sound waves reach the trough, then the particles will fall to the trough and no longer move. If the sound comes from two parallel acoustic sources, the troughs will form a line or a series of lines. And if the sound source is perpendicular to each other, the troughs will form evenly spaced lines or columns like a chessboard. Also, these particles will also be pushed to places where the sound no longer moves.

Because of its versatility, low energy consumption, simple technology, miniaturization and so on, acoustic tweezers show obvious advantages. Acoustic tweezers will be used as a tool for (biological) tissue engineering, cell research, drug screening and so on.

2 Research on separation principle

The particles in the channel, as shown below, are affected by 4 forces, there are gravity, buoyancy, surface acoustic wave (SAW) and medium viscous force. In the vertical direction, gravity and buoyancy are balanced, so as to the motion of particles, we only need to consider the surface acoustic wave and viscous force in the horizontal direction.

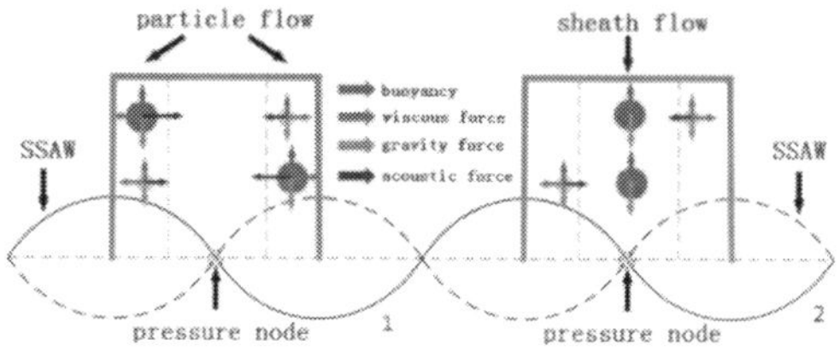

Fig.1 Force analysis of particles in channel

The acoustic radiation force comes from the interaction between particles and sound waves; the expression of the acoustic radiation force of particles by the acoustic surface standing wave produced by interdigital transducer is given below [4-6]:

$$F_r = -\left(\frac{\pi p_0^2 V_p \beta_m}{2\lambda}\right)\phi(\beta,\rho)\sin(2kx) \tag{1}$$

$$\phi = \frac{5\rho_p - 2\rho_m}{2\rho_p + \rho_m} - \frac{\beta_p}{\beta_m} \tag{2}$$

Equation (1) is the calculation of surface acoustic wave(F_r), therefore, without considering the environmental factors, the main factors affecting the surface acoustic wave(F_r) are wavelength(λ), particle volume(V_p)， ϕ function and sin(2kx). Obviously, the larger the wavelength(λ) is, the smaller the surface acoustic wave (F_r) is. When the wavelength(λ), particle and medium is constant, the surface acoustic wave depends on sin(2kx), k represents the wave vector, x is the distance between the particle and the node. According to the picture above, when the particles are between the antinode and the node of the wave, the farther the particles are from the wave node, the larger the surface acoustic wave (F_r) is, so as the other way. So the particles will move towards the pressure node under the effect of surface acoustic wave (F_r), after reaching the pressure node, the force of the particles will disappear, and then stop at the node position.

To different particles in the same position，the surface acoustic wave (F_r) depends on the particle volume （V_p） and ϕ function. Equation (2) is the calculation of ϕ function, where ρ_p and ρ_m correspond to density of the particles and density of medium, β_p and β_m correspond to compressibility of particles and compressibility of medium. So in the same medium, when the compressibility of different particles is very small, the volume of particles(V_p) determines the surface acoustic wave(F_r).

Equation (3) is the calculation of viscous force (F_v) [7], so for particles of different sizes, viscous force (F_v) depends on the radius of the particle(r).

$$F_v = -6\pi\eta r v \tag{3}$$

So when the particle size increases, the increase of surface acoustic wave (F_r) will be more than the increase of viscous force (F_v). In other words, the larger the particle, the larger the resultant force of F_r and F_v, and then particles move faster and more clearly. So the larger particles move faster than the smaller ones, which cause the separation of large and small particles.

3 The design of interdigital transducer

The chip includes interdigital transducer and PDMS microchannel, select the appropriate substrate, the PDMS microchannel is processed in the middle of the substrate and the interdigital transducers are processed on both sides.

3.1 Interdigital transducer(IDT)

In the whole system, interdigital transducers convert electrical energy into acoustic energy, and then acoustic energy is used to drive particles. In the design of interdigital transducers, we consider the following conditions. First of all, the interdigital transducers produce surface acoustic wave standing wave[8], to achieve SAW with a high coupling coefficient, we choose Y+128° X-propagation lithium niobate ($LiNbO_3$) wafer (500 mm thick) as the substrate. For the design of interdigital transducers, acoustic wave length should be 300um, so the width and pitch of the IDTs were 75 mm and 150 mm[9]. The acoustic aperture is 16mm, there are 20 pairs of IDT, and the distance between one pair of IDTs is 8mm. As shown in the following picture:

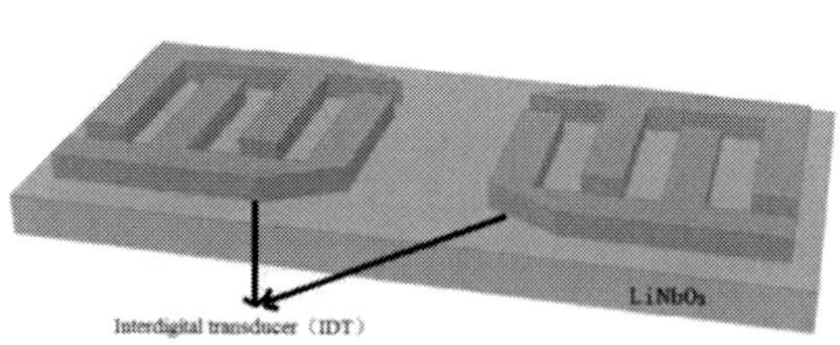

DOI: 10.1051/matecconf/201713900087

Fig.2 Structure of IDTs

The relation between acoustic wave frequency and wave velocity is as follows:

$$f = \frac{v}{\lambda} \tag{4}$$

So our signal frequency is set at 12.6 MHz.

3.2 PDMS microchannel

In the design of microchannel, considering the particles move to the pressure node or antinode under the action of SAW, we set the width of the microchannel half a wavelength (150um), so we can guarantee the separation position.

Fig.3 Structure of microchannel

1,2,3 is the entrance, 1 and 3 are injected the sheath fluid, the inlet 2 is injected liquid with particles, so when the liquid enters the channel together, it forms the laminar flow as shown in the figure below:

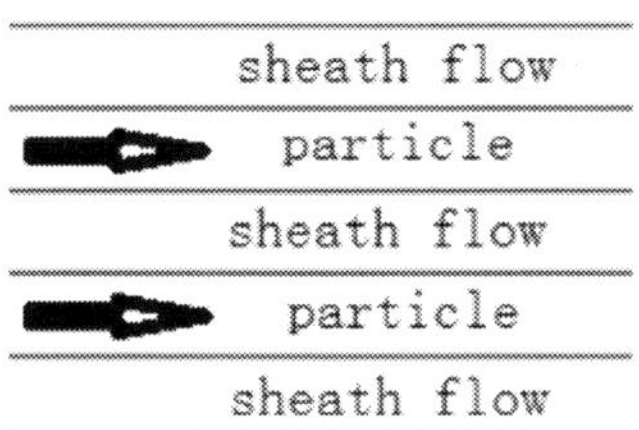

Fig.4 the laminar flow

3.3 Connection of inlets and outlets of micro channel

We choose polytetrafluoroethylene (PTFE) tubes to connect inlets and outlets. The inner diameter of the tube is 0.3mm and the external diameter is 0.6mm. We insert the tubes into PDMS and spread PMDS prepolymer on the connection, then dry it. This ensures the strength of the connection, avoids leakage and other phenomena. The model is as follows:

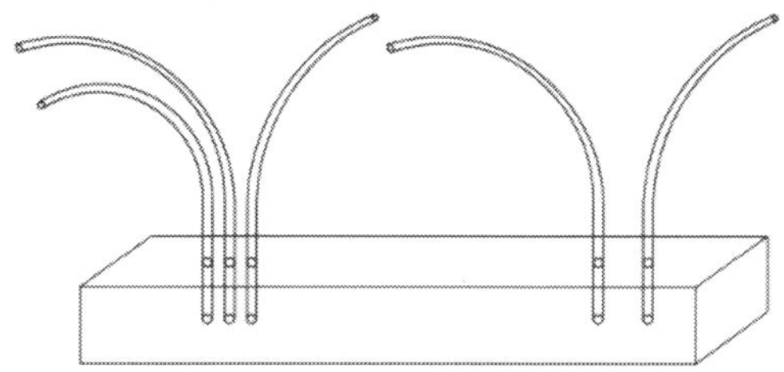

Fig.5 Connection of inlets and outlets of microchannel

4 Results and discussion

4.1 Simulation of microchannel

Through the front design, we basically determine the structure of the micro channel. But considering the flow of the fluid outlet, we need to get concrete results through simulation. So, in order to ensure the normal flow of outlets, we carried out the streamline simulation on export by COMSOL. We set the width of the middle channel 60um, bilateral symmetrical, 45 degrees tilt angles. The result of the simulation is as follows:

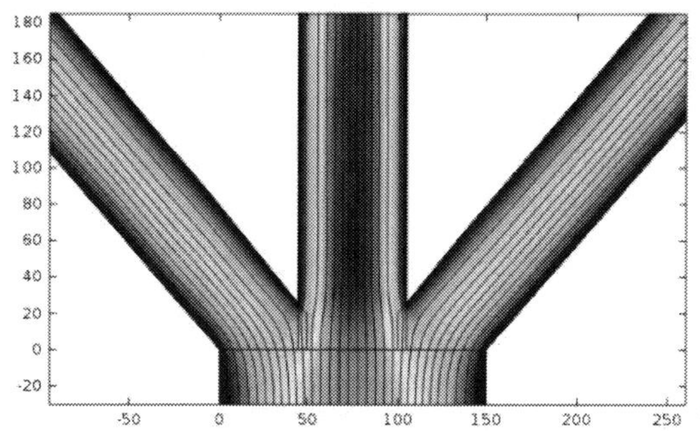

Fig.6 Simulation of outlet streamlines

It can be seen from the figure, there are 9 streamlines on both sides of the channel, 12 streamlines in the midde channel. For the design of the channel, on the one hand, this result can ensure the normal outlet liquid of the middle channel, on the other hand, it can also let fluid production of exports of both sides not be too much.

4.2 experiment of particle manipulation

In the experiment, we inject liquid by injection pump. As to particles, we choose fluorescent polystyrene microspheres with diameter of 5um and non-fluorescent polystyrene microspheres with diameter of 1um, the concentration was 1mg/ml. In order to prevent clusters of

DOI: 10.1051/matecconf/201713900087

particles, we use TWEEN-20 for the pretreatment of polystyrene microspheres. We use deionized water as the sheath and set the rate of flow 2.5ul/min. In the experiment, we clearly see the movement of fluorescent particles in the chip by fluorescence microscope. Also, we observed that fluorescent particles can be separated from non-fluorescent particles under the drive of acoustic tweezers.

5 Summary

Based on the unique advantages of acoustic tweezers, a particle sorting chip is designed in this paper. It produces surface acoustic wave by applying the pumping signal to IDTs, then we can observe the separation of particles by different diameters under the force of SAW. The experiment proves that fluorescent polystyrene microspheres with diameter of 5um and non-fluorescent polystyrene microspheres with diameter of 1um can separate from each other. This design is expected to be used in biomedical research.

References

1. N. Mu, The advent of Multifunctional micro acoustic tweezers[J], Technology development of enterprise:2, 2009, 16 (11): 55-55.
2. J. Xiao, Multifunctional micro acoustic tweezers[J], Technology and market, 2009, 16(11):88-89.
3. C. R. P. Courtney, C. E. M. Demore, H. Wu, et al, Independent trapping and manipulation of microparticles using dexterous acoustic tweezers[J], Applied Physics Letters, 2014, 104(15):492-660.
4. X. Ding, P. Li, S. C. Lin, et al, Surface acoustic wave microfluidics.[J], Lab on A Chip, 2013, 13(18):3626-3649.
5. G. H. Du, Z. M. Zhu, X. F. Gong, Basic of Acoustics[M], Nanjing University Press, 2012.
6. J. C. Cheng, Principle of Acoustics[M], Science Press, 2012.
7. J. Shi, H. Huang, Z. Stratton, et al, Continuous particle separation in a microfluidic channel via standing surface acoustic waves (SSAW)[J], Lab on A Chip, 2009, 9(23):3354-3359.
8. W. Zhou, Experimental Study on Mechanical Properties in Acoustic Streaming[D], Nanjing University, 2016.
9. J. F. Ping, Z. C. Le, Research and Design of Surface Acoustic Wave Interdigital Transducer[J], Application of Photoelectric Technology, 2011, 26(4):49-52.

Expert Database of Unified Parameter Adjustment Based on Long-step Calibration and Cubic Spline Interpolation

Liao Tianfa[*], Xue Jiaxiang
School of Mechanical and Automotive Engineering, South China University of Technology, 510640 Guangzhou,China

Abstract. There are many kinds of aluminum alloys suitable for double-pulse MIG welding, however, the welding parameters are much more and they cannot be easily adjusted. Thus, it is necessary to establish the expert database for the welding source. In order to improve the adjustment performance to precisely control the welding current, this paper proposes a kind of expert database of unified parameter adjustment based on the long-step calibration and cubic spline interpolation to make the welding parameters "continuously adjusted" and "precisely matched" as well as make the test results verified.

1 Introduction

The double-pulse MIG welding involves a large amount of control parameters, moreover, it would directly affect the stability and joint performance in the welding process whether the parameters selected are suitable or not. Therefore, it is an important content for the welding study to establish an expert database including welding conditions, methods and parameters. However, there is much less study on the expert database of double-pulse MIG welding of aluminium alloy at home. This paper establishes an expert database to make the welding parameters "continuously adjusted" and "precisely matched" by a specific algorithm based on the welding process test [1,2], which has a very important significance to study the high-end double-pulse MIG welding source of aluminium alloy.

2 Unified Database

It is the key for double-pulse MIG welding of aluminium alloy that the wire feed rate matches well with the current during the welding process. However, the difficulties to establish the double-pulse MIG welding expert database of aluminium alloy are as follows. (1) There are many kinds of aluminium alloys, as well as different wire feed rates are needed according to the different welding process requirements. So the database establishment would cost much more manpower, time and money. (2) The aluminium alloy has better conductive and heat-conducting properties. When the aluminium alloys are welded with the same welding parameters at different temperatures, the welding results are significantly different. This results in worse repeatability of welding data and worse consistency [3].
The original data are needed to establish the expert database, and they present as some discrete points in a two-dimensional plane. Then these discrete points are processed into a line by the interpolation algorithm or curve fitting approximation, thus obtaining the complete continuous curve. Finally, another parameter is found out to match with some parameter to establish a unified database [4,5].

There are two methods to build the continuous model, namely, interpolation method and curve fitting. The interpolation method requires that the interpolation function curve must go through these discrete points, but the curve fitting only requires that the fitting function curve approaches the known discrete points as far as possible as well as the fitting function curve can accord with the distribution profile of known discrete points [5].

It is assumed that the continuous function is in the form of $f \in [a,b]$. The simple function $\varphi \in \Phi$ in the finite-dimensional linear subspace Φ (or polynomial space P_n) is close to the continuous function f, and two kinds of measurement standards for the approximating function φ are proposed as follows.

(1) Best square approximation

For $f \in [a,b], \varphi^* \in \Phi = Span\{\varphi_0, \varphi_1, \dots, \varphi_n\}$ should be worked out to make

$$\int_a^b [f(x) - \varphi^*(x)]^2 dx = \inf_{\varphi \in \Phi} \int_a^b [f(x) - \varphi(x)]^2 dx$$

Where inf is the infimum sign.

(2) Best uniform approximation

For $f \in C[a,b], p \in p_n = Span\{1, x, x^2, \dots, x^n\}$ should be worked out to make

$$\max_{a \le x \le b} |f(x) - p^*(x)| = \inf_{p \in p_n} [\max_{a \le x \le b} |f(x) - p(x)|]$$

Then the best uniform approximation is also called as Chebyshev approximation.

3 Long-step calibration method

At present, the intelligent welding source with the parameter storage only has a certain automatic control functions, and they store much more welding parameters by a large amount of welding procedure tests. However, in the practical welding process, the existing welding parameters can only conduct the read operations rather than the write operations. This would result in two problems, as follows. (1) The huge number of data are needed, and the welding parameters which are not stored cannot be used for welding. But if all the welding data are stored, the cost is greatly high; (2) There possibility exist the "global risk", for example, if there exists local

[*] liaotianfa@163.com

difference in a type of welding sources, all the welding parameters are invalid so that the welding fails. In order to solve these problems, the new generation of double-pulse MIG welding source of aluminium alloy not only can store the welding parameters but also has the self-learning function.

So this paper proposes a kind of parameter self-learning method based on the long-step calibration, and then the expert database which only has read operation has the dynamically flexible self-adjusting function. This lays a foundation for studying the new generation of double-pulse MIG welding source of aluminium alloy.

The long-step calibration refers to the larger step among the data, and it only calibrates some typical parameters. For example, the scope of welding current for the double-pulse MIG welding of aluminum welding wire with a diameter of 1.6mm is [40A], so the step is set as 40A, then the welding parameter to be calibrated is 40A. If there exists some special parameters such as inflection point, and so on, the refinement calibration is needed near the special points.

4 Cubic spline interpolation algorithm

Equations should be centred and should be numbered with the number on the right-hand side.

In order to overcome the Runge's phenomenon of Lagrange interpolation curve as well as improve the smoothness of Newton interpolation in the nodes, this paper uses the cubic spline interpolation algorithm.

The principle of cubic spline interpolation algorithm is as follows. There are interpolation nodes in the interval [a,b], and they meet $a = x_0 < x_1 < \ldots < x_n < x_{n+1} = b$. $f(x)$ is the function in the interval [a,b], and if $S(x)$ meets the following three conditions:

(1) it goes through (n+1) interpolation points, namely, $S(x_i) = f(x_i) = y_i (i = 0,1,2,\ldots,n)$;

(2) it is second-order continuous, namely, $S(x) \in C^2[a,b]$;

(3) It is cubic subsection, namely, namely, the function $S(x)$ is all the cubic polynomial in each interval $[a, x_1], [x_1, x_2], \ldots, [x_n, x_{n+1}]$. $f(x)$ as well as its first and second derivatives are continuous for $x_i (i = 1,2,\ldots,n)$, then $S(x)$ is called as cubic spline interpolation function of $f(x)$.

Let $M_i = S''(x_i)$, then after twice integrals for the interpolation condition $S(x_i) = y_i, (i = 0,1,2,\ldots,n)$ and $S''(x) = \frac{x - x_i}{-h_{i-1}} M_{i-1} + \frac{x - x_{i-1}}{h_{i-1}} M_i (i = 1,2,\ldots,n)$, the expression of cubic spline interpolation function $S(x_i) = y_i, (i = 0,1,2,\ldots,n)$ could be obtained.

$$S(x) = \frac{M_{i-1}}{6h_{i-1}}(x_i - x)^3 + \frac{M_i}{6h_{i-1}}(x - x_{i-1})^3 + \left[\frac{y_{i-1}}{h_{i-1}} - \frac{M_{i-1}}{6}h_{i-1}\right](x_i - x) + \left[\frac{y_{i-1}}{h_{i-1}} - \frac{M_{i-1}}{6}h_{i-1}\right](x - x_{i-1})$$

$$, \quad x_{i-1} \le x \le x_i, (i = 1,2,\ldots n)$$

Where $h_{i-1} = x_i - x_{i-1}$.

Because the function $S(x)$ has the second derivative in the sampling point x_i , the natural boundary conditions $S''(x_0) = y'_0 = 0$ and $S'(x_n) = y''_n = 0$ are increased according to the cubic natural spline interpolation.

The equation set is obtained as follows.

$$\begin{bmatrix} 2 & u_0 & & & \\ \lambda_1 & 2 & u_1 & & \\ & \ddots & \ddots & \ddots & \\ & & \lambda_{n-1} & 2 & u_{n-1} \\ & & & \lambda_n & 2 \end{bmatrix} \circ \begin{bmatrix} M_0 \\ M_1 \\ \cdots \\ M_{n-1} \\ M_n \end{bmatrix} = \begin{bmatrix} U_0 \\ U_1 \\ \cdots \\ U_{n-1} \\ U_n \end{bmatrix}$$

Where $\begin{cases} u_i = \dfrac{h_{i+1}}{h_i + h_{i+1}} \\ \lambda_i = 1 - u \\ U_i = \dfrac{6}{h_i + h_{i+1}}\left[\dfrac{y_{i+1} - y_i}{h_{i+1}} - \dfrac{y_1 + y_{i-1}}{h_i}\right] \end{cases}$;

$i = 1,2,\ldots,n-1$; $u_{o=0}$; $U_0 = 0$; $U_n = 0$; $\lambda_n = 0$。

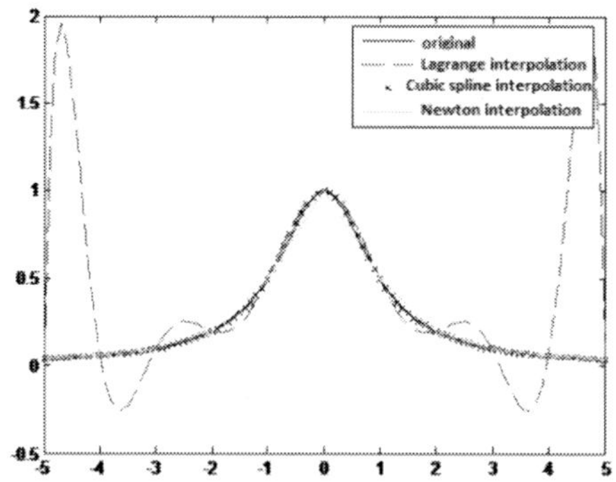

(a) The graph of cubic spline interpolation curve

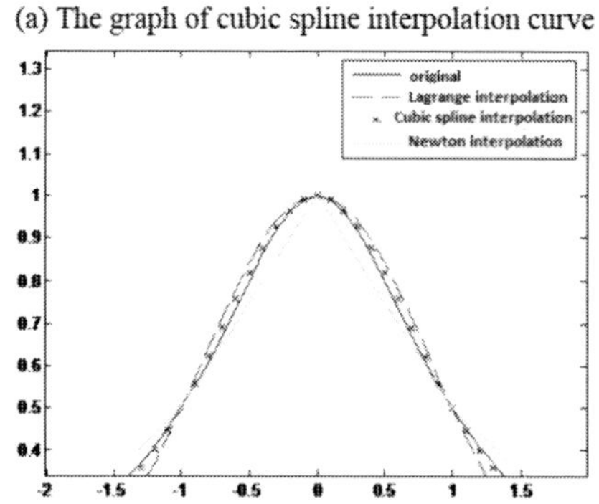

(b) The partial enlargement graph of Node

Fig.1 Graph of cubic spline interpolation algorithm

Figure 2 is the flow chart of cubic spline interpolation algorithm.

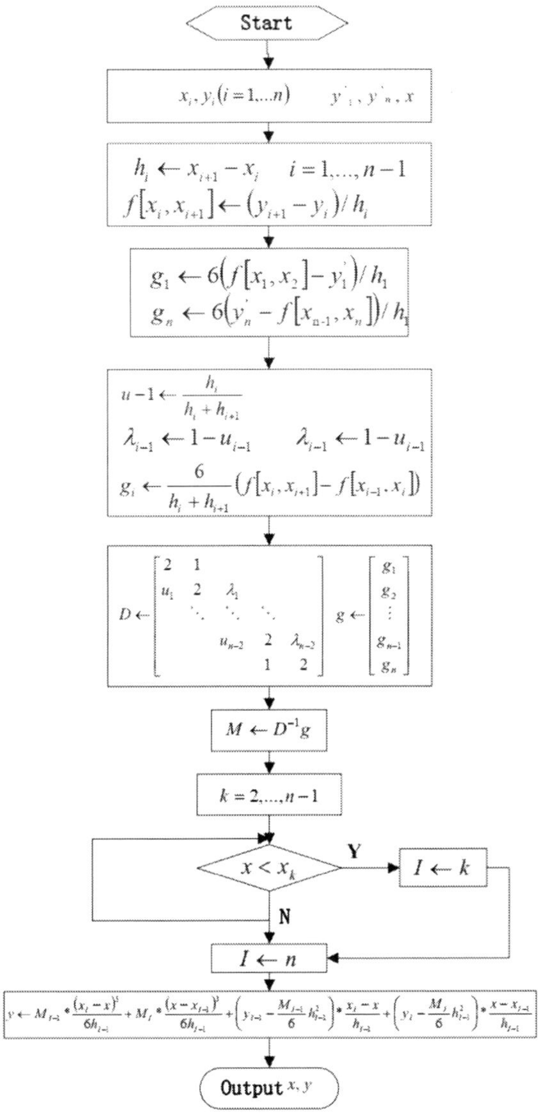

Fig.2 Flow chart of cubic spline interpolation algorithm

The equation set is solved to figure out $M_i(i=0,1,2,\ldots,n)$, then the solved $M_i(i=0,1,2,\ldots,n)$ is substituted into $S(x)$. Thus the cubic natural spline interpolation function in each subarea $[x_{i-1},x_i](i=1,2,\ldots,n)$ can be obtained.

During the process of actual programming, if the values of $x_i, y_i(i=1,\ldots,n)$ and y', y'_n, x are known, when the cubic spline interpolation algorithm is used to figure out x and y values, the solution is shown as the flow diagram in Figure 1.

The Matlab software is used to obtain the cubic natural spline interpolation curve of function $f(x)=\frac{1}{1+x^2}, x\in[-5,5]$, as shown in figure 1. From Figure 1, the cubic natural spline interpolation algorithm well overcomes the Runge's phenomenon of Lagrange interpolation algorithm, roughness in the nodes of Newton interpolation algorithm, and so on, the interpolation curve drawn is close to the original curve, as well as the interpolation curve is very smooth.

5 Result

After the text edit has been completed, the paper is ready for the template. Duplicate the template file by using the Save As command, and use the naming convention prescribed by your conference for the name of your paper. In this newly created file, highlight all of the contents and import your prepared text file. You are now ready to style your paper.

Figure 3 is the expert database established by the cubic natural spline interpolation algorithm, and the parameters selected from the expert database are used for the welding line of aluminium alloy by the double-pulse MIG welding. Moreover, the original instantaneous current and voltage waveforms are collected by the dynamic wavelet analyzer in the welding process, and then the U-I diagram is analyzed.

Figure 3a and b are the original instantaneous current and voltage waveforms in the double-pulse MIG welding of aluminium alloy, respectively. Clearly, the pulse waveforms dispaly the sine-wave modulation, and the instantaneous current and voltage waveforms can directly reflect the stability of welding process.

Figure 3c shows the U-I diagram after the wavelet filter. The edge line group keeps in order, their distribution is very concentrated, and the U-I diagram presents the high repeatability. Thus the welding quality and welding process control are stable and reliable.

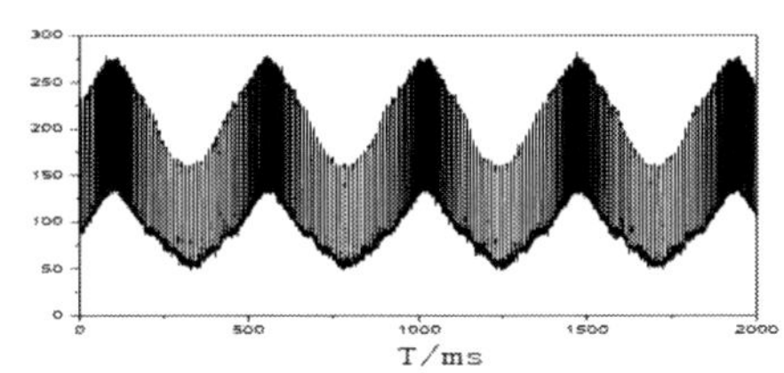

(a) The instantaneous waveform of original welding current

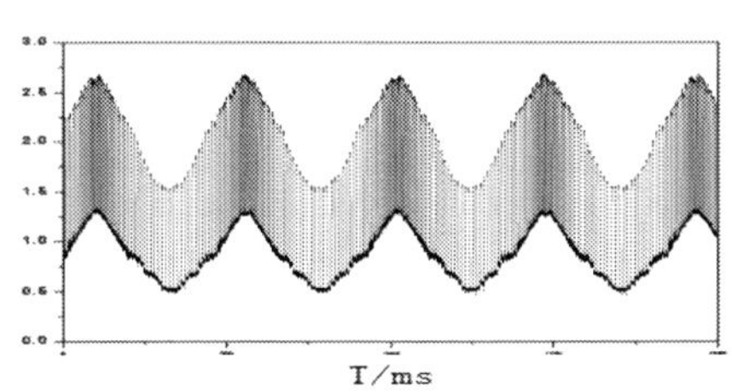

(b) The instantaneous waveform of original welding voltage

DOI: 10.1051/matecconf/201713900088

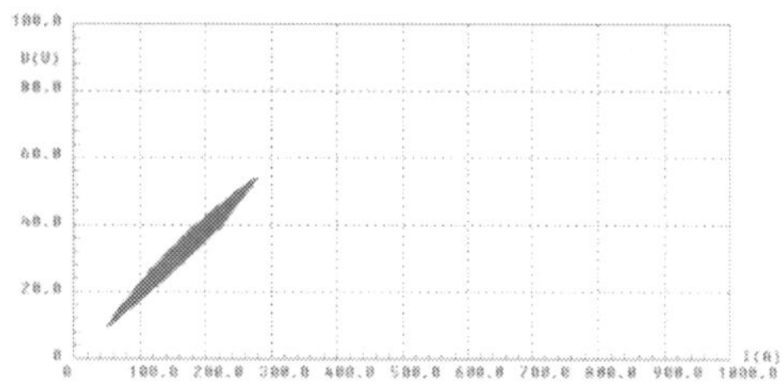

(c) The instantaneous U-I figure of welding

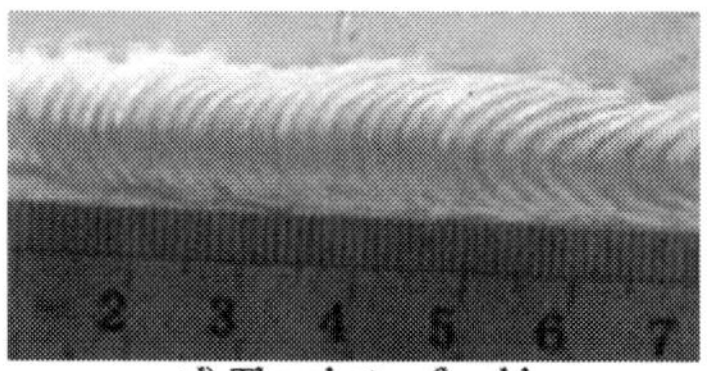

(d) The photo of weld

Fig.3 Welding test results based on cubic spline interpolation algorithm

Figure 3d shows the welding lines photos, in which the welding lines with clear fish-scale pattern corresponding to the sine waveform are displayed.

It is suggested by the test process and the results that the welding parameters obtained by the cubic spline interpolation algorithm are stable and reliable in the double-pulse MIG welding of aluminium alloy, and the welding lines are in fish-scale patterns. So these parameters can meet the welding requirement of aluminium alloy.

6 Conclusion

The cubic spline interpolation algorithm can well overcome the disadvantages such as Runge's phenomenon, roughness in the nodes, and so on. In this paper, the expert database of unified parameter adjustment based on the long-step calibration and cubic spline interpolation is proposed. The long-step calibration can make the welding parameters "continuously adjusted", the cubic spline interpolation can make the welding parameters"precisely matched", and thus the welding source has the unified adjustment function. This greatly decreases the repetitive process adjustment steps and meets the actual demand in the welding process.

References

1. Yao Ping, Xue Jiaxiang, Meng Wanjun, etc. The process parameters of aluminium alloy double pulse weld forming [J].The influence of welding journal, 2009, 30(3):69-72.
2. Xi Shengfeng. The research and application of data fitting program [J].Journal of Zhejiang University of Technology, 2003, 31(5):587-590.
3. Zheng Xianyi, Yao Yangxin, Lei Xiuren, etc. The application of numerical analysis [M].Guangzhou: South China University of Technology Press, 2008.
4. Chen Wenliu, Wang Ziyang. Fitting of cubic spline interpolation in engineering application [J]. Journal of China normal university: natural science edition, 2004, 38(4):418-422.
5. Yang Y U, Yuan J H, Qian J, et. al. Cubic Spline Interpolation with New Boundary Conditions[J]. Computer Engineering & Software, 2016.

MATEC Web of Conferences **139**, 00089 (2017)
ICMITE 2017 DOI: 10.1051/matecconf/201713900089

Dynamics of a two-species model with transitions in population interactions

Yaping Wang* and *Jingyu* Xiao

School of Science, Tianjin University of Technology and Education Tianjin 300222, P. R. China

Abstract. A two-species model with variable population interactions and harvesting on one of the species is studied. Existence and stability of equilibria and existence of periodic solutions are established, existence of some bifurcation phenomena are analytically and numerically studied, explicit threshold values are computed to determine the kind of interaction (mutualism, competition, host-parasite) between the two species. A brief discussion on the influence of the harvesting function on the dynamics of the model is also included.

1 Introduction

The population models with static interactions, like competition, predator-prey, mutualism, is widely and well studied in [2], [6], [7], [11], [14]. However, there are some evidences [5], [12], [13], [14] show that it not adequate the case as the interspecific relationship may change depending on population densities and environmental parameters and so on. Several studies [5], [4], [15] have focus on condition interaction which is represented via nonconstant functions. Jorge Rebaza [10] studied the model incorporates linear α-functions representing the variable interaction between the species, and also considered harvesting on one of species by an external agent. For the first time, Hopf bifurcations and periodic solutions are found in this kind of models. The aim of this paper is to consider more general and rational α-functions representing the variable interaction between the species, and discuss the main influence of harvesting on the dynamics of the model.

In the present paper, we consider the following model:

$$\begin{cases} \dfrac{du}{dt} = u(r_1 - k_1 u) + \alpha_1(u,v)uv - \dfrac{hu}{e+u} \\ \dfrac{dv}{dt} = v(r_2 - k_2 v) + \alpha_2(u,v)uv \end{cases} \quad (1)$$

Where $u(t) \geq 0$, $u(t) \geq 0$ are the population densities of species 1 and 2 respectively, and all the parameters are positive. The constants r_1, r_2 are the corresponding intrinsic growth rates, k_1 and k_2 are the coefficients of intraspecific competition. The function $H(u) = \frac{hu}{e+u}$ is the harvesting function which is increasing smoothly with the size u of the population, h is the rate-of-harvesting limit, and e is the number of species 1 it take to reach one half of the maximum harvesting rate. We consider the functions α_1 , α_2 represent the interactions between both species, where

$$\alpha_1(u,v) = \frac{a_1 - b_1 v}{c_1 + d_1 v} \ , \ \alpha_2(u,v) = \frac{a_2 - b_2 u}{c_2 + d_2 u} \quad (2)$$

The parameters a_i , b_i , c_i , d_i ($i = 1,2$) representing changes in environmental conditions. So α_1 , α_2 can take positive or negative values, and thus the interspecific interactions are not fixed but vary with the environmental parameters and the system state, and can take positive or negative values. Classically, one considers α_1 , α_2 as fixed parameters, and their signs determine the interspecific interaction. In particular, if $c_1 = c_2 = 1$, and $d_1 = d_2 = 0$, then the model reduces to the one studied in [1], [10].

The paper is organized as follows. In section 2, we studied the existence and local stability of the equilibrium. The existence of certain bifurcations both analytically and numerically is resolved in section 3. In section 4, the impact of harvesting on one of the species is discussed.

2 Local stability of equilibria

The system (1), (2) determines the following equilibria:

$$P_0 = (0,0), P_1 = (u_\pm, 0), P_2 = \left(0, \frac{r_2}{k_2}\right), P_3 = (u_3, v_3).$$

Here

$$u_\pm = \frac{(r_1 - k_1 e) \pm \sqrt{(r_1 - k_1 e)^2 + 4k_1(r_1 e - h)}}{2k_1} = \frac{(r_1 - k_1 e) \pm \sqrt{(r_1 + k_1 e)^2 - 4k_1 h}}{2k_1}. \quad (3)$$

Thus $u_\pm$ is real if $h \leq \frac{(r_1 + k_1 e)^2}{4k_1}$. This means that if the rate-of-harvesting limit above the threshold value in (3), there are no equilibria with species 1 only.

From (3) we can see that: (a) if $(r_1 e - h) < 0$ and $r_1 - k_1 e > 0$, then $u_\pm > 0$; (b) if $(r_1 e - h) > 0$, then only

* Corresponding author: ypwang1001@163.com

$u_+ > 0$; (c) if $(r_1e-h) < 0$ and $r_1 - k_1e < 0$, then $u_\pm < 0$; So if h is above the value r_1e , then at the same time the intrinsic growth rate r_1 of species 1, has to be above the value k_1e , to have a well defined equilibrium $(u_\pm,0)$.

For $u_3 > 0$, $v_3 > 0$, the coexistence equilibrium $P_3 = (u_3, v_3)$ is solution of

$$\begin{cases} r_1 - k_1u + \dfrac{a_1 - b_1v}{c_1 + d_1v}v - \dfrac{h}{e+u} = 0, \\ r_2 - k_2v + \dfrac{a_2 - b_2u}{c_2 + d_2u} = 0. \end{cases} \quad (4)$$

From the second nullcline in (4), we get v is a convex curve about u . So we can establish conditions under which there is at least one solution of (4), with $u_3 > 0$, $v_3 > 0$. For instance, denote by v0 the v-intercept of the first nullcline in (4), and by v^* its minimum value. Let $\frac{r_2}{k_2} < v^* < \frac{a_1}{b_1}$ and $h \geq r_1e$. For this nullcline to intersect at least once with the second nullcline in (4), we impose a negative slope on the first nullcline at the v-intercept, that is $h > r_1e^2$.

The general Jacobian A_0 of (1):

$$\begin{pmatrix} r_1 - 2k_1u + \dfrac{a_1 - b_1v}{c_1 + d_1v}v - \dfrac{h}{(e+u)^2} & \dfrac{(a_1c_1 - 2b_1c_1v - b_1d_1v^2)}{(c_1 + d_1v)^2}u \\ \dfrac{(a_2c_2 - 2b_2c_2u - b_2d_2u^2)}{(c_2 + d_2u)^2}v & r_2 - 2k_2v + \dfrac{a_2 - b_2u}{c_2 + d_2u}u \end{pmatrix}$$

In the theorem below, we denote

$$h^* = r_1e\left[1 + \frac{k_2a_2 - b_1r_2}{k_2(k_2c_1 + d_1r_2)}\right].$$

Theorem 2.1. The boundary equilibrium points of (1) have the following local stability properties:

(a) $P_0 = (0,0)$ is an unstable node if $h < r_1e$ and a saddle if $h > r_1e$. P_0 is never a focus nor of center type.

(b) $P_1 = (u_\pm,0)$ is a saddle or a stable/unstable node according to

1. if $h = r_1e$ and

$$\frac{r_1 - k_1e}{k_1} < \frac{(r_2d_2 + a_2) + \sqrt{(r_2d_2 + a_2)^2 + 4b_2r_2c_2}}{2b_2} \text{ then}$$

P_1 is a saddle, and a stable node if the inequality is reversed.

2. if $h < r_1e$; $r_1 - k_1e > r_2d_2 + a_2$ and $\frac{b_2r_2c_2}{r_1e-h} \leq k_1 \leq b_2$; then P_1 is a stable node, and a saddle if all the inequalities but the first one are reversed.

3. if $h > r_1e$, $b_2 < k_1 < \frac{1}{2}r_1$ and $h + 4k_1e < r_1 - k_1e < r_2d_2 + a_2$ then P_1 is a saddle, and an unstable node if $h > r_1e$, $k_1 \geq \max\left\{b_2, \frac{r_1}{2}\right\}$, $r_1 - k_1e < \min\{h - r_1e, r_2d_2 + a_2\}$.

4. P_1 is never a focus nor of center type.

(c) $P_2 = \left(0, \frac{r_2}{k_2}\right)$ is a saddle if $h < h^*$, and a stable node if $h > h^*$. P_2 is never a focus nor of center type.

Proof : (a) At P_0 the eigenvalues are $\lambda_1 = r_1 - \frac{h}{e}$ and $\lambda_2 = r_2$, which proves the claim. (b) At P_1 the eigenvalues of Jacobian are $\lambda_1 = \frac{e(h-r_1e)+2(h-r_1e)u - r_1u^2}{(e+u)^2}$, $\lambda_2 = \frac{-b_2u^2 + (r_2d_2 + a_2)u + r_2c_2}{c_2 + d_2u}$. Therefore, if $h \leq r_1e$, $\lambda_1 < 0$ .when $h > r_1e$, $\lambda_1 > 0$ if $0 < u < u'$ and $\lambda_1 < 0$ if $u > u'$,where $u' = \frac{1}{r_1}\left[h - r_1e + \sqrt{(h-r_1e)^2 + er_1(h-r_1e)}\right]$ is the positive root of the parabola. By comparing u' with $u = u_\pm$ in (3), we see that if $r_1 < 2k_1$ and $r_1 - k_1e < h - r_1e$ then $\lambda_1 > 0$,if $r_1 > 2k_1$ and $r_1 - k_1e > h - r_1e$ then $\lambda_1 < 0$. In the same way, we get if $h \geq r_1e$, $b_2 < k_1$ and $r_1 - k_1e < (r_2d_2 + a_2)$ then $\lambda_2 > 0$. If $h < r_1e$, $b_2 \leq k_1 \leq \frac{b_2r_2c_2}{r_1e-h}$ and $r_1 - k_1e < r_2d_2 + a_2$, then $\lambda_2 > 0$ and $\lambda_2 < 0$ if all the inequalities are reversed but the first one. Combining the inequalities above, the first three statements follow. In addition, P_1 is never a focus nor of center type, as none of the eigenvalues is complex.

(c). At P_2 , the eigenvalues of the Jacobian are $\lambda_1 = r_1 + \frac{(k_2a_2 - b_1r_2)r_2}{k_2(k_2c_1 + d_1r_2)} - \frac{h}{e}$, and $\lambda_2 = -r_2 < 0$ .Thus it is clear that $\lambda_1 > 0$ if $h < h^*$ and $\lambda_1 < 0$ if $h > h^*$ .Clearly, P_2 is never a focus nor of center type.

Interspecific relationships

For the coexistence equilibrium $P_3 = (u_3, v_3)$, with $u_3 > 0$, $v_3 > 0$, the sign the slope of the nullcline (4) at the corresponding equilibrium point indicates the type of the relationship: denoting with m_1, m_2 the slope of u and v nullcline in (4) respectively. We have mutualism if $m_1 > 0$ and $m_2 > 0$, competition if $m_1 < 0$ and $m_2 < 0$, host-parasite if $m_1 > 0$ and $m_2 < 0$ and $m_1 < 0$ and $m_2 > 0$. From (4), we get

$$m_1 = \frac{[k_1(e+u)^2 - h](c_1 + d_1v)^2}{(e+u)^2(a_1c_1 - 2b_1c_1v - b_1d_1v^2)},$$

$$m_2 = \frac{(a_2c_2 - 2b_2c_2u - b_2d_2u^2)}{(c_2 + d_2u)^2k_2}$$

For convenience we denote $D_1 = a_1c_1 - 2b_1c_1v - b_1d_1v^2$

$D_2 = a_2c_2 - 2b_2c_2u - b_2d_2u^2$, $E_1 = \sqrt{4b_1c_1(b_1c_1 + a_1d_1)}$, $E_2 = \sqrt{4b_2c_2(b_2c_2 + a_2d_2)}$, and $u^+ = \dfrac{E_1}{2b_1d_1} - \dfrac{c_1}{d_1}$, $v^+ = \dfrac{E_2}{2b_2d_2} - \dfrac{c_2}{d_2}$. Then if $e > \sqrt{\dfrac{h}{k_1}}$, $k_1(e+u)^2 - h > 0$, there is mutualism at P_3 when $D_1 > 0$ and $D_2 > 0$, that is when $0 < v < v^+$ and $0 < u < u^+$; there is competition at P_3 when $u > u^+$ and $v > v^+$ and host-parasite at P_3 when either $u > u^+$ and $0 < v < v^+$, or $0 < u < u^+$ and $v > v^+$. If instead $e < \sqrt{\dfrac{h}{k_1}}$, then $k_1(e+u)^2 - h > 0$ if $u > u_1$ ($u_1 = \dfrac{-k_1e + \sqrt{k_1h}}{k_1}$ and $k_1(e+u)^2 - h < 0$ if $0 < u < u_1$. Then coexistence equilibrium P_3 represents :

Mutualism when $e > \sqrt{\dfrac{h}{k_1}}$, $0 < v < v^+$, $0 < u < u^+$ or

$e < \sqrt{\dfrac{h}{k_1}}$, $v > v^+$, $0 < u < u^+$.

Competition when $e > \sqrt{\dfrac{h}{k_1}}$, $v > v^+$, $u > u^+$ or

$e < \sqrt{\dfrac{h}{k_1}}$, $0 < v < v^+$, $u > u^+$.

Host-parasite when $e > \sqrt{\dfrac{h}{k_1}}$, $0 < u < u^+$, $v > v^+$

($u > u^+$, $0 < v < v^+$) or

$e < \sqrt{\dfrac{h}{k_1}}$, $0 < u < u^+$, $0 < v < v^+$,

($u > u^+$, $v > v^+$).

Remark. These inequalities provide with threshold value that determine the kind of relationship between the species.

In the theorem below, we denote

$$B = \frac{k_1(e+u)^2 - h}{(e+u)^2}$$

Theorem 2.2 The coexistence equilibrium $P_3 = (u_3, v_3)$ of (1) has the following stability properties:

(a) It is locally asymptotically stable if $B > 0$ and $\dfrac{m_2}{m_1} < 1$ or $B > 0$, $Bu + k_2v > 0$, $\dfrac{m_2}{m_1} > 1$ and $(Bu + k_2v)^2 \le 4Bk_2\left(1 - \dfrac{m_2}{m_1}\right)uv$.

(b) It is locally asymptotically unstable if $B < 0$, $Bu + k_2v < 0$, $\dfrac{m_2}{m_1} > 1$ and $(Bu + k_2v)^2 \le 4Bk_2\left(1 - \dfrac{m_2}{m_1}\right)uv$.

(c). It is a saddle if $B > 0$, $\dfrac{m_2}{m_1} > 1$, or $B < 0$, $\dfrac{m_2}{m_1} < 1$ and $Bu + k_2v \ne 0$.

(d). It is a focus if $(Bu + k_2v)^2 \le 4Bk_2\left(1 - \dfrac{m_2}{m_1}\right)uv$. It is stable if $Bu + k_2v > 0$ and unstable if the last inequality is reversed.

(e). It is of center-type if $Bu + k_2v = 0$ and

$$\left[\left(h < k_1(e+u)^2\right)\frac{m_2}{m_1} < 1 \text{ or } \left(h > k_1(e+u)^2\right)\frac{m_2}{m_1} > 1\right].$$

Proof. Using the nullcline equations (4), Jacobian at P_3 can be written as

$$J(u_3, v_3) = \begin{pmatrix} -Bu & \dfrac{Bu}{m_1} \\ m_2k_2v & -k_2v \end{pmatrix}.$$

Thus, the eigenvalues are

$$\lambda_{1,2} = \frac{-(bu + k_2v) \pm \sqrt{(bu + k_2v)^2 - 4Bk_2\left(1 - \dfrac{m_2}{m_1}\right)uv}}{2}.$$

Through analysis symbols of λ_1 and λ_2, we confirmed the above conclusions.

3 Bifurcations

Equations should be centred and should be numbered with the number on the right-hand side.

Theorem 3.1. The system (1) undergoes a transcritical bifurcation at $P_0 = (0,0)$ when $h = r_1e$ and $r_1 \ne k_1e$, and at $P_2 = \left(0, \dfrac{r_2}{k_2}\right)$ when $h = h^*$, $r_1 > k_1e$ and $k_2 > k_2^* = \dfrac{b_1c_1r_2 + \sqrt{b_1^2c_1^2r_2^2 + a_1c_1b_1d_1r_2^2}}{a_1c_1}$.

Proof. At P_0, the Jacobian has one zero eigenvalue if and only if $h = r_1e$. When $\lambda_1 = 0$, an eigenvector is $x^T = [1; 0]$, and a left eigenvector is also $y^T = [1; 0]$. If we write the right hand side of (1) as $f(u,v)$, then one gets $y^T f_h(P_0) = 0$, $y^T Df_h(P_0)x = -\dfrac{1}{e} \ne 0$, $y^T D^2 f_h(P_0)(x,x) = \dfrac{2}{e}(r_1 - k_1e) \ne 0$,Thus , using Sotomayor's theorem one concludes that under the condition above, the system (1) undergoes a transcritical bifurcation at P_0.

Similarly, at P_2:

$\lambda_1 = r_1 + \dfrac{(k_2a_2 - b_1r_2)r_2}{k_2(k_2c_1 + d_1r_2)} - \dfrac{h}{e}$, when $h = h^*$, $\lambda_1 = 0$.

$\lambda_2 = -r_2 < 0$. An eigenvector is $x^T = [1, \frac{a_2}{c_2 k_2}]$, and a left eigenvector is also $y^T = [1;\ 0]$. So $y^T f_h(P_2) = 0$,

$y^T Df_h(P_2)x = -\frac{1}{e} \neq 0$,

$$y^T D^2 f_h(P_2)(x,x) = -2k_1 + \frac{2h}{e^2} + \frac{2a_2}{c_2 k_2} \frac{k_2^2 a_1 c_1 - 2b_1 c_1 r_2 k_2 - b_1 d_1 r_2^2}{(c_1 k_2 + d_1 r_2)^2}.$$

With $h = h^*$, we get

$$y^T D^2 f_h(P_2)(x,x) = -2k_1 + \frac{2r_1}{e} + \frac{2r_2}{ek_2} \frac{a_1 k_2 - b_1 r_2}{(c_1 k_2 + d_1 r_2)} + \frac{2a_2}{c_2 k_2} \frac{k_2^2 a_1 c_1 - 2b_1 c_1 r_2 k_2 - b_1 d_1 r_2^2}{(c_1 k_2 + d_1 r_2)^2}$$

Thus, $y^T D^2 f_h(P_2)(x,x) \neq 0$ when $-2k_1 + \frac{2r_1}{e} > 0$, $a_1 k_2 - b_1 r_2 > 0$, and $k_2^2 a_1 c_1 - 2b_1 c_1 r_2 k_2 - b_1 d_1 r_2^2 > 0$ which is true when $k_2 > k_2^*$ and $r_1 > k_1 e$.

Hence, under the given condition, there is a transcritical bifurcation at P_2.

Theorem 3.2. The system (1) undergoes a saddle-node bifurcation at

1. $P_1 = (u, 0)$, if $h = k_1(e+u)^2$ and $r_2 + \frac{(a_2 - b_2 u)u}{c_2 + d_2 u} \neq 0$.
2. $P_3 = (u_3, v_3)$, if any of these cases hold:
   (a). $h = k_1(e+u)^2$, and [($m_2 > 0,\ v > v^+$), or ($m_2 < 0,\ 0 < v < v^+$)].
   (b). $h < k_1(e+u)^2$, $m_1 = m_2 > 0$.
   (c). $h > k_1(e+u)^2$, $m_1 < 0$, $Bu + k_2 v \neq 0$.

Proof. 1. The general Jacobian at P_1 reduces to

$$J = \begin{pmatrix} -Bu & \frac{Bu}{m_1} \\ 0 & r_2 + \frac{u(a_2 - b_2 u)}{c_2 + d_2 u} \end{pmatrix}.$$

there is a simple zero eigenvalue when $B = 0$ and $r_2 + \frac{u(a_2 - b_2 u)}{c_2 + d_2 u} \neq 0$. $B = 0$ if and only if $h = k_1(e+u)^2$. For the zero eigenvalue, one can choose right and left eigenvectors as $x^T = y^T = [1;\ 0]$, and therefore $y^T f_h(P_1) = -\frac{u}{e+u} \neq 0$, $y^T D^2 f_h(P_1)(x,x) =$

$$(1,0)\begin{pmatrix} -\frac{2k_1 u}{e+u} \\ * \end{pmatrix} = \frac{-2k_1 u}{e+u} \neq 0$$

.Thus, sotomayor's theorem implies the existence of a saddle-node bifurcation.

2.(a).If $B = 0$, the general Jacobian at P_3 reduces to

$$J(u_3, v_3) = \begin{pmatrix} 0 & 0 \\ m_2 k_2 v & -k_2 v \end{pmatrix}$$

An eigenvector of the zero eigenvalue is $x^T = [1, m_2]$, and a left eigenvector is $y^T = [1;\ 0]$. Thus $y^T f_h(P_3) = -\frac{u}{e+u} \neq 0$ and $y^T D^2 f_h(P_3)(x,x) = -\frac{-2k_1 u}{e+u} + \frac{2m_2 D_1}{(c_1 + d_1 v)^3} - \frac{2m_2^2 c_1 u(b_1 c_1 + a_1 d_1)}{(c_1 + d_1 v)^3}$ .Then, if $m_2 > 0$, $v > v^+$ or ($m_2 < 0$, $0 < v < v^+$), $y^T D^2 f_h(P_3)(x,x) \neq 0$, if $m_2 > 0$ and $0 < v < v^+$,or ($m_2 < 0$, $v > v^+$), we also need $\frac{c_1 + d_1 v}{u} < \frac{m_2 c_1(b_1 c_1 + a_1 d_1)}{D_1}$.

(b)(c). If $B > 0$ and $m_1 = m_2 > 0$, or ($B < 0$ and $m_1 = m_2 > 0$, $Bu + k_2 v \neq 0$), an eigenvector of the zero eigenvalue is $x^T = [1, m_1]$, a left eigenvector is

$y^T = \left[1, \frac{Bu}{m_1 k_2 v}\right]$. Thus $y^T f_h(P_3) = -\frac{u}{e+u} \neq 0$,

$$y^T D^2 f_h(P_3)(x,x) = -\frac{hu}{(e+u)^3} + \frac{2m_2^2 c_1 u(b_1 c_1 + a_1 d_1)}{(c_1 + d_1 v)^3} - \frac{2Bc_2 uv(b_2 c_2 + a_2 d_2)}{m_1 k_2 v(c_1 + d_1 v)^3}$$

.Then, if $B > 0$ and $m_1 = m_2 > 0$, or ($B < 0$ and $m_1 = m_2 > 0$, $Bu + k_2 v \neq 0$), $y^T D^2 f_h(P_3)(x,x) \neq 0$. Sotomayor's theorem implies the existence of a saddle-node bifurcation for (1).

Theorem 3.3. The system (1) undergoes Hopf bifurcate-ions when P_3 satisfies conditions to be of center-type, and $v_3 \neq \frac{a}{2b}$.

Proof. Though shifting the equilibrium P_3 to the origin, we obtain the corresponding power series expansions:

$$u = a_{10}u + a_{01}v + a_{20}u^2 + a_{11}uv + a_{02}v^2 + a_{30}u^3 + a_{12}uv^2 + O_1\left(|(u,v)|^4\right)$$

$$v = b_{10}u + b_{01}v + b_{20}u^2 + b_{11}uv + b_{02}v^2 + a_{21}u^2 v + O_2\left(|(u,v)|^4\right)$$

where a_{ij} and b_{ij} are the coefficients of the expansions. From the assumptions, we know $D = a_{10} - b_{01} > 0$ and $a_{10} + b_{01} = 0$. The Lyapunov number is given by

$$\sigma = \frac{-3\pi R}{2a_{01}D^{\frac{3}{2}}} = \frac{-3b\pi R}{2u_3(a - 2bv_3)D^{\frac{3}{2}}}.$$

where R is a long polynomial expression in the terms of the parameters. There are parameter values so that P_3 satisfies Theorem 2.2(e). For instance, $r_1 = 0.6$; $r_2 = 0.4$; $k_1 = 0.337$; $k_2 = 0.5$; $a_1 = d_1 = d_2 = 1$; $b_1 = 1.3$; $a_2 = 0.4$; $b_2 = 1.5$; c1 = c2 = 0; h = 0.574805; e = 0.5, for which _ > 0, and turn $r_1 = 0.9$; $r_2 = 0.6$; $b_1 = 0.3$ and others keep constant, we have σ < 0. Thus , the system(1) experiments supercritical and subcritical Hopf bifurcations.

4 The impact of harvesting.

Here we discuss the influence of the harvesting function on the dynamics of the model. We compare our model (1) with the classical model with no harvesting.

Boundary equilibria. If the model with no harvesting, the boundary equilibria are $P_0 = (0,0)$, $P_1 = \left(\frac{r_1}{k_1}, 0\right)$, $P_2 = \left(0, \frac{r_2}{k_2}\right)$. Comparing with our model, there are not one but up to two equilibria $(u_\pm, 0)$ on the horizontal axis when $h \le \frac{(r_1 + k_1 e)^2}{4k_1}$. With no harvesting, P_0 is always an unstable node, and P_1 is a saddle or a stable node. Whereas with harvesting, P_0 can also be a saddle, P_1 can also be unstable node.

Coexistence equilibria. The coexistence equilibria were found to be nodes, saddles or focus. By including harvesting, the coexistence equilibria in our model can be nodes, focus, or even of center type.

Bifurcations. The system with no harvesting undergoes some bifurcations: transcritical, saddle-node bifurcations. Our model has even richer dynamics, not only those bifurcations, but more importantly, Hopf bifurcations.

5 Acknowledgements.

The authors gratefully acknowledge the support of the National Natural Science Fundamental of China(NNSFC) though Grant Nos. 11526157, 11526154 , Tianjin Research Program of Application Foundation and Advanced Technology though Grant Nos. 15JCQNJC01600 and Science and Technology Development Foundation of Tianjin University of Technology and Education though Nos. KJ1431. The authors declare that there is no conflict of interest regarding the publication of this paper.

References

1. B. Zhang, Z.Zhang, Z.Li and Y.Tao, Stability analysis of a twospecies model with transitions between population interactions, J.Theoret.Biol., 248(2007), 145-153.
2. B.Leard and J.Rebaza, Analysis of predator-prey models with continuous threshold harvesting, Appl.Math.Math.Comput, 217(2011), 5265-5278.
3. Z.Li, M.Han and F.D.Chen, Influence of feedback controls on an autonomous Lotka-Volterra competitive system with infinite delays, Nonliear Anal.RWA, 14(2013), 402-413.
4. T.Lara and J.Rebaza, Dynamics of transitions in population interactions, Nonliear Anal.RWA, 13(2012), 1268-1277.
5. M.J.Hernadez, Spatiotemporal dynamics in variable population interaction with density-dependent interaction coefficients, Ecol.Modeling, 214(2008), 3-16.
6. L.Ji and C.Wu, Qualitive analysis of a predator-prey model with constant-rate prey harvesting incorporating a constant prey refuge, Nonl. Anal: Real World Appl., 11(2010), 2285–2295.
7. K.Lan and C.Zhu, Phase portraits of predator-prey systems with harvesting rates, Disc.Cont. Dynam. Syst, 32(2012), 901-933.
8. S.Lefschetz, Stability of Nonlinear Control Systems, Mathematical Biosciences and Engineering, Academic Press, New York, (1965).
9. J.Hale, Ordinary Differential Euqations, J.Wiley and Sons (1980).
10. J.Rebaza, Bifurcations and periodic orbits in variable population interactions, Communication on pure and Applied analysis, 12(2013), 2997-3012.
11. C,Ji and D.Jiang, Persistence and non-persistence of a mutualism system with stochastic perturbation,, Disc. Cont.Dynam, 32(2012), 867-889.
12. M.Hernandez and I.Barradas, Variation in the outcome of population interactions: bifurcations and catastrophes , Math. Biol, 46(2003), 671-594.
13. T. Peschak, Currents of Contrast: Life in South Africa's Two Oceans, Struik Publ,(2006).
14. J.H, Cushman and J.F.Addicot, Conditional interzctions in antplant-herbivore mutualism, in:C.R.Huxley,D.F.Cutler, Ant Plant Interactions, Oxford University Press,(1991),92-103.
15. J.Holland and D.DeAngelis, Consumer-resource theory predicts dynamic transitions between outcomes of iinteractions, Ecol.Letters, 12(2009), 1357-1366.

DOI: 10.1051/matecconf/201713900090

Characteristic Analysis of Ambient Air Pollutants during Summer Season in Shenyang City

Cheng Liu, Zhonghu Yuan, Yingkui Du, Chunhe Shi, Yutong Gao, Xiaoxuan Qi, Nan Wang, and Xiaowei Han*

Shenyang University, School of Information Engineering, Key Lab of Equipment Manufacturing Comprehensive Automation.

110044 Shenyang, China

Abstract. Shenyang City was the political, economic and cultural centre of the northeast China and was also a heavily polluted industrial city. The understanding of the distribution characteristics of air pollutants concentration and changes was still lacked. To reduce the impact of disturbing factors such as firecrackers in the traditional Chinese festivals, the observation period of the monitoring data was selected from 17th May to 21st July in 2016. The data sources were picked from eight national monitoring stations and the daily average concentration of the main air pollutants that included $PM_{2.5}$, PM_{10}, SO_2, NO_2 and O_3. The overall analysis of distribution characteristics of the air pollutant was shown that the principal pollutants with highest frequency were O_3 and PM_{10}, the average proportion was 74.1% and 20.8% respectively.

1 Page layout

Shenyang was a representative manufacturing industrial city with industry with heavy air pollution. The city was the political, economic and cultural center of the northeastern China and was located in the heart of the Northeast Asian economic circle and the Bohai Sea economic circle.

Shenyang City located in central region of Liaoning Province. The terrain was mainly consisted of flat plain. The average altitude was about 50 meters and the east region was lower than the west. The north of Shenyang City was close to the south Horqin desert of the Inner Mongolia Autonomous Province and was threatened by sandstorm. Shenyang belongs to the northern temperate monsoon-influenced semi-humid continental climate.

The industry of Shenyang City was mainly included the traditional heavy industry such as coal, metallurgy, power, chemical and machinery manufacturing. With the process of urbanization and the rapid economic development, serious environmental problems especially the compound air pollution were arisen[1]. As the Shenyang Environmental Status communique in 2014[2] shown, the annual concentration of PM_{10} and $PM_{2.5}$ has reached 124μg/m³ and 74μg/m³ respectively that were the 3.1 and 4.9 times of the Chinese ambient air quality standard respectively. The annual good quality air quality was only 191 days. The atmospheric environment qualities of Shenyang City were ranked the last 15th and the last 11th during the 74 key cities in 2013 and 2014 years.

The understanding of the distribution characteristics of air pollutant concentration and changes was still lacked[1,7]. This main contribution was to analyze the characteristics of primary air pollutions of the $PM_{2.5}$, PM_{10}, SO_2, NO_2 and O_3 during summer season in Shenyang that was driven by the data of daily average concentration released by Shenyang environmental monitoring stations.

2 Data and methods

The data referred to in this paper were the monitoring data of air pollution during the heating season of 8 Shenyang environmental air quality monitoring stations. The time period was from 17 May to 21 July in 2016, 65 days total.

The data mainly include the daily average concentrations of $PM_{2.5}$, PM_{10}, SO_2, NO_2 and O_3 of the eight national monitoring stations which were released by Dongling Road, Jing Shen Street, Xiaoheyan Road, Culture Road, Xinxiu Street, Hunnan East Road, Lingdong Street and Taiyuan Street.

The R language was selected as the tool to realize the analysis and graphing the results. The obvious advantages of the R language were that it was open source and free with Built-in utility function with a large number of dedicated for statisticians.

Statistical methods were utilized for the analysis of the contribution characteristics of ambient air pollution. The occurrence proportion was defined as the proportion of the number of days of air pollutant occurred in the total number of days of the sample data to evaluate the occurrence frequency of the primary air pollutions. The

* Corresponding author: Yingkui Du, syu_dyk@163.com

distribution characteristics of the main air pollutants were plotted and shown by the boxplot tools of the R language.

3 Results and discussion

3.1. Overall distribution characteristics

The daily average concentrations of $PM_{2.5}$, PM_{10}, SO_2, NO_2 and O_3 air pollutants in Shenyang summer period were shown in Fig 1. The proportions of the number of days of the primary air pollutants occurred in the total number of days were calculated at the beginning and the distribution of the main pollutants as the primary pollutant was analyzed.

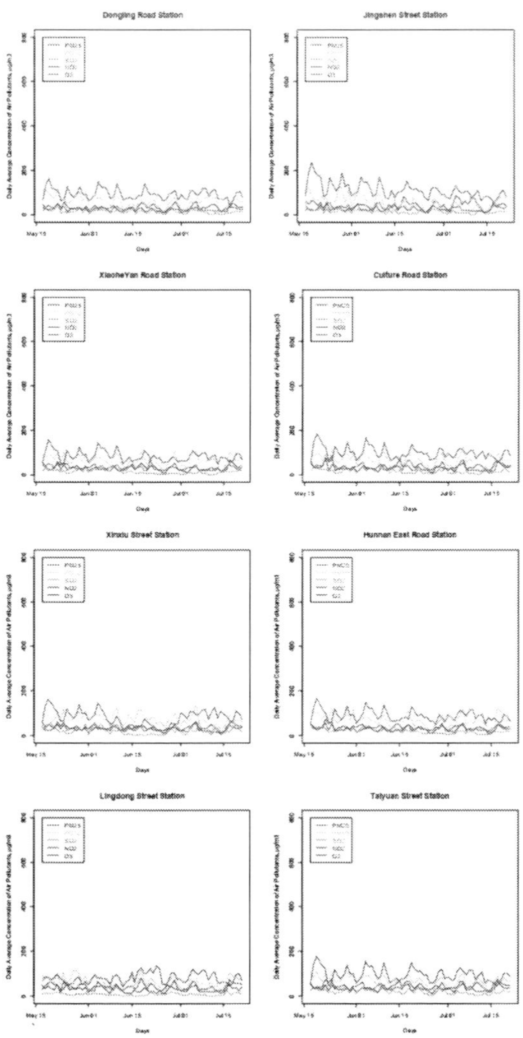

Fig. 1. Daily average concentration of pollutants during summer season

For the quantitative analysis that was shown in Table.1, the proportion of the number of days of occurrence of primary air pollutants includes two types: the independent proportion of each monitoring point and the average proportion. The principal pollutants with highest frequency were O_3 and PM_{10}, the average proportion was 74.1% and 20.8% respectively. The number of days that O_3 appears as the primary pollutant were 58, 63, 49, 57, 31, 51, 22 and 57respectively, and the proportions were 89.3%, 95.5%, 74.2%, 87.8%, 48.4%, 78.5%, 32.9% and 86.4% respectively, the mean was74.1%.

The PM_{10} was the 2nd pollutant in the sample data of 8 national monitoring stations The number of days was 7, 2, 17, 7, 34, 14, 20 and 9, the proportion was 10.7%, 3.0%, 25.8%, 10.7%, 51.6%, 21.5%, 29.8% and 13.6% respectively.

Table 3. The total proportion of the primary air pollutant

Station	$PM_{2.5}$		PM_{10}		SO_2		NO_2		O_3	
	D	P	D	P	D	P	D	P	D	P
Dongling Road	—	—	7	10.7%	—	—	—	—	58	89.3%
Jingshen Street	1	1.5%	2	3.0%	—	—	—	—	63	95.5%
Xiaoheyan Road	—	—	17	25.8%	—	—	—	—	49	74.2%
Culture Road	—	—	7	10.7%	—	—	1	1.5%	57	87.8%
Xinxiu Street	—	—	34	51.6%	—	—	—	—	31	48.4%
Hunnan East Road	—	—	14	21.5%	—	—	—	—	51	78.5%
Lingdong Street	—	—	20	29.8%	—	—	25	37.3%	22	32.9%
Taiyuan Street	—	—	9	13.6%	—	—	—	—	57	86.4%

*The D denotes the total number of days; the P denotes the proportion of each pollutant occurred.

According to the above analysis, it can be seen that the O_3 was the most important air pollutant with the highest occurrence proportion during the summer season. According to the average proportion of the main air pollutants as the primary pollutants, the PM_{10} Particle matter was ranked as the second one.

It was necessary to carry out special case that the average daily concentration of NO_2 as the primary pollutant during the summer season was 25 days in the data of Lingdong Street national monitoring station.

3.2 Fluctuation characteristics of main air pollutants

As shown in Fig 2, the fluctuation characteristics of the main air pollutants was displayed and analyzed using the box diagram method. The $PM_{2.5}$, PM_{10}, SO_2 and NO_2 fluctuation characteristics were regular and stable.

From the Fig 2, it was obviously can be seen that the region that nearby the road with larger traffic flow was with the lower fluctuation of the daily average concentration of O_3, such as the stations of the Jing Shen Street, Culture Road, Hunnan East Road and Taiyuan Street. On the contrary, lower traffic flow was corresponding to higher concentration in the regions of Dongling Road, Xiaoheyan road, Xinxiu Street and

DOI: 10.1051/matecconf/201713900090

Lingdong Street. As were shown in Table 2. To understanding the characteristics of the spatial distribution, the comparative analysis of the data of the 8 national stations will be done.

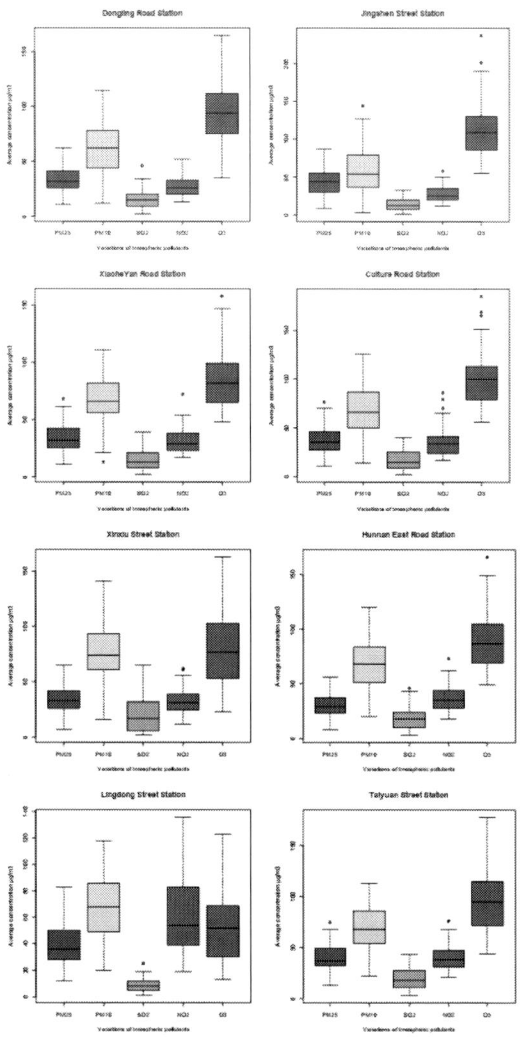

Fig. 2. Fluctuation characteristics of major air pollutants

Table 2. The statistical analysis in boxplots

	Dong ling Road	Jing shen Street	Xiao heyan Road	Culture Road	Xin xiu Street	Hunnan East Road	Ling dong Street	Tai yuan Street
1st Qu	75	86	65	79	53	69	30	72
Median	94	109	82	100	77	87	52	95
3rd Qu	112	130	99	113	103	105	69	115

3.3 Locations of the monitoring Station and corresponding urban areas

The locations of the 8 national monitoring stations involved in this paper and the corresponding urban area were shown in Table 3. The geographical location of each urban area was different, so its air pollution sources were largely different inevitably.

According to the results of the data analysis, the urban areas were divided into two types by the traffic flow that were denoted as the relative larger traffic flow urban area and the relative lower traffic flow urban area. The two types of urban area were referred to as larger flow area and lower flow area.

Table 3. The classification of urban corresponding to different monitoring station

Station	Urban Area	Location
Dongling Road	Qipanshan Development Dwastrict	3rd cyclic
Jing Shen Street	Shenbei New Dwastrict	3rd cyclic
Xiaoheyan road	Dadong Dwastrict	1st cyclic
Culture Road	Heping Dwastrict	1st cyclic
Xinxiu Street	Hunnan New Dwastrict	3rd cyclic
Hunnan East Road	Hunnan New Dwastrict	3rd cyclic
Lingdong Street	Huanggu Dwastrict	1st cyclic
Taiyuan Street	Heping Dwastrict	1st cyclic

3.4 Distribution characteristics of O_3 pollutant

As were shown in Table 4, the maximal daily average concentration of O_3 was 237μg/m³ in Jing Shen Street station, the minimum daily average concentration was 13μg / m³ in Lingdong Street station.

The maximal mean value of the daily average concentration of O_3 was 114.8μg/m³ in Jing Shen Street station, the minimum mean value was 54.6μg / m³ in Lingdong Street station. The mean value of the daily average concentration of O_3 in Jing Shen Street station was twice more than the mean value of Lingdong Street site.

Table 4. Primary air pollution distribution of the core area

National Monitoring Station	Daily Average Concentration of O_3 (μg/m³)		
	Max	Mean	Min
Dongling Road	165	96	35
Jing Shen Street	237	114.8	55
Xiaoheyan road	158	85.6	48
Culture Road	185	101.3	56
Xinxiu Street	163	80.3	23
Hunnan East Road	166	89.9	49
Lingdong Street	123	54.6	13

From the Table 4, it was obviously can be seen that the larger flow area was with the higher daily average concentration of O_3, such as the national monitoring stations of the Jing Shen Street, Culture Road, Hunnan East Road and Taiyuan Street.

On the contrary, lower flow area corresponding to the lower concentration. This type regions include Dongling Road, Xiaoheyan road, Xinxiu Street and Lingdong Street.

The mean value of the mean daily average concentration of O_3 in larger traffic flow region was 100.7μg/m^3 and the mean value of the lower traffic flow region was 79.1μg/m^3, which was lower about 21.4%.

4 Summary

The distribution characteristics of the major air pollutions in Shenyang City during the summer season were analyzed from different aspects. The overall analyzes of distribution characteristics of the air pollutant was shown that the principal pollutants with highest frequency were O_3 and PM_{10}. The mean proportion was 74.1% and 20.8% respectively. The regions that nearby the road with larger traffic flow can be confirmed with the lower fluctuation of the average concentration of O_3. It has active significance for the government decision making by providing necessary scientific basis and data support.

Acknowledgment

This work was supported by Science and Technology Program of Shenyang City (F16-155-9-00, 17-180-9-00) and Science and enterprise competitive selection project of Shenyang City (Study on the risk management and remediation technology of urban ecological environment).

References

1. Qiu Weiguang, Zhang Qingxin, Chen Zongjiao, Zu Biao, Tai Shanshan, Bai Lu, Zhang Xiaofeng. Characteristics and Cause Analysis of One Typical Air Pollution in the Winter of Shenyang City. Environmental Protection Science, 2016, 42(4) : 106-109.
2. http://www.shenyang.gov.cn/zwgk/system/2015/05/26/010114621.shtml.
3. Jiang, J., Zhou, W., Cheng, Z., Wang, S., He, K., & Hao, J. Particulate matter distributions in China during a winter period with frequent pollution episodes. Aerosol Air Qual. Res, 15(2): 494-503.
4. Zhang J, Zhu T, Kipen H, et al. Cardiorespiratory biomarker responses in healthy young adults to drastic air quality changes surrounding the 2008 Beijing Olympics. Research Report, 2013, pp:5-174.
5. Zhang X Y, Wang Y Q, Niu T, et al. Atmospheric aerosol compositions in China: spatial/temporal variability, chemical signature, regional haze distribution and comparisons with global aerosols. Atmospheric Chemistry and Physics, 2012, 12(2): 779-799.
6. Yan Wenlian, Zhou Deping, Wang Yangfeng, et al.Studies on the Concentration and Scale Distribution of Inhalable Particulate Matter in Shenyang and Winter. Journal of Applied Meteorology, 2008, 19 (4): 435-442.

MATEC Web of Conferences **139**, 00091 (2017) DOI: 10.1051/matecconf/201713900091
ICMITE 2017

Approximation properties of the modified Lupas-Kantorovich type operators

Bo-yong Lian ⋆

[1] *Department of Mathematics, Yang-En University, Quanzhou 362014, P.R.China*

Abstract. In this paper, the author introduce a class of modified Lupas-Kantorovich type operators which preserve constant and linear functions. By using modulus of continuity, modulus of smooth, K-functional and lipschitz class, the rate of convergence of these operators are derived. Finally, the author present a voronovskaya-type asymptotic formula.

1 Introduction

In the year 1987, Lupas [1] introduced Polya-Bernstein operators

$$P_n^{(1/n)}(f,x)=\sum_{k=0}^{n} f(\frac{k}{n})p_{n,k}^{(1/n)}(x), \qquad (1)$$

where $f\in C[0,1]$, $p_{n,k}^{(1/n)}(x)=\frac{2(n!)}{(2n)!}\binom{n}{k}(nx)_k(n-nx)_{n-k}$, $(n)_k=n(n+1)\cdots(n+k-1)$.

In [2], Miclaus studied some approximation properties of Bernstein-Stancu type operators based on Polya distribution. Recently, Gupta and Rassias [3] introduced the Durrmeyer variant of the operators (1) as follows:

$$D_n^{(1/n)}(f,x)=(n+1)\sum_{k=0}^{n} p_{n,k}^{(1/n)}(x)\int_0^1 p_{n,k}(t)f(t)dt, \qquad (2)$$

where $p_{n,k}(t)=\binom{n}{k}t^k(1-t)^{n-k}$. In[4], to approximate Lebesgue integrable functions, Agrawal introduced the following integral modification of the operators (1):

$$D_n^{*(1/n)}(f,x)=(n+1)\sum_{k=0}^{n} p_{n,k}^{(1/n)}(x)\int_{\frac{k}{n+1}}^{\frac{k+1}{n+1}} f(t)dt. \qquad (3)$$

In 2003, King [5] introduced the modified Bernstein operators. Later, this idea was applied to some other well-known approximating operators,such as, the Szasz-Mirakjan operators [6], the Baskakov operators [7], the q-operators [8]. Obviously, these operators have a better rate of convergence than the classical operators.

Inspired by the idea of King, we propose the modified variant of the operators (3) in the following way:

$$\widehat{D}_n^{(1/n)}(f,x)=(n+1)\sum_{k=0}^{n} p_{n,k}^{(1/n)}(r(x))\int_{\frac{k}{n+1}}^{\frac{k+1}{n+1}} f(t)dt, \qquad (4)$$

where $r(x)=x+\frac{2x-1}{2n}$, $x\in[\frac{1}{4},\frac{3}{4}]$.

The purpose of this paper is to study local approximation, pointwise estimates and global approximation results. Finally, we present a voronovskaya-type asymptotic formula.

We get some definition as follows:

Let $f(x)\in C[0,1]$, $\delta>0$,

Definition 1 (modulus of continuity)

$$\omega(f,\delta)=\sup_{0<h\le\delta}\ \sup_{x,x+h\in[0,1]}|f(x+h)-f(x)|,$$

⋆e-mail: lianboyong@163.com

Definition 2 (the second order modulus of smoothness)

$$\omega_2(f,\sqrt{\delta}) = \sup_{0<h\leq\sqrt{\delta}} \sup_{x,x+h,x+2h\in[0,1]} |f(x+2h)-2f(x+h)+f(x)|,$$

Definition 3 (K-functional)

$$K_2(f,\delta) = \inf_{g\in W^2}\{\|f-g\|+\delta\|g''\|\},$$

where $W^2 = \{g \in C[0,1] : g', g'' \in C[0,1]\}$.

Definition 4 (Lipschitz class)

$$Lip_M(\beta) = \{f|f \in C[0,1], |f(t)-f(x)| \leq M|t-x|^{\beta}\},$$

where $0 < \beta \leq 1, M > 0$.

Let $\phi(x) = \sqrt{x(1-x)}$ and $f \in C[0,1]$, the first order modulus of smoothness is given by [9]

$$\omega_\phi(f,t) = \sup_{0<h\leq t}\left|f(x+\frac{h\phi(x)}{2}) - f(x-\frac{h\phi(x)}{2})\right|,$$

where $x \pm \frac{h\phi(x)}{2} \in [0,1]$.

Further, the corresponding K-functional to $\omega_\phi(f,t)$ is defined by

$$K_\phi(f,t) = \inf_{g\in W_\phi[0,1]}\{\|f-g\|+t\|\phi g'\|\}(t>0),$$

where $W_\phi[0,1] = \{g : g \in AC[0,1], \|\phi g'\| < \infty\}$ and $AC[0,1]$ is the class of all absolutely continuous functions on $[0,1]$.

2 Some Lemmas

We start this section with the following uesful lemmas, which will be used in the sequel.

Lemma 1(see [2]) For $e_i = t^i, i = 0,1,2,$ we have

$$P_n^{(1/n)}(e_0,x) = 1, P_n^{(1/n)}(e_1,x) = x,$$

$$P_n^{(1/n)}(e_2,x) = x^2 + \frac{2x(1-x)}{n+1}.$$

Lemma 2(see [3]) For $e_i = t^i, i = 0,1,2,$ we have

$$D_n^{*(1/n)}(e_0,x) = 1, D_n^{*(1/n)}(e_1,x) = \frac{2nx+1}{2n+2},$$

$$D_n^{*(1/n)}(e_2,x) = \frac{3n^3x^2+9n^2x-3n^2x^2+3nx+n+1}{3(n+1)^3}.$$

Using the same method to lemma 2, we obtain

Lemma 3 For $e_i = t^i, i = 0,1,2, x \in [\frac{1}{4},\frac{3}{4}]$, we have

$$\widehat{D}_n^{(1/n)}(e_0,x) = 1, \widehat{D}_n^{(1/n)}(e_1,x) = x,$$

$$\widehat{D}_n^{(1/n)}(e_2,x) = \frac{3n^3r^2(x)+3n^2r(x)(3-r(x))+3nr(x)}{3(n+1)^3} + \frac{1}{3(n+1)^2}.$$

Remark 1 By simple applications of Lemma 3, we get

$$\widehat{D}_n^{(1/n)}(t-x,x) = 0, \tag{5}$$

$$\widehat{D}_n^{(1/n)}((t-x)^2,x) = \frac{2x(1-x)}{n} + O(n^{-2})(n\to\infty). \tag{6}$$

$$\widehat{D}_n^{(1/n)}((t-x)^4,x) = O(n^{-2})(n\to\infty). \tag{7}$$

Remark 2 When n sufficient large, we have

$$\widehat{D}_n^{(1/n)}((t-x)^2,x) \leq \frac{3x(1-x)}{n}. \tag{8}$$

Lemma 4 For $f \in C[0,1], x \in [\frac{1}{4},\frac{3}{4}]$, we have

$$\|\widehat{D}_n^{(1/n)}(f,x)\| \leq \|f\|.$$

Proof From the definiton (4) and Lemma 3, we have

$$|\widehat{D}_n^{(1/n)}(f,x)| \leq (n+1)\sum_{k=0}^{n} p_{n,k}^{(1/n)}(r(x))\int_{\frac{k}{n+1}}^{\frac{k+1}{n+1}}|f(t)|dt$$

$$\leq \|f\|\widehat{D}_n^{(1/n)}(1,x) = \|f\|.$$

Lemma 5(see[10]) For $f(x) \in C[0,1]$, then exists a constant $C > 0$ such that

$$K_2(f,\delta) \leq C\omega_2(f,\sqrt{\delta}).$$

Lemma 6(see[9]) For $f(x) \in C[0,1]$, then exists a constant $C > 0$ such that

$$K_\phi(f,t) \leq C\omega_\phi(f,t).$$

3 Conclusion

Theorem 1 For $f \in C[0,1]$, $x \in [\frac{1}{4}, \frac{3}{4}]$, we have

$$\lim_{n\to\infty} \widehat{D}_n^{(1/n)}(f,x) = f(x).$$

Proof From the Lemma 3, we get

$$\lim_{n\to\infty} \widehat{D}_n^{(1/n)}(e_i,x) = x^i, i = 0,1,2.$$

By Korovkin theorem, it follows that $\widehat{D}_n^{(1/n)}(f,x) \to f(x)$ as $n \to \infty$.

Theorem 2 For $f, f'' \in C[0,1]$, $x \in [\frac{1}{4}, \frac{3}{4}]$, then

$$\lim_{n\to\infty} n[\widehat{D}_n^{(1/n)}(f,x) - f(x)] = x(1-x)f''(x).$$

Proof By Taylor's expansion, we may write

$$f(t) = f(x) + f'(x)(t-x) + \frac{1}{2}f''(x)(t-x)^2$$
$$+\phi(t;x)(t-x)^2,$$

where $\phi(t;x)$ is the Peano form of the remainder, $\phi(t;x) \in C[0,1]$ and $\lim_{t\to x}\phi(t;x) = 0$.
By applying the operators $\widehat{D}_n^{(1/n)}(f,x)$ to the above relation, we obtain

$$n[\widehat{D}_n^{(1/n)}(f,x) - f(x)] = nf'(x)\widehat{D}_n^{(1/n)}(t-x,x)+$$
$$\frac{n}{2}f''(x)\widehat{D}_n^{(1/n)}((t-x)^2,x)+n\widehat{D}_n^{(1/n)}(\phi(t;x)(t-x)^2,x). \tag{9}$$

By the Cauchy-Schwartz inequality, we have

$$\widehat{D}_n^{(1/n)}(\phi(t;x)(t-x)^2,x)$$
$$\leq \sqrt{\widehat{D}_n^{(1/n)}(\phi^2(t;x),x)} \cdot \sqrt{\widehat{D}_n^{(1/n)}((t-x)^4,x)}. \tag{10}$$

Observe that $\phi^2(x;x) = 0$, $\phi^2(t;x) \in C[0,1]$, then it follows from Theorem 1 that

$$\lim_{n\to\infty} \widehat{D}_n^{(1/n)}(\phi^2(t;x),x) = \phi^2(x;x) = 0. \tag{11}$$

Now from (5), (6), (7), (10) and (11), we get immediately

$$\lim_{n\to\infty} nf'(x)\widehat{D}_n^{(1/n)}(t-x,x) = 0. \tag{12}$$

$$\lim_{n\to\infty} n\widehat{D}_n^{(1/n)}(\phi(t;x)(t-x)^2,x) = 0. \tag{13}$$

$$\lim_{n\to\infty} \frac{n}{2}f''(x)\widehat{D}_n^{(1/n)}((t-x)^2,x) = x(1-x)f''(x). \tag{14}$$

Theorem 2 now follows from (9), (12), (13), (14) immediately.

Theorem 3 For $f \in C[0,1]$, $x \in [\frac{1}{4}, \frac{3}{4}]$, when n sufficient large, we have

$$|\widehat{D}_n^{(1/n)}(f,x) - f(x)| \leq 2\omega\left(f, \sqrt{\frac{3x(1-x)}{n}}\right). \tag{15}$$

Proof Using linearity and monotonicity of $\widehat{D}_n^{(1/n)}(f,x)$ we easily get, for every $\lambda > 0$, that

$$|\widehat{D}_n^{(1/n)}(f,x) - f(x)| \leq \left|\widehat{D}_n^{(1/n)}(|f(t) - f(x)|, x)\right|$$

$$\leq \omega(f,\lambda)\left(1 + \frac{1}{\lambda}\widehat{D}_n^{(1/n)}(|t-x|,x)\right)$$
$$\leq \omega(f,\lambda)\left(1 + \frac{1}{\lambda}\sqrt{\widehat{D}_n^{(1/n)}((t-x)^2,x)}\right).$$

Choosing $\lambda = \sqrt{\widehat{D}_n^{(1/n)}((t-x)^2,x)}$, then (15) follows from (8) immediately.

Theorem 4 For $f \in C[0,1]$, $x \in [\frac{1}{4}, \frac{3}{4}]$, when n sufficient large, there exists a constant $C > 0$ such that

$$|\widehat{D}_n^{(1/n)}(f,x) - f(x)| \leq C\omega_2\left(f, \sqrt{\frac{3x(1-x)}{2n}}\right).$$

Proof Let $g \in W^2$. By Taylor's expansion, we get

$$g(t) = g(x) + g'(x)(t-x) + \int_x^t (t-u)g''(u)du.$$

From Lemma 3, we have

$$\widehat{D}_n^{(1/n)}(g,x) = g(x) + \widehat{D}_n^{(1/n)}\left(\int_x^t (t-u)g''(u)du, x\right),$$

Hence

$$|\widehat{D}_n^{(1/n)}(g,x) - g(x)|$$

$$= \left|\widehat{D}_n^{(1/n)}(\int_x^t (t-u)g''(u)du, x)\right|$$
$$\leq \widehat{D}_n^{(1/n)}\left(\int_x^t (t-u)|g''(u)|du, x\right)$$
$$\leq \widehat{D}_n^{(1/n)}((t-x)^2, x)\|g''\|$$
$$\leq \frac{3x(1-x)}{n}\|g''\|.$$

Thus

$$|\widehat{D}_n^{(1/n)}(f,x) - f(x)|$$
$$\leq |\widehat{D}_n^{(1/n)}(f-g,x)| + |f-g| + |\widehat{D}_n^{(1/n)}(g,x) - g(x)|$$
$$\leq 2\|f-g\| + \frac{3x(1-x)}{n}\|g''\|.$$

Taking the infimum on the right hand side over all $g \in W^2$, we obtain

$$|\widehat{D}_n^{(1/n)}(f,x) - f(x)| \leq 2K_2\left(f, \frac{3x(1-x)}{2n}\right),$$

By Lemma 5, we get

$$|\widehat{D}_n^{(1/n)}(f,x) - f(x)| \leq C\omega_2\left(f, \sqrt{\frac{3x(1-x)}{2n}}\right).$$

Theorem 5 For $f \in C[0,1]$, $x \in [\frac{1}{4}, \frac{3}{4}]$ and $\phi(x) = \sqrt{x(1-x)}$, when n sufficient large, there exists a constant $C > 0$ such that

$$|\widehat{D}_n^{(1/n)}(f,x) - f(x)| \leq C\omega_\phi(f, \sqrt{\frac{6}{n}}).$$

Proof Using the representation

$$g(t) = g(x) + \int_x^t g'(u)du,$$

we get

$$\widehat{D}_n^{(1/n)}(g,x) = g(x) + \widehat{D}_n^{(1/n)}(\int_x^t g'(u)du, x),$$

For any $x, t \in (0,1)$, we find that

$$\left|\int_x^t g'(u)du\right| \leq \|\phi g'\|\left|\int_x^t \frac{1}{\phi(u)}du\right|.$$

But

$$\left|\int_x^t \frac{1}{\phi(u)}du\right| = \left|\int_x^t \frac{1}{\sqrt{u(1-u)}}du\right|$$
$$\leq \left|\int_x^t \left(\frac{1}{\sqrt{u}} + \frac{1}{\sqrt{1-u}}\right)du\right|$$
$$\leq 2\left(|\sqrt{t} - \sqrt{x}| + |\sqrt{1-t} - \sqrt{1-x}|\right)$$
$$= 2|t-x|\left(\frac{1}{\sqrt{t}+\sqrt{x}} + \frac{1}{\sqrt{1-t}+\sqrt{1-x}}\right)$$
$$\leq 2|t-x|\left(\frac{1}{\sqrt{x}} + \frac{1}{\sqrt{1-x}}\right) \leq \frac{2\sqrt{2}|t-x|}{\phi(x)}.$$

Using Cauchy-Schwarz inequality, we obtain

$$|\widehat{D}_n^{(1/n)}(g,x) - g(x)|$$
$$\leq 2\sqrt{2}\|\phi g'\|\phi^{-1}(x)\widehat{D}_n^{(1/n)}(|t-x|, x)$$
$$\leq 2\sqrt{2}\|\phi g'\|\phi^{-1}(x)\left(\widehat{D}_n^{(1/n)}((t-x)^2, x)\right)^{1/2}$$
$$\leq \frac{2\sqrt{6}\|\phi g'\|}{\sqrt{n}}.$$

Thus

$$|\widehat{D}_n^{(1/n)}(f,x) - f(x)|$$
$$\leq |\widehat{D}_n^{(1/n)}(f-g,x)| + |f-g| + |\widehat{D}_n^{(1/n)}(g,x) - g(x)|$$
$$\leq 2\|f-g\| + \frac{2\sqrt{6}\|\phi g'\|}{\sqrt{n}}.$$

Taking the infimum on the right hand side over all $g \in W_\phi[0,1]$, we obtain

$$|\widehat{D}_n^{(1/n)}(f,x) - f(x)| \leq 2K_\phi\left(f, \sqrt{\frac{6}{n}}\right).$$

By Lemma 6, we get

$$|\widehat{D}_n^{(1/n)}(f,x) - f(x)| \leq C\omega_\phi(f, \sqrt{\frac{6}{n}}).$$

Theorem 6 For $f \in Lip_M(\beta)$, $x \in [\frac{1}{4}, \frac{3}{4}]$, when n sufficient large, we have

$$|\widehat{D}_n^{(1/n)}(f,x) - f(x)| \leq M\left(\frac{3x(1-x)}{n}\right)^{\beta/2}.$$

Proof By Hölder's inequality with $p = \frac{2}{\beta}, q = \frac{2}{2-\beta}$, in view of Remark 2 we have

$$|\widehat{D}_n^{(1/n)}(f,x) - f(x)|$$

$$
\begin{aligned}
&\leq \widehat{D}_n^{(1/n)}(|f(t) - f(x)|, x) \\
&\leq M\widehat{D}_n^{(1/n)}(|t - x|^{\beta}, x) \\
&\leq M\left(\widehat{D}_n^{(1/n)}((t - x)^2, x)\right)^{\beta/2} \\
&\leq M\left(\frac{3x(1 - x)}{n}\right)^{\beta/2}.
\end{aligned}
$$

Acknowledgment

This work is supported by the Program for Outstanding Youth Scientific Research Talents in Fujian Province University(2016).

References

[1] L. Lupas and A. Lupas, *Polynomials of binomial type and approximation operators*, Stud Univ Babes-Bolyai Math, 1987, 32(4): 61-69.

[2] D. Miclaus, *The revision of some results for Bernstein-Stancu type operators*, Carpathian J Math, 2012, 28(2):289-300.

[3] V. Gupta and T.M. Rassias, *Lupas-Durrmeyer operators based on polya distribution*, Banach J Math Anal, 2014, 8(2): 146-155.

[4] P.N. Agrawal, N. Ispir and A. Kajla, *Approximation properties of Lupas-Kantorovich operators based on polya distribution*, Rendiconti del Circolo Matematico di Palermo Series 2, 2016, 65(2):185-208.

[5] J.P. King, *Positive linear operators which preserve x^2*, Acta Math Hungar, 2003, 99(3): 203-208.

[6] O. Duman and M.A. Ozarslan, *Szasz-Mirakjan type operators providing a better error estimation*, Appl Math Lett, 2007, 20(12): 1184-1188.

[7] L. Rempulska and K. Tomczak, *Approximation by certain linear operators preserving x^2*, Turkish J Math, 2009, 33(3): 273-281.

[8] A. Aral and V. Gupta, *Applications of q-calculus in operator theory*, Springer, New York, 2013.

[9] Z. Ditzian and V. Totik, *Moduli of Smoothness*, Springer, New York, 1987.

[10] R.A. Devore and G.G. Lorentz, *Construtive Approximation*, Springer-Verlag, Berlin, 1993.

The Analysis and Design of Automatic Beverage Vending System based on UML

Yindi Dong

School of Information Engineering, Chongqing City Management College, Chongqing

Abstract. UML is a standard modeling language, which is easy to express, powerful and universal. At the same time, it can support the software development process from the demand analysis. Using UML modeling mechanism, which can be visual representation and construction of system model for all the development process of the application software system based on UML. We design the use case diagram, class diagram, timing diagram, etc.

1 Introduction

In recent years, the automatic beverage vending system get the majority of consumers as a kind of new intelligent products, with simple operation, convenient service. It has developed rapidly in the United States, Japan and other countries since the listing, and now has been widely spread in 45 countries. At the same time, it is developing rapidly as an emerging industry in China. Today, in the airport, subway, shopping malls, parks and other places, it is not difficult to find the automatic beverage vending system. If the vending system market is fully developed, its prospects are very impressive, when China's automatic beverage vending system will develop into a huge industry. In this paper, we have the requirement analysis and modeling by UML modeling language for automatic beverage vending system. UML modeling tools with high efficiency model, which can be designed to facilitate change, strong expansibility and easy maintenance of automatic beverage vending system.

UML (Unified Modeling Language) is used to describe the structure of software system and business modeling graphical language, which is a standard language with description, visualization, and documentation of object-oriented product development system. It has obtained excellent software engineering method, which is recognized in the field of modeling of complex systems with new ideas and new methods and the new technology into the field of software engineering and by amending the found problems in the practical application. It has become a recognized industry standard object oriented technology. Its scope is not only limited to object oriented analysis and design, but also supports the whole process of software development. IBM Rational Rose and MS Visio are UML tools. UML provides 9 kinds of rich and powerful model for the design and development of software system: use case diagram, activity diagram, class diagram, object diagram, state diagram, sequence diagram, collaboration diagram, component diagram and deployment diagram, which introduces the system model using the static model and dynamic model. It describes the development projects from different perspectives and different stages of software development, and can be a good understanding.

2 The modeling of automatic beverage vending system based on UML

2.1 System requirements analysis phase

Requirements analysis includes the establishment of business models and user requirements analysis in the problem domain. First of all, we need to get the requirements, the use cases that describe the system functions, business processes or clear text, which are observed from the outside of the system. In UML, the use case diagram and some simple class diagrams can be used to describe the user's functional requirements. The use case diagram is used to describe a type of document, which shows participants who complete a process using a system when the time sequence. The use case diagram displays a set of case system, the relationship between the use case and participants.

The function description of automatic beverage vending system: to provide customers with the purchase of goods; to verify the legitimacy of coins; to prompt abnormal situation (goods sold, the change inappropriate, etc.); to find change. Specific analysis is as follows:

The main purpose of the beverage vending system is to allow customers to buy drinks. Customers insert coins into it and choose what they want. If not, the vending system will give customers a message that there is not this beverage. Ideally, customers see this message, then the display will show other drink immediately, and the system must also be provided to retrieve money for the customer. It should give customers two options: let customers choose other drink and provide the drink to customers or refund money. If the change in the storage is run out, a message should be displayed to tell the user that the appropriate change is needed. The main work for the supplier and the payee is as follows: to supply goods timely and provide the conditions for handling abnormal situation. The supplier opened the machine and front desk shelf, then filled with various kinds of beverage on the shelves and put the change in it. Finally he put the front shelf and locked it. The pre-condition of the use case is the passing of a time interval, and the post-condition is that the supplier placed a new drink for sale.

2.2 System use case diagram

The use case model of a system is composed of a plurality of use case diagram, use case diagram of the main components are use cases, actors and system. The simple icon elements describe the actors and the system interaction and the function of the system. Use cases drive the method of system analysis and system design, which has become the mainstream of object-oriented method. First of all, we should analyze and determine the activity and the case in order to establish the case model during the development of a software system. Use case model is the link of developers to communicate with customers.

Buying drink module is part of the beverage vending system, which is mainly responsible for providing customers with the purchase of goods. From the previous analysis, we know that there are three actors for the system, customers, suppliers and payee. We could get the actors from the above requirements analysis. Customer: get the item you want to buy from a vending system. Supplier: responsible for the sales of beverage system. Payee: responsible for collecting the money from the vending system regularly.

Through the above analysis, we can get the system use case diagram as shown in Fig.1.

2.3 System class diagram

Class diagram is used to express the relationships between classes in a system. The model element that is allowed in the class diagram is only the class and its relationship. There are several classes of graphs in a system. The steps of constructing class diagram: research and analysis the problems, determine the system requirements; determine the class, make clear the meaning and responsibility of class, make sure the attribute and operation; express the relationships between classes with the association, generalization, aggregation, combination; adjust and refine the relationships, such as naming conflicts, repetitive function problems; draw class diagram and add the appropriate instructions. There are many classes involved in the automatic beverage vending system, and there is a cooperative relationship between them. We show the relationship between classes with the beverage vending system, the front desk, the coin record, the distributor and the transaction record. As shown in Fig.2.

2.4 System dynamic model

The dynamic model is used to describe the dynamic behavior of the system, and shows the dynamic interaction between the object and the system at different times during the system operation. In the UML, the dynamic model is established by using the time sequence diagram, collaboration diagram, state diagram and activity diagram. During the system analysis and design, we draw these graphics for the main use cases and objects in order to analyze the dynamic characteristics and behavior of the system.

Sequence diagram is used to describe the interactions between objects, and reveals the interaction of a specific scene. It is suitable for describing the time characteristics and time constraints in real-time systems, and the message is sent by the time. Collaboration diagram is also used to describe the interaction between system objects. Sequence diagram emphasizes time sequence of the operation between objects, and collaboration diagram emphasizes the relationship between the set of objects. Collaboration diagram and sequence diagram can be transformed into each other. The sequence diagram of beverage vending system is shown as Fig.3.

3 System construction and testing

In the construction or design phase, we transform the design phase of the class into some kind of object-oriented programming language code, and then complete the design class diagram of beverage vending system during the development cycle. We will get enough detail information about the generation of object definition field. When the design is mapped to the code, we write the definition and method of the class with the object-oriented programming language. In the program design phase, we still have to make many decisions to change design and continuous survey. Ideally, the whole structure and main design decisions should be completed before the encoding stage. There are many system testing methods, and different groups will take different UML diagram as the basic unit. The testing will be used the class diagrams and specifications, and the system test use case diagram to confirm the behavior of the system in line with the definition. For the automatic beverage vending system, the UML system modeling is the core work, the interface and the maintenance are the same important.

4 Conclusion

The above is a brief description of how to use UML modeling to develop the automatic beverage vending system. UML can provide flexibility and readability for the entire development process, and shows its strong function. UML combines the advantages of object-oriented modeling languages with a broad application prospects. Through the analysis, we can know that understanding the user needs and system design with the standard modeling language and tools can greatly improve the efficiency of application development. We hope that the wide application of automatic beverage vending system will provide more convenient and efficient service for people.

References

[1]Larman. The application of UML and model [M]. China Machine Press. 2016.5

[2]Meliir Page-Jones. UML object-oriented design. The Posts and Telecommunications Press. 2012.5

[3]Hu Hefen. UML object-oriented analysis and design. Tsinghua University press. 2012.5

[4]Xie Xing. UML foundation and Rose modeling practical guide. Tsinghua University press. 2011.1

[5]Grady Booch. James Rumbaugh. Ivar Jacobson. Unified Modeling Language User Guide. The Posts and Telecommunications Press. 2013.1

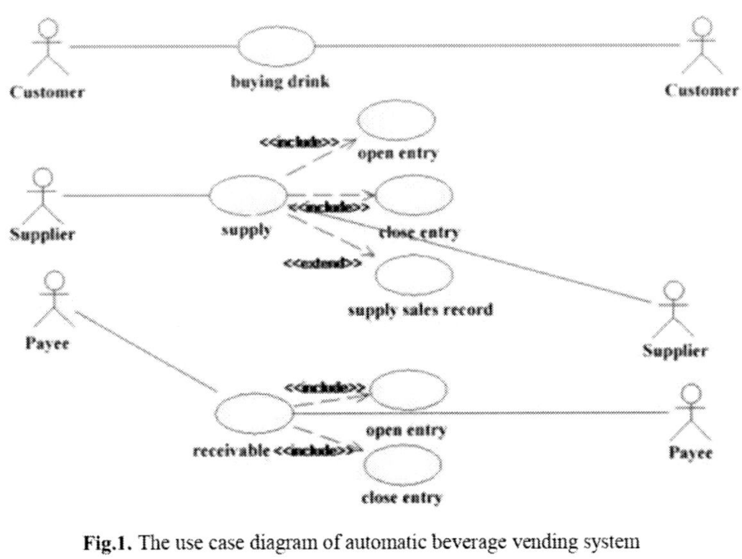

Fig.1. The use case diagram of automatic beverage vending system

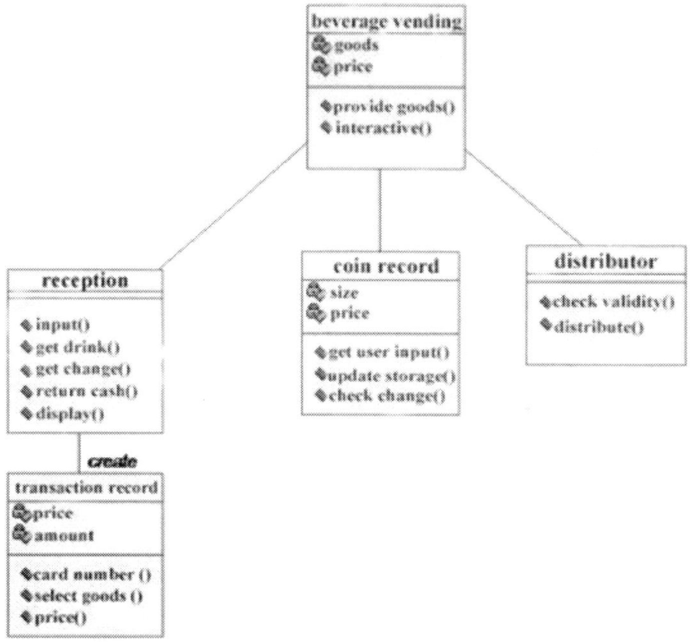

Fig.2. System diagram

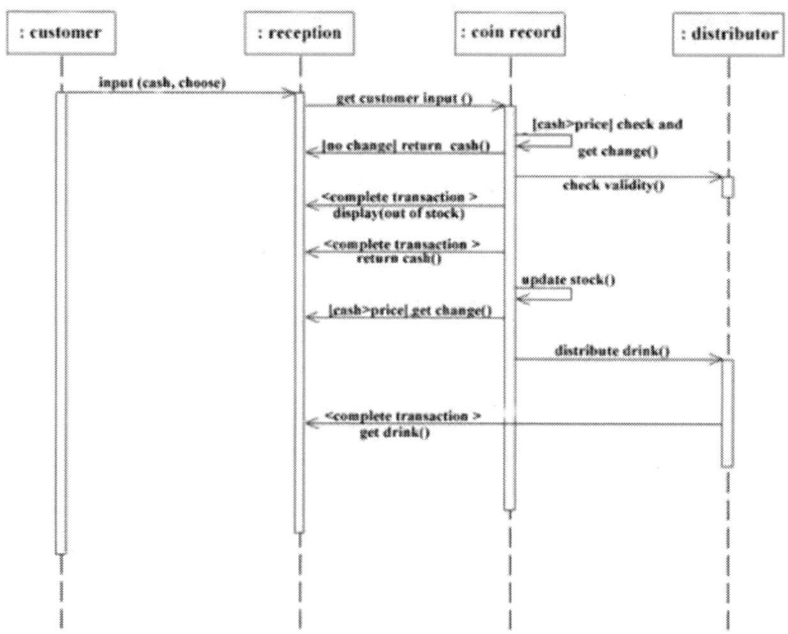

Fig.3. Sequence diagram of beverage vending system

Research on the Flow Field and Structure Optimization in Cyclone Separator with Downward Exhaust Gas

Weiwei Wang[1], *Linjiang* Zou[1*], *Jun* Li[1] and *Zhiming* Zhao[1]

[1]School of Energy and Environment, Anhui University of Technology, Ma'an shan, 243002, China

Abstract. A numerical software analysis of the turbulent and strongly swirling flow field of a cyclone separator with downward exhaust gas and its performances is described. The ANSYS 14.0 simulations based on DPM model are also used in the investigation. A new set of geometrical design has been optimized to achieve minimum pressure drop and maximum separation efficiency. A comparison of numerical simulation of the new design confirm the superior performance of the new design compared to the conventional design. The influence of the structure parameters such as the length of the guide pipe, the shape of the guide, the inlet shape on the separation performance was analyzed in this research. This research result has certain reference value for cyclone separator design and performance optimization.

1 Introduction

Gas cyclone separators are widely used in industries to separate dust from gas due to their geometrical simplicity and relative economy in power consumption[1]. Currently, with the speeding up of the exhaustion of global energy resources, environmental pollution has become the focus of world[2]. Cyclone separator with downward exhaust gas has some advantages such as small flow resistance, low pressure loss and relatively low cost; And it is able to meet the requirements of ∏ type in the circulating fluidized bed boiler; while separation efficiency and ensure lower pressure loss at the same time, further study of the fine particles separation mechanism with the separator and the structure optimization design are needed[3].

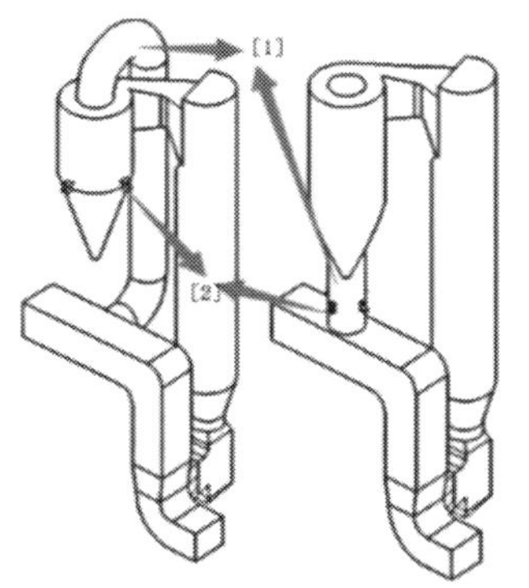

[1] Exhaust pipe [2] Supporting position

Figure 1. Comparison of cyclone separator with the downward and upward exhaust gas

As we can see from Fig.1, It is the arrangement of the circulating fluidized bed (CFB) boiler on flue gas dust removal device with semi dry desulfurization. It is shown that three elbows are less than the upward exhaust pipe of cyclone separator, and the length of the exhaust pipe is

*Conrresponding author: wahut@sina.com

much longer than that of the downward exhaust pipe. Therefore, the overall resistance losses will be much greater than the downward exhaust pipe, and based on the structure features of the exhaust separator, it's the overall layout also more flexible[4]. However, due to the vortex change of the flow field, particle entrainment phenomenon is relatively serious in the downward exhaust gas, and the separation efficiency is smaller than that of the upward exhaust gas[5]. Thus to obtain the best separation performance, and meet the needs of the actual operation, it is necessary to optimize the design of the structure in this paper.

2 Numerical models and boundary conditions

2.1 Physical model

The Fig.2 shows the schematic of the reference cyclone separator which is applied to some $400t/h$ CFB boiler. The power plant uses coal gangue as fuel, and CFB boiler adopting the cyclone separator with downward exhaust gas.

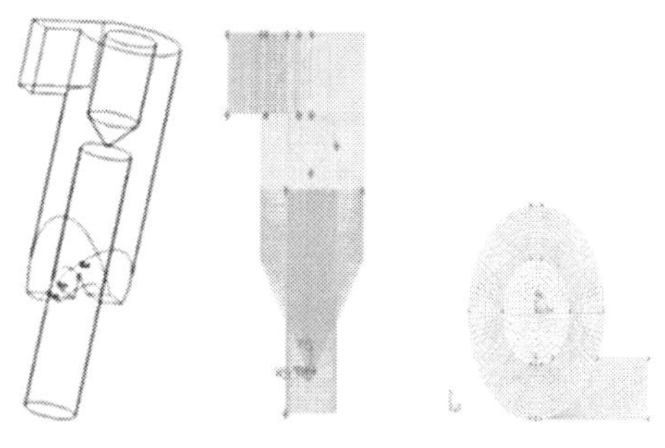

Figure 2. Schematic diagram of reference cyclone structure and grids of filter vessel.

2.2 Simulation condition

Due to the gas flow of the cyclone separator is a three dimensional turbulent flow, so simulation process is more complex. The research proposed that the RNG $k-\varepsilon$ model closed equations directly for the two order correlation of fluctuating velocity, and the determination of particle flow is ensured by the use of a DPM mode[6]. Because of there is the thin particles in the cyclone separator, the phase coupling is not considered, and the influence of the turbulent fluctuation is only considered. Meanwhile, the volume averaged conservation equation is discretized by the control volume finite element method, ANSYS 14 uses the isolation method to solve. The SIMPLE algorithm is adopted in the solution process, and the QUICK scheme is used to the convection term in the discrete scheme. Finally, this model shows the computational domain, containing 6682651 cells. The whole computational domain is divided by structured hexahedron grids.

3 NUMERICAL SIMULATION RESULTS AND ANALYSIS

In order to eliminate the influence of other operating factors, the simulation of each structure is carried out under the same operating parameters. Inlet flow velocity remains constant, which is $16.3m/s$, the particle size of the incident particles was Rosin-Rammler distribution. Particle size range is from to $0.075\ \mu m \sim 0.25\ \mu m$.

Table 1. Separation efficiency and pressure loss

Structure of cyclone separator	Separation efficiency (%)	Increase ratio of separation efficiency (%)	Pressure loss (Pa)	Increase ratio of pressure loss (%)
Reference structure	95.2	-------	925.6	-------
h_c =0.7a	96.1	0.945378	1485.2	60.4586
h_c = a	93.9	-1.36555	1058.8	14.3855
h_c = 1.3 a	93.2	-2.10084	960.6	3.7771
h_c = 1.6 a	92.3	-3.04622	779.5	15.7893
Cylindrical guide pipe	85.3	-10.3992	1127.6	21.8181
Entrance of square type	94.7	-0.52521	620.9	32.9186

As shown in Table 1, reference structure is tapered entrance, and the length of the guide pipe is $5000mm$, with the entrance height ratio is $1.25(5000/4000=1.25)$, Outlet pipe diameter is $3000mm$. In the Table 1, pressure loss is the flow resistance loss of the square inlet section to the exit section of the exhaust pipe. Meanwhile, the increase ratio of pressure loss and separation efficiency is based on the numerical simulation results of the reference structure.

3.1 Effect of inlet shape on separation efficiency and pressure loss

Research shows that pressure drop and separation efficiency are the principal parameters which characterize the performances of a cyclone separator. We only study the influence of tangential normal inlet and tangential tapered inlet structure, while their entrance area and initial velocity of entry ($v_0 = 16.3 m/s$) are the same. So the treatment of flue gas is equal.

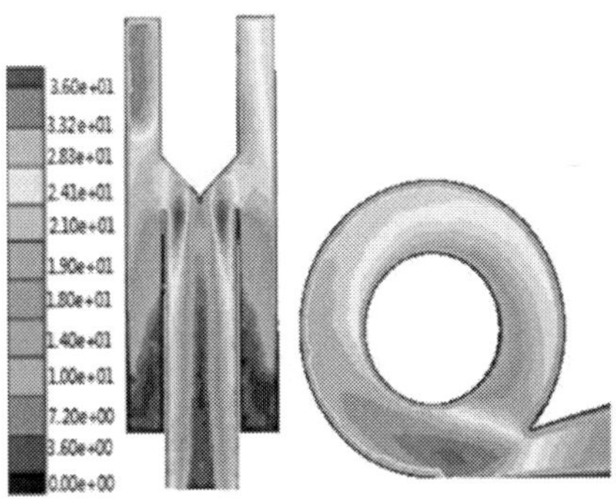

(a) Square cut converging entrance

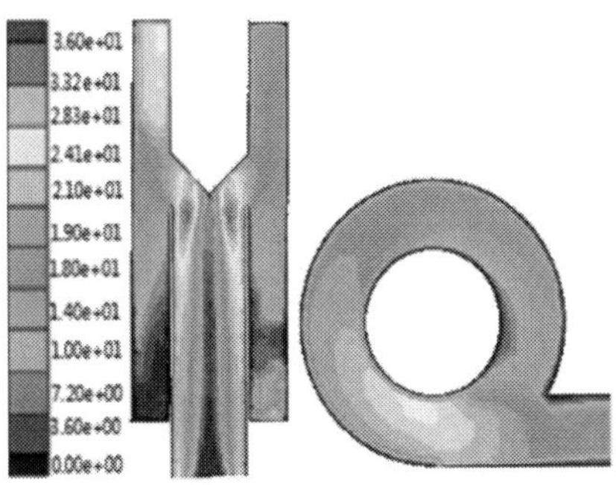

(b) Square cut normal type entrance

Figure 3. Comparison of velocity distribution under different entrance

As shown is Fig.3, when cyclone Separator is converging inlet, the speed of the area above the exhaust pipe is obviously increased. This is because the gas and solid mixture enters the tapered pipe, the inlet section is gradually reduced, which cause to the velocity of gas is increased. And the penetration ability of particles is also enhanced. More particles can pass through the vortex field to the wall of the separator. Meanwhile, the separation efficiency of the separator can be improved. Thus the above factors are taken into consideration: this article does not require the use of converging entrance structure.

3.2 Effect of guide pipe on separation efficiency and pressure loss

As shown is Fig.4, it is the numerical simulation results of two kinds of cyclone separators with different structures. Compared with a conical guide tube, the inlet velocity of the cylindrical guide pipe is larger, and a strong vortex is generated. Meanwhile, the cross section of the exhaust pipe suddenly reduced, resulting in a sudden increase in the speed of air whirling, which leads to the increase of pressure loss.

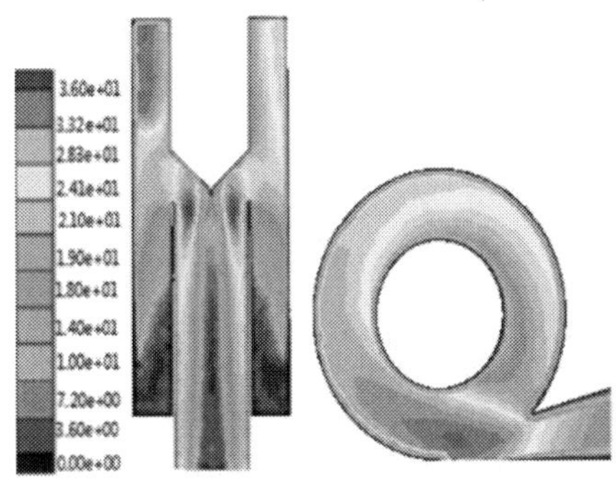

(a) The conical guide tube

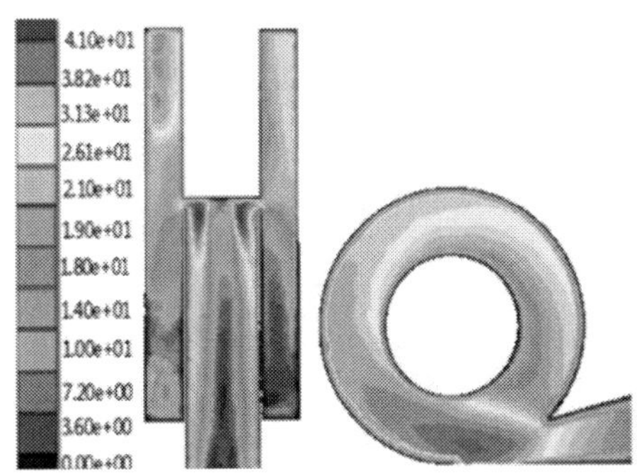

(b) The cylindrical guide pipe

Figure 4. Distribution of velocity

Can be seen from the Fig.5, Nearly all of the static pressure of the cylindrical guide pipe is larger than that of the conical guide pipe. Through the comprehensive analysis, the separation performance of the conical guide pipe is better than that of the cylinder guide pipe.

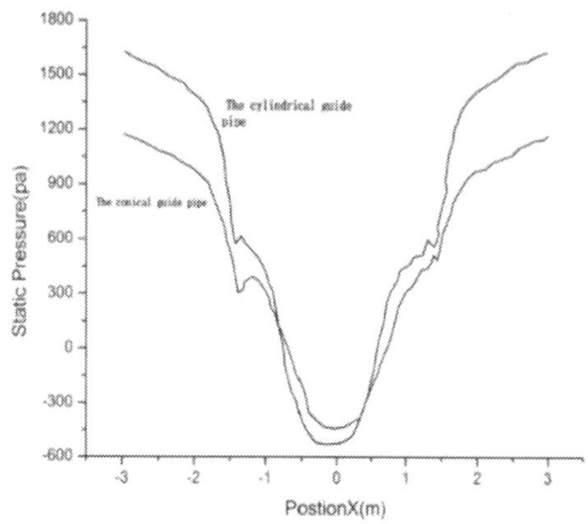

Figure 5. Comparison of static pressure distribution on the inlet diameter of exhaust pipe

3.3 Effect of the guide pipe length (h_c) on separation efficiency and pressure loss

As shown is Fig.6, the velocity field change of the two cyclone is consistent. But the main difference is that strong swirl of exhaust pipe began to grow with h_c increases, and longitudinal symmetry also increases. Thus we can draw a conclusion that the effect of h_c closer to the 1.3a will be more obvious. Through the comprehensive analysis, this paper selects $h_c = 1.25a$ as the most reasonable structure.

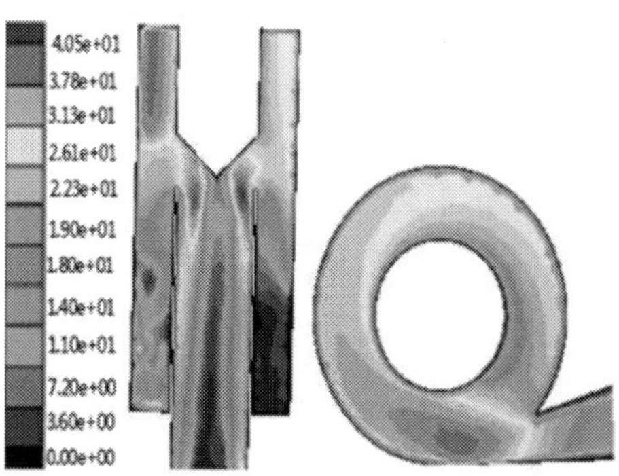

(a) h_c=a

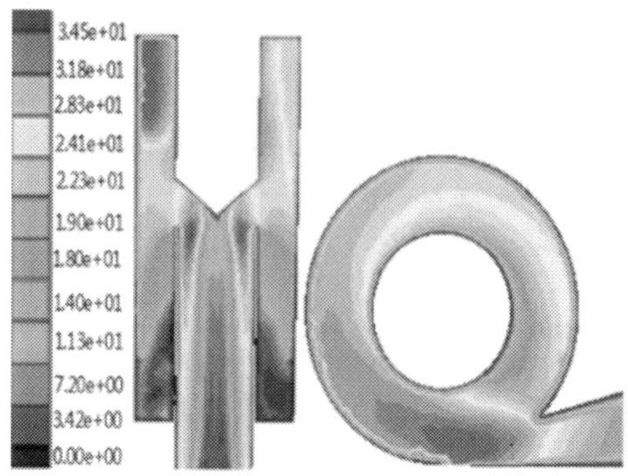

(b) h_c=$1.3a$

Figure 6. Velocity distribution of h_c=a and h_c=$1.3a$

4 CONCLUSIONS

The DPM model provided by ANSYS software was applied to study flow field of a cyclone separator used in some circulating fluidized bed (CFB). Thus compared different simulation results with reference model, separator are optimized to provide reasonable theory guidance for the optimization of industrial design.

When the square cut converging inlet is employed, the separation efficiency of the cyclone separator increases as the velocity increases, while the vortex flow also increases and the pressure loss increases obviously.

Compared with conical guide pipe, the cylindrical guide pipe entrance has a higher velocity and a stronger vortex.

With a certain range, the selection of nozzle length needs a comprehensive consideration of the influence of the separation efficiency and pressure loss.

References

[1] K.W. Chu, B. Wang, D.L. Xu, et al. CFD–DEM simulation of the gas–solid flow in a cyclone separator. Chemical Engineering Science, 2011, 66(5): 834-847.

[2] K. Elsayed, C. Lacor. Optimization of the cyclone separator geometry for minimum pressure drop using mathematical models and CFD simulations. Chemical Engineering Science, 2010, 65(22): 6048-6058.

[3] J. Gimbun, T.G. Chuah, A. Fakhru'l-Razi, et al. The influence of temperature and inlet velocity on cyclone pressure drop: a CFD study. Chemical Engineering and

DOI: 10.1051/matecconf/201713900093

Processing: Process Intensification, 2005, 44(1): 7-12.
[4] Oh J, Choi S, Kim J. Numerical simulation of an internal flow field in a uniflow cyclone separator. Powder Technology, 2015, 274:135-145.
[5] I. Mokni, H. Dhaouadi, H. Mhiri, et al. CFD Simulation Of The Flow Field In A Uniflow Cyclone Separator. Engineering Letters, 2009, 17(3): 173-177.
[6] P. González-Tello, F. Camacho, J.M. Vicaria, et al. A modified Nukiyama–Tanasawa distribution function and a Rosin–Rammler model for the particle-size-distribution analysis. Powder Technology, 2008, 186(3):278-281.

MATEC Web of Conferences 139, 00094 (2017) DOI: 10.1051/matecconf/201713900094
ICMITE 2017

Implementation of news module for news client based on ApiCloud

Xin Fu[1,a], Yu Liang[2,b], Sanxing Cao[2,c], Hongbo Gu[3,d]

[1] Electronic and Communication Engineering, Communication Unversity of China, Beijing, China
[2] New Media Institute, Communication Unversity of China, Beijing, China
[3] Information center, State Forestry Administration of China, Beijing, China

Abstract. With the development of new media technology, news client has become the main battlefield of news browsing. Based on the ApiCloud hybrid development platform, this paper uses HTML, JavaScript and other technologies to develop the mobile client news module, and uses WAMP integrated development environment to build a news publishing system on the server side.

1 INTRODUCTION

The traditional native development of App needs to use a platform-specific language for development. iOS platform using Objective-C language, Android platform using Java language, to generate a specific client to adapt to the platform. Web applications uses the embedded terminal browser, front-end uses HTML, CSS, interactive response through the JavaScript to achieve mobile application

Mobile clients are mainly divided into three types: native, web and hybrid development. ApiCloud consists of two parts: cloud API and port API, which can help developers quickly realize the whole lifecycle management of mobile applications, including development, testing, publishing, management and operation. The overall architecture of the ApiCloud is shown in figure 1.

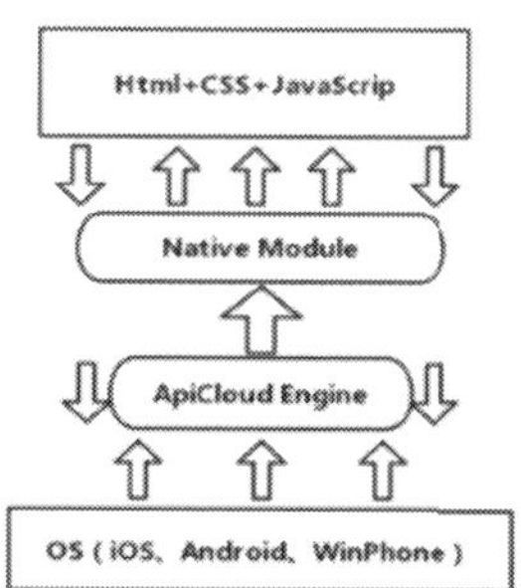

Fig. 1. Overall architecture of the ApiCloud

2.2 Apache server

development. The hybrid application development program adopted in this paper is the ApiCloud development platform, which implements the functions of news uploading, news channel setting and other functional modules.

2 INTRODUCTION OF RELATED TECHNICAL

2.1 APICLOUD

Apache HTTP Server (Apache) is a Web server software, its utilization rate ranked first in the world. In almost all the computer platform, Apache can run well. It is fast and reliable and has a powerful characteristics of crossing platform. It performs stably and it supports Web server applications which applied crossing platform, and it is widely used by many developers and websites.

2.3 PHP

PHP (Hypertext Preprocessor, hypertext preprocessor) is a universal open source scripting language, the file suffix called .php, which combines the C language, Perl and Java features. You can access the database using PHP. PHP is suitable for the field of Web development, it can carry out dynamic webpage faster, more efficient than other programming languages. Due to the Web system is to embed the program into HTML, and it almost matches all of the current database, such as MySQL, Oracle, Sysbase, SQLServer, dBase etc.

2.4 MySQL database

MySQL is an association database management system

developed by Swedish MySQL AB. Instead of setting all the data together, but the data would be stored in different tables, so as to speed up the access speed, and clear the database structure design, improve the flexibility of database structure. MySQL has the characteristics of open source, small size and fast speed. It is widely used in all kinds of systems under the Internet environment. It can match Apache and PHP to form a development environment with excellent performance at the same time.

2.5 Apache+PHP+MySQL

In this paper, the B/S architecture is used to implement the system. In the Browser, based on ApiCloud, the client interfaces display and user interaction functions. Server uses PHP to complete the implementation of functional components, in order to complete most of the function in logic. And the Server uses easily developed environment XAMPP as the operating environment. XAMPP (Apache+MySQL+PHP) is a full-featured integrated development environment in Web, it can run on the platform of Windows, OSX, and Linux, cross platform superior. News client is based on Windows APP background Server 2012 as the operating system, using Apache as Web server, MySQL database, development language is PHP, which can control the connection with database and ApiCloud client package.

3 Demand analysis for news client

3.1 Demand of front-end interface

In the front end of the news client, the home page is the most important and complex page. The news client designed in this paper contains the top and bottom navigation bars. The navigation bar contains many categories, such as headline, hot spot, Beijing and so on. The middle part of the home page is a thumbnail display of graphic news and video news. News is divided into graphic news and video news. Click on the news list, access to landing news page. At the bottom of the atlas are news sources, times, comments, and information.
At the same time, the user's personal center page is also an indispensable demand in the news client. The client's personal center page draws on the design of the NetEase news client. Including user login, user registration, news collection and other functions.

3.2 News release system

The client's news release system, news shown on the front end all need to edit the news materials uploaded to the database for client to read, display and use. The news includes pictures, news and video news. User reviews management is also the function of this system, which can delete the user's unqualified comments. These are the core components of server-side code.
In this news release system, the administrator logins in, inputs the administrator name and password to enter into the editing system, then they can edit text news, video news. And the system will store the operation information of administrator. The detailed development module will be described in the fourth chapter. The overall requirements architecture relationship of this news client is shown in figure 2.

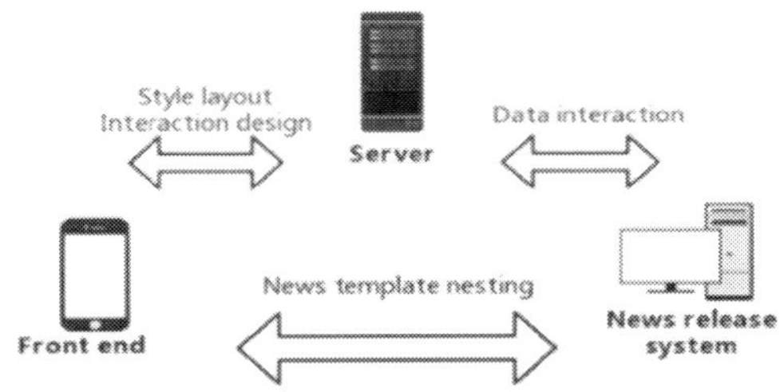

Fig. 2. Total demand framework

4 the design and implementation of news client module

The core of the news client is the news release system, and all the news content displayed at the front end is edited and uploaded through the background, and finally appeared at the front end interface. The background system of the news release system for editors includes uploading news, adding, deleting, modifying, user reviews management and other functions. The back-end server mainly handles various requests sent by the mobile client, and realizes the front presentation of the data.
The news client is divided into 4 modules, each module is responsible for different functions.

4.1 Channel settings module

The news client news is designed in different channels. The news channel setting system includes three parts: the setting of the channel, the modification of the channel name and the deletion of the channel.
The web page transmits the form through the post and passes the data to the addCate function in the doAminAction.php file, where the addCate function receives the data and writes the data into the database, form transfers code. Create article table storage uploaded news according to the demand, which sets field ID (int) as the primary key, the video thumbnails and memory address. Using ',' connected them into a string, and then stor tag (varchar) said that this news is hot or recommended, title (varchar) said the news' headlines, content (text) said the

theme of the news' text, imgs (varchar) said the news' thumbnails and video address, type (varchar) said the news' category, editdate (varchar) said the news' editor of the date, channel (varchar) said that belongs to which channel, editor (varchar) said for authors and editors, imgsarticle (varchar) said the thumbnails and video upload. While imgs and imgsarticle all represent the thumbnail, but their contents are not the same, the former is filtered. The latter is extracted, APP reads directly when the client request, and imgsarticle is the address uploaded through local without filtering. The specific settings are shown in Table1.

Table 1. Setting of database

Attribute name	**Type**	**Allow for empty**	**Auto_INCREMENT**
id	int(8)	no	Yes
tag	varchar(10)	no	Yes
title	varchar(50)	no	Yes
content	text	no	Yes
imgs	text	no	Yes
type	varchar(10)	no	Yes
editdate	date	no	Yes
channel	varchar(20)	no	Yes
editor	varchar(10)	no	Yes
imgsarticle	text	no	Yes

4.2 News upload module

The news upload module is the core of background editing, background administrator can upload news by setting the account after landing good news channel, including graphic news and video news. Different news can be related to different channels. Graphic news uploading includes thumbnails of pictures, news headlines, text editors, editors, authors, date settings, marking news for hot spots, and recommendations. Video news uploading includes video thumbnails, news headlines, text editors, editors, authors, date settings, marking news for hot spots, and recommendations.

The data uploaded by video news achieves the transmission by the form, and uploads the video to the specified folder and rename through the KindEditor. And choosing which channel the video is stored on, needs to retrieve the data stored in the database in channel, traverse it, and then store it in the select tag for selection. And accepting form data addPro1 function, accepting thumbnails and video data needs to extract the two from the KindEditor raw data, rearrange them, and then deposit them into the database, which is applied to regular expressions. By using regular expressions to extract the memory address of the uploaded thumbnails and video, and then deposit the string connected by ',' into the field of imgs. The client reading the field is only need to use the "split" function, it can separate thumbnail address and video address.

Graphic news uploads and video news uploads are about the same, but there are differences, thumbnails. Only need to upload one thumbnail news video can, and graphic information in the client has two display modes, one thumbnail mode and three thumbnail mode, which requires the need to determine the number of thumbnail stored in the database. More than three pieces is automatically selected three picture mode. The extraction of image address is through regular expressions, while extracted array automatically extracted data using the "," to connect. Single array directly transfer into a string and stored in the imgs field. The client first split the string in the imgs field by 'split' function when request comes, while the array number is greater than 2, it comes to the three display mode, the number of 1 or 2 when using only the first array value as a thumbnail display address.

4.3 User comments management module

After registered users logging in the news client, you can use four functions including comments, modifying personal information, collection news, sharing news. For some users' inappropriate comments, the system administrator from the background landing system has the right to delete user' comments, but did not permit to modify the users' comments.

4.4 Server response module

The server is mainly used to store CSS style files and JavaScript interactive files designed at the front end, as well as news and databases. Under normal circumstances, using AJAX for cross domain requests is prohibited because of security problems, so you should make a change in the server file. Each PHP file needs to have header ('Access-Control-Allow-Origin: *) statement, the statement that allows for cross domain requests.

Each entity in the database is associated with the overall E-R diagram, as shown in Figure 3 below.

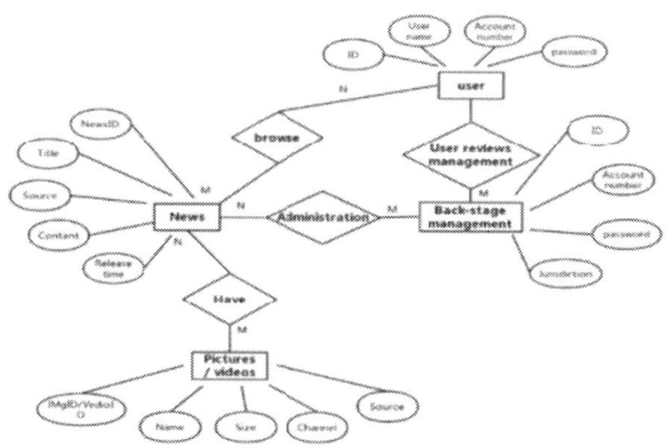

Fig. 3. Database entity association E-R diagram

5 Conclusion

The news client implemented in this paper is a new APP designed for iOS and Android smart phones and other intelligent mobile terminal user users. In the design aspect, using the flat design conforms to the handset user's operation custom. In terms of technology, the use of advanced HTML5+CSS3+JS technology sends mobile phone users to a light, smooth operation experience. In terms of function, compared to the PC terminal, the content structure is rearranged and designed to make it more concise and meticulous, more focused on content reading, more suitable for touch screen operation.

6 Acknowledgements

The work is sponsored by China Forestry Mobile Internet Development Strategy Research Project (2016 — R23) granted by State Forestry Administration Information Center and New Media Research Institute of Communication University of China.

REFERENCE

1. Wei Yabin. Design and implementation of mobile game news system based on HTML5[D]. Zhejiang University of Technology, 2015.
2. Guo Jundong. Design and implementation of recommender system based on wireless news client [D]. Shandong University, 2016.
3. Hu Yonghua, Dai Lin. Design and implementation of mobile campus news client [J]. information communication, 2015 (1): 110-111.
4. Li Yuan, Nie Yubao. IPhone platform network news client time module based on [J]. Journal of Shaoguan University, 2012, 33 (6): 18-22.
5. Li Ting. Design and implementation of science and technology news recommendation system based on iOS platform [D]. Dalian University of Technology, 2015.
6. Ding Nan. Research on mobile products based on HTML5 and CSS3's NetEase WAP touch screen version [D]. Huazhong University of Science and Technology, 2012.
7. Lane D. Web Database Applications with Php and Mysql[M]// Web database applications with php and mysql, second edition. O'Reilly Media, Inc. 2004.
8. Ding S Y, Cheng Z Q. Android Client Development based on Moodle Platform[J]. International Journal of Advanced Science & Technology, 2016,87

Research on Blended Teaching Mode Based on Gaoxiaobang Platform

Haoyan Shao[1,*], *Na Qiao*[1], and *Xiaohui Zhang*[2]

[1]School of Distance and Continuing Education, Dalian University of Technology
[2]School of Computer and Information Technology, Liaoning Normal University

Abstract. With the gradually deeper development of Higher education informationization and online open educational pattern, blended teaching mode has become a common concern. We practically use Gaoxiaobang course platform to carry out blended teaching mode in the course of Piano art and classics enjoy, and research teaching case to provide reference for blended teaching mode. **Keywords—MOOC.**Blended teaching mode; Gaoxiaobang course platform; Piano art and classics enjoy

1 Introduction

With the fast development of network, educational circles are full of teaching mode based on MOOC, and online open educational pattern makes most people of educational field looking forward to use MOOC to replace traditional classroom teaching. But the MOOC itself exists deficiency, and need face-to-face education to supplement. [1]Blended teaching mode combines advantages of traditional classroom teaching and online learning better. To carry out blended learning successfully, we need a network course platform to support learning on which course teacher can share teaching videos they made related with the course content, and upload auxiliary teaching resources to expand students' knowledge range such as PPT、PDF documents, etc. Teachers hope that the teaching platform has modules to discuss and communicate, and teachers and students can freely express their opinions. Furthermore, teachers hope to observe the students learning process and learning situation to evaluate students comprehensively by the platform. This paper takes the case for Cross-school course of Piano art and classics enjoy in Liaoning province of Dalian University of Technology open course teaching centre, to research blended teaching mode.

2 MOOC and blended teaching overview

2.1 MOOC Overview

2.1.1 The Concept of MOOC

MOOC（Massive open online course） is massive open online courses, and it often referred to as "Mu Ke" in Chinese. Three course providers represented by Coursera、Udacity and edX provide a platform for the development of MOOC. [2~3] In recent years, universities at home and abroad have joined MOOC,and developed rapidly. MOOC develop so rapidly by reason of the potential of enriched traditional classroom. MOOC meanwhile has many shortages, such as lack of communication between teachers and students and low attendance, etc. Therefore, we should treat MOOC dialectically.

2.1.2 The Characteristics of MOOC

MOOC has the following characteristics: ①massive. Massive means plenty of enrollment and abundant curriculum resource; ②open. Open means. learning space and learning resource is open in MOOC,and all the courses students interested in can be registered to study; ③online. All the learning process is done through the internet, such as teacher teaching、students learning 、 teacher-students/ student-students interaction 、 homework finishing and homework correcting; ④course. Course represents that the content provided by MOOC is complete course, and it is a progression that the MOOC relatives to open education resources. Open education resources provide many excellent resources, but all the learning process is incomplete because of the lack of teachers' guidance and companion's mutual help, therefore excellent educational resources are failed to make full use. Then the MOOC appears in the form of course with specific classes begin time and classes over time, and it can improve the finishing rate and the use value of education resources greatly.

2.2 Blended Teaching Overview

* Corresponding author: network4@dlut.edu.cn

Blended teaching is not only a kind of teaching form, is also a kind of teaching strategy, and the definition of it at home and abroad experiences from broad to narrow.

In a broad sense, blend in blended teaching includes: blend of learning theory、blend of learning resources、blend of learning resources、blend of learning style and blend of learning style。[4~6]This explanation led to confusion of the blended teaching concept, because of the excessive contents it contains, and blended teaching lost the significance to be a independent concept due to this definition.

In narrow sense, with the constantly understanding and in-depth research of blended teaching theory, the definition of blended teaching has gradually been refined internationally, and the combination of face to face traditional teaching mode and blended teaching mode has been focused on. Bonk and Graham have defined blended learning in the "blended learning handbook", they believe that blended learning is a new learning form which gathers the feature of Face-to-face teaching and many technology medium teaching. India NIIT Company describes blended learning as a new learning form including entity classroom teaching, online learning and learners' active learning in the book of "blended learning white paper".

Blended teaching is a new teaching mode which combines the advantages of traditional classroom teaching and online learning by the view of teaching designers, complement each other to promote the teaching goal better achieved. Blended teaching combines traditional classroom teaching and the latest information media means on the teaching method, adopts online and offline form in the teaching form, takes full advantage of teachers' domination and students' learning autonomy in the relationship between teachers and students. It establishes democracy and equality relationship on the relationship between teachers and students, and adopts the evaluation method which combines formative assessment with summative assessment.

3 Blended teaching mode of Piano art and classics enjoy based on Gaoxiaobang platform

3.1 Introduce of Cross-school course in Liaoning province

Cross-school credit study project of Piano art and classics enjoy is sponsored Education department of Liaoning province which is established in the form of educational reform project. Cross-school credit study course carries through teaching for full-time recruitment undergraduate level student, and course credit is recognized together by office of academic affairs and teaching teacher who use the course、video building teacher, the final result is depended on which office of academic affairs have recognized and published Cross-school credit study course is different from traditional classroom teaching mode, and newly increase a medium of network platform on the basis of the interaction between teachers and students of traditional teaching. The teaching of cross-school credit study course is a blended teaching mode based on online education, hope to change the deficiency existing in the traditional classroom teaching which mainly includes lack of teaching content novelty、teacher's guidance after class and learning resources.

The exploration of blended teaching mode on Gaoxiaobang course platform is divided into three types, as shown in the table 1.

Table 1. The exploration of three types of blended teaching mode on Gaoxiaobang course platform

Teaching Mode	Students Task	Teacher in Class	Video Building Teacher	Characteristic
Flipped Classroom	1、Complete knowledge study at home 2、teacher-students、student-students interaction in classroom	1、teacher-student interaction , question answer and discussion in classroom	face to face teaching teacher is video building teacher	Grading assignment is accomplished by video building teacher. Can make up the problem of weakness of teaching staff in some subjects in universities.
Half-flipped Classroom	1、Complete knowledge study at home 2、listen to the teaching content not existed in class	1、add the missing teaching content to video courseware 2、teacher-students interaction in classroom	face to face teaching teacher is the teacher of the school which use the course	Grading assignment is accomplished by the teacher of the school which use the course
complete auxiliary	1、Normal classroom listening 2、use video to supplement or revision offline	Normal classroom teaching	face to face teaching teacher and video building teacher is the same person or from the same school	Grading assignment is accomplished by teaching teacher

3.2 Introduce of Gaoxiaobang course platform

Arising of MOOC 、 micro-Course and flipped Classroom is impacting traditional education. The development of online education has become the world trend. In continuing education, the MOOC can be quality-oriented education, and it can also be applied in curricula education. In order to enhance open course resource construction and education work, promote quality education teaching resources sharing, use information technology to promote the teaching level, promote teaching reform and improve the quality of teaching, Dalian University of Technology founds open course teaching centre in October 2013 to enhance open course resource construction, and school of distance and continuing education takes charge of the centre' specific operations.

With Dalian University of Technology open course teaching centre work being carried out, a platform named Gaoxiaobang generates under the deployment of the college.

Open course teaching work takes Gaoxiaobang course platform as a online education medium, professional teachers for guidance, unite colleges and universities teacher to carry out teaching and explore all kinds of new type of teaching mode together. Figure 1 is Gaoxiaobang course platform login page.

Fig. 1. Gaoxiaobang course platform login page.

Gaoxiaobang course platform can scan code to download mobile phone APP for mobile learning, and it helps the students make full use of fragmentary time, break through space constraints, and learn the course at any time and any place. It is easy to obtain and has flexible schedules, so the students are inspired to participate in courses actively.

Gaoxiaobang course platform is the medium of network teaching, and teachers and students complete course teaching and learning on it. It has three function modules after logging in the platform, and they are courses square, my study and administrators entrance. The courses square collects 282 high-quality teaching course resources including 50 Liaoning province classic shared classes,112 Liaoning province Cross-school credit study course,11 Graduate student course,6 undergraduate course.

3.3 Blended teaching mode practice of Piano art and classics enjoy based on Gaoxiaobang course platform

Since the spring of 2015, the exploration of blended teaching mode of Cross-school course of Piano art and classics enjoy in Liaoning province is carried out. The platform invites teaching team of HuangCen teacher from Northeastern University recording video and using the related teaching resources they provided, and Dongbei University of Finance and Economics has carried out blended teaching of that course in the mode of flipped classroom for teaching attempt.

According to the requirement of the teacher given course, Gaoxiaobang course platform realizes all the traditional classroom teaching link, including video of classroom teaching, homework assignment, student work submission, student work review, the interaction between teachers and students in the teaching process, and final examination results generated and export.

The course of Gaoxiaobang platform is build by remedial teacher, and platform interface of remedial teacher as shown in figure 2. The remedial teacher enters into by administrators entrance, and starts a new course construction from clicking on course creation, then add the content course teaching needs of video,PPT, homework, tests, course introduction, teacher information to the platform. The teacher issues the course after finishing the course construction, then builds new class and adjusts the related content of teaching and examination according to the requirements of different semester. The course of piano art and classics enjoy has rich teaching resources, and the remedial teacher build course and maintain platform, meanwhile the video production teacher HuangCen carries out face to face teaching to the class of Dongbei University of Finance and Economics which uses the platform course. She also interacts with students, course answering for students, carries out the discussion, and assigns homework.

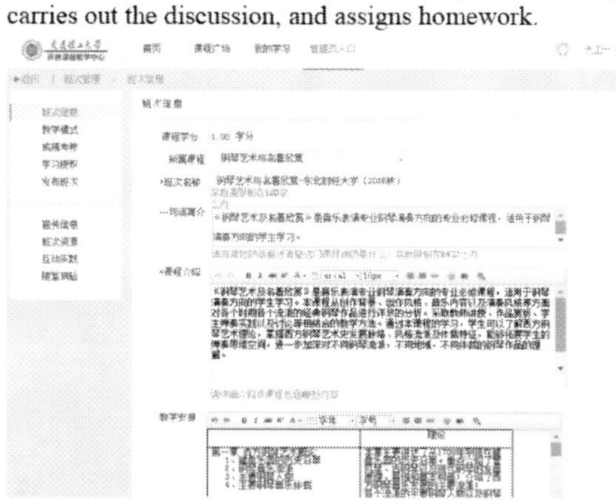

Fig. 2. Gaoxiaobang platform interface of remedial teacher.

* Corresponding author: network4@dlut.edu.cn

The remedial teachers get involved in the capacity of platform technical support and Course guidance teachers in the whole teaching process, help the teaching teacher strengthen the teaching management by multi-channel communication and supervise with the teachers and student. The remedial teachers' work task is shown in figure 3.

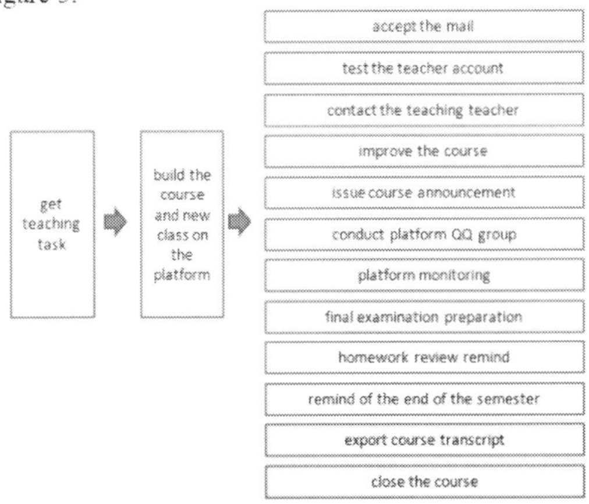

Fig. 3. Gaoxiaobang course platform remedial teacher work flow diagram.

The course examination form is composed of video, homework, tests, discussion, examination five parts according to the percentage composition, after the communication with the teaching teacher, the course final examination form percentage composition is: platform video record 20%, platform test 20%, classroom test 40%, furthermore, homework and discussion are used as self-test assisted learning which are not count to the final record. In the final stage, the remedial teacher derives course transcript from the platform, and deletes the list which is not count to the final record, then compiles course transcript to the teaching teacher.

4 Blended teaching mode practice and the analysis result

We carry out the blended teaching mode practice of Piano art and classics enjoy in undergraduate students of Dongbei University of Finance and Economics, and have successfully completed three semesters teaching task. In order to understand student perceptions of blended teaching mode, we carry out questionnaires and interviews to the students of three semesters, and the interviews mainly get through network, QQ chat, E-mail and other forms.

The results show that the blended teaching mode has great effect on improving students' interest and teaching effect.

(1) Most of the students said they would like to study on Gaoxiaobang course platform, and blended teaching has made up for the inadequacy of traditional classroom teaching mode.

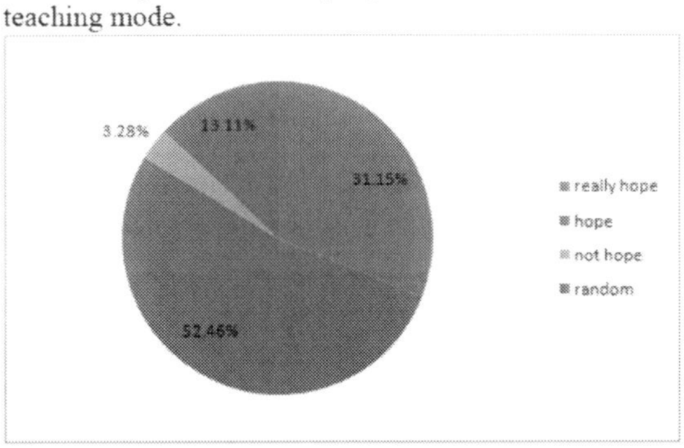

Fig. 4. Whether students hope to study by this mode.

(2) Students said, the diversification of teaching forms increases the students' interest in learning piano art and classic appreciation course, more teaching resources and expand resources on the platform are used to facilitate students' learning and improve the students' classroom participation.

(3) Students also said that the teaching mode expands the autonomous learning space, and their mastery of knowledge level has improved significantly by specific learning according to their knowledge.

(4) Many students said that this mode increased their interaction with the teacher and the students communicate more frequently, and they present original ideas and opinions broaden their thinking in the discussion on the platform. The course learning opens the test limits, and better explores the course content.

At the same time, the writer also gathers the defects in the blended learning of teachers' and students' feedback in the process of interviews, the content is as follows:

Table 2. The main deficiencies and suggestions for improvement of blended teaching mode get in the interviews

	Problems	Improvement Suggestions
teacher	(1) platform operation is not enough skillful (2) teaching practice increase teachers' workload (3) homework sets weeks ahead, individual students can't keep up with the rhythm.	(1) compile platform operation facilitation manual, teachers and students are more easy to master (2) students to do voluntary teaching assistant, set mutual evaluation of homework, reduce homework marking workload. (3) extend the effective time of network learning resources and homework, convenient the teachers and students.
student	(1) time to complete the task is limited (2) some students' learning efficiency is low, and not willing to try new things.	(1) the teacher should understand the idea of students actively and change the teaching design at any time, then select the most suitable teaching design for students' development.

	(3) individual module on platform difficult to use, password forget, network problems, etc.	(2) Teachers increase humanistic care to students. (3) further improve the platform function module, make the interface more friendly.

5 Conclusions

Blended teaching ensures that students get systematic knowledge in the traditional classroom, on the basis of advantages of teacher guiding role and good study atmosphere, the learners can learn any knowledge in any time and any place with the infinite extension of time and space, and it satisfies the learners' desire for learning. [7~8]The practice of blended teaching mode of Piano art and classics enjoy based on Gaoxiaobang course platform concludes that it has a great help to improve the teaching quality and promote students' individual development. It is a successful attempt that blended teaching mode applies to Piano art and classics enjoy course teaching, and blended teaching mode based on Gaoxiaobang course platform will play a greater role in schools teaching of various levels while the platform function further improved and the network education resources increased.

References

1. Ping Wang, New Development and Application of Massive Open Online Course:From CMOOC to x MOOC[J]. Modern Distance Education Research,2013,(3):56-62.
2. Zaihong He. Teaching Research of C lLanguage and Program Design Course Based on Blended Learning[D]. Changsha: Hunan University,2008,(5),15.
3. Li Yang,Dongsheng Zhao, Research of Blended Learning Based on Moodle Platform [J]. Journal of Capital Normal University(Natural Science) ,2010,(1):6-9.
4. Jide Wang,Yingying Feng.Ying Wang. Cold Think behind the Hot MOOC[J]. Education research, 2014, 9.
5. Kekang He, Look at the New Development of Education Technology Theory from Blended Learning[J]. E-education Research， 2003(3)： 6—10.
6. Apostolos,K., Rebecca J. Hogue. How to Succeed in a MOOC- Massive Online Open Course.Learning Solutions Magazine, October 8,2012.
7. Laura Pappand. The year of the MOOC. The New York Times. November 2, 2012.
8. Laura Pappand. The year of the MOOC. The New York Times. November 2, 2012.

Design of Technical Support System for Retail Company Based on Cloud

Ping Shao[1,2,*] ,*Fei* Ye[1,2] ,*Xiangyang* Wang[3] ,*Yanmin Guo*[1,2] ,*Guoning* Gao[3] , *Jian* Geng[1,2]

[1]China Electric Power Research Institute, 210003, China
[2]Electricity Marketing Operation Technology Laboratory of SGCC, Nanjing, 21003, China
[3]Zhejiang Electric Power Trading Center Co., Ltd, Hangzhou, 310037, China

Abstract. With the retail side of the market in China, the sale of electricity companies as a new source of power retail, they participate in the electricity market business. National and local governments subsequently introduced the corresponding policies and rules, the technical support system becomes one of the necessary conditions for the access of the retail company. Retail electricity companies have started the system construction, but has not yet formed a standardized, complete architecture. This paper analyzes the business and data interaction requirements of retail electricity companies, and then designs the functional architecture based on basic application, advanced application and value-added application, and the technical architecture based on "cloud". On this basis, the paper discusses the selection of private cloud, public cloud and mixed cloud model, and the rationalization suggestion of system construction. Which can provide reference for the construction of the technical support system of the domestic retail enterprises.
Keywords. power market; cloud; retail electric company; technical support system; system structure

1 Introduction

2015, "China Development and Reform Commission on the 9th document" established the current stage of China's power reform direction, the document clearly put forward the "steady promotion of retail electricity side reform, orderly release of retail electricity business to social capital" [1]. In November the same year, the National Development and Reform Commission, Energy Bureau issued a supporting document "on the promotion of retail electricity side of the implementation of the views of reform", in October 2016, issued a document "retail electricity company access and exit management rules", it further clarified the responsibilities and orientation of market members. Requiring the retail company "should have a fixed scale of business with the fixed place of business and power market technical support system and customer service platform" [2]. "Guangdong Province, the implementation of the pilot side of the implementation of the pilot reform program," clearly requires retail companies with user management, trading, billing and other functions of the technical support system [3]. As a result, it becomes one of the most important jobs for retail companies to enter the market that building technical support systems to meet market requirements.

Retail electricity company technical support system is different from the traditional power information system, one is the use of various types of small and medium-sized power users and retail electricity companies internal users; Second, China's electricity market has just opened soon, the business is still in the development stage, Its technical support system is also the side of the construction side to explore, it did not accumulate enough technology; the third is most of the retail electricity companies do not have a strong information technology and operation and maintenance team conditions. Simple to only through the rental cloud system, end users can use including PC, laptop or a smart phone, they can complete the power for the user information exchange and internal operations and other tasks.

"Cloud" appears, is to meet the latest Internet technology development direction, with high reliability, data processing capacity, flexible and scalable and high utilization of equipment advantages [4,5]. Cloud technology has matured in the power industry scheduling, marketing and other specialized systems [6-9], while the retail electrical company technical support system is a completely new information system for the Internet. The majority of small and medium-sized retail electricity companies with cloud technology, they do not bother to support the storage, Internet, servers and other information resources. Energy Internet and large data applications is the future development trend of retail electricity companies [10], only the Internet-facing cloud system architecture have the ability to achieve the future of large data integration, analysis, they are possible to provide users with better energy-packages and energy-solutions. At the same

* Corresponding author: shaoping@epri.sgcc.com.cn

time, the cloud architecture is also an information system of operational outsourcing [11,12], it can effectively reduce system construction and operating costs.

This paper analyzes the business requirements of the retail electricity company, and then designs the technical support system architecture of the retail company based on "cloud", and describes in detail the functional architecture, technical architecture and security measures of the system, and finally discusses the "cloud" architecture of the selection, this paper has the characteristics of low cost, high reliability, flexible and easy expansion, and the design idea has been applied in the construction practice of some retail electrician technical support system.

2 Requirements analysis

2.1 Business needs

Retail electricity market has been more mature in foreign countries, power users, distributed energy owners have been able to participate in the retail electricity market bidding decision, energy efficiency analysis, demand response business [13,14], the retail business is the core needs of the company in China. Electricity customers are the terminal service objects of the retail electricity companies, the electricity as the service carrier, the technical support system as the interactive tool, around the customer's electricity consumption to carry out various services.

China's power purchase transactions mainly in the annual long association and monthly bid trading, retail companies must understand the rules of transaction bidding in order to seize the initiative in the market. It is not only required to manage, track and analyze the purchase contract, but also to support the analysis of the results of the successful bidding of the historical transaction, to automatically generate the next round of market bidding through the auxiliary decision function to achieve the best cost of purchasing power.

The retail business is a power retail business for electricity customers. Customers are the core resources of retail electricity companies, in the fierce competition in the electricity market, the better customer service, increase customer viscosity, which is highly concerned about the content of retail electricity companies. In addition, after further liberalization of the market, retail companies must also face a series of problems caused by the explosive growth of power customers, how to effectively maintain customer information, rapid response to customer needs and improve efficiency, will become a retail company must consider problem.

The new round of power reform not only revitalize the retail electricity market, but also to promote integration with other related industries, retail companies to provide diversified value-added services to open the door. Energy efficiency control, demand side management, electrician on behalf of Victoria and so on are value-added services, these value-added services will be launched around the sale of electricity business, reflecting the value of retail professional services [10].

Therefore, the technical support system should be through information technology means to analyze the customer's electrical characteristics, rich design of electricity packages, to provide users with energy-saving recommendations such as personalized care service. These seemingly trivial features can greatly enhance the power of customers electricity experience and retail electricity service quality.

2.2 Data interaction requirements

To meet business needs, retail companies need to interact with third-party systems. Such as from the power trading center to obtain open transaction data, to the power trading center to provide quotations, retail contract reporting, etc. To obtain the power acquisition data from the distribution system; to obtain the grid model data from the dispatch system for bid decision, etc., these data will effectively support the normal operation of the retail electricity company business.

The current situation, regardless of the State Grid or the Southern Power Grid within the scope of the power exchange center at all levels failed to provide for these data needs of the interactive interface. It is reported that the relevant units have been in the lead in the development of trading platform and retail electricity company technical support system information industry standards. The mechanism will not only meet the basic needs of transaction settlement in data quality and timing, but also will help improve the overall quality of service in the retail electricity market [15]. Therefore, when the conditions are met in the external policy and automation, the system should provide the ability to dock data with the trading platform to enhance the efficiency of power purchase and sale of electricity business.

3. System function architecture

The technical support system mainly supports the retail electric companies to carry out bid trading and electricity sales service, at the same time for its terminal power customers to provide service window, to achieve the two-business interaction. System function architecture shown in Figure 1, according to the user's classification of business needs, can be divided into basic applications, advanced applications and value-added services in two parts, a total of 10 functional modules.

3.1 Basic application

3.1.1 Customer file management

The customer model is the basic information for the sale of electricity. China's power customers to enter the market is gradually released in accordance with the voltage level, so the minimum size of the business should be "electric card number + voltage level" of the particle size. In addition, the retail electricity companies are also involved in the wholesale market power purchase business, the establishment of trading institutions, power generation companies, transmission and distribution companies such as the market model for signing contracts, settlement and so on. This part of the model should inherit the transaction platform IEC 62325 model standard [16].

3.1.2 Customer service

There is a lot of information interaction and business interaction between retail electricity companies and power customers. Should use the Internet mature technology, through the online business hall, mobile APP, WeChat and other means for customers to facilitate and efficient means of interaction.

Customer service includes public and private information. Public information includes corporate announcements, market information releases, policies and regulations, price-packages, etc., which are open to all visitors. Private information includes historical transactions, electricity, electricity, contract signing, electricity demand, information subscription, etc., these are the customer's own private information.

Web page should pay attention to the comprehensiveness of information, the convenience of operation, through visual analysis, report display, etc., it provides users with intuitive information display, and support users to download and export, mobile APP and WeChat public number is to provide users with similar Simplified version of the function, user-friendly time to grasp the market information and their own electricity data.

3.1.3 Retail electricity management

It is responsible for the management and power customers related to the contract, electricity data, electricity billing and so on. The electricity contract is the legal agreement between the retail electricity company and the electricity customer, which is the basis of the electricity settlement. Specific features include: contract editor, annual contract automatically assigned to the month, the contract text management, electronic contract generation, contract changes and so on. Contract implementation process, combined with the actual customer electricity situation, the system provides the implementation of the contract tracking, deviation from the warning, risk analysis and other information, to give customers a comprehensive information to remind. The system provides monthly settlement function, according to the energy meter count or electricity customers to fill the actual electricity consumption, combined with the deviation assessment requirements, and then electricity calculation, and submitted to the power trading platform.

3.1.4 Power management

Another important function of the retail electricity company is to carry out electricity purchase activities in the wholesale market. Features include purchase contract management, power purchase plan, power purchase settlement and earnings analysis.

Purchase contract management: including power generation companies purchase contracts, transfer contracts, transmission and distribution service contracts, including bilateral and centralized, and so on. Specific features include contract editing, contract decomposition to month, contract generation documents, electronic signatures, etc., it is similar to the sale of electricity contract, but the nature of the contract is concerned about the difference.

Transaction plan management: for the new round of annual agreements, monthly concentration of transactions, as well as the day-ahead and real-time market transactions, through the collection of electricity demand data and forecast electricity data, and finally the formation of transaction plan management.

Power purchase settlement and benefit analysis: After the implementation of the monthly, according to the actual electricity data, combined with the deviation assessment requirements, the purchase side of the contract to settle, and with the electricity trading center settlement data to check. At the same time carry out profit analysis, statistical balance of sales, used to guide the next power transaction quotes.

3.1.5 Price-Packages

Electricity price is the most important means for retail customers to attract customers. The system should support the management of various types of price-packages, including fixed price model, divided into models, fixed price plus points and other mixed mode. With the rules of the market, the supply and demand situation often changes, so the functional design needs to consider its scalability.

In addition to electricity prices, the system should also provide the transmission and distribution price, capacity price, collection price and other catalog price maintenance, in order to help customers, complete the accounting of various types of electricity.

3.1.6 Distribution management

For the distribution and retail electricity in one company, according to No. 5 views, once the power distribution business, you need to bear the measurement, charges and other corresponding functions. Therefore, for such retail companies, the system should provide electricity collection, electricity billing, charge management, equipment operation and maintenance, asset management, and other original power companies provided by the support services. Due to the high real-time requirements of distribution network monitoring applications, it is recommended to adopt the independent system to realize the data integration with the system. Once the equipment failure affects the electricity, the information can inform the customers in time. In addition, the system can also provide electrical services, new equipment, industry expansion and other functions, the system provides value-added services for power customers, agents and all customer-related business and activities.

3.2 Advanced applications and value - added service applications

3.2.1 Auxiliary decision

The market will run for some time will gradually mature, the rules are relatively stable, but also accumulated some historical data, which support the auxiliary decision-making applications. Features include: load forecasting, bid-assisted decision making, power-assisted decision making, and so on.

Load Forecasting. Based on the customer's history of electricity, forecast the future period of electricity data. The system can be predicted for a single customer or full caliber. Time covers the annual and monthly, the future can be extended to a few days ago or even to the hour. Load factors cover data closely related to customers, such as production planning, order status, maintenance plans, industry factors, holiday settings and so on.

Bid support decision. In the submission of a new round of power demand before the need to combine the power purchase contract has been signed, the future of electricity forecast, the system automatically calculates the electricity gap. At the same time, according to the electricity sales contract price, the company operating costs and estimated profits, the system automatically calculate the upper limit of the purchase price. The measured data is used as a reference limit for the price declaration. In addition, the system through the analysis and prediction of other market members of the offer behavior, to carry out business simulation, or directly predict the market marginal price and so on.

Retail Electricity Assist Decision. In the power user side, through the customer history data mining analysis, the formation of the customer's typical electricity curve, and have similar business habits of customers clustering, the formation of customer classification. By analyzing the different customer categories and price-packages, the system to help retail companies to develop a better price-packages.

3.2.2 Micro-services

In the future, retail companies can provide the price reduction trend gradually convergence, its competitiveness will be more and more reflected in the value-added services. Features include micro-network equipment management, measurement and settlement, energy storage equipment management. The operation of the microgrid can be achieved by a separate system, but requires data integration with the system.

3.2.3 Demand response

With the advancement of the spot market, demand response through real-time electricity prices to guide the market members of the electricity behavior. Features include resource management, project management, customer participation willingness collection, DR instruction issuance, real-time load monitoring, revenue calculation, demand response assessment. Retail electricity companies to guide customers through the short-term electricity behavior, to reduce the power load for a certain period of time, it can guarantee the stable operation of the grid and inhibit the rise in electricity prices, and ultimately to achieve revenue sharing with the user.

3.2.4 Energy efficiency services

Retail companies can enhance customer stickiness through energy efficiency services. The system can provide electricity monitoring, energy efficiency analysis and benchmarking, energy efficiency potential analysis, analysis, etc., the system can help retail companies understand the entire park and the various equipment within the enterprise energy situation, and put forward energy-saving recommendations.

4 Technology Architecture

The technical architecture is the underlying architecture in the TPS technical support system, which describes how to build an IT runtime environment to support data and functional architectures to ensure that the business is working properly. This section will focus on the technical requirements of the retail company to support the software system and propose an adaptive technology architecture.

In order to solve these problems and challenges, the use of virtualization technology, load balancing, hot backup redundancy and network storage and distributed technology [17-19], designed based on "cloud" technology architecture technical support system. As shown in Figure 2, the technical architecture is divided into four layers: IaaS (Infrastructure as a Service), PaaS (Platform as a Service) SaaS (Software as a Service) And presentation layers.

Infrastructure layer IaaS, implemented by a third-party cloud computing platform or a retail company's existing cloud resources for virtualization. Through the virtualization platform, the host, storage devices and network equipment and other hardware resources for unified management. The PaaS platform layer is a collection of platform resources that are versatile and reusable. The platform layer provides software development kits (SDKs) and application programming interfaces (APIs) for the upper-tier power-hungry applications to develop a test environment that forms the operating environment for the sales application business. SaaS software application layer is a collection of application software on "cloud". All applications run as micro service in the service platform, according to the degree of hot and cold interface of the gateway. The first three layers are essentially the integration of resources, respectively, IT infrastructure, development and operation platform, application software as a service, the user on-demand use [20]. The top layer of the presentation layer to provide Web, App, WeChat and multi-carrier display terminal, etc., they are through the Restful-API and back-end interaction, to achieve front and back separation.

"Cloud" architecture reduce the difficulty of enterprise deployment, saving hardware costs, improve system scalability and other advantages, but in the deployment, there are some concerns in the power customer electricity information and enterprise core data leakage and system operation security. the system needs to use such as system isolation and authority control, identity authentication, communication and data encryption [21] and other technical measures to protect the security of the system. Specific measures are as follows:

(1) system isolation and permission control. The SaaS layer of the electricity application and the system operating space of the PaaS layer of the platform are isolated, and the IaaS layer virtual resource monitoring software is used for real-time monitoring analysis. [22] The customer account of the retail electricity company can only access its internal application data table, the power user's access account can only access the application data for the power user, the two applications through the interactive service synchronization data. Through the logical isolation of two relatively independent WEB services, and ultimately to ensure the security of the system through the logical security boundaries.

(2) identity authentication. According to the different use of the system, we will be divided into three categories: the first category is the retail electricity company agent power users, the main access to the power of the user's online business hall and APP applications; the second category is the retail company's ordinary users , The main access to the system purchase and sale of electricity-related basic application functions; the third category is the retail electricity company's core users, the main use of bidding decision-making and operating KPI indicators analysis applications. The first two types of users using the account password + mobile phone dynamic password mainstream authentication, user-friendly quick access to the system. The third type of users because of the confidentiality of business functions related to access, then the use of triple joint authentication, including CA digital certificate + account password + SMS password.

(3) communication security. Retail companies and their agents all need to access the Internet through the Internet technical support system, in order to protect the stability and security of the network, the system uses at least two of the Internet access channel. At the same time, the website using Hyper Text Transfer Protocol over Secure Socket Layer protocol to prevent data in the transmission process will not be stolen, change, to ensure data integrity. In addition, the system can resist DDos (Distributed Denial of Service) services through the purchase of network hardware devices or third-party services provided, it can ensure that the system communication and service is stable and reliable.

(4) data encryption. The system uses the PaaS layer to deploy Transparent Data Encryption [23]. The system performs real-time I/O encryption and decryption on application data. It ensures that the entire database is protected by writing encryption, reading and decrypting, protecting data files and logs file. In addition, the system can also encrypt the sensitive data of the retail company through the hardware encryption machine or software encryption algorithm, such as the system of auction transactions, contract price, settlement of electricity and other confidential data using a variety of encryption algorithms to encrypt operations [24].

5 Discussion on Architecture Selection and System Construction

In a variety of "cloud" architectures, such as private clouds, public clouds and hybrid clouds, they have advantages and disadvantages in terms of data privacy, flexibility, load capacity, technical difficulty, and operational costs. As shown in Table 1, private cloud applications and operational data in the self-built system, the privacy of its data has a higher protection, but because it can only be deployed on its own server, not only the existence of higher investment costs, And the system expansion is also limited; the public cloud has a better scale effect by constantly leasing new equipment to deal with uncertain load requirements, the underlying system and network operation and maintenance can be provided by the public cloud manufacturers, investment and operating costs lower; Mix-Cloud is a private cloud + public cloud mixed architecture that has higher data privacy and flexibility than public clouds. [25,26] The system offers highly scalable, low-cost, high-security services. When there is a high workload demand, the company can share the high load by renting the public cloud.

Table 1. Features comparison between three Clouds Structur

Characteristic	Private cloud	Public cloud	Mixed cloud
Data privacy	Higher	Medium	High
Flexibility	Medium	Higher	Higher
Reliability	High	High	Higher
Load Capacity	Medium	Higher	High
Technical Realization	Medium	Low	High
Cost of Investment	Higher	Low	High
Operating Costs	Higher	Low	High

Therefore, the selection of the "cloud" architecture of the retail technical support system needs to be considered from its own scale, system flexibility, technical difficulty and operating cost. Large-scale retail electricity companies generally have a strong economic and technical strength, the proposed hybrid cloud architecture system; medium-scale retail electricity companies and economic strength is good, high privacy requirements of the data, the proposed use of private cloud architecture; other Small-scale retail electricity companies recommend using public cloud to build their own technical support system.

6 Conclusion

This paper analyzes the business requirements of the retail technical support system and designs the technical support system architecture of the retail company based on "cloud", which can be flexibly selected according to the demand. The architecture of the technology to achieve operational flexibility, to meet the current market requirements and future development needs. At the same time, this paper analyzes the characteristics of cloud, public cloud and mixed cloud, and so on, according to the different requirements of various types of retail electric companies in terms of economy, technology and their own scale, and discuss the rationalization proposal of the system step by step construction. The technical support system architecture of the retail electricity company based on the cloud is a positive attempt to cater to the Internet +, the research results can provide reference for the design of the technical support system of the retail companies.

Acknowledgment: This work was supported by the Science and Technology Project of State Grid Corporation of China under Grant DZ71-17-019.

References

1. The CPC Central Committee and State Council. Zhong Fa [2015] No. **9** on further deepening the power system reform a number of opinions [Z] .2015-3-15
2. China National Development and Reform Commission National Energy Board. Development and Reform [2016] No. **2120** No. 0 Recycling and withdrawal management of retail companies [Z] .2016-10-11
3. Guangdong Provincial Development and Reform Commission Guangdong Provincial Economic and Information Commission National Energy Bureau of the South Authority. Guangdong Development and Reform Energy [2017] No. **48** Guangdong Province, the implementation of electricity sales reform pilot program [Z] .2017-1-19
4. WANG Dewen , SONG Yaqi , ZHU Yongli.Information Platform of Smart Grid Based on Cloud Computing[J].Automation of Electric Power Systems,2010,**34(22)**:7-12.
5. WANG Dewen,LIU Yang.A Task Scheduling Strategy of Cloud Data Center in Electric Power Corporation[J].Automation of Electric Power Systems,2014,**38(8)**:61-66.
6. YAO Jianguo,YANG Shengchun,SHAN Maohua.Reflections on Operation Supporting System Architecture for Future Interconnected Power Grid[J].Automation of Electric Power Systems,2013,**37(21)**:52-59.
7. ZHAO Junhua1 , WEN Fushuan1 , XUE Yusheng2 , LIN Zhenzhi1.Cloud Computing: Implementing an Essential Computing Platform for Future Power Systems[J].Automation of Electric Power Systems,2010,**34(15)**:1-8.
8. QU Zhaoyang,CHEN Shuai,YANG Fan, et al.An Attribute Reducing Method for Electric Power Big Data Preprocessing Based on Cloud Computing Technology[J].Automation of Electric Power Systems,2014,**38(8)**:67-71.
9. Li Dan, Shen Hao, Sun Wei. Exploration and practice of"Internet plus electric power marketing"mode [J]. Power demand side management, 2016, **18** (z1).
10. Zang Ningning. Power System Reform Under The Sale Of Electricity Companies How To Build A Marketing System [J]. China Energy, 2016, **38 (4)**: 44-47.
11. Zheng Guangcheng. An Soa Cloud Service Platform Architecture Research And Application [J]. Computer Applications and Software, 2011, **28(12)**:228-231.
12. Liang Shuang. Design and realization of cloud computing framework model based on SOA [J]. Computer Engineering and Applications, 2011, **47 (35)**: 92-94
13. MA Li,ZHANG,Xiaoxuan,WEI Zhe. Experience and Enlightenment of Power Sales Side Market Liberalization in France [J]. Southern Power System Technology, 2015, **9(8)**:9-12.
14. ZHANG Xiaoxuan,XUE Song,YANG Su, et al.International Experience and Lessons in Power Sales Side Market Liberalization[J].Automation of Electric Power Systems,2016,**40(9)**:1-8.
15. BAI Yang,XIE Le,XIA Qing, et al.Institutional Design of Chinese Retail Electricity Market Reform and Related Suggestions[J].Automation of Electric Power Systems,2015,**39(14)**:1-7.
16. ZHENG Yaxian,YANG Zhenglin,XUE Bike, et al.Latest Development of International Electricity Market Standards IEC 62325[J].Automation of Electric Power Systems,2015,**39(15)**:1-8.
17. He Hengjing, Zhao Wei, Huang Songling, Wang Qing. Research on the application of cloud computing in power user electric energy data acquisition system [J]. Electrical Measurement & Instrumentation, 2016, **53(1)**: 1-7.
18. Armbrust M, Fox A, Griffith R, et al. A view of cloud computing[J]. Communications of the ACM, 2010, 53(4): 50-58.
19. Rittinghouse J W, Ransome J F. Cloud computing: implementation, management, and security[M]. CRC press, 2016.
20. WANG Dewen , SONG Yaqi , ZHU Yongli.Information Platform of Smart Grid Based on Cloud Computing[J].Automation of Electric Power Systems,2010,**34(22)**:7-12.
21. FENG Chao-Sheng, QIN Zhi-Guang[2], YUAN Ding[3]. Techniques of Secure Storage for Cloud Data [J]. Chinese Journal of Computers, 2015, **38(1)**: 150-163.

22. LONG Feng, WEN Zhonghua, WANG Quan, WANG Jinzong. User privacy protection against providers in public cloud computing [J]. Computer Engineering and Applications, 2015, **51(23)**:106-111.
23. Deshmukh D, Pasha A, Qureshi D. Transparent Data Encryption--Solution for Security of Database Contents[J]. arXiv preprint arXiv:**1303.0418**, 2013.
24. Ali cloud. Ali cloud data security white paper [EB / OL]. [2016-08-10]. Https://help.aliyun.com/knowledge_detail/42566.html.
25. Zhang Zi-lang. The Research on Architecture of Enterprise Information Platform Based on Hybrid Cloud [J]. Information Security and Technology,2014,**(04)**:60-65.
26. Huang Dongmei, Du Yanling, He Qi. Migration Algorithm for Big Marine Data in Hybrid Cloud Storage [J]. Journal of Computer Research and Development,2014,**(01)**:199-205.

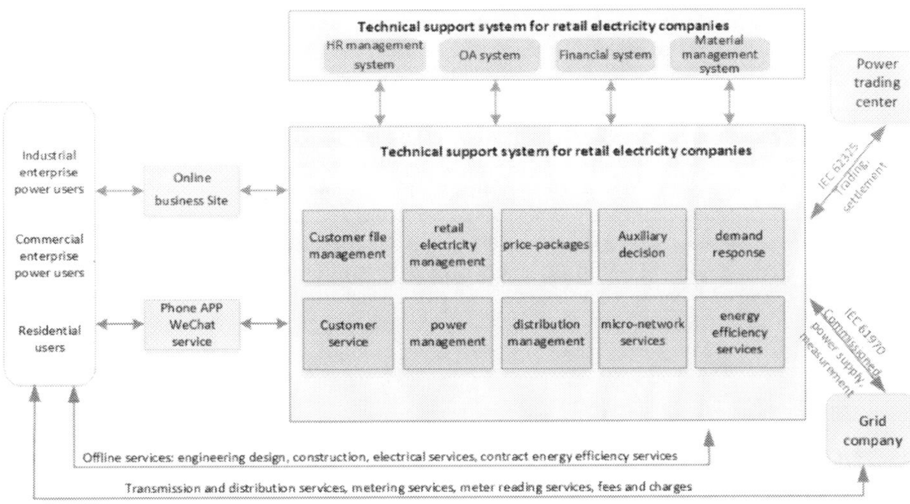

Fig. 1. System functions architecture of retail electricity company

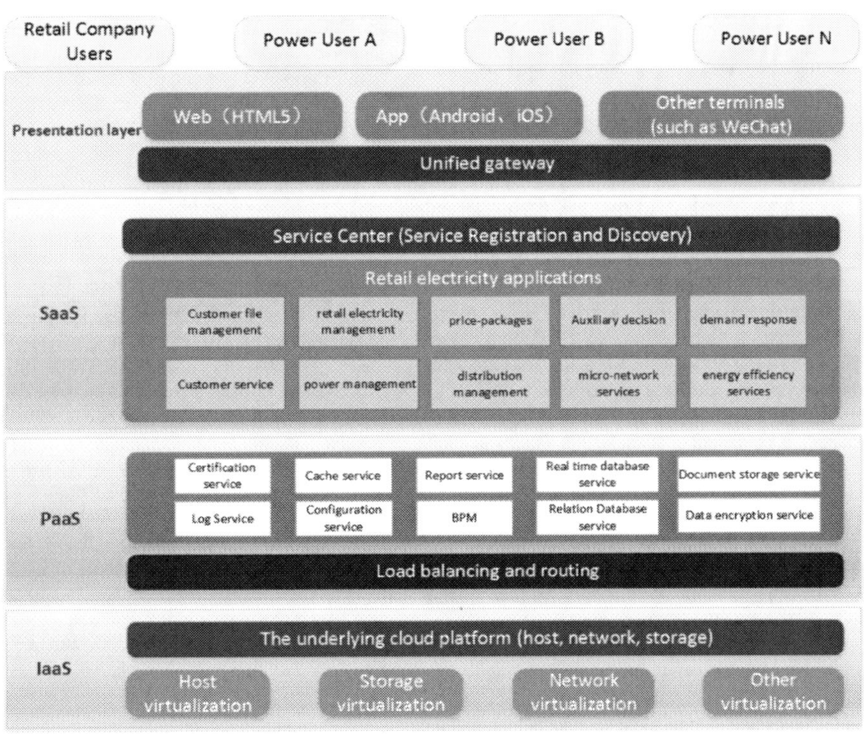

Fig. 2. System technical architecture of retail electricity company

The Design Concept and Realization of University Library's Online Lecture System

Linbo Duan[1,*], Ping Qin[1], Lingfei Qian[1], Ting Xie[1], Qingfeng Duan[2]
[1]Library,Nanjing University of Aeronautics and Astronautics,Nanjing, China
[2]State Grid Corporation of China,Luohe power supply company,Luohe,China

Abstract: The traditional library lectures are based on the physical space environment, which is problematic in terms of user needs analysis, time selection, and resource utilization, so that the value of the library unexploited.University library's online lecture system adopts the three-tier architecture design pattern, which based on ASP.NET as the development platform and SQL Server as the back-end database, designing the system to solve the problems in the traditional library service.The design and implementation of the University library's online lecture system broke the traditional library lecture mode, and further optimized the library service concept, improve the quality of service to meet the diverse needs of the user groups. Integrating online test, interactive question and answer, feedback mechanism to the system, the system has a good user experience, visual way to show the audience's learning situation and other information, optimize the content and quality of services to make comments and suggestions.

1 INTRODUCTION

University libraries as the literature information centers and resource sharing platforms, play an important role in teaching and scientific research.A series of lectures carried out by university libraries are important parts of college education and teaching. At the same time these lectures also play an important role in the aspect of educational functions.The traditional library services mainly include the construction of information retrieval work, retrieval of documents and databases ,manual retrieval of documents and so on.The digital library pay more attention to service mobility and flexibility, and they focus on the development of disciplines in colleges and universities with the application of network and information technology.The duties and responsibilities of university libraries are carrying out service work about literature and information according to different user groups .In the form of giving lectures in libraries, making user groups have a thorough understanding of the library resources present situation and the application method, and learn more about literature delivery, better serving the teaching and scientific research.

The traditional library service model has been unable to meet the diverse needs of the user groups, and the informational lecture mode is expected to change the current situation.Professor Su Xinning at Nanjing University point out in the article [1] "large data age digital library facing opportunities and challenges"that:Future digital library want to jump out of the traditional framework for resource development, turn the passive wait into a positive activity, establish a comprehensive service concept and service mode and the form of resource dissemination. digital librarians should be determined to become data analysts.Wu Weici, professor at Peking University points out[2], in the sixth session of the Chinese Library Association report that:The development of modern library should use the new information technology as much as possible, and make the library development towards the direction of network and information.In the "Library Development Trend Research Report: Environmental Analysis and Main Strategy", Chu Jingli and Wu Dongman pointed out that:Library development should change, take the initiative to respond, analyze user needs and environmental changes.Research and develop a sustained and dynamic development strategy.Therefore, in the ever-changing external environment, Service concept and service form of libraries should also change with the external environment changes.

2 The Necessity of Systematic Design of University Library 's Online Lecture

According to the different needs of different user groups for library resources Construction and development of modern digital library.The design and implementation of online lecture system in University Libraries will solve some problems in the process of Library Construction.At the same time,the function of online evaluation of the system will contribute to perfecting the function of University Library,improving service quality,enhancing user experience,further enhancing the information literacy of user groups,and realizing the extension and expansion of library service mode and service scope in the "Internet plus" era.

*Corresponding author:linboduan@nuaa.edu.cn

DOI: 10.1051/matecconf/201713900097

2.1 Meet multi-level needs and make value of lectures Survivability Structure Model of Large-scale Network

The service tenet of the library must be people-oriented,and the service form is bound to develop in the direction of diversification and hierarchy.The library user groups are characterized by diversity.Based on the teachers and students in the school and the related scientific research personnel, there are some differences between the different user groups in the demand level of the University library.Liang Xin, Guo Shiming [4] pointed out:At present, the service mode of the library is relatively simple,the training of the literacy and skills is lagging behind, the service mode is compounded, the user experience specialization, the deepening of the skills and the literacy education will become the main trend of the future of the library.Therefore, in order to meet the needs of the user groups, the content of the online lecture system is divided into different levels.

The service tenet of the library must be people-oriented,and the service form is bound to develop in the direction of diversification and hierarchy.Based on the teachers and students in the school and the related scientific research personnel, there are some differences between the different user groups in the demand level of the University library.Although the activities carried out by the library aims to serve the teachers and students as the core, but these activities tend to focus on the basic application, and they can not meet the needs of different user groups hierarchical, and the targeted is not strong, so the effect is often not ideal. Liang Xin, Guo Shiming [4] pointed out:At present, the service mode of the library is relatively simple,the training of the literacy and skills is lagging behind, the service mode is compounded, the user experience specialization, the deepening of the skills and the literacy education will become the main trend of the future of the library.Therefore, in order to meet the needs of the user groups, the content of the online lecture system is divided into different levels.

2.2 Optimize the time node, improve the lecture efficiency

The diversity of the library user group determines the time planning of the user group.As the majority of the user groups of university library lectures are teachers and students,the selection of the time node of the library lectures will often integrate various factors,as for the collective interests, overall schedule, avoid conflicts in the time node, creating opportunities for each student to attend the lecture.

The university library online lecture system based on the ASP.NET technology development will effectively break the impact of time node selection.With the help of Internet technology,we can realize the transformation from traditional service to digital library.After user groups loging in the system they can repeat learn a lot of times.This form can effectively avoid the impact of time conflict, at the same time teachers do not need to repeat the teaching.This formcan also reduce the workload of teachers, improving the ratio of input and output.

2.3 Save space resources and reduce the cost of activities

The lecture activities in the physical space environment need to occupy the site, audio equipment and so on.In addition to a certain amount of cost, the service life of the device will also produce losses.Under the academic environment of lacking funds for scientific research and advocating the recycling of resources, it is no doubt that it is the best choice to adopt the digital teaching mode.The design and implementation of the online lecture system of university library enables users to study and study under the normal network environment, and will not be restricted by the geographical environment and other factors.

2.4 Establish evaluation mechanism, optimize service content

Huwe, University of California Library [5] in the context the construction of digital library, points out that Digital library construction should be combined with big data.Through the library platform to strengthen the collection of user behavior information,then constantly optimize the service level of digital library.Therefore, in order to judging the quality of university library lectures and the results of user groups on the lectures, and we need to use certain indicators or data to measure. The traditional lecture mode generally uses the form of a live examination, and then manually corrected papers. This form not only takes a lot of time, but also having a low efficiency.The online examination function in the online lecture system of the university library can replace this task by replacing it manually. At the same time, the module can also realize the analysis and visualization of the assessment data.

3 System Design Concept

3.1 System design goals

The design goal of the online lecture system is to start from the actual needs of the user community. Using the help of internet,the system can simplify the lecture process and improve the service quality,it can also rationally utilize the resources, optimize the lectures service, enhance the user experience, change the traditional lecture mode from multiple angles,and build a new digital library online lecture platform.

3.1.1 Meet the needs of different groups

Wuhan University Chen Chuanfu [6] in the "large data age digital library construction research,"analyzes the problems faced by digital libraries,he put forward alot of opinions and suggestions about the transformation of ideas, the development of knowledge services, the improvement of financial input mechanism, and the improvement of the quality of librarians.Therefore, the user groups combined with library lectures are complicated. The diversity of the lecture content, the lecture pattern with characteristic of hierarchical, multi type, Diversification can meet the needs of the multi factor conditions.The online lecture system of university library based on the ASP.NET technology will meet the

 DOI: 10.1051/matecconf/201713900097

diverse needs of the user groups, reducing the time cost of user groups, and enhancing the experience of the user groups.

3.1.2 Optimize service mode, simplify service flow

The service object of the library has a variety of disciplines. In order to meet the diverse needs of the user groups, experts and scholars in various fields are invited to participate in the Library Lecture Service.With the help of online lecture system in university library, not only can we get rid of the dependence on the physical space, reduce the economic cost of the library, but also can increase the user groups to obtain knowledge and education opportunities, optimize the experts and user groups all the time, achieve the win-win purpose.

3.1.3 Establish information feedback mechanism and improve service quality

The establishment of good communication channels and information feedback mechanism in the network environment is also an important way to enhance the library service.Brition and others with large data technology at Massachusetts Institute of Technology participating in the library service user groups to conduct a deep excavation of the behavior of students and the association analysis[7].After the end of the traditional lecture,there is a short interactive quiz.This step can only grasp the information about the user groups on the content of the lecture,and it also can not quantify the quality of lectures.However,the online lecture system will effectively make up for this deficiency, and achieve the function like SPSS Modeler, Hadoop and so on.The system will be automatically transferred to the online evaluation module after the completion of the learning task of the user group, and then the system will assess the effect of learning of user groups.

3.2 System design principles

3.2.1 Focus on improving service quality

As the learning resource service center of the University, the library provides the best service for the majority of users.In the era of big data,network academic resources construction developed very quickly.Under such circumstances,the construction of digital library should establish a comprehensive service concept and expand the service mode and knowledge transmission form.Change passive to active, personalized, and new forms of service model will be more welcomed by the user groups.

3.2.2 Simple and user-friendly system operation

In the era of big data, the information literacy of the user group is getting higher and higher, and the information service level of the library is also put forward newer and higher requirements.University library lectures online system based on ASP.NET technology will have better man-machine interactive functions and friendly interface. The system function modules in the humanized design reflects the perfect combination with users.In order to meet the increasing needs of the user groups in the future, system structure with flexibility, scalability and the ability of easy to maintain the system provides a basic guarantee for the upgrading and optimization of the latter system.

3.2.3 Diversified information services

IFLA stated in the IFLA Digital Library Declaration [8]:University libraries should use their own advantages to provide useful information to users, and use the library's own role in the dissemination of digital media resources.In addition to providing basic online lectures , the diversified information service content makes the service strategy of the university library change in the subtle influence ,allowing the system to be recognized by more user groups. Addition to online evaluation function, the system also designed interactive communication, online Q & A, message boards, comments box and other functional modules.

4 System Design Flow

4.1 Development tools and related technologies

Online lecture system is built on the platform of ASP.NET and SQL Server database. It runs on the server of the Internet, the user groups accessing by B/S. ASP.NET technology support C#, Java and other programming languages. It also provides the convenience for the development of the online lecture system, and provides the conditions for the daily management of the system.

SQL Server database as a background database, the system has the advantages that customs application allows the use of Visual Studio development, and make the library with the safety, reliability and high scalability to run the most critical task. The system can reduce the time and cost of development and management data infrastructure, and ensure the security and stability of the system.

4.2 System structure design

The online lecture system of university library is a three layer structure, which mainly includes UI layer, business logic layer and data layer. The structure is one of software abstraction hierarchy and a kind of vertical segmentation of complex software. The type of operation is performed at each level in order to segment the code based on its mission to reduce the complexity of the software and improve the maintainability of the system.

ASP.NET works as shown in Figure 1:

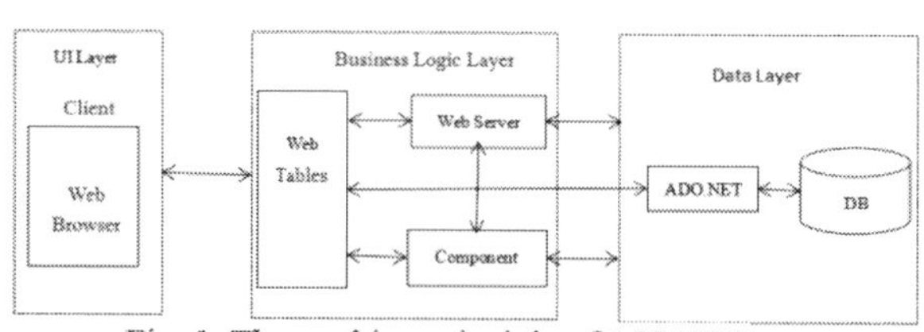

Fig .1. The working principle of ASP.NET

DOI: 10.1051/matecconf/201713900097

4.3 Database Design and Access

4.3.1 Database design

Database is all of the information resources storage warehouse. All the information of users will directly impact on the database. Establishing an efficient and secure database system is an important guarantee of normal operation of the whole system. Using SQL Server database to build a online meeting database, it contains five tables, each table contains the corresponding field,as shown in table 1:

TABLE I. Online meeting database contains tables and fields in the table

Tables	Fields
Administrator	aName, aPwd
Teachers	tName, tPwd
Users	uName, uPwd
Source	ID、Theme、Path
Tests	ID、Contents、Type、Answers、Score, GetScore

4.3.2 Database access

ADO.NET provides consistently access to SQL Server data sources. Data sharing application enables to connect data sources by using ADO.NET to find, update, and modify the internal data of the tables in the online meeting database. ASP.NET provides efficient components to achieve the storage of the database. The introduction of Data Set ADO.NET data set to achieve high-speed data buffer.

Some of the core code for the system to access the database is as follows:

```
SqlParameter[] pms=new SqlParameter[]{
  New      SqlParameter("@ID",SqlDbType.varchar,50)
      {value=txtID.Text.Trim()},
      new SqlParameter("@Pwd",SqlDbType.varchar,50)
      {value=txtPwd.Text};
      cmd.Parameters.AddRangs(pms);
       Con.open();
       int r=cmd.ExecuteNonquery();
}
```

4.4 System function design

The structure of the online lecture system is shown in Figure 2:

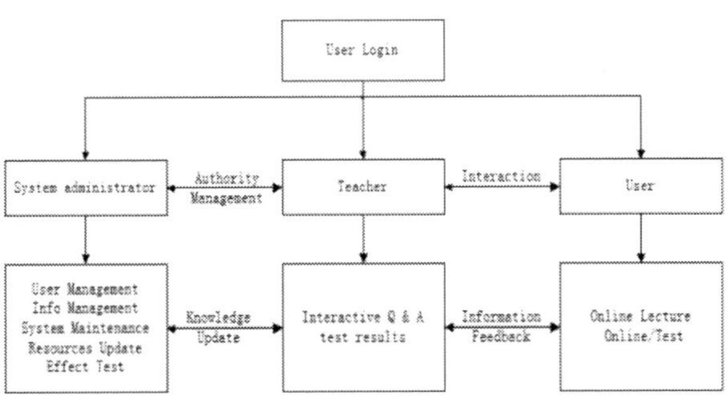

Fig.2. The structure and function of the system

4.4.1 User information management

The system administrator has the highest authority, and mainly responsible for user information management, resource update, system maintenance, etc. System administrators make different types of user groups to have a distinct level of authority and within the scope of the exercise of authority. For different types of lectures, the system can update resources in a timely manner. The combination of digital library construction and library service concept to meet the growing user needs of functional groups, optimize system functions, to achieve the continuous upgrading of the system.

4.4.2 Information resource management

After the administrator of the various types of user groups to complete the corresponding authorization, the user groups can be within the scope of authority to exercise the appropriate permissions.The teacher will be prepared in advance of teaching resources, examination questions and other information uploaded to the system.When the system is open to the outside world, the user group can go through the system front end, with the user name and login password into the system to learn.At the same time, accessing to the user group can be downloaded in the system corresponding to the courseware resources.

4.4.3 Online lecture

After user groups logged into the online lecture system, they can start a targeted learning task.As the learning resources in the system stored for a long time, the user community can avoid the time and space constraints, select the appropriate learning nodes, repeat learning, effective grasp of learning initiative, to avoid the impact of traditional lectures model.The design concept of the system pays more attention to the dissemination and reproduction of knowledge, emphasizes the ability of autonomous learning and training of user groups.

4.4.4 Lecture evaluation module

The examination module is the core of the university library online lecture system, and the innovation and uniqueness of the system.The design of the evaluation module combines the needs of the user community and the service concept of the digital library.Through the specific assessment data, the teacher timely and effective grasp of the user group learning situation.In view of the shortcomings in the course of teaching, teachers can adjust the teaching strategies in a timely manner, and constantly optimize the quality of library services.As part of the library to undertake some of the teaching tasks, the user group to master the content of the lectures is the basis for the quality of teacher assessment.

4.4.5 Interactive quiz module

The interactive Q & A module is designed to effectively address short boards present in traditional lecture modes.The survey found that most of the user groups are more inclined to real-time interactive mode of communication.The design of the interactive Q & A module not only facilitates the user community to ask questions, quickly receives a reply, but also helps the learning and understanding of the course content to a new level for the user groups that are not aware of the problem.At the same time, the interactive Q & A module design can continue to cultivate the interaction and sharing of user groups.For the problems posed by the user group, the system can be classified analysis, and provide the basis and direction for the library to optimize the service quality.

4.5 Comparison of the characteristics of 3.5 online lecture system

4.5.1 System Design Cost Economy

System features is the service of the majority of user groups, applications, knowledge sharing.
The biggest feature of the system is the system resources free of charge to the university user groups.The design of the system is simple and simple design style, reducing the time and cost of system development.The system shipped in the campus of the server, to facilitate the daily management of the system, management and upgrading.

4.5.2 System resource specialization

The learning resources of the online lecture system are professional and targeted. The learning resources within the system are designed for different subject backgrounds, identity categories and hierarchical needs of the user groups in colleges and universities. At the same time, the content of the teaching is also taught by senior experts in the field, to ensure the professional content of the teaching.In order to improve the learning efficiency, the system can update the learning content and shorten the learning cycle according to the learning situation of the user group.

4.5.3 Autonomous learning of user groups

Compared with the traditional face-to-face courses, online lectures system teaching mode can inspire the enthusiasm of the user community to learn independently and cultivate students hands-on and practical ability.Flexible teaching methods are also constantly encourage and guide the user groups from passive learning to active learning changes, to further increase the user group participation, to meet the information access needs and improve the user group integrated information literacy.

5 Conclusion

The university library online lecture system is based on the development of ASP.NET technology.The design and implementation of the system not only solve the problems existing in the traditional service model, but also provide the scientific basis for improving the work of the library.However, the system also has some shortcomings.The data obtained from the online evaluation module can be used for shallow analysis, presented in a simple visual form, and shallow data mining.Today, large data theory and technology are increasingly mature, in the enterprise has been widely used.In the future, the system should also introduce the corresponding data mining algorithm to realize the deep analysis and mining of the background data, mastering the development trend of college students' information literacy, giving full play to the data value, providing scientific data for the research of related fields.

REFERENCES

1. Su Xinning. Opportunities and Challenges for Digital Library in Large Data Times [J]. Journal of Library Science, 2015,06: 4-12.
2. Wu Weici. The development of Chinese library science and the mission of the new generation of library scholars - Report on the establishment of the eighth academic research committee of the China Library Association [J]. Journal of Library Science, 2009, 06: 66-69.
3. Chu Jingli, Wu Dongman. Library development trend research report: environmental analysis and the main strategy [J]. National Library Journal, 2010,01: 3-11.
4. Liang Xin, Guo Shiming. China Mobile Library Service Model Status and Development Trends [J]. Information Work, 2013,05: 98-102.
5. Huwe T K. Building digital libraries:big data and the library:a natural fit[J]. Computers in Libtaries,2014,34(2):17-18.
6. Chen Chuanfu, Qian Ou, Dai Yuzhu.Study on Digital Library Construction in Large Data Age [J]. Library and Information Work, 2014 (7): 40-45.
7. Renaud J,Brition S,Wang D,et al. Mining library and university data to understand library use patterns[J]. The Electronic Library,2015,33(3):355-372.
8. Zhang Xiaolin. Let the digital library to drive the library service innovation and development: read the "IFLA digital library declaration" [J]. Journal of Library Science, 2010 (3): 73-76.

Partial, Composite Fractional Operators, and Their Properties and Applications

Kaiying Deng [1,2*], *Jingwei Deng* [2], and *Suduo Li* [1]

[1] College of Mathematics and Computer Science, Northwest Minzu University, Lanzhou 730124, China
[2] Key Lab of Streaming Data Computing Technologies and Applications, Lanzhou 730030, China

Abstract. The paper discusses the properties of the partial fractional integrals, the partial fractional derivatives, and the composite fractional integrals and derivatives. Some basic formulas are derived; and the relations between the Riemann-Liouville and Caputo ones are also presented. Finally, as the specific application, a two dimensional Abel integral equation is solved.

1 Introduction

Fractional calculus, as a natural extension of the classical calculus, has the history of more than 300 years [3,4,7,8,9]. In fact, since the beginning of the theory of differential and integral calculus, the mathematicians began their investigations of the calculation of non-integer order derivatives and integrals. Most of the researchers of this topic cite 30th September 1695 as the birthday of the so called 'fractional calculus'. In a letter, dated 30th September 1695, L'Hopital wrote to Leibniz asking him a particular notation that he had used in his publication for the n-th derivative of a function $\frac{D^n f(k)}{Dx^n}$; L'Hopital posed the question to Leibniz, what would the result be if $n = 1/2$; Leibniz's response: "An apparent paradox, from which one day useful consequences will be drawn."

Following L'Hopital's and Liebniz's first inquisition, fractional calculus was primarily a study reserved for the best minds in mathematics. Fourier, Euler, Laplace are among the many that dabbled with fractional calculus and the mathematical consequences [7]. Many proposed, using their own notation and methodology, the definitions that fit the concept of non-integer order integral or derivative. The most famous of these definitions that have been popularized in the world of fractional calculus (not yet the world as a whole) are the Riemann-Liouville and Grunwald-Letnikov derivatives [6].

Most of the mathematical theories applicable to the study of fractional calculus were developed prior to the turn of the 20th century. However in the past 100 years, the widely engineering and scientific applications have been found. In some cases, the mathematics has to be made some changes to meet the requirements of physical reality. Caputo reformulated the more 'classic' definition of the Riemann-Liouville fractional derivative in order to conveniently specify the initial conditions. And the so called Caputo derivative was presented[2,3,4]. More recently, the space-time fractional operator was discussed [1].

In this paper, we introduce the partial fractional integral, partial fractional derivative, their compositions, and the relations between their Riemann-Liouville and Caputo versions. The properties of these operators are detailedly discussed; and a specific application of solving a two dimensional Abel integral equation is proposed. For easy of presentation, we restrict the left end point of the interval of the definition of fractional operators at '0'; and the technique of Laplace transform is used [5].

2 Partial Fractional Integrals And Their Compositions

In our notation the Cauchy formula reads

$$J^n f(t) := f_n(t) = \frac{1}{(n-1)!}\int_0^t (t-\tau)^{n-1} f(\tau)\,d\tau,\ t > 0, n \in \mathbb{N}, \quad (1)$$

where $\mathbb{N}$ is the set of positive integers. From this definition, it can be noted that $f_n(t)$ vanishes at t=0 with its derivatives of order $1, 2, \cdots, n-1$. Noting that $(n-1)! = \Gamma(n)$, and introducing the arbitrary positive real number α, one defines the fractional integral of order $\alpha>0$:

$$J^\alpha f(t) := \frac{1}{\Gamma(\alpha)}\int_0^t (t-\tau)^{\alpha-1} f(\tau)\,d\tau,\ t > 0, \alpha \in R^+. \quad (2)$$

Starting from the one dimensional case (2), we can naturally define the partial integral of the function $f(x, y)$:

$$J_x^\alpha f(x,y) := \frac{1}{\Gamma(\alpha)}\int_0^x (x-\tau)^{\alpha-1} f(\tau, y)\,d\tau,\ x > 0, \alpha \in R^+, \quad (3)$$

$$J_y^\beta f(x,y) := \frac{1}{\Gamma(\beta)}\int_0^y (y-\xi)^{\beta-1} f(x, \xi)\,d\xi,\ y > 0, \beta \in R^+, \quad (4)$$

For the one-dimensional case (2), the semigroup property can be easily derived as

* Corresponding author: dengky0420@163.com

$$J^{\alpha}J^{\beta} = J^{\alpha+\beta}, \alpha, \beta \geq 0. \quad (5)$$

So,

$$J_x^{\alpha 1}J_x^{\alpha 2} = J_x^{\alpha 1+\alpha 2}, \alpha_1 + \alpha_2 \geq 0; \quad (6)$$

$$J_y^{\beta 1}J_y^{\beta 2} = J_y^{\beta 1+\beta 2}, \beta_1 + \beta_2 \geq 0; \quad (7)$$

It can be noted that the fractional integral can be considered as the Laplace convolution between $\Phi_\alpha(x)$ and $f(x, y)$, i.e.,

$$J_x^{\alpha}f(x, y) = \Phi_\alpha(x) * f(x, y), \alpha > 0. \quad (8)$$

Similarly, there exists

$$J_y^{\beta}f(x, y) = \Phi_\beta(x) * f(x, y), \beta > 0; \quad (9)$$

$$J_x^{\alpha}J_y^{\beta}f(x, y) = \Phi_\alpha(x) * \{\Phi_\beta(y) * f(x, y)\}$$

$$= \Phi_\alpha(x) * \Phi_\beta(y) * f(x, y), \alpha, \beta > 0; \quad (10)$$

and

$$J_x^{\alpha}J_y^{\beta}f(x, y) = \Phi_\beta(y) * \{\Phi_\alpha(y) * f(x, y)\}$$

$$= \Phi_\beta(y) * \Phi_\alpha(x) * f(x, y), \alpha, \beta > 0. \quad (11)$$

For the two-fold Laplace transforms, we have

$$\mathcal{L}_y\mathcal{L}_x\{f(x, y)\} = \mathcal{L}_y\{\mathcal{L}_x f(x, y)\} = \tilde{f}^{yx}(s_1, s_2), s_1, s_2 \in C; \quad (12)$$

$$\mathcal{L}_x\mathcal{L}_y\{f(x, y)\} = \mathcal{L}_x\{\mathcal{L}_y f(x, y)\} = \tilde{f}^{xy}(s_1, s_2), s_1, s_2 \in C; \quad (13)$$

Noting that

$$\mathcal{L}_x\{\Phi_\alpha(x)\} = \frac{1}{s_1^{\alpha}}, s_1 \in C; \quad (14)$$

and

$$\mathcal{L}_y\{\Phi_\beta(y)\} = \frac{1}{s_2^{\beta}}, s_2 \in C, \quad (15)$$

see e.g., [5], and recalling the convolution theorems for Laplace transforms, we get the following rules for the Laplace transforms of the fractional integrals,

$$\mathcal{L}_x\{J_x^{\alpha}f(x, y)\} = \mathcal{L}_x\{\Phi_\alpha(x) * f(x, y)\} = \frac{\tilde{f}^{x}(s_1, y)}{s_1^{\alpha}}, \alpha > 0; \quad (16)$$

$$\mathcal{L}_y\{J_y^{\beta}f(x, y)\} = \mathcal{L}_y\{\Phi_\beta(y) * f(x, y)\} = \frac{\tilde{f}^{y}(x, s_2)}{s_2^{\beta}}, \beta > 0, \quad (17)$$

which are the straightforward generalization of the case with a n-fold repeated integral ($\alpha = n$).

3 Partial Fractional Derivatives

We note that

$$J_x^{n}D_x^{n}f(x, y) = f(x, y) - \sum_{k=0}^{n-1} f^{(k)}(0^+, y)\frac{x^k}{k!}, x > 0, \quad (18)$$

$$J_y^{n}D_y^{n}f(x, y) = f(x, y) - \sum_{k=0}^{n-1} f^{(k)}(x, 0^+)\frac{y^k}{k!}, y > 0. \quad (19)$$

One defines the partial fractional derivative

$$D_x^{\alpha}f(x, y) := \begin{cases} \frac{\partial^m}{\partial x^m}\left[\frac{1}{\Gamma(m-\alpha)}\int_0^x \frac{f(\tau, y)}{(x-\tau)^{\alpha+1-m}}d\tau\right], m-1 < \alpha < m, \\ \frac{\partial^m}{\partial x^m} f(x, y), \alpha = m; \end{cases} \quad (20)$$

$$D_y^{\beta}f(x, y) := \begin{cases} \frac{\partial^n}{\partial y^n}\left[\frac{1}{\Gamma(n-\beta)}\int_0^y \frac{f(x, \xi)}{(y-\xi)^{\beta+1-n}}d\xi\right], n-1 < \beta < n, \\ \frac{\partial^n}{\partial y^n} f(x, y), \beta = n. \end{cases} \quad (21)$$

In fact, $D_x^0 = J_x^0 = I$ and $D_y^0 = J_y^0 = I$. Then we can easily recognize that

$$D_x^{\alpha}J_x^{\alpha} = I, \alpha \geq 0; \quad (22)$$

$$D_y^{\beta}J_y^{\beta} = I, \beta \geq 0; \quad (23)$$

$$J_x^{\alpha}D_x^{\alpha} \neq I, \alpha > 0; \quad (24)$$

$$J_y^{\beta}D_y^{\beta} \neq I, \beta > 0; \quad (25)$$

And

$$D_x^{\alpha}x^{\gamma} = \frac{\Gamma(\gamma+1)}{\Gamma(\gamma+1-\alpha)}x^{\gamma-\alpha}, \alpha > 0, \gamma > -1, t > 0; \quad (26)$$

$$D_y^{\beta}y^{\gamma} = \frac{\Gamma(\gamma+1)}{\Gamma(\gamma+1-\beta)}y^{\gamma-\beta}, \beta > 0, \gamma > -1, t > 0. \quad (27)$$

We can get the Caputo partial fractional derivatives:

$$D_x^{\alpha}f(x, y) := \begin{cases} \frac{1}{\Gamma(m-\alpha)}\int_0^t \frac{f_x^{(m)}(\tau, y)}{(x-\tau)^{\alpha+1-m}}d\tau, m-1 < \alpha < m, \\ \frac{\partial^m}{\partial x^m} f(x, y), \alpha = m; \end{cases} \quad (28)$$

and

$$D_y^{\beta}f(x, y) := \begin{cases} \frac{1}{\Gamma(n-\beta)}\int_0^t \frac{f_x^{(n)}(x, \xi)}{(y-\xi)^{\beta+1-n}}d\xi], n-1 < \beta < n, \\ \frac{\partial^n}{\partial y^n} f(x, y), \beta = n, \end{cases} \quad (29)$$

where $f_x^{(m)}f(x, y) := \frac{\partial^m}{\partial x^m} f(x, y)|_{x=\tau}$ and $f_y^{(n)}f(x, \xi) := \frac{\partial^n}{\partial y^n} f(x, y)|_{y=\xi}$, respectively. These definitions are of course restrictive; since they require the absolute integrability of the partial derivative of order m and n, respectively. Whenever using the operators D_x^{α} and D_y^{β}, we implicitly assume that these conditions are met. It can be easily recognized that in general

$$D_x^{\alpha}f(x, y) := D_x^{m}J_x^{m-\alpha}f(x, y) \neq J_x^{m-\alpha}D_x^{m}f(x, y) := D_x^{\alpha}f(x, y) \quad (30)$$

$$D_y^{\beta}f(x, y) := D_y^{n}J_y^{n-\beta}f(x, y) \neq J_y^{n-\beta}D_y^{n}f(x, y) := D_y^{\beta}f(x, y); \quad (31)$$

unless the function $f(x, y)$ along with its first $m-1$ partial derivatives about xvanishes at $x = 0^+$and its first $n-1$ partial derivatives about y vanishes at $y = 0^+$, respectively. Furthermore, for $m-1 < \alpha < m$ and $x > 0$,

$$D_x^{\alpha}f(x, y) = D_x^{\alpha}f(x, y) + \sum_{k=0}^{m-1}\frac{x^{k-\alpha}}{\Gamma(k-\alpha+1)}f_x^{(k)}(0^+, y). \quad (32)$$

Similarly, we get

$$D_y^{\beta}f(x,y) = D_y^{\beta}f(x,y) + \sum_{k=0}^{n-1} \frac{y^{k-\beta}}{\Gamma(k-\beta+1)} f_y^{(k)}(x,0^+); \quad (33)$$

and therefore, recalling the partial fractional derivative of the power functions (26-27),

$$D_x^{\alpha}f(x,y) - \sum_{k=0}^{m-1} \frac{x^k}{k!} f_x^{(k)}(0^+,y)) = D_x^{\alpha}f(x,y) \quad (34)$$

$$D_y^{\beta}f(x,y) - \sum_{k=0}^{n-1} \frac{y^k}{k!} f_y^{(k)}(x,0^+)) = D_y^{\beta}f(x,y) \quad (35)$$

In particular, according to these definitions, it can be noted that the partial Caputo fractional derivative of a constant is zero, i.e.,

$$D_x^{\alpha}g_1(y) = g_1(y)D_x^{\alpha}1 \equiv 0, \alpha > 0; \quad (36)$$

$$D_y^{\beta}g_2(x) = g_2(x)D_y^{\beta}1 \equiv 0, \beta > 0. \quad (37)$$

We now explore the differences between the two kinds of the partial fractional derivatives, the Riemann-Liouville fractional partial derivative, and (28)-(29), the Caputo fractional partial derivative. Based on (26)-(27), we can derive that

$$D_x^{\alpha}x^{\alpha-1}g_1(y) = g_1(y)D_x^{\alpha}x^{\alpha-1} \equiv 0, \alpha > 0, t > 0; \quad (38)$$

$$D_y^{\beta}g_2(x)y^{\beta-1} = g_2(x)D_y^{\beta}y^{\beta-1} \equiv 0, \beta > 0, t > 0. \quad (39)$$

with $x > 0, y > 0$ and $m-1 < \alpha \le m, n-1 < \beta \le n, m \in \mathbb{N}, n \in \mathbb{N}$,

$$D_x^{\alpha}f(x,y) = D_x^{\alpha}g(x,y) \Leftrightarrow f(x,y) = g(x,y) + \sum_{i=1}^{m} c_i(y)\, x^{\alpha-i}; \quad (40)$$

$$D_y^{\beta}f(x,y) = D_y^{\beta}g(x,y) \Leftrightarrow f(x,y) = g(x,y) + \sum_{j=1}^{n} d_j(x)\, y^{\beta-j}; \quad (41)$$

$$D_x^{\alpha}f(x,y) = D_x^{\alpha}g(x,y) \Leftrightarrow f(x,y) = g(x,y) + \sum_{i=1}^{m} c_i(y)\, x^{m-\alpha}; \quad (42)$$

$$D_y^{\beta}f(x,y) = D_y^{\beta}g(x,y) \Leftrightarrow f(x,y) = g(x,y) + \sum_{j=1}^{n} d_j(x)\, y^{n-\beta}. \quad (43)$$

Incidentally, we note that (38)-(39) provides an instructive example to show why D^{α} is not the right-inverse to J^{α},

$$J_x^{\alpha}D_x^{\alpha}x^{\alpha-1} \equiv 0, \text{but } D_x^{\alpha}J_x^{\alpha}x^{\alpha-1} = x^{\alpha-1}, \alpha > 0, x > 0, \quad (44)$$

$$J_y^{\beta}D_y^{\beta}y^{\beta-1} \equiv 0, \text{but } D_y^{\beta}J_y^{\beta}y^{\beta-1} = y^{\beta-1}, \beta > 0, y > 0. \quad (45)$$

For the two definitions we also note a difference with respect to the formal limit as $\alpha \to (m-1)^+$. From (36)-(37) and (33)-(34), there exist

$$\lim_{\alpha\to(m-1)^+} D_x^{\alpha}f(x,y) = D_x^{m}J_x^{1}f(x,y) = D_x^{m-1}f(x,y); \quad (46)$$

$$\lim_{\beta\to(n-1)^+} D_y^{\beta}f(x,y) = D_y^{n}J_y^{1}f(x,y) = D_y^{n-1}f(x,y); \quad (47)$$

$$\lim_{\alpha\to(m-1)^+} D_x^{\alpha}f(x,y) = J_x^{1}D_x^{m}f(x,y) = D_x^{m-1}f(x,y) - f_x^{(m-1)} \quad (48)$$

$$\lim_{\beta\to(n-1)^+} D_y^{\beta}f(x,y) = J_y^{1}D_y^{n}f(x,y) = D_y^{n-1}f(x,y) - f_y^{(n-1)}(x,0^+). \quad (49)$$

For the standard partial fractional derivative D_x^{α} or D_y^{β}, the Laplace transform, assumed to exist, requires the knowledge of the (bounded) initial values of the partial fractional integral $J^{m-\alpha}$ and of its integer derivatives of order $k = 1,2, m-1$ or $n-1$ [8]. The corresponding rule reads

$$\mathcal{L}_x\{D_x^{\alpha}f(x,y)\} = s_1^{\alpha}\tilde{f}_x(s_1,y) - \sum_{k=0}^{m-1} D_x^k J_x^{(m-\alpha)} f(0^+,y)s_1^{m-1-k},$$
$$m-1 < \alpha \le m; \quad (50)$$

$$\mathcal{L}_y\{D_y^{\beta}f(x,y)\} = s_2^{\beta}\tilde{f}_y(x,s_2) - \sum_{k=0}^{n-1} D_y^k J_y^{(n-\beta)} f(x,0^+)s_2^{n-1-k},$$
$$n-1 < \beta \le n. \quad (51)$$

In fact,

$$J_x^{\alpha}D_x^{\alpha}f(x,y) = f(x,y) - \sum_{k=0}^{m-1} f_x^{(k)}(0^+,y)\frac{x^k}{k!}; \quad (52)$$

$$J_y^{\beta}D_y^{\beta}f(x,y) = f(x,y) - \sum_{k=0}^{n-1} f_y^{(k)}(x,0^+)\frac{y^k}{k!}; \quad (53)$$

We can prove the following rule for the Laplace transform,

$$\mathcal{L}_x\{D_x^{\alpha}f(x,y)\} = s_1^{\alpha}\tilde{f}^x(x,y) - \sum_{k=0}^{m-1} f_x^{(k)}(0^+,y)s_1^{\alpha-1-k}, \quad (54)$$
$$m-1 < \alpha \le m;$$

$$\mathcal{L}_y\{D_y^{\beta}f(x,y)\} = s_2^{\beta}\tilde{f}^y(x,y) - \sum_{k=0}^{n-1} f_y^{(k)}(x,0^+)s_2^{\beta-1-k}, \quad (55)$$
$$n-1 < \beta \le n.$$

4 Composite Fractional Operators

In general, both the operators of fractional differentiation, D_x^{α}, D_y^{β} and D_x^{α}, D_y^{β}, do not satisfy either the semigroup property, or the (weaker) commutative property. To show how the Law of Exponents does not necessarily hold for the standard fractional derivative, we provide two simple example (with power functions) for which

$$D_x^{\alpha 1}D_x^{\alpha 2} = D_x^{\alpha 2}D_x^{\alpha 1} \neq D_x^{\alpha 1+\alpha 2} \quad (56)$$

$$D_x^{\alpha 1}D_x^{\alpha 2} \neq D_x^{\alpha 2}D_x^{\alpha 1} = D_x^{\alpha 1+\alpha 2}. \quad (57)$$

For example, let us take $f(x,y) = x^{-1/2}$ and $\alpha_1 = \alpha_2 = 1/2$. Then, using (31), we get $D_x^{1/2}f(x,y) \equiv 0, D_x^{1/2}D_x^{1/2}f(x,y) \equiv 0,$ but $D_x^{1/2+1/2}f(x+y) = Df(x,y) = -x^{-3/2}/2$. Another example, let us take $g(x,y) = x^{1/2}$ and $\alpha_1 = 1/2, \alpha_2 = 3/2$. Then, again using (42), we get $D_x^{1/2}g(x,y) = \sqrt{\pi}/2, D_x^{3/2}g(x,y) \equiv 0,$ but $D_x^{1/2}D_x^{3/2}g(x,y) \equiv 0, D_x^{3/2}D_x^{1/2}g(x,y) = -t^{3/2}/4$ and $D_x^{1/2+3/2}g(x,y) = D_x^2 g(x,y) = -x^{3/2}/4$. By the same way, we can obtain

$$D_y^{\beta 1}D_y^{\beta 2} \neq D_y^{\beta 2}D_y^{\beta 1}, \beta_1, \beta_2 > 0. \quad (58)$$

it can be easily recognized that

$$D_x^{m1}D_x^{m2} = D_x^{m2}D_x^{m1}, D_y^{n1}D_y^{n2} = D_y^{n2}D_y^{n1}, n_1, n_2, m_1, m_2 \in \mathbb{N}; \quad (59)$$

$$J_x^{\alpha}J_y^{\beta} = J_y^{\beta}J_x^{\alpha}, \alpha, \beta \geq 0; \quad (60)$$

$$D_y^{n}J_x^{\alpha} = J_x^{\alpha}D_y^{n}, \alpha \geq 0, n \in \mathbb{N}; \quad (61)$$

$$D_x^{m}J_y^{\beta} = J_y^{\beta}D_x^{m}, \beta \geq 0, m \in \mathbb{N}. \quad (62)$$

Therefore, we have

$$D_y^{\beta}J_x^{\alpha} = D_y^{n}J_y^{n-\beta}J_x^{\alpha} = J_x^{\alpha}D_y^{\beta}, \alpha, \beta \geq 0; \quad (63)$$

$$D_x^{\alpha}J_y^{\beta} = D_x^{m}J_x^{m-\alpha}J_y^{\beta} = J_y^{\beta}D_x^{\alpha}, \alpha, \beta \geq 0. \quad (64)$$

Hence, there are the following commutative properties for the fractional partial derivatives

$$D_y^{\beta}D_x^{\alpha} = D_y^{n}J_y^{n-\beta}D_x^{m}J_x^{m-\alpha} = D_x^{\alpha}D_y^{\beta}, \alpha, \beta \geq 0 \quad (65)$$

$$D_y^{\beta}D_x^{\alpha} = J_y^{n-\beta}D_y^{n}J_x^{m-\alpha}D_x^{m} = D_x^{\alpha}D_y^{\beta}, \alpha, \beta \geq 0; \quad (66)$$

Considering that the Laplace transform is an integral transformation and noting (4)-(5), the following expression is valid,

$$\mathcal{L}_y\mathcal{L}_x = \mathcal{L}_x\mathcal{L}_y, \quad (67)$$

i.e., $\tilde{f}^{yx}(s_1,s_2) = \tilde{f}^{xy}(s_1,s_2)$. What is more, according to (8)-(9) and the previous expressions, we can get

$$\mathcal{L}_y\mathcal{L}_x J_x^{\alpha}J_y^{\beta}\{f(x,y)\} = \mathcal{L}_y\left\{\frac{1}{s_1^{\alpha}}\mathcal{L}_x\{J_y^{\beta}f(x,y)\}\right\}$$
$$= \frac{1}{s_1^{\alpha}s_2^{\beta}}\tilde{f}^{yx}(s_1,s_2), s_1, s_2 \in \mathbb{C}; \quad (68)$$

$$\mathcal{L}_x\mathcal{L}_y J_y^{\beta}J_x^{\alpha}\{f(x,y)\} = \mathcal{L}_y\mathcal{L}_x J_x^{\alpha}J_y^{\beta}\{f(x,y)\}$$
$$= \frac{1}{s_1^{\alpha}s_2^{\beta}}\tilde{f}^{xy}(s_1,s_2), s_1, s_2 \in \mathbb{C}. \quad (69)$$

5 Solving the Abel Integral Equation

Here we limit ourselves to put some emphasis on the method of the Laplace transforms; and use the results given in the above sections to solve the equation. Let us consider the two dimensional Abel integral equation

$$\frac{1}{\Gamma(\alpha)}\frac{1}{\Gamma(\beta)}\int_0^x \frac{1}{(x-\tau)^{1-\alpha}}d\tau \int_0^y \frac{u(\tau,\xi)}{(y-\xi)^{1-\beta}}d\xi = f(x,y), \alpha > 0, \beta < 1, \quad (70)$$

where $f(x,y)$ is a given function. By the notations of fractional integrals, there exists

$$J_x^{\alpha}J_y^{\beta}u(x,y) = f(x,y), \alpha > 0, \beta < 1; \quad (71)$$

and consequently from the properties of partial fractional calculus we can get

$$u(x,y) = D_y^{\beta}D_x^{\alpha}f(x,y). \quad (72)$$

We can also solve (70) by using the Laplace transforms. Noting from (68), there exists

$$\mathcal{L}_y\mathcal{L}_x\{J_x^{\alpha}J_y^{\beta}u(x,y)\} = \mathcal{L}_y\mathcal{L}_x\{f(x,y)\}, \quad (73)$$

i.e.,

$$\frac{1}{s_1^{\alpha}s_2^{\beta}}\tilde{u}^{yx}(s_1,s_2) = \tilde{f}^{yx}(s_1,s_2). \quad (74)$$

Now we can choose two different ways to get the inverse Laplace transform from (74). Writing (74) as

$$\frac{1}{s_1}\tilde{u}^{yx}(s_1,s_2) = s_1^{\alpha-1}s_2^{\beta}\tilde{f}^{yx}(s_1,s_2);$$

$$J_x^{1}\tilde{u}^{y}(x,s_2) = J_x^{1-\alpha}s_2^{\beta}\tilde{f}^{y}(x,s_2); \quad (75)$$

$$\tilde{u}^{y}(x,s_2) = D_x^{1}J_x^{1-\alpha}J_x^{1-\alpha}s_2^{\beta}\tilde{f}^{y}(x,s_2) = s_2^{\beta}D_x^{\alpha}\tilde{f}^{y}(x,s_2); \quad (76)$$

$$J_y^{1}u(x,y) = J_y^{1-\beta}D_x^{\alpha}f(x,y); \quad (77)$$

we obtain

$$u(x,y) = D_y^{\beta}D_x^{\alpha}f(x,y) = D_y^{1}J_y^{1-\beta}D_x^{\alpha}f(x,y). \quad (78)$$

On the other hand, writing (95) as

$$\tilde{u}^{yx}(s_1,s_2) = \frac{1}{s_1^{1-\alpha}}\left[\begin{array}{c} s_1\{s_2^{\beta}\tilde{f}^{yx}(s_1,s_2)\} - \\ \{s_2^{\beta}\tilde{f}^{y}(0^+,s_2)\} \end{array}\right] + \frac{s_2^{\beta}\tilde{f}^{y}(0^+,s_2)}{s_1^{1-\alpha}}; \quad (79)$$

$$\mathcal{L}_x\tilde{u}^{y}(x,s_2) = \mathcal{L}_x\{\Phi_{1-\alpha}(x) * D_x^{1}s_2^{\beta}\tilde{f}^{y}(x,s_2)\}$$
$$+\mathcal{L}_x\{\tilde{f}^{y}(0^+,s_2)\Phi_{1-\alpha}(x)\}; \quad (80)$$

leads to

$$\mathcal{L}_y u(x,y) = D_x^{\alpha}\{s_2^{\beta}\tilde{f}^{y}(x,s_2)\} + \tilde{f}^{y}(0^+,s_2)\Phi_{1-\alpha}(x). \quad (81)$$

Hence, we can ge

$$u(x,y) = D_y^{\beta}D_x^{\alpha}f(x,y) + \Phi_{1-\beta}(y)D_x^{\alpha}f(x,0^+)$$
$$+\Phi_{1-\alpha}(x)D_y^{\beta}f(0^+,y)\Phi_{1-\alpha}(x)\Phi_{1-\beta}(y)f(0^+,0^+); \quad (82)$$

$$D_x^{\alpha}f(x,y) = D_x^{\alpha}f(x,y) + \sum_{k=0}^{m-1}\frac{x^{k-\alpha}}{\Gamma(k-\alpha+1)}f_x^{(k)}(0^+,y); \quad (83)$$

$$D_y^{\beta}f(x,y) = D_y^{\beta}f(x,y) + \sum_{k=0}^{n-1}\frac{y^{k-\beta}}{\Gamma(k-\beta+1)}f_y^{(k)}(x,0^+). \quad (84)$$

As a specific example, for $0 < \alpha < 1$, there exists

$$D_x^{\alpha}f(x,y) = D_x^{\alpha}f(x,y) + \Phi_{1-\alpha}(x)f(0^+,y); \quad (85)$$

$$D_y^{\beta}f(x,y) = D_y^{\beta}f(x,y) + \Phi_{1-\beta}(y)f(x,0^+); \quad (86)$$

where we have used (52) with $m = 1$ and (53) with $n = 1$.

$$D_y^{\beta}D_x^{\alpha}f(x,y) = D_y^{\beta}D_x^{\alpha}f(x,y) + \Phi_{1-\beta}(y)D_x^{\alpha}f(x,0^+)$$
$$+\Phi_{1-\alpha}(x)D_y^{\beta}f(0^+,y) + \Phi_{1-\alpha}(x)\Phi_{1-\beta}(y)f(0^+,0^+). \quad (87)$$

Thus, the solutions (78) and (82) are expressed in terms of the partial fractional derivatives $D_y^{\beta}D_x^{\alpha}$ and $D_y^{\beta}D_x^{\alpha}$. In fact, many of the above conclusions can be directly generalized to those of n-dimensional case, where $n \geq 2$.

Conclusions

Since the beginning of the theory of differential and integral calculus, the mathematicians began their investigations of the calculation of non-integer order derivatives and integrals. In this paper, we introduce the properties of the partial fractional integrals, the partial fractional derivatives, and the composite fractional integrals and derivatives, and a specific application of solving a two dimensional Abel integral equation is proposed.

Acknowledgments

This work was supported by the Fundamental Research Funds for the Central Universities (Northwest Minzu University) under Grant No. 31920150039, the Northwest Minzu University Talent Research Funds (No.xbmuyjrc201632) and the Youth Science and Technology Foundation of Gansu Province (No. 1610RJYA109) and Fundamental Research Funds for the Central Universities (NO.31920170001, 31920170009, 31920170143).

References

1. B. Baeumer, M. Meerschaert, J. Mortensen, Proceedings of the American Mathematical Society, pp. 2273-2282, (2005)
2. M. Caputo, Elasticit à e Dissipazione, Zanichelli, Bologna, (1969)
3. K. Diethelm, Springer-Verlag, New York, (2010)
4. K. Diethelm, N. J. Ford, J. Math. Anal. Appl. **265** pp. 229-248,(2002)
5. G. Doetsch, Berlin, Germany, (1974)
6. R. Gorenflo, F. Mainardi, Mathematics, **49**(2), pp. 277-290, (2008)
7. K. Nishimoto, An essence of Nishimoto's Fractional Calculus, (Calculus in the 21st Century), Descartes Press Company, (1991)
8. K. B. Oldham, J. Spanier, New York, USA, (1974)
9. G. Samko, A. Kilbas, I. Marchev, Gordon and Breach Science Publishers, Singapore, (1993)

Design of Mobile Operation and Maintenance System Based on Power GIS and GPS

TAO *Jing* [1,2,*], WU *Liwen* [3], ZHANG *Yanlei* [3], ZHANG *Jiaojiao* [3] and YAO *Jiming* i[1,2]

[1]Global Energy Interconnection Research institute, Nanjing, China
[2]State Grid Laboratory of Electric Power Communication Network Technology, Beijing, China
[3]Information & Te1ecommunication Company, State Grid Jibei Electric Power Co., Ltd, Beijing, China

Abstract. In order to solve the low efficiency of operation and maintenance and the problems of management coordination caused by the coexistence of off-line operation and maintenance and multi system coexistence in current power telecommunication operation and maintenance, this paper proposes a mobile operation and maintenance system based on power GIS and GPS location technology. The system is based on the C/S architecture, and the mobile operation and maintenance terminal is connected with the operation and maintenance management terminal through wireless and wired network. In addition to meeting basic business requirements, the system integrates power GIS functions to visually demonstrate communication resources, personnel, and failures. The staff can use the wireless mobile terminal which integrated GPS module for to operate and maintain, and can locate accurately in real time.

1 Introduction

The operation and maintenance of power communication is a series of comprehensive management in order to ensure the stable operation of power system. Mainly include: (1)Communication scheduling (such as system operations monitoring, fault management, communication resource management, etc.), (2) The operation of communication (analysis of power communication data, technical management, safety management, etc.), (3) Maintenance of communication (such as daily information collection, regular inspection, troubleshooting, etc.).

Currently, the following problems exist in the implementation and management of power communication and maintenance operations:

•Lack of unified management platform. The management modes, software, hardware systems and media forms adopted by different provinces and cities are different. Multi mode coexistence does not facilitate centralized management of all communication systems under the state Power Grid Corp, and discrete data can easily lead to information silos,

•Part of the communication operation and communication service is still off-line. First, the management system distributes the work order to the operation and maintenance operator, and then the operator carries the printed work order to inspects and records simultaneous, and finally enters the check result into the system. This method reduces the efficiency of the operation and maintenance operator, and increases the verification workload of the management administrator. And the description of the fault or the hidden danger depends on the personal written expression ability of the operation and maintenance personnel,

•Lack of intuitive map display. Localization of communication resources, fault and maintenance personnel patrol record respectively depend on management accounting, fault report written description and work logs. When needed, the communication resources, failures and maintenance personnel can't be quickly located.

To solve the above problems, a mobile operation and maintenance system based on power GIS, GPS positioning and IOT identification is designed and applied to the field of power communication operation and maintenance.

2 System design

2.1. System overall design

* Corresponding author: taojing@geiri.sgcc.com.cn

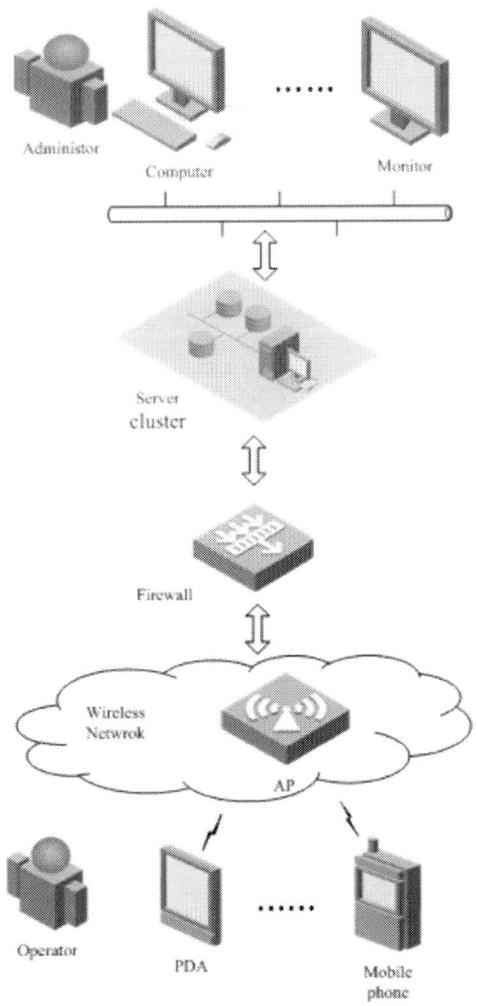

Fig. 1. Design of mobile operation and maintenance system.

Operators on the spot use mobile terminals for operation and maintenance, and mobile terminals use wireless VPN to access the wireless network via AP. Wireless networks connect to the VPN firewall. Depending on the security area which operation and maintenance data belongs, the data will be connected to the server cluster after the firewall of a specific security zone. Server cluster includes mobile operation and maintenance server, data server and so on. Management terminals connect to the server through network devices such as switches.

2.1. Function of Mobile Terminal

The mobile terminal contains the following hardware modules: GPS positioning module, wireless communication module, button module, processor module, power module, multimedia module and memory module. The processing module implements the software function of the mobile terminal, including terminal management, signal processing, data encryption, log management, security management, streaming media and information push. As shown in Fig 2:

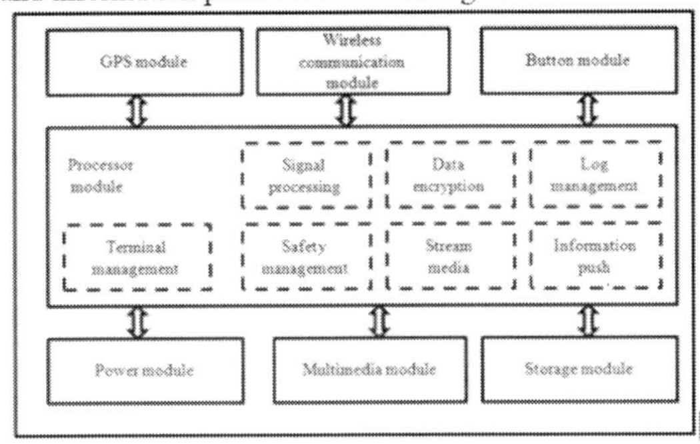

Fig. 2. Design of mobile terminal

Operators receive the work orders distributed by the management terminal through the mobile terminal, and respond accordingly. Operators performs maintenance operations on the spot according to the pushed checklist, safety notices and operation instructions issued in the mobile terminal, and reports communication equipment, circuit status and other information in real time. When the operation and maintenance site is found to be out of order or need to be dealt with online, it can be communicated with other operators or managers through the online call and online video module of the mobile terminal. As shown in the following figure:

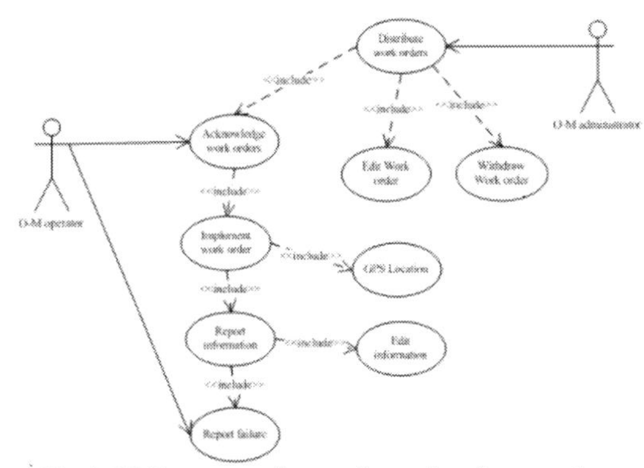

Fig. 3. UML use case diagram for work order execution.

2.3. Function of Management Terminal

The management terminal includes the following hardware modules: additional function interface, additional data interface, communication module, processor module, power module, multimedia module, data server. The processing module implements the software function of the management terminal , including signal processing, data encryption, network management, personnel management, Yun Wei plan management, report management, fault management, resource management, communication management, streaming media information release, GIS management, Yun Wei system management, terminal management, log management and security management. As shown in figure 4:

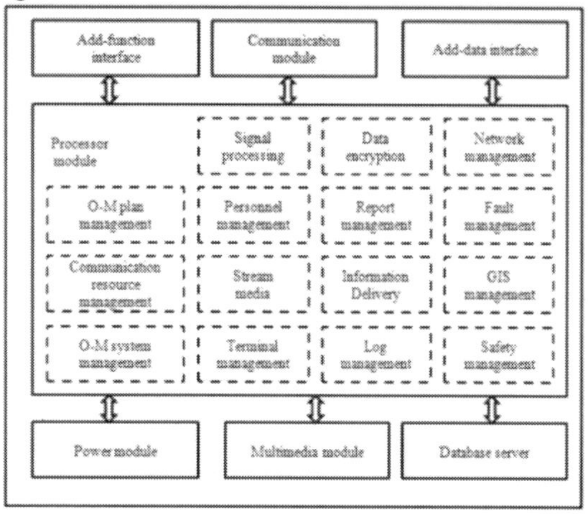

Fig. 4. Design of Management terminal.

In the overall coordination of scheduling: by using the management of terminals, the administrator can plan the operation and maintenance and manage all communication resources in the system, and also be able to control the failure and line status, generate corresponding reports according to different levels and requirements.

In management of a single task: the administrator distribute the work order to operator. And then, the administrator can inquire the current position of the personnel and the implementation of the plan through the personnel management module and the operation and maintenance plan management module.

3 Application of key technologies in the system

3.1. System Problems and Solutions

In order to solve the problems faced by the current operation and maintenance in the introduction and improve the intelligence level of the equipment, the mobile operation and maintenance system not only realizes the basic functions, but also focuses on the research and implementation of the following functions:

•Visual display. This function helps to improve the readability of all types of reports. In the event of a failure, the affected area and line can be viewed intuitively through the management terminal, and the operators can be dispatched to the scene as soon as possible. The operator can quickly query to the fault location through the mobile side. And ultimately reduce the duration of the failure.

•The high-precision positioning function of mobile terminal. This function helps to locate the fault quickly, manage the communication resources, and check the working discipline of the operators.

•Intelligent communication resource identification and marking. Using more efficient and intelligent equipment or line identification technology to avoid errors caused by manual registration. And after that, the status of the current resources can be recorded online, and the historical information can accessed in real time, so as to improve the efficiency of communication resource management.

In order to achieve the above functions, the system uses the power GIS, GPS positioning, Internet of things equipment identification technology.

3.2. Application of Key Technologies

3.2.1 Power GIS

GIS (Geographic Information System) is a data management system that can centralize, store, query, edit, process and display geographic information. Power GIS is an example of GIS in the application of power system, can integrate and display the roads, ordinary buildings, power system construction (such as substations and power plants etc.), tower, transmission line, communication resources (such as communication equipment, communication lines etc.) with realistic geographical environment.

In the mobile operation and maintenance system, both the server management and the mobile terminal can perform the visual fault tracking and resource scheduling through the power GIS function. As shown in the following figure:

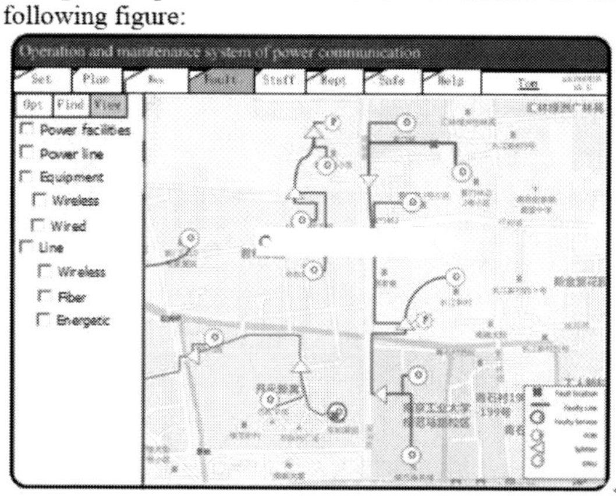

Fig. 5. The GIS view of the fault on Management Terminal.

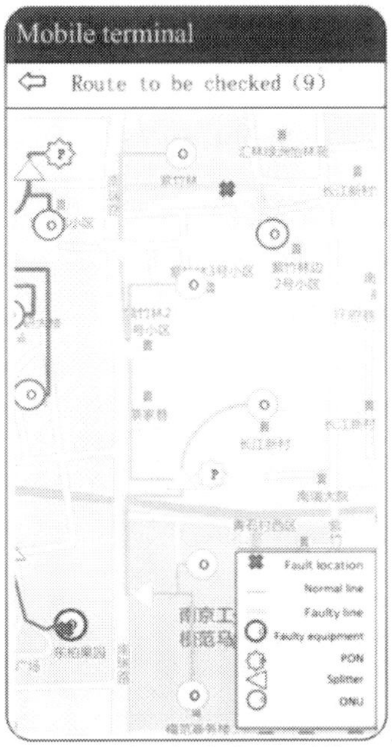

Fig. 6.Routes to be checked on Mobile Terminal.

3.2.2 GPS

In order to realize the high precision and real-time positioning of mobile terminals, several mainstream localization technologies are investigated. The advantages and disadvantages are shown in the following table:

Table 1. Comparison Of Several Positioning Techniques.

Positioning Technology	Virtues	Flaws
GPS	Technology is mature and module prices are low.	Positioning takes a long time (about a few minutes), civilian accuracy is low (about 10m), indoor positioning effect is poor.
A-GPS	Quick positioning, moderate civilian accuracy (5m or so).	Wireless cellular base station networks are needed, the use of A-GPS requires a fee.
Wifi AP	It is economical and practical without adding hardware modules	Clients are required to connect to the wireless network
FRID	Indoor positioning can be achieved	The location accuracy is low and the workload of the deployment location network is large

In consideration of : 1) most of the power communication operation and maintenance work is carried out on the ground, and the technology of supporting outdoor positioning can be selected; 2) the operation and maintenance work has its own duration, so rapid positioning is not necessary; 3) combined with the power GIS, the accuracy of 10m has met the use requirements; 4) convenience and economy. Therefore, this system finally selects the GPS positioning technology. Figure 7 shows the interface when the operators fill in the patrol record:

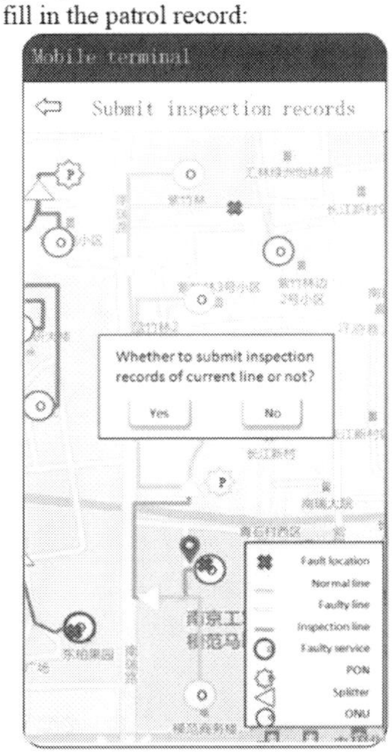

Fig. 7. Record submission

3.2.3 IOT Device Identification Technology

The Internet of things identification technology can query and edit the communication resource information rapidly on the mobile terminal, and can maintain all the historical information of the resource through the database at the management server. The current mainstream equipment identification technologies, including bar code, two-dimensional code and radio frequency identification technology, are investigated. Considering the difficulty and cost of the implementation, the bar code and two-dimensional code identification methods are selected.

The bar code capacity on the fixed assets label of the existing communication equipment and circuit can be expanded or replaced into two-dimensional codes. At the operation and maintenance site, the operator scans the bar code or two-dimensional code through the camera assembled by the mobile terminal and enters the information base of the resource. Operators update the corresponding attributes of the resources according to the actual situation at the scene. The management terminal updates the information to the data server to avoid the emergence of information islands.

4 Summary

The design can solve the current power communication operation and maintenance encountered in a series of problems. In order to ensure the stability of power grid and communication network operation, the design proposed in this paper adopts the mature technology. The extended function and data interface are reserved in the software and hardware modules of the terminal, so as to improve the adaptability of the device.

In addition, the design of mobile terminal and management terminal in this article is based on independent device, so it contains hardware module and software. In actual R & D and production, a large number of investigations are needed to determine the specific forms of implementation. Such as mobile side installs applications installed in Android, IOS and other system terminals, the management side installs executable programs in Windows, Linux and other systems.

Acknowledgment

This work was supported by the Science and Technology Foundation of the State Grid Corporation of China: Research and demonstration application of intelligent operation and maintenance for power communication field(SGJBXT00DDJS1700067).

References

1. YANG Guoqi. Consideration on construction of power telecommunication operation and maintenance system[J]. Communication world, 2017, 10:154-155.
2. YANG Zhimin, HUANG Yu, LEI Yun, etc. Power communication, operation and maintenance, mobile system and power communication operation and maintenance method:, CN 103685215 A[P]. 2014.
3. CHEN Dong, TU Li, LI Huimin, etc. Design of comprehensive operations automation system of power transmission and transformation equipment[J], Electrical Measurement & Instrumentation,2016, 53(2):1-9.
4. GAo Xiaoming. Design and Research of Power Equipment Inspection System Based on Mobile Internet[D].South China University of Technology,2013
5. SHI Wenbo, YU Junjie, LI Xiaoxiao. Design and realization of electric power GIS system[J]. Modern Electronics Technique, 2015, 38(21):155-158.
6. HU Shengqing. Research on fast positioning of optical fiber cable fault in power system based on GIS[J]. China High-tech Enterprises, 2016(35):7-8.
7. BI Jingxue, ZHEN Jie, GUO Ying. Accuracy of GPS and A-GPS Positioning on Android Phone[J]. Bulletin of Surveying and Mapping, 2016(7):10-13.
8. Shen Li, Bi Wei, Shu Kaiqi. Design of GPS and GP R S based remote synchronization detector[J] Electrical Measurement & Instrumentation, 2016, 53(21):124-128.
9. HUANG Hongbing, LIU Junyi. Management, technology, double pipe drive to improve the quality of power communication operation and maintenance[J]. Enterprise Management, 2016(S1):270-271.
10. KUANG Hong-hai, ZHANG Shu-yun, ZENG Li-qiong, etc. Design of rural intelligent distribution network remote monitoring system based on GPRS and GPS[J]. Advanced Technology of Electrical Engineering and Energy, 2017(4):83-88.
11. ZHANG Jing, ZHANG Qingwei, WANG Kaiyu, etc. Research on RFID positioning technology applied to Internet of Things devices[J]. Modern Electronics Technique, 2017, 40(5):29-32.
12. JIANG Shu. Comparative study on the application of bar code, two-dimensional code and RFID in archives management[J]. Beijing archives, 2015(8):36-37.
13. XIAO Qingwang, WANG Jinhua, ZHU Yixiang. Intelligent terminal device identification method of internet of things[J]. Telecommunications Science, 2017, 33(2):3-8.

DOI: 10.1051/matecconf/201713900100

Talking Aboat WIFI's New Security

Ling He[1,*], *Yangxin Teng*[1], *Mingxuan Li*[1], *Qingrui Guo*[1], *Feng Li*[1], *Cailing Wang*[2], and *Jizhao Yi*[3,] *Jinshan Gao*[2]

[1]Information and communication technology center, State Grid Xinjiang Electric Power Research Institute, Urumqi, China
[2]Urumqi railway bureau,Urumqi, China
[3]Tianjin Science and Technology University,Tianjin, China

Abstract. Wireless network technology's widely used brings the web's users huge convenience and flexibility, and also brings great challenges for the security of network.This paper introduces the MAC address authentication, Wired Equivalent Privacy (WEP), Wi-Fi Protected Access (WPA), WiFi Protected Access 2 (WPA2) and 802.1X and so on, giving priority to analysing their pros and cons. A new model will be used to cope with the security issues of WIFI based on it.

1 Introduction

WIFI means wireless local area network based on IEEE802.11b. As a new generation of wireless network technology, WIFI can get ethernet network performance, rate and availability for wireless network users, and also seamlessly combine variety technology of LAN. In recent years, the number of wireless AP is growing rapidly, on account of wireless network's convenience and efficiency can make it spread quickly[1].

At the same time, Open nature of the wireless network determines it is more vulnerable than wired network in safety. Accordingly, when we enjoy the convenience and flexible features which the wireless network brings, it will be of great use to take technical measures to improve security.

2 Development Prospects

WIFI is made up of access point and wireless network card. WIFI' s network structure[2] is shown in Fig 1.

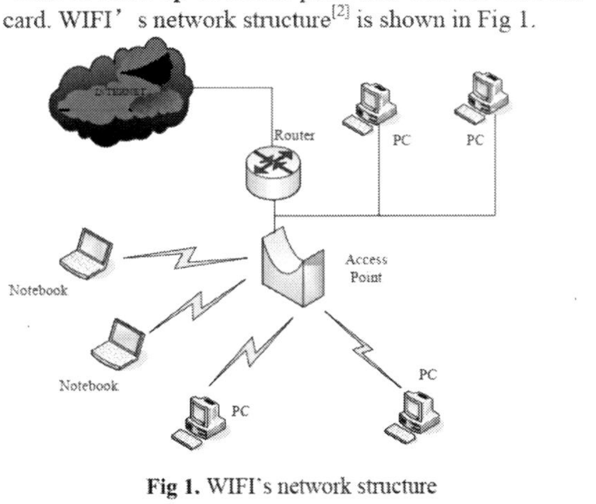

Fig 1. WIFI's network structure

The wi-fi technology is short distance in wireless information technology. It has widely used both in the office family. It usually uses 2.4 GHz frequency bands [3].

1) The radio waves used in wifi waves which covers a wide range. As an example, the Bluetooth radio waves can encompass a radius about 15 meters, at the same time the radio coverage of wireless technology can reach 100 meters. It makes the wireless technology to better meet the various needs of diverse types of users. With the continuous development of technological innovation, the effective coverage range will make further efforts to increase.

2) The high transmission speed. Depending upon the different criteria of wireless card are used, the transmission rate becomes different. For instance, the maximum number of IEEE802.11b can reach 11Mbps, the 802.11a and others can achieve 54Mbps. Comparing with the Bluetooth technology and others, the high-speed of WiFi technology can meet the requirements of the development of the information-based society.

3) The low application threshold. In China, wireless technology is growing rapidly. There are many public places use it, such as airports, stations, restaurants and business halls and so on, where set up wifi hotspots that we can use it by means of high-speed lines which access the Internet. On condition that wireless customers use laptops that support LAN pours into this area can easily access the network.

Owing to wireless networks have many virtues such as low cost and convenient use, which has become such an integral part of our daily life, at the same time, the most severe cases is that whether the security identification system of wifi technology is reliable. The following are comparative analyses about the security measures of wifi.

3 Compared

3.1. MAC address authentication

* Corresponding author: author@e-mail.org

Wireless AP checks whether the address is a legitimate user by checking the 48 bit MAC address contained in the data frame header sent by the user.

The drawback of this approach lies in that MAC address authentication is unreliable. Illegal invaders can track and detect by means of a sniffer software and then their phones can verify the MAC address of legal machines through modifying their own MAC address, on this count they can be disguised as legitimate users to connect the network.

3.2. Wired Equivalent Privacy

WEP[4] is a security protocol, which can be applied to wireless local area network (WLAN) of 802.11b standard. It is a kind of basic encryption measure for wireless devices. WEP uses the RC4 encryption algorithm based on 40/104-bit to share encryption keys and it uses this approach to encrypt all data transmitted by wireless network.

With the upgrade of wireless security, WEP encryption has occurred 100% crack methods. Linear CRC encryption algorithm, static keys as well as unspecified key management and vulnerability authentication mechanisms, which can make hackers disrupt confidential matters about WEP if they can get enough data packets by a network sniffer. For instance, AirSnortor or WEPCrack, they can try to break out of the 2.45*109 keys per second, which means that the 40-bit encrypted data can rank 240 keys will be deciphered provided that these tools collect enough data packets in 5min.

3.3. WPA and TKIP

Due to the vulnerability of WEP, before the new standards produced, WIFI alliance solutions develop a transitional solution — WPA. It will limit hacker attacks by adopting 48 bit IV, building mechanisms with each pack key and periodic alternating release keys as well as information coding. WEP uses the same symmetric encryption algorithm as RC4, which fixes quite a few dilemmas including short IVs in WEP and fixed keys etc. The protocol can be realized with software upgrades in the existing WEP so as to achieve the existing resources can obtain the maximum security.

Nevertheless, WPA still has many serious problems. Because of the weak encryption algorithm, it still may be cracked : hackers just listen to enough data packets with powerful computing equipment even under the protection agreement.

3.4. WPA2 and AES

In June 2004, IEEE 802.11i standard set up, wireless cable association re-release wireless security solutions— WPA2, which based on the latest IEEE 802.11i standard. WPA2 is a way to protect data security that is based on advanced Encryption Standard (AES). AES is a type of symmetrical block encryption technology, using 12bit block encrypting of data, and providing a more advanced encryption performance than RC4. WPA2 uses 48 bit IV and IV sequence rules to send the initial key, meanwhile, send out MIC using CCMP key tamper detection. CCMP adopts the cipher block chaining mode to encrypt the data into 128 bits, and provides information integrity detection via the MAC layer. Nevertheless, due to the high requirements of AES hardware, CCMP can not be achieved on the basis of existing equipment upgrades.

Currently, WPA2 encryption mode was once thought to be 100% safe but now it is considered to be extremely unsafe at present. Back when the original 2009, two Japanese security experts said that they had been developed a way that made use of wireless router to break WPA encryption system in a minute. Through the dictionary and PIN code we can easily solve almost 60% WPA2 secrets.

3.5. 802.1X

802.1x is a kind of port based network aceess control protocol [5]. It defines three important components: Supplicant's System, Authentication's System and authentication's Server's System. The following figure is the relationship between the three and communication between each other.

802.1x authentication protocol is layer protocol, its service message is directly loaded on the normal second layer message. In the course of the process of authentication, 802.1X doesn't need encapsulate frames into the Ethernet, so the process is relatively efficient. And at present, most hardware devices support this protocol. The performance requirements of the devices are not high and relatively simple to achieve. Therefore, 802.1X is supported by many hardware and software vendors.

But 802.1X is not overwhelmingly safe in reality. A new study indicates that 02.1X is easily attacked by Session hold—It's means that after attackers wait for legitimate users to pass authentication, on the one hand they through various forms of Dos to prevent legitimate users and servers' connection. On the other hand, they pretend to be legal users to establish a connection with the servers, which has been challenging the security of wireless network.

4 Propose a New Model to Improve Security

In view of existing situation, there are some latest attack ways: sniffing, disguising, pretending, utilizing existing vulnerabilities to attack wireless network with brute-force, cracking by dictionary and grasp the package for analyse. These attack patterns utilize a variety of vulnerabilities to attack, for instance, authentication attack, forged disconnect, forged AP, probing AP information and guessing ESSID and so on. All the above of vulnerabilities will interact with that data with wireless devices. In the process of interaction, MAC address must be copied by wireless devices, but provided that you want to find hidden troubles through enhancements, you must login management interface. It is unable to real-time monitor the security performance of the wireless devices and unable to find and solve cases. Many times, we don' t even notice it until it' s too late. How to better protect our safe router and display the status of wireless router to users in real time, which that needed to be addressed as a matter of urgency. We propose a new model based on this view that reinforce general wireless routers so as to quickly find the wireless network attacked and these issues can be received our immediate attention.

First we have to count client numbers who are connected by a wireless router and MAC address so as to form a standard document and lock (except public wireless routers). In order to obtain linked MAC addresses, we have to establish a session with a local wireless router. Comparing the two, once there are abnormal and connective MAC address records, it will prompt information to the desktop terminal instantly and inform the local administrator that there has been set up an illegal link. As shown in Fig 2.

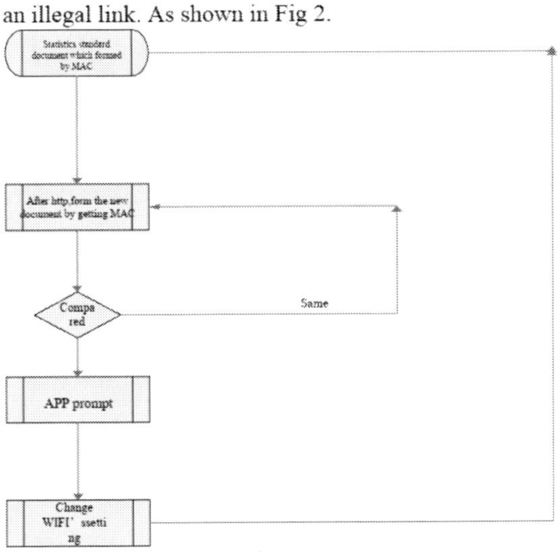

Fig 2. Security reinforce procedure

We need to find a cross-platform tool to realize the function that we can record the MAC visits and prompt authorized users and unauthorized users who get the relevant contents by visiting routers. We choose the Python to programme, it is a pure and free software, and source code and interpreter follow the GPL protocol. The advantages are clear grammars and rich and powerful databases,

The following we will take the TP-Link router as an example (the main noted type is TL-WR720N, the firmware version is TL-WR720N V3_120620 Standard Edition) Using Python as the platform to enhance the security of this router.

5 Solutions

1)Establishing a session named urllib2 with any wireless routers. We can simulate various kinds of loggings that meet the requirements through urllib2' s various attributes. We obtain wireless routers' relevant information with GET:

```
ip = '192.168.1.253'
login_user = 'admin'
login_pwd = 'admin'
url = 'http://' + ip + '/userRpm/WlanStationRpm.htm?Page=1'
auth = 'Basic ' + base64.b64encode(login_user+':'+login_pwd)
heads = { 'Referer' : 'http://' + ip + '/userRpm/WlanStationRpm.htm','Authorization' :
auth }
request = urllib2.Request(url,None,heads)
response = urllib2.urlopen(request)
html = response.read()
```

Once we get the wireless router ' s landing permissions, we can read the special message from special webs for our subsequent queries and uses.

2)We transcode the information which we obtain from the html through a type of component — BeautifulSoup. In this article, we just get linking MAC addresses as following:

```
soup = BeautifulSoup(html)
res = soup.find_all('script')[1]
cont = res.prettify()
reg1 = re.compile("<[^>]*>")
res1 = reg1.sub('',cont)
mac = re.findall('..-..-..-..-..-..', res1)
i = len(mac)
```

We will be able to carry out various types of operations through the linking MAC addresses. You can add a router' s white lists, and enable MAC filtering mechanism, etc.

3)Comparing the new MAC files with the standard files by continuous testing and then we can record the new MAC addresses.

```
file1 = open(r'tmp-b.txt','r')
file2 = open(r'tmp-c.txt','r')
file3 = open(r'tmp-ba.txt','w')
found = 0
l1 = file1.readline()
while l1:
  print
  print l1
  file2.seek(0,0)
  found =0
  l2 = file2.readline()
  while l2:
    print
    print l2
    if (l1==l2):
        found =1
      break
    l2 = file2.readline()
  if found == 0:
    file3.write(l1)
  l1=file1.readline()
file1.close()
file2.close()
file3.close()
```

We can identify some new MAC addresses through comparing and querying.

4)We can display the specific MAC addresses to users' desktop with Tkinter, which can remind the local users who have new MAC addresses to assess the wireless routers.

```
def show():
  bubb = Tkinter.Tk()
```

```
  f =open(r'tmp-ba.txt','r')
  mess = f.read()
  s = Label(bubb,text=mess,bg='blue',width='40',height='3')
  s.pack()

def creatmac():
  root = Tkinter.Tk()
  bub = Tkinter.Button(root,text="MAC",command=show)
  bub.pack()
  root.mainloop()
creatmac()
```

5)We can realize the real-time monitoring of local terminal by while loop and time components. The terminal can be a mobile phone or a computer that you only need to install the Python platform, and specific codes are as follows:

```
def main():
 while True :
   checkmac （）
     else :
       os.path.exists(r'tmp-c.txt')==True:
       break
   time.sleep(3)
```

It can achieve real-time monitoring of the local wireless routers’ MAC address transforms from the above five steps, and it will make a reminder through the unfamiliar MAC addresses access so as to improve the protective capability of wireless routers. The real-time monitoring of the running state of the wireless device can effectively prevent the wireless device from being attacked, and to a certain extent, it can improve the security of wireless devices. But we should be aware of the daily management of wireless devices. And do the followings：

Enhancing password strength, which means to it contains 8 or else more bits complex passwords and pieces of letters, numbers and symbols. For example, Bu/eB7@2.

Hiding SSID, which makes the SSID of wireless network no longer be broadcast can greatly reduce the probability of being hacked

MAC binding. Like the way to hide SSID, MAC binding can greatly increase the difficulty of hacking network. The additional complexity of the operation to make the majority of the hackers involved.

Turn off wireless routers’ QSS and WDS functions.

Try to avoid connecting your device to a public Wi-Fi. Even if connected to public wireless, you can’t use applications involving personal privacy.

Provided that you feel the internet speed slower significantly, you maybe suffered a session hijacking. At this point, you should immediately cancel the exit account and clear cookies and immediately modify the Wi-Fi password.

Do both from the technique and management, we can better improve the safety performance of the wireless network.

6 Conclusions

With the pace of life speeds up and further popularization of wifi, the increasing demands of people for wifi. At least technically, we need to strengthen the security mechanism of wireless devices, the safety consciousness and safety management measures in order to keep the WiFi is reliable.

Acknowledgment

The authors would like to thank the editor and the anonymous referees for their valuable comments, and thank the members who had participated in"wireless communications security in the State Grid Xinjiang power company technical application research.

References

1. Li xiaoyang.WIFI Technology And its Application And Development[J].Information Technology,2012,2:196-198
2. Lu yan,Mao xu.Compare And Analysis Based on WIFI’s wireless Network Security[J].Guangdong Communication Technology,2007，3:25-29
3. Yu zhenquan.A Wireless Network Security Protection System[J].Electronic Science and Technology,2006,9:1-9
4. Zhangyun.Analysis of Wireless Network Technology[J] .Science&Technology Information,2008,23:57-58
5. Lai yu,Zhang huajie.Information Network Security Measures to Explore Based on Wireless Network[J],Coal Technology,2013,10:234-235.

The Role Definition and Competency Model Building of the Teaching Assistant of the MOOCS

Wang Dongmei[1], *Yang Liyuan*[1], *Zhang Xiaohui*[2]

[1]*School of Distance and Continuing Education, Dalian University of Technology, Dalian, China*

[2]*School of Computer and Information Technology, Liaoning Normal University, Dalian, China*

Abstract. The teaching assistant has been the key element of the MOOCS. Their competence and ability affect the study support service quality of the MOOCS. Based on the connotation and function of the teaching assistant defined, combining the onion competence model theory and the study of relevant literature at home and abroad, this paper constructs the role and competency model of the teaching assistant of MOOCS, and explores the model application in the construction of teacher staff, in order to make effort to provide support for China's MOOCS.

1 Introduction

Massive open online courses is a new type learning mode of current education. As the "giant" of scale, the "diversity" of students and the "people-oriented" of the teaching mode, MOOCS has caused extensive concern of the society[1].Teachers of MOOCS come from the top universities all over the world, which have great attraction for the learners. On the other hand, many teachers also become the "stars" through MOOCS course quickly. The "stars" have won the "word of mouth" for the colleges and universities, which in turn have greatly promoted MOOCS institutions. MOOCS, however, are lack of tutors, and without the construction of faculty development support system. MOOCS increase the influence of the course by "the school, the teacher and the class", and collaborative works of "college joined, unified platform, unified production". On the other hand, lack of effective learning support services and complete the teaching management system, MOOCS safeguards the teaching quality by the prestigious reputation of the school only[2].Therefore, determine how to perfect learning support service and teaching management system, so as to promote the development of MOOCS? First, we need to clear the requirement of the teaching assistants in the learning support services. This paper will emphasis on this, namely the responsibilities and competency of the teaching assistants of MOOCS.

2 The connotation and function defining of the teaching assistant of the MOOCS

The lecturer of the MOOCS will be only one or a few people, and the course cannot be done by the lecturer itself. Before the course online, a lot of curriculum resources need to be prepared, including the content of curriculum, the reference materials, assignments and exams, etc., and the course video need to be recorded and made, which are completed by the production team coordinating with each other. Then the course will be designed to be better presented on the platform. Next, the course will be tested so to complete the operation. After the course online, the technical and content support will be needed and teachers also need to grasp the curriculum situation and students' learning dynamic condition timely. So the teaching team of MOOCS should consists of the lecturer, the curriculum production team and the other personnel, which we call "teaching assistant". In this paper, the teaching assistants will complete all the other work except lecturer's and the curriculum production teams', including curriculum designing, curriculum testing, perfecting the course resources and providing technical support services and so on. The role sharing and functions defining of MOOCS course mode is shown in Fig.1.

3 The theory of onion

The quality model theory was first originated by David McClell in America in the 1970 s. As the harvard university professor, in the paper of "testing competence rather than intelligence", David McClell first proposed the concept of "quality", namely the individual condition and behaviour characteristics which impacted on the job performance. In his paper, good performance on the job depends on personality traits, motive and values, which hidden in brain, in addition to the necessary knowledge and skills with work. Mcclelland put forward a new human resource evaluation analysis technology which is named the ability competence model, the model refine the core quality and ability of the performance outstanding personnel. Based on this theory, Richard Boyatzis from American put forward the "onion competence model "[3],which show the core elements of the quality, and illustrates the characteristics of the components can be observed and measured ,as is shown in fig. 2.

The characteristics and motives are the core of the onion model, and then the self-image, the social roles, the attitudes and values, the knowledge and skills expanding outward in turn. The elements which are more to the outer layer, more easy to be cultivated and evaluated; and more to the inner layer, more difficult to be evaluated and acquainted.

4 The relevant literature at home and abroad

Retrieving with the keywords of "teaching assistants + MOOCS" and "tutors + MOOCS" from the CNKI, Chinese digital periodical group, the Google academic, the Springer Online Journals, only 3 relevant articles are found. As the work of teaching assistant is similar to the distance counselling teachers, network guide teachers, network teachers, and e - tutoring, e - facilitating, e - mentoring, online teacher, online instructor. In this paper, further expanding the search scope, investigates and combs the relevant literature about the online education tutors and network guide teachers at home and abroad with the above keyword to provide the basis for subsequent research.

4.1 The relevant literature abroad

Combing foreign literature study (As shown in Table 1),we find foreign scholars about the quality of the teaching assistant focused on their ability, such as the teaching ability, the technology, the management and social dimension, which are get higher consensus. In addition, the ability of design, the research and evaluation dimensions are also got more mentioned.

Dimensions of the teaching ability obtained identity by all the literature, which was to provide support and guide for the students. Dimensions of the teaching ability focused on the students grasp the course content, such as the content catalysts, the processes catalysts, the solution of the problem, and to promote the students understanding of the courses, to organize and guide the learning intercourse, to answer students' questions about the course learning, and to motivate learning motivation through the teaching strategy, etc.

Ability of technology dimensions has also been mentioned in all documents, which referred to the technology integration and support capabilities of the teaching assistants, including selecting guidance technology, teaching design according to different technical environment, and technical feedback and support for the learning process. Teaching assistants don't need to be technical experts, but inform technical support for the students with difficulties.

Ability of social dimension also received high recognition. It referred to the ability of community construction and the maintenance of teaching assistants under the network. Ability of social dimension had nothing to do with students' learning tasks and learning content. It created the sense of belonging for students to eliminate cognitive obstacles and interference, including building positive network corporation, motivating students to participate in the network society, grasping the network community development direction and maintaining community cohesion.

The ability of management dimension is mentioned more. It referred to the ability of teaching organization and regulation of the teaching assistant. In the network autonomous learning, most learners may lack the ability to learn and self management. Therefore, an important responsibility of the assistant teaching is to organize the network learning activities and processes. So the teaching assistants must have the ability of management, in particular, guiding and controlling ability, stimulating learning motivation and enthusiasm ability, continuous attention and timely warning ability, handling teaching event ability and so on.

The ability of teaching design dimensions in the literature referred to the ability of the development and planning of teaching activities, including tutoring programming and planning online as well as the designing of resources based on the technology of interaction.

The ability of study dimensions in the literature referred to the research and innovation ability of the teaching assistant about the course guidance, including studying and providing learning strategy according to the psychological characteristics of adult learners, studying the frontier disciplines knowledge, innovation as well as promoting the open and distance education, etc.

The ability of evaluation dimension of the teaching assistant concentrated the monitoring and feedback ability which required in the tutoring work, including the teaching evaluation and demand forecast, monitoring and evaluation on the learning strategies and learning process as well as the evaluation of the inspection.

4.2 The relevant literature at home

The relevant literature at home about the teaching assistant mainly focused on the ability condition and role playing.

4.2.1 From the view of ability condition

Related the research abroad, the study at home also pointed out that the teaching assistants should possess the abilities, including teaching, evaluation, social, management, technology, design, etc.. Although every ability was mentioned individually, the views of domestic scholars are accordance with the foreign research(As is shown in Table 2).

4.2.2 From the view of role playing

Contrasting with the research about ability, the other domestic literature mainly discussed on the roles of teaching assistants. Jiang Yulian and Yao Wenjian [19] pointed out that an excellent teaching assistant was not only the tutor, but also the partnership, and the changer to drive the distance learning, as well as the service experts for the students' autonomous learning. Liu Ming[20] defined the teaching assistant as the navigator, the information provider of learning, the learning facilitator, and the director of the teaching goal. Ren Hong [21] indicated the teaching assistant should be the " guides" of the learning mode and the "instructor" of learning process, the " counsellor" of the learning content, as well as the "supervisor" of the learning effect and the "researcher" of the theory, Xie Qun etc. [22] pointed out that the teaching assistant may be the instructor for guiding students to effective learning, the servicer that provided circumspect support, and the researcher promoted the pen and distance education. Hu Liqiang [23] orientated the teaching assistant as twelve kinds of roles, such as the organizer of the learning process, the designers of the curriculum resources , etc..

Above the researches, this paper argues that the role of a good teaching assistant is not only a teacher to provide quality learning content and lead students to complete the task, but also a friend that observes students to learning situation and encourages students to study actively, and a support servicer provides circumspect service as well as a researcher that guarantees students to effective learning and promotes the development of remote teaching.

5 The role and competency model building of the teaching assistant of the MOOCS teaching mode

 DOI: 10.1051/matecconf/201713900101

The teaching assistant mode of the MOOCS is built based on the onion competence model theory and the relevant literature at home and abroad. The mode is composed by three dimension, as "for the student, the role-playing and the ability demanded. As is shown in Fig. 3.

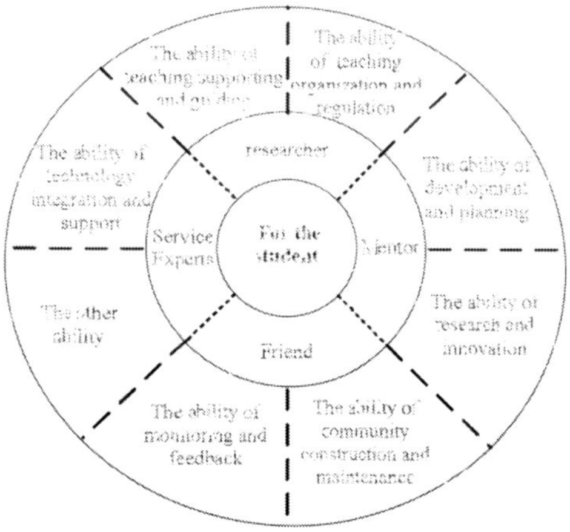

Fig.3 The competency model building of the teaching assistant of the MOOCs teaching mode based on the theory of onion

"For the Student" is the power quality of the teaching assistant, which requires teaching assistant should have the service consciousness, starting with students' demand , guiding and inspiring the right motivation, cultivating the students' interest, and help them solve psychological barriers, and improve autonomous learning ability, as well as provide learning methods and strategies, etc. The dedication and service consciousness is the power personality of the teaching assistant. It embodies the personality traits of the teaching assistant unity of the internal motivation on the basis of interest and the external motivation driving by the ideal.

The role-playing constitutes with the mentors, the friends, and the service of experts and the researchers, as the social role dimensions of the teaching assistant, which is described specifically above.

The dimensions of ability demanded constitutes of the ability of teaching supporting and guiding, the ability of community construction and maintenance, the ability of technology integration and support, the ability of teaching organization and regulation, the ability of development and planning, the ability of research and innovation, the ability of monitoring and feedback ,and the other ability. The other ability refers to the teaching assistants should provide students with the career and course selection consulting. The dimensions of ability demanded is the core competence of the teaching assistant, which belongs to the necessary condition for the teaching assistant.

The entire model embodies the competence factors of the teaching assistant as a professional in the sustainable and professional development. It also shows the teaching assistant should gradually from counselling daily to promoting students' effective learning, and improve the teaching support service quality and for the benign development of MOOCS. The Model clearly demonstrates the service consciousness and the social role of the teaching assistant, describes the necessary abilities to produce excellent performance of the teaching assistant, which help the teaching management system of MOOCS understand the competence level and performance improvement for the teaching assistant . on the other hand, the teaching assistant can promote their professionalism and specialization development through understanding and practical the competence model. The competence model also provides the policy basis and reference for the teaching staff construction of MOOCS in our country.

References

1. Li Man-li. Features and design principles of MOOCS. Tsinghua university education research.2013(4):13-21.
2. Yao Yuan, Han Xibin, Liu Yingqun, Cheng Jian. comparative study on the operation mechanism of MOOCS and remote education. Journal of distance education ,2013,6:3-10.
3. BOYATZIS A R. The Component Manager: A Model for Effective Performance[M].New York:John Wiley,1982.
4. Berge, Z.L. and Collins, M.P. (Eds.) Computer Mediated Communication and the online classroom[M]. Cresskill, NJ:Hampton Press. 1995, 1-3.
5. Shepherd, C.(2003). E-learning' s greatest hits[M]. Brighton:Above and Beyond Ltd,2003.

6. J Bawane, JM Spector.Prioritization of online instructor roles: implications for competency - based teacher education programs[J]. Distance Education ,2009,30(3):383-397.

7. Banzato, M. and Constantino, G.D. Competence Based Tutoring Online. Learning to Live in the Knowledge Society[J]. IFIP International Federation for Information Processing, 2008(281):209-216.

8. Peter Goodyear, Gilly Salmon, J.Michael Spector, Christine Steeples,Sue Tickner. Competences for Online Teaching[J]. Educational Technology Research and Development, 2001, 49(1): 65-72.

9. Dennis,B.,Watland,P.,Pirotte,S.,&Verday,N.Roleand Competencies of the E- tutor[DB/OL]. http://orbi.ulg.ac.be/bitstream/2268/12722/1/DENIS_WATLAND_PIROTTE_VERDAY_Roles_and_competencies_of_the_tutor_30_03_2009.pdf,2014-07-05.

10. Aydin,CH.Turkish mentors,perception of roles,competencies and resources for onlie learning[J].Turkish Online Journal of Distance Education-TOJDE,2005,6(3):58-80.

11. Sanjaya Mishra. Roles and competencies of academic counsellors in distance education[J]. Open Learning, 2005, 20(2):147–159.

12. From the athabasca university internal information

13. Teresa Guasch, Ibis Alvarez and Anna Espasa. Roles and Domains to Teach in Online Learning Environments: Educational ICT Competency Framework for University Teachers[A]. Changing Cultures in Higher Education[C]. Heidelberg:Springer Berlin Heidelberg Inc., 2010. 339-353.

14. Li Shuang and Zhang Yanxia. An Empirical Study on the Role Defining and Functional Transform of Course Tutor of Open Universities in Online Education Era[J]. China Educational Technology,2014,(9): 50-58.

15. Huang Xun, ChenWei, Zhou Chihe. empirical research on the ability framework of Online tutoring teachers [J].Journal of Guangxi Open University,2015,(3):11-17.

16. Li shuang, Chen Li. job analysis research of key professional roles of China's remote education [J]. China Educational Technology,2008,(12) :38-43.

17. HongQingGen. research of the role and responsibilities of Online tutor [J]. China Educational Technology,2010,(5):47-50.

18. Chen li, Feng xiaoying. research on the role ability condition of the Counseling teachers in the network [J]. China Educational Technology, 2012(7):58-63.

19. Jiang Yulian, Yao Wenjian. research on quality model building and application of the Distance education teachers [J]. China Educational Technology.2009(12):62-65.

20. Liu Min. thinking on Teachers' role in open education teaching [J]. Journal of adult education,2012,(7):51-52.

21. Ren Hong. the role change and relocation of the teacher in the Open University [J]. China's school education: in late,2012,(4): 68-68.

22. Xie Qun, Zhou Shaofeng, Liu Dingbang. the role evolution of the tutors of TV university course in the distance open education environment [J]. China distance education,2003,(5):36-39.

23. Hu Liqiang. study of the role of the counseling teachers Under the distance education[J]. Journal of education research,2009,(2): 90-91.

 DOI: 10.1051/matecconf/201713900101

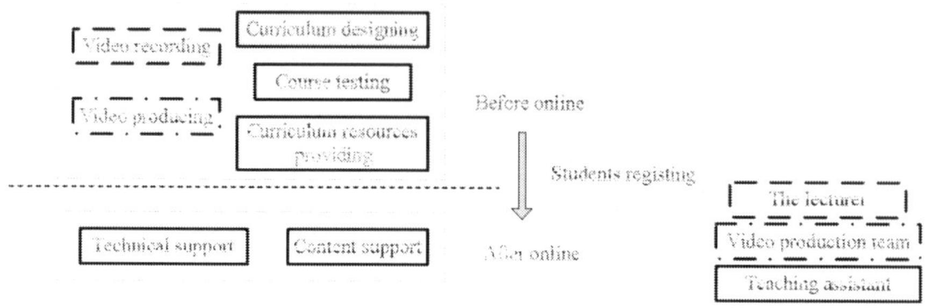

Fig. 1 The role sharing and functions defining of the MOOCs course mode

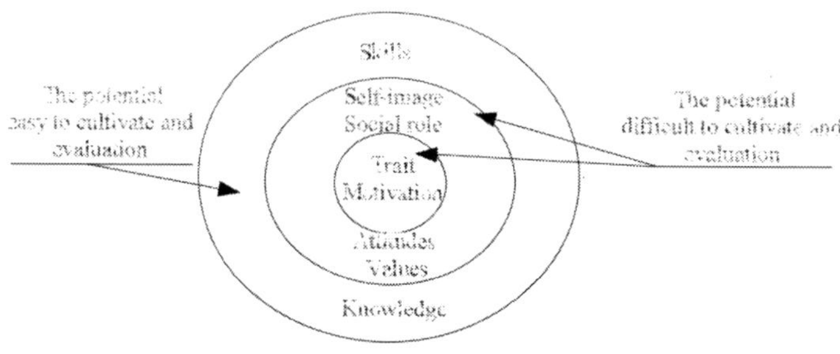

Fig. 2 The onion competence model

Table 1. The research summary abroad about the teaching assistant competence

The Researchers	Teaching	Technology	Social	Management	Design	Study	Evaluation	Consulting
Berge[4]	√	√	√	√				
Shepherd[5]	√	√			√			
Bawane[6]	√	√	√	√		√	√	√
Banzato[7]	√	√	√	√	√	√		
Goodyear et al[8]	√	√	√	√	√	√	√	
Dennis[9]	√	√	√	√	√	√	√	
Aydin[10]	√	√	√	√	√	√	√	
Sanjaya[11]	√	√	√	√		√	√	
Ally[12]	√	√	√	√			√	
Guasch[13]	√	√	√	√	√			
Total	10	10	9	9	6	6	6	1

Table 2. The research summary about the teaching assistant competence at home

The Researchers	Teaching	Evaluation	Social	Management	Technology	Design
LiShuang and ZhangYanxia[14]	√	√	√	√	√	
HuangXun[15]	√	√		√		√
LiShuang and ChenLi [16]	√	√	√			
Hong Qinggen [17]	√	√	√			
Chen Liand FengXiaoying [18]	√		√	√	√	√
Total	5	4	4	3	2	2

Research on Navigation Safety Evolution of Arctic Route Based on Entropy Theory

Liu Tongchao, *Li* Zhenfu

College of Transportation Management, Dalian Maritime University, Dalian 116026, China

Abstract: In recent years, with global warming, the development and utilization of the Arctic route has become possible. At present, the Arctic route navigation safety environment is the main factor that affects the passage of the Arctic route, so this paper mainly analyzes the development of the navigation safety environment of the Arctic route. First of all, using the fuzzy analytic hierarchy process to analyze the influence factors of the navigation safety of the Arctic route, the structure diagram of the navigation safety system of the Arctic route is divided. Secondly, according to the characteristics of navigation safety system of the Arctic route, entropy theory and fuzzy analytic hierarchy process are used to calculate the entropy values of navigation safety of Arctic routes, and the evolvement trend of navigation safety of Arctic routes is obtained. Finally, through the analysis of the trend of the Arctic route navigation security, some management countermeasures are put forward for the navigation safety of the Arctic route. The research results can provide reference for the development and utilization of the Arctic route.

1 Introduction

With the development of global warming and science and technology, human activities in the Arctic area are increasing day by day, and ships are sailing more frequently in the arctic. Due to the complex environment in the Arctic, the safety of ships in the Arctic region has attracted wide attention. In order to better understand the security of Arctic routes, it is necessary to analyze the evolution of the Arctic routes.

In recent years, the economic value of Arctic routes has been studied more than [1,2,3]. But little research for the Arctic route navigation environment, mainly Li Zhenfu respectively using the navigation environment model to explain the structure of Arctic route analysis and comprehensive safety evaluation method of Arctic route navigation environment by fuzzy comprehensive evaluation [4,5]. The existing assessment of the Arctic sea route safety environment mostly focuses on the evaluation of the navigation environment of the Arctic route without taking into consideration the factors of the Arctic route and the participation of management factors. The data show that 80% of the accidents are caused by human factors, 80% of which are caused by improper management. Therefore, although the geographical environment of the Arctic route is very important for the assessment of the security environment of the Arctic routes, human factors and management factors are also indispensable.

The navigation safety of the Arctic route is a systematic project, and the navigation safety of the Arctic route should be analyzed from the point of view of the system. Therefore this paper on Arctic route navigation safety evolution analysis method and safety entropy theory combining fuzzy analytic hierarchy process, and the use of safety system management method for the demonstration of the evolution trend, put forward the Arctic route navigation safety management countermeasures.

2. Theoretical

2.1 FAHP

AHP is widely used in multi attribute decision making method, but does not take into account the ambiguity of the people when they make decisions accordingly, fuzzy set theory and AHP combined with Fuzzy AHP method is proposed to fully considering the fuzziness of expert decision-making, and effectively reduce the AHP the inconsistency. Fuzzy analytic hierarchy process can analyze the system, divide the subsystems of different levels, and calculate the weights of different attributes [6].

2.2 Safety entropy theory

Entropy was put forward by T.Clausius in 1854. It is mainly used to describe the state of a system, and the order of the system state is represented by the entropy value. Originally, entropy was mainly applied to statistical physics and thermodynamics, and later expanded to biology, information theory and astrophysics, and then extended to various fields of natural science and some fields of social science. Entropy theory belongs to the initial stage in the application of security field. In this paper, we will expand the application of the safety entropy to analyze the evolution mechanism of the navigation safety system of the Arctic route.

2.3 Safety system management

Security system management is a comprehensive approach to ensure the safe operation of the system, mainly by using the laws and principles of safety science, system science and management science [7]. The main method of analysis is to analyze the state of the security system first, secondly, analyze the hidden security problems in the security system, and finally put forward some management countermeasures to the security system by using management methods.

3. Computing method

Firstly, the fuzzy analytic hierarchy process is used to analyze the system. Fuzzy AHP is used to analyze the hierarchical structure of the problem, and the complex problem is decomposed into several relatively simple sub structures. And through the fuzzy comparison of the 22 elements, and then obtain the weight of each element, and then get the relative important order. Fuzzy analytic hierarchy process (AHP) is a widely used method for partitioning the weights between different attributes.
The definition of [8] set matrix is an interval judgment matrix:

$$M=\begin{bmatrix} 1 & [l_{12},u_{12}] & \cdots & [l_{1n},u_{1n}] \\ [l_{21},u_{22}] & 1 & \cdots & [l_{2n},u_{2n}] \\ \vdots & \vdots & \ddots & \vdots \\ [l_{n1},u_{n1}] & [l_{n2},u_{n2}] & \cdots & 1 \end{bmatrix}$$

Where the upper and lower bounds of the interval values are represented respectively. If satisfied,
The interval judgment matrix is the interval reciprocal judgment matrix. The calculation of matrix, [9] Islam's method, this method can not only solve the consistency of interval reciprocal judgment matrix, but also can not satisfy the consistency for interval complementary judgment matrix is obtained by the weight vector is calculated as follows:

$$\begin{aligned} \min \ & \sum_{i=1}^{n-1}\sum_{j=i+1}^{n}\left(p_{ij}+q_{ij}\right) \\ s.t.\ & w_i - l_{ij}w_j + q_{ij} \geq 0, i=1,2,\ldots,n-1; j=i+1,i+2,\ldots,n \\ & w_i - u_{ij}w_j - p_{ij} \leq 0, i=1,2,\ldots,n-1; j=i+1,i+2,\ldots,n \\ & \sum_{i=1}^{n} w_i = 1, j=1,2,\ldots,n \\ & w_i, p_{ij}, q_{ij} \geq 0, i,j=1,2,\ldots,n \end{aligned} \quad (1)$$

It represents the offset distance that deviates from the lower bound of the interval; represents the deviation distance that deviates from the upper bound of the interval; it is the weight of different attributes.
Secondly, the utility function is used to calculate each attribute value, and different kinds of data are calculated. The concrete formula is as follows:

$$u_i = 1+\min(x_i)/\max(x_i) - x_i/\max(x_i) \quad (2)$$

$$u_i = x_i/\max(x_i) \quad (3)$$

Thirdly, the entropy theory is used to calculate the entropy of each attribute of the security system.

$$S(x_i)=k\log 1/P(x_i)=-k\log P(x_i)(i=1,2,\cdots n) \quad (4)$$

The article is a factor of safety; safety is the safety factor of the entropy is a factor of safety; safety; undetermined coefficient is, is an undetermined coefficient in order to facilitate and other combination of theory, which can be assigned to 1. The total entropy of safety can be weighted by the sum of the entropy of different attributes, and the concrete formula is as follows:

$$S=-\sum w_i \log P(x_i)(i=1,2,\cdots n) \quad (5)$$

Among them is the weight of the first attribute, which can be obtained by formula (1).

Only when the total entropy of the security system S=0, the security system in a most secure state, the system can be safe and orderly development of stability. The probability of occurrence of a safety accident is at its lowest level. Since the minimum value S is 0, the greater the value in any case $S \geq 0$, the more insecure the system state is S. The following analysis of the safety system changes according to different situations.
Suppose scenario 1, when $S_t = a, (a \geq 0)$, $S_{t+1} > a$, that is $dS/dt > 0$, represents the security system in the direction of instability, the system is increasingly insecure.
When $d^2S/dt^2 > 0$, the state of the safety system was shown to move towards a unstable state at a faster rate.
When $d^2S/dt^2 = 0$, the state of the safety system was shown to move to a steady state at a steady rate. In this case, there are not many in the real world
When $d^2S/dt^2 < 0$, the state of the safety system was shown to move towards a state of instability at a slower rate.

Suppose scenario 2, when $S_t = a, (a \geq 0)$, $S_{t+1} < a$,the value S is reduced. $dS/dt < 0$ represents the development of a security system toward a stable approach, and the system is becoming less secure.

When $d^2S/dt^2 > 0$, the state of the safety system was shown to move towards a unstable state at a faster rate.

When $d^2S/dt^2 = 0$, the state of the safety system was shown to move to a steady state at a steady rate.

When $d^2S/dt^2 < 0$, the state of the safety system was shown to move towards a state of instability at a slower rate.

To sum up, for the analysis of system security evolution, the steps are as follows:

Step 1 calculate the attribute weight of the security system according to formula (1)

Step 2 normalize the attribute values of the security system according to the formula (2)

step 3, according to formula (3) and (4), each attribute value of the security system is calculated by entropy

Step 4 calculate the total entropy of the safety system according to formula (5)

Step 5, according to different system security change scenarios, make different security management countermeasures.

4 Calculation and analysis of navigation safety system evolution of Arctic route

4.1 evolution calculation of navigation safety system of Arctic route

The security environment of the Arctic route plays an important role in the navigation and exploitation of the Arctic route. Therefore, it is very necessary to evaluate the security environment of the Arctic route. There are many factors affecting the safe navigation of the Arctic route. This paper uses fuzzy analytic hierarchy process (AHP) to systematically analyze the security status of the Arctic routes from three aspects, such as human, machine and environment. From the analysis of human factors, the main factors included safety awareness, safety operation, emergency response, risk identification and improper behavior. From the factor analysis of aircraft, the main factors included ice breaker, port infrastructure, navigation instruments, communication equipment and rescue equipment. From the analysis of environmental factors, the factors included ice, visibility, wind, channel width and water depth, channel obstacles, tides and currents, polar front and cyclones. The detailed structure is shown in figure 1. The Arctic route number of ship navigation data mainly from the northern sea route official information (http://www.arctic-lio.com/nsr_transits), ice area data mainly comes from the National Snow and Ice Data Center (https://nsidc.org/data).

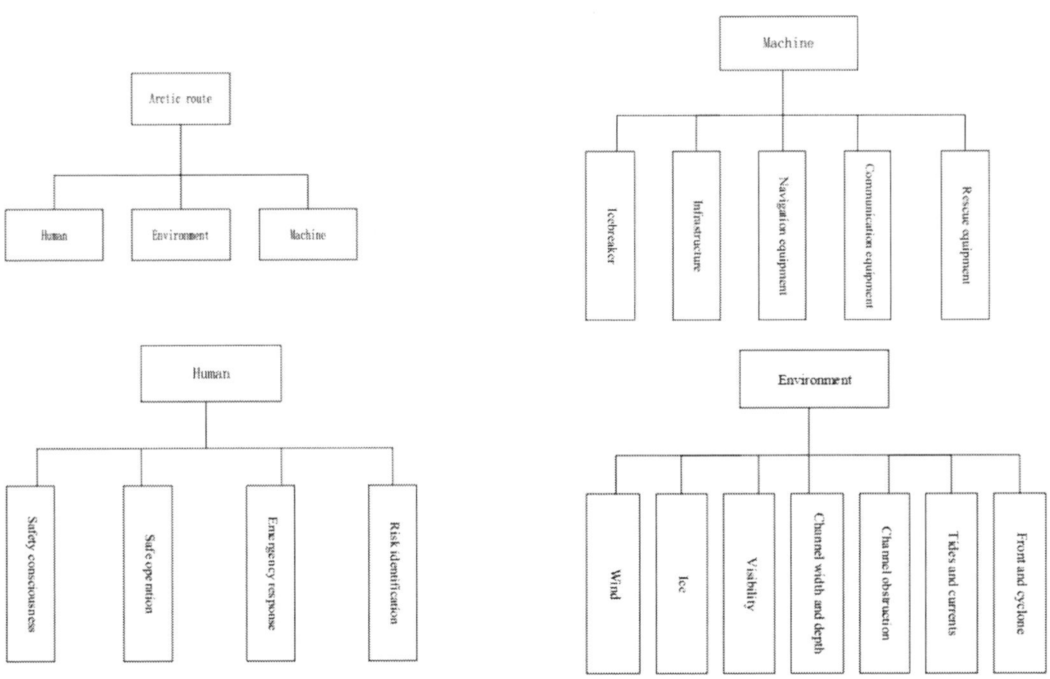

Fig.1. Arctic route safety structure

DOI: 10.1051/matecconf/201713900102

Table 1. Fixed value of human factors

	safety consciousness	Safe operation	Emergency response	Risk identification
Utility value	0.84	0.86	0.76	0.88

Table 2. Fixed value of machine factors

	Icebreaker	Infrastructure	Navigation equipment	Communication equipment	Rescue equipment
Utility value	0.77	0.65	0.75	0.68	0.70

Table 3. Fixed value of environmental factors

	wind	visibility	Channel width and depth	Channel obstruction	Tides and currents	Front and cyclone
Utility value	0.68	0.64	0.80	0.76	0.74	0.70

Table 4. Average area of ice per year

Year	Ice area
2003	11. 640
2004	11. 443
2005	11. 186
2006	11. 041
2007	10. 753
2008	11. 262
2009	11. 233
2010	10. 968
2011	10. 732
2012	10. 695
2013	11. 150
2014	11. 047
2015	10. 823
2016	10. 384

Table 5. Number of shipping traffic per year

Year	Ship number
2009	2
2010	4
2011	41
2012	46
2013	71
2014	31
2015	18

Table 6. Ice area standardization

Year	Ice area
2003	0.30
2004	0.31
2005	0.33
2006	0.34
2007	0.37
2008	0.33
2009	0.33
2010	0.35
2011	0.37
2012	0.37
2013	0.34
2014	0.34
2015	0.36
2016	0.39

Table 7. Ship quantity standardization

Year	Ship quantity
2009	1
2010	0.99
2011	0.95
2012	0.95
2013	0.91
2014	0.96
2015	0.98

Table 8. The total entropy of safety system in different years

Year	Safety entorpy
2002	0.158
2003	0.158
2004	0.158
2005	0.153
2006	0.152
2007	0.150
2008	0.154
2009	0.153
2010	0.151
2011	0.149
2012	0.150
2013	0.154
2014	0.154
2015	0.151
2016	0.147

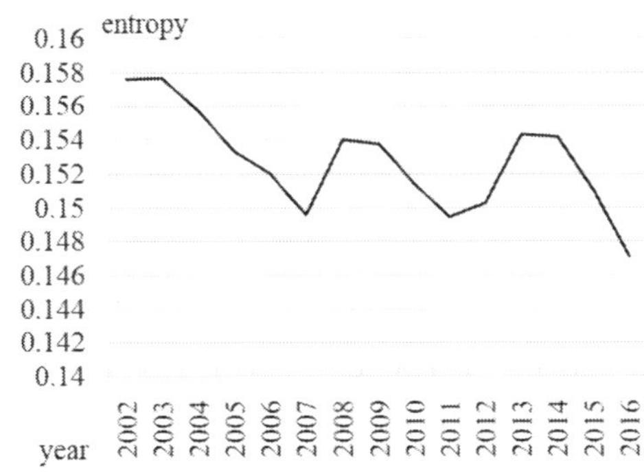

Fig.2. The evolution of the Arctic route navigation safety system

4.2 evolution analysis of navigation safety system of Arctic route

You can see from Figure 2 in the Arctic route navigation safety system in general is gradually to the safe state, but can be seen from the figure figure the evolution trend of the Arctic route navigation safety system has two fluctuations, data can be seen from the first wave is affected by the ice changes the whole Arctic ice area is decreasing, but the first time in 2008 and 2009 when the fluctuation of ice area more than a few years ago the ice area is large, so the Arctic route navigation safety system to the entropy fluctuations, but after the ice area declined, the Arctic route navigation safety system entropy correspondingly decline. For the second time in terms of fluctuation is mainly affected by the number of ship of the Arctic route, the number of ship navigation from 2009 to 2013 the Arctic route has increased year by year, so the Arctic route navigation safety system to the entropy fluctuations, but because the number of ship navigation and ice area decreased, while the Arctic route navigation safety system and entropy a downward trend. At present, with the melting of the Arctic ice layer, the navigation environment of the Arctic route will be more and more safe. However, with the increase of navigation ships, the navigation safety system of the Arctic route will inevitably be affected.

The analysis in Figure 2 except fluctuations in the region outside of the Arctic route navigation safety system entropy change, from Figure 2 can be found, decreased from 2002 to 2007, the Arctic route navigation safety system entropy, and from the figure can be found in the Arctic route navigation safety system based on the linear entropy decrease, namely, that the Arctic route navigation safety system the Arctic ice area of straight down. From 2014 to 2016 declined, the Arctic route navigation safety system entropy, and from the figure can be found in the Arctic route navigation safety system entropy basic straight down, that is, during this period the Arctic route navigation safety system from the Arctic ice area is reduced and the ship reduced the number of double effect, the Arctic route navigation safety system entropy appears sharp drop.

Overall, the Arctic route navigation safety system state change rate is not large, the main reason is the ice changes in the Arctic route navigation safety system occupies the proportion is 0.09, impact on the Arctic route navigation safety system is limited. There are many factors in the navigation safety system of the Arctic route, and various factors influence the change of the security system of the Arctic route synthetically. You can see from Figure 2 that the Arctic route navigation safety system is in the transition phase to a safe condition, according to the analysis of scenarios, for the Arctic route navigation safety system, this should continue to implement the management of Arctic route area, and according to some problems on the management of repair.

4.3 Management Countermeasures of navigation safety system in Arctic route

Although with the melting of Arctic ice, Arctic route navigation safety system is in the transition phase to a safe condition, but the Arctic route navigation safety system there are still many problems, such as the Arctic route along the port infrastructure is seriously insufficient, the sovereignty issue, such as lack of communication equipment. In view of these problems, the following suggestions are put forward.

First, we should strengthen infrastructure construction along the Arctic route and improve the transportation system of the Arctic route. At present, the port infrastructure along the Arctic route can not meet the safety of commercial vessel docking, and the infrastructure of the ports along the Arctic route should be strengthened. Second, strengthen the legal construction of the Arctic route and improve the legal system of the Arctic route. At present, there are very many laws about the Arctic routes, but these laws vary in quality and even contradict each other. Third, we should strengthen the monitoring of natural environment in the Arctic region and improve the accuracy of arctic weather prediction. For ship navigation safety in the Arctic, the biggest security threat is from the natural environment of the Arctic, the Arctic climate, be the most changeful, with great uncertainty, strengthen environmental monitoring can effectively reduce the number of ships sailing in the extreme weather or the ship in advance to take preventive measures to cope with the extreme weather, and safeguard people's safety and property safety.

5 conclusion

Based on the entropy theory, the application of existing entropy theory in the field of security development, put forward the analysis method of security system evolution, and using this method the evolution trend of Arctic route navigation safety system are analyzed, the results found that the Arctic route navigation safety system is in the transition phase to a safe condition. However, there are still some hidden dangers in the navigation safety system of the Arctic route. Therefore, this paper puts forward some management countermeasures for the navigation safety system of the Arctic route. In addition, the evolution of the navigation safety of the Arctic route in space is worthy of further attention.

References

1. Wang N, Yan B, Wu N, et al. Transport. Res. Part A Pol. Pract., 2016.
2. Lasserre F. Transport. Res. Part A Pol. Pract., 2016.
3. Lasserre F. Transport. Res. Part A Pol. Pract. 66(1):144-161. 2014.
4. LI Z F, LI Y, YU S Q. World Reg Stu, 22(2):11-17. 2013.
5. LI Z F, YAN L, XU M Q, et al. Comput Eng Appli, , 49(1):249-253. 2013
6. Tavana M, Zareinejad M, Caprio D D, et al. Appl Soft Comput, 40:544-557. 2016.
7. ZHANG S, SHI X Z, W C.Chin.Saf.Sci.J. 06:9-16. 2010.
8. Salo A A, Hämäläinen R P.. Eur.J.Oper.Res.,82(3): 458–475.1995.
9. Islam R, Biswal M, Alam S. Eur.J.Oper.Res, 97(1): 53–62. 1997.

Applicability Analysis of Bidding Strategy in Electricity Market

Suyan Zhou[1,*], Fei Chen[2], Yahui Qiao[3], Wenzhe Zhang[4], Kaifeng Zhang[1], Kun Yuan[1], Xuemei Dai[1]

[1]Key Laboratory of Measurement and Control of Complex Systems of Engineering, Southeast University, Nanjing, 210096, PR China
[2]Shanxi Electric Power Company, Xi'an, 710082, PR China
[3]Hohai University, Nanjing, 210098, PR China
[4]Chongqing Electric Power Company, Chongqing, 404100, PR China

Abstract. With the development of the electricity market, competition has been introduced in the generation side. It is the overall development trend of the electricity market reformulation to optimize the allocation of different resources through bidding. Therefore, it is significant to research the bidding strategies of the generation companies and the large consumers. This paper reviews the existing research methods of bidding strategy. According to the different market mechanisms, the market participators will choose different bidding strategies based on their own cases. We analyze the applicability of bidding strategies under the different conditions, and give suggestions on how to select bidding strategy for the different market participators under different conditions.

1 Introduction

With the development of the electricity market, competition has been introduced in the generation side and retail side. It is the overall development trend of the electricity market reformulation to optimize the allocation of different resources through bidding. Different from the planned power generation under monopoly mode, power generation companies need to achieve their own power generation and profit through bidding in the electricity market.

According to the theory of the economics, generation companies must quote the marginal cost of the unit if they want to maximize their expected profits in a perfectly competitive market. And in this case the market will have the highest operating efficiency [1]. However, the real electricity market is closely to oligopolistic market. This is mainly due to the fact that there are usually only a few dominant firms producing power in an area. In oligopolistic markets, generation companies try to maximize their expected profits by bidding strategically instead of bidding on marginal cost. In this case, the bidding strategies adopted by the generation companies in the market have a direct impact on their profitability.

With the thorough reform of the electric power system in China, the direct power-purchase for large consumers is one of the key points of this reform. So it is significant to research the bidding strategy of the generation companies and the large consumers in the electricity market.

Strategic bidding in the electricity market has been studied by many researchers using the different methods. A remarkable review is presented in [2], in which several different research methods on generation companies' bidding strategy before the year of 2000 are summarized. In [3], some new bidding strategies based on evolutionary game theory are presented.

Previous reviews on bidding strategies in the electricity market have mostly embarked from the perspective of optimal strategy, discussing how to construct the optimal strategy for generation companies. So the most common method based on power generation cost analysis has been often ignored. On the other hand, with the development of the electricity market, some new problems and approaches emerge, which have not been summarized in previous study works. Also the applicability of various bidding strategies under different market conditions are not discussed in previous reviews.

To summarize, the main contributions of this paper are as follows:

To sort and classify the existing research methods on bidding strategies in the electricity market from a more comprehensive perspective. It means that the classification is not from the perspective of how to construct optimal strategy but includes the common method based on power generation cost analysis in practice.

To analyze the applicability of various bidding strategies under the different market conditions and give some suggestions on how to select bidding strategy for the different market participators according to the different market conditions.

The rest of the paper is constructed as follows. In Section II, the existing methods of studying bidding strategy in the electricity market are summarized. In Section III, we analyze the applicability of bidding strategy under the different conditions and give suggestions on how to select bidding strategy for the different market participators under different conditions.

* Corresponding author: zhousuyan@seu.edu.cn

Conclusions and expectations are presented in Section Ⅳ.

2 The research methods of bidding strategy in electricity market

Most previous studies on bidding strategy in the electricity market have mostly embarked from the perspective of optimal strategy, discussing how to construct the optimal strategy for generation companies. The most common method based on power generation cost analysis has been ignored. However, it is a practical approach in real market that is easily grasped for most generation companies. In some cases, this method may be sufficient and more effective for generation companies. So in order to understand the research methods of bidding strategy more comprehensively, we sort out the common research methods and compare their advantages and disadvantages.

The methods used to construct bidding strategies include 5 types: method based on power generation cost analysis, method based on market clearing price forecasting, method based on guess of opponents' bidding strategies, method based on game theory and method based on evolutionary game theory. The following describes each of these.

2.1 The method based on power generation cost analysis

It is the most fundamental method of constructing bidding strategy for the generation companies [5,6]. In some cases, this method may be sufficient and more effective for generation companies than other methods. The principle of the method is simple and it is easily grasped for most generation companies. However, it is hard for generation companies to maximize their own expected profit by this method because the influence of other opponents' bidding strategies are not considered [4].

2.2 The method based on market clearing price forecasting

In the electricity market with unified clearing mechanism, this method is commonly used. Such method need to predict the market clearing price based on historical data. If the market price can be accurately estimated and the price is higher than the cost, the generation companies only need to bid slightly lower than the predicted price.

In fact, it is difficult to predict the market clearing price accurately. This is mainly because the accurate prediction need the sufficient historical data and reasonable forecast model. But for the developing electricity market, the historical data is not enough due to the market structure and rules are still being adjusted. Although in a relatively mature market, the accuracy of electricity price forecasting is difficult to guarantee because of the speculation in the market. But it does not mean that prediction is unnecessary. With the development of electricity market, the prediction is still an important reference when generation companies bid. At present, there are many methods to forecast the market price [7~14], such as time series prediction method [7], gray system prediction method, regression analysis [8], neural network [8~12], combination forecast method, wavelet analysis method [12] and so on.

2.3 The method based on guess of opponents' bidding strategies

In this method, the opponents' bidding strategies are estimated by probability or fuzzy method. On this basis, the optimal bidding model is established to seek the optimal bidding strategy.

In order to estimate the bidding behavior of the opponent accurately, it is necessary to know the historical bidding data of the competitors. On this basis, the statistical method [15] is used to estimate the bidding behavior of competitors. However, because a lot of bidding data in the market is confidential and the behaviors of the market participants are speculative, it is difficult to grasp the bidding laws of the competitors.

Most of the studies assume that the bidding laws of the competitors are known, such as the assumption of rival offer obeying the discrete probability distribution [16,17], uniform distribution [18] and normal distribution [19~21,23,26,27], or regarding it as a fuzzy set [22,24].

2.4 The method based on game theory

The research methods based on classic game theory are mainly divided into the following two categories. The first type is based on the matrix game model. This model is very difficult to apply in practice because it needs to know all the other competitors' strategies. In fact, it is hard to realize. And when the number of the market members is high, the game matrix will be very large and the solution is complicated. Therefore, this method can only analyze small systems and is far from practical applications.

The second type is based on the oligopoly game model, which mainly includes Cournot model, Stackelberg model and supply function model [31~33]. Many simplified assumptions are made in the application of these models, and the equilibrium point obtained in this case may not have much significance for the construction of bidding strategies. Generally, the methods based on the classic game theory are more suitable for rough analysis of the market participants' bidding behaviors.

2.5 The method based on evolutionary game theory

Compared with the method based on classic game theory, the evolutionary game theory seeks the equilibrium point through continuous learning in the environment [3].

Based on the intelligent agent technology, each market participant is modeled as an independent agent. Each agent knows only its personal information and the public information of the market. But it does not know the private information of other agents. This assumption conforms to the incomplete information game in the real market. At present, this method is widely used in power market simulation research [36~38].

3 The applicability of bidding strategy in electricity market

The operation modes of electricity markets in distinct countries are different, and the bidding rules and clearing rules adopted by different markets are also different. According to the different market mechanisms and management, market participators will choose different bidding strategies based on their own cases. In this section, we analyze the applicability of bidding strategy under the different conditions, and give suggestions on how to select bidding strategy for the different market participators under different conditions.

3.1 The bidding strategies in different electricity market

According to the time division, the electricity market system can be divided into the medium and long-term contract market, day-ahead market and real-time market. The generation companies can participant in different markets at the same time. Considering the generation companies' objectives in these markets, the bidding strategies adopted in different markets need to be adjusted.

For example, in order to avoid the risk of price fluctuations in the spot market, the goal of most power generation companies is to gain as much power as possible in the medium and long-term contract market.

Therefore, the generation companies usually bid low to improve the probability of bid winning. In this case, it is more appropriate for them to bid based on power generation cost in the medium and long-term contract market. This is mainly because the method based on cost analysis has less risk and makes generation companies win the bid more easily than other methods.

The purpose of most power companies participating in the spot market is to obtain additional economic benefits. Therefore, for these generation companies, they try to achieve the maximum expected profit in the spot market. But if the generation companies bid only based on their own cost, without considering the impact of the opponent's bidding in the spot market, it is not conductive to achieve the goal of maximizing profits. Obviously, it is more beneficial for some generation companies which are familiar with the competitors' information and have sufficient historical data in the spot market. These generation companies can estimate the opponents' bidding behaviors or predict the market clearing price based on these historical data for their reference. However, when the generation companies use these bidding strategies to gain high profits, they will also face different risks, such as market price risk, power generation risk and so on.

3.2 The bidding strategies of the generation companies with different scales

In terms of the generation company's scale and market share in the electricity market, it will compete as a price-taker or a price-maker. The greater the generation company's market share, the greater the ability to influence the market. In the oligopoly market, such large-scale generation companies can indirectly determine the market price [18]. In this case, some small and medium generation companies with high cost become price-taker due to lack of competitiveness in the market. So the generation companies with different scales should take the different bidding strategies to get more profits and enhance competitiveness.

Under uniform clearing pricing rule, the market price is mainly determined by a few large-scale generation companies in this area. So for the small and medium companies, they can bid based on cost to improve the probability of winning the bid. Once the companies win the bid and the marginal clearing price is higher than their bid, such companies will be able to obtain additional profits. In this case, it is wiser to choose the bidding strategy based on cost.

On the contrary, in order to achieve the maximum expected profits, the large-scale power companies, as the price-maker, need to consider the bidding behaviors of a few main competitors who have almost the same strength with them. For such companies, if they can collect more historical data about the market and other competitors, it is more beneficial to bid based on forecast price and estimated competitors' bidding strategies.

3.3 Different types of generation companies' bidding strategies

Most of the researches on generation companies' bidding strategies in the market have been carried out around thermal power plants. However, the characteristics of hydropower plants are different from the thermal power plants. This leads to different bidding strategies between them.

The studies on hydropower plants' bidding strategies are mostly from a price-taker perspective [38,39]. For the thermal power plant, the most common bidding strategy is to bid based on the fuel cost, which is modeled as a quadratic function of its output power. And fixed cost elements are ignored. On the contrary, the hydropower plants hardly consume fuel after they are completed. The cost of hydropower plants is mainly fixed cost and fuel cost can be neglected. Therefore, the original biding strategy based on cost needs to be adjusted when applied to the hydropower plants.

The bidding strategy based on clearing price forecasting is also commonly used in the electricity market. However, it is difficult to guarantee the accuracy of prediction due to the speculation or collusion in the market. This situation is extremely risky for hydropower plants with water constraints. Because the forecast value is slightly higher than the actual market price in a certain period, the hydropower plant will lose the chance of generating electricity. As a result, it is difficult to meet the constraints of water plan [40,41]. As a result, it is likely to abandon water in the future because of the limited regulation and storage capacity of the reservoir. And this will cause huge losses. But only by lowering the price to ensure the electricity is obviously uneconomical. Therefore, these original bidding strategies are difficult to meet the hydropower plants' demands and need to be adjusted according to practical experience. The hydropower plants' bidding strategies need for further study and improve.

3.4 The bidding strategies in the direct power-purchase for large consumers

With the thorough reform of the electric power system in China, the direct power-purchase for large consumers is one of the key points of this reform. So it is significant to research the bidding strategies of the generation companies and the large consumers in the electricity market.

In this section, we mainly discuss the generation companies' and larger consumers' bidding strategies under the different market clearing mechanisms. Under uniform clearing pricing rule, large-scale generation companies can adjust the bidding strategy according to their own situation and the market conditions. In this rule, the small and medium generation companies can bid based on fuel cost to improve the probability of winning the bid.

Different from uniform pricing, in high-low matching pricing rule, each generation company is paid at the average price of generation company's bid and matched large consumer's bid. It is difficult for generation companies to form a coalition in the electricity market adopting high-low matching clearing mechanism.

Under the circumstance of supply exceeding demand in electricity market, the small and medium generation companies can bid low to ensure a certain proportion of electricity or the maximum bidding electricity cleared. For these companies, bidding based on cost is a more effective and less risky strategy. But for some large companies, they have the ability to collect more historical data about the market and other competitors. It is more beneficial for them to bid based on forecast price and estimated competitors' bidding strategies.

However, large consumers must ensure the bided volume of electricity in order to meet the production demands. So they pay more attention to the bided volume of electricity rather than price in most cases. But the market clearing price is certainly the lower the better for large consumers. So in the electricity market some large consumers will bid the tariff for most of electricity to ensure the bided volume to satisfy production needs. And the offer of small amount of electricity should be as low as possible. This strategy can make large consumers sacrifice the minimum amount of electricity to obtain the lowest clearing price.

4 Conclusions

There are many factors that influence the choice of bidding strategies, including bidding rules, market clearing mechanism, and so on. The market participants should choose the suitable strategy according to the market environment and their own situation. At present, there have been many research results of bidding strategies for generation companies, but there are few researches on bidding strategies for large consumers. The bidding strategy for large consumers is worthy of further study.

Acknowledgement

This work was supported by State Grid Science Project (DZN17201500036).

References

1. Gans J, King S, Stonecash R, et al. Principles of economics[M]. Cengage Learning, 2011.
2. Wen F, David A K. Bidding strategies in electricity markets[J]. Automation of Electric Power Systems, 2000, 24(14): 1-6.
3. Wang Shuai. A review of bidding strategies for generation companies[J]. Moderm Economic Information, 2009 (23): 107-110.
4. Wang Xifan, Wang Xiuli, Chen Haoyong. Power market theory[M]. Xi'an Jiaotong University Press, 2003.
5. Li Tao, Jiang Chuanwen, Hou Zhijian. A cost offer model on the generation side of power market[J]. Information Journal Hydroelectric Energy, 2001, 19(3): 62-63.
6. Zhang Shaohua, Li Yuzeng. Probabilistic forecasting of short-run marginal cost of power generation under electricity market[J]. Proceedings of the CSEE, 2000, 20(10): 19-22.
7. Nogales F J, Contreras J, Conejo A J, et al. Forecasting next-day electricity prices by time series models[J]. IEEE Transactions on power systems, 2002, 17(2): 342-348.
8. Huang Rixing, Kang Chongqing, Xia Qing. System marginal price forecasting in electricity market[J]. Automation of Electric Power Systems, 2000, 24(22): 9-12.
9. Li Caihua, Wang Zhiwei. Short-term marginal price forecasting with hybrid model[J]. Automation of Electric Power Systems, 2000, 24(22): 9-12.
10. Liu Zhengguo, Gao Fuying. Forecasting market-clearing price in Zhejiang generation market using neural network[J]. Automation of Electric Power Systems, 2002, 26(9): 49-52.

11. Yang Li, Qiu Jiaju, Jiang Daozhuo. BPN based day-ahead unconstrained market clearing price forecasting model[J]. Automation of Electric Power Systems, 2001, 25(19): 11-14.
12. Song chao. Application of wavelet analysis and neural network in electricity price forecasting[D]. Zhejiang University, 2002.
13. Conejo A J, Nogales F J, Arroyo J M. Price-taker bidding strategy under price uncertainty[J]. IEEE Transactions on Power Systems, 2002, 17(4): 1081-1088.
14. Ni E, Luh P B, Rourke S. Optimal integrated generation bidding and scheduling with risk management under a deregulated power market[J]. IEEE Transactions on Power Systems, 2004, 19(1): 600-609.
15. Bialek W, Callan C G, Strong S P. Field theories for learning probability distributions[J]. Physical review letters, 1996, 77(23): 4693.
16. Zhang D, Wang Y, Luh P B. Optimization based bidding strategies in the deregulated market[C]//Power Industry Computer Applications, 1999. PICA'99. Proceedings of the 21st 1999 IEEE International Conference. IEEE, 1999: 63-68.
17. Song H, Liu C C, Lawarree J. Decision making of an electricity supplier's bid in a spot market[C]//Power Engineering Society Summer Meeting, 1999. IEEE. IEEE, 1999, 2: 692-696.
18. David A K. Competitive bidding in electricity supply[C]//IEE Proceedings C-Generation, Transmission and Distribution. IET, 1993, 140(5): 421-426.
19. Wen F, David A K. Optimal bidding strategies and modeling of imperfect information among competitive generators[J]. IEEE transactions on power systems, 2001, 16(1): 15-21.
20. Wen F, David A K. Coordination of bidding strategies in day-ahead energy and spinning reserve markets[J]. International Journal of Electrical Power & Energy Systems, 2002, 24(4): 251-261.
21. Ma Li, Wen Fushuan. A preliminary investigation on bidding strategies employing step-wise bidding rules[J]. Automation of Electric Power Systems, 2002, 26(9): 16-19.
22. Widjaja M, Sugianto L F, Morrison R E. Fuzzy model of generator bidding system in competitive electricity markets[C]//Fuzzy Systems, 2001. The 10th IEEE International Conference on. IEEE, 2001, 3: 1396-1399.
23. Lamont J W, Rajan S. Strategic bidding in an energy brokerage[J]. IEEE Transactions on Power Systems, 1997, 12(4): 1729-1733.
24. Yang Li, Wen Fushuan. A possibility theory based approach for building optimal bidding strategies in electricity markets[J]. Automation of Electric Power Systems, 2002, 26(23): 12-17.
25. Wen F, David A K. Strategic bidding in reserve market[C]//Advances in Power System Control, Operation and Management, 2000. APSCOM-00. 2000 International Conference on. IET, 2000, 1: 80-85.
26. Wen F, David A K. Optimal bidding strategies for competitive generators and large consumers[J]. International Journal of Electrical Power & Energy Systems, 2001, 23(1): 37-43.
27. Hao S. A study of basic bidding strategy in clearing pricing auctions[C]//Power Industry Computer Applications, 1999. PICA'99. Proceedings of the 21st 1999 IEEE International Conference. IEEE, 1999: 55-60.
28. Diao Qinhua, Lin Jikeng, NI Yixin, et al. Game theory and its applications in power markets[J]. Automation of Electric Power Systems, 2001, 25(1): 19-23.
29. Wu Zhiyong, Kang Chongqing, Xia Qing, et al.Strategic bidding with application of game theory[J]. Automation of Electric Power Systems, 2002, 26(9): 7-11.
30. Contreras J, Candiles O, de la Fuente J I, et al. A cobweb bidding model for competitive electricity markets[J]. IEEE Transactions on Power Systems, 2002, 17(1): 148-153.
31. Song Yiqun, Ni Yixin, Hou Zhijian, et al. A novel model of Gencos strategic behaviors based on conjectured supply function equilibrium[J]. Automation of Electric Power Systems, 2003, 27(13): 15-18.
32. Chattopadhyay D. Multicommodity spatial Cournot model for generator bidding analysis[J]. IEEE Transactions on power systems, 2004, 19(1): 267-275.
33. Xian W, Yuzeng L, Shaohua Z. Oligopolistic equilibrium analysis for electricity markets: a nonlinear complementarity approach[J]. IEEE Transactions on Power Systems, 2004, 19(3): 1348-1355.
34. Xie Shiyu. Evolutionary game theory under bounded rationality[J]. Journal of Shanghai University of Finance and Economics, 2001, 3(5): 3-9.
35. Gao Jie, Sheng Zhaohan. Elementary groping for evolutionary game theory and its application in electricity market[J]. Automation of Electric Power Systems, 2003, 27(18): 18-21.
36. Fujii Y, Okamura T, Inagaki K, et al. Basic analysis of the pricing processes in modeled electricity markets with multi-agent simulation[C]//Electric Utility Deregulation, Restructuring and Power Technologies, 2004.(DRPT 2004). Proceedings of the 2004 IEEE International Conference on. IEEE, 2004, 1: 257-261.
37. Xiong G, Hashiyama T, Okuma S. An electricity supplier bidding strategy through Q-Learning[C]//Power Engineering Society Summer Meeting, 2002 IEEE. IEEE, 2002, 3: 1516-1521.

38. Wang Yeping, Yang Yan, Jing Zhaoxia, et al. Application of intelligent Agent-based simulation to electricity market[J]. Journal of South China University of Technology(Natural Science Edition) , 2010, 38(3).
39. Chen Zhixu, Zhang Lizi, Shu Jun, et al. Bidding strategy research for price-taker in electricity market[J]. Electric Power, 2007, 40(6): 1-5.
40. He Li. Research on Bidding Problems for Hydropower Plant in Electricity Market[D]. Huazhong University of Science & Technology, 2007
41. Shi Xiaojun. Bidding Strategy Research for Hydropower Plant in Electric Power Market[D]. Xi'an University of Technology, 2005.

 DOI: 10.1051/matecconf/201713900104

Comprehensive Evaluation and Analysis on Automobiles Performance Considering Objective Weights

Jian Zhang*, Xuexing Jiang, Liping Wang, Li Zou
School of Transportation and Automobile Engineering, Panzhihua University, Panzhihua 617000, China

Abstract. A fuzzy comprehensive evaluation method for vehicle performance based on variation coefficient method was proposed. The evaluation index system was constructed and the mathematical models of fuzzy comprehensive evaluation about vehicle performance were established. The method can cause evaluation bias because of the subjective weights of factors in traditional fuzzy comprehensive evaluation. In this paper, variation coefficient method which is objective weights was introduced to integrate with subjective weights from experts. In the end, the numerical examples were given, compared and evaluated the evaluation results with practical sales volume. The practical application showed that the evaluation method was effective and had good practicality.

1 Introduction

Automobiles gradually become daily-life necessities, and its vehicle performance has become key point for massive consumers, as a result, the scientific vehicle performance evaluation methods have become many scholars' goal [1]. Currently, there are many scholars have studied every major vehicle performance. It mainly concentrates on structure performance including handling stability, engine performance, automobile ride comfort, etc[2-3]. A variety of evaluation models are proposed and different evaluation methods are used to study vehicle each-part performance. However, fewer studies on vehicle performance. Due to multifarious and fuzzy vehicle performance indicators, it is an effective method to apply fuzzy comprehensive method to assess vehicle performance. When fuzzy mathematics theory is applied on comprehensive evaluation, the weights have a direct influence on the evaluation results [4-5]. Vehicle evaluation performance index parameters have been determined as soon as the vehicle get out of the car factory, which means that the effect of weights is objective, So objective weighting method should be applied. However, everyone's feeling vary from person preference to person on vehicle body aspect, it's appropriate to apply subjective weighting method. Integrating subjective and objective weights for evaluation index, finally more scientific vehicle comprehensive evaluation results were gained through fuzzy evaluation theory, which promoted manufacturers innovation of technology and provide reference for consumers.

2. The determination of vehicle performance indicators systems and all indicators weights

* Corresponding author: zhangjianpzh@126.com

2.1 The vehicle-performance indicators systems

In identifying evaluation index, choosing important, intuitive and specific indicators is convenient to compare differences among vehicles performance. The three indicators, space, configuration, process quality what consumers attach an importance to are chosen in the vehicle body. For the power assembly, what the most direct way is to compare power performance, acceleration and fuel economy. The car is not simply walking implement. Therefore its security, passing ability, practicality and operational economy are an increasing importance when the usability index is determined. According to evaluation system principle, establish valuation index system as shown in Fig.1.

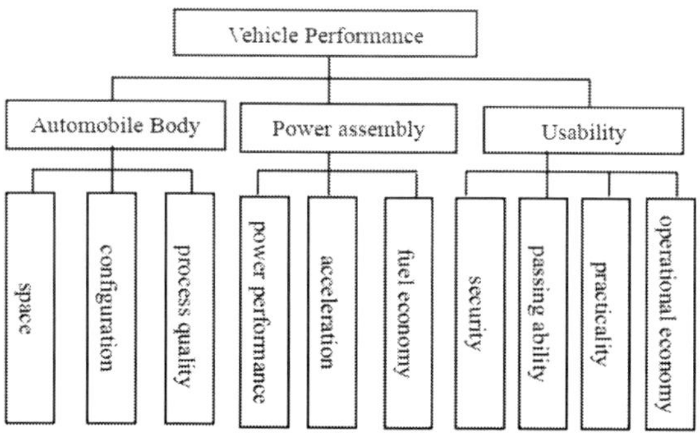

Fig.1. Automobile performance evaluation system structure

2.2 Determine weights

The determination of the weights can be divided into objective and subjective weighting method. Objective weighting method is a method which can directly obtain weights from each-index original information after a certain mathematics deal. Subjective weighting method refers to determining each-index importance weights after subjective analysis and study by professional knowledge and practical experience. Variation coefficient method is an objective weighting method which determine evaluation index weights according to variation degree of each evaluation index weights. Vehicle performance indicators are reflected by objective parameters. The index data that intuitive and variation degree is larger are suitable for Variation coefficient method. According to the characteristics of the performance evaluation index system and data statistics, power assembly and usability choose the variation coefficient method to determine weights. Vehicle body choose analytic hierarchy process(AHP) of subjective weighting method to determine weights.

The basic steps of variation coefficient method are:

(1) Calculate the average of index $\overline{x_f}$ and variance S_f^2

We suppose that there are q objects to evaluate, and among them have p indicators $x_1, x_2, \cdots, x_p$ to describe. Then, calculate the average of indicators $\overline{x_f}$ and variance S_f^2.

$$\overline{x_f} = \frac{1}{q}\sum_{g=1}^{q} x_{gf} \quad (1)$$

$$S_f^2 = \frac{1}{q}\sum_{g=1}^{q} (x_{gf} - \overline{x_f})^2 \quad (2)$$

(2) Calculate the each-index variation coefficient v_f.

$$v_f = \frac{S_f}{\overline{x_f}} \quad (3)$$

(3) Normalized processing for variation coefficient v_f, get each-index weight.

$$\omega_f = \frac{v_f}{\sum_{f=1}^{p} v_f} \quad (4)$$

3 The fuzzy evaluation method of vehicle performance

Fuzzy evaluation is a method of using fuzzy mathematics theory to analyze and judge fuzzy phenomena. Fuzzy comprehensive evaluation judges fuzzy items in fuzzy environment for a particular purpose and takes into account multiple influence factors. Vehicle performance evaluation belongs to fuzzy problem for it involves multiple aspects, reflects more performance evaluation indicators, has complex influence factors. Thus fuzzy comprehensive evaluation method can be used to establish a multi-aspect ,multi-level and scientific system that can reflect the vehicle performance.

For vehicle performance fuzzy comprehensive evaluation problem, the mathematical model can be established as follows. Suppose the known factors affecting things set:

$$X = \{x_1, x_2, \cdots, x_n\} \quad (5)$$

X expressed as vehicle-performance various evaluation indicators set, is a single evaluation index. Evaluation set of thing is :

$$U = \{u_1, u_2, \cdots, u_m\} \quad (6)$$

U expressed as vehicle-performance various evaluation results set, is each-evaluation results. Statistics experiment or experts score are applied to establish distribution matrix of the weights between each factors.

3.1 Establish the evaluation index weights set

When we evaluate vehicle performance, each-evaluation index is given weight to fully reflect its important degree. The weights which consists of each weight coefficient are:

$$W = \{w_1, w_2, \cdots, w_n\} \quad (7)$$

In the formula (7), $\sum_{i=1}^{n} w_i = 1$, Each weight should meet negative requirements.

3.2 Construct comprehensive evaluation matrix of vehicle performance

Determine the degree of membership of the evaluation object to each element in the evaluation set for the vehicles to be evaluated and fuzzy comprehensive evaluation matrix is constructed.

$$R = \begin{bmatrix} r_{11} & r_{12} & \cdots & r_{1m} \\ r_{21} & r_{22} & \cdots & r_{2m} \\ \vdots & \vdots & \cdots & \vdots \\ r_{n1} & r_{n2} & \cdots & r_{nm} \end{bmatrix} \quad (8)$$

3.3 Vehicle-performance fuzzy evaluation

After W and R are determined, fuzzy decision integration B is obtained by fuzzy synthesis operation B = W· R = (b1, b2,..., bm) , where symbol "· "is the operator notation. This is single factor evaluation result for the first level evaluation index. If those evaluation indicators are given weights W' again, then secondary level comprehensive evaluation C = W'· R' can be carried out. Similarly more hierarchy comprehensive evaluation can be continued. According to the specific results, the final evaluation conclusions are gained.

4 The calculation and analysis of examples

4.1 Choose evaluation objects

In establishing evaluation indicators, the seven secondary indicators including power performance, acceleration and fuel economy, as well as security, passing ability, practicality and operational economy can use corresponding data to reflect except the three secondary indicators of vehicle body. After dealing with and analyzing the statistical date of Autohome[6], finally engine horsepower values (ps), 0-100 km of per car acceleration time (s), various car models 100km fuel consumption (L), the minimum ground clearance (cm), braking distance (cm)of speed 100-0 km, trunk volume (L), each routine maintenance costs are regarded respectively as judgment basis of seven indicators--power performance, acceleration, fuel economy, safety, passing ability, practicality, operational economy. Then get power assembly and usability, as shown in Table 1.

4.2 Calculate weights

According to the formula (1), (2), (3), (4), we gain corresponding secondary level indicators, as shown in Table 2.W_{21}={0.41 0.24 0.35},W_{22}= {0.34 0.04 0.09 0.53}.

Three secondary indicators, space, configuration, process quality, corresponding to vehicle body are analyzed and calculated by AHP, and importance degree between them are compared, which can get weights W_{23} = {0.15, 0.56, 0.29}. When consider three primary indicators, vehicle body, power assembly and usability performance, a plenty of investigations and researches are performed, and finished the results. The weights W '= {0.34, 0.33, 0.33}.

4.3 The example of vehicle performance

Now, take the joint venture vehicle models including Ashkenazi, U.S. and Japanese for example based on comprehensive evaluation of six different brand with same price, well-sold vehicle in the market. Six case on assessment are set up respectively expressed with D = {D1, D2, D3, D4, D5, D6}. According to former evaluation index system and weights, experts are hired to score for each index. The score is shown as Table 3.

4.4 The first level evaluation

(1) Vehicle body

Through evaluating three indicators, space, configuration, process quality, vehicle body comprehensive evaluation are:

$$b_{11} = w_{21}\cdot\begin{bmatrix} b_{21} \\ b_{22} \\ b_{23} \end{bmatrix} = (0.15,0.56,0.29)\cdot\begin{bmatrix} 90 & 85 & 95 & 90 & 90 & 95 \\ 85 & 95 & 95 & 90 & 80 & 85 \\ 85 & 80 & 95 & 95 & 85 & 85 \end{bmatrix} \quad (9)$$
$$= (85.75,89.15,95,91.45,82.95,86.5)$$

(2) Power assembly

Through evaluating each-vehicle three indicators, power performance, acceleration and fuel economy, power assembly comprehensive evaluation is:

$$b_{12} = w_{22}\cdot\begin{bmatrix} b_{24} \\ b_{25} \\ b_{26} \end{bmatrix} = (0.41,0.24,0.35)\cdot\begin{bmatrix} 85 & 95 & 85 & 90 & 80 & 90 \\ 85 & 85 & 85 & 90 & 80 & 85 \\ 85 & 80 & 90 & 100 & 90 & 85 \end{bmatrix}$$
$$= (85,87.35,86.75,93.5,83.5,87.05) \quad (10)$$

(3) Usability Performance

Through evaluating each-vehicle four indicators, security, passing ability, practicality, operational economy, the comprehensive evaluation is:

$$b_{13} = w_{23}\cdot\begin{bmatrix} b_{27} \\ b_{28} \\ b_{29} \\ b_{30} \end{bmatrix} = (0.34,0.04,0.09,0.53)\cdot\begin{bmatrix} 85 & 90 & 80 & 85 & 80 & 85 \\ 85 & 90 & 85 & 85 & 80 & 85 \\ 80 & 85 & 100 & 90 & 95 & 75 \\ 80 & 85 & 80 & 80 & 85 & 90 \end{bmatrix}$$
$$= (81.9,86.9,82,82.8,84,86.75) \quad (11)$$

4.5 The second level evaluation

According to formula (9), (10), (11), the second level fuzzy comprehensive evaluation matrix of overall performance is:

$$R' = B = \begin{bmatrix} 85.75 & 89.15 & 95 & 91.45 & 82.95 & 86.5 \\ 85 & 87.35 & 86.75 & 93.5 & 83.5 & 87.05 \\ 81.9 & 86.9 & 82 & 82.8 & 84 & 86.75 \end{bmatrix} \quad (12)$$

Combined with three primary indicators weights W = {0.34, 0.33, 0.33}, the secondary level evaluation result is:

$$C = W'\cdot R' = (0.34,0.33,0.33)\cdot\begin{bmatrix} 85.75 & 89.15 & 95 & 91.45 & 82.95 & 86.5 \\ 85 & 87.35 & 86.75 & 93.5 & 83.5 & 87.05 \\ 81.9 & 86.9 & 82 & 82.8 & 84 & 86.75 \end{bmatrix} \quad (13)$$
$$= (84.232,87.8135,87.9875,89.272,83.478,86.764)$$

4.6 Discussion of results

By analyzing fuzzy comprehensive evaluation results of vehicle performance we can get following conclusions:

From formula(13), we can conclude that the overall evaluation order of the six vehicle models is D4,D3,D2,D6,D1,D5. Thus comprehensive performance of D4 is the best, D3, D2 are basically equal, D6 is a little worse, and D1, D5 are the worst. D1, D5 should further improve in all aspects to achieve better quality.

At the same time, from the formula (9), (10), (11), the primary level evaluation results can be analyzed. It is easy to know which individual index is more outstanding or deficient among each-model body, power assembly and usability, which provide a reference for the modification of manufacturers.

4.7 The analysis of comparing evaluation results with the actual sales

Chinese domestic sales rank respectively was D1,D6,D3,D5,D2,D4 in 2015 by looking up the annual sales of six vehicle models in China. The sales of six vehicle models in North America were also looked up in 2015. Because data on D4,D5,D6 models were lacking in North America, for the another three models, the rank respectively was D2,D3,D1. To compare test performance results whether is positively related to actual sales, the Fig.2 was made.

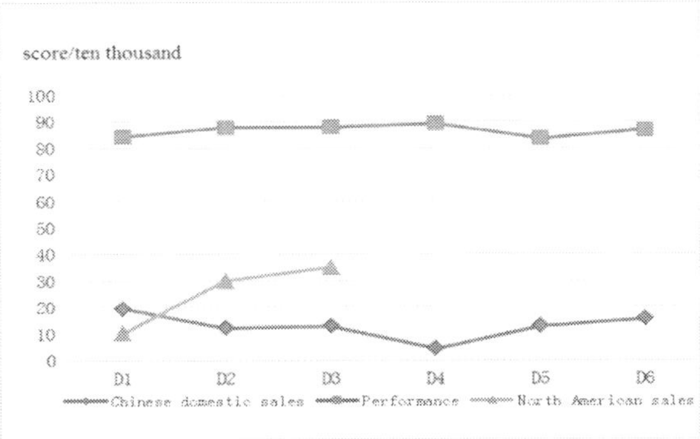

Fig.2. Comparison figure of evaluation and sales

As the Fig.2 shown, we can find out sales in North America is basically coherent with performance assessment rank, which proved that evaluation methods are effective and practical. But from the point of view of Chinese domestic sales, the pros and cons of the evaluation result is not positively related to actual sales except vehicle comprehensive performance factors. That is because of brands, products and personal preferences, etc. For example, the Ashkenazim joints venture vehicle models accumulated long-time public praise in Chinese automobile market, won high popularity. Its good image, fashion and generous appearance, accord with Chinese aesthetics, therefore, it is more praised by Chinese consumers. And partial Chinese consumers can not choose Japanese joints venture models because of personal preferences. Above factors affect all kinds of car sales in China, so the evaluation result is not positive correlation with the Chinese domestic actual sales. It's state that the domestic automobile consumption in China is not yet mature at present, and subjective factors of the car buyers have a greater impact.

5 Conclusions

A more scientific and rigorous vehicle performance evaluation of fuzzy comprehensive evaluation method based on variation coefficient was proposed in this paper. This method combined qualitative analysis with quantitative analysis, and considered comprehensively the influence of each evaluation index on vehicle performances. In order to solve subjectivity of determining weights in fuzzy comprehensive evaluation, we proposed the method that combined objective weight with the traditional subjective weight to determine weights, and testified its feasibility, effectiveness by practical examples. Its operation was simple, easy to achieve. Therefore, this method had good application value, not only provided a more actual new method for the automobile performances, also provided manufacturers with basis of improving and promoting automobiles performances.

Acknowledgment

In this paper, the research was sponsored by 2014 Research Program of Education Office of Sichuan Province (Project No. 14ZB0409)

References

1. Daodong Liu, Tie Wang, Yi Gao, Hailin Li. Research on Evaluation Method of Vehicle Performance[J]. Mechanical Science and Technology for Aerospace Engineering, 2012 (3): 40-44.
2.Zhiqiang Zhang, Bin Xu, Yongling He, ect. Engine Performance Evaluation Based on Analytic Hierarchy Process[J]. ACTA ARMAMENTARII, 2008(5):115-118.
3. Gang Chen, Weigong Zhang, Zongyang Gong, ect. Evaluation Method for Vehicle Shift feel Based on Evidence Theory and ANFIS[J]. Automotive Engineering, 2009(4): 19-23.
4.Jian Zhang, Xiaoying Liu, Jingdong Zhang, ect. Agriculture machinery safety and risk evaluation based on extension method[J]. Journal of Chinese Agriculture Mechanization, 2014(3):57-60, 66.
5. Caiyun Mao, Muchun Wu, Huazhong Lu. Fuzzy Synthesis Evaluation Method on Performance of Automobile Engine[J]. Journal of South China Agriculture University, 2010(3): 115-118.
6.Information on http://www. autohome. com. cn

DOI: 10.1051/matecconf/201713900104

Table 1. parameters statistic of some meddle car table

data	Power assembly				practicality		
	Power performance	acceleration	Fuel economy	Passing ability	security	practicality	Operational economy
Passat	180	8.5	6.8	105	4133	490	798
Mondeo	203	8.1	7.9	124	3912	516	625
Accord	186	8.9	7.7	100	3887	512	307
Atenza	192	8.3	7.1	169	3896	483	543
Camry	184	9.1	7.8	159	3985	506	599
Magotan	160	8.9	7.8	114	4046	565	957

Table 2. The calculation results of variation coefficient

data	Power assembly				practicality		
	Power performance	Acceleration	Fuel economy	Passing ability	security	practicality	Operational economy
average	184	8.6	7.5	129	3977	512	638
standard deviation	14.3	0.4	0.5	28.9	98	28.9	222.5
variation coefficient	0.0777	0.0465	0.0667	0.2240	0.0246	0.0564	0.3487
weight	0.41	0.24	035	0.34	0.04	0.09	0.53

Table 3. The score and the weights of vehicle-performance evaluation indicators system

evaluation index system and weights		weights	D1 score	D2 score	D3 score	D4 score	D5 score	D6 score
Vehicle body	Space	0.15	90	85	95	90	90	95
0.34	Configuration	0.56	85	95	95	90	80	85
	Process quality	0.29	85	80	95	95	85	85
Power assembly	Power performance	0.41	85	95	85	90	80	90
0.33	Acceleration	0.24	85	85	85	90	80	85
	Fuel economy	0.35	85	80	90	100	90	85
	Passing ability	0.34	85	90	80	85	80	85
Usability performance	Security	0.04	85	90	85	85	80	85
0.33	Practicality	0.09	80	85	100	90	95	75
	Operational economy	0.53	80	85	80	80	85	90

Business Process Modelling based on Petri nets

Jianglong Qin[1,2], Na Zhao[1,2], Zhongwen Xie[1,2] and Qi Mo[1,2]
[1]School of Software Yunnan University Kunming, China
[2]Key Laboratory in Software Engineering of Yunnan Province Kunming, China
e-mail:king8414@163.com, zhaonayx@126.com, xiezw56@126.com, moqiyueyang@163.com

Abstract. Business process modelling is the way business processes are expressed. Business process modelling is the foundation of business process analysis, reengineering, reorganization and optimization. It can not only help enterprises to achieve internal information system integration and reuse, but also help enterprises to achieve with the external collaboration. Based on the prototype Petri net, this paper adds time and cost factors to form an extended generalized stochastic Petri net. It is a formal description of the business process. The semi-formalized business process modelling algorithm based on Petri nets is proposed. Finally, The case from a logistics company proved that the modelling algorithm is correct and effective.

Keywords-Business Process Modelling; Basic Block; Modelling Algorithm; Petri nets

1 INTRODUCTION

Business process everywhere, to improve the business process management and processing capabilities has very important significance to improve work efficiency for the government and to improve business performance for the enterprise. With the development of information technology, it creates the technical foundation for a substantial increase for business process management capabilities and processing power [1].

Business process modelling is the way business processes are expressed. Business process modelling is the foundation of business process analysis, reengineering, reorganization and optimization. The common modelling languages include UML[2],EPC[3],Workflow Graph[4],BPMN[5], Pi-Calculus[6] and Pet nets[7]. Petri is not only intuitive, but also a formal definition. Based on the prototype Petri net, this paper adds time and cost factors to form an extended generalized stochastic Petri net. It is a formal description of the business process. The semi-formalized business process modelling algorithm based on Petri nets is proposed.

2 DEFINITION OF BUSINESS PROCESS

Definition 2.1[7] A triple N=(C, A, F) is called a net if

(1)$C \cup A \neq \Phi$;

(2)$C \cap A = \Phi$;

(3)$F \subseteq (C \times A) \cup (A \times C)$ is a binary relation, the flow relation of N;

(4) $dom(F) \cup cod(F) = C \cup A$.

C which is called the condition set of N, A is called the activity set, F is called arc set. $X = C \cup A$ is called the N element set. Elements in C are called the condition or element C, Elements in A are called activity or element A. The Φ is empty. $\times$ is the two-set (space) of the Cartesian product operation, so F is the set of ordered pair composed of a C element and a A element. dom (F) is the order dual of F contains the first element (starting) of the set, cod (F) is the second element (the end) of the collection: dom (F) = {x | y: (x, y) $\in$ F}; cod (F) = {x | y: (y, x) $\in$ F}. Net standard graphics is that circle represents the condition, with the box that transition s from x to y by a directed arc (arrow) that ordered dual (x, y). (hence (x, y) $\in$F), it is also known as directed arcs).

Definition 2.2 Satisfy the following six conditions set Σ = (C, A, F, M, λ, g) is called a business process system:

(1) (C, A, F) is a net;

(2) where T is divided into time transition Set At and instantaneous transition set Ai 2 sub-sets. The implementation of the time transition will take some time, the instantaneous transition implementation time is negligible, that is, $A = At \cup Ai$, $At \cap Ai \neq \Phi$;

(3) F in the permitted inhibitor arcs, inhibitor arcs exist only from the position to transition the arc. Inhibitor arcs can be implemented by the conditions of the original link becomes unenforceable (disable) the conditions, the original non-implementation of the conditions can be implemented into the conditions, and transition in the linked implementation, not marked out from the connected position;

DOI: 10.1051/matecconf/201713900105

(4) λ ={ λ 1, λ 2, ... λ m}, λ i is the average implementation rate of the transition ai (ai ∈ A), which represents average implementation times in the case of the implementation in unit time. The unit is times/hour;
(5) g={g1,g2,...,gm}, gi is the average implementation costs of the transition ai (ai ∈ A), which represents implemented cost in the case of implementation per unit time, The unit is the yuan/hour.

Where each delay of the transition 1/ λ ={1/ λ 1, 1/λ2, ...,1/λm} are exponentially distributed, the cost of each transition g = {g1, g2, ..., gm} are exponentially distributed .

Definition 2.3 Let Σ = (C, A, F, Mλ , g) represents a business process system. Let M0 ∈ M (M0⊆C), p = (C, A, F, M0, λ , g). M0 is called the initial marking of p and p is called a business process. If confusing, then, C, A, F, M0, λ ,g respectively p.C, p.A, p.F, p.M0, p. λ , p.g to represent.

Definition 2.4 Let p = (C, A, F, M0, λ , g) represents a business process. For p in any one activity a ∈ A, Inflow (a) = {(x, a) | (x, a) ∈ F, x ∈ C}, Inflow (a) is called the input stream for the activity of a, Outflow (a) = {(a, y) | (a, y) ∈ F, y ∈ C}, Outflow (a) is called the output stream for the activity of a[3].

Definition 2.5[7] Let N={C, A, F} as a net. For x ∈ C ∪ A, .X={y | y ∪ A ∧ (y, x) ∈ F}, x. = {y | y ∪ A∧ (x, y) ∈F} .X is called the front set or the input set of x, x. is called the back set or the input set of x.

Definition 2.6 Let N = {C, A, F} as a net, ∀c1∈C, ∀c2∈C , c1≠c2,if and only if c1.∩.c2≠Φ∨ .c1 ∩c2. ≠Φ, c1 and c2 are called the adjacent condition, denoted by border (c1, c2). Otherwise c1 and c2 are called non-adjacent condition, denoted by¬border(c1,c2).

Definition 2.7 Let N = {C, A, F} as a net, StartPlace (N) = {c ∈ C | a ∈ A: (a, c) ∉ F}, StartPlace (N) is called the initial condition of net ; EndPlace (N) = {c ∈ C | a ∈ A: (c, a) ∉ F}, EndPlace (N) is called termination condition of net; SpecialPlace (N) = StartPlace (N) ∪ EndPlace (N), SpecialPlace (N) is called special condition of the net.

Definition 2.8[8] Sequence block, concurrency bock, selection block and iteration block are called basic block. A basic block can be described as a 5-tuple b=(C, A, F, Ae, Ax) where

(1)C, A and F are called the condition set, the activity set and the arc set respectively.

(2)Ae, Ax⊆A are called the entrance and the exit of b respectively.

Basic blocks, which are enclosed by dotted lines in figures, are described as follows

Where each ei (i=1,2,...) denotes an activity.

Sequence block: It describes activities ei and ej are executed sequentially, shown in Figure 1. Formally, C={c}, A={ei, ej}, F={(ei, c), (c, ej)}, Ae={ei}, Ax={ej}

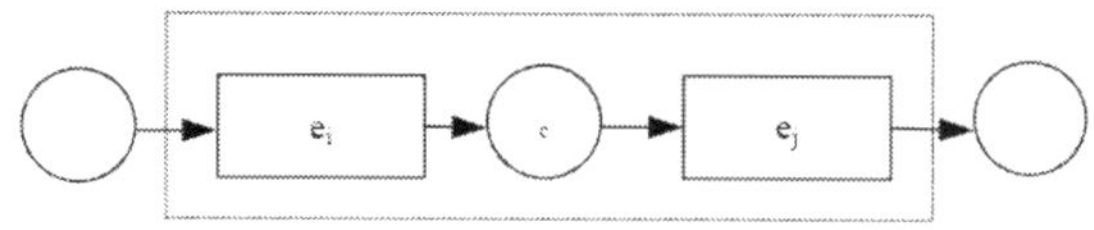

Figure 1. Sequence Block

Concurrency bock: It describes activities ei and ej are executed concurrently, shown
in Figure 2. Formally, C={c1, c2, c3, c4}, At={ ei, ej}, Ai={e0,en}, A={e0, ei, ej, en}, F={(e0, c1), (e0, c2), (c1, ei) ,(c2, ej), (ei, c3), (ej, c4), (c3, en), (c4, en)}, Ae={e0}, Ax={en}.

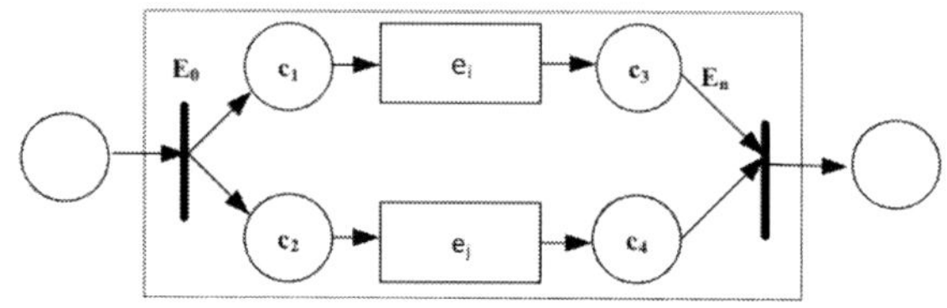

Figure 2. Concurrency Block

Selection block: It describes activities ei and ej are executed selectively, shown in Figure 3. Formally, C={ }, A={ei, ej}, F={ }, Ae={ei, ej }, Ax={ei, ej}.

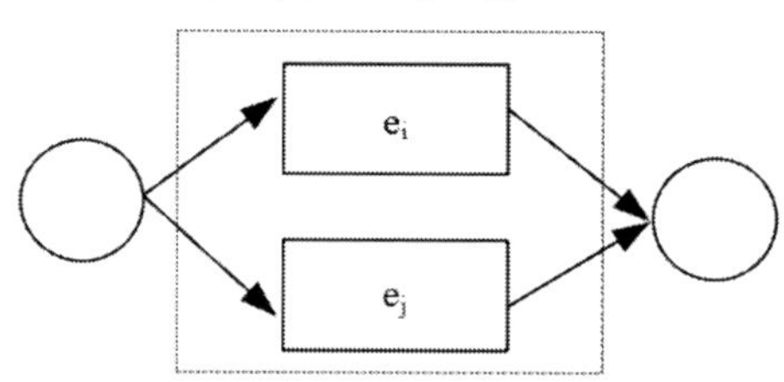

Figure 3. Selection Block

Iteration block: It describes activities ei and ej are executed repeatedly, shown in Figure 4. Formally, C={c1, c2}, At= { ei, ej,en },Ai= {e0 } A={e0, ei, ej, en}, F={(e0, c1), (c1, ei), (ei, c2), (c2, ej),(ej, c1), (c2, en)}, Ae={e0}, Ax={en}.

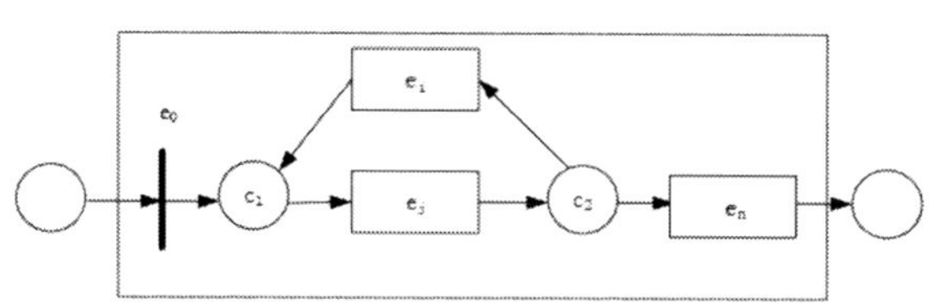

DOI: 10.1051/matecconf/201713900105

Figure 4. Iteration Block

3 BUSINESS PROCESS MODELLING

Business process modelling method is from top to bottom, and gradually decomposed. The first step in any business process is to create an initial block and the definition of the initial block is shown in Figure 5. Business process by layer refinement, the modeler can get a business process to meet their needs model. The modelling of business processes is described in Function 3.1. Two sub-functions are called in the process of business process modelling. Function 3.2 is used for business process layer by layer decomposition and refinement. Function 3.3 is used to establish a link between non-adjacent libraries in a business process.

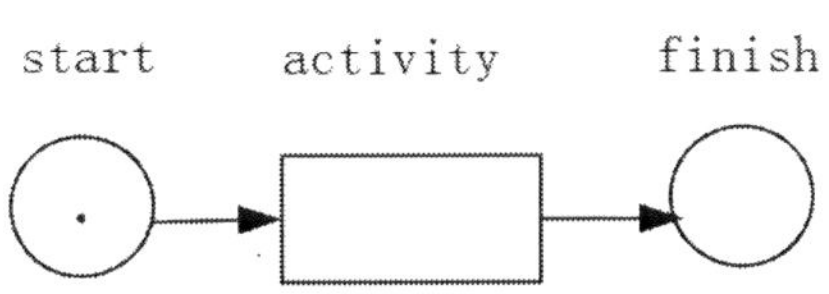

Figure 5. Initial block

```
Function 3.1(Business process modelling method)
 p Modelling_Business_Process( )
 {
/* I represents the number of business processes    */
 int i;
 /*  Max represents the maximum number of stages   */
 p(0), …, p(MAX);
 Define p (0) as an initial block;
 If(The initial identity of the initial block must be
determined by other conditions)
 {
/* p(0).M0 represents the initial identity of the business
 process p (0) */
p(0).M0=Φ;
}
else
{
    p(0).M0={start};
}
i=0;
/* Call Function 3.2*/
p(0)=decompose(i);
/* Call Function 3.3*/
p(0)=Modelling_NoBorderCondition_Relation( p(0) );
    λ : p(0).Ai -> λ i
   g : p(0).Ai ->gi
return p(0);
}
Function 3.2
p decompose(int i)
{
p(i+1)=p(i);
for(a : p(i).A)
{
 Analyze activity a;
 if（Activity a needs to be broken down into basic
blocks）
  {
   Build a basic block b={C, A, F, Ae , Ax}
p(i+1).C= p(i+1).C ∪ b.C;
p(i+1).A= p(i+1).A-{a} ∪ b.A;
p(i+1).F= p(i+1).F-Inflow(a)-Outflow(a) ∪ b.F
∪ {(x, y)|(x, a) ∈ Inflow(a) ∧ y ∈ b.Ae}
∪ {(x, y)|x ∈ b.Ax ∧ (a, y) ∈ Outflow(a)};
}
}
If( p(i+1) need to continue refinement refinement)
{
   decompose(i+1);
}
else
{
    return p(i+1);
}
}
```

```
Function 3.3
p Modelling_NOBorderCondition(p p1)
{
p p2;
p2=p1;
for(c1 ; p1.C; c1 ∉ SpecialPlace(N))
{
for(c2 ;p1.C, c2 ∉ SpecialPlace(N))
{
if( ¬border(c1,c2) && There is some basic block between
c1 and c2)
{
Build a basic block d={C, A, F, Ae , Ax};
p2.C= p1.C ∪ d.C;
p2.A= p1.A ∪ d.A;
if( (C1 is connected to d.Ae) && (C2 is connected to
d.Ax) )
{
p2.F= p1.F ∪ d.F ∪ {( c1,y)| y∈b.Ae}
∪ {( y, c2)| y∈b.Ax};
}
else if( (C1 is connected to d.Ax ) && (C2 is connected to
d.A e) )
{
p2.F= p1.F ∪ d.F ∪ {( c2,y)| y∈b.Ae}
∪ {( y, c1)| y∈b.Ax};
}
}
}
}
return p2;
}
```

4 CASE STUDY

This paper combines the application examples of an enterprise to illustrate the modelling of extended generalized stochastic Petri nets. Fig.6 is a logistics company's operating process.

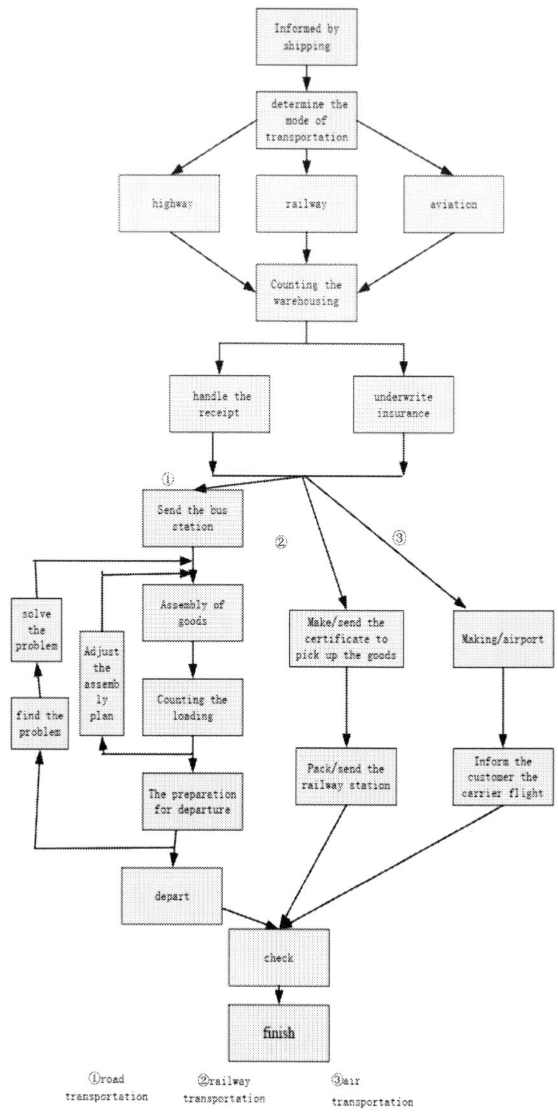

Figure 6. The flow of a logistics company

According to the business process modelling algorithm(Function 3.1), get a formal description of business process modelling. The description is as follow.

A={a1,a2,a3,a4,a5,a6,a7,a8,a9,a10,a11,a12,a13,a14,a15,a16,a17,a18,a19,a20,a21,a22,a23};

Ai ={a6,a9,a11};

At={a1,a2,a3,a4,a5,a7,a8, a10,

a12,a13,a14,a15,a16,a17,a18,a19,a20,a21,a22,a23};

C={c1,c2,c3,c4,c5,c6,c7,c8,c9,c10,

c11,c12,c13,c14,c15,c16,c17,c18,c19};

F={(c1,a1),(a1,c2),(c2,a2),(c2,a3),(c2,a4),(a2,c3),(a3,c3),(a4,c3),(c3,a5),(a5,c4),(c4,a6),(a6,c5),(a6,c6),(c5,a7),(c6,a8),(a7,c7),(a8,c8),(c7,a9),(c8,a9),(a9,c9),(c9,a10),(a10,c10),(c10,a11),(a11,c11),(c11,a12),(a12,c12),(c12,a13),(a13,c13),(c13,a14),(a14,c11),(c13,a15),(a15,c14),(c14,a16),(a16,c15),(c15,a17),(a17,c10),(c14,a18),(a18,c18),(c9,a19),(a19,c16),(c16,a20),(a20,c18),(c9,a21),(a21,c17),(c17,a22),(a22,c18),(c18,a23),(a23,c19)};

M0={c1};

Finally, we get the extended generalized stochastic Petri nets representation of business processes. It shown in Figure 7. According to Figure 7, we can quantify and optimize business processes of the logistics company.

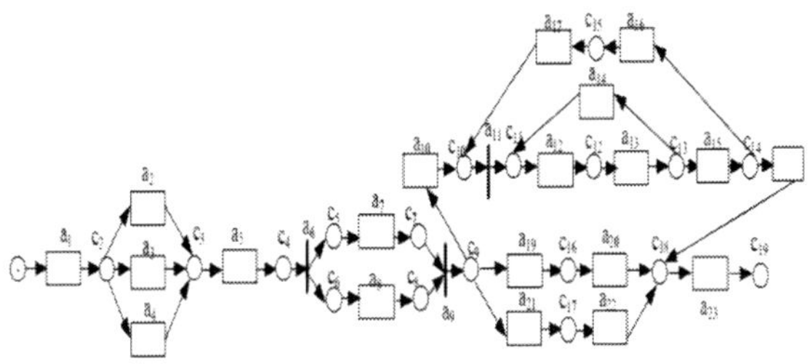

Figure 7. Extended generalized stochastic Petri net representation of business processes

5 CONCLUSIONS

In this paper, the concept of cost is introduced to the generalized stochastic Petri nets. It is called it extended generalized stochastic Petri nets. At the same time, basic block is introduced to business process modelling. Business process is got described with the extended stochastic Petri net. The semi-formalized business process modelling algorithm based on Petri nets is proposed. Finally, the case from a logistics company proved that the modelling algorithm is correct and effective.

Acknowledgment

This work has been supported by the National Science Foundation of China under Grant No. 61379032.

References

1. HammerM, James C. Reengineering the Corporation: A Manifesto for Business Revolution [M]. New York: HarperCollins Publishers Inc. ,1993.
2. OMG. Unified Modelling Language (UML). http://www.omg.org/cgi-bin/doc?ptc /2004-10-02 (8/8/05).
3. van der Aalst W.M.P. Formalization and Verification of Event-driven Process Chains. Information and Software Technology, 1999, 41(10):639-650.
4. Sadiq W., Orlowska M.E. Analyzing Process Models Using Graph Reduction Techniques. Information Systems, 2000, 25(2):117-134.
5. White S.A. Business Process Modelling Notation (BPMN). Version 1.0. http:// www.bpmn.org/Documents/BPMNV1-0May2004.pdf.
6. Milner Robin. Communicating and Mobile Systems: the Pi-calculus. Fifth Printing, USA: Cambridge Press, 2004.
7. Wu Zhe Hui. Introduction to Petri nets. Beijing: Machinery Industry Press, 2006.(In Chinese)
8. Li Tong. An Approach to Modelling Software Evolution Processes. Springer-Verlag and Tsinghua University

MATEC Web of Conferences 139, 00106 (2017) DOI: 10.1051/matecconf/201713900106
ICMITE 2017

Study on numerical calculation method for hydrodynamic parameters of WEC

Shen Lijiao
National Key Laboratory on Ship Vibration & Noise, China Ship Development and Design Center, Wuhan 430064, China

Abstract: For the effect of hydrodynamic parameters on the dynamic performance of wave energy devices is very significant, these parameters must be considered carefully when adjusting dynamic characteristics of devices. On the other hand calculating hydrodynamic parameter of devices accurately can guarantee rational dynamic property parameter adjustment. By using CFD technique and considering the definition of hydrodynamic parameters, the phase relationship between added mass and damp as well as the equation of forces, one new calculation method of hydrodynamic parameter was presented. Finally one example demonstrated the effectiveness of the new analysis method presented in this paper.

Keyword: wave energy extraction, hydrodynamic parameters, CFD, resonance dynamic characteristics

1 Introduction

Wave engine covert (WEC) is one of the main methods to harvest wave engine, which belongs to hot issue of using wave engine at present. Existing experiment result and theory indicates that the efficiency of WEC is the highest when WEC is in resonance with the wave. Therefore using the resonance method is one high-efficiency way to convert the wave engine. The effect of hydrodynamic parameters on the dynamic performance of wave energy devices is very significant, and it is important to calculate the hydrodynamic parameters of WEC accurately. Hence, to work out a new method to calculate hydrodynamic is necessary, not only for WEC, but also for the ocean construction and ship.

If the shape of the construction is simple and regular, such as sphere, cylinder etc, the hydrodynamic parameters is easy to be solved by the analysis equation. However if the shape of the construction is complete, it is hard to get the hydrodynamic parameters. Therefore, many calculating methods for the hydrodynamic parameters are put forward at home and aboard. Slice theory as well as the slender body theory has been developed based on the potential flow theory, which is appropriate for the hydrodynamic parameters of the ship whose shape is simple. However, the results of these methods need to be modified by the empirical equation gotten from experiment, and the veracity is weak. Another kind of method is Green function method, which belongs to Hess-Smith method. This method supposes the source density of every element is constant, so that velocity potential can be calculated using source sink and doublet distribution on wet surface. This method is more accurate than the slender body theory, but it is hard to calculate the growth characteristics, high frequency oscillation and singularity accurately. According to the correlational research of ISSC and ITTC, it is found that results of different program based on the same theory are inconsistent.

With the development of the computer technology, computational fluid mechanics is more and more mature. Because of the advantages such as the low computational expense and comprehensive analysis factor, the method based on CFD theory has become one of the most important methods to study the hydrodynamic parameters. More and more methods to calculate hydrodynamic parameters are put forward.

The shape of the WEC proposed by ref.[5] is complete. Hydrodynamic parameters have an effect on adjusting the dynamic characteristics of the device. In order to ensure that the device keeps the resonant mode, it is important to calculate the hydrodynamic parameters of the device accurately. Therefore, according to the definition of the hydrodynamic parameters, this paper put forward a computing method based on CFD. The ideal of this method is: Firstly, a numerical wave tank and couple modal of device and water is built. Secondly, the resultant force on the device is calculated at every time step while the device is on forced harmonic vibration. Finally, the hydrodynamic parameters are gotten by decomposing the resultant force. The resultant force includes added mass and add damping dynamic pressure, inertia force, hydrostatic restoring force and so on. Add damping could be decomposed from dynamic pressure by using triangular decomposition. Then, based on the equilibrium equation, the dynamic pressure caused by added mass could be separated from remain pressure. Hydrodynamic parameters are calculated by separated pressure.

2 Method

Floating body radiates wave in the process of movement, the wave induces a pressure field in fluid domain. According to the potential flow theory, a part of the dynamic pressure field has the same phase as the acceleration of floating body. Meanwhile another part has the same phase as the velocity acceleration of floating body. The dynamic resultant force can be gotten by dynamic pressure integration on the wet surface of floating body. the resultant force is:

$$M_{ij}\frac{\mathrm{d}^2\eta}{\mathrm{d}t^2}+C_{ij}\frac{\mathrm{d}\eta}{\mathrm{d}t}=\iint_S p_d\boldsymbol{n}\mathrm{d}s \quad (1)$$

$$M_{ij}\frac{\mathrm{d}^2\eta}{\mathrm{d}t^2}+C_{ij}\frac{\mathrm{d}\eta}{\mathrm{d}t}=\iint_S p_d(\boldsymbol{r}\times\boldsymbol{n})\mathrm{d}s \quad (2)$$

The pressure filed caused by the floating body can be divided into two part, one has the same phase with the acceleration of floating body, another has the same phase with the velocity acceleration of floating body like this:

$$M_{ij}\frac{\mathrm{d}^2\eta}{\mathrm{d}t^2}=\iint_S p_{\mathrm{dm}}\boldsymbol{n}\mathrm{d}s \quad (3)$$

$$C_{ij}\frac{\mathrm{d}\eta}{\mathrm{d}t}=\iint_S p_{\mathrm{dc}}\boldsymbol{n}\mathrm{d}s \quad (4)$$

Added mass and added dumping are defined by equation (3) and equation (4). Where, M_{ij} is added mass. C_{ij} is added dumping, η is displacement of floating body, t is time, p_d is dynamic pressure caused by floating body, S is the wet surface of the floating body.

Added mass and added dumping are related to the floating body shape, when the floating body shape is complete, it is hard to get the hydrodynamic parameters by analysis equation. Otherwise, the hydrodynamic parameters are functions of frequency. Equation (3) and function(4) is interesting, they are the product of the frequency domain function and time domain function. if $m_{ij}(t)$ is the time domain function of added mass, then the dynamic pressure induced by added dumping can be calculated in the same way:

$$M_{ij}(\omega)\frac{\mathrm{d}^2\eta}{\mathrm{d}t^2}=m_{ij}(t)*\frac{\mathrm{d}^2\eta}{\mathrm{d}t^2} \quad (5)$$

From the above function, it can be found that the radiation pressure caused by floating body movement appears as the convolution type. Hence, it is hard to calculate the hydrodynamic parameters by equation (3) and equation (4).

When the floating body moves in a harmonic motion, its displacement is a harmonic function, which shows as a δ function in frequency domain. As a result, the convolution is a harmonic function in time domain, and the corresponding added mass and added dumping is the value under the harmonic movement frequency. if the accelerate of the floating body is as follows:

$$\frac{\mathrm{d}^2\eta}{\mathrm{d}t^2}=A_m\sin(\omega t) \quad (6)$$

$$F^{-1}(M_{ij}(\omega)\cdot A_m\pi/2(\delta(\lambda+\omega)-\delta(\lambda-\omega)i)=M_{ij,\omega}A_m\sin(\omega t) \quad (7)$$

where $m_{i,j}$ is added mass with the frequency being ω,the added dumping can be calculated in the same way.

According to the equation(7),when the floating body moves as a constant frequency, the added mass and added dumping could be calculated by the steady hydrodynamic pressure or moment on the wet surface of the floating body. Because hydrodynamic pressure induced by the added mass and the added dumping has the same phases with the accelerate and the velocity of the floating body, the phases difference between the hydrodynamic pressure induced by the added mass and the added dumping is $\pi/2$.

Without incident wave, floating body is forced to move in a harmonic motion η_j, the equilibrium equation on the floating body is as follows:

$$F_i=-M_{ij,\omega}\frac{d^2\eta}{dt^2}-C_{ij,\omega}\frac{d\eta}{dt}+f_i \quad (8)$$

Where, F_i is the resultant force on the floating body, f_i is the statics force on the floating body at this direction, including restoring force or moment on the floating body and the force(moment) of gravity at this direction.

$$f_i=K_w\eta_j+G_i \quad (9)$$

Known from equation (8), the resultant force on the floating body is the sum of the radiation pressure caused by the added mass and the added dumping as well as the statics force. In practice, it is easy to calculate the resultant on the floating body, but it is hard to calculate the pressure separately, thus the added mass and the added dumping can not be calculated by the resultant force directly.

The displacement of the floating body can be taken as a harmonic function as follows:

$$\eta_i=A_i\sin(\omega t+\varphi) \quad (10)$$

Substitute equation (10) into equation (8), then

$$F_i=\left(-M_{ij,\omega}\omega^2+K_i\right)A_i\sin(\omega t+\varphi)-\omega C_{ij,\omega}A_i\cos(\omega t+\varphi) \quad (11)$$

Known from the above equation, the resultant force can be divided into two parts, of which the phase difference is $\pi/2$. So by taking the time of equation(10) as a standard, based on equation(11), the resultant force can be divided into two part by using triangular decomposition. And noticed that the second part is the force induced by added dumping, then, the added dumping of float body can be calculated by the equation (4).

The first part of equation (11) also can be divided into two parts: restoring force and the force caused by the added mass. For restoring force can be calculated directly by the shape of floating body, the force induced by the added mass can be calculated after substituting the restoring stiffness into equation (11), then added mass with the frequency being ω can be calculated.

Based on CFD, the couple model of floating body and fluid field should be built, the floating body is forced to move as a harmonic motion, and the resultant force on the floating body is calculated at every time step. Then the hydrodynamic parameters of the floating body can be calculated by the method proposed in this paper. Because the water viscidity can be considered in CFD, the degree of accuracy is higher.

3 Example

Based on the software FLUENT, a circular wave tank is established for the incident wave is not needed. The boundary of the tank should meet the Sommerfeld radiation condition, thus the source term wave absorbing method is employed. Wave absorbing field is near the boundary and the wave energy covert is at the center of the tank. The field around the device is dynamic grid field. The radius of the tank is 2.5 times of the wave length induced by the device. The height of the tank is equal to the wave length induced by the device. The field above the device model is full of air, and the field under the device is full of water. The free surface is traced by the method of VOF, and control functions are solved by RANS.

The wave energy convert shape of reference 5 is shown in fig 2. The upper part of the device is cylindricality, of which the diameter is 3.5m, with the height of the device being 3.3m. Because the device has two disymmetric faces, surging, heaving and pitching mode of the device is not coupled with the swaying, rolling and yawing mode . And for the

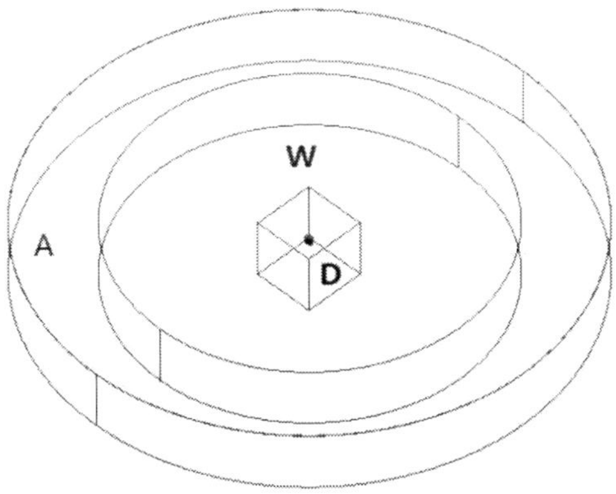

Figure 1. Numerical flume model

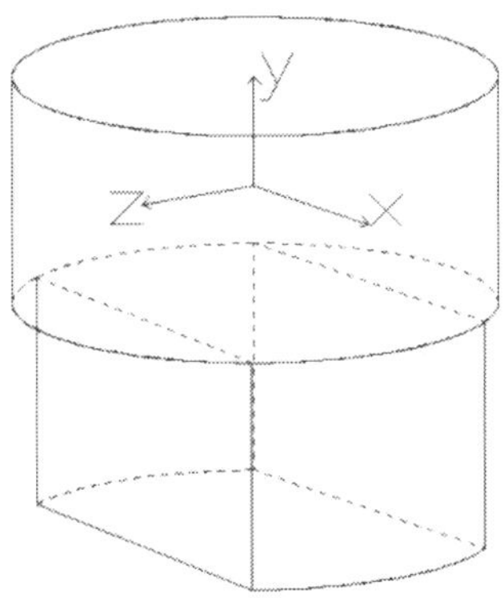

Figure 2. Device outside diagram

device harvesting the wave energy is mainly through heaving and pitching direction, the hydrodynamic parameters of the device at the direction of surging, heaving and pitching are concerned. On account of the relationship of heaving and pitch, this paper only calculates the hydrodynamic parameters of the device at the direction of surging and heaving.

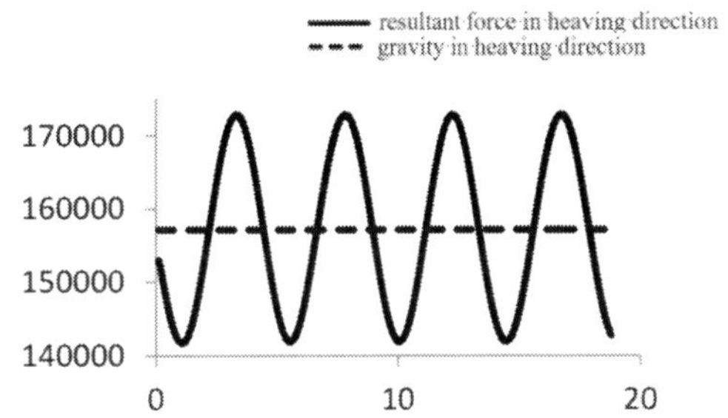

Figure 3. Force on the heaving direction

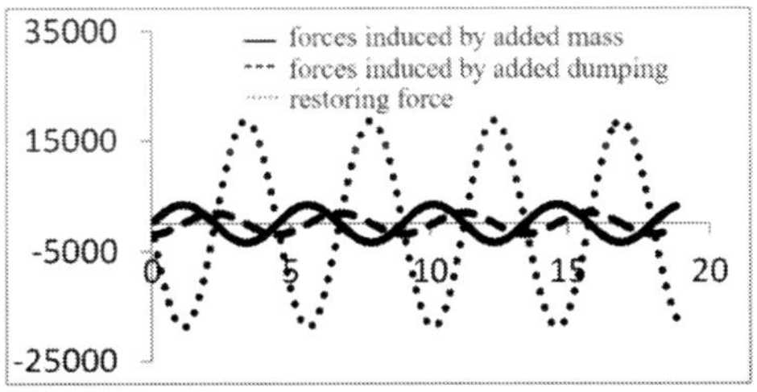

Figure 4. Force component on the heaving direction

In the process of calculation, the device is forced to be a rigid motion of which the displacement is a sine function. At every time step, the force at every direction on the wet surface of the device is calculated. The displacement amplitude is 0.2m, and the period is 4.5s. The force at the direction of heaving on the wet surface of the device changes over time is shown in fig 3. From the figure it can be found that when the device is forced to be a rigid harmonic motion, the force on the device comes into a stable state quickly, and conform to the harmonic rule. Thus it can be seen that the wave absorbing performance of the numerical tank is fine. Only two or three periods data is needed to realize the triangular decomposition, thus this method can save a lot of calculation time.

The calculated resultant force is divided by the method proposed in this paper, the curves of components are shown in figure 4. From the figure, it can be found out that the phase difference between forces induced by the added mass and added dumping is $\pi/2$, and the phase difference between the force induced by the added mass and the restoring force is π

The wet surface has a great effect on the hydrodynamic parameters, in order to confirm the method proposed in this paper, different periods and amplitudes are employed to calculate the hydrodynamic parameters.

The results of calculated added mass and added dumping are modified by dimensionless treatment. The results are shown in table 1, from the table, it can be found that the effect of the amplitude on the hydrodynamic parameters are small.

Table. 1. Relationship between the hydrodynamic coefficient and vibration state

Vibration peirod	4.5s	4.5s	5.5s	5.5s
amplitude	0.2	0.3	0.2	0.3
added mass factor μ_{ij}	0.55985	0.55940	0.77568	0.78060
added dumping factor λ_{ij}	0.1227	0.1393	0.1361	0.1361

If the amplitude is 0.2m, the periods is 3s~7s, curves of the added mass and added dumping are shown in figure 5, Obviously, the relationship between the period and the hydrodynamic parameters is complete.

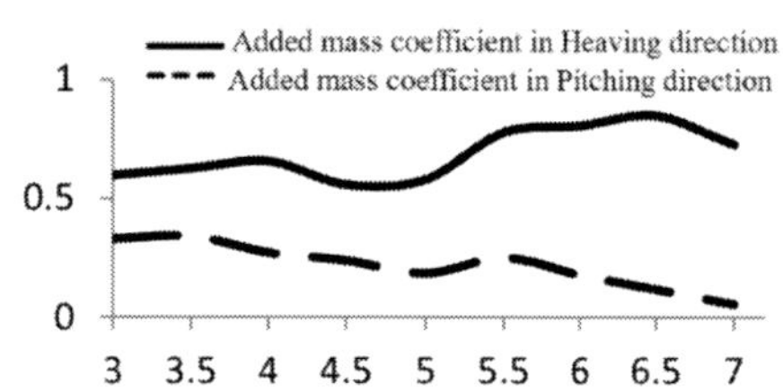

(a)Added mass coefficient

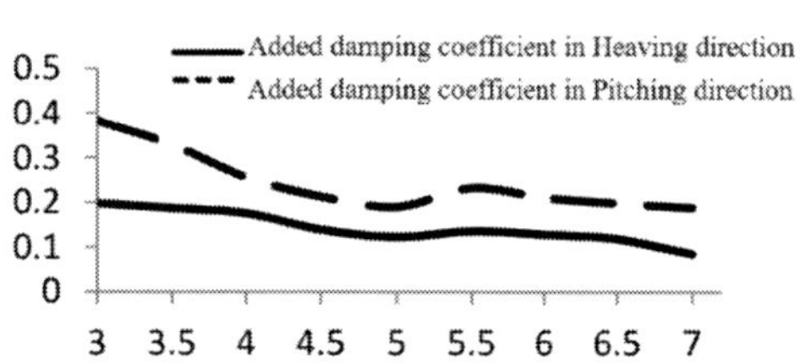

(b)Added damping coefficient

Figure 5. Relationship between the hydrodynamic coefficient and vibration period

4. Conclusion

It is important to calculate the hydrodynamic parameters for the WEC which harvests the wave energy by resonance stage with the wave. The method proposed in this paper can get the hydrodynamic parameters quickly and actually. In this method, the effect of the amplitude on the hydrodynamic parameters is small, but the relationship between the period and the hydrodynamic parameters is complete.

Reference

[1] Cai Yuanqi,Li Ling. Study on Classification Methods of Wave Power Generation echnology[A]//Proceedings 2012 International Conference on Material Renewable Energy and Environment[C].2012,711~715.

[2] Yu Zhi. The application of wave theory in the ocean wave energy utilization[J]. Chinese Journal of Nature,1987,(10): 736-739.

[3] Li Yunwu. Ocean energy development [M].Beijing: The Ocean Publishing Company, 2008.

[4] Haren.P, Mei.C.C. Rafts for Absorbing Wave Power [A].13th Naval Hydrodynamics symposium[C], Preprint, Tokyo,1980: 31-36.

[5] Cai Yuanqi. Resonant wave power device[P]. China:201110233308, 2011-08-16.

[6] Cai Yuanqi, Yang Shengbang. Adjustable film type stiffness regulator and using method [P]. China: 201110231566.1, 2011-08-12.

[7] Zhu Renchuan, Guo Haiqiang, Miao Guoping, Yu Jianwei. A Computational Method for Evaluation of Added Mass and Damping of Ship Based on CFD Theory [J]. Joutnal Of Shanghai Jiaotong University. 2009(43)2:198-203.

[8] Dai Yishan, Duan Wenyang. Potential flow theory of ship motions in waves. National Defence Industry Press. 2008:28-35 .

[9] O. M. Faltinsen. Sea loads on ships and offshore structures. Shanghai Jiaotong University Press, 2008.

[10] Fu Huipin, Li Jie. Numerical Studies of Added Mass Based on The CFD Method[J]. Journal of Harbin engineering university. 2011(32)2:148-152.

[11] Ma Ye, Shan Xuexiong. A New Numerical Computation Method for Added Masses of Complicated Object[J]. Computer simulation. 2007(24)5:75-79, 113

MATEC Web of Conferences 139, 00107 (2017)
ICMITE 2017 DOI: 10.1051/matecconf/201713900107

Design and Implementation of Track Record System Based on Android Platform

Jiachen Zhang[1,*], Qizhong Cai[1]

[1]College of Electrical and Information Engineering, Guangxi University of Science and Technology Liuzhou 545006,P.R.China

Abstract. For the problem of losing and missing of vulnerable groups, a track record system is designed.The mobile terminal Android system is used as a platform,with the help of Auto Navi Map Android SDK positioning function,realize the positioning data acquisition of mobile terminals;using Apache Tomcat Server and MySQL database to build a Server which haves C/S(the client and the server) server architecture.The mobile terminal interacts with the server through the JSON data transmission mode based on the HTTP protocol,and the server saves the relevant information provided by the mobile terminal through the JDBC to the corresponding table in the database. It can be used to monitor the trace of the family and friends,compared with the PC terminal, it is not only more flexible,convenient and fast,but also has the characteristics of real-time and high efficiency.Through the test,all functions can be used normally.

1 Introduction

Android is based on Linux as the core of open source mobile terminal application software platform [1]. According to statistics, the year 2016 Android platform mobile phone global market share has reached 85%. This article to Android mobile terminal as a platform is also because of its openness, cross-platform and the use of breadth based on the choice.

In China, the annual missing children according to incomplete statistics there are about 200,000, and get them back in less than 0.1%, which does not include the lost old man [2]. The annual female college students and other vulnerable groups of missing events are frequently appear in our lives.This is the original purpose of the track record system in this article.The tracking system provides the location information of the stalker in real time with the SDK location function provided by Auto Navi Map.Use the Tomcat server and MySQL database to build the server side. Records, saves and calls the user's account information, the location information of the stalker, and the corresponding relationship of the user. It helps users monitor the movements of the elderly and children in the home, and then it helps prevent the loss of the vulnerable population.The track record system can be extended to all mobile intelligent terminal devices covered by the Android system, and it can be recorded in a fast, real-time and convenient location.

2 Track record system model

2.1 The overall structure of the track record system

In simple terms, track record system is the process of data collection and processing.It's a track record of the movements of the moving application, and the system is the type of C/S software.Compared with the traditional browser-based C/S software system type, we can avoid the fixed and complexity of traditional methods and make full use of mobile, real-time and flexible mobile devices.

Figure 1 shows the overall structure of the track record system,and the server is the skeleton and core of the system,all data on the mobile terminal must interact and store data through the server side.For the mobile terminal,need to provide the information of objects, namely, the sending end will actually when positioning information saved to the server through the server database;then we can use the receiving end to read the stored data from the server to the sending end.The data from the system is based primarily on the mobile location.

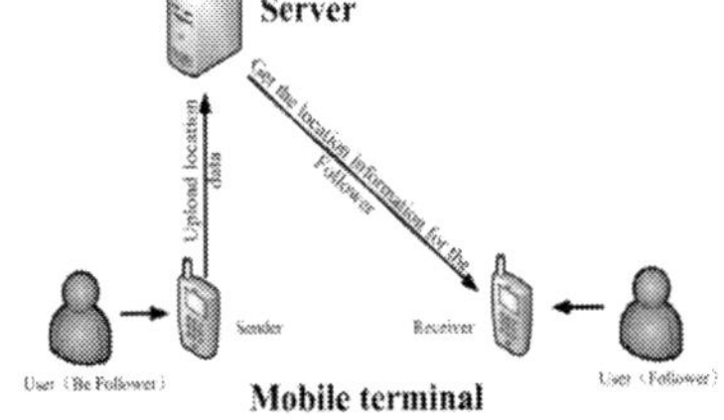

Figure 1. Track record system structure

2.2 Function design

According to the above system functional requirements analysis, the system mainly includes the following functions:registration/login, self-positioning, location

* Corresponding author: jiachen0718@126.com

information upload, receiver follow /unfollow sender, receiver select time period view the sender's track, lost prompt.As shown in Figure 2.

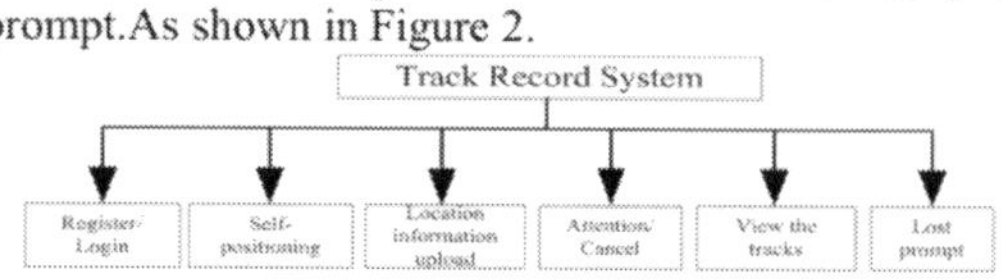

Figure 2. System main function

The positioning function of the track record system is mainly realized through the android SDK provided by the open platform of Auto Navi Map.In order to improve the positioning precision of the system, adopted based on the hybrid precise positioning of GPS + base + WiFi.The hybrid localization way has the advantage of millisecond response of milliamperes power, accuracy within 10 meters, is not only dependent on the GPS, break through the traditional mode of GPS positioning to the success of high power consumption, regional requirements open defect. No matter in the city or in the valley, skyscrapers can achieve more accurate positioning.

3 Service terminal design

3.1 The overall service terminal design

The server is mainly composed of Apache server and MySQL database, and the peripherals include Android phone and PC [3].Overall framework of the server shown in Figure 3,the sender sends location data to the Apache server and stores the data through the MySQL database,receiver through port 80 to monitor the server, the data through the JSON format data interaction between them, the system administrator can through the PC to server maintenance.

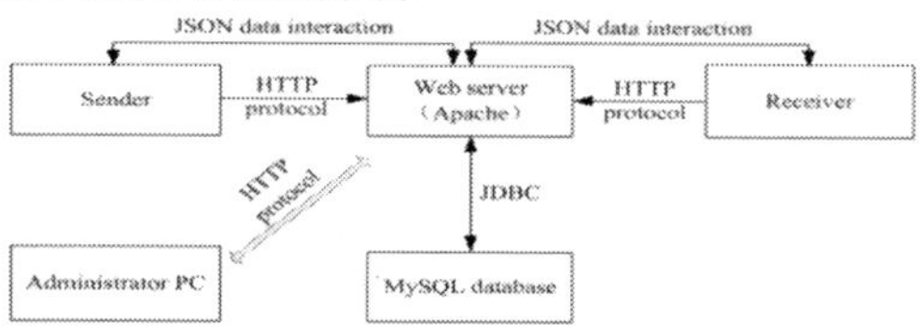

Figure 3. Service terminal overall architecture

3.2 The service terminal database structure design

Database is the support of data information for the entire design.According to the system's implementation, three tables are required in the database.The first table is used to save the user registration information, name it User. The main function is to record the user's account password, user type and phone number, and independent number for the above information.The second table is used to store information uploaded on the sending end,name it Path. The main function is to record the sender to upload the location of the longitude and latitude, name, phone remaining power and upload time.The last table is used to store the user relationship information, named Userinfo, mainly on the status of attention to real-time records, to avoid repeated attention.

Table 1. User.

Field name	Data type	Remark
userid	smallint(6)	non-null,auto increment
usertype	smallint(6)	non-null,default 0
username	char(50)	non-null
userpass	char(20)	non-null
phonenumber	char(11)	non-null
status	smallint(6)	non-null,default 0
createtime	datatime	non-null,default current time
updatetime	datatime	non-null

Table 2. Path.

Field name	Data type	Remark
pathid	smallint(6)	non-null,auto increment
pathlongitude	char(50)	non-null
pathlatitude	char(50)	non-null
pathname	char(50)	non-null
electric	char(50)	non-null
status	smallint(6)	non-null,default 0
pathuserid	smallint(6)	non-null
createtime	datatime	non-null,default current time

Table 3. Userinfo.

Field name	Data type	Description	Remark
userinfoid	smallint(6)	Relationship ID	non-null,auto increment
userid1	smallint(6)	Follower ID	non-null
userid2	smallint(6)	Be follower ID	non-null
status	smallint(6)	Attention status	non-null,default 0
createtime	datatime	createtime	non-null,default current time
updatetime	datatime	updatetime	non-null,default current time

Take the user table as an example,the main create code as follows, PRIMARY KEY and UNIQUE KEY are constraints category.There is only one PRIMARY KEY in a table, which cannot be repeated, and is used in logical design as the record label,UNIQUE KEY can have multiple, nullable, mainly to ensure the uniqueness of domain/domain groups.

```
DROP TABLE IF EXISTS `user`;
CREATE TABLE `user` (
`userid` smallint(6) NOT NULL AUTO_INCREMENT,
`usertype` smallint(6) NOT NULL DEFAULT '0',
`username` char(50) NOT NULL,
`userpass` char(20) NOT NULL,
`phonenumber` char(11) NOT NULL,
`status` smallint(6) NOT NULL DEFAULT '0',
`createtime` datetime NOT NULL DEFAULT CURRENT_TIMESTAMP,
`updatetime` datetime NOT NULL DEFAULT
CURRENT_TIMESTAMP ON UPDATE CURRENT_TIMESTAMP,
PRIMARY KEY (`userid`),
UNIQUE KEY `username` (`username`),
UNIQUE KEY `phonenumber` (`phonenumber`)
ENGINE=InnoDB AUTO_INCREMENT=1 DEFAULT CHARSET=utf8;
```

The other two tables create the same code, just slightly modified on the constraints.

3.3 The service terminal data preservation and interface design

The data is primarily saved through the service terminal, and the server is mainly used to interact with the data on the mobile terminal through the interface.The code is as follows:

Registration interface main code:

```
String sql="insert into user(usertype,username,userpass,phonenumber) values (?,?,?,?)";
```

Login interface main code:

```
hashMap = (HashMap<String, Object>) jdbcUtil.findSimpleResult("select * from user where username=" + params.get(0), null);
```

Submit location information interface main code:

```
String sql= "insert into path (pathlongitude,pathlatitude, pathname,electricquantity, pathuserid ) values (?,?,?,?,?)";
```

Query user's trajectory interface main code:

```
hashMapList=jdbcUtil.findModeResult("select*from path where pathuserid = ? and createtime BETWEEN ? AND ? order by pathid desc", params);
```

Add user interface main code:

```
String sql = " insert into userinfo(userid1,userid2) values (?,?)";
```

Remove user interface main code:

```
String sql="update userinfo SET status = 1 WHERE userid1 = ? and userid2 = ?";
```

To acquire the binding user list interface main code:

```
hashMapList1=jdbcUtil.findModeResult("select userid2 from userinfo where status = 0 and userid1=" + params.get(0), null);
```

All of the updated methods:

```
jdbcUtil.updateByPreparedStatement(sql, params);
Update method main code:
pstmt = connection.prepareStatement(sql);
    int index = 1;
    if (params != null && !params.isEmpty()) {
      for (Object param : params) {
        pstmt.setObject(index++, param);
      }
    }
    result = pstmt.executeUpdate();
```

The service terminal interaction with the mobile terminal in addition to the commonly used hypertext transfer protocol HTTP,JDBC, which is based on Java's API, is an important part in the database that the service terminal brings to the database through a network request.As shown in Figure 4,SQL statements can be easily transmitted to almost any database using JDBC,programs written in Java can automatically send SQL statements to the appropriate database server.This can be achieved through the addition and deletion of SQL statements to achieve service terminal interface and database data interaction process.So,the service terminal is mostly interacting with the mobile data through HTTP, through the JDBC implementation of the data interaction with the database,save the data collected on the mobile end to the database.

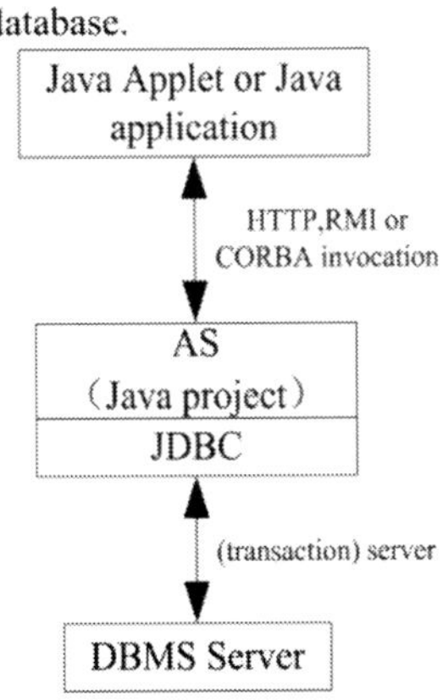

Figure 4. Access the database via JDBC

4 Mobile terminal software design

The mobile terminal of the track record system, namely the sender and receiver, need to be implemented on the intelligent terminal device, APP form implantation of the Android system.Considering the limitations of mobile terminal hardware and network factors, the complex data mainly processing realized through the service terminal.The main function of the mobile terminal is to locate data collection and transmission, user interface drawing and simple user data non-null verification.The main function is shown in figure 5 as follows:

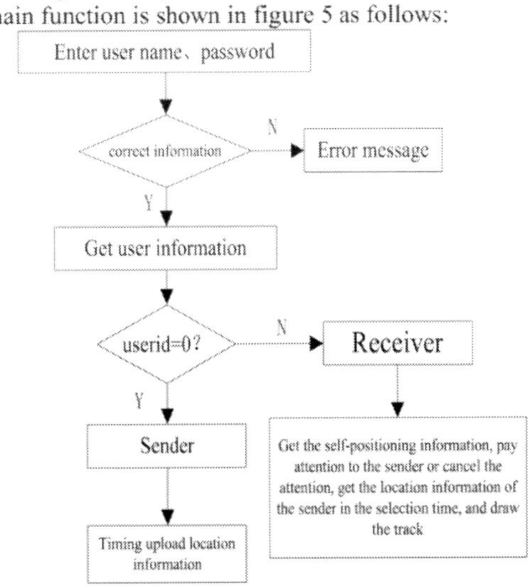

Figure 5. Android mobile terminal software design

The function of locating data on the sender is based on the long connection method based on the HTTP protocol [4].Because all functions in the design of the amount of data transferred is less, and the data structure is relatively simple,so achieve the unity of data so all transmit and receive data via JSON data format [5,6].

Android mobile terminals are divided into two roles, mainly by marking different user type to differentiate the functions required by different users.So the interface and the functionality that the sender and the receiver user logged in to see is slightly different.The mobile interface main code is as follows:

```
registerReceiver(batteryReceiver, new IntentFilter(Intent.ACTION_BATTERY_CHANGED));
    id_iv_back.setVisibility(View.GONE);
    id_btn_right1.setText("My attention");
    id_btn_right2.setText("Attention");
    ......
    if (BaseConfig.CURRENT_USER.getUsertype() == 1) {
      id_btn_right1.setVisibility(View.VISIBLE);
      id_btn_right2.setVisibility(View.VISIBLE);
      id_layout_choice.setVisibility(View.VISIBLE);
    } else {
      id_btn_right1.setVisibility(View.GONE);
      id_btn_right2.setVisibility(View.GONE);
      id_layout_choice.setVisibility(View.GONE);
    }
    if (aMap == null) {
      aMap = id_map.getMap();
    }
  myLocationStyle = new MyLocationStyle();
  myLocationStyle.myLocationType(MyLocationStyle.LOCATION_TYPE_LOCATION_ROTATE);
  myLocationStyle.interval(10 * 60 * 1000);
  aMap.setMyLocationStyle(myLocationStyle);
  aMap.setMyLocationEnabled(true);
```

```
aMap.setOnMyLocationChangeListener(new
 AMap.OnMyLocationChangeListener() {@Override
 public void onMyLocationChange(android.location.Location location) {
if (BaseConfig.CURRENT_USER.getUsertype() == 0) {locateMe(); }}});
```

In addition to the interface, the main function of the sender is positioning which is the cornerstone of the system.The implementation of the function is also mentioned in the previous section,it is mainly based on the open platform of Auto Navi Map, on this basis, the method provided by Auto Navi Map was encapsulated into a tool class,process and returning the results of its return.The positioning function main code is as follows:

```
LocationManager locationManager = new LocationManager(context);
locationManager.startLocation(new
LocationManager.LocationManagerListener()
 { @Override
  public void onLocationChanged(Location location) {
  if (location.getErrorCode() != 0) {
  LogUtil.d("ErrorCode=" + location.getErrorCode() + "ErrorInfo="+
location.getErrorInfo() + "LocationDetail=" +location.getLocationDetail());
  return; }
  id_tv_title.setText(location.getCity());
  submitLocation(location); }});
```

The main function of the receiver is to look at the trace of the sender and draw it on the map.This aspect of the treatment is mainly through the interface to get a sender whereabouts trajectory data set, and then by iterating through the data collection for each point data at a time, and tracing points and trajectory drawing method to draw on the map.The track mapping main code is as follows:

```
for (int i = 0; i < jsonArray.length(); i++) {
 Location location = JSON.parseObject(jsonArray.optString(i),
Location.class);
 latLngList.add(new LatLng(location.getPathlatitude(),
location.getPathlongitude()));
 locationList.add(location);
 if (i == 0) {aMap.addMarker(new
MarkerOptions().position(latLngList.get(i)).title(location.getPathname() +
"Remaining battery: " + location.getElectricquantity() + "%").snippet("Last
position"));}
 else {aMap.addMarker(new
MarkerOptions().position(latLngList.get(i)).title(location.getPathname() +
"Remaining battery: " + location.getElectricquantity() + "%"));}
}
 Polyline polyline = aMap.addPolyline(new PolylineOptions().
 addAll(latLngList).width(getResources().getDimensionPixelSize(R.dim
en.dp1)).color(getResources().getColor(R.color.blue_12b7f5)));
```

5 Test

5.1 Registration function test

Open the phone client, enter the registration interface, fill in the registration information, and submit it, and then click back to the login screen.Then enter the MySQL database to see the registration information has been written in the user table and whether the registration information is incorrect.The test results are shown in figure 6.Registered function after the test is available.

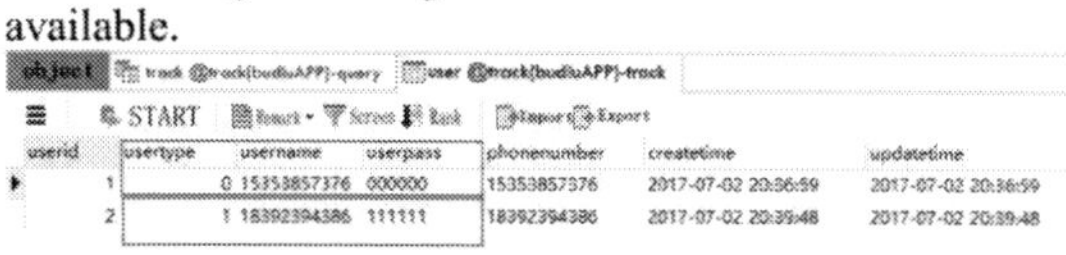

userid	usertype	username	userpass	phonenumber	createtime	updatetime
1	0	15353857376	000000	15353857376	2017-07-02 20:36:59	2017-07-02 20:36:59
2	1	18392394386	111111	18392394386	2017-07-02 20:39:48	2017-07-02 20:39:48

Figure 6. Registration function test

5.2 Login function test

If enter the wrong username or password, the system cannot login,as shown in figure 7(a).If enter the correct username and password, login successfully and jump to the location map interface,as shown in figure 7(b).After testing, the login function can be used normally.

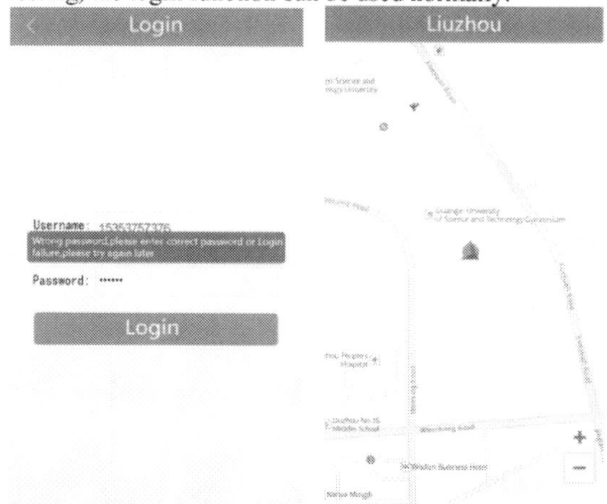

(a)Login failure (b) Login successful

Figure 7. Login function test

5.3 Location function test

Login the sender account, open the MySQL database, and see the latitude and longitude data is successfully written or not.As shown in figure 8,the real-time latitude and longitude data have been written into the database,the location function is normal.

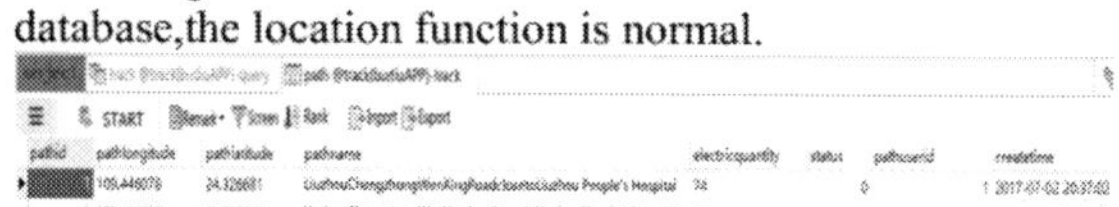

Figure 8. Location function test

5.4 Attention function test

The receiver can be attention/cancel attention to the sender, and can view the track of the sender at the selected time period.Login to the receiver account, click the "Attention" button in the upper right corner of the interface to enter the phone number who needs attention.If enter the wrong information or the phone was not registered,will receive a system prompt.Only when you enter the correct phone number, it will shows "success!",as shown in figure 9(a).Click "My attention" to list the phone number that has been concerned,click on the corresponding phone number and select the start and end times that you want to view,then user's trajectory will appear,as shown in figure 9(b).If you need to cancel attention, touch and hold the phone number, as shown in figure 9(c). After testing, the attention function can be used normally.

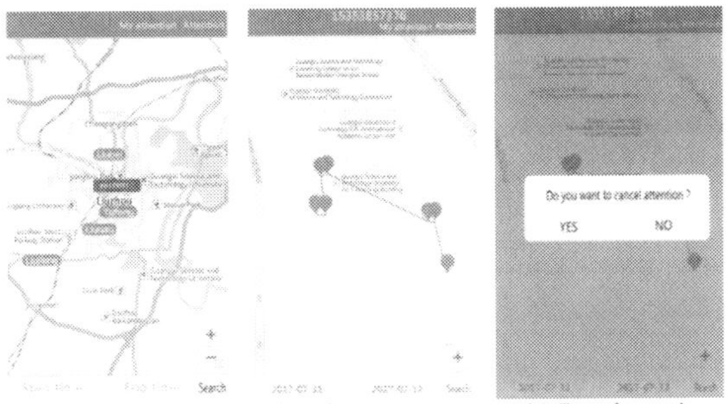

(a) Attention successful (b) Select time view (c) Cancel attention

Figure 9. Attention function test

5.5 Lost prompt function test

Shut off the sender phone APP for 24 hours,that is,24 hours did not upload location data.View on the receiver,the system prompts the user is in a lost state,and show the user's last occurrence of the location and the remaining battery power,as shown in Figure 10.After testing, Lost prompt function can be used normally.

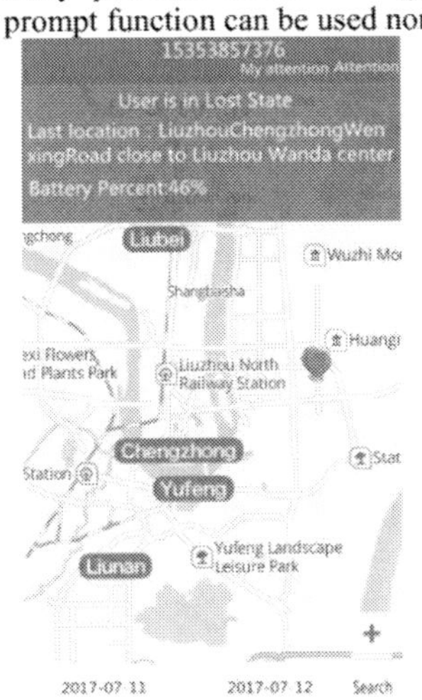

Figure 10. Lost prompt function test

6 Epilogue

This paper presents a model of the inbound record system, which is embedded into mobile devices based on the Android platform.Achieve a location function of the mobile terminal through the Auto Navi Map open platform.Using Apache Tomcat server and MySQL database to build service terminal.System realized many functions such as the sender data preservation ,the receiver attention to the sender,data calling and view the sender's track. It can be used to monitor the trace of the family and friends,designed to protect children, the elderly and other vulnerable groups.Through the test,all functions can be used normally. Compared with the PC terminal, it is not only more flexible,convenient and fast,but also more real-time and high efficiency.

Acknowledgments

This work was supported by the Guangxi province Natural Science Foundation of China Grant No. 2014GXNSFAA118392 and Educational Commission of Guangxi province Grant No. YB2014209.

References

1. Liu Dan.Method of voice locatable image acquisition at mobile terminal based on Android platform.Journal of Modern Electronics Technique, 2016,39:53-55.
2. Liu Xiaobing.The design and implementation of APP intelligent tracing and guardian base on Android platform.Nanjing, Southeast University,2016.
3. Zhang Yangyang.Location-based Services System Design and Prototype Implementation under the Environment of the Mobile Internet.Xi'an,Chang'an University,2015.
4. Li Dong,Guo Shi-ming.Research on the Current Status of LBS Service of Mobile Libraries.Journal of INFORMATION SCIENCE,2014,10:80-85.
5. Wen Yu.Software architecture design programmers need to transition to the architect.Beijing,Publishing House of Electronics Industry,2012.
6. Zhu Rui,Che MIN.Analysis of server program based on HTTP protocol.Journal of Modern Electronics Technique,2014, 35:117-119.

How to Improve the Quality of Teaching in the Course of Network Education

Na Qiao[1] ,Haoyan Shao[1] ,Xiaohui Zhang[2]

[1]Dalian University of Technology, Dalian,116023,PR China
[2]School of Computer and Information Technology, Dalian,116023,PR China

Abstract. This paper expounds the characteristics and advantages of modern distance education. On this basis, this paper puts forward the limitations and deficiencies of the current educational practice, and analyzes the reasons that affect the quality of education. On the basis of analyzing the reasons, solutions to the problems in the course of work are put forward according to working experience and practice. The realization of the virtual experiment will ease difficulties and pressures effectively, which the various modern distance education in colleges and universities generally faced in funds, venues and equipment and other aspects. So carrying out the virtual experimental teaching is timeless and beyond the constraint of space, and help to improve the quality of teaching. Therefore, the virtual experiment will become an inevitable trend in the practice teaching of modern distance education.

With the development of modern information technology, modern distance education is a new form of education, which is the main means to build a lifelong learning system in the era of knowledge economy. It can exploit the advantages of various educational resources to the full, and provide strong support for improving the quality of education, and provide a convenient, fast and wide range of educational services to different people. With the advent of the information age, the era of knowledge economy, the rapid development of science and technology, and emerging industries continue to emerge and develop, people must continue to learn, lifelong learning. Therefore, modern distance education is responsible for improving the quality of the whole people and promoting the development of economy and society. In the new situation, we must repositioning of the modern distance education, reunderstanding of modern distance education, and reunderstanding of the practice of modern distance education, to explore the teaching mode of modern remote education.

1 Characteristics and advantages of modern distance education

1.1 Maximizing the use of resources

A variety of educational resources across the space constraints through the network, so that the school education can become more open to the wider area of radiation than the open education. Schools can give full play to their own strengths and advantages of educational resources, the best teachers, and spread the best teaching results through the network to all directions.

1.2 Autonomous learning behavior

The application of network technology to distance education is characterized by: any person, any time, any place, starting from any chapter, learning any course. Network education is convenient and flexible, "any of the five", is the most direct manifestation of the active learning characteristics, and meet the needs of modern education and lifelong education fully.

1.3 Interactive learning

It can carry out a full range of communication through the network, narrow the psychological distance ,and increase the exchange of opportunities and scope between teachers and students, or students and students. We can carry on the statistical analysis about question type, the number, the frequency through the computer to the student, cause the teacher to understand the students' questions, the difficulty and the main question in the study , more targeted to instruct the student.

1.4 Personalized teaching

In distance education, we can apply the information database management technology and two-way interactive function in computer network lock. On the one hand, tracking system for each network, the character information, learning process and stage can achieve complete, on the other hand, the system of teaching and learning service system can propose personalized learning suggestions according to the records of personal data and different students.

* Corresponding author: network5@dlut.edu.cn

1.5 Teaching management automation

The computer network teaching management platform has the function of automatic management and remote interactive processing.
Remote students' counseling, registration, payment, course selection, query, school management, operation and examination management, etc., can be completed through the network remote interaction.

2 The modern distance education in the lack of time and analysis of reasons

Practice is the essence of modern educational thought, and it is widely used in various forms of education, especially for the cultivation of applied talents in vocational education and adult education. Practice is the basic way to carry out the practical education thought, to cultivate the learning ability, practical ability and innovation ability, which plays an important role in the teaching process. But there are still many problems in the practice of education, the existence of such problems, makes a lot of problems still exist in current education practice link, the existence of such problems, the current practice of distance education link requirements vary greatly, the task is far from complete, which seriously restrict the improvement of the quality of education. In general, the impact of the practice of the reasons for the completion of the following points:

2.1 Distance education object is mainly the general public on the job

Now the social conditions, due to various social reasons, many people study is still relatively backward to stay in to get a diploma, lack of awareness of the importance of various abilities, training does not attach importance to the ability of learning is the exam, only by rote. As a result, we do not attach importance to the practice of training ability does not , do not participate in the practice and implementation of the impact of the implementation of the effect.

2.2 Difficult to organize the school

First, the distance education is a joint form of education, a variety of forces to form a joint force more difficult. In addition, there are still some problems in running a school, such as: College away from students, etc. Austrian education provides only a learning platform, its strength is relatively weak..

2.3 Practice is difficult

Students do not have too much time, learning center staff is less, lack of manpower, no base, lack of material conditions. The work is not mature and fixed mode, cannot be quantified, unable to assess and manage.

2.4 Concept problem

The social and educational groups are not aware of the importance of the practice education, have not realized the important influence of the link to the teaching effect, and have not enough understanding of the importance of improving the quality of teaching. So there is no need, no pressure, no confidence and courage, no initiative to overcome the difficulties of enthusiasm.

3 The solution of modern distance education

The practice of engineering education is indispensable part of the teaching process, but also a weak link in the development of modern distance education, now, it has become a constraint weak link in the development of distance education, which has become the bottleneck of current distance education engineering professional development. In the modern distance education practice teaching, in addition to the traditional practice teaching mode must be used, it should give full play to the advantages of modern distance education technology network advantages and modern management advantages, combined with the characteristics of education objects, continuous innovation and appropriate new practical teaching mode. In modern distance education, the feasibility of small students personally from universities in the real laboratory experiments, so the creation of online virtual experiment course has become an inevitable trend to carry out the experiment teaching in modern distance education..
The development and application of the virtual laboratory on the Internet has played an important role in making up the resources and enhancing the teaching effect. After several years of exploration and practice, the school of continuing education of Dalian University of Technology has introduced a virtual experiment model to solve the problem of difficult to realize in practice. Virtual laboratory can save the classroom practice of virtual experiment, practice less funding, less material, easy preparation, high safety, can be more intuitive to let students understand the knowledge, enhance the perceptual knowledge, can be very good to attract the attention of students, so as to achieve better teaching effect; the vivid image, can be flexible, can effectively improve the students' interest in learning, which is helpful for students to understand and master the knowledge; virtual experiment teaching can be closely linked, is conducive to the breakthrough teaching difficulty, help to validate the concepts and knowledge, is not easy to understand in the textbook problem; in addition, the virtual experiment for teacher edge demonstration while explaining, to guide the students to grasp the key, break through the difficulty, consolidation points.
In the practice of the introduction of virtual experiments, for civil engineering, road and bridge engineering, water conservancy projects, as well as marine and marine engineering, such as the implementation of the customer

DOI: 10.1051/matecconf/201713900108

based professional practice teaching is of great significance.

In this paper, a virtual experiment of soil mechanics experiment base on network environment is used to illustrate the important role of virtual laboratory in modern education. It is difficult for students to remember the experimental conditions and precautions only from the learning materials, and it is difficult to form a complete and clear process in mind, theory and practice very hard in the process of learning is very good with. In view of this reason, the real environment can be recorded in the experimental or operational process into a video, so that students can understand the specific operation of the experiment by watching the video to complete the experimental report. Figure 1~4 is a fragment of the experiment.

Fig. 1. Experimental instrumentations of soil mechanics experiment

Fig. 2. Sketch of soil compression curve

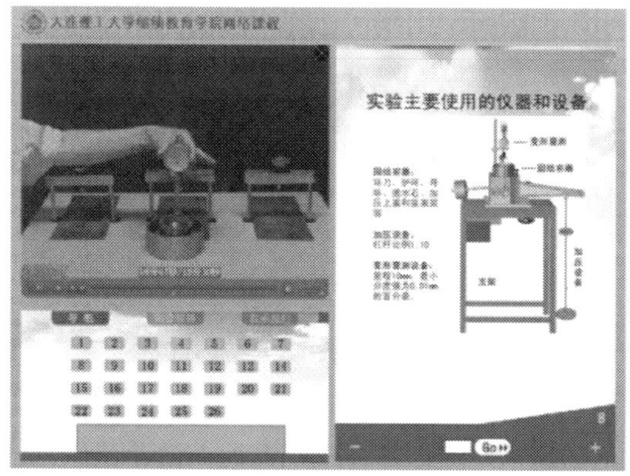

Fig. 3. Visualizations of experimental instrumentations

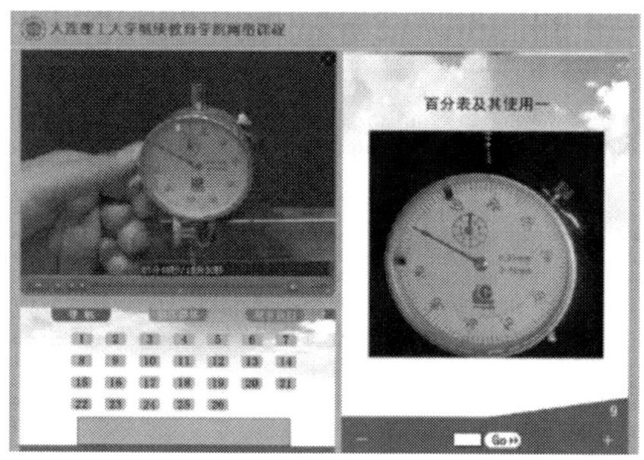

Fig. 4. Example of percentmeter

4 Conclusion

Based on the characteristics of modern distance education, this paper puts forward the problems in practice, and puts forward the method of making use of virtual experiment on the Internet to make up for the defects of the link. The practice and experiment combination can make full use of distance education multimedia experimental video resources, low cost, short production cycle, although unable to achieve the same results and real experiments, but the learners can still pass the video has a profound perceptual knowledge of experimental course, it can greatly improve the network education the effect has become indispensable tools in modern distance education mode.

References

1. Shuqin Lou, Xinzhi Sheng. Design and development of Digital electronic technology foundation base on network course [J]. Chinese distance education, 2002(3). (In Chinese)
2. Qibiao Deng. Advantages and characteristics of modern distance education [J]. Journal of Fujian Radio & TV University, 2002(4). (In Chinese)
3. Xiaoliang Zhao, Xinfeng Cui. Analysis of the concept、characteristics and development trends of modern distance education [J]. Techno-economic market, 2010(4). (In Chinese)
4. Hong Wei,Wei Hui. Construction and Practice of Distance Education in Engineering Practice Instructional Modes [J]. Technology and Education, 2010(6). (In Chinese)

Research on the Influence of Social Media Information on Consumers' Purchase Intention

Yanqun Qin

School of Management, Shanghai University, No.99 Road, China

Abstract. In addition to the channels of information, the information itself will also affect the intention of consumers to purchase. This paper extends para-social interaction (PSI) theory to the social media environment, and studies the influence of the information itself through it. Through the data analysis, we found that openness and interactivity of information have a positive impact on the purchase intention of consumers, but interactivity of information has a direct negative impact on consumers' purchase intention. The result has great significance on the theory and strategy of marketing.

1 Introduction

Advancement and innovation of technology promotes the continuous development of Web 2.0 technology, promoting the establishment and improvement of social network, making social media arises at the historic moment. The way that people communicate, collaborate, and linked to others also changed, and more and more marketers, media professionals, scholars and enterprises have gradually realized that social media have a major role on consumers. Social media allows us to liberate from the traditional books, newspapers, television, radio and outdoor billboards and other "push media", of course, this does not mean "push media" has been nothing, but their impact is gradually reduced . Social media can be seen as a 'pull media', which makes direct interaction possible. Companies and consumers can communicate or share information with each other, so there is a higher autonomy and relevance. This paper introduces the theory of para-social interaction in the social media environment, to carry out the impact of social media information on consumer purchase intention. Study will extend the theory of para-social interaction from traditional mass media to social media, widening the scope of the theory, providing new perspectives for the social media marketing theory. Through the study of social media information to identify the positive factors on consumer purchase intention, and that factors how to affect consumers' attitudes and behaviors, to put forward feasible proposals for social media marketing strategy, which has great significance on theory and application, based on the relevant theories and models, as well as its marketing strengths, weaknesses and prospects for developments. Basis of relevant theories

2 Basis of relevant theories

Social media in the "What is Social Media" is considered to give the user a great space to participate in the new online media. It is a medium of exchange of information, ideas and opinions between people, has a very high participation degree, high interaction, and breaks the media by changing the traditional physical media internet to personal media internet and entirely dividing the audiences and communicators, so that audiences and media became a truly organic integration. In this context, social media is translated from "Social Media". Although social media is a new concept in our country, appearing in recent years, but more and more experts and scholars recognize its important role. Currently, social media is to occupy the mainstream position in the Internet .Alexa's statistics show that half of the top ten large sites are social media sites.

2.1 Para-social Interaction Theory

Para-social interaction theory has different translations in Chinese. It was proposed firstly by American psychologist Horton and Whol in the article "Psychiatry magazine" in 1956. Horton and Whol defined para-social interaction as a unilateral relationship that is similar to the relationship between the media personalities and the audience in the real world, and it is not easy to develop bilaterally. Horton and Wohl believes that media figures have initiatives on the development of this relationship, they can use their own image or remarks to influence the audience perception of their relationship. But for the development of this relationship, the audience is relatively passive, they can only continue or stop watch a program to control the relationship between the media personalities.

Para-social interaction theory has the following three characteristics: unidirectional, indirect, similarity. Unidirectional means that media personalities always release information, audiences can only receive information from beginning to end; indirect means the occurrence of para-social interaction is based on various media as relations intermediary. These intermediaries may be a conventional television, film, radio, telephone but can also be new media, such as micro-channel, microblogging, forums, communities, and so on; similarity means para-social interaction relationship and real relationships are very similar. Their roots are nature of reliance on others, and they look forward to gain certain satisfy from others.

2.2 Six Degrees of Separation Theory

Six degrees of separation theory was firstly proposed by the American social psychologist Milgram in the 1960s, "The interval between you and anyone who is a stranger will not more than six, that is, you will be able to recognize any up to six individual."

Microsoft researchers have applied this theory to social media, Jurc Lskovcc and Eric Horvitz randomly selected messages of a month on Microsoft's MSN. By filtrating and comparing more than 300 million communications information form 240 million MSN users, the results largely proved the theory. 48% users can associate with each other less than 6 times, and 78% of users can associate with each other less than 7 times. According to the theory of six degrees of separation, interrelated relationship network of social media audience ultimately form a large social relationship network. In this network, any point may associate with strange point, if the firm release information at any point, the information would have the possibility to be passed to every consumer.

2.3 Strong Ties and Weak Ties Theory

American sociologist Granovetter believes that the Interpersonal networks can be classified as two networks, strong ties and weak ties. Strong relationship refers to the Interpersonal network has more homogeneities and more closely relations, and this close relationship mainly sustains by emotional factors. Weak relationship refers to the Interpersonal network has greater qualitative differences. In weak ties, there are extensive Individuals, so it's impossible to have an close relationship and obviously emotional maintain.

3 Study decign

3.1 Openness of Information

In para-social interaction theory, the para-social interaction is similar to the friendship between friends. Message' senders send relevant information to users to make users establishing a sense of intimacy and trust on senders. In this article, the para-social interaction is that senders abandon her / his / its originally remote or virtual character for users, to disclose information about themselves to the audience. If users feel they obtained information about the senders of the various aspects, along with it of para-social interaction emotion was established. During the course of para-social interaction, users are eager to understand all aspects of the message and private information of senders, as though it were the real world relationships. In communication, the key to establish para-social interaction emotion is to send out the suitable message, such as openness of information: the information is timely issuance and can be easily access and use; information including all details users want to know; information may relate to intimate details of the sender. Therefore, we propose the following hypothesis:

H1a: openness of information is positively correlated with para-social interaction emotion.

3.2 Interaction of Information

Interactive network environment are seen as an important feature in a lot of literature, but there is not a consensus definition. Some researchers define interaction based on technical factors, such as the ability to provide feedback, the speed of website. In this paper, interaction is only related with the information itself. Message that make users feel sender is listening and responding, that is, users realized that he is communicate with sender, is seen as interaction of information in this article. Therefore, we propose the following hypothesis:

H1b: interaction of information is positively correlated with para-social interaction emotion.

3.3 Consumers' Purchase Intention

Schiffma and Kanuk consider consumers' willingness is a measure of the possibility of buying a product. Zhixian Zhu noted that consumers' willingness is to buy certain commodity suiting their needs, is a prelude to buying behavior and reflects consumers' psychology. Consumers' willingness is an important mental activity before purchase decision, directly affecting the final buying behavior to a large extent. Previous studies demonstrated that, in general, formed consumers' can ultimately translate into purchasing behavior. Therefore, we propose the following hypothesis:

H1c: para-social interaction emotion is positively correlated with Consumers' Purchase Intention.

3.4 Para-social Interaction Emotion

Prior to the establishment of real-life relationships between the sender and the user, relationships between the two is para-social interaction, recipients establish intimate feelings and trust on sender is para-social interaction emotion. Para-social interaction emotion and real relationships are similar. The current studies of para-social interaction support this view. For example, in a variety of social media, when the sender send information through the microblogging, Renren, WeChat and other social networking platform to share their experiences, emotions and feelings, it's the same as that in real life. Interested in sent information lead to the generation of para-social interaction emotion; and then those who with para-social interaction emotion would like to prove a close relationship with the sender by focusing on information, increasing participation, buying or recommending related products and so on. Details information and constant feedback improve the persuasiveness and credibility of the information, it can also increase the trust of the audience, and directly impact attitudes and behavior of the audience. Therefore, we propose the following hypothesis:

H2a: Openness of information is positively correlated with Consumers' Purchase Intention.

H2b: Interaction of information is positively correlated with Consumers' Purchase Intention.

3.5 Construction of the Model

This paper discusses the impact of social media information on Consumers' Purchase Intention, and the ultimate aim is to identify factors affecting consumer purchase intention. All Things Considered, there were many factors, but in this article, we limit the choice of content in social media information itself. So the openness and interaction of information described above are two very important factors. Based para-social interaction theory, the model constructed in this paper is shown here.

Figure 1. Model

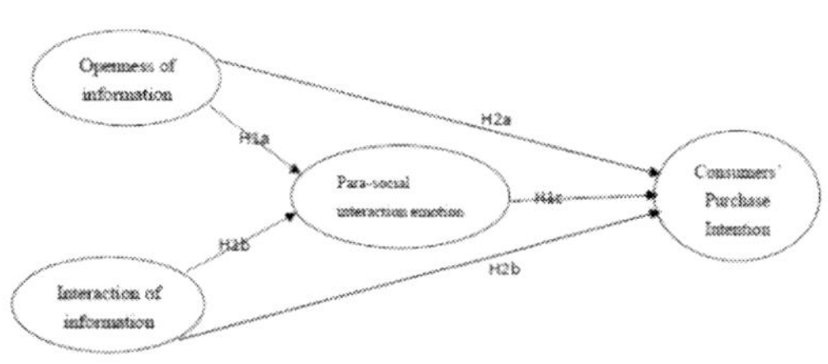

Before testees fill out the questionnaire, testees will be asked whether to surf the Internet, whether to use the social platforms. Those whose answer is no don't participate in questionnaires.

4 Hypothesis testing analysis

Table 1. the correlation coefficient and its significance of openness of information and consumer's purchase intention

Items		openness of information	para-social interaction emotion	Consumers' Purchase Intention
openness of information	Correlation	1	—	—
	Significant	—	—	—
para-social interaction emotion	Correlation	0.370(**)	1	—
	Significant	0.000	—	—
Consumers' Purchase Intention	Correlation	0.120(**)	0.790(**)	1
	Significant	0.000	0.000	—

** : Significant correlation at the 0.01 level (bilateral)

We can see from table 4.5, the correlation coefficient of openness of information and para-social interaction emotion is 0.370, and its correlation is positive significant, the data show openness of information has a greater positive on para-social interaction emotion. The correlation coefficient of openness of information and Consumers' Purchase Intention is 0.120, the data show openness of information has a positive impact on Consumers' Purchase Intention, and the impact is significant. The correlation para-social interaction emotion and Consumers' Purchase Intention is up to 0.790, the data show para-social interaction emotion has a greater positive on Consumers' Purchase Intention.

Table 2. the correlation coefficient and its significance of interaction of information and consumer's purchase intention

Items		interaction of information	para-social interaction emotion	Consumers' Purchase Intention
interaction of information	Correlation	1	—	—
	Significant	—	—	—
para-social interaction emotion	Correlation	0.430(**)	1	—
	Significant	0.000	—	—
Consumers' Purchase Intention	Correlation	-0.070	0.790(**)	1
	Significant	0.057	0.000	—

** : Significant correlation at the 0.01 level (bilateral)

We can see from table 4.6, the correlation coefficient of interaction of information and para-social interaction emotion is 0.430, and its correlation is positive significant, the data show interaction of information has a greater positive on para-social interaction emotion. The correlation coefficient of interaction of information and Consumers' Purchase Intention is -0.070, the data show interaction of information has a negative impact on Consumers' Purchase Intention, and the impact is not significant.

Based on survey data analysis, openness and interaction of information has a significant positive correlation on para-social interaction emotion, analysis results also show that para-social interaction emotion was positively related to Consumers' Purchase Intention. Three hypothesis of para-social interaction: H1a, H1b and H1c have been verified.

Openness and interaction of Information has direct influence on Consumers' Purchase Intention, and there is indirect impact through para-social interaction emotion as an intervening variable. Openness of information' direct impact on Consumers' Purchase Intention is path .12. Openness of Information' direct influence on Consumers' Purchase Intention isn't positive correlation, it's path -.07. Based on the above data, H2a has been verified, but the analysis of the data do not support H2b.

From the above data analysis, we can see that social media information has a significant positive impact on Consumers' Purchase Intention, especially the form of para-social interaction emotion has a very important impact on Consumers' Purchase Intention. Because the indirect effects of openness and interaction of information on Consumers' Purchase Intention is significantly greater than the direct impact, and information's indirect effect Consumers' Purchase Intention use para-social interaction emotion as an intervening variable. In the current era of information explosion, simple information issuance is hard to move consumers, so that only by making consumers generate para-social interaction emotion, making them think they are also a part of it, focusing on information and even providing information, can it bring long-term positive impact for product or company. In this study, the key to generate para-social interaction emotion, the most important is to issue what audience interested in and effective social media information. Openness and interaction of information are worthy of our attention.

Mentioned above, between any two strangers are not spaced more than six, so we can issue information to a wider audience and broader coverage. But in the current era of Web.2.0, social media platform provides us with a convenient interrelated platform, we can regard these platforms and the audience of the platforms as space between strangers. And fifth question in questionnaire, testees use which platform pay an attention to brands also shows that social media platform is a very important factor. The relationship of these platforms are not black or white, but may produce a synergistic effect. Most marketers are concerned about how to let more audience access to information, compared to choose a single platform to find a wider audience, selecting a number of social media platforms to send information is more convenient, and the effect will be more noteworthy. No matter how big the base of a platform, it's less than superposition of many platforms. In addition, there is not a social media platforms whose audiences that is far more than other platforms.

Indirect effects of openness and interaction of information on Consumers' Purchase Intention was significantly greater than the direct impact. Its indirect impact use para-social interaction emotion as an intervening variable, and their relationship based on the emotion. So the indirect effects of openness and interaction of information on c Consumers' Purchase Intention can be seen as strong relationship, and a direct impact of openness and interaction of information on Consumers' Purchase Intention can be seen as weak relationship. Thus, the establishment of para-social interaction emotion is especially important.

In summary, the marketers can use appropriate para-social interaction theory, integrating strong ties, weak ties theory, six degrees of separation theory and so on, issuing effective information in social media platforms as many as possible, such as information with openness and interaction can generate a positive impact on Consumers' Purchase Intention to a large extent. In addition to the concern about social media information, social platforms can't be ignored.

5 Conclusions

We can see form the results of this study that para-social interaction has a positive impact on Consumers' Purchase Intention. There two dependent variable and independent variable in above study model. Analysis of research data verify the most of hypothesis: First, openness and interaction of information is positively correlated to para-social interaction emotion, and para-social interaction emotion is also positively correlated to Consumers' Purchase Intention. Second, openness of information is positively correlated para-social interaction emotion, but interaction of information is negatively correlated para-social interaction emotion. Third, openness of information and Consumers' Purchase Intention may be directly positive relationship, it can also be a positive correlation by using para-social interaction emotion as an intervening variable. While positive relationship between interaction of information and Consumers' Purchase Intention may be directly positive relationship can only use para-social interaction emotion as an intervening variable. In general, para-social interaction emotion is helpful for the form of Consumers' Purchase Intention. In the social media environment, social media platforms are convenient available resources for marketers, and marketers can design information content itself and the way to issue to generate para-social interaction emotion.

Results of this study has a positive meaning on marketing in social media environment, but there are still some deficiencies and limitations during the course of the study. This article did not study the relationship between openness of information and interaction of information. For example, the highly interactive nature of information may involve openness of information, in addition to help enhance the interactivity level, may leading to improve openness. Mutual influence between openness of Information and interaction of Information also worth our inquiry. The accuracy of the data is one of the key factors in the study, so future research will pay more attention on the accuracy of the data. Besides using questionnaire, future research might select data already on the social platforms directly to collected real data under non-experimental conditions.

References

1. Antony Mayfield, What is Social Media, Spannerworks, vol.1, pp.3-6, 2007.

2. Toni Ahlqvist, Asta B ck, and Minna Halonen, Social Media Roadmaps Exploring the futures triggered by social media, VTT research notes, vol.23, pp. 5-15, 2008.

3. Horton Donald, R Richard Wohl, Mass Communication and Para-social Interraction: Observations on Intimacy at a Distance, Psychiatry, vol.19, pp.215-229, 1956.

4. Schiffma L G, Kanuk L F.Consumer behavior,Prentice Hall, vol.7, pp.188~190, 2000

5. Zhao, Xinshu, John G Lynch Jr, and Qime Chen, Reconsidering Baron and Kenny: Myths and Truths About Mediation Analysis, Journal of Consumer Research, vol.37:2, pp.197-206, 2010.

6. Jinpin Ge, Research on the New Progress of Para-social Interaction in the West, Journal of Zhejiang University of Media and Communications, vol.20:1, pp.96-100, 2013

Research on Construction of Road Network Database Based on Video Retrieval Technology

Fengling Wang[1]

[1]Hezhou University, School of Mathematics and Computer Hezhou Guangxi 542899, China

Abstract. Based on the characteristics of the video database and the basic structure of the video database and several typical video data models, the segmentation-based multi-level data model is used to describe the landscape information video database, the network database model and the road network management database system. Landscape information management system detailed design and implementation of a detailed preparation.

1 Introduction

Video database in the field of distance education, telemedicine, e-commerce, digital library, video surveillance and other fields have a wide range of applications. It has become one of the hot spots in multimedia technology research. As the video database technology theory and technology although there are some video database prototype, but in this area there is no complete theoretical system. In many ways, there is no proven way to mature, and there is no widely used commercial video database software platform. It is still in the laboratory research and promotion stage.

2 Video Database Features

Video data refers to the storage of audio and video information is a very special data, it is far greater than the transmission of static images and text information, in general, video data and text data has the following characteristics:

2.1 Video Data has a High Information Resolution.

The so-called information resolution refers to the number of details provided by the media, such as the description of the crime scene of the video data, we can from crime sites, backgrounds, criminals, criminal tools and criminal means as well as other details.

2.2 The relationship between video data is complex, is unstructured data organization.

In the video segment, the time between the video segment on the object relationship, as well as the relationship between the space, in addition, the video data also has a specific application areas.

2.3 The diversity and ambiguity of video data interpretation.

It does not have a complete objective explanation of the character numerical data, and there are often personal subjective factors. Because of the ambiguity of video data, when we conduct content-based queries, we can not query specific records as traditional database searches, so we can only query similarity [1].

3 The Basic Structure of the Video Database

According to the characteristics of the video database, in order to achieve the insertion and retrieval of video data, the structure of the video database has its own characteristics. However, there is no standard video database architecture, the existing design is mainly for specific applications, the following video database structure to do a simple introduction [2].

3.1 The hierarchical structure of the video database

The first layer of the video database is the media support layer, which is the main video segmentation and identifies the original information and other operations. Segmentation means that the video stream is divided into meaningful cells and is identified by identifying the video description information. The second layer is called the data model, which mainly completes the logical

* Corresponding author: wf1232983@163.com

storage of the physical storage of the video data to the mapping, which is defined in the logical content definition, as well as the data and data relationships determined by the data model. This layer is actually an implementation of the conceptual model of the database. The third layer is the video user interface layer, which completes the video query description and query results on the user information.

3.2 The basic structure of video database system

The establishment, use and maintenance of the video database requires the support of the video data management system. Video data management system based on the operating system, unified management and control database, the user through the video data management system to insert and retrieve database commands, such as implementation. At present, the video data management system is not a unified structure, which is due to the current video database management does not exist a complete theoretical system, technically immature, the researchers are still in the exploratory stage, the proposed video data management system is also often for a specific category s application.

The video data management system consists of the following components.

3.2.1 Video Database

The video database (VDB) is a physical storage of video data that is often stored on VDB in compressed digital video. Data storage format and system used by the file system. VDB should use an efficient, easy-to-access structure that makes it easy to provide video material.

3.2.2 Database System

The video data management system should have a dedicated database system (DBS) that stores a target pattern for video management, which may be used to associate the logical location of the data with the physical location. The structure of the database system and the application of the system and query type support [4].

3.2.3 Query Processing Unit

The basic function of the unit is to enable the user to query the video data in a formatted manner. The query formatted by this format can be converted to the target mode in the database system. In the actual video database system, the mode usually describes the video. The attribute's attribute representation matches the existing pattern in the system to achieve the physical location of the query video. The user's query request can be represented using a specially designed video data query language, and sometimes an example of a query can also be used. Obviously, the unit is an important module for the user to retrieve video data.

3.2.4 Insert the Module

Insert module is one of the important modules of video data management system. Use this module to insert the original video data into the video database. Due to the large amount of video data, the module can be automatically or semi-automatic insertion, especially to improve efficiency. The main function of the module is to segment the video stream and extract the video from the target model stored in the database system. The target model is implemented from manual extraction to fully automated [5].

3.2.5 User Interface

The user interface in the video data management system has an important role in the user interface, so that the user can easily enter the formatted query request, while the query results visual visual expression. The user interface can also be used to complete the interactive video query task.

4 Typical Video Data Model

4.1 Data Model Based on Video Segmentation

This data model is presented by Dr. Hampapur. By studying the application and generation of different video data, the model can be used as the basis of video data model design. The data model mainly divides the video data through two segmentation criteria: grammar segmentation criteria and semantic segmentation criteria. The video segment is defined by the start frame tb and the end frame te and has n different characteristics (F1, F2, ... Fn), and one feature type (W1, W2, ... Wn) for each feature. The feature is based on the principle of content dependency, time expansibility and label requirements [6].

This model introduces elements such as time relationships, images and video features, which can more fully reflect the temporal and spatial relationships and characteristics of video data. It has great reference value for video editing and video database, but this model does not give complete video data The presentation and description of the entity and its relationship can only be used as the main representation of the video data model.

4.2 Data model based on hierarchical identification

Due to the lack of flexibility in the data segmentation based on video segmentation, Swanberg et al. Proposed hierarchical model. The bottom of the model is a lens, and the content-based lens describes the characteristics of the lens (e.g., image, sound, etc.) taken from the lens itself. Once the appropriate algorithm has been determined, and because of the characteristics from the video itself, it has a more complete representation. Depending on the temporal and spatial relevance of the different shots, the scene can be constructed semantically and multiple scenes with internal relationships are constructed, and the associated sequence forms a composite unit. This pyramid-based video data model has a reference value for the application of data from small to coarse requirements.

4.3 Time description class model

An important feature of video data is the two-dimensional structure of time and space, the time dimension model [7]. Breiteneder proposed a time-based model in which there are four types of entities: a cinema entity representing the structure of video information composed of multiple tracks; and a channel entity from the media entity. In addition, each channel is derived from the media. In addition, the model allows the information to be grouped into a set of intermediate entities. The entity layer supports channel combining and reuse. The advantage of the timeline model is that the video user provides a very clear and straightforward expression, from the timeline can clearly see the lens partition, play time. The disadvantage of the timeline model is the use of absolute time coordinates, thus increasing the complexity of video editing. This model can not be considered strictly in the form of a complete video data model that can only be viewed vividly in the main form of video data time series and relationships. This model does not introduce advanced abstractions, contains very few semantics, does not take into account the logical structure of video and video between hierarchies, nor does it include the operation of the video operating mechanism, can not share video material and generate new video streams.

4.4 Algebraic video data model

Duda presents an algebraic video model that introduces the hierarchical relationship between video segments and video algebra operations and generates new video streams from the original video segment through video algebra operations. The algebraic data model is a nestable hierarchical structure whose basic elements are algebraic video nodes and some operations contained in nodes. The algebraic video node contains the name of the child node, which may be an algebraic video node or an original video segment. The original video segment can be represented by the interval between the original video name and the original video. The algebraic video node also contains descriptive information about its contents and a combination and presentation information describing the playback characteristics of all its child nodes. Algebraic video model is characterized by video synthesis, reuse, organization, search and browse the basic functions. Video algebra can be used to simulate the complex nested logic of video. Video algebra is the core of this model, the contents of the video, the logical structure and the spatial composition of the access; the basic function of the algebraic video model in the algebraic operation based on the model of the level of unity; model node carrying video synthesis description, representation, query and browse Information, and represent the hierarchical and nested structure of the video at the model level [8].

It can be seen from the above characteristics that the algebraic video data model not only includes the segmentation and temporal relationship of video data, but also provides feature information describing the content of the video segment, supports content-based retrieval, and defines the video performance characteristics, which is a more perfect model.

4.5 Universal video data frame model

Hjelsvold proposed a general video data frame model, through the use of physical connection diagram to establish a clearer conceptual model. It provides core concepts and modules. Video segmentation is used to define the structure of a video file. Level abstract structure, support scene, sequence and composite unit of multi-level abstraction. Object - oriented technology for model building. Each video object has a unique object identifier. In the model, the concept of video data context is introduced. Through the context, the original video can be synthesized as a new video, and can generate new semantics. The video query algebra is defined in the model, and the video data can be easily manipulated [9].

The starting point of the generic video data frame model is to try to build a generic video data model. VideoSTAR is based on this model of video storage and retrieval experiment system. One of the shortcomings of the model is that the structure is complex and its use is limited. In addition, the model only proposes a framework that is not discussed in terms of encapsulation, inheritance, sharing, and index generation.

5 Design of Landscape Information Database Model

Based on the analysis of several typical video data models, the segmented multi-level data model is used to describe the landscape information video database according to the characteristics of video data.

This is because in the video stream there are two inherent abstractions: the entire video stream and a single frame. For most applications, the entire video stream is too rough for the abstraction layer; on the other hand, a single frame is rarely a unit of interest because even a short video has a large number of frames. Naturally presenting other intermediate levels, such as a lens, a scene, a sequence, a composite unit, and the like. Hjelsvold and Midtstraum define the scene as a collection of space and time-related scenes, with the specified meaning of the scene composition sequence, the relevant sequence of the composite unit. Because different users access different levels of video information, we use a segmented multi-level data model to describe the video database. In order to facilitate the realization of the original video stream through the lens split into the lens as the basic unit of the video stream, these basic units stored in the media to form a storage media library. Plus the function and annotation to form the lens library. With the lens library, you can create a video library as needed to form a user view. The video document may be a scene, a sequence, a composite unit, or the like. Its purpose is to achieve the independence of video data, in order to facilitate the sharing of different users of video data and reuse [10]. Figure 3 shows the organization of the video database.

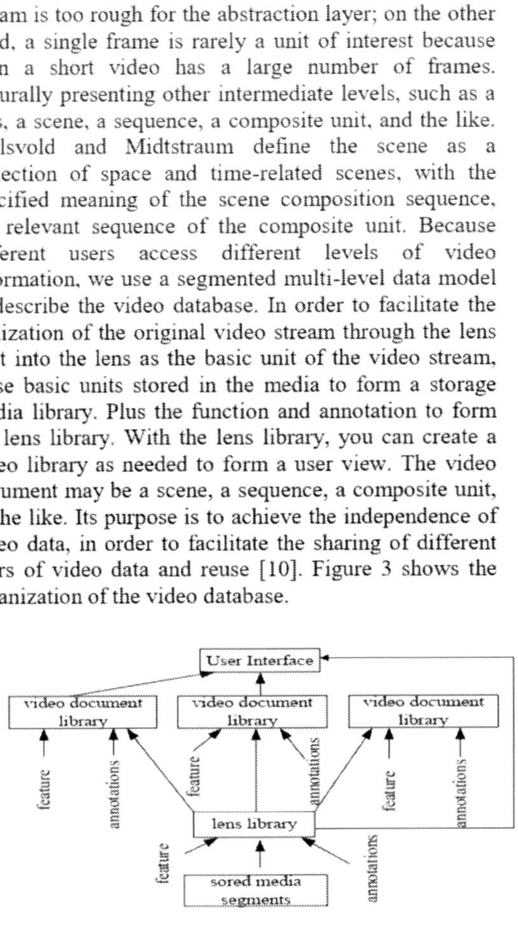

Fig 1. The organization of the video database

The lens library and video library can use the following data model:

V: video interval: (oid, tb, te)

Features: (F1, F2, ..., Fn)

Note: (A1, A2, ..., An)

Here, oid denotes an interval identifier, and tbte denotes a start frame and a stop frame, respectively. For lenses, this feature is unique to each lens, and the other lenses are not critical. It can be called context-free; also comment on a single shot. For video document libraries, this feature takes into account the relationship between the video intervals, leaving the video interval of the environment, features no longer exist. The note of the video interval depends on the context. This only reflects the time dimension of the video data.

Summary

Due to the complexity of the video data itself and the wide application, it is still a long way to go to build a common model that can be widely used in many fields. However, as the latest technology for database technology development, content-based retrieval of video database systems is only a beginning, but its prospects are brilliant. With the gradual resolution of the problem, the video database system will play an important role in all areas of the information society.

Acknowledgment

The author thanks to the support from Hezhou University 2016 Professor of scientific research start fund project under Grant No.HZUJS201615.

References

1. x.s. Zhou, T. S. Hunag, Exploring the Nature and Variants of Relevance Feedback,in proc. IEEE. Workshop on Content-Based Access of Image and Video Libraries, Hawaii,Dec.2001.
2. Liu w z,Huang J.Multimedia Content Analysis Using Audio and Visual Information.IEEE Signal Processing Magazine,2000,17(6):12-36.
3. Hisashi Miyamori, Shun-ichi lisaku ,video annotation for content-based retrieval using human behavior analysis and domain knowledge, Proceedings of the Fourth IEEE International Conference on Automatic Face and Gesture Recognition 2000.
4. Yu Zhong, Hongji ang Zhang, Anil K. Jain , automatic caption localization in compressed video, IEEE Transactions on Pattern Analysis and Machine Intelligence, Volume 22, Issue 4,April, 2000, pp. 385-392.
5. B. Shahraray, Scene Detection and Content-Based Sampling of Video Sequences, Digital Video Compression: Algorithms and Technologies, Vol. SPIE 2419,February 1995.2-13.
6. Wolf W. Key Frame Selection by Motion Analysis. In: Proc. IEEE Int.Conf. Acoust, Speech, and Signal Proc., 1996
7. Rune Hjelsvold and Roger Roger Midtstrum, Modeling and querying Vedio Data, In Proceedings of the 20th International Conference on Very Large Data Bases, September 1994.
8. Zhu Yingying, Zhou Dongru. A method of extracting key frames from compressed video streams. Computer Engineering and Applications, 2003,18.
9. Meng Qian. Research on video database data model based on content retrieval. Journal of Xuzhou

Normal University (Natural Science Edition), 2003,21 (4).

Analysis and Design of the Logistics System for Rope Manufacturing Plant

Xue Sun[1,2,*], *Chao-Chin* Wu[3], and *Liang-Rui* Chen[2]

[1] College of Urban Rail Transit and Logistics, Beijing Union University, Beijing, China
[2] Department of Electrical Engineering, National Changhua University of Education, Changhua, Taiwan
[3] Department of Computer Science and Information Engineering, National Changhua University of Education, Changhua, Taiwan

Abstract. In order to promote logistics system for manufacturing plant, this paper proposed a new design for the logistics system of a rope manufacturing plant. Through the analysis in the aspects of workshop facility layout, material handling and inventory management, the original logistics system of the plant is optimized. According to the comparison of the simulation results between original and optimized design, the optimized model has the higher productive efficiency. This can provide the references for the other manufacturing plant in analysis and design of the logistics system to improve plant efficiency.

1 Introduction

With the rapid development of technologies, in manufacturing industry original single variety of mass production is transformed into low-volume and multi-species production. The new production method is more adaptable to the faster changes of the market. In the process of manufacturing, it spends more than 90% of the time in the logistics process such as transport, handling, and packaging. However, some manufacturing enterprises especially small and medium enterprises only perform simple analysis by experience of practice, which results in contradiction between the backward enterprise logistics system and improvement of production becoming increasingly prominent. Therefore, making a design of good internal logistics system can not only reduce the production costs, but also help enterprises to improve production efficiency, thereby enhance the competitiveness of enterprises.

This paper takes a certain rope manufacturing plant as the research object. As it is a new plant, the entire logistics systems within the plant have a lot of imperfections. According to the investigation of the plant, there are unreasonable facility layout of the production, not high efficiency of logistics handling, low space utilization and other issues. This research is to analyze and optimize the plant logistics system on workshop facility layout, handling and inventory management. Finally, the rational design of the optimization system is verified by simulation.

2 Factory logistics principle

Factory logistics is from the beginning of raw materials procurement into the plant. After certain manufacturing processes, formulate the semi-finished products, and then assemble them into a certain value of the finished products. Ultimately store the finished products in the warehouse or transport them to the users. In this process, materials have experienced a series of circulation from the procurement into the storage to manufacturing and finished goods shipping. Generally speaking, the factory logistics refers to the flow of goods within the factory.

The main research to factory logistics system is as follows:

(1) Survey data collection

Collect the raw data of the plant in a systematic way, including production operations and related data, to analyze the existing problems of the plant.

(2) Plant facility layout

Facility layout looks at physical allocation of space for economic activities in the plant. The main objective of the facility layout planning is to design effective workflow as to make equipments and workers more productive.

(3) Material handling

Material handling in the plant mainly involves short-distance movement within the confines of a building or between a building and a transportation vehicle [1]. Material handling also affects facility layout to some extent, and the layout can be modified according to the efficiency of the material handling.

(4) Warehouse and inventory management

Warehouse and inventory management in a plant is the practice overseeing and controlling of the ordering, storage and use of components that a manufacturing enterprise uses in the production of the items it sells.

In the following sections, we choose a manufacturing plant called Plant L, which produces ropes for hunting, carrying, lifting and climbing, to analyze and design the plant logistics system in plant facility layout, material handling system and inventory management.

* Corresponding author: sunsnow11@163.com

 DOI: 10.1051/matecconf/201713900111

3 Facility layout of rope manufacturing plant

Plant L is a high-tech manufacturing enterprise which is mainly produces many kinds of ropes. This plant has a better sales performance, that the monthly total sales of various products are generally around one hundred tons. According to the actual investigation, the plant facility layout was almost depended on experiences of managers and there was no systematic analysis, resulting in low logistics efficiency and not high space utilization. To solve the problems, we adopted Systematic Layout Planning (SLP) to analyze and design the plant facility layout. Through systematic analysis of SLP, we got three alternative options. Furthermore we used Analytic Hierarchy Process (AHP) to select the best scheme.

3.1 Overview of SLP

The SLP is a tool used to arrange a workplace in a plant by locating areas with high frequency and logical relationships close to each other [2,3]. There are four levels of detail in plant layout design what are site layout, block layout, detailed layout and workstation layout. Five major elements of a SLP methodology are P (Products), Q (Quantity), R (Routing), S (Service), and T (Timing). The basic steps of SLP are:

- Prepare original data

The preparation of the original data includes clarification of five major elements and division of the plant operating units. This information is the prerequisite and key for system layout design.

- Analyze logistics relationship

Logistics analysis is an important part of the facility layout. According to process flow diagram, some factors about logistics between two departments such as flow costs can be quantified. Such relationships can be described by using REL diagrams.

- Analyze non-logistics relationship

Non-logistics relationship is the key element of facility layout design. The factors of non-logistics are such as noise, safety and pollution, the continuity of the production process, and convenience of material handling. The non-logistics relationship between two departments is also quantified by using REL diagrams.

- Analyze comprehensive correlation and draw related table

In the actual, score of comprehensive correlation can be obtained by the weighting of logistics analysis and non-logistics analysis, and then comprehensive correlation table can be drawn.

- Draw space relationship diagram

In this part, substitute in the actual area on each department and fit the departments into the available space.

- Put in the constrains

Finally, other existing constrains are employed to cut down number of feasible solutions, to result in a small set of solutions. From among these, direct comparison can be used to rank, eliminate, or select optimum design.

3.2 Application of SLP in Plant L

There are mainly nine raw materials which can be divided into two categories, one need for drawbench and the other need not. According to the collected information, daily demand of raw materials can be obtained which is shown in Table 1. The production process of Plant L includes inspecting, wire drawing, twisted strand, stranded rope, and packaging. Semi-productions produced by wire drawing, twisted strand and stranded rope must be inspected in order to ensure the product quality. Draw the process flow diagram for plant L is shown in Fig.1. Ten departments are separated, and the required area for each department is calculated and presented in Table 2.

Table 1. Daily demand of raw materials.

ID	Raw material name	Weight(kg/day)
1	Polyester fiber	1360
2	Polypropylene fiber	1190
3	Nylon	170
4	HMPE	170
5	Sisal	136
6	Cotton	272
7	Kevlar fiber	103
8	PP	933
9	PE particle	234

Table 2. List of department and required space.

No.	Department	Size m²
1	Raw Material Storage	795
2	Inspecting Department	270
3	Wire Drawing Department	2072
4	Twisted Strand Department	5325
5	Strand Rope Department	2560
6	Waste Department	135
7	Semi-product Department	576
8	Finished Product Department	672
9	Packaging Department	784
10	Office Department	135

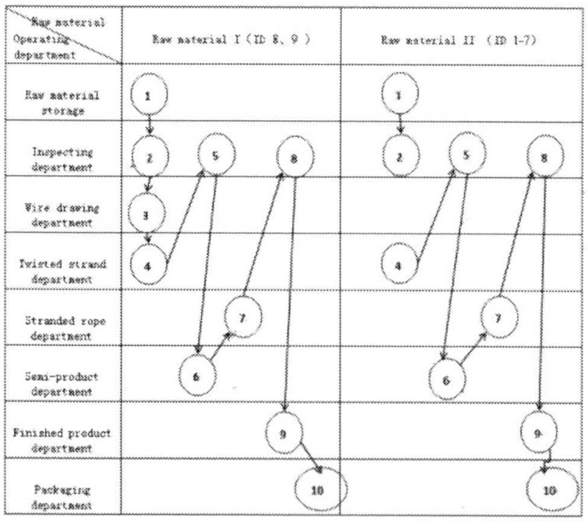

Fig. 1. Process flow diagram.

In SLP technique, to the rope plant L, after analysis of logistics and non-logistics, REL chart is required to show the importance of adjacency between pair of departments. Fig.2 is the REL chart where have A, E, I, O, U to represent absolutely necessary, especially important, important, ordinary closeness okay, unimportant, and undesirable. REL chart is used to develop the relationship diagram shown in Fig.3, which has three different alternatives to be generated for plant L.

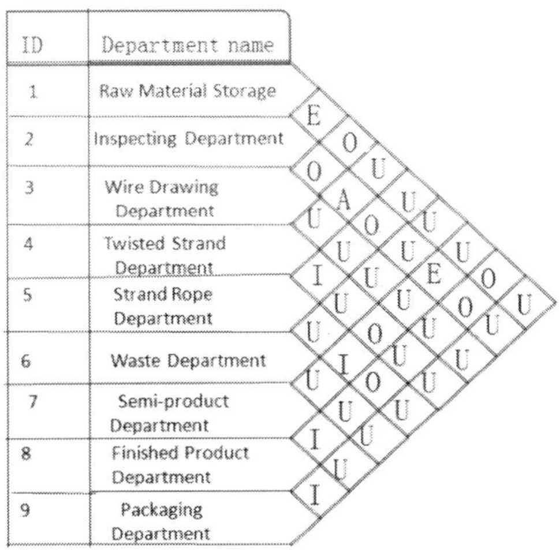

Fig. 2. REL chart.

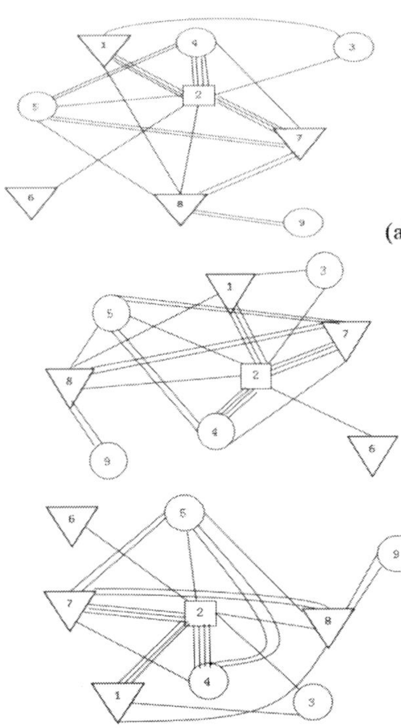

(a) scheme A

(b) scheme B

(c) scheme C

Fig. 3. Relationship diagrams.

3.3 Evaluate alternatives for Plant L with AHP method

Three alternatives of space relationship diagrams are produced by SLP method. One of the best among them should be obtained as a decision. The analytic hierarchy process (AHP) method can be used. AHP method is a structured technique for analyzing and organizing complex decisions based on psychology and mathematics [4]. In accordance with the production status of L plant, five evaluation factors were chosen for study, what are safety and comfort of the working environment, the space utilization in the workshop, the convenience of the management, the efficiency of production logistics and the convenience to the warehouse.

Let U be the five evaluation factors, and set V as the three types of the relationships:

U = { safety and comfort of the working environment ($u1$),
the space utilization in the workshop ($u2$),
the convenience of management ($u3$),
the efficiency of production logistics ($u4$),
the convenience to the warehouse ($u5$) }

And

$$V = \{ \text{REL 1 } (v1),\ \text{REL 2 } (v2),\ \text{REL 3 } (v3) \}$$

Based on the enterprise managers and experts scoring for statistics to form the single factor evaluation matrix, collect the scoring and statistics and use formula for processing the statistical results. Through the test of consistency of judgement matrix, get the weights of the influence factors as:

$$\omega = (0.114206 \quad 0.126946 \quad 0.097889 \quad 0.370940 \quad 0.290019)^T$$

And then get the results of comprehensive evaluation is:

$$[\omega A \quad \omega B \quad \omega C] = [0.1897 \quad 0.2360 \quad 0.5743]$$

Because the weight value accounted for 0.5473 is the largest, the plant L should choose the scheme C of the space relationship digrams. The facility layout for plant L is shown in Fig.4.

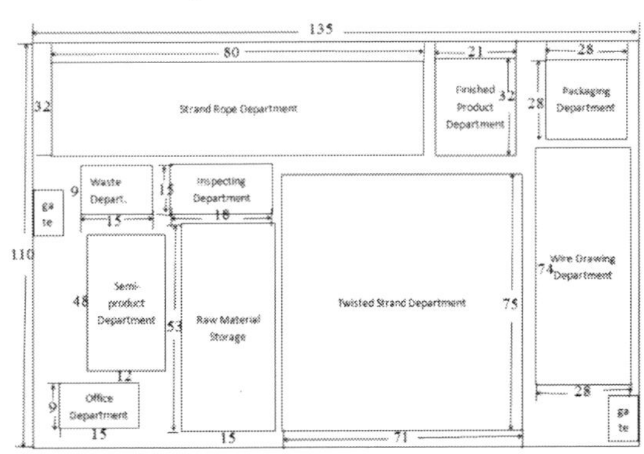

Fig. 4. Facility layout of plant L.

DOI: 10.1051/matecconf/201713900111

4 Analysis of material handling system

Material handling is the movement, protection, storage and control of materials and products throughout manufacturing, warehousing, distribution, consumption and disposal. Material handling incorporates a wide range of manual, semi-automated and automated equipment and systems that support logistics and make the supply chain work.

For plant L, after analysis and design of the facility layout, the handling analysis can be carried out in conjunction with F-D diagram. Calculate the distance between the pairs of the department having the handling contact, and draw the F-D diagram shown in Fig.5.

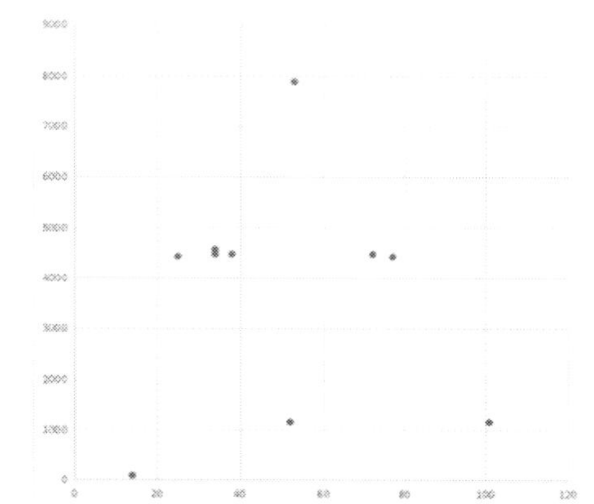

Fig. 5. F-D diagram for plant L

From the F-D diagram for plant L, most of the handling is concentrated in the lower part of the figure which is for a reasonable handling. The main handling tools of plant L are forklifts. It is suitable for the handling needs. The scheme *C* of the layout of the handling for plant L is feasible.

5 Analysis of inventory management system

In manufacturing enterprises, good inventory management is critical in ensuring the normal production and the demand. ABC classification is an effective method for managing the products. ABC analysis is also called Pareto analysis, which is a method of classifying events, items, or activities according to their relative importance [5].

Through the investigation of the plant L, there are about thirty kinds of products, involving a number of application areas, such as outdoor leisure, rescue, protection, safety, fishery, yachts, agriculature, animal husbandry, garden, and other daily necessities. We have ABC classification for the products, the specific steps are:

(1) According to the product price and inventory level to caculate monthly inventory amount.

(2) Order the monthly inventory amount from high to low order of the products.

(3) Calculated the accumulated monthly inventory based on the order of the products after sorting.

(4) Calculate the percentage of accumulated monthly inventory for each kind of product and classify it by ABC classification.

After the steps, products can be divided into three classification. Products in class A need to be focused on management, the raw materials for this type of product are regularly ordered and ordering cycle should be minimized. Class B is in general management in the use of quantitative orders, which adjusts the inventory level according to the business strategy. Class C products can be mass-produced due to the small amount of captial occupied.

5 Modelling and simulation

We adopted simulation software Flexsim 7.5 to model and simulate our analysis and design which we called optimized workshop. Build up Flexsim model for original and optimized layout of plant L according to our analysis and design. Set the simulation runtime of the working time that is eight hours or 28800 seconds a day. The summary reports of simulation results are shown in Figure 6 and Figure 7. Figure 6 is a report for original layout, and Figure 7 is a report for optimized layout.

Flexsim Summary Report
Time: 28800

Object	Class	stats_contentavg	stats_input	stats_output	stats_staytimemin	stats_staytimemax	stats_staytimeavg
Raw material II	Source	1	0	862	0	125.908279	26.410489
Raw material I	Source	1	0	276	0	728.99176	101.894826
Waste products	Queue	16.63206	29	0	0	0	0
Inspection I	Processor	0.999875	276	275	10	731.086792	104.423408
Semi-products	Queue	298.69816	1106	503	12.162463	15647.97365	7798.21258
Drawbench	Processor	0.999527	264	365	20	731.086792	108.332404
Stranding machine I	Processor	0.99885	265	264	30	731.086792	108.666997
Finished products	Queue	0	500	500	0	0	0
Packaging machine	Processor	0.698421	500	499	40	40	40
Closing machine	Processor	0.610219	503	502	35	35	35
Inspection II	Processor	0.994834	862	861	10	127.059332	33.268178
Inspection III	Processor	0.611319	1116	1115	10	36.649323	15.783728
Inspection IV	Processor	0.174398	502	502	10	10	10
Sink	Sink	0	499	0	0	0	0
Forklift	Transporter	0.104279	503	503	5.969196	5.969196	5.969196
Stranding machine II	Processor	0.997208	853	852	30	127.059332	33.694598

Fig. 6. A report for original layout

Flexsim Summary Report
Time: 28800

Object	Class	stats_contentavg	stats_input	stats_output	stats_staytimemin	stats_staytimemax	stats_staytimeavg
Raw material II	Source	1	0	866	0	114.206032	26.2408
Raw material I	Source	1	0	303	0	504.498077	92.5157
Waste products	Queue	13.862098	30	0	0	0	0
Inspection IV	Processor	0.200096	576	576	10	10	10
Semi-products	Queue	284.582912	1139	577	6.878744	14412.41211	7196.601
Closing machine	Processor	0.70024	577	576	35	35	35
Stranding machine I	Processor	0.998831	298	297	30	504.828843	96.63407
Finished products	Queue	0	870	570	0	0	0
Packaging machine	Processor	0.790687	570	569	40	40	40
Drawbench	Processor	0.999527	299	298	20	504.828843	98.37691
Inspection III	Processor	0.612285	1153	1152	10	29.123381	15.29008
Inspection II	Processor	0.985205	866	865	10	120.346291	33.11848
Inspection I	Processor	0.999875	303	302	10	504.828843	95.13551
Sink	Sink	0	569	0	0	0	0
stranding machine II	Processor	0.999181	857	856	30	124.123382	33.60037
Forklift	Transporter	0.141884	577	577	7.07949	7.07949	7.07949

Fig. 7. A report for optimized layout

From the comparison of the two reports, for the semi-finished products, the capacity of the optimized workshop is smaller than that of the original workshop. Optimized workshop can save the space area of semi-finished products, and the waiting time of them for processing is shorter, further to reduce the waste of time.

Overall, the sum of the average staytime of the optimized workshop is shorter than that of original workshop, that indicates the handling efficiency of the system is improved. In the same period of time, outputs of the optimized workshop are 569, which is significantly higher than the outputs of the original workshop 499. The productive efficiency is increased by nearly 14%. Therefore, the optimized solution (Scheme C) of the logisitics system for plant L is better than the original design.

6 Conclusions

The purpose of this paper is to develop a new design for analyzing and designing of the logistics system for manufacturing plant. This research took a certain rope manufacturing plant as an example, through analysis of current production and logistics in the plant, provide the logistics design in workshop facility layout, material handling and inventory management. For workshop facility layout, SLP and AHP method can be combined for the best layout selection. With F-D diagram, analyze the material handling to adjust the workshop layout. For inventory management, use ABC classification to analyze the plant inventory. Finally, compare the optimized design to the original design with the simulation software. The results indicate our design of logistics system for the plant is better than the original design, and it can improve the production efficiency of the plant.

This work was supported by the 2017 Project supported by China Society of Logistics under Grant No. 2017CSLKT3-163.

References

1. Coyle, J.J., "Management of Business Logistics", Mason, OH: South-Western. p. 308 (1992)
2. Muther, R., "Systematic layout planning", Cahners Books (1984)
3. Ojaghi Y, Khademi A, Yusof N M, et al, "Production Layout Optimization for Small and Medium Scale Food Industry", Procedia Cirp, 26:247-251 (2015)
4. Sun X, "E-Commerce Logistics Distribution Mode Analysis and Decision", Applied Mechanics & Materials, 744-746:1873-1877 (2015)
5. Sun X, "Simulations on the Spare Parts Inventory Management of Equipment Maintenance in Manufacturing Enterprise Based on Flexsim", Advanced Materials Research, 715:3181-3186 (2013)

Factor Analysis of E-learning Services Quality

Xinglong Xie*
Economics &Management Institute, Xi'an University of Technology Xi'an, China, P.Q.710048

Abstract- This paper aims to perform a factor analysis of e-learning service quality for providing valuable references and proposing strategies, which would advance successful e-training operations. This paper delineates services features and overviews current literature on e-education quality assessment. From e-learning attributes and the outlook of SSCE (services science, management, and engineering), the paper designs an indicator structure consisting of 10 criteria. Based on questionnaires, each indicator is given a certain value, and e-views is employed to calculate principal components. The paper finds six factors are crucial to measure of e-learning services quality for representing a 82.5% proportion in the total variance. Top three factors are ease of use, security, and reliability, indicating that network facilities, e-safety protection, and high-levels of instructing and punctual answering learner–raised questions after class, are leading prerequisites of maintaining superior e-educational quality. E-education suppliers should put efforts into these dimensions for providing quality e-training services.

1 Introduction

The rapid evolution of information technology has created new applications, such as e-banking, e-learning and e-health [1]. E-learning is a common application that is widely used in the educational sector [2]. The main objective of ICT is to reduce the limitations on time and location, in the context of higher education. It allows access to desired information, without limitations [3]. So, it fills the gap of traditional educational modes. Recently, this learning mode is taking on a swift upward trend, mainly driven by strong demand for new technology, knowledge and academic degrees of on-the-job individuals, increasing recognition of online diplomas, building of occupational and adult education networks.

*Corresponding author:long86886@aliyun.com

It is estimated that e-education for China will grow at annual 150% rate, averaging US$ 1014.5US$ in expenditures per capita, with an annual expanding capacity of market reaching US$0.6087 Billion. Under this backdrop, quality for e-learning is increasingly emphasized as learners are required to become more active and self-constrained due to diminishing of instructors' dominating role, which exists in the traditional teaching activity. Otherwise e-education would fail to make the expected contribution to society.
Traditional instruction wields restrictions a learner's time and space and indeed robs learning of the convenience of choosing the time to learn, which can even cause learners to lose the motivation to learn [4]. Web-based e-learning has been promoted in government institutions to improve

national digital competitiveness. Additionally, some enterprises have successively introduced web-based e-learning as a platform for training employees, creating a shared learning environment and strengthening the professionalism and information-related competencies of employees [5].

Under swift development of e-learning, quality has triggered a great deal of attention from scholars, government and public as it is recognized as a lifeline of e-education. Large quantities of service quality evaluations are conducted on the foot indicators arising from experts, authorized organizations and academic institutes. To name but few, European Treasure Browser Quality Research Project, Development of a European service for Information on Research and Education, E-learning Courseware certification Standards, and Enforcing Regulating of E-colleges and Improving Instructing Quality(China). Regarding reference to determinants exercising effects on e-learning services quality, there is no universal conclusion but there exist generalized several schools of viewpoints: bi-factors referring to deliberated courses and superior learning resources for trainees, tri-factors meaning instructors, curriculums and education management, four elements covering teaching resources, technological facilities, educators and learners, and multiple-elements including objective and subjective dimensions. Thus, there have been no criteria over gauging of e-learning services quality inside China. It remains controversial on what factors should fall into the structure of services quality measurement for e-education. Determining these contributors is extremely significant and pressing to safeguarding e-training quality and driving the education into a scientific development track.

The aim of this paper is to contribute to the growing body of literature on e-learning by dealing with principal factors affecting e-learning services quality based on e-learners' perceived value. This paper can alleviate pitfalls existing in traditional teaching services quality assessment such as overlooking the role of trainees. Our analysis employs data stemming from 1052 questionnaires, which gives our study a uniqueness to other similar studies dedicated to examining e-educational services. We develop use of principal component and factor analysis to seek those factors which play a leading role in the e-learning services quality. Those factors could be a reference for other investigations measuring services quality related to e-education and act as guidance for companies or individuals that are engaged in e-teaching business. One contribution is to facilitate the current complex measurement of e-educational services by streamlining the structure bracing more variables and alleviating defects from linear regression like variable correlation.

2 Literature review

2.1 Services and services science

A service is an activity which has some element of intangibility associated with it, which involves some interaction with customers or with property in their possession, and does not result in a transfer of ownership. A change in condition may occur and production of the service may or may not be closely associated with a physical product. Services mainly features intangibility, heterogeneity, inseparability, and perishability.

Distinctive perspectives are used by scholars to explain services science (services science, management, and engineering (SSME)): (1) from the discipline perspective. SSME is a cross–multi discipline of computer science, operational research, industrial engineering, math, management science, decision-making, and law science. Spohrer defines SSME as a science which applies theories of science, management and engineering into jointly completing a specific lucrative task by a person, or an institution or a system with another person, another institution or another system. (2)From the systematic viewpoint, serving system is a major researched object in SSME. Spohrer points out that SSME resolves complicated practical problems through synthesizing various analysis methods from disciplines such as service science, management and engineering. (3) From the value-creating outlook. The role that SSME plays in services innovation and optimization during the course of value creating is brought to light. Bitner believes that SSME, through examining basic science, models, theories and their usage, resolves the problems of innovation, competition, and quality arising from serving courses to achieve value creation. (4)From contents in services, innovation and capability are underscored. Abe holds that SSME, through efficiently projecting production efficiency, quality, and performance of services, developing and repetitively using knowledge to optimize services, promotes innovation and productivity of services. IBM are credited with putting the SSME acronym together representing the cross disciplinary approach of Services Science, Management and Engineering while Cambridges Institute for Manufacturing and IBM also discussed development in the field in 2007.To sum up, services science is defined as an emerging discipline that aims to combine fundamental science and engineering theories, models and applications with facets of the management field, particularly knowledge, supply chain and change management, for and advancing service innovation and is becoming a key driver of socio-economic growth.

2.2 Importance regarding e-learning

As a learning model unrestricted by time or space, e-learning is a type of remote learning. In addition to integrating learning and the Internet, it also enables learners to develop their professional knowledge, to select the information they need, and to set the pace of their own learning. Learners can use the Internet to obtain educational content, and it is possible to extend e-learning via e-learning assistance and community knowledge-sharing systems. Through the Internet, e-learning allows digitized instructional materials to be quickly and efficiently transmitted, lowers production costs, and greatly lowers the time needed for learners to search for the educational content needed to engage in self-directed study. E-learning acts as the most cost-effective educational and training tool, enabling the learner to obtain abundant and diverse content quickly [6].Obstacles that are encountered during the process of e-learning can be discussed with teachers or peers through the Internet in a process of cooperative online learning, which enhances the interactive aspects of learning. Furthermore, the Internet accelerates the speed of updating educational content, enabling the learner to access new knowledge faster. Thus, e-learning has been viewed as an efficient learning technique.

2.3 E-learning Service Quality Assessment

For e-training, services quality assessment has aroused concern from academicians and produced much fruit. Quite representative are America's ASCI, Euripe's ECSI,SCCTB from Swede, and CNESI proposed by Chinese experts. Additionally, some scholars claimed that strong commitment fulfillment capacity and instructors' high qualification exert a great impact on the confidence and trust of learners. According to Charles and Paula, five dimensions stemming from SERVQUAL, are utilized to measure perceived quality of e-learning services. Having conducted an empirical analysis of learners' satisfaction degree, Qian(2009) found that perceptive services quality serves as a crucial determinant[7]. It is imperative that China need put efforts into promote web education systems in ease of use, practicality and stability. Jin, Shi and Zhen claimed that deficiency in learning time wields negative effects on customer satisfaction [8]. E-education services are reliant on five criteria including responsiveness and indicators for services quality differences are employed to investigate e-education analysis.

3 Empirical analysis

3.1 Indicator framework

Designing an indicator framework for assessing e-learning is aimed at satisfying online education 's specialized features, deriving learners of becoming armchair strategists, getting involved in various practical management activities, and access to data collection, for cultivating their management technique and skills, and improving their ability to analyze and tackle real problems. Hence, an indicator structure needs to give prominence to e-education's property, contemplate user needs, and serve as a reference to rating instructing quality at the same time. In view of deep exploration from scholars and achievements concerning e-education, following "objectivity, operability, supervisoryness, stability and popularization" , this study, sets an indicator frame gauging e-learning services quality(table 1).

Table 1. Indicator structure

Criteria/symbol	Measure of services prior sales
	Sub-criteria

 DOI: 10.1051/matecconf/201713900112

Ease of Use/EU	Swift Website speed	Convenience concerning course inquiry and retracing	Simplicity and easiness of transaction procedures	Explicit course information and knowledge
Tangibility/TA	Superior, perceived image and concise name		Rational page layout	Audio-visional and trouble-free steering and service center
Interaction /INC	Clients interact with learners to offer help.			
Security/SE	Safe deal	Privacy is preserved	Accountable data in the website	
Reliability/RE	Clients are trustworthy	Accurate course contents		
Branding/BR	Credibility and prestige	Public praise		
	Services after sales			
Reliability/RE)	Complete curriculum resources	Regularly update schooling resources	Answer questions at the reasonable time	Play E-lecturing documents for audiences affluently and stably
Validity for schooling resources/VSR	Teacher competence meets educational demand	High quality textbooks and supporting materials	Contents are accurate, without errors or redundant messages	Lessons can be showed online or offline
Interest/INT	Contents are delightful, productive to improving creativity and imagination	Designing of learning stages is interesting, beneficial to raising academic records	Website functions are vividly designed, facilitating self-tests or looking up data	
Responsiveness/RES	Customer services providers are ready to tackle problems for students.	Amicably deal with complaints lodged by learners	Assist trainees to carry out matters such as dropping out and canceling a course	Correct homework and respond to questions from learners
Solicitude/SO	Offer examination-related information including reminding services	Provide online-technological support to take up problems emerging during the lesson-listening course	Can supply trainees with customized services	Offer a complaint mechanism to unwind issues confronting customers.

Table 2. Profile of questionnaire respondents

Classification	Sample distribution
gender	Males accounting for 62.9%;female accounting for 37.1%
age distribution (%)	Samples between 20 and 29 years 58.1%; 30-39 years 33.4%; 40-49 years 8.0%;50-59 years 0.5%; 60 years or above 0.1%
education background	Undergraduate 5.7%; bachelor 41.4% (four years); bachelor(starting from undergraduate) 50.4%; professional master 0.4%; second bachelor 1.5%; academic master 0.4%; doctor o.2%
e-education time	One year below 21.4%, 2-3years 70.9%; 4-5years 6.4%; 6 years above 1.3%

Table 3. Indicator value

If an indicator is	very unimportant	unimportant	uncertain	important	absolutely important
Its value	1	2	3	4	5

Table 4. Reliability of questionnaire

Criteria	EU	TA	INC	SE	REL	BR	VSR	INT	RES	SO
Cronbach Alpha	0.91	0.87	0.90	0.91	0.90	0.90	0.90	0.90	0.86	0.95

Table 5. KMO and Bartlett testing

Kaiser-meyer-Olkin		0.80
Bartlett	Approximate chi square	226.5
	Sig.	0.000

Table 6. Results of principal analysis

Explained total variance									
Component	Initial eigenvalue			Extracting sum of squares loading			Rotating sum of squares loading		
	Total	Variance	Accumulative	Total	Variance	Accumulative	Total	Variance	Accumulative
F1	4.08	40.8%	40.8%	4.08	40.8%	40.8%	2.32	23.2%	23.2%
F2	2.08	20.8%	61.6%	2.08	20.8%	61.6%	1.25	12.5%	35.7%
F3	0.91	9.1%	70.7%	0.91	9.1%	70.7%	1.24	12.4%	48.1%
F4	0.68	6.8%	77.5%	0.68	6.8%	77.5%	1.23	12.3%	60.4%
F5	0.6	6.0%	83.5%	0.6	6.0%	83.5%	1.11	11.1%	71.5%
F6	0.51	5.1%	88.6%	0.51	5.1%	88.6%	1.10	11.0%	82.5%

Table 7. Matrix for factor loadings

Criteria	Principal Component					
	F1	F2	F3	F4	F5	F6
EU	**0.79**	0.1	0.63	-0.06	-0.11	0.08
TA	0.63	0.21	-0.01	0.40	0.23	0.23
INC	0.05	0.03	0.16	0.26	0.81	0.22
SE	0.33	**0.96**	0.15	0.02	0.02	0.01
REL	0.1	0.1	**0.83**	0.03	0.05	0.17
VSR	0.02	0.43	0.1	**0.89**	0.05	0.09
RES	0.11	0.29	-0.05	0.22	**0.89**	-0.03
SO	0.15	0.02	0.16	0.21	0.08	**0.65**
INT	0.13	0.01	0.17	0.69	0.26	0.21
BR	0.42	0.12	0.17	0.12	0.09	0.05

Table 8. Factor naming

Factor	1	2	3	4	5	6
Title	easy utilization	security	reliability	validity	responsiveness	solicitude

3.2 Data availability

We made a survey of service quality to customers in March, April, October, and December of 2016, issuing 1200 questionnaires(see table 2), in which questions are formulated thanks to a 5-level Likert scale(see table 3), and getting back 1052. EXCEL is adopted for calculation. According to questionnaires, we summarize, process, and transform information into data calculating the value of an indicator. By means of e-views 6.0, the paper conducts factor analysis, which principal component analysis mixes a group of certain related initial variables into a set of irrelevant indicators through mathematical transformation with a view to obtaining linear combinations, which retain information as much as included in the primary variables. The primary step is to test variable correlation, then followed by principal analysis, and finally factors are determined which play critical part in e-study services quality measurement. Note that table Tables 4 and 5 register the result concerning questionnaire reliability and KMO and Bartlett testing respectively. Coefficient matrix eigenvalue, contribution and accumulative contribution are reported in table 6, the factor loadings matrix table 7, and factor names in table 8.

3.3 Discussion

Table 4 demonstrates that the Cronbach Alpha coefficient for each criterion, the maximum topping 0.91 and the minimum at 0.87, all, is in excess of 0.8, implying that there is a high reliability for the designed questionnaire. Validity test must be conducted to confirm whether the factor approach is fit for the question in this study. To this end, we resort to the exploratory factor, voicing that the higher KMO, the more common factors among variables and the more suitable the factor methodology will be. If KMO exceeds 0.7, the approach of factor analysis will be reviewed as appropriate. It can be found from table V that the relationship among variables is suitable to be examined through the way with the value for KMO above 0.80 and the significant linear connection between indicators implied by Barlett testing.

Inter-variable relevance must be investigated to certify whether these variables have a certain dependency relationship, whereby finding out their correlation both degree and direction. In order to achieve this aim, various ways have been utilized in literature, in which Pearson correlation test receives wide-spread attention and usage. In this study, we also depend on Pearson to examine the correlation among ten indicators of gauging e-study services quality. The results show that there exists a significant positive linkage between two dimensions at the 5% level, for example, ease of use and tangibility, or security or reliability, validity of learning resources and solicitude. All this dictates that principal component analysis is suitable to be applied into dealing with services quality regarding e-learning.

Table 6 reveals that the cumulative variance of the cumulative eigenvalue for six former factors making up 82.5% of total variance, thus they are identified as principal components. As a result, ten original indicators can be shrunken into six main factors. Table 7 exhibits factor loadings after Kaiser rotation. Obviously, ease of use has the largest loading (0.79) in the first common factor, so the first factor is termed as the easy utilization factor. In the same token, factors are named security, reliability, learning resources' validity, responsiveness, and solicitude due to the largest loading in its own principal component, accordingly. Each of these factors represents a more definite economics meaning. Hence, gauge of e-learning services quality can be generalized into six major factors (see table 8). The loading for interaction comes in at 0.05, 0.03, 0.16, 0.26, 0.81, and 0.22 in the six main factors, separately, and this displays interaction is mainly explained by easy responsiveness and solicitude factors, especially dominant effects by timely settlement of problems for learners.

4 Conclusion and proposal

The contribution of this paper is to provide an evaluation of services quality in e-learning. We extend existing literature via dealing with the principal components that mainly affect e-education services and thereby identifying the most significant factors shaping the services quality of e-education. Data are calculated grounded on answers from 1052 respondents, allowing us glean a broad range of new insights into e-learning services quality.

- The most important factor confirmed by its loading of 0.79 is easy utilization for e-learning services, demoting unhindered page linkage and strong searching functions, vigorously driving learners to carry out studying and related activities such as consulting curse-connected contents, instruction purposes, instructor resources, and supplement services. Thus, network facilities must be superior, assuring fresh information technology of being putting into use in the whole e-training process, updating teaching contents and supplying frontier knowledge dynamics.
- Factor security has evolved into anther critical element for e-educational quality with the loading reaching 0.96 and this suggests that one company would obtain access to the market if it provides safety and privacy guaranteeing for e-customers technologically and legally.

• With reference to factor reliability, it plays an important role in improving e-training quality for its loading hitting 0.83. Customers may not materially be benefited from purchasing an educational service commodity, for which production and consumption go side by side and this requires learner involvement and lecturer-student interaction. Otherwise, there would be no expected performance to be achieved. So, instructors should be equipped with ability to exactly fulfill commitments made. Additionally, strategies such as course updating and punctual offering answers to student-raised problems need to be put into effects for promoting e-learning quality improvement.

• Factor validity with a 0.89 loading displays professional competence and working spirit of instructors are the primary concern of e-learners. The loading of factor responsiveness at 0.89 shows e-students' learning process supervision from e-organizations and assuring trainees of obtaining access to answers to questions stand out for e-education services quality betterment. Concerning the solicitude factor, it also plays an important part in the measurement of e-learning services quality with its loading valued at 0.65. It can be taken as the ability to willingly offer aid and enable to furnish just-in-time services for customers, for example, customized training. Thus, it is necessary and imperative for e-schools to build an effective system of supporting learning, bringing in a more meticulous and benevolent e-education, in which learners are centered.

Our results provide valuable insights into building and maintaining superior e-learning services:

• Initially, e-education suppliers ought to focus attention on web technology improvement as high qualify services depend on information telecommunications facilities and lay higher requirements on system environments, schooling platforms, software instrument and various management mechanisms. These technical infrastructures must be benign enough to guarantee successful teaching fulfillment, which can provide all prerequisites including independent and sufficient IP addresses and names, valid schooling information, managerial functions of users, and stability in net-work system operation. Furthermore, web technology, instructors' teaching techniques, and media communications should be integrated into a whole one to achieve the optimal result.

• e-educational schools must develop an advantage in the branch of learning or cooperate with those intuitions, which have an advantage in the regarding, or co-develop e-courses along with these organizations. As for launching a new major, it should be based on preponderant specialties. Moreover, lifelong-education must be the primary concern for e-lecturers themselves with substantial innovative skills and knowledge emerging by constant development of information technology, to allow their lectured knowledge to satisfy economic reality.

Acknowledgment

The authors are thankful to the financial aid from Shaanxi's fund program No.2015KRMO98.

References

[1] A. Y. Alsabawy, A.Cater-Steel, J.Soar, Determinants of perceived usefulness of e- learning systems, Computers in Human Behavior, 64, 2016,843-858.

[2]A. K. M. N. Islam, E-learning system use and its outcomes: moderating role of perceived compatibility, Telematics and Informatics,33, 2016,48-55,2016.

[3] A.Althunibat, Determining the factors influencing students' intention to use m-learning in Jordan higher education, Computers in Human Behavior, 52, 2015,65-71, 2015.

[4] H.R.Chen, H.L.Huang, User acceptance of mobile knowledge management learning system:design and analysis, Educational Technology & Society, 13, 2010,70–77.

[5]S. Sridhar, S.E-government – a proactive participant for e-learning in higher education, Journal of American Academy of Business, vol.7, 2005, 258–268.

[6] A. Powell, M. Barbour, An examination of government policies for e-learning in New Zealand's secondary schools, Journal of Open, Flexible and Distance Learning, 5, 2011,75–89.

[7] X.Q. Qian, Empirical analysis of students' degree of satisfaction of e-education services quality, Distance Education in China, 7, 2009, 57-60.

[8]L.W. Jin, Zh.P. Shi, Q.H. Zhen, On factors affecting learners' degree of e-education satisfaction, Beijing Radio and TV University Journal, 2,2015, 49-52.

The Realization and Test Research of Standardized Mobile Learning Courseware Based on SCORM

Ruan Lanjuan
Department of traffic information engineering, Guangxi Vocational and Technical College of Communications, Guangxi, 530023, P.R. China

Abstract: High-tech education is the new trend of the development of modern education. This article attempts to use the modern educational development model to conduct a reasonable attempt of a more scientific and more authoritative new teaching model. The research on the realization and test of standardized mobile learning material based on SCORM is carried out on the continuous Chinese education reform. The article makes a brief analysis on the overall technical framework of SCORM, and points out the specific issues the SCORM has in standardized mobile courseware learning. At the same time, this article studies and analyzes the main produces and steps in the courseware design and development.

1 Introduction

The contemporary society is an economical construction society, with the development of science and technology as its core. The cultivation of talents and the strategy of rejuvenating China through science and education are main direction of national development. Education will help the cultivation of comprehensive quality of talents on the background of social development. Therefore, as for educational development, the reform of educational forms and ways is the main target of scientific educational development. The development of modern technology and the rapid development of internet provide a strong foundation to the continuous reform of educational science. In modern education, diversified talent training model has become a new pattern of educational development, and mobile learning based on online courseware, as a new learning method, is welcomed by the new generation of educated. On the conditions that education has a scientific and multi-directional development, mobile teaching and mobile learning becomes an innovative and scientific form of educational development. Developing this model is in favor of education and teaching take place anytime, anywhere, and is in favor of a comprehensive ideological education and scientific and cultural education, and promote long-term development of harmonious education and social security. Therefore, the implementation of standardized mobile learning courseware based on the shared content reference model (hereinafter referred to as SCORM) will be a major development trend of education. We should assess the situation, study this kind of education model deeply, scientifically and rationally, so as to help it develop scientifically and create great opportunities for the development of China's education and long-term effectiveness of quality education.

2 The analysis of existing problems in the development of information education

With the modernization of education and the extension of internet, people have a great change of their thinking way and a growing awareness of the information highway. However, under the same conditions of education development, the application of the Internet in the education process needs to be improved. Particularly, in the same educational time, different regions in China has unequal educational conditions, which makes the Internet plays defective functions under the same conditions of education, and leads to the lagging development of education in most regions in our country.

The arrival of the information technology era marks that the human civilization steps into a new stage of development. Internet plays a very important role in all walks of life and promotes the rapid development of all walks of life. Internet has made great differences in promoting the reform of education. Firstly, it promotes the realization of multimedia teaching. Secondly, it promotes the formation and development of mobile courseware. The development of human civilization has great changes both in time and in space, which promotes the continuous development of knowledge. Advanced and forward-looking as the mobile education mode is, it has a broad development prospect. However, anything in the development process must first go through a difficult and tortuous process, mobile teaching and learning courseware is no exception. It will be confronted with many difficulties under the immature conditions of development.

China's mobile communications equipment applications on the current stage, has reached a comprehensive development trend. Whether it is adult or adolescent, and even child, nearly everyone has a mobile phone. It will lay a solid foundation for the comprehensive development of standardized mobile learning courseware based on SCORM. When we study and develop this education mode, we should take full account of the problems that we will meet in early stage. To be specific, the development of this learning method will be slow in early stage, mainly by the impact of China's traditional education model and the relevant education system constraints, and because of the lack of attention in the social class. Therefore, this learning mode plays only a preliminary auxiliary role in the education development, can only achieve its service but is difficult to achieve its educational leadership role. For example, when it comes to the role of mobile phone in education, teachers seldom consider its application for extracurricular learning, and rarely let students learn courseware through mobile devices. To a large extent the educational resources that would have been devoted to education is wasted. Although in our education, courseware through internet is designed in multimedia teaching equipment, but the possibility of the realization that courseware is used in multi-platform in the form of SCORM is really low, and even zero in some places. Thus, to achieve the full development, this teaching method still needs a considerable period of development, and needs the support from policy and education systems.

3 The existing problem in the development of SCORM standardized mobile courseware in current network.

3.1 Network courseware sharing resources is relatively lacking, it is difficult to meet the needs of the current education groups

At present, the network courseware is chaotic, quite mixed, the quality of network courseware is different, and the system resources is lack of effective integration. In terms of format and structure, there is also a clear lack of unity. In terms of network sharing, it is difficult to meet the needs of people to share network resources. In terms of network sharing, it is difficult to meet the needs of people to share network resources. Each system document format is different, courseware design form is also very different, and each take the road, there is a chaotic phase of network courseware development. The courseware system and the student information system developed by different organizations are difficult to realize the mutual sharing and the exchange of educational methods, and form their own way of development. This situation is not good for the development and sharing of more resources, thus not good for the application of online resources in education and the development of modernized information education. Even within its own system, courseware content updates are not strong. Even for a long period of time, even if the courseware designers want to update it, they have to go through layers of change, and succeed to update after a long period of time. Throughout the content of educational courseware, most content are electronic design of the paper teaching materials. There are not much education innovation, and the innovation and development efforts are not strong. In terms of knowledge structure, the content is profound, the universality of the courseware is not obvious, and it is not suitable for the education group with the basic cultural ability. From a number of angles to think, most courseware lack of dialectical teaching practice and play little role on actual teaching. Most of the learning groups are from different social levels. Their ability of learning understanding is different, and thus the level of knowledge they can accept is also different. Therefore, the unsuitable courseware will lead to weariness in the learning process, so that the role of such teaching services greatly reduced, failed to achieve the actual effect of courseware design. For the practical role of teaching, the teaching courseware each school needs is different. For different learners, learning strategies are also easy to lead to different learning effects, and the degree of application of network courseware will produce different results. In a word, the development of online courseware at this stage presents repeated and low level of content, it is difficult to independently form a new system of the scientific development of standardized and automated network courseware.

3.2 The development of network courseware mainly to meet the requirements of today's education system, the majority of the form of development is formalistic.

The online courseware was originated in the United States, developed in the United States, and then continue to spread to the vast majority of the world. The development of network courseware in our country belongs to the development stage of formal requirements, and it is in the initial stage of development in many cases. Many online courseware designers are easy to confuse the online mobile courseware with the traditional multimedia pages, and make it develop in accordance with in accordance with the multimedia web production thinking. This kind of courseware attracts readers mainly by the screen color matching and dynamic color changes to strengthen their learning interests. However, the designers often take little of courseware content, the degree of education, courseware framework, courseware development process and even courseware transfer mode, so they often neglect them in the process of design. When learners use the shared online courseware to learn, it is difficult for them to know what content they really need in the face of a wide range of content. Because the difficulty to identify the authenticity and effectiveness of the contents of the network courseware, many learners waste a lot of time in the process of blindly learning. Although they really want to learn, but they get half the result with twice the effort. Many online courseware is just a simple combination of some teaching content, its teaching depth and the breadth of digging is not enough. It reduces the learning effect. For example, when learners use online courseware, they cannot check their learning effect as the courseware lack of the systems of learning time, learning page, test scores and so on. Which led to the learners cannot further understand their own learning state, and cannot monitor their own learning process. Besides, this will also lead to teachers cannot know more about students' learning situation in the teaching process, so the process of teaching and learning will be chaotic and the learning effect will be reduced.

3.3 At present, there is not yet a unified technical standard for mobile learning courseware design

At present, China's domestic development of SCORM standardized mobile learning courseware is lagging behind, cannot keep up with the pace of research in the United States, and to a large extent almost copy the traditional model of others, research efforts are surprisingly weak. In term of the education system, due to the lack of attention from the education system, the design of network courseware and the building of research aspects is still in the initial stage of development. In terms of technical processing, it is difficult to form a major technical development force, the development of SCORM in a long period of time is in a semi-stagnant state, and technological development is poor. In the ever-changing wave of network resources, if SCORM standardized mobile learning courseware cannot strengthen itself timely with the technology as its core competitiveness, it will inevitably be forced to withdraw from the stage of historical development. We should find ways to regulate the relevant management system of network courseware technological development to promote its close integration with the way of education. However, due to the constraints of development and educational thinking, coupled with the impact of China's social development and the development of market economy, the production and design of online courseware cannot adapt to the reality, thereby reducing the efficiency of education. The network space is quite broad, but the design and dissemination of network courseware presents chaotic situation. There are many blind spots in the system, and a standardized management system is not achieved. Technology development, development and research of market are also weak, which is not conducive to the scientific advance of network courseware design and development.

4 Design, development and test research of SCORM standardized mobile courseware

4.1 Design and development of SCORM standardized mobile courseware

The traditional network courseware design and the process of actual operation almost without any control mechanism carried out. SCORM standardized courseware divide the teaching materials into a number of small points of knowledge, make it many learning components with the unit as a small unit. In the SCORM standardized design, the macro framework design of teaching content is achieved according to the curriculum teaching objectives, and the design and planning of the size of each learning element is realized in the micro level. Teaching cases with innovative ideas and teaching values are ported to learning components to achieve a flexible combination of knowledge to facilitate learner use and learning. In the design of the whole courseware, a number of learning elements combine into a set of learning units, and then a number of learning units combine into a learning unit courseware package, so as to facilitate learners in the mobile way to effectively learn.

4.2 The specific design steps of SCORM courseware.

4.2.1 Follow the SCORM concept to standardize the division of curriculum unit. SCO is the most basic courseware unit for courseware editing, and the smallest unit of platform organization and tracking. ①The SCO is composed of one or more materials, and is designed from the chapters of text contents or a variety of knowledge points according to specific conditions and requirements. In addition, its size is not strictly limited. This division meets the "miniaturization, granulation" requirements of the fragmented mobile learning on the courseware knowledge group. However, SCO in the design process requires the inclusion of at least one learning goal that can provide the function of the unit being recorded, since the SCO is a learning resource that enables the SCO to be able to track the granularity of the IMS under the SCORM run-time conditions, thereby SCO can independently divide the context of the content, form a clear individual content and exist in the courseware. Besides, under normal circumstances, the static and dynamic graphic material in the courseware is mostly completed in the traditional way of network education courseware.

4.2.2 Standardize the construction of the metadata file framework refer to SCORM design ideas. SCORM standardized mobile courseware contains nine major categories of metadata, which perform their duties, to achieve the integrated development, and can accurately achieve the rapid processing of data. They can do coordinated work of searching, managing, sharing and reorganizing the information, quickly realize the operation of metadata. However, in many cases, SCORM does not use metadata files. Metadata files only serve to retrieve and provide basic information in many cases in practical applications. In view of this, the rational construction of metadata files is directly related to the effective use value of courseware and contribution to resources sharing.

In the design of SCORM courseware, several metadata making tools are often used, but mainly the reload editor can achieve flexibility and efficiency in use and can describe and save the relevant information in xml file. ②In contrast to the entire courseware, the order of work for each xml file inside it is orderly, and their order of work is the same. They can work orderly in this high-speed work environment, achieve the rapid transmission of information and provide the basic feasible conditions for modern mobile learning.

4.2.3 Standardize the encapsulation of courseware content refer to the SCORM standard. In the process of design and completion of SCORM standardized mobile learning courseware, the main idea performed in the development and application is the idea of component work. That is to say, a complete course of the network courseware is composed of multiple groups of interactive courseware, and each unit courseware is composed of a plurality of unit courseware packages that can realize the mutual influence, thereby realizing that every layers act in concert with each other and work with interaction. The individual content consists of two parts, which are used to describe the content inventory file IMSMANIFEST.XMI of the learning resource organization structure and the entity file of the learning resource.

4.2.4 Test based on SCORM normalized learning software. Conformance Test is mainly to test whether the production content is consistent with the requirements of SCORM. Regardless of the storm-based courseware and learning management system developed with any tool, you can use ADL's validation tools to test, debug and complete standard verification. ③This test can clearly detect the matching effect of SCORM in the running process, providing reliable data monitoring security for the operation quality of entire network courseware. It will monitor its own quality and improve itself continuously in the work process, which is really helpful for the improvement of its own quality. In the future operation of SCORM, the achievement of the continuous innovation of research methods, will be more dependent on the continuous operation of such tests. Thus, it can be seen that such tests play a very important role in the testing of SCORM standardized learning software.

In addition, the work test is the test that can achieve the multi-effect the network courseware can give modern education and teaching in the mobile environment. It is divided into quantitative testing and job performance testing. But at this stage due to the development of educational facilities and courseware are subject to different constraints, the effectiveness of testing is not high, and the test is not developed well.

5 Conclusions

In summary, the development of network courseware and its contribution in education, and its rapid update but not enough follow-up changes and inadequate supplement conditions in its operation process, all presents that at this stage SCORM standardized mobile learning courseware has many shortcomings in the development, and need to be improved and improved. In addition, there are fewer successful cases of network teaching applications, its influence is still in the initial stage. Commercial use in public application platform and resource sharing is more serious, which is not conducive to the sharing and rapid dissemination of public education resources.

SCORM contributed a lot to the education industry, it can effectively promote the continuous development of modern education, and promote the continuous advance of network education resources. In the course of modern education and teaching, SCORM has set a brilliant example for network teaching, so that the network education and teaching can develop rapidly. In the process of multimedia teaching, the new situations that courseware developed comprehensively is realized due to the continuous contributions of SCORM. The effect of education and teaching is constantly improving, and the mutual influence and mutual penetration of knowledge at all levels are realized in the educational development platform. In its development process we should continue to stimulate developers and network courseware makers to study and innovate, and constantly tap the potential of the role of SCORM in education, so as to make it contribute more to the modern information education.

Acknowledgements

The project of 2016 to enhance the basic ability of the young teachers in Colleges and universities in Guangxi: *Design and Application Research of Mobile Micro - learning Support System Based on Micro – course.* Item Number: KY2016YB668 Project Host: Ruan Lanjuan

DOI: 10.1051/matecconf/201713900113

References

[1]. Yang. W. Research on Standardized Mobile Learning Courseware Based on SCORM. East China Normal University. 32(2009)

[2]. Rui. L, Xiuling. L, Lingling. Z. Research and Implementation of Network Courseware Design Based on SCORM Standard. China Education Technology & Equipment. **11**, 39(2012)

[3]. Yutong. C. Design and Implementation of E - Learning Solution for Catering Enterprises Based on Social Learning. Tianjin University. 34(2014)

AUTHOR INDEX

AUTHOR INDEX

AUTHOR INDEX

AUTHOR INDEX

AUTHOR INDEX

AUTHOR INDEX

AUTHOR INDEX

EDP Sciences – Web of Conferences
17, Avenue du Hoggar
Parc d'Activité de Courtabouf
BP 112
F-91944 Les Ulis Cedex A
France

ISBN 978-1-5108-5350-8

CURRAN ASSOCIATES INC.
proceedings.com

3rd International Conference on Mechanical, Electronic and Information Technology Engineering (ICMITE 2017)

MATEC Web of Conferences Volume 139 (2017)

Chengdu, China
16 - 17 December 2017

Part 2 of 2

Editors:

Bing Xu
Yinong Chen

ISBN: 978-1-5108-5350-8

Printed from e-media with permission by:

Curran Associates, Inc.
57 Morehouse Lane
Red Hook, NY 12571

proceedings.com

Some format issues inherent in the e-media version may also appear in this print version.

Printed by Curran Associates, Inc. (2017)

For additional information, please contact EDP Sciences – Web of Conferences at the address below.

EDP Sciences – Web of Conferences
17, Avenue du Hoggar
Parc d'Activité de Courtabœuf
BP 112
F-91944 Les Ulis Cedex A
France

Phone: +33 (0) 1 69 18 75 75
Fax: +33 (0) 1 69 28 84 91

contact-edps@webofconferences.org

Additional copies of this publication are available from:

Curran Associates, Inc.
57 Morehouse Lane
Red Hook, NY 12571 USA
Phone: 845-758-0400
Fax: 845-758-2633
Email: curran@proceedings.com
Web: www.proceedings.com

TABLE OF CONTENTS

PART 1

PART 2

A Multilevel Association Model for IT Employees' Life Stress and Job Satisfaction: An Information Technology (IT) Industry Case Study

Khalid Mehmood[1*], *Amir* Hussain[2]

[1] Department of Organization Behavior and Human Resource Management, Beijing Institute of Technology, Beijing, China
[2] School of Natural Sciences, University of Stirling, Stirling, FK9 4LA, Scotland, UK

Abstract. The aim of this research was to investigate the association among IT employees' life stress and job satisfaction in information technology (IT) firms. Data on 250 IT employees' in 30 working groups was obtained from 10 Information Technology (IT) Chinese firms from Beijing, and analyzed using hierarchical linear modeling (HLM). Results found momentous association among life stress of IT employees' and their job satisfaction at an individual-level and group-level in IT firms. Furthermore, life stress in Beijing at group-level moderates the association among job satisfaction and IT employees' life stress at an individual-level. Finally, limitations and implications of the present study are also discussed.

1 Introduction

Over the past couple of decades, stress has been turning into an inexorably significant issue in IT firms. Stress is a constructive condition of chance, demand, or assets related to an employee's wants and the outcomes are viewed as obscure and important [1]. According to [2], stress is an unfavourable response which occurs when an individual applies huge pressure or different sorts of desires upon themselves. This was delivered when individuals feared not being able to deal with stress during work. Life Stress has a hopeful upshot on employees' of IT firms, however the degree to which an IT employee can deal with it, by and large surpasses bearable limits and has a negative outcome on IT employees'[3]. IT employees' countenance pressures in family life, include salary, working hours and housing problems in Beijing etc.

In every IT firm and at each juncture of IT management, IT executives and specialized IT professionals, a customary level of stress is to be found, fundamentally influencing the job satisfaction of employees [4]. The job satisfaction of an IT employee may be an indispensable variable related to stress during work. Researchers deem that job satisfaction relies upon substance and circumstances outside of themselves, and in addition inside objectives, discernments and wishes [5]. As per the viewing of [6], job satisfaction alludes to the "gratifying enthusiastic stipulation approaching from the evaluation of one's job as accomplishing or encouraging one's job standards". The relationship among uneasiness stress with job satisfaction in IT firms and the execution of IT employees' had been inspected, and discovered that lower stress advances the performance of IT employees' which they premeditated across various IT administrative levels of the firm [3]. On the foundation of literature, the accompanying hypotheses were estimated:

Hypothesis 1: IT employees' life stress pessimistically influences on job satisfaction at individual level in IT firms.

Hypothesis 2: IT employees' life stress pessimistically influence on job satisfaction at group level in IT firms.

Hypothesis 3: IT firms' group level life stress moderates the association among IT employees' life stress and job satisfaction at individual level.

2 Methodology

2.1 Research sample

Participants of the study were employees in ten indigenous Chinese firms based in Beijing, capital of P.R. China and the sample was taken randomly from Beijing based organization. Among all, ten firms were private owned. The number of employees range from 50 to 200 in these IT firms. Prior to data collection, members of the research team obtained consent from the HR departments of every IT firm. They clarified the objective and purpose of the research and ensured confidentiality and voluntary participation of employees. Employees from all departments voluntarily participated in the survey. All participants gathered in the main hall of their respective organizations. In order to minimize potential common method bias, one of the research team members explains the objective of the study and also clarifies that there is no right and wrong answer and that they should give answer as honestly as possible. Moreover, an ample distance was observed between the participants to reduce eye contact. During a survey research members seated outside of the hall and within were free to fill in questionnaires. All participants returned their feedback in an anonymous envelope. We distributed 300 questionnaires, recovered 270 questionnaires and after removal of outliers and incomplete information were left with 250 for further analysis, providing a response rate of 83%. The participants average age was 30 years (SD=10.47). The

* Corresponding author: khalidmir@bit.edu.cn

employees had an average workday of 8.5 hours. All the respondents had at least an undergraduate university education, with 66.7% having Master's degrees.

2.2 Measures

2.2.1 Life stress

The 10 items tool of [7] is used in this study to measure the life stress of IT employees' in the IT firm. The tool is associated with standard family life, including salary, working hours and housing problems and etc. The items range for measure stress is commencing from (no stress= 1 to strongly stress= 5) from respondents in IT firms. The coefficient alpha for this scale was 0.92.

2.2.1 Job satisfaction

The single item tool [4] is used in this study to measure job satisfaction of IT employees'. The tool is associated how fulfilled they were with their jobs in IT firms, taking everything together: "Everything well thought-out, how fulfilled are you with your job as a whole these days? It would be an ideal if you utilize this card to help your answer [1= disappointed to 5= satisfied]". The (ICC1) intraclass correlation for this variable [8] was 0.15.

3 Analysis stratergy

While our model was multilevel on the whole, we choose an appropriate regression analysis for each hypothesis according to the level involved. We analyzed our multilevel study by using hierarchical linear modeling (HLM) [9]. Hierarchical linear modeling (multilevel analysis) is an iterative process. The objective of multilevel modeling is to permit more precise prediction of dependent variable based on a function of predictor variables, at more than one level.

In our multilevel study, we used the individual and group level into three models. If life stress variable at group level is associated "variability of the intercept", which shows that the variable at group-level is forthwith associated to job satisfaction as dependent variable and also variable at group- level is forthwith associated to the "variability of the slope", it gives indication of cross level moderating effect. IT employees' job satisfaction was predicted in Model 1 as a null model where and it satisfies the hierarchical linear modeling first condition. Model 2 included only the random intercept for the group that is individual level main effect among IT employees' life stress and IT employees' job satisfaction was anticipated. This provides test of hypothesis 1. Life stress variable was analyzed as group level variable at Model 3. Also, in model 3 we analyzed individual level interaction among life stress of group level, job satisfaction and individual level IT employees' life stress. This model 3 provides assay of hypothesis 2 and 3. The multilevel analysis permissible the researcher to establish at what level these predictive interactions was operative.

4 Results

4.1 Correlation analysis

Summary statistics of the variables and the individual level correlations showed in table I. job satisfaction of employees' and IT employees' life stress has negative significant correlation($r=-0.52$, $p<0.01$). Significant negatively correlation among life stress and age was established ($r=-0.40$, $p<0.01$) and also positively correlation is established between job satisfaction of IT employees' and age ($r=0.30$, $p<0.01$). Represented gender as (1= male, 2= female) and also gender and job satisfaction of IT employees' were negatively correlated.

Table 1. Descriptive statistics and correlations

Descriptive Statistics and Correlation				
Variables	**1**	**2**	**3**	**4**
1) Life stress	1			
2) Job satisfaction	-.52**	1		
3) Age	-.40**	.30**	1	
4) Gender	.17	-.15**	-.31**	1
Mean	4.10	4.19	30.12	1.56
S.D.	.77	.94	10.47	.57
**P<0.01(two tailed test); *P<0.05(two tailed test)				

4.2 Multilevel Moderation Effect and Main Effect

Results of job satisfaction of IT employees' were provided in table II for model 1 and model 2. It showed that hypothesis 1 has significant support in this study. In table II, Υ_{00} showed that either departments of IT in firms as group can be eminent from one another by their IT employees' job satisfaction. Job satisfaction of IT employees' might vary at group level according to significant results of table II. There is abundant significance for hierarchical linear modeling. IT employees' life stress has a pessimistic significance association with job satisfaction of IT employees' in table II, and showed that ($\Upsilon_{10} = -0.43$, $p<0.05$). The senior oldest males of IT firms are most likely to be content with their jobs at IT firms ($\Upsilon_{20}= 0.19$, $p<0.05$; $\Upsilon_{30}= -0.11$, $p<0.05$).

Moderating effects findings for life stress showed in table III and also suggested significant validation for hypotheses 2 and 3. Association among individual level IT employees' life stress and employees' job satisfaction moderated by life stress at group-level ($\Upsilon_{11} = 0.21$, $p<0.05$) showed in table III. Furthermore, according to results life stress at group-level had a negative momentous association with job satisfaction at individual level ($\Upsilon_{01} = -0.19$, $p<0.05$) showed in table III. Moreover, model I, model II and model III value of σ^2 progressively decreased from ($0.89 < 0.72 < 0.68$).

5 Conclusions

This multilevel study concentrated on the interaction among the IT employees' life stress and job satisfaction in IT firms. The outcome of this study uncovers that IT employees' life stress is pessimistically allied with job satisfaction in IT firms at an individual level which supports hypothesis 1. Life stress at group level was adversely related with job satisfaction and supports hypothesis 2. Group- level life stress is also moderated by association among life stress of IT employees' and job satisfaction on an individual-level, which supports hypothesis 3.

Table 2. Hierarchical linear models of Model 1 and Model 2

	Model 1	Model 2
	β (S.E.)	β (S.E.)
Fixed effect		
job satisfaction (β0)		
Intercept (ϒ00)	3.43** (0.09)	3.41** (0.09)
Life stress (β1)		
Intercept (ϒ10)		-0.43* (0.06)
Age (β2)		
Intercept(ϒ20)		0.19* (0.03)
Gender(β3)		
Intercept (ϒ30)		-0.11*
Variance component		
Group Level		
Average intercept (τ00)	0.19*	0.20*
Slope (τ11)		
Individual Level (σ2)	0.89	0.72
**P<0.01(two tailed test); *P<0.05(two tailed test); "ϒ" as intercept'; " τ" as slope		

Table 3. Hierarchical linear models of Model 3

	Model 3
Factors	*β (S.E.)*
Fixed effect	
job satisfaction (β0)	
Intercept (ϒ00)	2.12* (0.13)
Slope (ϒ01)	-0.19*(0.09)
Life stress (β1)	
Intercept (ϒ10)	-0.29* (0.19)
Slope (ϒ11)	0.21*(0.11)
Variance component	
Level 2	
Average Intercept (τ00)	0.15*
Slope (τ11)	0.11*
Level 1 (σ2)	0.68
**p<0.01(two tailed); *p<0.05(two tailed)	

The result in table II of job satisfaction and IT employees' life stress being adversely correlated shows that minor stress increases performance during the job in IT firms. IT firms management bolster helps in diminishing worry in employees, authoritative help and management boost acts during work as a lagging which acts decidedly in diminishing work related stress in employees' [10]. According to [11], Job satisfaction of IT employees' was adversely associated with work related anxiety or stress and as a result the employees' encounter lower job satisfaction due to high stress during work. Furthermore, employees' encounter lower job satisfaction during work not only due to work related stress, but higher life stress.

This multilevel study was intended to analyze at cross level, the moderating impact of life stress at group-level on associations among IT employees' life stress at individual level as well as employees' job satisfaction in IT firms. Many studies have been conducted on job satisfaction of employees [12], but the association between individual level job satisfaction and group level stress had not been examined. Consequently, this research suggested life stress implicit the contextual variable role in group level in IT firms.

In table III, life stress at group level strongly moderated the association among individual-level IT employees' life stress and employees' job satisfaction in IT firms (Υ_{11} =0.21, $p<0.05$). Also, it showed that the general understanding of group level life stress would emphatically encourage the connection among individual-level IT employees' life stress and employees' job satisfaction. Findings of this research uncovered that amid accentuation on the association among IT employees' life stress and job satisfaction, basic function of the well IT firms working conditions and customs climate in developing low sentiment of stress for IT employees' ought to be given more vigilant consideration. According to [1], employees in socioeconomic status firms tend to have mediocre job satisfaction compared to those in privileged socioeconomic status firms. We instigate that male IT employees were more happy and content with their jobs than females IT employees in IT firms in Beijing. The results from table II showed that older IT employees were more happy and content with their job than younger IT employees' (Υ_{20} = 0.19, $p<0.05$).

Finally, this study has some limitations which can provide further study roadmaps for careful consideration. First, the sample size of this study is relatively small (the total population), which may limit the effectiveness of this study. The respondents in this study came from Beijing, China. Since the results of this study are based on a sample of Beijing based IT employees', the universality of the findings is limited. Furthermore, to evaluate job satisfaction of IT employees' using single item might be not imitate this construct as utterly as would the exploit of multi items. In fact, according to [13], single item measures illustrate good reliability and good contemporaneous validity in surveys of IT firms. Future studies needs to replicate this study in different environments or use cross-cultural data to determine the universality of our findings. Unstructured feedback from employees, in the form of natural language text, could also be collated and analyzed using advanced sentiment and opinion mining techniques [14]. The findings could be correlated where appropriate, with analytical outcomes from conventional structured studies reported in this paper, and also elsewhere.

References

1. U. Bashir and M. I. Ramay, "Impact of stress on employees job performance: A study on banking sector of Pakistan," J. Mark., **2**, 122–126, (2010)
2. S. E. Anderson, B. S. Coffey, and R. T. Byerly, "Formal organizational initiatives and informal workplace practices : links to work – family conflict

and job-related outcomes," J. Manage., **28**, 787–810, (2002)

3. J. M. Ivancevich and J. H. Donnelly, "Relation of organizational structure to job satisfaction, anxiety-stress, and performance," Adm. Sci. Q., **20**, 272–281, (1975)
4. M. Rose, "Good deal,baddeal? Job satisfaction in occupations," Work Employ. Soc., **17**, 503–530, (2003)
5. R. G. Netemeyer, J. G. Maxham, and C. Pullig, "Conflicts in the work–family interface: links to job stress, customer service employee performance, and customer purchase intent," J. Mark., **69**, 130–143, (2005)
6. C. L. Ho and W. T. Au, "Teaching satisfaction scale:measuring job satisfaction of techers," Educ. Psychol. Meas., **66**, 172–185, (2006)
7. R. P. Chaplain, "Stress and job satisfaction: a study of english primary school teachers," Educ. Psychol., **15**, 473–489, (1995)
8. L. R. James, R. G. Demaree, and G. Wolf, "Rwg: An assessment of within-group interrater agreement," J. Appl. Psychol., **78**, 306–309, (1993)
9. A. S. Bryk and S. W. Raudenbush, *Hierarchical Linear Models for Social and Behavioral Research: Applications and Data Analysis Methods* (Newbury Park, CA: Sage, 1992)
10. C. L. Stamper and M. C. Johlke, "The impact of perceived organizational support on the relationship between boundary spanner role stress and work outcomes," J. Manage., **29**, 569–588, (2003)
11. H. M. Weiss, "Deconstructing job satisfaction: Separating evaluations, beliefs and affective experiences," Hum. Resour. Manag. Rev., **12**, 173–194, (2002)
12. J. L. Romano and K. Wahlstrom, "Professional stress and well-being of K-12 teachers in alternative educational settings: a leadership agenda," Int. J. Leadersh. Educ., **3**, 121–135, (2010)
13. A. M. Abdel-Khalek, "Measuring happiness with a single-item scale," Soc. Behav. Pers., **34**, 139–150, (2006)
14. Poria, S., Cambria, E., Bajpai, R. and Hussain, A. "A review of affective computing: From unimodal analysis to multimodal fusion". Information Fusion, **37**, 98-125, (2017)

Comparative study of prediction methods of gel point of mixed crude oil

Chaofei Nie[1], Huanhuan Wang[2], Hao Lan[1], Bo Xu[1], Feng Yan[1], Qifu Li[1], Yu Diao[1]

[1] PetroChina Pipeline Research and Development Center, Lang fang, China
[2] Southwest Oil an Gas Field Company Northwest of Sichuan Gas Production District, Mianyang, China

Abstract:Gel point is an important parameter of the safe pipeline transportation of the waxy crude oil, and predicting the gel point of the mixed crude oil quickly and precisely has a certain practical significance for this kind of mixed transportation. By adopting the scientific evaluation method of the prediction model, the adaptation of the Model Li Chuangwen, Model Liu Tianyou I, Model Liu Tianyou II, and Model Chen Jun to predicting the gel point of mixed crude oil transported in a certain pipeline is analyzed and compared, by which the most appropriate model for the prediction of the mixed crude oil is confirmed. It is found that Model Liu Tianyou I is most suitable for the mixed crude oil. Therefore, from the engineering point of view, for the mixed crude oil mentioned in this text, the gel point calculated by Model Liu Tianyou I has a certain reference value.

Keywords:mixed crude oil; gel point; prediction model; evaluation method.

I. INTRODUCTION

The gel point is the highest temperature of crude oil losing fluidity under the test conditions, and it is an important parameter to guide the safe pipeline transportation of waxy crude oil [1]. When two components crude oil of different gel points are mixed, the gel point of mixed crude oil is usually between that of two components crude oil, and does not take a simple linear relation with the mix ratio of the two components [2]. The ratio of the two components crude oil is subject to changes with practical crude oil production and transportation. While the gel point can be measured by experiment, the testing process is relatively complicated. Therefore, a fast and accurate calculation method of the gel point of mixed crude oil has a certain practical significance for the pipeline process calculation.

II. COMMON GEL POINT PREDICTION MODEL

The existing calculation model of mixed crude oil gel point is the empirical or semi-empirical formula derived mainly by the users from inductive experimental data. The existing calculation model is often of good calculation precision for the oil sample used in modeling, but not for the other crude oil. When these models are used, it is necessary to evaluate the applicability of the models first, and select the model of higher accuracy specific to a certain research object.

The common gel point calculation models are Model Li Chuangwen [3-4], Model Liu Tianyou I, Model Liu Tianyou II [5-6] and Model Chen Jun [7]. Model Li Chuangwen and Model Liu Tianyou are calculation formula gained by amending the linear model through the experimental data of crude oil and mixed crude oil in the representative single well of the various Xinjiang oil fields. When these two models are used, not only the gel point of each component crude oil but also of each two-component crude oil mixed in the same ratio need to be known. In the analysis of the data of gel point of mixed crude oil, Chen Jun found that the absolute value of the difference between the measured value of and the linear weighted method calculated value of the gel point of the two-component mixed crude oil correlates properly with the absolute value of the difference between the gel points of each component crude oil. Therefore, a new method of calculating mixed crude oil gel point is put forward where the gel point of components crude oil mixed in various ratios can be predicted only when the gel point of each component crude oil is known. In this paper, a model suitable for predicting the gel point of the mixed crude oil in a pipeline is selected from the above four models.

The calculation formula of each model is shown in Table I.

 DOI: 10.1051/matecconf/201713900115

Table I Calculation Model of Gel point of Mixed Crude Oil

Model	Calculation Formula
Li	$T_G=\sum_{i=1}^{n}(X_iT_i)+\sum_{j=1}^{n-1}\sum_{k=j+1}^{n}\left(B_{jk}C_{jk}X_jX_k\right)$ $B_{jk}=\left[\log(100X_j)/\log(100X_k)\right]^{\mathrm{sgn}(C_{jk})}$ $C_{jk}=2(2T_{jk}-T_j-T_k)$
Liu I	$T_G=\sum_{i=1}^{n}(X_iT_i)+\sum_{j=1}^{n-1}\sum_{k=j+1}^{n}\left(B_{jk}C_{jk}X_jX_k\right)$ $B_{jk}=1-\frac{X_k}{2}+\frac{X_j}{2}$ $C_{jk}=2(2T_{jk}-T_j-T_k)$
Liu II	$T_G=\sum_{i=1}^{n}(X_iT_i)+n^nX_1X_2\cdots X_n(T_{1/n}-\frac{1}{n}\sum_{j=1}^{n}T_j)$
Chen	$T_G=\sum_{i=1}^{n}(X_iT_i)+\sum_{j=1}^{n-1}\sum_{k=j+1}^{n}\left(B_{jk}C_{jk}X_jX_k\right)$ $B_{jk}=1-\frac{X_k}{2}+\frac{X_j}{2}$ $C_{jk}=\pm0.698\lvert T_j-T_k\rvert^{1.1456}$

Note: In the table, T_G stands for the gel point of mixed crude oil, X_i stands for the mix proportion of component oil i, T_i the gel point of the component oil i, T_{jk} the gel point of components oil j and k mixed in the same ratio, n the number of component oil type, $T_{1/n}$ the gel point of n kinds of components oil mixed in the same ratio.

As can be seen from the table, all the methods above have adopted a hybrid rule method. The main difference is that Bij, modified parameter of the ratio and Cij, modified parameter of mixed oil properties have different calculation methods. If Model Li Chuangwen, Model Liu Tianyou I and Liu Model Tianyou II are used, not only the gel point and the ratio of each component oil but also the gel point of any two components oil mixed in the same ratio need to be known. Nevertheless, under the condition of Model Chen Jun, the gel point of the mixed crude oil can be calculated only when the gel point and the ratio of each component oil are known.

III. Prediction Model Evaluation Method

A. Evaluation index

For the screening of gel point models of mixed crude oil, the following evaluation indexes were selected.

① Absolute Average Deviation（AAD）

$$AAD=\frac{1}{n}\sum_{i=1}^{n}\lvert GPD_i\rvert \quad (1)$$

$$GPD=T_p-T_m \quad (2)$$

In the formula above, T_p is the measured value of gel point and T_m is the calculated value of gel point.

GPD is the absolute deviation of the measured value and the calculated value of gel point. ADD, Absolute Average Deviation, is the average value of the absolute deviations calculated in n kinds of gauging points with the given model.

② Maximum Absolute Deviation（MAD）

$$MAD=\max\lvert GPD_i\rvert \quad (3)$$

MAD, Maximum Absolute Deviation, is maximum value of the absolute deviations calculated in n kinds of gauging points with the given model.

③ Absolute Deviation Distribution（Da）

$$Da=\frac{n_a}{n} \quad (4)$$

Da, Absolute deviation distribution, is the ratio of the number of measuring points whose deviation exceeds a certain limit to the total number of measuring points. It is stipulated in SY / T0541-2009 "Measurement Method of Gel point of Crude Oil" that if the same operator using the same equipment in the same lab measures the same oil sample twice under the same experimental conditions, the difference between the two measuring results should not exceed 2 ℃. Therefore, na takes the measuring points where the calculation deviation of the model exceeds 2 ℃.

The above three evaluation indexes are inverse indexes, that is, the smaller the index value, the better the model.

B. Data Standardized Process

For evaluation process of the multi-index evaluation, the measurement units and the magnitude of the indexes are not the same, and the significance degree of the comprehensive evaluation of the accuracy of the models is not the same for different indexes, so the statistical index values cannot simply be added. In order to make each index value comparable, the data should be processed first, and then the appropriate method is selected to integrate data.

Data standardization is also the indexation of statistical data. The indexes of different properties directly added cannot correctly reflect the comprehensive results of different forces. Therefore changing the property of the inverse index data has to be considered first so that all the indexes have the same chemotaxis on the evaluation scheme, and then the correct result can be gained by adding the indexes. The data dimensionless processing is mainly to solve the problem of data comparability. After the chemotaxis and dimensionless processing of the data, the original data are converted into the measured values of the dimensionless indexes, that is, the index values are in the same level of magnitude, where a comprehensive evaluation analysis can be conducted.

There are many methods of data standardization. For the same chemotaxis treatment of inverse index, the threshold value method is used to deal with the data matrix formed by each model index to be evaluated to change the index property. That is,

$$X = (x_{ij})_{m\times n} \xrightarrow{\text{阈值法}} Y = (y_{ij})_{m\times n} \quad (5)$$

In the formula:

X—data matrix made up of the index values;

Y—data matrix of index values after dimensionless treatment;

m—the number of models;

n—the number of indexes.

Among them, the conversion rule of the threshold value method matrix is,

$$y_{ij} = \frac{\max\limits_{j=1,2\cdots m} x_{ij} - x_{ij}}{\max\limits_{j=1,2\cdots m} x_{ij} - \min\limits_{j=1,2\cdots m} x_{ij}} \quad (i=1,2,\cdots k) \quad (6)$$

In the formula, i and j represent the jth index of the ith model; m and k represent the number of indexes and models respectively.

The evaluation value y_{ij} calculated by the formula 6 decreases as x_{ij} increases in the interval [0,1]. After the treatment of threshold value method, the three indexes, which are originally the inverse indexes, can be converted into the positive index evaluation values, and the original data can be transformed into the dimensionless index evaluation values, which can be processed by the comprehensive evaluation analysis.

C. Determining the weight of the indexes

The significance of each evaluation index varies from model evaluation to model evaluation. To reflect the difference of its function, it is necessary to make the scientific weight adjustment to the valuation indexes to ensure the rationality of the comprehensive evaluation result. The method of determining the weight of indexes mainly includes subjective weighting method and objective weighting method. The subjective weighting method adopts qualitative method, where weights are obtained by experts making subjective judgments according to experience, and then the indexes are evaluated synthetically, such as Analytic Hierarchy Process (AHP), Expert Survey (Delphi), Fuzzy Analysis, Binomial Coefficient and so forth. The advantages of these methods are that experts can determine the sequencing of the indexes more reasonably based on practical problems. The objective weighting method starts with the independent data itself, and the comprehensive evaluation is undertaken based on the correlation between the research indexes of historical data or between the indexes and the evaluation results. These methods include Maximum Entropy Technique, Principal Component Analysis, Multi-Objective Planning Method, Standard Deviation method, etc. Although the evaluation results of these methods have a strong mathematical theoretical basis, the method depends heavily on the actual problem domain, versatility and participation of decision-makers are poor, and the calculation methods are more cumbersome.

Considering that the three evaluation indexes selected in this paper do not accord with the characteristics of mutual independence, Expert Survey Method in the subjective weighting method are selected to determine the index weight. It is the Absolute Deviation Distribution (Da) among the three indexes that has the greatest effect on the model. Second only to Absolute Deviation Distribution (Da), is the Absolute Average Deviation (AAD), and the Maximum Absolute Deviation (MAD) has the smallest effect. The weights of the three indexes are ω_{Da}, ω_{AAD} and ω_{MAD} respectively, then

$$\omega_{Da} + \omega_{AAD} + \omega_{MAD} = 1 \quad (7)$$

$$\omega_{Da} > \omega_{AAD} > \omega_{MAD} \quad (8)$$

It is assumed that the weight of the index value that has the greatest effect on model accuracy is five times that of the least significant one, then

$$\omega_{Da} = 5\omega_{MAD} \quad (9)$$

Combine the formula 7 and 8 to determine the weights of the three indexes were $\omega_{Da}=0.5$, $\omega_{AAD}=0.4$, $\omega_{MAD}=0.1$ respectively.

D. Multi-index comprehensive evaluation method

The Weighted Average Method is adopted to synthesize the evaluation values of the three indexes of each model. That is, the comprehensive evaluation matrix E is obtained by multiplying the data matrix Y after the index transformation and the weight matrix ω T, seen in formula 10. The descending sequencing of the comprehensive evaluation values of the matrix is accordance with the relative superiority and inferiority (sequencing) of the model accuracy.

$$Y \cdot \omega^T = E \quad (10)$$

IV. Prediction Model Comparison

In order to verify the calculation accuracy of the above models, gel points of 111 sets of mixed crude oil of different components and mix ratio for are tested, and the experimental results and the calculated values of the gel points of the four models are compared. The comparison results are shown in Fig.1 to Fig.4, and the statistical results seen in Table II.

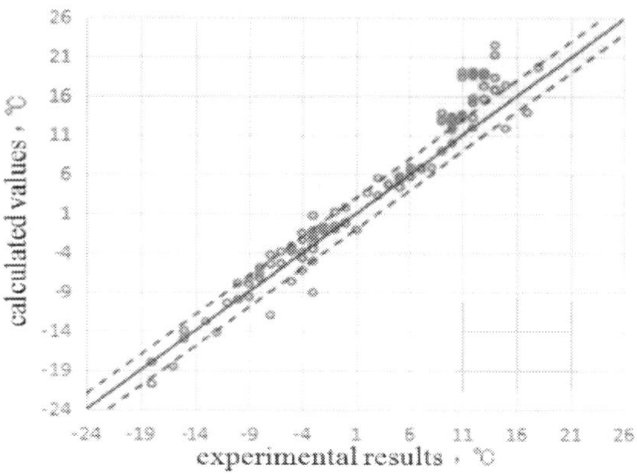

Figure.1 Comparison Between Calculated Value and Measured Value of Model Li Chuangwen

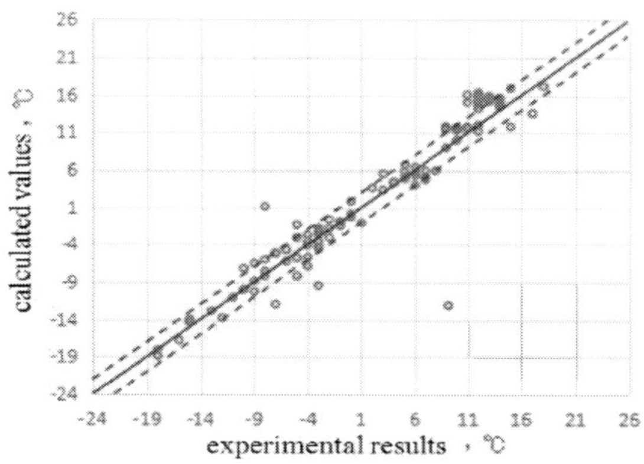

Figure.2 Comparison Between Calculated Value and Measured Value of Model Liu Tianyou I

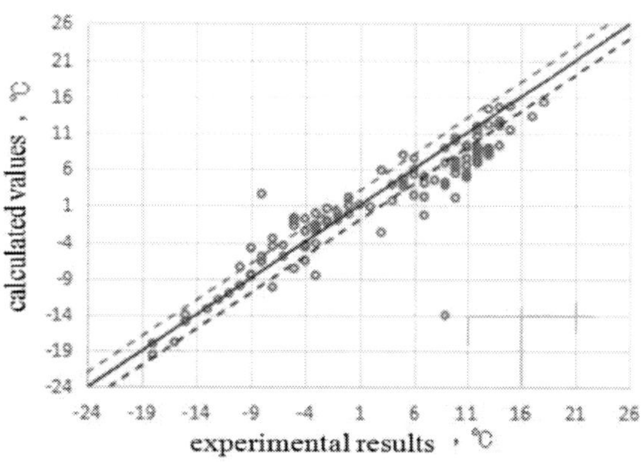

Figure.3 Comparison Between Calculated Value and Measured Value of Model Liu Tianyou II

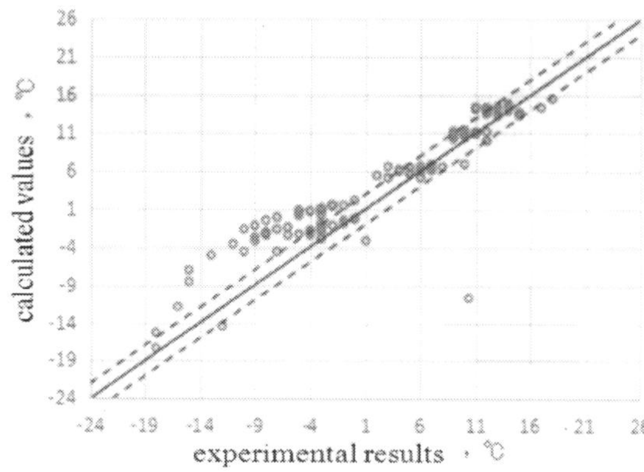

Figure.4 Comparison Between Calculated Value and Measured Value of Model Chen Jun

Table II Statistical Result of Calculation Value of the Four Models

Evaluation Index	Statistical Analysis of Calculation Results of the Four Models			
	Model Li	Model Liu I	Model Liu II	Model Chen Jun
AAD	2.57	2.61	2.82	4.32
MAD	3.10	3.40	3.68	7.98
Da	32.35	17.65	17.65	55.88

The contrastive analysis results indicate that there is no significant difference in the Absolute Average Deviation of calculation results between Model Li Chuangwen, and Model Liu Tianyou Modified II and that Model Liu Tianyou Modified I and Model Liu Tianyou Modified II are nearly equivalent in the distribution of both Absolute Average Deviation and Absolute Deviation and that the accuracy of Model Chen Jun is lower than the former three.

According to the above-mentioned model evaluation methods and through scientific calculation, the comprehensive evaluation value of model accuracy is shown in Table III.

Table III Comprehensive Evaluation Results of Gel point Calculation Models of Mixed Crude Oil

Model	X			Y			E
	AAD	MAD	Da	AAD	MAD	Da	
Liu I	1.65	9.13	0.3051	1	0.66	1	0.97
Li	2.40	8.55	0.4831	0.18	0.92	0	0.14
Chen	2.5	8.38	0.4746	0	1	0.0478	0.13
Liu II	2.37	10.58	0.4746	0.15	0	0.0478	0.09

V. Conclusions

In summary, based on the comprehensive evaluation results, when the gel point of each component crude oil and of two components crude oil mixed in the same ratio are available, it is recommended to use Model Liu Tianyou Modified I to calculate the gel point of mixed crude oil. Model Li Chuangwen equals Model Chen Jun in the comprehensive evaluation results. Of the four models, Model Liu Tianyou Modified II is least suitable for calculating the gel point of mixed crude oil.

References

[1] Luo Tanghu. Rheological Properties and Pipeline Transportation Of Waxy Crude Oil [M]. Petroleum Industry Press, 1991.

[2] Yang Xiaoheng, Zhang Guozhong. Design and Management of Oil Pipeline [M]. Petroleum University Press, 1996.

[3] Li Chuangwen, Yan Dafan, Zhang Jinjun. Change of Congealing Points of Crudes and Blends of Xin-jiang Oil Field [J]. Petroleum Planning and Design, 1996 (06):24-27.

[4] Li Chuangwen. A Study of Rheology of Mixed Crude Oil and Its Compatibility Regularity [D]. Beijing: School of Petroleum and Natural Gas Engineering of Petroleum University, 1992.

[5] Liu Tianyou, Zhang Xiujie, Xu Cheng. Calculation of Pour Point of Xinjiang Mixed Oil [J] .Oil and Gas Storage and Transportation, 1993 (2): 37-45.

[6] Liu Tianyou, Sun Weizhong, Gao Yanqing, Xucheng. Study on the Ordinary Temperature Process of Multi-Blended Crude [J]. Oil and Gas Storage and Transportation, 1999(09):1-6.

[7] Chen Jun, Zhang Jinjun, Zhang Fan. A New Method for Determining Gel Point of Mixed Crude [J]. Journal of the University of Petroleum (Natural Science Edition), 2003 (05):76-80.

Research on Connection and Function Reliability of the Oil&Gas Pipeline System

Bo Xu[1], *LinJie Duan*[1,*], *XiangDong Xue* [1], *Hao Lan*[1], *Kaizhi Chen*[2]

[1]Petro China Pipeline Research and Development Center, Lang fang, China
[2]Petro China West-to-East Gas Pipeline Company Wuhan Management Department, Wuhan, China

Abstract. Pipeline transportation is the optimal way for energy delivery in terms of safety, efficiency and environmental protection. Because of the complexity of pipeline external system including geological hazards, social and cultural influence, it is a great challenge to operate the pipeline safely and reliable. Therefore, the pipeline reliability becomes an important issue. Based on the classical reliability theory, the analysis of pipeline system is carried out, then the reliability model of the pipeline system is built, and the calculation is addressed thereafter. Further the connection and function reliability model is applied to a practical active pipeline system, with the use of the proposed methodology of the pipeline system; the connection reliability and function reliability are obtained. This paper firstly presented to considerate the connection and function reliability separately and obtain significant contribution to establish the mathematical reliability model of pipeline system, hence provide fundamental groundwork for the pipeline reliability research in the future.

1 Introduction

Pipeline transport is one of the five main transportation modes in the world, and it is the most suitable way for crudes, refine products and natural gas transportation. Because the pipeline system has the advantages of continuous operation, large capacity and low cost, and it is no limited by the climate and other ground. Taken the natural gas pipeline system within China as an example, it supplies more than 600 cities, feed more than 0.8 billion local citizens each day; it has become a lifeline to secure energy supply in China.

Because the mileage distance of China's natural gas resources and markets is somehow dramatically long, it has many hazards such as geological hazards along the pipeline network, fragile ecological environment and complex humanity environment, with potential to threat the safety and reliable operation of the pipeline. Considering all these factors together, China's pipeline network system becomes the most complicated and the toughest transport task [1-2].

During the period of service, oil&gas Pipeline pipe often under challenges such as various load, corrosion, and third party damages, which could lead to pipeline system failure, cause unscheduled shutdown. Some hazards, such as pipeline leak will directly cause substantial threat to the natural environment and social security [3-4].

How to effectively assess the risk of oil & gas pipeline network system, work out the weakest point of the system, implement a preventive management for the large scale oil&gas pipeline network in its whole life cycle, to ensure the safe operation safety and achieve efficient transportation is the key issue in today's pipeline research area. All these are major problems to be urgently solved in pipeline industry [5].

Therefore, it is important to use a well-accepted tool such as the reliability technology to analyse the pipeline system, to satisfy the demands which are presented by the pipeline industry to ensure that the pipeline system could be settled in a safe and reliable state for its life time.

Reliability is a fundamental attribute for the safe operation of any modern technological system. Focusing on safety, reliability analysis aims at the quantification of the probability of failure for the system and its protective barriers [6-7]. It refers to that the ability or possibilities of system to complete the required function within the stipulated time and working conditions.

These barriers are intended to protect the system from failures of any of its components, hardware, and software, human and organizational. These all need to be addressed by the system reliability analysis in a comprehensive and integrated manner [8].

In this paper, the reliability of the pipeline system is calculated and analysed. Through the reliability modelling of the oil and gas pipeline system, the connection reliability and function reliability of the system are analysed separately. The system failure is divided into its physical function failure and process condition failure, which can provide substantial reference for the further study of pipeline system reliability.

* Corresponding author: 270426450@qq.com

DOI: 10.1051/matecconf/201713900116

2 Pipeline system reliability model

In the pipeline industry, reliability modeling method is not well established. Therefore it need to be started from a simple case study, and then on this basis, to optimization the pipeline system reliability mode and method gradually. According to ISO/TR 12489 which describing pipeline reliability modeling method, it includes the statistical method, Boolean model method and reliability of simulation method for modeling reliability.

Based on the common reliability modeling, reliability block diagram method is relatively straightforward, which would establish a simple physical model for the pipeline system, as shown in figure 1.

Due to the complexity of station structure, the paper firstly presents a typical station structure for the gas station as shown in Fig. 2 in order to make the complex structure simpler. Further the station valves, compressors and other equipment within station needs to be initialized with its own reliability model.

3 Pipeline system reliability calculation methods

According to the established pipeline system reliability model, system reliability is calculated. It is known that if m compressor units in the n stations are available, that meets the production requirement. In this paper, pipeline failure mode can be divided into connection failure and function failure.

Connection failure describe the physical structure in the pipeline system to a fault, by any means that cannot form effective pathway, cause physical truncation pipeline system; Function failure is the piping system under the condition of can mutually connected to form pathways, due to the technological equipment but cannot meet the required by the process conditions such as transporting the required amount of volume's

When calculate system reliability, different failure modes should be taken into account. Connection and function reliability of system is calculated respectively. Given compressed station as an example, if it is categories according to the different failure mode, then all of the devices within the station will be divided into function devices and connectivity devices, and then the compressor unit will be regarded as a function device.

3.1. Pipeline system connectioned reliability calculation

By the system reliability block diagram, pipeline system is series system that consists of station, pipe and valve chamber. The pipeline system reliability formula can be written as:

$$R_{SC} = R_{Csti} \cdot R_{Pi} \cdot R_{vsi} \tag{1}$$

Where, R_{SC} is system connectioned reliability, R_{Csti} is the i-th connectioned reliability of station i, R_{Pi} is the pipe segment reliability, R_{vsi} is the valve chamber reliability.

Pipe segment inside the pipeline system can be divided into pipes in the station and pipes outside the station. For the pipes in station section, due to the short distance, inspections in a timely manner and the maintenance level is high, the reliability is relatively high, close to 1,when calculating the pipe reliability is not considering. Pipes outside the station is used to connection all kinds of pipe between the station and valve chamber. Based on section reliability of literature at home and abroad, found that pipe fault time obey to the exponential distribution, indicators for the constant failure rate. The available section failure rate:

$$\lambda_i = \frac{1}{MTBF_i} \tag{2}$$

According to the index distribution function, pipe reliability calculation formula can be written as:

$$\mathrm{R_{Pi}} = \exp(-\lambda_i \mathrm{t}) \tag{3}$$

To analsye the station shown in Fig 2， compressor unit equipped with a bypass circuit, and station has the bypass station processes. So compressor unit is considered to keep confectioned all the time when calculating the station connect reliability. The compressor reliability is 1. The Station connection reliability formula can be written as:

$$R_{CSTi} = \left[1 - \left(1 - R_{filt}R_{bypass}\right) . \left(1 - R_{trans}\right)\right] . R_{pullin} . R_{output} \tag{4}$$

Where, R_{inpart} is pull in part reliability, R_{bypass} is by pass valve reliability, R_{filt} is filtering part reliability, R_{trans} is the trans station part reliability, R_{output} is in put part reliability.

3.2 Pipeline System Function Reliability Calculation

Under the condition of meet the component form interconnection pathways, pipeline system may still fail. This is mainly due to the process equipment cannot reach the required process conditions. This kind of failure is called the function failure. It is need to calculation and analysis of system function reliability.

For pipeline system, satisfy the technological conditions of n in any m stage compressor in the available state that meet the production requirements. The pipeline system reliability function can be written as:

$$R_{SF} = R_{Fsti} \cdot R_{Pi} \cdot R_{vsi} \tag{5}$$

Where, R_{SF} is system function reliability, R_{Fsti} is the i-th function reliability of station i, R_{Pi} is the pipe segment reliability, R_{vsi} is the valve chamber reliability.

The compressor and its corollary equipment are the main function components of the station. In the process of calculation, for possible to simplify the model, other provisions different compressor station used are the same type, performance indicators. Because the station function realization based on system connection, the station function reliability can be defined as:

$$R_{Fst} = R_{Cst} \cdot \sum_{k=m}^{n} C_n^k \cdot R_{com}^k \cdot \left(1 - R_{com}\right)^{n-k} \tag{6}$$

Where, R_{Fst} is station function reliability, R_{Cst} is station connection reliability, R_{com} is compressor unit reliability.

According to the vendors to determine parameters and historical operation data, compressors and its components reliability index is determined.

4 Application

According to the above method, combining the reality of an active pipeline, selection of calculation and analysis a particular operating conditions. The pipeline system consists of four stand five valve chamber and confectioned section of each length of pipe section list as shown in table I. First (ST1) and station 3 (ST3) for compressor station, station is equipped with three the same type compressors. Under the selection of operation condition, the compressor work plan for 2 in 1 case.

Table 1. The length of each pipe

pipe	length (km)	pipe	length (km)
1	21.2	5	22.8
2	27.6	6	18.6
3	29.7	7	18.7
4	7.2	8	14.7

The pipeline system physical model is selected from the SCADA system, as shown in figure 3. Modeling method based on reliability, reliability modeling was carried out on the pipeline, selection of pipeline reliability model as shown in figure 4.

The pipeline system connectioned reliability is calculated based on the above method.

$$\begin{aligned} R_{SC} &= R_{CST1} \cdot R_{P1} \cdot R_{VS1} \cdot R_{P2} \cdot R_{VS2} \cdot R_{P3} \cdot \\ &R_{CST2} \cdot R_{P4} \cdot R_{VS3} \cdot R_{P5} \cdot R_{VS4} \cdot R_{P6} \cdot R_{CST3} \\ &\cdot R_{P7} \cdot R_{VS5} \cdot R_{P8} \cdot R_{CST4} \end{aligned} \tag{7}$$

Station connectioned reliability:

$$\begin{aligned} R_{CST1} &= R_{CST3} = R_{output} . R_{pullin} \cdot \\ &\left[1 - \left(1 - R_{filt} R_{bypass}\right) . \left(1 - R_{trans}\right)\right] \\ &\cdot R_{P7} \cdot R_{VS5} \cdot R_{P8} \cdot R_{CST4} \end{aligned} \tag{8}$$

Pipeline System Function Reliability:

$$\begin{aligned} R_{SF} &= R_{FST1} \cdot R_{P1} \cdot R_{VS1} \cdot R_{P2} \cdot R_{VS2} \\ &\cdot R_{P3} \cdot R_{FST2} \cdot R_{P4} \cdot R_{VS3} \cdot R_{P5} \cdot R_{VS4} \\ &\cdot R_{P6} \cdot R_{FST3} \cdot R_{P7} \cdot R_{VS5} \cdot R_{P8} \cdot R_{FST4} \end{aligned} \tag{9}$$

Station Function Reliability:

$$R_{FST1,3} = \left[R_{com}^3 + C_3^2 \cdot R_{com}^2 \cdot \left(1 - R_{com}\right)\right] \cdot R_{CST1} \tag{10}$$

Equipment unit reliability data information was got by historical statistics, as shown in table II.

Table 2. The length of each pipe Equipment unit reliability data information

name	Reliability
valve	0.9902
Compressor	0.9775
natural gas filter	0.906
Trans station valve	0.999
Pipe failure rate	0.00032time/km•year

Pipeline System connection and function Reliability is 0.85223 and 0.84968 respectively.

$$R_{SC} = 0.85223$$

$$R_{SF} = 0.84968$$

5 Conclusions

This article simplifies the extreme complicated oil and gas pipeline system and it is the first paper to establish the reliability model of pipeline system based on its typical structure. Based on these achievements, the system connection reliability and function reliability calculation method are presented. This paper can provide significant reference for the future research in the area of pipeline system reliability. However it did make some assumptions, as follows:

First, the pipeline system described in the paper is inherent with the real pipeline system to establish a simplified virtual pipeline system, but the modeling process involves only the key components, ignore other trial components.

Second, in this paper, the calculation is based on the relatively normal working condition; it did not include all kinds of operation conditions, where the computation will be much more difficult. But the strategy and the methodology would be the same, and it can be dialed with the model and algorithm proposed in the paper.

References

1. Huang Weihe. Reliability of large-scale natural gas pipeline network [J]. Acta Petrolei Sinica, 2013 Vol. 34 (2): 401-404.
2. Ai Muyang. Discussion on issues regarding the reliability of large-scale oil and gas pipeline network systems [J]. oil & gas storage and transportation, 2013 Vol. 32 (12): 1265-1270.
3. Kariyawasam S, Peterson W. Revised corrosion management with reliability based excavation criteria. In: Proceedings of IPC 2008, IPC2008-64536. Cal-gary: ASME; 2008.
4. Stephens M, Nessim M. A comprehensive approach to corrosion management based on structural reliability methods. In: Proceedings of 6th International Pipeline Conference, IPC2006-10458. Calgary, Canada: ASME; 2006.
5. Huang Weihe. Transportation technigue for oil&gas pipoeline[M] . Beijing : Petroleum Industry Press，2012：9-11.
6. Reason J. Managing the risk s of organisational accidents. Aldershot: Ashgate Publishing Ltd.; 1998.
7. Baraldi P, Zio E. A combined Monte Carlo and possibilistic approach to uncertainty propagation in event tree analysis, risk analysis; 2008.
8. Flage R, Aven T, Zio E. Alternative representations of uncertainty in system risk and reliability analysis: review and discussion. In: Proceedings of ESREL2008, Valencia Spain, 22–25 September 2008.

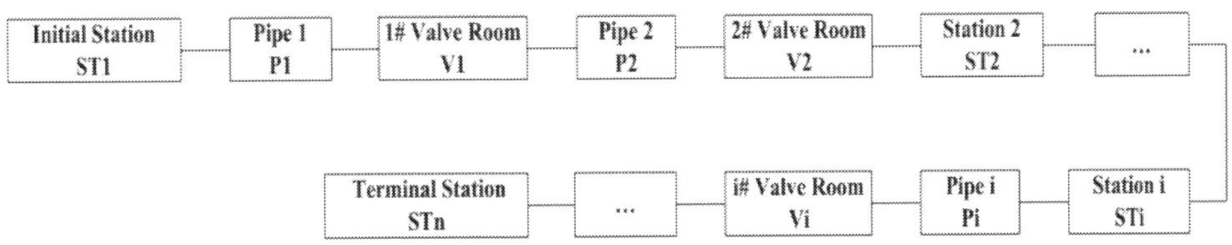

Fig. 1. Brief diagram for Oil & gas pipeline system reliability model.

Fig. 1. .

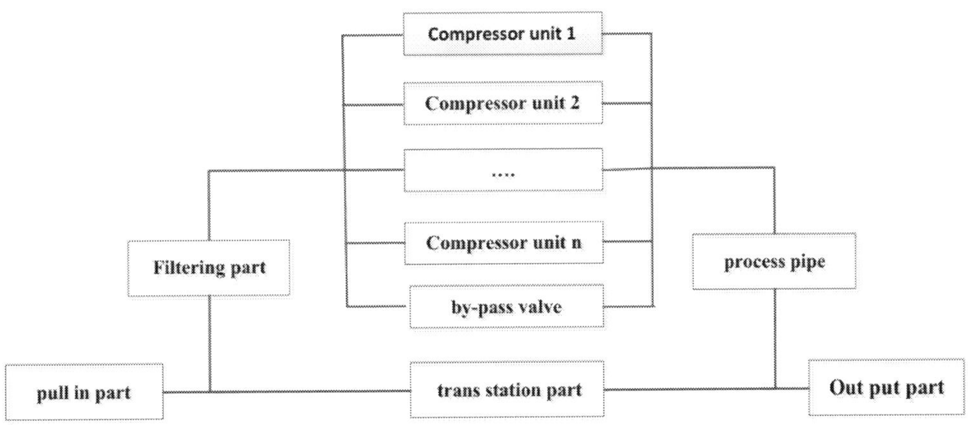

Fig. 2. The compressed air station reliability model.

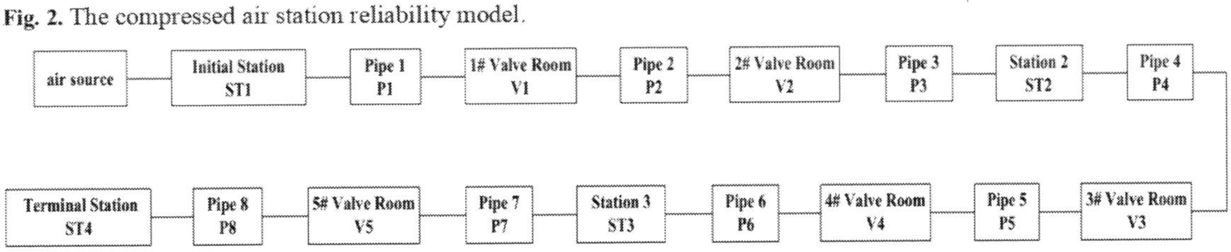

Fig. 3. The pipeline system physical model.

DOI: 10.1051/matecconf/201713900116

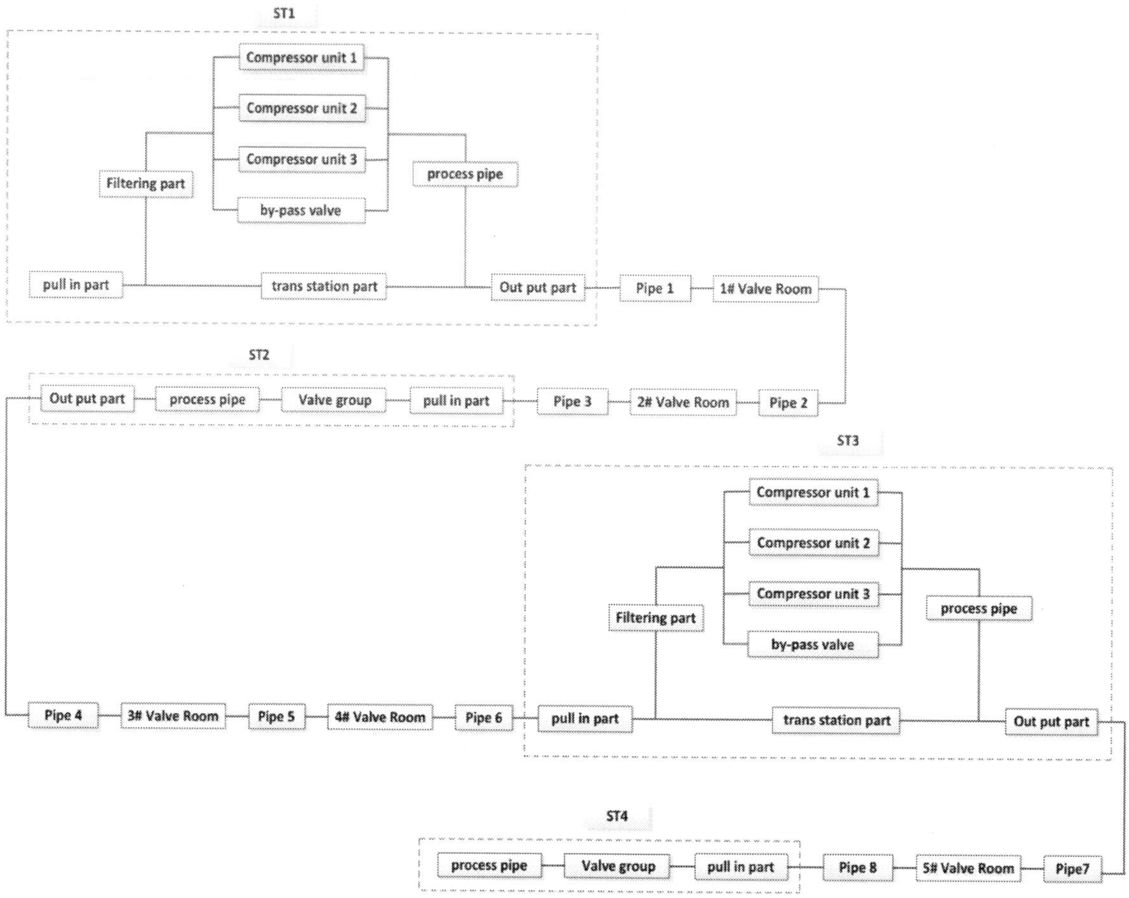

Fig. 3. The pipeline system reliability block diagram.

* Corresponding author: 270426450@qq.com

Display Technology of General Production Report on Cement Enterprise

Yiling Shao [1], and *Shaohong Jing* [2]

[1] School of Electical Engineering, University of Jinan, Jinan, China
[2] School of Electical Engineering, University of Jinan, Jinan, China

Abstract. Production report plays a significant role for the adjustment of industrial structure on cement enterprises. However, types of production reports are various, using general report technology can accelerate development progress and simplify report development process. This paper designs a kind of display technology of general production report based on datagrid which unifies report style, designs fixed template, realizes display function of header area and detail area on production report, and completes information management of cement enterprises

1 Introduction

Production reports play a vital role in the cement industry. Through the production report, cement enterprises understand the drawbacks of their own energy consumption structure, adjust the industrial structure, avoid overcapacity, and achieve energy conservation and emissions reduction. But because types of production reports are numerous, such as xx cement group, the main production reports are divided into two types: basic statistics and comprehensive statistics. Basic statistical reports are segmented into electricity statistics statements, basic production statistics reports, basic electricity statistics statements, basic power consumption statistics statements and other small categories; Similarly, the consolidated statistical statements are also divided into energy-related comprehensive statistical statements, comprehensive power consumption statistical report of clinker line, comprehensive power consumption statistics statements of cement line and other small categories. Two types of production reports have more than 400 sheets in total, and the production report styles of xx cement group headquarters and subsidiaries are different. If taking the traditional way to create a production report for each branch separately, the workload is unpredictable. The use of general production report technology speeds up the development progress and streamlines development process of production report.

At present, there are two commonly used development methods in the common reporting system: ① the use of third-party components available in the market, such as Crystal reports, and so on. Though its operation is simple and powerful, the price of this report system software is higher, especially for small and medium-sized enterprises. ② Install the report system on the server. The software need to be constantly updated and is not conducive to be put into use in bulk [1]. In order to reduce procurement costs and improve work efficiency, this paper designs a kind of display technology of general production report based on datagrid. The development platform is Visual Studio 2012, database is SQL Server 2012.

2 Overall design

Report is universally segmented into header area, detail area, and footer area. Header area occurs at the beginning of a report, and is usually used to display field captions and other information; Detail area is the core of the report and generally shows the actual information of the report body; Requirements on the footer for cement enterprises are not high, not too much discussion. Database stores information of header area and detail area, then the database engine executes SQL queries, through a series of processing, the browser finally shows production report, as shown in Fig. 1.

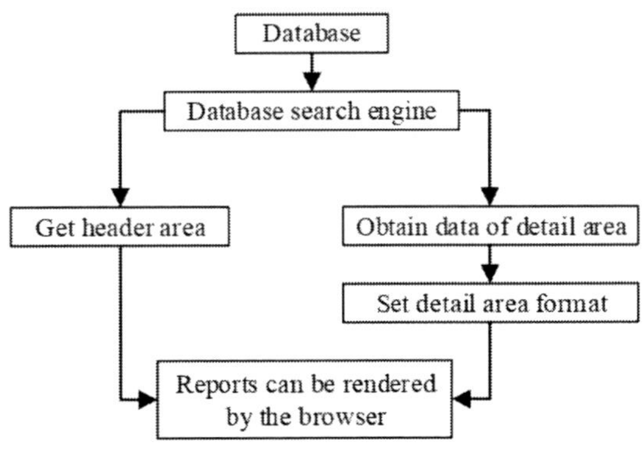

Fig. 1. Overall structural design

* e-mail:cse_jsh@163.com

3 Related technology

JQuery EasyUI is a collection of jQuery-based UI plug-ins, the goal is to help web developers more easily create a feature-rich and beautiful UI interface. Developers don't need to write complex JavaScript or have a deep understanding of CSS styles, and just know a few simple HTML tags. JQuery EasyUI provides easy-to-use components that enable Web developers to quickly build program pages on popular jQuery kernels and HTML5.

Datagrid is a set of plugins of jQuery EasyUI which extends from $.fn.panel.defaults. It displays data in tabular format and provides rich support for selecting, sorting, grouping, and editing data. It is designed to reduce development time, and does not require the developer have specific knowledge. It is lightweight, but multifunctional which its characteristics include merging of cells, multi-column header, frozen columns and footers, and so on.

Datagrid has a variety of methods that can be created by using existing table elements, <table> tags, or Javascript. Use Javascript to create datagrid, code is simple, therefore this article takes the way to write code.

AJAX("Asynchronous Javascript And XML") refers to a web page development technology for creating interactive web applications. By making a small amount of data exchange with the server in the background, AJAX can make the page asynchronously updated. This means that you can update a portion of the page without reloading the entire page. However, if the traditional web page (does not use AJAX) needs to update the content, you must reload the entire page.

C# is a secure, stable, simple, elegant, and object-oriented programming language which is derived from C and C++. It inherits C and C++ powerful features while removing some of their complex features (such as no macro and does not allow multiple inheritance).C# combines simple visualization of VB and high operating efficiency of C++, and is the preferred language for .NET development with powerful operating power, elegant grammar style, innovative language features and convenient component-oriented programming support.

SQL is the abbreviation for Structured Query Language. SQL is designed specifically for the database and a full-featured database language. When you use it, you only need to issue a "do what" command, "how to do" is not considered by users. SQL is powerful, easy to learn and use. It has become the basis for database operations, and now almost all of databases support SQL.

JSON (JavaScript Object Notation) is a lightweight data exchange format. JSON uses a completely language- independent text format that makes JSON become an ideal data exchange language. It is easy to read and write, but also easy to machine analysis and generation.

4 Get header area

The flowchart of getting header area is shown in Fig. 2.

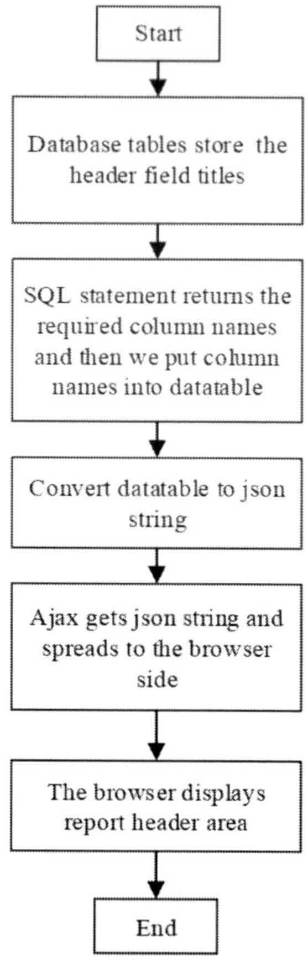

Fig. 2. The flowchart of getting header area

Specific process requires the following steps:

(1)Original data of general production reports are stored in a database table and field titles of these data are mostly named after English. Such as statistics on coal consumption of xx cement group, create a new database table called system_TitleName, and respectively set four fields which are ID, titleName, chineseName and tableName. ID is as an identifier, and automatically generated by the system; The other three fields store English names of field titles, the corresponding Chinese names and the production report name in turn.

(2)After building the system_TitleName table, C # connects to the database. If the connection is successful, you need take out English names of field titles and the corresponding Chinese names from the above table and put it into datatable, conversely return error information.

(3)In order to facilitate the browser-side language Javascript's processing, you need to convert data source formats such as datatable which is queried to JSON string. When transmitting the same data, JSON is smaller than xml. First determine whether there is content in the datatable, no content to return "\" rows \ ": []", on the contrary, you need to follow the JSON string format to return to the contents of datatable. Data format of JSON string is generally '{"name": "Zhang three", "age": "30"}'. JSON holds objects with curly braces which can contain multiple name / value pairs.

(4)AJAX requests JSON data and Javascript parses with it. AJAX have type, url, contentType, dataType, success, and so on several common parameters. If the request succeeds, the JSON data are invoked in the success method. This method has two parameters: the data returned by the server, and the state returned. The format of the call is as follows:

```
function (data, textStatus) {
    // data could be xmlDoc, JSONObj, html, text, etc...
    this; // the options for this AJAX request
}
```

The key code is as follows:

```
success: function (msg) {
var m_msg = JSON.parse(msg.d);
//Convert the JSON string into a Javascript object
  if (m_msg.total == 0) {
  alert("error!");
  }
  else
  {
   var titnames = [];
   for (var i = 0; i < m_msg.rows.length; i++){
   titnames.push({ filed: m_msg.rows[i].titleName,
                 title: m_msg.rows[i].chineseName,
                 width:100});
        }
   $("#Windows_Report").datagrid({
   striped: true,
   rownumbers: true,
   singleSelect: true,
   fit: true,
   pagination: true,
   pageSize: 50,
   toolbar:'#toolbar_ReportTemplate',
   columns: [titnames]
     });
   }
}
```

Datagrid columns are the column configuration objects of the data grid (datagrid) whose properties are objects. To dynamically add the columns property, you must assign values strictly in accordance with its attributes format. Create an array called titnames, push titleName, chineseName fields into it one by one according to the attributes of columns, and assign values to the datagrid columns. The field titles in a web page can be displayed in Chinese correctly and easily.

5 Detail area's content acquisition and format settings

5.1 Detail area obtains production data

Details' data are the core part of the cement production report which can represent the current production levels of the cement enterprises in some extent and provide objective basis for the enterprise production target in the future.

Obtaining the content of the detail area is similar to that of the header field, specific process is as follows:

1)Connect to the database

2)SQL statement query and return the report production data

3)Put the queried report production data into datatable

4)Transfer datatable into JSON string

5)AJAX gets and deals with JSON string

6)The web page shows the production data of detail area

5.2 Detail area sets format

Types of production data are roughly divided into two categories, the numerical and the non-numerical. Determines the data type, the numeric is right justified, the non-numerical is left-aligned.

Many of production data content are more. Setting the width adaptively is easy to appear scroll bars which are likely to affect the appearance of the web page. Set the fixed width, if the production data exceed this width, then the excess will be displayed in the form of ellipses; When the mouse touches the excess, the browser will show the entire production data.

Use the datagrid's formatter property to format the detail. The core code is as follows:

```
function formatCellTooltip(value) {
var reg = new RegExp("^[-+]?[0-9]+(\.[0-9]+)?$");
if (reg.test(value)) {
   return '<span style="float: right ">' + value + '</span>';
      }
return "<span title='"+ value + "'>" + value +"</span>";
  }
```

6 Conclusion

This paper presents a kind of general display design idea for the production report of cement enterprises. This paper, by using database to store the data of the table names and subsidiary area in the production report, and then through the SQL statements, c# datatable, JSON and Javascript and so on a series of processing, eventually makes production reports have common display function. How do I make the header area with multilayer structure, detail area's format more variety as well as display functions of general production report more efficient, which is the next step to be solved.

Acknowledgment

This work is supported by The Research Fund for the Taishan Scholar Project of Shandong Province of China, Independent Innovation and Achievements Conversion of Shandong Province under Grant 2015ZDXX0101F01 and Independent Innovation and Achievements

Conversion of Shandong Province under Grant 2014CGZH0601.

References

1. Baidong JIN. Design of Universal Report Based on Self-defined Template. Journal of computer age, 2009;10 :38-39.
2. Xiaojuan Zhou. Design of General Report System in Web. Railway Computer Application. 2004; **13**(5) :43-44.
3. Weihu Kang. Excel reporting design and implementation based on the. NET. Software Guide. 2016; **15**(5):101-103.
4. Fengxia Cong. Research on Automatic Generation Technology of General Crystal Report. Computer technology and development, 2013;**23**(5): 54-57+62
5. Xiaochun Yu. General Report Programming. Microcomputer Applications, 2000; **16**(10): 61-54

Data Extraction Based on Page Structure Analysis

Yichao Ren[1], *and Jiayin Tian*[2].

[1]College of Information Engineering, Communication University of China, Beijing, China
[2] College of Information Engineering, Communication University of China, Beijing, China

Abstract. The information we need has some confusing problems such as dispersion and different organizational structure. In addition, because of the existence of unstructured data like natural language and images, extracting local content pages is extremely difficult. In the light of of the problems above, this article will apply a method combined with page structure analysis algorithm and page data extraction algorithm to accomplish the gathering of network data. In this way, the problem that traditional complex extraction model behave poorly when dealing with large-scale data is perfectly solved and the page data extraction efficiency is also boosted to a new level. In the meantime, the article will also make a comparison about pages and content of different types between the methods of DOM structure based on the page and HTML regularities of distribution. After all of those, we may find a more efficient extract method.
Keywords. Page Structure Analysis, Data Extraction, JSON

1 Introduction

The content on Internet is all half-structured data, which contains structural data and non-structural data like natural language and images. Web has already developed to a giant source of information. What's more, Web information extraction with optimistic application prospect has certainly became a research hotspot these years. Exactly because of the manifold information on the network and half-structured web data, it is difficult for us human to search useful information among the sea of data. Not even to think about popular application programs, for they are unable to analysis diverse information on the Internet directly. So as to enhance the availability of network data, web information extraction shows up. The technology makes it easy to extract the information on the web page in a more structured way. These useful information could be helpful in data analysis system, batch query system and so on.

The information we need has some confusing problems such as dispersion and different organizational structure. In addition, because of the existence of unstructured data like natural language and images, extracting local content pages is extremely difficult. In the light of the problems above, this article will apply a method combined with page structure analysis algorithm and page data extraction algorithm to accomplish the gathering of network data[1]. In this way, the problem that traditional complex extraction model behave poorly when dealing with large-scale data is perfectly solved and the page data extraction efficiency is also boosted to a new level. In the meantime, the article will also make a comparison about pages and content of different types between the methods of DOM structure based on the page and HTML regularities of distribution[2]. After all of those, we may find a more efficient extract method[3].

2 Data Extraction Based on Page DOM Structure Analysis

The technology uses hierarchical relation of label to combine the content in webpage.As a result, an orderly structure is built up. Then, the technology gets useful data by matching rule of regular express[4].

If the language matching the structure of data could not be described by context-free grammar of global consistent query accurately, and there is no locally uniformly context-free grammar of graduation, such as natural language and image, could be obtained through any ordered partition about data, then the data is non-structured. If the language matching data structure could not be described by context-free grammar of global consistent query accurately, but there is a well-organized division of data in which the local consistency of context-free grammar of Partition No. i+1 could be concluded by the clue of the semantic information of the foregoing partitions, then the data is semi-structured[5]. Directly, data extraction is a process during which we find the useful content from non-structured or semi-structured data and then extract the data to make it structured and clear.

The algorithm based on the structure of tree depends on the structure of label in text. However, standard has little requirements on grammar, and different webpages that come from different designers do not share a common executable power of standard. For example, labels often do not match the content. Now, almost all of Web programming regards HTML of webpage as a document tree. From the observation of DOM document tree, we know that many labels can be deleted in advance and there is not a bit influence on main data content of HTML. DOM document tree is the structure tree of HTML which defines many normal connectors that have specific access to webpage labels. DOM is a connector that is completely independent of different languages and platforms. Different languages and platforms could also have the chance to access and operate the structure, data and even layout of DOM text. Here comes several frequently-used points of HTML DOM:

Get Element By Id(id): obtain node(element) with specified ID

Append Child(node):insert new child node(element)

Remove Child(node):delete child node(element)

Several frequently-used properties of HTML DOM:

Inner HTML：textual value of node(element)

Parent Node：parent node of node(element)

Child Nodes：child node of node(element)

Attributes：attribute node of node(element)

After normalizing, HTML source code is parsed to DOM tree structure. Each label of HTML corresponds to each node of DOM tree, and the property of node is exactly the property of label. HTML node of DOM tree has two child node: Head and Body. Head label defines the head of document and it is the container of all head elements[6].

Element in the head could be adopted in script, indicate browser to find style sheet, provide meta-information and so on. The head of document describes property and information of the file, including title, location in the web and connection with other documents. The data contained in the head would not be revealed as normal content in most documents.

Element of body defines subject of the whole document. Body element contains all content of the file, such as text, hyperlink, image, form, list and so on. The analysis of DOM tree is mainly in body elements and all child node of it. Structurally, the child tree of DOM tree that body lies contains all kinds of structure information, for example, XPath property of label A, class property of DIV label, label FROM and the like.

HTML label is of nested feature, so HTML webpage has a hierarchical structure. Many methods parse HTML document through DOM tree, operating every single node(HTML label) to extract text or other feature of the node. If a leaf node does not contain text, then we delete the node expediently. Analyzing the link between regions in the webpage and semantic connotation between hierarchical structure by page structure, then independent meaning unit is determined.

Applying algorithm to extract webpage content on DOM parse tree. Below is specific description of the algorithm:

The algorithm expresses the node which contains 'w'-the number of information, remarked 'n' , remarked 'p' on the parent node by the form of tetrad-(s,w,n,p). Gaining the end of extracting information of content by finding a set of nodes sequence in tree T.

The tetrad-(s,w,n,p) works like this. Location number-s shows the location sequence of nodes, counting from left to right, up to down. As a result, the 's' value of root node is 0. The quantity of information-'w' shows the number of characters in both Chinese and English that included in the nodes of the tree. The flag value of the exact node is represented by 'n', while the flag value of parent node of the node in the above is represented by 'p'.

We are searching for a set of nodes sequence in which node tag-'n' and parent tag-'p' of every node array must equal to each other and the distance of any two nodes-'Xi' and 'Xj' in the tree equals to different value of node-'j-i'.What's more, 'i' and 'j' mean the location number of node.

Then, we handle the sequence by getting set like the following steps:

Do away with node sequence remarked 'a' or node sequence of which node information-'w' is less than 'q'. What calls for special attention is that the value of 'q' is 16 in our system.

Combine or filter the sequences with only one node in the sequence set. Inspect parent node of the node in all single node sequence with algorithm, then combine the nodes that the parent node of which comply with the requirements for algorithm to generate a new queue. If the node that conforms to the condition is not node Max, then do the step of filtering and deleting. If the node is node Max, then reserve it. When the filtration is over, export all node information of the remaining nodes sequence. What we get is exactly the content information of extracted webpage.

The label of HTML could be devided two types based on the difference of function of webpage label. One type is structure label aiming to build frame of webpage, while the other is layout label specifically for modifying text of webpage. We only care about the structural distribution of webpage, so we need to filter the layout label that modifys webpage text, such as font label<font>, text layout label<b>,<em>,<small>,<big>,<strong>,<l>,<b>,<u>, text style label<style> and so on. Then, we only consider structure label when we remark node with quadruple notation, such as <p>,<div>,<table>,<tr>,<td>,<li>,<span> and so on.

Subject data extraction based on webpage mainly filters the webpage in the process of judging webpage and subject. However, the difference of webpage and traditional text information is the webpage is semi-structured. Webpage expresses different meaning of text through different labels, and what we need is the useful information in the page while analyzing webpage and subject semantics, such as text, title, link and so on. So, we have to extract these information from webpage and represent text with the meaning of label when analyzing the relevant semantics[7].

3 Data Extraction Based on HTML Regularities of Distribution

The main idea of the method is to extract body content of the webpage step by step. At first, webpages of common similar structure need to be blocked. In other word, the whole webpage will be divided to two parts-head and body. Then, useless label elements and the related content will be deleted by analyzing HTML label semantic in the two domain blocks[8].

In webpage structure generated by HTML, XHTML, XML auto format, there are <head></head>and <body></body> which mean head block and body block.

Take out the content of body block. Body block format is as below:

<body>

The content of body block:

(1) The content of body block:

(2) Webpage text content

(3) Others

</body>

Delete useless information in the body block. Body blocks of movie reviews page in Douban consists of text in which contains series of property with controlling power like size, color, font and so on. Then, delete content irrelevant to text content in body block.

Delete information irrelevant to text content in body block. It always occurs that there are some word links and image links which are irrelevant to the text content. In order to extract text without irrelevant links, all irrelevant links need to be cleared.

We could get webpage text content smoothly through the method of deleting HTML element based on regional block. It satisfies the need of the article, and hit the mark of extracting text content.

 DOI: 10.1051/matecconf/201713900118

4 Text Extraction Experiment

Similar with commonly used data extraction, data extraction of user leading webpages, such as post bar, forum, Weibo, has to follow the design principle of common data extraction like flexibility and robustness. Because of special need of system, data extraction part is different from common data extraction.

We take user reviews information in Douban for example to carry on the experiment. When users are surfing on the movie section, they can choose an interested movie section. Then, simply click the area to enter the page of the selected section. What shows up is relevant content about the section. When users choose a specific movie label, the page changes to a page of the exact movie. So users could comment in the page. Posts page is regarded as leaf node of forum.

The experiment extracts text in HTML source code by regular expression. Analyzing the code with beautifulsoup in Python.

Another easier way is to match text content in original HTML source code by regular expression directly. The rule of replacement is '<(\s|.)*?>'. '/s' means blank character which contains blank character, CRLF and tag symbol. '|' means the front one or the latter one is chosen. '.' means any symbol, so '\s|.' means that match all arbitrary character. '*' matches none or multiple symbol that has showed up before. Thus, '(\s|.)*?' means that any symbol shows up any times which equals to any alphabetic string. So, we can conclude that regular expression -'<(\s|.)*?>' means any alphabetic string that starts with '<' and ends with '>'. The alphabetic string is exactly HTML label. We replace the text according to matching rule to filter the HTML label in HTML source code.

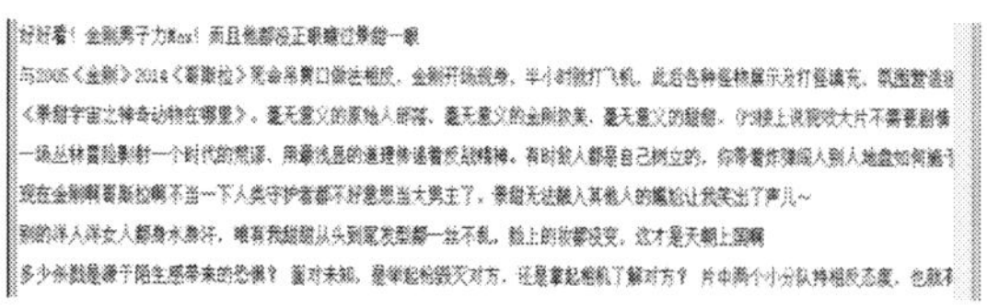

Fig.1. Data Etraction

5 The Text Comment Is Put Out As Data of JSON Format

JSON as a standard module could accept Python data structure and turn them to a representation of string. This process is called serialize. The types that JSON code supports contains none, bool, int, float and str. There is no doubt that the types also contains lists, tuples and dictionary that of basic type. As for dictionary, key is assumed as string. It is necessary to note that any non-string key would be transferred to string during coding process. To confirm to JSON standard, only Python list and dictionary should be coded.

In different languages, a collection of name/value pairs is unscrambled as object, record, structure, dictionary, hash table, keyed list or associative array. In most languages, an ordered list of values is unscrambled as array.

These are usual data structure. In fact, most modern computer languages support them in some form which makes the exchange of a data structure in programming language based on these structure in the same way possible.

The simple data types that put out through encode and repr() which represents the original way share a common result. However, several types of data change during the process. For example, the tuples mentioned in the above are converted to list. During the programming process of JSON, there is a conversion between Python which is the original type and JSON. Some basic types of Python such as tuple turn to list. When the object is turned back to python type, it doesn't come with the change that list type turns to tuple one. In the meantime, the format of the code changes to unicode.

The method of json.dumps() reverse a object back which is of str type and called encode json.

encodedjson = json.dumps(obj)

An object of python could not convert to json type directly. The only way to finish the conversion is to change the object to dictionary first, and then change it to json. As for json, the conversion acts as the same way that the target is changed to dictionary as a transfer station[9].

As for the text comments that we have already got through the methods above, we transfer them to json.

First of all, we create a method, analyze data, rewrite default method of JSONEncoder, and convert object to dict.

```
class userEncoder(json.JSONEncoder):
    def default(self, obj):
        if isinstance(obj, user):
            return {
                        obj.comment
            }
        return json.JSONEncoder.default(obj)
        }
    else:
        return obj
```

The Encode process is to convert python object to json object. There are two frequently-used functions, one is dumps and the other is dump.The only difference between them is that a document flow of fp is generated during the conversion of dump while during the conversion of dumps, a string is generated instead.

When JSON deals with Chinese string, the string is showed as unicode. As a result, unicode usually acts as intermediate code during a conversion. The process works as this: decode string in other code to unicode, and then, transfer unicode to another code. The role of decode is to convert strings in other code to unicode. For example,

str1.decode('gb2312') means convert str1 coded by gb2312 to unicode. The rule of encode is convert string in unicode to strings in other codes. Here comes another example, str2.encode('gb2312') means convert str2 to string coded by gb2312. Therefore, we have to make it clear before we start conversion.

We define the variable, convert dictionary to object in JSON, put out data of JSON. The result is as below.


```
\u6f02\u4eae\uff0c\u7279\u6548\u534e\u4e3d\u3002\u57fa\u672c\u4e0a\uf
91cd\u8981\u7684\u662f\uff0c\u7ea2\u6843\u7687\u540e\u6709\u4e00\u989
\u662f\u4e00\u5207\u4f20\u5947\u6587\u5b66\u7684\u4e00\u4e2a\u7cbe\u5
8b8a\u597d\u770b\uff0c\u96d6\u7136\u6709\u5e7e\u5e55\u6232\u7684 3D \u
\u4e24\u767e\u4e2a\u60f3\u8981\u780d\u6b7b\u7684\u4eba\uff0c\u2014\u2
38\u8fdc\u662f\u7f8e\u597d\u7684\n", "69/100\u3002\u6211\u5728\u60f3\u
u8c61\u6700\u6df1\u5c45\u7136\u662f\u7247\u5c3e\u7684\u82b1\u5f00\uff
u5fd2\u4e3b\u65cb\u5f8b\u4e86\uff0c\u8fdc\u4e0d\u53ca\u53f2\u4e91\u68
206\uff0c\u4ee5\u663e\u793a\u81ea\u5df1\u7684\u54c1\u5473\u4e0e\u4f17
\u3002\u914d\u4e50\u8d5e\uff0c\u753b\u9762\u534e\u4e3d\uff0c\u6700\u6
\u4e66\u91cc\u9762\u5173\u4e8ejabberwocky\u8fd9\u4e2a\u5c0f\u7247\u6b
0c\u60c5\u8282\u5355\u8584\uff0c\u4fd7\u5957\uff0c\u70c2\u3002\u7efc\u
1\u5931\u671b\u8fc7 \u5305\u62ec\u7531\u4ed6\u6267\u5bfc\u7684\n", "
```


Fig.2.JSON Transmission Format

The reason responsible for the result is that json.dumps generates matching character encoding instead of original character towards non-ascii character in default occasion.

If the original character is required, just handle as below:

```
print json.dumps(js, ensure_ascii=False)
```

6 Conclusion

HTML language is semi-structured. It lacks of semantic to describe information. So, when it comes to analyzing webpage structure by DOM tree, most of the methods are studying how to use the semantic information to design data waiting to be extracted by regular expression with directed attention. Then, the detail data could be extracted from the webpage.

Web information extraction method based on HTML regularities of distribution or the relationship between the tags is appropriate for simple webpages with regular structure. The method does good at this kind of webpage. And the method could recognize text content in unconventional web structure correctly. Then, the exact text content could be extracted.

JSON is a simple format of text same as XML. Relative to XML, JSON is more readable and easier to check by users. At the grammatical level, the difference between JSON and other formats is characters that separate data. Characters that separate data in JSON are confined to single quotes, parenthesis, bracket, brace, colon and comma. At first glance, the advantage of using data delimiter in JSON is not obvious enough. However, there is an essential reason that makes JSON special. Data delimiters in JSON simplify data access which will open an easier access path than DOM. Another advantage of JSON is its feature of not long. In XML, it is necessary to open and close the tag which leads to the satisfaction of compliance of tag. But in JSON, all these requirements would be satisfied only through a simple comma. In the process of data exchange that contains hundreds of fields, traditional XML tag would lengthen the time of data exchanging.

Acknowledgment

The work on this paper was supported by the open project foundation of Key Laboratory of Performing Art Equipment System Technology of Ministry of Culture, P. R. China.

DOI: 10.1051/matecconf/201713900118

References

[1] Hong, Haikun, et al. "Web Data Extraction Based on Tree Structure Analysis and Template Generation." International Conference on E-Product E-Service and E-Entertainment IEEE, 2010:1-5.E.
[2] Ma, Anxiang, et al. "Web Data Extraction Based on Structure Feature".Applied Informatics and Communication. Springer Berlin Heidelberg, 2011.
[3] Hongyao Chang, "Research on The Body of Web Pages Based on Labelling Analysis". Chongqing University, 2015.
[4] Wen Zhao. " Research on Web Page Structure Mining for Information Extraction". Artificial Intelligence and Recognition Technology, 2008, 1(1):126-129.
[5] Jiang Liu. "A survey on sparse subspace clustering." Harbin Institute of Technology,2012.
[6] Wang, Jiying, and F. H. Lochovsky. "Data extraction and label assignment for web databases." International Conference on World Wide Web ACM, 2003:187-196.
[7] Cormier, Michael. "Computer vision-based analysis of web page structure for assistive interfaces." Web for All Conference ACM, 2016:24.
[8] Chau, Michael, and H. Chen. "A machine learning approach to web page filtering using content and structure analysis." Decision Support Systems44.2(2):482-494.
[9] Cesarini, Francesca, et al."Page Classification for Meta-data Extraction from Digital Collections." International Conference on Database and Expert Systems Applications Springer-Verlag, 2015:82-91.

Design of intelligent power consumption optimization and visualization management platform for large buildings based on internet of things

Shulan Gong[1,a], *Yuling* Wang[1], *Mingyu* Zhang[1]

[1] School of Information & Electrical Engineering , Shandong Jianzhu University

Abstract. The buildings provide a significant contribution to total energy consumption and CO2 emission. It has been estimated that the development of an intelligent power consumption monitor and control system will result in about 30% savings in energy consumption. This design innovatively integrates the advanced technologies such as the internet of things, the internet, intelligent buildings and intelligent electricity which can offer open, efficient, convenient energy consumption detection platform in demand side and visual management demonstration application platform in power enterprises side. The system was created to maximize the effective and efficient the use of energy resource. It was development around sensor networks and intelligent gateway and the monitoring center software. This will realize the highly integration and comprehensive application in energy and information to meet the needs with intelligent buildings

1 Introduction

Globally, buildings account for almost 40% of the total end use of energy, as well as 24% of total CO2 emissions.[1] Domestic energy shortage， short of conventional sustainable energy supply and extensive development of energy resource led to the rapid growth of China's energy demand. At present , the energy consumption of buildings ,the industrial energy consumption and the traffic energy consumption are the three main energy consumers[2], the building energy-conservation is extremely urgent. The energy consumption statistical commission of CABEE(China Association of Building Energy Efficiency) released China's energy research report in 2016. It showed that the energy consumption of civil buildings in China accounts for 20%-25% of the total energy consumption .In 2014 the data was 19.12%[3]. In the nearly 20 million square meters of the new building every year , only 1% can be regarded as energy-efficient buildings[4]. Building energy consumption in china will continue to rise with people's living standard improved. Building energy consumption will be more than 3 times now[5]. The building energy-conservation will be huge challenge. conservation in developed countries is that the new buildings energy-conservation will reach the targets of zero energy consumption or near zero energy consumption in Euro-American countries in 2020-2030. Current situation of building energy-conservation in china is that the proportion from the new technologies and new products to applications is low .The use of a power consumption optimization and visualization management platform is designed in substantial savings in both hardware components and software.[6]The smart grid with the comprehensive energy consumption monitoring and visual management demonstration platform in the large buildings can play an important role in energy conservation.

The main contents in the article contain (1) To design intelligent sensors based on Internet of things technology to gain information parameters of power about people position and environment parameters such as temperature, humidity, illumination and power consumption by the methods of electricity information monitoring technology and detection technology. (2) To design intelligent user control terminals, including the smart socket, intelligent switch, intelligent gateway with wireless communication which can control air conditioning, lighting and other electrical appliances.(3)To design the system architecture of intelligent electricity interactive supporting platform on large buildings by components and extensible technology and massive intelligent power information mining and large scale concurrent user access processing technology .(4) To design the system of demand beside response analysis control and energy efficiency optimization to monitor and control for lighting, electrical appliances, energy management and other aspects.(5) To realize electricity visualization platform, the data of power consumption is sent to the data center for statistical purposes through the network which can provide many measuring techniques based on different needs.

* Corresponding author: gongsl2008@sdjzu.edu.cn

2 The summary of the whole structure of the system.

Intelligent power consumption optimization system and visual management platform of large buildings is a set of monitoring and control system based on Internet of things. It is set in the monitoring center of the power industry which will manage intelligent electricity and energy monitoring information of the buildings and further change the light switch. The framework of the whole system network is a sensor network connected to Internet. The monitoring center of the building can be responsible for information monitoring and related equipment control in the building which will interact information with the monitoring center of the power industry. The whole structure of the system is shown as Fig.1

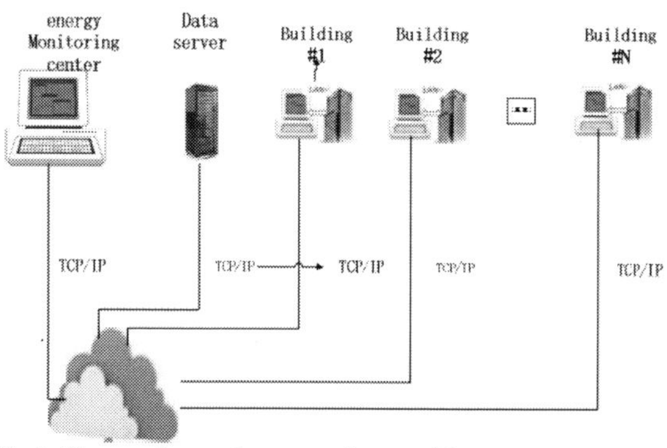

Fig.1. The overall network structure diagram of the system

The system in single building includes intelligent power consumption optimization system , visual management platform , data server ,multimedia gateway in each floor ,room controller , intelligent sensors ,intelligent lighting system ,etc. The system structure of a single large building is shown as Fig.2.

3 The intelligent sensor

The emerging technology of wireless sensor network(WSN) has become an increasingly feasible approach to realize control and management of building environment[7,8,9].

Intelligent sensors are responsible for environmental information monitoring in intelligent power system .The single-chip MSP430F149 is used as the controller of intelligent sensors which will accomplish a remote smart home system through WIFI and Internet. Intelligen t sensors in this system include intelligent temperature and humidity sensors, intelligent illumination sensor, smart meter and intelligent air quality sensors . The environment variables will be converted into analogy variables through intelligent sensors, then the analogy variables will be converted into digital variables through the single-chip MSP430F149. The digital variables will be translated to the room controller through WIFI network. The room controller will exchange information with the multimedia gateway through WIFI network. The multimedia gateway has the function of inverting each other between the serial port data and the Ethernet data which will be translated to monitoring center by LAN bus. The people will observe environmental parameters and electricity consumption of each room and then realize the remote intelligent control through control command and control strategy.

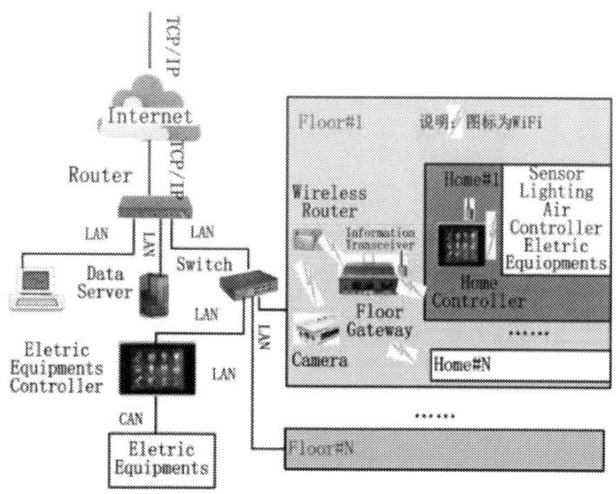

Fig.2. Tthe system structure of a single large building

3.1The design of intelligent temperature and humidity sensor

The hardware part of intelligent temperature [10]and humidity sensors mainly include: MCU and its peripheral circuit, digital display circuit, sensor circuit, serial communication circuit, nonvolatile memory circuit, real-time clock circuit, A/D conversion circuit. MSP430F149 is used as MCU. DS18B20 is used as temperature sensor, HS1101 is used as humidity sensor.

MSP430F149 reads DS18B20 temperature value by I2C bus, and then converts the value to Celsius degree which is displayed through the liquid crystal displays.In the software design of the humidity sensor module, the method scanning port is used to start the measurement at intervals of 8s. The measuring time length is 1s. The controller will count the number of pulses per unit time, determine the range of the humidity value, and display the humidity value through the LCD.

3.2 Intelligent illumination sensor

Intelligent illumination sensor adopts light detector with higher sensitivity and have high precision linear amplifying circuit. Its shell is designed to wall-hanging Style with nice appearance. MSP430F149 is used as MCU. The master clock circuit is supplied by 8MHz external crystal oscillator. BH1750FVI is used as intelligent illumination sensor.

3.3 Smart meter

The measurement of electric current, voltage and power in the each room can be received through the smart meter with the energy measuring chip RN8209 which will communicate with the host by SPI interface.

The hardware of the smart meter includes power module, measurement module with RN8209 chip , MCU Mega16 , WIFI module ESP8226, LCD display module, external E2PROM

3.4 The intelligent air quality sensor

The hardware of the Intelligent air quality sensor model includes power module, measurement module with Infrared absorption sensor 6004 produced by Telaire company , MCU MSP430F149 , WIFI module ESP8226, LCD display module, clock model DS1302 , AD conversion module and DA conversion module. CO2 will be monitored by the intelligent air quality sensor in this design. Depending on the actual occupancy rate (CO2 concentration), the ventilation rate is adjusted to maintain indoor air quality and the comfort of the users in order to conserve energy.

4 The Intelligent Control of Electric Equipment

4.1 The intelligent switch

The intelligent switch is used to control the lamp by manual touch or wireless control. The intelligent switch equipped with electricity metering unit is used to measure the voltage, current, power of lamp in the room. The data will be translated to room controller by WIFI. The electricity information can be viewed by user login from the remote through mobile phone. The on-off of the intelligent switch is controlled by the thyristor circuit with photo-electric coupling method. The intelligent switch circuit model mainly includes MCU module AtmelMEGA8, measurement module with RN8209, SCR control module MOC3041 and WIFI module.

The red arrow indicates the direction of power supply, the thin blue arrow indicates the flow of information, the thick blue arrow indicates the strong current of the 220V in the Fig.3.

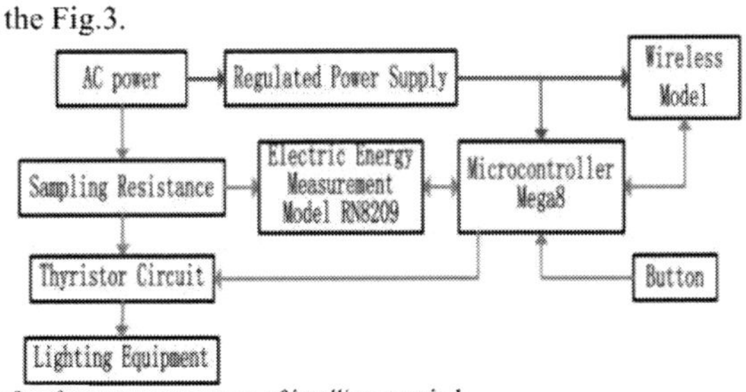

Fig.3. the system structure of intelligent switch

4.2 The Smart Socket

The Smart sockets are used to control household electric appliances. The user can manually control the Smart sockets through the room controller , or implement remote control it by landing room controller through the phone to control the household electric appliances on-off according to the intelligent control strategy. The Smart sockets equipped with electricity metering unit are used to measure the voltage, current, power of load in the room. Smart socket is designed in accordance with 86 boxes with the shape of the ordinary socket. The relay is used as the on-off control device of the smart socket. The triode is used as switch of the smart socket..

5 The Room Controller

The Room controller is designed with touch-sensitive panel and android operating system .The communication module is designed to use Socket communication technology with Java programming language. It can perform remote monitoring of the electricity consumption, power consumption and remote control of the electrical devices. The client of the room controller communicates with server by WLAN network between the sensors collecting surroundings information in the room. The data reading and assigning of the communication model is realized by JSON data-interchange format on the android platform.

The room controller can send out the communication and control instruction with simple display interface. There are command buttons on TV, lighting, air conditioner and the display interface to view temperature, humidity, light intensity and a refresh button and so on. There are two switching modes for manual mode and automatic mode. We can get the latest information about environment and the state of household appliances by the refresh button. The intelligent room controller can realize to monitor and collect the environment information in the room and then send the information to the monitoring center and the mobile phone.The room controller interface is shown as Fig.4

6 The Smart Gateway

The smart gateway is the core of the whole building's internal functions realizing and the hub of internal communication. Floor multimedia gateway is consisted of the wireless router and the floor gateway. The wireless router is mainly responsible for video accessing with WIFI camera and it can upload video by LAN network. The floor gateway can collect the data information from the controlled devices of each room by WIFI communication protocol and upload the data by LAN network. The information and data from LAN network is translated to the controlled devices in the room by the smart gateway and room controller. The structure of the smart gateway is showed as Fig.5. The workflow chart of the smart gateway is showed as Fig.6.

 DOI: 10.1051/matecconf/201713900119

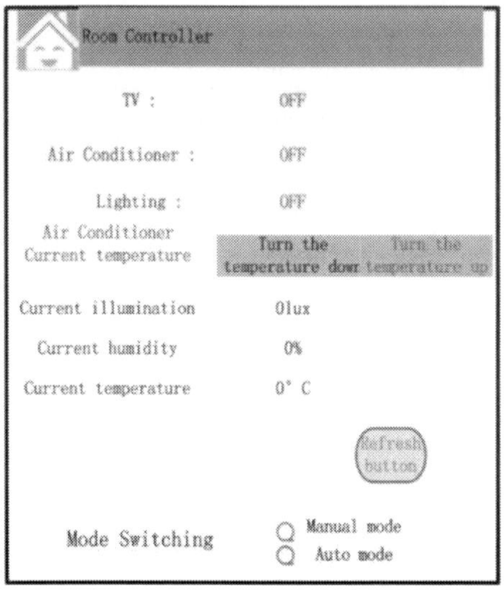

Fig.4. the room controller interface

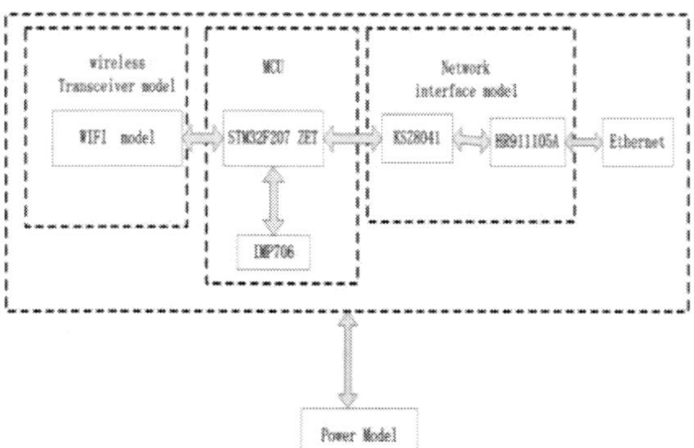

Fig.5. the structure chart of the smart gateway

7 The Implementation of The Monitoring Center

The monitoring center is the upper layer software of electrical power information collection and energy saving control system of the building. The monitoring center is responsible for the monitoring, controlling and data storage of the whole building. The server of the monitoring center receives and analyses the information of energy monitoring information and environment parameters from the floor gateway through TCP communication protocol. These data will be written into database at the same time. It can send control commands through the floor gateway and room controller to the controlled devices.

The main functions of the monitoring center is to implement

(1) The center monitoring software design of building energy consumption detection and control system. It includes the receiving and processing of power information, the sending of control instruction and the connection of database.

(2) Design communication protocol and algorithm to implement the network connection between the monitoring center and the smart gateway.

(3) The database design of building energy consumption detection and control system.

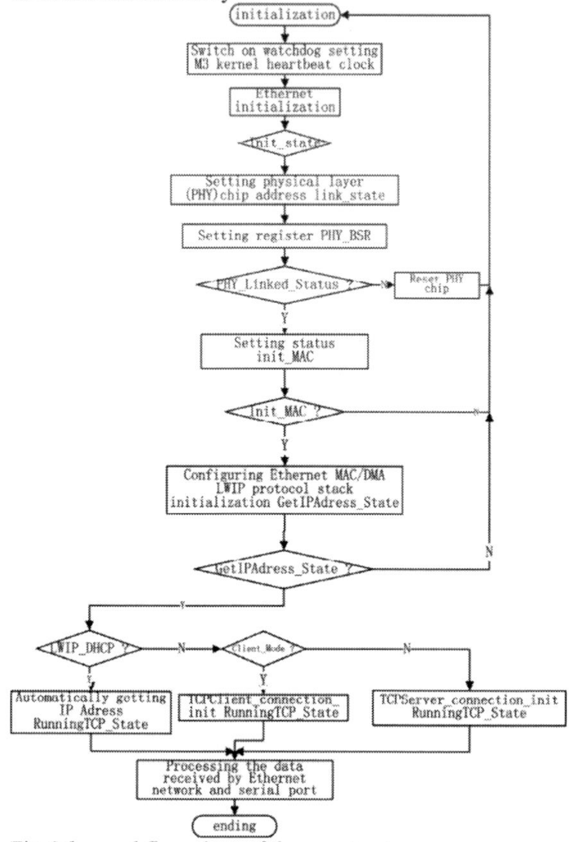

Fig.6The workflow chart of the smart gateway

According to the function of the monitoring center, the interface design of the monitoring center platform is showed as Fig.7.

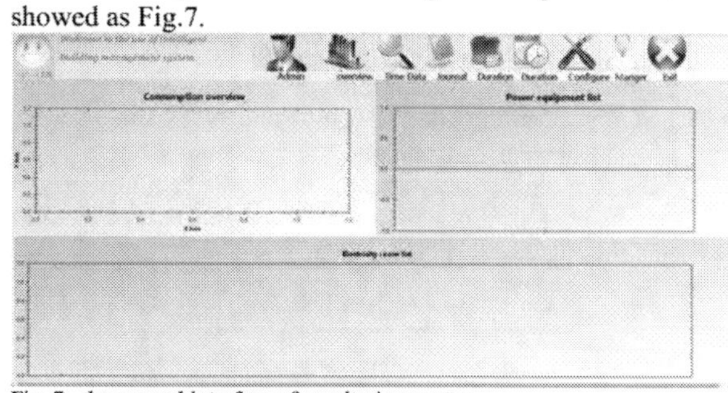

Fig. 7. the general interface of monitoring center

When the real time data button is clicked , you can look at the real energy distribution of the specific floor , it will be showed as Fig.8. When the historical time data button is clicked , you can look at the historical energy distribution of the specific floor , it will be showed as figure Fig.9.

 DOI: 10.1051/matecconf/201713900119

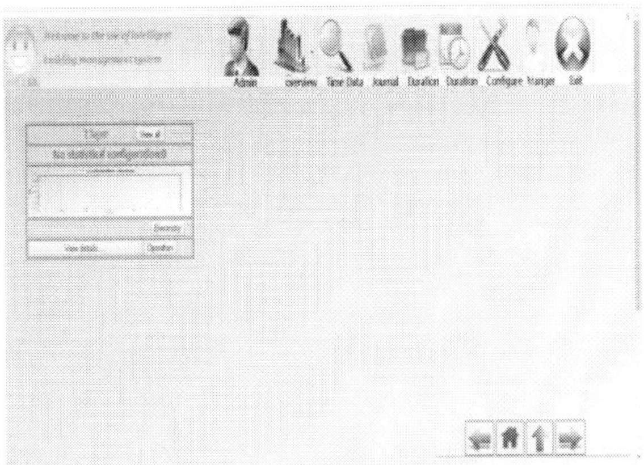

Fig. 8. the real energy distribution of the specific floor

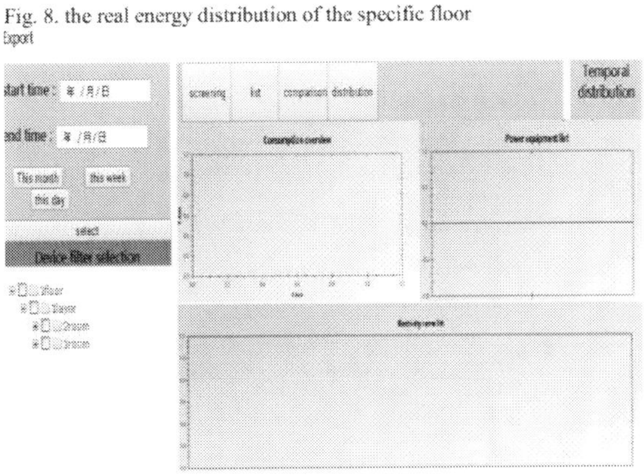

Fig.9. the historical energy distribution of the specific floor

8 Conclusions

The intelligent power consumption optimization and visualization management platform for large buildings based on internet of things is designed in this paper. The real-time data can be gotten by sensor network and socket communication. The parameters of air conditioner, lighting, electrical appliance equipments can be controlled by the control strategy through the remote monitoring center. The proposed system will have potential prospect in large buildings.

Acknowledgment

The work is partially supported by the key Science and Technology Planning Project of Shandong Province under Grant No.2014GGX101017 , No.2013GGX10131, No.2015GGX101024, The National Natural Science Foundation of China No.61273326. Thanks to all members of our task group . As a representative of our group, I express my sincere appreciation here to editors and specialist reviewer for your instructions and helps to the article.

References

1. Park,J.and Hong, T. Maintenance management process for reducing CO2 emission in shopping mall complexes .Energy and Buildings,2011,43:894-904M. Ben Rabha, M.F. Boujmil, M. Saadoun, B. Bessaïs, Eur. Phys. J. Appl. Phys. (to be published)
2. Lin Jianping . Building energy consumption is too high for comfort[N]. China Information News, 2007,11.
3. Development of China's building energy saving industry in 2016. Cnii ,2016,05.
4. Jiang Yi. Building energy consumption situation and key energy saving in China[J]. Building technology, 2007,05:26-29.
5. Guan Zhineng. Building- Maximum energy consumption black hole[N]. Economic Daily, 2007,11.
6. Trejo-Perea,M.,et al. Local predictor for air conditioning systems using artificial neural network models for intelligent buildings. Nova Science Publishers, Hauppauge, NY, 2009a:337-351.
7. Tessa,D.,Elena,G.,James,B. Wireless sensor networks to enable the passive house-deployment experiences. European Conference on Smart Sensing Systems and Context[C], 2009:177-192.
8. Karsten,M.,Dirk,P.,Brendan,F.,etc.Towards a wireless sensor platform for energy efficient building operation. Tsinghua Science and Technology[J],2008(13):381-386.
9. Chiara ,B.,Alberto,F.,Roberto,V. An IEEE 802.15.4 wireless sensor network for energy efficient buildings. In The Internet of Things[J],2010;329-338.
10. Charles,C.,Jeffrey,F.,Asad,D.,and Ruei,C.(2009). Temperature control framework using wireless sensor networks and geostatistical analysis for total spatial awareness. 10th International Symposium on Pervasive System, Algorithms, and Networks,717-721.

Role - based Access Control in Educational Administration System

LIU Dongdong[1], XU Shiliang[1] ,ZHANG Yan[1] ,TAN Fuxiao[1],NIU Lei[1],ZHAO Jia[1]
[1] School of Computer and Information Engineering, Fuyang Normal University Fuyang, China

Abstract. In the 21st century, the network of teaching management not only improves its quality and efficiency but also brings convenience to teachers and students. However, as a network application system, it also faces a variety of security issues. In order to improve the system security , the widely-used RBAC control method is introduced in this paper. Based on the refinement of system privilege and user role, this paper puts forward the security management model of "user classification, role authorization , Unified management ", which is more suited to the structure of multi-level applications by controlling the data range accessible to users, and ultimately achieves the purpose of strengthening the security of the system.

1 Introduction

In the 21st century, along with the rapid development of computers and the widespread use of the network, the growing size of the school, are making the school's educational administration is extremely important, educational management network model not only to teachers and students to bring convenient But also improve the quality and efficiency of teaching management. Combine the characteristics of the current school development, make full use of the existing campus network resources, development in line with the future development of the school educational management system, and further optimize and improve educational management methods and processes for better teaching services, has important practical Meaning, but also can promote the school's information technology and digital campus construction. As an important subsystem of the digital campus, the educational administration system is also very important. In order to solve the traditional security problem in educational administration system, this paper puts forward the RBAC-based control method which is widely used in the system[1]. Through the refinement of system privilege and user role, the control data can be accessed, So that the model is more suitable for multi-level application structure. The paper puts forward the security management mode of "user classification, role authorization, unified management", and finally achieves the purpose of strengthening the security access of the system.

2 RBAC

Role-based access control (RBAC) is the role through the contact users and permissions, by directly to the role of authorization to control the ownership of the user's access to the system operation. What is "role", in fact, the user can operate in the system permissions set, the definition of the role, the system can give users different roles to obtain the appropriate access. Users have many-to-many relationships with roles, and roles and permissions are also many-to-many. This mechanism allows the user to have privileges only through the role. In the basic structure of RBAC, each user in the landing system will establish a session (Session), and through the session to activate the role of the user and activate the role contains all the privileges (as shown in Fig1). In the RBAC model, role can exist inheritance relationship, that is, upper level role can be part or all of the inheritance of the role of the next level of authority, this role inheritance relationship to form the role of the hierarchy.

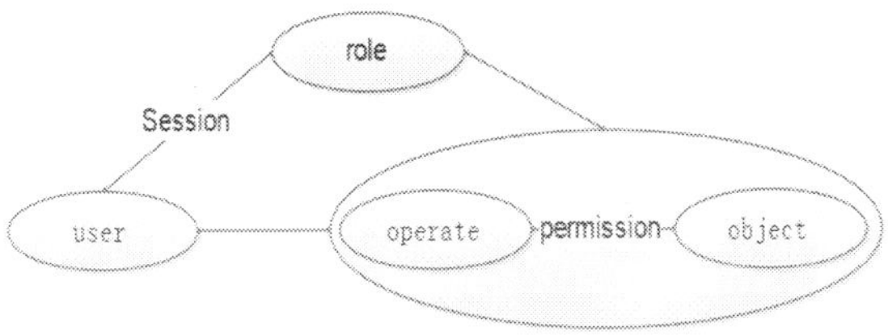

Fig1 Basic structure of RBAC

The hierarchical structure of the role can correspond to the hierarchical structure within the organization, the different roles corresponding to different functions or positions, which can be based on the user's position to grant the role. So that once the responsibility of the user position changes or add new applications in the system, the system as long as relatively simple to change the user's role, or in the role of adding new permissions to modify the user's permissions, in addition, you can revoke user roles Or role permissions.

3 System security analysis

The security scheme of the educational administration system is based on the role-based access control under the premise of identity authentication, so the system security requirements mainly involve the following aspects:

1) users and rights management

The system administrator is responsible for managing each user and authority in the system. Its specific management functions mainly include creating users and roles and defining and assigning various operation rights of the system, and according to the user's level and responsibility. Role of the grant and recycling, to ensure that each user can only be authorized within the scope of legitimate operations, while the system administrator also has the appropriate resources to manage the operation of the corresponding functional data responsibilities.

2) user authentication

School students, teachers and managers at all levels use their own account number (student number, employee number, etc.) for system login. Only users with identity and password authentication during logon can perform their own privileges in the system.

3) Access control

WEB server, database server are deployed in the campus network platform, the system must strictly control the various users can only perform their own operating authority to ensure system security.

3.1 User role analysis

RBAC model in the management of user rights is based on the role of the general role can be given to multiple users with a role, but also can give a variety of roles to a single user. Any user of a system has a unique symbol (UserID, user name, employee number, student number, etc.)[2] that can identify itself in the system. The user must first authenticate the identity through the authentication module when logging in the system. Of users can log on the system, and the system also need to determine whether the user is already in the login state of the function, to prevent duplication of registration caused by various information fraud. The following are two requirements for the roles in the educational administration system:

1) The role of the educational administration system created by the administrator, specifically according to the school's organizational structure and the actual business functions set, the system provides a special module to manage users and roles.

2) Each role in the academic management system also has a unique number to identify their own, and save the role of the relevant institutional information (such as college, department number), which can facilitate the role and authority to manage.

3.2 Object privilege analysis

Objects in the system include the various visible objects in the system window, such as menus, buttons, etc., but also invisible objects in the database, such as records, fields, and so on. Educational Administration System itself consists of various functional modules and a variety of system window menu command together. Each user or role can have its own functional modules and menus, through the role and user authorization to achieve the control of the object operation. Specific practices are as follows:

1) the establishment of the role of the authority table,

in which the role of each role in the system records of the operation of the system resources. At the same time, this table also reflects the system's various business functions.

2) The module function of the system can be expressed in the form of the function menu, which can facilitate the control of users with certain roles to add, delete, change and so on the function modules in the system.

3) As the role of the size of the problem of partition, each role has a corresponding set of permissions, and when a user has more than one role, then the user should have access to all the roles of its authority is a role of authority And set, in this case, the union may appear duplication of the same user rights, but this does not affect the user has the total permissions.

4) The administrator in the user function module for the distribution of authority, the main strategy is to set the user in the function of the various components of the object permissions, such as menu, field properties.

In the basic structure of the RBAC model, the user will establish a session with the system upon login, which is responsible for activating all the roles granted by the user. The user can have the role of the authority during this session, the user during the establishment of the process of access to the system role of the specific process as shown in Fig2.

After the user logs in to the system and the authentication succeeds, the user obtains the role and the corresponding authority, and determines whether to allow the access request resource according to the role privilege[3]. Fig5 is a state diagram of the user login system, reflecting the process of user login system status.The security scheme of the educational administration system is role-based access control under the premise of identity authentication, so only the users who succeed in authentication can have the role privilege and access the corresponding system resources. The authentication function of the educational administration system is realized by Java security programming interface The certification process is roughly described as follows:

When the user logs in, the system creates the LoginContext object. The LoginContext object is then used to query the configuration file login.config, and the LoginModule object is loaded according to the configuration file. The user is authenticated by running login () and the CallbackHandle object is created at the same time..

Principal is used to represent the user's access rights, thus completing the verification step.Fig3 shows how the authentication framework processes the Web page request.

4 RBAC-based access control model design

4.1 Design of role privilege management scheme

In the overall scheme of access control, granting and reclaiming various rights is the key point of the authorization process, and in order to increase the security and flexibility, we

plan to manage the role categories and functions independently. According to the specific situation of the school, the specific plan of RBAC in the educational administration system is to classify the school personnel, to determine various roles, to restrict role based on role, and to unify management roles and rights: "user classification, role authorization and unified management".

1) Role definition

Due to the fact that there are a lot of users in academic institutions and the different levels of identities, levels and levels in the secondary school, there are many types of roles that need to be divided in the system. Often, Of the staff will have a lower level of operational authority, so the system through the introduction of inheritance between different roles to reflect the relationship between the different levels of human rights include (for example: in charge of teaching authority will include teaching in charge of two Level role of the Dean of the authority), role inheritance is the use of object-oriented programming thinking inheritance between the characteristics of the specific approach is to expand and classify existing roles to derive new roles, the benefits of doing so is to avoid duplication of definition, This is in line with the teaching management in the flexible management and level management.

2) permission to configure

In RBAC, the role is essentially a user collection,

and because the permissions are granted roles, so the role is the permission set. And can grant a user multiple roles, but also can grant a role more than one authority, of course, can also grant a role to more than one role. Permission configuration is actually the role of configuration permissions, because only the role has permissions, can be granted to the user through the role.

- Divide staff roles

In the educational administration system, it is the main core idea of RBAC that authority and user are connected by role. In view of the division of personnel roles, this paper takes the design idea as follows: starting from the specific needs, analyzing all kinds of personnel roles in the school, on the basis of the division and division of good user roles, and then the role of classification, and finally assigned to the hierarchical authority Character.

In fact, in any unit, people or positions are divided into different levels of hierarchy, different positions have different roles, and therefore have different permissions. Although different positions are not exactly the same as the role of division, but the role of the system as the level of distinction between the level of the same job[4]. For example, in the system, in charge of teaching school principals and system administrators, the role of the system administrator is actually more important than the principal, so the system administrator should have greater system privileges, which in the role of the division to reflect the different Level of the hierarchy.

In the school, the general use of educational administration system users mainly include: system administrators, departmental administrators, module administrators, teaching management personnel and a series of management personnel and general teachers and students. When we divide roles into these users, the strategy is to first split the roles into two different levels:

1)low-level roles both general role: including ordinary teachers and students, the role of the class have some common ground: the authority is generally fixed limited, in the role of such authorization, the general can be granted directly to the fixed authority .

2)senior roles both special roles, including system administrators, module administrators and various types of teaching administrators, the role of the general level of ownership of more authority in the role of such authority may be granted directly, but also Granting other roles, that is to say other roles become their sub-roles, so that sub-role can be granted through the indirect grant. For example, in charge of teaching the role of school principals and in charge of the teaching of the secondary college dean role is clearly a role in the inheritance relationship. Fig4 is the role of educational management system in two types of division.

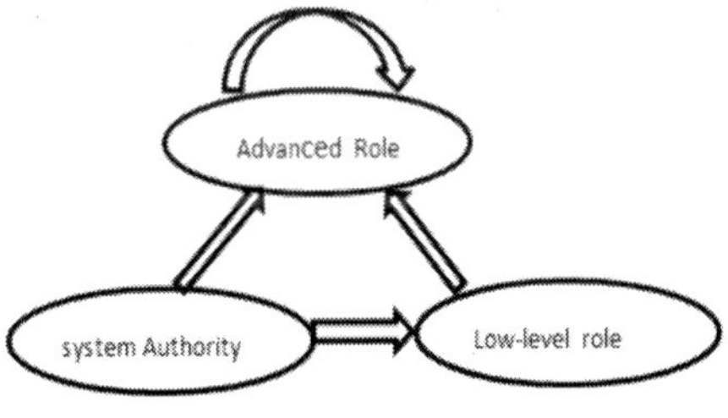

Fig 4 Two types of role classifications

As you know, a privilege can be granted both a high-level role and a low-level role, but there are differences between the two types of roles in granting permissions: the lower-level role can be granted directly, and the higher-level role can Inheritance of low-level role or other high-level roles, the role of the parent role of the inheritance of sub-role, the role of the high-level authorization is more flexible, in addition to its authority can be granted directly, but also inherit other roles permissions.[5]

Follow the above rules to classify and classify the various roles of the educational administration system (low-level role, senior role), the specific division is as follows:

1) Low-level role: students, general teachers, counselors.

2) senior role: Senate Secretary, Department (College) administrator, school leadership, super administrator.

- Assign system privileges

RBAC design program includes two key technologies: one is the role, we have previously described the division and hierarchical roles; the other is the permissions, and role similar to the authority also need to divide the RBAC RBAC management system in this paper design, The plan divides authority into three

DOI: 10.1051/matecconf/201713900120

categories[6]:

1) Departmental authority: This kind of authority generally belongs to the departmental role. For example, the academic secretary of the School of Economics and Management only has the authority of the college's resources in the operating system, and can not operate the resources of other colleges.

2) Function Permission: This kind of authority decides which system function module the user can use and can not use which function module.

3) data permissions: This type of permissions are mainly used to limit the operation of the data in the database, used to determine the different data objects (data tables, views, etc.) for what the permissions (search, entry, change, delete, etc.) .

The above types of permissions through the role of the system with the operating system resources to limit access to the purpose of protecting the security of the system. In the specific application, in the role of grant permissions often need to combine three types of permissions, therefore, the division of authority requires fine-grained division.

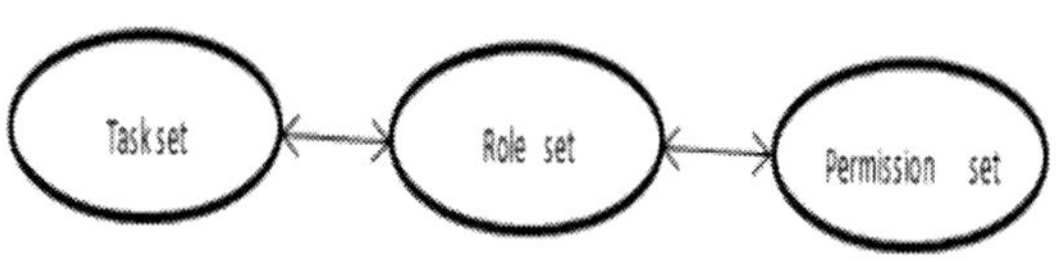

Fig 5 System authority division

Fig5 depicts the division of authority in the educational administration system, explained as follows:

1) Task Set: task division is actually the decomposition of the system functions, and function refers to the need to perform the task, task set is a collection of all functional tasks.

2) Role set: the role corresponds to the number of tasks to perform, so according to the task set the role, and the permissions are granted to the role. A character set refers to a collection of roles.

3) permission set: role with the system permissions are determined by the role of the corresponding set of tasks, the role of the corresponding set of permissions on the various modules of the call.

4.2 Model structure design

Due to the large number of staff, the number of users who use the educational administration system at the same time is also very large. . In order to ensure the security of educational administration system, it is very important to design a set of secure and flexible security access control module. Based on the traditional RBAC model, this module introduces the security principle of separation of users and responsibilities, Authorization, unified management "objectives, the design of educational administration management system security access control model structure shown in

Therefore, the security management system is composed of two parts: identity authentication controller and access controller. Among them, the premise is the identity authentication controller, the core is the access control server, access control server consists of four parts: the user role server, role permissions server, permissions library, and the role of libraries.

5 Summary

In the educational administration system, we introduce the commonly used relatively secure and flexible role-based access control strategy, and RBAC model in the role and permissions to be graded and refined, and the design of the
RBAC model program applied in the educational administration system. From the actual results, this RBAC program can meet the system security requirements, and has better control access capability, and effectively protect the educational administration system in operation security.

ACKNOWLEDGE

This work was supported by the following project:
(1) National Natural Science oundation(61673117,61401101)
(2) Natural Science Foundation of Anhui Provincial Education Department(KJ2016A551, KJ2016A549,KJ2015A295, 2015KJ014,2015KJ007)
(3) National College Students' innovative projects (201610371028, 201610371029)

(4)Fuyang Normal University Foundation (2013ZYSD05,2013KJFH05,2013FSKJ14,2015KJFH02, 2015FSKJ09, 2016PPJY30, rcxm201710)
(5)Anhui Province Quality Project(2014zdjy083, 2014zdjy080,2014zy048)

. Reference

[1] Jia, Z., Pang, L., Luo, S. S., Zhang, J. Y., & Xin, Y.. A privacy-preserving access control protocol for database as a service. In2012 International Conference on Computer Science & Service System (CSSS) (pp. 849–854). IEEE. (2012)

[2] Jia, Z., Pang, L., Luo, S. S., Zhang, J. Y., & Xin, Y.. A privacy-preserving access control protocol for database as a service. In2012 International Conference on Computer Science & Service System (CSSS) (pp. 849–854). IEEE. (2012)

[3] Yaish, H., & Goyal, M. Multi-tenant database access control. In2013 IEEE 16th International Conference on Computational Science and Engineering (CSE) (pp. 870–877). IEEE. (2013)

[4] Xinchao Song, Yishuang Geng, Distributed community detection optimization algorithm for complex networks, Journal of Networks, 9(10), 2758-2765, Jan. (2014)

[5] Jiang D, Ying X, Han Y. Collaborative multi-hop routing in cognitive wireless networks[J]. Wireless Personal Communications: 1-23. (2015)

[6] Yaish, H., Goyal, M., & Feuerlicht, G. An elastic multi-tenant database schema for software as a service. In IEEE Ninth International Conference on Dependable, Autonomic and Secure Computing (Dasc), 2011 (pp. 737–743). IEEE.(2011)

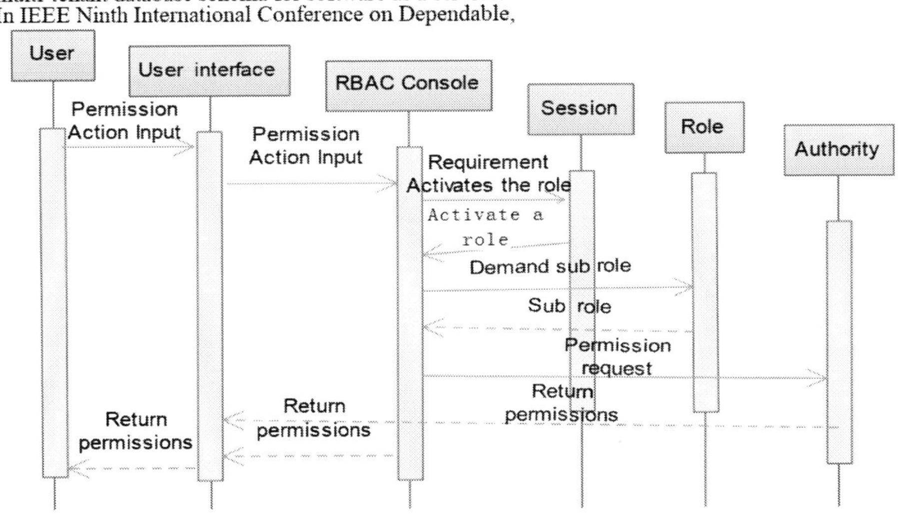

Fig 2 Collaboration diagram of the RBAC session

DOI: 10.1051/matecconf/201713900120

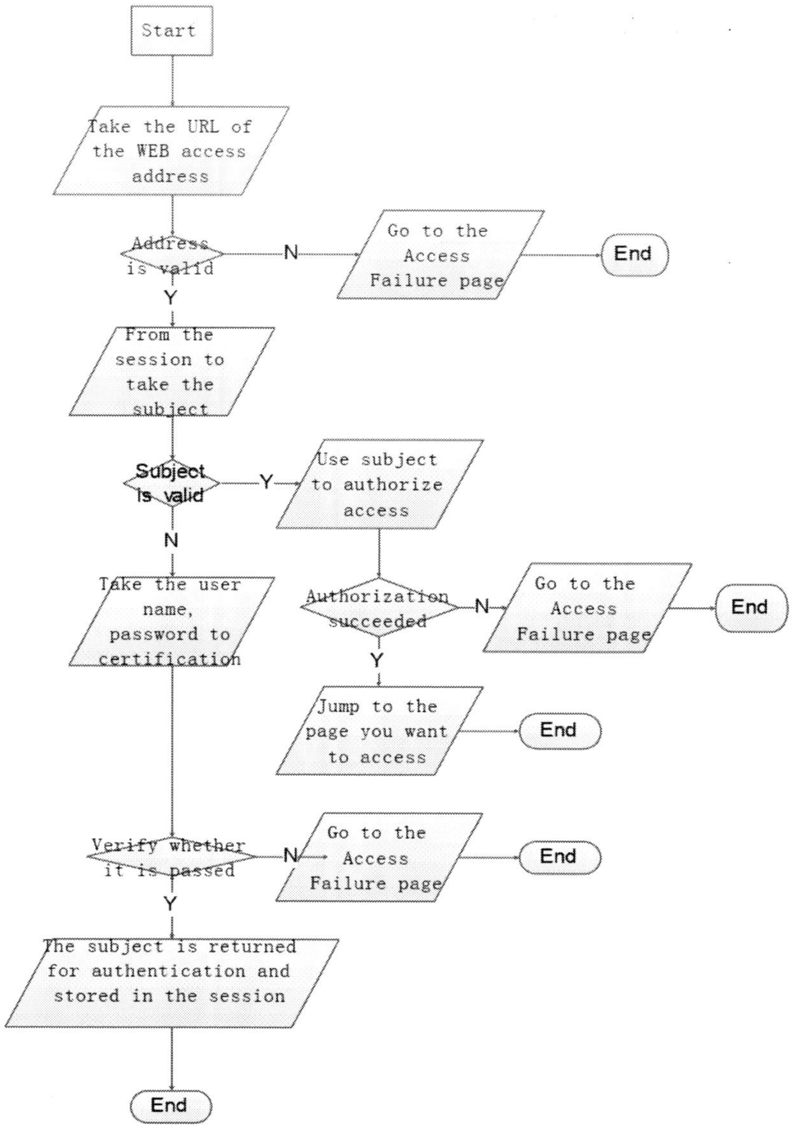

Fig 3 Authentication framework to handle WEB page requests

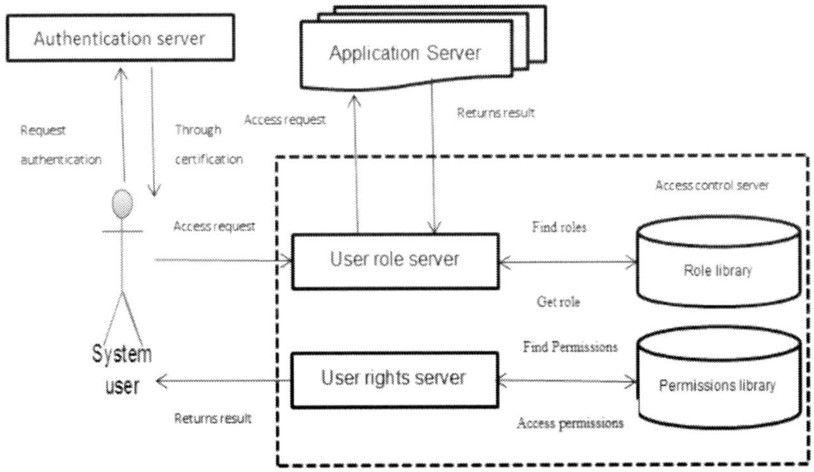

Fig6 Role-based access control system architecture

MATEC Web of Conferences 139, 00121 (2017) DOI: 10.1051/matecconf/201713900121
ICMITE 2017

Research on Innovative Ability of Computer Majors Based on Discipline Contest

LIU Dongdong[1,*], XU Shiliang[1] ,ZHANG Yan[1] ,TAN Fuxiao[1],NIU Lei1,Wang hao[1]
[1] School of Computer and Information Engineering, Fuyang Normal University Fuyang, China

Abstract. Based on classroom teaching, the subject competition is used to stimulate students to link theory with practice and develop their ability of independent working. It consists of a series of activities that can help students find problems and solve them in practice, which further enhances their self-confidence in learning and working. The actual situation of promoting the reform of curriculum construction with the subject competition to improve students' innovative ability in practice varies in different universities and disciplines. This paper takes the computer science of Fuyang Normal University as an example to study how the subject competition promotes students' innovative ability in practice and their comprehensive quality as well.

1 Introduction

The subject competition is based on the close combination of classroom teaching, to competition methods, to stimulate student theory with practical and independent work ability, through practice to find problems, solve problems, enhance student learning and work self-confidence series of activities. Combined with the national long-term education reform and development plan (2010-2020) mentioned in the "to improve the quality of higher education and the quality of personnel training," the current major colleges and universities on the subject competition is very important[1]. Because the subject competition is an effective

means and an important carrier to cultivate the comprehensive quality and innovative spirit of college students, it is necessary to promote the construction of school style style, cultivate students 'innovative spirit, team consciousness and practical ability to stimulate students' interest and potential for creating a good atmosphere of innovation education Important role; for the training of applied innovative talents, improve the quality of personnel training has a very important significance.

From the current situation of domestic and foreign research, college students innovation and practical ability training is already a wide range of issues of concern, subject competition for the promotion of innovative ability to enhance the role of college students have also been recognized. However, how to use the subject competition as the carrier, to promote the reform of curriculum construction, and thus promote the cultivation of innovative ability of college students in each university, each professional is not the same specific situation[2]. This thesis takes the computer major of Fuyang Teachers College as an example to study how the computer specialty can enhance the students' innovative practical ability and comprehensive quality through the subject competition.

2 Promoting the Cultivation of Innovative practice Ability of Computer Majors through Discipline Contest

"Participate in the competition, regardless of the results, can fully mobilize the students' subjective initiative, encourage them to hands, innovation, collaboration, proactive, apply their knowledge.
Therefore, should give full play to the competition in the training of innovative talents in the important role. The effect of subject competition on the cultivation of innovative practical ability of computer students is mainly reflected in:

2.1 To promote the professional curriculum system and teaching content reform

Deepening the reform of curriculum system and teaching content, changing teaching methods and teaching methods is an important content of the construction of teaching quality in colleges and universities. The instructor can participate in the whole process of the subject competition, including the selection of the players, the training to the pre-selection guide, the game command and the post-match summary, can find the problems in the daily teaching, the curriculum system, teaching content and teaching methods To improve. Such as students to participate in the Chinese University Students Computer Design Competition requires a strong analytical capacity, innovation, practical ability, and thus in the latest version of the computer (software engineering) professional training program added C language programming, database principles, Java programming , Linux operating system, Jsp project development and other five curriculum design; increase teaching practice in the proportion of teaching, while encouraging students to participate in innovative business projects; encourage students to participate in teaching and research projects, research projects[3]. Through the theoretical knowledge of learning and practical ability, independent thinking ability and psychological quality and other aspects of training, to cultivate students' innovative consciousness and innovative practical ability to promote teachers teaching, research activities, improve the quality of teaching purposes.

2.2 To cultivate students' innovative consciousness and innovative thinking ability

90 after the students are very strong personality, but also very willing to participate in some activities to highlight their own personality. And the school's conventional teaching is difficult to take care of each student's personality differences, coupled with the current teaching methods are relatively simple, resulting in students awareness of innovation and innovation and lack of creative thinking ability. To carry out the subject competition, you can according to the needs of the content and students to accept the ability to hobbies of different treatment, for some better, flexible thinking, handsome students to create certain conditions, so that they are strengthened in the innovation activities Exercise, and further improve the thinking of innovation and scientific research and innovation ability.

2.3 To help students find their own learning defects, improve learning methods

Daily final exam, memory content more, it is difficult for students to find their own learning defects. By participating in the competition, students can understand their knowledge of the degree of knowledge and the lack of use, so that they continue to improve learning methods, from passive to take the exam to take the initiative to learn, after the active reading of reference materials, Research literature, research and research technology, open up professional field of vision, in order to achieve "innovation in the competition, learning in innovation." Applicants have been surveyed in the class, many students are willing to participate in the subject competition, but because the school or secondary level of publicity is not in place, many students are unable to get the relevant information, resulting in the teacher had just called a few Students to participate, not to promote the purpose of the game.

2.4 To enhance students' sense of collaborative combat and competition

Students who participate in the subject competition may not come from a professional, a grade, a class, before each other do not know. By participating in the

DOI: 10.1051/matecconf/201713900121

subject competition, the original players are not familiar with each other to learn from each other, work together, is conducive to training students to organize, coordinate and enhance the sense of collaborative combat. At the same time the competition itself is a kind of suffering, can test the perseverance of students. The game itself is cruel, there will be win or lose. Through the competition

This can enhance the students brave the first, unyielding competition in the sense of competition, so that students learn to compete, to adapt to competition in the competition for innovation and development[4]. Participate in the game regardless of the outcome of how the players are a valuable experience and wealth, for the future to participate in work, to adapt to the community to lay a solid foundation.

2.5 To build a communication platform between institutions

Participate in the competition, through comparison and learning can be found in the school teaching mode, competition training, competition and other aspects of the lack of security. Through communication with the brothers and institutions, can better promote the understanding of both sides, at the same time for the future study and exchange to provide a good platform.

3 Development Strategy of Innovative Practice Ability of Computer Students Based on Discipline Contest

At present, there are teachers in the university, the pressure of scientific research is big, and the discipline competition is not rewarded. The students' learning initiative is not high and they are not willing to participate in the discipline competition. The disciplines, Which restricts the improvement of the competition level and the improvement of students' innovation ability. Based on the experience of participating in the competition in recent years, the paper summarizes the relationship between the subject competition and the students' innovative practice ability.[5]

3.1 To establish "discipline competition + scientific research project" training mode

"Discipline competition + research project" training model is the school's innovative talents training objectives for the purpose, and strive to improve the practical ability of computer students to innovate. Through the subject competition, scientific research project training, etc., to cultivate a group of students with strong practical ability to practice innovation and entrepreneurship activities, part of the competition or

research project content really hatched into a practical practice of innovative business activities, from the real meaning To improve students' practical ability to innovate. At the same time, students will be in the practice of innovation and entrepreneurship activities accumulated experience to further nurture and enrich the content of subject competitions and research projects to make it richer, more practical, this "subject competition + research project" training model (Figure 2) and innovation and practice to form a mutually reinforcing and mutually reinforcing relationship, so as to jointly improve students' practical and innovative ability.

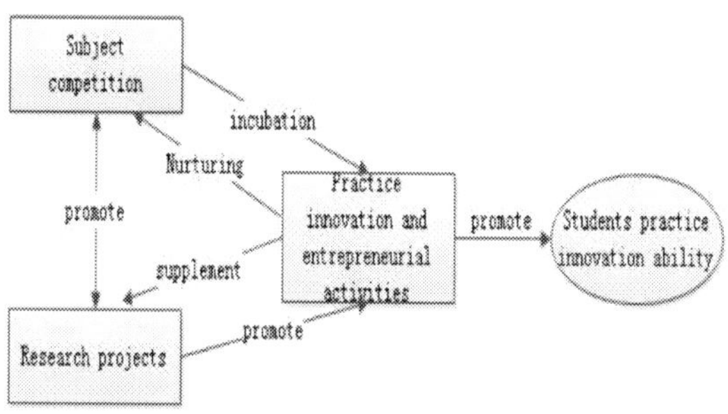

Fig.3. "four years of competition" of the subject competition learning mechanism

3.2 Constructing the Learning Mechanism of "Four - year Competition" in Discipline Contest

In order to make the subject competition become an effective way to cultivate high-quality innovative talents, we can construct the "four-year competition" subject competition learning mechanism (shown in Figure 3), form a series of disciplines, Competition will focus on the "national science and technology competition - to raise the level, to honor the honor; school-level science and technology competition - to expand the students benefit, strong ability" to carry out the work of the practice of teaching and competition combined.

The first year is the basic course of study. After the students are enrolled in the school, the instructor will instill the contents of the relevant competition in the course of the student's course and consciously conduct the guidance and coaching. Organize new students to watch the contests of senior students, to promote past competition results, to stimulate new interest in science and technology competition.

The second year is a school-level competition. In the sophomore stage, the teacher will team up according to the student's hobbies and assign a instructor and a number of learning assistants (each of the experienced senior students) to each team. Teachers in accordance with the interests of students in different types of competition content counseling, regularly carry out the relevant knowledge of the course discussion, learning experience and social surveys, in a variety of ways to stimulate students 'interest in learning and improve students' ability to practice innovation.

The third year is the provincial and ministerial level competition. Students who perform well in school competition are selected into a high level competition team to prepare for a provincial or national science and technology competition. Will be selected out of the outstanding students team, to strengthen training, team members work together. This can not only cultivate students 'team spirit but also improve the students' practical ability and innovation consciousness, and finally can cultivate a high level, all-round,

innovative and practical talents.The fourth year "pass, help, with". Senior students in the professional knowledge and experience are relatively rich. With strong ability of students as the core construction project team, senior and junior students, old members and new members of the "pass, help, with" to promote the rapid growth of team members. And finally to promote the development of competition to promote the practice of students to promote the cultivation of innovative ability[6].

3.3 Curriculum construction and reform of teaching methods

Establish the daily competition "competition view", according to the subject competition and the students to practice the relationship between innovation and practice ability to revise the professional personnel training program. First, the establishment of innovative practice credits, set up an independent competition elective courses to join the school's interdisciplinary elective courses for students to choose, and in accordance with the needs of the competition to improve the curriculum teaching; the second is to reform classroom teaching methods to class discussion, group debate The ability to imitate to the innovative thinking, innovation ability, improve the overall quality of the direction of transformation; Third, the formation of interest groups, so that students can learn from the simple knowledge of knowledge, Starting from the freshman start, strengthen the database and database construction, pay attention to the usual training for the discipline competition reserve talent.

4 Summary

The article combines the traditional teaching, the college students subject competition and scientific research projects and other innovative activities into the cultivation of practical innovation ability. Through the "subject competition + scientific research project" model, the "four years of competition" learning mechanism and other strategies to cultivate college students innovative thinking and practical problems to solve practical problems, while the subject competition to promote the professional curriculum system and teaching content reform, Professional curriculum construction and teaching methods reform, to cultivate students' innovative consciousness and innovative practical ability, to promote teachers teaching, research activities, improve the quality of teaching purposes

ACKNOWLEDGE

This work was supported by the following project:
(1) National Natural Science Foundation (61673117, 61401101)
(2) Natural Science Foundation of Anhui Provincial EducationDepartment(KJ2016A551, KJ2016A549,No.KJ2015A295, 2015KJ014,2015KJ007)
(3) National College Students' innovative projects (201610371028, 201610371029)
(4) The national statistical science project plan research(2014LZ32)
(5)Fuyang Normal University Quality Project (2016PPZY09,2016JYXM37,2016PPJY30, 2015KJFH02,2015FSKJ09, 2016PPJY30, rcxm201710)
(6)Anhui Province Quality Project(2014zdjy083, 2014zdjy080,2014zy048)

. Reference

[1]Xue Yanru .Based on "subject competition" to improve the innovation of college students' ability to cultivate research [J]. Young writers ,24(15): 197 (2011)

[2] Wang Yun lu. Relying on subject competition, training innovative engineering talents [J]. Popular Technology, 08(20): 227 ~ 228 (2011)

[3]Zhang Shuqin, Dong Yuejun, Dong Zhiyong.Study on the Cultivation of Innovative Practice Ability of Computer Majors Based on Science and Technology Competition [J]. Computer Education, 17(22): 14 ~ 16 (2010)

[4]Wang Feng, Wan Zhiping, Ye Shitong, Mo Weijian, Lu Zhimin. Construction of Innovation Team under Scientific Research Project and Subject Competition [J]. Journal of Ningbo University (Educational Science Edition) ,01(15): 57 ~ 60 (2014)

[5]Wang Yicheng, Wang Fangyin, Xiang Wenjiang, Qin Bo Lian, Yu Yougui, Ma Wenyin, Xie Bing.Study and Practice of Cultivating Sci-tech Competition and Innovative Ability of College Students [J]. China Electric Power Education, 10(18):187 ~ 189 (20131)

[6]Zhang Jie. Based on scientific research projects to cultivate independent college students scientific research and innovation ability [J]. Cultural and educational materials,,09(16): 188 ~ 189 (2010)

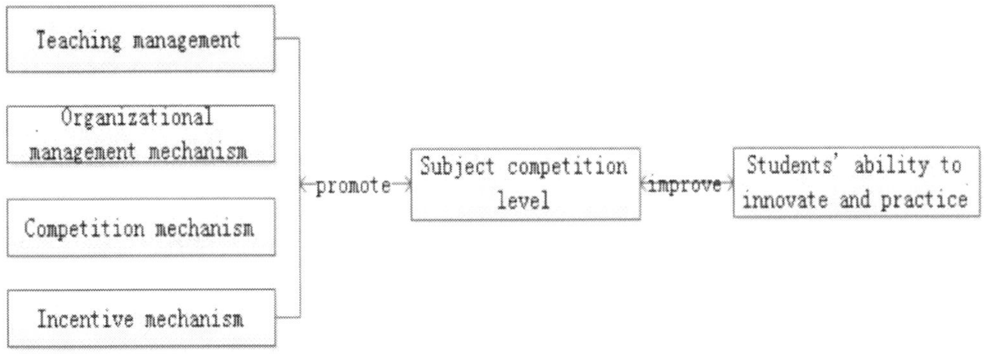

Fig.1. The relationship between the subject competition and the student's ability to practice innovation

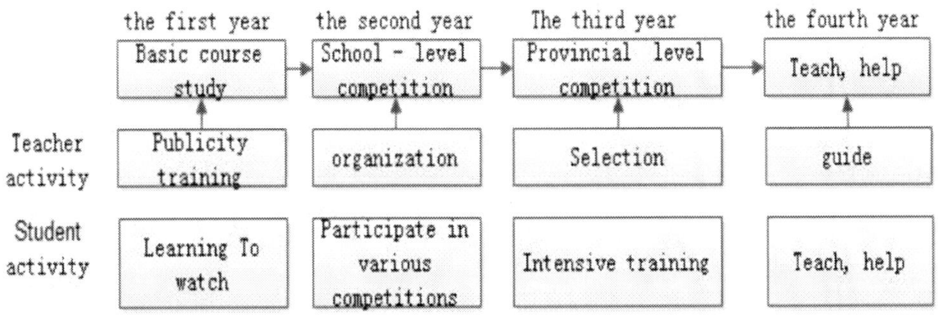

Fig. 2."discipline contest + research project" training mode

Dynamic characteristics of dual-rotor system with coupling faults of misalignment and rub-impact

Zhiwei Huang*

China Ship Development and Design Center, Wuhan 430064, China

Abstract. According to fault problems of the rotor system local rubbing caused by mass eccentricity, a dynamic model for the dual-rotor system with coupling faults of misalignment and rub-impact was established. The dynamic behaviours of this system were investigated by using numerical integral method, as parallel and angular misalignment varied. Various nonlinear phenomena compressing periodic, three-periodic and quasi-periodic motions are observed. The results reveal that the process of the rotor rub-impact is extremely complex and has some frequencies with large amplitude, especially at the 1/3X component. Meanwhile, quasi-periodic regions exhibit different configurations of attractors and the phenomenon of beat vibration.

1 Introduction

Rotor misalignment is one of the most common difficulties in the operation of rotating machinery, yet it remains incompletely understood. Despite the rapid increase in understanding of rotor dynamics, no satisfactory analysis explains the range of observed phenomena. At a time when the rotor dynamics may be regarded as a mature technology, misalignment remains as an outstanding area, where basic understanding is somewhat lacking [1]. Part of the difficulty may arise because the characteristics of misalignment cover a range of situations and the resulting behaviour arises from a combination of physical process. Under the misaligned condition, a series of negative dynamic effects such as wear of bearing, bending deformation of shafts, rotor rub-impact and so on [2-4], would be induced during the motion process of equipment. The rotor unbalance contributed to the misalignment of the journal bearing is always existent [5]. To achieve maximum efficiency in rotating machinery, it is essential to minimize the clearance between the rotor and stator. As a result, the rub phenomenon is also one of the main problems [6].

There are many types of industrial couplings and some of which are of rigid, gear and flexible type. The couplings' function is to transmit torque from the driver to the driven machine, There are two types of coupling misalignments: parallel and angular. A combination of parallel and angular misalignment is common in the industry [7]. In the past several years, efforts have been dedicated on studying and understanding the nonlinear phenomena in rotating machinery with the rotor misalignment or rub-impact. According to the characteristics of the coupling and the present situation in researching on the mechanism of misalignment both at home and abroad, the misalignment was classified in detail by Xia [8]. Hu [9] studied a rig for testing lateral misalignment effects in a flexible rotor supported on three or more hydrodynamic journal bearings. The kinetic mechanism of the rotor system connected by the gear coupling and its faults physical properties were researched by Li [2] and Han [10] form the view point of kinematics when they are of misalignment. The dynamic and characteristics of vibration malfunction of various kinds of shaft misalignment were discussed by Wang [11]. Lees [1] reported some early studies on rotor misalignment and clearly the work had been concentrated on a single form of misalignment. Recent analytical and numerical simulations by Al-Hussain [12] and Bouaziz [13] show the influence of angular misalignment on the dynamic behaviour of a rotor system. From experimental investigations on misalignment rotor systems, frequency spectrum with series of harmonics of running speed was reported by Saavedra and Ramirez [14]. It was also reported [15] that steady-state vibration response at integer fraction of the first bending natural frequency is investigated. Analysis on complicated characteristics of the rotor system with rub-impact has been studied comprehensively [16-18]. Many practical analytical methods were also developed to analyze and interpret the results. Despite a significant number of theoretical and experiment investigations were conducted so far, most of the previous research was focused on gear coupling and horizontal unit with the parallel misalignment. Moreover, little attention was given to the rotor rub-impact due to rigid coupling misalignment that contains both parallel misalignment and angular misalignment. Dynamic behaviours of the rotor system with coupling faults are more complication than that with only one fault. Hence, intensive on the mechanism and phenomena of nonlinear vibration

* Corresponding author: hzwhust@aliyun.com

coupled faults with parallel misalignment and rub-impact in the rotor system has become one of the key issues for accurate fault diagnosis, dynamic design and control.

To make a profound understanding of the dynamic characteristics and malfunctions of rotating machinery with coupling faults of misalignment and rub-impact, the paper presents a model of the dual-rotor system with coupling faults and analyzes its dynamic characteristics by using modern nonlinear dynamics and rotor dynamics theories. For different vibration types of the rotor system, it shows different phenomena and dynamic behaviours, such as periodic, period-three and quasi-periodic. It is indicated that this study can contribute to a further understanding of the non-linear dynamics of such a rotor system with coupling faults of misalignment and rub-impact.

2 Dynamic model of the dual-rotor system

Rigid coupling by which the rotor 1 and rotor 2 are connected together cannot compensate the relative displacement of the two shafts and force the rotor 1 to follow the rotor 2. The motion process of the rotor misalignment and rub-impact is very complicated, involving a lot of factors. In order to make the main problem outstanding, the mechanical model can be simplified as shown in Fig.1. O_1 and O_2 are the center of the rotor 1 and rotor 2, and m_1 and m_2 with eccentricity e_1 and e_2 are the masses of the rotor 1 and rotor 2, respectively. Their radial displacements are assumed as (x_1, y_1) and (x_2, y_2), respectively.

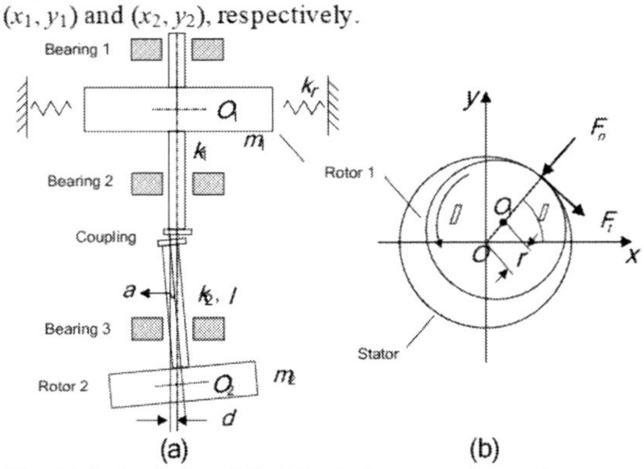

Fig. 1. Mechanical model of the dual-rotor system and schematic diagram of rotor rub-impact

To analyze above the dual-rotor system, the following assumptions are made:

(a) the masses of the two rotors are lumped at mid-point;

(b) the rotors, shafts and bearings are rigid and symmetric;

(c) axial and torsional vibration are negligible;

(d) the mass of shaft and the torque of rotor disk are negligible.

It is assumed that there is an initial clearance of δ between the rotor 1 and stator. Compared with one complete period of rotating, the time during rub-impact is so short that the contact between the rotor1 and stator can be regarded as elastic impact. Under these assumptions, the rub-impact model is illustrated in Fig.1(b). When rubbing happens, the radial impacts force F_N and the tangential rub force F_T can thus be expressed [5] as

$$\begin{cases} F_n = (r-\delta)k_r \\ F_t = fF_n \end{cases} \quad (r \geq \delta) \tag{1}$$

where k_r is the radial stiffness of the stator, f is the friction coefficient between the rotor 1 and stator and $r=\sqrt{x_1^2+y_1^2}$ is the radial displacement of the rotor 1. In xoy reference frame, the forces F_x and F_y are represented as

$$\begin{bmatrix} F_x \\ F_y \end{bmatrix} = -H(r-\delta)\frac{(r-\delta)k_r}{r}\begin{bmatrix} 1 & -f \\ f & 1 \end{bmatrix}\begin{bmatrix} x_1 \\ y_1 \end{bmatrix} \tag{2}$$

where $\boldsymbol{H}$ is the Heaviside function, can be defined as

$$H(x) = \begin{cases} 0 & x \leq 0 \\ 1 & x > 0 \end{cases}$$

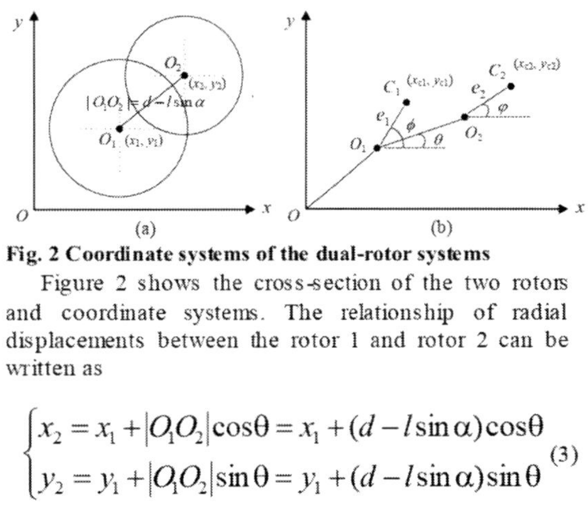

Fig. 2 Coordinate systems of the dual-rotor systems

Figure 2 shows the cross-section of the two rotors and coordinate systems. The relationship of radial displacements between the rotor 1 and rotor 2 can be written as

$$\begin{cases} x_2 = x_1 + |O_1O_2|\cos\theta = x_1 + (d - l\sin\alpha)\cos\theta \\ y_2 = y_1 + |O_1O_2|\sin\theta = y_1 + (d - l\sin\alpha)\sin\theta \end{cases} \tag{3}$$

where d is the values of parallel misalignment, θ is the angle of rotor 2 rotating around rotor 1, $\theta=\omega t+\theta_0$.

The centroid displacements of the rotor 1 and rotor 2 can be respectively calculated as (see Fig.2b)

$$\begin{cases} x_{c1} = x_1 + e_1\cos\phi \\ y_{c1} = y_1 + e_1\sin\phi \end{cases} \tag{4}$$

$$\begin{cases} x_{c2} = x_2 + e_2\cos\varphi\cos\alpha \\ y_{c2} = y_2 + e_2\sin\varphi\cos\alpha \end{cases} \tag{5}$$

where (x_{c1}, y_{c1}), (x_{c2}, y_{c2}) are the centroid coordinates of the rotor 1 and rotor 2, ϕ, φ are the angle of the two rotors rotating respectively, $\phi=\omega t+\phi_0$, $\varphi=\omega t+\varphi_0$.

The total kinetic energy T is constituted of driver kinetic energy T_G and the driven kinetic energy T_r. The system kinetic energy can be written as

$$T = T_G + T_r \tag{6}$$

in which

$$\begin{aligned} T_G &= \frac{1}{2} m_1 v_{c1}^2 + \frac{1}{2} m_2 v_{c2}^2 \\ &= \frac{1}{2} m_1 (\dot{x}_1^2 + \dot{y}_1^2 + e_1^2 \dot{\phi}^2 - 2 e_1 \dot{x}_1 \dot{\phi} \sin\phi \\ &+ 2 e_1 \dot{y}_1 \dot{\phi} \cos\phi) + \frac{1}{2} m_2 [\dot{x}_2^2 + \dot{y}_2^2 + (e_2 \cos\alpha)^2 \dot{\varphi}^2 \\ &- 2(e_2 \cos\alpha) \dot{x}_2 \dot{\varphi} \sin\varphi + 2(e_2 \cos\alpha) \dot{y}_2 \dot{\varphi} \cos\varphi] \end{aligned} \tag{7}$$

$$T_r = \frac{1}{2}(J_1 + m_1 e_1)\dot{\phi}^2 + \frac{1}{2}(J_2 + m_2 e_2)\dot{\varphi}^2 \tag{8}$$

where J_1, J_2 are the inertia moment of the rotor 1 and rotor 2 respectively.

The elastic potential energy U of the system is defined as

$$\begin{aligned} U &= \frac{1}{2} k_1 |OO_1|^2 + \frac{1}{2} k_2 |OO_2|^2 \\ &= \frac{1}{2} k_1 (x_1^2 + y_1^2) + \frac{1}{2} k_2 (x_2^2 + y_2^2) \end{aligned} \tag{9}$$

where k_1, k_2 are the bearing stiffness of the rotor 1 and rotor 2 respectively.

It is assumed that only the damping c and the external exciting forces of the rotor 1 are not considered. So the generalized force for the dual-rotor system can be written as

$$\begin{cases} Q_x = -c\dot{x}_1 + F_x \\ Q_y = -c\dot{y}_1 + F_y \end{cases} \tag{10}$$

Upon substituting the kinetic and potential energy expressions into the Lagrange equation, performing partial differentiation and manipulation, if $x=x_1$, $y=y_1$, the system equation of motion is found by Equation (1) ~ Equation (10) as

$$\begin{cases} (m_1 + m_2)\ddot{x} + c\dot{x} + (k_1 + k_2)x = m_1 e_1 \omega^2 \cos\phi + \\ m_2 e_2 \omega^2 \cos\alpha \cos\varphi + m_2 (d - l \sin\alpha)\omega^2 \cos\theta \\ -k_2 (d - l \sin\alpha)\cos\theta + F_x \\ (m_1 + m_2)\ddot{y} + c\dot{y} + (k_1 + k_2)y = m_1 e_1 \omega^2 \sin\phi + \\ m_2 e_2 \omega^2 \cos\alpha \sin\varphi + m_2 (d - l \sin\alpha)\omega^2 \sin\theta \\ -k_2 (d - l \sin\alpha)\sin\theta + F_y \end{cases} \tag{11}$$

In right side of the equation(11), the preceding two items are respectively centrifugal inertial force of the rotor 1 and rotor 2 as the shaft is rotating, and the centrifugal inertial force of the rotor 2 caused by parallel and angular misalignment is given in the third item, the fourth is the force exerted by the bearing and the last item is the rub-impact force mentioned above. It can be concluded clearly that the mass eccentricity, parallel misalignment and stiffness are the main factors on the stability of the rotor system.

For the convenience of calculating composition and results analysis, the dimensionless transformations are given as follows:

$$X = \frac{x}{\delta}, \quad Y = \frac{y}{\delta}, \quad D = \frac{d}{\delta} \tag{12}$$

3 Simulation result and analysis

Due to the strongly nonlinear characteristics of the rub-impacting rotor system, the vibration responses are quite complicated. Numerical analysis is carried out by the fourth-order Runge-Kutta method. During the calculation a smaller integration step has to be chosen to ensure that a stable solution will be obtained as well as to avoid the numerical divergence at the point where derivatives of the rub-impact forces are discontinuous.

The values for the computational parameters are as follows:

m_1=3.5×10^{5}kg, m_2=3.0×10^{5}kg, c=2.1×10^{6}N·s/m, k_1=1.0×10^{7}N/m, k_2=1.2×10^{7}N/m, k_r=2.5×10^8 N/m, δ=2.0 mm, f=0.01, ϕ_0=0.5rad, φ_0=0.3rad, θ_0=0.1rad, ω=35rad/s.

3.1. Influence of parallel misalignment

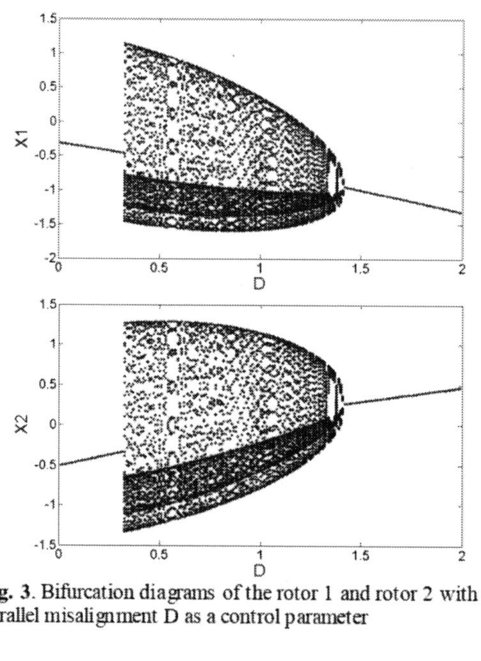

Fig. 3. Bifurcation diagrams of the rotor 1 and rotor 2 with the parallel misalignment D as a control parameter

Figure 3 shows the bifurcation diagrams of the rotor 1 and rotor 2 using the parallel misalignment D as the control parameter with mass eccentricity e_1=0.8 mm, e_2=1.0 mm and α=0.0001rad. It can be seen that the rotor responses contain periodic and quasi-periodic motions at D=0~2. To illustrate the various process of motion, Fig.4 shows the nonlinear characteristics of the dual-rotor system with the plots of Poincaré maps, axis orbits, time histories and amplitude spectrum diagrams at different D. When parallel misalignment is small, the rub phenomenon does not occur, the motion is synchronous with period-one, and only one point is correspondingly displayed in the bifurcation diagram. It can also be clearly seen form the spectrum that a single spectrum line can be observed at the 1X component and zero at other positions. Fig. 4(a) shows the dynamic characteristics at D=0.2. With the increase of the parallel misalignment, rub does occur between the rotor 1 and the stator. When D>0.32, the rub-impact happens and the rotor motion becomes quasi-periodic. At D=0.57, there are some isolated points in the Poincaré map and periodic motions in the axis orbit. The beat vibration is observed on this time history and the amplitude of the 1/3X component is bigger to that of 1X component as shown in Fig. 4(b). The more greatly the parallel misalignment increases, some changes can be seen when D=1.36 as shown in Fig. 4(c). In this case there is a closed curve in the Poincaré map and the vibration waveforms in some periods are not so regular which indicates that the rub-impact has become more serious. As D exceeds 1.41, the motion comes into period-one again because the dual-rotor system reaches dynamic balance. Although the two rotors are rigidly coupled by the shaft, dynamic characteristics of the rotor 1 with the rub-impact force are more complicated than those of the rotor 2 such as axis orbits and time histories.

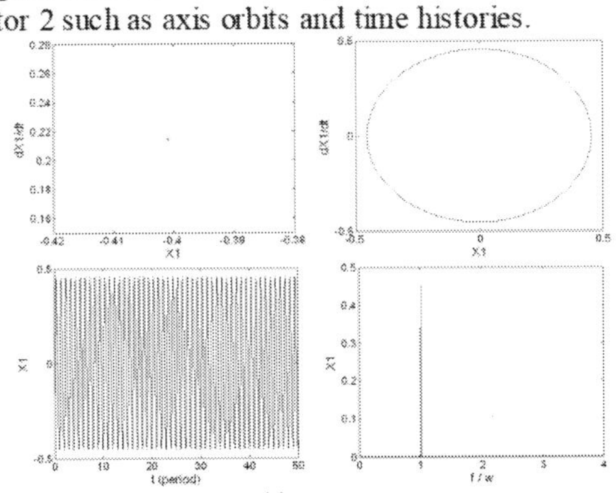

(a) D=0.2

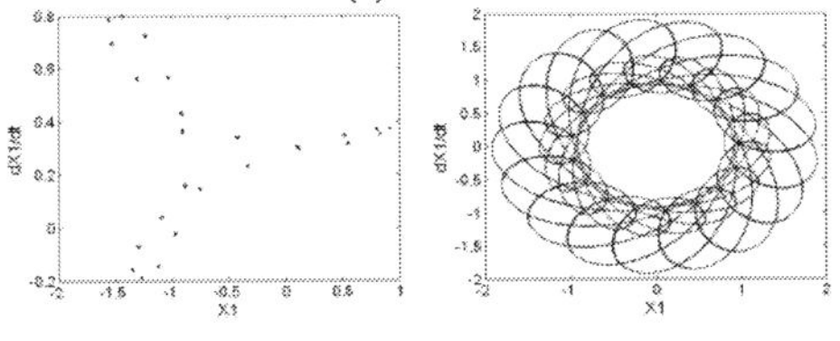

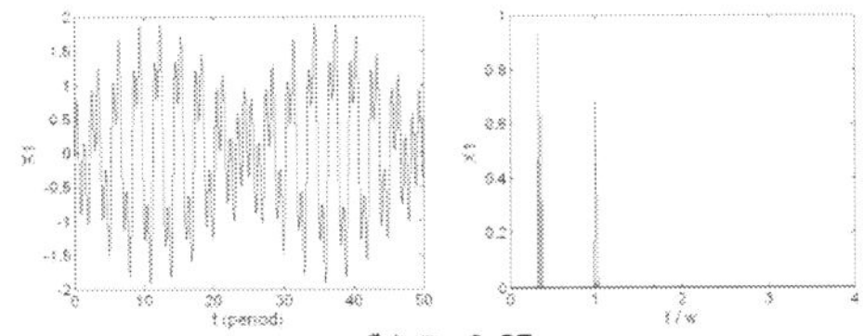

(b) D=0.57

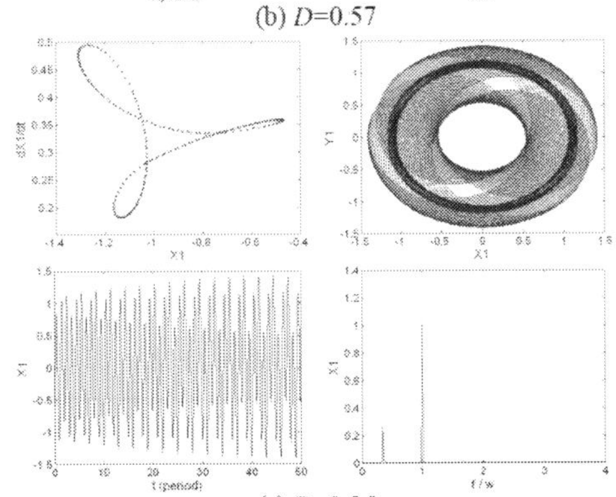

(c) D=1.36

Fig. 4. Poincaré maps, axis orbits, time-histories and amplitude spectrum diagrams on the response of the rotor 1 for different parallel misalignment

From amplitude spectrums of Fig. 4(a)-(d), we find some low frequencies with large amplitude besides 1X component, especially at 1/3X component. Amplitude of the rotor 1 at 1X component is increasing constantly. At the same time, amplitudes of low frequencies of the two rotors are decreasing. Responses of sub-harmonic for the dual-rotor system caused by the coupling faults with misalignment and rub-impact make amplitudes of low frequencies bigger than that at 1X component in some conditions. The phenomenon can be defined as a typical feature for the coupling faults. In addition, it also could provide important theory basis to diagnose the kind of fault.

2.2 Influence of angular misalignment

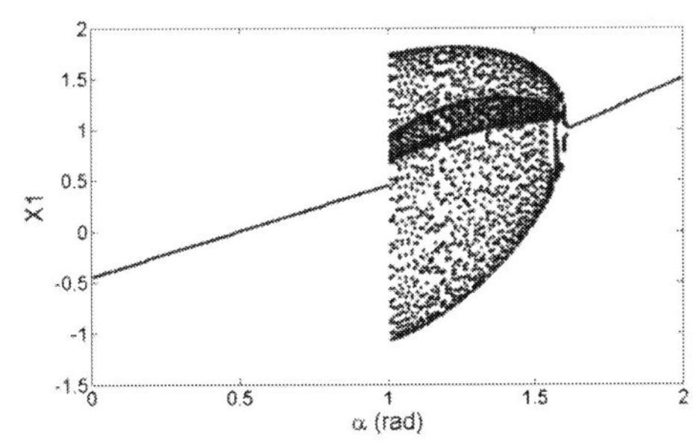

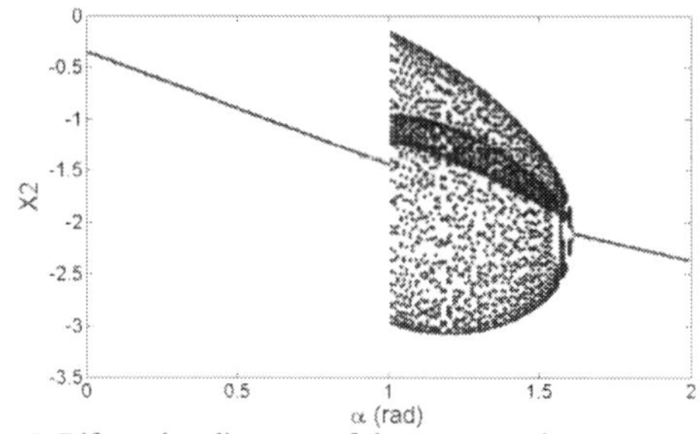

Fig. 5. Bifurcation diagrams of the rotor 1 and rotor 2 response with the angular misalignment D as a control parameter

Fig. 5 shows the bifurcation diagrams and its magnification of angular misalignment α on the response of the rotor 1 and rotor 2 with mass eccentricity e_1=0.8 mm, e_2=1.0 mm and D=0.1. It can be seen that system responses exhibit periodic, there periodic and quasi-periodic motions at the interval of (0<α<0.002) rad. At α<1.0×10^{-3} rad, the system response displays steady-sate synchronous motion with period-one as shown in Fig. 6(a), the vibration amplitude of the dual-rotor system is small and rub phenomenon generally does not occur. As the imbalance increases, the vibration amplitude of the dual-rotor system becomes bigger and the rub phenomenon occurs between the rotor 1 and stator. At α>1.0×10^{-3} rad, the system response comes into quasi-periodic motion and exhibits a wider range of quasi-periodic motion as the control parameters vary. Fig. 6(b) illustrate the dynamic characteristics at α=1.4×10^{-3} rad. It can be seen from that the closed from of the attractor is decomposed in the Poincaré map and two discrete frequency components arise in the amplitude spectrum. All of these give proof of that the motion is quasi-periodic. The more greatly the angular misalignment increases, the motion becomes a sub-synchronous vibration with period-three and its attractor is three isolated points, exhibiting three different circular rings in the axis orbit and amplitudes in time history correspondingly as shown in Fig. 6(c). 1/3X component can be clearly seen from the spectrum except the 1X component. At α>1.61×10^{-3} rad, the motion comes into period-one again.

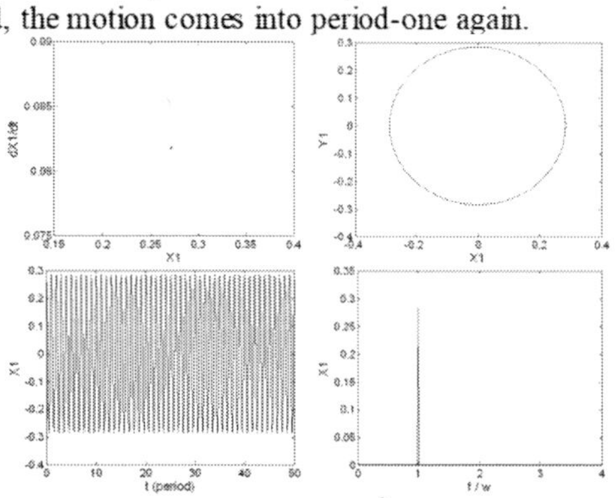

(a) α=0.8×10^{-3}rad

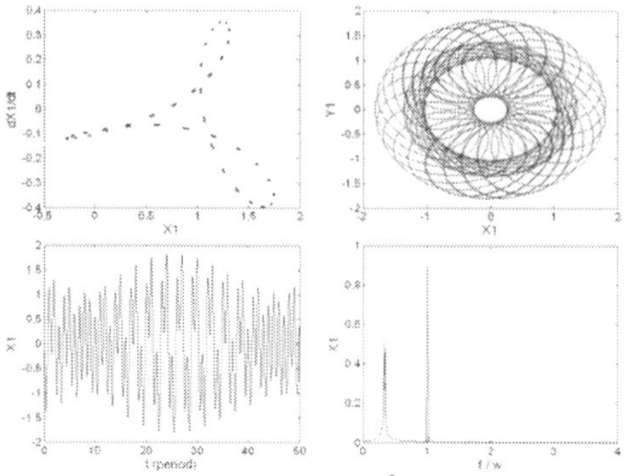

(b) α=1.4×10^{-3}rad

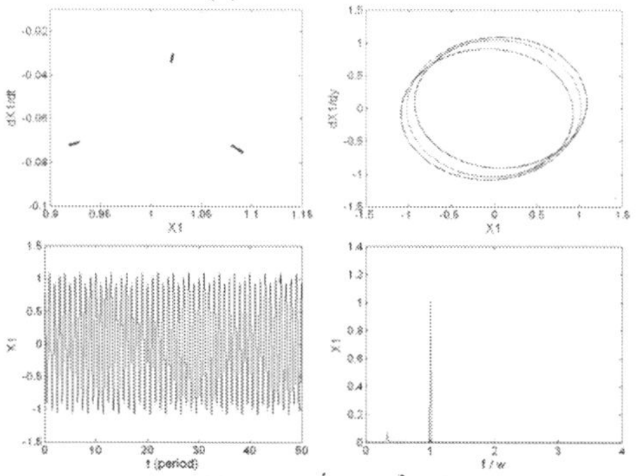

(c) α=1.61×10^{-3}rad

Fig. 6. Poincaré maps, axis orbits, time-histories and amplitude spectrum diagrams on the response of the rotor 1 for different angular misalignment

Comparison between Figure 3 and Figure 5 reveals that dynamic characteristics of the dual-rotor system are very similar as the increase of parallel misalignment and angular misalignment. However, the changes of angular misalignment can effect the centrifugal inertial of the rotor 1, so there are some distinctive responses of the dual-rotor system such as axis orbits and amplitude spectrum. It can be also observed that the influence of parallel misalignment is stronger than that of angular misalignment.

4 Conclusions

In this study, according to fault problems of the dual-rotor system local rubbing caused by mass eccentricity, a model of the dual-rotor system with coupling faults of misalignment and rub-impact has been developed. Some non-linear analysis methods were employed to explore dynamic behaviors of the discussed system, including Poincaré maps, axis orbits, time histories, amplitude spectrum and bifurcation diagrams. The influence of parameters such as parallel misalignment, angular misalignment and mass eccentricity on the dynamic characteristics of the dual-rotor system is studied theoretically in detail. It can be observed that the system exhibits various forms of periodic, three-periodic and quasi-periodic. The simulation results show that there are some low-frequency components with large amplitude in the amplitude spectrum, especially at the 1/3X component and beat vibration phenomenon occurs in the time history for quasi-periodic. Although the rotor 1 and rotor 2 are rigid coupled, dynamic characteristics of two rotors are some distinctive. It is found the mass eccentricity can obviously change radial vibration of the dual-rotor system. With the increase of mass eccentricity, system responses mainly alternate between periodic and quasi-periodic motions. It will be seen from this that the performances of the dual-rotor system are significantly influenced by the characteristics of the misalignment. It is imperative to understand the nonlinear behavior of the dual-rotor system with the coupling faults. The purpose of the further research is effectively to diagnose the coupling faults, reduce the failure and optimize the design of the dual-rotor r system.

Acknowledgments

This work has been supported by Project of Program of National Natural Science Foundation of China (No.50539140).

References

1. Lees AW. Misalignment in rigidly coupled rotors. Journal of Sound and vibration, 2007, 305: 261-271.
2. Li M, Yu L. Analysis of the coupled lateral torsional vibration of a rotor-bearing system with a misalignment gear coupling. Journal of Sound and vibration, 2001, 243(2): 283-300.
3. Hamdi T, Selcuk E, Ibrahim U. Experimental analysis on fault detection for coupled rotor-bearing system. Measurement, 2013, 46(1): 336-344.
4. Li QK, Liao MF, Jiang YF. The vibration features analysis of twin spool rotor with misalignment fault. Mechanical Science and Technology for Aerospace Engineering, 2015, 33(12): 1916-1920.
5. Li ZG, Li M. Non-linear dynamics of a flexible multi-rotor bearing system with a fault of parallel misalignment. Applied Mechanics and Materials, 2012, 138-139: 104-110.
6. Wan C, Jian C, Chen CK. Chaos of rub-impact rotor supported by bearings with nonlinear suspension. Tribology International, 2009, 42: 426-439.
7. Al-Hussain KM, Redmond I. Dynamic response of two rotors connected by rigid mechanical coupling with parallel misalignment. Journal of Sound and vibration, 2002, 249(3): 483-498.
8. Xia SB, Zhang XJ, Liu ZS, et al. A survey of research on misalignment of rotary machinery. Journal of vibration, Measurement and Diagnosis, 1998, 18(3): 157-160.
9. Hu W, Miah H, Feng NS, et al. A rig for testing lateral misalignment effects in a flexible rotor supported on three or more hydrodynamic journal bearings. Tribology International, 2000, 33: 197-204.
10. Han J, Shi LD. Study on kinematic mechanism of misalignment fault of rotor system connected by gear coupling. Journal of Vibration Engineering, 2004, 17(4): 416-420.
11. Wang YB. Shaft misalignment induced vibration in high capacity steam-turbine generator units. Power Engineering, 2004, 24(6): 768-774.
12. Al-Hussain KM. Dynamic stability of two rigid rotors connected by a flexible coupling with angular misalignment. Journal of Sound and Vibration, 2003, 266: 217-234.
13. Bouaziz S, Hili MA, Mataar M, et al. Dynamic behaviour of hydrodynamic journal bearings in presence of rotor spatial angular misalignment. Mechanism and Machine Theory, 2009, 44: 1548-1559.
14. Saavedra PN, Ramirez DE. Vibration analysis of rotors for the identification of shaft misalignment Part Ⅱ: Experimental validation. Journal of Mechanical Engineering Science, 2004, 218(c): 987-999.
15. Patel TH, Darpe AK. Experimental investigations on vibration response of misalignment rotors. Mechanical System and Signal Processing, 2009, 23: 2236-2252.
16. Liu Y, Tai XY, Yao HL, et al. Study on misalignment-rubbing coupling fault of dual-disk rotor-bearing system including the impact of motor coupling. Journal of Vibration, Measurement and Diagnosis, 2013; 33(5): 819-823.
17. Zhu YY, Zhou QZ, Wang DS. Vibration characteristics of a rub-impact rotor system with an angular misalignment under the unsteady oil-film forces. Journal of Wuhan University (Natural Science Edition), 2005, 283: 621-643.
18. Shen XY, Jia JH, Zhao M. Effect of parameters on the rubbing condition of an unbalanced rotor system with initial permanent deflection. Archive of Applied Mechanics, 2007, 77(12): 883-892.

OceanSense-A general application development framework for marine environment monitoring with multi-sensors Conferences

Bin Zhao[1],Yu Yu[1]*, Honghai Zhu[1], and Li Yang[1]

1、Institute of Oceangraphic Instrumentation, Shandong Academy of Sciences ZheJiang Road 28, 266001, Qingdao, China

Abstract. This paper presents a component-based framework OceanSense which supports rapid development of marine environment monitoring application with multi-sensors. By decomposing systems into relatively independent, mature and reusable modules, component-based programming provides stability, reliability and extensibility compared to other programming methods. The framework covers kernel components for marine monitoring application, which are data management component, data transmission component, electronic nautical chart component, sensor communication component. OceanSense supports commonly used marine environment monitoring sensors, which increases its availability. A real project is carried on to validate the framework and result shows its availability.

Keywords.marine environment monitoring; electronic nautical chart;development framework

1 Introduction

With the rapid development of society and economy, an increasing number of human activities have gradually destroyed the marine environment. Maine environment monitoring is a vital problem and has increasingly attracted agreat deal of research and development attention. During the past few decades, various marine environment monitoring systems have been developed [1-6]. Base on decades of experiences on marine monitoring, a component-based framework for rapid development of marine environment monitoring application is proposed. Leveraging on large amount of related projects, commonly used sensors of different interfaces and manufacturers are summarized and catalogued. Multi-sensor supporting gives flexibility and extensibility of OceanSense.

Rest of this paper is organized as follows. The framework architecture of OceanSense is presented. Then, three kernel components of OceanSense are described. After that, an application example is presented. Finally, conclusion and the future work are given.

2 Framework architecture

As demonstrated in Fig. 1, the framework is composed by Electronic nautical chart component, data management component, data transmission component and sensor communication component. Those components can be imported into visual studio 2010. The properties of those components can be set by visual interface in visual studio 2010 and the interfaces of those components can be called by using C#. Each component will be further described as the following.

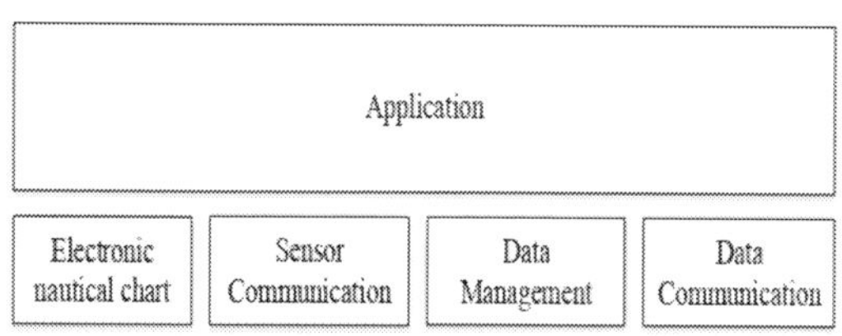

Fig. 1. System architecture

2.1 Data management component

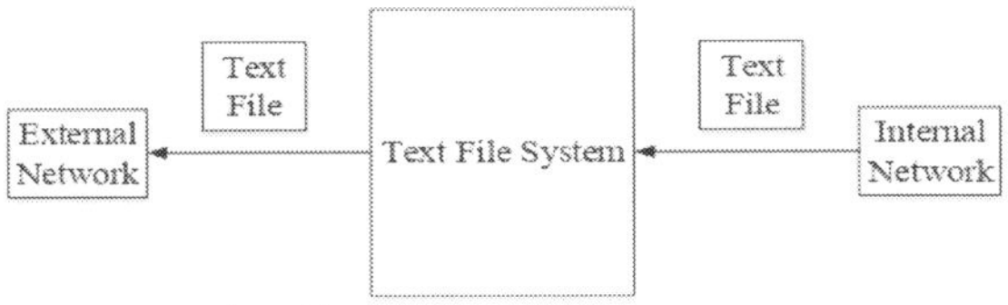

Fig. 2. Network architecture.

Ocean observation data are usually sensitive data, which has high requirements for safety. A typical network architecture is described in Figure 2. Data transmission from internal network to external network can only be achieved through transmission of text files. Applications from external network periodically scan dedicated text files in the mediator system. Then, contents in those text files are compared with last time scanning contents to decide whether contents is updated by applications in the internal network. XML is preferred for its advantages in data storage and data sharing. A typical observation data in XML text files is shown in Figure 3.

```xml
<observations>
  <location timezone="8" distance="0.00" longitude="120.31" latitude="36.56" >
    <observation utcTime="2017-05-21T10:20:00.00+01:00" elevation="31.00">
      <temperature>26.00</temperature>
      <humidity>87</humidity>
      <windSpeed>5.56</windSpeed>
      <windDirection>30</windDirection>
      <barometerPressure>1016.26</barometerPressure>
      <visibility>7.00</visibility>
    </observation>
  </location>
</observations>
```

Fig. 3. Observation in XML

2.2 Data transmission

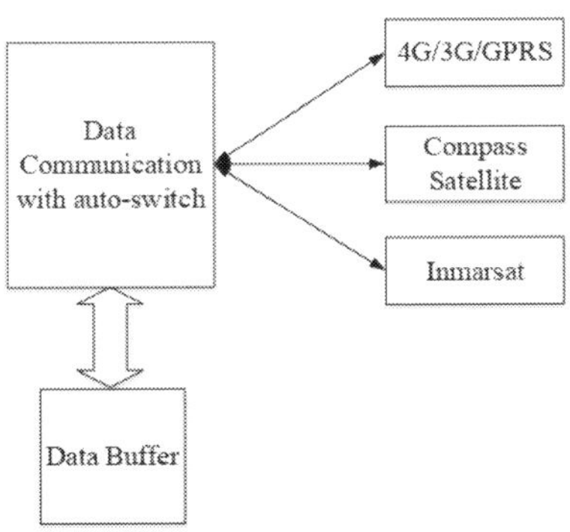

Fig. 4. Data communication architecture

Observation data transmission are usually transmitted via satellite to the ground station. In china coastal water and the Asia-Pacific region, Compass short messages provide a better way than Inmarsat for its high efficiency and low price (See Figure 5). In the near land waters, high speed data transmission with 3G, 4G or GPRS signal coverage is feasible. In the design of data communication component, communication modes are switched automatically according to different communication strategy.

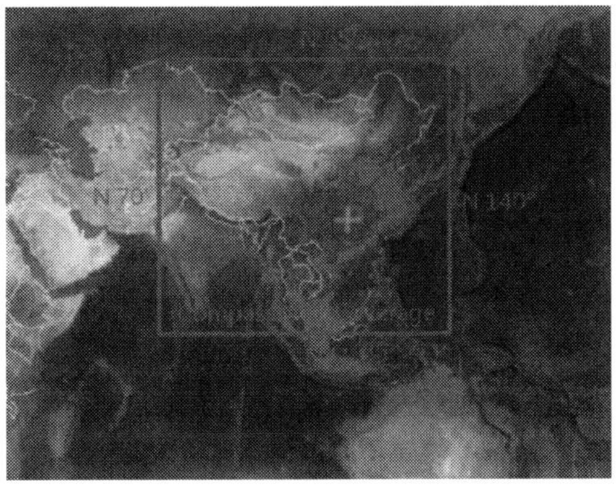

Fig. 5. Compass No1 satellites coverage area.

2.3 Electronic nautical chart component

Electronic nautical chart (ENC) component plays an important part in marine monitoring data demonstration. It provides direct connections between monitoring data and its geographical features. The extensibilities of ENC component is through supporting of IHO S-57 format. The S-57 file with .000 extensions will be translated to proprietary format, in order to ensure efficiency of S-57 objects rendering (See Fig. 6).

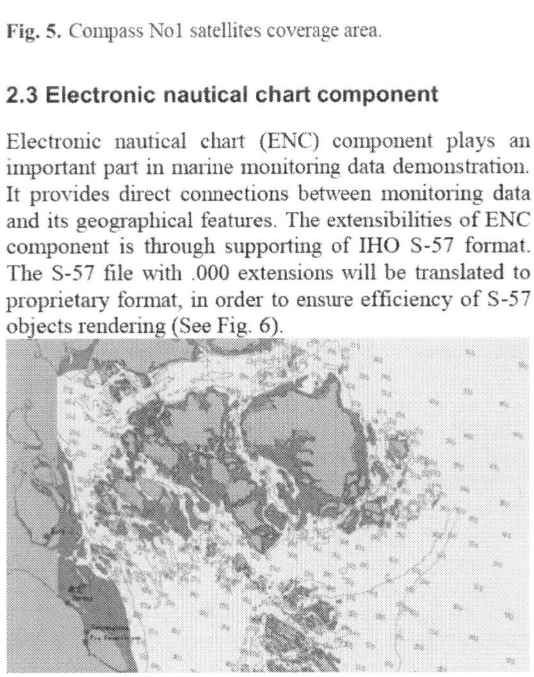

Fig. 6. ENC display.

2.4 Sensor communication component

Sensor communication component encapsulates the details of communications between seaMass and the instruments. A base abstract class 'Command' is provided to encapsulate the structure of communication command (see Fig. 7).

Based on communication direction between instrument and application, two abstract classes 'OutCommand' and 'InCommand' inherited from 'Command' are further given. The 'OutCommand' class provides 'ToHex' function which transfers the instantiated class object into raw hex array. Meanwhile, the 'InCommand' class provides 'ValidateCheckSum' function which calculates and validates Checksum of the instantiated class object. New command can be added by inheriting from either 'InCommand' or 'OutCommand' classes. The

'Command' classes' layout of inheritance is further depicted in Fig 8.

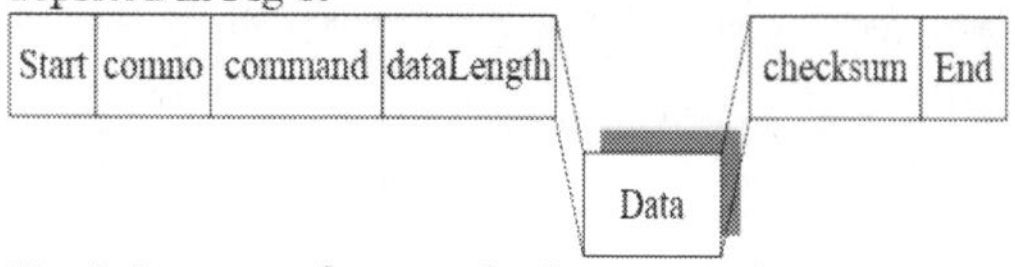

Fig. 7. Structure of communication command

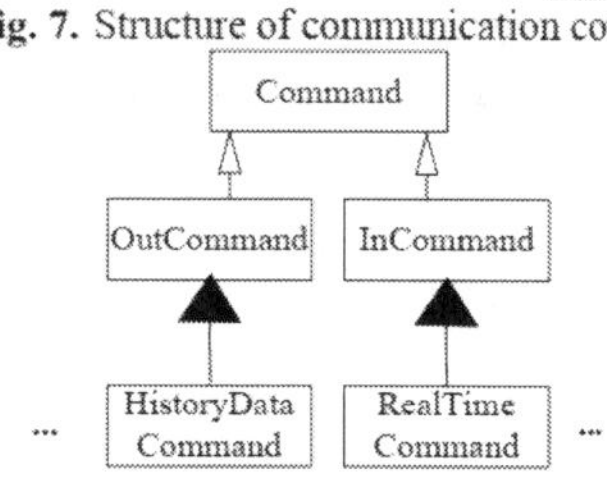

Fig. 8. Command classes layout of inheritance.

Another key interface for the instrument communication component is 'Interpreter'. The 'translate' function in 'Interpreter' interface translates the raw byte array received from serial port into concrete class object. The pseudo-code of 'Interpreter' is listed as follows:

```
If (buffer[2] == RealTime)
RealTimeCmd rtCmd = new RealTimeCmd(…);
else if(buffer[2] == HistoryData)
HistoryDataCmd hdCmd = new HistoryDataCmd(…);
else if(buffer[2] == SysStatus)
SysStatusCmd sysCmd = new SysStatusCmd(…);
…
```

The concrete class object follows a plain ordinary java object (POJO) format. The low level details of serial port communication are encapsulated by 'SerialPortOperation' class, which mainly responses for serial data income event handling.

The abstract classes and interface described above are the fundamentals for our automatic 'Command' handling mechanism. In the automatic 'Command' handling mechanism, 'Command' is expressed by XML with schemas. Taking the 'RealTimeCommand' for example, its corresponding XML definition is shown in Fig. 9.

Besides that, the 'Command' can also be validated against its corresponding XML Schema before sending out (See Fig. 10), which greatly improves system reliability.

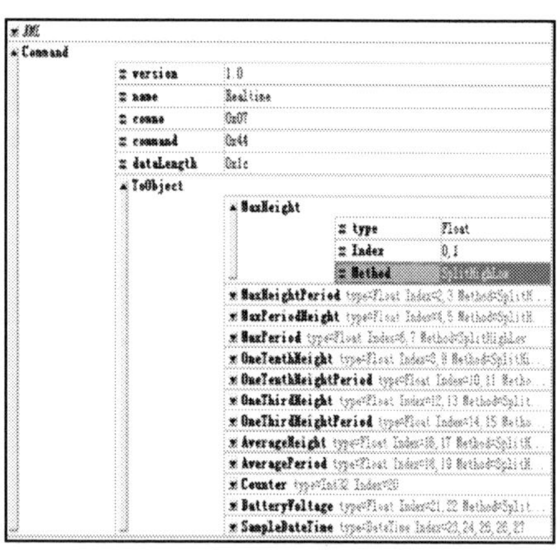

Fig. 9. Xml definition of 'RealTimeCommand'

3 Application example

In order to validate the development framework, it has been used in volunteer boat marine observation (See Fig. 11), which is dedicatedly designed for monitoring seawater temperature and salt, wind speed and direction, tidal, rainfall, and etc.

4 Conclusion

In OceanSense, kernel components for marine environment monitoring application development are provided. Yet, those components with .net runtime are proved to be reliable, stable and easy to use. In future, development and refinement on those kernel components will be continued and an open source project will be founded.

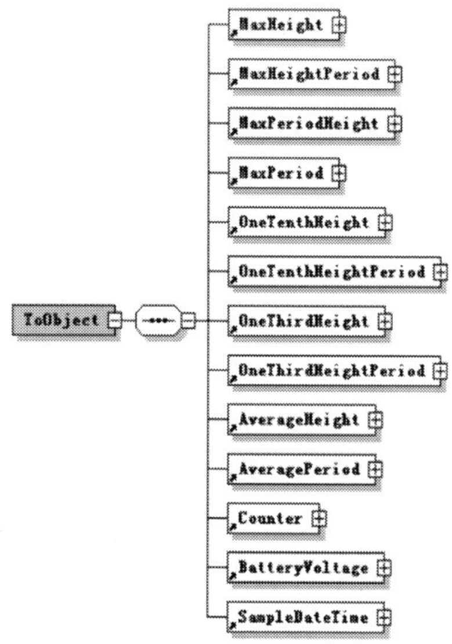

Fig. 10. Xml schema for 'RealtimeCommand'.

Acknowledgment

This work is supported by Shandong Academy of Sciences for Young Scientists Fund (grant no. Y15-304), Key Research and Development program of Shandong Province (grant no. 2015GSF115011), Qingdao City Southern District Science and Technology Development Fund (grant no. 2016-2-012-ZH).

Reference

1. Xu G, Shen W, Wang X. Applications of wireless sensor networks in marine environment monitoring: A survey[J]. Sensors, 2014, 14(9): 16932-16954.
2. De Jonge V N, Elliott M, Brauer V S. Marine monitoring: its shortcomings and mismatch with the EU Water Framework Directive's objectives[J]. Marine Pollution Bulletin, 2006, 53(1): 5-19.

 DOI: 10.1051/matecconf/201713900123

3. Borja Á, Elliott M, Carstensen J, et al. Marine management–towards an integrated implementation of the European Marine Strategy Framework and the Water Framework Directives[J]. Marine Pollution Bulletin, 2010, 60(12): 2175-2186.
4. Rogers S I, Greenaway B. A UK perspective on the development of marine ecosystem indicators[J]. Marine Pollution Bulletin, 2005, 50(1): 9-19.
5. Imperial M T. Institutional analysis and ecosystem-based management: the institutional analysis and development framework[J]. Environmental management, 1999, 24(4): 449-465.
6. Rudd M A. An institutional framework for designing and monitoring ecosystem-based fisheries management policy experiments[J]. Ecological Economics, 2004, 48(1): 109-124.

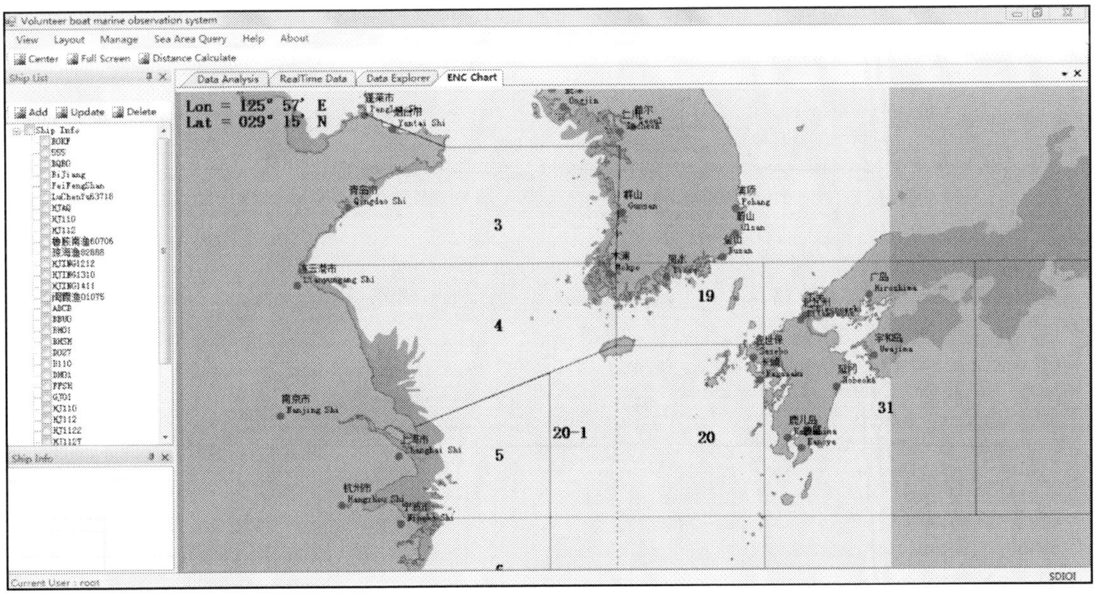

Fig.11. volunteer boat marine observation system

Design for The Indoor Visible Light Communication Application System Based on LED Visible Light

Wenyu Lian[1,2] ,Jia Liu[1,2] ,Shigang Cui[1,2] ,Lin He[1,2] ,Yunxi Zhang[1,2, a]

[1]Tianjin Key Laboratory of Information Sensing and Intelligent Control, Tianjin University of Technology and Education, Tianjin 300222, P.R.China
[2]School of Automation and Electrical Engineering, Tianjin University of Technology and Education, Tianjin 300222, P.R.China.

Abstract. This paper designs an indoor visible light communication application system based on LED. The system can modulate the original signal one or more times, move to a specific frequency band, transmit on the power line, in the LED terminal use this module to decode, restore the Ethernet signals. This design is applicable to the simplicity of the LED visible light communication applications, which provide the premise and guarantee for the construction of smart home network.
Keywords. LED; visible light communication; Ethernet signal; smart home network

1 Introduction

Visible light communications technology is a new type of wireless communication developing rapidly in the past ten years, through the increase in public infrastructure lighting auxiliary function of data transmission, combining communication with indoor lighting Light source, can build indoor Visible Light wireless communication network, information from the server to the client's wireless transmission [1]. Due to visible light is closely related to daily life, office, family, and many of the public equipment is visible light, such as office lights, road lights, the square of the display screen, etc, so the indoor visible light communication technology can meet the users of mobile communication network of wide coverage, high link quality requirements, for indoor users with the convenience of data services anytime and anywhere. At the same time, the indoor visible light communication technology using visible light transmission medium which can ensure the security of mobile communication network, including network data security and user safety and health, the user information privacy protection, avoid malicious intercepted by other users, even for a long time in the mobile communication network, will not affect their health.

At present, the main mode of transmission for wireless communication system is radio frequency communication. Compared with the radio frequency communication technology, indoor visible light communication technology has many advantages:

First of all, indoor visible light communication technology can provide a large number of potential available bandwidth, and the bandwidth is not limited, so it does not need to obtain the authorization of the management agency [2]. On the contrary, the use of radio frequency band, especially the widespread use of mainstream products is lower than that of the 30GHz band, the electromagnetic wave spectrum is a limited resource, the frequency of strict control by the authorities, leading to the use and maintenance is expensive.

Secondly, indoor visible light communication does not exist electromagnetic radiation [3], is not susceptible to external electromagnetic interference, which can be widely used in sensitive to electromagnetic interference, even to eliminate electromagnetic interference on special occasions [4], such as hospitals, aircraft, gas station and chemical factory, etc. On the contrary, the penetration of radio frequency signals is likely to cause interference between electronic devices, so the areas sensitive to electromagnetic waves are limited to mobile devices.

Thirdly, the indoor visible light communication technology has built a higher network security [5]. the technology using the transmission media is visible light cannot penetrate walls and other obstructions in limiting transmission of users within visual range, which means that the transmission of network information is confined to one room (assuming no windows or transparent transmission medium, even if some words can be used to remove the physical block), such as to effectively avoid the transmission of information by malicious interception, ensure the security of the information.

Finally, the indoor visible light communication technology supports the rapid construction of wireless networks, which can facilitate the establishment of

* Corresponding author: zyunx2004@mail.nankai.edu.cn

flexible temporary network and communication links, reduce the use and maintenance costs of network [6]. Radio frequency signals such as subways and tunnels cover blind areas, and if radio frequency communication is used, high cost is required to establish base stations and pay expensive maintenance costs. Indoor visible light communication technology can use its indoor lighting source as base station, combined with power line communication, to provide users with convenient indoor wireless communication services.

Visible light communication technology is a new type of communication developed with lighting support high speed switch [7], can effectively alleviate the current problem of the shortage of radio frequency communication band, provides a new method for short distance wireless communication [8,9]. Indoor visible light communication can meet the requirements of users for security, stability, speediness and environmental protection of communication links, and provides a new broadband access mode for indoor wireless communication networks. The application of visible light communication technology based lighting in the next generation wireless communication technology is feasible, is an alternative construction of indoor wireless communication system, is an effective complement to the existing wireless radio frequency communication technology.

Based on this, this paper researches the key technology of the application of visible light communication system under indoor channel environments, in order to apply simple LED visible light communication, in indoor applications without re laying of cables, the design of a indoor visible light communication system based on the LED will be one or more original signal modulation, moving to the specific frequency band for transmission on power line. At the end of the LED lamp, this module is used to decode and restore the Ethernet signal.

2 Overall Design Scheme

The utility model is based on the USB light cat network card module and the light cat module, and establishes a smart home hardware and software integration system of a wireless local area network based on visible light communication, so as to build an intelligent home network. The main research and development of portable visible light communication interface module, can be directly embedded in all kinds of electrical appliances and network access equipment, realize low cost and high rate of local area network (LAN) networking technologies, household appliances can be used to realize intelligent.

Indoor visible light communication based on LED application systems, including USB light cat network card module (1), its characteristic is that the USB light cat network card module (1) connected to the light cat module (2) , and the light cat module (2) connected to the photosensitive detector (3), and the photosensitive detector (3) connected to the LED drive modulation circuit (4), and the LED drive modulation circuit (4) connected to the LED lamp (5), and connected to the terminal (6). As shown in Fig. 1.

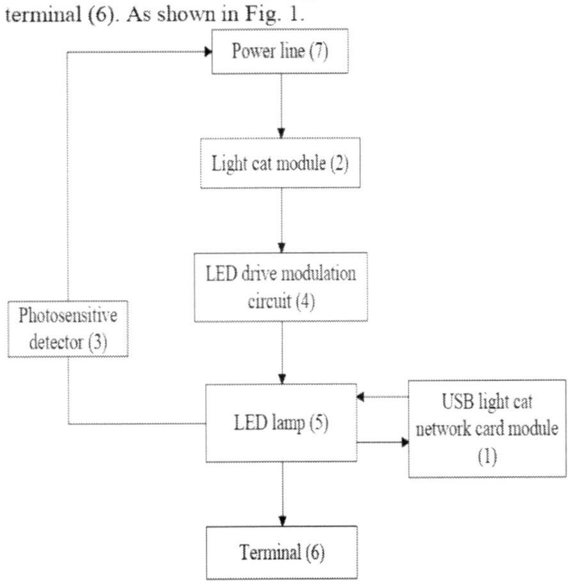

Fig. 1. A block diagram of an indoor visible light communication application system based on LED

(1)USB light cat network card module; (2) light cat module; (3) photosensitive detector; (4) LED drive modulation circuit; (5) LED lamp; (6) terminal.

3 Key Technology Research

3.1 Design of USB Light Cat Network Card Module

This design carries on the LED modulation signal to receive, and send it to the PC machine, and carries on the PC signal similarly by the LED modulation way to the LED lamp. Realize full duplex data transmission of LED visible light communication. The design scheme adopts the USB network card chip AX88772B produced by Taiwan AsiaInfo Engineering Department, and the light cat module is still used in ML6652 chip.

AX88772B is a highly integrated high-performance and low-power USB 2.0 10/100M fast Ethernet control chip, can increase the low cost, small package, plug and play fast Ethernet networking characteristics for all kinds of applications, can be used for desktop computer, notebook computer, ultra portable computer, the bracket / port replicator / dock, game machine, intelligent home furnishing and any of the USB interface of embedded devices. This chip is made up of USB to Ethernet bridge, voltage regulator, crystal oscillator and PLL, SRAM, memory arbitrator, Ethernet MAC, Ethernet PHY, general input and output circuit and interface. AX88772B is used to connect the USB host side and conform to the USB 1.1/2.0 standard USB port, support the 10/100M Ethernet functionality based on the IEEE802.3 and IEEE802.3u specification and built-in network packet buffer SRAM, and integrated of

DOI: 10.1051/matecconf/201713900124

10/100M Ethernet PHY, thus simplifies the system design.

AX88772B can cooperate contains a USB Host microcontrollers, support twisted pair cable or through a fiber optic PECL interface to connect the external 100BASE-FX optical transceiver module, the ability to provide embedded system Ethernet connection, network data transceiver and remote control. AX88772B has the following characteristics:

(1) USB 2 turns RMII, supports HomePNA and Homeplug PHY.

(2) On the USB interface, the AX88772B integrates the USB 2.0 transceiver and serial interface engine, which conforms to the USB 1.1/2.0 specification; Support USB full speed and high speed mode, power drive capability, support bus power supply mode and self powered mode; Support 4~6 programmable USB bus endpoints; Support isolated power saving mode, automatically remove the USB host when it detects the removal of the network line; With unique burst transmission mechanism, it can make the cover transmission on USB bus more efficient.

(3) AX88772B has advanced power management function which can effectively reduce power consumption.

(4) Support dynamic power management to save power consumption under no load or light load; Support for AutoDetach power saving mode, which will automatically remove the USB host when the network line is detected.

(5) Universal extended media interface. AX88772B optional RMII interface to support the MAC working mode, for connecting AX88772B external 100Bease-FM or Homeplug PHY; Optional Reverse-RMII interface to support the PHY working mode, support MAC-to-MAC seamless connection.

(6) Extremely low power sleep mode that supports waiting for network events to wake up.

The LED dimming circuit adopts the MAX3967 chip, MAX3967 is a programmable LED driver that can be used for the highest 270 Mb/s data rate optical transmitter. The circuit includes a high speed current driver with programmable temperature coefficients, a LED pre bias voltage regulator, and a disable function circuit. The circuit accepts the PECL data input and operates at the +2.97V to +5.5V supply voltage. The SFP LED driver can cut the current of up to 100mA into a typical high speed light-emitting diode. With the increase of temperature, the device of the modulation current can be according to the temperature coefficient increases, the temperature coefficient of programmable range: 2500ppm/°C to 12000ppm/°C. The modulation current is set by an external resistor. The pre bias voltage of MAX3967 can be programmed in 400mV to 925mV. The pre bias circuit generates a peaking current to improve the switching speed of the LED. The compensation current output helps maintain a constant supply current and reduces the EMI and power noise generated by the transmit module. The MAX3967 is provided in the form of a bare chip or provided in a 4mm*4mm 24 pin and a thin QFN package.

3.2 Design of Light Cat Module

In this paper, the function of the light cat module is to convert the Ethernet signal (from the power carrier module) into the level signal modulated by LED. This design adopts ML6652 chip. This chip is a low cost / low LED current drive single chip transceiver, which provides double stranded wire and fiber optical Ethernet technology between 10 Mb/s and 100 Mb/s. The device supports 10 Mb/s and 100 Mb/s operating data rates with Auto-Negotiation using 850nm or 1300nm optics. One or two interfaces of fiber optic and double stranded wires can be interfaced to industry standard miniature fiber optic components or Physical Media Dependent (PMD) modules using Positive Emitter Coupled Logic / Low Voltage Positive Emitter Coupled Logic (PECL / LVPECL) compatible modes. ML6652 has the following characteristics:

(1) Fully implement the media interface between optical fiber and twisted pair.

(2) 3.3V low voltage power supply.

(3) Support ISO/IEC802.3, IEEE802.3, and TIA/EIA785 industrial standards.

(4) Six optional working modes: mandatory 10Mb/s, mandatory 100Mb/s, automatic negotiation of non transparent half duplex, automatic negotiation of non transparent full / half duplex, automatic negotiation of transparent 10 / 100Mb/s, non transparent special case.

(5) The control mode is flexible, which can be controlled by hardware and software. The hardware configuration is achieved by setting the level of the pins in the chip; The software control mode is realized by the corresponding bits in the operation control register.

(6) Small 44 pin Tiny Quad Flat Package (TQFP) and Leadless Plastic Chip Carrier (LPCC) / Quad Flat No-lead (QFN) package.

10M/100M light cat module, as shown in Fig 2.

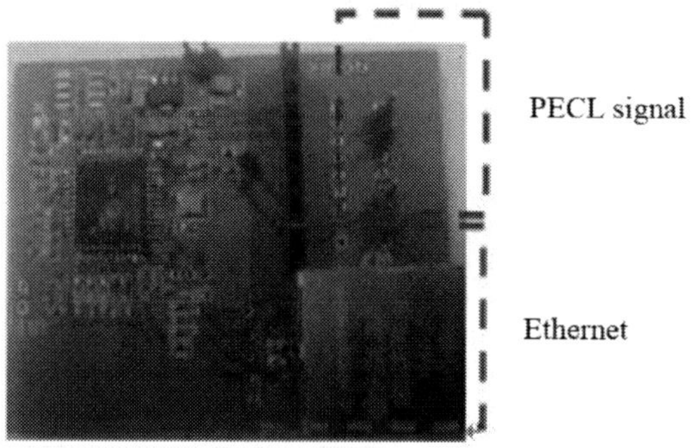

Fig. 2. 10M/100M light cat module.

3.3 Design of LED Driver Modulation Circuit

The input of LED driver chip is PECL level, and the output circuit of the photosensitive element is also the PECL level. Through the light cat module, the PECL level can be converted to the Ethernet level. ML6652 chip is used to convert PECL level signal into Ethernet network signal.

The USB card AX88772B and 10M / 100M light cat module ML6652, optical detection and optical emission module integrated in a 5.6 cm * 2.2 cm of small PCB board, through the USB interface is connected to a computer, to realize the miniaturization of optical communication module, the practical application.

3.4 Other Instructions

In the design of this system, the photosensitive detector can receive PECL level signal and convert it to an Ethernet signal. This design uses LED light as the signal source, as long as the light can be obtained where the communication can be carried out, there is no communication blind spots, convenient and quick. Terminal for PC, printer, smart phone and other communication equipment.

Designed based on LED indoor visible light communication application system, can be directly embedded into all kinds of electric equipment and network access equipment, realize low cost and high rate of local area network networking technologies, commonly used household appliances can realize intelligence.

4 Conclusion

On the basis of key technology of USB light cat network card module and light cat module, a smart home hardware and software integration system based on visible light communication wireless local area network is established, and the smart home network is constructed. The main research and development of portable visible light communication interface module, can be directly embedded into all kinds of electrical appliances and network access equipment, realize low cost and high rate of local area network networking technologies, commonly used household appliances can realize intelligence.

This work was supported by the National High Technology Research and Development Program (863 Program) of China under Grant No. 2015AA033303, the National Natural Science Foundation of China under Grant Nos. 61503164, 61573197, 61403227, the National Natural Science Foundation of Tianjin under Grant No.15JCQNJC04200, the Natural Science Foundation of Jiangsu Province under Grant No. BK20140241, Tianjin Higher School Science and Technology Development Fund Project under Grant No. JWK1614 and the Talent Fund Project of Tianjin University of Technology and Education under Grant Nos. KYQD14032, KJ15-21..

References

1. T. Komine, M. Nakagawa, IEEE Transactions on Consumer Electronic, **50**, (2004), 100-107.
2. T. Chen, L. Liu, Zhongxing communication technology, **19**, (2013), 49-52,.
3. J.Y. Wang, Z.Y. Xu, opto electronics, lasers, **23**, (2012), 1314-1318.
4. C.Z. Wu, modern transmission, **03**, (2012), 8-19.
5. N. Xu, D.T. XU, Chinese Journal of quantum electronics, **29**, (2012), 629-636.
6. N. Kumar, N. Lourenco, Journal of Engineering and Applied Sciences, **5**, (2010), 296-307.
7. H.T. Luo, C.Y. Chen, optical communication technology, **02**, (2011), 32-36.
8. H. Zhang, Z.X. Song, Changchun Polytechnic University Journal of science and technology, **33**, (2010), 70-73.
9. J.B. Wang, X.X. Xie, optics and precision engineering, **20**, (2012), 24-30.

The Research and Application of Electric Massage Chair with Residual Current Protection Function

Rui Zhang[1,*], Jungang Zhou[2], and Zhaoxia Shang[2]

[1]Shandong Institute of medicine and health information, Jinan Shangdong, China
[2]Shandong Product Quality Inspection Research Institute, Jinan Shangdong, China

Abstract. The electric massage chair is a common instrument in family health care.In addition to health function, to provide safety for the users is also very important. Comparing with the current safety protection technology of electric massage chair, this paper designs a residual current detection function to increase the safety performance,which provides a new way for this field.When the residual current is generated,the power is disconnected in time.It not only can protect the user's personal safety,but also can reduce the occurrence rate of fire.

1 Introduction

At present, the composition of a massage chair generally includes a mechanical part and an electrical part which are made of a metallic material with electric conductivity. The two parts are connected by a wire when the power source is used.If the electrical insulation is damaged,the two parts will become exposed conductive body,which leads potential danger.If the electrical insulation of the massage chair is damaged and the power cannot be disconnected in time,the user's life will be dangerous with touching the exposed conductor.If the bare conductor contact with the earth directly,it will produce a larger current and at the same time release much heat, which will likely ignite combustible materials and then cause a fire.So the electricity safety protection is very important to the massage chair. From the perspective of electricity safety analysis,the lack of residual current protection will increase the danger of electricity safety,so it has very practical significance to design a residual current protection (RCP).

According to GB4706.1 and GB 4706.10[1],the scope of massage apparatus is "single-phase appliances for rated voltage not exceeding 250V, other appliances for rated voltage not exceeding 480V".The normal power source of massage apparatus with larger work power is generally from the electric motor, which is used at home and similar places adopting the wall power directly.The general work power is 220V alternating current.For example,in the use condition of 220V,if the user touches the power accidentally,then residual current is produced,it will cause a great risk of injury for user.If the residual current caused by damaged insulation,the heat will be produced continuously,and it is possible to cause fire,so it becomes a hidden danger of fire.

At present, the electric massage chair use 3 to 4 sets of 15-30W permanent magnet DC motor to drive the movement of kneading,knock shaft and also drive the movement up and down.Electric massage is used the RC-260,RC-280 and other small motors to generate vibration.The maximum power of the motor output is limited,if the motor power is designed too small, the load may exceed the nominal output power,the motor may be overloaded.The overload will make the motor heating,vibration,speed decreases,sound abnormalities and other phenomena.When the overload is very serious, the motor will burn.According to 6.1 in GB4706.1,the motor protection of the standing massage chair is not only dependent on basic insulation but also includes an accessory safety precaution which links the accessible conductive parts to the grounded protective conductor in the fixed wiring.Some non-portable appliances with a weight greater than or equal to 18 kg cannot be grounded,it rely on double insulation or fortified insulation,and at the same time safety special low voltage power supply to achieve anti-shock protection.

So the existing safety precautions of electric massage chair may still produce residual current and cause the risk of electric shock,resulting in excess heat,leading to security incidents.In order to compensate for the shortcomings of the existing technology,this paper puts forward and realizes a massage chair with residual current protection(RCP)function improving the safety of electricity.

2 The RCP mechanism

2.1 The Design of RCP

This research present the residual current protection technical scheme is summarized as follows:the motor portion connected to a mechanical linkage portion and an air compression device portion connecting an airbag portion through a PVC pipe,and the electronic integrated control remote controller being connected to a motor portion,an air compression unit portion are all connected in parallel with the outgoing end of the residual current detection module through the wire.And the input terminal of the residual current detection module is connected in series with the working power source(see figure.2).

The motor part is composed of motors which belong to each functional unit having an automatic forward

* Corresponding author:Rui Zhang, E-mail: zr2352_cn@sina.com.This work was supported by the Innovation Project of Shandong Academy of Medical Sciences.

rotation,a reverse rotation and a limit function.The forward and reverse rotation of the motor drives the mechanical linkage device realizing the lifting and descent of the functional unit in the spatial position,the forward and backward,the front and the rear, and so on.

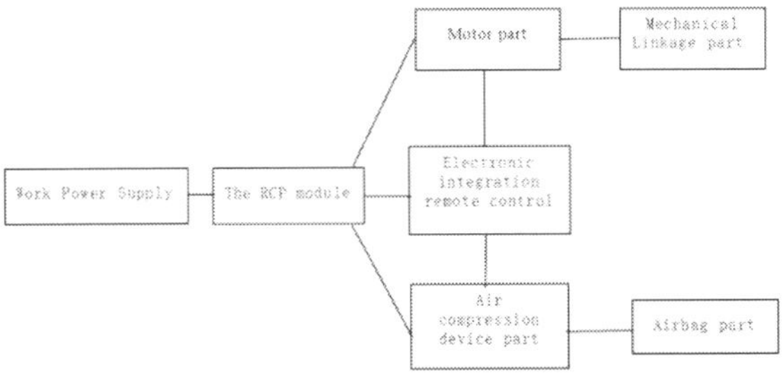

Fig. 1. The frame of RCP

The mechanical linkage part is composed of connecting rod, bearing,push rod and other mechanical parts achieving varieties of mechanical functions through the mutual cooperation of the above device[4].The air compression device is composed of an air compressor and solenoid valves corresponding to each functional unit,which is connected with each other through a PVC pipe.The control of each solenoid valve enables the each functional unit inflation and aeration.The compressed gas is passed through the PVC pipe to the various functional units, as the driving force of massage.The airbag portion is composed of an airbag of each functional unit.When the airbag is filled with the gas,the body parts of the user are subjected to fasten flexibly.The electronic integrated control remote controller integrates the functions by writing a corresponding program,and the motor part and the air compression device can be operated according to the predetermined program realizing the corresponding function.The working power supply described in the paper is 220V single-phase AC power supply with fire line L and zero line N.

2.2The connection of RCP module

The residual current detection module described in the paper is made up of a switch installed on the connection circuit of the incoming and outgoing terminals(see figure.2).

The switch is connected in parallel with the automatic identification unit,the voice alarm and the reset instruction button.The automatic identification unit is connected in series with the voice alarm.The automatic identification unit is used for indicating the presence of residual current.The switch is used to close or connect the connection between the incoming and outgoing terminals of the electric circuit.The voice alarm beeps to indicate the presence of residual current.The resetting indicator button upsprings when the residual current is detected.The reset indicator button and switch is interlocked with each other.When the reset button recovery original position,the switch can be closed and disconnected properly,otherwise the switch cannot be closed.

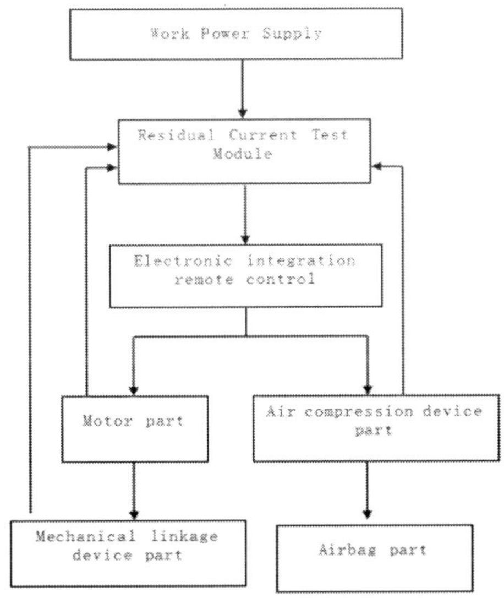

Fig. 2. The Schematic diagram of circuit connection.

When the residual current is generated,the automatic identification unit will identify it automatically.The switch is disconnected within 0.1s automatically,the voice alarm beeps,the reset indicator button bounce.The switch mentioned above is respectively connected with the working power source, the motor part,the electronic integrated remote control and the air compression device part.

2.3 The work flow of the RCP module

Work power supply provides power to the residual current detection module,the motor part,the air compression device and electronic integrated remote control.

The electronic integrated remote control makes the motor part,the air compression device part,the mechanical linkage part and the airbag part cooperate complete the corresponding action.The residual current detection module starts at the time when the working power is turned on,and detects whether there is residual current at real-time.When the work power supply is connected,the automatic identification unit monitors whether there is residual current at real-time.Assuming the current which come from the power line L and go into the current detection module is recorded as I_1.The current which is come from the zero line N and go into the detection module current is recorded as I_2.The residual current is recorded as I_3.when the residual current I_3 is generated,$I=I_1-I_2$,$I_1>I_2$.The automatic identification unit records the I_3 value.When $I_3 \geq 10mA$,the voice alarm beep,the reset instructions button bounce,the switch is disconnected within 0.1s automatically,the massage chair stops working;When $I_3<10mA$,the switch, the voice alarm,the reset instructions button is in normal working condition,the massage chair works properly.(see figure.3)

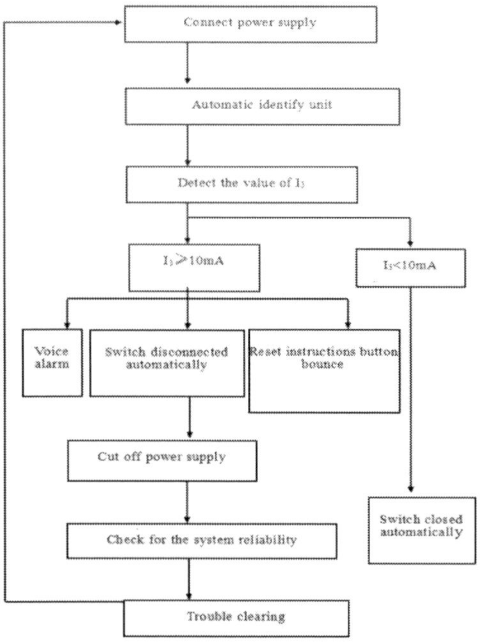

Figure.3. The working principle diagram of residual current detection

2.4 The actual case

Case1:As shown in Figure 2,work process of the residual current test module:Connect power supply→Automatic identify unit→Detect the value of I_3→$I_3 \geq 10mA$→Voice alarm, Switch disconnected automatically,Reset instructions button bounce → Check for the system reliability → Trouble clearing → Connect power supply.Connect power supply→Automatic identify unit →Detect the value of I_3→$I_3<10mA$→Switch closed automatically.More specifically,as show in Figure.4. Residual current test module is made up of four parts such as automatic identify unit 8, switch 9,voice alarm 10 and reset instructions button 11.The four parts cooperate with each other to complete the function of residual current test the residual current module.

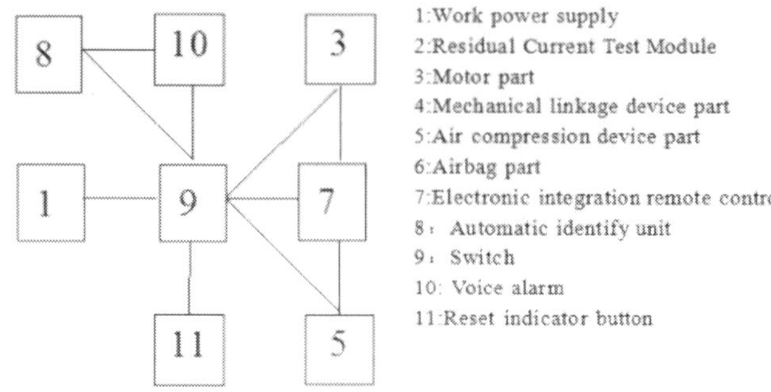

Figure.4. The relation between each part of case1

First turn on the power supply 1,then residual current test module began to work, automatic identify unit 8 detect the value of residual current I_3.In fact, there are two cases:the first case is when detect $I_3 \geq 10mA$, voice alarm 10 just beeps to indicate residual current fault occur,switch 9 automatically disconnected within 0.1s to ensure that the existence of the residual current won't cause dangerous accident in a short time.Reset instructions button 11 bounce indicates the presence of residual current.Maintenance personnel cut off power supply and check for the above mentioned faults,after ruling out possible faults connect power supply 1,the automatic identify unit 8 in residual current test module 2 began to detect residual current and restart the workflow.The second case is when $I_3 < 10mA$,switch 9 close automatically,the massage armchair work properly.Switch 9 and reset instructions button 11 is interlocked.When reset instructions button 11 bounce,press to restore manually,so the switch 9 can close automatically when $I_3 < 10mA$ and vice versa.

Case2:As shown in Figure 5, the work process of massage armchair module: Work Power Supply → Residual Current Test Module → Electronic integration remote control→Motor part→Mechanical linkage device part; Electronic integration remote control → Air compression device part → Airbag part; Motor part, Mechanical linkage device part, Air compression device part→Residual Current Test Module.

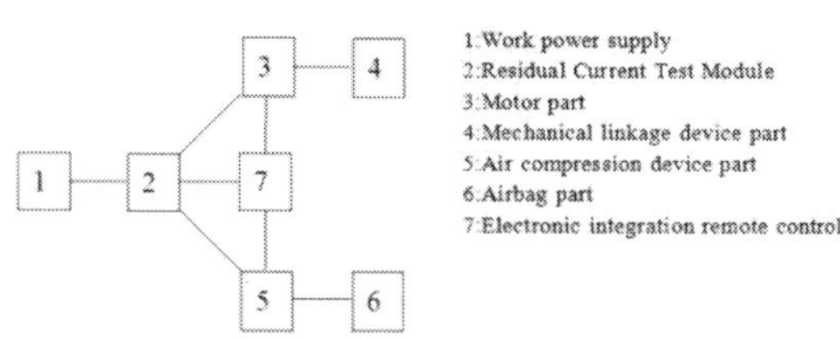

Figure.5.The relation between each part of case2

More specifically,as show in Figure 5,the connection from power supply 1 to single-phase 220V alternating current power supply makes the massage Chair can work properly;Switch inside residual current test module 2 closed automatically;Electronic integration remote control 7 is used to select the corresponding function. Motor part 3 and air compression device part 5 began to work according to the functional requirements.Motor part 3 drives mechanical linkage device part 4;Air compression devices part 5 drives airbag part 6;In the normal work,the failures of electrical insulation damage might happen on motor part 3,mechanical linkage device part 4, and air compression devices part 5, this can lead to residual current;Residual Current Test Module 2 obtains the feedback of residual current value, so residual current test module 2 can play the corresponding protection function according to the residual current value.Throughout the course of the work, it achieves closed cycle automatic control for residual current test of massage chair, and further achieves protection function for residual current.

3 Conclusion

Compared with the prior technology[5][6][7],our research has the residual current protection function, which increases the electrical safety performance of the massage chair.When the residual current is generated,the power supply can be disconnected in order to protect the user's personal safety and can also prevent the occurrence of fire.When the residual current is generated,the voice alarm beeps prompting the existence of residual current,it's easy to find the problem and repair.When the residual current accident is not resolved,the reset indicator button bounces.And the reset indicator button interlock with the switch of residual current detection module making the massage chair cannot work.The residual current detection module will eliminate the existing safety risks and make the massage chairs work with strong security.

References

1. GB4706.1.Safety of household and similar electrical appliances[S].General Administration of Quality Supervision.Inspection and Quarantine of the People's Republic of China, China National Standardization Administration Committee,2008.
2. Xudong Zhang.Massage equipment national standards Q & A[J].Appliances & Technology,2010 (5):52-53.
3. Yu Song.Dynamic model and simulation of four-link electric chair motor[J].Journal of Mechanical Engineering,2015(10):47-50.
4. Xudong Zhang.Application of Small Motor in Electric Massage Equipment[J].Micro and special motor,2007(12):55-58.
5. Fanfang Kong,Junji Wu,Xueyong Ye,Wei Yang,Junfang Zhang.Simulation study of adaptive current protection scheme[J].Power system protection and control,2008 , 36 (4) :10-14.
6. Chenglian Luo,New principle and technology of relay protection and automation[M].Huazhong University of Science and Technology press, 2005.
7. Yaozhong Ge,The principle and technology of new type of relay[M].Xi'an Jiao Tong University press,1996.

 DOI: 10.1051/matecconf/201713900126

The Present Situation and Trend of Wearable Equipment in Electric Field Application

Shao Guangwei, *Ji Zhongjun*, and *Liu Bing*
Economic and Technological Research Institute, Liaoning Electric Power Co., Ltd. State Grid (Liaoning Electric Power Construction Supervision Co., Ltd), Shenyang, China

Abstract. With the innovation of mobile internet technology and intelligent terminal technology, the wearable equipment has been gradually applied in various industries. Through the typical application of wearable equipment, the characteristics of wearable equipment technology and application are summarized. Based on the analysis of power field business application and technology, combined with the demand of electric field management, the design situation of the wearable equipment in electric field service is designed, and the future development direction of electric field wear can be forecasted.

1 Introduction

The wearable equipment refers to the use of wear technology for daily wear intelligent configuration of the equipment. All kinds of sensing, identification, connectivity and cloud services and other technologies are integrated into people's glasses, rings, watches, bracelets, clothing and footwear and other daily wear equipment in order to achieve the user five sense of ability to expand, life steward, social entertainment, health monitoring and other functions. The general appearance of the equipment is more beautiful and easy to wear, with a certain degree of computing power and have a dedicated application and function and so on.

As a portable and wearable computing device, the wearable equipment has miniaturization, portable, small size and mobility and other characteristics. Therefore, being different from the general computing equipment or smart devices in the human-computer interaction, the wearable equipment is a kind of human machine directly seamless, fully connected to the interactive way; its main features include single (double) hand release, voice interaction, perception enhancement, tactile interaction, consciousness interaction and so on.

1.1 Perception

The wearable equipment can promote people's perception of the world, others, and us. We can perceive the world through wearable equipment. For example, smart glasses can help us to identify the animals and plants, smart chopsticks can help us to perceive the food contains nutrients, even through the wearable equipment to read the meaning of other people's micro-expression, understand the language of their pets, the wearable equipment provides human beings with the infinite possibilities of perceiving the world[1].

1.2 Management

The wearable equipment can help people achieve the control of other electronic equipment, achieve the location of the records and allow us to visually see a good line planned by us through the sensors. The progress of the project and the control of the own act can also be achieved. The wearable equipment can not only achieve the control of other equipment, but also achieve the control functions move to the person's body from the original mobile phone or computer. So that the function is more intuitive and life is simpler.

1.3 Expression

Because the wearable equipment is portable, so the expression is more convenient, more image and more accurate in the process of communication achieved by projection, display, three-dimensional imaging and file transfer, etc.

2 Wearable equipment applications

The development and application of wearable technology puts more demands on the portability and mobility of the device. As an important basis for field management and means, the wearable equipment effectively complete the real-time recording and online control combined with the special nature of industry and professional, fully playing portability and real-time characteristics[2].

The wearable equipment directly wearing by the human body determines its three main features:

• Human relevant data can be obtained by it, powerful and small size.

* Corresponding author: sy_liangqian@sina.com

• Solve the most direct needs, simple and easy to use, strong affinity.
• Beautiful, small, comfortable and suitable.

Wearable equipment is mainly used in personal health, medical services, environmental protection, fitness, recreation and other industries. The use of its own high portability, high mobility and high timeliness, the excellent application result has been accessed[3].

2.1 Medical field

In the medical field, the wearable equipment will be in vitro data collection and cross-analysis through data collection to help users to manage important physiological activities, to make the monitoring of various health indicators. Wearing medical care allows patients to manage their own diseases.

The current application of wearing equipment in the medical field is mainly reflected in two aspects: one is in vitro data collection, mainly through the G-sensor with three-dimensional motion sensor or GPS to obtain movement distance and exercise, to help users to exercise and manage sleep, while introduce the concept of social platform for users to measure and contrast the data between different users to promote the user's continue behavior and to change the health of users. The other is to help users manage important physiological activities by means of sign data such as heart rate, pulse rate, respiratory rate, body temperature, heat consumption, blood pressure, blood glucose and blood oxygen, hormones and BMI index, body fat content. The results of these numerical cross analyses can be used to analyze the user's current physical condition and health risk assessment, combined with data the several key physiological activities can be given: sleep, diet, exercise and medication personalized improvement recommendations, allowing users to stay in a stable physical conditions [4,5].

2.2 Sports field

In the field of sports, the wearable equipment will be used in the field of physical health management and professional sports applications, including the application of ski, swimming, mountain climbing, running and other special sports areas.

At present, the wearable equipment based on sports and fitness is mainly used in two aspects: on the one hand applied to the health aspects of such products are mainly concerned with the movement of fitness crowd, they can record their daily real-time exercise, consumption of calories, food Into the amount of sleep and thus effectively urge themselves to increase the amount of exercise to maintain good health; On the other hand, applied to professional sports applications, such as skiing, swimming, mountain climbing, running and other special sports areas, for example, the wearable equipment is applied in the ski field to confirm their position in the snow, location, height, taxi speed and is applied in swimming to confirm their data such as heart rate, calorie consumption and maximum heart bear ability when athlete does not stop endurance exercise.

2.3 Entertainment field

In the field of entertainment, the wearable equipment will be in social, games, photography, reading, Internet, APP and other aspects; it will have a wider range of applications.

Entertainment applications of the wearable equipment become part of today's social life and the inseparable part of the mobile terminal has become the main tool for users of fragmented life entertainment, similar to smart phones, PAD, wearable equipment in order to obtain public recognition. Entertainment and social applications become another key application for its future development. In the future, the wearable equipment in social, games, photography, reading, Internet, APP and other aspects will have a wider range of applications. And now the wearable equipment entertainment social field applications has become a positive development focus of the manufacturers, Google Glass can achieve unique glasses perspective of the photo and video, Samsung Galaxy Gear take pictures of photography, and support for Internet access and social information access. Japan Telepathy has introduced a wearable device that can create real-time access to social networking and cloud services, equipped with cameras, projection units and wireless communication modules, etc. The wearers can use the device to broadcast their own views and so on. Replay launched the first social function of the smart jeans products, the biggest feature is to support the Bluetooth function, jeans and smart phones and other equipment can be connected.

At present, the types and how to wear of equipment mainly include:

• Headwear: glasses, headband.

• Hand wear class: bracelet, watch, gloves.

• Clothing categories: belts, bandages, clothing.

3 Power scene business situations

Power site is mainly reflected in the construction, construction supervision, operation and maintenance, customer service site. In the field work the field equipment, device information must be collected and the work of the scene need be recorded, at the same time management personnel can complete the work of the scene and quality supervision and control.

At present, on the professional power scene the typical application of the wearable equipment has not been seen. Electricity major professional complete the power construction, production maintenance, marketing services, communications and other professional business data entry and collection are mainly by mobile terminals, handheld equipment and other equipment. But the power of the majority of professional business information and data processing, still rely on the PC side to complete.

Power field business mobile devices mainly include:

- Handheld Terminal: Based on commercial PDA terminals, their applications are in the power industry, mainly in power distribution and transmission line inspection applications. Its advantages are wireless communication and interactive operation of distribution line, courts information marking and collection, as well as transmission line inspection and recording.

- Measuring instrument: In the power construction, the need to measure the site construction, construction, access to site, equipment data, Electric work personnel, the use of special measuring instruments to complete the measurement operation.
- Infrared thermal imager: In the power production site, operation and maintenance personnel regularly check the state of power equipment operation, through the equipment parameters to determine the equipment status and operating situation. The surface of the node thermal images and ensuring accurate measurement of temperature data is achieved by the use of infrared thermometer, shooting equipment.
- Meter reading device: Mainly used for power marketing on-site meter reading use. Solve the problem of manual recording of electric energy meter reading table. As the functional requirements are relatively simple, you can record the number of electricity can be. Meter reader terminal performance is relatively low, storage space is small, and the display is black and white LED.
- GPS locator: With GPS and GPRS wireless communication capabilities, mainly used in transmission, distribution professional, in the line inspection, with the power operators to complete the positioning and data upload operation.
- Digital Camera: In the construction of electric power construction site, supervision site, maintenance site, most of the need to collect on-site work of the image information, as well as equipment, personnel related images. So the power of personnel digital cameras, power scene image records.

Through the power of the professional and on-site mobile terminal equipment classification and application can be seen, the ultimate goal is to complete the power field control and business information records, thereby enhance the power business site control capabilities. The power field applications and business needs are shown in Table 1.

Table 1. Power field applications and business needs

Profession	Main job content	Business needs
Infrastructure	Foundation construction	Site control, personnel control
Supervision	Engineering supervision, quality management	Site control, personnel control
Substation	Substation construction, operation and maintenance	Site control, personnel control, equipment status records
Transmission	Line construction, inspection line, operation and maintenance	Site control, personnel control, equipment status records
Power distribution	Distribution lines, Substation collection	Site control, personnel control, equipment status records
Marketing	Customer service, inspection, electricity management	Site control, personnel control
Communication	Communication construction, operation and maintenance	Site control, personnel, equipment status records

Power construction site and the number of operating sites, scattered widely, carrying work inconvenient, and thus suitable for wearable equipment applications. With the strengthening of the power field business management, as well as the improvement of data accuracy requirements, wearable equipment is small, easy to portable, and gradually requires the ability to enter more comprehensive information and content. Support a richer business, assist in the completion of on-site business control, and enhance the real-time control of power business. The wearable equipment is bound to be used in the electricity field, to replace the original handheld, portable and other equipment used to become the necessary work on the power site.

4 The application trends

The wearable equipment in the power industry applications are still in the initial stage, the main application of handheld terminal equipment, in the field for recording use. In the infrastructure, substation, transmission, supervision, distribution, scheduling, marketing and other management professionals, are lack of business in the field of terminal equipment. But for the power industry is an urgent need to address the field business processing and control, wearing tools and devices has not yet applied. With the power of professional management and business applications continue to deepen the demand for wearable equipment will become increasingly urgent, it can be seen the wearing equipment in the power of professional applications with considerable prospects.
The wearable equipment applications are mainly concentrated in the following areas.

• On-site Supervision

Electric field construction, construction operations, operation and maintenance, customer service, complete the field process records and monitoring, improve the quality of work, improve quality service. The wearable equipment on-site automatic recording, does not affect the normal business processing.

• Personnel Control

Personnel control, mainly infrastructure, substation, transmission site, the need for real-time monitoring of operational safety, real-time perception of the physical condition of workers, to avoid access to dangerous areas. So through the wearable equipment, the physical condition of workers is monitored real-time and the personal safety management of electricity workers are effectively strengthen to avoid casualties and supervise the scene at the same time.

• Equipment Control

In the power scene the operation of the equipment is often needed to monitor and the scene operating equipment is needed to record. The wearable equipment can operate at the same time, real-time, accurate monitor of equipment running, does not affect the normal work.

The wearable equipment performance and functional technology innovation and development will promote the power industry wearable equipment business applications and the original power professional business and management further extend. Aiming at the visualization demand of electric field, combined with the typical application scene of wearable equipment, the technological innovation and application trend of the wearable equipment in power field is summarized.

The wearable equipment types, wear patterns and application are shown in Table 2.

Table 2. Power field applications and business needs

Types	Pattern	Application
Wearing class	Glasses, headband	Site control, status records, on-site control
Hand wear class	Bracelet, watch, glove	Personnel control, safety monitoring, equipment records, on-site control
Clothing category	Belts, bandages, clothing	Personnel control, safety monitoring, site control,

4.1 Controllable wearable equipment

The embedded design, the formation of small equipment are used, easy to install and wear in the helmets, shoulders, eyes, arms and other parts. To meet the needs of power site safety and control, it is facilitated the work of electric power workers in the field to carry and use. The controllable wearable device integrates high-definition video capture, image processing, video storage, voice recording, communication function, efficient supervision and control of electric field personnel and equipment, and realizes power scene management and equipment management.

4.2 Guard proof wearable equipment

Mainly for power workers to carry out a full range of state monitoring, as well as operating personnel on-site security protection. Structural design more inclined to bracelet, watches, wristbands and other types, easy to wear at the same time, more importantly, the signs and status of the operator can accurately be recorded. So the wearing equipment must be more closely with the operator. The wearable equipment assimilated into the pulse, blood pressure, sleep, body temperature and other precision sensors monitor the physical signs of workers all the time. At the same time the electromagnetic, current, voltage sensor sensors are integrated to prevent the operation of personnel into the high pressure area and operate the full range of electrical workers effectively.

4.3 Recording wearable equipment

Mainly used for electric field information recording and collection, replace the PDA and other mobile terminals to complete the work of the scene. Recordable wearable equipment can be worn on the operator's arm, chest and other positions with a key operation function. According to the power scene the power scene, equipment and work conditions are recorded. At any place in the field electricity on-site work and the situation are collected all-weather real-time. The wearable equipment support the power operation personnel to record the scene situation, complete the power scene multi-dimensional data, images, data records, collection and storage.

The wearable equipment, full support for wireless communication and storage backup, complete the power field work control, record upload and storage. Configuration of large capacity rechargeable lithium battery, to ensure that wearing equipment can work longer than 12 hours. Provide both active and passive power supply, to meet the power of various field applications, to complete the wearable device in the passive environment of the power supply.

4 Conclusions

Power field control is an important guarantee for the stable supply of electricity. Targeted to strengthen the power site management, to meet the technical means and equipment needs, the control means are needed to strengthen. To meet the management needs of the scene and the full complete the personnel, on-site, work and equipment monitoring, and to deal with inconvenience and the not timely recording, complex operation and other issues, with its upgrading and innovation the wearable equipment is bound to promote the use of the power industry. The wearable equipment provide a more comprehensive supervision and control, strengthen the protection of workers and equipment to enhance the quality of the quality of live operations for the power field business.

References

1. Yi Geng, Hui An, Yang Li, Hua Jiang. Present Status and Future Perspectives of the Wearable Devices [J]. Electronic Science and Technology, 2014, 02: 238-245.
2. Gatzoulis L, Iakovidis I. Wearable and portable eheahh systems. Technological issues and opportunities for personalized care [J]. IEEE Eng Med Biol Mag, 2007, 26(5): 51-56.
3. Rutherford JJ. Wearable technology health-care solutions for a growing global population [J]. IEEE Eng Med Biol Mag, 2010, 29 (3): 19-22.
4. Appelboom G, Camacho E, Abraham M, et ol. Smart wearable body sensors for patient self-assessment and monitoring [J]. Arch Public Health, 2014, 72(1): 28-28.
5. Pandian PS, Mohanavelu K, Safeer KP, et al. Smart vest: wear- able multi-parameter remote physiological monitoring system [J]. Med Eng Phys, 2008, 30(4): 466-477.

Establishment of Visual Supervision Mechanism for on - site Supervision of Electric Power Engineering

Zhang Yunshan, Liu Guofu, and *Zhang Lei*
Economic and Technological Research Institute, Liaoning Electric Power Co., Ltd. State Grid (Liaoning Electric Power Construction Supervision Co., Ltd), Shenyang, China

Abstract. The quality of power engineering is an important guarantee for the power supply. The existing engineering supervision mainly relies on manual completion and lacks effective supervision mechanism. From the perspective of power engineering site supervision, the function and significance of visualization supervision mechanism are introduced, the application of visualization of power industry is enumerated, the necessary elements of visualization supervision of power engineering is summed up, and the visual supervision mechanism establishing for power engineering is discussed. The establishment of visualization regulatory mechanisms has a role in upgrading for power engineering site supervision and quality control.

1 Introduction

In order to ensure the quality and progress of the project, the supervisor should supervise and manage the construction progress, and directly determines the quality of the project. With the development of society and economy, the demand for electricity increased, the number of power supply enterprises in power engineering construction projects increased year by year. The supervision tasks corresponding increase in the power engineering. At present, the number of electric power projects in transmission and substation is relatively large, and the field is scattered in various cities. The management means are relatively simple, mainly through manual feedback on the situation, the information content mainly relies on the text, the photo records, and the provincial and the city company can not grasp the project in real time as well as the supervision situation. The effective supervision and control mechanism is lack for the project site supervision, and the supervision process problems can not be found. The visual data files are lack in power engineering supervision and management, and the query and trace technical means are more backward[1].

In the project site supervision work, the supervisors often record the situation on the spot, including supervision, rectification matters, problems and the next stage of work and so on. At present in the project site, the supervisor mainly relies on the manual record and the photograph way, records the project supervision key and the data[2]. Manual notes use text records; descript the scene supervision and problems. Photo recording methods mainly use digital cameras or mobile phones shoot construction site related images. The supervision staff can not fully show the scene supervision situation, such as the project quality, the progress of the problem and the lack of intuitive presentation, failed upload images and information in the field. The hidden dangers of electrical engineering safety and quality aspects are easily leaded. Therefore, for the on-site supervision of power engineering, the establishment of centralized visualization on-site supervision mechanism and real-time understanding of the project site situation are of great significance to strengthen the project supervision and control and improve the quality of the project[3].

2 Power visualization applications

Power industry visualization applications came out and developed with the visual technology, and on going construction and upgrading. Visualization technology applications are mainly used to solve the power of the professional security protection, equipment monitoring, real-time recording, service monitoring, remote command and other issues. Visual technology in the power industry site applications includes: substation monitoring, transmission line monitoring, and distribution network operation monitoring, emergency command, power supply business hall monitoring. Visualization application is realized by the video surveillance system; the real-time collection of video files will be timely transmission to the monitoring center, and real-time dynamic report the situation of monitoring point, find problems in times and process it[4].

The power professional visual construction and application are as follows[5].

2.1 Substation security monitoring

With unattended substation gradually increased, in order to ensure the safe operation of power grid and solve the

* Corresponding author: sy_liangqian@sina.com

problem of unattended substation security, power supply company has gradually implemented the substation remote video surveillance system. Substation video surveillance system is designed to achieve substation safety protection and operation of remote visual management. Improve the safety and reliability of substation operation and maintenance. Substation install a the camera in the indoor and outdoor, it used to complete image acquisition of the important areas, camera images collected to the station hard disk recorders, use streaming media service module the video stream to the request to connect the workstation through the LAN forwarding.

2.2 Transmission line monitoring

In the process of transmission line operation, the line conductor position, sag, ice and other running status are needed to regularly check. Transmission video image monitoring system is installed on the line, so that the transmission line is running in the visual monitoring, the scene surroundings are real-time monitored by digital video compression technology, low power technology, GPRS / CDMA / 3G / 4G wireless communication technology, solar power technology. Line managers can remotely view the line monitoring images or regularly upload image information, in order to achieve real-time monitoring of the transmission line all-weather.

2.3 Distribution network operation visualization

At present, the automation of the distribution network has realized the running data of the distribution line, the running status of the switching equipment and the related data, and transmitted to the background monitoring center through the optical fiber communication and GPRS communication. The background monitoring center can carry on the remote control operation to the remote equipment and tuning settings. Distribution automation increase "remote" on the basis of "four remote" to achieve remote real-time video surveillance distribution line switching equipment operation. "Remote" terminal video surveillance equipment using network high-definition camera, the camera itself directly to the video acquisition of analog signals into digital models, and directly transfer to the system background by the wireless router .

2.4 Emergency Command Visualization

Power dispatching, safety supervision and other professional in the field operations or troubleshooting need remote command site personnel view the fault node, and analyze the cause of the failure, quickly deal with field failure. Emergency command video system is installed in the vehicle, the camera is installed in the roof position; it can rotate 360 degrees and cover the scene panorama. The camera is connected to the in-car video server, and the video server transmits the video signal back to the dispatching command center through the wireless website. So that experts in the command center can view the job or accident scene to buy the number.

2.5 Business hall video surveillance

Business hall as a power supply office of the external service window, its quality of service on the power supply bureau has a direct impact on the image. In order to ensure the normal operation of the business hall and improve the quality of service power supply bureau must strengthen the management of the business hall. Complete real-time video on the basis of improving the quality of business services, to ensure the safe and orderly operation of the business hall. In the business hall window, counter and other locations, there are many real-time video surveillance management systems; all install a high-definition camera, real-time record of customer and customer service status, to ensure to archive the whole process of customer service video.

3 Power engineering visualization elements

Through the power industry visualization application and development power engineering visualization are mostly fixed place application, camera fixed shooting and easy to move after installation can be seen in the substation, transmission, distribution, marketing and other professionals, at the same time the scene has a stable power supply and network supply, the equipment supply and communication transmission problems are not regarded. This makes the difficulty of visual construction and cost greatly reduced. Camera equipments are installed in the indoor and outdoor areas, there are no restrictions on space and requirements, and so in the visual system to build the process does not need to consider the size of the camera equipment. There are a variety of camera types, including gun-type cameras and dome cameras and infrared cameras.

Table 1. Power visualization monitoring system content

Application Environment	Camera type	Network	Intelligent identification
Substation	Simulation, high definition	Wired LAN	Yes
Transmission line	Simulation, high definition	Wired LAN	No
Distribution lines	Simulation, high definition	Wired LAN	No
Emergency command	high definition	Wireless LAN	No
Business hall monitoring	Simulation, high definition	Wired LAN	No

Outdoor monitoring commonly used standard definition camera, resolution 720P, indoor high-definition camera, resolution up to 1080P. In the substation, transmission line visual monitoring has increased the intelligent behavior analysis function, through the video image recognition, the substation, the line operation abnormal situation, carries on the automatic recognition discrimination. The power visualization monitoring system building content is shown in Table 1.

There is a clear difference between the power engineering site and power substation, transmission, distribution, safety supervision, marketing and other business. Power engineering site are under construction in the substation project and transmission line project, and their construction period are always several years. The visualization systems and equipment are only used during construction and will no longer be used on site after completion of the project for the power engineering sites. While other professional visualization systems will be running on-site after the completion of the project.

Before the construction of the electric power is completed, the temporary power or generators is used to provide power for the project, therefore a stable power supply can not be guaranteed. Most power engineering sites do not provide power supply for visualization systems and equipment. As the site construction site and the environment often change, the installation of camera equipment can not be in the ideal site. The lack of full-angle and real-time monitoring mechanism can not guarantee the construction of power engineering supervision in place and can not be intuitive and accurate response to the project site happening.

There is no communication equipment at the power engineering site, can not provide a fixed network communication environment. In the video transmission of the camera, the use of traditional wired connection, can only be used as temporary during the project; line laying costs are high, and the construction period is long, the construction process and after completion will be removed. Construction site environment and other physical factors also increased the difficulty of erection of cables. Project site camera head repeatedly disassembling likely cause damage to the camera, affect the camera life, and the camera is larger and lack of flexibility to increase the supervision and construction workers burden.

In the power engineering construction site supervision work, the construction progress and the scene situation are needed to check by going to the various construction sites, once the supervision is not timely, supervision lack of important basis and information is caused easily. Even if the visualization system be used in the project site, only a single point of management is existed, the lack of centralized unified monitoring and management reduce the efficiency of the visualization system to increase the cost of supervision and management work.

In order to realize the establishment and application of visualization supervision of power engineering, it is necessary to solve the problems such as camera fixed, power supply stability and communication network. The key elements of visualization supervision of power engineering site supervision are as follows.

• Convenient for supervisors to carry and use on site.
• Small size, easy to move, easy to operate.
• Be used for a long time without power.
• With wireless communication and data transmission function.
• A number of project visual centralized management.

4 Power scene visualization establishment methods

Power engineering construction involves transmission, substation and other engineering sites, is affected by the environmental conditions of the project site, resulting in more difficult visualization supervision, and thus the video-related device equipment can not be installed. Therefore, the visualization and mobility of the visualization device and equipment is the key to realize the visual supervision of power engineering site supervision. In the field of electric power engineering, the comprehensive utilization of portable, mobile and fixed visualization devices, they have high-definition image acquisition, video storage, voice recording and transmission and communication functions, can real-time record the project supervision progress and record full-angle and multi-dimensional situation. Visualization device has good portability and mobility, is suitable for all kinds of power engineering site supervision applications, can complete the scene image, video recording, collection, transmission, and is easy to carry and use the supervision staff, do not increase the on-site staff burden, and deploy and install at the supervision site quickly.

The wearable visual collection device due to the use of embedded design, it can be easily installed on the helmet, shoulders, chest, easy to carry and use the officers. The wearable visual collection device integrates high-definition video capture, image processing, video storage, voice recording, communication function, complete project supervision site visualization records; supports wireless communication and storage backup mode, completes on-site supervision visualization upload. To ensure that the wearing device can work longer than 12 hours the large capacity lithium battery is used. The on-site use of mobile acquisition devices are suitable and easy to carry. Built-in high-definition camera, high-definition camera ball machine, storage devices, communication devices, fixed bracket, it can quickly be installed in the supervision of the site. According to the supervision of the scene situation, the mobile visualization device can use the fixed bracket, be placed in the field of any location, and real-time collect all-weather supervision site work content and circumstances, and assist the crew to record the scene to complete the supervision of the scene image and video recording. Both active and passive power supply for visualization equipment is to meet a variety of engineering site environment applications. The use of solar panels and batteries can complete the visualization of the device under the passive environment of the power supply. The visual acquisition device with active environment and passive environment can all-weather real-time collect the on-site supervision work of the situation.

The difficulty of on-site monitoring is increasing because power engineering site coverage is relatively wide and the project time is long. In order to ensure the supervision and control of the power engineering site supervision business, it is necessary to realize the whole process monitoring of the project site. The full coverage visualization device application is the basis and guarantee of power engineering site supervision. In the power engineering site, the power engineering supervision on-site visualization can be achieved by the use of wireless or wired communication to support portable visualization devices, mobile visualization devices or fixed visualization device access applications, covering the power engineering supervision site and supervision of the various processes. In the project supervision site, the application of wireless transmission technology and wireless communication network, the formation of transmission network connection and completing the supervision of the scene multi-channel visual video wireless collection and finishing, all-round displaying the supervision site can meet all kinds needs of engineering site environment applications. In the power engineering supervision site, the use of 4G communication network, the establishment of wireless transmission network and wireless hotspot AP, and visualization system platform can establish a network connection and upload supervision site visualization images. Because of the strong scalability, the visualization regulatory mechanisms can be flexibly and easily installed and deployed to provide high reliability and ensure uninterrupted visual monitoring.

Power engineering site visual supervision needs to rely on the system platform to real-time check the view of the project supervision site and the process of retrospective project records. The centralized management of visualization process management has been achieved after power engineering site supervision visualization resources collection and uploaded to the unified system platform. It can monitor the scene of the various power engineering supervision of the unified data management, focus on the various supervision of the scene, the progress of the project, the construction of the management staff to master the work of supervision work, sum up project supervision process visualization files; real-time check the supervision site visualization of the image, achieve on-site visual remote call, command and coordinate on-site supervisors work; establish the project supervision visualization process archives, in the process of project supervise and display project site supervision of image and video data, enrich the project supervision process visual query trace function, check playback power engineering supervision live situation.

5 Conclusions

In view of the problems and requirements of on-site supervision of electric power engineering, the application of visualization technology is innovated, and a visual supervisory mechanism is established at the power engineering site. The establishment of visualization regulatory mechanisms can focus on the management of power engineering site, achieve full supervision of power engineering site supervision, and be convenient and quick complete supervision on-site inspection and supervision; and can improve the efficiency of the scene, reduce the strength of the supervision staff, timely detect the problems in the construction process, enrich the site supervision and control technology to further enhance the quality and level of power engineering site management.

References

1. XIAO Donghui, LIN Li. Solution Scheme of Integrated Video Surveillance Platform of Power System[J]. Automation of Electric Power Systems, 2013, 5: 74-80.
2. XIAO Xing-quan, LI Fu-xiang. Discussion on the Interconnection of Sichuan Power Video Monitoring System Platforms[J]. Telecommunications for Electric Power System, 2010, 7: 30-35.
3. CUI Yanming, LIU Xiaoxian, MA Chao, CHEN Jianning. Standards of Grid Video Surveillance System and Interface Technology[J]. Automation of Electric Power Systems, 2010, 20: 13-16.
4. ZHANG Jinzhi. Real-time Video Monitoring and Management System of Power Supply Business Hall Based on Local Area Network Transmission[J]. Guangdong Electric Power, 2009, 2: 58-62.
5. CHEN Xinhe, SONG Wei, LU Ying. Information Integration Demand Analysis and Data Modeling for Interactive Electric Power Business Hall[J]. Automation of Electric Power Systems, 2013, 7: 77-81.

Examining the Internet-Based Free Talk in College English Classes from the Motivation Perspective

Ming Li
School of Foreign Languages, Sias International University
Xinzheng, Henan, China
bahalucy@163.com

Abstract: Free Talk is recognized as an effective approach to teaching college English in China to improve students' English speaking. With the popularity of Internet around the world, the Internet-based Free Talk demonstrates more advantages in motivating students to engage in English learning. In this paper, the author compares the main features of the Internet-based Free Talk with the five components of the MUSIC Model of Motivation synthesized from current research and theory in the field of motivation. Furthermore, the author illustrates how Internet facilitates Free Talk through online writing service system and online QQ community. The comparison reveals that the success of the Internet-based Free Talk is consistent with the key motivation principles. This paper indicates that professors and researchers in higher education could design and evaluate their instruction according to the components of the MUSIC model of motivation.
Key words: The Internet-based Free Talk; college English class; the MUSIC model of motivation

1 Introduction

Free Talk is recognized as an effective approach to college English classes in China to improve students' English speaking. With the popularity of Internet around the world, the Internet-based Free Talk demonstrates more advantages in motivating students to engage in English learning. The main features of the Internet-based Free Talk program include: two or three students are the focus in class as active speakers during the first few minutes of the class period; the Free Talk topics are based on a survey regarding students' background such as hometown, hobbies, ideal scenic places, the short term goal and the long term goal, teachers allow students to make decisions on the topics; an online writing service system and an online QQ community are provided for the writing and video of the Free Talk; the multiple choice of the prescribed higher level reading materials for the Free Talk.

From the interviews with the students participated in the Internet-based Free Talk, the author found that students are interested in such an activity in their English learning process. And as a result students are motivated to engage in their college English courses. The purpose of this research is to identify to what extent the Internet-based Free Talk program relates to the motivation principles. From this comparison the author hopes that professors and researchers in higher education can apply such principles to their instructional design to improve student motivation and engagement.

In the realm of educational psychology, educators and researchers have conducted research on academic motivation since the 1940s [1]. Up to now, there are a lot of motivation theories such as the self-determination theory [2][3], the self-efficacy theory[4], the expectancy-value theory[5][6], the four-phase interest theory[7]and theories concerning interpersonal relationship[8][9]. Instructors outside the motivation field are usually overwhelmed by these terms and have difficulty in implementing them in their classes. Based on such problems, Jones synthesized the current research and theories of motivation, grouped them into five components, and called them the MUSIC model of motivation[10].

MUSIC is an acronym of the key principles of motivation related to eMpowerment, Usefulness, Success, Interest and Caring. Jones explains that students will be more motivated when they perceive that: (1) they have control over the coursework, (2) the content is useful for their short-term or long-term goals, (3) they can be successful if they put forth appropriate effort, (4) they are interest in the content and class activities, and (5) they feel cared for by teachers and peers in the learning environment[10][11]. Research has shown that the MUSIC model components are correlated, yet distinct in samples of elementary schools [12], secondary and middle schools [13] and postsecondary schools [14][15].The main body of this paper compares the Internet-based Free Talk program with the five components of the MUSIC model. Then we discuss how it is possible for college professors to implement the

MUSIC model in the Internet environment to increase their students' motivation and engagement.

2 Comparing the Internet-Based Free Talk to the music model

2.1 Students are the Focus of Free Talk who have Opportunites to Choose topics and Make Decisions.

The component of empowerment in the MUSIC model refers to the extent to which students feel that they have control over some parts of their coursework [11]. This component is mainly based on the self-determination theory and the individuals' need for autonomy[2][3]. Deci and Ryan held that if individuals have choices toward a task, they will be more likely to spontaneously engage in an activity that interests them[2]. They also argued that autonomy, competence, and relatedness are a human's innate psychological needs[2]. Therefore, it will be possible for instructors to empower their students when they provide students with a variety of opportunities regarding an assignment and allow them to make decisions [16].

In the Internet-based Free Talk program, students are always the focus in the first few minutes in the English class period and teachers provide them with opportunities of the topics and allow them to make decisions. First, at the very beginning of the class in the new semester, the college English teacher gives students an assignment of free talk and asks each of them to give a one minute free talk in front of the class. During each class period, there will be two or three students for the free talk in the first few minutes of the class. Second, the teacher provides students with a list of topics that students can choose one or two as their free talk topics. Third, the students have the power to decide when to give a free talk in class. In other words, students are not forced to give a free talk at a prescribed time. All of these enhance students' sense of empowerment because they have control over the topic and they are able to make decision when they will give the free talk. At the

Fig. 1 The first year college students are discussing the list of the topics for their free talk.

Fig. 2 The first year college student is presenting her free talk

same time, these strategies motivate students in respect to other components of the MUSIC model such as interest, success, and usefulness. Figure 1 and Figure 2 show that the first year college students are discussing the list of the topics with their classmates and then choose one as their own free talk topic.

2.2 The Usefulness of the Free Talk for Students' English proficiency.

The component of usefulness in the MUSIC model means that instructors use strategies to help students understand why the course they are learning is useful for their short- and long-term goals [11]. This component is consistent with the future time perspective theory and the expectancy-value theory. The former theory argues that students will be more motivated if they could see that their present task or coursework are useful for their future goals [17]. The latter theory holds that if students can see the values of a task or the coursework, they will put forth more effort and persistence on the task or the coursework [6]. Jones holds that instructors should explain the usefulness of the course to students' daily lives or help students to see the usefulness of the content by designing some activities themselves [11]

In the Internet-based Free Talk program, the teacher uses three strategies to demonstrate the usefulness of the free talk activities. The fundamental strategy is that the teacher illustrates the video samples of English interview programs in class at the very beginning of the semester. The teacher also picks up a small amount of the genuine conversation scripts from the National College English Test Band Four (CET 4). After the demonstration of these samples, the teacher uploads them to the online QQ community built up specifically for the Free Talk program. In so doing, students can apparently see the usefulness of the free talk both for their English speaking proficiency and for the certificate contributing to their long-term goal. Second, the teacher provide students with the opportunities of free talk after class with foreign experts on campus. Students can go and talk with the foreign experts before or after their free talk in class. Third, the teacher invites the upper class students from the same campus as guest presenters to class who have high English speaking proficiency and share their experience of the usefulness of English speaking for graduate exam or job hunting. The teacher uploads the upper class students' free talk text to the online QQ community in case the students want to read them again after the guests' presentation. The English interview videos and the CET 4 conversations, the free talk with the foreign experts, and the guest presenters help the students realize the usefulness of the Free Talk program.

2.3 The Success that Students can Perceive When They Prepare and Present their Free Talk.

The success component in the MUSIC model indicates that students are more likely to put forth more time and effort if they believe they can succeed in their coursework [10][11]. This component is based on the self-efficacy theory [4] and the expectancy of success aspect of the expectancy-value theory [6]. These two theories hold that an individual's beliefs of their capabilities to be successful in a task decide to what extent they will persist, put forth time and effort, and choose cognitive self-regulation strategies [4][6]. Jones also argued that it is necessary for the instructors to help students believe that they can succeed if they put forth the effort required[11]. One implication of the success component is that teachers should consider students' prior knowledge and connect the difficulty level of the class activities and assignments with students' existing capabilities; another implication is that teachers should regularly provide students with honest and specific feedback [11].

The college English teacher in the Internet-based Free Talk program improves students' sense of success by providing the different level topics for the free talk, the quick response to students' questions, the regular feedback on students' free talk text, and the uploading of the individual's free talk video to the online QQ community. In the first college semester, the teacher provides students with the lower level daily life topics such as the student's hometown, hobbies, favorite food, best friend, and the ideal scenic places around the world. Then the teacher provides students with the higher level topics derived from the book entitled A concise course of American literature. For example, students will talk about his or her favorite American writer based the list of the writers in the book. In this way the students can perceive their success in speaking and writing. Second, the teacher asks the student who will give a free talk the next day to submit his or her free talk text to the online writing service system and then revise the text until there is no mistake about grammar or spelling. Therefore, when the student is giving a free talk in class, the text on the big screen in the front of the class will be one without any mistake. The teacher is approachable and will praise the students online or offline when students correct the mistakes in their free talk text. During the Free Talk presentation, both teacher and peers listen to the student's free talk attentively. Teachers and peers will applaud when the individual's free talk is over. In addition, the teacher records the student's free talk performance as video in class and then upload the video to the online QQ community in a few hours after class. All these strategies will enhance students' sense of success and lead students to believe that they can succeed if they put forth appropriate time and effort on the free talk assignment.

2.4 The Interest that Students can Perceive from the Free Talk Program.

The interest component in the MUSIC model means both the content and the class activities should interest students and teachers should help students to enjoy their learning experiences[11]. This component is consistent with Hidi and Renninger's interest theory and they state that "the potential for interest is in the person but the content and the environment define the direction of interest and contribute to its development" (p. 112) [7]. Teachers can increase students' interest by using novelty and various instructional activities to catch and hold student attention[11].

The Internet-based Free Talk program interests students in many ways. First, students have choice about what to talk in class. In contrast to the prescribed textbook, the topics of the Free Talk are interesting. Second, the teacher adds relevant pictures or words to students' free talk text before the free talk time, and then show the revised text in class. These pictures or words provide students with more interests about the free talk. Third, the teacher uploads the relevant movies based on American literature to the online QQ community, and then students can enjoy them to deepen their understanding of the American writers and their novels, poems, or fictions. In contrast to the paper-and-pen test, the free talk, a part of students' class presentation score, is more interesting due to the colorful pictures, vivid movies, and the diversity of topics.

2.5 The Caring that Students can Perceive from Teachers and Peers.

The component of caring in the MUSIC model indicates that students will be more motivated when they believe that the teachers and peers care about their learning and about them as a person [11]This component is mainly derived from research in attachment, relatedness, caring, and sense of community [2][18]. Bergin and Bergin and Wentzel argue that students tend to be more active in their coursework and perform better academically when they have close relationships with their teachers and classmates[18][19]. Noddings holds that besides academic achievement, the main aim of education should also include the moral perspective fostering students to be "competent, caring, loving and lovable" (p. ix)[18]. One implication is that teachers should be available to students when they need help. Another implication is that teachers should respect students and ensure that students in the class respect one another [11]

The Internet-based Free Talk program demonstrates four aspects of caring between the teacher and the students and among the peers. First, at the very beginning of the semester, the teacher builds up two online communities; one is the online community of the writing service system and the other is online QQ community. In the online writing system, the teacher provides students with the free talk writing assignment and asks students to upload and revise their free talk text based on the guidance of the writing service system. The teacher can assign the students to pair work or three to five peers group work randomly in this community so that the peers can help each other and witness each other's progress. The teacher is approachable when students meet problems that the writing system cannot solve. Second, in the online QQ community, all the students upload their revised free talk text so that each of them can share their text with each other. After the free talk presentation in class, the teacher will upload the free talker's video to the QQ community. The teacher and some of the other students send symbolic pictures such as applause, gifts, and smile faces to the free talkers in the community. Also, some students upload the relevant picture and videos to the QQ community. Third, during the free talk period in class, both the teacher and the students listen to the free talk attentively. After the free talk, the teacher and the students applauds the free talker's good job and the teacher usually spends a few seconds commenting the free talk briefly. In addition, the teacher is available to students inside and outside the two communities. For example, students can also send messages to the teacher via WeChat or QQ personally. The teacher gives quick, honest and specific feedback to the students whenever they need help.

3 CONCLUSION

This paper provides evidence that the Internet-based Free Talk program is consistent with current motivation research and theory. Furthermore, this paper demonstrates how the MUSIC model of motivation serves as a lens examining the effectiveness of the Internet-based Free Talk program from five key components (empowerment, usefulness, success, interest, and caring). In addition, the online writing service system and the online QQ community play an important role in promoting the power of the Free Talk program. This paper concludes that the Internet-based Free Talk program is successful in college English classes in China because of the following reasons. It empowers students by providing a lot of opportunities and decision making regarding free talk topics, when and with whom to discuss and give the free talk. The various interview videos, the conversations in the listening section of the CET 4 and the guest presentations by the upper class students contribute to students' enhanced sense of usefulness toward this free talk program; the quick and honest feedback from the teachers and peers, the use of the online writing service system, and the free talk performance in the front of the classroom result in students' strong sense of success. The list of diverse daily life topics and the different writers in the American literature, and the uploaded free talk videos bring a lot of interests to students. Finally, the online communities of writing service system and QQ chatroom provide teachers and students the sense of family because they care about each other both in English learning and their life. The teacher's quick and honest feedback on students' questions strengthens students' sense of caring. As a result, other instructional design programs can use the key components of the MUSIC model of motivation to determine whether the design is consistent with the current motivation research and theory and then the program really motivate students to engage in their learning.

Acknowledgment

This work was supported by Sias International University with the grant No. 2016JGZD03.

References

[1] B.Weiner. History of Motivational Research in Education. Journal of Educational Psychology, *82*(4), 6-16,1990.

[2] E.l.Deci, R.M,Ryan, R. M. A Motivational Approach to Self: Integration in Personality. In University of Nebraska (Lincoln campus). Department of Psychology. Nebraska Symposium on Motivation, 38, 237-288,1991.

[3] R.M.Ryan, E.L.Deci. Self-determination Theory and the Facilitation of Intrinsic Motivation, Social Development, and Well-being. American Psychologist, 55(1), 68,2000.

[4] A.Bandura. Social Foundations of Thought and Action: A Social Cognitive Theory. Englewood Cliffs, N.J: Prentice-Hall.1986.

[5] J.S.Eccles, A.Wigfield, A. In the Mind of the Actor: The Structure of Adolescents' Achievement Task Values and Expectancy-related Beliefs. Personality and Social Psychology Bulletin [H.W. Wilson - SSA], 21(3), 215-225,1995.

[6] A.Wigfield, J.S. Eccles, J. S. Expectancy–value Theory of Achievement Motivation. Contemporary Educational Psychology, 25(1), 68-81,2000.

[7] S.Hidi, K.A.Renninger. The Four-phase Model of Interest Development. Educational

[8] Psychologist, 41(2), 111-127.2006. Doi: 10.1207/s15326985ep4102 4
[9] T.M.Freeman, L.H.Anderman, & J.M.Jensen. Sense of Belonging in College Freshmen at the Classroom and Campus Levels. The Journal of Experimental Education, 75(3), 203-220,2007.
[10] J.Reeve.Motivating others: Nurturing Inner Motivational Resources. Boston: Allyn and Bacon,1996.
[11] B.D. Jones.Motivating Students to Engage in Learning: The MUSIC Model of Academic Motivation. International Journal of Teaching and Learning in Higher Education, 21(2), 272-285,2009.
[12] B.D. Jones. Motivating Students by Design: Practical Strategies for Professors. Charleston, SC: CreateSpace,2015.
[13] B.D. Jones. M.L.Sigmon. Validation Evidence for the Elementary School Version of the MUSIC® Model of Academic Motivation Inventory. Electronic Journal of Research in Educational Psychology, 14(1), 155-174,2016. Retrieved from http://dx.doi.org/10.14204/ejrep.38.15081
[14] K.Parkes, B.D. Jones.& J.Wilkins. Assessing Music Students' Motivation Using the MUSIC Model of Academic Motivation Inventory. UPDATE: Applications of Research in Music Education. Advance online publication,2015.
[15] B.D. Jones.C.M. Epler, P. Mokri, L.H. Bryant, & M.C. Paretti, M. C. The Effects of a Collaborative Problem-based Learning Experience on Students' Motivation in Engineering Capstone Courses. Interdisciplinary Journal of Problem-based Learning, 7(2), 2.2013
[16] B.D. Jones, G.E.Skaggs. Measuring Students' Motivation: Validity Evidence for the MUSIC Model of Academic Motivation Inventory. International Journal for the Scholarship of Teaching and Learning, 10(1),2016. Retrieved from http://digitalcommons.georgiasouthern.edu/ij-sotl/vol10/iss1/7
[17] A.F.Gardner, B.D. Jones. Examining the Reggio Emilia Approach: Keys to Understanding Why it Motivates Students. Electric Journal of Research in Educational Psychology, 14(3), 602-625, 2016.
[18] D.F.Kauffman., J. Husman. Effects of Time Perspective on Student Motivation: Introduction to a Special Issue. Educational Psychology Review, 16(1), 1-7, 2004.
[19] N.Noddings.The Challenge to Care in Schools: An Alternative Approach to Education. New York: Teachers College Press,1992.
[20] C.Bergin, D. Bergin, Attachment in the Classroom. Educational Psychology Review, 21(2), 141-170,2009.

How group-buying servicescape affect consumers' purchase intention, the regulating effects of price discount and time pressure

Lingying Zhang[1], *Chunling Zhang*[1,*], *Cui Li*[1] and Kang Wang[1]

[1]College of Management, Shenzhen University, Shenzhen 518060, PR China

Abstract. This paper constructs a conceptual model of how group-buying servicescape affect consumers' purchase intention and discusses the regulating effects of price discount and time pressure. This research uses a survey approach to collect data, as a result, we collected 506 valid questionnaires. The results show that aesthetic appeal, information exchange, interpersonal interaction and perceived security have a positive effect on positive emotion; layout and functionality, information exchange and interpersonal interaction have a positive effect on virtual touch. Positive emotion and virtual touch can promote the consumer's purchase intention, and time pressure and price discount play a regulatory function. The results of this study have an important reference value for group-buying operators to use servicescape to attract and retain consumers.

Keywords—group-buying servicescape, virtual touch, positive emotion, group-buying intention.

1 Introduction

Group-buying as a new e-commerce model, through the consumer self-group and professional group-buying web site to enhance the bargaining power of customers and businesses. China Group-buying rise later than abroad, group-buying is in the budding period in the year of 2005 and 2006, and enter the stage of vigorous development until in the year 2010. After several years of development, there are many third-party group-buying web site, such as meituan.com, lashou.com, nuomi.com, manzuo.com, 55.com, etc. According to iResearch released data showed that in the first half of 2016, China group-buying market turnover reached 77.01 billion yuan, compared with the same period in 2015 increased by 167.7%. Nevertheless, (Lloyd C Harris & Goode, 2004) [1] study said that commentator had found that some group-buying e-tailers fail to make profits. Considering the universality of such views, it is perhaps not surprising to find a growing body of research that is identifying and exploring the myriad of factors that may account for the attitudes and behaviors of online shoppers. Therefore, to explore the impact of group-buying servicescape on consumer's purchase intention, will help operators to create a good shopping environment, so as to attract and retain more consumers.

This research belongs to the literature on aspects of servicescapes. This paper have three objectives. Firstly, to determine whether online group-buying servicescape (physical service environment, social interaction and financial security) has significant relationship with emotional cognition (positive emotion and virtual touch). Secondly, to explore the relationship between emotional cognition (positive emotion and virtual touch) and purchase intention. Thirdly, to study the regulating role of time pressure and price discount on emotional cognition and purchase intention.

2 Literature review

2.1 Group-buying servicescape

The context of service encounters has been conceptualized a number of scholars using various terms. For example, (Turley & Milliman, 2000) [2] defined it as marketing environment, (Tai & C., 2003) [3] defined the online shopping as store environment. The term "servicescape" is created by (Bitner, 1992) [4]. (Lloyd C Harris & Ezeh, 2008)[5] proposed that the concept of a network servicescape and defined it as all environmental factors in the process of network service delivery (from pre-purchase to post-purchase), which was widely accepted and became a generic term. There are two main aspects of the research on the network servicescape, one aspect is the division of the dimension, another aspect is about the influence on the online behavior of the consumers.

Different people have different perspectives about the division of servicescape. Traditional store servicescape not only includes the physical factors of visualization in the service scenario, but also includes the social factors of interpersonal interaction. (Baker, Grewal, & Parasuraman, 1994) [6] proposed that traditional store servicescape consists of three factors, including atmosphere factors, design factors and social

* Corresponding author: 1715167698@qq.com

factors. In contrast, (Bitner, 1992) [4] argued that store servicescape is made of ambient conditions, layout and functionality, and signs, symbols, and artifacts. Later scholars adopt this concept in many offline stores study. For example, (Wakefield & Blodgett, 2013) [7] apply this concept into their study about the servicescape of leisure service settings. However, the advent of Internet and information technology, making a lot of scholars begin to study online network servicescape. (Eroglu, Machleit, & Davis, 2003) [8] divided the online store atmospherics into high-mission-related factors and low-mission-related factors based on media richness theory. (Kim, Kim, & Kandampully, 2009) [9] divided the online store servicescape into six dimensions: convenience, customization, informality, interaction, the appearance of the site and the value of entertainment. (Lloyd C. Harris & Ezeh, 2008) [5] argued that the online servicescape should be consist of three dimensions, including aesthetic appeal, functional layout and financial security.

There are two paths to the study of the influence of online servicescape on consumer behavior. One path is based on S-O-R. It is believed that emotional reaction is the intermediary of online servicescape and consumer reaction. With the deepening of research, many scholars realize that the relationship between emotional stimuli and consumer behavior is explained partly. Therefore, some scholars introduce customer cognition as another rese arch path. (Kim et al., 2009) [9] proved that the online servicescape positively affects customer perceived service quality and satisfaction, which leads to customer buying behavior.

2.2 Group-buying interaction

Online group-buying is a novel business model. Based on the number of units a group buys, consumers have a motivation to regroup for unit price reduction. With the purchasing amount increases, consumers will receive lower product prices. (Matsuo, 2009)'s [10] research showed that group-buying enables consumers to gain volume discounts and discounted group prices when consumers aggregate the requested group size. At present, there are many third-party group-buying web site, such as meituan.com, lashou.com, nuomi.com, manzuo.com, 55.com, etc. It can be seen, the trend of group-buying socialization has been strengthened, the consumer interaction in group-buying has become an inevitable trend, but the existing group-buying literature less concerned about the interaction between users.

(Florenthal & Shoham, 2010) [11] defined group-buying interaction as the degree to which one or more individuals respond to the behavior of a particular information provider. (Bonfield, 1974) [12] pointed that the members of group-buying often share the purchase experience. (Shan, Luo, & Zhang, 2013) [13] put forward that the interaction between customers is divided into two types: information exchange and social interaction, the purpose of information exchange is to obtain certain information to solve the problem, social interaction is in order to obtain emotional satisfaction through chatting on the Internet.

2.3 Virtual touch

In recent years, sensory research on the field of consumers has begun to receive attention in the world, especially tactile research is on the rise. (Peck, Barger, & Webb, 2013)[14] thought that tactile is a direct experience of the consumer on the goods. (Reb & Connolly, 2007)'s [15] research showed that the touch can narrow the emotional distance between the two sides and thus affect the consumer's attitude to the goods and purchase behavior. (Grohmann, Spangenberg, & Sprott, 2007)[16] argued that the tactile elements of a commodity can inspire consumers' favor sense and emotional involvement to change buying decisions. (Citrin, Jr, Spangenberg, & Clark, 2003) [17] proposed that a shortcoming of Internet-based retailing efforts is consumers' inability to touch products during their purchase decision-making processes, the results show that women have a higher need for tactile input compared to men in making product evaluations. The Internet and the World Wide Web (WWW) are predicted to become the center of the commercial universe. Customers can touch the product in offline stores, but if they cannot touch the product when making purchase online. (Peck & Childers, 2003)[18] found that product's intrinsic cues such as tactility, color or smell (cues which, if changed would produce a resultant change in the physical product itself) are more important in a product's perceived quality than its extrinsic cues.

There are many literatures on tactile at present, but there is relatively little literature on virtual touch. Through the integration of a small number of foreign literature, we can clear the formation of virtual tactile mechanism. Because of the interactive function of human organs, enabling the customers produce a tactile experience through the sense of visual and auditory, and creating a telepresence from the overall scene, thus evoking the historical tactile memory, and thus produce tactile perception through the association.

3 A conceptual model

3.1 The relationship between group-buying servicescape and positive emotions

According to the literature review, the group-buying servicescape can be divided into physical dimension, social dimension and financial security. (Lloyd C. Harris & Goode, 2004) [1] divided the physical dimension into aesthetic appeal and layout and functionality based on the characteristics of group-buying online. Aesthetic appeal refers to the atmosphere of the attractiveness of the group-buying web site, layout and functionality refers to the organization of the web site, organization, layout and adaptability. Social interaction is divided into information exchange and interpersonal interaction.

Many scholars have confirmed that the online shopping physical environment has an impact on consumer sentiment. In terms of aesthetic appeal, (Kawaf & Tagg, 2012)'s [19] research showed that the text, picture, and music atmosphere in the online shopping environment can significantly affect the consumer's emotion. (Wang, Minor, & Wei, 2011)[20] suggested that site with a rich media tools, can make the store a better atmosphere, which leading to customer pleasure mood. (Shobeiri, Laroche, & Mazaheri, 2013)'s [21] study found that the aesthetic factors in the online shopping environment help consumers to immerse them and promote the purchase. In terms of layout and functionality, (Lloyd C. Harris & Ezeh, 2008)[5] study showed that if shopping sites can be customized, and allow customers to personalize their settings, the customer's satisfaction will increase significantly.

(Rose, Clark, Samouel, & Hair, 2012)'s [22] study confirmed that the usefulness and customization of the online stores positively affects the customer's pleasure. When consumers browse the group-buying site, physical elements can attract the user's eye and stimulate the user to produce a happy mood. Therefore, the hypothesis 1a, 1b is proposed.

H1a: Aesthetic appeal is a positive impact on the group-buying consumer's positive emotions.

H1b: Layout and functionality is a positive impact on the group-buying consumer's positive emotions.

Social media allows consumers who coming from different space to gather together on the Internet, the consumers can communicate with each other. So, some scholars began to focus on the impact of interaction on consumer emotional response. (Lin & Liang, 2015)[23] believed that the information conveyed by consumer interaction contains some useful clues that give the audience a pleasant feeling. (Fiore & Jin, 2003)[24] proposed that online interactivity position perceived interactivity as a central driver of consumers' interpretations and behavior. Group-buying users in the virtual community, QQ group and WeChat and other social media interaction group, not only exchange product-related information, but also free communicate personal experience and emotions. So we propose the hypothesis 2a, 2b.

H2a: Information exchange is a positive impact on the group-buying consumer's positive emotions.

H2b: Interpersonal interaction is a positive impact on the group-buying consumer's positive emotions

Online financial security refers to the extent to which consumers perceive the payment processes and general policies of a web site as secure or safe. (Fusaro, Charron, & Theoret, 2003)[25] thought that perceived security is a crucial aspect of online group-buying servicescape. (Szymanski & Hise, 2000)'s [26] study showed that perceived safety has a significant impact on customer satisfaction. Safe payment environment can promote the positive emotions of consumers. Therefore:

H3: Perceived security is a positive impact on the group-buying consumer's positive emotions.

3.2 The relationship between group-buying servicescape and the virtual touch

With the development of virtual reality technology, the sensory experience of online shopping began to receive attention. (Li, Daugherty, & Biocca, 2003)[27] proposed that virtual experience is a psychological state which consumers interacts with the commodity in the virtual environment. (Marzocchi, Pizzi, & Scarpi, 2016)'s [28] study showed that the realization of consumer virtual experience is affected by objective factors, sensory factors, perceptual factors, stimulating response factors, cognitive and affective factors, encounter factors. Commodity quality, price and other marketing factors play a role through the consumer's sensory process. (Mccabe & Nowlis, 2003)'s [29] research showed that media tools that provide a wealth of product information can effectively compensate for the disadvantages that the goods cannot be touched when shopping online. Therefore, the development of technology allows the group-buying operators to display the product vividly and create a harmonious and comfortable atmosphere of online shopping, which making consumers feel like in the store. Accordingly the hypothesis 4a, 4b are putted forward.

H4a: Aesthetic appeal is a positive impact on the group-buying consumer's virtual touch.

H4b: Layout and functionality is a positive impact on the group-buying consumer's virtual touch.

Although the development of virtual reality technology can make up for the lack of consumer online shopping sensory experience, but the display of goods and the layout of the site are from the online stores. For consumers, the product information obtained from other customers is more objective and real. Therefore, many consumers may choose to communicate with other consumers online to understand the product. (Skadberg & Kimmel, 2004)'s [30] study found that the higher the level of interaction, the higher the virtual experience of the consumer. (Zhao, Cai, Shan, University, & University, 2014)'s [31] study indicate that network marketing tools such as communities, real-time chat, etc., contribute to the generation of high-quality online interactions that inspire the consumer's touch experience. The interaction between consumers can enable consumers to get more information about product and create a good social atmosphere, so that consumers have a sense of presence. So we propose the hypothesis 5a, 5b.

H5a: Information exchange is a positive impact on the group-buying consumer's virtual touch.

H5b: Interpersonal interaction is a positive impact on the group-buying consumer's virtual touch.

3.3 The relationship between emotional cognition and group-buying intention

The relationship between positive emotions and purchase intention has been proved by many scholars. (Rose et al., 2012)[22] analyzed that how price perception and service awareness awake consumer positive emotions and then affect the final shopping process. But there is relatively little research on the relationship between virtual tactile and purchase intention, (Peck & Johnson, 2011)'s [32] research showed that touch cannot make

DOI: 10.1051/matecconf/201713900129

consumers get enough product information, but can change the consumer's attitude about product and enhance the purchase intention through the sensory experience. (Zhao et al., 2014)'s [31] empirical study showed that online commodity display and online interaction can directly and indirectly affect the consumer impulse to buy, virtual touch plays an indirect role in the incomplete intermediary variables. On the online purchase scenario, although the biggest problem facing consumers is difficult to touch the product, many consumers can not touch the product to give up the final purchase. The maturity of virtual technology can make up for this shortcoming. Visual experience and online interaction can trigger a tactile experience, and consumers get a virtual touch. So this paper think that the virtual touch can affect the purchase intention as the same as the real touch.

H6a: Positive emotions have a positive impact on consumer's group-buying intention

H6b: Virtual touch has a positive impact on consumer's group-buying intention.

3.4 The regulating role of time pressure and price discount

(Bonfield, 1974)[12] argued that time pressure refers to the pressure response that consumers are caused by completing shopping decisions within a limited time. In the e-commerce platform, promotional methods of limited time are widely used, namely, with the passage of time and the closing date of the approaching, consumers perceive the lack of time to produce emotional reactions. Although time is an important factor in the decision-making of consumers to buy, less scholars pay attention to the role of time pressure. (Dhar & Nowlis, 1999)'s [33] study show that when time pressure is large, consumers are more likely to reduce their risk expectations and impulse consumption. (Sepp, 2014)'s [34] research show that time pressure in the situation of group-buying has a positive impact on attitude and purchase intension, in the case of large time pressure, the effect of attitude on purchasing intention is stronger than less time pressure. Limited time purchase is the difference between group-buying and ordinary online shopping, the customers is affected by group-buying activities and group-buying atmosphere and more inclined to transfer the positive perception into the purchase intention. So we propose the hypothesis 7a, 7b.

H7a: Time pressure have a regulatory role between positive emotions and group-buying intention.

H7b: Time pressure have a regulatory role between virtual touch and group-buying intention.

Price is one of the most likely to stimulate consumer desire to purchase, different from ordinary online shopping, the characteristics of group-buying are gathering a certain number of customers to get a low prices. (Rudolph, Bauer, & Steiner, 2010)'s [35] research shows that the discount rate of advertising will strengthen the impact of consumer's perception on the purchase intention.

(Moon & Lee, 2013)[36] pointed out that the price discount can adjust the impact of user perception on the purchase decision, the higher the price discount, the higher the customer's perceived value, and then stimulating the purchase behavior of customer. To sum up, in the situation of group-buying, when the consumer has a positive perception of the environment created by online stores, price discounts will stimulate consumers to grasp the chance to buy.

H8a: Price discount has a regulatory role between positive emotions and group-buying intention.

H8b: Price discount has a regulatory role between virtual touch and group-buying intention.

The conceptual model of this paper is showed in Figure1.

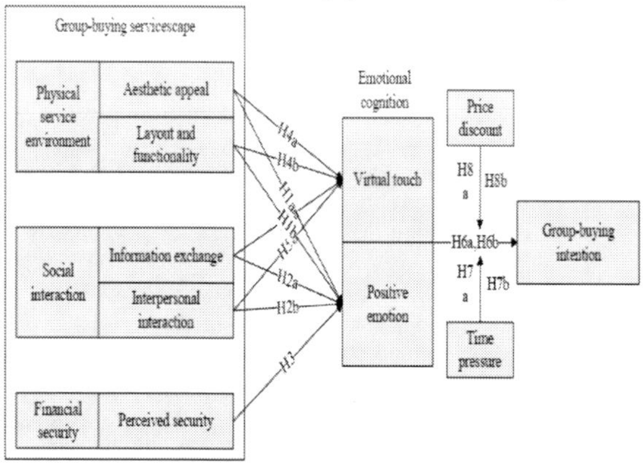

Fig. 1. The conceptual model.

4 Methodology

4.1 Data Collection

This paper used the form of a questionnaire to verify the hypothesis. We designed an online questionnaire, which is used to collect data. We created a questionnaire link and share it with people through WeChat, e-mail and QQ etc. As a result, we collected 506 valid questionnaires. In terms of gender, there are 271 females and 235 males; in terms of age, in the 15-20 years old accounted for 5.73 percent, in the 21-25 years old accounted for 29.45 percent, in the 26-30 years old accounted for 21.74 percent, in the 31-35 years old accounted for 39.13 percent, more than 35 years old accounted for 3.95 percent; in terms of education, bachelor degree had 318 people, master's degree had 35 people.

4.2 Measures

The measurement items in this paper were from previous studies. The items of online survey in this research used 1-5 Likert scales, on which 1-point indicated "strongly disagree" and 5-point "strongly agree". In this paper, we use the first letter of each word in variables plus the number to represent the indicators. The indicator of aesthetic appeal was described as AA1, AA2 and AA3. The same way to define layout and functionality (LF), information exchange (IE), interpersonal interaction (II), perceived security (PS), positive emotion (PE), virtual touch (VT), group-buying intention (GI), price discount (PD), price discount (PD).

Table 1. The result of principal component analysis.

	Components									
	1	2	3	4	5	6	7	8	9	10
AP1	**.641**	.105	.236	.108	.205	-.002	.163	-.091	-.131	-.010
AP2	**.537**	.187	.349	.284	.025	-.021	.125	.006	-.014	.039
AP3	**.518**	.313	.268	-.014	.096	-.136	-.029	.086	.072	.043
AP4	**.609**	.290	.345	.243	-.131	.010	.010	.007	.007	.064
LF1	.477	.310	.168	**.564**	-.022	-.015	.153	.037	.052	-.201
LF2	.399	.315	.376	**.634**	-.212	.171	.306	.088	.060	-.073
LF3	.186	.283	.202	**.680**	.132	-.128	-.086	-.128	.087	-.059
LF4	.414	.279	.404	**.675**	-.090	.045	.279	-.006	-.044	.045
LF5	.354	.279	.237	**.571**	.138	.051	-.018	.195	.105	-.041
LF6	.343	.399	.379	**.594**	-.202	-.063	.209	.073	-.143	.114
LF7	.577	.263	.252	**.547**	.045	.110	.019	-.015	.017	-.092
IE1	.243	.393	**.657**	-.028	.120	.027	.081	.179	.173	-.072
IE2	.413	.307	**.546**	-.053	.017	-.001	-.163	.030	.017	.127
IE3	.186	.316	**.518**	.143	-.135	-.031	.096	.082	.046	-.026
IE4	.334	.359	**.676**	-.163	-.037	-.024	-.079	-.024	.002	-.065
II1	.200	-0.35	.385	.479	.086	-.076	.183	-.001	**.634**	.017
II2	.387	.084	.218	.445	.145	.137	-.017	.052	**.578**	.067
II3	.334	.025	.416	.465	.007	-.051	.094	.020	**.628**	-.014
II4	.209	.032	.265	.365	-.085	-.019	.073	.036	**.682**	-.144
PS1	.042	.054	.012	.003	.052	**.954**	-.042	.443	.454	.055
PS2	-.062	-.029	.017	.036	.103	**.838**	-.097	-.002	-.024	.025
PS3	-.014	.018	.024	.092	-.115	**.637**	-.794	.043	.054	.901
PS4	.023	-.019	.033	.037	.079	**.907**	-.034	-.027	-.023	.055
PS5	.067	-.007	.077	.075	.449	**.679**	.096	.044	.068	-.116
PE1	.472	-.028	.410	-.014	.000	-.048	-.058	**.588**	-.017	.060
PE2	.338	.093	.159	-.058	-.027	.074	.030	**.609**	.056	-.008
PE3	.420	.101	.349	-.020	.027	-.062	.036	**.623**	.038	.047
PE4	.148	-.063	.233	.188	-.034	-.009	-.165	**.729**	-.012	.179
PE5	.316	.085	.290	-.090	**.560**	.160	.003	.068	.063	-.105
VT1	.295	-.218	.254	-.039	**.607**	-.103	-.007	-.099	-.072	.007
VT2	.127	.195	.407	-.085	**.569**	.097	.133	-.076	-.087	-.006
GI1	.347	**.675**	.081	.034	-.077	.247	.054	.086	.047	-.041
GI2	.274	**.814**	.368	.042	-.127	.096	-.087	.121	.121	.072
GI3	.395	**.558**	.057	.022	-.097	.017	.064	.038	.066	-.052
TP1	.195	.249	**.694**	.025	.172	-.012	-.094	-.019	-.014	.015
TP2	.324	.331	**.531**	-.124	.169	.029	.088	.055	-.019	-.045
TP3	.194	.251	**.670**	-.002	.128	.063	-.035	-.083	-.091	-.045
PD1	-.009	.451	.282	.120	-.102	.075	.025	.084	-.091	**.526**
PD2	.083	.442	.316	-.029	.219	.040	-.069	.064	.060	**.617**
PD3	-.067	.242	.392	-.081	-.206	.173	-.096	.025	.035	**.510**
PD4	.069	.322	.237	.157	.309	-.166	-.036	.054	.037	**.529**

5 Data analysis

5.1 Reliability and validity analysis

Reliability analysis is an effective method to analyze whether the comprehensive evaluation system has a certain stability and reliability, mainly test the stability and consistency of measurement results and measurement tools. This paper used SPSS22.0 software and the widely used Cronbach's alpha to analyze the reliability. The result is showed in Table 2, from Table 2, we can see that the Cronbach's alpha coefficient is all great than 0.70 which indicates that the sample is quite good and commonly acceptable level for explanatory research. Table 1 shows that all the indicators have a high load (all above 0.5) on the respective factors and low on other factors, indicating that the scale has a good structural validity.

Table 2. Analysis of scale reliability.

Factors	**Cronbach's Alpha**
AA	0.751
LF	0.830
IE	0.751
II	0.735
PS	0.800
PE	0.780
VT	0.749
GI	0.751
TP	0.721
PD	0.747

5.2 Regression analysis

According to the conceptual model and research hypothesis, we need multiple regression analysis of three sets of variables. Three sets of variables are group-buying servicescase and emotional cognition, emotional cognition and group-buying intention and the regulating role of time pressure and price discount.

5.2.1 Group-buying servicescase and emotional cognition

This paper use spss22.0 software to stepwise regression, the result of regression analysis of the group-buying servicescape and emotional cognitive is shown in Table 3 and Table 4.

In all dimensions of group-buying, the results of regression analysis show that the positive effect of aesthetic appeal, information exchange, interpersonal interaction, perceived security on positive emotion is supported, but layout and functionality failed to enter the positive emotion of this model, so its positive role is not supported. The positive effect of layout and functionality, information exchange and interpersonal interaction on virtual touch get support, however, aesthetic appeal did not enter the virtual touch of the model, so their positive role is not supported.

Table 3. The results of regression analysis of group-buying servicescape and positive emotion.

Independent Variables	Normalization coefficient	T Value	Significance level	VIF	D.W Value
AA	0.398	5.758	0.000	1.048	1.803
IE	0.311	4.646	0.000		
II	0.283	3.603	0.001		
PS	0.374	5.036	0.000		

Table 4. The results of regression analysis of group-buying servicescape and virtual touch.

Independent Variables	Normalization coefficient	T Value	Significance level	VIF	D.W Value
LF	0.119	1.988	0.000	1.003	1.766
IE	0.291	4.879	0.000		
II	0.408	6.793	0.004		

5.2.2 Emotional cognition and group-buying

The results of regression analysis of emotional recognition and group-buying intention is shown in Table 5. The regression analysis results show that the two factors of the emotional cognition are entered into the model, and the positive effects of positive emotion and virtual touch on the group-buying intention are supported.

Table 5. The results of regression analysis of emotional recognition and group-buying intention.

Independent Variables	Normalization coefficient	T Value	Significance level	VIF	D.W Value
PE	0.710	7.065	0.000	1.341	2.097
VT	0.673	2.758	0.006		

5.2.3 The regulatory effect of time pressure and price discount

The effect of time pressure and price discount on positive emotion and group-buying intention. In this paper, the positive emotion, time pressure, price discount, the product of positive emotion multiply time pressure and positive emotion multiply price discount are taken as independent variable; group-buying intention is taken as dependent variable. Multi-layer regression analysis is the main method to analyze the regulatory effects of time pressure and price discount. The results are shown in Table 6.

According to the regression analysis results in Table 6, it was found that the negative correlation coefficient R^2 was strongly changed (0.018) after the time pressure was added, and its significance was 0.001, which was significant at the <0.05 level, indicating that the time pressure was positively correlated with the positive emotion and group-buying intention. In model 3, the negative correlation coefficient of the model was changed (0.011) after adding the price discount factor, and it was also significant at the <0.05 level, indicating that the price discount had effect on the positive emotion and the group-buying intention.

Table 6. Time pressure and price discount on positive emotions and group-buying intention regression model coefficient.

Model	Model Summary[d]			
	R	*R²*	*ΔR²*	*Sig.F*
1	0.751[a]	0.564	0.564	0.000
2	0.763[b]	0.582	0.018	0.001
3	0.770[c]	0.593	0.011	0.000

Note: a predictor variables are (constant), time pressure, positive emotion, price discount; b predictor variables are (constant), time pressure, positive emotion, price discount, the product of positive emotion multiply time pressure; c predictor variables are (constant), time pressure, positive emotion, price discount, the product of positive emotion multiply time pressure and positive emotion multiple price discount; d as group-buying intention presents dependent variable.

The effect of time pressure and price discount on virtual touch and group-buying intention. In this paper, virtual touch, time pressure, price discount, the product of virtual touch multiply time pressure and virtual tactile multiply price discount are used as independent variable, group-buying intention is taken as dependent variable. Multi-layer regression analysis is the main method to analyze the regulatory effects of time pressure and price discount. The results are shown in Table 7.

Table 7. Time pressure and price discount on the virtual touch and group-buying intention of the regression model coefficient.

Model	Model Summary[d]			
	R	*R²*	*ΔR²*	*Sig.F*
1	0.715[a]	0.511	0.511	0.000
2	0.723[b]	0.523	0.012	0.000
3	0.730[c]	0.533	0.010	0.004

Note: a predictor variables are (constant), time pressure, virtual touch, price discount; b predictor variables are (constant), time pressure, virtual touch, price discount, the product of virtual touch multiply time pressure; c predictor variables are (constant), time pressure, virtual touch, price discount, the product of virtual touch multiply time pressure and virtual touch multiple price discount; d as group-buying intention presents dependent variable.

According to the regression analysis results in Table 7, it was found that the negative correlation coefficient of the model was changed (0.012) after adding the time pressure factor, and its significance was 0.000, which was significant at the <0.05 level, indicating that the time pressure has regulatory effect on the virtual touch and group-buying intention. In model 3, after adding the price discount element, the model negative correlation coefficient R^2 is changed (0.010), and significant at the <0.05 level, indicating that the price discount on the virtual touch and group-buying intention to have a regulatory role.

6 Discussion and conclusion

This paper used empirical research methods to explore the impact of group-buying servicescape on consumer purchasing intention. Most of the assumptions have been proved to be true, only H1b and H4a are not supported.

The research indicates that aesthetic appeal, information exchange, interpersonal interaction and perceived security have a positive effect on positive emotion; layout and functionality, information exchange and interpersonal interaction have a positive effect on virtual touch. Positive emotion and virtual touch can promote the consumer's purchase intention, and time pressure and price discount play a regulatory function. The results of this study have an important reference value for group-buying operators to use servicescape to attract and retain consumers.

H1a is supported, this means that pictures, colors, text, sound and video display is conducive to mobilize the consumer's emotions. Unexpectedly, H1b are not supported, the layout and functionality does not affect the positive emotions, the reason may be that consumers are more concerned about the product itself, customization of online group-buying may is a waste of time. H2a and H2b are supported, the result shows that the social interaction are related to the consumer's emotions. By obtaining real and objective information to enable consumers to get a good mood. And the exchanges of value of life and daily of life also can help they get a good mood. H3 is also supported, a safe environment of group-buying web site will make consumers have a good mood. H4b is supported, but H4a is not supported, the one reason may be that consumers are more concerned about the inherent nature of the product and not concerned about the aesthetic elements of the site, or the other reason may be that consumers bought this product before, they already know the quality of the product, now they want to buy this product again, only the price can cause their attention. H5a and H5b are supported, that proves that in the servicescape of group-buying, when the content of the consumer discussion is about the purchase of products, the description of others can objectively realistic display of the actual situation of the product, which contains the product touch thread than the business description easier to let consumers feel the product real existence. H6a and H6b are supported, the result shows that in the e-servicescape of group-buying, positive emotions and virtual tactile perception can promote the willingness of consumer's group-buying. On the one side, there exists a positive impact on the purchase intention when they are in a good mood. On the other side, one of the important factors of many consumers give up online group-buying is cannot touch the product, and physical dimensions and social dimensions produced by the visual display as a compensation mechanism can make consumers have a virtual touch feel, thereby enhancing the consumer's desire to buy. H7a and H7b are supported, this indicates that group-buying limited time make the consumer have no sufficient time to fully search for information, but in the state of positive emotions and perceived touch experience, the pressure from time led to consumers to make a decision to grasp promotions in time. H8a and H8b are supported, indicating that the price discount play a regulatory role in the emotional cognition and group-buying.

At last, we provide some advice to the operators of online group-buying. Firstly, as the development of virtual reality technology, operators can use more advanced technical to display products, and enhance the display effect of products, such as to take 3D display, set the virtual dressing room and real models. Secondly, China's current group-buying web sites are mostly belong to third-party web site which lacking of consumer interaction platform, according to the results of this study, operators can set up a special information exchange platform on the page of group-buying products, then consumers can communicate with other people about the product. At the same time, operators set up a dedicated emotional exchange section on the website, for consumers to engage in social communication to enhance their sense of belonging. Thirdly, due to time pressure and price discount has a significant regulatory role, operators can continue to retain the practice and further strengthen it. For example, setting the countdown time on the buy page. Timekeeping and supplemented by the practice of ticking voice, making customers feel the passage of time through visual and auditory, which can make consumers are more concerned about the factor of time.

Acknowledgments

This work is partially supported by National Natural Science Foundation of China under Grant No. 71572115; Major Program of Social Science Foundation of Guangdong under Grant No. 2016WZDXM005; National Science Foundation of SZU under Grant No. 836.

References

1. L. C. Harris and M. M. H. Goode, "The four levels of loyalty and the pivotal role of trust: a study of online service dynamics," Journal of Retailing, vol. 80, pp. 139-158, 2004.
2. L. W. Turley and R. E. Milliman, "Atmospheric Effects on Shopping Experimental Evidence," 2000.
3. A. R. Tai and S. T. C., "Store Environment and Shopping Behavior," Journal of International Consumer Marketing, vol. 15, pp. 71-99, 2003.
4. M. J. Bitner, "Servicescapes: The impact of physical surroundings on customers and employees," Journal of Marketing, vol. 56, pp. 57-71, 1992.
5. L. C. Harris and C. Ezeh, "Servicescape and loyalty intentions: an empirical investigation," European Journal of Marketing, vol. 42, pp. 390-422, 2008.
6. J. Baker, D. Grewal, and A. Parasuraman, "The influence of store environment on quality inferences and store image," Journal of the Academy of Marketing Science, vol. 22, pp. 328-339, 1994.
7. K. L. Wakefield and J. G. Blodgett, "The effect of the servicescape on customers' behavioral intentions in leisure service settings," Journal of Services Marketing, vol. 10, pp. 45-61, 2013.
8. S. A. Eroglu, K. A. Machleit, and L. M. Davis, "Empirical testing of a model of online store

atmospherics and shopper responses," Psychology & Marketing, vol. 20, pp. 139–150, 2003.

9. J. H. Kim, M. Kim, and J. Kandampully, "Buying environment characteristics in the context of e - service," in Meetings on Acoustics 1 Meeting Acoustical Society of America, 2009, pp. 2155-2167.

10. T. Matsuo, "A REASSURING MECHANISM DESIGN FOR TRADERS IN ELECTRONIC GROUP BUYING," vol. 23, pp. 1-15, 2009.

11. B. Florenthal and A. Shoham, "Four - mode channel interactivity concept and channel preferences," Journal of Services Marketing, vol. 24, pp. 29-41, 2010.

12. E. H. Bonfield, "Attitude, Social Influence, Personal Norm, and Intention Interactions as Related to Brand Purchase Behavior," Journal of Marketing Research, vol. 11, pp. 379-389, 1974.

13. M. Y. Shan, Y. H. Luo, and R. L. Zhang, An Analysis and Empirical Study of the Impact of Customer Relationship Quality on Repurchase Intention in Online Shopping Context: Springer Berlin Heidelberg, 2013.

14. J. Peck, V. A. Barger, and A. Webb, "In search of a surrogate for touch: The effect of haptic imagery on perceived ownership," Journal of Consumer Psychology, vol. 23, pp. 189-196, 2013.

15. J. Reb and T. Connolly, "Possession, feelings of ownership and the endowment effect," Judgment & Decision Making, vol. 2, pp. 107-114, 2007.

16. B. Grohmann, E. R. Spangenberg, and D. E. Sprott, "The influence of tactile input on the evaluation of retail product offerings," Journal of Retailing, vol. 83, pp. 237-245, 2007.

17. A. V. Citrin, D. E. S. Jr, E. R. Spangenberg, and M. J. Clark, "Consumer need for tactile input : An internet retailing challenge," Journal of Business Research, vol. 56, pp. 915-922, 2003.

18. J. Peck and T. L. Childers, "Individual Differences in Haptic Information Processing: The “Need for Touch” Scale," Journal of Consumer Research, vol. 30, pp. 430-442, 2003.

19. F. Kawaf and S. Tagg, "Online shopping environments in fashion shopping: An S-O-R based review," Marketing Review, vol. 12, pp. 161-180, 2012.

20. Y. J. Wang, M. S. Minor, and J. Wei, "Aesthetics and the online shopping environment: Understanding consumer responses," Journal of Retailing, vol. 87, pp. 46-58, 2011.

21. S. Shobeiri, M. Laroche, and E. Mazaheri, "Shaping e-retailer's website personality: The importance of experiential marketing," Journal of Retailing & Consumer Services, vol. 20, pp. 102-110, 2013.

22. S. Rose, M. Clark, P. Samouel, and N. Hair, "Online Customer Experience in e-Retailing: An empirical model of Antecedents and Outcomes," Journal of Retailing, vol. 88, pp. 308-322, 2012.

23. J. S. C. Lin and H. Y. Liang, "The influence of service environments on customer emotion and service outcomes," Journal of Service Theory & Practice, vol. 21, pp. 350-372, 2015.

24. A. M. Fiore and H. J. Jin, "Influence of image interactivity on approach responses towards an online retailer," Internet Research, vol. 13, pp. 38-48, 2003.

25. M. Fusaro, C. Y. Charron, and Y. Theoret, Generating Trust in Online Business: From Theory to Practice: Isabelle Quentin Editeur, 2003.

26. D. M. Szymanski and R. T. Hise, "E-satisfaction: an initial examination," Journal of Retailing, vol. 76, pp. 309-322, 2000.

27. H. Li, T. Daugherty, and F. Biocca, "The Role of Virtual Experience in Consumer Learning," Journal of Consumer Psychology, vol. 13, pp. 395-407, 2003.

28. G. L. Marzocchi, G. Pizzi, and D. Scarpi, "When a picture's worth a thousand words: The effects of visual construal priming on information acquisition and choice," Marketing Letters, vol. 27, pp. 1-12, 2016.

29. D. B. Mccabe and S. M. Nowlis, "The Effect of Examining Actual Products or Product Descriptions on Consumer Preference," Journal of Consumer Psychology, vol. 13, pp. 431-439, 2003.

30. Y. X. Skadberg and J. R. Kimmel, "Visitors' flow experience while browsing a Web site: its measurement, contributing factors and consequences," Computers in Human Behavior, vol. 20, pp. 403-422, 2004.

31. H. Zhao, Z. Cai, H. E. Shan, B. University, and L. T. University, "The Relationship between Online Merchandise Displaying,Online Interaction and Impulsive Buying Based on Virtual Tactility," Chinese Journal of Management, 2014.

32. J. Peck and J. W. Johnson, "Autotelic need for touch, haptics, and persuasion: The role of involvement †," Psychology & Marketing, vol. 28, pp. 222–239, 2011.

33. R. Dhar and S. M. Nowlis, "The effect of time pressure on consumer choice deferral," Journal of Consumer Research, vol. 25, pp. 369-384, 1999.

34. K. Sepp, "Shoppers' motivations at the airport; The impact of impulse buying tendency and time-pressure," 2014.

35. T. Rudolph, J. C. Bauer, and W. J. Steiner, "Price-promotion-framing—a state-of-the-art review," Zeitschrift Für Betriebswirtschaft, vol. 80, pp. 285-327, 2010.

36. J. Y. Moon and K. H. Lee, "Influence of time pressure on the purchase decision making process in apparel shopping," Research Journal of the Costume Culture, 2013.

MATEC Web of Conferences 139, 00130 (2017) DOI: 10.1051/matecconf/201713900130
ICMITE 2017

Consistency Analysis of Port Industry Development and Chinese Economic Development

Jianping Sun,Zhenfu Li,Yunxiao Guan,Tongchao Liu

College of shipping economics and management,Dalian Maritime University, Dalian 116026, China

Abstract. As a function in organization and management, coordination is used to adjust the interrelation of the individuals in the organization in order to achieve the ultimate goal. Coordination can make individuals groups in a whole, reduce the contradiction and conflict between each other and promote the realization of the overall goal. The research on the coordination of relationship between the port industry of China and the growth of economy plays an important guiding role for the port industry and healthy development of national economy. According to the correlation model between port industry and economic growth, the correlation factor between port industry and economic growth is extracted, and the coordination indicator system of the relationship between the port industry and economic growth is established. Then, this article makes correlation analysis of the standardization of indicators, removes the bigger indicators of correlation coefficient and eliminates the impact of the duplication of information between indicators in order to simplify the indicator evaluation system. The actual results by data calculation show that there were six obvious changes in the coordination degree of the relationship between the port industry and economic growth in China during the period from 1990 to 2014.

1 Introduction

Coordination in the organization and management is a function to achieve the ultimate goal by regulating the relationship between the individual organizations, it can make these individuals in a whole, to reduce the contradiction between each other, to promote the realization of the overall goal [1]. The study of the coordination relationship between port industry and economic growth plays an important role in guiding the healthy development of port industry and national economy. In order to effectively measure the relationship between the two, this paper introduces the coordination index to assess the relationship between port industry and economic growth in China. In view of the lack of research on the relationship between port industry and economic growth, and taking into account the important role of the coordination relationship between port industry and economic growth on the development of national economy, this paper focuses on the coordination degree of the relationship between the two industries. Through the investigation of the correlation pattern between the two, the coordination degree evaluation model is established, and the coordination level of the two is quantitatively analyzed to explore the coordination degree between the port industry development and economic growth in China, and then guide the healthy development of China's port industry.

2 A Study on the Relevance Model of Port Industry and Economic Growth

As a demand industry which is derived from the process of national economic development, the port industry plays an important role in the development of national economy. It combines the production, distribution, exchange and consumption of domestic and foreign trade and economic activities organically. It plays an important pillar role in the development of the national economy. At the same time, shipping as a carrier of import and export trade, for the import and export trade in the world to provide the possibility of the port as an important interface to maritime transport development has a great role in promoting. Therefore, it is clear that the correlation between port industry and economic growth contributes to the further innovation of the port industry and can also serve as an important basis for the development of national development strategies.

2.1 Contribution of Port Industry to Economic Growth

The first one is the impact of a region's economy. Combined with the development of domestic and international port industry, can be found in the development of the port industry has a strong effect of aggregation, according to Hong Kong City, Hong Kong and cities linked to the economic development model more common. Port as the core of the development of the city, because it has close to the transport node, low transport costs, the advantages of a single volume, often become the best area to attract the relevant industries gathered to form a port industrial clusters, these industrial clusters Attracting talent, technology and other elements of a large number of inflows, the formation of a strong competitive industrial clusters, and thus promote the

development of the city, its direct performance is the region's GDP growth. And as a comprehensive reflection of the regional economic development indicators, GDP can clearly show the role of port industry in promoting economic growth.
The second one is the promotion of labor and employment role [2]. Expressed ad the performance of the port industry in promoting the employment, each additional unit of output value can directly use or mobilize the number of labor personnel situation in relevant industry. When the port industry participates in economic activities, it has a great correlation with other industries. The port industry, as a derivative industry, provides services such as loading and nloading and transit for the economic activities of other sectors. At the same time, the port industry needs to develop manufacturing, trade and other industries to provide services and products, and in the process of labor and employment generated by the relevant areas to meet the needs of the employment of workers in the region to promote the employment of workers in order to create more social production value and promote economic growth.

2.2 The Effect of Economic Growth on the Development of Port Industry

One is to provide direct financial support for the development and construction of the port industry. In the course of economic development in a country or region, the accumulation of capital can provide financial support for the development of the national economic sector. As the economy continues to grow, the region's capital stock will increase, in order to promote the port industry and related industries to upgrade the allocation of resources for the port industry to provide a strong scale of protection. When the infrastructure construction of the port industry reaches a certain level, it can speed up the development of the port industry and enhance the service level of the port industry, so as to better serve the economic construction. Similarly, when the development of the financial system of an area's port industry is relatively complete, it can attract relevant industries such as industry, trade and logistics, and bring together the resources, talents, technology and other resources in the region to form a whole industrial chain, so as to enhance the quality of the development of the port industry in the region, the region's port industry to increase the core competitiveness.
The other one is to provide indirect trade logistics needs for the development of the port industry. The purpose of the port industry is to meet the logistics needs, so as to promote economic and social development. Economic growth, as the driving force behind the demand for logistics, can provide constant demand for the development of the port industry. When a region's economy continues to grow, it will stimulate the continuous development of various industrial sectors, resulting in a huge trade logistics needs, and shipping with large volume, low cost, low energy consumption advantages, in terms of foreign trade, Country or region of the main transport tasks. Port as a hub for maritime transport, in the realization of maritime missions play an important role, therefore, by the economic growth caused by the demand for trade logistics will be completed by means of the port industry, which indirectly contributed to the development of the port industry.

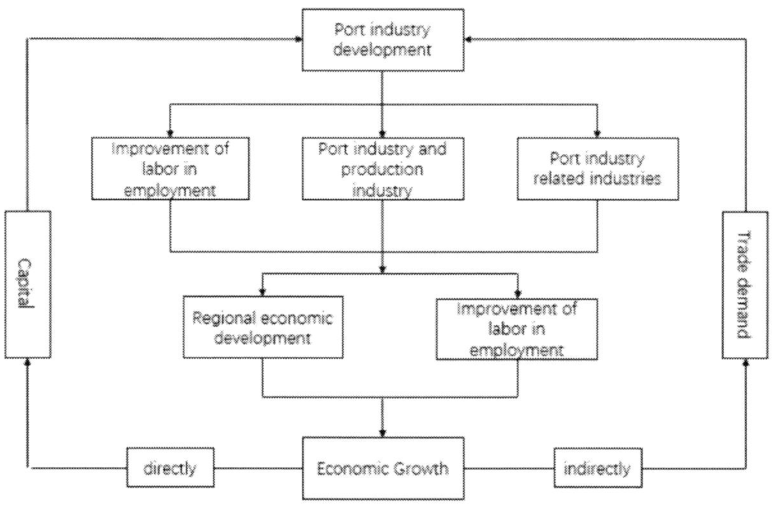

Fig 1. The correlation model between port industry and economic growth

3 Coordination Model of Port Industry and Economic Growth

The purpose of coordination is "harmony and consistency, with appropriate" [3]. From the system point of view, coordination refers to the system or between the various levels of the system, the various parameters of the system between the harmony, with the appropriate, structure and function of the relationship between the integration of coordination through the system can reduce the negative Effect, and improve the overall output function and synergistic effect of the system. The coordination degree of a system can be expressed as the degree of harmony between the elements of the system or the system, which can reflect the trend and degree of the system from disorderly to orderly development, and can effectively measure the internal factors of the system or system The level of coordination between the level. Thus, coordination is a measure of the coordination of the system.

Port industry and economic growth belong to two systems. The development of port industry is affected by factors such as economic trade, shipbuilding industry and government's legal policy. Similarly, economic growth will be affected by the accumulation of capital, labor input, technological progress and other factors The coordination between the port industry and economic growth is to deal with the interrelationship between these two parameters in order to improve the overall output function of the two systems and to promote the continuous development of the world economy. And coordination is used to measure the relationship between the two indicators, is a quantitative measure of this relationship.

3.1 Index System

In order to be able to measure the degree of coordination between the two industries, considering to reflect, monitor, compare, evaluate and predict the principle of the analysis of the object, the paper analyzes the relationship between the port industry and the economic growth, Port industry development and economic growth indicators to select the index system, in order to reflect the development of China's port industry and economic growth of the respective characteristics of the characteristics and factors. Due to the large number of factors influencing the port industry and economic growth, and there is a great correlation between the different indicators, through the analysis of the index system, the indicators that have significant influence on the evaluation results are selected as the analysis of the port industry and economic growth Indicators of covariance measurement.

To select indicators [4] influencing the development of the port industry, respectively, from the port-related industries development level, maritime fleet, economic and trade, talent and technology, ports, government investment in six aspects to measure. As shown in Table 1.

As for indicators [5] selected affecting the economic growth. Specifically, from the following seven aspects such as the accumulation of capital, resource consumption, economic development system, labor costs, science and technology education level, urbanization rate and degree of industrialization to make judgments.

3.2 Coordination Model and Type

The relationship between port industry and economic growth is a dynamic process. There may be a big gap between the two stages. However, there is a reasonable interval between the port industry and the economic growth. As long as the coordination of the two in the threshold range, not enough to affect the overall relationship between the two, this development model can be recognized, of course, the overall development of the two to maintain the optimal state is the best. For this idea, we define the formula for the degree of coordination:

$$C_{XY} = 1 - \delta/\mu \quad (1)$$

$$\mu = \frac{X+Y}{2} \quad (2)$$

$$\delta = \sqrt{\frac{(X-\mu)^2+(Y-\mu)^2}{2}} \quad (3)$$

In the above formula, X is the comprehensive index of the development of port industry in China, Y is the comprehensive index of China's economic growth, C_{XY} is the coordination degree of port industry and economic growth, $C_{XY} \in [0,\ 1]$. C_{XY} is determined by the variables X and Y, and when X and Y are both positive and the values are equal, C_{XY} takes the maximum value and the value is 1. According to the value of X and Y, the coordination degree of the two will change, and the degree of coordination will be divided into the following categories according to the value of C_{XY}.

3.3 Calculation Method

3.3.1 Standardization of data

Due to the existence of different dimensions of the original data for the different indicators in the system, and the size of the order of magnitude is not the same. Data for different indicators could not be compared and calculated directly. In order to eliminate the impact of these factors, evaluation index of different dimensions need to be transformed into dimensionless standardized indicators, through an appropriate way. That is, the standardization of indicators. Considering that the covariance matrix of each index data processed by the mean method can reflect the difference in the degree of variation of each index in the original data and the information of the degree of mutual influence of each index, this paper adopts the mean method to carry out the index standardized processing.

Let $\hat{x}_{ij}$ be the normalized value of the i-th-evaluation index at the j-th time point, $\hat{x}_{ij}$ is the original value of the i-th evaluation index at the jth time point, and m is the total length of the time series. Then

Standardization formula for port industry development indicators:

$$\hat{x}_{ij} = \frac{x_{ij}}{\bar{x}_i} \quad (4)$$

$$\bar{x}_i = \frac{1}{m}\sum_{j=1}^{m} x_{ij} \quad (5)$$

Among them, $\bar{x}_i$ is the average of the original indicators of the development of the port industry;

Standardization formula for economic growth indicators:

$$\hat{y}_{ij} = \frac{y_{ij}}{\bar{y}_i} \quad (6)$$

$$\bar{y}_i = \frac{1}{m}\sum_{j=1}^{m} y_{ij} \quad (7)$$

Among them, $\bar{y}_i$ is the average of the original indicators of economic growth;

Table 1. The indicator system of the effects in the port industry's development

Target layer	Criteria layer	Index layer
The indicator system that affects the development of port industry	Port - related industry development level	Shipbuilding industry development level
		Maritime trade service industry development level
		Average return on capital of shipping enterprises
		Enterprise fleet market share
	Sea fleet	China's maritime fleet capacity scale (100t dwt)
		The proportion of domestic imports and exports of goods transported by their own ships
		Fleet structure
		Fleet age structure
		Route density
	Economic and Trade	China's import and export trade volume (ten thousand US dollars)
		Sea freight
		The degree of development of the financial industry
	Talent and Technology	Sailing senior personnel training
		Training of professional personnel in shipping enterprises
		Maritime information service platform
		Maritime research investment
	Port	The natural conditions of the port
		Port throughput Port infrastructure improvement
		Port logistics service level
	Government investment	The state of the domestic shipping industry policy force
		Government 's financial support for the development of maritime industry

3.3.2 Relevance analysis for indexing

The basic principle of correlation analysis is to calculate the correlation coefficient between the two evaluation indexes, and to remove the set of indexes with large correlation coefficient, so as to eliminate the influence of information repetition and simplify the index evaluation system [6].

Let r_{ij} be the correlation coefficient between the i-th index and the j-th index, Z_{ki} denote the value of the k-th time period of the i-th index, $\bar{Z}_i$ denote the expected value of the time series of the i-th index, and n be the length of the time series.

According to the calculation of the correlation coefficient, then

$$r_{ij} = \frac{\sum_{k=1}^{n}(z_{ki}-\bar{z}_i)(z_{kj}-\bar{z}_j)}{\sqrt{\sum_{k=1}^{n}(z_{ki}-\bar{z}_i)^2(z_{kj}-\bar{z}_j)^2}} \quad (8)$$

In order to be able to effectively select a representative evaluation index, we need to define the value of r_{ij}, specify the critical value M (0 <M <1), if $|r_{ij}|$ >M, you can delete one of the two evaluation Index, leave another, if $|r_{ij}|$ <M, that is to say there is little correlation between the two indicators ,and both are retained.

Through the relevance of the index screening, the impact of repeated information on the evaluation can be effectively removed, and the evaluation of the index system being simplified.

Table 2. The indicator system of the effects in the economic growth

Target layer	Criteria layer	Index layer
The indicator system that affects economic growth	Capital accumulation level	GDP
		Foreign investment level
		Foreign exchange reserves
	Economic development system	The proportion of public ownership economy
		Taxation accounts for the proportion of the national economy
	Labor input	Employment rate
		Per capita income of urban residents
		Per capita income of rural residents
	Urbanization level	Proportion of urban and rural population
		Urban land
		Urban population aggregation degree
	Level of science and technology education	Accept the number of higher education
		Per capita education level
		Education and Research Funds Proportion of National Economy
		Number of national patents
	Degree of industrialization	Three types of industrial structure than
		Industry coordination degree
		High - tech industrialization rate
	Resource utilization	Energy consumption elasticity coefficient
		Resource consumption
		Renewable energy utilization

3.3.3 Calculation of Comprehensive Evaluation Index Affecting Port Industry Development and Economic Growth

Because the factor influencing the development of the port industry and the economic growth is the comprehensive index composed of different indicators, it is a multi-index statistical analysis method. Therefore, this paper chooses the correlation-principal component analysis [7] to measure the indicators. The essence of the principal component analysis is the linear combination of the observed indicators. The calculation model of principal component analysis is:

$$x_i = \alpha_{i1}\hat{x}_{i1} + \alpha_{i2}\hat{x}_{i2} + \cdots + \alpha_{in}\hat{x}_{in} \quad (9)$$

$$X = \sum \omega_i\hat{x}_i$$

$$y_j = \beta_{j1}\hat{y}_{j1} + \beta_{j2}\hat{y}_{j2} + \cdots + \beta_{jn}\hat{y}_{jn} \quad (10)$$

$$Y = \sum v_j\hat{y}_j \quad (11)$$

In the formula, x_i and y_j are the sovereign factors after the standardization of port industry and economic growth, and ω_i and v_j are the contribution rates of the sovereign factor eigenvalues of port industry and economic growth. $\hat{x}_{ij}$ is the j-th index of the i-th sovereign factor which influence the port industry. $\hat{y}_{ji}$ is the i-th index of the j-th sovereign factor which influence the economic growth. X and Y are the comprehensive indexes of port industry development and economic growth respectively, n is the number of indicators.

The relevant steps [7] of the principal component analysis is, take the calculation method of x_i as an example:

① Standardize the index data processing, unified dimension;

② Calculate the normalization of the index data after the correlation coefficient, obtain the results of correlation coefficient matrix $R_{n\times n}$;

③ Calculate the eigenvalues $\lambda_i(i = 1,2 \ldots n)$ of the matrix $R_{n\times n}$, λ_i represents the total variance of the data of the original index explained by the i-th sovereign factor x_i;

④ Determine the value of k, make the utilization rate

of information reaching 85% or more, k indicates the number of indicators that affect the sovereign factor after screening;

$$\frac{\sum_{i=1}^{k}\lambda_i}{\sum_{i=1}^{n}\lambda_i} \geq 0.85 \quad (12)$$

⑤ The contribution rate ω_i of the eigenvalue of the sovereign factor x_i is:

$$\omega_i = \frac{\lambda_i}{\sum_{i=2}^{n}\lambda_i} \quad (13)$$

⑥ The weight of the index α_{ij} is determined by the expert consultation method [8].

Through the correlation analysis and principal component analysis, we can eliminate some non-quantifiable and high correlation index with the known indicators, select the indicators that have higher impact on the evaluation object as the calculation factor of the coordination degree, so as to analyze The relationship between port industry development and economic growth.

4 The Empirical Analysis and Evaluation of Coordination Degree between Port Industry and Economic Growth

4.1 Selection of Samples and Data Sources

In view of the index system which affects China's economic growth and the development of China's port industry, this paper chooses the data from 1990 to 2014 as the time series to analyze the relationship between China's economic growth and port industry. The data for the indicators used for coordination are from the website of the National Bureau of Statistics of the People's Republic of China, the website of the Ministry of Transport, the China Yearbook of Transportation, and the website of the China Shipbuilding Industry Association.

Table 3. The classification of coordination degree between the port industry and economic growth

Coordination type	C_{XY}	Port industry and economic growth of the comprehensive indicators X, Y value	Coordination characteristics
Class I	$0.8 \leq C_{XY} \leq 1$	$X \geq 0, Y \geq 0$	Very coordinated, the port industry and economic growth are optimized development, the system was the best state
Class II	$0.6 \leq C_{XY} < 0.8$	$X \geq 0, Y \geq 0$	Coordination, the development of port industry and economic growth are relatively high level, the overall system to the good direction
Class III	$0.5 \leq C_{XY} < 0.6$	$X \geq 0, Y \geq 0$	Relatively coordinated, the port industry development and economic growth in one of the comprehensive level higher than the other, but the system as a whole is still relatively coordinated
Class IV	$0.4 \leq C_{XY} < 0.5$	$X \geq 0, Y \geq 0$	Not coordinated, economic growth and port industry, the level of imbalance, the development of the system instability
Class V	$0.2 \leq C_{XY} < 0.4$	$X \geq 0, Y \geq 0$	Uncoordinated, the development of the port industry and the level of economic growth are relatively low, and the development of the two is very uneven, the overall system of instability
Class VI	$0 \leq C_{XY} < 0.2$	$X \geq 0, Y \geq 0$	Very uncoordinated, port industry and economic growth in the development of a comprehensive level are in a low state, the overall system was in recession

Table 4. The coordination degree of the port industry's development and economy growth in China (1990-2014)

Time	Port Industry Development Index	Economic growth index	Coordination degree	Coordination type
1990	0.2294	0.6786	0.3004	Class V
1991	0.2294	0.6863	0.2945	Class V
1992	0.2249	0.6489	0.3138	Class V
1993	0.2306	0.6739	0.3069	Class V
1994	0.2563	0.6745	0.3646	Class V
1995	0.2618	0.7148	0.3439	Class V
1996	0.2680	0.7246	0.3495	Class V
1997	0.2900	0.7336	0.3871	Class V
1998	0.3021	0.7273	0.4157	ClassIV
1999	0.3426	0.7201	0.4976	ClassIV
2000	0.4071	0.8109	0.5312	ClassIII
2001	0.4548	0.8750	0.5531	ClassIII
2002	0.5195	0.9557	0.5818	ClassIII
2003	0.6198	1.0871	0.6128	Class II
2004	0.7503	1.1281	0.7156	Class II
2005	0.8860	1.0872	0.8558	Class I
2006	1.0235	1.0595	0.9755	Class I
2007	1.1761	1.1001	0.9528	Class I
2008	1.3481	1.1056	0.8602	Class I
2009	1.4005	1.1813	0.8799	Class I
2010	1.7948	1.2881	0.7676	Class II
2011	2.0476	1.3941	0.7315	Class II
2012	1.9988	1.4209	0.7610	Class II
2013	1.9695	1.4652	0.7923	Class II
2014	1.9685	1.4838	0.8014	Class I

Source: 《China Statistical Yearbook》, 《China Transportation Yearbook》 and China Shipbuilding Industry Association website

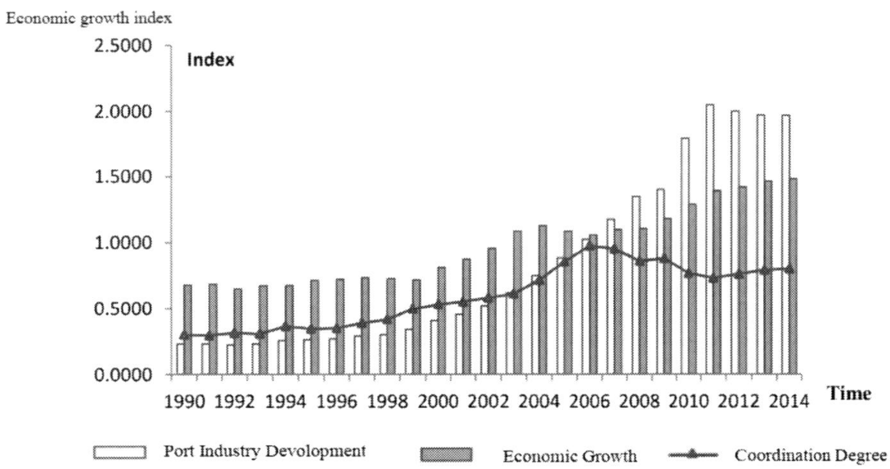

Fig 2. The changing curve of coordination degree between the port industry's development and economic growth in China

4.2 Analysis of Results

After extracting the evaluation index by the correlation-principal component analysis, the index weight determined by the expert consultation method is brought into the calculation model of the coordination degree, and the coordination table between China's port industry and economic growth can be obtained. As shown in Table 4.

It can be seen from Table 4 and Figure 2, there have been six significant changes of the coordination of the relationship between China's port industry and economic growth. The first stage was in the early 1990s (1990-1991). Although China's reform and opening-up policy was effective, economic growth reached a certain scale, but because of the poor economic base in China, the structure of economic development is irrational and the economic growth is instability, resulting in a series of problems in China's economic growth, economic development, the level of comprehensive development in a low state. At the same time, the rapid development of economy and foreign trade brought by reform and opening-up policy has stimulated the huge demand for import and export goods for maritime transport. However, due to the limited ship capacity and relatively backward construction of port infrastructure, serious problem such as the pressure port phenomenon has been appeared in the development of the port industry.

The second stage is in 1992-1997, when China's port industry development and economic growth is in a state of incongruity. In 1992, Comrade Deng Xiaoping's "Southern Tour Speech" pointed out the direction of China's economic development. The upsurge of reform and opening up continued in our country. The foreign capital entered China's territory, and there was an investment boom in China. However, due to China's economic development system reform in the early stages, lacking effective policy guidance and macroeconomic regulation and control, which caused a lot of duplication of the project, a serious waste of resources, leading directly to China's economic growth reached a climax in 1992-1993, causing the phenomenon of overheating. Although economic growth maintained a high rate in this period, the economic development of the structural problems also highlighted. Therefore, the state in the 1993-1996 stage implemented the macro-control, which successfully controlled the momentum of economic growth and the level of economic growth has also been improved. Similarly, in this stage of development, the state has further expanded the scope of coastal openness, which greatly promoted the development of the coastal port industry. During that time, Ministry of Communications put fully effort to carry out the long-term planning of traffic infrastructure construction which named "three main one support" [9], made a series of planning and constructing achievements, not only in the supporting system, but the main water channel and the main port as well. At the same time, it brought the initial formation of the coal, oil, iron ore, grain-based bulk cargo terminals and container specialized terminals. However, due to the late development of China's port industry, infrastructure is still weak, so a short period of time the level of development will not be qualitative improvement, the overall level of development is still in a low state.

The third stage is in 1998-1999, when China's port industry and economic growth is in a state of less coordinated. During this period, China's economic growth rate is lower than the 9.5% growth rate since the reform and opening up. Economic growth failed to maintain the momentum of development, partly from the Southeast Asian financial turmoil in the second half of 1997. Although the financial turmoil did not have a direct impact on the Chinese mainland economy, coincided with Hong Kong and Macao return to the motherland, Hong Kong has been greatly affected as a financial center in Asia in the face of this financial turmoil. At the same time, Southeast Asia as an important area of China's foreign trade, China has also encountered some problems in foreign trade and foreign investment, which also led to some limits of the development of China's port industry at this stage.

The fourth stage is in 2000-2004, when China's port industry development and economic growth transformed from a relatively coordinated state to a coordinated state. From this stage we can see that China's port industry development system and economic growth are in the rising stage, and the development model is maturing, one of the important turning point is China's successful accession to the World Trade Organization in November 2001, which has produced a great impact on the Chinese economy growth. A significant change is that China's economic growth rate being substantial rebound. At the same time, with the formation of China's all-round, multi-level and wide-ranging pattern, China has made certain achievements in economy, politics, science and technology, culture and security. The relationship between China's economy and the world economy has been strengthened, and further reform of China 's economic system has also been accelerated. China has begun to play an increasingly important role in the world stage. However, in the early years of China's accession to the WTO, China was still in a passive and one-way state in relations with various WTO members [10], especially developed countries. Therefore, China has always been at the disadvantage in the foreign trade, which brought certain restrictions in implementing the strategy of "going out". As an important carrier of foreign trade, the port industry also ushered in a new era of development after China's accession to the WTO. As an important manufacturing country in the world, a large number of product that made in China began to enter international markets, which promote further innovation in China's port industry.

The fifth stage is from 2005 to 2009, during which China's economic development as a whole showed a continuous wave of heat, China played an increasingly important role in international economic affairs. As the world's largest developing country, China's contribution rate to the world economy during this period of time reached 15%, which has made great contributions to the development of the world economy. At the same time, China successfully held the 2008 Beijing Summer

Olympics during this period, which not only shown the great achievements since China's reform and opening up to the world, but also enhance China's international status. The development of China's port industry has experienced a climax. China has grown as the world's maritime power. As 《2009 China's Shipping Development Report》 shows, which issued by the Ministry of Transport, the national port to complete the cargo throughput of 7.657 billion tons. In the top 10 ports of global cargo throughput, China has occupied eight seats and has become the world's largest country of port throughput. Therefore, at this stage, China's economic growth and the development of the port industry are in a period of rapid development of the golden period, the coordinated development of the relationship between the two are in good condition.

The sixth stage is from 2010 to 2014, the development between port industry and economic growth is in a coordinated state. However, suffering the 2008-2009 US "subprime mortgage crisis", the world economy declined at this stage as a whole, which directly reflected by the substantial reduction in global trade demand. World's economic development is in trouble. Similarly, China's economic development in the face of this crisis has not escaped the tragedy of decline, China's economic growth rate has declined for five consecutive years. The reason for this can be divided into two kinds. One reason is that the volume of China's economic increased gradually, which reduced the pace of development. The other reason is that China's foreign trade development has been facing the challenge of global economic downturn. For this situation, China's government has introduced a series of policies to slow down the impact of the financial crisis on China, optimize the industrial structure, promote the supply side of the reform, China's economy also grew in a high speed in face of this crisis. As for port industry, after a boom of economic development in the last stage, China's port industry booming, has made unprecedented achievements, the comprehensive level of China's port industry development has reached a new height. But the resulting shortcomings of the airport empty ship is also fully exposed at this stage, the development of China's port industry experienced a cliff-style decline. However, the crisis also contributed to China's port industry, making it continue to develop in a more reasonable direction.

5 Conclusion

Based on the systematic review of the co-ordination model, this paper analyzes the various factors related to the growth of the port industry and economic and economic factors. By using the correlation analysis and the principal component analysis, the indicators are screened and the indicators with large correlation coefficients are removed, the impact of information duplication was also reduced, which makes the indicators strong representative, independent of each other. Through the quantitative measurement of the coordination degree, we can accurately identify constraints restricting the development of the port industry, in order to provide an important basis for formulating the strategy of the port industry developing in the future, and then guide the healthy development of the port industry.

The actual data calculation results show that there had been six significant changes in the relationship between the port industry and economic growth in China during the period from 1990 to 2014: ① in the early 1990s (1990-1991), the rapid development of economic and foreign trade caused by the reform and opening up policy brought about a series of pressure ship pressure port phenomenon; ② in 1992-1997, China's port industry development and economic growth were in a state of incongruity; ③ in 1998-1999, China's port industry and economic growth was in a state of less coordinated; ④ in 2000-2004, China's port industry development and economic growth were transforming from a coordinated state to a coordinated state; ⑤ in 2005-2009, the growth of China's economy and the development of the port industry were in a period of rapid development, and the coordinated development of the two was in a good state; ⑥ During the period from 2010 to 2014, the development of the port industry and economic growth is in a state of coordination. These changes are closely related to the background and policy environment of the time, in which the coordination degree of the two reached a high level in 2005-2009.

This work was supported by the major projects of the National Social Science Fund (13 & ZD170), the New Century Excellent Talents Support Program of the Ministry of Education (NCET-13-0701) and the "Thirteen Five" Research Project of Dalian Maritime University (3132016361)

References

1. Zhangjingjing,Chenzhengjiang,Yangdegang. Construction and Application of Evaluation Model of Urban and Rural Coordination Degree [J]. Journal of Arid Land Resources and Environment,2007,21 (2) :85-94
2. Liming. Water transport on the contribution of China's national economy [D]. Dalian Maritime University master's degree thesis,2004
3. Zhangxiaodong,Chitianhe. Analysis on the Coordination of Regional Economy and Environment in China in the 1990s [J]. Geography Research,2001,20 (4) :506-515
4. Sunyabo,Fanhouming,Liuyiying,Lizhenfu. Construction of Evaluation Index System of Shipping Power Based on Reliability and Validity Analysis [J]. Journal of Shanghai Maritime University, 2014,35(4):26-31
5. Wangjunlei,Wangzhaokai,Yangxiaoming. Economic Growth Quality Evaluation Model Based on Analytic Hierarchy Process [J]. Statistics and Decision,2007
6. Chiguoqin,Caotingting,Zhangkun. Based on the correlation-principal component analysis of human

comprehensive development evaluation index system construction [J].Systems Engineering Theory & Practice,2012,32（1）:111-119

7. Lihongxi. Research on Port Logistics Evaluation Based on Correlation - Principal Component Analysis [D]. Dalian University of Technology Master of Engineering degree thesis,2013
8. Hanxiaohai,Zhangzhaihui,Sunfujun,Wangshaohua. Method of Determining Weight of Index Based on Principal Component Analysis [J]. Journal of Sichuan Ordnance,2012,33(10):124-126
9. Tongxinchun. On the Development and Changes of the Maritime Industry in New China [J]. Research on China 's Economic History,2012,2:127-137
10. Chentaifeng,Huangang. The Influence of China 's Entry into WTO on China：Post-evaluation based on international returns [J]. National situation report,2009,12:1-34
11. Zhangxiaodong,Chitianhe. Analysis on the Coordination of Regional Economy and Environment in China in the 1990s [J]. Geography Research,2001,20（4）:506-515
12. Liuqiang，Chenyixian. A Summary of China 's Economic Growth since the 1990s [J]. Journal of Zhengzhou Institute of Aeronautical Industry Management, 2004,22（1）:8-11
13. Panwenyan，Wangzongjun. Research on Evaluation Index System of Low Carbon Competitiveness Based on Coordination Model [J]. Information magazine，2012,31（10）:76-82
14. Liuyaobin ， Songxuefeng. Analysis on the Coordination Degree between China 's Industrialization and Urbanization since the Reform and Opening - up [J]. Science and Technology Herald，2005，23（2）：48-51
15. Zhangxiaodong，Zhudehai. Forecast and Analysis of China 's Regional Economy and Environment Coordination [J]. Resource science,2003, 25(2):1-6
16. Ministry of Communications,Ministry of Commerce. Provisions on the Administration of Foreign Investment in International Maritime Industry.〔2004〕Order 1
17. B Büttner. Effectiveness versus Efficiency: Growth Accelerating Policies in a Model of Growth without Scale Effect，German Economic Review，2006, 7 (3), 297-316.
18. GLASS,A. and X.WU. "Intellectual Property Rights and Quality Improvement". Journal of Development Economics, 2007, 82: 393-415
19. Maddison A. Chinese economic performance in the long run[M].Organization for Economic,1998
20. Jones, C.R&D-based Models of Economic Growth[J].Journal of Political Economy,1995,6（3）:286-297

MATEC Web of Conferences 139, 00131 (2017)
ICMITE 2017 DOI: 10.1051/matecconf/201713900131

Application of 802.11ah wireless technology in Global Energy Interconnection

Wang Liang[1,*], *Zheng Zhe*[2]

[1] Global Energy Interconnection Research Institute, Beijing China
[2] State Grid Key Laboratory of Information & Network Security, Beijing China

Abstract. The Global Energy Interconnection (GEI) will be the foundation of future clean energy, and it has drawn great challenges for communication technology. GEI has significant need for low power-consuming, long-range, medium-rate sensing and measuring method which can't be achieved by current technology. Hence, in this article, applying 802.11ah in GEI sensing and measuring is proposed; key features in 802.11ah are introduced; the DCF mode base-band circuit is built, and the correctness is verified by experiment.

1 Introduction

The Global Energy Interconnection(GEI) is a smart grid characterized by the global interconnection and ultra-high voltage network. It delivers clean energy to the world. To form a new pattern of energy development in the power grid connectivity and global power optimization in various continents; To establish new energy production and consumption patterns featuring clean energy and high electrification[1].

2 The needs and challenges of global energy interconnection in communication technology

The concept of GEI brings new demands and challenges to communication technology.

The current challenges are: transnational grid interconnection and scheduling, remote energy base management, global clean energy trading.

We can see from the information demand of the GEI that the global demand side, dealing with cross-border transmission scheduling and remote energy base management requirements in operation, transaction and monitoring[2].

Regional demand side, focus on the domestic level, and the distribution network to clean energy allotment, energy trading system, giving full play to the advantages of wide coverage of information, interactive and convenient, to realize regional services.

The GEI will form a unified information architecture that integrates distribution and concentration, tight and loose logic to support the global energy interconnection. The overall architecture of GEI information is shown in figure 1.

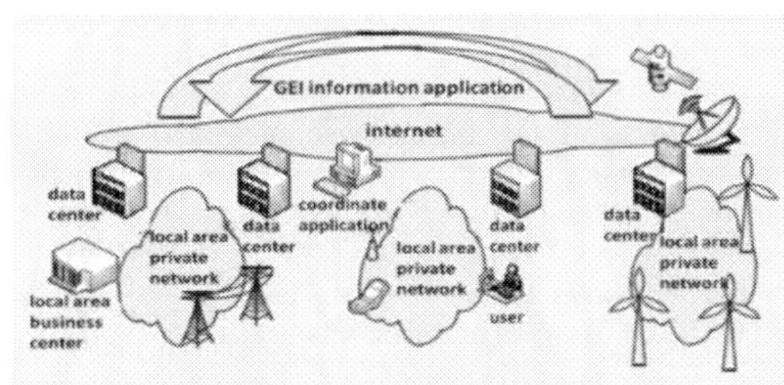

Figure 1.GEI information architecture

Internet of Things (IOT) uniquely identifies equipments, connect any articles according to certain protocol and through the network, makes them exchange information and communication, realizes intelligent positioning, identifing, interaction. In the GEI, the research on the IOT is still at progress, the standard is not accomplished, some key technologies are still under research.

3 The application of intelligent sensing technology in energy Interconnection

The communication between equipments is pervasive in the GEI. Measurement monitoring is the basis of GEI operation, which reflects the overall performance of it.

The IOT sensors used for power grid have been extensively studied, such as routing protocols that can help packets avoiding congested areas, and security issues. The sensor network is the basis of intelligent measurement and perception and will be widely used in the energy interconnection. The sensor network will evolve towards self-organization, high bandwidth utilization, self-energized and low power consumption.

The main applications of intelligent sensor technology are:

(1)intelligent power Transmission and substation

* Corresponding author: zhiyanyuanwl@126.com

The new generation of smart substations will be widely used in the energy Interconnection to improve the level and quality of information interaction between substations. This requires the online monitoring and early warning of power transmission and transformation equipment, and through data analysis and big data process, to realize the interaction between devices, human and data, to obtain accurate perception of equipment's state, to provide support for intelligent maintenance.

(2)Intelligent distribution

The basis of intelligent decision-making and optimization of energy Interconnection is the acquisition and processing of power distribution information. This application is oriented towards improving the success rate and calculation efficiency of the distribution data acquisition. Comprehensive utilization of information acquired from all process of energy interconnection, realize the power grid and the user's real-time interaction, so as to improve the user's satisfaction, promote the lean marketing management level of power grid.

(3)Big data analysis and prediction

The application of big data analysis and prediction technology in power is mainly focused on the analysis of power load main components and the main factors that affect the power load, so as to facilitate the modeling of load prediction.

We can also see that in order to adapt to the development of the GEI, development based on IOT sensor network information acquisition, data mining, safety protection is very necessary to improve the processing speed and accuracy of huge amounts of data, it electric power dispatching and trading globally. And all of this is based on a sensor network[3].

4 wifi and 802.11

A. WiFi introduction

WiFi uses 2.4 GHz and 5 GHz band radio frequency (RF) technology, replaces the wired LAN, provides all functions of the traditional wired LAN does. It is easy to install, expand and with high mobility, strong confidentiality[4].

The 802.11 series are the standard set by IEEE for WiFi. It involves the physical enhancement technology, QoS guarantees, safety control mechanism, networking mode, network management, co-existence with other protocols, all together to build a set of WLAN standard system.

However, the existing IEEE 802.11 series have some problems as follow:

- The existing 802.11 protocol does not cover the low power consuming considerations of each network node.
- 2.4GHz and 5GHz are the two primary bands used in WiFi, but those two bands are not suitable for long-range transmission. To meet the needs of IOT, IEEE has launched the sub-1GHz 802.11ah standard develop.

802.11ah can be the protocol standard for IOT. It uses OFDM technology in the physical layer, it has enhanced IEEE802.11 MAC layer technology to match its physical layer technology, it provides the corresponding mechanisms to support its coexistence with ZigBee or Bluetooth technology. Using the sub-1GHz band, it can be widely used in energy Interconnection, industrial automation, building automation, public environmental monitoring and other fields. The 802.11ah final standard is released in 2016. In 2018, chips and systems based on 802.11ah will be put into the market[5].

B. 802.11ah key technology

The advantages of 802.11ah are low power, long-range, massive nodes(up to 6000). This is achieved by the key technologies below:

1)Relay

IOT needs wide coverage, so 802.11ah uses relay technology. The relay is a function entity, includes relay-AP and relay-STA. On the bottom, relay connects with STA, plays as AP. On the up side, relay connects with AP, plays as a STA. In the Fig.2, relay can transfer data packets to make a wider coverage.

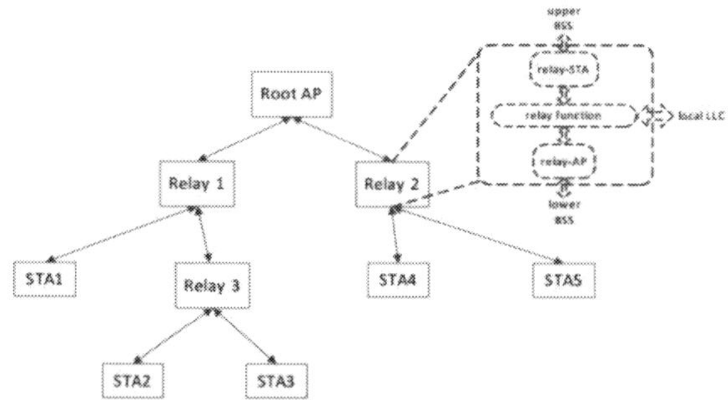

Figure 2. relay

2)Traffic indication map(TIM)

To reduce the power consuming, nodes will enter sleep mode. AP sends the TIM element in the beacon packet to inform the sleeping nodes whether there are buffered data for them to receive, as is illustrated in Fig.3. traditional TIM in WLAN only supports 2000 nodes, but IOT needs more, so 802.11ah uses hierarchical TIM. It slices TIM into many pages, and then many blocks for a page, and many sub-blocks for a block. AP sends these pages in the interval of beacons to support massive nodes.

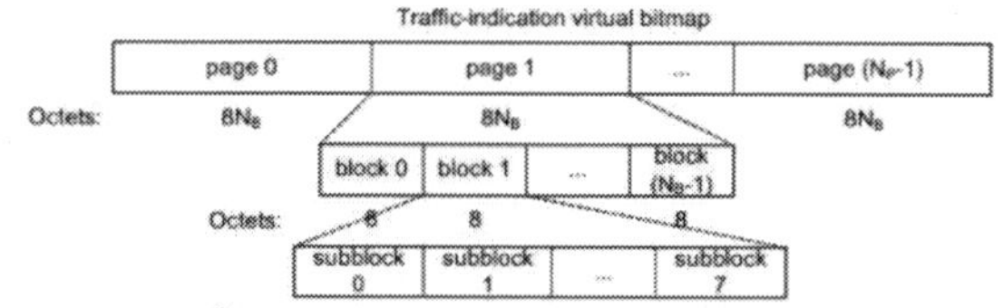

Figure 3. TIM

3)Target wake time(TIM)

In order to achieve low power consumption, node needs to shut down the Tx/Rx module as much as possible to enter a sleep mode, and wake up only when there is data to send or to receive. 802.11ah uses target wake time mechanism, allowing AP to manage the wake up time of subordinate nodes, make them wake up in different time, this reduces the competition, increases the efficiency of the system, eventually reduces the power consumption of the node.

There are two kinds of TWT: active TWT and passive TWT. In the active TWT, AP will tell the next TWT wake-up time to node in packet; And in the passive TWT, nodes need calculate the wakeup time by the TWT information obtained before.

4)Restricted access window(RAW)

To ensure low power consuming and support a large number of nodes, as well as increase the efficiency of the system at the same time, 802.11ah introduces RAW. RAW keeps the nodes into different groups, and make them transmit signal in specified time period. This makes sure each node have equal chance to transmit, and reduces conflicts, improves the system efficiency.

In the Access stage, the STA wakes up in the specified time, and began to listen to beacon frames, which will carry the RAW information, such as start time, time duration, the number of time slots, and which RAW will the node be assigned to. After receiving such information, STA will know the access time allotted by AP, and before this time the STA enters sleep mode. When the access time arrives, STA wakes up for channel access. As shown in Fig.4, the STA enters sleep mode in the first 4 slots, wakes up for data transmission in the fifth slot.

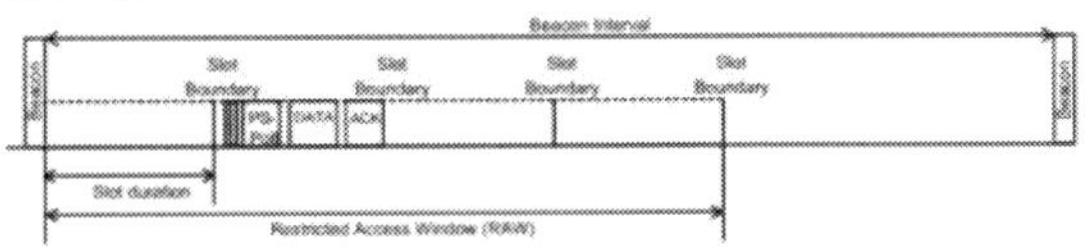

Figure 4. RAW

5 implementation of 802.11ah

A. architecture of the implementation

This article realizes the 802.11 baseband system, as shown in Fig.5. The main device is XILINX ZYNQ7000 FPGA. The parameters of this FPGA is below:

- 36kb dual-port RAM, 72bit wide maximum, can be configured as two 18kb RAM.
- 18x25 multiply operation, 48 bit adder/accumulator. 25 bit pre-adder.
- 16 transceivers maximum, speed up to 12.5Gb/s.
- 350k logic elements, 218600 look up tables, 437200 F-Fs, 19.1Mb RAM, 900 DSP blocks, supports AES and SHA 256.

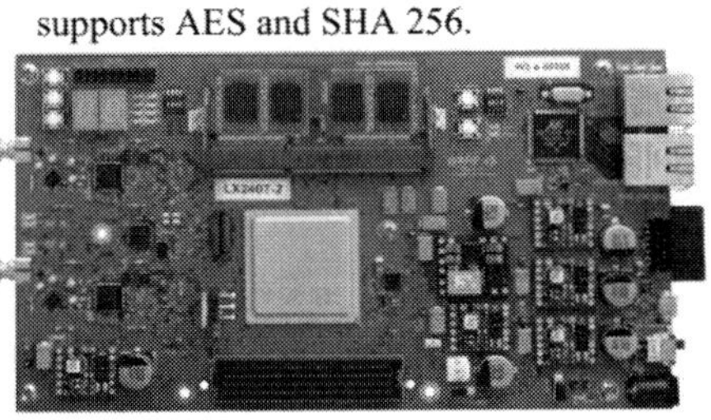
Figure 5 baseband circuit

The architecture of software and hardware is Fig.6 and Fig.7.

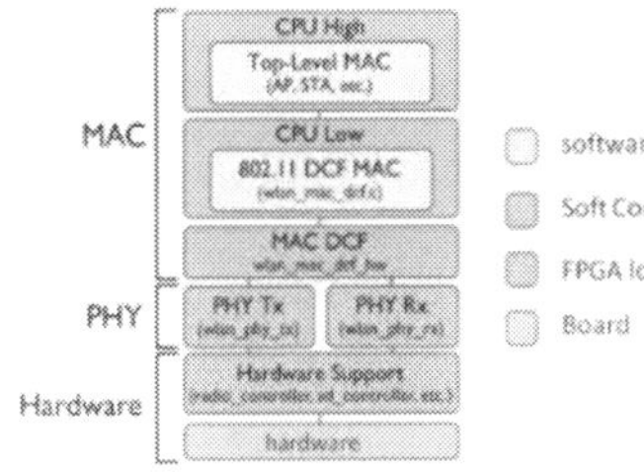

Figure 6. software

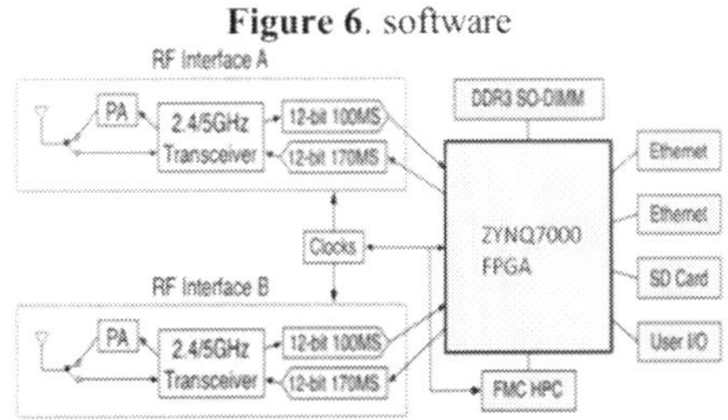

Figure 7. hardware

The physical layer is shown in Fig.8, there are OFDM Tx, OFDM Rx, and AGC in the physical layer.

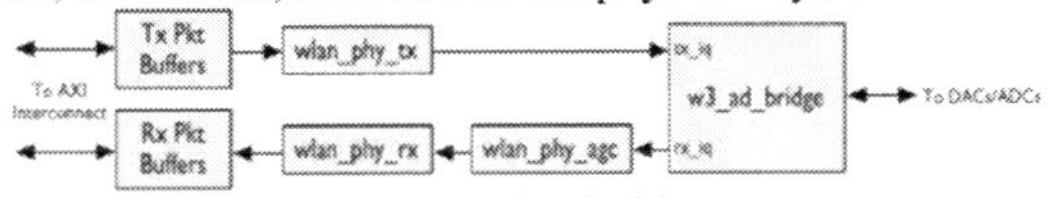

Figure 8. physical layer

The specifications of the PHY are:

- Clock frequency: 160MHz.
- Bandwidth : 10MHz,20MHz.
- OFDM format: 64 sub-carrier,16-sample cyclic prefix.
- Frame Format: As specified in 802.11ah, SIGNAL field as first OFDM symbol
- Data rate: Each data rate is realized for every MCS.

The architecture of the receiver and transmitter of FPGA core is illustrated in Fig.9 and Fig.10. The packet detection module implements two packet detection schemes: simple energy detection based on RSSI and auto-correlation of the I/Q samples searching for the preamble STS. The antenna selection is automatic. A dual-port circular sample buffer records all incoming samples. A complex channel coefficient is calculated for each non-zero subcarrier to estimate channels. The de-interleaved soft values are decoded using a standard Viterbi decoder.

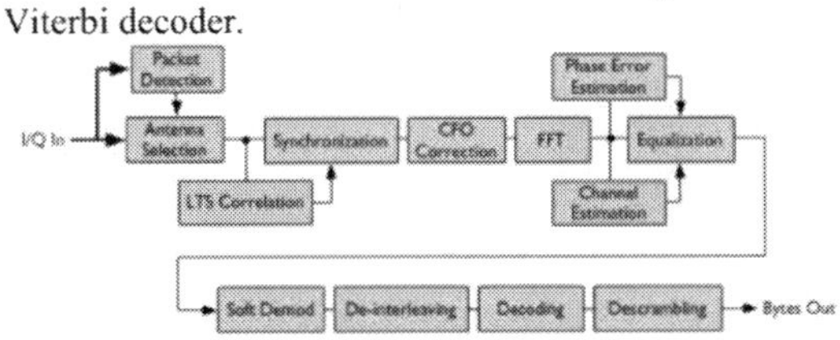

Figure 9. receiver

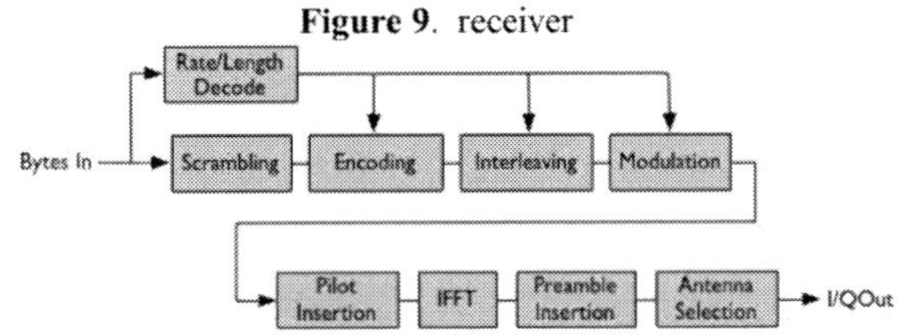

Figure 10. transmitter

 DOI: 10.1051/matecconf/201713900131

B. Experiment and conclusions

An experiment is carried for DCF mode bi-directional communication, as is shown in Fig.11. A 802.11ah network is formed with 3 nodes, 1 for AP and 2 for STA. The characteristics of this experiment is below:

- Packet Length: 1400 byte payloads (1428 byte MPDUs, with MAC header and FCS).
- PHY Rate: 18 Mbps (QPSK, code rate 3/4).
- Tx Power: -10 dBm.
- Trial Duration: 300 seconds.
- 2.4GHz channel 1.
- Physical Carrier Sensing Threshold: approximately -70 dBm, when enabled.

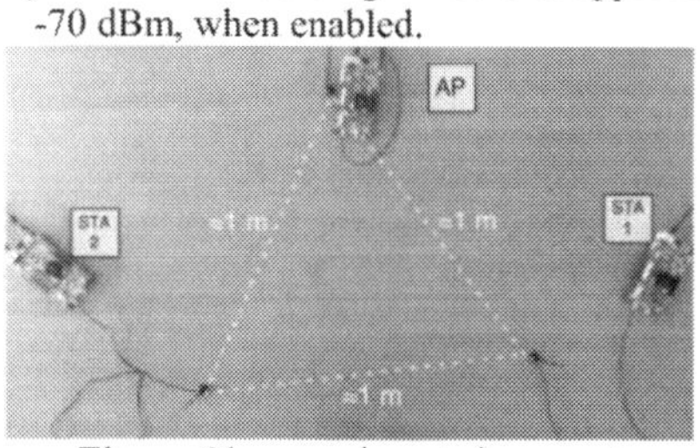

Figure 11. experiment placement

We use 4 traffic flows in our experiments, as is shown in Fig.12

Figure 12. traffic flow

The experiments framework include a flexible logging system that runs in real-time at every node. This system keeps a record of every Tx and Rx event at the node. For Tx events the log includes both the high-level MPDU Tx (as implemented in CPU High) and the low-level PHY Tx events for each MPDU. Each low-level Tx record includes the timestamp of the actual PHY transmission plus MAC parameters for the transmission (number of backoff slots, current contention window, re-transmission count, etc.). By retrieving the logs from every node following the experiment and analyzing them together, we can gain significant understanding of the behaviors of the nodes and how various parameters impact performance on short and long timescales.

Fig.13 and Fig.14 plots the received power of every packet for each node, grouped by traffic flow. These plots clearly demonstrate the low-mobility, mid-to-high SNR propagation environment for our experiments. And, as expected, the physical carrier sensing threshold has no substantial impact on Rx power.

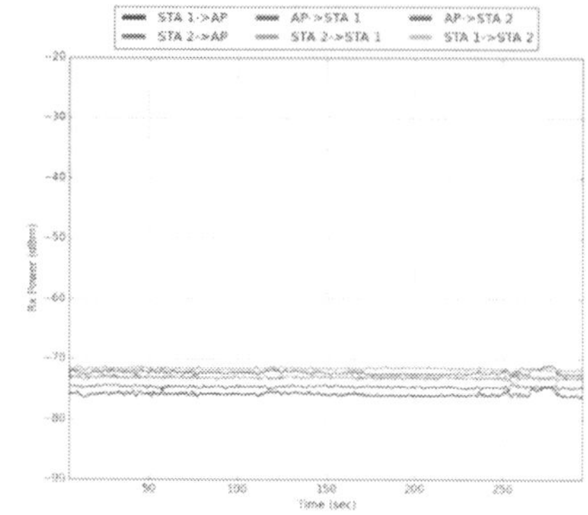

Figure 13. physical carrier sensing enabled

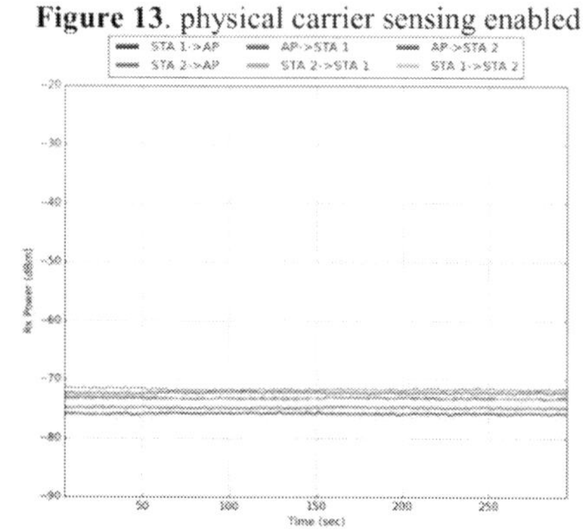

Figure 14. physical carrier sensing disabled

Fig.15 and Fig.16 show the throughput of each flow as a function of time. Each throughput curve is computed as a 1 second rolling window of bits received by each node for each flow. This 1 second window is rolled over the full 300 second log in steps of 1 msec.

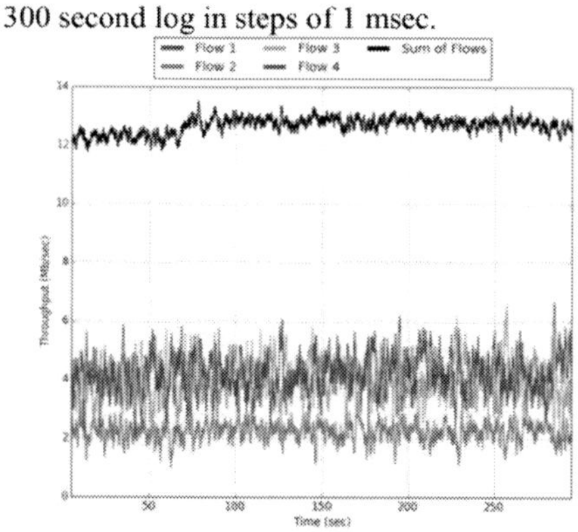

Figure 15. physical carrier sensing enabled

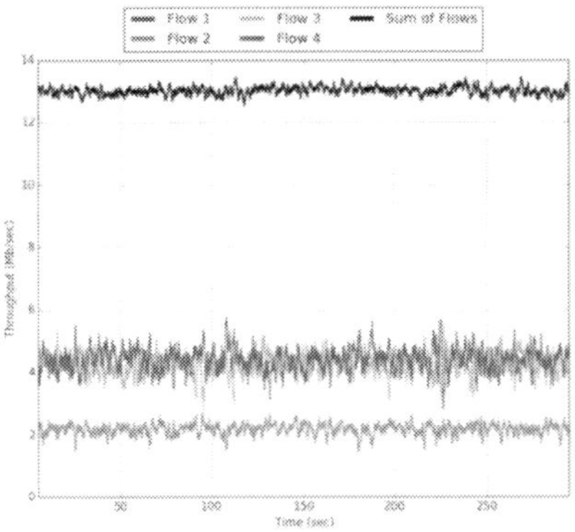

Figure 16. physical carrier sensing disabled

The 802.11 DCF contention window determines the distribution of available slot counts when selecting a random back-off period. The contention window grows with each failed attempt to transmit a packet. The contention window growth is exponential, starting at a width of 16 and doubling with each re-transmission, up to a maximum width of 1024. Ideally a node's contention window would never increase from 15. In reality packet collisions and losses require re-transmissions, leading to inevitable contention window increases. The contention window resets to its minimum following a successful transmission.

Fig.17 and Fig.18 show the actual contention window value of every node at the time of every PHY transmission. Note, the same rolling average technique used in the above throughput graphs is being employed here. The impact of physical carrier sensing on contention windows is dramatic. As explained above, more collisions lead to more transmission failures, leading to higher expected contention window values, resulting in longer back-off periods and lower throughput.

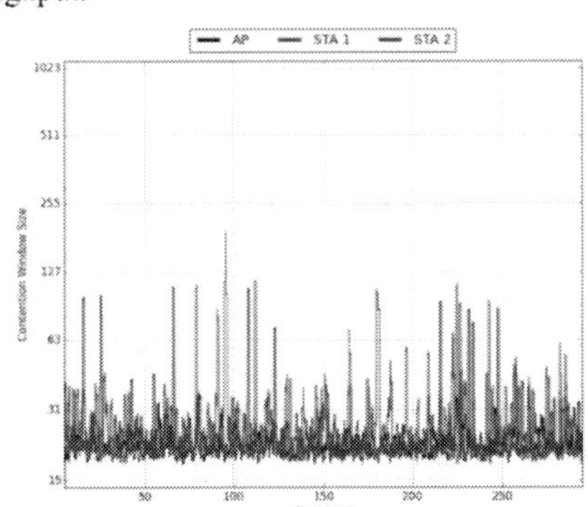

Figure 17. physical carrier sensing enabled

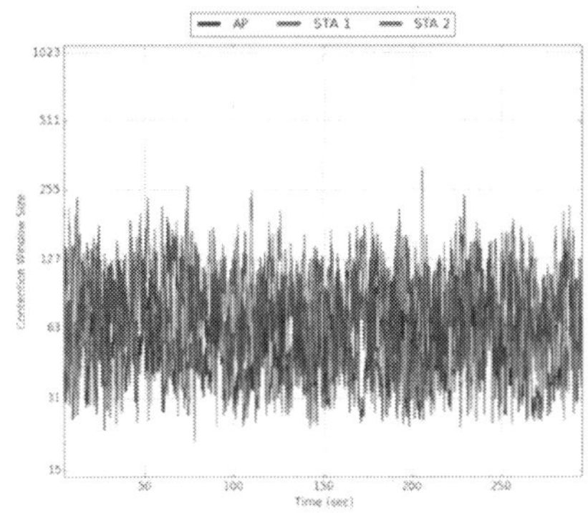

Figure 18. physical carrier sensing disabled

6 summary

As a new kind of WiFi technology, 802.11ah will be widely used in the field of GEI and IOT for data acquisition. This paper realizes the MAC protocol in 802.11ah using FPGA hardware platform, which is a bi-directional communication. The result is exactly the same as specified in the 802.11 ah standard.

References

1. Liu Zhen Ya, Global Energy Interconnection. Beijing: China Electric Power Publish House, 2015.
2. Quan Nan, Zhang Ya Ping, Si Jing Xin, "The Information Top-Level Architecture of Global Energy Interconnection" Electric Power Information Technology, vol. 14, Mar. 2016.
3. Guo Yun Fei, Liang Yun,Huang Feng, "Research on Power Business Model Based on Information and Communication Fusion" Electric Power Information Technology, vol. 13, Feb. 2016.
4. Weiping Sun, Munhwan Choi and Sunghyun Choi, "IEEE 802.11ah: A Long Range 802.11 WLAN at Sub 1 GHz".Journal of ICT Standardization, Vol. 1, 83–108.
5. IEEE std. IEEE 802.11ah/D5.0. Part 11: wireless LAN medium access control (MAC) and physical layer (PHY) specifications, March. 2015.

MATEC Web of Conferences **139**, 00132 (2017) DOI: 10.1051/matecconf/201713900132
ICMITE 2017

Improvement on Cleaning System of SIROX Tobacco Expander

Guiping Wu[1,*], *Wei* Ding[2], *Mingxing* Li[3], *Kaihua* Yang[4], *Feng* Xie[5], *Qisheng* Chen[6], and *Hanying* Li[7]

[1,2,3,4,5,6,7]China Tobacco Zhejiang Industrial Co., Ltd. Hangzhou Cigarettes Factory. No. 118, Kehai Road, Hangzhou City, Zhejiang Province, China

Abstract. Based on analyzing SIROX the tobacco expander of its pitfalls on cleaning system like poor drainage, valve leakage, cleaning time consuming, etc., this article proposes an improvement of its cleaning system as well as an optimization of its control mode. Furthermore, this article presents a new design of automatic drainage system of the cleaning pipe as well as a real- time monitoring system. This set of system not only addresses quality issues like the over temperature of cylinder wall of the tobacco dryer and the abnormal fluctuation of the tobacco moisture after dryer, but also visualizes the valve leakage thus helps to improve the prompt maintenance of these devices. Observing the two-year operation in HCF, this system proves a high application value.

Warming and humidifying process of tobacco producing process is to warm and humidify the cut leaves and stems to make the temperature and moisture content of the leaves or the stems meet certain requirements, then speedy drying after cylinder drying or air drying, so as to increase the filling value of leaves or stems[1-3]. As a key link of tobacco processing, warming and humidifying process seems extremely significant.

Hangzhou cigarette factory introduced the SIROX tobacco expander produced by German HAUNI company during "the 11th Five-Year" technical transformation, this tobacco expander has the characteristics of large steam flow range, high temperature control and moisture control ability. Because of its good performance and simple operation, it is widely used in various large and medium-sized factories [4]. In the work process, SIROX tobacco expander will use much saturated vapor, while the internal environment of high temperature and humidity leads to serious fouling at the end of the equipment. Due to the high tightness of equipment, artificial cleaning seriously restricts production efficiency. In this paper, the improvement of the cleaning system and the optimization scheme of the control mode are put forward.

1 Working principle of SIROX

The SIROX tobacco expander consists of feeding airlock, expansion unit, compressed-air system, vapour system, cleaning system, drainage system and vibration slot [5]. As shown in figure 1, the working principle is as follows: tobacco in the front section flows to rotary rake (1-4) via charging conduit (1-2) and air lock (1-3). The rotary rake shaft is hollow, the hollow pin is screwed in on the shaft, the cylindrical periphery and end face of the pin is provided with a hole of 1.3 mm. Saturated steam is injected through the hollow pin into the tobacco, thereby expanding and humidifying the tobacco. After warming and humidifying, the tobacco falls onto a covered discharging trough (1-6) to enter the next process. While the excess exhaust gas from feed trough and outlet of material vibration trough is exhausted via exhaust fan (1-7) and exhaust pipe (1-8) [6].

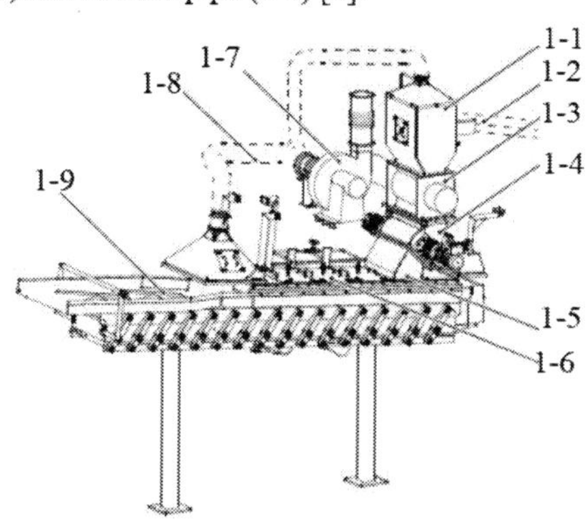

Fig. 1. Structure of SIROX tobacco expander

2 Problem analysis of SIROX cleaning system

As shown in figure 2, 5- way clean water is designed on the hydraulic main (2-1) of the SIROX tobacco expander, the role it to clean feed housing (2-2), air lock box (2-3), vibrating tank cover plate (2-4), discharging trough (2-5), and rotary rake (2-6). After automatic cleaning program starts up, each pipeline will clean the key parts in discontinuity based on setup settings, but after running for a period of time, the system has found such problems as poor drainage, leakage from valves and long cleaning time.

* Corresponding author: wuguiping@zjtobacco.com

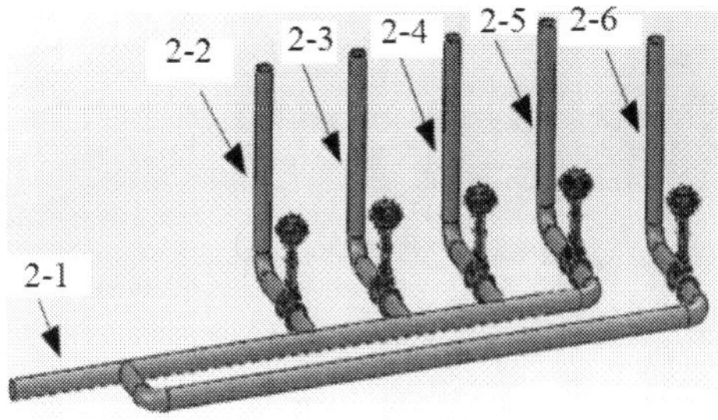

Fig. 2. Automatic cleaning pipeline of SIROX

2.1 Poor drainage

Terminal exit of the equipment cleaning pipeline is located above the vibrating slot, with a 2.5 m drop to pneumatic diaphragm stop valve, each time the automatic cleaning is finished, the valve closes automatically, and a section of water column will be formed on the end of 5-way cleaning pipeline. After the device starts running, the steam flows to the equipment via the shared pipe (2-6) of clean rotary rake to warm and humidify the tobacco. In the production process, water in the remaining 4 road cleaning pipeline will flow to the logistics channel with the vibration of equipment, resulting in the formation of water cured tobacco to affect product quality.

2.2 Leakage from valve

Pneumatic diaphragm stop valve is adopted in the cleaning pipeline of the equipment, the spring in the valve seals the spool via guide rod loading fillings. The repeated action of the valve will cause minor wear of the packing and sealing surface and cause water leakage in the stop valve, meanwhile, the pressure difference of 6 bar around the valve will deteriorate the leakage of the valve to a certain extent.

Once valve leakage, the clean water will overflow from the end nozzle to the tobacco production line under the role of pressure difference, which will result in high partial moisture in tobacco, ultra high wall temperature of post process dryer cylinder, abnormal moisture fluctuation after drying, and other problems. As there is no effective means to detect the equipment's valve conditions and leakage degree currently, it will lead to delaying maintenance, and there is a big hidden danger of product quality.

2.3 Long cleaning time

According to the technological requirements of our factory, the production and maintenance mode of "1 time cleaning for every production of 2 batches of tobacco" of SIROX tobacco expander. At present, SIROX automatic cleaning process takes about a total of 20 min, and 2 times of artificial cleaning take about 10 min, each cleaning takes about 30 min. Under current production mode and cleaning frequency, it takes about 1.5 h per day for SIROX equipment cleaning, which greatly prolongs the batch change time in the production section, seriously affects the continuity of production and reduces the production efficiency.

3 Improvement measures and effects

In view of the above 3 problems, transformation and optimization have been made on the SIROX cleaning pipeline, valve control system, automatic cleaning procedures.

3.1 Improvement on cleaning pipeline

As shown in Figure 3, one drain pipe line (3-1) is designed on the pneumatic diaphragm stop valve outlet of feed housing, vibrating slot cover plate, discharging trough and rotary rake, the drain pipe line is provided with an air operated ball valve (3-2). And, the pneumatic diaphragm stop valve (3-3) is normally closed, the pneumatic ball valve is normally open, and these two have opposite states of opening and closing: that is, when the pneumatic diaphragm stop valve (3-3) is opened, the new pneumatic ball valve (3-2) is closed; when the pneumatic diaphragm stop valve closes, the pneumatic ball valve opens. The design has the following advantages:

(1)Solve the problem of cleaning water retention in pipe line. When the automatic cleaning program operates, the pneumatic diaphragm stop valve opens and the pneumatic ball valve closes, the clean water cleans the equipment through the original cleaning pipeline; When the automatic cleaning procedure is finished, the pneumatic diaphragm stop valve closes and the pneumatic ball valve opens, the retained water in pneumatic diaphragm stop valve to the end nozzle of clean pipe flows from the new pipe (3-1) under gravity.

(2)Eliminate the hidden trouble on product quality caused from the leakage of pneumatic diaphragm stop valve. If the pneumatic diaphragm stop valve leaks, the leaking water will flow out of the new pipeline and will not overflow the material on the production line, thus eliminating the hidden danger of the product quality.

(3)Realize the visualization inspection of valve leakage. Equipment maintenance personnel can observe whether there is water flow on the production process to determine whether the corresponding pneumatic diaphragm stop valve leaks to conduct timely maintenance after evaluation.

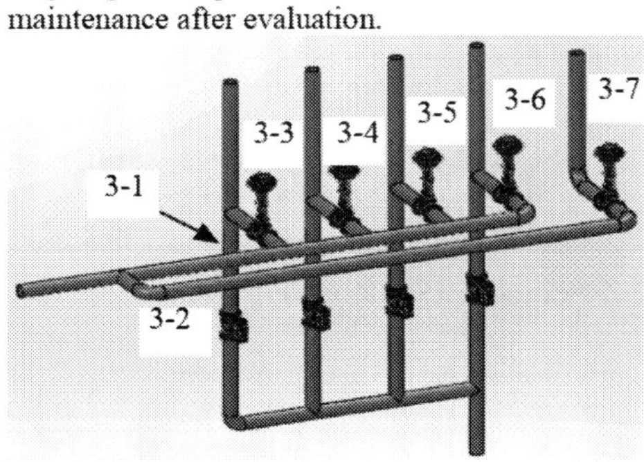

Fig. 3. Improvement on cleaning pipeline

3.2 Improvement on valve control system

The opening and closing of the additional pneumatic ball valve shall be controlled by the pressure air line, while the pressure air valve island of the SIROX does not have enough air pressure modules to be reserved, and after the addition of the module, reprogramming is needed. Therefore, according to the logic relation between the opening and closing of the valve, the existing air pressure pipeline has been reformed in this paper. A 8-mm T interface is added on the control air pressure pipeline of the pneumatic diaphragm stop valve of above 4 way water pipe, while the pneumatic diaphragm stop valve and pneumatic ball valve on other 2-way pipeline are controlled by by-pass. This added T-interface makes the pneumatic diaphragm stop valve and pneumatic ball valve form an interlocking relationship, a single signal controlled dual actuator is thus implemented: when the corresponding branch valve island module has a signal, the pneumatic diaphragm stop valve opens and the pneumatic ball valve closes; When the module has no signal, the pneumatic diaphragm stop valve closes and the pneumatic ball valve opens.

3.3 Optimization on automatic cleaning program

SIROX automatic cleaning process carries out 5 cycles, each cycle is divided into 8 stages, the 5-way cleaning pipeline in each stage cleans the corresponding parts according to procedures.

After repeated tests by the project team, the opening of each valve has been determined, as shown in table 1.

Table 1. Cleaning valve opening at all stages in automatic cleaning process

stage	valve 2-4 (3-5)	valve 2-5 (3-6)	valve 2-3 (3-4)	valve 2-2 (3-3)	valve 2-6 (3-7)
Stage 1	Off	On	On	On	On
Stage 2	Off	On	On	On	Off
Stage 3	Off	On	On	On	On
Stage 4	On	Off	Off	On	Off
Stage 5	On	Off	Off	On	On
Stage 6	On	Off	Off	On	Off
Stage 7	On	Off	Off	On	On
Stage 8	On	Off	Off	On	On

After completing the program improvement, the project team will evaluate the cleaning effectiveness. The team has carried out 4 cleaning effect evaluation successively, the evaluation results are consistent. Table 2 shows the fourth evaluation record.

Table 2. Cleaning effect evaluation in each cleaning stage of SIORX

Cleaning stages	One cycle	Two cycles	Three cycles	Four cycles	Five cycles
Feed housing	Not clean	Not clean	Clean	Clean	Clean
Airlock box	Not clean	Not clean	Clean	Clean	Clean
Rotary rake	Not clean	Clean	Clean	Clean	Clean
Vibrating tank cover plate	Not clean	Not clean	Clean	Clean	Clean
Discharging trough	Not clean	Not clean	Clean	Clean	Clean

Based on the 4 cleaning effect evaluation, the optimized SIROX tobacco expander has reached the goal of thoroughly cleaning the key parts after the third cleaning cycle.

4 SUMMARY

In this paper, the problem of poor drainage is solved by the retained clean water in linkage relation evacuation line of pneumatic ball valve and pneumatic diaphragm stop valve; and the hinder danger of flowing into the production line due to leakage of pneumatic diaphragm stop valve is eliminated by the normally opening drain pipeline, which guarantees product quality. And, the leakage condition of the pneumatic diaphragm stop valve is quantified by observing the effluent situation of the emptying pipeline, which improves the timeliness of equipment maintenance. Meanwhile, by optimizing the cleaning process and the number of cycles, the cleaning time is reduced, and the water resource is saved, and the stability of the SIROX tobacco expander and the efficiency of the production are further enhanced. Therefore, the project has achieved remarkable results in the application of Hangzhou cigarette factory, which has high application value.

References

1. Chen-zhe Song, Guang-zhou Liu, Zhe Jin, etc. Study on tobacco expansion effect using steam expansion technology. Journal of Zhengzhou University of Light Industry (Natural Science Edition), **12**,29(2014).
2. Zhi-yan Chen. Effects on the Characteristics of Cut Stems of Flue-cured Tobacco in the Process of the Expansion Drying. Journal Of Anhui Agricultural Sciences, **38**, 7262-7263(2010).
3. Yong-ren Zhang, Zhe-kun Li, Yan-hong Song, etc. The Research of SIROX Warming and Humidification Equipment. New Technology & New Process, **10**, 6–8(2015).
4. Quan-dao Wang, Optimization of technological parameters of tunnel-type expanded processing of cut stem, Acta Agriculturae Jiangxi, **23**, 100–103(2011).

DOI: 10.1051/matecconf/201713900132

5. Liang Zhou, China tobacco commercial enterprise development direction. Journal of hunan university of arts and sciences(natural science edition), **6**, 92-94(2011).

6. Zhi-hui He, Jia Luo, Wen-liu Lian, etc. The Fourier transform of cigarette tar is near infrared measurement technology. Journal of hunan university of arts and sciences(natural science edition), **9**, 59-67(2009).

Design and analysis of disc conveyor of cutlery polishing machine

Min Fu[1,2] , *Yuanshuang* Chen[1,2,*] , and *Loulou* Zhang[1]

[1] College of Mechanical Engineering, Tianjin University of Science & Technology, Tianjin, 300222, P.R. China
[2] Tianjin Key Laboratory of Integrated Design and Online Monitoring for Light Industry & Food Machinery and Equipment, Tianjin, 300222, P.R. China

Abstract. The paper designs a new disc conveyor of stainless steel cutlery polishing machine. The disc conveyor is mainly composed of transmission mechanism, locating mechanism and clamp mechanism. The paper makes finite element analysis on rotating disc and improves the structure design according to analysis result in order to enhance statics characteristic of rotating disc structure. Through modal analysis, the paper verified the rationality of the design of rotating disc structure. The prototype was put into enterprise and the result showed that this system could reliably operate.

1 Introduction

The processing of stainless steel cutlery needs to experience more than ten working procedure, of which, polishing is an important link in the production process of stainless steel cutlery. Polishing of stainless steel cutlery can enhance corrosion resistance and smoothness of cutlery, and obtain decorative appearance [1]. The polishing of cutlery is a very complicated technological process and the polishing quality affected the yield of product directly.

Cutlery polishing technology developed earlier in foreign cutlery manufacturers. Most of them adopt full-automatic polishing machine with high production efficiency and good product quality. Domestic research on cutlery polishing technology starts late. Some small and middle-sized enterprises mainly introduce outdated and obsolete stainless steel cutlery polishing machine from foreign countries, repair and improve it to realize semi-auto production. Large enterprises mainly purchase foreign advanced automatic polisher. However, they fail to master core technology, so the repair and subsequent maintenance cost are high.

With the increasing of labour cost, traditional manual polishing cannot meet the development of enterprise. Domestic cutlery manufacturers increase research spending of automatic polisher. A set of high-automation stainless steel cutlery polishing machine could save labour, reduce the cost and significantly enhance productivity.

According to technological requirement of stainless steel cutlery manufacturers for cutlery polishing, the paper designs a high-automation stainless steel cutlery disc conveyor.

There is a set of five-station cutlery polishing conveyor which is applied to rough grinding of cutlery. Because of fewer configuration and lower production efficiency, which cannot meet the requirement of manufacturers.

According to the disadvantage of polishing conveyor at present, the paper increases number of station according to the program and requirement of manufacturer. We can achieve rough grinding and medium grinding by using one polishing conveyor. The setting of cutlery overturning station and cooling station make the handle and handlebar of cutlery polishing by using a set of conveyor.

2 The design of cutlery polishing system

2.1 General structure layout of cutlery polishing machine

Cutlery polishing machine is composed of cutlery conveyor that can realize the cutlery intermittent rotational motion, cooling mechanism, cutlery overturning mechanism and layout in the periphery of the eight polishing wax institutions. The transmission mechanism of cutlery polishing machine mainly includes index plate and telescopic cutlery clamping device, Index plate under the drive of the motor to realize the intermittent rotation. It drives extensible clamp mechanism to be transmitted among 12 stations. The function of clamp mechanism is to hold stainless steel cutlery. The polishing and waxing mechanism realizes feeding in transverse and longitudinal direction in the

* Corresponding author: 2461216748@qq.com

 DOI: 10.1051/matecconf/201713900133

polishing process so as to complete cutlery polishing and polishing paste painting in the polishing process. After cutlery polished in four stations, the cooling mechanism cools the cutlery, which avoids the cutlery temperature is exorbitant and appears insufficient clamping force. Cutlery overturning mechanism realizes conversion among cutlery handle and handlebar. The general layout diagram is shown in Fig. 1.

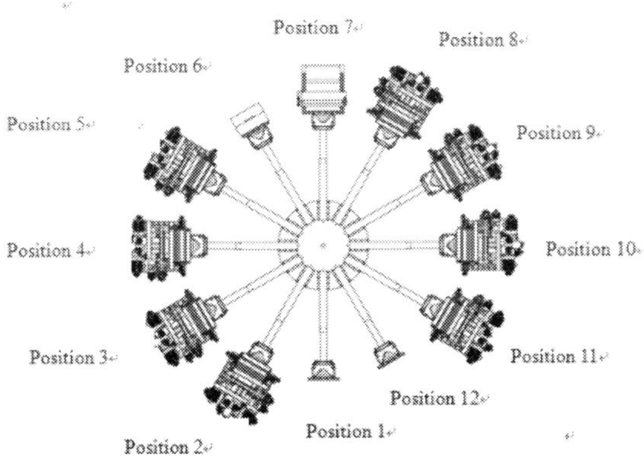

Fig. 1. General structure layout of cutlery polishing machine.

The control action of the whole operation process of cutlery polishing machine is realized by PLC control system. In the actual processing course, the manufacturers can adapt to processing requirement of different kinds of stainless steel cutlery through adjusting transverse and longitudinal feed rate of polishing and waxing mechanism and cutlery polishing wax institutions and the position of the clamping mechanism. The polishing wax agencies in 2 to 9 position of polishing time can be set in advance. The whole set of device has emergency stop control button, to avoid the unexpected incidents in the process of actual production and processing bring unnecessary losses. Only need one worker to preset the parameter and feed the workpiece of cutlery to be processed in the manufacturing process.

2.2 Structure design of disc conveyor

Disc conveyor is mainly composed of bottom plate, shore, transmission mechanism, locating mechanism and clamping mechanism and its structure is shown in Fig. 2.

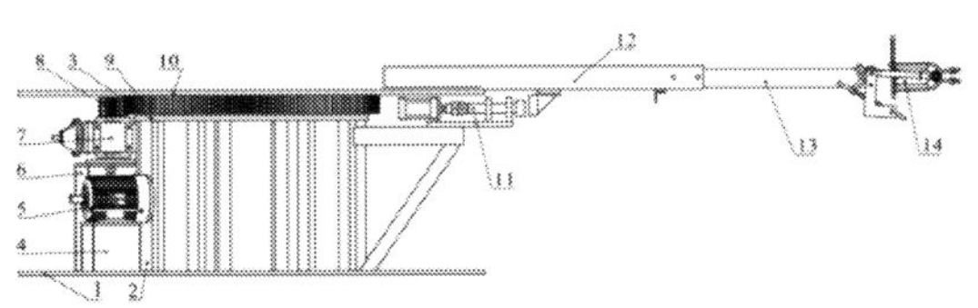

1. Baseplate; 2. Shore; 3. Flange; 4. Motor support; 5. Motor; 6. Reducer support; 7. Reducer; 8. Small gear; 9. Slewing bearing; 10. Rotary disc; 11. Locating mechanism; 12. Big rectangular steel tube; 13. Small rectangular steel tube; 14. Clamping mechanism

Fig. 2. General structure layout of cutlery polishing machine.

Bottom plate is fixed on the ground using the anchor bolts. The one end of shore on the bottom plate is welded to bottom plate, and the other end is connected to flange. A driving device is fixed on the flange, and the rotating disk is connected with the driving device, and twelve extensible cutlery clamping devices are evenly distributed on the rotating disk, and the fixing frame of the positioning device is connected with the upright post.

2.2.1 Intermittent motion mechanism

The core component of the cutlery transmission mechanism is intermittent movement mechanism, and the quality of the intermittent movement mechanism influences the advantages and disadvantages of the cutlery transmission mechanism directly. Many mechanisms can achieve intermittent movement, such as indexing cam mechanism, Maltese mechanism, ratchet mechanism, incomplete gear mechanism, pin wheel mechanism, gear rod compound mechanism, etc. [2]. Through combining technological requirements of actual production, the main transmission system of polisher adopts YVP variable-frequency and variable-speed motor, drives helical gear-spiral bevel gear reducer by belt drive, and passes through small gear of reducer so as to drive rotation of slewing bearing and the rotation of index plate. According to the bearing (1,000 Kg) and maximum rotating speed (2r/min) of rotary disc and combining with the design experience, the motor power is 2.2KW and rated speed is 1440r/min.

2.2.2 Pedestal module

The pedestal of cutlery polishing machine conveyor, on one hand, supports the rotary disc, on the other hand, it is connected to transmission parts and provides power for the rotary disc. The lowest part of the pedestal is baseplate with the diameter of 2,200mm. There are eight rectangular hollow steel tubes on the baseplate for supporting. The upper part of the steel tube is flange with the thickness of 20mm and on the flange is a slewing support. Index plate is connected to rotation support through bolt. Motor drives the rotation of reducer through belt drive. Reducer is connected to small gear, and small gear and rotation bearing realize mesh transmission so as to drives the rotation of rotary disc. Fixed plate of transmission system and fixed plate of locating mechanism are mounted on the pedestal. The structure is shown in Fig. 3.

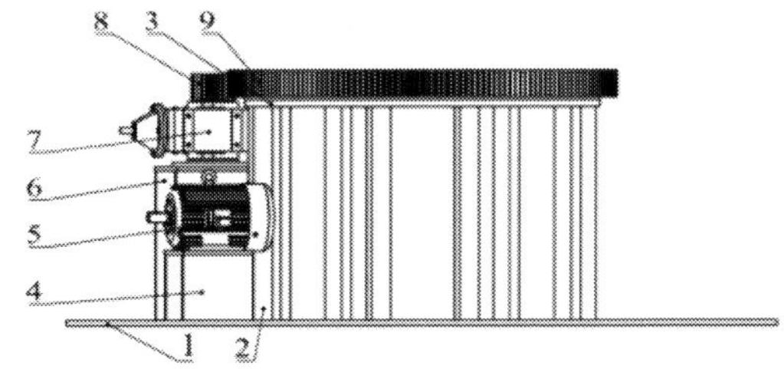

1. Baseplate; 2. Shore; 3. Flange; 4. Motor support; 5. Motor; 6. reducer support; 7. Reducer; 8. Small gear; 9. Slewing bearing.

Fig. 3. Pedestal components.

2.2.3 Locating mechanism

When the big disc is stopped, the location may be inaccurate owing to inertia, there is a certain deviation. And the polishing machine will have a certain force on the meal in the grinding process of cutlery, the force of the telescopic clamping mechanism by conduction to the chassis causing the chassis vibrating in the grinding station, thereby affecting the polishing quality of cutlery. This paper has designed a set of locating mechanism composed of tip and dead head. On one hand, the mechanism can provide combined force of tip and dead head in horizontal direction; on the other hand, it can ensure tip to keep good guiding orientation in the working process, and its structure is shown in Fig. 4.

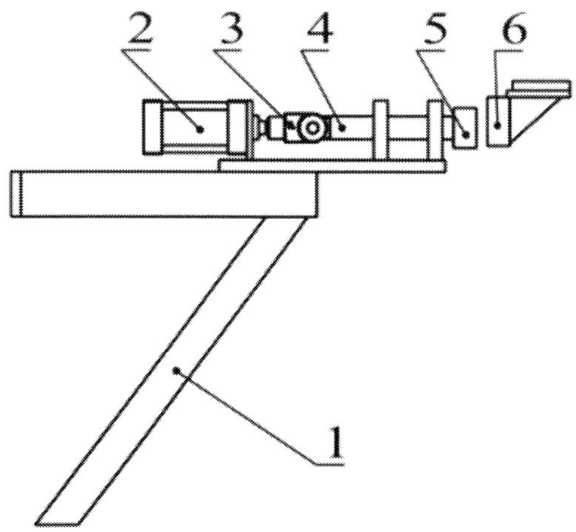

1. support; 2. Air cylinder; 3. Air cylinder contact; 4. Guide shaft; 5. Tip; 6. Dead head.

Fig. 4. Locating mechanism.

2.2.4 Extensible clamp mechanism

The cutlery clamp mechanism is an important component of cutlery disc conveyor. When polishing the cutlery, the cutlery often flies off and causes personnel injury because of the high revolving speed of polishing and the low frictional force. The clamp force of pneumatic clamping at the end of such clamp mechanism is provided by rubber air spring. After test analysis, the clamp mechanism can provide enough clamp force and avoid flying-off of cutlery when polishing. Clamp fixture is connected to rotary disc through dead plate, the big rectangular steel tube is welded on the fixed plate, the small rectangular steel tube is installed on the big rectangular steel tube and the small rectangular steel tube is flexible and regulated relative to big rectangular steel tube, the end of small rectangular steel tube is connected to the pneumatic clamp. The structure is shown in Fig. 5.

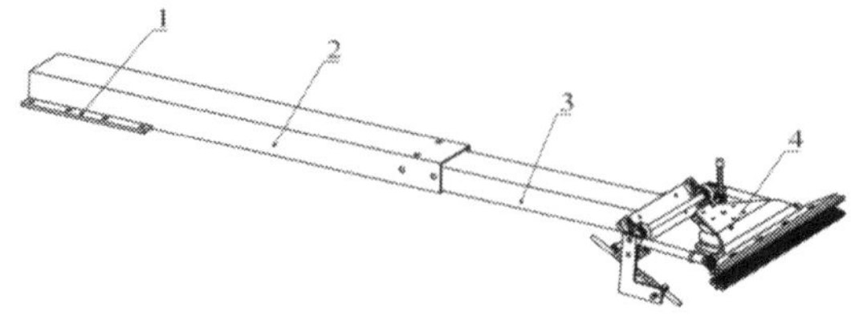

1. Dead plate; 2. Big rectangular steel tube; 3. Small rectangular steel tube; 4. Pneumatic clamping

Fig. 5. Extensible clamp mechanism.

3 Finite element analysis and structure improvement of rotating disc

3.1 Finite element analysis

The support stiffness and operational stability of rotary disc are directly related to the polishing quality of cutlery. Therefore, the finite element analysis of rotating disc structure is needed to determine the deformation and stress of the rotating disc under the condition of force.

3.1.1 Model pretreatment

The three-dimensional model is built by using 3D software Solidworks, and the model established by Solidworks is saved as x_t format file, and then the file is imported into ANSYS Workbench to analyze the linear statics structure. The material is Q235 structural steel, whose modulus of elasticity is 2.0×1011Pa, poisson's ratio is 0.3. The model is divided into 81534 nodes and 18520 units by using the method of automatic meshing and the default cell size. Finite element model is shown in Fig. 6.

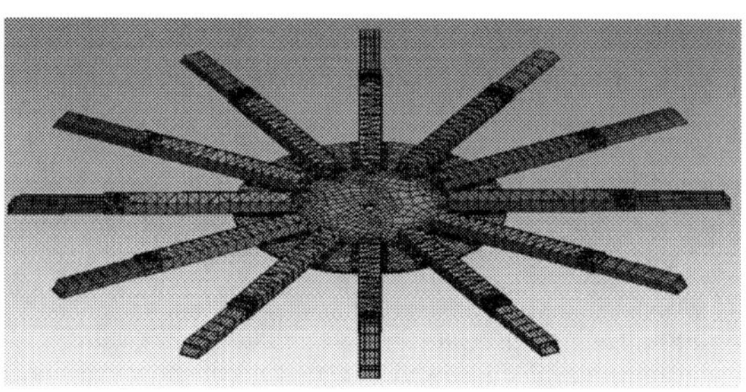

Fig. 6. Finite element model of rotary disc structure.

3.1.2 Apply load and boundary condition

Rotary disc and rotary bearing are connected through bolts, so a constraint condition is applied to the 12 bolt holes on the rotating disk, so that the freedom degree is completely restrained. The load applied to the rotary disc mainly originates from the weight of pneumatic clamping and its own weight. After measurement, the weight of each set of pneumatic weight is determined to be about 30Kg; self-weight load mainly includes rotary disc, large rectangular steel tube, and small rectangular steel tube. When making finite element analysis, the Standard Earth Gravity is selected, and the uniformly distributed load is applied on each unit at the gravity acceleration of 9.8066m/s2 in ANSYS Workbench [3]. After applying load, the model is shown in Fig. 7.

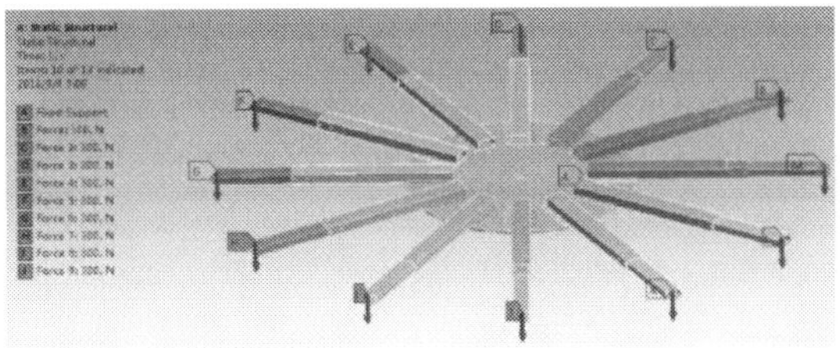

Fig. 7. Schematic diagram of impose restriction and load.

3.1.3 Result analysis

The overall deformation diagram and stress nephogram of the rotating disk structure are obtained by solving the finite element model as is shown in Fig. 8.

According to Fig. 8, the deformation between pneumatic clamp and small rectangular steel tube is the largest. The maximum deformation is 7.5032mm, and the maximum stress is at the connecting bolt hole of rotary disc and rotary bearing. The maximum stress is 288.92Mpa.

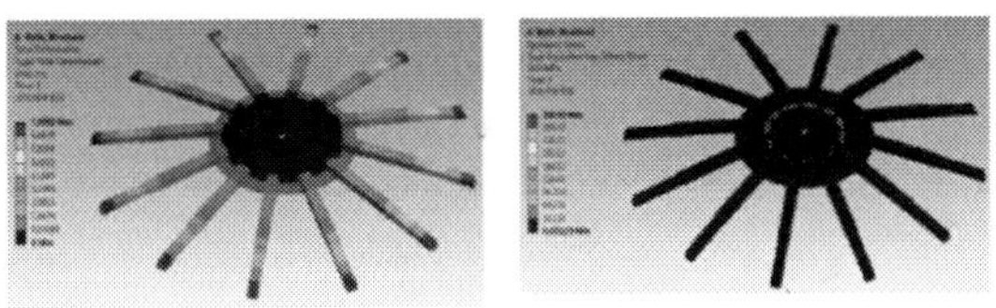

Fig. 8. Overall deformation diagram and stress nephogram of rotary disc structure.

3.2 The design of structure improvement

As the basis of conveyor work, the overall deformation of rotary disc structure will affect the polishing quality of cutlery. To reduce overall deformation of rotary disc structure and improve the stress between rotary disc and rotary support connecting bolt hole and enhance the safety factor of equipment, the paper makes structure improvement design of rotary disc structure.

Referring to the design of cable-stayed bridge, the paper determines to set up a cylinder with the diameter of 500mm, height of 800mm and thickness of 10mm on a basis of existing structure. The two section 30 × 30 × 3mm square tube is taken as a cable-stayed component. One end of the inclined member is connected with the cylinder, and the other end is connected with the end of the large rectangular steel tube. Two sections of cable-strayed members are connected using regulating threaded rod, and the rotation direction of thread one both ends of threaded rod is opposite so as to ensure movement together of these two sections of cable-strayed members. The modified structure is shown in Fig. 9.

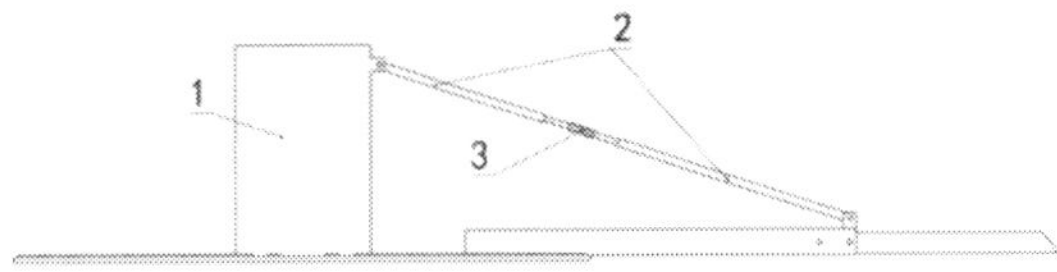

Fig. 9. Schematic diagram of cable-stayed structure.

Static analysis of the improved rotating disk structure model is carried out by using ANSYS Workbench, and the overall deformation and stress nephogram are obtained, as is shown in Fig. 10.

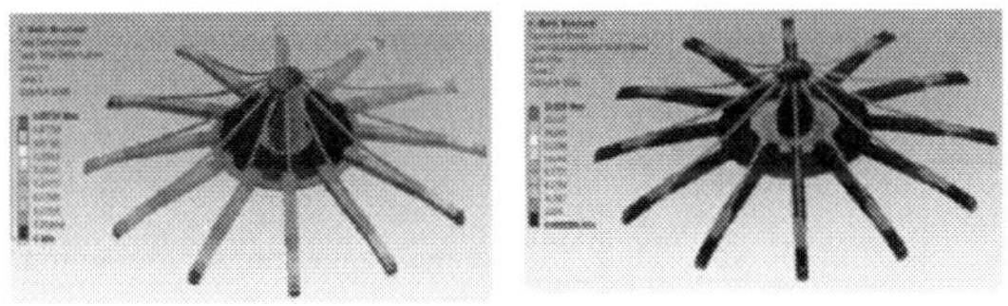

Fig. 10. Overall deformation diagram and stress nephogram after modificaiton.

According to analysis result, it is known the maximum deformation of the modified rotary disc structure is 0.53724mm, the maximum stress is 18.829Mpa and the safety factor is 13.728; the maximum deformation and maximum stress decline obviously, safety factor is enhanced greatly, stress concentration is improved and stress of rotary disc is distributed around the cylinder evenly. Therefore, the static characteristics of the rotating disc are greatly improved by the improved structure design.

In the process of cutlery polishing, rotary disc structure will produce a certain vibration. During modal analysis, for the one with prestressing force, the inherent frequency is greater than that without prestressing force, and it is more realistic than the actual situation. Thus, the modal analysis of the rotating disc structure is carried out to determine the inherent frequency and vibration mode, so that the external excitation to avoid its natural frequency, to avoid the resonance phenomenon, to ensure stable operation of equipment [4].

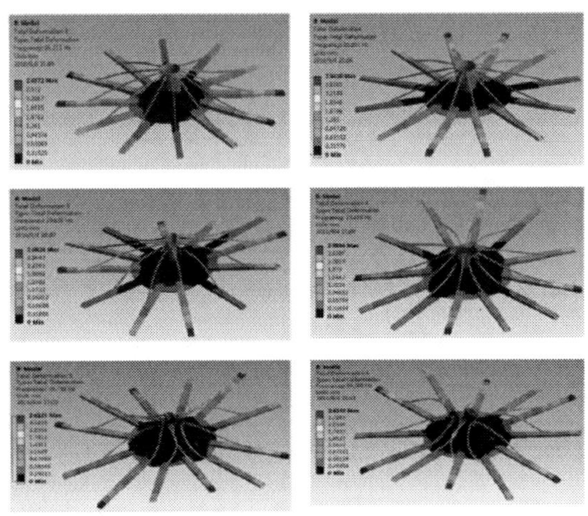

Fig. 11. Mode of vibration and frequency of the first six orders.

On the general conditions, the high-order mode has little effect on the vibration of the structure, usually only consider the low-order mode [5, 6]. Here, ANSYS Workbench is used to analyze the inherent frequency and mode of vibration of the six orders of rotary disc structure. As shown in Fig. 11.

The modal frequency value and mode of vibration of rotary disc structure at each order as are shown in Table 1.

Table 1. Modal frequency and mode of vibration of the first six orders of rotary disc structure.

Order	1	2	3	4	5	6
Inherent frequency (Hz)	26.097	26.111	29.606	29.619	39.752	39.766
Mode of vibration	Swing around X axis	Swing around Y axis	Rotate around Y axis	Rotate around X axis	Rotate around Z axis	Rotate around Z axis

According to modal analysis result, the first six order frequency of rotary disc approximates, showing the symmetry of the rotating disk structure. The mode of first six order vibration involves in the swinging and torsion of the overall structure. The inherent frequency range is 26.097～39.766Hz, 5Hz higher than the drive frequency of motor and much lower than the polished roller grinding vibration frequency of 180Hz. Thus, in the process of cutlery transmission and polishing, the equipment will works safely and reliable which without resonance phenomenon.

4 Conclusions

The paper designs a new stainless steel cutlery polishing machine disc conveyor, introduces the mechanical structure composition and work process. The new design has been put into production and used by enterprise. It can polish the predetermined daily 20,000 pieces of the task. The system operates stably and good in polishing effect. It can be popularized and applied to stainless steel cutlery polishing manufacturing industry.

Through modified design of rotary disc structure, the statistical analysis of modified structure by ANSYS Workbench software shows that the rotary disc structure are enhanced greatly in rigidity and strength, also increased the stability of cutlery transmission device which improved the cutlery polishing quality. Moreover, through modal analysis, which verify the rationality of rotary disc structure so as to ensure the equipment operates safely and reliable.

References

1. J.Y. Hu, J.X. Zhang, M. Fu, Journal of Tianjin University of Sciences & Technology, **30**, 6, 61-65 (2015)
2. M. Fu, L.F. Lin, Machinery Design & Manufacture, **4**, 4, 94-97 (2016)
3. B. Li, Z.J. He, X.F. Chen, *ANSYS Workbench design, simulation and optimization* (Tsinghua University Press, Beijing, 2013)
4. M. Fu, X. Tang, L.Y. Jin, W. Zhang, Machine Design and Research, **3**, 6, 58-61 (2015)
5. Y.X. Wu, J. Han, J.J. Liu, Machinery Design & Manufacture, **4**, 4, 130-132 (2016)
6. H.C. Zhang, *Theoretical analysis and engineering application* (China Machine Press, Beijing, 2013)

Strength analysis and optimization of writing mechanism of steel billet marking machine

Min Fu[1,2] , *Yan* Li[1,2,*] , and *Liang* Zhao[1]

[1] College of Mechanical Engineering, Tianjin University of Science & Technology, Tianjin, 300222, P.R. China
[2] Tianjin Key Laboratory of Integrated Design and Online Monitoring for Light Industry & Food Machinery and Equipment, Tianjin, 300222, P.R. China

Abstract. According to steel billet marking theory of plasma arc nicking, the paper designs a dual laser ranging marking machine against online marking of special steel billet and realizes multi-character marking of the end face of hot steel billet. Writing mechanism bases on the rectangular coordinates marking form, Z axis adopts cantilever structure. It completes the overall marking task utilizing the synergy of KK module in X axis, Y axis and Z axis. It makes modal analysis on the writing mechanism model established by Pro/Enginner utilizing ANSYS Workbench at the position of X1Y1Z1, and obtains the first six order modal frequency and analyzes the vibration in the writing process. Moreover, the paper analyzes the static structure of the cantilever of writing mechanism, computes its maximum stress and total deformation. To make the writing mechanism reach the target of light weight, the paper optimizes Z-axis cantilever of writing mechanism. According to the analysis, it is known that the optimized Z-axis cantilever of the writing mechanism still meets the strength and rigidity requirement and total mass declines approximately 30%.

1 Introduction

Steel billet marking machine is one of important machines for the production of steel enterprise. To facilitate management of production and quality of steel works, the steel enterprise usually marks the steel billet product and marks the sole number composed of related information on the surface of steel billet. If quality problem occurs, the continuous casting machine number, furnace number, serial number, time and other important information of steel billet can be traced according to the mark number so as to discover and solve the machine problem earlier [1]. Hereinto, writing mechanism is an important component of steel billet marking machine. It provides movement basis for the steel billet marking machine, so the position accuracy of writing mechanism must be guaranteed. The steel billet marking machine on the continuous casting line of special steel billet designed in the paper utilizes dual laser ranging system to prevent the mobile box colliding with the steel billet of writing mechanism. Writing mechanism is installed in the mobile box. It applies PLC to control servo motor and drive KK module and makes the movement of writing mechanism to complete the marking work. Kinetics and strength feature of writing mechanism affects the effect and stability of steel billet marking machine and prevents serious vibration of writing mechanism in the operation. Thus, it is necessary to analyse the static structure and modality of writing mechanism [2]. Meanwhile, considering economy and performance requirement, the paper makes optimal design of writing mechanism on the premise of meeting the strength and use requirement.

2 Determine writing mechanism structure and build 3D model using Pro/E

Marking device needs to mark on the end face of hot steel billet, defines the length direction of steel billet as Z axis, the plane vertical to it is XY plane (namely writing plane). Servo motor and uniaxial robot are installed in XYZ axes to drive the movement of spearhead, and one or two ancillary slide ways are installed in XYZ axes to support and assist the movement, which enhances the rigidity of writing mechanism.

The quality consistence of end surface of steel billet after cutting is poor, and there is a certain cutting slope (about 1.5°~2.5°). The movement of Z-axis KK module is used for compensating the distance deviation caused by cutting inclination. Z axis adopts cantilever structure. Plasma spearhead and spearhead protection mechanism are hung on the Z-axis support through rectangular beam and rectangular structure to make the component which is easily affected by the temperature be far away from the writing position of spearhead. The paper makes 3D

* Corresponding author: 1632366977@qq.com

modeling using Pro/E and the writing mechanism is shown in Fig. 1.

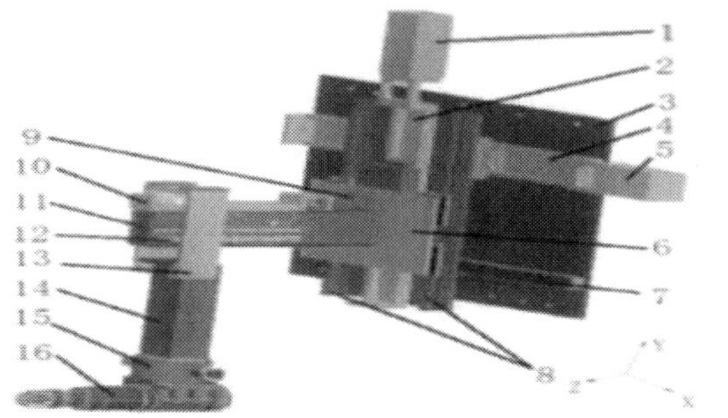

1. Y-axis motor 2. Y- single axis robot 3. X-baseplate 4. X-single axis robot 5. X-axis motor 6. Z-baseplate 7. X-ancillary guideway 8. Y- ancillary guideway 9. Z-axis motor 10. A-single robot 11. Z-support 12. Z-ancillary guideway 13. Rectangular structure 14. Vertical beam 15. Spearhead protection mechanism 16. Plasma spearhead

Fig. 1. Structure chart of writing mechanism.

3 Modal analysis of writing mechanism

The working of writing mechanism is that servo motors in the axis of XYZ drive KK module to realize coordination movement in these three directions. A certain vibration occurs in the writing process. To prevent the motivation of writing mechanism and servo motor causing low-order frequency resonance and affecting marking, or even damaging spearhead, it is necessary to analyse the inherent frequency and mode of vibration of writing mechanism [3, 4]. The stroke of KK module is controlled through proximity switch. When the sliding block of writing mechanism is located in the middle point of stroke, recording it as X1Y1Z1 position. The paper utilizes ANSYS Workbench to make modal analysis of writing mechanism in the position of X1Y1Z1, analyses, solves and extracts the first 6-order modality, as shown in Fig. 2. Table 1 is the description of inherent frequency and mode of vibration of various order modalities.

According to the following table, it is known, the inherent frequency 60.01Hz of second-order modality approximates to inherent frequency 62.79Hz of third-order modality, but the maximum amplitude occurs in different positions. Maximum amplitude of the second-order modality occurs at the Y-axis motor end, but maximum amplitude of the third-order modality occurs at the front end of spearhead. According to mode of vibration diagram of modal analysis, it is known, the nearer the centre of gravity of the writing mechanism is to the distance of fixed end and that of symcenter of structure, the better the rigidity of the whole system is and the higher the inherent frequency is. Thus, during marking, Z-axis sliding block should close to the drive motor, and sliding block in X axis and Y axis should close to the midpoint of stroke so as to increase system rigidity of writing process and reduce the possibility of resonance occurrence. The inherent frequency of minimum first order of writing mechanism is 43Hz, but the maximum of exciting force frequency of motor when writing is only 11.1Hz. It indicates that the rigidity of writing mechanism is very good and the possibility of system producing resonance in the marking process is very small.

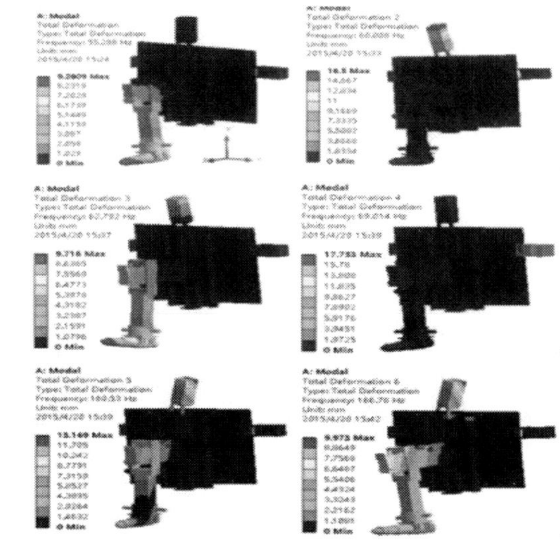

Fig. 2. The modal analysis results of location X1Y1Z1.

Table 1. Modal frequencies and mode shapes of location X1Y1Z1.

Order	1	2	3	4	5	6
Inherent frequency HZ	55.29	6 0.01	6 2.79	6 9.01	160.53	166.78
Mode of vibration	The cantilever of spearhead swing around Z axis	The cantilever of spearhead and motor end in Y axis swings around X axis	The cantilever of spearhead and motor end in Y axis swing around X axis	The motor end in X axis bends toward Y axis	The cantilever of spearhead and motor end in Y axis swings around Y axis	The cantilever of spearhead and motor end in Y axis swings around and bend toward Z axis

4 Static structure analysis of cantilever of writing mechanism

4.1 Establishment of Z-axis cantilever structure of writing mechanism

Z-axis cantilever of writing mechanism is composed by welding seven pieces of plates. All the plates in the initial design is made from Q235 steel, the boundary dimension of cantilever structure is 474×60×100mm, the thickness of rib plate is 3mm and the thickness of other plates is 8mm. Z-axis cantilever structure is shown as Fig. 3. (The figure is 1/2 structure.)

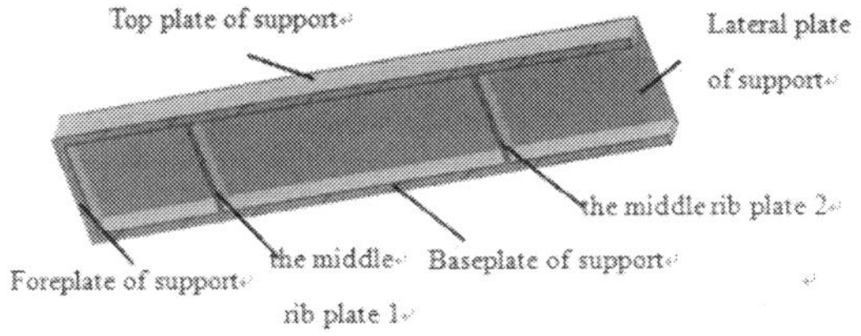

Fig. 3. The structure diagram Z- cantilever beam.

4.2 Characteristic parameter of material

Cantilever structure is processed using spot welding and molding. The characteristic parameter of material is shown in Table 2.

Table 2. Z-axis cantilever material parameter.

Grade	Elastic modulus / MPa	Poisson ratio	Density / (kg/mm^3)	Yield strength / MPa	Tensile strength / MPa
Q235	2×10^5	0.3	7.85×10^{-6}	235	375-460

4.3 Finite element analysis and computation

4.3.1 Apply load and boundary condition

The front end of Z-axis cantilever is connected to spearhead structure, so the axial load is applied to the front end of cantilever, the measured axial load is F1=468N. In addition, the self-gravity is F2=312N. The back end of Z-axis cantilever is fixed on the Z-axis support, so fixed constraint is applied to the back end.

4.3.2 Compute the stress and deformation of Z-axis cantilever of writing mechanism

The horizontal plate, lateral plate and rib plate of writing mechanism is made of Q235 steel plate through spot welding and molding method. The paper applies load and constraint condition so as to compute the equivalent stress and total deformation nephogram, as shown in Fig. 4.

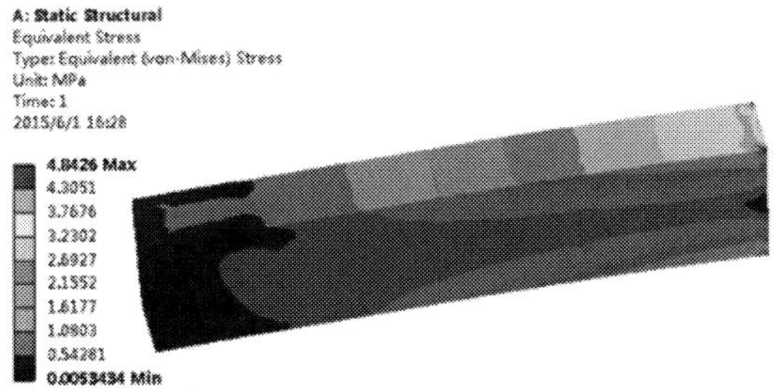

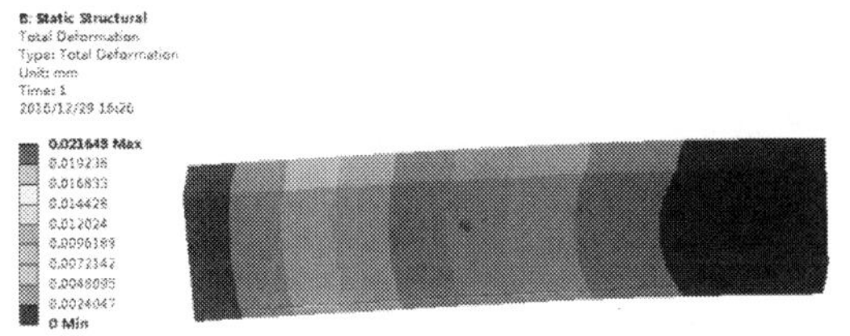

Fig. 4. Deformation and equivalent stress nephogram of Z-axis cantilever.

4.3.3 Result analysis

According to equivalent stress and total deformation nephogram, it is known the maximal deformation and maximal equivalent stress of Z-axis cantilever of writing mechanism. The maximal deformation of Z-axis cantilever of writing mechanism occurs at the front end of cantilever, the maximum deformation is δ max= 0.0216mm, the maximum stress occurs at the joint of back end of cantilever and maximum stress is σ max=4.84Mpa. Considering that the spearhead must be accurate when writing mechanism works, so the deformation cannot be too big. Thus, the design structure meets the requirement of maximum deformation. Combining with the above factors and actual cases, the cantilever made of Q235 steel plate using spot welding and molding method has good performance. We take safety factor of yield strength as n=2, and allowable stress as [σ]= σ s/n=235/2=117.5 MPa >4.84 MPa, so the strength and rigidity of cantilever made of Q235 steel plate suing spot welding and molding method conforms to the requirement and it has great safety margin, so it only needs to ensure the welding quality. The design has great design margin, so it is necessary to make optimal design.

5 Optimal design of cantilever of writing mechanism

5.1 Theoretical basis of optimal design

The structure size of cantilever of writing mechanism is optimized using Design Explorer in ANSYS Workbench and multi-objective optimization function. Hereinto, the optimal design made by Design Explorer is achieved through response surface (line) and the fitting of response surface (line) is completed through design point [5, 6]. Through determining the establishment of multi-objective optimization function, the paper calls in goal driven optimization for Design Explorer optimization design of structure size of connecting plate.

5.2 Establishment of optimal design

5.2.1 Establishment of multi-target optimization function

Other components of writing mechanism is standard component and the design margin of the height of cantilever and thickness of plate is great, so optimizing the cantilever of writing mechanism can reduce the weight. The weight can be reduced from boundary dimension of cantilever and the thickness of the used plate. The width and length of cantilever are related with the corresponding KK module, so the paper selects height of cantilever (P2) and thickness of plate (P1) as the input parameters and takes mass P6 and equivalent stress P4 as the optimized output parameter so as to reduce the mass of cantilever of writing mechanism and establish multi-objective optimization function, as shown in Formula (1):

$$\begin{aligned} &\text{Min } f(X)=P6 \\ &X=[x1, x2]=[P1,P2] \\ &\text{s.t. } \sigma <= P4 = 117.5\text{Mpa} \\ &5\text{mm}<=P1<=8\text{mm} \\ &70\text{mm}<=P2<=100\text{mm} \end{aligned} \quad (1)$$

In the formula, P1 is the thickness of cantilever plate; P2 is the height of cantilever; P4(σ max) is the equivalent stress; p6(m) is mass (the below is the same).

5.2.2 Call in and compute goal driven optimization

The paper optimizes parameter using Response Surface in the optimization module so as to observe the impact of the change in each design point on the optimal target. Through 2D curve and 3D curve, it is seen the response of each design point to the optimal target. The optimization target and importance of each parameter can be set up according to the impact of input parameter on output parameter. The impact of the plate thickness P1 and height P2 of cantilever of writing mechanism on mass P6 and maximum stress of cantilever of writing mechanism is shown as the generated response surface in Fig. 5.

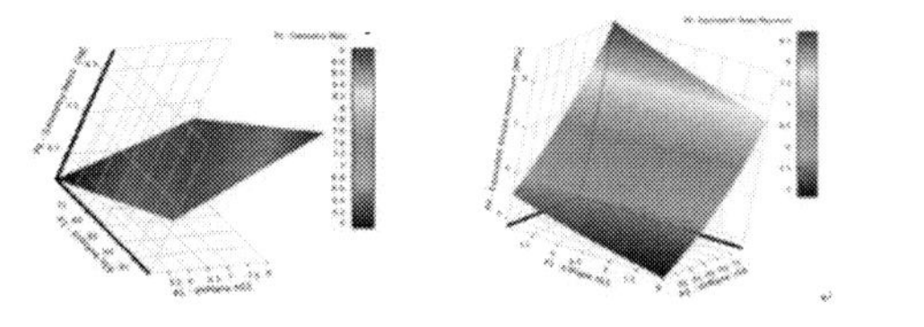

The varies of mass p6 with plate thickness p1 and height p2. The varies of maximum stress p6 with plate thickness p1 and height p2.

Fig. 5. Response surface of parameters.

Through analyzing response surface, it is known, the structural mass P6 increases with the increasing of height P1 and plate thickness p2 of cantilever, but the corresponding equivalent stress P4 decreases accordingly. So does the actual situation. With the increasing of the plate thickness and section of cantilever, the bearing capacity increases, structural mass also increases. When bearing the same load, the stress value and deformation also decrease. The impact of plate thickness and height on the cantilever is almost the same. The impact of height of cantilever on the equivalent stress is greater than the plate thickness on it. That is, the impact of P1 and P2 on P4 and P6 is P1<P2.

Goal driven optimization result ensures minimum mass and minimum maximal deformation and updates new response point after optimization. After analysis and comparison, the paper selects response point as the final optimization result, namely taking P1=6mm and P2=70mm to generate model and analyze its maximum stress. The parameters comparison before and after optimization are shown in Table 3.

Table 3. Parameter change comparison of cantilever of writing mechanism before and after optimization.

Parameter name	Thickness of cantilever beam P1/mm	Height of cantilever beam P2/mm	Maximum mass P2/kg	Maximum stress P4/MPa
Before optimization	8	100	31.2	4.84
After optimization	6	70	21.8	8. 50

6 Conclusions

The marking effect chart of cantilever of writing mechanism before and after optimization is shown as Fig. 6 when the whole steel billet marking machine is produced, processed and assembled at the factory. From the figure, it is seen, the character written by the writing mechanism before optimization has obvious "shaking" phenomenon, and the character line is not smooth. Writing mechanism can complete the marking content, but the strokes of character it marks loses stability seriously and the marking effect is not good; but the character stroke marked by the writing mechanism after optimization is neat and smooth and the marking effect is improved greatly, which proves that the optimization effect is obvious.

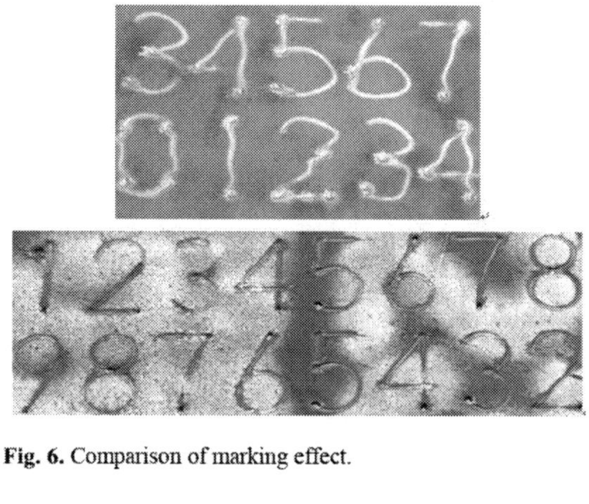

Fig. 6. Comparison of marking effect.

The paper analyzes the inherent modality of writing mechanism in common working position X1Y1Z1 and the writing mechanism will not have strong vibration phenomenon in the marking processes. To make mass of the structure lightest and most rational and ensure the rigidity and strength of structure, the paper makes optimal design of the size and maximum stress of Z-axis cantilever structure of writing mechanism made from Q235 steel plate through welding and molding:

(1) After finite element static structure analysis, it is found, the deformation of Z-axis cantilever structure of writing mechanism made from Q235 steel plate through welding and molding is small. It meets the strength and rigidity requirements and has great safety margin.

(2) After optimal design, the plate thickness of cantilever of writing mechanism is reduced and the height is reduced greatly, which reduces the mass of cantilever of writing mechanism from 31.2kg to 21.8kg, about 30%. On the condition of meeting requirement, maximum stress increases little and has a certain safety margin. It achieves the target of lightest and most rational structural mass and reduces the cost.

References

1. X.Y. Jiang, Z.Q. Xiang, J.X. Liang, Industrial Design, **29**, 4, 191-194 (2001)
2. M. Fu, Q.X. Song, Journal of Tianjin University of Science & Technology, **4**, 54-59 (2014)
3. M. Fu, L.F. Lin, Y. Liu, Manufacturing Automation, **11**, 6, 49-51 (2015)
4. X. M. Song, J. J. Cui, X. L. Li, Machinery Design & Manufacture, **3**, 103-106 (2016)
5. B. Li, Z.J. He, X.F. Chen, *ANSYS Workbench design, simulation and optimization* (Tsinghua university Press, Beijing, 2013)
6. J.B. Chen, Y. Yang, *Method of Optimization* (China Sinopec Press, Beijing, 2010)

Based on the Hardware Resources Configurable Shanke PLC Building Energy Consumption Detection System

Guanghe Cheng[1,2], *Shuai Zhao*[1,a], *Maoli Wang*[2], *Rui-rui Sun*[2], *Xiang Sun*[2]

[1]Qilu University of Technology, 250353,Jinan, Shandong China

[2]Shandong Computer Science Center(National Supercomputer Center in Jinan), Shandong Provincial Key Laboratory of Computer Networks,250014,Jinan,Shandong,China

Abstract. According to the actual situation of the comprehensive office building and the functional requirements of the building energy consumption monitoring and management system, the office building energy consumption monitoring and management system is designed by using the hardware resource configurable Shanke PLC(SKPLC) as the data collector. The system uses data bus technology and field data acquisition technology to achieve the building energy consumption data acquisition and management. Practice has proved that energy-saving effect is good.

1 Introduction

At present, in China's large public buildings, excessive power consumption is a prominent issue in building energy consumption. According to statistics, public building energy consumption of 63.35 million tons of standard coal, accounting for 3.6% of the total energy consumption. The energy cost of spending growth rate is much higher than the national energy consumption growth rate. Unit area power consumption is 5 times to the ordinary public buildings, while it is Europe, Japan and other developed countries similar to the building 1.5 to 2 times. Therefore, researching a set of practical building energy consumption monitoring and management system for real-time monitoring and analysis of energy consumption data statistics, it is necessary for the relevant departments to provide energy efficiency assessment basis.

Based on the actual situation of the office building of the eastern district of Shandong Academy of Sciences and the technical characteristics of SKPLC whose hardware resources can be configured. This paper designs a set of monitoring and management system of office building energy consumption based on data bus technology and field data acquisition technology[1]. One of the main equipment data acquisition use the Shandong province Computing Center independently developed hardware resources configurable PLC[4-5]. The system mainly completes the classification of

[a]Corresponding author: 760825602@qq.com

large-scale buildings, sub-energy data collection and processing, to achieve the comprehensive office building energy consumption dynamic monitoring and management.

2 System Composition

The system includes three types of hardware devices:metering terminal, data acquisition control terminal and data collector; two software parts:data transfer station and data center, web information distribution system. The system architecture is shown in Figure 1.

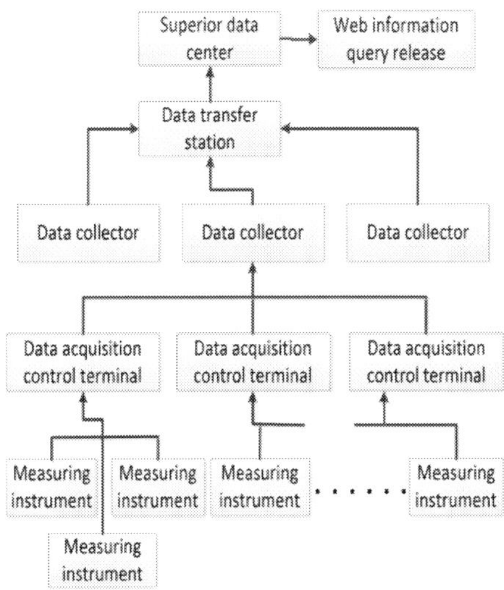

Figure 1. Building energy consumption monitoring and management system architecture

The metering **instrument** is responsible for energy, water, gas and other classifications for energy consumption data measurement. The data acquisition and control terminal converts the collected meter data into a digital signal. The data collector collects data from the RS485 bus network to collect and control the digital signals of the terminals, and the data acquisition and control terminal controls sends data to the data center or transit station by the LAN and TCP/IP protocol. After data processing, the data center is provided to the web information publishing system for users to access and query the building energy consumption monitoring data through the LAN and browser. In this system, a total of 40 power metering terminals, 6 water flow meters and 8 data collectors were installed.

The data collector is installed in the data acquisition and control terminal centralized location, mainly to complete the acquisition of different types of energy consumption to collect terminal equipment and sub-energy equipment data and upload data to the host computer such as data center server, which is the most complex in this system of the hardware equipment. In the integrated office building building energy consumption monitoring and management system, we use the SKPLC that the hardware resources can be configured as the core of the mountain equipment to set up a data acquisition device.

3 Realization of Data Acquisition Device of SKPLC Based on Hardware Resource Configuration

In this system, the data collector collects data of 6 remote water flow meters, 40 power station capable of measuring terminals and 2 central air conditioning equipment, etc., and uploads data to the data center server through the LAN. The hardware resource configurable SKPLC, which is independently developed by Shandong provincial computer center, targets the Siemens S7300 series[2-3], whose hardware resources can be configured, supporting RS85, RS232 and Ethernet interface, having the characteristics of the large capacity I/O channel expansion, modular design, simple programming, reliable performance. After the logic programming of SKPLC, the function of data collector is realized, which is the core equipment of building energy consumption monitoring management system.

3.1 Power metering terminal data acquisition

Because the power metering terminal uses the RS485 custom agreement, which is same as the data collector and data center protocol, so the data collector sends the data acquisition command from the data center, directly from the RJ45 forward to connect multiple metering terminal on the RS485 bus, the arrival of a receiving electricity metering terminal in the RS485 interface, Ethernet can be forwarded to RS485. Collect the metering terminal program flow in Figure 2.

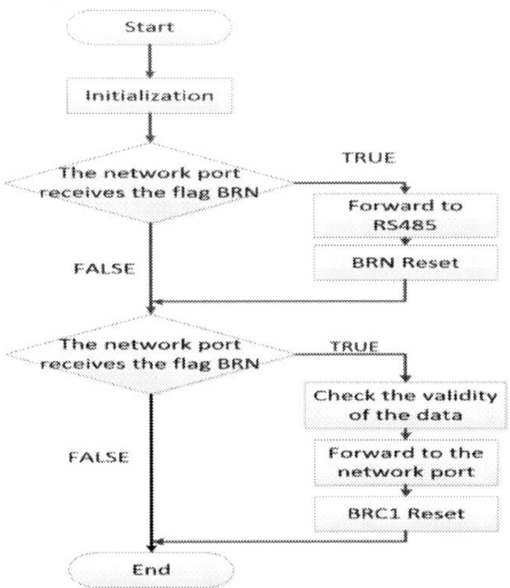

Figure 2. Collect the flow chart of the metering terminal

3.2 Tap water flow meter data collection

Because the tap water flow meter uses the third party product, its communication agreement uses the factory FE FE 68 19 as the guide character customization agreement. The data collector uses the regular poll flow meter data, checks after assembling the data frame waiting to send, responds to the data center Patrol command to send data frames. Acquisition of three flow meter program shown in Figure 3.

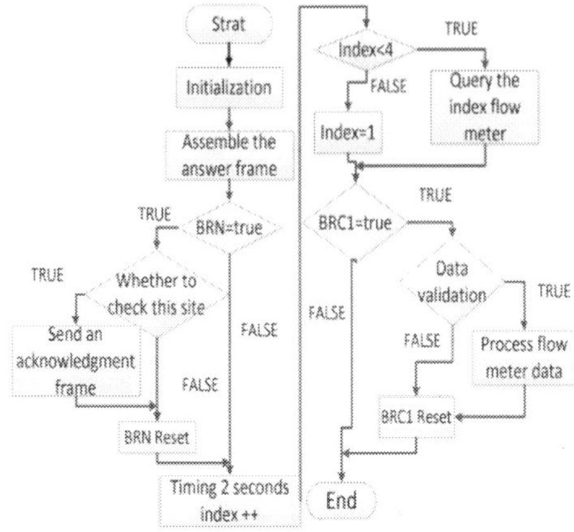

Figure 3. Flow chart of collecting flow meter program

3.3 Central air conditioning data collection

The integrated office building of the eastern of the Shandong Academy of Sciences used Central air-conditioning system, which reserved MODBUS data remote interface. The data collector collect central air-conditioning operation data in accordance with MODBUS agreement. The Central air-conditioning data acquisition process flow shown in Figure 4.

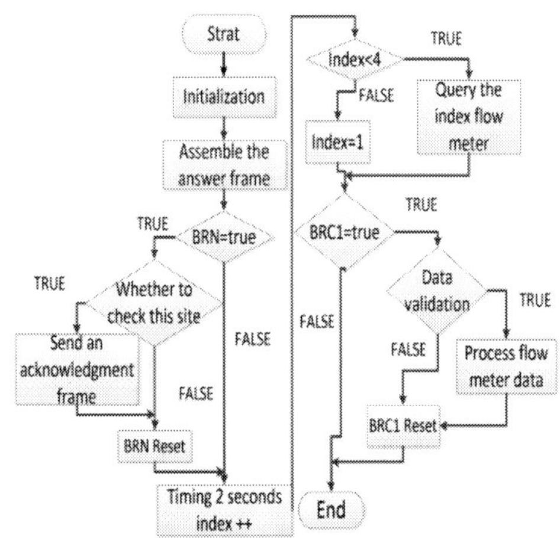

Figure 4. Flow chart of collecting central air conditioning program

4 Conclusions

Through the operation of the past six months, the energy consumption monitoring and management system of the comprehensive office building of the eastern district of Shandong Academy of Sciences is stable and reliable. The relevant leadership and management departments can grasp the energy consumption of the buildings through real-time dynamic building energy monitoring. Query analysis of building energy consumption rules, energy consumption trends and other information to improve the level of energy consumption management. Before the implementation of the system, the total cost of energy consumption per year of the office building is about 2.2 million yuan. After the implementation of the system, the total cost of the system is reduced to 1.98 million yuan, which is achieved by saving 10% of the energy consumption. It has good economic benefits and social benefit.

Acknowledgement

This work has been supported by the project of the Shandong Province independent innovation and the results of special conversion (2014CGZH0804), Major Projects of Science and Technology in Shandong Province (2015ZDZX10002,2015ZDXX0101G05) and Shandong Province key research and development plan (2017GGX30117) and Technology research(GG201703090037).

References

1. National Ministry of Housing and Urban and Rural Development. Technical Guidelines for Data Acquisition of Energy Consumption in State Office Building and Large Public Building Energy Consumption Monitoring System.S.
2. ChangChu Miao. The Programming Method and Engineering Application of Programmable Logic Controller. M. ChongQing: Chongqing University Press, 2001.
3. YunXin Gong, HouYu Zhao, Benzhi Qi Editor. PLC technology and application: based on Siemens57-200. Tsinghua University Press. 2009,9.
4. Feng Zhao, Guanghe Cheng, Xiang Sun, Xuan Luo, XuXai Ren. Journal of electronic measurement and instrument . 237-239, (2009).
5. HuiQiang Li, Guanghe Cheng, LingYan Han, KeYing Li, MaoLi Wang. Automation & Instrumentation. 1, (2010), 17-20.

Joint Equipment Support Task Assignment Model and Simulation Based on Genetic Algorithm

Hai ying Ma[1], *Bo Zhang*[1]*, *Dongyan Li*[1], *Bo Bao*[1]

[1]Air and Missile Defense College, Air Force Engineering University, Xi'an Shanxi 710051, China

Abstract: With the transformation of the war pattern under the information conditions, the corresponding object of equipment support also changes from "equipment-oriented" to "warfare-oriented", and the equipment implement the joint support. Based on the characteristics of the equipment support task under the system combat conditions, the Joint Equipment Support task assignment model with the shortest transition time is established, and a model algorithm based on genetic algorithm is proposed. The validity and accuracy of the model and the algorithm are verified by the example simulation.

1 Introduction

At present, in order to adapt to the trend of the change of war pattern, and to improve combat effectiveness, the world army carry out the adjustment and reform of the military system[1][2]. This adjustment and reform shows that the future pattern of warfare will be transformed from single soldier operations to joint operations and system operations, and the object of equipment protection closely related to it will be changed from "equipment-oriented" to "warfare-oriented", and implementation Joint Equipment Support.

Joint Equipment Support is to deploy all the safeguarding forces and resources in the combat area by the Joint Support Command, the support forces deploy adopt "modular combination" method to form a temporary support team to implement equipment support according to support task. At present, there are many studies on the allocation of equipment support tasks and the ensemble of support forces [3][4], but most of the models are not suitable for the practical application of the joint equipment support of regional operations and system operations under modern high-tech conditions. Therefore, arming at the characteristics of the joint equipment support in theater, how to establish the support force model which fit the support task demand from the numerous support forces is the key for commander of wartime equipment support to make the accurate guarantee decision-making.

2 Equipment Support Task Assignments Model

2.1 Basic conditions and assumptions of equipment support tasks

Considering the real conditions that a multi-combat unit needs to be supported at the same time, and each combat unit has many professional equipment or parts needs to be supported, the support personal belonging to a single support task can be divided into multiple

* Corresponding author: zhb8706@163.com

 DOI: 10.1051/matecconf/201713900136

support groups to implement according to the different organizations, levels and professional skills, the proposed definition and assumptions of air defense and anti-missile equipment support force grouping problem are as follows:

(1)Support tasks B_{ij}: All safeguards for maintenance, electrical inspection, troubleshting and replacement of parts of a weapon unit shall be called a safeguard task. Support tasks of different operational direction have different priority, and the support task of key combat direction should be considered as priority.

(2)Support Team T_i: The minimum professional equipment support unit which is composed of different levels of professional members that taken from the safeguard module of the security force system, known as the protection group. The support group can complete a variety of maintenance tasks, different support groups will take different time for the same support tasks.

(3)There is a sequence of operations among all the support tasks of any combat unit, and there is no sequence of support tasks between different combat units.

(4)Each support group can only operate one support task at any time. Each support task of any combat unit can be operated by only one support group at any time of protection.

(5)Without loss of generality, it is assumed that each support task of each fire unit, once started, cannot be interrupted.

2.2 Problem description

In joint operation, there may be multiple combat directions operations at the same time, so all the combat units of different direction are necessary to be support, and there is a prior act based on the priority of the different combat direction, when a support group finished a security task, can turn to support another combat unit, and ultimately complete all the tasks. Therefore, the problem of equipment support task assignment is to find the most scientific matching scheme for support force and support task, including each combat unit for each security task to select the most appropriate protection group, each group needs to determine the best sequence for the support task set, and once each support task start, it should complete all the tasks of combat units in the shortest possible time.

2.3 Support task allocation model

According to the above description of the equipment support task and the description of the problem of the task assignment, the task assignment model is established as follows:

Objective function:

$$\min\left(\max_{i\in[1,\dots,I]}(t_i)\right) \quad (1)$$

Which is subject to

$$\sum_{i=1}^{I} x_{ij}^k = 1, j=1,2,\dots,J, k=1,2,\dots,h_j \quad (2)$$

$$x_{ij}^k = \begin{cases} 1, & s_{ij}^k \neq 0 \\ 0, & s_{ij}^k = 0 \end{cases} \quad (3)$$

$$0 \le s_{ij}^k \le \sum_{j=1}^{J} h_j \quad (4)$$

$$s_{ij}^k \in Z \text{ , } Z \text{ is a natural number} \quad (5)$$

The objective function of Eq. (1) represents the shortest time to complete the task. Eq.(2) indicates that at any one time, each support group can only perform a support task; Eq.(3) shows the order of the k-th task in the task sequence of support task group T_i. Eq.(4) indicates that the task sequence number of the support group cannot be greater than the total sequence number of the guaranteed task.

In the objective function, t_i is the total time for support group T_i to finish the support task. The priority weights of the different combat directions is expressed, and is calculated within the guarantee time, the support time in the calculation formula also includes the transition time of the support unit from a certain place to the supported unit.

$$t_i = at_{ij_1} + \sum_{n=1}^{m} st_{ij_n}^{k_n} + \sum_{n=2}^{m} zt_{j_{n-1}j_n} \tag{6}$$

Where, at_{ij} represents the arrival time of the support team T_i from its support base to the combat unit B_j. st_{ik}^{j} represents the support time of support task B_{jk} that the support group T_i on the combat unit B_{jk}. zt_{ij} represents the time of combat unit Bi transport to B_j.

3 Genetic algorithm for support task assignment model

As a widely used intelligent optimization method, Genetic algorithm (GA) is applied to solve the discrete combinatorial optimization problems [5-7]. Based on the genetic mutation between individuals and the natural selection mechanism of the tournament, the paper designs the coding/decoding and fitness function for the grouping problem of support forces, and obtains the most scientific task assignment scheme.

3.1 Algorithm flow

Based on the evolutionary mechanism of genetic algorithm and iterative optimization, this paper designs the algorithm for solving the task assignment optimization problem of the air defense and anti-missile equipment maintenance team as shown in Fig.1.

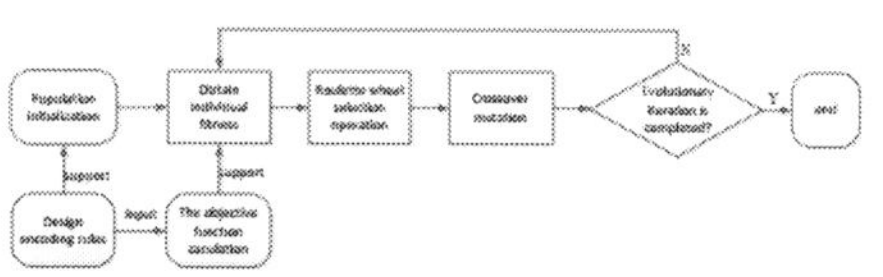

Fig.1. Algorithm process of support task assignment

The key points of the algorithm: First, through great coding and decoding rules, building a seamless transition between the construction of the support power grouping scheme and the feasible solution of the genetic evolution operation. Second, design the simple and scientific calculation method of the objective function, generate the fitness of each individual in the population genetic evolution operation by the support force grouping scheme.

3.2 Encoding and decoding rules

According to the characteristics of the equipment support task assignment problem, we assume that the input of support task is shown in Table.1.

Table 1. Input of maintenance support task problem

task	B_{11}	B_{12}	B_{13}	B_{21}	B_{22}	B_{23}
T_1	1.5	2	∞	∞	1.5	∞
T_2	2	∞	3	2.5	∞	3
T_3	1	1.5	∞	2	2	∞
T_4	∞	∞	2	∞	2.5	2

A non-zero number in the table indicates the time that the support team performed the corresponding support task, and "∞" indicates that the corresponding task cannot be performed.

The design code is to be able to clear the only refers to the support group corresponding to all maintenance tasks in the sequence of operations. And the design code is to be able to refer to the only support group for each task.

Therefore, according to the characteristics of the task assignment problem of equipment support, this paper proposes a two-level coding rule based on the maintenance support task and the maintenance support group with the natural number serial number coding, and is illustrated by a simple example, as shown in Table.2.

Table 2. Feasible solution of support group scheme

Task	B_1	B_1	B_1	B_2	B_2	B_2
First layer code	1	3	4	2	1	2
Second layer code	2	1	1	1	1	2

The first layer code is used to refer to the support group label corresponding to each safeguard task; the second layer code is the task sequence number for the support task in the corresponding support group.

Decoding rules: Confirm the support group corresponding to the support task force according to the security group label, and then, confirm the order of implementation according to the support task sequence number of support group. In this way, we can know the tasks of the support group T_1 and its order $B_{22} \rightarrow B_{11}$, the support group T_2 and its order B_{21}, the support group T3 and its order B_{12}, the support group T_4 and its order B_{13}.

3.3 Fitness value calculation

The purpose of support force marshaling is to minimize the total support time, so the Eq. (1) is used as the fitness value of the chromosome to realize the individual evaluation and ranking.

3.4 Select, crossover and mutate operations

Genetic Algorithm(GA) is selected by the roulette wheel operation, select the next generation of iterative evolution of the population of individuals, the selected individuals which participate in the next generation of evolution must be a large fitness of individuals, and the greater the fitness, the larger the chance, the size of the area represents this chosen opportunity.

Crossover operation is the most important step in GA. The new generation can be obtained by crossing, which can combine with the excellent gene of the parent individual. Because of the method with double layer encoding and sequence in this paper, the genetic crossover operation adopts the same level independent crossover method, the genetic cross-operation using the same level of independent crossover method, as shown in Fig.2.

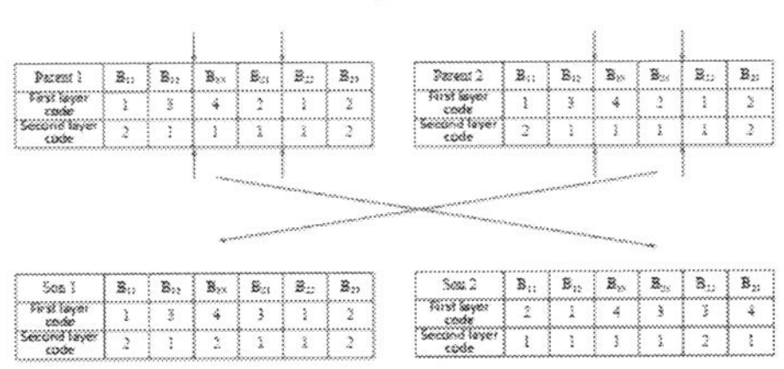

Fig.2. Double layer independent crossover

In Fig.2, the position of the arrow is the crossover point. Here two-point crossover method is selected.

After the new individuals are obtained by crossover and mutation operations, check its feasibility through the Eq. (2) - (4) and the initial input, if not satisfied, repeating the selection, crossover and mutation operations.

After several iterations, we can obtain the optimal support task assignment scheme and the order and time for different teams to accomplish different tasks.

4 Simulation analysis

4.1 Simulation case

Suppose that there has a support group consisting of 10 support units in the combat area, there are 6 combat units need to be supported. In order to verify the reasonableness of the algorithm, set 6 support tasks for each combat unit, and each support task optional maintenance group set and the time of maintenance group to complete the corresponding maintenance task as shown in Table.3.

In Table.3, combat unit 3 and combat unit 4 as the main combat direction that supported by priority. Combat units 1,2,5,6 as the non-combat direction combat unit. The maintenance support tasks are represented by the subscript of combat units, such as the subscript digit "12" of "B_{12}" means the second support task of combat unit 1, but even the second digit of the different combat units subscript are same, does not mean that the same type of support tasks. Each support group and the corresponding digital of maintenance task in the table represent the time of the support group to complete the task, "∞" means that the support group can't complete the maintenance task. The difficulty, quota size of the task and the ability of the support group to support different tasks are indicated by different maintenance time.

Table 3. Input data of equipment maintenance support task example

Task	support group set (time of complete the repair task)									
	G1	G2	G3	G4	G5	G6	G7	G8	G9	G10
B_{11}	∞	∞	1.5	∞	∞	∞	∞	∞	∞	2.5
B_{12}	5	∞	∞	∞	∞	∞	∞	∞	∞	∞
B_{13}	∞	4.5	∞	∞	∞	∞	∞	∞	∞	∞
B_{14}	∞	∞	∞	2.5	∞	∞	2	∞	∞	∞
B_{15}	∞	∞	∞	∞	∞	1.5	∞	1.5	∞	∞
B_{16}	∞	∞	∞	∞	5	∞	∞	∞	∞	∞
B_{21}	∞	3	∞	∞	∞	∞	∞	∞	∞	∞
B_{22}	∞	∞	4	∞	∞	∞	∞	∞	∞	∞
B_{23}	∞	∞	∞	∞	0.5	∞	∞	2	∞	∞
B_{24}	∞	∞	∞	∞	∞	2.5	3	∞	∞	∞
B_{25}	1.5	∞	∞	∞	∞	∞	∞	∞	∞	1.5
B_{26}	∞	∞	∞	1.5	∞	∞	∞	∞	∞	1.5
B_{31}	∞	∞	0.5	∞	∞	∞	∞	∞	2	∞
B_{32}	∞	∞	∞	2.5	∞	∞	3.5	∞	∞	∞
B_{33}	∞	∞	∞	∞	∞	2.5	∞	3	∞	∞
B_{34}	2.5	∞	∞	∞	∞	∞	∞	∞	∞	∞
B_{35}	∞	4.5	∞	∞	∞	∞	∞	∞	∞	5.5
B_{36}	∞	∞	∞	∞	0.5	∞	∞	∞	∞	∞
B_{41}	∞	∞	∞	3.5	∞	∞	∞	∞	∞	∞
B_{42}	2	∞	∞	∞	∞	∞	∞	∞	1.5	∞
B_{43}	∞	∞	2	∞	∞	∞	3	∞	∞	∞
B_{44}	∞	1.5	∞	∞	∞	∞	∞	2.5	∞	∞
B_{45}	∞	∞	∞	∞	0.5	∞	∞	∞	∞	∞
B_{46}	∞	∞	∞	∞	∞	1.5	∞	∞	∞	∞
B_{51}	∞	∞	∞	∞	3	∞	∞	∞	∞	∞
B_{52}	∞	5	∞	∞	∞	∞	6	∞	∞	∞
B_{53}	∞	∞	3.5	∞	∞	∞	∞	∞	∞	4.5
B_{54}	∞	∞	∞	∞	∞	4	∞	∞	4	∞
B_{55}	2.5	∞	∞	∞	∞	∞	∞	∞	∞	∞
B_{56}	∞	∞	∞	2	∞	∞	∞	3.5	∞	∞
B_{61}	∞	1	∞	∞	∞	∞	∞	∞	∞	∞
B_{62}	∞	∞	∞	2	∞	∞	3.5	∞	∞	∞
B_{63}	∞	∞	∞	∞	∞	3	∞	∞	4.5	∞
B_{64}	0.5	∞	∞	∞	∞	∞	∞	∞	∞	∞
B_{65}	∞	∞	∞	∞	2.5	∞	∞	4	∞	∞
B_{66}	∞	∞	1.5	∞	∞	∞	∞	∞	∞	∞

The arrival time from the support base to each combat unit is indicated by the vector $\boldsymbol{AT}$.

$$\boldsymbol{AT} = [1.5\ \ 2\ \ 1.5\ \ 2.5\ \ 1.5\ \ 2]$$

The transition time matrix $\boldsymbol{ZT}$ between combat units is

$$\boldsymbol{ZT} = \begin{bmatrix} 0 & 1 & 1.5 & 2 & 1 & 1.5 \\ & 0 & 2 & 2.5 & 1.5 & 1 \\ & & 0 & 1.5 & 1 & 1.5 \\ & & & 0 & 2 & 2.5 \\ & & & & 0 & 1.5 \\ & & & & & 0 \end{bmatrix}$$

According to the optimization model and solving algorithm, the simulation results are shown in Fig.3 and Table.4.

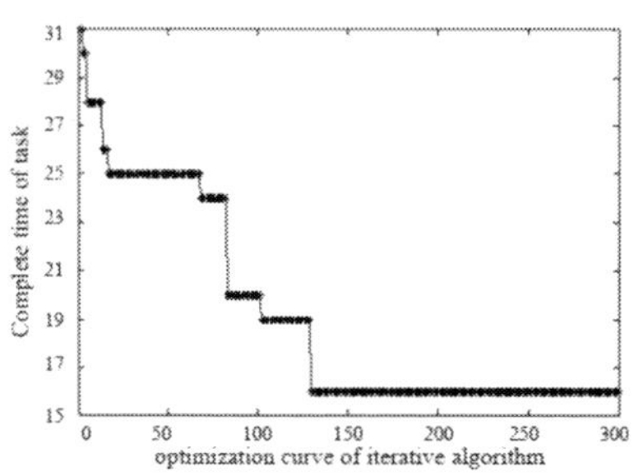

Fig.3. Optimization curve of iterative algorithm

Table.4 Maintenance task assignment results

Combat Unit No.	Task	Support Group	Task Sequence	Task Start Time	Task End Time
Unit 1	B_{11}	T_{10}	1	1.5	4
	B_{12}	T_1	1	1.5	6.5
	B_{13}	T_2	3	8	12.5
	B_{14}	T_7	3	14.5	16.5
	B_{15}	T_6	2	5.5	7
	B_{16}	T_5	4	10	15
Unit 2	B_{21}	T_2	4	13.5	16.5
	B_{22}	T_3	2	4	5
	B_{23}	T_8	1	2	4
	B_{24}	T_6	3	7	9.5
	B_{25}	T_{10}	3	13	14.5
	B_{26}	T_{10}	4	14.5	15
Unit 3	B_{31}	T_3	1	1.5	2
	B_{32}	T_4	1	1.5	4
	B_{33}	T_6	1	1.5	4
	B_{34}	T_1	2	8	10.5
	B_{35}	T_{10}	2	5.5	11
	B_{36}	T_5	2	5.5	6
Unit 4	B_{41}	T_4	2	5.5	9
	B_{42}	T_9	1	2.5	4
	B_{43}	T_7	1	2.5	5.5
	B_{44}	T_2	1	2.5	4
	B_{45}	T_5	3	7.5	8
	B_{46}	T_6	4	12	13.5
Unit 5	B_{51}	T_5	1	1.5	4.5
	B_{52}	T_7	2	7.5	13.5
	B_{53}	T_3	3	9.5	13
	B_{54}	T_9	3	12.5	16.5
	B_{55}	T_1	4	14.5	16.5
	B_{56}	T_8	3	10.5	14
Unit 6	B_{61}	T_2	2	6	7
	B_{62}	T_4	3	11.5	13.5
	B_{63}	T_9	2	6.5	11
	B_{64}	T_1	3	12.5	13
	B_{65}	T_8	2	5	9
	B_{66}	T_3	4	14.5	16

4.2 Simulation results and discussions

According to the optimization algorithm, we obtain the optimal maintenance support power grouping scheme of the study example.

In this study, the order and the total time for each support group to complete the maintenance tasks are:

Support Group T_1: $B_{12} \rightarrow B_{34} \rightarrow B_{64} \rightarrow B_{55}$, task completion time: 16.5.

Support Group T_2: $B_{44} \rightarrow B_{61} \rightarrow B_{13} \rightarrow B_{21}$, task completion time: 16.5.

Support Group T_3: $B_{31} \rightarrow B_{22} \rightarrow B_{53} \rightarrow B_{66}$, task completion time: 16.

Support Group T_4: $B_{32} \rightarrow B_{41} \rightarrow B_{62}$, task completion time: 13.5.

Support Group T_5: $B_{51} \rightarrow B_{36} \rightarrow B_{45} \rightarrow B_{16}$, task completion time: 15.

Support Group T_6: $B_{33} \rightarrow B_{15} \rightarrow B_{24} \rightarrow B_{46}$, task completion time: 13.5.

Support Group T_7: $B_{43} \rightarrow B_{52} \rightarrow B_{14}$, task completion time: 16.5.

Support Group T_8: $B_{23} \rightarrow B_{65} \rightarrow B_{56}$, task completion time: 14.

Support Group T_9: $B_{42} \rightarrow B_{63} \rightarrow B_{54}$, task completion time: 16.5.

Support Group T_{10}: $B_{11} \rightarrow B_{35} \rightarrow B_{25} \rightarrow B_{26}$, task completion time: 15.

It can be seen that the latest task completion time of all the support groups in this support power is 16.5, and the earliest task completion time is 13.5.

Through the analysis of the optimal support force marshaling program, the maintenance task latest completion time of combat unit 3 in the important direction is 11, the maintenance task completion time of combat unit 4 is 13.5. The maintenance task latest completion time of combat unit 1 in non-important direction is 16.5. The maintenance task latest completion time of combat unit 2 is 16.5. The maintenance task last completion time of combat unit 5 is 16.5. The maintenance task latest completion time of combat unit 6 is 16. In this example, the maintenance task latest completion time of combat unit 3 and 4 in important direction are shorter than the combat units in non-important combat direction, which shows that the task assignment scheme is meets the requirement that the mission priority of the important direction and the non-important direction comes second. And the feasibility and validity of the proposed model and algorithm are verified.

5 Conclusion

The support task allocation model is based on the fact that the regional support forces are all set up in a support base, and the total support time is the support group from the support base to the supported combat units, also include the transition time from one operational unit to another when a task was finished. The model takes into account the importance of the combat units in different combat directions, and the difficulty, quota size of the task and the ability of the support group to support different tasks are indicated by different maintenance time. The factors considered in the model are similar to those of the actual support forces, and the model is practicality.

References

1. Chad C Serena. A Revolution in Military Adaption: *The US Army in Iraq War*. (Washington, D.C: Georgetown University Press, 2011)
2. Harvey Sapolsky. *US military Innovation since the cold War: Creation without Destruction*. (New York: Roultledge Press, 2009)
3. F Y Zhang, Q Gao, P He. Research on task assignment model and algorithm of wartime equipment maintenance. Operations research and management science, **15** (1): 62-65(2006)
4. H F Ling, X Z Zhou, X L Jiang, et al. Study on equipment maintenance task assignment based on multi-objectives particle swarm optimization. Application Research of Computers, **29** (6): 2090-2092(2012)
5. Q H Zhang. *Artificial intelligence technology and application*. (Beijing: People's Liberation Army Publishing Housing, 2012)
6. Y P Shao, S Y Chen, Y Liu, et al. Research on Dispatching Problem of War Zone Vehicle Equipment Emergency Support Force. Ordnance Industry Automation, **32** (6): 20-22(2013)
7. An Equipment Support Task Discrimination Model Based on GA Algorithm. System Simulation Technology and application, **8**:193-196(2014)

A Grid Voltage Feed-forward Control Strategy Based on The Virtual Impedance

Chu Wang[1], Jin Wang[1], Tian Qin[1], Xuhui Liu[1], Xiang Wu[*]
[1]School of Mechatronics Engineering, University of Electronic Science and Technology of China, Chengdu 611731, China

Abstract. This paper initially investigates the conventional LCL filter grid-connected inverts system, which introduces the control strategy with inverter-side current feedback, and analyzes the deficiencies within the strategy. The stability of the controller can be improved by adopting the active damping scheme. However, the output current of the inverter is severely influenced by the external grid voltage distortion and the grid low-order harmonic voltage. In order to reduce this impact and gain high-quality power, a novel control strategy is proposed in this paper, in which the virtual complex impedance is introduced to feed grid voltage forward. By using the proposed control strategy, the total harmonic distortion can be greatly decreased, allowing the grid current to synchronize the reference current well. In addition, the modulation of the virtual complex impedance by controller can also quantificationally regulate the equivalent output impedance, which brings great convenience for inverters system designers to calculating and devising. Simulation results are presented to prove the validity and improvements achieved by the proposed controller.

I Introduction

With the acceleration of the industrialization process, the pollution problems caused by the use of fossil fuels and problems of the energy shortage have attracted people's attention, which also promotes the development of renewable energy technologies, such as photovoltaic power generation technology[1-2], wind power generation technology[3-4]and tidal power generation technology[5-6]. In order to meet the demands of users, the power generated by these renewable energy power generation devices needs to be connected into grid through inverters[7]. Hence, stable operation of the inverter can greatly improve the overall capacity and the stability of the grid.

In the case of grid connection, the total harmonic distortion(THD) of the grid current and the power factor of the generation are two important technical indicators[8]. Under the condition in which inverters are required to work both independently and corporately, the inverter normally adopts an LCL filter to inhibit harmonics and control the THD of the grid current in a reasonable range. The distortion of the inverter can also meet the distortion limits as recommend in IEEE Std[9]. However, the LCL filter increases the order of the grid-connected inverter, thereby reducing the system's stability. To address this problem, many kinds of strategies are presented, including double-closed loop control strategy with active damping [10-13],closed-loop control in inverter-side current[14],Split capacitance in closed-loop control[15-16]. In active damping control schemes, the closed-loop control in inverter-side current and the split capacitance in closed-loop control are the most widely used, and compared with other control strategies, the power factor of grid is lower. However, these two strategies have both a stable system and only single

[*] Corresponding author: wx121314@sina.com

current loop which can make the design of parameters easy to achieve. But all these control methods have a common deficiency that in some areas, the grid current can be affected by the quality of the grid voltage because of the severe distortion of grid voltage in these areas.

The control strategy based on the virtual impedance adopts the virtual impedance feedforward branch, which can regulate the output impedance quantificationally and decouple the active and reactive power, bringing convenience to the design of the control algorithm. This control strategy which balances the active and reactive power is widely applied in the parallel inverters system.

In this paper, the analysis of conventional LCL grid-connected inverter system is presented, and furthermore, the module of the system is built. Nevertheless, while applying the control strategy of the inverter-side current feedback, according to the Routh criterion, the quadratic coefficient of the characteristic equation of the system is zero, which means the system is unstable. To stabilize the system, the active damping is introduced, adding a quadratic coefficient and enhancing the stability of the system. Under this control strategy, however, the grid harmonic current is still influenced by the grid harmonic voltage and the grid voltage distortion. To compensate this impact and improve the quality of the grid current, a novel control strategy with the adoption of the virtual complex impedance is proposed in this paper. Applying this controller, the impact can be greatly reduced, and the controller enables the grid current to synchronize with the reference current signal.

II Theoretical Analysis of Conventional Inverter

A. Conventional Inverter Modeling

Full-bridge inverter circuit, consisting of four IGBTs in parallel with the freewheeling diodes, is one of the most widely used circuits in the single phase voltage inverter system [17]. At the same time, the LCL filter needs to be connected to the post-stage circuit because the post-stage circuit contains a great amount of high-order harmonic voltage. The typical structure of the inverter circuit is shown in Fig. 1.

This paper adopts the PWM control mode to get the equivalent SPWM wave. In fact, the on-off action of the IGBTs may have an adverse impact on the system, but this impact is limited. In order to establish the mathematical model of the PWM inverter and the LCL filter, for convenience, this part of impact can be ignored and the PWM inverter bridge can be regarded as a gain K_{pwm}[18]:

$$K_{pwm} = \frac{V_{dc}}{V_{tri}} \tag{1}$$

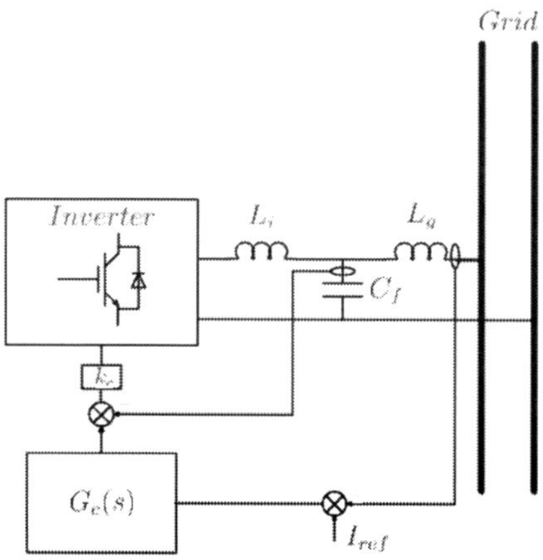

Fig. 1. Power stage of grid-connected inverter using LCL filter

where V_{tri} is the triangular carrier peak; $V_{dc}$is the input voltage of inverter.

At the same time, the complex equations of the LCL filter voltage and current in the output stage can be got, ignoring the filter inductance and stray capacitance. The complex equations are derived as follows:

$$\begin{cases} sL_iI_i(s) = V_i(s) - V_{cf}(s) \\ sL_gI_g(s) = V_{cf}(s) - V_g(s) \\ I_c(s) = I_i(s) - I_g(s) \\ sC_fV_{cf}(s) = I_c(s) \end{cases} \tag{2}$$

where $L_i$is the inductance in inverter side ; I_i is the inverter output current ; V_i is the inverter output voltage ; V_{cf} is the filter-capacitor voltage ; I_C is the filter-capacitor current;C_f is the filter capacitor ; L_g is the inductance in the grid side ; I_g is the grid current ; V_g is the grid voltage. According to the voltage and current equations, we can draw the block diagram of LCL filters. The block diagram is shown in Fig.2:

DOI: 10.1051/matecconf/201713900137

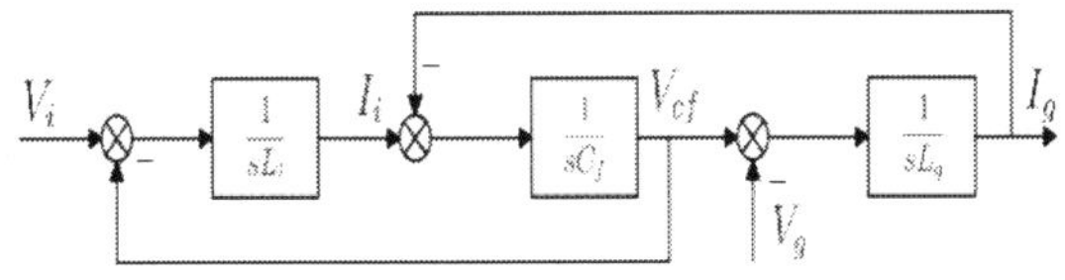

Fig. 2. Block diagram of LCL filter

From the block diagram, we can get the transfer function from the inverter output voltage to the grid current:

$$G_{gi}(s) = \frac{1}{s^3 L_i L_g C_f + s(L_i + L_g)} \quad (3)$$

B. Active Damping

According to the transfer function (3), when the control scheme of the current loop is proportion, the system is a third order system and the system's characteristic equation is $D(s) = a_0 s^3 + a_1 s^2 + a_2 s + a_3$. This equation indicates the instability of the system for the second-order term is 0. To stabilize this system, an active damping control scheme was adopted [17]. The system chart of damped system is depicted in Fig.3.

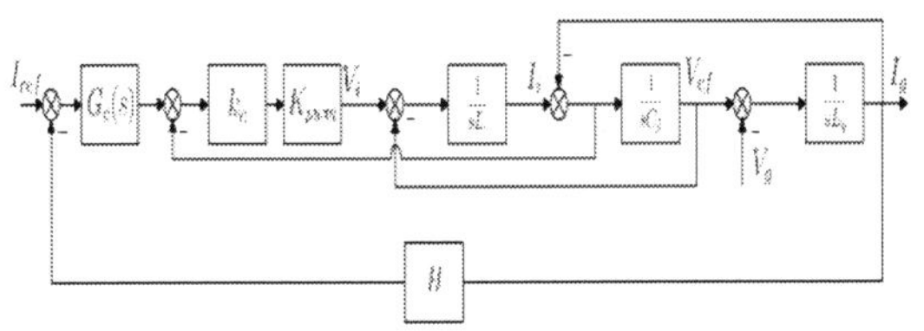

Fig. 3. Block diagram of an inverter with the active damping

Where $G_c$is the transform function of the controller, k_c is the coefficient of active damp , H is the gain of current feedback stage, I_{ref} is the reference current.

The mathematical model of the system can be established and the transform function of this system is as follow:

$$G_{gi}(s) = \frac{G_c K}{s^3 L_i L_g C_f + s^2 K L_g C_f + s(L_g + L_i) + K G_c H} \quad (4)$$

where the$K = K_c K_{pwm}$. According to this function, with the active damp adopted, a second order term is added to the characteristic equation, thereby stabilizing the system.

C. The Impact of Grid Voltage on The Output Current

The output stage of the inverter is connected to the grid through a LCL filter and the output current is influenced by the grid voltage. Whereas the voltage can be affected by a lot of factors such as power-generation quality, nonlinearity of loads, etc. and there are much high order harmonics in the grid voltage. The transform function of grid voltage to output current is derived as follows:

$$I_{gv} = -\frac{s^2 L_i C_f + sKC_f + 1}{s^3 L_i L_g C_f + s^2 K L_g C_f + s(L_g + L_i) + KG_c H} V_g \quad (5)$$

III Virtual Impedance Controller

The grid connected inverter can be modeled as a current source and the equivalent circuit is shown in Fig.4.

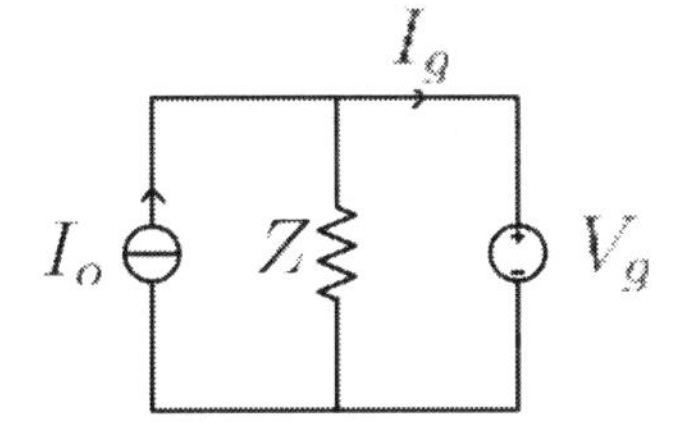

Fig. 4. Equivalent circuit of the grid connected inverter

The output current is expressed as:

$$I_g = I_o - \frac{V_g}{Z} \quad (6)$$

Assuming that the output impedance $Z = Z_0 \angle \theta$, output current$I_o = I_0 \angle \gamma$, grid voltage $V_g = V_0 \angle 0°$, the vector diagram that illustrates the relationship between the output current and the grid voltage can be depicted in Fig.5.

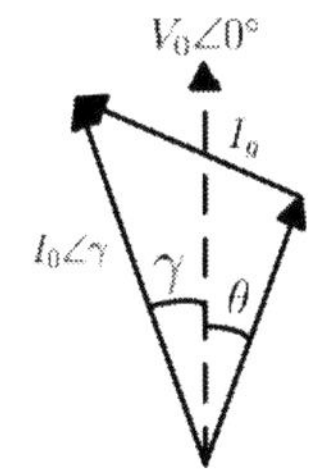

Fig. 5. Vector diagram of equivalent circuit

The impact of grid voltage on the output current depends on two factors: (1) Value of inverter output impedance; (2) The output impedance angle. The impact of grid voltage on the output current is reduced with the increase of the output impedance.

Fig.6 shows a series of bode plots of output impedance for four values of K, varying from 10 to 40, and other parameters are listed in Tab.1. The coefficient of the PI controller are$k_p = 1, k_i = 200$.

With the decrease of K, the output impedance decreases, which enhances the impact of grid voltage. In addition, the impedance in medium frequency is tiny, making this situation even worse.

By implementing a fast control loop, the virtual

impedance was adopted to change the output impedance of the inverter. Virtual impedance can be implemented by drooping the reference current proportional to the grid voltage and virtual capacitor can be implemented by subtracting the reference current proportional to the time derivative of the grid voltage. Virtual impedance was adopted in UPS design, which provides the following advantages: 1)System damping;2)automatic harmonic current sharing; and 3) active sharing that is barely affected by phase errors[18]. In this paper, a new virtual impedance module was proposed to eliminate the impact of grid voltage on the output current of inverter connected to grid.

Tab 1. Parameters Used for Drawing Bode Plot

L_i/mH	1
L_g/mH	1
C_f/mF	1
k_c	0.1

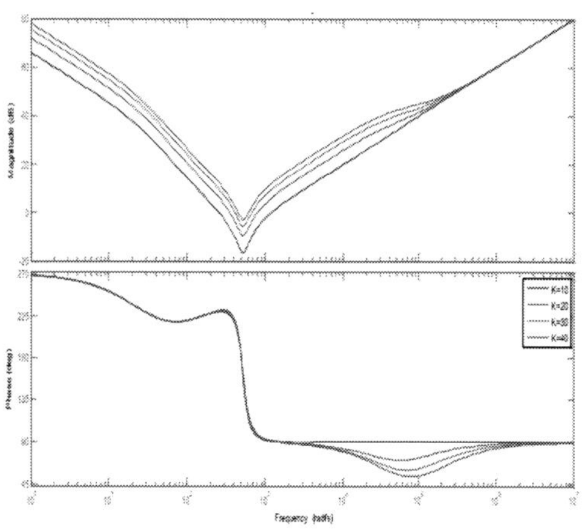

Fig. 6. Bode plot of output impedance

The system chart with a virtual impedance module is depicted in Fig.6

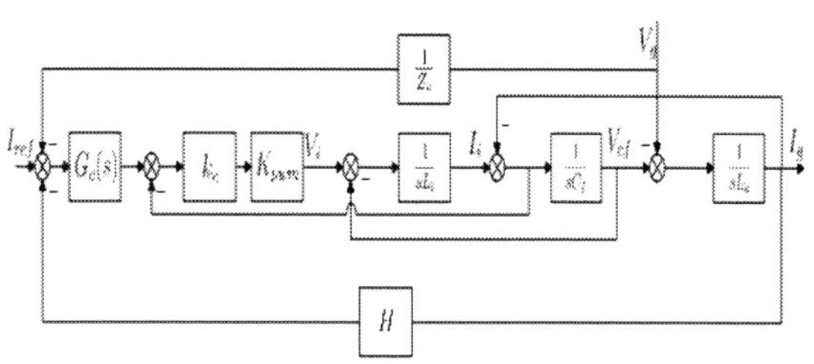

Fig. 7. Block diagram of an inverter with the virtual impedance

where Z_v is virtual impedance. With this module, the transform function from output current to the reference current can be described as follows:

$$G_{gi}(s)=\frac{G_cK}{s^3L_iL_gC_f+s^2KL_gC_f+s(L_g+L_i)+KG_cH} \quad (7)$$

The transform function from output current to the grid voltage is:

$$G_{gv}(s)=-\frac{\frac{G_cK}{Z_v}+(s^2L_iC_f+sKC_f+1)}{s^3L_iL_gC_f+s^2KL_gC_f+s(L_g+L_i)+KG_cH} \quad (8)$$

So the output current can be described as follows:

$$\begin{aligned} i_g &= G_{gi}(s)I_{ref}+G_{gv}(s)V_g \\ &= \frac{G_cK}{s^3L_iL_gC_f+s^2KL_gC_f+s(L_g+L_i)+KG_cH}I_{ref} \\ &- \frac{\frac{G_cK}{Z_v}+(s^2L_iC_f+sKC_f+1)}{s^3L_iL_gC_f+s^2KL_gC_f+s(L_g+L_i)+KG_cH}V_g \end{aligned} \quad (9)$$

The PI controller was adopted to track the reference current. Assume that the transform function of controller is

$$G_c(s)=k_p+\frac{k_i}{s} \quad (10)$$

The characteristic equation of the proposed system is

$$\begin{aligned} D(s) &= s^4L_iL_gC_f+s^3k_cK_{pwm}L_gC_f+s^2(L_g+L_i)+ \\ & sk_cK_{pwm}K_pH+k_cK_{pwm}K_iH \end{aligned} \quad (11)$$

According to the Routh criterion, the Routh table can be described as in Tab.2.

To stabilize the system, each term in the first column of Routh table must be greater than 0, which indicates that:

$$\begin{cases} L_i+L_g-k_pHL_i>0 \\ (L_i+L_g-k_pHL_i)Kk_pH-k_c^2k_{pwm}^2L_gC_fk_iH>0 \end{cases} \quad (12)$$

In order to eliminate the effects of grid voltage harmonics, the controller should be designed to make $G_{gv}(s)=0$. So, the expression of virtual impedance controller can be derived as follows:

$$H_{zv}(s)=\frac{1}{Z_v}=-\frac{s^2L_iC_f+sKC_f+1}{G_cK} \quad (13)$$

where $G_c(s)$ is the PI controller mentioned above:

$$G_c(s)=k_p+\frac{k_i}{s} \quad (14)$$

Therefore:

$$H_{zv0}(s)=-\frac{s^3L_iC_f+s^2KC_f+s}{K(k_ps+k_i)} \quad (15)$$

Since the order of the numerator is higher than denominator, this controller cannot be realized. A second order low pass filter should be adopted and the transform function of the low pass filter is:

$$H_{LP}(s) = \frac{1}{(1+\frac{s}{\omega_n})^2} \quad (16)$$

Where the ω_n is the cut-off frequency. Then, the controller function can be derived as follows:

$$H_{zv}(s) = -\frac{s^3 L_i C_f + s^2 K C_f + s}{K(k_p s + k_i)} \frac{1}{(1+\frac{s}{\omega_n})^2} \quad (17)$$

With the ω_n set high enough, $H_{zv}(s)$ will be equivalent to $H_{zv0}(s)$. In this way, the effect of grid voltage harmonics can be eliminated.

Tab 2. Routh Table for Active Damping Closed Loop System

s^4	$L_i L_g C_f$
s^3	$k_c K_{pwm} L_g C_f$
s^2	$L_i + L_g - k_p H L_i$
s^1	$\frac{(L_i+L_g-k_p H L_i) k_c K_{pwm} k_p H - k_c^2 k_{pwm}^2 L_g C_f k_i H}{L_i+L_g-k_p H L_i}$
s^0	$k_c K_{pwm} k_i H$

IV Simulation and Data Analysis

After using the control strategy according to the control block diagram shown in Fig. 6,the grid-connected inverter model of the LCL filter can be established. Here, the PI control coefficient are $k_p = 5, k_i = 100$. The grid voltage contains a serial of harmonic voltage, including the fundamental wave 220V(50HZ),the 3rd harmonic voltage 73.7V, the 5th harmonic voltage 44V , and the 7th,9th,11th harmonic voltage 5V.

Figure 8 (a) and (b) depict the reference current and grid current waveform under the corresponding control schemes and the spectrum analysis of is shown in(c), (d).

It can be concluded from the simulation results of Fig. 8 (a) and (c) that when the virtual complex impedance is not added, the grid current contains a large amount of low-order harmonics (mainly 3 and 5 harmonic waves).Besides, the phase position is lagged behind the reference phase and the current distortion rate THD is 14.16%.The waveform has obvious distortion and this issue has not been properly treated in the conventional techniques.

In contrast, from the simulation results in Fig. 8 (b) and (d), it can be seen that after adding the virtual complex impedance, the low-order harmonics contained in the grid current have been obviously inhibited and the phenomenon of phase lagging is improved. In addition, the distortion is not very obvious, with the THD turning into 5.43%, and the 3rd, 5th harmonics amplitudes have reduced significantly.

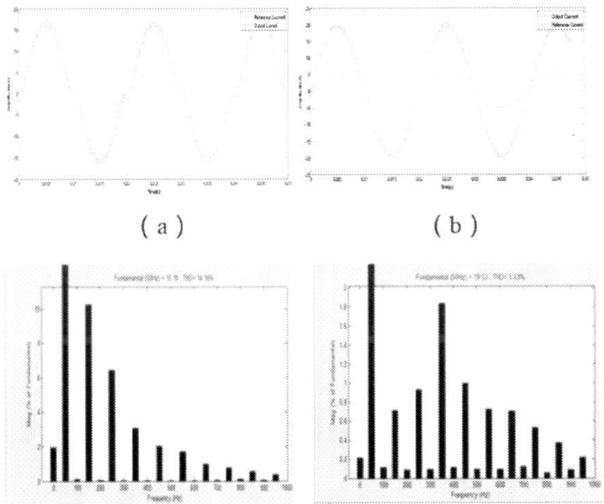

(a) (b)

$k_c K_{pwm} k_i H$ () (c) (d)

Fig. 8. LCL grid-connected in verter simulation waveform and spectrum analysis(a) the current waveform without virtual com plex impedance; (b) the curre nt waveform with the virtual complex impedance; (c) the current spectrum without virtual complex impedance; (d) the current spectrum with the virtual complex impedance;

It is clear that the current quality has been improved significantly after adding the feedforward branch. However, due to the impact of the low-pass filter, the high-order signal is not sent into the feedforward branch, this part of the high-order signal may result in waveform distortion. As a consequence, the improvements in the impacts of the 7th,9th,11th harmonics are not very obvious. Moreover, the LCL capacitor inductance value selection, the controller temperature drift and the inverter DC power supply voltage U_d can increase the voltage distortion rate. Nonetheless, compared with the distortion caused by the high-order harmonics of the grid voltage, these factors only have few impacts. In the real power grid, there are only few harmonics, whose orders are higher than 5. So the differences between the real situation and the simulation results are not huge. Therefore, it is also proven that the virtual complex impedance can minimize the harmonic components of the output current and the output current can track the reference signal well. The feedforward loop has effectively inhibited the output current distortion.

V Conclusion

This paper is based on the analysis of the conventional grid-connected inverter system with a LCL filter, using the inverter-side current feedback control strategy, and points out its inherent deficiencies. To fix those deficiencies, a novel control strategy for grid-connected inverter was presented. Applying the proposed controller, grid-connected inverters can deliver high-quality power, even when there is voltage distortion in the grid. In the paper, the active damping module was adopted to improve the deficiency of the conventional LCL inverters' control strategy and the module increases the stability of the system. Besides, the relationship among the grid voltage, the grid current and the reference current are presented and well investigated as well. With the introduction of the virtual complex impedance module, the impact of the grid voltage distortion and harmonic grid voltage on the grid current can be substantially reduced.

A series of simulations was carried out to test the performance of the proposed controller. According to the results, by adopting the control scheme, the impact of the grid harmonic voltage on grid current has been effectively minimized, and the grid current could be synchronized with the reference current. Simulation results substantiated the validity and improvement of the proposed control scheme.

References

1. Zhou, X. S., Shen, X. B., & Ma, Y. J. (2012). The review of photovoltaic power generation technology. Advanced Materials Research, 609, 118-121.
2. Lapointe, F., & Co/F. (2009). Photovoltaic power generation technology and applications maximum power point tracking technology. Power Electronics.
3. Bao, E. (2004). Current situation of development of wind power generation technology. Renewable Energy.
4. Chen, M., & Yang, G. (2008). The wind power generation technology and its development prospect. Journal of Electric Power.
5. Elghali, S. E. B., Benbouzid, M. E. H., & Charpentier, J. F. (2007). Marine Tidal Current Electric Power Generation Technology: State of the Art and Current Status. Electric Machines & Drives Conference, 2007. IEMDC '07. IEEE International (Vol.2, pp.1407-1412). IEEE.
6. Shiono, M., Suzuki, K., & Kiho, S. (2000). An experimental study of the characteristics of a darrieus turbine for tidal power generation. Electrical Engineering in Japan, 132(3), 38-47.
7. Nigim, K. A. 1. (2006). Development and opportunities of distributed generation fuelled by renewable energy sources. International Journal of Global Energy Issues,26(3-4).
8. Shmilovitz, D. (2005). On the definition of total harmonic distortion and its effect on measurement interpretation. IEEE Transactions on Power Delivery, 20(1), 526-528.
9. Sponsor. (2000). 929-2000 - ieee recommended practice for utility interface of photovoltaic (pv) systems.
10. Malinowski, M., & Bernet, S. (2008). A simple voltage sensorless active damping scheme for three-phase pwm converters with an , filter.Industrial Electronics IEEE Transactions on, 55(4), 1876-1880.
11. Twining, E., & Holmes, D. G. (2003). Grid current regulation of a three-phase voltage source inverter with an lcl input filter. IEEE Transactions on Power Electronics, 18(3), 888-895.
12. Xu, Z. Y., Xu, A. G., & Xie, S. J. (2009). Dual-loop grid current control technique for grid-connected inverter using an lcl filter. Proceedings of the Csee,29(27), 36-41.
13. Xu, A. J., Xie, B. S., Kan, C. J., & Ji, D. L. (2015). An improved inverter-side current feedback control for grid-connected inverters with LCL filters. International Conference on Power Electronics and Ecce Asia (pp.984-989).
14. Abeyasekera, T., Johnson, C. M., Atkinson, D. J., & Armstrong, M. (2005). Suppression of line voltage related distortion in current controlled grid connected inverters. IEEE Transactions on Power Electronics,20(6), 1393-1401.
15. Shen, G., Xu, D., Xi, D., & Yuan, X. (2006). An improved control strategy for grid-connected voltage source inverters with a LCL filter. IEEE (Vol.23, pp.7 pp.). IEEE.
16. Li, Q., & Wolfs, P. (2008). A Review of the Single Phase Photovoltaic Module Integrated Converter Topologies With Three Different DC Link Configurations. IEEE Transactions on Power Electronics, 23(3), 1320-1333.
17. Laaksonen, H., Saari, P., & Komulainen, R. (2005). Voltage and frequency control of inverter based weak LV

network microgrid. International Conference on Future Power Systems (pp.6 pp.-6). IEEE.

18. Tuladhar, A., Jin, H., Unger, T., & Mauch, K. (2000). Control of parallel inverters in distributed ac power systems with consideration of line impedance effect. Industry Applications IEEE Transactions on, 36(1), 131-138.

Research on simulated devices for Solar photovoltaic grid-connected generation system

Zhang quan-zhu, Ge yan-xiang, Deng yong-hong
College of electronic and information engineering, North China Institute of Science and Technology, 065201

Abstract: On the standpoint of energy conservation and emission reduction, one device simulated photovoltaic grid-connected generation system based on SPWM was designed in the paper. And DC/AC inverter could transduce efficiently direct current to alternating current. The MCU(Micro-Control-Unit), in this system could achieve the control method for maximum-power-point and tracking for frequency and phase. Moreover, the MCU could implement PWM (Plus-Width Modulating) through programming. The system showed clearly the whole photovoltaic grid-connected generation system using simulated methods and ways.

0 Introduction

With the rapid development of modern science and technology, the depletion of petrochemical energy and the impact on the environment are becoming increasingly serious, looking for renewable clean energy has become increasingly urgent global research topics. At present, the solar energy inexhaustible clean energy has been favored. The radiant energy of the sun can be converted into heat, which is a common use of solar energy. Photovoltaic power generation is the use of the energy directly into electrical energy the sun radiation, that is to say, light radiation can be converted into electrical energy by solar cells, after energy storage, control and transformation, which become direct or alternating current energy that can be used directly by people. The so-called grid connected, simply speaking, when voltage, phase, phase sequence and frequency are same, the principle of parallel operation can be achieved [1,2].

1 System structure and analysis

As shown in Figure 1, Schematic diagram of photovoltaic grid connected analog device. The voltage U_s generated by the solar panel, through the DC-AC power conversion, then the sine pulse width modulation wave (SPWM) is obtained, through the L-C consisting of sine wave filters, the filter to filter out high frequency harmonics, so that the sine wave fundamental voltage u_{o1} is finally obtained, through the step-up grid connected transformer coupling, and provide voltage u_o to the load. In the figure, R_s is equivalent to the resistance of solar panels, u_f is network synchronization voltage, that is a input of the control circuit coupled with the grid connected transformer, u_{REF} is a given voltage for control circuit, and also is a instruction for the DC-AC converter when adjusting the voltage.

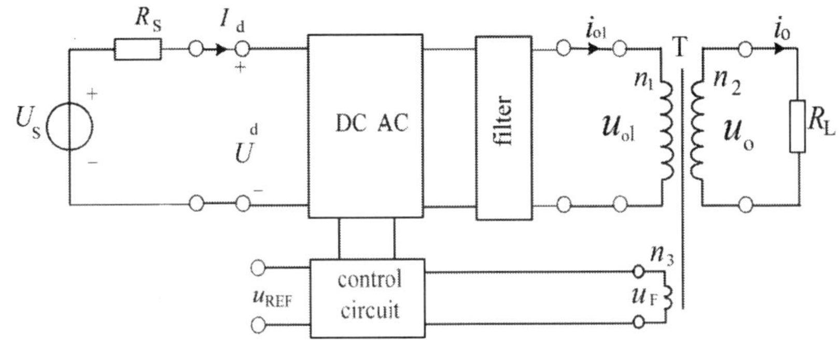

Fig. 1 Schematic diagram of photovoltaic grid connected analog device

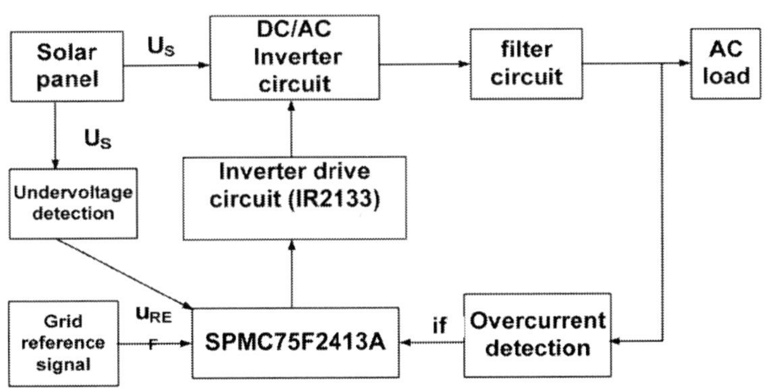

Fig. 2 the schematic diagram of the simulated devices for Solar photovoltaic grid-connected generation system

Figure 2 shows the schematic diagram of the simulated devices for Solar photovoltaic grid-connected generation system. The DC voltage Us which is output by the solar panel, through the DC/AC inverter circuit composed of IGBT, the SPWM wave is formed. L-C sinusoidal filter is used to filter the output waveform, and finally formed a relatively stable sine wave, then through the Boost-Type Grid-Connected transformer, the simulated devices be operated under load. Thereinto, the Sunplus 16bit microcontroller (SPMC75F2413A) as the core constitutes a control circuit, while completing the function of system protection, Photovoltaic Maximum Power Point Tracking. Through sampling the input voltage and output current of the system, the microcontroller forms the protection signal which is under voltage and over-current. At the same time, the maximum power point tracking is completed by sampling the input voltage; The reference voltage of the network is also input into the microcontroller through the detection circuit. As the reference voltage of the grid connected control, the frequency and phase tracking of the power grid is realized. The simulation scheme of the system is complete, the control and drive functions are complete, and the simulation operation can be carried out steadily and reliably.

2 Inverter and filter main circuit

Figure 3 shows the PV inverter simulation device and main filter circuit, inverter main circuit adopts the single-phase voltage type inverter circuit, two bridge arm consisting of four IGBT IGBT, Infineon produced by IKA15N60T, the IGBT with anti parallel diode, the main circuit connection more simple. In addition, the maximum emitter voltage of the IGBT is 600V, the maximum collector current is 15A, and the maximum junction temperature is 175 DEG C, which meets the application requirements. Through the drive circuit of the 4 wave SPWM wave drive to its grid, to achieve the DC/AC transform, output SPWM wave, through the SPWM wave duty cycle D and frequency f control to realize the output voltage and frequency adjustment [3,4].

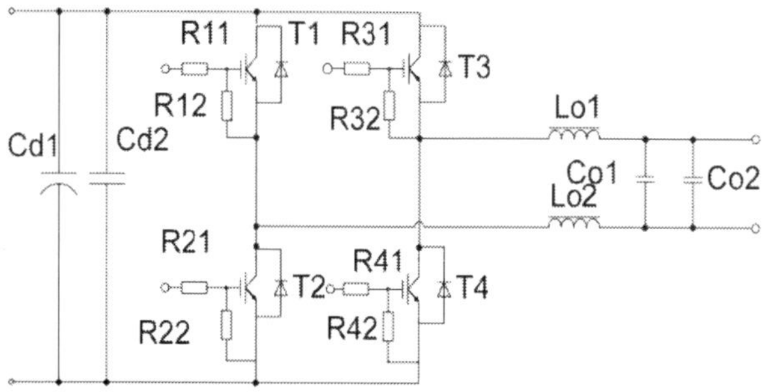

Fig. 3 inverter and filter main circuit of photovoltaic grid connected analog device

3 drive circuit

In this design, we use IR (International, Rectifier) company's IR2133 chip as the driving circuit to drive IGBT (IKA15N60T). IR2133 is designed for high voltage, high speed power switching devices MOSFET and IGBT, which has the advantages of simple and convenient, especially suitable for three-phase Variable frequency transformer circuit.

3.1 IR2133 drive circuit

IR2133 is a driver chip, when it needs PWM pulse input, it will produce pulse output, IR2133 input pulse is provided by 16bit microcontroller SPMC75F2413A. The input signal of the IR2133 is compatible with the output signal of the CMOS or TTL circuit of 5V, and the maximum voltage of the output signal can reach 20V, so that the driving ability of the driving signal is obviously enhanced, and the purpose of driving the IGBT reliably is achieved. In addition, the input control circuit also provides the dead time for high-end devices and low-end devices with one arm in the chip, in order to avoid the phenomenon of simultaneous misoperation that the two powers element is driven with one arm during the switching transition. Thus, the reliability of the drive pulse is improved. drive circuit based on IR2133, which outputs 6 pulse signals, is used to drive the IGBT three-phase inverter.IR2133 based IGBT inverter drive circuit. The output of the systemic inverter circuit is a single phase AC voltage, only 4 drive signals are enough, and the redundant 2 signals can be suspended or grounded through the 10kΩ resistor. If the protection function is not needed during the application of IR2133, each protective terminal can not be processed, which is equivalent to that the IR2133 has been kept in the normal working state.

4 system detection circuit

The design of the detection circuit is very important, and it directly affects the accuracy of the data acquisition by the microcontroller. The system has five detection circuits, which respectively realize the functions of frequency tracking, maximum power point tracking, under voltage protection, overcurrent protection and phase sampling .The frequency tracking detecting circuit transforms the sine wave into the rectangular wave of the same frequency, input the rectangular wave to the microcontroller, and the time difference is obtained by the zero crossing of the detecting waveform. Thus, the frequency of the voltage signal is obtained, and frequency tracking is realized. The sampling circuit of maximum power point tracking is based on the features of maximum power output, When the load equivalent resistance is equal to the load internal resistance, the power output power is the maximum. use the matching method of the sampling voltage values, If the supply voltage 60V is constant (output voltage of solar panels),after the load is connected, the full bridge type inverter voltage regulation is used, in order to make the load equivalent resistance equal to the internal resistance of the power supply, then the Load obtain the output voltage U_d, should be equal to half of the power supply voltage, to realize the function of maximum power point tracking (MPPT).

 DOI: 10.1051/matecconf/201713900138

4.1 frequency detection circuit

As shown in Figure 4 for the frequency detection circuit, the detected AC input voltage, through the differential amplifier circuit the which consists of integrated operational amplifier (TL082), according to the resistance in the figure, calculated the difference signal magnification is 50, the output voltage of the operational amplifier is $50(u_1\text{-}u_2)$. As the supply voltage of TL082 is: ±15V, the maximum output voltage of the amplifier is: ±15V. When the input peak-to-peak voltage is twice the grid reference voltage U_{ref}, the output voltage signal is turned off and the actual output voltage is a square wave. After the square wave passes through the diode IN4148, the negative half-cycle of the original waveform is eliminated, only the positive half-cycle rectangular wave is output. which laid the foundation for the Sunplus single chip microcomputer accurately collecting the zero crossing of waveform. Thus, the frequency of the original sine wave is detected. Sunplus MCU regulates the output AC frequency of the inverter circuit by adjusting the output SPWM wave. which realize that the frequency of the output voltage is equal to the frequency of the reference signal of the power grid, and the voltage frequency of the power grid can be automatic tracked .

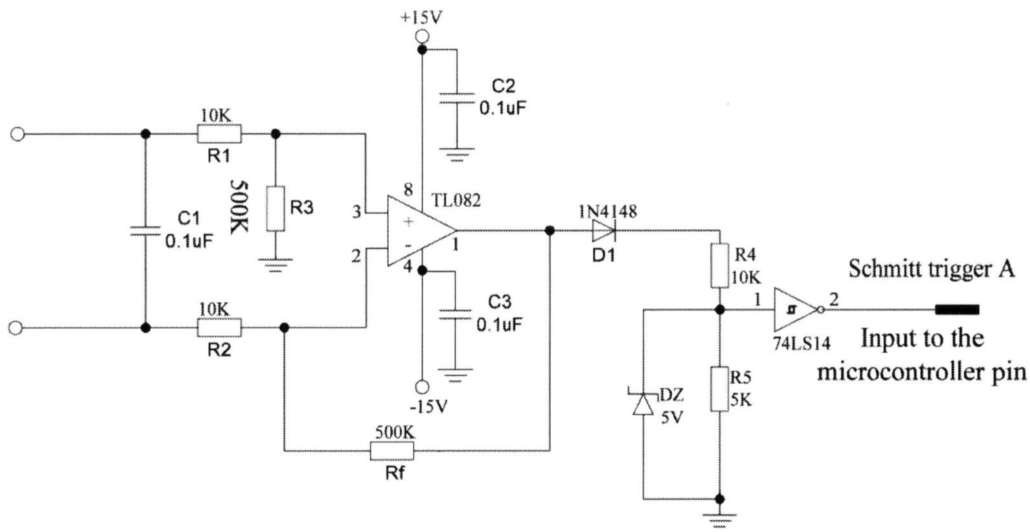

Fig.4 frequency detection circuit

4.2 voltage and current detection circuit.

The detection of voltage and current is almost similar. The voltage detection circuit is shown in figure 5. volts, where the use of the proportional operation is the buck operation, the voltage signal is reduced to less than 5V, so that the microcontroller can accurately sample. The maximum power point tracking (MPPT) sampling circuit is basically the same as the voltage detection circuit, which has been described previously.

The current detection circuit is that a small resistor is connected in the current detecting circuit, such as milli ohm resistance, using the same differential amplifying circuit, the voltage signals at both ends of the ohm resistor are detected, adjusted and input to the microcontroller, after calculation of microcomputer, corresponding to calculate the current value.

As shown in Figure 6, the current detection circuit. The operational amplifier uses the independent operational amplifier of TL082, which can avoid mutual interference between the sampling circuits, so that the output signal is accurate and anti-interference is strong. It is worth noting that the sampling circuit must be repeatedly debugged, to make the amplitude, stability and smoothness of the waveform determine, only in this way can the detection signal be input to the microcontroller to ensure the accuracy of the signal.

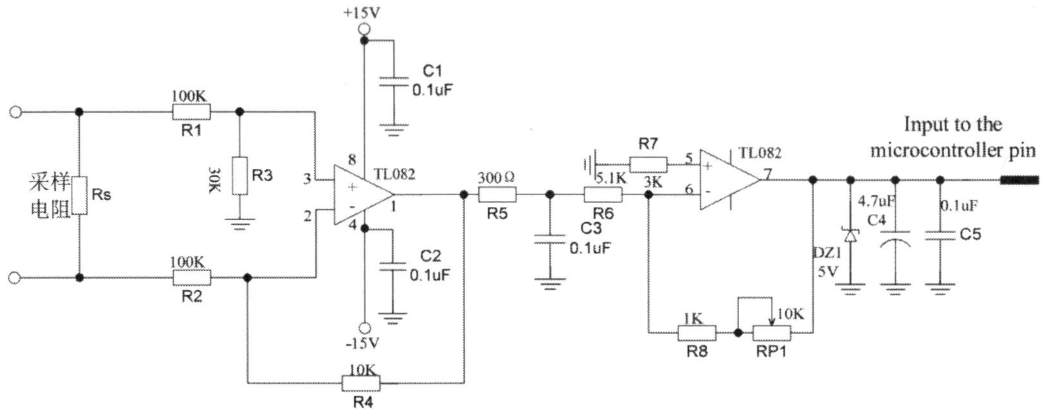

Fig.5 voltage detection circuit

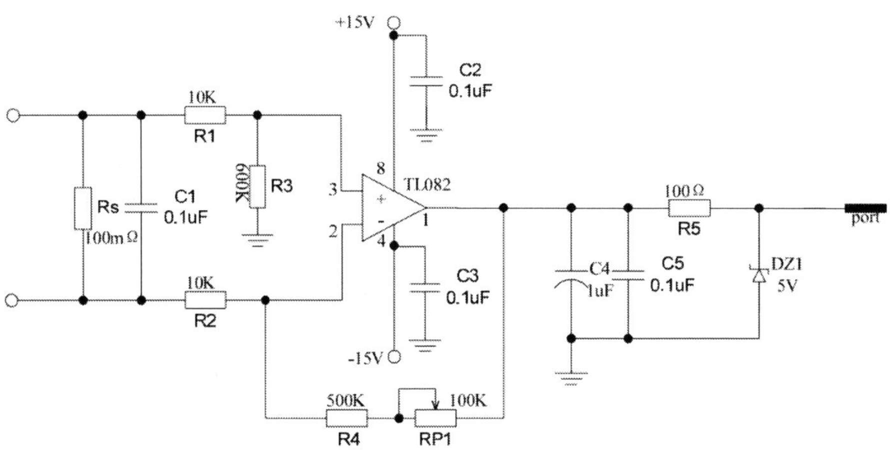

Fig.6 current detection circuit

5 Conclusion

The device is a simulation device, which uses the voltage source to simulate the photovoltaic solar cell, so some places need to be perfected. For example, the actual maximum power point tracking (MPPT) is more complex, need to consider many factors of light intensity, temperature, humidity. but in the simulation device, Just focusing on simulating the intensity of the light (such as when the maximum output power is reached, the load resistance value is equal to the resistance value in the power supply), and other factors is ignored. We are still actively studying and have achieved more new results.

references:

[1] Dong Mi, Luo an. Design and control method of inverters in grid connected photovoltaic systems [J]. automation of electric power systems, 2006 (20): 10-13;

[2] Zhang Zhi-qiang, Ma Qin, Cheng Da-zhang. research on the control technology of solar photovoltaic power generation system [J]. Low voltage apparatus, 2008. (12):28-31;

[3] Zhao Zhen-ming. Solar photovoltaic power generation and application [M], Science Press, 2005,10;

[4] Liu Dao-chun, solar photovoltaic technology and its development prospects [J], public power, 2009 (03): 16-18.

DOI: 10.1051/matecconf/201713900138

[5], Hou Guo-qing, Chen Shi-hua. 20 kW grid connected photovoltaic power generation system [J], solar energy, 2008 (11): 44 – 46.

Foundation item: the project by science and Technology Department of Hebei province (1627460316214408); Chinese Coal Industry Association (MTKJ2016-35, MTKJ2016-329); fundamental research funds for the central universities such as funding (3142016022, 3142017044)

Design of the control system of the bidirectional DC/DC converter for the storage and regeneration

Deng Yonghong , Ge Yanxiang, Qian Huifa
College of electronic and information engineering, North China Institute of Science and Technology, 065201

Abstract: In view of the phenomena of energy waste, environmental pollution and increase of operating cost caused by the braking of the rail transport vehicle, the storage and regeneration of bidirectional DC/DC converter control system is designed, It takes TMS320F28035 DSP as the core control circuit, and adopts voltage and current double closed loop PI control. The system realizes the storage and regeneration of energy during operation. The experiment shows that the system has high reliability and anti-interference ability, so it can be popularized and applied.

0 Introduction

The accumulator locomotive is powered by the storage battery, so it is suitable for transportation in the tunnel with more gas and coal dust accumulation, safety-type battery locomotives can also operate safely in the main ventilation tunnel of super gas mines. As the traditional traction power supply system can not reverse recovery energy, the energy generated by the regenerative braking of the traction tractor is not only wasted, but also increased the burden of other electrical equipment operation, and increasing the ventilation system load. Similarly, urban rail vehicles also have this problem. This paper studies energy storage and regeneration technology of rail vehicles, and make reasonable use of the energy produced by braking, to achieved the purpose of saving energy, improving carrying capacity and stabilizing the supply voltage of a storage battery or a power supply network. Using super capacitor bank as the scheme of accumulator. At present, there is no mature product in the domestic application, the regenerative braking energy of super capacitor adopt in domestic research has just started, and it has just started in urban rail transportation. In coal mine, it is of great practical significance to study the storage and regenerative bidirectional DC/DC converter control system in track operation.

1 Constitution and principle of control system

In orbit, the structure of a power supply system incorporating a bidirectional DC-DC converter is shown in figure 1. The system consists of a storage battery or a power supply network, a bidirectional DC-DC converter, a super capacitor set, an inverter and a He Yongci synchronous machine. In the process of vehicle braking charging, the average charging current is determined according to the regenerative braking power and the capacitor voltage of the locomotive, and the charging process can be automatically ended when the super capacitor voltage reaches the highest charging voltage. In the discharge process, the discharge current is determined according to the starting power of the vehicle and the voltage of the super capacitor, and the bidirectional DC-DC converter is turned off when the voltage of the ultracapacitor is reduced to the minimum operating voltage.Whether the converter operating condition is the charging state or the discharging state, It is determined by the conditions of the vehicle operation, the voltage of the power supply network and the super capacitor voltage. In the case of running conditions, controling the control signal of the main circuit of the corresponding contactor controlchip and a certain duty cycle PWM pulse,which were sent by high-speed output port of control chip , and by the drive circuit to isolate the signal after the drive, the main circuit switching device IGBT Work according to the required duty cycle. thus achieve the recovery of braking energy and reuse.

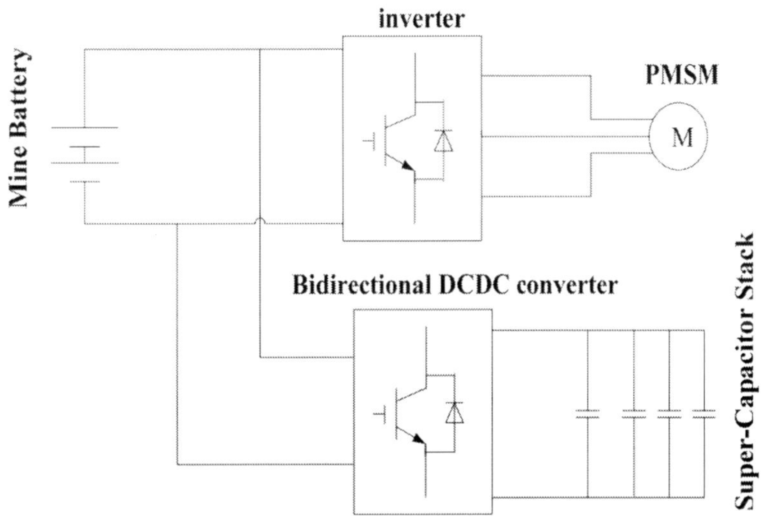

Figure 1 Mine electrical locomotive power supply system structure

1.1 Analysis of operation principle of bidirectional DC/DC converter

As shown in Fig. 2, The bidirectional DC/DC converter consists of a three-phase three chopper circuit, the difference of control pulse is 120 degrees, and its working principle is as follows:

(1) when the vehicles brake again , resulting the power grid voltage increased, bidirectional DC/DC converter circuit in Buck circuit model, and the super capacitor energy storage system connected, for charging. The ultracapacitor group absorbs regenerative braking energy while reducing the voltage of the power supply network.

(2) when the vehicle is accelerating or starting, power supply network voltage will be lower, then the controller of bi-directional DC/DC converter working in Boost circuit mode, super capacitor energy feedback through bidirectional converter. Under this condition, the super capacitor energy storage system is put into use, which reduces the requirement of the peak power of the power supply system, and makes the voltage drop of the power supply network be eased.

(3) when the running vehicle is idle, the controller can control the energy storage state of the super capacitor set by the main circuit of the bidirectional DC/DC converter according to the capacitance of the super capacitor group.

To sum up the analysis can be seen, according to the different operation condition of the vehicle and power supply voltage, the controller control bidirectional DC/DC converter to achieve super capacitor energy storage system and inverter energy exchange, and achieve energy storage utilization.

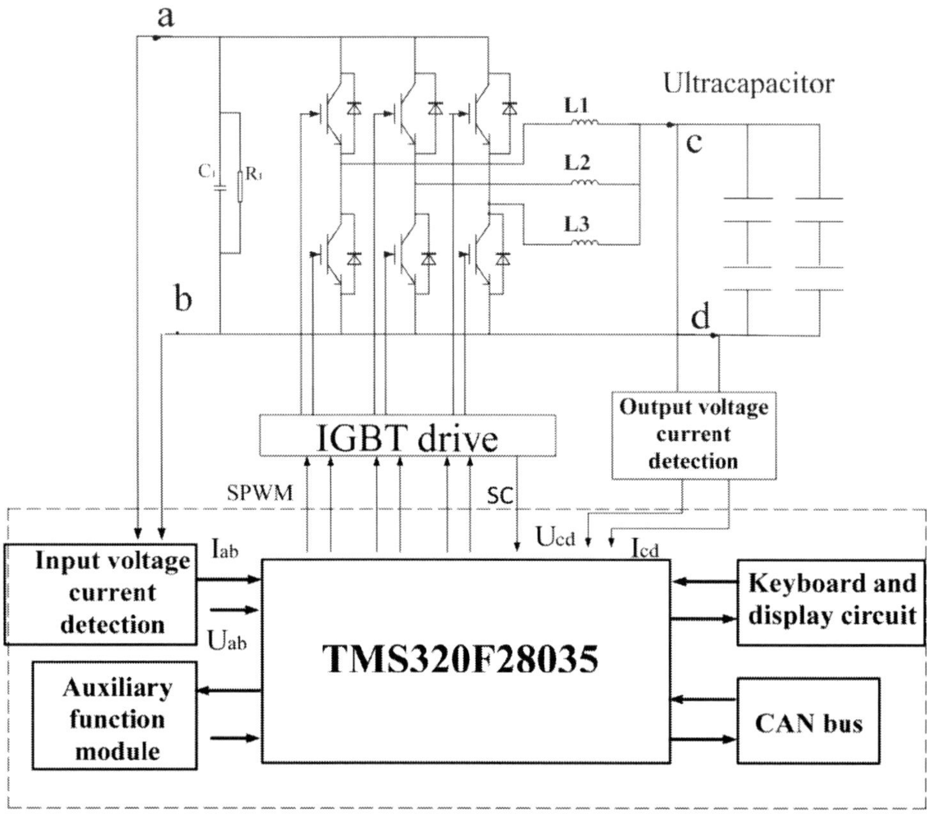

Fig. 2 overall structure diagram of the system

2 DSP control block diagram

The TMS320F28035 control circuit is the core part of the system, and its block diagram is shown in figure 3. The system consists of five parts, such as DSP digital system, touch sensitive interface circuit, IGBT drive circuit, CAN communication circuit and keyboard display circuit. The functions of each part are as follows:

(1) the main task of DSP digital system is receiving the upper control computer or run vehicle bus from the given conditions according to the given information and instructions, Then according to the given, To compare and calculate the physical quantities such as voltage, current with setting value ,the output PWM switch control pulse and the signal control system;

(2) analog analog interface circuit converts the strong current signal into weak current signal for digital sampling of DSP;

(3) the IGBT drive circuit converts the DSP system's signal into the pulse signal that can drive IGBT, and meanwhile, when the IGBT has short circuit or over-current fault, it sends back the fault signal to the DSP system;

(4) the function of the keyboard display circuit is to set various instructions and parameters through the keyboard, display the various states of the system operation, and record the system history running fault.

 DOI: 10.1051/matecconf/201713900139

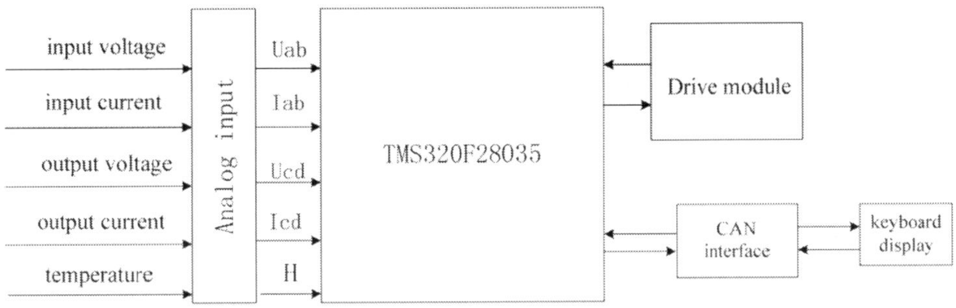

Fig. 3 block diagram of control system

3 Control Algorithm and Software Design

3.1 control algorithm analysis

Block diagram of control algorithm is used in the system, as shown in Figure 4 (a), its working principle is as follows: the feedback value of output voltage U_1 of the power supply system is different from the given reference value U_{ref}, U_1 through the PI regulator as the given current signal output. The current command I_{ref} is different from the inductance current i_L, and finally get the PWM signal of Power tube VT_1 of the Upper bridge arm after PI adjustment; maximum U_{2max} and minimum U_{2min} of allowable voltage compared to the voltage U_2 of super capacitor, the voltage amplitude limiting closed loop of ultra capacitor side is formed, closed limit the highest limit voltage loop and minimum voltage respectively .Comparing the current i_{ref1} of regulator output with the Limiting amplitude of Maximum current I_{max} and minimum current I_{min} , the current command i_{ref1} gets the I_{ref} after the clipping closed loop, and the I_{ref} is used as the input value of the current closed loop.

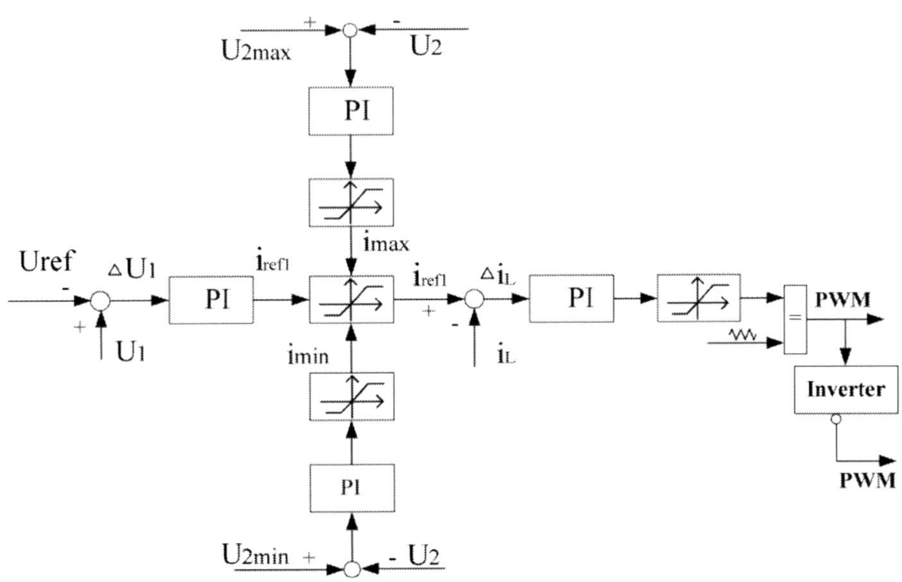

Fig. 4 (a) block diagram of bidirectional DC-DC control algorithm

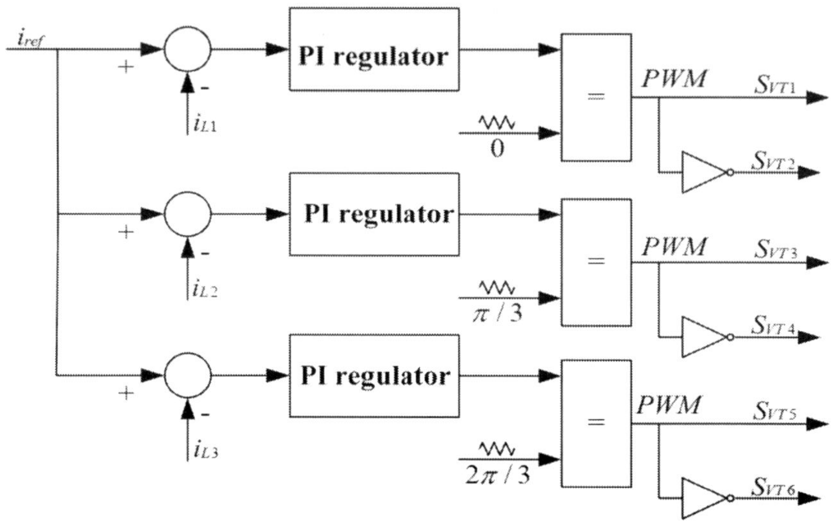

Figure 4 (b) closed-loop current sharing control diagram

Figure 4 shows that the voltage and current double PI closed loop control algorithm, through the DSP to calculate the value of the PI regulator, the output difference of 120 degrees of PWM wave control, three-phase three - way bi-directional DC-DC converter.

3.2 software design

As shown in Figure 5, the software program diagram of the system is composed of main program, interrupt service program, CAN interface program and A/D interface program.

Then the interrupt get into circulation, waiting for DSP initialization, clock circuit configuration, the peripheral circuit and software initialization variables in the definition and assignment, system interrupt configuration of system initialization task; then open communication interrupt, accept the control signal of the upper computer or the inverter controller carries out data transmission, To enable the system to get boot instructions or super capacitor charge and discharge instructions.

The interrupt service program completes all control algorithms, such as current and voltage detection, A/D converter, closed loop PI control and PWM occur, The program sets up three interrupts: interrupts, reads, sampling results, interrupts, and fault protection interrupts, As mentioned in the preceding paragraph, Compare interrupt is used to generate control cycles periodically, and A/D sampling completes sampling results after reading.

The CAN interface program simply passes the status data and control instructions. The basic steps include initialization of the CAN module, sending data, receiving data, and checking the correctness of the data.

The A/D interface program completes the frequency used by AD and sets the sampling mode.

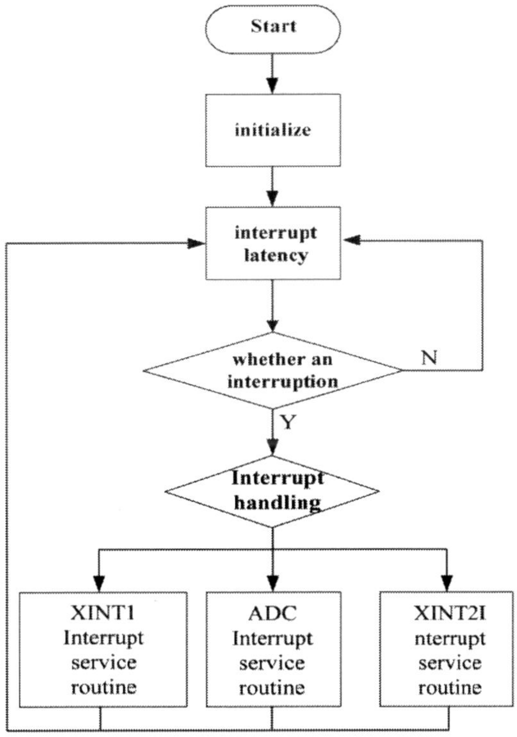

Fig. 5 program flow chart

4 experiment debugging and result analysis

The bidirectional DC-DC converter system has been successfully implemented. The main parameters of the system are as follows: the voltage range of the power grid is between DC163.2V and DC288V, the actual test was DC210V at the time of the experiment, rated power of permanent magnet synchronous shock motor is 22kW * 2. Bidirectional DC-DC converter parameter: The value of the inductor is 2.8mH.The ultracapacitor group is 60V/125F, two series connected, set the PWM frequency of the control system to 5K. Carry on load test, charge the super capacitor when the vehicle slows down or stop. Through the buck chopper circuit, the constant current charging of the bidirectional DC-DC converter. When the motor vehicle accelerates or starts, the super capacitor is supplied to the inverter through a boost chopper circuit, and the actual voltage is 195V. Figure 6 shows the DSP output PWM debug waveform, and the bi-directional DC-DC converter system is stable.

5 Conclusion

Energy storage and regeneration bidirectional DC-DC converter system is Realized for rail vehicles, and the current and voltage double closed loop PI regulator control system has been successfully developed based on DSP. The hardware of the system is simple, modular strong, reliable operation, the recovery and reuse of brake energy are realized. The system has better engineering practicability and popularization value.

DOI: 10.1051/matecconf/201713900139

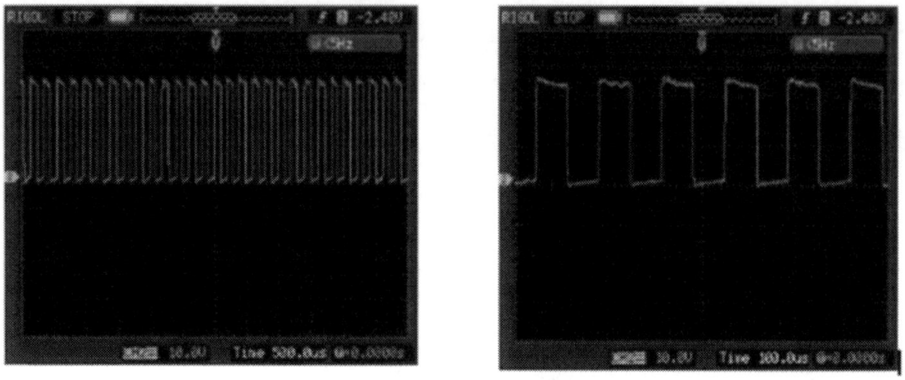

Figure 6 PWM debug waveform

Reference:

[1] Wang Zhaoan, Huang Jun. Power electronic technology. Fourth edition. Beijing: China Machine Press, 2006

[2], Pan Xinmin, Wang Yanfang. Practical tutorial of microcomputer control technology. First edition. Beijing: Publishing House of electronics industry, 2010

[3] Liu Lingshun, high bright, Zhang Shutuan, Wang Fang.TMS320F28335DSP principle and development programming. Beijing: Beihang University press, 2011

[4] Chen Boshi. Electric drive automatic control system. Third edition. Beijing: Mechanical Industry Press, 2011

[5] Hu Shousong. The principle of automatic control. Fifth edition. Beijing: Science Press, 2007

[6] Xu Dehong. Modeling and control of power electronic systems. First edition. Beijing: China Machine Press, 2006

[7], Wang Xinggui, Zou Yingwei, Liu Jinlong. Modeling and control of full bridge DC/DC switching power. Power electronics, 2007.7, 41 (7): 86~88

[8] Cui steel. Research and development of PWM switching power supply based on DSP control: dissertation. Dalian: Dalian University of Technology, 2006.6.1

[9] Deng Yonghong, Zhang Quanzhu, Ren Xuejun, Zhi Qiang. Research on the power supply mode of the class driven offshore oil logging device of motor.

Power technology. 2011, 35 (2): 202-206.

[10] Wang Xuedi. Research on super capacitor energy storage system based on urban rail transit. [Dissertation], Beijing: Beijing Jiaotong University, 2008.6

Foundation item: the project by science and Technology Department of Hebei province (1627460316214408); Chinese Coal Industry Association (MTKJ2016-35, MTKJ2016-329); fundamental research funds for the central universities such as funding (3142016022, 3142017044)

Intelligent Shutter Speech Control System Based on DSP

Deng Yonghong, Ge Yanxiang ,Zhao Zimei,
College of electronic and information engineering, North China Institute of Science and Technology, 065201

Abstract: Based on TMS320F28035 DSP, this paper designed a smart shutters voice control system, which realized the functions of opening and closing shutters, intelligent switching of lighting mode and solar power supply through voice control. The traditional control mode is converted to voice control at the same time with automatic lighting and solar power supply function. In the convenience of people's lives at the same time more satisfied with today's people on the intelligent and environmental protection of the two concepts of the pursuit. The whole system is simple, low cost, safe and reliable.

Key words: blind ,DSP , voice control , solar power supply ,automatic chasing light, intelligent home ,environmental protection

1 Ask questions

With the rapid development of today's society and the continuous improvement of people's living needs, the traditional home and residential has become increasingly unable to meet people's demand for intelligence, environmental protection, convenience, fast and comfortable. At present, life mainly rely on manual or electric to open and close the blinds, when the outdoor light changes also need to adjust the location of the blinds so that indoor light to maintain comfort. The blinds controlled by this mode will inevitably bring inconvenience to the user, especially those with reduced mobility. Waste people's time at the same time also delayed people's work. Therefore, this paper combines the current market demand, combined with engineering and practical application, designed this set of voice control, automatic chasing light, solar power supply three functions in one and can well meet the needs of people living green blinds.

2 System composition

2.1 target function

(1) Through the voice to control the opening and closing shutters

(2) When the outdoor light changes, the blinds will automatically adjust the current chase light mode, rotate the corresponding angle to achieve the desired effect

(3) In the shading at the same time the largest use of resources, through the fan on the solar panels to the battery charge

2.2 System frame structure

Figure 1 shows that the DSP-based intelligent shutters voice control system consists of a battery charging module, a stepper motor drive module, a voice module, a photoresistor sampling module and a TMS320F28035 main control board.

(1) battery charging module: mainly for the entire system to provide power support. The voltage on the solar panel is converted to the required voltage to power the system.

(2) stepper motor drive module: the main system to provide dynamic support.

(3) voice module: the main responsibility for the entire system of information feedback.

(4) photoresistor sampling module: is mainly responsible for the collection of external light signals and information back to the main control board.

(5) DSP main control panel.

Each module with each other, coordination of work, constitute a complete intelligent shutters voice control system.

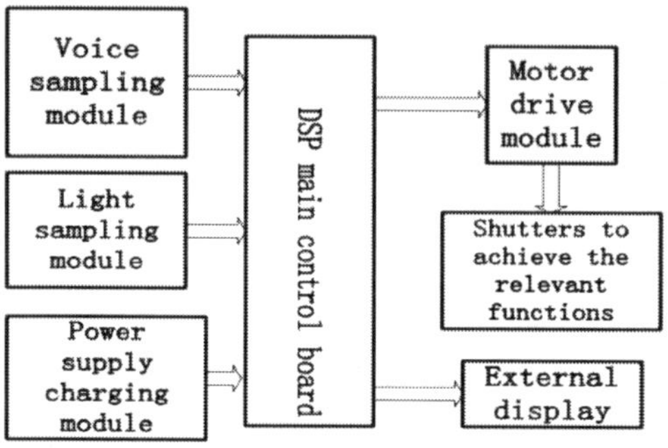

Figure 1 system overall frame structure

3 Hardware circuit design

3.1 voice recognition module

The voice module circuit is mainly based on LD3320 chip-free programming non-specific speech recognition module. The module will provide a specific return value to the main control board when a specific voice signal is received. The main control board according to the different statements of different return value issued a different order so that the blinds to achieve different functions.

Speech recognition consists of two phases: training phase and recognition stage.

(1) training phase: through a number of voice input after preprocessing and feature extraction feature parameters obtained in order to establish a voice reference model library.

(2) identification stage: the input voice parameters and reference model library model similarity comparison, and the highest degree of similarity as a result of the output.

LD3320 chip internal voice recognition principle: first of the input sound spectrum analysis, and then extract the voice feature to get the keyword information, and then the voice reference model of the data characteristics of the comparison, and ultimately find the best keywords and As a result of the output. Schematic diagram shown in Figure 2:

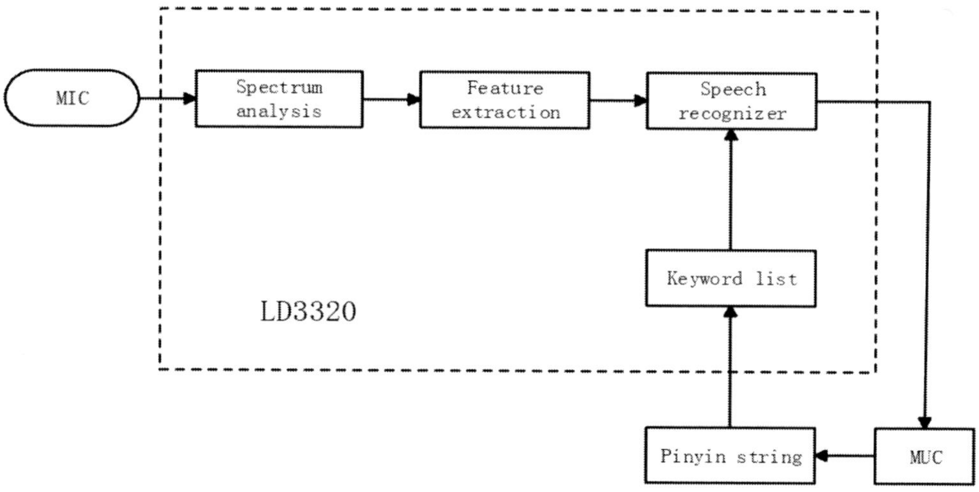

Figure 2 speech recognition module

3.2 Motor drive

The motor drive module is mainly for the entire system to provide power support, is to complete the motor action command executor. In order to ensure the use of shutters, we use the more precise step stepper motor as the drive motor, and ULN2003 chip composition shown below the motor drive circuit to drive, when the main control board to receive the signal will be implemented after the corresponding Action to ensure the realization of the various functions.

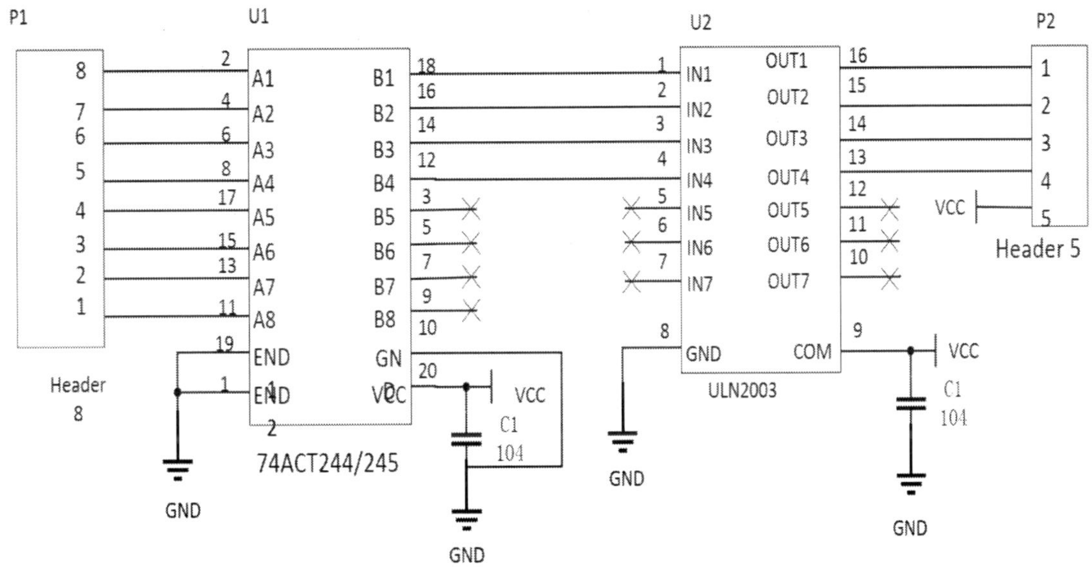

Figure 4 Schematic diagram of stepper motor drive

Mainly for the entire system to provide power support. To meet the needs we set the following two power supply.

(1) mains power supply: DC 5V power supply or with a 5V transformer to convert the mains of electricity to 5V DC power supply for the entire system. Mainly used when the battery is fully charged or the solar power supply system can not be used normally.

(2) solar power: solar energy supply and environmental protection, the principle is mainly the use of fan blades on the solar panels to collect solar energy, and then use the CN3722 chip composed of solar cells with the maximum power tracking function and can effectively charge a number of lithium batteries Management of the circuit, to achieve the purpose of energy saving and environmental protection. The circuit diagram is as follows.

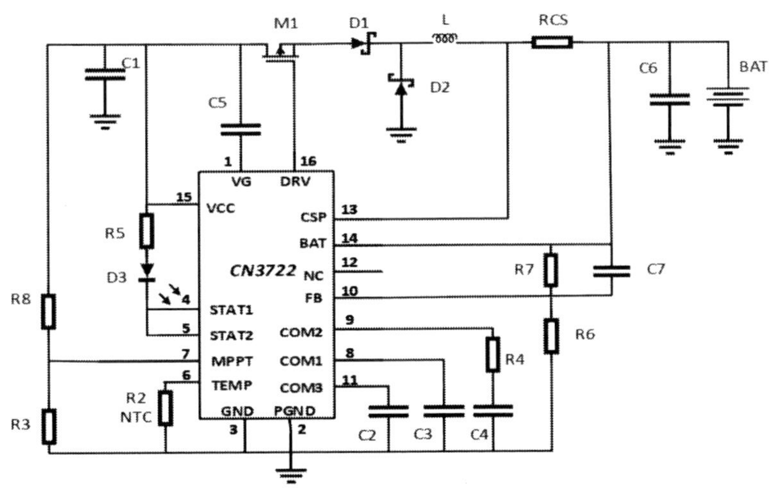

Figure 5 CN3722 based on the battery charging circuit schematic

We use the CN3722 is a can use the solar power supply PWM decompression mode charge management integrated circuit, with solar cell maximum power point tracking function. CN3722 is very suitable for single or multi-cell lithium battery or lithium iron phosphate battery charge management, with a small package, small external components and the advantages of simple use. CN3722 with constant current and constant voltage charging mode,

DOI: 10.1051/matecconf/201713900140

very suitable for lithium batteries or lithium iron phosphate battery charging. In constant voltage charging mode, the constant voltage charging voltage is set by the external resistor divider network; in the constant current charging mode, the charging current is set by an external resistor. For deep discharge batteries, the CN3722 charges the battery with 15% of the constant current charge current set when the battery voltage is less than 66.7% of the set constant voltage charge voltage. In the constant-voltage charging stage, the charging current is gradually reduced, the charging process continues until the charging current is reduced to zero, so that even if the solar cell output a very small power, but also the battery charge; because the battery terminal BAT pin voltage The same, so the battery will not be overcharged. When the input power is lost or the input voltage is lower than the battery voltage, the CN3722 automatically enters the low-power sleep mode. Other features include input low voltage latch, battery temperature monitoring, battery side overvoltage protection and charge status indication.

CN3722 uses constant pressure to track the maximum power point of solar cells. In the voltammetric characteristic curve of the solar cell, when the ambient temperature is constant, the output voltage corresponding to the output of the maximum power is basically the same at different sunshine intensities, that is, as long as the output voltage of the solar cell is kept constant, It is possible to ensure that the solar cell can output the maximum power when the light intensity is different at this temperature. When the ambient temperature changes, the voltage corresponding to the maximum power point of the solar cell varies with the temperature in accordance with the temperature coefficient of -0.4% / ℃, Formula to calculate the maximum power point. This maximum power point tracking method is very suitable for the four major temperature difference is relatively large or relatively large temperature difference between the situation, and our blinds are very consistent with the situation.

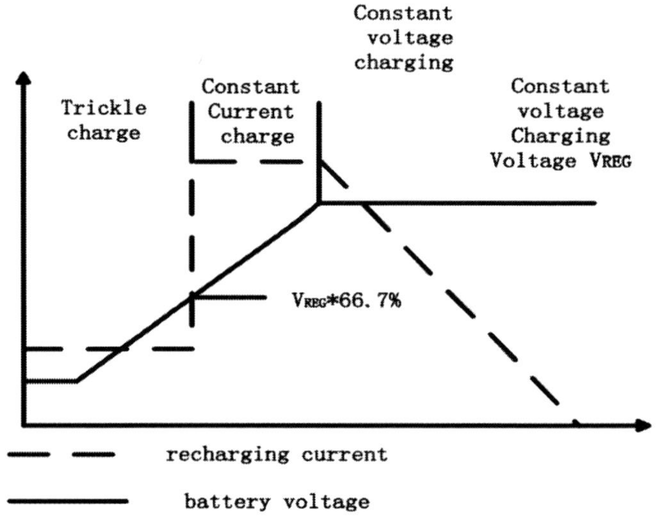

Figure 6 Schematic diagram of the charging process

4 Software implementation

The entire system to TMS320F28035 microcontroller as the control core, multi-sensor information fusion, the composition of the detection unit; click on the module composed of the implementation unit, according to the acquisition of voice and light information, effective, complete and reliable completion of the blinds open and automatic chasing light and other actions. The use of single-chip C voice programming and ccs compiler software, modular design ideas, is conducive to the commissioning procedures and follow-up maintenance upgrades. Design ideas as shown in the Figure 7 System program flow chart . Through the voice module to receive voice signals and 2.4-inch LCD serial port touch screen display the current working state. When a specific voice signal is received, the louver will perform the corresponding action until the next voice command is monitored.

5 System debugging

5.1 Speech recognition static test

LD3320 speech recognition chip in the following two cases to give recognition results

(1) external into the scheduled time of the voice data, the chip after the analysis of these voice data analysis gives the results.

(2) external voice data stream, the voice recognition chip through the endpoint detection technology to detect the user to stop talking, and the user from the beginning to speak to stop the voice of the time period of the voice data analysis analysis gives the identification results.

The system designed in this paper is mainly used in voice control shutters, in the practical application of the main focus on voice recognition rate and real-time, so we use the second identification method. In order to better detect the recognition effect, we selected a number of different sound of the staff in the experimental environment several times to test. The results show that the average speech recognition rate is as high as 85% in the experimental environment.

5.2 Test of the operation of the blinds in different power supply modes

This experiment tested the effect of two different power supplies on the working condition of the blinds. Experimental results show that, regardless of the type of power supply, the blinds can work properly and complete the instructions

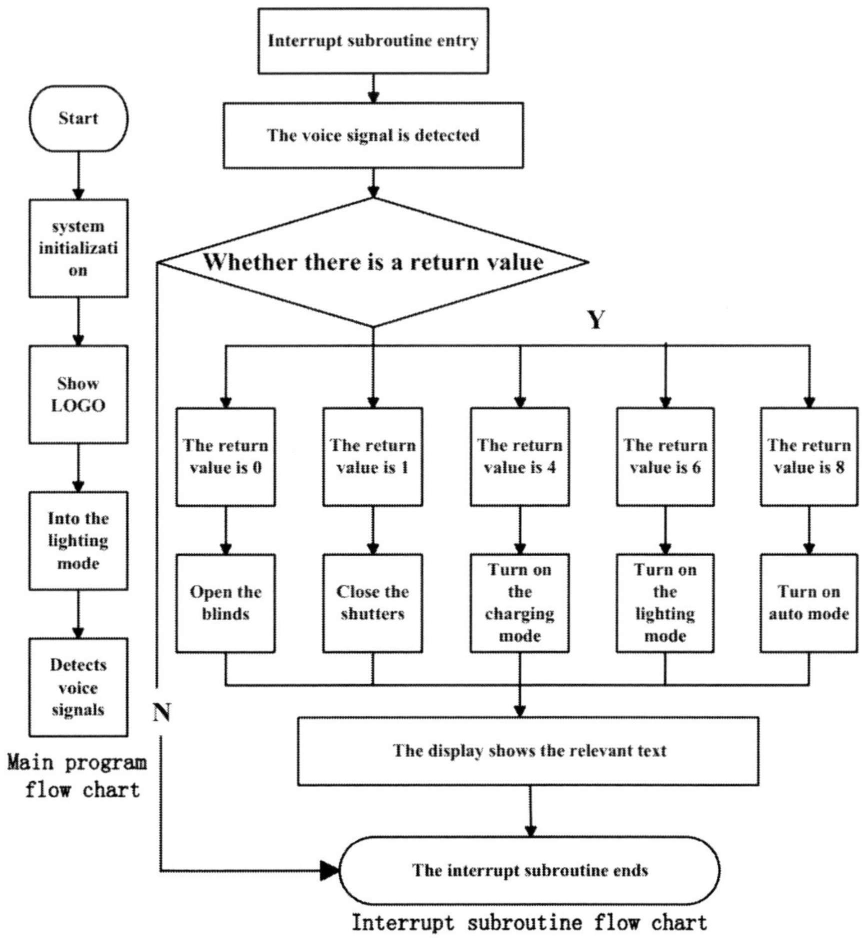

Figure 7 System program flow chart

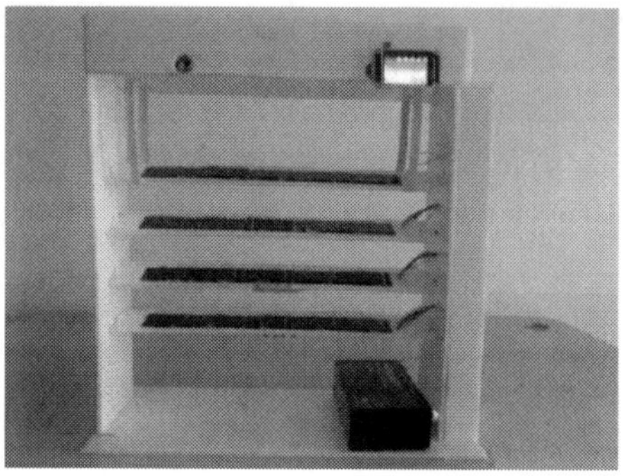

Figure 8 system test chart

6. Conclusion

Natural voice recognition, voice control and the rational and efficient use of natural resources is the focus of current research. This paper designs the DSP-based intelligent shutters voice control system to TMS320F28035 microcontroller as the control core, combined with the use of LD3320 chip and the use of CN3722 chip, completed the voice control and solar charging module design. Compared with the traditional blinds, we designed the intelligent blinds more in line with the current era of home intelligent and energy saving and environmental protection requirements, to facilitate people's lives. At present, there are still some shortcomings in the works, such as voice module recognition rate is very susceptible to external interference, but also need to further improve and deal with.

references

[1] Chen Xiaoyan, Pang Tao, Lian Ruoxin. Based on MCU multi-machine communication intelligent curtain design [J]. Measurement and Control Technology, 2012,31 (6): 76-78.

[2] Zhang Kairi, Deng Yanmei, Gao Shi. Photovoltaic power smart blinds system [Papers]. "Science and Technology Information", 2014, 21:34-37;

[3] Zheng Zhenjie, Jiang Yanxuan, Yu Dezhi. Single-chip microcomputer combined with ULN2003 driven stepper motor. "Motor Technology", 2011,31:27-32.

[4] Chen Boshi. Electric drive automatic control system. Third edition. Beijing: Mechanical Industry Press, 2011

[5] Hu Shousong. The principle of automatic control. Fifth edition. Beijing: Science Press, 2007

[6] Xu Dehong. Modeling and control of power electronic systems. First edition. Beijing: China Machine Press, 2006

Foundation item: the project by science and Technology Department of Hebei province (1627460316214408); Chinese Coal Industry Association (MTKJ2016-35, MTKJ2016-329); fundamental research funds for the central universities such as funding (3142013101, 3142016022)

Backstepping based integral sliding mode control with neural network for ship steering control

Renqiang Wang[a], Hua Deng, Keyin Miao, Yue Zhao, Jiabao Du

Navigation College, Jiangsu Maritime Institute, Nanjing 211170, PR China

Abstract. An integral sliding mode controller with neural network was investigated for ship navigation, which realizes the accurate and stable tracking of the ship in the steering process, on the basis of backstepping method. First of all, an integrator sliding surface were designed with the sliding mode variable structure control technology. Secondly, radial basis function neural network was applied to approximate the system nonlinear function and uncertain parameters. Furthermore, a nonlinear damping law was introduced to overcome the bounded outside interference. Finally, on the basis of the above, the system control law was deduced by using the backstepping method. The simulation results show that the neural network can accurately approximate the nonlinear function and uncertain parameters of the ship, and the controller output is smooth and the heading output is not sensitive to perturbation of parameters and external disturbance, and the controller has strong robustness.

1 Introduction

The factors that affect the performance of ship motion control are nonlinearity of ship motion system function, uncertainty of ship model parameters and unknown disturbance.

It is well known that the most advantage of sliding mode control [1] is able to overcome uncertainty of the system, which has highly robustness for uncertain dynamic system with outside disturbance, especially effectiveness for control of modelless system. It is an important method for suppressing high frequency switching that causes chattering of control input. And the variable structure control system algorithm is simple, fast response, and robustness to external noise and parameter perturbation. Therefore, it is useful for ship steering control. The integral link [2] joined into the algorithm of sliding mode control can also effectively eliminate the bounded interference outside so that the original controller has a stronger anti-jamming capability.

With the rapid development of computer technology, the performance of intelligent control [3], as the representative of neural network, has been improved greatly. The nonlinear function and the uncertain parameters can be arbitrarily identified by neural network with a universal approximation performance. The neural network [4] was used to infinitely approach and compensate the interference outside and the uncertainties of controlled system can reduce buffeting; The controller parameters were optimized online by genetic optimization method, so that the buffeting is reduced. A ship course RBF neural network sliding mode control algorithm is devised in literature [5], where sliding mode control based on the design of a hyperbolic tangent function for new reaching law can speed up the approaching speed and RBF neural network solves the problem of parameter variation and external disturbance. Global finite-time heading control of surface vehicles is designed in literature [6], which uses continuous control law to ensure the stability of the whole finite time closed-loop control system and provide accurate ability of heading control with excellent anti-jamming capability. In literature [7], taking into consideration of the uncertainty of ship model parameter, a control algorithm based on RBF neurol network which approximates the uncertainty has been derived by using backstepping method.

Ship course response model has significant nonlinear characteristics, and backstepping [2] is a very effective method used to deal with the problem about nonlinear system controller design. It uses the system's structural properties, recursively constructs the entire system Lyapunov function, and makes the controller being more flexible and robust, but also reduces the difficulty and energy. Backstepping refers to the control system consists of a number of integrators in series, the state feedback controller is pushed through the integrator to the beginning of the control input, by way of coordinate transformation, systematic structure step by step and stabilization control Lyapunov function, so as to make the system asymptotically stable equilibrium point. Then, ship course sliding mode controller is designed in this paper by using backstepping method..

[a] Corresponding author: wangrenqiang2009@126.com

2 Problem description

Considering the problem of ship maneuvering. the Bech [3] nonlinear model of ship motion nonlinear response was selected in paper, and its mathematical expression is:

$$T_1T_2\dddot{\varphi}+(T_1+T_2)\ddot{\varphi}+H(\dot{\varphi})=K(T_3\dot{\delta}+\delta)+\omega_0 \quad (1)$$

Where, T_1 , T_2 and T_3 represent ship follow index, K represents ship turning index, φ represents ship course, δ represents rudder angle, ω_0 represents ship external interference. And $H(\dot{\varphi})$ is nonlinear function with regard to $\dot{\varphi}$, which can be approximated as follows, $H(\dot{\varphi})=\alpha_0+\alpha_1\dot{\varphi}+\alpha_2\dot{\varphi}^2+\alpha_3\dot{\varphi}^3$.

Make the following settings:

$x_1=\varphi$, $x_2=\dot{x}_1=\dot{\varphi}=r$, $x_3=\dot{x}_2=\ddot{\varphi}=\dot{r}$,

$a=\frac{1}{T_1}+\frac{1}{T_2}$, $b=\frac{1}{T_1T_2}$, $c=\frac{K}{T_1T_2}$, $u=T_3\dot{\delta}+\delta$.

Obviously, it is difficult to solve the system control input u which contains derivative items $T_3\dot{\delta}$, therefore, make the following setting $\omega=\omega_0+\frac{K}{T_1T_2}T_3\dot{\delta}$.

Therefore, formula (1) can be converted into the following format:

$$\begin{cases}\dot{x}_1 = x_2\\ \dot{x}_2 = x_3\\ \dot{x}_3 = f(x)+cu+\omega\\ y = x_1\end{cases} \quad (2)$$

Where, x_1 , x_2 and x_3 are system variable, y is the actual output of the system, ω represents ship external interference, u is the control input of system, $f(x)$ is system function, $f(x)$=-b($\alpha_0+\alpha_1x_2+\alpha_2x_2^2+\alpha_3x_2^3$)-a$x_3$.

Based on above, the goal of this article is to seek control law u which is able to make the system output y asymptotically tracking expected course φ_r , the tracking error will be infinitely close to zero, that is $e=y-\varphi_r\to 0$, and the stabilization time will be reduced as much as possible.

3 Dteering control design based on backstepping

The design steps of the ship course controller contain the following steps.

(1) According to the characteristics of system (2), the following sliding surfaces are defined.

$$\begin{cases}\dot{\xi} = z_3\\ z_1 = x_1-\varphi_r\\ z_2 = x_2-\sigma_1\\ z_3 = x_3-\sigma_2\end{cases} \quad (3)$$

Where, φ_r is the expected course, σ_1 and σ_2 are virtual stabilization function, z_1 is the heading error. ξ is the integral term which can be able to eliminate the static error caused by the uncertain interference term in the control process.

(2) The second step, Lyapunov function is constructed to prove the asymptotic stability control system by using backstepping method. The first Lyapunov functionis constructed.

$$V_1=\frac{1}{2}z_1^2 \quad (4)$$

According to (3), it is obtained (5).

$$\dot{z}_1=z_2+\sigma_1-\dot{\varphi}_r \quad (5)$$

It can be inferred formula (6) by derivation formula (4).

$$\dot{V}_1=z_1\dot{z}_1=z_1\left[z_2+\sigma_1-\dot{\varphi}_r\right] \quad (6)$$

It can be obtained (7) , according to (5) and (6).

$$\dot{V}_1=-k_1z_1^2+z_1z_2 \quad (7)$$

Based on the above, it can be obtained (8) by considering formula (5) and (7).

$$\dot{z}_1=k_1z_1+z_2 \quad (8)$$

Therefore, it can be obtained (9).

$$\dot{\sigma}_1=-k_1(-k_1z_1+z_2)+\ddot{\varphi}_r \quad (9)$$

Similarly, according to (8), it is obtained (10).

$$\dot{z}_2=\dot{x}_2-\dot{\sigma}_1=z_3+\sigma_2-\dot{\sigma}_1 \quad (10)$$

The second Lyapunov function is constructed.

$$V_2=V_1+\frac{1}{2}z_2^2 \quad (11)$$

It can be inferred formula (12) by derivation formula (11).

$$\begin{aligned}\dot{V}_2&=\dot{V}_1+z_2\dot{z}_2\\ &=-k_1z_1^2+z_2z_3+z_2(z_1+\sigma_2-\dot{\sigma}_1)\end{aligned} \quad (12)$$

Making the following setting.

$$\sigma_2=-k_2z_2-z_1+\dot{\sigma}_1 \quad (13)$$

It can be inferred formula (14) by (13) converted into (12).

$$\dot{V}_2=-k_1z_1^2-k_2z_2^2+z_2z_3 \quad (14)$$

Therefore, it can be obtained (15) by considering formula (10) and (13).

$$\dot{z}_2=-k_2z_2-z_1+z_3 \quad (15)$$

It is also acquired (16) by considering formula (13) and (15).

$$\dot{\sigma}_2=(k_1^2-1)\dot{z}_1-(k_1+k_2)\dot{z}_2+\dddot{\varphi}_r \quad (16)$$

And the third Lyapunov function is constructed.

$$V_3 = V_2 + \frac{1}{2}z_3^2 + \frac{\lambda}{2}\xi^2 \tag{17}$$

It is easy obtained (18) by formula (3).

$$\dot{z}_3 = f(x) + cu + \omega - \dot{\sigma}_2 \tag{18}$$

It can be inferred formula (19) by derivation formula (17).

$$\begin{aligned}\dot{V}_3 &= \dot{V}_2 + z_3\dot{z}_3 + \lambda\xi z_3 \\ &= -k_1 z_1^2 - k_2 z_2^2 + z_3\left(z_2 + \dot{z}_3 + \lambda\xi\right) \\ &= -k_1 z_1^2 - k_2 z_2^2 + z_2 z_3 + \lambda\xi z_3 \\ &\quad + z_3[f(x) + cu + \omega - \dot{\sigma}_2]\end{aligned} \tag{19}$$

(3) The RBF network is used to adapt the system function $f(x)$, and its algorithm is (20).

$$\begin{cases} h_j = \exp\left(-\dfrac{\left\|x - c_j\right\|^2}{2b_j^2}\right), \ j = 1, 2, \cdots, m \\ f(x) = W^{*T}\mathbf{h}(x) + \varepsilon \end{cases} \tag{20}$$

Where $\boldsymbol{x}$ is the network input, i represents one of the first input network input layer, j is a hidden layer of the network the first network input, $\boldsymbol{h} = \left[h_j\right]^{\mathrm{T}}$ is high Gaussian function, $\boldsymbol{W}^*$ is the ideal for weighting value, ε is ideal for neural network approximation error.

$$\hat{f}(x_2) = \hat{W}^T\mathbf{h}(\mathbf{x}) \tag{21}$$

Where, $\tilde{W}$ is network connection weight adjustment error, there is $\tilde{W} = \hat{W} - W^*$.

Therefore, the compensation control law is described as formula (22).

$$u = \frac{1}{c}\left[-\hat{f}(x) - \omega - \lambda\xi + \dot{\sigma}_2 - z_2 - k_3 z_3\right] \tag{22}$$

The global Lyapunov function is constructed.

$$V = V_3 + \frac{1}{2}\gamma\tilde{W}^T\tilde{W} \tag{23}$$

It can be inferred formula (24) by derivation formula (23).

$$\dot{V} = \dot{V}_3 + \gamma\tilde{W}^T\dot{\hat{W}} \tag{24}$$

It can be obtained formula (25) by considering formula (19) and (22).

$$\begin{aligned}\dot{V} &= -k_1 z_1^2 - k_2 z_2^2 - k_3 z_3^2 \\ &\quad + \left[f(x) - \hat{f}(x)\right] z_3 + \gamma\tilde{W}^T\dot{\hat{W}} \\ &= -k_1 z_1^2 - k_2 z_2^2 - k_3 z_3^2 + \tilde{W}^T(\gamma\dot{\hat{W}} - \mathrm{h}z_3)\end{aligned} \tag{25}$$

In order to ensure system asymptotically stable, the following adaptive law is designed.

$$\dot{\hat{W}} = \frac{1}{\gamma}\mathbf{h}z_3 \tag{26}$$

(4) In order to eliminate the uncertain interference term in equation (19), the nonlinear damping law [9] was designed.

$$u = \frac{1}{c}\begin{bmatrix} -f(x) - \lambda\xi + \dot{\sigma}_2 \\ -z_2 - \left(k_3 + \eta\right)z_3 \end{bmatrix} \tag{27}$$

It can be inferred formula (28) by considering formula (25) and (27).

$$\begin{aligned}\dot{V} &= -k_1 z_1^2 - k_2 z_2^2 - k_3 z_3^2 - \eta z_3^2 \\ &\quad + \left(\omega + \varepsilon\right) z_3 + \tilde{W}^T(\gamma\dot{\hat{W}} - \mathrm{h}z_3)\end{aligned} \tag{28}$$

Therefore, it is established that $\dot{V} \le 0$, and the system (2) is asymptotically stable.

It is easy obtained formula (29) by theorem "$xy \le \eta x^2 + \frac{1}{4\eta}y^2$".

$$\varpi z_3 \le \eta z_3^2 + \frac{1}{4\eta}\left\|\varpi\right\|_\infty^2 \tag{29}$$

Where, $\varpi = \omega + \varepsilon$.

The fitness law of the network weights is still adopted formula (26), therefore, the formula (30) can be acquired.

$$\begin{aligned}\dot{V} &= -k_1 z_1^2 - k_2 z_2^2 - k_3 z_3^2 - \eta z_3^2 + \varpi z_3 \\ &\le -k_1 z_1^2 - k_2 z_2^2 - k_3 z_3^2 + \frac{1}{4\eta}\left\|\varpi\right\|_\infty^2 \\ &\le -k_2 z_2^2 + \frac{1}{4\eta}\left\|\varpi\right\|_\infty^2\end{aligned} \tag{30}$$

In order to ensure that the system (4) is asymptotically stable, it is only need to prove the establishment which is $k_2 z_2^2 \ge \frac{1}{4\eta}\left\|\omega\right\|_\infty^2$. Proof slightly, see the literature [8].

4 Simulation

Ship course continuous tracking control simulation is carried out by applying the above control algorithm.

4.1 Simulation Object and Conditions Setting

Effectiveness of control algorithm is verified through simulation of training vessel "YUKUN". It is known that the parameters of the ship are: $T_1 \cdot T_2 = 9.238$, $T_1 + T_2 = 23.944$, $K = 7.926$, $T_3 = 10.064$, $\alpha_0 = 0$, $\alpha_1 = 1$, $\alpha_2 = 0$, $\alpha_3 = 30$, $a_1 = 1$, $a_2 = 30$, $K = 0.2$, $T = 64$ .Outside interference is approximately substituted by white noise driven waves pope [9].

A required mathematical model which can represent ship's actual performance [10] is selected as following

$$\begin{aligned}&\phi_m(k+2) + 0.1\phi_m(k+1) + 0.0025\phi_m(k) \\ &= 0.0025\phi_r(k)\end{aligned} \tag{31}$$

Where, $\phi_m(k)$ represents the characterize performance of ship heading. $\phi_r(k)$ is a processed input signal, and $\phi_r(k) = (sign(\sin(\pi k / 500)) + 1)\pi / 12$, whose range is from 0 ° to 30 ° and period is 500s.

4.2 Simulation Results and Analysis

Simulation was executed by Matlab/ Simulink, whose results were shown in figure 1 and figure 2.

Figure 1 depicts the curves of ship course and desired course, although the system parameter is changing, it rarely affects the tracking performances, which primarily affected by outside interference completely inhibited in the paper. Through the simulation demonstrated in Figure 1, it can be concluded that the time complexity of the algorithm presented in the paper is substantially approximately linear order $O(n)$, and the proposed algorithm frequency time $T(n)$ is approximately $51n$.

Figure 2 depicts the curves of control rudder angle, which is smooth and no buffeting and met the actual needs of the project.

Overall, the control system is stable, reliable, and which has strong robustness.

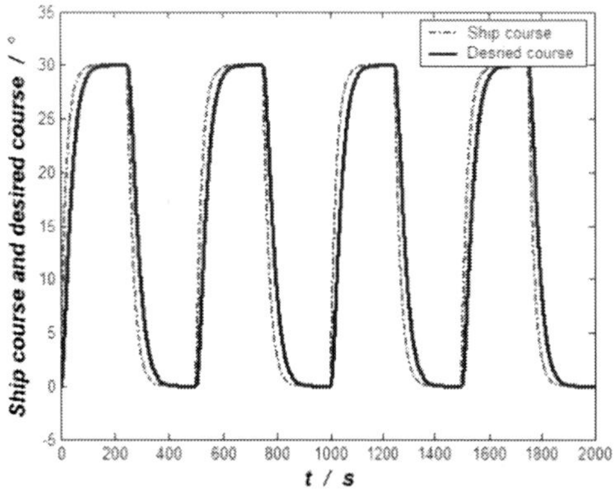

Fig.1. Curves of ship course and desired course

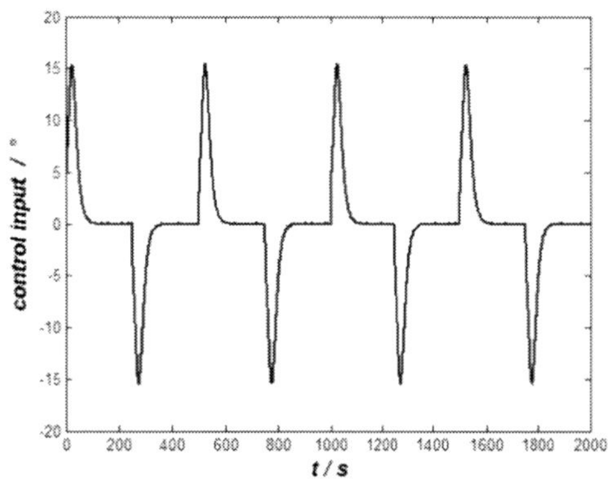

Fig.2. Curve of control rudder angle

5 Conclusion

A ship steering neural network controller based on integral sliding mode was designed by using the backstepping algorithm, based on the Bech ship motion mathematical model, for ship 's large scale maneuvering problem. The results show that the output of the controller is smooth and no jitter, and the heading output is not sensitive to the perturbation of the ship parameters and the external disturbance. Therefore, it has strong robustness.

Acknowledgment

This work was supported by the Foundation of Jiangsu Maritime Institute under Grant No.XR1501 and No.XR1504, and Jiangsu University Brand Professional Construction Project under Grant No. PPZY2015B177.

References

1. F. J Lin, W. D Chou, An Induction Motor Servo Drive using Sliding mode Control with Genetic Algorithem, Electric Power System Research, 2003, 64(2): 93-108.
2. Zhang Xianku.Ship motion simple and robust control, Beijing:Science Press, 2012.
3. Renqiang Wang, Yuelin Zhao, Keyin Miao, Application of Neural Network Minimum Parameter Learning Algorithm in Ship's Heading Tracking Control, Proceedings of 9th International Symposium on Computational Intelligence and Design, 2016, 12: 135-139.
4. Wang Renqiang, Zhao Yuelin, Sun Jianming, Application of Optimized RBF Neural Network in Ship's Autopilot Design, Proceedings of IEEE Advanced Information Management, Communicates, Electronic and Automation Control Conference, 2016, 10: 1642-1646.
5. Wang Renqiang, Gong Jianyun, Zhao Yuelin, Neural Network Sliding Mode Control under New Reaching Law and Application, Proceedings of Advanced Information Technology, Electronic and Automation Control Conference, 2015, 12: 911-917.
6. Wang Ning, Lv Shuailin, Liu Zhongzhong, Global finite-time heading control of surface vehicles, Neurocomputing, 2016, 175(106): 662-666.
7. Yang Wang, Chen Guo, Fuchun Sun. Dynamic neural fuzzified adaptive control of ship course with parametric modelling uncertainties, Int. J. of Modelling, Identification and Control, 2015, 13(4): 251-258.
8. Benaskeur A. R, Desbines A. Backstepping based adaptive PID control, Proceedings of Control Theory and Applications, 2002, 149(1): 54-59.
9. Guanshan Hu, Yingbing Zhou, Application of Fuzzy Neural Network in Ship Course Control, Applied Mechanics and Materials, 2012, 1511(135): 309-315.
10. Xin Wang, Tieshan Li, Weilin Luo, Direct Adaptive Neural Network Control for a Class of Ship Course Uncertain Discrete-time Nonlinear Systems, Marine Engineering Frontiers, 2013, 37(1): 42-48.

Numerical simulation of transverse jet flow field under supersonic inflow

LI Qian[1,*], WANG Dian-kai[1], YANG Yin-jun[2], FANG Juan[3]

[1]State Key Laboratory of Laser Propulsion & Application, Space Engineering University, Beijing 101416, China
[2]Xichang Satellite Launch Center of China, Xichang Sichuan, 615606
[3]Jiuquan Satellite Launch Center of China, Jiuquan Gansu, 732750

Abstract. Transverse jet flow field under supersonic inflow is simulated numerically for studying the characteristic of fuel transverse jet and fuel mixing in scramjet combustion chamber. Comparison is performed between simulated results and the results of references and experiments. Results indicate that the CFD code in this paper is applicable for simulation of transverse jut flow field under supersonic inflow, but in order to providing more effective numerical predictive method, CFD code should be modified through increasing mesh density and adding LES module.

Key words. supersonic inflow; transverse jet; fuel mixing; numerical simulation

1 Introduction

With the development of aerospace technology, higher flight velocity is purchased by people, so the study of scramjet is becoming one of the key problems in the research field of hypersonic vehicle. Compared with other ramjet engines, problems exist in the mixing of air and fuel in scramjet: (1) When the inflow velocity achieved supersonic speed, residence time of fuel in the combustor is ~ms, and the mixing efficiency of fuel and air descends rapidly with the increasing of Mach number [1]. (2) In supersonic air flow, compression and stability of the mixed layer informed by mixing of fuel and air are very strong. At the same main inflow and density ratio, diffusion coefficient of compressible mixed layer is 1/3 of incompressible one [2]. (3) Because the energy density of H_2 is very low, hydrocarbon fuel should be used for realization of high specific impulse propulsion [3].

Experimental method is usually used in the study on mixing of fuel and air of supersonic vehicle. However, for the limitation of costs, etc., superiority of numerical simulation becomes more and more outstanding gradually. Koji[4] compared the reliability of tetrahedron mesh, optimized mesh and structured mesh with CFD software. Separated shear layer can be captured distinctly through using optimized mesh and structured mesh. For the capture of shocks, optimized mesh is more adaptive. Xiao[5] computed the transverse jet flow field under supersonic inflow with SST turbulent model, and N-S equations are solved with AUSMPW scheme. He[6] added the component transport equations into the controlling equations, selected standard k-2 model as the turbulent model, then computed 2-D transverse side jet flow field under supersonic inflow. Sparkman[7] optimized the fuel array spray design of scramjet with JET-PEN analysis software based on former experimental data.

In summary, as favorable supplement of experiments, study of numerical predictive method is very necessary, so as to provide references for engineering. Based on the needs, aimed at the problem of fuel jet and supersonic mixing in the combustor, transverse jet flow field under supersonic inflow is simulated numerically. In the further study, the method will be modified, and provide references for experiments.

2 Computational method and conditions

Computational zone and boundary conditions are shown in Figure 1. The number of mesh is 900, 000. Left edge is supersonic inflow, right edge is outflow, continuous line is no-slip solid surface, and imaginary line is symmetry boundary condition. The Mach number of inflow is 2.5, static pressure is 5.8×10^4Pa, and static temperature is 130K. The diameter of jet is 2mm, total pressure of is 1Mpa, and total temperature is 300K.

* Corresponding author: 49225312@qq.com

DOI: 10.1051/matecconf/201713900142

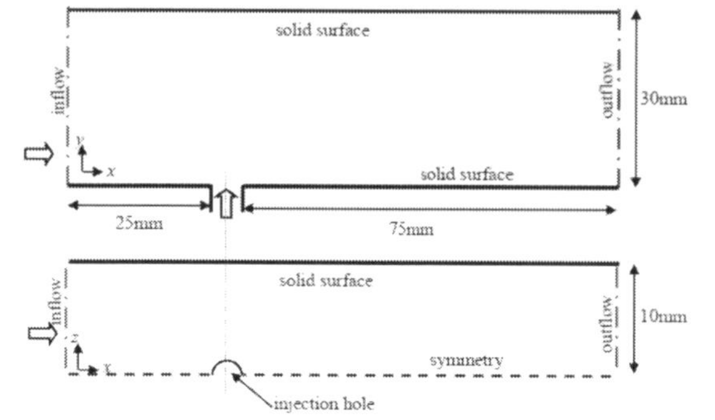

Figure 1. Computational zone and boundary conditions

Control equations of the CFD code programmed by us are 3-D unsteady compressible Reynolds average Navier-Stokes equations. Numerical method is implicit (dual-time step) finite volume method. Computation of inviscid flux is upwind scheme based on rotational Riemann resolver. MUSCL differential method is adopted, and the space accuracy achieved second-order. Second-order implicit scheme is used in time divergence, and the turbulent model is k-ω.

3 Results and Discussion

3.1 Simulated results

Density gradients of equal spaced sections parallel to the symmetry are shown in Figure2. Separated shocks, trigeminal shocks and bow shocks can all be observed. From Figure 2, it is clear that with the increasing of z value, the positions of separated shocks, trigeminal shocks and bow shocks moved backward. According to the space distributions of each section, separated shocks, trigeminal shocks and bow shocks are all around the jet, which agrees well with the 3-D configuration in Figure 3.

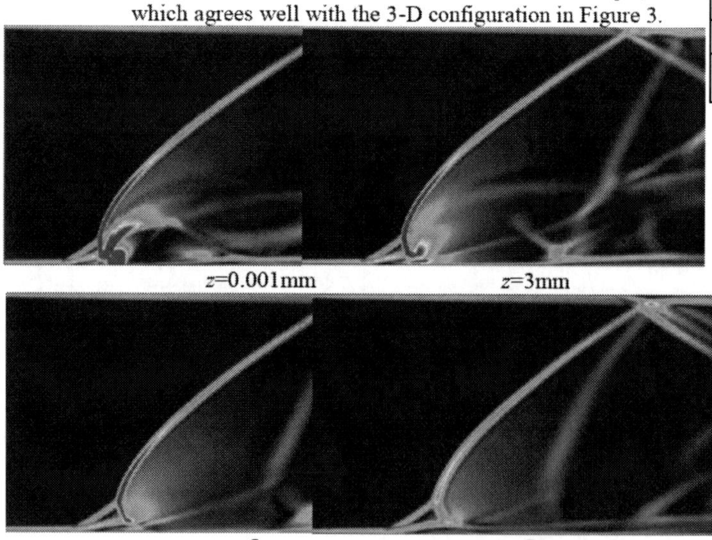

z=0.001mm z=3mm

z=5mm z=7mm

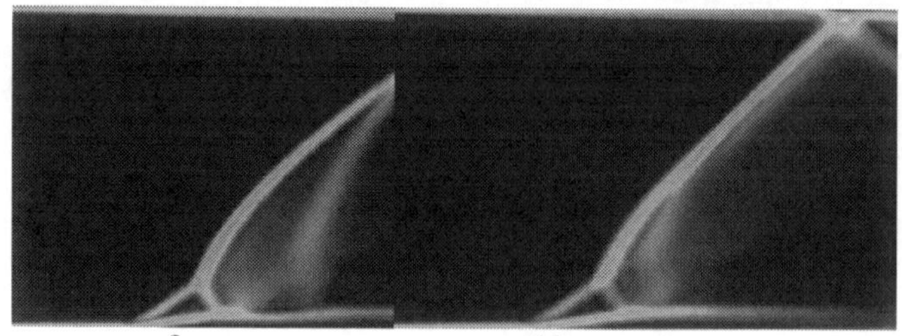

z=9mm z=12mm

Figure 2. Density gradients of equal spaced sections parallel to the symmetry

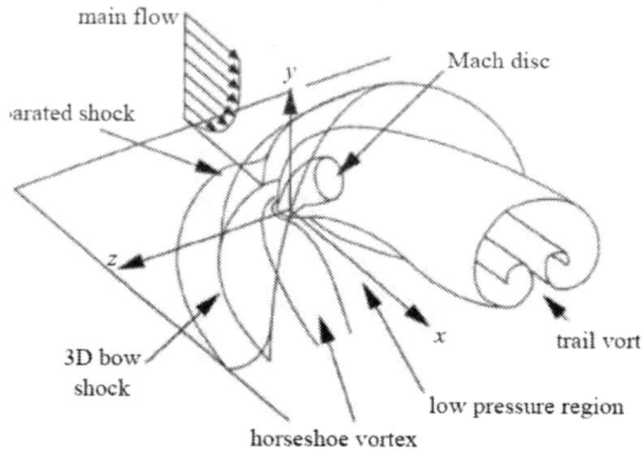

Figure 3. 3-D configuration of the transverse jet flow field under supersonic inflow[8]

3.2 Comparison of references

Simulated conditions of this paper and Cranfield University[9] are shown in Table 1. Mach number contours of both are shown in Figure 4 and Figure 5. It should be noticed that the simulated conditions is the same with that of experiments.

Table 1. Simulated conditions

	M_∞	P_0/kPa	T_0/K	d/mm	Mesh number
This paper	2.5	58	130	2	90W
Cranfield University[9]	1.6	241	295	4	914W

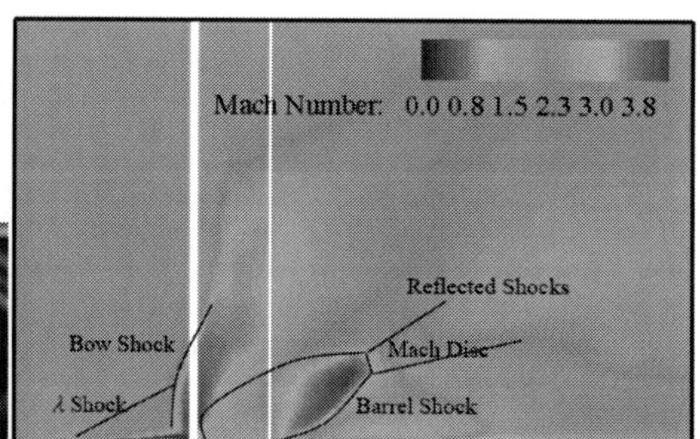

Figure 4. Mach number contours(this paper)

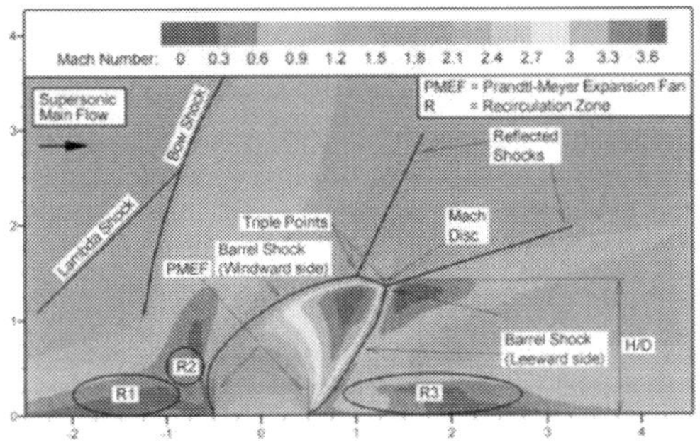

Figure 5. Mach number contours(Cranfield University) [9]

λ shock, Mach disc and separated region can be observed in Figure 4, but the bow shock is not simulated ideally, and large scale eddies configuration is not shown. The reason is that the mesh is rare, especially the mesh far from the wall. There is no LES turbulent model in the program, too. Bow shock, Mach disc, λ shock and separated region are all be shown in Figure 5. Although LES is adopted, large scale eddies is not simulated.

3.3 Comparison of experiments

Simulated result and schlieren photos of the flow field are shown in Figure 6. Figure 6 (a) is integral result of the density along z axis. Figure 6 (b), (c) and (d) are photos of monochrome schlieren, continuous xenon lamp colored schlieren and strobe light colored schlieren respectively. The experimental model, inflow conditions and jet conditions are all same with that of simulation.

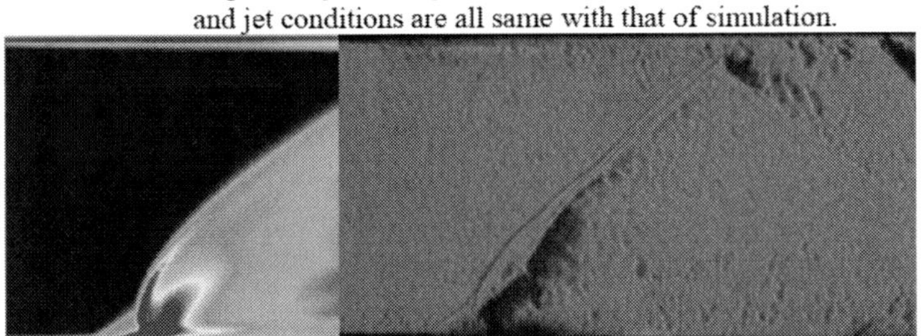

(a) simulated result (b) monochrome schlieren photo

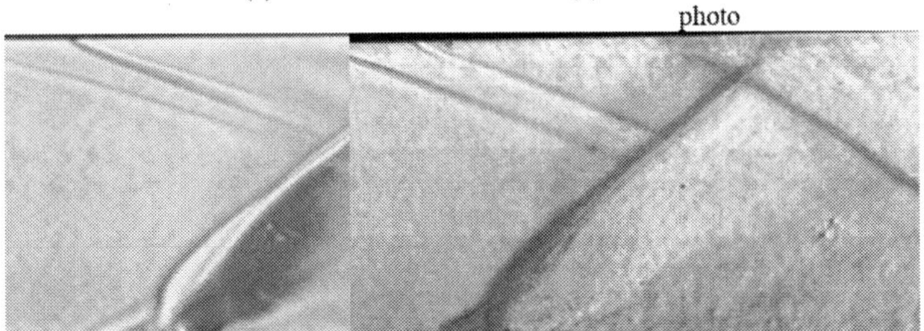

(c) continuous xenon lamp colored schlieren photo (d) strobo light colored schlieren photo

Figure 6. Simulated result and schlieren photos of the flow field

From Figure 6) it can be observed that the simulated results agree well with that of experiments if the oblique shock induced by the rough upper wall is neglected. Mach disc, barred shock, bow shock, trigeminal shock and reflected shock can all be shown clearly. Compared with colored schlieren, information is less in monochrome schlieren, and the sensitivity is lower. Even so, if the cutting edge is deep enough, that is, the sensitivity of monochrome schlieren is high enough, barred shock and rear mixing zone can both be observed. Otherwise, distinction between bow shock and mixing layer boundary can be seen obviously through monochrome schlieren, while through the colored schlieren with high sensitivity, it is not easy to be districted. Photos taken by long time exposure can display wave configuration. Exposure time of several ns is suitable for the display of turbulence, especially the disordered situation of turbulence can be observed through colored schlieren.

4 Conclusion

From the comparison of research results, references and experiments, barred shock and Mach disc can be simulated, while λ shock and back flow zone in transverse jet can be computed, too, which indicates that the computational method and scheme is suitable for research of this kind of problem. However, the deficiency is that detailed vortexes are not be displayed. The main reasons are: (1) the number of mesh is few; (2) there is no LES module in CFD code; (3) affection on transverse jet of boundary layer development is not considered. In the further study, modification should be made aimed at the above problems, so more accurate numerical predictive method would be provided.

Acknowledgement

This research was supported by the Natural Science Foundation of China (11372356, 11372357).The authors would like to acknowledge all discussions and technical help provided by the former and current members of the State Key Laboratory of Laser Propulsion and Application.

References

1. John M, Seiner S M. Historical Survey on Enhanced Mixing in Scramjet Engines[R]. Journal of Propulsion and Power, 1998, 4(4): 1273-1286.
2. Georgiadis N J, Alexander J L D, Reshotko E. Investigation of a compressible turbulent mixing layer using a Hybrid RANS-LES method[R]. AIAA 2001-2563.
3. Law C K. Combustion Physics[D]. Cambridge University Press, 2006.
4. Koji M, Yoshiyuki K. Effect of Grid Resolutions and Injectant Species in Computations of Supersonic Mixing[R]. AIAA 2009-3946.
5. Xiao Yinli, Song Wenyan, Zhao Zhi. Numerical study of injection angle effect on fuel mixing characteristics in supersonic flow[J]. World Sci-tech R&D, 2008, 30(4), 436-439.

6. He Bo, Nie Wansheng, Feng Songjiang. Numerical simulation of underexpand heterogeneous lateral jet into supersonic air freestream[J]. Journal of the Academy of Equipment Command & Technology, 2008, 19(3), 74-79.
7. Sparkman B T, Chrissis J W, Gruber M R, et al. Optimization of a Scramjet Fuel Injection Array: An Application of Mixed Variable Generalized Pattern Search With Kriging Surrogates[R]. AIAA 2008-5861.
8. Gruber M R, Nejad A S, Dutton J C. Circular and Elliptical Transverse Injection into a Supersonic Crossflow-The Role of Large-Scale Structure [R]. AIAA 1995-2150.
9. Rana Z A, Thornber B, Drikakis D. An ILES analysis of transverse jet injection in to supersonic cross-flow with synthetic turbulent boundary layer[R]. AIAA 2011-231.

RBF network based integral backstepping sliding mode control for USV

Renqiang Wang[a], Hua Deng, Keyin Miao, Yue Zhao, Jiabao Du

Navigation College, Jiangsu Maritime Institute, Nanjing 211170, PR China

Abstract. A kind of USV course RBF network control algorithm is putted forward, which is on the basis of integral backstepping sliding mode. First of all, an integrator sliding surface were designed with the sliding mode variable structure control technology. Secondly, radial basis function neural network was applied to approximate the system nonlinear function and uncertain parameters. Furthermore, a nonlinear damping law was introduced to overcome the bounded outside interference. Finally, on the basis of the above, the system control law was deduced by using the backstepping method. The simulation results show that the neural network can accurately approximate the nonlinear function and uncertain parameters, and the controller output is smooth and the output is not sensitive to perturbation of parameters. Therefore, the proposed algorithm is effective for USV course control.

1 Introduction

Maritime shipping and production activities are becoming more and more busy, the Unmanned Surface Vehicle (USV) due to the advantages of high nautical rate, high efficiency, low cost, severe sea condition that can instead people to the various activities has become an important development trend of the future of the ship. Therefore, USV is playing increasing roles in commercial, scientific and military applications [1]. Due to the disturbances of wind, wave, current and other marine environment, the USV will inevitably deviate from the original given course when it is sailing. Precise course control is a prerequisite to solve the trajectory tracking, autonomous navigation and collision avoidance and other issues. The design of ship course controller is the issues of common concern for the control theory and control engineering [2].

With the development of computer technology and modern control theory, a variety of new control algorithms, such as the Lyapunov's direct method [3], backstepping control [4-6], sliding mode control [7-9], fuzzy control [10-12] and neural network control [13-16] have been applied to the research of ship course control. It is well known that the most advantage of sliding mode control is able to overcome uncertainty of the system, which has highly robustness for uncertain dynamic system with outside disturbance, especially effectiveness for control of modelless system. It is an important method for suppressing high frequency switching that causes chattering of control input. And the variable structure control system algorithm is simple, fast response, and robustness to external noise and parameter perturbation. Therefore, it is useful for ship steering control. The integral link joined into the algorithm of sliding mode control can also effectively eliminate the bounded interference outside so that the original controller has a stronger anti-jamming capability.

In this paper, by taking the "Lanxin" USV of Dalian maritime university as the research object, aiming at its motion control system, the nonlinear mathematical model for the USV planar motion is established by the method of responsive integrated backstepping modeling. Based on the Nomoto mathematical model and the combination of integral backstepping control, sliding mode control and neural network control, a kind of USV course RBF network control algorithm of USV course control is proposed. With the disturbances of wind, wave and current, simulation results show that the designed course controller can be properly adopted to the "Lanxin" USV course keeping with good effectiveness.

2 Problem description

In actual voyages, ship movements usually exhibit a non-linear state. Therefore, in the mathematical model of USV plane motion, the nonlinear model is considered. Based on the linear mathematical model of plane three degrees of freedom, the nonlinear model is described by adding nonlinear term as following.

$$\ddot{\varphi}+\frac{1}{T}H(\dot{\varphi})=\frac{K}{T}\delta+d(t) \qquad (1)$$

Where, K and T represent ship turning and follow index,

[a] Corresponding author: wangrenqiang2009@126.com

φ represents ship course, δ represents rudder angle, $d(t)$ represents ship external interference. And $H(\dot{\varphi})$ is nonlinear function with regard to $\dot{\varphi}$, $H(\dot{\varphi})=\alpha_1\dot{\varphi}+\alpha_2\dot{\varphi}^3+\alpha_3\dot{\varphi}^5+\cdots$,where, $\alpha_{i(i=1,2,3,\cdots)}$ is the real value of the constant.

Make the following settings, $x_1=\varphi$, $x_2=\dot{\varphi}$, $u=\delta$. Then formula (1) can be converted into the following format:

$$\begin{cases}\dot{x}_1=x_2\\ \dot{x}_2=f(x_2)+g\cdot u+d(t)\\ y=x_1\end{cases} \tag{2}$$

Where, $f(x_2)=-\frac{1}{T}H(\dot{\varphi})$, $g=\frac{K}{T}$, x_1 and x_2 are system variable, y is the actual output of the system, $f(x_2)$ is unknown function of system, g is known gain of control input, $d(t)$ represents ship external interference, and u is the control input of system.

Based on above, the goal of this article is to seek control law u which is able to make the system output y asymptotically tracking expected course φ_r, the tracking error will be infinitely close to zero, that is $e=y-\varphi_r\to 0$, and the stabilization time will be reduced as much as possible.

3 Dteering control design based on backstepping

The design steps of the ship course controller contain the following steps.

Step 1. According to the characteristics of system (2), the following sliding surfaces are defined.

$$\begin{cases}\dot{\xi}=z_2\\ z_1=x_1-\varphi_r\\ z_2=x_2-\sigma_1\end{cases} \tag{3}$$

Where, φ_r is the expected course, σ_1 is virtual stabilization, z_1 is the heading error. ξ is the integral term which can be able to eliminate the static error caused by the uncertain interference term in the control process.

Step 2. Lyapunov function is constructed to prove the asymptotic stability control system by using backstepping method. The first Lyapunov functionis constructed.

$$V_1=\frac{1}{2}z_1^2 \tag{4}$$

According to (3), it is obtained (5).

$$\dot{z}_1=z_2+\sigma_1-\dot{\varphi}_r \tag{5}$$

It can be inferred formula (6) by formula (4).

$$\dot{V}_1=z_1\dot{z}_1=z_1\left[z_2+\sigma_1-\dot{\varphi}_r\right] \tag{6}$$

The virtual stabilization σ_1 is set as formula (7)

$$\sigma_1=-k_1z_1+\dot{\varphi}_r \tag{7}$$

Therefore, it can be obtained (8).

$$\dot{\sigma}_1=-k_1\left(-k_1z_1+z_2\right)+\ddot{\varphi}_r \tag{8}$$

It can be obtained (9) , according to (6) and (7).

$$\dot{V}_1=-k_1z_1^2+z_1z_2 \tag{9}$$

Similarly, it is obtained (10) according to (3).

$$\dot{z}_2=\dot{x}_2-\dot{\sigma}_1=f(x_2)+g\cdot u+d(t)-\dot{\sigma}_1 \tag{10}$$

The second Lyapunov function is constructed.

$$V_2=V_1+\frac{1}{2}z_2^2 \tag{11}$$

It can be inferred formula (12) by formula (11).

$$\begin{aligned}\dot{V}_2&=\dot{V}_1+z_2\dot{z}_2=-k_1z_1^2+z_1z_2\\ &+z_2\left(f(x_2)+g\cdot u+d(t)-\dot{\sigma}_1\right)\end{aligned} \tag{12}$$

And the third Lyapunov function is constructed.

$$V_3=V_2+\frac{\lambda}{2}\xi^2 \tag{13}$$

It can be inferred formula (14) by formula (13).

$$\begin{aligned}\dot{V}_3&=\dot{V}_2+\lambda\xi\dot{\xi}=-k_1z_1^2+z_1z_2+\lambda\xi z_2\\ &+z_2\left(f(x_2)+g\cdot u+d(t)-\dot{\sigma}_1\right)\\ &=-k_1z_1^2+z_2\left(z_1+f(x_2)+g\cdot u+d(t)-\dot{\sigma}_1+\lambda\xi\right)\end{aligned} \tag{14}$$

Step 3. The RBF network is used to adapt the system function $f(x)$, and its algorithm is (15).

$$\begin{cases}h_j=\exp\left(-\frac{\|x-c_j\|^2}{2b_j^2}\right),\ j=1,2,\cdots,m\\ f(x)=W^T\mathbf{h}(x)+\varepsilon\end{cases} \tag{15}$$

Where $\boldsymbol{x}$ is the network input, i represents one of the first input network input layer, j is a hidden layer of the network the first network input, $\boldsymbol{h}=\left[h_j\right]^{\mathrm{T}}$ is high

Gaussian function, W^* is the ideal for weighting value, ε is ideal for neural network approximation error.

$$\hat{f}(x_2) = \hat{W}^T \mathbf{h}(\mathbf{x}) \tag{16}$$

Where, $\tilde{W} = \hat{W} - W^*$.

The global Lyapunov function is constructed.

$$L = V_3 + \frac{1}{2}\gamma \tilde{W}^T \tilde{W} \tag{17}$$

It can be inferred formula (18) by formula (17).

$$\dot{L} = \dot{V}_3 + \gamma \tilde{W}^T \dot{\hat{W}} \tag{18}$$

It can be obtained formula (19) by considering formula (14) and (16).

$$\begin{aligned}\dot{L} =& -k_1 z_1^2 + \gamma \tilde{W}^T \dot{\hat{W}} \\ &+ z_2 \begin{bmatrix} z_1 + f(x_2) + g \cdot u \\ + d(t) - \dot{\sigma}_1 + \lambda\xi \end{bmatrix}\end{aligned} \tag{19}$$

Step 4. In order to eliminate the uncertain interference term in equation (19), the nonlinear damping law[9] was designed.

$$u = \frac{1}{g}\left[-z_1 - \hat{f}(x_2) - (k_2 + \eta) z_2 - \lambda\xi + \dot{\sigma}_1\right] \tag{20}$$

It can be inferred formula (21) by considering formula (19) and (20).

$$\begin{aligned}\dot{L} =& -k_1 z_1^2 + \gamma \tilde{W}^T \dot{\hat{W}} \\ &+ z_2\left[f(x_2) - \hat{f}(x_2) + d(t) - z_2(k_2 + \eta)\right] \\ =& -k_1 z_1^2 - k_2 z_2^2 - \eta z_2^2 + \tilde{W}^T(\gamma \dot{\hat{W}} - \mathbf{h} z_2) \\ &+ \left[d(t) + \varepsilon\right] z_2\end{aligned} \tag{21}$$

In order to ensure system asymptotically stable, the following adaptive law is designed.

$$\dot{\hat{W}} = \frac{1}{\gamma}\mathbf{h} z_2 \tag{22}$$

It is easy obtained formula (23) by theorem "$xy \le \eta x^2 + \frac{1}{4\eta} y^2$".

$$\varpi z_2 \le \eta z_2^2 + \frac{1}{4\eta}\|\varpi\|_\infty^2 \tag{23}$$

Where, $\varpi = d(t) + \varepsilon$.

The fitness law of the network weights is still adopted formula (22), therefore, the formula (24) can be acquired.

$$\begin{aligned}\dot{L} =& -k_1 z_1^2 - k_2 z_2^2 - \eta z_2^2 + \varpi z_2 \\ \le& -k_1 z_1^2 - k_2 z_2^2 + \frac{1}{4\eta}\|\varpi\|_\infty^2 \\ \le& -k_2 z_2^2 + \frac{1}{4\eta}\|\varpi\|_\infty^2\end{aligned} \tag{24}$$

In order to ensure that the system (2) is asymptotically stable, it is only need to prove the establishment which is $k_2 z_2^2 \ge \frac{1}{4\eta}\|\omega\|_\infty^2$. Proof slightly, see the literature [4].

4 Simulation

USV course tracking control simulation is excuted by applying the above control algorithm.

4.1 Simulation Object and Conditions Setting

The establishment of USV planar motion mathematical model requires 8 known USV parameters, as shown in the table 1.

Table 1. The "LanXin" USV Parameters

Length Between Perpendiculars	7.02 m
Breadth	2.60 m
Speed	<=35 kn
Draft (full load)	0.32 m
Block Coefficient	0.6976
Displacement (full load)	2.73 m^3
Rudder Area	0.2091 m^2
Distance Between Barycenter and Center	0.35 m

In this paper, the planer motion mathematical model for the "Lanxin" USV adopts Nomoto model from the nonlinear mathematical model of formula (1).

There is the formula of matrix parameters of the mathematical model of the ship to calculate the above parameters, When the speed is 8.5kn, the gain constant and time constant are obtained through MATLAB programming, an $K = 1.50226$, $T = 0.90503$, $\alpha_1 = 0.001$, $\alpha_{i(i=2,3,\ldots)} = 0$. They are the Nomoto model parameters of the "Lanxin" USV [17]. In the actual simulation, the influence of wind and wave for USV motion is obtained with white noise and a second-order wave ' s transfer function [18, 19]. The formula can be written as

$$y(s) = h(s)\, w(s) \tag{25}$$

Where $w(s)$ is the zero-mean Gaussian white noise, the power spectral density is 0.1, $h(s)$ is a second-order wave transfer function, and $h(s)$ can be written as formula (26).

$$\begin{cases} h(s) = \dfrac{K_\omega s}{s^2 + 2\xi\omega_0 s + \omega_0^2} \\ K_\omega = 2\xi\omega_0\sigma_m \end{cases} \quad (26).$$

Where ω_0 is the dominant frequency of sea waves, ξ is the damping coefficient, K_ω is the gain constant. σ_m is the constant describing wave intensity.

Based on USV motion model, the controller adds the model of the rudder servo system as formula (27).

$$\dot{\delta} = -\frac{1}{T_r}\delta + \frac{1}{T_r}\delta_r \quad (27).$$

Where T_r is time constant, for USV is generally about 0.2s. δ_r is Command rudder angle. There is the actually limit of rudder angle that is the formula (28).

$$|\delta| \le 35^\circ \quad (28).$$

4.2 Simulation Results and Analysis

4.2.1 Course changes experiment

(1) experiment of 20°change of course

Protocol initial heading as 000°, expected course 020°, the simulation results without external interference and model parameter perturb are shown in Fig.1.

It can be seen from Figure 1, the control effect is good,and the output of course will reach the expected value faster, costing 30s without overshoot nearly; and rudder control horn is reasonable which can meet the characteristics requirements of steering system.

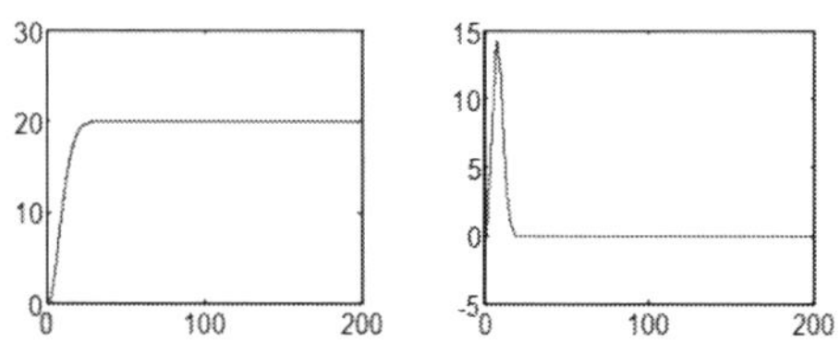

(a) course angle (b) rudder angle

Fig.1. the output of course and rudder angle

(2) experiment of continuous change of course

Conduct experiment of tracking square wave, set the cycle as 100s, the change of course as 030°, the experimental results will be shown in Fig.2.

It can be seen from Figure 2, in the case of desired track changes, the new controller can do better in steering and easing the helm counter steering and output the right rudder angle according to ship characteristics and sea condition at the right time, and also conduct without overshoot in Fig 2.

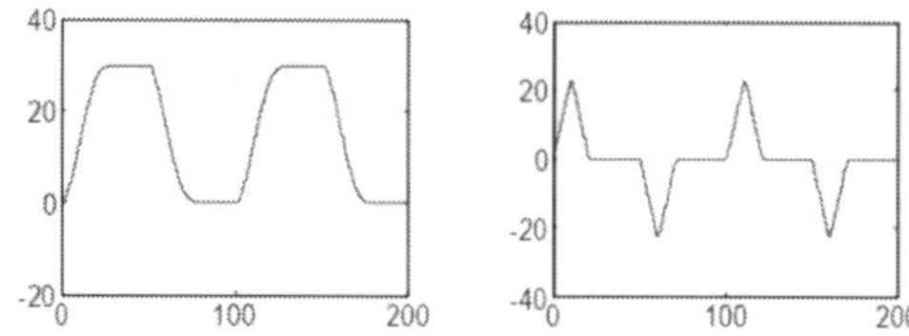

(a) course angle (b) rudder angle

Fig.2. the output of course and rudder angle when expected course continuous change

4.2.2 Interference experiment

(1) experiments of white noise interference

Protocol initial heading as 000°, desired track as 030°, apply amplitude as 0.1 white noiseinterference to course angle, the experiment results are shown in Fig.3.

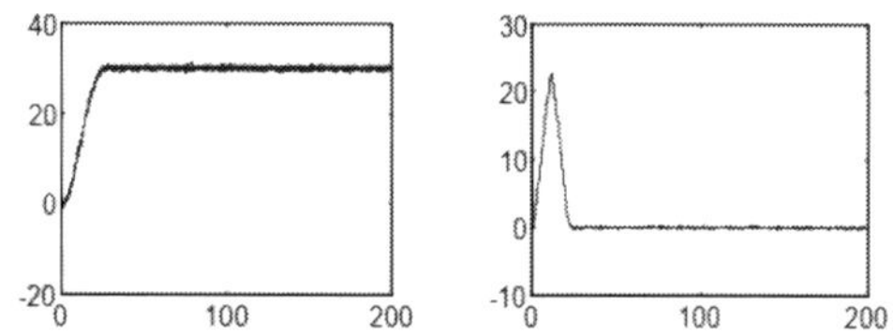

(a) course angle (b) rudder angle

Fig.3. the output of course and rudder angle under white noise interference

It can be seen from Figure 3, when the ship was disturbed by the outside world, the designed course controller will overcome outside interference to a certain degree, and will track the expected course with no shock of rudder controlling.

(2) experiments of ship parameter perturbation and white noise interference

Setting the protocol initial heading as 000°, desired track as 030°, propose maneuvering ship indices K、T perturb at 40%， apply amplitude as 0.1 and make white noise interference to the course, the experiment results are shown in Fig 4.

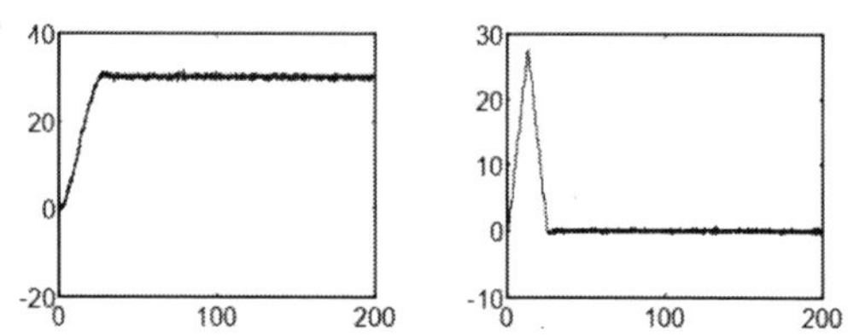

(a) course angle (b) rudder angle

Fig.4. the output of course and rudder angle under model parameter perturb and white noise interference

It can be seen from Figure 4, when the ship model parameters occur perturbat with interference from the outside world, the course output from the new controller has smaller fluctuation around expect course, and has smaller shock amplitude of rudder controlling, thereby, it is illustrated that the anti-interference ability of the new controller is stronger.

5 Conclusion

An integral sliding mode controller with neural network was investigated for USV navigation, which realizes the accurate and stable tracking of the ship in the steering process, on the basis of backstepping method. With disturbances of wind, wave and current, the RBF network based integral backstepping sliding mode control system is simulated and the results of simulation show that the proposed algorithm is effective for USV course control.

Acknowledgment

This work was supported by the Foundation of Jiangsu Maritime Institute under Grant No.XR1501 and No.XR1504 , and Jiangsu University Brand Professional Construction Project under Grant No. PPZY2015B177.

References

1. B. Volker, Unmanned Surface Vehicles - A Survey, Skibsteknisk Selskab, Copenhagen, Denmark, 2008.
2. S. Bennett. Nicolas Minorsky and the automatic steering of ships, IEEE Control Systems Magazine, 1984, 4(4) : 10-15.
3. S. Kumarawadu, K. Kumara, On the speed control for automated surface vessel operation, in Proceedings of the 2007 third international conference on information and automation for sustainability, Melbourne, Australia, 2007: 135-140.
4. Benaskeur A.R, Desbines A. Backstepping based adaptive PID control[C]. Proceedings of Control Theory and Applications. 2002, 149 (1): 54-59.
5. Y. Pan, D. Huang, Z. Sun, Backstepping adaptive fuzzy control for track-keeping of underactuated surface vessels, Control Theory and Applications, 2011, 7: 907-914.
6. Renqiang Wang, Yuelin Zhao, Keyin Miao. Application of Neural Network Minimum Parameter Learning Algorithm in Ship's Heading Tracking Control[C]. Proceedings of 9th International Symposium on Computational Intelligence and Design, 2016, 12: 135-139.
7. Wang Renqiang, Gong Jianyun, Zhao Yuelin.Neural Network Sliding Mode Control under New Reaching Law and Application[C], Proceedings of Advanced Information Technology, Electronic and Automation Control Conference, 2015, 12 : 911-917.
8. X. Bao, K. Nonami, Z. Yu, Combined yaw and roll control of an autonomous boat, in Proceedings of the 2009 IEEE International Conference on Robotics and Automation, Kobe, Japan, 2009: 188-193.
9. Wang Ning, Lv Shuailin, Liu Zhongzhong. Global finite-time heading control of surface vehicles[J], Neurocomputing, 2016, 175 (106) : 662-666.
10. K. Do, Z. Jiang, J. Pan, Underactuated ship global tracking under relaxed conditions, IEEE Trans. on Automatic Control, 47(9): 1592–1536, 2002.
11. L. Yuan, H. Wu, Multiple slidfing mode adaptive fuzzy controller for nonlinear marine autopilot systems, CAAI Transactions on Intelligent Systems, 2010, 4: 308-312.
12. T. Li, Y. Yang, B. Hong, et al, Robust adaptive fuzzy design for ships track-keeping control, Control Theory and Applications, 2007, 3: 445-448.
13. Guanshan Hu, Yingbing Zhou. Application of Fuzzy Neural Network in Ship Course Control[J], Applied Mechanics and Materials, 2012, 1511(135) : 309-315.
14. Xin Wang, Tieshan Li, Weilin Luo. Direct Adaptive Neural Network Control for a Class of Ship Course Uncertain Discrete-time Nonlinear Systems[J], Marine Engineering Frontiers, 2013, 1(3) : 42-48.
15. Yang Wang, Chen Guo, Fuchun Sun. Dynamic neural fuzzified adaptive control of ship course with parametric modelling uncertainties[J], Int. J. of Modelling, Identification and Control, 2015, 13(4) : 251-258.
16. Wang Renqiang, Zhao Yuelin, Sun Jianming.Application of Optimized RBF Neural Network in Ship's Autopilot Design[C], Proceedings of IEEE Advanced Information Management, Communicates, Electronic and Automation Control Conference, 2016, 10: 1642-1646.
17. FAN Yunsheng, SUN Xiaojie, WANG Guofeng, GUO Chen. On Fuzzy Self-adaptive PID Control for USV Course[C], Proceedings of the 34th Chinese Control Conference, 2015, 7: 8472-8478.
18. T. Fossen, Guidance and control of ocean vehicles. England: John Wiley & Sons LTD., 1994.
19. C. Yang, X. Jia, Y. Bi, Rudder Stabilization of Ships and its Robust Control. Dalian: Dalian Maritime University Press, 2011.

Research on a detection algorithm for abnormal state of load

Dan Yu[1,2*], *Yubo* Wang[1,2], *Jun* Du[1,2], *Lizong* Liu[1,2], *Haojie* Tian[3], *Yan* Guan[3] and *Shengfang* Li[1,2]

[1]State Grid Key Laboratory of Power Industrial Chip Design and Analysis Technology, Beijing Smart-Chip Microelectronics Technology Co., Ltd. Beijing 100192, China
[2]Beijing Engineering Research Center of High-reliability IC with Power Industrial Grade, Beijing Smart-Chip Microelectronics Technology Co., Ltd. Beijing 100192, China
[3]State Grid Liaoning Electric Power Supply Co., Ltd. Liaoning 110006, China

Abstract. This paper describes the development status and main problems of load monitoring, introduces the key technologies of load monitoring, and by using a load state monitoring system, emphatically illustrates a detection algorithm for abnormal state of the secondary load of the current transformer on the three-phase line of the power grid. The algorithm mainly achieves the real-time detection of abnormal state such as disconnection, short connection and series connected semiconductor. In the light of this algorithm, the working principle is explained, the model formula is worked out, and the state criterion is given. The load condition monitoring system is debugged, put into operation and tested in the pilot operation, and the results show that the algorithm has a good effect.

1 Introduction

With the development of national economy, electric energy has become an indispensable energy source in daily production and life. However, abnormal state of the power load in our country is common and serious, which brings huge economic losses.

The combination of key technical measures such as network technology, sensor technology, communication technology, and information technology makes load monitoring of our country change from conventional passive way to all-weather, all-round, real-time active and intelligence way.

At present, load monitoring is divided into two types of intrusive and non-intrusive [1-2]. In the past, intrusive load monitoring was used, that is, sensors and other information acquisition devices which are installed at the load to obtain load information. As the market of load monitoring is increasingly expanding, non-intrusive load monitoring (NILM [3]) is proposed and widely concerned which is compared with the traditional intrusive [4-5]. The installation of information acquisition devices is at the entrance of the monitored power area to obtain load information, such as current and power. Based on these, the situation of each load in the area is analyzed. Compared with intrusive load monitoring, non-intrusive load monitoring has the advantages of simple installation, good economy, high reliability and easy to popularize [1, 6], and is suitable for independent user installation mode [1, 7].

In order to discover and deal with abnormal state of the secondary load of the current transformer on the three-phase line of the power grid (usually including disconnection, short connection and series connected semiconductor), the load state monitoring system is developed to make up for the lack of existing measures. The load state monitoring system ensures the normal operation of the power grid, so that non-intrusive load monitoring is achieved smoothly.

According to the requirement of load state monitoring system, a detection algorithm based on analysis of self oscillation and AC current harmonics is studied to identify the abnormal state of secondary load by using FFT signal processing technology and impedance model of the secondary load. The algorithm is small in complexity, the identification state is accurate and the validity is verified by practice, so that the load state monitoring system achieves real-time, accurate, intuitive and comprehensive monitoring of the secondary load state.

2 Description of the detection algorithm

2.1 Detection principle of disconnection, short connection

A resonant circuit is constructed by using isolated magnetic couplers and capacitors, which is incorporated into the secondary load circuit for forming self-excited oscillation.

Self-excited oscillation is the basis for detecting abnormal disconnection and short connection. These abnormal states of the secondary load will show the abrupt change of the secondary load impedance, and this change will be directly expressed as the change of self-excited oscillation frequency. This self-excited

* Corresponding author: ewebbac@163.com

oscillation frequency becomes high in short circuit and it becomes low in open circuit.

If the measurement accuracy of this frequency is strictly required, the detection of the abnormal disconnection will become a problem. Therefore, it is necessary to inject an AC signal with the appropriate frequency band to assist identification, and the corresponding signal pickup circuit is also needed.

2.2 Analysis of temperature compensation during open circuit and short circuit

In the practical application, it is found that the isolated magnetic coupler for self-excited oscillations has a temperature characteristic which affects the self-excited oscillation frequency, because the permeability of isolated magnetic coupler core material will change greatly when the ambient temperature changes.

In this case, the temperature experiment of the high-precision liquid bath was completed for the isolated magnetic coupler composed of MnZn ferrite and ultra-microcrystalline (nanocrystalline). In this experiment, the temperature was kept stable for half an hour at different temperatures, and changes in the parameters was measured by the standard LCR tester. LCR tester measurement model: Ls-Q (series equivalent), the test level is 1V.

The parameter of the corresponding magnetic core inductance is leakage inductance in short circuit, and it is self inductance in open circuit.

When abnormal disconnection or short connection is happened in the secondary load circuit, the temperature characteristics of ferrite material parameters are shown in Figure 1 or Figure 2. The curves of 5, 15, 20 and 30 in Figure 1 and Figure 2 show the values measured at the test frequencies of the LCR tester (5kHz, 15kHz, 20kHz, and 30kHz, respectively).

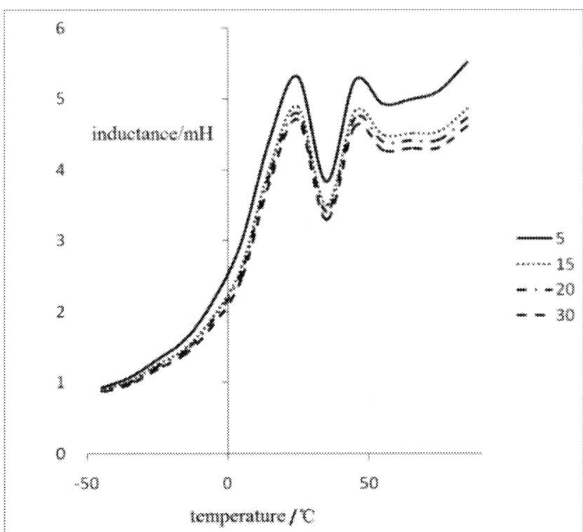

Fig. 1. The self inductive parameters of ferrite vary with temperature.

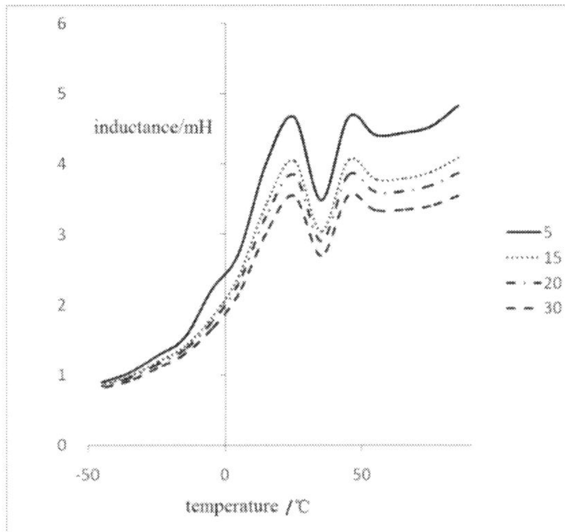

Fig. 2. The leakage inductive parameters of ferrite vary with temperature.

The characteristics of ferrite material parameters are shown in Figure 1 and Figure 2:

- Large leakage inductance (self inductance and leakage inductance are not much difference);
- Occurrence of the diving fluctuation, 25℃~45℃;
- Basic stability (there is a certain upward trend), above 50℃;
- The downward trend of the approximate quadratic function curve, below 25 ℃.

When abnormal disconnection or short connection is happened in the secondary load circuit, the temperature characteristics of ultra-microcrystalline material parameters are shown in Figure 3 or Figure 4. The curves of 5, 10, 15 20, 25 and 30 in Figure 3 and Figure 4 show the values measured at the test frequencies of the LCR tester (5kHz, 10kHz, 15kHz, 20kHz, 25kHz and 30kHz, respectively).

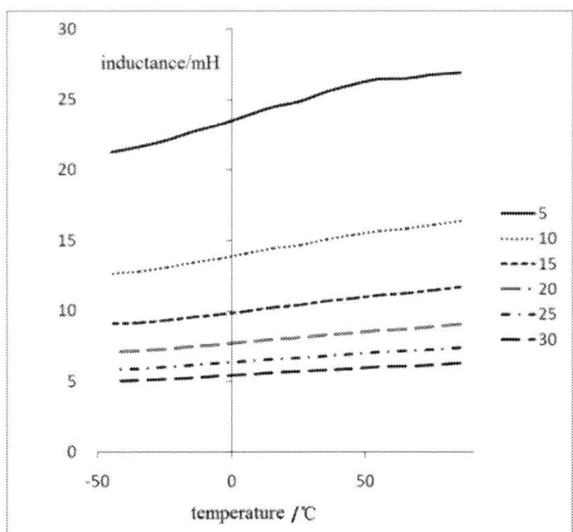

Fig. 3. The self inductive parameters of ultra-microcrystalline vary with temperature.

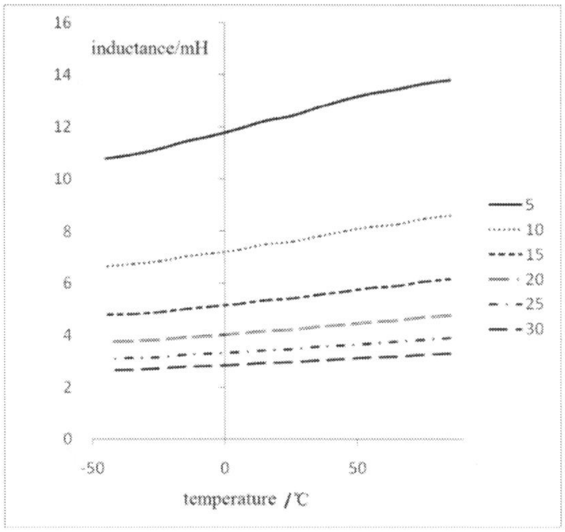

Fig. 4. The leakage inductive parameters of ultra-microcrystalline vary with temperature.

The characteristics of ultra-microcrystalline material parameters are shown in Figure 3 and Figure 4:

- high permeability, small leakage inductance;
- almost linear variation with temperature;
- the slope with the change of temperature is small (relatively gently).

Thus, different temperature compensation methods should be given for different coupling element material characteristics, and for ferrite should be segmented compensation, whereas for nanocrystalline only linear compensation is needed.

The temperature compensation is realized by the curve fitting method based on the least square principle, The regression analysis model is constructed by linear equation and polynomial.

2.3 Detection principle of series connected semiconductor

The semiconductor described in this paper refers to the diode. Due to the unidirectional conductivity of the diode, the load current changes from the pure AC signal to the pulsating DC signal after the semiconductor is connected in the secondary load circuit.

The key to the detection algorithm of series connected semiconductor is how to calculate and analyze the waveform characteristics of the secondary load current in the abnormal state by using sampled series current data.

Fourier spectrum analysis is the key to the analysis of the load current waveform in the abnormal state of series connected semiconductor [8].

According to the Fourier transform [9]:

$$\hat{f}(\xi) = \int_0^{\frac{1}{\xi}} f(t) e^{-i2\pi\xi t} dt \tag{1}$$

In formula (1): $f(t)$ - time domain function; $\hat{f}(\xi)$ - frequency domain function.

In the generalized the Fourier transform, the upper and lower bounds of the integral cycle should be -∞ ~ + ∞, but only $1/\xi$ is used, because it is no necessary to consider the effects of subharmonics and non-integer harmonics, and the calculation is convenient. $1/\xi$ is the minimum integral cycle, of course, you can also take an integer multiple of $1/\xi$.

Further solving vector ($\hat{f}(\xi)$): according to Euler's formula, put $e^{-i2\pi\xi t} = \cos 2\pi\xi t - i \sin 2\pi\xi t$ into formula (1), then formula (2) is formed.

In order to express the relationship between science between quantity and degree, and convenient program portability, engineering calculation will usually be normalized to the number of numerical and harmonic content into a normalized value, then processing.

$$\hat{f}(\xi) = \int_0^{\frac{1}{\xi}} f(t) \cos 2\pi\xi t dt - i \int_0^{\frac{1}{\xi}} f(t) \sin 2\pi\xi t dt \tag{2}$$

In formula (2): $\int_0^{1/\xi} f(t)\cos 2\pi\xi t dt$ - cosine component of harmonic frequency; $\int_0^{1/\xi} f(t)\sin 2\pi\xi t dt$ - sinusoidal component of harmonic frequency.

In order to scientifically demonstrate the relationship between quantity and degree, and facilitate the transplant of the program, many numerical values are usually normalized in engineering calculation, that is, the above harmonic content is converted to the per-unit value, and then processed.

Due to the similar "fingerprint" feature of such harmonics is fixed at the ratio of the even harmonic spectral density (except the fundamental wave), the ratio of the adjacent even harmonic spectral density can be used to confirm "Fingerprint" features of current harmonics in the abnormal state of series connected semiconductor. Since the higher harmonic content is usually very small, and the cost of the system sampling rate and calculation is high, the lower harmonics are selected to be solved. Because of the limitation of space, the solution process is omitted. The following are three criteria for series connected semiconductor:

$$R_{21} = \frac{|U_{2b}|}{|U_b|} = 42.44\% \tag{3}$$

$$R_{42} = \frac{|U_{4b}|}{|U_{2b}|} = 20.03\% \tag{4}$$

$$R_{64} = \frac{|U_{6b}|}{|U_{4b}|} = 42.82\% \tag{5}$$

In formula (3), (4), (5): U_b - Fundamental value; U_{2b} - Second harmonic value ; U_{4b} - Fourth harmonic value ; U_{6b} - Sixth harmonic value.

If the formula (1), (2), (3) are true, the abnormal state of series connected semiconductor is detected.

If the formula (2), (3) are true, and the equation (1) is false, it is detected that there is the half-wave direct current component in the primary current of the current transformer on the three-phase line of the power grid.

Note: In formula (1), 42.44% is the maximum value. When there is a part of the DC component, the value of this item will be small or even close to zero.

2.4 Saturation analysis of magnetic circuit in series connected semiconductor

In the practical application, it is found that when the secondary load circuit is greater than a certain value, the core of the current transformer on the three-phase line of the power grid enters the saturation state; when the secondary load circuit is greater than a larger value, then it enters the deep saturation state.

After the semiconductor is connected in the secondary load circuit, the effect of magnetic saturation caused by the DC bias is shown in Figure 5. At this point, parameters of the current transformer on the three-phase line of the power grid: the change ratio is 150: 5; the maximum secondary load is 5VA; the accuracy level is 0.2S.

The measured current in Figure 5 is the measured value obtained by CPU processing after sampling the secondary load current with an analog-to-digital converter. The input current is the actual value of the secondary load current.

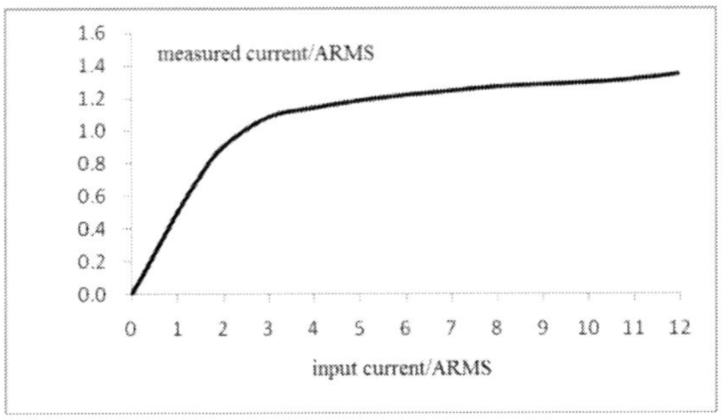

Fig. 5. The relationship between measured current and input current.

After the semiconductor is connected in the secondary load circuit, the current waveform varies with the secondary load current as shown in Figure 6, Figure 7, Figure 8, and Figure 9.

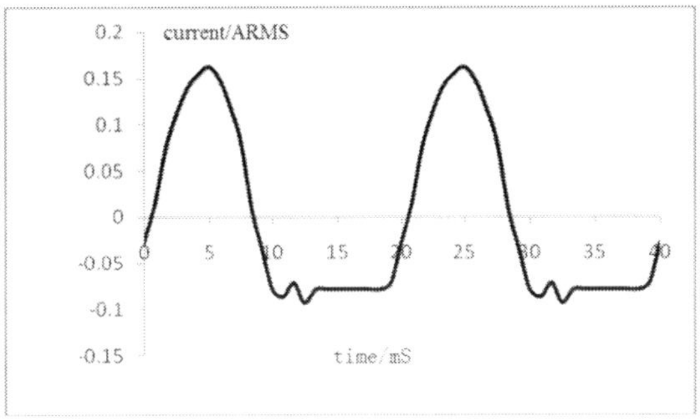

Fig. 6. In the half wave excitation, the waveform of the input current varies. (The input current is 200mA)

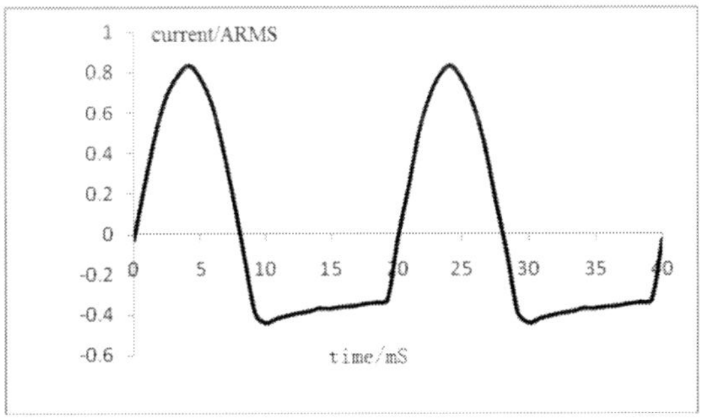

Fig. 7. In the half wave excitation, the waveform of the input current varies. (The input current is 1A)

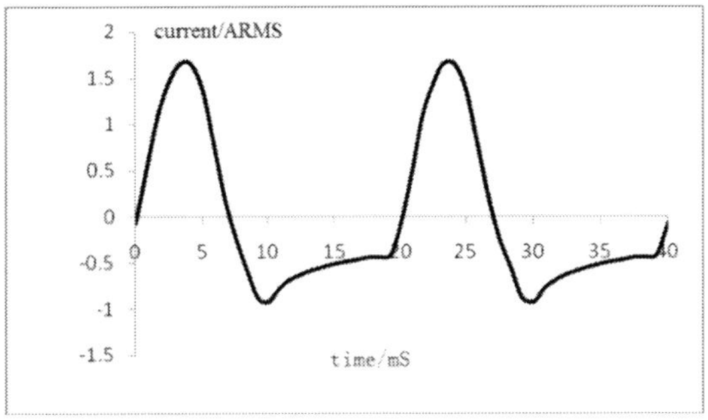

Fig. 8. In the half wave excitation, the waveform of the input current varies. (The input current is 2A)

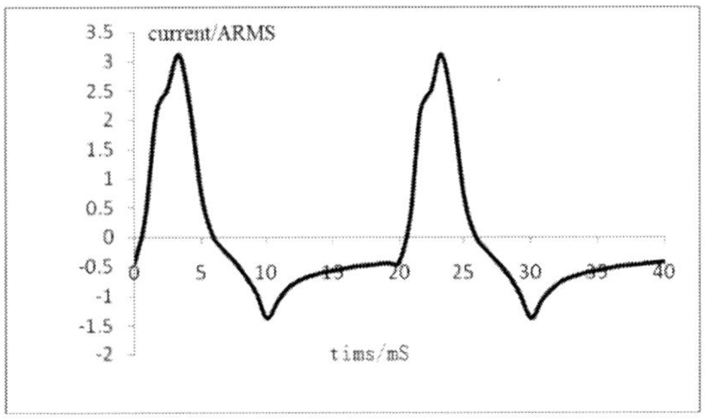

Fig. 9. In the half wave excitation, the waveform of the input current varies. (The input current is 4A)

Analysis and comparison of the waveform changes of the secondary load current:

- When the current is less than 2A, the current waveform basically conforms to the ideal half wave model. Among them, the half wave in Figure 6 should be the closest ideal (the bottom of the half-wave is flat), the bottom of the half wave in Figure 7 appears linearly tilted, and the bottom of the half wave in Figure 8 appears nonlinear tilt.

- When the current is greater than 2A, the distortion of the current waveform is exacerbated, as shown in Figure 9.

- Because the saturation of magnetic circuit leads to the nonlinear transmission of signals, the current waveform is no longer the superposition of a DC component and even harmonic components, in which odd harmonic components are added.

In order to balance the function and cost, the core saturation of the current transformer on the three-phase line of the power grid is difficult to avoid. In this way, the criterion for detecting the abnormal state of series connected semiconductor can not simply depend on the ratio of adjacent even harmonics. It needs to be combined with the distortion of a certain number of harmonic components.

3 Conclusion

This paper studies a detection algorithm for abnormal state of the secondary load, as the core part of the load condition monitoring system to ensure the normal operation of the intrusive load monitoring.

The advantages of the algorithm are as follows:

- The algorithm provides a more practical criterion, and the corrective measures are given in combination with the effects of ambient temperature and magnetic saturation.

- The algorithm has low computational complexity and short calculation time, so it can meet the requirement of real-time.

In addition, the algorithm is simple, effective, stable and easy to maintain.

References

1. Yu Yixin, Liu Bo, Luan Wenpeng. Nonintrusive Residential Load Monitoring and Decomposition Technology. Southern power system technology, **7**, 4 (2013)
2. Liu Bo, Luan Wenpeng. Conceptual Cloud Solution Architecture and Application Scenarios of Power Consumption Data Based on Load Disaggregation. Power System Technology, **40**, 3 (2016)
3. Hart G W. Nonintrusive appliance load monitoring. Proceedings of the IEEE, **80**, 12 (1992)
4. Drenker S, Kader A. Nonintrusive monitoring of electric loads. IEEE Computer Applications in Power, **12**, 4 (1999)
5. Li Peng. Non-intrusive Method for Power Load Disaggregation and Monitoring. Tianjin University Ph.D. Thesis (2009)
6. Gao Yun, Yang Honggeng. Household Load Measurement by Classification Based on Multi-scale Wavelet Decomposition. Proceedings of the CSU-EPSA, **27**, 7 (2015)
7. Niu Lulu , Jia Hongjie. Transient Event Detection Algorithm for Non-intrusive Load Monitoring. Automation of Electric Power System, **35**, 9 (2011)
8. GB/T 17215.321-2008 Electricity metering equipment (a.c.) – Particular requirements – Part 21: Static meters for active energy (class 1 and 2). (IEC 62053-21:2003, IDT)
9. E.Chu, A.George. Inside the FFT Black Box: Serial and Parallel Fast Fourier Transform Algorithms. Canada: CRC Press LLC (2000)

Design and Test of Sensor for Aircraft Engine Gas Path Debris Electrostatic Monitoring

Wei Liu [1,2,*], Manguo Huang [1,2], Defeng Liu [1,2], Yanshan Wang [1,2], Meiju Zhang [1,2]

[1]Beijing Changcheng Aeronautic Measurement and Control Technology Research Institute, Beijing 101111, China
[2]Aviation Key Laboratory of Science and Technology on Special Condition Monitoring Sensor Technology, Beijing 101111, China

Abstract. The research and design of sensor for aircraft engine gas path debris electrostatic monitoring are given in this paper. The mechanism of charged debris and principle of electrostatic induction of sensor are researched firstly, and when the charged material near or keep away from the electrostatic sensor, the induced charge of opposite polarity will be induced on the sensor's probe. And then, the structure of electrostatic monitoring sensor is designed based on the principle of electrostatic induction. Electrostatic sensor is constituted by sensitive element, dielectric, internal circuit and grounding shell. Due to the particularity of working environment of electrostatic sensor, various factors such as performance requirements and using occasions should be considered in designed process. Finally, the effectiveness of electrostatic monitoring sensor is verified using experimental system. From the experimental results we can know when the charged particles through the electrostatic sensor, the sensing signal amplitude and the activity level are change obviously. These works provide theoretical method for the early fault warning of aircraft engine gas path components, and the monitoring and management of aircraft engine working condition.

1 Introduction

In 1970s, the U.S. air force institute of technology found that when the aircraft engine gas path components fault occurs, the electrical activity in the gas path channel will increase [1]. When aircraft engine combustor jet exhaust, meanwhile, with a lot of electronics, ions and charged particles including the state information of aircraft engine gas path components generated [2]. Using the electrostatic sensor to detect these charged particles can obtain status information of the engine gas path components, which can provide early warning information for engine. In the 1990s, Powrie and Fisher et al studied the technology and method of using the state of the electrostatic sensors to monitor engine. This method has been used successfully in the U. S. joint fighter F-35 [3-4]. The current domestic research on gas path electrostatic sensor monitoring technology just started, a lot of researched results of the engine gas path simulated test [5-6] and engine's gas path electrostatic monitoring signal [7-8] has been achieved, but the current domestic public research mainly based on the ground of a certain type of test-platform of aircraft engine. As the main research target, the engine exhaust gas electrostatic signal was studied, but the research of electrostatic sensor that is used to monitor charged particle is less public. In this paper, the electrostatic sensor for aircraft engine gas path debris detection have been researched, the principle of electrostatic monitoring was analyzed, the model and the design of structure are given, and the effectiveness of the proposed sensor was verified by the simulated research finally.

2 The Principle of Electrostatic sensor

2.1. The mechanism of charged particles

Apart from carbon dioxide, nitrogen, water vapor, hydrocarbon, carbon monoxide, sulfur oxides and produced by incomplete combustion of carbon particles and other normal particles, the exhaust of aircraft engine may also include abnormal particles because of the fault of engine's gas path components. There are a lot of reasons for the abnormal particles appeared in the exhaust gas, for example, the suction inlet of an alien, the falling blade caused by local instantaneous high temperature, disbonding, the loss of leaf material caused by cyclical stress under the local environment of high temperature, excessive carbon smoke particles in chamber caused by local rich oil, as well as the rubbing between blade and casing or sealing ring.

In aircraft engine combustion chamber, the high temperature from the burning fuel oil makes the gas molecules to occur the chemical ionization reaction, so there are many charged ions owing to the gassy chemical ionization reaction in the gas path. When the carbon smoke particles formed in the rich oil region pass through the high-temperature flame zone, interaction between the free electrons and ions formed by chemical

* Corresponding author: lw880623@126.com

ionization make the particles charged [9]. The smoke from the burning of particle collide with the random thermal motional ions under the environment of high temperature, which can produce electric charge, the process can be expressed as:

$$q_p = \frac{d_p kT}{2e^2}\ln\left(1+\frac{d_p c\pi e^2 Nt}{2kT}\right) \tag{1}$$

In the equation, q_p is the particle charge, d_p is the particle diameter, k is the Boltzmann constant, T is temperature, e is the electronic charge, c is the average speed of particles, N is particle concentration, t is time.

2.2 Principle of electrostatic sensor

When a charged object closes to or keep far away from a non-charged conductor, there will be induced charge on the surface of conductor and the phenomenon is called electrostatic induction. The basic principle of electrostatic induction is that the charged objects can produce electric field in space, and the charge will redistribute within the electric force of the conductor and then induced current generated inside the sensor. Electrostatic sensor principle diagram is shown in Fig. 1. When the charged material near or keep away from the electrostatic sensor, the induced charge of opposite polarity will be induced on the sensor's probe.

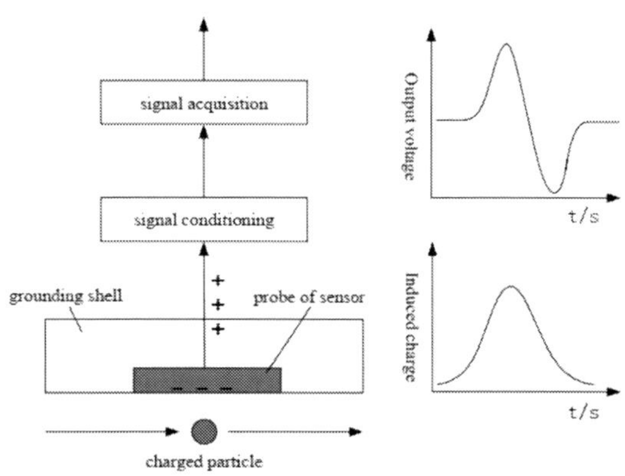

Figure 1. Schematic diagram of induced charge and current.

Assuming that the induced charge on the probe is $q(t)$, the measurement of the output voltage can be represented by equivalent circuit shown in Fig. 2.

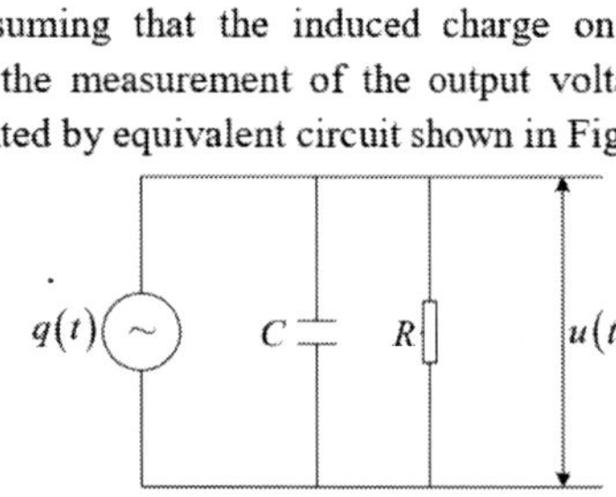

Figure 2. Measuring equivalent circuit.

In Fig. 2, R represents the equivalent resistance measuring circuit, C represents the equivalent capacitance measuring circuit. Assuming that the initial state of the circuit are zero, then the relationship between the Laplace transform $U(S)$ of output voltage and the Laplace transform $Q(S)$of induced charge are:

$$U(S)=\frac{RSQ(S)}{1+RCS} \tag{2}$$

When RCS<<1, the measurement of model equals to:

$$u(t)=Rq'(t) \tag{3}$$

The equation reflects the changes of the induced charge.

3 Design of the Electrostatic Sensor

3.1. Constitution of electrostatic sensor

Electrostatic sensor is constituted by sensitive element, dielectric, internal circuit and grounding shell, as shown in Fig. 3. When charged particles flow through the measuring electrode, it will induce the charge equal in size, symbol instead on the proximal and the distal. The induced amount of charge is all the sum of the induced charge in electrostatic sensor. You can take single charged particles as point charge here. The induced characteristic will emerge when the point charge pass through the measuring electrode, reflecting their own sensing characteristics of the measuring electrode also. The mathematical models of electrostatic sensor are established on the basis of the point charge in this paper.

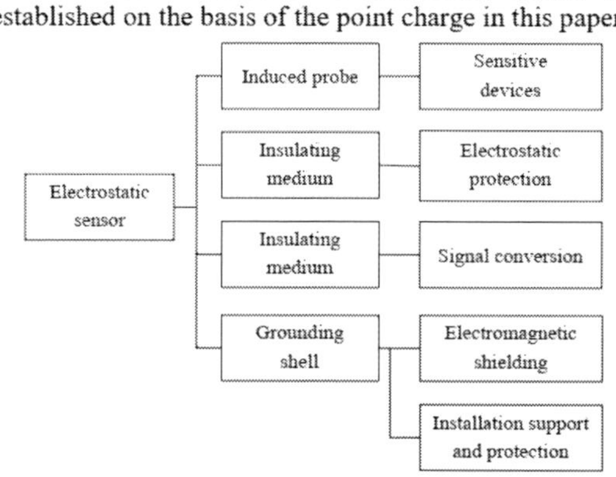

Figure 3. Structure and function model diagram of electrostatic sensor.

3.2 Design of electrostatic sensor

Due to the particularity of working environment of electrostatic sensor, various factors such as performance requirements and using occasions should be considered in designed process. The probe is put in the nozzle directly. Not only the materials of probe should satisfy the sensitive induction electric field but also the high temperature characteristic. The isolation materials between probe with grounding shell material behave conductive not easily and generate electrostatic charge not easily, and have the characteristics of high temperature resistant as same as the mat.

The design of electrostatic sensor is shown in Fig. 4. In this figure, 1 is the stainless steel plug, 2 is the stainless steel shell, 3 is the ceramic plate, 4 is ceramic sleeve, and 5 is nickel base alloy probe.

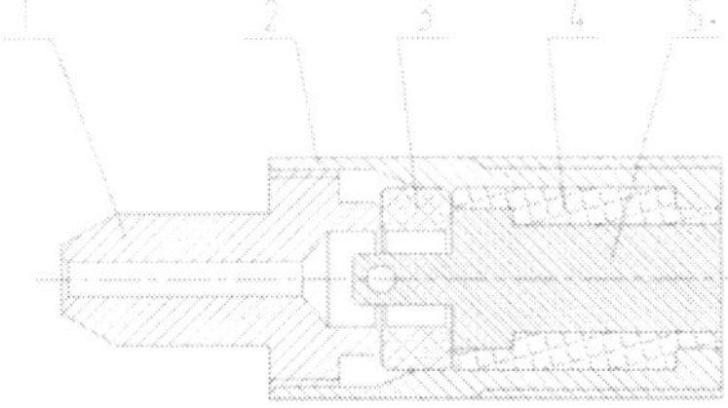

Figure 4. Structure of electrostatic sensor.

The plug and the shell are made by stainless steel with the characteristic of high temperature resistance. The shell mainly has the effect of external electromagnetic interference shielding, protecting the internal circuit structure at the same time, and grounded well. Cover and sleeve are made by ceramic material with good insulated performance and high temperature resistance, the main effect of cover and sleeve is to prevent the interference of external static and the loss of internal static signal. The probe is made by conductive easily nickel base alloy material with the characteristic of high temperature resistance, ensuring the good conductivity. The extended wire of electrostatic sensor is made by single core shielded wire.

4 Analysis of Experiment

To verify the effectiveness of the designed electrostatic sensor, the electrostatic monitoring experiment of simulated engine pipeline particles was done, the experimental system is shown in Fig. 5.

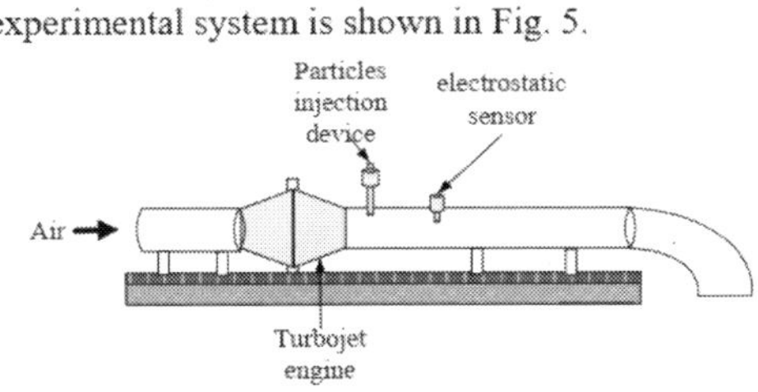

Figure 5. Simulation experiment system of electrostatic sensor.

The whole tested system consists of particles injection device, electrostatic sensor, small turbojet engine system, exhaust pipe, exhaust device, installation of platform. The electrostatic sensor installed on the exhaust pipe. The particles injection device is used to inject simulated particles to exhaust pipe, the small turbojet engine system is used to simulate the high-temperature environment.

In the process of experiment, the turbojet engine was opened firstly, and the iron powder particles were injected into the pipeline periodically. The particles sampled signal using electrostatic sensor is shown in Fig. 6.From this figure we can see that the sampled signal amplitude has an obvious change when the iron powder particles were injected, and the peak to peak value is increased obviously.

The activity level (AL) of sampled signal is shown in Fig. 7. From this figure we can know that the activity level of particles sampled signal is higher than the background signal, which indicates the electrostatic sensor can detect the charged particles effectively.

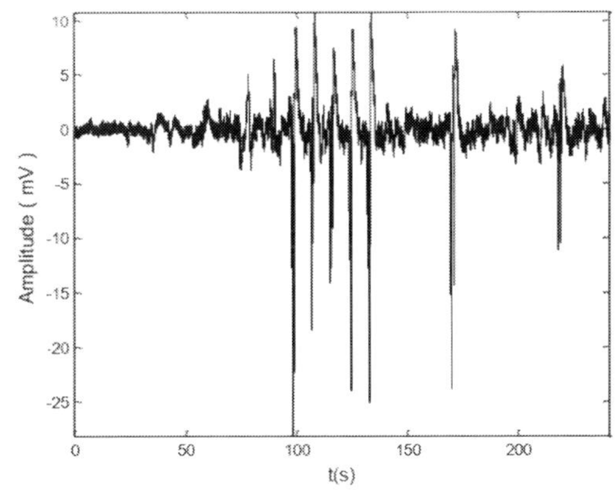

Figure 6. Particles sampled signal.

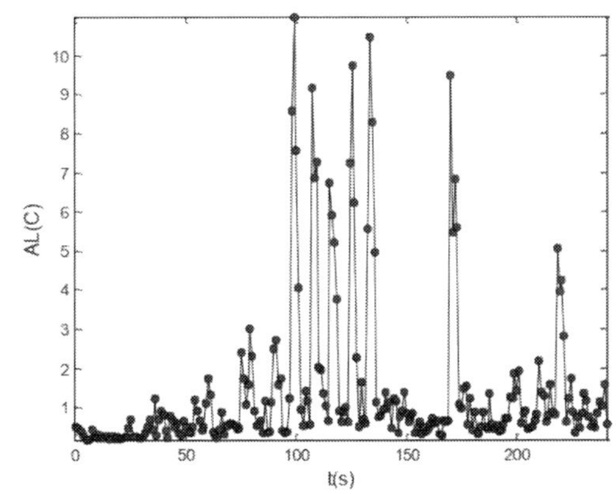

Figure 7. Activity level.

The activity level (AL) is defined as:

$$AL(t)=\sqrt{\frac{1}{T}\int_0^T Q_f^2(t)dt}=\sqrt{\frac{1}{N}\sum_{n=1}^{N}Q_{f_n}^2} \tag{4}$$

In the equation, Q_f is the electrostatic charge signal, T is the intercepted signal length, usually takes for 1s, N

is the sampling points in T after discrete sampling for Q_f ,and Q_{f_n} is the electrostatic charge signal after discrete sampling for Q_f .

5 Conclusions

Based on the principle of electrostatic induction sensor, the design of electrostatic sensor is given, and the electrostatic monitoring experiment was processed. From these analyses, we can obtain the following conclusion: When the charged particles through the electrostatic sensor, the sensing signal amplitude and the activity level are change obviously, which verifies the effectiveness of the electrostatic sensor designed in this paper.

Acknowledgment

The author thanks all those who helped her during the process of writing this manuscript and during her research. This work was supported by the Aeronautical Science Fund 2014ZD34005.

References

1. H. E. G. Powrie , K.Mcnicholas, Gas path condition monitoring during accelerated mission testing of a demonstrator engine, AIAA paper, 972904 (1997)
2. A. Sorokin , F. Arnold, Electrically charged small soot particles in the exhaust of an aircraft gas - turbine engine combustor: comparison of the model and experiment, Journal of Atmospheric Environment, 38 , 17 (2004)
3. H. E. G. Powrie , A Novis. Gas path debris monitoring for F - 35 to be strike fighter propulsion system PHM, Aerospace Conference Proceedings, 2006 Montana, USA, 11 (2006)
4. C. E. Fisher, Gas path debris monitoring - a 21 st century PHM tool, Aerospace Conference Proceedings, 2006 Montana, USA, 6 (2006)
5. Li Yaohua, Zuo Hongfu, Wen Zhenhua, the Simulated Experiment of Aircraft Engine Gas Path Debris Monitoring Technology, ACTA AERONAUTICA ET AS TRONAUTICA SINICA, , 30 , 4 (2009)
6. Li Yaohua , Zuo Hongfu, Method for Rub Fault Detection Based on Electrostat ic Technique : Model and Simulated Experiments, ACTA AERONAUTICA ET ASTRONAUTICA SINICA, 31, 6 (2010)
7. Liu Pengpeng, Zuo Hongfu, Fu Yu, Sun Jianzhong. Exhaust gas electrostatic monitoring and gas path fault feature for turbojet engine, Journal of Aerospace Power, 28, 2 (2013)
8. Sun Jianzhong, Zuo Hongfu, Liu Pengpeng, Yu Youfeng, and Chen Zhixiong, "Analysis and application of baseline model of aero-engine exhaust gas electrostatic monitoring signals," Journal of Aerospace Power, 28, 3 (2013)
9. Ren Jun, Shen Jian, Lu Shouci. The science and technology of particle dispersion , Beijing: chemical industry press, (2005)

3 tons pure electric vehicles power system design based on Cruise

LIU Xinyu, TIAN Houjie

Tianjin Transportation Vocational College, Tianjin300110, China

Abstract: The pure electric minivan is different from electric car. Combined with a given vehicle, vehicle simulation model established in Cruise software, complete simulation by setting tasks for the selected models designed drivetrain. Simulation results show that: The design of the transmission ratio can best meet the performance requirements of the matching target power analysis and simulation of electric minivan provides a new way, with practical guidance.

Key words: pure electric minivan; simulation model; Cruise; match

INTRODUCTION

With the increasingly prominent environmental protection and energy problems, pure electric vehicles become the development direction of many large car enterprises and research institutes because of their advantages such as zero emission and low noise. The pure electric mini truck usually travels in the city or the suburbs. By optimizing the power system, the system can reduce the system loss, prolong the mileage and improve the efficiency of the whole vehicle. Before the battery problem, how to choose the reasonable parts and related parameters, the optimal matching conditions, in the same battery, better meet the requirements of vehicle power and increases the range, researchers always pursue the goal of this paper is [1], the main purpose of the research.

In this paper, a model of pure electric vehicle of Tang Jun company is taken as an example to design the transmission system parameters of the whole vehicle. Because the motor itself has the function of starting with load and has the function of speed regulation, the transmission structure adopted should be simpler. The structural parameters and target dynamic performance of the vehicle are as follows:

Table 1 structural parameters of whole vehicle

Mass /kg	Full load mass /kg	Windward area /m^2	drag coefficient	Wheelbase/mm
1500	3000	1.65	0.35	2000
Wheel rolling radius/m		Final reduction ratio	Coefficient of rolling resistance	transmission efficiency
0.265		4.4	0.0165	0.92

Table 2 dynamic performance requirements

No-load maximum speed Km/h	Full load maximum speed km/h	Maximum climb%
80	65	20
Acceleration time s (0~40km/h)	Continued mileage km	
10	85	

1 MATCHING OF PURE ELECTRIC VEHICLE TRANSMISSION SYSTEM

1.1 determination of motor parameters

Pure electric micro trucks often work in the suburbs, and run in the starting, acceleration, deceleration, climbing, braking and other conditions, so the drive motor to meet the needs of various operating conditions.

Power selection of 1.1.1 motor

The power size of the motor is directly related to the power performance of the electric vehicle. The motor power is, the better the acceleration and climbing performance of electric vehicles, but the size and quality of the motor will also increased, at the same time the motor can not always keep working in high efficient area, reduced the rate and the vehicle mileage by using the energy of pure electric vehicles. Therefore, the selection of motor power plays a vital role in the dynamic performance of the whole vehicle [2].

1) to meet the maximum speed requirements, the power required by the motor is:

$$P_1 \geq \frac{v_{max}}{3600\eta_T}\left(mgf + \frac{C_D A v_{max}^2}{21.15}\right) \quad (1)$$

The corresponding parameters are taken into account: P_1=7.32kw

2) to meet the requirements of climbing performance, the power required by the motor is:

$$P_2 \geq \frac{v}{3600\eta_T}\left(mgf\cos(\arctan\alpha) + mg\sin(\arctan\alpha) + \frac{C_D A v^2}{21.15}\right) \quad (2)$$

The corresponding parameters are taken into account: P_2=8.65kw

3) the total power required by the vehicle acceleration performance index is calculated. Assuming that the output characteristic of the motor is equal to the power, the [3] can be deduced from the dynamic equilibrium equation:

$$v = v_m\left(\frac{t}{t_m}\right)^{0.5} \quad (3)$$

Vm - the final speed of the acceleration process;

TM - the time required to accelerate the speed from 0 to Vm

According to the hypothesis, the total power of the power source in the whole acceleration process is constant, and the average power and the output speed of the power output in the whole acceleration process are the same when the power source output power is the maximum. The total power required for the acceleration process is derived as follows:

$$P_3 \geq \frac{1}{3600 t_m \eta_T}\left(mgf\frac{v_m t_m}{1.5} + \frac{C_D A v_m^3}{21.15 * 2.5} t_m + \delta m \frac{v_m^2}{2}\right) \quad (4)$$

The corresponding parameters are taken into account: P_3=18.45kw

In order to ensure the design of the vehicle can travel at the maximum speed is constant, and meet the climbing performance and accelerating performance requirements, the design of motor rated power should be greater than P1, in the range of overload permitted (overload 3-5 times), to meet the power requirements and accelerate climbing.

Determination of torque of 1.1.2 motor

The main reduction ratio of pure electric vehicles is Io. When the speed reaches the highest speed, it is usually on the direct gear $n > i_g i_{ov}/0.377r$ to meet the Maximum speed of motor. In the motor overload allowed range, select the motor rated speed. Then, the rated torque and maximum torque of the motor are determined by the rated power and the rated speed.

The motor type is selected according to the basic motor parameters determined above.

1.2 Design of transmission ratio of transmission

The principle of transmission ratio and gear position determination: two main power and resistance balance points need to be considered in motor drive. One is the torque balance of the conventional speed, and the two is the torque balance point at the highest speed [4]. According to the operating characteristic curve of the motor, the power balance diagram of the motor under the maximum speed and the basic frequency is plotted.

 DOI: 10.1051/matecconf/201713900146

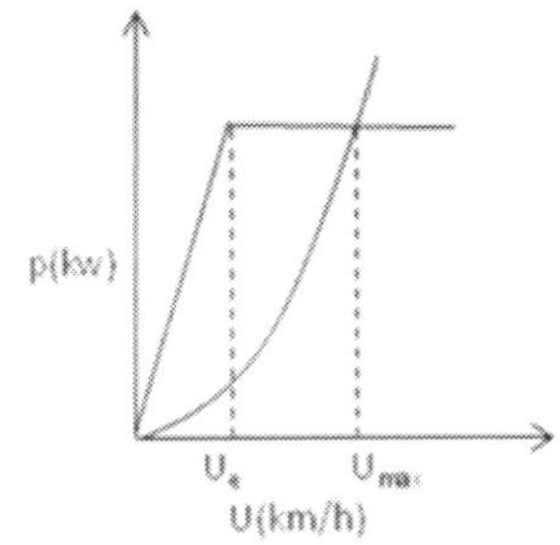

Fig. 1 power balance diagram of pure electric vehicle of $n_{max}/n_e > 2.5$

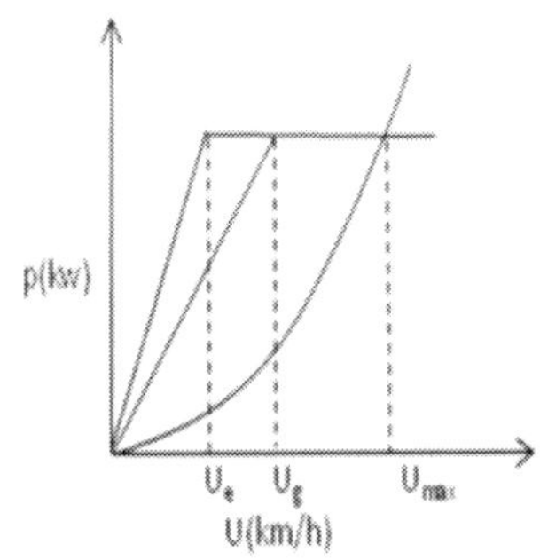

Fig. 2 power balance diagram of pure electric vehicle of $n_{max}/n_e \leq 2.5$

1) as shown in Figure 1: when the maximum speed of the motor and the basic speed ratio of >2.5, the motor speed range from the basic frequency is large enough to select a stall to meet the requirements. But this kind of situation requires higher motor, it is difficult to guarantee that the inverter always works in high efficiency area.

2) as shown in Figure 2: maximum motor speed and pitch speed ratio is less than 2.5, the motor from the baseband to speed range is not wide enough, should consider to add a file.

At the same time, pure electric mini truck driving at top speed, with minimum transmission ratio of driving gear; the maximum gradient, the maximum transmission ratio of driving gear, and the gear ratio of known io=4.6, the largest, small transmission range from the following two type available:

$$ig_{max} \geq \frac{mg(f\cos(\arctan\alpha)+\sin(\arctan\alpha))r}{T_{max} i_o \eta_T} \quad (5)$$

$$ig_{min} \leq \frac{0.377 r n_{max}}{v_{max} i_o} \quad (6)$$

Within the allowable range of the transmission ratio, the smaller the transmission ratio, the less the gear set, and the simpler the structure. However, smaller transmission ratios may result in a decrease in climbing performance. Therefore, the number of gears is determined according to the maximum speed of the motor and the speed ratio of the rated motor. Combining the range of transmission ratio obtained by type 5 and 6, the preset transmission ratio, the simulation model is established, and the simulation results are optimized continuously.

1.3 power battery type selection

The operation of pure electric vehicles rely entirely on battery energy, battery can provide timely backup power in the vehicle load conditions, regenerative braking can absorb energy, thus requiring battery with high energy and high power and long service life cycle. In addition, the utility model also has good use safety performance and self discharge performance. Currently available for electric vehicles, power batteries are mainly lead-acid batteries, lithium-ion batteries, Ni-MH batteries, super capacitors and so on. Because of the advantages of Ni-MH cell such as high energy density, high power density, fast charge and low cost, we use Ni-MH cell in this design.

In order to meet the design goals of the dynamic performance and driving range requirements, in accordance with the calculation of the vehicle speed is set to the level of wind resistance: good road, calculate the driving force required in the work condition, and according to the relationship between power, force and distance from the stroke energy required for [5]. Ni-MH battery is selected as the object, full of power, calculated according to SOC 100%, discharge 90%, the rated voltage of the battery is determined by the selected motor, and the rated capacity of the battery is calculated according to the energy balance.

The battery capacity mainly guarantees the power demand of the electric vehicle and the need of the continuous mileage, and the battery number can be selected according to the maximum power and the driving range.

The number of cells determined by the maximum power is the number of cells:

$$N = \frac{P_{e\max}}{P_{bmax}\eta_e\eta_{ec}} \quad (7)$$

Pemax is motor, maximum power is kW;

P_{bmax} is the maximum output power of single cell kW;

η_e for motor efficiency;

η_{ec} is the working efficiency of the motor controller.

Number of battery cells determined by continued mileage

The number of battery cells determined by the continued mileage shall be:

$$N = \frac{1000LW}{C_i U_i} \quad (8)$$

Type (8): L is the driving range of electric vehicle km;

Ui is the single cell voltage V;

Ci is a single cell rated capacity A * h;

W consumes energy per unit mileage of an electric vehicle kJ.

The larger value of N in formula (7) and formula (8) is the required number of battery monomers .

2 EXAMPLE BASED DESIGN PROCESS

In order to improve the development of pure electric vehicles of a certain type of Tang Jun company as an example, according to the design requirements of the power, in accordance with the design process mentioned above, the optimized matching transmission system of vehicle, and simulation model is established by using cruise software. Through further analysis of the simulation results, optimize the transmission system, so as to achieve the power requirements while improving transmission efficiency. For the example of a Tang Jun model of a pure electric truck, the basic parameters of the motor and battery are selected as shown in table 3:

Table 3 basic parameters of selected motors and batteries

Motor type		Rated / maximum power Kw	Rated / maximum speed r/min
Permanent magnet brushless DC motor		7.5/18.5	2000/3500
Rated / maximum torque N.m	Battery type	Rated voltage of battery V	Rated capacity of battery Ah
36/60	Ni-MH	120	200

3 SIMULATION MODELING AND RESULT ANALYSIS

3.1 modeling and calibration of pure electric vehicles based on AVL CRUISE

AVL Cruise software is used for the study of vehicle dynamics, fuel economy, emission performance and braking performance of advanced simulation analysis software. The software can be used in powertrain matching, vehicle performance prediction, and modeling, simulation and performance simulation of hybrid vehicles and electric vehicles.

Based on the Cruise platform, a miniature pure electric truck model is built up as shown in figure 3. The model mainly includes wheel model, main reducer model, transmission model, motor model, battery model and driver model.

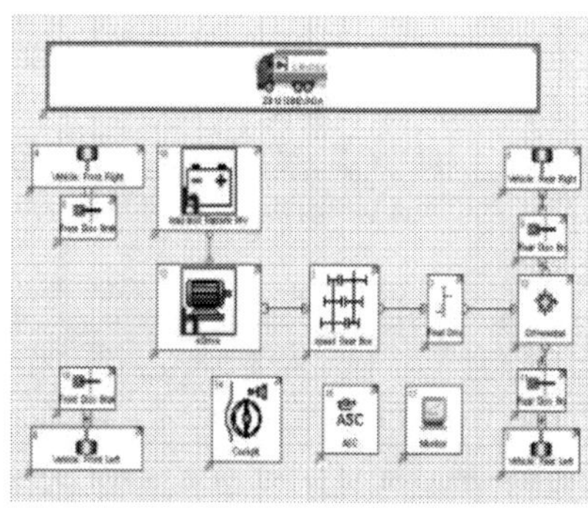

Figure 3 vehicle model

After the vehicle model is completed, the parameters of each module are set up, and the physical connection between the modules is established and the signal is connected. By adding relevant calculation tasks, the pure electric vehicle is operated in a given set of working conditions. According to the traffic conditions of the city, added Climbing Performance, Constant Drive, Full load Acceleration, Cycle run four computing tasks, respectively to verify the design of vehicle climbing performance, maximum speed, acceleration, mileage.

Due to the objective conditions, the parameters of the whole vehicle and parts needed for the accurate model of Cruise can not be obtained. In order to make the model precise, the model is calibrated by using the existing data. In this design, we use the parameters of the original car and the experimental data for calibration. The parameters of the original car into the model, the simulation results are compared with experimental results, to adjust the parameters in the model can not be accurately obtained according to the error, until the error of simulation results and experimental results in the set range, calibration [7].

When the model is built and calibrated, the model is checked to see if there are any problems such as module definition, information connection, data input error and so on. After checking, the model can be run for task calculation, and the result of simulation data will be automatically generated in file form at Result Manger.

3.2 simulation results analysis

By adding a model to the cruise software, a calculation task is added to the transmission ratio. The simulation results show that in the design phase, through the use of Cruise for vehicle optimization matching, not only improve the performance of the whole vehicle, but also shorten the design cycle, reduce design costs and shorten the feedback chain. Among them, ig1=2.55, ig2=1.8, simulation results are as follows:

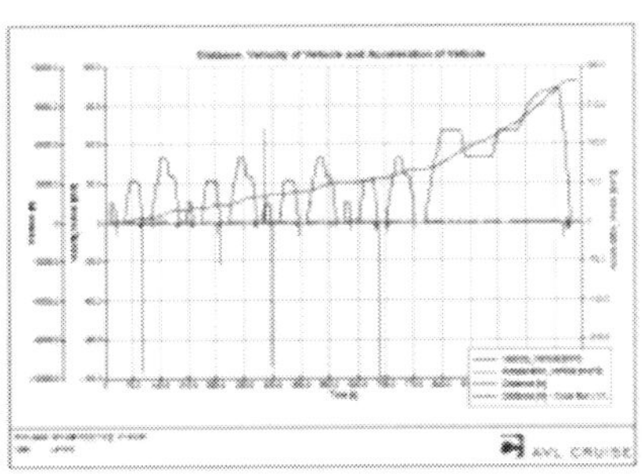

Figure 4 no gear mode, continuous mileage under one NEDC condition

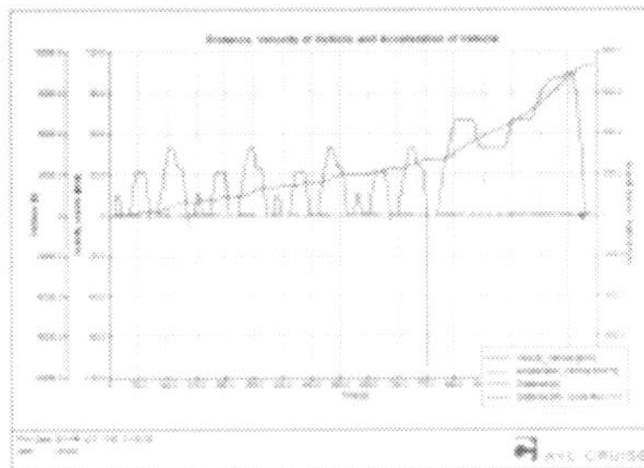

Figure 5 two stall mode, once NEDC conditions, continued mileage

As can be seen from Figure 4 and figure 5, in the case of a NEDC, the speed of the pure electric vehicle and the continuous mileage of the two models are not very different. Compared with the two gear mode, the model is more ideal, but it has little advantage. This is because the required speed, acceleration performance, no gear and two gears are not very different from each other under this working condition. Therefore, as to the NEDC working condition, which of these two gears can be chosen, all of them can meet the requirements.

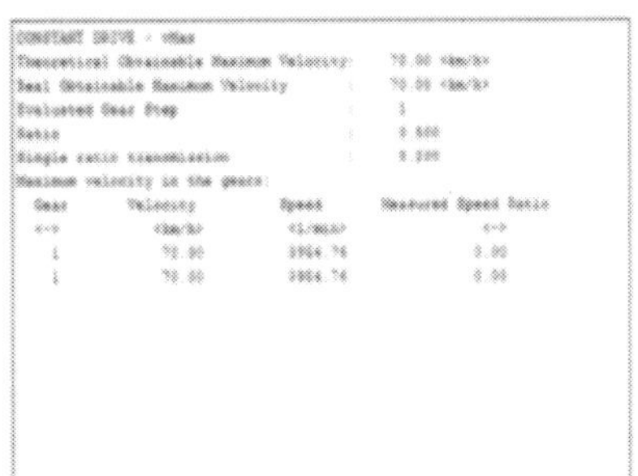

Figure 6 maximum speed without stall mode

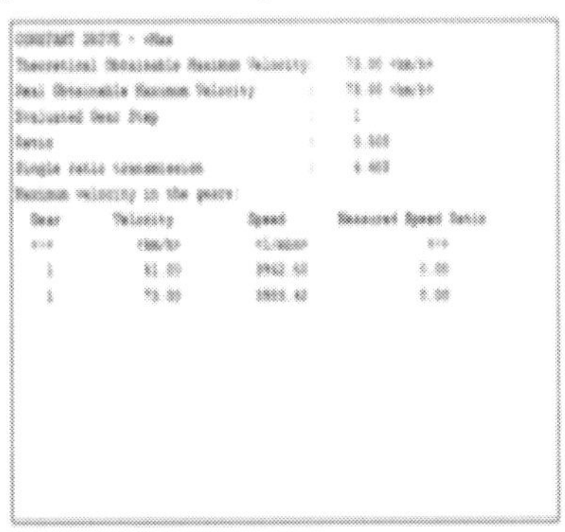

Figure 7 two gear mode top speed

From Figure 6 and Figure 7 shows, two stalls under the mode of the maximum speed of pure electric vehicles is the highest, reaching 72km/h, there is relatively low but the two little difference between the maximum speed gear mode, and pure electric vehicles 3T total power quality requirements are relatively high, without excessive speed. This is because the two stalls will change the transmission mode than the transmission system for high speed, sacrificing some performance, so pure electric vehicles two gear mode the highest speed is relatively high, but the pure electric vehicles running in urban areas of major truck speed is not high. Thus, the two models are available for maximum speed.

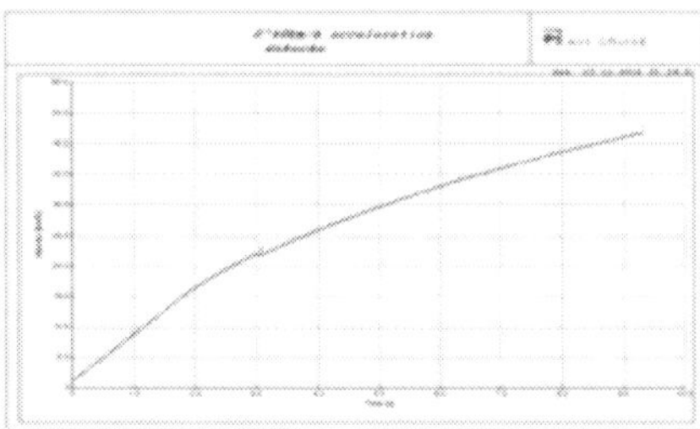

Figure 8 no stall mode 0~40km/h acceleration time

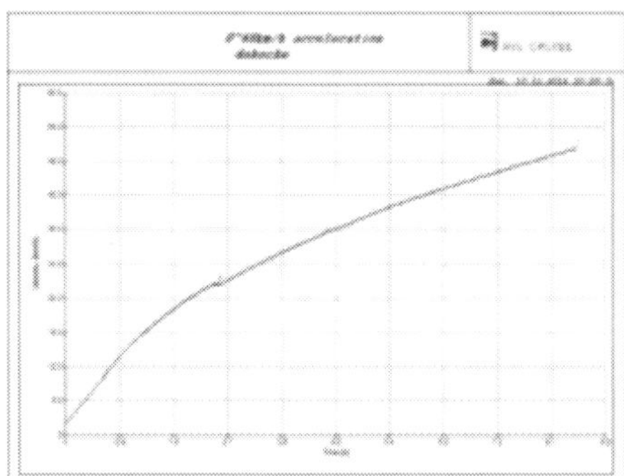

Figure 9 no stall mode 0~40km/h acceleration time

From figure 8 and Figure 9 shows, without shift mode, pure electric vehicles speed from 0 up to 40 km/h to about 5.5s; while the two gear mode, electric truck speed from 0 up to 40 km/h only 5S, shorter time. This is because the two gear mode drive system can sacrifice more energy for acceleration, so the acceleration capability is better. Therefore, in terms of acceleration time, the 3T pure electric vehicle chooses the two gear mode best.

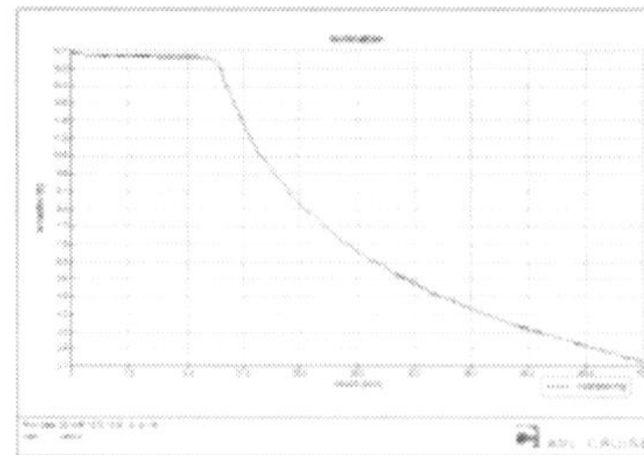

Fig. 10 maximum climbing grade without stall mode

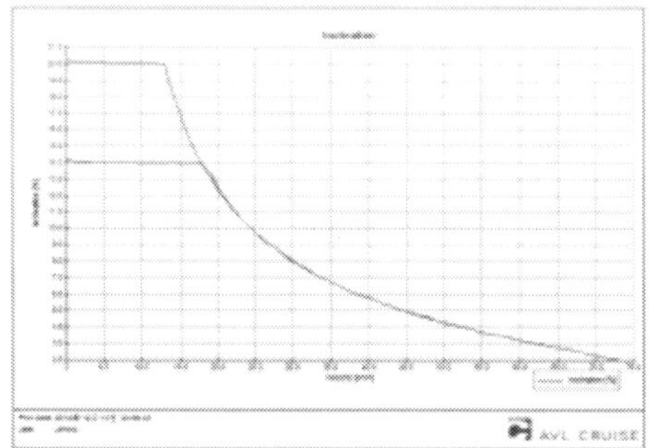

Fig. 11 maximum climbing grade of two stall mode

From Figure 10 and Figure 11 shows that the pure electric vehicle maximum climbablegradient maximum reached 20 under the two gear mode, this is because the two gear mode can choose 1 gear low gear, the speed is very low at this time, the same power for climbing more energy, so the maximum climbablegradient greater. Therefore, the two gear mode is the best for the maximum climbing grade.

4 CONCLUSION:

Power system is the core component of a car, which is particularly important for a pure electric vehicle. This design uses the manual gearbox with the main reducer, transmission power, the mini truck from the mini car, dynamic performance of the manual file can quickly improve the automobile, expand the scope of work in high efficiency area of the motor, save climbing, acceleration time, extended mileage.

In this design, an example of a pure electric truck of Tang Jun company is analyzed, the motor and battery are selected according to the dynamic performance index, and the range of the drive ratio is set. Combined with AVL Cruise software, the whole vehicle model is set up, and the calculation task is established by setting up the matrix operation of the transmission ratio. The simulation results show that this design and simulation have strong feasibility and rationality for matching the power system of pure electric vehicle, and greatly reduce the complexity of the traditional design and verify the optimization process.

REFERENCE:

[1] Yu Zhisheng car theory [M]. 5th edition. Beijing Machinery Industry Press,

[2] Zhai Li, Sun Fengchun. Electric car matching calculation and simulation of electromechanical drive system. [J]. Beijing University of Technology, 2007,3(10):869-873

[3] Dai Tianlu. pure electric car drive system design and matching parameters [C]. Symposium Chinese Society of Automotive Engineers, 2011, 2 (1): 1283-1286

[4] Ceng Hu, Huang Juhua. Pure electric vehicle motor and transmission matching [J]. Equipment Manufacturing Technology Journal, 2010,12 (2): 40-42

[5] Zhu Zhengli, Yin Chengliang, Zhang Jianwu. Pure electric vehicle powertrain optimization parameters based on genetic algorithm [J] Shanghai Jiaotong University, 2004, 38 (11): 1907-1972.

[6] Chang Lv. Pure electric mini-car powertrain design and dynamic simulation parameters [J]. Mechanical design and manufacturing, 2010(6):43-45

[7] WANG Guo-rong, ZHAO Fu-quan. Study on Matching Optimization of Automotive Power Train Based on Cruise [J]. Automotive Engineer, 2013(4):29-31.

Corresponding author: Liu Xinyu, (1972-) male, Tianjin, master degree in automotive, electrical and electronic technology.1617596880@qq.com .

MATEC Web of Conferences 139, 00147 (2017) DOI: 10.1051/matecconf/201713900147
ICMITE 2017

Dynamic Establishment of Weight of Enterprise's Internal Function Based on AHP and QFD

Pei HE, Yue-Xiong GONG, Li-Xing ZHANG, Li XIE

China Tobacco Zhejiang Industrial Company Limited, Hangzhou, 310024, China

Abstract: Modern enterprise management strives to become effective and pragmatic, usually, KPI is the most essential key indicator for the enterprise examination department. However, because the supporting and bearing roles of KPI for different departments are different, no a rather scientific quantification for the weight between them is available, unbalanced examination weights between indicators and departments usually exists. Here, it is with balanced score card to breakdown enterprise's key tasks, card the performance indicators supporting enterprise's key tasks, and utilize AHP and QFD method to dynamically establish a standard for measuring each department's indicators so as to provide a basis for enterprise's scientific examination.

1.Integral Frame of the Project

Here, with the strategy of Hangzhou Cigarettes Factory as a start point, the key performance indicators of current year on basis of enterprise's tasks of the said year are carded, the broken-down indicators are sorted with KJ method, and the weight of the key indicators are determined with AHP method. Comparison with industry advanced level is made with key indicators-functional department matrix, and finally the weight supporting each department's key indicator is obtained. The integral frame is shown in Fig. 1.

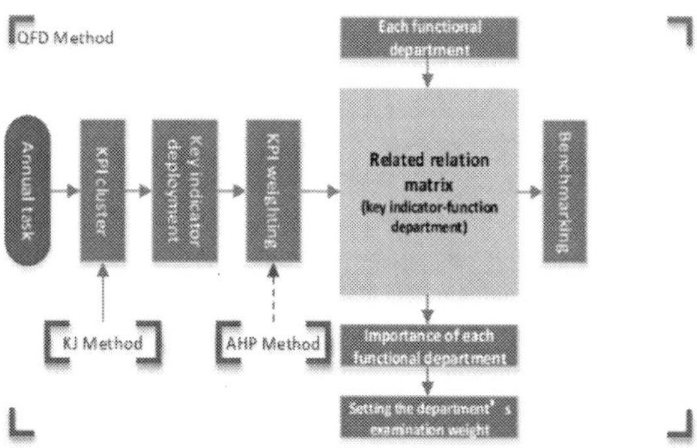

Fig. 1 Diagram of Integral Study Frame

2.Breaking Down of Annual Key Performance Indicators

Some enterprise's annual tasks are rather definite, while some tasks are hidden in the high-level documents or laws or regulations of external origins, usually, the enterprise conducts an overall carding to ensure comprehensiveness of its annual tasks. Generally, an enterprise may be carded from four aspects (finance, customers, internal operation, and learning and growth) of the balanced score card. After breaking down of tasks, each supporting key indicator is re-divided. Taking the carding of year-2015 key performance indicators of Hangzhou Cigarettes Factory as example, the carding of Hangzhou Cigarettes Factory in 2015 in four aspects determines the annual tasks, as shown in Table 1.

Tab. 1 Breaking Down Table of Key Performance Indicators of Hangzhou Cigarettes Factory in 2015

Level-I indicators	Level-II indicators (Balanced Score Card)	Level-III performance indicators
	Finance	Controllable management fee of a single box

Rational setting of enterprise key performance indicators		Cut filler rate
		Roll paper consumption of a single box
		Composite energy consumption of 10000 pieces of cigarette
	Customers	Consumers' satisfaction on product quality
		Satisfaction on inter-department cooperation
	Internal operation	Mean productivity per on-post employee
		Sigma level of production process
		Acceptance ratio of random inspection of Level-I and Level-II station
		Complaint and claiming rate of product quality
		Outage rate of tobacco primary processing equipment
		Run efficiency of packing equipment
		Run efficiency of rolling equipment
	Learning and growth	Standard compliance ratio of training hours
		Ratio of middle/senior technical title

3.Ordering of Importance of Key Performance Indicators

After determining the key performance indicators, the project group invites the personnel of functional departments to compare relative importance of level-I, level-II and level-III performance indicators with a pair-wise method, and analyze the data with yaaha analytic hierarchy process software. The calculated coefficient of test of concordance is below 0.1 and meets requirements. Due to limited space, here, merely the ordering and test of concordance of level-II indicators and total weight distribution of level-III indicators are cited, as shown in Table 2 and Table 3.

Tab. 2 Distribution of Weight of Level-II Indicators

	Setting of key performance indicators : 0.0544: setting of weight of "key performance indicators: 1.0000: λmax: 4.1454				
	Finance	Internal operation	Learning and growth	Customers	wi
Finance	1.0000	0.5000	4.0000	0.3333	0.1903
Internal operation	2.0000	1.0000	4.0000	0.5000	0.2836
Learning and growth	0.2500	0.2500	1.0000	0.2500	0.0759
Customers	3.0000	2.0000	4.0000	1.0000	0.4503

Tab. 3 Distribution of Weight of Level-III Indicators

Scheme	Weight	Scheme	Weight	Scheme	Weight
Consumers' satisfaction on product quality	0.4052	Mean productivity per on-post employee	0.0211	Run efficiency of rolling equipment	0.0297
Composite energy consumption of 10000 pieces of cigarette	0.0432	Sigma level of production process	0.0282	Run efficiency of packing equipment	0.0335
Roll paper consumption of a single box	0.0233	Acceptance ratio of random inspection of Level-I and Level-II station	0.0613	Standard compliance ratio of training hours	0.0379
Cut filler rate	0.0805	Complaint and claiming rate of product quality	0.0842	Ratio of middle/senior technical title	0.0379
Controllable management fee of a single box	0.0432	Outage rate of tobacco primary processing equipment	0.0256	Satisfaction on inter-department cooperation	0.0450

4.Building of House of Quality of Key Performance Indicators-functional Department

Competitive Assessment of Key Performance Indicators

In order to find out the difference between indicators of us and other benchmarking enterprises, our indicators are compared with other enterprise's indicators in five grades (score 1-5) on basis of data on the excellence and standard-compliance of year-2015 cigarette industry, and in combination with features (highlights) of our key performance indicators, their absolute importance is obtained, as shown in Fig. 2.

Level-I	Level-II	Level-III (Customers' demands \ Quality/Technical characteristics)	Competitive assessment: Hangzhou Cigarettes Factory	Competitive assessment: Domestic benchmarking enterprise	Planned target: Planned quality	Planned target: Improvement ratio	Planned target: Product characteristics	Weight: Absolute importance	Weight: Relative importance
Rational setting of enterprise key performance indicators	Finance	Controllable management fee of a single box	2	3	5	2.50	1.0	10.8	6.686
		Cut filler rate	3	4	5	1.67	1.0	13.4	4.294
		Roll paper consumption of a single box	2	5	5	2.50	1.5	5.8	3.948
		Composite energy consumption of 10000 pieces of cigarette	3	4	5	5.00	1.2	13.8	6.408
	Customers	Consumers' satisfaction on product quality	4	5	5	1.25	1.5	50.7	18.885
	Internal operation	Mean productivity per on-post employee	5	5	5	1	1.0	2.1	3.958
		Sigma level of production process	5	4	5	1	1.0	4.7	2.058
		Acceptance ratio of random inspection of Level-I and Level-II station	4	5	5	1	1.0	7.7	5.408
		Complaint and claiming rate of product quality	3	3	5	1.67	1.2	16.8	7.881
		Outage rate of tobacco primary processing equipment	3	5	5	1.67	1.0	4.3	3.144
		Run efficiency of packing equipment	2	5	5	2.50	1.2	10.1	5.360
		Run efficiency of rolling equipment	2	5	5	2.50	1.2	8.9	4.822
	Learning and growth	Standard compliance ratio of training hours	3	5	5	1.67	1.0	8.3	3.388
		Ratio of middle-senior technical title	3	4	5	2	1.0	7.2	6.425

Fig. 2 Competitive Assessment Table of Key Performance Indicators

5.Matrix Analysis of Key Technical Indicators-functional Department Relation

After completion of the fore-phase work, the project group determines the matrix of "key performance indicators-each department's function" two-dimensional relation matrix, which is used to express the complex relation between each key performance indicator item and each functional department. Here, figure 6 expresses the "Strong" relation, i.e. the improvement of a key performance is strongly related with the compliance of corresponding functional department; 3 expresses "medium" relation, i.e. the improvement of a key performance is medium related with the compliance of corresponding functional department; 1 expresses the "weak" relation, i.e. the improvement of a key performance is weakly related with the compliance of corresponding functional department; and the blank expresses the no direct relation. When the selected product performance indicators are closely related with the compliance of corresponding functional department, for embodying the correlation, figure 9 is used, as shown in Table 3.

Level-I	Level-II	Level-III (Customers' demands \ Quality/Technical characteristics)	General Office	Production Technology Department	Tobacco Primary Processing Shop	Packing Shop	Power Shop	Tooling Department	Trade Union	Political Education Office	Information Center	Administrative Affairs Department	Supervision Section	Personnel and Labor Department	Technical Reformation Office	Safety and Security Section
Rational setting of enterprise key performance indicators	Finance	Controllable management fee of a single box		3	9	9		6			1	3				
		Cut filler rate		6	6											
		Roll paper consumption of a single box				9		3			1					
		Composite energy consumption of 10000 pieces of cigarette		1	6	9	9	3			1					
	Customers	Consumers' satisfaction on product quality	1	3	3	6										
	Internal operation	Mean productivity per on-post employee		3	9	9	3	3			1	1		3		
		Sigma level of production process		3	6	6	3	1			1					
		Acceptance ratio of random inspection of Level-I and Level-II station		3	6	6		1			1					
		Complaint and claiming rate of product quality		1	3	9	1	1			1					
		Outage rate of tobacco primary processing equipment			9			3			1					
		Run efficiency of packing equipment				9		6			3					
		Run efficiency of rolling equipment				9		6			3					
	Learning and growth	Standard compliance ratio of training hours	1	1	3	3	3	1	1	1	1	1	1	9	1	1
		Ratio of middle-senior technical title	1	1	3	9	3	1	1	1	1	3	1	9	1	3

Fig. 3 Matrix Analysis of Key Performance Indicators-Functional Department Relation

6.Building of House of Quality for Key Performance Indicators and Functional Departments

As per the principle and structure of the house of quality, roof structure should exist, however, which is omitted, as shown in Fig. 4, since the relation between enterprise's functional departments is positive, instead of being negative.

Level-I	Level-II	Level-III (Customers' demands \ Quality/Technical characteristics)	General Office	Production Technology Department	Tobacco Primary Processing Shop	Packing Shop	Power Shop	Tooling Department	Trade Union	Political Education Office	Information Center	Administrative Affairs Department	Supervision Section	Personnel and Labor Department	Technical Reformation Office	Safety and Security Section	Weight: Absolute importance	Weight: Relative importance
Rational setting of enterprise key performance indicators	Finance	Controllable management fee of a single box		3	9	9		6			1	3				1	10.8	6.686
		Cut filler rate		6	6												13.4	4.294
		Roll paper consumption of a single box				9		3			1						5.8	3.948
		Composite energy consumption of 10000 pieces of cigarette		1	6	9	9	3			1						13.8	6.408
	Customers	Consumers' satisfaction on product quality	1	3	3	6											50.7	18.885
	Internal operation	Mean productivity per on-post employee		3	9	9	3	3			1	1		3			2.1	3.958
		Sigma level of production process		3	6	6	3	1			1						4.7	2.058
		Acceptance ratio of random inspection of Level-I and Level-II station		3	6	6		1			1						7.7	5.408
		Complaint and claiming rate of product quality		1	3	9	1	1			1						16.8	7.881
		Outage rate of tobacco primary processing equipment			9			3			1						4.3	3.144
		Run efficiency of packing equipment				9		6			3						10.1	5.360
		Run efficiency of rolling equipment				9		6			3						8.9	4.822
	Learning and growth	Standard compliance ratio of training hours	1	1	3	3	3	1	1	1	1	1	1	9	1	1	8.3	3.388
		Ratio of middle-senior technical title	1	1	3	9	3	1	1	1	1	3	1	9	1	3	7.2	6.425
		Importance of quality/technical characteristics	68	382	805	809	389	307	34	18	306	93	13	551	18	96	162.6	100.00
		Importance of quality/technical characteristics	[illegible]	[illegible]	[illegible]	[illegible]	[illegible]	[illegible]	[illegible]	[illegible]	[illegible]	[illegible]	[illegible]	[illegible]	[illegible]	[illegible]		

Fig. 4 House of Quality for Key Performance Indicators and Functional Departments

From the above-mentioned analysis of house of quality, we concluded that if type I department is scored above 600, type II department is scored 100-599, and type III department is scored below 99, the Packing Shop and Tobacco Primary Processing Shop are type I department, the Tooling Department, Production and Technology Department, Power Shop, Information Center, and Personnel and Labor Department are type II department, and the General Office, Trade Union, Political Education Office, Supervision Section, Technical Reformation Office, and Safety And Security Section are type III department.

7.Conclusion

The enterprise's work tasks of current year are carded from its strategy, the key performance indicators of the current year are broken down through comprehensive consideration with balanced score card mode, the supporting and contribution of each functional department to the key performance indicators is transformed with QFD method, and finally the types of various departments are scientifically decided, which provides the support to examine each enterprise department's weight and the objective data basis to set each department's basic wage and bonus. Dynamic adjustment with this method will dynamically set the weight of each department's function.

References

[1] Xiong Wei, Quality Innovation-QFD-Based System Method [M]. China Zhijian Publishing House, 2015: 47-89

[2] Xiong Wei, Quality-Function-Development [M]. Beijing: Chemical Industry Press, 2005: 43-95

[3] Chen Honggen, Liang Gongqian Quality Assurance Model of QFD-based Postgraduate Culture [J]. Academic Degrees and Graduate Education, 2007 (2): 14- 1 [6]

[4] Sun Yuanyuan, Liu Fei, Li Li, Determination of Attribute Final Importance for Personalized Product Development Using Kano Model and QFD Integration Method [J]. Computer Integrated Manufacturing Systems, 2014, 20 (11): 2697-2703

[5] Jin Gang, Li Ming, Analysis and Research of Customer's Requirement Based on Quality Function Configuration, [J]. Manufacturing Technology & Machine Tool, 2009 (8): 116 -126

[6] Wang Shuang, Yin Guofu, He Zhongxiu, Research on Customer Requirements' Target System Based on Kano Model [J] Packaging Engineering, 2006, 27 (4): 209-213

MATEC Web of Conferences 139, 00148 (2017) DOI: 10.1051/matecconf/201713900148
ICMITE 2017

Research on Cigarettes Customer Needs Importance Algorithm Based on KJ / RAHP / KANO

Xiong-Jun NI[1], Pei HE[1], Wen-Xi XU[1], Yue-Xiong GONG[1], Qiang ZHU[1], Wei-Zhong HUANG[1], Juan-Li WANG[2]

[a] China Tobacco Zhejiang Industrial Company Limited, Hangzhou, 310024, China.

[b] Zhejiang University of Water Resources and Electric Power, Hangzhou, 310018, China

gongyuexiong@zjtobacco.com

Abstract: To express the ambiguity and uncertainty of customer needs importance, an algorithm was proposed. It integrated KJ method, rough analytic Hierarchy Process and KANO model. It calculated the customer needs importance in rough set. A case study of cigarettes customer needs importance illustrated the feasibility and validity of the algorithm.

Research Background and Significance

Customer needs importance information is an important input of QFD (Quality Function Deployment). Most existing researches on customer needs importance algorithm based on QFD focuses on combining fuzzy set and AHP to calculate customer needs importance. Although fuzzy set can solve uncertainty and ambiguity of customer needs information, customer needs algorithm based on fuzzy sets is subjective because fuzzy set needs pre-set membership function of fuzzy numbers. However, rough set needs not any pre-set parameters, the data can "speak by themselves", which has more advantages in solving uncertainty and ambiguity of customer needs information compared with fuzzy sets. In addition, although AHP (Analytic Hierarchy Process) has advantages in calculating relative importance of customer needs sub-items under the same parent item, but there may be a relatively large difference in the number of customer needs sub-items under different parent items, which will, to some extent, affect AHP effectiveness. As for solving relative importance of customer needs sub-items which are at the same level but belong to different parent items, KANO model has a comparative advantage. Therefore, it is necessary to use KANO model to further adjust customer needs information processed by AHP, QFD provides the possibility of integration between KANO model and AHP.

Customer Needs Analysis framework Based on KJ / RAHP / KANO

The framework of customer needs analysis based on KJ / RAHP / KANO is in Figure 1.

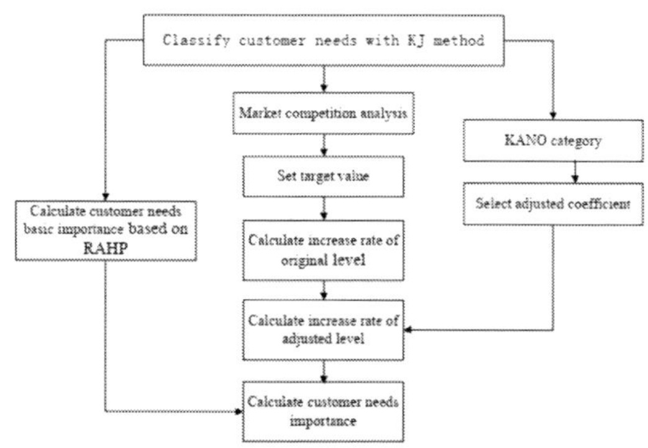

Fig. 1 Customer needs analysis framework based on KJ / RAHP / KANO

Source: This study.

Calculate customer needs with KJ method

Customer needs information obtained through market research is quite complex, and the same meaning may be expressed in many different ways; some customer needs information are broad, while some are very narrow; some customer needs information contains other customer needs information. Therefore, it is necessary to classify and simplify customer needs information by using Affinity Diagram, KJ method to give an analysis of customer needs information. KJ method will be used to clarify the customer needs in this paper.

Determination method of customer needs importance based on RAHP

Customer needs information with hierarchical structure can be clarified with KJ method, indicating subordination and containing the relationship between customer needs concept, which plays an important role in expressing and understanding complex customer needs information (Cheng,

Y., Leu, S., 2011). Due to the complexity of customer needs information and limitation of KJ legal analysis method, customer needs information which is got from KJ method cannot express relative importance between customer needs, and there are still multi-meanings, uncertainty and ambiguity. So it needs to further tap customer needs information and solve its multi-meanings, uncertainty and ambiguity by combining quantitative analysis method. Customer needs importance determining method based on RAHP (Rough Analytic Hierarchy Process) provides an effective solution to expressing relative importance of customer needs and solving multi-meaning, uncertainty and ambiguity of customer needs information.

RAHP-related concepts and definitions

AHP method can completely reflect the system analysis of system engineering and system integration idea. It takes a complex problem as a system, and in system analysis phase, according to membership function between factors within the system, transforms a complex problem into a methodical and hierarchal problem, and directly reflects the relationship between factors within the system in the form of hierarchical diagram. This decomposes complex system into relatively simple subsystems, and then integrates them step by step (in system integration phase).

Calculation steps of customer needs importance based on RAHP

AHP is easy to operate, especially for more complex customer needs analysis, but AHP itself is unable to cope with ambiguity and subjectivity of customer language, and is also unable to cope with judgment from many decision-makers. Application of rough number and rough boundary interval on AHP analysis can enable customer needs analysis to better cope with ambiguity of information evaluation from customer, while gathering judgment information from several survey participants, making judgment of customer needs importance more representative. Existing rough sets and research combining with AHP mostly discusses how to reduce evaluation guidelines by rough sets, namely, reducing the number of items in sub-hierarchy (Jain, VK, Sobek, DK, 2006; Zongjun Wang, Maoguang Chu, 2009). Customer needs determining method based on rough AHP proposed by Xiaotun Wang (2010) provides the idea of calculating customer needs importance with rough number and rough interval. But with this calculation method, customer needs importance expressed with rough number is transformed into the numerical form in the fifth step, to some extent; it weakens the advantage of rough number uncertainty and ambiguity (Xiaotun Wang, 2010). In this paper, customer needs importance determining method based on rough AHP proposed by Xiaotun Wang (2010) was amended, proposing that customer needs importance is always be expressed in the form of rough number.

KANO model and customer needs weight adjustment method

The concept of multidimensional quality property

According to Kano (2001), quality property can change over time, that is to say, a successful property will follow the life cycle of from indifference to an attractive, to become one-dimensional single item, and eventually become a must. For example, a property may be indifferent by customers when it first enters the market. But with the development of market, the property may become attractive, that is to say, it makes customers feel satisfied, even if the product does not contain the property, customers will think it is normal. However, if the property disappears, customers who frequently use this property will feel quite dissatisfied. After frequent use, the perception of this property might become one-dimensional and customers will satisfy or dissatisfy it depending on the satisfaction extent of this property. As time passes, one-dimensional property value will be accepted by consumers, and gradually become an item taken for granted (Kano, 2001).

KANO classification and customer need weighting correction coefficient

Relationship of customer satisfaction existing in KANO model and product (service) performance can be expressed approximately with a function $s = f(k,p)$, s means customer satisfaction, p is product (service) performance, k is adjustment coefficient of KANO classification.

Obviously, a good product (service) performance can bring high customer satisfaction, and the information brought by KANO model is much more. KANO model tells us that contribution of each customer satisfaction property to customer satisfaction is not equal. Not only each customer satisfaction property is different in customer satisfaction importance, but also contribution way of each customer satisfaction property to total customer satisfaction is different (Kano, 1984).

In order to obtain maximum customer satisfaction with the most economical cost, QFD practitioners not only need to be clear what customers need most, but also need to know the cost of each customer need to obtain desired customer satisfaction level. KANO adjustment coefficient and conversion function provides useful information.

Customer need classification information based on KANO model through market competition analysis combines with customer need basic importance algorithm based on amended RAHP constitutes customer need importance algorithm based on RAHP and KANO models.

Case study on Cigarette Customer Needs

To illustrate the feasibility and effectiveness of customer needs importance algorithm based on KANO and RAHP models, I selected cigarettes as an example, to explain customer needs importance algorithm and verify its effectiveness based on RAHP and KANO models.

Cigarette customer needs analysis

When customers express their needs information in the voice of customer (VOC), the form is usually various, including comments, complaints, evaluation, and hope, so original data collected not only has complex content but also has mixed concepts. KJ method is used to clarify and reduce customer needs information. KJ method to process cigarette customer needs, which are good taste, strong taste, no throat hurting, steady quality, and good package.

Cigarette customer needs importance determining method based on RAHP

(1) AHP questionnaire and consistency inspection

By inviting customers to fill in AHP questionnaire, conduct pair comparison of importance of cigarette customer needs importance, forming AHP judgment matrix. In this case, seven customers are invited to judge cigarettes customer need importance and calculate consistency ratio of judgment matrix according to formula (1). Judgment matrix and its consistency ratio.

(2) Constitute rough group decision-making matrix, to obtain rough pair comparison matrix

The above five AHP judgments matrix are expressed in the form of rough group decision-making matrix:

$$P_i^* = \begin{bmatrix} 1,1,1,1,1 & 1/5,1/4,1/2,3,2 & 1/9,1/4,1,1/3,1/2 & 1/5,1/5,1,1/2,1/2 & 1/6,1/6,1,2,2 \\ 5,4,2,1/3,1/2, & 1,1,1,1,1 & 2,2,1/2,1/6,1/2 & 1/2,1/2,1/2,1/4,1/4 & 1,1,1,1,1/2 \\ 9,4,1,3,2 & 1/2,1/2,2,6,2 & 1,1,1,1,1 & 1,1/2,1,1,1/2 & 2,1,2,4,2 \\ 5,5,1,2,2 & 2,2,2,4,4 & 1,2,1,1,2 & 1,1,1,1,1 & 2,1/3,1,3,3 \\ 6,6,1,1/2,1/2 & 1,1,1,1,2 & 1/2,1,1/2,1/4,1/2 & 1/2,3,1,1/3,1/3 & 1,1,1,1,1 \end{bmatrix}$$

According to algorithm, the average rough interval of p_{12} can be obtained:

$$RN(p_{12}^*) = [0.765, 1.992]$$

Similarly, according to the formula, rough number of all elements and average rough interval of group decision-making matrix can be obtained, and rough pair comparison matrix can be obtainable:

$$P^* = \begin{bmatrix} [1.000,1.000] & [0.765,1.992] & [0.310,0.664] & [0.372,0.772] & [0.602,1.519] \\ [1.154,3.682] & [1.000,1.000] & [0.463,1.380] & [0.313,0.438] & [0.625,0.875] \\ [2.160,5.827] & [1.528,4.278] & [1.000,1.000] & [0.625,0.875] & [1.611,3.111] \\ [1.722,3.722] & [2.500,3.500] & [1.250,1.750] & [1.000,1.000] & [1.093,2.426] \\ [1.250,4.000] & [1.250,1.750] & [0.403,0.778] & [0.676,2.009] & [1.000,1.000] \end{bmatrix}$$

(3) Calculate customer needs importance

Cigarettes customer needs importance can be obtainable from the formula (22)

$$RN(Q^*) = \{[0.556,1.092],[0.636,1.142],[1.271,2.324],[1.425,2.231],[0.843,1.614]\}$$

Cigarette customer needs basic importance is in Table 1.

Tab. 1 Cigarette customer needs basic importance

Customer needs items	Customer needs basic importance $RN(d_i)$
Good taste	$[0.556, 1.092]$
Strong taste	$[0.636, 1.142]$
No throat hurting	$[1.271, 2.324]$
Steady quality	$[1.425, 2.231]$
Good package	$[0.843, 1.614]$

Source: This study.

Cigarette customer needs importance adjustment coefficient based on KANO model

(1) Customer needs classification based on KANO model

According to the definition of multi-dimensional quality needs based on KANO model and KANO questionnaire template, I designed a KANO questionnaire on cigarette customer needs. 144 consumers filled in the KANO questionnaire, 95 valid questionnaires were recovered. With the classification of customer needs based on percentage, data processing results are shown in Table 2.

Tab. 2 KANO classification on cigarette customer needs

No.	Customer needs items	A	M	O	R	Q	I	Total(%)	KANO classification
1	Good taste	22.11	26.32	45.26	0.00	2.11	4.21	100	O
2	Strong taste	17.89	28.42	25.26	2.11	0.00	26.32	100	M
3	No throat hurting	14.74	30.53	45.26	0.00	1.05	8.42	100	O
4	Steady quality	6.38	44.68	38.3	0.00	1.06	9.57	100	O

				0					
5		2		2		1.	1	100	
		9.	28	3.	1.	0	6.		A
	Good	1	.1	9	0	4	6		
	package	7	3	6	4		7		

Source: This study.

(2) Market competition analysis and customer importance adjustment coefficient based on KANO needs classification

According to conventional steps of market competition analysis, the customers compare and evaluate cigarette products and products of their competitors. Experts group from cross-functional departments of certain company compare and evaluate the result, and set product's target quality value and product characteristics. As shown in Table 3.

Tab. 3 Evaluation table on cigarette market competition

No.	Customer needs	KANO classification	This company	Competitor 1	Competitor 2	Target quality	Level increase ratio	adjustment	Product characteristics
1	Good taste	O	4	4	3	5	1.25	1.25	1.20
2	Strong taste	M	3	3	3	3	1.00	1.00	1.00
3	No throat hurting	O	4	4	3	5	1.25	1.25	1.50
4	Steady quality	O	4	4	4	5	1.25	1.25	1.00
5	Good package	A	4	4	3	5	1.25	1.56	1.20

Source: This study.

(3) Calculation of customer needs importance

According to the calculation result of customer needs basic importance in Table 1, customer needs classification result in Table 2, and market competition evaluation result in Table 3, customer needs importance can be calculated, and the results are in Table 4.

Tab. 4 Customer needs importance calculation

No.	Customer needs	Customer needs basic importance	Level increase ratio after adjustment	Product characteristics	Customer needs importance
1	Good taste	$[0.556, 1.092]$	1.25	1.20	$[0.834, 1.638]$
2	Strong taste	$[0.636, 1.142]$	1.00	1.00	$[0.636, 1.142]$
3	No throat hurting	$[1.271, 2.324]$	1.25	1.50	$[2.384, 4.358]$
4	Steady quality	$[1.425, 2.231]$	1.25	1.00	$[1.782, 2.789]$
5	Good package	$[0.843, 1.614]$	1.56	1.20	$[1.578, 3.021]$

Source: This study.

According to the calculation results of customer needs importance, comparison drawing of customer needs importance is shown in Figure 2.

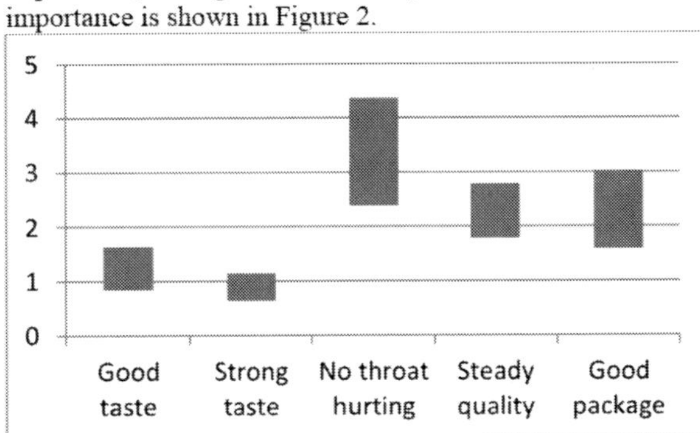

Fig. 2 Cigarettes customer needs importance

According figure 2, the algorithm that this paper proposed calculate the cigarette customer needs importance in rough set. By comparing with traditional customer needs importance research, it could express the ambiguity and uncertainty more validly.

Conclusion

In this paper, KJ method, rough sets theory, AHP method and KANO theory were integrated to analyze customer needs, constituting customer needs analysis method based on KJ / RAHP / KANO, and put forward customer needs importance algorithm based on RAHP and KANO models. This method gave a comprehensive and deep analysis of "voice of the customer" (VOC). KJ method was used to clarify, analyze and reduce customer needs information, obtaining customer needs information in the form of language, which provided a basis for understanding customer needs information. To further tap customer needs information, combination of rough sets, AHP, and KANO methods were used to give a quantitative analysis of relative importance of customer needs and put forward customer needs importance algorithm based on KANO and RAHP models. This algorithm used rough number and rough interval to deal with uncertainty and ambiguity of AHP customer needs importance judgment information. Besides, KANO model was used to conduct rich and detailed explanation to customer needs information, supplementing disadvantage of AHP customer needs importance judgment method caused by difference in the number of customer needs items. Meanwhile, I took cigarettes customer needs as example to explain customer needs analysis process based on QFD / RAHP / KANO and verified the validity of customer needs importance algorithm based on RAHP and KANO models.

Reference

[1] Cheng, Y.& Leu, S. Integrating data mining with KJ method to classify bridge construction defects. Expert Systems with Applications, 2011, 38(6): 7143-7150

[2] Cornelis, C., De Cock, M.& Eikerre, E. Intuitionistic fuzzy rough sets: At the crossroads of imperfect knowledge. Expert Systems, 2003, 20(5): 260-270

[3] Jackson, A. G., Pawlak, Z.& LeClair, S. R. Rough sets applied to the discovery of materials knowledge. Journal of Alloys and Compounds, 1998, 279(1): 14-21

[4] Jain, V. K.& Sobek II, D. K. Linking design process to customer satisfaction through virtual design of experiments. Research in Engineering Design, 2006, 17(2): 59-71

[14]

[5] Kano, N. Life cycle and creation of attractive quality. Linkoping, Sweden, 2001

[6] Kano, N., Seraku, N.& Takahashi, F. Attractive quality and must-be quality. Journal of the Japanese Society for Quality Control, 1984, 14(2): 39-48

[7] Kwong, C. K.& Bai, H. Determining the Importance Weights for the Customer Requirements in QFD Using a Fuzzy AHP with an Extent Analysis Approach. IIE Transactions, 2003, 35(7): 619-626

[8] Nepal, B., Yadav, O. P.& Murat, A. A fuzzy-AHP approach to prioritization of CS attributes in target planning for automotive product development. Expert Systems with Applications, 2010, 37(10): 6775-6786

[9] Pawlak, Z.& Skowron, A. Rough sets: Some extensions. Information Sciences, 2007, 177(1): 28-40

[10] Saaty, T. L. The analytic hierarchy process: Planning, priority setting and resource allocation. New York: McGraw-Hill, 1980: 196-203

[11] Tan, K. C., Shen, X. X. Integrating Kano's model in the planning matrix of quality function deployment. Total Quality Management, 2000, 11(8): 1141-1151

[12] Tan, K. C., Xie, M.& Shen, X. X. Development of innovative products using Kano's model quality function deployment. International Journal of Innovation Management, 1999, 3(3): 271-286

[13] Wang, X. T. Research on quality function deployment under uncertain circumstances. Doctoral Dissertation of Zhejiang University, 2011.

Research on Fault Prediction of Distribution Network Based on Large Data

Zhou Jinglong[1,a], Han Changzhan[1] ,Wu Xiangming[1],An Jiakun[2],He Chunguang[2] and Han Jinglin[2]

[1]State Grid Hebei Electric Power Company,No.32, fuqiang street, shijiazhuang, hebei province, China
[2]State Grid Hebei Economic Research Institute, fuqiang street, shijiazhuang, hebei province, China

Abstract. With the continuous development of information technology and the improvement of distribution automation level. Especially, the amount of on-line monitoring and statistical data is increasing, and large data is used data distribution system, describes the technology to collect, data analysis and data processing of the data distribution system. The artificial neural network mining algorithm and the large data are researched in the fault diagnosis and prediction of the distribution network.

Big data refers to the conventional software tools in a certain range of time capture, management and processing of data, is the need for new processing mode to have a stronger decision-making ability, insight discovery and process optimization capabilities of massive, high rates of growth and diversification of information assets.

Big data has four characteristics: massive data scale, fast data transfer, various data types and low value density. With the rapid development of computer information technology and intelligent distribution network technology, the application of large data in fault prediction and diagnosis of the distribution network is of great significance to improve the degree of automation of the distribution network.

1 Big data of distribution

Distribution network is composed of overhead lines, cables, towers, distribution transformers, isolating switches, reactive power compensator, and some ancillary facilities, which play an important role in the distribution of electrical energy.

The research and application of the data distribution system can improve the network operation and management level. Therefore, the application of data distribution system has a broad market prospect. The distribution network system uses the methods of data collection, retrieval, analysis and storage to provide reliable information technology services for the distribution network.

The distribution network including big data distribution automation system, distribution management and scheduling system, power demand side management system, power quality monitoring system, production management system, geographic information system, electricity information collection system and load monitoring system[1]. Among them, the distribution automation and management system is an important part of the distribution network monitoring, operation and management. The distribution automation and management system highly integrates information of dispatching automation system, production management information system, power marketing system and metering automation system.The main component of data distribution system as shown in Figure 1.

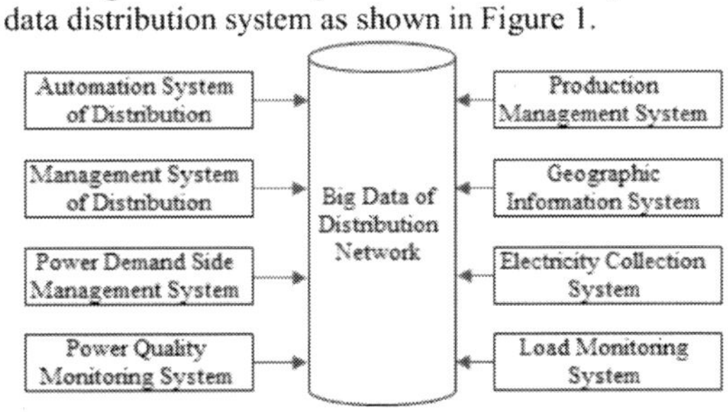

Figure 1. The main component of data distribution system

According to the data structure, data distribution system is mainly divided into structured data and unstructured data. Structured data stored in the database, most of the data in the distribution network is the kind of form, and with distributed energy, electric vehicles and its facilities appear in a large number of active distribution network, this type of data will continue to grow. Unstructured data is the data that cannot be expressed by two-dimensional logic table[2]. This part of the data mainly includes the monitoring of the lines and

[a] Corresponding author: zhoujinglong2017@sohu.com

equipment, pictures and video, equipment maintenance management, log information and so on.

2 Analysis and processing of large data

Large data analysis, including data analysis and interpretation. Large data analysis refers to the process of in-depth research and analysis of a large number of types of data, to find the implicit relationship between data and potentially useful data value. Big data interpretation refers to a deeper analysis of the process and results of data analysis and wood. Large data with multi dimension form of display, and big data analysis results will be converted to a specific industry issues.

Data visualization is the database of each data item as a single element, in order to construct a time dimension, space dimension and logic dimension of image data and other dimensions, so that from the different dimensions of the data were observed and analyzed deeply. Through from the time, space and logic and other dimensions of the data centralized management with visualization of multidimensional data, can get the stereo view[3].

Data mining is a work of extracting potentially useful information and knowledge from large amounts of incomplete data. Data mining consists of many steps, and each step is closely linked. In the process of data mining, data mining needs to be adjusted iteratively according to the knowledge acquired by mining data. Data mining can predict the unknown data according to the potential rules. The data mining process can be generalized into four parts: determination task, data preparation, mining modeling and result analysis application.The process of data mining is as shown in Figure 2.

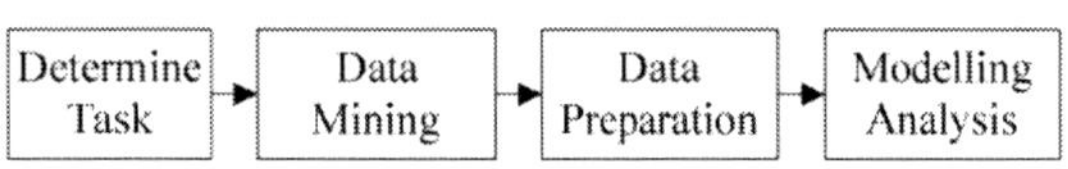

Figure 2. The process of data mining

Associated with a lot of data mining technology and data warehouse technology, data mining is processed on data mining, so data mining data preprocessing, data preprocessing, including data cleaning, data transformation and data integration process.

Because of the huge amount of large data, only after processing can we obtain useful information in these large amounts of data. At present, many fields of power system involve large data applications. The basic attributes of large data include quantity, speed, diversity, and so on, so the method of big data analysis is the key factor to determine whether information is valuable or not. Large data analysis methods generally have the following basic aspects:

Large data analysis techniques are applied to different types of users. Visual analysis is the basic requirement of users using large data analysis techniques. Visual analysis can display its characteristics through intuitive data graphics, and users can understand the complex relationship existing behind the data by comparative information[4].

The core of large data analysis is data mining. The algorithms of data mining are usually founded on data types and formats and apply accepted algorithms to dig into the data and dig out useful information. Because of the large amount of data in large data, these data mining algorithms give large data processing faster solutions. If the algorithm takes a long time to reach a conclusion, then the big data is of no value and significance.

Data warehouse to complete data preprocessing provides a platform for data mining, thus eliminating the complicated process of data preparation. In addition, in the process of data warehouse construction, a comprehensive data processing and data analysis infrastructure including data access, integration, merging, database conversion and WEB access is set up around the data warehouse. The Structure of typical data mining system is as shown in Figure 3.

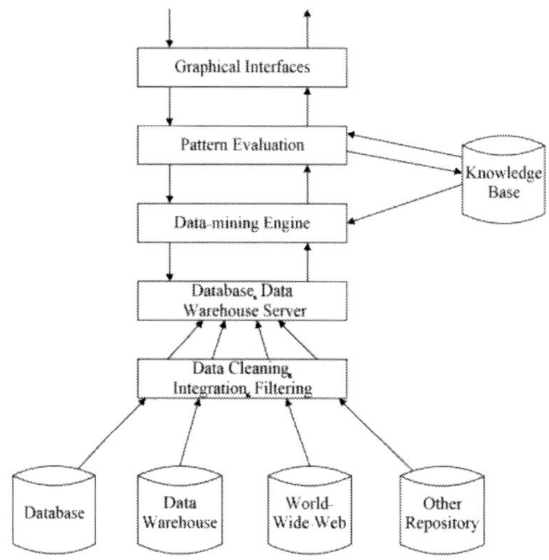

Figure 3. Structure of typical data mining system

3 Design of data mining algorithm

Large data mining methods are commonly used classification, regression analysis, neural network method, Web data mining, etc.. The methods of data mining from different angles, using the method of neural network technology to mine the data distribution system, through the analysis of data mining, prediction and fault diagnosis in the distribution network.

Artificial neural network (ANN) is a widely used technique in data mining. The data mining method of neural networks is to train the learning data sets repeatedly by imitating the human nervous system, and finds the model for prediction and classification from the data sets to be analyzed.

Neuron is the basic processing unit of artificial neural network. It is a nonlinear component with multiple inputs and single output. Besides the input signal, the output of neuron is also influenced by other factors in neuron. In neural modeling, an additional input signal is often added to become a deviation or threshold.

Neurons are the foundation of work in a neural network. Its structure is quite simple, so the processing capacity is simple. A large number of neurons constitute a neuronal network that has many superior characteristics. The processing of information by neural networks is accomplished by a large number of neurons. It is a collection of functions that have access to information, distributed storage, and associative memory.

The neural network consists of different levels of node set, and each layer is output to the next level node. The output value is magnified, attenuated or suppressed due to the connection weights between nodes.

The structure of artificial neural networks is as shown Figure 4.

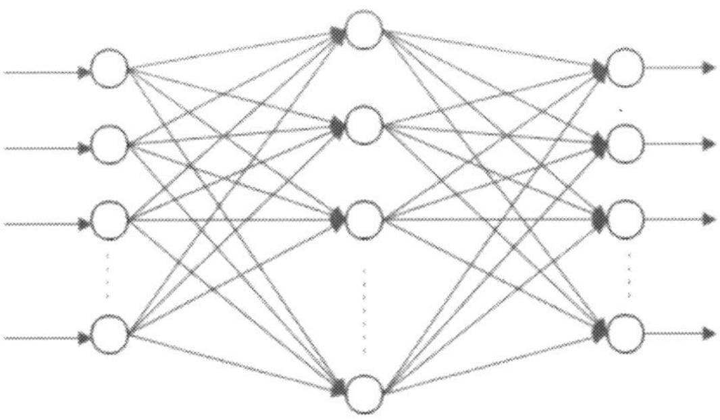

Figure 4. Structure of artificial neural networks

Feedforward networks have feedback signals in the training process, and in the classification process, data can only be transmitted forward until the output layer is reached, and there is no backward feedback signal between the layers. Perceptron and BP neural networks belong to feedforward networks. Figure 4 is a 3 layer feed forward neural network, in which the first layer is the input unit, the second layer is called the hidden layer, and the third layer is called the output layer[5].

The choice of activation function is an important step in the construction of neural networks. The activation function used in this paper is introduced below.

The main difference between the bipolar S shape function and the *S* shape function lies in the range of function, and the range of the bipolar S shape function is (-1,1), and the range of the *S* shape function is (0,1).Since the S shape functions and the bipolar S shape functions are derivable, the two activation functions are suitable for use in the BP neural network.

The double S type function is as formula （1）

$$f(x)=\frac{2}{(1+e^{-ax})}-1(-1<f(X)<1) \qquad (1)$$

The derivatives of double *S* type function is as formula （2）

$$f'(x)=\frac{2ae^{-ax}}{(1+e^{-ax})^2}=\frac{a[1-f(x)^2]}{2} \qquad (2)$$

For single neuron information, the information from other neurons is X_i, and their interaction strength with the neuron, that is, the connection weight is W_i, i=0,1,... n-1, the internal threshold of the processing unit is the θ.The input of the neuron X is as formula （3）

$$X=\sum_{i=0}^{n-1} w_i x_i \qquad (3)$$

The output of neurons is as formula （4）

$$Y=f(\sum_{i=0}^{n-1} w_i x_i-\theta) \qquad (4)$$

4 Fault prediction and of distribution network

Distribution network is the last link of the power system to supply power to users. Distribution network equipment is complex, many users, wide coverage, diverse geographical conditions, and affected by external conditions. Fault diagnosis of the distribution network is an important part of distribution automation. The realization of the high automation level of distribution network fault diagnosis is an essential guarantee for the economic development and the improvement of people's quality of life.

Distribution network fault prediction is based on distribution network fault data and fault influence factor data. Mining the relationship between fault and data of distribution network, constructing fault prediction model by using the correlation method, and forecasting the fault in a certain time area. The distribution network fault prediction is usually based on the actual demand of the power company for the fault prediction results, and the time scale and the region of the fault prediction of the distribution network are determined according to the optimization principle of the prediction model[6,7].

When a fault occurs in the distribution network, fault diagnosis system of large data analysis and judgment according to the real-time fault monitoring system to obtain the information on the distribution network based on the proposed regional, correct and effective power recovery strategy, fault isolation region. After years of operation, the distribution network has accumulated a large amount of sufficient fault data.

Data mining technology has powerful analysis function, can find out the cause of the malfunction analysis technique using its integrated, and formulate corresponding rules of decision-making, to take measures to reduce the failure probability, reliability and economic operation of the distribution network benefit.

4 Summary

With the continuous development of intelligent information system in distribution network, a rapid increase in the operation of the distribution network data collection, storage, analysis and processing of the effective data distribution system, provides important data for the stable operation of the distribution network. This paper mainly introduces the distribution network data, and the typical application of data distribution system analysis.

The application of large data in distribution network improves the management level of distribution network. Through the analysis of large data, it is of great significance to promote the construction of intelligent distribution network by quickly predicting and diagnosing the fault of distribution network.

References

1. Zhang Dongxia, Miao Xin, Liu Liping, Zhang Yan, Liu Keyan.Research on Development Strategy for Smart Grid Bid Data[J].Proceedines of the CSEE,2015,35(1):2-11.
2. Tao Shun,Xiao Xiangning.Infrastructure of the power quality assessment system of power systems[J].Transactions of China Electrotechnical Society,2010,25(4): 171 — 175.
3. Research of Distribution Network's Fault Diagnosis Based on Data[D].Guangxi University,2007.
4. Zhang Pei,Wu Xiaoyu,He Jinghan.Review on Big Data Technology Applied in Active Distribution Network[J].Electric Power Construction,2015,36 (1):52-59.
5. Lu Yuzi.Research on Distribution Network Outage Prediction Based on Data Mining Technology[D]. Beijing Jiao Tong University,2017.
6. Zhang Tiefeng,Liang Sibo,Gu Jiangwei.Overview of the distribution and utilization big data application[J]. Electrical Measurement &Instrument, 2017,54 (2):92-98.
7. Wang Qian.Significance of the study methodology Big Data Analytics[D].Wuhan University of Science and Technology,2015.

Development Of An DDF Unit For Power Communication

Lin Bing-hua
Sanming Power Supply Bureau of Fujian Electric Power Co.1td.Sanming 365000.Fujian.China

Abstract:DDF unit is an importantpart in power communication field. It is the connection and connection equipment between digital multiplex equipment, digital multiplex equipment and program-controlled switching equipment or data service equipment and other professional equipment. However, the traditional DDF unit needs to be tested by 2M error code, which is inconvenient to carry and will not be used if it is not for the communication staff. Especially in the remote substation, substation operation and maintenance personnel at the scene of the failure, but can not match. This paper provides a DDF communication unit, to overcome the shortcomings of existing technology, substation operation and maintenance personnel can visually see the connectivity, distance to deal with the fault, shorten the fault time, improve work efficiency.

Key words: DDF unit; Power communication; detection device

DDF (Digital Distribution Frame) unit is an important part in the field of electric power communication。It is suitable for the transmission rate is 2 Mb/s ~ 155 Mb/s digital terminal equipment or SPC digital signal wiring and switching circuit, switching and scheduling, with test function, convenient maintenance, flexible circuit line of business change and adjustment. Therefore, the equipment has a wide range of application, and it is the necessary transmission equipment for the digital transmission room and the program-controlled switchboard room. The position of the digital distribution frame is shown in figure 1.

1 Background Technology

DDF (Digital Distribution Frame) unit, also called high-frequency distribution frame, has more and more advantages in digital communications. The DDF unit can make the digital stream digital communication equipment connected as a whole, is between the wiring, digital multiplex equipment of digital multiplex equipment and program-controlled switching equipment or non business connections between devices with equipment, wiring, jumper, switching and testing function flexible.

The DDF unit consists of several pairs of 2M terminals, carrying the important business such as relay protection and dispatching automation. There are two reasons for business failure, one is terminal equipment failure, the other is channel fault. The traditional detection method is 2M error code, the disadvantage is that the instrument must be carried anytime, and if it is not communication staff will not be used. Especially in remote sites, substation operators at the scene, but noway to match.

2 DDF Unit Structure Of Power Communication

The aim of the invention is to overcome the shortcomings of existing technology, provides a new DDF unit, substation operation and maintenance personnel can visually see the connectivity, distance to deal with the fault, shorten the fault time, improve work

 DOI: 10.1051/matecconf/201713900150

efficiency.

The utility model relates to an DDF unit, and the structure of the utility model is characterized in that the utility model comprises metal plate body, inscription strip, connecting terminal and LED lamp.

The appearance of the DDF unit is gray, can be turned, and the rotation angle is 180 degrees to facilitate the operation of the connector plug. A DDF unit is a unit board 20 system, divided into two rows, each row on the 10.

The metal plate is made of full copper material and has good corrosion resistance.

The inscription for labeling, length of the unit plate length corresponding to the outside with a layer of transparent plastic, inside a long strip of paper, up and down each row.

The connecting terminal is an unbalanced coaxial connector with 75 ohm, the surface is plated with gold, the contact resistance is low, and the connection is reliable. 20 a unit board of a system with 4 connecting terminals. In general, the connecting terminal 3 is connected with the service side, and the connecting terminal 5 is connected with the service side, and the connecting terminal 4 is connected with the hair of the equipment side, and the connecting terminal 6 is connected with the side of the device.

The LED light is a system of 4, connectivity LED lamp 7 indicates connection terminals 5 and 6, LED lamps 8 indicates connection terminal connectivity 3 and 4 LED, indicating lamp 9 connection terminals 3 and 5 connectivity, LED connectivity indicator lamp 10 connection terminals 4 and 6. For lights, no lights.

In normal use, as shown in Figure 3, the vertical plug connection terminals 3 and 4 connected to the circuit in Figure 2, R1, D1, diode D1 lights, LED lights lit 8; connection terminals 5 and 6 connected, R2, D2, diode D2 lights, LED lights, 7 lights. LED lights 9 and 10 are not lit.

When the loopback test is shown, as shown in Figure 4, that is, the horizontal plug is connected with the connecting terminals 3 and 5, the R3 and the D3 work, the diode D3 is bright, and the LED lamp 9 is bright; when the connecting terminals 4 and 6 are connected, the R4 and the D4 work, and the LED lamp 10 is bright. LED lights 7 and 8 are not lit.

In summary, the present invention has the following advantages over prior art:

1.Simple and intuitive, not communication staff, but also through the LED light directly observe the channel, the initial judgment is equipment problems or channel problems, reduce misoperation, misoperation probability.

2.When the communication personnel software loop back test, on-site substation operators can check the ring back through the LED lamp is wrong, remote processing failure.

Brief description of drawings

Figure 2 is a schematic diagram of a panel of the present invention. Label description 1 metal plates; 2 Ming bars; 3, 4, 5, 6 connecting terminals; 7, 8, 9, and 10 LED lights.

Figure 3 is an internal circuit diagram of the present invention.

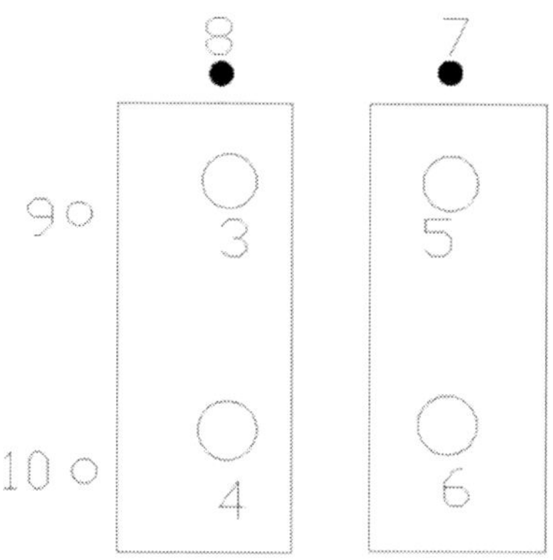

Figure 4 A connection diagram when the connection terminal is used.

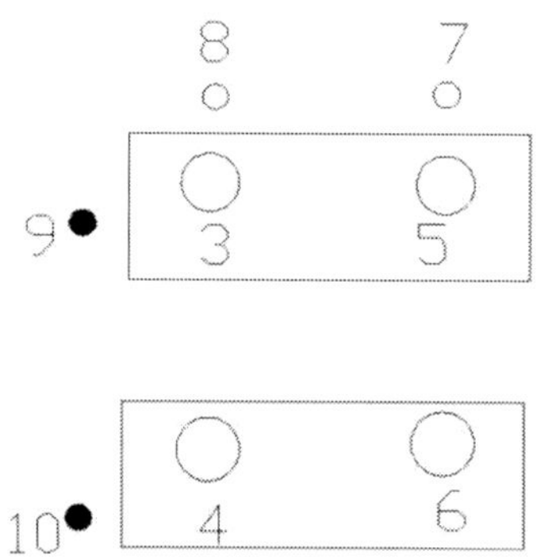

Figure 5 The connection diagram of the connection terminal ring back.

3 application example analysis

Embodiment 1

The utility model relates to an electric communication DDF unit, which comprises a metal plate body, a inscription strip, a connecting terminal and a LED lamp. The appearance of the DDF unit is gray, can be turned, and the rotation angle is 180 degrees to facilitate the operation of the connector plug. A DDF unit is the unit board of the 20 system. It is divided into two rows and 10 pairs on each row. The metal plate is made of full copper material and has good corrosion resistance. The inscription for labeling, length of the unit plate length corresponding to the outside with a layer of transparent plastic, inside a long strip of paper, up and down each row. The connecting terminal is an unbalanced coaxial connector with 75 ohm, the surface is plated with gold, the contact resistance is low, and the connection is reliable. 20 a unit board of a system with 4 connecting terminals. In general, the connecting terminal 3 is connected with the service side, and the connecting terminal 5 is connected with the service side, and the connecting terminal 4 is connected with the hair of the equipment side, and the connecting terminal 6 is connected with the side of the device. The LED light is a system of 4, connectivity LED lamp 7 indicates connection terminals 5 and 6, LED lamps 8 indicates connection terminal connectivity 3 and 4 LED, indicating lamp 9 connection terminals 3 and 5 connectivity, LED connectivity indicator lamp 10 connection terminals 4 and 6. For lights, no lights.

In normal use, as shown in Figure 3, the vertical plug connection terminals 3 and 4 connected to the circuit in Figure 2, R1, D1, diode D1 lights, LED lights lit 8; connection terminals 5 and 6 connected, R2, D2, diode D2 lights, LED lights, 7 lights. LED lights 9 and 10 are not lit. When the loopback test is shown, as shown in Figure 4, that is, the horizontal plug is connected with the connecting terminals 3 and 5, the R3 and the D3 work, the diode D3 is bright, and the LED lamp 9 is bright; when the connecting terminals 4 and 6 are connected, the R4 and the D4 work, and the LED lamp 10 is bright. LED lights 7 and 8 are not lit. The LED light 7 in Fig. 1 corresponds to the diode D1 in Fig. 2, and the LED light 8 in Fig. 1 corresponds to diode D2 in Fig. 2, and the LED light 9 in Fig. 1 corresponds to diode D3 in Fig. 2, and the LED light 10 in Fig. 1 corresponds to diode D4 in Fig. 2.

4 Conclusion

The DDF (Digital Distribution Frame) unit is a wiring and connection equipment between the digital multiplex equipment, the digital multiplex equipment and the program-controlled switching equipment or the data service equipment and other specialized equipment.

The invention can directly observe the channel condition through the LED lamp, and preliminarily judge whether the equipment is a problem or a passage problem, and reduce the probability of misoperation and misoperation, and is simple and direct. At the same time, when the communication personnel software loop back test, on-site substation operators can check the ring back through the LED lights are wrong, remote processing of faults, shorten the failure time, improve work efficiency.

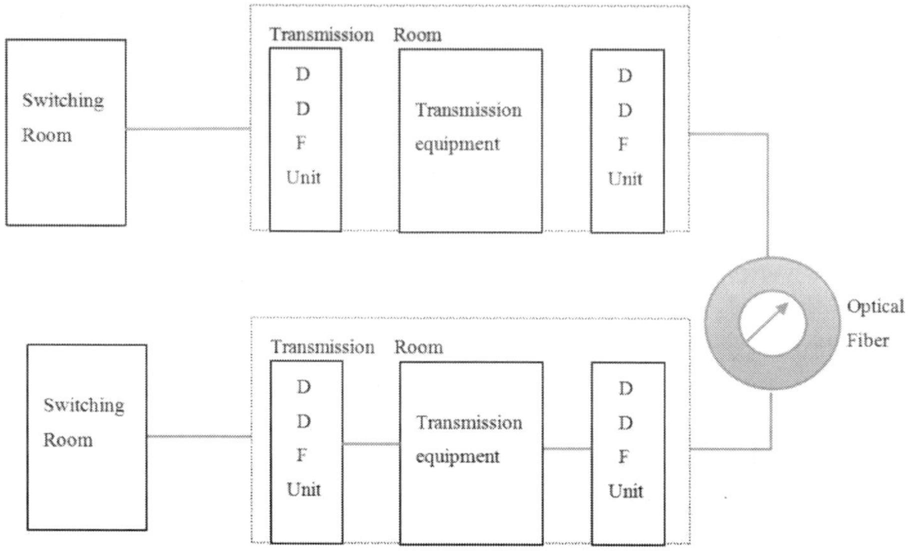

Figure 1 Location diagram of the DDF unit in the transmission

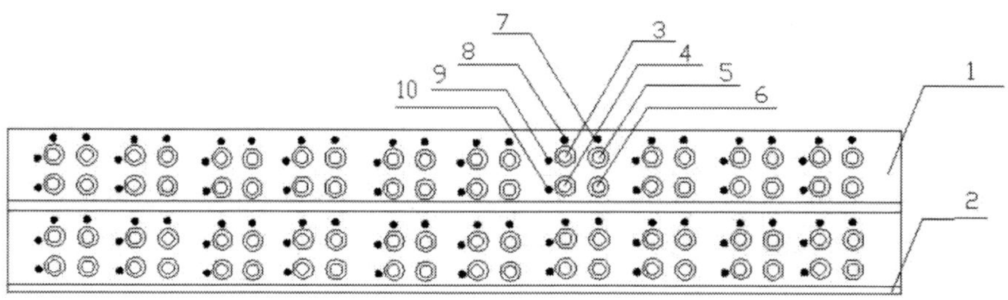

Figure 2 A schematic diagram of a panel of the present invention

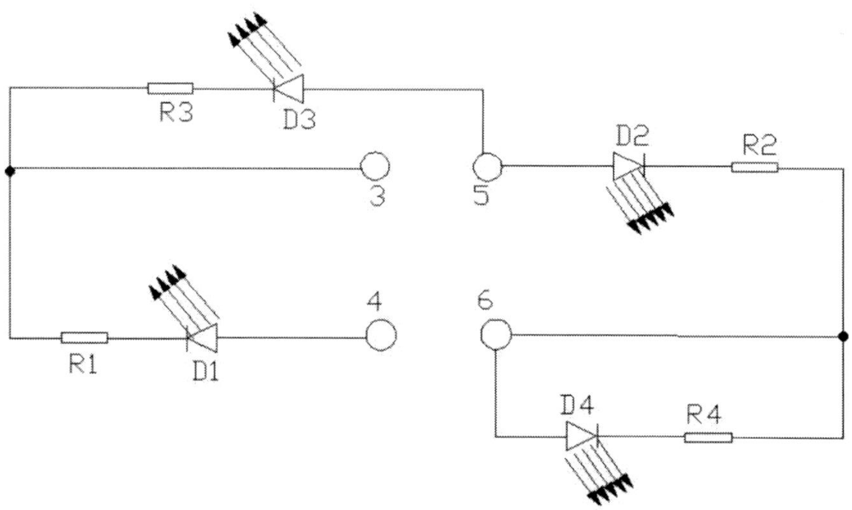

Figure 3 An internal circuit diagram of the present invention

A Trusted Computing Architecture of Embedded System Based on Improved TPM

Xiaosheng Wang [1,1*], *Gaochao Xu* [2], *Yongfei Han* [1,2], *and Yanchun Yang* [1,3]

[1] School of Information Technology, Shandong Women's University, Jinan 250300, P.R. China
[2] School of Computer Science and Technology School, Jilin University, Changchun 130022, P.R. China

Abstract. The Trusted Platform Module (TPM) currently used by PCs is not suitable for embedded systems, it is necessary to improve existing TPM. The paper proposes a trusted computing architecture with new TPM and the cryptographic system developed by China for the embedded system. The improved TPM consists of the Embedded System Trusted Cryptography Module (eTCM) and the Embedded System Trusted Platform Control Module (eTPCM), which are combined and implemented the TPM's autonomous control, active defense, high-speed encryption/decryption and other function through its internal bus arbitration module and symmetric and asymmetric cryptographic engines to effectively protect the security of embedded system. In our improved TPM, a trusted measurement method with chain model and star type model is used. Finally, the improved TPM is designed by FPGA, and it is used to a trusted PDA to carry out experimental verification. Experiments show that the trusted architecture of the embedded system based on the improved TPM is efficient, reliable and secure.

1 Introduction

At present, some of traditional information security technologies used by the information system are lack of active defense function [1] to lead to the majority of security problems to lag solving without a unified security system. With the development of Internet of Things (IoT), a variety of embedded systems applications continue to grow such as sensor networks, smart cards, mobile devices, medical, avionics, automotive and smart grid control systems etc. [2]. On one hand, the embedded system is open to the Internet/mobile Internet and it has external attack channels in the IoT. On the other hand, the architecture of embedded systems is too simple and its resource permissions can be arbitrarily used and its executable program is easily included malicious code to control system. So the embedded system may bring security problems. Solving the security of embedded systems is a new challenge in the field of embedded application in the IoT era.

To solve the security of embedded systems in the IoT, it is necessary to improve the architecture and increase the active monitoring and prevention mechanism for the virus or malicious code, and enhance the security of the whole system.

2 Related works

The Trusted Computing Organization (TCG) has proposed a set of trusted computing concepts and methods for the PC information security problem [3], which requires a trusted platform to include the TPM chip as the root of trust for measurement, and then build a trusted chain level by level to extend the trust relationship to the entire PC system to achieve a trusted security computing environment on the entire platform to guarantee the operation of the computing to execute normally.

At present, the trusted concept for expanding embedded and mobile terminal with TPM chip also is proposed [4-6] to solve the growing trusted computing requirement in the field of embedded information security, such as Apostolos et al propose an embedded system hardware architecture capable of providing security and trust along with physical attack protection using trust zone separation [6]. This paper draws lessons from the trusted PC architecture against its shortcomings and improves the original TPM to propose a trusted architecture for the embedded system based on embedded trusted cryptography module (eTCM) with the Chinese cryptosystem, and creates FPGA hardware platform to verify the prototype system of the trusted architecture.

3Trusted embedded system architecture based on improved TPM

Since the embedded system has strict requirements for function, reliability, cost, size and energy consumption, and the original TPM made by TCG has defects which lacks master capability and complexly calculates cryptography (using only RSA but not the use of symmetric cryptography) and so on. Thus, the original

* Corresponding author: xswang@sdwu.edu.cn

PTM is not entirely suitable for embedded systems, it is necessary to improve the TPM. In addition to the original functions, the improved TPM should have some new features for embedded systems.

In this paper, we present a trusted embedded computing architecture based on the improved TPM and the Chinese cryptosystem as shown in Figure 1 (a kind of hierarchy). In the architecture, we use a cryptosystem combined with symmetric and asymmetric cryptography to form an embedded trusted cryptographic module (eTCM), and propose the embedded trusted platform control module (eTPCM) based on the cryptosystem as the trust root, and the trust chain of the embedded information system is established to ensure the security of the applications and the critical information.

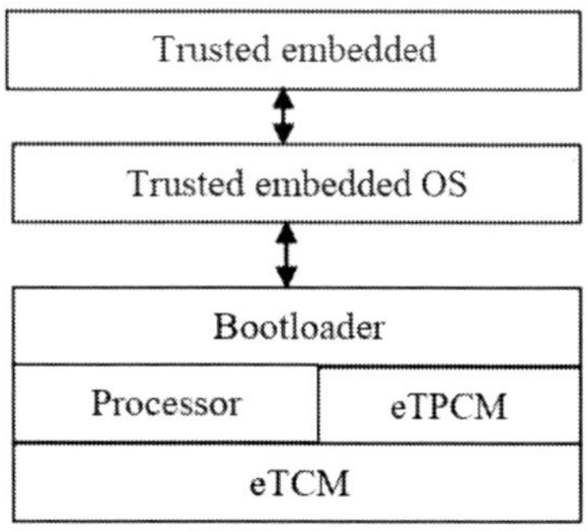

Fig.1. The trusted embedded system architecture based on Chinese cryptosystem.

The eTCM included the cryptosystem is a necessary and critical basic components in the trusted computing platform which provides a support of independent cryptography algorithm. The eTCM includes hardware and firmware that can be integrated with eTPCM in an IP core to serve as an improved TPM.

4 Embedded system trusted platform module

In this paper, an embedded system trusted platform module is designed for embedded systems combining the original TPM and Chinese cryptosystem. The module consisting of the eTCM and the eTPCM is an improvement of the original TPM. The improved TPM has the ability of active control, faster speed of the symmetric encryption and decryption, enhanced the reliability of the system, and supports to achieve chain and star type trust measurement model as a trusted root to effectively reduce the decay and time in the trust transfer. The architecture of the improved embedded system trusted platform module is shown in Figure 2.

According to Chinese trusted computing cryptosystem, the elliptic curve cryptography algorithm used in asymmetric cryptography in eTCM includes three sub-algorithms: digital signature algorithm SM2-1, key exchange protocol SM2-2, public key encryption algorithm SM2-3. The SM4 is used in symmetric cipher algorithm. The SM3 is used in hash algorithm. The cryptographic mechanism uses to protect sensitive data in the system and user sensitive data. The eTPCM is used as a trusted root to provide a series of trusted computing functions such as integrity measurement, safely store, trusted reports, and cryptography services etc. In the eTPCM, a combination of symmetric and non-symmetric cryptography is adopted to improve security and efficiency and active control and measurement are achieved. The bus arbitration module is used to control the main processor in embedded system and the improved TPM access to external memory. i.e. the eTPCM is in the active position due to control the read and write to the external memory and can verify each component in the system separately. A backup memory physically protected is added on the basis of the original TPM to store the boot program and key data of the OS, during they are verified, if they are tampered, it can be easily restored to the external main memory, thus, the reliability of the system is enhanced.

5 The eTPCM trusted measurement protocol

In the paper, a security trust measurement protocol is built with eTPCM as a source of trust for embedded system to achieve the entire hardware and software modules trusted level by level. In the circuit design, the eTPCM is required to start before the embedded main processor so that holds control to the entire circuit system (changing traditional ideas that previous TPM is a passive device). Trusted chain transfer process shown in Figure 3. The specific steps are as follows.

Step 1: Bootloader code is read by the core (RTM) code of the trusted measurement root in the eTPCM, and it is measured, and the report is transferred to the eTPCM. If successful, the control is transferred to Bootloader, go to Step 2. Otherwise, Bootloader code is not trusted, it need to be recovered from the backup memory to be re-measured.

Step 2: Bootloader executes the trusted measurement for embedded OS core code and reports the measurement value to eTPCM. If successful, the OS is loaded and the control is transferred to the OS, go to step 3. Otherwise the OS core code is not trusted and needs to be restored and re-measured.

Step 3: The embedded OS completes the trusted measurement for the trusted service manager, and reports the measurement value to the eTPCM. If successful, the trusted service manager is started. Otherwise, it needs to be restored and re-measured.

Step 4: The trusted service manager completes the trusted measurement of multiple applications that require trusted authentication.

Finally, the trusted service manager resides in the system memory to perform an active measure that measures trust of the applications on runtime from time to time to ensure these applications to be detected timely and remedied after modified by viruses or malicious code.

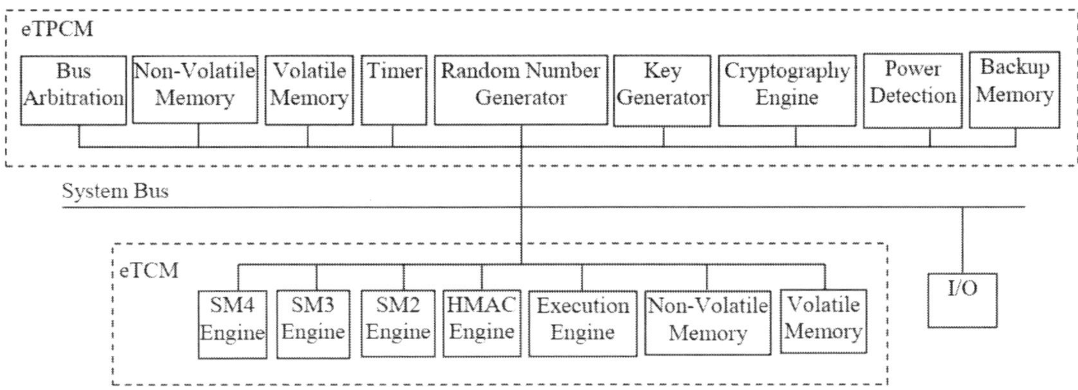

Fig.2. Trusted platform module architecture based on improved TPM for embedded systems.

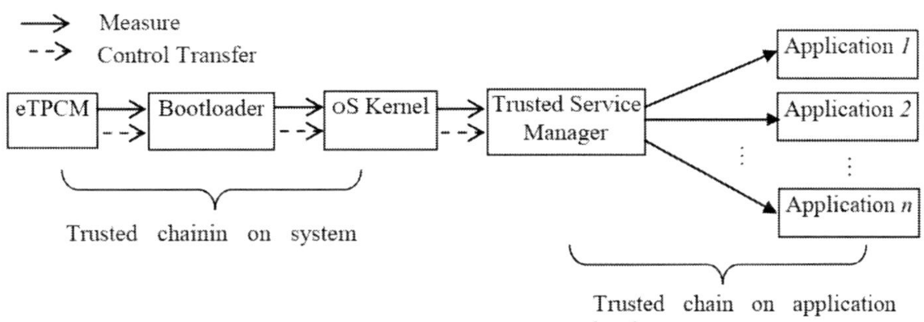

Fig.3. Trusted measurement model of the eTPCM

In the trusted measurement protocol, the chain of trust transfer is combined with chain and star type model. The chain type trusted measurement model is used in the system level and the star type model is used in the application level, which overcomes the defect that only the star type model is used to result in a long measurement due to the eTPCM processor typically performs less. The application is measured by trusted service manager in the user kernel, which is computed by the embedded system host processor, greatly reducing the measurement time.

6 Experimental test and verification

6.1 Verification platform building

In order to validate and evaluate the feasibility and security of the embedded system trusted computing architecture, Altera's Stratix IV E series EP4SE530H40 FPGA is used to perform the improved TPM functionality and apply it to a trusted PDA with ARM processor to complete the test, and a good result is achieved. The prototype system verification platform is shown in Figure 4.

In the start of the trusted PDA, paying attention to the following points:

(1) The improved TPM must be run before power-on and the operation of the ARM processor is controlled by it.

(2) Since the eTPCM to read the external memory data for reliability verification, there are both the ARM processor and the eTPCM need to access the external memory, so the external memory bus needs to be arbitrated.

(3) If you want to upgrade the Bootloader, it must be authorized by eTPCM.

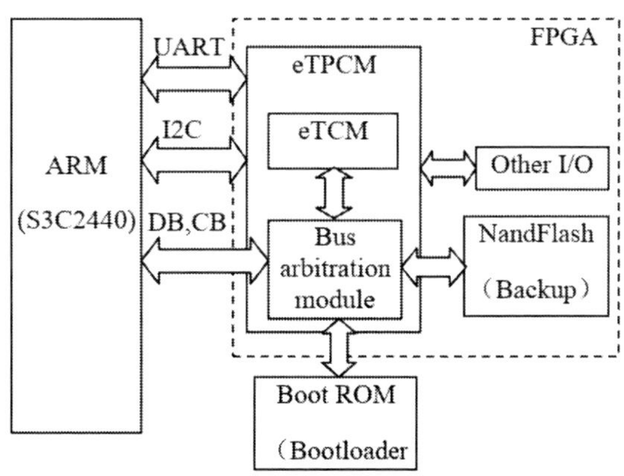

Fig.4. Schematic diagram of improved TPM and trusted PDA prototype verification system

* Corresponding author: xswang@sdwu.edu.cn

In the trusted PDA, the platform acts as a slave and the start-up process of the computer system is controlled by the improved TPM. Before the platform starts, the boot program, the operating system were measured on integrity by the improved TPM and the measurement results of integrity is compared with the values stored previously in the improved TPM to determine whether they are trusted, only the programs that are determined to be trusted can run. If there is an error in the integrity verification process, the improved TPM will automatically call the backup-recovery module to implement the system recovery.

After the system passes the integrity verification, improved TPM allows the ARM processor to run, trusted PDA normal start up to work. During the system executing, the improved TPM is still in the monitoring state, and once the exception is detected it is possible to terminate the trusted PDA reading or writing to the external memory at any time.

From the trusted start-up process of the PDA and the use of the encryption and decryption engine, the improved TPM with the bus arbitration module and the backup- recovery has greatly improved control capability and speeds up the symmetric encryption and decryption.

6.2 Analysis of experimental results

In this experiment, eTPCM needs to perform a trust measurement for the underlying configuration in the starting, and the measurement process can't take too long, if is, although the verification can be completed, but the meaning of the application is lost.

(1) Verifying the embedded system boot program and OS kernel: The verified program size is about 1MB in the experiment, in which boot program is about 60KB and OS kernel is about 800KB. Through the actual verification, entire starting process of experimental system takes less than 0.2s, and the verification speed is nearly 10MB/s, which can meet the requirements of embedded applications.

(2) Testing the speed of the symmetric cryptography encryption and decryption: The symmetric cryptography encryption and decryption engine SM4 is implemented in the hardware of improved TPM, and the execution speed of the engine is up to 7.2MB/s by test. Compared with the 74.6KB/s of the SM4 algorithm that we implemented by software in the PDA, the result of using hardware symmetric encryption and decryption engine greatly improves the encryption and decryption speed.

7Conclusion

This paper proposes a trusted architecture for embedded system from the hardware point of view, and proposes an embedded trusted cryptographic module (eTCM) and an embedded trusted platform control module (eTPCM) based on the Chinese cryptosystem for the architecture. They are combined with each other to complete the improvement of the original TPM. The improved TPM not only conforms to the TCG specification, but also its control over embedded systems is enhanced and it is more compliant with embedded system features. Experiment analysis shows that the design of improved TPM is very effective for embedded systems.

The paper focuses on the improvement of TPM and the functional design in the embedded system environment. The next step is to study the hardware security of improved trusted platform modules.

Acknowledgements

This work has been supported by Project of the Program of The Key Research and Development of Shandong Province, China, under the Grant Numbers 2015GGD01007.

References

1. C.X. Shen, B. Gong. The innovation of trusted computing based on the domestic cryptography. Journal of Cryptologic Research, 2015, 2(5): 381-389.
2. S.L. Keoh, S.S. Kumar, H. Tschofenig. Securing the Internet of Things: A Standardization Perspective. IEEE Internet of Things Journal, 2014, 1(3): 265-275.
3. TCG PC Client Specific TPM Interface Specification (TIS), Specification Version 1.3, 2013.
4. K. Tang, X. Xu, C. Guo. The Secure Boot of Embedded System Based on Mobile Trusted Module, Proceedings of the Second International Conference on Intelligent System Design and Engineering Application, January 6-7, 2012, Sanya: China, IEEE, 2012, 1331-1334.
5. TPM 2.0 Mobile Reference Architecture, Level 00 Revision 142, 2014.
6. Apostolos P. Fournaris, Nicolas Sklavos. Secure Embedded System Hardware Design - A Flexible Security and Trust Enhanced Approach. Computers and Electrical Engineering, 2014, 40(1):121-133.

New Algebraic Groups Produced By Graphical Passwords Based On Colorings And Labellings

Hui Sun[1], *Xiaohui* Zhang[1], *Meimei* Zhao[3] and *Bing* Yao[1,2,*]

[1] College of Mathematics and Statistics, Northwest of Normal University, Lanzhou 730070, CHINA
[2] School of Electronic and Information Engineering, Lanzhou Jiaotong University, Lanzhou 730070, CHINA
[3] College of Science, Gansu Agricultural University, Lanzhou 730070, CHINA

Abstract. Safety of plain text passwords has been questioned in current researching information passwords. Graphical passwords are another way for alternative text-based passwords and to improve the user account security. As we are constructing Topsnut-graphical passwords that can be traced to an idea of "Graph structure plus the number theory" proposed first by Hongyu Wang with her colleagues, we find that some of Topsnut-graphical passwords can be composed of algebraic groups under the principle of Abelian additive finite group. We apply the odd-elegant labelling of graph theory to produce Topsnut-graphical passwords, and verify our Topsnut-graphical passwords can form algebraic groups, called *labelling graphical groups*. Our results can provide those users who have business in two or more banks, and our methods are easily transformed into algorithms with polynomial times.

1 Introduction

Everyone in today's society has at least two or three passwords. Your password can not only ensure the safety of personal privacy, but also provide a sense of security for the user. The password is stolen and lost it, which is often seen in daily life. Safety of plain text password has been questioned. A graphical password is one of the ways to alternative for text-based passwords and to improve the user account security, because of graphical passwords are easy for users and difficult to be broken by attackers [1]. New graphical passwords can be made by the idea of "Graph structure plus the number theory" due to Wang *et al.* in [2] and [3], we call such new graphical passwords as *Topsnut-graphical passwords* for distinguishing with other graphical passwords. Studies have shown that humans remember pictures more than texts, and psychological research supports the hypothesis [4]. In addition, Topsnut-graphical passwords differ from those graphical passwords mentioned in [1], since they need less storage space and can be realize in communication quickly.

We show an example to introduce Topsnut-graphical pass-words simply. Alina goes to a bank for her business. First of all, she is given six topological structures (also, six graphs) shown in Fig.1, and is asked for selecting a graph for making a key. Secondly, she select the graph shown in Fig.1 (e) and Fig.2 (1) as the base of her key. Next, she labels each small circle of the graph shown in Fig.2 (2) with a positive integer, continuously, she labels each *edge* that joins two small circles with a positive integer, these two small circles are called the *ends* (also, vertices in graph theory) of the edge. Finally, she obtain her key shown in Fig.2 (3), this key is called a opsnut-graphical password. Observe the well-labelled graph Fig.2 (3), we can find that three numbers x,y,z on an edge labelled with number y and its two ends labelled with numbers x,z hold $x+z=y \pmod{16}$, that is, $x+z=y$ if $x+z<16$, and $x+z-16=y$ if $x+z\geq 16$.

Naturally, from the view of theoretical investigation and practice, we want to solve some basic problems: (i) Find all graphical passwords for the graph Fig.2(1)? (ii) Find connections among graphical passwords of the graph if the problem (i) are determined; (iii) Find all of graphs having 7 vertices and 8 edges. As known, many graph labellings can be used to other applications out of graph theory. We will apply operations, colorings and labellings of graph theory ([5], [6], [7]) to answer partially these three problems. In the following discussion, we will show an application of the odd-elegant labelling of graph theory.

* Corresponding author: yybb918@163.com

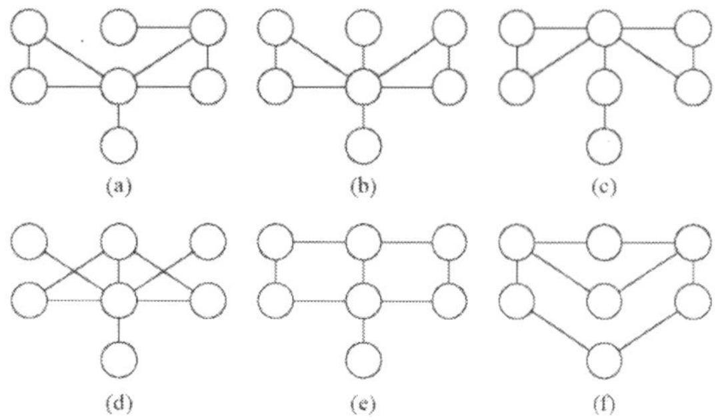

Fig. 1. Six topological structures.

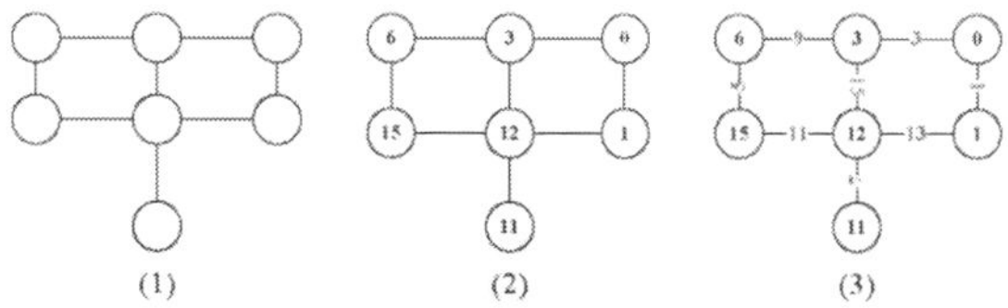

Fig. 2. Alina makes a graphical password.

2 Graphical groups

2.1 Preliminary

The shorthand notation $[m,n]$ to indicate a set $\{m,m+1,\dots,n\}$ with integers m,n respect to $0\leq m<n$; the symbol $[s,t]^o$ denotes an odd number set $\{s,s+2,\dots,t\}$ with odd numbers s,t keeping $1\leq s<t$; the notation $[\alpha,\beta]^e$ stands up an even number set $\{\alpha,\alpha+2,\dots,\beta\}$ with even numbers α, β falling into $2\leq\alpha<\beta$; $|X|$ is the cardinality of elements of a set X. All graphs mentioned here are undirected, finite, no multiple-edges and no loops, and we call them *simple graphs*. A (p,q)-graph G has its vertex set $V(G)$ of cardinality p and edge set $E(G)$ of cardinality q. Another two notations will be used in this article are $N(x)=\{v:v\in V(G)\}$ (the set of all vertices adjacent to a vertex v in G, called the *neighborhood* of x) and $N_e(x)=\{xv:v\in N(x)\}$ (called the *edge-neighborhood* of x). The symbol $\deg(x)$ stands for the degree of a vertex x, also, $\deg(x)=|N(x)|$.

Definition 1. Let X be a non-empty set, and let $\odot$ be a 2-element operation on X. We call the algebraic structure $\Gamma=(X,\odot)$ as a group if it holds:

(1) *Uniqueness and Closure*: $x\odot y\in X$ holds for any pair of elements $x,y\in X$;

(2) *Unit element* (*zero element*): There exists an element $e\in X$ such that $e\odot x=x\odot e=x$ for each element $x\in X$;

(3) *Inverse element*: Any element $x\in X$ corresponds another element $y\in X$ such that $x\odot y=y\odot x=x$, we call y the inverse element of x, denoted as $y=x^{-1}$;

(4) *Associative law*: Any triple $x,y,z\in X$ holds $(x\odot y)\odot z=x\odot(y\odot z)$.

Furthermore, $\Gamma=(X,\odot)$ is called an *Abelian group* (a *commutative group*) if the 2-element operation $\odot$ holds the commutative law true.

3 Graphical groups based on graph labellings

3.1 A graphical group obtained from the graph odd-elegant labelling

Let $f_1: V(H_1)\rightarrow[0,q]$ be an odd-elegant labelling such that $f_1(E(H_1))=[1,2q-1]$ for a (p,q)-graph H_1. Define a *modular labelling* $f_k(x)=f_1(x)+(k-1)\pmod{2q}$ with $x\in V(G)$ and $k\in[1,2q]$. Notice that $F(H_{2q+1})=F(H_1)$. As $k\in[2,2q]$, each copy of the (p,q)-graph H_1 is written as H_k, and $F(H_k)$ is the graph labelled the vertices of H_k by the labelling f_k. Define a 2-element operation $F(H_i)\oplus_k F(H_j)$ on a set $O_{dde}(H)=\{F(H_k): k\in[1,2q]\}$ by $[f_i(x)+f_j(x)-f_k(x)]_{(\bmod 2q)}$ with $x\in V(G)=V(H_i)=V(H_j)=V(H_k)$, where $[f_i(x)+f_j(x)-f_k(x)]_{(\bmod 2q)}$ means that

$$[f_i(x)+f_j(x)-f_k(x)]_{(\bmod 2q)}=2q-[f_i(x)+f_j(x)-f_k(x)] \quad (1)$$

if $f_i(x)+f_j(x)-f_k(x)<0$ or $f_i(x)+f_j(x)-f_k(x)\geq 2q$, otherwise,

$$[f_i(x)+f_j(x)-f_k(x)]_{(\bmod 2q)}=f_i(x)+f_j(x)-f_k(x) \quad (2)$$

We call this particular operation as the *odd-elegant additive-sum* of graphs.

Theorem 1. *The set $O_{dde}(H)$ based on a (p,q)-graph H_1 forms an Abelian group, we call $O_{dde}(H)$ a labelling graphical group (labelling graphical Abelian group) based on the graph G hereafter. Furthermore, $f_k(uv)=f_1(uv)+2(k-1)$ with $k\in[1,2q]$.*

Proof. First of all, the equation

$$F(H_i)\oplus_k F(H_j)=F(H_{i+j-k}) \quad (3)$$

with $i+j-k\pmod{2q}\in[1,2q]$ follows the equation

$$[f_i(x)+f_j(x)-f_k(x)]_{(\bmod 2q)}$$
$$=f_1(x)+(i+j-k-1)\pmod{2q}=f_{i+j-k}(x)$$

with $x\in V(G)=V(H_i)=V(H_j)=V(H_k)$. We are ready to show five principles for a standard group in the following.

(1) *Zero element*. The zero element of $O_{dde}(H)$ is $F(H_1)$ as $k=1$ in the equation (3).

(2) *Inverse element*. The inverse of each element $H_i\in O_{dde}(H)$ is the element $H_{2q+2-i}\in O_{dde}(H)$. In fact,

$$F(H_i)\oplus_1 F(H_{2q+2-i})=F(H_{2q+1})=F(H_1).$$

(3) *Uniqueness and Closure*. $F(H_i)\oplus_k F(H_j)\in O_{dde}(H)$ follows the equation (3).

(4) *Associative law.*

$[F(H_i) \oplus_k F(H_j)] \oplus_k F(H_l) = F(H_i) \oplus_k [F(H_j) \oplus_k F(H_l)]$.

Since

$[F(H_i) \oplus_k F(H_j)] \oplus_k F(H_l) = F(H_{i+j-k}) \oplus_k F(H_l) = F(H_{i+j+l-2k})$, and

$F(H_i) \oplus_k [F(H_j) \oplus_k F(H_l)] = F(H_i) \oplus_k F(H_{j+l-k}) = F(H_{i+j+l-2k})$,

so the associative law holds true.

(5) *Commutative law.*

$$F(H_i) \oplus_k F(H_j) = F(H_j) \oplus_k F(H_i).$$

Since $F(H_i) \oplus_k F(H_j) = F(H_{i+j-k}) = F(H_{j+i-k}) = F(H_j) \oplus_k F(H_i)$.

We conclude that $O_{dde}(H)$ is an odd-elegant graphical group. Next, we determine a connection between $f_1(uv)$ and $f_k(uv)$ with $k \in [1,2q]$ in the following.

Notice that $f_1(E(H_1)) = [1,q]$. Consider the labelled graph H_k, take an edge $uv \in E(H_k)$, we have

$f_k(uv) = f_k(u) + f_k(v) = [f_1(u) + (k-1)] \pmod{2q} + [f_1(v) + (k-1)] \pmod{2q}$,

and there are the following cases:

$$\begin{aligned} f_k(uv) &= f_k(u) + f_k(v) \\ &= [f_1(u) + (k-1)] + [f_1(v) + (k-1)] \\ &= f_1(u) + f_1(v) + 2(k-1) \\ &= f_1(uv) + 2(k-1); \end{aligned}$$

The proof of this theorem is complete.

A *dual labelling* $\overline{f_k}$ of a modular labelling f_k is defined by $\overline{f_k}(x) = 2q - f_k(x)$ with $x \in V(G)$. Let $\overline{O}_{dde}(H)$ be the set of graphs H_k, where each H_k is a labelled H by the dual labelling $\overline{f_k}$ *of* the modular labelling $f_k(x)$ with $x \in V(H)$ and $k \in [1,2q]$. We have

Corollary 2. *The set* $\overline{O}_{dde}(H)$ *forms a labelling graphical group (labelling graphical Abelian group) based on the* (p,q)*-graph* H.

3.2 Examples for illustrating Theorem 1

We show an example in Fig.3 for illustrating Theorem 1.

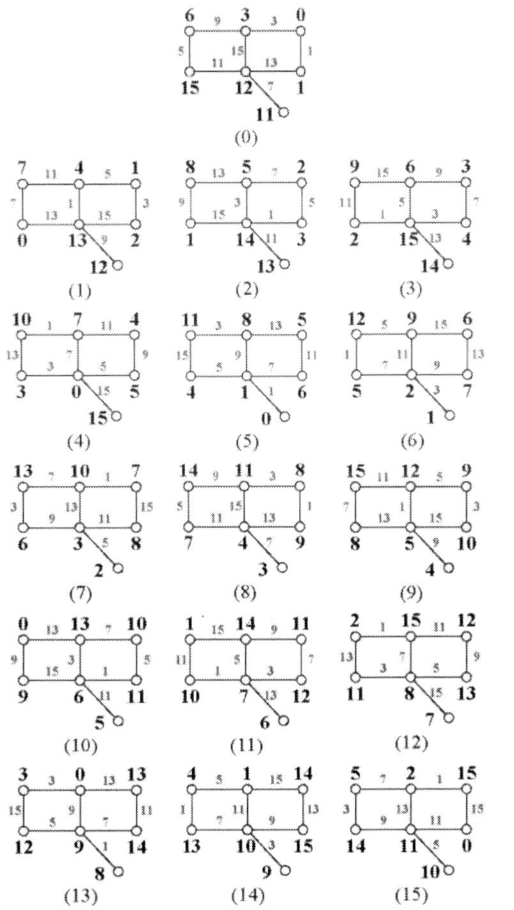

Fig. 3. A labelling graphical group for illustrating Theorem 1.

3.3 A new graph labelling

We, based on the previous results and examples, define a new labelling of graphs as follows:

Definition 2. A labelling h of a (p,q)-graph G is called a $2q$-***modular odd-elegant labelling*** if $h : V(G) \rightarrow [0,2q-1]$, and $h(E(G)) = \{[h(u) + h(v)] \pmod{2q} : uv \in E(G)\} = [1,2q-1]^o$.

According to Definition 2 we can define a negative-labelling (see Fig.4), or a mixed-labelling for a graph G (see Fig.3 and Fig.4).

3.4 A mixed labelling graphical group

In fact, we can generalize our coloring/ labelling/operation graphical quasi-groups/groups. A labelling graphical group Hroup is formed by two collections A and B shown in Fig.3 and 4. It is not hard to define a 2-element operation on Hroup defined by

$[g_i(x)+g_j(x)-g_k(x)]_{(\bmod 2q)}$, $[h_i(x)+h_j(x)-h_k(x)]_{(\bmod 2q)}$, $[g_i(x)+g_j(x)-h_k(x)]_{(\bmod 2q)}$, $[h_i(x)+h_j(x)-g_k(x)]_{(\bmod 2q)}$, $[h_i(x)+g_j(x)-h_k(x)]_{(\bmod 2q)}$ and $[g_i(x)+h_j(x)-g_k(x)]_{(\bmod 2q)}$.

This example implies a new graph labelling, called a mixed odd-elegant labelling (see g_1 shown in Fig.3 and h_1 shown in Fig.4).

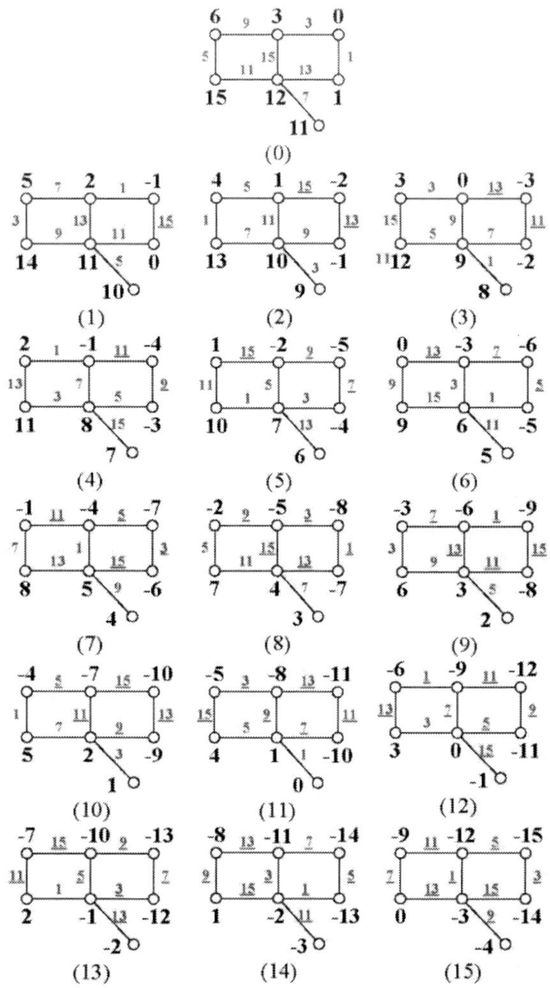

Fig. 4. A negative-labelling graphical group opposite with the group shown in Fig. 3.

Theorem 3. *The set $MO_{dde}(H)$ based on a (p,q)-graph H_1 forms an Abelian group, we call $MO_{dde}(H)$ a labelling graphical group (labelling graphical Abelian group) based on the graph G hereafter.*

Proof: A labelling graphical group H_{roup} is formed by two collections A and B. Define a 2-element operation on H_{roup} defined by

(i) $[g_i(x)+g_j(x)-g_k(x)]_{(\bmod 2q)}$,

(ii) $[h_i(x)+h_j(x)-h_k(x)]_{(\bmod 2q)}$,

(iii) $[g_i(x)+g_j(x)-h_k(x)]_{(\bmod 2q)}$,

(iv) $[h_i(x)+h_j(x)-g_k(x)]_{(\bmod 2q)}$,

(v) $[h_i(x)+g_j(x)-h_k(x)]_{(\bmod 2q)}$

(vi) $[g_i(x)+h_j(x)-g_k(x)]_{(\bmod 2q)}$.

It is not hard to verify $g_i(x) = h_{2q-i}(x) \pmod{2q}$ for $i\in[1,2q-1]$, and $g_0(x) = h_0(x)$. So, this theorem holds true according to Theorem 1.

4 Conclusion

Based on the new definition of graph coloring / labeling operations, we found the graph coloring / label constitute a algebraic group. In Theorem 1, we found that set $O_{dde}(H)$ and 2-element operation $F(H_i) \oplus_k F(H_j)$ constitute a algebraic group for which we are in the same class label, converted from one label to another label provides a feasible method. Our results show that there are groups in which any element can be defined as the zero element by the 2-element operation $\oplus_{k_0}$ in [14].

We already know that there is a label f of graph G, then it must have the corresponding dual label $\overline{f}$. That is to say, the original label and its dual label always appear in pairs 2. According to Definition 2 we can define a negative- labelling, and a mixed-labelling for a graph G. In theorem 3, We find a way to extend the two groups into a new group.

The mixed odd-elegant labelling, mentioned the above, can leads to some mathematical conjectures , such as: *There is a mixed odd-elegant labelling of every tree.* Furthermore, we can transplant this new labelling to other graph labellings, for instance, felicitous labelling in [8], (generalized) edge-magic total labelling in [9], [10] and [12], strongly graceful labelling in [11], even other labellings in [13] and so on, which means that exploring new graphical passwords will bring more new mathematical subjects and problems. Moreover, the mixed odd-graceful labeling has been studied in [14]. Exploring the topological structure of some kind of graphs, finding its more intrinsic properties, and providing theoretical help for new graphical cryptography are indispensable. [15] and [16] explore the equivalent definitions of cactus graphs and euler graphs respectively. [17] and [18] explore odd-graceful labeling and odd-elegant labeling on ring computer networks. Finally, we will explore other new 2-element operations and design new graphical passwords for determining graphical quasi-groups/groups in practice.

Acknowledgements.

The work has been supported by the National Natural Science Foundation of China under grants No. 61163054, No. 61363060, and No. 616620660.

References

1. Xiaoyuan Suo, Ying Zhu, G. Scott. Owen, Graphical Passwords: A Survey. *In Annual Computer Security Applications Conference* (ACSAC), December 2005: 463-472. DOI: 10.1109/CSAC.2005.27 Source: DBLP
2. Hongyu Wang, Jin Xu, Bing Yao. The Key-models And Their Lock-models For Designing New Labellings Of Networks. *Proceedings of 2016 IEEE Advanced Information Management, Communicates, Electronic and Automation Control Conference* (IMCEC 2016) pp565-5568.
3. Hongyu Wang, Jin Xu,Bing Yao. Exploring New Cryptographical Construction Of Complex Network Data. *IEEE First International Conference on Data Science in Cyberspace*. IEEE Computer Society, (2016):155-160.
4. R. N. Shepard, Recognition memory for words, sentences, and pictures, *Journal of Verbal Learning and Verbal Behavior* **6**, pp 156-163, (1967).
5. Bela Bollobás, The Modern Graph Theory, *Springer-Verlag*, New Tork, Inc. (1998)
6. Joseph A. Gallian, A Dynamic Survey of Graph Labelling. *The Electronic Journal of Combinatorics*, **14** (2013), #DS6.
7. Douglas B. West. Introduction to Graph Theory (Second Edition). *Mechanical Industry Press*, Bei Jing (2006).
8. Zhang J, Yao B, Wang Z, et al. Felicitous Labellings of Some Network Models. *Journal of Software Engineering and Applications* (2013) **6** (3):29-32.
9. Hongyu Wang, Bing Yao, Chao Yang, Sihua Yang, Xiang-en Chen, Ming Yao, Zhenxue Zhao. Edge-Magic Total Labellings Of Some Network Models. *Proceeding of ISCCCA13, Applied Mechanics and Materials* Vols. 347-350 (2013) pp 2752-2757.
10. Hongyu Wang, Bing Yao, Chao Yang, Sihua Yang, Xiang'En Chen. Labelling Properties Of Models Related with Complex Networks Based On Constructible Structures. *Advanced Materials Research* Vols.765-767 (2013) pp 1118-1123. DOI:10.4028/www.scientific.net/AMR.765-767.1118
11. Bing Yao, Hui Cheng, Ming Yao and Meimei Zhao, A Note on Strongly Graceful Trees. *Ars Combinatoria* **92** (2009), 155-169.
12. Hongyu Wang, Bing Yao, Ming Yao. Generalized Edge-Magic Total Labellings Of Models from researching Networks. *Information Sciences* **279** (2014) 460-467. DOI:10.1016/j.ins.2014.03.132
13. Bing Yao, Xia. Liu and Ming Yao. Connections Between Labellings Of Trees. *Bulletin of the Iranian Mathematical Society*, Vol. **43** (2017), No. 2, pp. 275-283.
14. Bing Yao, Hui Sun, Meimei Zhao, Jingwen Li and Guanghui Yan. On Coloring/Labelling Graphical Groups For Creating New Graphical Passwords. 2017 submitted. in Press.
15. Hui Sun, Bing Yao. On Equivalent Definitions Of Cactus Graphs. 2017 submitted.
16. Hui Sun, Bing Yao. Exploring equivalent definitions of Euler graphs. 2017 submitted.
17. Hui Sun, Bing Yao. Exploring the Odd Gracefulness of Cyclic-Dragon Graphs. 2017 submitted.
18. Hui Sun, Bing Yao. On the odd-elegant quality of cyclic-dragon graphs. 2017 submitted.

MATEC Web of Conferences **139**, 00153 (2017)
ICMITE 2017

DOI: 10.1051/matecconf/201713900153

Genetic Algorithm's Application for Optimization of PID Parameters in Dynamic Positioning Vessel

Liu gai ye ; Hou Yuan-bin ; Luo yan ; Li dong

School of Electric and Control Engineering, Xi'an University of Science and Technology, Xi'an 710054,China

Abstract. Genetic algorithm as intelligent optimization ,There are important applications in many field,The genetic algorithm is applied to the tuning of PID parameters in the dynamic positioning of ship,First,In this paper, a model of three degree of freedom model of ship motion and the ocean environment disturbance model of wind, wave and current are established,Secondly,PID control in dynamic positioning system is,Difficulties in debugging PID parameters,This paper focuses on the use of genetic algorithm to optimize the PID parameters,To achieve the purpose of optimizing the PID,Last,The optimization algorithm is simulated,Through the analysis of simulation results,Come to conclusion,In ship dynamic positioning,When using PID controller,The genetic algorithm can be used to find the optimized PID parameter group.

1 introduction

With the development of the ocean and the expansion of human activities, Marine resources have become a strategic resource of fierce competition among countries. Humans of ocean exploration activities will need to ship and platform structures locate a specific location in the ocean, and traditional way of anchorage due to factors such as water depth, working time, precision requirements for the restrictions cannot meet the requirements of relevant offshore operation. With the development of electronic computer and automatic control theory, the ship control technology of dynamic positioning is generated. The so-called dynamic positioning refers to the ability of the ship's own thrusters to realize its position or preset track in the complex Marine environment automatically.

The control algorithm of ship dynamic positioning is generally based on the classic PID control, PID control has many advantages and has a strong robustness. However, the parameters of the PID control is currently rely mainly on pool test setting, when a multi-channel PID control, there will be a lot of PID parameters, setting the PID parameter together only depend on the test will cost a lot of time and energy. Aiming at this problem, this paper proposes a method of PID control parameter setting based on genetic algorithm, and the optimization of PID parameters in dynamic positioning control is optimized by using genetic algorithm.

2Ship dynamic positioning model

2.1 kinematics model of ship

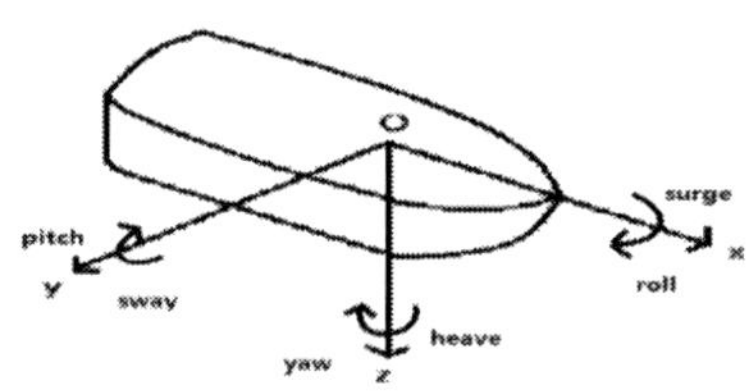

Fig. 1. schematic diagram of ship motion six degrees of freedom

The ship kinematics includes linear motion and angular motion model. According to the literature [1], the ship motion six degrees of freedom can be directly written:

$$\eta = J(\eta)v \tag{1}$$

It is equivalent to:

$$\begin{bmatrix} P^n \\ \Theta \end{bmatrix} = \begin{bmatrix} R_b^n(\Theta) & 0_{3\times3} \\ 0_{3\times3} & T_\Theta(\Theta) \end{bmatrix} \begin{bmatrix} U_o^b \\ \Omega_{nb}^b \end{bmatrix} \tag{2}$$

Type in the: $\eta = [x\ y\ z\ \phi\ \theta\ \psi]^T$, It's the position and the attitude vector; $v = [u\ v\ w\ p\ q\ r]^T$, It's the velocity and the angular velocity vector; $J(\eta)$ It's the jacobian transformation matrix.

2.2 ship dynamics model

Ship dynamics is the study of ship movement position and attitude change when hull is affected by

* Corresponding author: liugaiye 454592459@qq.com.org

force and torque. According to the literature [1], the ship dynamics model can be directly written:

$$M\dot{v}+C(v)v+D(v)v+g(\eta)=\tau+g_o+w \qquad (3)$$

On the right side of the equation are the various torque items which are applied to the ship, including the control moment and the interference moment.

In order to simplify the problem, this paper studies the ship motion at a lower speed, which can be assumed that the damping is only linear damping [2], namely

$$D(v)=D_L+D_{NL}\approx D_L \qquad (4)$$

D_L、D_{NL} (v) It represents the damping matrix, $C(v)v$ The quadratic velocity is negligible, The quadratic velocity is negligible,

$$C(v)\approx 0 \qquad (5)$$

In this paper, the six-degree-of-freedom model is simplified to three degrees of freedom, swinging and rolling. Its simplified model is:

$$\dot{\eta}=R(\psi)v \qquad (6)$$

$$M\dot{v}+D(v)v=\tau+w \qquad (7)$$

3 Genetic algorithm for optimization of PID parameters of DP ship and simulation

3.1 the PID control

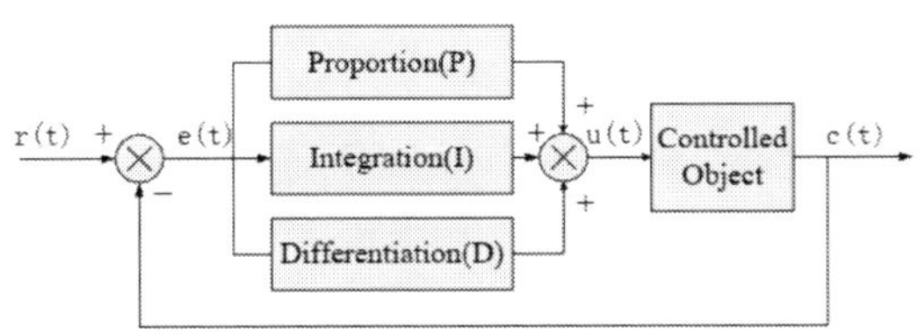

Fig. 2. PID control flow chart The control rules of PID control are expressed as:

$$u(t)=K_p\left[e(t)+\frac{1}{T_I}\int_0^t e(t)dt+\frac{T_D de(t)}{dt}\right] \qquad (8)$$

Type, K_p Proportional gain, T_I Is the integral time constant, T_D For the differential time constant, $e(t)$ For the deviation, $u(t)$ To control the amount of。

PID The functions of each link can be summarized as follows:

(1) proportion : proportional amplification deviation signal. Once the deviation is generated, the controller generates corresponding control to reduce the deviation.

(2) integration : it is mainly to eliminate the static difference of the system and improve the degree of the system. The larger the integral time constant, the weaker the integral action, and the stronger the other.

(3) differential link: the change trend of the deviation is reflected, and the early correction signal can be introduced in the system in advance, so as to speed up the response speed of the system, reduce the shock, and reduce the adjustment time.

3.2 genetic algorithm

Genetic Algorithm [3] (based Algorithm) study began in the late 1960 s to the early 70 s, mainly by the American professor John Holland and his colleagues at the university of Michigan, the students formed a relatively complete theory and method.

The genetic algorithm begins with a population of possible potential solutions for the problem, and a population is composed of a certain number of individuals that are encoded by genes. Original population carried out in accordance with the fitness of the selection, each subsequent generation evolution produces better approximate solution, in each generation, according to the problems of individual fitness (fitness) size selection (selection) and by means of genetic crosses genetic operators (crossover and mutation (mutation), on behalf of the new solution set of the population. The optimal individual in the last population is decoding, which can be the approximate optimal solution for the problem. The operation process of genetic algorithm can be explained by the following image:

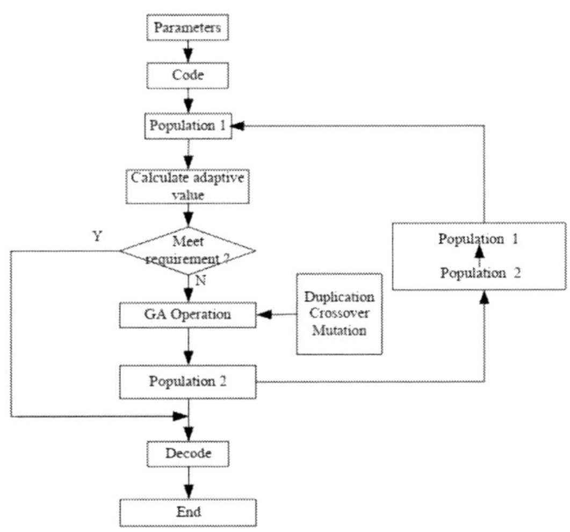

Fig. 3.the operation process of genetic algorithm .

3.3 the tuning of PID parameters by genetic algorithm

The PID parameter optimization of dynamic positioning system is to find suitable parameters to achieve the minimum value of [4-6]. For dynamic positioning ship three degrees of freedom, need nine parameters optimization, so many variables, if set a performance index function, affects the optimization precision of nine parameters, in order to solve this problem, after trying, we can independently optimization on three degrees of freedom, rather than the overall optimization, so we set the three performance index function index function matrix, this article selects a linear combination of the rise time and quantity of the deviation, control as the performance of the system control [7] :

$$J=\int_0^t (w_1\,|e(t)|+w_2\sqrt{|u(t)|})dt+w_3t_u \tag{9}$$

$$J=\begin{bmatrix} J\_X \\ J\_Y \\ J\_\psi \end{bmatrix}=\begin{bmatrix} \int_0^t (w_{1x}\,|e_x(t)|+w_{2x}\sqrt{|\tau_X|})dt+w_{3x}t_x \\ \int_0^t (w_{1y}\,|e_y(t)|+w_{2y}\sqrt{|\tau_Y|})dt+w_{3y}t_y \\ \int_0^t (w_{1\psi}\,|e_\psi(t)|+w_{2\psi}\sqrt{|\tau_N|})dt+w_{3\psi}t_\psi \end{bmatrix} \tag{10}$$

Based on the above analysis, the objective function will be taken from bottom, get fit function of genetic algorithm we need matrix, the genetic algorithm with adaptive function to evaluate group of PID parameters, in order to get the optimal PID parameters [8-9]

In this paper, the Matlab simulation is used to verify the overall flow chart of the design.

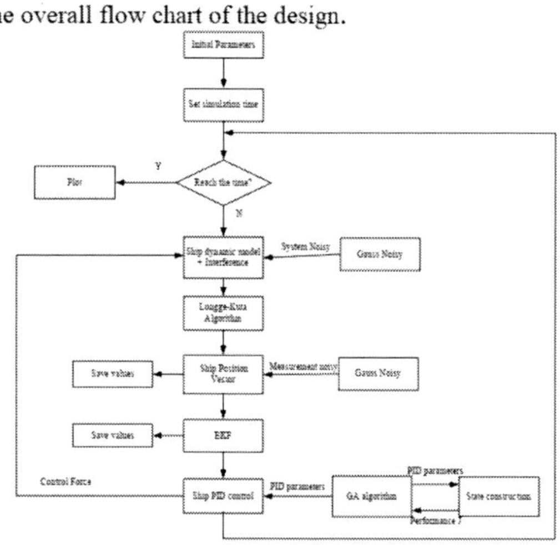

Fig. 4.system simulation flow chart .

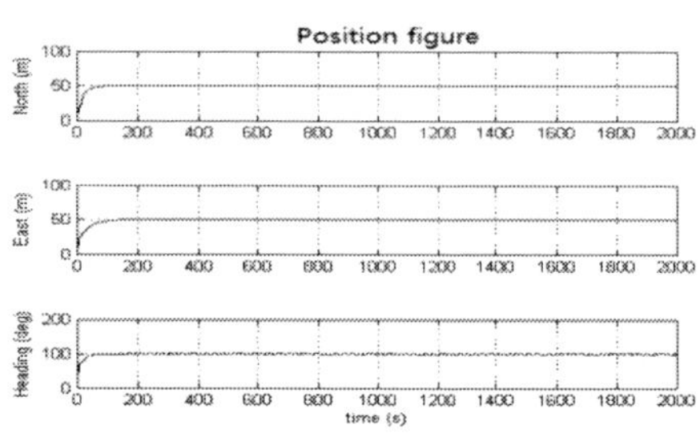

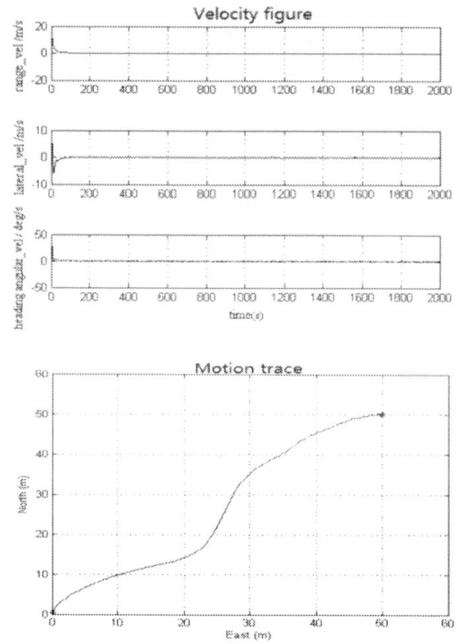

Fig. 5. control results of PID control parameters obtained by genetic algorithm

The change of performance indicator function and adaptation function in the optimization process of genetic algorithm is shown in figure6

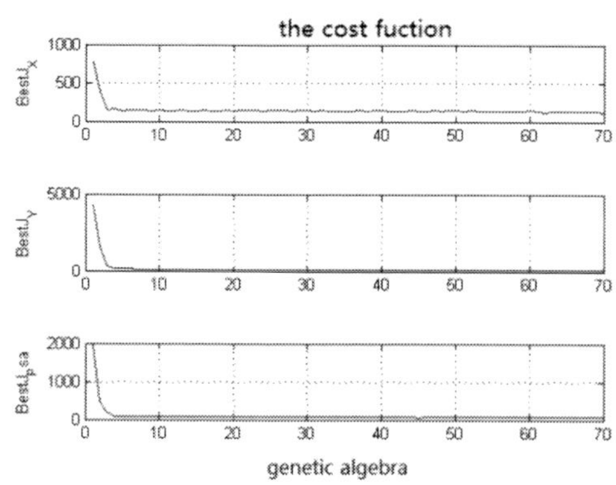

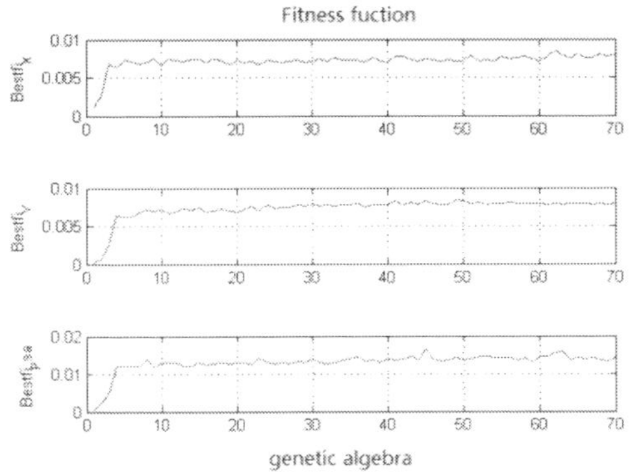

Fig. 6. shows the variation of performance indicator function and adaptation function in optimization of genetic algorithm

The control result of FIG. 5 shows that the PID parameters optimized by genetic algorithm achieve a good control effect. From FIG. 6, we can see that the convergence of genetic algorithm is very strong.

From the obtained effect picture, whether the location image or the speed image [10], the optimized parameters are satisfied with the required accuracy.

In addition, through the simulation experiment for many times, found that the genetic algorithm in optimization of this case, there will be a good optimization results, I think it may be as intelligent algorithm, genetic algorithm itself will have certain randomness, is adopted in this paper can also be offline optimization, under the circumstance of state for refactoring may exist near the optimal solution under the local optimal value, have certain does not conform to the state of refactoring and the actual .

4 conclusion

From the simulation results, the PID parameter with genetic algorithm can also achieve the desired control effect. Gather together test through engineering personnel constantly experimental debugging, select a set of good group of PID parameters, gather together test relies heavily on the experience of the engineers, and will waste a lot of time and energy for engineers [9]. When the genetic algorithm is used to optimize the PID parameters, the simulation results show that the genetic algorithm can achieve the optimal control effect by optimizing the group of PID parameters. Thus, it is difficult to overcome the difficulty in adjusting PID parameters by engineering personnel.

References

1. boundary letter qian, fu mingyu, wang yuanhui. Ship dynamic positioning [M]. Beijing: science press, 2011:1-13.
2. jia xinle, Yang yansheng. Mathematical model of ship motion mathematics - mechanism modeling and identification modeling. Dalian: dalian maritime university press. 1999:294-359.
3. chung. Linear systems theory [M]. Beijing: tsinghua university press, 2005.
4. A study on dynamic positioning of ships under mooring conditions [D]. Harbin: Harbin engineering university. 2013
5. Research on Marine dynamic positioning simulation technology [D] : China ship research institute; 2011.
6. Research on modeling and characteristics of ship movement [D] : wuhan university of technology; 2010.
7. Research on the dynamic positioning control of ships under single point mooring [D] : Harbin engineering university; 2013.
8. Math Works Inc. Genetic Algorithm and Direct Search Toolbox-MATLAB® version 2.4.1,User's guide,2009.
9. Sefrioui M,Periaux J. Nash Genetic Algorithms: examples and applications[J].In:2000 Congress on Evolutionary Computation,volume1,IEEE Service Center,Piscataway, 2000-07:509~516.
10. Zhang Y-h, Jiang J-g, Jiang Y-j.Dynamic Positioning Nonlinear Control System of the Dredger Based on PROFINET IO[J].Journal of Computers,2013,8(2).

DOI: 10.1051/matecconf/201713900154

Rock Granularity Analysis by Deep Belief Network

Guo Jiancheng[1], *Wenhui Guo*[2,*]

[1]School of Computer Sciences, Computer Department, 71006 Xi'an Shiyou University, China
[1]School of Computer Sciences, Computer Department, 71006 Xi'an Shiyou University, China

Abstract. Granularity analysis is one of the most essential issues in authenticate. To improve the identification accuracy, a Deep Belief Network (DBN) based method is proposed in this paper. DBN can extract features from image samples automatically.4800 rock images from Ordos Basin are used in this paper, 1200 of them are used to test the model accuracy and the result achieves 94.75%. According to theoretical and actual data test, the method proposed in this paper is efficiency and accuracy.

1 Introduction

There are many problems in the traditional methods for granularity analysis, such as time-consuming and low accuracy. With the development of computer and digital image processing technology, many researchers have done researches in the analysis of rock images, such as, Song Xiangfa and Li Jiaocheng proposed multi - instance image classification based on sparse representation[1]. Amir Mollajan et. improved porosity identification of rock thin images by the integration of multi-kNN, RBF and SVM[2]. The methods above all improved the accuracy of rock image classification, but there still are many problems. Firstly, to extract image features is a manual work, it means that which features will be taken to be the accordance for the classification is depended on the researcher. So the essential features may not be extracted because of the complex of rock image. Secondly, the classification accuracy still limits researches on authenticate.

In view of the above deficiencies, a new method based on Deep Belief Network (DBN) proposed in this paper to improve the accuracy and efficiency of rock granularity analysis. DBN is a representative algorithm in deep learning. It can extract features automatically for classification and has be successfully applied into all walks of life. Many home and abroad researcher have done researches on it, such as, Shi Xin and Zhu Yongli proposed DBN based transformer fault classification method by the analysis of dissolved gas in power transformer oil using DBN[3]. Tang Xielun and Zhou Jialin proposed a EEG signal identification method using DBN[4]. Shi Xugan and Zhang Shiqing proposed a DBN based new method for face expression identification by the combination of DBN and KNN[5]. Mohamed Elleuch, Olivier Pauplin use DBN to identify Handwritten Arabic[6,7] and so on. In view of the advantages of DBN in pattern recognition, this paper applies it to the rock granularity analysis to improve the identification accuracy by extracting features automatically.

2 ROCK IMAGES

It is usually determined by the professional for type and structural parameters of rock after identification under polarized light microscopy. The rock images used in this paper are all feldspar sandstones collected from an oil field in Ordos, they are divided into three types: 1) Coarse feldspar sandstones, as is shown in figure 1(a). 2) Medium granular feldspar sandstone, as is shown in figure 1(b). 3) Fine feldspar sandstone, as is shown in figure 1(c).

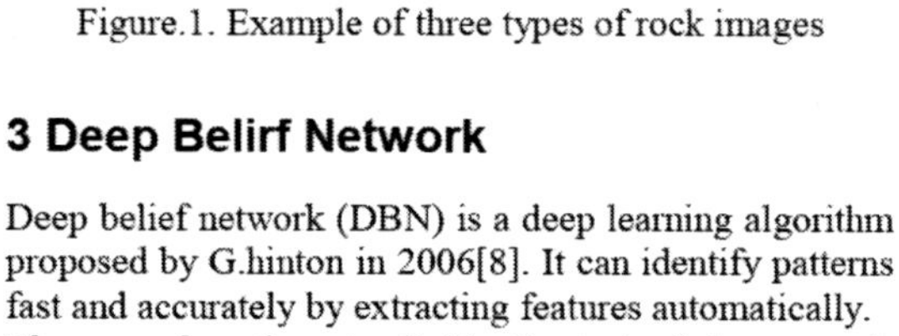

(a) Coarse feldspar sandstone
(b) Medium granular feldspar sandstone
(c) Fine feldspar sandstone
Figure.1. Example of three types of rock images

3 Deep Belirf Network

Deep belief network (DBN) is a deep learning algorithm proposed by G.hinton in 2006[8]. It can identify patterns fast and accurately by extracting features automatically. The use of sections to divide the text of the paper is optional and left as a decision for the author. Where the author wishes to divide the paper into sections the formatting shown in Table 2 should be used.

3.1 Restrict Boltzmann Machine

* Corresponding author: gjcheng@xsyu.edu.cn，,WenhGuo@163.com

Restrict Boltzmann Machine (RBM) can extract image features. A network consists of multilayer RBM has a strong ability to extract features. RBM can be seen as a simple neural network that consists of visual layers and hidden layers, and its node value is 0 or 1.The structure of RBM is shown as figure 2, where v and h respectively represent visual layer and hidden layer, w represents connection weight, b and c respectively represent bias of visual layer and hidden layer.

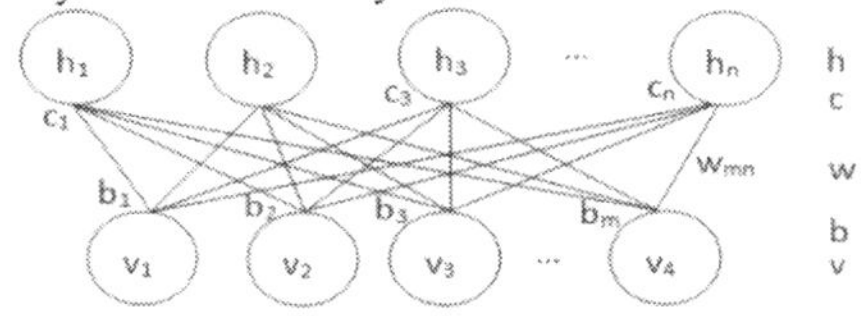

Fig 2. RBM structure

Usually, describing a system by an energy function. The smaller the system energy, the more ordered the system or the more concentrated probability distribution. On the contrary, it means the more disorderly or the probability distribution of the system becomes more uniform. For a RBM, its energy function is shown as formula 1.

$$E(v,h) = -\sum_{i\in v} b_i v_i - \sum_{j\in h} c_j h_j - \sum_{i,j} v_i h_j w_{ij} \qquad (1)$$

According to formula 1, the probability of visual layers and hidden layers is deduced as formula 2.

$$p(v\,|\,h) = \frac{e^{-E(v,h)}}{\sum_h e^{-E(v,h)}} \quad , \quad p(h\,|\,v) = \frac{e^{-E(v,h)}}{\sum_v e^{-E(v,h)}} \qquad (2)$$

So, the joint probability between visual layer and hidden layer can be deduced as formula 3,

$$p(v,h) = \frac{e^{-E(v,h)}}{Z(\theta)} \qquad (3)$$

Where represents the target parameter , is the normalization factor.According to formula 1, formula 3 can be deduced as formula 4.

$$P(v,h) = \frac{1}{Z(\theta)} \exp\left(-\sum_{i=1}^{n}\sum_{j=1}^{m} w_{ij} h_i v_j - \sum_{j=1}^{m} b_j v_j - \sum_{i=1}^{n} c_i h_i \right) \qquad (4)$$

The target of training RBM is to minimize formula 4. Gibbs is the best way to train RBM, but it needs to sample many more times. Contrast divergence algorithm is usually used that is an optimization of Gibbs.

3.2 DBN

DBN can be seen as multilayer RBM. If a DBN consists of two layers of RBM (RBM1 and RBM2) and a classifier, the hidden layer of RBM1 can be seen as the visual layer of RBM2 and the output of RBM1 is the input of RBM2. Steps of training a DBN is shown as follows:

Step 1. Start from the input layer, train RBM1 using the input data and fix the parameters.

Step 2. Train RBM2 taking the output of RBM1 as its input and fix its parameters.

Step 3. Train the remaining layers as the way of training RBM2 until RBMn has be trained(n represents number of RBM in DBN that is training).

Step 4. Classify the original image by a classifier with the input of the combination of the sample labels and the output of RBMn.

Step 5. Fine-tuning the whole network by back propagation algorithm.

4 Experiments and results

4.1 Experiments

4.1.1 Data set

In this paper, 4800 rock images in RGB colour space are selected as experimental samples, normalize the size of them to 224*224. There are 1600 coarse-grained rock images, 1600 medium-grained rock images and 1600 fine-grained rock images. And 1200 rock images of each type are used as the training set, the remaining 400 are used as the test set.

4.1.2 DBN structure

The DBN structure designed in this paper is shown as figure 3 which is a 5-layer network. The activation function of the first four layers is sigmoid, the last layer uses softmax as a classifier. Training it as steps in 2.2 can achieve images classification.

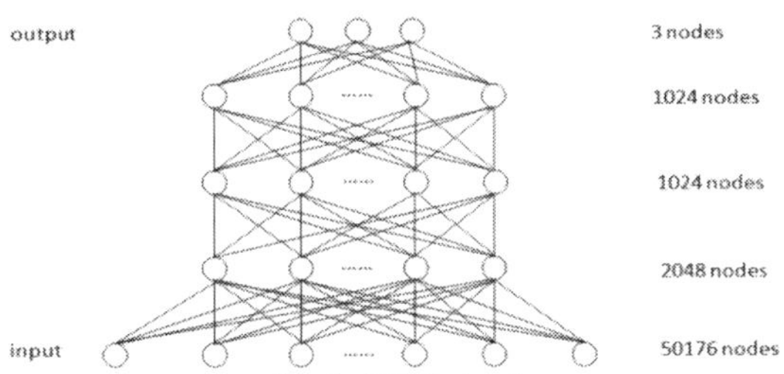

Fig 3. DBN structure

4.2 Architecture adjustment

In this paper, the influences of layers, classifier and dropout algorithm to classification accuracy are studied. The experimental data is loaded into memory by batch whose size is 100.

4.2.1 DBN structure

DBN extracts features layer by layer, so the hidden layers number has a great influence to its classification accuracy. For different hidden layers number, the identification results of the DBN designed in thispaper are shown as table 1 and figure 4.

Table 1. Error rate of different number of hidden layers

Number of hidden layers	5	4	3	2
Error rate	39.53%	10.85%	6.92%	15.74%

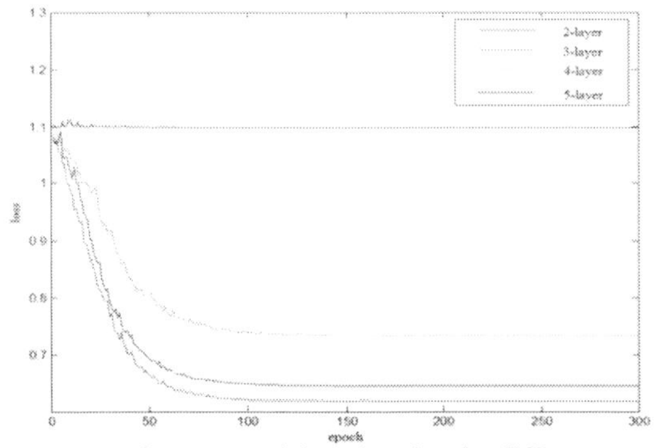

Fig4. Network loss function for different convolution layers

It can be seen in table 1 and figure 4 that the network designed in this paper is over-fitting when setting its hidden layers number to 5. When its hidden layers number is respectively 2, 3 or 4, the loss functions all can converge and the epoch times are roughly the same. But it is obviously that when the number is 4, the loss function and error rate are all high while they are all the lowest when the hidden layers number is 3. Therefore, the hidden layers number of DBN designed in this paper is set to 3.

4.2.2 Classifier

The choice of classifier determines the accuracy of the final classification of the network to a large extent. In this paper, a DBN structure with 3 hidden layers whose hidden nodes assigned to 2048-1024-1024 is studied with Logistic classifier, Linear classifier and Softmax classifier respectively. The error rates are shown in table 2.

Table 2. Error rate of different classifiers

Classifier	Logistic	Linear	Softmax
Error rate	26.17%	35.61%	5.64%

It can be seen in table 2 that the error rate is the highest when using Linear classifier and it is the smallest when using Softmax classifier. So, the network designed in this paper using Softmax as its classifier.

4.2.3 Dropout

To prevent the network from over-fitting, the dropout algorithm is added and its influence is studied in this paper. The thought of it is abandoning some nodes for a certain percentage which is set as 50. Figure 5 is the error rate of the network designed in this paper with or without the dropout algorithm. It can been seen clearly that the error rate curve is lower and becomes stable earlier when dropout algorithm is added.

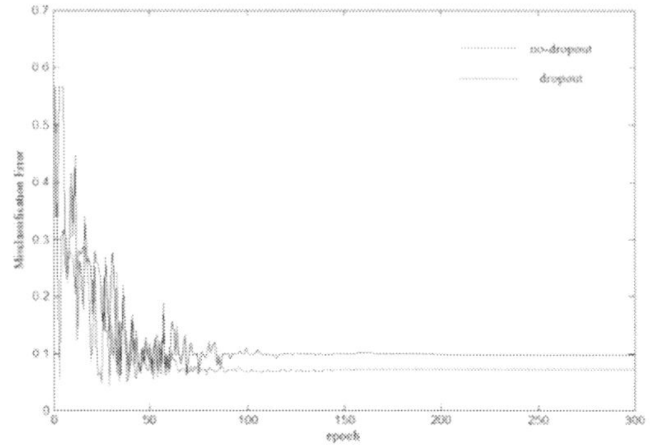

Fig 5. Network error rate for the network with or without dropout

4.3 The results

The experiment results in this paper are shown in figure 6 ,figure 7 and table 3. Figure 6 is the loss function curves and figure 7 is the error rate curves. It is clearly that either loss function curve or error rate curve is similarly on the test dataset and the training dataset. And they are going to convergence when Iterates about 90 times, it means that the network does not over-fitting, and it is obvious that the error rate is low.

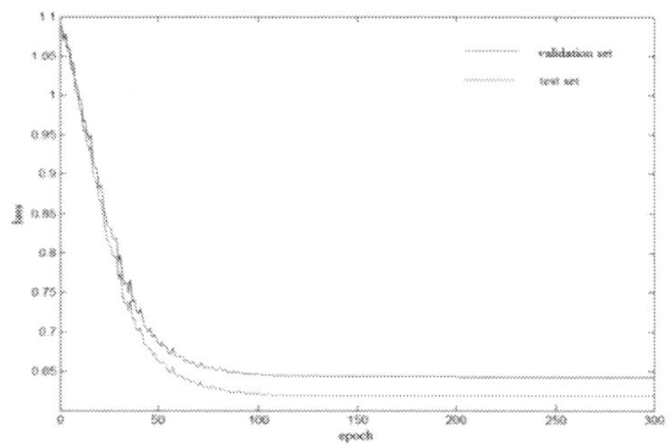

Fig 6. Loss function

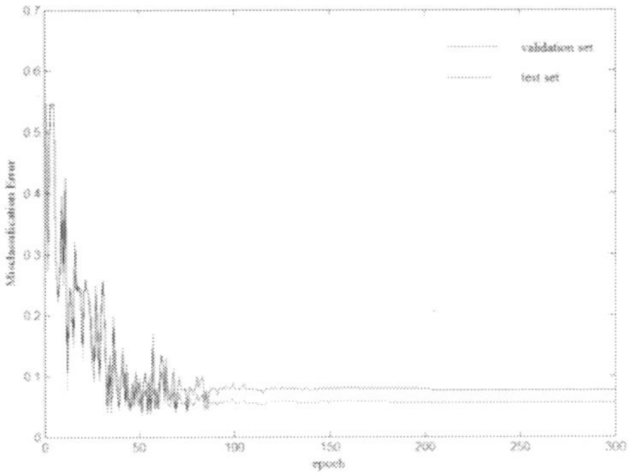

Fig 7. Error rate

Table 3 is the classification results. It shows that it is effective to classify rock images by the method proposed in this paper. In 400 images of type I , 21 are error classified as type ii and 5 are error classified as type iii. In 400 images of type II, 16 are error classified as type i and 6 are error classified as type iii. And in 400 images of type III, 6 are error classified as type I and 9 are error classified as type ii. The whole error rate is 5.25%.

Table 3. Classification results

Result of DBN	Real result		
	I	II	II
I	374	21	5
II	16	378	6
III	6	9	358

5 Conclusion

The experiment results in this paper prove that using DBN to classify rock images greatly improve its accuracy and effectiveness and can provide references in authenticate. However, the experimental results still have deviation which may be caused by the complex of rock images and the use of single polarized images. The next research can be start on the two aspects to improve classification accuracy. Firstly, the multi-polarized light can be used in the rock image. Secondly, comparing the classification accuracy of rock granularity under multiple colour spaces and select the one with the highest classification accuracy.

Acknowledgments

The main project is supported by the National Science and Technology (No.2011ZX05044) and Shaanxi Province industrial science and technology (No.2015GY104).

References

[1] Song Xiangfa, Jiaoli Cheng, Classification of Multi-instance Image Based on Sparse Reprsentation[J], Computer Science, Volume 42(1), Jan 2015, Pages 2292-2297.

[2] Amir Mollajan, Javad Ghiasi-Freez, Hossein Memarian, Improving pore type identification from thin section images using an integrated fuzzy fusion of multiple classifiers[J], Journal of Natural Gas Science and Engineering 31 (2016) Pages396-404.

[3] Shi Xin, Zhu Yongli et. Power transformer fault classifying model based on deep belief network[J], Power System Protection and Control, Vol.44(1) Jan. 2016, ,Pages 71-76.

[4] Tang Xianlun, ZHOU Jialin et, Recognition of Motor Imagery EEG Based on Deep Belief Network[J], Volume 44(6),2015,Pages 717-721.

[5] Shi Xu-gan, ZHANG Shi-qing, ZHAO Xiao-ming, Facial Expression Recognition by Integrating Deep Belief Networks with Multi-layer Perceptron[J], Vol.36(7),2015,Pages 1629-1632.

[6] Mohamed Elleuch, Najiba Tagougui, Monji Kherallah, Optimization of DBN using Regularization Methods Applied for Recognizing Arabic Handwritten Script, Procedia Computer Science, Volume 108, 2017, Pages 2292-2297

[7] Olivier Pauplin, Jianmin Jiang, DBN-based structural learning and optimisation for automated handwritten character recognition, Pattern Recognition Letters, Volume 33, Issue 6, 15 April 2012, Pages 685-692

[8] Hinton G E, SALAKHUTDINOV R. Reducing the dimensionality of data with neural networks[J].Science,2006,313(5786):504-507.

A Multi-objective Optimization Behavior Fusion Avoidance Method for Snake-like Rescue Robot

Li Hongyan [1,*], *Hou Yuanbin* [1]

[1] Xi'an University of Science and Technology, College of Electrical and Control Engineering, 710054 Shaanxi Xi'an, China

Abstract. The obstacle avoidance path planning of rescue robot has been the key problem of autonomous navigation for rescue robot. Most of the existing path planning algorithms is aimed at the optimization of a particular target, often resulting in deadlock or oscillation in order to obtain the optimal solution or the optimal path. Therefore, using the multi-objective optimization theory, this paper presents a multi-objective optimization dynamic obstacle avoidance algorithm based on interval weights, the fusion process of obstacle avoidance process is divided into three seed action, by giving the behavior function of different weights, the output rate of real-time dynamic change of different action. This is not to obtain the optimal solution or the optimal path at the current moment, but to obtain only the most efficient solution or the most satisfactory path at the current moment. Experiments show that the algorithm can effectively improve the flexibility and security of the obstacle avoidance process under the premise of ensuring the real-time and robustness of the search and rescue robot obstacle avoidance process.

1 Introduction

Snake-like robot is one of the branches of bionic robot in recent years. Unlike robots that traditionally use wheels, legs, or tracks, the snake-like robot can imitate the movements of snake, and realize the no limb movement through the twisting and stretching of the body. It has the advantages of good stability, small cross sectional area and high flexibility.

Autonomous obstacle avoidance is an important part of autonomous navigation for snake-like rescue robot. Nowadays, scholars have put forward many different obstacle avoidance theories according to different navigation environment. The typical methods include Curvature rate method[1],Vector field histogram method[2-3], Dynamic window method[4-5], Neighborhood histogram method[6], Limit cycle method[7], these methods can improve the robustness of obstacle avoidance in different degrees, but they all need time to calculate the optimal path or the optimal solution, it will sacrifice some time. Even deadlock or shock phenomenon, affect the real-time robot obstacle avoidance.

Therefore, this paper proposes a real-time obstacle avoidance algorithm based on multi-objective optimization for snake-like rescue robot. This algorithm does not need to calculate the current optimization path, but only the most efficient solution or the most satisfactory path is needed. Under the premise of safe obstacle avoidance, the real-time and robustness of the obstacle avoidance process are effectively improved.

2 Obstacle avoidance process behavior function

The obstacle avoidance process of snake-like rescue robot can be divided into three sub behaviors: fast forward, keep target direction and real-time obstacle avoidance. In the process of obstacle avoidance navigation, the robot determines the most appropriate behavior function at the current moment according to the weights of each sub behavior and the priority parameters, so as to ensure the security and stability of the obstacle avoidance process.

2.1 Move forward quickly

The purpose of this behavior function is to make the snake-like rescue robot move to the target as fast as possible. Its objective function is a linear function about the translational speed, moving forward rapidly, and the sub function can be expressed as $b_{mf}(v,\omega)=a(v/v_{\max})+b$ .In the formula, V is the velocity, and ω is angular velocity. When the empirical constant a=1, b=0, the range of the target function varies between [0, 1], and is directly proportional to the V value, irrespective of the angular velocity ω.

2.2 Keep the target direction

Keeping the target direction behavior function requires the robot to keep the same direction as the target in the

* Corresponding author: lihongyan@xust.edu.cn

process of obstacle avoidance so as to ensure that the robot can accomplish the task in the shortest distance. The higher the value of the behavior function, the more likely the robot will be in the same direction with the target at the current moment. Conversely, the less likely the same direction is. Thus, the behavior function that maintains the target direction selects a Π function as the behavior map, that is, $\prod(x) = 1/\left\{1+\left[(x-\gamma)/\beta\right]^2\right\}$. In the formula, the angle between the forward direction and the target direction of the robot is determined by γ, which determines the extremum of the function; β is the empirical constant, which determines the size of the function graph window. In the function curve shown in figure 1, $\gamma = 0^o, \beta = 45^o$.

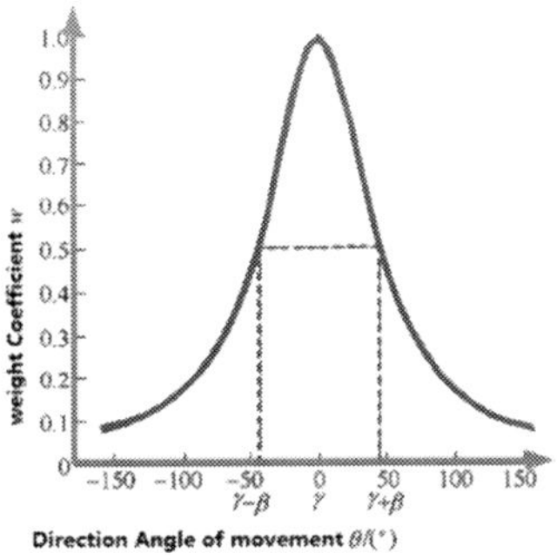

Fig. 1. Plot of the Π function

According to the target position, the snake-like rescue robot determines the moving direction of each time toward the target, and its behavior function can be expressed as $b_{mh}(v,\omega) = \prod(\theta(v,\omega)) = 1/\left\{1+\left[(\theta(v,\omega)-\theta_{target})/\beta\right]^2\right\}$.

In the formula, $\theta = \theta_0 + \omega T$ indicates that the robot is expected to move in the direction of (v,ω) control, θ_0 is the initial direction of motion, T is the control instruction cycle, and θ_{target} is the direction of the robot when it reaches the target point. If the higher the objective function value is, the more likely the robot will be in the same direction as the target points at the current moment; otherwise, the smaller the likelihood is.

2.3 Real-time obstacle avoidance

The purpose of obstacle avoidance for snake-like rescue robot is to make the robot bypass the obstacle safely and steadily. Figure2 illustrates the obstacle avoidance process of a robot passing through any set of static obstacles.

A snake-like rescue robot can be viewed as a cylinder in which R is the radius of the robot cylinder; $A(x_0, y_0)$ is the center of the robot arc motion; φ is the angle between the A point to the coordinate origin line and the A point to the barrier $P_i\ (x,\ y)$; d_i is the barrier spacing, and the D is the actual distance from the robot to the obstacle. The obstacle avoidance behavior function can determine a set of control parameters (v,ω) with the shortest safe path for a robot according to the minimum distance of a set of distance obstacles. The obstacle avoidance behavior function can be defined as

$$b_{ao}(v,\omega) = \min\left\{D(v,\omega,d_i)\middle| i \in N\right\}$$

Considering the size of the robot and the position of the obstacle, the obstacle avoidance function can be expressed by piecewise function:

$$D = \begin{cases} (v/\omega)\varphi \quad \omega \neq 0 (v/\omega - r) \leq \sqrt{(x-x_0)^2+(y-y_0)^2} \\ \leq (v/\omega + r), \max(d_i) > 2r \,|\, i \in N \\ \sqrt{x^2+y^2} \quad \omega = 0; |x| \leq r \\ \infty \quad \omega \neq 0; \max(d_i) < 2r \,|\, i \in N \end{cases}$$

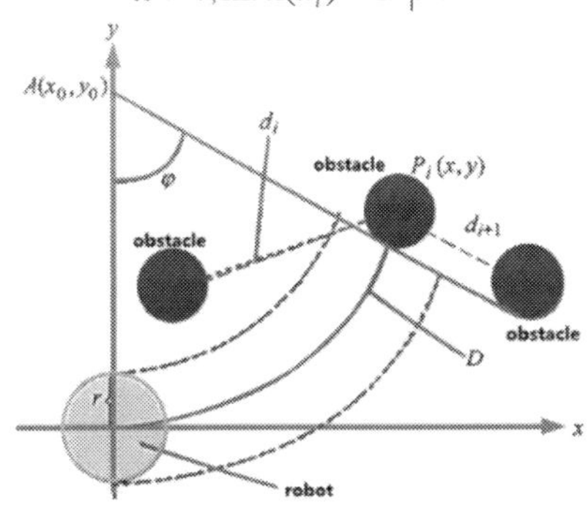

Fig. 2. Plot of obstacle avoidance process

The obstacle avoidance trajectory of snake-like rescue robot is related to V, ω, obstacle position P, and location and size of robot itself, Therefore, the segmentation function of robot obstacle avoidance function is discussed.

The first segment value in the segmentation function represents the arc path of the robot to make the arc motion. At this point, the robot moves in an arc to the front of the obstacle, sensing the distance information of the obstacle in front of the airborne infrared sensor, and adopting the appropriate obstacle avoidance strategy;

The second segment indicates that the robot makes a uniform rectilinear motion trajectory. At this point, the obstacle is very close to the robot, and the robot's actual moving distance is the distance between the obstacle and the coordinate origin;

The third segment indicates that the robot airborne sensor detects very small distances between the obstacles and the robot body cannot pass through the barrier.

3 Multi-objective optimization problem implementation

The simultaneous optimization problem with multiple objectives (or indexes) is called a multi-objective optimization problem or a vector extremum problem. From the point of view of mathematical theory, the multi-objective optimization problem can be defined as follows

$$\max[o_1(x), o_2(x), \cdots o_n(x)] \quad x = (x_1, x_2, \cdots x_N) \in R^N$$

In the formula, x is a set of feasible action sets, and $o_n(x)$ is the action set credibility. According to the multi-objective optimization method, the navigation process of an autonomous mobile robot specifying the starting and ending points can be decomposed into three sub actions, namely, obstacle avoidance, keeping the target direction and moving rapidly toward the target. As shown in figure 3.

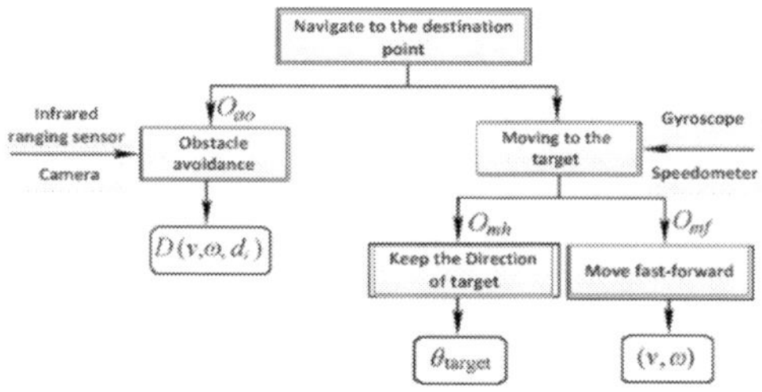

Fig. 3. specifies the navigation process behavior decomposition of the target

According to the decision theory, the weights, priorities and other factors are taken into account. A weighted objective programming method with dynamic interval weight is proposed to obtain the satisfactory solution of the system at present:

$$(v^*,\omega^*)=\arg\max_{(v,\omega)\in R^N}[w_{mf}b_{mf}(v,\omega)+w_{mh}b_{mh}(v,\omega)+w_{ao}b_{ao}(v,\omega)]$$

$$w_{mf}+w_{mh}+w_{ao}=1,\ w_{mf}\in[o,0.1],\ w_{mh}\in[o,0.4],\ w_{ao}\in[o,1].$$

In the formula, w_{mf}, w_{mh}, w_{ao} respectively represent fast forward, keep the target direction and the obstacle avoidance behavior of three sub function weights.

The weighted values of the sub behaviors of the algorithm can be adjusted dynamically within a given interval, so as to ensure that the mobile robot can obtain the most efficient solution in real time.

4 Reliability analysis of multi output obstacle avoidance system

Definition1(multiple behavior output system).If a control system with n behavioral parameters has at least k successful output, and then the system is called $k-out-of-n$ system [8]. If the outputs of each subsystem in the system are independent of each other, the reliability function of the system can be expressed as:

$$R(p_1,\cdots,p_n)=P\{X_i=1|i=k,\cdots,n\}=\sum_{l=k}^{n}\sum_{x\in\Omega(l,n)}\{\psi(x)|k=1,2,\cdots,n\} \quad (1)$$

$$=\sum_{l=k}^{n}\sum_{x\in\Omega(l,n)}\{\prod_{i=1}^{n}\Delta_i(x)|k=1,2,\cdots,n\}$$

In the formula, $\psi(x)$ represents the probability that $x\in\Omega(k,n)$ is satisfied. $\Omega(k,n)$ represents all subsets of k behaviors that are successfully output in n behavior. If the system succeeds in output, then the k behavior value is 1, and the n-k behavior value is 0.

For example, $\Omega(2,3)=\{(1,1,0),(1,0,1),(0,1,1)\}$, the function $\Delta:\{0,1\}^n\to[0,1]$ is defined as

$$\Delta_i(x)=\begin{cases}P_i & x(i)=1\\ 1-P_i & otherwise\end{cases}$$

Aiming at the snake-like rescue robot obstacle avoidance system (2-out-3) with 3 sub outputs, the independence of obstacle avoidance behavior is analyzed, if the behavior of reliability for P_1, P_2, P_3, the formula (1) can be used to calculate the system reliability[9-10].

$$R(p_1,p_2,p_3)=\sum_{l=2}^{3}\sum_{b\in\Omega(l,3)}\prod_{i=1}^{3}\Delta_i(b)=\sum_{b\in\Omega(2,3)}\prod_{i=1}^{3}\Delta_i(b)+\sum_{b\in\Omega(3,3)}\prod_{i=1}^{3}\Delta_i(b)$$

$$=\sum_{b\in\{(1,1,0),(1,0,1),(0,1,1)\}}\prod_{i=1}^{3}\Delta_i(b)+\sum_{b\in\{(1,1,1)\}}\prod_{i=1}^{3}\Delta_i(b)$$

$$=\prod_{i=1}^{3}\Delta_i(1,1,0)+\prod_{i=1}^{3}\Delta_i(1,0,1)+\prod_{i=1}^{3}\Delta_i(0,1,1)+\prod_{i=1}^{3}\Delta_i(1,1,1)$$

$$=p_1p_2(1-p_3)+p_1(1-p_2)p_3+(1-p_1)p_2p_3+p_1p_2p_3$$

If X represents the successful output of child behavior in n behavior, the reliability of the multi output system can be expressed as

$$R(p_1,\cdots,p_n)=\sum_{l=k}^{n}P\{X=l\} \quad (2)$$

Which contains the parameter $(n,\ p)$ binomial random variable function p the probability form can be expressed as $p(i)=P\{X=i\}=\binom{n}{i}p^i(1-p)^{n-i}$, $i=0,1,\cdots,n$.

When $p=p_1=p_2=\cdots=p_n$, a binomial random variable with parameters $(n,\ p)$ can describe the reliability of a multi output system.

Theorem 1: for a given behavior set, the reliability of $k-out-of-n$ systems has the following properties: $R(k)\ge R(k+1)$, $k=0,1,\cdots,n$.

Proof: formula (1) can be written in recursive form $R(k)=R(k+1)+\sum_{x\in\Omega(k,n)}\psi(x)$. Since $\sum_{x\in\Omega(k,n)}\psi(x)\ge 0$, then $R(k)\ge R(k+1)$ is proved.

So far, it can be seen that the reliability of the obstacle avoidance system can be expressed as the mean function of the reliability of each sub member. Therefore, the reliability of the whole system can be reflected by changing the reliability of the sub behaviors in the system.

5 System debugging and analysis

5.1 On-board sensor accuracy test

In order to verify the effectiveness and real-time of snake-like rescue robot obstacle avoidance algorithm based on multi-objective optimization. Using the snake-like rescue robot DK-Snake-II developed by the embedded Laboratory of Xi'an University of Science And Technology, the obstacle avoidance navigation test has been carried out. The snake-like rescue robot head equipped with infrared sensors and visual sensors to obtain obstacle distance information within 80cm by infrared sensor[11], on the basis of the obstacle avoidance behavior conduct dynamic adjustment.Figure 4 shows the obstacle avoidance process, a set of infrared sensor ranging error statistics. The maximum distance of infrared sensor is 80cm, the effective range is 25~ 45cm, and the range error is 5.62cm.The range of small range error can ensure that the rescue robot avoids collisions due to odometer or visual error, and improves the security and robustness of obstacle avoidance.

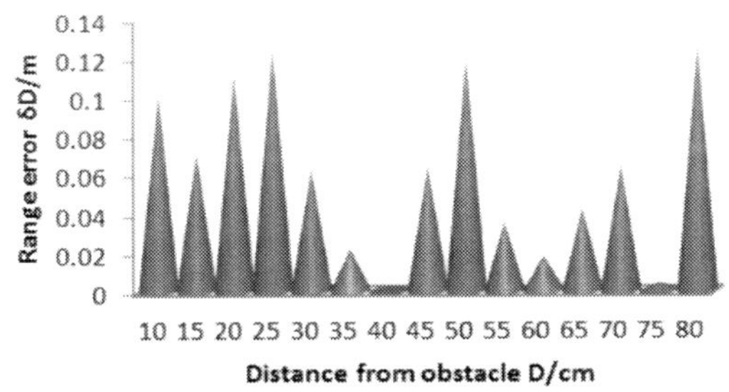

Fig. 4. Infrared ranging error distribution map

5.2 Obstacle avoidance process analysis based on different decision algorithms

Snake-like rescue robots use four different decision methods to avoid static obstacles, as shown in figure 5.

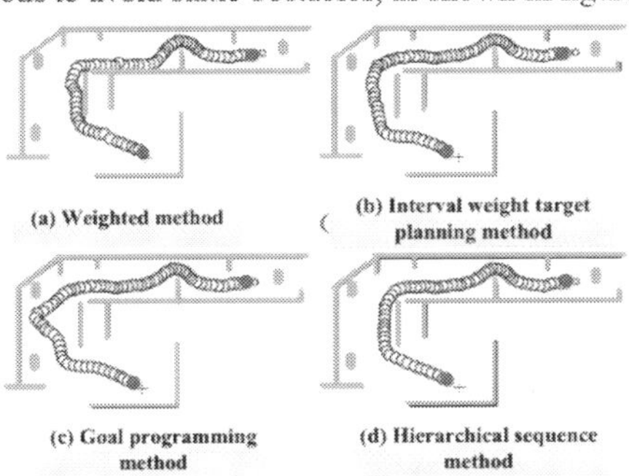

Fig. 5. Different decision-making methods obstacle avoidance algorithm to avoid static obstacles

From the figure, the interval weight goal programming method and hierarchical sequence method of obstacle avoidance trajectory is smooth, but the hierarchical method and the weighted method of obstacle avoidance radius is too small, the process of moving the robot velocity and angular velocity are not easy to control, easy to collide, While the interval weight goal programming method has long moving distance, but the trajectory is smooth and the safety factor is high. It is easy for the fast moving robot to adjust the body posture.

In view of the four multi-objective decision making methods, multiple experiments are carried out in different environments with different obstacle avoidance complexity and different obstacles placed. Table 1 presents the comprehensive performance index of the multi objective optimization obstacle avoidance method. In this table: A_{bz} stands for obstacle avoidance safety; V_{bz} for obstacle avoidance; robot movement speed; t represents avoidance obstacle response time; C_{bz} for obstacle avoidance success; ϕ is trajectory smoothing angle.

Table 1. Comprehensive performance of obstacle avoidance based on multiple object optimization.

Decision Method	A_{bz} /cm	γ/(°)	V_{bz}/(cm·s^{-1})	t/s	C_{bz}/%	ϕ/(°)
Weighted method	38.60	33.40	21.70	0.23	77	3.1
Hierarchical sequence	36.50	44.30	29.30	0.24	83	3.2
Interval weighted goal programming	41.20	37.60	24.00	0.36	86	3.3
Goal programming	43.50	47.90	28.00	0.25	82	3.7
Average Value	39.95	40.80	25.75	0.27	82	3.325

5.3 Obstacle avoidance test analysis

Figure 6 shows the process of obstacle avoidance snake-like rescue robot based on multi-objective optimization.

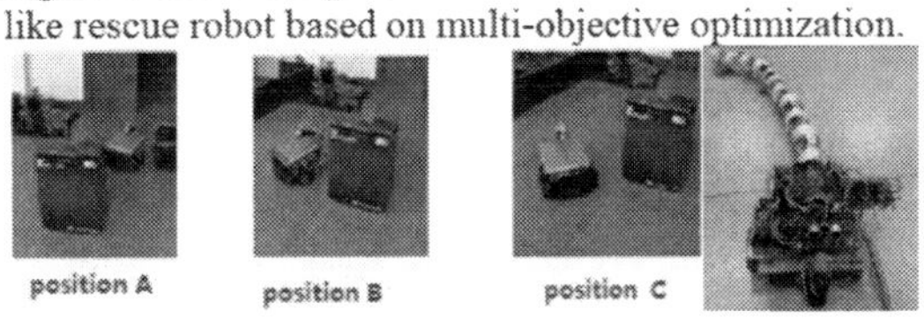

Fig. 6. Rescue robot head obstacle avoidance process

During the obstacle avoidance of the snake-like robot head starting from the initial position of A, bypassing the obstacle position B, enter the safe position of C.

Under the control of three sub functions, the robot selects a path which is most suitable for avoiding obstacles and adjusts its pose to the target position.

The position of the three positions relative to the target point, the slip angle and the sampling period are as follows:

Location A indicates $\theta_0 - \theta_{target} = 0.879\pi, T = 0.045$

Location B indicates $\theta_0 - \theta_{target} = 0.007\pi, T = 0.052$

Location C indicates $\theta_0 - \theta_{target} = 0.896\pi, T = 0.039$

5.4 Reliability analysis of multi-objective optimization output system

The simulation test of autonomous obstacle avoidance system for snake-like rescue robot with three output behaviors is carried out. Figure7 gives the mean function curve of system reliability at k=5 and k=10 respectively. Among them, the reliability of each subsystem is equal. That is $P(i) = 0.5, i = 0,1,\cdots,n$.

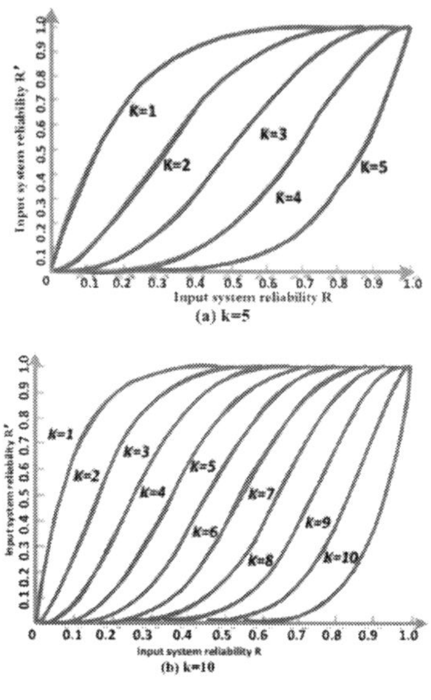

Fig. 7. Multi-behavior output system reliability relation of input and output system

As can be seen in figure 7, after multi-behavior fusion output system reliability can be represented as a function of the input system reliability behavior means, and increases with the increase of behavior reliability mean, or input system with low reliability behavior can be constructed with high reliability of the system output. The larger the K value is, the lower the reliability of the output system is. At the same mean time, the reliability of the system $k-out-of-n$ is higher than that of the $k+1-out-of-n$ system when $k=0,1,\cdots,n$.

Therefore, in the obstacle avoidance system based on multi behavior fusion, the proper number of sub behaviors should be determined. It is necessary to ensure the high reliability of the fusion system, and to ensure that multiple sub behaviors cannot bring heavy burden to the system.

6 Conclusion

In this paper, the autonomous navigation system for mobile robot due to too much time to seek the optimal path or the optimal solution of deadlock or shock phenomenon, put forward the snake-like rescue robot behavior of a dynamic weighted fusion algorithm of obstacle avoidance.

The algorithm uses the multi-objective optimization method to obtain the most efficient solution of the rescue robot, and divides the mobile robot's obstacle avoidance navigation process into three subsystems. By dynamically changing the weights and priorities of the sub behavior function, the most satisfactory path or the most effective path at the current time is obtained in real time.Experimental results show that the algorithm can effectively improve the security and smoothness of obstacle avoidance navigation, and improve the reliability of the system, while ensuring the robustness of the obstacle avoidance process.

References

1. SIMMONS R. The curvature velocity method for local obstacle avoidance[C] Proceedings of the IEEE International Conference on Robotics and Automation,Minnesota,Minneapolis:IEEE CNF,1996:3375-3382.
2. SHI Dongqing,COLLINS E G,DUNLAP D. Robot navigation in cluttered 3-D environments using preference-based fuzzy behaviors[J]. IEEE Transactions on Systems,Man., and Cybernetics, Part B,2008,37(6) :1486-1499.
3. KIRBY R, SIMMONS R, FORLIZZI J. Companion: A constraint optimizing method for person-acceptable navigation[C]ROMAN2009,Toyama,Japan:IEEE,2009:607-612.
4. SEDER M,PETROVIC I. Dynamic window based approach to mobile robot motion control in the presence of moving obstacles[C] Proceedings of the 1996 IEEE International Conference on Robotics and Automation,Roma,Italy:IEEE CNF,2007:1986-1992.
5. SCHRÖTER C, HÖCHEMER M, ROSS H M. A particle filter for the dynamic window approach to mobile robot control[C] Proceedings 52nd Int. Scientific Colloquium (IWK), Ilmenau,Germany:Universitätsverlag,2007:425-430.
6. NUNKAEW W,PHRUKSAPHANRAT B. A fuzzy multiple objective decision making model for solving a multi-depot distribution problem[C] Proceedings of the International Multi Conference of Engineers and Computer Scientists,Hong Kong,China:IMECS,2010:17-19.
7. SADJADI S J,HABIBIAN M,KHALEDI V.A multi-objective decision making approach for solving quadratic multiple response surface problems[J]. Int. J. Contemp. Math. Sciences, 2008,32(3) :1595-1606.
8. PIRJANIAN P. Reliable reaction [mobile robot navigation][C].IEEE/SICE/RSJ International Conference on Multisensor Fusion and Integration for Intelligent Systems,1996:158-165.
9. Fu Chao, Zhang Lefeng. Research on autonomous obstacle avoidance for indoor mobile robots [J]. industrial control computer, 2015,28 (07): 44-46+49.
10. Yongyue, Li Shuqin. Research on multi robotic fish formation strategy planning based on [J].Computer simulation, 2013, 30 (11): 369-373.
11. Li Hongyan, Hou Yuanbin. Design of actuator and control system of snake robot based on LPC2132 [J].Journal of Xi'an University of Science And Technology, 2016,36 (02): 265-270..

The Algorithm of Habitat Discovery in Bird Migration

Zhengzheng Wei[1,*], *Zengzhen* Shao[1], *Dong* Chen[1], and *Yancong* Li[1]

[1]School of Information Science and Engineering, Shandong Normal University, Jinan, 250300, P.R.China

Abstract. Bird migration has attracted an increasing attention. The study of habitats has played a vital role in the birds migratory. Previous researches, however, have encountered many problems, such as great limitations on research methods, low data utilization rate, statistics-focused and ineffective data processing and analysis methods. In this paper, the algorithm of habitat discovery is put forward by using computer's data-mining technology based on the spatio-temporal characteristics of bird-watching data. First the algorithm detects and eliminates duplicate data to guarantee data standardization. Then density-based clustering algorithms are used to identify habitats where birds gathered. Finally the habitats of birds migratory are discovered.

1 Introduction

All groups of birds are special in that a large number of species migrate annually between their breeding and non-breeding areas [1]. The migration of birds has a great impact on the environment and production life of human beings. Studying the migration of birds can help people prevent the spread of epidemics and maintain species diversity. Therefore the study of migration habitat is crucial for people to protect the birds and natural environment and maintain species diversity.

Various methods are adopted and developed to carry out research on migratory birds from different aspects by domestic and foreign researchers in order to understand the migration patterns of birds. Among these methods, the fixed-point investigation is the earliest bird migration research method. The most common and most popular method to study the bird migration is bird-banding[2]. It can be easily implemented and widely applied. But its monitoring cycle is long and the data recycling is quite complicated [3]. Next method is the satellite positioning method. Its accurately collected data can achieve continuous tracking to the individual bird. However it is not suitable for small birds with high cost and difficult popularization as well as limited amount of data. There are also some other methods including radar monitoring, sensitive geographical location and others. But they have low precision, difficulty in popularizing, limited data and other issues. In addition, the usage and analysis of the collected migration data of birds has also attracted the attention of researchers at home and abroad. It is analysed by the early bird data that only through the track point marked in GIS by biologists or the distribution points gotten by artificial statistics can migratory lands and migration routes be acquired [4]. In 2004, the Japanese scientist Shimazaki proposed to deal with bird flight data through using the method of ISODATA clustering [5]. In this method the migratory state of birds is determined in accordance with their flight speed and further the migratory location is obtained. Nonetheless the migratory routes of birds still need to be manually marked and further the spatial location information is unable to be processed with this method. In 2010, Zhou Yuanchun and others found the birds gathering land by using the density-based hierarchical clustering algorithm to cluster bird GPS, the association of the aggregation rules and further migratory routes of birds by virtue of the Apriori algorithm or GSP algorithm [6]. However some defects that the changes of the habitats and migratory routes of birds cannot be found still exist due to small amount of data, less number of birds and short time span of the data. In 2012 Li Xueyan and others established a bird-watching database using China Birding Report and displayed the changes of bird distribution of recent years by GIS [7]. But it only relies on artificially statistical methods to mark bird discovery sites in the GIS, neither using the "quantitative observation" to analyse the data deeply, nor dealing with the problems of repeated sampling and uneven distribution for the bird-watching data.

From the above there are still a lot of problems that need to be solved in the above researches which are embodied in: 1) the collected data shows the following problems such as incomprehensive, imprecise, and limited amount; 2) there is no much work on data standardization of source data; 3) the amount of data for analysis and study is relatively small; 4) the hidden knowledge in the data fails to be dug out.

In this paper, the problem of traditional biology is abstracted as a computational problem. A feasible, efficient and general method is sought to solve the above problems, to achieve the effective treatment and utilization of bird-watching data, to find the habitats of migratory bird. This method can make up the shortcomings of the research of bird migration in China.

DOI: 10.1051/matecconf/201713900156

2 Data source

As the important supplementary information of traditional bird distributions, Chinese bird-watching data is comprehensive and reflects Chinese bird watching achievements accurately. These data comes from three aspects. The first is from the network such as Bird Report (www.birdreport.cn) [8] and China Bird Watching Network (www.chinabirdnet.org) [9] etc. The second comes from ornithological books and literature such as China Bird Report 2003-2007 [10], China Coastal Waterbird Census Report 2005–2011 [11]and *A Checklist and Distribution of the birds in Shandong* [12]etc. The third is provided by many ornithologists led by Prof. Sai Daojian. Total 189350 bird watching records have been verified by ornithologists that insure the authority of these data.

2.1 Data characteristics

The spatial-temporal information of Chinese bird-watching data records including species, date, location, number and observer in detail as shown in Fig. 1. The following "Number" records the number of birds observed in each bird watching.

Fig. 1. Chinese Bird-Watching Data

2.2 Data Statistics

The 239350 Chinese bird-watching records involve 24 Orders, 100 Families, and 1230 Species accounting for 85.7% of China's existing bird species [13]. The range of these records covers 34 provinces, municipalities and autonomous regions, including Hong Kong, Macao and Taiwan approximately 46 years ranging from 1970 to the present. Seeing from Chinn's bird-watching records, the number of bird-watching records is more in the east than that in the west, and more in the south than that in the north. The proportions of bird-watching records also vary with years. The number of records from 2001 to 2016 accounts for more than half of the total. In addition, among the whole bird-watching records, there are 142078 records of migratory bird which accounts for 59.36% of the total. The remaining records are for non-migratory bird.

Although China's bird-watching records are distributed in the species, time and space, the advantages of bird-watching data are obvious such as low cost, bulk information, easy access, various species coverage, substantial data, long time span, high accuracy and convenient spatial-temporal data analysis [5].

3 Algorithm

3.1 Overview of Algorithm

Bird migration is a relatively long and complex process which reflects in spatial and temporal changes. The study of habitats has played a vital role in the birds migratory. In this paper Chinese bird-watching data is used as the data source. First the important attribute --"Number" (Fig 1) is introduced to solve the problems of repeated sampling and uneven sampling distribution based on the data characteristics of bird-watching records. The quality of the data will be improved. Then according to the temporal and spatial structure of the migration trajectory, the time and space attributions of the bird-watching records are dealt with separately. The potential information of the spatial -temporal data of bird is discovered. As a consequence, the quality of data mining will also be improved to some extent. This study will lead to more efficient utilization, processing and analysis of bird-watching data and provide new perspectives and new ideas for the study of bird migration in China.

Moreover migratory birds are employed as the study objects. This Algorithm is used to achieve two goals of the research: 1) to solve the problems of duplicate sampling and uneven distribution of sampling data; 2) to identify the habitats of migratory birds; These steps of the Algorithm are described in the following:

1) the bird-watching data is digitized and stored in the database as the form of GPS track point; 2) the temporal and spatial distance is calculated between the points to discover implicit duplicate data and the special points instead of the duplicated data are used to produce a standardized data set; 3) the preprocessed new trajectory point collection is clustered by the density-based clustering algorithm to obtain the high density area of the migratory activity that is used as a habitat during the migration.

The relevant definitions and steps of the Algorithm are described as follows in more detail.

3.2 Digitizing the Text into GPS Trajectory Points

Each bird-watching record contains its unique spatial-temporal data information. In order to better show the distribution and migration of birds each bird-watching record is abstracted into coordinate points with times before the mining and analysis steps. Each record corresponds to a coordinate point. This tracing point that can stand for an individual or a group is used to display the bird distribution and migration. The textual information of "bird-watching site" in each bird-watching record needs to be converted into the latitude and longitude coordinates of GPS in order to facilitate compare calculation and presentation. API interface of Baidu map with high accuracy is applied to the textual information of "bird-watching site" in each bird-watching record. The coordinates of Baidu Map can be accurate after the 10 decimal point. By this Algorithm, the Migratory birds' corresponding coordinates are retrieved.

Definition1. Trajectory point of moving object is to describe the moving object sampling including three parts of latitude, longitude and timestamp and moving object identification which is expressed as pi (ej) = <(xi, yi), ti, ej>, pi (ej) ∈ P (ej), ej ∈ E, j ∈ [1, J], i ∈ [1, n] where (xi, yi) is the track point position component. For example, ej = Hirundo rustica. A trajectory point is expressed as pi(ej) = <(xi, yi), ti, ej> where (xi, yi) are coordinates and ti is a time stamp. pi(ej)∈P(ej), j∈[1,J], i∈[1,n].

Definition2. Migratory trajectory of migratory bird: a sequence of spatial locations with time stamps is called a migratory trajectory of birds. A migratory trajectory can be expressed as: P(ej) = {p1(ej), p2(ej),..., pi(ej),..., pn(ej)} where pi(ej) is a sampling point of the trajectory and n is the number of sampling points. Ej is the moving object (event), ej ∈ E, E = {e1, e2, ..., ej, ..., eJ} is the set of moving objects (events), j ∈ [1, J] and J is the moving object Event).

3.3 Selection of Feature Point

There is a considerable duplication in the original bird-watching data. If these duplicate data could not be removed, it would affect the quality of data mining. Therefore the original trajectory points of migratory bird should be pre-processed and removed duplicates before analysis of the data. The problem of uneven distribution of bird-watching records is solved initially, which lays the solid foundation for further analyzing of the data.

There are two types of duplicate data in bird-watching data. The first type is explicit duplicate data manifesting the same time and place, namely pi(ej)=pi+1(ej). For such duplicates simple merge processing is needed. The second type is implicit duplicate data which is not easy to find. When a kind of bird repeatedly sampled over a continuous interval and a small regional range, its sample can be considered repeated. They manifest multiple sampling points of a bird species with similar temporal and spatial characteristics.

Definition3. Duplicate Points Set (DPS): the set of the tracking point of a certain type of moving object P'(ej) = {pu(ej), pu+1(ej),...,pi(ej), ..., pv(ej)}, j ∈ [1,J], i ∈ [u,v]. If Dist(pu, pi)≤ 2 θ r, Dist(pu, pv+1)>2 θ r, and tv - tu< θ t, the set P'(ej) is called a Duplicate Points Set (DPS) marking asP'(ej)={pu'(ej), pu+1'(ej), ··· ,pv'(ej)}. The sequence {P1'(ej), P2'(ej),...,Pk'(ej),...,PK'(ej)} is called the set of DPSs where K is the number of DPS and k ∈ [1,K]. The tracking points of the DPS are implicit duplicate data. Fig 2 shows the DPSs.

Definition4. Feature Point (FP): In a DPS, a single point P'(ej) that can be used to replace P'(ej) is called a Feature Point (FP). The FP can be expressed as: sk(ej) = <(xk, yk), tu, tv, ej> where tu is the starting time of the DPS, tv is the ending time and (xk, yk) is the coordinate of the FP, k∈[1,K].

The steps of the selection of Feature Point are as follows:

Step1: Assign the values of θr and θt;

Step2: A point pi (ej) is chose as a center arbitrarily in the set P(ej) = {p1(ej), p2(ej),..., pi(ej), ..., pn(ej)};

Step3: Calculate the distances of the remaining points except the central point of the set P(ej) to pi(ej);

Step4: If Distance (pother, pi)≤ θr && |tother – ti|< θt, then the point pi(ej) is added to the DPS P'(ej);

Step5: Output P'(ej);

Step6: The k-mediods algorithm (Han et al. 2012) is used in the DPSs P'(ej). The cluster of k-mediods algorithm is set to 1. The central point of cluster sk (ej is used as the feature point of pi (ej), sk (ej) = <(xk, yk), tu, tv, ej>;

Step7: Calculate the weighted average of points in P'(ej). The weight average of a point is the value of "Number" that is used as a new weight of sk (ej);

Step8: Repeat Step2-Step7 for each point in P'(ej) until all FPs are output;

Step9: Rearrange the feature points and the tracking points that are not in the DPSs. Then a set of new trajectory points about ej is got, S(ej) = {s1(ej), s2(ej),..., sk(ej),...,sm(ej)}, m is the number of the new trajectory points;

Step10: Output S(ej).

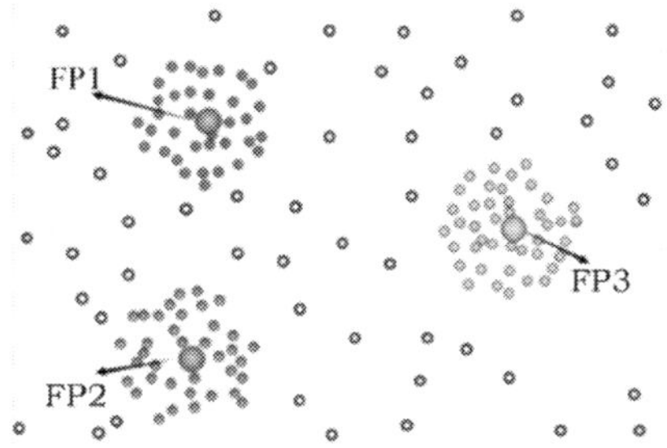

Fig. 2. Duplicate Points Sets (DPS)

3.4 The Discovery of Habitats

There are many stopover sites in the route of birds migrant which consist the habitants with wintering ground and breeding place [14][15]. Within the habitat, the number of bird populations is usually larger than that of the other regions [16][17].That is a place that has larger amount and higher density of birds is much more important to birds themselves which is more likely to become the potential habitats for birds. Following this idea, the area where birds are densely distributed must be found after the pre-processing of the new trajectory points. A density-based clustering algorithm is employed in the new trajectory points S(ej) = {s1(ej), s2 (ej),..., sk(ej),...,sm(ej)} to find the high density areas for bird activity. And the value of "Number" is introduced to calculate the values of each cluster and outlier. Then high-density and high-volume areas are screened out as habitats for migratory bird. Fig 3 shows various types of clusters.

 DOI: 10.1051/matecconf/201713900156

Definition5. Heat: The importance of a point or region is called the Heat Degree (HD). The "number of birds" is used as the weight of the point which is called the Heat Degree Point (HDP). If it has higher weight, the heat range will be larger. The greater the sum of HDPs is in a region, the greater the HD of the region will be.

The steps of the discovery of habitats are as follows:

Step1： The DBSCAN algorithm (Han et al. 2012) is used in the new trajectory points sequence of ej, S(ej) = {s1(ej), s2(ej),…,sm(ej)} to get various shapes of clusters, denoted as C(ej)={c1(ej),c2(ej),…,cl(ej),…,cL(ej)}, l ∈ [1,L], L is the number of the clusters;

Step2： The "number of birds" is used as the weight to calculate the HD of each cluster: the sum of multiplication of all points and the weights in a cluster cl(ej). The HD of a cluster is called the HDC;

Step3： Calculate the HDP of each outlier point;

Step4： Arrange all HDPs and HDCs in ascending order. If HD>MinHeat (the value of MinHeat can be set by the user) the point or cluster will be output that is used as the habitat of migratory bird ej: D(ej)={d1(ej),d2(ej),…,dl(ej),…,dL(ej)}, l∈[1,L];

Step5： Repeat Step1 to Step4 for E ={e1, e2,…, ej,…, eJ} until all habitats are output.

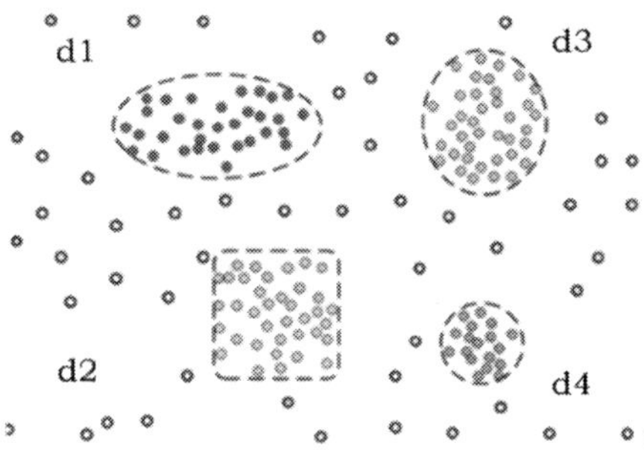

Fig. 3. Clusters of Arbitrary Shape

3.5 Time complexity of Algorithm

There are two stages in this Algorithm. The time complexity of each stage will be separately analysed.

The k-mediods algorithm is used when selecting Feature Points (FP). The time complexity of the k-mediods algorithm is O (k (n-k) 2). In this paper setting k=1, n is the total number of trajectory points, S is the total number of points in the Duplicate Points Set (DPS) and S < n. The time complexity of algorithm is O ((S/K) 2) when executing once and K is the number of Duplicate Points Sets (DPSs). This needs to be performed K times and K<<n. Therefore the time complexity of this stage is O (S2/K) < O (n2).

Then in the discovery of habitats the improved DBSCAN algorithm was executed once. Hence its time complexity is O (m2) where m is the number of input trajectory points and m<n (n is the total number of trajectory points).

In conclusion the time complexity of the Algorithm is: O (S2/K) + O (m2) < O (n3). This shows that this Algorithm can be completed in a relatively short time.

4 Experiment

4.1 Experimental environment

Bird-watching data is used for studying migratory bird in China thus the algorithm of habitat discovery is proposed. In the environment barn swallows (Hirundo rustica) is used as examples to explore their habitats. The feasibility and effectiveness of the algorithm is verified by comparing with the results of authoritative ornithological literatures *A Field Guide to the Birds of China*[18] and *A Checklist on the Classification and Distribution of the Birds of China[19]*.

The operating environment of this experiment is windows7 operating system and C # language is used to write the algorithm. The software development environment is Microsoft Visual Studio 2010 and SQL server 2010.

4.2Experiment One: Selection of Feature Point

The authors adopt the Feature Point (FP) to replace duplicate sampling points in the trajectory data of migratory bird.

Fig 4 shows partial data of a Duplicate Points Set (DPS) of Hirundo rustica from May 2, 2008 to May 9, 2008.

BirdsChinName	ActionTime	BirdsX	BirdsY
Hirundo rustica	2008/5/2 0:00:00	118.823466	31.97004
Hirundo rustica	2008/5/2 0:00:00	118.836165	31.072
Hirundo rustica	2008/5/2 0:00:00	118.781166	31.09214
Hirundo rustica	2008/5/3 0:00:00	118.836111	31.07212
Hirundo rustica	2008/5/3 0:00:00	118.856166	31.05534
Hirundo rustica	2008/5/3 0:00:00	118.800006	31.07215
Hirundo rustica	2008/5/4 0:00:00	118.8361	31.06214
Hirundo rustica	2008/5/4 0:00:00	118.743656	31.987115
Hirundo rustica	2008/5/4 0:00:00	118.801935	31.016251
Hirundo rustica	2008/5/4 0:00:00	118.685015	31.985155
Hirundo rustica	2008/5/5 0:00:00	118.713761	31.027857
Hirundo rustica	2008/5/5 0:00:00	118.809197	31.939081
Hirundo rustica	2008/5/5 0:00:00	118.78735	31.936139
Hirundo rustica	2008/5/6 0:00:00	118.779405	31.943003
Hirundo rustica	2008/5/6 0:00:00	118.781601	31.996915
Hirundo rustica	2008/5/6 0:00:00	118.757454	31.043939
Hirundo rustica	2008/5/6 0:00:00	118.760904	31.091918
Hirundo rustica	2008/5/8 0:00:00	118.698813	31.005794
Hirundo rustica	2008/5/8 0:00:00	118.752855	31.986915
Hirundo rustica	2008/5/9 0:00:00	118.763203	31.927214
Hirundo rustica	2008/5/9 0:00:00	118.843692	31.992015
Hirundo rustica	2008/5/9 0:00:00	118.794249	31.952808
Hirundo rustica	2008/5/9 0:00:00	118.765865	31.956729

Fig. 4. Partial data of a DPS (Hirundo rustica)

In the light of the bird-watching data and spherical distances based on latitude and longitude coordinates the following parameters are shown in Table1. After experimental tests the clustering results are best when θr=6 km and θt=10 days.

Table1. Parameters of DPS

NO.	Event E	θr(km)	θt(day)
1	ej	6	10
2	ej	6	15

3	ej	10	10
4	ej	10	15
5	ej	3	10
6	ej	3	15

When ej= Hirundo rustica the trajectory points in Fig 4 (on May 2, 2008 to May 9, 2008, near Nanjing, Jiangsu Province) are marked on the map shown in Fig 5a. After clustering the canter of the cluster – FP is gotten (Fig 5b) which is used to replace all DPSs by this method. After data pre-processing, the trajectory points become fewer (shown in Fig 5d) than that of the original (shown in Fig 5c).

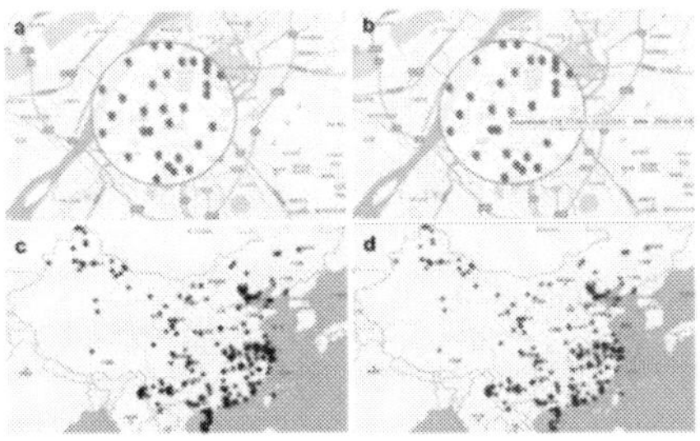

Fig.5. Data pre-processing of the Trajectory Points (Hirundo rustica).
a Duplicate Points Set. b Selecting the Feature Point. c The Original Trajectory Points d New Trajectory Points after Data Pre-processing

4.3 Experiment Two: The discovery of habitats

For the new trajectory points S(ej) after preprocessing the density-based clustering algorithm and Heat Degree (HD) formula is applied to get habitats for avian migration.

When ej= Hirundo rustica 25 clusters C(Hirundo rustica) can be gotten through DBSCAN clustering that is shown on the map Fig 6a. The HD of each point is then calculated and shown on the heat map Fig 6b. Finally 11 habitats are output. D (Hirundo rustica)= {d1, d2, d3, d4, d5, d6, d7, d8, d9, d10, d11}, their HD={369, 1287, 1885, 1221, 287, 155, 177, 159, 231, 189, 122} (Fig 6c).

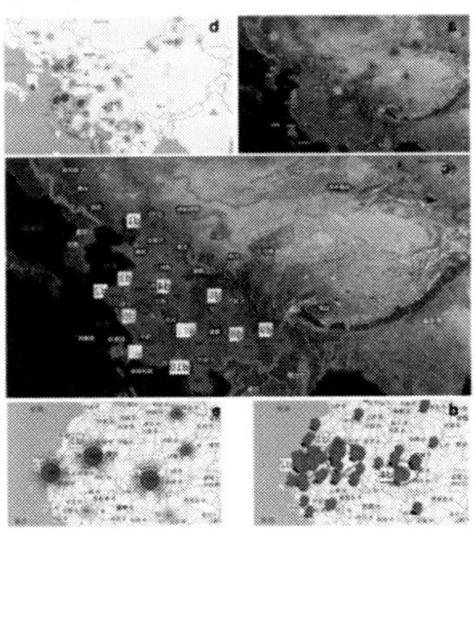

Fig .6. The Discovery of Habitats (Hirundo rustica)
a The Clusters of Trajectory Points after DBSCAN. b The Distribution Diagram of the Heat Degree c Habitats d Trajectory Points within the Habitat. e Part of the Habitat

5 Discussion

5.1 Social and scientific value

This Algorithm is general, practical and convenient. Ideally it can be applied to all bird-watching data of migratory bird in China. However the algorithm depends on a certain scale of bird-watching data. If the quantity of data is greater, the results of data mining may be more precise. If the data sample is too small, the accuracy of the algorithm will decline and the distribution and migration of migratory bird cannot be reflected truly. These results are confirmed by the experiments. Therefore it is important to collect bird-watching data constantly and adequately data pre-processing is necessary before data analysis and mining. On the one hand, the algorithm solves the repeated sampling of data sets and ensures the accuracy of data mining. On the other hand, massive and redundant data is compressed by this process. Thus the efficiency of data analysis is also improved.

5.2 Limitations and shortcomings

1) Bird-watching data cannot track individual bird. Therefore it is not enough to just rely on bird-watching data itself to verify the accuracy of the results. 2) There are spatial and temporal discontinuities in bird-watching data. Hence it is difficult to solely rely on such data to analysis changes of bird migration's habitats over the years. Moreover this may insert a negative impact on the predicted results of bird migration. 3) The uneven distribution of sampling still exists which prevents deeper data analysis and mining activities.

5.3 Future directions

First bird-watching data will be collected continuously. Additional migratory bird data such as satellite-tracking data and bird-banding data will be gradually introduced to supplement further mining and analyses. Second meteorological data will be added to our study. The authors will further study the effects of climate and environment on bird migratory habitats.

Conclusion

Bird watching activity has been developing rapidly in China in recent years. This activity will help people to understand the distributions of birds and population dynamics etc. This paper analyses the problems of bird migration in-depth from the perspective of mining data. Based on Chinese bird-watching records, the algorithm of habitat discovery is proposed and used for the selection

 DOI: 10.1051/matecconf/201713900156

and discovery of habitats during the migratory process of birds. Taking Hirundo rustica as example, maps and GIS demonstrate the feasibility of the algorithm. The time complexity of the algorithm is small, resulting in its high efficiency. The migration routes and habitats of the birds derived by this work are compared with that of the authoritative ornithological literature which shows more accurate and real results and verifies the accuracy of the algorithm.

References

1. Bock WJ. Presidential Address: Three Centuries of International Ornithology. Acta Zoologica Sinica. 2004, 50(6): 880-912
2. Wang Y, Zhang ZW, Zheng GM, Li JQ, Xu JL, Ma ZJ. Ornithological Research: Review of the Past Twenty Years and Future Perspectives in China. Biodiversity Science. 2012, 20: 119-137
3. Ma ZJ. The Research Methods and Study Advances of Bird Migration. Bulletin of Biology. 2009, 44:5-9
4. Bousquet O, Luxburg UV, Rätsch G. Advanced Lectures on Machine Learning. Springer-Verlag, 2003.
5. Shimazaki H, Tamura M, Higuchi H. Migration Routes and Important Stopover Sites of Endangered Oriental White Storks (Ciconia boyciana) as Revealed by Satellite Tracking. Revista Enfermeria Del Trabajo. 2004, 3:39-43
6. Zhou YC, Tang MJ, Cui P. Research and Implementation of Data Mining Algorithm for Bird Migratory Behavior in Qinghai Lake. E-Science Technology and Application. 2010, 1:38-50
7. Li XY, Liang LY, Gong P, Liu Y, Liang FF. Revealing Bird Distribution Changes of Bird Watching in China. Chinese Science Bulletin. 2012, 57: 2956-2963.
8. Bird Report. http://www.birdreport.cn/. 2016. Accessed 23 Apr 2016.
9. Bird Watching Network of China. http://www.chinabirdnet.org/indexc.html. 2016. Accessed 1 Apr 2016.
10. China Ornithological Society. China Bird Report 2007 (in Chinese). Beijing: China Ornithological Society, 2008
11. China Coastal Waterbird Census Team. China Coastal Waterbird Census Report (in Chinese). Hong Kong: Bird Watching Society, 2011
12. Sai DJ, Sun YG. A Checklist and Distribution of the birds in Shandong. Beijing: Science Press, 2013
13. Liu Y, Wei Q, Dong L. Updated New Bird Records in China Recently. Chinese Journal of Zoology. 2013, 48:750-758
14. Ma ZJ, Li B, Chen JK. Study on the Utilization of Stopover Sites and Migration Strategies of Migratory bird. Acta Ecologica Sinica. 2005, 25:1404-1412.
15. Ma ZJ, Li B, Chen JK. Physiological Ecology of Migratory bird during the Stopover Periods. Acta Ecologica Sinica. 2005, 25:3067-3075.
16. Yang Y, Wen JB, Hu DF. A Review on Avian Habitat Research. Scientia Silvae Sinicae. 2011,47:172-180.
17. Munster VJ, Baas C, Lexmond P, Wallensten A, Fransson T, Rimmelzwaan GF, et al. Temporal, and Species Variation in Prevalence of Influenza A Viruses in Wild Migratory bird. Plos Pathogens. 2007, 3:630-638.
18. MacKinnon J R, Phillipps K and He F. The Field Guide to the Birds of China. Changsha: Hunan Education Press, 2010.
19. Zheng GM. The Checklist of the Classification and Distribution of the Birds of China (Second Edition). Beijing: Science Press, 2011.

Whale optimization algorithm with nonlinear control parameter

Minghui Zhong[1], and Wen Long[2*]

[1]*College of Information Engineering, Guangxi University of Foreign Language, Nanning 530222, PR China*
[2]*Key Laboratory of Economics System Simulation, Guizhou University of Finance & Economics, Guiyang 550025, PR China*

Abstract. Whale optimization algorithm (WOA) is a relatively novel intelligence optimization technique which has been shown to be competitive to other population-based algorithms. However, the control parameter a is a major factor to affect the algorithm's convergence precision and speed. At present, few of them are aiming at control parameter setting in WOA algorithm. This paper proposes corresponding improved WOA algorithm with different nonlinear adjustment strategy of control parameter a by adopting sinusoid, cosine, tangential, logarithmic and quadratic curves. The experimental results for six benchmark test functions show that the proposed nonlinear adjustment strategies are superior to the classical linear strategy.

Keywords: whale optimization algorithm; control parameter; nonlinear; benchmark test function

1 Introduction

Whale optimization algorithm (WOA) is a relatively new meta-heuristic optimization technique proposed by Mirjalili and Lewis [1], which inspires the bubble-net hunting behaviour of humpback whales. WOA easy to implement and has few adjustment parameters, which make it superior than particle swarm optimization (PSO), grey wolf optimizer (GWO) algorithm, and gravitational search algorithm (GSA), etc. As a result, WOA has attracted much attention and has been applied to handle many practical engineering application problems, such as training multi-layer perceptron in neural network [2], tracking MPP of photovoltaic system [3], optimizing the active and reactive power dispatch problem [4], feature selection [5], parameter estimation of photovoltaic cells [6], multilevel thresholding image segmentation [7], and so on.

Similarly to other meta-heuristic algorithm, as the growth of the search space dimension, WOA is also easily trapped in the local optimum and provides a poor convergence [8]. Consequently, researchers increasingly are paying close attention to improve the performance of WOA. For instance, Trivedi et al. [8] proposed an improved version of WOA (called AWOA) based on modified position-updated equation. Ling et al. [9] developed an improved version of WOA (called LWOA) with Levy flight trajectory to make faster convergence and avoid premature convergence. Hu et al. [10] presented an improved version of WOA (called IWOA) based on inertia weight.

In the standard WOA, the distance control parameter a is an important parameter to balance the abilities of exploration and exploitation. However, the distance control parameter a is linearly decreased from 2 to 0 over the course of iterations [1]. Thus, in this paper, we introduce the different distance control parameter strategies to WOA to design the improved whale optimization algorithms (IWOAs). The proposed IWOAs are tested with six benchmark functions and the results show that the proposed algorithm is effective in most cases. The structure of the rest of this paper is as follows. In Section 2, the standard WOA is summarized. The improved version of WOA is proposed in Section 3. In Section 4, six benchmark functions are introduced to test the performance of proposed algorithm. Finally, Section 5 summarizes the conclusion of this paper.

2 Whale optimization algorithm

Fig.1 shows the bubble-net hunting behaviour of humpback whales. Humpback whales prefer to hunt school of krill or small fishes close to the surface. Humpback whales swim around the prey within a shrinking circle and along a spiral-shaped path simultaneously to create distinctive bubbles along a circle or "9"-shaped path [1].

Figure 1. Bubble-net feeding behaviour of humpback whales.

In 2016, Mirjalili and Lewis [1] developed a novel meta-heuristic optimization algorithm, WOA, which mimics the bubble-net hunting technique of humpback whales. To simulate this technique in WOA, there is a probability of 50% to choose between the shrinking encircling mechanism and the spiral model to update the position of whales during optimization.

* Corresponding author: lw227@mail.gufe.edu.cn

2.1 Shrinking encircling prey

In WOA, to mathematically model the encircling mechanism, the following formulas are introduced [1]:

$$\vec{X}(t+1)=\vec{X}^*(t)-A\cdot|C\cdot\vec{X}^*(t)-\vec{X}(t)| \quad (1)$$

$$A=2\cdot a\cdot r-a \quad (2)$$

$$C=2\cdot r \quad (3)$$

where $\vec{X}$ is the position vector, $\vec{X}^*$ is the historically best position, t indicates the current iteration, r is a random number in [0,1], a is called distance control parameter and is linearly decreased from 2 to 0 over the course of iterations [1]:

$$a(t)=2-2\frac{t}{t_{\max}} \quad (4)$$

where $t_{\max}$ indicates the maximum number of iterations.

2.2 Spiral bubble-net feeding strategy

A spiral equation is used between the position of whale and prey to mimic the helix-shaped movement of humpback whales as follows:

$$\vec{X}(t+1)=\vec{X}^*(t)+e^{bl}\cdot\cos(2\pi l)\cdot|\vec{X}^*(t)-\vec{X}(t)| \quad (5)$$

where b is a constant for defining the shape of the logarithmic spiral and l is a random number in $[-1,1]$.

2.3 Search of prey

In order to have a global optimizer, when A is greater than 1 or less than −1, the search agent is update according to a randomly chosen search agent instead of the best search agent:

$$\vec{X}(t+1)=\vec{X}_{rand}-A\cdot|C\cdot\vec{X}_{rand}-\vec{X}(t)| \quad (6)$$

where $\vec{X}_{rand}$ is a random position vector chosen from the current population.

For further details, the reader may refer to [1].

3 Improved WOA algorithm

As we know, meta-heuristic optimization algorithm must has a good balance between exploration and exploitation. From [1] and Eq.(1), exploration and exploitation abilities of WOA are guaranteed by the adaptive values of distance control parameter a. But the values of distance control parameter a are linearly decreased from 2 to 0 over the course of iterations.

Although the linearly distance control parameter strategy shows faster convergence and high accuracy in the early stage for most cases by empirical investigations with some well-known problems [10]. However, the actual search process of WOA is nonlinear and much more complicated, so it can not be truly reflected by the linearly decreasing adaptive parameter a. Therefore, inspired by PSO algorithm, this work proposes some improved versions of WOA by introducing non-linear strategies of distance control parameter a :

$$a(t)=(a_{\max}-a_{\min})\times\sin(\mu\cdot\frac{t}{t_{\max}}\pi) \quad (7)$$

$$a(t)=(a_{\max}-a_{\min})\times\cos(\mu\cdot\frac{t}{t_{\max}}\pi) \quad (8)$$

$$a(t)=(a_{\max}-a_{\min})\times\tan(\mu\cdot\frac{t}{t_{\max}}\pi) \quad (9)$$

$$a(t)=(a_{\max}-a_{\min})\times\log(0.5+(e-1)\times(\frac{t}{t_{\max}})) \quad (10)$$

$$a(t)=(a_{\max}-a_{\min})\times(\frac{t}{t_{\max}})^2 \quad (11)$$

where t indicates the current iteration, $t_{\max}$ indicates the maximum number of iterations, μ is the nonlinearly adjust factor, $a_{\max}$ and $a_{\min}$ are the maximum and minimum values of the distance control parameter a, respectively.

Based on the above considerations, the flow chart of improved version of WOA is showed in Fig.2.

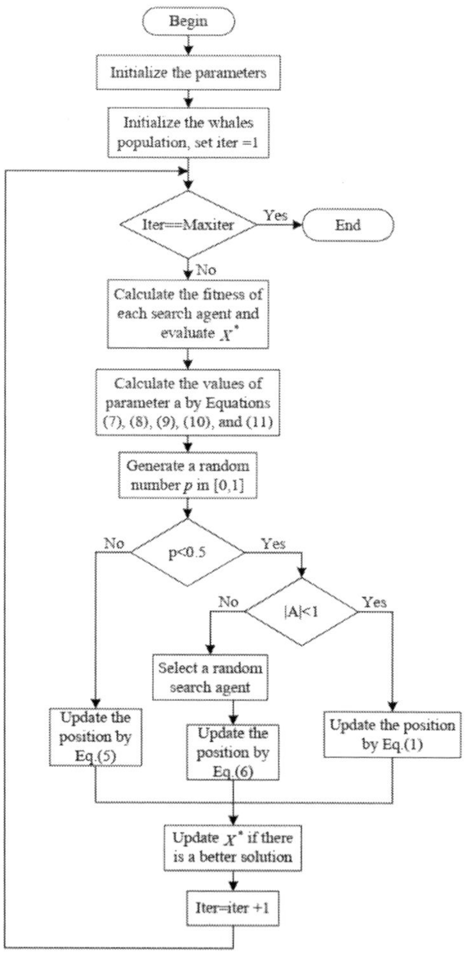

Figure 2. The flow chart of improved WOA algorithm.

Thus four kinds of improved WOAs are obtained as follows:

(1) WOA with nonlinear distance control parameter strategy shown Eq.(7) (donated as SinWOA).

(2) WOA with nonlinear distance control parameter strategy shown Eq.(8) (called as CosWOA).

(3) WOA with nonlinear distance control parameter strategy shown Eq.(9) (called as TanWOA).

(4) WOA with nonlinear distance control parameter strategy shown Eq.(10) (called as LogWOA).

(5) WOA with nonlinear distance control parameter strategy shown Eq.(11) (called as SquareWOA).

4 Simulation and comparison

To test the performance of the proposed algorithm, we select six benchmark functions from literature [1]. These functions are summarized in Table 1.

Table 1. Benchmark test functions.

Function	Search range
$f_1(x)=\sum_{i=1}^{n} x_i^2$	[−100, 100]
$f_2(x)=\sum_{i=1}^{n}\lvert x_i\rvert+\prod_{i=1}^{n}\lvert x_i\rvert$	[−10,10]
$f_3(x)=\sum_{i=1}^{n}(\lfloor x_i+0.5\rfloor)^2$	[−100, 100]
$f_4(x)=\sum_{i=1}^{n} ix_i^4+random[0,1)$	[−1.28,1.28]
$f_5(x)=\sum_{i=1}^{d}[x_i^2-10\cos(2\pi x_i)+10]$	[−5.12, 5.12]
$f_6(x)=\frac{1}{4000}\sum_{i=1}^{d} x_i^2-\prod_{i=1}^{d}\cos\left(\frac{x_i}{\sqrt{i}}\right)+1$	[−600, 600]

The set of experiments test on six functions are executed. For a fair comparison among six algorithms, they are tested using the same setting of the parameters, that is, the population size is equal to 30, the maximum iteration is equal to 500, and the function dimension is set to 30. The other parameters of our algorithm are set as follows: a_{max}=2, a_{min}=0, μ=7. Table 2 shows the best result (Best), the mean result (Mean), and the standard deviation (Std) result. All source codes are written in MATLAB 7.11.0 (win64) and executed in an Apple iMac computer with 2.8 GHz Intel Core i5, 8 GB 1867 MHz DDR3 under MAC OS operation system. All results reported are obtained based on 30 independent runs. For clarity, the results of the best algorithms are marked in boldface.

Table 2. Benchmark test functions.

Function	Result	WOA	SinWOA	CosWOA	TanWOA	LogWOA	SquareWOA
	Best	2.94E−083	4.67E−112	**3.64E−128**	5.15E−115	5.62E−113	7.62E−091
f_1	Mean	5.51E−075	4.34E−100	**2.38E−122**	3.18E−108	4.46E−102	1.49E−082
	Std	1.23E−076	1.94E−100	**4.10E−122**	5.00E−108	3.10E−102	1.29E−082
	Best	3.97E−056	1.15E−075	**3.23E−077**	2.05E−076	4.41E−070	1.95E−062
f_2	Mean	6.42E−053	2.64E−069	**2.51E−073**	3.58E−071	1.63E−066	1.43E−059
	Std	8.65E−053	7.36E−069	**3.45E−073**	7.52E−071	1.19E−065	1.09E−058
	Best	**0**	**0**	**0**	**0**	**0**	**0**
f_3	Mean	**0**	**0**	**0**	**0**	**0**	**0**
	Std	**0**	**0**	**0**	**0**	**0**	**0**
	Best	4.85E−004	2.18E−003	**6.27E−005**	6.06E−004	3.55E−004	4.16E−004
f_4	Mean	5.43E−003	8.91E−003	**1.13E−003**	8.37E−003	4.35E−003	4.79E−003
	Std	5.80E−003	8.45E−003	**1.59E−003**	6.71E−003	3.17E−003	4.76E−003
	Best	**0**	**0**	**0**	**0**	**0**	**0**
f_5	Mean	**0**	**0**	**0**	**0**	**0**	**0**
	Std	**0**	**0**	**0**	**0**	**0**	**0**
	Best	**0**	**0**	**0**	**0**	**0**	**0**
f_6	Mean	**0**	**0**	**0**	**0**	**0**	**0**
	Std	**0**	**0**	**0**	**0**	**0**	**0**

From Table 2, an interesting result is that the six WOA-based algorithms are able to find the global optima value (0) on three test functions (f_3, f_5, and f_6) with fixed number of iterations. For test functions f_1 and f_2, the mean values obtained by six algorithms are highly near global optimum. With respect to WOA algorithm, SinWOA and TanWOA can obtain better results on two test functions (f_1 and f_2) and similar results on three test functions (f_3, f_5, and f_6). However, for function f_4, WOA can find better results. Compare with WOA algorithm, CosWOA, LogWOA, and SquareWOA are able to provide better results for test functions f_1 and f_2. For the rest of test functions, four algorithms find similar results. In general, the performance of the CosWOA algorithm is the best and the standard WOA algorithm is the worst among the six compared algorithms in a statistically significant fashion.

Fig.3 illustrates the convergence curves of fitness values with respect to the number of iterations for the six benchmark test functions with D=30. It can be observed from Fig.3 that the CosWOA algorithm is faster than the standard WOA and the other four improved WOA algorithms on all the benchmark test functions.

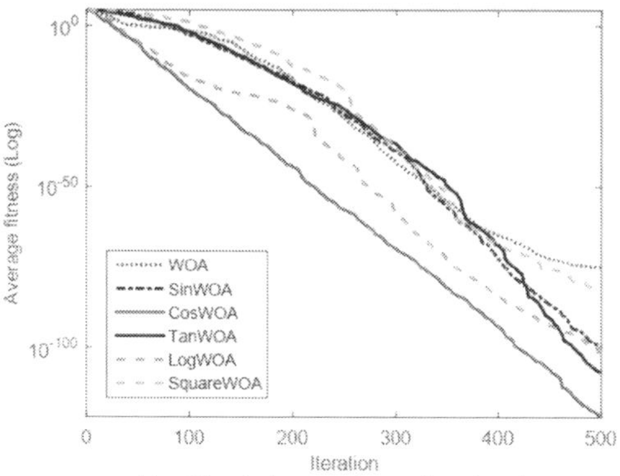

(a) Simulation results on function f_1

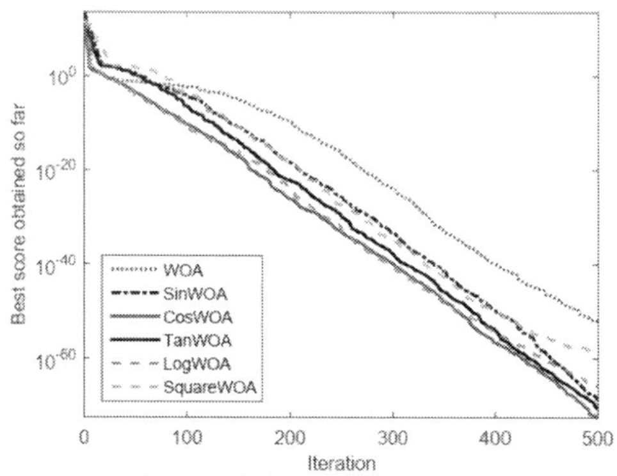

(b) Simulation results on function f_2

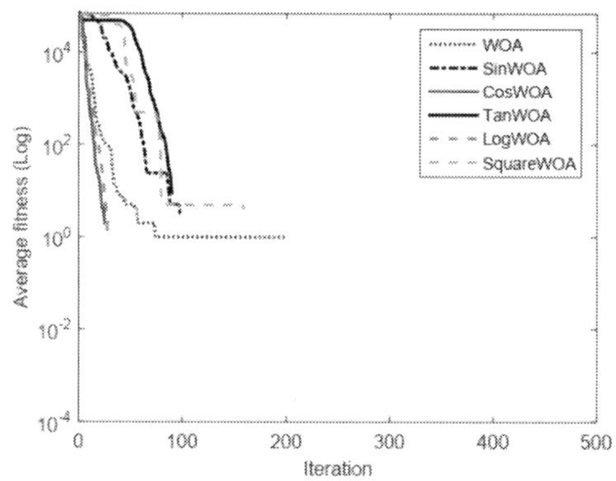

(c) Simulation results on function f_3

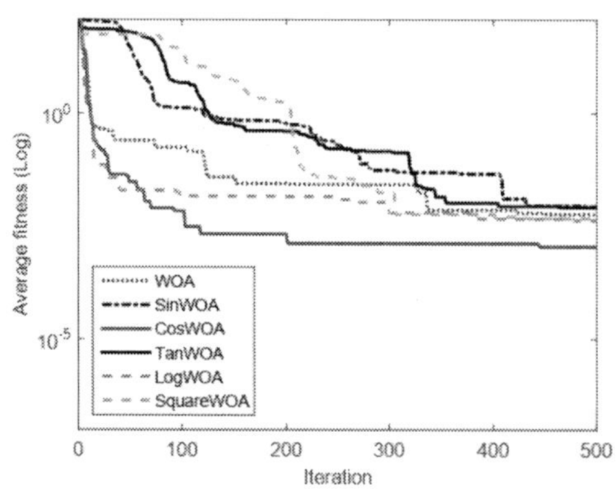

(d) Simulation results on function f_4

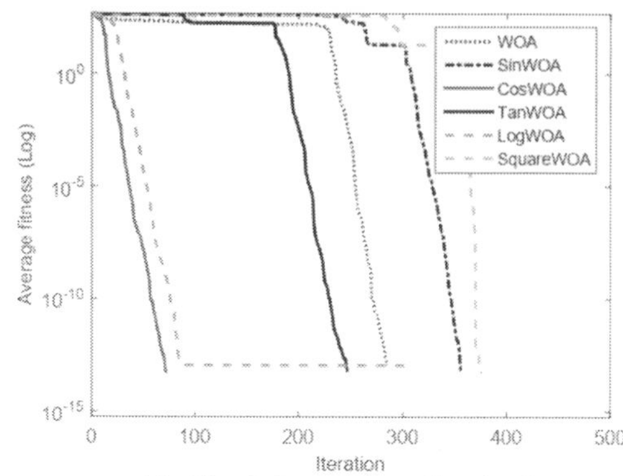

(e) Simulation results on function f_5

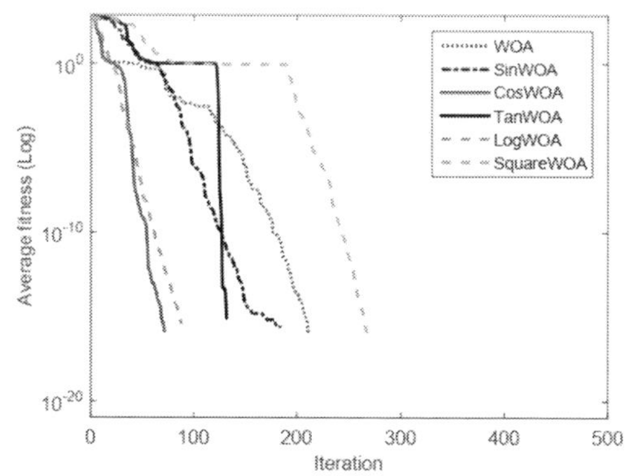

(f) Simulation results on function f_6

Figure 3. Convergence curves of six functions.

5. Conclusion

Whale optimization algorithm (WOA) is a novel meta-heuristic optimization technique inspired by the bubble-net hunting behaviour of humpback whales. Exploration and exploitation abilities of WOA are guaranteed by the adaptive values of a. This paper introduces nonlinearly distance control parameter strategies to design improved WOA algorithm. Six benchmark test functions were employed to verify the performance of the proposed improved WOA algorithm. The experimental results show that the proposed improved WOA with nonlinearly distance control strategies can provide highly competitive results.

Acknowledgements

This work is supported in part by the National Natural Science Foundation of China under the Grant Number 61463009, and supported in part by the Science and Technology Foundation of Guizhou Province under the Grant Number [2016]1022.

References

1. S. Mirjalili, A. Lewis, Advances in Engineering Software, **95**, (2016), 51-67.

2. R.H. Bhesdadiya, P. Jangir, N. Jangir, I.N. Trivedi, D. Ladumor, Indian Journal of Science and Technology, **7**, (2016), 1-15.
3. C.H.S. Kumar, R.S. Rao, International Journal of Renewable Energy Development, **5**, (2016), 225-232.
4. R.H. Bhesdadiya, S.A. Parmar, I.N. Trivedi, P. Jangir, M. Bhoye, N. Jangir, Indian Journal of Science and Technology, **7**, (2016), 1-12.
5. M.M. Mafarja, S. Mirjalili, Neurocomputing, **260**, (2017), 302-312.
6. D. Oliva, M.A.E. Aziz, A.E. Hassanien, Applied Energy, **200**, (2017), 141-154.
7. M.A.E. Aziz, A.A. Ewees, A.E. Hassanien, Expert Systems with Applications, **83**, (2017), 242-256.
8. I.N. Trivedi, P. Jangir, N. Jangir, A. Kumar, D. Ladumor, Indian Journal of Science and Technology, **9**, (2016), 1-6.
9. Y. Ling, Y.Q. Zhou, Q.F. Luo, IEEE Access, (to be publised)
10. H.P. Hu, Y.P. Bai, T. Xu, International Journal of Circuits, Systems and Signal Processing, **11**, (2017), 12-26.

Study on the Method of Face Detection Based on Chaos Genetic Algorithm Optimization AdaBoost Algorithm

Yu Chai[1], *Yang Yang*[2,*]

[1,2]School of Electrical and Control Engineering, Xi'an University of Science and Technology, Xi'an, 710054, China

Abstract. Aiming at the problems that the traditional AdaBoost algorithm has complex feature computation, long training time and low detection rate, a method of face detection based on chaos genetic algorithm optimization adaBoost algorithm was proposed. Firstly, this algorithm uses the image color segmentation for coarse screening on the face image, in order to determine the human skin area. Secondly, the adaptive median filtering is applied to denoise the face image to improve the quality of the face image. Finally, the chaotic genetic algorithm is used to optimize the AdaBoost algorithm to achieve higher detection rate and detection speed. Compared with the traditional AdaBoost algorithm, the experimental results showed that the face detection method based on chaos genetic algorithm optimization AdaBoost algorithm proposed in this paper has a significant improvement in detection rate and detection speed.

1 Introduction

Face detection uses a certain strategy or a set of image sequences to search for any given image to determine the position and region of the face, to detect the presence of a face in a variety of different images or image sequences, and to determine the number of faces and the process of spatial distribution [1]. Face detection is a key technology of face information processing which has a very broad application [2], such as: face recognition; face tracking; pose estimation; facial expression recognition; image retrieval and digital video etc [3].

At present, face detection methods are mainly based on the feature detection method and statistical based detection method [4]. Among them, the feature-based detection method is intuitive and easy to implement, but there are great limitations. However, the statistical detection method is not easy to implement, but it has better robustness and universality [5]. In this paper, the AdaBoost algorithm based on statistical detection method is used as the core algorithm of face detection.

Firstly, the skin color segmentation is used to screen the face image. Secondly, the adaptive median filter is used to denoise the face image. Finally, the chaotic genetic algorithm is used to optimize the AdaBoost algorithm. The method reduces the detection time and improves the detection efficiency.

2 Face detection

2.1 Image color segmentation

In order to reduce the search time of AdaBoost algorithm, this paper adopts the color segmentation of the image to rough screen the face image, and then determine the skin area of human body, thus it can avoid the traversal search of AdaBoost algorithm.

Because the skin color has better clustering characteristics in YCbCr color space, the conversion relation with RGB color space is simple, and there is no singular point [6], so YCbCr color space is chosen as the skin segmentation model. In the YCbCr color space, the Y component represents brightness information, and the Cb component represents the blue chromaticity information, and the Cr component represents the red chromaticity information. The transformation formula of RGB color space and YCbCr color space:

$$\begin{bmatrix} Y \\ Cb \\ Cr \end{bmatrix} = \begin{bmatrix} 0.299 & 0.587 & 0.114 \\ -0.168 & -0.331 & 0.500 \\ 0.500 & -0.419 & -0.0813 \end{bmatrix} \begin{bmatrix} R \\ G \\ B \end{bmatrix} \tag{1}$$

According to the distribution statistics:

$$\begin{cases} 77 \le Cb \le 127 \\ 133 \le Cr \le 173 \end{cases} \tag{2}$$

If the pixels fall within this range, the pixels of the skin color are considered. The raw image and skin segmentation results are shown in Figure 1 (a) and (b).

2.2 Image denoising

Normally, the image information obtained by the image acquisition device contains various noises and distortions [7]. Therefore, after the segmentation of the skin color, we need to denoise the face image so as to improve the quality of the face image.

Because of noise and distortion, the image is distorted and degraded, so denoising is the basic step in

* Corresponding author: 18792906858@126.com

DOI: 10.1051/matecconf/201713900158

image processing. In the denoising process, filtering is necessary in face detection [8]. The commonly used image filtering methods include smoothing filtering and sharpening filtering, etc. smoothing filtering is more widely used in face image denoising processing. In this paper, an adaptive median filtering algorithm is used to denoise the image. The denoising results of the face image are shown in Figure 1 (c).

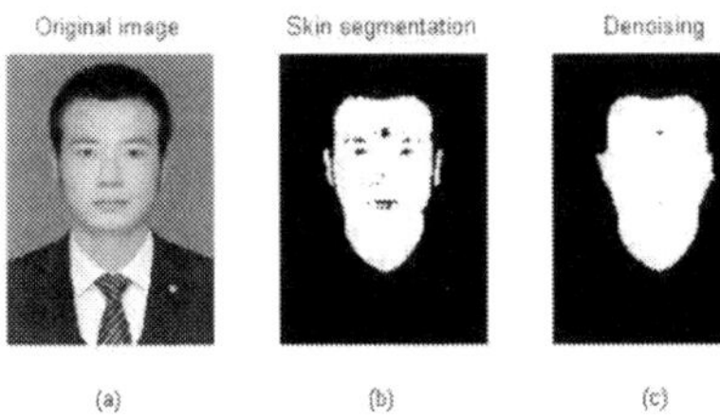

Fig. 1. Skin egmentation and denoising of face images

2.3 AdaBoost algorithm

Freund and Schapire improved the Boosting algorithm in 1995, named it AdaBoost. The algorithm is efficient and does not need to know any prior knowledge about the weak learning algorithm. Paul Viola and Michael Jones in 2001 proposed a face detection algorithm based on AdaBoost, this method combines several strong classifiers into cascade classifiers. When the sub-window is recognized, the sub-window is sent to the subsequent classifier to continue processing when the classifier of the previous stage is judged to be positive , otherwise it is judged as negative directly, and the sub-window will be judged to be positive only if all classifiers are judged to be positive [9-10].

AdaBoost algorithm specific description follows [11]:

1) The input is N training samples.

$(x_1, y_1), \cdots, (x_i, y_i), \cdots, (x_N, y_N)$, where $i = 1, 2, \cdots, N$, $x_i \in X$, $y_i \in Y = \{-1, 1\}$ are non-face samples and face samples respectively.

2) Initialize weights, $\omega_{1,i} = \frac{1}{2m}, \frac{1}{2l}$, where m, l are training sample set of non-face samples and face samples respectively.

3) Start iteration loop, $t = 1, 2, \cdots, T$.

a. Normalized weight $\bar{\omega}_{t,i}$.

$$\bar{\omega}_{t,i} = \frac{\omega_{t,i}}{\sum_{j=1}^{n} \omega_{t,j}} \tag{3}$$

where n is the characteristic number.

b. For each feature j of the weak classifier h_j , the error rate of the current weight is calculated, $\varepsilon_j = \sum_{j=1}^{n} \bar{\omega}_{t,j} \left| h_j(x_i) - y_i \right|$.

c. The decision function h_t corresponding to the minimum error rate ε_t is used as the weak classifier obtained by this iteration.

d. Judge the weight of each training samples according to the decision function h_t , that is $\omega_{t+1,i} = \omega_{t,i} \beta_t^{1-e_i}$. Among, if h_t judge the sample i correctly, $e_i = 0$; otherwise, $e_i = 1$, $\beta_t = \frac{\varepsilon_t}{1-\varepsilon_t}$.

4) Generating a strong classifier.

$$h(x) = \begin{cases} 1 & \sum_{t=1}^{T} \alpha_t h_t(x) \geq 0.5 \sum_{t=1}^{T} \alpha_t \\ 0 & Others \end{cases} \tag{4}$$

where $\alpha_t = \log_2 \frac{1}{\beta_t}$.

3 Optimization of AdaBoost algorithm based on chaos genetic algorithm

3.1 Improved genetic algorithm

Genetic algorithm is an adaptive global optimization probability search algorithm [12], which mimics the genetic and evolutionary processes of organisms in the natural environment. The idea of solving the genetic algorithm is to map the solution space of the problem into the coding space, and use the fitness function value of the individual as a measure of the quality of the individual. In the process of evolution, the new optimal individuals are generated by continuous selection, crossover and mutation, and finally the global optimal solution is decoded [13].

Genetic algorithm is widely used in many fields, because of its relatively simple and low demand characteristics of the objective function, but the search for the optimal individual is more difficult, the slow speed of the search in the vicinity of the global optimal solution, and may fall into the local optimal solution, slow convergence speed and premature convergence is also related to the lack of. Therefore, in order to improve the above shortcomings of genetic algorithm, a new algorithm combining genetic algorithm with other algorithms is proposed.

The theoretical support of chaotic optimization algorithm is the ergodic property of chaos, so it is a global optimization algorithm. Chaos is a class of random phenomena which appear to be chaotic and exist fine rules in a definite system [14]. Because chaos has the ergodic property, it is better to use chaotic variables to optimize the search than blind random search, and prevent the evolutionary algorithm from being trapped into the local optimum. Chaos optimization algorithm and genetic algorithm are fused here.

3.2 Chaotic genetic algorithm design

3.2.1 Encoding mode

For function optimization, real number coding is the most effective, so we choose real coding to determine individual encoding.

3.2.2 Fitness function

Fitness function is the only basis for natural selection, and it is the criterion used to judge the quality of individuals in a group, which is usually transformed by objective function [15]. In this paper, the decision function h_t in AdaBoost algorithm should have the minimum error rate ε_t. That is, the objective function is

$$\min \varepsilon_j = \sum_{i=1}^{n} \bar{\omega}_{t,j} \left| h_j(x_i) - y_i \right| \tag{5}$$

The objective function is to find the minimum error rate ε_j, that is, the reciprocal of the objective function is used to judge the fitness value of the individual. Individuals with small function values have large fitness values and are individually superior. Therefore, the fitness function is chosen as

$$F(\varepsilon_j) = \frac{1}{\varepsilon_j} \tag{6}$$

3.2.3 Initial population

In this paper, Logistic mapping is adopted, and its mathematical model is as follows:

$$x_{k+1} = \mu x_k (1 - x_k),\ k = 1,2,\cdots \tag{7}$$

Where μ is control parameter, x_k is chaotic variable, k is the ordinal number of the chaotic variable. When $\mu = 4$, $0 < x_0 < 1$, the system has no stable solution and is in chaotic state. It is a full mapping process, so it can improve the genetic algorithm to initialize population diversity.

3.2.4 Selection

It is assumed that there are N populations that produce N offspring individuals after crossing and mutation. According to the fitness function value, from large to young, sub parent generation of $2N$ groups for sorting, take the start of $N/2$ individuals, from $2N$ individuals last selected $N/2$ individuals were matched. In this way, the population is kept converging to the optimal solution, and the diversity of the population is reduced slowly.

3.2.5 Cross variation

In order to simplify the experimental process, the single point crossover operator as the crossover operator in individual encoding string random sets a crossover point, then part of gene transformation; the basic bit mutation as the mutation operator, the mutation probability of individual gene encoding string on the base of value of gene mutation operation.

In this paper, the flowchart of the chaotic genetic algorithm is shown in Figure 2.

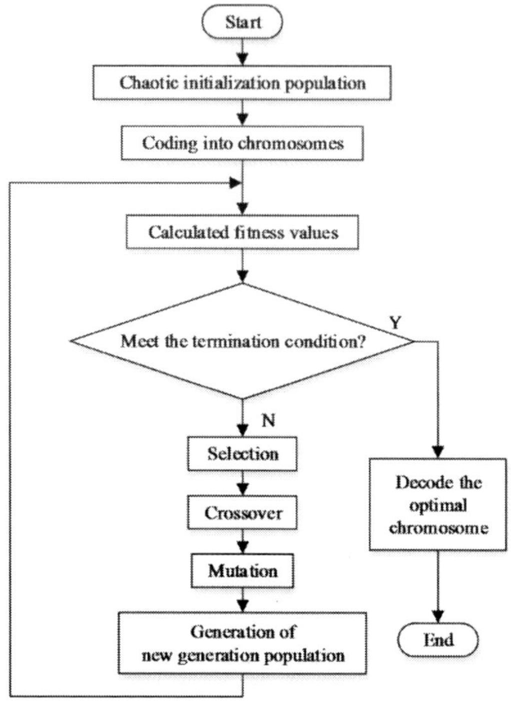

Fig. 2. Flowchart of the chaotic genetic algorithm

3.3 Optimization algorithm flow

The optimization of AdaBoost algorithm using chaotic genetic algorithm is as follows:

1) The input is N training samples.

$(x_1, y_1), \cdots, (x_i, y_i), \cdots, (x_N, y_N)$, where $i = 1, 2, \cdots, N$, $x_i \in X$, $y_i \in Y = \{-1, 1\}$ are non-face samples and face samples respectively.

2) Initialize weights, $\omega_{1,i} = \frac{1}{2m}, \frac{1}{2l}$, where m, l are training sample set of non-face samples and face samples respectively.

3) Start iteration loop, $t = 1, 2, \cdots, T$.

a. Normalized weight $\bar{\omega}_{t,i}$.

$$\bar{\omega}_{t,i} = \frac{\omega_{t,i}}{\sum_{j=1}^{n} \omega_{t,j}} \tag{8}$$

where n is the characteristic number.

b. The genetic algorithm population x_i is initialized using chaotic variables and encoded into chromosomes by real coded encoding.

c. The fitness value of each individual is calculated according to the fitness function of the individual $F(\varepsilon_j)$.

d. Determine whether or not the condition has been met. If the condition is reached, the chromosome is decoded so that the decision function h_t corresponding to the minimum error rate ε_t can be output. If the conditions are not met, the selection, crossover and mutation operations are performed to generate a new generation of population and return to step c to continue searching.

4) Judge the weight of each training samples according to the decision function h_t, that is $\omega_{t+1,i} = \omega_{t,i}\beta_t^{1-e_i}$. Among, if h_t judge the sample i correctly, $e_i = 0$; otherwise, $e_i = 1$, $\beta_t = \frac{\varepsilon_t}{1-\varepsilon_t}$.

5) Generating a strong classifier.

$$h(x) = \begin{cases} 1 & \sum_{t=1}^{T}\alpha_t h_t(x) \geq 0.5\sum_{t=1}^{T}\alpha_t \\ 0 & Others \end{cases} \tag{9}$$

where $\alpha_t = \log_2 \frac{1}{\beta_t}$.

4 Experimental result

4.1 Experimental environment and data set

Experimental environment: 2.50GHz, 4GB memory, VS2010+OpenCV. In this paper, MIT face sample library is selected. The training sample consists of 2000 face samples and 5000 non-face samples. The test set consists of 1000 face samples and 2500 non-face samples. All the samples are 20 × 20 pixels. Some of the samples are shown in Figure 3.

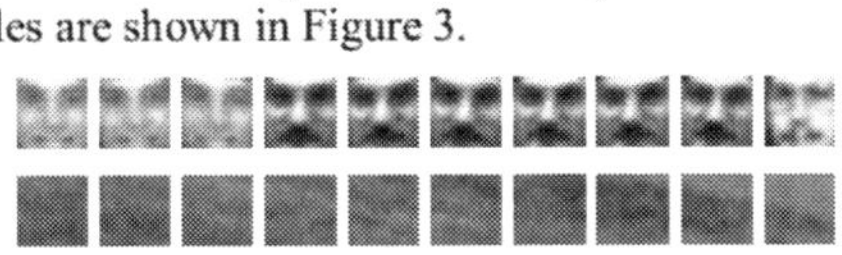

Fig. 3. Partial face and non-face samples

4.2 experimental result

The related parameters of the algorithm are as follows: population size 25, maximum reproduction algebra 500, cross probability 0.85, variation probability 0.02. In order to verify the effectiveness of this algorithm, the author uses the traditional AdaBoost algorithm, the AdaBoost algorithm based on skin segmentation and this algorithm three algorithms for comparative analysis.

In this experiment, a strong classifier is trained, in which the training data include the average number of features and the training time required by the classifier. Table 1 presents the training comparison results of the three algorithms.

Table 1. The average number of features andtraining time comparison

Method	Average characteristic number / peace	Training time / h
Traditional AdaBoost algorithm	225	25.7
the AdaBoost algorithm based on skin segmentation	187	8.3
this algorithm	146	4.9

You can get it from table 1, this algorithm is compared with the traditional AdaBoost algorithm and the AdaBoost algorithm based on skin segmentation, the average number of features required in the training process is much less, the training time is reduced a lot.

In the detection process, a trained strong classifier is used to test the sample set, and the statistical results are checked. Table 2 lists the test results of the test set.

Table 2. Comparison of test results

Method	Detection rate / %	Detection error rate / %
Traditional AdaBoost algorithm	94.67	5.85
the AdaBoost algorithm based on skin segmentation	96.15	3.04
this algorithm	99.26	1.29

From the comparison of the detection results, we can see that the detection performance of the proposed algorithm is improved compared with the other two algorithms. The reason is that this paper uses chaotic genetic algorithm to optimize the AdaBoost algorithm, which has more powerful feature search ability when constructing weak classifiers, so as to achieve higher detection performance and detection speed.

The picture uses the author's class photo, a total of 34 people. The actual effect of the traditional AdaBoost algorithm for face detection is shown in Figure 4, detecting a total of 27 faces, time-consuming 1134ms. The actual effect of this algorithm for face detection is shown in Figure 5, and all 34 faces are detected, consuming time 817ms. This shows that the algorithm is better in detection speed and detection accuracy.

Fig. 4. The actual effect of the traditional AdaBoost algorithm for face detection

Fig. 5. The actual effect of this algorithm for face detection

5 Conclusion

In this paper, a method of face detection based on chaos genetic algorithm optimization adaBoost algorithm is proposed. Firstly, this algorithm uses the image color segmentation for coarse screening on the face image, to determine the human skin area. Secondly, the adaptive median filtering is applied to denoise the face image to improve the quality of the face image. Finally, the chaotic genetic algorithm is used to optimize the AdaBoost algorithm to achieve higher detection rate and detection speed. Experimental results show that this method can improve the detection speed and ensure the accuracy of detection. The next step can aim at the real-time, robustness and generalization of the algorithm. Further optimization of the AdaBoost algorithm, in order to reduce the effects of illumination, facial expression and gesture on face detection.

References

1. Zhang Ya-Nan; Zhang Zuo. Survey of human face detection and tracking technology[J]. Computers and modernization,2012,12(2):60.
2. Xu Qian; Zhao De-An; Zhao Jian Bo. Face detection and localization based on improved AdaBoost algorithm[J]. Transducer and microsystem technologies,2010,29(1):94.
3. You Qing-Qing; Shen Hai-Yun; Luo Jun et al. Survey of face detection technology[J]. Wireless internet technology,2017,5(10):137
4. Zhang Yun-Long; Xie Ze-Qi; Zhang Hui-Min et al. A method of face detection in complex background[J]. Transducer and microsystem technologies,2011,30(6):42
5. Liang Lu-Hong; Ai Hai-Zhou; Xu Guang-You et al. Survey of face detection[J]. Journal of Computer Science,2002,25(5):449.
6. Liu Yu-Sheng; Feng Rui. A face detection algorithm based on AdaBoost[J]. Computer engineering and Applications,2016,52(11):209.
7. Liu Zhuo-Ya. Review of image denoising techniques[J]. Science and technology information,2013,15:317
8. Tomasz Marciniak; Agata Chmielewska; Radoslaw Weychan et al. Influence of low resolution of images on reliability of face detection and recognition[J]. Multimed Tools Appl,2015,74: 4329.
9. Zhang Hua-Xun. A classifier training method for face detection based on AdaBoost[C] //Proceedings of the International Conference on Transportation,Mechanical and Electrical Engineering,2011:731.
10. Chen Shi-Gang; Ma Xiao-Hu; Zhang Shu-Kui. AdaBoost face based on haar-like intensity features and multithreshold features[C] //Proceedings of the International Conference on Multimedia and Signal Processing,2011: 251.
11. Zhang Jun-Yi; Zhang Ze. Face detection based on improved AdaBoost algorithm[J]. Computer simulation,2011,28(7):240.
12. Pratibha Sukhija; Sunny Behal,Pritpal Singh. Face Recognition System Using Genetic Algorithm[C] //International Conference on Computational Modeling and Security(CMS 2016),2016,85:410.
13. Zhou Xian-Fu. The principle and application of genetic algorithm[J]. Science and technology,2017,03:265
14. Petra Snaselova; Frantisek Zboril. Genetic Algorithm using Theory of Chaos[C] //ICCS 2015 International Conference On Computational Science,2015,51:316
15. Xuan Deng-Ying; Wang Fu-Lin; Gao Min-Hui et al. A genetic algorithm with improved fitness function[J]. Mathematics in practice and theory,2015,16:232

Electromagnetic Pulse Coupling Analysis of Electronic Equipment

Lei Hong[1], LI Qingying[2]
[1]Aviation Industry Corporation of China, Shenyang Aircraft Design Institute, Shenyang, China
[2] Electronic Information Engineering, Shenyang Aerospace University, Shenyang, China

Abstract High-intensity nuclear explosion caused by high-altitude nuclear electromagnetic pulse through the antenna, metal cables, holes and other channels, coupled with very high energy into the electronic device, and cause serious threats. In this paper, the mechanism, waveform, coupling path and damage effect of nuclear electromagnetic pulse is analyzed, and the coupling mechanism of nuclear electromagnetic pulse is studied.

Keywords Nuclear electromagnetic pulse Coupling Cable Hole seam

1 Electromagnetic Pulse

The electromagnetic pulse (EMP) has a short duration and a bandwidth. Electromagnetic pulse is usually caused by the explosion (especially nuclear explosion) and other conditions of electromagnetic radiation, acting on electronic materials, blasting equipment or the surrounding medium of electromagnetic shock wave. Any form of nuclear explosion can produce electromagnetic pulse, but the form is different, high altitude nuclear explosion caused by electromagnetic pulse greater damage, nuclear electromagnetic pulse intensity, high frequency components rich, covering a wide area. For more than 100km high altitude nuclear explosion, the resulting high field strength in the ground coverage of up to thousands of kilometers. Therefore, this paper mainly discusses the high-altitude nuclear explosion electromagnetic pulse (HEMP). Electromagnetic radiation environment produced by high altitude nuclear explosion, early (0 <t <1μs), medium (1μs <t <1s) and late (t> 1s) from time to time. Which early nuclear electromagnetic pulse on the destruction of electronic systems plays a major role. At present, there are many kinds of description of HEMP radiation environment at home and abroad, mostly in the form of double exponent as the mathematical expression of radiation waveform, the time domain expression is [1]:

$$E(t) = kE_0 (e^{-\alpha t} - e^{-\beta t}) \quad (1)$$

Where k is the correction coefficient, E_0 is the peak field strength, generally $50kV/m$, β, α is the characterization of the pulse before and after the parameters. The figure1 shows the waveform of the electromagnetic pulse when the parameter is $k = 1.3$, $E_0 = 50kV/m$, $\alpha = 4 \times 10^7$, $\beta = 6 \times 10^8$.

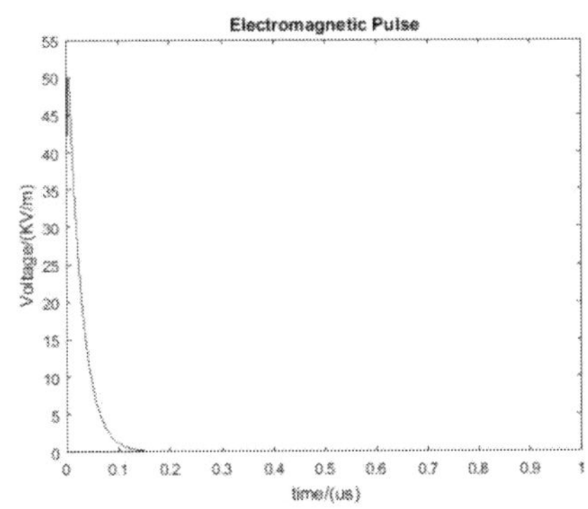

Figure 1 electromagnetic pulse waveform

2 Damage effect of electromagnetic pulse

Electromagnetic pulse coupling into the electronic system or equipment will affect the electronic components, lines and the entire system or equipment, so that the bit error rate, electronic components performance degradation or even damage, resulting in communication system performance degradation or failure, known as electromagnetic pulse effect. This effect may be temporary or permanent. The sensitivity of the system is the response of the system to the transient electromagnetic shock of the coupled electromagnetic pulse. The system caused by the impact of electromagnetic pulse temporary work disorders called work failure, and permanent damage is called functional damage.

Electromagnetic pulse no obvious damage to humans and non-electronic devices. The damage caused by electromagnetic pulse is related to its field strength and pulse width, which is related to nuclear explosion equivalent, burst height and distance. The effect of electromagnetic pulses on electronic systems or devices can be divided into four levels, namely: ineffective, Interference, disturbance and damage, depending on the size of the coupling energy entering the target, the coupling energy representing the incoming electronic system or device and subsequently generating The power to enter the target electronics depends on the power that can be generated by the electromagnetic pulse, the distance from the target, and the characteristics (frequency, pulse rate, pulse duration, etc.) of the electromagnetic pulse radiation, as well as the protection capability of the target. Electromagnetic pulse mainly through the energy of the conduction coupling, radiation coupling mode, the damage mode can be summarized as the following four aspects:

2.1 Thermal effect

The thermal effect of electromagnetic pulses is usually done in nanoseconds or microseconds, and is an adiabatic process. This effect can be used as ignition source and detonating source, instantaneous cause flammable, explosive gas and other items burning explosion; Can also make the communication system in the microelectronic devices, electromagnetic sensitive circuit overheating or input protection resistance burned, resulting in local thermal damage, resulting in deterioration of circuit performance or failure, and even lead to inventory burning explosion.

2.2 Interference and "surge" effects

The transient over-voltages or "surge" effects of electromagnetic pulses can also cause hard damage to the communication system. The performance of the device or circuit may deteriorate or completely fail, mainly for the semiconductor device open circuit, short circuit, current gain drop, leakage current increases, metal melting, and oxide breakdown; May also form a cumulative effect, buried the potential harm, so that the reliability of the circuit or equipment to reduce, resulting in malfunction.

2.3 Strong electric field effect

The strong electric field formed by the electromagnetic pulse source can not only cause the dielectric breakdown between the gate oxide layer and the metallization line of the metal oxide semiconductor (MOS) circuit in the communication device, so that the circuit fails, and the system self-test instrument and the sensitive device.

2.4 Magnetic effect

The strong current caused by the electromagnetic pulse can produce a strong magnetic field, so that the electromagnetic energy directly coupled to the system, interfere with the normal work of the communication system.

3 Coupling Path of Electromagnetic Pulse

Electromagnetic pulses can be damaged by the front door coupling and back door coupling to the electronic system. Front door coupling: The communication system must transmit the baseband signal to the transmit antenna before it can be effectively propagated outward, and vice versa. When the communication system is irradiated with electromagnetic pulses (especially when the main valve is aligned), the pulse signal is received by the antenna and transmitted to the RF port. RF module contains a band-pass filter, the purpose is to select the useful signal, abandon the out-of-band signal. But the face of the electromagnetic pulse is too large amplitude, the ordinary band-pass filter simply cannot withstand, the risk of being burned out.

Backdoor Coupling: Electromagnetic pulses cause large transient currents or electrical standing waves through the fixed electrical connections of the device and the interconnected cables of the device, which can cause damage to the circuit inside the device. The back door coupling of the communication system is mainly through the power line, the signal line, the cavity hole coupling to the inside of the system [2].

3.1 Front door coupling

Electromagnetic pulse coupling to the electronic device communication system transceiver antenna will produce the induced voltage or current, of course, but also consider the antenna band response and gain, the frequency is very high in a very short antenna can sense a large voltage. These voltages or currents are transmitted through the cable to the inside of the system to affect the normal operation of the system, which can damage the electronic circuits and their components in severe cases.

Electromagnetic pulse through the electronic equipment of the power line, the signal line input and output, coupled with electromagnetic pulse energy to produce a certain strength of the voltage or current signal, and then in a conductive way into the electronic device to go. Even if the current is not introduced from the leader of the transmission cable core, but in the outer shield on the induction, through the transfer impedance will be coupled to the core, directly into the electronic device system [3].

There must be more than one module within a system; each module may not be under the same mechanical structure under the shield. The transmission of power, control and signal between different modules is mostly done by cable. Cables tend to be long and exposed to the air. When the electromagnetic pulse is irradiated, the cable receives the electromagnetic wave through the antenna effect, especially when the cable length and the wavelength of the electromagnetic wave can be compared, the coupling is particularly serious.

The coupling of the EMP to the cable is generally on the order of 100 degrees and increases with the length of the cable and the height of the cable, which increases as the load resistance decreases across the cable.

Antenna system is the main carrier of the transmission and reception of the electronic equipment. The electromagnetic pulse energy is easy to enter the communication system of the electronic equipment through the "front door" coupling of the antenna, thus destroying the sensitive devices inside the electronic equipment. The center-fed dipole antenna shown in Fig. 2 is taken as an example.

 DOI: 10.1051/matecconf/201713900159

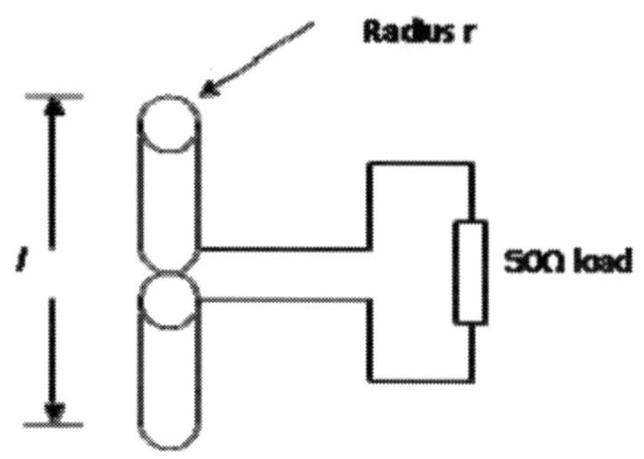

Figure 2 Center feed dipole

Its structural parameters are:

$$\chi=2ln\left(\frac{l}{r}\right) \qquad (2)$$

Where: χ for the antenna structure parameters, l for the antenna length (m), r antenna radius (m).
The resonant frequency of the antenna is:

$$f_c=\frac{c}{2l} \qquad (3)$$

Where: f_c for the antenna resonant frequency (Hz), c for the electromagnetic wave propagation velocity in vacuum (m/s), l for the antenna length (m).
50Ω load on the flow through the transient induced current mathematical expression is:

$$I_L(t)=kI_p e^{\left(\frac{-\pi f_c t}{Q}\right)} sin(2\pi f_c t) \quad t\geq 0 \qquad (4)$$

Where: $I_L(t)$ is the antenna load induced current (A) I_p is the current peak (A), Q is the damping coefficient.
The damping coefficient of the center feed dipole antenna $Q=\chi/3.6$. According to, the maximum induced current of the antenna in free space is [4]:

$$I_{max}=Hl \qquad (5)$$

Where: I_{max} is the maximum induced current (A), H is the incident wave magnetic field (A/m).
From the above formula, the induced current waveform for the attenuation of oscillation, the cycle is about $\frac{2l}{c}$, where l is the length of the wire. c is the speed of light in free space. As the length of the wire increases, the magnitude of the induced current increases [5].

3.2 Backdoor coupling

In order to facilitate the monitoring of the electronic system to open the ventilation system, in addition to the electronic chassis structure on the surface due to the existence of surface roughness, resulting in a random gap. These leaks provide coupling channels for electromagnetic pulses. When the size of the hole is greater than half the wavelength of the electromagnetic wave, the electromagnetic wave can enter the chassis, greatly reducing the shielding effectiveness of the chassis. Once the electromagnetic pulse into the box inside, bare Internet and printed circuit board can receive all kinds of electromagnetic pulse, into the sensitive circuit will damage the components, affecting the entire system. Take the seam as an example:

$$V(\omega)=I(\omega)\,Z_T(\omega) \qquad (5)$$

Where ω is the electromagnetic wave angular frequency(rad/s), $V(\omega)$ is the seam's induced voltage (V), $I(\omega)$ is the current flowing through the joint (A), $Z_T(\omega)$ is the transfer impedance of the joint (Ω).
The transfer impedance of the joint is related to the direction of the surface current. When the current is perpendicular to the joint, the transfer impedance is the largest, and when the surface current is parallel to the joint, the transfer impedance is very small [6]. According to the literature, the coupling electric field perpendicular to the joint direction is larger than the coupling electric field parallel to the joint direction [4].

4 Concluding remarks

In summary, the coupling of electromagnetic pulse is mainly for front door coupling and backdoor coupling. The electromagnetic pulse enters into the interior of the electronic device by coupling and causes a degree of damage to it. To study how the electromagnetic pulse on the

DOI: 10.1051/matecconf/201713900159

electronic equipment, electrical system damage, so as to study the corresponding protection methods, we must first clear the electromagnetic pulse coupling path. Different electromagnetic pulse coupling path due to different frequencies, coupling, the way is also different. Electromagnetic pulse through the front of the electronic equipment coupling or back door coupling into the system will cause serious damage to electronic equipment.

References

1. GAO Guangguang, GUO Xiaoqiang, ZHOU Hui, XIE Yanzhao. Double Exponential Function Fitting Method for Simulated Waveform of High-altitude Nuclear Electromagnetic Pulse [J]. Strong laser and particle beam. 2004, 14 (3): 336—340.
2. LU Xinke. Effect and Protection of Electromagnetic Pulse [D]. Xi'an: Xi'an University of Electronic Science and Technology master's degree thesis, 2009.
3. ZHANG Xin. Study on Coupling Simulation of Electromagnetic Pulse on Cable [D]. Xi'an: Xi'an University of Electronic Science and Technology master's degree, 2014.
4. Xiong Run. Study on Coupling of Electromagnetic Pulse on Equipment Chassis Lap Joint [J]. Modern Physics , 2015, 5, 88-92.
5. ZHU Lijie, SHEN Weidong, WANG Jianli, WU Honglin, GAO Guige, SHI Zhiqiang. Electrical pulse coupling and protection analysis of communication vehicle [J]. Mobile power and vehicles. 2012, (3): 24—29.
6. Xia Nenghong, Yi Xueqin, Song Wenwu, Huang Songgao. Study on Coupling Characteristics of Metal Chassis with Punching under EMP Excitation [J]. Equipment Environmental Engineering. 2009, 6 (3): 21-24

Design of four axis aircraft based on Android control

Linglong Tan ,Qing Wang , Zexin Li

Electronic communication engineering college , Anhui Xinhua University, Anhui Hefei, Hefei,China

Abstract .A four axle aircraft is designed with the open source firmware MWC is used as the flight control board .The MPU6050 gyro sensor and BMP085 pressure sensor are used to acquire the attitude data of aircraft in real time. Combining the Calman filtering and mean filtering, the attitude data is filtered and fused to obtain more accurate attitude information. In the control algorithm, the PID control algorithm is adopted to control the speed of the motor, which ensures that the aircraft can quickly and stably change the attitude of the flight. The Android handset is connected to an aircraft by using the WiFi signal which is launched on the four axle aircraft. The user can control the aircraft by using the appropriate App. Experimental results show that: The four axle aircraft can realize the functions of attitude, heading and hover under the control of App , and the system has friendly interface, flexible operation, stable and reliable.

1 Introduction

The four axis aircraft is becoming more and more popular. Its common control is to use radio remote control [1].There are three main types of remote control transmitters for four axis aircraft at present: The first one is a portable lever type remote control transmitter, the second one is a hand-held gun remote control transmitter, and another one is a small hand-held remote control transmitter for cassette keys. Gun type transmitter is the most popular remote control system at present, The gun type transmitter is a runner and a similar trigger lever to control the direction and speed respectively, remote control system of switch type analog circuit, the general model of ordinary toy remote control will use the remote control system, operation instruction is relatively simple, only the "on" and "off" of the two states, Performance and remote control distance is low. The four axis aircraft is controlled by Android phones. Using Android mobile phones to control four axis aircraft removes the traditional handheld wireless remote controller and the operation is simple.

2 Introduction of four axis aircraft

Four axis aircraft using MWC open source firmware for flight control board, and has a digital pressure sensor BMP085, MPU6050 digital three axis accelerometer and three axis gyro sensors and HMC5883L three axis digital magnetoresistive sensor, can help determine the four axis aircraft flying state with the sensors, and can adjust the flight attitude according to the parameters of each sensor. Therefore, the open source firmware is suitable for the development of four axis aircraft. In addition, the firmware has reserved other interface pins, such as I2C interface and serial port interface, which makes the ports for other sensors to be added later. The four axis flight control module consists of the following components:

1. master chip

Master chip is used ATmega328P microcontroller, with 1Kb EEPROM and 32Kb Flash, clock frequency is 16MHz. Arduino IDE can be used to edit and burn programs [2].

2. MPU6050 sensor

MPU6050 integrates 3 axis gyroscope and 3 axis accelerometer, and an extensible digital motion processor Digital Motion Processor (DMP), digital sensor and connected to a third party through the I2C interface. After expansion, the MPU6050 can output a 9 axis digital signal with its I2C or SPI interface. The MPU6050 sensor uses three 16 bit ADC for gyroscopes and accelerometers to convert the analog measurements into digital quantities. In use, in order to be able to accurately track the equipment fast and slow motion, the measuring range can be controlled, the range is + 250, gyroscope can measure the + 500, + 1000, + 2000 DEG / sec, the scope for accelerometer can be measured as + 2, + 4, + 8, + 16g.

3. HMC5883L

Highly integrated HMC5883L module is a weak magnetic sensor chip with digital interface, which can be used in the field of magnetic field testing. HMC5883L has an integrated circuit of Honeywell patents, including amplifier circuits, bias calibration circuits, and automatic degaussing drives. HMC5883L uses Honeywell anisotropic magnetoresistive technology, which is characterized by high linearity, high accuracy and high sensitivity. This advantage is not realized by other sensor technologies. The sensor has a solid structure of low sensitivity to the orthogonal axis, and the direction and magnitude of the earth's magnetic field can be measured using this structure, with measurements ranging from Gauss to 8 Gauss.

4. BMP085 sensor

Four axis aircraft in flight, if you need to lock height, you need to use the pressure sensor, BMP085 is a low-power, high-precision pressure sensor, can be applied to the pressure measurement equipment. Its absolute accuracy at the lowest value can reach 0.03hPa, at the same time, its power consumption is very low to 3uA. The sensor can be mounted on the four axle to provide pressure data for the four axle to cause pressure changes due to height changes, and to transfer the data to the master chip for processing. On the package, the BMP085 features an ultra thin package of 8-pin ceramic leadless chip, which looks very delicate. Its communication mode is to transmit data by using I2C bus and control chip.

3 Introduction of control algorithms

If a gyroscope or accelerometer is used alone to detect the attitude of the flight, it is impossible to provide reliable data for the balance of the aircraft. According to the characteristics of both, if the method used for filtering, fusion of gyroscope and accelerometer output signal to compensate for the cumulative error, gyro drift error and dynamic error of the acceleration sensor, can make the two mutual calibration, which can play the advantages, can also avoid their disadvantages, so the measurement can get a dip a better approximation, thus more accurate attitude measurement.

The design of the MPU6050 sensor output data using Calman filtering and mean filtering to filter and fusion, in order to obtain more accurate attitude information. Using the PID control algorithm to control the speed of the motor, to maintain the four axis aircraft in changing the attitude of the flight can be fast and stable.

3.1 Calman filtering algorithm

Calman filter[3] can measure a series of signals that do not contain noise completely, and can estimate the state of dynamic system optimally. It is an efficient recursive filter [4].

It is assumed that a linear stochastic differential equation is used to describe a discrete control process system:

$$X(k) = AX(k-1) + BU(k) + W(k) \quad (1)$$

The measured values of the system are:

$$Z(k) = HX(k) + V(k) \quad (2)$$

$X(k)$ is the system state of the k , and u is the amount of control that is added to the system at the k moment. In formulas, A and B are intrinsic parameters of the system, and when the system is multiple modules, they are matrices. k time measurement value is $Z(k)$, when it is in the measurement system, parameter measurement system is H matrix. W and V respectively represent system processes and system measurements of noise. Assuming that the Gauss white noise does not vary with the system state, then the covariance of $W(k)$ and $V(k)$ is Q and R. The Calman filter is the best processor for the process satisfying the above conditions.

First use system process model to predict the next state:

$$X(k, k-1) = AX(k-1, k-1) + BU(k) \quad (3)$$

$X(k, k-1)$ is the value predicted by the previous state. $X(k-1, k-1)$ is the optimal result of the previous state, and $U(k)$ is the control of the current state.

The covariance equation corresponding to $X(k-1, k-1)$ is:

$$P(k, k-1) = AP(k-1, k-1)A^T + Q \quad (4)$$

$P(k, k-1)$ is the value of the covariance of the $X(k, k-1)$, $P(k-1, k-1)$ is the covariance of the $X(k-1, k-1)$, and A^T is the transpose of the A , and the value of Q is the covariance of the system process.

DOI: 10.1051/matecconf/201713900160

The current state measurement value of the system is collected, and the optimal estimation value of the current state (k) can be obtained by fusing the predicted value and the measured value $X(k, k)$:

$$X(k, k) = X(k, k-1) + K_g(k)\left[Z(k) - HX(k, k-1)\right]$$

The K_g is Calman gain in the formula:

$$K_g = \frac{P(k, k-1)H^T}{HP(k, k-1)H^T + R} \quad (5)$$

In order to ensure that the Calman filter can continue running, the covariance of the $X(k, k)$ in the k state is also updated:

I is the unit matrix in the formula and I=1 is in the single model single measurement system. When the system is in the $k+1$ state, the $P(k, k)$ becomes the $P(k-1, k-1)$ at this point. In this way, we can do the regression calculation and realize the optimal estimation.

The average filter takes the average value of N values as the measured value, so that the measured value is close to the actual value, and can reduce the random error and weaken the influence of the external disturbance on the control process.

3.2 PID control algorithm

In the control of four axis aircraft, because there are many unknown disturbances and disturbances that affect the flight of an aircraft, the PID control [5] is adopted in this design.

The controller based on the ratio of deviation, integral and differential is called the PID controller. This automatic control law is widely used in the control system[6,7].

As shown in Fig.1, $r(t)$ is the system set value, $u(t)$ is the control quantity, and $c(t)$ is the actual output value

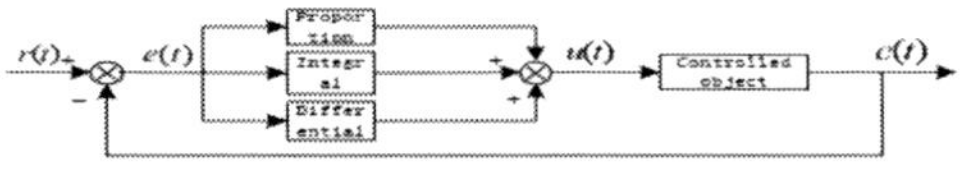

Fig. 1. Schematic diagram of PID system control

The control formula of analog PID controller is:

$$u(t) = K_P\left[e(t) + \frac{1}{T_I}\int_0^t e(t)dt + T_D\frac{de(t)}{dt}\right] \quad (6)$$

In the formula, $e(t) = r(t) - c(t)$ is the system deviation; K_P is the coefficient of proportionality; T_I is the integral time constant; T_D is the differential time constant.

In the four axis flight vehicle system, the four axis can be adjusted quickly by the PID control algorithm, so that the flight parameters can be adjusted rapidly, and the stability of the four axis flight state can be maintained.

4 Design of the system

4.1 overall design ideas

The design uses Android mobile phone as the remote control, through WiFi to send control instructions to the four axis aircraft, to achieve the lifting, pitching, steering, rotating and other flying posture.After setting up the WiFi connection with the four axis, the mobile phone sends the relevant control data by judging the user gesture operation. The overall design flow chart of this system is shown in fig.2.

4.2 App software design

Open the App, first stop 2S in the interface, and then enter the squared unlock page, if successful unlock mobile phone will automatically determine the current WiFi connection status, when success is connected with the four shaft, start recording connection time, while users can manipulate the two button to move on the screen, and the calculated coordinate information button control different data, sent through the network to the WiFi module on the four axis aircraft.

The flow chart of the App software design is shown in fig.3.

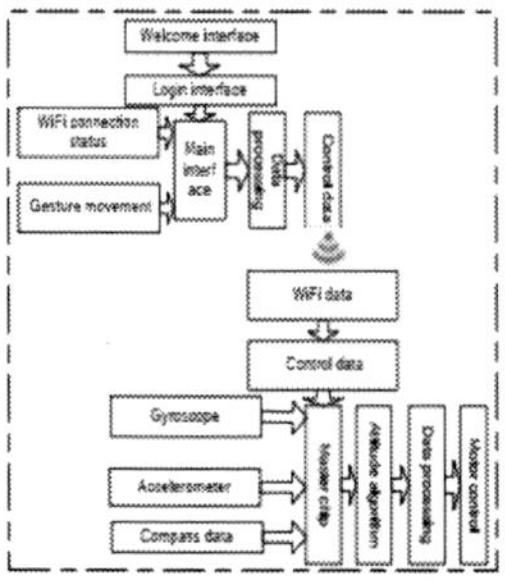

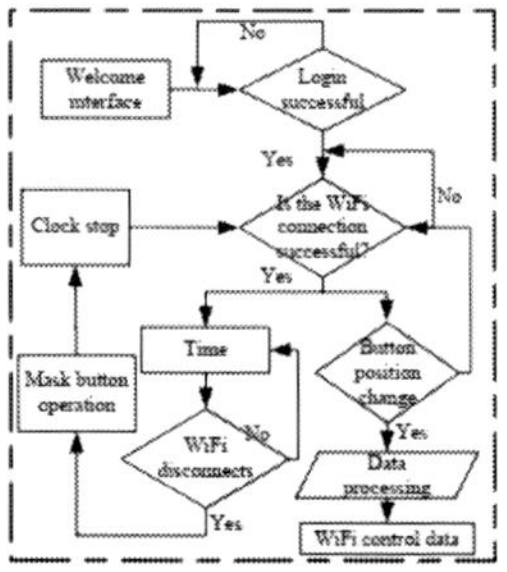

Fig.2. Overall framework of the system

Fig.3. App design flow chart

 DOI: 10.1051/matecconf/201713900160

4. 3 four axis aircraft design

The four axis aircraft uses MWC open source firmware as the flight control board, and the master chip determines the flying posture of the aircraft through related sensors. After processing the WiFi data sent from the Android mobile terminal, the PPM control signal is transmitted to the master chip, and then processed into a signal to control the motor to control the attitude of the four axis flying. The design flow chart of the four axis aircraft is shown in fig.4.

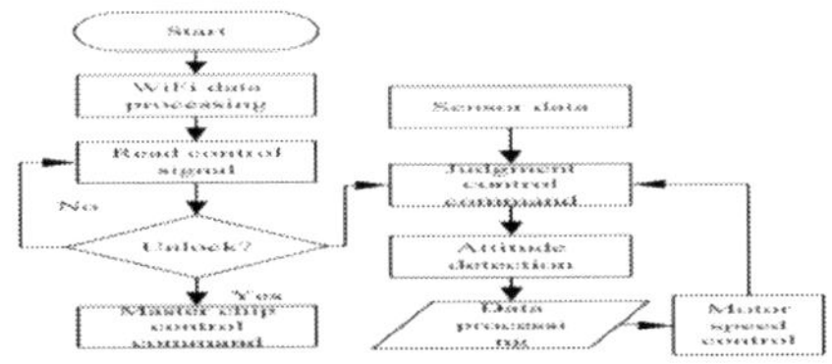

Fig.4. four axis design flow chart

5 Design and implementation of App software

5.1 App main interface

5.1.1 layout of main interface

The interface layout file uses the Frame Layout. Frame Layout is a layer layout, and all controls are placed in the upper left corner of the screen. The defined controls are superimposed up and down in sequence, and the controls that are placed will be displayed directly on the previous control. The previous control is blocked when the previous control size is smaller than the later control.

In the main interface, each side has a circle with a larger radius, which is the range that can be moved when the user moves the button, and the button can only move freely within the range. The upper left corner of the screen is the text displayed by the wifi connection state, and the user can see whether the current connection is wifi and the name of the connected wifi. The upper right corner of the recording connection time App and four aircraft, when disconnected, stop time, can help the user to observe the stability of the connection. There are 4 vertical bar in the middle of the screen, with the graphical representation of the current amount of rudder throttle, direction, pitch and aileron of the 4 channel value, users can easily visualize the operation of each channel.

In the main interface layout, in addition to the system controls, but also custom control, on both sides of the screen painted a radius of 350px circle, the circle is the range of movement of the control button. Inside the circle, two intersecting straight lines provide the button position reference function for the user. The main interface measured as shown in figure 5.

5.1.2 App detects the WiFi connection state program

Detection of wifi connection state in App through the radio to listen to the wifi i connection state, when the program into the main interface, the main interface has been through the radio to listen to the wifi connection state, when the Android mobile phone connected to the wifi, the program will automatically try to connect four aircraft, and the bomb box prompts the user is working with four axis connection. Until the connection is successful, the frame closes and starts recording the connection time, then the user can manipulate the button and send a control command to the four axis aircraft. Interface measurements when connected to the WiFi are shown in fig.6.

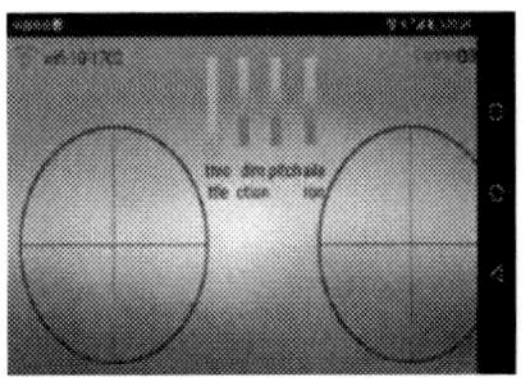 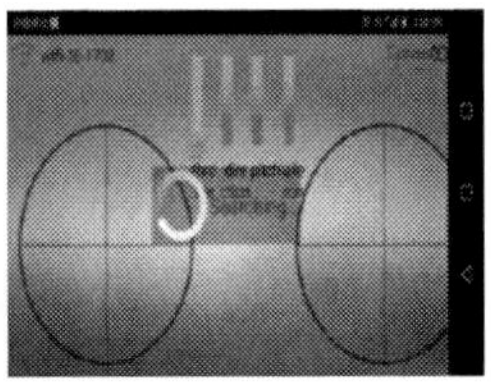

Fig.5. App master interface measurement

Fig.6. App connection four axis measurement

When the program through the radio monitoring to wifi is disconnected, the first button is set to click, is unable to send instructions to the four axis, at the same time, stop the clock connection time, and pop-up dialog box prompts the user, if you set the wifi connection. Click Cancel, the frame disappears, and click OK to jump to the system wifi settings page. When the wifi is disconnected, the main interface is shown in fig.7.

5.1.3 App connect four axis method

When the main interface is connected to the wifi, the method is called to open a thread, and the program has been in the connection loop since it came in, trying to connect to the IP address of the wifi module on the four axis aircraft. If no connection is successful within a specified time, the connection exception is indicated. If the connection is successful, you will jump out of the loop, send the connection success message to the main thread, close the connection prompt box, set the button to clickable status, and start recording the connection time. The actual measurements of the App and four axis connections are shown in fig.8.

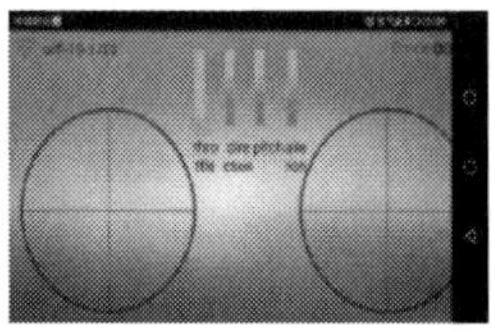

Fig.7. WiFi cut off the shot frame

Fig.8. App connection successfully measured

5.1.4 App send and manipulate instructions

After a successful connection, the user through the mobile button on the left and right sides to control four axis aircraft, when the user presses the button, the program will capture the current screen touch points in the X and Y axis coordinate point, when the finger moves in the circle area, the program will change X and Y axis control with respect to the amount of acquisition and, to control set new coordinates, so that the control can be moved with the movement of the finger. In the process of moving, the program will take the measured coordinate value and coordinate value changes as parameters, calculate the control data, and then send the control data to the four axis aircraft through the network. Then, the data is received by the WiFi module on the four axis aircraft and transmitted to the master chip through the serial port.

6 Summary

The four axis aircraft designed in this paper is a portable aircraft with a small volume, which is powered by 4 brushless motors, with high flexibility and flexibility. The whole machine is fully electronic stabilized and easy to operate. It has a wide range of applications in many fields. Four axis aircraft flight control system based on Android design in this paper, the Android mobile phone is connected with the four axis aircraft by launching four aircraft wifi, off user with the appropriate App can control the four axis aircraft, landing and other attitude change. Not only the operation is simple, but also eliminates the traditional handheld wireless remote control, increasing the user's operating experience.

ACKNOWLEDGMENT

This work was supported by the Scientific Research Project of Anhui Xinhua College No. 2017zr003 and Natural Science Project of Anhui province No.KJ2017A618.

References

1.Zheng Xu, Ji Zhijian, Pan Zhenzhen. Control and development of APP client for four axis aircraft based on Android [J]. industrial control computer,6(2017)

2.Yang Shubin, Huang Junrong. Design of small four axis aircraft based on real-time operating system [J]. Journal of Wuhan Institute of Technology, 1(2017)

3.Chen Peng exhibition, Ru Yan, Chen Xiaoyue. Simulation study on Optimization Strategy of control system of four axis aircraft [J]. computer simulation, 12(2016)

4.Zhong Yongxi, Wu Liming, Zhang Shanshan,.WiFi system design of Micro Four Axis aircraft, [J]. single chip microcomputer and embedded system application, 11(2016)

5.Ma Zhenghua, Sun Bin. Design of short range low altitude transportation system based on four axis aircraft [J]. computer measurement and control, 10(2016)

6.Han Ping, Sang Wilin, Shi Qing research. A new nonlinear Calman filtering method [J]. Chinese Journal of scientific instrument, 3(2015)

7.Liu Haopeng, Long Changjiang River, Wan Peng, Wang Xiaoyi, Hu Ben. Fuzzy PID control of plant protection four axis aircraft [J]. Chinese Journal of agricultural engineering,1(2015)

FM jamming cancellation suppression based on noise floor estimation with dynamic threshold

Jie Bai[1] , Guohong Wang[1] ,Min Wu[2] ,Shichao Liu[1] ,Yang Lin[1],Xin Tingting[1]

[1] Naval Aeronautics University , Yantai, China
[2] 75841 troop, Changsha, China

Abstract. Technique of FM jamming time domain cancellation suppression based on noise floor estimation with dynamic threshold is researched, by using the colored noise NLR filter estimation algorithm instead of the narrowband filtering algorithm, the problem that the target echo signal is lost in the process of jamming amplitude estimation is solved. Through the phase analysis of the jamming signal, the problem of jamming signal amplitude estimation in the background of echo signal is transformed into the problem of useful signals spectral estimation in colored noise background. Combined with the characteristics of the low frequency components of the useful signal, the traditional NLR filtering algorithm is improved by adding the dynamic threshold, so as to achieve the purpose of colored noise's non-invasive estimation, and then to obtain more accurate jamming signal amplitude estimation. The simulation results show that compared with the traditional jamming cancellation algorithm, this algorithm can improve the SJR of the echo signal better and the anti - jamming ability is stronger.

1 Introduction

FM jamming has advantages such as variable bandwidth and small amplitude dynamic range, because of which it has a wide range of applications in the field of radar confrontation. At present, scholars have done a lot of research on FM jamming. Literature [1] analyzes the signal, spectrum and the interference effect on pulse compression radar of the FM jamming; literature [2] introduces random differential theory, and studies the interference effect the FM jamming has on pulse compression filter. In the literature [3], the radar signal containing the FM jamming signal is decomposed in different frequency bands by multi-scale time-frequency analysis and reconstruction of the wavelet packet, and then the original signal is reconstructed to realize the interference suppression.

In those literatures, the generation and suppression of FM jamming are studied respectively, but the suppression algorithm of strong FM jamming is less. In response to this problem, the literature [4] proposed a noise cancellation interference suppression algorithm based on time domain cancellation, the mathematical model and the main parameters of FM jamming are discussed. The interference amplitude is estimated by logarithmic transformation and the jamming suppression is accomplished by reconstruction. However, when the jamming signal could not enter the receiver totally, the constant signal characteristics of the jamming signal is destroyed, thereby reducing the effectiveness of the jamming suppression algorithm. The literature [5] proposed the amplitude disturbance of the FM jamming suppression algorithm, through the echo signal phase demodulation and narrowband filtering to estimate the disturbance, and then complete the time domain cancellation. However, this algorithm has lost in the target echo signal, and is not conducive to the detection of the target. To solve this problem, in this paper, the jamming reduction algorithm is improved by combining with the noise floor estimation algorithm.

2 Suppression of FM Jamming Based on Time Domain Cancellation

2.1. Jamming Cancellation Algorithm

In the background of FM jamming, the radar received echo signal model expresses as:

$$x(t) = s_r(t) + J_{FM}(t) + n(t) \tag{1}$$

In formula (1), $s_r(t)$ and $n(t)$ represent target echo signal and Gaussian white noise. FM jamming with amplitude disturbance expresses as:

$$J_{FM}(t) = \left(A_j + \zeta(t)\right)\exp\left(j\phi_j(t)\right) \tag{2}$$

In formula (2), $\zeta(t)$ represents the amplitude disturbance term, suppose that the $\zeta(t)$ is a narrowband low frequency Gaussian random noise with mean of 0. $\phi_j(t)$ is the instantaneous phase, subjecting to 0 mean random distribution, $\zeta(t)$ and $\phi_j(t)$ are independent of each other.

* Corresponding author: 1541753296@qq.com

In strong interference environment, the target echo signal and the noise signal are much weaker than the jamming signal, therefore, put the target echo signal and process noise into one, formula (2) can be simplified as:

$$\begin{aligned} x(t) &= s'(t) + J_{FM}(t) \\ &= A'(t)\exp(j\phi'(t)) + (A_j + \zeta(t))\exp(j\phi_j(t)) \end{aligned} \tag{3}$$

In formula (3), $A'(t)$ and $\phi'(t)$ represent the instantaneous amplitude and instantaneous phase of $s'(t)$. A' and A_j have the following relationship:

$$A'(t) << A_j \tag{4}$$

For further transformation, formula (3) can be expressed as:

$$x(t) = J_{FM}(t)(1 + s_1(t)) \tag{5}$$

Where,

$$\begin{aligned} s_1(t) &= \frac{s'(t)}{J_{FM}(t)} \\ &= \frac{A'(t)}{A_j + \zeta(t)}\exp(j(\phi'(t) - \phi_j(t))) \end{aligned} \tag{6}$$

Logarithmic transformation to both sides of formula(5), and use first order approximation of Taylor expansion, the following formula is got:

$$\begin{aligned} \ln(x(t)) &= \ln(J_{FM}(t)) + \ln(1 + s_1(t)) \\ &= \ln(A_j + \zeta(t)) + j(\phi_j(t) - 2k\pi) + s_1(t) \end{aligned} \tag{7}$$

Where, $k = \mathrm{mod}(\phi_j(t), 2\pi)$.

The imaginary part of (7) is:

$$I(t) = \mathrm{Im}[\ln(x(t))] = \phi_j(t) - 2k\pi + \mathrm{Im}[s_1(t)] \tag{8}$$

For $|s_1(t)| << 1$, $I(t) = \mathrm{Im}[\ln(x(t))] \approx \phi_j(t) - 2k\pi$ can be got from formula (8), which is the phase estimation of jamming signal. The phase demodulation of the echo signal is carried out according to the interference instantaneous phase estimation:

$$\begin{aligned} y(t) &= x(t)\exp(-j\,\mathrm{Im}(\ln(x(t)))) \\ &= [s'(t) + J_{FM}(t)]\exp(-j\,\mathrm{Im}(\ln(x(t)))) \\ &= s'(t)\exp(-j\,\mathrm{Im}(\ln(x(t)))) + (A_j + \zeta(t))\exp[-j\,\mathrm{Im}(s_1(t))] \end{aligned} \tag{9}$$

The jamming item is indicated by $\vartheta(t)$:

$$\begin{aligned} \vartheta(t) &= (A_j + \zeta(t))\exp[-j\,\mathrm{Im}(s_1(t))] \\ &= (A_j + \zeta(t))\exp[-j\kappa(t)] \end{aligned} \tag{10}$$

Where, $\kappa(t)$ is a random process that satisfies the following formula:

$$|\kappa(t)| << 1 \tag{11}$$

Do approximation of Taylor expansion to formula (11):

$$\begin{aligned} \vartheta(t) &= (A_j + \zeta(t))\exp[-j\kappa(t)] \\ &= (A_j + \zeta(t))\left[1 - j\kappa(t) - \frac{\kappa^2(t)}{2!} + \cdots\right] \\ &= (A_j + \zeta(t)) + (A_j + \zeta(t))o(\kappa(t)) \end{aligned} \tag{12}$$

Because $|\kappa(t)| << 1$, the following formula holds:

$$\begin{aligned} \vartheta(t) &= (A_j + \zeta(t))\exp[-j\kappa(t)] \\ &\approx (A_j + \zeta(t)) \end{aligned} \tag{13}$$

Therefore,

$$\begin{aligned} y(t) &= x(t)\exp(-j\,\mathrm{Im}(\ln(x(t)))) \\ &\approx s'(t)\exp(-j\,\mathrm{Im}(\ln(x(t)))) + (A_j + \zeta(t)) \end{aligned} \tag{14}$$

For $y(t)$, it can be seen as the sum of weak echo signal entry and strong jamming signal entry, among which, the echo signal is a broadband signal, while the jamming signal is a narrowband signal under the assumption. At the same time, because the echo signal and jamming signal are generated in different ways, so the power rate changes differently. Based on the above analysis, the echo signal term is regarded as non-Gaussian colored noise, and the jamming signal term is regarded as the target signal. The estimation problem of the jamming signal item can be converted into the detection problem of the target signal in the background of colored noise. To solve the problem: the noise floor of the input signal is first estimated and then the power spectrum of the target signal is corrected according to the estimation result.

2.2 Noise Floor Estimation Algorithm Based on NLR Filter

Let $S(n)$ be the input signal power spectrum, $Y(n)$ be the estimated noise base, the noise base estimation process based on NLR filtering is represented by the following recursion:

$$Y(k) = Y(k-1) + e(k) \tag{15}$$

$$e(k) = \begin{cases} S(k) - Y(k-1) & |S(k) - Y(k-1)| < \lambda \\ \lambda\,\mathrm{sgn}(S(k) - Y(k-1)) & else \end{cases} \tag{16}$$

Where, λ is the threshold, $\mathrm{sgn}(x)$ is the symbolic function. When $x < 0$, $\mathrm{sgn}(x) = -1$, when $x > 0$, $\mathrm{sgn}(x) = 1$.

It can be seen from formula (16), when the power spectrum of the input signal changes slowly, the power spectrum difference between adjacent two moments is smaller than the threshold value, and the power spectrum value at the next time is directly equal to the input signal power spectrum, while when the power spectrum of the input signal changes rapidly, the power spectrum difference value of the adjacent two moments is greater than the threshold value, and the power spectrum value at the next time is increased by the smaller step size. It can be seen that the NLR filtering effect is closely related to the threshold, and the threshold changes with the bandwidth of the input signals. When the input signal's bandwidth is wide or includes signals in different bandwidth, the threshold selection of the NLR filtering algorithm will be contradictory. The threshold setting is too large, which will cause the noise power spectrum estimation value to rise faster, resulting in the estimation of the noise power spectrum. If the threshold setting is small, the noise base estimation The change step is small, resulting in noise power spectrum estimation.

2.3 NLR Filtering Algorithm Based on Dynamic Threshold

From the formula (14), it can be seen that in the background of this paper, the input signal is composed of the weaker broadband noise and the strong narrowband low frequency target signal, so the low frequency target signal in the input signal is very prominent, it is easy to get the start and end locations of the power spectrum of target signal. In order to achieve the lossless estimation of the noise base in the no-target signal band, it is necessary to select a large threshold to match the rate of change of the noise floor fluctuation. In the target signal band, improve the Traditional NLR Filtering Algorithm: the adjustment factors a and b are used to dynamically adjust the threshold value, （$a<1$， $b>1$）, so that in the target signal frequency band, the noise base can be more accurate non-destructive estimation, so as to obtain a more accurate target signal power spectrum correction results.

In the target signal band, the adjustment factor on the threshold of the correction process is as follows: when the value of $S(k)-Y(k-1)$ is greater than 0 for multiple times, it can be considered that this location is the start point of the target signal power spectrum, In this case, the threshold λ is corrected to be:

$$\lambda=\lambda a^{n}, a<1 \tag{17}$$

Where, n is the number of times that $S(k)-Y(k-1)>0$ appears continually. Using the adjustment coefficient a to gradually reduce the threshold to avoid that the noise floor is estimated higher than the true value. When the value of $S(k)-Y(k-1)$ is smaller than 0 for multiple times, it can be considered that this location is the end point of the target signal power spectrum, In this case, the threshold λ is corrected to be:

$$\lambda=\lambda b^{n}, b>1 \tag{18}$$

Where, n is the number of times that $S(k)-Y(k-1)<0$ appears continually. Using the adjustment factor b to gradually increase the threshold, the noise base estimate is rapidly decreased, approaching the true noise base as soon as possible.

The introduction of the adjustment coefficient solves the problem of noise base over-estimation and under-estimation caused by the fixed threshold. but at the same time, to make the filtering effect better in the whole frequency, the filtering threshold of the target signal band λ_s , the initial threshold value λ_j and the adjustment coefficient a and b at the target signal band need determining

For the threshold λ_s at the target signal band, in order to make the estimation algorithm match the change rate of the noise floor, select the maximum value of the adjacent time power spectrum difference as the threshold value:

$$\lambda_s=\max\left[S_s(k+1)-S_s(k)\right], k=1,2\cdots \tag{19}$$

Where, $S_s(k)$ is the band of no target signal.

Since the NLR filtering algorithm is based on the difference between the input signal power spectrum and the noise floor estimation, the relationships between the power spectrum and the initial threshold, as well as the adjustment coefficient are analyzed at the target signal band. The initial threshold λ_j is determined by using the average difference of the adjacent time of the input signal power spectrum, and the value of the adjustment coefficient a , b are determined according to the selected initial threshold and the noise floor fluctuation.

It can be seen from a large number of simulation experiments that, take half of the average difference in the power spectrum of the adjacent time as the initial threshold, the estimated value of the noise base can be matched with the true value. So the calculation of the initial threshold is as follows:

$$\lambda_j=\frac{1}{2}\frac{1}{N-1}\sum_{k=2}^{N}d(k) \tag{20}$$

Where, $d(k)=S_j(k)-S_j(k-1), k=1,2\cdots$, $S_j(k)$ indicates the target signal band.

Under the action of the adjustment coefficient a , the noise base estimation value gradually increases and eventually stabilizes. Due to the adjustment of the coefficient a , the maximum estimated value of the noise base is:

$$\lambda+a\lambda+a^{2}\lambda+\cdots \approx \lambda\frac{1}{1-a} \tag{21}$$

Let the noise floor fluctuation of input signal may reach be m dB, then make

$$\lambda\frac{1}{1-a}=m \tag{22}$$

The value of the available adjustment factor a is $1-\lambda/m$.

Since the value of the maximum noise base fluctuation is not known precisely, or the value of the noise floor fluctuation changes with time due to the time-varying characteristics of the channel, the noise floor fluctuation is deviated from the real situation. When the estimated value is less than the true value, it may cause the estimated value of the noise floor from the true value. Therefore, if the undulation of the real noise could not be predicted, the noise floor can be selected to ensure no loss of the noise floor estimated value.

Under the action of the adjustment coefficient b , the noise floor estimate is rapidly decreased at the end of the power spectrum of the target signal to approximate the true value of the noise floor as soon as possible, which has little effect on the estimation of the noise base, and any optional adjustment coefficient that satisfies $b>1$ is ok.

At the same time, in order to solve the problem of asymmetry of target signal power spectrum caused by adjustment coefficients a and b in different values. The input signal power spectrum is arranged in reverse order and the dynamic threshold NLR filter is re-estimated, get the average of both sides. By means of the mean value, the spectral peak shape of the target signal can be substantially preserved, and the signal-to-noise ratio loss caused by the excessive noise floor estimation can be reduced. The estimated value of the power spectrum of the target signal can be obtained by subtracting the estimated value of the noise spectrum of the input signal power spectrum.

NLR filtering algorithm based on dynamic threshold makes it possible to estimate the power spectrum of the interfering signal more accurately. Compared with the narrowband filter estimation results of the interfering signal in the literature [5], this algorithm not only can get more accurate interference signal estimation, and can effectively reduce the loss of the echo signal. The estimation of the interference signal term $\varsigma(t)$ can be obtained by NLR filtering, reference formula (2), the reconstruction of FM jamming signal with amplitude disturbance is expressed as followed:

$$\hat{J}_{FM}(t)=\varsigma(t)\exp(jI(t)) \tag{23}$$

Where $I(t)$ is the instantaneous phase estimation of the interfering signal obtained from formula (8). Eliminate the reconstructed interference signal from the echo signal $x(t)$:

$$\begin{aligned}\hat{s}'(t)&=x(t)-\hat{J}_{FM}(t)\\&=s'(t)+J_{FM}(t)-\hat{J}_{FM}(t)\end{aligned} \tag{24}$$

Through the time domain cancellation algorithm, in the more complete preservation of the echo signal under the premise of most of the FM jamming signal energy is eliminated, which is conducive to the subsequent detection of the target.

3 Simulation results and analysis

Simulation parameter settings: LFM signal pulse width $T_p=30\mu s$, the carrier frequency $f_0=4MHz$, bandwidth $B=2MHz$. The carrier frequency of FM jamming signal $f_j=4MHz$, modulation index $m_{fe}=1000$, the process noise is Gaussian white noise. Radar sampling frequency $f_s=20MHz$, FM jamming signal bandwidth $B_j=0.3$ (normalized frequency), SJR is $-30dB$, SNR is $-10dB$. Under this simulation condition, the validity of the algorithm is verified.

3.1 Verification of Dynamic Threshold NLR Filtering Algorithm

Determine the threshold λ_s, the initial threshold λ_j, and the value of the adjustment factor a according to formula (23), (24) and (26), adjustment factor $b=1.5$. In the band with the jamming signal, the threshold λ is dynamically adjusted according to formula (21) and (22). Figure 1 shows the results of the power spectrum estimation of the echo signal of the traditional NLR filter algorithm and the correction of the power spectrum of the interference signal under the condition of the simulation parameters. Figure 2 shows the results of the power spectrum estimation of the echo signal of the dynamic threshold NLR filter algorithm and the correction result of the power spectrum of the interference signal under the simulation parameters.

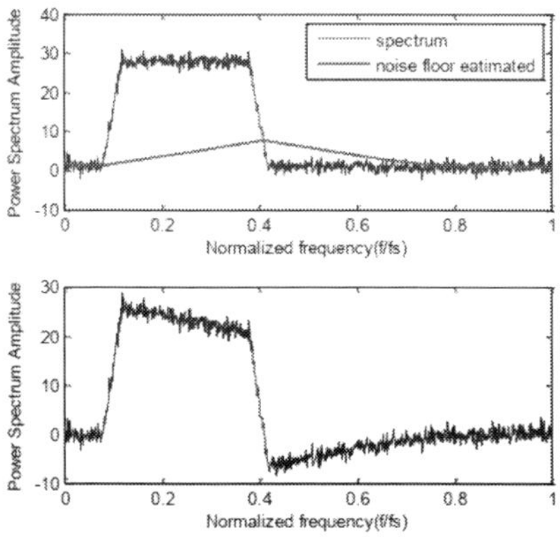

Figure.1 Traditional NLR filtering algorithm

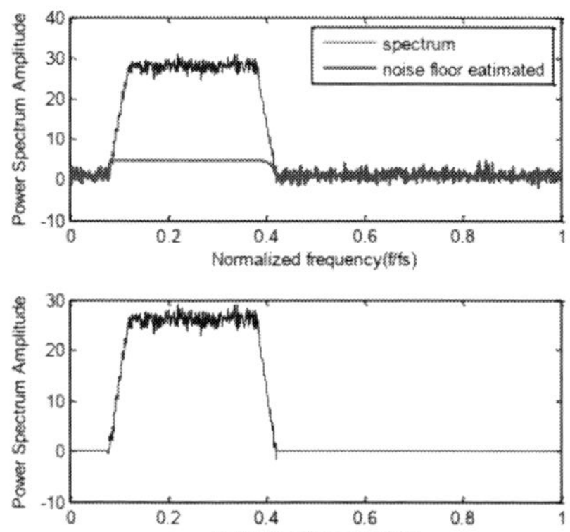

Figure.2 NLR filtering algorithm with dynamic threshold

In figure 1 and figure 2, the abscissa indicates the normalized frequency, and the ordinate represents the power spectral amplitude value. The red lines in figure 1 (a) and figure 2 (a) show the power spectrum estimation of the echo signal term. It can be seen from the comparison of figure 1 and figure 2 that the traditional NLR filtering algorithm has a large error in the power spectrum estimation result of the interference signal. At the end of the power spectrum of the interfering signal, the estimation result is difficult to converge quickly and the dynamic threshold NLR filtering algorithm can realize the accurate estimation of the power spectrum of the interference signal in the whole frequency band, and it will not cause the loss of the echo signal term.

3.2 Verification of validity of cancellation algorithm

According to the dynamic threshold NLR filtering results, the interference signal is reconstructed and canceled in the time domain. Figure 3 and Figure 4 show the pulse compression results of the echo signal before and after the cancellation.

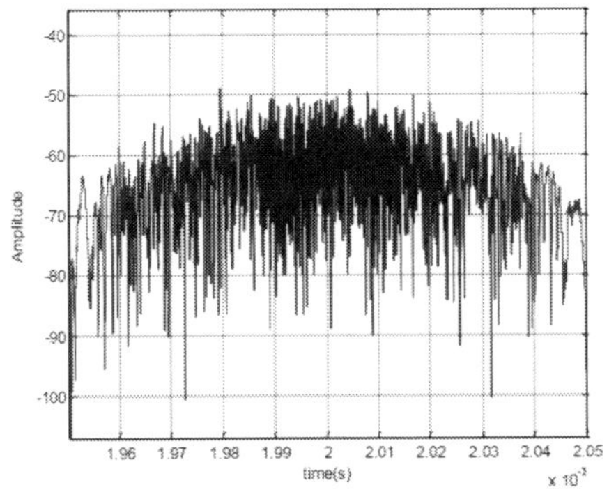

Figure.3 The pulse compression before the interference cancellation

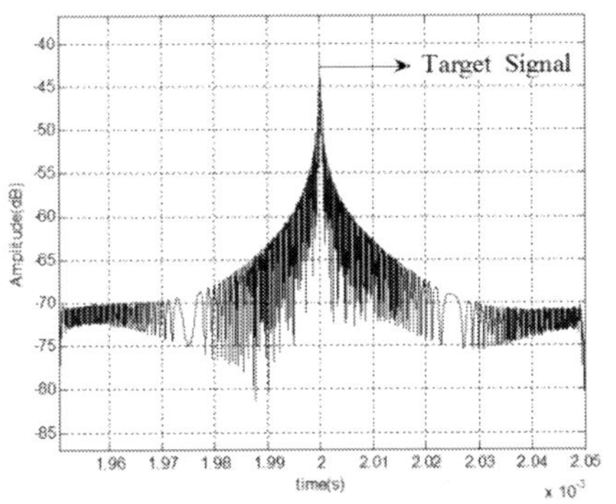

Figure.4 The pulse compression before and after the interference cancellation

The abscissa represents the time and the ordinate represents the signal amplitude in figure 3 and figure 4. As can be seen from Figure 3 and Figure 4, the time domain interference cancellation algorithm eliminates most of the noise Energy, in order to achieve a strong FM jamming environment for the effective detection of the target signal.

3.3 Algorithm comparison

Change the FM jamming amplitude disturbance signal bandwidth range of 0 to 0.5 (normalized frequency), where the bandwidth of 0 means that the interference amplitude is not received disturbance . Figure 5 and figure 6 show the signal-to-interference ratio (SJR) curves of the literature [5] and the algorithm of this paper in different perturbation signal bandwidth.

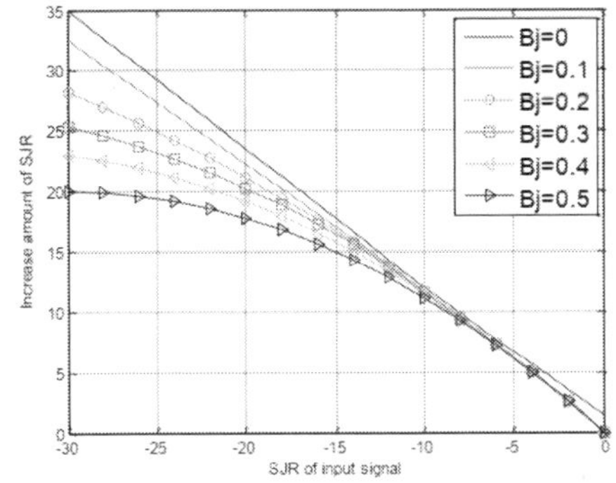

Figure.5 SJR increased amount of literature [5]

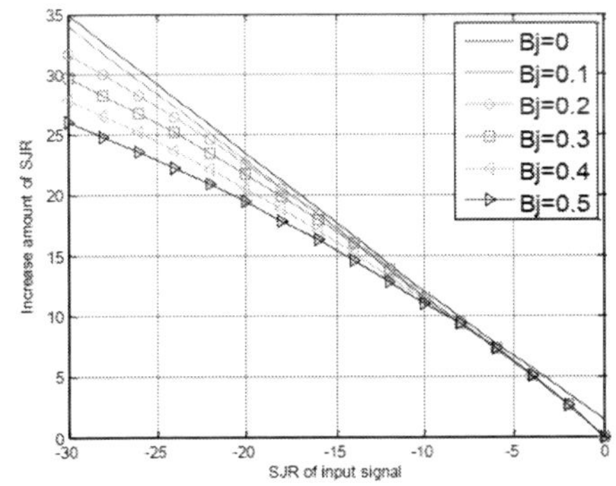

Figure.6 SJR increased amount of this paper

The abscissa indicates the signal-to-noise ratio of the echo signal, and the ordinate represents the increase in the signal-to-noise ratio of the output signal in figure 5 and figure 6. It can be seen from figure 5 that the narrowband filtering algorithm in literature [5] not only could not achieve the accurate estimation of the interference signal term, but also causes the loss of the target echo signal, and the greater the bandwidth of the disturbance signal, therefore, the SJR increases with the widening of the disturbance signal bandwidth under the same conditions. As can be seen from figure 6, since the algorithm proposed in this chapter does not cause loss to the target echo signal, the change in SJR increases with the disturbance signal bandwidth compared with the literature [5], due to the interference signal Estimate of the error. Through the comparison of figure 5 and figure 6, we can see that the algorithm in this paper solves the shortcoming of the loss of echo signal in literature [5] algorithm. In the strong FM jamming environment, the improvement of the signal-to-interference ratio is superior to that of the literature [5].

Conclusions

In this paper, the problem of loss of the target echo is solved by using the NLR filtering algorithm instead of the narrowband filter algorithm. According to the application background of NLR filtering, this paper combines the low frequency components of the input signal, The traditional NLR filtering algorithm is improved to complete the nondestructive estimation of the echo signal term, and the estimation of the

 DOI: 10.1051/matecconf/201713900161

interference signal term is corrected to achieve the better reconstruction effect of the interference signal. It can be seen from the simulation results that this algorithm has better interference suppression ability in strong noise FM interference environment.

References

1. Y.S. Li, H.Lv, and Y.J.Liu. Research on Noise FM Jamming and its Realization on FPGA, Electronic Science and Technology, 28,(2015), 139-141.
2. Q.Q. Sun, K.Zheng, X.Y.Cao, Jamming Mechanism of Wideband Noise FM Signals to Pulise Compression, RADAR & ECM, 36,(2016), 30-34.
3. Z.G. Liu, W.Wang. J.Y.Zhang., Research on shipboard radar anti noise frequency modulation jamming based on envelope filtering, SHIP SCIENCE AND TECHNOLOGY, 31,(2009),70-72.
4. H.L. Dong, S.F. Li, D.L.Luo. Simulation Analysis of Anti-noise FM Jamming Algorithm Based on Time-domain Cancellation, Journal of Air Force Early Warning Academy, 4,(2013), 249-251.
5. D.P.Du, Research on Radar Suppression Interference Suppression Algorithm. Chengdu: University of Electronic Science and Technology, 2008, 83-87.

Research on MPPT Algorithm Based on Mixed Strategy

Yu Chai [1], *Xu Fang*[2]

[1,2]Xi'an University of Science and Technology, Xi'an 710054, China

Abstract: Based on the mathematical model of PV array and its output characteristics, a hybrid control strategy combining fixed step method and variable coefficient approximation method is proposed. The method is followed by a fixed step in the search of the maximum power point. When the slope of the PU curve changes for the first time, the search interval is fixed in an initial step, and then the search is gradually approximated to maximum power point in the section. The method is analyzed theoretically and modeled in Matlab / Simulink environment, and compared with the variable step perturbation observation method. The simulation results show that the proposed method can quickly and accurately track the maximum power point under the standard condition and the sudden change of light intensity, and has faster tracking speed.
Keywords: PV array; hybrid strategy; fixed step; progressive approximation method; maximum power point

1 Introduction

Traditional fossil energy consumption brought about environmental pollution, greenhouse effect and many other environmental problems, it has been a great threat to human survival and development. Solar energy is a clean, safe, reliable and well-stocked renewable energy source that has become a major alternative to traditional fossil fuels. The output voltage and output current of the PV array vary with the temperature, light intensity and other external environment changes, and at a particular voltage in order to make the output power value to the maximum. Therefore, it is necessary to add the maximum power tracking module to the PV array to stabilize its output power at maximum. The variable perturbation observation method is a kind of control algorithm that easy to implement and widely used, but the algorithm itself has some defects, including the optimization speed is slow, also easy to oscillate at the maximum power point.

In this paper, based on the analysis of variable step perturbation observation method, a hybrid control strategy combining fixed step and variable coefficient approximation is proposed. The method can be divided into two steps, first to a fixed step to search. According to the P-U characteristic curve of the PV array, it is known that when the slope is changed for the first time during the search, the search range can be fixed within an initial step range. In this range, the gradually approximation of the variable coefficients is performed. Compared with the perturbation observation method with slope as variable step size, the method has its advantages both in tracking speed and eliminating oscillation.

2 The output curve of photovoltaic cells

Under standard test conditions(1.illumination 1000W/㎡;2.temperature 25℃; 3.spectral characteristics AM1.5 standard spectrum), according to I_{sc}, U_{oc}, I_m, U_m parameters can get the output curve of the current.

$$I_L = I_{sc}[1 - C_1(e^{\frac{U_m}{C_2 U_{oc}}} - 1)] \tag{1}$$

$$C_1 = (1 - \frac{I_m}{I_{sc}})e^{\frac{-U_m}{C_2 U_{oc}}} \tag{2}$$

$$C_2 = (\frac{U_m}{U_{oc}} - 1)[\ln(1 - \frac{I_m}{I_{sc}})]^{-1} \tag{3}$$

The current output curve in the above formula is the standard test condition, and when the temperature and light intensity change does not equal to the reference temperature and the reference light intensity, this formula is not applicable, it is need to amend to describe the new characteristic curve. First, the temperature and light intensity are corrected.

$$\Delta T = T - T_{ref}\ ,\quad \Delta S = \frac{S}{S_{ref}} - 1 \tag{4}$$

* Corresponding author: fangxu152@sina.com

T_{ref} (Reference temperature)

S_{ref} (reference light intensity)

The parameters are optimized under temperature and light intensity correction.

$$I_{sc}' = I_{sc}\frac{S}{S_{ref}}(1+\alpha\Delta T) \quad (5)$$

$$U_{oc}' = U_{oc}(1-\gamma\Delta T)\ln(1+\beta\Delta S) \quad (6)$$

$$I_{m}' = I_{m}\frac{S}{S_{ref}}(1+\alpha\Delta T) \quad (7)$$

$$U_{m}' = U_{m}(1-\gamma\Delta T)\ln(1+\beta\Delta S) \quad (8)$$

Typical value: α =0.0025/°C, β =0.5,

γ =0.00288/°C.

The optimized parameters are brought into the equation (1) can get the curve with different temperature and light conditions.

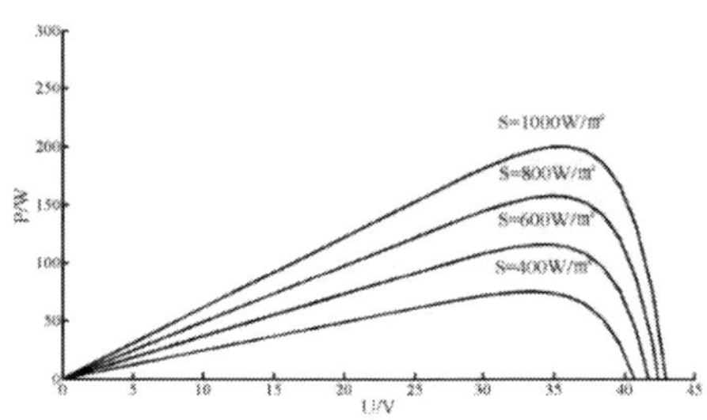

Fig.1. P-U characteristic curves of PV array for different insolations (25°C)

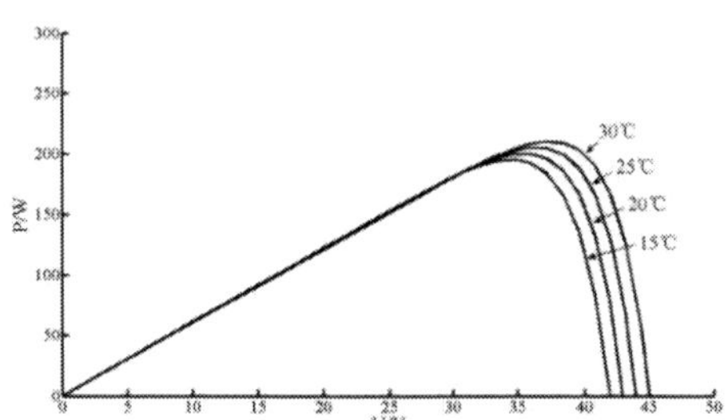

Fig.2. P-U characteristic curves of PV array for different temperature (1000W/m²)

Figure 1 and Figure 2 shows that when the light intensity and temperature conditions change, the output curve will be offset. In a particular light intensity or temperature conditions under the curve, the voltage increases when the output power first monotonically increasing and then monotonically decreasing, at a certain voltage output power to reach the maximum. The search process of the maximum power point is essentially a singular optimization process.

2 Research on Maximum Power Point Algorithm

2.1 variable step perturbation observation method

Perturbation observation method is a control method that widely used in practice, the method by applying a disturbance to the system to determine the next step. If the power increases, then continue with the direction of disturbance, if the power is reduced, then change the direction of disturbance, until the step or power difference reaches the set threshold, then found that the maximum power point has been reached at this time. The traditional disturbance observation method takes a fixed step tracking. If the step size is too small, the tracking speed is slow; if the step size is too long, it will oscillate near the maximum power point, resulting in energy loss so that can not track the maximum power point.

The variable step perturbation observation method is optimized for the slope, with the slope as the variable step size coefficient. When away from the maximum power point, the slope $|k| = dP/dU > 0$. At maximum power point, $|k| = dP/dU = 0$,step size $D_k = kD_{k-1}$. Through the slope to optimize the step size, real-time change step size. When away from the maximum power point, the long size step is used to track quickly. When close to the maximum power point, the small size step is used to track stably. In order to quickly and accurately track, the process of finding the maximum power point need to constantly change the step size. Compared with the fixed-step perturbation method, the control effect of the variable-step perturbation observation method is more rapid and accurate.

2.2 Mixed control strategy

Based on the variable perturbation observation method, this paper proposes a hybrid control method combining fixed step method and variable coefficient continuous approximation method.

As shown in Fig. 3, a fixed step is selected to start the search with the PV array's P-U characteristic curve as a reference. The search process is performed from the left side of the curve to the right, and the slope of the curve is positive at the beginning of the search, and the slope of the curve is negative at the right of the maximum power point.

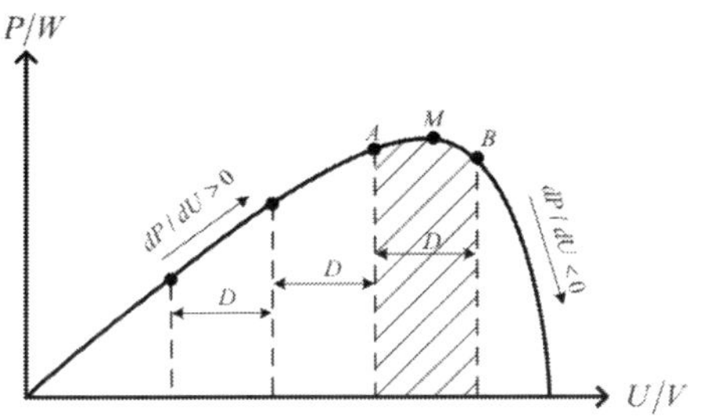

Fig.3. The first curve of search

In the stepwise search process, when the slope changes, that is, the first time from positive to negative (Figure 3, point A to point B), indicating that at this time has crossed the maximum power point, the maximum power point should be in a fixed initial step within the area. Then change the search strategy, reverse to γD step to search, γ range is from 0.5 to 1(γ is 0.618 in this paper), so that within a certain range does not appear multiple searches.

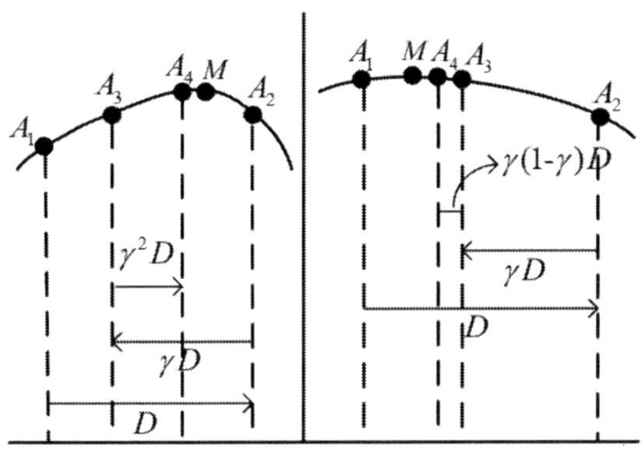

Fig.4. The second curve of search

If the slope does not change, indicating that the maximum power point has not been exceeded, continue with the direction of search, the maximum power point should be in the remaining $(1-\gamma)D$ range, select the step $\gamma(1-\gamma)D$ for the same direction search. If the slope changes, indicating that the maximum power point has been crossed, the maximum power point should be in the γD range, then the search direction should be changed to search in step $\gamma^2 D$. The search interval is changed by the coefficient γ and $1-\gamma$, and each search will divide the interval. The maximum power point in the large side of the interval, select the larger coefficient γ to optimize the next search; in the small side of the interval, then select a smaller coefficient $1-\gamma$ to optimize the next search, making the search interval more rapid convergence. If the slope changes, that is, over the maximum power point, the step multiply γ, the opposite direction search. If the slope has not change ,the step multiply $1-\gamma$, the same direction to continue search, the algorithm flow chart shown in Figure 5.

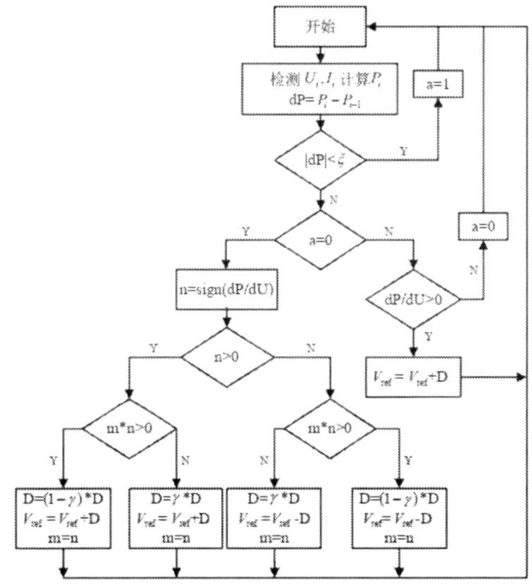

Fig.5. Flow chart of the algorithm

In the flow chart, a is a numeric variable, the variable value can only be **0** or **1**, the initial value is **1**. When the slope is changed for the first time, that is, when the search process crosses the maximum power point, assign a to **0**, and the second step strategy is the variable coefficient approximation method. Until the power difference is less than the set threshold, then it is determined that the maximum power point has been reached, let a be **1**, and proceed to the next sampling. m, n in the flow chart is a symbol variable, used to determine the last and the sampling slope of the positive and negative. m or n is greater than the value **0** indicates that the search is on the left side of the maximum power point, less than **0** indicates that the search for the maximum power point on the right.

3 Simulink Modeling of Maximum Power Tracking System

In order to verify the validity of the fixed step method combined with the variable coefficient approximation method proposed in this paper, building a model based on Matlab / Simulink. As shown in Figure 6, the model includes a photovoltaic cell module, an MPPT algorithm module, a typical boost circuit model, and a PWM module. Photovoltaic battery parameters selected a domestic production of the battery panel parameters (U_{oc} =200V, I_{sc} =6.1A, U_m =36V, I_m =5.55A, T=25℃). The MPPT module is programmed using the S function.

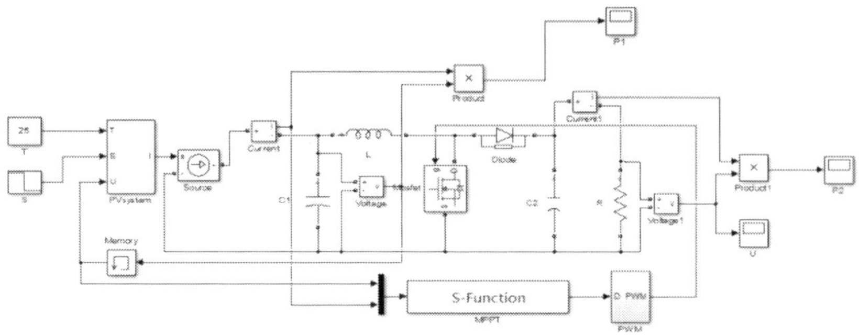

Fig.6. MPPT Simulation model of PV system

4 Simulation results analysis

The parameters of each component(C_1 =400 μF , C_2 =100 μF , L =5 mH , R =18 Ω , P_m =200W). The simulation time is set to 0.2s. The simulation results shown in Figure 7,8,9. Figure 7 is the voltage time curve under standard conditions, Figure 8 for the standard conditions of the power time curve, Figure 9 for the sudden change in the power time curve.

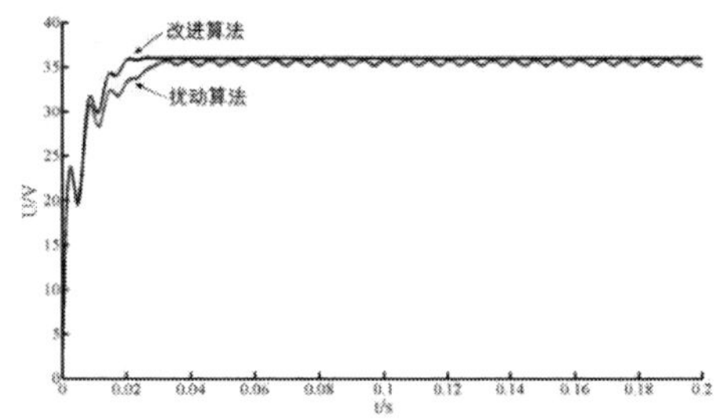

Fig.7. U-t curve under standard conditions(1000W/㎡,25°C)

Figure 7 shows the output voltage and time of the PV panel. The theoretical maximum power point voltage of the panel is 36V. Under the standard conditions of light intensity of 1000W / ㎡ and temperature of 25℃. The perturbation algorithm enters the maximum power point voltage oscillation interval around 0.03 s, it can not accurately track the maximum power point voltage. The improved algorithm accurately tracks the maximum power point voltage at 0.02s, and there is no obvious oscillation. Compared with the traditional disturbance observation method, the improved algorithm has obvious advantages in eliminating oscillation.

Figure 8 shows the comparison between the improved algorithm and the variable step perturbation method. Under the standard conditions of light intensity 1000W/ ㎡ and temperature 25 ℃, the two algorithms can quickly search for the maximum power point and have good stability. And the improved algorithm has faster tracking speed than the variable step perturbation observation method. The variable-length perturbation method is stable at 0.03s, while the hybrid algorithm in this paper traces the maximum power point at 0.02s, and the velocity is greatly improved.

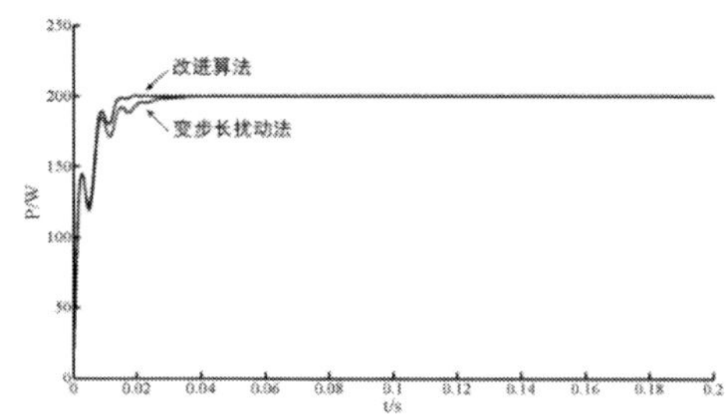

Fig.8. MPPT curve under standard conditions(1000W/㎡,25°C)

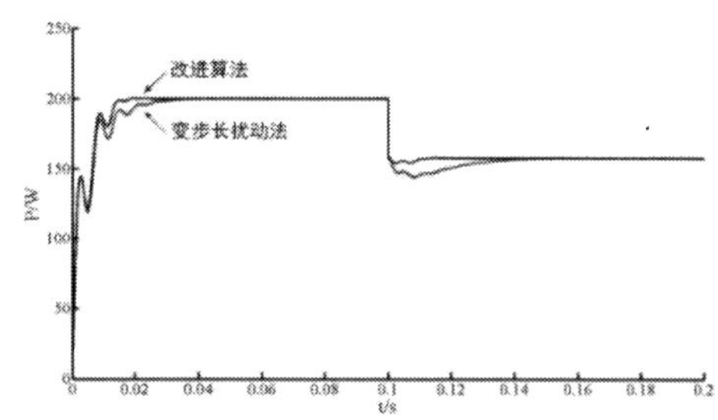

Fig.9. MPPT curve with the change of light intensity

In Fig. 9, the light intensity changes from 1000W/ ㎡ to 800W/ ㎡ at 0.1s when the light intensity is abruptly changed. The variable step perturbation method tracks the maximum power point again at 0.14s, and the improved algorithm tracts the maximum power point again at 0.11s. The improved algorithm still has a faster tracking rate in the event of a sudden change in light intensity.

* Corresponding author: fangxu152@sina.com

5 Conclusion

In order to track the maximum output power point of the photovoltaic cell accurately and quickly, on the basis of the variable step perturbation observation method, this paper proposes a hybrid control method combining the fixed step method and the variable coefficient approximation method.

The simulation results of Matlab / Simulink are used to compare the simulation results of different algorithms. The simulation results show that the fixed step method combined with the variable coefficient approximation method has its advantages both in the stability condition and in the light mutation condition, both in tracking speed and eliminating the oscillation.

References

1. Xiong Yuansheng ,Yu Li, Xu Jianming , MPPT con trol of photovoltaic generation system combini-ng c onstant voltage method with P&O method [J]. Elect ric Power Automation Equipment,**29(6)**:85-88,(200 9).

2. ZHU Tuofei,CHEN Guoding,Survey of MPPT for photovoltaic power generation[J].Chinese Journal of Power Sources,**35(10)**:1322-1324,(2011).

3. LI Xingshuo , WEN Huiqing , Research on an imp-roved β-based variable step MPPT algorithm [J] . Power System Protection and Control,**44(17)**:58-63, (2016).

4. XIE Yiwen,ZHU Jinrong,CONG Fangzhou,SHEN Deming,The Control of MPPT on the Photovoltaic System Based on the Improved Variable Step Pert-urbation Algorithm [J]. Power Electronics , **51(1)**:5 3-55,(2017).

5. PAN Chao, MENG Tao, CAI Guowei,YIN Hang , Multiobjective optimization planning of generalize power in active distribution network[J]. Advanced Technology of Electrical Engineering and Energy , **35(3)**:69-75,(2016).

6. Huang Zhipeng, Pan Sanbo,MPPT control of model predictive control combined with golden section method [J]. Renewable Energy Resources , **34(12)**: 1766-1771,(2016).

7. LIU Fangrui, DUAN Shanxu , LIU Fei . Avariable step size INC MPPT method for PV systems [J] . IEEE Transactions on Industrial Electronics ,**55(7)**: 2622-2628,(2008).

8. Piegari L, Rizzo R. Adaptive perturb and observe algorithm for photovoltaic maximum power point tracking[J].IET Renewable Power Generation,**4(4)**: 317-328,(2010).

9. K.L.Lian, J.H.Jhang, I.S.Tian, A maximum power point tracking method based on perturband observe combined with particle swarm optimization , IEEE, **vol4**,pp. 626-633,(2014).

10. Chen Kun. Photovoltaic power generation systems MPPT control algorithm research [D.Chongqing, Chongqing University,(2013).

11. Zhu Minglian, Li Chensong, Chen Xin. A variable step size P&O MPPT method for PV systems [J]. Power Electronics,**44(1)**:20-22,(2010).

12. Wang Shuzheng , Li Xianyun , Application of a n-ovel adaptive perturbation and observation method in MPPT strategy for photovoltatic power system [J]. Acta Energiae Solaris Sinica,**37(9)**:2393-2396, (2016).

Research on the Standardization of Drug Test Data

Weiqiang Lin[1], Guojiang Xu[2], Xiaomin Guo[1], Shuiping Wu[2], Jun Xiao[1], Senlian Ye[2]

[1] Guangdong Institute for Drug Control, Guangzhou, China
[2] Guangdong Wisdom Source Information Technology Limited Company, Guangzhou, China

Abstract. In order to solve the problems that test data of the drug control institution is not standardized and low quality, the data exchange and information sharing are realized, the data value is mined and the information level is improved. Method: combined with the business of control institution and information practice, refer to the practice of relevant standard development; carry out the research on standard development work from the content, principle and process. Result and conclusion: this research completes development of the local standard of Guangdong drug test data, which can provide reference for the development of similar standards in future.

Keywords. Drug inspection, data, standards, compilation

1 Introduction

Standardization is technology support and important foundation that comprehensively promotes the information; and it is one of the six key factors that form national information [1]. "Guiding Opinions on further strengthening information construction of food and drug supervision delivered by China Food and Drug Administration " (China Food and Drug Administration [2013] No.32) clearly put forward: standardization system of food and drug supervision information is the key construction task, and it is the basis that food and drug supervision information resources external link and service, it is an important protection that improves the overall level of food and drug supervision information construction.

In order to strengthen the food and drug supervision information construction, Information Office of China Food and Drug Administration published, "Food and Drug Supervision Information Standard System" in February 2014 [2], Food and Drug Administration in Guangdong Province and some provinces have released data standards system for drug supervision field inside the province [3]. However, these standards are primarily based on data on drug supervision work and cannot meet the requirements of drug test information for data standards.

In 1996, the Guangdong Provincial Institute for Drug Control independently developed the first set of Windows-based drug testing business management system, and in the province to promote the use of municipal drug testing; 2007, Guangdong Province, the drug testing in the domestic drug testing institutions First Applied Laboratory Information Management System (LIMS). After more than 20 years of development, Guangdong province's drug testing system in the use of information technology to improve the quality and efficiency of the test at the same time accumulated a large amount of test data, but since the current domestic drug testing data has not yet established standards, the data generated by various agencies is not unified, the data content is not standardized, low data quality, it is difficult to achieve effective data sharing and large data analysis. Drug testing agencies in other provinces in the country are also facing the same problem.

Therefore, planning, development, application and management requirement based on drug control information resources, combined with scientific standardization methods, drug control data standard research and preparation is carried out, improve the standardized system of drug data, strengthen the information work of drug control, achieve drug control data exchange and information sharing, improve the groundwork for of utilization level of drug control data, it has the importance and urgency. This paper integrates practice of local standard research and compiling work of drug test data in Guangdong Province, the content, principle and process of standard for drug test data are discussed.

2 Compilation Of Content

2.1. Standardized Framework

The standardized framework is the general description of the standard preparation of the entire drug test data, including the composition of standard, logical structure, the sub-item description of the standard system. The standard preparation of drug test data difficult and complex system engineering, which is a top-down normative formulation process. The research and implementation of data standards must be based on a certain standard and standardized framework, and use scientific methods guide it.

"Food and Drug Supervision Information Standard System" includes the overall standards, network infrastructure standards, information resources standards, application support standards, application standards, information security standards, information management standards seven parts. Because the drug test information is closer to drug supervision in the network infrastructure, application support, application, information security and other aspects, so the corresponding drug supervision information standards can be used. Therefore, based on information standard system framework, this topic mainly studies what are directly related to drug test data and the part which is different from drug supervision information standards, the standard of drug test data includes the standard framework, business data standard, data interface standard, standard management method.

2.2. Business Data Standard

The business data standard is the core content of the data standard preparation work; it is the standard that verifies business process data of various drugs, the business process includes the whole process of business commission, sample acceptance, sample inspection, report review and report release, the business types include sampling inspection, registration inspection, import inspection, approval of biological products, commissioning, etc..

Business data standard is by analyzing drug test information system needs and scope, the business data preparation terminology and standard standards involved in the drug test industry, data standard, data classification and coding standards, database design specifications. The elaboration scope of basic data information items include the internal identifier, the data element identifier, the data element name, the definition, the data type of the data element value, the representation format, the permission value of data element and so on.

2.3. Data Interface Standard

The drug test information system is not an isolated system, in addition to internal management data of drug test; it also involves data exchange among drug test agencies, drug supervision and other departments. Therefore, the drug test data interface does not only include the city institute for drug control, as well as higher-level drug supervision authorities, port drug administration and other institutions. The data interface standard mainly includes the interface specification, the execution order, and dependency relation and so on.

2.4. Management Method of Standard

The data standard content of drug test is complex, wide range, long time span and high requirement, it is necessary to standardize the procedure of compiling, implementation and revision, then ensure the formality of the preparation process, the authority of the prepared product, the feasibility of the standard implementation, and normalization of standard revision. Standard management method is authoritatively describe the standard planning, preparation, approval, release, record, implementation, revision and other stages, and data management, data permissions, data storage and other aspects are carried out standardized management.

3 Compilation Principles

3.1. Principle of Authority

The compiled data standards, first of all, must ensure its authority, namely, the content must comply with relevant laws and regulations of drug supervision, drug test and laboratory management, and the form should meet the requirements of standardization. Business rules and regulations are the guidelines that are jointly followed by the inspection institution, the data standards conform to these laws and regulations, in addition to authority, it more tally with the actual situation of the inspection institutions and reduce the implementation difficulty.

In addition, the rules and regulations are sorted out in accordance with the authority, application scope, the issued departments, and solve the inconsistent problems of different laws and regulations, norms. Some laws and regulations, norms have been improved and refined in the implementation process, the preparation of data standards should also be used as a basis. Data standard preparation work should also develop their own norms, namely, criterion of standard preparation work, avoid similar problems uneven or contradictory, so as to ensure the authority of the standard.

3.2. Principle of Practicality

The carrier of information technology standard is the document, if the practicality is poor, and it ultimately will become a pile of waste paper locked in the filing cabinet. The development of data standards is absolutely not a simple standardized professional work. If only the standardized technical personnel carry out data collection, document retrieval, document filing, it will become the mechanical handing of text, no matter what the framework, logical relations, content of data standards are out of line with business, and it ultimately cannot be implemented. Therefore, there must be staff who are familiar with the business fully participate in the preparation of data standards, standardized technical staff should also fully understand the business, the relevant regulations and norms of business are regard as the yardstick to meet the construction requirements for management system of drug test. At the same time, we must also fully consider the applied practice of existing information systems and historical data status, and follow the standards and norms of information technology.

Drug test data standards involve a large number of laws, standards and norms, its range is wide and field is broad, overall planning are carried out in accordance with the ontology, system analysis methods, project management and other methods in the standard preparation, these standards and norms are teased out, analyze the internal link its test data is analyzed to ensure the scientific nature, systematic nature, practical applicability universality of standard preparation.

3.3. Principle of Traceability

Standard compilation process takes " from inside to outside, bottom up" advance gradually. "From the bottom up", namely start from existing business applications the provincial institute for drug control, the internal management system documents are combed, refer to the existing standards, verify the relevant laws and regulations; "from inside to outside" start from the provincial institute for drug control, representative city institute for drug control are surveyed step by step, and refer to the situation of other provincial institute for drug control. In this process, the definition of data element is cross-referenced, constantly changing, and constantly strive for perfection. Because this work is long span, the number of participants is many, the data is large amount, the process is complex, need technology means to protect the data element of data standard, with the continuous progress of work, the evolution process of data element definition and other contents are effectively recorded, the process changes can be traced, and can back to the source. So the authority and practical applicability of preparation standard can be better ensured.

4 Compilation Process

Construction of a set of highly operational, forward-looking standards must follow a certain process and methods, in-depth research and research, the standard preparation is used to carry out top design, norms, and guide opening and share of drug test information resources, and ultimately achieve scientific applications of data. The process of data standard preparation includes three stages: business research stage, standard preparation stage, review and report stage, the specific process is shown in Figure 1[7].

4.1. Business Research Stage

The business research stage mainly use document retrieval, investigation of drug test institution on the spot, business expert' consultation collect the relevant business rules, specifications and software system tables, report forms, interfaces, databases, and extract the corresponding data fields, then analyze and arrange the formed data sets and code sets, the final data element, data element range code will be obtained from these field information, the standard situation of drug test data, user requirements and technical requirements should comprehensively be analyzed, the scope and target of the standard, the construction idea of the conceptual model, and classification research method of the related information resource should be cleared [8].

4.2. Standard Compilation Stage

This phase first the actuality of drug test field, according to information theory of conceptual model, information model and data model to analyze systematic function model and application framework, then determine the standard research methods of drug test data. Then, through a series of analysis of process, table, data flow, interface, etc., then develop standardized business function and data resource design framework. Finally, the collected information in business research phase is used, research methods which is identified in earlier stage is utilized, according to the proposed framework, the specific content of data standards of drug test, eventually it is converted to data standard document in line with the standard.

In this process, through the application combing of drug test business, the relevant business applications are sorted out, and tease out existing business information systems (especially the laboratory information management system) business and related forms, sort out the data element, data dictionary; and check the data element and data dictionary by reference to management system documentation within the organization, and study relevant laws and regulations, management norms, calibrate related data and data dictionary, and find the authoritative definition of relevant data element; It is more important to repeatedly communicate and confirm with professional business sector experts. In the field of information technology, according to the relevant standards and norms of information technology and resources, comb the data element data sets, databases and other standard suitable for the test data standards, the overall compilation ideas is shown in Figure 2.

4.3. Standard Review and Report Stage

The compilation of standard is a spiral iterative process. After the team forms the standard document, the team should invite the relevant experts in the field of drug test and the experts in the standard to carry out the periodical standard review, special subject review, mid-term review and final report review, and constantly adjust and improve the standard document in accordance with the expert's opinion. The review mainly includes internal self-examination and external expert review.

After the above process completes the compilation of drug test data standards, it can be used as internal

DOI: 10.1051/matecconf/201713900163

standards, also further declare it to quality and technology supervision departments as national, local or group standards.

5 Conclusion

This research through the combing and analysis of drug test business, combining the current situation of drug test institution information, the compiled content of drug test data standard are summarized from the system framework, data standards, interface standards and management method four aspects. According to the specification of standard preparation, combining the reality of drug test institution, the principle of authority, practicality and traceability should be followed in the standard preparation of drug test data, and analyze the standard preparation process and the work methods and the problems that should be paid attention to. The scientific, systematic and advanced data standards of prepare ideas and methods in this research is used to complete the first domestic drug test data standard-preparation of Guangdong province pharmaceutical inspection data local standard, and it can provide reference for the next similar type of standard preparation.

Acknowledgements

Fund project: special fund project for public interest research and capacity building in Guangdong province (2014B010103005).

References

1. The information leadership group office of the Ministry of Water Resources, et al. Information Guidance for Water Conservancy [M], Beijing: China Water&Power Press, 2003
2. CFDAB/T 0101-2014, Information Standard System of Food and Drug Administration [S]
3. Q/GDYJ 5001-2007, Data Maintenance and Management Standard of Guangdong Food and Drug Administration Industry Standard [S]
4. Lin Weiqiang. Use Computer Technology to Strengthen the Business Management of Institute for Drug Control [J]. Chinese Pharmaceutical Affairs, 1998, 12(2):89-90
5. Lin Weiqiang. Application of LIMS in Guangdong Province Institute for Drug Control [J], Pharmacy Today, 2008, 18(4):88-90
6. Hu Bing, Huang Xiaoqin, Sun Jiaming, Xu Kuidong, Zhang Fan. Research on the Framework of Shanghai Sanitary Supervision Data Standard [J], Chinese Journal of Health Inspection, 2010, 17(3):225-230
7. Xu Zhan, Wang Hezhen, Chen Jiangling. Research on the Standardization Line of High-End New Electronic Information Industry in Guangdong Province [J], Standard Science, 2013, (04):59-61
8. Hu Bing, Huang Xiaoqin, Feng Donglei, Chang Bo, Sun Jiamin. Research on Developing Method of Shanghai Sanitary Supervision Information Data Standard [J], Computer Applications and Software, 2010, 27(12):299-301

DOI: 10.1051/matecconf/201713900163

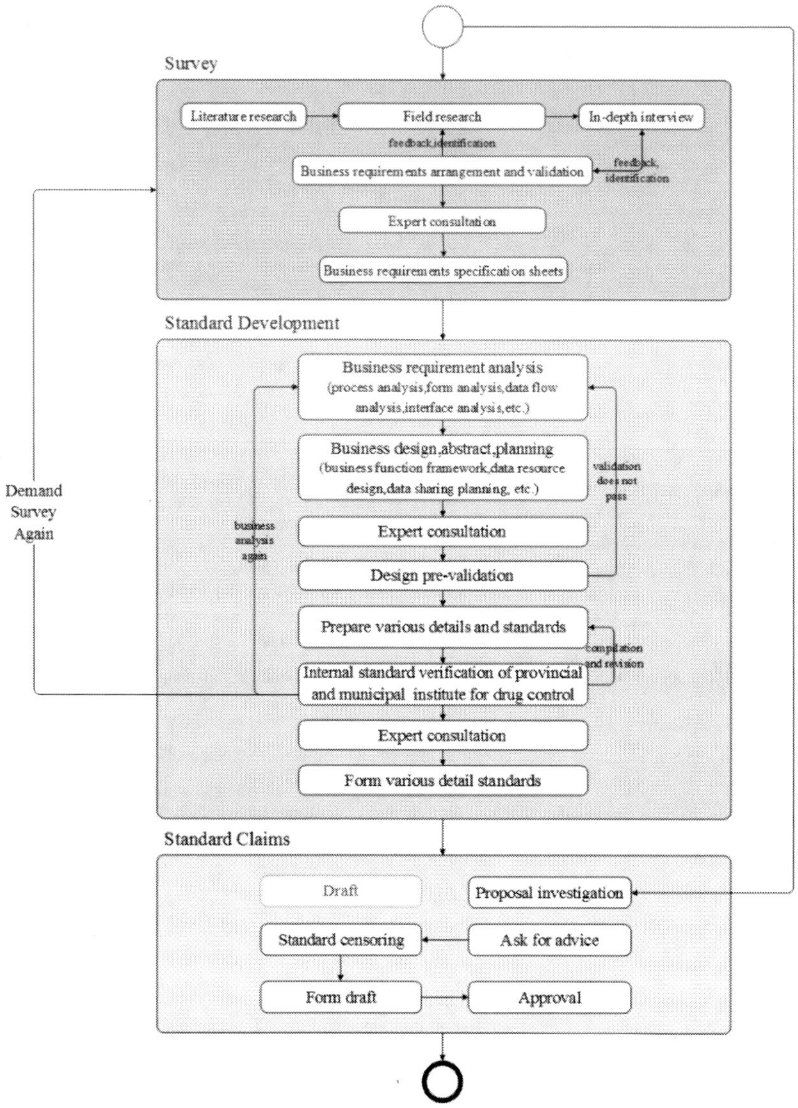

Figure 1. Compilation process of data standard of drug test

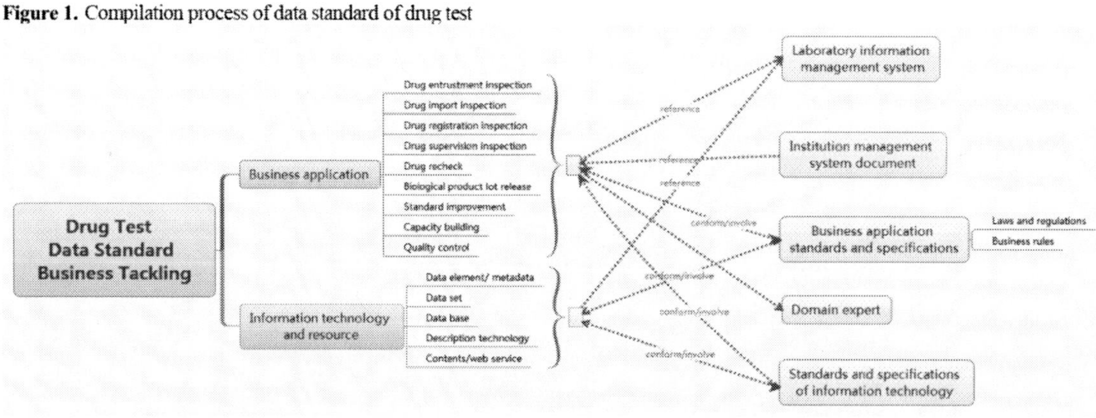

Figure 2. Framework diagram of business tackling of data standard of drug test

Research of Penetration Overload Signals Processing Method Based on EEMD and WT

Guo Yan[1,*], *Zhao* Hai-feng[1,2]

[1] Faculty of Mechanical and Electrical Engineering, Nanjing College of Information Technology, Jiangsu, 210023,China
[2] Faculty of Mechanical and Electrical Engineering, North University of China, Taiyuan, 030051,China

Abstract. Hard Target Penetration is a very complex dynamic problem, and penetration signals contain axial de-acceleration signals, vibration signals and other weakly noise signals. It is difficult to obtain penetration feature signals through some methods filtering unwanted vibration signals and noise signals. As such, we propose a joint filtering method based on EEMD (Ensemble Empirical Mode Decomposition) and WT (Wavelet Transform) to solve this problem. This method consists of four main steps: (a) penetration signals decomposing via EEMD method, this gets the IMF (Intrinsic Mode Function) components, (b)then we calculate the Whole BURG power spectrum of the original signals and each component ._after that compare the power spectrum of each component with the original signals, this draws the original signals EEMD decomposition scale, (c) the high-frequency components of IMF filtering based on the WT threshold, (d) signals reconstruct by using low-frequency IMF components which reflect signals characteristic and high-frequency IMF components through wavelet filtered. Experiment show that proposed method can effectively determine the penetration signals decomposition scale, eliminate vibration and noise signals of penetration process, avoid the losing of the high-frequency components when using a single wavelet filtering method. The result of the proposed method can get better SNR than WT, the velocity and depth are in good agreement with the results of experiment.

1. INTRODUCTION

Recently, hard target penetration overload signals processing has become increasingly important in national defense research. Projectile important parameters (such as the penetration depth, time, layers of medium, etc.) in the process of penetration could be obtained through analysis of measured overload signals. And projectile parameters provide an important basis for the design of projectile structural strength, explosive stability and fuse reliability [1]. When penetration occurring, the overload signals generally contain two major components: one is the de-acceleration signal that has been formed by encountered resistance in the process of penetrate the target medium, this signal spreads in the manner of tress wave; Another is vibration signal in the process of penetration, which included projectile horizontal and vertical vibration[2]. The key of penetration signal processing is find that the right filtering cut-off frequency, then remove the vibration signal during penetration process (including projectile horizontal and vertical vibration signal), weakly interference signal, finally reserve de-acceleration signal that reacts projectile penetrating resistance.

At present, domestic and foreign scholars have carried out many studies about penetration signals processing technology. For example, American Sandia Laboratories had finished 3KHZ frequency filtering wave of the measured data in the 1990s[3].Domestic Nanjing University of Science and Technology measured acceleration curve and made an analysis comparison of 3KHZ and 10KHZ filtering wave in the early 2000[4]. Beijing Institute of Space Long aircraft applied the first-order cut off frequency filtered wave. These all three methods are hardware threshold filtering. Those characteristics are given based on specific conditions of a fixed frequency threshold. When the frequency-domain transform is greater than the threshold value, the original value has been retained, otherwise has been set to zero. This filtering method is a single threshold function processing, and has a general de-noising effect. With the development of signal theory, Donoho et al. proposed wavelet threshold de-noising method [5]. Many people started this extensive research [6] .This method decomposes of the signals by selecting the appropriate wavelet, and filters noise from high-frequency component after wavelet decomposition by using reasonable threshold function, then gets the de-noising signals by wavelet reconstruction. Wavelet can better analyze the details and characteristics of the signals than traditional methods, and can remove Gaussian noise, get good SNR. Currently, wavelet decomposition de-noising method based on different thresholds also been applied to the field of penetration acceleration signal processing . However, the choice of wavelet and threshold function needs further analysis in actual signals processing. EEMD is a signal decomposition method which pluses auxiliary noise on the basis of the Empirical Mode Decomposition. And it solves the model mixing problem when signals adopt EMD decomposition method in the complex and abnormal noise environment . This method has certain advantages for the complex noise signal's separation and processing.

We propose a joint filtering method based on EEMD and WT for the non-stationary random vibration

* Corresponding author: guoyan@njcit . cn

acceleration signals in the hard target penetration processing.

2. EEMD and penetration acceleration signals analysis

2.1 EEMD Principe

Empirical Mode Decomposition has been proposed as an adaptive time-frequency data analysis method by Huang et al. of National Aeronautics and Space Administration (NASA) in 1998 . It has been proved quite versatile in extracting useful signal from these non-stationary and non-linear signals .

The essence of EMD is obtained intrinsic mode functions (IMF) by using time scale characteristic of signals. The procedure of extracting an IMF is called sifting. The sifting process is as follows:

1) Identify all the local extrema in the original data S (t).
2) Connect all the local maxima by a cubic spline line as the upper envelope.
3) Repeat the procedure for the local minima to produce the lower envelope.

The upper and lower envelopes should cover all the data between them. Their mean is M_1. The difference between the data and M_1 is the first component H_1:

$$S(t)-M_1=H_1 \quad (1\text{-}1)$$

Ideally, H_1 should satisfy the definition of an IMF, for the construction of H_1 described above should have made it symmetric and having all maxima positive and all minima negative. After the first round of sifting, a crest may become a local maximum. New extreme point generated in this way actually reveals the proper modes lost in the initial examination. In the subsequent sifting process, H_1 can only be treated as a proto-IMF. In the next step, it is treated as the data, then

$$H_1-M_{11}=H_{11} \quad (1\text{-}2)$$

After repeated sifting up to k times, H1 becomes an IMF, that is

$$H_{1(k-1)}-M_{1k}=H_{1k} \quad (1\text{-}3)$$

Then, it is designated as the first IMF component from the data:

$$C_1=H_{1k} \quad (1\text{-}4)$$

Overall, C_1 should contain the finest scale or the shortest period component of the signal. We can, then, separate C_1 from the rest of the data by

$$R_1=X(t)-C_1 \quad (1\text{-}5)$$

Since the residue, R_1, still contains longer period variations in the data, it is treated as the new data and subjected to the same sifting process as described above. This procedure can be repeated to all the subsequent and several indecomposable IMF C_i can be obtained until C_n.

From the above equations, we can induce that the original signal sequence as:

$$S(t)=C_1+C_2+\ldots C_i+\ldots+C_n+R_n \quad (1\text{-}6)$$

C_i represents the first i intrinsic mode function; R_n becomes a monotonic function from which no more IMF can be extracted.

In the above method, obtaining reasonable IMF component depends on the distribution of signal extreme points. If the signal extreme points are uneven in distribution, the case of mode mixing will appear. Huang proposed EEMD method through adding white Gaussian noise on the basis of EMD. The decomposition process is similar. The only difference is that N different sets Gaussian white noise is added to the original signal before EMD decomposition , and form a new whole signal sequence S1 (t), S2 (t), ... , Sn (t). Then different scales intrinsic mode functions IMF_{i1}, IMF_{i2}, ... , IMF_{in} has been got by decomposing the new signal sequence S1 (t), S2 (t), ... ,Sn (t). Finally, the whole signal's IMF_1, IMF_2, ... , IMF_n has been obtained after meaning of results of decomposition. Noise will be inter-neutralization because of added noise has zero mean characteristics [14].

2.2 EEMD decomposition of Penetration acceleration signal

EEMD can resolve the mode mixing problem which happening in using EMD method to decompose signals. EEMD can clearly decompose the natural frequency of the penetration signals. The decomposition process show in Figure 1:

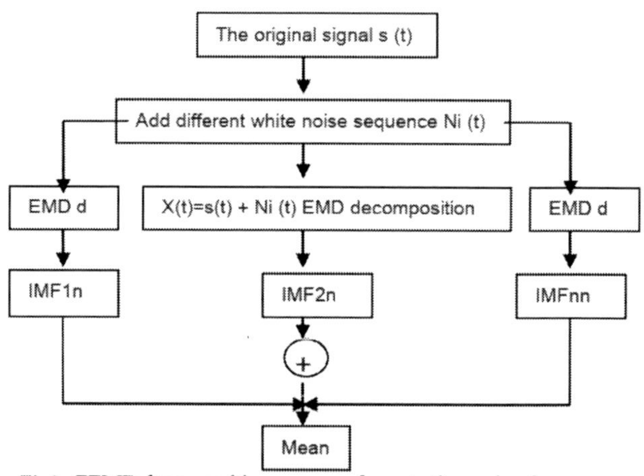

Fig1. EEMD decomposition process of penetration signals

Penetration acceleration signals curve has been obtained when 100mm diameter projectile penetrates unlimited concrete target, which is shown in Figure 2.

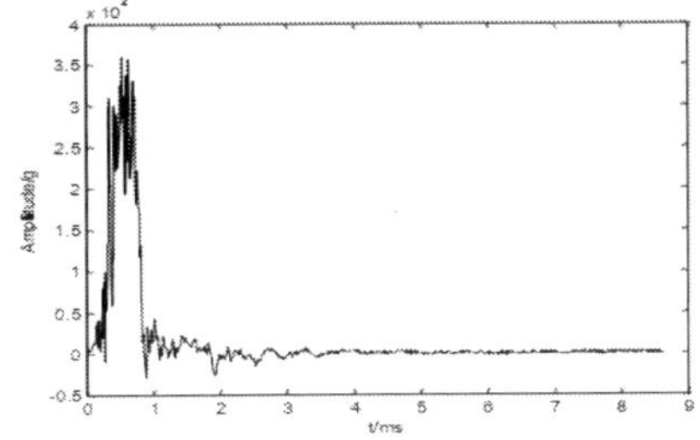

Fig2. Penetration acceleration signals curve

The Whole BURG Power Spectrum has been calculated before signal decomposition by using EEMD. The result is shown in Figure 3. We know the whole signal energy concentrated in the vicinity of the frequency1083HZ and 3813HZ from Figure. Then projectile energy gradually decline with increasing of penetrate target's depth, and finally drop to zero. And this result are very similar with the signal frequency domain analysis in Figure 4, the main frequency is 1071Hz and 3836Hz, 1071Hz is the first-order natural frequency of projectile that reflects corresponding intrinsic vibration frequency of penetration process, 3836Hz may be the two or multiple harmonic frequency of projectile.

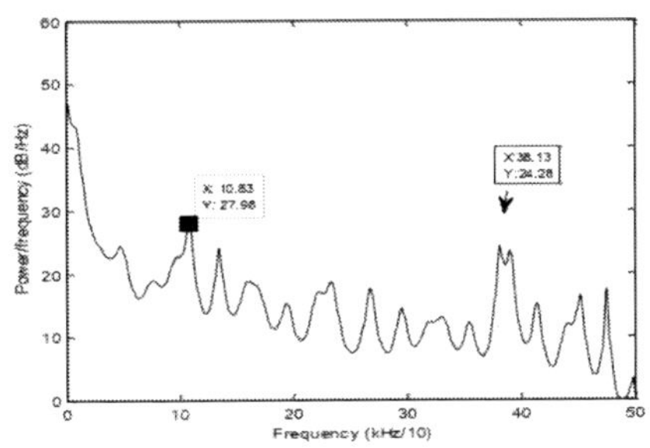

Fig3. The Whole BURG Power Spectral of Penetration acceleration signals

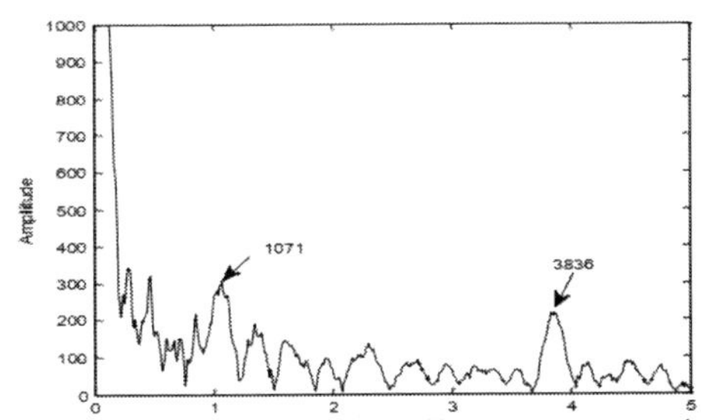

Fig4. Frequency analysis of Penetration acceleration signals

In order to remove the noise signal from acceleration curve of penetration process, we decompose penetration signals according to the above EEMD decomposition process after calculating of whole projectile energy . The IMF_i components and the rest of the data R have been got after EEMD decomposition, and it is show in Figure 5.

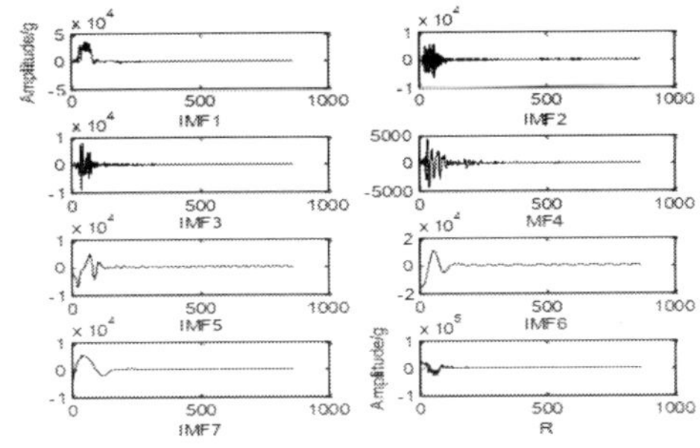

Fig5. IMF components of penetration signal after EEMD decomposition

We can find IMF_1, IMF_2, IMF_3 has small amplitude and high frequency feature from Figure 5. These characteristics reflect the high-frequency components in the penetration process, which represent horizontal and vertical vibration signals of the projectile and high-frequency vibration signals causing by accelerometers and external factors. If continue to decompose signals, we discover IMF_4 and IMF_5 component is boundary of the high frequency and low frequency, and IMF_6 and IMF_7 component is completely similar, which shows signals don't need to continually decompose after the 6-order decomposition. At this time the measured signals have reached the trend term. But what order is the optimal decomposition order that need to further analysis. Paper presents to calculate the decomposed BURG Power Spectrum of each order IMF method to analysis. From Figure 6, we find IMF_4 Power Spectrum is very similar with the previous obtained Whole Signal Power Spectrum. The energy of IMF_4 Power Spectrum concentrated at the 1083HZ, and the energy of IMF_2 Power Spectrum concentrated at the 3871HZ, then the energy began to decline sharply after IMF_4 component. Furthermore, the second harmonic response or multiple harmonic response frequency—3813Hz high frequency has been decomposed after EEMD the fourth-order decomposition, that is to say the fourth-order decomposition has separated the high-frequency vibration signals which generate in the penetration process. So we can draw a conclusion that IMF4 is the optimal EEMD decomposition, which reflects the response acceleration signal of penetration process.

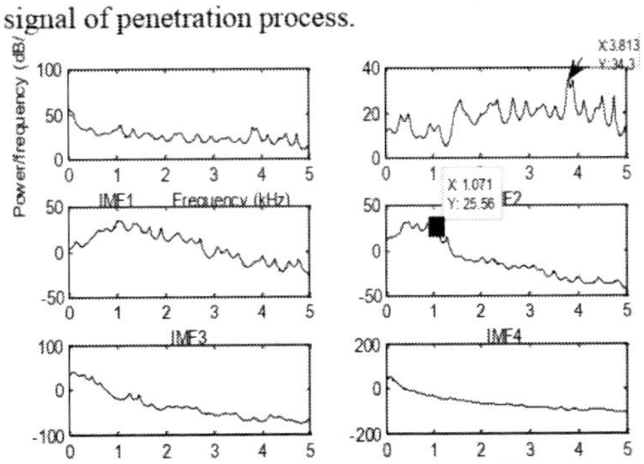

Fig6. The BURG Power Spectral of each IMF component

3 Wavelet Multi-resolution Analysis and Penetration acceleration signal filtering

3.1 Wavelet Analysis

Morel proposed wavelet analysis when he was analyzing the local characteristics of seismic wave in 1984. The wavelet analysis is different with traditional Fourier transform. With the development of the various wavelet basis and proposal of the multi-resolution thinking, the wavelet analysis is widely used for non-stationary signal analysis and filtering processing because of good localization features in the time-frequency domain.

It is defined as: function $\psi(t)\in L^2(R)$, its Fourier transform is $\hat{\psi}(w)$. Meet the following local conditions:

$$C_\psi = \int_{-\infty}^{\infty} \frac{|\hat{\psi}(w)|^2}{|w|} dw < \infty \tag{1-7}$$

That is C_ψ bounded, $\psi(t)$ has been called a basic wavelet or mother wavelet. Scaling and shifting the mother wavelet function, a as its stretching factor, b as shift factor, translation function $\psi_{a,b}(t)$ has been got after stretching and shifting:

$$\psi_{a,b}(t)=\frac{1}{\sqrt{a}}\psi\left(\frac{t-b}{a}\right) \quad a,b\in R \quad a\neq 0 \tag{1-8}$$

So $\psi_{a,b}(t)$ has been called associated wavelet basis function with parameters a and b. Due to continuous transformational values of a and b, it is also called continuous wavelet basis function. These series functions were got through the same mother function after scaling and translating.

When the analytical signal $f(t)\in L^2(R)$, Wavelet transform is it's expand on this wavelet function, marked:

$$WT_f(a,b)=<f,\psi_{a,b}>=\frac{1}{\sqrt{|a|}}\int_R f(t)\cdot\overline{\psi\left(\frac{t-b}{a}\right)}dt \tag{1-9}$$

As such, a time function can be projected onto 2-Dimensional time-scales plane, and the same time the local characteristics of the signal were obtained in the time-frequency domains.

3.1 Multi-resolution Analysis

MRA (Multi-resolution Analysis) is an effective analytical method which had been proposed by Mallet and Meyer on the constructing wavelet basis in 1986. The main idea is that use different time resolution to approximate the analyzed signal f (t) after continuous signal are disserted. This approach decomposes the signal step by step on the orthogonal space of $L^{2(R)}$. Each level of the input signal can decompose into two parts: the high frequency detail signals and the low frequency approximation signals. Sample frequency halves step by step. If the frequency space of signal f (t) is A_0 which was decomposed into two sub-spaces in the first stage: low frequency A_1 and high frequency D_1. A_1 will continue to break down into low frequency A_2 and high frequency D_2. This decomposition process continues until no useful information could decompose. Its decomposition process can be noted as:

$$A_0=A_1+D_1;\ A1=A_2+D_2;\ A_{i-1}=A_i+D_i \quad (1\text{-}10)$$

$\oplus$ represents two sub-spaces "orthogonal sum", A_i represents multi-resolution analysis sub-space corresponding to the resolution 2^{-j}. This result indicates that the signal f (t) could be infinitely approximated by 2^{-j} multi-resolution analysis. Signal f (t) can be expressed as:

$$F(t)=A_0(f)=A_1(f)+D_1(f)=A_2(f)+D_1(f)+D_2(f)=\ldots \quad (1\text{-}11)$$

3.2 Penetration acceleration signals Wavelet analysis

Wavelet analysis of signals are also widely used in recent years, We summarized wavelet decomposing of penetration acceleration signals in the following three steps:

1) The penetration acceleration signals wavelet decomposition. Select the appropriate wavelet and determine the layers N of filtering, then the signal N layers wavelet decomposition;

2) The high-frequency signals of wavelet decomposition filtering noise. Choose reasonable threshold filtering High-frequency coefficients of each layer from 1 to layer N;

3) The signal reconstructing. The signals wavelet reconstructs according to the low-frequency coefficients N-layer of wavelet decomposition and high-frequency coefficients from 1 to N after filtering.

The WT process of penetration acceleration signals is following figure 7.

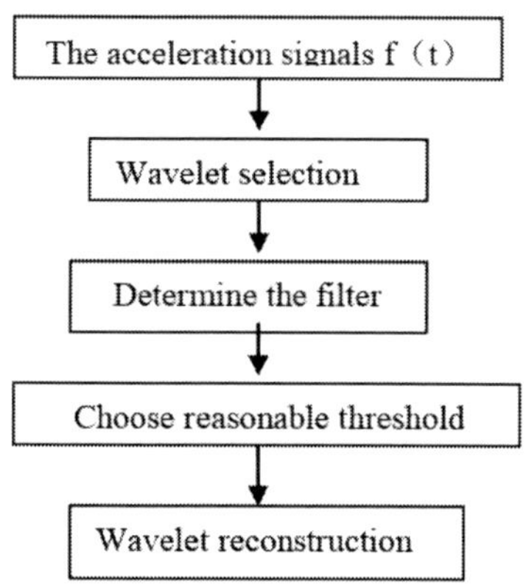

Fig7. Wavelet de-noising process

The main factors that determine the filtering effects to penetration acceleration signals wavelet analysis are:

1) Wavelet Type. Because different type of wavelet have different result of the decomposition, So paper decomposes signal by selecting the Daubechies II wavelet as mother wavelet due to penetration overload signals characteristics and previous experience.

DOI: 10.1051/matecconf/201713900164

Daubechies II wavelet has the following characteristics: Proximity with penetration measured characteristic curve, compactness, orthogonality etc. The main components of the measured signal can be obtained by Daubechies II wavelet decomposition;

2) Wavelet decomposition layers. How many layers have been decomposed which depends on the low-frequency signal could not appear obvious peak extinction phenomenon after WT decomposition. Paper decomposes the penetration acceleration signals by using Daubechies II wavelet, and finds 3 layers of decomposition is better than others. "Daubechies II wavelet decomposition coefficients" is following diagram 8.

3) The choice of threshold function. Different threshold function determines the different filtering effect of the high-frequency coefficients of wavelet decomposition, and also determines the effect of the reconstruction signal after signal filtering. Paper uses two different threshold methods filtering: One is high-frequency coefficients compulsory de-noising, and this method can obtain relatively smooth curve, but it filters out the useful components of signals; Another method is default threshold de-noising, which retains some of the high-frequency vibration signals and has the better details of the signal and SNR than compulsory de-noising. Different filtering result has been shown in figure8, SNR value is following table 1.

Table1. SNR of two filtering methods

filtering methods	Compulsory de-noising	Default threshold de-noising
SNR	29.69	45.28

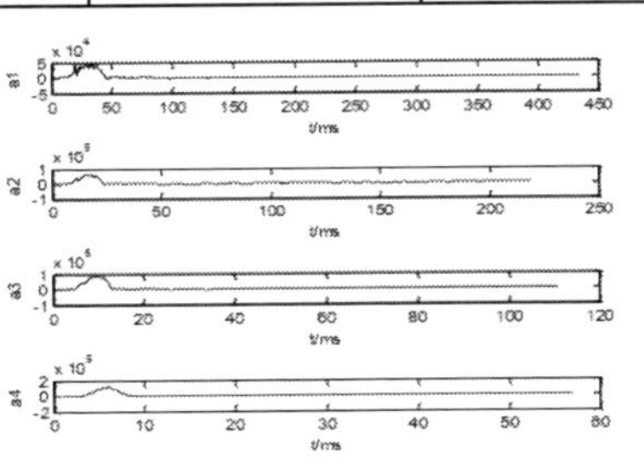

a. Low-frequency coefficients

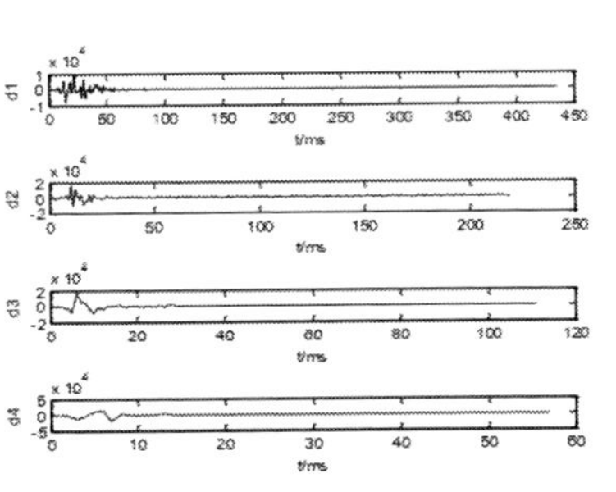

b. High-frequency coefficients

Fig8. Coefficients of the db2 wavelet decomposition

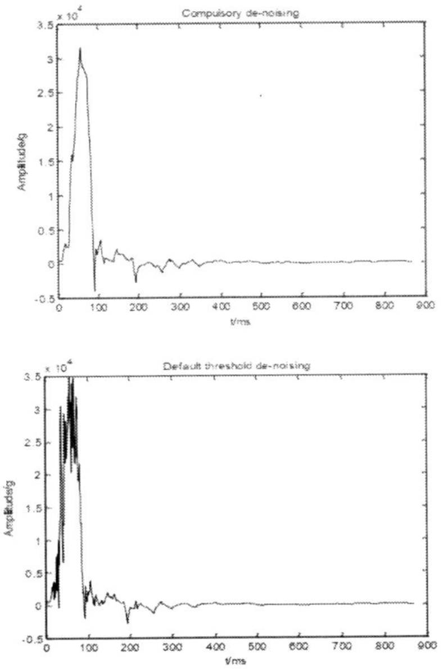

Fig9. Different threshold Wavelet de-nosing

Although WT filtering has been widely used in penetration signals process, but there are two difficulties.

1) The choice of Wavelet has not strictly theoretical foundation. Many Wavelets has been choice depending on continuous experiments and previous experience. It brings a lot of work of signals filtering and don't obtains the optimal result.

2) Wavelet decomposition layers depends on the low-frequency signals could not appear obvious peak extinction phenomenon after WT decomposition. This conclusion always been drawn by direct observation, there is no strict criteria, which may bring bias of decomposition levels.

Based on the above advantages and disadvantages of EEMD and WT, we propose a joint filtering method based on EEMD and WT to analysis complex penetration acceleration signals.

4 EEMD and WT joint filtering analysis

4.1 Filtering method and process

Firstly, the penetration acceleration signals have been decomposed by using EEMD method. This gets the series IMF components, and calculates the power spectrum of each component. Secondly, then we compares the BURG Power Spectrum between original signals and IMF components, obtains decomposition scale of original signals, and distinguish the high and low frequency boundaries of signals decomposition. At same time the natural frequency of the projectile overload bodies can be precisely drawn in this process. And next we filter projectile vibration and environmental noises from high-frequency components of IMF by using the different Wavelet (WT) threshold. Finally, we reconstruct the signal by using low-frequency IMF components which reflect signal characteristic and high-frequency IMF components which have been filtered by wavelet. The process of EEMD and WT joint filtering analysis has been shown in figure 10.

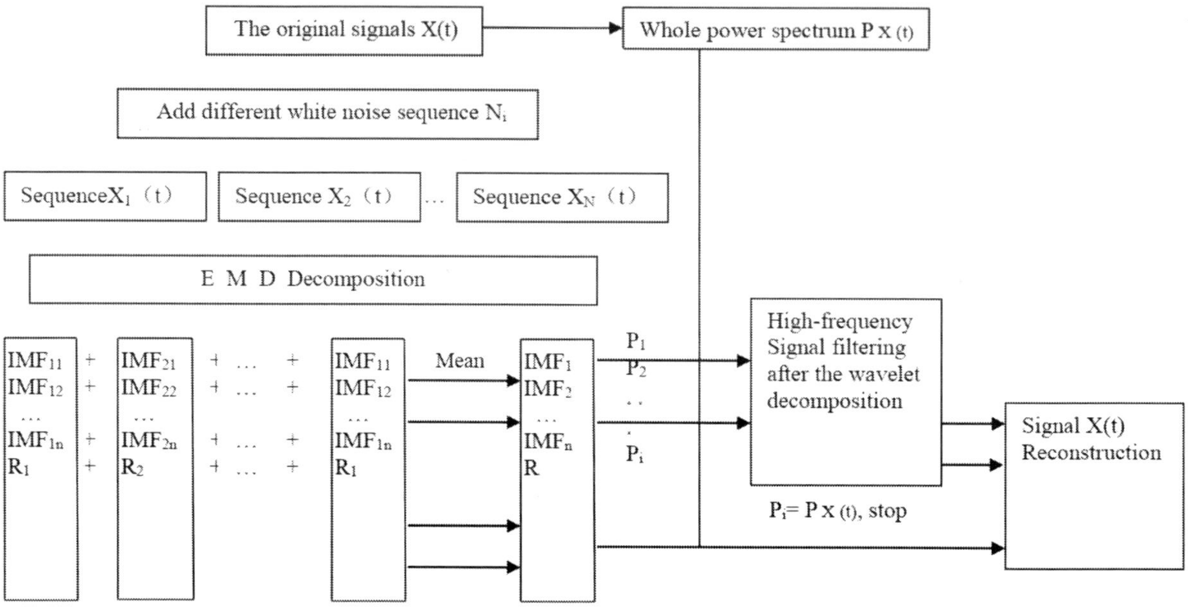

Fig10. EEMD and WT joint filtering process

4.2. Experiment

We can draw a conclusion that the optimal decomposition scale is 4 levels by comparing the BURG Power Spectrum between original signals and decomposed IMF components, because IMF_4 Power Spectrum is similar with original signal whole power spectrum. Peak value of power spectrum reflects the first order nature frequency of projectile in the penetration process. Then we filters high-frequency component from IMF_1、IMF_2 and IMF_3 by using Daubechies II wavelet. Finally, signal has been reconstructed by using low-frequency IMF_4 component which reflects the signal characteristic and high-frequency IMF_1、IMF_2 and IMF_3 components after wavelet filtering. And we get the desired signal through EEMD and WT joint filtering. The result is show in figure11.

Figure a) shows that the signal has been reconstructed by using wavelet compulsory de-noising method after EEMD decomposed high-frequency. Figure b) shows that the signal has been reconstructed by using wavelet default threshold de-noising after EEMD decomposed high-frequency. SNR values of the two methods are shown in Table 2.

Table2. SNR of two WT filtering methods after EEMD decomposition

filtering methods	Compulsory de-noising	Default threshold de-noising
SNR	31.13	54.79

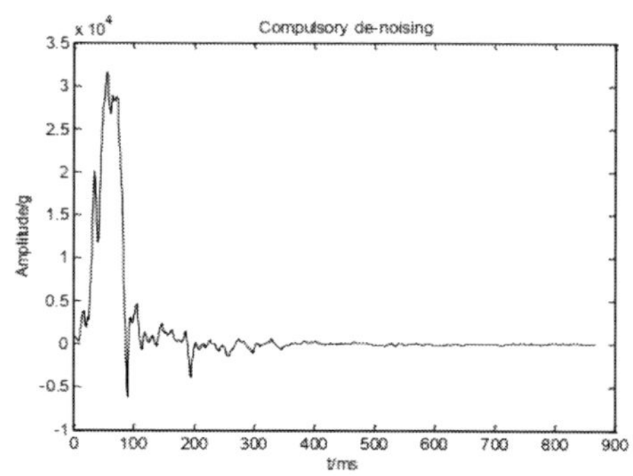

a. Compulsory de-noising

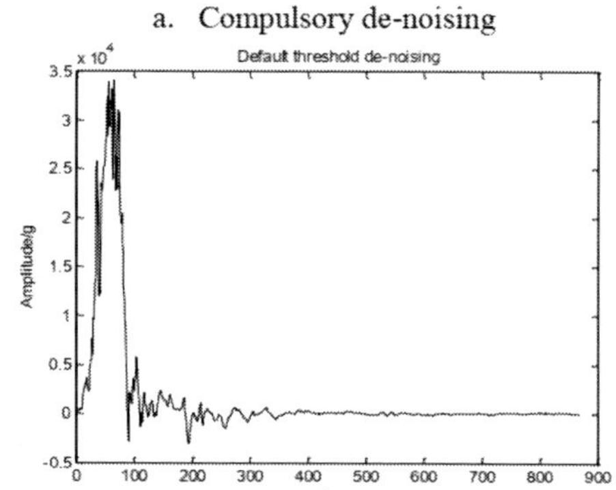

b. Default threshold de-noising

Fig11. EEMD and WT joint filtering signal

Table 2 shows the paper proposed method has better de-noising capability than single wavelet filtering method. Further comparing the signal curves a and b, we find signal curve a is more smooth than curves b. This is because signal curve a filter noise by using compulsory de-noising method, and the method filter out part of the signal which reacts projectile characteristic in penetration process. Signal curve b maintains projectile vibration detail, and has better SNR than signal curve a.

In order to further verify the correctness of the paper propose signal processing method, the Velocity-time curve and Displacement-time curve of projectile penetration process have been got by integrating the acceleration signals b after EEMD and WT joint filtering, and they are shown in Figure12 and Figure13. These indicated that the striking velocity was 424.2m/s and the displacement during the penetration was 1.003 m. Table 3 summarized the results of the paper proposed method and experiment results, and it indicated that the experimental results were in good agreement with results of paper proposed method .

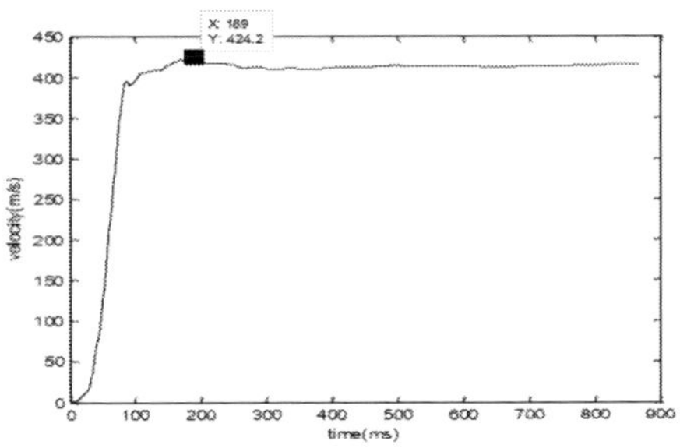

Fig12. Velocity -time curve of projectile penetration process

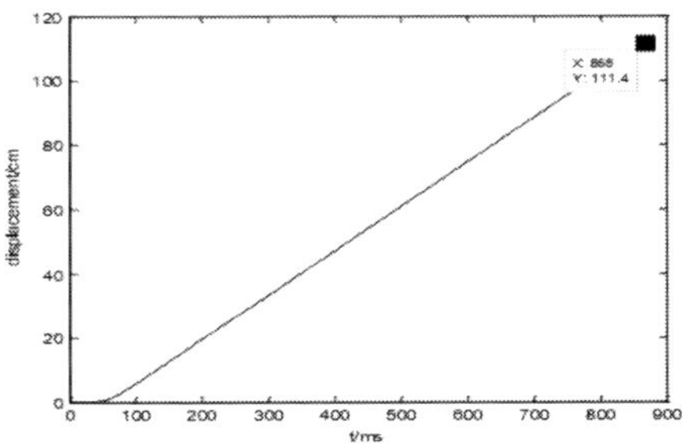

Fig13. Displacement-time curve of projectile penetration process

Table3. The results of the paper proposed method and experiment results

Result	Striking velocity (m/s)	Residual velocity (m/s)	Displacement of penetration (m)
Experiment	478	0	1.005
Proposed method	424.2	0	1.114

5 Conclusion

Through comprising the result of different filtering analysis methods that paper proposes that EEMD and WT joint filtering analysis is better than others. The advantages mainly reflect in four aspects:

1) The penetration signals decomposition scale is easy determined through comprising of the BURG power spectrum between the whole signals and IMF components after EEMD decomposition. It solves the problem that wavelet decomposition scale is not easy to determine. And the same time projectile vibration natural frequency could be obtained.

2) This method can eliminate mixed projectile vibration noise and void the loss of itself high-frequency components in the penetration process.

3) High SNR.

4) Penetration velocity and depth are in good agreement with results of expriment.

Overall, paper provides a better filtering method for the research of typical non-stationary random vibration signals filtering processing, and has certain significant guide for high-g penetration signals filtering. However, more detailed analysis of the penetration process and the environments are need to known because some penetration process is relatively complex.

Acknowledgment

This work is supported in part by the Scientific research of Nanjing College of Information Technology (YK20140201) , and the "Overseas research and Training Program of Outstanding Middle-aged University Teachers and Principals in Jiangsu Province". Their supports are greatly appreciated.

[1] Xu Peng, Zu jing, Fan Jinbiao, "Research Development to High-g Penetration Acceleration Test and Its Correlative Technology," J. Acta Armamentarii, vol. **32**,2011,pp739-744.

[2] FAN Jinbiao, ZU Jing, XU Peng, WANG Yan, "Impact Deceleration Signal Processing for Concrete Target Penetration," J. Journal of Detection & Control, vol.**34**,2012,pp1-6.

[3] Xu Peng, FAN Jinbiao, ZU Jing, "Study on Acceleration Test Technique of High Velocity Kinetic Energy Projectile Penetrating into Hard Target,"J.Journal of Vibration and Shock, vol.**26**,2007,pp 118－122.

[4] Zhang Zhian. "Research on Hard Target Penetration Fuze Semi-physical Simulation Technology,"D. Nanjing:Nanjing University of Science and Technology，2007.

[5] D.L.Donoho."De-noising by Soft-Thresholding,"J. IEEE.Transactions on Information Theory, 1995,41(3):612-627.

[6] Berkner, K. and R.O. Wells, "Smoothness estimates for soft-threshold denoising via translation-invariant wavelet transforms,"J. Applied and Computational Harmonic Analysis,2002,12(1):1-24.

Method of Geometric Connected Disk Cover Problem for UAV realy network deployment

Liu Chuang, Lv Na, Chen Ke-fan, Zhang Bu-shuo,Cao Fang-bo
School of informational and navigation, Air Force Engineering university, Xi'an 710077.China

***Abstract*—Aiming at the problem of the effective connectivity of a large number of mobile combat units in the future aeronautic swarm operation, this paper proposes an idea of using UAV(Unmanned Aerial Vehicle) to build, and studies the deployment of the network. User coverage and network connectivity are important for a relay network planning which are studied separately in traditional ways. In order to effectively combine these two factors while the network's survivability is taken into account. Firstly, the concept of node aggregation degree is proposed. Secondly, a performance evaluation parameter for UAV relay network is proposed based on node aggregation degree, then analyzes the lack of deterministic deployment and presents one a PSO (VFA-PSO) deployment algorithm based on virtual force. Finally, compared with the existing algorithms, the validity and stability of the algorithm are verified. The experimental results show that the VFA-PSO algorithm can effectively improve the network coverage and the survivability of the network under the premise of ensuring the network connectivity, and has better deployment effect.**

Keywords*—aeronautic swarm; unmanned aerial vehicle relay; network deployment; invulnerability; particle swarm optimization;

I. INTRODUCTION

With the deepening of the digital battlefield, a small number of combat platform to complete a single combat mission combat mode has been unable to meet the complex and changing battlefield environment. In order to make the air combat platform collaborative and efficient completion of operations, integrated system to realize multi platform, multi task combat, researchers will biological swarm concept applied to the field of air combat and put forward the concept of aeronautic swarm [1~3]. There are usually many types of combat platforms in aeronautic swarm, such as early warning aircraft, fighters, reconnaissance aircraft, unmanned aerial vehicles, etc., and the scale is large, generally reach dozens, hundreds, and platform mobility, wide distribution. In order to solve the different operational tasks of OODA (Oberve-Orient - Decide-Attack), the different types of platforms are based on the dynamic combination of task requirements, and the OODA complete combat process is realized on a continuous, continuous and closed loop basis on the basis of the coordinated operation of OODA .

In the aeronautic swarm combat environment, the platform position changes with the target location, the platform combination with the task requirements change, so that the distribution of the aeronautic swarm platform is random, discrete and wide area characteristics. The distribution of the platform matches the operational and operational objectives, but may not match the communication topology, that is, some platforms may become isolated nodes of the communication, resulting in the information can not be transmitted to each other, the swarm collaborative task can not be achieved. Therefore, how to achieve a large number of mobile aircraft in the mobile, decentralized combat units between the effective connectivity is the need to study the problem. The use of UAV relay can effectively solve the above problems. UAV relay platform with a wide range of coverage, easy deployment, the use of flexible features, and can achieve over-the-horizon communication [4].

In this paper, we present a UAV relay network to support aeronautic swarm operations, and study the deployment of UAV relay network. This paper is organized as follows. In paper Section II we review related work .In Section III we formulate the problem, proposed the node aggregation degree and the deployment evaluation function. A random deployment algorithm based on particle swarm is designed in Section IV. In Section V we evaluate the algorithms via simulation. We summarize the results in Section VI.

II. RELATED WORK

The problem of network deployment is a pre-planning problem of the network. At present, the related research on UAV relay network is divided into two categories. One is the dynamic maintenance of UAVs. This kind of research mainly focuses on the low speed users to study the optimal flight path and network performance optimization of relay UAV[5-6]. Because this paper mainly focuses on the background of aviation cluster operations, UAVs have lower speeds compared to combat aircraft, can not be dynamically maintained. Therefore, this paper studies relatively static network deployment. This type of research is divided into deterministic deployments [7] and random deployment [8].

Deterministic deployment refers to the fact that the nodes are placed at the specified location according to a certain geometrical structure in order to obtain the minimum or optimal connectivity and coverage of the nodes. The deterministic deployment approach is applicable to situations where the environment is in good condition and the fixed-point deployment costs are low. The advantage is that the number, density, location or proximity of the nodes in the network can be calculated and arranged according to different connectivity radius and coverage radius to obtain the optimal network deployment effect with the least number of nodes or the connectivity coverage. The disadvantage is that there is a boundary area resource redundancy, flexibility is poor, does not apply to irregular areas.

The location information of the nodes in the random deployment is unpredictable. The optimization problem is constructed according to the known conditions, and the location points of the deployment are obtained by intelligent algorithms such as genetic, virtual force and particle swarm. It has strong flexibility and more practicality. At present, the study of coverage and connectivity problems is usually independent of them to optimize, how to combine the two is the focus of many scholars. [9] designed a greedy algorithm, the algorithm first to complete the area coverage, in the realization of network connectivity, and finally delete the redundant deployment node to amend. However, there is a certain blindness in the greedy algorithm, and separate consideration and coverage will also affect the final outcome. In this method, the nodes are initially moved to the center of the region by gravity, and when the nodes are gathered to a certain extent, the network is connected, and the nodes are uniformly spread through the repulsive force to realize the maximum coverage of the nodes. However, this method is easy to cause nodes to always converge in the central part of the region, can not effectively achieve the edge of the area near the coverage. The use of particle swarm optimization algorithm for network deployment usually assumes that the network is connected, in the case of a large number of nodes. Does not consider the number of nodes in the case of limited coverage and connectivity to deploy.

III. SCENES AND MODELS

A. *Scene Description*

The discussion of this paper is based on the swarm combat application scenario shown in Figure 1, where a number of types of combat aircraft are scatred, with single isolated aircraft nodes and formation subnets, which are dynamically organized according to different combat missions. The distribution of these aeronautical swarm platforms presents randomness, discretization and wide-area characteristics. Therefore, this paper transforms the effective connectivity problem of decentralized maneuvering unit into the coverage problem of task area. The task area is divided according to the specific combat missions and scheduled routes in advance, can be a regular task area, it can be irregular. Through the rational deployment of multiple trunk UAVto achieve the stability of the task area coverage for the entire operational process to provide communication support.

B. *The geometric model of relaying unmanned aerial vehicles*

In general, the coverage area of the relay UAV is determined by the altitude of the flight, the height of the coverage area and the maximum propagation distance of the relay signal, and the maximum propagation distance of the relay signal is related to the status of the relay link channel. In this paper, considering the relay of the combat aircraft, assuming that the maximum propagation distance of the relay signal is R, the deployment height of the relay unmanned aerial vehicle is H, and for the aerial area with height h, if the transmission power and the antenna beam limitation factor are not taken into account, The maximum coverage radius of the relay signal is

$$r(h) = \sqrt{R^2 - (H-h)^2} \tag{1}$$

As the height difference between the different air platforms is tens of meters to several kilometers, and the maximum transmission distance of the relay signal can reach hundreds of kilometers or more, the height difference between the platforms and the horizontal distance between each other can be almost ignore. Therefore, in order to facilitate modeling, this paper assumes that the relay UAV has the following characteristics:

(1)on the task area of the coverage radius and relay UAV connected between the same radius, are set to R;

(2) Consider deployment issues on a two-dimensional plane;

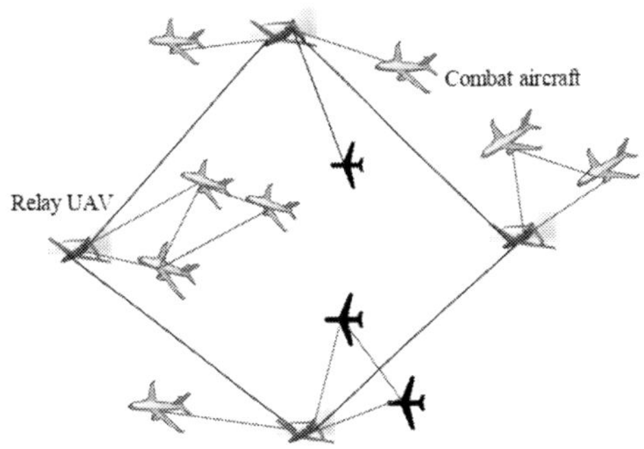

Figure.1. Schematic diagram of the scene

IV. RELATED DEFINITIONS

Definition 1. The degree of aggregation

For a network, the invulnerability of the network is to measure the ability of the network topology to remain connected when the attack fails. At present, the survivability optimization model focused on topology optimization, mainly discusses two problems: to meet the requirements of network connectivity, the lowest total cost or at a given cost, the topology optimization design of the largest network connectivity. In this paper the UAV relay network deployment issues, as the relay UAV number, radius of communication is limited, how to construct a under this condition should not only satisfy the coverage and connectivity requirements, at the same time, the largest network of survivable deployment mode is the main problem considered in this paper.

In a relay network, each relay node not only wants to relay users within its own coverage area, but also consider forwarding information for neighbor nodes. Therefore, according to the problem of unmanned aerial vehicle relay network deployment, combined with the degree of cohesion and the function of the relay node itself, the evaluation parameter of the node aggregation network resource is defined, that is, the aggregation degree β_i of the relay node i:

$$\beta_i = t_1 \frac{n-1}{2\sum_{i,j\in V} d_{ij}} + t_2 \frac{\sum N_i(S) - \sum_{n\neq i}(S)}{N_0(S)} \tag{2}$$

In order to evaluate the network coverage performance effectively, this paper uses the grid method to divide the task area into grid. Assume that the number of pixels in the task area is $m\times n$. The area coverage ratio $R_{area}(S)$ of the node set S is the ratio of the sum of the total number of pixel points of the node set S to the total pixel of the task area. Where n is the number of nodes in the network, d_{ij} is the shortest distance between any two points i and j in the network, and V is the set of all the nodes in the network. $N_0(S)$ is the total number of pixels that a single node covers. And t_1, t_2 is the weight coefficient, in order to balance effect of network topology and coverage changes on the degree of aggregation.

V. Design Of deployment algorithm for UAV relay network

At present, there are two characteristics of the research on connectivity and coverage problems by means of intelligent algorithms such as genetic, virtual force and particle swarm. ①There are fixed nodes and mobile nodes in the network, and the location of the mobile nodes is optimized by the intelligent algorithm. ②The connectivity radius of the node is much larger than the coverage radius of the node, and the number of nodes is relatively dense. The above characteristics make the study of network deployment problems usually based on the default network is connected to optimize. In the context of this paper, there is no fixed node, and the task area is large, the number of relay unmanned aerial vehicles, the communication radius is limited, the particle search results are prone to the interconnection between the network nodes can not judge the quality of the solution, To a certain extent limit the use of PSO algorithm.

A. Deterministic deployment model analysis

Strip, triangular, quadrilateral, hexagonal, and so on because of the simple deployment of the deployment of wireless sensor network deployment and terrestrial emergency wireless communication relay network deployment issues. There are a lot of literature from the perspective of coverage and connectivity of these four models to study, so do not do too much. In this paper, based on the coverage of connectivity, the aggregation degree of nodes is analyzed. The deployment mode of several existing deployment modes is analyzed and the deployment mode satisfying the requirements of this paper is studied.

Figure 3 shows the four deployment modes that the size of the mission area is $500km\times 400km$ and the communication radius R is 100km. Table I gives a performance comparison of the four deployment patterns.

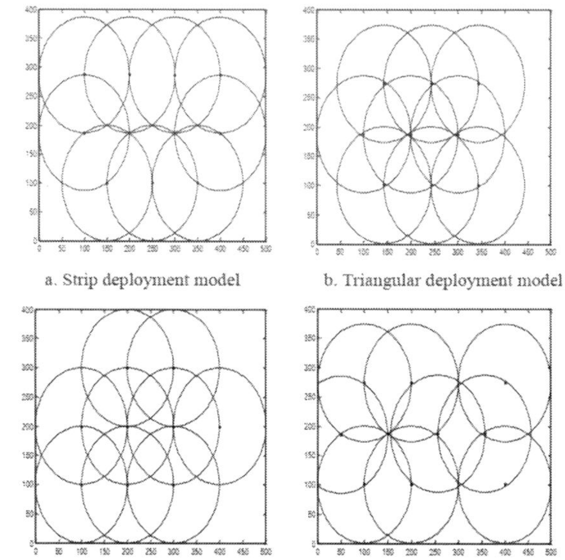

a. Strip deployment model b. Triangular deployment model

c. Quadrilateral deployment model d. Hexagonal deployment model

Figure 3. schematic diagram of four deterministic deployment mode

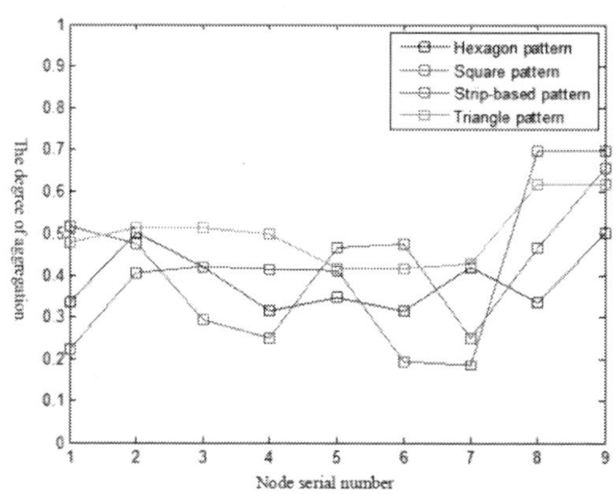

Figure 4. The diagram of the variation of the node aggregation degree

TABLE I. THE PERFORMANCE COMPARISON OF THE FOUR DEPLOYMENT PATTERNS

PATTERNS	COVERAGE RATE	AVERAGE DEGREE	VARIANCE
STRIP	84.60%	0.3780	0.0057
TRIANGULAR	70.91%	0.5022	0.0060
QUADRILATERAL	76.70%	0.4278	0.0186
HEXAGONAL	83.12%	0.4051	0.0371

It can be seen from Table I that in the same number of relay nodes, the deployment coverage ratio is the highest. Contrast stripe pattern and hexagonal pattern can be seen that although both coverage are smaller, but the hexagon mode node concentration variance is bigger, each node in the network "importance" gap is bigger, easy to cause this topology network anti-destroying ability is poor. And the pattern of stripe, although the node is distributed more evenly, and the coverage is higher, but the highest connection degree of the node is 2, the network robustness is poor, unsuitable for practical application. Triangle and quadrilateral model can provide a better for network connectivity, but the triangle pattern redundant nodes is bigger, the coverage is low, can lead to a number of relay nodes use too much, not suitable for limited number of relay UAV communication relay network deployment.

Result in practical application, the task area will become irregular shape, fixed deployment model can produce covered hole in the region boundary, the ground base station or emergency relay network is mostly through the border region put node to "fix" connect the weak area of coverage. Such repairs, while filling the gaps in the cover, will actually cause the area near the border to be repeatedly covered, causing considerable redundancy. This paper proposes an algorithm for boundary region deployment - based on improved particle swarm algorithm.

B. *PSO algorithm principle*

Particle Swarm Optimization (PSO) is a new evolutionary computing technology based on group intelligence, which is widely used in various optimization fields. Each particle in the PSO algorithm represents a potential optimal solution, and each particle perturbs its flight trajectory according to its own current information, its own historical information and population information, so as to find the optimal solution in the search space. Let the population of m particles search the Q-dimensional space (ie, the dimension of each particle). The position of particle i is expressed as $X_i = (x_{i1}, x_{i2}, \ldots x_{iQ})$. The velocity represented by particle i is expressed as $V_i = (v_{i1}, v_{i2}, \ldots v_{iQ})$. The basic PSO algorithm is updated as follows:

$$v_{iQ}(k+1) = w \times v_{iQ}(k) + c_1 \times \gamma_1 \times [p(k) - x_{iQ}(k)] + c_2 \times \gamma_2 \times [g(k) - x_{iQ}(k)] \quad (4)$$

The location update formula for the particle in the search process is:

$$x_{iQ}(k+1) = x_{iQ}(k) + v_{iQ}(k+1) \quad (5)$$

In the formulas (4) and (5), k represents the number of iterations; w is the inertia weighting factor; c_1 and c_2 are the learning factor; γ_1 and γ_2 are the random number within the (0,1) interval; $v(k) \in [-V_{\min}, V_{\max}]$ represents the velocity range of the particle; $x(k) \in [L_{\min}, L_{\max}]$ represents the search space; $p(k)$ is the historical value of the particle itself; $g(k)$ is the historical optimal value of the whole population; Particle velocity and position update can directly determine the quality of the solution and the efficiency of the solution. Although the update of the particle position is too fast, the particle will jump out of the local optimal solution to a certain extent. However, due to the limitation of the connectivity between the relay nodes, the destination location update will cause the relay nodes to not communicate with each other. In view of this, in order to consider the network coverage and connectivity to deploy the node, and to avoid the situation can not calculate the value of the situation because the node is not directly connected, adding virtual force to affect the location of the node update after each iteration.

In a two-dimensional plane, if the particle swarm algorithm is used to deploy the N relay nodes, the search space dimension of the algorithm is Q=2N. In the process of particle motion, the quality of the solution depends on the fitness function. In this paper, the reciprocal of the deployment evaluation function is taken as the fitness function Fitness of the PSO algorithm. According to the above rules, design PSO deployment algorithm.

VI. Simulation Analysis

In a clocked at 3GHZ, the memory is 4GB on the PC, the proposed algorithm is validated based on the MATLAB environment. The area of the task area is set $500km \times 400km$ so that the communication radius R of the relay unmanned aerial vehicle is 120 km. Learning factors c_1 and c_2 are both 2. $w_{\max} = 0.9$, $w_{\min} = 0.4$ and $|F_{th}| = 5$, c_3=0.5 , w_a=5 and $w_r = 0.2$, Maximum displacement distance $\max step = 30km$.

In order to compare the performance of VFA-PSO algorithm, the basic PSO algorithm and two improved PSO algorithms are used in the simulation experiment. One is based on the speed update factor improved with gravitational and repulsive PSO (combined gravitational and repulsion PSO, CAR-PSO) algorithm. Another PSO (explorative capability enhancement PSO, ECE-PSO) algorithm is based on improved particle search capabilities. The group size is set to 10 and the maximum number of iterations is 100. For these three algorithms in the iterative process, due to the connection between nodes can not calculate the fitness value, we do not take the particle fitness value comparison.

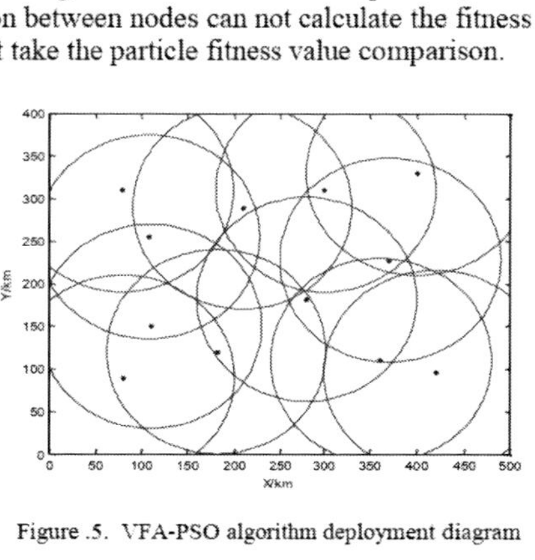

Figure .5. VFA-PSO algorithm deployment diagram

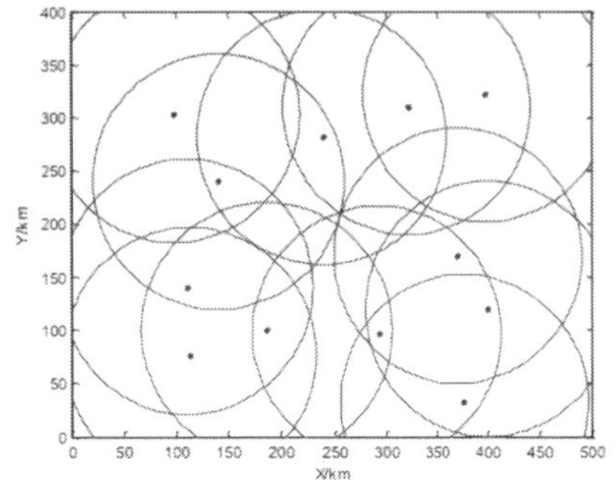

Figure.6. CRC-PSO algorithm deployment diagram

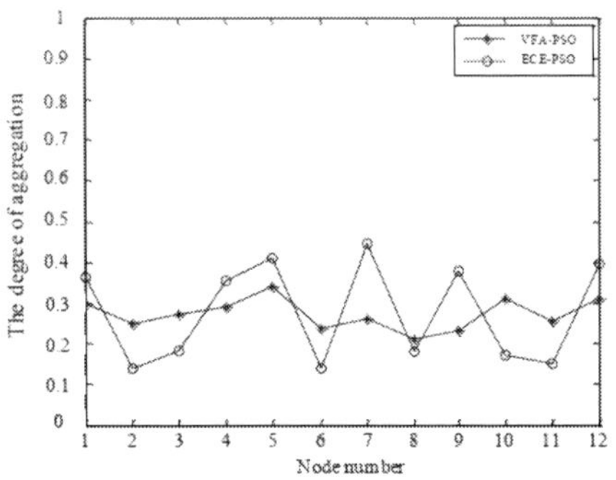

Figure.7. The curve of node aggregation degree change in two deployment situations

First, in order to verify the effectiveness of node aggregation degree for network survivability evaluation, 12 nodes are deployed in the task area. Figure 5 shows the results of the VFA-PSO algorithm deployment with a coverage of 98.7%. Figure 6 shows the coverage of the CAR-PSO algorithm. The coverage rate is 97.3%. Figure 7 shows the node aggregation degree of the two deployment results. The coordinates of the nodes are sorted according to the x-axis coordinates from small to large. However, as can be seen from Figure 7, the variance of the network node aggregation in Figure 4 is significantly smaller than that of Figure 6. From the perspective of the survivability of the network, we can see that, for example, the number of nodes with the corresponding coordinates (290, 97) in Figure 6 is the largest, and the node is removed. The coverage does not change, but the network will not be connected. In the deployment result of Fig. 1 (b), the node distribution is more "scattered". It can be clearly seen in Figure 4 and Figure 6 that each node in the deployment network of Figure 6 has a large gap in "action", resulting in poor distribution of nodes with uneven distribution of nodes. This verifies the accuracy of the node aggregation degree definition and can effectively evaluate the role of each node in the relay network.

VII. CONCLUSIONS

In this paper, we puts forward an idea of using UAV to construct relay network and study the deployment of the network based on the background of aeronautic swarm operations. In order to consider the deployment of combat unit coverage, network connectivity and survivability, the concept of node aggregation degree is put forward under the constraints of the number of relay UAV and the communication radius. And designs a VFA-PSO deployment algorithm based on virtual force adjustment. The simulation results show that the VFA-PSO algorithm can effectively improve the network coverage and the survivability of the network under the premise of ensuring the network connectivity, and has better deployment effect.

ACKNOWLEDGMENT

This work was supported by the National Natural Science Foundation of China under Grant No.61472443 and Research on the Application of Swarm Intelligent Induced Emergence in Aeronautic Swarm.

REFERENCES

[1] Duan H B, Shao S, Su B W. New development thoughts on the bio-inspired intelligence based control for unmanned aircraft system flight plan.USA:Washington DC: 2009-2047[R].2010.

[2] Huo D J.Operational of network swarms [M].Beijing : National Defense University Press,2013.

[3] Chaumette S, Laplace R, Mazel C, et al. CARUS, an operational retasking application for a swarm of autonomous UAVs: First return on experience[C]// Military Communications Conference, 2011 - Milcom. IEEE, 2012:2003-2010.

[4] Duan X H, Wang J W, Zhao Y P, et al. Design and Implementation of UAV in Communication Relay[J]. Sci-tech Innovation and Productivity, 2016.

[5] Marinho M A M, Freitas E P D, Costa J P C L D, et al. Using cooperative MIMO techniques and UAV relay networks to support connectivity in sparse Wireless Sensor Networks[C]// International Conference on Computing, Management and Telecommunications. IEEE, 2013:49-54.

[6] Journal of Tsinghua University (Science and Technology), 2011 (2): 150-155 .Department of Computer Science and Technology, Tsinghua University, Beijing 100084, China); A Multi-UAV Relay Network Supporting Mobile Ad Hoc Networks [J].

[7] Khoufi I, Minet P, Laouiti A, et al. Survey of Deployment Algorithms in Wireless Sensor Networks: Coverage and Connectivity Issues and Challenges[J]. International Journal of Autonomous & Adaptive Communications Systems, 2014.

[8] Zhou F B, Xue S L. Advanced Sequence-Based Localization Algorithm in Wireless Sensor Networks[J]. Applied Mechanics & Materials, 2014, 651-653(1):387-390.

Liu Chuang is a Master student at Air Force Engineering university, He received his bachelor of engineering in Chongqing University. His research interests are on the topics of Military aviation communication.

Lv Na is a Professor, doctoral supervisor at Air Force Engineering university, His research interests are on the topics of Military aviation communication.

Shaanxi Province, Xi'an Lianhu District Feng Gao Road No. 1, Phone number 13289396256.

Program Design of JSON to Structured Data Conversion

Zhongyue Da[1,a], *Wenjie Yang*[1], *Peipei Ran*[1], *Yuke Huo*[1]

[1]School of Printing & Packaging Engineering, Beijing Institute of Graphic Communication, Beijing 102600, PR China

Abstract. With the maturity of structured data storage technology and the increasing demand for enterprise information development, some enterprises need to convert the data storage model of application system from JSON to structured data. A file management system is token as an example to explain how to convert. In this article, it is achieved through the following three parts. First, the structure diagram of data transformation is built. Then, JSON model and structured data model are built. Finally, the conversion of data from JSON to MySQL is realized by using java language.

1 Introduction

With the development of relational database technology, relational database systems are becoming more stable and efficient. Relational data are studied to store semi-structured data in many enterprises, whose goal is to share data between multiple different system environments [1]. And it is beneficial to the diversified development and progress of the application system.

In non-database environments, semi structured data is common, such as XML and JSON [2]. The latter is discussed in this paper. JSON is easy to read and write, and is easily parsed and produced by machines. JSON format is not only language independent text format, but also conforms to the habit of C language family [3].

In the actual application of enterprises, different data storage technologies are used in different network systems. For example, in a file management system, two sets of storage methods are used, JSON semi-structured storage and MySQL database storage. To facilitate the application and rapid query, it is necessary to convert and merge the data storage mode. In this paper, the file management system is taken as an example to discuss and implement the conversion from JSON to relational database.

2 Program Design

2.1 Demand analysis

In this paper, the data conversion in a file management system is taken as an example to illustrate the process, which needs to get date from JSON file and store it into MySQL database. JSON file is composed of JSON [4] and unstructured binary data. Before storing the JSON data, the original files should be pre-processed to achieve the purpose of splitting binary data and JSON data. After it, semi-structured data is converted into JSON documents, while binary data is stored as pictures.

2.2 The overall design

The whole process of data conversion includes pre-processing, mode analysis, mode conversion, data conversion, data storage and so on. The specific process is shown in figure 1.

[a] Corresponding author: dazhongyue@qq.com

 DOI: 10.1051/matecconf/201713900166

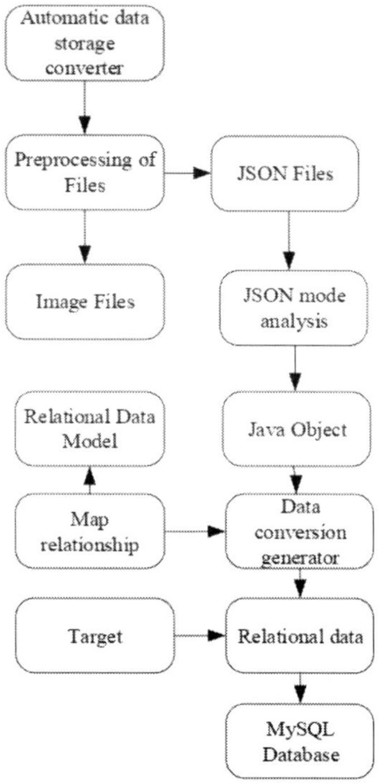

Fig. 1. Schematic diagram of the structure

3 Model Analysis

Schema transformation is the foundation of the whole process. The task needs to be accomplished is to build JSON semi-structured data model and generate corresponding relational data table model, then process JSON semi-structured data into structured data and store it in database.

3.1 Model Analysis

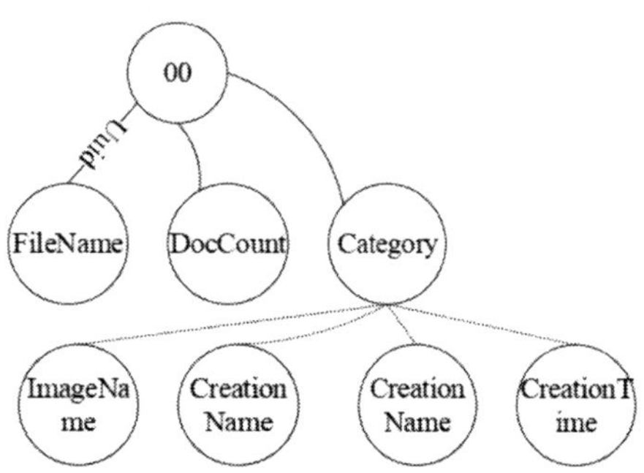

Fig.2. JSON Model

As shown in Figure two, the JSON data is deeply parsed, and then the structure model diagram is built, which is a simplified JSON OEM model diagram [5]. First, the types of nodes are partitioned, the node Category of the pattern division is found. Because Its child nodes are only one layer and the number of child nodes is appropriate. Other nodes are considered as data nodes.

Second, based on the previous results, two tables, info and file, were created. All the data nodes are put in the nodes of the model they belong to. According to the actual situation, attributes are added to type.

Third, according to the subordination in the OEM model [5], a master-slave table relation is generated, and a foreign key index is established between the two tables. Info is the primary table, and File is a child table. The foreign key Mainno in the table corresponds to FileName in JSON.

After the above process, the JSON tree structure in Figure 2 is converted to the table structure of Table 1 and Table 2.

Table 1. Information table(info)

Properties	Data type	Meaning
Mainno	VARCHAR	Primary key
Count	INT	File number
CreateName	VARCHAR	Record man
CreateTime	DATETIME	Record time

Table 2. File storage table (file)

Properties	Data type	Meaning
ID	INT	Primary key
Mainno	VARCHAR	Foreign key
imageName	VARCHAR	File name
dir	VARCHAR	File directory
remark	VARCHAR	Remark

3.2 Model implementation

The JSON document storage and MySQL database storage model is described in figure 3. The following is a description of the data model:

In the JSON document storage model, FileName is the name of the document, RecordDate is the document creation date, count is the number of images that the document contains, and imageName is the name of the picture. Remark is a note.

The attributes of the table in the chart are shown in Table 1 and Table 2.

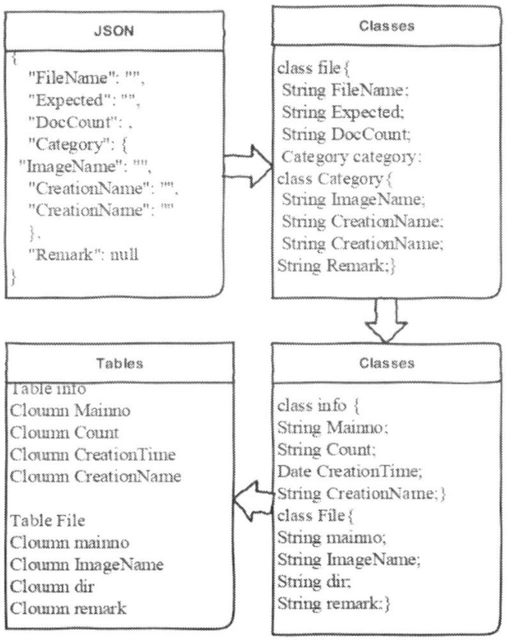

Fig.3 JSON model transformation diagram

4 The implementation of conversion

4.1 Data analysis

When the schema transformation is completed, the application data can be transformed into the target mode according to its corresponding relationship, and the basic process is divided into 3 steps:

(1) Read the JSON document data, get JSON format data;

(2) Translate from the original data to the target model of data according to the corresponding relationship between the data in Figure 2;

(3) Write the second step of the data to the MySQL database.

4.2 Program implementation

The following are classes implemented by the conversion program, as shown in figure 4.

(1) Doc class represents a data model class of JSON, and other data models are not described

(2) UpdateAction class belongs to the database connection class

(3) JSONAysis and AysisRead class, belonging to the JSON object read and model conversion class

(4) Read class is the core class, used to call other classes and complete all the steps of a file.

(5) The Account class is a target model class, where setter and getter represent the get and set functions for all attributes. Other data models are not described.

(6) FileOut class is the picture file output class, which converts binary data into JPG images.

4.3 Matters need attention

In the process of technical implementation, we should pay attention to the following aspects.

(1) In the aspect of time type transformation, the time expression format in JSON is dependent on the environment because JSON does not contain a variable type. First, the program reads data in a string, then determines the type, converts the format, and stores it in the database. If not, the system can't work. For example, "2017-03-27T16:10:11.626398+08:00", "2017-03-27", "2017-03-27T16:10:11.626+08:00" and other expressions are translated into "2017-03-27"".

(2) When reading multiple byte numbers (such as 16-bit integers), we should pay attention to the problems of high byte and low byte order. For example, when IO is written to a file, the order of bytes written by the C++ is from low to high, while Java is from high to low.

(3) The following code is added to avoid the running error when the JSON is serialized. Because the JSON object and the JSON structure are not one-to-one.

```
objectMapper.configure(DeserializationFeature.F
AIL_ON_UNKNOWN_PROPERTIES, false);
```

5 Conclusions

In this paper, it discusses process from design model to the source code and gradually studies the storage of semi-structured data. First, the flow chart is used to describe the whole process of program implementation. Then the storage model of semi-structured data is constructed by using OEM model [5]. Next, according to the mapping rules, the JSON storage model is mapped to structed data, and data conversion rules from JSON to MySQL are implemented. In addition, the program has been successfully applied to the file management system, which has realized the data conversion and achieves good results. The next step is to achieve dynamic transformation between semi-structured data and structured data

Acknowledgments: The study is supported by the 'Development and application of micron level vision quality detector (2013YQ140517)'.

6 References

1. Wang Wei. Zhu Dong. Peng Kaiyuan. From file system to database system: automatic conversion method and practice of application software. The Tenth National Academic Conference on Database, **608-612**

2. Xiaoyue Xin. Design and Research Study on Data Model of Document Database. Sun Yat-sen University (2015).

3. Fellipe Freire, Crishane Freire and Damires Souza Academic Unit of Informatics, Federal Institute of Education, Science and Technology of Paraiba, João Pessoa, Brazil Enhancing JSON to RDF Data Conversion with Entity Type Recognition (2017) .

4. Huyin Zhang, Qiansong Qu, Ruiyun Hu. Data exchange model based on JSON, Computer engineering and design.36(2015), **3380-3384**

5. HangYujia1, SuZhongbin, WuHuarui. ZhuHuaji. TaoYong. RESEARCH ON DYNAMIC TREE STORAGE MODEL OFSEMI-STRUCTURED DATA, Computer Applications and Software, **28,** (2011), **86-90.**

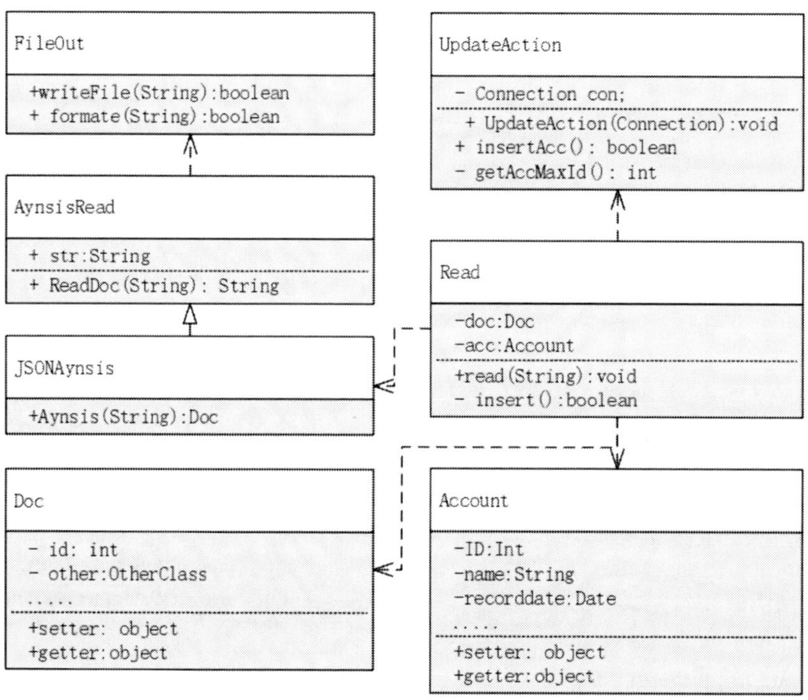

Fig. 4. Class Diagram of Program Implementation

Preference Selection Index Method for Machine Selection in a Flexible Manufacturing Cell

Shaoyong Jian[1], Shi Ying[*]

[1]Department of Mathematics and Computer Sciences , Xinyu University , Xinyu,338004 , China

[*]Department of fundamental courses ,Wuhan Technology and Business University Wuhan, Hubei,430065,China

Abstract. The selection of a desirable machine is an important concern for the manufacturing firm. The selection process contains some critical selection attributes and the task of this process is to choose the desirable machine from a number of candidate machines. Then the machine selection problem is actually a multi-attribute decision making problem. This paper will develop a preference selection index method to solve the problem of machine selection in a flexible manufacturing cell. A case study is used to demonstrate that the proposed method is effective and feasible.

1 Introduction

Manufacturing companies worldwide are forced to undergo transformation processes in order to improve their ability to succeed with their products on extremely competitive international markets. In this perspective, an adequate selection of the appropriate machine tools if often crucial but very difficult to achieve [1]. The paper [1] developed the TOPSIS method for solving packaging machine selection problem under intuitionistic fuzzy (IF) environment. The paper [2] solved the machine tool selection problem using a 0–1 integer programming combining with genetic algorithm. The paper [3] proposed the AHP approach to the material tool selection problem. The paper [4] used the fuzzy AHP method to solve the machine selection problem, in which the evaluation attribute values are expressed with triangular fuzzy numbers. The paper [5] put forward a novel 0-1 integer programming model for solving a problem of dynamic machine-tool selection and operation allocatio -n with part and tool movement policies in a flexible manufacturing system(FMS) environment. Flexible manufacturing cell (FMC) is a group of machines, working together to perform a set of functions on a particular part or product, with the added capability of being conveniently changeable to other parts or products [6]. Flexible manufacturing cells (FMCs) have received great attention in today's dynamic manufacturing environment. The evaluation attributes of machine selection in a FMC include many influence factors, such as purchasing cost, machine type, number of machines required, productivity, production output requirements, product quality, task and operating preference, interrelation among operating processes, type of control and accuracy of the machine, number of available AGVs, etc. [6,7]. Many authors studied the machine selection problems in a flexible manufacturing cell (FMC). The paper [8] developed the AHP method to the problem of machine tool selection. The paper [9] developed an expert system for manufacturing systems machine selection. The paper [10] and the paper [11] put forward a fuzzy goal-programming approach to solve the machine tool selection problem. The paper [12] proposed a fuzzy AHP method for evaluating machine tool alternatives. The paper [13] developed the digraph and matrix methods for the machine group selection in a FMC. The paper [14] suggested a decision support system based on fuzzy AHP and fuzzy TOPSIS for the selection of machining centre. This paper will develop preference selection index method to solve the problem of machine selection in a flexible manufacturing cell. The rest of this paper is organized as follows. In Section 2, the MADM model for machine selection problem is firstly established, and then the new decision method is proposed. In Section 3, a machine selection in a flexible manufacturing cell example is given to demonstrate the feasibility and effectiveness of the new method. Finally, Section 4 gives the conclusion of this paper.

2 Preference selection index (PSI) method

Preference selection index (PSI) method is firstly proposed by Maniya and Bhatt in 2010, which is not necessary to assign relative importance between attributes, but overall preference value of attributes are calculated using concept of statistics [15]. This method is specifically effective when there is conflict in deciding the relative importance between attributes and that is the advantage of PSI method. Using overall preference value, PSI for each alternative is calculated and alternative with higher value of PSI is selected as the best alternative. The detail calculation steps of PSI method are given as follows [15]:

Step 1. Identify the goal. Find out all possible the candidate machines (alternatives), selection attributes and its measures for the given application.

Step 2. Establish the MADM decision matrix. The solving each MADM problem begins with constructing decision matrix. Let $X=\{x_1,x_2, \ldots ,x_m\}$ be a set of alternative, $O=\{o_1,o_2, \ldots ,o_n\}$ be a set of decision attributes or criteria, a_{ij} is the performance of alternative x_i on the attribute o_j. Then the machine selection problem can be expressed with the decision matrix form $A=(a_{ij})_{m\times n}$

Step 3. Normalize the decision matrix A into normalization decision matrix $R=(r_{ij})_{m\times n}$.

The process of transforming attributes value into a range of [0,1] is called normalization and it is required in MADM methods to transform performance rating with different data measurement unit in a decision matrix into a compatible unit[15]. The normalization method adopted from the paper [16], and the formulas are given as follows: If the the j-attribute is benefit attribute, then $r_{ij}=a_{ij}/\max_{1\le i\le m}\{a_{ij}\}$; If the j-attribute is cost attribute attribute, then $r_{ij}=\min_{1\le i\le m}\{a_{ij}\}/a_{ij}$.

Step 4. Compute preference variation value (PV_j), which is defined as: $PV_j=\sum_{i=1}^{m}(r_{ij}-\bar{r}_j)^2$, where $\bar{r}_j=\sum_{i=1}^{m}r_{ij}/m$ is the mean of normalized value of attribute j.

Step 5. Determine overall preference value (Ψ_j) for each attribute. To get the overall preference value, it is required to find deviation (Φ_j) in preference value (PV_j) and the deviation in preference value for each attribute is determined using the following equation: $\Phi_j=1-PV_j$, and overall preference value (Ψ_j) is determined using following equation: $\Psi_j=\Phi_j/\sum_{j=1}^{n}\Phi_j$ .The total overall preference value of all the attributes should beone, i.e. $\sum_{j=1}^{n}\Psi_j=1$.

Step 6. Calculate the PSI (I_i), where $I_i=\sum_{j=1}^{n}r_{ij}\times\Psi_j$, for each alternative.

Step7. Rank all the alternatives according to the PSI. After calculation of the PSI (I_i), alternatives are ranked according to descending or ascending order to facilitate the managerial interpretation of the results, i.e. an alternative is ranked/selected first whose PSI (I_i) is highest and an alternative is ranked/selected last whose PSI (I_i) is the lowest and so on.

3 Case Study

This example is taken from the case study conducted by Wang et al. [7]. Wang et al. [7] had presented a real case of a FMC including two CNC milling machines, one CNC lathe and one robot for material handling. In this problem the factory manager had decided to purchase some machine facilities after sufficient discussion and complete evaluation. The two CNC milling machines, one CNC lathe and a robot are composed into a possible alternative for the FMC, that is, all are put together and considered as a machine group required for FMC. After putting different purchasing constraints on the total purchasing cost, and the specifications of milling machine, lathe machine, and the robot, suitable machines of FMC were composed into ten possible alternatives. Evaluation attributes areTotal purchasing cost (dollars) (o_1), Total floor space (m2) (o_2), Total machine number (o_3), Productivity*(mm/min) (o_4). Here, the value of Productivity corresponds to the machine with the slowest feedrate in FMC.

Table 1. Decision matrix and normalized decision matrix for machine group selection in an FMC [7].

No.	Evaluation attribute values			
	o_1	o_2	o_3	o_4
1	581818	54.49	3	5500
2	595454	49.73	3	4500
3	586060	51.24	3	5000
4	522727	45.71	3	5800
5	561818	52.66	3	5200
6	543030	74.46	4	5600
7	522727	75.42	4	5800
8	486970	62.62	4	5600
9	509394	65.87	4	6400
10	513333	70.67	4	6000

No.	Evaluation attribute values			
	o_1	o_2	o_3	o_4
1	0.8370	0.8389	1.0000	0.8594
2	0.8178	0.9192	1.0000	0.7031
3	0.8309	0.8921	1.0000	0.7813
4	0.9316	1.0000	1.0000	0.9063
5	0.8668	0.8680	1.0000	0.8125
6	0.8968	0.6139	0.7500	0.8750
7	0.9316	0.6061	0.7500	0.9063
8	1.0000	0.7300	0.7500	0.8750
9	0.9560	0.6939	0.7500	1.0000
10	0.9486	0.6468	0.7500	0.9375

The specific calculation steps of the proposed method are given as follows:

Step 1. The normalized attribute values are reported in Table1;

Step 2. The vectors of preference variation value (PV_j), deviation (Φ_j), and overall preference variation value (Ψ_j) are respectively obtain as:

PV =(0.0379, 0.1967, 0.1736, 0.0701);

Φ =(0.9621,0.8033,0.8264,0.9299);

Ψ =(0.2732,0.2281,0.2347,0.2640).

Step 3. The preference selection index (I_i) are calculated as I =(0.8816, 0.8534, 0.8714, 0.9566, 0.8840, 0.7921, 0.8080, 0.8467, 0.8595, 0.8302). Then the ranking order is 4 - 5 - 1 - 3 - 9 - 2 - 8 - 7 - 10 - 6, which is almost the same as the one obtained in paper [6]: 4 - 5 - 1 - 3 - 2 - 9 - 8 - 10 - 7 - 6. The desirable machine is the machine No. 4.

4 Conclusion

For the machine selection problem, we first construct a decision matrix, and then develop a new decision making method named preference selection index method for solving it. This method is specifically effective when there is conflict in deciding the relative importance between attributes. Finally, a case study is use to demonstrate and validate the application of the proposed method. The proposed method can also extend to other application to robot selection, investment selection and project selection.

References

1 D. Aloini, R. Dulmin and V. Mininno. A peer IF-TOPSIS based decision support system for packaging machine selection. Expert Systems with Applications, Vol. 41 (2014), p. 2157–2165.

2 C. Moon, M. Lee, Y. Seo and Y. H. Lee. Integrated machine tool selection and operation sequencing with capacity and precedence constraints using genetic algorithm. Computers & Industrial Engineering, Vol. 43 (2002) ,p. 605–621.

3 M. Yurdakul. AHP as a strategic decision-making tool to justify machine tool selection. Journal of Materials Processing Technology, Vol. 146 (2004), p.365–376.

4 O. Durán and J. Aguilo. Computer-aided machine-tool selection based on a Fuzzy-AHP approach. Expert Systems with Applications, Vol. 34 (2008), p. 1787–1794.

5 M. H. M. A. Jahromi and R. T. Moghaddam. A novel 0-1 linear integer programming model for dynamic machine-tool selection and operation allocation in a flexible manufacturing system. Journal of Manufacturing Systems, Vol. 31 (2012), p.224–231.

6 R. V. Rao: *Decision Making in Manufacturing Environment Using Graph Theory and Fuzzy Multiple Attribute Decision Making Methods*. (Springer-Verlag, London 2013).

7 T. Y. Wang, C. F. Shaw and Y. L. Chen. Machine selection in flexible manufacturing cell: a fuzzy multiple attribute decision making approach. Inter J Prod Res, Vol. 38 (2000), p. 2079-2097.

8 Z. Taha and S. Rostam. A hybrid fuzzy AHP-PROMETHEE decision support system for machine tool selection in flexible manufacturing cell. J Intell Manuf, Vol. 23 (2012) p. 2137-2149

9 H. Chtourou, W. Masmoudi and A. Maalej. An expert system for manufacturing systems machine selection. Expert Syst Appl, Vol. 28 (2005), p. 461-467

10 F. T. S. Chan and R. Swarnkar. Ant colony optimization approach to a fuzzy goal programming model for a machine tool selection and operation allocation problem in an FMS. Robot Comput Integr Manuf, Vol. 22 (2006), p. 353-362

11 F. T. S. Chan, R. Swarnkar and M. K. Tiwari. A fuzzy goal-programming model with an artificial immune system (AIS) approach for a machine tool selection and operation allocation problem in a flexible manufacturing system. Inter J Prod Res, Vol. 43 (2005), p. 4147-4163

12 Z. Ayag and R. G. Ozdemir. A fuzzy AHP approach to evaluating machine tool alternatives. J Intell Manuf, Vol. 17 (2006), p. 179–190

13 R. V. Rao. Machine group selection in a flexible manufacturing cell using digraph and matrix methods. Inter J Ind Syst Eng, Vol. 1 (2006), p.502–518

14 Y. Tansel and M.Yurdakul. Development of a decision support system for machining center selection. Expert Syst Appl, Vol. 36 (2009), p. 3505-3513

15 K. Maniya and M. G. Bhatt. A selection of material using a novel type decision-making method: Preference selection index method, Materials and Design, Vol. 31 (2010), p.1785-1789

A Convolutional Neural Networks Approach to Devise Controller

Xiangdi Liu[1,2,*] and Yunlong Dong[1,2,**]

[1] Huazhong University of Science and Technology
[2] School of Automation

Abstract. PID controller is widely used in many fields. The input sequence and output sequence of a well-parameterized PID controller are transformed into a matrix presented in images in this paper, and a Every-Timestep data augmentation algorithm is operated.This paper propose a image to image Convolutional Neural Networks(CNN) structure to learn the PID controller, which incorporates the successful methods in computer vision and deep learning to control field. The simulation is performed using Matlab/Simulink and Keras[1]. The simulation demonstrates that the CNN controller is capable in performance.
Keywords: PID controller, CNN, Deep learning, Control.

1 Introduction

In practice PID controller is the most widely used common control algorithm in many fields. PID controller has very simple structure and if parameters tuned well can achieve high performance. Neural Network is an information processing structure abstracted by the biological structure, consisting of processing element interconnected together with unidirectional signal channels called connections.As Neural Networks have three advantages, which have been widely used in pattern classification, system identification, image processing and other fields. First, Neural Networks are data-driven self-adaptive methods. Second, they are universal function approximator[2]. Third, Neural Networks are powerful nonlinear models, which make them flexible in handling real world relationships. Over the last few years, machine learning algorithms have led to more efficient methods and traditional neural networks have been developed into different structures for different problems. Neural Network(NN) was proposed to compose a PID controller[3–5]. Cheon[6] proposed a deep learning method to learn the PID controller.

This paper combines the deep learning algorithms with the control filed, and propose a CNN controller to replace the PID controller. The simulation is performed using Matlab/Simulink and the details of traditional PID controller and CNN controller were discussed.This paper is organized as follows. CNN is described in section II. In section III, Devise Controller is explained. The comparison details between the proposed CNN controller and traditional PID controller presented with the simulation results are shown in section IV. Finally, the conclusion follows in section V.

*e-mail: lxd@hust.edu.cn
**e-mail: dyl@hust.edu.cn

2 Convolutional Neural Networks

Convolutional Neural Network(CNN) is a kind of structural variation of traditional neural network, utilizing layers with convolutional filters that are applied to extract features . After Alex Krizhevsky used Deep Convolutional Neural Networks in ImageNet classification problem[7], machine learning researcher reach a consensus that general CNN structure composed of convolution layer, pooling layer and fully connected layer. The structure is shown in Fig 1[7].

Originally invented for computer version, CNN models have subsequently been shown to be effective for many different problems including, Simultaneous Localization and Mapping(SLAM)[8], Decision Making[9] and Automatic Driving[10].

The Convolutional Neural Network(CNN) plays a significant role in image classification[7], image caption[11], image semantic segmentation[12] and so on. CNN can directly learn from raw images and automatically extract the feature which usually outperform handcrafted feature engineering. CNN uses filters to extract feature from high dimensional sensory data(especially images) and have reached enormous success in many fields.

3 Devise Controller

3.1 Traditional PID controller

To devise a novel controller, a traditional PID controller was first proposed to be the baseline. A well-parameterized PID controller was the target we want to mimic. The PID controller calculates the output in the following rule(continous time) (1), where the k_p, k_i, k_d are the parameter to be settled in practice, e stands for the error in feedback control loop. The control scheme is showed in Fig 2 .

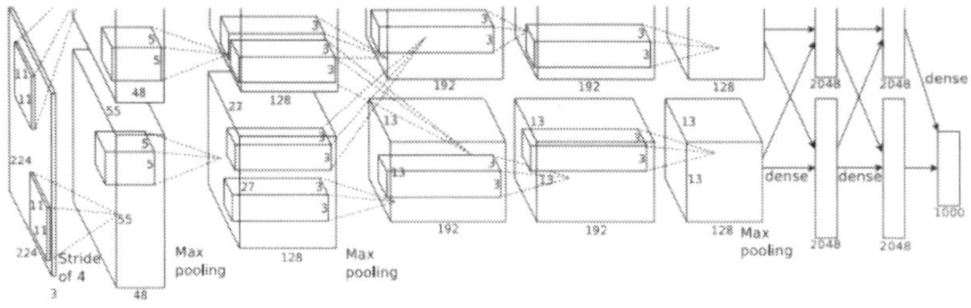

Figure 1. CNN structure

$$u = k_p \times e + k_i \times \int e dt + k_d \times \frac{de}{dt} \quad (1)$$

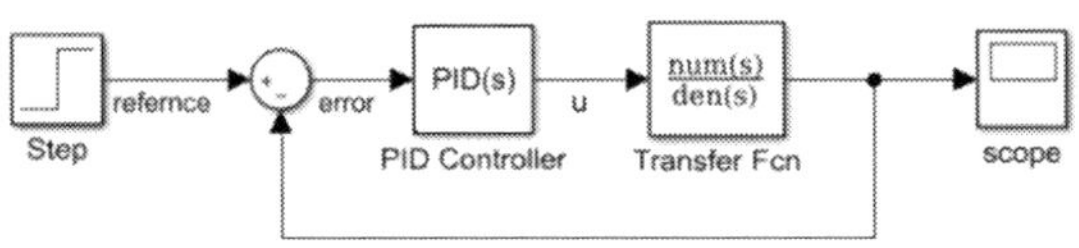

Figure 2. PID control scheme

The PID controller calculates the output based on the error sequence $\{e(0), e(1), e(2), ..., e(T)\}$ in the rule(Discrete time) (2). Kangbeom Cheon[6] used a Deep Believe Network to learn the PID controller's input and output, but they do not analysis the underlying principle in PID controller. The PID controller in fact consists of three filter, proportional, integrate, differential. Using the powerful efficiency of Deep learning methods, the map between PID controller input and output can be easily learned via supervising learning. The learning scheme is shown in Fig 3.

$$u(k) = k_p \times e(k) + k_i \times \Delta t \sum_{i=0}^{K} e(i) + k_d \times \frac{e(k) - e(k-1)}{\Delta t} \quad (2)$$

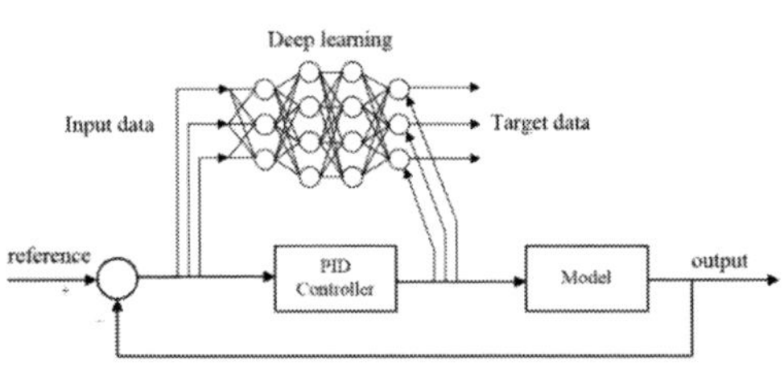

Figure 3. learning scheme

3.2 Transfer to Image

The error sequence $\{e(0), e(1), e(2), ..., e(T)\}$ is the input of our learning algorithm(Deep learning methods). Instead of directly dropping the Sequence to the model, We analysis the sequence first. The sequence is obviously a time sequence which implies the system response in time domain. The RNN[13] like models are capable of dealing with such sequence. But the time sequence is just a One-Dimensional relationship, we want further exploration. So we transform the error sequence to a matrix which has a spatial structure, which can contain more information. We samples 20s time(the default terminated time) in error with sample time 0.01s to get a 2000 length sequence. Then the Sequence is transformed into a 40×50 matrix called **error plane**. The same process was done in the output of PID controller called **u plane**. For visualization convenience, The matrix(One channel) are shown with colorful format in Fig 4.

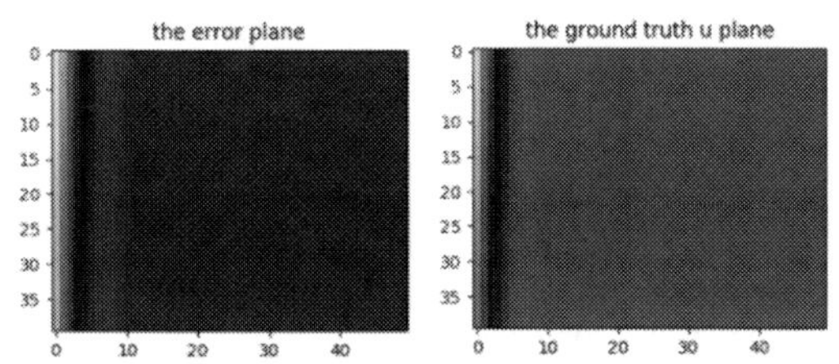

Figure 4. error plane(left), u plane(right)

As can be clearly seen, there is a pattern in the error plane and u plane. The u plane is like the blurry error plane. We can treat the sequence like image. There are many well-developed theory, models and algorithm processing images. Now we transfer the learning controller problem to a image to image supervising learning problem.

3.3 Devise CNN controller

The architecture used to devise the controller is shown in Fig 5. The architecture has purely convolutional layers. The output of the CNN controller should have the same size with the input, so there no no pooling layers and Flattern layers . The first layer has 32 3×3 filters with relu

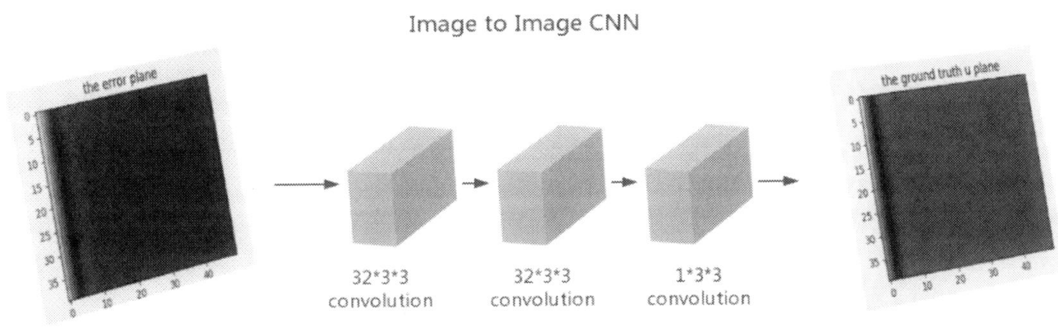

Figure 5. The CNN controller architecture.

activation function, the second layer has the same filters as the first layer, and the last layer have 1 3×3 filter with relu activation function to make the output of CNN has size 40×50.

Two-Dimentional convolutional operation can be written in eqn (3), where K is the filter with size $k_r \times k_c$, $In(s,t), out(s,t)$ are the input and output of the convolutional operation with size $N_c \times N_r$ respectively. Through convolutional operation, the CNN is able to capture the relationship and maps between error plane and u plane. Semantic, we want the filter to learn the pid caclulating rule eqn (2).

$$Out(s,t) = \sum_{m=0}^{k_r-1}\sum_{n=0}^{k_c-1} K(m,n) \times In(s-m, t-n)$$
$$s.t. \; 0 \leq s \leq N_r - 1, \; 0 \leq t \leq N - c - 1 \qquad (3)$$

3.4 Training

In practice the whole length of error sequence is not able to obtain, cause the length of error increases as the controlling time passing. In general, in time step $k, k < 2000$, the error sequence is $\{(e(0), e(1), e(2), ..., e(k))\}$. We fullfill the sequence to length of 2000 with zeros to obtain $\{(e(0), e(1), e(2), ..., e(k), 0, 0, ...)\}$. In training we only gather the full-length error plane and u plane, in other words, only one pair of data is available. In controlling loop, every time step should obey the rule(???), so the error plane and u plane can be split in every time step to generate a called **Every-Timestep error-u** dataset. The detail goes in **Algorithm 1**. $error, u$ are the error and u sequence with length T(default terminated time). Once the augmented dataset has been obtained. A mini-batch Stochastic Gradient Descent opimizer with $lerning\_rate = 0.1, batch\_size = 32, momentum = 0.9, epochs = 10$ is used to train the CNN. Keras [1] was used to achieve these stuff. The results are shown in Fig 6. We use 70% of dataset to train and the rest 30% to test, and proving that overfitting did not appear. The visualization of training peroformance is shown in Fig 7.

Algorithm 1 Every-Timestep($error, u, T$)

$error\_gen = Zeros[T, T]$
$u\_gen = Zeros[T, T]$
for $i = 1$ to T **do**
  $error\_gen[i, 1:i] = error[1:i]$
  $u\_gen[i, 1:i] = u[1:i]$
end for
return $\{error\_gen, u\_gen\}$

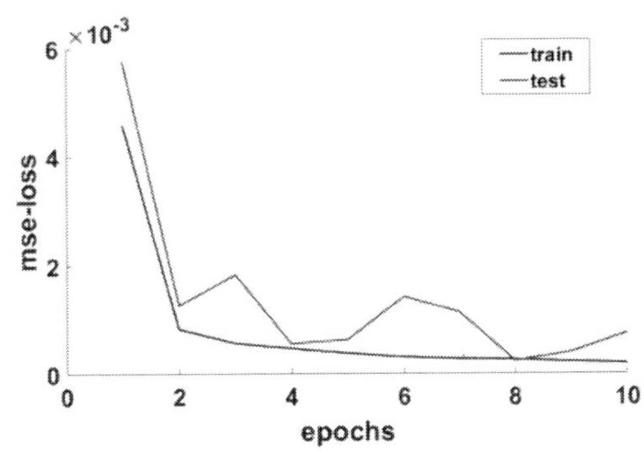

Figure 6. training results

4 Evaluation

In evaluation, we first parameterized a traditional PID controller in Simulink to simulate. The closed loop is shown in Fig 8. Then we replace the PID controller with our proposed CNN controller. Simulating in 2s. The comparing results of step response are shown in Fig 9.The CNN controller perform almost the same with PID controller even a little more smooth.

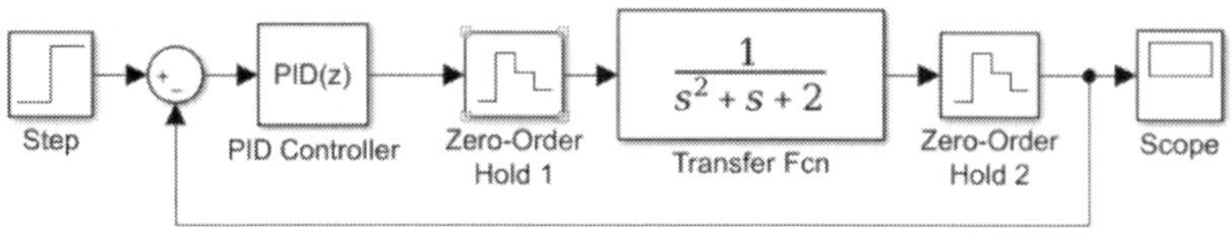

Figure 8. simulink model

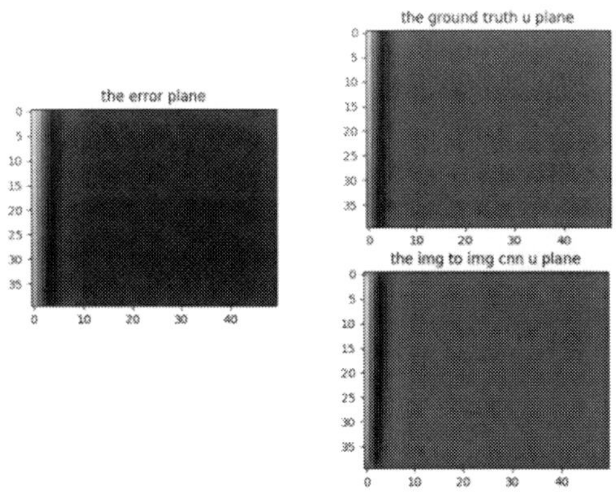

Figure 7. training visualization

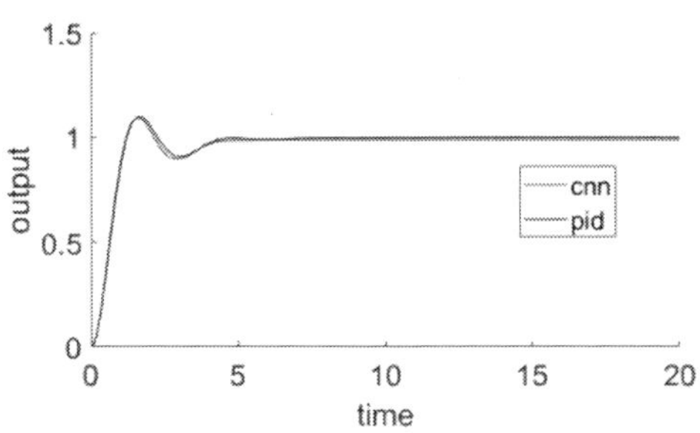

Figure 9. results

5 Conclusion

In this paper, a Convolutional Neuron Networks controller was devised as an approach to replace the PID controller. So far this paper is the first to combine images with control, incorporating the successful mothods in Computer Vision to control field, and propose a image to image CNN to learn the controller. The simulation demonstrates that the CNN controller is capable.

References

[1] F. Chollet, *keras*, https://github.com/fchollet/keras (2015)

[2] K. Hornik, M. Stinchcombe, H. White, Neural Networks **2**, 359 (1989)

[3] F. Shahraki, M. Fanaei, A. Arjomandzadeh, Chemical Engineering Transactions pp. 1395–1400 (2009)

[4] H. Shu, Y. Pi, Computers & Chemical Engineering **24**, 859 (2000)

[5] H. Shu, Y. Pi, networks **1**, 8 (2005)

[6] K. Cheon, J. Kim, M. Hamadache, D. Lee, Journal of Automation and Control Engineering Vol **3** (2015)

[7] A. Krizhevsky, I. Sutskever, G.E. Hinton, in *Advances in Neural Information Processing Systems 25*, edited by F. Pereira, C.J.C. Burges, L. Bottou, K.Q. Weinberger (Curran Associates, Inc., 2012), pp. 1097–1105

[8] T. Naseer, M. Ruhnke, C. Stachniss, L. Spinello, W. Burgard, *Robust visual SLAM across seasons*, in *Intelligent Robots and Systems (IROS), 2015 IEEE/RSJ International Conference on* (IEEE, 2015), pp. 2529–2535

[9] A. Giusti, J. Guzzi, D.C. Cireşan, F.L. He, J.P. Rodríguez, F. Fontana, M. Faessler, C. Forster, J. Schmidhuber, G. Di Caro et al., IEEE Robotics and Automation Letters **1**, 661 (2016)

[10] M. Bojarski, D. Del Testa, D. Dworakowski, B. Firner, B. Flepp, P. Goyal, L.D. Jackel, M. Monfort, U. Muller, J. Zhang et al., arXiv preprint arXiv:1604.07316 (2016)

[11] K. Xu, J. Ba, R. Kiros, K. Cho, A. Courville, R. Salakhudinov, R. Zemel, Y. Bengio, *Show, attend and tell: Neural image caption generation with visual attention*, in *International Conference on Machine Learning* (2015), pp. 2048–2057

[12] J. Long, E. Shelhamer, T. Darrell, *Fully convolutional networks for semantic segmentation*, in *Proceedings of the IEEE Conference on Computer Vision and Pattern Recognition* (2015), pp. 3431–3440

[13] T. Mikolov, M. Karafiát, L. Burget, J. Cernockỳ, S. Khudanpur, *Recurrent neural network based language model.*, in *Interspeech* (2010), Vol. 2, p. 3

The influence of the opening on the capacity of the community based on matlab software

Daoming Feng

College of Mathematics and Computer Science,Xinyu University ,Xinyu 338000, China

Abstract. This paper mainly studies the influence of community opening on road traffic. Combined with the reality of life, from the lane of the influencing factors, to find out several important factors, and accordingly the establishment of mathematical models for the community to put forward rationalization proposals.Firstly, the influence of different factors on the same index is selected by using the found data and analogy method. Through the analytic hierarchy process, the weight vector of each index is calculated by using MATLAB, and an appropriate evaluation index system is established.This paper establishes a mathematical model based on the equal relationship between traffic volume, traffic capacity and saturation before and after the opening of the district, and studies the influence of the opening of the district on the surrounding road traffic according to the basic traffic capacity and the crossing capacity of the access road.

1 PROBLEM DESCRIPTION

In recent years, with the national economic development level gradually increased, people have the ability to buy their own cars and rooms. Now most people are willing to go to the district to buy a house, first of all that the environment is better and close to the developed city, and that the surrounding area surrounded by the security of the enclosure is better from outside, but now because of the construction of the district more and more The greater the area occupied by the city, so that the road to building traffic less, causing frequent road congestion, so the public on the issue of whether the opening of the district can ease the traffic pressure to start a discussion, if the district can be opened by the The road inside and out of the district, but although the increase in the number of roads, but also led to other problems, such as will cause confusion within the district, the increase in traffic led to traffic congestion and traffic accidents increased probability.

In addition to security, the opening of the district is really open to optimize the road network structure, the following open to the surrounding roads on the impact of the following four questions:

Question 1: To evaluate the impact of open area on the surrounding road access to select the appropriate evaluation index system.

Question 2: to establish a mathematical model on the passage of vehicles, to study the opening of the district on the impact of the surrounding roads.

2 Problem analysis

For the first question: First, understand the changes before and after the opening of the district, the use of comparative law, the more likely to affect the traffic capacity of the indicators, select the impact of relatively large, due to the impact of factors are more common, for example, we can ignore external factors (Such as weather) and other unnecessary factors, the use of control variables to extract variables, residential open to the surrounding factors on the impact of the surrounding area through the study of our district road structure, community space size, traffic flow, residential parking, residential road structure between the data changes, The above factors are set up as the criterion layer and the target layer is the appropriate evaluation index system for the community. Therefore, we establish the model of the analytic hierarchy process, and then compare each other to determine the weight of the target layer. Finally, we use MATLAB to calculate the weight vector of each index , It will be drawn which indicators will affect the opening of the district and the impact of the size of the extent of the impact.

For the second question: First, the number of road around the road several cases of traffic analysis, the number of lanes is usually divided into two-way street, three-way street, four-way street, and analysis of residential road structure, residential road is a one-way street, and then we According to each road saturation is similar to the calculation of the district before and after the opening of the road saturation changes. There are T-shaped, Y-shaped, and round-shaped intersections at the intersection of the entrance and exit of the main road on the main road. However, we can only discuss the T-shaped, and then the three roads cross the road together, analyze the three cases Of the traffic density (hourly traffic) due to the main road, the traffic on the road will be different, assuming that the choice of road vehicles will not form a blockage, the initial determination of its saturation unchanged, the vehicle can run normally, and finally Compare the saturation of these three roads to determine the size of the impact on the surrounding roads, which is best.

3 General assumptions and definitions

3.1 Assumptions

DOI: 10.1051/matecconf/201713900169

- Assume that there is no vehicle failure or accident on the road.
- Suppose the road is flat, regardless of the slope effect.
- Assume that road vehicles are cars.
- Assume that the district is connected to the surrounding roads without traffic lights.
- Assuming that the speed of the surrounding road vehicle is constant, it is 30km / h.

3.2 Terms and definitions

Table 1. Terms and Definitions

B	Cell size
C	District surrounding road structure
Q	Traffic flow
D	District parking volume
S	District road structure
W	Weight vector
h	Indicates the number of lanes
N	Road capacity
M	Road saturation
K	Road reduction factor
W_B	The weight vector of the cell space
W_C	The Weight Vector of Residential Road Structure
W_Q	Traffic volume vector
W_D	The right vector of parking volume
W_S	Weight vector of district road structure
λ_{max}	The largest feature root
E	Unit matrix
C_i, C_j	The i, j elements of the criterion layer (i, j = 1, 2, 3, 4, 5)

4 Establishment and solution of the model

4.1 The establishment of the hierarchy

The decision-making problem is divided into two levels. According to the influence of the opening of the district on the surrounding roads, the corresponding evaluation indexes are set up. According to the information, the spatial size of the district B, the surrounding road structure (economic development area) Flow Q, parking volume D, road structure S The main five factors affect the passage of the surrounding roads. The above factors are used as the second level of the level, and the pairwise comparison matrix of 5 × 5 is established according to the degree of the magnitude relationship of the influence of the second level on the first level. Finally, the pairwise comparison matrix is normalized The maximum eigenvector of each factor is calculated. If the matrix is consistent, then the maximum weight vector is used as the weight vector, and the evaluation index system with the first three effects is selected according to the order of the ratio. The fuzzy hierarchy is established according to the above method Analysis model one, as shown below:

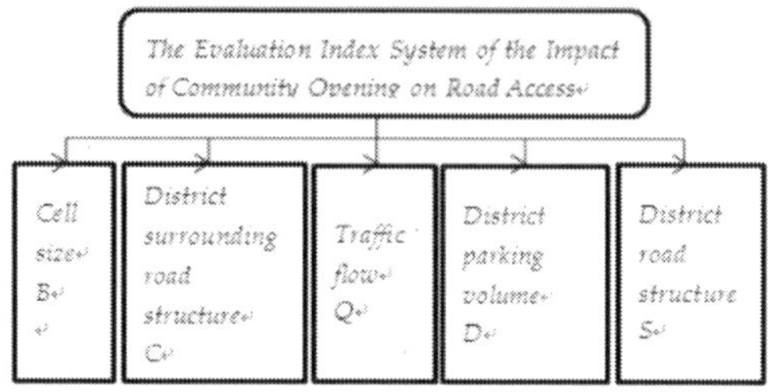

Fig. 1. Hierarchical model

4.1.1 Determination of pairs of matrices

With the development of the economy, people's living standards continue to improve, the number of residential quarters also will increase, the traffic volume also has an impact, in the decision-making factors, we can refer to the corresponding size of the data, from five factors were removed two A factor out of comparison. Use a_{ij} to represent the ratio of the i-th and the j-th factor, regardless of the influence of the other three factors, and so on.

As shown in Figure 1: Compare the impact of the second layer of five factors （Cell space size C1、Community surrounding road structure C2、Traffic flow C3、Residential parking volume C4、Road structure C5）on the opening of the first layer of the district. From the second layer to select A_5^2 times two different factors to compare, when $C_i = C_j$ means that the selected for the same two factors for comparison. According to the proportion of reflexivity 1 ,that is $a_{ii} = 1, i = 1,2,3,4,5$ are always established,used to indicate the impact on the factors of the previous level,The proportional result can be used as a comparison matrix

$$A = (a_{ij})_{5\times5}, a_{ij} > 0, a_{ji} = 1 / a_{ij} \quad (1)$$

According to the comparison of the factors affecting the surrounding roads, we can form a pair of matrices,and let $a_{ij} = \frac{w_i}{w_j}$,then

$$A = \begin{bmatrix} \frac{W_B}{W_B} & \frac{W_B}{W_C} & \frac{W_B}{W_Q} & \frac{W_B}{W_D} & \frac{W_B}{W_S} \\ \frac{W_C}{W_D} & \frac{W_C}{W_C} & \frac{W_C}{W_Q} & \frac{W_C}{W_D} & \frac{W_C}{W_S} \\ \frac{W_Q}{W_B} & \frac{W_Q}{W_C} & \frac{W_Q}{W_Q} & \frac{W_Q}{W_D} & \frac{W_Q}{W_S} \\ \frac{W_D}{W_B} & \frac{W_D}{W_C} & \frac{W_D}{W_Q} & \frac{W_D}{W_D} & \frac{W_D}{W_S} \\ \frac{W_S}{W_B} & \frac{W_S}{W_C} & \frac{W_S}{W_Q} & \frac{W_S}{W_D} & \frac{W_S}{W_S} \end{bmatrix} \quad (2)$$

The pairwise comparison matrix is obtained by substituting the law and the proportional relationship:

$$A=\begin{bmatrix} 1 & \frac{1}{3} & \frac{1}{5} & \frac{4}{7} & \frac{5}{8} \\ 3 & 1 & \frac{4}{3} & \frac{5}{4} & \frac{7}{3} \\ 5 & \frac{3}{4} & 1 & \frac{9}{5} & \frac{7}{5} \\ \frac{7}{4} & \frac{4}{5} & \frac{5}{9} & 1 & \frac{3}{5} \\ \frac{8}{5} & \frac{3}{7} & \frac{7}{5} & \frac{5}{3} & 1 \end{bmatrix} \quad (3)$$

4.1.2 Compare scale determined

(1) compare the two different nature of the indicators and come to a conclusion, we need to set a comparison scale to determine whether these indicators are realistic, according to the information and multi-party discussion, the indicators used to compare the use of two levels Analysis, determination of indicators:

Table 2. The meaning of each factor scale of Judgment matrix

Scale aij	Meaning
Similar to 1	C_i has the same effect as C_j
Similar to 3	C_i and C_j slightly stronger
Approximately 5	The influence of C_i and C_j is strong
Similar to 7	The effect of C_i and C_j is obviously enhanced
Similar to 9	The influence of C_i and C_j is absolutely strong
Similar to 2,4,6,8	The ratio of C_i to C_j is between the two adjacent levels
Similar to the reciprocal of the above numbers	C_i / C_j becomes C_j / C_i

(2) The proportion of the factors in the criterion layer is displayed according to the data of the search and can be fully accorded with the 1-9 scale. Therefore, it is only necessary to approximate the corresponding scale, for example, in matrix A The index is 7/4 to about 1/2, to determine its 1/2 of the indicators, which can be derived from the other.

4.1.3 Solve the weight vector and the maximum eigenvalue step

According to the goal of the solution process, the use of step by step approach, starting from the bottom, according to the degree of correlation between the indicators listed in order, the specific steps are as follows: Calculate the weight vector W of each element on the target layer of the criterion layer.

According to the influence of the second level on the first level goal, the proportion of the two pairs is compared to the comparison matrix.

Find the eigenvalues W_i (i = 1, 2, 3, 4, 5) under the eigenvalues are substituted by the formula

$$(\qquad |\lambda E - A| = 0 \qquad (4)$$

The eigenvectors of pairs of comparison matrices are calculated by MATLAB software.

The normalization of each eigenvector and the determination of the weight vector are as follows: the largest eigenvector (normalized) is chosen as the weight vector, and the normalized vector sum is 1 for accuracy. On the basis of this, we use the knowledge of MATLAB to calculate the maximum eigenvector as the weight vector $W = [0.1911 \quad 0.651 \quad 0.4777 \quad 0.3384 \quad 0.44367]^T$ as the weight vector of each factor.

Calculate the maximum eigenvalue λ_{max} of each element of the criterion layer for the target layer.

The eigenvalue is calculated from（4）, and the largest eigenvalue of the eigenvector is selected from the calculated eigenvalue. The algorithm of the maximum eigenvalue can be realized by MATLAB, and λ_{max} = 5.0979.

4.Consistency test passed

According to the search data, the maximum eigenvector $\lambda \geq n$ of the n-th pair of the comparison matrix A is shown, and A is a consistent matrix when $\lambda = n$. But the calculation is inevitable error, in order to better measure the consistency of this measure to establish a measure to determine the scope of consistency, when the characteristic root λ than the order of the n-order matrix is much larger, to determine the degree of A serious degree of inconsistencies, so With the normalized eigenvector as the weight vector, the algorithm will be as follows

$$CI = \frac{\lambda_{max}}{n-1} \qquad (5)$$

As an indicator of consistency,when the CI is similar to 0, it is determined that A is a consistency matrix;The greater the degree of inconsistency of CI, the more serious.In order to study the scope of the inconsistency,we need to find the maximum amount of measure to measure the inconsistency of A,Substituting the matrix into the above formula,using the MATLAB to calculate the maximum feature vector λ_{max} = 5.0979 into the above consistency index formula,substituting in (5) to get CI = 0.02448.To verify the consistency of the test,Here we introduce the randomness of the indicators RI,specific data as shown in Table III:

Table 3. Random consistency indicator RI values

n	1	2	3	4	5	6	7	8	9	10	11
RI	0	0	0.58	0.90	1.12	1.24	1.32	1.41	1.45	1.49	1.51

The ratio of the random consistency index RI is called the consistency ratio CR, that is, the consistency ratio CR

$$CR = CI / RI < 0.1 \quad (6)$$

It can be seen that the maximum eigenvector is within the allowable range, and the eigenvector can be used as the eigenvector after the eigenvector is the normalized vector of the eigenvector $\lambda = 5.0979$

$$W = [0.1911 \ \ 0.651 \ \ 0.4777 \ \ 0.3384 \ \ 0.44367\ ]^T$$

And then by (6),find the

$$CI = (5.0979\text{-}5)/4 = 0.2448$$

Since the established pairing matrix is 5 by 5 for the matrix in Table II, RI = 1.12,use the formula (6) compute put

$$CR = CI / RI = 0.02448 / 1.12 = 0.02186 < 0.1$$

Conform to the set range, and consistency verification is passed as a weight vector.

A comprehensive evaluation of the indicators with the right vector,the basic steps are shown in Figure 2:

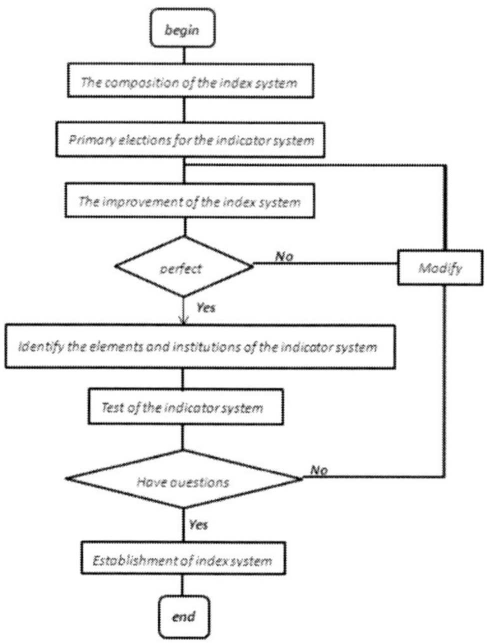

Fig. 2. Evaluation flow chart

Community open to the impact of road access we selected five evaluation index system,The weight vector of the district road structure is

$$W = [0.1911 \ \ 0.651 \ \ 0.4777 \ \ 0.3384 \ \ 0.44367\ ]^T$$

In order to correspond to the above-mentioned index system, WC= 0.6511, WQ=0.4777, WS=0.4436, WD =0.3384, WB=0.1911, which can be arranged from large to small.

It is concluded that the influence of the opening of the district on the traffic volume is arranged from large to small area for the surrounding road structure, traffic volume, community road structure, community parking volume and cell space size.

4.2 Analyze the structure of the surrounding roads of the district to determine the road type of the vehicle

As the structure of the plot of a variety of styles, where we take the road around the hospital for example, as shown in Figure 3: the district has a number of roads around, and each road structure is not the same, the road points For the one-way street, two-lane, three lanes, four lanes and other forms, and in the road junctions have to study the plane crossroads in the form of cross, T shaped, Y-shaped, round cross, multi-channel cross, In order to facilitate the study of the choice of T-word intersection, so we mainly consider the road type and the surrounding road structure and traffic flow of the three main factors, its open area on the surrounding road traffic will cause what impact and the intersection capacity The impact of the situation.

Based on the previous settings and information, we assume that the surrounding roads are composed of two different types. And assuming that there are h1 lane in n1 roads around the area, and each road traffic is Q1; there are h2 lane in n2 roads, and each road traffic flow is Q2. You can create a model:

$$\begin{cases} \dfrac{Q_1}{N_1(k_1+\cdots+k_{h1})} = \dfrac{Q_2}{N_1(k_1+\cdots+k_{h2})} \\ n_1Q_1+n_2Q_2=Q \end{cases} \quad (7)$$

In order to make the calculation more convenient, we can make:

$$(k_1+\cdots+k_{h1}) = r_1, (k_1+\cdots+k_{h2}) = r_2$$

The equation can be reduced to :

$$\begin{cases} \dfrac{Q_1}{N_1r_1} = \dfrac{Q_2}{N_1r_2} \\ n_1Q_1+n_2Q_2=Q \end{cases} \quad (8)$$

From the (8) Can get

$$\begin{cases} Q_1 = \dfrac{Qr_1}{n_1r_1+n_2r_2} \\ Q_2 = \dfrac{Qr_2}{n_1r_1+n_2r_2} \end{cases} \quad (9)$$

According to the obtained Q1, Q2, the saturation can be obtained

$$M_1 = \frac{Q}{N_1(n_1r_1+n_2r_2)} \quad (10)$$

N1 is the maximum traffic volume for a lane,M1 for road saturation, $k_1,\cdots,k_h$ is the ratio reduction coefficient of the road.

Because the road in the district is mainly two-way one-way road, but the number of single-street road number is different, so we can set the district has n3 two-way one-way street, if the open cell, the district road traffic for each road Q3, Each road flow becomes Q4, and the traffic volume of each car in the h2 lane becomes Q5. You can build the model from this:

$$\begin{cases} \dfrac{Q_3}{N_1} = \dfrac{Q_4}{N_1r_1} = \dfrac{Q_5}{N_1r_2} \\ n_3Q_3+n_1Q_4+n_2Q_5=Q \end{cases} \quad (11)$$

By the above equation together to obtain

$$M_2 = \frac{Q}{N_1(n_1 r_1 + n_2 r_2 + n_3)} \quad (12)$$

$$M_3 = \frac{Q_3 + Q_4}{N_2} \quad (13)$$

$$M_4 = \frac{Q_3 + Q_5}{N_3} \quad (14)$$

According to this model, the impact of the road on the surrounding roads can be studied by comparing the road saturation before and after the opening of the district and the change of the traffic volume of the main road and the intersection.

5 Result analysis

We get three conclusions:

(1) When the cell structure and the surrounding road structure does not change, the greater the traffic flow, the community before and after the saturation change the greater the change. But the greater the traffic flow at the same time, the community will make the intersection of the higher degree of saturation.

(2) when the cell structure and traffic flow does not change, the less the surrounding roads, residential area before and after the saturation changes the greater the change. But the less the surrounding roads, residential open will make the intersection of the higher saturation.

(3) When the surrounding road structure and traffic flow does not change, the more the internal road, the district before and after the saturation change the greater the change. And found the impact of the internal roads on the intersection.

Therefore,we have made the following recommendations based on these three points: we can give priority to open the surrounding roads within the road more of the district, because the number of internal roads on the intersection of the intersection will reduce the number of problems in the three tables have many types of district open Although the road saturation , But also will increase the burden of each intersection, when M> 1, the traffic will be in a blocked state.

For the traffic flow of the district, the surrounding roads less of the district should be carefully considered whether open, although they can be a good way to alleviate the surrounding road traffic, but also likely to lead to crossing the block, so that all the surrounding roads blocked.

At the same time, we can rationalize the surrounding roads to improve the road capacity. Mitigate road traffic conditions.

References

1. Xiao zheng, zhu jiamin, xia Ming, building kang. Quantitative analysis on the influence of the opening of the community on the traffic capacity of surrounding roads[J].Journal of jiamusi university (natural science edition), 2015, 35(02):336-339.
2. Zhang heng, sun zhonggui, wu huimin. Mathematical modeling of the impact of the opening of the community on road traffic capacity [J]. Electronic technology and software engineering, 2017,(04):-189. (2017-03-07).
3. Xiao jiqing. Evaluation study on the impact of the opening of the community on the impact of road capacity[J].Technology horizon, 2016,(26):198-+207.
4. Li xiaoqing, zhu cunbin, li dongmei, Chen kai. Research on the influence of the opening of the community on the traffic capacity of surrounding roads[J].Journal of northern college of hebei (natural science edition), 2013, 33(09):39-43.
5. Liu Yang, zhang xueya, Yang zongyue, gong tenant. The impact model of open community on road traffic capacity[J].Technology innovation guide, 2015, 14(01):93-95.
6. Wang chai ping. Research on traffic network optimization in wuchang urban area based on Matlab simulation[D].Hubei university of technology, 2016.
7. Liu jiang. Selection of enterprise transportation based on analytic hierarchy process[D].University of international business and economics, 2006.

Research on taxi software policy based on big data

Daoming Feng

College of Mathematics and Computer Science.Xinyu University ,Xinyu 338000, China

Abstract. Through big data analysis, statistical analysis of a large number of factors affect the establishment of the rally car index set,By establishing a mathematical model to analyze the different space-time taxi resource "to match supply and demand" degree.combined with intelligent deployment to solve the "taxi difficult" this hot social issues. This article takes Shanghai as an example,the central park, Lu Xun park, century park three areas as the object of study.From the "sky drops fast travel intelligence platform" big data,Extracted passenger demand and the number of taxi Kongshi data.Then demand and supply of taxis to establish indicators matrix,get the degree of matching supply needs of the region.Then through the big data relevant policies of each taxi company.Using the method of cluster analysis,to find the decisive role of the three aspects of the factors,using principal component analysis,compare the advantages and disadvantages of the existing company's programs.Finally, according to the above research to develop a reasonable taxi software related policies.

1 PROBLEM DESCRIPTION

Taxi is one of the important means of transport for the public,with the "Internet +" era,the popularity of mobile internet,there are a number of companies rely on the mobile Internet to build a taxi software service platform,to achieve the passenger and taxi drivers between the information exchange,making people in the daily life of the taxi to become more efficient and convenient,changed the way the traditional travel taxi,and thus to the general public's life caused a certain impact,However, on the other hand the taxi driver complained that labor intensity,income is relatively low,even the occurrence of a taxi driver strike,this reflects the taxi market management there are some problems,The existing taxi pricing is not reasonable, the speed is too high,leading to the entire taxi industry downturn,The long run will affect social stability,worthy of attention.China's cities in the next period of time,The scale will continue to expand, the population will continue to grow, people's living standards will continue to improve.The demand for a taxi will also change.How according to the city's population and travel intensity and the share of the taxi,To develop effective and practical programs in line with the strategic objectives of urban development,Plan a reasonable number of car rental,to maximize the people to meet the travel needs,at the same time, according to reduce environmental pollution and resource consumption.Coordinate the interests of all walks of life,is worthy of further study.

Taxi is one of the important vehicles of public travel, " Take a taxi difficult" is a hot topic of social concern. With the era of "Internet +", there are a number of companies rely on mobile Internet to establish a taxi service platform software, and information exchange between passengers and taxi drivers, together with a variety of taxi subsidy scheme.

Through data analysis, research mathematical model the following questions:

(1) try to establish a reasonable indicators, and analyze the different space-time taxi resource "to match supply and demand" degree.

(2) analysis of each company whether the taxi subsidy program "to alleviate taxi difficult" helpful?

(3) to a taxi service software platform for the newly created design a reasonable subsidy program.

2 Problem analysis

2.1 Analysis of the first question

For the first question, through the establishment of a city to study the G region of the supply and demand matching, the number is $(D_1, D_2, \cdots, D_G)$, the public daily in 24 hours a day unit of the demand for indicators of the taxi $(Q_1, Q_2, \cdots, Q_{24})$ and taxi supply index $(V_1, V_2, \cdots, V_{24})$. Use the formula

$$D_i = \frac{V_{i,j} - Q_{i,j}}{Q_{i,j}} \quad i = 1,2,3\cdots G; j = 1,2,3\cdots 24 \tag{1}$$

the supply and demand matching model of a certain area in different time periods is established by analyzing the degree of "supply and demand matching" of taxi resources in different time and space.Through the analysis of supply demand data about time series, The image more intuitively illustrates the change in the supply demand relationship in each region, for further results analysis. But also through the past few years the demand for taxis.

2.2 Analysis of the second question

For the second question, through the Baidu search can understand the taxi company to ease the taxi and their own profits promulgated by some of the policies (here we listed nine more representative of the policy), we can through clustering analysis These nine policies are divided into three categories, namely:

(1)taxi promotion policy for software;

(2)taxi company incentives for non-peak car customers;

(3)the taxi driver's reward;

And then through the questionnaire survey and expert interviews, respectively, the three main factors of fuzzy comprehensive evaluation, the establishment of models for analysis.Through these three programs as a program layer to explore the distribution of indicators of weight, and then get the level of comprehensive sort, and then compare Get the merits of the existing company's programs.

2.3 Analysis of question three

On the basis of the third question and the first and the second question, the author puts forward a relatively reasonable scheme in combination with the relevant actual situation and related regulations in combination with the weight of the three main decision-making schemes.

3 General assumptions and definitions

3.1 Assumptions

•Assuming that the demand of passengers every day,like the law does not consider the impact of holidays,

•Assuming that the taxi company has a reasonableallocation of taxis,

•Assuming that the latitude and longitude within a certain range can well represent the vicinity of a region,

•Assumptions consistent commuting time on the ground of general corporate or enterprise,

•Assuming that the results which useing the big data analysis can effectively reflect the supply demand situation in the region.

3.2 Terms and definitions

Table 1. Terms and Definitions

$D_1, D_2, D_3 \cdots D_G$	1 to G regions of the matching degree
$Q_1, Q_2, Q_3 \cdots Q_{24}$	The demand matrix for different time periods in a given day
$V_1, V_2, V_3 \cdots V_{24}$	Taxis supply index matrix for different periods of time in a day
X_{ij}	Sample matrix
R_i	Policy factors
W_i	Weights

4 Establishment and solution of the model

4.1 Problem one model is established and solved

4.1.1 The establishment of evaluation indicators

Taxi traffic has different eigenvalues as time changes and location displacements, and is analyzed on a day-by-day basis. According to the number of months, the day is divided into $T_1, T_2, T_3 \cdots T_{24}$, a total of 24 time units, assuming that the day's passenger demand for Q, then 24 hours a day in the unit of passenger demand corresponding to $Q_1, Q_2, Q_3 \cdots Q_{24}$,to meet: $Q_1 + Q_2 + Q_3 + \cdots + Q_{24} = Q$; The number of empty taxi is V, the number of empty taxis in a certain period of time is marked, assuming that it is $V_1, V_2, V_3 \cdots V_{24}$, then: $V_1 + V_2 + V_3 + \cdots + V_{24} = V$; Assuming that the supply and demand of G regions are matched, then the number is $D_1, D_2, D_3 \cdots D_G$, for G regions , The demand Q for a day in the area is a matrix and the number of empty taxi numbers V (ie supply number) is a matrix, respectively:

$$\begin{bmatrix} V_{1,1} & V_{1,2} & \cdots & V_{1,24} \\ V_{2,1} & V_{2,2} & \cdots & V_{2,24} \\ \vdots & \vdots & \ddots & \vdots \\ V_{G,1} & V_{G,2} & \cdots & V_{G,24} \end{bmatrix} \begin{bmatrix} Q_{1,1} & Q_{1,2} & \cdots & Q_{1,24} \\ Q_{2,1} & Q_{2,2} & \cdots & Q_{2,24} \\ \vdots & \vdots & \ddots & \vdots \\ Q_{G,1} & Q_{G,2} & \cdots & Q_{G,24} \end{bmatrix} \quad (2)$$

And then use the formula:

$$D_i = \frac{V_{i,j} - Q_{i,j}}{Q_{i,j}} \quad i = 1,2,3 \cdots G; j = 1,2,3 \cdots 24 \quad (3)$$

you can find the matching of different regions j supply and demand. If the absolute value of D_i is closer to zero, it means that the matching degree of supply and demand is better. If the absolute value of D_i is larger, the greater the matching degree of supply and demand is. If D_i is greater than zero, the supply is oversupply, D_i is less than zero, the amount.

4.1.2 Data collection and processing

This problem is mainly research and analysis In the "Internet +" era under the influence of some taxi software with the Internet to build a taxi software service platform to achieve the passenger and taxi drivers between the information exchange, influence and change the traditional way of taxi Circumstances, the taxi industry in different time and space regions of the impact of supply and demand matching. But also in the existing “didi taxi”, “fast taxi”, “haha carpool” and other software, the biggest difference in the market share of the taxi, so the data from the title of "sky dripping fast intelligent travel platform," the large data, Out of Shanghai within a day of passenger demand and the number of vacant taxi part of the data. See Appendix 1 for specific data. (In Appendix 1, each data is composed of a square bracket which consists of four parts separated by semicolons, such as: ["", 121.4861, 31.2269, 861], the first data is double quotation marks , Is estimated to be a code, the second third, respectively, said latitude and longitude, the fourth represents the latitude and longitude of the needs and supply).

In this paper, by selecting the Shanghai Central Park (longitude: 121.46906000000001 latitude: 31.23169), and then within a certain range of the central park (ie, longitude: 121.46 latitude: 31.23) within a day of travel and demand data, City Lu Xun Park (longitude: 31.27, latitude: 121.49) Century Park (longitude: 31.21, latitude: 121.55) The specific data are as follows:

Table 2. Center Park Taxi Supply Demand Table

Time	Demand	Supply	Time	Demand	Supply
0:00	54	30	12:00	299	199
1:00	60	35	13:00	274	224
2:00	66	43	14:00	271	211
3:00	77	50	15:00	204	156
4:00	104	66	16:00	195	105
5:00	137	90	17:00	392	235
6:00	171	150	18:00	221	202
7:00	259	203	19:00	206	152
8:00	322	300	20:00	249	186
9:00	291	250	21:00	216	143
10:00	266	203	22:00	204	125
11:00	235	171	23:00	67	21

Table 3. Lu Xun Park Park Taxi Supply Demand Table

Time	Demand	Supply	Time	Demand	Supply
0:00	151	45	12:00	352	266
1:00	158	63	13:00	341	231
2:00	163	71	14:00	285	227
3:00	181	76	15:00	257	189
4:00	209	93	16:00	171	146
5:00	215	114	17:00	198	175
6:00	221	166	18:00	353	211
7:00	289	215	19:00	335	231
8:00	360	265	20:00	351	254
9:00	563	261	21:00	309	218
10:00	426	235	22:00	260	154
11:00	311	213	23:00	169	75

Table 4. Century Park Taxi Supply Demand Table

Time	Demand	Supply	Time	Demand	Supply
0:00	59	23	12:00	141	83
1:00	72	21	13:00	137	62
2:00	68	26	14:00	131	50
3:00	71	30	15:00	123	47
4:00	85	34	16:00	120	42
5:00	93	39	17:00	123	63
6:00	91	42	18:00	134	78
7:00	101	57	19:00	145	92
8:00	124	83	20:00	153	101
9:00	130	89	21:00	113	92
10:00	266	203	22:00	204	125
11:00	235	171	23:00	67	21

4.1.3 Explanation and explanation of the phenomenon

Through the above data table can be obtained $V_{i,j}, Q_{i,j}$ supply and demand matrix, through

$$D_i = \frac{V_{i,j} - Q_{i,j}}{Q_{i,j}} (i = 1,2,3\cdots G; j = 1,2,3\cdots 24) \quad (4)$$

Calculated The supply matching degree of the central park $D_1 = 0.3132$;

Lu Xun Park supply demand matching degree $D_2 = 0.5226$;

Century Park supply demand matching degree $D_3 = 0.7373$;

Further, as long as we have obtained the supply demand data for each region, we can use a similar approach to find the supply demand matching relationship in the corresponding region.

Through the calculation results can be drawn: $D_1 < D_2 < D_3$,

That is: the center of the park around the scope of the supply needs of the best matching, the scope of the century around the park supply demand matching the worst degree; and D_1, D_2, D_3 are greater than zero, so you can know the three areas of the supply demand of the taxi are Is the supply is greater than the demand, indicating that the taxi industry there is a greater competitive edge, on the other hand can also show that customers can hit the car in the more time, to meet the requirements of customer taxi!

Through the above data form, and further, we through the excel to draw the supply chain of each region needs to change the map, you can more intuitive image of the supply needs of the changes observed, the specific image below:

Fig. 1. Center Park Taxi Supply Demand

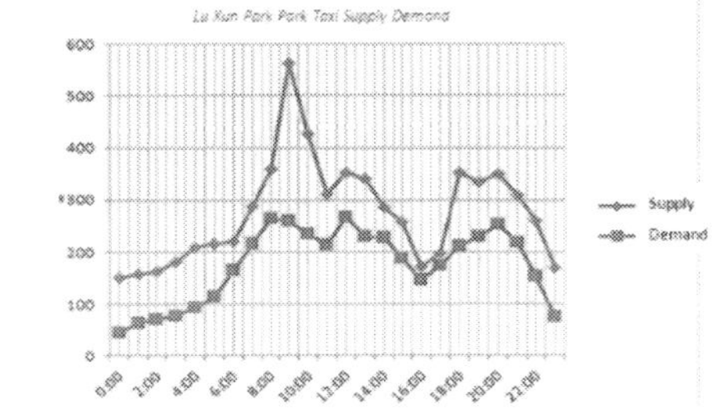

Fig. 2. Lu Xun Park Park Taxi Supply Demand

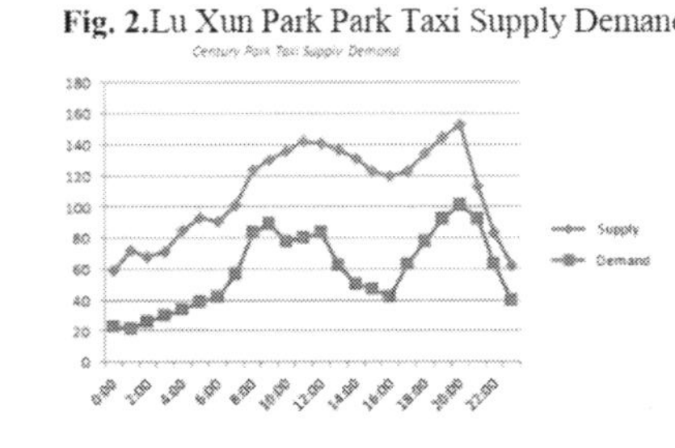

Fig. 3. Century Park Taxi Supply Demand

Through the image, you can intuitively know: the supply is greater than the demand, and generally in the morning 7: 00-9: 00 and 18: 00-20: 00, these times the supply and demand of the smallest difference between that In the case of a taxi company taxi supply certain circumstances, 7: 00-9: 00:00 and 18: 00-20: 00 These two time periods are commuting to the peak of commuting, so the passengers are relatively more; In other times, people flow is relatively small, relatively less car customers, and some customers will choose the subway or bus, which caused a certain impact on the taxi industry.

4.2 Problm 2 model building and solving

4.2.1 Taxi company related subsidy policy

February 17, 2014 Di tick hit the car, and its Wechat to pay the third round of marketing activities officially opened, to restore subsidies and has been strengthened. Beijing, Shanghai, Shenzhen, Hangzhou drivers with Wechat to pay the collection fee, 10 yuan per single, 10 per day, the other city drivers every day The first 5 single prize 5 yuan each, after 5 single 10 yuan per single prize. It is noteworthy that the Di Di car extra reward new users: passenger first single by 15 yuan, the first driver of the first prize 50 yuan.

4.2.2 Utilization of Big Data on subsidy programs analysed

This paper is to study the various regions of Shanghai taxi supply matches extent,Through the network platform data on the Shanghai area of the public taxi situation and the impact of the subsidy policy,Converted to the main factor of the final total score U_i,and then through the formula.

$$W_i = \frac{U_i}{\sum_{i=1}^{3} U_i} \tag{5}$$

to determine the weight of each factor.Finally, by summing up the statistics, the final score of each factor is obtained .

Table 5. Policy weight score

R_i \ Index value	Total score U_i	Specific gravity W_i
Taxi software promotion R_1	654	31．53%
Each time period Subsidy for car customers R_2	723	34．86%
The reward for the driver R_3	697	33．61%

From the table, we can see that the proportion of subsidies to non-peak passenger customers is the largest, which reflects the common problems of taxi companies and the general public. Secondly, the proportion of incentives to drivers is the largest, indicating that the taxi drivers Subsidies to make the driver can more actively meet the needs of customers; the last is the promotion of taxi software policy favored by customers, but the result is relatively low. The reason is based on the analysis because some older and some younger people are less accustomed to using taxi software to spend.

4.2.3 The solution to the problem three

From the second analysis of the results we can see:
When the taxi company's decision-makers in the development of policies, we should give priority to the factors from different time to consider the improvement of customer satisfaction,and then combined with the driver's reward and punishment system.Finally, in combination with the promotion of taxi software policy,making the relevant policies can not only reflect the broad masses of customers the taxi consumption needs,but also to maximize the profitability of the taxi company to achieve a win-win situation.Thereby promoting attracting more consumer spending;Finally, through the careful consideration of the company's leadership, the following three policies have been drawn:
(1) every Monday to Thursday 06: 00-10: 00, 20 yuan per subsidy;
Every Monday to Thursday 16: 00-22: 00, 10 yuan per subsidy.
Every Friday 06: 00-22: 00, 30 yuan per subsidy.
Every Friday 16: 00-24: 00, 25 yuan per subsidy.
Every Saturday, Sunday 0: 00-2: 00, each compensation 10 yuan.
Every Saturday, Sunday 11: 00-24: 00, 10 yuan per subsidy.
(2) the incentive for the driver according to the weekly travel volume and the customer under the number of equal weight and reward:On the completion of 40 weekly or more weekly travel more than 1,500 km of the driver, an additional reward of 200 yuan;On the completion of 50 per week or more than the weekly

travel volume of more than 1,800 km of the driver, an additional reward of 500 yuan;On the weekly completion of 80 single or more weeks more than 2200 km trip driver, an additional reward of 850 yuan;On the completion of 100 per week or more than the weekly travel volume of more than 2500 km of drivers, additional awards 1,200 yuan;
(3) the use of taxi software then subsidize customers ranging from 8-20 yuan.Daily subsidy three single, and for seven consecutive days using taxi software more than ten times, to carry out the activities of sending calls. In every day among all the use of taxi software, customers randomly selected part of the 5-10 yuan per person to pay as a subsidy.

5 Result analysis

5.1 This article is worth learning from

In this paper, large data analysis, with persuasive;The model of this paper directly through the taxi company's supply and demand relationship to illustrate the degree of matching analysis, the theory is conducive to the realization;In this paper, clustering analysis, with a wide range of applicability;The establishment of the planning model with universal, combined with local data can be directly promoted; Due to the limitation of the data span, we can only analyze the data given in the topic, and the model is rather rough and omitted some cases.

5.2 Need to be improved

Data processing needs to be further precise;The collection of inquiries on the policy is not perfect.

References

1. Zhiqiang Qu, jing Qiao.Effects of Big Data Processing on E-commerce [J]. Chinese Business Theory, 2015, (16): 61-63..
2. Jie Jia,Zongzhe Sun,Ming-ming Mao,Yubin Bao,Research on MapReduce Large Data Processing Platform and Algorithm [J]. Journal of Software, 2017,28 (03): 514-543.
3. Li xiaofei. Research on big data processing system based on cloud computing technology [J]. Journal of changchun engineering college (natural science edition), 2014,15(01).
4. Juntong Wang, Shitong Wang, Zhaohong Deng.There are Some Problems in Cluster Analysis [J] .Control and Decision, 2012,27 (03): 321-328.
5. Deng xue,Jiaming Li,Chen Junyang, jun-feng zhao. Ahp weight calculation method for the analysis and application study [J]. Journal of mathematics practice and understanding, and 2012 (7) : 93-93. [2017-09-28]..
6. Liu xiao. Research on development strategy of didi taxi company [D]. Guangdong university of finance and economics, 2016.
7. Li Juan. Research on the optimization of traditional taxi return and low speed driving cost [D]. South China university of technology, 2016.

Fatigue life test of reinforced grinding bearings

Zhuan Zhao[1, 2],Xiaochu Liu[1, 2, *],Zhe Qin[1, 2], Youcheng Chen[1, 2]

1Department of Mechannical and Electrical Engineering，Guangzhou University，Guangzhou 510006， China
2.Key Laboratory of high-peformance Metal Materials Reinforced Grinding Machining，Guangzhou University，Guangzhou 510006，China

Abstract：There are 4 groups of car wheel bearings, in which the inner ring of the 2 groups of bearings were reinforced by grinding process while the other 2 groups of bearings were not. After finishing grinding, then following by superfinishing to make them fit with steel and cage. When it was completed, 4 groups of bearings on the bearing fatigue testing machine were able to test the change of surface quality, operation temperature, vibration and clearance. The result shows that the surface quality, operating temperature, vibration and clearance changes of bearings are better than those untreated after reinforced grinding.

1.Introduction

Bearing is an important and basic component, which is widely used in military and civilian areas. With the implementation of 2015 China's manufacturing, intelligent, automation of mechanized equipment improve continuously, a new challenge about the strength, reliability and fatigue life of the bearing is emerging. In terms of fatigue life of bearings, many scholars have done a lot of researches. P.C. Bastias, S.T. Hadfield [1, 2] et al believe that the fatigue failure of the bearing is divided into two stages. The first stage is the plastic deformation of the material, and cracks are formed in the plastic deformation zone. This stage is relatively short, accounting for about 10% of the damage process, but it creates the condition for the formation, expansion and spalling of crack in the second stage. Therefore, the plastic deformation of metal is the precondition of crack generation. G.R. Miller [3] et al studies and suggests that rolling contact fatigue is mainly in two forms: initial rolling contact fatigue of the subsurface and of the near-surface. Li Xunjiao [4] et al thinks that the rolling bearing contact fatigue is under the action of alternating cyclic contact stress, local permanent cumulative damage is occurred on the contact surface and experienced the process of crack generation , expansion, and resulted in small pieces or small pieces of metal peeling . Wanchang Sen and G.. John [5, 6] at al introduce that under the condition of normal use, the materials of bearing surface are subjected to repeated action of high contact stress, crack initiation on the surface or subsurface, and ultimately lead to fatigue failure of rolling bearings.

This paper will make a fatigue life test on reinforced grinding bears and untreated bearings. The reliability and feasibility of reinforced grinding process are revealed by comparing different aspects. It is beneficial to improve the fatigue life of bearings.

2.Reinforced grinding experiments

*Corresponding author:XiaoChu Liu.gdliuxiaochu@163.com

 DOI: 10.1051/matecconf/201713900171

2.1 The principle of reinforced grinding

Reinforced grinding is a processing method based on composite fatigue resistance, corrosion resistance and wear precision machining technology of metal materials. This kind of processing technology can help obtain bearing ring with residual stress. The principle is shown in Figure 1 and 2. We fused strengthened abrasive material with high pressured gas and used injection system to produce mixed high-pressure gas liquid solid three-phase jet flow. As the collision of ring surface happened randomly [1], thereby grinding on the workpiece surface was strengthened.

Figure 1. Reinforced grinding machine

Figure 2. Electromagnetic centerless fixture

2.2 Experimental sample

We selected 4 groups of bearings and numbered them as 1~4， the inner ring of the No.1 and No.2 bearings were reinforced by grinding process while the other 2 groups of bearings were not. The bearing sample was automobile hub bearing with model DAC34640037B, the sample was removed from the production line after grinding the bearings (no super fine), the main size was shown in Table 1. The bearing was a double row angular contact ball bearing. The inner and outer ring physical map is shown in figure 3.

Table 1. Main dimensions of DAC34640037B

Bearing code	d	D	B	C	Seal up	Net weight
DAC34640037B	34	64	37	37	B	0.43kg

Figure 3.Inner and outer rings

2.3 Process parameters of Reinforced grinding

The process parameters are set up and then processed, process parameters are shown in table2 and table3.

Table 2. compounding proportion of abrasive materials

Bearing steel ball with diameter of 2mm	1400gram
Abrasive powder（Brown corundum）	400gram
Triethanolamine solution	600gram
30 pecent of AES solution	200gram
Water	400gram

Table 3.other process parameters

diameter of nozzle（mm）	10
Jet angle（°）	45
Processing time（min）	3
Workpiece speed（r/min）	100
Jet pressure（Mpa）	0.6

3.Rolling contact fatigue life test

After grinding, the raceway surface of 4 groups of bearings is machined to superfinish. The processing capacity is 10μm after superfinish. The No.1 and No.2 bearing are loaded with steel ball and cage, so that they can be assembled completely and keep good lubricity, and put into the bearing fatigue test machine bearings, which is shown in Figure 4. Because the bearings are automotive wheel hub bearings, therefore they require simulation of vehicles in general driving conditions. Each bearing runs 65 hours, the speed is 999r/min, the radial load is 8kN, the test machine will record the running condition of the bearing every 10s, including the temperature, vibration and clearance of bearings.

a

b

Figure 4. Bearing fatigue life testing machine

4 Results analysis

Figure 5 and 6 show the conditions of inner ring after finishing test. Comparing the surface of the bearing inner raceway with the reinforced grinding process or without the process, it can be seen that after running through the 65h load, the bearing rings surface that after grinding were still bright, while the untreated bearing rings surface began to become yellow and purple because of the high temperature. So the bearing's running temperature is closely related to the force, clearance, vibration of bearings.

Figure5. Bearing inner ring after test

Figure 6. No. 2 and No. 3

Figure 7 shows the variation of the temperature of bearings along with time during the test. In the first 3 hours, the temperature of the 4 sets of bearings increased sharply to about 87°C. After that, the temperature remained between 85~90°C. Within 35 hours of the test, the No.1 and No.2 bearing were almost the same temperature relative to the other 2 groups, but after 35 hours, the temperature of No.1 and No.2 bearing was significantly lower than No.3 and NO.4. It is precisely because of high temperature, the No.3 and No.4 bearings become yellow, purple as a result. Figure 8 shows the variation of vibration of bearings with time during the test. The vibration of the bearing is closely related to the lubrication characteristics and processing quality of the surface during operation.

The standard deviations of the 4 groups of bearings were: $\sigma_1 = 0.07630$, $\sigma_2 = 0.07633$, $\sigma_3 = 0.1000$, $\sigma_4 = 0.15426$

The standard deviation of No.1 and No.2 bearing is smaller than No.3 and No.4 bearing, which indicates that the stability of No.1 and No.2 bearing is better than No.3 and No.4.

DOI: 10.1051/matecconf/201713900171

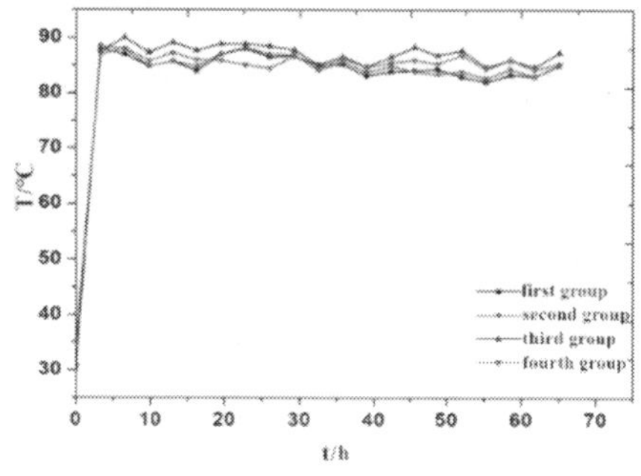

Figure6. Bearing temperature variation

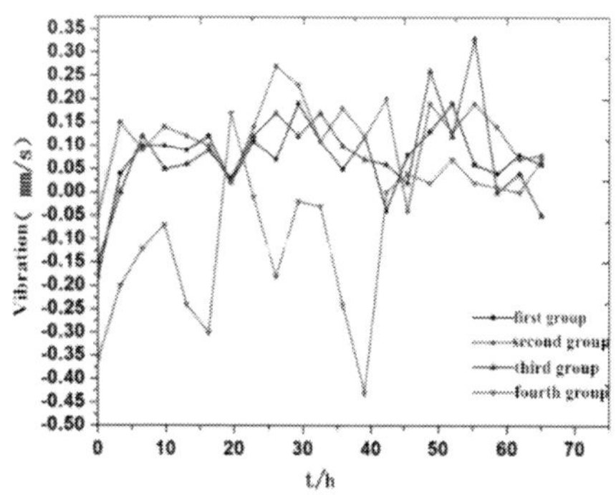

Figure 7. Bearing vibration change

Figure 9 shows the clearance change of bearings with time during the test. The clearance change range of No.1 and No.2 bearing was obviously smaller than No.3 and No.4. No.4 bearing had the biggest change of clearance. It also confirms that No.4 bearing has the worst running stability. It also shows that the temperature and vibration performance of the latter two groups are lower than those of the first two groups.

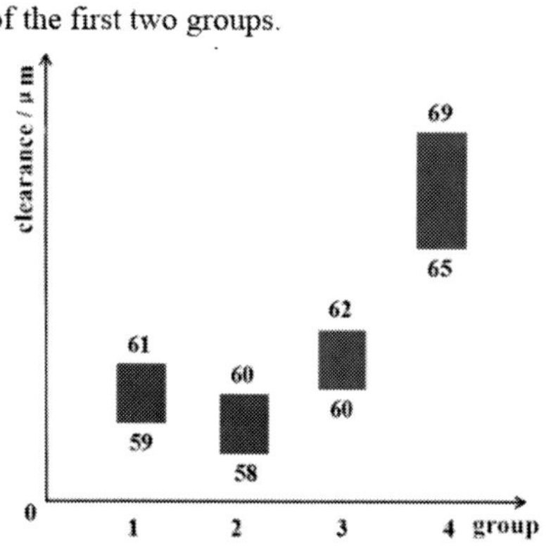

Figure 9.Bearing clearance variation

5 Conclusion

(1) The bearing will be yellow and purple after running for a certain period of time under the load. If the bearing raceway is subjected to reinforce grinding, this will improve the situation, at least the deterioration time of raceway surface will be prolonged

(2) Compared with the bearings without the raceway surface treatment, the bearings with reinforced grinding had better running temperature, stability and clearance change.

References

1. Bastias P C, Hahn G T, Rubin C A, V Gupta,X Leng. Wear, **171**,(1994),169–178.
2. Hadfield M, Tobe S, Stolarski T A. Tribolog Interna-ational, **27**,(1994),359-367.
3. Miller G R, Keer L M, Cheng H S. Proceedings of the Royal Society A,**397**,(1985),197-209.
4. Eryu Shao, Xunjiao Li. Journal of Henan University of Science and Technology(Natural Science),**1**,(198 6),51-60.(In Chinese)
5. Changsen Wan. *Method of analysis of rollingbearin-gs*. (China Machine Press, 1987). （In Chinese）
6. Eleftherakis J G,Shelton D C.Lubrication Engineeri-ng, **51**,(1995), 205-208.

 DOI: 10.1051/matecconf/201713900172

Brushless DC motor speed control strategy of simulation research

Xiang WEN [1,*] ,*Zhen-qiang LI* [2]
[1, 2]College of Electrical and Information Engineering, Guangxi University of Science and Technology, Liuzhou Guangxi 545006, China

Abstract. In view of the brushless DC motor speed regulation problem, an ideal control strategy is designed. Through the model and analysis of Brushless DC motor, the mathematical model of the brushless DC motor is obtained. By comparing three control strategies of PID control strategy, fuzzy control strategy and fuzzy PID control strategy, PID controller, fuzzy controller and fuzzy PID controller are designed respectively for simulation test. The simulation results show that the fuzzy PID controller has good control effect.

Keywords. Brushless DC motor; Fuzzy controller; Control strategy

0. Introduction

The DC brushless motor is divided into two types because of the different types of its winding drive current waveforms [1]: One is the square wave permanent magnet synchronous Motor, because the armature driver current is the square wave (trapezoidal wave), which is called Brushless DC Motor. The other is the permanent magnet synchronous motor, which is called the permanent magnet synchronous motor because the armature driver current is a sine wave. Compared with the sinusoidal permanent magnet synchronous motor, brushless DC motor has obvious superiority, feedback device is simpler, the power density is higher [2], the output torque is bigger, the control structure is simpler, make the motor and inverter fully play their own potential. Therefore, the application and research of brushless DC motor have been paid more and more attention. In many cases, other kinds of motors have been replaced by their technological advantages.

Compared with other type motor, brushless DC motor in the form of square wave excitation brushless DC motor, improve the utilization rate of the permanent magnetic material and reduced the volume of motor, increasing the motor output, with high efficiency, high reliability characteristics. This paper adopts 22 conduction method [3], rotor each corner 60° electrical degrees, Inverter bridge is a converter, the stator magnetic state is a corresponding change. However, the principle and process of anti-electromotive force and electromagnetic torque generation are similar to those of traditional brushless dc motors, so the analysis process is similar.

In order to ensure the stability of the motor speed, the motor speed closed loop control [4,5,6] is realized by measuring the motor speed. In this paper, three control strategies, PID control strategy, fuzzy control strategy and fuzzy PID control strategy, are used to simulate the brushless DC motor.

PID control algorithm has better control effect on linear time invariant system [7]. Fuzzy control [8,9] can solve the system difficult to model the control problem, because it does not need the exact model of the controlled object, and often adjust the time is short. However, the fuzzy controller is difficult to eliminate the steady-state error because it does not have integral links. Fuzzy control PID control algorithm for nonlinear time-varying systems can achieve good control effect.

* Corresponding author: 920592971@qq.com

Fuzzy PID control is a combination of fuzzy control and PID control, which takes full advantage of the advantages of both [9,10]. At the same time, it proves that it has better control effect.

1. Working principle and model analysis of Brushless DC motor

1.1 Working principle

In order to realize the commutation without mechanical contact, the brushless DC motor cancels the brush, and the armature winding and permanent magnet steel are placed on the stator and rotor side respectively, thus becoming the "DC -DC motor" mechanism. In order to control the speed and the rotation direction of the motor, the brushless DC motor must be realized by the rotor position sensor, the control circuit and the power inverter bridge. The power drive mode is divided into [3]: half bridge type, full bridge type (1. 22 conduction mode, 2. 33 conduction mode), C-Dump type, H bridge type, Four switch type.

1.2 Brushless DC motor, mathematical model

The transfer function is derived from the three-phase full bridge drive and the 22 winding mode of stator winding. At this point, each phase of the stator winding leads to a 60 degree electrical angle, and each electrical cycle of the motor undergoes a 6 commutation. In the commutation process, the three-phase winding has current flow because of the diode freewheeling. When the shutdown phase current is reduced to 0, the freewheeling diode stops, the commutation process is completed, the new motor in two-phase conduction state. Commutation torque ripple caused by commutation is an important dynamic process of Brushless DC motor. However, due to its short duration, it has little effect on the effective value of the electrical quantity. Therefore, in the steady state analysis and the transfer function deduction, it is neglected to simplify the calculation. It is assumed that only two phase windings are connected and their currents are equal and opposite. Take A and B phase winding conduction as an example.

$$i_A = -i_B = i \tag{1.1}$$

The linear voltage equation of Brushless DC motor is expressed as

$$\begin{bmatrix} u_{AB} \\ u_{BC} \\ u_{CA} \end{bmatrix} = \begin{bmatrix} R & -R & 0 \\ 0 & R & -R \\ -R & 0 & R \end{bmatrix} \begin{bmatrix} i_A \\ i_B \\ i_C \end{bmatrix} + \begin{bmatrix} L-M & M-L & 0 \\ 0 & L-M & M-L \\ M-L & 0 & L-M \end{bmatrix} \frac{d}{dt} \begin{bmatrix} i_A \\ i_B \\ i_C \end{bmatrix} + \begin{bmatrix} e_A - e_B \\ e_B - e_C \\ e_C - e_A \end{bmatrix} \tag{1.2}$$

By (1.2) available

$$u_{AB} = 2Ri + 2(L-M)\frac{di}{dt} + (e_A - e_B) \tag{1.3}$$

Irrespective of the transient process of commutation, that is, when the A and B phases are steady conduction, the size of e_A and e_B are equal and the sign is opposite, and the formula (1.3) can be rewritten as

$$u_{AB} = U_d = 2Ri + 2(L-M)\frac{di}{dt} + 2e_A$$

$$= r_a i + L_a \frac{di}{dt} + k_e \Omega \tag{1.4}$$

Type: U_d -DC bus voltage;

r_a - winding wire resistance, $r_a = 2R$;

L_a - winding equivalent line inductance, $L_a = 2(L-M)$;

k_e - Line back EMF coefficient, $k_e = 2P\Psi_m = 4PNSB_m$.

Similar to other motors, the power and torque of Brushless DC motor can be analysed from the point of view of energy transfer. The electric power from the power supply when the motor is running, the power of small part into the copper loss and iron loss, most of the air gap magnetic field to transfer torque to the rotor of the permanent magnet rotor, this power is part of the electromagnetic power, instead it is equal to the electromotive force of the three-phase winding and the phase current and the product of that.

$$P_e = e_A i_A + e_B i_B + e_B i_B \tag{1.5}$$

Regardless of mechanical losses and stray losses of the rotor, the electromagnetic power is converted into rotor kinetic energy, so

$$P_e = T_e \Omega \tag{1.6}$$

Type: T_e - Electromagnetic torque;

Ω - Mechanical angular speed of motor.

By formula (1.5) and (1.6)

$$T_e = \frac{e_A i_A + e_A i_A + e_A i_A}{\Omega} \quad (1.7)$$

$$e_A = 2\mathrm{NS}\omega B_m f_{A(\theta)} = \omega\Psi_m f_A(\theta) \quad (1.8)$$

Type: B_m - the maximum value of the air gap flux density distribution of the rotor permanent magnet;

Ψ_m - The maximum value is _m per phase winding chain, $\Psi_m = 2NSB_m$;

$f_A(\theta)$ - A contrary to the waveform function of electromotive force;

e_A - A contrary electromotive force.

By substituting (1.8) for (1.7), another form of the torque equation can be obtained

$$T_e = \mathrm{p}[\Psi_m f_A(\theta) i_A + \Psi_m f_B(\theta) i_B + \Psi_m f_B(\theta) i_B] \quad (1.9)$$

Type: P - motor pole count.

When the brushless DC motor running in 120°conduction mode, without considering the phase change of transient process, so the type (1.9) can be simplified to

$$T_e = 2\mathrm{p}\Psi_m i_A = K_T i \quad (1.10)$$

Type: K_T - motor torque coefficient;

i - winding phase current in steady state.

Motor equation

$$T_e - T_L = J\frac{d\Omega}{dt} + B_v\Omega \quad (1.11)$$

Type: T_L - load torque;

J - Rotor moment of inertia;

B_v - Coefficient of viscous friction.

In formula (1.4), the relation between the bus voltage and the angular speed can be obtained by representing the current with angular velocity, and then the motor transfer function is introduced. It will be (1.10) in substitution (1.11)

$$K_T - T_L = J\frac{d\Omega}{dt} + B_v\Omega \quad (1.12)$$

The transfer function of the load will be discussed later, considering the no-load condition at which the armature current is

$$\mathrm{i} = \frac{J}{K_T}\frac{d\Omega}{dt} + \frac{B_v}{K_T}\Omega \quad (1.13)$$

Replace (1.13) with (1.4)

$$U_d = r_a\left(\frac{J}{K_T}\frac{d\Omega}{dt} + \frac{B_v}{K_T}\Omega\right) + L_a\frac{d}{dt}\left(\frac{J}{K_T}\frac{d\Omega}{dt} + \frac{B_v}{K_T}\Omega\right) + k_e\Omega \quad (1.14)$$

Therefore

$$U_d = \frac{L_a J}{K_T}\frac{d^2\Omega}{dt^2} + \frac{r_a J + L_a B_v}{K_T}\frac{d\Omega}{dt} + \frac{r_a B_v + k_e K_T}{K_T}\Omega \quad (1.15)$$

The transfer function of the brushless DC motor is obtained by Laplace transformation and sorting (1.15)

$$G_u(s) = \frac{\Omega(s)}{U_d(s)} = \frac{K_T}{L_a J s^2 + (r_a J + L_a B_v)s + (r_a B_v + k_e K_T)} \quad (1.16)$$

The transfer function shown in equation (1.16) is a two order system and is sorted into the canonical form of the two order system

$$G_u(s) = \frac{K_T}{r_a B_v + k_e K_T}\frac{\omega_n{}^2}{s^2 + 2\xi\omega_n s + \omega_n{}^2} \quad (1.17)$$

Type: $\omega_n = \sqrt{\frac{r_a B_v + k_e K_T}{L_a J}}$ - Natural frequencies of two order systems;

$\xi = \frac{1}{2}\frac{r_a J + L_a B_v}{\sqrt{L_a J}\sqrt{r_a B_v + k_e K_T}}$ - Damping ratio of two order system.

When the load is not zero, it can be regarded as the input of the system, and the system structure is shown as shown.

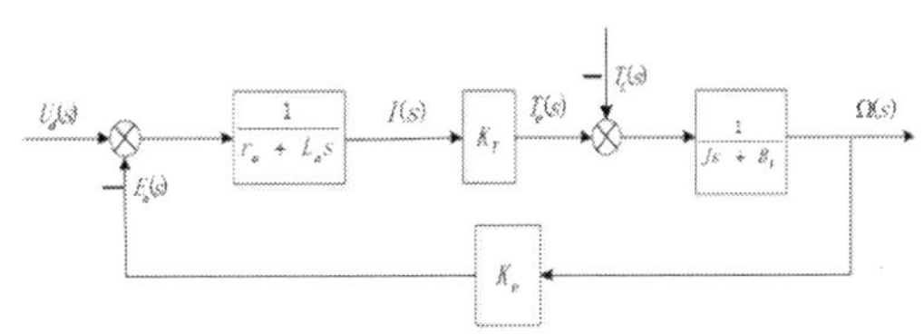

According to the superposition principle, at this time the output equals the sum of the responses of $U_d(s)$ and $T_L(s)$. When $U_d(s)$ =0 is shown in Fig.

$$\left[-k_e\frac{1}{r_a + L_a s}K_T\Omega(s) - T_L(s)\right]\frac{1}{Js + B_v} = \Omega(s) \quad (1.18)$$

That is

$$\Omega_L(s)\left[\frac{(r_a + L_a s)(Js + B_v) + k_e K_T}{(r_a + L_a s)}\right] = -T_L(s) \quad (1.19)$$

At this point, the transfer function between the load torque and the speed is

$$G_L(s) = \frac{\Omega(s)}{T_L(s)} = -\frac{r_a + L_a s}{L_a J s^2 + (r_a J + L_a B_v)s + (r_a B_v + k_e K_T)} \quad (1.20)$$

Therefore, the speed response of Brushless DC motor under the combined action of voltage and load torque is

DOI: 10.1051/matecconf/201713900172

$$\Omega(s) = G_u(s)U_d(s) + G_L(s)T_L(s)$$

$$= \frac{K_T U_d(s) - (r_a + L_a s)T_L(s)}{L_a J s^2 + (r_a J + L_a B_v)s + (r_a B_v + k_e K_T)} \quad (1.21)$$

2. Control strategy design

The transfer function G_{dv} between the known input voltage and the output speed is considered. On this basis, the block diagram of the PID controller, fuzzy controller and fuzzy PID controller are conceived respectively, as shown in figure 4.

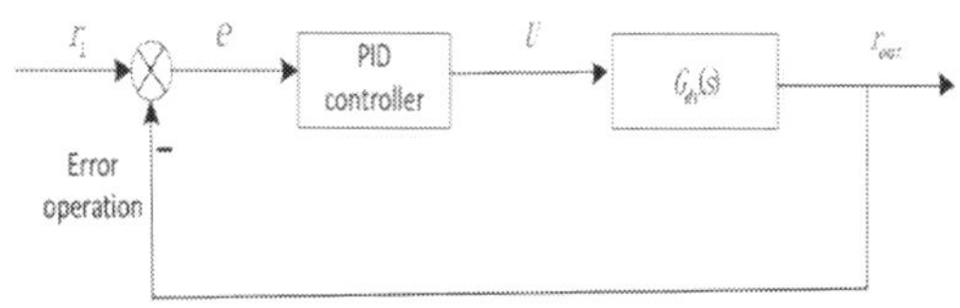

(a).PID controller

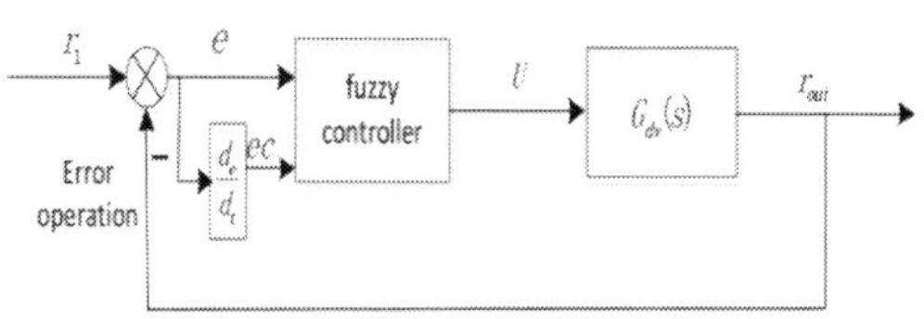

(b).fuzzy controller

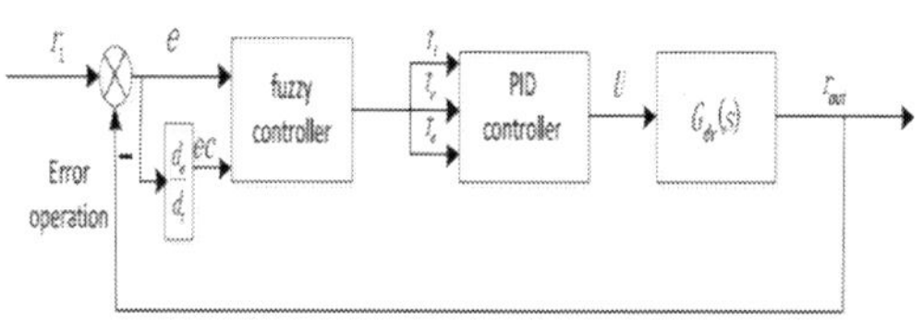

(c).fuzzy PID controller

Figure 4 three structural block diagrams

Take the main circuit parameters, respectively

$U = 120V$,

$r_a = 14m\Omega$,

$L_a = 0.0001H$

$T_T = 0.0275$

$J = 0.0002kgm^2$

$B_V = 0.004586wb$

$K_e = 0.0275$,

$T_L = 10Nm$.

2.1 PID controller design

A block diagram of the system consisting of the PID controller is shown in Figure 4 (a). The main circuit parameter is substituted into G_{dv}

$$G_{dv}(s) = \frac{4.19 \times 10^4 - 1.22s}{2.44 \times 10^{-5}s^2 + 5.58 \times 10^{-4}s + 1}$$

After the addition of PI control, the transfer function of the PI action is set to $T(s) = 1 + 1/s$, and the open-loop transfer function of the system becomes:

$$G(s) = G_{dv}(s) \cdot T(s) = \frac{4.19 \times 10^4 - 1.22s}{2.44 \times 10^{-5}s^3 + 5.58 \times 10^{-4}s^2 + 1}$$

2.2 fuzzy controller design

A block diagram of the system consisting of a fuzzy controller is shown in Figure 4 (b).

The specific design process of fuzzy controller is as follows:

1) Determine the input language variables for error e and error change ec, the output variable is r.

$e(k) = r_{ref} - r_o(k)$, $ec(k) = e(k) - e(k-1)$

2) Select fuzzy subsets e,ec, u={NB (negative large), NM (negative middle), NS (negative small), ZE (zero), PS (positive small), PM (positive middle), PB (positive large)}.

3) Determine fuzzy control rules.

4) By resolving the ambiguity, the output control R, and then after processing, the output voltage.

2.3 fuzzy PID controller design

A block diagram of the system consisting of a fuzzy PID controller is shown in figure 4(c).

The fuzzy PID controller is combined with fuzzy controller and PID controller, error e and error change ec as input, using the fuzzy control rules to modify the PID parameters to meet the different moments of e and ec on PID parameters.

The design process is similar to the fuzzy controller. The input linguistic variables are error e and error variation ec, but the output variables become proportional, integral and differential parameters, namely, kp、ki and kd.

At run time, the control system completes the on-line adjustment of PID parameters by processing the fuzzy logic rules, look-up table and operation.

3. simulation results and analysis

Table 1 The simulation results of the three kinds of control scheme

simulated effect	PID controller	Fuzzy controller	Fuzzy PID controller
overshoot	big	small	small
debug time	long	short	middle
steady-state error	middle	big	small

To verify the effectiveness of the above three control strategies. In the MATLAB/Simulink software, the simulation models of the three schemes are built, in which the initial PID tuning parameters are consistent. The setting speed is $\Omega_i = 1500r/min$ as follows. The simulation model is shown in Figure 5, and the simulation results are shown in Figure 6

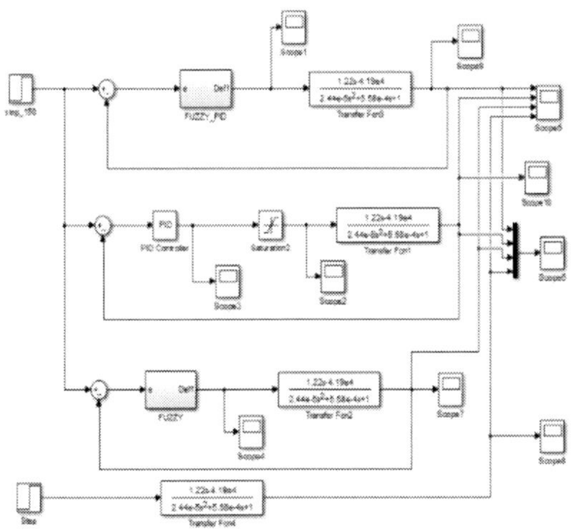

Figure 5 Simulink simulation model

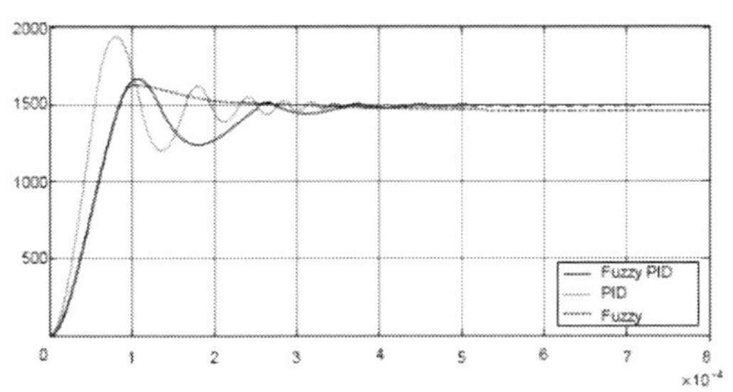

Figure. 6 Comparison of the simulation results of the three control schemes

Through the simulation, the PID overshoot is relatively large, the adjustment time is relatively long; fuzzy PID controller, the overshoot is small, the adjustment time is slightly shorter than the PID controller, the steady-state error is almost zero; fuzzy controller overshoot, and adjust the shortest time, but there is a large steady-state error. Details are shown in table 1.

4. Conclusion

In this paper, three different control strategies simulation of Brushless DC motor, through the experimental results, the fuzzy PID controller combined with fuzzy control and PID small overshoot and small steady-state error, good simulation effect.

5. Reference

1. Zhang Yan. DSP brushless DC motor speed control system of high performance research based on [D]. Xi'an Electronic and Science University, 2007.
2. Yin Yunhua. Design and Simulation of Brushless DC motor control system based on DSP [D]. North Central University, 2007.
3. Xia Changliang. Brushless DC motor control system [M]. Science Press, 2007..
4. Du Juan, Guo Zhonghua, Ma Hua jie. Speed control system of Brushless DC motor based on adaptive fuzzy PID control [J]. electronic technology and software engineering, 2016, (03): 139-141.
5. Lian QinJian. Design and Simulation of Brushless DC motor drive system of electric vehicle based on DSP [D]. Xi'an University Of Architecture And Technology, 2014.
6. Wang Ling, Liu Weiguo. Simulation of Brushless DC motor speed control system based on fuzzy PI control [J]. computer simulation, 2009, (10): 186-189.
7. Wei Hua, Li Qun, Chen Debao. A new nonlinear parameter fuzzy PID control method of [J]. computer technology and development, 2008,18 (2): 237-239.
8. Cheng Lirong, Sun Changzhi. Simulation of three phase brushless DC motor speed regulation system [J]. microcomputer information, 2007, (04): 251-252.
9. Zhang Jinhuan. PID control system and fuzzy adaptive PID control system research and comparison [J]. Journal of Wuhan University of Technology, 2005,27 (5): 286-290.

 DOI: 10.1051/matecconf/201713900172

10. Wang Shuyan, Shi Yu, Feng Zhongxu. Research on control method based on fuzzy PID controller [J]. mechanical science and technology, 2011,30 (1): 166-172.

Analysis on the influence of fuze warhead distance to the fuze warhead coordination

Shaoyong Jian[1],Shi Ying *

[1]Department of Mathematics and Computer Sciences, Xinyu University, Xinyu,338004 ,China

* Department of fundamental courses ,Wuhan Technology and Business University Wuhan, Hubei,430065,China

Abstract. For anti-aircraft missiles to intercept the small size of the object in a small miss distance condition cited war with the problem, analyze the warhead to Fuze distance interception prejudice to such objectives, through theory and simulation analysis, the distance is small miss distance to intercept small size of the target with the effect of the lead battle influence evident in the actual engineering design, we need to consider the impact of the distance to lead the war with the design of the proposed boundary distance constraints.

1.Introduction

Past air defense missile guidance deviation is bigger, in the design and analysis of fuze-warhead coordination, warhead and fuze is overlap, this assumption in the miss distance is reasonable, is largely due to the fuze trigger delay time of flight (t_f) greater than the fragment (t_{v0} , flying fragment to a point of time) with the inherent delay fuze and (τ_0), so as to get the best fuze delay (r) is larger, the warhead and fuze distance (D_{x1}) in the direction of relative velocity (V_r) ($\Delta\tau=D_{x1}/V_r$) is the best time delay error caused by the delay of proportion is very small, and thus has little influence on fuze average starting position.With more and more small deviation guided anti-aircraft missiles, the presence of trigger delay fuze fragments smaller than the inherent delay of flight time and the circumstances, and if you do not consider the impact on the optimal delay will result in less than optimal delay 0, average starting position relative to the fuze warhead fragments scattered region lagging behind, unable to achieve the desired effect with the lead battle, therefore, a small amount off target intercept targets need to consider the impact of the issue of war with the lead.

2.issues analysis

For a short-range interceptor weapon system, select the feature points (see Table 1) with the analysis cited war, warfare systems cited basic parameters are as follows: First Fragment velocity 2000m/ s, broken piece of the main scattering direction angle of 90 °, inclination fuze antenna 50 °, the inherent latency of 1.0ms, fuze warhead to a distance of 0.8m, the more typical trajectory without considering the lead to war under warhead fuze with effect from the condition shown in Figure 1.

Table 1 typical trajectory points

No.	Target speed (m/s)	Height (m)	Target slant range (m)	Goal Route Links (m)	Relative vel	Misalignme
1	300	500	3000	0	719	15
2	300	500	3000	1450	710	17

From the lead battle with the effects of simulation map view, without considering the distance fuze to warhead, fuze average starting position at 1m off target amount condition lags Warhead dynamic scattering area, that broken pieces scattering area can not cover the target key points, features point lead with ineffective war, after analysis of the optimal delay models, that the main reason is the distance to the warhead fuze impact on the battle with the effects of lead.

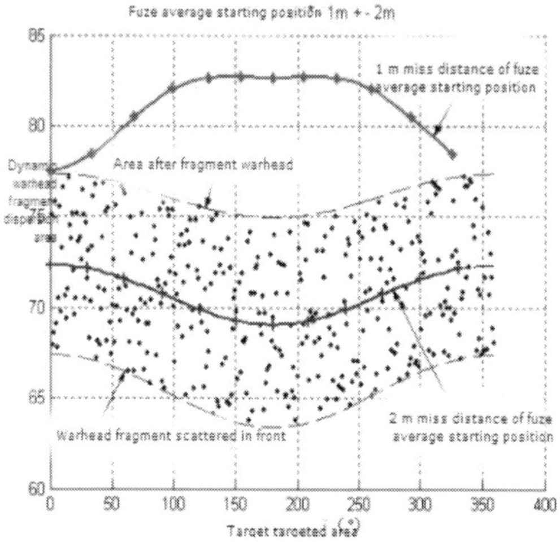

(1#)

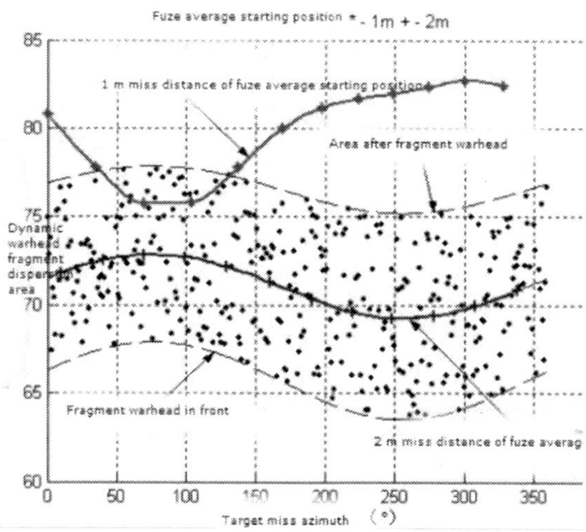

(2#)

Figure 1 does not consider the conditions lead to the warhead a distance of 1 # and 2 # ballistic cited war with renderings

2.1 Quantitative analysis of the impact of best delay D_{x1}

In order to further analyze the effect of D_{x1} on the lead and the delay time delay, with the following delay from the simplified calculation model lead war starting, quantitative analysis of the issue, in the missile and the target is on or near the intersection opposite the intersec -tion (misalignment angle less than 15°) conditions, elastic mesh intersection model and delay formulas are as follows:

$$\tau = t_f - t_{f0} - \tau_0 = \frac{\Delta L + \frac{\rho}{\tan(\Omega_f)} + D_{x1} - \rho\tan(\delta)}{V_r} - \frac{\frac{\rho}{\cos(\delta)}}{V_0} - \tau_0 \quad (1)$$

Type: τ for Fuze optimal burst delay time; ρ for the target miss distance; Ω_f for the tilt angle of antenna peak; ΔL for a target trigger point to the target key center distance (ΔL = 0 that target trigger point is the

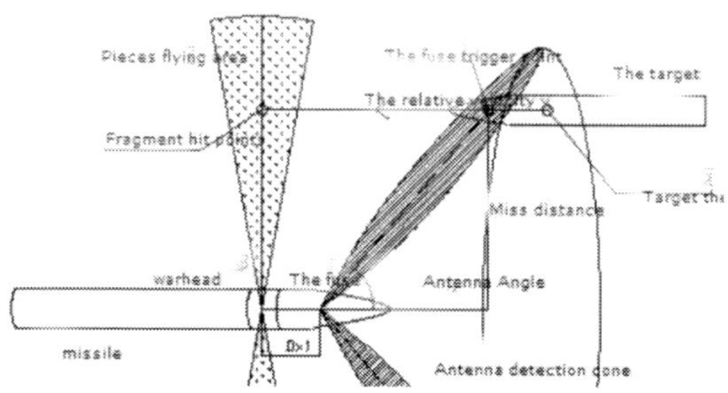

Figure 2 Missile project is on the intersection of computational geometry model best delay

target key points, V_r for missile target relative velocity; V_0 as the initial velocity of the fragments δ fragment dispersion main direction angle of anteversion or retroversion angle (forward positive and backward tilt is negative)[1].

Assume that the target off target amount of change in the range of 0m ~ 30m, fuze antenna peak inclination of 50 °, D_{x1} is 0.8m, the relative speed of 800m/ s, early fragment velocity of 2000m/ s, broken piece of the main scattering direction is 90 °, fuse inherent delay as 1.0ms, based on the above assumptions, analysis $\Delta\tau$ of the fuze optimal initiating delay time τ influence, under condition of different target volume $\Delta\tau$ of the best delay τ accounts for the proportion ($\Delta\tau/\tau$) as shown in Figure 3 shows.

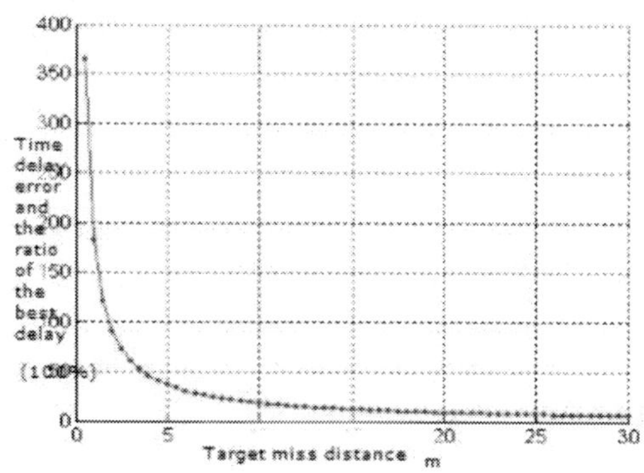

Figure 3 Delay error caused delays accounted for

optimum ratio (100%)

From the map view, at 0.5m, 1m, 2m and 3m respectively at the proportion of 364%, 182%, 91% and 61%, while in the 10m, 20m and 30m $\Delta\tau$ accounted for τ the ratio was 18%, 9% and 6%, When the miss distance is smaller, $\Delta\tau$ to τ the proportion of larger, the fuze average starting position is greater, the greater the impact on the final fuze warhead coordinate on effect.

At the time of fuze warhead coordination analysis, in order to ensure the fuze average starting position not outside the fragment dynamic scattering section, theory, caused by $\Delta\tau$ the fuze average starting position change is less than 1/2 the fragment dynamic scattering section width, namely:

$$\Delta\tau \times V_{r\max} \le \frac{\rho\varphi}{2} \quad (2)$$

Where ϕ is the width of fragmentation dynamic scattering region.

2.2 pairs D_{x1} boundary conditions constraint

For formula 1, in the best delay is not less than zero conditions, in order to ensure fuze average starting position is not lagging behind the fragmentation dynamic scattering area at this time, when the miss distance tends to 0($\Delta L = 0$), the presence of the following formula :

$$\lim_{\rho\to 0}\left[\frac{\frac{\rho}{\tan(\Omega_f)}+D_{x1}+\Delta L}{V_r}-\frac{\rho}{V_0\sin(\theta)}-\tau_0\right]=\frac{D_{x1}}{V_r}-\tau_0\ge 0 \quad (3)$$

Where by D_{x1} need to satisfy the following formula:

$$D_{x1}\ge V_{r\max}\times\tau_0 \quad (4)$$

Where D_{x1} is the distance to warhead fuze, $V_{r\max}$ is the largest bomb target relative veloci-ty, τ_0 is the inherent delay fuse. Actual engineering design, you need to consider the trade-off relationship D_{x1} , $V_{r\max}$ and τ_0 between reasonable design specifications, equipment layout after preliminary determination on the bomb, cited warfare systems need to focus on the inherent time delay fuse indicators, for example, the maximum relative speed 800m/s, the distance to the warhead fuze is under 0.8m condition, the inherent delay fuse not greater than 1m/s.

In D_{x1} can not satisfy Formula 3 missile to intercept the target's lead war with Figure 1, in a small miss distance, the fuze average starting position outside the fragmentation dynamic scattering area range, lead to the removal of the target amount warhead can not hit the target key point.

3 Model Improvement

Best Start Delay Fuze engineering calculation model, the distance to the warhead fuze fuze trigger delay mainly affects the calculation has no impact on fragments flying time, the inherent delay and off target azimuth correction time, consider the following fuze to fight the influence of the distance portion of the original model to improve the trigger delay.

3.1 Coordinate system definition

This paper relates to the coordinate system of the ground, elastomers and relative velocity coordinates, the concept of indirect references to other coordinate systems, see the relevant single shot kill probability models to calculate the model file [2].

Ground we coordinate system: $ox_dy_dz_d$ ground coordina -te system represented by qwer, which is defined as: the origin 0 is located in the missile launch point, oz_d instantaneous axis of missile target flight speed parallel projection on the ground, and taking the opposite direction is positive, oy_d axis vertical upward is positive, oz_d axis and oy_d and ox_d right-handed coordinate system constituted.

Body coordinates: With $ox_1y_1z_1$ indicates the origin of the coordinate system projectile fuze is located in the center of missiles along the longitudinal axis of the shaft ox_1 forward, oy_1 to take in the missile body axis symmetry plane, constitute oz_1 right-handed coordinate system.

Relative velocity coordinates: With $oxyz$ indicates, o take the origin of coordinates in the target centroid, oz axis in the opposite direction to take missiles opposite direction of the target velocity vector V_r is positive, as oy axis perpendicular to the axis oz, ox axis constituted by right-handed rule.

From the body coordinate system to the coordinate

MATEC Web of Conferences **139**, 00173 (2017) DOI: 10.1051/matecconf/201713900173
ICMITE 2017

system relative velocity transformation matrix:

$$A_{1x} = A_{dx} \times A_{1d} = M_x[\varepsilon_r]M_y[\beta_r - 90\ ] \times M_y[-\psi]M_z[-\upsilon]M_x[-\gamma] \quad (5)$$

Where, A_{dx} is ground coordinate system to convert the coordinate system relative velocity matrix, ε_r is the relative trajectory angle, β_r is the relative trajectory yaw angle; A_{1d} is body coordinates to the ground coordinate system transformation matrix, ψ, υ and γ respectively projectile yaw angle, pitch angle and projectile missile body roll angle.

3.2 trigger delay calculation model

In the relative velocity coordinates, to establish the trigger delay calculation geometry shown in Figure 4, the figure, O points for the target organism center, A point relative velocity coordinate the play off target coordinates ($\rho\cos\theta$, $\rho\sin\theta$, 0) , B point for the target trigger points relative velocity coordinate system position coordinates (x_i, y_i, z_i) , N point fuze center position coordinates relative velocity, the distance off the target for $V_r \times t_i$, t_i to trigger time, M point for the location warhead fuze center relative to the center of the relative speed of the coordinate system, the position of its center is relatively fuze (D_{rx1}、D_{ry1}, D_{rz1}) , which is calculated, see equation (14).

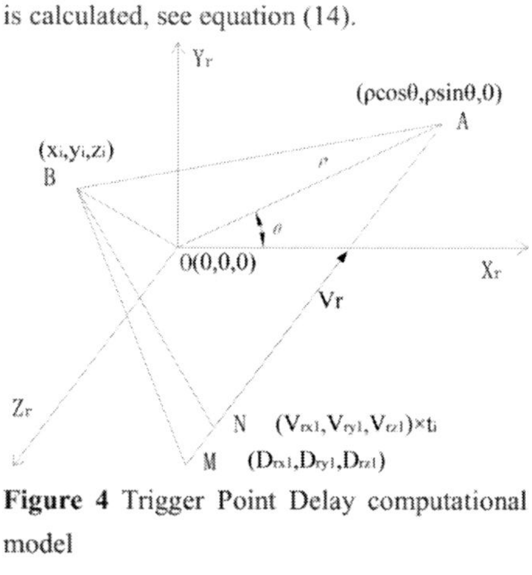

Figure 4 Trigger Point Delay computational geometry model

In Figure 4, the coordinate system relative velocity, AB is represented as follows [3]:

$$OA=(\rho\cos\theta,\rho\sin\theta,0) \quad (6)$$

$$OB=(x_i,y_i,z_i) \quad (7)$$

$$AB=OB-OA=\begin{bmatrix} x_i-\rho\cos\theta \\ y_i-\rho\sin\theta \\ z_i \end{bmatrix} \quad (8)$$

When playing the system, consider the warhead center position vector with respect to the center of the fuse $\overrightarrow{MN}$, and $\overrightarrow{AB}$ converted from the relative velocity vector coordinates to the missile body coordinates, the position of the trigger point in playing the system i have the following relationship:

$$MN=(D_{x1},D_{y1},D_{z1}) \quad (9)$$

$$NA=(V_{rx1},V_{ry1},V_{rz1})\times t_i \quad (10)$$

$$MA=MN+NA=\begin{bmatrix} D_{x1} \\ D_{y1} \\ D_{z1} \end{bmatrix} + \begin{bmatrix} V_{rx1} \\ V_{ry1} \\ V_{rz1} \end{bmatrix} \times t_i \quad (11)$$

$$\overrightarrow{MB}=\overrightarrow{MA}+\overrightarrow{AB}=\begin{bmatrix} x_{1i} \\ y_{1i} \\ z_{1i} \end{bmatrix} = A_{x1} \times \begin{bmatrix} x_i-\rho\cos\theta \\ y_i-\rho\sin\theta \\ z_i \end{bmatrix} + \begin{bmatrix} D_{x1} \\ D_{y1} \\ D_{z1} \end{bmatrix} + \begin{bmatrix} V_{rx1} \\ V_{ry1} \\ V_{rz1} \end{bmatrix} \times t_i \quad (12)$$

In the formula, D_{x1}, D_{y1} and D_{z1} warhead missile system in the center position with respect to the fuse center.

According to the actual engineering computing needs, location of the point system will trigger bomb formula 12 is converted to the following formula:

$$\begin{bmatrix} x_{1i} \\ y_{1i} \\ z_{1i} \end{bmatrix} = A_{x1} \times \begin{bmatrix} x_i-\rho\cos\theta+D_{rx1} \\ y_i-\rho\sin\theta+D_{ry1} \\ z_i+D_{rz1} \end{bmatrix} + \begin{bmatrix} V_{rx1} \\ V_{ry1} \\ V_{rz1} \end{bmatrix} \times t_i \quad (13)$$

$$\begin{bmatrix} D_{rx1} \\ D_{ry1} \\ D_{rz1} \end{bmatrix} = A_{1x} \times \begin{bmatrix} D_{x1} \\ D_{y1} \\ D_{z1} \end{bmatrix} \quad (14)$$

Equation (13) is considered relatively fuze warhead fuze position after the trigger delay calculation model, using ground test trigger line fuse data, combined with the above formula iterative solver get fuze trigger delay, and then calculate the best start delay fuse, be analysis of lead battle with effect[4].

4 Simulation Analysis

Through the delay calculation model improvement, re-lead war with the simulation analysis, Table 1, the two feature points lead battle simulation with the effect shown in Figure 5.

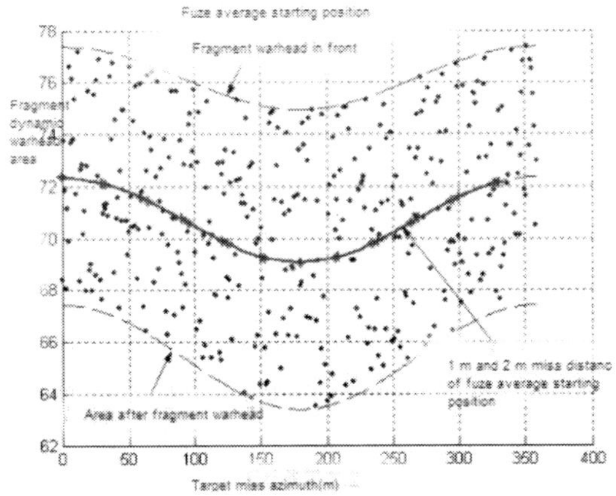

(3#)

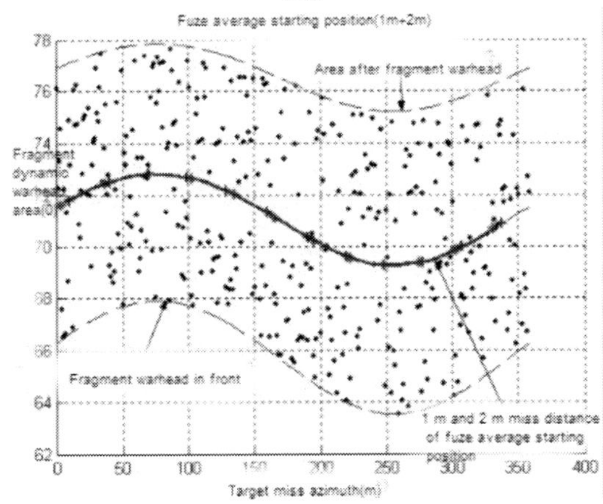

(4#)

Figure 5 The improved design 3# and 4# ballistic cited war with effect simulation map

From Figure 5, the consideration of the impact fuze distance to warhead, improved trigger delay calculation model, citing the war on the orientation of each off-target with good effect, trigger delay calculation model improvements achieved to improve the effective -ness of combining the purpose of the lead battle, to meet the target of high-precision missile defense requirements[5].

5 Conclusion

With the presence of small off-target for the amount of lead war to intercept targets poor problem, through theoretical analysis and model to improve that when a small amount off target interception target, you need to consider the impact of distance warhead fuze to war with the effects of lead, and the distance and delays inherent in the relationship between the fuze given constraint equation, cited improved model with war simulation results show that the missile intercept targets with a good primer war.

References

1 Zhihong Zhang air defense missile and warhead fuze andwarhead design with efficiency Beijing: Astronautics Publishing House .(1994)

2 Tongwen Liang antiaircraft missile fuze design and simulation technology. Beijing: Astronautics Publishing House .(1994)

3 Bozhen Cao winged missile warhead and fuze Beijing: Astronautics Publishing House. (1995)

4 Patrick Girard, Nicola Nicolici, Xiaoqing Wen Power-Aware Testing and Test Strategies for Low Power Devices,[M] Springer Science & Business Media, Technology & Engineering, pp.297-300, 2010 .

5 X. Lin and Y. Huang, "Scan shift power reduction by freezing power sensitive scan cells," Journal of Electronic Testing, vol. 24, no. 4, pp. 327 - 334, 2008.

Modeling and Predicting Cross-sectional Surface Roughness for the Grinding Process

Qiang Li [1,a], Peilian Tan [1,b *]
[1]School of Mechanical Engineering, Hunan Industry Polytechnic, Changsha 410082, China

Abstract. This paper proposes a new algorithm for predicting the cross-sectionalsurface roughnesst of the workpiece for the grinding process, which reduces three-dimensional modeling to two-dimensional modeling by virtue of the cross-sectional area of the undeformed grinding chip. Besides, this paper develops a search technique to figure out systematically the surface roughness in order to determine the final cross-sectional surface roughness produced by thousands of grinding wheel grains with randomly distributed protrusion heights. The simulated results are consistent with the measurement data,which proves the effectiveness of the proposed algorithm.

1 Introduction

The face gear transmission performance, vibration, noisy and working life are decided by tooth surface in a great extent. In Engineering, the surface roughness is usually measured perpendicular to the grinding direction. The typical sampling length for the grinding surface roughness measurement is 0.8 mm, and usually the evaluation length should contain 5 consecutive sampling lengths. Current methods, however, aim at three-dimensional area simulation which is carried out on a rather small scale [1-5]. Much more time will be consumed if the simulation scale is increased. In this paper, a new algorithm is developed for predicting the cross-sectional surface roughness of the workpiece for the grinding process. It can simulate an arbitrary cross-section of the surface topography, and take into consideration the random distribution of the grain protrusion heights. First, the principle of the simulation technique is introduced by virtue of the cross-sectional area of the undeformed grinding chip to reduce three-dimensional modeling to two-dimensional modeling. Then, the interaction of grain with cross section is given and the projection of the undeformed grinding chip on the cross-section is studied. Still, a search algorithm is developed to obtain the final surface roughness. Finally the simulation results are given and verified in part of characteristic in comparison with experimental data.

2 Simulation Technique

The schematic diagram of surface grinding is shown in Fig.1. A $o-xyz$ coordinate system is set with its origin O fixed on the grinding wheel spindle. The cross-section I is any work-section perpendicular to the grinding direction. The abrasive grains in the grinding wheel are represented with polar coordinates (y , ρ , ϕ).

In order to simplify the grinding kinematic analysis, the assumptions adopted in the present work for predicting the cross-sectional surface roughness are as follows: 1) the abrasive grains are conical in shape; 2) the distribution of the grain intervals is uniform, and the grain interval $L\,(mm)$ can be determined as $L = 137.9M^{-1.4}\sqrt[3]{\frac{\pi}{32-S}}$ [6], where M is the grit number and S is the structure number of the wheel; 3) the grain protrusion height h is distributed with mean value μ and standard deviation σ . μ is given by $68M^{-1.4}$, and σ can be estimated as $(15.2M^{-1} - 68M^{-1.4})/3$.

During the grinding process, the grains on the surface of the wheel pass through the workpiece and cut a portion of the workpiece material. The final topography of cross-section I is the surface shape of the material remaining on the cross section after the passage of all the engaged grains through the cross-section. The generation of the cross-sectional surface is sketched in Fig.2, where the numbers represent the grains in one row. According to the assumptions, the grain protrusion heights are subject to Gaussian distribution, in j row, not all the grains will leave marks on the cross-section and those that do will be the ones trajectory with low intersection with the cross-section.

The simulation comprises three steps: (i) the intersecting point of the arbitrarily distributed grains in j row with the cross-section is calculated, and subsequently, the lowest point is calculated in j row; (ii) the grain protrusion height associated with the lowest point is confirmed and the intersection of the two adjacent triangles is calculated; (iii) surface generation is simulated geometrically with reference to the shape of the projection triangle.

* Corresponding author: 715847062@qq.com

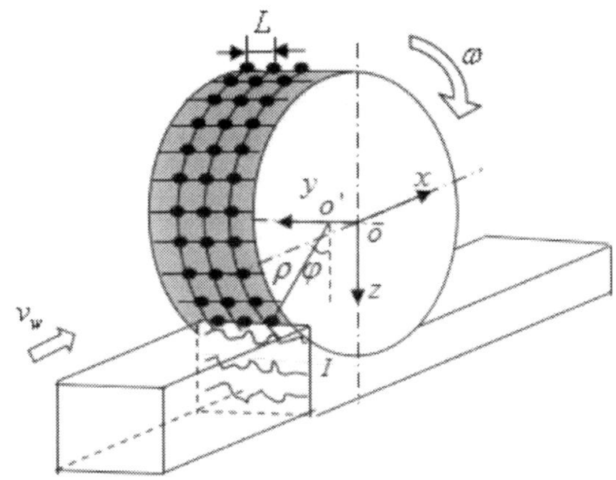

Fig.1 Kinematic relation between wheel and workpiece.

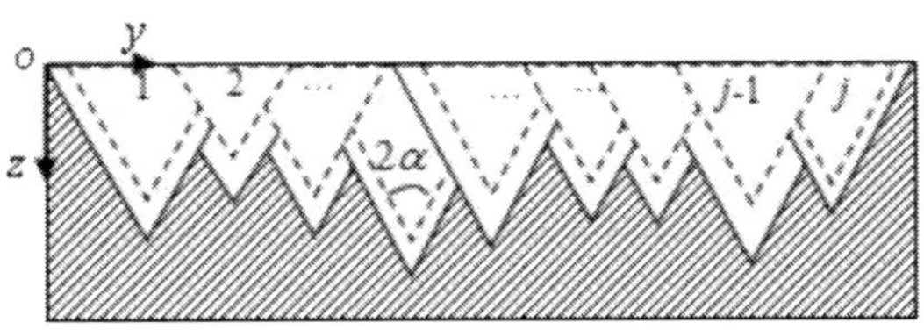

Fig.2 Schematic of the cross-section generation during grinding.

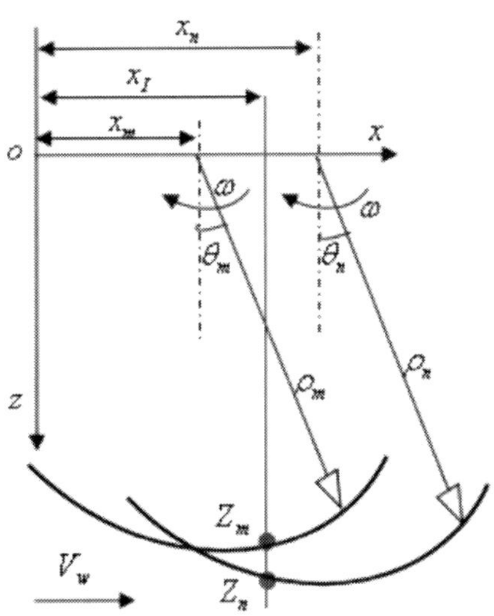

Fig.3 The intersecting point of the arbitrarily distributed grains in one row with the cross-section.

2.1 Intersecting Point of the Arbitrarily Distributed Grains in one Row with the Cross-section

The intersecting points of the arbitrarily distributed grains m and n in one row with the cross-section are is shown in Fig.3. In the $x-z$ coordinate system, the trajectory of grain m can be expressed as

$$\begin{cases} x = x_m + \rho_m \sin\theta_m \\ z = r_m \cos\theta_m \end{cases} \tag{1}$$

where x_m is the moving distance of the wheel center when grain m becomes involved in grinding, and it can be expressed as

$$x_m = \frac{v_w}{\omega}\phi_m \tag{2}$$

Since θ_m is small, it can be approximated that $\theta_m \approx \sin\theta_m$. Substitute it into Eq. (1) and the equation can be simplified to

$$\begin{cases} x = x_m + \rho_m\theta_m \\ z = \rho_m(1 - \dfrac{\theta_m^2}{2}) \end{cases} \tag{3}$$

Substitute Eq. (2) into Eq. (3) to eliminate parameter θ_m, then the equation can be expressed as

$$z = \frac{2\rho_m^2 - (x - \frac{v_w}{\omega}\phi_m)^2}{2\rho_m} \tag{4}$$

Set x_I as the distance between the cross-section I and the origin of the coordinate, and then the intersecting point of grain m with the cross-section I can be expressed as

$$z_m = \frac{2\rho_m^2 - (x_I - \frac{v_w}{\omega}\phi_m)^2}{2\rho_m} \tag{5}$$

Similarly, the intersecting point of grain n with the cross-section I can be expressed as

$$z_n = \frac{2\rho_n^2 - (x_I - \frac{v_w}{\omega}\phi_n)^2}{2\rho_n} \tag{6}$$

The smaller value of z_m and z_n is chosen as the final intersection which will leave marks on the cross-section. Set a_e as the depth of cut and $\rho_{\min}$ as the minimum radius in one row, and then the limit value of the intersection of the grain with the cross-section can be expressed as

$$\rho_{\min} - a_e = \frac{2\rho_{\min}^2 - (x_I - \frac{v_w}{\omega}\phi)^2}{2\rho_{\min}} \tag{7}$$

Along each row, number the grains engaged with the cross-section I as $\{1,2,...i-1,i\}$, and then the grain in j row which will leave marks on the cross-section can be determined as

DOI: 10.1051/matecconf/201713900174

$$z^{(j)} = \min\{z_{j1}, z_{j2}, ..., z_{j(i-1)}, z_{ji}\} \quad (8)$$

2.2 Prediction of Surface Roughness

The generation of the cross-section during grinding is shown in Fig.4. Set $B_1, B_2, ..., B_{j-1}, B_j$ as the grains associated with the lowest intersection point in each row in the $y-z$ plane. The point B_1 is the origin in y direction coordinate, and the value in Z direction coordinate which can be given according to Eq. (8) can be written as (0, z^1). Linear equation of B_1D_1 can be expressed as

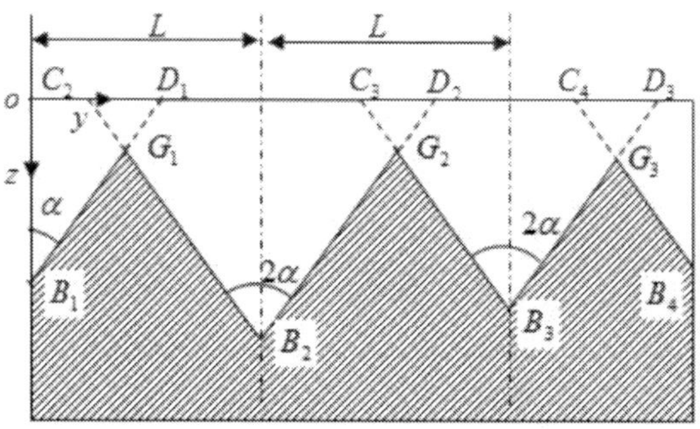

Fig.4 Generation of the cross-section during grinding.

$$z - z^{(1)} = \frac{1}{\tan\alpha} y \quad (9)$$

According to the basic assumption, the uniform grain interval is L, so the coordinate of point B_2 can be written as (L, z^2). Linear equation of B_2C_2 can be expressed as

$$z - z^{(2)} = -\frac{1}{\tan\alpha}(y - L) \quad (10)$$

The intersection G_1 between these lines can be solved by using Eq. (11) and Eq. (12) as

$$\begin{cases} y = \dfrac{\tan\alpha}{2}(z^{(2)} - z^{(1)} + \dfrac{L}{\tan\alpha}) \\ z = \dfrac{1}{2}(z^{(2)} + z^{(1)} + \dfrac{L}{\tan\alpha}) \end{cases} \quad (11)$$

If there is no intersection between line B_1D_1 and B_2C_2, it indicates that only one grain which has smaller intersection with the cross-section will leave marks on the final profile. Select the smaller value of z_1 and z_2 as point B_1, and then calculate the intersection with the adjacent grains.

Similarly, for the arbitrary intersections B_p and B_q in the $y-z$ plane, the linear equations of B_pD_p and B_qC_q can be expressed respectively as

$$z - z^{(p)} = \frac{1}{\tan\alpha}[y - (p-1)L] \quad (12)$$

$$z - z^{(q)} = -\frac{1}{\tan\alpha}[y - (q-1)L] \quad (13)$$

The intersection G_p between these lines can be solved by using Eq. (12) and Eq. (13) as

$$\begin{cases} y = \dfrac{\tan\alpha}{2}(z^{(q)} - z^{(p)} + (p+q-2)\dfrac{L}{\tan\alpha}) \\ z = \dfrac{1}{2}(z^{(p)} + z^{(q)} + (q-1)\dfrac{L}{\tan\alpha}) \end{cases} \quad (14)$$

The final surface profile of the cross-section is formed by connecting successive points B_1, G_1, B_2, G_2, …, B_j and G_j, which can be solved by using Eq. (12) to Eq. (14).

3 Simulation

The proposed method is applied to simulate the cross-sectional surface roughness of the experiment data in [7]. Table 1 lists the grinding and wheel parameters. The surface roughness on the cross section is measured with a cut-off length of 0.8 mm. Since the evaluation length should contain 5 consecutive sampling lengths, the simulation length in the cross-sectional direction is taken as 5 mm. Table 2 shows the roughness results compared with the measurement and references.

From table 2, we can see the simulation result is consistent with the experiment data. The main simulation error caused by this method is attributed to the difference between the assumption and the process in the distribution and the shape of the grains. In practice, the grain interval is random and the grain shapes are complex basic geometries, such as ellipsoid, tetrahedron, cuboids and octahedron. In simulation, the undeformed chip cross section is assumed to be triangular in shape. However, the undeformed chip cross section for a dulled flattened pyramidal-shaped grit having a wear flat at its tip of dimension leads to a trapezoidal cross section. For all that, the simulation data proposed in this paper is closer to the experiment data compared with references [8]. Since only one of the cross-sections is simulated in this method, a lot of time is saved and the simulation scale is expanded.

Table 1. Wheel and grinding parameters

Nominal wheel diameter (mm)	Grain mesh size	Structure number	Wheel spindle speed (rpm)	Table speed (m/s)
203	180	8	4500	0.06

Table 2. Surface roughness compared with the experiment

	Simulated data(this paper)	Simulated data [7]	Experiment data [8]
R_a	2.0	2.73	1.5

4 Conclusion

(1) A new algorithm for predicting the cross-sectional surface roughness of the workpiece for the grinding process is developed in this paper. It takes into consideration the random distribution of the grain protrusion heights and reduces three-dimensional modeling to two-dimensional modeling by introducing the cross-sectional area of the undeformed grinding chip, which not only saves a lot of time but also expands the simulation scale.

(2) The simulation shows that the proposed method gives the results that are consistent with the measurement ones, thus the effectiveness of the algorithm is guaranteed.

This research did foundation work for the modeling and predicting surface roughness of the face gear grinding process.

Acknowledgement

This Project is supported by Scientific Research Fund of Hunan Provincial Education Department(14B052)

References

1. Y.Liu, A.Warkentin, R.Bauer, et al: Precis Eng Vol. 37, 758-764(2013).
2. E.Brinksmeier, J.C.Aurich, E.Govekar, et al: Annals of the CIRP Vol. 55,667-696(2006).
3. X.Chen, W.B.Rowe:Int J Mach Tool Manu Vol. 36,871-882,(1996) .
4. X.Chen, W.B.Rowe: Int J Mach Tool Manu Vol. 36,883-896,(1996).
5. P.Koshy, L.K.Ives, S.Jahanmir: Int J Mach Tool Manu Vol.39, 1451-1470,(1999).
6. S.Agarwal, P.Venkateswara Rao: Int J Mach Tool Manu Vol. 50,1065-1076,(2010)..
7. X.Zhou, F.Xi: Int J Mach Tool Manu Vol.42, 969-977,(2002).
8. T.W.Hwang., C.J.Evens., S.Malkin:J Manuf Sci E-T ASME Vol. 122,42-50 (2000).

A Review of Risk Assessment Methods for Power System

Liang Zhao[1*], *Tianyang* Mao[1], *Wen* Xu[2], *Jingzhao* Luan[2], *Jiangning* Wu[2], *Guangyuan* Qi[2].

[1]Faculty of Electronic Information and Electrical Engineering, Dalian University of Technology, Dalian 116024, China
[2]State Grid Dalian Electric Power Supply Company, Dalian 116021, China

Abstract. In view of the current social background of power system becoming more and more complicated, and the current research situation of related fields, this paper discusses the common methods of power system risk assessment, including the method of hierarchy analysis, Markov chain model, Monte Carlo method, fuzzy comprehensive evaluation and related research progress. The advantages and disadvantages of the four methods are enumerated respectively. It is considered that the power system should be evaluated synthetically in the practical application and the effective risk assessment method should be adopted rationally.
Keywords.power system; risk assessment; analytic hierarchy process; Markov chain model; Monte Carlo method; fuzzy comprehensive evaluation

1 INTRODUCTION

With the extensive application and popularization of information technology, the information system has gradually penetrated into all walks of life, and has played a significant role in promoting social development. The safe operation of information system has become a necessary guarantee for the normal development of society. And as an important social infrastructure, the normal operation of the power system is the primary condition of sustainable development of social economy. On the other hand, in the power system to achieve great networking benefits at the same time, also had to bear a greater potential risk. Especially with the advancement of power market reform, the change of management institutions and the participation of new members in the market, the uncertain factors that are difficult to control and its influence on the power grid are more profound, making the planning, operation and maintenance of power system and asset management face a great challenge. Therefore, the importance of power system risk research is becoming more and more obvious.

An et al (2009) [1] analyzes the current situation of power system operation risk, power system equipment risk, power market risk and power system security risk, and points out that electric power workers in the system operation, equipment management, safety assessment and market analysis of these areas Research ideas change from deterministic research to uncertainty research. Zhang et al (2014) [2] analyzes the current situation of the research on the vulnerability of power system, and thinks that the current evaluation method can not meet the needs of modern power system, and it needs a lot of in-depth research. Liu et al (2016) [3] introduces the management system, database structure, index establishment principle and classification method of the reliability evaluation system of the large power system based on the operational data in North America, and explains the calculation and analysis method of the index.

Based on the present research situation of power system risk assessment at home and abroad, this paper introduces several risk assessment methods commonly used in power system, and analyzes their advantages and disadvantages.

2 COMMON RISK ASSESSMENT METHOD FOR POWER SYSTEM

2.1 Analytic Hierarchy Process

The analytic hierarchy process (AHP) is used to decompose the decision problem into different hierarchical structures according to the overall goal, the subordinate goals, the evaluation criteria and the specific alternatives. Then, the method of solving the matrix vector is obtained. The final weight of each element of each level is given to the priority of the previous level, and the final weighting method is the final weight of the candidate scheme to the total target. The final weight is the optimal scheme. The analytic hierarchy process is more suitable for the target system with hierarchical evaluation index, and the target value is difficult to quantitatively describe the decision problem. Its use is to construct the judgment matrix, find its maximum eigenvalue. And its corresponding eigenvector W, after normalization, is a certain level of indicators for the previous level of a relative importance of the relative importance of indicators.

Based on the multi - factor analysis of large - scale blackout events, Wang et al (2011) [4] set up a set of safety and risk assessment system of power system with objectivity, practicability and applicability by using fault tree analysis and analytic hierarchy process. And the evaluation system is used to evaluate the power system of the provincial power system. The evaluation results validate the feasibility of the assessment methodology and the assessment indicators. Feng et al (2013) [5] applied 3 scale analytic hierarchy process assessment model for the establishment of the state of the transformer. Case analysis proved that assessment of power transformer condition based on Analytic Hierarchy Process method is reasonable and effective. Zhang et al (2013) [6] uses the 3-scale AHP method to derive the comprehensive risk of components through over-voltage risk, low-voltage risk, line overload risk, transformer overload risk and loss-of-load risk. And through the IEEE RTS-79 system to verify the risk based on the power system component risk assessment approach is feasible. Yang et al (2013) [7] use the analytic hierarchy process and the fuzzy comprehensive rating method to recalculate the expert evaluation weight for the current situation of the subjective factors of the expert in the risk assessment of the current power information system. The method is used to evaluate the components in the power information system. The results show that the method can reduce the subjective judgment factors. He (2016) [8] analyzes the current situation and existing problems of power information system construction, and uses the analytic hierarchy process to identify the main risks of power information system construction project. The research found that the risk of construction project of power enterprise information system mainly focused on the project stage and the realization stage, which provided the help for the future construction. In view of the shortcomings of the information security risk assessment mechanism in traditional power systems, such as poor accuracy, lack of perfection and low efficiency, Liang et al (2017) [9] introduce the analytic hierarchy process into the risk assessment mechanism according to the specific application of the power system. They use the fuzzy mathematics knowledge in the process of risk calculation to design a new model of information security risk assessment, namely AF-RA model, and elaborate and analyze the model.

2.2 Markov Chain Model

Markov chain is a stochastic process of discrete events with Markovian properties in mathematics. In the process, in the case of a given current knowledge or information, the past is irrelevant for the prediction of the future. The Markov chain describes a state sequence whose state values depend on the previous finite state. The Markov chain is a sequence of random variables with Markovian properties. The range of these variables, that is, their all possible sets of values, is called "state space" and the value of Xn is in the state of time n. If Xn+1 is Xn function of the conditional probability distribution for the past state, then:

$$P(X_{n+1}=x|X_1=x_1,X_2=x_2,...,X_n=x_n)=P(X_{n+1}=x|X_n=x_n) \quad (1)$$

Where x is a state in the process. The above identity can be regarded as the Markovian nature.

Guo et al (2012) [10] define an improved Markov chain, the peak-type Markov chain, for the defects of traditional Markov chains in some special cases of power systems, and prospects its application in the power system. Magnus et al (2012) [11] proposed a new method for generating samples of the power at all nodes given a set of transfers through specified corridors of the power system. In the reliability analysis of power system, the Monte Carlo simulation method requires a large number of computational resources and the slow convergence speed. From the operational point of view, based on the Markov chain, Wang et al (2013) [12] proposed a model for rapid analysis of power system reliability. The validity of the study is proved by the analysis of the examples. Tang et al (2013) [13] proposed a refinement pattern recognition method for transient stability assessment of power systems based on a two-stage parallel Hidden Markov model. The simulation and analysis of the actual regional grid environment verify the effectiveness and accuracy of the method. In the case of recognition accuracy, this method is less than the training samples required for commonly used artificial intelligence methods (such as ANN, SVM, etc.) and converges faster. In the reliability analysis of power system, the shortcomings of long time residence time in the same state are often simulated by using the discrete Markov chain model, based on the continuous time Markov chain, Jin et al (2015) [14] propose a model and give an example of the proposed research method. Based on the hidden Markov model, Ma et al (2016) [15] improve the existing static risk assessment algorithm based on the influence of entity behavior on system risk. This paper analyzes the system assets, threats and vulnerabilities, and presents an improved risk calculation method. Theoretical analysis and experimental results show that the improved method improves the reliability and timeliness of the evaluation results. Mindaugas et al (2016) [16] proposed stochastic automata networks to allow reducing the size of state space of Markov chain model and simplifying system specification. Modelling results showed that the implementation of Markov chain model by using SAN method is a relatively easy task.

2.3 Monte Carlo Method

Monte Carlo method, also known as statistical simulation method, random sampling technology, is a stochastic simulation method, based on probability and statistical theory method is a calculation method, is the use of random numbers to solve a lot of computing problems. The problem is solved by a certain probability model, and the statistical simulation or sampling is realized by an electronic computer to obtain the approximate solution of the problem.

Bie et al (1997) [17] introduce the basic principles of Monte Carlo method and its application in power system reliability assessment. Through the comparison and analysis of Monte Carlo method and analytic method, the characteristics of Monte Carlo method and its application range are expounded. Then, the application status of Monte Carlo method in power system reliability evaluation is introduced from the aspects of improving the convergence speed of Monte Carlo method, the combination of Monte Carlo method and analytic method, pointing out the future research direction. In order to comprehensively evaluate the economics and reliability of wind power generation, Wu et al (2004) [18] proposed a reliability model of wind turbine based on sequential Monte Carlo simulation, and gave a detailed algorithm, taking the random characteristics of wind energy and the forced outage of wind turbines into account. The model is used to simulate the wind turbine and the time series load model is used to calculate the system reliability index and economic index under different design schemes. Simona et al (2007) [19] presented the development of a reliability simulation program for electrical power stations based on Monte Carlo method.

The program can help in equipment and system condition evaluation and monitoring. Based on the power system operational risk assessment system, based on the analysis and evaluation of the critical system of power system operation risk index set, Xia et al (2014) [20] use non-sequential Monte Carlo simulation technology quantitative analysis and comprehensive consideration of equipment failure probability and consequences of the risk index. They through the various control measures after the implementation of the changes in the risk indicators to assist the system operator decision-making, which has important engineering application value. Bao et al (2014) [21] analyzed the differences in the application of the two conventional sequential Monte Carlo simulations in the power system reliability assessment of the state duration sampling (SDS) and the time interval sampling (TIS). The validity of the proposed method is verified by the simulation of IEEE-RTS79 test system, which indicates that the modified TIS method has more obvious advantage when the system failure rate increases. Chen (2016) [22] first outlined the concept and basic theory of power system risk assessment, and then details the most commonly used Monte Carlo method in risk assessment, which is mainly divided into sequential Monte Carlo method and non-sequential Monte Carlo method. This paper uses the data of Hainan power grid to carry out the risk assessment analysis. The risk assessment results verify the validity and practicability of the power system risk analysis in the power system.

2.4 Fuzzy Comprehensive Evaluation

Fuzzy comprehensive evaluation method is a comprehensive evaluation method based on fuzzy mathematics. The comprehensive evaluation method is based on the membership degree theory of fuzzy mathematics to transform the qualitative evaluation into quantitative evaluation, that is, using fuzzy mathematics to make a general evaluation of things or objects subject to various factors. It can better solve the problem which is fuzzy, difficult to quantify, suitable for a variety of non-deterministic problems.

Daniela et al (2001) [23] presented practical applications of fuzzy sets methods to power systems control, design, operation and management, and demonstrated that fuzzy methods can solve many difficult problems. Liang et al (2009) [24] put forward a comprehensive evaluation model and method of information system security risk based on fuzzy comprehensive evaluation theory, and realized the goal of quantifying the security risk of information system. Based on the set of safety risk factors, the set of weights and the weight coefficient of the information system, the fuzzy risk comprehensive evaluation matrix is established and applied to the security risk assessment of the Web components of the power information system. Wu et al (2011) [25] proposed the concept of "relevant fuzzy number" and "independent fuzzy number" for the deficiency of fuzzy operation defined by expansion theorem, and an improved method suitable for fuzzy operation is proposed in combination with the principle of expansion, and the rationality of the method is proved. The fuzzy uncertainty of equipment outage and load level under different operating conditions of power system is analyzed, and the fuzzy model of equipment outage and multi - level load level is established. The risk assessment procedure and risk index of power system based on improved fuzzy operation are given. Finally, taking the IEEE - RTS 24 - node test system as an example, the improved fuzzy operation method is used to calculate the specific system risk index, and the results are compared and analyzed. The correctness and rationality of the method are verified. Based on the research of the theory of information fusion system and the security characteristics of power system, Song (2014) [26] designs the quantitative risk assessment method of power information physical fusion system by introducing the comprehensive fuzzy risk assessment method. Using the fuzzy set theory, the triangular fuzzy number is used to carry out the risk data transformation, and the parameters are integrated to calculate the overall risk level of the system. Tan (2014) [27] studied the fault probability grading algorithm based on fuzzy mean clustering method, and proposed the grading algorithm of fault consequence severity based on fuzzy reasoning principle, and established the safety warning level by fuzzy operation. The wind speed prediction model based on time series state transition method and the output model and outage model of wind turbine with consideration of wake effect are established. Based on fuzzy synthetic evaluation method, Wu et al (2017) [28] assessed the risks in public-private partnership straw-based power generation projects in China. This study expands the knowledge and contributes to the literature.

3 ANALYSIS ON ADVANTAGES AND DISADVANTAGES OF POWER SYSTEM EVALUATION METHOD

The analytic hierarchy process has the advantages of systematic analysis method, simple and practical decision-making method and less quantitative data information. But also can not provide new solutions for decision-making, less quantitative data, more qualitative components, not convincing, too many indicators when the data statistics, and the weight is difficult to determine the eigenvalues and eigenvectors of the exact method of more complex shortcomings.

The Markov chain model has a good effect on the state prediction of the process, and can be used to predict the production of dangerous state, but it is not suitable for long-term prediction.

Monte Carlo method can simulate the characteristics of things with random properties and physical experimental process, subject to the constraints of geometric conditions is small, the convergence rate has nothing to do with the dimension of the problem, in dealing with multi-dimensional problems have more obvious advantages , With the ability to simultaneously calculate multiple solutions with multiple unknowns, the error is easy to determine, the program structure is simple, easy to use computer to achieve the advantages. At the same time, it has the shortcomings of slow convergence and error.

Fuzzy evaluation through accurate digital means to deal with fuzzy evaluation of the object, can be hidden information on the information presented by the more scientific, reasonable, close to the actual quantitative evaluation. And the evaluation result is a vector, rather than a point value, contains a wealth of information, both can accurately describe the object to be evaluated, but also can be further processed to get reference information. But the calculation is complex, and the subjective weight of the index weight vector is strong.

4 CONCLUSION

The above article introduces five common methods of risk assessment of power systems and compares their advantages and disadvantages. It can be seen that each method has to be in its appropriate environment to play the greatest effect. In the actual production, according to the actual situation, a reasonable choice of assessment methods, if necessary, can be used in conjunction with multiple methods to achieve rapid and accurate risk assessment.

Acknowledgment

The present research is supported by the national science and technology Power Grid Corp (2017YF-36).

Reference

1. T.-Y. An, Z.-Y. Wang, X.-Z. Jin, J.-N. Zhang and H. Hu, Power System and Clean Energy **9**, pp.4-10, September(2009)
2. X.-D. Zhang, R.-K. Sun, P. He and S. Zhang, The Proceedings of the 8th Annual Conference of Science and Technology of Jilin Province, China. (2014)
3. W.-X. Liu, X.-D. Shan, W.-M. Jiang, Z.-Q. Wang and B.-L. Xia, Power System Technology **40**, pp. 3811-3817, December(2016).
4. B. Wang, D.-H. You, X.-G. Yin and Q.-Q. Chen, Power System Technology **35**, pp.40-45, January (2011)
5. Y. Feng, Z.-H. Zhang and L.-Z. He, Advanced Materials Research **805-806**, pp. 793-800(2013)
6. Y.-M. Zhang, Z.-H. Zhang, F. Yao, L.-Z. He and W. Xu, Power System Protection and Control **41**, pp.73-78, December(2013)
7. H. Yang, Y. Wang, X.-G. Pei and S. Chen, Journal of Shanghai University of Electric Power **29**, pp.518-522, December(2013)
8. Y.-M. He, North China Electric Power University, (2016)
9. Z.-Q. Liang and D.-S. Lin, Netinfo Security **4**,

pp.86-90(2017)

10. H.-K. Guo and S.-K. Heng, Journal of Electric Power **27**, pp.96-98, April(2012)
11. M. Perninge and L. Soder, European Transactions on Electrical Power **22**, pp.140-151, March(2012)
12. Y. Wang, X.-S. Han, Y. Ding and J. Sheng, Power System Technology **37**, pp.405-410, February(2013)
13. F. Tang, B. Wang, X.-M. Zha, Z.-H. Ma and Y.-N. Shao, Proceedings of the CSEE **33**, pp.90-97, April(2013)
14. S.-H. Jin, J.-G. Lu, Y.-P. Wan, S.-G. Sun and J.-P. Liu, Fifth International Conference on Reliability and Electrical Contact for Electrical Products, China(2014)
15. Z.-C. Ma, P. Yang, X.-F. Zhang and L.-T. Ding, Modern Electronic Technique **39**, pp.165-170, September(2016)
16. M. Snipas, V. Radziukynas and E. Valakevicius, Reliability Engineering and System Safety **157**, pp.13-22, January(2017)
17. C.-H. Bie and X.-F. Wang, Automation of Electric Power Systems **22**, pp.68-75, June(1997)
18. Y.-C. Wu and M. Ding, Electric Power Automation Equipment **24**, pp.70-73, December(2004)
19. S. Dzitac, T. Vesselenyi, I.Dzitac and E. Valeanu, The 4th International Federation of Automatic Control Conference on Management and Control of Production and Logistics, September(2007)
20. T. Xia, T.-G. Zhou, X. Zhang, Y. Wang and G.-J. Liu, Journal of State Grid Technology College **17**, pp.1-21(2014)
21. Y.-K. Bao, Y. Wang, J.-X. Tang, C.-X. Guo, X. Wang and H. Shi, Power System Technology **38**, pp.1189-1195, May(2014)
22. Z. Chen, Technology Innovation and Application. pp.9-10(2016)
23. D. Andone, R. Dobrescu and S. S. Iliescu, IFAC Proceedings Volumes **34**, pp.151-156, July(2001)
24. D.-X. Liang and X. Chen, Power System Protection and Control **37**, pp.61-64, March(2009)
25. H.-B. Wu and J. Hu, Journal of North China Electric Power University **38**, pp.34-39, November(2011)
26. C.-Y. Song, Electronic Technology & Software Engineering. pp.261(2016)
27. S.-M. Tan, North China Electric Power University, (2014)
28. Y.-N. Wu, L.-W.-Y. Li, R.-H. Xu, K.-F. Chen and Y. Hu, Journal of Cleaner Production **161**, pp.977-990, September(2017)

Analysis and prediction of the summer cooling load characteristics

XU Hao[1], CHEN Lipin[1],YE Qiang[1] , REN ZhiChao[1], LIANG Yuan[2]
[1]State Grid Sichuan Economic Research Institute SGSERI Chengdu, China
[2]College of Management Science, Chengdu University of Technology
Chengdu, China

Abstract. Nowadays, cooling load is mainly attributable to upgrade of a new high of power grid load. An accurate calculation and prediction of cooling load can better meet power demands at peak hours in summer. Based on the historical summer cooling load and temperature data of Sichuan province, the characteristics of summer cooling load and temperature in Sichuan are examined firstly. Next, the random and nonlinear relations of power load are studied. Finally, the BP neural network model optimized by the genetic algorithm (GA) is employed to achieve accurate prediction of summer cooling load in Sichuan province. Results suggest that cooling load is closely connected with temperature and air-conditioner ownership, and that the BP neural network optimized by the GA is effective in predicting summer cooling load. To sum up, this research can provide solid basis for power load scheduling.

1 Introduction

Cooling load refers to rapid increase of the power grid network within a short period of time due to launch of cooling equipment, notably the air-conditioner [1]. In recent years, along with rapid development of economy, continuous improvement of people's living standards, and worsening of the "heat island" phenomenon, cooling load is increasing on an annual basis [2][3]. Nowadays, cooling load has been a major factor attributable to continuous upgrade of a new high of the power grid load in summer. This has not only changed the original load characteristics [4-6], but also brought negative effects on the system voltage and frequency stability [7].

The special landform of the Sichuan Basin has led to a moist and scorching hot summer in most parts of Sichuan. To cool down the temperature, Sichuan has a high requirement of cooling load in summer, which has a significant influence on power load. In order to increase accuracy of load prediction at different time scales and to ensure stability, safety and economical operation of Sichuan's power grid in summer, it is imperative to measure and analyze cooling load of Sichuan's power grid.

Calculation of cooling load should take into account not only the local climate conditions, but also the local use habits of cooling equipment (e.g. air-conditioners), national energy-saving policies [8], and perception of physical comfort [9]. Since it is no easy task to collect all the necessary data, accuracy of cooling load calculation has long remained a challenge. Traditional algorithms for power load prediction include time series, multiple linear regression and Fourier component. Every algorithm, with respective advantages and disadvantages, is applicable to different occasions. Since power load changes along with weather conditions and human activities, a series of random and nonlinear relations has formed, making it hard to improve the power load prediction accuracy [10]. As a universal approximator, the artificial neural network can recognize systems, which are nonlinear in essence. Therefore, the artificial neural network can be regarded as a highly effective load prediction technology. Characterized by large-scale parallel, distributed storage and processing, self-organization, self-adaption and self-learning ability, the neural network is especially suitable for inaccurate, fuzzy information processing with many factors and conditions to take into consider.

In optimizing the back propagation (BP) neural network, the genetic algorithm (GA) can improve efficiency and robustness of global searching, and has shown outstanding performance in practical application. After the neural network is globally optimized by the GA, the BP algorithm can be adopted for fine tuning to accelerate the network convergence speed and reduce the probability of training failure [12].

This paper collects and processes the power load and temperature data in June, July and August of 2009 to 2016. Then, statistical analysis is adopted to sum up rules and influencing factors of power load changes. General changes of power load are roughly predicted. Finally, the BP neural network model optimized via the GA is employed to achieve precise prediction of cooling. It is hoped that this research can contribute to contribute to the power load prediction and scheduling in Sichuan province.

* Corresponding author: luckfunny_wang@163.com

DOI: 10.1051/matecconf/201713900176

2 Analysis of summer cooling load characteristics in Sichuan province

The power load and temperature historical data of Sichuan province in June, July and August from 2009 to 2016 are collected, analyzed, and used for load prediction. Results show October and April are the two months with the most comfortable temperature in Sichuan in that almost no cooling equipment is required. Hence, the mean value of power grid in October and April is adopted as a reference load for calculation of cooling load. The maximum daily cooling load is calculated as:

ΔL= Daily maximum power load-(Daily maximum power load in October+ Daily maximum power load in April)/2

2.1. Consistency between daily maximum cooling load and daily highest temperature changes

As shown in Fig. 1 and Fig. 2, the daily maximum cooling load changes via the daily highest temperature, and the change trend of the two is consistent with each other. To be specific, with increase of the daily highest temperature, the daily maximum cooling load also rises. On the contrary, when the daily highest temperature decreases, the daily maximum cooling falls.

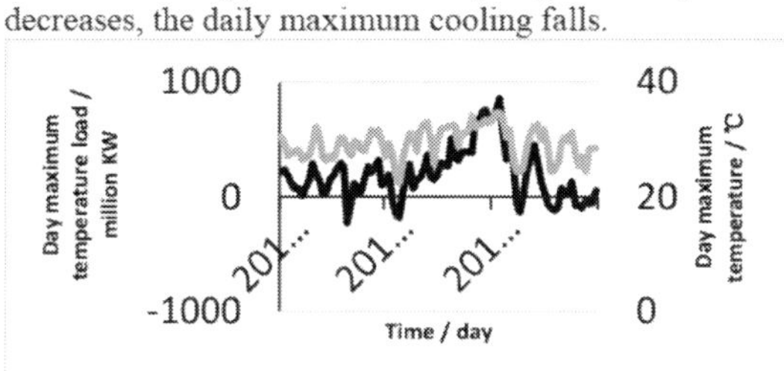

Figure 1. Change trend of the daily maximum cooling load and the daily highest temperature from June to August 2015

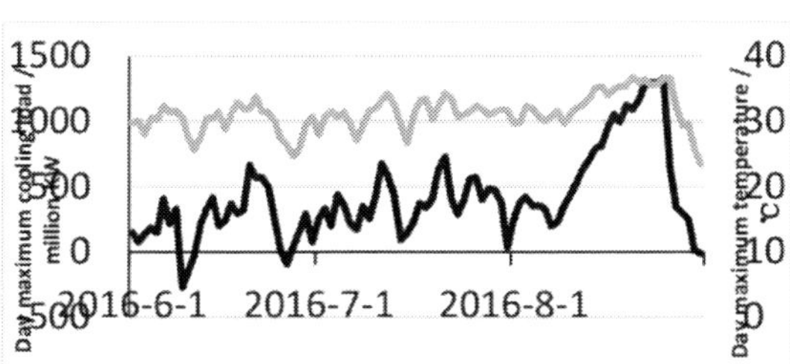

Figure 2. Change trend of the daily maximum cooling load and the daily highest temperature from June to August 2016

The daily highest temperature has an accumulative effect on the daily maximum cooling load. From Fig. 1 and Fig. 2, it can be seen that, after days of high temperature above 30 ℃, the power load increases rapidly, because the human body has an upper limit for continuous high temperature. (The dark broken line below represents the load; while the light broken line above represents the temperature.)

2.2 Influence of GDP on the daily maximum cooling load long-term change trend

As one notices in Fig. 3, the daily maximum cooling load of Sichuan's power grid experiences continuous growth from 2009 to 2009, and the percentage of the maximum cooling load in the daily load is also on an upward trend.

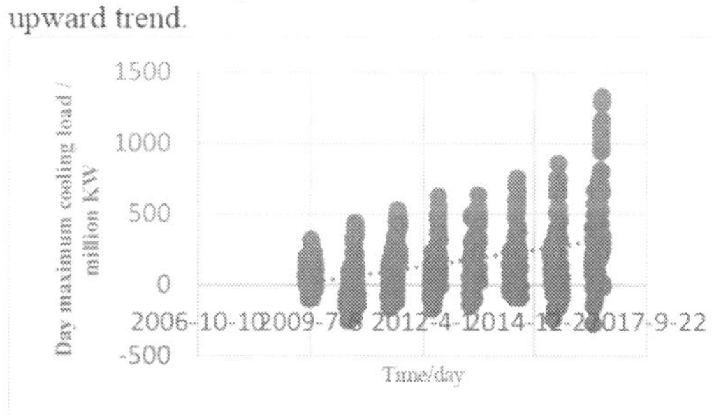

Figure 3. Percentage and linear change trend of the daily maximum cooling load from June to August of 2009 to 2016

As one observes in Fig. 4, the annual summer daily highest temperature shows no dramatic changes from 2009 to 2016. This provides solid basis for the conclusion that the daily maximum cooling load is positively correlated with the daily highest temperature. The daily maximum cooling load and the daily highest temperature changes represent changes in the short term. The annual changes of the daily maximum cooling load are changes in the long term, and are affected by the annual GDP of Sichuan province. The change trend of the GDP is shown in Fig. 5. (Unit: 100 million yuan) It is widely agreed that growth of GDP can contribute to improvement of people's living standards. With basic living demands satiated, people pay more attention to comfortability of their living and production environment. To the end, quantity and use frequency of cooling equipment both increase, thus leading to rise of the daily maximum cooling load.

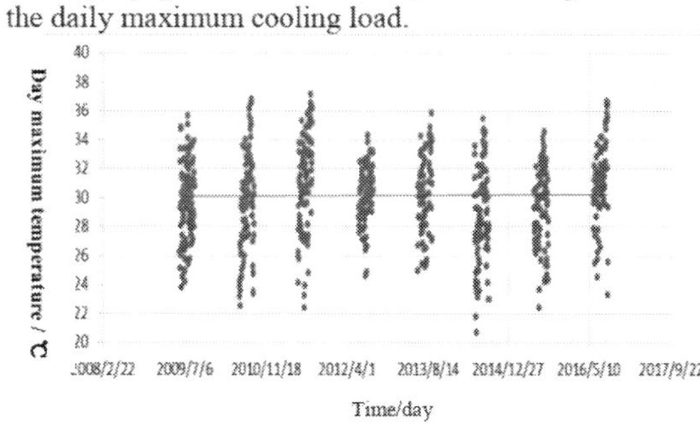

Figure 4. Changes and linear change trend of the daily highest temperature from June to August of 2009 to 2016

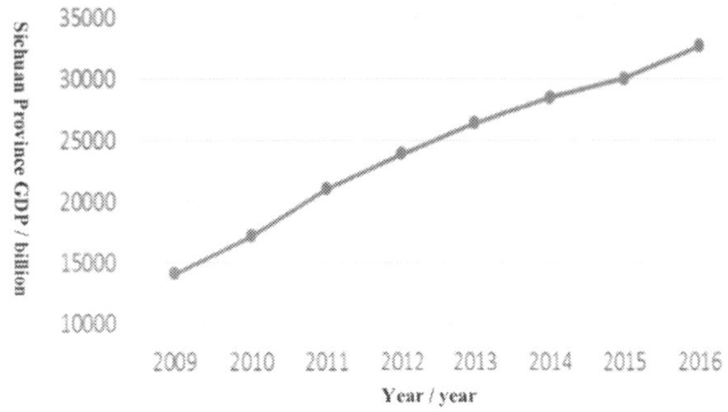

Figure 5. Sichuan provincial GDP level from 2009 to 2016

2.3 Influence of small hydropower stations on the daily maximum cooling load

A negative maximum cooling load is found in each year from 2009 to 2016. When the maximum cooling load is negative, it suggests the total load of the current day is lower than the reference load of the current year. One possible reason for the phenomenon is that there might be other forms of energy supply or that the load required by production activities decreases.

Through statistical analysis of annual data, it can be found that the maximum cooling load concentrates in July, seldom appearing in June and August. Historical data show that July is a month with abundant precipitation. Besides, there are many small hydropower stations in Sichuan, which serve as an important supplementation to the national power grid. Therefore, though data of small hydropower stations are incomplete, this paper still believes that the negative maximum cooling load results from ample power supply from small hydropower stations during the period of abundant precipitation period. On the Dragon Boat Day, the maximum cooling load is mostly negative. This represents a typical decrease of load required by production activities.

3 Summer cooling load prediction

3.1 Prediction model and algorithm

In this paper, the BP neural network model optimized via the genetic algorithm (GA) is adopted for prediction of summer cooling load in Sichuan province. The GA is an algorithm for global optimization and random search. On the basis of individual gene representative, the genetic operator is adopted to simulate various phenomena occurring during the process of inheritance, including replication, intersection and variation, and pick the superior individuals from the population. Finally, the optimal individuals can be obtained. In this paper, the improved GA and BP neural network are combined to form a BP neural network prediction model optimized by the GA algorithm.

The model consists of the following three parts:

(1) Determining the BP neural network structure: The BP neural network topological structure is determined according to the chaotic time series input and output of parameters. Based on that, the length of the GA individuals can be determined.

(2) Optimizing the BP neural network weight and threshold value using the GA: The linear interpolation function is adopted to generate the population individuals. Every individual in the population contains a BP neural network weight and threshold value. The degree of individual fitness is calculated using the fitness function. The GA search the corresponding individual with the optimal fitness value via a series of operations, including selection, intersection and variation.

(3) Predicting the BP neural network: The optimal individual obtained via the GA is used to assign the initial weight and threshold value of the BP neural network. Then, the BP neural network is adopted to conduct local optimum-seeking so as to obtain the predicted value with the global optimal solution.

Below are steps of the model:

Step 1: Set the population size as p;

Step 2: Confirm the individual evaluation function;

Step 3: Use the roulette method to choose the operator; (Select the chromosome from every generation of population based on the percentage of the fitness degree.)

Step 4: Adopt the real number cross method considering the real number coding of individuals;

Step 5: Choose the j gene from the i individual for variation;

Step 6: Decompose the optimal individual obtained by the GA into the connecting weight and threshold value of the BP neural network;

Step 7: Use the BP algorithm to train the BP neural network prediction model and work out the optimal solution of the chaotic time series prediction.

Here, the feedforward neural network (FNN), empirical mode decomposition (EMD), support vehicle machine (SVM), BP neural network and GA-optimized BP neural network are used for prediction. Their prediction performance is compared. Due to limit of the paragraph, the model theory is not further expounded.

3.2 Prediction results and performance

The daily maximum cooling load and daily highest temperature historical data from June to August of 2009 to 2016 are adopted as research data. There are 604*7 sample data in total, and 92 data in 2016 are predicted. It is found out that the variation between the predicted value and the actual value is small when the GA-optimized BP neural network is adopted for prediction. Indexes to verify the prediction results include MSE,

MAE, rMAE, MAPE and RMSPE, and the index data are presented in Table 1.

Table 1. Evaluation indexes of prediction performance

	MSE	MAE	rMAE	MAPE	RMSPE
FNN	31501	135.7970	0.3649	1.7608	6.4412
EMD_FNN	121340	256.9908	0.6578	5.8897	46.5189
SVM	171630	1495.2	0.0404	4.1540	1.1767
BP neural network	1344.1	12.7733	0.0316	0.0226	0.0514
Genetic algorithm optimization BP neural network	**16.8606**	**1.4747**	**0.0036**	**0.0042**	**0.0077**

According to the index data and evaluation criteria in Table 1, it is obvious that the GA-optimized BP neural network model is the best, and can meet the prediction requirements.

4 Conclusions

Through descriptive statistics of the daily maximum cooling load and daily highest temperature data, the following conclusions can be reached:

(1) The change trend of the daily maximum cooling load and the daily highest temperature generally coincides with each other, but the latter has a cumulative effect on the former;

(2) Temperature is a major factor influencing the daily maximum cooling load in the short term, while GDP is a long-term influencing factor of the daily maximum cooling load;

(3) A major cause of the negative daily cooling load is the large number of small hydropower stations in the period of abundant precipitation.

Analysis of the daily maximum cooling load prediction results and performance shows that the GA-optimized BP neural network model outperforms other prediction models in terms of every prediction index. Therefore, the GA-optimized BP neural network algorithm can be combined with the local weather forecasts to help predict the future maximum cooling load in June, July and August of Sichuan province.

References

1. Yao Wei, Cheng Shijie, et al. *Application of HVDC technology in grid - connected offshore wind farm* CEP, **40** (2007)
2. Liu Lin, Ge Xu Bo, and so on. *China 's offshore wind power development status and analysis*.ETE, **24**(2012)
3. Wang Jingquan, Cheng Jiansheng, Li Feng. *On the construction of offshore wind farms in China* .CES, **12**(2010)
4. Yin Yufen. *Influence of temperature on regional load characteristics*. GEPr, **24**(2011)
5. Li Canbing, Shang Jincheng, Zhu Shouzhen, and so on. *Analysis of Energy Consumption Caused by Cumulative Effect of Temperature on Temperature Regulation Load*. Power System Automation, **34** (2010): 30-33.
6. KYUNG B S, SEONG K H, JUNG W P, et al. *Hybrid load forecasting method with analysis of temperature sensitivities*. IEEE Trans Oil Power Systems, **21**(2006): 869—876.
7. Zhang Limin, Fu Hongjun, Li Yuyan, et al. *Estimation of Cooling Load Ratio of Air Conditioning and Its Influence on Voltage Stability*. Relays, **33** (2005): 33-36.
8. TONG Shulin, WEN Fushuan. *Calculation and analysis of the annual maximum high-temperature related load in the energy saving and emission reduction environment in Guangdong Province*.Journal of North China Electric Power University, **37**(2010):32-37.
9. TAO Yong, SHEN Ying. *Influence of summer weather condition on regional air conditioning loads* . East ChinaElectric Power, **34**(2006):29-30.
10. Moazzami M, Khodabakhshian A. *Hooshmand R. A new hybrid day-ahead peak load forecasting method for Iran's National Grid.* Appl Energy **101**(2013):489-501.
11. Lawan S, Abidin W, Masri T, Chai W, Baharun A. *Wind power generation via ground wind station and topographical feedforward neural network (T-FFNN) model for small-scale applications*. Journal Of Cleaner Production**143**(2017) :1246-1259.
12. WANG Jun-tao.*Study on Optimized BP Neural Network Algorithm Based on Genetic Algorithm*. Small and Medium Enterprises Management and Technology, **4** (2017).
13. Wang J, Shang L, Chen S, et al. *Application of fuzzy classification by evolutionary neural network in incipient fault detection of power transformer*.IEEE International Joint Conference on Neural Networks, (2004)
14. Han Liqun. *Artificial neural network theory, design and application*. Beijing. Chemical Industry Press .(2007)
15. Huang N E, Shen Z, Long S R, et al. *The Empirical Mode Decomposition and the Hilbert Spectrum for Nonlinear and Non-Stationary Time Series Analysis*. Proceedings Mathematical Physical & Engineering Sciences, **454** (1998):903-995.
16. Vapnik V N. *The nature of statistical learning theory*. New York Springer, (1995).
17. Mukherjee S, Osuna E, Girosi F. *Nonlinear prediction of chaotic time series using support vector machines*. Proceedings of IEEE NNSP ,(1997)
18. Yan Pingfan, Zhang Changshui. *Artificial neural network and simulated evolutionary computation*. Beijing: Tsinghua University Press, (2005)

Feature extraction and influence analysis for the structural performance of complex component

Haiyan Hua [1,*], *Zhenhui* Shen [2], and *Yangmei* Zhang[2]

[1]School of Mechanical & Automotive Engineering, FuJian University of Technology, Fuzhou 350118, P.R. China
[2]Engineering College, Fujian Jiangxia University, Fuzhou 350108, P.R. China

Abstract. To explore the structural performance of complex component deeply, a new method of feature extraction and influence analysis is proposed and its automatic mechanism is established. Using this mechanism, the complex component can be divided into several sub-regions, and the maximum stress in each sub-region can be extracted. After evaluating the danger situation for all sub-regions, the absolute influence of structural parameters for the characteristic stresses in feature regions are carried out based on Spearman rank correlation analysis. And then, the feature knowledge of complex component is acquired. Finally, the excavator boom is taken as an example, which demonstrates that the useful knowledge can be extracted from the structural performance of complex component automatically and effectively by this method.
Keywords. Feature extraction; Influence analysis; Structural performance; Spearman rank correlation coefficient; Significant sensitive

1 Introduction

High strength and lightweight are two important but contradictory targets for modern mechanical design[1], which are more difficult to be coordinated for complex component. Exploring the stress feature is one of the most important work for realizing reasonable lightweight design and improving the comprehensive structural performance of complex component [2,3]. In recent years, many feasible strategies of extracting stress features have been proposed in some researches. For example, the welding regions have been taken as stress features by Uzer[4], but the influence of complex geometry for the structure performance is still not be solved. The regions of the bus with different shapes and sizes have been explored and optimization by Zhang, but the curved beam of the bus needs to be simplified to straight beam[5]. However, using existing method, the stress features of complex component cannot be fully extracted because of the continuous change of geometry. And the influence for stress features is also difficult to be acquired. For complex component, it is necessary to seek a more effective method to extract the stress feature.

Therefore, a new method of automatic extracting stress features based on stress survey is proposed in this paper. And then, the influence analysis for stress features based on Spearman rank correlation analysis is discussed.

2 An automatic mechanism for feature extraction and influence analysis

To explore the structural performance of complex component deeply, an automatic mechanism for feature extraction and influence analysis (FE&IA Mechanism) is constructed and shown in Fig.1. This mechanism includes three major modules, which are illustrated as follows.

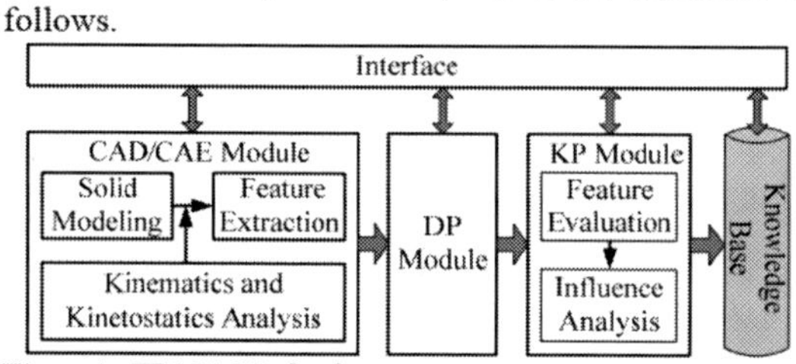

Figure1. FE&IA Mechanism

CAD/CAE module. This module is constructed to carry out the stress survey for model samples and extract the stress features from samples, which includes three sub-modules, i.e., the solid modeling sub-module based on Pro/Engineering, the kinematics and kinetostatics analysis sub-module, as well as the feature extraction sub-module based on ANSYS. In this module, the efficiency of modeling and analysis can be improved greatly by connecting VC++ with Pro/Engineering and ANSYS.

Data processing (DP) module. In this module, the data of stress survey for all samples can be handled, and the statistics analysis of danger situation for stress features in all samples can be realized.

Knowledge processing (KP) module. KP module is constructed to evaluate the stress features of complex

* Corresponding author: hhy810608@163.com

component and acquire the absolute influence of structural parameters for every stress feature based on Monte Carlo simulation and Spearman rank correlation analysis method.

These three modules can be operated and contacted each other by the interface, and the results of feature extraction and influence analysis can be saved in knowledge base.

3 The new strategy of exploring the structural performance for complex component

3.1. Feature extraction for complex component

The complexity of stress distribution is determined by the complexity of component. Feature extraction for the structural performance of complex component has always been a bottleneck problem. In order to explore the stress distribution and extract stress features for the complex component, a new method based on stress survey is proposed, which can be realized automatically by the CAD/CAE module.

Some model samples with different values of structural parameters should be constructed and analyzed. The main steps of feature extraction are shown in Fig.2 and illustrated as follows.

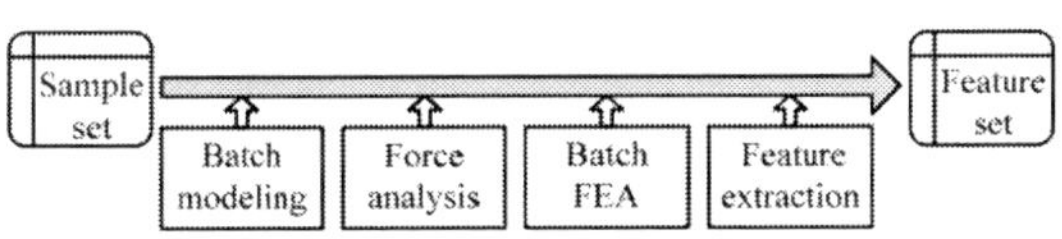

Figure 2. The main steps of feature extraction

In the beginning, some samples are taken by the method of Latin Hypercube Sampling. The value ranges of parameters are determined according to the series components.

• Step1: Batch modeling. The batch modeling of component is realized by the solid modeling sub-module. All the solid model samples must be evaluated and some of geometric deformity models should be taken out.

• Step2: Force analysis. The forces of every model are calculated by the kinematics and kinetostatics analysis sub-module, which will be utilized in the loading of finite element analysis(FEA).

• Step3: Batch FEA. Stress survey based on model samples are carried out by the batch FEA, which are realized by the feature extraction sub-module.

• Step4: Feature extraction. In the post-processing of FEA, each model is divided into several sub-regions and the maximum stress in every sub-region is extracted, which is taken as the feature of stress distribution and called characteristic stress (CS).

In the end, the feature set, i.e., the set saved the CS of all samples, is transmitted to DP module.

3.2 Influence analysis for complex component

The influence analysis for complex component is realized by DP module and KP module. The main steps of influence analysis are shown in Fig.3 and illustrated as follows.

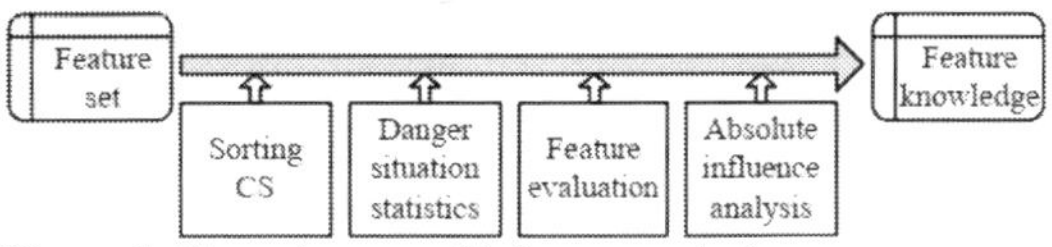

Figure 3. The main steps of influence analysis

•Step1: Sorting CS. After the feature extraction, the feature set can be acquired. For every sample, the CS of all sub-regions should be sorted firstly. According to the result of sorting, the danger regions in every sample can be identified.

•Step2: Danger situation statistics. The danger situation in different sub-region may be different. Therefore, it is necessary to count the times of occurring danger for every sub-region in all samples.

•Step3: Feature evaluation. According to the danger situation, the danger grade of all sub-regions can be identified. And then, the sub-regions can be divided into several classes. The danger sub-region, in which the number of times occurred danger is much higher than other regions, is called feature regions.

• Step4: Absolute influence analysis. In order to extract feature knowledge, the absolute influence analysis for the CS of feature regions is carried out. The Monte Carlo simulation technique and the method based on Spearman rank correlation analysis are utilized in this paper. The Spearman rank correlation models between structure parameters and CS are constructed. The Spearman correlation coefficients are taken as the absolute influence for CS. After that, by using rule reasoning, the feature knowledge is acquired.

4 Application

The gooseneck type boom of excavator is taken as an example, in which the structural performance under four working conditions[6] is explored in this paper.

4.1. Feature extraction for excavator boom

Firstly, 500 boom samples are generated by the method of Latin Hypercube Sampling, and their solid models are built by the solid modeling sub-module. After evaluating geometry, only 247 boom models are taken as the qualified samples of feature extraction. Secondly, the force analysis for every boom model under four working conditions is achieved, and then the stress survey is carried out.

This stress survey is a comprehensive survey for stress distribution. In order to improve the commonality of the feature extraction sub-module, the sub-regions are acquired by segmenting the line of BC and the line of BF equally. In this example, both the line of BC and the line of BF are divided into ten equal parts, and the segmentation plane is perpendicular to BC or BF. According to this division method, a boom model is divided into 23 sub-regions. The division of one of boom sample is shown in Fig.4 as an example. By using the method of feature extraction illustrated in section 3.1, the CS of every sub-region for all boom models are extracted and the feature set is generated.

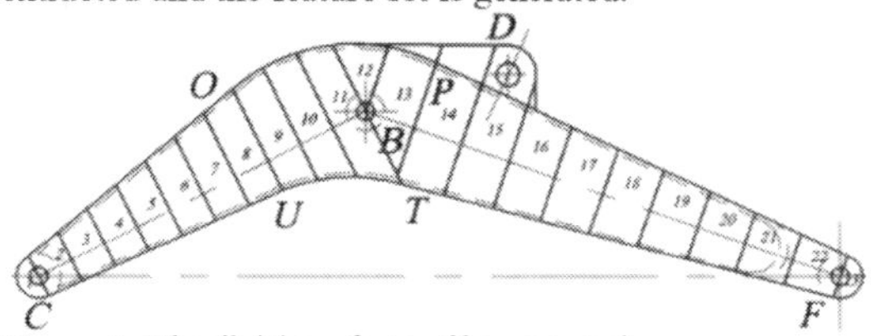

Figure 4. The division of one of boom sample

4.2 Influence analysis for excavator boom

After acquiring the feature set, for every sample, the CS of all sub-regions are sorted. And then, the evaluation of danger situation for sub-regions is realized. Danger grades are identified according to the sorting of CS in all sub-regions, i.e., from danger grade 1 to danger grade 23. Danger grade 1 means that the maximum stress in sub-region is the global maximum value among 23 sub-regions. Similarly, danger grade 2 means that the maximum stress in sub-region is the second maximum value among 23 sub-regions, and so on. In this example, the evaluation of danger situation is carried out by counting the times from danger grade 1 to danger grade 5. The total times of occurring danger for every sub-region under four working conditions is shown in Fig.5.

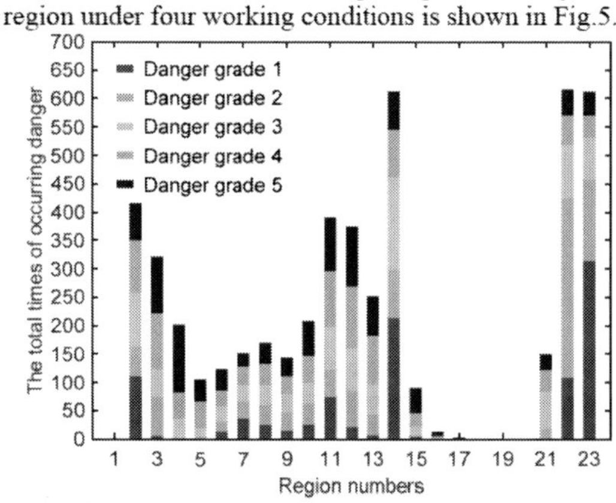

Figure 5. The total times of occurring danger

By this evaluation, the sub-regions can be divided into four classes in this paper, which is shown in Table 1. The four classes are the frequent danger regions, the potential danger regions, the occasional danger regions and the safe regions. Table 1 shows that four regions, i.e., No.2, No.14, No.22 and No.23 region, are the frequent danger regions. The total times of occurring danger in these four regions are significantly higher than other regions. The times of occuring danger in potential danger regions are only second to those in frequent danger regions. Fig.5 also shows that among all potential danger regions, the times of occuring danger in No.11 region is higher than other potential danger regions. In this example, the frequent danger regions and the potential danger regions are taken as feature regions of boom. As the occasional danger regions and the safe regions rarely occur danger, they will no longer be further researched in the following.

Table 1. The classification of sub-regions

Class No.	Name	Region No.
Class 1	Frequent danger	2,14,22,23
Class 2	Potential danger	6,7,8,9,10,11,12
Class 3	Occasional danger	3,13,15
Class 4	Safe	1,4,5,16,17,18,19,20,21

Because the values of maximum stress in the frequent danger regions are the most needed to be controlled, the absolute influence of structural parameters for CS in these four regions are analyzed firstly. And the results are shown in the following four figures, i.e., from Fig.6 to Fig.9. The parameters model of boom is shown in Fig.10 and the explanation of structural parameters is shown in Table 2.

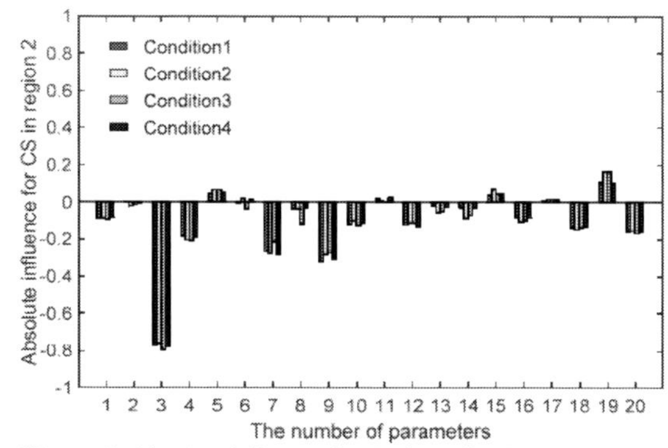

Figure 6. Absolute influence for CS in region 2

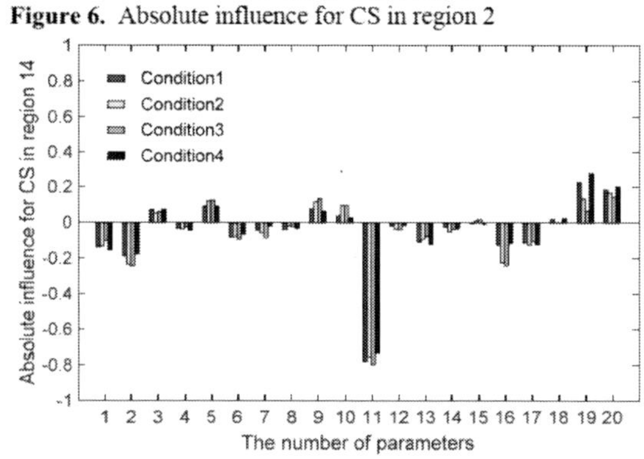

Figure 7. Absolute influence for CS in region 14

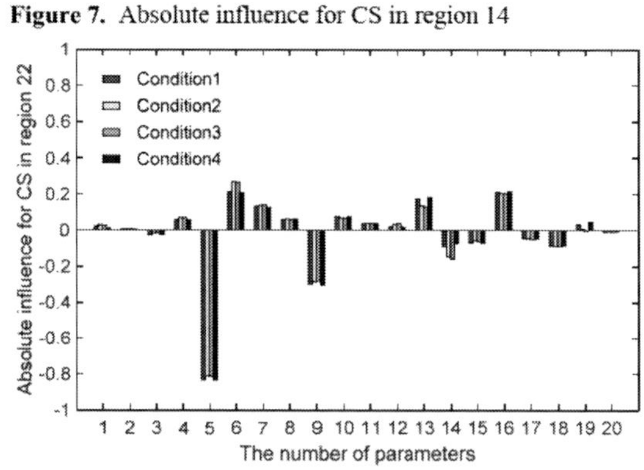

Figure 8. Absolute influence for CS in region 22

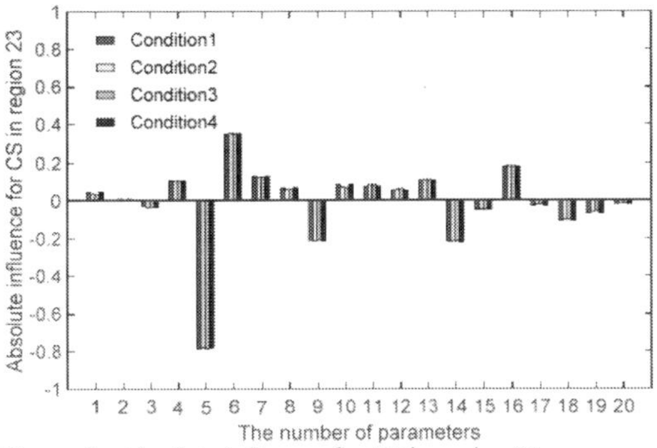

Figure 9. Absolute influence for CS in region 23

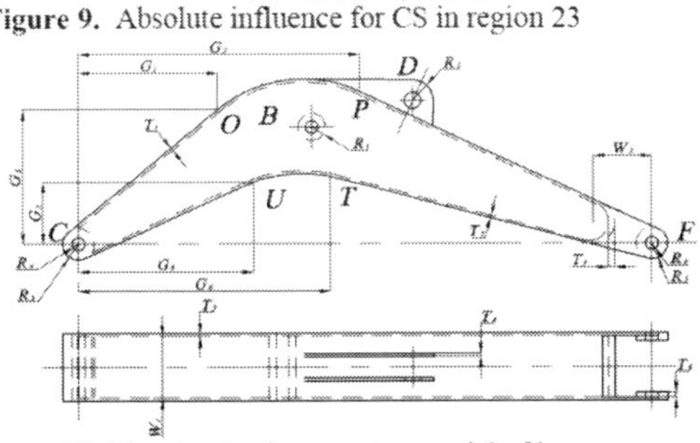

Figure 10. The structural parameters model of boom

Table 2. The explanation of structural parameters

NO.	Name	Implication	Type
1	R_1	External dimeter of reamed hole B	local
2	R_2	External dimeter of reamed hole D	local
3	R_3	External dimeter of reamed hole C	local
4	R_4	Internal dimeter of reamed hole C	local
5	R_5	External dimeter of reamed hole F	local
6	R_6	Internal dimeter of reamed hole F	local
7	T_1	Thickness of upper cover plate	global
8	T_2	Thickness of web plate	global
9	T_3	Thickness of side plate	global
10	W_1	The distance of side plates	global
11	T_4	Thickness of ear plate	local
12	T_5	Thickness of bending plate	local
13	W_2	The distance between the center of bending plate and hole F	local
14	T_6	Thickness of lug boss	local
15	G_1	Lateral distance between O and C	geometry
16	G_2	Portrait distance between O and C	geometry
17	G_3	Lateral distance between P and C	geometry
18	G_4	Lateral distance between U and C	geometry
19	G_5	Portrait distance between U and C	geometry
20	G_6	Lateral distance between T and C	geometry

Comparing all of frequent danger regions, the significantly sensitive parameters for CS in No.2, No.14 and No.22 region are different. But the sensitivity of each structural parameter for CS in No.22 region and No.23 region are similar. Observing the Fig.4 and the results of FEA, it can be found that the danger often occurs in the lug boss, which causes the similar absolute influence for CS in No.22 region and No.23 region. Comparing the absolute influence of each structural parameter for the same CS in Fig.6 to Fig.9, it is shown that the absolute influence of most paremeters under different conditions are similarly. Only a few parameters, such as G_2, G_5, the absolute influence under different conditions are relatively different.

According to the absolute influence under four working conditions, the classification of structural parameters for frequent danger regions are realized. For each region, all the structural parameters are divided into four types, i.e., significant sensitive, high sensitive, medium sensitive and low sensitive. The number of parameters in first three classes are shown in Table 3, and the others belong to the low sensitive class are not listed in this paper.

Table 3. The classification of structural parameters for frequent danger regions

	Significant sensitive	High sensitive	Medium sensitive
No.2	3	7,9	4,19,20
No.14	11	2,16,19	20
No.22	5	6,9,16	7,13,14
No.23	5	6,9,14,16	7

Table 3 shows that No.3 parameter, i.e., the external dimeter of reamed hole C, is the significant sensitive parameter for No.2 region ; No.11 parameter, i.e., the thickness of ear plate, is the significant sensitive parameter for No. 14 region ; and No.5 parameter, i.e., the external dimeter of reamed hole F, is the significant sensitive parameter for No. 22 and No. 23 region. It is indicated that these four regions are the local regions of occurring danger and easily to be controlled, because the significant sensitive parameters for these regions are local parameters of boom.

Table 1 also shows that seven regions, i.e., from No.6 region to No.12 region, are the potential danger regions. Because of the similarly among different conditions, the absolute influence of structural parameters for CS in No.6 region to No.9 region combining four working conditions is shown in Fig.11.

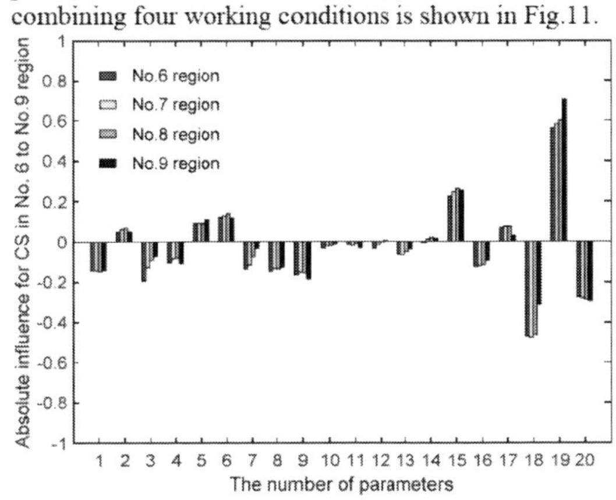

Figure 11. Absolute influence for CS in No.6 region to No.9 region conbining four working conditions

In Fig.11, for most of parameters, the absolute influence of the same parameter for the CS in No.6 to No.9 region are similar. Only a few parameters, such as R_3, G_4 and G_5, the absolute influence are relatively different.

The absolute influence of structural parameters for CS in No.10 to No.12 region combining four working conditions is shown in Fig.12. For many parameters, the absolute influence for the CS in these three regions are different.

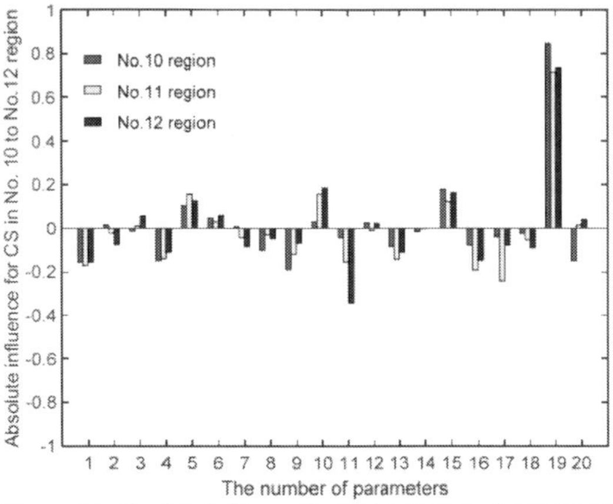

Figure 12. Absolute influence for CS in No.10 region to No.12 region conbining four working conditions

According to Fig.11 and Fig.12, the classification of structural parameters for the regions from No.6 to No.12 is acquired. The number of parameters in three classes are shown in Table 4 and the others belong to the low sensititive class are not listed in this paper.

Table 4. The classification of structural parameters for No.6 to No.12 region

	Significant sensitive	High sensitive	Medium sensitive
No.6	18,19	15, 20	3,9
No.7	18,19	15, 20	9
No.8	18,19	15, 20	9
No.9	18,19	15, 20	9
No.10	19	9	1,15,20
No.11	19	16,17	1,10
No.12	11,19	10	1,15,16

Table 4 shows that No.18 and No.19 parameters are the significant sensitive parameters for CS in No. 6 to No. 9 region; No.15 and No.20 parameters are the high sensitive parameters in No. 6 to No. 9 region. It is indicated that these four regions are the global regions of occurring danger and can be controlled by geometry parameters. Table 4 also shows that No.19 parameter is the significant sensitive parameters for CS in No. 10 to No. 12 region. It means that these three regions also can be controlled by geometry parameters. But the influence of global parameters for there three regions is higher than that for No. 6 to No. 9 regions.

5 Conclusions

In this paper, a new method of feature extraction and influence analysis for complex component is proposed. By verifying the proposed method in the example of excavator boom, the conclusions are extracted as follows:

(1) The danger regions of complex component may be changed with the change of the values for structural parameters and the working conditions, but it is always concentrated in the certain regions more frequently. It is feasible to take the CS in these regions as the feature of structural performance for complex component.

(2)For each feature region, the influence of every structural paremeter for CS may be different, but at least one significant sensitive or high sensitive parameter can be identified.

(3)According the absolute influence of structural parameters for CS as well as the classification of structural parameters, the controlling parameters for feature regions can be identified. All these result can be utilized to guide the structural optimization of complex component.

Acknowledgements

This work is supported by the Education Scientific Research Project for the Young Teacher of Fujian Province, China (Grant No. JAT170374), and the Scientific Research Foundation Project of Fujian University of Technology, China (Grant No. GY-Z14075).

References

1. G.H. Shi, Y. Chen, Y.Z. Yang, et al. Journal of Mechanical Engineering. **48**,8,(2012),110-114.
2. J. París, F. Navarrina, I. Colominas, et al. Computer Methods in Applied Mechanics and Engineering. **199**, (2010), 2110-2122.
3. C.Y. Li, W.C.Wang, Y.Ma. China Mechanical Engineering. **22**, 4,(2011), 397-402.
4. C. C. Uzer. *Shape optimization of an excavator boom by using genetic algorithm.* (Middle East Technical University, Turkey, 2008).
5. D.Q. Zhang, T.X. Zhang, G.S. Zhang, et al. Journal of Mechanical Strength, **33**,6, (2011), 913-920.
6. Tianjin Mechanical Engineering Research Institute. *GB9141-88*, Standardization Administration of the People's Republic of China, (1988).

Analysis and Experimental Study on Perforation Abnormity of Perforating Gun

Yu Han [1,*], *Ma* Tao [2]

[1]China Oilfield Services Limited (COSL) Tianjin Branch, No.1581, Hai Chuan Road, Tanggu Marine Hi-tech Development Area, Binhai New District, Tianjin, China
[2]Xi'an Wuhua Juneng Blasting Equipment Co., Ltd,20th Zhuque Avenue, Xi'an, China

Abstract. In the oilfield development, the perforation process is known as the "goal", therefore, the perforating operation is the most important process in the well completion process. However, in the perforating process, the perforating gun is working in the down hole; the operator cannot observe on the ground. And the perforating gun itself, and the perforating gun assemble process produced some unpredictable situation, such as broken blasting, gun broken, etc. These problems will influent the well completion process; even make the whole well scrapped. In this thesis, we will analysis one unpredictable situation in Bohai oilfield, through a variety of experiments to find and analyze the cause of the abnormal cases, to avoid the same problems in the future work.

1 INTRODUCTION

The Bohai oilfield in recent years due to anti-sand and fracturing, used 60 shots per meter (18SPF), this type of gun has the advantages of ultra-high density and big hole size, can satisfy the requirements of perforation process in fracture sandstone heavy oil well completion process, to effectively reduce the pressure near well, and production differential pressure in the same production quantity, to achieve stable sand control purpose. But in the process, due to the increased quantity of shaped charge, in the moment of perforating, the gun body also will have a lot of pressure, if in the assemble process have a problem, is likely to cause serious accidents of perforating gun[1][2]. This well the first TCP process on February 8th, 2014, after the process of running gun, depth correction and pressure firing, when lifting the gun, the operator found gun seized, after 180Klb, the gun released; on the ground, the operator found the 9# perforating gun occurred abnormal phenomenon (See Fig1,2) the rest of perforating gun perforated normal[3].

2 Figures and tables

2.1 Quality review

Through the quality review of detonating cord in the process on this well (Type: 80RDXXHV, Lots No.: M1303), results: the material (powder), manufacture, testing and acceptance process were not found abnormal, the production process is under the quality control.
The Lot M1303 80RDXXHV detonate cord produced in December of 2013, this lot produced 47110.5 meters, at 7th of March of 2014, delivered 39072.5 meters, 20 boxes deliver to COSL. Others delivered to CPL Tuha, Bohai drilling, Shengli Logging, SPT Chongqing, GWDC Logging, etc. (see Table I). All users have not feedback abnormal information.

Fig1 9# perforating gun lifting out of wellbore

Fig2 entry hole of upper and lower side on perforating gun

* Corresponding author: mingmingzj418@163.com

 DOI: 10.1051/matecconf/201713900178

Table I Lot M1303 80RDXXHV delivery report

Product	*Type*	*Lots No.*	*Quantity*	*Date*	*Delivered to*	*Delivered qty*
Detonating cord	80RDXXHV	M1303	47110.5m	201312	COSL	3060m
					CPLTuha	19890m
					Bohai Drilling	2448m
					Shengli Logging	3000m
					SPT Chongqing	7767.5m
					GWDC Logging	2907m

2.2 Establishment of fault tree

Using failure mode analysis method, to analyze the possible cause of abnormal perforation of 9# perforating gun in this well. (see Fig 3).
The failure mode analysis (FAT) is using to find all causes of system failure, and combine the environments and state, after excluding the impossible cause[4], to diagnose the final cause of the system failure.

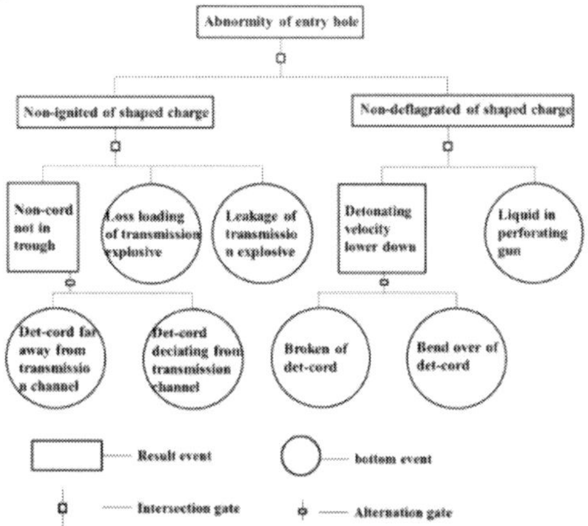

Fig 3 FAT of abnormity of entry hole

2.3 Colour illustrations

You are free to use colour illustrations for the online version of the proceedings but any print version will be printed in black and white unless special arrangements have been made with the conference organiser. Please check with the conference organiser whether or not this is the case. If the print version will be black and white only, you should check your figure captions carefully and remove any reference to colour in the illustration and text. In addition, some colour figures will degrade or suffer loss of information when converted to black and white, and this should be taken into account when preparing them.

3 EQUIPMENT ANALYSIS

3.1 Shaped Charge

If the shaped charge during the production process occurs: Booster missing load and booster leakage and other quality problems, caused reduce the sensitivity of shaped charge, when the shaped charge igniting caused detonating or burning, and shaped charge cannot shaped normal, resulting the abnormal perforation in perforating gun. The production of shaped charge is production line, if in production occurs missing load or leakage, the performance is discrete distribution. However, in the process[5], only 9# perforating gun occurs abnormal perforation, and in the 9# perforating gun occurs continuous 2.2 meters abnormal perforations, which can eliminate the shaped charge cause abnormal perforation.

3.2 Detonating cord

If the density of detonating cord is not enough, it may cause the lateral output shock wave to be insufficient to detonate the shaped charge, cause the shaped charge detonating or burning,The detonating cord is made up by powder, high strength Kevlar weaving layer, and nylon skin (see Fig 4), the main parameter affecting by the powder density.

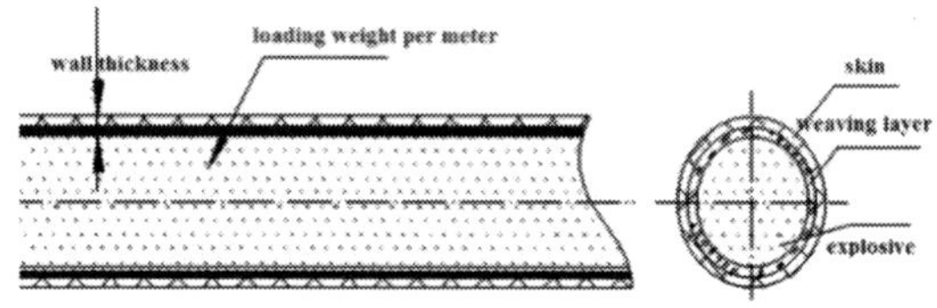

Fig4 Structure of detonating cord

The detonating cord production is product line, every mixing, knitting, extrusion and rolling will produce 800 meters, and because of ultra-high velocity detonating cord must be rolled to increase the density of powder. (see Fig 5). Now COSL using the detonating cord was produced by the product line which imported from Switzerland, this product line is automatic and can detect the density of detonating cord all the time, if occurs density abnormal in the product process, the equipment will alarm and stop automatically.

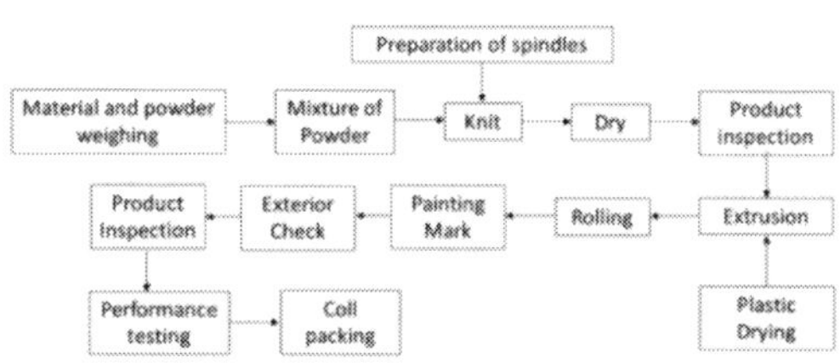

Fig5 manufacturing process of detonating cord

From the production process and acceptance of the detonating cord, no abnormality is found, and each detonating cord needs to be extracted to be test. The Lot M1303 80RDXXHV detonating cord accepted 128 units, the exterior check, outside diameter, density detection (design requirements: ≥ 17g/m, the measure value: 18.36~22.76g/m), detonating velocity test (design requirements: ≥ 7800m/s, the measure value: 7831~8146m/s,) etc., and all satisfy the design requirements.

In Summary: Detonating cord density cause this problem in the operation is low.

3.3 Some Common Mistakes

3.3.1 Detonating cord damage

During the detonating cord assemble, if the detonating cord was damage, cause the density of powder reduce, the lateral output shock wave is insufficient to detonate the shaped charge, cause shaped charge detonating or burning[6].

The detonating is protected by package during the delivery, is not easy to be damaged. In the loading gun process, is easy to be squeezed and bent by shaped charge which leads the change the density of the detonating cord. In the design of 60shots per meter perforator, the detonating cord deviates from the centerline of the perforation gun 2.3mm, the distance of two adjacent shaped charges is 16.7mm, the detonating cord is repeated bending (60 times bending per meter)[7]. Under this condition, the detonating cord may be damaged and in extreme cases may lead to decrease the detonating velocity cause the detonation of detonating cord, lead the problem in this process. (see Fig 6,7)

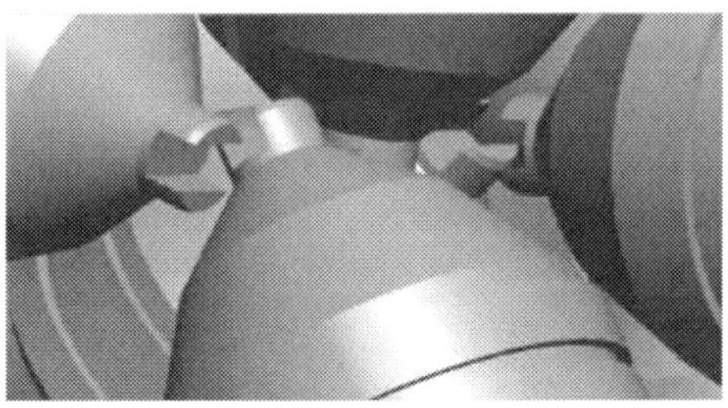

Fig 6 60 shot/m 127shaped charge loading simulation

Fig 7 60 shot/m 127shaped charge loading in reality

3.3.2 Detonating cord assemble

During the assemble of detonating cord, such as: the detonating cord is not normally installed in the slot of the shaped charge, the bottom of shaped charge is on the detonating cord, or detonating cord is bypassed from the side of shaped charge[8]. May caused the detonation energy of detonating cord is lower, caused shaped charge detonating or burning.

3.4 Perforating gun water seepage

The liquid of well goes into the gun, it has a very big influence on the normal operation of the explosive device. Although in this process, the downhole pressures is not very high, but no eliminate the possibility of the seal is not good enough to lead the gun water seepage.

4.TEST VERIFICATION

According to the above analysis, the tests of the detonating cord and shaped charge under different conditions and verify.

4.1The experiments of broke outside and velocity in bend condition

4.1.1 The experiment method

Harm the outside plastic layer of the detonating cord, and bend in the loading gun, at the same time of detonating, to test the velocity on input and output, to compare (see Fig 8 to 10)

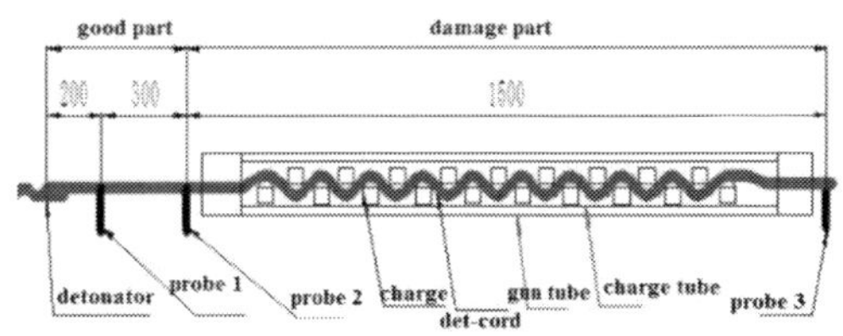

Fig 8 Test plan of broken skin and bend speed of detonating cord

Fig 9 Status after broken

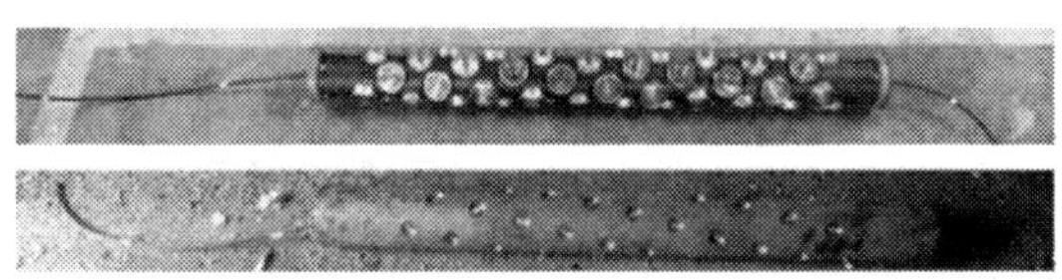

Fig 10 Status after loading

4.1.2 The experiment result

The detonating cord complete part, the velocity is 7913m/s, the harm outside and bend detonating cord, the velocity is 8215m/s.

4.1.3The experiment conclusion

DOI: 10.1051/matecconf/201713900178

The detonating velocity of the 2 states is all in the high detonation velocity range, and it is not necessary to have problems in the current configuration of equipment, and also confirms the condition of use.

4.2 The detonating cord away and deviated from blasting hole test

4.2.1 The experiment method

In the assemble of shaped charge and detonating cord, according the normal installation, the bottom of shaped charge against the detonating cord, and detonating cord deviated the shaped charge, and record the each blind hole position[9], observed the blind hole after detonating. The test quantity is 5. (see Fig 11)

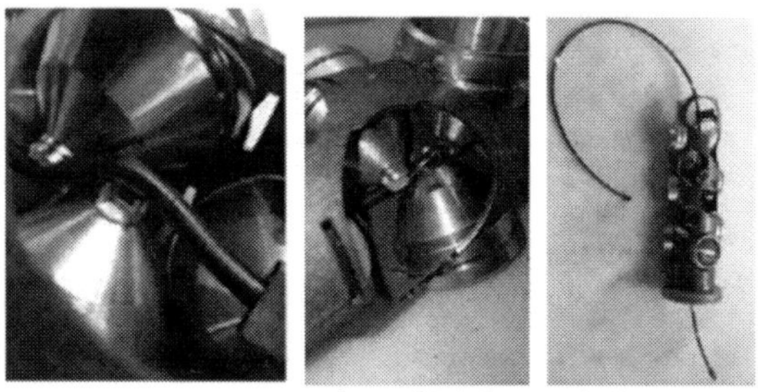

bottom of shaped charge against the detonating cord — detonating cord deviated the shaped charge — Perforating gun after load

Fig 11 Test Course

4.2.2 The experiment result

The normal assemble and shaped charge bottom against the detonating cord the performance is normal to perforating, the detonating cord deviated the shaped charge is not perforating. (see Fig 12)

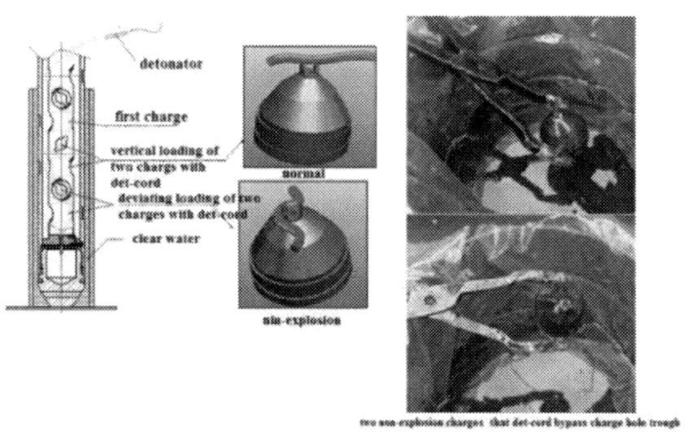

Fig 12 Test assembling and result

4.2.3 The experiment conclusion

When the detonating cord is slightly away from the detonation hole, the shaped charge can be detonated normal. No eliminate the possibility that the detonating cord deviated the detonation hole and caused shaped charge is not detonated.

4.3 Simulation test of perforating gun water seepage

4.3.1 The experiment method

According the normal loading gun condition to load 5 shaped charge, and put the 3 shaped charge of bottom side into the water after 3 hours, and detonate the perforating gun, to observe the gun body. (See Fig 13 to15)

Fig 13 three charges at bottom end inwater in 3h

Fig 14 Test status

4.3.2 Test Result

perforating charges penetrated normally in air. Three perforating charges penetrated abnormally in water.(See Fig 16 to19).

4.3.3 The experiment result

2 shaped charge in the air perforating normal, 3 shaped charge in the water perforating normal.

Fig 15 Status after test

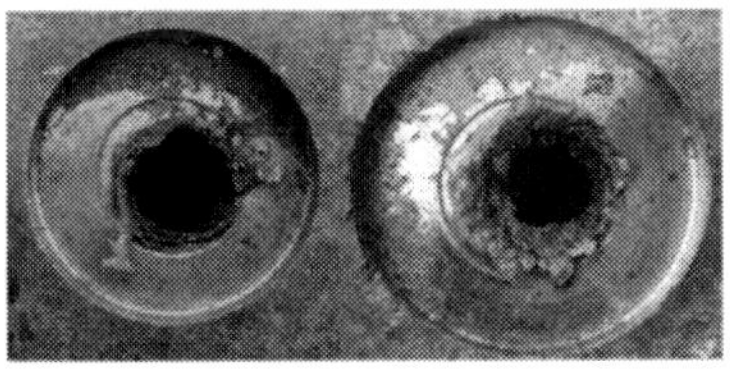

Fig 16 2 charges' normal entry hole

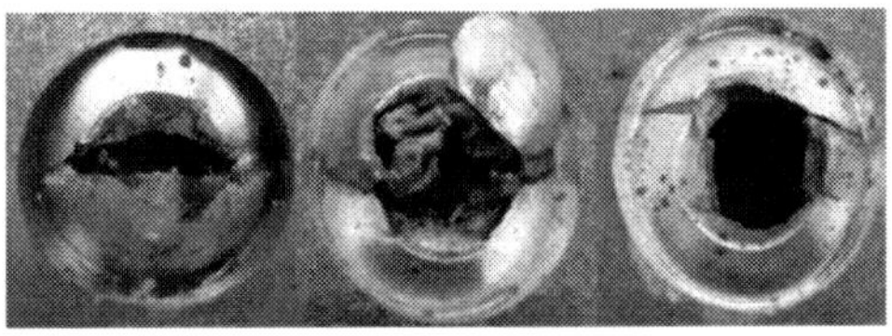

Fig 17 4 charges' un-normal entry hole

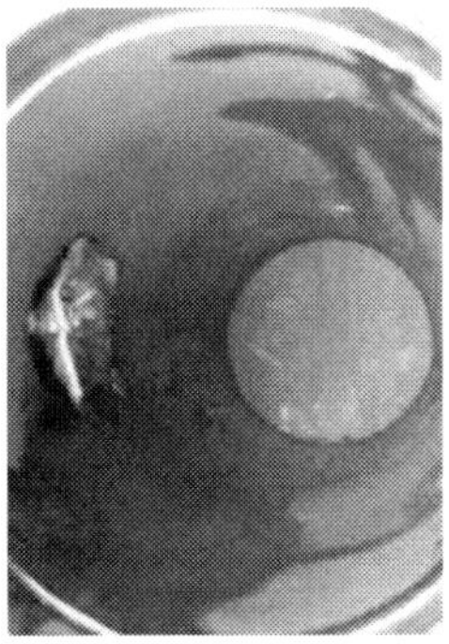

Fig 18 charge case being close to entry channel in un-normal condition

4.3.4 The experiment conclusion

If the perforating gun is water seepage, the shaped charge soaked in water after certain time, the water will goes into the powder, and shaped charge performance occurs abnormal phenomenon.

Fig 19 charge trough against to det-cord

4.4 Initiation test of whole perforating gun without target

4.4.1 The experiment method

1. When the gun loaded, the detonating cord is not processed and keeps the original condition, corresponding to the 27 shaped charges of top side
2. In the bottom side, the 26 shaped charges corresponding detonating cord, using cutter to break the outside skin, to expose the knitting layer, however ensure the power is not exposed, the damage is quite the same as the most serious damage when the gun is loading in the oilfield.
3. The rest of detonating cord keeps complete.
4. To select the 1st, 2nd, 19th, 20th, 38th, 39th shaped charge form the igniting side, to vertical assemble the detonating cord and shaped charge[10].

4.4.2 The experiment result

All perforations are normal. (See Fig 20 to 23)

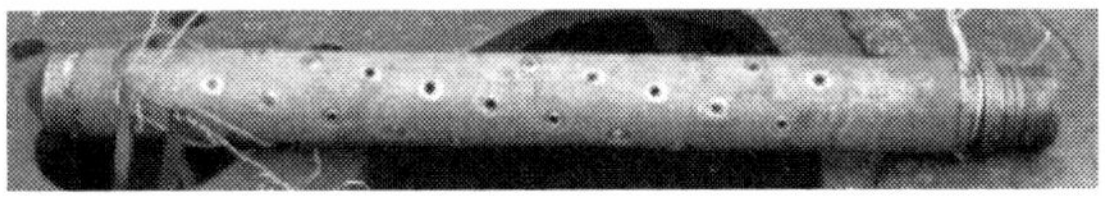

Fig 20 gun after perforating

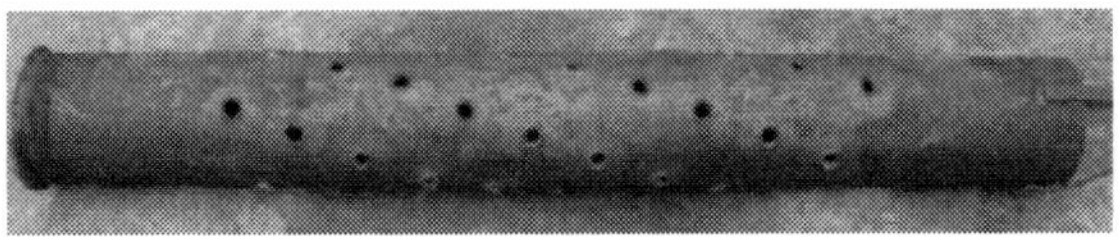

Fig 21 casing after perforating

Fig 22 entry hole on gun tube after test

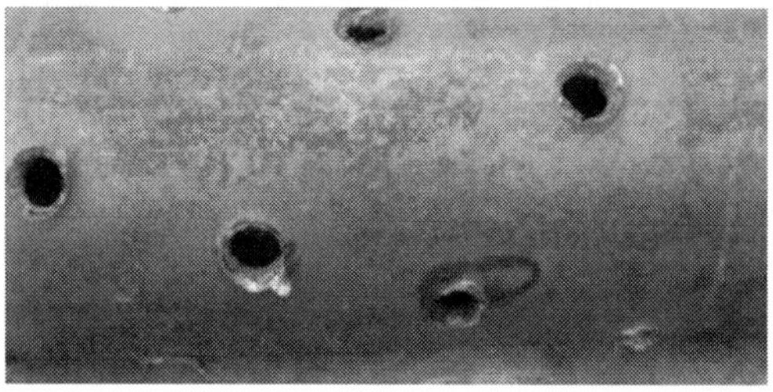

Fig 23 entry hole on casing after test

4.4.3The experiment conclusion

The experiment did not simulate process problems.

5 CONCLUSIONS

1. According the fault tree analysis, the possible cause of gun bulge and no perforation mainly is four aspects: the quality of product, the bend damage of detonating cord, shaped charge abnormal load, and well liquid influence.
2. After quality review, we can eliminate the quality of product caused the problems.
3. Test the detonating cord under damage and bend condition, the explosive velocity is not decreased, although the test quantity is not big, but the result this reason caused the problem is less.
4. The abnormal installation of detonating cord and shaped charge could be a great influence the reliability of shaped charge.
5. Perforating the shaped charge that soaked in water, could cause gun bulged, not perforation, is similar to this process, cannot eliminate and also cannot be fully determined the water seepage cause the problem in this process.
6. Through statistical analysis, since 2011, the 60 shot per meter 127 perforator the operation quantity is about 4000 meters, almost 200 thousands shaped charges, this problem is the first found, should be caused by accidental factors.

References

[1] Song SJ,Chen XB,Niu XX & Yang YX,High Efficiency Composite Perforating Technology, Explosion and Shock Waves. 21,8-9 (2001).

[2] Hu MM, Zhu X,Present situation and trend of compound perforating technology, Explosion Materials. 31 (2017) 30-31.

[3] S.N.Domenico,HARNESSING THE GHOST, Abstracts of Beijing(89) International Symposium on Exploration Geophysics,.01 (1989) 20-23.

[4] Zhao ZX, Formulation development and characterization of cellulose acetate nitrate based propellants for improved insensitive munitions properties, Defence Technology. 02 (2014) 08-11.

[5] Cui JJ,Wang DY,Shi QQ, Structural Design Method on Elastic-plastic of Oil Perforating Guns, Proceedings of 2010 International Conference on Computer,Mechatronics, Control and Electronic Engineering (CMCE 2010) Volume 3. 08 (2010) 12-15.

[6] Zhang HQ;Sun XP;Wang YZ;Yin JL:Wang CY Dynamic Numerical Simulation of Detonation Wave of Perforating Bullet and Research of Its Measuring Method, Proceedings of 2010 International Conference on Computer,Mechatronics, Control and Electronic Engineering (CMCE 2010) Volume 2. 08 (2010) 28-32.

[7] T ManningJeffrey WK KlingamanViral PE RozumovJohn ,BM Wang, YS Patel, Enhanced propellant performance via environmentally friendly curable surface coating. Defence Technology. 06 (2017) 18-22.

[8] Wang,ZG, Application of Separable Perforating-Fracturing Combined Bullet to the Test of Oil and Gas Wells, Natural Gas Industry. 03 (1996) 01-05.

[9] Yang LL, Song DL, Shi X Study on Titanium Alloy TC4 Ballistic Penetration Resistance Part I:Ballistic Impact Tests, Chinese Journal of Aeronautics. 06 (2012) 12-16.

[10] Wu BY ; PU YouZhe ; YU XiaoYang ; SU WanHua ; Experimental Study on the Feasibility of Methane Drainage in Coal Seams with Compound Technique of Perforating and Fracturing, Engineering Sciences. 03 (2007) 09-12.

MATEC Web of Conferences 139, 00179 (2017) DOI: 10.1051/matecconf/201713900179
ICMITE 2017

Research on Whipstock Packer Oriented Perforating Technology

Xing Rendong[1,*], *Wang* Feng[2]

[1]China Oilfield Services Limited (COSL) Tianjin Branch,No.1581, Hai Chuan Road, Tanggu Marine Hi-tech Development Area, Binhai New District, Tianjin, China
[2]Xi'an Wuhua Juneng Blasting Equipment Co., Ltd 20th Zhuque Avenue, Xi'an, China

Abstract. Multilateral well drilling and completion technology reduces well development cost and increases effectively the benefit and productivity of oil & gas well, which is widely used and expanded in recent years at home and abroad. Whipstock packer oriented perforating technology, a critical step in multilateral well drilling and completion, is for perforating whipstock packer and communicating lower branch wellbore. This technology meets the higher requirement of orientation and accuracy of penetration, perforator being able to perforate whipstock packer but not perforate outer casing of whipstock, and also meets complicated operation process.

1 Introduction

As for petroleum exploitation, multilateral well technology contributes to set a reasonable development program, so as to develop multilateral well effectively with lower cost. Considering from drilling, all branch wells have a common wellbore and upper well section, so this technology will greatly reduce drilling cost and land occupation, and also contribute to environment protection. Besides reducing cost, the multilateral well technology can also increase production rate and ultimate recovery, improve oil drainage mode and increase reserves. It has a great development prospect. In multilateral well completion technology, whipstock oriented perforating has a particularly effect on communicating upper and lower branch wellbore. That is, lower branch wellbore's drilling and completion proceeds firstly, after that lowering down the whipstock packer at the upper branch wellbore depth, then proceeding upper branch wellbore's drilling and completion; after that lowering down perforator and perforating upon whipstock packer, communicating lower branch wellbore and lowering down production tube, so as to accomplish joint exploitation of multilateral well[1].

Whipstock packer oriented perforating is a synthesis technology, with a high technical difficulty and strict process requirement, which designs every technical parameters based on well condition and determines perforating parameters by ground simulation test. This technology meets the higher requirement of orientation, accuracy of penetration and strictness of entry hole, and also meets complicated operation process[2].

Use a two-column format, and set the spacing between the columns at 8 mm. Do not add any page numbers.

2 Wihpstock Structure and Requirement of Perforating Parameter

2.1 A. Multilateral Well Completion Technology

Multilateral well is that drilling two or more branch wellbore (second stage wellbore) in an original wellbore. The original wellbore could be vertical well, oriented well or horizontal well. Multilateral well has various types, such as overlay type, reversed type, etc., which can be optimized selected according to well reservoir.

In 1997, technical experts on multilateral well from the world major oil company and service company had a communication. They appointed a multilateral well system, that is, TAML (Technology Advancement Multi Laterals) stages. TAML evaluates the technology and stages of multilateral well based on its three features: Connectivity, Isolation and Accessibility. Completions of multilateral well completion is divided into 1～6S as follows.

(1) Grade 1 completion: Original wellbore and branch wellbores are all uncased. During well completion, different reservoir will not be separated and pressure difference among reservoir will not be disposed.

(2) Grade 2 completion: Original wellbore is cased and cemented; branch wellbores are uncased or only completed by sieve tube, not cemented.

(3) Grade 3 completion: Original wellbore and branch wellbores are all cased; original wellbore is cemented, but branch wellbores are not.

(4) Grade 4 completion: Original wellbore and branch wellbores are all cased and cemented at joint position.

(5) Grade 5 completion: Having the joint technology features of Grade 3 and Grade 4 branch wellbore, it also adds a completion machine for pressure seal at joint position of branch wellbore's screen pipe and original

* Corresponding author: mingmingzj418@163.com

 DOI: 10.1051/matecconf/201713900179

casing; original wellbore is totally cased and joint is hydraulic isolated.

(6) Grade 6 completion: Pressure at joint and wellbore pressure is the same which is an entirety pressure. It can be acquired by lowering down casing, does not need downhole completion tools.

(7) Grade 6S completion: Using downhole divider or downhole wellhead device; mainly using a downhole double-casing head as the wellhead to divide a large diameter original wellbore into two small equant branch wellbore[3].

Currently, whipstock packer oriented perforating technology is mainly used in Grade 5 completion. The wellbore structure of Grade 5 completion is showed as Fig 2. Taking Grade 5 completion multilateral well's whipstock as the example, this paper will discuss the perforating parameter and operation technology.

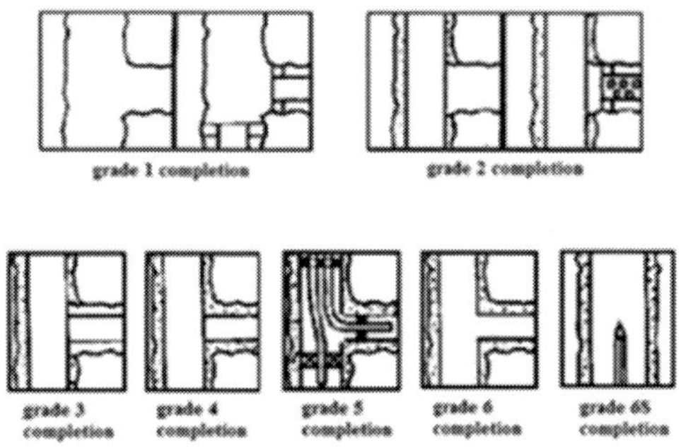

Fig1 Multilateral well TAML grade diagram

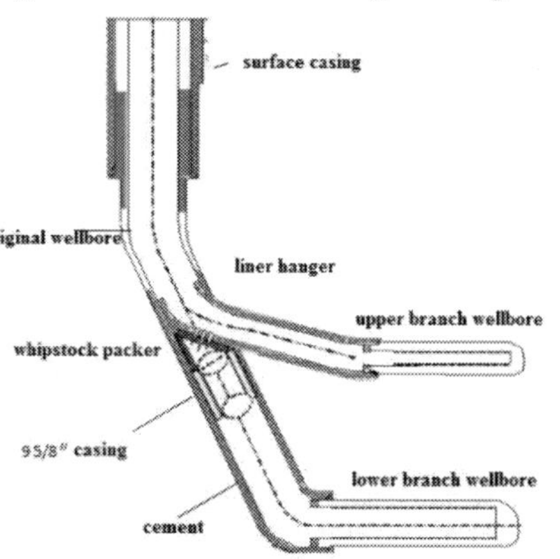

Fig2 Wellbore Structwre of the Grade5 Completiong

2.2. Structure of Whipstock Packer

In branch wellbore, the mostly used whipstock packer is hollow whipstock. Its structure is as Fig 3. This type of whipstock connects direction sub (UBHO Sub / MWD Sub) on ground, lowering down by drilling rig to preset depth. After adjusting orientation, operator pressurizes on ground and sets packer, meanwhile, releases loading tool and raising drilling rig upon whipstock. On account of the restriction of whipstock structure, the perforating area could only be between C and D. Radioactive isotope mark could be loaded at both ends of the area before perfortor running down, so as to accomplish an accurate positioning.

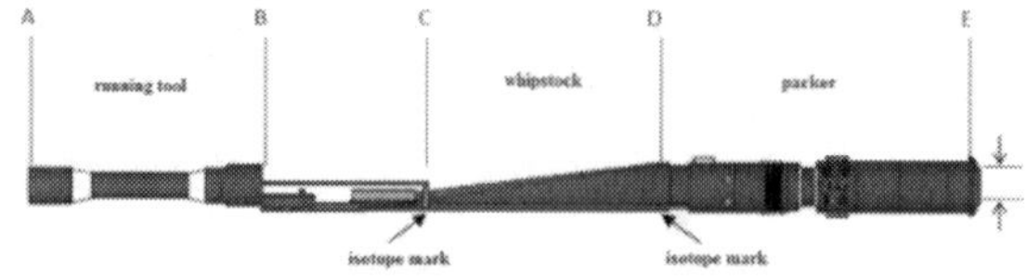

Fig3 Whipstock Packer

2.3 Requirement of Perforating Design

(1)To communicate branch wellbore and original wellbore, it is required that perforating gun, branch wellbore casing, cementing and whipstock cover board shall be perforated successively, and to prevent wellbore losing, perforating cannot damage the original wellbore casing of outer whipstock.

(2)The perforator shall be positioned accurately in whipstock perforating area and perforating azimuth error at tool surface shall be ±10°.

(3)To increase oil recovery, the total area of effective perforating entry hole on whipstock cover board shall be more than the area of minimum inner diameter of lowered down production tube.

(4)Because functional dielectric material is different and whipstock changes gradually along angular surface, penetration of perforating charge shall be considered and simulation test of practical well condition is needed.

3.Selections of Perforator And Simulation Test

Based on multilateral well' s structure and setting status of whipstock packer, it is required that perforator shall perforate 7″ branch wellbore casing, 9 5/8″ cover board of whipstock packer and cementing ring between them, but it shall not perforate 9 5/8″ casing of original wellbore, besides that, the effective perforating area (unit entry hole*hole quantity) on whipstock cover board shall be more than the area of minimum inner diameter (2.313″) of production tube in original wellbore.

3.1 Perforating Gun

Based on casing dimension of branch wellbore, the 5″ perforating gun is selected, 2.3m, with a 2m effective perforating length, phasing 0°, density 16 shot/m, charge distance 62.5mm, can be loaded with 32 pieces of perforating charge.

3.2 Perforating Charge

Type 127 Big Hole perforating charge is used, loading with normal temperature RDX and 39g explosive weight. Shape structure is as Fig 4 and loading structure is as Fig 5.

3.3 Simulation Test of Single Perforating Charge and Result

Based on practical well condition, the simulation test device for single perforating charge is designed as Fig 6. Targets for perforating successively are scallop thickness of perforating gun, water in the gap between perforating gun and casing, 7″ casing thickness, cement and whipstock cover board and whipstock back board. Therein, the gap of whipstock cover board and side wall is between 0.769″ (19.53mm) and 4.26″ (108.20mm), and the thickness of whipstock back board is 1″ (25.4mm).

(1)Casing gap (gap spacer 1): simulating liquid thickness between scallop of perforating gun and 7″ casing, setting it to be 6.5mm.

(2)ID of whipstock (gap spacer 2): simulating liquid thickness between scallop of whipstock cover board and whipstock side wall, setting it to be 25.9mm.

(3)Cement: Simulating the thickness of cement, setting it to be 25.0mm, 35.5mm, 61.0mm and 40.0mm.

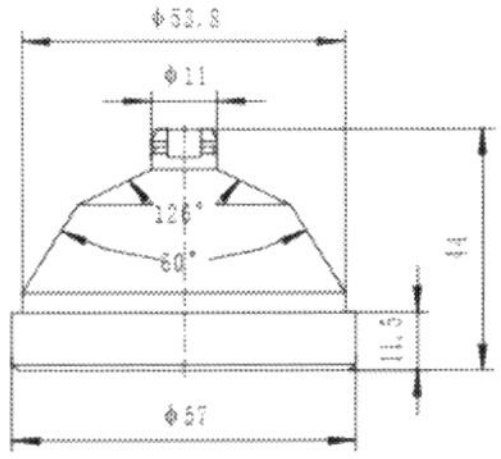

Fig 4 Shape structure of perforating charge

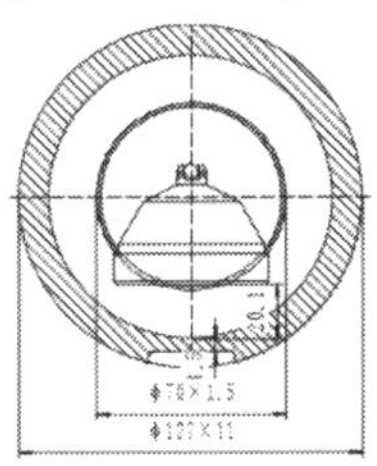

Fig 5 Loading structure of perforating charge

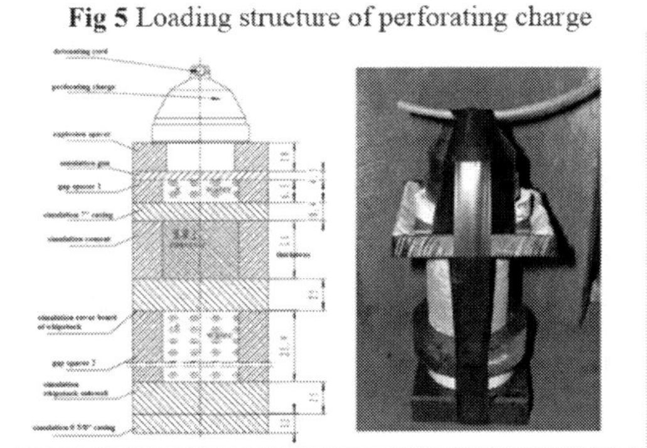

Fig6 Target Test (minimum gap)

3.4 Test Data and Result

Based on functional target analysis of perforator and practical perforating status, simulation test under different status is performed, showing as Table 1. Test result based on Table 1's simulation test is as Table 2.

Table 1 Status Parameter of Simulation Test

Status group	No	Explosion spacer (mm)	Simulation thickness of gun (mm)	Simulation casing gap (mm)	Simulation 7" casing thickness (mm)	Simulation cement thickness (mm)	Simulation thickness of whipstock cover board (mm)	Whip stock ID (mm)	Simulation thickness of whipstock side wall (mm)	Simulation 9 5/8" casing thickness (mm)
I	1	18	4.5	6.5	10.4	35.5	25.4	20	25.4	12
	2									
II	3					61				
	4									
III	5					40				
	6									
IV	7					25				
	8									

Table 2 Test Result

Group	No	7" casing entry hole size (thickness: 10.4mm) (mm)			Whipstock cover board entry hole size (thickness25.4mm)						Damage of simulation whipstock sidewall (thickness 25.4mm)(mm)		Damage of simulation 9 5/8"casing (mm)
					Entrance side (mm)			Exit side (mm)					
		long axis	minor axis	average	long axis	minor axis	average	long axis	minor axis	average	entry hole	penetration	
I	1	21.98	21.3	21.64	11.84	11.55	11.7	14.77	13.24	14	7	10.8	none
	2	22.78	22.5	22.64	12.11	10.82	11.47	14.68	13.86	14.26	4.8	8.4	none
II	3	20.01	18.57	19.29	13.76	12.86	13.31	9.71	9.82	9.67	none	none	none
	4	20.06	19.18	19.62	11.86	10.76	11.31	9.55	9.07	9.31	none	none	none
III	5	24.87	23.71	24.29	11.87	10.24	11.11	10.65	10.56	10.61	none	none	none
	6	21.04	20.5	20.77	15.42	15.42	15.42	10.87	10.31	10.59	5	8.71	none
IV	7	25.58	22.37	23.98	16.03	16.03	16.03	12.66	12.23	12.45	3	2.77	none
	8	24.3	22.15	23.23	14.02	13.9	13.96	13.96	13.02	13.5	5	12	none
Average				21.93			12.98			11.81			none

Calculating the test result, it is concluded that the effective perforating area (S1) on whipstock cover board is larger than the flowing area (S2) of original wellbore production tube.

$$\frac{S1}{S2} = \frac{\pi * 11.81^2}{\pi * (2.313 * 25.4)^2} > 1$$

4.Operation Process

4.1 Introduction of Operation Process

TCP (tubing conveyed perforating) oriented perforating technology is used in whipstock packer perforating. In this perforating technology, pressure actuated firing head is connected at the top of perforating gun, and tubing short circuit, X-over, UBHO Sub, 3 1/2″ drilling pipe, isotope sub (depth sub) and drilling rig are connected to well mouth. All downhole tools must be connected reliably before running down perforating string. Included angle between perforator and orientation key of oriented sub must be measured accurately. It cannot rotate between them. The distance between the first perforating charge at bottom gun and isotope sub should be measured and will not change in wellbore[4]-[6].

TCP oriented perforating technology runs down perforator into preset depth (Fig 8) based on perforating string (Fig 7). After that, running down wireline correlation apparatus to correct depth; calculating the distance (Z) between isotope sub and isotope mark on the top of whipstock packer by calculating isotope sub's depth, meanwhile, adjusting amount (Y-Z-X) of perforating string is concluded. The perforator shall be located at perforating depth and error shall be controlled in 0.2m. Then running down a Gyro with guide device by wireline; the orientation of orientation key should be measured when the guide device of Gyro inserting in the orientation key of orientation sub and being fixed; adjusting perforating orientation by rotating tube strings on ground, and pressurizing to initiate perforator after the perforating orientation being confirmed[7].

 DOI: 10.1051/matecconf/201713900179

NO.	DESCRIPTION	SUP.	Box	Pin	ID"	OD"	L(m.)	DEPTH(m.)
11	5" Drill pipe							
10	X-over 4-1/2"IF(box)×3-1/2"IF(pin)	CCLT	[illegible]	3-1/2"IF	[illegible]	[illegible]		
9	3-1/2" drilling pipe	CCLT	3-1/2"IF	3-1/2"IF	2.75	5.000		
8	Isotope sub	COSL	3-1/2"IF		2.75	5.000	[illegible]	
	Isotope mark	COSL		3-1/2"IF	2.75	5.000	[illegible]	
7	3-1/2" single drill pipe	CCLT	3-1/2"IF	3-1/2"IF	2.75	5.000	[illegible]	
6	UBHO	COSL	3-1/2"IF	3-1/2"IF	2.000	4.750	[illegible]	
5	X-over 3-1/2"IF(box)×2-7/8"EUE(pin)	COSL	3-1/2"IF	2-7/8"EUE	2.441	5.000	[illegible]	
4	2-7/8" EUE Tubing	COSL	2-7/8"EUE	2-7/8"EUE	[illegible]	[illegible]	[illegible]	
3	Pressure Actuated Vent Firing Head	COSL	[illegible]	[illegible]		3.750	[illegible]	
	Hollow perforating gun zone and blind zone	COSL	[illegible]	[illegible]		5.000	[illegible]	
	First charge at top							
2	5" perforating gun	COSL	[illegible]	[illegible]		5.000	[illegible]	
	First charge at bottom							
	Blind zone	COSL	[illegible]	[illegible]		5.000	[illegible]	
1	Bottom Sub						[illegible]	
REMARK:								

Fig 7 Combination of perforating string

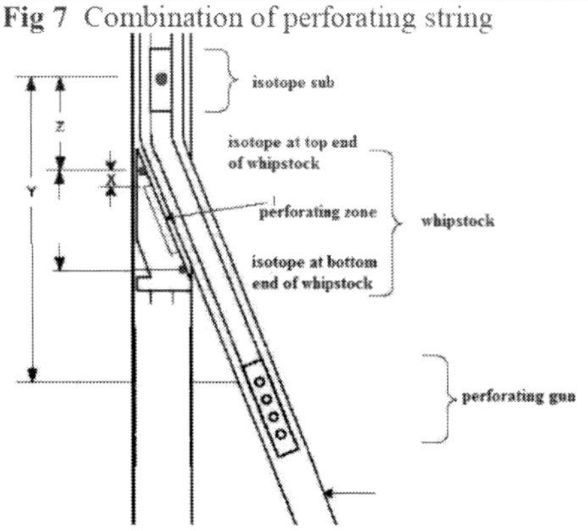

Fig8 Perforating String before depth correction

In that operation process, the function of pressure actuated firing head is to initiate perforator by pressurizing on ground and also to open circulation hole for circulation well killing after communicated lower branch wellbore; the function of tubing short circuit is to remain enough space at the bottom after the guide device of Gyro inserted in the orientation key of UBHO sub[8]; the function of UBHO sub is to guarantee the accuracy of perforating orientation by measuring included angle between perforator orientation key on ground and measuring orientation of the orientation key in downhole; the function of isotope sub is to accurately locate the depth of perforating string.

3.2 Main Steps of Operation Process

(1)Running down perforating strings: locking phasing mark of whipstock oriented perforating system with oriented sub, then running down it to desired perforating zone by tubing.

(2)Wireline depth correction: When the perforating string is run down to desired depth, GR/CCL combination is run down for wireline depth correction. Perforating gun needs to be below the whipstock; isotope sub must not overlap with isotope of whipstock; CBL/GR/CCL logging Fig or open-hole GR logging material shall be supplied by the owner.

(3)Adjusting perforating string to designed depth: Adjusting perforating string to designed depth based on measuring result.

(4)Orientation measuring: Running down Gyro to measure orientation after adjusted perforating string[9].

(5)Orientation adjusting: Based on measuring result of Gyro, rotating tubing string repeatedly to transfer torque to perforator, so as to adjust orientation of string.

(6)Orientation confirmation: After adjusted perforating orientation, the Gyro shall be run down again to measure orientation. It meets requirement if the orientation error is with the scope of +/-10; it is required to adjust orientation again if the orientation error is more than 10°.

(7)Pressurizing and initiating: Pressurizing from well mouth to initiate perforator under the command of supervisor.

(8)Lifting perforating string and releasing perforating gun.

3.3 Announcements during Operation

To guarantee the preparation and safety of operation, the announcements below shall be considered.

(1)Perforating string shall run down after drifting, flushing and confirmation of stabilization of well loss.

(2)Tubing strings must be drift diameter checked. Measuring apparatus and Gyro shall be reliably run down to designed depth.

(3)It needs to be guaranteed that all downhole tools must be connected reliably, preventing screw loose at connection and dropping down.

(4)Speed of tripping must be smooth and steady. Drastic running or cease during tripping is forbidden.

(5)Any object needs be prevented from dropping down into wellbore. If any object dropped down, operator shall report to supervisor to avoid bigger loss[10].

(6)Load fluid need to be poured every three tubes' running. After full poured, it continues to dripping, avoiding over value of pressure difference from inside and outside tubing.

(7)The distance between first perforating charge of top perforator and isotope mark must be accurately measured. It should confirmed by perforating engineer and supervisor.

(8)It requires that lifting and running tubing strings repeatedly to transfer torque to perforator when rotating tubing string on ground for orientation adjustment, and running down Gyro again to measure perforating orientation if it is under allowable error scope.

3 Conclusion

Whipstock packer oriented perforating technology is currently a new operation technology in China, with little operation experience. Requiring high techniques, with complicated process and high risk, this technology is acknowledged as a highly difficult technology in industry. Based on field operation experience accumulation cooperated with international technology, this paper comes to a conclusion that according to test data of perforating charge and optimizing of perforating string, this technology will reduce influence factors of orientation measurement, enhance operation efficiency and lower operation risk.

References

[1] Ya JS,Huang JH,Phase Behavior of Sphere-Forming Triblock Copolymers in Films, Proceedings of 2011 World Congress on Engineering and Technology(CET 2011) VOL04. 19 (2011) 6-9.

[2] C SRINIVASANN,Variation of Flow Field Around Twin Cylinders with and Without the Outer Perforated Cylinder , Numerical Study. 3(2016) 3-4.

[3] ZHAO X, Yang JU, Yang Y, Sun SU, Gong WB,Impact of hydraulic perforation on fracture initiation and propagation in shale rocks, Science China (Technological Sciences),.02(2016) 10-13.

[4] Li JF,Eric Diggins.A proposed multilateral welt classification matrix. World Oil, 1997 : 107.

[5] Mic D.Burnaman,S Smith Shale Gas Well Completions and Maximizing Gas Recoveries. China Petroleum Exploration , 03(2009),12-15.

[6] Li Y,Wang XN,Zhang Dj,Control strategies for aircraft airframe noise reduction. Chinese Journal of Aeronautics , 01(2013),6-10.

[7] Modeling natural convection heat transfer from perforated plates. Journal of Zhejiang University-Science A(Applied Physics & Engineering),03(2009),12-15.

[8] Qiu WJ,Li SB, Yong-yun Ji; Lin Xi Zhang SELF-ASSEMBLY OF LINEAR ROD-COIL MULTIBLOCK COPOLYMERS. Chinese Journal of Polymer Science,09(2013),40-45.

[9] Lin LM,Zhong XF,WU YX Effect of Peak Perforation on Flow Past A Conic Cylinder at Re=100:Drag, Lift and Strouhal Number. China Ocean Engineering,06(2017),4-8.

[10] Modified Atmosphere Packaging Enhances Postharvest Green Celery Quality. Collection of Extent Abstracts of 2004 CIGR International Conference(Volume.1),11(2004),10-14.

A new UAV PTZ Controlling System with Target Localization

Qiaojin Guo [*], *Zhongyan* Liang, *Jian* Xu, and *Jie* Hu

Nanjing Research Institute of Electronic Engineering, Nanjing 210000, China

Abstract. In recent years, with the continuous exposure of unmanned aerial vehicles in the local battlefield, as well as the rapid development of UAV technology, UAV's military application values, especially the advantages in the low-altitude reconnaissance have been the world's attention. Unmanned reconnaissance aircraft is becoming very promising new equipment in technology battlefield and an important information combat front. However, most of the current UAV controlling systems only provide manually PTZ controlling functions, which is inconvenient for the user to control the sensor to monitor targets on demand. This paper propose a new UAV PTZ controlling system, which enables user to keep target in the view filed by simply selecting a position from the digital map. The proposed system is also able to do real-time target localization for any pixel of the video. Experimental results show that our system is easy for PTZ controlling and the localization error is small.

1 Introduction

Unmanned Aerial Vehicle (UAV) [1,2], as new modern military equipment, with its unique advantage, is fully compatible with the modern war concepts of "no contact" and "zero casualties". UAVs can carry out tasks such as reconnaissance and surveillance, laser guidance, electronic jamming, communication relay, target location, combat assessment, and precision strike by carrying different equipment. UAVs can also be used in civil areas. It can be applied to field monitoring, weather detection, road inspection, exploration and mapping, flood monitoring, aerial photography, traffic management, power line inquiries, forest fire prevention and so on.

Optical sensor as one of the most common sensors carried by UAVs can be used for real-time collection and processing. The real-time battlefield video information, which is the most basic and most important information, plays an important role in flight controlling, battlefield commanding, reconnaissance, striking, and so on.

The sensor movement controlling, also known as PTZ (Pan/Tile/Zoom) controlling [3,4], is an important problem for continuous monitoring. At present, most UAVs only support the artificial control of the yaw, pitch and roll of the sensors, as shown in Figure 1, which is difficult to keep targets in the view field, especially when the UAV is moving. It's also inconvenient for the user to control the sensor to monitor multiple targets on demand. To solve this problem, we propose a new sensor controlling scheme, which allow users to control the sensor by directly selecting targets on the map.

In order to control sensors automatically, we need to know which place the sensor is now watching, which is known as target localization [5,6]. The main goal of the target localization with the UAV output video is to match the pixel position of the target in the video image with the coordinates of the object in the real scene, such as GPS coordinates. The main method is based on the navigation information and equipment parameters to calculate the spatial coordinate system, which builds the conversion system between UAV carrier coordinate system and the geocentric coordinate system.

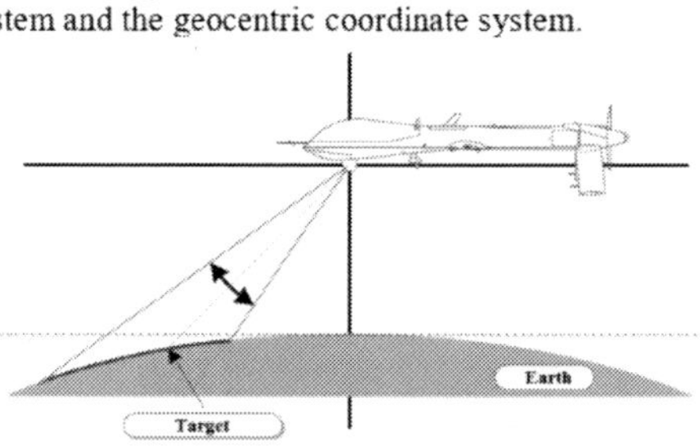

Fig. 1. Example of UAV PTZ controlling for keeping target in the center of view field.

At present, the localization technology based on UAV video images mainly focus on how to match the image between real-time images and the reference images [7-10], such as satellite images. The basic image matching algorithm is already very mature, and the key is how to match the images of different sources and how to improve the reliability and speed of matching. Based on the image matching, the algorithm uses multiple points on a pair of images to describe the correspondences [11,12]. However, if the view field optical sensor contains a limited number of landmarks, the matching performance on the images cannot be guaranteed, and the algorithm will fail. In addition, the algorithm performs image matching is complex and time-consuming, which is not able to do real-time localization [13,14]. This paper proposes a real-time

[*] Corresponding author: guoqiaojin@163.com

DOI: 10.1051/matecconf/201713900180

target localization method based on UAV's video, which is able to calculate the localization of every pixel in the image on the fly.

The rest of this paper is organized as follows. In the next section, we describe the main components of proposed UAV PTZ controlling system. In section 3, we detail the real-time target localization method and the PTZ controlling method will be described in section 4. The prototype system and the experimental results are shown in section 5.

2 System Overview

The proposed system comprises a metadata processing module, an elevation data management module, a target localization module, a PTZ auxiliary control module and a GIS module, which is shown in Figure 2.

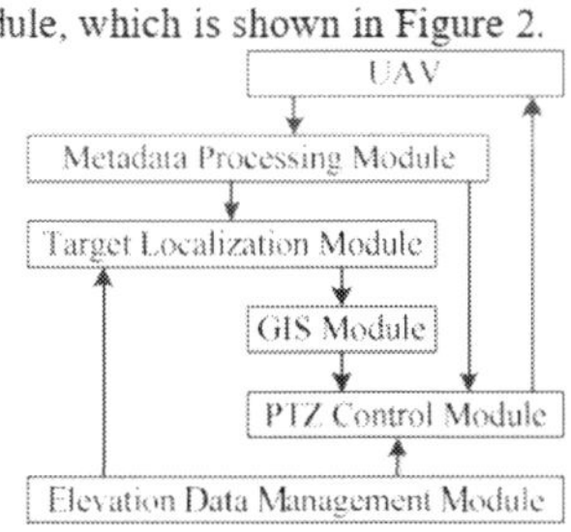

Fig. 2. Components of proposed automatic UAV PTZ controlling system.

The metadata processing module is responsible for receiving the UAV Flight data and sensor status data, and extracted useful information such as latitude, longitude, height, pitch angle, azimuth angle, roll angle of UAV and sensor, horizontal field angle, vertical field angle and video resolution.

The elevation data management module is responsible for storing and accessing the elevation data, converting the divided DEM data to the geocentric coordinate system, dynamically loading and buffering the target according to the UAV's flight trajectory.

Table 1. Descriptions of Metadata.

Metadata	Description
α_{lon}	Longitude of UAV
α_{lat}	Latitude of UAV
α_{alt}	Height of UAV
β_y	Yaw angle of UAV
β_p	Pitch angle of UAV
β_r	Roll angle of UAV
λ_y	Yaw angle of sensor
λ_p	Pitch angle of sensor
W	Width of video frame (pixels)
H	Height of video frame(pixels)
ψ_h	horizontal field angle of sensor
ψ_v	vertical field angle of sensor

The target localization module calculates the angles of the center of view field according to the received metadata and performs the coordinate transformation, and calculates the intersection of view sight with the elevation data in the geocentric coordinate system. Thus the latitude, the latitude and the height of the center of the view field are obtained.

The GIS module displays the real-time view filed of sensor and enables user to select the position of target for monitoring, which is send to PTZ control module for further processing.

The working flow is described as follows. First, the metadata processing module receives the flight data and load data returned by the UAV in real time which is shown in Table 1. Then the cached metadata is sent to target localization module and PTZ control module. The target localization module uses the metadata and elevation data as input and outputs the geospatial location of the image pixels. And the PTZ control module use received metadata, elevation data and selected target position to calculate derivate angles and send controlling commands back UAV. The details of proposed target localization method and PTZ controlling method will be described in the next sections.

3 Proposed Target Localization Method

The received metadata information is buffered in Redis (a memory-based efficient key-value database) for quick access, and the cache invalidation time is set to 20 seconds. For each received video frame, the target localization module queries the metadata before and after the video frame from Redis according to its timestamp and interpolates to obtain the metadata information corresponding to the current frame, and then calculates the longitudes and latitudes of the video center and the four corners. The details are described as follows.

1. Calculate the distance between CCD and the lens which denotes as $L=W/(2\tan(\psi_h/2))$, where W is the width of video frame, and ψ_h is the sensor horizontal view field angle;
2. Calculate the rotation matrix $R_\sigma=R(V_c, V_p)$ between the visual sight $V_c=<w_c, D_c>$ and the target line of sight $V_p=<w_c, P_c>$, where $w_c=(W/2,H/2,L)$, $D_c=(W/2,H/2,0)$, $P_c=(x,y,0)$, W is the height of video frame, H is the height of video frame and (x,y) is the pixel coordinates of target;
3. Calculate the rotation of UAV in geocentric system which denotes as R_{lon} and R_α, where $R_\alpha=R(<(0,0,0),(0,0,1)>,<\alpha_0, \alpha_1>)$, $\alpha_0=T(\alpha_{lon}, \alpha_{lat}, \alpha_{alt}-100)$, $\alpha_1=T(\alpha_{lon}, \alpha_{lat}, \alpha_{alt})$ and function T transform longitude, latitude and height to geocentric coordinates;
4. Calculate the rotation matrix between UAV and the warp which records as R_β based on the yaw, pitch and roll angles of UAV;
5. Calculate the rotation matrix between the sensor sight axis and the UAV body axis which notes as R_λ based on the yaw angle and pitch angle of sensor;
6. obtain the UAV flight area elevation data, convert it to the geocentric coordinate system, and denote the elevation surface as S;
7. Define a straight line in the geocentric coordinate system $V_t=<(0,0,0),(0,0,-1)>$, and then rotate this

line with the rotation matrixes $V_{tr} = V_t * R_\lambda * R_\beta * R_{lon} * R_\alpha * R_\sigma$;

8. Calculate the intersection point of line V_{tr} and the elevation surface S, and the geocentric coordinates of the point is denoted as $P_s = (P_x, P_y, P_z)$;
9. Transform P_s to longitude, latitude and altitude, $P_a = (P_{lon}, P_{lat}, P_{alt})$, which is the target localization result.

4 Proposed PTZ Controlling Method

The PTZ auxiliary control module obtains the target elevation according to the latitude and longitude selected from GIS map, and then calculates the azimuth and pitch angle according to the UAV real-time metadata information. Finally, the module calculates the relative yaw and pitch angle which need to be changed and send commands to UAV for actions. The details are described as follows.

1. Select target in the GIS map, and get the longitude and latitude (P_{lon}, P_{lat});
2. Get the elevation data P_{lat} with the longitude and latitude of selected target, and transform to geocentric coordinates where the target is at $P = (P_x, P_y, P_z)$;
3. Calculate the rotation of UAV in geocentric system which denotes as R_{lon} and R_α, where $R_\alpha = R(<(0,0,0),(0,0,1)>, <\alpha_0, \alpha_1>)$, $\alpha_0 = T(\alpha_{lon}, \alpha_{lat}, \alpha_{alt} - 100)$, $\alpha_1 = T(\alpha_{lon}, \alpha_{lat}, \alpha_{alt})$ and function T transform longitude, latitude and height to geocentric coordinates;
4. Calculate the rotation matrix between UAV and the warp which records as R_β based on the yaw, pitch and roll angles of UAV;
5. Define a straight line in the geocentric coordinate system $V_t = <(0,0,0),(0,0,-1)>$, and then rotate this line with the rotation matrixes $V_{tr0} = V_t * R_\beta * R_{lon} * R_\alpha$;
6. Calculate the deviation azimuth angle K_y and the deviation pitch angle K_p between V_{tr0} and β, where β is the geocentric coordinates of UAV;
7. Send the rotation command to UAV and rotate the sensor with K_y and K_p, the target will be displayed in the view field.

5 Experimental Results

Both target localization and PTZ controlling module need elevation data for calculation. In this paper, we use SRTM [15], which covers an area of more than 119 million square kilometres between 60 degrees north latitude and 56 degrees south latitude, covering more than 80% of the global land surface. The original data is more than 90G and cannot all be loaded into memory. In order to obtain the elevation information of the specified latitude and longitude in real time, the system divide the elevation data into the small pieces and dynamically load those pieces nearby the position of UAV. During the flying of UAV, the system will automatically load new pieces and unload historical pieces, thus using a limited memory space to ensure the 100% hitting rate of getting elevation information of specified latitude and longitude with limited flight speed.

Fig. 3. Examples of target localization results, (a1) and (a2) are video frames from real-time video and selected target (red rectangle), (b1) and (b2) are localization results shown on Google map.

The target localization module can locate the target of any pixel in the video without laser ranging, and the time of single pixel positioning takes less than 50ms. In order to verify the effectiveness of the method, the method of the invention uses the UAV's historical metadata information and video data to locate and compares with the satellite map. The experimental results show that the positioning error of the method is up to 9 to 300 meters. Figure 3 shows the examples of target localization results, (a1) and (a2) are video frames from real-time video and selected target (red rectangle), (b1) and (b2) are localization results shown on Google map.

During the UAV flight, the proposed system can display the position of the view field on the satellite map in real time, and the existing UAV operating system, such as DJI, can only display the UAV and the direction of camera. Our system provides a more convenient way for aerial reconnaissance. We also evaluate performance the proposed PTZ controlling method. We manually select different targets from GIS map, and find the corresponding target in the video after controlling command are executed. The result is shown in Figure 3, where the rectangle denotes as the video area and the markers are the selected targets. We select 50 different targets during the flight and all of them were successfully monitored from the video after the automatic rotations. Figure 4 shows the results of PTZ controlling, where (a) is the selected target from GIS map, and (b) is the video frame after rotation. Our proposed PTZ controlling method successfully rotated the pan and the selected target from Google map is captured in the video.

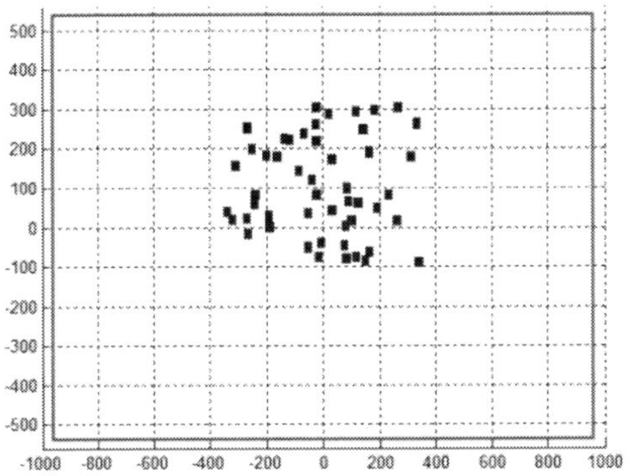

Fig. 3. Results of PTZ controlling method, where the red rectangle is area of video and the blue markers are the positions of selected targets in video frames.

6 Conclusion

UAV's military application values, especially the advantages in the low-altitude reconnaissance have been the world's attention. However, most of the current UAV controlling systems only provide manually PTZ controlling functions, which is inconvenient for the user to control the sensor to monitor targets on demand. This paper propose a new UAV PTZ controlling system, which enables user to keep target in the view filed by simply selecting a position from the digital map. The proposed system is also able to do real-time target localization for any pixel of the video. Experimental results show that our system is easy for PTZ controlling and the localization error is small. In the future work, we will improve the localization accuracy and we will focus on developing new PTZ controlling method which supports focusing on selected targets when the UAV is moving.

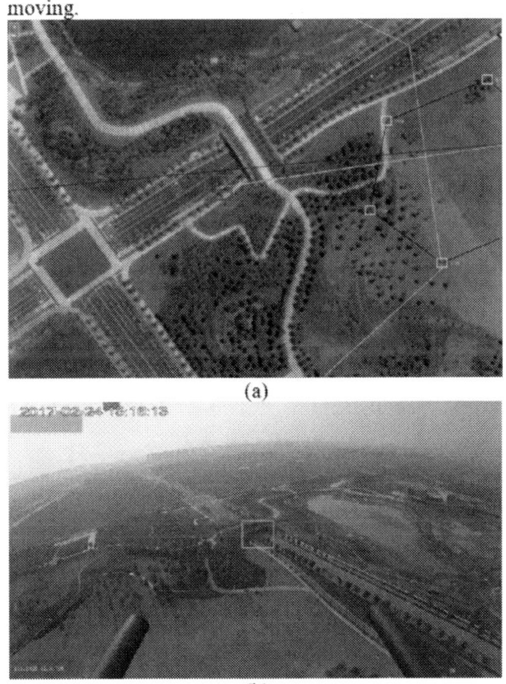

(a)

(b)

Fig. 4. Result of PTZ controlling, where (a) is the selected target from GIS map, and (b) is the video frame after rotation.

References

1. S.R. Herwitz, L.F. Johonson, S.E. Dunagan, et al. *Computers & Electronics in Agriculture*, **44**, (2004) , 49-61.
2. O. Burdakov, P. Doherty, K. Holmberg, et al. *International Journal of Robotics Research*, **29**, (2010) , 1069-1087.
3. S. Hrabar, P. Corke, V. Hilsenstein. *Proceedings of Robotics & Automation*, **1**, (2011), 240-247..
4. M. Tarhan, E. Altug. *Journal of Intelligent & Robotic Systems*, **61**, (2011), 119-134.
5. H. Kwon, D.J. Pack. *Optimization Letters*, **6**, (2012), 1707-1717.
6. S. Sohn, B. Lee, J. Kim, et al. *IEEE Transactions on Aerospace & Electronics Systems*, **44**, (2009), 1391-1401
7. L. Yang, B. Xiao, Y. Zhou, et al. *Proceedings of Cyber Technology in Automation*, **1**, (2016), 321-326.
8. T. Suzuki, Y. Amano, T. Hashizume. *Proceedings of SICE*, **1**, (2010), 2960-2964

9. G. Zhou. *IEEE Transactions on Geoscience & Remote Sensing*, **4**, (2009), 739-747.
10. G. Zhou, D. Wei. *Proceedings of Geoscience & Remote Sensing Symposium*, **5**, (2009), 510-513
11. M. Brown, D.G. Lowe. *International Journal of Computer Vision*, **74**, (2007), 59-73
12. Z.L. Yang, B.L. Guo. *Proceedings of IIH-MSP*, **1**, (2008), 1422-1425
13. H. Kwon, R. Sharma, J. Yoder, et al. *Proceedings of ICC*, **1**, (2012), 454-458
14. M. Quigley, M.A. Goodrich, S. Griffiths, et al. *Proceedings of ICRA*, **1**, (2005), 2600-2605.
15. E. Rodriguez, C.S. Morris, E.J. Belz. *Photogrammetric Engineering & Remote Sensing*, **72**, (2006) , 249-260.

A Study on the Motif Pattern of Dark-Cloud Cover in the Securities

Jing Long[1], Wen-Gang Che[1], Ren Yu[1], Zhi-Yuan Zhou[1]

[1] *Faculty of Information Engineering and Automation*
Kunming University of Science and Technology Kunming, P.R. China

Abstract: Morphological analysis is the analysis and mining of the graphics formed of the securities price changes. Investors need to forecast the trend of future before buying and selling points, which can avoid great loss. Therefore, the analysis of motif pattern of K-line in the form of futures investment technology analysis is very significant. Based on the thoughts of short-term trend clustering, this paper proposes a method of detecting the motif pattern of Dark-Cloud Cover in stock time series by analysing stock historic data and K-line shape, in order to predict the stock market trends. And we prove the effectiveness and practicality of the method by a series of experimental analysis.

1 Introduction

The fluctuation of the stock market is a typical time series. In other words, the stock price will change with time. At the point of view of mathematical statistics, it is a series of discrete random data. This orderly random data (such as X_t, t=1,2,3,...,n) in chronological order is called the time series[1]. The stock time series is a sequence related to the data points, including continuous measurement of the stock price on the trading day, that is, the opening price, the closing price, the highest price, the lowest price, the volume and the rate of return. As we all know, the stock market is a huge system that is affected by many factors. It has complex law. And market prices are also changing rapidly. As a comprehensive external form of expression, the financial time series data of stock market contains many objective laws and information. So it is significant to dig useful information from these data, which could help us to better know, master and utilize the law to make forecast, decision and risk management of stocks investment.

At present, morphological analysis is the main method of reversal forecasting of stock price in financial market.[2] It analyses and mines the graphics formed by changes in the securities price, and regard the morphology as best. When the tendency chart of the security price repeated emerge in some morphology, the price will change according to the prior trend, which could predict the trend of future price. Zhou Ding-bin (2013)[3] has published a study of the stock trading operation in the K-line analysis, which used Guizhou 21 stocks as an example, And it illustrated that the methods of morphological analysis impacted the practicality and accuracy of k-line morphology and the stocks trading. Zhang Dan(2012)[1] has published Reversal Pattern Discovery in Financial Time Series Based on Fuzzy Candlestick Lines, which used a variety of k-line morphology as an example for morphological analysis. It proves the validity of morphological analysis of reversal pattern of mining fuzzy k-line. This paper is based on the ideas that these scholars propose morphological analysis of the stock time series, by combining with short-term tendency theory, we study the motif pattern of Dark-Cloud Cover morphology in the stock time series.

In this paper, the experiment is divided into two parts: the first part is motif pattern that dig Dark-Cloud Cover morphology from the data; the second part is to verify the rewards of investment of Dark-Cloud Cover morphology in the financial market. The experimental data are derived from Qianlong database. Numerical calculation and motif pattern mining of Dark-Cloud Cover morphology are completed by MATLAB software. The main purpose of the study is to dig out the hidden motif pattern from the financial time series, find out the inflection point of the stock price tendency, and evaluate the effectiveness of the motif pattern for the actual investment in the financial market. This could provide reference for investors when they choose stocks buying and selling point. And by assessing the effect of investment income of reversal morphology in the financial market, we can estimate whether or not the research of this kind of reversal morphology is valuable in the stocks time series.

2 The related definitions

The models, commonly used by traditional financial time series analysis are mostly a comprehensive description

* Corresponding author. Email address: wgche@yahoo.com (W.-G. Che)

of the data. For example, an autoregressive model is used to examine the time series. If the time series satisfied the autoregressive model, every time point of the time series applies to the autoregressive relational expression. However, the motif pattern is a local concept. And it reflects more information about a certain aspect of the data, which is about some interesting information fragments on the part of the data. Reversal, as a common language of stock market, refers to that the stock price moves to the opposite direction of the original trend. Specifically, it means that the stock price goes from the bull market to short market or from the short market to bull market. The purpose of this paper is finding out the reversal morphology that the financial market analysts and other investors are interested through the method of data mining, which is also defined as singularity in the financial time series.

Definition 1: Stock Time Series: Given a stock time seriesX, denoted by$X(t_i, o_i, c_i, h_i, l_i)$,which o_i, c_i, h_i, l_i are used to represent the opening price, the closing price, the highest price and the lowest price of a stock time series at the trade day of t_i.

Definition 2: Motif pattern: If there are two or more similar segments of $S_k = X_k[t_{i-1}, t_{i+3}]$ for k=1,2,3,...n within stock time series of X_k.This similar segment having the same short-term tendency pattern is defined as time series pattern of short-term tendency; that is motif pattern.

$$S_k = X_k[t_{i-1}, t_{i+1}] \quad k = 1,2,3, \dots, n \qquad (1)$$

Definition 3: Reversal morphology: The Dark-Cloud Cover morphology is a kind of top reversal morphology of the motif pattern, which describes the external force that impacts the reversal of the securities price changed trend when the stock price rises to high-order after a certain period of time. The morphology generally appears after the rising trend, and in some cases it may also appear at the top of the horizontal consolidation range.

Definition 4: Reversal point: Taking the closing price of the historical trading data of the stock as the object of investigation and the closing price of the stock will be linked into a line in the short-term tendency, there will be a trading day t of the Dark-Cloud Cover morphology as the reversal point. The significance of the reversal point t is to give investors an early warning that they can sell their stocks to maintain the proceeds or to effectively stop losses when the stock price is turned downward.

Definition 5: Short-term average rate of return: Based on the idea of short-term tendency, a cycle is selected, such as: 3 days, 5 days, 7 days etc. The average price of the closing price of this cycle is computed, and the formula is:

$$R_n = \frac{\sum_{i=1}^{n} close(i)}{n} \quad (i = 1,2,3, \dots, n) \qquad (2)$$

3 Dark-Cloud Cover detection and algorithms

The method of motif pattern detection in Dark-Cloud Cover morphology is based on the short-term tendency clustering theory of time series, and combines with the feature of the Dark-Cloud Cover morphology. The numerical calculation and pattern mining are completed by MATLAB software. In conjunction with definition 3, the following algorithm is available:

Algorithm1: Short-term growth tendency algorithms

Step 1: Take a stock time series of $X(t_i, o_i, c_i, h_i, l_i)$, from any trade day of i, compare the closing price of day i and day $i-1$, if $c_{i-1} > c_i$, go to step 2; otherwise increase $i$to next day and go back to step 1.

Step 2: Compare the stock prices of day $i+1$and day i, if $c_{i+1} > c_i$, go to step 3; otherwise increase $i$to next day and go back to step 1.

Step 3: Compare the stock prices of day $i+2$and day $i+1$, if $c_{i+2} > c_{i+1}$, go to step 4; otherwise increase $i$to next day and go back to step 1.

Step 4: Compare the stock prices of day $i+3$ and day $i+2$, if $c_{i+3} > c_{i+2}$, go to step 5; otherwise increase i to next day and go back to step 1.

Step 5: A short-term growth tendency is detected. To determine whether there is a Dark-Cloud Cover morphology after growth trend; combining algorithm 2.

Algorithm2: Dark-Cloud Cover morphology detection algorithm

Take a stock time series of $X(t_i, o_i, c_i, h_i, l_i)$, from any trade day of i.

Step 1: respectively compare the opening price o_i of day i with the closing price $c_{i-1}$and the highest price h_{i-1} of day$i-1$, if $o_i > c_{i-1}$ or$o_i > h_{i-1}$, go to step 2; otherwise increase i to next day and go back to step 1.

Step 2: respectively compare the closing price c_i of day i with the closing price c_{i-1} and the opening price o_{i-1} of day $i-1$, if $c_i > o_{i-1}$&&$c_i < c_{i-1}$ or $c_i < \frac{c_{i-1}+o_{i-1}}{2}$, go to step 3; otherwise increase i to next day and go back to step 1.

Step 3: If the selected time series transaction day i satisfies step1and satisfies step 2, it is judged that the Dark-Cloud Cover morphology is detected.

Algorithm3: Short-term average rate of return of Dark-Cloud Cover morphology

Calculating the short-term average rate of return is to determine the stock trend that will continue to rise, down or sideways when the Dark-Cloud Cover morphology appears. This gives a warning that the investors should throw or continue to follow up.

First, investors have already held positions for 5 days before Dark-Cloud Cover morphology (a cycle of short-term tendency in this article). We assume that the Dark-Cloud Cover morphology appears on the trading day$i$, then the formula that calculates rate of return of day i and day $i-1$ is recorded asR', that is:

$$R' = \frac{close(i)+close(i-1)}{2} \qquad (3)$$

And later 5 days after day i as a cycle of short-term tendency, calculate the average return of 5 days trading day after day$i$, denoted byR'':

$$R'' = \frac{\sum_{i=1}^{5} close(i)}{5} \quad (i = 1,2, \dots, 5) \qquad (4)$$

Finally, given a standard rangeγ, it is determined the stock trend, which is rising, down or sideways after the Dark-Cloud Cover morphology. That is:

DOI: 10.1051/matecconf/201713900181

$$\gamma = \frac{R''-R'}{R'} \quad (5)$$

Algorithm 4: Motif pattern

Generally, k-line morphology of time series is a pattern that frequently appears in a time sequence, which is very useful to forecast the stock short-term tendencies and prices. In this section, we propose an algorithm that is used to find those motifs of short-term tendency in stock time series.

Step 1: Take a stock time series of$X(t_i, y_i)$, detect the short-term tendency and remember the time duration.

Step 2: Take another stock time series of $X(t_i, y_i)$, detect the short-term tendency and compare it with former. Then determine whether there are similar segment of time series during same time period. If yes, got to step 3, otherwise go to step 2.

Step 3: Remember the number of similar segments that is detected and then got back to step 2.

Step 4: The motifs of the stock time series are detected.

Taking the Shanghai A-stocks as an example, we randomly selected n stocks, and found the motif pattern of Dark-Cloud Cover morphology in the stock time series by the above algorithm, as shown in "Fig.1, Fig.2, and Fig.3".

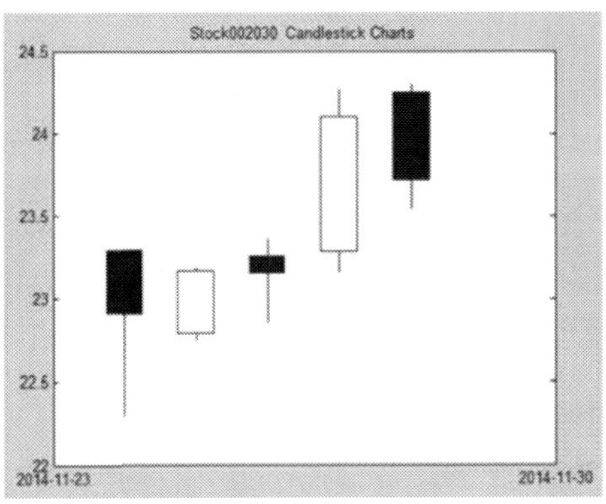

Figure1. Periodicity Dark-Cloud Cover morphology

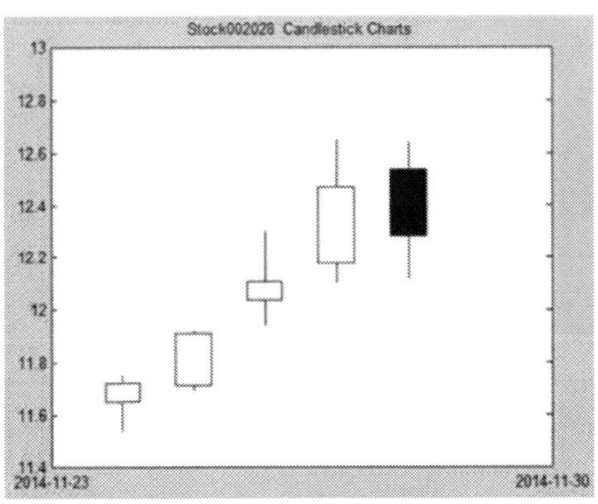

Figure2. High Dark-Cloud Cover morphology

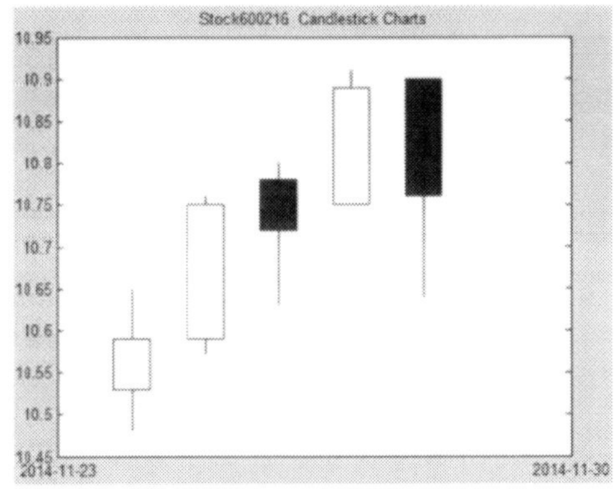

Figure3. Unreversed Dark-Cloud Cover morphology

4 Empirical Analyses

Combined with other scholars' research on the factors that influence the fluctuation of securities prices, it is found that there is a tendency in the process of securities price changes. The direction of securities price changes is determined by the current trend. Unless there is an external force that stops or reverses the current trend, the trend will continue. This paper, based on this idea, proposes a meaningful detection method based on motif pattern to detect motif pattern of Dark-Cloud Cover morphology in stock time series. Combining with the short-term tendency theory, and citing the short-term average rate of return, we analysis the short-term trend of Dark-Cloud Cover morphology in stock time series, in order to predict the next trend of the stock market, and to give investors an early warning.

In this paper, we selected 200 stocks of Shanghai A-stocks from January 2014 to December 2016 for a period of three consecutive years as the experimental data, which were detected 338 segments of the Dark-cloud cover morphology. This experiment is classified with two quarters as standard of the time period (three months is a quarter), which is divided into a total of six time periods. The data table is shown in Table 1; the time series is shown in "Fig.4"; the moving average is shown in "Fig.5", and the dark is turning point in the Fig.5. In addition, day i and day $i-1$ constitute Dark-cloud Cover morphology.

Table1. The distribution of the number of Dark-Cloud Cover morphology

time period	1	2	3	4	5	6
Dark-Cloud Cover morphology	42	80	72	49	43	52

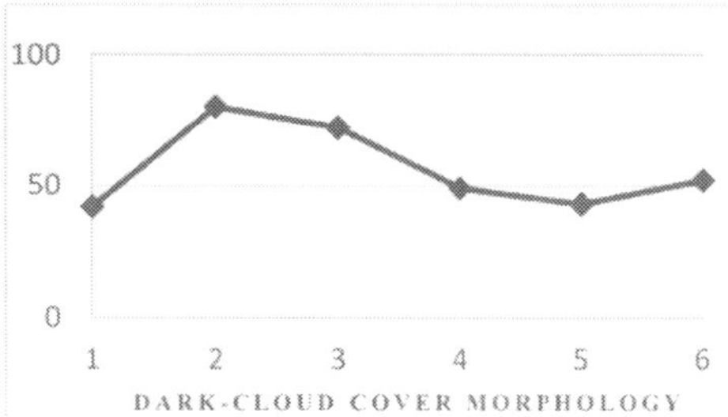

Figure4. Time series diagram

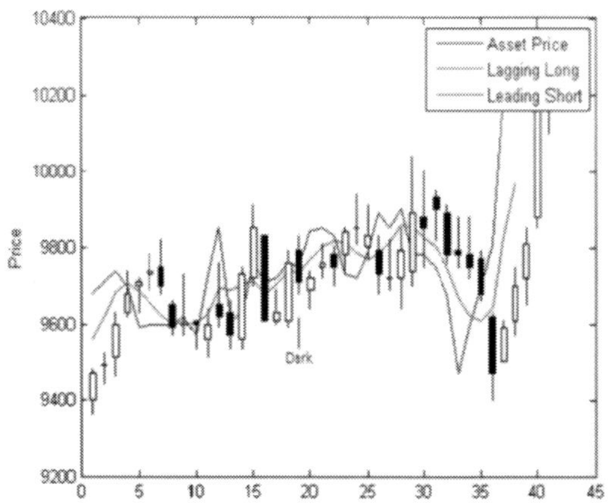

Figure4. The moving average diagram

We get the number and frequency of Dark-Cloud Cover morphology in the Shanghai A shares in January 2014 to December 2016 for three consecutive years by analyzing the data table. However, if we want to better to determine the follow-up trend of the stock by using Dark-Cloud Cover morphology, we need to combine short-term Trends, with calculating the short-term average returns and ranges. This experiment collects the closing price of trading day of Dark-Cloud Cover morphology and the closing price of the next five days after Dark-Cloud Cover morphology. The results calculated by the formula (4) are combined with the collected data. In this paper, the standard range γ of the stock trend in the late stage of Dark-Cloud Cover morphology is setting up as 2%. The data statistics are shown in Tables 2 and 3 (Table 2 shows the distribution of the late trend of the stock; Table 3 shows the ratio of the stock overall trend)

Table2. The distribution of the late trend of the stock

Time period	1	2	3	4	5	6
rise	2	7	6	2	2	2
down	35	48	44	35	34	47
sideways	5	25	22	12	7	3

Table3. The ratio of the stock overall trend

Stock trend	rise	down	sideways
gross ratio	21	243	74
	6.21%	71.90%	21.89%

By analyzing the data in Table 3, it can be concluded that the number of stocks down accounted for more than 71.90%, sideways trend accounted for 21.89%, and the rise trend that was the lowest, accounted for only 6.21% of the gross ratio. So the possibility of the stock down is largest when Dark-Cloud Cover morphology appears, which provides reference for investors to sell the timing, and reminds investors should not blindly chase high to avoid inflection point of stock price in a certain period of time. Besides, it is more effective to protect the vested income of investors.

5 Conclusions

Based on the idea of short-term trend, this paper proposes a method to detect motif pattern of singular morphological fragments in stock time series. Through the detection and analysis of Shanghai A stocks data, the results show that we can achieve better results by using the method to analysis the morphological of stock time series and the late trend. Finally, though the analysis of the closing price corresponding to the motif pattern of singularity, the validity of the method proposed in this paper is further confirmed. Of course, there are many factors that affect the stock price, this article only analyses the closing price. In the later period, we will analyze the volume of the stock market, which hope to achieve the expected results.

References

1. Qiu-jun Lan, Dan Zhang, Long-ling Xiong. Reversal Pattern Discovery in Financial Time Series Based on Fuzzy Candlestick Lines. In: International Symposium on Complex Systems Theory & Applications. Changsha, 2011
2. De-hong Liu. Technical analysis of stock investment. Beijing: Economic management press.2009,pp-78-146(in Chinese)
3. Ding-bin Zhou, Jie Wang. study of the stock trading operation in the K-line analysis. Economics and management science 2013, (10) (in Chinese).
4. Engle RF. Autoregressive conditional heteroscedasticity with estimates of the variance of United Kingdom inflation. Econometrical

Journal of the Econometric Society, 1982(6),pp-987-1007

5. Lee C-H L, Liu A. Pattern discovery of Fuzzy Time Series for financial prediction. IEEE Transactions on Knowledge and Data Engineering, 2006, 18(5),pp-613-624
6. Xiao-di Liu, Wen-gang Che, and Qing-jiang Zhao, "Analysis of outliers and public information arrivals using wavelet transform modulus Maximum," The 2nd IEEE International Conference on Information and Financial Engineering, September 2010, pp-176-179.
7. Qiang Fu, Wen-gang Che, "Study on isolated singularity data mining algorithm based on sliding window," JINGXI SCIENCE, vol. 29, pp-273-276, April 2011(in Chinese).
8. Yue Zhang, Jie Liu, and Hang Li. An outlier detection algorithm based on clustering analysis, The 1st International Conference on Pervasive Computing, Signal Processing and Applications, 2012,pp-1126-1128.

ACT-R Model for Addition Under Emotional Stimulus of Major Depressive Disorder and Control

Xiaojing Yang [1,2*], Yiting Wang[1]
1. College of Computer and information engineering, Anyang Normal University, Anyang 455000
2. International WIC Institute, Beijing University of Technology, Beijing 100124

Abstract. Neural correlate of major depressive disorder (MDD) and control of addition under emotional stimulus is still unclear. Addition under positive emotional stimulus (AP) and addition under negative emotional stimulus (AN) are two typical different cognitive processing, and with a common core component of rule induction. Previous studies have demonstrated that different processing is adopted in AP and AN completion tasks; even the underlying rules are identical. In the present study, we examined cortical activation as a function of two different machining processing for solving addition calculation under different emotional completion tasks. In AP completion tasks, involves direct retrieving of arithmetic knowledge to get the relations between items. In AN completion tasks, requires counting a certain number of times to detect the relations linking two items. The two processes require essentially the equivalent cognitive processes, but they have different working memory demands. In AN completion tasks produced significant greater activity involved in memory retrieval (dorsolateral prefrontal cortex, DLPFC) and mental representation/maintenance (posterior parietal cortex, PPC). An ACT-R model of the tasks successfully predicted the behavioral performance and BOLD response. The present findings positive emotional stimulus has promoting effect on addition cognitive calculate. Negative emotional stimulus has inhibiting effect on addition cognitive calculate.

1 Introduction

Major depressive disorder (MDD) [1-2]is a significant and lasting low mood as the main mark, and with lower self-esteem, lack of pleasure and a series of symptoms of affective disorders [3]. It has the characteristics of chronic deferment, high incidence and easy recurrence, the suicide rate and high morbidity, work, study and daily life of the patients with many aspects, such as, interpersonal communication has a serious negative emotion [4]. Over the past 20 years, the development of neuroimaging provides a new train of thought to explore the neural mechanism of brain, which has become the indispensable important tool of disease physiological basis [5-6]. Through the comparative analysis of the normal people and MDD can be found, it may provide indicators for the diagnosis of the disease to reveal the development rule of disease, provide the basis for relevant treatment in the future. Researchers believe that these brain regions in the process of emotional and cognitive form regulation loop [7-9]. Researchers assumed that the structure and function of patients with depression anomalies mainly related to emotion recognition loop [10]. Depression of the frontal cortex, knee cortical volume decreases [11-12]. Task state fMRI experiments study found that patients with depression and emotional, cognitive processing part of the brain associated with memory function impairment [13-14]. Researchers found that MDD of the dorsolateral prefrontal cortex (dlPFC) has the abnormal activation of significant in cognitive processing, and were positively associated with the degree of cognitive load [15]. Relevant research shows that negative emotional stimulus has attention retention for cognitive task [16-18].

Digital computing is one of the basic human brain cognitive functions [19]. Simple arithmetic natural mating, the simplest and most commonly used and has some related [20]. Dehaene and Riveral believes that a simple single digits addition and multiplication events (additive formula, my multiplication) in the process of calculating extraction depend on the participation of the basal ganglia [21]. Division operation and carry and abdication of addition and subtraction are complex, there have been many studies about digital cognition theory and experimental psychology, neuroimaging that a particular area of the brain with different digital processing function. The PFC and PPC are considered to be one of the most important areas [22]. This study is to choose a non-carry addition calculation equation under different emotional picture stimuli as the experimental paradigm, in order to studying the effect of different emotional stimuli to addition calculation, the positive, neutral, negative emotional stimuli pictures appear before an addition equation present. Experimental design is shown in figure 1, the participants only need to judge the two digits addition equation right and wrong under the emotional stimulus pictures if the result is right to press the left key, or else press right key.

ACT - R can simulate the behavior process and the characteristics of the brain functional areas in the process of physiological activity, can predict fine granularity time information processing (e.g., 0.2 s) and the active model of ROI [23]. The shortcoming of fMRI technology is can't link brain activity with specific information processing, the shortage of the ACT-R simulation is unable to confirm the rationality of the forecast results,

* Corresponding author: yangxj84@163.com

ACT - R meets fMRI can be complementary advantages to avoid shortage. Zhong Ning et al. combined with artificial Intelligence, cognitive science and neuroscience research contents, advocate using BI (Brain Informatics) research to development of WI (Web Intelligence) applications [24]. John Anderson, Yinlin Qin et al study on Hanoi tower, solution of the equation, geometric proof et al. In their study, eight brain regions corresponding to different cognitive component, the parietal lobes activities reflect the state of problem space representation, left prefrontal activities reflect the declarative memory retrieval, basal ganglia activities reflect to extract the production rule, the area before the cingulate activities reflect the target status updates, etc. the BOLD effect of these areas has a high degree of consistency in different experiments. Research results have been integrated into the ACT - R cognitive system, make the ACT - R can effectively forecast the BOLD response of these regions[21-23].

In this paper, adopt non-carry addition calculation under different emotional stimulus for experimental paradigm, using ACT-R meets fMRI to study the non-carry addition calculation under the different emotional pictures. According to the function of the eight ACT - R modules and synergy work principle, because of ACT - R without emotion module by adjusting the length of time to indirect reaction the degree of emotional impact on cognitive computing. Experimental results based on the questionnaire survey and the behavior hypothesis model to establish the hypothesis model, model and simulate the cognitive process of the addition calculation under different emotional pictures stimulus, the experimental results verify the validity of the behavior experiment data

2 Materials and Methods

2.1 Subjects

MDD: 23 (male:15 female:8) average aged (32.76 ± 2.48), average education (10.37 ± 1.62) subjects were in depressive episodes come from Anding Hospital, Capital Medical University, accord with Chinese classification and diagnostic criteria by mental disorders in version 3 (CCMD - 3), ruled out with organic mental disorder or mental active substance and the addiction caused by depression, serious body disease, severe side effects of psychotropic drugs, alcohol and substance abuse.

Control: From the same region, according to the group of the patient's gender, age, level of education, such as indicators, workers and their families in Beijing university of technology, such as student volunteers in matching selection, selection of standard is: without mental illness, neurological diseases, serious body disease history and voluntary cooperation. 23subjects (male:15 female:8), average aged (31.72±2.63), average education (9.88±1.58), as shown in Tab.1.

In the two groups had no age (t = 0. 41, P = 0. 59), level of education (t = 0. 03, P = 0. 76) had no significant differences.

Table 1 . General demographic information of MDD and control

	Control	MDD
Gender(male/female)	15/8	15/8
age	31. 72±2. 63	32. 76±1. 48
education	9. 88±1. 58	10. 37±0. 62

2.2 Stimuli and Experimental Design

An experimental procedure compiled and run in E - prime 2.0 (Psychological Software Tools, Inc., Sharpsburg, PA, USA). The stimulate content is presented to subjects by MRI machine by vision systems, subjects may react by button box. The reaction time and accuracy is recorded automatically by program. Before formal experiment, using unified sequence of instructions and practice to train participants to ensure that the participants understand the experiment content and operation method.

Experiment tasks using the addition calculate equation under different emotional stimuli as experimental paradigm, control: addition calculate equation under positive emotional stimuli (CP), addition calculate equation under neutral emotional stimuli (CNe), addition calculate equation under negative emotional stimuli (CN); MDD: addition calculate equation under positive emotional stimuli (MP), addition calculate equation under neutral emotional stimuli (MNe), addition calculate equation under negative emotional stimuli (MN). Firstly present an emotional picture (2s), then present an addition calculation equation (4s) and judge the equation right or wrong. It asked participants to judge the equation as soon as possible as shown in Tab.2. Using experimental block design, the experiment contains 18 blocks, and each block contains three trials, each trial contains one emotional stimuli picture and one equation, the picture and equation alternate rendering as shown in Fig.1, Whole experiment duration 486s.

Table 2. Examples of experimental tasks

	Task	Options		Answer
positive	12+14=26	A.right	B.wrong	A
neutral	13+15=38	A.right	B.wrong	B
negative	14+13=27	A.right	B.wrong	A

2.3 MR Data Acquisition

Scanning was performed on a 3.0 T MRI system (Siemens Trio Tim; Siemens Medical System, Erlangen, Germany) with a 12-channel phased array head coil. Foam paddings and headphones were used to limit head motion and reduce scanning noise. Structural images were obtained by using a three-dimensional magnetization prepared rapid acquisition gradient echo (3-D MPRAGE) sequence with the following parameters:

 DOI: 10.1051/matecconf/201713900182

repetition time (TR)/echo time (TE) = 1600/2.25 ms, flip angle = 9°, 192 sagittal slices, voxel size = 1 × 1 × 1 mm3. Functional data were acquired using a T2* gradient echo-planar imaging (EPI) pulse sequence (TR/TE = 2000/31 ms, 30 axial slices, voxel size = 3.75 × 3.75 × 4.0 mm3, 0.8 mm inter-slice gap, 90° flip angle, 64 × 64 matrix size in 240 × 240 mm2 field of view [FOV]). This acquisition sequence generated 364 volumes for each session. The presentation of every trial was synchronized with the scanner.

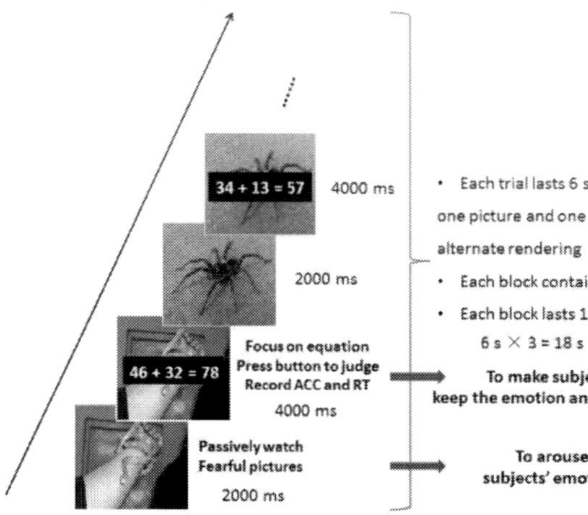

Fig.1. Paradigm of stimuli presentation

2.4 Data Preprocessing

Data were analyzed using SPM8 software (http://www.fil.ion.ucl.ac.uk). The first four images for each session were discarded to allow for T1 equilibration effects. The remaining fMRI images were first corrected for within-scan acquisition time differences between slices and then realigned to the first volume to correct for inter-scan head motions (Head movement was b 2 mm and b 2° in all cases). The structural image was co-registered to the mean functional image created from the realigned images using a linear transformation. The transformed structural images were then segmented into gray matter (GM), white matter (WM) and cerebrospinal fluid (CSF) by using a unified segmentation algorithm. The realigned functional volumes were spatially normalized to the Montreal Neurological Institute (MNI) space and re-sampled to 3 mm isotropic voxels using the normalization parameters estimated during unified segmentation. The registration of the functional data to the template was checked for each individual subject. Subsequently, the functional images were spatially smoothed with a Gaussian kernel of 8 × 8 × 8 mm3 full width at half maximum (FWHM) to decrease spatial noise.

2.5 fMRI Analysis

Only the correct response trials were included in data analysis (wrong response trials and interferential trials were not included). The epoch of interest is the duration between the presentations of the first equation to the first button response. The BOLD[21] signal was modeled using the canonical HRF with time derivative and the RT as duration trial by trial. Condition effects at each voxel were estimated according to the general linear model and regionally specific effects were compared using linear contrasts. Each contrast produced a statistical parametric map of the t-statistic, which was subsequently transformed to a unit normal z-distribution. The contrast images of each subject were then used in random effect analysis to determine what regions were the most consistently activated across subjects using a one sample t-test. For confirmatory analysis, region of interest (ROI) analyses focused on two predefined functional regions in ACT-R, LPFC (TAL., − 43, 23, 24) and PPC (TAL., − 23, − 63, 40) [25-27].

The statistical results were based on the beta-values (of general linear model implemented in SPM8) averaged within the two ROIs. In order to relate our results to previous studies, we reported the contrast associated with two simple main effects of Task (CP–CN) and (MP–MN), but are primarily interested in the activations common to both strategies [(CP–CN) and (MP–MN)]. The common activation map was achieved by the following two steps: The map of (CP–CN) was first outputted as a mask, and then (MP– MN) was examined within the mask. The Task by Content interaction comparisons [(CP–CN)–(MP–MN)] and [(MP–MN)–(CP– CN)]) were also executed to reveal the areas specific to each strategy. The activation reported survived a voxel-level intensity threshold of p=0.05 corrected for false discovery rate (FDR) for the whole brain volume with a minimum cluster size of 10 contiguous voxels.

2.6 ACT-R Modeling

Having now reviewed the result of the experiment, we come to the question whether we can understand them in the framework of the ACT-R theory.

The model we constructed primarily depends on the visual module to perceive the stimuli, the manual module to respond, the retrieval module to retrieve a fact from memory, and the imaginal module to encode and update its stored representation. Communication among these modules is achieved through a procedural module. For example, the following might be a production rule for producing of the addition under neural emotional stimulus of control:

IF the goal is to solve the addition calculation problem and the problem is of the form "Number1 Number2 Number3" and "Number1 + Number2 = Number3" has been retrieved

THEN the answer of the problem is Number3.

As the fronto-parietal network plays an important role in additional calculate process, the retrieval and imaginal module are of special interest in the present study. In the following, we would illustrate the prediction of BOLD responses in the two modules. Fig.2 and Fig.3 illustrates the sequences of activities in the four relevant buffers of ACT-R, by taking addition

("23+35=58") under different emotional stimulus of control and addition ("23+35=58") under different emotional stimulus of MDD as examples. Each box reflects a period of time during which that buffer is active, and the vertical length of the box reflects the duration that the buffer is active. With the processes in Fig.2 and Fig.3 and their durations set, we will test whether the BOLD responses in the fronto-parietal regions can be predicted.

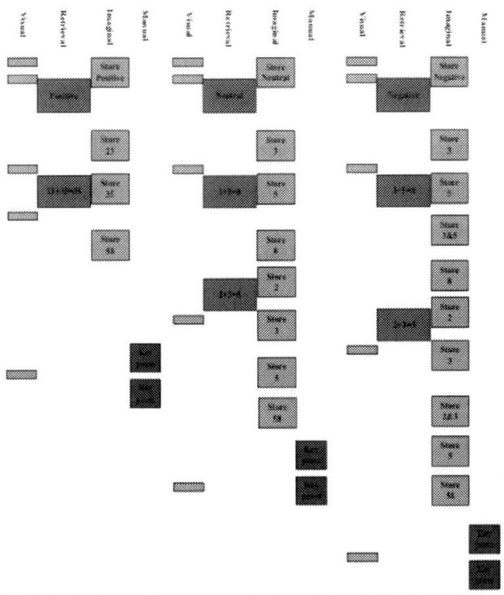

Fig.2. A schematic representation of the ACT-R model's solution to addition under different emotions of control

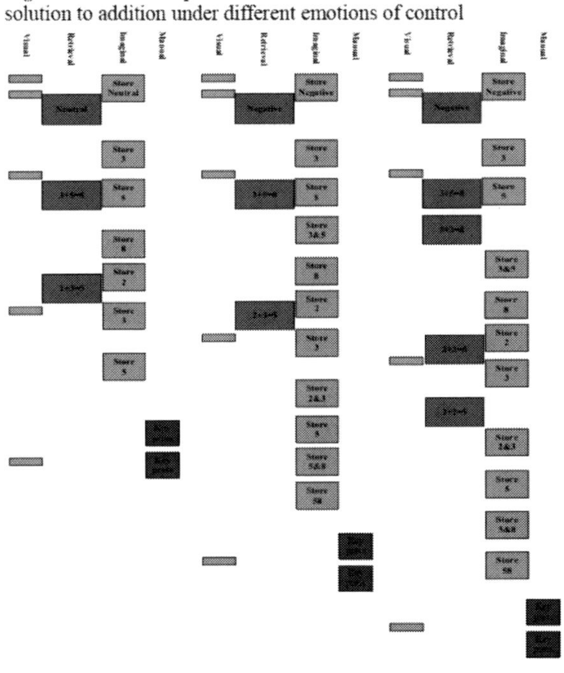

Fig.3. A schematic representation of the ACT-R model's solution to addition under different emotions of MDD

3 Results

3.1 Behavioral Performances

We carried out analyses of variance for two factors: Content (number vs. letter) and Task (induction vs. baseline) on both RT and accuracy. Behavioral scores indicated that subjects performed the task in the expected manner (Tab.3).

Table 3. Behavioral scores

	RT(ms)		
	positive	neutral	negative
control	1861 ± 372.36	2092 ± 719.61	2283 ± 928.05
MDD	2348 ± 569.42	2481 ± 983.08	2704 ± 762.42

	Accuracy(%)		
	positive	neutral	negative
control	96 ± 7.34	93 ± 4.86	89 ± 11.04
MDD	91 ± 4.15	88 ± 3.87	86 ± 5.03

3.2 Results of ACT-R Modeling

The fit of the predictions of the model to the RT and ACC data are presented in Fig.7. The parameters were estimated to fit the behavioral data: a factor that scaled the time to retrieve a declarative memory fact (0.4 s) and the time to modify the contents of the imaginal module (i.e., 0.2 s). We fixed α to be 2 and s to be 2 s. Once the timings of buffer actions are all set, we can predict the BOLD functions by estimating the magnitude parameter m from the BOLD response in each brain region[22-24]. The estimates of these parameters and measurement of the quality of the prediction are given in Tab.4 and Tab.5. Fig.8 and Fig.9 displays percent change of BOLD response (relative to the baseline defined by the average BOLD response of the first two scans and the last two scans), along with the prediction of the ACT-R model to be presented.

Table 4. Parameters and the quality of the BOLD function prediction of control

	Imaginal	Retrieval
Exponent(α)	1.8	1.8
Scale(s)	1.8	1.8
Magnitude		
M $\Gamma(\alpha+1)$	5.96	3.57
Correlation(r)	96	93

*This is a more meaningful measure because the height of the function is determined by the exponent as well as M.

Table 5. Parameters and the quality of the BOLD function prediction of MDD

	Imaginal	Retrieval
Exponent(α)	2.1	2.1
Scale(s)	2.1	2.1
Magnitude		
M $\Gamma(\alpha+1)$	6.24	4.92
Correlation(r)	97	94

 DOI: 10.1051/matecconf/201713900182

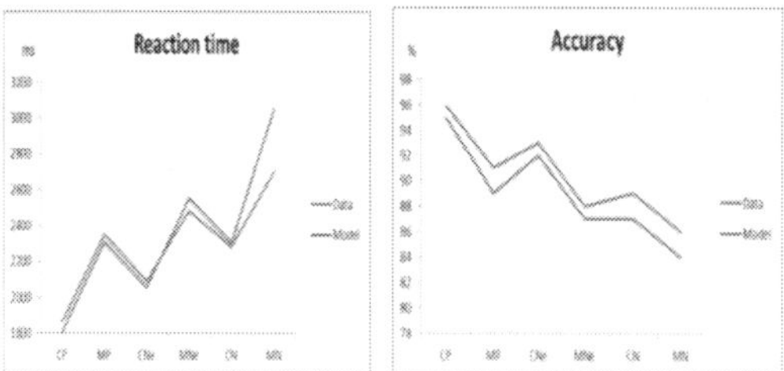

Fig.7 Data (blue lines) and Model(red lines) for addition under different emotional stimulus of control and MDD
(**NOTE:** C- denotes control, M- is MDD)

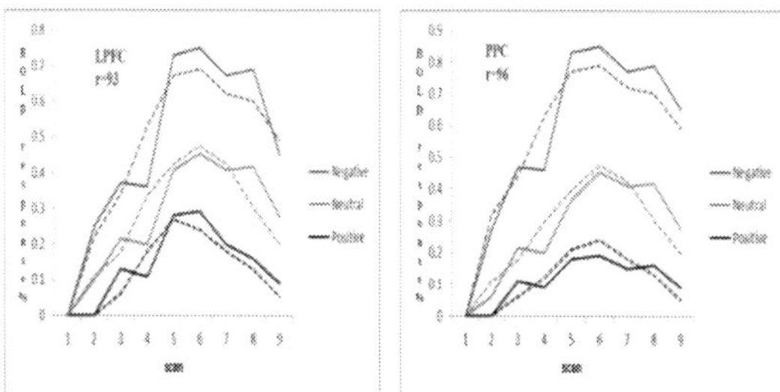

Fig.8 BOLD functions obtained for the prefrontal and parietal regions for addition under different emotional stimulus of control
(**NOTE:** ——— denotes real data, ·········· is model data)

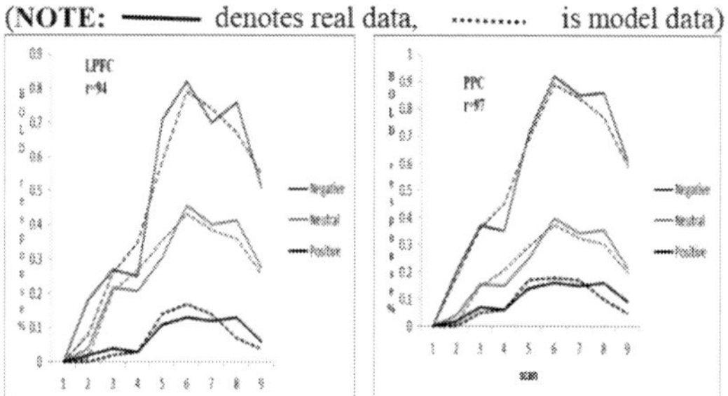

Fig.9 BOLD functions obtained for the prefrontal and parietal regions for addition under different emotional stimulus of MDD

(NOTE: ——— denotes real data, ·········· is model data)

4 Conclusion

MDD emotional active areas and cognitive function areas may have damage[9]. So reaction time of control for addition in the same emotional stimuli is shorter than MDD, at the same have higher accuracy; it has a retention phenomenon due to the negative sentiment[13-15]. So in the same group, reaction time for addition under positive emotional stimuli is shorter than negative emotions, and the accuracy is higher. The ACT-R model and simulation results for addition under different emotional stimuli verify the hypothesis.

References

[1] Neugebauer R. Mind matters:the importance of mental disorders in public health' s 21st century mission [J]. Am J Pub Health, 1999, 89: 1309-1311.

[2] Mayberg H S, Lozano A M, VoonV, et al. Deep brain stimulation for treatment-resistant depression [J]. Neuron, 2005, 45: 651-660.

[3] Baref B R. Introduction to functional magnetic resonance imaging:principles & techniques[M]. London: Cambridge University Press, 2002:272.

[4] Zhang D, Raichle M E. Disease and the brain's dark energy [J].Nat. Rev. Neurol., 2010, 6: 15-28.

[5] Bower G H, Forgas J P. Affect, memory, and social cognition. In: Eich E, Kihlstrom J F, Bower G H, et al., eds. Cognition and Emotion. New York: Oxford University Press, 2000. 87-168.

[6] Langeslag S J E, Franken I H A, van Strien J W. Dissociating love-related attention from task-related attention: An event-related potential oddball study [J]. Neurosci Lett, 2008, 431: 236-240.

[7] Fox M D, Raichle M E. Spontaneous fluctuations in brain activity observed with functional magnetic resonance imaging [J]. Nat.Rev. Neurosci., 2007, 8: 700-711.

[8] Raichle M E.The brain's dark energy [J]. Sci. Am., 2010, 302: 44-49.

[9] Sharp S C, Hellings J A. Efficacy and safety of selective serotonin reuptake inhibitors in the treatment of depression in children and adolescents: practitioner review [J]. Clin Drug Investig, 2006, 26(5): 247-255.

[10] Murray C J L, Lopez A D. Alternative projections of mortality and disability by cause 1990-2020: Global Burden of Disease Study [J]. Lancet, 1997, 349: 1498-1504.

[11] Pessoa L. On the relationship between cognition and emotion [J]. Nat Rev Neurosci, 2008, 9: 148-158.

[12] Lyons W. The philosophy of cognition and emotion. In: Dalgleish T, Power M, eds. Handbook of Cognition and Emotion. Chichester: Wiley, 1999. 21-44.

[13] Ochsner K N, Phelps E. Emerging perspectives on emotion-cognition interactions [J]. Trends Cogn Sci, 2007, 11: 317-318.

[14] Dolan R. Emotion, cognition, and behavior [J]. Science, 2002, 298: 1191-1194.

[15] Barrett L F. Emotions as natural kinds? [J]. Perspect Psychol Sci, 2006, 1: 28-58.

[16] Pessoa L. How do emotion and motivation direct executive control? [J]. Trends Cogn Sci, 2009, 13: 160-166.

[17] Barrett L F, Mesquita B, Ochsner K N, et al. The experience of emotion [J]. Annu Rev Psychol, 2007, 58: 373-403.

[18] Izard C E. Emotion Theory and research: Highlights, unanswered questions, and emerging issues [J]. Annu Rev Psychol, 2009, 60: 1-25.

[19] Ekman P. Handbook of Cognition and Emotion [M]. Chichester: Wiley, 1999.

[20] Mehrabian A. Framework for a comprehensive description and measurement of emotional states [J]. Genet Soc Gen Psychol Monogr, 1995, 3: 339-361.

[21] Martino B D, Kumaran D, Seymour B, et al. Frames, biases, and rational decision-making in the human brain [J]. Science, 2006, 313:684-687.

[22] Irrational Decisions Driven By Emotions, ScienceDaily. Retrieved June 29, 2009, from http://www.sciencedaily.com/releases/ 2006/08/060803171138.htm.

[23] Clore G L, Storbeck J. Affect as information about liking, efficacy, and importance. In: Forgas J, ed. Affect in Social Thinking and Behavior [J]. New York and Hove: Psychology Press, 2006.104: 123-142.

[24] Buchanan T W. Retrieval of emotional memories [J]. Psychol Bull, 2007, 133: 761-779

Optimum topology design for the concentrated force diffusion structure of strap-on launch vehicle

Yong Mei[1*], Guoqiang Liu[1] and Zebin Shen[2]

[1]The Third Engineer Scientific Research Institute, Luoyang 471023, China
[2]PLA Foreign Language University, Luoyang 471023, China

Abstract. The thrust from the booster of strap-on launch vehicle is transmitted to the core via the strap-on linkage device, so the reinforced structure to diffusion the concentrated force should be employed in the installation site of this device. To improve the bearing-force characteristics of the concentrated force diffusion structure in strap-on linkage section and realize the lightweight design requirements, topology optimization under multiple load cases is conducted for the concentrated force diffusion structure in this study. The optimal configuration finally obtained can achieve 17.7% reduction in total weight of the structure. Meanwhile, results of strength analysis under standard load cases show the stress level of this design scheme of the concentrated force diffusion structure meet design requirements and the proposed topology optimization method is suitable for the design of the concentrated force diffusion structure in concept design phase.

1 INTRODUCTION

Aerospace structures are often assembled from multiple sections. The concentrated force is likely to impact in the installation site of mechanical joints during the external load transferring through the internal structure. The linkage device of the strap-on launch vehicle coupling the core with the booster, and transmits the thrust from the booster to the core. Due to the core and booster employ thin-walled skin structure, which structural forms have small stiffness that is usually necessary to design concentrated force diffusion structure in the installation area of the joints. The force diffusion structure can meet the reliability requirements and transmit the axial or lateral concentrated load from the booster to the core.

The large or heavy launch vehicle with high thrust becoming the focus of research under the development of space technology and demand of the commercial launch, manned lunar landing, deep space exploration [1]. It is reported that a new generation of high-thrust strap-on launch vehicle's booster thrust will reach more than 400% the maximum thrust of the existing model and the strap-on launch vehicle's core size specification will also be more than twice the existing structure [2]. The traditional strap-on launch vehicle compartment using the skin structure strengthening method which is difficult to bear the large concentration force transferring from the large booster. At the same time that the structure form of solid booster technology may be used in the future changing greatly compared to the existing liquid booster system, so the existing strap-on linkage device structure scheme needs to be redesigned. In addition, the redesigned form should meet the structural stiffness and strength requirements, as far as possible to reduce the mass of concentrated force diffusion structure to realize the lightweight design requirements.

Topology optimization aims to find the best possible layout for a structure within a specified design domain under a set of loads and boundary conditions, which can achieve the optimal performance by changing the topology relation under the premise of satisfying the constraint, and can provide the conceptual design of the structure with the optimal force path in the initial stage of engineering design [3]. Topology optimization has become a very important research in the engineering sector with the tendency of engineering field pursuing lower manufacturing costs, lighter structural weight and better structural performance [4]. It is also concerned with the structural topology with the best performance indexes under the constraints of stress, deformation and other constraints with the deepening of topology optimization study. To this end, carrying out the topology optimization research of concentrated force diffusion surround the typical linkage components bearing-force characteristics in the aerospace structure, especially in the launch vehicle structure is very urgent.

This study carries out the conceptual design for the concentrated force diffusion structure based on the load index and structure specification of a certain type high-thrust strap-on launch vehicle. Firstly, the finite element model of the initial structure design is built by using MSC.PATRAN/ NASTRAN, and the simulation analysis is carried out under the standard load case to

* Corresponding author: meiyong1990@qq.com

determine the topology optimization area. Then, based on the FE-DESIGN company's TOSCA software platform, the topology optimization for the concentrated force diffusion structure is carried out by using the minimizing weighted compliance of the multiple load cases as the objective function, the manufacturing constraints and the volume ratio as the constraint conditions. Finally, a new type of concentrated force diffusion structure is designed and the intensity checking is carried out according to the topology optimization results.

2 MATHEMATICAL MODEL OF TOPOLOGY OPTIMIZATION PROBLEM UNDER MULTIPLE LOAD CASES

Solid isotropic material with penalization (referred to as SIMP) is the most commonly used continuum structure topology optimization method. The relative density of the element is usually designated as the design variable and generally (0, 1) distributed in the initial design area after topology optimization [5]. The interpolation model of SIMP algorithm mostly uses density-stiffness exponential model which assumes that the elastic modulus of the material and its density have the following relation:

$$\begin{cases} \rho(x) = x_e \rho_0 \\ E(x) = x_e^p E_0 \end{cases} \quad (1)$$

Where x_e is the relative density of the element, $\rho(x)$ is the design variable of topology optimization, ρ_0 is the initial density of the element in the design region, E(x) is the elastic modulus after optimization, E_0 is the initial elastic modulus, p is the penalty factor.

Before the separation of the strap-on launch vehicle booster and the core, it needs to focus on the load case in several typical states of the concentrated force diffusion structure. The minimum compliance of the structure under different load cases is different for the topology optimization problem with the objective of minimizing the compliance. In order to ensure that the compliance of the structure is as small as possible under multiple load cases, Gong [6] uses the weighted average of the structural compliance under each load case as the optimal compliance of the structure, and the original optimization problem is transformed to multi-objective optimization problem. Thus, the topological optimization model of multiple load cases can be established based on the SIMP algorithm with the objective of minimizing the structural weighted average of mean compliance C and the volume ratio α as the constraint:

$$\min_{\rho_1,\cdots,\rho_2} \boldsymbol{C} = \sum_{i=1}^{l} \eta_i \boldsymbol{C}^i = \sum_{i=1}^{l} \eta_i \boldsymbol{U}_i^T \boldsymbol{K} \boldsymbol{U}_i$$

$$subject\ to \begin{cases} \boldsymbol{F} = \boldsymbol{KU} \\ \sum_{j=1}^{n} V_j \rho_j - V^* \le 0 \quad V^* = \alpha V_0 \\ 0 < \rho_{\min} \le \rho_j \le 1 \quad j = 1,2,\cdots,n \end{cases} \quad (2)$$

Where l is the total number of load cases, η_i is the weight factor of the ith condition, $\boldsymbol{C}^i$ is the compliance of the ith condition, $\boldsymbol{C}$ is the weighted compliance of the multiple load cases, $\boldsymbol{K}$ is the total stiffness matrix , $\boldsymbol{F}$ is the load vector, $\boldsymbol{U}$ is the displacement vector, α is the volume ratio, V_0 is the initial volume of the design domain material, V^* is the volume of the material allowed, V_j is the volume of the element, ρ is the density of the element, ρ_{min} is the lower limit value of the design variable and generally in order to avoid the total stiffness matrix singular, take $\rho_{\min} = 0.001$, n is the number of elements.

3 FINITE ELEMENT MODELING AND ANALYSIS OF CONCENTRATED FORCE DIFFUSION STRUCTURE

3.1 Introduction of concentrated force diffusion structure

For the strap-on launch vehicles, the booster's gravity and the engine thrust will be transmitted to the core through the strap-on linkage device and the installation area of the linkage device is the part of concentrated force applied. The concentrated force diffusion structure of the shell is usually designed as the main structure for the bearing capacity, which using reinforced bearing and joint structure to diffusion concentration corresponding to the concentrated force of the part.

Concentrated force diffusion structure consists of the front and rear frame components, three intermediate frame, skin, composite beam, vice beam, reinforced bearing, joint and stringer (the design of concentrated force diffusion structure in this study does not consider the import and export hole of cable). The whole structure is symmetrical by the quadrant. Where reinforced bearing and joint structure using 7A09 aluminum alloy, the rest of the structure using 2A14 aluminum alloy material, the relevant material parameters shown in Table 1.

Table 1. Material property parameter

Material	Elastic modulus	Poisson's ratio	Yield limit	Strength limit
2A14	70560 MPa	0.3	—	355 MPa
7A09	70560 MPa	0.3	315 MPa	500 MPa

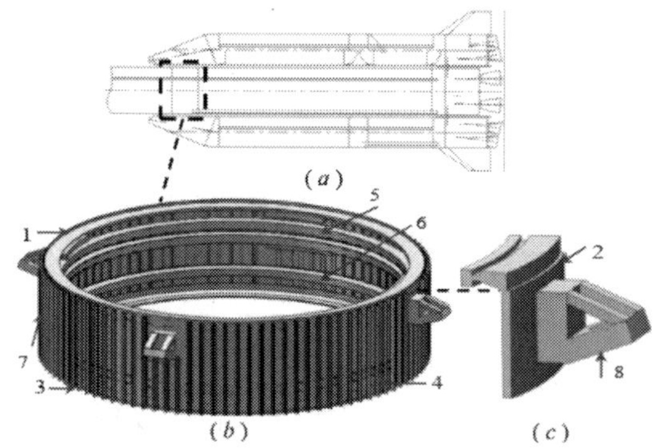

1. front frame, 2. reinforced bearing, 3. rear frame, 4. stringer, 5. front intermediate frame, 6. rear intermediate frame, 7. skin, 8. joint

Fig. 1. The concentrated force diffusion structure

3.2 Finite element model of concentrated force diffusion structure

The load cases and the corresponding load value of the concentrated force diffusion structure in mechanics analysis are determined according to the analysis results of the early strap-on launch vehicle transmission path. Since the boosters are generally symmetrically mounted around the core, 1/4 finite element model of concentrated force diffusion structure built based on MSC.PATRAN can be used for the analysis. In this study, the front and rear ends of concentrated force diffusion structure connects with an extended section, which is set to be elastic supporting region as non-design domain considering the connection stiffness of the structure in engineering practice. The boundary conditions on the top of front extension and bottom of rear extension are set to be fixed, and the extension section is connected to the main structure by means of the nodes equivalent. In addition, the symmetry boundary conditions in the cylindrical coordinate system are applied to both left and right sides of the concentrated force diffusion structure. The finite element model shown in Figure.2 (a) is modeled by Hex8 entity element with a total of 260504 elements, 345223 nodes, and the influence of the skin is not considered during the analysis.

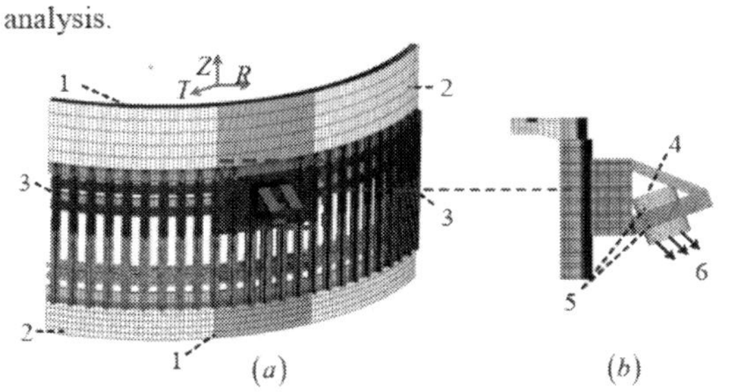

1. fixed support, 2. extended section, 3. symmetric boundary condition, 4. strap-on linkage device, 5. contact interface, 6. load

Fig. 2. The FEA model of the concentrated force diffusion structure

In this study, the actual assembly relation between the strap-on linkage device and the concentrated force diffusion structure is simplified, which is equivalent to contact and the contact type between the device and the joint defined as flexible contact, the load transmitted by the booster is loaded on the end of joint on the strap-on linkage device, other boundary conditions processing methods are shown in Figure 2 (b). The strength analysis of the concentrating force diffusion structure on strap-on launch vehicle under multiple load cases is carried out using the linear contact solver 101 of NASTRAN. The strength analysis in the conceptual design phase requires consideration of the 4 standard conditions shown in Table 2, requiring that the maximum Von Mises stress of the structure does not exceed the strength limit of the associated material and the maximum deformation is less than 15 mm.

Table 2. Analysis load cases

Load case	time/ s	load / kN			remarks
		Z	*R*	*T*	
1	-5.0	235	-296	0	Loaded with fuel
2	0+	688	-89	0	Take off
3	149	1860	-94	0	Reach the maximum acceleration
4	150	-214	27	0	Turn off the booster engine

3.3 Finite element analysis results of concentrated force diffusion structure

The strength analysis results of concentration force diffusion structure under multiple load cases are shown in Table 3. The results show that the initial design configuration satisfies the structural stiffness and strength design requirements, and the maximum Von Mises stress is much smaller than the material strength limit, so in this study will carry out topology optimization based on this design configuration.

Table 3. The structural static response under the typical load cases

Load case	Maximum Von Mises Stress/ MPa	Maximum deformation/ mm
1	55	0.5
2	131	3.0
3	375	8.6
4	41	0.9

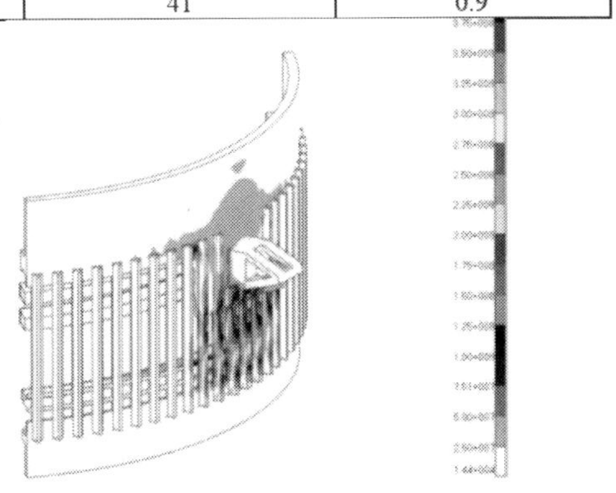

Fig. 3. Stress nephogram of the concentrated force diffusion structure

It can control the optimization and non-optimization parts of the structure by specifying the design and non-design domain in the topology optimization. Figure.3 shows the Von Mises stress nephogram of the concentrated force diffusion structural of strength analysis under load case 3. The results of the stress distribution show that the maximum stress concentrated in the reinforced bearing and joint (the conclusion of 1, 2, 4 load cases is the same as that described in case 3). In addition, in order to strengthen the reinforced bearing and joint using a large number of materials in the initial design which quality accounts for 33% of the total weight. There is an urgent need for weight reduces in the realization of lightweight design while changing the original concentration form of force diffusion. Therefore, the front frame, intermediate frame, rear frame, etc. are designated as non-design domain and reinforced bearing, joint are set to be optimize design domain. At the same time that the manufacturing constraints of the symmetrical distribution are defined based on the centerline of the structure.

4 OPTIMUM DESIGN AND RESULT CHECK OF CONCENTRATED FORCE DIFFUSION STRUCTURE

4.1 Optimum design of concentrated force diffusion structure

Based on Tosca software platform and integrate Nastran solver for optimum topology design of the concentrated force diffusion structure, and the SIMP algorithm is carried out by using the method of moving asymptotes algorithm (referred to as MMA) to iterative solution of the optimization model. In the conceptual design stage set 4 conditions are equally important and the weight factor η_i in the formula (2) is processed according to the linear weighted method. In TOSCA system set the compliance weight factor of each condition is 25% to the total compliance, so as to obtain the weighted average structure compliance under multiple load cases.

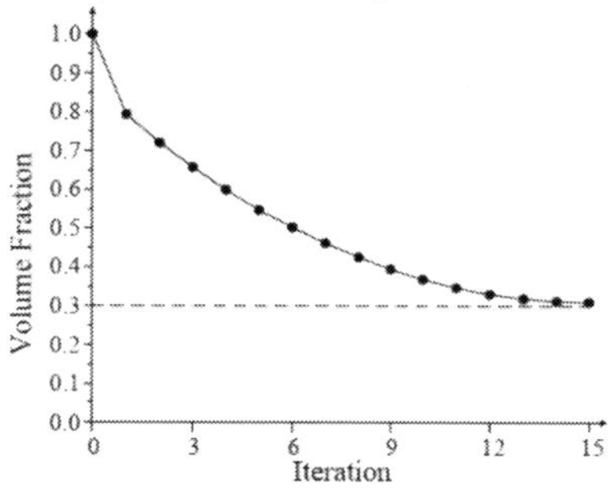

Fig. 4. Evolution history of volume fraction

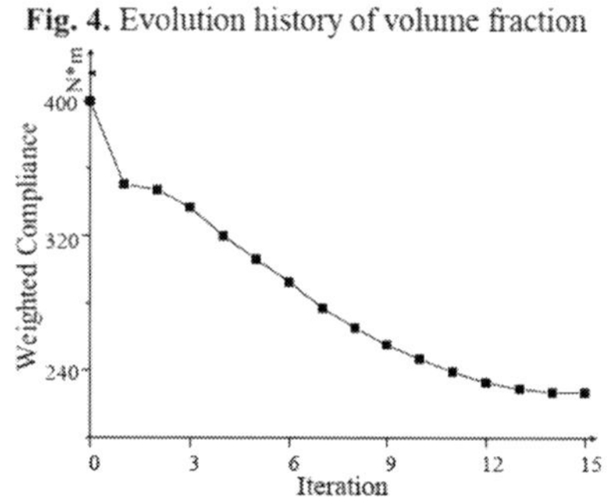

Fig. 5. Variable trend of the weighted compliance

The optimum topology design of the structure is carried out by taking the minimum weighted compliance for the target volume of 30% of the initial design domain as the objective, considering the performance of concentrated force diffusion and the stiffest design as the constraints based on the SIMP algorithm. Figure 4 shows the evolution history of volume fraction for the reinforced bearing and joint structure under 4 load cases. It can be seen that a stable material removal was carried out through the SIMP optimization process after 15 iterations and the volume ratio is approximately 30% of the initial design domain. The iteration history of the weighted compliance is shown in Figure 5.

The optimal material distribution results of the reinforced bearing and joint structure after topology optimization and boundary smoothing within the design domain are shown in Figure. 6. The density of the whole topology optimization result is basically symmetrical distribution and achieving a smooth solution.

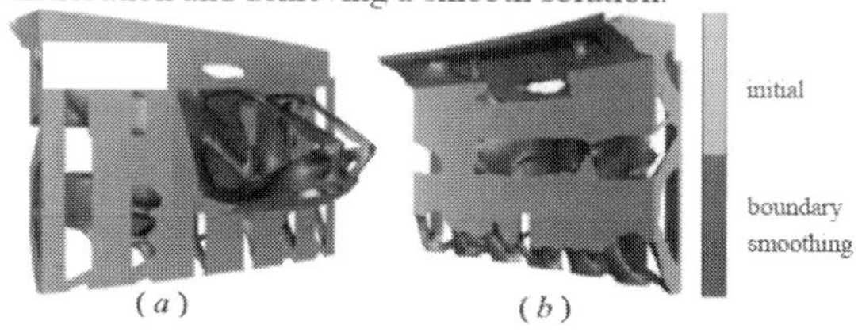

Fig. 6. Optimal topological structure of support and joint

4.2 Re-modeling and result check of the new concentrated force diffusion structure

It is necessary to redesign the reinforced bearing and joint structure on the basis of the engineering practice according to the optimum topology configuration. The main considerations are as follows: (1) If a density based method, like SIMP, is employed for topology optimization, there is a need to threshold at an arbitrary density to achieve a discrete solution which contains some parts of intermediate relative density material that is not present in the actual structure and should be modified to improve the optimality of the solution; (2) the shape of the material-filled region boundary is more complex or even discontinuous in the material distribution of the optimum topology design, which is inconsistent with the actual structure design rules at the same time increasing the processing difficulty. So it is necessary to simplify the optimal topological configuration and require additional post-processing, such as smoothing and shape optimization, to generate a manufaturable design with smooth and clearly defined boundaries. The redesigned reinforced bearing and joint structure are shown in Figure 7 out of consideration of these circumstances, the results showed a significant improvement in boundary representation and structural performance of the solutions. Compared with the initial configuration shown in Figure.1 (c), the material at the top of the reinforced bearing and joint is partially cut. The whole structure of new reinforced bearing and joint is integrally trapezoidal. The new structure will result in a 17.7% reduction in the overall concentrated force diffusion structure and achieve the goal of lightweight design.

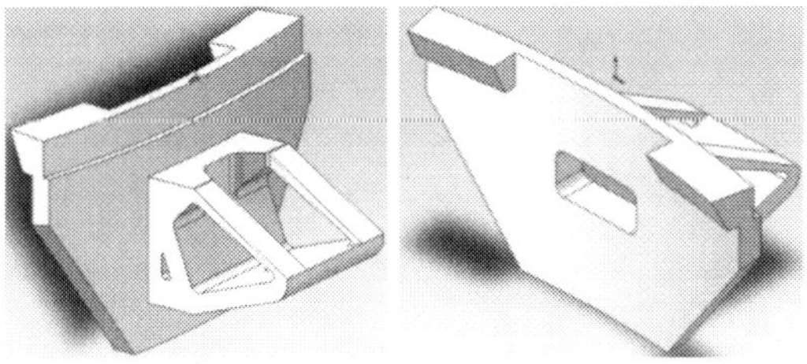

(*a*) (*b*)

Fig. 7. The new support and joint according to the optimum topology design

In order to verify whether the designed newly concentrated force diffusion structure can meet the requirements of structural strength and stiffness, the newly designed structures are discretized and meshed by finite element replacing the original structures which have defined material parameters and element properties. Define the contact boundary conditions and deformation constraints between the various components, and perform the strength analysis under the standard load cases as with the previous analysis.

The calculation results of the maximum Von Mises stress and deformation under 4 typical load cases are shown in Table 4. The maximum stress and deformation appear in the case 3, the corresponding maximum stress is 466MPa which is less than the material strength limit, and the maximum deformation is 11mm which is still less than the stiffness requirement in the concept design. Therefore, the maximum stress and the deformation value of the redesigned concentrated force diffusion structure according to the topology optimization results conform to the design index of the stiffness and strength in the conceptual design stage. So the structure design scheme is reasonable.

Table 4. Optimized design results

Load case	Maximum Von Mises Stress / MPa	Maximum deformation / mm
1	84	0.9
2	158	3.9
3	466	11.0
4	55	1.2

5 CONCLUSIONS

In this study, take the minimum weighted compliance of the structure as the objective function, considering the performance of concentrated force diffusion and the stiffest design for the target volume of 30% of the initial design domain as constraint, the optimum topology design of the concentrated force diffusion structure is carried out based on the SIMP algorithm in the optimization software TOSCA integrating finite element analysis software MSC.PATRAN/ NASTRAN.

The additive manufacturing design of the structure is carried out according to the topology optimization results. The weight of new concentrated force diffusion structure configuration is 17.7% less than the original ones, and the maximum stress and deformation of the structure under specific multiple load cases are lower than those in the conceptual design stage. Finally, the goal of lightweight design is realized.

Consider the fact that this study only considers the structural optimization under static load cases. The results can be more in line with the engineering requirements if the topology optimization taking into consideration under dynamic response of the flight process.

References

1. L.H. Long, X.J. Wang, Y. Rong, The development prospect of expendable launch vehicle in China. Chinese science E: technical science, **39**, (2009), 460-463.
2. W. He, W. Liu, G. Li, Heavy launch vehicle and its application. Missile and space vehicle, **1**, (2011), 1-5.
3. X.H. Wang, H.F. Zheng, H.P. Liu, et al, Structural optimization of folding wing based on HyperWorks. Journal of solid rocket technology, **6**, (2012), 795-798.
4. A.R. Yildiz, Comparison of evolutionary- based optimization algorithms for structural design optimization Engineering applications of artificial intelligence, **1**, (2013), 327-333.
5. M.P. Bendsoe, O. Igmund, Material interpolation schemes in topology optimization, Archive of Applied Mechanics, **69**, (1999), 635-654.
6. S.G. Gong, X. Liu, G.L. Xie, et al, Topology optimization under multi-load cases based on element-free Galerkin method, Journal of mechanical engineering, **12**, (2009), 137-14

Research on the Network Security Strategy for Digital Distance Education Platform

Minzhu Zhang[1,*] and *Xiaofei* Dong[2]

[1]Teaching Affairs Department, Suzhou Chien-Shiung Institute of Technology, 215411 Taicang, China
[2]School of Mecheanical Engineering, Yancheng Institute of Technology, 224051 Yancheng, China

Abstract. Distance education has been an important development tendency and learning platform with the emphasis of lifelong learning of the society. Networked learning and teaching is a main characteristic of distance education, which inevitably needs to transmit large magnitude of private data among students, teachers and the education platform. To protect the security of data transmission and storage, a networked security strategy was proposed. The security strategy is based on the technologies of intrusion detection and digital signature. An intrusion detection model was established in accordance to the main tasks of distance education platform. The encryption process of digital signature was illustrated along with the information flow of the distance education platform. The paper offers an effective reference for solving security problems of distance education platforms.

1 Introduction

Distance education, also called distance learning, is defined as an education model that lesson giver and accepter are not face-to-face but through media such as networked computer. Distance education is not a fixed concept but a developing one, which is based on the development of computer network. Current trends in the field of distance education indicate a shift in pedagogical perspectives and theoretical frameworks, with student interaction at the heart of learner-centered constructivist environments [1]. It has been said that correspondence study represents not only as historical legacy but a continuing dominance of distance education practice. And assumptions regarding learning are implicit in designs of instruction and education [2]. With the development of distance education, many researches about networked technology of distance education have been studied. For example, Hillman et al [3] have found that the interaction occurred between the learner and the technologies used to deliver instruction has not been well considered. Therefore they presented the concept of learner-interface interaction and recommends instructional design strategies that facilitate students' acquisition of the skills needed to participate effectively in the electronic classroom. Berge [4] has list the roles, tasks, and functions of a computer conference moderator used to promote students' learning skills.

Modern distance education mainly relies on digital satellite transmission and computer network, adopting multi-media resource and methods, to form a personalized learning style, which is suitable for different students, especially for those who are not convenient assessable to school. Distance education has become an expectation with students who have grown up with technology as well as nontraditional and working students. Most universities now offer a wide range of online courses and degree programs to meet these needs [5]. The education department of many countries, such as America, British, German, and China, have taken distance education as an important tendency of developing lifelong learning. Distance education has many special merits that traditional education cannot achieve, such as owning broad coverage over students of different ages, low cost, not limited to the basic infrastructures, and not limited to teachers and other teaching materials, which has been widely accepted by mass.

However, with the rapid development of Internet, network security has been a more and more severe problem. And web-based distance education is also facing the same problem, such as virus infection, illegally access to database, account and PIN code theft, information leakage, distance attack, which often lead to teaching activities not carried normally. Unsafe environment will bring lots of incontinence to students. Lin et al [6] focused on how tool development can combine with security to provide a safe, efficient, and effective learning environment, and put forward a number of distance education applications developed in our lab and show how security and privacy may be integrated into these applications. Stallings [7] proposed there are two requirements for secure use of symmetric encryption: a strong encryption algorithm and a secret key known only to senders or receivers. Hoque et al [8] adopted Genetic Algorithm to implement the intrusion detection system. The evolution theory to information evolution is applied in this paper to filter the traffic data

* Corresponding author: 310897265@qq.com

and thus reduce the complexity. Rao et al [9] presented a support vector machine-based intrusion detection system, and the paper has given a comparison of detection ability between different detection method and the method proposed in this paper. Bauer and Koblentz [10] have described a knowledge-based prototype network intrusion detection expert system, which combines knowledge describing the target system, history profiles of users' past activities, and intrusion detection heuristics from a knowledge-based system capable of detecting specific violations that occur on the target system. Patel et al [11] identified a list of germane requirements by considering the desired characteristics of intrusion detection system and cloud computing systems and provided an appropriate set of all possible solutions and a layered taxonomy of intrusion detection system. Johnson et al [12] has described the digital signature technology called the elliptic curve digital signature algorithm and discusses related security, implementation, and interoperability issues. Courtois et al [13] has studied an approach of achieving McEliece-based digital signature scheme, which is based on one of the oldest known public key crptosystems.

In this paper, the strategy network security management will be studied in accordance to the requirement of distance education system.

2 Safety requirements of distance education system

The network system of distance education has distinct deficiency of existing network, operation system and serve leakages, the distance education platform sometimes will receive active and inactive attacks from Internet visitors. The adoptable security measures for distance education mainly includes following sides.

- The security of Internet serve is mainly assured through setting management of password and user access limitation, visiting control of networked resources. Feasible and useful measurements of maintaining Internet serve is the best method of defending outside attacks.
- Establishing physical security transmission media and encrypting transmitted data are two different ways of ensuring transmission security of Internet platform of distance education. For transmitting teaching resources such as students' test paper or grades, an encrypting algorithm will be applied to the communication of data link and end-to-end. Thus even the illegal invader has got the data, he still cannot read the data, and the security of the network is safeguarded.
- The firewall is the primary security technology of computer network system, which is assembled between outside Internet and the innernet of distance education platform. The firewall is currently one of the most effective security measures to prevent illegal invasion. As the blocking and control point, firewall can greatly reduce the risk of network system through filtering unsafe and uncertain services.

Detection technique is divided into anomaly detection and misuse detection. Pattern matching technology belongs to misuse detection, which is the most commonly used. A data detection technology is the most practical and widely used technology. Only deepen research on intrusion detection method can make faster and better development of intrusion detection technology. The intrusion detection method of the core will be very meaningful and effective. The efficiency of detection can be improved by integrating more flexible and effective into the intrusion detection system.

3 Intrusion detection of distance education system

In a practical intrusion detection system, the most important part is the detection model, which determines what kind of technology will be used for the intrusion detection. Intrusion detection system provided in this paper aims on transferring static protection to dynamic protection, through consistency inspection, traffic statistics, and pattern matching and application. Intrusion detection system of distance education platform is based on common intrusion detection framework, of which the basic framework is shown in Figure 1. The intrusion detection system is mainly made up of the following four pats. The system need to analyse the data known as events, and the events is based on network data packet, which can also be based on a host system log. The main function of event generator is to acquire events from the whole computer environment and offer the events to other parts of the system. The analysing results will be sent to the event analyser. Responding element will make out suitable and repaid responds to results sent by event generator. The responding element can make drastic responds such cutting off linkage and revising file attributes. Event database is a place in the system where the intermediate and final data are stored.

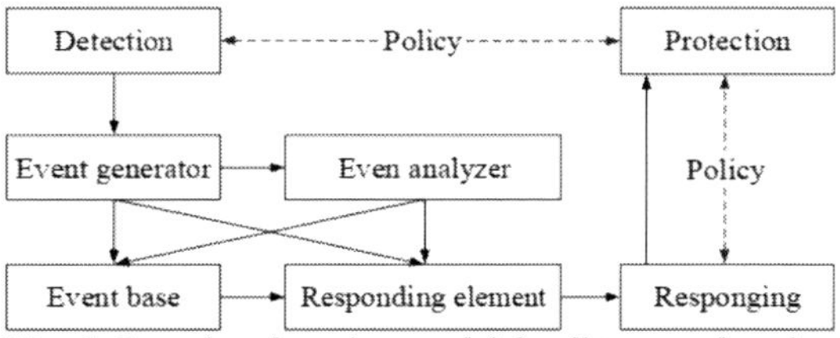

Fig. 1. Intrusion detection model for distance education system

The technology of intrusion detection system can be simply divided into two categories: feature based detection and anomaly based detection. The feature based detection mainly includes protocol analysis and pattern matching. Network intrusion detection method based on information system and pattern matching is known and will be collected from the misuse of pattern knowledge were compared to find the intrusion. The protocol analysis with respect to the pattern matching technology is a kind of intrusion update the detection technology. It captures data packets, and then analyze the data packets, including network protocol and command parsing, quickly detect some attack characteristics in the parsing code. Anomaly detection technology based on a lot of factors, such as the use of statistical model, expert system technology. It is the first important thing to make a statistics on the behavior of the system, the statistical performance of normal use of the system, such as the number of visits, operation failure times and delay. The statistical performance is used to compare with the network, the behavior of the system, when the observed values in the normal range, the intrusion detection system will determine the invasion.

3.1 Main tasks of intrusion detection system of distance education platform

The main task of intrusion detection system for distance educations platform has its own specialties, which are mainly generalized as the following items: 1) monitoring, analysis of user and system activity of the distance education platform; 2) proceeding statistical analysis of abnormal behavior patterns and issue intrusion regulations; 3) checking the correctness of configuration and security vulnerabilities of the distance education system, and prompting the administrator to patch the vulnerability; 4) responding in real-time to the intrusion behaviors; 5) assessing the integrity of key resources and data files of the system; 6) tracking audit management of the operating system, and identifying the breach of security strategy. Though firewall and intrusion detection system are both helpful to the security of the distance education system, but there still exists distinctive difference. System firewall is an important defense, which no longer do any operation on the data that has passed through the firewall. However real-time intrusion detection will respond immediately in case of new invasions, which is an important completion of the defense of firewall. Firewall allows some inner hosts can be accessed externally, while intrusion detection does not have these functions, but monitoring and analyzing the activities of the users (students, teachers, and local teaching spots) and the system of distance education.

3.2 Framework of intrusion detection system

A complete intrusion detection system is typically made of data collecting model, data analyzing model, data management model, data preprocessing model, communication model, data storing model, and data responding model. Data from Internet are collected by the data collecting model and preprocessed by data preprocessing model, and transferred to data analysis model. Data management model provides information overflow with users through user interface model. Data preprocessing model includes many functions such as data filtering, data classifying, data merging and data converting. A typical structure of an intrusion detection system can be illustrated by the following figure (as shown in Figure 2).

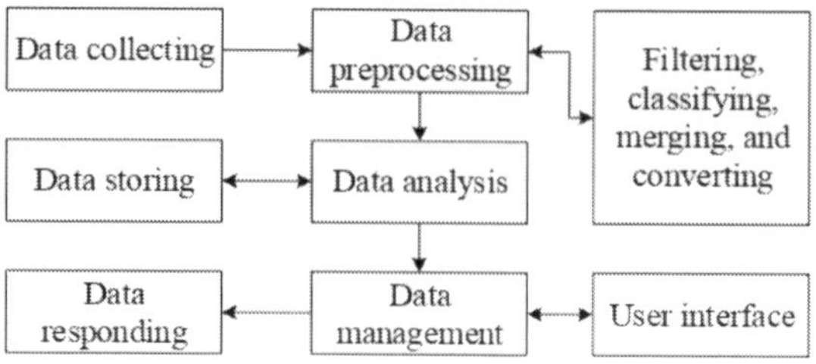

Fig. 2. The basic structure of an intrusion detection system

3.3 Strategy of digital signature for distance education platform

With the rapid growth of Internet, a great need has appeared for a mechanism capable of protecting the ownership of video or image authors and distant learning resource providers. A classical means to protect distant learning material is known as "Digital Signature", where the video or image-data material is slightly modified ("signed") in order to embed a number or image or other information, which can later be retrieved [14][15]. With the application of digital signature, the teaching resources and private information of learners are well protected. Digital signature is a kind of technology that utilize sending port to produce an encoding algorithm which is difficult to be enciphered.

The digital signature has a unique specialty of the signer, which is difficult to be stolen and revised by illegal attackers. The encoding, transmission and decoding process of the digital signature can be expressed as the following three steps: firstly, the learner encodes the information by encoding key which only be mastered by himself or herself; secondly, the learner transmits the encoded information to the other person who he or she want to send to; finally, the accepter decoding the information by secret key and calculate the characteristic value through a HASH function, and compare the characteristic value with digital signature. If the characteristic value is not changed that means the information has been safely transmitted and accepted.

For distance education platform, the strategy of digital signature needs to meet following requirements. The digital signature needs to be pre-versified. The distal signature can keep the intact of the file which is transmitted between the learner and teacher through Internet. The application model of digital signature used between teachers and learners for distance education plater is illustrated by Figure 3.

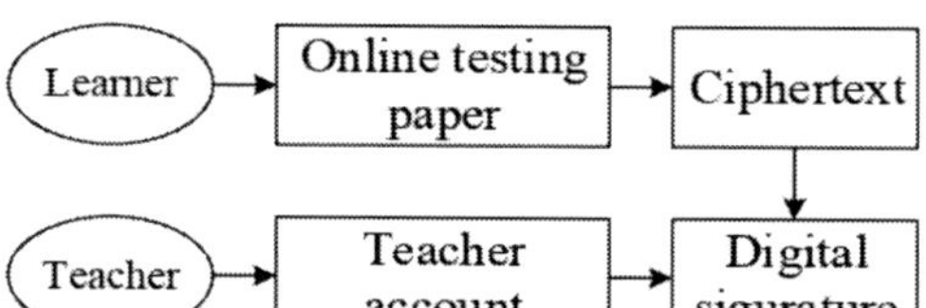

Fig. 3. The basic structure of digital signature for distance education platform

3.4 Strategy of digital signature for distance education platform

Through distance education platform, students carried a series of learning activities, which are encrypted into protective Class-One File by Key One, then encrypted into Class-Two File by Key Two, next into Class-Three File by Key One, finally applying for an public key of a certain curriculum from the distance education platform. The Class-Three File will be finally encrypted into a Class-Four File by the public key, thus the whole encrypting process is complete. When students or local learning spot received a secret file from education management department through the distance education platform, a private key need to be applied from the private key generator to decrypt the secret file into a readable file. The process of applying digital signature on the platform of distance education can be modelled by the Figure 4.

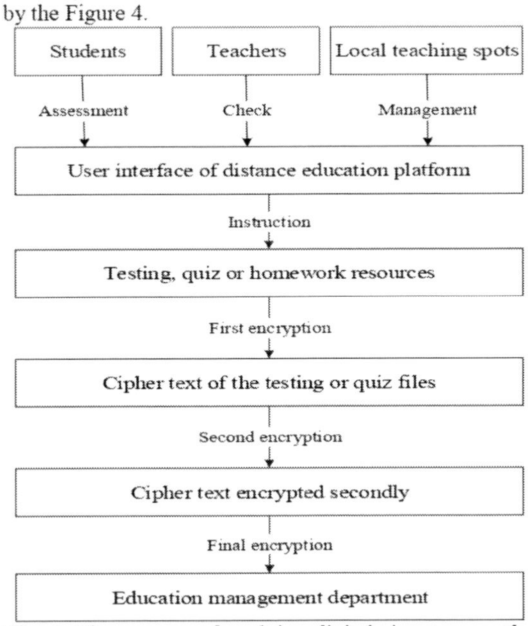

Fig. 4. The process of applying digital signature on the platform of distance education

4 Conclusions

With the development of the concept of lifelong learning, more and more people, especially adults, have resorted to Internet to acquire the knowledge they want. Under the requirement of convenient accessing learning resources and guidance of teachers, lots of distance education platforms have been developed and applied on the purpose of mass learning. However, information security is an important task that needs to be dealt with by the developers of the distance education platforms. Learning and teaching via user interface is a main characteristic of distance education, which inevitably needs to transmit large magnitude of private data (such as delivering test paper, individual homework and quizzes, handing over finished paper, and correcting and grading the test paper) among students, teachers and the education platform. A strategy of intrusion detection and digital signature is very useful to protect the security of data transmission and storage through distance education platform. An intrusion detection model was established in accordance to the main tasks of distance education platform. The encryption process of digital signature was illustrated along with the information flow of the distance education platform.

6 Acknowledgments

The authors gratefully acknowledge the support by Taicang Key Project of Research and Development Plan (Industry Perspective and Common Key Technologies) under Grant No. TC2016GY07 and Jiangsu Provincial Policy Guidance Program (Enterprise-University-Research Institute Cooperation)-Perspective Joint Research Program under Grant No. BY2016065-52.

References

1. Y. Beldarrain. *Distance education*. 27, 2 (2006). *Distance education*. 27, 2 (2006)
2. D.R. Garrison. *Distance education*. 14, 2 (1993).
3. D.C. Hillman, D.J. Willis, and C.N. Gunawardena. *American Journal of Distance Education*. 8, 2 (1994)
4. Z.L. Berge. *Educational Technology-Saddle Brook NJ-*, 35, 22-22 (1995).
5. S.W. Tabor. *Quarterly Review of Distance Education*. 8, 1 (2007)
6. N.H. Lin, L. Korba, G. Yee, T.K. Shih, and H.W. Lin. *In Advanced Information Networking and Applications, 2004. AINA 2004. 18th International Conference on* IEEE. 1 (2004)
7. W. Stallings. *Pearson Education India*. (2006)
8. M.S. Hoque, M. Mukit, M. Bikas, and A. Naser. *arXiv preprint arXiv*. 1204.1336 (2012).
9. X. Rao, C.X. Dong, and S.Q. Yang. *Journal of Software*. 14, 4 (2003): 798-803.
10. D.S. Bauer, and M.E. Koblentz. *In Computer Networking Symposium, Proceedings of the IEEE*. (1988) 98-106
11. A. Patel, M. Taghavi, K. Bakhtiyari and J.C. JúNior. *Journal of network and computer applications*, 36, 1 (2013) 25-41
12. D. Johnson, A. Menezes, and S. Vanstone. *International Journal of Information Security*. 1, 1 (2001) 36-63.
13. N. Courtois, M. Finiasz, N. Sendrier. *Advances in Cryptology-Asiacrypt*. 2248 (2001) 157-174.
14. T. Vynne, and F. Jordan, Cray Research, Inc., Embedding a digital signature in a video sequence. *U.S. Patent 5,960,081*. (1999)
15. M. Abdalla, and L. Reyzin. *In International Conference on the Theory and Application of Cryptology and Information Security. Springer Berlin Heidelberg* (2000)

Research and Design of Dynamic Migration Access Control Technology Based on Heterogeneous Network

Feng Wang, Weichuan Ni, Shaojiang Liu, Jia Lu, Zhiping Wan
Xinhua College of Sun Yat-sen University, Guangzhou, 510520, PR China

Abstract: With the continuous development of wireless networks, the amount of privacy services in heterogeneous mobile networks is increasing, such as information storage, user access, and so on. Access control security issues for heterogeneous mobile radio network, this paper proposes a dynamic migration access control technology based on heterogeneous network. Through the system architecture of the mutual trust system, we can understand the real-time mobile node failure or abnormal state. To make the service can be terminated for the node. And adopt the 802.1X authentication way to improve the security of the system. Finally, it by combining the actual running test data, the trust update algorithm of the system is optimized to reduce the actual security threats in the environment. Experiments show that the system's anti-attack, the success rate of access, bit error rate is in line with the expected results. This system can effectively reduce the system authentication information is illegally obtained after the network security protection mechanism failure and reduce the risk of user data leakage.

Keywords: Heterogeneous network; wireless network; mutual trust system; trust update algorithm

0. Introduction

The diversification of Internet services and the spread of communication technology services have led to the rapid development of mobile devices. This makes wireless network service demand more than ever before. Heterogeneous mobile wireless network access and network type are showing diversity. Different network implementation technology, transmission mechanism, organization, control mechanism there is a huge difference. Access Control Technology for Heterogeneous Mobile Wireless Networks, which need to request the user's network status based on the current visit. According to the user's location and business in a different way to smooth between the accesses that can to provide reliable access to support services.

Through different access network landing personal account information to access the service system on the mobile device, compared to a single access network to authenticate the way more convenient. As a result, more authentication processes are done on heterogeneous mobile wireless networks [1-2]. Heterogeneous mobile networks involved in the security-critical business, which the need for strong security access control technology as a support. To ensure that the only legitimate user account and other information will not be illegal access. Today's heterogeneous mobile wireless networks offer a variety of security services. On this basis, the relevant security strategies are: encryption, authentication, access control, key management, certificate management, etc. [3-4]. Among them, the most widely used authentication mechanism for network security-sensitive services is authentication methods based on personal information such as ID / password and digital certificates. Personal information will be due to the user's own negligence, but also by other illegal means of invasion of technology, such as the loss or theft of smart cards, access to information is illegal tracking and collection. Even if the application layer and network layer at the same time the implementation of security certification, but due to the dependence between the network layer and the application layer. The application layer's security risks still exist [5]. Because heterogeneous network services are the interaction between users and service delivery systems, service security is based on mutual trust between them. The user terminal can access the service providing system simultaneously through a different access network. When the authentication information of the only legitimate user is illegally stolen by another person, the service providing system cannot distinguish whether the object to be accessed is a legitimate user. As a result, there is a security issue for the mutual authentication system of authentication between the user terminal and the service providing system.

This paper studies the security of heterogeneous mobile

 DOI: 10.1051/matecconf/201713900185

wireless network access control. We proposed architecture consists of a client and a security service system installed on a trusted user node. It can make the security service system can obtain the real-time status of the mobile node. Make the system can be real-time understanding of mobile node failure or abnormal state, which can terminate the service for the node. And this paper is building the modeling method to analysis and comparison with safe access methods. It make the system finding a safer, more efficient access method. And we are use 802.1X authentication to improve the security of the system. And in the authentication process, the user's identity is used as a public key, so that the key management easier, and do not need to perform the key revocation of the certificate. Node are using a non-interactive computing a symmetric key, which can reduce communication overhead. Finally, after the actual operation of this test, the trust update algorithm to optimize the system to reduce the actual operation of the security threats in the environment.

1. Heterogeneous Network

Security issues as wireless heterogeneous network development process must pay attention to an important issue. As the expanding range of network applications and the diversification of access methods, it make a variety of means of attack growing, security plays a vital role in all key technical issues heterogeneous networks. As in all aspects of heterogeneous network routing, authentication, billing, node cooperation, intrusion detection, there are security vulnerabilities. At present, people have carried out relevant research work on some aspects of heterogeneous network security and achieved certain results. However, due to the extremely complex nature of heterogeneous networks, the need to address the security issues is still more.

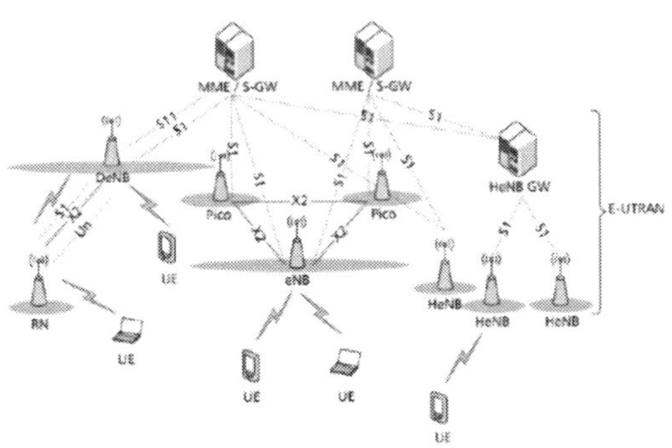

Figure 1. isomorphic network overall architecture map

2. Article Algorithm

2.1 System Architecture Based on Mutual Trust System

Our proposed architecture consists of a client and a security service system installed on a trusted user node. The security service system realizes the user and assigns unique identifiers to each user. The client can manage multiple network interfaces. And the establishment of a mutual trust system to establish and maintain network links based on security services. And it can use security service system to send mobile conditions under the changing network status information, which including the connection status, network address dynamic changes.

The study of the mutual trust system in the system is the key to the design of the whole system architecture. The mutual trust system enables the secure service system to obtain the real-time status of the mobile node. It can make the system understand the fault or abnormal state of the mobile node. If the illegal node chooses not to send the state data to the security service system in order to hide its own abnormal state information, the mutual trust rule will be broken, and the security service system may terminate the service for the node.

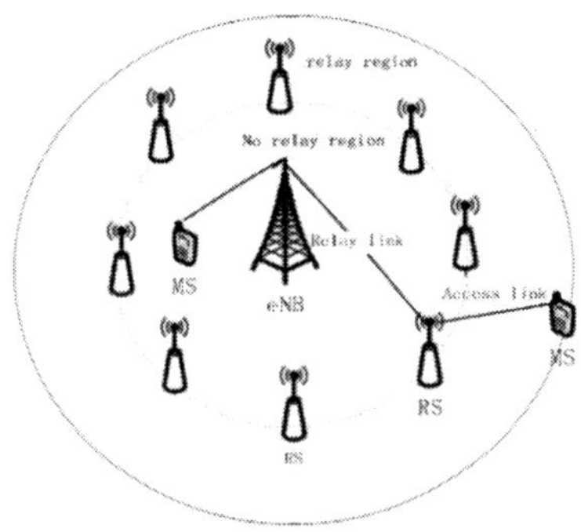

Figure 2. Enhanced cell coverage model

Trusted Verification Security Formula 1: The system state v is to satisfy the trusted authentication security formula. iff for any allowed（s,o,A,T）=true. The following formula holds:

（1）当 A=r 时，fc_c(s)>=fo_c(o) ∧ fc_w(o)>=fo_w(s) ∧ Time(trans,s)<=T;

（2）当 A=w 时，fc_c(s)<=fo_c(o) ∧ fc_w(o)<=fo_w(s) ∧ Time(trans,s)<=T;

Trusted Verification Security Formula 2: system∑（R,D,W,v_0）is a security system。Each state of the system（v_0，v_1，......，v_n）are safe.

Here s is the set of subjects in the system. o is the guest set. A={r,w,a,e} indicates that the subject requests access. They represent the read, write, append and execution. Trusted state set Ts={trusty,untrusty,wtrusty}, which represents trusted entities, untrusted entities, and integrity trusted entities, respectively. And it corresponds to a trusted area, an untrusted area, and a controlled area in the TNC structure.

2.2 Secure Access Model for Heterogeneous Mobile Networks

Through the modeling method [6], which can make all aspects of the system behavior in the early stages of development, is predicted, and thus easier to solve the problem. Therefore, we need to abstract the security access problem and establish the theoretical model at first. Through the model of the method to compare a variety of secure access methods to find a more secure, more efficient access method.

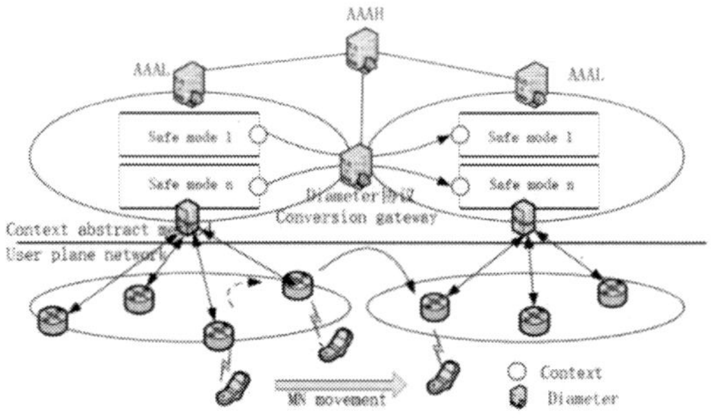

Figure 3. Secure access model for heterogeneous mobile networks

The network is abstracted into two levels. One for the user network plane, it is mainly represents the heterogeneous network topology and the behavior of the user in the network. The other for the abstract level of the context, it is mainly used to ensure the security of the transfer of context. There are multiple Diameter nodes in the fig.3.it is including AAA home server (AAAH), AAA local server (AAAL) and protocol conversion gateway. Among them, if the mobile node is only in the local mobile, then only need to pass AAAL certification, or in the context of the transfer process need to go through AAAH processing.

2.3 Secure Access Authentication Method for Heterogeneous Mobile Networks

In this paper, the network access authentication process, mainly including three levels:

Access control layer, is to control access [7]. Which allows users to access, when to allow access, to allow access to where and so on. It mainly adopts the way of Web authentication and 802.1X authentication.

The authentication control layer, which is the interactive rule that controls the authentication message [8-9]. Based on the key mechanism of the service system, the user responds to the authentication requirements of the service system according to the authentication method and its parameters needed by the negotiation.

Authentication implementation layer, which completion of authentication based on in the selected access and authentication control. At present, commonly used authentication methods include account number / password, CHAP, AKA, etc. Authentication control and authentication implementations are usually combined to complete access authentication.

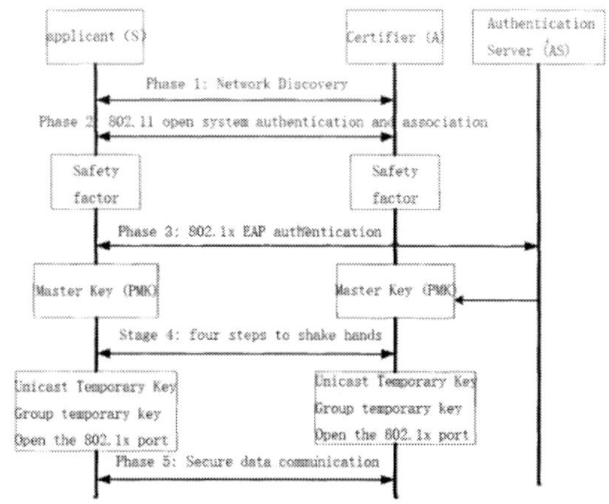

Figure 4. Identity encryption key mechanism

2.4 Key mechanism based on identity encryption

n the identity encryption mechanism, the user's identity is used as a public key [10-11]. This makes it easier to manage keys and does not need to perform a certificate for key revocation. The node uses noninteractive computing symmetric keys to reduce communication overhead. This model of minimizing PKC overhead is very appealing For mobile, resource-constrained heterogeneous mobile networks.

Compared with the traditional encryption algorithm, this article uses a symmetric key cryptography system. The elliptic curve encryption method based on signature (El-Jamal type) is used on the decryption key to avoid tampering with unauthorized attackers.

According to its use of the key only one, both parties use this key to encrypt and decrypt the data. This requires the decryption party to know the encryption key in advance. Symmetric encryption algorithm is characterized by open algorithm, small computation, fast encryption, high encryption efficiency.

2.5 Trust update algorithm

Through the system in the actual operating environment for a large number of security scenes testing, the average number of times the security analysis information is obtained is q_Δ. Here Δ is a placeholder. Respectively, with BSD, HSD and RSD instead of the respective indicators of information on the average number of replacement. The frequency of each security is f_Δ[12-13].

$$f_\Delta = q_\Delta/T$$

The system's total update frequency can be calculated by the above formula:

$$f = f_{BSD} + f_{HSD} + f_{RSD} - f_{BSD} * f_{HSD} * f_{RSD}$$

In this paper, the network security threats are simply divided into three types: warnings, errors and attacks. The threat frequency is expressed as R_y . Here y = 1, 2, 3, which representing the three threat levels. The resulting attack frequency function is g(f).

The replacement period T of the system can be calculated and the following equation is satisfied:

$$T = 1/f^*$$

Here f^* satisfies $\lim_{f \to f^*} g(f) = f^*$.

3. Simulation Experiment

The research group will build a heterogeneous mobile network platform based on NS2. And we using the MATLAB simulation platform to attack the system, the success rate of access, bit error rate and other parameters to simulate. By comparing the data with the traditional system and the literature [14], we can verify the anti-attack effect, the success rate and the data fidelity ability of the algorithm.

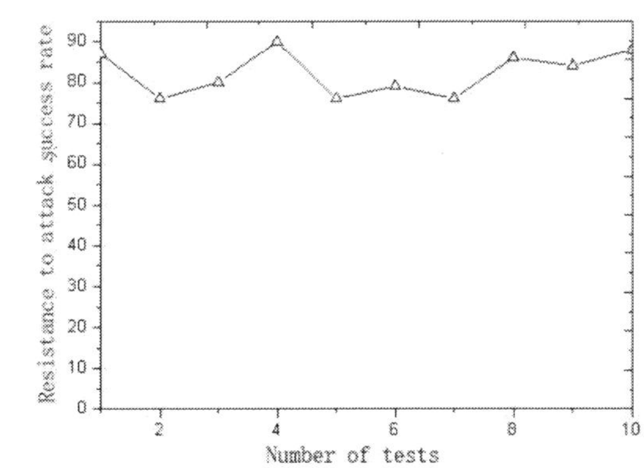

Figure 5. system of anti-attack data curve

In this paper using 10 groups of commonly used network attacks (including message tampering, MitM attacks, etc.) in the actual network environment. And each group we carried out 100 attacks. The experiment to obtain FIG. 5 system against attacks graph data, we can see that the system can effectively detect and block these network attacks. It can guarantee the safety of the system, and each group of anti-attack success rate of more than 75%.

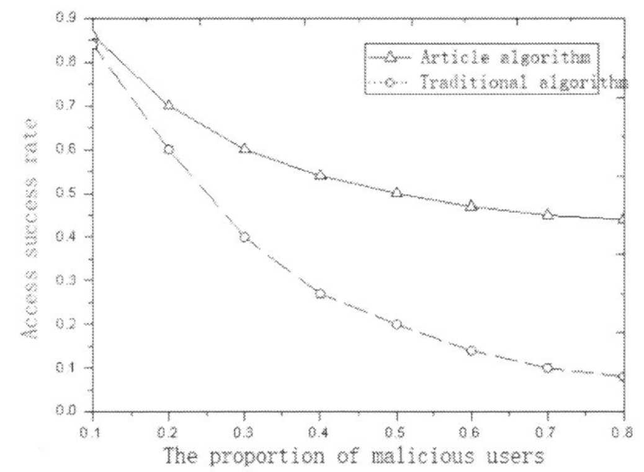

Figure 6. System access success rate change data curve

By observing the data rate of the success rate is not difficult to find that The success rate of its authorized users to access gradually become smaller with the system of malicious users continue to increase. But compared with the traditional algorithm, we can that success rate of this algorithm is always guaranteed to be above the traditional algorithm data curve. Article System Access success rate gradually stabilized with the gradual increase of malicious users. It can effectively prove the stability of the system.

In order to better verify the accuracy of the algorithm. We compare the error rate between the algorithms in the equalizer in the same environment. The data is shown below.

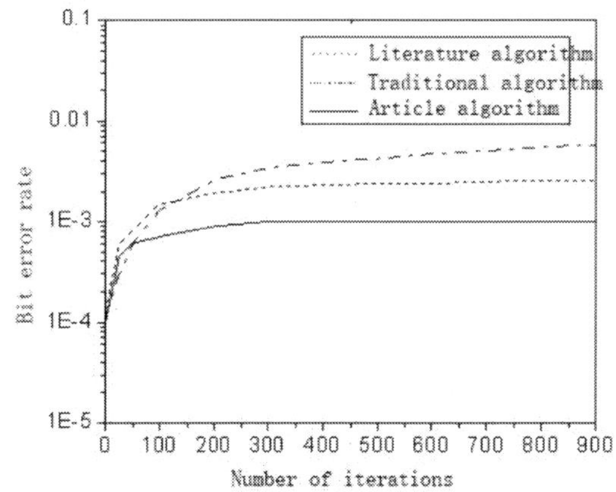

Figure 7. error rate curve of each algorithm

Looking at Figure 7 we can see that the error rate of this algorithm has high stability. The bit error rate remains essentially unchanged, after the number of iterations is 300. But the traditional algorithm has been growing from beginning to end. By comparison, although the algorithm can have a more stable bit error rate, But the steady-state error rate is higher than this algorithm. Therefore, it can be proved that the algorithm has high accuracy.

4. Conclusion

This article proposed a system architecture based on mutual trust system. The service is terminated for the node by detecting the fault or abnormal state of the mobile node in real time. In the process of system authentication using 802.1X authentication to improve the security of the system. And it combines the actual running test data to optimize the trust update algorithm. It will be reduce the actual operational security threats in the environment. The experimental results show that the experimental results are in accordance with the expected results by testing the system's anti - attack, the success rate of access and the bit error rate. It shows a good anti-jamming capability with data fidelity. And it can effectively reduce the risk of user data leakage, with a certain practical value.

Acknowledgements

The paper is subjected by the following project funding:

Guangdong Provincial Department of Education of young creative talents of Natural Science Fund Project (2015KQNCX227)

References

[1] Zhu L, Zhan Z. A random key management scheme for heterogeneous wireless sensor network[C]// Cyber Security of Smart Cities, Industrial Control System and Communications (SSIC), 2015 International Conference on. IEEE, 2015:1-5.

[2] Kumar P, Ylianttila M, Gurtov A, et al. An Efficient and Adaptive Mutual Authentication Framework for Heterogeneous Wireless Sensor Network-Based Applications[J]. Sensors, 2014, 14(2):2732-2755.

[3] Yu X. Security Key Management Scheme in Heterogeneous Wireless Sensor Networks Based on Clustering Strategy[J]. Journal of Networks, 2013, 8(9): 2114-2120.

[4] Turkanović M, Brumen B, Hölbl M. A novel user authentication and key agreement scheme for heterogeneous ad hoc wireless sensor networks, based on the Internet of Things notion[J]. Ad Hoc Networks, 2014, 20(2):96–112.

[5] Deebak B D. Secure and Efficient Mutual Adaptive User Authentication Scheme for Heterogeneous Wireless Sensor Networks Using Multimedia Client–Server Systems[J]. Wireless Personal Communications, 2015:1-23.

[6] Jiang JM, Peng MG, Zhang KC, Li L. Energy-efficient resource allocation in heterogeneous network with cross-tier interference constraint[J]. IEEE International Symposium on Personal, Indoor and Mobile Radio Communications, PIMRC. 2013.9:168-172

[7] Samuel H, Zhuang W. Preventing Unauthorized Messages and Achieving End-to-End Security in Delay Tolerant Heterogeneous Wireless Networks[J]. Journal of Communications, 2010, 5(2):152-163.

[8] Mahshid M K, Eslamipoor R. An efficient and secure authentication for inter-roaming in wireless heterogeneous network [J]. Social Network Analysis & Mining, 2014, 4(2):1-10.

[9] Hou H F, Xin-Sheng J I, Liu G Q. Identity-based anonymity authentication protocol in the heterogeneous wireless network[J]. Journal on Communications, 2011, 32(5):153-161.

[10] Liu G Q, Hou H F. Research on access authentication mechanisms for 3G and WLAN and integrated wireless network between them[J]. Application Research of Computers, 2009, 26(11):4023-4026.

[11] Shrestha AP, Choi DY, Kwon GR, Han SJ. Kerberos based authentication for inter-domain roaming in wireless heterogeneous network[J]. Computers and Mathematics with Applications. 2010,60(2):245-255.

[12] Deng Y, Wang G, Cao J. Trust-Based Fast Inter-Domain Secure Handoff over Heterogeneous Wireless Networks[C]// Wireless Communications, Networking and Mobile Computing (WiCOM), 2011 7th International Conference on. IEEE, 2011:1-5.

[13] Durresi A, Durresi M, Barolli L. Secure authentication in heterogeneous wireless networks.[J]. Mobile Information Systems, 2008, 4(2):119-130.

[14] Yoon SHY, Lee SS, Kim, SH. Robust mutual trust architecture for safety critical service in heterogeneous mobile network environment[J]. Telecommunication Systems. 2015,60(2): 275-284.

Improved adaptive unscented Kalman filter algorithm for target tracking

Chunyao Han[1,*], *Jiajun* Xiong[1], *Kai* Zhang[1]
[1]Air Force Early Warning Academy, 430019 Wuhan, China

Abstract: An adaptive unscented Kalman filter (AUKF) algorithm is proposed to solve the problem that the statistical characteristics of the process noise are unknown in the target tracking, which leads to filter divergence or low filtering precision. The improved Sage-Husa estimator is used to estimate the statistical characteristics of the unknown process noise in the filtering process, and to judge and suppress the filtering divergence, which effectively improves the numerical stability of the filtering and reduces the error of the state estimation. The simulation results show that the improved AUKF algorithm not only keeps convergence but also improves the accuracy and stability of the target tracking under the condition of unknown time-varying process noise statistic, compared with the standard UKF algorithm.
Keywords: adaptive unscented Kalman filter; time-varying process noise statistic estimator; target tracking; motion model;

1 Introduction

Target tracking is a basic problem in the military and civil fields. Its task is to use the measured information obtained by the sensor to estimate the current state of the target and the future state of the target. In the tracking of high maneuvering targets, such as fighter, ballistic missile and reentry vehicle, the state estimation is confronted with three problems: one is the nonlinearity of the system, including the nonlinearity of the motion model and the measurement model, the second is that the process noise can't be accurately described, the third is that the measurement noise is usually not accurate.

The most widely used method to deal with nonlinear estimation is the extended Kalman filter (EKF). This method has the obvious drawback that the linearization error is large and the Jacobian matrix is difficult to be calculated. In the case of high nonlinearity, the filter is unstable, and the estimation error will be large or even divergent. Unscented Kalman filter (UKF) is a nonlinear filtering method, and it does not need to calculate the Jacobian matrix of the nonlinear state equation and the nonlinear measurement equation, and the unscented transformation (UT) can theoretically be approximated to the posterior mean and covariance of the state at least the third-order Taylor precision, regardless of the degree of the nonlinearity system[1][2]. UKF has the advantages of simple realization, high filtering precision and good convergence. Therefore, UKF has received extensive attention from domestic and foreign scholars.

UKF filtering algorithm applied to target tracking need to know the mathematical model of the object to be studied and the priori knowledge of noise statistical characteristics. If the filter is designed based on inaccurate mathematical model and inaccurate noise statistical characteristics, it may lead to large state estimation error and even filter divergence. In general, the statistical properties of the measurement noise can be obtained in advance by the physical characteristics of the sensor. However, for tracking systems, it is difficult to represent process noise with an accurate statistical characteristics due to external disturbances, physical characteristics of acceleration, and maneuvering[3]. Therefore, its statistical properties are often unknown and time-varying.

To solve that problem, adaptive filtering algorithm[4][5][6] came into being. The adaptive filtering algorithm can be divided into three categories: scaling state covariance matrix (P-matrix), multi-model adaptive estimation and adaptive stochastic model modeling. [7] proposed a filtering algorithm of scaling the process noise covariance matrix Q. It is believed that the tuning parameters are less and the robustness is better than the algorithm of scaling the state covariance matrix. But admitting is different from the general adaptive random Model modeling algorithm, can not adjust the variance of a single noise. In recent years, the UKF has been widely used in estimation of state and parameters, control, target tracking, navigation and fault diagnosis[8][9]. Many scholars have tried to improve the UKF in terms of its numerical stability and computational speed, e.g., a decrease sigma points UKF and square root decomposition of the UKF[10].

The Sage-Husa algorithm is a suboptimal unbiased maximum posteriori noise estimator for the uncertain process noise and measurement noise.

In this paper, an AUKF algorithm is proposed. The algorithm uses the improved Sage-Husa algorithm to estimate the unknown noise, and

Corresponding author: hanchunyaoc@163.com; tel: 15072389460

estimates and corrects the process noise statistical characteristics in real time to reduce the state estimation error. This method adopts a method based on covariance matching criterion to judge the filter divergence trend, introduces the adaptive attenuation factor to correct the prediction covariance, and adjusts the filter gain matrix to eliminate the divergence phenomenon in the adaptive filtering. Both theoretical analysis and experimental results show that the proposed AUKF algorithm is superior to the standard UKF algorithm in estimating performance and stability, which greatly improves the ability of real-time target tracking.

2 Mathematical description of state estimation

Consider a nonlinear discrete system as follows:

$$x_k = f_{k-1}(x_{k-1}) + v_{k-1} \tag{1}$$

$$z_k = h_k(x_k) + w_k \tag{2}$$

where x_k is n-dimensional state vector, z_k is m-dimensional measurement vector, v_k is n-dimensional process noise vector, w_k is m-dimensional measurement noise vector, $f_k(\cdot)$ represents non-linear state equations and $h_k(\cdot)$ represents non-linear measurement equations.

Hypothesis1: v_k and w_k are uncorrelated Gaussian white noise and have the following statistical properties:

$$\begin{aligned} &E(v_k) = q_k, \operatorname{cov}(v_k, v_j^T) = Q\delta_{kj} \\ &E(w_k) = 0, \operatorname{cov}(w_k, w_j^T) = R\delta_{kj} \\ &\operatorname{cov}(v_k, w_j^T) = 0 \end{aligned} \tag{3}$$

where Q is non-negative definite symmetric matrix, R is positive definite symmetric $x \sim N(\hat{x}, P_x)$ matrix, δ_{kj} is Kronecker–δ function.

Hypothesis2: the initial state x_0 is not related with v_k and w_k, and obeys the Gaussian normal distribution. The mean and covariance matrix of the initial state are

$$\begin{aligned} \hat{x}_0 &= E(x_0) \\ P_0 &= \operatorname{cov}(x_0, x_0^T) \end{aligned} \tag{4}$$

The problem is to design an adaptive UKF filtering algorithm with noise statistic estimator based on the measurement $Z_k = \{z_1, z_2, \cdots, z_k\}$ when the statistical characteristics of the process noise are unknown or time varying.

3 Adaptive UKF

3.1 Unscented transformation

UKF uses a deterministic sampling strategy to approximate the nonlinear distribution. Its core idea is to use the unscented transformation to estimate the distribution of random variables through a nonlinear system, and still use the standard Kalman filtering framework.

Let n-dimensional variable, m-dimensional variable z is a non-linear function of x

$$z = f(x). \tag{5}$$

The main idea of UT is to design a series of sigma points $\xi_i, i = 0,1,\cdots,L$ to approximate the distribution of random variables under the premise of the sampling mean $\hat{x}$ and covariance P_x, then calculate the value γ_i, $i = 0,1,\cdots,L$ of the sigma point after passing through the nonlinear system, and calculate the distribution of $z \sim (\hat{z}, P_z)$.

In the UT algorithm, the most important thing is to determine the sigma sampling strategy, which is to determine the number, the location and the corresponding weights of sigma points. There are several typical sigma sampling strategy, symmetric sampling, single sampling, third-order moment skew sampling and fourth-order moment symmetric sampling of Gaussian distribution.

The UT implementation is described as follows:

StepA1: According to the selected sampling strategy, calculate the sigma sampling points and their weight coefficients on the basis of the statistical characteristics of $x \sim N(\hat{x}, P_x)$. Let the corresponding weight of sigma points $\xi_i, i = 0,1,\cdots,L$ are $W_i^{(m)}$ and $W_i^{(c)}$, which are the weight coefficients of the first and second order statistical properties respectively. The sampling strategy given in is given below:

$$\begin{cases} \xi_0 = \hat{x} \\ \xi_i = \hat{x} + \sqrt{n+\lambda}\left[\sqrt{P_x}\right]_i, \quad i = 1,\cdots,n \\ \xi_{n+i} = \hat{x} - \sqrt{n+\lambda}\left[\sqrt{P_x}\right]_i \end{cases} \tag{6}$$

$$\begin{cases} W_0^{(m)} = \lambda/(n+\lambda) \\ W_0^{(c)} = \lambda/(n+\lambda) + (1-\alpha^2+\beta), i = 1,\cdots,2n \\ W_i^{(m)} = W_i^{(c)} = 1/2(n+\lambda) \end{cases} \tag{7}$$

$$\lambda = \alpha^2(n+\kappa) - n \tag{8}$$

where $\left[\sqrt{P_x}\right]_i$ is the i th row of the matrix square root of P_x, holds $\sqrt{P_x}^T\sqrt{P_x} = P_x$, W_i is the weight associated with the i th point, λ is a scaling parameter determining the spread of the sigma points around $\hat{x}$, n is the dimension of x. The parameter α is usually set to a small positive value, $(10^{-4} \le \alpha < 1)$. κ is a secondary scaling parameter, and usually set to 0 or $3-n$. β is utilized to combine prior knowledge of the distribution of x, and for Gaussian distributions, $\beta = 2$ is optimal.

StepA2: calculate the propagation of the sigma points through a nonlinear function:

$$\gamma_i = f(\xi_i),\ i = 0,1,\cdots,L \tag{9}$$

StepA3: calculate the mean $\hat{z}$ and covariance P_z of the random vector x after passing through the nonlinear function $f(\cdot)$.

$$\hat{z} = \sum_{i=0}^{L} W_i^m \gamma_i \tag{10}$$

$$P_z = \sum_{i=0}^{L} W_i^c (\gamma_i - \hat{z})(\gamma_i - \hat{z})^T \tag{11}$$

3.2 Standard UKF

On the basis of the standard KF, the mean and covariance of the state vector through the nonlinear system are calculated by means of the UT, hence the UKF. UKF algorithm for the process:

StepB1: initialize the state $\hat{x}_0$, state covariance P_0, process noise covariance Q and measurement noise covariance R.

StepB2: select the sigma point sampling strategy.

Select the sigma points $\xi_{i,k-1}$ according to the estimated mean and covariance matrix of the current state, and evaluate their corresponding weights $W_i^{(m)}$ and $W_i^{(c)}$.

StepB3: time update.

$$\boldsymbol{\xi}_{i,k|k-1} = f(\boldsymbol{\xi}_{i,k-1}), i = 0,1,\cdots,L \tag{12}$$

The mean of the transformed set of sigma points $\hat{x}_{k|k-1}$ is calculated by

$$\hat{x}_{k|k-1} = \sum_{i=0}^{L} W_i^{(m)} \xi_{i,k|k-1} \tag{13}$$

And the predicted covariance is

$$\begin{aligned} P_{k|k-1} &= E\left[(x_k - \hat{x}_{k|k-1})(x_k - \hat{x}_{k|k-1})^T\right] \\ &\sum_{i=0}^{L} W_i^{(c)} (\xi_{i,k|k-1} - \hat{x}_{k|k-1})(\xi_{i,k|k-1} - \hat{x}_{k|k-1})^T + Q \end{aligned} \tag{14}$$

The predicted observation is

$$\zeta_{i,k|k-1} = h(\xi_{i,k|k-1}), i = 0,\cdots,L \tag{15}$$

$$\hat{z}_{k/k-1} = \sum_{i=0}^{L} W_i^{(m)} \zeta_{i,k|k-1} \tag{16}$$

It should be noted that $\xi_{i,k|k-1}$ can also be obtained by resampling the state prediction value $\hat{x}_{k|k-1}$ as stepA1.

StepB4: measurement update.

Calculate the covariance and the cross correlation matrix by

$$\begin{aligned} P_{z_k} &= E\left[(z_k - \hat{z}_{k/k-1})(z_k - \hat{z}_{k/k-1})^T\right] \\ &= \sum_{i=0}^{L} W_i^{(c)} (\zeta_{i,k|k-1} - \hat{z}_{k/k-1})(\zeta_{i,k|k-1} - \hat{z}_{k/k-1})^T + R \end{aligned} \tag{17}$$

$$\begin{aligned} P_{x_k z_k} &= E\left[(x_k - \hat{x}_{k/k-1})(z_k - \hat{z}_{k/k-1})^T\right] \\ &= \sum_{i=0}^{L} W_i^{(c)} (\xi_{i,k|k-1} - \hat{x}_{k/k-1})(\zeta_{i,k|k-1} - \hat{z}_{k/k-1})^T \end{aligned} \tag{18}$$

$$K_k = P_{x_k z_k} (P_{z_k})^{-1} \tag{19}$$

$$\varepsilon_k = z_k - \hat{z}_{k|k-1} \tag{20}$$

where K_k is the filter gain matrix. Then update the mean and covariance of state vector x.

$$\hat{x}_k = \hat{x}_{k|k-1} + K_k \varepsilon_k \tag{21}$$

$$P_k = P_{k|k-1} - K_k P_{z_k} K_k^T \tag{22}$$

The standard UKF has a good performance in the estimation of general nonlinear systems, but it has a poor performance in tracking mutation parameters and states.

3.2 Time-varying noise estimator

Incorrect mathematical models and inaccurate noise statistic properties often lead the filter to divergence. In order to solve this problem, adaptive filtering technology has been widely studied. The Sage-Husa algorithm is a noise estimator based on the suboptimal unbiased maximum a posteriori (MAP), which is simple in calculation and clear in principle, and can estimate the first and second moments of noise at the same time.

For the unknown process noise in Eq.(1), we use the Sage-Husa algorithm to estimate q_k and Q_k.

$$\hat{q}_k = \frac{1}{k} \sum_{j=1}^{k} \left[\hat{x}_{j|k} - \sum_{i=1}^{L} W_i^{(m)} f_{j-1}(\xi_{i,j-1|k}) \right] \tag{23}$$

$$\begin{aligned} \hat{Q}_k = \frac{1}{k} \sum_{j=1}^{k} \Bigg\{ &\left[\hat{x}_{j|k} - \sum_{i=1}^{L} W_i^{(m)} f_{j-1}(\xi_{i,j-1|k}) - \hat{q}_k \right] \\ &\left[\hat{x}_{j|k} - \sum_{i=1}^{L} W_i^{(m)} f_{j-1}(\xi_{i,j-1|k}) - \hat{q}_k \right]^T \Bigg\} \end{aligned} \tag{24}$$

The suboptimal Sage-Husa estimator can be obtained by replacing the smoothing estimate $\hat{x}_{j|k}$ with the estimation value $\hat{x}_j$ approximately. For time varying noise, the role of recent data should be emphasized, and the role of old data should be gradually forgotten and disappeared. Therefore, we use fade-out memory weighting method, that each term of Eq.(23) and Eq.(24) is multiplied by a different weighting factor [12]. The suboptimal Sage-Husa estimator is combined with the KF to obtain an adaptive filter for linear systems with time-varying noise characteristics [11] [12], and can be combined with

EKF to generalize to nonlinear systems. Then $\boldsymbol{q}_k$ and $\boldsymbol{Q}_k$ can be recursively obtained by the time-varying noise estimator:

$$\hat{\boldsymbol{q}}_k = (1-d_{k-1})\hat{\boldsymbol{q}}_{k-1} + d_k\left[\hat{\boldsymbol{x}}_k - \sum_{i=0}^{L} W_i^{(m)} f_{k-1}(\xi_{i,k-1})\right] \quad (25)$$

.. (26)

$$d_{k-1} = (1-b)/(1-b^k) \quad (27)$$

where b is the forgetting factor, the value range is $0.95<b<0.99$. The forgetting factor b can be used to limit the memory length of the filter. The larger b, the greater the weight of the new measurement data. When the noise statistical characteristics change quickly, b should take a large value; when the change is slow, it should be smaller.

The sub-optimal Sage-Husa estimator is introduced into the UKF algorithm to estimate and correct the statistical properties of the process noise in real time, which reduces the model error, suppresses the filtering divergence and improves the filtering accuracy. The estimated mean, estimation variance, and filter gain in Eq.(25) and Eq.(26) can be obtained during UKF filtering.

3.4 Suppression of filtering divergence

The suboptimal Sage-Husa algorithm will be divergent sometimes. In this paper, a method based on covariance matching criterion is proposed to judge the filter divergence trend.

$$\tilde{\boldsymbol{z}}_k^T \tilde{\boldsymbol{z}}_k \le S\mathrm{tr}\left[E\left(\tilde{\boldsymbol{z}}_k \tilde{\boldsymbol{z}}_k^T\right)\right], S \ge 1 \quad (28)$$

where S is adjustable coefficient, $\tilde{\boldsymbol{z}}_k$ is the residual sequence, i.e. $\tilde{\boldsymbol{z}}_k = \boldsymbol{z}_k - \hat{\boldsymbol{z}}_{k|k-1}$. If Eq.(28) does not hold, correct one-step prediction covariance matrix as follows:

$$P_{k|k-1} = \lambda_k \sum_{i=0}^{L} W_i^{(c)} \left(\xi_{i,k|k-1} - \hat{\boldsymbol{x}}_{k|k-1}\right)\left(\xi_{i,k|k-1} - \hat{\boldsymbol{x}}_{k|k-1}\right)^T + \hat{\boldsymbol{Q}}_{k-1} \quad (29)$$

When the convergence condition is satisfied, the Sage-Husa algorithm described above is used. When the filter is divergent, the adaptive weighting coefficient is computed by Eq.(30). And then use λ_k to amend $P_{k|k-1}$, thereby increasing the role of the current observation to suppress the filter divergence. The adaptive weighting factor λ_k is calculated as follows:

$$\lambda_k = \begin{cases} \lambda_0, & \lambda_0 \ge 1 \\ 1, & \lambda_0 < 1 \end{cases} \quad (30)$$

$$\lambda_0 = \frac{\mathrm{tr}\left(C_{0,k} - R\right)^T}{\mathrm{tr}\left[\sum_{i=0}^{L} W_i^{(c)} \left(\xi_{i,k|k-1} - \hat{\boldsymbol{x}}_{k|k-1}\right)\left(\xi_{i,k|k-1} - \hat{\boldsymbol{x}}_{k|k-1}\right)^T\right]} \quad (31)$$

$$C_{0,k} = \begin{cases} \tilde{\boldsymbol{z}}_k^T \tilde{\boldsymbol{z}}_k, & k=1 \\ \dfrac{C_{0,k-1} + \tilde{\boldsymbol{z}}_k^T \tilde{\boldsymbol{z}}_k}{1+\rho}, & k>1, 0<\rho\le 1 \end{cases} \quad (32)$$

where ρ is the attenuation coefficient. The attenuation coefficient can further improve the fast tracking ability of the filter. Usually, the value of ρ is about 0.95. The larger the value, the smaller the proportion of information before current time, and the more effect of current residual vector. This filter algorithm has strong ability to track the mutation state, and keeps the tracking ability for the gradual change and the mutation state when the filtering reaches the steady state.

3.5 Implementation of the improved AUKF algorithm

In order to solve the problem of target tracking described by Eq.(1) and Eq.(2), the proposed AUKF algorithm can realize the recursive estimation of the unknown process noise statistic and the UKF filtering of the nonlinear system. The concrete steps are as follows:

StepC1: initialize the state , state covariance $\boldsymbol{P}_0$, process noise mean $\hat{\boldsymbol{q}}_0$ and covariance $\hat{\boldsymbol{Q}}_0$, measurement noise covariance $\boldsymbol{R}$.

StepC2: for a given $\hat{\boldsymbol{x}}_{k-1}$ and $\boldsymbol{P}_{k-1}$, calculate the state prediction $\hat{\boldsymbol{x}}_{k|k-1}$ and prediction covariance $P_{k|k-1}$. The predicted observation is the same as standard UKF.

$$\hat{\boldsymbol{x}}_{k|k-1} = \sum_{i=0}^{L} W_i^{(m)} \xi_{i,k|k-1} + \hat{\boldsymbol{q}}_k \quad (33)$$

$$P_{k|k-1} = E\left[\left(\boldsymbol{x}_k - \hat{\boldsymbol{x}}_{k|k-1}\right)\left(\boldsymbol{x}_k - \hat{\boldsymbol{x}}_{k|k-1}\right)^T\right]$$
$$\sum_{i=0}^{L} W_i^{(c)} \left(\xi_{i,k|k-1} - \hat{\boldsymbol{x}}_{k|k-1}\right)\left(\xi_{i,k|k-1} - \hat{\boldsymbol{x}}_{k|k-1}\right)^T + \hat{\boldsymbol{Q}}_{k-1} \quad (34)$$

StepC3: divergence judgment.

According to Eq.(28) to judge the divergence of the filter. If divergence, modify covariance .. according to Eq.(29), otherwise go to the next step.

StepC4: measurement update.

Calculate the covariance $\boldsymbol{P}_{\tilde{z}_k}$, the cross correlation matrix $\boldsymbol{P}_{x_k z_k}$, the filter gain matrix $\boldsymbol{K}_k$, and then update the state mean and covariance according to Eq.(17) to Eq.(22).

StepC5: Update the mean and covariance of the process noise according to Eq.(25) to Eq.(26).

4 Application in target tracking

In order to verify the filtering performance of the AUKF algorithm in this paper, the standard UKF algorithm and the AUKF algorithm are used to estimate the target state when the statistical characteristics of the process noise are unknown. In the case of the ballistic target tracking, the target state is $X(t)=\left[x(t),y(t),z(t),v_x(t),v_y(t),v_z(t)\right]^T$ in the ENU coordinate system. In the reentry stage, the motion model of the ballistic target can be expressed as:

$$\begin{aligned}
\dot{x}(t)&=v_x(t)\\
\dot{y}(t)&=v_y(t)\\
\dot{z}(t)&=v_z(t)\\
\dot{v}_x(t)&=C_G(t)x(t)+C_D(t)v_x(t)\\
\dot{v}_y(t)&=C_G(t)y(t)+C_D(t)v_y(t)\\
\dot{v}_z(t)&=C_G(t)z(t)+C_D(t)v_z(t)
\end{aligned} \tag{35}$$

where

$$C_G(t)=-\mu/R^3(t) \tag{36}$$

$$C_D(t)=-\rho v(t)/2\beta \tag{37}$$

μ is the gravitational constant of the earth, $R(t)$ is the distance from the target to the center of the earth at time t, holds $R(t)=\sqrt{x^2(t)+y^2(t)+\left(z(t)+R_e\right)^2}$, R_e is the radius of the earth, $v(t)$ is the velocity of the target, holds $v(t)=\sqrt{v_x^2(t)+v_y^2(t)+v_z^2(t)}$, β is the mass-to-resistance ratio, which is assumed to be known, $\beta=6\,000\ \text{kg/m}^2$. When filtering, it can be used to predict the state of the target at the next moment by means of discretization of differential equation or numerical integration, and will not be repeated here. The state data of the ballistic missile during the reentry process are obtained by numerical integration of the accurate ballistic motion equations.

Z_k is the measurement vector, and its components are slanting distance R, azimuth A and pitch angle E. Radar distance measurement standard deviation is $\sigma_R=100m$, angle standard deviation is $\sigma_A=\sigma_E=10^{-3}\text{rad}$, sampling period is 1s. The measurement noise covariance matrix is also a diagonal matrix. When the measurement noise is zero mean Gaussian distribution, the diagonal elements can be taken as 1/9 times the square of the angular accuracy.

Process noise and measurement noise are independent zero mean Gaussian white noise. The measurement noise is known, the statistical characteristics of the process noise is unknown, and is time varying.

In the initial stage of filtering, the error of the target state vector is large. By adjusting the process noise Q, the divergence of the filter can be effectively avoided. The standard UKF process noise covariance matrix is set to:

$$Q=\begin{cases} diag\left(\left[10^4 \quad 10^4 \quad 10^4 \quad 10^4 \quad 10^4 \quad 10^4\right]\right) & k\le 10\\ diag\left(\left[10^{-5} \quad 10^{-5} \quad 10^{-5} \quad 10^{-5} \quad 10^{-5} \quad 10^{-5}\right]\right) & k>10\end{cases} \tag{38}$$

The mean of the process noise for AUKF is $\hat{q}_1=\left[0 \quad 0 \quad 0 \quad 0 \quad 0 \quad 0\right]^T$, and the initial process noise covariance for AUKF is the same as standard UKF.

The initial state and its covariance matrix are calculated according to the first two points of the measurement. Assuming that the first two points of measurement data in the ENU coordinate system are $\left[x_0 \quad y_0 \quad z_0\right]^T,\left[x_1 \quad y_1 \quad z_1\right]^T$, the initial state is set to:

$$X(1)=\left[x_1 \quad y_1 \quad z_1 \quad \frac{x_1-x_0}{T} \quad \frac{y_1-y_0}{T} \quad \frac{z_1-z_0}{T}\right]^T \tag{39}$$

The initial state covariance is a diagonal matrix, and the square of the maximum error of the initial state filter value can be chosen as the initial value. The initial state covariance is set to

$$P(1|1)=\text{diag}\left(10^6 \quad 10^6 \quad 10^6 \quad 10^7 \quad 10^7 \quad 10^7\right) \tag{40}$$

The two filtering methods are compared taking the root mean square error (RMSE) of the filtering method as an index. RMSE is defined as:

$$RMSE=\sqrt{\frac{1}{N}\sum_{i=1}^{N}\left\|\hat{X}_i(k)-X_i(k)\right\|_2} \tag{41}$$

where N is the Monte Carlo simulation times.

In an independent experiment, the true trajectory of the target is shown in Fig.1, and the velocity versus time is shown as Fig.2. Fig.3 to Fig.4 compared the RMSE of the two filtering algorithms under unknown time varying process noise. Tab.1 is the mean RMSE of the two filtering algorithms at different times.

The AUKF algorithm proposed in this paper uses the Sage-Husa estimator to estimate the statistical characteristics of the process noise in real-time. According to the innovation, the filtering divergence trend is judged and the attenuation factor is introduced to restrain the divergence, which can not only guarantee the positive definite and nonnegative definite of the process noise variance matrix, but also effectively solve the filter instability problem caused by unknown statistical characteristic of the process noise, and improve the convergence speed and the estimation precision of the filter.

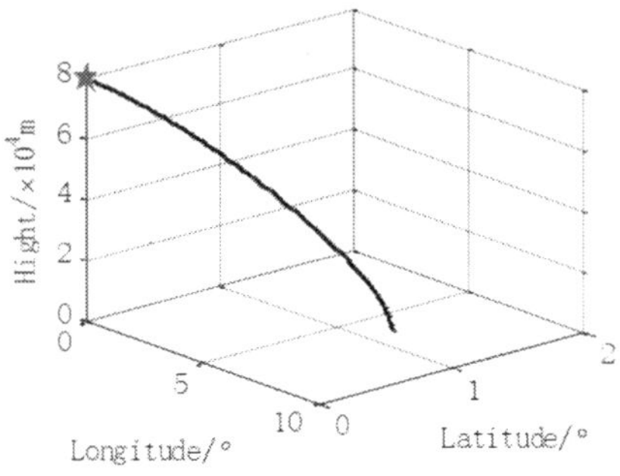

Figure 1. Height of BM

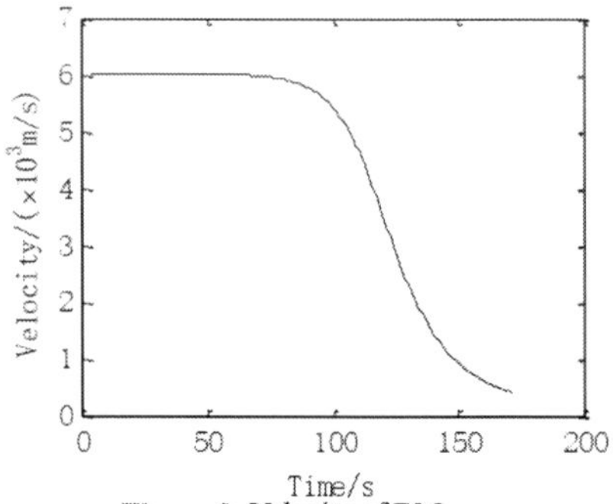

Figure 2. Velocity of BM

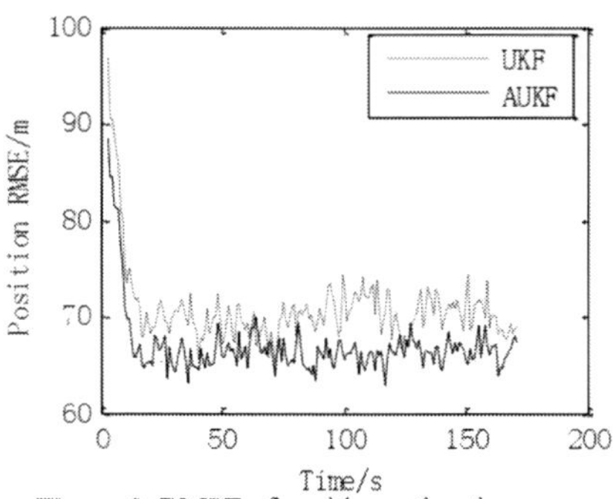

Figure 3. RMSE of position estimation

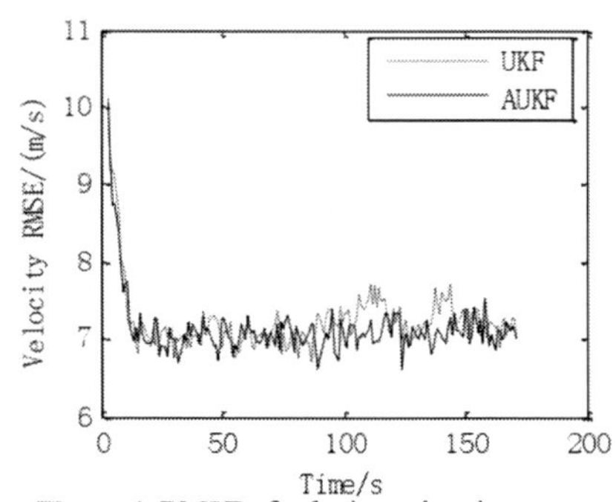

Figure 4. RMSE of velocity estimation

Table 1. RMSE of the two filtering algorithms

Algorithm	Position variance (m)	Velocity variance (m/s)
UKF	71.1938	7.203
AUKF	67.8747	6.985

It can be seen from Fig.3 and Fig.4 that the algorithm has good filtering effect under the condition of unknown time varying noise, and the RMSE of position and velocity is obviously smaller than that of the standard UKF algorithm. Simulation results show that the AUKF algorithm is superior to the standard UKF algorithm in filtering precision and stability, and the state estimation is closer to the real state.

5 Conclusions

A new AUKF algorithm which adaptively adjusts the variance of noise based on innovation is proposed for non-linear target tracking problem with unknown process noise. In the filtering process, we adopt the Sage-Husa estimator to estimate and correct the process noise covariance, and take measures to suppress the possible divergence phenomenon. The new algorithm improves the real-time tracking ability and the numerical stability of the filter. The AUKF algorithm proposed in this paper is much better than the standard UKF algorithm, showing higher filtering precision and stronger anti-noise performance.

References

[1] S. C. Zhang, G. D. Hu. *Acta Automatica Sinica*, **33**, 1220 (2007).
[2] K. Xiong, C. L. Wei, L. D. Liu. *Asian Journal of Control*, **12**, 426 (2010).
[3] C. Liu, P. Shui, S. Li. *Journal of Systems Engineering and Electronics*, **22**, 188 (2011).
[4] L. Zhao, X. X. Wang, M. Sun, et al. *Acta Automatica Sinica*, **36**, 1007 (2010).
[5] Y. Shi, C. Z. Han. *Acta Automatica Sinica*, **37**, 755 (2011).
[6] F. Deng, J. Chen, C. Chen. *Journal of Systems Engineering and Electronics*, **24**, 655 (2013).
[7] Y. Wang, S. Sun, L. Li. *Journal of Guidance Control and Dynamics*, **37**, 1696 (2014).
[8] G Valverde, V Terzija. *IET Generation, Transmission and Distribution*, 5, 29 (2011).
[9] J. Liu, J Ma, J W Tian. *Journal of Systems Engineering and Electronics,* **21**, 675(2010).
[10] J S Kim, D R Shin, E Serpedin. *Digital Signal Processing*, **19**, 504 (2009)

Design of Mine Locomotive System Based on CAN Bus

Yuanhong Li, Quanzhu Zhang and Wenshan Zhang

North China Institute of Science and Technology, Information and Control Technology Institute Beijing East Yanjiao 101601

Abstract: Based on CAN bus, this paper studies the system control and management system of locomotive in mine, analyzes the working principle of locomotive system, gives the CAN bus scheme, hardware circuit design and CAN communication protocol,and implements long-distance, high-reliability communication function and remote monitoring function . Experiments show that the auxiliary system based on CAN bus control easier, operation more secure, as well as improving the control performance and service life of the electric locomotive .

0 Introduction

With the development of power electronics technology, CAN bus technology to use, coal mine electric locomotive now the signal transmission system can't meet the requirements of the development of the times [1].Electric locomotive now the auxiliary converter power supply has some shortcomings: Each device on the locomotive (transformers, air conditioning units, etc.) must individually lead a line to control the start and stop of the device, failure and all the feedback of the signal, and transmitted to the console, and automatic or manual control.As shown in Figure 1, the system for the auxiliary converter working principle, various devices, contactors, etc,with the key switch to control, then this is not only complicated wiring, maintenance inconvenience, the most important is still very easy to cause dangerous accidents.But we use the auxiliary converter system instead of the original system, combined with CAN bus technology, it is very convenient to achieve the above control, both to improve the efficiency, while the system also take into account advanced science and technology, and has a good use of value, there is a broad application market [2-3].

1 The overall structure of the system

Figure 1 shows a block diagram of the system-assisted converter, which is composed of phase-controlled rectifiers, four inverters, and various electrical equipment (brake fan, traction machine, compressor, oil pump).And each device is controlled by a contactor (and a spare contactor).This system will first AC794 voltage through the phase-controlled rectifier into DC540V, and then DC540V voltage inverter, inverted into AC380V supply equipment needed to use the device, in order to ensure normal operation of equipment.

In conjunction with Figure 1, we draw the CAN bus block diagram.As shown in picture 2,there are the start stop button and the display screen of each device on the driver's console, which can display the real-time status of each device.Before starting, the driver console receives phase fault rectifier fault signal and 4 inverter fault signal, if there is no fault, then start the phase-controlled rectifier, and then as needed to open the various inverter operation.

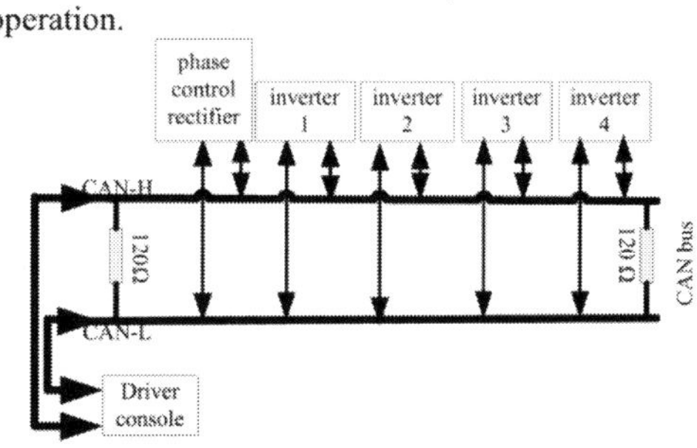

Figure 2 CAN bus system architecture

2 Hardware design

Figure 3 shows the CAN circuit block diagram, which phase control rectifier, four inverters are used a STM32F103ZET6 chip to control.STM32F103 chip comes with CAN controller used to achieve data transmission.bxCAN is STM32F1 comes with CAN controller, which supports CAN protocol 2.0A and CAN2.0B.This controller uses the smallest CPU load to

DOI: 10.1051/matecconf/201713900187

handle a large number of received packets, but also supports the priority of packet transmission [4].

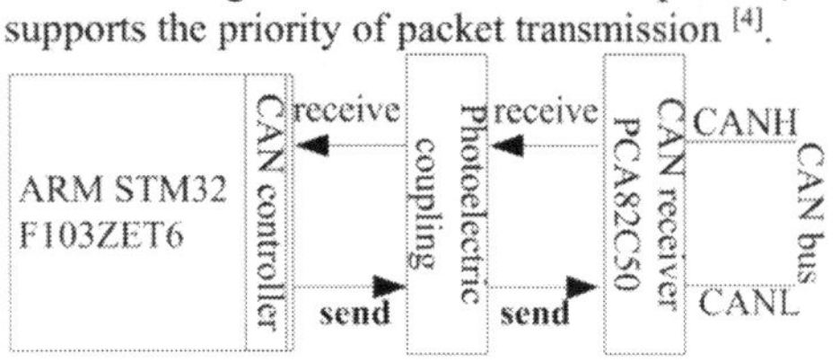

Figure 3 CAN circuit block diagram

2.1 CAN circuit design

Figure 4 for the separate power supply module, the use of "F0505D-1W" DC-DC converter, The 5V voltage is converted into a 5V power isolation, improve the noise immunity of the bus, isolated power supply CAN optocoupler isolation and transceiver ,and increased CAN bus stability and anti-jamming capability.

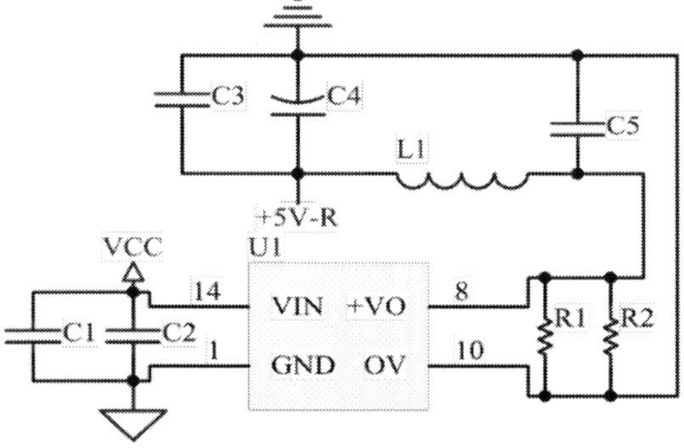

Figure 4 power supply module

Figure5 shows the optical isolation circuit, we use high-speed optocoupler 6N137, the chip sent by the signal CANTXD and received signal CANR, with optocoupler isolation, so that the speed of the CAN network can be achieved and the speed of the previous network drive the same,so that the signal transmission more timely, stable, reliable, the most important can improve the anti-jamming.

Figure 6 shows the CAN transceiver circuit, we use the PCA82C50 chip, the chip can provide differential transmission capacity of the bus, CAN controller to provide differential reception capability, and the chip to CAN in high-speed mode, so that the transmitter output stage transistors at the fastest possible opening and closing speed, improve the efficiency of the CAN. The chip in the circuit accepts the CANT, CANR signal from the optocoupler, will become a stable differential signal, after the follow-up filter circuit, the protection circuit sent to the CAN bus, which can improve the signal anti-jamming capability.

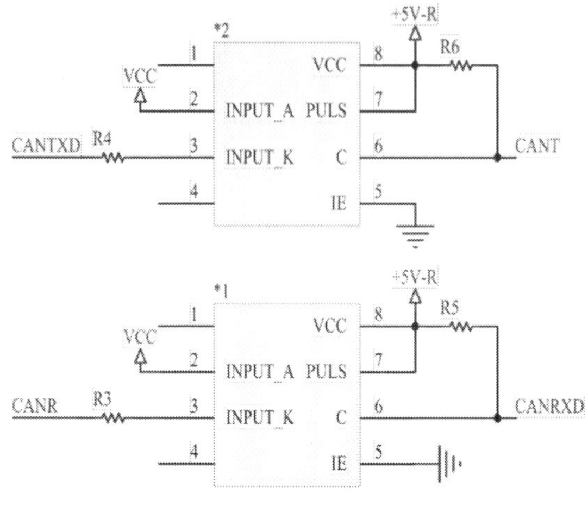

Figure 5 Optoelectronic isolation circuit

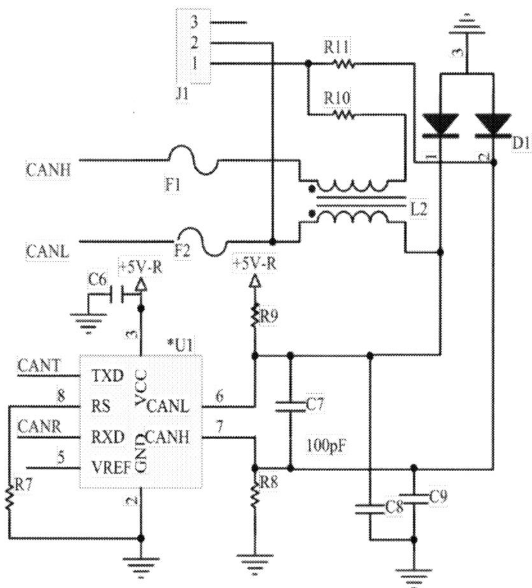

Figure 6 CAN Receiver PCA82C50 wiring diagram

3 CAN communication protocol

For CAN (Controller Area Network), it is a serial data communication protocol, with high performance, high reliability, easy to develop and low-cost field bus.It consists of CANH and CANL lines, the transmission medium can be unshielded twisted pair, cable and fiber, the communication rate up to 1Mb / S, and its transmission distance up to 10Km. [5-8]

In this system, set up the control station, phase control rectifier, the inverter's own frame number.The control station sends the rectifier, the inverter frame number is 0X80 instruction frame, receives the real-time data of each electrical equipment and sends the control command signal.The frame number sent by the phase control rectifier to the control station is 180H, and the frame numbers sent by the four inverters to the console are 181H, 182H, 183H and 184H respectively.As the end of the Table 1, 2, 3 is the byte definition of each module

4 Experimental waveform

For different devices and contactors connected to them, we will be 24 hours continuous monitoring, and the parameters of each device for real-time collection.By testing our system can run continuously for 24 hours without interruption.The following figure is our test in the CAN bus data waveform and CAN bus data frame waveform, as shown in Figure 7,8.

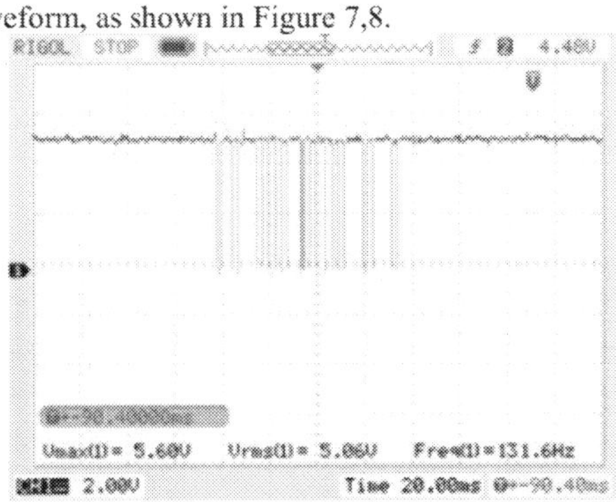

Figure 7 CAN bus data waveform

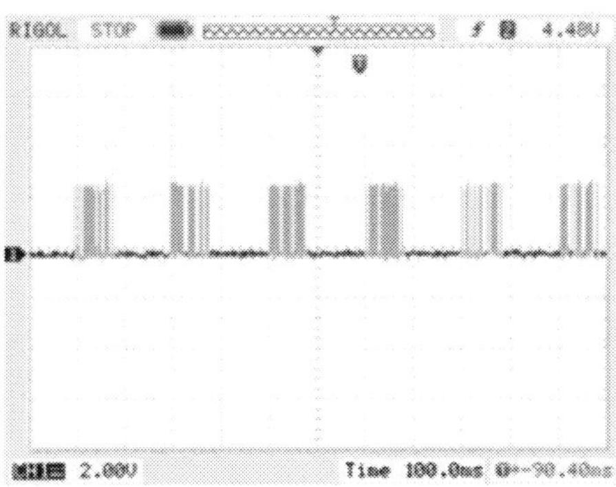

Figure 8 CAN bus data frame waveform

After 24 hours of uninterrupted monitoring of the console display, we found that the four frequency converter through the CAN bus transmission voltage, current, frequency, fault and other information more timely and effective, and information feedback more stable and reliable.

5 Conclusion

Aiming at the auxiliary converter used in the locomotive, this paper designs the information transmission system which uses CAN bus technology for real-time communication. CAN bus system is a good solution to the locomotive power equipment operating frequency, current, voltage, fault and other information transmission problems, and managers can timely and accurate through the console display to get the operating parameters of each electrical equipment , with fast and accurate, high reliability characteristics. Through the installation and commissioning and operation of the locomotive, the system of seismic, electromagnetic compatibility, reliability and other aspects of the design can meet the practical requirements. This paper provides a reliable basis for the use of CAN bus technology in locomotives, and has good marketing value.

Fund Project:

Hebei Province Science and Technology Support Project (16274603,16214408)
Central University of basic scientific research business fees (3142017045,3142013101,3142016022)

References

[1] A.J. YUAN ,X.M. Tang, etc. *A generalized software design technique for CAN fieldbus instrument* [J]. Computer Engineering, **27** (2001), 144-146.
[2] G.J. Bao , M.M. Ji, L. Zhang , T.L. Wang . *CAN bus technology, system implementation and development trend*. Journal of Zhejiang University of Technology, Feb. (2003).
[3] Y.T. Rao, J.J Zou , Y.Y Zheng . *Fieldbus CAN principle and application technology* [M]. Beijing: Beijing University of Aeronautics and Astronautics Press, (2003).
[4]N. Li. *MDK-based STM32 processor development and application* [M]. Beijing: Beijing University of Aeronautics and Astronautics Press .(2008).
[5]F.C. Xiong. *Automobile CAN bus communication technology research* [D]. Wuhan University of Science and Technology .(2009).
[6]X. Hong,W.C. Kuang,Y.P. Xu.*Marway. remote monitoring system based on CAN bus* [J]. Microcomputer Information,**24** (2008) ,73-74,97.
[7] D.B. Liu , Y.J. Chen , H.W. Mao , ect. *Remote monitoring system of coal mine underground substation based on CAN bus and Ethernet* [f]. Industrial and mining automation, **35** (2009), 11-15.
[8] C.Y. Huang ,Y.H. Deng ,Q.Z. Zhang . *A new type of CAN bus dust monitoring system* [J]. Coal Mine Safety, **43** (2012) ,95-97.

DOI: 10.1051/matecconf/201713900187

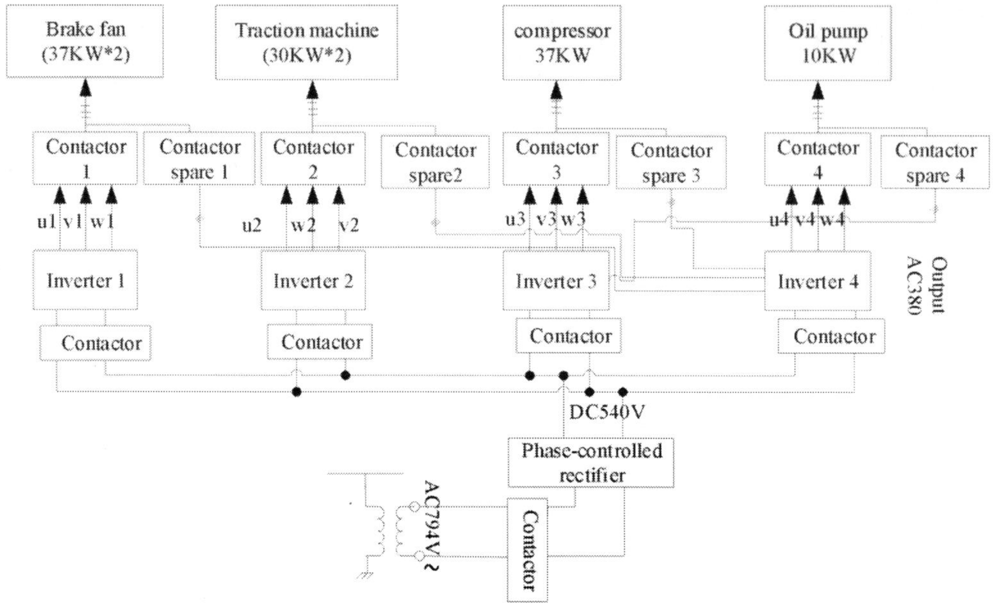

Figure 1 Auxiliary converter works

Table 1 Receive: Division console ID: 80

	Bit7	Bit6	Bit5	Bit4	Bit3	Bit2	Bit1	Bit0
Byte1	80H							
Byte2	08H							
Byte3	inverter1							run 1, stop 0
Byte4	inverter2							run 1, stop 0
Byte5	inverter3							run 1, stop 0
Byte6	inverter4							run 1, stop 0
Byte7	rectifier							run 1, stop 0
Byte8	reserved	reserved	reserved	reserved	reserved	reserved	reserved	normal 0 failure 1

Table 2 Send: Rectifier ID: 180H

	Bit7	Bit6	Bit5	Bit4	Bit3	Bit2	Bit1	Bit0
Byte1	180H							
Byte2	08H							
Byte3	voltage low byte							Normal 0 overcurrent 1
Byte4	voltage high byte							
Byte5	current							
Byte6	temperature							
Byte7	rectifier							
Byte8	reserved	reserved	reserved	reserved	reserved	reserved	reserved	normal 0 failure 1

Table 3 Send: inverter1/2/3/4 ID:181H/182H/183H/184H
inverter1 ID: 181H

DOI: 10.1051/matecconf/201713900187

	Bit7	Bit6	Bit5	Bit4	Bit3	Bit2	Bit1	Bit0
Byte1	181H							
Byte2	08H							
Byte3	voltage low byte							normal 0 overcurrent 1
Byte4	voltage high byte							
Byte5	current							
Byte6	temperature							
Byte7	rectifier							
Byte8	reserved	reserved	reserved	reserved	reserved	reserved	reserved	normal 0 failure 1

About the author: Li Yuanhong (1994-), male, Shanxi Yuncheng, undergraduate, graduate student of North China Institute of Science and Technology. Tel: 15732632524, E-mail: 15732632524@163.com

Design and Experiment Study of High Power Automobile Water Pump Based on CFD

Zhanlong Cao[1,a], Qunpo Zhang[2]

[1] School of Mechanical Engineering, Hubei University of Automotive Technology, Shiyan, China
[2] Dongfeng Motor Pump Company Limited, Shiyan, China

Abstract. A high power automobile cooling water pump is designed and developed based on the theory of centrifugal pump design and computational fluid dynamics (CFD). The structure of the automobile cooling water pump adopts a curved guide rib structure, so as to avoid the dry friction of the automobile cooling water seal of the pump. The structural elements and flow field of the automobile cooling water pump assembly are optimized by simulation analysis. The peak efficiency of the automobile cooling water pump assembly is more than 51%, the automobile cooling water pump performance has been significantly improved. The effective methods and means of product verification have been developed to ensure the high reliability and durability of the automobile cooling pump assembly.

1 Introduction

The working principle of an automobile engine is to promote the movement of the crankshaft through the compression and expansion of the gas. However, in this process, a lot of heat is harmful to the parts. Therefore, this part of the heat required to use the pump to pump cooling water away, this pump is the engine cooling water pump [1, 2]. The cooling water pump is the key component of the engine cooling water system, and it is a kind of engine assembly with high reliability, stability and matching requirement. Its performance directly determines the operation of the engine. With the domestic and international heavy truck gradually to the heavy, high power development trend, the engine power will bring the necessary parts of the upgrade, at the same time high-power engine cooling water pump also need to be matched. Therefore, it is necessary to design a new type of cooling water pump which can meet the requirements of high power engine, high efficiency, high reliability and long life. The pump is designed to reduce the pump internal volume loss and hydraulic loss as the main target, considering the key pump between the key parameters of the flow components of the interdependence and restriction relations, take the "small range selection, optimization analysis method of fixed value" CFD simulation. The water pump is used to bend the reinforcement technology, at the same time; it has been a unique product verification methods and means.

2 Design parameters of automobile water pump

The pump is designed to apply the engine displacement is 13L, power range is 405.5~552.6 metric horsepower. Pump performance index: Engine speed is 3702rpm, pump flow is 398l/min, and pump lift is not less than 19.7m, pump peak efficiency is more than 50%, the service life is 1000000 km or 10 years, double SIC water seal is used, the reliability of the water seal is to provide evidence to prove, pump development process must be scientific.

3 Overall parameters and structure design of automobile water pump

The pump is designed based on the basic theory of centrifugal pump design. At the same time, the existing pump structure is referenced. After the calculation of the relevant parameters of the pump design, the final arc blade, the control blade inlet and outlet area ratio of 1.12 (The recommended value is 1.0~1.3) is selected. In order to effectively reduce the hydraulic loss of the pump, the hydraulic radius of the impeller is 24mm * 23mm. In order to reduce the volume loss and the friction loss of the water in the pump, the outer diameter of the impeller D2 is reduced reasonably, the larger blade outlet angle 2 is selected, the impeller outlet width B2 and the volute throat area are increased. The pump is mainly composed of pump casing, pump shaft, seal, pump impeller, belt pulley, and pump cover. The specific structure is shown in Figure 1.

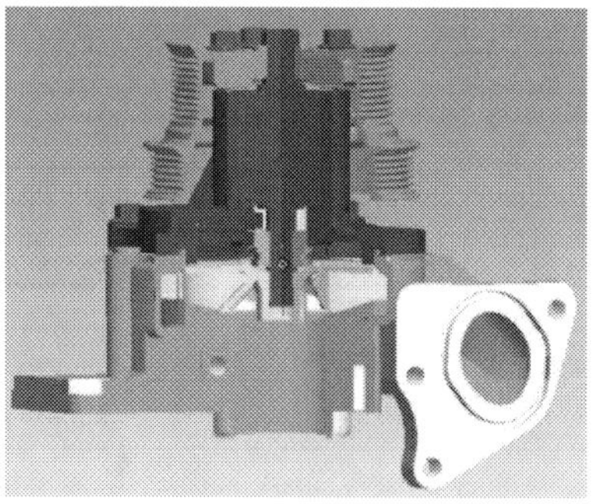

Fig1. Overall structure of automobile water pump

* Corresponding author: caozl@mail.ustc.edu.cn

4 Design of automobile water pump

4.1 CFD optimization design

Numerical analysis of flow field is one of the most widely used methods in pump design. Based on the theory of computational fluid dynamics (CFD), the flow field of the cooling water pump is analyzed by using the flow field analysis software, which can be used to predict the performance of the cooling water pump [3]. Performance results are obtained, which can optimize the pump.

The automobile cooling water pump assembly is optimized by using CFD technique [4, 5, 6]. Taking the cross section of the volute shell as an example, the overall optimization of the vortex shell is analyzed. The results show that the primary section of the vortex chamber is 1015 mm2 (mean section velocity 6.54m/s) and the isobaric distribution in the volute is shown in figure 2.

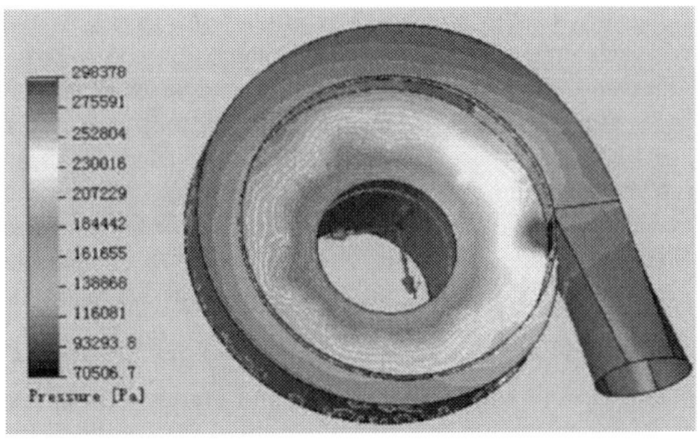

Fig2. Isobaric distribution in a vortex shell with a cross section of 1015 mm2

When the cross section is 1189 mm2 (mean section velocity 5.58m/s), the isobaric distribution in the volute is shown in Figure 3.

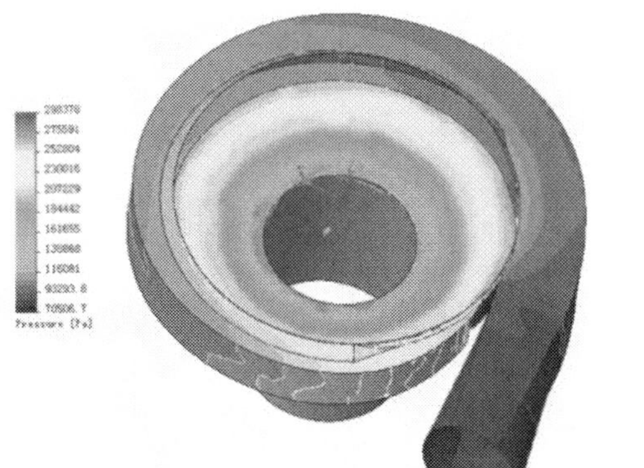

Fig3. Isobaric distribution in a vortex shell with a cross section of 1189 mm2

As can be seen from the chart, the section is 1189 mm2, the coolant flow from the vortex chamber starting end to the diffusion tube inlet section, the vortex distribution curve of indoor pressure gradually increased, change of uniform, fully illustrated the vortex chamber can be high energy liquid from the impeller flow fully into high potential and the required loss is small, and the formation of a pump required pressure in the pump outlet, the pump complies with the design requirements. Therefore, the cross section of the constituency is in line with the requirements, using a cross-section of 1189 mm2.

The pressure line final optimization of the pump assembly, as shown in Figure 4, it can be seen from the figure: pump internal flow from the pump inlet to the pump outlet uniform gradient, the pressure gradually increased, such as the pressure of hypertension did not appear abnormal trajectories, consistent with the operating characteristics of hydraulic.

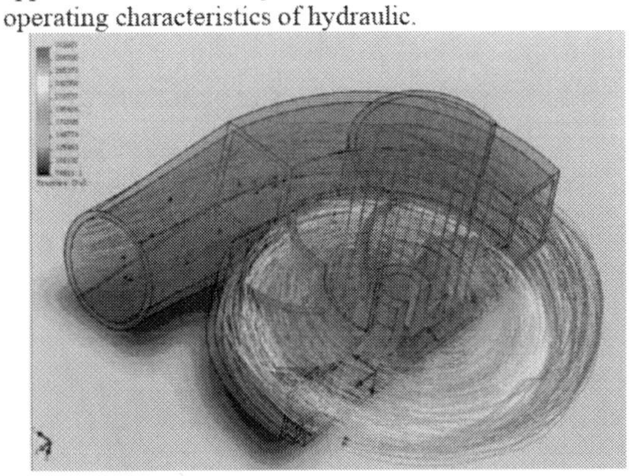

Fig4. Hydraulic trajectory of automobile cooling water pump optimization

4.2 Selection of automobile water pump seal

Water seal failure will directly lead to the failure of the automobile water pump, so the choice of water seal is very important. Therefore, specific "theoretical calculation test" unique selection mode, the selection mode not only solves the matching of the antifreeze liquid rust water seal and water pump assembly and use, but also to avoid the blindness of seal selection. 3 preliminary meet the requirements and technical maturity of the seal was analyzed: Eng40089 type seal for John Crane, and KACO 91470400 (60N) and 93301000 (45N), and to analyze the pump working environment as the basis, through the calculation of key parameters of water seal rigorous theoretical data, see table I. It can be seen that the 3 kinds of water seals are more balanced in the water seal and the back pressure coefficient λ, and the Eng40089 is more dominant in the balance coefficient β. While the two KACO seal by increasing the contact area of the effective movement ring also avoid the shortcomings in the area, and in the end the most important friction two side PV pressure K and end face seal 93301000 than the other two kinds of water sealing requirements to meet the national standard in the numerical values are more scattered and minimum, meet the pump assembly need. Therefore, the comprehensive performance of the KACO 93301000 type water seal is relatively good compared with other parameters, and the KACO 93301000 model water seal is tested. The test results show that it meets the requirements.

Table 1. Water seal calculation data

Item/ Seal type	Eng40089	91470400	93301000
Compliance of water seal(time /min)	≥8000	≥8000	≥8000
Equilibrium coefficientβ	0.1837	-0.0963	-0.0963
Back pressure coefficientλ	0.667	0.667	0.667
Face specific pressure K（Kpa）	982.8~1965.2	1050.4~1530.5	690.3~1170.4
PV value of end face friction pairs（Kpa•m/s）	4676.8~9351.6	5654.8~8239.4	3716.3~6300.9

4.3 Structural design of curved guide bar

The water flow around the water seal is taken as the optimization objective, and it is optimized by three aspects, such as the velocity of water flow, the pressure of water flow and the increase of water flow rate, so as to increase the design of the structure of the curved guide bar, as shown in Figure 5. Through this design to ensure that the water seal around the pump flow stability, adequate, no cavitations' occurs.

Fig5. Structure of curved guide bar

The flow path of the pump assembly is not optimized and the optimized flow path of the pump assembly is shown in Figure 6, Figure 7. After the flow trajectory (shown in yellow for the comparison can be seen through the impeller bottom), volute and impeller clearance and reach the water flow around the impeller bottom and water increased significantly, reaching the bottom water seal impeller optimization is expected to improve the surrounding water status.

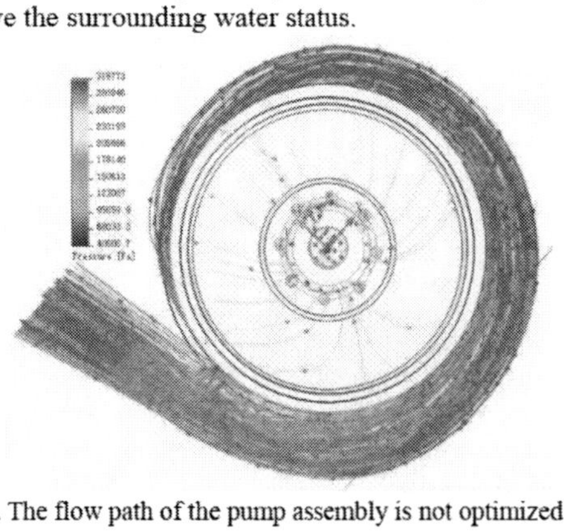

Fig6. The flow path of the pump assembly is not optimized

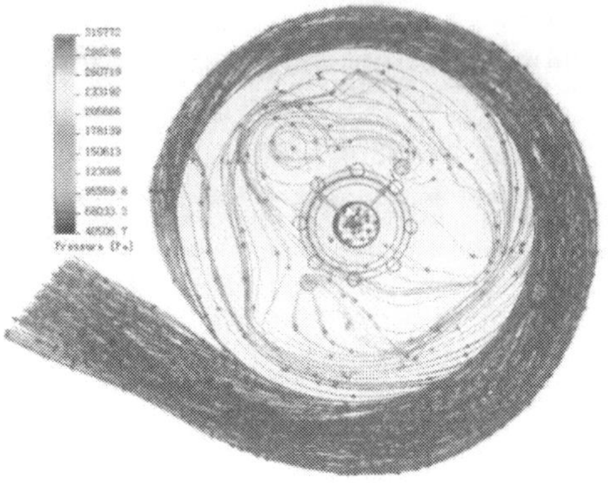

Fig7. The flow path of the pump assembly is optimized

4.4 Design of automobile water pump life

In order to ensure the reliability of the pump life, put aside the previous design method of the whole and part of the pump, update the design concept, and take the reliability of the overall matching of the pump as the research object. In the process of development in order to ensure the reliability of sealing and bearing life to meet the requirements on the basis of the pump cover two key parts installation hole shape and position tolerance and strict requirements, and select the appropriate seal assembly process and bearing, ensure the pump assembly to ensure matching, to ensure the high life, high pump reliability. For the selection of water seal and bearing, through the combination of theoretical calculation and experimental verification, select the two key parts suppliers and product types. The bearing according to the theoretical calculation for reliability coefficient was 0.21 and 99%, according to the checking speed and stress value calculation, get life in different conditions of the engine under the bearing 21171.93 hours for the comprehensive life value in the pump rated speed, and meet the design life of the pump.

4.5 Manufacturing and assembling technology of automobile water pump

In order to ensure the high reliability of water pump assembly, in the assembly, also need to avoid ever in the assembly of the water seal, adversely affect the hidden failure of the bearing and housing, to ensure that key parts of the selection of the best use.

The bearings in the assembly process, there may be too much pressure on the pump cover a hidden cracks and other issues, bearing assembly with hot charging equipment, through high temperature heating rod of the bearing bore of bearing hole producing thermal expansion, and then directly to the bearing in the bearing hole, through natural cooling the bearing hole to eliminate the hard pressed and natural problems in the implementation process of the process, through a number of tests, and heating temperature on the pump cover shell are calibrated to avoid high temperature caused damage to the pump cover, as shown in Figure 8, the red circle as part of the pump cover.

Fig8. Hot charging equipment for automobile water pump

Stealth fragmentation problems may arise in the process of pressing water, attached have been designed, the water seal flip directly with the pressing head contact, avoid pressing water when the pressure head mounted on the lower part of the water seal static ring momentum, as shown in Figure 9, the red circle section for the seal installation position.

Fig9. New water seal pressing method

Through the improvement of the above methods, there is no problem that the water pump cover is broken or the water seal leaks in the sealing test.

5 Verification and experiment of automobile water pump

The automobile water pump assembly needs to be subjected to the routine test[7, 8], and the GOC, Debris Coolant Test, water seal surface temperature test and torque destructive test should be carried out at the same time. Therefore the comprehensive performance of automobile water pump test bench was developed. The advanced comprehensive performance test rig, three test stations, all test automation, through the design of special together meet the full flow test to verify the vast various types of automobile water pump. The sealing performance test of 2 sets of automobile water pump is made by itself, which can meet the sealing test of automobile water pump.

Test medium: anti rust and anti rust liquid , temperature: : 90℃±5℃, inlet pressure greater than 0 Kpa.

The specific tests are as follows: leak-test, index point performance test, cavitations' test, evaporation test, Seal Face Temperature Measurement, reliability test. The experimental results show that the parameters of the automobile water pump are in line with the requirements, see Table 2.

Table **2.** Experimental project and result of water pump

Test name	**Test objective**	**Test requirements**	**Test result**
Leak-test	Check pump leakage	Leakage: cover total: 2.5ml/min; assembly: 6ml/min	Pass
Index point performance test	Performance meets requirements	Performance to achieve the target point requirements	Pass
Cavitations' test	Performance meets requirements	No cavitations' occurs when importing 0Kpa	Pass
Evaporation test	Comprehensive performance of water seal	No more than 300mg/hr	Pass
Seal Face Temperature Measurement	Matching of water seal and assembly	The difference between the water surface temperature and the inlet temperature is less than 6℃	Pass
Reliability test	Reliability	More than 1000 hours	Pass

Automobile water pump was carried out a variety of different engine output speed performance test data are shown in Table 3. When the rated speed is 3072rpm, the test data flow of the automobile water pump is displayed as 698.71l/min, the head is 19.85m, the peak efficiency is 51.03%, and the hump phenomenon is not appeared. The performance indexes of automobile water pump have reached the design requirements.

Table 3. Test data of performance indexes of automobile water pump at different speeds

Speed r/min	**Flow l/min**	**Lift m**	**Torque N*m**	**Shaft power KW**	**Peak efficiency %**
3455	780.82	24.91	16.61	6.01	51.44
3072	698.71	19.85	13.46	4.33	51.03
2685	603.01	15.33	10.51	2.96	49.97
2304	510.98	11.53	8.01	1.93	48.43

6 Conclusions

In accordance with the requirements of the relevant parameters, a high-power automobile water pump has been designed. When the water pump is designed, the key components are selected, the assembly process is improved, and the advanced simulation technology is applied to the integrated technology. The flexure type diversion technology was created for the first time. In order to improve the operation environment of water seal, the dry friction problem of high elastic water seal is solved. Unique product validation methods were carried out, the peak efficiency of the automotive water pump reached 51.03%, its efficiency has been greatly improved. The design and development of the automobile cooling water pump product has been laid the foundation.

You are free to use colour illustrations for the online version of the proceedings but any print version will be printed in black and white unless special arrangements have been made with the conference organiser. Please check with the conference organiser whether or not this is the case. If the print version will be black and white only, you should check your figure captions carefully and remove any reference to colour in the illustration and text. In addition, some colour figures will degrade or suffer loss of information when converted to black and white, and this should be taken into account when preparing them.

Acknowledgments

This work was supported by Science and technology research project of Hubei Provincial Department of Education（B2017086）.

References

[1] Shouqi Yuan,Tingting Zhang,Jinfeng Zhang,Jianrui Liu,Bing Pei, "Optimization design and test study on automobile cooling pumps," Journal of Drainage and Irrigation Machinery Engineering, vol. 32, pp. 93–96, February 2014.

[2] Jie Wu ,Qian TangG,Yuanxun Zhang ,Qixiang Fengi, "Impeller Design and Experiment for Automobile Water Pump Based on CFD," Mechanical Research and Application, vol. 26, pp. 89–91, February 2013.

[3] Jie Wu, "Flow field simulation and design of water pump for automobile cooling system," Chongqing University, pp. 34–60, April 2013.

[4] Fujun Wang, "The theory and application of computational fluid dynamics CFD software," Tsinghua university press, 2004.

[5] Dongping Shen,Hongbao Zhang,Ming Yang etc, "Application of CFD numerical simulation in automobile water pump product development," ANSYS.CFD proceedings of China, pp. 205–208, 2007.

[6] Voorde J V, Dick E, Vierendeels J, Serbruyns S, "Performance prediction of centrifugal pumps with steady and unsteady CFD," Advances in Fluid Mechanics, pp. 559–568, 2002.

[7] "Experimental method of cooling water pump for automobile engine," QC/T 288.2-2001.

[8] "Test method for performance of internal combustion engine cooling pumps," JB/T 8126.2-1999.

MATEC Web of Conferences **139**, 00189 (2017)
ICMITE 2017

DOI: 10.1051/matecconf/201713900189

Design of OBDH Subsystem for Remote Sensing Satellite based on Onboard Route Architecture

WANG Wenping, *WANG* Xianghui, *HAN* Bo, *ZHANG* Yahang, and *LEI* Yong

Beijing Institute of Spacecraft System Engineering, Beijing 100094, China

Abstract. This paper describes the On-Board Data Handling Subsystem of remote sensing satellite, which plays nowadays a relevant role to support satellite assembling, integration and testing activities in the implementing complex system-level requirements. Firstly, the architecture and functionalities of On-Board Data Handling Subsystem are described. Secondly, layered On-Board Data Handling Subsystem flight software architecture based on onboard route as well as software maintenance and tolerance designs are depicted. In particular, we present the redundancies and fault processing means for ensuring the correctness of On-Board Data Handling Subsystem.

1 Introduction

In this paper, we present the architectural approach of the On-Board Data Handling Subsystem (OBDH S/S) running on the China-Brazil Earth Resource Satellite onboard computer. OBDH S/S is one of the subsystems, which belongs to the service platform of remote sensing satellite. Satellite OBDH S/S can be regarded as a system-level entity, which plays a relevant role in implementing some complex requirements, and is the main interface for the ground operators. OBDH S/S integrates the TT&C video frequency functions into a system based on microprocessor equipment to perform the telecommand(TC), telemetry(TM), program control, auto-control onboard, housekeeping functions for the satellite and generate on-board time reference, etc. The OBDH S/S life cycle and reliability requirements are to be defined at the very beginning of any satellite project [1].

Satellites are becoming increasingly software-dependent, together with other complex systems [2]. The OBDH flight software (OBDH F/S) provides all the necessary functions for nominal and emergency operations of the satellite and enables the ground operators to control and supervise the various the state of all sub-systems for the whole satellite mission.

This paper is organized as follows. Section 2 describes its layered architecture. Section 3 outlines the OBDH S/S functionality. Section 4 presents the flight software components based architecture as well as some software maintenance and tolerance designs. Section 5 focuses on the redundancies and fault processing means to fulfill the need of having a robust, reusable, modular OBDH S/S. Finally, section 6 concludes the paper.

2 Architecture descriptions

The deployment architecture diagram of OBDH S/S running on the on-board computer is shown in Fig.1. The OBDH S/S shall be configured with two-level distribution structure by means of a data bus with the core of the fault tolerant computer. It comprises the following units: a Central Terminal Unit(CTU), several Remote Terminal Units(RTUs), a Telecommand Unit(TCU), a Command Decode Unit(CDU), a Ultra-Stable Oscillator(USO) and a Data Recording and Processing unit(DRP). Flight software are mounted on the different intelligent units.

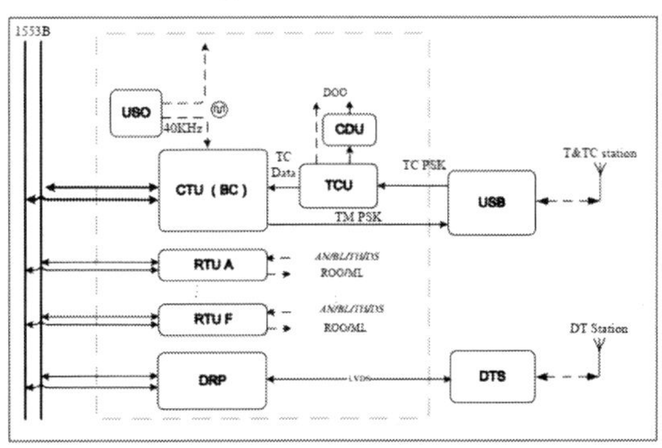

Fig. 1. OBDH Subsystem topology

2.1 Telecommand unit–TCU

The TCU of OBDH S/S is an internal hot redundant unit, which performs the functions of receiving the TC video signal from S-band receiver, demodulating synchronization of the TC signal, decoding and validating the TC frame(including the Address and Synchronization Word (ASW), mode and message codes), distributing and executing the direct ON/OFF command to the satellite subsystem, sending the TC

* Corresponding author: wangwenping304@163.com

message code to the CDU or CTU through the Serial Load Command (SLC) interface according to the mode, and controlling the data injection channels configuration.

2.2 Command decode unit–CDU

The CDU of OBDH S/S is also an internal hot redundant unit, which has the ability of receiving a serial message from the TCU and generating the corresponding direct ON/OFF command to the users.

2.3 Central terminal unit–CTU

The CTU is the core unit of OBDH S/S, which shall be cold redundancy in one box and shall perform the following functions.

The CTU is the central terminal unit of satellite, which plays the relevant role to process and distribute the routed commands & data and on-board commands. Routed commands and data are received from the ground station and on-board commands are generated by on-board computer based on some parameter measurements and a control algorithm. The types of the routed commands and data include real-time/time-tagged ON/OFF commands, real-time/time-tagged Memory Load Commands (MLC), OBDH housekeeping data, Attitude and Orbit Control Sub-system (AOCS) control data and payload control data, etc. In addition, the CTU performs the functions of executing program-control commands, command group and time-tagged commands, and deleting time-tagged commands (groups).

2.4 Remote terminal unit–RTU

The OBDH S/S owns several RTUs, which are identical and are distributed among the different positions in the satellite to provide the direct TM/TC interfaces to the satellite subsystems. Each RTU is identified by the SDB remote terminal address. The microprocessor-based RTU provides the ability of TM acquisition, command distribution and some data processing. The operation of the RTU is controlled by the CTU via the SDB communication.

The RTU provides four kinds of channels for the acquisition of telemetry data: analog channels (AN), digital bi-level channels (BL), serial digital channels (DS) and thermal channels (TH). An analog channel is sampled by the RTU and then converted into an 8 bit binary word. A digital bi-level channel is two-level digital information delivered on a single line. A serial digital word is transferred from user to the RTU with the control of a sampling signal and an acquisition clock signal. A thermal channel is the same as the analog channel.

There are two kinds of interface channels to distribute and execute of commands and data to users: ON/OFF command channels and Memory Load Command (MLC) channels. The RTU also provide the function of self-testing, which includes the tests of CPU, RAM, PROM, AN, DS, ON/OFF command channel and MLC channel.

2.5 Ultra-stable oscillator–USO

Ultra-Stable Oscillator is an internal cold redundant unit outputting high-stability 40 kHz and 5MHz clock; 40KHz clock to CTU, which is utilized to generate the on-board time reference and 5MHz clock to GPS.

2.6 Data recording and processing unit–DRP

Data recording and processing unit is an internal cold redundant unit, which receives satellite TM or housekeeping data from CTU through SDB bus and playbacks the data from TM downlink channel (on-orbit inquire mode)and Data Travelling Subsystem channel(DTS playback mode). DRP has the ability of saving the important data of OBDH, AOCS, and etc.

In On-orbit inquire mode, the ground controller can downlink eight frames of specially selected TM data packets each time, according to the packet APID, or the packet saved time scale uploaded.

In DTS playback mode, the ground controller can upload the time-tagged command to playback all the TM data from DTS.

3 Functional requirements

In order to support the platform and payload operations, the OBDH S/S shall provide the following functions.

For the TC routing, execution and report generation, the OBDH S/S should be able to receive, demodulate and decode the telecommand messages from ground and distribute telecommands to all the satellite subsystems (see Fig.2). Usually, it shall handle three types of commands, the direct ON/OFF commands, routed commands and uplink data injection. In addition, the capability of performing program-control commands, on-board commands, command groups and time-tagged commands should be provided.

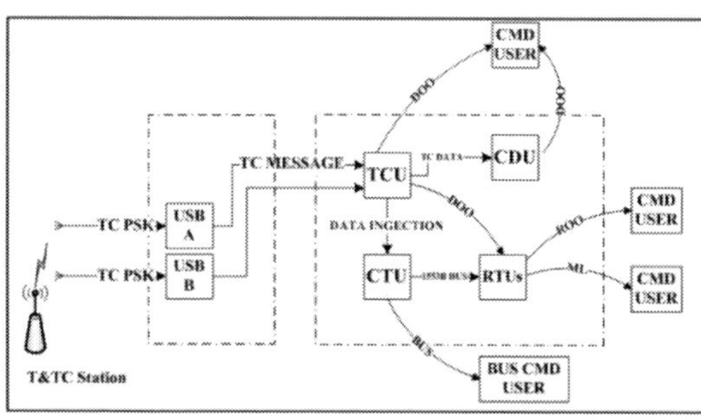

Fig. 2. Satellite telecommand receiving, routing and execution

In order to report relevant satellite parameters values and on-board operational status, the OBDH S/S regularly gathers and encodes the satellite telemetry data, and sends TM packets to the ground in real-time via S-band telemetry transmitter or stored in the on-board memory for later down-link (see Fig.3). The CTU generates the OBDH S/S important data, and stores or restores the important data between the OBDH S/S(CTU, RTU, DRP) and AOCS. The CTU also provides the ability of dumping all the time-tagged command queues (On/Off, MLC and SDB) by using one single telecommand from the ground station.

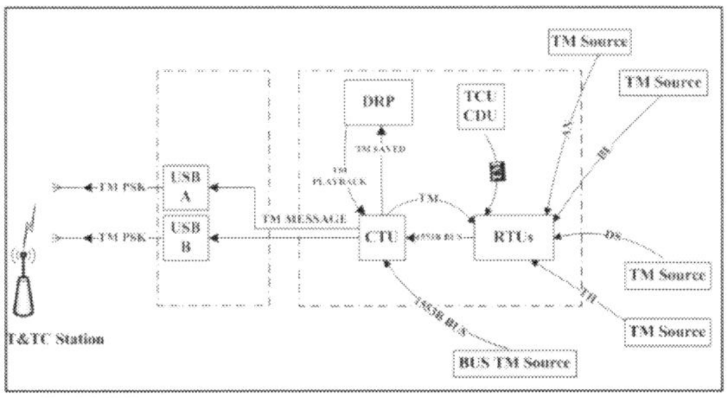

Fig. 3. Satellite telemetry data gathering, encoding and down-link

Generally, all the satellite data, which come from RTUs, AOCS, GPS receiver, thermal controller etc, are gathered and formatted into Advanced Orbiting System (AOS) frame [3-5]. Fig.4 describes how the OBDH handles the TM data based on the AOS. In addition, the capability of changing telemetry scheduling period should be provided.

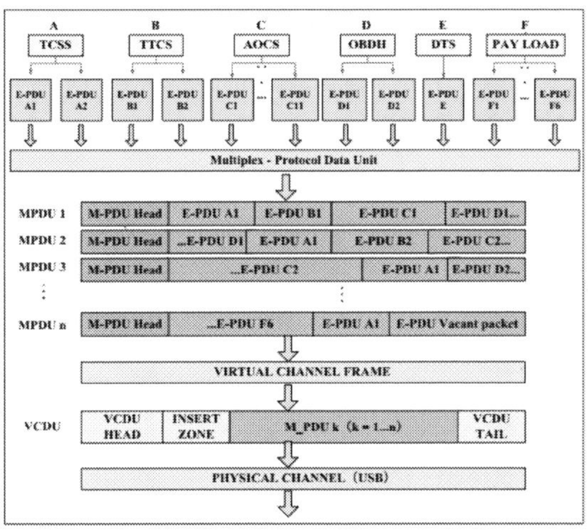

Fig. 4. TM data transmitting based on AOS

On-board time reference and timing signal are generated and distributed to satellite subsystems in-need. The OBDH S/S has the ability of generating the on-board time reference and distributing timing signals to the spacecraft subsystems. The on-board time reference shall be updated periodically according to the GPS clock or ground clock reference. The ways of concentrating time calibration and uniform time calibration should also be provided. Time reference is transferred to remote terminals via serial data bus There are three ways to correct the on board time, they are GPS time self-calibration mode, GPS time coercive calibration for one time mode and TC calibration mode (concentrated time calibration and uniform time correction).

In addition, OBDH S/S also provides the capability of housekeeping processing including system configuration, status data and event report collection, fault diagnosis, isolation and processing. Especially, OBDH S/S provides the capability of auto-control onboard, user-dedicated processing and data formatting. The serial data bus communication management is carried out by the OBDH S/S. Housekeeping functions of the CTU include system configuration, self-diagnosis and fault handling, telemetry scheduling mode switching, acquiring channel reconfiguration, and execution of the uplink control data, as well as the collection of the status data and historic event reports. Moreover, the CTU provides the functions of auto-control onboard, and user-dedicated data generation, processing and distribution. The CTU is specifically designed to prevent the memory and components from SEU and SEL, and to provide the capability of memory write-protection.

4 Flights software

The OBDH F/S plays a relevant role in the implementation of the following functionality: management of the satellite AOCS, thermal control, spacecraft health management (also referred to as Failure Detection, Isolation, and Recovery (FDIR)), onboard autonomy, and spacecraft mode management. The OBDH F/S is composed of the CTU software, the RTU software, DRP software and TCU software. The CTU software is defined as management software, the RTU software is defined as execution software, DRP software is defined as data processing software and TCU software is defined as uplink data processing software.

The software of the CTU is the most important part in the OBDH, which is based on real-time operating system (RTOS). There are two hierarchical layers: one is nucleus which provides routines to serve all interfaces, and the other is application processes that implement the CTU tasks necessary. The CTU software is loaded in the PROM and run in the RAM.

The RTU software provides the functions of communication with the CTU through the SDB, telemetry acquisition (digital, analog and serial), command generation (on/off and serial) and internal OBDH processing.

The software architecture of the RTU, DRP & TCU are sequential, which are loaded and executed in the PROM.

4.1 Software Framework

The development of the on board software should follow the software engineering standards and consider carefully the various aspects, for example, development methodology, quality assurance, configuration management and analysis/design methodology and tools. Furthermore, the capability should be provided by the flight software to support the unit testing of the OBDH S/S.

During the process of development of data management sub-system on traditional spacecrafts, the management of protocol data form and data transaction take plenty time, because of the meticulous analysis of all kinds of transport protocols. A new design of spacecraft OBDH software framework based on onboard route is introduced in this article (see Fig.5). Using this framework, data link layer protocols, such as AosLink, TcLink and 1553Blink, are handled by network layer route, which works as a component. When application layer software needs to send/receive data, the sending/receiving interfaces of the network layer are utilized, and the background running route will choose the destination and service type needed for the data automatically, according to the route table. The data flow between ground, OBDH and other on board sub-systems is thus formed. This framework separates functions of the application layer from commutation on the bottom layer totally, which makes the application layer software design a lot more convenient [6-8].

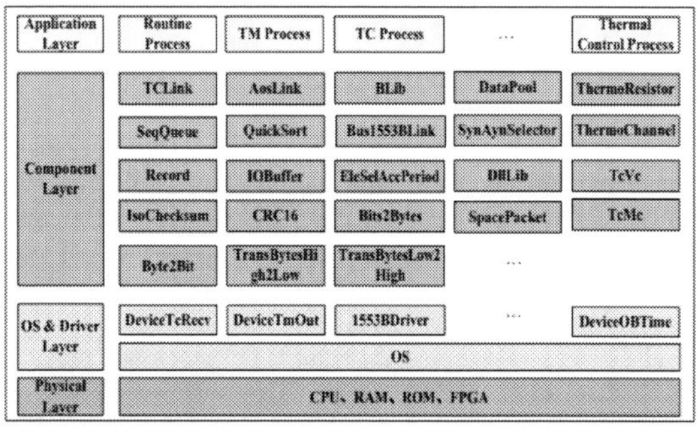

Fig. 5. OBDH software framework based on onboard route

4.2 Software Maintenance

The software maintenance in orbit is to fix and update CTU software through the uplink TC block to correct, replace some wrong software module, or to find a new application in order to satisfy new user requirements. Memory dump, program load, and process creation/activation/inactivation functions should be provided by the flight software maintenance in orbit.

4.3 Software Tolerance

4.3.1 Important Data Store/Restore

The important data should be able to store and to restore between the OBDH S/S and other subsystems. OBDH important data presents the OBDH operational state, interval of uniform correction; flag of sending safety switch-off command group, thermal auto-control, temperature range, payload initialize data, main/redundant table, antenna bias data, etc. Its purpose is that the previous working state makes a comeback when CTU resetting or switching.

OBDH important data shall be stored respectively in AOCC, RTU, and DRP while it updates. The OBDH important data and on-board time should be firstly restored from AOCC or RTUs, DRP to perform setup, CTU should also require AOCC to transmit AOCS important data to store when CTU power-on, resetting or switching.

4.3.2 Software Auto-control

There are several cases which could be able to cause CTU resetting by software (namely soft-reset), for example, Non-masked interrupt (watch-dog), RAM write-protect, RAM Hamming code checkout error, Process overtime and etc.

Under the permission of ground station, the CTU may be auto-switched by software normally, when the soft-reset occurs some times, or the CTU fails to communicate with all RTs.

The CTU unconditional auto-switch caused by software may occur at the instance of the failing of Memory self-test when CTU powered on, the wrong testing of TM bit rate, and the stop of On-board time.

4.3.3 Data Estimation

Some important TM parameters, for example, which participate in the controlling of satellite-launcher separation signal, should be judged at least three times continuously. If the testing results are identical, OBDH can make sure some event occurred. Considering the security of satellite, all data acquired by the RTU channels for satellite-launcher separation signal should be judged.

4.3.4 Self-Testing and Self-Diagnosis

OBDH software should provide the capability of self-testing and diagnosis for some critical interface such as TM video interface, and clock interface etc depending on the hardware. Error Detection and Correction Circuit (EDAC) should be adopted to correct the single-bit error and to detect the double-bit error for the memory. The memory should be refreshed periodically by using "read" operation to reduce the probability of piling up the single-bit error and to prevent from double-bit error, which has been proved to be an effective measure to prevent Signal Event Upsets (SEU).

5 Redundancy and fault processing

The following redundancies shall be provided by the OBDH subsystem. The TCU and CDU are dual redundant devices of "hot backup". The CTU, DRP and USO are designed to be dual redundant with "cold backup" spare unit. The serial data bus SDB is dual redundant and the second bus is in "hot backup" status. The routed TC redundancy shall be provided using two interconnected RTUs. The RTU is not required to be redundant for TMs. TM channels redundancy is implemented by routing TM to two RTUs.

The OBDH S/S provides several fault tolerance techniques for failure protection, isolation and recovery. The criterion to determine the position when fault occurred is that, TM value used for testing, comparing or judging must be made a decision during a specified time limit. If the diagnosis result is "true" for three times continuously, the fault is determined to be happen.

In order to judge the fault occurrence, the OBDH S/S provides the capability of self-testing and self-diagnosis including CPU, RAM, PROM, analogue channel, serial digital channel, ON/OFF command channel, Memory Load command channel, TM/TC interfaces and on-board time, etc. Also, it provides the capability of collecting and recording the working status and event reports of the satellite.

Incorrect-write access protection memory, watch dog timer and the code selection filled in spare area of the memory are designed to prevent the program from dead-loop and going astray. The OBDH S/S provides the capability of detection of error telecommand frame and important data, detection of application running overtime, and auto-selecting of telemetry data acquired by main and backup channels onboard, which gives the first priority to ground station to fill in the TM frame.

6 Conclusions

This paper has described the design of OBDH Subsystem for the China-Brazil Earth Resource Satellite based on onboard route. As described, the OBDH is composed by different units in order to provide data handling service to all the equipment of the satellite. The main units and functionalities of OBDH have been presented. An innovative flight software framework based on onboard route has been reported. In particular, attention has been paid to software maintenance and software tolerance. Finally, this paper has stressed the redundancy design and fault processing means provided by OBDH for ensuring the correctness.

Acknowledgements

This work has been supported by Project of China-Brazil Earth Resource Satellite.

References

1. Tan Weichi , Gu Yingqi . Space data system[M] . Beijing : China Science and Technology Press, 2004 (in Chinese).
2. Ralf Wenker, Cedric Legendre, etc. On-Board Software Architecture in MTG Satellite. The 4th IEEE International Workshop on Metrology for AeroSpace. 2017.
3. WANG Xianghui, WANG Tonghuan, LI Ningning, et al. An Efficient Scheduling Algorithm of Multiplexing TM Service Based on the AOS [J]. Beijing: SPACECRAFT ENGINEERING, 2011(in Chinese).
4. CCSDS, AOS Data Link Protocol. Recommendation for space data system standards, CCSDS 732.0-B-2. Blue Book[S]. Issue 1. Washington D.C.: CCSDS, 2006.
5. Space Packet Protocol. Recommendation for Space Data System Standards, CCSDS 133.0-B-1. Blue Book. Issue 1. Washington, D.C.: CCSDS, July 2003.
6. Garlan, D., Shaw, M. An introduction to software architecture. Technique Report, CMU/SEI-94-TR-21, Carnegie Mellon University, 1994.
7. ZHANG Yahang, YUAN Jun, GUO Jian. Design of Reconfigurable Genral TM Based on Software Component [J]. Beijing: SPACECRAFT ENGINEERING, 2013(in Chinese).
8. ZHANG Yahang, ZHANG Meng, YuanJun, et al. Research of Spacecraft Software Components Based On C Language. The 1st Conference on High Graphics Earth Observation Technology. 2012.12(in Chinese).

Research on the management system of lithium battary

Wenqiang Xie[1,*]

[1]Xi'an Research Institute of China Coal Technology & Engineering Group Corp

Abstract. Lithium battery is the first choice for the energy source of explosion-proof electrical equipment because of its small size, light weight and large energy. In order to ensure the safety, improve the battery life and prolong the service life of the lithium battery, it is necessary to study the lithium battery management system. For this reason, the article uses the modular design idea to design a lithium battery intelligent management system. The system takes the microcontroller as the core to control the current voltage signal sampling, power indicator, serial communication and other management functions, and combined with the software process, the implementation of the system control algorithm is described in detail. Finally, combined with the system software and hardware function, and through repeated experiments and continuous improvement and improvement, it can be run safely and stably.

1 Introduction

The rapid growth of the application portable electronic equipment, the application of lithium battery has opened up a broad prospect, especially for the use of lithium batteries for instruments and meters in coal mines. It is well known that the power use of lithium battery requires the battery series and parallel connection to increase the voltage and capacity, due to the inconsistency of the battery in the battery pack, if one of the batteries has problems, the whole battery cannot continue to work for safety reasons. When the battery appears overcharge, excessive short circuit, temperature over limit and so on, the battery management system will judge according to the abnormal data collected, and then send out the state signal, and through the controller to the charge and discharge protection, temperature protection, equalization management circuit and other corresponding control signals, real-time monitoring its use status, timely protection. And the various operating data of the battery to store backup, establish the use of archives, in order to provide reference for the safe use and maintenance of batteries in the future. Therefore, according to the standards of GB3836-2010 and MT1051-2007 series standards, a lithium battery intelligent management system is designed to meet the safety performance management of lithium ion batteries in coal mine.

2 System design and composition

The management system of lithium battery mainly includes the intelligent charging, battery pack equalization and other protection circuit, the supporting management device realizes the hardware and software protection and intelligent energy management. The management device collects all kinds of battery data, and uses the software set rules to protect and optimize all the cell batteries in the battery pack. The battery management system design the main tasks, as shown in table 1:

Table 1. Main tasks of battery management system design.

Task	Content	Execution unit
Prevent overcharging	Voltage limiting current limit	Charging control circuit
Prevention of over - discharge	Voltage limiting current limit	Discharge control circuit
Temperature control	Refrigerating plant	Heating and cooling part
Status information	Voltage, current, temperature	Digital display
Balanced management	Performance matching	Equalizer circuit

In short, the battery management system is the core of the safe use of lithium battery, it can collect all kinds of data in real time, deal with the data, and issue control signal to protect and balance the whole battery pack, calculate the residual capacity of the battery, ensure the safety of lithium battery, improve the battery life and prolong the service life of lithium battery.

3 System hardware design

The system hardware design uses the microcontroller control scheme, real-time detection of battery current, voltage and temperature parameters, and then carries on the data processing, calculates the residual battery power of the battery pack by the appropriate algorithm, and sends the battery voltage, current, temperature and residual capacity into the display module, real-time display, in order to ensure the timely grasp of the battery

* Corresponding author: xwq7323182@163.com

state parameters, ensure the use of safety. The hardware control scheme structure diagram, as shown in figure 1:

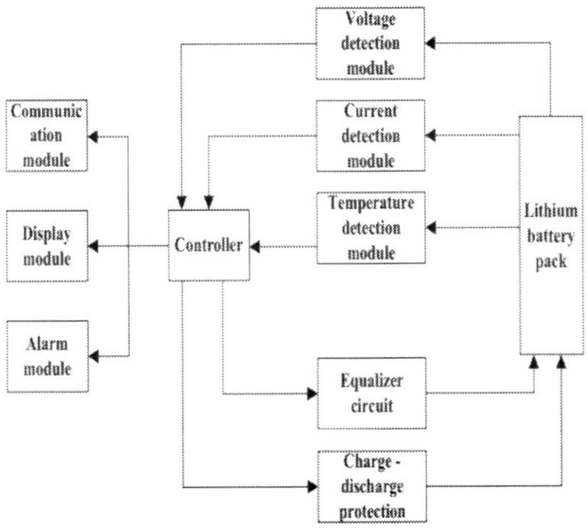

Fig. 1. Structural diagram of hardware control scheme

3.1 Voltage measurement module

The system takes the direct measurement method, and takes the voltage of the two ends of the monomer cell into the differential amplifier circuit for sampling. In order to eliminate the effect of load change on the output voltage, we connect a voltage follower at the output end, so that the input resistance tends to infinity, and the output resistance tends to zero. After the voltage follower output, and then use the analog switch to select the channel to sample, directly into the microcontroller of the 10 - bit ad converter. Since it is an integrated analog to digital converter in the chip, it is not necessary to add the sampling and holding circuit to measure the voltage of any cell.

3.2 Current measurement module

The current is a key parameter for estimating the capacity of the battery, and it is an important reference factor for the over-current protection circuit of the battery charging and discharging process. Therefore, the current sampling accuracy, anti-interference ability and linearity are very high in the lithium battery intelligent management system. The system uses the hall current sensor as the current detection device, and realizes the isolation with the main circuit.

3.3 Temperature measurement module

The battery will produce heat in the process of charge and discharge, the temperature of the battery has a very important relationship with the performance and working status of the battery. The temperature will greatly affect the efficiency of the battery charging, and even cause serious damage to the battery. The temperature difference between the batteries will increase the imbalance between the battery cells, which greatly affects the service life of the battery. Therefore, the temperature measurement is also very important in the system. The temperature measurement in the system is based on the temperature sensor produced by Dallas company.

3.4 Charging and discharging protection module

The normal voltage range of lithium battery is 4.2v ~ 2.7v, 4.2v is the protection value of charging voltage, and 2.7v is the voltage protection value of discharge. The Provision and Guo delay time are all 1s, and the Guo delay time is determined according to the current size, the more current the current is, the smaller the current delay time. When the voltage of all batteries is between 4.2v and 2.7v, the battery is in the normal charge and discharge state; When the voltage of any section of the battery exceeds 4.2v and the duration is more than 1s, the control signal is given to turn off the MOSFET tube and the charging circuit is disconnected; When the voltage of the battery is less than 2.7 v and the duration is more than 1s, the control signal is issued to shut off the MOSFET and disconnect the discharge circuit. When the current sampling circuit is converted to a voltage greater than the short-circuit protection reference voltage, the protection circuit will immediately shut off the control MOSFET and disconnect the discharge circuit to prevent the flow or short circuit.

3.5 Equalizing circuit module

The realization of equalization circuit is through the collection of lithium battery terminal voltage, internal resistance, current, temperature and other information, and then through the SOC estimation algorithm to calculate the state of charge of a monomer cell, in order to obtain a reasonable equilibrium strategy. Therefore, the preparation of these external parameters is crucial. After the data acquisition, the analysis of the specific control scheme, through the control circuit to control the operation status of each balanced module, and ultimately complete the balance of the whole system. The system is realized by active equalization method.

3.6 Hardware anti - interference measures

The battery management system contains a variety of electronic devices and different types of lines, relatively complex. The interference between the various transmission channels can affect the work of the controller, or even cause damage. The battery management system is in the electric vehicle, the working environment is complex, there is noise interference around. In order to ensure the system can operate stably in complex operating conditions, the following measures are introduced to the relative stability and accuracy of the collected data.

The wiring design of PCB is also very important. Pcb provides the electrical connection between the circuit

 DOI: 10.1051/matecconf/201713900190

components. the design of the circuit has great influence on the anti-jamming of the system. the unreasonable design of the circuit board will lead to the serious distortion of the data acquisition and even the system can not work. Therefore, the anti-interference measures are particularly important in the PCB design. When arranging components, try not to use jumpers, generally use a 45-degree corner to reduce the wire impedance and distributed capacitance inductance. At the same time, according to the function of the circuit layout, so as to prevent the different regions of RF current coupling caused interference. The different power supply must be separate wiring, such as the system has + 5v, + 12v and so on different voltage must pay attention to separate; On the ground wire, the digital ground and analog ground should be separated, and the width of the power line and ground wire should be larger than the general signal line, so as to improve the anti - interference.

4 System software design

The main task of the lithium battery management system is to measure the voltage, current and temperature of the battery, through the analysis of the measurement parameters, the battery charge and discharge protection system to achieve effective management. Parameter measurement is the basic part of the whole battery management system, the battery charge and discharge protection and SOC estimation are based on this part. When the measured value exceeds the upper limit set value, the battery charge and discharge protection circuit shall be protected by the system.

Because of the inconsistency of the single cell, such as operating temperature, humidity, chemical composition and other characteristics, when the battery is charged with the same power supply, the battery charge or discharge will be different, when the battery needs to be effectively balanced, otherwise it will greatly shorten the life of the battery. For the estimation of residual power, the corresponding relationship between open circuit voltage and battery power is estimated, and the current value of detection is integrated by ampere - hour integral method in charge and discharge, which can be realized by timer interrupt of single chip microcomputer.

CAN communication is the intermediary between the battery management system and the central control system. the battery management system sends the information of the battery voltage, current, temperature and SOC to the central control system through the CAN bus, and displays the battery pack status. In the system, the central control system sends data signals to different modules, gets the data information needed by the system, and then summarizes the data and makes corresponding processing and display. The main program flow diagram of the system, as shown in figure 2:

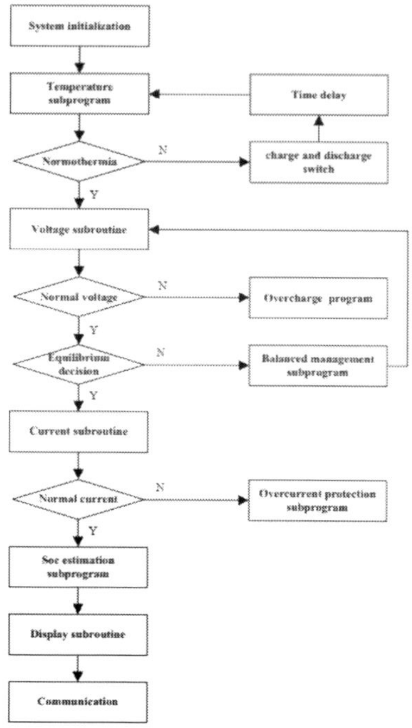

Fig. 2. System master program flow chart

4.1 Flow chart of measurement of voltage and current

Fig. 3 is a flow chart of the voltage measurement subroutine. The current measurement is basically consistent with the voltage measurement in the software implementation, the current measurement is only in the hardware circuit, the measured circuit is converted into voltage, and then sent to the microcontroller ad converter.

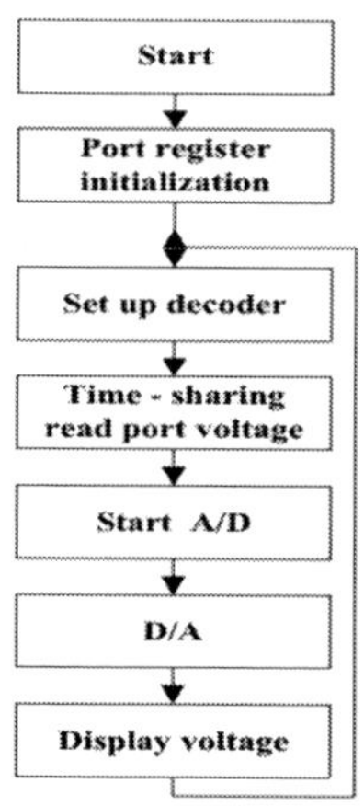

Fig. 3. Process flow chart of voltage measurement subprogram

4.2 Balanced control subprogram chart

The equalization strategy adopted by the system is to take the active and balanced way in the battery charging,

 DOI: 10.1051/matecconf/201713900190

which can improve the charging efficiency, save the power consumption, and can effectively balance the battery pack. The balanced control subroutine diagram is shown in figure 4:

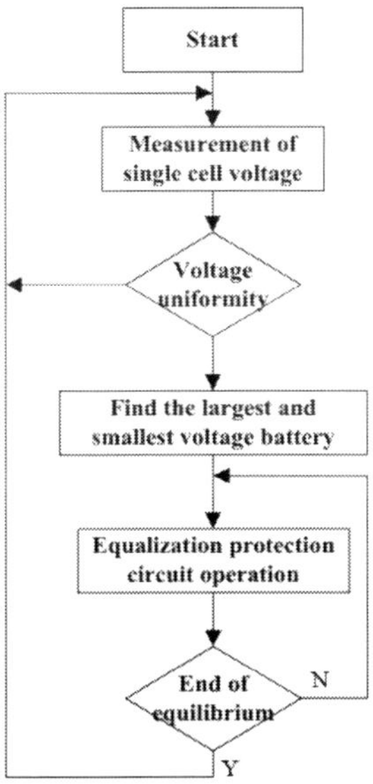

Fig. 4. Flow chart of balanced control subprogram

5 System test

Using the system to collect the voltage and temperature data of 24 series lithium battery, and the voltage balance of lithium battery. The monitoring data of 10 series lithium batteries are compared, and the results are shown in table 2. It can be seen that before the voltage equalization of the lithium battery, the maximum pressure difference between the 10 series lithium battery is 0.289V, the maximum pressure difference after adopting the system is only 0.052V.

Table 2. Pre-and post-equilibrium voltage comparison of lithium battery

Number	1	2	3	4	5
Equalizing front voltage（V）	3.412	3.386	3.316	3.285	3.136
Equalizing post voltage（V）	3.356	3.347	3.366	3.352	3.319
Number	6	7	8	9	10
Equalizing front voltage（V）	3.359	3.219	3.227	3.401	3.123
Equalizing post voltage（V）	3.322	3.314	3.321	3.326	3.329

6 Conclusion

In order to meet the safety performance management of lithium ion battery used in coal mine, a lithium battery intelligent management system is designed according to GB3836-2010 and MT1051-2007 series standards. From the system principle method, hardware functional circuit, software design and implementation of three aspects comprehensive analysis, lithium battery charging, discharge characteristics, voltage acquisition, current collection, protection circuit design and the balance management of the battery were studied and tested, and the test results were given. Through the actual test, the system can improve the consistency of lithium battery, effectively achieve the battery equalization, which is conducive to prolong the battery life and battery life, greatly reducing the workload of on-site maintenance.

References

1. Cui Yufeng, Yang Qing, Zhang Linshan, Research on the development status and charging technology of electric vehicles at home and abroad[J]. *Yunnan power technology*. 2010.
2. Xie Bin, Research on SOC identification and equilibrium management of electric vehicle lithium battery group [D], *Guangzhou: south China university of technology*. 2013.
3. Yueren Chao. Research on battery management system [D]. *Beijing: Beijing jiaotong university*. 2010.
4. Liu wenhao, Chen Jingyuan, Shen Wei, Design of Li - ion power battery protection circuit based on pic16 [J]. *Journal of Zhejiang university of technology*, 2008.
5. Wang Wei, Luo Guangyi, qianzhaoming. An improved lossless equalization charge circuit for VRLA battery [J]. *power electronics technology*. 2003.
6. Liu Hao, Xie Hua, jiangjiuchun, Research on SOC estimation scheme for Li - ion batteries for pure electric vehicles [J]. *Electrical technology*. 2010.
7. Yoon-Ho Kim. Design of Interface Circuits With Electrical Battery Models [J]. *IEEE Transactions on Industrial Electronics*. 2013.

An improved reconstruction algorithm based on multi-user detection for uplink grant-free NOMA

Chengyan Hou[1], *Xiaohui* Li[1]

[1]School of Electronics and Information Engineering, Anhui University, Hefei ,230000, P.R. China

Abstract. For the traditional orthogonal matching pursuit(OMP) algorithm in multi-user detection(MUD) for uplink grant-free NOMA, here is a poor BER performance, so in this paper we propose an temporal-correlation orthogonal matching pursuit algorithm(TOMP) to realize muli-user detection. The core idea of the TOMP is to use the time correlation of the active user sets to achieve user activity and data detection in a number of continuous time slots. We use the estimated active user set in the current time slot as a priori information to estimate the active user sets for the next slot. By maintaining the active user set $\hat{T}^l$ of size K(K is the number of users),but modified in each iteration. Specifically, active user set is believed to be reliable in one iteration but shown error in another iteration, can be added to the set path delay $\hat{T}^l$ or removed from it. Theoretical analysis of the improved algorithm provide a guarantee that the multi-user can be successfully detected with a high probability. The simulation results show that the proposed scheme can achieve better bit error rate (BER) performance in the uplink grant-free NOMA system.

1 Introduction

With the rapid development of mobile communication, the spectrum resources become increasingly scarce, in order to meet the needs of future 5G in spectrum efficiency and the number of connections; Non-orthogonal multiple access (NOMA) as a candidate for 5G technology has been widely concerned [1].The future 5G wireless communication network will meet the wireless networking needs of a variety for network equipment, especially for the scene of 5G Massive Machine Type Communication(MMTC), the sporadic overload systems are some small transmission packets, continue along by orthogonal multiple access (OMA) will be bound to cause the waste of resources and excessive scheduling overhead, therefore, in order to avoid the waste of resources, we need to study the new non-orthogonal multiple access(NOMA) technology to improve the efficiency of small packet transmission.The existing multi-user detection of NOMA system is based on the premise that all users are active, however, not all users are active at the same time in the system, a large number of statistical results show that even in the busy time, active users generally do not exceed 10% of the total users, which means that although there are massive user connections in 5G, access to the system is only a small part of the user at the same time, so active users are sparse [2].The base station can detect the sparse active users and data simultaneously by means of joint detection. Since the base station does not need to know the active user status information, the user does not need

[a] Corresponding author: houchengyan520@163.com

to interact with the base station in advance, therefore, grant free transmission is highly expected in uplink NOMA systems, which greatly reduces signaling overhead and delay, and realizes user overload. It meets the requirements of future 5G mass connection, high spectrum utilization and low latency.

2 Compressive sensing

The premise of compressive sensing (CS) [3] is that the signal is sparse, but, most of the natural signal $f \in R^N$ is not sparse in the time domain, but f can be sparse by a sparse basis $\mathbf{\Psi}$ transform, where $\mathbf{\Psi} = (\Psi_1, \Psi_2, ..., \Psi_N)$, so the signal f can be expressed as:

$$f = \sum_{i=1}^{N} \Psi_i x_i = \Psi \mathbf{x} \quad (1)$$

The sparse vector $\mathbf{x}$ is k-order sparse, so $\| x \|_0 \leq k$, indicates that the number of nonzero values are equal to $k (k < N)$ .If we apply a corresponding linear transformation to the coefficient $\mathbf{x}$ with measuring matrix Φ: $M \times N (M << N)$ which is not related to the sparse base $\mathbf{\Psi}$, get measurements $y : M \times 1$:

$$\mathbf{y} = \mathbf{\Phi} f = \mathbf{\Phi \Psi x} = \mathbf{Ax} \quad (2)$$

The vector $\mathbf{x}$ is k-order sparse, so the initial signal can be accurately recovered from the linear measurement set using the optimal method, the observation matrix $\mathbf{\Phi}$ and the sparse base $\mathbf{\Psi}$ follow the constraint isometric (RIP) condition, so reconstruct the signal $\mathbf{x}$, i.e., to find the minimum ℓ_0 norm [4]:

$$\hat{x} = \min_{x \in R^n} \| x \|_0 \;\; s.t. \;\; \mathbf{Ax} = \mathbf{y} \quad (3)$$

where $\mathbf{A} = \mathbf{\Phi \Psi}, \mathbf{A} \in \mathbf{R}^{M \times N} (M << N)$,for the general $\mathbf{A}$ and $\mathbf{y}$, there are nondeterministic polynomials for the equivalent constraint problem of ℓ_0, so the following method is proposed:

$$\min_{x \in R^n} \| x \|_1 \;\; s.t. \;\; \mathbf{y} = \mathbf{Ax} \quad (\ell_1)$$

The optimal sparse solution requires a very sparse premise, and the perceived matrix $\mathbf{A}$ must satisfies the incoherent condition,however, in practice, these conditions cannot be satisfied, ℓ_1 cannot find the optimal solution, we must turn to other nonconvex problems, which needs weaker conditions to guarantee the accuracy of the successful recovery [5]:

$$\min_{x \in R^n} \| x \|_p^p \;\; s.t. \;\; \mathbf{y} = \mathbf{Ax} \quad (\ell_p)$$

where $0 < p < 1$, it can be seen from the above formula that when p is close to zero. The solution of ℓ_p will be closer to the solution of ℓ_0.

3 System model

This paper considers a typical uplink NOMA system with one base station(BS) and K users,here the single antenna scheme is considered without loss of generality.The transmitted symbol x_k for user k is modulated onto a spreading sequence $\mathbf{s_k}$ of length N, in particular,we consider the case of $N < K$,that is, the overloaded system, where the number of users can be larger than the length of spreading sequences; After that, signals from all active users are superimposed,and then are transmitted over N orthogonal OFDM subcarriers [6].The received signal on subcarrier n at the BS is denoted as:

$$y_n = \sum_{k=1}^{K} g_{nk} s_{nk} x_k + v_n, n = 1,2,..., N \quad (4)$$

where s_{nk} is the nth component of spreading sequence $\mathbf{s_k}$, and $v_\mathbf{n}$ is the Gaussian noise on subcarrier n with zero mean and variance σ^2, g_{nk} is the channel gain of user k on the nth subcarrier, and all of them are independent and identically distributed complex Gaussian variables with zero mean and unit variance [6-8], we combine the received signals over all N subcarriers, and then the received signal vector $\mathbf{y} = [y_1, y_2, ..., y_N]^T$ is denoted as:

$$\mathbf{y} = \mathbf{Hx} + v \quad (5)$$

where $\mathbf{x} = [x_1, x_2, ..., x_K]^T$ denotes the vector of transmitted signal for all K users, $\mathbf{H}$ is the measurement matrix of size $N \times K$, its value h_{nk} in the nth row and the kth column equals to $g_{nk} s_{nk}$, $\mathbf{v} = [v_1, v_2, ..., v_N]^T$ is the noise vector that follows the distribution of $CN(0, \sigma^2 I_N)$ [8].

4 TOMP algorithm

The core idea of the TOMP is to use the time correlation of the active user sets to achieve user activity and data detection in a number of continuous time slots.This is the biggest difference with OMP. The main steps of the algorithm are as follows:

Input:

Measurements : $\mathbf{Y} = [y^{(1)}, y^{(2)}, ..., y^{(t)}]$;

Measurement matrix: $\mathbf{\Phi} = [\phi_1, \phi_2, ..., \phi_N]$;

Maximum sparsity level: K.

Initialization:

1) The initial support is set as $\hat{T}_0 = \phi$, residual signal $\mathbf{R} = \mathbf{Y}$;

Iteration:

2) if $k > K$, stop, otherwise continue.

3) The estimated signal:

$$\mathbf{Z} = \mathbf{\Phi}^{\mathbf{H}} \mathbf{R} \tag{6}$$

4) User estimated support:

$$\hat{T}_0 = \hat{T}_0 \bigcup \arg\max_i \lambda_i = \sum_j | Z_{i,j} | \tag{7}$$

5) Residual: The number of elements in the estimated support do not exceed the sparsity level K:

$$R = Y - \Phi_{\hat{T}_0} \Phi_{\hat{T}_0}^{+} Y \tag{8}$$

6) The estimate of the support set at the end of the iteration by $\hat{T}_0$,which also serves as the initial estimate support for the following operation:

$$\hat{T}^l = \hat{T}_0 \tag{9}$$

7) Search matching users: $l = l + 1$,searching for the most relevant atoms and indexs to join set $\hat{T}^{l-1}$,in each iteration , the user that correlates best with the residual signal will be included in the active user set at first.So, the active user set can be calculated by:

$$\tilde{T}^l = \hat{T}^{l-1} \bigcup \arg maX_i \beta_i = \phi_i^H r^{l-1} \tag{10}$$

where r^{l-1} is the residual signal.

8) Update: The sparse transmitted signal vector in this lth time slot can be updated as follows:

$$\tilde{x}^l = \Phi_{\tilde{T}^l}^{+} y \tag{11}$$

9) Modify the estimated support: Estimated support set is constantly being modified in each iteration. For example, in an iteration, the active user set is reliable, but the active user set is not accurate at the next iteration,and then it should be removed from the $\hat{T}^l$. The estimated support is revised by:

$$\hat{T}^l = \{K \text{ indices with largest magnitude in } \tilde{x}^l\} \tag{12}$$

10) Update the residual signal:

$$r^l = y - \Phi_{\hat{T}^l} \Phi_{\hat{T}^l}^{+} y \tag{13}$$

11) When the energy of the residual error signal does not decrease,the iteration will be terminated with the following condition:

$$\| r^l \|_2 \geq \| r^{l-1} \|_2 \tag{14}$$

12) Output：After calculating the active user set in each time slot, the sparse transmission signal vector can be restored in the tth time.

$$\hat{x}^{(t)} = \Phi_{\hat{T}}^{+} y^{(t)} \tag{15}$$

5 Simulation results

The simulation platform is the MATLAB R2014a version. Let's assume that the number of active users are already known at the BS, with QPSK modulation;Especially,the prerequisites for the simulation system are that the number of users is $K = 200$, active users is $k = 20$ and orthogonal subcarriers is $N = 100$,so the overloading factor is 200%. The simulation results are shown in the following Fig.1 and Table.1:

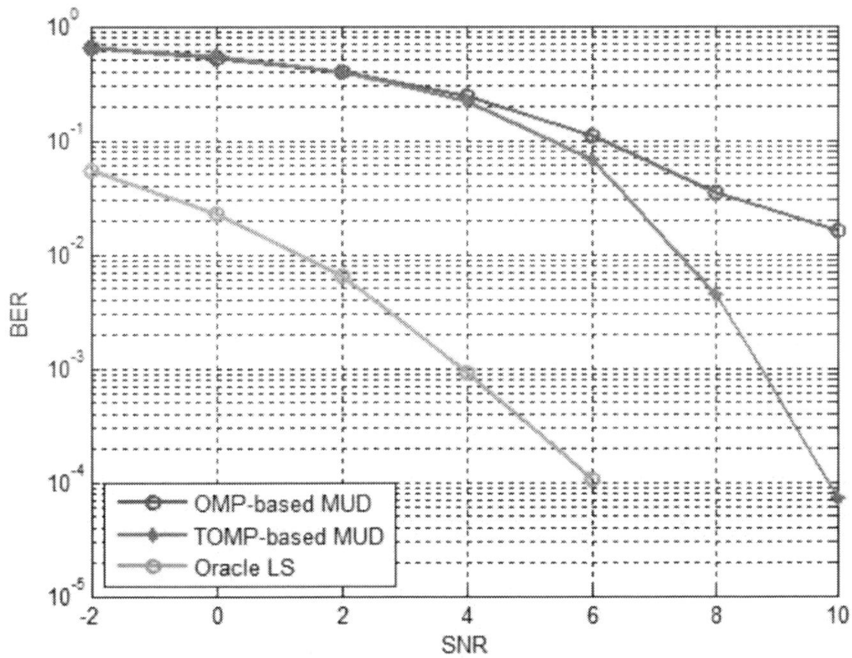

Fig.1. Comparison of performance between BER and SNR in uplink grant-free NOMA system

Table 1. Experimental data

BER	SNR(dB)		
	OMP-based MUD	TOMP-based MUD	Oracle ls
10^{-1}	—	5.4	6
10^{-2}	10.4	7.4	1.4
10^{-4}	—	9.8	6

In summary, it can be clearly seen from the above Fig.1 and Table.1 that when the SNR is less than 4 dB, the BER performance of the multi-user detection based on the traditional OMP algorithm is approximately the same as the BER performance of the improved algorithm. However, when SNR increases gradually, the BER performance of the multi-user detection based on the improved algorithm is superior to the traditional OMP algorithm.For example, when SNR is 10^{-2} , the multi-user detection based on the improved algorithm is 3 dB SNR gains over the traditional OMP. This is due to the temporal correlation of active user sets in several continuous time slots. In addition, we can also see that, compared to the oracle LS algorithm, when the SNR is 10^{-4} , the TOMP-based MUD suffers from about 4dB performance loss compared to the Oracle LS. This is due to the uncertainty support estimation of the greedy recovery algorithm in CS when the SNR is lower.

6 Conclusions

In this paper, a novel optimization algorithm is proposed based on the MUD scheme of traditional CS classical orthogonal matching pursuit (OMP) in NOMA systems. Simulation results show that the proposed scheme can achieve better BER performance in uplink grant-free NOMA systems, a novel solution for MUD is provided.

DOI: 10.1051/matecconf/201713900191

References

1. L. Dai, B. Wang, Y. Yuan, S. Han, Z. Wang. *Nonorthogonal multiple access for 5G: Solutions, challenges, opportunities,and future research trends,* IEEE Commun. Mag.,**53**, (2015), 74-81.
2. Z. Ding, L. Dai, H. V. Poor. *MIMO-NOMA design for small packet transmission in the Internet of Things*,IEEE Access,**4**, (2016), 1393-1405.
3. A. T. Abebe, C. G. Kang. *Iterative order recursive least square estimation for exploiting frame-wise sparsity in compressive sensingbased MTC*,IEEE Commun. Lett.,**20**, (2016), 1018-1021.
4. E. J. Candes , T. Tao. *Near optimal signal recovery from random projections:Universal encoding strategies*,IEEE Transactions on information theory, **52**, (2006), 5406-5425.
5. L.Gan. *Block compressed sensing of natural images,*Proceedings of 15th International Conference on Digital Signal Processing, (2007), 403-406.
6. B. Wang, L. Dai, T. Mir, et al,*Dynamic compressive sensing based multi-user detection for uplink grant-free NOMA,*IEEE Communications Letters,**20**, (2016), 2320-2323.
7. B. Shim, B. Song, *Multiuser detection via compressive sensing,*IEEE Commun. Lett.,**16**, (2012), 972-974.
8. J. P. Hong, W. Choi, B. D. Rao. *Sparsity controlled random multiple access with compressed sensing*,IEEE Trans. Wireless Commun.,**14**, (2015), 998-1010.

Design and implementation of fingerprint access control system based on ZigBee wireless network

Jin Zhang[1],Shaoyong Jian[*]

[1]Changzhou Vocational Institute of Engineering.Changzhou,213000,China

[*]Department of Mathematics and Computer Sciences, Xinyu University, Xinyu,338004 ,China

Abstract: this paper mainly introduces a kind of fingerprint access control system based on ZigBee wireless network design and implementation method, specify when passengers should fingerprints collected and stored in the system at the information desk PC, and then assign rooms;Guests entered the room according to the fingerprint module, the system will be fingerprint information through ZigBee coordinator node module and network to transmit the fingerprint characteristic value to management system and the fingerprint information stored, if consistent with the electromagnetic lock open, allowed to enter the room, does not conform to the issued a warning sound.Additional access control system can control the corresponding fingerprint information storage room and remove, such as is required for a security incident broke in situation is special open mode can be set up.System software design consists of two aspects of the bottom and upper machine.

1 The introduction

Hotels are a place where people stay short and change quickly, which brings convenience and safety hazards to the people who provide them. Often in the news, the suspect in the news of the suspect in the fake hotel guest to open the door of the room to steal, although there are surveillance equipment and see the suspect, but solving the case is not easy. Even if the amount of loss is small, but some id CARDS, credit CARDS and other things, it is troublesome to apply for a replacement.

Hotels access control system from the original key to the current popular induction card, the goal of the implementation is to be safe, convenient. However, in these systems, there is a fundamentally unsolvable problem where the key can be copied, the sensor card can be stolen, or the user identity cannot be identified. Therefore, the ultimate goal of the development of the gated system can only be biometric identification. At present, the common biometric identification system has fingerprint, iris, etc. Relatively, the fingerprint identification is simpler, and there is more choice. Therefore, if the hotel and hotel access control system is changed to the fingerprint access system, the thief will have nothing to do and greatly improve the security of housing.

But one problem that needs to be addressed is how to

DOI: 10.1051/matecconf/201713900192

design the minimum cost of access control system transformation program. The network transmission scheme of fingerprint eigenvalue information is usually transmitted by GPRS, Wifi, ZigBee and other modes. Look from the actual application effect, the wireless mode sensing module installation is simple, the equipment maintenance workload small, the advantages of the engineering costs are relatively cheap makes it have a huge impact to the traditional cable transmission mode.

2 System structure and design thinking

Fingerprint access control system based on ZigBee wireless network is divided into fingerprint input module and fingerprinting module, fingerprint input module is mainly to collect the fingerprint characteristic value of passenger and entry to PC system, interface designed by QT. Fingerprinting module mainly by the fingerprint acquisition module, ZigBee terminal module and ZigBee coordinator of the fingerprint characteristic value of the package and sent to ZigBee terminal, through the ZigBee real-time calibration coordinator and the upper machine, than success, return to unlock signal, open the electromagnetic lock. If the fingerprint is not successful, the fingerprint information will be collected. If the fingerprint information is not successful, the alarm signal will be generated and the access control function will be realized.

The design idea mainly considers in the following aspects:

2.1 Biometric identification to prevent counterfeiting;

When checking in, the guest needs to register the fingerprint information and store it on the system's upper computer. The terminal will collect the fingerprint information and compare it with the upper computer via the ZigBee wireless network. The match will open the electromagnetic lock, otherwise the error will be prompted, and a number of errors will generate an alarm will not be opened again.

signal. After the passenger has checked out, the fingerprint information will be deleted and the door lock

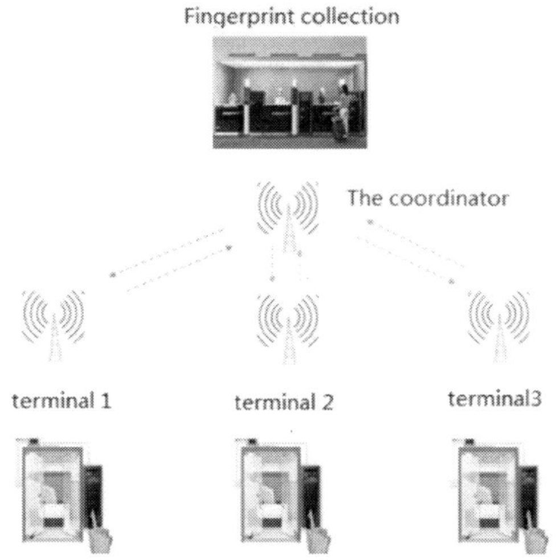

FIG. 1 schematic diagram of the system structure

2.2 Can quickly add authorized person, delete authorized person, record user and other operations;

2.3 Based on ZigBee's wireless network, the fingerprint information can be transmitted safely without upgrading the existing network.

Fingerprint information needs to be sent to the server for validation, using traditional lines of cable transmission will produce a great burden, so first consider adopting wireless fingerprint characteristic value in the form of transfer.

Despite the selection of various wireless networks, the ZigBee wireless sensor network has considerable advantages over the GPRS and other communications networks and WiFi wireless networks. The use of wireless networks such as GPRS or CDMA should be considered as the first consideration of the cost. Combined with the monthly fee, it will be difficult to accept. The advantage of WiFi is that the technology r&d threshold is low and the product cost is low. But the biggest problem is low security and poor wireless stability. Besides, the power consumption of WiFi is also very weak, and the number of devices that the application system can connect is very limited, so the future development space is limited.

ZigBee technology has the advantages of high security, low power consumption and strong networking ability, compared with ZigBee wireless transmission fingerprint characteristic value.

3 System hardware design

The main components of ZigBee's wireless fingerprint access terminal are divided into three main components:

3.1 CC2530 core board

The separation of the core plate and the function board is more convenient for the maintenance and testing of the equipment. It has 21 universal I/O pins and two 20mA current absorbing or current supply capacity. The working band is 2.4gHZ. Low power consumption, receiving current <30mA, emission current < 50mA. Use 2. 4G full - to antenna, reliable transmission distance of 250 meters. Automatic reconnect is as high as 110 meters.

3.2 ZigBee node module.

The ZigBee node module has a size of 15*10 cm, and its power supply mode is square port USB, DC2. 1 power base (5V). Serial port communication is performed by using the USB transfer serial port function (pl-2303). With the Debug interface, it is compatible with TI standard simulation tool, which brings out all IO ports, commonly used serial port pins and 5V/3. 3V pins and other functional interfaces.

3.3 TFS-M711 fingerprint module.

TFS-M711 and ZigBee node module and CC2530 core board are connected by serial port and finally communicate with the upper computer.

4 System software design

4.1 Overall structure of system software

ZigBee wireless fingerprint access control system software development mainly divided into two areas: one is in the PC QT management control interface input fingerprint information, room number and distribution, at the same time save the fingerprint information used for comparison. The second is to collect the fingerprint information on the fingerprint module of the guest room, and send it to the management system by the ZigBee module and the coordinator to check the fingerprint information and return the judgment result to the guest room. Use the lock correctly, otherwise the alarm signal will be generated. The details are shown below:

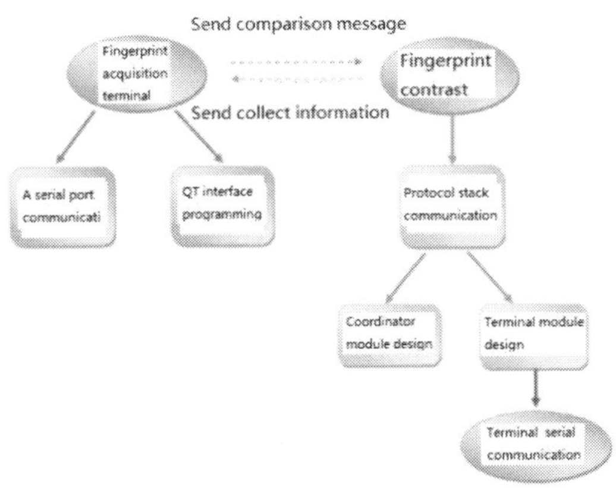

FIG. 2 overall architecture of the software

4.2 Upper computer software design

The upper computer software is divided into two modules: QT serial port design and QT interface design respectively. The serial port design module of QT mainly includes the serial port design of the fingerprint input module, the serial port design of the ZigBee coordinator and the sending and receiving design of the serial port. The interface design of QT mainly includes graphical module design of interface, signal slot design of interface and initializing design of constructors.

DOI: 10.1051/matecconf/201713900192

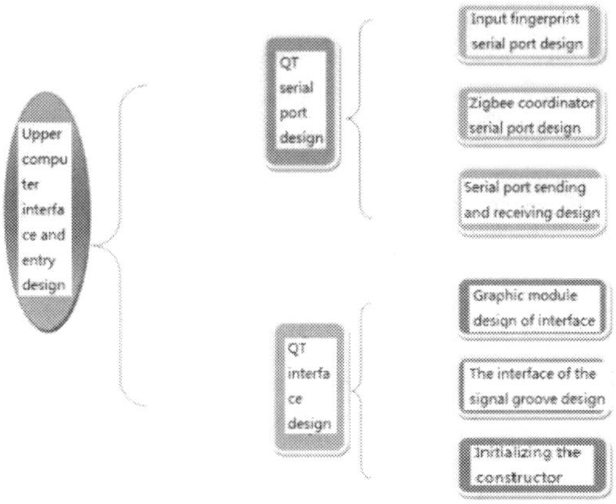

FIG. 3 upper computer programming structure

User input interface design is shown as follows:

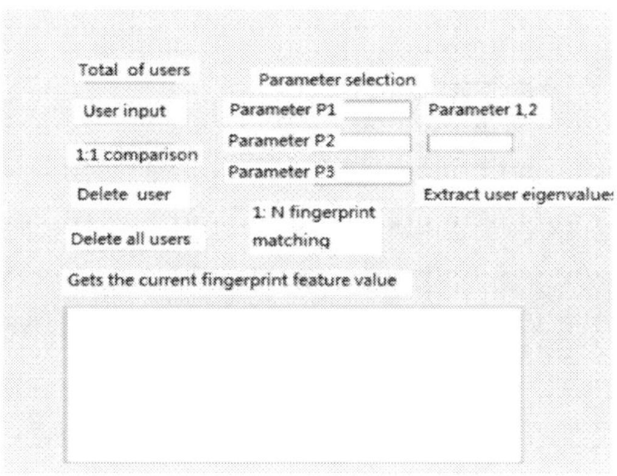

FIG. 4 user input QT interface design

5 The conclusion

The tenant can no longer worry about room card theft, or falsely use another identity into the room, his fingerprint is the key to the room, at the same time of convenience for passengers, further improve the performance of the hotel safe.

5.1 Centralized management control of the reception desk.

When checked in at the customer service desk management computer can directly input fingerprint information entry, when the check-out is removed from the entrance guard management system fingerprint information of the customers, and be able to access to any fingerprints for remote control. At the same time, the system can record the opening time of the fingerprint and send it to the front desk management system, and it can generate customer billing information and make it easier to manage.

5.2 The supervision room authorizes manual intervention function.

When some special events, such as: fire, kidnapping, catch the thief, etc., the monitoring center can be directly authorized to control the switch of access control points, so that the flexible handle emergency events, ensure users get more time and reduce the loss, etc.

Because the identity information, the corresponding relationship between the fingerprint of the fake id hotel also has a huge deterrent effect, has a poor record of customer if identity information does not accord with the fingerprint will automatically generate alarm signal.

References:

1. Guo xie chao, "information security" （2009.1）, based on ZigBee network's wireless fingerprint identification system
2. Pan wei, "science and technology and innovation", based on ZigBee network's wireless fingerprint identification system design and implementation
3. Network bee, ZigBee combat drills v3.1 experimental manual
4. Wang xiaoqiang, ouyang jun. Design and implementation of ZigBee wireless sensor network, chemical industry press
5. Lu wenzhou, Qt5 development and examples (2nd edition), electronic industry press

Impact of Laser Additive Enhanced Powder Feeding Methods on Melt Pool Temperature Field

Ren *Yuan*[1,*], Li *Wen-long*[1], Ge *Hai-long*[1], Wang *Wen-tao*[1], Zhang *Xian-hu*[2] and Cheng *Wei*[1]
[1]Laser Institute of Shandong Academy of Sciences, Ji Nan, 250014, China
[2]Remanufacturing Branch of Yan kuang Dong hua Heavy Industry Co.,Ltd. Zhou Cheng, 273500, China

Abstract: The surface of Q345 steel are coated with the Fe62 alloy powder by the synchronous powder feeding method and the prepositioned powder feeding method. The impacts of the two methods on the laser pool forming rate and the temperature field are tested and analyzed with a Charge-Coupled Device (CCD) colorimetric thermometer. The results indicate that the laser pool has faster forming rate, more stable and homogeneous temperature field with the synchronous powder feeding method than the prepositioned powder feeding method under the same conditions. The outcomes are of great significant to optimize the process of the laser additive enhancement technology.
Keywords: Laser Additive Enhancement; Temperature Detection; Synchronous Powder Feeding; Prepositioned Powder Feeding

1.Introduction

Laser additive enhancement is a new surface modification technology emerging with the development of high-power lasers in the 1970s. During the laser additive enhancement, inside the melt pool lie high energy, momentum, and mass transfer, which resulted in heat transmission, convection, and mass transfer. These directly affect the metallurgical performance and surface quality of laser for remanufacturing [1]. By detecting the distribution of the laser melt pool temperature field, we analyzed the energy, momentum, and mass transfer in the process of additive enhancement. The analysis is of great significance to the development and application of laser additive manufacturing (LAM) [2].

Laser additive enhanced power feeding methods usually refer to synchronous powder feeding and prepositioned powder feeding. The former is to use a gas-borne powder feeder, through which laser cladding powder can be directly delivered into a spot and form a cladding layer as moving along with the spot on the workpiece surface. The latter is to place the power on the path of the laser spot in advance. With moving of the spot, the prepositioned powder together with the substrate is melted and hardened to have a laser formation layer. The two methods have direct impacts on the quality of the laser additive enhancement formation layer due to different powder melting processes [3].

Based on the experiment that Q345 steel was used as a substrate, Fe62 as powder, and laser additive enhancement was carried out by two methods of synchronous powder feeding and prepositioned powder feeding, this paper studied the distribution of the laser melt pool temperature field and analyzed the melt pool formation and the change of the temperature field of two powder feeding methods. The experiment indicated that under the same condition, the synchronous powder feeding method could achieve faster melt pool stabilization and more uniform distribution of the temperature field as opposed to the prepositioned powder feeding. At the same time, the study results of this paper lay a theoretical and experimental foundation for the optimization of the laser additive enhancement in the application process [4].

2 Test Materials and Methods

Q345 steel is used as a substrate, cladding powder is Fe62, and the powder particle size is in the range of 80-115μm. A high-power optical fiber laser is used as a laser source in the experiment, the laser power is 2,000W, the spot diameter is 5mm, the focal length is 200mm, and the movement speed is 3mm/s.

Before the test, we closed up the melt pool temperature field to monitor the dynamic balance state of a CCD camera and adjusted the installation angle and position and focal length of a lens. The change of the melt pool temperature field in the process of laser additive enhancement is observed via the synchronous and prepositioned powder feeding methods. Pneumatic powder feeder is adopted to feed powder and prepositioned powder feeding thickness is 3mm.

DOI: 10.1051/matecconf/201713900193

3 Experiment Results and Analysis

3.1 Dynamic Process and Analysis of Melt Pool Temperature Field in Laser Additive Enhancement of Synchronous Powder Feeding

Figure 1 shows the melt pool temperature field at different moments when laser additive enhanced feeding of Fe62 metal powder. The laser is turned off at 19.82s. It can be noted from the figure that as slag and powder generated by melting Fe62 powder fall in, the heat radiation and transmission of the laser melt pool is remarkably decreased, and the heat radiation brightness at the central part of the melt pool turns dark. However, the temperature of the laser melt pool doesn't see any decrease.

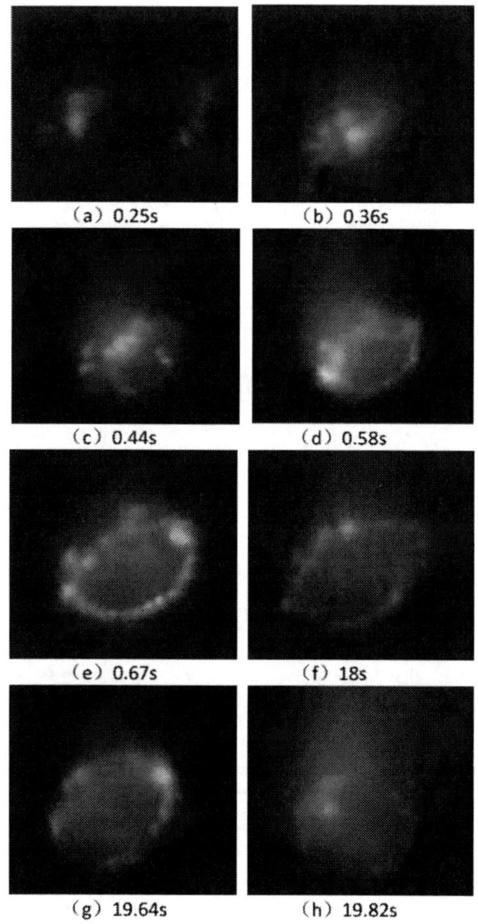

Figure 1 Heat Radiation Imaging of Lateral Powder Feeding Melt Pool

When laser radiates at 0.25s, a part of powder is melted. At this time, the temperature of the melting part is about 1600K and heat rises at the speed of 8×10^3 K/s. As time passes, melted powder is increasing fast, and the size of the formed laser melt pool turns larger fast. At 0.36s, the size is about 1.5mm. At 0.44s, it is about 2.0mm while at 0.58s, it is increased to 3mm. The result shows that in the initial process, the melt pool area has extremely high heating speed. Furthermore, melting starts from the center of the laser spot, then the size of the melt pool increases rapidly. Until 0.67s, the melt pool enters the basically stable state. The average temperature of the melt pool remains at 1800K, and the size of the melt pool is about 3mm.

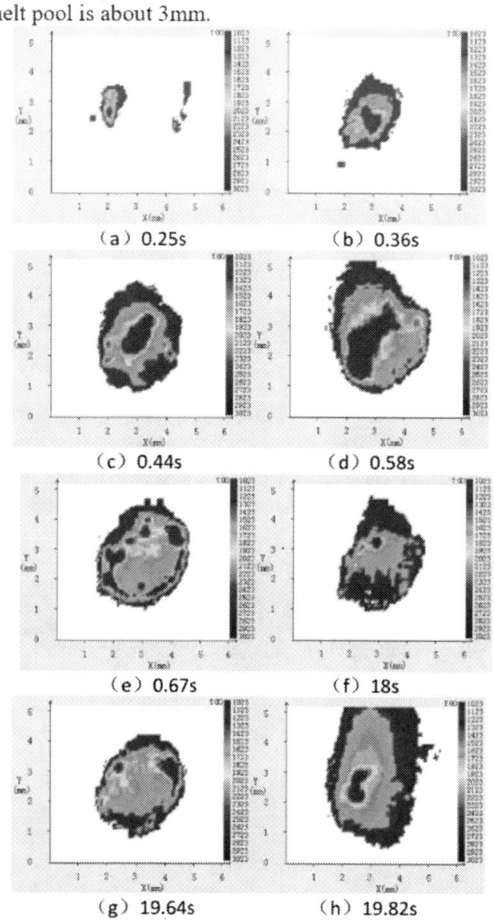

Figure 2 Temperature Field Distribution of Lateral Powder Feeding Melt Pool

Under the condition that powder feed, laser power, and processing speed remain constant, the temperature distribution of the melt pool is stable, and so is the size of the melt pool, which is about 3mm. At 19.64s since the laser was turned off, the temperature of the laser melt pool goes down rapidly, the melt pool shows obvious gradient distribution of temperature. These indicated that when the laser heat input was cut off, heat transmission of the workpiece is interrupted and the temperature of the melt pool rapidly decreases. At 19.82s, no heat radiation images can be seen [5].

3.2 Dynamic Process and Analysis of Melt Pool Temperature Field in Laser Additive Enhancement of Prepositioned Powder Feeding

Figure 3 illustrates the heat radiation of the laser melt pool at different moments when powder is prepositioned. At 0.25s, the melt pool generally becomes stable and complete. At 19.82 the laser is turned off, the heat radiation of the laser melt pool is gone fast.

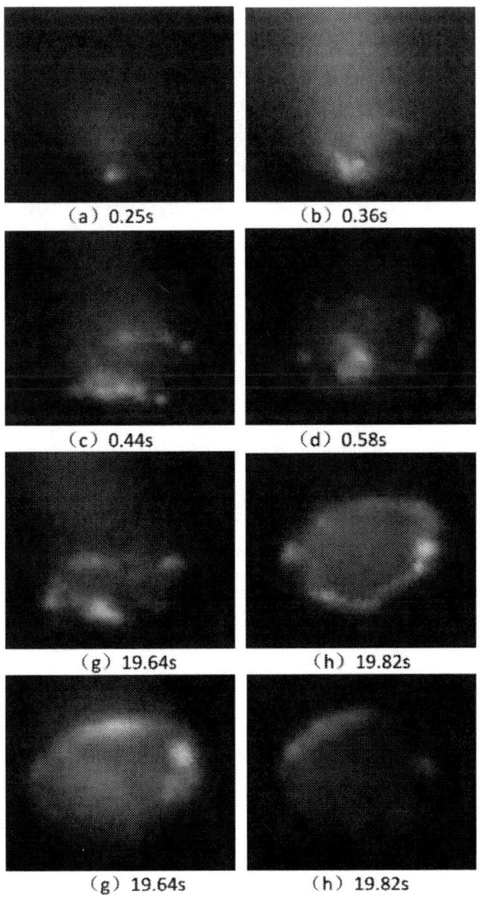

Figure 3 Heat Radiation Imaging of Dynamic Process of Powder Prepositioned Laser Cladding

At the very beginning, an impurity was burned up and it can be seen that the flame is substantially symmetric to the center of the melt pool. From the beginning to 0.25s, only a part of powder was not melted, so that an incomplete melt pool was formed. At 0.25s, a complete melt pool is basically formed and starts turning stable. It can be seen from the photos of the temperature distribution between 0.67s and 19.64s, the laser melt pool presents a stable distribution.

Figure 4 shows the temperature field distribution of the laser additive enhancement melt pool when powder is prepositioned. With the prepositioned powder feeding method, the melting point of Fe62 powder is about 1100°C, lower than the substrate (the melting point of 45 steel is at 1433°C). A laser beam firstly radiates the powder, and the powder can be melted fast to be a melt pool due to its high absorptivity [6].

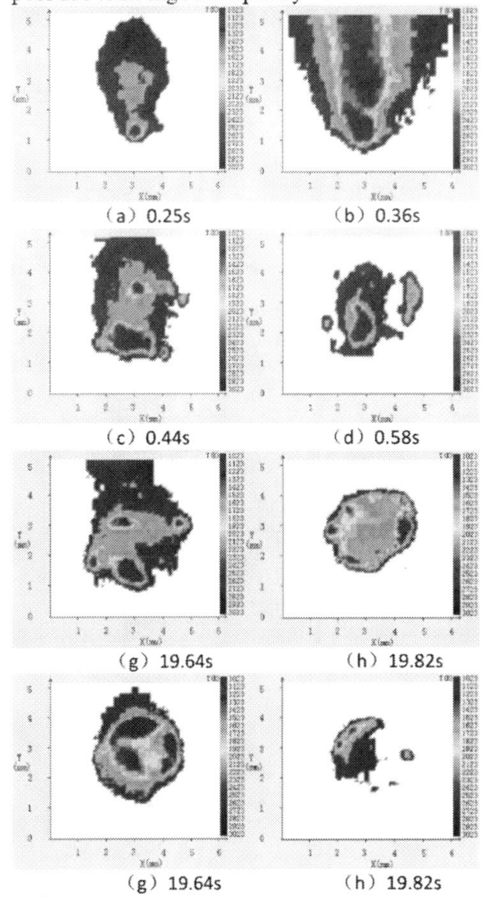

Figure 4 Temperature Field Distribution of Powder Prepositioned Laser Cladding

By comparing the changes of the melt pool temperature fields caused by the laser additive enhanced synchronous and prepositioned powder feeding methods, we found that the laser melt pool can be stable at fast and the temperature distribution more uniform under the powder feeding circumference [7]. Through analysis, during the laser additive enhancement, it can be seen that the powder of the synchronous powder feeding technology is heated uniformly, the heating area is the laser spot area, and the heat of the powder is transferred too insufficient to melt the powder. By contrast, the prepositioned powder feeding technology will transfer a part of heat to the powder around during the powder melting, and powder performance is exposed to oxygen and nitrogen in the air during heating, so the formation speed of the melt pool is slower and the formation quality is not as good as that of the synchronous powder feeding technology. However, when the melt pools formed by the prepositioned powder feeding and synchronous powder feeding methods remain stable, the

DOI: 10.1051/matecconf/201713900193

average temperature of the melt pool formed by the former is higher than the melt pool formed by the latter. This result plays a good guidance role in the laser additive enhancement technology [8-10].

4 Conclusions

1) In the process of laser additive enhancement, the quality of the formation layer can be controlled by the temperature field distribution and the size of the melt pool. It takes some time for the melt pool to be stable. According to the time and processing speed, the processing volume can be calculated in advance. In the stable state, the size of the melt pool may vary to some extent and the variation has to be taken into consideration during processing.

2) Also in this process, the average temperature before the formation of the melt pool is higher than that of the stable melt pool. This indicates that laser can generate a melt pool at instant high temperature, when the temperature is stable, so is the melt pool.

3) The analysis on the changes of the temperature fields from the synchronous and prepositioned powder feeding method of laser additive enhancement indicated that by the former method, powder is stably heated and the temperature distribution of the laser melt pool is uniform, and correspondingly the formation quality is better than that of the latter method.

Acknowledgement

This work has been supported by Major Research Project of Shan Dong Province under the project numbers 2016GGX103017 and 2016ZDJS02A09.

References

[1] Claudia H. Second Time Around and Around-Remanufacturing is Gaining Ground in Corporate America[N]. The New York Times, 1998-07-14

[2] W.Bloehs, B.Grunenwald, F.dausinger, H. Hugel. Recent progress in laser surface treatment.II. Adopted Processing for High Efficiency and Quality[J]. Journal of Laser Application, 1996, 8(1): 65~77

[3] M.Simon, T.Dowie. Disassembly process planning[C]. Proceeding of 30th International MATADOR Conference, Manchester, UK, 1993

[4] Yang Xi-chen, Wang Yun-shan, Lei Jian-bo et al.. Laser re-fabricating of expensive parts of the equipment in oil-field on the sea[C]. Lasers in Material Processing and Manufacturing II, Proceedings of SPIE(2005), 2005.1, v5629: 23~31

[5] Yang Xi-Chen, Zhao Xin, Wang Yun-shan. New Development of Laser Cladding System on Large Area for Industrial Application[C]. The Proceedings of SPIE, Nov.6-7, 1996, Vol.2888:14-20

[6] Sankaranarayanan Srikanth, Guo Wen, Kar Aravinda. Characteristics of laser-fabricated metal structures[J]. Materials and Manufacturing Processes, 1998, 13(4): 537-554

[7] R.H.Hayes. Laser powder fusion/an innovative technology for the gas turbine component repair industry[C]. World Aviation Gas Turbine Engine Overhaul and Repair'97, 1997

[8] E.Beyer. Laser macro processing –Today and tomorrow[C]. Proceedings of 4th International Congress on Laser Advanced Materials Processing-LAMP2006, May 16-19,2006, Kyoto, Japan, #06-1

[9] Ohba Hironori, Shibata Takemasa. Temperature profiles on liquid metal surface during electron beam evaporation[J]. Japanese Journal of Applied Physics, Part 1: Regular Papers & Short Notes & Review Papers, 1995, 34(8A): 4253~4257

[10] Kaplinsky Michael B., Kosonocky Walter F., McCaffrey Nathaniel J., et al.. Muti-wavelength imaging pyrometer for non-contact temperature sensing[C]. Proceedings of the 1995 IEEE International Symposium on Industrial Electronics, ISIE'95. Part 1 (of 2), 1995, 1: 199~204

A new type of biomimetic fish with oscillatory propulsion method

Zhang Lei, Deng Yongchao*, Li Kunlun, Qiu Linchuan
School of Physics and Electronics, Henan University

Abstract. The traditional multi joint mechanical serial parallel structure biomimetic fish has lower propulsive efficiency, more energy transmission loss, more complicated mechanical design and control algorithms because of more mechanical joints. While the flexible body has infinite degrees of freedom, and the energy in it is stored by elasticity and inertia, it is more viscous than that of multi joints. This project studies the principle of carangidae fish in carangidae carp, takes it as the bionic object, and designs a biomimetic fish with a flexible body tail structure to improve propulsive efficiency. This study explores the biomimetic fish propulsion speed and caudal aspect ratio, caudal oscillation frequency, modal distribution curve, and optimizes a parameter set of fish tail that can achieve effective promotion. So it is good for the exploration of the development of high speed, high efficiency, high maneuverability of bionic propeller swing on theoretical basis and practical aspects.

Keywords: bionic propulsion; oscillating propulsion; fluid-structure interaction system; propulsive performance

1 Introduction

Underwater vehicle plays an important role in marine resources exploration, underwater target search and rescue and military investigation. However, the traditional thruster based on propeller mechanical structure has the defects of high noise, poor maneuverability and low propulsive efficiency[1]. In addition, the propeller is easily entangled by debris, resulting in the loss of propulsion power. Therefore, it is significant to study the propulsion of new propulsion methods. After millions of years of natural selection and evolution, fishes form unique patterns and movements to suit for the complex underwater environment. These movements are mainly divided into the body / caudal fin propulsion mode and the central fin / fin propulsion model[2]. The propeller based on BCF type of fish with exercise using caudal and pectoral not only has high mobility but also has very good propulsive efficiency. For instance, the speed of BCF type of tuna in still water can be 80km/h, promoting the efficiency almost to 90%, while the carangidae mode of carp startup fast, acceleration - close to 50m/s[3]. Due to the excellent underwater movement ability of fish, many propulsion models of biomimetic propulsion have been studied by many researchers, and a series of important achievements have been achieved[4]. With the development of bionics[5], fluid mechanics and material science, it can be predicted that the biomimetic fish like robot has the advantages of high efficiency, high concealment and high maneuverability compared with the propeller.

2. Biomimetic fish prototype design and manufacture

2.1. An external 3D model of biomimetic fish and caudal fin design

This design based on the theoretical and experimental study of CARANGOID fish body / tail fin propulsion mode[6], is to improve the biomimetic fish propulsion efficiency. The main source of fish propulsion comes from its body that 1/3 after caudal part. So in the design of the fish, the front 2/3 part of the fish body is designed as a rigid structure[7], while the 1/3 part uses silica gel mold into a flexible structure with a certain rigidity, and the tail is set to be mixed with flexible structure.

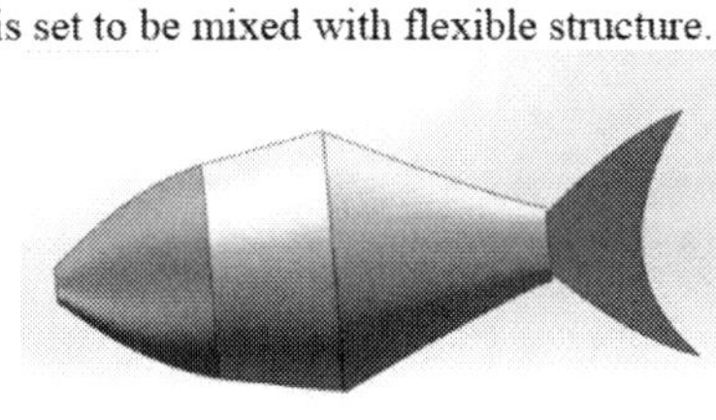

Fig.1. external structure of biomimetic fish

After hundreds of millions years of evolution, fishes have different tail fins of different shapes and flexibility[8], and the shape of the caudal fin is often represented by a formula (1):

$$P = L / S \tag{1}$$

Where P is the aspect ratio and L is the tail fin extension, and S is the caudal fin area. The results show that the speed of bionic tuna propulsion is high[9], but not necessarily the best propulsion efficiency. This shows that the influence of the ratio of the tail to the chord on the speed and propulsive efficiency of the fish is not a simple linear relationship. In this case, five different aspect ratio fins are designed as a reference. Through the actual measurement of carp body proportion parameters,

DOI: 10.1051/matecconf/201713900194

when the fish body length was 230mm, the average length of the caudal fin show is about 70mm. The caudal fin models with the aspect ratios 2.3, 3.4, 3.6, 4.2, and 6.3. Comparing with the corresponding real caudal fin, as Fig. 2.

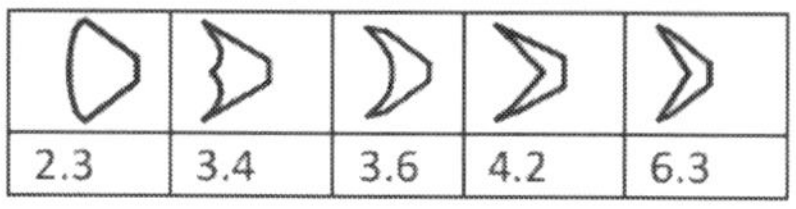

Fig. 2. Shape of caudal fin with different aspect ratio

Flexible tail fin structure. In this study, the flexible body fishtail is made of silica gel, and the tail handle is made of rigid PLV material. The two components constitute a flexible and rigid tail structure: (1) ensure that the hardness of the tail of the flexible body can handle the pressure underwater without deformation in movement; (2) ensure the flexible body has a certain degree of hardness, and not hinder the movement of the structure of swing.

2.2. Biomimetic fish internal power unit and transmission structure

According to the principle of electromagnetic induction, the conductor will produce inductive magnetic field when it is energized[10]. The magnet produces a strong magnetic field, the hollow coil is fixed in the structure with freedom in the horizontal direction only. When the current passes through the hollow coil, the coil is deflected by the torque in the horizontal direction. So we design the swing structure of the magnet and the hollow coil according to above principles, with two magnetic poles being opposite, hollow coil in the magnet middle. When the magnitude and direction of the current in the coil are changed, the coil can swing the swinging rod horizontally in the horizontal direction.

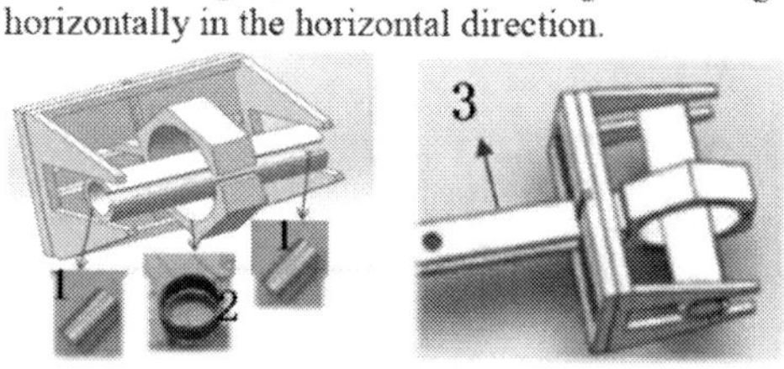

1:magnet; 2: hollow coil; 3: swinging rod

Fig. 3. Schematic diagram of oscillating device

For a single joint body / tail fin like biomimetic fish, the tail fin swing translation complex motion is the first consideration in the design of biomimetic fish. Therefore, according to the relevant experimental conclusions, the following parameters are set for flexible tail fin:

(1) Caudal to the maximum angle of attack: $\alpha_{max} \in$ (15°, 25°)

(2) The aspect ratio of caudal fin is: D=2.3, D= 4.2, D=6.3.

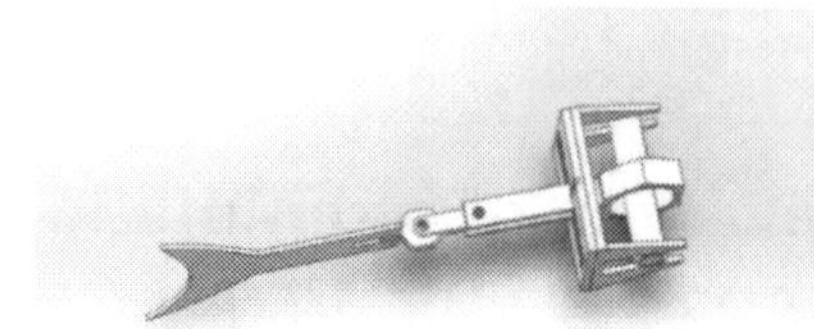

Fig. 4. Schematic diagram of driving structure of tail fin double pendulum lever

Fig. 5. The final biomimetic fish model

In Fig. 4, three micro bearings are respectively placed in three circular through holes. The tail fin and the frame part are fixed by silica gel to form two fixing points, and the connecting rod between the two bearings is a connecting rod, and the whole structure is composed of a double pendulum rod structure[11]. When the coil is energized, the coil will deflect under the action of the magnetic field force to drive the oscillating coil pendulum rod, and then drive the tail fin swinging rod through the connecting rod to realize the transmission. In this process, the coil and swing rod belong to the active rod, and the tail fin and pendulum rod belong to the driven rod; on the contrary, when the coil power belongs to the active flexible tail fin, pendulum rod coil pendulum belongs to the follower rod. The double pendulum transmission structure makes full use of the flexible body to store and dissipate energy and improve energy efficiency. Figure 5 shows the final biomimetic fish model

3. Control system design

The driving device of the tail fin is to control the magnitude and direction of the current in the coil by PWM signal. Therefore, the essence of the control of the biomimetic fish is the control of the duty ratio and the frequency of the PWM signal. Biomimetic fish control system is a relatively complex system, which needs to control the output of the signal, data transmission and data analysis simultaneously. So it requires close integration of multiple functional modules and execution in accordance with certain processes. The control system adopts the top-down design idea of multiple modules.

The flow chart is shown in Figure 6.

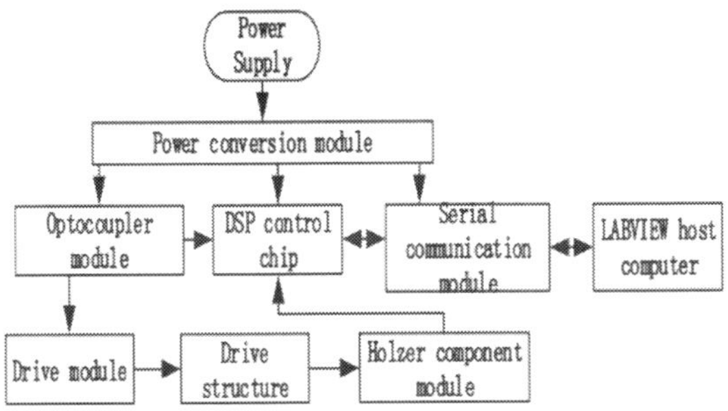

Fig. 6. Flow chart of biomimetic fish control system

4. Experiment

According to the characteristics of the biomimetic fish swimming, in order to ensure the independence movement of the biomimetic fish and reduce experimental device caused on the result of the measurement error, the experimental design uses the image resolution of 640*480 camera to shoot biomimetic fish swimming under water. The image processing method is used to intercept the image in the computer. We identify and track the biomimetic fish through target recognition and template matching, and record the position coordination of the biomimetic fish, and then draw the trajectory curve. Finally, we calculate the swimming speed of the biomimetic fish.

5. Results and discussion

5.1. The relationship between the oscillating frequency and the velocity of the caudal fin

According to the characteristics of carangidae fish tail fin, we select features for maximum swing of 25 degrees and aspect ratio were 4.2 and 2.3, the flexible fin structure 6.3. In this experiment, two cycles of PWM wave are used, one of which is duty cycle S=50%,T=64s and the other duty cycle is S=90%, T=64s The tail fin oscillation frequency is 1-9hz.

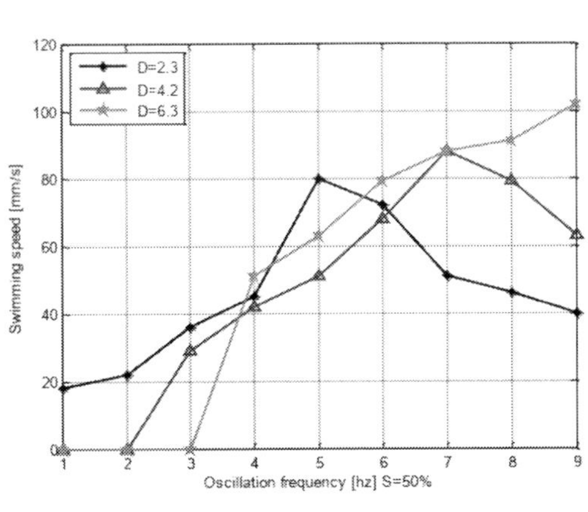

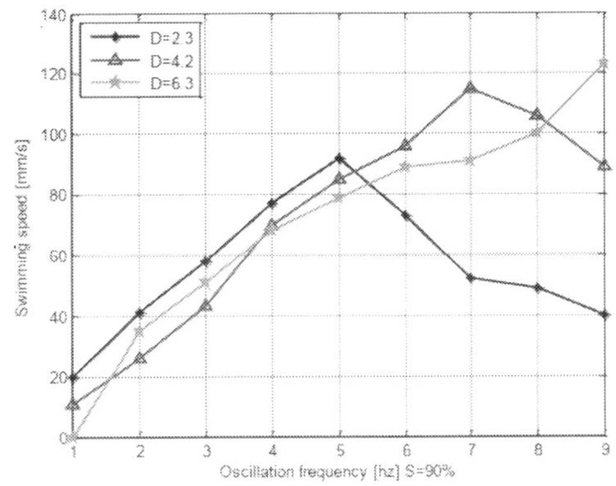

Fig. 7. Relation curve between tail fin oscillation frequency and fish speed

As shown in Figure 7 above:

(1). For a tail with arbitrary aspect ratio, its moving speed follows a two curve relationship with the oscillation frequency, and reaches its maximum near a certain frequency.

(2). When the aspect ratio is relatively small, the motion velocity following the frequency change gently and distributes evenly.

(3). When the aspect ratio is relatively large, the motion speed following the frequency change rapidly, but the overall change range is small.

5.2. The curvilinear relation between input power and speed

In this measurement experiment, the input voltage of the system is E=6.5v, the tail fin oscillation frequency is 6hz, the PWM duty cycle of the control signal is from 10% to 90%, and the duty cycle is increased by 10% each time, and the experiment is carried out once. The relationship between the input power and the swimming speed is shown in Figure 8.

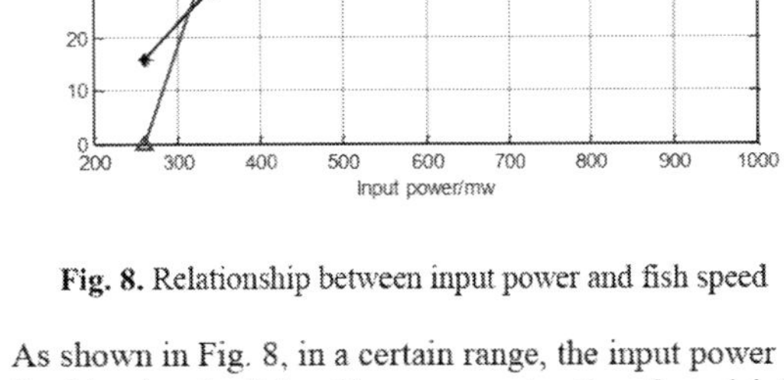

Fig. 8. Relationship between input power and fish speed

As shown in Fig. 8, in a certain range, the input power of the biomimetic fish with any aspect ratio of caudal fin has a positive correlation with the swimming speed of fish. When the input power increases, the swimming speed of fish will increase, and the smaller the aspect ratio, the greater the influence of input power on the speed of biomimetic fish swimming.

5.3. The relationship between the oscillating frequency of the caudal fin and the propulsive efficiency

The propulsive efficiency of biomimetic fish is determined by various parameters. But in the same experimental environment[8], the coefficient of fluid resistance, the density of fluid and the surface area of fish are small, so they can be regarded as constants[9]. Now the product of c is the product of the three. The propulsive efficiency of the robotic fish can be expressed by expressions (1), (2) and (3):

$$P = \frac{1}{2}\rho C_d S v^2 \bullet v \quad (1)$$

$$c = \frac{1}{2}\rho C_d S \quad (2)$$

$$\eta = \frac{P}{UI} = \frac{cv^3}{UI} \times 100\% \quad (3)$$

Where P is the effective power, ρ is the density of water, and C_d is the drag coefficient. Where S is the cross section, U is the voltage at both ends of the oscillating device, and I is the current in the hollow coil. Where η is the efficiency of converting electric energy into kinetic energy on the biomimetic fish, v is the biomimetic fish swimming speed. In this experiment, the driving signal PWM duty cycle is 90%, the corresponding system input voltage is 5.6v, the system input current is 168mA, and the system input power is 940.8mW. The aspect ratio of flexible biomimetic fish fin is D=2.3, 4.2 and 6.3 respectively.

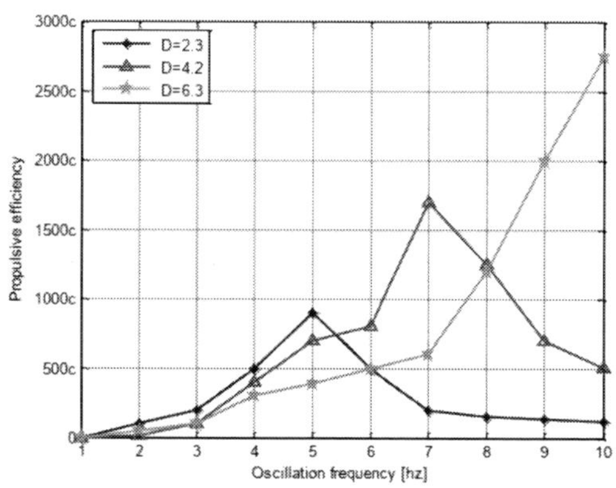

Fig. 9. Relationship curve of tail oscillation frequency and propulsive efficiency

As can be seen from Fig.9, (1) biomimetic fish propulsion efficiency with oscillating frequency increases and then decreases, the propulsion efficiency reaches the maximum value at a certain frequency. (2) under the same condition, when the ratio of the fin to tail ratio affects the propulsive efficiency of the biomimetic fish, the oscillation frequency of the caudal fin and the ratio of the aspect ratio of the tail fin are greater, and the oscillation frequency of the fish when the fish reaches the maximum propulsive efficiency is greater. (3) within a certain range, the greater the tail fin aspect ratio, the greater the propulsive efficiency of the biomimetic fish.

6. Conclusions

In this study, a biomimetic fish with flexible fish tail was designed to reproduce the complex motion of fish, and the influence of parameters of different flexible caudal fin on propulsion speed and propulsive efficiency was discussed. For a tail with arbitrary aspect ratio, its moving speed follows a two curve relationship with the oscillation frequency, and reaches its maximum near a certain frequency. In a certain range, the input power of the biomimetic fish with any aspect ratio of caudal fin has a positive correlation with the swimming speed of fish. For flexible fish body system, the effect of increasing the swing frequency or simply increasing the input power on the propulsion speed is not complete. In order to get the best propulsion efficiency, we should balance the tail fin ratio, swing frequency, input power and other factors, to maximize the speed of fish movement, in order to achieve high performance propulsion.

Acknowledgements

This work has been supported by the Foundation of Key science and technology research projects of Education Department of Henan Province （CX0001F01158）.

References

1. J. Edward Colgate, Kevin M. Lynch, Mechanics and Control of Swimming: A Review, IEEE Journal of Oceanic Engineering, 2004, 29(3): pp. 660-673.
2. M. Sfakiotakis, D. M. Lane, and J. B. C. Davies. Review of fish swimming, modes foraquatic locomotion. IEEE Journal of Oceanic Engineering, 1999, 24(2): pp. 237–252.
3. J. Deng, X. M. Shao, Z. S. Yu. Hydrodynamic Studies on Two Traveling Wavy Foils in Tandem Arrangement. Physics of Fluids. 2007, 19 : pp. 1-10.
4. J.D.Liu, H.S.Hu.Mimicry of Sharp Turning Behaviours in a Robotics Fish. Proceedings of the 2005 IEEE, International Conference on Robotics and Automation, Barcelona, Spain, April ,2005.
5. J.E dward Colgate, Kevin M.Lynch.Mechanics and Control of Swimming:A Review. IEEE Journal of Oceanic Engineering.2004,29(3):660～669.
6. G. Buccino, et al.Action Observation Activities Premotor and Parietal Areas in a Somatotopic MannerEuropean Journal of Neuroscience 2001(13):400～404 .
7. Shinjo N, Swain GW.Use of a Shape Memory Alloy for the Design of an Oscillatory Propulsion System[J]. 2004 IEEE Journal of Oceanic

Engineering, 2004, 29(3): 750~755.
8. B. E. Flammang, G. V. Lauder, D. R. Troolin and T. Strand. Volumetric imaging of shark tail hydrodynamics reveals a three-dimensional dual-ring vortex wake structure. Proceedings of the Royal Society, Biological Sciences, 2011,5 :pp. 1-9.
9. Valdivia y Alvarado, P. and K. Youcef-Toumi, Design of Machines with Compliant Bodies for Biomimetic Locomotion in Liquid Environments, ASME Journal of Dynamic Systems, Measurement, and Control 128, March 2006: pp. 3-13.
10. Tayebi A, Mcgilvray S, Roberts A, et al. Attitude estimation and stabilization of a rigid body using low-cost sensors[C]. Decision and Control, 2007, IEEE Conference on. IEEE, 2008:6424-6429.
11. Bouabdallah S, Becker M, Siegwart R. Autonomous miniature flying robots: coming soon - Research, Development, and Results[J]. Robotics & Automation Magazine IEEE, 2007, 14(3):88-98.

DOI: 10.1051/matecconf/201713900195

Research on Centralized Data-Sharing Model Based on Master Data Management

Hong Dandan[1], *Zhang* Yajuan[1], *Luo* Junfeng[1,*], *Liu* Chen[1], *Xu* Mo[1], *Suo* Zhihai[1]

[1]Network and Information Center of Xi'an Jiaotong University, Xi'an, Shaanxi, 710049, China

Abstract. This research aims to analyse the data-sharing model widely used in information centres of Chinese universities. A university centralized data-sharing model (UCDSM) based on master data management system was proposed to provide unified data standards, exchange services, data-quality control and comprehensive data-sharing services for universities in big data environment. Then the core implementation methods of this data-sharing model were analysed and implemented. Finally, the application results show that the UCDSM can support large-scale data sharing for digital campus and all the data transmission processes can be monitored and managed under a unified management platform.

1 Introduction

With the rapid development of digital campus, many colleges and universities in China have built tremendous independent systems providing diverse services for different user groups in the campus [1][2].

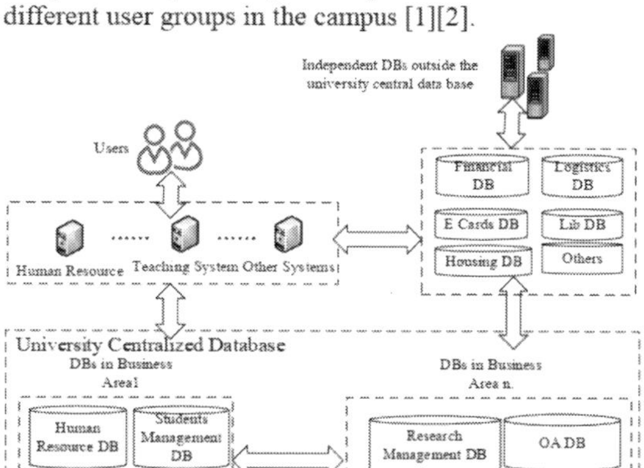

Figure 1. The traditional data sharing model in universities

As shown in figure 1, the campus database model commonly used in Chinese universities is consisted of two major parts. One is the university centralized database platform (UCDP) managed by the university network department, and the other is the independent databases managed by university business departments themselves [3].

On the university centralized database platform (UCDP), various independent databases belonging to different business systems are hosted by the network department centrally to reduce the pressure of database management in each business department. These model brought benefits to the whole university information construction:

- The UCDP provides a unified hardware platform for business departments, so that they do not need to buy any database hardware servers or software.
- The UCDP provides central database management services to business systems including database operation monitoring, user accounts management, database security control, etc. So that the business departments can focus on the software development of their systems.

Due to these advantages of UCDP, the construction of digital campus has been highly improved in the last years [4]. But with the rapid developments of university information systems, the data and objects are stored in a large amount of repetitions in UCDP. This phenomenon brings storage pressure to UCDP, and affects the data sharing efficiency. Besides, the same type of data is usually stored in different encoding formats. If one business department wants to change some information of one object in its database or add new information to an object, the databases in other third-party systems associated with it cannot be updated synchronously. This imposes a major obstacle to data analysis in large data environments, because data analysts need to spend much time in cleaning and converting heterogeneous data in different formats [5].

To solve these problems, the work in [6] proposed to use MDM (Master Data Management) to resolve the issues of data duplication, inaccuracy and inconsistency. Baghi designed an architecture for master data application toward a decision model and it verified that the master data application architecture is better at the data quality aspect [7]. Gomede presented an approach to improving the data quality of a data warehouse based on MDM [8]. Jaksic analysed a common integration architecture using MDM [9]. These researches focused

* Corresponding author: luojf@xjtu.edu.cn

on building central master data management systems, which are crucial to improve the efficiency of data-sharing and data analysis.

Motivated by these related work, the university centralized data-sharing model (UCDSM) based on master data management system is proposed in this paper. The structure of our proposed UCDSM and the difference between UCDP and UCDSM will be discussed and analysed in section 2. Three major modules of our UCDSM including MDMS, master database and master data warehouse will also be explained in this section. Section 3 will describe the core steps of the data-sharing implementation in UCDSM. Section 4 will summarize this work and propose the future work.

2 Structure of UCDSM based on MDMS

The UCDSM proposed by this paper is constructed based on the UCDP described in section 1. To improve the efficiency of data exchange and data sharing, the UCDSM integrated MDMS, master database and data warehouse, therefore the UCDSM can integrate all the advantages of these three parts. The structure of the UCDSM is shown in figure 2.

As shown in figure 2, four main modules of UCDSM system includes master data management system (MDMS), master database, master data warehouse and the public database group. The four modules work together to implement the data sharing for services in university.

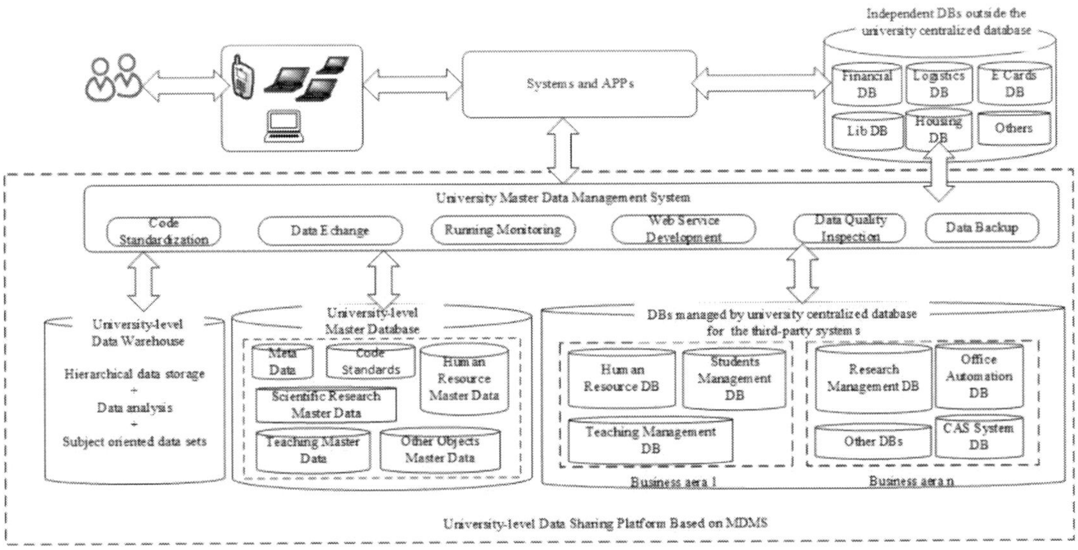

Figure 2. The university data sharing model based on MDMS

2.1 Master Data Management System

The Master data management system (MDMS) is the core part of the UCDSM. Its main functions are shown as follow.

Data standardization: it helps university establish, manage, publish and share its official data standards. The data collected from various data sources except master database must be formatted into one official standard of the university, therefore it can be shared to other systems requiring this data in digital campus. The detail process of data standardization is explained in section 3.2.

Data exchange: this module uses its ETL tools to implement the data exchange between the master database and the databases applied by business systems. The detail process is also described in section 3.2, along with data standardization.

Operation monitoring: this module includes system operation monitoring management, data integration monitoring, database operation monitoring and data flow management. It helps the UCDSM managers to detect system exception and security threats in time.

Web service publishing: this module is responsible for interface development, management and authorization. According to the comprehensive data sharing scheme defined in section 3.2, it is necessary to develop web service interfaces for data sharing, for example the interfaces in JSON and XML formats. The systems that requiring data from UCDMS can get the shared data via the authorised interfaces.

Data quality inspection: this module is responsible for the data quality check, including metadata quality, master data quality. Whenever the abnormal data is found, such as inconsistent code, inconsistent data field in master database, then it gives alarms and rehabilitate suggestions to UCDSM manager in time.

Data backup: this module helps the master data warehouse to retain the historical data of university master data and reproduce the daily data, which provides an important method for data analysis in the time dimension.

DOI: 10.1051/matecconf/201713900195

2.2 Master Database

As shown in figure 2, the master database stores all kinds of master data managed by the MDMS, such as metadata, data encoding standards, the master data objects abstracted from each business database and the activity information of the objects. For example, master data of scientific research projects, human resource master data (HR master data), students master data, teaching master data and the other management data relating to them.

The shared data that required by related systems are provided in unified format by master database via the ETL tools of UCDSM.

2.3 Master Data Warehouse

The master data warehouse provides various data subsets for different analysis topics. For example, the research project dataset is used to analyse the scientific research achievements of the university. The teaching dataset is used to analyse the most popular lectures.

Meanwhile, under the help of OLAP (On-Line Analytical Processing), BI (Business Intelligence) and SDAS (Sale Decision Analysis System), the master data warehouse also supports hierarchical data storage and provides more valuable analysis results for university administrators.

3 Data Sharing Implementation based on MDMS

In this section, we define the university itself as the first level master object. The design pattern of its master data is discussed in section 3.1. The human resource master data is designed and discussed as its child object. Meanwhile, the core data-sharing implementation (Data standardization and data exchange) is also described in section 3.2.

3.1 Design Pattern of a University Master Data

Relying on the 2012 edition of the Chinese Ministry of education information standards and characteristics of universities, the university itself is defined as the first-level master object (Figure 3), all the other master data objects inherit it and extend its child objects and properties.

For example, the human resource (HR) master data, education resource master data and the other business master data as shown in figure 3 are defined as the second-level university master data.

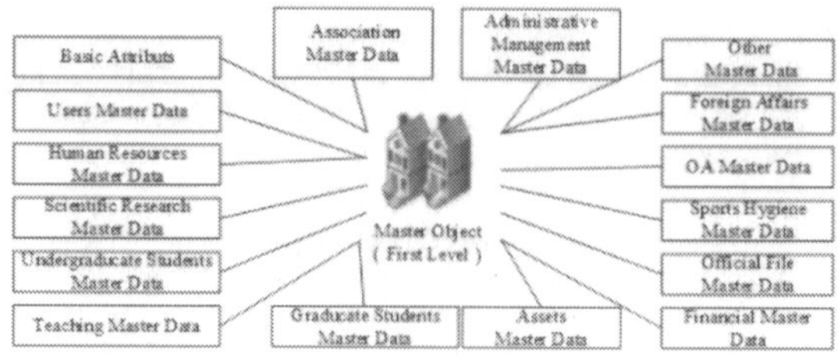

Figure 3. Design Pattern of the First Level Master Data (Only the first and second levels are displayed)

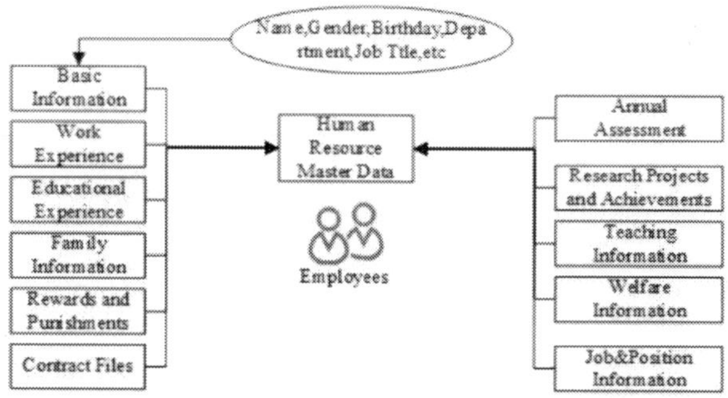

Figure 4. Design Pattern of the HR Master Data

As shown in figure 4, the HR master data model of the university is consisted of 11 child objects currently, namely basic information, work experience, education experience, family information, rewards and punishments, contract files, annual assessment, research projects and achievements, teaching information, welfare information, job&position information. Besides them, the model can be enriched synchronously with the development of the master data construction.

3.2 Implementation of Data Standardization and Exchange

As mentioned in section 2, data standardization and data exchange are the core steps of the whole data sharing process, which consists of three steps: data integration, data cleaning and data sharing (figure 5). These steps work together to implement the data sharing of the UCDSM in high efficiency and help universities conduct more detailed and standardized data analysis in big data environment.

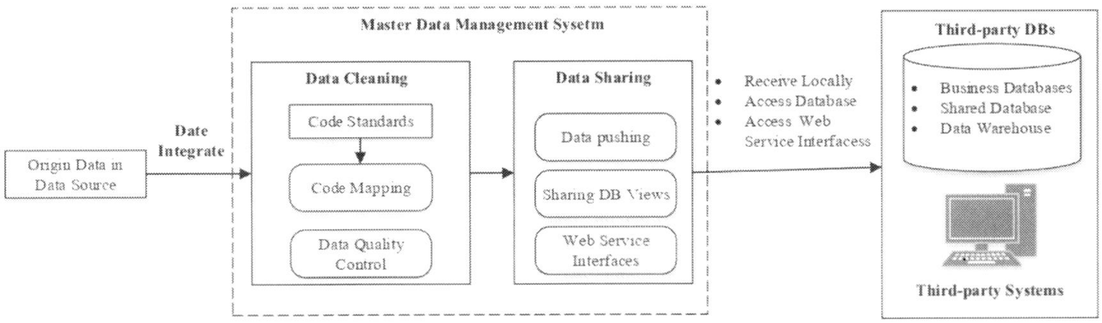

Figure 5. Data Standardization and Exchange Process

3.2.1 Data Integration

MDMS firstly finds information about the data source from the meta database stored in the master database. Then MDMS sends an access request to the data source and extract the data it needs.

To ensure data extraction efficiency, there are two approaches to extract data from databases as follows.

• The whole data extraction

MDMS extracts all the data it needs from the data sources into the temporary area of the master database. The advantage of this approach is that the retrieved data is a whole copy of the data stored in its original database, so it can keep in line with it in three aspects, namely integrity, quantity and timeliness.

The disadvantage is that if the amount of data is large (for example, all the employees' photos in Oracle blob format), the data integration process will spend much time for data extraction. Besides, if the data transfer process is accidentally terminated for some reason, the MDMS must restart this data extraction procedure no matter if the same data has been already extracted.

• The incremental data extraction

This approach only extracts the data that has been changed in its source system after the last extraction. The advantage of this approach is that it can greatly reduce the amount of data to extract and improve the efficiency of data extraction. Meanwhile, the data timeliness and data integrity can also be guaranteed.

Based on the above analysis, the second extraction (incremental data extraction) approach is adopted in UCDSM proposed in this paper.

3.2.2 Data Cleaning

Data cleaning consists of two parts: data quality control and code mapping.

• Data quality control

Data quality control is responsible for two tasks. One is the data format conversion. The other is data quality monitoring. Data format conversion aims to unify different formats of the same data into standard data formats. For example, convert different types of time formats (e.g., 2017-9-17 12:30,2017/9/17 12:30:00, etc.) into the unified time format (e.g., 2017-09-17 12:30:45).

Data quality monitoring aims to check the contents, standards and code formats of the retrieved data. If these indicators do not meet the requirements of the MDMS, alarms will be shown in the MDSM.

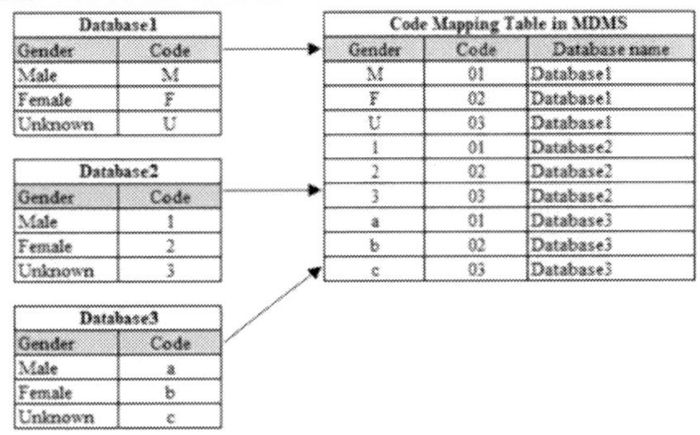

Database1	
Gender	Code
Male	M
Female	F
Unknown	U

Database2	
Gender	Code
Male	1
Female	2
Unknown	3

Database3	
Gender	Code
Male	a
Female	b
Unknown	c

Code Mapping Table in MDMS		
Gender	Code	Database name
M	01	Database1
F	02	Database1
U	03	Database1
1	01	Database2
2	02	Database2
3	03	Database2
a	01	Database3
b	02	Database3
c	03	Database3

Figure 6. Code Mapping Table in MDMS

• Data code mapping

The data code mapping creates a code mapping table in the master database and establishes the mapping relationships between the code stored in the original database and the master code defined in the master database.

As shown in figure 6, the gender code of an employee, in database1 (DB1), "M, F, U" are encoded for male, female, and uncertainty respectively. In database2 (DB2), "1" is for "male", "2" is for "female", and "3" is for uncertainty. Under this scenario, the MDMS needs to unify these data encoding rules according to the same standard. For example, all the codes extracted from different systems for "male", "female" and "uncertainty"' are re-encoded as "01", "02", and "03" respectively.

3.2.3 Data Sharing

Data Sharing model provides three approaches to help the third-party systems get the standard data from university master database.

• Data pushing

The third-party system provides the JDBC remote connection authority to the MDMS. Then the MDMS uses a data-exchange tool (e.g., Oracle ODI tool) to push

* Corresponding author: luojf@xjtu.edu.cn

DOI: 10.1051/matecconf/201713900195

the standard data directly into the database defined in the JDBC URL.

• Sharing DB views/tables

In this approach, the MDMS creates a data-sharing view (or table) in the master database, then the shared view can be authorized to multiple systems. And the third-party systems decide if the shared data should be copied and stored in its own database.

• Web service interfaces

In this approach, MDMS creates web service interfaces (JSON, XML) for data sharing. If the third-party systems need to get these data, MDMS authorizes the third-party systems to access the interfaces according to the roles of the third-party systems.

All the advantages and disadvantages of these three approaches are compared and summarized in the following table 1.

Table 1. Advantages and disadvantages of the three data-sharing methods in UCDSM.

Method	Advantages	Disadvantages
Data pushing	• The shared data are stored in the third-party system, and the third-party system does not need to connect to the master database frequently, so it can reduce the access pressure of the master database. • MDMS is responsible for data synchronization at regular time. The third-party systems need not to maintain the shared data, so the working pressure is reduced.	• The same type of shared data are stored repeatedly in many third-party systems within the campus. • MDMS must create several data-pushing procedures (for example, oracle ODI procedures) to update the shared data stored in various third-party databases.
Sharing DB views(tables)	• MDMS does not need to create data-exchange interfaces, and all the granted third-party systems can access to this shared view directly to get their requiring data. • The same shared data are only stored in one copy in the university master database. The third-party systems decide whether it should be copied locally.	• The number of the third-party systems requiring to access the master database directly will continue increasing. This brings pressure to the master database. • There are too many users accessing the same shared view, which brings risk to system security and data security of the master database.
Web service interfaces	• All the shared data are encapsulated from the web service interfaces. The third-party system can not access to the master database directly. • The data in the interface are strictly consistent with the master database	• There are too many users accessing the same shared web service interfaces, which brings system access pressure to master database.

Through these comparisons, a comprehensive scheme is proposed as follows.

- If the thirty-party systems have highly frequent access which will brings access pressure to the master database, the first way is highly recommended.
- If the shared data are changed frequently, and there are less third-party systems requiring these data, the third way is highly recommended.
- The second way is not recommended. Because the DBA of the master database in UCDSM must manage a large amount of DB accounts, the relationships between the accounts and the DB views. This will bring pressures to DBA and database management. Meanwhile, the data security is also threatened if several accounts access to the same shared DB view or DB table.

4 Application and Conclusion

The university data sharing model based on MDMS designed in this paper has already been applied in Xi'an Jiao Tong University (XJTU).

As shown in figure 7(a), 2332981 records have already been stored in the university master database. And 468991 records are stored as payment master data whose parent is the finance master data. 267735 records are stored as curriculum schedule master data whose parent is the teaching master data.

Figure 7(b) shows that there are more than 14 types of code standards managed by MDMS to support the code mapping and data exchanging between the data sources and the third-party systems. The public code standards, teaching affairs code standards and scientific research code standards are the top 3 standards in the MDMS (statistics based on code quantity).

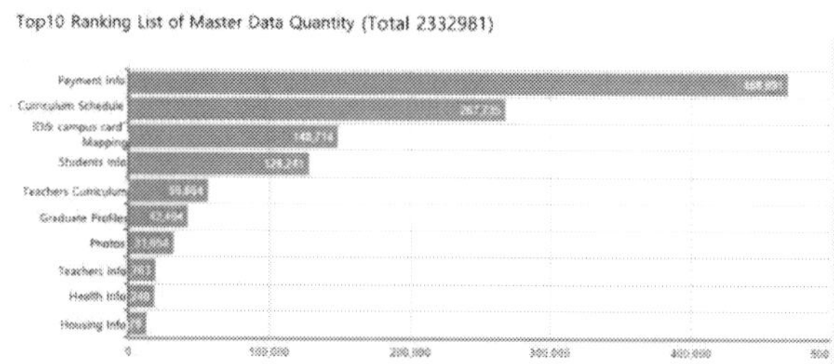

Figure 7 (a)

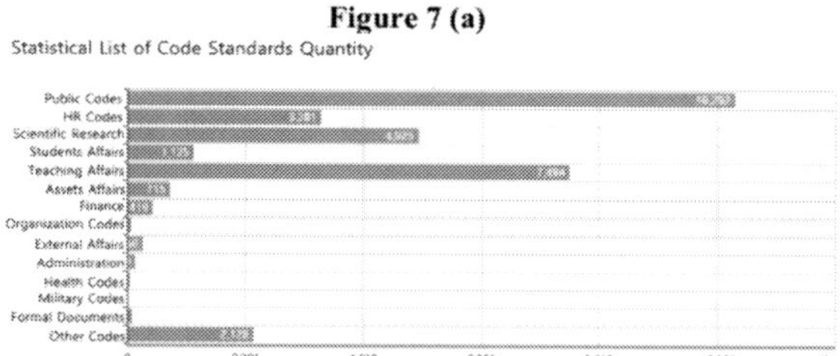

Figure 7 (b)

Figure 7(c) shows that there are more than 25 data sharing interfaces integrated into the MDMS, which are distributed in various systems, such as teaching management system, organization management system, human resource management system and so on.

Figure 7(d) shows that there are more than 373 thousands data records shared among 5 systems. For example, the all-in-one card system contributes the most shared master data in the university campus.

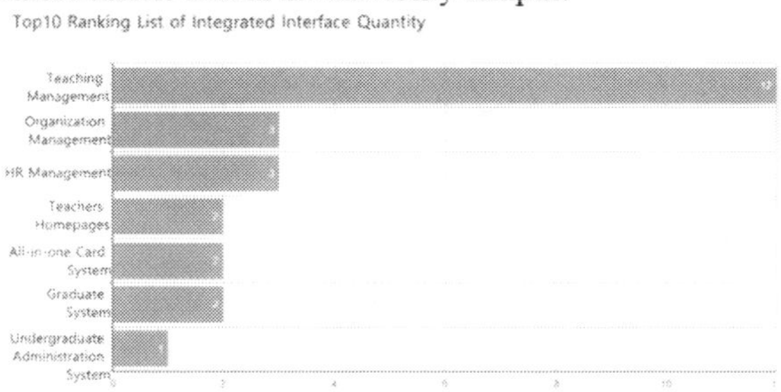

Figure 7 (c)

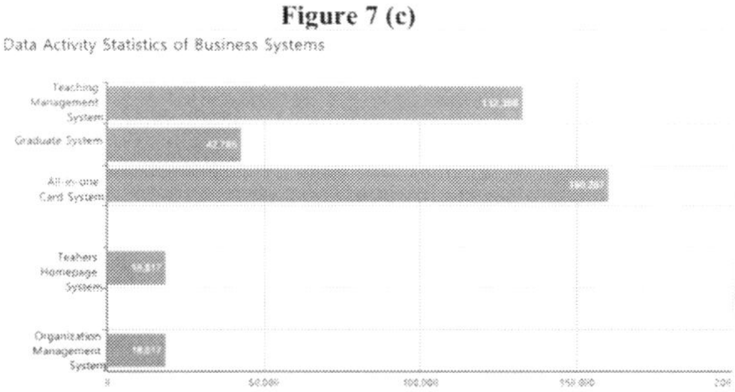

Figure 7 (d)

Figure 7. The application results of UCDSM in XJTU

In summary, the application result shows that this model can satisfy the requirements of large-scale data sharing management in digital campus. The joint analysis of different data in XJTU has been greatly improved with the help of unified data standard, unified data quality control in the big data environment.

In the future, more effort will be devoted to research on data mining of multiple data in university campus, e.g., the analysis of education data [10], research on providing accurate service based on users' behaviour [11] [12]. Meanwhile, the university master data warehouse will also be concerned to support more official and authoritative decision analysis [13] [14].

References

1. Li M J. The Digital Campus Information System Integration Research Based on SOA[C]// International Conference on Intelligent Computation Technology and Automation. IEEE, 2015:781-784.
2. Zhu R, Chen Y, Zhuang C. Experimental Center Construction of Teaching Management Platform under the Background of Digital Campus[C]// Eighth International Conference on Measuring Technology and Mechatronics Automation. IEEE, 2016:286-289.
3. Yang M, Zhang X. The research of resource sharing platform for universities using virtual database[C]// Advanced Research and Technology in Industry Applications. IEEE, 2014:690-693.
4. Zhu R, Chen Y, Zhuang C. Experimental Center Construction of Teaching Management Platform under the Background of Digital Campus[C]// Eighth International Conference on Measuring Technology and Mechatronics Automation. IEEE, 2016:286-289.
5. Xu Q, Dazhi. The application of heterogeneous database general query in digital campus[C]// Cross Strait Quad-Regional Radio Science and Wireless Technology Conference. IEEE, 2011:1193-1195.
6. Haneem F, Ali R, Kama N, et al. Resolving data duplication, inaccuracy and inconsistency issues using Master Data Management[C]// International Conference on Research and Innovation in Information Systems. 2017:1-6.
7. Baghi E, Schlosser S, Ebner V, et al. Toward a Decision Model for Master Data Application Architecture[J]. 2014:3827-3836.
8. Gomede E, Barros R M. Master Data Management and Data Warehouse: An architectural approach for improved decision-making[C]// Information Systems and Technologies. IEEE, 2013:1-4.
9. Jaksic D, Jovanovic V, Poscic P. Integrating evolving MDM and EDW systems by data vault based system catalog[J]. Proceedings of Jubilee International Convention on Information & Communication Technology Electronics & Microelectronics, 2017:1401-1406.
10. Shu J, Wang X, Wang L, et al. Exploration on college education big data open service platform[C]// IEEE, International Conference on Cloud Computing and Big Data Analysis. IEEE, 2017:161-165.
11. Guivarch V, Paz J F D, Villarrubia G, et al. Hybrid system to analyze user's behaviour[C]// Computational Intelligence. IEEE, 2017:1-5.
12. Yu C. Sale Decision Analysis System Design Based on Data Warehouse[C]// International Conference on Intelligent Systems Design & Engineering Applications. IEEE Computer Society, 2015:712-718.
13. Khanna L, Singh S N, Alam M. Educational data mining and its role in determining factors affecting students academic performance: A systematic review[C]// India International Conference on Information Processing. 2016:1-7.
14. Singh R P, Singh K. Design and Research of Data Analysis System for Student Education Improvement (Case Study: Student Progression System in University)[C]// International Conference on Micro-Electronics and Telecommunication Engineering. IEEE, 2017.

Research on I / Q Unbalanced Channel Reciprocity Calibration Algorithm in Massive MIMO System

Jinming Cui[1], *Liang* Yin[1, *], *Yue* Ma[2], *Shufang* Li[1]

[1]Beijing Key Laboratory of Network System Architecture and Convergence, Beijing University of Posts and Telecommunications, 100876, Beijing, China

[2]School of Computer Science, Beijing University of Posts and Telecommunications, 100876, Beijing, China

Abstract. Massive MIMO system uses the transmitter channel state information (CSI) to preprocess the transmitted data, which can effectively reduce the interference and enhance the Massive MIMO system capacity. In the actual transmission process, reciprocity is subject to many factors, resulting in application difficulties. In this paper, we mainly study the influence of RF device's noncardiosis on channel reciprocity of Massive MIMO system. Combined with channel estimation based on pilot signal and FRRH subspace tracking algorithm, this paper proposes a channel reciprocity calibration method for Massive MIMO. Finally, the compensation effect of the compensation algorithm is verified by simulation.

1 Introduction

Massive MIMO system uses the transmitter channel state information (CSI) to pre-process the transmitted data, which can effectively reduce the interference and enhance the Massive MIMO system capacity [1]. In the TDD system, the uplink and downlink data transmission uses the same frequency, and the uplink and downlink wireless channel environments are the same. Therefore, the base station can get the downlink channel CSI from the uplink channel CSI obtained by the uplink channel estimation, that is, the uplink and downlink channels are reciprocal [2, 3].

In the actual transmission process, reciprocity is subject to many factors, resulting in application difficulties. Channel estimation error, channel time-varying, non-ideality of radio frequency devices and asymmetry of uplink and downlink interference modes will lead to a certain degree of loss of channel reciprocity, resulting in that the base station can not directly obatain the downlink channel CSI from uplink channel CSI [4, 5]. In the case where the channel reciprocity is affected, it is necessary to use the channel estimation to compensate for the lack of reciprocity of the channel [6]. This paper mainly studies the influence of the non-ideality of RF devices on the channel reciprocity of Massive MIMO systems.

I / Q imbalance is caused by the non-ideal characteristics of the wireless transceiver circuit, which is present in the process of upconversion and down-conversion. The result is, the amplitude gain of the I / Q branch of the signal is not equal or the phase is not orthogonal [7, 8]. In the presence of the I / Q imbalance, the channel will lose its reciprocity even if the radio channel itself remains unchanged during an uplink and downlink subframe interaction.

[9] proposed a calibration method based on mutual coupling of time-division duplex massive MIMO system. [10] considered the channel estimation error to simulate and analyze the effect on TDD multi-user massive MIMO system linear precoding caused by RF mismatch. [11] analyzes the bidirectional signaling method of time division duplex reciprocal calibration (RC) in multi-user beamforming system. [12] and [13] proposed a new closed-loop reciprocal calibration method. Compared with the existing method, the proposed method can be more accurate to support online calibration.

In this paper, based on the above research, combined with the channel prediction, the impact of the RF devices' non-ideal channel reciprocity and compensation algorithm are simulated and researched.

2 System model

The system studied in this paper contains L cells, where each base station (BS) is configured with M antennas and has N single antenna users (UTs).

Foundation items: National Natural Science Foundation of China, China (No. 61427801)

* Corresponding author: yinl@bupt.edu.cn

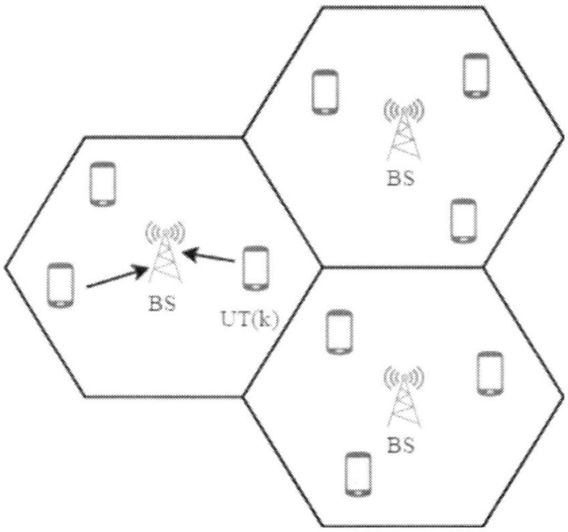

Fig. 1. System model diagram.

The SISO system model with I / Q unbalanced receiver is shown in Fig. 2.

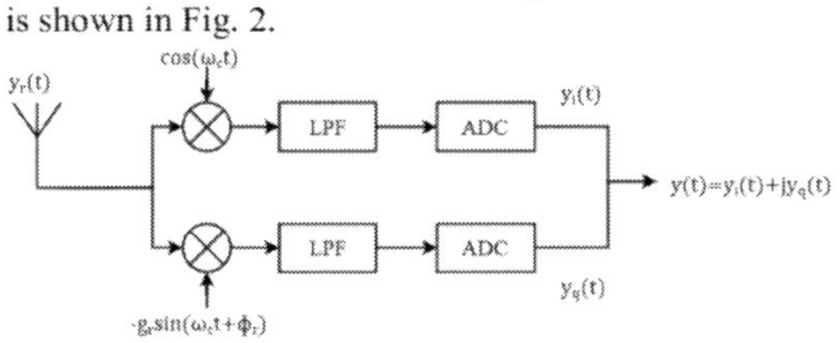

Fig. 2. I / Q unbalanced SISO system model.

The signal from the user to the base station (uplink baseband signal) can be expressed as:

$$Y_i = K_{Bi} H K_{Ui} X + N_W \tag{1}$$

Where i = 1, ..., L.

The transmit signal is:

$$X = [\mathrm{x}_{1I}, \mathrm{x}_{1Q}, \cdots \mathrm{x}_{NI}, \mathrm{x}_{NQ}] \tag{2}$$

Wherein the elements represent the transmission signals from the first to the N-th UT to the BS in the i-th cell.

K in the expression represents the I / Q unbalance matrix between the BS and the UT. The I / Q unbalance matrix at the BS side can be expressed as:

$$K_{Bi} = \begin{pmatrix} K_{Bi1} & 0 & 0 \\ 0 & \ddots & 0 \\ 0 & 0 & K_{BiM} \end{pmatrix} \tag{3}$$

Where the matrix elements are the I / Q unbalance parameters of the first to the M-th BS antennas at the BS in the i-th cell, respectively, and can be expressed as:

$$K_{Bik} = \begin{pmatrix} \alpha_{Bik}(1+\beta_{Bik}) & -\alpha_{Bik}\sin\theta_{Bik} \\ 0 & \alpha_{Bik}\cos\theta_{Bik} \end{pmatrix} \tag{4}$$

Where k = 1, ..., M.

The UT side I / Q unbalance matrix can be expressed as:

$$K_{Ui} = \begin{pmatrix} K_{Ui1} & 0 & 0 \\ 0 & \ddots & 0 \\ 0 & 0 & K_{UiN} \end{pmatrix} \tag{5}$$

Where the matrix elements are the I / Q unbalance parameters of the first to N-th UTs in the i-th cell, respectively, and can be expressed as:

$$K_{Uil} = \begin{pmatrix} \alpha_{Uil}(1+\beta_{Uil}) & -\alpha_{Uil}\sin\theta_{Uil} \\ 0 & \alpha_{Uil}\cos\theta_{Uil} \end{pmatrix} \tag{6}$$

Where l = 1, ..., N.

α in the I / Q unbalanced parameter is the power holding factor, β is the difference ratio of I / Q signal, θ is the phase difference of I / Q signal.

In the expression, **H** is the channel environment between the BS and the UT, and can be expressed as:

$$\mathrm{H} = \begin{pmatrix} \mathrm{h}_{11I} & -\mathrm{h}_{11Q} & \cdots & \mathrm{h}_{1NI} & -\mathrm{h}_{1NQ} \\ \mathrm{h}_{11Q} & \mathrm{h}_{11I} & \cdots & \mathrm{h}_{1NQ} & \mathrm{h}_{1NI} \\ \vdots & \vdots & \ddots & \vdots & \vdots \\ \mathrm{h}_{M1I} & -\mathrm{h}_{M1Q} & \cdots & \mathrm{h}_{MNI} & -\mathrm{h}_{MNQ} \\ \mathrm{h}_{M1Q} & \mathrm{h}_{M1I} & \cdots & \mathrm{h}_{MNQ} & \mathrm{h}_{MNI} \end{pmatrix} \tag{7}$$

$\mathbf{N_w}$ in the expression is noise, subject to the Gaussian distribution and is independent of the user channel.

3 Compensation algorithm

3.1. Channel estimation and compensation

Using pilot signals:

$$\mathbf{Y}_\mathbf{p}(t) = \mathbf{H}\mathbf{S}_\mathbf{p} + \mathbf{n}_\mathbf{p}(t) + \mathbf{n}_\mathbf{p}^*(t) \tag{8}$$

where:

$$\mathbf{H} = \left(\mathbf{h}(\mathrm{t}) \quad \mathbf{h}^*(\mathrm{t})\right) \tag{9}$$

$$\mathbf{S}_\mathbf{p} = \left(\mathbf{x}_\mathbf{p}(\mathrm{t}) \quad \mathbf{x}_\mathbf{p}^*(\mathrm{t})\right)^T \tag{10}$$

The channel estimation result of the LS estimation is:

$$\mathbf{H} = \mathbf{Y}_\mathbf{p}(t)\mathbf{S}_\mathbf{p}^H(\mathbf{S}_\mathbf{p}\mathbf{S}_\mathbf{p}^H)^{-1} \tag{11}$$

The mean square error (MSE) is:

$$\begin{aligned} \sigma_\Omega^2 &= E[\left\|\mathbf{H}-\mathbf{H}\right\|_F^2] \\ &= E[\left\|(\mathbf{n}_\mathbf{p}(t)+\mathbf{n}_\mathbf{p}^*(t))\mathbf{S}_\mathbf{p}^H(\mathbf{S}_\mathbf{p}\mathbf{S}_\mathbf{p}^H)^{-1}\right\|_F^2] \\ &= 2\,\mathrm{n_R}\,\mathrm{N_0}\mathrm{tr}[(\mathbf{S}_\mathbf{p}\mathbf{S}_\mathbf{p}^H)^{-1}] \end{aligned} \tag{12}$$

To minimize the MSE, the rows of the matrix SP should be orthogonal, and the equivalent channel estimates can be further obtained:

$$\mathbf{H}_{eq} = \begin{pmatrix} y_p \\ y_p^* \end{pmatrix} \left(\begin{pmatrix} x_p \\ x_p^* \end{pmatrix}^H \left(\begin{pmatrix} x_p \\ x_p^* \end{pmatrix} \begin{pmatrix} x_p \\ x_p^* \end{pmatrix}^H \right)^{-1} \right) \quad (13)$$

3.2. FRRH Channel estimation algorithm

In the Massive MIMO system, due to the large number of antennas and the large scale, the traditional channel estimation method is not efficienct enough. Therefore, a low-complexity semi-blind channel estimation method based on subspace tracking algorithm is introduced: FRRH (Fast Recursive Row-Householder) channel estimation algorithm.

Using the FRRH algorithm, the fuzzy matrix is estimated as follows:

$$\hat{\Theta}(n) = [(U^S(n))^H U^S(n)]^{-1} (U^S(n))^H \hat{H}_{eq} \quad (14)$$

Where U^S is the signal subspace of the receiver pilot signal.

The channel estimation expression is:

$$\begin{aligned} \hat{H}(n) &= U^S(n)\hat{\Theta}(n) \\ &= U^S(n)[(U^S(n))^H U^S(n)]^{-1} (U^S(n))^H \hat{H}_{eq} \end{aligned} \quad (15)$$

Through the recursive channel estimation, the computational efficiency can be greatly improved, and the accurate channel estimation result can be obtained, and the channel reciprocity calibration is carried out.

4 Simulation analysis

4.1. Simulation environment

It is assumed that the base station (BS) in each cell is configured with 64 antennas and has 8 single antenna users (UTs).

The antenna parameters are set as follows:

Table 1. 64 × 8 antenna parameter settings.

β_{Bik}	θ_{Bik}	β_{Uil}	θ_{Uil}
1.15	5°	1.05	10°

Note: k = 1, ···, 64; l = 1, ···, 8

It is assumed that the base station (BS) in each cell is configured with 128 antennas and has 16 single antenna users (UTs).

The antenna parameters are set as follows:

Table 2. 128 × 16 antenna parameter settings.

β_{Bik}	θ_{Bik}	β_{Uil}	θ_{Uil}
1.15	5°	1.05	10°

Note: k = 1, …, 128; l = 1, …, 16

In this paper, the bit error rate and channel capacity are used as the parameters to measure the effect and compensation effect. In this paper, the water capacity algorithm is used to calculate the channel capacity:

$$C = \frac{1}{2} E\left\{ \sum_{i=1}^{M} \log_2 \left(1 + \frac{P_{Si} |\beta_{i1}|^2}{P_N + \sum_{j=1, j\neq i}^{M} P_{Sj} |\beta_{ij}|^2}\right) \right\} \quad (16)$$

4.2 Simulation results

The channel capacity comparison result of 64 × 8 antenna-scale Massive MIMO system's ideal I / Q channel case and imbalanced I / Q channel case is shown in Figure 3:

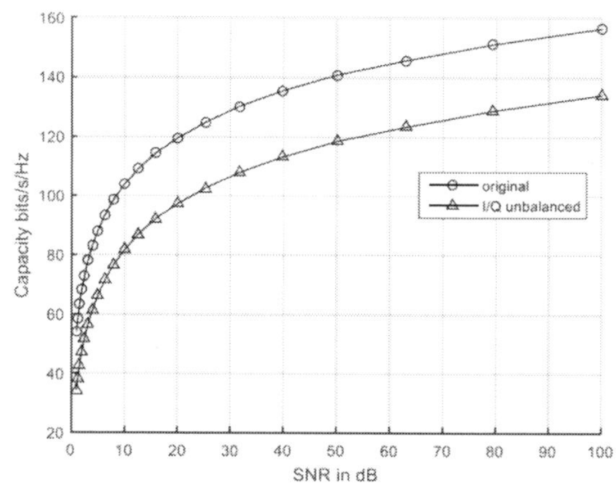

Fig. 3. Channel capacity comparison result.

Using the channel analysis of the unbalanced I / Q channel compensation method, the channel reciprocity can be compensated. Figure 4 shows the channel capacity before and after the compensation:

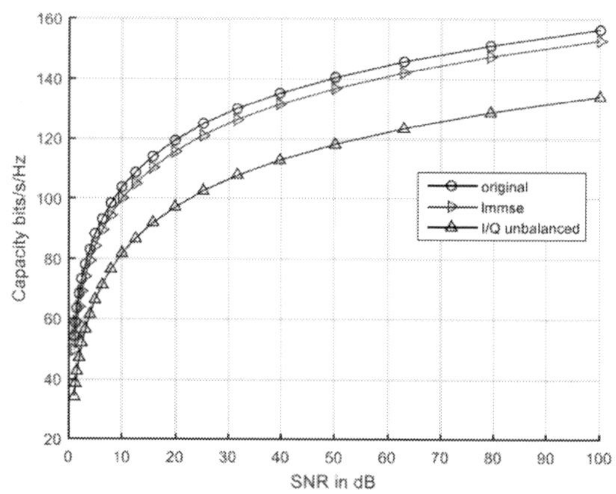

Fig. 4. Channel capacity before and after compensation.

Figure 5 shows the error rate before and after the compensation:

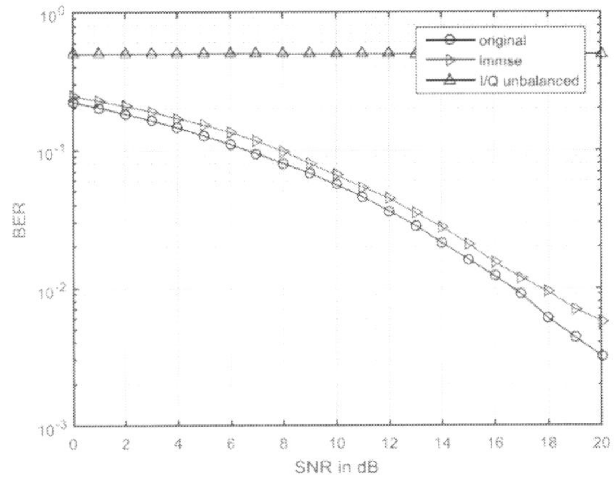

Fig. 5. Error rate before and after the compensation.

For the Massive MIMO system with an antenna scale of 128 × 16, Figure 6 shows the comparison of the channel capacity before and after compensation:

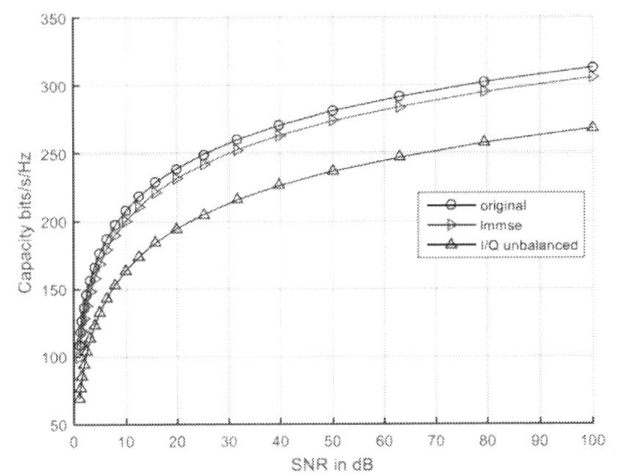

Fig. 6. Channel capacity comparison result (128×16 scale).

Figure 7 shows the error rate before and after the compensation:

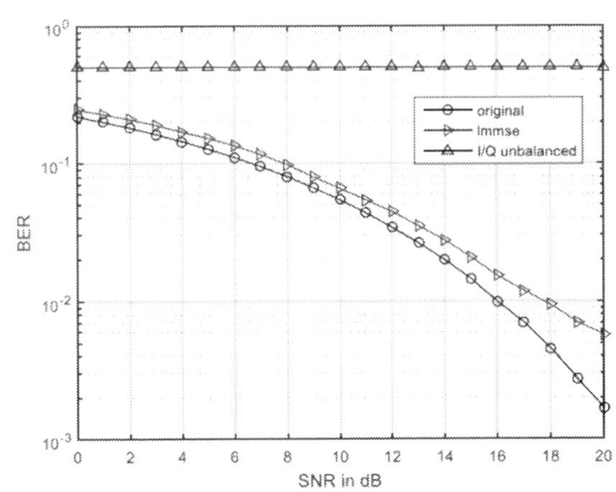

Fig. 7. Error rate before and after the compensation (128×16 scale).

From the simulation analysis, it can be seen that the non-ideality of the RF devices in the Massive MIMO system with 64 × 8 antenna scale has a great influence on the channel reciprocity both in terms of channel capacity and bit error rate. With the compensation algorithm proposed, the channel capacity and bit error rate can be nearly restored to the ideal RF device situation, to achieve a good compensation effect. While, for the Massive MIMO system with an antenna scale of 128 × 16, the compensation algorithm proposed in this paper can also achieve great compensation effect.

5 Conclusions

Based on the research of Massive MIMO channel estimation, the paper researches on the channel reciprocity of TDD Massive MIMO system, mainly on the effect of non-ideal RF device and the compensation method. The simulation results show that the compensation method proposed in this paper has great compensation effect on the channel reciprocity of Massive MIMO system. The further studies could be the verification of the compensation method in the semi-physical simulation platform, and lower complexity of the algorithm to reduce frequency pollution effects.

References

1. Marzetta T L. Noncooperative cellular wireless with unlimited numbers of base station antennas[J]. IEEE Transactions on Wireless Communications, 2010, 9(11):3590-3600.
2. Fu Xiaowei. Channel estimation and pilot scheduling of Massive MIMO systems under pilot pollution [J]; Journal of Southwest Jiaotong University;
3. Liu Yijing. TDD system channel reciprocity analysis and application of [D]. Beijing University of Posts and Telecommunications, 2014.
4. He Fuli. TDD MIMO channel non-reciprocal compensation algorithm [D]. Xidian University, 2011.
5. Gao Q, Qin F, Sun S. Utilization of channel reciprocity in advanced MIMO system[C]// Communications and Networking in China (CHINACOM), 2010 5th International ICST Conference on. IEEE, 2010:1 - 5.
6. Bourdoux A, Come B, Khaled N. Non-reciprocal transceivers in OFDM/SDMA systems: impact and mitigation[C]// Radio and Wireless Conference, 2003. RAWCON '03. Proceedings. IEEE, 2003:183-186.
7. Björnson E, Matthaiou M, Debbah M. Massive MIMO with Non-Ideal Arbitrary Arrays: Hardware Scaling Laws and Circuit-Aware Design[J]. IEEE Transactions on Wireless Communications, 2014, 14(8):4353-4368.

8. Zarei S, Gerstacker W H, Aulin J, et al. I / Q Imbalance Aware Widely-Linear Receiver for Uplink Multi-Cell Massive MIMO Systems: Design and Sum Rate Analysis[J]. IEEE Transactions on Wireless Communications, 2016, 15(5):3393-3408.
9. Vieira J, Rusek F, Edfors O, et al. Reciprocity Calibration for Massive MIMO: Proposal, Modeling and Validation[J]. 2016.
10. Mi D, Dianati M, Zhang L, et al. Massive MIMO Performance with Imperfect Channel Reciprocity and Channel Estimation Error[J]. IEEE Transactions on Communications, PP(99):1-1.
11. Liu D, Ma W, Shao S, et al. Performance Analysis of TDD Reciprocity Calibration for Massive MU-MIMO Systems With ZF Beamforming[J]. IEEE Communications Letters, 2016, 20(1):113-116.
12. Luo H, Zhang Y, Huang L K, et al. A new reciprocity calibration method for massive MIMO systems[C]// IEEE International Symposium on Broadband Multimedia Systems and Broadcasting. IEEE, 2016:1-4.
13. Luo H, Zhang Y, Huang L K, et al. A Closed-Loop Reciprocity Calibration Method for Massive MIMO in Terrestrial Broadcasting Systems[J]. IEEE Transactions on Broadcasting, 2016, PP(99):1-9.

Design and implementation of a user- extensible network packet generator

Guozhen Shi [1*], Yaoming Pan [1], Feng Yang [1], Shengyu Chen [2], Kai Zhang [3]
[1]School of Information Security, Beijing Electronic Science and Technology Institute, Beijing 100070, China
[2]School of Telecommunications Engineering, Xidian University, Xi'an 710071, Shanxi, China
[3]Xi'an center new software Co., Ltd,Xi'an 710071, Shanxi, China

Abstract. The security problems of network equipment are becoming more and more important with the fast development of the Internet. Aiming at the security, stability and protocol consistency of network security equipment, a user extensible network security device testing framework is proposed, and a user extensible network packet generator is designed based on Libnet and Libpcap library. The system uses the extensibility of XML document and metadata-based reflection technology to design a set of extensible protocol template, which is composed of user-defined data and data length to construct a specific network data packet to meet the specific network security equipment testing. The security, stability and protocol consistency of the network security equipment are tested by the test method through the test in the LAN simulation environment to ensure the security of the network equipment and the network environment.

1. Introduction

Internet services are increasingly rich, online banking, online shopping, mobile office and other increasingly popular, at the same time a growing number of hidden security risks, viruses, worms, and other malicious attacks get endless. IN the actual network environment, network security devices, such as firewall, virtual private network (VPN), intrusion detection system (IDS) and various encryption machines are all essential parts of the network, however, if the network security equipment itself is attacked, it may cause the entire network paralyzed. Therefore, the security of network security equipment is an important guarantee for network system security [1].

During the safety equipment leave factory or security overhaul, the equipment reliability and security of targeted testing is needed. The network packet generator is an important means of network devices permeability tests to find potential safety problems. At the same time, in order to ensure that network security equipment from external virus attacks, the network protocol used in network security equipment is usually opaque, or of protocol customization. Although the traditional network packet generator can send TCP /IP data packets, it cannot meet so many requirements of network security equipment security, stability testing and protocol conformance testing. So this article designed a user-expandable network packet generator for the above problems. The test results show that the network packet generator can not only send TCP /IP packet simulation attacks, but also can send scalable custom protocol packets, in order to ensure the security, stability and protocol consistency of network security devices.

DOI: 10.1051/matecconf/201713900197

2 Related works

The network packet generator, the traffic generator, is a tool for detecting devices in the access network, by generating protocol packets through the network packet generator, the real network environment is simulated and the network equipment is tested for security protocol consistency, equipment reliability and performance stability. At present, there are many commonly used tools to generate IP packets, the mainstream tools are sendip, Nessus and sniffer [3].

Sendip is relatively short and more complete, more suitable for use in daily testing. Sniffer tool is easy to use; you can send any possible data packets. The function of Nessus tool is more comprehensive, but in terms of contract awarding sendip. At present, many users develop the network packet generator with automatic simulation attack function, which can tailor the attack time, quantity and so on for a few attacks. However, these test tools, with single function and limited application scope, can only send fixed-protocol packets. They are powerless for application layer protocol data packets and user-defined protocol packets and cannot make consistency tests for a variety of security device protocol, and the network equipment is not suitable for complex test requirements. Furthermore, most of these tools are based on Socket with the inefficient code, also affected by other tasks, the traffic rate and attack effects produced greatly reduced as well. The use of Libnet library can not only simplify programming, and can directly construct the packet and send it to the network, do not need server/client mode [4].Therefore; this paper designs an extensible the network packet generator based on Libnet open source library on the embedded development board platform.

3. Introduction to technical theory

3.1 Introduction to scalability technology

To implement software scalability, techniques used include: Extensible mark-up Language (XML),plug-in, code automatically generated, dynamic compilation, metadata-based reflection technology and so on[5,6],.The extensible network packet generator designed in this paper achieves scalable design by using extensible mark-up language, JAXB technology, and metadata-based reflection techniques.

3.1.1 XML extensible mark-up language

A mark-up language used to mark an electronic file to make it structural; XML (Extensible Mark-up Language) is an extensible mark-up language, a subset of the standard general mark-up language. Having self-descriptive, scalability, content and display separation and cross-platform features and advantages, XML is designed to transfer data [7].This paper designs the extensible protocol template by using the extensibility of XML language. Users can generate custom data packets according to the test template and change the test data in real time.

3.1.2 JAXB technology

JAXB (JAVA Architecture for XML Binding, JAVA XML Binding Architecture) is an industry standard that can generate JAVA classes based on XML Schema. The core of JAXB is the bi-directional mapping technology for XML Schema and Java classes and interfaces,[8].JAXB technology not only avoids the traditional way to manually parse XML documents, but also can generate Java class objects automatically. The extensible network packet generator designed in this paper uses JAXB technology to reverse line the XML document into Java objects, which solves the problem of writing specific Java class object code and makes the program have extended features.

3.1.3 Flective technology based on metadata

The basic meaning of Metadata is "data about data". In the field of software configuration, metadata is defined as an object in the program is not being processed. Reflective technology is in the program running state, for any class or object; you can know all the attributes and methods of this class, this dynamic access to information and dynamic call object method technology called reflection technology [9].

In this paper, through the reflection technology based on the metadata, you can easily access all attributes and methods of the running class or object, and then use the acquired attributes and methods for dynamic scalable development.

3.2 Libnet open source library introduction

Libnet is a small interface library implemented by the C language that provides the construction, processing and sending of low-level network packets. The purpose of Libnet's development is to create a simple and unified network programming interface to block the differences in the underlying network programming of different operating systems, allowing programmers to focus on solving key issues [10, 11].Based on the reusable and highly portable Libnet library, this paper generates network packets at a constant rate on the embedded development platform to detect the device security, stability and protocol consistency.

4 Design and implementation of system function

The network packet generator which can be expanded by users using the upper and lower machine running mode, the host computer is responsible for interactive control, the slave computer to achieve the contract function to ordinary PC and embedded board hardware platform, on which the software part running is developed by JAVA language and C Language correspondingly, the core part are the construction of the scalability of the design of the upper computer and the network protocol development package based on Libnet of the lower computer. The system integrates the general contractor's contracting capabilities with scalable service in a test system, to realize automatic test, multi-function testing and scenario testing, can detect including but not limited to, the IDS, encryption machine and other network security equipment.

4.1 Functional module design

The network packet generator, which can be expanded by user, consists of the upper computer interaction module and the lower computer simulation contract module,

(1) The upper computer interaction module includes the subcontract module input sub-module, the interactive command set construction and sending sub-module, in which the sub-module can be subdivided into interface definition input and XML definition input, but the interface definition input only can complete the simple TCP / IP contract.

The sub-module is the core of the upper computer, mainly to complete the network security equipment online detection structure of the contract command interface display, XML document upload and the corresponding parameter settings.XML document includes network security device custom protocol template document and custom protocol test document, the former document is used to make the system to resolve the agreement corresponding to the POJO class, and the latter document parses the attribute values of the test protocol based on the template class to construct the interactive command set.

The interactive command set construction and delivery sub-module is used to convert multiple groups of packet commands into byte stream group, and send it to the lower machine through the upper and lower machine custom communication protocol.

(2) Lower machine simulation contract module includes control command set processing sub-module, network packet generation sub-module and constant rate output sub-module.

Control command set processing sub-module is mainly for processing the command set sent by the upper computer, including command set split, command recognition and other operations.

Network packet generation sub-module is the core of the lower machine, mainly based on TCP/IP network protocol and custom protocol, network data packets are generated, and the data packet function of the specified command is implemented on the base of Libnet library.

Constant rate output is an indicator of the performance of the network packet generator, through the constant rate output algorithm, making the lower machine in accordance with the upper computer to send network packets according to a constant rate.

4.2 Scalability design

In order to achieve the universality of custom protocol testing for network security devices, you need to

DOI: 10.1051/matecconf/201713900197

use the advantage of XML configuration language since the descriptive and extensibility, define a set of test templates that conform too many custom protocols. Users only need to change the test data in real time according to the test template. The scalable packetizer can automatically obtain the data of the tag in the template through the metadata-based reflection technology to generate the specified test data packet to realize the system scalability.

The custom protocol XML template uses the XML element tree structure model to determine the attributes of the tag element, the number of tags, and the length of the data, and so on in the attribute list of the root element. The extensible design is implemented in the tag element. The custom protocol XML template tree structure model is shown in Figure1:

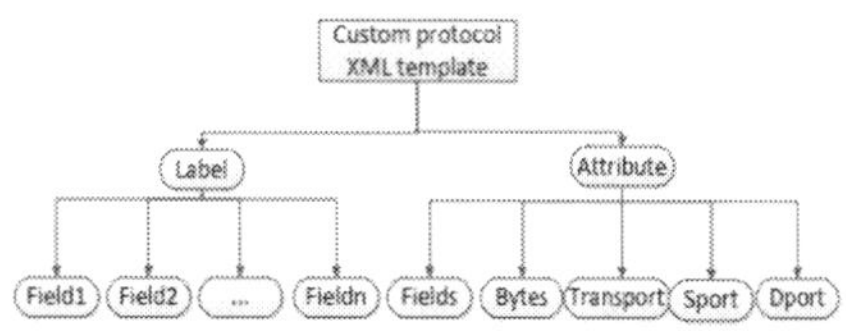

Figure 1. Custom protocol XML template tree structure model

The network protocol application layer header contains several fields, each field has a different data length, and the application layer protocol is based on the TCP/UDP transport layer protocol, so the custom protocol requires the field name, field length, and the transport layer of the protocol to be marked in the template file. The custom protocol can specify 1024 to 65535 any port number, but the standard network protocol already specified has a specified port number, for example, the DNS port number is 53 and the HTTP port number is 80, so the source port number and destination port number should be specified in the template file. In addition, the default value of each field also needs to be listed in the template file, and the system will determine the field attribute data type and create a custom protocol POJO class based on the default value. In short, the XML template file is the field specification for each upper layer protocol and the default padding. The attribute field and field description are shown in Table 1:

Table 1 XML attribute field and description.

XML Label	XML attribute field	Description
Protocol name (root node)	Fields	This is the enumeration of each protocol field, the order of each protocol field corresponds to the order of occurrence, and the middle of each field is separated by commas.
	Bytes	This is an enumeration of the number of bytes in the protocol header of each protocol field. The sequence of each protocol corresponds to the order in which the order appears. The integers are separated by commas.
	Transport	This is the description of the transport layer corresponding to the application layer, including "TCP" and "UDP"".
	Sport	This is the source port description of the sender, because a specific protocol has a specific port number. If the protocol constructed by the user is not a standard protocol, any number in the 1024~65535 can be filled here.
	Dport	This is the destination port description of the receiver, and this port description is the same as the sport instruction.
Each protocol field	Null	This is the field description of extensible protocols, and also the key to stitching scalable protocols.

4.3 Network packet generation module design

The network packet generation sub-module is the core of the lower computer, and it is also one of the core of the whole scalable network packet generator. The generation of network packets mainly uses the functions of the construction data provided by the Libnet open source library. This system not only realizes the function of sending the TCP/IP layer network packet by the ordinary sender, but also realizes the application layer network packet which can be extended to send to user.

The specific process of the sub-module is:

(1) First, use the function libnet_init () for memory initialization and environment establishment. What is supposed to be noted here is if the command structure provides the content in the underlying protocol header field, parameter injection type use LIBNET_LINK. If the protocol is above the IP layer and the command structure only provides the application layer protocol header. The parameter injection type uses LIBNET_RAW4;

(2) Packet structure. The principle is from the upper to the lower, the contract unit network packets can be divided into four categories: the link layer packets (MAC, LLC, the SNAP packets),network layer, data packets (IP,ARP packets/RARP),packet transport layer (ICMP,TCP,UDP packets) and application layer packet packets (custom).The concrete structure is to construct the load first, then construct the upper layer protocol header and the lower layer packet header, then assemble each part into a byte array to complete the structure of the packet.

(3) After constructing, use the function Libnet_write () to send the packet through the network card;

(4) Then use the function libnet_destroy () to destroy the applied memory space and initialize the environment.

The specific process design shown in Figure 2:

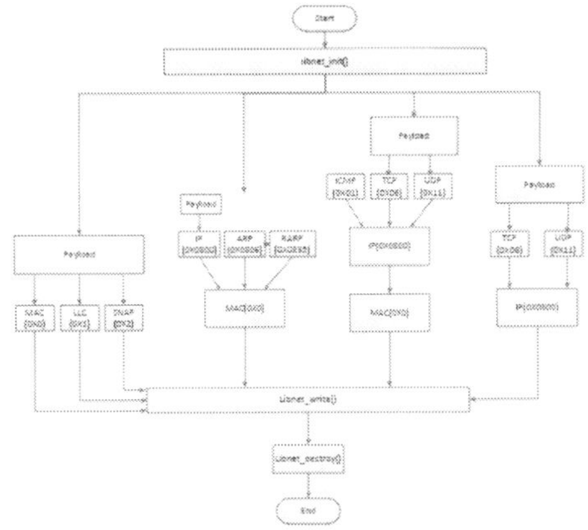

Figure 2. The network data package sub module constructs the packet flow

The table of concrete protocol constructs and protocols as follows:

Table 2.Protocol headers and constructors

protocols	Libnet concrete protocol constructs
MAC	libnet_build_ethernet ()
802.2LLC	libnet_build_802_2 ()
802.2SNAP	libnet_build_802_2snap ()
IPV4	libnet_build_ipv4 ()
ARP	libnet_build_arp ()
RARP	libnet_build_arp ()
ICMP	libnet_build_icmpv4_echo ()
TCP	libnet_build_tcp ()
UDP	libnet_build_udp ()

For the application layer protocol, Libnet only provides some of the protocol head of the construction methods such as DNS, and the use of existing protocol header construction method also limits the scalability of the scaler. In order to meet the scalability of the scaler, this paper chooses the LIBNET_RAW4 primitive socket injection type so that it will be from the network layer to start building data packets without having to fill the Ethernet head, The transport layer structure function of the application layer can be set up according to the transport layer protocol and port number obtained from the control command processing sub-module, and then all the contents of the command structure as a data load, placed in the transport layer protocol load location, and the transmission function complete the custom data

DOI: 10.1051/matecconf/201713900197

packet transmission.

5 System test results and analysis

5.1 Test the environment

In the test of network security equipment, very few devices placed in the actual network, because the actual network environment is uncontrollable, and the actual network environment is too strong, it is difficult to accurately test the device system, therefore, a special network test environment is constructed.

The test of network equipment is guided by black box testing, use active test, through test and online and offline testing combined method [12] to test the security, stability and reliability of network devices and the consistency of device protocols. Test environment consists of the upper computer, network security equipment, the next crew, three switches and monitoring host. The upper computer sends control commands to the lower machine, and receives the feedback from the lower machine. The lower computer sends the network test packet through the LAN, and the three-layer interaction machine mirrors the test packet port to the monitoring host, and forwards the test packet to the network security device. After crossing the network security device, the packet is processed by the lower computer and the data packet is returned to the upper computer, through comparing the packets captured by the upper computer Wireshark and monitor the host Wireshark and analysing the network security equipment custom network protocol security effect, achieve a goal of the detection of the network equipment security, stability and reliability.

The test environment of the network security device detection tool is shown in Figure 3:

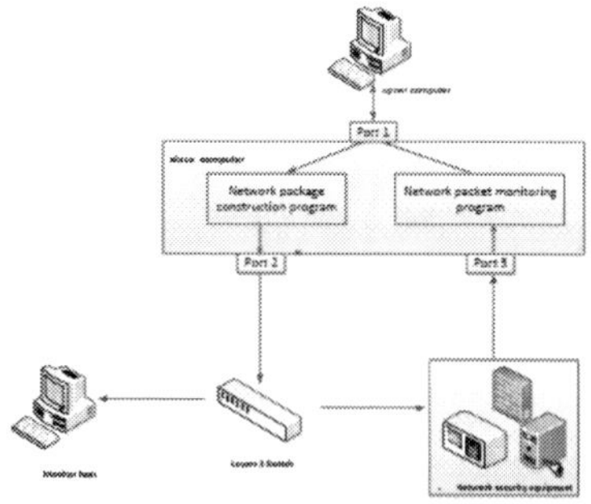

Fig 3.Testing environment for network security device testing tools

Based on the campus network as a local network, the scalability of the network transmitter is tested, and the computer configuration and tools are shown in Table 3:

Table 3.Computer configuration and tools

Experimental equipment	Configuration
Monitor host	AMD A10, 3.5GHz CPU, 4G Memory
Monitor host	AMD A10, 3.5GHz CPU, 4G Memory
Slave computer	MC-AM335X Development board,800MHz clock speeds, 256MB DDR3 Memory dual Gigabit Ethernet port
Layers 3 Switch	BDCOM S2210
Network security computer	IDS、Firewall

5.2 Test results and analysis

5.2.1 TCP / IP packet authentication

This network packet generator implements basic contracting capabilities and can simulate network attacks such as Syn Flood, UDP Flood, and Land Attack to detect against network devices.

Extensible network packet generator implementation the function of sending TCP/IP packet, the packet interface shown in Figure 4:

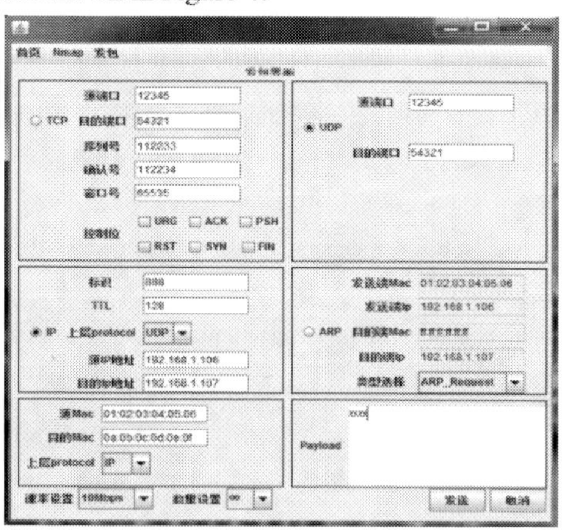

Figure 4.TCP/IP packet sending interface

The network packet generator can send any protocol packets in the TCP / IP protocol stack and send the generated packet command for the above figure. The experimental results are shown in Figure5 below:

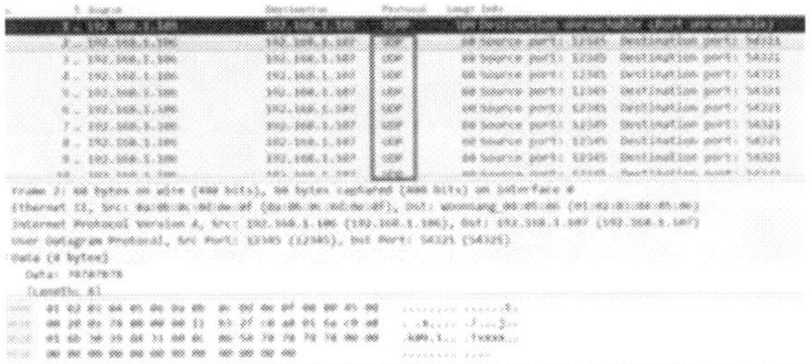

Figure 5. Test results generated by TCP/IP packets

(2) Extensible packet authentication

The network packet generator implements an application layer protocol such as SSL, SSH, DNS, and OICQ, as well as any custom network protocol. In order to verify the scalability of the scaler, the experiment selects the alarm protocol of the SSL protocol and the unencrypted SSH protocol to do the extensibility test.The XML test files are shown in Table 4 and Table 5 respectively.

Table 4.SSL alarm protocol XML file

```
<? Xml version="1.0" encoding="UTF-8"?>
<SSL
fields="contentType,version,length,alertMlever,alertMdescrip" bytes="1,2,2,1,1" transport="tcp" sport="443" dport="54321">
    <!-- TLS 记录层 -->
    <contentType>22</contentType>
    <version>771</version>
    <length>2</length>
    <!-- TLS 报警协议 -->
    <alertMlever>1</alertMlever>
    <alertMdescrip>112</alertMdescrip>
</SSL>
```

Table 5.SSH protocol XML file

```
    <?xml version="1.0" encoding="UTF-8"?>
    <SSH fields="length,encryptedPacket"
    bytes="4,0" transport="tcp" sport="22" dport="54321">
    <length>18</length>
    <encryptedPack-
et>encrypedPacketencrypedPacketencrypedPacket</encryptedP
acket>
</SSH>
```

The network packet generator can send any protocol for a given protocol template, construct a custom data packet for Table 4, and Table 5 XML file templates. The Wireshark capture packet results are shown in Figure 6 and Figure 7:

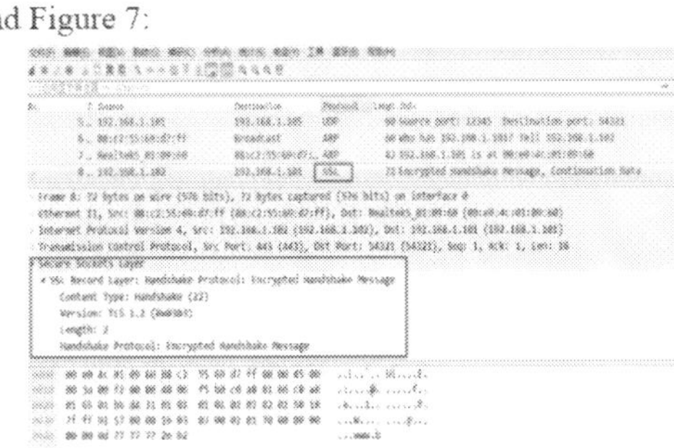

Figure 6.SSL protocol for custom protocol testing

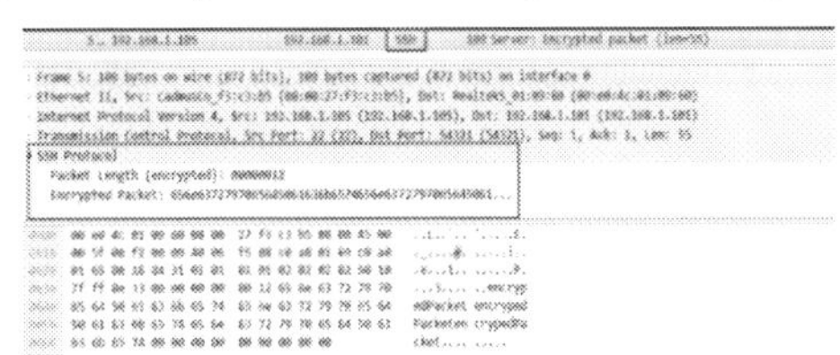

Figure 7. SSH protocol for custom protocol testing

6 Conclusions

This paper designs and develops a user-scalable network packet generator which can detect the security of general network security equipment. Through the concrete programming and project implementation, it is proved that the system can effectively and automatically send various network test packets, so as to assess the security of network security equipment. The system will increase the late password component evaluation module. Thus, the encryption and authentication of security equipment are tested professionally. The system has the advantages of strong expansibility and wide application range and it is easy to use. It is an effective tool for detecting network security equipment and has good practical value.

Acknowledgement

This work was supported by the Beijing Natural Science Foundation (4152048) and National Key Research Program of China (2016YFB0800304).

References

1. Wei Sun, Chuang Bao. Current situation and trends of international network security products market [J]. National Defense Science and technology, 2016, 37(2):**59-64**.
2. Hong Zhang. Design and implementation of online security detection system for network equipment [D]. Beijing University of Posts and Telecommunications, 2012.
3. Yiqiang Wang. Research and implementation of IP traffic generator [D]. Chengdu University of Technology, 2010.
4. Jordan P, Patten C V, Peterson G, et al. Distributed PowerShell load generator (D-PLG): A new tool for dynamically generating network traffic[C]// International Conference on Simulation and Modeling Methodologies, Technologies and Applications. IEEE, 2017.
5. Oquendo F, Leite J, Batista T. Designing Scalability in Software Architectures [M]// Software Architecture in Action. Springer International Publishing, 2016.
6. Jindal A, Kumar J, Kasralikar R S, et al. Systems and methods for increasing the scalability of software-defined networks [J]. 2016.
7. Xiangyi Geng, Yueping Zhang. XML Basic course, The Second Edition [J]. 2012.
8. Cover R. Java Architecture for XML Binding (JAXB) [J].
9. Deng X, He G, Chen Y, et al. CIM lead-in based on java reflection mechanism in AEMS of Shanghai power grid [J]. Automation of Electric Power Systems, 2007, 31(18):21-25.
10. Quan D, Pei C, Zhu C, et al. Design and Implementation of Network Traffic Generator Based on Libnet [J]. Modern Electronic Technique, 2005. Wang L X, Wen-Wei L I. Active Measurement Tool Development Kit [J]. Computer Systems & Applications, 2014.
11. Meyer R A, Klassen M. Methods, systems, and computer readable media for controlling processor card power consumption in a network test equipment chassis that includes a plurality of processor cards [J]. 2017.

Research on Handling Stability of Steering-by-wire System

Ying Yuan[1,*], Guangxing Tan[1]

[1]College of Electrical and Information Engineering, Guangxi University of science and technology, Liuzhou 545006 China

Abstract. The main function of steer-by-wire (SBW) system are improving steering characteristics, security and stability of the vehicle. In this paper, the variable steering ratio of SBW system is analyzed, and the method of steering ratio based on fuzzy control and neural network are researched. In order to solve the actual working condition, the wheel angle may not reach the expected value, this paper establishes a two-degree-of-freedom (2-DOF) vehicle model, and a Matlab/Simulink simulation model, in which a control strategy based on PID controller is put forward to control the front wheel steering angle. Simulation results show that proposed control strategy based on fuzzy neural network can effectively reduce lateral deviation and improve the handling stability and comfortability of the vehicle.

1 Introduction

The steering system is the link between driver and vehicle, which plays an important role in the automobile active safety technology[1]. SBW system is a new technology of steering system, which improves the handling stability of the vehicle and reduces the burden on the driver. When driving, we mainly consider improving the handling stability of the vehicle[2]. The maneuverability means that the car can be driven in accordance with the driver's steering intention and not deviate from the intended trajectory. Stability refers to the ability of the car to maintain stable driving when disturbed by the outside world[3]. The centroid side-slip angle reflects the trajectory retention of vehicle, and the yaw rate is focused on the oversteer or understeer[4]. This paper first design the ideal steering ratio to guarantee yaw rate gain is a constant. And then the steering ratio based on fuzzy control and fuzzy neural network were discussed. Next, 2-DOF vehicle based on PID controller model is established, Finally, the vehicle output parameters are analyzed and compared to ensure the stability of the yaw rate and centroid side-slip angle.

2 2-DOF vehicle dynamics model

2-DOF vehicle handling stability model includes lateral motion, yaw motion and roll motion. Use the front wheel angle as input to observe the automobile steering movement. In the state of automobile movement, the output observation values are yaw rate γ , lateral acceleration a_r, and centroid side-slip angle β. Ignores the air resistance and tire's non-linear, the car model is established[5]. According to the lateral force balance and the balance of the moment around the centre of gravity:

$$\begin{cases} mV(\dot{\beta}+r)=-(P_{Y1}+P_{Y2}) \\ I\dot{r}=bP_{Y2}-\alpha P_{Y1} \end{cases} \quad (1)$$

Along with:

$$\begin{cases} P_{Y1}=K_1\delta_1=K_1(\beta+\alpha r/V-\delta) \\ P_{Y2}=K_2\delta_2=K_2(\beta-br/V) \end{cases} \quad (2)$$

In the formula, we can simplify it as follow:

$$\begin{cases} mV(\dot{\beta}+r)=-(K_1+K_2)\beta+(bK_2-\alpha K_1)r/V+K_1\delta \\ I\dot{r}=(bK_2-\alpha K_1)\beta-(\alpha^2K_1+b^2K_2)r/V+\alpha K_1\delta \\ a_r=V(\beta+\dot{\gamma}) \end{cases} \quad (3)$$

Above equation is subjected to a Laplace transform, that is, $r\to r(s)$, $\beta\to\beta(s)$, $\dot{r}\to sr(s)$, $\dot{\beta}\to s\beta(s)$, and make initial condition is 0, then the transfer function is:

$$\begin{cases} r/\delta(s)=(mV\alpha K_1s+lK_1K_2)/\{\Omega+\Phi s+\Psi\} \\ \beta/\delta(s)=(IK_1s+blK_1K_2/V-mV\alpha K_1)/\{\Omega+\Phi s+\Psi\} \end{cases} \quad (4)$$

Among them:

$$\begin{cases} \Omega=mVIs^2 \\ \Phi=[I(K_1+K_2)+m(\alpha^2K_1+b^2K_2)] \\ \Psi=K_1K_2l^2(1+KV^2)/V \end{cases} \quad (5)$$

The s is the Laplace variable in the formula, when written as standard transfer function form:

$$\begin{cases} r/\delta(s)=G_r(\tau_1s+1)/(T_2s^2+T_1s+1) \\ \beta/\delta(s)=G_\beta(\tau_1's+1)/(T_2s^2+T_1s+1) \end{cases} \quad (6)$$

Among them, G_r and G_β are steady-state gain:

$$\begin{cases} G_r=r/\delta(s)|_{\delta=0}=V/l(1+KV^2) \\ G_\beta=\beta/\delta(s)|_{\delta=0}=(b-V/C^2)/l(1+KV^2) \end{cases} \quad (7)$$

Among them:

$$\begin{cases} K=K_1/mb-K_2/m\alpha \\ T_1=V(b+\eta\alpha mb/K_1l+\alpha+\eta bm\alpha/K_2l)/l(1+KV^2) \\ T_2=IV^2K_1K_2l^2/m^3\alpha^2b^2(1+KV^2) \\ \tau_1=VK_2l/m\alpha \\ \tau_1'=\eta/(K_2l/m\alpha V-V/b) \end{cases} \quad (8)$$

* Corresponding author: yrlsx99@126.com

 DOI: 10.1051/matecconf/201713900198

Table 1. Vehicle system parameter.

Parameters/Unit	The physical meaning of parameters	Value
M/(Kg)	Vehicle weight	1359.8
α/(m)	Distance from mass center to front axle	1.06282
l/(m)	Wheelbase	2.548
K1/(N/rad)	Front tire cornering stiffness	52480
I/(Kg·m2)	Moment of inertia	1992.54
b/(m)	Distance from mass center to rear axle	1.305
K2/(N/rad)	Rear tire cornering stiffness	88416
K	Stability factor	0.0034
η	Mass distribution coefficient	0.9283

3 Design of ideal steering ratio

In the research of steering characteristics, the steering ratio under steady-state steering gain is the most important. As the steering ratio of the SBW is variable, in order to reduce the burden of the driver, it can be designed as follows: The steering ratio is small when the car is driving at low speed, so the driver turning a small steering handle angle can have a greater steering wheel angle[6]. When the speed is high, the steering ratio is large to ensure that the steering stability in the process of steering. Thereby simplifying driver's steering operation and improving the active safety of automotive SBW[7].

The steering ratio of SBW is determined by the front wheel angle δ and the steering wheel angle δ_h, which can be expressed as i:

$$\delta = \delta_h / i \tag{9}$$

Two steering gains are defined in the vehicle dynamics. One of steering gain refers to the ratio of front wheel angle δ and yaw rate ω_r, represented by $G_\delta^{\omega_r}$. And the steady-state conditions are as follows:

$$G_\delta^{\omega_r} = \omega_r / \delta \tag{10}$$

Another steering gain $G_{\delta_h}^{\omega_r}$ refers to steering wheel angle δ_h and yaw rate ω_r, under the steady-state conditions:

$$G_{\delta_h}^{\omega_r} = \omega_r / \delta_h \tag{11}$$

If the steering wheel angle is linearly related to the road travel angle, that is, $G_\delta^{\omega_r} G_s = K_s$ is a constant, Where $G_s = 1/i$. Then the steering operation can be simplified and the driver's burden can be reduced. The ideal steering ratio is variable and yaw rate gain K_s is constant. The steering characteristics of ideal steering ratio are easily to be mastered by the driver.

$$i = G_\delta^{\omega_r} / K_s = v / K_s L(1 + Kv^2) \tag{12}$$

In above formula: $K = m(a/k_2 - b/k_1)/L^2, L = a + b$.

When the car is low speed or static steering, the larger static friction force makes the steering resistance torque become large, in order to make the steering "light", transmission ratio should be smaller and constant. Substituting the data of the Volvo 264 model[8], the maximum steering wheel angle of car is 420°, the maximum of front wheel angle is 45°. The transition speed v_0=*30 km/h* is selected from "Fig. 1", so the minimum value of steering ratio i:

$$i_{\min} = \delta_{h\max} / \delta_{\max} = 420/45 \approx 9.333 \tag{13}$$

Therefore, the ideal steering ratio based on the yaw rate gain K_s constant should be:

$$i = \begin{cases} 9.333 & (v < 30 \quad km/h) \\ v / K_s L(1 + Kv^2) & (v \geq 30 \quad km/h) \end{cases} \tag{14}$$

In order to select a reasonable yaw rate gain K_s, the values of K_s are 0.19,0.27,0.32,0.45, and the following curves are obtained.

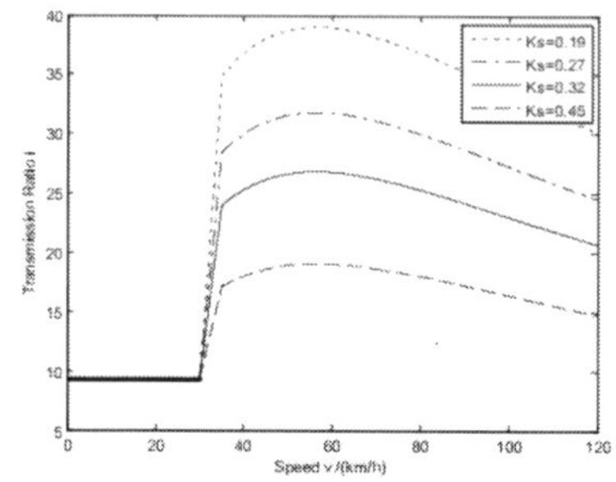

Fig. 1. Ideal steering ratio curves with different speed.

The figure shows that when the car driving slowly, it is possible to improve the steering sensitivity of SBW by using a constant and smaller steering ratio. In the high speed driving, the driver can avoid the fault operation. It is advantageous to improve the safety and stability of car.

4 Design of fuzzy steering ratio

Generally, the steering ratio should be kept between 15:1 and 20:1. From the above figure, the ideal steering ratio does not meet the actual requirements of the steering system[9]. Considering the influence of vehicle speed and steering wheel angle, this paper use fuzzy control algorithm to establish a more reasonable steering ratio of SBW system. Based on fuzzy control toolbox of Matlab, the selected fuzzy subset language is {NB, NM , NS , 0 , PS, PM , PB}. Fuzzy rules are designed as follows:

v= *{NB , NM , NS , 0 , PS, PM , PB}*, the language values are {0 , 20 , 40 , 60 , 80 , 100 , 120}.

δ_h= *{NB , NM , NS , 0 , PS, PM , PB}*, the language values are {0 , 70 , 140 , 210, 280 , 350 , 420}.

i= *{NB , NM , NS , 0 , PS, PM , PB}*, the language values are {3 , 6 , 9 , 12 , 15 , 18 , 21}.

The fuzzy controller is double input and single output, and membership function graphs are Gaussian curves. Based on the driver's driving behavior and expert knowledge, the fuzzy steering ratio control table is established as follows:

Table 2. Rule of steering ratio fuzzy control.

v	δh						
	NB	***NM***	***NS***	***Z***	***PS***	***PM***	***PB***
NB	NB	NB	NB	NB	NB	NB	NB
NM	NB	NB	NB	NB	NB	NB	NB
NS	NM	NS	NS	Z	NS	NS	NM
Z	Z	PS	PS	PM	PS	PS	Z
PS	PS	PM	PM	P	PM	PM	PS
PM	PM	PM	PM	PB	PM	PM	PM
PB	PB	PB	PB	PB	PB	PB	PB

According to the fuzzy rule table, the steering ratio curve surface diagram as follows:

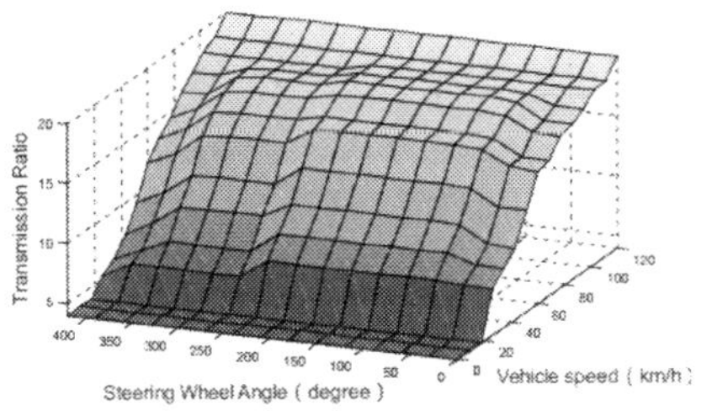

Fig. 2. Relationship between i,v and δ_h.

It can be seen from the figure, as the speed increases, the steering ratio increases gradually. So as to effectively improve the vehicle handling stability and reduce the incidence of traffic accidents. But the fuzzy controller output curve is not smooth enough.

5 Fuzzy neural network makes better steering ratio

Because of fuzzy control relies too much on expertise, so used the fuzzy neural network algorithm to improve the design. In order to use fuzzy neural network to adjust membership rules and functions of fuzzy control, the steering ratio of two inputs and an output fuzzy neural network model is established[10]. Based on the fuzzy control system model, the fuzzy network structure is based on the extended radial basis function (RBF) neural network, which functions is a TS fuzzy inference system.

There are 5 layers of network structure:

The first layer: input layer, the number of nodes is 2, that is, steering wheel angle and speed.

$$f(X) = [x_1, x_2] \tag{15}$$

The second layer: fuzzy layer, 14 nodes, the role is to complete the input of fuzzy.

$$\mu_i^j(x_i) = \exp\left[-\left(x_i - ci_j\right)^2 / \sigma_j^2\right] \tag{16}$$

The third layer: fuzzy inference layer, 49 nodes, the role is to achieve conditional fuzzy set operations.

$$\varphi_j = \mu_1^{i_1}\mu_2^{i_2}, i_1 = 1,2,...m_1, i_1 = 1,2,...m_2, m = m_1 m_2 \tag{17}$$

Fourth layer: normalization calculation.

$$a_j = \varphi_j / \sum_{k=1}^{\varphi_j} \varphi_k \tag{18}$$

The fifth layer: output layer, that is transmission ratio.

$$y(X) = \sum_{j=1}^{m} a_j \omega_j, \omega_j = p_{j0} + p_{j1}x_1 + p_{j2}x_2 \tag{19}$$

Given a typical speed and a different steering wheel angle as input, the fuzzy control will produce the corresponding steering ratio. After repeated operation, save vehicle data in various circumstances and establish a database. And then some training data as follows.

Table 3. A part of the training data.

v	δh	i	v	δh	i	v	δh	i
30	-210	8.10	70	-210	12.32	110	-210	15.48
30	-140	8.78	70	-140	13.76	110	-140	15.98
30	-70	9.20	70	-70	14.14	110	-70	16.68
30	0	9.88	70	0	15.21	110	0	17.32
30	70	9.23	70	70	13.98	110	70	16.57
30	140	8.84	70	140	13.10	110	140	15.80
30	210	8.12	70	210	12.43	110	210	15.41

The membership functions of the input variables are optimized as follows: from the figure we can see that the central value and width of the membership functions have a certain degree of adjustment.

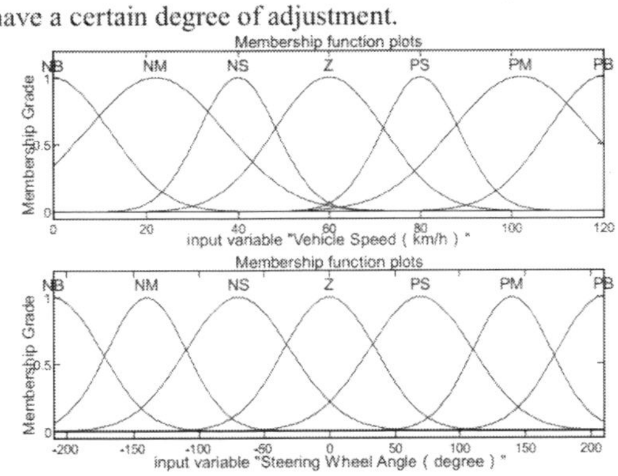

Fig. 3. Membership function of input variable after optimization.

6 Design of PID controller for two degree-of-freedom vehicle model

The driver turns the steering wheel according to the speed of the vehicle under different conditions, and the corresponding steering wheel angle expectation value is obtained by designing the variable transmission ratio model. Therefore, in this paper, the yaw rate and centroid side angle of the steering vehicle are taken as the control target, and the control stability of the vehicle is analyzed by PID control. The system input is front wheel angle determined by fuzzy neural network ratio of SBW system. Using the steering response of SBW system to better track the vehicle model and the smaller error evaluation index set controller: The conventional PID controller input is the difference $e(t)$ between the expected value $r(t)$ and the actual output value $y(t)$. Used proportional operation, integral operation and derivative operation for the $e(t)$, and a proper control quantity $u(t)$ is obtained by summation.

The PID control transfer function is:

$$G(s) = K_p + K_i / s + K_d s \tag{18}$$

The mathematical equation of the PID controller is:

$$y(X) = \sum_{j=1}^{m} a_j \omega_j, \omega_j = p_{j0} + p_{j1}x_1 + p_{j2}x_2 \tag{19}$$

In which, $e(t)=u(t)-y(t)$, K_P is the coefficient of proportionality, $T_i=K_P/K_i$ is integral time constant, $T_d=K_d/K_P$ is differential time constant. Based on Matlab/Simulink software, build a simulation model as follow:

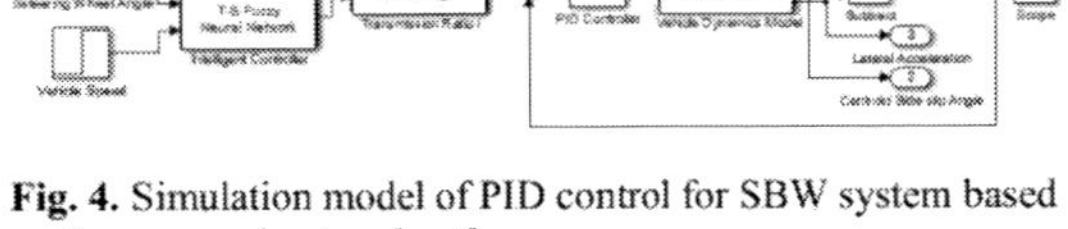

Fig. 4. Simulation model of PID control for SBW system based on fuzzy neural network ratio.

First set the initial value of the three parameters of the PID, generally the value of K_P is given first, and the

K_i and K_d are set to 0, which can avoid overshoot and improve the response speed. In this paper, the system input is a single lane change. K_d can improve the dynamic characteristics of the system, restrain the controlled variable, shorten the adjustment time and reduce the steady-state error. Through several simulation comparisons, the output parameters of the automotive SBW system are simulated as follows:

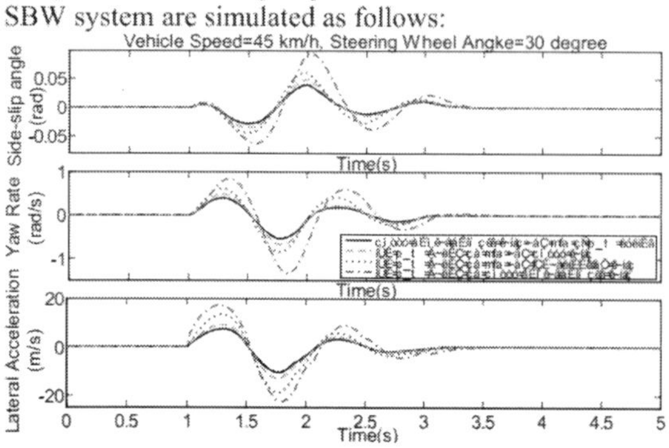

Fig. 5. Curve of parameter change in vehicle steering process.

Table 4. Simulation result comparison analysis.

The vehicle output parameters		Fuzzy neural network ratio and PID of SBW system	Fuzzy ratio and PID of SBW system	Ideal ratio and PID of SBW system	Fuzzy neural network ratio of SBW system
Side-slip angle	*Maximum (rad/s)*	0.0392	0.0487	0.0579	0.0933
	Settling time(s)	3.3639	3.4139	3.3639	3.7639
Yaw rate	*Minimum (rad/s)*	-0.5538	-0.6884	-0.8589	-1.3468
	Settling time(s)	3.2139	3.3639	3.2639	3.9639
Lateral acceleration	*Minimum (rad/s)*	-10.4499	-12.8488	-18.7831	-22.6140
	Settling time(s)	3.6639	3.7639	3.7639	3.9639

The graphs show the different algorithms of steering ratio on SBW system PID control model. Since the yaw rate and lateral acceleration mainly focus on describing vehicle stability, the centroid side-slip angle reflects vehicle trajectory maintenance. Therefore, the vehicle stabilization time is shorter, indicating that the vehicle has good handling stability. From graphs data show that the output parameters of SBW system based on PID and fuzzy neural network control are lower than others. Without PID controller, the SBW system has the longest stability time, and the curve of lateral acceleration is not smooth. The results proved that the PID control model of automobile SBW system based on fuzzy neural network steering ratio can improved the steering characteristics and the handling stability, reduce the burden on driver.

7 Conclusion

As can be seen from this paper, the ideal transmission ratio is difficult to meet the actual requirements. To solve this problem, this paper establishes a two-degree-of-freedom vehicle model, and the fuzzy neural network control algorithm is used to optimize the steering ratio. The system can improve steering sensitivity of vehicle at low speed, and in the high speed driving, the driver can be prevented from fault operation, thereby improve the handling stability. The simulation results show that use the PID controller to adjust SBW system of fuzzy neural network steering ratio, the vehicle output parameters can be stabilized in a short time, so as to meet the requirements of vehicle steering sensitivity at different speeds. This system reduces burden of the driver, improve the vehicle handling stability and driving safety.

Acknowledgments

Project supported by the Major Program of the National Natural Science Foundation of China (61563005) and Innovation Project of GuangXi University of Science and Technology Graduate Education (GKYC201709).

References

1. C. Wang, T. Cui, W. Zhao, et al. Active front wheel steering control based on ideal transmission ratio. Transactions of the Chinese Society of Agricultural Engineering, **31**(4):85-90(6),(2014)
2. Q. Zhang. Research on Modeling and Control Strategy of Electric Forklift Steer-by-wire System. HeFei University of Technology, (2015)
3. S. Tang. Research on Automobile Steer by Wire System Based on Fuzzy Adaptive PID Control. Shandong University of Technology, (2011)
4. Y. Liu. Research on Modeling and Active Steering Control Algorithm for Electric Forklift Steer-by-Wire System. MECS, **8**(11):70-79, (2016)
5. L. Yu. On Lateral Stability Analysis of Steering-by-Wire System Based on Fuzzy PID Control. Journal of Shandong Jiaotong University, (2014)
6. G. Wang. Study on Variable Steering Ratio and Active Steering Control Strategy of Steer-by-wire System. Liaoning University of Technology, (2015)
7. Q. Cao, G. Tan. Research on front wheel of steer by wire system based on fuzzy Immune PID. Guangxi University of Science and Technology, (2014).
8. J. Tian. Research on Dynamic Analysis and Control Method of Steer-by-wire System for Vehicle. Jiangsu University, (2011)
9. K. Zhang, S. Cui, J. Wang. Path Tracking Control of Intelligent Vehicle Based on Fuzzy Neural Network. Automotive Engineering, **37**(1):38-42 and 77, (2015)
10. N. Wang, J. Shi, G. Li. Study on Active Steering Control Strategy for Steer-by-Wire System. Automobile Applied Technology, **1**:72-73, (2016).

 DOI: 10.1051/matecconf/201713900199

Air Compressor Pressure Control System Based On Gearshift Integral PID Controller

Chunyue PAN

Wenzhou Vocational & Technical College, Zhejiang, China, 325035

Abstract. The application of gearshift integral PID controller to air compressor pressure control system is introduced,Its kernel is single chip microcomputer PIC16F877.The design of hardware and solfware are introduced too. Practical application shows that this system has many advantages.

Keywords. Gearshift Integral , Pressure control system , Single chip microcomputer , PID control .

1 Introduction

In modern industrial production ,compressed air is an indispensable power source. Compressed air, electricity, water and steam are called the four major energy sources in industry, As a compressed air generator, air compressor system is widely used in metallurgy, chemical, tobacco, machinery, medicine, food and other industries, This paper introduces the pressure control system Based Gearshift Integral PID control algorithm, with PIC16F877 as the core device.

2 System Composition

2.1 Pressure control principle of air compressor

Shown in Figure 1, The pressure sensor senses the pressure of the compressed air and sends the signal into the single chip microcomputer. Single chip microcomputer calculates the pressure difference between the setting value and measurement value preset .

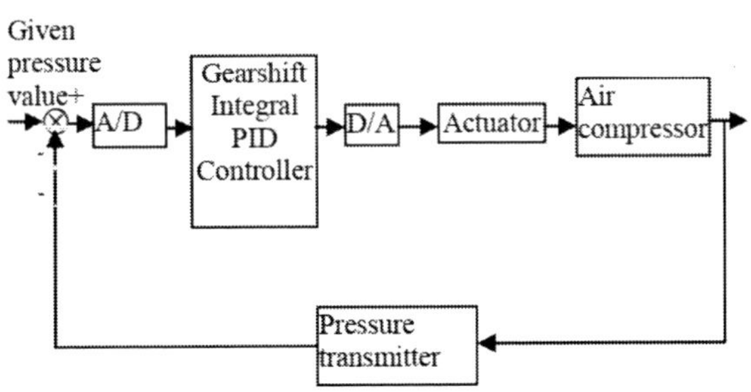

Fig. 1. variable integration pressure control schematic diagram

2.2 Hardware circuit of pressure control system

The hardware of the pressure closed-loop control system consists of pressure sensor transmitter, inverter, air compressor, microcontroller and related control circuit. The schematic diagram is shown in figure 2. The pressure sensor transmitter sends the pressure signal to the single chip microcomputer, and the MCU outputs the signal according to the control rules to control the output frequency of the inverter, than adjust the speed of the air compressor motor, the exhaust volume of the air compressor changes with the change of the gas consumption, so as to ensure the constant pressure of the pipe network.

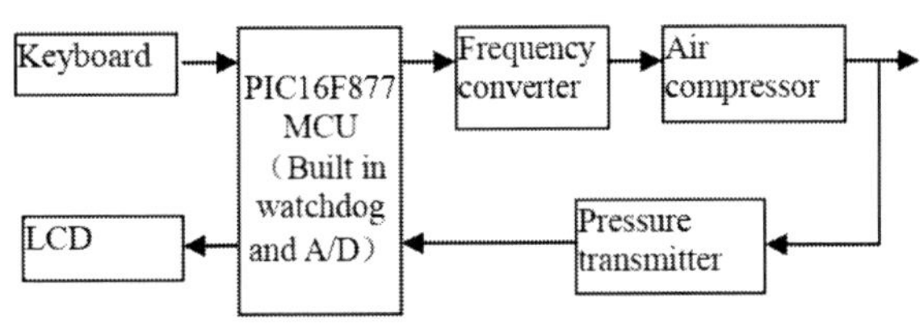

Fig. 2. hardware structure of pressure control system

The pressure transmitter includes two parts: pressure detection and signal processing. YTZ-150 remote pressure gauge is used for pressure measurement. Its measuring range is 0 ~ 1.0MPa, and the output is a resistance signal. The range of resistance is 0~400 ohms. The remote pressure gauge is cheaper and the signal processing circuit design is simple, but the consistency of different pressure sensors is poor and debugging is troublesome.

* Corresponding author: 603199516@qq.com

The pressure detection and signal processing circuit consists of operational amplifier, reference voltage source and some resistors to form constant current source. A constant current of 1mA is provided to the output resistance of the remote pressure gauge, Convert the resistance value of the pressure gauge to the DC voltage signal, After in-phase amplification by the high sensitivity integrated operational amplifier, output DC voltage signal of 0~5V is sent to A/D conversion built in single chip microcomputer. Due to the poor consistency of the remote pressure sensor, the zero and full scale of the amplifiers of different sensors must be individually adjusted.

The operating environment of air compressor is abominable,such as the electromagnetic interference caused by high power motor, AC variable frequency speed regulator , power supply voltage fluctuation is more serious. So the detection instrument and wiring should avoid the strong magnetic field as far as possible, and the sensor should be powered by high precision power supply.

3 System Composition

3.1 Control Algorithm

PID control algorithm is simple, good applicability, widely used, But the integral gain K_I is constant in the traditional PID algorithm, and remains constant in the adjustment process,so it is better to the small deviation control .When the system is disturbed and the pressure fluctuation range is relatively large, it will cause larger overshoot and longer transition time.Therefore, the requirements of the pressure control system are: When the deviation is large, the integral action should be weakened so as to avoid overshoot and even integral saturation; when the deviation is small, the integral action should be strengthened, otherwise it can not meet the requirements of accuracy. The introduction of variable speed integral PID control algorithm can make the control performance satisfied. For this reason, the coefficient $f[E(k)]$ is set, which is a function of the deviation E(k); when [E(k)] increases, $f[E(k)]$ decreases; otherwise, it increases. After each sampling, multiply E(k) with $f[E(k)]$, and then accumulate:

$$P_I(k)=K_I\{\sum_{j=0}^{n-1} E(j)+f[E(k)]\,E(k)\} \qquad (1)$$

In the formula: $P_{I(k)}$ represents the output value of the variable integration.

The relation between $f[E(k)]$ and E (k) can be expressed as:

1) when $|E(k)|\leq B, f[E(k)]=1$,

2) when $B<|E(k)|\leq A+B$, $f[E(k)]=\dfrac{A-|E(k)|+B}{A}$

3) when $|E(k)|>A+B$, $f[E(k)]=0$

$f[E(k)]$ changes in the range of [0, 1],when| E(k)| is greater than a given interval A+B then $f[E(k)]=0$,and the integral coefficient is 0.When | E(k)| less than or equal to B then $f[E(k)]=1$, and the integral coefficient is K_I .When| E(k)| in the interval [B,A+B], $f[E(k)]$ changes with|E(k)|.

$P_I(k)$ is substituted into the PID formula:

$$P(k)=K_PE(k)+K_I\{\sum_{j=0}^{k-1} E(j)+f[E(k)]\,E(k)\}+K_D[E(k)-E(k-1)] \qquad (2)$$

According to formula (2), the Flow chart of variable speed integral PID control algorithm is worked out, as shown in figure 3.

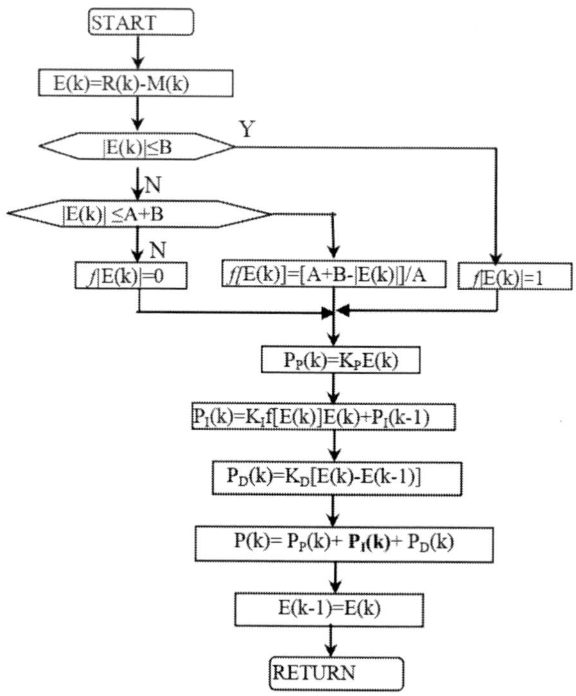

Fig.3. Flow chart of variable speed integral PID control algorithm

In this system, a simple variable step integral PID control is adopted, taking A=10, B=2. The experimental result is better.

3.2 Optimization of PID parameters

The system uses genetic algorithm off-line to optimize PID parameters.

Genetic algorithm (GA) is an adaptive global optimization probability search algorithm that simulates the genetic and evolutionary process of organisms in the natural environment, It will be the problem to be solved into the evolution of groups made up of individuals and the operation of the group of a group of genetic operators, the whole system in accordance with the "survival of the fittest in natural selection principle, the survival of the fittest" introduced such as reproduction, crossover and mutation method, generation - Evaluation - Selection - evolution operation is repeated until the search optimal solution.

This system uses GA algorithm to optimize the PID parameters K_P, K_I and K_D in PID discretization expression, so that the performance of the system is optimal. The flow chart of the algorithm is shown in figure 4.

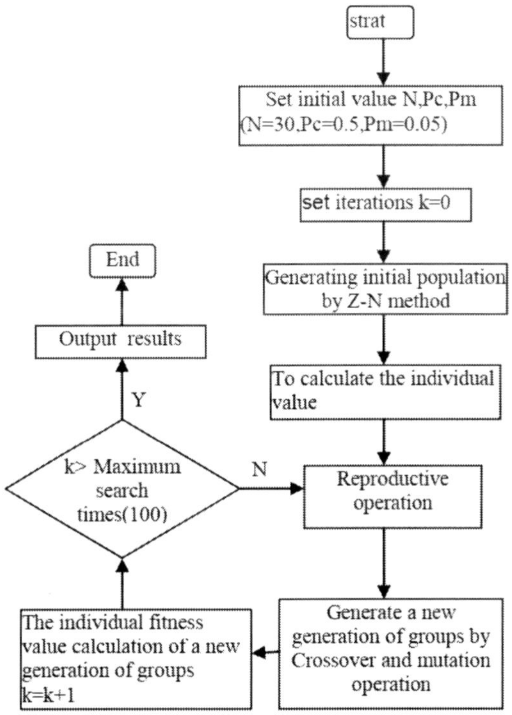

Fig.4. Flow chart of genetic algorithm

3.3 Master Program Design of System

Design of system software based on PIC16F877 MCU,the main program of single chip microcomputer includes initialization, display panel and other subroutine. The implementation of pressure signal acquisition, digital filtering, scale conversion, pressure display, variable speed integral PID control algorithm and other functions are completed by each subroutine. The software also includes the protection of the system. The software flow chart is shown in figure 5.

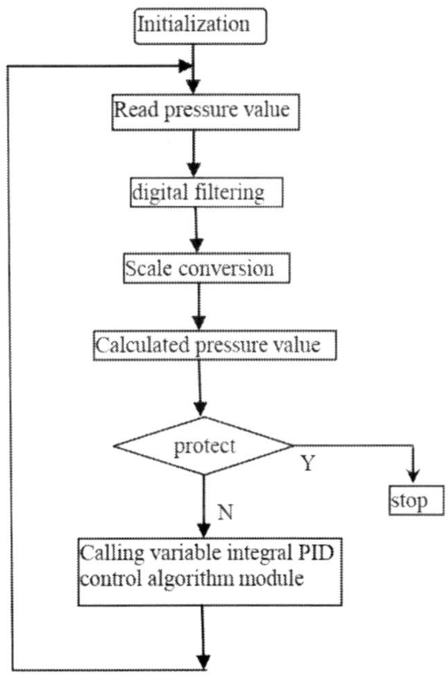

Fig.5. software flow chart of pressure control system

3.3 Master Program Design of System

Design of system software based on PIC16F877 MCU,the main program of single chip microcomputer includes initialization, display panel and other subroutine. The implementation of pressure signal acquisition, digital filtering, scale conversion, pressure display, variable speed integral PID control algorithm and other functions are completed by each subroutine. The software also includes the protection of the system. The software flow chart is shown in figure 5.

4 Experimental Results and Analysis

The pressure of the gas tank increased from 0.2Mpa to 0.5Mpa in the experiment,and maintain the pressure with the pressure value fluctuates up and down ± 0.02Mpa. When the traditional PID control algorithm is adopted ,the dynamic characteristic curve of the system is shown in figure 6,its dynamic performance is as follows: delay time t_d =160s, overshoot σ =3.9%, rise time t_r =420s, regulating time t_c=920s.When,

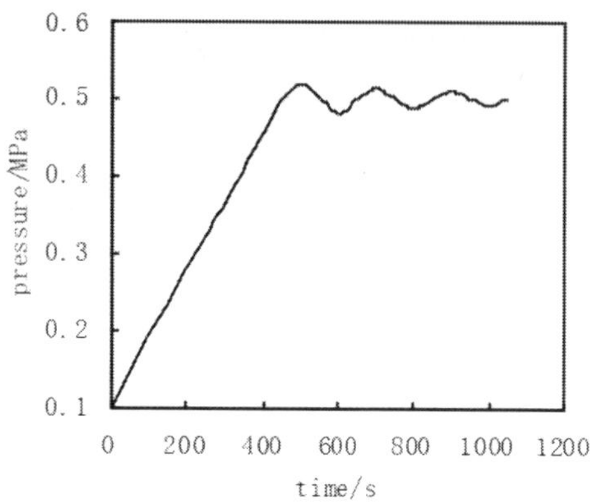

Fig.6. system dynamic characteristic curve(traditional PID)

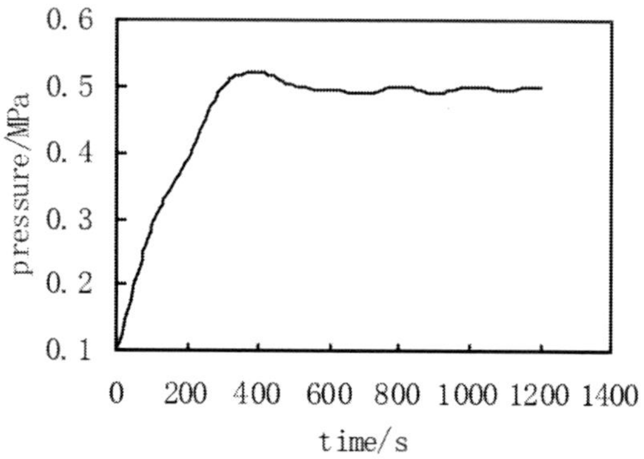

Fig.7. system dynamic characteristic curve(variable speed integral PID control)

5 Conclusion

In the air compressor pressure control system of variable speed integral PID control algorithm, large deviation, accumulated slowly, integral effect is weak.When small deviation, the integral accumulation speed is fast, and the integral action is strong. Using genetic algorithm off-line to optimize PID parameters,then the system has the characteristics of fast dynamic response, good control steady state performance, and can quickly become stable.

References

1. HE Xicai.Sensors and their applications [M]. Beijing: National Defense Industry Press,2001.
2. LI Shaoyuan,WANG Jingchen. intelligent control [M]. Beijing: Machinery Industry Press .2005,158-168
3. ZENG Rong. Design of Experimental Simulation Platform for Pressure Control System of Gas Collector [J], Experimental Science and Technology,2016(8)
4. ZHU Qingzhi, SHEN Yige. Research on Fuzzy Control of Boiler Pressure Control System [J], Precision Manufacturing and Automation,2017(2)
5. WU Jing-hua, Parameter Solving and Identification of Hydraulic Pressure Control System Based on Improved GA[J], Hydraulic and pneumatic,2010.(3)
6. WANG Shu-xia,ZHU San-yuan,TU Jun-ying,Parameter Solving of a Hydraulic Pressure Control System and the Simulation Based on Improved GA[J], Computer engineering and Science,2010(7)
7. FU Long-fei,TIAN Guang-lai,LIANG Bo,SHI Qi, Real Time Modeling and Simulation of Aircraft Cabin Pressure Control system[J],computer simulation,2016(2)

Fiber fabry-pero micro pressure sensor

Xiaojing Yang[1,2], Yiting Wang[1,3]

1.College of Computer and information engineering, Anyang Normal University, Anyang 455000

2. International WIC Institute, Beijing University of Technology, Beijing 100124

Abstract: A new fiber fabry-perot pressure sensor is designed. It is made by photolithography, silicon wafer etching, anode bonding, microelectromechanical system technology and so on. It is suitable for the pressure measurement of micro-pressure environment under harsh environment and small space. The structure of the transmitter and the way to make it are described in detail. The design smartly uses the Fiber optic flange, the optical fiber end face is parallel to the sensitive film, thus a high-quality Fabry-Pero-cavity is formed. The structure also contributes to the stability of the initial cavity, the sensor error is reduced. The experimental demodulation system is established, the pressure, temperature and other characteristics are tested in detail. The experiment result shows that within the pressure range of 0~0.1MPa, the sensor has a good linearity, high repeatability and a sensitivity of 61.6 μ m/MPa.

Keywords: sensors; micro-pressure sensor; fiber optics sensor; Fabry-Perot; micro-electromechanical system technology

1 Introduction

Micropressure sensor is the most commonly used pressure sensor in industrial practice,widely used in various industrial automatic control environment, it is involved in oil pipelines, water and electricity, railway transportation, intelligent building, production of automatic control and machine tools and other industries[1.2]. But in the industrial security zone, against bad conditions, micro pressure leak in confined Spaces and test requirements for pipeline microflow have not been satisfied, need small size, high anti-interference, high sensitivity, micro - pressure sensor products with wide range and not lose the precision. In recent years, the development direction of micropressure sensor is starting to shift to new materials, new mechanism and new structure, the process is being developed from silicon micropressure sensor [3,4] to optical micropressure sensor.

Meanwhile, fiber fabry-perot(F-P) sensor is developing rapidly, it has been applied to severe electromagnetic interference and high temperature and other harsh conditions.The optical fiber F-P interferometric pressure sensor structure is simple, low cost, it is the most commonly used interferometric fiber pressure sensor. It not only has the characteristics of general optical fiber sensor good reliability, anti-electromagnetic interference, corrosion resistance, etc ,but also has the characteristics of high measure precision, high dynamic range, linearity and so on.

This paper designs and develops a fiber F-P micro - pressure sensor, make full use of the advantages of optical fiber sensor, the disadvantages of traditional micro pressure sensor in small space and strong electromagnetic environment are overcome, and linearity is good, the sensitivity is high. The sensor is made of microelectromechanical system(MEMS) technology and suitable for mass serialproduction.

2 sensor design

The structure of the sensor is shown in figure 1, among them a is the fiber optic flange, b is the fiber and ceramic inserts, c is the glass ring,d is monocrystalline silicon piece, also called the pressure sensitive membrane of the sensor. The light travels vertically through the fiber, part of it is reflected off the fiber end, the other part passes through the air cavity, the silicon sensitive film is reflected back into the optical fiber, two beams of light interfere in the optical fiber, the F-P interference cavity was formed under the surface of the fiber optic end and the sensitive film. When external pressure changes, the sensitive membrane changes under the action of pressure, that changes the length of the F-P cavity, the phase of the reflection spectrum is changed. By means of the demodulation of its reflection spectra, you get the length of the F-P cavity.

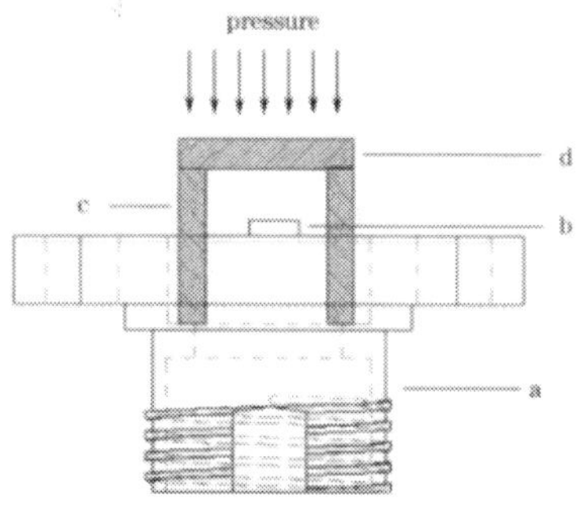

Figure 1 Schematic diagram of the structure

MEMS technology is used in the structure of sensor, the main features include lithography, silicon wafer etching and anode bonding, See figure 2. steps of:1) photoetching, is the geometric figure on the mask board, through photochemical reactions, transfer the pattern to a photoreceptor coated on a wafer. The process of coating, prebaking, exposure, developing and postbaking,ect, forming a photoresist pattern on a silicon wafer, as shown in figure2(a),(b);2) wet corrosion of silicon wafers coated with photoresist protection layer, use a certain concentration of KOH solution[9-10], the (100) wafers to silicon is etched into the desired shape and thickness, as shown in figure2(c), and degumming and surface cleaning processing, as shown in figure2(d);3) the obtained silicon wafers and pre-optical polishing, the long - precision controlled glass ring is used for the anode bonding, as shown in figure2(e), the bonding temperature selection is the key. The temperature is too low to reach the glass melting point, the temperature is too high because of the different thermal expansion coefficient, The stable work of the sensor is affected;4) the bonded silicon and glass rings are fixed on the FOCI fiber flange, finally insert the ceramic pin. The whole sensor is low cost and suitable for mass production.

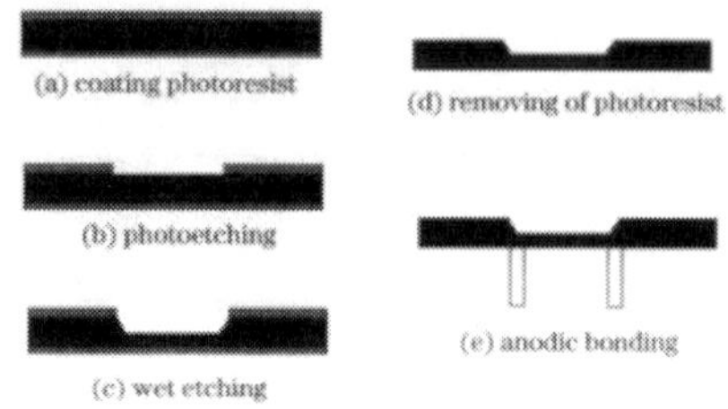

Figure 2 Sensor production process

In this sensor, the mechanical model of silicon sensitive film with glass ring bonding can be abstracted as radius r, a round plate of surrounding solids, the surface is subjected to uniformly distributed pressure P, then the relationship between the center deflection of silicon membrane and pressure P is

$$\frac{Pr^4}{Eh^4} = \frac{16}{3(1-v^2)}\left(\frac{Y}{h}\right) + \frac{7-V}{3(1-V)}\left(\frac{Y}{h}\right)^3 \quad (1)$$

The h is sensitive film thickness, y is the sensitive film center deflection. In general(y<h/2),(1) can be approximated to A linear equation:

$$y = \frac{3pr^4(1-v^2)}{16Eh^3} \quad (2)$$

P is the external pressure, r is the sensitive membrane radius, E is young's modulus, v is poisson's ratio.For monocrystalline materials, E is 160GPa,v is 0.22. The assignment $r = 3800\mu m$, $h = 200\mu m$, apply pressure of 0.1MPa, use software to simulate, the results are shown in figure 3,different colors represent different degrees of deflection,the specific values are given by the ruler in the figure, the deflection of the circular film center is $5.9\mu m$.

3 Analysis of experimental results

The pressure test system of micropressure sensor is shown in figure 4,the broadband source with the optical fiber sensor Si720 is used as the light source of the sensor. The light passes through the coupler and divides into two equal paths, all the way as the incident light enter fiber F-P micro pressure sensor, the light reflected by the sensor passes through the same coupler into the sensing analyzer. The spectral signal of the reflected light is collected by the sensor analyzer to find the cavity length under certain pressure. Due to the membrane deflection and pressure are linear relationships, therefore, the corresponding pressure can be determined according to the change of lumen length.

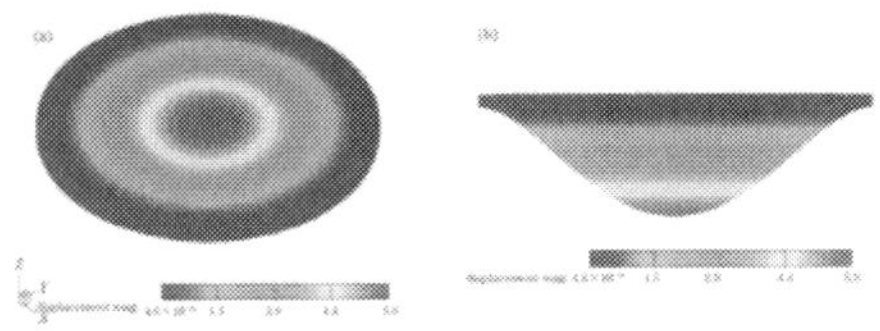

Figure 3 Simulation results of the sensitive membrance. (a)Front view; (b)side view

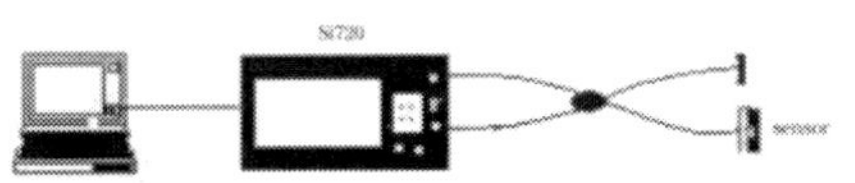

Figure 4 Pressure test system of the micro sensor

The reflectance spectra of the sensing analyzer are shown in fig.5.

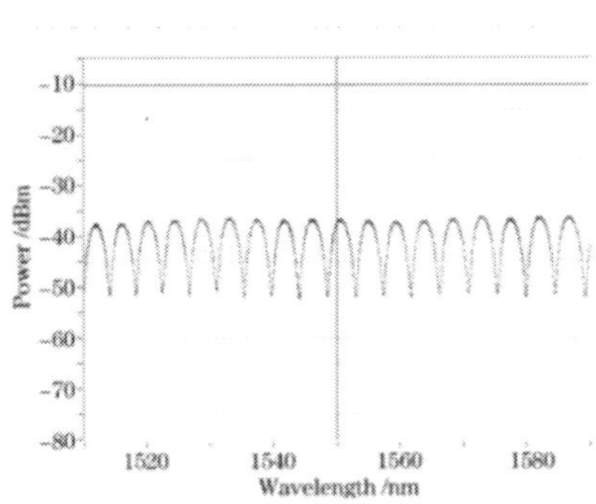

Figure 5 Reflectance spectra of micro pressure sensor

For the adjacent levels in the spectrum m,m+1, its spectral peak corresponds to the wavelength λ_m, λ_{m+1} and cavity length L. The following relationships exist:

$$\frac{4\pi L}{\lambda_m} = (2m + 1)\pi \qquad (3)$$

$$\frac{4\pi L}{\lambda_{m+1}} = [2(m + 1) + 1]\pi \qquad (4)$$

F-P cavity length can be solved by (3), (4), We get the change in the cavity length, so let's figure out what the external pressure is.

The finished micro pressure sensor is marked with the YS-2.5 piston type pressure gauge, the pressure is uniformly increasing from 0 to 0.1 MPa. The initial cavity length of the micropressure sensor is $208.36\mu m$, the length of the lumen is calculated using Matlab software, the results are shown in figure 6, the lumen length fitting equation G=208.4+61.79P, error of fitting 0.9999, root-mean-square error 0.02033. The sensitivity of the sensor is $61.6\mu m/MPa$. The simulation results are shown in figure 3, external pressure change is 0.1 MPa, deflection change of the circle membrane center is $5.9\mu m$, the experiment changed $6.1\mu m$, the experimental results are consistent with the simulation results.

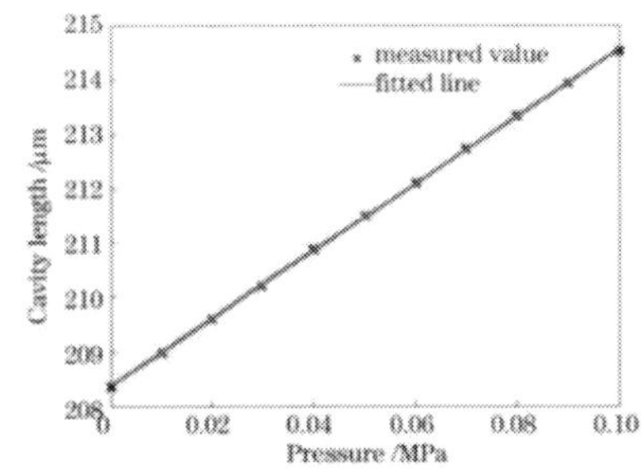

Figure 6 Relationship between pressure and cavity length of the micro pressure sensor

To test the delay performance of the sensor, increase the pressure from 0 to 0.1MPa, and then it's uniformly down to zero, The contrast of lumen length in the process of lifting and lowering is shown in fig. 7. It can be seen that the delay effect of micro pressure sensor is very weak, the rise and fall curves almost coincide.

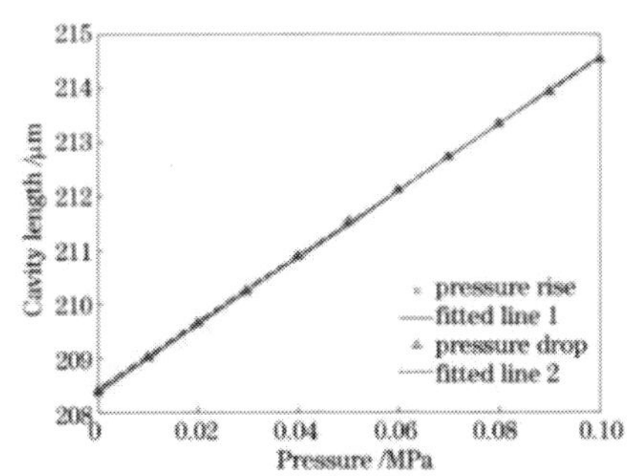

Figure 7 Cavity length changes during pressure rising and dropping

The experimental group also tested the repeatability of

pressure sensors, the sensor was tested for four days in a row, the results are shown in table 1, the mean relative error of the calculation is 0.7%,sensitivity to the sensor is minimal.

In addition, the temperature characteristic of the micro - pressure transmitter is studied. Place a finished sensor in a high temperature test chamber, control the temperature rise from 10 ℃ to 65 ℃, the length of the lumen curve is shown in figure 8. It can be seen that the cavity length of the sensor is linear with temperature change, the temperature sensitivity calculation value is 80 nm / ℃. In stress tests, a good temperature linearity also facilitates temperature compensation for the sensor, to get accurate test results.

Table 1 Repeatability of the sensor

Pressure /MPa	0	0.01	0.02	0.03	0.04	0.05	0.06	0.07	0.08	0.09	0.10
1st day	−0.0002	0.0101	0.0201	0.0298	0.0405	0.0508	0.0606	0.0706	0.0806	0.0903	0.1000
2nd day	0.0005	0.0106	0.0209	0.0305	0.0409	0.0508	0.0608	0.0707	0.0806	0.0901	0.0999
3rd day	−0.0002	0.0099	0.0199	0.0298	0.0403	0.0507	0.0606	0.0705	0.0805	0.0900	0.0997
4th day	0.0000	0.0101	0.0202	0.0299	0.0403	0.0506	0.0606	0.0706	0.0807	0.0899	0.0996

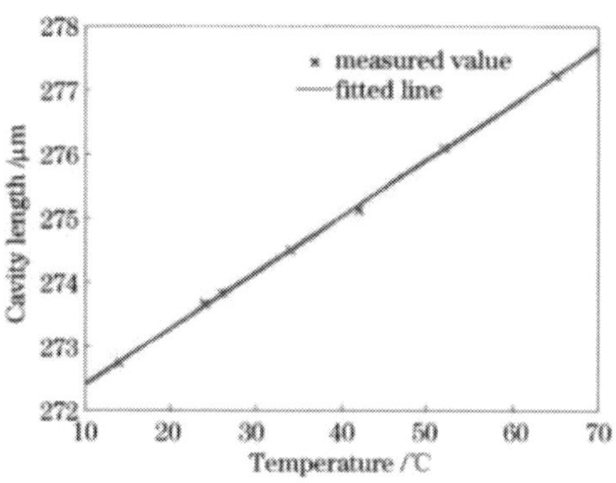

Figure 8 Temperature characteristic of the sensor

4 Conclusions

The design and manufacture of optical fiber F-P micro pressure sensor, a number of tests have been carried out. The results show that the micro-pressure sensor is stable, good hysteresis, high sensitivity, and it is easy to draw materials, Compact construction, the advantages of low production cost, suitable for mass production.

References:

1 Gregory T A Kovacs.Micromachined Transducers:Sourcebook [M].Boston:WCB,1998.

2 Minxin Zhou,Qingan Huang,Ming Qin.Modeling,design and fabrication of a triple-layered capacitive pressure sensor[J].Sensors and Actuators A:physical,2005,117(1):71-81.

3 E G Bakhoum,M H Cheng.Capacitive pressure sensor with very large dynamic range [J].IEEE Transactions on Components and Packaging,2010,33,79-83.

4 Zhenhai Chen,R C Luo.Design and implementation of capacitive proximity sensor using microelectromechanical systems technology[J].IEEE Transactions on Industrial Electronics,2002,45:886-894.

5 Yang Chundi,Wang Ming,Ge Yixian,et al..A miniature extrinsic fiber Fabry-Perot pressure sensor[J].Acta Optica Sinica,2011,30(5):1358-1361.

6 Wang Lihua,Wang Ming,Rong Hua,et al..Design and simulation of miniature shear stress sensor based on Fabry-Perot interference[J].Acta Optica Sinica,2012.32(sl):sl28003.

7 Ni Xiaoqi,Wang Ming,Chen Xunxing,et al..Wavelength multiplexing of optical fiber microelectromechanical system Fabry-Perot pressure sensor[J].Acta Optica Sinica,2007,27(5):0776-0780.

8 Ke Tao,Zhu Tao,Rao Yunjiang,et al..Accelerometer based on all-fiber Fabry-Perot interferometer formed by hollow-core photonic crystal fiber[J].Chinese J Lasers,2010,37(1):171-175.

9 Yanfeng Jiang,Qingan Huang.A physical model for silicon anisotropic chemical etching[J].Seminconductor Science and Technology,2005,20(6):524-531.

10Nacer Eddine Demagh,Assia Guessoum,Hind Aissat,Chemical etching of concave cone fibre ends for core fibre alignment[J].Meas Sci Technol,2006,17:119-122.

11Gregory T A Kovacs.Micromachined Transducers Sourcebook[M].Zhang Wendong et al.Transl.Beijing:Science Press,2003.181.

Analysis and design on airport safety information management system

Lin Yan

School of Air Transportation, Shanghai University of Engineering Science, China

Abstract. Airport safety information management system is the foundation of implementing safety operation, risk control, safety performance monitor, and safety management decision for the airport. The paper puts forward the architecture of airport safety information management system based on B/S model, focuses on safety information processing flow, designs the functional modules and proposes the supporting conditions for system operation. The system construction is helpful to perfecting the long effect mechanism driven by safety information, continually increasing airport safety management level and control proficiency.

1 Introduction

Safety is the eternal theme of civil aviation. Ensuring aviation safety is the focus of the world aviation industry. In the "13th Five-Year" period, China will construct 74 new airports. Till 2020, the total number of airport in China will be about 260. The increase in the number of airports and the promotion of the scale will inevitably make the airport safety management face a huge challenge. High quality safety information is the basic guarantee of airport safety management. A good airport safety information management system is of great significance to the sustainable and safe development of the airport.

2 System requirements analysis

At present, existing aviation safety information management system research focuses on the whole civil aviation or an individual company. The related research on airport is very few. The existing airport safety information management systems have some disadvantages, such as insufficient function, poor operability, difficult to expand and so on. A good airport safety information management system should be able to: (1) accumulate a large number of basic original safety information; (2) make investigation and treatment timely on unsafe incidents; (3) find the useful information from the safety information to support the decision and improve safety management level of airport; (4) facilitate information retrieval of different authorized person and provide statistical reports, security reports and so on; (5) provide materials for relevant personnel to learn systematically and draw lessons from typical cases; (6) establish information sharing and exchange between airport subsidiary companies, airports and resident units.

The paper focuses on the overall design of a good airport safety information management system. With the system, the safety information can be collected comprehensively, the information transmission process is smooth, the information can be fully utilized, and the airport safety management level can be improved.

3 System's data sources

Airport business can be divided into aviation and non-aviation business. The units providing service at the airport include airport authorities, customs, border inspection, entry-exit inspection and quarantine, the airlines, the ground services company, trading companies, cleaning companies, security companies and so on, therefore, the sources of airport safety information are very wide.

Besides collecting the safety information requested by Civil Aviation Administration of China (CCAR-396-R3), the airport safety information management system should also collect the information that affects passenger safety and airport operation. The specific types of safety information are divided into: people's unsafe behaviours, unsafe conditions of devices, unsafe environment, unsafe production process, other potential risks and so on.

From the perspective of information processing, system's data sources are divided into: mandatory reporting, voluntary reporting, security report, complaint, danger or hidden danger report, safety performance monitor, operation deviation report, industry's accident and incident investigation etc.

Corresponding author: lylucy@163.com

4 System architecture and business process

4.1 System architecture

Group Company is the management mode of most of airports in China. An airport group company may manage multiple airports and related business. Information communication between the subsidiaries is carried out through the network. Therefore, with the design principle of practicality, maintainability, security and friendly operability, airport safety information management system can be designed based on B/S mode, using Oracle as database management system, under the SS2H framework of the JavaEE platform. Client terminal uses the browser to access server's web page or web interface. Server uses a multitier architecture, including a presentation tier, a business logic tier, and a data storage tier. The logical structure diagram of the system is shown in figure 1.

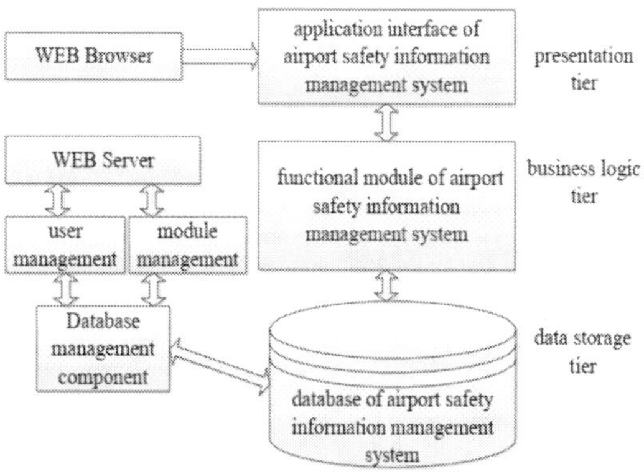

Fig. 1. Logical structure of the system

4.2 System's business process

The business process of airport safety information management system is divided into: information collection and collation, information analysis and evaluation, information release and sharing, information utilization, as shown in figure 2.

4.2.1 Information collection and collation

With the means of reports, records, networks, media, telephones, etc., People can submit various safety information of airport. Safety information personnel will categorize them into the system.

4.2.2 Information analysis and evaluation

Some safety information will trigger the incident investigation procedure, according to the results of the investigation, whether to further trigger the procedure of correction and prevention, risk control, or emergency rescue will be determined. The whole process of safety management activities will be recorded and monitored. At the same time, the airport safety situation is statistically analysed; the safety trends and laws are reflected, providing reference for leadership decision-making.

4.2.3 Information release and sharing

According to the content, nature and influence degree, the airport safety information which has been treated by certain procedures, can be published, spread and shared in different ways and scopes, such as safety meeting, internal propaganda, alarm information, etc. Typical cases can be chosen for safety education and training, to improve the safety awareness of employees and create a good safety culture atmosphere.

4.2.4 Information utilization

On the basis of information analysis and evaluation, information release and sharing, safety information can be further used by airport safety operation quality monitor, airport safety performance management, department safety assessment, airport regulation and standards revision, and so on.

5 System's functional modules

Based on the business process of the system, the functional modules of the system include safety information reporting module, safety supervision and audit module, incident investigation module, correction and prevention module, risk control module, emergency rescue module, data analysis module, safety notice module, safety education and training module, system management. They correspond to the input, processing and output of information management. The structure of functional modules is shown in Figure 3.

5.1 Safety information reporting

Safety information reporting module collects safety information from mandatory reporting, voluntary reporting, security report or complaint, danger or hidden danger report, industry's accident and incident investigation. The module provides special forms of information, and also provides a general form. Airport staffs or resident units can input information through the network. The information entered by the general form must be checked and classified by safety information personnel. Safety information can also be submitted by telephone, fax, mail, etc., then input by safety information personnel. The module can also query, modify and delete the information.

5.2 Safety supervision and audit

According to the airport safety situation and goal, safety supervision and audit module help to establish the plan of safety supervision and audit, completely record information of safety performance monitor and technical operation deviation, and automatically generate the safety supervision and audit report.

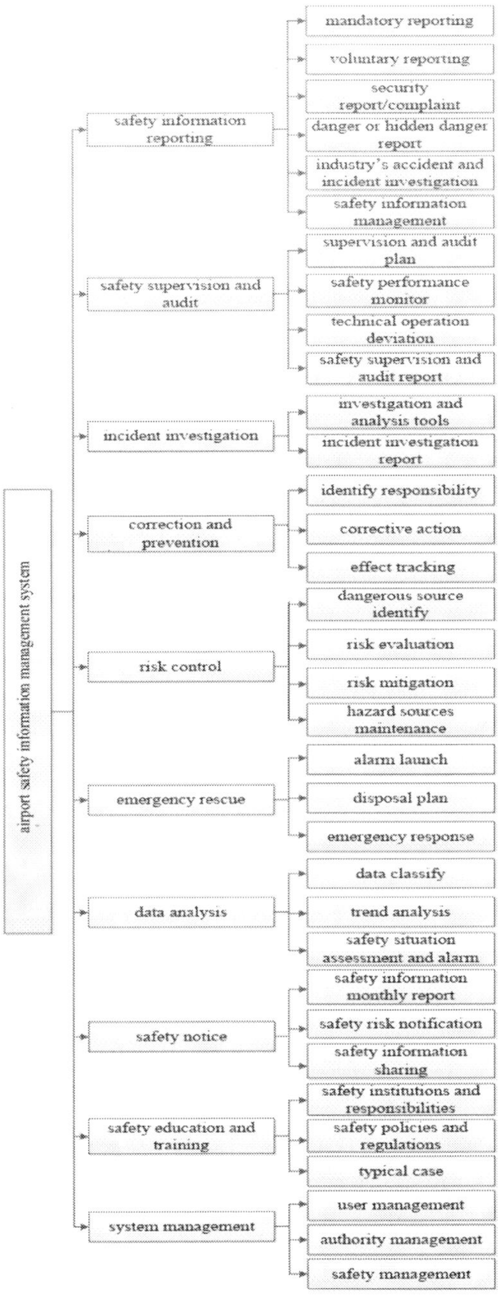

Fig. 3. Structure of functional modules of the system

5.3 Incident investigation

Incident investigation module will start, when unsafe events or risks are found through safety information analysis. Incident investigation module has functions of creating, editing, inquiring and automatically generating report. It can provide investigation and analysis tools to help safety experts to carry out event analysis and evaluation, and provide suggestions and measures. With approval by the leaders, it can trigger other safety management modules such as correction and prevention, risk control, emergency rescue, safety supervision and audit, safety education and training.

5.4 Correction and prevention

For the hidden danger which is classified as low hazard level by incident investigation module, corrective and preventive module can identify the responsibility, and issue the corrective notice to the relevant business departments. The relevant business departments should establish corrective actions and submit them to the module for verification. The module also has the function of track the completion of corrective actions.

5.5 Risk control

Risk control module carries out information maintenance of hazard sources. When the incident investigation module finds serious hazards, risk control module will launched to identify the dangerous source, using fuzzy clustering, artificial neural network, grey system and other methods to further assess the severity of the hazard, determine the risk level. The module can propose risk mitigation measures according to the evaluation and airport safety management requirements, and track the effectiveness of risk mitigation measures.

5.6 Emergency rescue

When a sudden and destructive emergency happens in the airport, emergency rescue module will star. The module alarms the relevant departments and staffs according to the event type. The module provides the disposal plan to help to make reasonable decisions on the rescue plan and resources. In the rescue process, with the module, the scene information can timely be fed back, so that the leader can grasp the progress of the rescue in real time. The module can also record the various decision-making and progress information in the emergency rescue process, so as to summarize the experience and lessons afterwards.

5.7 Data statistics

Data statistics module can classify the data according to the time interval, the severity of the incident, the event type and so on. It can also perform trend analysis. This can help decision-makers to find out the cause of event correlation, analyse safety trends, excavate safety laws, and obtain airport safety situation. When the airport reaches a certain safety threshold, the module can automatically alarm, and the relevant experts or leaders can judge and take countermeasures accordingly.

5.8 Safety notice

After certain treatment, the safety information can form the safety information monthly report and the safety risk notification, which is convenient for the communication of various departments of the airport. The module provides an open interface. Airport subsidiary can submit a safety information report to the group company. Information exchange can also be easily carried out between airport subsidiaries. Other business enterprises at the airport can also voluntarily join the airport information sharing.

5.9 Safety education and training

Safety education and training module includes airport safety institutions and responsibilities, safety operation policies and regulations, typical cases and so on. The module has full text browsing, keyword search and other functions, which can provide training materials for airport safety education and training, and contribute to the improvement of the overall safety level of the airport.

5.10 System management

System management module includes user management, authority management, safety management and other functions. Through it, the system administrator can set which users can use the system, which functions can be used. In the process of user's use, the module automatically records its operation track and forms a log.

6 Supporting conditions for system operation

Airport safety information management system can improve the level of airport safety management, but its effective operation also requires the airport to strengthen the following two aspects of construction.

(1) Enhance the ability of safety information personnel

With more and more safety information collected by airport safety information management system, the role of safety information personnel will become more and more important, and their ability requirements are also getting higher and higher. But the present situation is: safety information personnel lack of professional background or lack of training and experience of data mining. Data mining ability is weak, resulting that the analysis quality of information is not high, and it is difficult to make a convincing evaluation. Therefore, in order to effectively collect and apply safety information, the ability of safety information personnel should be enhanced.

(2) Build a positive safety information culture

Safety information culture is a culture that respects facts, praises rationality and emphasizes precision. Respecting data and respecting objective facts is the foundation of giving play to safety information. The airport should carry out the education on information report honesty, strengthen the consciousness of honest report, and improve the information report quality of staffs.

7 Conclusions

Airport safety information management system can realize the effective integration and maximum sharing of safety information, deepen the analysis and utilization of safety information, and provide decision support for airport safety. In this paper, the overall framework of airport safety information management system based on B/S model is constructed, and the supporting conditions of system operation are proposed. However, the construction of the system still has a lot of work to do, the design of the system function needs to be further refined, the system module algorithm needs to be implemented. In addition, because airport has more than one computer management system, how to integrate the airport safety information management system and other systems needs further study.

References

1. Y. Q. Wang, analysis and design of unsafe event management system in civil airport, Journal of Safety Science and Technology, **6**, (2010),163-166.
2. Y. Chen, research on airport-oriented aviation safety information system, Computer Applications and Software, **29**, (2012),46-49.
3. Y. J. Shi, Y. Q. Chen, problems and countermeasures on aviation safety information management, Journal of Safety Science and Technology, **6**, (2010),116-120.
4. M. Zhan, acquisition, integration and mining of aviation safety production data, Science and Technology for Development, **8**, (2012),36-39.
5. R. Zhang, R. Cheng, Y. Fan, design and research of risk management system for air traffic management, Intelligent Computer and Applications, **6**, (2016),111-112.
6. X. Y. Qi, L. L. Liu, S. Y. Bao, on the evaluation of the safety information management and control capability with the civil aviation system, Journal of Safety and Environment, **2**,(2017),565-572.

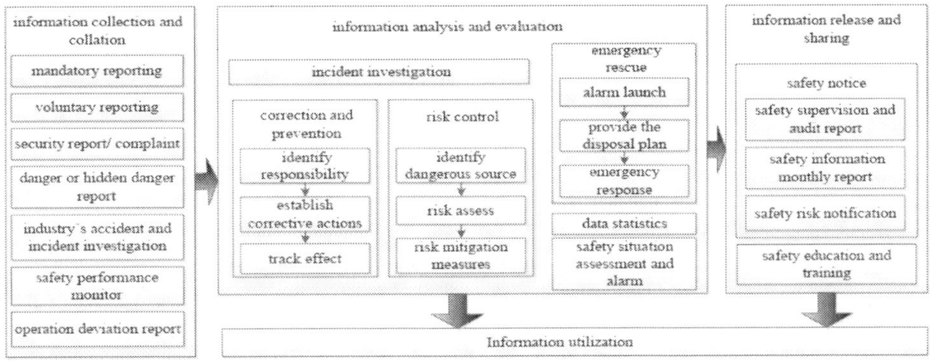

Fig. 2. Business process of the system

A Research on the Electrical Test Fault Diagnostic and Data Mining of a Manned Spacecraft

Feng Yang[1, 2], *Ye* Lv [3, 2], and *Wenting* Li[4]

[1] Department of Aeronautics and Astronautics Engineering, Nanjing University of Aeronautics and Astronautics, Nanjing, 210016, China
[2] Institute of Manned Space System Engineering, China Academy of Space Technology, Beijing, 100094, China
[3] School of information system and management, National University of Defense Technology, Changsha, 410073, China
[4] Beijing Aerospace Automatic Control Institute, Beijing, 100039, China

Abstract. The paper introduces the modeling method and modeling tool for the fault diagnosis of manned spacecraft, the multi-signal flow graph model of a manned space equipment was established using this method; the framework of the fault detection and diagnosis system of manned spacecraft is proposed, the function of ground system and function of the spacecraft are clearly defined. The structure of the functional module is given separately; finally, the tool builds the fault detection and diagnosis system, the application of fault diagnosis method for manned spacecraft is used for reference.

1 INTRODUCTION

Manned spacecraft is a large complex system with the latest cutting-edge achievements in the field of machinery, electronics, materials, control, energy, communication and computer technology. Because of the design level, manufacturing technology, the influence of such factors as cost limits, the manned spacecraft reliability of each subsystem, component is reduced, and thus prone to failure, affect the flight mission, threatened the personal safety, and manned spacecraft to reliable operation in the task of thousands years' time. Therefore, the design of manned spacecraft must be capable of fault monitoring and diagnosis, which is an extremely important auxiliary means to improve the reliability of manned spacecraft [1~3].

In recent years, the research on spacecraft fault diagnosis method has been paid more and more attention. Various fault diagnosis methods have advantages and disadvantages, and the fault diagnosis of spacecraft is a complicated problem. In general, it is not only expected to be able to diagnose faults in individual subsystems and components, but also to diagnose at the system level. By comparing various fault diagnosis methods, it can be seen that each method has its own advantages and disadvantages, and different research objects apply different diagnostic methods. Based on the analysis of the subjects and the advantages and disadvantages of various diagnostic methods, this paper selects the model based method to study the fault diagnosis of manned spacecraft. [4~6].

2 Fault Monitoring and Diagnosis Key Technology Research

The model-based fault monitoring and diagnosis method mainly involves the modelling, fault monitoring and diagnosis system architecture, monitoring and diagnosis system construction and other key technologies.

2.1. Modelling Methods and Tools

Currently in use within the territory of the fault signal modelling methods mainly include relational database modelling method, the information flow model modelling method and the model of multiple signal flow graph modelling method, through the comparison, this paper select multiple signal flow graph model modelling method.

2.1.1. Model Modelling Method

Conceptually, the multi-signal flow graph model covers multiple information flow models, and the model is closer to the physical structure of the system. In addition, "signal" corresponds to the independent unit in the transmission function of the system or the unique attribute that constitutes the function of the system [3]. Because the model is very similar to the real system structure, the model integration and verification are greatly simplified and effectively improve the efficiency of the model builders.

Formally speaking, the multi-signal flow graph model consists of the following elements:

- A finite system constitutes a set of components, $C=\{c_1,c_2,\cdots,c_L\}$;
- System - related independent signal sets $V=\{v_1,v_2,\cdots,v_k\}$;
- N - dimensional test set $T=\{t_1,t_2,\cdots,t_n\}$;
- n_p dimensions can be used to test point sets $TP=\{TP_1,TP_2,\cdots,TP_{n_p}\}$;

* Corresponding author: yangfeng501@163.com

- Each test point TP_p corresponds to a set of tests $SP(TP_p)$;
- Each element c_i affects a set of signals $SC(c_i)$;
- Each test t_j detects a set of signals $ST(t_j)$;

Directed graph $DG = \{C, TP, E\}$, there are the physical connections to the diagram to represent the system.

The multi-signal flow graph model has advantages in obtaining the important knowledge of system fault diagnosis, and the multi-signal flow graph modelling method does not require the system component to specify the failure mode knowledge. In the modelling process of multi-signal flow graph, the failure mode influence analysis is carried out through the failure mode description of the relevant signals of each part. The multi-signal flow graph modelling method represents the fault space of the system. Therefore, in the multi-signal flow graph method, only the propagation process of system component failure mode is modelled. Compared with quantitative modelling and qualitative modelling methods, the cost and difficulty of modelling are greatly reduced.

2.1.2. Modelling Tools

Modelling method has been supported by software tools, the TEAMS of both companies in the United States is a set of software testability, maintainability and system status monitoring function of the commercial software products, complex object can be static, dynamic testability analysis, mainly for defence and aviation, space systems, to provide users with advanced, high-quality health diagnosis, prediction and system management products and solutions. At the same time, the real-time status monitoring and remote fault diagnosis of complex system can be realized through the supporting modules of TEAMS-RDS and TEAMS-RT.

The TEAMS-RT software is a supporting tool for online and real-time diagnosis. It is based on the team test model of the established system, as well as the 0/1 state of each test point in the model (passed/not passed). Such a team -RT software can identify known good, bad, and suspect components.

The TEAMS-RT is designed for remote monitoring/diagnosis. Using this technique, users can upload sensors or test data to the RDS using the Internet, and they can be diagnosed in a few seconds by intelligent automatic processing based on the test case design model.

TEAMS software for testability analysis and evaluation of complex product, the real-time condition monitoring and remote diagnosis provides a good solution, can be on the basis of the test model to build real-time monitoring and diagnosis system of running status of manned spacecraft, and improves the automation of manned spacecraft, management level and management level.

2.1.3. Modelling Steps

During the study, follow these steps.

The first step, input structure model, schematic diagram model or concept block diagram. In the TEAMS, the structural model can be automatically generated or entered directly through the graphical user interface through VHDL structure model, EDIF wire table, etc.

Second, load the signal to the module and test point. A signal set can be found from a functional or pass function independent variable. For example, the signal characteristics of the power amplifier will include output distortion, harmonic distortion and power output. Generally any unique attribute has a relevant signal. For example, a bus with multiple independent addressable devices is used as a signal for each device's address.

Step 3: modify the model under special circumstances. When the following special situations occur in the modelling process (not necessarily all of them), the necessary corrections can be made. In the TEAMS, all of these fixes are available.

- If system with redundant structure composition (Both A and B have to be invalidated), redundant components to make use of the AND node set.
- If the system has a different mode of operation, the use of modeling the SWITCH node.
- If the system can contain parts in the test machine, the TEAMS should indicate in the modeling. It is particularly noteworthy that in-machine self-testing can be modeled with special types of signal names.
- If the system has a replaceable digital integrated circuit components, can be modeled by their equivalent model. The equivalent model is to simplify the chip model and obtain the necessary correlation information to detect faults in the chip (without fault isolation).
- In special circumstances, the system has a counter the correlation between each other, must identify and eliminate such correlation.

Step 4: Test the Model. The multi-signal model is tested according to the test data and physical model.

2.2. Fault Monitoring and Diagnosis Technology of Manned Spacecraft

2.2.1. Overall Structure and Plan

The manned spacecraft is based on the system structure of the fault monitoring and control system of the earth and earth. The fault monitoring and diagnosis system of world integration is mainly composed of monitoring and diagnosis system, ground monitoring and diagnosis system. On monitoring and diagnosis system is responsible for the spacecraft status monitoring, fault diagnosis and emergency treatment, the ground monitoring and diagnosis system is responsible for fault prediction, fault diagnosis and system reconfiguration, troubleshooting and manned spacecraft health status evaluation. The system architecture of fault monitoring and diagnosis system is shown in Fig.1.

Among them on monitoring and diagnosis system has fault management architecture of data acquisition, data processing, condition monitoring (abnormal monitoring), health assessment, fault diagnosis and prediction) and decision support (fault disposal), and other functions.

The display and alarm system has status monitoring, health assessment support and decision support function. The display and warning system can simultaneously receive data from the monitoring and diagnosis system and ground monitoring and diagnostic system status monitoring results data, health assessment module output fault diagnosis data. In addition, the health assessment support and decision support module also provides an area of expertise in human-computer interaction.

Ground monitoring and diagnosis system for data processing, condition monitoring, health assessment, health, forecast and decision support function, under the support of experts in the field, with more accurate ability of health assessment and decision support.

Based on the existing state monitoring and diagnosis technology, the real-time status monitoring and fault location of manned spacecraft is realized based on the telemetry data and remote control instructions of manned spacecraft. Using Teams testability analysis software system model is set up, and to evaluate the testability of manned spacecraft, the output correlation matrix and diagnosis strategy, monitoring and diagnosis system based on Teams - RT on real-time condition monitoring and diagnosis, the ground to realize fault detection and diagnosis system combined with domain experts' further reasoning to locate the fault. The overall scheme of fault monitoring and diagnosis system for the integration of heaven and earth is shown in Fig.2.

2.2.2. On Board Monitoring and Diagnosis System

Shown on the monitor and diagnosis system structure in Fig. 4. IN figure 3 upper part, for the development of manned spacecraft system real-time monitoring and diagnosis system, firstly according to the need of multi signal flow graph modelling method mentioned in the introduction, the establishment of multi signal flow model of the diagnosed object using the TEAMS software, and then based on the interface file generation and RT TEAMS model, the interface file in the TEAMS-Designer environment can directly generate the suffix were INP, PRM, MAP file, and then according to the characteristics of manned spacecraft system of real-time fault diagnosis reasoning program. In order to facilitate the development of real-time fault diagnosis system, QSI provides a real time diagnostic reasoning application program interface -- TEAMS-RT API. In TEAMS-RT API, the input and output data format, the function code, and the returned diagnostic data format are provided for the interaction with the TEAMS dynamic link library. Among them, the core code of real-time diagnosis reasoning is encapsulated in TEAMS dynamic link library. Real time diagnosis is the use of TEAMS-RT API development for the application of the system, the real-time file regularly read the test results and the model, called TEAMS.dll for diagnostic reasoning and the diagnosis result output process.

2.2.3. Ground monitoring and diagnosis system

He structure of the ground monitoring and diagnosis system is shown in Fig. 3, below half part, the ground monitoring and diagnosis system based on Browser / server (B/S) architecture, the server real-time diagnostic procedures and database management system, the user can receive authorization in computer network browsing real-time monitoring results.

The ground monitoring and diagnosis system is mainly composed of three parts: ground control system, diagnosis centre server and remote diagnostic resource (expert). The ground control system receives manned spacecraft telemetry parameters, send control commands, receive the telemetry parameters; diagnosis centre server is the core logic of diagnosis system, it mainly includes the data processing module, monitoring module, fault diagnosis module, remote expert consultation platform and based on Web database and knowledge base system, provides intelligent diagnosis and maintenance support service for the manned spacecraft; remote diagnosis resources (expert) is to provide the technical support of remote real-time diagnosis system of manned spacecraft, solve the difficult fault.

The ground monitoring and diagnosis system of manned spacecraft system has the following processing steps:

- Data acquisition: manned spacecraft operation data management system for acquisition, processing and storage of manned spacecraft operation data, the data is formatted in the packaging process, the channel encoding and subcarrier modulation after telemetry downlink RF telemetry channel is sent to the center.
- Comprehensive diagnosis: telemetry center sends data to the diagnostic server for processing after being readable. The diagnostic center server in real-time monitoring and fault reasoning module of manned spacecraft system into a multi signal model based on the parameters obtained on the basis of inference, can quickly isolate the fault, and according to the prior knowledge base generating maintenance suggestions. Manned spacecraft system complex if the automatic fault isolation tools cannot real-time situation, immediately set up interactive diagnosis task and expert system, the expert system in the browser can access to Web diagnosis center server, access to the existing state of manned spacecraft, and according to the information of the remote expert consultation.
- Maintenance suggestions: system fault isolation, fault maintenance suggestions generated through the browser interface, feedback to the transportation management center of manned

spacecraft guidance and technical personnel, repair tasks.

3 Intelligent mining technology of test data

In spacecraft testing, data interpretation is an important part. At present, the main data interpretation system is based on the division system, and the main interpretation methods are:

- Draw the curve of the parameter test data to the time, and judge whether the jump point is reasonable.
- The criterion interval of the correct value of the given parameter, and the transfinite data is determined to be abnormal.

The main disadvantages of current interpretation methods are:

- Unable to interpret the change trend of parameters in a large period of time. Using the current interpretation method, the interpretation interval is concentrated on a certain day or a test stage, and is not compared with the previous stage, the previous model test data, cannot see the changing trend of data.
- It is impossible to interpret the possible differences in the status of multiple similar products. The spacecraft has multiple telemetry parameters of the same or similar function, which characterize the states of the same kind of products. The current interpretation method does not focus on the specific interpretation of such parameters, and cannot find the differences between similar products.
- Neglecting the connection between parameters. The current interpretation method uses single parameter as the unit to interpret the data, ignoring the specific relationship between some parameters. Using the current interpretation method, easy to cause the following situation, lead to leakage judgment: in a number of correlation parameters, the test data of each parameter are in the normal range, but the parameters do not conform to the curve.

In view of the above advantages and disadvantages analysis, we have carried on the data intelligent mining analysis research. Data intelligent mining analysis refers to the automatic data query and processing through the database, to achieve the same parameter test data longitudinal ratio, the same parameter test data transverse ratio, correlation parameter test data joint ratio.

Users can query the historical test data of multiple spacecraft by software, and then pre-process the data and compare the data. The work flow is shown in figure 4. There are two ways to deal with data: image processing and statistical processing. Image processing is to display one or more test data visually on the image, the horizontal axis can be time, data serial number, can also be a data. Statistical processing refers to the selection of different parameters in a certain period of time, to facilitate users to find outliers.

4 Conclusions

This paper studied by monitoring and fault diagnosis method of manned spacecraft, a test model of a typical aerospace equipment, construction of the monitoring and fault diagnosis system of the system, realize real-time monitoring, remote client access and interactive expert diagnosis, provides a reference scheme of monitoring and fault diagnosis of manned spacecraft. The results show that:

- Through the research of system level testability design method, master the multi signal flow chart modeling method and modeling tools, and provide reference for the system level testability design of spacecraft;
- The thought of fault monitoring and diagnosis system is put forward, the construction monitoring and diagnosis system for spacecraft integrated, and ground monitoring and fault location reasoning of the spacecraft fault at the same time, the two complement each other, the thought is of great significance for the space station and other large spacecraft fault diagnosis;
- The real-time algorithm of spacecraft fault monitoring and diagnosis is introduced, and the real-time diagnosis algorithm is optimized according to the characteristics of manned spacecraft, and the optimized algorithm flow is given;

Finally, according to the method studied in this paper, using the "three ratio" test data intelligent mining method, the test data is further analyzed and verified.

References

1. Y. Yan, H.X. Wu, X.L. Wang, Z.B. Li. Overview of Spacecraft Fault Diagnosis and Fault Tolerant Control Technology. Journal of Astronautics, 2013, 24(3): 221-226.
2. B. Long. The Research of Multi-signal Flow Modelling and Fault Diagnosis Method used in Spacecraft. Harbin Industry Technology, 2005: Harbin.
3. J.Y. Shi, X. Zhang, T.G. Zhou. Application of Multi-signal Modelling and Diagnosis Strategy Design Technology. Systems Engineering and Electronics, 2011, 33(4):811-815, 2011.
4. R. Sakthivel, S. Selvi, K. Mathiyalagan Fault Tolerant Sampled Data Control of Flexible Spacecraft with Probabilistic Time Delays. Nonlinear Dynamics, 2014, 79(3):1835-1846.
5. Y. Shen, X.L. Zhang, Z.H. Wang. Spacecraft Fault Diagnosis Based on Empirical Mode Decomposition and Directed Factor Graph. Journal of Harbin Institute of Technology, 2013, 45(1):19-24.
6. H. Kato, J. Ota. Real-time Spacecraft Actuator Fault Diagnosis with State Segmented Particle Filtering. Advanced Robotics, 2014, 28(19): 1265-1276.

DOI: 10.1051/matecconf/201713900202

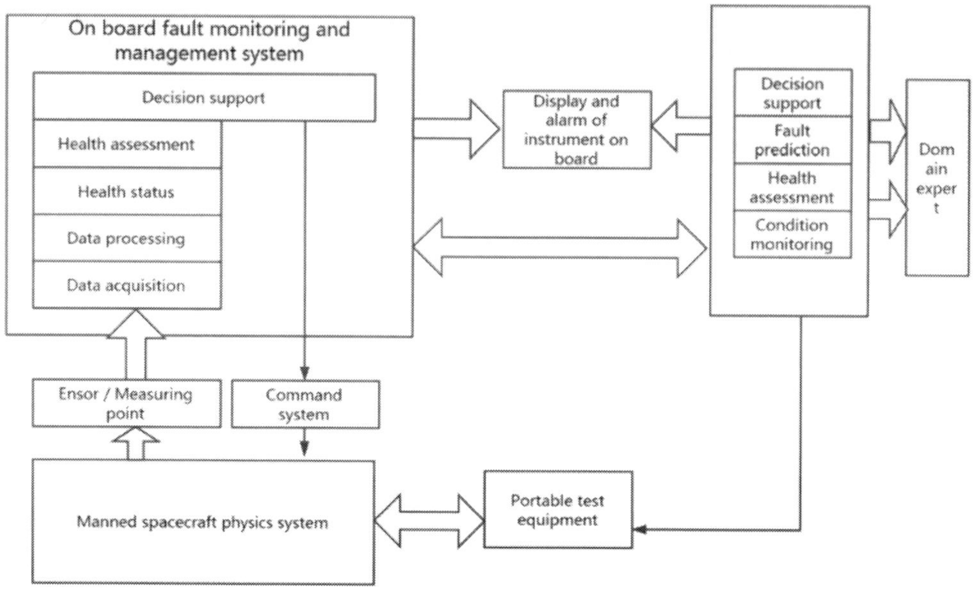

Fig 1. Fault monitoring and diagnosis system architecture based on integration of earth and sky

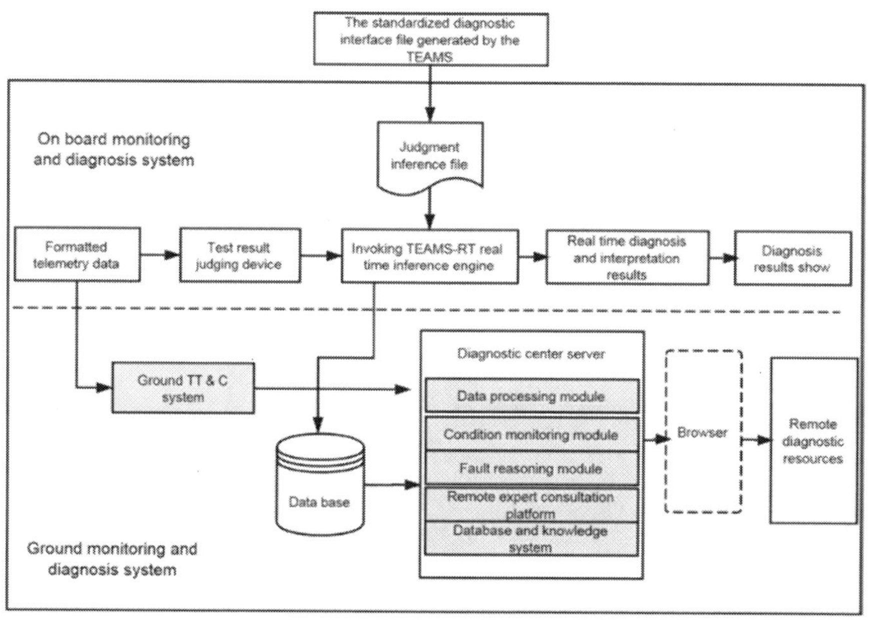

Fig 2. Overall scheme of fault monitoring and diagnosis for integration of heaven and earth

* Corresponding author: yangfeng501@163.com

Research on Automotive Dynamic Weighing Method Based on Piezoelectric Sensor

Wei ZHANG, Chun-li LI, Xiao-feng DI, Mi CHEN, Sheng TAO

1. China Academy of Transportation Sciences, Beijing 100029, Chin)

Abstract: In order to effectively measure the dynamic axle load of vehicles in motion, the dynamic weighing method of vehicles based on piezoelectric sensor was studied. Firstly, the influencing factors of the measurement accuracy in the dynamic weighing process were analyzed systematically, and the impacts of road irregularities and dynamic weighing system vibration on measurement error were discussed. On the basis of the analysis, the arithmetic mean filter method was used in the software algorithm to filter out the periodic interference added in the sensor signal, the most suitable n value was selected to get the better filtering result by simulation comparison. Then, the dynamic axle load calculation model of high speed vehicles was studied deeply, based on the theoretical response curve of the sensor, the dynamic axle load calculation method based on frequency reconstruction was established according to actual measurement signals of sensors and the analysis from time domain and frequency domain, also the least square method was used to realize the identification of temperature correction coefficient. A large amount of data that covered the usual vehicle weighing range was collected by experiment. The results show that the dynamic weighing signal system identification error all controlled within 10% at the same temperature and 60% of the vehicle data error can be controlled within 7%. The temperature correction coefficient and the correction formula at different temperatures ranges are well adapted to ensure that the vehicle temperature error at different temperatures can also be controlled within 10% and 70% of the vehicle data error within 7%. Furthermore, the weighing results remain stable regardless of the speed of the vehicle which meets the requirements for high-speed dynamic weighing.

Key words: Road engineering; Dynamic weighing; Frequency reconstruction; Piezoelectric-film sensor; Parameter identification

Classification No.: U492.8 **Literature code**: A

1 Introduction

Traffic stream axle load information is essential for road safety and road service life. Statistical data show that 70% of road safety accidents are caused by oversize and overload of vehicles [1]. Axle load information can be used to automatically identify overload vehicles, so as to prevent from damage of bridge and road surface caused by overload vehicles. Additionally, axle load information can be used to realize long-term accumulation of road service performance data. Axle load measurement when automotive is normally driving in road is a global problem. Among the common dynamic weighing detectors, piezoelectric sensor has increasingly become the research direction of automotive dynamic technology due to its advantages such as small size, small pavement incision, low maintenance workload, concealing incision, etc.

When dynamic weighing is conducted, the precision of dynamic weighed results can be affected seriously by such randomly uncertain factors as very short action time (within dozens of ms) of tyre on piezoelectric sensor, the force of axle weight on piezoelectric sensor, as well as disturbing force caused by many factors including vehicle speed, vehicle vibration, road disturbance, tyre force, etc. [2-3].

In this article, we firstly systematically analyzed factors affecting measuring accuracy in the course of automotive dynamic weighing, and deeply studied the calculation model of high-speed vehicle dynamic axle load. Additionally, analysis was conducted in the aspects of time domain and frequency domain based on the theoretical response curve and actual measurement

This work was supported by the Scientific Research Project of China Academy of Transportation Sciences (20161203).
* Corresponding author: dixiaofeng@catsti.com

signal of sensor, so as to establish calculation method based on frequency reconstruction, and to realize identification of system parameters by a multitude of actual measurement data and by means of least square method.

2 Analysis of factors affecting the measuring accuracy in the course of automotive dynamic weighing

Vehicle dynamic weighing refers to automotive weighing when automobile keeps the current motion status. When dynamic weighing is conducted, vehicle which moves through weighing sensor at a certain speed not only can be weighed truly due to the complexity of automotive motion, but also the true weight of vehicle can be easily to be covered by various interference caused by complexity of automotive motion. Automobile is the main body of dynamic vehicle weighing system. It has quite complex motion; especially the motion in the course of dynamic weighing is more complex. The main reasons that caused complex motion are as follows:

(1) Automobile is a complex multivariant power system. The motion status and power characteristics of automobile in the course of dynamic weighing can be objectively and comprehensively described through at least more than 20 degrees of freedom. Therefore, simplified model is generally used in actual analysis.
(2) The reasons of automobile vibration is quite complex. The vibration is the primary motion form affecting weighing in the course of dynamic weighing. While the reason resulting in automobile vibration is quite complex, for example, automobile motor, tyre elasticity, uneven road surface, etc. are all the reasons which result in automobile vibration.
(3) The forms of vibration load are diverse. The complexity of automobile vibration also causes diversity of ground load from automobile. Ground load from automobile mainly covers three kinds of forms including stable load, random load and impact load. The generation of all of these loads comes with great uncertainties.

In a word, the measurement error occurred in automobile dynamic measurement are mainly associated with all kinds of vibrations.

2.1 Measurement error of automobile forced vibration resulting from unevenness of road surface

The unevenness of road surface is the main factor resulting in automobile vibration. Therefore, firstly the vehicle random vibration caused by the unevenness of road surface shall be analyzed to determine the influence of the unevenness of road surface on measurement error. Given the condition that the connection form between cargo and vehicle body is not considered, the automobile vibration model can be simplified as shown in figure 1:

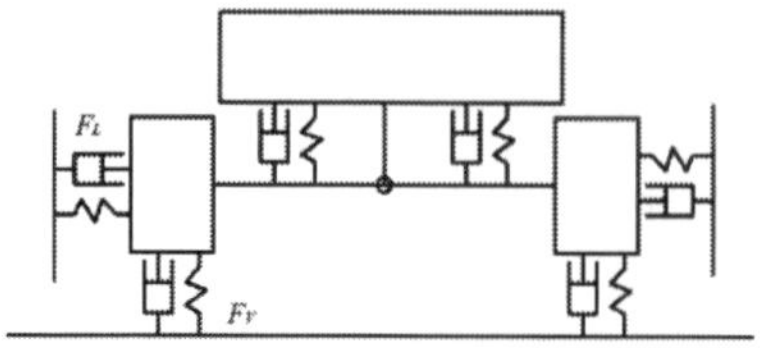

Fig .1 Simplified Model of Automobile Vibration

Where FV refers to the vertical force of tyre, FL refers to the lateral force of tyre. Where, the formula of the vertical force of tyre, is as follows:

$$F_V = \begin{cases} K_V \cdot \delta^e + C_V \cdot \overset{o}{\delta}, \delta > 0, \\ 0, \delta <= 0 \end{cases} \quad (1)$$

Where, K_V, C_V and e refer to constant, δ refers to vertical deformation of tyre. Generally, e is more than 1, that is tyre is featured by characteristics of hard spring. Additionally:

$$K_L = \min(K_L \cdot \lambda + C_L \cdot \overset{o}{\lambda}, \varphi \cdot F_V) \quad (2)$$

Where, K_L, C_L refers to lateral rigidity and damping of tyre respectively, φ refers to lateral friction coefficient of tyre, λ refers to lateral deformation of tyre. When the lateral force resulted by lateral deformation of tyre is more than lateral adhesive force of tyre, sideslip of tyre occurs, and the lateral force of tyre is equal to the lateral friction force.

The damping of vibration damper is non-linear. Generally, stretch stroke damping is more than compression stroke damping, namely:

$$F_D = \begin{cases} C_{D1} \cdot \overset{o}{v}, \overset{o}{v} \geq 0 \\ C_{D2} \cdot \overset{o}{v}, \overset{o}{v} < 0 \end{cases} \quad (3)$$

Where, C_{D1}, C_{D2} refer to stretching damping and compression damping respectively, and $C_{D1} > C_{D2}$; $\overset{o}{v}$ refers to the relative speed of vibration damper.

From the above two formulas, it can be concluded that the unevenness of road surface will result in relatively serious forced vibration of vehicle, so that error will be conveyed to dynamic weighing system.

2.2 Measurement error resulting from vibration of dynamic weighing system

The measurement error mainly refers to error caused by elastic deformation of dynamic weighing system. The system model is always necessarily simplified as required in actual analysis [14], as is shown in figure 2:

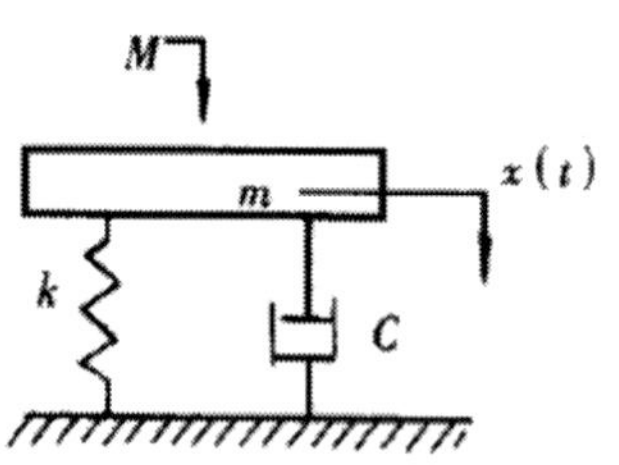

Fig .2 Simplified Model of Dynamic Weighing System

In the figure, k, C refer to rigidity and damping coefficient of weighing system, M refers to vehicle mass, m refers to mass of weighing table, x refers to displacement of weighing system. Then, the kinetic model of dynamic weighing system can be represented as follows:

$$(M+m)\cdot\ddot{x}+C\cdot\dot{x}+k\cdot x=Mg\cdot u(t) \tag{4}$$

Where, $u(t)=\begin{cases}0, t<0\\1, t\geq 0\end{cases}$, the load of weighing sensor can be obtained as follows:

$$F(t)=Mg\cdot\left[1-\frac{e^{-\varsigma\omega_n t}}{\sqrt{1-\varsigma^2}}\right]\sin(\omega_d+t+\varphi) \tag{5}$$

Where, undamped natural frequency $\omega_n=\sqrt{\frac{k}{M+m}}$, damping coefficient $\varsigma=\frac{c}{2\sqrt{(M+m)\cdot k}}$, damped oscillation frequency $\omega_d=\omega_n\sqrt{1-\varsigma^2}$, phase angle $\phi=\arctan\frac{\sqrt{1-\varsigma^2}}{\varsigma}$.

Consequently, the difference between the sensor output resulted from vibration and true value is as follows:

$$e=\frac{Mg\cdot e^{-\varsigma\omega_n t}}{\sqrt{1-\varsigma^2}}\cdot\sin(\omega_d t+\varphi) \tag{6}$$

In accordance with mode of integral or average of repeated sampling generally adopted in dynamic weighing, the absolute error can be represented as follows:

$$e=\frac{1}{\tau}\int_0^\tau e dt \tag{7}$$

Where, $\tau=l/v$ refers to sampling time, l refers to length of sensor, v refers to vehicle speed. If the worst phase is considered, namely $\phi=90^\circ$, the relative error will be as follows:

$$\delta=\frac{e}{Mg}=\frac{ve^{-\varsigma\omega_n t}}{l\omega_d\sqrt{1-\varsigma^2}}\sin(\omega_d\cdot l/v) \tag{8}$$

From the above formula, it can be concluded as follows:

(1) With the increase of speed, dynamically affected relative error gradually increases.

(2) Installation site of piezoelectric sensor and rigidity difference asphalt pavement materials result in not large error, but the evenness of road surface greatly affects dynamic relative error.

(3) As for mechanical vibration error of weighing system, the relative error is associated with vehicle weight. The larger the vehicle weight is, the larger the relative error is.

3 Research on calculation model of high-speed vehicle dynamic axle load

For calculation of vehicle dynamic axle load, axle load signal shall be required to be analyzed, interference factors shall be removed, so as to conduct mechanism modeling, and identify model parameters. The vehicle weight will be obtained, and the results will be analyzed and verified.

3.1 System model

When piezoelectric material is forced by external force in some direction, deformation occurs, which results in the relative displacement of positive and negative charge center inside the material to generate polarization, so as to result in positive and negative bound charge occurring on the relative surface of the material. When the external force is small, piezoelectric material restores to the uncharged status, which is called piezoelectricity or direct piezoelectric effect. The sensor manufactured according to piezoelectricity is called piezoelectric sensor. As vehicle moves through piezoelectric sensor, sensor generates charge signal due to tyre impact. The charge signal is converted into voltage signal after it is amplified, and the amplitude variation of voltage signal is directly proportional to the pressure applied by tyre on the sensor, the circle of signal is correlated to time of tyre staying on the sensor.

The width of signal waveform is correlated to vehicle speed, the faster the speed is, the shorter the time of tyre staying on the sensor is, and the narrower the signal waveform (as is shown in figure 5), the smaller the area between signal waveform and transverse axle. But the product of the area and the corresponding speed does not change, that is, the value is a constant directly proportional to vehicle load.

$$W=A\cdot v\cdot C \tag{9}$$

Where, A refers to area of hook face, v refers to vehicle speed, C refers to adjustment coefficient. They can be determined by test that the vehicle whose weight is known moves through sensor. The vehicle weight can be obtained through multiplying the three.

Furthermore, it can be seen that speed measurement error is directly proportional to axle load measurement error. The more stable the vehicle speed in driving is, the more accurate the measured axle load is.

 DOI: 10.1051/matecconf/201713900203

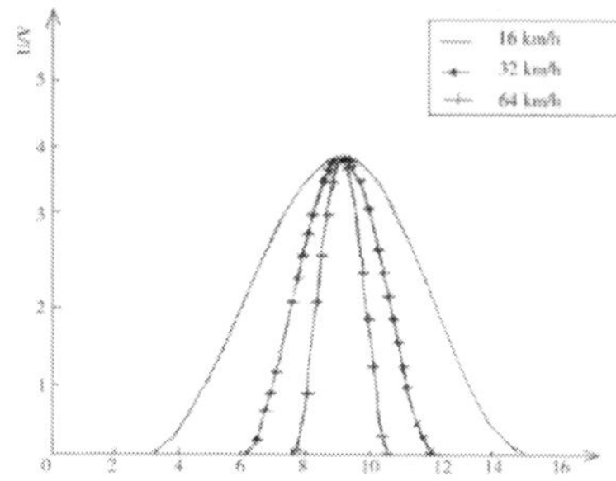

Fig .3 Calculating Schematic Diagram of Charge Dynamic Weighing System

Pretreatment of filtering wave for signal has been conducted in the previous research to basically eliminate the influence of periodic interference. Then the random error interference caused by such factors as filtering vehicle vibration, filtering unevenness of road surface, etc. will be researched.

3.2 Time domain analysis

The actual action of vehicle weight on sensor is the effect of rectangular pulse overlapping triangle pulse, as is shown in figure 7:

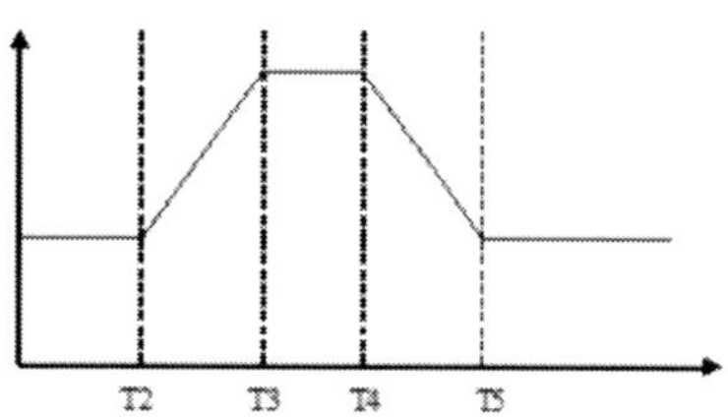

Fig .4 Effect Diagram of Vehicle Weight

Where, the process from T2 to T3 represents the whole width of sensor which is gradually covered by tyre, the time difference is W/v, where W refers to sensor width, v refers to vehicle speed. The process from T3 to T4 represents the process from tyre totally covering sensor to tyre which is about to start to leave sensor, the time difference is $(L-W)/v$, where L refers to the touchdown width of tyre. The process from T4 to T5 is opposite to the process from T2 to T3. The actually selected sensor width in the project is 6.6mm, while generally the action area of tyre is 60~100mm. Therefore, theoretically, it is analyzed that the time width of table period shall be about 5 times as long as the tilt period

However, actually the measured output signal of piezoelectric sensor is shown in figure 8. Where blue curve refers to signal when vehicle passes by, red curve refers to signal curve when none of vehicle passes by. It can be seen that the table period theoretically anticipated does not occur in the middle part of blue curve, and the base electrical level of signal curve decreases before and after tyre moves through sensor. The reason is that the natural frequency of piezoelectric sensor and the measurement system composed of the successive amplifying circuit are not infinite, it can be regarded as a band-pass filter. Therefore, the vehicle weight action signal is expanded by measuring system in time domain. While decrease of base electrical level of curve refers to charge signal generated from tangential deformation of sensor resulted from tyre's friction on ground.

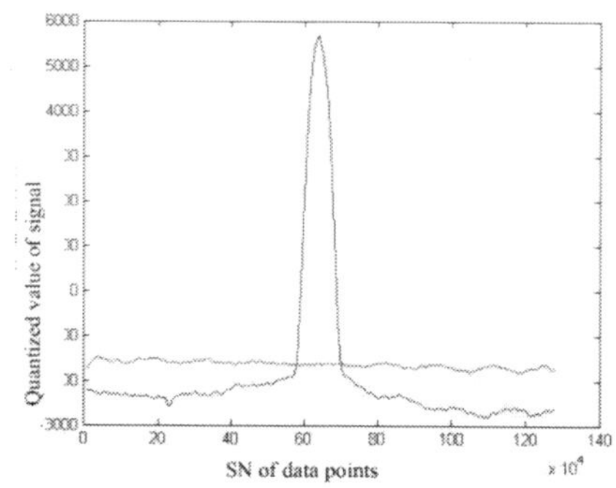

Fig .5 Signal of Piezoelectric Sensor

The flat period of signal can not be directly measured, but it can be discovered from careful observation of signal chart that there is a apparent point before signal sharply ascends and after signal sharply declines respectively. Compared driving speed of vehicle with tyre touchdown width, it can be discovered that the distance vehicle moves through in the period is basically in line with theoretical anticipation. Therefore, the interval between the two points includes three stages of T2, T3 and T4, while the total weight of vehicle shall be the sum of acting force of the three stages of T2, T3 and T4.

3.3 Frequency domain analysis

Before eliminating basic offset of frequency domain, frequency spectrum of piezoelectric sensor signal before and after vehicle moves through piezoelectric sensor is firstly analyzed as shown in figure 9. It can be seen that the basic signal is basically thought as direct current quantity, while frequency response of vehicle axle load is also centralized in low-frequency area of less than 300HZ, and direct current component in frequency spectrum of axle load signal is equivalent to that in the basic signal. Therefore, before treating all the frequency spectrum of sensor signal, firstly signal frequency must subtract frequency spectrum of basic signal in frequency domain, so as to filter basic signal.

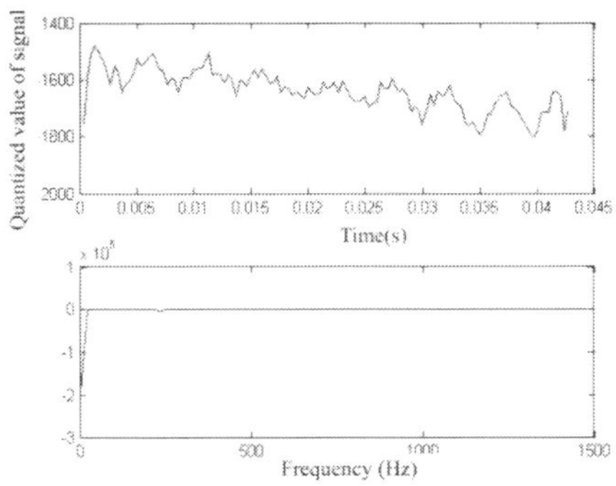

(a) Vehicle does not pass by

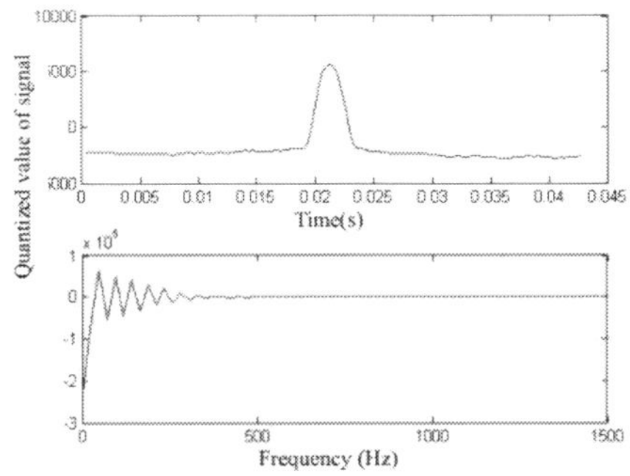

(b When vehicle passing by

Fig .6 Time-Domain and Frequency-Domain Graphs before/after Passing

It can be known from time domain analysis that vehicle weight action can be regarded as overlap of rectangle pulse and triangle wave pulse. The expression form of the two kinds of pulses in frequency domain is shown in figure 10:

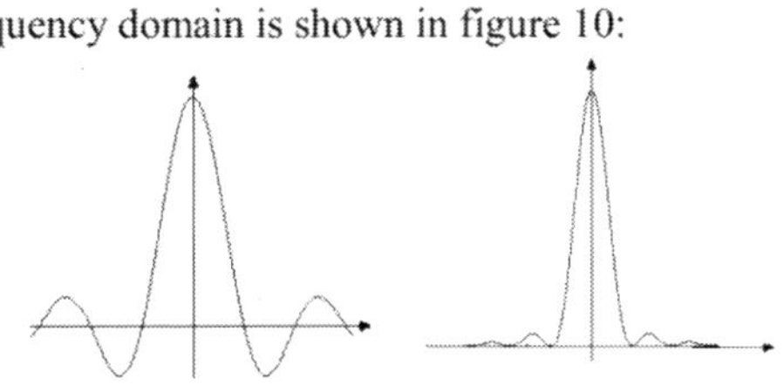

(a) Rectangle pulse (b) Triangle wave pulse

Fig .7 Pulse Spectrum Diagram

Therefore, output signal spectrogram in frequency domain of Piezoelectric Sensor should be similar with Fig.10. A signal spectrum of good original data result can be compared with that of poor original data result, as shown in Fig.11.

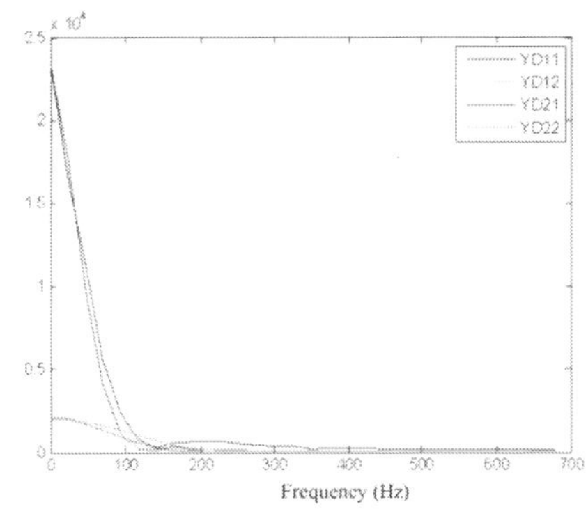

(a) Error -8% signal

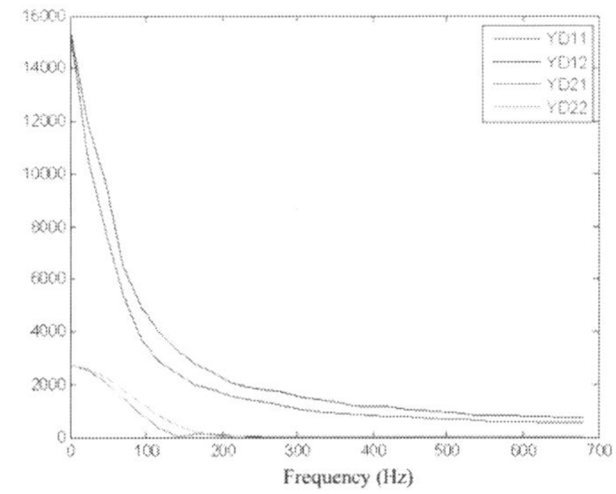

(b) Error 23% signal

Fig .8 Original Signal Spectrum Diagram

It can be seen from Fig.11 that when the real original signal curve spectrum gets relatively close to the theoretical spectrum, the error is relatively small, while when the real original signal curve spectrum gets relatively far from the theoretical spectrum, the error is relatively big. The reason is that: high frequency errors such as vehicle vibration mingle in low frequency signals of vehicle weight, which thus lead to the variation of signal spectrum. The processing method proposed by former scholars is to process the spectrum of variation through wavelet analysis and neural network scheduling algorithm so as to obtain the most real vehicle weight signals.

3.4 Dynamic Axle Load Calculation Algorithm Based on Frequency Reconstruction

It's found through big quantity of data spectrum diagrams that the difference between real signal spectrum and theoretical spectrum can clearly reflect the magnitude of errors. Therefore, if the real spectrum can be directly corrected based on theoretical spectrum, the errors may likely be reduced. The Dynamic Axle Load Calculation Algorithm Based on Frequency Reconstruction is proposed based on this and as shown in Fig.12, the algorithm description is as follows:

DOI: 10.1051/matecconf/201713900203

(1). Elimination of basic offset. It can be seen from figure 11 that there is a apparent basic offset on blue and black curve in figure (b), which is most different from theoretical frequency spectrum. Therefore, basic offset is firstly eliminated. The first zero crossing point of signal frequency spectrum can be estimated from vehicle speed. And all the calculated values subtract the value of the zero-crossing point within frequency domain, so that basic offset returns to zero.

(2). Data after the first zero crossing point are sorted, so that the situation of gradually declining in theory is presented in the sequence. If "warped-tail phenomenon" occurs, basic offset will be eliminated according to the second zero crossing point, that is all the calculated values after the first zero crossing point subtract the value of the zero-crossing point.

(3). Repeat the above steps, until the frequency of the zero-crossing point is more than 300HZ.

(4). Data are converted to time domain to be treated by

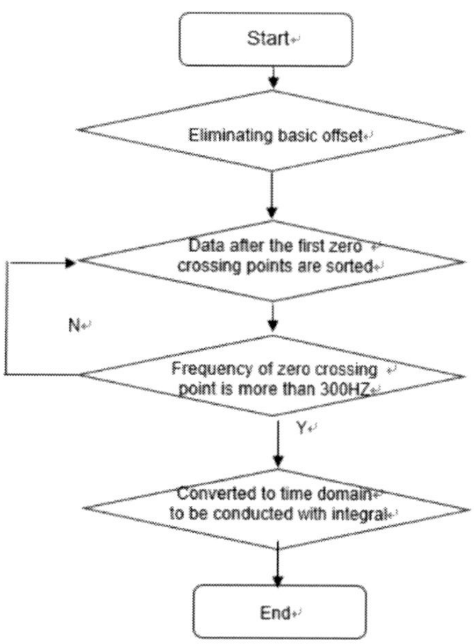

Fig .9 Dynamic Axle Load Calculation Algorithm Based on Frequency Reconstruction

The result of the processed data of Fig.11(b) Original Signal Spectrum Diagram following the above steps is as shown in Fig.13 with the error of -2.38%.

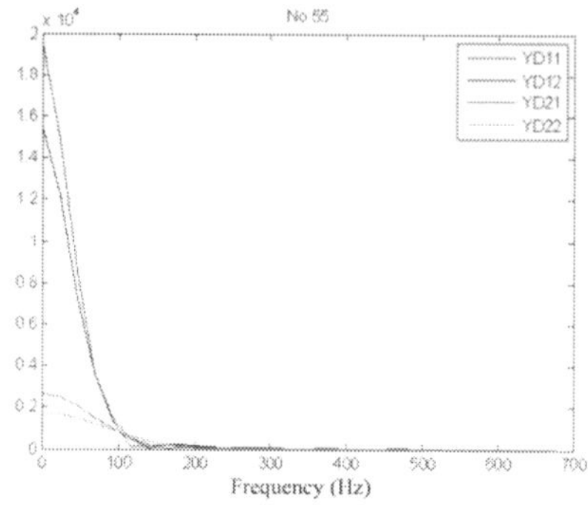

Fig .10 Data after Processing

3.5 System identification

Objective function has been obtained in the aforesaid $W=A\cdot v\cdot C$, therefore the identification issue turns to a least squares issue that solves adjustment parameter C.

Select the least square method for system identification and assume a SISO (single input/output) system, as shown in Fig.14:

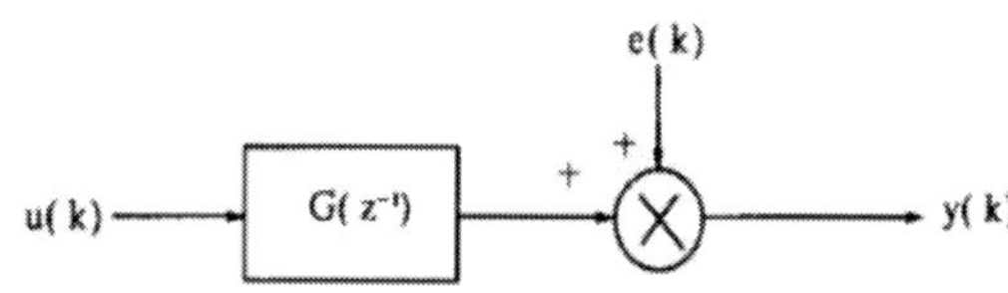

Fig .11 SISO System Structure Diagram

Discrete transfer function is:

$$G(z^{-1})=\frac{B(z^{-1})}{A(z^{-1})}=\frac{b_1z^{-1}+b_2z^{-2}+\ldots+b_nz^{-n}}{1+a_1z^{-1}+a_2z^{-2}+\ldots+a_nz^{-n}} \quad (10)$$

Relation between input and output is:

$$u(k)\cdot G(z^{-1})+e(k)=y(k) \quad (11)$$

It can be further obtained:

$$y(k)\cdot A(z^{-1})=u(k)\cdot B(z^{-1})+e(k) \quad (12)$$

In the formula, disturbance quantity $e(k)$ is an average value, the white noise unrelated with 0.

The above formula is converted into the form of difference equations:

$$\begin{aligned}y(k)&=-a_1y(k-1)-a_2y(k-2)-\ldots-a_ny(k-n)\\&+b_1u(k-1)+b_2u(k-2)+\ldots b_nu(k-n)+e(k)\end{aligned} \quad (13)$$

Assuming:

$$\begin{cases}\phi(k)=[-y(k-1)\ \ldots\ -y(k-n)\ \ u(k-1)\ \ldots\ u(k-n)]^T\\ \theta=[a_1\ \ a_2\ \ldots\ a_n\ \ b_1\ \ b_2\ \ldots\ b_n]\end{cases}$$

The formula can be:

$$y(k) = \phi^T(k)\theta + e(k) \quad (14)$$

Extending to N inputs and outputs observed value $\{u(k), y(k)\}(k=1,2...N)$. Substituting into the formula to write in matrix form as:

$$Y = \Phi\theta + e \quad (15)$$

Where,
$$\begin{cases} Y = [y(n+1) \quad y(n+2) \quad \dots \quad y(n+N)]^T \\ e = [e(n+1) \quad e(n+2) \quad \dots \quad e(n+N)]^T \end{cases}$$

$$\Phi = \begin{bmatrix} -y(n) & \dots & -y(1) & u(n) & \dots & u(1) \\ -y(n-1) & \dots & -y(2) & u(n+1) & \dots & u(2) \\ \vdots & \dots & \vdots & \vdots & \dots & \vdots \\ -y(n+N) & \dots & -y(N) & u(n+N) & \dots & u(N) \end{bmatrix}$$

Functional $J(\theta)$ is:

$$J(\theta) = \sum_{i=1}^{N}(Y - \Phi\theta)^2 = \sum e^2(n+i) = e^T \cdot e = (Y - \Phi\theta)^T(Y - \Phi\theta) \quad (16)$$

The principle of the least square method is to minimize $J(\theta)$ to evaluate extreme value and obtain:

$$\frac{\partial J}{\partial \theta} = \frac{\partial}{\partial \theta}[(Y - \Phi\theta)^T(Y - \Phi\theta)] = 0 \quad (17)$$

The least squares value of system can be obtained in such way.

$$\theta = (\Phi^T\Phi)^T\Phi^T Y \quad ()$$

The least squares value of system parameter is obtained in such way.

4 Analysis on system identification result

Parameter identification on dynamic weighing data of vehicles with different weights under the state of natural vehicle stream was carried out based on a large number of experimental data. The result is as shown in Tab.1. Full data is not given due to the big data volume. Weights of vehicles cover a range from 2 ton to 55 ton. True value, measured value and error results are as shown in Fig.15.

Tab .1 System Identification Results at the Same Temperature

Plate No.	Test time	Tem pera ture	(kg) Measu red value	True value (kg)	Error (%)
Yu G738xx	2013/6/22 13:59	23	55531	52160	6.46
YuA1N8xx	2013/6/22 16:16	23	2756	2740	0.58
YuG288xx	2013/6/22 19:31	23	28451	27060	5.14
YuJ768xx	2013/12/10 7:55	-4	51470	51980	0.98
Yu F121xx	2013/12/10 3:56	-3	56760	57560	1.39
Yu J629xx	2013/6/28 12:41	44	59372	54020	9.91
Yu G208xx	2013/6/28 13:04	45	16296	17560	7.19
...	...	...	...	...	...

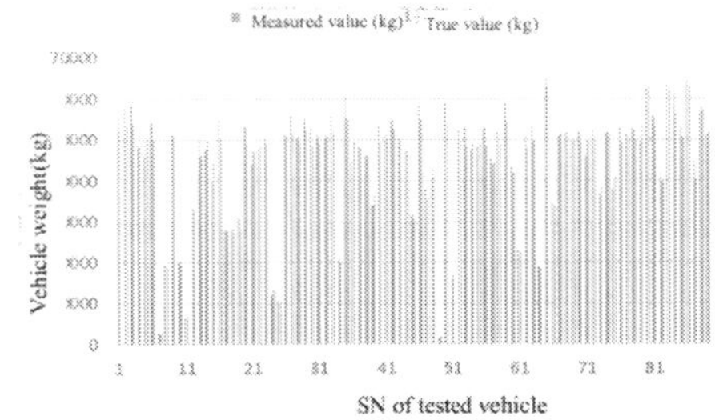

(a) Measured value and true value

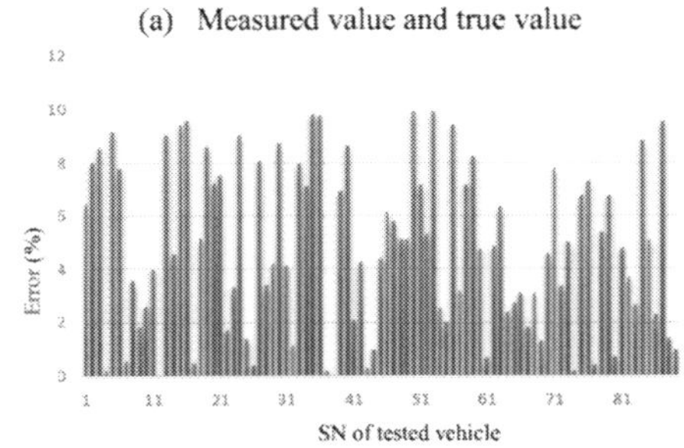

(b) Absolute measurement error

Fig .12 Dynamic Weighing Results at the Same Temperature

Results show that all of the errors of the vehicle data under the state of natural vehicle stream after system identification are controlled within 10% and 70% of vehicle data errors are within 7%. No matter how fast or slow the vehicle passes, the weighting results remain stable regardless of the speed of the vehicle which meets the requirements for high-speed dynamic weighing.

5 Conclusion

The automotive dynamic weighing technology based on piezoelectric sensor was researched in this thesis. Analysis was performed in the aspects of time domain and frequency domain based on theoretical response curve and actual measured signal of sensor to establish calculation method on the basis of frequency reconstruction. The identification of correction coefficient was finally realized by means of least square method through a multitude of actual measured data. The results of actual collected data showed that frequency reconstruction algorithm and system parameter

DOI: 10.1051/matecconf/201713900203

identification propose in this article can meet the requirement of high-speed dynamic weighing.

References:

[1]. HE Shu-xin. History, Current Situation and Prospect of Vehicle Dynamic Weighing Technology[J]. Journal of China & Foreign Highway, 2004,24(**6**):104- 108.

[2]. CEBON D, COORDINATOR B J. Weigh-in-motion of axles and vehicles for Europe (WAVE) [R]. General Report, 2001.

[3]. FAN Li-hui. Vehicle Dynamic Weighing Technology[J]. Middle South Automotive Transportation, 1998(2): 5~7.

[4]. FENG Xiu-li, LI Xiao-lin, YANG Hong-kou. Influence of the Road Roughness and the Speed of Car on Dynamic Car Scale Weighing[J]. Sciencepaper Online, 2012, (11):857-861.

[5]. SHI Chang-yan. State of Arts and Trends for the Dynamic Weighing and Force Measuring Technology[J]. Acta Metrologi- ca Sinica, 2001, 22 (3): 201~205.

[6]. LI Jia-lin, SU Xiao-lou. Research on Dynamic Electronic Track Scale without Foundation[J]. Railway Quality Control, 1990(5):35-38.

[7]. [7] Ono, T.etal., A dynamical Measurement Device of Mass Utilizing a Micro-computer, System and control, 1980, 24(3).

[8]. [8] Ono, T. and shimaoka.H. On the Dynamical Mass-estimation Algorithm of Displacement and velocity Sensing Type. Transactions of the society of Instrument and control Engineers, 1983,19(4).

[9]. GUO Lan-ying, LIANG Bo, DONG An-guo. New Technology for Weigh-in-motion of Vehicles[J]. Journal of Chang'an University (Natural Science Edition), 2009, 29(2):98- 100.

[10]. WANG Li-li. High Degree of Accuracy Research on Dynamic Car-weighting System on Freeway[D]. Liaoning University of Engineering and Technology, 2008.

[11]. TAN Shuang, LI Li-hong. Vehicle Dynamic Weighing Data Processing Based on Grey Neural Network[J]. Chinese Journal of Sensors and Actuators, 2016, 29(8):1205-1209.

[12]. ZHOU Zhi-feng, CAI Ping, CHEN Ri- xing, et al. A New Method of Processing Data of Weigh-in-Motion of Vehicles Based on Nonlinear Curve-Fitting[J]. Journal of Shanghai Jiaotong University, 2006, 40(5):709-712.

[13]. PAN Ruo-yu LI Lei. Study of Algorithm for Dynamic Weigh System Based on Singular Spectrum Analysis[J]. Journal of Highway and Transportation Research and Development, 2010, 27(12):128-133.

[14]. YU Zhi-sheng. Automobile Theory[M]. China Machine Press, 1995(2).

[15]. LIU Jiu-qing. The design & Calculation of Disk Load cell[J]. Weighing Apparatus, 2002,31(6):10-13.

Strongly (k,d)-Graphical Labellings For Designing Graphical Passwords In Communication

Hui Sun[1], *Xiaohui* Zhang[1], and *Bing* Yao[1,2,⋆]

[1]*College of Mathematics and Statistics, Northwest of Normal University, Lanzhou 730070, CHINA*
[2]*School of Electronic and Information Engineering, Lanzhou Jiaotong University, Lanzhou 730070, CHINA*

Abstract. The safety of the traditional text passwords is questioned by many users in the present information research. Graphical passwords become new research object of many researchers, who hope they can replace the existing text passwords and improve user account security. We are studying Topsnut-graphical passwords, which can be traced back to an idea of "topological structure plus number theory" proposed first by Hongyu Wang with her colleagues, and we found the equivalence between with Topsnut-graphics passwords of the trees having perfect matching. Through designing new parametric graphical labellings, we hope to enrich the variety of Topsnut-graphical passwords, and study the relationship between new parametric graphic labellings and existing labellings on trees having perfect matching. In addition, we also found, by two kinds of special operations, trees having perfect matching can be converted into paths perfect matching, and the nature of the original passwords remains unchanged. Our results can be provided to two or more banking users, and our method can easily be converted into polynomial time algorithms.

1 Introduction

Password have become an important part of human daily life, whether it is shopping or chatting and playing games, people use passwords. Your password can not only ensure the safety of personal privacy, but also provides a sense of security for the user.However, attacks to passwords appear in our daily life, and more by more. The safety of the traditional text passwords is questioned by many users in the present information research. Graphical passwords become new research object of many researchers, who hope they can replace the existing text passwords and improve user account security, because of graphical passwords are easy for users and difficult to be broken by attackers [1]. We call new graphical passwords as *Topsnut-graphical passwords* , denoted by Topsnut-GPWs, for distinguishing with other graphical passwords, since they are made by an idea of "topological structure plus number theory" (Topsnut) proposed firstly by Wang *et al.* in [2] and [3]. Studies have shown that humans remember pictures more than texts, and psychological research supports the hypothesis [4]. However, one of the biggest drawbacks in the existing graphical passwords in [1] is that they occupy large storage space, slow communication speed and lack of practicability.

We show an example for illustrating Topsnut-GPWs simply. Olaf goes to a bank to establish an account for his deposit. At first, Olaf are provided for making a password based

⋆e-mail: yybb918@163.com

 DOI: 10.1051/matecconf/201713900204

on six given small circles shown in Fig.1(a). At second, he connected by a straight line between small circles and drew a Chinese character with six small circles and seven sides – "day" in Fig.1(b). Next, he label numbers in these circles (see Fig.1(c)). Finally, label seven lines with particular numbers (see Fig.1(d)). We use $f(u)$ to indicate the label of a circle, and $f(uv)$ to be the label of an edge uv joining two circles u, v as its ends. Thus, $|f(u) - f(v)| = f(uv)$ holds true for each line uv. Thereby, we have made the desired password, called a *Topsnut-GPWs* in which the graphical structure is a graph without numbers labelled to the vertices and edges of the graph. If Olaf establishes multiple bank accounts at the same time, it is clear that it is not safe to share one password in multiple accounts, and it is not easy to memorize many different passwords. At this point, he can set up a parametric password to deal with this situation(see Fig.1(e)). When $k = 1, d = 1$, it is Fig.1(d). We will design more complicated parametric graphical passwords in this article.

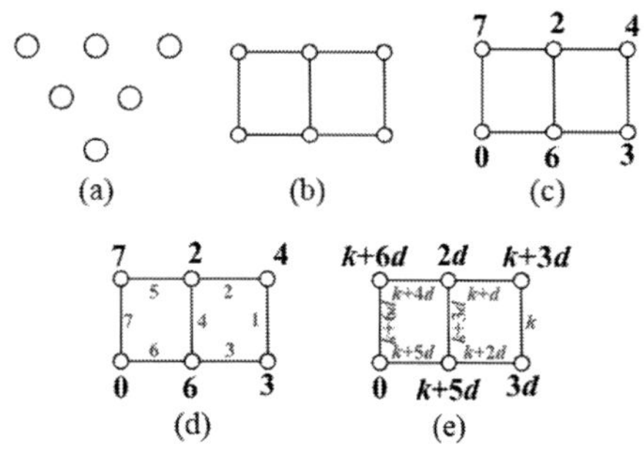

Figure 1. Olaf makes a graphical password.

2 Graphical passwords

2.1 Preliminary

All graphs mentioned here are undirected, finite, and no multiple-edges and loops, and they are called *simple graphs*. We use standard terminology and notation of graph theory, other undefined concepts or definitions can be found in [5] and [6]. Additional notations are: the symbol $[m, n]$ indicates a set $\{m, m+1, \dots, n\}$ with integers m, n respect to $0 \leq m < n$; the $[s, t]^o$ denotes an *odd-integer set* $\{s, s+2, \dots, t\}$ with odd integers s, t maintaining $1 \leq s < t$; and the notation $[a, b]^e$ stands up an *even-integer set* $\{a, a+2, \dots, b\}$ with even integers a, b falling into $2 \leq a < b$. $|X|$ is the *cardinality* of elements of a set X. A (p, q)-graph G has p vertices and q edges. The set of all vertices adjacent to a vertex v in G is denoted as $N(x) = \{v : xv \in E(G)\}$, called the *neighborhood* of the vertex x. And the *degree* $\deg(u)$ of a vertex u is equal to $\deg(u) = |N(u)|$. A vertex of degree one is called a *leaf* here. We restate the definitions of well-known graph labellings as follows:

Definition 1 *[5, 7] Suppose that a connected (p, q)-graph G admits a mapping $\psi : V(G) \rightarrow \{0, 1, 2, \dots\}$. The induced edge label is defined as $\psi(xy) = |\psi(x) - \psi(y)|$ for each edge $xy \in E(G)$. Write $\psi(V(G)) = \{\psi(u) : u \in V(G)\}$, $\psi(E(G)) = \{\psi(xy) : xy \in E(G)\}$. There are the following constraints:*

(1) $|\psi(V(G))| = p$.

(2) $|\psi(E(G))| = q$.

(3) $\psi(V(G)) \subseteq [0, q]$, $\min \psi(V(G)) = 0$.

(4) $\psi(V(G)) \subset [0, 2q-1]$, $\min \psi(V(G)) = 0$.

(5) $\psi(E(G)) = \{\psi(xy) : xy \in E(G)\} = [1, q]$.

(6) $\psi(E(G)) = \{\psi(xy) : xy \in E(G)\} = \{1, 3, 5, \dots, 2q-1\}$.

(7) G is a bipartite graph with the bipartition (X, Y) such that $\max\{\psi(x) : x \in X\} < \min\{\psi(y) : y \in Y\}$ ($\psi(X) < \psi(Y)$ for short).

(8) G is a tree having a perfect matching M such that $\psi(x) + \psi(y) = q$ for each matching edge $xy \in M$.

(9) G is a tree having a perfect matching M such that $\psi(x) + \psi(y) = 2q - 1$ for each matching edge $xy \in M$.

A graceful labelling ψ *holds (1), (3) and (5) true; a* set-ordered graceful labelling ψ *holds (1), (3), (5) and (7) true; a* strongly graceful labelling ψ *satisfies (1), (3), (5) and (8); a* strongly set-ordered graceful labelling ψ *holds (1), (3), (5), (7) and (8) true. An* odd-graceful labelling ψ *holds (1), (4) and (6) true; a* set-ordered odd-graceful labelling ψ *holds (1), (4), (6) and (7) true; a* strongly odd-graceful labelling ψ *satisfies (1), (4), (6) and (9); a* strongly set-ordered *odd-graceful labelling ψ holds (1), (4), (6), (7) and (9) true, simultaneously.*

In [8], the authors show: *A connected bipartite graph H admits a (strongly) set-ordered graceful labelling if and only if H admits a (strongly) set-ordered odd-graceful labelling.* We will use the graph labellings defined in Definition 2 in this article.

Definition 2 *Let G be a (p, q)-graph, and let $S_{k,d} = \{k, k+d, \dots, k+(q-1)d\}$ for integers $k \geq 1$ and $d \geq 1$.*

(1) [9] A (k, d)-graceful labelling *f of G holds $f(V(G)) \subseteq [0, k+(q-1)d]$, $f(x) \neq f(y)$ for distinct $x, y \in V(G)$ and $f(E(G)) = \{|f(u) - f(v)|;\ uv \in E(G)\} = S_{k,d}$. Especially, a $(k, 1)$-graceful labelling is also a k-graceful labelling.*

(2) [5] A felicitous labelling *f of G holds $f(V(G)) \subseteq [0, q]$, $f(x) \neq f(y)$ for distinct $x, y \in V(G)$ and $f(E(G)) = \{f(uv) = f(u) + f(v) \pmod q : uv \in E(G)\} = [0, q-1]$; and furthermore, f is* super *if $f(V(G)) = [0, p-1]$.*

(3) [5] An edge-magic total labelling *f of G holds $f(V(G) \cup E(G)) = [1, p+q]$ such that for any edge $uv \in E(G)$ and $f(u) + f(v) + f(uv) = c$, where the magic constant c is a fixed integer; and furthermore f is* super *if $f(V(G)) = [1, p]$.*

(4) [5] An edge-magic graceful labelling *f of G holds $f(V(G) \cup E(G)) = [1, p+q]$ such that for any edge $uv \in E(G)$, $| f(u) + f(v) - f(uv) = c |$ where the magic constant c is a fixed integer; and furthermore f is* super *if $f(V(G)) = [1, p]$.*

(5) [10] A labeling f of G is said to be (k, d)-arithmetic *if $f(V(G)) \subseteq [0, k+(q-1)d]$, $f(x) \neq f(y)$ for distinct $x, y \in V(G)$ and $\{f(uv) = f(u) + f(v) : uv \in E(G)\} = S_{k,d}$.*

(6) [11] An odd-elegant labelling *f of G holds $f(V(G)) \subset [0, 2q-1]$, $f(u) \neq f(v)$ for distinct $u, v \in V(G)$, and $f(E(G)) = \{f(uv) = f(u) + f(v) \pmod{2q} : uv \in E(G)\} = \{1, 3, 5, \dots, 2q-1\}$.*

By the definition of a (k, d)-graceful labelling, we know that the graph having a (k, d)-graceful labelling must be bipartite, and this property can be obtained by odd-graceful graphs [8]. Guo *et al.* [12] defined a *strongly (k, d)-graceful labelling* as: *Let a (p, q)-graph G has a (k, d)-graceful labelling f and a perfect matching M. If $f(u) + f(v) = k + (q-1)d$ for each edge $uv \in M$, then f is called a strongly (k, d)-graceful labelling of G, and G is called a strongly (k, d)-graceful graph.* When $(k, d) = (1, 1)$, f is just a strongly graceful labelling of G; as $(k, d) = (1, 2)$, f is a strongly odd-graceful labelling of G [5]. We show the examples shown in Fig.2 for understanding Definition 1,2. Three new labellings are defined in Definition 3.

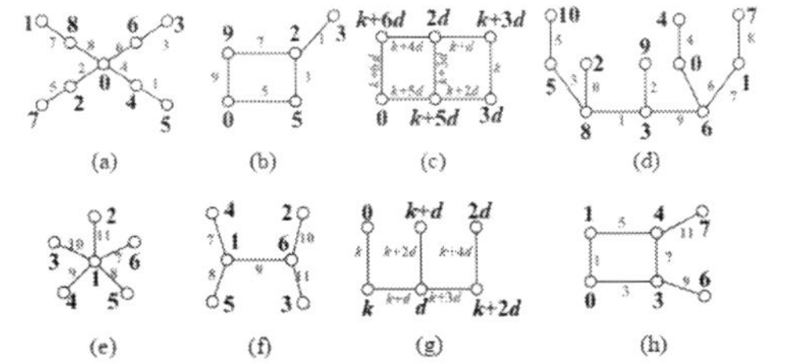

Figure 2. An example for illustrating Definition 1,2 (a) A strongly graceful labelling; (b) a set-ordered odd-graceful labelling; (c) a (k, d)-graceful labelling; (d) a felicitous labelling; (e) a super edge-total graceful labelling; (f) a super edge-magic graceful labelling; (g) a (k, d)-arithmetic labelling; (h) an odd-elegant labelling.

2.2 New graphical passwords

Definition 3 *Let T be a (p, q)-tree having a perfect matching M, and let $S_{k,d} = \{k, k+d, \dots, k+(q-1)d\}$, $S'_{k,d} = \{k, k+2d, \dots, k+2(q-1)d\}$, $S''_{k,d} = \{k, k+d, \dots, k+(p+q-1)d\}$ for integers $k \geq 1$ and $d \geq 1$. Suppose that each leaf of T is an end of some matching edge of M, so $M(L)$ is the set of all leaves of T, and furthermore the graph $T - M(L)$ obtained by deleting all leaves of T is a tree having $c = |V(T - M(L))|$ vertices. AM is the abbreviation of arithmetic matching.*

(1) An AM-(k, d)-totally odd/even graceful labelling *f of G holds $f(V(G)) \subseteq [0, k+2(q-1)d]$, $f(x) \neq f(y)$ for distinct $x, y \in V(G)$ and $f(E(G)) = \{|f(u) - f(v)| : uv \in E(G)\} = S'_{k,d}$. Meanwhile, $f(u) + f(v) = k + 2(q-1)d$ for each edge $uv \in M$.*

(2) An AM-(k, d)-felicitous labelling *f of G holds $f(V(G)) \subseteq [0, k+(q-1)d]$, $f(x) \neq f(y)$ for distinct $x, y \in V(G)$ and $f(E(G)) = \{f(uv) = f(u) + f(v) \pmod{qd} : uv \in E(G)\} = S_{k,d}$; and*

furthermore, f is super *if* $f(V(G)) = [0, (\frac{p}{2} - 1)d] \cup [k + \frac{p}{2}d, k + (p-1)d]$. *Meanwhile,* $|f(u) - f(v)| = k + cd$ *for each edge* $uv \in M$.

(3) An AM-(k,d)-edge-magic total labelling *f of G holds* $f(V(G) \cup E(G)) \subseteq [0, k+(p+q-1)d]$ *such that for any edge* $uv \in E(G)$ *and* $f(u) + f(v) + f(uv) = a$, *where the magic constant a is a fixed integer; and furthermore, f is* super *if* $f(V(G)) = [1, \frac{p}{2}d] \cup [k + \frac{p}{2}d, k+qd]$. *Meanwhile,* $|f(u) - f(v)| = k + (c-1)d$ *for each edge* $uv \in M$ *and the elements of* $f(E(T))$ *yield an arithmetic progression having* $|M|$ *terms and the first term* $k + 2(p-1)d$ *and the common difference* $-2d$.

(4) An AM-(k,d)-edge-magic graceful labelling *f of G holds* $f(V(G) \cup E(G)) \subseteq [0, k + (p+q-1)d]$ *such that for any edge* $uv \in E(G)$, $|f(u) + f(v) - f(uv)| = a$ *where the magic constant a is a fixed integer; and furthermore, f is* super *if* $f(V(G)) = [1, \frac{p}{2}d] \cup [k + \frac{p}{2}d, k + qd]$. *Meanwhile,* $|f(u) - f(v)| = k + (c-1)d$ *for each edge* $uv \in M$ *and the elements of* $f(E(T))$ *yield an arithmetic progression having* $|M|$ *terms and the first term* $k + pd$ *and the common difference* $2d$.

(5) Suppose that T has a (k,d)-arithmetic f. If $|f(u) - f(v)| = k$ *for each edge* $uv \in M$, *and the elements of* $f(E(T))$ *produce an arithmetic progression having* $|M|$ *terms and the first term k and the common difference $2d$. Then we call f an* AM-(k,d)-arithmetic labelling *of T.*

(6) An AM-(k,d)-totally odd/even-elegant labelling *f of G holds* $f(V(G)) \subset [0, k + 2(q-1)d]$, $f(u) \neq f(v)$ *for distinct* $u, v \in V(G)$, *and* $f(E(G)) = \{f(uv) = f(u) + f(v) (\bmod\ k + 2qd) : uv \in E(G)\} = S'_{k,d}$. *Meanwhile,* $|f(u) - f(v)| = k$ *for each edge* $uv \in M$ *and the elements of* $f(E(T))$ *yield an arithmetic progression having* $|M|$ *terms and the first term k and the common difference $4d$.*

Remark. The parity of AM-(k,d)-totally odd/even graceful labelling is related to the value of k, if $k = odd/even$, then it is AM-(k,d)-totally odd/even graceful labelling.

Eight examples shown in Fig.3 are for understanding Definition 1,2 and 3.

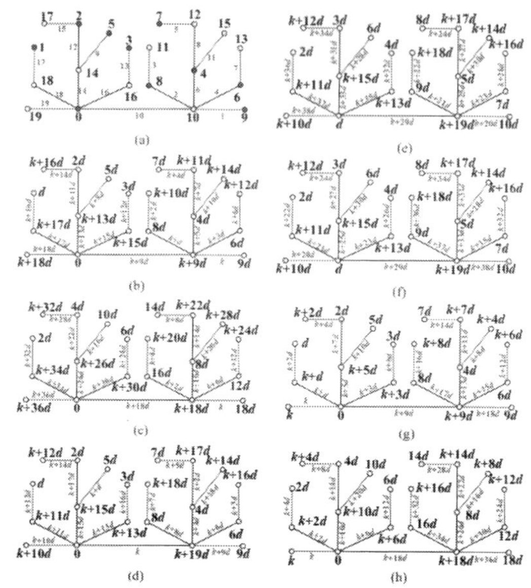

Figure 3. Eight examples for illustrating Definition 1,2 and 3. (a) A strongly set-ordered graceful labelling; (b) a strongly (k,d)-graceful labelling; (c) an AM-(k,d)-totally odd/even graceful labelling; (d) an AM-(k,d)-super felicitous labelling; (e) an AM-(k,d)-super edge-magic total labelling; (f) an AM-(k,d)-super edge-magic graceful labelling; (g) an AM-(k,d)-arithmetic labelling (h) an AM-(k,d)-totally odd/even-elegant labelling.

3 Main results on trees having perfect matching

3.1 Connections of strongly (k,d)-graphical labellings on trees having perfect matching

Lemma 1 *Let T be a tree having n (≥ 4) vertices and a perfect matching M, and let (X, Y) be its bipartition. Let* $S_{k,d} = \{k, k+d, \ldots, k + (n-2)d\}$ *for integers $k \geq 1$ and $d \geq 1$. If each edge of the perfect matching M has an end of degree one, then we have the following pairwise equivalent assertions:*

(1) T admits a strongly set-ordered graceful labelling;

(2) T admits a strongly (k,d)-graceful labelling;

(3) T admits an AM-(k,d)-totally odd/even graceful labelling;

 DOI: 10.1051/matecconf/201713900204

(4) T *admits an AM-(k,d)-super felicitous labelling;*

(5) T *admits an AM-(k,d)-super edge-magic total labelling;*

(6) T *admits an AM-(k,d)-super edge-magic graceful labelling;*

(7) T *admits an AM-(k,d)-arithmetic labelling;*

(8) T *admits an AM-(k,d)-totally odd/even-elegant labelling.*

Lemma 2 *Suppose that a tree T containing a perfect matching $M(T)$ admits a strongly graceful labelling f. Then*

(i) *(Edge-exchanging operation) T admits another strongly graceful labelling g defined as $g(u) = f(v)$ and $g(v) = f(u)$ for each edge $uv \in M(T)$.*

(ii) *(Leaf-decreasing operation) Suppose that an edge $uv \in M(T)$ holds $\deg_T(v) = 1$ and $N(u) = \{v, u_i : i \in [1, d_u - 1]\}$, where $d_u = \deg_T(u) \geq 3$. The tree H obtained by deleting edges uu_j and joining v with u_j for $j \in [s+1, d_u - 1]$, where $0 \leq s < d_u - 1$, is strongly graceful too, and H has a perfect matching $M(H)$ such that $|L(T) \cap V(M(T))| - 1 = |L(H) \cap V(M(H))|$.*

(iii) *(Leaf-increasing operation) Suppose that an edge $uv \in M(T)$ holds $\deg_T(u) \geq 2$ and $\deg_T(v) \geq 2$. We have a strongly graceful tree G obtained by deleting the edge uv from T, identifying u with v into one vertex, denoted as u_0, and joining a new vertex v_0 with u_0, and furthermore G has a perfect matching $M(G) = \{u_0v_0\} \cup (M(T) \setminus \{uv\})$ such that $|L(T) \cap V(M(T))| + 1 = |L(G) \cap V(M(G))|$.*

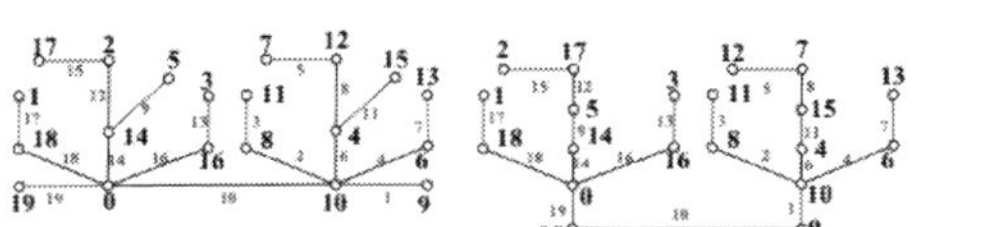

Figure 4. An example for understanding the results of Lemma 2. The strongly graceful tree (a) has ten leaves greater than that of the strongly graceful tree (b) obtained by Lemma 2.

Fig.4 is for understanding Lemma 2. By Lemmas 1 and 2, we get the main result of this article as follows:

Theorem 1 *The eight assertions in Lemma 1 on a tree having a perfect matching are equivalent to each other.*

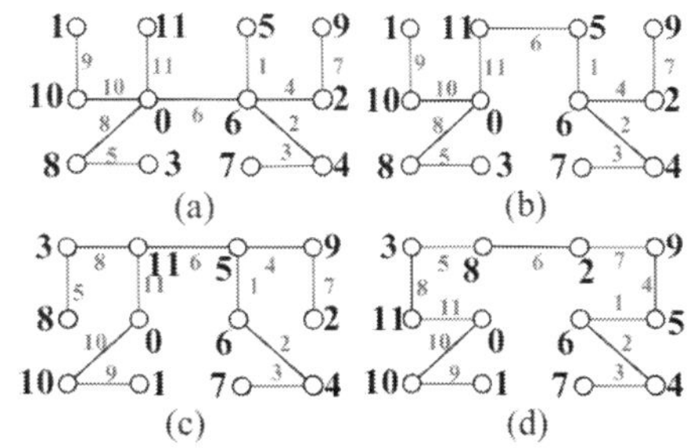

Figure 5. From (a) to (d) is the process trees having perfect matching can be converted into paths perfect matching in turn.

3.2 Trees having perfect matching can be converted into paths perfect matching

By studying the three operations of lemma 2, we define the following operation:

Definition 4 *Suppose that a (p,q)-graph G having perfect matching M have a strongly labelling f, if $f(xy) = f(uv), uv \in M, xy \notin E(G)$, then delete the edge uv and add the edge xy, this operation is called* mismatched edge transfer operation.

By mismatched edge transfer operation and operations of lemma 2, trees having perfect matching will be converted into paths having perfect matchings.

4 Conclusion

We have introduced six new graph labellings for trees having perfect matchings discovered by our working on researching graphical passwords, and show the equivalencies of eight graph labellings of trees having perfect matchings in lemma 1 and theorem 1. Moreover, by mismatched edge transfer operation and operations of lemma 2, trees having perfect matching will be converted into paths perfect matching, and the nature of the original passwords remains unchanged. Clearly, our results can help designers to produce Topsnut-GPWs quickly. On the other hands, our methods in the proofs of our rusults can be easily transformed into feasible and effective algorithms having polynomial complex. Thereby, our methods on dealing with Topsnut-GPWs can yield Topsnut-GPWs having high security level by polynomial algorithms. Moreover, by our experience working on graph labellings, we propose the following conjecture for further studying Topsnut-GPWs: *Trees with perfect matchings admit each one of six new labellings defined in Definition 3.*

Exploring the topological structure of some kind of graphs, finding its more intrinsic properties, and providing theoretical help for new graphical cryptography are indispensable. [13] and [14] explore the Equivalent Definitions Of Cactus Graphs and Euler graphs respectively. [15] and [16] explore odd-graceful labeling and odd-elegant labeling on ring computer networks.

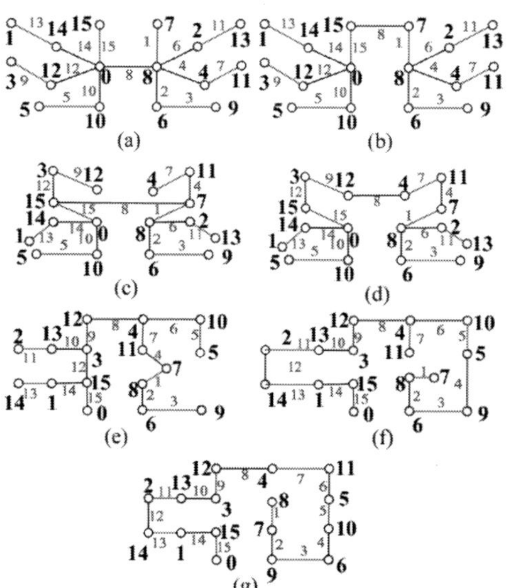

Figure 6. From (a) to (g) is the process trees having perfect matching can be converted into paths perfect matching in turn.

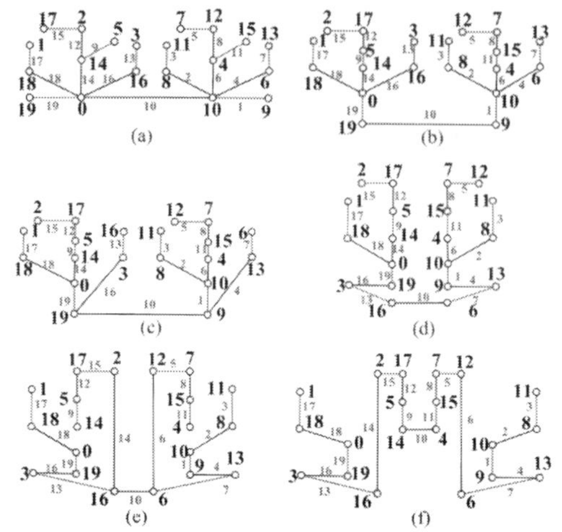

Figure 7. From (a) to (f) is the process trees having perfect matching can be converted into paths perfect matching in turn.

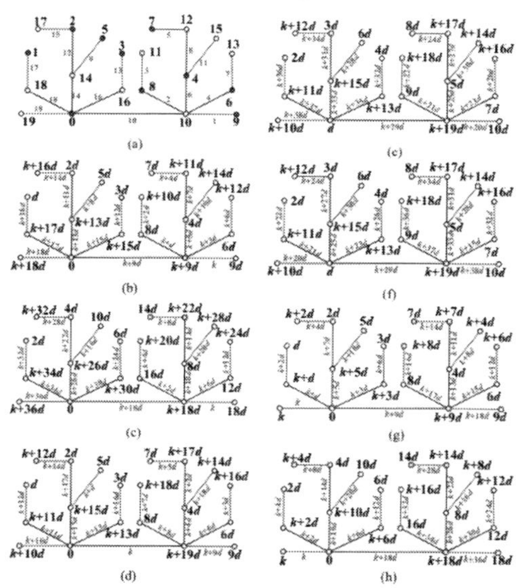

Figure 8. Eight examples for illustrating Definition 1,2 and 3. (a) A strongly set-ordered graceful labelling; (b) a strongly (k, d)-graceful labelling; (c) an AM-(k, d)-totally odd/even graceful labelling; (d) an AM-(k, d)-super felicitous labelling; (e) an AM-(k, d)-super edge-magic total labelling; (f) an AM-(k, d)-super edge-magic graceful labelling; (g) an AM-(k, d)-arithmetic labelling (h) an AM-(k, d)-totally odd/even-elegant labelling.

Acknowledgment

This project was supported by the National Natural Science Foundation of China under grant No. 61163054, No. 61363060 and No. 61662066.

References

[1] Xiaoyuan Suo, Ying Zhu, G. Scott. Owen, Graphical Passwords: A Survey. In Annual Computer Security Applications Conference (ACSAC), December 2005: 463-472. DOI: 10.1109/CSAC.2005.27 Source: DBLP

[2] Hongyu Wang, Jin Xu,Bing Yao. The Key-models And Their Lock-models For Designing New Labellings Of Networks.Proceedings of 2016 IEEE Advanced Information Management, Communicates,Electronic and Automation Control Conference (IMCEC 2016) pp565-5568.

[3] Hongyu Wang, Jin Xu,Bing Yao. Exploring New Cryptographical Construction Of Complex Network Data. IEEE First International Conference on Data Science in Cyberspace. IEEE Computer Society, (2016):155-160.

[4] R. N. Shepard, Recognition memory for words, sentences, and pictures, Journal of Verbal Learning and Verbal Behavior, vol. 6, pp. 156-163, 1967.

[5] Joseph A. Gallian. A Dynamic Survey of Graph Labeling. The Electronic Journal Of Combinatorics, 17 (2014), # DS6. (308 pages, 1693 reference papers)

[6] Bondy J. A. and Murty U. S. R. Graph Theory. Springer, ISBN: 978-1-84628-969-9, e-ISBN: 978-1-84628-970-5, DOI: 10.1007/978-1-84628-970-5, 2008.

[7] Bing Yao, Xia Liu and Ming Yao. Connections between labellings of trees. Bulletin of the Iranian Mathematical Society, ISSN: 1017-060X (Print) ISSN: 1735-8515 (Online), Vol. **43** (2017): 275-283.

[8] Xiangqian Zhou, Bing Yao, Xiang-en Chen and Haixia Tao. A proof to the odd-gracefulness of all lobsters. Ars Combinatoria **103** (2012), 13-18.

[9] S.M. Hegde. On (k,d)-graceful graphs. Journal of Combinatorics, Information and System Sciences, Vol. **25** (1-4) (2000), 255-265.

[10] B. D. Acharya and S. M. Hegde. Arithmetic graphs. *J. Graph Theory*, **14** (1990), 275-299.

[11] Xiangqian Zhou, Bing Yao, Xiang'en Chen. Every Lobster Is Odd-elegant. Information Processing Letters **113** (2013), 30-33.

[12] Jinxia Guo, Bing Yao, Jiajuan Zhang. On Generalized Gracefulness Of Edge-Symmetri Trees. Journal Of Southwestern University (in Chinese), 2013, **35** (2): 62-68.

[13] Hui Sun, Bing Yao. On Equivalent Definitions Of Cactus Graphs. 2017 submitted.

[14] Hui Sun, Bing Yao. Exploring equivalent definitions of Euler graphs. 2017 submitted.

[15] Hui Sun, Bing Yao. Exploring the Odd Gracefulness of Cyclic-Dragon Graphs. 2017 submitted.

[16] Hui Sun, Bing Yao. On the odd-elegant quality of cyclic-dragon graphs. 2017 submitted.

Exploring Topsnut-Graphical Passwords by Twin Odd-elegant Trees

Hong-yu WANG[1], *Jin* XU[1,⋆], and *Bing* Yao[2,3]

[1] *School of Electronics Engineering and Computer Science, Peking University, 100871, China*
[2] *College of Mathematics and Statistics, Northwest Normal University, Lanzhou, 730070, China*
[3] *College of Electronic and Information Engineering, Lanzhou Jiaotong University, Lanzhou 730070, China*

Abstract. Graphical passwords are facing a good opportunity as 2-dimension codes are accepted by many people, since it has been applied in mobile devices, electronic equipments with touch screen, and so on. QR codes can be considered as a type of graphical passwords. Topsnut-graphical password differs from the existing graphical passwords, and has been investigated and developed. In this article, a new type of Topsnut-graphical passwords has been designed by technique of graph theory, called twin odd-elegant labelling. We make the twin odd-elegant graphs for one-key vs two or more locks (conversely, one-lock vs two or more keys). These Topsnut-GPWs show perfect matching characteristics of locks (TOE-lock-models) and keys (TOE-key-models). We show examples for testing our methods which can be easily transformed into effective algorithms.

1 Introduction

Since its introduction in 1994, the QR Code (Quick Response codes, QR-codes, they are referred as a *printable computer language*) has gained wide acceptance in such diverse industries as manufacturing, warehousing and logistics, retailing, healthcare, life sciences, transportation and office automation [2]. QR codes can be considered as a type of graphical passwords that are used widely, since it has been applied in mobile devices, electronic equipments with touch screen, and so on. QR code (abbreviated from Quick Response Code) is the trademark for a type of matrix barcode (or two-dimensional barcode) first designed for the automotive industry in Japan. A barcode is a machine-readable optical label that contains information about the item to which it is attached. A QR code uses four standardized encoding modes (numeric, alphanumeric, byte/binary, and kanji) to efficiently store data; extensions may also be used [1]. However, QR codes can not be used to those departments required with high-level security. We have known that passwords in the security community should be easy to remember, and the user authentication protocol should be executable quickly and easily by humans. And passwords should be secure, i.e. they should look random and should be hard to guess; they should be changed frequently, and should be different on different accounts of the same user; they should not be written down or stored in plain text. On the other hands, no report tells us that other graphical passwords were applied to business and were accepted by people (Ref. [7], [3], [5]). Wiedenbeck *et al.* [12] asked for two principal research questions:

⋆e-mail: jxu@pku.edu.cn

Question 1: Are graphical passwords a viable alternative to alphanumeric passwords in terms of security, as well as password creation, learning, performance, and retention?

Question 2: Are users' perceptions of graphical passwords different from those of alphanumeric passwords?

An idea of "graphical structure plus number theory" (Topsnut) for creating new type of graphical passwords was proposed firstly by Wang *et al.* in [9, 10]. And in making new Topsnut graphical passwords (Topsnut-GPWs), Wang *et al.* [11] have designed some Topsnut-GPWs by knowledge of graph theory for answering the following problem of number theory:

Problem of odd-graceful KL-matching pairs: Let $[0, 2q] = \{0, 1, 2, \dots, 2q\}$ and $[1, 2q-1]^o = \{1, 3, 5, \dots, 2q-1\}$ with $q \geq 2$. If there are two subsets X_1, X_2 of $[0, 2q]$ hold: (i) $X_1 \cup X_2 = [0, 2q]$ and $\{0\} \subseteq X_1 \cap X_2 \subseteq [0, 2q]$; (2) $\{|a-b| : a \in X_i^{odd}, b \in X_i^{even}\} = [1, 2q-1]^o$ where X_i^{odd} is an odd-number set and X_i^{even} is an even-number set of X_i such that $X_i^{odd} \cup \in X_i^{even} = X_i$ and $X_i^{odd} \cap \in X_i^{even} = \emptyset$ with $i = 1, 2$. We say X_1 and X_2 to be an odd-graceful KL-matching pair. Find all odd-graceful KL-matching pairs for any given set $[0, 2q]$ with $q \geq 2$.

Clearly, one can use odd-graceful KL-matching pairs to design more complex Topsnut-GPWs. We, in Section 2, will introduce several techniques by knowledge of graph theory. Our main works are to show new Topsnut-GPWs made by the techniques, which also are the part solutions of the problem of odd-elegant KL-matching pairs proposed in the last section. On the other hands, our methods for dealing with the new Topsnut-GPWs can be easily transformed into algorithms.

2 Definitions for techniques

We use standard notation and terminology of graph theory [4]. Graphs mentioned are loopless, no multiple edges, undirected, connected and finite, unless otherwise specified. A (p, q)-graph G is one with p vertices and q edges. The shorthand symbol $[m, n]$ stands for an integer set $\{m, m+1, \dots, n\}$, where m and n are integers with $0 \leq m < n$; the notation $[s, t]^o$ indicates an odd-set $\{s, s+2, \dots, t\}$, where s and t both are odd integers with $1 \leq s < t$; and the notation $[k, \ell]^e$ indicates an even-set $\{k, k+2, \dots, \ell\}$, where k and ℓ both are even integers with $0 \leq k < \ell$.

Definition 1 *[14] Suppose that a (p, q)-graph G admits a mapping $f : V(G) \to [0, 2q-1]$ such that $f(u) \neq f(v)$ for distinct $u, v \in V(G)$, and the label $f(uv)$ of every edge $uv \in E(G)$ is defined as $f(uv) = f(u) + f(v) \pmod{2q}$ and the set of all edge labels is equal to $[1, 2q-1]^o$, we call f an* odd-elegant labelling *and G to be* odd-elegant.

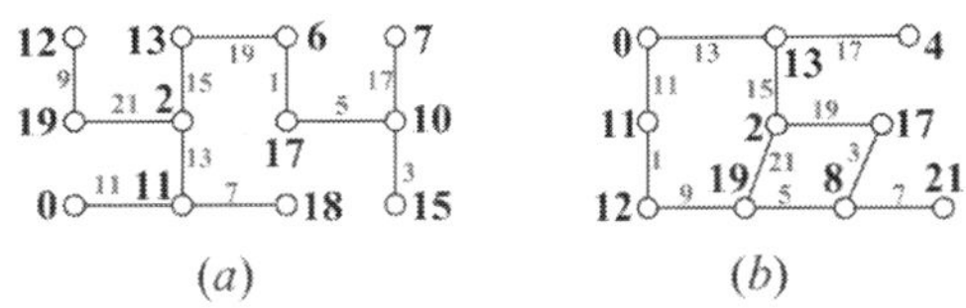

Figure 1. (a) An odd-elegant tree; (b) An odd-elegant graph.

Definition 2 *[13] Suppose that a bipartite graph G admits a labelling f such that $\max\{f(x) : x \in X\} < \min\{f(y) : y \in Y\}$, where (X, Y) is the bipartition of vertex set $V(G)$ of G. We call f a set-ordered labelling (an So-labelling for short).*

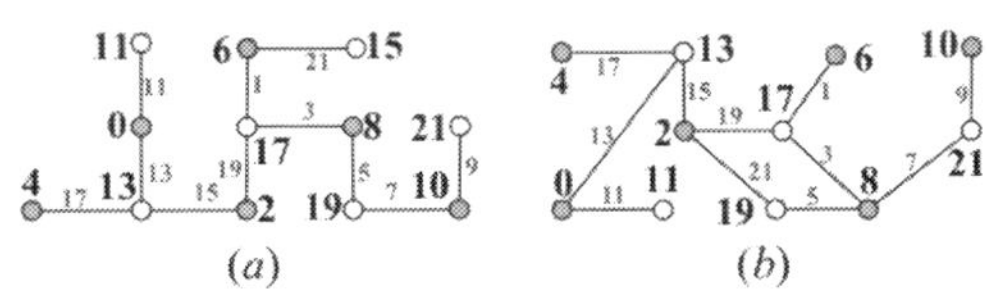

Figure 2. (a) A set-ordered odd-elegant tree; (b) A set-ordered odd-elegant graph.

Let G_j be a (p_j, q_j)-graph with $j = 1, 2$. The graph G obtained by identifying two certain vertices $x_{i,1}$ of G_1 with two certain vertices $x_{i,2}$ of G_2 into one vertex x_i, where i=1,2, x_1 and x_2 is called the *identified-vertex*, respectively. is denoted as $G = G_1 \circ_2 G_2$, called an *double identifying graph*. So, the graph $G = G_1 \circ_2 G_2$ has $p_1 + p_2 - 2$ vertices and $q_1 + q_2$ edges.

Suppose that two vertices w_1 and w_2 of G are two identified-vertices of identifying $x_{i,1}$ with

$x_{i,2}$, where $i = 1, 2$. Then we can split the identified-vertices w_i into two vertices $x_{i,1}$ and $x_{i,2}$ (called the *splitting-vertices*), so G can be split into two parts G_1 and G_2.

The above process of producing G from G_1 and G_2 is called an *identifying operation*, another process of splitting G into two parts G_1 and G_2 is called a *splitting operation*.

Definition 3 *Let two connected (p_i, q)-graphs G_i with $i = 1, 2$, and let $p = p_1 + p_2 - 2$. If the identifying (p, q)-graph $G = G_1 \circ_2 G_2$ has a mapping f: $V(G) \rightarrow [0, q-1]$ such that (i) $f(x) \neq f(y)$ for any pair of vertices $, y \in V(G)$; (ii) f is an odd-elegant labelling of G_i with $i = 1, 2$. Then we say G to be a* twin odd-elegant graph *(a TOE-graph), f a* TOE-labelling. *G_1 a* TOE-source graph, *and G_2 a* TOE-associated graph, *vice versa.*

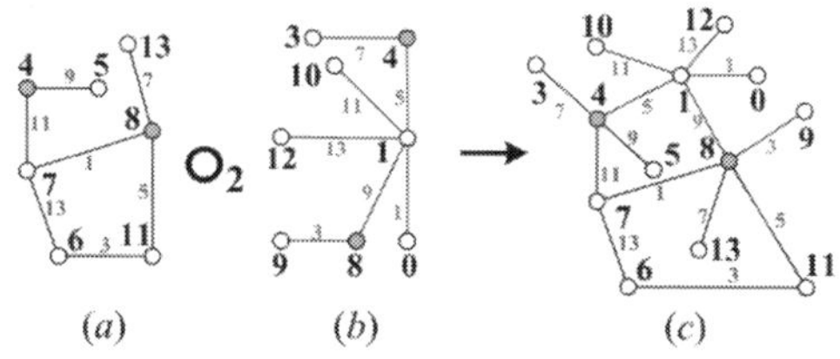

Figure 3. A twin odd-elegant graph $G = G_1 \circ_2 G_2$ with its OE-source graph G_1 and OE-associated graph G_2 by an identifying operation.

Furthermore, if each G_i with $i = 1, 2$ is a connected graph in Definition 3, and exist TOE-source G_1 is a bipartite connected graph having its own bipartition (X_1, Y_1) satisfy Definition 2, we say that the identifying (p, q)-graph $G = G_1 \circ_2 G_2$ is a *set-ordered twin odd-elegant graph*, f a *set-ordered twin odd-elegant labelling*. Here, the source graph G_1 is a set-ordered odd-elegant graph by Definition 1 and 2.

Similarly, we present the definition of a twin odd-elegant tree as follows.

Definition 4 *Let T_i be a tree of order n with $i = 1, 2$. If the identifying tree $G_T = T_1 \circ_2 T_2$ has a mapping $f : V(G_T) \rightarrow [0, 2n-3]$ such that (i) $f(x) \neq f(y)$ for any pair of vertices $x, y \in V(G_T)$; (ii) f is an odd-elegant labelling of T_i, where $i = 1, 2$. Then G_T is called a* twin odd-elegant tree *(a TOE-tree), f a* TOE-labelling, *and T_1 a* TOE-source tree, *and T_2 a* TOE-associated tree *of T_1. Furthermore, we say T_1 a* TOE-source self-associated tree *if T_1 is isomorphic to T_2.*

Two examples for illustrating Definition 4 are show in Figure 4 and Figure 10. Let T_1 has its bipartition (X_1, Y_1) in Definition 2; and if the TOE-source tree T_1 has a set-ordered odd-elegant labelling, then we call G_T a *set-ordered twin odd-elegant tree*(an So-TOE tree for short), and f is *set-ordered twin odd-elegant labelling*(an So-TOE labelling for short).

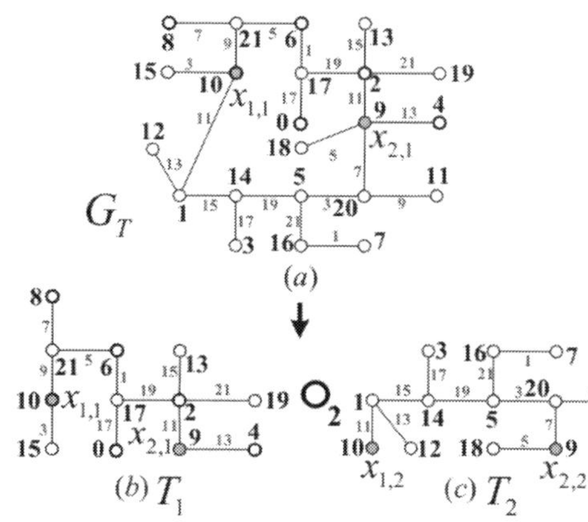

Figure 4. A TOE tree G_T has its TOE-source tree T_1 and TOE-associated tree T_2, and two split-vertices $x_{i,1}$ and $x_{i,2}$ labelled with numbers 9 and 10, where $i = 1, 2$, by a splitting operation.

In our design of the new Topsnut-GPWs, a source graph/tree and its associated graph/tree in Definition 3 or Definition 4 are called a *TOE-lock-model* and a *TOE-key-model*, respectively.

3 Theoretical analysis and guarantee

Lemma 1 *Any star $K_{1,n}$ is a TOE-source tree of a So-TOE-tree G_T.*

Proof. Let $V(K_{1,n}) = \{x, y_1, y_2, \ldots, y_n\}$ and $E(K_{1,n}) = \{xy_i : i \in [1, n]\}$ be the sets of vertices and edges of $K_{1,n}$, respectively. We can define an odd-elegant labelling f with $f(x) = 0$,

$f(y_i) = 2i - 1$ with $i \in [1, n]$, which induce $f(xy_i) = f(x) + f(y_i) \pmod{2n}$ for $i \in [1, n]$. So, $f(E(K_{1,n}) = [1, 2n - 1]^o$. i.e. $K_{1,n} = T_1$

Let T_2 be a copy of $K_{1,n}$. We have the sets of vertices and edges of T_2 as $V(T_2) = \{x', y'_1, y'_2, \dots, y'_n\}$ and $E(T_2) = \{x'y'_i : i \in [1, n]\}$, respectively. Next, we define a labelling f' in the way $f'(x') = 1$, $f'(y'_i) = 2i-2$ with $i \in [1, n]$. Clearly, $f'(x'y'_i) = f'(x') + f'(y'_i) \pmod{2n}$ for $i \in [1, n]$. We have that $f'(E(T_2) = [1, 2n - 1]^o$.

Since $f(y_1) = f'(x') = 1$ and $f(x) = f'(y'_1) = 0$, we identify the vertex y_1 of $K_{1,n}$ with the vertex x' of T_2 into one w_1 and the vertex x of $K_{1,n}$ with the vertex y'_1 of T_2 into one w_2 for forming the identifying tree $G_T = K_{1,n} \circ_2 T_2$. It is not hard to provide a labelling f_{G_T} of G_T as: $f_{G_T}(u) = f(u)$ for $u \in V(K_{1,n})$, and $f_{G_T}(v) = f'(v)$ for $v \in V(T_2)$. Immediately, $\{f(xy_i) : xy_i \in E(K_{1,n})\} = [1, 2n - 1]^o$ and $\{f(x'y'_i) : x'y'_i \in E(T_2)\} = [1, 2n - 1]^o$. Therefor, we claim that G_T is a TOE-tree since f_{G_T} is a TOE-labelling of G_T by Definition 4, which means that $K_{1,n}$ is an So-TOE source tree of G_T, and T_2 is a TOE-associated tree of $K_{1,n}$.

It is noticeable, $T_2 = K_{1,n}$ means that the TOE-source tree $K_{1,n}$ is isomorphic to its associated tree T_2 in T. So, $K_{1,n}$ is a *TOE-source self-associated tree*.

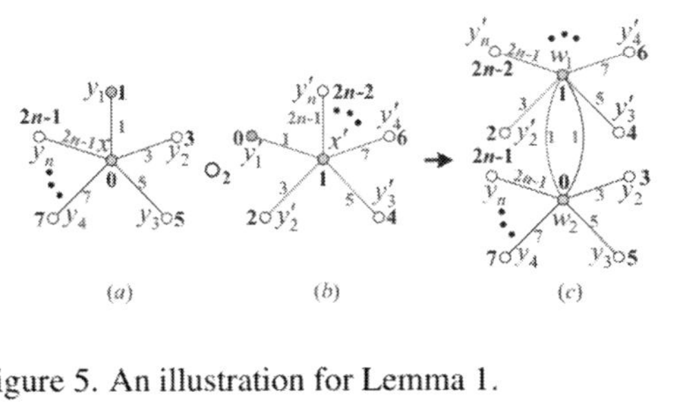

Figure 5. An illustration for Lemma 1.

Theorem 1 *Any set-ordered odd-elegant tree being not a star is an So-TOE-source tree of at least two So-TOE trees.*

Proof. Suppose that T_1 is a tree with n vertices, its vertex bipartition is (X, Y), where $X = \{x_i : i \in [1, s]\}$, $Y = \{y_j : j \in [1, t]\}$, $s + t = n$ and $\min\{s, t\} \geq 2$. By the hypothesis of the theorem, T_1 has a set-ordered odd-elegant labelling f_1 defined by $f_1(x_i) = 2(i - 1)$, $i \in [1, s]$; $f_1(y_j) = 2(j + s - 1) - 1$, $i \in [1, t]$; and $f_1(x_iy_j) = f_1(y_j) + f_1(x_i) = 2(s + i + j - 2) - 1 \pmod{2(n - 1)}$ for each edge $x_iy_j \in E(T_1)$.

We can observe $f_1(V(T_1)) = \{0, 2, \dots, 2s - 2, 2s - 1, 2s + 1, \dots, 2n - 3\}$ and $f_1(E(T_1)) = [1, 2n - 3]^o$.

Case 1. We construct a labelling f_2 of another tree T_2 by the labelling f_1 such that $f_2(V(T_2)) = \{1, 3, \dots, 2s - 3, 2s - 2, 2s - 1, 2s, \dots, 2(n - 2)\}$, and $f_2(E(T_2)) = \{f_2(uv) = f_2(u) + f_2(v) \pmod{2(n - 1)} : uv \in E(T_2)\} = [1, 2n - 3]^o$, and furthermore, $f_2(u) \neq f_2(v)$ for any $u, v \in V(T_2)$. This tree T_2 can be built up in the way: A bipartition (U_1, V_1) with $U_1 = \{u_i : i \in [1, s]\}$ and $V_1 = \{v_j : j \in [1, t]\}$, such that $f_2(u_i) = 2i - 1$, $i \in [1, s]$; $f_2(v_j) = 2(s - 2 + j)$, $j \in [1, t]$. Any edge $u_iv_j \in E(T_2)$ satisfy $f_2(u_iv_j) = f_2(u_i) + f_2(v_j) \pmod{2(n - 1)}$ with $i \in [1, s]$ and $j \in [1, t]$. We construct the edge set of T_2 as $\{u_1v_j, u_iv_t : i \in [1, s], j \in [1, t - 1]\}$ such that the edge labels are $f_2(u_iv_t) = 2i - 3$, $f_2(u_1v_j) = 2j+2s-3 \pmod{2(n-1)}$ for $i \in [1, s]$ and $j \in [1, t - 1]$. Observe that $f_2(E(T_2)) = [1, 2n - 3]^o$, $f_1(y_1) = f_2(u_s)$ and $f_1(x_s) = f_2(v_1)$.

Now, we can identify the vertex x_s and y_1 of T_1 with the vertex v_1 and u_s of T_2 into one (the identified vertex) w_1 and w_2, respectively, so we obtain the desired tree $G_T = T_1 \circ_2 T_2$. And G_T has a labelling f defined as: $f(x_i) = f_1(x_i)$, $i \in [1, s - 1]$; $f(y_j) = f_1(y_j)$, $i \in [2, t]$; $f(u_k) = f_2(u_k)$, $k \in [1, s - 1]$, $f(v_l) = f_2(v_l)$, $l \in [2, t]$, $f(w_1) = 2s - 2$ and $f(w_2) = 2s - 1$. Clearly, any two vertices of G_T are assigned different numbers. According to Definition 4, G_T is an So TOE-tree having the source tree T_1.

Case 2. Similarly to Case 1, we can get the following results: Let $f_2(V(T'_2)) = \{1, 3, \dots, 2s-3, 2s - 2, 2s, 2(s + 1), \dots, 2n - 4\} \cup \{0\}$, and $f_2(E(T'_2)) = [1, 2n-3]^o$, and furthermore $f_2(u) \neq f_2(v)$ for $u, v \in V(T'_2)$. This tree T'_2 can be built up in the way: A bipartition (U_2, V_2) with $U_2 = \{u_i : i \in [1, s - 1]\}$ and $V_2 = \{v_j : j \in [1, t + 1]\}$, such that $f_2(u_i) = 2i - 1$, $i \in [1, s - 1]$; $f_2(v_j) = 2(s - 2 + j)$, $j \in [1, t]$, $f_2(v_{t+1}) = 0$. Any edge $u_iv_j \in E(T'_2)$ satisfies $f_2(u_iv_j) = f_2(u_i) + f_2(v_j) \pmod{2(n - 1)}$ with $i \in [1, s - 1]$ and $j \in [1, t + 1]$. We construct the edge set of T'_2 as $\{u_1v_j, u_iv_{t+1} : i \in [2, s], j \in [1, t]\}$

such that the edge labels are $f_2(u_i v_{t+1}) = 2i - 1$, $f_2(u_1 v_j) = 2(s + j) - 3$ for $i \in [1, s-1]$ and $j \in [1, t]$. Observe that $f_2(E(T_2')) = [1, 2n-3]^o$, $f_1(x_1) = f_2(v_{t+1})$ and $f_1(x_s) = f_2(v_1)$.

Now, we can identify the vertex x_1 and x_s of T_1 with the vertex v_{t+1} and v_1 of T_2' into one (the identified vertex) w_1 and w_2, so we obtain the desired tree $G_{T'} = T_1 \circ_2 T_2'$. And $G_{T'}$ has a labelling f defined as: $f(x_i) = f_1(x_i)$, $i \in [2, s-1]$; $f(y_j) = f_1(y_j)$, $i \in [1, t]$; $f(u_k) = f_2(u_k)$, $k \in [1, s-1]$, $f(v_l) = f_2(v_l)$, $l \in [2, t]$, $f(w_1) = 0$ and $f(w_2) = 2s - 2$. Clearly, any two vertices of $G_{T'}$ are assigned different numbers. According to Definition 4, $G_{T'}$ is an So-TOE tree having the source tree T_1.

Corollary 1 *(1) Every So-TOE-source tree is a TOE-source self-associated tree.*

(2) Not every TOE-tree consisting of TOE-source self-associated tree has an odd-elegant labelling.

Theorem 2 *If the TOE-source tree T_1 is a set-ordered odd-elegant tree, then the simple So-TOE graph $G_T = T_1 \circ_2 T_2$ has an odd-elegant labelling.*

4 Test and simultaneous

Experiment 1 *We show two examples for illustrating the results of Theorem 1. Our methods appeared in the proof of Theorem 1 are easily applied the design of Topsnut-GPws, since the methods are constructive. We illustrate theorem 2 with Figure 6(c) and Figure 9.*

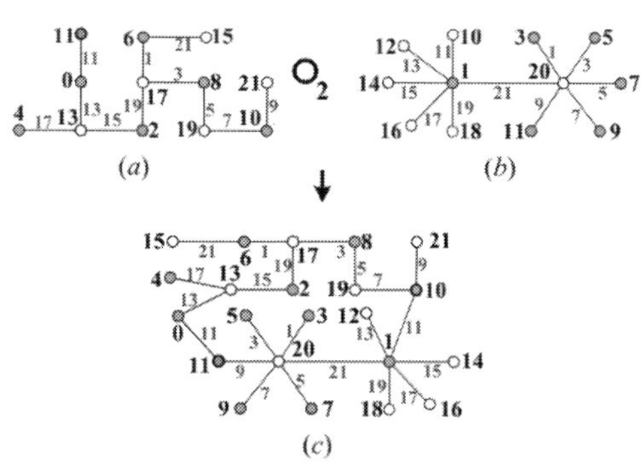

Figure 6. An example for illustrating Case 1 of Theorem 1.

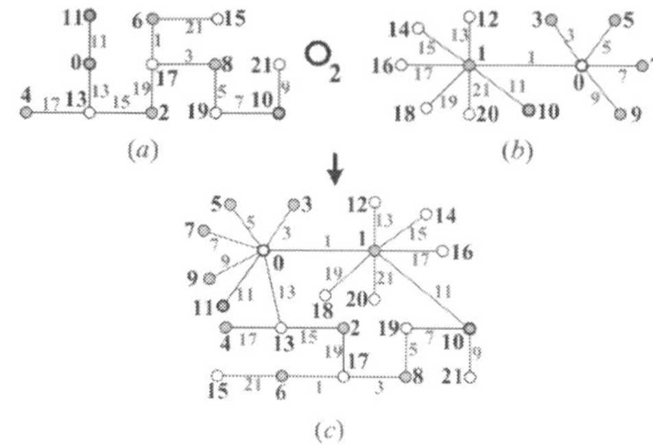

Figure 7. An example for illustrating Case 2 of Theorem 1.

Experiment 2 *Based on the technique provided by Corollary 1, we make a Topsnut-GPW shown in Figure 8(c) for the purpose of simultaneous.*

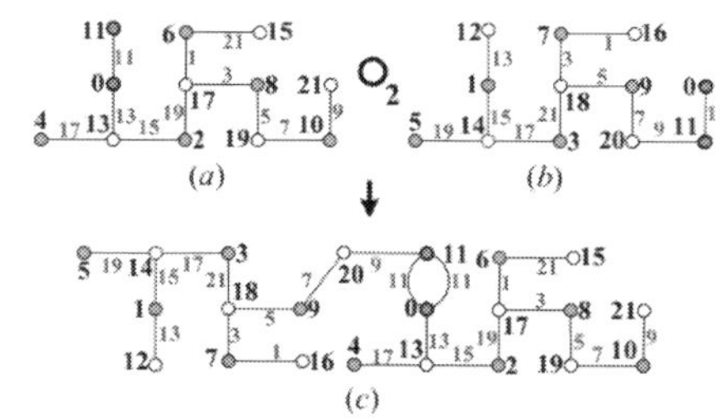

Figure 8. An example for illustrating the technique of Corollary 1.

Experiment 3 *It is is very important to get the property P of large scale of topological structure for smaller topological structures having a property P. We test the results of Theorem 2 by the scheme shown in Figure 9. Here, we can confirm that our method for proving Theorem 2 can be transformed into an effective algorithm.*

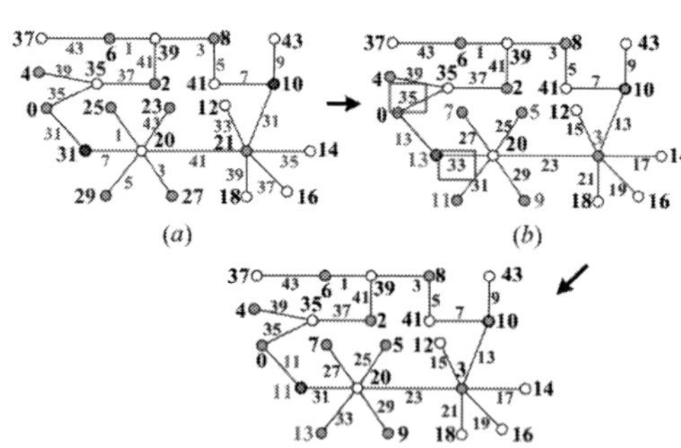

Figure 9. A conversion process from an So-TOE-tree to odd-elegant graph for illustrating Theorem 2.

Experiment 4 *We have the following results* (1) A TOE-source graphs may have different TOE-associated graphs, and vice versa.

(2) A TOE-graph may contain two or more groups of TOE-source and TOE-associated graphs.

(3) Two groups of TOE-source and TOE-associated graphs may produce a TOE-graph.

As shown in Figure 11, there are eight different TOE-key-models for Figure 10 (b).

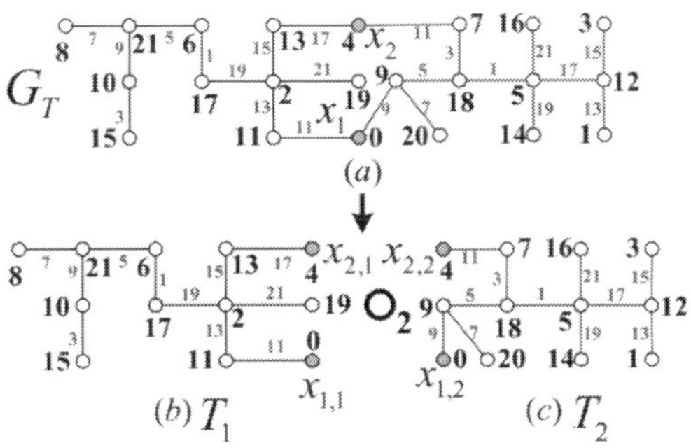

Figure 10. An So-TOE tree G_T has its TOE-source tree T_1 and TOE-associated tree T_2, and two splitting-vertices $x_{i,1}$ and $x_{i,2}$ labelled with numbers 0 and 4, where $i = 1, 2$, by a splitting operation.

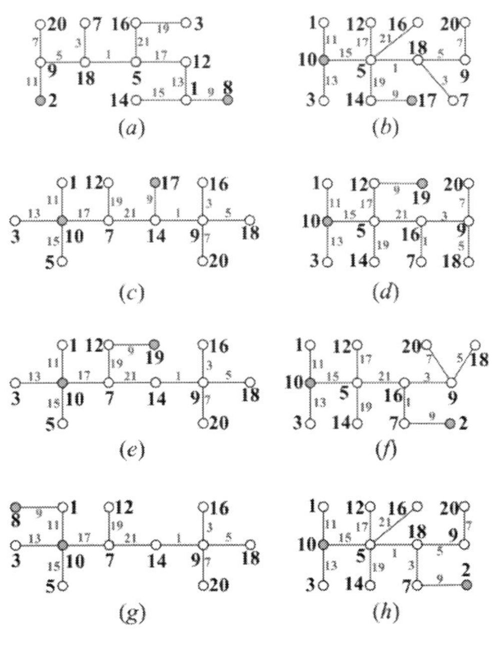

Figure 11.(a)-(h) Eight TOE-key-models of the TOE-lock-model for Figure 10(b) (eight TOE-associated trees of the TOE-source tree T_1).

5 Conclusion and researching problems

We have designed a new technique, called *(set-ordered) twin odd-elegant labelling*, by the idea of graph labellings. Furthermore, we have found some themods/algorithms to realize Topsnut-GPWs based our new graph labellings. These Topsnut-GPWs show perfect matching characteristics of locks (TOE-lock-models) and keys (TOE-key-models), which enable us to produce one-key vs two or more locks (conversely, one-lock vs two or more keys). We have many graph colorings were introduced and investigated by algorithmic methods [4], and have more graph labellings shown in a large survey due to Gallian [6], the survey collected over 1400 papers. Thereby, Topsnut-GPWs are supported by graph colorings/labellings. Every thing can become a password, and t is just beginning for Topsnut-GPWs. By our experience, we propose:

Conjecture 1 *Any tree is the TOE-source tree of a certain twin odd-elegant tree.*

Conjecture 2 *Any twin odd-elegant tree is simple having an odd-elegant labelling.*

Furthermore, we, motivated from problem of odd-graceful KL-matching pairs, propose the following problem of number theory:

Odd-elegant KL-matching pairs' problem. Let $[0, 2q-1] = \{0, 1, 2, \dots, 2q-1\}$ and $[1, 2q-1]^o = \{1, 3, 5, \dots, 2q-1\}$ with integer $q \geq 2$. If there are two subsets Y_1 and Y_2 of $[0, 2q-1]$ such that

(I) $Y_1 \cup Y_2 = [0, 2q-1]$ and $\{0, k\} = Y_1 \cap Y_2 \subseteq [0, 2q-1]$ for some $k \in [1, 2q-1]$; and

(II) the set $\{a + b \pmod{2q} : a \in Y_i^{odd}, b \in Y_i^{even}\} = [1, 2q-1]^o$, where Y_i^{odd} is an odd-number set and Y_i^{even} is an even-number set of Y_i holding $Y_i^{odd} \cup Y_i^{even} = Y_i$ and $Y_i^{odd} \cap Y_i^{even} = \emptyset$ with $i = 1, 2$.

We call these two sets Y_1 and Y_2 to be an odd-elegant KL-matching pair, and want to find all possible odd-elegant KL-matching pairs.

Acknowledgment

This work is supported by National Key R&D Program of China (No. 2016YFB0800700), National 973 Program of China (2013CB329600) and the National Natural Science Foundation of China (Nos. 61572046, 61502012, 61672050, 61363060 and 61662066).

References

[1] "QR Code features". Denso-Wave. Archived from the original on 2013-01-29. Retrieved 3 October 2011.

[2] "QR Code Essentials". Denso ADC. 2011. Retrieved 12 March 2013.

[3] R. Biddle, S. Chiasson, P. C. V. Oorschot. Graphical passwords:Learning from the First Twelve Years. Acm Computing Surveys, 44,4 (2012):1-41.

[4] J. A. Bondy, U. S. R. Murty. Graph Theory. Springer London, 2008.

[5] H. Gao, W. Jia, F. Ye, et al. A survey on the use of graphical passwords in security . Journal of Software, 8, 7(2013):320-329.

[6] J. A. Gallian. A Dynamic Survey of Graph Labeling. The electronic journal of combinatorics, 17 (2014), # DS6.

[7] X. Suo, Y. Zhu , G. S. Owen. Graphical Passwords: A Survey. Computer Security Applications Conference. DBLP, (2005):463-472.

[8] X. Suo . A Study of Graphical Password for Mobile Devices. International Conference on Mobile Computing, Applications, and Services. Springer, Cham, (2013): 202-214.

[9] H. Y. Wang, J. Xu , B. Yao. Exploring New Cryptographical Construction of Complex Network Data. IEEE International Conference on Data Science in Cyberspace. IEEE, (2016):155-160.

[10] H. Y. Wang, J. Xu , B. Yao. The key-models and their lock-models for designing new labellings of networks. Advanced Information Management, Communicates, Electronic and Automation Control Conference. IEEE,(2016):565-568.

[11] H. Y. Wang, J. Xu , B. Yao. Twin Odd-graceful Trees Towards Information Security. Procedia Computer Science 107 (2017):15-20, Elsevier Science Publishers B. V. 2017, DOI: 10.1016/j.procs.2017.03.050

[12] S. Wiedenbeck, J. Waters , J. C. Birget, et al. PassPoints: design and longitudinal evaluation of a graphical password system[J]. International Journal of Human-Computer Studies, 63, (1-2)(2005):102-127.

[13] B. Yao, H. Cheng, M. Yao, et al. A note on strongly graceful trees. Ars Combinatoria, 92 (2009):155-169.

[14] X. Q. Zhou, B. Yao, X. E. Chen. Every lobster is odd-elegant. Information Processing Letters, 113,1-2(2013):30-33.

MATEC Web of Conferences 139, 00206 (2017) DOI: 10.1051/matecconf/201713900206
ICMITE 2017

Graph Theory Towards Module-K Odd-Elegant Labelling Of Graphical Passwords

Xiaohui Zhang[1], *Hui* Sun[1], and *Bing* Yao[1,2,⋆]

[1] *College of Mathematics and Statistics, Northwest Normal University, Lanzhou, 730070 CHINA*
[2] *School of Electronic and Information Engineering, Lanzhou Jiaotong University, Lanzhou 730070, CHINA*

Abstract. Graph labellings have been applied in many areas of science and engineering, such as in the development of redundant arrays of independent disks which incorporate redundancy utilizing erasure codes, algorithms, design of highly accurate optical gauging systems for use on automatic drilling machines, design of angular synchronization codes, design of optimal component layouts for certain circuit-board geometries, and determining configurations of simple resistor networks which can be used to supply any of a specified set of resistance values. Based on the idea of of "topological structure plus number theory" and magic type of various labellings for solving network transfer delay by using other type of graphical passwords. We define a new graph labelling, called *Module-K odd-elegant labelling*, and find some network models that admit our new labelling, and furthermore our methods can be transformed into effective algorithms.

1 Introduction

No report tells us that much of graphical passwords were applied to business and practice (Ref. [1], [2], [3]). An idea of "topological structure plus number theory" (Topsnut) as an alternative the existing graphical passwords is proposed in [6], which can be realized by graph labellings, and this idea is related with many mathematical conjectures, such as "Every tree is graceful" due to Rosa [4]. The survey article [5] has collected over 1700 papers on research of graph labellings. Motivated from the some graph labellings we define a new labeling, called Module-K odd-elegant labelling. The Module-K odd-elegant labelling can induces some mathematical problems, such as: *Every tree admits a Module-K odd-elegant labelling*. Furthermore, we can plant this new labelling to other graph labellings, see [7], strongly graceful labelling in [10], odd-graceful labelling in [11], odd-elegant labelling in [18], seven other labellings in [13] and so on, which mean that exploring new graphical passwords will bring more new mathematical subjects and new problems. An idea of "topological structure plus number theory" for creating new type of graphical passwords was proposed firstly by Wang *et al.* in [14] and [15]. In comparing with those graphical passwords mentioned in [16], the new graphical passwords obtained by Topsnut needs less storage and implements quickly in network communication.

2 Preliminary and new labellings

We use standard notation and terminology of graph theory. Graphs mentioned are loopless, no multiple edges, undirected, connected and finite, unless otherwise specified. A (p, q)-graph

⋆e-mail: yybb918@163.com

G is one with p vertices and q edges. The shorthand symbol $[m, n]$ stands for an integer set $\{m, m+1, \dots, n\}$, where m and n are integers with $0 \leq m < n$; the notation $[s, t]^o$ indicates an odd-set $\{s, s+2, \dots, t\}$, where s and t both are odd integers with $1 \leq s < t$; and the notation $[k, \ell]^e$ indicates an even-set $\{k, k+2, \dots, \ell\}$, where k and ℓ both are even integers with $0 \leq k < \ell$. Graphical password will be abbreviated as GPW, and "Topsnut-GPW" is the abbreviation of "graphical passwords based on the idea of topological structure plus number theory".

Definition 1 *[5] If a (p, q)-graph G admits a mapping f: $V(T) \to [0, q]$ such that $f(u) \neq f(v)$ for distinct $u, v \in V(G)$, and the edge labelling set $\{f(uv) = |f(u) - f(v)| : uv \in E(G)\} = [1, q]$, then f is called a graceful labelling, and G is called a graceful graph.*

We write the vertex labelling set $f(V(G)) = \{f(u) : u \in V(G)\}$ and the edge labelling set $f(E(G)) = \{f(uv) : uv \in E(G)\}$ hereafter. Suppose that a bipartite graph G admits a graceful labelling f such that $\max\{f(x) : x \in X\} < \min\{f(y) : y \in Y\}$, where (X, Y) is the bipartition of $V(G)$, we call f a set-ordered graceful labelling, and this case is denoted as $f(X) < f(Y)$ for short ([17, 18]).

Definition 2 *[9] Suppose that a (p, q)-graph G admits a mapping $f : V(G) \to [0, 2q - 1]$ such that $f(u) \neq f(v)$ for distinct $u, v \in V(G)$, and the label $f(uv)$ of every edge $uv \in E(G)$ is defined as $f(uv) = f(u) + f(v) (\text{mod } 2q)$ and the set of all edge labels is equal to $[1, 2q - 1]^o$, we call f an* odd-elegant labelling *and G to be* odd-elegant.

Definition 3 *[9] Let (V_1, V_2) be the bipartition of a bipartite (p, q)-graph G. If G admits an odd-elegant labelling f such that* $\max\{f(u) : u \in V_1\} < \min\{f(v) : v \in V_2\}$, *then we call f a* set-ordered odd-elegant labelling, *and write this case as $f(V_1) < f(V_2)$.*

Motivated from the above graph labellings we define a new Module-K odd-elegant labelling as follows.

Definition 4 *Suppose that a (p, q)-graph G admits a mapping $f : V(G) \to [0, 2q - 1]$ such that $f(u) \neq f(v)$ for distinct $u, v \in V(G)$, and the label $f(uv)$ of every edge $uv \in E(G)$ is defined as $f(uv) = f(u) + f(v) + K (\text{mod } 2q)$ for some given K, and the set of all edge labels is equal to $[1, 2q - 1]^o$ or $[2, 2q]^e$, we call f a* Module-K odd-elegant labelling *and G to be* Module-K odd-elegant. *Moreover, if (V_1, V_2) is the bipartition of G, and the Module-K odd-elegant labelling f obeys* $\max\{f(u) : u \in V_1\} < \min\{f(v) : v \in V_2\}$, *then we call f a* set-ordered Module-K odd-elegant labelling, *and write this case as $f(V_1) < f(V_2)$.*

In the current paper, we make the following contributions to the literature on password security:

(1) be used conveniently in usually;

(2) with strong security, that is, it is difficult to be broken;

(3) there are enough graphs and labellings for making desired keys and locks.

For answering the above problems, we prove the series cryptographical graphs have good properties, and show the guarantee for constructing large scale of cryptographical trees from smaller cryptographical trees. The methods used for constructing the desired cryptographical graphs can be transformed into efficient algorithms.

3 Main results and their proofs

Theorem 1 *A connected bipartite graph G is set-ordered graceful if and only if there exists a K such that G admits a Module-K odd-elegant labelling.*

Proof. *Necessity.* Suppose that a bipartite graph G admits a graceful labelling f such that $max\{f(x) : x \in X\} < \min\{f(y) : y \in Y\}$, where (X, Y), we define $X = \{x_i : i \in [1, s]\}$ and $Y = \{y_j : j \in [1, t]\}$ with $s + t = p$ and $x_i, y_j \in V(G)$. Let G have a set-ordered graceful labelling f with $f(x_i) = i - 1$ for $i \in [1, s]$ and$f(y_j) = s - 1 + j$ for $j \in [1, t]$, and $f(x_i y_j) = f(x_i) - f(y_j) = s + j - i - 2$ for each edge

$x_i y_j \in E(T)$. Let G have a labelling g define as: $g(x_i) = 2f(x_i)$ for $i \in [1, s]$, $g(y_j) = 2p - 1 - 2f(y_j)$ for $i \in [1, t]$. for each edge $x_i y_j \in E(G)$, For some given $K \geq 0$, $g(x_i) + g(y_j) + K = 2f(x_i)+2p-1-2f(y_j)+K = 2(i+p+j-s)+K-1$, and $g(x_i y_j) = g(x_i) + g(y_j) + K(\text{mod } 2q)$, which implies $g(x_i y_j) \in [1, 2p-1]^o$ or $[2, 2q]^e$, such that g is Module-K odd-elegant labelling with for each edge $x_i y_j \in E(G)$.

Sufficiency. Suppose that G admits a Module-K odd-elegant labelling h such that $h(x_i) = 2(i-1), i \in [1, s]; h(y_j) = 2p - 1 - 2(s - 1 + j), j \in [1, t]$. $h(x_i y_j) = 2(i + p - s - j) - 1 + K(\text{mod } 2q)$ for $i \in [1, s]$ and $j \in [1, t]$, and $h(x_i y_j) \in [1, 2q-1]^o$ or $[2, 2q]^e$. We define a labelling f_1 of G as: $f_1(x_i) = \frac{1}{2}h(x_i)$ for $i \in [1, s]$; $f_1(y_j) = h(y_j) - 2p + 1$ for $j \in [1, t]$; $f_1(x_i y_j) = h(x_i y_j) - K$ for $i \in [1, s]$ and $j \in [1, t]$. which produces $f_1(V(G)) = [1, q]$ and $f_1(x_i y_j) \in [0, q]$. Thereby, f_1 is a set-ordered graceful labelling of G.

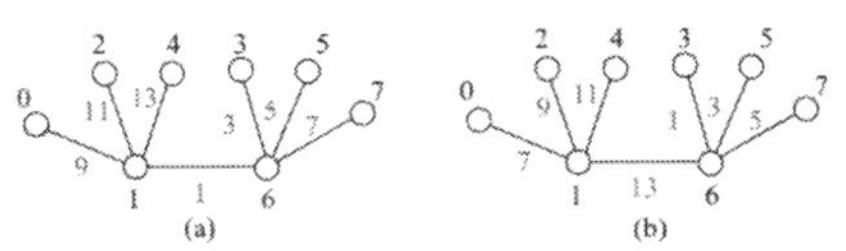

Figure 1. (a) A Module-8 odd-elegant labelling; (b) a Module-6 odd-elegant labelling.

A union graphical quasi-group $\bigcup_i(G)$ of order n is defined as: $\bigcup_i(G) = \{G : V(G) = 1, 2, \ldots, n\}$, the operation between elements of $\bigcup_i(G)$ is the graph union operation "$\bigcup$", and $G \bigcup H$ produces a simple graph of order n.

(*i*) (Zero element) the zero element of $\bigcup_i(G)$ is the unique complete graph K_n, since $G \bigcup K_n = K_n$.

(*ii*) (Inverse element) the complementary G of each element $G \in \bigcup_i(G)$ is the inverse element of G, since $G \bigcup G = K_n$.

(*iii*) Uniqueness and closure. $G \bigcup H$ is unique and $G \bigcup H \in \bigcup_i(G)$ for any element $H \in \bigcup_i(G)$.

(*iv*) (Associated law) $(G_1 \bigcup G_2) \bigcup G_3 = G_1 \bigcup (G_2 \bigcup G_3)$.

A graphical group obtained from the graph Module-K odd-elegant labelling. Let $f_1 : V(G_k) \to [0, 2q-1]$ be a Module-K odd-elegant labelling such that the set of all edge labels is equal to $[1, 2q-1]^o$ or $[2, 2q]^e$ for a (p, q)-graph G_k. Define every edge $uv \in E(G_k)$ is defined as $f(uv) = f(u) + f(v) + K(\text{mod } 2q)$ with some given K. Notice that $T_i (i \in [1, n])$ has identical vertex labelling of G_k. Each copy of the (p, q)-graph G_k is written as $T_i (i \in [1, n])$. Define a 3-element operation $T_i \oplus T_j \ominus T_k = T_{i+j-k \ (\text{mod } K)}$ We call this particular operation as the Module-K odd-elegant additive-sum of graphs.

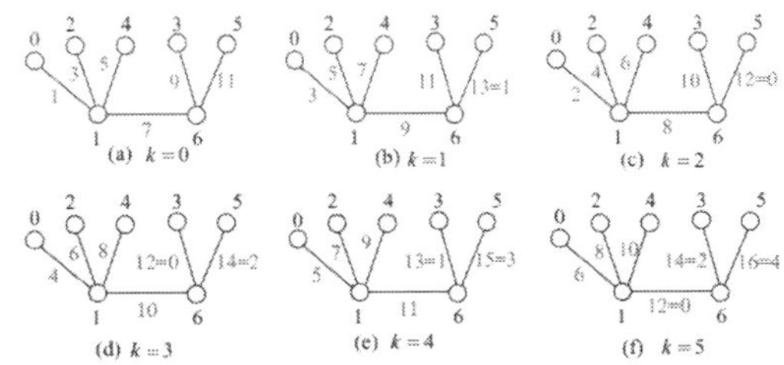

Figure 2. $T_i (i \in [1, 4])$.

Corollary 1 *The set Oele(G) based on a (p, q)-graph G_1 forms an Abelian additive group, we call Oele(G) a labelling graphical group based on the graph G hereafter. Furthermore, $f(uv) = f(u) + f(v) + K(\text{mod } 2q)$ with some given K.*

Proof. First of all, the equation $F(T_i) \oplus F(T_j) \ominus F(T_k) = F(T_{i+j-k \ (\text{mod } 2q)})$ with $f(uv)(\text{mod } 2q) \in [1, 2q-1]^o$ or $[2, 2q]^e$ $f(uv) = f(u) + f(v) + K(\text{mod } 2q)$ follows the equation $[f_i(x) + f_j(x) + K](\text{mod } 2q)$ with $V(G) = V(G_i) = V(G_j) = V(G_k)$. We are ready to show four principles for a standard group in the following.

(1) Zero element. The zero element of Oele(G) is $F(T_1)$ as $k = 0$ in the equation.

(2) Inverse element. The inverse of each element $T_i \in$ Oele(G). In fact, $F(T_i) \oplus F(T_j) \ominus F(T_k) = F(T_{i+j-k \ (\text{mod } 2q)})$.

(3) Uniqueness and closure. $F(T_i) \oplus F(T_j) \in$ Oele(G).

(4) Associative law. Since $[F(T_i) \oplus F(T_j)] \oplus F(T_l) = F(T_{i+j-l \ (\text{mod } 2q)})$, so the associative law holds true.

We conclude that Oele(G) is a Module-K odd-elegant graphical group. Next, we determine a connection between $f_1(uv) = f_1(u) +$

$f_1(v) + K(\text{mod } 2q)$ and $f_k(u) + f_k(v) + K(\text{mod } 2q)$ with some given K in the following. Notice that $f_1(E(T_1)) = [1, 2q-1]^o$ or $[2, 2q]^e$. Consider the labelled graph T_k, take an edge $uv \in E(T_k)$, we have $f_k(uv) = f_k(u) + f_k(v) + K(\text{mod } 2q) = [f_1(u) + f_1(v) + (k-1)](\text{mod } 2q)$, and there are the following cases: $f_k(u)+f_k(v) = f_1(u)+f_1(v)$. The proof of this theorem is complete.

Theorem 2 *Let $G_1, G_2, \ldots, G_m$ be m bipartite graphs having Module-K odd-elegant labellings and (X_i, Y_i) be the bipartition of each $V(G_i)$ with $i \in [1, m]$. The graph G is obtained by joining a vertex of G_i with a vertex of G_{i+1} for $i \in [1, m]$. When m is odd, Then G has a Module-K odd-elegant labelling.*

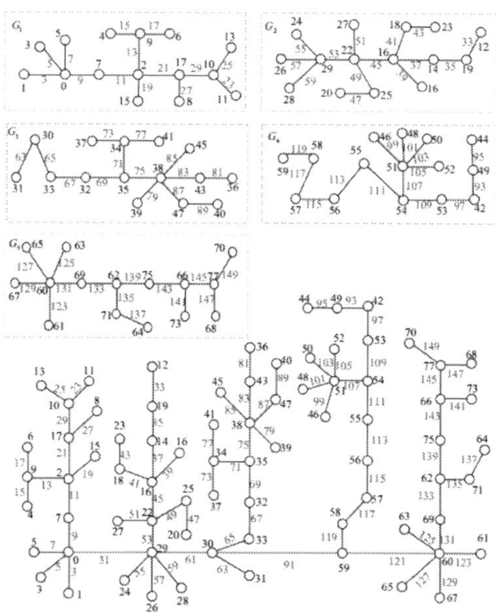

Figure 3. A Module-2 odd-elegant labelling for illustrating Theorem 2.

A dual labelling f_k of a modular labelling f_k is defined by $f_k(uv) \in [1, 2q-1]^o$ or $[2, 2q]^e$. Let Oele(G) be the set of the graphs H_k, where each H_k is a labelled Oele(G) by the dual labelling f_k of the modular labelling f_k with $x \in V(G)$ with some given K. We propose two new graphical subject to some con-straint sets as: When $G_1, G_2, \ldots, G_m$ be m bipartite graphs and neither m is odd or even with the Module-K odd-elegant labelling g induced by the vertex set-labelling, we show that every simple, connected (p, q)-graph G admits a Module-K odd-elegant labelling, and any tree having a Module-K odd-elegant labelling. We propose the following problems for further researching new graphical passwords:

Theorem 3 *Let $G_1, G_2, \ldots, G_m$ be m bipartite graphs having Module-K odd-elegant labellings and (X_i, Y_i) be the bipartition of each $V(G_i)$ with $i \in [1, m]$. The graph G is obtained by joining a vertex of G_i with a vertex of G_{i+1} for $i \in [1, m]$. Then G has a Module-K odd-elegant labelling.*

Theorem 4 *Let $G_1, G_2, \ldots, G_m$ be m bipartite graphs having Module-K odd-elegant labellings and (X_i, Yi) be the bipartition of each $V(G_i)$ with $i \in [1, m]$. The graph G is obtained by joining a vertex of G_i with a vertex of G_{i+1} for $i \in [1, m]$. If m is even, then G has a Module-K odd-elegant labelling.*

4 Conclusion and further researches

We are working on new type of graphical passwords based on the idea of "topological structure plus number theory", vividly speaking, new graphical passwords can be called "mathematical fingerprints". And we are exploring transforming human faces into mathematical models in order to recognizing and comparing human faces through mathematical models rather than image recognition among vast of human faces, such works can be considered as "mathematical faces". However, we will make effort to let mathematical fingerprints approximating to human fingerprints. By our experience on working Topsnut-GPWs, we propose the following conjectures:

(i) *Every bipartite graph admits a Module-K odd-elegant labelling.*

(ii) *A graph is odd-graceful if and only if it admits a Module-K odd-elegant labelling.*

(iii) *Any tree T has a Module-K odd-elegant labelling.*

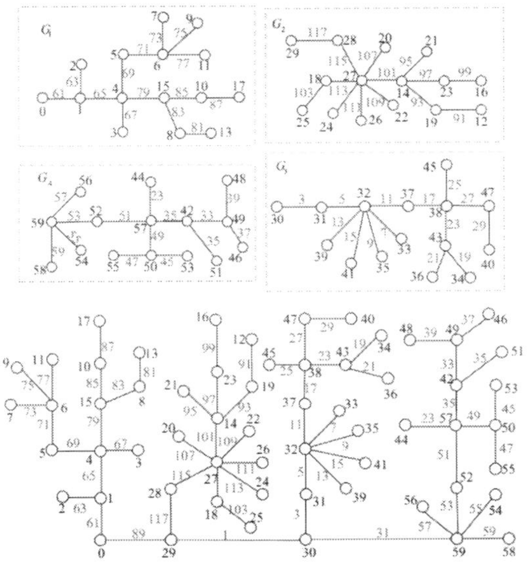

Figure 4. A Module-60 odd-elegant labelling for illustrating Theorem 4

Some of recent works have demonstrated that textual passwords are particularly vulnerable to targeted online guessing. Textual passwords may be influenced by personal information and textual passwords may be reused, both of which contribute to the vulnerability of textual passwords to online guessing. "Topological structure plus number theory" can be realized by graph labellings, future tests could be designed to answer this question.

Acknowledgment

This research was supported by the National Natural Science Foundation of China under grants No. 61163054, No. 61662066 and No. 61363060.

References

[1] Xiaoyuan Suo, Ying Zhu, G. Scott. Owen. Graphical Password: A Survey. In: Proceedings of Annual Computer Security Applications Conference (ACSAC), Tucson, Arizona. IEEE (2005) 463-472.

[2] R. Biddle, S. Chiasson, and P.C. van Oorschot. Graphical passwords: Learning from the First Twelve Years. ACM Computing Surveys, 44(4), Article 19:1-41. Technical Report TR-09-09, School of Computer Science, Carleton University, Ottawa, Canada. 2009.

[3] Haichang Gao, Wei Jia, Fei Ye and Licheng Ma. A Survey on the Use of Graphical Passwords in Security. Journal Of Software, Vol. 8, No. 7, July 2013, 1678-1698.

[4] Rosa A. On certain valuations of the vertices of a graph. Theory of Graphs, 1967:349-355.

[5] Joseph A. Gallian. A Dynamic Survey of Graph Labeling. *The electronic journal of combinatorics*, 17 (2014), # DS6.

[6] Hongyu Wang, Jin Xu, Bing Yao. Exploring New Cryptographical Construction Of Complex Network Data. IEEE First International Conference on Data Science in Cyberspace. IEEE Computer Society, (2016):155-160.

[7] Hongyu Wang, Bing Yao, Chao Yang, Sihua Yang, Xiang-en Chen, MingYao, Zhenxue Zhao. Edge-Magic Total Labellings Of Some Network Models. (Proceeding of ISCCCA13, Published by Atlantis Press, Paris,France, 2013, 0491-0494.) Applied Mechanics and Materials Vols. 347-350 (2013) pp 2752-2757.

[8] Hongyu Wang, Bing YAO, Chao YANG, Sihua YANG, Xiang'en CHEN. Labelling Properties Of Models Related with Complex Networks Based On Constructible Structures. Advanced Materials Research Vols. 765-767 (2013) pp 1118-1123. DOI:10.4028/www.scientific.net/AMR.765-767.1118

[9] Hongyu Wang, Bing Yao, Ming Yao. Generalized Edge-Magic Total Labellings Of Models from researching Networks. Information Sciences 279 (2014) 460-467. DOI:10.1016/j.ins.2014.03.132

[10] Bing Yao, Hui Cheng, Ming Yao and Meimei Zhao, A Note on Strongly Graceful Trees. Ars Combinatoria 92 (2009), 155-169.

[11] Xiangqian Zhou, Bing Yao, Xiang-en Chen and Haixia Tao. A proof to the odd-

gracefulness of all lobsters. Ars Combinatoria 103 (2012), 13-18.

[12] Xiangqian Zhou, Bing Yao, Xiang'en Chen. Every Lobster Is Odd-elegant. nInformation Processing Letters, 113 (2013), 30-33.

[13] B. YAO, X. LIU and M. YAO. Connections Between Labellings Of Trees. Bulletin of the Iranian Mathematical Society, Vol. 43, No.2 (2017): 275-283.

[14] Hongyu Wang, Jin Xu,Bing Yao. The Key-models And Their Lock-models For Designing New Labellings Of Networks.Proceedings of 2016 IEEE Advanced Information Management, Communicates,Electronic and Automation Control Conference (IMCEC 2016): 565-5568.

[15] Hongyu Wang, Jin Xu,Bing Yao. Exploring New Cryptographical Con-struction Of Complex Network Data. IEEE First International Conference on Data Science in Cyberspace. IEEE Computer Society, (2016):155-160.

[16] Xiaoyuan Suo, Ying Zhu, G. Scott. Owen, Graphical Passwords: A Survey. In Annual Computer Security Applications Conference (ACSAC), December 2005: 463-472. DOI: 10.1109/CSAC.2005.27 Source: DBLP

[17] Bing Yao, Hui Cheng, Ming Yao, Meimei Zhao, A note on strongly graceful trees,Ars Combinator. 92 (2009) 155-169.

[18] Xiangqian Zhou, Bing Yao, Xiang'en Chen, Haixia Tao, A proof to the odd-gracefulness of all lobsters, Ars Combinator. 103 (2012) 13-18.

L-band 4×4 Full-redundant Matrix Based on Microwave Multilayer Technique

Xiaoming Li[a], *Qinghui* Song, and *Jiangkun* Du

The 54th Research Institute of CETC, Shijiazhuang 050081, Hebei, P.R. China

Abstract. A planar RF 4×4 microwave switch matrix has been proposed to reduce the size of traditional RF matrix at L-band, where multilayer microwave technique based on PCB is employed to integrate embedded power divider and RF interconnections. Buried components are simulated and optimized using 3D EM software to obtain good circuit performance. A prototype of the matrix is designed and fabricated on Arlon "CLTE-XT" substrate with "25N" prepreg. For all ports of the matrix and among the working band from 950MHz to 2150MHz, both input and output VSWR are less than 1.7, transmission gains are between 6.8 and 8, while ON-OFF isolations are better than 50 dB. The dimension of the matrix is not greater than 70mm×120mm×20mm, which is reduced by 2 order of magnitude compare with traditional design.

1 Introduction

Microwave switch matrix (MSM) is a frequently used device in modern electric systems, especially in multi-beam systems [1]. Performance of MSM playing an important role in system behaviour, it got continuously attention in both academic and industry fields. François Assal in COMSAT Lab. and Xavier Rozec in Thomson Corp. proposed a lightweight, low-power-consuming 8×8 MSM [2] in 1979, it has implemented cross-bar configuration to provide full interconnect flexibility, and use push-pull connectors to ease the assembly procedure. Although the matrix of François and Xavier seems bulky in nowadays, the structure is still popular due to its advantage on modulization and expanding ability. R. K. Gupta in COMSAT proposed another type of 4×4 quasi-planar MSM in 1988 [3], it is composed of input plane, output plane, metal structure between them and connections. Gupta's structure provides much smaller size compare with the former one, but the arrangement of its input and output port is not convenient for combination. Recent research on matrix are concentrates on new integration platform like SIP or SOC, but none of them are fully-redundant [4, 5].

With rapidly development of MIC and MMIC, more and more attentions are transferred to planar circuit in RF design, which can reduce system size and cost tremendously. It is obviously that the bottleneck of planar circuit miniaturization is component mounting area, which is equal to top and bottom layer area in traditional design. Microwave multilayer technique [6, 7] is developed to deal with the problem, it is focused on use of inner layer of PCB, LTCC and other planar circuit platforms. As many microwave components like filter, power divider, and coupler are realized via transmission line, it is convenient to integrate in planar circuit platform directly. In this paper, design method of microwave multilayer technique based on PCB is discussed, a 4×4 Full-redundant MSM work in L-band is designed using the method, which exhibit good RF performance and occupy very small circuit area. In addition, the proposed MSM have the ability to rearrange RF port conveniently.

2 System design of MSM

2.1. Topology

In order to meet the requirement of full-redundant, topology in figure 1 is frequently used for N×M MSM. Typical MSM is composed of N M-way divider module, M N-way switch module and N×M connection RF cables between them. The input of M-way power divider is connected to each input port of the matrix, while M outputs of divider are connected to corresponding switch module. The output of N-way switch module is connected to each output port of the matrix. In practice, switch module usually use two or more stage of MMIC switch to improve isolation between different channels.

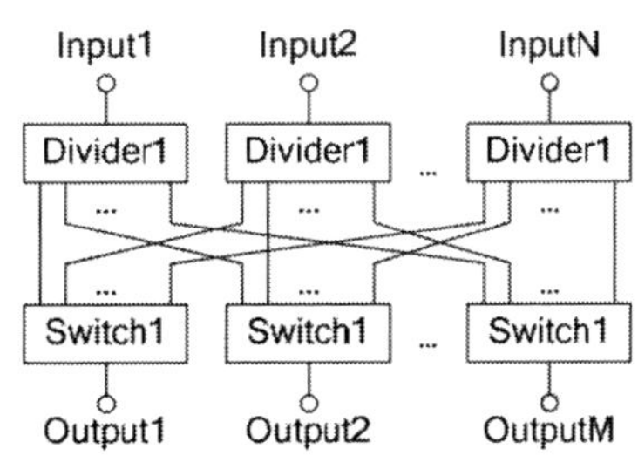

Fig. 1. Topology of full-redundant MSM

[a] Corresponding author: 624910659@qq.com

2.2 Structure

PCB platform is selected for multilayer layout to reduce costs. Symmetry for both input and output is chosen so unitary divider and switch module are employed. Four buried Wilkinson 4-way equal divider is used to provide isolation between outputs connected to same input. Two stage MMIC switch is cascaded to construct 4-way switch module. Sketch map of the proposed structure is illustrated in figure 2. four Wilkinson 4-way dividers are stacked in the inside of the PCB board, top and bottom layer of the board are used to mounting Dc-Dc module, control unit, switch module, and amplifier modules. Buried divider occupied no surface area and can provide good VSWR and insertion loss compare with chip divider. These characters make superiority of the proposed MSM both on size and RF performance.

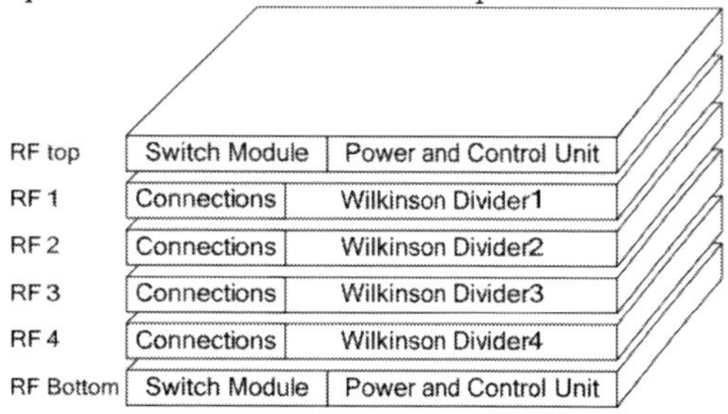

Fig. 2. Structure sketch map of proposed MSM

In figure 2, four buried RF layers and two component layers are stacked together, so six RF layers in all are designated.

3 Material system and laminates design

3.1 Material system

Based on comprehension consideration of cost, size, performance, and the ability of buried resistor integration, the laminate material "CLTE-XT" and prepreg material "25N" produced by Arlon Corp. are selected. "CLTE-XT" has a relative dielectric constant of 2.94, loss tangent of 0.0012, x and y axis CTE of 8, z axis CTE of 20. "25N" has a relative dielectric constant of 3.38, loss tangent of 0.0025, x and y axis CTE of 15, z axis CTE of 50. Two laminates and one prepreg composite a single buried RF layer illustrated in figure2.

3.2 Laminates design

As well known, for each buried RF layer or component layer, thicker laminates and prepreg can provide more impedance precise under fabrication, but it also introduced larger circuit size. Considered that impedance precise can influence the RF performance, miniaturization and RF performance forms a trade-off on circuit design. Another restriction on laminate design is the total thickness of PCB board, fabrication of 6 mm thick or above PCB usually causes an unacceptable price. Under these considerations, the thickness of the laminate is chosen to be 0.254 mm, while the thickness of prepreg is set to be 0.1 mm. PCB stack scheme of the proposed MSM is illustrated in figure 3. Totally 10 laminates and 9 prepregs are used to form a 20 layer PCB structure. Thickness of copper foil cladding is 1/2 oz for inner layers and 1 oz for component layers. 9 types of via are used in the design which can be degenerated to 5 types in process flow because its symmetry. Totally thickness of the board is below 4 mm, which is economic applicable in fabricate.

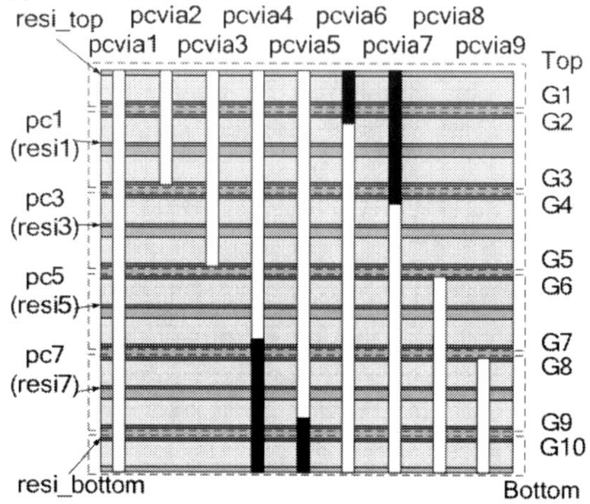

Fig. 3. PCB stack scheme of proposed MSM

4 Design of buried transmission line and passive components

4.1 Buried transmission line

Transmission line (TL) is the most basic element of RF circuits, different components are connected by TL and some RF components are composed of TLs it self. Typical buried TL is somewhat like strip line (SL) or eccentric strip line (ESL) but not the same. The key distinction between buried TL and SL or ESL is that the dielectric around signal line is inhomogeneous, which is illustrated in figure 4.

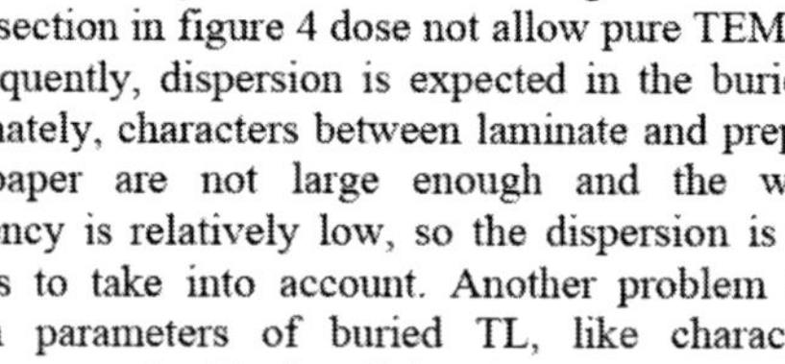

Fig. 4. Cross-section of buried transmission line

The boundary condition of inhomogeneous dielectric cross-section in figure 4 dose not allow pure TEM mode. Consequently, dispersion is expected in the buried TL. Fortunately, characters between laminate and prepreg in this paper are not large enough and the working frequency is relatively low, so the dispersion is not as bad as to take into account. Another problem is that design parameters of buried TL, like characteristic impedance and effective dielectric constant can not be calculated via expressions in existence. An effective way to get buried TL design parameters is to solve the Maxwell equations on the cross-section by numerical method, several commercial EM softwares can be used to do this. This paper use ANSYS EM to compute characteristic impedance and effective dielectric constant vary with strip width w_0, simulation results are shown in figure 5(a) and (b) respectively. The intersection point of 50 ohm grid and impedance curve is marketed on figure 5(a), x coordinate of the point is the width of 50 ohm TL. Using the same method, effective dielectric constant of 50 ohm buried TL can be read from figure 5(b), which is about 3.03.

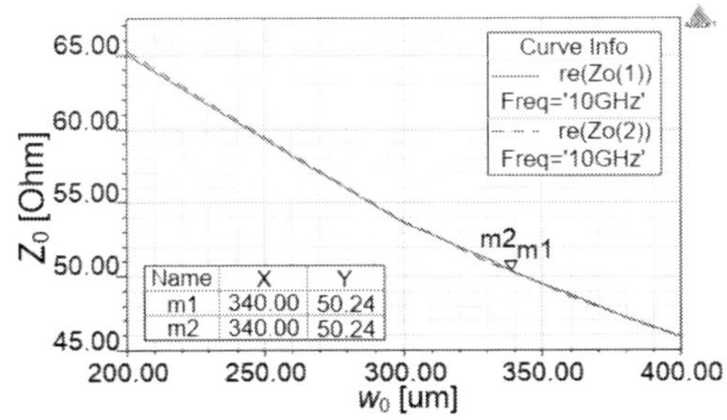

(a) Characteristic impedance with w_0

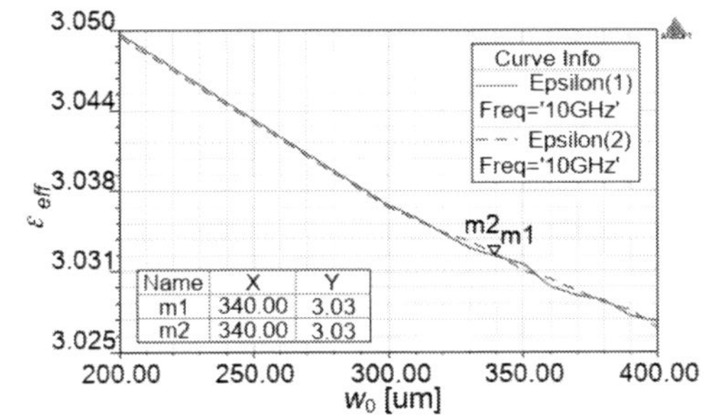

(b) Effective dielectric constant with w_0

Fig. 5. Simulation results of buried TL using ANSYS EM

4.2 Buried Wilkinson divider

Buried 4-way Wilkinson divider is composed of three 2-way divider, witch model is illustrated in figure 6 (a). The simulation of 2-way divider is also made by ANSYS EM, simulation results are shown in figure 6(b)-(d).

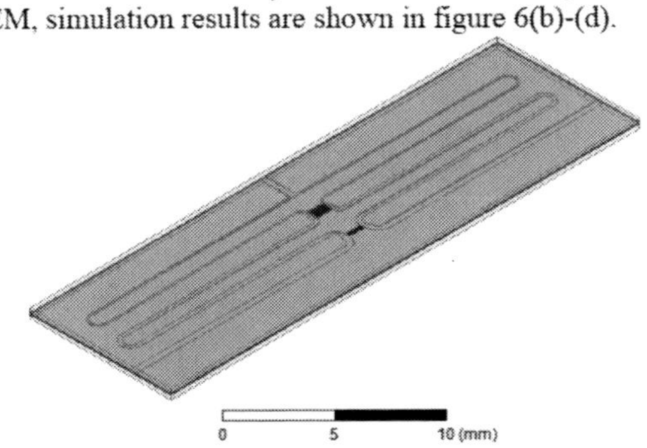

(a) Simulation model of 2-way divider

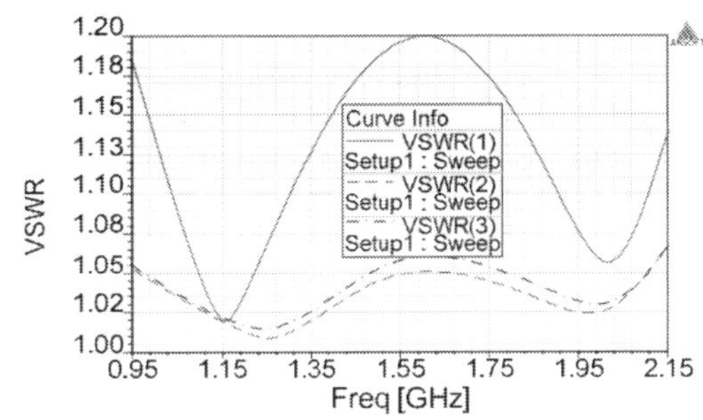

(b) VSWR of 2-way divider

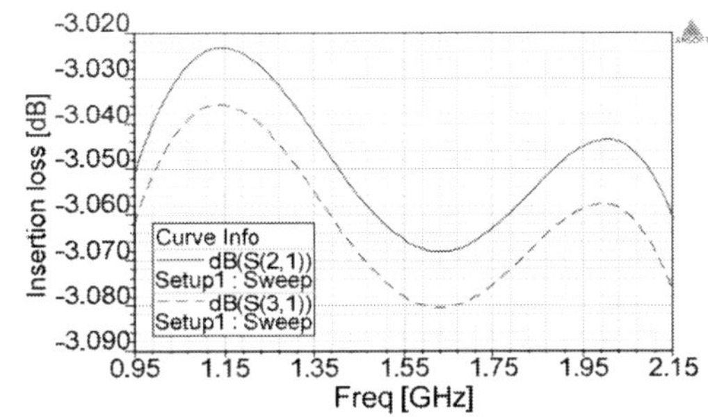

(c) Insertion loss of 2-way divider

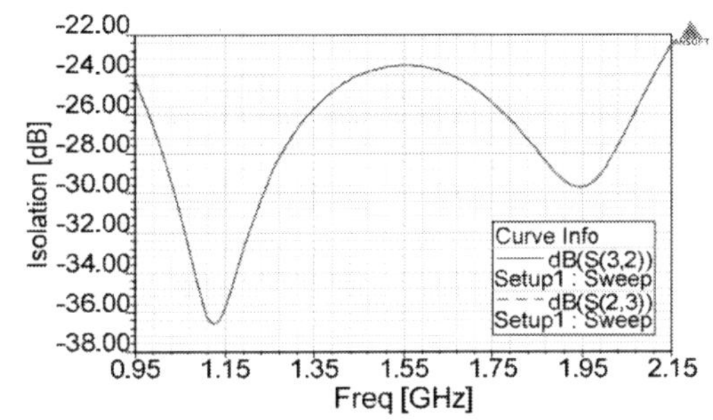

(d) Isolation of outputs of 2-way divider

Fig. 6. Model and simulation results of 2-way Wilkinson divider

5 Fabrication and measurement

A prototype of proposed MSM is fabricated based on former designs. The pictures of its mother board and MSM itself are given in figure 7(a) and (b). The matrix is measured by Agilent vector network analyser N5230A, which is calibrated by 85052D calibration kit. Measurement results are in figure 7(c)-(f). It can be seen that the VSWR of both input and output is less than 1.7; the transmission gain of the on-state is between 6.8 and 8. The off-state insertion loss is greater than 45 dB, which leads a on-off isolation (on-state transmission gain plus off-state insertion loss) more than 50 dB.

 DOI: 10.1051/matecconf/201713900207

(a) Mother board of MSM (b) Control environment

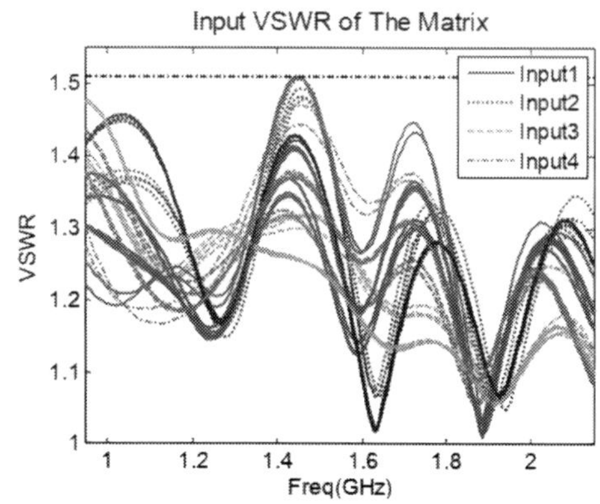

(c) VSWR of the input ports

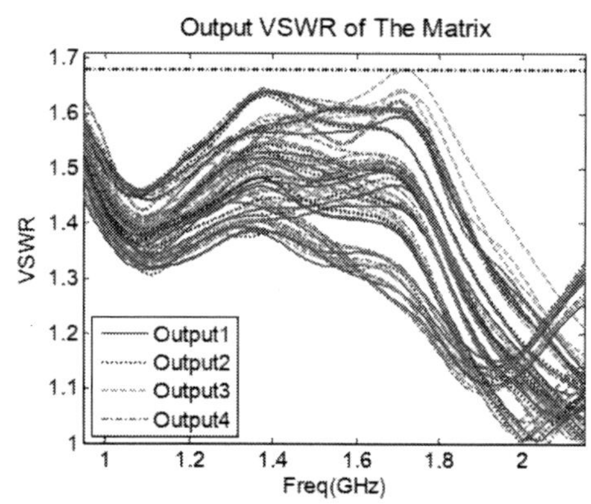

(d) VSWR of the input ports

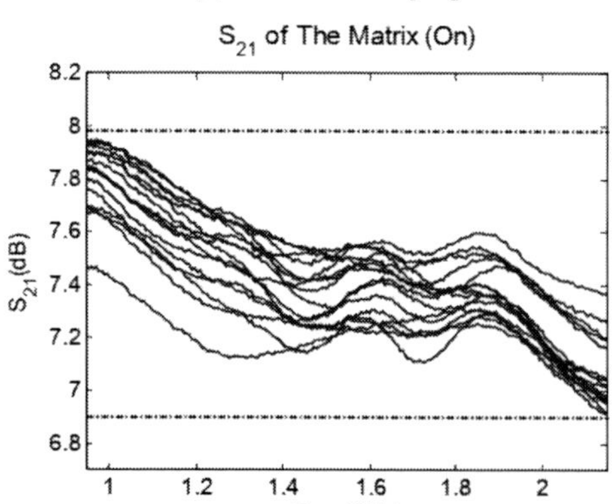

(e) Insertion loss of the on-state

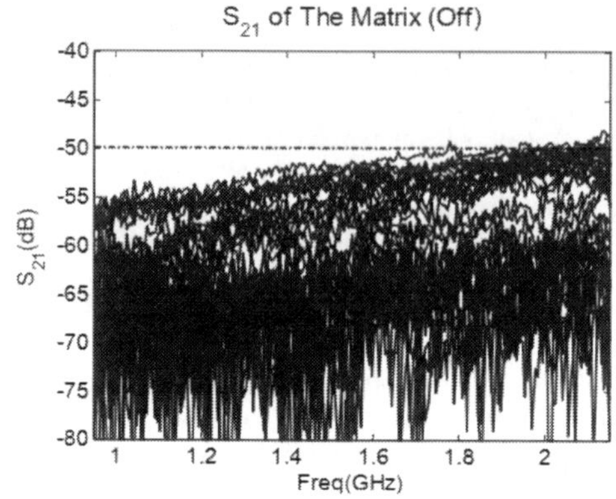

(f) Off-state insertion loss of the MSM

Fig. 7. Picture and measurements of the proposed MSM

6 Conclusion

In this paper, a novel L band planar 2×2 MSM based on microwave multilayer technique is proposed. Prototype of the MSM is designed and realized via PCB technique. The measurements of the MSM shown good performance on VSWR, transmission gain and on-off isolations, while a tremendous size reduce is achieved compare with traditional way. Benefits from mature PCB technique, the proposed MSM also have the character of low cost. It is obviously that the proposed MSM have potential application prospects in various communication systems, especially in space borne environment.

References

1. L. Qinghua, L. Pei, L. Xiao, etc. A prototype of feed subsystem for a mutilple-beam array-fed reflector antenna. ISAP (2015), 1-3.
2. F. Assal, X. Ro. Fast, Fully-Redundant, 4 GHz, 8x8 Microwave Switch Matrix for Communications Satellites. 9th EMC (1979), 218-222.
3. Gupta R. K., Gerson H. I., Ross P. B., etc. Design and packaging approach for MMIC insertion in a broadband 4×4 microwave switch matrix. GaAs IC Symposium (1988), 261-264.
4. S. Kaleem, S. Rentsch, S. Humbla, etc. Highly integrated reconfigurable microwave switch matrix module for geostationary satellites. 44th EMC (2014), 992-995.
5. A. Ebert, S. Kaleem, J. Müller. An industry-level implementation of a compact microwave diode switch matrix for flexible input multiplexing if a geo-stationary satellite payload. COMCAS (2015), 1-4.
6. Metzen P L. Globalstar Satellite Phased Array Antennas. ICPAT (2000), 207-210.
7. Gustafasson A, Malmqvist R, Pettersson Lz. A Very Thin and Compact Smart Skin X-Band Digital Beamforming Antenna. European Radar Conference (2004), 313-316.

Research on vehicle speed detection technology based on micro-magnetoresistive sensing equipment

Wei Zhang, *Li* Wang*, *Chunli* Li, *Mi* Chen, and *Sheng* Tao

China Academy of Transportation Sciences, Beijing 100029, China

Abstract. In order to effectively detect road traffic to achieve intelligent road traffic management, the vehicle speed detection algorithm based on micro-magnetoresistive sensing equipment was studied. A new pairwise differential eigenvalue vehicle speed calculation algorithm was proposed to solve the problem that selecting one eigenvalue on the magnetoresistive curve may be influenced by the actual condition that the data curves of the two sensors and the relative base voltage cannot be exactly the same. The results of the application on a certain expressway indicate that speed detection accuracy can be 95.4%, besides, both the integrity and the validity of the data can meet the requirements.

Key words. Road engineering; Vehicle speed detection; Pairwise differential eigenvalues; Micro-magnetoresistive sensor

1 Introduction

Real-time road traffic information collection is the basis of road administration department to make road monitoring management, traffic information analysis and relevant research and decision. Enhancing research on traffic information sensing technology and equipment is the premise to achieve intelligent road traffic management.

As a kind of shallow-embedded vehicle detection equipment, micro-magnetoresistive vehicle detector has the characteristic such as no reaction to non-ferromagnetic objects, informative, small and light and easy to be installed. It has outstanding effect in the application of dynamic and static vehicle detection, estimate of vehicle speed and length as well as classification of vehicle type[1,2,3,4].

In recent years, many relevant scientific research institutes and enterprises at home and abroad have been conducting research on magnetoresistive vehicle detection technology. How to separate vehicle information from continuous signal sequence is the problem that needs to be solved by magnetoresistive sensor vehicle recognition algorithm. Sing[5] introduced a fixed threshold value detection algorithm, but it doesn't solve the problem that the sensor is liable to have datum drift and the distortion. In order to address the defect of threshold value judgment, Knaian[6] proposed state machine detection method, which does not only judge the existing state of vehicles, but also judges the information such as coming, existing and leaving of vehicles. The subdivision of the state enhances the robustness of detection algorithm and improves algorithm recognition rate, but the drift of datum value of sensor still exists. In a word, detection algorithm is mainly affected by noise, discontinuity of detection signal and the drift of datum value. At present, there is no good solution for drift of datum value and the commonly used method is re-calibration of datum value regularly.

This paper conducted research on vehicle speed detection algorithm based on micro-magnetoresistive sensing equipment and proposed a pairwise differential eigenvalue vehicle speed calculation algorithm, which reduced the error of vehicle speed detection caused by drift of datum value and effectively improved the vehicle speed detection accuracy of micro-magnetoresistive sensing equipment.

2 Research on vehicle speed detection algorithm based on micro-magnetoresistive sensing equipment

This work was supported by the Scientific Research Project of China Academy of Transportation Sciences (20161203).
* Corresponding author: wangli@catsic.com

DOI: 10.1051/matecconf/201713900208

2.1 Vehicle speed detection principle

The principle of vehicle speed detection with micro-magnetoresistive sensing equipment is: when a vehicle comes close to magnetoresistive equipment, signal of magnetoresistive 1 and 2 will exceed threshold value successively and produce similar changing Curve. As shown in Figure 1, under ideal circumstance, signal curve of magnetoresistance 2 could overlap with that of magnetoresistance 1 when making time-domain translation Δt, that is, a corresponding similar point A_2 could be found on magnetoresistance 2 if A_1 is taken randomly on the curve of magnetoresistance 1. The time difference between the two points is the time delay Δt required for calculating speed. Thus, as long as the accurate time delay is got, the accuracy of vehicle speed calculation can be ensured. Supposing serial number of sampling point of A_1 is N_1 while that of sampling point of A_2 is N_2, sampling frequency is f_{sp} and the distance between two sensors is l, vehicle speed could be expressed as:

$$v = l \cdot f_{sp} / (N_2 - N_1) \tag{1}$$

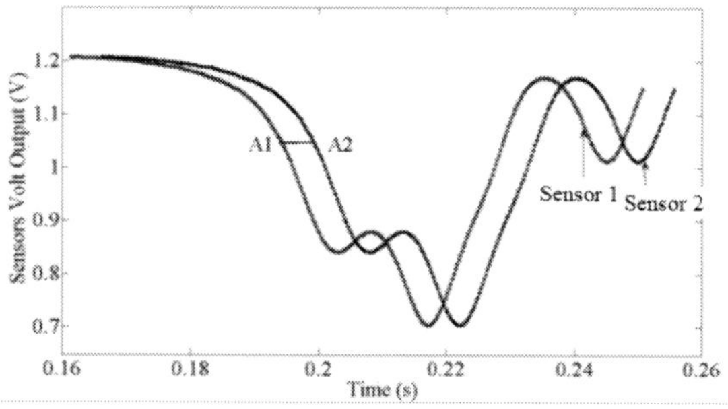

Fig.1. Schematic of Speed Detection

However, as the static output voltage of the two magnetoresistive sensors is different in practical application, there is certain difference between the magnification times of signal in module. Besides, in case the driving direction of vehicle has certain included angle with axial direction of magnetoresistive equipment, the part of automobile passing two magnetoresistances will also differ, which, combining with random jitter of vehicle and road roughness, will result in different signal sensed by the two sensors. Therefore, under the influence of many factors, the curve of signal sensed by two magnetoresistances cannot completely overlap in practice. Thus, we cannot select two similar points on magnetoresistive curve randomly, but shall select appropriate eigenvalue to calculate time delay. The research and practical result indicates selecting the peak value of the two magnetoresistive curves or the threshold value triggered by the coming of vehicles as eigenvalue to calculate time delay could guarantee the accuracy of time delay calculation.

2.2 Vehicle detection error caused by signal amplitude difference

It is found in practical application that the data curves of the two sensors and the relative base voltage cannot be exactly the same, thus calculation with one eigenvalue may not get better effect. As shown by measurement signal in Figure 2, Figure 2 (b) is the local amplification of Figure 2 (a).

In Figure 2, two signals have been subtracted with base signal. After careful observation of Figure 2(b), it can be found although the two curves have similar shape, their amplitude varies, which result in: in case single eigenvalue of one time is used to calculate speed, there will be speed measurement error as the signal amplitude of the two sensors is different.

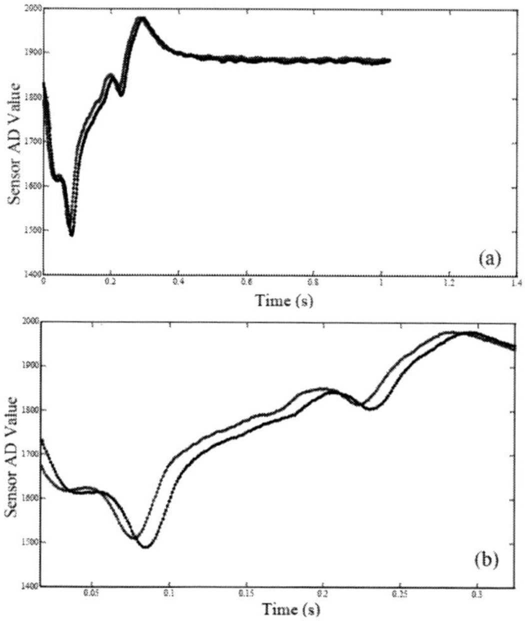

Fig.2. Filtered Signal of Magnetoresistive Sensor (a) global curve (b) local magnified curve

The principle of the measurement error caused by signal amplitude differences is shown in Figure 3, obviously, $t_2 > t_1$ in the figure, which means the speed calculated with difference between two times has error. Supposing actual time difference is t_0, t_1 could be expressed as $t_1 = t_0 - dt$ while t_2 as $t_2 = t_0 + dt$.

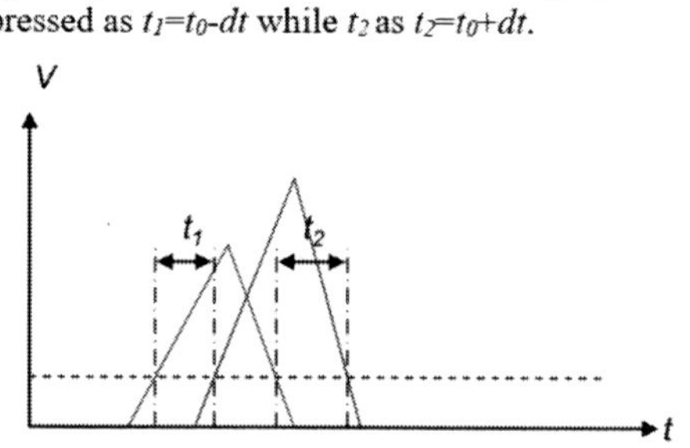

Fig.3. Schematic Diagram of Speed Measurement Error Caused by Signal Amplitude Differences

DOI: 10.1051/matecconf/201713900208

2.3 Vehicle speed detection algorithm based on pairwise differential eigenvalue

In order to solve the problem of vehicle speed detection error caused by signal amplitude difference, previous method is to make signal normalization processing by using hardware or software, but this method needs to do lots of test and the consistency of signal value cannot be absolutely guaranteed. It can be seen from aforesaid analysis that the average value of t_2 and t_1 is true value t_0. Based on this method, this paper proposed a kind of method of calculating average time with pairwise differential eigenvalue. Its schematic diagram is described as follows.

First of all, set threshold value T and difference value d, and then make binarization processing of sensor signal S. The method is as shown in formula (2).

$$S_n' = \begin{cases} 1, S > (T-d), S_{n-1}' = 1 \\ 1, S > (T+d), S_{n-1}' = 0 \\ 0, S < (T+d), S_{n-1}' = 0 \\ 0, S < (T-d), S_{n-1}' = 1 \end{cases}, S_0' = 0 \quad (2)$$

Where, T-d and T+d are referred to as a pair of pairwise differential eigenvalue.

At last, make XOR of binarization sequence signal of the two sensors to get time difference sequence $\{T_d\}$, and take even-even time difference from the sequence, calculate its average value and use it as actual vehicle passing time t to calculate final speed.

This method could effectively reduce the speed measurement error caused by the difference of sensor signal amplitude. In addition, the hysteresis effect caused by $2d$ could effectively reduce the influence of fluctuation of signal in a small range.

3 Installation of micro-magnetoresistive vehicle detection equipment

In Figure 4, equipment arrangement diagram is provided with a unidirectional three-lane highway as an example. The arrangement method of bidirectional multi-lane is similar. Magnetoresistive equipment is a cylindrical one and single magnetoresistive equipment also has directionality, which is referred to as axial direction of equipment in this research. The arrow in the figure represents the driving direction, that is, the direction from magnetoresistance 1 to 2 within the equipment. It means when a vehicle arrives, it will be sensed by magnetoresistance 1 and 2 successively.

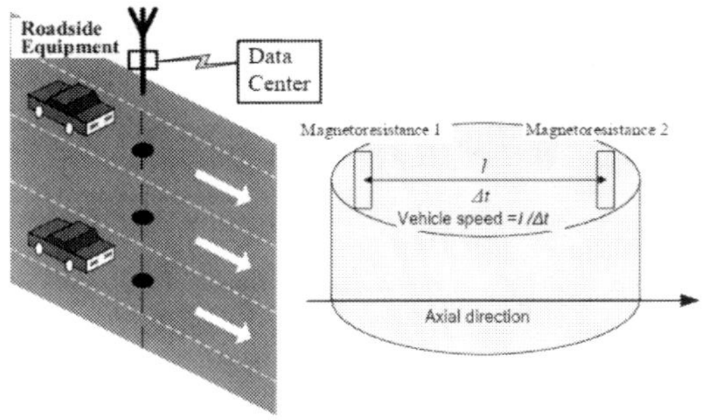

Fig .4. General Scheme of Magnetoresistive Equipment

Magnetoresistive sensing equipment is installed at the center of every lane of the road and is kept to be at a straight-line vertical to the driving direction. One magnetoresistive equipment can only be used to detect vehicles passing such lane. By adjustment of threshold value, the magnetoresistive equipment located at the center of some lane can only be triggered by the vehicle passing right over or above, thus there is no interference from adjacent lanes.

In case magnetoresistive equipment is buried under the pavement, the plane of its top shall be parallel to and be slightly lower than the pavement, and be sealed and covered with proper road repair material. In case it is installed on pavement, the axial direction of magnetoresistive equipment shall be parallel to the driving direction of vehicles and the arrow shall point to driving direction. In view of the microminiaturization of magnetoresistive equipment, the pavement drilling diameter shall be about 150mm. Thus, the distance between two magnetoresistive sensors shall be set as l=120mm. The vehicle speed $v = l \,/\, \Delta t$ can be calculated by measuring the time delay of sensing of vehicles by the two magnetoresistive equipment.

4 Application of micro-magnetoresistive vehicle detection equipment

In order to inspect the practical application effect, 54 sets of magnetoresistive sensing equipment and 14 sets of roadside antenna receiving equipment are installed at certain expressway, which can provide average speed and flow of traffic stream at 14 cross sections. Figure 5 is the actual installation picture of micro-magnetoresistive vehicle detection equipment on an interchangeable cross section. This cross section is composed of two lanes and every lane is installed with micro- magnetoresistive vehicle detection equipment in the center.

Fig.5. Installation Diagram of Expressway Interchangeable Magnetoresistive

In order to verify the influence of pairwise differential eigenvalue on vehicle speed detection accuracy proposed, several traffic cross sections installed with micro-magnetoresistive vehicle detection equipment are randomly selected as test points and hand-held microwave vehicle speed detector with detection accuracy as 1% are used to detect and record the speed of vehicles passing such cross section, which is used to compare with the vehicle speed measured by magnetoresistive vehicle detector. The number of sample vehicle detected is 100 and the speed range of test sample is 20~120km/h. The two algorithms used in verification could be downloaded to vehicle detection equipment with wireless communication method.

First of all, the simple eigenvalue vehicle detection algorithm with peak eigenvalue is tested. Through test and sample arrangement calculation, 89.1% expectation of vehicle speed detection is got.

Secondly, vehicle detection algorithm with pairwise differential eigenvalue is tested. Through test and sample arrangement calculation, 95.4% expectation of vehicle speed detection is got. Vehicle speed detection accuracy is greatly improved.

5 Conclusion

This paper mainly conducted research on vehicle speed detection algorithm of micro-magnetoresistive vehicle detection equipment and developed a kind of magnetoresistive traffic flow collection equipment with low power consumption and small size based on the algorithm which s relatively higher practical value.

In the research of vehicle speed measurement algorithm, the pairwise differential eigenvalue vehicle speed calculation algorithm is created with the aim of reducing the defect of existing algorithm in practical application, and the speed measurement error caused by the difference of signal amplitude of sensor is calibrated. The result of application on a certain expressway shows the vehicle speed detection accuracy of this equipment is above 95%. This research creates condition for building road traffic flow monitoring system.

References

1. S. Y. Cheung, S. C. Ergen, and P. Varaiya, "Traffic Surveillance with Wireless Magnetic Sensors", ITS World Congress, 2005:648-649.
2. M. H. Kang, B. W. Choi, K. C. Koh, et al., Experimental study of a vehicle detector with an AMR sensor, Sensors & Actuators A Physical, 2005, **118**(2):278-284.
3. M. J. Caruso, C. H. Smith, T. Bratland, et al., A New Perspective on Magnetic Field Sensing, Sensors, 1998, **15**(12).
4. X. Li, G. Yu, Application of AMR sensors to vehicle detection, Journal of University of Science and Technology Beijing, 2006, **28**(6):587-590.
5. S. Y. Cheung, S. Coleri, B. Dundar, et al., Traffic measurement and vehicle classification with single magnetic sensor, Transportation Research Record, 2005, 10.
6. A. N. Knaian, J. Paradiso, A. C. Smith, A Wireless Sensor Network for Smart Roadbeds and Intelligent Transportation Systems, Massachusetts Institute of Technology, 2000.

Application research of power allocation based on Buck circuit in DC microgrid

Zihao Wang, Mingyu Zhou
College of Electrical Engineering, Guangxi University, Nanning, China

Abstract: In a traditional DC microgrid, the power sharing control strategy has been always used in the distributed power converters, resulting in not making outer power allocation arbitrarily. In order to solve the power output allocation problem of wind power in DC microgrid, the intelligent Buck circuit based on PI algorithm and the load current feed-forward method was used to realize the arbitrary regulation of the output power of the wind power in the DC microgrid system. Compared with traditional distributed generators power-sharing method, the simulation and experimental results show the proposed method can realize arbitrary power outputting from distributed generators. Finally, the simulation and experimental results prove the validity and effectiveness of the control method.

1 Introduction

Due to the energy crisis in the new era and the pressure of environmental protection, the demand for new energy power generation technology is becoming increasingly urgent. The microgrid system consisting of new energy DG(distributed generations) has attracted much attention and support in society [1].Modern household appliances are mostly low voltage and small power DC loads. Thus, the small low-voltage DC microgrid receives more and more attention.

In traditional DC microgrid, the equal power sharing model with equal voltage and equal current is used. Literature [2] shows an average current model making the DC/DC converters outputting equal current; The active DC/DC module is studied in document[3]. However, these method can only achieve equal power outputting of each distribution power supply. These method cannot realize arbitrary power allocation from distributed power sources in DC microgrid system.

In this paper, we study two terminal low voltage DC microgrid based on wind power and batteries. The wind power simulator is connected with the new intelligent Buck in series and connected to the DC bus, and then connected to the resistive load. The structure of this paper is organized as follows: Section 2 presents structure and control method of two terminal DC microgrid and new intelligent Buck circuit. Section 3 establishes the simulation and experimental results. Section 4 concludes the paper.

2 The Structure and control strategy of dc microgrid

2.1 The Structure of Two-terminal DC Microgrid

Nowadays, due to the use of a large number of low-voltage DC household appliances, low voltage DC micro grid has received more and more attention. Figue.1 shows a structure of a low-voltage two-terminal DC microgrid. The distributed generator DG connects to the DC bus trough a DC/DC converter. The wind power, always a AC power supply, rectifies into DC after trough rectifier and connects to DC bus trough the intelligent Buck circuit. Generally, DG is used as DC bus regulator, and a stable constant voltage source such as battery is always used to stabilize DC bus voltage. The system connects to a steady resistive load.

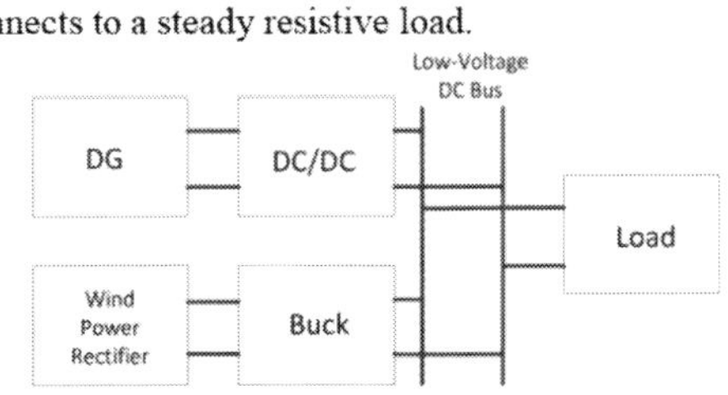

Figue.1. Structure of Two-terminal DC Microgrid

 DOI: 10.1051/matecconf/201713900209

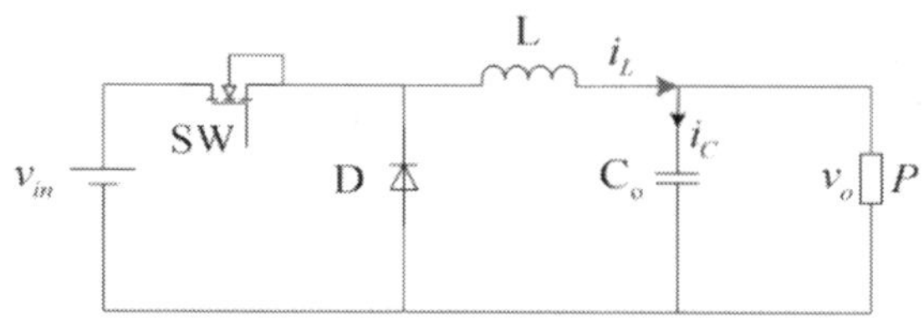

Figue. 2. Structure of new intelligent Buck converter

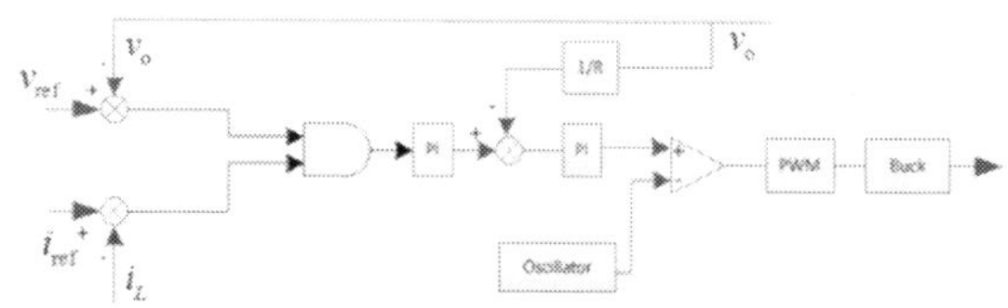

Figue. 3. Control structure of a new intelligent Buck circuit

2.2 The Control of New intelligent Buck circuit

The buck converter connecting the wind power rectifier in Figue.2 is used to supply a constant power and current .The state－space equations of the buck converter when the switch is on (i.e., $0<t<dT$) and off (i.e., $dT<t<T$) are given by (1) and (2), respectively [3].

When $0<t<dT$:

$$\begin{cases} \dfrac{di_L}{dt} = \dfrac{1}{L}(v_{in} - v_o) \\ \dfrac{dv_o}{dt} = \dfrac{1}{C}(i_L - \dfrac{P}{v_o}) \end{cases} \tag{1}$$

When $dT<t<T$:

$$\begin{cases} \dfrac{di_L}{dt} = \dfrac{1}{L}(-v_o) \\ \dfrac{dv_o}{dt} = \dfrac{1}{C}(\dfrac{P}{v_o} - i_L) \end{cases} \tag{2}$$

where d and T are the duty ratio and switching period of the converter, respectively. Using the state－space averaging method the dynamic model of the buck converter can be written as

$$\begin{cases} \dfrac{di_L}{dt} = \dfrac{1}{L}(v_{in}d - v_o) \\ \dfrac{dv_o}{dt} = \dfrac{1}{C}(i_L - \dfrac{P}{v_o}) \end{cases} \tag{3}$$

Consider small perturbations in small-signal model[4]:

$$\begin{cases} v_{in} = V_{in} + \tilde{v}_{in} \\ d = D + \tilde{d} \\ v_o = V_o + \tilde{v}_o \\ i_L = I_L + \tilde{i}_L \end{cases} \tag{4}$$

Where V_{in}, D, V_o, and I_L are the average values of v_{in}, d, v_o and i_L, respectively. Substitute (4) into (3), the dynamic model of the buck converter becomes

$$\begin{cases} \dfrac{d\tilde{i}_L}{dt} = \dfrac{1}{L}(V_{in}\tilde{d} + D\tilde{v}_{in} - \tilde{v}_o) \\ \dfrac{d\tilde{v}_o}{dt} = \dfrac{1}{C}(\tilde{i}_L - \dfrac{P\tilde{v}_o}{V_o^2}) \end{cases} \tag{5}$$

Note considering the fact that $V_o \gg \tilde{v}_o$:

$$\begin{cases} I_L - \dfrac{P}{v_o} = \dfrac{P}{V_o} - \dfrac{P}{V_o + \tilde{v}_o} = \dfrac{P\tilde{v}_o}{V_o(V_o + \tilde{v}_o)} = \dfrac{P\tilde{v}_o}{V_o^2} \\ I_L = \dfrac{P\tilde{v}_o}{V_o^2} + \dfrac{P}{v_o} \end{cases} \tag{6}$$

The transfer functions of the system can be obtained from (4) as follows:

$$\begin{cases} H_1(s) = \dfrac{\tilde{v}_o(s)}{\tilde{d}(s)} = \dfrac{\dfrac{V_{in}}{LC}}{s^2 - (\dfrac{P}{CV_o^2})s + \dfrac{1}{LC}} \\ H_2(s) = \dfrac{\tilde{v}_o(s)}{\tilde{v}_{in}(s)} = \dfrac{\dfrac{D}{LC}}{s^2 - (\dfrac{P}{CV_o^2})s + \dfrac{1}{LC}} \end{cases} \tag{7}$$

Considering the steady constant load based on a linearized small-signal model, it can expand into a big-signal model[5]. The transfer function of the buck converter with a steady load can be obtained as:

$$H_3(s) = \frac{\tilde{v}_o(s)}{\tilde{d}(s)} = \frac{\dfrac{V_{in}}{LC}}{s^2 + \dfrac{1}{RC}s + \dfrac{1}{LC}} \tag{8}$$

where the two poles are in the left half plane. Therefore, the system is stable.

The control structure of a new intelligent Buck circuit is shown in Figue.3.

The current source voltage and current control method is used in this paper. This method compares the load voltage v_o and the reference voltage v_{ref} to prevent external voltage suddenly increasing that results in a sudden increase in load over voltage damaging and system collapsing. Linear PI controllers can be designed to stabilize the system around a specific operating point based on a linearized small-signal model. And the system is stable from (8).

3.Simulation and experiment results

3.1 The Simulation results

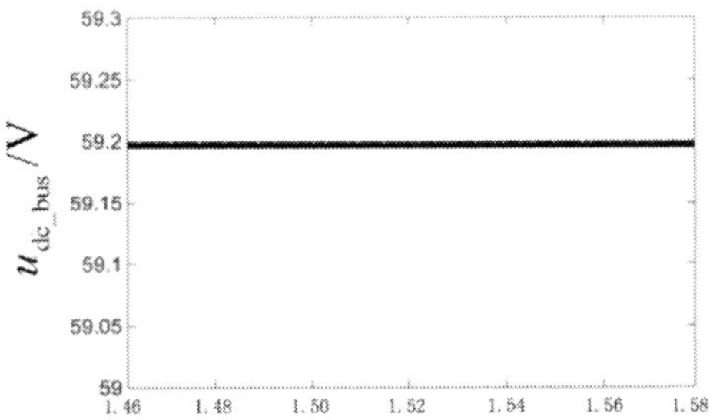

Figue.4. DC bus voltage when inductor outputting 4A

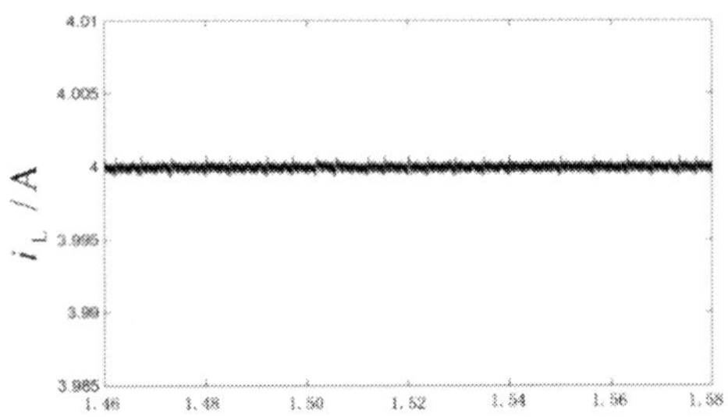

Figue.5. Inductor current when inductor outputting 4A

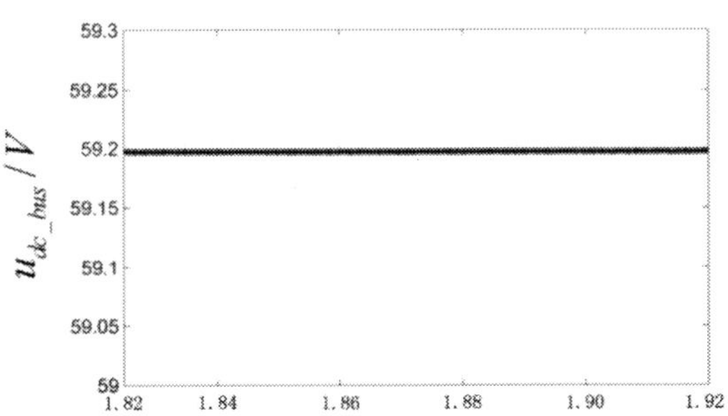

Figue.6. DC bus voltage when inductor outputting 6A

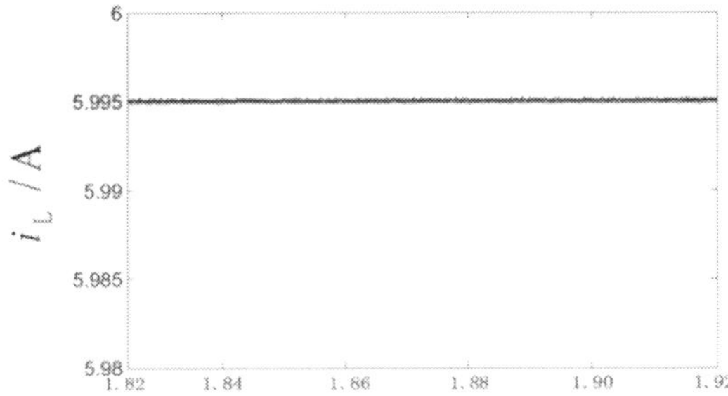

Figue.7. Inductor current when inductor outputting 6A

To verify validity and effectiveness of the control method in this study, a simulation model is built in MATLAB/Simulink using a two-terminal DC microgrid model. A 60V battery is used as the DC microgrid bus voltage regulator. Wind power simulator fluctuates at 100V-120V according to the simulation curve and connects to the intelligent Buck circuit after rectified through a uncontrolled full bridge rectifier. As for the intelligent Buck circuit, the inductor is 0.1mH,the switching frequency is 50kHz, and the constant resistive load is 10Ω.The simulations results are shown in Figue.4-Figue.7.As shown in Figue.4 and Figue.5,the DC bus voltage is 59.2V and inductor outputs 4A current .Then we can know that the wind power outputs 4A and battery outputs 6A for the load. As shown in Figue.6 and Figue.7,the DC bus voltage is 59.2V and inductor outputs 6A current .Then we can know that the wind power outputs 6A and battery outputs 4A to the load.

As shown in Figue.6 and Figue.7, the proposed algorithm can realize the free adjustment of the output power of the distributed power supply. In Figue.5, the inductance current is 4A. The Figue.4-Figue.7 show that when the output current of wind power rectifier is changed, Figue.4 and Figue.6 show DC microgrid bus voltage is constant 59.2V, and the inductor current changed from 4A of Figue.5 to 6A of Figue.7.The simulations show that this algorithm can realize power outputting arbitrarily.

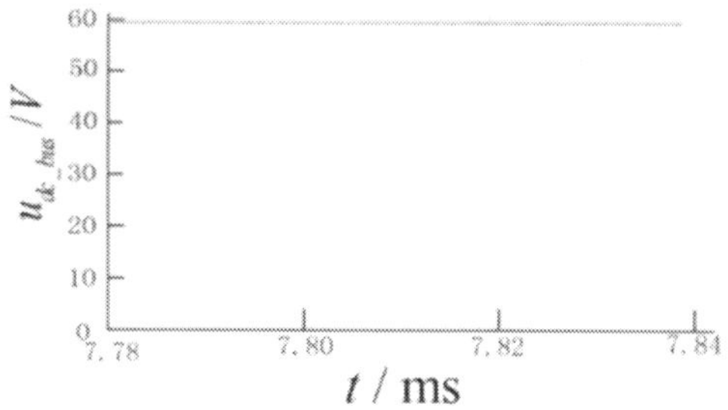

Fig.8. DC bus voltage 1

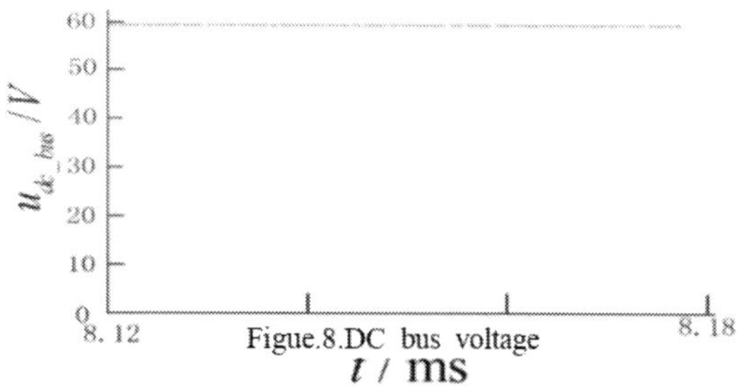

Fig.10 DC bus voltage 2

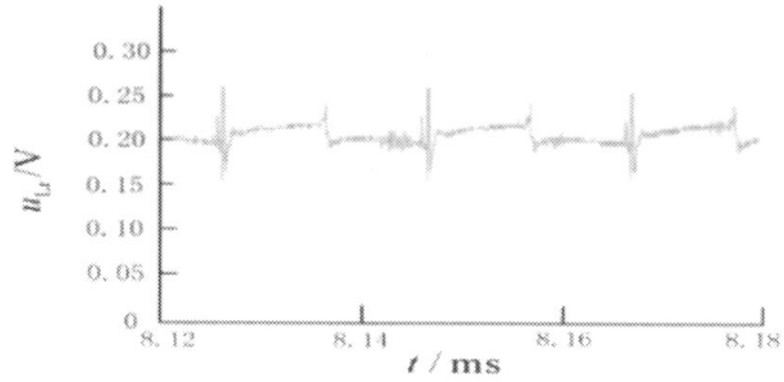

Fig.11. u_{Lr} voltage 2

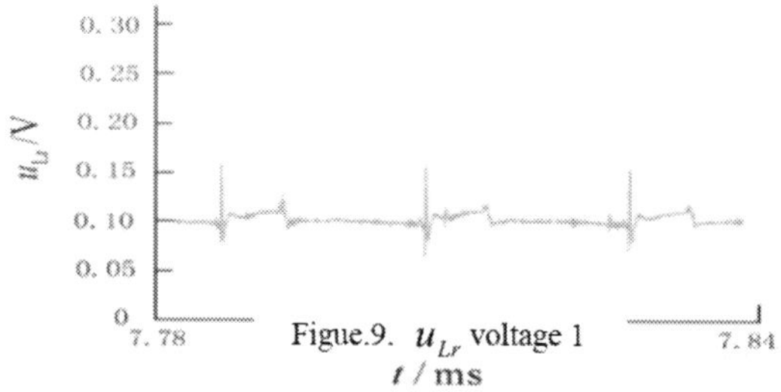

Fig.9. u_{Lr} voltage 1

3.2 The Experiment results

A 60V battery is used as the DC microgrid bus voltage regulator. Wind power simulator fluctuates at 100V-120V according to the simulation curve and connects to the intelligent Buck circuit after rectified through a uncontrolled full bridge rectifier. The intelligent Buck circuit selects a 0.1mH inductor,50kHz switching frequency and DSP28335 as a controller to realize the control algorithm. The constant resistive load is 100Ω.A 0.1Ω copper wire is connected with the load of the Buck circuit, and the signal u_{Lr} is amplified by LM324 shown in Figue.9 and Figue.11 .

Figue.10.DC bus voltage

The DC bus is 59V shown in 8 when two-terminal works. As shown in , u_{Lr} is 0.10 V so the output current of the inductor is 0.2A by the corresponding amplification factor. We can know that battery output 0.4A so the power allocation rate is about 1:2 .The DC bus is 59V shown in Figue.10 when two-terminal working. As shown in Figue.11, u_{Lr} is 0.2V so the output current of the inductor is 0.4A by the corresponding amplification factor. We can know that battery output 0.2A so the power allocation rate is about 2:1 .

As can be seen from Figue.8 to Figue.11,the DC bus is constant 59V based on the proposed algorithm when wind power output arbitrarily. The results are the same as the simulation shown in Figue.4 to Fig7.

4 Conclusions

In this paper, we clarify the space-state of Buck circuit and propose a power allocation algorithm based on PI and load current feed-back. Then we built a two-terminal DC microgrid in MATLAB/Simulink and validate the effectiveness of the algorithm in a real battery and wind power two-terminal DC system. The simulation and experiment results prove the validity and effectiveness of the control method.

Figue.11. u_{Lr} voltage 2

References

[1] N. H. Doerry, “Next generation integrated power systems for the future fleet,” in Proc. IEEE Elect. Ship Technol. Symp., Apr. 20–22, 2009, pp. 1–27.

[2] N.H.Doerry and J. Amy, “Functional decomposition of amediumvoltage DC integrated power system,” in Proc. ASNE Symp. Shipbuilding Support Global War Terrorism, Apr. 14–17, 2008, pp. 1–21.

[3] A. Griffo, J. Wang, and D. Howe, “Large signal stability analysis of DC power systems with constant power loads,” in Proc. IEEE Veh. Power Propulsion Conf., Sep. 3–5, 2008, pp. 1–6.

[4] R.D. Middlebrook and S. Cuk, “A general unified approach to modeling switching converter power stages,” in Proc. IEEE Power Electron. Spec.Conf., Jun. 8–10, 1976, pp. 18–34.

[5] C.N. Onwuchekwa and A. Kwasinski, “Analysis of boundary control for buck converters with instantaneous constant-power loads,” IEEE

DOI: 10.1051/matecconf/201713900209

Transactions on Power Electronics, vol. 25, no. 8, pp. 2018–2032,Aug 2010.

Development and Application of the Rail-type Inspection Robot used in Substation Rooms

Jiang Keqiang*, Sun Zhizhou, Liu Yongcheng, Sui Jichao, Fu Chongguang, Li Yunchang

Shandong Luneng Intelligence Technology Co, LTD., Jinan 250101, China

Abstract. The outdoor inspection robots are widely used in substation at present. At the same time, the requirements of inspection robot used in substation room are put forward. First, the background of substation indoor equipments was analyzed in this paper, and the inspection content requirements of robot used for these equipments were given. Then, the overall structure of intelligent inspection robot was designed. The key components, such as the platform of horizontal movement, motion control system, and PD (Partial Discharge) detection device were closely designed. Finally, the practical application example of inspection robot was given. The application results show that the structure of rail type inspection robot used in substation room is reasonable and its operation is reliable. The inspection robot can work well instead of staff.

1 Preface

The reliable operation of equipments used in substation room (such as relay protection room、GIS room、switchgear room, etc.) is very important for power grid safety. At present, the inspection and operation condition evaluation work of equipments used in substation room is usually finished by the staff[1, 2]. However, manual inspection and test has some disadvantages such as security risk, low efficiency and high test probability of those equipments no need to be tested together[3]. In order to make the power grid run safely and reliably, the state grid company of China plans to set up strong power grid and intelligent substation[4], and put forward requirements of unmanned inspection and early warning of equipment fault. So it is very important to research and develop rail type inspection robot used for equipments in substation room.

The rail type inspection robot can detect the equipments automatically, show the inspection data and give the potential failure by data analysis. In order to satisfy unmanned inspection requirement, the intelligent inspection robot can be designed together in the design stage of new intelligent substations and set up for the existing common substations.

2 Introduction of intelligent inspection robot

2.1 Functional analysis

The rail type intelligent inspection robot used in subway station equipment room should have the following functions.

(1)Motion navigation and obstacle avoidance function

The layout of the indoor equipments in the substation is different. In order to improve flexibility and adaptability of inspection robot, the track that will be installed on the ceiling or side wall is used to provide the function of support and navigation. The track can be linear, S-type or ring depending on the position of detected equipments.

The lift mechanism which can move together with the detection component on inspection robot body has a large stretch ratio. It can effectively expand the scope of inspection. At the same time, the robot has the obstacle recognition function. The robot can stop running automatically when it detects an obstacle around it and continue to run when the obstacle is removed.

(2)Multiple detection functions

According to the different contents of the inspection, the robot can be equipped with different detecting equipments, such as infrared thermal imager, optical camera and PD (Partial Discharge) detector, which can be used to check the appearance, temperature and partial discharge level of switchgear and other substation indoor equipments.

(3)Comprehensive analysis and early warning function

The background system can automatically analyze and deal with different inspection data, for example, it can automatically identify the state of all types of switches, pressure plate, pilot lamp and the value of instruments. At the same time, historical data curves and inspection reports can be automatically generated.

* Corresponding author: jkq_py@163.com

 DOI: 10.1051/matecconf/201713900210

When the inspection data is greater than the default alert value, the background system will automatically pop up the window for alarm. It also can be used to perform horizontal and vertical comparison of various types of data. Combined with the failure model, the robot can predict some potential faults or hidden dangers of inspected equipments.

(4)Equipment linkage control function

Inspection robots can be used to control some other devices automatically. The indoor lights can be turned on and the automated curtains can be opened by the robot to improve inspection results when then the inspection robot start. The robot can work with the fire automatic alarm system. When a fire occurs, the robot can arrive at the fire position and sent the scene of the fire to control centre automatically.

(5)Auxiliary monitoring function

The inspection robot is equipped with two-way voice intercom equipment, and the auxiliary detection module detects temperature, humidity, ozone and smoke content in real time. These detection contents can be provided to field staff to make accurate judgment for the operation status of substation.

2.2 System architecture design

The architecture of indoor rail type intelligent inspection robot system is network distributed, and it includes three parts: the background layer, communication layer and intelligent terminal layer. The terminal layer is located in equipment room, and it is composed of track system, inspection robot body and auxiliary monitoring device. Terminal layer performs inspection tasks and achieve auxiliary monitoring function. The background layer is composed of server, display monitor and intercom equipment, and it is in control of the terminal layer .The background software is used to make the analysis of inspection data.

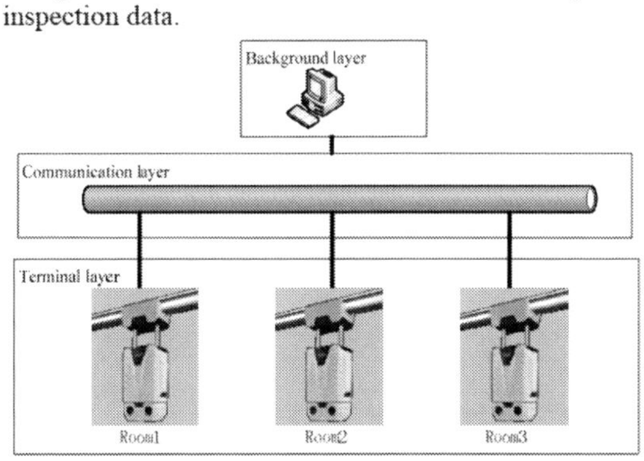

Figure 1. System architecture of inspection robot

Communication layer consists of network switches, photoelectric converters, power line carrier modem and other communications equipment, and it is responsible for the establishment of the network channel between the back-end and intelligent terminal layer. Communication medium between the terminal layer and the background layer can be cable or fiber depending on the distance between the background layer and the terminal layer.

The system supports centralized control mode. Multiple robots can be centrally controlled remotely.

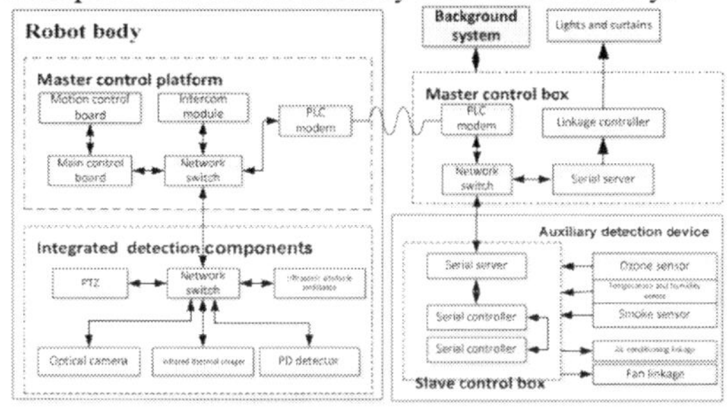

Figure 2. Device structure of inspection robot

2.3 Robot body design

Robot movement body moves along aluminum track which is installed on the ceiling or side wall of equipment room. The inspection robot body consists of horizontal sports platform、master control platform、lift movement component、PD detection component and PTZ detection component. PTZ detection component consists of optical camera、infrared thermal imager and PTZ. PTZ can take these detection equipments achieve horizontal and pitch rotation.

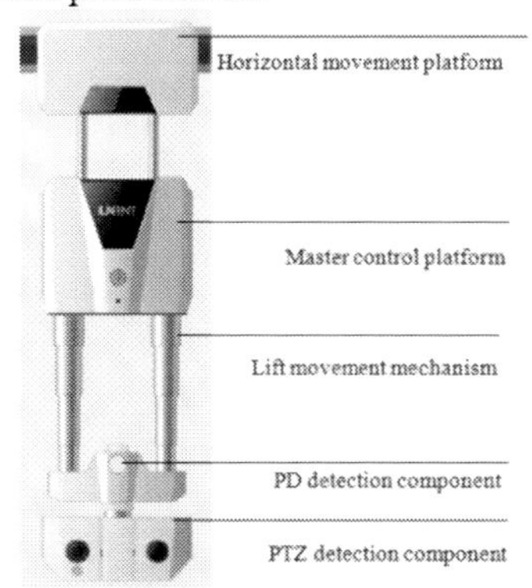

Figure 3. Structure of the robot body

Horizontal movement platform also plays a role of carrier. It is driven by a DC servo motor. The driving force comes from the friction between the wheel installed on horizontal movement platform and the track. The platform can turn around among curved track. The robot is equipped with lifting device which can take detection devices move up and down.

3 Key technologies of intelligent inspection robot

3.1 Design of horizontal turning movement platform

The detected equipments layout in different rooms is different. In order to expand the scope of the inspection as much as possible, the robot moves along the installed track. The track can be flexibly mounted on the ceiling where the wheeled robot can not reach. The track is used to support and navigate the robot, and it acts as the mounting carrier of sliding contact power line and RFID(Radio Frequency Identification). The sliding contact power line is used to provide power for the robot and it act as communication line by power line carrier technique. The RFID is used for motion positioning for motion control.

The track are made of aluminum alloy. The two sides of the upper wing are used to guide, and the upper surfaces of the lower wing are used to support robot. Both sides of the track's rib are used to install other devices such as RFID.

In order to meet environmental and functional requirements, the carrying capacity of the track should be greater than 100 kg and its life should be greater than 10 years. The design of the track should be easy to be processed and installed and it can be arranged along the s-type. The cross-section of the material would be deformed when bended, so the curve track is made of some machined parts.

The horizontal movement platform is mounted on the upper part of the robot. It is used to support and take all other parts move. There are two bogies on the front and rear ends of the platform, and they can flexibly rotate. There are two guide wheels on the upper and lower side of each bogie. The bogies and guide wheels make the movement of platform smooth.

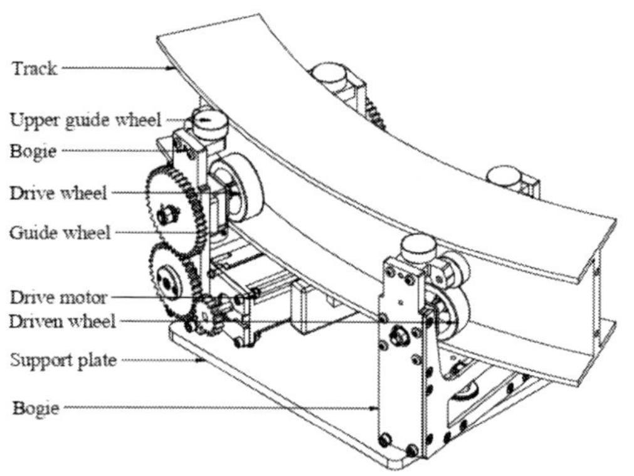

Figure 4. Structure design of horizontal movement platform

Two drive wheels are mounted on both sides of the front bogie, and two driven wheels are mounted on both sides of the rear bogie. The driving force of the robot movement comes from the friction between the driving wheel and the track surface. The drive wheel is driven by a DC servo motor.

3.2 Design of motion control system

The basic movement of the robot is divided into horizontal movement and lifting movement. In order to meet the precise identification requirements, the positioning error of each direction should be control within 5mm.

There are two ways chosen to achieve the positioning of horizontal movement. The RFID installed on the side of the track is used for absolute positioning, and the number of pulses emitted by DC servo motor driver is used for incremental positioning. In autorun mode, the location information sent by the host computer contains the RFID position and the number of pulses which are given by the debugging staff in site. This combined positioning method reduced positioning error compared with the relative positioning method, and improved positioning accuracy in the case of cost reduction.

A proximity switch mounted at the top of the lifting mechanism plays a role of the zero position in the vertical direction. The number of pulses emitted by DC servo motor driver is used to make the positioning of vertical movement. The number of pulses is proportional to the lift distance. The lift stroke is 2m, so the cumulative error of the lift movement is very small and. Vertical positioning error actual measured is less than 2mm when the robot runs at the speed of 0.1m/s in the vertical direction.

The inspection robot will call the PTZ preset position after the robot reaches the specified position to complete the detection work, or call the PD detection component to perform the partial discharge detection task.

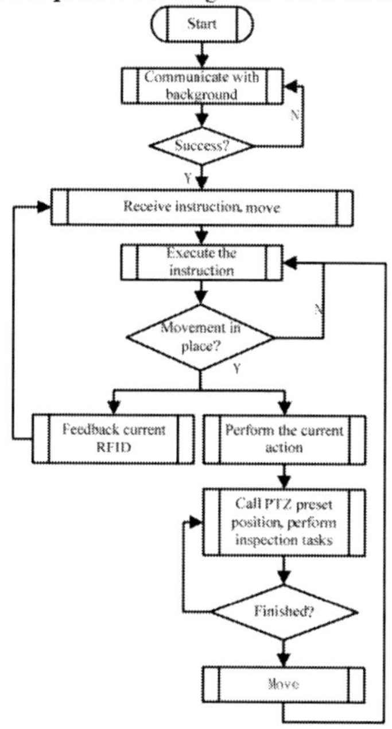

Figure 5. Flow chart of motion control

3.3 The key technologies of PD detection

3.3.1 PD detection component design

There are two ways used to detect the PD level of the switchgear, named direct measurement method and indirect measurement method[5]. The TEV (Transient Earth Voltages) measurement method is an indirect measurement method. TEV technology has high detection reliability and sensitivity, so it is very suitable for PD level detection of switchgear[6]. The TEV sensor should be tightly close to the surface of the detected cabinet when the PD is detected. So the PD detection sensor used to carried out PD detection and the method used to make TEV sensor close to switchgear should be considered at the same time.

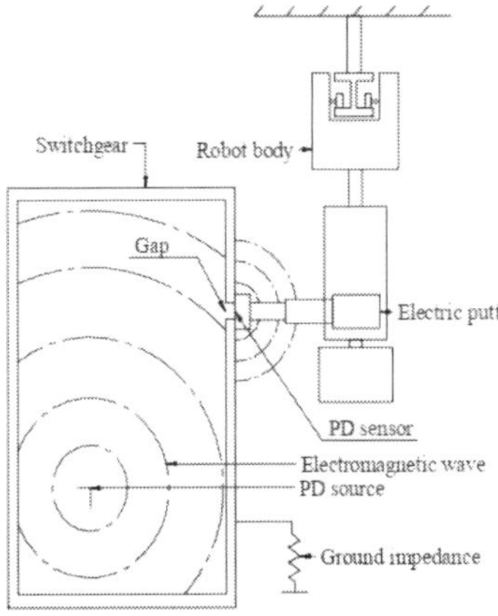

Figure 6. Partial detection diagram

The PD detection component consisted of PD sensor and electric putter is integrated on the robot body. PD sensor can be taken tightly close to the surface of the cabinet by the electric putter which can extend 280mm. The robot track is installed near the cabinet and the distance between the track and the cabinet should be large enough to ensure barrier-free passage of the robot.

3.3.2 Data process of PD detection

PD detection of the switchgear is significantly different from other electrical equipment in terms of the structure and spectral characteristics of the switchgear. First, the discharge part is enclosed in the metal shell, and the sensor of the detection device is difficult to penetrate the inside of the device, so the detection process is difficult to eliminate the influence of environmental electromagnetic noise. Secondly, the switchgear mainly use air insulation or epoxy resin solid insulation, so the dielectric strength is weak, and the electromagnetic discharge spectrum is low, basically with the ambient electromagnetic noise band coincidence, so TEV measurement needs to follow certain process.

The background noise should be detected and recorded before transient voltage detection. The background noise should be measured repeatedly from time to time considering the fluctuation of background noise, then the switchgear transient voltage data is detected following normal process. The detected values of background noise and PD were analyzed synthetically according to certain threshold criteria, and the actual partial discharge data of the switchgear is evaluated finally.

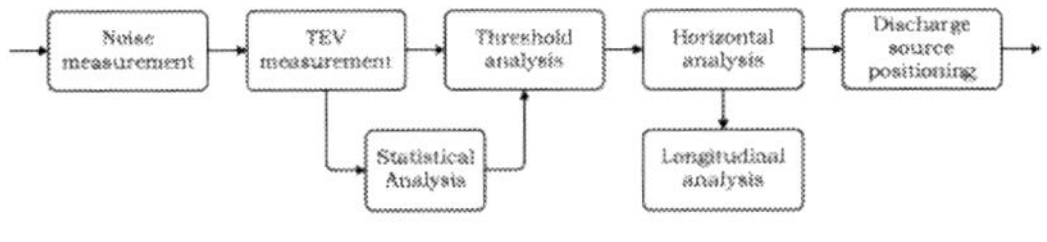

Figure 7. TEV detection data process

There are four methods to analyze the switchgear PD data: threshold analysis, horizontal analysis, longitudinal analysis and statistical analysis. If the simple threshold analysis can not show the information of switchgear PD level, especially when switchgear PD is weak, the horizontal analysis method can be used to judge the PD level of the switchgear by comparing the detection data of the same type of devices.

Compared with the threshold analysis and the lateral analysis technique, the statistical analysis can be used to analyze and discover the development and evolution of the partial discharge state of the switchboard of the power enterprise from the macroscopic point of view, and it can help the operation and maintenance staff to formulate the correct maintenance strategy.

Longitudinal analysis is a kind of analysis method than can be used to analyze the data of the same device at different time. By comparing the data at different times, it can help the staff find the potential defects of the equipments.

4 Field application

The robot started work in April 2017, and it has been running for more than six months and got a lot of data. The inspection data were analyzed by horizontal and longitudinal as follow.

Four detected points, two named point 1-1 and 1-2 on No.1 switchgear and the other two named point 2-1 and 2-2 on No.2 switchgear were selected. The No.1 and No.2 switchgears are the same type cabinet. The point 1-1 and point 2-1 are in the same position on No.1 and No.2 switchgears, and the point 1-2 and 2-2 are the same. The detection data of the four points were analyzed and following analysis results were obtained.

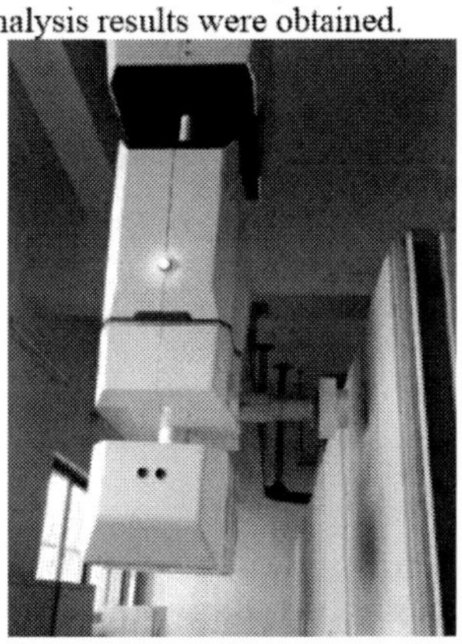

Figure 8. Robot application picture

Point 1-1 and 2-1 are located on the outer upper surfaces of the two switchgears. The PD amplitude of upper part was generally below 15dB through horizontal comparative analysis. It showed that the PD level is weak, and the two switchgears run properly. Points 1-2 and 2-2 are located on the back of the bottom of the switchgears, where the cable connectors are inside. The PD level of points 1-2 and 2-2 was higher than that of the points 1-1 and 2-1, and individual data exceeded 20dB.

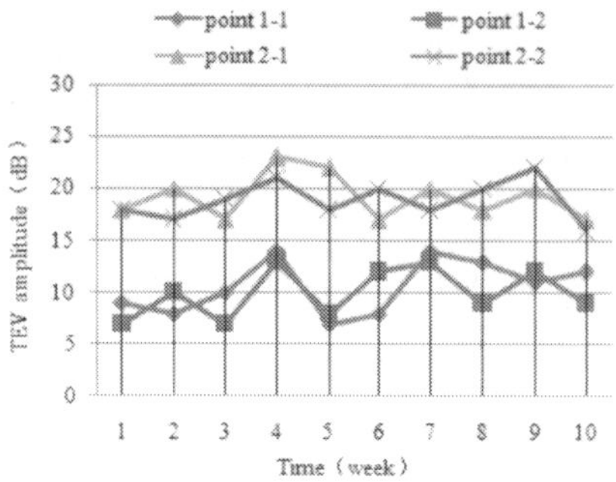

Figure 9. History Curve of PD Detection Data

The PD level of the same point at different time is relatively stable through longitudinal analysis. It showed that the two switchgears run stably and PD level was weak. But the PD amplitude of both two switchgears was bigger than 20dB at some same time. Relative humidity information at the same time was analyzed, and it showed that the PD level was generally high when the relative humidity was high. So the relative humidity of the air affects the PD level of the equipments. Comprehensive comparative analysis is helpful for the operation and maintenance staff to make accurate judgment.

5 Conclusions

Intelligent inspection means should be used to improve the intelligent level of substation, and rail type inspection robot can be used to realize the automatic inspection of the indoor equipment. Field application shows that the rail type inspection robot used in substation room is reliable. It can work without interruption and the inspection data is more objective and real. So the inspection robot is significant to improve the intelligent level of substation and ensure the safe operation of power equipment. Next the operation function of the robot could be researched in order to expand the scope of robot application and serve the power operation and maintenance better.

References

1. Cao nan, Li Gang, et al. Key technologies and construction methods of smart substation[J].Power system protection and control, 2011,39(5):63-68
2. Zhou Lihui, Zhang Yongsheng, et al. Development and Application of Equipment Inspection Robot for Smart Substations[J].Automation of Electric Power Systems, 2011,35(19):85-88
3. Lu Shouyin, Qian Qinglin, et al. Development of a Mobile Robot for substation Equipment Inspection[J]. Automation of Electric Power Systems, 2006(13):94-98
4. State Grid. Q/GDW 1168-2013 Regulations of condition-based maintenance & test for electric equipment[S].Beijing: China electric power press,2014
5. Li Haide. Research on partial discharge detection methods of HV switchgear based on TEV and Ultrasonic wave[D]. North China Electric Power University,2013.3
6. Ren Ming, Peng Huadong, Chen Xiaoqing, et al. Comprehensive Detection of Partial Discharge in Switchgear Using TEV[J]. HIGH VOLTAGE ENGINEERING,2010,36(10):2460-2466

Line Features of Detail Enhancement Thermal Infrared Image for SLAM

Fan Li[*], *XueZhao Zhang, YaoHui Zhang and HeLun Song*
Suzhou Institute of Nano-Tech and Nano-Bionics Chinese Academy of Sciences Suzhou, China

Abstract. To solve the problem of simultaneous localization and mapping of mobile robot vision navigation using infrared thermal imaging. A point-line feature extraction and matching algorithm after infrared image enhancement is proposed for visual odometry. We first used a guided filter to smooth the input image and separate it into the base layer and the detail layer. Then constraining the gradient of the detail will be used gain mask to enhance it. Finally, the two parts of the image combined with weighted coefficients will be exported into the second guided filter. The output image will use LSD to extract line feature, then using ORB algorithm will get point feature extraction and matching from line feature image . The results show the it can effectively improve the defect of infrared image blurred with similar background temperature and difficult to extract image features for mobile robot visual SLAM.

1 Introduction

Monocular visual odometry for its rich visual information and the computation is relatively small, and become an important branch of Visual-Simultaneous localization and mapping(V-SLAM)[1].

The properties of the infrared thermal imaging technology make it not only visually display the object surface temperature field in the form of pictures, but also enable it to show the environment and identify targets highly robustness to changing lighting conditions and other environmental effects. It is a relatively new field to exploit the thermal infrared imaging as a mobile robot sensing modality, such as being used for mobile robot object Identification [2],or SLAM.

However, compared with the visible light camera, the context of infrared images in the video-based SLAM system has some most serious challenges:

- Limited resolution
- High noise
- Low contrast
- Poor texture distribution

The four listed disadvantages of infrared thermal imaging have a great influence to achieve a large number of reliably tracked features between frames in a thermal-infrared video sequence. Therefore, infrared image pretreatment method is meaningful and necessary for robot vision.

Point feature is the most widely used feature in visual SLAM, and it is mature in feature extraction, matching and representation. However, the point feature is more dependent on the environment and does not perform well in scenes such as missing textures. In the artificial structured environment, there are abundant line features, which play a complementary role with the point features in the image. At the same time, the line feature is a higher level feature, and the environment map constructed by line segment feature has more intuitive geometric information. Because of the thermal infrared images characteristic defects, extracting line features is more needed.

The usual points are specialized as corner points or spots in images, and specific detection algorithms are needed to extract the features of these images Historically, researchers have proposed many image features. The SIFT is a classic, has good scale, rotation, perspective and illumination invariance, but the large amount of calculation and time-consuming, so far, PC CPU was unable to compute SIFT features. Other features are considered to reduce the accuracy and robustness, and to improve the speed of calculation ORB feature is a representative real-time image feature. It improves the problem that the FAST detector does not have directionality, and uses the fast binary descriptor BRIEF to accelerate the feature extraction of the whole image

At present, there are many visual odometry and visual SLAM based on line segment features too. Georg Klein [3] Used point and line segment features to constrcut a SLAM system based on key reconnaissance optimization, Guoxuan Zhang[4] proposed the pure line features of SLSLAM(Stereo Line-based SLAM), framework based on the binocular vision SLAM system, including the construction of line feature map of the scene, closed loop detection and so on.

According to using infrared image for VSLAM and comprehensive efficiency and cost, this paper proposed an algorithm which firstly doing image enhancement and denoising and then deal with point and line feature extraction. This pretreatment process can effectively improve the infrared visual navigation robot localization accuracy and robustness.

[*] Corresponding author: fli2012@sinano.ac.cn

2 The principle of the proposed algorithm

2.1. Guided Image Filtering

Guided Image filter[5] smooths an input image p under the guided image I. Local linear model assumes that filtered output q is expressed by a linear function of I in a window ωk centered at pixel k:

$$qi = akIi + bk, \forall i \in \omega k \quad (1)$$

where (ak, bk) are some linear coefficients assumed to be constant in window ωk. A square window is used in the original formula of 2D guided filtering. We can determine the linear coefficients by minimizing the difference between q and p.

$$E(a_k, b_k) = \sum_{i \in w_k}\left((akIi + bk - pi)2 + \epsilon a_k^2\right) \quad (2)$$

The least squares solution of (2) is given as:

$$a_k = \frac{\frac{1}{|w|}\sum_{i \in w_k} I_i p_i - \mu_k \overline{p_k}}{\sigma_k^2 + \epsilon} \quad (3)$$

$$b_k = \overline{p_k} - a_k \mu_k \quad (4)$$

where μ_k and σ_k^2 are the mean and variance of I in w_k, $|w|$ is the number of pixels in w_k, and $\overline{p_k}$ is the mean of p in w_k. Because each pixel lies in the overlapping windows, then a pixel's smooth result is the average of multiple estimates, we rewrite the final Equation:

$$q_i = \overline{a_i} I_i + \overline{b_i} \quad (5)$$

where $\overline{a_i} = \frac{1}{|w|}\sum_{k \in w_i} a_k$ and $\overline{b_i} = \frac{1}{|w|}\sum_{k \in w_i} b_k$ are the average coefficients of all windows overlapping i.

2.2 Gain mask enhancement

The edge-preserving filtering property of guided filter explained intuitively help us to design the gain mask.[4] Consider the case where I = p.

Case 1: "High variance" $a_k \approx 1$.

Case 2: "Flat patch" $a_k \approx 0$.

We simplify the gain mask as follows:

$$T = (KH - KL)\overline{a_i} + KL \quad (12)$$

The parameter $KH(\text{the max}), KL: (\text{the min}$)can modify the masking function, which makes the detail enhancement more controllable. With this function we are able to enhance the detail region as much as we need, and directly using $\overline{a_i}$ save a number of computation.

2.3 ORB feature extraction

ORB(Oriented FAST and Rotated BRIEF)[6], the algorithm is improved based on the BRIEF feature description algorithm and the FAST corner extraction algorithm.

FAST is defined that if a pixel is located in different regions with enough pixels in its surrounding area, the pixel may be an angular point. Compared with other corner detection algorithms, FAST only needs to compare the pixel brightness size. When a discrete Bresenham circle with a radius equal to 3 pixels is taken as the center of the pixel, there are 16 pixels on the boundary of the circle.Fig.1.

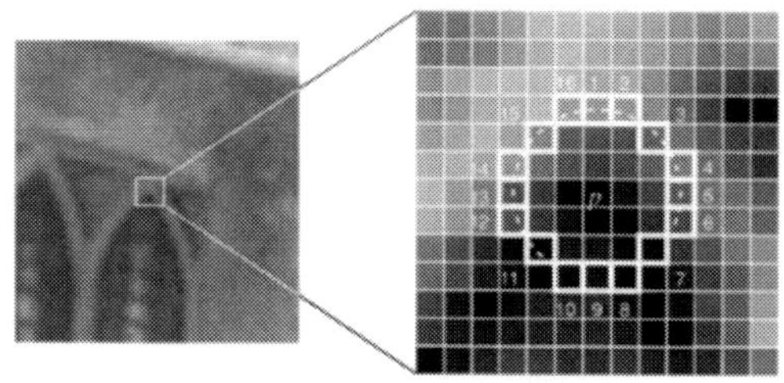

Figure 1. FAST Characteristic point

To select a pixel P from the image, we first set the brightness value to Ip. Set a proper threshold T (for example, Ip's 20%). If there are n continuous pixels in the size of 16 pixels on a circle, or their pixel values are larger than those of Ip+t, or smaller then Ip-t, then it is a corner.
In view of the fact that the FAST corner has no directionality and scale weakness, the gray centroid method is proposed. Connecting the geometric center O and centroid C of the image block, then the angle between the feature point and the centroid is defined as the direction of the FAST feature point. FAST corner has the description of scale and rotation, which greatly improves the robustness of the expression between images. So in ORB, the improved FAST becomes Oriented FAST.

BRIEF, which provides a binary descriptor which consists of many 0 and 1, does not need to compute a feature descriptor similar to SIFT. It needs to smooth the image first, then select a Patch around the feature point, and pick a pair of n points in the Patch by a selected method. Then, for each pair of points (p, q), we compare the brightness values of these two points, and if I (p) >I (q), then this point pair. BRIEF uses a randomly selected, very fast, and the use of the binary expression, storage is also very convenient, generally speaking, BRIEF is suitable for feature extraction of a very high efficiency in real-time image matching method.

2.4 Line Segment Detector

LSD(Line Segment Detector)[7] is an image processing method for fast detection of line segments in images. It was Published by Rafael Grompone von Gioi et al in

2012. It is a local extraction line segment which can obtain sub-pixel precision, and the output of LSD results is the coordinates and attributes of the line segment, including the starting point, the end point, the thickness of the line, and so on.
The first step of LSD is to scale the input image to 80% of its size. Then, the gradient magnitude and gradient direction of the image are calculated. The gradient field LSD is formed by all pixel gradients. Firstly, the angle between each pixel and the level-line is calculated to form a gradient phase field (level-line Field). Merge the pixels with approximately the same direction in the field, so that a series of regions can be obtained, these regions are called line support regions. Fig2.

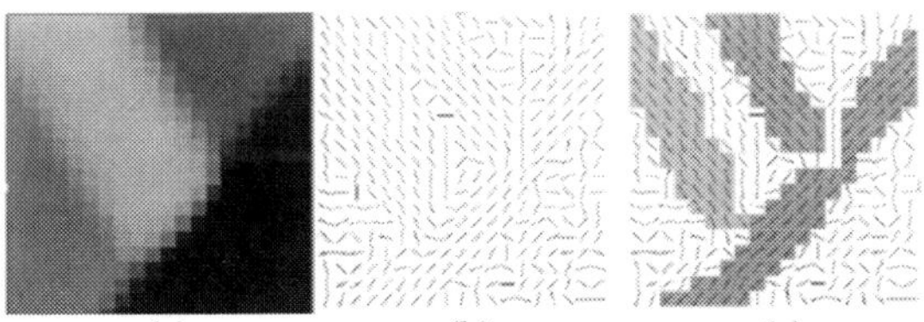

(a) (b) (c)

Figure 2. LSD line detection. (a) gray image. (b) level-line Field. (c) line support regions.

Finally, to verify each rectangular area, the main direction of a support line of pixels in the area of the level-line angle and the minimum bounding rectangle of the angle difference in tolerance (tolerance) in 2 tau, then this point is called aligned point". The minimum number of pixels in the circumscribed rectangle and the number of aligned points in it are used to determine whether the line support region is a straight line segment. When the ratio is greater than or equal to the threshold D, the Number of False Alarms function NFA is calculated. If the NFA is less than or equal to the internal parameters, the line segment is identified as the line segment.

3 Experimental

3.1 The process of two Guided filter enhancement method

The workflow of the method displays in Fig.3. In Fig. 3(a) is the original IR image. Fig. 3(b) and (d) are the base layer and detail layer after filtering original image by GF. Then, we first project histogram the base layer's gray levels distribution and get Fig. 3(c), and it can be noticed that the brightness of Fig. 3(c) is brighter than Fig. 3(b). Meanwhile, we process the detail layer. As shown in the detail layer Fig. 3(d), there is very dark and blurry of the details because the temperature of objects in original IR image is very small. It is necessary to make sure that the detail derivatives and the input derivatives have the same sign. After applying the technique removal of halos artifacts, the correction result in Fig. 3(f). Then the gain mask enhancement of Fig. 3(f) is shown in Fig. 3(g). After dealing with the base layer and detail layer, Fig. 3(e) is the combination of them. Finally, Fig. 3(h) is the output of the second GF of Fig. 3(e).

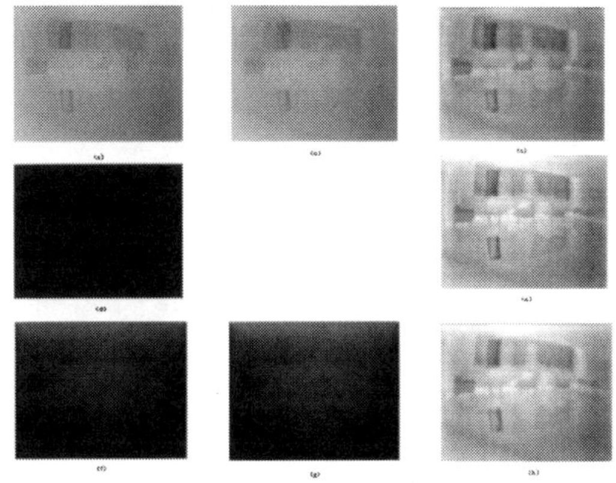

Figure 3. The proposed method flow chart on test image . (a) Original IR image. (b) The base layer by GF. (d) The detail layer by GF. (c) The projection histogram of base layer. (f) The halo removal of detail layer. (g) The Gain mask enhancement of (f). (e) The combination of (c) and (g). (h) The second GF of (e).

The bright of the enhancement could see the details clearly. The contrast is better than original images, and the local details are perceived well. The parameters setting are as follow: the twice of GF is same as: r = 4; eps = 0.5^2. The Gain Mask is KL =1; KH=3;

3.2 Feature point extraction and matching of ORB

As shown in figure 4~5, the experiment shows that the ORB feature point matching has good stability in the image with obvious heat source, and is not weaker than that of the visible image with rich feature points. However, if the intensity of image enhancement is not enough, the possible feature points are scarce in the background environment with similar temperature. It is easy to lose when running ORB-SLAM, aiming at this problem. A strengthening of image enhancement is not

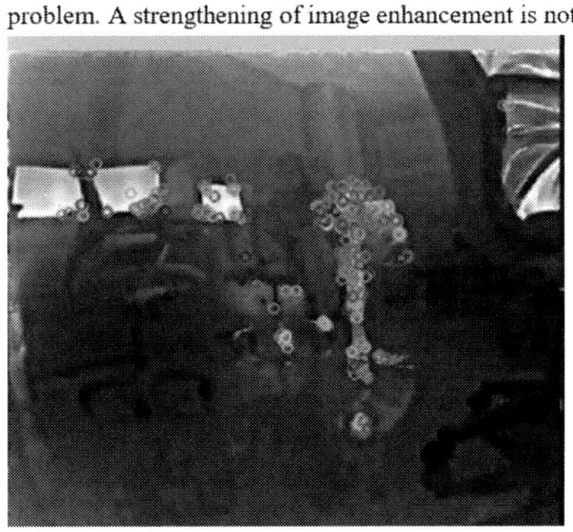

Figure 4. Extracting ORB feature points from infrared images with obvious heat source

Figure 5. feature points matching selection

the same, there will be a lot of noise also increases, there will be a large number of false feature points, and also easy to frame loss. So the stability of feature information needs to be improved from other aspects, such as the development of new feature lines or semantic maps.

3.3 LSD line extraction

LSD algorithm is used to extract the line features of infrared image. For after the extraction of line features in the actual process, can extract the matching points with ORB algorithm, convergence to the ORB-SLAM system, aiming at the basic background temperature as cause extremely lack of scene texture

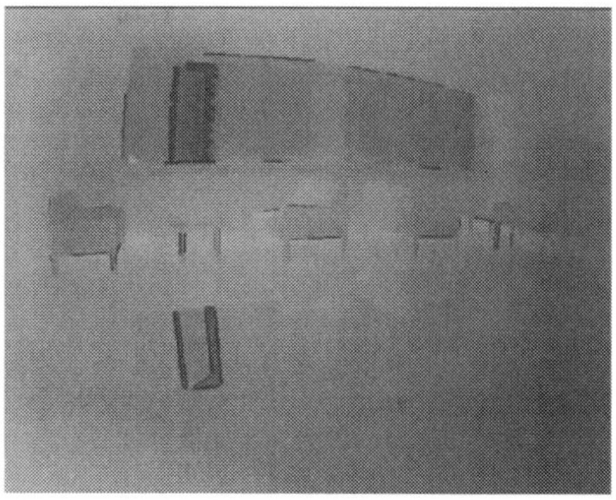

Figure 6. using LSD to extract line features

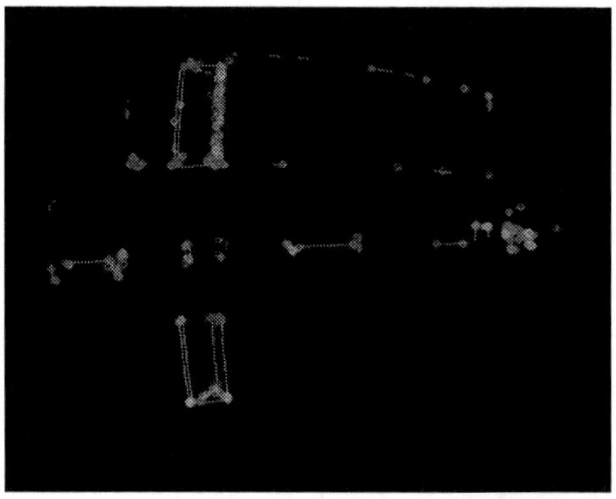

Figure 7. Extracting ORB feature points from line feature image

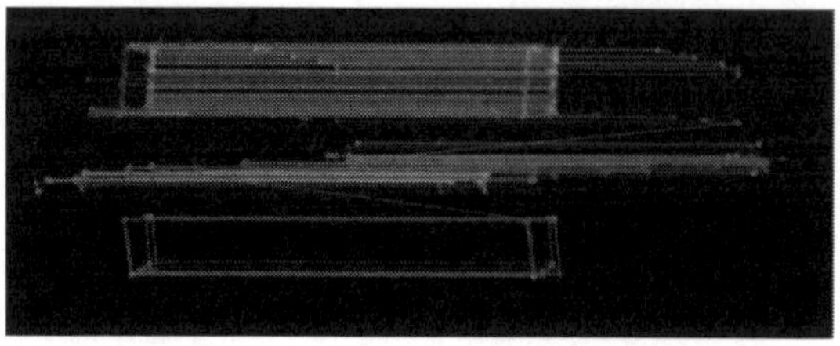

Figure 8. feature points matching selection in line feature image

4 Conclusions

This paper studies how to develop and use the infrared thermal imaging for mobile robot localization and mapping as visual vision sensor, dig the characteristics of infrared image, infrared image with respect to the defects of the visible image denoising, feature extraction, enhancement, increasing the line features aspects of strengthening the robustness and meet the real-time constraint down achieve the aim of study through the images, line features can make up for the lack of feature point feature scenes, especially for the infrared image background temperature often very similar and very lack of texture, to further enhance the robustness of the front end of SLAM. Improvement the effect of each module is verified by experimental test.

References

1. G. Klein and D. Murray. In International Symposium on Mixed and Augmented Reality (ISMAR), (2007)
2. Fehlman Ii, William L., and M. K. Hinders. Journal of Field Robotics 1211.1:2076-2083(2010)
3. Klein G, Murray D. European Conference on Computer Vision, Marseille, France, October 12-18 (2008)
4. Zhang G, Lee J H, Lim J, et al. IEEE Transactions on Robotics, 2015:1-14
5. K. He, J. Sun, X. Tang, IEEE Trans. Pattern Anal. Mach. Intell.35 (6) 1397–1409 (2013)
6. Rublee E, Rabaud V, Konolige K, et al. IEEE International Conference on Computer Vision. IEEE, 2564-2571(2012).
7. Gioi R G V, Jakubowicz J, Morel J M, et al. Image Processing on Line, 2(4):35-55(2012)

Research on Travel Control System of Hydrostatic Transmission Chassis

Liang Zhao [1,2,a], Jin Wang [2] and Zhengwu Zhang [2]

1 Chang'an University, School of Construction Machinery, 710064 Xi'an ,China
2 Shaanxi University of Technology, School of Machinery Engineering, 723001 Hanzhong , China

Abstract. Aiming at the control problem of driving system of hydrostatic transmission chassis, the composition of the control system of hydrostatic transmission chassis is introduced and the control method of dual engine is solved. According to the number of driving axles in driving process. The external characteristic curve of the engine controls the variable hydraulic pump by one parameter, controls the rotational speed of the variable hydraulic motor according to the change of the vehicle speed, and introduces the control flow of the brake system. It provides a reference for the design of driving control system of multi-axis hydrostatic transmission chassis.

1 Introduction

With the development of the national economy, more and more high-rise buildings in the city, and its accompanying construction and fire-fighting machinery and equipment are also toward large-scale development in order to meet the construction and fire safety requirements. For large-scale construction machinery and equipment to develop high load capacity and through the hydrostatic hydraulic chassis, in order to make it get better power and economy, the study of multi-axis hydrostatic drive chassis drive system in the control Some methods can solve the problem of lack of theory in the driving system of multi-axis hydrostatic drive chassis and provide a solution for the speed control of the hydraulically driven chassis at high speed.

Large-scale machinery and equipment need to have a high carrying capacity of the chassis, and to be able to facilitate the operation, to achieve the vehicle's micro-driving and high-speed requirements. Hydrostatic transmission technology has a compact structure, wide speed range, easy to control, you can achieve the vehicle stepless speed regulation and freestyle, in the construction machinery walking system has been a lot of applications. At present, hydrostatic transmission technology is mainly used in low-speed operation of engineering vehicles, in multi-axis drive and high-speed vehicles in the application of research is still blank, speed control problems in driving lack of theoretical support.

2 Composition and working principle

The hydrodynamic drive vehicle has four hydraulic pumps and 14 drive motors, which are driven in a fully driven manner. Hydraulic drive vehicle unilateral driving hydraulic and control system works as shown in Figure 1, including the controller, the engine 1, the variable hydraulic pump 2, the pressure sensor 3, the accelerator pedal 4, the handle 5, cartridge valve 6 , An electromagnetic directional valve 7, a safety relief valve 8, a variable hydraulic motor 9, a speed sensor 10, and a wheel reducer 11.

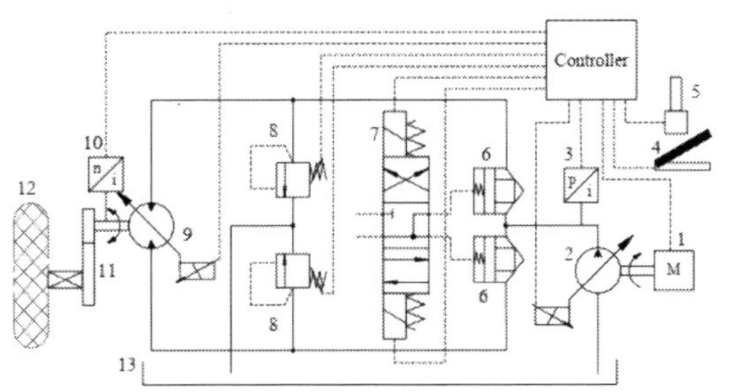

Figure 1. Schematic diagram of single-sided driving hydraulic system

The engine 1 and the variable hydraulic pump 2 are connected. The controller controls the engine 1 to drive the variable hydraulic pump by detecting the signal of the accelerator pedal. The hydraulic system is supplied to the travel system. The hydraulic motor 9 is driven by hydrostatic pressure. The wheel is driven by the wheel reducer. The electromagnetic directional control valve 7 controls the flow of the hydraulic oil to change the steering of the variable hydraulic motor and realize the function of advancing and retreating the vehicle. The hydraulic pump uses constant power control with pressure cut off, adjusts the displacement of the variable hydraulic pump according to the external characteristic curve of the engine. The variable hydraulic motor 9 is controlled by the electric proportional control. The controller detects the gear signal of the gear position and

[a] Corresponding author: zhaoliangslg@163.com

DOI: 10.1051/matecconf/201713900212

the pressure of the hydraulic system. The magnitude of the hydraulic motor coil is controlled to control the displacement of the variable hydraulic motor.

3 Driving principle

Control of the drive section of the multi-axis hydrostatic drive chassis drive includes control of the engine, variable displacement hydraulic pump, variable displacement hydraulic motor and associated solenoid valves. The control schematic shown in Figure 2. By adjusting the accelerator pedal stroke to adjust the engine throttle opening to adjust the engine speed, according to the engine's external characteristics curve to determine the maximum power of the engine at each speed, combined with the hydraulic system to determine the variable displacement hydraulic pump target displacement. Through the gear lever gear signal and hydraulic system pressure variable displacement hydraulic motor displacement, change the variable speed hydraulic motor to change the vehicle speed.

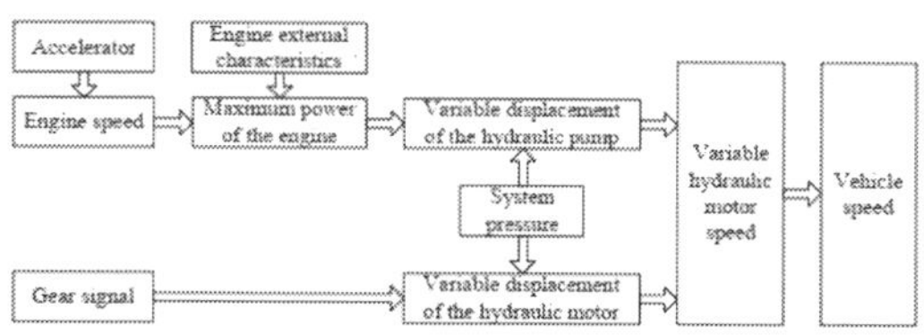

Figure 2. Schematic diagram of the operation of the hydrostatic transmission vehicle

3.1 Power control system

Power control system is mainly on the engine control, the engine has two engines, when traveling need to use two engines; in the operation of the load when the need to use an engine, priority use of the main engine, if there is a fault available from Engine; two engines can be a separate start-up, downtime; two engines share a single accelerator pedal while driving. As shown below. The engine speed is directly controlled by the accelerator pedal in the cab. Both the master and slave engines are equipped with KSM, and the bodywork can separately control the engine speed regulation.

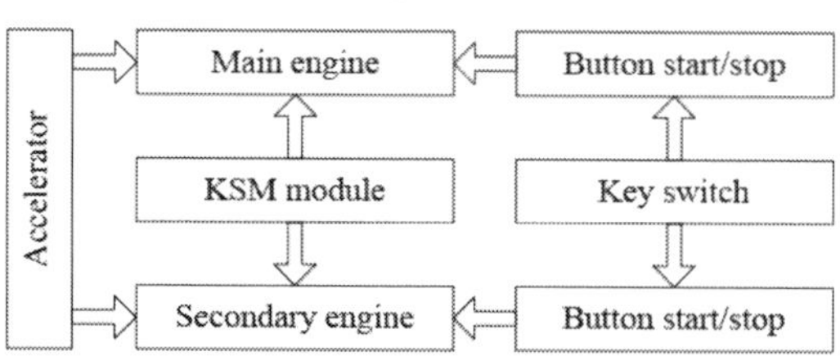

Figure 3. Power control system schematic

3.2 Hydraulic pump control

The hydraulic pump is directly connected to the engine, the input speed, torque and engine output speed, torque and power of the hydraulic pump are equal. The displacement of the hydraulic pump for a single parameter control, according to the engine changes set the pump displacement, pump displacement and the engine speed n .

$$q_b = f(n) \tag{1}$$

Where q_b is the displacement of the hydraulic pump; n is the speed of the engine.
The relationship is determined by the following figure:

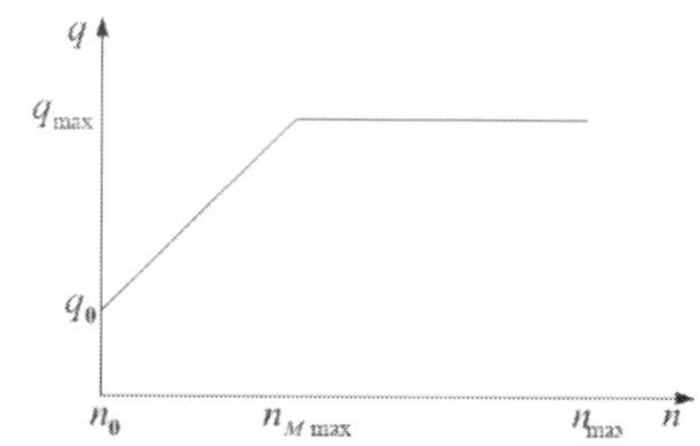

Figure 4. Relationship between the displacement of pump and engine speed

Where q_0 is the minimum displacement of the hydraulic pump, q_{max} is the maximum displacement of the hydraulic pump, n_0 is the engine idling, $n_{M\max}$ is the maximum torque speed of the engine, n_{max} is the maximum engine speed.
Hydraulic pump control strategy shown in Figure 5.

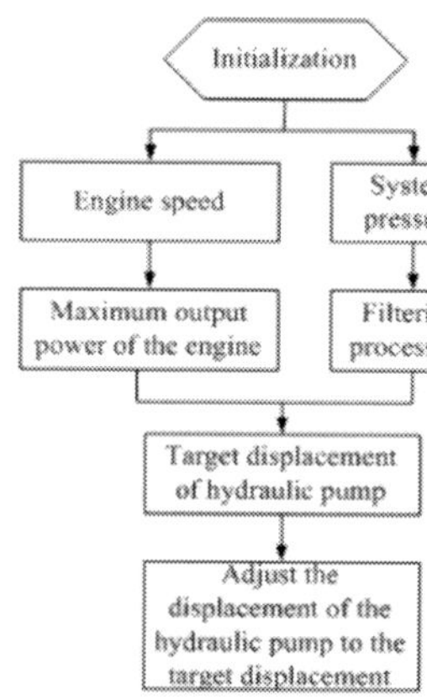

Figure 5. Hydraulic pump control strategy

3.3 Hydraulic motor control

Hydraulic motor displacement According to gear, speed and system pressure to calculate the displacement of the motor at any time, the speed and force into a hyperbola similar to the current vehicle speed calculated according to the target displacement is:

$$q_m = \frac{q_{\min} \cdot v_{\max}}{v} \tag{2}$$

Where q_m is the displacement of the hydraulic motor; v is the speed of the vehicle; $v_{\max}$ is the maximum speed per file ; q_{min} is the minimum displacement corresponding to each gear of the motor.

First, the motor displacement opening to the maximum, and then according to the highest speed of each file to determine the motor displacement, the motor displacement from the maximum transfer to the maximum speed corresponding to the displacement, and then automatically adjust the motor displacement according to the system pressure.

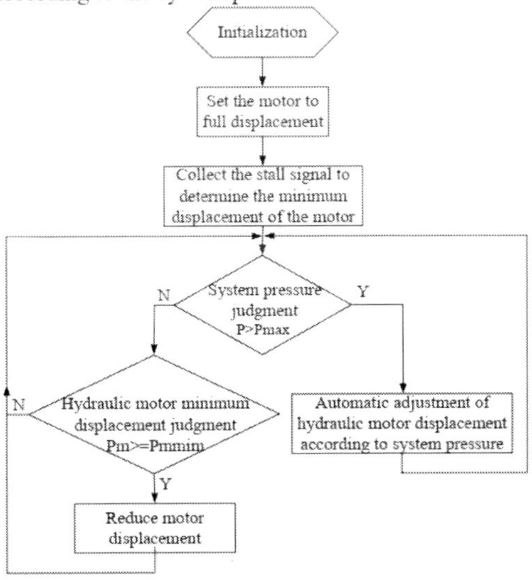

Figure 6. Control strategy for hydraulic motors

3.4 Division of travel vehicles

Multi-axis hydrostatic transmission vehicles in the transmission system without gearbox, the gear design is based on the number of drive axles to determine the number. When the vehicle is moving forward, the number of axles is changed from 7 to 3, and the number of axles is divided into five gears according to the number of drive axles. Can be manual and automatic gear switch, when the need to drive the number of less when the corresponding turn off the displacement of the hydraulic motor, let it in a zero displacement state, to improve the speed of other drive hydraulic motor3.

4 Hydraulic pump control strategy

Braking system can be divided into three brake: brake brake, parking brake and emergency brake. Brake brake: mining brake pedal, the valve is closed, brake relief valve and motor displacement by the brake pedal travel, slope determined. Parking Brake: On flat road or on the slope, the detection speed is zero and the parking brake is started. This button does not really work. Emergency brake: press the stop button, the engine flameout, the pump displacement minimum, maximum motor displacement, the valve closed, the pressure relief valve has been the largest, parking brake start. Control flow chart is as follows:

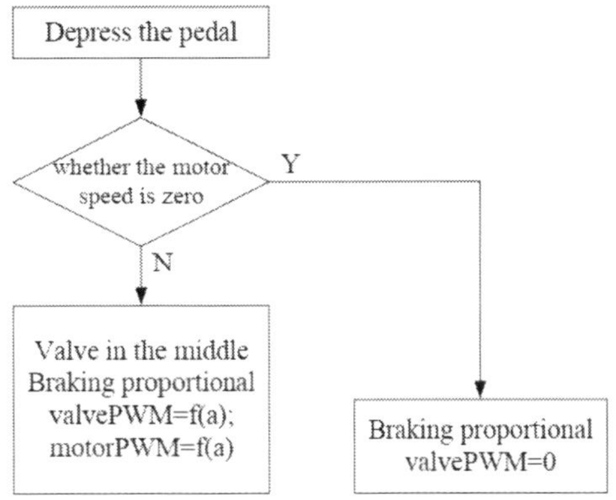

Figure 7. Brake control flow chart

5 Conclusion

Based on the composition and working principle of hydrostatic transmission chassis, this paper introduces the control methods of hydraulic pump, hydraulic motor and brake system.

(1)According to the engine's external characteristics of the variable displacement of the hydraulic pump control, the engine fuel consumption is the best;

(2)According to the driving conditions of the vehicle to control the displacement of the variable hydraulic motor, the torque required at the time of starting is large, and the variable hydraulic control of the variable hydraulic motor is used to maximize the displacement of the hydraulic motor. Constant power control, regulation of hydraulic motor displacement;

(3) Depending on the number of axle of the gear we have been divided.

Acknowledgement

The article was supported by Special research project of Shaanxi Provincial Department of Education (Grant No. 14JK1134),Shaanxi Province Natural Science Basic Research Project(Grant No. 2017JM5087).

References

1. Liu Peng, Zhang Runli, Liu Xinfu, etc. "Research on Matching Method of Hydrostatic Transmission Engine and Variable Pump Power". Engineering Machinery Abstracts, 2012（4）,43-45
2. LIU Yu-hui, JIANG Ji-hai. "Performance of static-hydraulic flow coupled system with secondary regulation". Journal of Jilin University (Engineering and Technology Edition), 2008,38（5）,1095-1100.
3. Feng Kai-lin. "Study on the Electro-control Method of Hydrostatic Driven Multipurpose Mobil". Hydraulic and pneumatic, 2002（8）,21-23.
4. Landers, Kirk. "The Operator-designed Motor Grader". Better Roads.2006,76(6):50-55.
5. IZQUIERDO J, IGLESIAS P L. "Mathematical Modelling of Hydraulic Transients in Simple

 DOI: 10.1051/matecconf/201713900212

Systems".Mathematical and Computer Modelling, 2002, (7),801-812.
6. Mohamed Saber Ahmed Ibrahim-Sokar.Verlag.Investigation of Hydraulic Transmissions for Passenger Cars. Diss. RWTH Aachen University,2011
7. Zhang Qingyong. "Study On the Hydraulic System of Hydraulic Hybrid Vehicles". Nanjing University of Science&Technology, 2009,10.
8. YI Xiaogang, WANG Xin. "Matching and control techniques for hydrostatic grader".Road Machinery & Construction Mechanization, 2008, （3）,18-21.
9. GENG Lingxin, ZHANG Lijuan,LI Xuefei. "Matching and Simulation to Hydraulic Transmission System of Motor Grader",Journal of China Construction Machinery Corporation，2003, （8）,86-88.
10. TIAN Jinyue,YU Ying. "Research on performance of hydrostatic transmissions of vehicles". Transactions of the Chinese Society for Agricultural Machinery, 2002, (4),32-34.

Design of plant growth information acquisition and management system based on WIFI

Shaowei Chen, *Jing* Le, *Jian* Wang, *Ze* Wang, *Xili* Duan

Mechanical and Precision Instrument Engineering,Xi'an University of Technology, Xi'an,*China*

Abstract. A portable plant growth information acquisition and management system based on WIFI was developed to monitor plant growth status by the non - picking leaves. It uses STM32 as main control chip. The chlorophyll characterization value SPAD and water characterization value WI were obtained with the transmittance of plant leaves under the light of 650, 940 and 1450nm, and the fluorescence images were obtained by 460nm excitation chlorophyll fluorescence to analyze the infection of the disease. The nitrogen content of the leaves can be retrieved with the fluorescence parameter R_{fd}. The software of the upper computer, which adopts the B/S architecture, is designed to manage and display the data based on the JAVA language. The stability and reliability of the designed system are verified by field application.

0 Introduction

Multi-sensor fusion, WSN, Internet of Things and other new technologies make the measurement system more comprehensive, more widely used, and more flexible to form, and they have also been applied in agricultural production and management[1,2]. The measurement systems mentioned above can overcome the shortcomings of single parameter detection to make a comprehensive and detailed judgment of the actual growth of plants. However, the network-based agricultural information monitoring systems currently is mainly to monitor the growth environment of plants[3]. The nutrition that the plant absorbed can not be grasped by them. The non-destructive testes which based on the geometric characteristics of the leaves, fluorescence parameters, etc, mainly detect the single parameter and remain a relatively large error[4]. Therefore, the trend of the comprehensive evaluation of plant growth condition is to carry out long-term tracking measurement of the leaf and multi-parameter detection directly to plant leaves, and the use of plant fluorescence information has the advantage of predicting disease and plant stresses.

A portable monitoring system based on chlorophyll fluorescence and infrared absorption characteristics of crop growth status is developed to achieve the prediction of the main stress in the early crop growth to reach the purpose of accurate irrigation and reasonable fertilization and avoiding the damage to crops. In this paper, the content of nitrogen fertilizer was determined by inducing chlorophyll fluorescence kinetics characteristics, and the distribution of fluorescence images and the change of light intensity were used to predict the degree of disease. WI and SPAD, which were considered to represent the values of moisture and chlorophyll, were measured by near infrared spectroscopy. To build a data acquisition, preservation, analysis and management platform based on Internet of Things, to provide a non-destructive, in-vivo tracking method for agricultural cultivation, a small multi-functional agricultural information detection system is configured to meet the requirements of the agricultural information age.

1 Design of measurement system

The functions of the measurement system include the acquisition, display, storage, processing, remote transmission of data, storage, display of web pages, query and remote control of plant growth signals. STM32F103 is used to manage the acquisition and processing of multiple parameters of plant leaves (all information collection are carried out in the dark room). Considering the factors such as transmission rate and measurement object, this system adopts WIFI as a wireless transmission tool,receive and manages data with PC to achieve long-term monitoring and analysis of plant growth. The system comes with lithium battery power, suitable for field monitoring.Below is a description of each component of the framework shown in Fig. 1.

According to the whole idea of the system, the design scheme is divided into several parts.

* Corresponding author: lejing@xaut.edu.cn

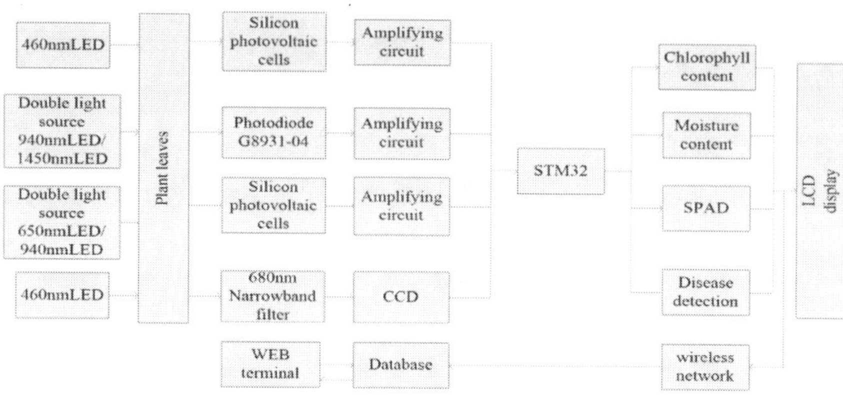

Fig. 1. Overall design block diagram

1.1 Design of the circuit

In the acquisition of plants measured signal, the sensor measurement program firstly is determined based on different measured parameters of the spectral characteristics.

1.1.1 Sensor selection and determination of measurement scheme

In view of the measurement of plant moisture, it is considered that in the near-infrared absorption region, near-infrared light of 1450nm is easily transmitted through the plant leaves, the near-infrared light of 980nm is not, and to reduce the thickness of the blade and the influence of transmission process irregular optical path on measurement results as far as possible. Therefore, in this measurement system, we select the high-power LED with a wavelength of 1450nm as the measuring light source, and select the 980nm high-power LED as the contrast light source.

In the photoelectric sensor, the photo-multiplier tube has poor noise resistance and high power requirements. This creates an obstacle at field measurements, and also the response frequency of the photocell can not meet these experiment requirements. Therefore, Hamamatsu G8931-04 indium gallium arsenide photo-diode is selected, which wavelength range detected is 800-1700nm.

Plant nitrogen fertilizer measurements can take advantage of a feature that nitrogen fertilizer content and chlorophyll content have a definite relationship. Plant nitrogen content can be inversion by experiment, chlorophyll content measurement can be used in the following two ways.

Chlorophyll is more sensitive to blue light (400-500nm) and red (600-700nm). Therefore, the principle of SPAD detection is selecting the red light area as the main band for measuring chlorophyll according to the absorption rate of chlorophyll spectrum. Similar to the measurement of moisture, to eliminate the thickness of the blade and other aspects of the measurement results in infrared light emission and reception, the 650nm, 940nm light-emitting diodes are selected as the excitation light source, the silicon photo-voltaic diode 2DU10 is used to detect.

The system can also obtain the chlorophyll content by using 460nm to induce chlorophyll fluorescence, then through a narrow band filter with a wavelength of 680nm and finally receiving fluorescence signal by the silicon photo-voltaic diode 2DU10.

The plant disease detection and early prediction can be achieved by using a CCD camera to capture the fluorescence images induced by 460nm LED.

After determining the measurement scheme and the sensor, it is necessary to carry out the circuit design and conditioning.

1.1.2 Signal process module

According to the design requirements, the signal acquisition circuit needs ±12V power supply. Selecting two 12V portable rechargeable batteries, the protection circuit designed to meet the system requirements and to avoid the damage that hardware circuit components suffered under the over-voltage, over-current conditions and other circumstances.

Analog signal conditioning circuit module is mainly to adjust the photoelectric signal meeting the acquisition voltage range of ADC, it is convenient for subsequent analysis and processing. The conditioning circuit mainly includes the I/V conversion circuit and the amplifier filter circuit.

As shown in Fig.2, AD795 is used as the preamplifier (maximum input bias current is 1pA and the maximum offset voltage is 250μV), and the second stage amplifier is AD620. The output signal is within 0-3.3V.

In order to use conveniently, ensuring the minimal impact of interference factors such as the outside light, the location of the leaves placed and so on, several independent photoelectric conversion probes are designed by ourselves that photoelectric conversion and follow-up circuit are installed into an organic whole.

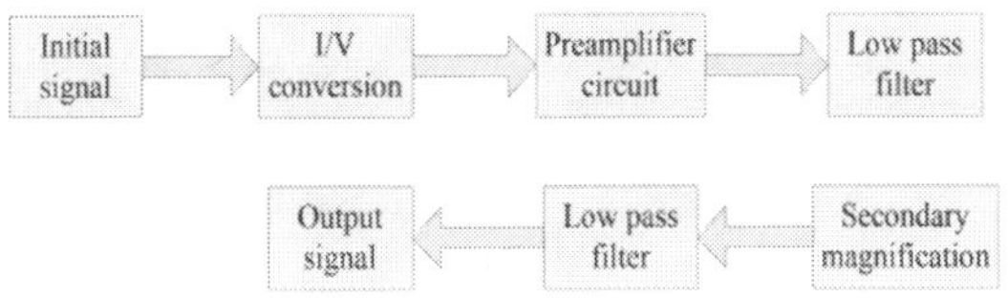

Fig. 2. Signal conditioning flow chart.

1.2 MCU module

Acquisition and control is achieved by STM32F103, which controls the acquisition and storage of plant growth information by pressing the key to control internal ADC, then the corresponding characterization parameters are calculated. Indicators are used to indicate the operation of the program, while the AT command is used to transfer data between MCU and the monitoring center.

For the acquisition of moisture in the leaves, it is necessary to alternately irradiate the photoelectric receivers with two states of no-load and placed leaves, then the water characterization value WI according to Bill-Lambert's law[5] is obtained by equation (1).

$$WI = K\lg\left\{\frac{U_2/U_1}{U_4/U_3}\right\} \tag{1}$$

(K is a constant, U_1 is the voltage corresponding to the light intensity of 980nm passing through the blade, U_2 is the voltage corresponding to the light intensity at no load; U_3 is the voltage corresponding to the light intensity of 1450nm passing through the blade, U_4 is the voltage corresponding to the light intensity at no load. This process needs to collect two channel signals.)

Similarly, the photoelectric receivers were alternately irradiated with no-load and placed leaves to obtain the chlorophyll characterization values SPAD:

$$SPAD = K\lg\left\{\frac{IR_t/IR_0}{R_t/R_0}\right\} \tag{2}$$

(K is a constant, IR_t is the light intensity of 940nm passing through the blade, IR_0 is the intensity of the emitted light, R_t is the received light intensity of 650nm passing through the blade, while R_0 is the intensity of the emitted red light. This process needs to collect two channel signals.)

In addition, the chlorophyll fluorescence induced by 460nm was used to collect the characteristic points of the fluorescence kinetics curve after 15 min dark adaptation. The fluorescence characteristic parameter R_{fd} was calculated to invert the nitrogen content of the leaves.

$$R_{fd} = (F_m - F_s)/F_s \tag{3}$$

(F_m is the maximum fluorescence value, F_s is the steady-state fluorescence yield, measured in the saturated pulse light 3min ~ 4min[6].)

So the MCU need to acquire the five channel signals with internal ADC. Below is a description of Signal acquisition and processing process shown in Fig. 3.

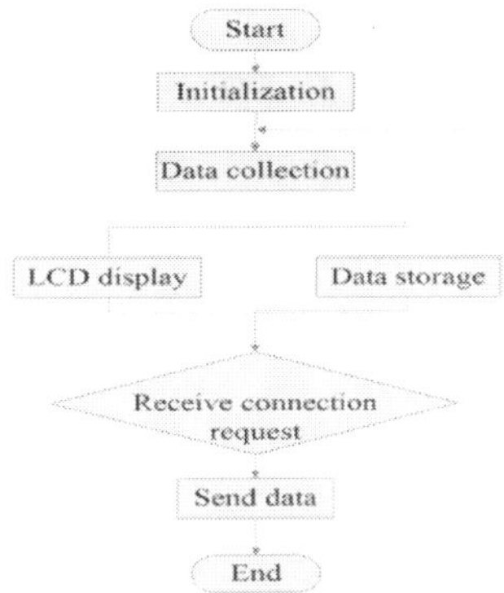

Fig. 3. Signal acquisition and processing process.

1.3 Wireless network coverage and access

The wireless network acts as a bridge between the nodes and the monitoring stations, which plays an important role in the whole measurement system. As a server, low-power ESP8266 is configured as AP, TCP / UDP Client transparent mode to achieve data transmission. When the module does not receive data, it will be moved into deep-sleep state. Its power consumption will be as low as 10μA.

After the MCU completes the data acquisition and processing, we can send the processed leaf water, fertilizer and chlorophyll information to the monitoring terminal through the AT command and generate the spreadsheet to facilitate the import of the database. The fluorescence images captured by the CCD are transmitted to the monitoring center through the map transmission module, and the images will be processed to obtain the R_{fd} distribution of each pixel point to observe the change of the disease area.

1.4 Information monitoring and management

1.4.1 The overall design of the monitoring center

For the monitoring and management part, we construct a plant information transmission and management system based on JAVA language. MySql technology was used to establish a database to store the transported data, JSP technology was used as a tool for building the system terminal page, and a connection between website and database was established by JDBC technology. The Web includes users management (registration, logout, logout, etc.), data management (plant location, time, information), and reference information (weather, agricultural advice, etc.), users can use the website to search for plant growth information.

As shown in Fig. 4, the farmland information management system is divided into three layers: view layer, control layer and model layer. Then the model layer is divided into two layers: one layer is the service layer, the other is the data access layer. The data access layer encapsulates the most basic database connection methods provided by JDBC. The service layer encapsulates the data access layer, which can improve the system's scalability.

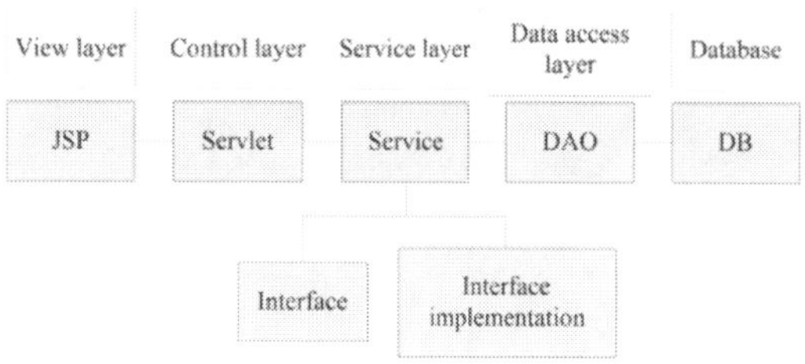

Fig. 4. System structure of information management

1.4.2 Interface and function design

The Login interface function is achieved by JSP technology. After the user enters the account number and password, the control layer Servle will obtain and verify the user information, the verification method provided by the model layer. After successful landing, users can enter the main interface. The management system includes four function modules: personnel management, crop data management, farmland data management and MCU management. Users can jump through the corresponding function interface by clicking these modules.

1) Personnel management: Users can add, delete, modify and query the staff;

2) MCU management: Controlling MCU uploads crop information, and import information into the database through the JAVA Socket class;

3) Crop data management: Users can query the corresponding crop of water, nitrogen, chlorophyll, disease information, and the information can be displayed through the form of digital curve, and imported into the spreadsheet;

4) Farmland data management: The different farmland can be increased, deleted, modified and queried. The information display is implemented by JSP hyperlink tags and JavaScript, which mainly provides some agricultural knowledge and real-time agricultural information. The software interface is shown in Fig. 5.

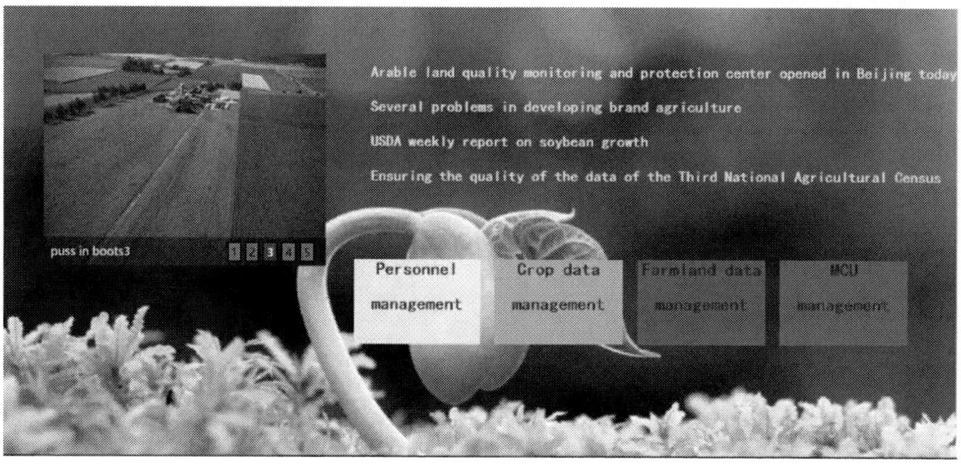

Fig. 5. Home page display.

2 System testing and analysis

The system can observe the growth status of plants under different conditions, such as water stress, fertilizer stress, early prediction of disease, etc. There are some experimental data to clarify the system's practicality and feasibility.

56 slices of green radish and 30 pieces of heather were selected in the experiment. The values of WI and SPAD and fluorescence intensity of the selected samples were measured by the designed system. The chlorophyll concentration and water content were obtained by spectrophotometry and drying method. It can be seen from Fig. 6 that the moisture content is in direct proportion to the value of WI, and the relative error of the mid-waist part of the plant leaves is relatively small, which can best reflect the real moisture content of the plant leaf. The relationship between chlorophyll content and the value of SPAD is similar to the quadratic polynomial model, what is the same as nitrogen fertilizer content and R_{fd}.

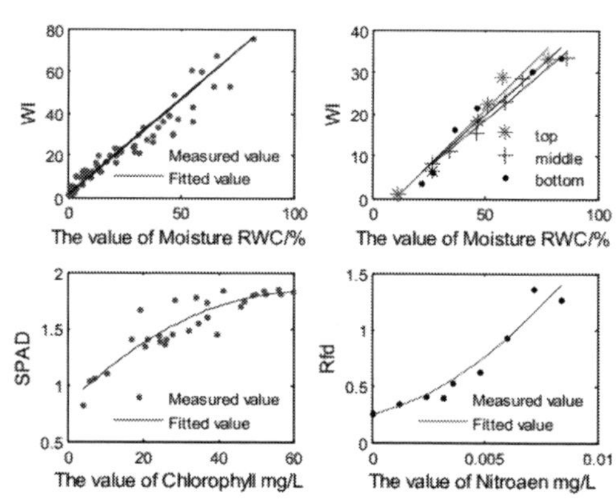

Fig. 6. The fitting curve of WI,SPAD and R_{fd}

The R_{fd} diagram of leaves was obtained by using C++

to analyze the fluorescence images of green radish and rose leaves. The relationship between the mean value of R_{fd} and the degree of infection was analyzed and the pseudo color images of green leaves[7] were shown in Fig. 7,8.

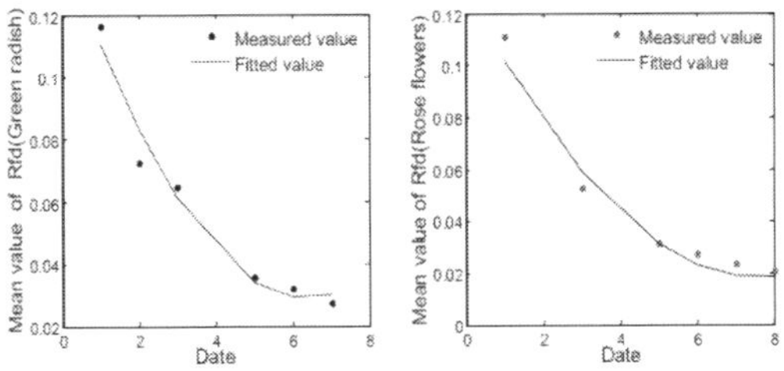

Fig. 7. Curve of infection degree and R_{fd}

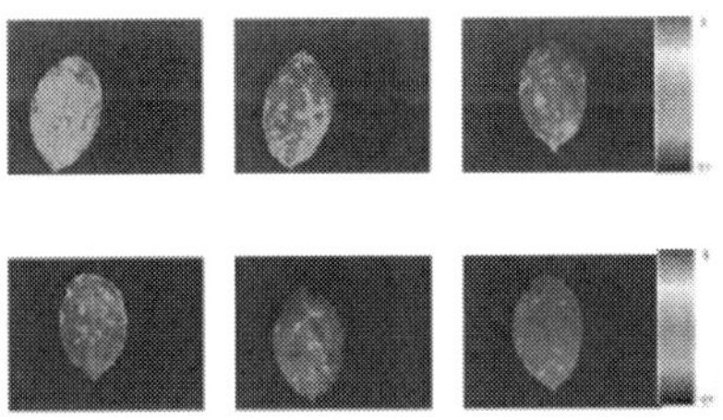

Fig. 8. Pseudo-color map of green radish leaves with increased degree of infection

The relationship between the degree of infection and R_{fd} value of green radish and rose is more consistent with quadratic polynomial model according to Fig. 7. With the increase of the degree of infection, the fluorescence characteristic parameter R_{fd} will be reduced to a certain stationary value.

The results of different experiments show that the measured value can reflect the real growth information of crops well. Therefore, the system designed in this paper has good reliability and can be used for inversion of crop water, fertilizer and chlorophyll.

3 Concluding

In this paper, the WIFI-based agricultural information acquisition and management system designed for crop growth problem is portable, a wide covering range of structure, simple and stable. This paper mainly presents a detailed design scheme including constructs of the hardware and software system, the measurement principle completed the experimental verification. This system can achieve the database establishment, invert the chlorophyll, water and nitrogen fertilizer content of the crop, and analyze the disease and the stress. Meanwhile using these parameters, the influence of the crop growth state can be analyzed by the multi-sensor information fusion technology. The research can provide some technical support to achieve the agricultural input-output optimization and the economic benefit maximization.

This work has been supported by the special project of Agricultural Machinery Administration of Shaanxi Province (Shaanxi Agricultural Machinery number [2017]4). Sincere thanks for the comments and contributions of anonymous reviewers and members of the editorial team.

References

1. Yong H, Pengcheng N, Fei L, Transactions of the Chinese Society for Agricultural Machinery, **44**, 10, 216-226, (2013).
2. Sokratis Barmpounakis , Alexandros Kaloxylos, Aggelos Groumas, et al, INFORMATION PROCESSING IN AGRICULTURE, **2**, 51–63, (2015).
3. Suryakant Sawant , Surya S. Durbha, Jagarlapudi Adinarayana, Computers and Electronics in Agriculture, **138**, 175–187, (2017).
4. Wenbo W , Dengxin H, Jing L, Meixia L, Ning C, Acta Physica Sinica, **62**, 19, 190601-1-6, (2013).
5. Chunyan Z, Dengxin H, Jing L,Spectroscopy and Spectral Analysis, **35**, 9, 2629-2633, (2015).
6. Claus Buschmann,Hartmut K.Lichtenthaler, Plant Physiology, **152**, 297-314, (1998).
7. Haibo L, Jing S, Zhenxun Y, *Detailed explanation of digital image processing technology of Visual C++* (Machinery Industry Press, 2014).

Analysis of dynamic meshing characteristic of planetary gear transmission in wind power increasing gearbox

Jungang Wang[1,*], Ruina Mo[2], Zhaoping Tang[1]

[1]School of Mechanotronics & Vehicle Engineering, East China Jiaotong University, Nanchang 330013, China
[2]School of Basic Science, East China Jiaotong University, Nanchang 330013, China

Abstract. Dynamic behavior of planetary gear's tooth contact surface in the different location can better conform operation condition comparing to the general gear pair. Nonlinear finite element algorithm was derived according to the basic control equation of contact dynamics. A finite element model of planetary gear transmission in wind power increasing gearbox was proposed considering different meshing locations based on nonlinear finite element solution. The characteristics of stress distribution at different meshing positions were analyzed. A simulation of the meshing process was conducted using finite element analysis. It was shown that node stresses of external meshing planetary gear varied significantly at different position. The analysis provides some useful insights into the performance of planetary gear's tooth contact surface.

Key words: Equivalent stress, planetary gear, finite element analysis, analysis of stress distribution

1 Introduction

Planetary gear transmission is used widely, which use more than one planet gear to support load on the structure of the average allocation. Compared with the general gear meshing transmission, planetary gear exists internal meshing and external meshing at the same time. There is a vibration and impact of planetary gears in the process of the actual work. It is very difficult to test the actual work condition of structure strength. With the help of simulation analysis, the danger zone load characteristics of planetary gears can be obtained. However, few papers are concerned with the internal meshing and external meshing planetary gear pair meshing process and the research is not enough in-depth.
The research on the contact problems of gear drives was reported by many investigators. Lin et al. [1] proposed a method for mesh generation to analyze meshing stiffness under static loading. A three-dimensional r model of helical gears was presented by Eritenel [2]. Cai et al. [3] investigated the dynamic behavior of a gear-bearing system considering the nonlinear gear meshing force effect. Chen et al. [4] carried out a study on the contact stress of gear pairs with non-parallel axes by means of finite element method. Zhang et al. [5] carried out a study on the bending fatigue failure of a bevel gear. Palermo et al. [6] proposed a contact element to simulate global dynamic behavior of gear assemblies. The tooth engagement behavior of beveloid gears was analyzed through formulating a pitch cone design theory by Zhu [7]. Tamarozzi et al. [8] analyzed flexible multibody model of two meshing gears by numerical simulation. Dong et al. [9] proposed a new method to study the contact fatigue life of gear-bearing system of a wind turbine. Li et al. [10] studied the influence of addendum on bending strength and tooth contact strength of spur gears. Khabou et al. [11] investigated the dynamic performance of a spur gear reducer in transient regime.

Few reports about stress distribution of external meshing gear tooth profiles in different mesh position using finite element model are concerned. Lumped parameter model and the finite element analysis (FEA) just for teeth were adopted in references [1-10] to analyze the dynamic characteristics of meshing contact transmission with parallel axis. However, the FEA just for teeth is not fully conformed to actual condition of planetary gear transmission in engineering application, leading to that the actual condition cannot be reflected in full scale. A finite element model with different meshing locations of planetary gear pairs is limitedly reported.
Differences of stress distribution on tooth profile at different meshing positions of planetary gear transmission in wind power increasing gearbox were investigated in this paper. The finite element model of entire gear with both external and internal meshing was established. The entire gears were built in order to simulate dynamic process of planetary gear pair more accurately. Boundary and initial conditions of the finite element model were decided. Load was applied and working parameters were set. The characteristics of

* Corresponding author: sduhys@163.com

 DOI: 10.1051/matecconf/201713900214

stress distribution of external meshing planetary gears were obtained. The stress distribution rule in the different areas of the gear tooth profile of the external gear planetary was studied in detail.

2 Nonlinear finite element solution to the contact problem

According to the basic control equation of contact dynamics, the nonlinear differential equation of the contact system can be derived from the Hamilton variation principle:

$$M\ddot{X} + D\dot{X} + r(X) = f(t) \quad (1)$$

Where D , M , $r(X)$ and are damping matrix, mass matrix and nonlinear matrix of the contact system, $\ddot{X}$, $\dot{X}$ and X are acceleration vector, velocity vector and displacement vector of the system, $f(t)$ is load vector.

Under the known initial conditions, convert equation (1) into the nonlinear algebraic equation set in the incremental form, and then use Newton-Raphson for iterative solution. Assuming that at the time of the t_n , all variable values are known, and determines the displacement $X^* = X_{t_n+\Delta t_n}$ at the time of $t_n + \Delta t_n$.

The unbalanced force of the system can be obtained from equation (1):

$$\tilde{f}(X^*) = M\ddot{X} + D\dot{X} + r(X) - f(t) \quad (2)$$

According to equation (2), the values at the time of $t_n + \Delta t_n$ and in the (i-1) step of iterative process can be represented by:

$$\tilde{f}(U^{(i-1)}_{t_n+\Delta t_n}) = M\ddot{U}^{(i-1)}_{t_n+\Delta t_n} + D\dot{U}^{(i-1)}_{t_n+\Delta t_n} + rU^{(i-1)}_{t_n+\Delta t_n} - f_{t_n+\Delta t_n} \quad (3)$$

The expansion of equation (4) can be obtained from Taylor series:

$$\tilde{f}(X^*) = \tilde{f}(X^{(i-1)}_{t_n+\Delta t_n}) + \left(\frac{\partial \tilde{f}}{\partial X}\bigg|_{X^{(i-1)}_{t_n+\Delta t_n}}\right)(X^* - X^{(i-1)}_{t_n+\Delta t_n}) + O\left(\left\|X^* - X^{(i-1)}_{t_n+\Delta t_n}\right\|^2\right) \quad (4)$$

The displacement increment is represented by:

$$\Delta X^{(i-1)} = X^{(i)}_{t_n+\Delta t_n} - X^{(i-1)}_{t_n+\Delta t_n} \quad (5)$$

From (5) and the second order small quantities in the ignored equation (4), the following expression can be obtained:

$$\Delta X^{(i-1)} = -\left(\frac{\partial \tilde{f}}{\partial X}\bigg|_{X^{(i-1)}_{t_n+\Delta t_n}}\right)^{-1} \tilde{f}(X^{(i-1)}_{t_n+\Delta t_n}) \quad (6)$$

From equations (5) and (6), the approximate value of the displacement X at the time of $t_n + \Delta t_n$ can be obtained:

$$X^{(i)}_{t_n+\Delta t_n} = X^{(i-1)}_{t_n+\Delta t_n} - \left(\frac{\partial \tilde{f}}{\partial X}\bigg|_{U^{(i-1)}_{t_n+\Delta t_n}}\right)^{-1} \tilde{f}(X^{(i-1)}_{t_n+\Delta t_n}) \qquad (i = 1,2,\cdots) \quad (7)$$

$\tilde{f}(X^{(i-1)}_{t_n+\Delta t_n})$ in the equation (7) can be directly calculated from equation (3).

Use Newmark to calculate Jacobian matrix $\partial \tilde{f} / \partial X$ in equation (7) :

$$\frac{\partial \tilde{f}(X)}{\partial U}\bigg|_{X^{(i-1)}_{t_n+\Delta t_n}} = M\frac{X}{\partial X}\bigg|_{X^{(i-1)}_{t_n+\Delta t_n}} + D\frac{X}{\partial X}\bigg|_{X^{(i-1)}_{t_n+\Delta t_n}} + \frac{\partial r(X)}{\partial X}\bigg|_{X^{(i-1)}_{t_n+\Delta t_n}} \quad (8)$$

In Newmark algorithm, the expressions of the increments of speed and acceleration:

$$\dot{X}_{i+1} = \dot{X}_i + (1-\gamma)\Delta t_i \ddot{X}_i + \gamma \Delta t_i \left(\frac{X_{i+1} - X_i - \dot{X}_i \Delta t_i}{\beta \Delta t_i^2} - \left(\frac{1}{2\beta} - 1\right)\ddot{X}_i\right) \quad (9)$$

$$\ddot{X}_{i+1} = \frac{X_{i+1} - X_i - \dot{X}_i \Delta t_i}{\beta \Delta t_i^2} - \left(\frac{1}{2\beta} - 1\right)\ddot{X}_i \quad (10)$$

Where: γ and β are calculation coefficients.

Substitute equations (9) and (10) into equation (8) and simplify it, and then obtain:

$$\frac{\partial \tilde{f}(X)}{\partial X}\bigg|_{X^{(i-1)}_{t_n+\Delta t_n}} = \frac{1}{\beta \Delta t_n^2} M + \frac{\gamma}{\beta \Delta t_n} D + \frac{\partial r(X)}{\partial X}\bigg|_{X^{(i-1)}_{t_n+\Delta t_n}} \quad (11)$$

Substitutes equations (3) and (11) into equation (6) and obtains:

$$\left(\frac{1}{\beta \Delta t_n^2} M + \frac{\gamma}{\beta \Delta t_n} D + \frac{\partial r(X)}{\partial X}\bigg|_{X^{(i-1)}_{t_n+\Delta t_n}}\right)\Delta X^{(i-1)} = f_{t_n+\Delta t_n} - M\ddot{X}^{(i-1)}_{t_n+\Delta t_n} - D\dot{X}^{(i-1)}_{t_n+\Delta t_n} - rX^{(i-1)}_{t_n+\Delta t_n} \quad (12)$$

$$\ddot{X}^{(i-1)}_{t_n+\Delta t_n} = \frac{X^{(i-1)}_{t_n+\Delta t_n} - X_{t_n} - \dot{X}_{t_n}\Delta t_n}{\beta \Delta t_n^2} - \left(\frac{1}{2\beta} - 1\right)\ddot{X}_{t_n} \quad (13)$$

$$\dot{X}^{(i-1)}_{t_n+\Delta t_n} = \dot{X}_{t_n} + (1-\gamma)\Delta t_n \ddot{X}_{t_n} + \gamma \Delta t_n \ddot{X}^{(i-1)}_{t_n+\Delta t_n} \quad (14)$$

Where $\ddot{X}_{t_n}$, $\dot{X}_{t_n}$ and X_{t_n} are the acceleration vector, the displacement vector, the speed vector and at the time of t_n . According to the displacement, velocity, acceleration and interpolation function of the calculated node point, physical quantities, such as internal displacement, strain, stress, can be obtained.

2 Simulation of dynamic meshing contact

The finite element model of the planetary gear pair was established, which had both the functions of inner and outer contact, as shown in Fig. 1. Boundary conditions parameters were specified. The finite element model with constraints setting was shown in Fig. 2.

DOI: 10.1051/matecconf/201713900214

Fig. 1 A planetary gearing finite element model

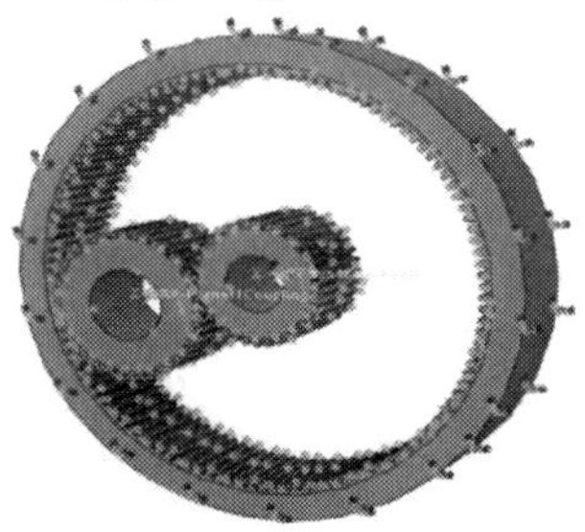

Fig. 2 Planetary gear pair with constraints

3.1 Analysis of stress distribution around addendum

The simulation of planetary gear pair was adopted to study dynamic behavior around addendum of external meshing planetary gear in meshing process. 5 nodes were chosen, numbered by 1, 2, 3, 4 and 5, respectively, as was shown in Fig.3 (a). Node 1 was on the meshing surface of addendum, and node 2 in the area near meshing surface. Node 3 was in the middle of addendum. Node 4 was in the area near non-meshing surface, while node 5 on the non-meshing surface of addendum. The time history curves of equivalent stresses of nodes shown in Fig.3 (a) were presented in Fig.3 (b), where it can be found that in meshing process the equivalent stress of each node increases first and then decreases, showing similar variation.

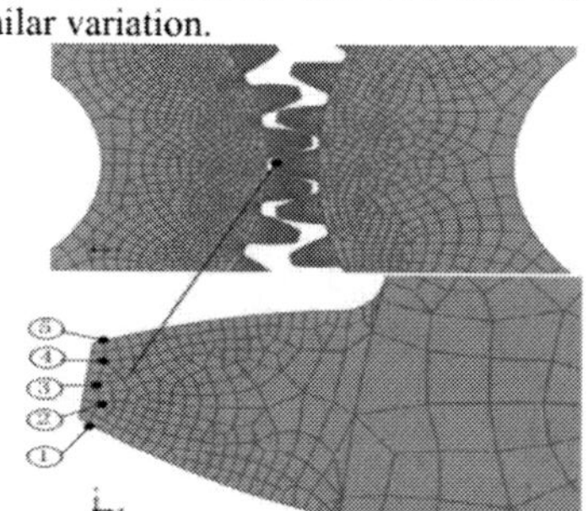

(a) Nodes around addendum of external meshing planetary gear

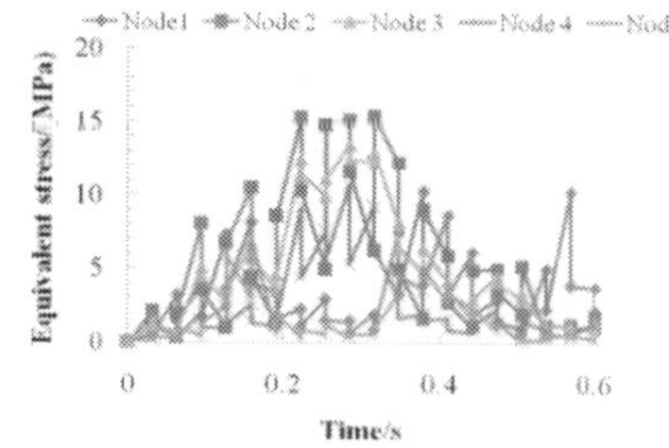

(b) Time history curves of equivalent stress of each node around addendum

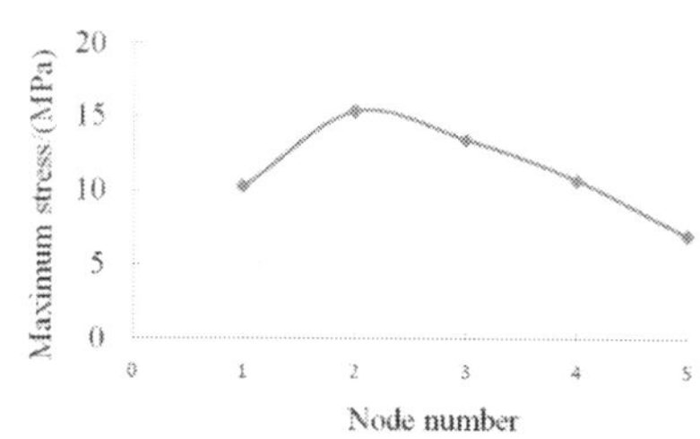

(c) Maximum equivalent stress of each node around addendum

Fig.3 Stress around addendum of external meshing planetary gear

The diagram was drawn taking each node shown in Figure 2(a) as abscissa, the maximum equivalent stress corresponding to each node as ordinate, as shown in Fig.3 (c). It could be known from both Fig.3 (c) and Fig.3 (a) that minimum stress was on the non-mating surface of the addendum. Less stress was in the non-engagement surface area close to the addendum. Greater stress was on the meshing surface of the addendum. Maximum stress was in the area close to the addendum mating surface. The maximum stress of node 1 was 33.06% smaller than that of node 2, the maximum stress of node 2 was 14.26% greater than that of node 3, the maximum stress of node 3 was 25.37% greater than that of node 4, the maximum stress of point 4 was 54.51% greater than that of node 5, and a difference between the minimum equivalent stress and the maximum was 54.82%.

3.2 Analysis of stress distribution around pitch circle

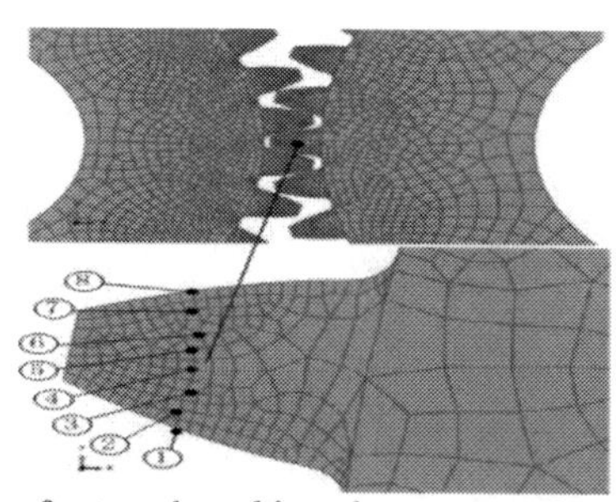

(a) Nodes of external meshing planetary gear's tooth profile around pitch circle

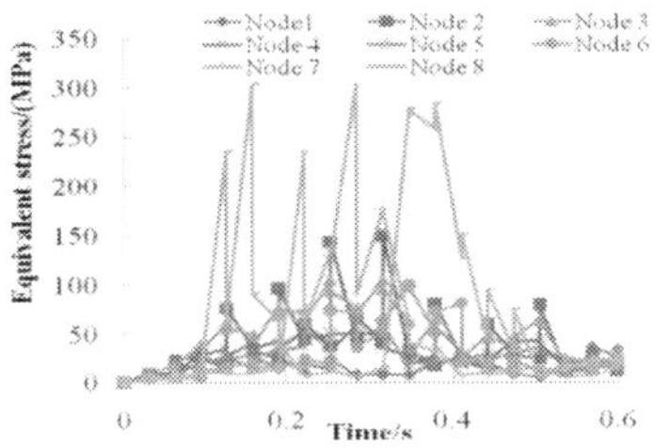

(b) Time history curves of equivalent stress

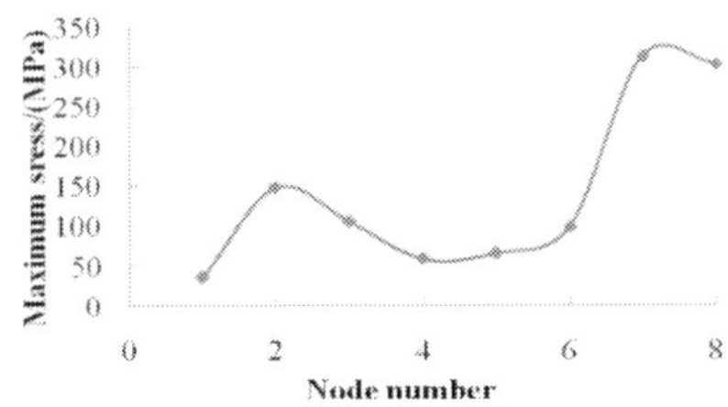

(c) Maximum equivalent stress of each node around pitch circle

Fig.4 Stress of external meshing planetary gear's tooth profile around pitch circle

The simulation of dynamic meshing process was adopted to study dynamic behavior around addendum of external meshing planetary gear. 8 nodes were chosen, numbered by 1, 2, 3, 4, 5, 6, 7 and 8, respectively, as was shown in Fig.4 (a). Node 1 was on the meshing surface of pitch circle, and node 2 in the area near meshing surface. Nodes 3~6 were in the middle of pitch circle. Node 7 was in the area near non-meshing surface, while node 8 was on the non-meshing surface of pitch circle. The time history curves of equivalent stresses of nodes shown in Fig.4 (a) were presented in Fig.4 (b), where it can be found that in meshing process the equivalent stress of each node increases first and then decreases, showing similar variation.

The maximum equivalent stress of each node around pitch circle was calculated, as was presented in Fig.4 (c), where abscissa was node number and ordinate was corresponding maximum equivalent stress. It can be shown in Fig.4 (c) combined with Fig.4 (a) that the maximum stress on the meshing surface of pitch circle is the smallest, while that in the area near non-meshing surface is the largest. The stresses in the middle of pitch circle are relatively small, and those on the non-meshing surface and in the area near meshing surface are larger. The maximum stress of node 1 is 75.82% less than that of node 2, which is 29.18% larger than that of node 3, while maximum stress of node 7 is 68.39% larger than that of node 6. The difference between minimum and maximum values of maximum equivalent stress is 88.53%. The stresses of nodes at different position of addendum of external stress planetary gear vary significantly.

3.3 Analysis of stress distribution around dedendum

Considering the different features between pitch circle and dedendum, 6 nodes were chosen to study the dynamic behavior around dedendum, instead of 8 nodes. The nodes were numbered by 1, 2, 3, 4, 5 and 6, successively from meshing surface to non-meshing surface, as was shown in Fig.5 (a). Time history curves of equivalent stresses of nodes in Fig.5 (a) were shown in Fig.5 (b) using finite element modeling. The equivalent stress around dedendum increases first, followed by decreasing.

The relationship between the number of node around dedendum (Fig.5 (a)) and corresponding maximum equivalent stress was obtained in Fig.5 (c). Significantly different from the law around pitch circle, the curve shows a "V" type. The maximum equivalent stress around dedendum decreases at first to 25.85 MPa at the position of node 3, from 56.51MPa at that of node 1, and then increases continuously to 93.56 MPa at the position of node 6, which is also the largest maximum equivalent stress.

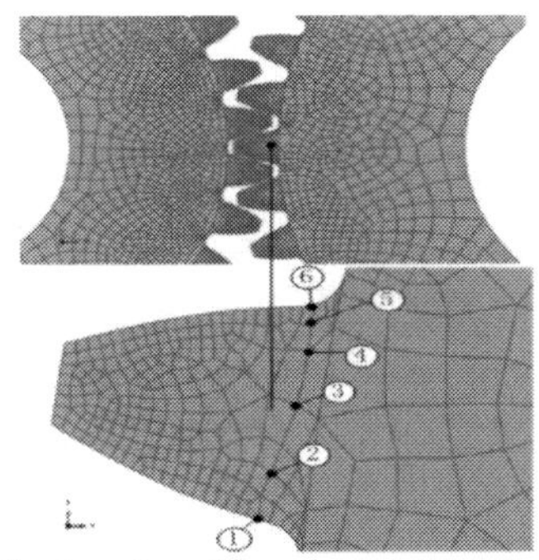

(a)Nodes of external meshing planetary gear's tooth profile around dedendum

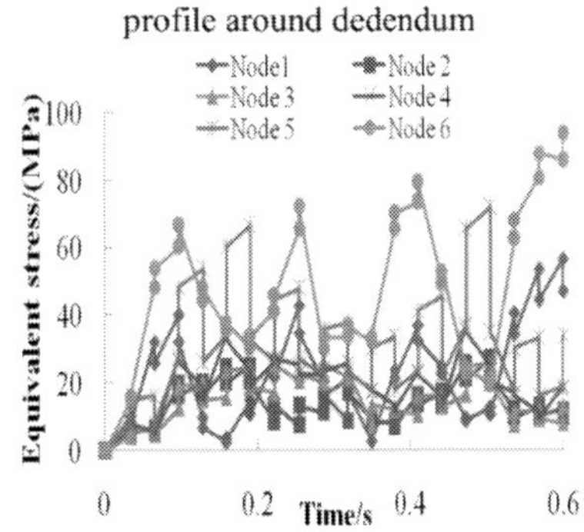

(b) Time history curves of equivalent stress

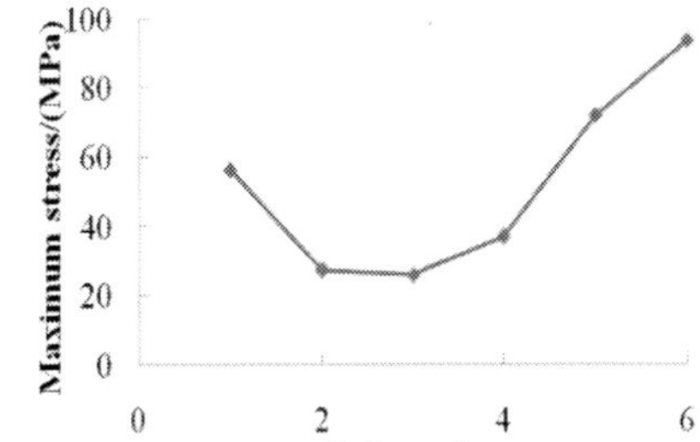

(c) Maximum equivalent stress

Fig.5 Stress of external meshing planetary gear's tooth profile around dedendum

4 Conclusion

Stress at each node of external meshing planetary gear has the similar change rule on the time course, which increases first and then decreases in the dynamic meshing process. Characteristics of stress distribution differ from one another in different meshing areas of the external meshing planetary gear's tooth profile. The difference of the equivalent stress around addendum, pitch circle and dedendum is relatively significant from the mating surface to the non-mating surface. The maximum stress of nodes at different position of external stress planetary gear varies significantly. Around dedendum, the maximum equivalent stress is obtained on the non-mating surface, while the smallest value of it is in the middle of dedendum.

5 Acknowledgement

The authors would like to thank anonymous reviewers for their helpful comments and suggestions to improve the manuscript. This research was supported by the Natural Science Foundation of Jiangxi Province (Grant No. 20161BAB206153), and the National Natural Science Foundation of China (Grant No.51765020), and the National Natural Science Foundation of China(Grant No.51765015).

References

[1] Lin T, Ou H, Li R F. A finite element method for 3D static and dynamic contact/impact analysis of gear drives . Computer Methods in Applied Mechanics and Engineering, 2007, 196(9): 1716-1728.
[2] Eritenel T, Parker R G. An investigation of tooth mesh nonlinearity and partial contact loss in gear pairs using a lumped-parameter model. Mechanism and Machine Theory, 2012, 56:28-51.
[3] Cai W, Chang J. Bifurcation and chaos analysis of the porous squeeze film damper mounted gear-bearing system. Computers & Mathematics with Applications, 2012, 64(5):798-812.
[4] Chen Y C, Liu C C. Contact stress analysis of concave conical involute gear pairs with non-parallel axes . Finite Elements in Analysis and Design, 2011, 47(4): 443-452.
[5] Chen Y C, Liu C C. Investigation on bending fatigue failure of a micro-gear through finite element analysis . Engineering Failure Analysis, 2013, 31: 225-235.
[6] Palermo A., Mundo D., Hadjit R., Desmet W. Multibody element for spur and helical gear meshing based on detailed three-dimensional contact calculations . Mechanism and Machine Theory, 2013, 62:13-30.
[7] Zhu C C, Song C S, Lim T C, Peng T. Pitch cone design and influence of misalignments on tooth contact behaviors of crossed beveloid gears . Mechanism and Machine Theory, 2013, 59:48-64.
[8] Tamarozzi T, Ziegler P., Eberhard P, Desmet W. Static modes switching in gear contact simulation. Mechanism and Machine Theory, 2013, 63:89-106.
[9] Dong W B, Xing Y H, Moan T, Gao Z. Time domain-based gear contact fatigue analysis of a wind turbine drivetrain under dynamic conditions. International Journal of Fatigue, 2013, 48:133-146.
[10] Li S. Effect of addendum on contact strength, bending strength and basic performance parameters of a pair of spur gears . Mechanism and Machine Theory, 2008, 43:1557-1584.
[11] Khabou M T, Bouchaala N, Chaari F, Fakhfakh T, Haddar M. Study of a spur gear dynamic behavior in transient regime . Mechanical Systems and Signal Processing, 2011, 25(8):3089-3101.

Research on Parameter Design of Multi - axis Hydrostatic Transmission Vehicle

Liang Zhao [1,2,a], Jin Wang [2] and Zheng-wu Zhang [2]

[1] Chang'an University, School of Construction Machinery, Xi'an ,China

[2] Shaanxi University of Technology, School of Machinery Engineering, Hanzhong , China

Abstract. In order to obtain reasonable parameters in the design of driving system of multi-axis hydrostatic transmission vehicle, the working principle of single-side drive of hydrostatic transmission vehicle is introduced. The matching and control of engine and hydraulic pump are analyzed. According to the driving equation of vehicle, The driving force required for driving system is determined, and the parameters of hydraulic motor, hydraulic pump, system working pressure and braking system are designed and calculated, which provides the parameter design for driving system of multi-axis hydrostatic transmission Reliable theoretical basis.

1 Introduce

It is a special chassis for operation, such as concrete mixer truck chassis, concrete pump truck chassis and large fire engines, etc., must be a second type of chassis to be modified to meet the vehicle after the conversion requirements, such as full power Output force device, and even install the transfer case and vice engine. Hydraulic transmission technology with a wide range of speed, easy to control, can achieve stepless speed and other characteristics, can achieve the vehicle's stepless speed regulation and freestyle, in the construction machinery walking system has been a lot of applications [1-2]. As the need for special chassis models are growing, engineering machinery and often in a certain work area for micro-operation. According to the characteristics of hydraulic transmission, the use of hydrostatic transmission of the way, so that the power transmission layout flexible and easy to achieve the vehicle's stepless speed.

2 Composition and working principle of hydrostatic transmission vehicle

The hydrodynamic drive vehicle has four hydraulic pumps and 14 drive motors, using a fully driven approach. The working principle of the hydraulic drive system is shown in Figure 1, which includes the engine 1, the hydraulic pump 2, the fuel tank 3, the cartridge valve 4, the electromagnetic directional valve 5, the relief valve 6, the hydraulic motor 7, Wheel reducer 8 and tire 9.

The engine 1 is connected to the hydraulic pump 2 to drive the hydraulic pump to rotate, to supply the hydraulic oil to the travel drive system, to drive the hydraulic motor 7 via the cartridge valve 4, to rotate the wheel 9 by the wheel speed reducer 8, The solenoid valve 5 controls the cartridge valve 4 to change the flow of the hydraulic oil so that the vehicle is advanced or retreated. The constant power control of the hydraulic pump with the pressure cut-off is used, and the displacement of the hydraulic pump varies depending on the rotational speed of the engine 1. The hydraulic motor 7 is controlled by the electrical ratio, and the displacement of the hydraulic motor is controlled by the magnitude of the current of the given hydraulic motor coil.

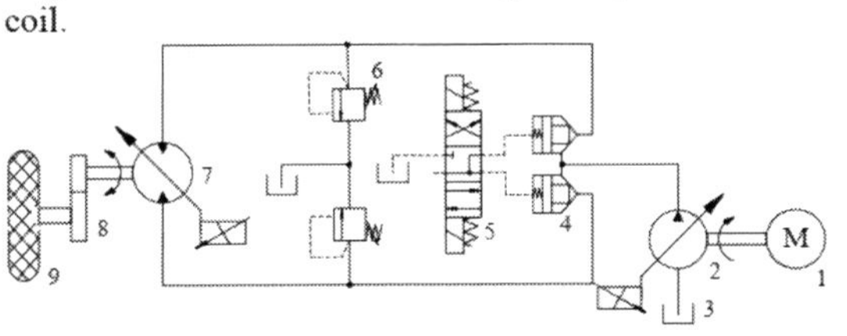

Figure 1. Schematic diagram of single-sided driving hydraulic system

3 Matching and control of engine and hydraulic pump

In order to make the engine in the vehicle running in the process of working in the best state, according to the external characteristics of the engine to control the displacement of the hydraulic pump, when the engine load rate of 90%, the engine fuel consumption is the best, so according to the engine The relationship between the

[a] Corresponding author: zhaoliangslg@163.com

engine output torque Me and the speed n is obtained by the least squares method: [3,4]

$$M_e = M(n) \tag{1}$$

Where $M(n)$ is the engine output torque at different speeds.

The external characteristics of the engine are shown in Figure 2, where n_1 is the idle speed of the engine; n_2 is the engine maximum output torque when the speed; n_3 is the engine maximum output torque when the speed.

Considering the economy of the engine, the maximum displacement control of the hydraulic pump at each speed is:[5]

$$q_b = \frac{0.9M_e 2\pi}{\Delta p} \tag{2}$$

Where q_b is the displacement of the hydraulic pump; Δp is the pressure difference between the outlet and inlet of the hydraulic pump.The displacement control curve of the hydraulic pump is controlled along the curve $\mathbf{1'-2-3}$ in Figure2.

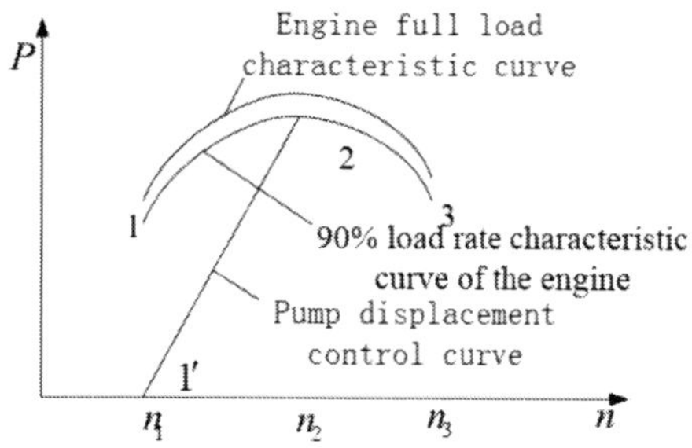

Figure 2. Maximum displacement control curve for hydraulic pumps

4 Calculation of vehicle driving force

Driving process, the chassis to provide the driving force and the resistance should be maintained in balance, the vehicle's driving equation is:

$$F_t = F_f + F_w + F_i + F_j \tag{3}$$

Where F_t is the driving force; F_f is the rolling resistance, $F_f = Mgf$; F_w is the wind resistance, $F_w = \frac{C_D A v}{21.15}$, C_D is the wind resistance coefficient, A is the windward area of the vehicle, v is the speed of the vehicle; F_i is the ramp resistance, $F_i = Mg\sin\alpha$; F_j is accelerated resistance.

4.1 Driving force of maximum speed

When the vehicle is traveling at the highest speed on the ground, The climb gradient $\alpha = 0$, so the ramp resistance $F_i = Mg\sin\alpha = 0$, The acceleration resistance $F_j = 0$. According to the design of the maximum speed, the wheel needs the driving force:

$$F_t = F_f + F_w \tag{4}$$

Given the parameters of the vehicle, you can determine the drive force required for the wheel,

Where $M = 84000\text{kg}$, $C_D = 0.8$, $f = 0.012$, $A = 5.5\text{m}^2$, $v = 80km/h$, the driving force required for the maximum speed $F_t = 11.2\text{kN}$.

4.2 Maximum driving force when climbing

When the vehicle is running on the maximum slope, the speed of travel is relatively low, can ignore the wind resistance, driving resistance only rolling resistance and ramp resistance, which:

$$F_t = F_f + F_i \tag{5}$$

According to 30% of the gradient, $\alpha = \arctan(0.3) = 16.67^\circ$, the driving force required for the wheel at the maximum slope is $F_t = 24.6\text{kN}$.

5 Parameter design of hydraulic Motor

In the hydraulic motor selection calculation, to make the hydraulic motor to meet the hydraulic drive vehicle power requirements and the maximum speed requirements, and can zero displacement. The vehicle has the greatest running resistance at the maximum slope, and the speed of the hydraulic motor is required to meet the requirements of the running speed when traveling on the ground.

The driving torque of the hydrostatic transmission vehicle:

$$T_m = \frac{F_t R_r}{n i_g \eta_g} \tag{6}$$

Where n is the number of drive wheels when driving, i_g is the gear ratio of the wheel reducer, η_g is the mechanical efficiency of the gear unit, R_r is the rolling radius of the wheel.

The hydraulic motor is also required to meet the maximum speed of the vehicle design, The maximum speed of the design is $v_{max} = 80km/h$. Calculate the maximum speed of the hydraulic motor according to the design maximum speed..

$$n_{m\max} = \frac{v_{max}60}{2\pi R_r 3.6} i_g \quad (7)$$

Where $n_{m\max}$ is the maximum speed of the hydraulic motor, i_g is the wheel reducer gear ratio, R_r is the wheel rolling radius.

6 System work pressure calculation

The working pressure of the hydraulic motor is determined by the running resistance of the hydrostatic transmission vehicle. The working pressure is calculated as:

$$\Delta P_m = \frac{100T_m}{1.59q_m\eta_{mh}} \quad (8)$$

Where ΔP_m is the working pressure at both ends of the hydraulic motor, T_m is the torque on the hydraulic motor, q_m is the displacement of the hydraulic motor, η_{mh} is the mechanical efficiency of the hydraulic motor.

7 Parameter design of hydraulic pump

Hydraulic pump in the calculation of parameters to meet the system's working pressure and hydraulic motor hydraulic flow requirements. Hydrostatic transmission vehicles in the ground to high speed, the hydraulic motor displacement to take 50ml / r, according to the number of drive the number of systems required flow:

$$Q_{max} = n_{max}q_m n \quad (9)$$

Where $n_{m\max}$ is the maximum speed of the hydraulic motor, q_m is the displacement of the hydraulic motor, n is the number of drive wheels.
Calculate the displacement of the hydraulic pump based on the maximum flow required by the system and the speed of the engine.

$$q_b = \frac{Q_{max}}{n_b n_p} \quad (10)$$

Where n_b is the number of hydraulic pumps, n_p is the maximum speed at which the engine is operating.

8 Braking system parameter calculation

8.1 Braking performance calculation

Braking operation, close the motor control valve, while maximizing the motor displacement, the use of the relief valve motor outlet pressure to slow down the motor, in order to achieve braking. The maximum pressure of the entire walking hydraulic system is the system relief valve settings.

The maximum hydraulic resistance moment:

$$M_{br} = 1.59\frac{P_{br}P_{m\max}}{10} i_g n \quad (11)$$

Where P_{br} is the brake relief pressure, the minimum is set to 20MPa, the maximum is 35MPa, determined by the brake pedal position, i_g is the wheel reducer gear ratio.

The maximum brake deceleration is:

$$a_{br} = \frac{M_{br}}{R_r M} \quad (12)$$

The full load brake the shortest distance:

$$s = \frac{v^2}{2a_{br}} + v(t_a + 0.5t_b) \quad (13)$$

Where v is the full speed, determined according to the current speed, t_a is the braking system reaction time, refers to the brake pedal when braking to overcome the free travel, hydraulic oil pressure transmission and other time required, the general hydraulic brake system reaction time is 0.015 - 0.03s, taken here 0.02s. t_b is deceleration deceleration time, the general hydraulic brake system is 0.15 - 0.3s, taken here 0.3s.

8.2 Braking performance calculation

According to the design requirements, the parking slope is 30%, about 16.7 ° slope.

The parking brake torque to be provided by the wheel reducer is:

$$T_t = \frac{Mg\sin\alpha R_s}{ni_g} \quad (14)$$

Where M is the vehicle quality, R_s is the wheel's static radius.

9 Engineering case analysis

According to the vehicle driving conditions, multi-axis hydrostatic transmission vehicle driving system hydraulic pump selection Rexroth A11VO190 variable pump, hydraulic motor selection A6VM200 variable motor, A11VO190 variable pump nominal pressure of 350bar, A6VM200 variable motor nominal pressure of 400bar。According to (6) and (8) to calculate the hydraulic motor working pressure range of 174bar -

318bar, can make the system pressure to get a better match. According to (11),(12) and (13),the running braking torque of the vehicle can be obtained as $M_{br} = 210\text{kN}\cdot\text{m}$,the maximum running braking deceleration of the vehicle can be obtained as $a_{br} = 3.77m/s^2$; According to (14), the parking brake torque of the vehicle can be obtained as $T_t = 795\text{N}\cdot\text{m}$

10 Conclusion

In this paper, the parameter matching of driving system of multi-axis hydrostatic transmission vehicle is researched. The parameter matching of running drive system, driving resistance, parameters calculation of hydraulic motor and hydraulic pump, and running brake system are analyzed. The parameter calculation of the hydraulic drive system is explained, which provides the basis for the parameter calculation of the hydrostatic drive vehicle.

Acknowledgement

The article was supported by Special research project of Shaanxi Provincial Department of Education (Grant No. 14JK1134),Shaanxi Province Natural Science Basic Research Project(Grant No. 2017JM5087), Natural Science Project of Shaanxi Provincial Science and Technology Department (Grant No. 2016JM5039).

References

1. YAO Huaixin. "Construction machinery chassis and its hydraulic drive theory", China Communications Press,2002
2. YAO Huaixin. "Rationale and Data Matching for Control of Vehicle's Hydraulic Driving Systems.ChinaJournal of Highway and Transport", 2002, 15(3):115-118
3. WANG Yongqi, SHAN Xinzhou. "Parameter matching and controlling of static hydraulic driving device in bulldozers". Construction mechanization, 2003（10）:34-53
4. Wang Xin, Zhang Chao, Yi Xiaogang." Research on Power - Load Adaptive Method for Full Hydraulic Leveling Machine", 2007,(38)
5. LONG Shuigen, YANG Xilong, REN Shijie. "Matc Parameters Calculation and Matching in the Traveling Driven System of Full Hydraulic Bulldozer", Hydraulic and pneumatic, 2007(12) :97-100
6. Zuo Delong, Yin Wenqing, Hu Fei . "Matching and Simulation of Engine and Hydrostatic Transmission", Mechanical transmission, 2010,(34)
7. ZHAO Liang. "Matching and control study on driving system of full hydraulically-driven chassis",Journal of China Construction Machinery Corporation，2013,（6）
8. LI Xiang sheng ， CHANG Siqin. "Matching Efficiency of Power Source System for Vehicle with Energy Accumulation Hydrostatic Drive", China Journal of Highway and Transport,2007 ， 20(1) :118-122
9. Yi Xiaogang, Jiao Shengjie, Liu Zhengfu." Research on Key Technical Parameters of Full Hydraulic Bulldozer" China Journal of Highway and Transport,2004(2)
10. LI Xiang sheng, CHANG Si qin."Design of the Power SourceSystem for the Vehicle of HydrostaticTransmission System "Journal of Nanjing Forestry University(Natural Sciences Edition,,2005,29(4) :65-68

Research on modulation transfer function of the linear-array CCD for hyperspectral imager

ZHANG Hong qiang[1], *WANG* Chun hong[1,*]
Suihua University, Suihua 152061, China

Abstract. As an image-spectrum merging technology, hyperspectral imaging has been used in battlefield reconnaissance rapidly. Based on response of CCD to the sine input signal, the period and extremal positions of output signal were analyzed and simulated. According to the definition of MTF based on contrast, three expressions of the average MTF of line-array CCD were presented and compared with those based on Fourier transform method. Finally, the average MTF of line-array CCD is confirmed. Accordingly, the characteristics of modulation transfer function of line-array CCD are studied. The simulation results indicate that the frequency increase, while at Nuquist frequency, the effect of the initial position, which becomes much stronger with frequency increase, while at other frequency, the effet of the initial position is weak. In the frequency range of $|\zeta|$>0.1%,the effect of the initial position on MTF can be ignored.

Key words. Hyperspectral Imager; Discrete imaging; Line-array CCD; MTF; Initial position

1 Introduction

With the rapid development of space technology and computer technology, the application of charge coupled device (CCD) in fields like hyper-spectral imaging is becoming more and more broad. As an internationally general method for evaluating the quality of imaging systems, modulation transfer function (MTF) can subjectively reflect the spatial frequency characteristics of an imaging system [1]. Thus, research on MTF characteristics of CCD and its influence on the imaging quality of the whole system are quite essential [2].

CCD is a discrete imaging device whose sampling of input signals is a linear process but not a complete spatial invariant process [3]. So far, Fourier method and contrast method are mainly applied to research MTF of CCD. Based on assumption of continuous scanning of a single image element, with application of contrast method, Kai M. Hock deduced the expression of average MTF of line-array CCD that is consistent with MTF of line-array CCD by Fourier method. Subsequently[4], O. Hadar proved the correctness that sinc function is used as the sampling function for arbitrary direction of two-dimensional sampling array[5]. The modulation transfer function is affected by the initial position. S.B.Campana pointed out in using contrast method (streak target) to measure MTF of CCD that if the phases of streak target and CCD change when the input signal frequency is near Nyquist frequency[6], MTF values of CCD fluctuate. S. K. Park proposed to use average MTF to evaluate the imaging quality of discrete imaging systems based on point spread function Fourier transform[7]. J.Feltz used a line-array CCD signal in the sinusoidal input signal period to output the maximum value and the minimum value according to the contrast definition of MTF so as to offer the line-array CCD and expression of phase-related MTF[8]. Song min experiments verified that the experiment results agree with Feltz's theoretical calculation[9]. And Jyrki uses the sinusoidal signal to measure the modulation transfer function of CCD (steps of 0.1*(pixel size) in the object plane). The simulation results show that the relative position has a strong effect on the modulation transfer function at the near Nuquist frequency. It has less influence on other frequencies. however, no theoretical analysis was made[10].

In this paper, based on specific analysis on response characteristics of line-array CCD to sinusoidal input signal, the mathematical model of MTF of line-array CCD is established according to the contrast definition of MTF so as to discuss the expression forms of average MTF of line-array CCD, based on which the influence of the initial position on MTF of line-array CCD. The research content in this paper can provide certain theoretical evidences for measurement of MTF of CCD.

2 Theoretical Analysis

2.1 The response characteristics of line-array CCD to sinusoidal input signal

* Corresponding author: wangcx2003@126.com

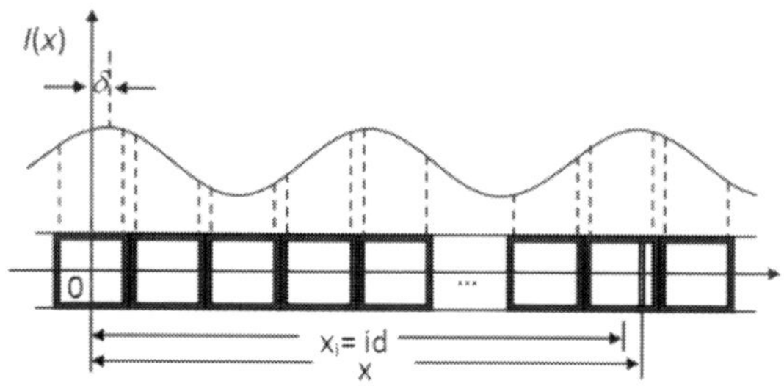

Figure1. Sketch map of line-array CCD sampling

As shown in Fig.1, the image element photosensitive unit width of line-array CCD is w and the spacing between image elements is d. The center of the first CCD image element is taken as the origin of coordinate and its serial number i is 0. $x_i=id$ refers to the coordinate of the center of No.i image element. If the signals that are irradiated on the line-array CCD are distributed as

$$I(x) = 1 + C_{in}(f)\cos(2\pi f(x+\delta)) \tag{1}$$

In the formula, f, $C_{in}(f)$ and δ separately refers to the spatial frequency, modulation degree and relative location (the relative spatial location of the input signal and No.0 CCD image element) of the input signal. During the sampling process, the signal output on each image element is obtained by accumulation of $I(x)$ from the interval between $(x_i\text{-}w/2)$ and $(x_i+w/2)$. thus, the signal output of No.i CCD image element is

$$\begin{aligned} I_i(f,\delta) &= 2\int_{x_i-w/2}^{x_i+w/2} w(1+\cos(2\pi f(x+\delta)))dx \\ &= 2w^2(1+\sin c(fw)\cos(2\pi f(x_i+\delta))) \end{aligned} \tag{2}$$

The following conclusion can be drawn from formula (2):

1.the position of signal output extremum

When $2\pi f(x_{i\max}+\delta)=2k\pi$ (k=0, ±1, ±2...), that is, $x_{i\,max}=k/f\text{–}\delta$, there are maximal values in continuously-sampling signal output. If $x_{i\ max}=i'_{\max}d$ and $round(x)$ refers to x after rounding. When the serial number of CCD image elements $i_{max}=round(i'_{\max})$, the maximal value in signal output $I_{max\ i}$. $|i'_{max}-i_{max}|$ reaches the minimum and there is the maximum value I_{max}. $|i'_{max}-i_{\max}|=0$ in signal output, the position of the maximum value of signal output coincides with the position of the minimum value of input signal and $I_{\max}=w^2(1+sinc(wf))$.

When $2\pi f(x_{i\ min}+\delta)=(2k+1)\pi$ (k=0,±1, ±2…) , that is, $x_{i\ min}=i'_{\min}d=(k+1/2)/f\text{-}\delta$, there are minimal values in continuously-sampling signal output. Assuming $x_{i\ min}=i'_{\min}d$, when the serial number of CCD image elements $i_{\min}=round(i'_{\min})$ and the minimal value in signal output $I_{max\ i}$. $|i'_{\max}-i_{\max}|$ reaches the minimum, there is minimum value $I_{\min}$. $|i'_{\min}-i_{\min}|=0$ in signal output. The position of the minimum value of signal output coincides with the position of the minimum value of input signal and $I_{\min}=w^2(1\text{-sinc}(wf))$.

2. the period of signal output

Because of periodic distribution of input signals, the output signal size of subsequent sampling points following a certain image element will repeat the result of its previous sampling point. Such as signal output unit is named as a signal output period. The period of input sinusoidal signal is $T=1/f$ and the period of corresponding signal output image element number is $T_i=1/(fd)$. $T'_i=rats(T_i)$ and $rats(x)$ numerically rationalize x. Because of discrete sampling of CCD image elements, the period of line-array CCD signal output image element number refers to the numerator part of numerically-rationalized T. Thus, the period of image element number is related to the value of f and d but not related to δ and w.

2.2 Average MTF of line-array CCD

It can be seen from formula (2) that when the input signal is a sinusoidal signal, the amplitude of output signal changes with the relative location δ whose changing period is d. Thus, MTF of line-array CCD is related to the relative location δ. The average MTF of line-array CCD is defined as:

$$\overline{MTF}(f) = \frac{\sum_{j=1}^{m} MTF(f,\delta_j)}{m} \tag{3}$$

Where $MTF(f, \delta_j)$ refers to MTF whose initial position is δ_j and m refers to the number of initial positions. According to definition of MTF, MTF refers to the ratio of the system's output modulation degree $C_{out}(f)$ and input modulation degree $C_{in}(f)$. When the input modulation degree $C_{in}(f)$ is 1,

$$MTF(f,\delta_j) \equiv \frac{C_{out}(f)}{C_{in}(f)} = \frac{I_{\max}(f,\delta_j)-I_{\min}(f,\delta_j)}{I_{\max}(f,\delta_j)+I_{\min}(f,\delta_j)}$$

On account of discrete sampling of CCD image elements, when the spatial frequency is f and the initial position is δ_j, there will be several maximal values $I_{max\ i}$ and minimal values $I_{min\ i}$ in signal output within the period of a CCD image element signal output and there are three expressions of MTF shown as below:

2.2.1 The method of maximum and minimum values of signal output

Through the maximum and minimum values, $I_{\max}$ and $I_{\min}$, of signal output within a signal output period of line-array CCD, $MTF''(f, \delta_j)$ whose initial position is δ_j can be directly obtained, that is:

$$MTF''(f,\delta_j) = \frac{I_{\max}(f,\delta_j)-I_{\min}(f,\delta_j)}{I_{\max}(f,\delta_j)+I_{\min}(f,\delta_j)} \tag{4}$$

Through substituting formula (4) into formula (3), $\overline{MTF''}(f)$ is obtained.

2.2.2 the averaging method of MTF

* Corresponding author: wangcx2003@126.com

n' MTFs that are obtained through maximal value $I_{max\,i}$ and minimal value $I_{min\,i}$ of signal output within No.i semi period of input sinusoidal signal are averaged to act as $MTF'(f, \delta_j)$ whose initial position is δ_j, that is:

$$MTF'(f,\delta_j) = \frac{1}{n'}\sum_{i=1}^{n'} \frac{I_{max\,i}(f,\delta_j) - I_{min\,i}(f,\delta_j)}{I_{max\,i}(f,\delta_j) + I_{min\,i}(f,\delta_j)} \qquad (5)$$

Through substituting formula (5) into formula (3), $\overline{MTF'(f)}$ is obtained.

2.2.3 the averaging method of maximal and minimal values of signal output

n maximal and minimal values of signal output within a signal output period are separately averaged to act as maximal value $I_{max\,i}$ and minimal value $I_{min\,i}$, through which $MTF(f, \delta_j)$ whose initial position is δ_j is obtained, that is:

$$MTF(f,\delta_j) = \frac{1}{n}\frac{\sum_{i=1}^{n} I_{max\,i}(f,\delta_j) - \sum_{i=1}^{n} I_{min\,i}(f,\delta_j)}{\sum_{i=1}^{n} I_{max\,i}(f,\delta_j) + \sum_{i=1}^{n} I_{min\,i}(f,\delta_j)} \qquad (6)$$

Through substituting formula (6) into formula (3), $\overline{MTF(f)}$ is obtained.

$$\overline{MTF(f)} = \frac{1}{mn}\sum_{j=1}^{m}\frac{\sum_{i=1}^{n} I_{max\,i}(f,\delta_j) - \sum_{i=1}^{n} I_{min\,i}(f,\delta_j)}{\sum_{i=1}^{n} I_{max\,i}(f,\delta_j) + \sum_{i=1}^{n} I_{min\,i}(f,\delta_j)} \qquad (7)$$

3 Simulation Analysis

Simulation calculation will be conducted on signal output characteristics of line-array CCD, expressions of average MTF and the influence of initial positions on MTF next. In detailed calculation, the spacing between image elements is taken as d=18μm and photosensitive unit width of image elements is w=16μm. Corresponding Nyquist frequency is f_N=1/(2d)=27.78 lp/mm.

3.1 Signal output characteristics

If f=18lp/mm, δ=(3d)/20=2.7μm, the detailed calculation is shown as below.

3.1.1 signal output period

T_f=1/(fd)=250/81, that is, the signal output period is 250 CCD image elements.

3.1.2 position of extremums

Position of maximal values: according to $i'_{max}d=k/f-\delta$, i'_{max}=(k−0.0486)/0.324, k=0, ±1, ±2, ±3⋯, That is, i'_{max}=2, 6.02, 9.11, 12.20... and the serial number of corresponding CCD image element is i_{max}=$round(i'_{max})$ =2, 6, 9, 12⋯.

Position of minimal values: according to $i'_{min}d=(k+1/2)/f-\delta$, i'_{min}=(2k+0.9514)/0.648, k=0, ±1, ±2, ±3⋯, That is, i'_{min} =1.47, 4.55, 7.64, 10.73... and the serial number of corresponding CCD image element is i_{min}=$round(i'_{min})$=1, 5, 8, 11⋯.

3.1.3 computer simulation

The signal output curve of No.0~100 CCD image elements is shown as Fig.2.

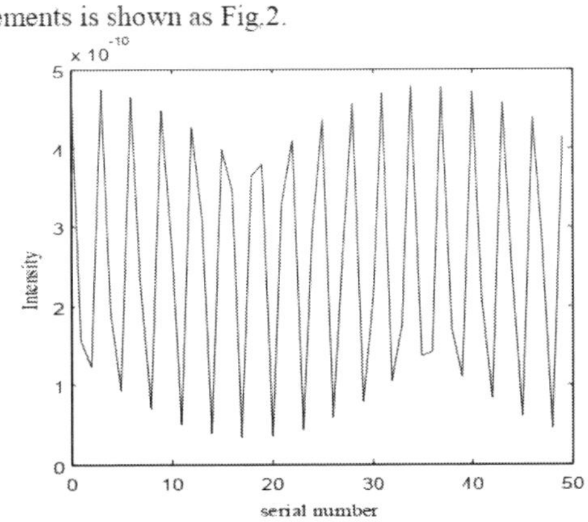

Figure2. Intensity distribution of output signal

3.2 Comparison between average MTF and MTF by Fourier method

Through Fourier method, MTF of line-array CCD is the product of Fourier transform of rectangle function $rect\,(x/w)$ and Fourier transform of sampling function $rect\,(x/d)$, that is:

$$MTF = \left|\sin c(fw)\sin c(fd)\right|$$

Values of input signal frequency are uniformly taken in Nyquist frequency. m=50, that is, step length is 18/50μm. $\overline{MTF''(f)}$ curve, $\overline{MTF'(f)}$ curve, $\overline{MTF(f)}$ curve and MTF obtained through Fourier transform are shown as Fig.3, in which Nyquist frequency is normalized as 1. It can be seen from Fig.3 that there is greater deviation between $\overline{MTF''(f)}$ curve and MTF curve, smaller deviation between $\overline{MTF'(f)}$ curve and MTF curve and $\overline{MTF(f)}$ curve coincides with MTF curve. It can be known from simulation comparison results that the expression formula of average MTF through contrast method is formula (7).

* Corresponding author: wangcx2003@126.com

 DOI: 10.1051/matecconf/201713900216

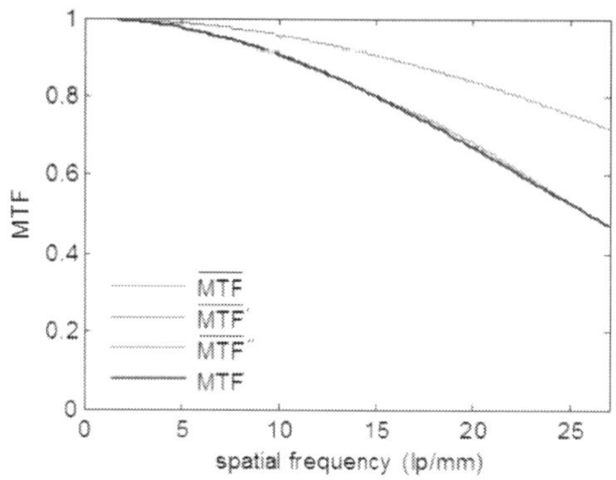

Figure3. Simulation and comparison of MTF

3.3 MTF characteristics in f_N frequency domain

In engineering design and tests, people usually pay more attention to the system's MTF of line-array CCD at Nyquist frequency. Thus, according to formula (7), we focally calculate Nyquist frequency and the changing principle of MTF near Nyquist frequency division with positions, during which CCD image element number n=1000, m=50.

f_N refers to Nyquist frequency. f_N / l (l=1, 2, 3, 4 ⋯) refers to Nyquist frequency division. l refers to Nyquist frequency division number. ζ refers to input signal frequency. f refers to the percentage deviation relative to Nyquist frequency division, that is,

$$\zeta = \frac{f - f_N / l}{f_N / l}$$

The deviation percentage ζ used in calculation is ±0.005%, ±0.02%, ±0.04%, ±0.06%, ±0.08%, ±0.1%.

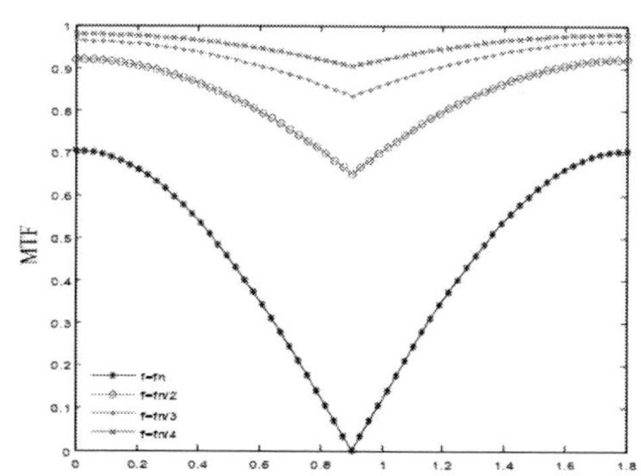

Figure4. Curve of MTF with initial positions

When input signal frequency $f=f_N / l$ (l=1, 2, 3, 4), the curve of MTF with initial positions is shown as Fig.4. When l=1 and l=2, under the deviation percentage mentioned above, the curve of MTF with initial positions (ζ<0) is shown as Fig.5and Fig.6.

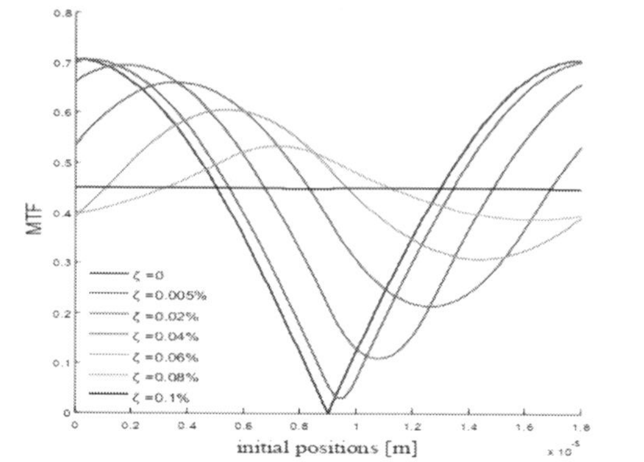

Figure5. Curve of MTF with initial positions at near Nyquist frequency division (l=1)

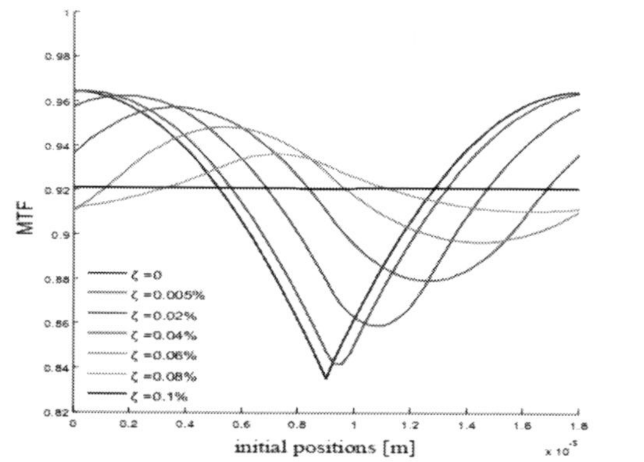

Figure6. Curve of MTF with initial positions at near Nyquist frequency division (l=3)

It can be indicated from comparison between Fig.4, Fig.5and Fig.6 that, at the Nyquist frequency division and near Nyquist frequency division, the influence of the initial position on MTF decreases as l increases. The maximum amplitude difference of MTF at $f_N / 4$ is 0.052 and MTF almost does not change with initial positions. Near Nyquist frequency division, the influence of the initial position on MTF decreases as the deviation percentage ζ increases. When ζ=±0.1%, the maximum amplitude difference of MTF is 0.007 and MTF no longer changes with initial positions.

4 Conclusion

◆ In this paper, the response characteristics of line-array CCD to sinusoidal input signal are analyzed and the theoretical analysis and numerical value simulation results of signal output period and extremum positions are provided.

◆ In this paper, starting from definition of MTF, MTF analytic expression formula of line-array CCD is deduced and the expression formula of average MTF of line-array CCD is obtained through simulation comparison, based on which MTF characteristics of line-

* Corresponding author: wangcx2003@126.com

array CCD are researched. Conclusion is described as below: the simulation results indicate that the frequency increase, while at Nuquist frequency, the effect of the initial position, which becomes much stronger with frequency increase, while at other frequency, the effet of the initial position is weak. In the frequency range of |ζ|>0.1%,the effect of the initial position on MTF can be ignored.

Acknowledgements

This work has been supported by Project of the Doctoral Scientific Research Foundation of Suihua University (Grant No. SB11001)

References

1 Christophe Latry, Vincent Despringre, et al. Automatic MTF measurement through a least square method. SPIE, 2004, 5570: 233-244.
2 Barnard K J, Anisimov I, Scheihing I E. Random laser speckle based modulation transfer function measurement of midwives infrared focal plane arrays. Optical Engineering, 2012, 51(8), 083601:1-8
3 Brian T. Teipen, Duncant L. MacFarlane. Liquid-crystal-display projector-based modulation transfer function measurements of charge-coupled-device video camera systems . Applied Optics, 2000, 39(4): 515-525.
4 K. M. Hock. Effect of oversampling in pixel arrays, Optical Engineering, 1995, 34(5): 1281-1288.
5 O. Hadar, A.Dogariu, G. D. Boreman. Angular dependence of sampling MTF . SPIE, 1998, 3110: 536-547.
6 S.B.Campana. Techniques for evaluating charge couple dimagers . Optical Engineering, 1977, 16(3): 267-274.
7 S. K. Park, R. A. Schowengerdt, M. A. Kaczynski. Modulation-Transfer-Function Analysis for Sampled Image Systems, Applied Optics, 1984, 23(15), 2572-2582.
8 J. Feltz, M. Karim. Modulation transfer function of charge coupled devices. Applied Optics, 1990, 29(5): 717-722.
9 Song Min, Kuai Xin kai, Zheng Ya ru. Influence of phase match between input signal and CCD pixel on the MTF. Otical Technique, 1999, 25(3):84-86.
10 Jyrki Laitinen. Method for evaluation of imaging in automated visual web inspection. Optical Engineering, 1997, 36(8): 2184-2196.

* Corresponding author: wangcx2003@126.com

Technical Research on the Electric Power Big Data Platform of Smart Grid

MA Ruiguang[1], WANG Haiyan[1], ZHANG Quanming [1], LIANG Yuan[2]

[1]State Grid Sichuan Economic Research Institute SGSERI Chengdu, China
[2]College of Management Science, Chengdu University of Technology, Chengdu, China

Abstract. Through elaborating on the associated relationship among electric power big data, cloud computing and smart grid, this paper put forward general framework of electric power big data platform based on the smart grid. The general framework of the platform is divided into five layers, namely data source layer, data integration and storage layer, data processing and scheduling layer, data analysis layer and application layer. This paper makes in-depth exploration and studies the integrated management technology based on big data, the index technology based on electric power big data, the analysis technology based on electric power big data and the excavation technology based on electric power big data, to achieve the breakthrough of key technologies of electric power big data. Finally, the typical case of big data applications shows that the key technology application of electric power big data platform will bring changes for the power enterprises, which has demonstration meaning and important reference meaning.

1 Introduction

With the advent of the Age of Big Data, countries around the world attach great importance to the development of big data. In 2012, the United States proposed the "Big data Research and Development Initiative", with the investment of more than 200 million funds of scientific research for the acquisition, organization and management of massive data, and authorized the government's six federal agencies to join, in to promote the application and development of big data related technologies. The Chinese government proposed the plan of "The Implementation of the National Big Data Strategy" in 2013 and issued *The Outline of The Action of Promoting the Development of Big Data*, to comprehensively promote the development of big data and accelerate the construction of great power of data. Big data has become a national basic strategic resources, and the "second petroleum". The rise of big data has whipped up huge waves in many fields of the country, and also to brought changes to a lot of industries, such as the telecommunications industry, power industry, consuming industry, which are based on big data. Of course, big data has also provided opportunities of development for many enterprises [1].

The energy industries, including electricity, petroleum, gas, new energy and other energy fields, generate massive data due to related technology applications of collection, processing, analysis and application, which form the big data of energy. At present, the main sources of big data are electricity, petroleum, gas, new energy, population, geography, meteorology and other data. The energy big data makes the energy production, consumption and relevant technical revolution deeply integrated into the big data philosophy, and promotes the thorough application of big data technology in the energy field, and also accelerates the development of energy industry and business model innovation.

In 2013, the issuance of White Paper for the Development of Chinese Electric Power Big Data aroused a hot trend of research on the electric power big data, which brought vitality to the traditional power industry, promoting the transformation of the power industry. Electricity mainly includes power generation, transmission, power transformation, power distribution, power consumption and scheduling, each producing a lot of data. The data are divided into several categories: the first one is the management data of the power enterprise; the second one is the power grid operation data; the third one is the electric power marketing data and the fourth one is the power grid monitoring data. Just like other big data, electric power big data is featured by 4V: (1) Volume: The automatic system of power enterprises collect, store and compute huge amount of data everyday; (2) Velocity: the data circulation and processing speed of power enterprises are very fast, which realizes real-time processing and analysis of a great deal of data in a fraction of a second; (3) Variety: The electric power big data have various types including structured data and unstructured data, and the sources are diversified, which may be relational data or text, pictures, audio and other data; (4) Very low value: There are not so much valuable data in the massive data, because the majority of data are normal. For the power monitoring data, the abnormalities of the data are the important data, because the abnormal data is the most important basis for condition based maintenance [2].

* Corresponding author: luckfunny_wang@163.com

With the in-depth advancing of big data and the intelligent transformation of the power grid, the electrical power system of each link has achieved digitization, informatization and intellectualization. The use of a variety of intelligent electric meters, sensors and information systems has injected a powerful driving force into the construction of electrical power system, which penetrate into the power generation, transmission, power transformation, power distribution, and power consumption and scheduling, and other links of production and management. The use of these devices brings more data sources for the electric power big data, and also provides continuous massive data for the power sector. Therefore, strengthening the relevant technical research on the electric power big data help promote the sustainable development of the power industry, and also promote the use of electric power big data.

2 Relation among the Electric Power Big Data, Cloud Computing and Smart Grid

The smart grid integrates information, computer, Internet and communication, and forms a new type of smart grid combined with devices of power transmission and distribution, which has high energy efficiency and power supply security. A mass of data generated by the smart grid carry out information interaction through the modern network and communication technology, and make information exchange with the power grid device, to ensure the exchange of information. The power grid completes the automated management of smart grid through automatic collection, automatic control, automatic monitoring and other functions. Automated management is based on the analysis of big data. It is just because of the existence of big data analysis that massive data analysis of smart grid is applied. The relationship between the cloud computing and big data is based on the powerful computing power of cloud computing, which can deal with massive data quickly, and also provide demanded service conveniently; the big data storage, management and analysis technologies are built on the basis of cloud computing, rooted in the cloud computing, which are also the basis of big data application [3-4]. The development of big data can promote the development of cloud computing, and the cloud computing can promote the realization of cloud computing application, both of which supplement each other.

The smart grid, big data and cloud computing are complementary to each other. The smart grid is the usage scenario of big data platform, and the cloud computing provides handling ability to the massive data storage and calculation of electrical power system. The platform which integrates the electric power big data is helpful for improving the on-line analysis ability and real-time control ability of the smart grid, and improving the reliability of the power supply and reducing the power loss of the electric energy of transmission grid, and alleviating the environmental impact and reducing the cost of the basic equipment [5].

3 Ease of Use General Framework of the Electric Power Big Data Platform

The general framework of electric power big data has good versatility, which can regulate and define the electric power big data platform. The general framework defines the functional procedure of electric power big data, which plays an exemplary role [6] for the power enterprise in the platform technology selection. As shown in Figure 1, based on other architecture platforms, with the integration of relevant characteristics of power enterprises, the five-layer framework is built on the basis of electric power big data, which is composed of the data sources, integration and storage, data processing and scheduling, data analysis, application. This framework integrates the data of different fields in the power sector, to form unified data resource pool; the data integration and storage layer and the data processing and scheduling layer are open and expandable, which can provide interface support to the data from other sources and data processing; the data analysis layer can provide mature and advanced big data analysis services; the application layer can provide personalized, secure and reliable application services, which provides public service to the society [7].

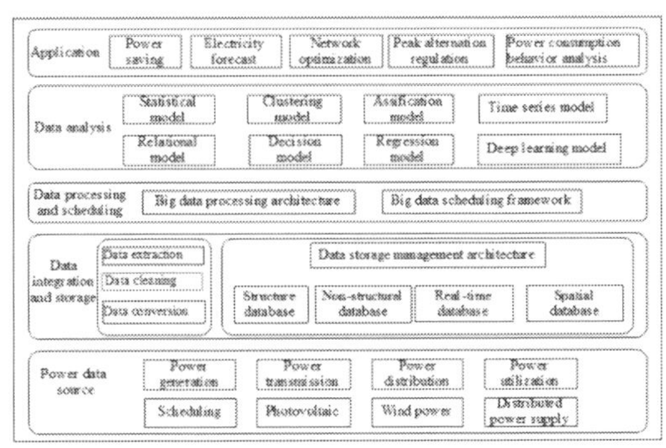

Figure 1. The General Framework Based on Electric Power Big Data

The power data source layer integrates the data sources of each links of power generation and management, such as power generation, transmission, power transformation, power distribution, and power consumption and scheduling, and also includes the photovoltaic and wind power, distributed power supply and other new energies, which provide ceaseless massive data for the power industry, laying the data basis for electric power big data [8].

At the data integration and storage layer, with the rise of cloud computing, the traditional centralized storage is no longer the mainstream storage architecture of the data center, but replaced by distributed storage. Distributed storage architecture is featured by retractility and scalability, which provides solutions to big data storage. As shown in Figure 2, the general business data uses relational database storage, and high-value, high-density data uses MPP database, low-value low-density structured, unstructured, semi-structured data use distributed file system. The architecture of sorted storing is conducive to improving the storage efficiency of data [9]. Distributed

DOI: 10.1051/matecconf/201713900217

file system has become the core module of the big data platform, which is used for storing all types of electric power big data.

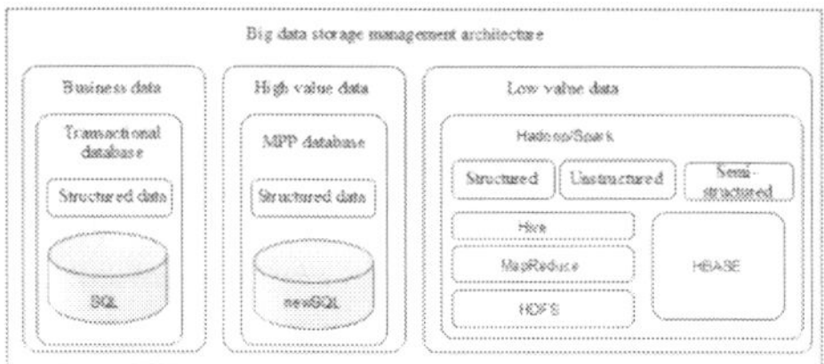

Figure 2. Big Data Storage and Management Architecture

The big data processing and scheduling layer, as shown in Figure 3, includes big data processing and big data scheduling and other related content. The big data processing includes distributed computing, high-performance computing, parallel computing, memory computing, stream computing, grid computing, etc. The big data scheduling, includes streamline scheduling, distributed assist, device, taskscheduling in WAN, which achieves big data organization and scheduling, thus providing the necessary conditions for analysis.

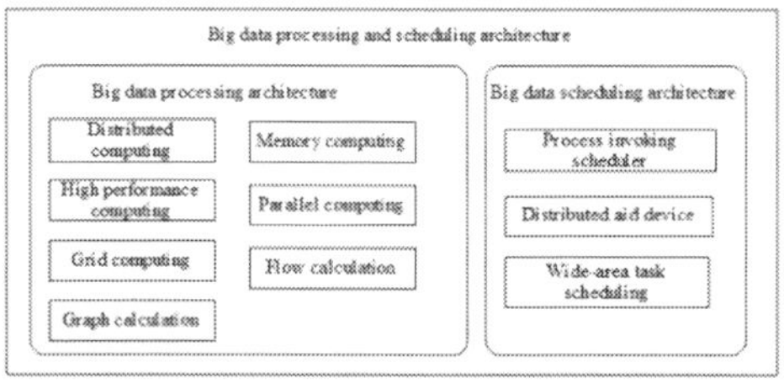

Figure 3. Big Data Processing and Scheduling Framework

Big data analysis layer achieves big data related analysis methods and models, such as statistical analysis, clustering analysis, classification analysis, time series analysis, correlation analysis, regression analysis, decision analysis, neural network analysis, etc. The big data platform can be used to carry out data analysis, and provide mature and advanced big data analysis service.

The application layer provides application service based on the analysis of big data, including energy saving, power consumption predication, grid structure optimization, taggering power consumption, power consumption behavior analysis and other application services.

4 Key Technology of Electric Power Big Data

A. Research on Integrated Management Technology Based on Electric Power Big Data

Power data synthesizes multiple links of data, involving different formats, features and properties of data that should be stored in the database organically and orderly. Therefore the big data platform should be able to handle multi-source, complex, and diverse data [10]. Therefore, the technical research on the integrated management based on the electric power big data aims to solve the problem of massive multi-source heterogeneous data storage in the power sector. Through data extraction, cleaning and transition, it extracts the entities and relationships and then achieves high efficiency storage to make the data for big data analysis stable and reliable. In addition, it supplies good expansibility and efficient data access ability.

B. Research on the Index Technology Based on Electric Power Big Data

In order to effectively manage massive and diverse power data and provide "demand-obtaining" services, the electric power big data platform must be extensible, usable, and capable of running in heterogeneous environment and provide efficient query [11]. Therefore, it is also one of the keys to strengthen the index technology research based on electric power big data and provide good index technology solutions [12]. Because the query performance will seriously affect the service quality, and good index technology can improve the query processing performance effectively, to reduce the CPU using time and disk reading frequency. It is the core technology of the big data platform and a research topic of great importance to build effective index structure and increase the query processing performance.

C. Research on the Analysis Technology Based on Electric Power Big Data

With the use of electric power big data analysis technology, potential mode and laws can be found in the massive data of electric power, which provides decision support to decision-makers after analysis and processing. The big data analysis technology includes statistical analysis, clustering analysis, classification analysis, time series analysis, correlation analysis, regression analysis, decision analysis, neural network analysis and other methods, which is mainly used in online analysis of power grid security, line state analysis, user power consumption behavior analysis, etc., which provides reliable data analysis [13-14] for the safe and stable operation of the electrical power system.

D. Research on the Mining Technology Based on Electric Power Big Data

The mining of electric power big data refers to extracting knowledge from complex data set to assist with decision-making mainly through using the association rules, classification, clustering and other mining methods. Electric power data mining includes data preparation, law searching and law representation. Data preparation refers to the data set used for excavation; the law searching refers to finding the law of data using different methods; the law representation refers to showing the laws found in a certain way or visual means.

E. Stealing Electricity and Electric Leakage Automatic Diagnosis Platform Based on Electric Power Big Data

With the popularization of SGCC smart grid and the improvement of data collection means, various relevant information about the user can be collected, which lays a data basis for the big data analysis. There are many applications based on electric power big data, so the stealing electricity and electric leakage automatic diagnosis platform is taken for example here to illustrate the application prospect of big data in the smart grid. The intelligent acquisition system collects power load data and power consumption abnormal data of current, voltage, power of each phase and other factors from the electric energy meter, voltage transformer, secondary metering circuit, electric power measurement bin and other relevant device, and stores it in the database system of big data platform. As shown in Figure 4, the stealing electricity and electric leakage automatic diagnosis platform based on electric power big data extracts relevant data from the database selectively through the data extraction module and read relevant data in real time, for cleaning and conversion treatment, and then provide effective, high-quality data to big data analysis module for analysis and processing, and the big data analysis module will use big data analysis method for analysis and processing, before sending the results to the user diagnosis and display module. Through building user identification model based on big data, automatic detection and automatic diagnosis, it identifies stealing electricity and electric leakage and relevant behaviors.

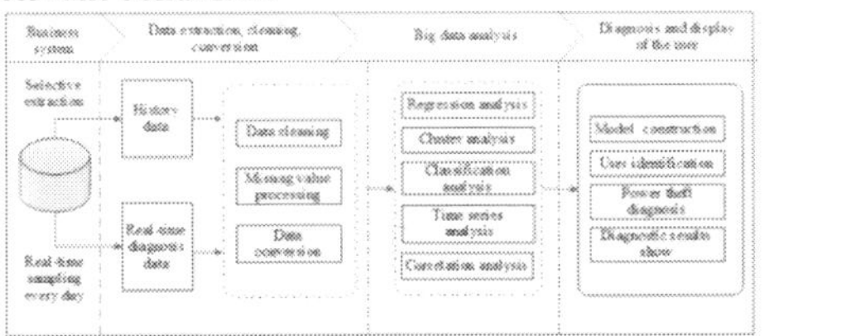

Figure 4. Stealing Electricity and Electric Leakage Automatic Diagnosis Platform Based on Electric Power Big Data

Through elaborating on the associated relationship among electric power big data, cloud computing and smart grid, this paper put forward general framework of electric power big data platform based on the smart grid. The general framework of the platform is divided into five layers, namely data source layer, data integration and storage layer, data processing and scheduling layer, data analysis layer and application layer. The framework points out the direction for the power enterprise informatization, which is of reference value to the promotion of the electric power big data platform. This paper also focuses on the integration of big data in the framework and storage architecture, and the big data processing and scheduling architecture, and then solves the problem of structured data and unstructured data storage in the way of distributed storage; and solves the calculation and processing problem through big data organization and scheduling. Among various key technologies, this paper focuses on the integrated management technology based on big data, the index technology based on big data, the analysis technology based on electric power big data and the excavation technology based on electric power big data, to achieve the breakthrough of key technologies of electric power big data. Finally, in terms of the typical case, it uses stealing electricity and electric leakage automatic diagnosis platform based on electric power big data as the study case, which has demonstration effect on power enterprises in the development of big data related researches.

References

1. Yao Wei, Cheng Shijie, et al. *Application of HVDC technology in grid - connected offshore wind farm* CEP, **40** (2007)
2. Yan Long-chuan, Li Ya-xi, Li Bin-chen, Zhao Zi-yan, Opportunity and Challenge of Big Data for the Power Industry, J. Electric Power Information and Communication Technology. 11(2013) 1-4.
3. Deng Chuan-hua, The Application Prospects and Challenges Research of Power Big Data, J. CA.Journal of the Hebei Academy of Sciences. 3(2015) 66-69.
4. Yan Long-chuan, Chen Liang, Zhao Zi-yan, Cloud computing data center architecture supporting power big data research, J. Distribution & Utilization. 8(2014) 37- 40.
5. Wu Kai-feng, Liu Wan-tao, Li Yan-hu, Su Yi- peng, Xiao Zheng, Pei Xu-bin, Hu Song lin, Study on the Impacts of Short-Term Demand Response on Generation Expansion Planning with Wind Power, J. CSCD.Electric Power.2(2015) 111-116 + 127.
6. Zhang Ming, Wang wei, Shi Jian-hua, Zhao De-wei, Optimization of Dispatching Cloud Based on Power Big Data, J.CSCD. Computer Simulation. 11(2014) 123-126 + 137.
7. Duan jun-hong, Zhang Nai-dan Zhao Bo, Yan Xiao-bin, Research on Basic System Architecture and Application of Electric Power Big Data,J. Electric Power Information and Communication Technology. 2(2015) 92-95.
8. Zhu Chao-yang, Wang ji-ye, Deng Chun-yu, Platform Research and Design of Electrical Big Data ,J. Electric Power Information and Communication Technology. 6(2015) 1-7.
9. Peng Xiao-sheng, Deng De-yuan, Cheng Shi-jie, Wen Jin-yu, Li Chao-yang, Niu Lin: Key Technologies of Electric Power Big Data and Its Application Prospects in Smart Grid , J. CSCD. Proceedings of the CSEE, 3(2015) 503-511.
10. Dan Tao, The key technology analysis of the Power Big Data for Smart Grid Applications [J]. ublic Communication of Science & Technology. 20(2015) 31 + 29.
11. Qiao Mao-bin, The Key Technology of the Power Big Data for Smart Grid Applications [J]. Resources Economization & Environmental Protection. 8(2015) 2-4.
12. Luan Kai-ning, Zheng Hai-yan, Ding Chen, Li Kun-ming,Dynamic Indexing Technology of Fast Combination Query for Power Large Data ,J. Electric Engineering,1(2015) 113-116 + 122.
13. Yang De-sheng, Chen Jiang-jiang, Zhang Ming,Power Big Data High speed Data Storage And Retrieval And Application Of Key Technologies ,J. Electronic Test. 3(2014) 62-63 + 61.
14. Shen Hao-bin, The Application Status of Power Big Data and Multi-Source Heterogeneous Data Analysis Technology J. Heilongjiang Science and Technology Information. 24(2015) 181-183.

The Design of Variable Stiffness Massage head Based on Magnetorheological Fluid

ZhiCheng Xu, *Xing* Ming, *YiKai* Shi*, and *WenDong* Wang

School of Mechanical Engineering, Northwestern Polytechnical University, China

Abstract. A controlled massage head is designed to vary the massage intensity rapidly. It bases on the property of magnetorheological fluid stiffness influenced by magnetic field. In this paper, we draw on relevant methods, design electromagnetic structures with numerical analysis software. The method of positive and negative current with permanent magnet compensation is presented to achieve a bigger range of magnetic field strength adjustment. Further, the massage head prototype is manufactured and tested. The experiment data are processed and indicates the design of the magnetic field strength and massage strength can meet the requirements.
Keywords magnetorheological fluid; massage instrument; electromagnetic simulation; structure design

1 Introduction

Massage has always been one of the effective means of relieving and treating chronic pain, approved and trusted by lots of people [1]. Traditional massage needs to be done by experienced masseurs, not only with high costs, but also the number of the masseurs is not enough comparing with public requirements. To solve this problem, many companies have produced a variety of massage machines on the market, such as massage hammer, massage chair, massage pillow, etc. However, these massage machine structure and functions are too simple with poor flexibility. They are unable to adjust the massage intensity in time for different massage situation also can not satisfy the personalized requirements. In view of the above disadvantages, we present a variable stiffness massage head based on magnetorheological fluid. MRF is a new controllable fluid material, its property can change within a few milliseconds from Newtonian fluid to structure which is similar to solid and controlled by magnetic field [2].At present, the magnetorheological fluid is mainly used in the design of semi-active shock absorbers and dampers, and some mature products have been come out [3]. Apart from this, there are many researchers committed to the application of controllable stiffness and deformation property of magnetorheological fluid. Yuki OKATANI from Kyushu Institute of Technology designed a MR fluid robot gripper, it is able to grab different shapes of objects weight 5kg approximately [4]; Jin Xing Kong put MRF into thin spherical shell enhancing its stiffness to improve process precision [5] Zhang XZ from University of Science and Technology of China analyzed the flexible fixture based on MRF, and carried out theoretical and experimental studies on the extrusion enhancement effect of MRF [6]. According to these features such as fast response to magnetic field, the stable state transition process and the controllable stiffness, the design idea is presented of using the magnetorheological fluid in semi-active control massage head.

2 The massage head electromagnetic structure design and methods

According to the theory of electromagnetism, the attenuation of the magnetic field is proportional to the square of the distance. Therefore, the circular massage head ensures the most efficient use of the magnetic field generated by the electromagnetic coil. Disposed outside of the coil is the yoke connect to the iron core, in order to reduce the air gap of the magnetic circuit and also suppress the magnetic flux leakage. The material of massage head surface is silicone. Silicone is non-toxic and odorless, the chemical properties are stable and most important it's insoluble in MRF base fluid [4]. We use MRF-140CG of LORD Company for massage head working stuff. The yield strength and B-H curve of the MRF are shown in figure 1 [7], the massage intensity designed for 0~30N. The yield strength of the MRF is designed about 10kPa. According to the data in Fig. 1, the magnetic field strength is initially designed to be 20kA/m. The MRF massage head structure schematic diagram is shown in figure 2. The coil connected to DC generat es a controlled magnetic field, and the massage head MRF influenced by the magnetic changes the characteristic. Adjust the current in the coil, affect and change the yield strength of the magnetorheological fluid, thus alter the contact stiffness with human, finally control the massage intensity.

* Corresponding author: ykshi@nwpu.edu.cn

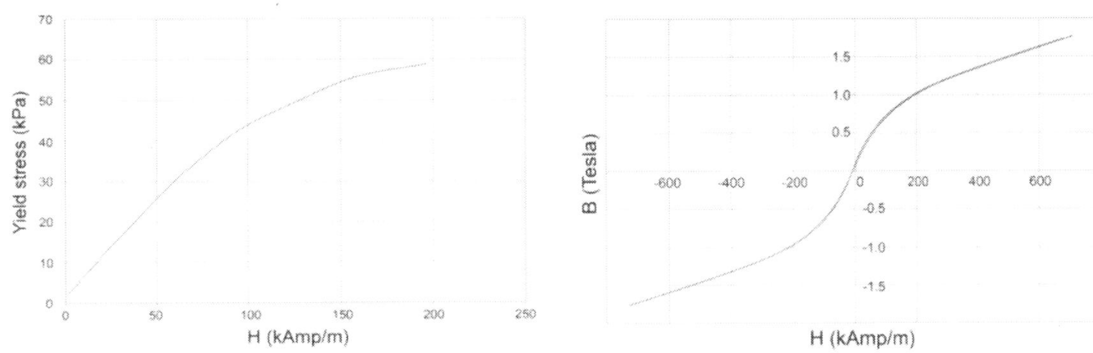

(a) (b)

Fig. 1. MRF-140-CG Yield stress - field correlation (a) and BH curve (b)

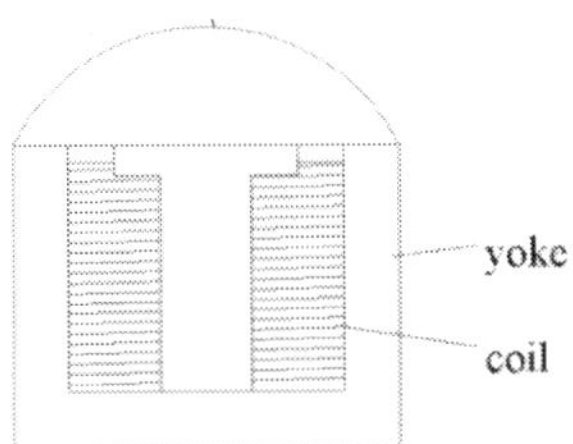

Fig. 2. The structure schematic diagram

The theoretical basis for the analysis of electromagnetic is the Maxwell Equations, as for the structure designed, ignore the propagation of electromagnetic waves, the current and the field change slowly with time, it meets the special case of Maxwell's equations:

$$\nabla \times \vec{E} = -\frac{\partial \vec{B}}{\partial t} \quad (1)$$

$$\nabla \times \vec{H} = \vec{J} \quad (2)$$

$$\nabla \vec{D} = 0 \quad (3)$$

$$\nabla \vec{B} = 0 \quad (4)$$

$$\nabla \vec{J} = 0 \quad (5)$$

Here E is the electric field intensity, B is the magnetic flux density, H is the magnetic field strength, J is the current density, D is the power density. Also:

$$\vec{D} = \varepsilon \vec{E};\ \vec{B} = \mu \vec{H}; \vec{J} = \sigma \vec{E} \quad (6)$$

ε is the dielectric constant of the material, μ is the permeability of the material, and σ is the conductivity of the material.

However, even with the boundary conditions, it is very difficult to solve the above equations. The traditional electromagnetic field calculation method is usually used mathematical and physical simulation method or approximate analysis method. These methods are not accurate, often need to be compensated according to the measured values after the completion of the design to meet the requirements of the field distribution [8]. Magnetic circuit method is also widely used to calculate magnetic field, it has the advantage of simplicity and the result is accurate when the air gap is small. But it can't be used for the open magnetic field calculation. Therefore, we refer to the empirical formula of electromagnet design [8]:

$$NI \approx \frac{BL_{air}}{4\pi \times 10^{-7}} \quad (7)$$

Here B is the magnetic induction of the target point, and L_{air} is the working air gap of the target point.

The preliminary calculation indicates a large magnetic flux about 3500Amp, But the size of massage head shouldn't be too large, we design using wire diameter 0.75mm, rated current 2A, 1500 turns of the DC coil, and add some permanent magnet as the magnetic field compensation to reach the demand of magnetic field. The compensation details are described in section 2.2.

3 Simulation and optimization

3.1 Ansoft Maxwell simulation

With the development of computer technology, the numerical analysis method of electromagnetic field calculation is widely used, including the finite difference method, the finite element method and the boundary element method. Based on these methods, some computer programs and software are designed for calculation, we can get the magnetic field distribution accurate results simply and fast. Ansoft Maxwell finite element numerical analysis software is an effective way to solve the problem of electromagnetism, it is widely used in the low-frequency electromagnetic field analysis. We use it to revise the preliminary results by finite element analysis and get the effective value. Maxwell's intuitive and convenient modeling interface and the abundant material database make structure model building efficiently. Establish a massage head mechanical structure model and select the material for different parts. According to the coil size, design the size and material of the yoke and the iron core: The core through the coil frame, the length is larger than the yoke height, choose DT4 electric iron because of its high permeability and coercivity with big magnetic saturation; Yoke slightly larger than the coil in diameter, 4mm thick, higher than the coil height of about 5mm for the placement of wire and the coil cover, the material choose No.10 steel for big magnetic saturation. There is no data of DT4 and MRF in Maxwell material library, we can add them to the project material library conveniently according to the B-H curve in figure1 and the DT4 data in reference [8].The data and progress are shown in figure 3.

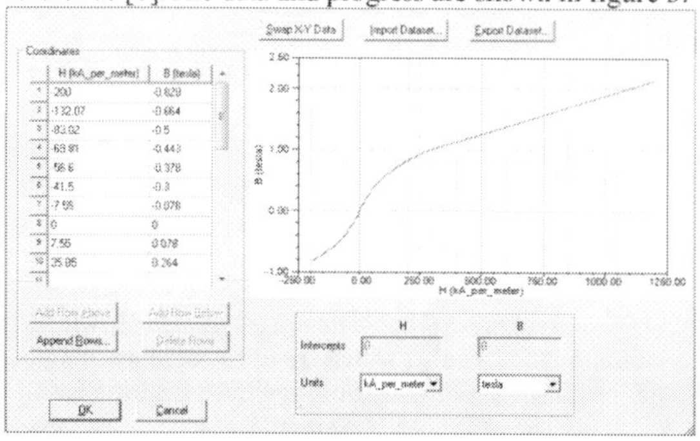

(a) The B-H curve of MRF

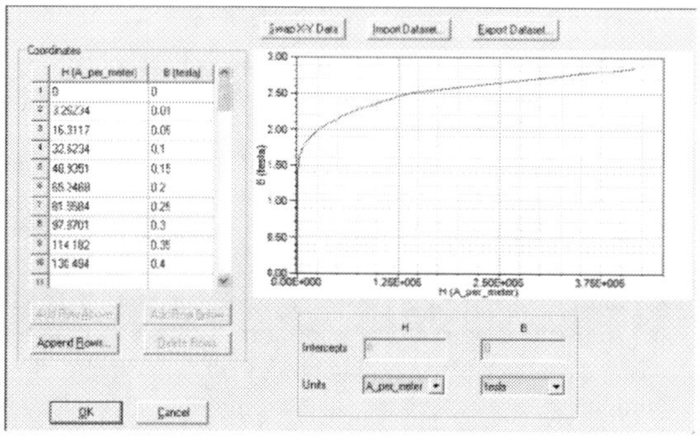

(b) The B-H curve of DT4

Fig. 3. Add property curve of DT4 and the MRF to Maxwell

3.2 magnetic field compensation

According to the information in Figure 1, the results of calculation and simulation (Fig.4) indicate that the magnetic field generated by the coil is not sufficient to make the magnetorheological fluid achieve the desired yield strength for the massage. Therefore, on the basis of the magnetic field superposition principle, it is necessary to place additional permanent magnets outside the core coil to ensure the magnetic field strength. We found that with the increase of the working point axial distance, the attenuation of magnetic field intensity is the main cause of the deficiency of the magnetic field. Common permanent magnets are disc-shaped, but it can't decrease the axial distance efficiently. Hemispherical permanent magnets are better for magnetic field compensation, however, the manufacture of hemispherical magnet is difficult, magnetizing is hard to be uniform, and the magnetic energy product is worse than the chip magnet. In order to ensure the effective massage volume of the massage head, we design with round-shaped NdFeB magnets as compensation, the simulation results are as follows.

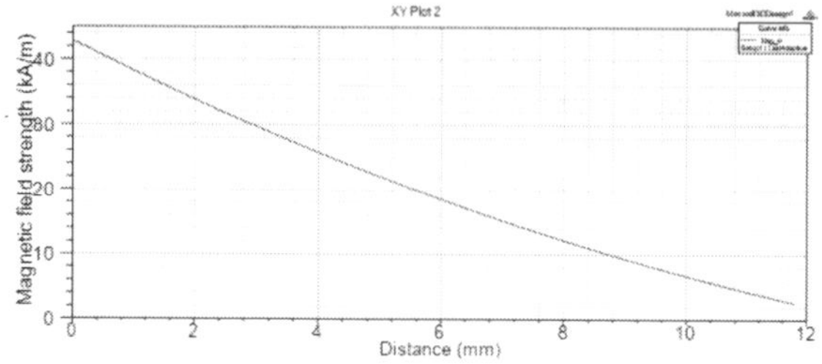

Fig. 4. Axial magnetic field simulation result without magnets

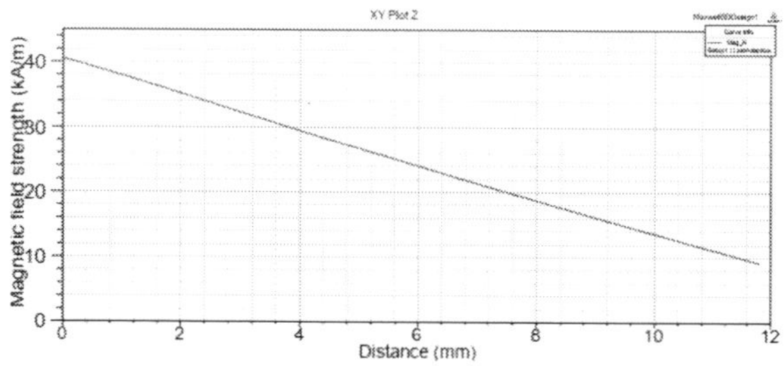

Fig. 5. Axial magnetic field simulation result with magnets

After the addition of permanent magnets, the magnetic field still maintains a high level when the current in the coil is very small. This is contrary to the original intention of our design, the stiffness of MRF cannot be well controlled. So we propose that reverse the current in the coil to offset some part of the magnets magnetic field in order to the bigger range of changes in the stiffness of the massage head. The intrinsic coercive force H_{cj} is a physical quantity that measures the anti-demagnetization ability of the magnet. It means the coercivity when the magnetization in the material is zero. It is an important index to judge whether the permanent magnet material can maintain its original magnetization. When the current reversed, the magnetic field strength generated by the coil is 129.3kA/m, calculated by Maxwell. It is far less than the intrinsic coercivity of the NdFeB magnets, so that a reasonable superposition of the magnetic field can be achieved. The MRF working magnetic field strength when the current is negative or positive are shown in Figure 6.

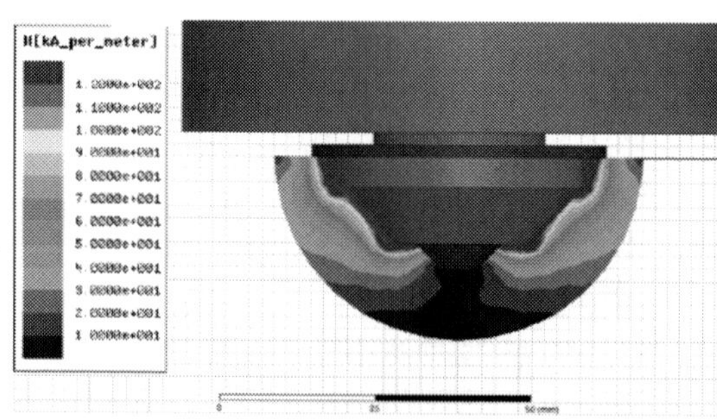

(a) The magnetic field strength in negative current

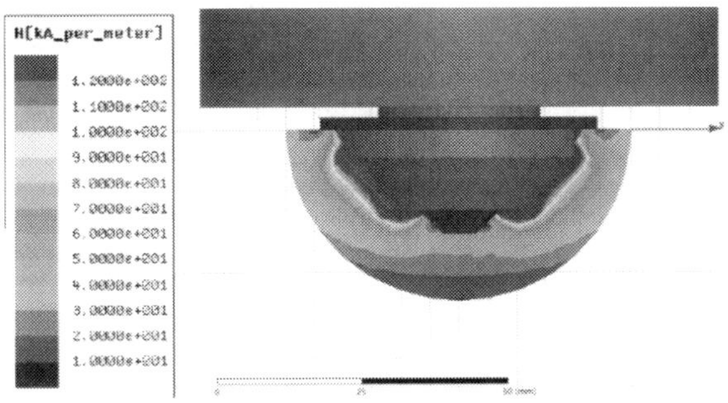

(b) The magnetic field strength in positive current

Fig. 6. The magnetic field strength situations

3.3 Sealed structure design and mechanical construction

Magnetorheological fluid is an oily liquid based material, it has fluidity without the influence of magnetic field. Leakage of the MRF may damage the power coil, also contaminate massage head surface affecting the experience of users. Therefore, the seal of the MRF is also an important part of the mechanical structure design. We design to seal the MRF in a hemispherical silicone massage head. The massage head edge apply squeeze-type seal, use the shell with a trench thinner than the silicone membrane to fix. The shell is bolted to connect the coil cover and the yoke to seal the edge of the silicone membrane. The NdFeB magnets have center through-holes, put a sealing ring into the notch of the coil cover on the large side of magnets, meanwhile the sealing ring separates the iron core from the magnets to reduce the influence of the magnetic saturation; Set a special low carbon steel bolt through the magnet center through-hole and the gasket over the small side of the magnet, tight them with the iron corn to seal the MRF. The final design massage head structure is shown in Fig.7

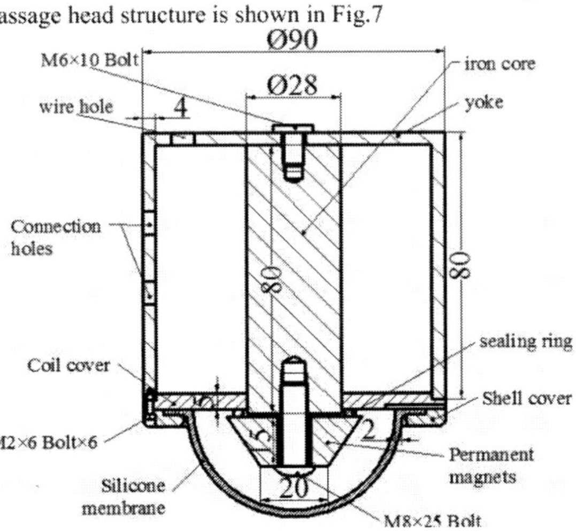

Fig. 7. The Massage Head structure

4 Experiment and data process

4.1 measure of the magnetic field

The stiffness of MRF is affected by the magnetic field, so whether the massage force can reach the expected aim or not depends on the magnetic field. To prove the simulation results are correct, we use a gauss meter to measure the magnetic field after the structure is designed and assembly. The gauss meter cannot be used in MRF, so we have to measure the field in the air. Therefore, the air is the magnetic medium in simulation, set "None model" for the MRF part in Maxwell, don't change any others parameter of the magnetic only the medium. If the simulation results match the measured data, it can be considered that the simulation results are correct in the MRF medium. We found there is significant disparity of the data in different times of experiment due to the manual measure and the high sensitivity of the gauss meter in milliTesla degree. Therefore, we use MATLAB to deal with it. The data are shown in table 1.

Table 1. The measure data of magnetic flux density in different distance

Distance from the magnet/mm	1	3	5	7	9
The magnetic flux density of measure points in positive current (mT)	254.3	193.8	153.8	126.9	104.9
	270.7	192.2	151.7	125.1	107.5
	262.0	199.2	157.6	129.1	104.7
	269.3	202.0	154.0	126.8	106.6
The magnetic flux density of measure points in negative current (mT)	104.1	89.0	74.3	55.3	44.7
	114.3	90.3	69.8	54.4	42.3
	112.4	83.8	67.2	53.9	42.7
	105.8	85.0	70.2	53.0	42.8

Matlab curve fitting tool has powerful curve fitting ability and provides various type of objective functions. The magnetic field of massage head is made up of 2 parts and the magnetic flux density decreases till 0 with the distance increases. Comparing different fitting curves we find that the exponential double terms function approximates the physical principles of magnetic and its R-square is 99.58%, so we choose this as our objective function. The compare between the fitting curve and Maxwell simulation curve are shown in Fig.8

It indicates that the measured data basically meet the magnetic flux density required for the yield strength of MRF, and the positive and negative current can make the magnetic field strength in a relatively large range of control. However, the simulation results and experimental measurements have a great diversity. As for the reason, maybe the special low-carbon steel bolt material is not standard, the magnetic field through its conduction is limited; or maybe the seal ring and the gasket are squeezed to cause the working distance from the magnet smaller compared with the simulation, which led to a deviation of the curve. Overall, the electromagnetic structure we

DOI: 10.1051/matecconf/201713900218

designed effectively controlled the massage head magnetic field

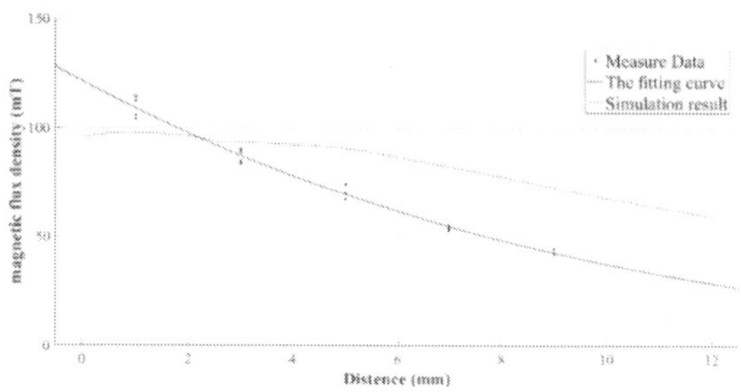

(a) The fitting curve of data in positive current

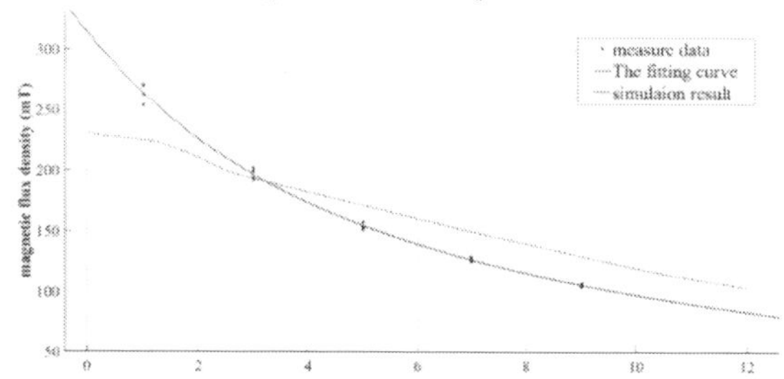

(b) The fitting curve of data in positive current

Fig. 8. The curve comparison between experiment and simulation

4.2 massage force measure

After filling and sealing the MRF, use the sensor to measure the massage intensity at different current. When the MRF head is observed to be obviously out of shape by pressure, record the sensor data. The data are shown in table 2.

Table 2. The massage head yield force in different current.

Current/A	-2	-1.5	-1	-0.5	0	0.5	1	1.5	2
Yield force/N	2.4	4.5	5.9	7.5	10.6	14.8	19.9	22.1	25.5

The result of the experiment shows that the massage head fulfills the demand of different massage force with fast response.

Fig. 9. The experiment of massage head yield force test

5 Conclusion.

We design a controllable massage head based on the MRF. First we analyzed the massage head requirements, and demonstrated the feasibility of variable stiffness. Next, design the electromagnetic structure of the massage head according to the prospective massage force and simulate in computer software. Moreover, design the permanent magnet compensation for the insufficient magnetic field drawing on the classic electromagnet design method, and present the idea of forward and reverse excitation current for better control. Then, design the structure and seal of the massage head, manufacture the prototype and test, deal with the experimental data by mathematical process. The results show that the variable stiffness massage head can effectively achieve the transformation of different massage intensity to meet the design requirements.

In future studies, the miniaturization of the massage head, the more freedom degrees of the massage machine, the optimization design of the magnetic field will be improved.

References

1. Huang, Y., Li, J., Huang, Q., & Souères, P. (2015). Anthropomorphic robotic arm with integrated elastic joints for TCM remedial massage. *Robotica*, **33(02)**, 348-365.
2. X. Zhang, F. Ma, F. Zhao, K. Guo, and M. Ahmadian, "Optimization of the Magnetic Property of a Magnetorheological Squeeze Mount," *The FISITA 2012 World Automotive Congress*, Beijing, (2012), p. 14.
3. F. Imaduddin, S. A. Mazlan, and H. Zamzuri, "A design and modelling review of rotary magnetorheological damper," *Materials & Design*, **vol. 51**, pp. 575-591, (Oct 2013).
4. Y. Okatani, T. Nishida, K. Tadakuma, and Ieee, "Development of Universal Robot Gripper Using MR alpha Fluid," *7th Scis and 15th Isis*, pp. 231-235, (2014).
5. J. X. Kong, Y. C. Zheng, and X. K. Wang, "Fixture Design Based on Magneto-rheological Fluids for Thin Wall Spherical Shell Precision Machining," in *Digital Design and Manufacturing Technology Ii*. **vol. 215**, C. Lu, Ed., ed, (2011), pp. 315-319.
6. X. Z. Zhang, Q. M. Wang, P. Q. Zhang. "The flexible fixture based on Magnetorheological Fluids," *Journal of Experimental Mechanics*, **No.2 vol. 18**, pp. 185-192, (2003).
7. A. Spaggiari and E. Dragoni, "Effect of Pressure on the Flow Properties of Magnetorheological Fluids," *Journal of Fluids Engineering-Transactions of the Asme*, **vol. 134**, (Sep 2012).
8. Q. F. Li, *Numerical Calculation of Electromagnetic Fields and Design of Electromagnets*. (TsingHua University, 2002).

Model-based safety analysis of a control system using Simulink and Simscape extended models

Nian Shao [1,2], *Shuguang Zhang*[1,2], and *Hui Liang*[1,2]

[1]School of Transportation Science and Engineering, Beihang University, Beijing, China
[2]Airworthiness Technology Research Centre, NLAA, Beijing, China

Abstract. The aircraft or system safety assessment process is an integral part of the overall aircraft development cycle. It is usually characterized by a very high timely and financial effort and can become a critical design driver in certain cases. Therefore, an increasing demand of effective methods to assist the safety assessment process arises within the aerospace community. One approach is the utilization of model-based technology, which is already well-established in the system development, for safety assessment purposes. This paper mainly describes a new tool for Model-Based Safety Analysis. A formal model for an example system is generated and enriched with extended models. Then, system safety analyses are performed on the model with the assistance of automation tools and compared to the results of a manual analysis. The objective of this paper is to improve the increasingly complex aircraft systems development process. This paper develops a new model-based analysis tool in Simulink/Simscape environment.

1 Introduction

Safety and reliability are essential for commercial aircraft design. The conventional safety analysis methods became very mature after long time research and development. However, with the increase of the complexity of aircraft systems, some limitations of these methods arise.

Fenelon [1] pointed out that the design process and the safety assessment process are usually not satisfactorily integrated. He used the expression "over the wall" to describe the gap between design and safety process. In addition, conventional safety analysis methods are faced with the problem of low efficiency.

With the problems of conventional safety analysis methods, an increasing demand of effective methods arises. One approach is the utilization of model-based technology. The ESACS [2] project and its following up project ISAAC developed some safety analysis methodology and tools in Simulink, SCADE, Statemate, NuSMV and AltaRica [3]. In the year 2005, Joshi et al. proposed the term "Model-Based Safety Analysis(MBSA)" for this kind of safety analysis method [4]. The European project MISSA [5] was dedicated to integrating the model-based security assessment into the system development process described by ARP4754A [6] and testing these processes in industrial examples. In 2016, Schallert [7] established safety analysis methods using multi-domain object-oriented models in Modelica.

MBSA becomes more and more popular and gradually gets the recognition from the authority. AC 20-115C [8] released by FAA in 2013 formally recognizes the model-based methodology.

2 System safety assessment process

2.1. Typical system safety assessment process

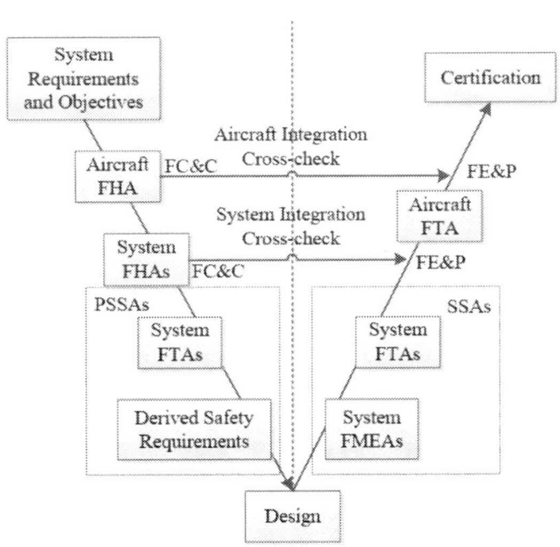

Fig. 1. System safety assessment process described in SAE ARP-4761.

Fig.1 shows the overall safety assessment process according to ARP-4761 [9]. The whole process includes the requirements validation and the products verification. An aircraft-level Functional Hazard Analysis (FHA) is

* Corresponding author: gnahz@buaa.edu.cn

initially conducted, followed by a System FHA, then Preliminary System Safety Assessment (PSSA). During PSSAs, safety requirements are derived using Fault Tree Analysis (FTA) and probability budgeting. This is usually an iterative process including the PSSA and the system design definition and refinement.

After the determination of the final design, the compliance of the implemented products with the safety requirements are verified during the System Safety Assessment (SSA). During this process, the system Failure Modes and Effects Analysis (FMEA) provides for the failure probabilities of the single items. Those are the inputs for the system-level FTA and the aircraft-level FTA, which are conducted to verify that the systems and the overall aircraft satisfy the top-level safety requirements.

2.2 Utilization of MBSA in typical safety process

Simulation models mentioned in this article can be the formal specification in a model-based development. It is also available for assistance of conventional safety processes. The integration work is enlighten by the qSafe project [10].

In the PSSA process, the objective of MBSA is to assist the validation of system requirements. The model describes the system architecture and functional properties. The models are based on Simscape language, which provide the possibility for computer assistance. MBSA is introduced to conventional safety assessment process to assist the FTA process. MBSA can traverse all possible failure mode connections of a system without omission in a short period of time. More importantly, MBSA can generate minimal cut sets for FTA.

In the SSA process, FMEA and bottom-up FTA are applied for the verification of system implementation. MBSA focus on the physical implementation and specific performance of components and systems. Information of failure modes and effects can be identified through simulation and analysis. MBSA is utilised into the ARP-4761 [9] safety process to assist FTA and FMEA of complex multi-domain systems in the PSSA and SSA processes.

2.3 Simscape modelling

The complete work described in the article is developed in the MATLAB environment.

The models are built and modified in Simulink environment. The formal system model is built with Simscape. Simscape enables rapidly creation of physical system models within the Simulink environment. With Simscape, physical component models based on physical connections that directly integrate with block diagrams and other modeling paradigms can be easily built. Modelling of systems like electric motors and hydraulic actuators can be achieved by assembling fundamental components into a schematic. Furthermore, as the principle of Simscape is physical modelling, the resulting models are intuitively comprehensible without detailed knowledge of numerical simulation methods and modelling paradigms, which benefits the interaction between design and safety analysis.

Simscape is one of the undirected numerical simulation models. Undirected numerical models describe systems as networks as functional elements that exchange energy through their ports. The component behaviour is encoded by differential algebraic equations. Every component is associated with some variables. Undirected numerical models distinguish protentional (voltage, velocity, pressure, etc.) and flow variables (current, force, flow rate, etc.).

3 A practicable MBSA modelling process using Simscape

3.1 Modelling of the exemplar control system

The example is a simplified control system model for a control surface (see also Fig. 2.). The function of the system is to control the position of the control surface (represented by the mass block m in Fig. 3.). There is a control circuit providing power and transmitting control signals. It includes a resistor, connected with a control voltage source and a fuse, as well as a motor driving the control surface to move. A sensor is connected to the motor to detect the angular displacement.

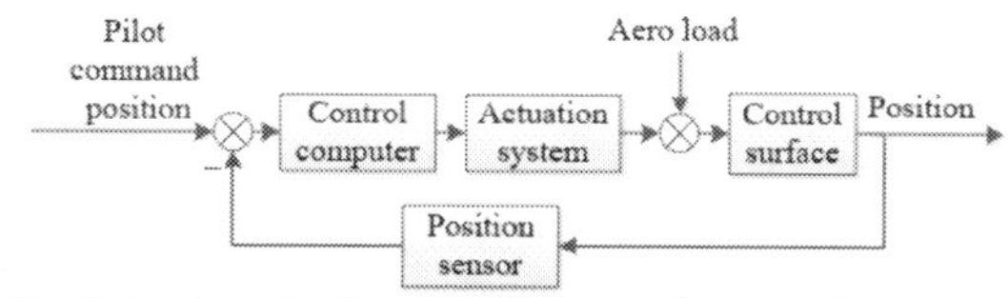

Fig. 2. A schematic diagram of the exemplar control system.

The system input are the pilot commands on the control column, which are simulated by a position command signal in the "Pos_cmd" block. Then the control signal is transmitted to the PI controller to control the voltage value of the voltage source. The voltage source drives the electric motor, and the motor generates a torque which is applied to the control surface.

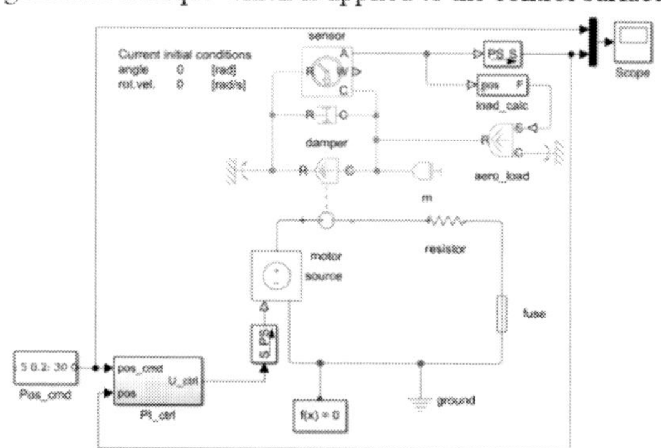

Fig. 3. Simscape model of the exemplar control system.

The resulting angular displacement is measured by the sensor. The purpose of the damper block is to roughly approximate the dynamic behaviour of a real motor. The sensor will transmit the angular displacement value to the controller as a negative feedback value used by the controlled to track the desired position signal: Once the command position is reached, the controller will give a constant value to the voltage source due to its integrating behaviour, and the position of the control surface will stay at the intended position. If the current in the circuit is too large, the fuse will cut off the circuit.

3.2 Fault modelling

3.2.1 Classification of component failures

For the effectiveness and usability of a simulation model, the failures should be modelled in a simple, intuitive way in case of very large system and numerous combinations of possible component states.

Schallert [7] establishes a catalogue of component failures for system safety analysis. This catalogue limits the possible failure modes of every component to three basic types of behaviour. This limitation on the failure numbers intends to avoid unnecessary modelling details and high computational effort during simulation, when combinations of failure modes are considered. A component will have 4 states in total.

Table 1. General component failure modes.

Mode	0	1	2	3
State	Normal function	Loss of function1	Loss of function2	Inadverten t function

Table 2. Components failure modes.

Mode	0	1	2	3
Sensor	Nominal	Loss of signal		Signal bias
Motor_m echanical	Nominal	Disconne ction	Jam	
Motor_el ectrical	Nominal	Open circuit	Short circuit	
Resistor	Nominal	Open circuit	Short circuit	
Fuse	Nominal	Stuck open		Stuck close
Power supply	Nominal	Loss of output		

Mode 0 means the normal function or intact state of a component or function.

Mode 1 means a type of function loss, where the transducing function of a component is lost. This can be an electrical O/C ($i \rightarrow 0$) or mechanical disconnection ($f \rightarrow 0$ or $\tau \rightarrow 0$). For a sensor, mode 1 represents "loss of signal".

Mode 2 is also a type of function loss, where a transducing component is stuck in a certain state It can be an electrical short to ground ($v \rightarrow 0$) or mechanical jamming ($v_e \rightarrow 0$ or $\omega \rightarrow 0$).

Mode 3 means some inadvertent activation of a component. This can be for example the "stuck close" failure mode of a switch or an erroneous signal given by a sensor (bias).

In this simplified system. There are five components with failure modes: the energy source, the resistor, the fuse, the motor and the sensor. For convenience and efficiency of simulation, the motor is split into a mechanical part and an electrical part.

3.2.2 Failure mode control

To exhaustively simulate all the combinations of failure modes, a way to control the failure mode must be provided. Simscape enables customized components, which means customer can modify library components or create components. Extended models are mostly modified from Simscape library models. Simscape models are based on equations written in Simscape language (source code). A variable named "failure mode" is added to the source code of each component. The value of the variable will be assigned according to the mode it's intended to fail. The value of "failure mode" changes the equations of each component accordingly.

For example, the source code of the Simscape component Resistor is:

```
component resistor
  R = { 1, 'Ohm' }; % Resistance
end
equations
  assert(R>=0);
  v == R*i;
end
end
```

A resistor can fail short or fail open. According to the classification stated in subsection 4.2.1, the open circuit state of a resistor is represented as "FM_R = 1". Similarly, "FM_R = 2" means short circuit of a resistor. The resistor is extended by modifying the source code. The following source code represents a new component extended from the original resistor. The mode of this new resistor can be controlled by importing different FM_R. In this way, the failure mode of the component is controlled through the modality of its parameter. The source code of the extended resistor equation is as follows:

```
if FM_R == 1
   R = Roc;
else if FM_R == 2
   R = Rsc;
  else if FM_R ==0
     R = R;
  else
     error('id:id','invalid FM_R');
  end
  end
end
```

In similar way, failure models of a set of several other commonly used components are built and stored in a failure model library.

DOI: 10.1051/matecconf/201713900219

3.3 Extension of the system model

In the extend system model, the original component models are replaced by fault models. There are mode input ports to control behaviour of all the concerned component models.

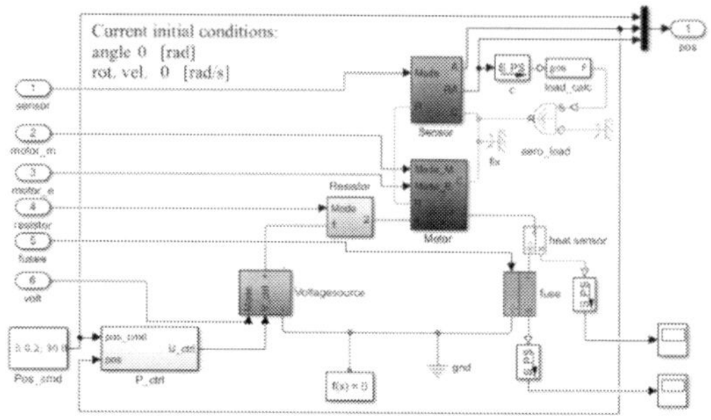

Fig. 4. Extended exemplar control system model.

The motor is controlled by two mode input port as it is represented by an electrical and a mechanical part internally. The sensor also gives an additional unbiased "real" signal to the scope for the convenience of observation. The "heat sensor" block is designed to take consideration of a special condition: short circuit of the resistor with the fuse stuck close. In this condition, the heat sensor will fall down from 1 to 0, which indicates potential burning down of the circuit.

3.4 Analysis of the exemplar control system through simulation

Partial automation of the safety analysis process can both reduce the cost and improve the quality of the results.

The result is a form containing all combination of failure mode numbers and corresponding FE No, which can be analysed to identify critical combinations of failures or even to construct the cut-sets leading to a certain effect.

3.4.2 FMEA

An FMEA intends to find out single faults and their local effects. This example model used here is quite basic and not structured hierarchically. Therefore, the FMEA local effects determined by the FMEA can be considered to be functional failure effects on system level.

The single failures can be simulated by simply applying the respective failure mode number of corresponding component.

The results of the simulation are shown in Fig. 5 and Fig. 6. The positions (rad) are shown on the y-axis and the time (s) is on the x-axis. There are 3 types of position signals depicted: command position given by the pilot (plotted in green), the position detected by the sensor (plotted in blue and dash) and the real position (plotted in red and plots).

At first, the control surface stays at position 0. After 5 seconds, a command is given to the system to move the control surface to position 0.2. At the 30th second, the position command is released.

Fig. 5 shows the normal function of the sample control system. The real position follows the control input well, and there is no bias between the detected position and the real position. Fig. 6 shows the condition where there the sensor signal is biased: The system tends to force the detected position to follow the command position and there is a steady bias between the real position and the command position.

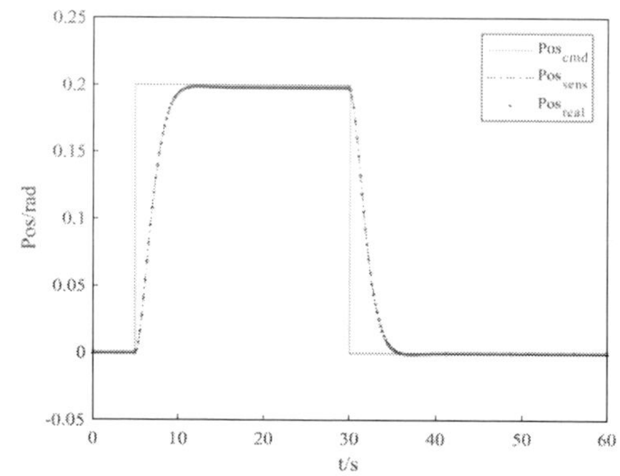

Fig. 5. Nominal function display.

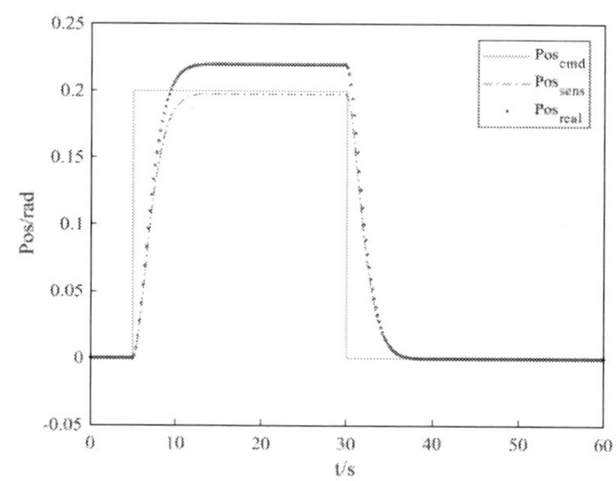

Fig. 6. Position bias display.

3.4.3 Combinations of failure modes

This analysis intends to find out all failure effects caused by the combinations of different failure modes (as there might be hidden failure effects, which cannot be detected by the single failure FMEA).

To automate the simulation, two basic steps are implemented. The first step is to generate a matrix contains all "combination vectors". Each vector represents a specific combination of the failure modes. For example, (1 0 0 0 2 0) means sensor "loss of signal" combined with fuse "stuck close". A total of 486 combinations is generated. After removing the conditions where more than 3 components fail at the same time, 62 combinations are taking into consideration. The second step is to automatically run the simulation of all the combinations. The values of mode input ports are set as specific elements of the vectors. The third step is to plot the results and save the pictures for the analysis reports.

Yet it is not enough to have only the results plotted as time traces. The next goal is to automatically analyse the simulation results, i.e. the determination of the effects resulting from different combinations. The analysis results can also help to find out the corresponding combinations leading to a specific top event (i.e. the cut-sets).

Table 3. Failure effect codes.

FE No.	0	1	2	3	4	5
Failure Effect	Nominal	Follow wind	Stuck	Position bias	Unstable	No effect

A set of expected failure effects can be received from the FMEA and the simulation results. A unique failure effect code is assigned to each failure effect (see Table 3.).

Fig. 6 shows the "Stuck" failure effect, which is the consequence of motor "Jam" failure.

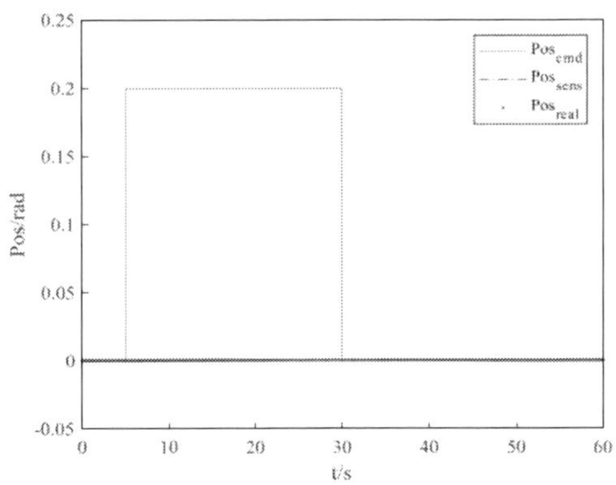

Fig. 6. Failure effect "Stuck".

With this form, the minimum cut sets of a specific top event can be identified, which is the essence of a fault tree.

Table 4. Combinations of failure modes and failure effects.

S	M_m	M_e	R	F	E	FE
0	0	0	0	0	0	0
1	0	0	0	0	0	4
2	0	0	0	0	0	3
0	1	0	0	0	0	1
1	1	0	0	0	0	1
2	1	0	0	0	0	1
2	2	0	0	0	0	2
...	...	...	...	...	...	...

4 Conclusions

This paper develops a new model-based analysis tool in Simulink/Simscape environment. The objective is to improve the increasingly complex aircraft systems development process. Extended Simscape model is utilized for the partial automation of typical safety assessment tasks.

During that process, some obvious advantages of model-based safety analysis became clear:

a) The ability to exhaustively explore all possible combinations of the components failure modes using automated analysis tools. The quality of the safety analysis will be therefore improved.
b) The ability to automatically generate analysis outcomes. These outcomes can be used to enhance or cross-check the manually created ones. Therefore, the method lightens the work load of safety analysts and reduce the costs, while simultaneously improving coverage and quality of the results.
c) The possibility to integrate the safety process and the design process

References

1. Fenelon, Peter, et al. "Towards integrated safety analysis and design." ACM SIGAPP Applied Computing Review 2.1 (1994): 21-32.
2. Bozzano, Marco, et al. "ESACS: an integrated methodology for design and safety analysis of complex systems." Proc. ESREL. (2003).
3. Akerlund, O., et al. "ISAAC, a framework for integrated safety analysis of functional, geometrical and human aspects." Proc. ERTS 2006 (2006): 1-11.
4. Joshi, Anjali, et al. "A proposal for model-based safety analysis." Digital Avionics Systems Conference, 2005. DASC 2005. (The 24th. **Vol. 2**. IEEE, 2005).
5. European Commission. MISSA More Integrated System Safety Assessment [EB/OL]. www.optics-project.eu/wp-content.
6. Society of Automotive Engineers, "ARP5754A: Guidelines for Development of Civil Aircraft and Systems," (SAE Aerospace, 2010).
7. Schallert, Christian. "Integrated safety and reliability analysis methods for aircraft system development using multi-domain object-oriented models." (2016).
8. Federal Aviation Administration, "Advisory Circular 20-115C: Airborne Software Assurance," (FAA, 2013).
9. Society of Automotive Engineers, "ARP4761: Guidelines and Methods for Conducting the Safety Assessment Process on Civil Airborne Systems and Equipment," SAE International, 1996.
10. Grigoleit, Florian, et al. "The qSafe Project–Developing a Model-based Tool for Current Practice in Functional Safety Analysis."(2016).

DOI: 10.1051/matecconf/201713900220

Application design of piezoelectric effect in cabin seat belt status indication

Lixia Gao, Hanlin Shen

Aviation Engineering Institute, Civil Aviation Flight University of China, Guanghan City, Sichuan Province, China

Abstract. During the special operation stage of the aircraft, if the passengers did not fasten their safety belts in accordance with the requirements, it may bring unexpected injury, resulting in unnecessary losses. So if there is a corresponding system can help the crew to accurately confirm the state of the safety belts of the passengers under special circumstances, what we said before can be avoided. The seat belt status indicator system is designed based on the piezoelectric effect can help the crew to know the state of the safety belt clearly, to ensure the effectiveness of their work.

1 Introduction

Because of the particularity of the operation of the aircraft, especially during the stage of take-off, landing and air turbulence encountered, the safety belt protection to passengers is particularly important. Under the existing circumstances, the plane crew need to repeat one by one in order to confirm repeatedly and remind constantly the passengers to use seat belts. Under some special circumstances, some of the passengers do not listen to the crew in time, so it easily leads to accidental injury. If the crew can make sure of the actual state of the safety belts of all passengers in the crew work area, it can reduce the workload of the crew, help the crew to give a timely reminder, ensure the effectiveness of the crew work, guarantee the safety of passengers.

2 Design concept

Piezoelectric material is a kind of material which is sensitive to machine, electricity, sound, light and heat. It is widely used in the field of the sensor, transducer, non-destructive testing and communication technology. The pressure can be caused by the action of the seat belt buckle. The piezoelectric material is installed in the corresponding position to accurately identify the safety belt buckle or not. The piezoelectric material is using to feel the pressure, and converts the pressure signal into corresponding electrical signals. The safety belt state display device is set in the service area. The induction electrical signal from the piezoelectric material will be sent to the corresponding display device. The crew can accurately identify with the safety state of all the passengers. A timely reminder will be given to the corresponding passengers.
The performance of piezoelectric materials is directly related to the accuracy of the acquisition signal. It will affect the correctness of the display. It can eventually determine whether the system is successful. The piezoelectric ceramic is a kind of important piezoelectric materials. It has excellent piezoelectric properties, so it is used in the application design.

3 Design scheme

The seat belt status indicator system is designed based on the piezoelectric effect. The hardware design includes power module, piezoelectric signal acquisition module, signal processing module, the main processor module and status indication module. Among them, the acquisition and processing of piezoelectric materials is the focus of the entire design. The basic hardware structure of system diagram as shown in figure 1.

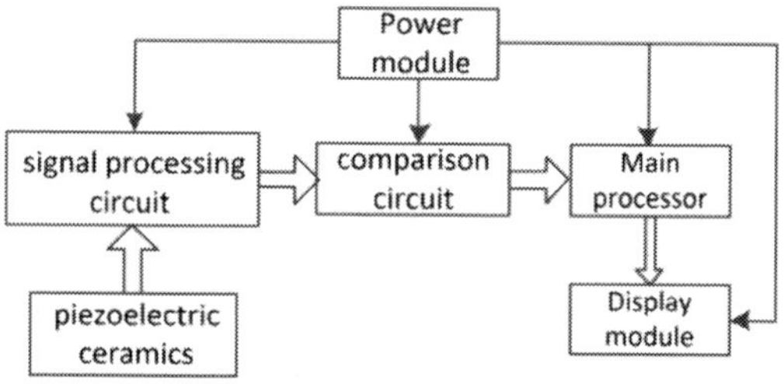

Figure 1. Basic hardware structure of the system

Considering the work environment characteristics and the demand of I/O, the TI MSP430 is chosen as the main processor. The power module can complete the conversion of electrical energy form. It supplies the power to the signal processing circuit, system voltage comparator, main processor and indication module. In this system, the main processor needs to be provided 3V working voltage, the indicating module needs to provided 5V voltage, the signal processing circuit and the comparison circuit need to be provided working voltage of 12V. Through a combination of considerations, an AC-DC step-down power module is chosen to meet the circuit power supply demand. It can provide a stable voltage source of 3V, 5V and 12V. The signal processing circuit finishes the signal processing from piezoelectric ceramics. The result is sent to the next level circuit for further processing. The main processor circuit completes collection the outputs form the comparison circuit and the collected signal is processed correspondingly. The processing result is sent to the corresponding indicating position to trigger the indicating system to work.

4 Piezoelectric signal obtaining and processing

4.1. piezoelectric signal analysis

With the different compression modes and different intensity of the situation, the piezoelectric ceramics generates the electric signals with different size and stability. In order to obtain the characteristics of electric signals generating under different conditions, and to complete the accurate judgment of the actual situation according to the characteristics, so to provide accurate signal to the processor to processor output to the indicator system clear signal, thus the performance test of piezoelectric ceramic of the selected type must be finished firstly. When the piezoelectric ceramic is triggered, the signal appeared in short time and each time the trigger is not the same. So in order to clarify the characteristics and laws of the signal, a large number of signals are tracking and acquisition by the oscilloscope. And the signal is statistically analyzed, to prepare for the subsequent design of the circuit.
According to the diameter of the selected piezoelectric sheet, specific test for different situations was done. Through a lot of tests and comparison, more clutter signal will be found out in the signal waveform when the pressure is far from the central zone of the piezoelectric plate. A waveform is generated by a lateral press as shown in Figure 2.

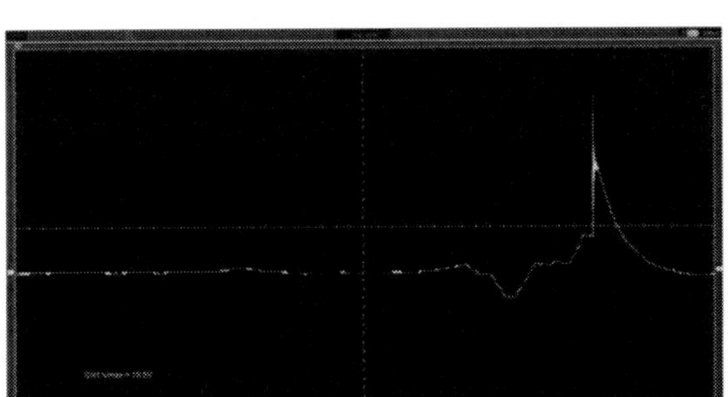

Figure 2. Waveform generated by a lateral press

If the pressure position is near the central zone of the piezoelectric plate, the clutter signal of the waveform is less. The waveform is complete and clear, convenient for further analysis, processing and use.
But if the pressure is different, there is a big difference between the voltage signals when the piezoelectric ceramic is pressed. One voltage waveform signal obtained by a light touch of the central zone of the piezoelectric plate is shown in Figure 3. The short-term peak voltage is 6.43V, and the voltage signal duration is short.

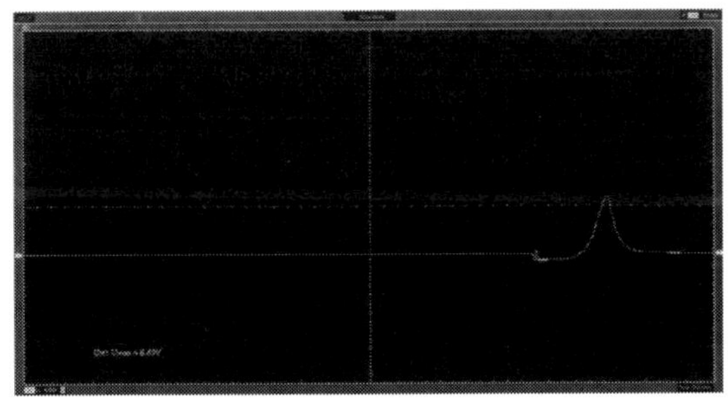

Figure 3. One voltage waveform from a light touch

One voltage waveform obtained by pressing the central zone of the piezoelectric plate is shown in Figure 4. The peak value of short-time voltage is 23.7V, and the voltage signal duration is longer.

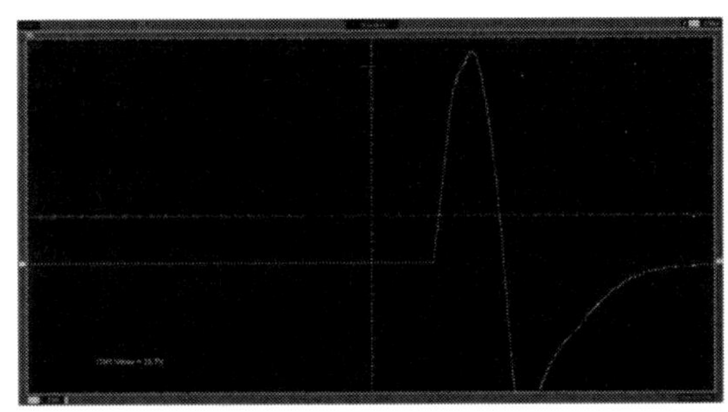

Figure 4. One voltage waveform from a normal touch

A large number output signals of the sensors are obtained by many tests of different forms and potency dimensions. Through the comparison of the output signals, it is found out that the short voltage value generated by the piezoelectric ceramic is more than 20V when the piezoelectric ceramic is under enough pressure. If the pressing pressure is insufficient or touch case, the short voltage values are below 12V.

4.2. Comparison circuit design

The comparison circuit design is used to identify the height of the piezoelectric ceramic output voltage. According to the large number of measurement results and their distribution characteristics, the state of the

corresponding device will be identified. It is triggered normally, or not triggered, or false triggered. It will help to avoid error indication.
Since the voltage range of the generated voltage is just confirmed to identify whether the device has been triggered or not, so the voltage comparator circuit is used to complete the corresponding function.
LM339 is the voltage comparator professional. It is a four differential comparator including four independent voltage comparators. The characteristics for it include fast switching speed, small delay time, a wide range of power supply voltage. The single power supply is 2 ~ 36V, and the double power supply voltage is + 1V ~ + 18V. It can be used for special occasions. In addition, its ratio differential input voltage range is large enough to be equal to the supply voltage. And the output potential can be selected flexibly and conveniently. So it is suitable for the voltage comparison circuit. The circuit principle diagram of the comparison circuit is shown in Figure 5.

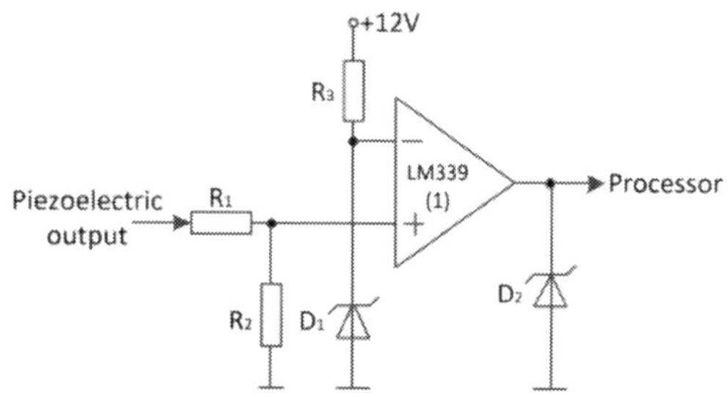

Figure 5. Comparison circuit principle diagram

According to the measuring result analysis, under the normal and abnormal pressure, the voltage value obtained is higher than 20V and lower than 12V. The voltage scope distinction is clear. In order to ensure the normal pressure is identified and the abnormal pressure is filtered out, the circuit hardware parameters are reasonably designed. R_1, R_2 and R_3 are respectively designed as 10 kΩ , 6 kΩ and 5 kΩ , D_1 as a 6V regulator.
The comparative results for the voltage comparator will be sent to the microcontroller to identify. So the voltage values must meet the processor voltage input range. In order to meet the requirements of signal, a stabilivolt with the corresponding parameters is set in the circuit, before the microcontroller and after comparison circuit. When a high voltage signal comes from the comparator, it can be limited to a fixed value, and this value is no high than the requirement of the microcontroller. Thus, a 4.5V regulator is used in the circuit after the output end of the comparator. When the comparator output is in high level, the signal sent to the main processor size does not exceed 4.5V, to meet the requirements of the size of the input signal processor.

4.3. Eliminate false triggering design

In order to reduce the possibility of the false triggering further to some extent, those voltage signals obtained by normal pressing, edge pressing and gent pressing are sent to the comparison circuit to deal with. Through a large number of tests, it is clear that the high level signal output can be stable for a long time. A certain measurement effect is shown in Figure 5.

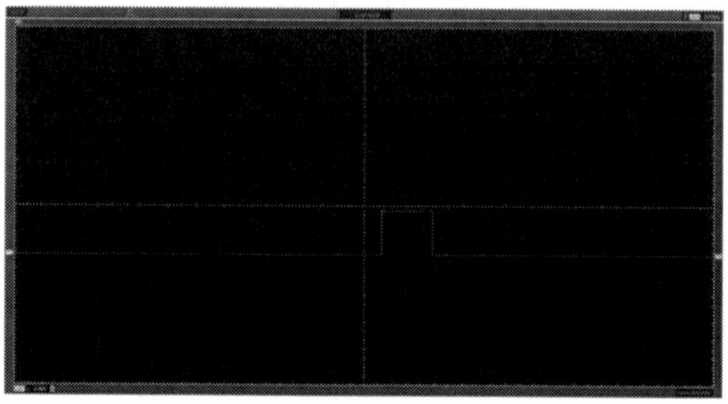

Figure 6. A certain measurement effect for normal pressure

In the case of false triggering, the comparator circuit even produces a high level output, and the output signal has a shorter duration, as shown in Figure 6.

Figure 7. A certain measurement effect for false triggering

According to a large number of repeated test statistics, there is an obvious difference in duration between the normal pressing and abnormal pressing. The time of duration of the abnormal is about no more than 100ms. Therefore, to reduce the possibility of false triggering further, during the system programming, the high level signal delay time is obtained and compared, and the value which is less than 100ms is believed to be from false triggering. So the error triggering signal is filtered further.

5 Status indicating

The display system of the design should be able to realize the explicit directions of each seat belt state. This function can be achieved by using LED lamp or LCD display. Initially, all seats corresponding state indicator lamp is lit. When the corresponding seat safety belts are buckled, the corresponding state indicator lamp is off. If the safety belt is released again, this kind of touch will make the indicator lamp light again. And the next safety belt buckle action will leads to the corresponding indicator lights go out again. The logic design is applied to the practical situation that the safety belt is buckled and unlocked many times during the operation of the aircraft.

6 Conclusions

This system has the advantages of simple structure, low cost, high reliability. And it has great practical value, not only can be used in the plane, but also can be used in high iron and some special entertainment facilities. The performance of the system is closely related to reliability of piezoelectric ceramics which is chosen, thus the performance improvement of the material will make the system stability and reliability to be further improved.

References

1. Xu Bin, Guo Xueyi, Chen Hongbing. Active Icing Monitoring for Wind Turbine Blade Models with PZT Technology. Piezoelectrics & Acoustooptics, **39**(2017).
2. Zhang Chuanjuan. Study of a compound piezoelectric energy harvester. Power Supply Technology and its Application, **19** (2015).
3. Yang Gaofeng. Preparation of Alarm Equipment PZT Piezoelectric Ceramics. Ceramics, **43**(2017)
4. Yuan Jiangbo. Power Generation and Pivotal Technique of Discal Energy Harvester (doctoral thesis) , Harbin Institute of Technology (2011).
5. Congjian Li. Research on Energy Harvester Used in Roads Based on Piezoelectric Effect (master's thesis). Harbin Institute of Technology, (2015).
6. Xian Yongbiao, Li Ketian. Design of a Piezoelectric Ceramic Driver. Mechanical and Electrical Engineering Technology, **46**(2017).

DOI: 10.1051/matecconf/201713900221

Research for Unmanned Aerial Vehicle components reliability evaluation model considering the influences of human factors

Yifang Tan[1], Ding Feng[2], Hanlin Shen[1]

1. Aviation engineering institute, Civil Aviation Flight University Of China, China

2. School of electrical engineering, Southwest Jiaotong University, China

Abstract—Human operating error is one of the important factors affecting the reliable operation of unmanned aerial vehicle (UAV) components, however, people always neglect human-operating-error of the existing unmanned aircraft systems and components reliability assessment methods. This article set the problem to the most important position to search, divided the scene of UAV components assignment to the Status debugging monitoring model and Preventive maintenance type. First, analysis of two scenarios UAV components made qualitative analysis of human error using the method of failure mode and effect analysis; and then put forward the two scenarios of human reliability assessment model, used to quantify the probability of human error. In example analysis, it considers the primary factors those affect humans reliability, to calculate and compare the proportion of human error at different situation. Results show that the proposed method can quantify human error probability, and quantitative characterization of human error can affect the reliability of the UAV components.

1 Introduction

At present, Whether the private company or professional unmanned aerial vehicle (UAV) management department for UAV components using the traditional way to checking and repairing, Maintenance system problems fundamentally have brought certain blindness and speculative to maintenance work, it's easy to cause excess maintenance or inadequate maintenance, which could lead to temporarily eliminated the risk factors or ruled out the fault on the surface, but buried a safety hazard for unmanned aerial vehicle (UAV) components follow-up operation, even the probability of failure is not reduce instead of increase. At the same time, all kinds of type of work staff due to the limited knowledge level, outside pressure interference and psychological quality, etc., may not be able to quickly and accurately finish operation, and create the serious consequences such as equipment damage, expanding failure. Therefore, because the participation of human activity will introduce artificial operation risks inevitably which affects the reliability of the UAV components.

There are existing methods for the reliability analysis of unmanned aerial vehicle (UAV) components, According to Bayes data fusion method, fault tree analysis method, the life distribution of key-module to establish the evaluation model, to reflect the reliability of the unmanned aerial vehicle (UAV) in literature [1]; Literature [2] is on the basis of the proportion distribution method and rate allocation method, through expert method to determine the weight coefficient, and got the UAV system reliability allocation scheme; Power system and the flight control system are important components of the unmanned aerial vehicle (UAV), therefore the system analysis of its failure mode will improve the reliability and reduce the risk of unmanned aerial vehicle (UAV) flight[3-7].It is very necessary to analysis failure mode and characteristic for UAV system failure, by studying the unmanned aircraft maintenance to improve the reliability of the unmanned aerial vehicle (UAV)[8-9]. Although the literature [10] using HFACS model for unmanned aerial vehicle (UAV) accident analysis, mined the key of accident cause, and put forward corresponding safety recommendations to made support for the establishment and perfection of the UAV airworthiness regulations and improve the level of industry safety management. The study of human factors on the unmanned aerial vehicle (UAV) reliability assessment is lack, so it's need to consider human factors of unmanned aerial vehicle (UAV) reliability to research and analysis.

This article according to the execution case and problem of UAV components, research and analysis the human error mechanism and human influence factors those influence the reliability of UAV parts. In these two scenarios, the reliability evaluation model of UAV parts is considered for human error, and the model is improved and improved based on the characteristics and influence of human factors. Set up scenarios and collect data, then, giving results are obtained by example. Verify the feasibility and applicability of the model. Thus, the reliability evaluation of UAV parts is realized. At the same time, improving the reliability of human operation has

* Corresponding author: yifang-t@foxmail.com

become the key to ensure the safe and reliable operation of UAV parts.

2 Reliability analysis of considering human factors

2.1 FMEA

FMEA (Failure Mode and Effect Analysis) often used to evaluate potential failure modes and their causes. From the system, it is a way of system safety analysis, based on the system's separable characteristics. It is a design method to improve the reliability of system and equipment according to the system's separable characteristics, the analysis of the components and elements of the system [11]. It analyzes possible failure modes through dividing the system into subsystems with different functions. classifies according to the severity of each potential failure mode. To put forward the possible planning, prevention and improvement methods, eliminate or reduce the possibility of failure.

Therefore, this article uses the method of FMEA to qualitative analysis the failure mode of failures and their causes and consequences of UAV components, it can provide the basis for scene classification of subsequent quantitative calculation of the reliability.

3 Quantitative analysis of reliability considering human factors

Human reliability, refers to within the prescribed time limit, if have time requirements), and any required in system operation stage, the probability of successfully completing job or task by humans [12]。Research has shown that UAV components failure part failure is caused by human errors in the process of debugging, operation, 10% ~ 15% of the total failure is caused directly by human improper operation [13]. It's enough to cause people's attention to research for unmanned aerial vehicle components reliability evaluation model considering the influences of human factors.

There are many types of UAV components, And all kinds of complicated parts operation condition, The influence of human error factors in different situations are different. Reasonable division of working condition of equipment scene type is the key to accurately quantify human reliability. Through the analysis of the characteristics of artificial operation and the working condition of equipment type,

In this paper, the operation condition of the UAV components scene can be divided into two categories: type: ESM (Equipment State Monitoring), EE&R (Equipment Examining and Repairing). On the basis of qualitative research of the FMEA forms, put forward the analysis method to quantitative modeling human error probability of the UAV components under different working scene.

Human reliability, refers to within the prescribed time limit, if have time requirements), and any required in system operation stage, the probability of successfully completing job or task by humans [12]。Research has shown that UAV components failure part failure is caused by human errors in the process of debugging, operation, 10% ~ 15% of the total failure is caused directly by human improper operation [13]. It's enough to cause people's attention to research for unmanned aerial vehicle components reliability evaluation model considering the influences of human factors.

There are many types of UAV components, And all kinds of complicated parts operation condition, The influence of human error factors in different situations are different. Reasonable division of working condition of equipment scene type is the key to accurately quantify human reliability. Through the analysis of the characteristics of artificial operation and the working condition of equipment type, In this paper, the operation condition of the UAV components scene can be divided into two categories: type: ESM (Equipment State Monitoring), EE&R (Equipment Examining and Repairing). On the basis of qualitative research of the FMEA forms, put forward the analysis method to quantitative modeling human error probability of the UAV components under different working scene.

3.1 Human reliability analysis of unmanned aerial vehicle (UAV) components in equipment state monitoring （ESM-HRA）

UAV components state surveillance scenes, means the task scenarios that need a long time for continuous inspection, monitoring and debugging of UAV components, at this point, the continuous working time is the dominant position, working long hours can make people appear low energy, such as fatigue state. At the beginning of the unmanned aerial vehicle (UAV) components debugging and monitoring work stage, the personnel on duty state is the best, human error probability HEP remain at low levels. But with the increase of personnel continuous operation time, the probability of human error HEP also will increase. In addition, HEP also will be affected by many other outside covariates including the cognitive level, the working environment, work intensity, physiological factors, psychological factors and the training level.

The human error probability calculation adopts the PHM (proportion hazard model) that was put forward firstly by D.R.Cox in the early 1970s [14]， The model considered a series of factors such as the equipment state parameters, work load and failure as the impact factors of the life[15]。Widely used in the field of reliability analysis. The failure rate function form:

$$h(t, Z) = h_0(t)\psi(Z) \quad (1)$$

Among them, $h_0(t)$ is hazard function ， said human errors change with working time.

The functional form had no special requirements, the application can choose the reliability parameter distribution according to actual systems or components, one of the most common is two parameter Weibull distribution [16]. Its function is:

$$h_0(t) = \beta t^{\beta-1} / \alpha^{\beta} \quad (2)$$

In the formula, α is the characteristic parameters of life; β is the shape parameter.

$\Psi(Z)$ is the connection function, it is the influence on human errors of 6 other collaborators variables (cognitive level, the working environment, work intensity, physiological factors, psychological factors and training level), each collaborators variable can be quantified for a particular measurement evaluation value. Z=[Z_1, Z_2, Z_3, Z_4, Z_5, Z_6], the influence index of each collaborators variable Z_i (i=1,2,3…6) can be{0,1,2}, The larger value indicates that the negative effect on the human operating reliability of the the variable is more obvious, and vice versa. γ is every collaborator variable weight coefficient. Its function is:

$$\psi(Z) = \exp(\sum \gamma \times Z) \quad (3)$$

If the installation personnel began to perform a task from zero moment, continue to perform to t moments don't have the probability of human error is the operation reliability of function, its expression is:

$$R_h(t) = \exp[-\int_0^t h(s,Z)ds] \quad (4)$$

The probability of human error HEP is:

$$HEP = 1 - R_h(t) \quad (5)$$

3.2 Human reliability analysis of unmanned aerial vehicle (UAV) components in equipment examining and repairing（EE&R-HRA）

UAV components preventive maintenance scenarios is according to certain standard and cycle, to regularly check of regulation of unmanned aerial vehicle (UAV) components, and ensure unmanned aerial vehicle (UAV) normal state. In this scenario, this paper uses the CREAM(Cognitive Reliability and Error Analysis Method) [16-17]. The main idea of this method is that human error is not random, but caused by the operators cognitive mode and the specific operation scene controlling and need according to certain process to complete. The direct statistics from the scene cannot be accurately and objectively reflect a series of factors change at any time including the external environment of operation, the operation staff psychological quality, human conditions and so on, but CREAM got a more specific, accurate, objective and quantitative human errors probability at the specific task scenario by CPC (Common performance conditions) impact factor repairing human error.

Table 1 is the relationship between general efficiency conditions of CPC and the expectation effect of the CREAM, shows the quantification of the influence on people's cognitive errors.

Table 1. Common performance conditions CPC

CPC name and number (i)	level	Expectation effect (Si)
1.Organization integrity	Very effective (9-10)	1
	Effective (6-8)	0
	Invalid (3-5)	0
	Effect poor (0-2)	-1
2.Working condition	Superior (8-10)	1
	Matching (4-7)	0
	Mismatching (0-3)	-1
3.Man-machine interface	Support (9-10)	1
	Adequate (6-8)	0
	Tolerable (3-5)	0
	Inaptitude (0-2)	-1
4. Plan availability	Appropriate (8-10)	1
	Acceptable (4-7)	0
	Impropriety (0-3)	-1
5. Target number at the same time	Within the ability (8-10)	1
	Conformity ability (4-7)	0
	Beyond the ability (0-3)	-1
6. Available time	Adequate (8-10)	1
	Temporary shortage (4-7)	0
	Lack of continuous (0-3)	-1
7. On duty time	Day (6-10)	0
	Night (0-5)	-1
8.Training experience adequacy	Experienced (8-10)	1
	Experience limited (4-7)	0
	Inadequacy (0-3)	-1
9. Quality of team cooperation	Very effective (9-10)	1
	Effective (6-8)	0
	Invalid (3-5)	0
	Poor effect (0-2)	-1

Model assumption:

(1) The task environment is continuous, and the human error probability of any point on the continuous space can be quantitative calculated by the human error probability prediction model.

(2)It can be described with a logarithmic relationship between the change of task environmental conditions with increasing or reducing of human respond ability [18]。

Quantification and weight set for HEP influence level with CPC:

CREAM method using level of CPC to describe the task environment, each CPC factor importance should be differentiated different in task environments, so it should determine the corresponding weights of CPC at the specific task environment firstly. It is W_i。

The integer 1, 0, 1 represent the three performance expectation effect, said the quantitative influence on people's cognitive errors. That determines the individual scores of CPC, as S_i. Specific processing: improved corresponding to 1, reduced corresponding 1, not significant is corresponding to 0. As shown in table 1 column 3.Set of two variables: I=∑improved score values, R=∑lower total value.

$$I = \sum_{i=1}^{9} W_i \times S_i (S_i = 1, i = 1, 2, ..., 9)$$

$$R=\sum_{i=1}^{9} W_i \times S_i \,(S_i=-1, i=1,2,...,9)$$

Operators task environment factors can be divided into two categories: improvement and reduction for human factors reliability. The first category: CPC that only cause improvement effect, the corresponding HEP (Human error aim-listed probability) is HEP_I, the second category: CPC that only cause reduction effect, the corresponding HEP is HEP_R. In addition to set up a basic environment: all the influence degree of the CPC to the human factors reliability is not significant,, that is S_i=0(i= 1, 2...9), at this point the corresponding HEP is HEP_0, and the total human error probability is HEP. The model hypothesis (2) can get the following expression:

$$\log(HEP_I / HEP_0)=k_1 I \quad (6)$$

$$\log(HEP_R / HEP_0)=k_2 R \quad (7)$$

Thus the relationship of HEP_I, HEP_R, HEP_0, I, R is established:

$$HEP_I = HEP_0 \times 10^{k_1 I} \quad (8)$$

$$HEP_R = HEP_0 \times 10^{k_2 R} \quad (9)$$

Table 2. The Basis of Failure Probability in Different Failure Modes in CREAM

Cognitive behavior	Failure Modes	Probability
Observation (P1)	Observation target error (P11)	0.0010
	Misidentification (P12)	0.0700
	Observation is not done (P13)	0.0700
Explanation (P2)	Diagnosis Failure (P21)	0.0200
	Decision Failure (P22)	0.0100
	Delay to explain (P23)	0.0100
Plan (P3)	Priority error (P31)	0.0100
	Improper planning (P32)	0.0100
Execution (P4)	Action mode error (P41)	0.0030
	Action time error (P42)	0.0030
	Action target error (P43)	0.0005
	Action sequence error (P44)	0.0030
	Action omission (P45)	0.0300

HEP_0 value can be calculated by the table 2, and then determine the values of parameters k_1, k_2, the ultimate human error probability HEP values as follows:

$$HEP=(HEP_I - HEP_0)+(HEP_R - HEP_0)+HEP_0 \quad (10)$$

$$HEP = HEP_0 \times (10^{-k_1 I}+10^{k_2 R}-1) \quad (11)$$

4 Example analysis

Select the example scenario is that the installation personnel state monitoring and debugging unmanned aerial vehicle (UAV) components of the management.

The Weibull parameters in the model can get through the historical data evaluation [12], α=200 、 β=3 ;

Selection of unmanned aerial vehicle (UAV) system6 collaborators under three different operating state variables influence index Z_i(i=1,2, 3, ...,6) values and through the analytic hierarchy process to get value of 6 collaborators variable weighting coefficient γ. Shown in the following table

Table 3. 6 covariates influence index and weight

Influence factor	Cognitive level	operating environment	operating intensity	physiological factors	Psychological factors	Training level
	Z_1	Z_2	Z_3	Z_4	Z_5	Z_6
Scene1	0	0	1	0	0	0
Scene2	1	1	1	1	1	1
Scene3	2	2	1	2	2	2
Scene4	2	0	0	0	0	0
Scene5	0	0	0	0	2	0
Scene6	0	0	0	0	0	2
weight	0.26	0.07	0.14	0.05	0.09	0.39

From the table above, the work intensity of the three scenarios are still in the general level (grade 1), the other five covariate effects on human error probability HEP. MATLAB simulation of the final diagram below:

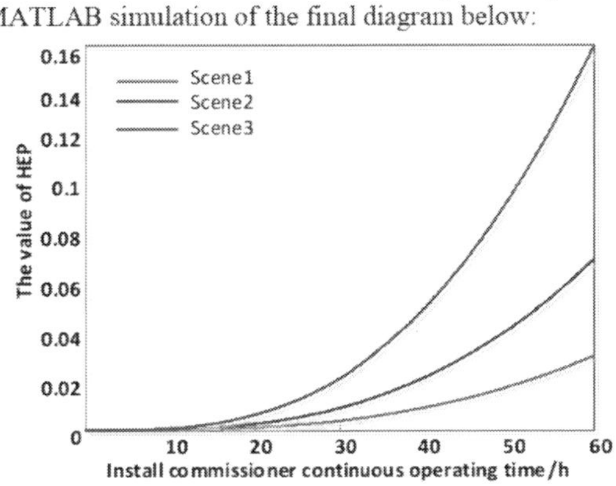

Figure 1. HEP with continuous working time curve

The last three kinds of circumstances in table 3, operating environment, work intensity, physiological factors that 3 collaborators variable values can promote the human operating reliability. In scene 4, the cognitive level is the most negative effect on the results, in scene 5, the psychological factors is the most negative effect on the results, in scene 6, the training level is the most negative effect on the results. The MATLAB simulation as shown in figure 2:

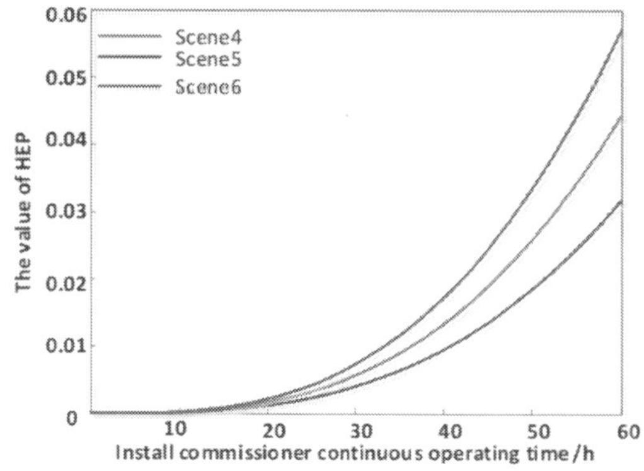

Figure 2. HEP with continuous working time curve

Results analysis:

1) The figure 1 shows that under the same work intensity, five other collaborators variables: cognitive level, working environment, physiological factors, psychological factors and training level have the lower value that the more significant promoting effect on the operating reliability, eventually the human error probability value is smaller, the operating reliability is higher. Thus, management department should make the operator with UAV components installation and debugging work in harmony, reduce such as fatigue, poor health, inattention, low technical level, lack of experience and operation technology is not skilled and so on unstable factors, strengthen the inspection of parts, confirm components state of technology, potential for may affect the safe operation of the UAV components timely report, handle in time.to improve the reliability of the UAV components running.

2) The figure 2 shows that, training level, cognitive level and psychological factors that 3 collaborators variable degree of influence on human error probability *HEP* from high to low arrangement. the 3 collaborators variable level are relatively easy to improve through artificially training. Therefore, managers should do a good job for staff training, strengthen the training of the personnel cognitive level and psychological quality assessment, to ensure operational reliability.

3) Combined with figure 1 and figure 2,in the first ten hours of continuous operation of personnel installation and debugging, Human errors HEP has maintained at the lower value, personnel operating reliability is higher at this stage. So the management department should perfect the related system, try to avoid same class staff long time continuous operation.

5 Conclusion

Human error is an important factor effecting the reliability of UAV components. This paper draws on research in the field of human reliability analysis, qualitative and quantitative analysis of the human error factors influents in the operating process of unmanned aerial vehicle (UAV) components. After considering the characteristics of its operating, involved the artificial operation scene into two classes, put forward the quantitative method of human error probability under different scenarios. The example analysis results show that, this method can consider the influence of personnel operation behavior, quantify the probability of human error and change trends. In the study of UAV system safe operation and accident prevention, according to the results of quantitative analysis, find out greater influence element of the human reliability, and put forward the targeted improvement measures, in order to reduce the occurrence probability of human error, to improve the overall reliability of the UAV system.

References

1. Wang Dawei, Wang Wei, Zhang Tiechun. Unmanned aerial vehicle (UAV) with a small aircraft engine reliability evaluation [J]. Journal of China civil aviation university, 2012, 30 (01) : 14-17 + 40..
2. Zhang Rui, Li Bo. UAV system reliability allocation method [J]. Journal of Nanjing university of aeronautics and astronautics, 2009, 9 (S1) : 106-108.
3. Feng Jun. Unmanned aerial vehicles based on FTF method of hydraulic and air conditioning system reliability analysis [D]. University of electronic science and technology, 2011.
4. Chen Yuan. Aircraft power system reliability analysis based on fuzzy FMECA method research [D]. University of electronic science and technology, 2011.
5. Shi Weiping, Han Mei. UAV servo mechanism with Angle sensor fault analysis and countermeasures [J]. Journal of quality and reliability, 2011 (2):50-52.
6. Ji Hui, Pan Shengbing, Zhang Xiaojing, etc. Unmanned aerial vehicle (UAV) fault tolerant flight control systems [J]. Journal of research and application of computer measurement and control, 2013, 21 (9).
7. JiHui, Pan sheng-bing, Zhang Xiao-lin. Flight control system fault diagnosis and reconstruction technology research [J]. Computer measurement and control, 2015, 23 (4): 1073-1074.
8. PuHai, Chen Zi-li. Reliability centered maintenance analysis technology and its use in unmanned aerial vehicle (UAV) [C]. Application of high and new technology to improve maintenance support capability conference. 2005.
9. Liu Jian-shan. Unmanned aerial vehicle (UAV) cooperative maintenance technology research [C]. Aviation equipment maintenance technology and application international symposium. 2015.
10. Lu Xiaojian, He Qiang. Human factors analysis of the UAV accident based on HFACS [J]. Science and technology wind, 2017, (17) : 245.
11. Yue Zhang, Chao Shi, Laihua Fang. Study on comprehensive analytical method and its application based on FMEA and HAZOP[J]. *Journal of Safety Science and Technology* , 2011,7(7):146-150.
12. BS Dillon (Canada), Mou Zhizhong, Xie Xiuling, Wu Fubang. The reliability of [M]. Shanghai: Shanghai science and Technology Press, 1990.
13. Hu Guo, Wenhua Chen, Weie Zhang, etal. Human Reliability Analysis of Operators under the Fluctuating Pressure [J]. *Transactions of The Chinese Society of Agricultural Machinery* .2002, 33(6): 107-109.
14. Cox D R. Regression models and life: tables, breakthroughs in statistics[M]. New York: Springer, 1992: 527-541.
15. Jiang S T, Landers T L, Rhoad T R. Assessment of Repairable-system Reliability Using Proportional Intensity Model: a Review [J] . *IEEE Transactions on Reliability*,2006.55 (2) : 328-336.
16. Hollnagel E. Cognitive reliability and error analysis method[M].
Amsterdam: Elsevier, 1998: 151-189.
17. Fujita Y, Hollnagel E. Failures without errors: quantification of context in HRA[J]. *Reliability Engineering & System Safety*, 2004, 83(2) : 145-151
18. Yao Wang, Xuhong He. Quantitative Analysis of Cognitive Errors

and Its Application in PSA [J]. *Nuclear Power Engineering.* 27(4) : 54-58.

AdaBoost typical Algorithm and its application research

TU Chengsheng , LIU Huacheng , XU Bing
Computer Science and Engineering college, Chongqing Three Gorges University ,ChongQing WanZhou 404000

Abstract. Boosting is popular algorithm in the field of machine learning. Adaboost is the most typical Algorithm in the Boosting family.This paper introduces Boosting and its research status briefly, and introduces the typical algorithms of each series respectively.

KeyWords: Machine learning; Adaboost Boosting algorithm; Text categorization; Combining prediction

1.Introduction

Boosting was suggested by Freund and Schapire in 1990 [1] .It is an efficient instrument for improving the predicting ability of learning system and a most typical method in coordinating learning. For Boosting, there are two main problems: how to adjust the training set to enable the weak classifier to conduct training on it, and how to combine the weak classifiers, gained through training, into a strong one. To solve the above problems, Freund and Schapire introduced Adaboost (adaptive boosting) algorithm in 1995, which works through adjusting weight without the need of any priori knowledge on learner learning. Algorithms of AdaBoost.M1 、 AdaBoost.M2 、 AdaBoost.R, etc. were suggested based on later innovation. These algorithms could change voting-weights and solved many practical problems of the early Boosting algorithm. [3] To solve the multi-class problem when there is a big category, Freund and Schapire introduced AdaBoost.OC algorithm [5] in 1997 based on the combination of AdaBoost.M2 与 ECOC[4]. A more popular AdaBoost algorithm was suggested in 1998 by them, and a confidence predication was introduced so as to improve the quality of Boosting. They also proposed the algorithms of AdaBoost.MH 、 AdaBoost.MR, which could solve multi-class and multi-label problems, and recommended the AdaBoost.MO [6] based on the combination of ECOC and AdaBoost.MH.

The study and application of AdaBoost algorithm mainly focus on classification problems. For example, the application of literature [7] ,mainly solves the two-class problem, multi-class single label problem, multi-classification and multi-label problem, class single label problem and regression problem.

Based on the application of AdaBoost in text categorization, this paper presents the basic idea and application of the two-class single label problem, and analyzes the algorithms of AdaBoost.M1and AdaBoost.M2 and their application in solving the multi-class single label problem, and the algorithms of AdaBoost.MR and AdaBoost.MH and their application in solving the multi-classification and multi-label problem.

2. AdaBoost algorithm and the two-class single label problem

Prepare training sets$(x_1, y_1), \ldots, (x_n, y_n)$. $x_i \in X$, X represents a certain domain or instance space, and each member is a training example with a label. $y_i \in Y=\{-1,+1\}$,proposes the two-class problem of learning. The weights of all the training examples are initially set to be 1/m equally. Adaboost conducts T times of iteration through repeatedly calling weak learning algorithm. When the *tth* $(t=1, 2, \ldots, T)$iteration was effected, the weight distributing on the sample $\{x_i, y_i\}$ was recorded as $D_t(i)$, and a set of distribution of weights are preserved in the training set after each iteration. The weight of each instance would increase while the training fails each time, so as to enable the weak learner to forcedly focus on the instance of failure in training. As per D_t distribution, the weak learner finds

 DOI: 10.1051/matecconf/201713900222

appropriate weak hypothesis h_t: $X \to R$; thus, predicting function sequence is gained. Each predicting function gives weight, and the weight of a learner with a better predicating effect is bigger. In the simplest case, if the scope of each h_t is two-valued {−1, +1}, the task of the learner is to minimize the error. Combining T- weak hypotheses, the final predicting function(hypothesis) H is gained after T times of circulation, with a weighted majority voting method. The learning accuracy rate of a single weak learner is not high enough; however, with the application of Boosting algorithm, the accuracy rate of the final result is to be improved. The AdaBoost algorithm is described in detail, as follows:

(1)With the given N-training sample set with labels: $(x_1,y_1),(x_2,y_2),......(x_n,y_n)$,among which $x_i \in X$, $y_i \in Y=\{-1,+1\}$.

(2)For t=1,2,……,T, initially set up weight D(i) and make D(i)=1/N.

(3)For t=1,2,……,T, T represents the maximum circulation times of training. For T-training, firstly, the weight distributing on the sample $\{X_i, Y_i\}$ is recorded as $D_t(i)$ while the Tth iteration happens, and as per the distribution D_t the weak learner finds its weak hypothesis h_t: $X \to \{+1, -1\}$ and adjusts distribution. Secondly, the error rate in computing h_t $\varepsilon_t = \sum_{i=1}^{N} D_t(x_i)\,[h_t(x_i) \neq y_i]$.

Thirdly, compute the weight of weak classifier based on the error rate: $\alpha_t=(1/2)ln((1-\varepsilon_t)/\varepsilon_t)$. Fourthly, update the sample weight:

$$D_{t+1}(x_i) = \frac{D_t(x_i)}{Z_t}\exp(-\alpha_t y_i h_t(x_i)),$$

among which,

$$Z_t = \sum_{i=1}^{n} D_t(x_i)\exp(-\alpha_t y_i h_t(x_i))$$

Z_t is a standardized factor that meets the probability distribution.

(4)T-weak classifiers are gained after T times of circulation, and a strong classifier $H(x) = sign(\sum_{t=1}^{T}\alpha_t h_t(x))$ is gained after adding to the updated weight.

3. AdaBoost.M1 and AdaBoost.M2 algorithms solve multi-class single label problems

Each sample to be classified is a single class within multiple category. To define $[\![\pi]\!]$, if π is true, $[\![\pi]\!]=1$, otherwise $[\![\pi]\!]=0$.

3.1 AdaBoost.M1 algorithm

Input sequence of *N*-examples (x_1, y_1), ⋯, (x_n, y_n), label $y_i \in Y=\{1,2, \cdots, k\}$, and set the distribution on instance is D , the weak learning algorithm WeakLearn, and the iteration T. In initialization, for $i=1,2,...,T$,set the weight vector $W_i^1 = D(i)$. For $t=1,2,......,T$, execute the following:

(1)the whole distribution $P^t = W^t / \sum_{i=1}^{N} W_i^t$. (2)call the WeakLearn and transmit the distribution P^t to it. (3) calculate the error rate of h_t $\varepsilon_t = \sum_{i=1}^{N} P_i^t [\![h_t(x_i) \neq y_i]\!]$. If $\varepsilon_t > 1/2$, set T=t−1 and exit a loop. (4) establish $\beta_t = \varepsilon_t/(1-\varepsilon_t)$. (5) Calculate the new weight:

$$W_i^{t+1} = W_i^t \beta_t^{1-[\![h_t(x_i) \neq y_i]\!]}$$

and gain the final hypothesis:

$$h_f(x) = \arg\max_{y \in Y} \sum (\log(1/\beta_t))[\![h_t(x) = y]\!].$$

When AdaBoost.M1calls the WeakLearn, the error rate produced is the hypothesis $\varepsilon_1, \cdots, \varepsilon_T$. Propose each $\varepsilon_t \leq 1/2$, the upper boundary of the error $\varepsilon = \Pr_{i \sim D}[h_t(x_i) \neq y_i]$ of the final hypothesis is $\varepsilon \leq 2^T \prod_{t=1}^{T}\sqrt{\varepsilon_t(1-\varepsilon_t)}$.

This is the most direct multiclass extension of AdaBoost. When the WeakLearn is strong enough to gain proper precision in the difficult distribution generated by

AdaBoost, it is enough to solve multiclass problems. Provided that the WeakLearn can not gain a precision of no less than 50%, the method fails. Therefore, multiclass problems are generally solved by simplifying them into multiple "two-value problems". This idea is the preferred choice for algorithms of AdaBoost.M2, AdaBoost.MH, AdaBoost.MR, etc. to solve multiclass problems.

3.2 AdaBoost.M2 algorithm

Input N-example sequence $(x_1, y_1), \dots, (x_n, y_n)$,label $y_i \in Y=\{1,2, \dots, k\}$, and set D as the distribution on the instance, WeakLearn as the weak learning method, and T as the iterative times. Initially, for $i=1,2,\dots,T$, set the weight vector $W^1_{i,y} = D(i)/(k-1), y \in Y-\{y_i\}$. For $t=1,2,\dots,T$, execute: (1) establishing $W^t_i = \sum_{y \neq y_i} W^t_{i,y}$, $q_t(i,y) = W^t_{i,y} / W^t_i$, and for $y \neq y_i$, setting up $D_t(i) = W^t_i / \sum_{i=1}^{N} W^t_i$; (2) calling WeakLearn, and transmitting distribution D_t and label weight function q_t to it; returning to hypothesis $h_t: X \times Y \rightarrow [0,1]$; (3) computing pseudo loss of h_t:

$$\varepsilon_t = \frac{1}{2}\sum_{i=1}^{N} D_t(i)\left(1-h_t(x_i,y_i)+\sum_{y \neq y_i} q_t(i,y)h_t(x_i,y)\right);$$

(4)set $\beta_t=\varepsilon_t/(1-\varepsilon_t)$; (5)computing the new weight: $W^{t+1}_{i,y} = W^t_{i,y}\beta_t^{\frac{1}{2}(1+h_t(x_i,y_i)-h_t(x_i,y))}$, and gain the final hypothesis: $h_f(x) = \arg\max_{y \in Y} \sum (\log(1/\beta_t)h_t(x,y)$.

AdaBoost.M2 is a special case of AdaBoost.MR. For each sample x_i *with correct label* y_i and each incorrect label y (y represents the other k-1 type except y_i), raise such a two-value question as : for x_i, the correct label is y or y_i? Here, the given h is used to answer the k-1 two-value questions, and separate out the correct label y_i from the incorrect label y. Propose h fetches its value from {0,1}, if $h(x_i, y)=0$ and $h(x_i, y_i)=1$, then the answer to the above question is y_i ; if $h(x_i, y)=1$ and $h(x_i, y_i)=0$, the answer to the said question is y. If $h(x_i, y)=h(x_i, y_i)$, then select any one from the two at random. When h fetch its value from [0,1], $h(x, y)$ can be interpreted as a random decision: select a random bit $b(x, y) \in \{0, 1\}$, the probability that the value is equal to 1 is $h(x, y)$, and the probability that the value is equal to 0 is $1-h(x, y)$. Hence, the probability to select incorrect answer y is:

$$\Pr[b(x_i, y_i) = 0 \wedge b(x_i, y) = 1] + \frac{1}{2}\Pr[b(x_i, y_i) = b(x_i, y)]$$

$$= \frac{1}{2}(1 - h(x_i, y_i) + h(x_i, y))$$

If answers of all the k-1 questions are equally important, the assumed loss is defined as average value:

$$(1 - h(x_i, y_i) + (\sum_{y \neq y_i} h(x_i, y)) / (k-1)) / 2$$

. However, various distinctions is very likely to have different importance in diverse situations. The method to attach different importance to diverse questions is to attach a weight to each question. Therefore, to specify weight $q(i,y)$ for every instance x_i and incorrect label $y \neq y_i$, it is linked to the differentiated label y and the correct label y_i. Put it into the above formula, and the result is referred to as the loss of q when h is on the training example i: $ploss_q(h,i)$

$$\approx \frac{1}{2}(1 - h(x_i, y_i) + \sum_{y \neq y_i} q(i, y)h(x_i, y)).$$

The function $q:\{1,\dots N\} \times Y \rightarrow [0,1]$ is termed as label weight function. For all the i, $\sum_{y \neq y_i} q(i,y) = 1$. The object of the WeakLearn is minimize the expected pseudo-loss of the distribution D and the weight function q: $ploss_{D,q}(h) := E_{i \sim D}[ploss_q(h,i)]$.

By controlling D and q, this algorithm effectively let the WeakLearn focus not only on difficult instances, but also on incorrect class labels which are most difficult to be eliminated. On the contrary, this kind of pseudo loss measurement may make the WeakLearn easily get weak dominance. AdaBoost.M2 is to produce a hypothesis with a pseudo loss of $\varepsilon_1, \dots, \varepsilon_T$ when it calls the WeakLearn. Thus, the upper boundary of

$\varepsilon = \Pr_{i\sim D}[h_t(i) \neq y_i]$, the error of the final hypothesis h_f, is $\varepsilon \leq (k-1)2^T \prod_{t=1}^{T} \sqrt{\varepsilon_t(1-\varepsilon_t)}$.

4. AdaBoost.MR and AdaBoost.MH algorithms solve multi-class and multi-label problems

Each object can belong to one or more of multi categories. A single label classification problem is a particular example of multi-label ones. Suppose γ is the finite set of a label or a class, and make $K=|\gamma|$. In the multi-label case, each instance $x_i \in X$ may belong to multi labels in γ; therefore, the example with a label is (x, Y) pairs, among which $Y \subseteq \gamma$ is a label set given to x. The purpose of learning is the need to look for a label hypothesis of a label set given to the instance by prediction, i.e., the purpose is to look for the probability $H{:}X \rightarrow \gamma$ of the minimized $H(x) \notin Y$ on a new type (x, Y). This type of measurement is referred to as one-error of hypothesis H since it measures a probability even with an incorrect label. The $one\text{-}error_D(H)$ is used to express the one-error about distribution D made by hypothesis H when observing on (x, Y), namely,

$$one-error_D(H) = \Pr_{(x,Y)\sim D}[H(x) \notin Y].$$

4.1 AdaBoost.MH algorithm based on the hamming loss

Input training sets $(x_1, y_1), \ldots, (x_m, y_m)$, among which $x_i \in X$, $Y_i \subseteq \gamma$. initially, $D_1(i,l)=1/(mk)$. For $t=1,2\ldots, T$ loops execution: (1)Distribution D_t trains WeakLearn; (2)obtain weak hypothesis h_t: $X \times \gamma \rightarrow R$; (3)choose $\alpha_t \in R$; (4)revise:

$$D_{t+1}(i, l) = (D_t(i, l) \exp(-\alpha_t Y_i[l] h_t(x_i, l))) / Z_t,$$

among which Z_t is a normalization factor, and then gain the final hypothesis:

$$H(x,l) = sign(\sum_{t=1}^{T} \alpha_t h_t(x,l)).$$

AdaBoost.MH operates by producing a set of two value problems to each sample x and each sample y, as follows : for sample x, its correct label is y or something else? Suppose the present target is only to predict all the correct labels, then learning algorithms generate the hypothesis of predictive label set, and the assumed loss depends on how large the difference b/w predictive set and observed result is. Consequently, the loss is $E_{(x,Y)\sim D}[h(x)\Delta Y]/k$, among which Δ represents symmetric difference, and is referred to as the hamming loss of H and expressed as $hloss_D(H)$. In order to minimize it, the problem is resolved into k-Orthogonal binary classification ones, namely, Y can be considered as the special value of k-binary labels (relying on whether a certain label y is included in the Y). Similarly, $h(x)$ can be considered as k-binary prediction. Thus, the loss can be regarded as an average error rate of h in k-binary problems. For $Y_i \subseteq \gamma, l \in \gamma, Y[l]$ is defined as $Y[l] = \begin{cases} +1 & if \quad l \in Y \\ -1 & if \quad l \notin Y \end{cases}$. To simplify the expression, the two- variables function of H: $X \times \gamma \rightarrow \{-1, +1\}$, defined with $H(x,l)=H(x)[l]$, is used to express the corresponding function of $H{:}X \rightarrow 2^{\gamma}$. This shows Boosting minimizes the hamming loss, and the main idea of the minimization is to replace each training example of (x_i, Y_i) with k-examples of $((x_i, l), Y_i[l])$ (for $l \in \gamma$). This minimization also lead to the select of the final hypothesis. It proves that the upper bound of the hamming loss is $hloss(H) \leq \prod_{t=1}^{T} Z_t$.

4.2 AdaBoost.MR based on ranking loss

Input training sets $(x_1, y_1), \ldots, (x_m, y_m)$, among which $x_i \in X$, $Y_i \subseteq \gamma$. Initially:

$$D_1(i, l_0, l_1) = \begin{cases} 1/(m \cdot |Y_i| \cdot |\gamma - Y_i|) & if \quad l_0 \notin Y_i \quad and \quad l_1 \in Y_i \\ 0 & others \end{cases}.$$

T loops execution: (1) use distribution D_t to train the WeakLearn; (2)gain weak hypothesis h_t: $X \times \gamma \rightarrow R$; (3) choose $\alpha_t \in R$; (4)revise:

$$D_{t+1}(i, l_0, l_1) = (D_t(i, l_0, l_1) \exp(\frac{1}{2} \alpha_t (ht(x_i, l_0) - ht(x_i, l_1)))) / Z_t,$$

among which Z_t is a normalization factor (make D_{t+1} as the distribution), and gain the final hypothesis: $f(x,l) = \sum_{t=1}^{T} \alpha_t h_t(x,l)$. The purpose of AdaBoost.MH is to find the hypothesis of the label set connecting with each instance, while the purpose of AdaBoost.MR is to find a hypothesis which sorts the labels, in the hope that the right labels can reach the highest level. To formally express, find a hypothesis shaped like $f: X\times\gamma\rightarrow R$, and interpret it as: to sort the labels in a given instance x, γ as per $f(x)$, that is to say, if $f(x, l_1)>f(x, l_2)$, the level of a label l_1 is considered to be more higher than that of l_2. For an observation (x, Y), care only about the relative sort relation of key pair of l_0, l_1 ($l_0 \notin Y$, $l_1 \in Y$). If $f(x, l_1)\leq f(x, l_0)$, then the key pair of l_0, l_1 can be considered as being wrongly sequenced by f. The target here is to look for a function f with only a small amount of error sorting, and let the labels in Y be sorted before those outside of Y. Therefore, the anticipated key pair with error sorting need to be minimized. Such a measure is called a ranking loss. A distribution D on a observation is defined as

$$E_{(x,Y)\sim D}\left[\frac{\left|\left\{(l_0, l_1)\in(\gamma-Y)\times Y : f(x,l_1)\leq f(x,l_0)\right\}\right|}{|Y||\gamma-Y|}\right],$$

and expressed by $rloss_D(f)$. It would not fail for any observation Y, and would not equal to γ; or otherwise, it needs no sorting. This algorithm maintains the distribution of D_t on $\{1,2, ...,m\}\times Y\times Y$. Only when l0, l1 are key pairs in relation to (x_i, Y_i), this distribution is non-zero in the triples (i, l_0, l_1).

4.3 Improved algorithm of AdaBoost.MR

Input training sets (x_1, y_1), ⋯, (x_m, y_m), among which $x_i\in X$, $Y_i\subseteq\gamma$. Initially:

$$v_1(i,l) = \left(m\cdot|Y_i|\cdot|\gamma-Y_i|\right)^{-\frac{1}{2}}$$

For $t=1,2,..., T$ loops, and execute: (1)training WeakLearn with the distribution D_t; (2)gaining a weak hypothesis $h_t: X\times\gamma\rightarrow R$; (3)selecting $\alpha_t\in R$; (4) revising:

$$v_{t+1}(i,l) = (v_t(i,l)\exp(-\frac{1}{2}\alpha_t Y_i[l]h_t(x_i,l)))/\sqrt{Z_t},$$

among which:

$$Z_t = \sum_i\left[\left(\sum_{l\notin Y_i} v_t(i,l)\exp\left(\frac{1}{2}\alpha_t h_t(x_i,l)\right)\right)\left(\sum_{l\in Y_i} v_t(i,l)\exp\left(-\frac{1}{2}\alpha_t h_t(x_i,l)\right)\right)\right],$$

and gaining the final hypothesis $f(x,l)=\sum_{t=1}^{T}\alpha_t h_t(x,l)$.

When the original AdaBoost.MR method has a lot of labels, the efficiency both in time and space is not high enough. $|Y_i|\cdot|\gamma- Y_i|$-weights need to be maintained for each training example(x_i, Y_i) and each weight needs to be revised each time. In reality, for the same algorithm, $O(mk^2)$ may be the difference between space complexity and time complexity that each iteration needs , *and only weight* v_t needs to be maintained in $\{1,2,\ldots,m\}\times\gamma$. Provided that l_0, l_1 is the key pair in relation to (x_i, Y_i), it is true that $D_t(i,l_0,l_1)=v_t(i,l_0).v_t(i,l_1)$. Thus, it can be seen that this algorithm is equal to the original AdaBoost.AR algorithm and the overhead is only $O(mk)$.

5. Conclusion

AdaBoost has the advantages of being quick in speed, simple in operation and easy to be programmed. There is no need to adjust parameters except for the number of iterations. It can be flexibly combined with any method to look for weak hypothesis without the need of any priori knowledge of WeakLearn. Given sufficient data and a WeakLearn with only reliable moderate accuracy, it can provide a set of theoretical guarantee of learning. It focuses on look for a weak learning method only better than random predication, instead of trying to design an accurate algorithm in the whole space. However, AdaBoost still has its own disadvantages. For example, its real performance in a particular problem clearly relies on data and WeakLearn; in the event of insufficient data set, it has a poor performance in too complex and too weak hypotheses; and it seems to be very sensitive to noise.

Acknowledgements

It is a project supported by the scientific fund assistance project from Chongqing Education Committee (NO: KJ131115).

References

[1] Robert E. Schapire. ***The strength of weak learnability.*** Machine Learning, 5(2):197-227, 1990

[2] Freund Y., Schapire R.E.A., ***Decision-Theoretic***

Generalization of OnLine Learning and an Application to Boosting.Journal of Computer and System Sciences,1997,55(1):119-139.

[3] Yoav Freund and Robert E. Schapire. ***A decision-theoretic generalization of online learning and an application to boosting***. Journal of Computer and System Science, 55(1):119-139, August 1997

[4] Thomas G.Dietterich and Ghulum Bakiri. ***Solving multiclass learning problems via error-correcting output codes***. Journal of Artificial Intelligence Research, 2:263-286, January 1995

[5] Robert E. Schapire and Yoram Singer. ***Using output codes to boost multiclass learning problems***. In Machine Learning: Proceedings of the Fourteenth International Conference, pp.313-321. 1997

[6] Robert E. Schapire and Yoram Singer. ***Improved boosting algorithms using confidence-related predictions***. In Proceedings of the eleventh Annual Conference on Computational Learning Theory, pp.80-91, 1998

[7] Diao Li-li,Hu ke–yun,Lu Yu-chang, ***Improved Stumps Combined by boosting for Text Categorization***.Journal of Software.2202,13(8).

[8] Yoav Freund. ***An adaptive version of the boost by majority algorithm***. In Proceedings of the Twelfth Annual Conference on Computational Learning Theory, 1999

A novel global optimization algorithm based on the smoothing function

Tian Jingjing[1,*] and *Yuan* Liu-yang[1,**]

[1]College of Science, Wuhan University of Science and Technology, Wuhan, Hubei, P.R. China, 430065

Abstract. In this paper, a new global optimization algorithm based on the smoothing function is suggested. Firstly, one of the minimizer x_1^* of the smooth function F is found by employing any local minimizer finder, then the function $m(x, x_1^*) := \min\{F(x), F(x_1^*)\}$ is considered instead of the objective function F and the minimizer x_2^* of the function m is searched. Because of the non-smooth of the function m, the smoothing function for m is used in order to make the local minimizer finding algorithms for smooth optimization available. Finally, a global optimization algorithm based on the smoothing function is presented. The implementation of the algorithm on several test problems is reported with numerical results.

1 Introduction

We consider the box constrained global optimization problem as follows:

$$\begin{aligned} \min \quad & F(x) \\ \text{s.t.} \quad & x \in D, \end{aligned} \tag{1}$$

here $F(x) : R^n \to R$ is a continuously differentiable function and D is a box. As $F : R^n \to R$ is a coercive function, i.e., $F(x) \to +\infty$ when $\|x\| \to +\infty$, then the following problem:

$$\min_{x \in R^n} F(x)$$

can be reformulated as the box constrained optimization problem in (1).

Global optimization has many applications in almost all fields of science and technology. Because a general non-convex objective function has many local minima, then it makes global optimization hard to solve ([2, 6, 13]). Currently, Global optimization makes great progresses on theories and practical applications, and there are many methods have been presented, for example the integration-level set methods [1], the branch and bound methods [6], the outer approximation methods [7], the tunneling function methods [8], the interval methods [12], the filled function methods [3–5, 9–11, 14–22]. Here, a novel global optimization algorithm based on the smoothing function is presented.

Most of global optimization algorithms must use local searches. The main difficulty for these types global optimization algorithms is to pass lower minimizers from the current one. Our minimization algorithm is also based on finding the lower local minimizer than the current local minimizer. Indeed, one of the minimizer x_1^* of the smooth function F is found by employing any local minimizer finder, then the function $m(x) := \min\{F(x), F(x_1^*)\}$ is considered instead of the objective function F and the minimizer x_2^* of the function $m(x)$ is searched. This loop proceeds until the global minimizer is found. It has to be stated that, the function $\min\{F(x), F(x_1^*)\}$ has the same minimizers with the function F except x_1^*. Therefore, one of the main difficulties of global optimization process is eliminated.

On the other hand, the function m can be non-smooth because of the presence of "min" operator. Therefore, we propose using the smoothing functions for m in order to make the local minimizer finding algorithms for smooth optimization available. After smoothing process, the next local minimizer is found by considering any of local minimizer finder for smooth optimization.

The structure of the paper is as follows. After this introduction, a smoothing function for the function $m(x, x_1^*) = \min\{F(x), F(x_1^*)\}$ is designed in section 2, and the properties of the smoothing function are analysed and discussed. Moreover, in section 3, a global optimization algorithm based on the smoothing function is proposed. Finally, the numerical experiment results are introduced in section 4. Compared with the numerical experiment results of [21], the results indicate that the global optimization algorithm in the paper is effective. Section 5 is the conclusion.

2 The smoothing function

Suppose that x_1^* is a current local minimizer of problem (1), then the smoothing function $m(x, x_1^*) := \min\{F(x), F(x_1^*)\}$ is considered in stead of the objective function F and the minimizer x_2^* of the function m is searched. It can be stated that, the function m has the same minimizers with the function F except x_1^*. However, the function m can be non-smooth because of the presence of "min" operator. Therefore, we propose using the smooth-

*e-mail: 646864258@qq.com
**Corresponding author, e-mail: yangly0601@126.com

ing functions for m in order to make the local minimizer algorithms for smooth optimization available.

For the function $m(x, x_1^*) := \min\{F(x), F(x_1^*)\}$, we have the following:

$$m(x, x_1^*) := F(x_1^*) + \min\{F(x) - F(x_1^*), 0\}.$$

Thus, The following smoothing function is designed:

$$S(x, x_1^*) := F(x_1^*) + q(F(x) - F(x_1^*), u),$$

where $u > 0$ and

$$q(t, u) = \begin{cases} t, & t \le -u, \\ -\frac{(t-u)^2}{4u}, & -u < t < u, \\ 0, & t \ge u. \end{cases}$$

Proposition 2.1. For any $(t, u) \in R^2$, we have the following results.

(1) $q(\cdot, \cdot)$ is continuously differentiable at any $(t, u) \in R^2$ with $u > 0$;

(2) $q(t, 0) = \min\{t, 0\}$;

(3) $1 \ge \frac{\partial}{\partial t} q(t, u) \ge 0$ at any $(t, u) \in R^2$ with $u > 0$.

Theorem 2.1. For every $x \in R^n$,

$$-\frac{u}{4} \le S(x, x_1^*) - m(x, x_1^*) \le 0$$

Proof. For every $x \in R^n$,

$$S(x, x_1^*) - m(x, x_1^*) = q(F(x) - F(x_1^*), u) - \min\{F(x) - F(x_1^*), 0\},$$

(1) When $|F(x) - F(x^*)| \ge u$,

$$S(x, x_1^*) - m(x, x_1^*) = 0.$$

(2) When $0 \le F(x) - F(x^*) \le u$,

$$S(x, x_1^*) - m(x, x_1^*) = -\frac{(F(x) - F(x_1^*) - u)^2}{4u}.$$

Therefore,

$$-\frac{u}{4} \le S(x, x_1^*) - m(x, x_1^*) \le 0.$$

(3) When $-u \le F(x) - F(x^*) \le 0$,

$$S(x, x_1^*) - m(x, x_1^*) = -\frac{(F(x) - F(x_1^*) - u)^2}{4u} - (F(x) - F(x_1^*)).$$

Therefore,

$$-\frac{u}{4} \le S(x, x_1^*) - m(x, x_1^*) \le 0.$$

From the theorem above, we can see that S is the smoothing function of m when u is sufficiently small.

3 Global optimization algorithm

In this part, a novel global optimization algorithm for solving problem (1) is proposed based on the smoothing function $S(x, x_1^*)$, and the algorithm can find a global optimal solution or an approximate global optimal solution of problem (1).

The novel global optimization algorithm for problem (1) is described as follows. we refer the algorithm as SMGP(the global optimization algorithm based on the smoothing function for problem (1)).

Algorithm SMGP

Step 0: Choose sufficiently small numbers $u_L > 0$, $\lambda_L > 0$. Choose a positive integer number I and directions e_i, $i = 1, \dots, I$. Choose a starting point $x_0 \in D$. Set $i := 0$;

Step 1: Seek for a local minimizer x_i^* of problem (1) by using local search methods starting from x_i;

Step 2: Let

$$S(x, x_k^*) := F(x_k^*) + q(F(x) - F(x_k^*), u),$$

where $u > 0$ and

$$q(t, u) = \begin{cases} t, & t \le -u, \\ -\frac{(t-u)^2}{4u}, & -u < t < u, \\ 0, & t \ge u. \end{cases}$$

Set $l = 1$ and $\lambda = 2$;

Step 3: (a)If $l \le I$, then go to 3(b); Otherwise, go to Step 5;

(b)If $\lambda \ge \lambda_L$, then let $y_i^l := x_i^* + \lambda e_l$, go to 3(c); otherwise, let $l := l + 1$, $\lambda = 2$, go to 3(a);

(c)If $y_i^l \in D$, then go to 3(d); otherwise, let $\lambda := \frac{\lambda}{2}$, go to 3(b);

(d)If $F(y_i^l) < F(x_i^*)$, then let $x_{i+1} := y_i^l$, $i := i + 1$, go to Step 1; otherwise, go to Step 4;

Step 4: Seek for a local minimizer for the following optimization problem from the starting point y_i^l:

$$\min_{x \in D} S(x, x_i^*), \tag{2}$$

Let y_i^* be a found local minimizer of problem (2). If y_i^* satisfies $F(y_i^*) < F(x_i^*)$, then let $x_{i+1} := y_i^*$, $i := i + 1$, go to Step 1; otherwise, let $l = l + 1$, $\lambda = 2$, go to Step 3(a);

Step 5: Let $x_s = x_i^*$ and stop.

Next, we give the explanation of the motivation and mechanism of the algorithm.

In Step 0, we choose the directions e_i, $i = 1, \dots, I$ as follows. For example, as $n \ge 2$, set $I = 2n$, for $i = 1, \dots, n$, and the i-th component of e_i is one, the other component of e_i is zero; for $i = n + 1, \dots, I$, the i-th component of e_i is -1, the other component of e_i is zero.

In Step 0, we choose a starting point. Then search for a local minimizer x_i^* of problem (1) starting from the initial point by using a local minimization method. If x_i^* is not a global minimizer, then to find deeper local minimizers of problem (1) is our main task.

In step 3, we try to choose an appropriate initial point to find a local minimizer of problem (2) by using a local minimization method. If the set of directions $\{e_i, i = 1, \ldots, l\}$ is large enough, no matter how to choose the initial points, local minimizers y_i^* of problem (2) will equal to x_i^*, and $S(y_i^*, x_i^*, u) = f(x_i^*)$, then we can say that the local minimizer x_i^* in Step 1 is an approximate global optimal solution of problem (1). When u is sufficiently small, the global optimal solution of problem (1) can be found.

4 Numerical experiments

In this part, several sets of numerical experiments are implemented and the results can show that Algorithm SMGP is effective. All the numerical experiments are implemented in Matlab2010b. We use The Matlab function 'fmincon' in the algorithm to search for local minimizers of problem (1) and problem (2).

The experiment results of each example are summarized in tables. The following symbols used in the tables are given:

i The iteration number in finding the i-th local minimizer;

x_i^0 The initial point in the i-th iteration for searching for the i-th local minimizer of problem (1);

x_i^* The i-th local minimizer for problem (1);

$F(x_i^*)$ The objective function value for x_i^*;

y_i^0 The initial point in the i-th iteration for searching the i-th local minimizer of problem (2);

y_i^* The i-th local minimizer for problem (2);

$F(y_i^*)$ The objective function value for y_i^*;

$CPU(s)$ The CPU time in seconds for the algorithms.

and throughout our computational experiments, we set the parameters in Algorithm SMGP as follows:

$$u_L = 10^{-6},\ \lambda_L = \frac{1}{2^3}.$$

Example 1.

$$\begin{array}{ll} \min & F(x) = \frac{x^6}{6} - \frac{5x^4}{4} + 2x^2 \\ \text{s.t.} & -2.5 \le x \le 2.5, \end{array}$$

This example has three local minimizer $x_1^* = -2$, $x_2^* = 0$, $x_3^* = 2$, where x_1^*, x_3^* is the global minimizer with $f(2) = f(-2) = -1.3333$. The computational results are given in Table I.

Example 2. (The Goldstein and Price function in [3, 22]).

$$\begin{array}{ll} \min & F(x) = g(x)h(x) \\ \text{s.t.} & -3 \le x_1, x_2 \le 3, \end{array}$$

where

$$\begin{aligned} g(x) &= 1+ \\ &(x_1 + x_2 + 1)^2(19 - 14x_1 + 3x_1^2 - 14x_2 + 6x_1x_2 + 3x_2^2), \\ h(x) &= 30+ \\ &(2x_1 - 3x_2)^2(18 - 32x_1 + 12x_1^2 + 48x_2 - 36x_1x_2 + 27x_2^2). \end{aligned}$$

The Example has the global optimal solution $x^* = (0, -1)^T$ and $f(x^*) = 3$. The computational results are summarized in Table II.

Example 3. (Six-hump back camel function in [3, 22]).

$$\begin{array}{ll} \min & f(x) = 4x_1^2 - 2.1x_1^4 + x_1^6/3 - x_1x_2 - 4x_2^2 + 4x_2^4, \\ \text{s.t.} & -3 \le x_1, x_2 \le 3. \end{array}$$

This problem has the global optimal solutions: $x^* = (0.0898, 0.7127)^T$ or $(-0.0898, -0.7127)^T$, where $f(x^*) = -1.0316$. The computational results are summarized in Table III.

5 Conclusion

In this paper, A new global optimization algorithm based on the smoothing function is proposed. Firstly, one of the minimizer x_1^* of the smooth function F is found by employing any local minimizer finder, then the function $m(x, x_1^*) := \min\{F(x), F(x_1^*)\}$ is considered instead of the objective function F and the minimizer x_2^* of the function m is searched. Because of the non-smooth of the function m, the smoothing function for m is used in order to make the local minimizer finding algorithms for smooth optimization available. Finally, a global optimization algorithm based on the smoothing function is proposed. The numerical experiments are carried out on several test problems. Compared with experiment results of [21], the results indicate Algorithm SMGP and Algorithm FFGP in [21] have the similar results, However the Cpu time of Algorithm SMGP is less than Algorithm FFGP in most case. Consequently Algorithm FFGP is effective. Nevertheless, the set of directions $\{e_i, i = 1, \ldots, l\}$ in Algorithm SMGP is crucial. so without appropriate directions, The local minimizers of problem (2) can not be found. Next, we will pay attention to how to modify it.

Acknowledgment

The research was financially supported by the National Natural Science Foundation of China (NSFC)(11401450 and 11401126).

References

[1] S. H. Chew and Q. Zheng, Integral Global Optimization, Volume 298 of Lecture Notes in Economics and Mathematical Systems, Springer–Verlag, 1988.

[2] L. C. W. Dixon, J. Gomulka and S. E. Herson, "Reflection on global optimization problems", in *Optimization in Action* (Dixon, L.C.W. etc.), Academic Press, New York, pp. 398-435, 1976.

[3] R. P. Ge, "A filled function method for finding a global minimizer of a function of several variables," Mathematical Programming, vol.46, pp. 191-204, 1990.
[4] R. P. Ge and Y. F. Qin, "A class of filled functions for finding a global minimizer of a function of several variables," Journal of Optimization Theory and Applications, vol.54, pp. 241-252, 1987.
[5] R. P. Ge and Y. F. Qin, "The globally convexized filled functions for global optimization," Applied Mathematics and Computation, vol.35, pp. 131-158, 1990.
[6] R. Horst, P. M. Pardalos and N. V. Thoai, Introduction to Global Optimization, 2nd edition, Kluwer Academic Publishers, Dordrecht, 2001.
[7] R. Horst, N. V. Thoai and H. Tuy, "Outer approximation by polyhedral convex sets," Operations Research Spektrum, vol.9, pp. 153-159, 1987.
[8] A. Levy and A. Montalvo, "The tunneling algorithm for the global minimization of functions," SIAM Journal on Scientific and Statistical Computing, vol.6, pp. 15-29, 1986.
[9] X. Liu, "Finding global minima with a computable filled function," Journal of Global Optimization, vol.19, pp. 151-161, 2001.
[10] H. W. Lin, Y. P. Wang, L. Fan and Y. L. Gao, "A new discrete filled function method for finding global minimizer of the integer programming," Applied Mathematics and Computation, vol.219, pp. 4371-4378, 2013.
[11] H. W. Lin, Y. L. Gao and Y. P. Wang, "A continuously differentiable filled function method for global optimization," Numerical Algorithms, vol.66, pp. 511-523, 2014.
[12] R. E. Moore, Enterbal Analysis, Prentice-Hall, NJ: Englewood Cliffs, 1966.
[13] P. M. Pardalos, H. E. Romeijn and H. Tuy, "Recent development and trends in global optimization," Journal of Computational and Applied Mathematics, vol.124, pp. 209-228, 2000.
[14] Z. Wan, L. Y. Yuan and J. W. Chen, "A filled function method for nonlinear systems of equalities and inequalities," Computational&Applied Mathematics, vol.31, pp. 391-405, 2012.
[15] W. X. Wang, Y. L. Shang, L. S. Zhang and Y. Zhang, "Global minimization of non-smooth unconstrained problems with filled function," Optimization Letters, vol.7, pp. 435-446, 2013.
[16] F. Wei and Y. P. Wang, "A new filled function method with one parameter for global optimization," Mathematical Problems in Engineering, vol.2013, pp. 1-12, 2013.
[17] F. Wei, Y. P. Wang and H. W. Lin, (2014), "A new filled function method with two parameters for global optimization," Journal of Optimization Theory and Applications, vol.163, pp. 510-527, 2014.
[18] Y. J. Yang and Y. L. Shang, "A new filled function method for unconstrained global optimization," Applied Mathematicas Computation, vol.173, pp. 501-512, 2006.
[19] Y. J. Yang, Z. Y. Wu and F. S. Bai, "A filled function method for constrained nonlinear integer programming," Journal of Industrial and Management Optimization, vol.4, pp. 353-362, 2008.
[20] L. Y. Yuan, Z. Wan, J. J. Zhang and B. Sun, "A filled function method for solving nonlinear complementarity problems," Journal of Industrial and Management Optimization, vol.5, pp. 911-928, 2009.
[21] L. Y. Yuan, Z. Wan and Q. H. Tang, "A criterion for an approximation global optimal solution based on the filled functions," Journal of Industrial and Management Optimization, vol.12, pp. 375-387, 2016.
[22] L. S. Zhang, C. NG, D. Li and W. Tian, "A new filled function method for global optimization," Journal of Global Optimization, vol.28, pp. 17-43, 2004.
[23] Q. Zheng and D. Zhuang, "Integral global minimization: Algorithms, implementations and numerical tests," Journal of Global Optimization, vol.7, pp. 421-454, 1995.

Table 1. Experiment results for Example 1

i	x_i^0	x_i^*	$F(x_i^*)$	CPU(s)
0	−0.5	0	0	
1	2	2	-1.3333	SMGP: 0.0156
0	−0.5	0	0	
1	2	2	-1.3333	FFGP: 0.0468

Table 2. Experiment results for Example 2

i	x_i^0	x_i^*	$F(x_i^*)$	CPU(s)
0	$\begin{pmatrix} -1 \\ -1 \end{pmatrix}$	$\begin{pmatrix} 0 \\ -1 \end{pmatrix}$	3	SMGP: 0.0156
0	$\begin{pmatrix} -1 \\ -1 \end{pmatrix}$	$\begin{pmatrix} 0 \\ -1 \end{pmatrix}$	3	FFGP: 0.0312

Table 3. Experiment results for Example 3

i	x_i^0	x_i^*	$F(x_i^*)$	y_i^0	y_i^*	$F(y_i^*)$
0	$\begin{pmatrix} 2 \\ -1 \end{pmatrix}$	$\begin{pmatrix} -1.7 \\ -0.8 \end{pmatrix}$	-0.2155	$\begin{pmatrix} 0.3 \\ -0.8 \end{pmatrix}$	$\begin{pmatrix} -0.09 \\ -0.71 \end{pmatrix}$	-1.0316
0	$\begin{pmatrix} -2 \\ 1 \end{pmatrix}$	$\begin{pmatrix} 1.7 \\ 0.8 \end{pmatrix}$	-0.2155	$\begin{pmatrix} -0.3 \\ 0.8 \end{pmatrix}$	$\begin{pmatrix} 0.09 \\ 0.71 \end{pmatrix}$	-1.0316
0	$\begin{pmatrix} 2 \\ -1 \end{pmatrix}$	$\begin{pmatrix} -1.7 \\ -0.8 \end{pmatrix}$	-0.2155	$\begin{pmatrix} 0.3 \\ -0.8 \end{pmatrix}$	$\begin{pmatrix} 0.3 \\ -0.8 \end{pmatrix}$	-0.3571
1	$\begin{pmatrix} 0.3 \\ -0.8 \end{pmatrix}$	$\begin{pmatrix} -0.09 \\ -0.71 \end{pmatrix}$	-1.0316	-	-	-

A Vegetation Selection Model for Ecological Bank Protection Works based on Improved TOPSIS Method and Its Application

Wei Xu*, *Zengchuan* Dong, *Jie* Ren, *Jianting* Wang, *Yue* Zhou and *Kedi* Ma

College of Hydrology and Water Resources, Hohai University, Nanjing, China

Abstract. To solve the complexity and difficulty of vegetation species selection for ecological bank protection works in a certain river effectively, an improved technique for order preference by similarity to ideal solution (TOPSIS) that using the multiple attribute decision making is proposed in river control planning. Besides precise number, interval number and triangle fuzzy number and their processing methods are introduced in TOPSIS. Through the improved TOPSIS, the order preference of the vegetation species could be recommended. The case study in Nenjiang Midstream shows that the proposed model is capable of selecting vegetation species for ecological bank protection works and could provide useful information for the regional integrated planning and management of river control.

1 Introduction

Most of the bank protection works along the river are often made of mortar or dry block stone and concrete. And concrete revetment has now been verified to lead to the degradation of ecology and environment, causing the continuity of hydrology destroyed and biodiversity loss. In recent years, ecological bank protection has become a sort of popular bank protection technique [1]. And a lot of studies have been carried out for analyzing the mechanism of plant bank protection technology [2]. First, when the flood passes through the vegetation fields, the flow velocity is greatly slowed down and the erosion of the soil surface by the flood flow is reduced due to the drag force of the plants. The effect of the waves on the bank are also reduced by the branches of the vegetation. Second, the developed root system can increase the mechanical strength of erosion resistance of soil. Then the collapse and erosion of the river bank is reduced. Third, the growth of roots, stems and leaves can improve the soil structure, increase the soil organic content and enhance the soil moisture content, thereby enhancing soil erosion resistance. Fourth, plants can also purify water quality. They can absorb pollutants such as organic matter, nitrogen and phosphorus in water.

Currently the ecological bank protection is more and more favored by stakeholders decision makers for its advantage in wave and energy dissipation, increasing the stability of the embankment, less investment and its fine harmonization with surroundings. Vetiveria zizanioides was planted on the bench land at the water side of the Jinjiang River embankment in Changsha City for its good bank protection effect and low cost [3]. Reed was adopted and planted in regulation project of Beijing-Hangzhou Canal Huaian-Huaiyin Ship Lock Section. The results showed that the reed planting had the best effect of wave absorption [4]. Calamus and narrow leaf bulrush were used to clean the Qinhuai River water, and both of the plants can clean the polluted water in deep autumn [5]. However, due to the characteristics of regional climatic and soil of the special planting area, growth adaption of the different vegetation species and their protection effect, how to select the optimal vegetation species for a given channel is a complex semi-quantificational systemic decision problem.

Nowadays building mathematical models has become an important approach to solve the optimum selection problems. And technique for order preference by similarity to ideal solution (TOPSIS) is a multi-objective decision making approach more commonly applied in the fields of hospital or university ranking [6], groundwater resources carrying capacity evaluation [7] and wetland ecosystem health evaluation [8]. And the greatest advantage is that this approach can make full use of data. The degree of suitability of the vegetation species for ecological bank protection works and its selection is a multi-objective concept, so the use of TOPSIS is feasible in practice.

In this paper, we study vegetation species suitability for ecological bank protection works, and construct an improved technique for order preference by similarity to ideal solution (TOPSIS) based on multiple attributes that include the conditions of precise number, interval number and triangle fuzzy number. The structure is organized as follows: Section 2 presents the classical TOPSIS and describes its deficiencies. Section 3 builds the vegetation species suitability index system, and describes the improved multiple attribute decision making. The interval number, triangle fuzzy number and their processing methods are introduced. Section 4

* Corresponding author: hhuxuwei@163.com

DOI: 10.1051/matecconf/201713900224

performs case study of vegetation species selection in Nenjiang Midstream. Finally, Section 5 concludes the paper.

2 Classical TOPSIS

TOPSIS is a very practical technique for dealing with multi-objective decision making problems. In TOPSIS, the concept that is relatively close to measurement is adopted. And steps of the classical TOPSIS are as follows [8]:

(1) Suppose there are m schemes to be selected, and each scheme has n assessment objectives, then the data matrix of the problem is:

$$X'=\begin{pmatrix} x_{11}' & x_{12}' & \cdots & x_{1n}' \\ x_{21}' & x_{22}' & \cdots & x_{2n}' \\ \vdots & \vdots & \vdots & \vdots \\ x_{m1}' & x_{m2}' & \cdots & x_{mn}' \end{pmatrix} \tag{1}$$

(2) Conduct normalization processing to the above matrix:

$$r_{ij} = \frac{x_{ij}'}{\sqrt{\sum_{i=1}^{m} x_{ij}'^2}} \tag{2}$$

Then the normalized matrix is obtained:

$$R=\begin{pmatrix} r_{11} & r_{12} & \cdots & r_{1n} \\ r_{21} & r_{22} & \cdots & r_{2n} \\ \vdots & \vdots & \vdots & \vdots \\ r_{m1} & r_{m2} & \cdots & r_{mn} \end{pmatrix} \tag{3}$$

(3) Determine the weights of all objectives, then the weight vector can be got as follows:

$$W = (w_1, w_2, \cdots, w_n) \tag{4}$$

(4) Multiply weight by normalized matrix to get the weighted decision matrix.

$$z_{ij} = w_j \bullet r_{ij} \tag{5}$$

(5) Seek for ideal solution A^* and negative ideal solution A^- of the problem. For the goals, the greater the better:

$$\begin{aligned} A^* &= \max_i z_{ij} = \left[a_1^*, a_2^*, \cdots, a_n^*\right] \\ A^- &= \min_i z_{ij} = \left[a_1^-, a_2^-, \cdots, a_n^-\right] \end{aligned} \tag{6}$$

For goals the smaller the better:

$$\begin{aligned} A^* &= \min_i z_{ij} = \left[a_1^*, a_2^*, \cdots, a_n^*\right] \\ A^- &= \max_i z_{ij} = \left[a_1^-, a_2^-, \cdots, a_n^-\right] \end{aligned} \tag{7}$$

(6) Calculate the Euclidean distance from all schemes to the ideal solution and Euclidean distance to the negative ideal solution:

$$d_i^* = \sqrt{\sum_{j=1}^{n} \left(z_{ij} - a_j^*\right)^2} \tag{8}$$

$$d_i^- = \sqrt{\sum_{j=1}^{n} \left(z_{ij} - a_j^-\right)^2} \tag{9}$$

Where a_j^* and a_j^- are the elements of ideal solution A^* and negative ideal solution A^- respectively.

(7) Calculate the approach degree C_i between schemes and ideal solutions:

$$C_i^* = \frac{d_i^-}{d_i^- + d_i^*} \tag{10}$$

The approach degree is 1 if problem is the same with the ideal solution, and the approach degree is 0 if the problem is the same with the negative ideal solution. Consequently, the closer the approach degree is to 1, the better the scheme will be. Rank C_i^* from the largest to the smallest. Then the scheme with the largest approach degree is the best choice.

From the computational steps of conventional TOPSIS above, we can find that the method mainly focuses on the element that is a precise number. But in many practical problems, interval number and triangle fuzzy number are also frequently involved. Therefore, there is a deficiency in the traditional TOPSIS. To expand the applications of the method, the interval number, triangle fuzzy number and their processing methods are introduced.

3 Vegetation selection model for ecological bank protection works based on improved TOPSIS

To realize vegetation species selection for ecological bank protection works effectively, a vegetation selection model based on the vegetation species suitability index system and the improved TOPSIS was built.

3.1 Index system and its weight

Much is involved in a vegetation species suitability index system for ecological bank protection. As a special vegetation planting at the water side of the embankment, developed root system, good flexibility of the branches, vegetation height are important factors that determine the bank protection effect when floods happen. Considering growing environment of plants, strong resistance to diseases and insect pests, regional climatic and soil characteristics are also taken into account. In addition, the vegetation must have some certain economic values and landscape values for the country wants to raise revenue especially for the people in those poor rural areas and construction of beautiful village is now a national strategy. Thus the vegetation species suitability index system was built and given in Table 1. From Table 1, the index system was classified into 3 groups: bank protection subsystem, auxiliary function subsystem and growth habits subsystem. Among them, some indexes may not be expressed as a precise number appropriately. For example, vegetation height can be expressed as a interval number and landscape value can be described qualitatively like excellent, good, medium, poor or very poor. And these five-grade method can be converted into a triangle fuzzy number. In this index system, the weights of each index were also given in Table 1. And indexes weights were determined by expert evaluation method [9].

Table 1. Vegetation species suitability index system for ecological bank protection works in Nenjiang midstream.

Subsystems	Items	Weights
Bank protection	Developed root systems	0.19
	Flexibility of the branches	0.18
	Vegetation height	0.11
Auxiliary function	Economic value	0.07
	Landscape value	0.07
Growth habits	Kinds of diseases and insect pests	0.13
	Cold endurance	0.13
	Soil adaptability	0.12

3.2 Improved TOPSIS

Among the indexes above, the index kinds of diseases and insect pests belongs to the cost type, and the other 7 indexes belong to the benefit type. Before normalization processing, invert the cost type index into the benefit type:

$$x_{ij} = 1 / x_{ij}' \tag{11}$$

For different attributes, there are corresponding normalization processes .

(a) For the index that is a precise number (x_{ij}), its normalization processes is the same as formula (2).

(b) For the index that is a interval number ($\left[x_{ij}^L, x_{ij}^U\right]$), its normalization processes is as follows:

$$r_{ij}^L = \frac{x_{ij}^L}{\sqrt{\sum_{i=1}^{m} x_{ij}^{U2}}} \qquad r_{ij}^U = \frac{x_{ij}^U}{\sqrt{\sum_{i=1}^{m} x_{ij}^{L2}}} \tag{12}$$

(c) For the indexes that is a triangle fuzzy number ($\left[x_{ij}^L, x_{ij}^M, x_{ij}^U\right]$), its normalization processes is as follows [10]:

$$r_{ij}^L = \frac{x_{ij}^L}{\sqrt{\sum_{i=1}^{m} x_{ij}^{U2}}} \qquad r_{ij}^M = \frac{x_{ij}^M}{\sqrt{\sum_{i=1}^{m} x_{ij}^{M2}}} \qquad r_{ij}^U = \frac{x_{ij}^U}{\sqrt{\sum_{i=1}^{m} x_{ij}^{L2}}} \tag{13}$$

For different attributes, there are also corresponding methods of determining ideal solution A^* and negative ideal solution A^- of the problem.

(a) For the index that is a precise number, see formula (6).

(b) For the index that is a interval number:

$$A^* = \left[\max z_{ij}^L, \max z_{ij}^U\right] \qquad A^- = \left[\min z_{ij}^L, \min z_{ij}^U\right] \tag{14}$$

(c) For the index that is a triangle fuzzy number:

$$A^* = \left[\max z_{ij}^L, \max z_{ij}^M, \max z_{ij}^U\right] \qquad A^- = \left[\min z_{ij}^L, \min z_{ij}^M, \min z_{ij}^U\right] \tag{15}$$

Calculate the Euclidean distance from all schemes to the ideal solution and Euclidean distance to the negative ideal solution:

$$d_i^* = \sqrt{\sum_{j=1}^{n} d_{ij}^{*2}} \tag{16}$$

$$d_i^- = \sqrt{\sum_{j=1}^{n} d_{ij}^{-2}} \tag{17}$$

Where

(a) For the index that is a precise number:

$$d_{ij}^* = z_{ij} - a_j^* \qquad d_{ij}^- = z_{ij} - a_j^- \tag{18}$$

(b) For the index that is a interval number:

$$d_{ij}^* = \sqrt{\left(z_{ij} - a_j^{L*}\right)^2 + \left(z_{ij} - a_j^{U*}\right)^2} \qquad d_{ij}^- = \sqrt{\left(z_{ij} - a_j^{L-}\right)^2 + \left(z_{ij} - a_j^{U-}\right)^2} \tag{19}$$

(c) For the index that is a triangle fuzzy number:

$$d_{ij}^* = \sqrt{\left(z_{ij} - a_j^{L*}\right)^2 + \left(z_{ij} - a_j^{M*}\right)^2 + \left(z_{ij} - a_j^{U*}\right)^2} \qquad d_{ij}^- = \sqrt{\left(z_{ij} - a_j^{L-}\right)^2 + \left(z_{ij} - a_j^{M-}\right)^2 + \left(z_{ij} - a_j^{U-}\right)^2} \tag{20}$$

Then the approach degree C_i can be calculated by formula (10).

4 Its application

Nenjiang River is the largest tributary of Songhuajiang River in North-eastern China. It originates from Greater Khingan Range, wanders through the southern semiarid region, passes through the western plain, and finally flows into Songhuajiang River. The total length of this river is about 1370 km and its drainage area is 2.97×10^8 km^2. The basin is characterized by a continental monsoon climate. The mean annual precipitation is about 500 mm and mainly concentrates between June and September, which accounts for 82% of total precipitation. The mean annual temperature is 3.2℃, and the extreme minimum temperature in history records is −39.5℃. The special climatic conditions determine that Nenjiang is a flood prone river [11]. At present, due to the bad foundation of embankment itself along Nenjiang Midstream, the waves do great damages to embankment in flood season. To reduce the wave disaster to embankment, it seems particularly important to build ecological bank protection works. Considering the cold weather and soil fertility condition, the index values of 8 vegetation species for ecological bank protection works that include Reed (R), Calamus (C), Thalia dealbata (TD), Giant reed (GR), Salix integra (SI), Typha (T) and Scirpus tabernaemontani (ST) were collected and given in Table 2.

Table 2. Index value of different vegetation species

Items	R	C	TD	GR	SI	T	ST
Developed root systems	Excellent	Poor	Very poor	Excellent	Excellent	Good	Excellent
Flexibility of the branches	Good	Medium	Good	Good	Excellent	Good	Medium
Vegetation height /m	[1,3]	[0.9,1.5]	[1.2,5]	[2,4]	[1,3]	[1.3,2]	[1,2]
Economic value	5	2	1	4	2	5	2
Landscape value	Good	Medium	Excellent	Very poor	Poor	Medium	Poor
Kinds of diseases and insect pests	2	8	1	1	3	2	2
Cold endurance	Excellent	Medium	Poor	Very poor	Excellent	Excellent	Excellent
Soil adaptability	Good	Medium	Good	Good	Poor	Good	Very poor

Obtain the data matrix of the problem according to the relationship between qualitative description and triangle fuzzy number.

$$\begin{bmatrix} [0.6,0.7,0.8] & [0.5,0.6,0.7] & [1.0,3.0] & 5 & [0.5,0.6,0.7] & 2 & [0.6,0.7,0.8] & [0.5,0.6,0.7] \\ [0.3,0.4,0.5] & [0.4,0.5,0.6] & [0.9,1.5] & 2 & [0.4,0.5,0.6] & 8 & [0.4,0.5,0.6] & [0.4,0.5,0.6] \\ [0.2,0.3,0.4] & [0.5,0.6,0.7] & [1.0,2.5] & 1 & [0.6,0.7,0.8] & 1 & [0.3,0.4,0.5] & [0.5,0.6,0.7] \\ [0.6,0.7,0.8] & [0.5,0.6,0.7] & [2.0,4.0] & 4 & [0.2,0.3,0.4] & 1 & [0.2,0.3,0.4] & [0.5,0.6,0.7] \\ [0.6,0.7,0.8] & [0.6,0.7,0.8] & [1.0,3.0] & 2 & [0.3,0.4,0.5] & 3 & [0.6,0.7,0.8] & [0.3,0.4,0.5] \\ [0.5,0.6,0.7] & [0.5,0.6,0.7] & [1.3,2.0] & 5 & [0.4,0.5,0.6] & 2 & [0.6,0.7,0.8] & [0.5,0.6,0.7] \\ [0.6,0.7,0.8] & [0.4,0.5,0.6] & [1.0,2.0] & 2 & [0.3,0.4,0.5] & 2 & [0.6,0.7,0.8] & [0.2,0.3,0.4] \end{bmatrix}$$

Then invert the cost type index kinds of diseases and insect pests into the benefit type by formula (10). Conduct normalization processing to the matrix and multiply weight by normalized matrix to get the weighted decision matrix.

$$\begin{bmatrix} [0.061,0.083,0.113] & [0.049,0.069,0.097] & [0.015,0.102] & 0.039 & [0.022,0.032,0.046] & 0.038 & [0.043,0.058,0.079] & [0.036,0.052,0.074] \\ [0.031,0.047,0.070] & [0.040,0.058,0.083] & [0.014,0.051] & 0.016 & [0.018,0.026,0.039] & 0.010 & [0.028,0.041,0.059] & [0.029,0.043,0.063] \\ [0.020,0.036,0.056] & [0.049,0.069,0.097] & [0.015,0.085] & 0.008 & [0.027,0.037,0.052] & 0.077 & [0.021,0.033,0.049] & [0.036,0.052,0.074] \\ [0.061,0.083,0.113] & [0.049,0.069,0.097] & [0.031,0.136] & 0.032 & [0.009,0.016,0.026] & 0.077 & [0.014,0.025,0.040] & [0.036,0.052,0.074] \\ [0.061,0.083,0.113] & [0.059,0.081,0.111] & [0.015,0.102] & 0.016 & [0.013,0.021,0.033] & 0.026 & [0.043,0.058,0.079] & [0.022,0.034,0.053] \\ [0.051,0.071,0.099] & [0.049,0.069,0.097] & [0.020,0.068] & 0.039 & [0.018,0.026,0.039] & 0.038 & [0.043,0.058,0.079] & [0.036,0.052,0.074] \\ [0.061,0.083,0.113] & [0.040,0.058,0.083] & [0.015,0.068] & 0.016 & [0.013,0.021,0.033] & 0.038 & [0.043,0.058,0.079] & [0.015,0.026,0.042] \end{bmatrix}$$

And ideal solution A^* and negative ideal solution A^- of the problem could be obtained.

$$A^* = ([0.061,0.083,0.113],[0.059,0.081,0.111],[0.031,0.136],0.039,[0.027,0.037,0.052],0.077,[0.043,0.058,0.079],[0.036,0.052,0.074])$$

$$A^- = ([0.020,0.036,0.056],[0.040,0.058,0.083],[0.014,0.051],0.008,[0.009,0.016,0.026],0.010,[0.014,0.025,0.040],[0.015,0.026,0.042])$$

Then, the approach degree of 7 vegetation species were calculated by formula (16) and (17) and were shown in Table 3. According to Table 3, we could find that the approach degree of Reed is the largest. It indicates that Reed is the most suitable vegetation species for ecological bank protection works in Nenjiang midstream. Besides, the approach degree of the species Giant reed, Salix integra and Typha are also larger relatively. But among them, Giant reed is not a native species in this region. According to the principle of giving priority to local species, the recommended order of vegetation species for ecological bank protection works in Nenjiang River midstream is the Reed, Salix integra and Typha.

Table 3. The approach degree of different species

Species	d^*	d^-	C^*
R	0.058	0.135	0.700
C	0.140	0.052	0.270
TD	0.115	0.099	0.463
GR	0.073	0.149	0.670
SI	0.080	0.125	0.610
T	0.086	0.112	0.565
ST	0.107	0.109	0.503

5 Conclusions

In this paper, we analyze the deficiencies of conventional TOPSIS and propose an improved method that includes the conditions of precise number, interval number and triangle fuzzy number. Then, we build a vegetation species selection model for ecological bank protection works using the multi-index system. Finally, we validate the effectiveness of the proposed model through the case study in Nenjiang midstream. The model could provide a useful tool for decision-makers in

DOI: 10.1051/matecconf/201713900224

aspects such as flood control works planning and ecological remediation.

Acknowledgements

This work was supported by the Applied Technology Research and Development Program of Heilongjiang Province under Grant No.GZ16B031 and No.GZ16B035 and the Water Resources Department of Shanxi Province under Grant No. 20165011212.

References

1. D.D. Liu, L. Zhu, J. Zhang, and J. Liang, Select the Optimal Ecological Bank-Protection Technique for Plain-Area Channel Using AHP. Journal of Convergence Information Technology, 2012, 7(1): 214-221.
2. H. Zhao, S.B. Shu, and J.L. Liu, Study on Application of Bank Protection Engineering Technology by Vegetation. Modern Agricultural Science and Technology, 2010, 17:248-250.
3. Q. Wang, Application of Vetiver Grass System in Regulation of Medium and Small Rivers. Hunan Hydro & Power, 2010, 6:73-75.
4. H.Q. Ge, Y.Q. Zhu, Application of Reed Ecological Bank Revetment in Regulation Project of Beijing-Hangzhou Canal Huaian-Huaiyin Ship Lock Section. Journal of Nantong University (Natural Science), 2009, 8(2): 62-64.
5. D.W. Yan, "Research on the Adaptability of Submerged Macrophytes Used by Ecological Riverbank in Qinhuai River," Environmental Science and Management, 2006, 31(9):154-156.
6. L. Guo, W.P. Liang, and Y. Han, Comprehensive Evaluation on Medical Quality of a Hospital by TOPSIS Method. Soft Science of Health, 2014, 4:219-221.
7. J.X. Xu, H. Fan, and X.T. Hu, The Model Combined Entropy Weight with Improved TOPSIS and Its Application in Groundwater Resources Carrying Capacity Evaluation. China Rural Water and Hydropower, 2012, 2:30-33.
8. L. Ren, J.T. Liu, J.J. Ni, and X.Y. Xiang, Health Evaluation of a Lake Wetland Ecosystem Based on the TOPSIS Method. Polish Journal of Environmental Studies, 2014, 23(6):2183-2190,.
9. W. Xu, Z.C. Dong, Z.C. Hao, D.Y. Li, and L. Ren, River Health Evaluation Based on the Fuzzy Matter-Element Extension Assessment Model. Polish Journal of Environmental Studies, 2017, 26(3):1353-1361.
10. Z.S. Xu, Study on Method for Triangular Fuzzy Number-Based Multi-Attribute Decision Making With Preference Information on Alternatives. System Engineering and Electronics, 2002, 8:9-12.
11. H.Y. Li, M. Xie, and S. Jiang, Recognition Method for Mid- to Long-Term Runoff Forecasting Factors based on Global Sensitivity Analysis in the Nenjiang River Basin. Hydrological Processes, 2012, 26(18):2827-2837

Design of New Test Function Model Based on Multi-objective Optimization Method

SHANG Zhaoxia [1,*], *ZHANG Rui* [2], *WANG Jiankang* [1], and *ZHANG Duo* [1]

[1] *Shandong Product Quality Inspection Research Institute, Jinan250102, Shandong, China.*

[2] *Shandong Institute of medicine and health information, Jinan250062, Shandong, China.*

Abstract. Space partitioning method, as a new algorism, has been applied to planning and decision-making of investment portfolio more and more often. But currently there are so few testing function for this algorism, which has greatly restrained its further development and application. An innovative test function model is designed in this paper and is used to test the algorism. It is proved that for evaluation of space partitioning method in certain applications, this test function has fairly obvious advantage.

Keywords: Multiple-objective optimization, test function, investment portfolio, model design, space partitioning method

1. Introduction

Test function is an important tool for algorism evaluation. Generally it evaluates an algorism's performance from 3 aspects – time complexity, distribution and convergence. Whitley[1] proposed a guideline for test function design in 1996. Deb[2] divided the design method of test function into three types. In addition, the test function proposed by literature[3] in 1999 is widely accepted. A test function of alterable decision variables dimension was proposed in the literature[4]. An index for distribution assessment proposed by Schott, et al[5] is the most popular parameter now. In 2010, Ishibuchi et al[6] proposed to evaluate the fitness of the individual using different Scalarizing Functions in MOEA/D at the same time. In recent years, new test functions are proposed one after another. Some of them are already well-known such as the test method based on non-dominated sorting proposed by Srinivas[7], the test method based on genetic algorithm proposed by Deb[8], and the scalable multi-objective test toolkit proposed by Huband[9].

* Corresponding author: gogogo255@163.com

DOI: 10.1051/matecconf/201713900225

The test function must be closely related to the problem solved by the algorism, in other words, individualized test function shall be adopted according to characteristics of the problem to make it worthy. A large portion of current functions are universal functions based on historical data and complex solution; there are quite few specific test functions facing certain type of problems. In view of the above limitation, a new test function for space partitioning application in investment portfolio as well as the performance measurement index is designed in this paper. Considering the requirement of portfolio on risk scattering, the design makes a point of test stability and distribution uniformity; considering the investor may want to control and intervene in the decision-making process, this test function is designed to be visualized and controllable.

Since the test function is designed for space partitioning method, distribution uniformity has to be the most important evaluation index. The solution set which distributes evenly can not only provide better options to the decision-maker, but also affect search capability of the algorism to a large degree[10]. There are multiple possible reasons for nonuniform distribution of population. For example, the target function is nonlinear or has multiple variables; parameters of different target functions are interrelated; nonlinear mapping exists between decision variables and parameters; the Pareto frontier is inconsecutive [11], etc. Therefore the test function proposed in this paper is designed not only to avoid that the population fall in the optimal zone of a single target function, but also to reflect restrictive relation between target function and solution through assessing the distribution uniformity. Most algorism just considers the impacts of the entirety on the individuals and does not consider the impacts between individuals. But in some practical cases， the affect is although small, but can't be ignored. Therefore, this article fully considers the influence between individuals, and also screens the running solution and performs maintenance on the external population by using test function.

The algorism proposed in this paper showed great superiority in the application field which have high request for dispensability. For example, in specific application of portfolio decision-making and planning, the investor values risk diversification of portfolio, avoid the combination plan is too single and close, and also avoid deviating from the actual. It is the major need for avoiding the interest rate risk. Correspondingly, the stability and distribution evenness were highly stressed during the designing of the test algorism.

2. Experimental

Firstly define the test function as:

$$\mathrm{TES}_i=\alpha^2+\beta\frac{K}{k} \quad (1)$$

i is the individual code; α is the number of individuals which fall in the grid where i stays; β is the number of individuals which fall in adjacent grids; K is the number of all possible grids in adjacent region; k is the number of grids in valid adjacent regions. Higher TES_i represents smaller fitness of individuals. Reason is that in a certain region, more densely the individuals are distributed, less contributions they make to the population. Accordingly these individuals have lower chance to be chosen. The evaluation of an algorism's performance shall also include deviation of its result to the actual result. Therefore the fuzzy expert evaluation system shall be set. First step is to integrate opinions of the experts and regulate them as Pro_i value, which is called as expert evaluation index. Then the index can be screened and tested as a decision factor after TES_i. The secondary fuzzy expression of the test function is:

$$\mathrm{TES}^*_i= \alpha^2+\beta\frac{K}{k}+\mathrm{Pro}_i \quad (2)$$

Then grant weight to two evaluation aspects; the secondary fuzzy expression changes to:

$$\mathrm{TES}^*_i=W_T(\alpha^2+\beta\frac{K}{k})+W_P\mathrm{Pro}_i \quad (3)$$

It shall satisfy:

$$W_T+W_P=1 \quad (4)$$

W_T is objective weight; W_P is subjective weight. The calculation procedure without expert evaluation index shows as following:

Step 1. Define scale of the population.

Step 2. Get TES_i value of the solutions sets.

Step 3. Re-rank --- according to TES_i.

Step 4. Compare the results of two ranking and evaluate the algorism.

Step 5. Move out or move in individuals from external population according to the result of step 4.

Step 6. Maintain the population according to Pro_i.

3. Simulation

The original test results generated by using method of literature [12] are shown in table 1. Then re-rank these results by TES_i value using the method proposed in this paper. The new results are shown in table 2.

Table 1 Original Test Results

No.	1	2	3	4
Individual	18,(4,2-1, 2.000000)	5,(4.2, 2.000000)	4,(5,2-1, 2.400000)	1,(5,2-2, 2.400000)
TES_i	7	7	7	7
No.	5	6	7	8
Individual	6,(3,1, 2.800000)	15,(6,2, 2.800000)	22,(3,4, 2.800000)	25,(8,1, 3.000000)
TES_i	6.3	6	6.3	6.3

No.	9	10	11	12
Individual	14,(4,4-1, 3.200000)	21,(4,4-2, 3.200000)	24,(7,2, 3.200000)	8,(5,4, 3.600000)
TES_i	7.2	7.2	7	5.8
No.	13	14	15	16
Individual	23,(8,2, 3.600000)	3,(7,3-1, 3.800000)	13,(7,3-2, 3.800000)	19,(6,4-1, 4.000000)
TES_i	10.8	11	11	7.4
No.	17	18	19	
Individual	11,(8,3, 4.200000)	20,(8,3, 4.300000)	17,(7,4, 4.400000)	
TES_i	12	12	9	

Table 2 New Test Results ranked by TES i value

No.	1	2	3	4
Individual	8,(5,4, 3.600000)	15,(6,2, 2.800000)	6,(3,1,2.8 00000)	22,(3,4, 2.800000)
TES_i	5.8	6	6.3	6.3
No.	5	6	7	8
Individual	25,(8,1, 3.000000)	18,(4,2-1, 2.000000)	5,(4,2, 2.000000)	4,(5,2-1, 2.400000)
TES_i	6.3	7	7	7
No.	9	10	11	12
Individual	1,(5,2-2, 2.400000)	24,(7,2, 3.200000)	14,(4,4-1, 3.200000)	21,(4,4-2, 3.200000)
TES_i	7	7	7.2	7.2
No.	13	14	15	16
Individual	19,(6,4-1, 4.000000)	17,(7,4, 4.400000)	23,(8,2, 3.600000)	3,(7,3-1, 3.800000)
TES_i	7.4	9	10.8	11
No.	17	18	19	
Individual	13,(7,3-2, 3.800000)	11,(8,3, 4.200000)	20,(8,3, 4.300000)	
TES_i	11	12	12	

Through comparison and analysis of the above two ranking results, the algorism can be evaluated. It shows that the algorism has comparatively low congestion and good distribution.

4. Conclusions

Through the above experimentation, it shows that the test function considered not only the whole's impact on the individuals, but also the individuals' impacts to each other. In the experimentation, the test function was used to screen data and maintain the external population; it showed the interactions among target functions and interactions among solutions. It not only tested the algorism, but also optimized distribution of the algorism.

Since the algorism tested is mainly applied in the investment portfolio area, higher stability is required. Therefore the algorism has adopted a selection mechanism integrating elite selection, random selection and test selection. Concerns over random issues, sensitivity analysis and uncertainty analysis of intervals are still to be studied in future practice; moreover how to apply the test function in more complex dynamic situation is also the target for future development.

References

1. Whitley D, Rana S, Dzubera J, et al. Evaluating evolutionary algorithms.Artificial Intelligence, 1996, 85(1-2): 245-276
2. Abraham A, Jain L, Goldberg R. Evolutionary Multi-objective Optimization: Theoretical Advances and Applications. London: Springer, 2005

3. Van Veldhuizen David A, Lamont G B. Multiobjective evolutionary algorithm test suites. In: Carroll J, et al. Editors, Proceedings of the 1999 ACM Symposium on Applied Computing, 1999: 351-357
4. Fonseca Carlos M, Peter J Fleming. Multiobjective genetic algorithms made easy: selection, s haring, and mating restriction. In: Proceedings of the First In ternational Conference on Genetic Algorithms in Engineering Systems: Innovations and Applications, 1995: 42-52
5. SCHOTT J R. Fault tolerant design using single and multicriteria genetic algorithm optimization. Cambridge: Massachusetts Institute of Technology, 1995
6. Ishibuchi, Hisao, et al. "Simultaneous Use of Different Scalarizing Functions in MOEA/D." Genetic and Evolutionary Computation Conference, GECCO 2010, Proceedings, Portland, Oregon, Usa, July 2010:519-526.
7. Srinivas N, Deb K. Multiobjective Optimization Using Nondominated Sorting in Genetic Algorithms. Evolutionary Computation, 1995, 2 (3): 221-248
8. Deb K, Prat a PA, Agarw al S, et al. A Fast and Elitist Multiobjective Genetic Algorithm: NSGA-II. IEEE Transactions on Evolutionary Computation, 2002, 6: 182-197
9. Huband S, Barone L, While L, et al. A Scalable Multi-objective Test ProblemToolkit. Proceedings of the Evolutionary Multi-criterion Optimization'05, Berlin, Germany: Springer, 2005: 280-295
10. Laumanns M, Thiele L, Deb k, et al. Combining convergence and diversity in evolutionary multiobjective optimization. Evolutionary Computation, 2002, 10(3): 263-282
11. Cheng P, Zhang Zili. Design and analysis of test problems of multi objective evolutionary algorithm. computer project, 2009, 35(14): 238-240(in Chinese)
12. Shang Z X, Liu H. A Multiple-Objective Algorism Design and Strategy Study Based on Target Space Partitioning. Advanced Materials Research, 2011, 1228: 219-220(in Chinese)

Analysis of the Relationship between Port Industry and Economic Growth Based on Cointegration Theory

Sun Jianping, *Li* Zhenfu, *Chen* Zhuo, and *Liu* Tongchao

College of shipping economics and management,Dalian Maritime University,Dalian,China

Abstract. The reciprocal covariance function between port Industry indicator and economic growth indicator is constructed according to time series, the cross spectral density between port industry indicator and economic growth indicator is obtained by Fourier transform after Yessentuki-Henning window truncation, and the correlation, coherence and leading lag relationship between port industry and economic growth were calculated by calculating the estimated values of mutual amplitude spectrum, coherence spectrum and phase spectrum. Then, the causality between port industry and economic growth is tested by cointegration theory, and the relationship between endogenous factors and exogenous variables is determined by impulse response and variance decomposition. The calculation results of the actual data show that the development of China's port industry and economic growth maintain a high degree of consistency in the length of 3 to 4 years and economic growth will lead the development of the port industry 0.8 years to 1.1 years. The results of dynamic econometric analysis show that China's port industry and economic growth are Granger reasons; China's port industry has made a negative response to economic growth. The economic growth has made a positive response to the impact of the development of the port industry, and the response has gradually increased with the passage of time; China's port industry and economic growth are affected by endogenous variables, and both of them affect each other in the course of development.

1 Introduction

In the process of economic development in a country or region, a very important characteristic is the export strategy of industrial products. Throughout the history of the western developed economies, from the beginning of the first industrial revolution to the present era of the information age, the import and export trade industry has played a significant role in promoting economic growth. As a carrier of import and export commodities, the port industry provides the possibility for the import and export trade in the world. Therefore, the relationship between port industry and economic growth is related through import and export trade, and the uncertainty of this relationship provides the direction for the research of this paper. At present, the research on the relationship between port industry and economic growth is lack of empirical evidence, or the empirical research is limited to simple data statistics, which does not fundamentally reveal the relationship between port industry and economic growth. Therefore, it is of great significance to clarify the relationship between port industry and economic growth.

Aiming at the uncertain relationship between port industry and economic growth, this paper uses spectrum analysis technology to decompose the time series of the two groups of indicators of port industry and economic growth into a series of frequencies representing the fixed-length cycle. The cross-spectrum is used to analyze and evaluate the correlation between the wavelengths of the two groups and the cross-spectrum analysis method between the port industry and the economic growth is established[2-4]to clarify the leading, lag and other related relations between the development of port industry and economic growth [3],at the same time, the intensity of the relationship between the port industry and economic growth is analyzed through the impulse response and variance decomposition, and the influence mechanism between port industry and economic growth is studied in depth.

2 Cross - spectrum Analysis of Port Industry and Economic Growth

2.1 Cross - Spectral Analysis Model of Port Industry and Economic Growth

For the different types of spectral analysis methods, the paper uses the cross-spectrum analysis method based on the Wiener-Khinchine theorem, which points out that the power spectral density of the wide stationary stochastic process is the Fourier transform of its autocorrelation function.

According to the theorem, it can be seen that the time series of cross spectral analysis based on Wiener-Khinchine theorem need to be stationary and stochastic, so it is necessary to smooth and randomize the time series.

For the time series y_t after the smooth processing, we give the observation value of n samples, then the mathematical expectation is

$$E(y_t) = \mu_y = \frac{1}{n}\sum_{t=1}^{n} y_t \quad (1)$$

The standard deviation is

$$s_y = \sqrt{\frac{1}{n-1}\sum_{i=1}^{n}(y_t - \mu_y)^2} \quad (2)$$

The autocorrelation coefficient of the sample is $C(\tau)$

$$C(\tau) = \frac{1}{n-k-1}\sum_{t=1}^{n-k}(y_t - \mu_y)(y_{t+k} - \mu_y) \quad (3)$$

The spectral density is obtained by Fourier transform of $C(\tau)$

$$Q_y(f) = \sum_{\tau=-M}^{M} C(\tau)e^{-j2\pi f\tau} \quad (4)$$

Then, the Euler formula is then taken into the spectral density formula, we can get

$$Q_y(f) = \sum_{\tau=-M}^{M} C(\tau)[\cos(2\pi f\tau) - j\sin(2\pi f\tau)] \quad (5)$$

By using the stable characteristic of covariance and the correlation property of trigonometric function, we can get

$$Q_y(f) = C(0) + 2\sum_{k=1}^{M} C(\tau)\cos(2\pi f\tau) \quad (6)$$

As the development of port industry and economic growth can be expressed by variables or parameters, by selecting the indicator variables that can represent the port industry and economic growth to form a binary sequence, spectrum analysis technology is used to decompose the time series of the two groups of indicators of port industry and economic growth into a series of frequencies representing the fixed-length cycle, and the correlation between the wavelengths of the two groups is analyzed by the cross- spectrum, and the cross-spectrum analysis method between the port industry and the economic growth is established.

(1)Correlation and Cross Spectrum of Binary Sequences

1) Mutual variance function and cross spectrum density function

First, the port industry indicator time series $\{X_t\}$ and Economic growth indicator time series$\{Y_t\}$ constitute a set of binary sequences, and excluding all the trend factors in the data, its autocorrelation coefficient$\hat{R}_{xy}(k)$, $\hat{R}_{yx}(k)$are formulae (7), (8):

$$\hat{R}_{xy}(k) = \frac{1}{N}\sum_{t=1}^{N-k}(x_t - \mu_x)\left(y_{t+k} - \mu_y\right), k = 0,1,2,\cdots,M \quad (7)$$

$$\hat{R}_{yx}(k) = \frac{1}{N}\sum_{t=1}^{N-k}(x_{t+k} - \mu_x)\left(y_t - \mu_y\right), k = 0,1,2,\cdots,M \quad (8)$$

$$\mu_x = \frac{1}{N}\sum_{t=1}^{N} x_t \quad (9)$$

$$\mu_y = \frac{1}{N}\sum_{t=1}^{N} y_t \quad (10)$$

Among them, μ_x , μ_y respectively represent the mathematical expectation of time series $\{X_t\}$ and $\{Y_t\}$, N is the length of time series, M is the maximum lag;

The correlation structure between the port industry indicator time series $\{X_t\}$ and the economic growth Indicator time series $\{Y_t\}$ is described by the reciprocal covariance function$R_{xy}(k)$, the formula is (11):

$$R_{xy}(k) = \mathrm{cov}(X_t, Y_{t+k}) = E(X_t - \mu_x)\left(Y_{t+k} - \mu_y\right) \quad (11)$$

Then, according to the correlation theory of spectral analysis, the spectral density function [4]$Q_{xy}(f)$ and the cross covariance function $R_{xy}(k)$ are subjected to the Fourier transformation of formula (12):

$$Q_{xy}(f) = \sum_{k=-\infty}^{\infty} R_{xy}(k)e^{-j2\pi fk}\ |f| \le \frac{1}{2} \quad (12)$$

$Q_{xy}(f)$is the cross-spectral density function, which represents the average of the product of the component with the frequency f both in the time series $\{X_t\}$ of the port industry indicator and the time series $\{Y_t\}$ of the economic growth indicator.

Thus,$Q_{xy}(f)$reflects the correlation of the phase of time series $\{X_t\}$ and $\{Y_t\}$ at f.Since $R_{xy}(k)$ is real and is obtainable, $Q_{xy}(f)$ is also present at$f\in\left[-\frac{1}{2},\frac{1}{2}\right]$.

2)Co-direction spectrum and orthogonal spectrum

For the binary stationary time series composed of the time series $\{X_t\}$of the port industry indicator and the time series $\{Y_t\}$ of the economic growth indicator, since $R_{xy}(k)$ is not an even function,so $Q_{xy}(f)$ is usually a complex function, then the true part and the imaginary part are respectively $\alpha_{xy}(f)$ and $\beta_{xy}(f)$, then $\alpha_{xy}(f)$ reacts to the relativity of the same direction frequency component of two time series, i.e., the co-direction spectra, $\beta_{xy}(f)$ reacts to the correlation of the heterogeneous frequency components of two time series, i.e. orthogonal spectra. The cross spectrum can be expressed as:

$$Q_{xy}(f) = \alpha_{xy}(f) - j\beta_{xy}(f) \quad (13)$$

When the time series of the port industry and economic growth are processed by Fourier transform, the relation between the whole time domain and the frequency domain is studied. But when using computer to process time series, it is impossible to measure and calculate infinite time series. So we usually select its limited time segment for analysis. In order to reduce the leakage of spectrum energy during the phase process, this paper uses Yessentuki-Henning window to truncate the time series of port industry and economic growth.

$$\lambda(k) = \begin{cases} \frac{1}{2}\left[1 - \cos\left(\frac{2\pi k}{M}\right)\right], & 0 \le k \le W \\ 0 \end{cases} \quad (14)$$

$\hat{R}_{xy}(k)$ and $\hat{R}_{yx}(k)$ are truncated by Henning window, and then the co-direction spectrum and orthogonal spectrum of $\{X_t\}$ and $\{Y_t\}$ are obtained by Fourier transform.

$$\alpha_{xy}(f) = \sum_{k=-\infty}^{\infty}\frac{1}{2}\left[\hat{R}_{xy}(k) + \hat{R}_{xy}(-k)\right]\cos(2\pi fk) = \frac{1}{2}\left[\hat{R}_{xy}(0) + \hat{R}_{xy}(0)\right] + \sum_{k=1}^{W}\lambda(k)\left[\hat{R}_{xy}(k) + \hat{R}_{yx}(k)\right]\cos(2\pi fk) \quad (15)$$

$$\beta_{xy}(f) = \sum_{k=-\infty}^{\infty}\frac{1}{2}\left[\hat{R}_{xy}(k) - \hat{R}_{xy}(-k)\right]\sin(2\pi fk) = \sum_{k=1}^{W}\lambda(k)\left[\hat{R}_{xy}(k) - \hat{R}_{yx}(k)\right]\sin(2\pi fk) \quad (16)$$

Where W represents the truncation point, the size of which affects the length of the selected time series.If W is larger, more sample data is included, will lead to the choice of K is too large, the curve peak is more, there are a lot of false peak, while the smaller W will lead to the choice of K is too small, the curve will be too smooth. So the selection range of the cut-off point is usually taken in$[2\sqrt{n}, 3\sqrt{n}]$, where n is the number of observation values.

(2)Mutual amplitude spectrum, coherence spectrum and phase spectrum.

For the representation of complex numbers, the cross-spectral density of the time series $\{X_t\}$ of the port industry indicator and the time series $\{Y_t\}$ of the economic growth indicator can also be expressed by the polar coordinate notation, in addition to the formula (13) for:

$$Q_{xy}(f) = \alpha_{xy}(f) - j\beta_{xy}(f) = A_{xy}(f)e^{j2\pi\phi_{xy}(f)} \quad (17)$$

Among them,

$$A_{xy}(f) = |Q_{xy}(f)| = \sqrt{\alpha_{xy}(f)^2 + \beta_{xy}(f)^2} \quad (18)$$

$$B_{xy}(f) = |Q_{xy}(f)|/\sqrt{Q_x Q_y} = \sqrt{\alpha_{xy}(f)^2 + \beta_{xy}(f)^2}/\sqrt{Q_x Q_y} \quad (19)$$

$$\phi_{xy}(f) = \tan^{-1}\left[-\frac{\beta_{xy}(f)}{\alpha_{xy}(f)}\right] \quad (20)$$

$A_{xy}(f)$ represents the modes of the co-direction spectrum and orthogonal spectrum and it is called the reciprocal amplitude spectra, which reflects the relation between the amplitudes of each frequency component of the time series of the port industry indicator $\{X_t\}$and the time series of the economic growth indicator$\{Y_t\}$.Due to the influence of the observed sequence values, the observed time series refers to the sequence of the different values of a certain indicator at different times, so the reciprocal spectral values of the two different time series are not comparable.$B_{xy}(f)$, which represents the coherence spectrum, is used to measure the consistency of the two sequences in frequency. Q_x and Q_y are the spectral density functions of the sequence X_t and Y_t respectively. The value of coherence spectral $B_{xy}(f)$ after coherence processing is in [0, 1], if the result is closer to 1, it indicates that the correlation of the two sequences is higher, the consistency performance is better, conversely, if the result is far away from 1, the correlation between the two sequences at this frequency is poor.

$\phi_{xy}(f)$ represents the phase spectrum of the cross-spectral density, which reflects the phase difference between the time series $\{X_t\}$ and $\{Y_t\}$ at each frequency, that is, the leading and lag relationship between the two. For the time series$\{X_t\}$ and the time series $\{Y_t\}$, it is limited to a certain interval, such as$[-\pi, \pi]$, by observing the phase spectrum value of the two sequences in this interval, we can determine the leading and lag relationship between them.If$\phi_{xy}(f)$ is greater than 0, then x is ahead of y and vice versa. By comparing the overall fluctuation of the two time series, the time difference between them can be well judged.

If the results of the observations of $\alpha_{xy}(f)$ and $\beta_{xy}(f)$are brought into the formula (18), (19), (20), we can obtain the mutual amplitude spectrum $A_{xy}(f)$,coherence spectrum $B_{xy}(f)$ and phase spectrum$\phi_{xy}(f)$, then the atlas can be processed and analyzed.

2.2 Implementation of Cross Spectral Analysis Method

(1)Index extraction and data processing

1) The selection of indicators

Port cargo throughput is an important quantitative indicator reflecting the results of port production and operation, its fluctuation is the direct manifestation of the cyclical fluctuation of economic development in the port industry, therefore, the selection of port cargo throughput represents the development indicator of China's port industry. GDP as the final result of production activities of all habitual units of a country or region in a certain period is the most important basis to measure the economic health of a region in a country, and its change can represent the development of economy in different time, so its time series can be used as the index to reflect the change of China's economic growth. The indicators are summarized as shown in table 1.

2) Basic methods of data processing

The use of the cross spectrum analysis method requires that the variable is a stationary time series, so the data collected will be tested for smoothness first, which is used in the ADF test [7] in this paper. If the data shown after the test is non-stationary, then the data collected will need to be processed to be stabilized to meet the requirements of the spectral analysis method. According to the characteristics of the data used and the characteristics of each kind of time series processing methods, this paper uses the Hodrick-prescott filter method (hereinafter called H-P filter method) [5-6]to separate the trend of time series, and extracts the factors which contain only fluctuation and periodic component to carry on the cross-spectrum analysis. The processed data is then tested again to verify whether the data meets the requirements or not.

Table 1 Indicator selection summary table

Indicator type	*Indicator name*	*Data unit*	*Data nature*	*Time quantum*
Indicator sequence of port industry	*Cargo throughput of major coastal ports in China*	*Million Tons*	*Year*	1985-2014
	Growth rate of cargo throughput of major coastal ports in China	*Dimensionless*	*Year*	
Indicator sequence of economic growth	*China's gross domestic product (GDP)*	*Billion Yuan*	*Year*	
	China's GDP growth rate	*Dimensionless*	*Year*	

Source: China Statistical Yearbook

In terms of the port cargo throughput indicator and the GDP growth rate indicator included in the indicator setting, the growth rate of existence itself does not have the trend component factors, but considering that under the planned economy system of our country, the indicator of growth rate may have some trend effect, and the filtered data will fluctuate around the 0 axis, so it is easy to compare with other indexes when using cross spectrum analysis. Therefore, in order to ensure the validity of the final data processing results, H-P filter processing is also done for the relative indicators of growth rate in this paper. The results of the ADF test of the indicator sequence after H-P filter processing are shown in Table 2.

Table 2. Statistics of ADF test results of the indicator sequences after H-P filtering

Indicator sequence	*t test*	1%	5%	10%	*Prob.*
Cargo throughput of major coastal ports in China	-1.989	-2.647	-1.953	-1.610	0.044
Growth rate of cargo throughput of major coastal ports in China	-4.540	-2.647	-1.953	-1.610	0.000
China's gross domestic product (GDP)	-3.610	-2.665	-1.956	-1.609	0.001
China's GDP growth rate	-3.386	-2.647	-1.953	-1.610	0.001

It can be seen from table 2 that the time series of each indicator after H-P filter processing passes the ADF test. Among them, the indicator of cargo throughput of major coastal ports in China rejects the original hypothesis at significant level$\alpha = 0.05$, indicating that the probability of existence of single association is small in the sequence, and the other time series index test results show that it rejects the original hypothesis at the significant level$\alpha = 0.01$, it can be considered that the filtered time series of indicators is in stable state and the cross-spectrum analysis can be done on them.

(2)Cross spectral analysis

The spectral analysis of the coherence spectrum and the phase spectrum between the port industry and the economic growth can be obtained by spectral analysis after filtering and passing through the time series after the stationary test, as shown in Figure 1 and Figure 2. According to the change of the graph, we can analyze the consistency and leading and lagging relationship between port industry and economic growth under different periods.

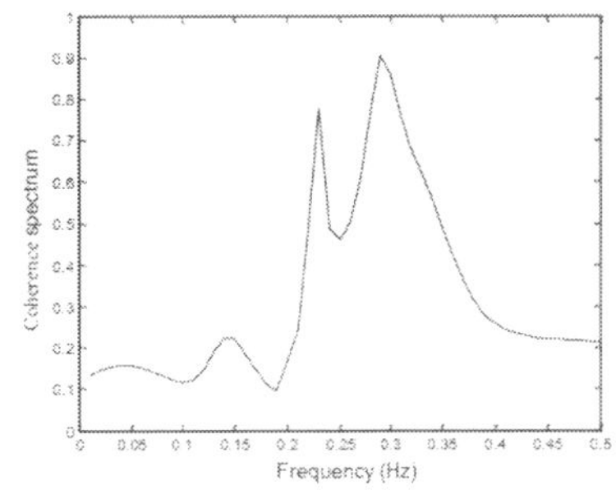

Fig. 1. Coherence spectrum of the throughput of major coastal ports and GDP of China

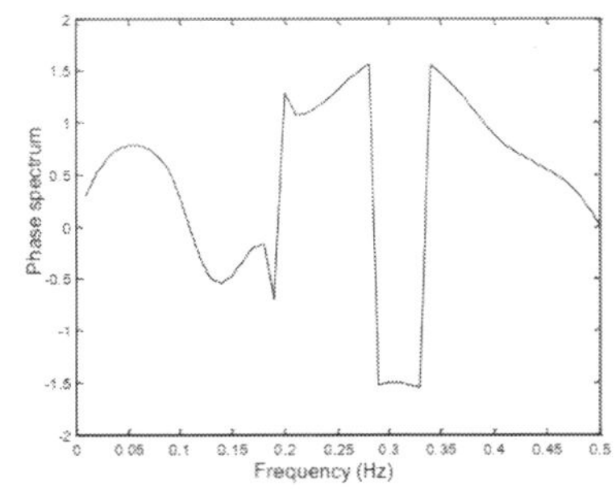

Fig.2. Phase spectrum of the through put of major coastal ports and GDP of China

According to the results shown in Figure 1, we can see that the coherence spectra value of the major coastal ports and GDP of China are higher at the frequencies of 0.23 and 0.28, and are greater than 0.75, so the consistency of the two sequences here is higher. As the reciprocal of the frequency indicates the length of the cycle, it can be analyzed and judged that the correlation between the development of China's port industry and the economic growth is high in the cycle of 3.57 and 4.45 years of length. The horizontal axis of the spectrogram shows the frequency, and the more the spectral graph is left, the smaller the frequency and the value of the coherence spectrum are, which can be used to analyze the long period of more than 10 years, the correlation between the development of port industry and the economic growth of China is small. Fig. 2 shows the phase spectrum of the throughput of the major coastal ports in China and of GDP. It can be seen that the values of the phase spectra at frequencies of 0.23 and 0.28 are 1.14 and 1.51, respectively, indicating that the GDP indicator representing economic growth is ahead of the throughput indicator representing the development of maritime industry. Among them, the period of length of 4.5 years and the period of the length of 3.7 years are respectively leading to 0.8 years and 0.86 years (time difference = phase spectrum $\div$ 2π $\times$ cycle length).From the coherence spectrum and the phase spectrum of the throughput of the major costal ports and GDP of China, it can be seen that there is a mutual influence and interaction between the development of port industry and economic growth of China.

Fig. 3 and Fig. 4 respectively indicate the coherence spectrum and phase spectrum between the growth rate of throughput of major coastal port and the growth rate of GDP of China.

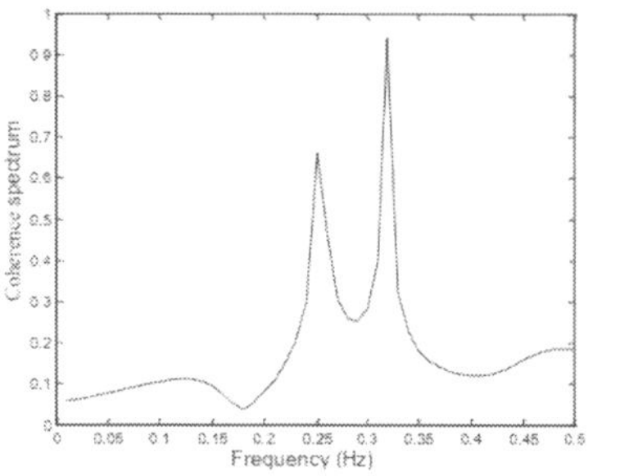

Fig. 3. Coherence spectrum of the growth rate of throughput of major coastal ports and of GDP of China

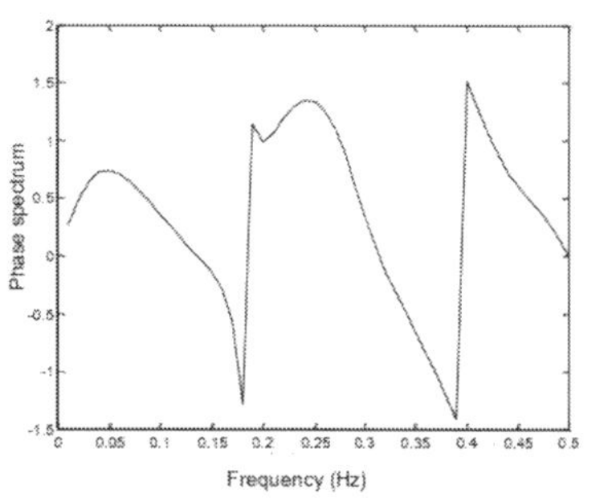

Fig. 4. Phase spectrum of the growth rate of throughput of major coastal ports and of GDP of China

The coherence spectra of Figure 3 show that the coherence spectra value of frequencies at 0.25 and 0.31 are significantly higher, that is, the growth rate of throughput of major coastal ports of China is higher than that of GDP in the period of 4 years and 3.13 years. But the frequency under 0.1 gradually tends to be gentle and the correlation degree is small, namely, the correlation degree between development of port industry and the economic growth of China is low in the period length over 10 years. It can be seen from the phase spectrum of Figure 4 that the phase spectra are 1.34 and 0.1 at frequencies 0.25 and 0.32, and the phase spectrum is positive, that is, the growth rate of China's GDP is ahead of the throughput of the major coastal ports, and the leading length of time is 0.85 years and 0.05 years respectively.

3 Dynamic Measurement Analysis of China 's Port Industry and Economic Growth

The method of cross-spectrum analysis can be used to analyze the consistency and phase relationship between the development of port industry and economic growth of China through the analysis of coherence spectrum and phase spectrum, but the influence between the two is not determined by the interaction or one-way effect, and whether there is a causal relationship between the two, what the causal relationship is like is not determined, therefore, after doing the cross spectrum analysis, we need to further study the relationship between the two. This paper further explains the specific relationship between time series of the development of port industry and the time series of the economic growth of China through dynamic econometric analysis.

3.1 Variable selection and data description

When using cross-spectral analysis to analyze the consistency and the phase relationship between the development of port industry and the economic growth of China, the throughput indicator of major coastal ports and the GDP indicator of China from 1985 to 2014 are used, which respectively represent the main indicator of the development of ports industry and economic growth of China, in order to further analyze the causality between the two and the response of endogenous variables and external variables, the paper also uses the original data of throughput of the major coastal ports of China and GDP of China in 1985-2014 years in determining the cointegration relationship and the analysis of the impulse response and the variance decomposition.

3.2 Cointegration and causality test of port industry and economic growth

The development of port industry and economic growth will be influenced by many factors, such as the volume of world economy and trade, the input of labor force, the level of shipbuilding industry, the status of Port construction, the State policy, etc. To more accurately describe the relationship between the port industry and economic growth, first of all, it is necessary to test the stability of the throughput indicator of port industry and GDP indicator of economic growth of China, and to determine the number of integrated order of the unstable time series, so as to judge the possibility of cointegration test[11].The test results are shown in Table 3.

The test results from table 3 indicate that CX and CY are second-order single Integer I (2), which may have cointegration relationship, so the cointegration test is needed. The results of cointegration regression statistics are shown in table4.

The regression equation can be obtained by Table 4:

$$CX_t = 1.795CX_{t-1} - 0.7795CX_{t-2} + 0.609CY_t + \varepsilon_t \quad (21)$$

The determination coefficient of the sample is $R^2 = 0.999016$, the final coefficient of the adjusted sample is $\bar{R}^2 = 0.998937$, and the D-W test statistic result is 1.722383, the regression test of the second order sequence correlation does not have the pseudo regression phenomenon.

The unit root test results of ε_t are shown in Table 5.

Table 3. Unit Root Test Results for Indicator Sequence of China's Port Industry Development (CX) and Economic Growth (CY)

Variable	*Type of test (c,t,n)*	*ADF test value*	1%	5%	10%	*Result*
CX	*(c,t,3)*	-2.071	-4.356	-3.595	-3.233	*Non-stationary*
D(CX,1)	*(c,t,3)*	-2.789	-4.324	-3.581	-3.225	*Non-stationary*
D(CX,2)	*(c,0,3)*	-6.580	-3.711	-2.981	-2.630	*Smooth*
CY	*(c,t,3)*	-0.833	-4.324	-3.581	-3.225	*Non-stationary*
D(CY,1)	*(c,t,3)*	-2.605	-4.324	-3.581	-3.225	*Non-stationary*
D(CY,2)	*(0,0,3)*	-5.420	-2.653	-1.954	-1.610	*Smooth*

Note: The test type (c, T, N) represents a constant entry, a time trend, and a lag order in the ADF test. The lag order is judged by the minimum value of test results of AIC and SC.

Table 4. Results of the cointegration regression of the development of China's port Industry (CX) and economic Growth (CY)

Variable	*Coefficient*	*Std. Error*	*t-Statistic*	*Prob.*
CY	0.609	0.178	3.428	0.0021
AR(1)	1.795	0.205	8.747	0.0000
AR(2)	-0.779	0.223	-3.491	0.0018

Note: CY is the independent variable, CX is the dependent variable

Table 5. Test results for unit root of ε_t

Variable	*Type of test(c,t,n)*	*ADF test value*	1%	5%	10%	*Result*
ε_t	*(c,0,3)*	-4.412	-3.700	-2.976	-2.627	*Smooth*

From the test results of Table 5, it can be seen that ε_t is I (0), that is ε_t is stationary. Therefore, it is assumed that CX and CY are cointegration, that is, CX and CY have long-term equilibrium relationship.

Then, the Grainger causality test is performed on CX and CY, and the results are shown in table 6.

Table 6. Grainger causality test of CX and CY

Null Hypothesis:	*Obs*	*F-Statistic*	*Prob.*
CY does not Granger Cause CX	28	4.051	0.031
CX does not Granger Cause CY		10.956	0.001

Note: The lag order is 2.

It can be seen from the test results of Table 6 that CY is the Granger cause of CX, that is, the economic growth of China is the cause of the development of the port industry, the same, CX is CY's Granger cause, that is, China's port industry development is the Granger cause for the economic growth.From the relationship between the two, we can see that the development of port industry and economic growth of China are interrelated and mutual causality.

3.3 Impulse Response and Variance Solution of Port Industry and Economic Growth

We can know the causal relationship between the port industry and economic growth through the cointegration analysis, but how the change of the unit variable affects the disturbance of the whole system through its intrinsic relation, and how the change of the response of each variable to these disturbances needs to establish the VAR model to do the impulse response analysis on the port industry and economic growth [13], finally the long-term relationship between variables can be determined. The autoregressive model of P-order vectors is:

$$y_t = A_1 y_{t-1} + A_2 y_{t-2} + \cdots + A_p y_{t-p} + B_1 x_{t-1} + \cdots + B_r x_{t-r} + \varepsilon_t \quad (22)$$

In the formula, y_t is m-dimensional endogenous variable vector, x_t is d-dimensional exogenous variable vector, p, r is respectively the delay order number of endogenous variable and exogenous variable, A_1, …, A_p and B_1, …, B_r is the parameter matrix to be estimated, and ε_t represents the error vector.

(1) VAR model test. Firstly, the data is preprocessed. Since the natural logarithmic transformation of the data does not change the original cointegration relationship and can linearize its trend, it eliminates the heteroscedasticity phenomenon in the time series. Therefore, the natural numerical value of all data is obtained to enhance the linearization trend of the data and eliminate the variance. The AR root test results of the VAR model are shown in Figure 5.

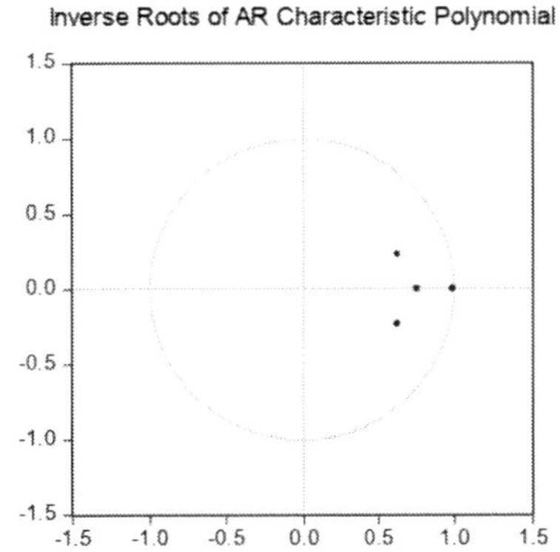

Fig. 5. AR root test result of VAR model

It can be seen from Fig. 5 that the AR root is all within the unit circle, that is, it can be concluded that the VAR model is stable and can carry out impulse response analysis on GDP and throughput of major coastal ports of China.

(2) Impulse response analysis. The impulse response analysis describes the magnitude of the influence of the current value of the model endogenous variable on the

DOI: 10.1051/matecconf/201713900226

perturbation term plus the impact of the unit standard deviation and the other endogenous variables through the dynamic structure of the model. GDP (CY) impulse response to major coastal port throughput (CX) of China is shown in Fig.6 and the impulse response of major coastal port throughput (CX) to GDP (CY) of China is shown in Fig.7.

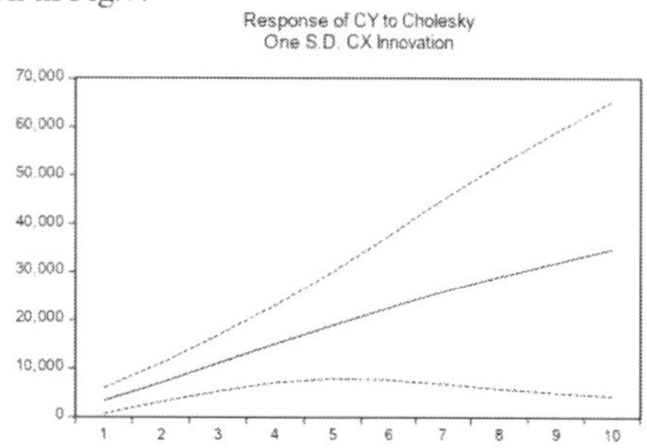

Fig. 6. Impulse response of GDP (CY) to major coastal port throughput (CX) of China

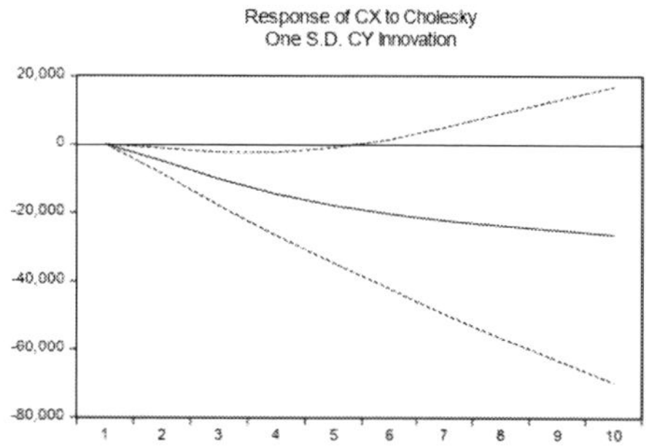

Fig. 7. Impulse response of major coastal port throughput (CX) to GDP (CY) of China

It can be seen from the results of impulse response that China's GDP has a positive shock response on the throughput of major coastal ports, and the response is increasing over time. The main port throughput of our country has a negative shock response on GDP, and the response of negative response will increase over time, but after the 7th period, it gradually becomes more balanced.

(3) Variance Decomposition. Variance Decomposition method [14] is to decompose the fluctuation of endogenous variables into the contribution of the variable shocks in the system, and then analyze the importance of different shocks. The variance decomposition of the throughput of the main coastal ports in China and GDP is done through Eviews8, and the results are shown in table 7.

Table 7. Variance Decomposition of throughput in major coastal ports in China (CX)

Variance Decomposition of CX:			
Period	*S.E.*	*CX (%)*	*CY (%)*
1	8497.317	100.000	0.000
2	18004.070	92.423	7.577
3	29002.770	84.668	15.332
4	40680.200	79.369	20.631
5	52507.380	75.999	24.000
6	64222.410	73.903	26.097
7	75743.080	72.623	27.375
8	87095.950	71.874	28.126
9	98368.580	71.464	28.536
10	109679.300	71.274	28.726

The result of variance decomposition in table 7 shows that the throughput of major coastal ports in our country is only affected by its own fluctuation in the 1th period, starting from the 2nd period, the GDP has an effect on the throughput of the main coastal ports, and the influence gradually increases over time, but at the beginning of the 7th period, it gradually tends to balance, affecting about 28%.It can be seen that the development of China's maritime industry is largely affected by economic growth. If the trend of China's economic growth can be effectively analyzed, it can guide the development of maritime industry.

The variance decomposition of China's GDP (CX) is shown in table 8.

Table8. Variance Decomposition of China's GDP(CX)

Variance Decomposition of CY:			
Period	*S.E.*	*CX*	*CY*
1	7424.40	20.215	79.785
2	11564.54	45.839	54.161
3	16092.19	71.177	28.823
4	22274.60	83.200	16.800
5	30139.08	85.090	14.915
6	39244.54	83.295	16.705
7	49167.87	80.857	19.143
8	59615.40	78.703	21.297
9	70413.62	77.011	22.989
10	81479.98	75.743	24.257

It can be seen from the result of variance decomposition of table 8 that the growth of GDP in China is affected by the throughput of major coastal ports in the 1th period, and the influence reaches 79.8%, which is greater than its own influence, however, the impact of the throughput of the major coastal ports was gradually reduced before the 7th period, and the 7th period was affected by 80.9%, and after the 8th period, the impact of the throughput of the major coastal ports began to increase. Thus, it can be judged that China's port industry and economic growth affect and promote each other in the process of development.

4 Conclusion and Analysis

From the analysis of cross-spectrum results, we can see that there is a certain correlation between the development of port industry and economic growth. In the coherence spectrum generated by time series, the correlation between the throughput and GDP of the main coastal ports in China is at a higher level in the period of 3 to 4 years, while the correlation between the two is at a lower level than in the period of 10 years. The result of this kind of data is inseparable from the development of our country's planned economy system, in formulating the economic growth plan, the state should make an adjustment in the 5 year cycle. Under the state macro control, China's economic development will develop in a long time and be in a good direction. At the same time, under the guidance of the national policy, the development of China's port industry will also be coordinated with economic development in a short period of time. Therefore, in a 5-year cycle, the correlation between the development of China's port industry and economic growth is higher, however, in the long period of port industry and economic growth in the development of the time axis, the development of port industry is less affected by endogenous variables and exogenous variables compared with the economic growth, while economic growth can be pulled through a variety of channels, except for the promotion of port industry. Therefore, in the long period of consistency analysis, the relationship between port industry development and economic growth will be in a low state. China's economic growth in the short cycle is ahead of the development of the port industry. In the phase spectrum formed by the throughput index of the main coastal ports in China and GDP index, the growth of our economy will lead to the development of the port industry from 0.8 years to 1.1 years in the short period of 3-4 years. However, due to the different response time of economic growth and port industry to external shocks, the development of China's port industry will be ahead of economic growth in a short period of time. The reasons for this result are closely related to the development characteristics of our country. In the long-term economic development of our country, due to the scarcity of capital, the development of many industries in our country is limited by the financial difficulties. The port industry as a capital investment, investment return cycle of a long industry, is also subject to a lot of restrictions. Therefore, in the whole fluctuation process of the development of China's port industry, the overall development of it will lag behind the growth of China's economy.

The results of dynamic econometric analysis show that China's economic growth and the development of port industry are Granger causes, which are interrelated and mutually reinforcing. It can be seen from the results of the impulse response that China's economic growth has made a positive response to the impact of the development of port industry. Meanwhile, the response gradually increases over time. However, China's port industry has made a negative impact on economic growth, and the negative response will increase over time, and eventually will gradually become more balanced. The results of variance decomposition show that China's port industry will be affected by greater economic growth in the long-term development process, and this effect will gradually tend to a stable state, the impact level is about 28%, meanwhile, the development of economic growth in the short term will be closely affected by the port industry, the impact of the port industry on the volume of foreign trade completion will outweigh the factors of economic growth, but in the long run, this effect will gradually weaken, and the influence of endogenous variables of economic growth will gradually increase. It can be judged that China's port industry and economic growth are in a state of mutual influence and promotion in the process of development.

Acknowledgment

This paper is supported by the major projects of the National Social Science Fund (13 & ZD170), the New Century Excellent Talents Support Program of the Ministry of Education (NCET-13-0701) and the "Thirteen Five" Research Project of Dalian Maritime University (3132016361)

References

1. Ma Zheng-bing.The Application of Fourier Series and Fourier Transform in the Derivation of Two Formulas [J]. Examweekly 5 (2015)
2. Gu Zu-kai.Study on the Relationship between Bulk Cargo Price and Freight Price Based on Cross Spectral Analysis [D]. Master 's Degree Thesis of Dalian Maritime University (2016)
3. Duan Kai-yu, Yin Shi-shu, Fu Ming .An Empirical Study on the Cross - Spectrum of the Price Discovery Function of Grain Futures in China [J]. Statistics& Decision 1 (2005)
4. Wu Xu, Wang Rui-jun, Hu Si-ji.Cross-spectra Analysis on Relationship between Railway Transportation and National Economy [J].China Soft Science 5 (2016)
5. Yu Jie, Han Xiao.Analysis on Price Volatility Cycle of Chinese Raw Milk Based on X12 Seasonal Adjustment Method and H-P Filter Model [J]. China Dairy Cattle 18 (2014)
6. Li Dong-sheng, Liu Xiao-le.New Characteristics of Pig Price Fluctuation in China - An Empirical Analysis Based on HP and BP Filtering [J].Rural Economy 6 (2015)
7. Nie Qiao-ping, Zhang Xiao-dong .Research on LM Statistics of Joint Hypotheses for ADF Unit Root Test [J] .Statistical Research 2 (2007)
8. Mo Da-long.Modeling time series with ADF validation [J].Times Finance 7 (2010)
9. Zhao Xue-yu .Research on the Relationship between World Trade and World Shipping Based on Co - integration Theory [D].Master 's Degree Thesis of Dalian Maritime University (2013)
10. Xu Guo-xiang.Statistical Forecasting and Decision Making (Fourth Edition)Shang Hai: Shanghai University of Finance & Economics Press (2012)

11. Kong Fan-wen, Cai Xu, Yu Miao.Analysis and Application of Granger 's Causality Test Model [J]. Journal of Shenyang Jianzhu University Natural Science 2 (2010)
12. Wu Wen-qing, Liu Chao .A Study on the Fluctuation of Purchase and Sale of State-owned Grain Enterprises in China - Based on Wavelet Analysis and Granger's Causality Test [J].Economic Theory and Business Management 10 (2001)
13. Zhong Xiao-jun.A Study on the Relationship between Direct Investment and China's Import and Export Trade - Impulse Response Function and Variance Decomposition Method Based on VAR Model [J].Statistical education 6 (2009)
14. Zhang Rong, Zhang Cai-qin.House price index, CPI and social consumption level -- An Interpretation Based on impulse response and variance decomposition [J].Academic Forum 10 (2015)
15. B Büttner.Effectiveness versus Efficiency: Growth Accelerating Policies in a Model of Growth without Scale Effect[J], German Economic Review 7 (3), 297-316 (2006)
16. GLASS,A., X.WU.Intellectual Property Rights and Quality Improvement[J]. Journal of Development Economics 82,393-415(2007)

Improved particle swarm optimization algorithm in professional skills identification management system

Yinghui Peng

Tianjin University of Technology and Education, Tianjin, China

Abstract. This paper adopt improved particle swarm algorithm to solve multi-objective optimization problem; so this project can determine vocational skills certification examination test of the different professions, different types and different levels of difficulty; the making test paper covers a wide range of knowledge and chapter score; and the difficult level of making test paper conform to normal distribution, meet different groups requirements of vocational skills certification examination.

1 Introduction

With the development of information technology and continuously improve the level of education, and educators intelligently put forward higher requirements for the speed and efficiency, after a test of test questions is not limited to a single subject, objective, requirements test questions are diverse, covering the wide range of knowledge, suitable degree of difficulty of the questions, More attention to test students ' knowledge and practical skills. In occupational skill testing questions management system, a single, slow group volumes long, remains a widespread problem, still cannot meet the needs of professional skill appraisal assessment. So I propose a realization policy of vocational skill identification bank management system based on improved Particle Swarm Optimization algorithm

2 Theoretical study on occupational skill testing questions intelligently

Vocational skills appraisal test each system has a certain role, each of which has a certain meaning, each index parameter is for examination paper analysis, automatic test, examination management services, these parameters are optimized, system based on parameters also has some optimizations. Intelligent test paper composition parameters are as follows

2.1 Test scores M

Test scores $\sum_{i=1}^{M} K(i)a_{i2} = M$, a_{i2} is a question I score. K(i)(1≤ i ≤ M) as I question the state ,variable k(I) take the value 0 or 1.

2.2 Exam time T

The elapsed time to complete all the questions, most students finish and check the problem need consuming time, item writers can adjust it according to the experience in teaching comprehensive objective information; $\sum_{i=1}^{M} K(i)a_{i3} = T$, a_{i3} is estimated time to finish the test I.

2.3 Chapter points

Course sections and the various types of questions in question bank exists relation, in order to facilitate assessment and management; $\sum_{i=1}^{M} K(i)a_{i4}C(j) = S_j$, Sj is chapter j score, aI4 is questions I belongs to section, c (j) take the value 0 or 1.

2.4 Knowledge points score

Syllabus requires assessment of knowledge; $\sum_{i=1}^{M} K(i)a_{i5}O(j) = D_j$, Dj is point j score, a_{i5} is a question I owned knowledge, o(j) take the value 0 or 1.Intelligent test paper composition of knowledge in selected to meet uniform and focus problems, through the creation of hierarchical tree structure of knowledge management chain, solve linear density of knowledge in the management of large or small problems, knowledge distribution, given in the hierarchy according to the weights of the weight calculation of two-level model knowledge point reference weights build classification knowledge points are evenly weighted selection algorithm.

2.5 Paper type fraction

Paper type change because of the different professions,

different types of user needs, usually set to 5 type: judgment questions, multiple choice, fill in questions, short answer, essay questions, etc; $\sum_{i=1}^{M} K(i)a_{i6}T(j) = F_j$, Fj is score in j type, ai6 belong to the types of questions I, t (j) take the value 0 or 1.

2.6 Ability Level score

$\sum_{i=1}^{M} K(i)a_{i7}A(j) = G_j$, Gj is j ability level score, a_{I7} questions I belong to ability level, a (j) take the value 0 or 1.

2.7 Test difficulty

Difficulty is a measure of test questions and test measure of the degree of difficulty. Difficulty is obtained by calculation of students' difficulty value of test. $[\,[\sum_{i=1}^{M} K(i)a_{i2}a_{i8}]/M = DIF$, DIF is general degree of difficulty, aI8 is the difficulty coefficient of test I, Fuzzy five levels listed in the following table I.

2.8 The degree of examination questions

Questions of discrimination is an important indicator of performance, is a measure of the difference between test ability, difficulty and discrimination are closely linked discrimination of the questions is Max at the time of 0.3-0.6 difficulty. Method of calculating distinction is that test scores of the candidates are arranged from the highest score to the lowest score, and the highest scores group is from 27% of total scores, and the lowest scores group is from 27% of total scores. D= (SH-SL)/S, SH and SL respectively are average scores of the highest and the lowest scores , S represents the full mark of the question, D represents the degree of examination questions, the greater of the D represent the greater of testers score differences. Degree evaluation criteria listed in the following table II. $[\sum_{i=1}^{M} K(i)a_{i2}a_{i9}]/M = DIS$, DIS is distinction of the papers, ai9 is the distinction of questions I.

The more difficult questions and the more easier questions should be fewer in proposition, the questions of the medium difficulty are more larger , Makes questions of average difficulty close to 0.5, difficulty and distinguish degrees is a relative indicators for groups, no absolute difficulty and distinguish degrees, if different capacity of test students groups are tested with same questions ,difficulty and distinguish degrees of questions also will different, according to students of knowledge capacity level situation general rendering distribution of features, proposition teachers determine meaning of initial value, then further amended according to students' questions exam, and set the basis for next item difficulty and discrimination.

3 Occupational skill testing questions intelligently mathematical modeling

In vocational skills appraisal test this algorithm, the strategy of intelligent test paper Bank system data source is a matrix. as shown in equation 3.1.

$$A_{M\times N} = \begin{pmatrix} a_{11} & \cdots & a_{1N} \\ \vdots & \ddots & \vdots \\ a_{M1} & \cdots & a_{MN} \end{pmatrix} \quad (3.1)$$

Where m is the total number of test questions, n is the number of attributes, N=9, attribute indicators for each question that is the n-dimensional vectors, including scores, item type, degree of distinction etc. 8 indicators. All properties of each question is the rows in the matrix, as column respects each property of all the questions. Generate test constraints up to n-1, each column is question number in the destination matrix.

Generating test paper process is analysis test multiple constraint parameters based on the purpose of the test, searching rows x=(x1,x2,x3...xp) of meeting the requirements in the matrix shown in 4.1.where p is the number of questions, Xi=(xi1,xi2,...xij) is vector composted of i type questions , $X_{i,j}$ is j question of I type.

3.1 Intelligent test paper composition function

Test paper problem is a multi-objective problem solving process, under normal circumstances, constraints above and generating test request error make objective function f ,when f minimum, generated the papers wanted by us, thus, multi-objective optimization problem transfer into single-objective optimization problem, the objective function such as

$$\text{Min } f=\sum_{j=1}^{2} d_j \times w_j+\sum_{i=1}^{h} d_{3i} \times w_3+\sum_{i=1}^{j} d_{4i} \times w_4 + \sum_{i=1}^{p} d_{5i} \times w_5+\sum_{i=1}^{q} d_{6i} \times w_6+\sum_{j=7}^{8} d_i \times w_j$$

$$\text{St } \sum_{j=1}^{M} K(j)a_{j2} + d_1=\text{M}$$

$$\sum_{j=1}^{M} K(j)a_{j3} + d_2=\text{T}$$

$$\sum_{i=1}^{h}(\sum_{j=1}^{M} K(j)a_{j2}\, C(i) + d_{3i})=\sum_{i=1}^{h} S_i$$

$$\sum_{i=1}^{j}(\sum_{j=1}^{M} K(j)a_{j2}\, O(i) + d_{4i})=\sum_{i=1}^{j} D_i$$

$$\sum_{i=1}^{p}(\sum_{j=1}^{M} K(j)a_{j2}\, T(i) + d_{5i})=\sum_{i=1}^{h} F_i$$

$$\sum_{i=1}^{q}(\sum_{j=1}^{M} K(j)a_{j2}\, A(i) + d_{3i})=\sum_{i=1}^{q} G_i$$

$$[\sum_{j=1}^{M} K(j)a_{j2}a_{j8}]/M + d_2=\text{DIF}$$

$$[\sum_{j=1}^{M} K(j)a_{j2}a_{j9}]/M + d_2=\text{DIS} \quad (3.2)$$

In the optimization of the objective function above, d is deviation in the intelligent test paper generation between question properties and evaluation requirements , wj (j=1, 2, 3, ..., 8) represent weight of the target group, the sum of weight is to 1.

3.2 The steps of generating test paper

(1) Determine subject knowledge points and scope of the examination, the examination content includes sections and scope. In the specified test range, clear analysis of knowledge content, focus, difficulties and a clear ability to assess students' ability level mastering of knowledge of each point.

(2) Determine the distinction and difficulty of paper

(3) Determine type of test papers. Generally include blanks, choice questions, judgment questions, short answer and essay questions.
(4) Determine the number of test questions and test time
(5) Determine the paper structure, namely, chapters of the proportion of scores in the paper.
(6) Generating test paper, according to papers information, form a set of paper as possible to meet the requirements of the papers.
(7) Output, adjust the layout and output the final test paper.

3.3 Basic idea of particle swarm optimization algorithm

Particle Swarm algorithm is inspired by the bird foraging behavior which is a new type of swarm intelligence algorithm after Ant Colony algorithm. Particle Swarm Optimization algorithm description: total of n particles, the position of I particles is to $Xi=(X_{i1}, X_{i2},..., Xin)$, each particle has a measurable quality XI's fitness; particle's velocity of the I-th particle (location change ratio) is expressed as $Vi=(V_{i1}, V_{i2},...,Vin)$
Particles passing through the individual extreme value XI*= (Xi1*, Xi2*, ., Xin*) and a global optimization g*=(g1,g2,g3...,gD) to update the velocity and position. The speed and position of the particle would be updated according to the following equation

$$V_i^{i+1}=V_i^{i}+\alpha r_1(X_i^{*}-X_i^{I})+\beta r2(g^{*}-X_i^{I}) \quad (3.3)$$

$$X_i^{j+}1=X_i^{j}+V_i^{j+1}\Delta t \quad (3.4)$$

Vi is the previous speed, αr1(xi*-xi I) called cognitive part, βr2(g*- xi I) affect the next behavior of particles. Xi and VI are the position and velocity of particle I, the searching optimal position of particle I is XI* and g* are searching optimal locations of current Particle Swarm optimal, w is inertia parameters and control the dynamics of particles flying, α and β are called parameters or constant acceleration.

In General, access α=β=2. I is the number of iterations, the R1 and R2 are random numbers in range of the [0, 1]. Random initial position and velocity of particle swarm, then iteration using the above equation until you find the optimal solution.

3.4 Algorithm steps

(1) Initialize the particle position

(2) Number of iterations I=I+1

(3) Initial value of i=0 and fitness evaluation

(4) Calculation groups best historical position

(5) Update the particle swarm location according to location update formulas

(6) I=I+1

(7) Determine if I is greater than n, if established, go to step 8,otherwise, go to step 3

(8) Determine whether the termination condition is met, if established, end program after the output. Otherwise, go to step 2

3.5 Improved Particle Swarm Optimization algorithm

In order to accelerate convergence, Accelerate the Particle Swarm Optimization algorithm has been proposed by a doctor of Cambridge University in 2008, speed updated is done by a simpler formula as 3.5 below.

$$V_i^{i+1}=V_i^{i}+\alpha r(t)+\beta r(g^{*}-X_i^{I}) \quad (3.5)$$

$$X_i^{j+1}=X_i^{j}+V_i^{j+1}\Delta t \quad (3.6)$$

Where r is a random number and can be obtained by normal distribution, in essence, updating formula is the same as formula (3.3), using formula (3.5) instead of the speed in formula (3.6), you can obtain a formula which can represent location update status, such as formula (3.7) shows.

$$X_i^{j+1}=(1-\beta)X_i^{j}+\alpha r+\beta g^{*} \quad (3.7)$$

Speed does not appear In equation (4.7) ,speed may not be updated in improved Particle Swarm algorithms, therefore the algorithm simpler and easier to implement. R is obtained by a probability distribution, and make the algorithm does not fall into any local optimal solutions. Under normal circumstances, α=0.7-0.9, β=0.2-0.7.

Under normal circumstances, standard acceleration particle swarm algorithm has strong ability to search exploration, and can be find the area of the best solution, but its mining solution is poor weak and cannot always guarantee convergence to the optimal solution. Now add a variation factors in the original acceleration Particle Swarm Optimization algorithm, so can greatly improve the speed of convergence of the algorithm, and solve more practical problems. The introduction of variation factors can accelerate the PSO, to increase the diversity of the population, improve search efficiency and accelerate convergence of optimal solutions,

$$X_i^{j+1}=x_{r1}^{j}+F^{*}(X_{r2}^{j}-X_{r3}^{j}) \quad (3.8)$$

Where f is the differential evolution variants, r1,r2,r3 is uniformly distributed random integer between 1 and NP, NP is the number of population. Using micro differentiation of variants to fine-tune the new solutions in order to increase diversity and improve search efficiency.

3.6 weight determination of the intelligent test paper

Weight is a number ranging from 0 to 1, and all top level indicator weights sum to 1, weight is scientific and oriented, weight of important need to increase, and weight of not important indicators need small. Scientific distribution of weight indicator is also an important step in composing test paper.

According to the required 8 constraints in the system to make the test object weight WJ (j=1, 2, 3, ..., 8), through the analysis of 8 constraints will break it into a tree structure, designed a NXN square matrix according to the constraints, while $At=\lambda t=0$, nonzero vector can be the eigenvectors of a, λ is the eigenvalues of the matrix A.

Get a relative value of aIJ, forming a square matrix A, suppose there is a normalized eigenvector $A\omega=\lambda_{max}\omega$, λ_{max} is the largest eigenvalue of square matrix A, ω is the normalized eigenvectors corresponding to the λ_{max}. ω is the weights factors. If matrix satisfy three conditions ① $a_{IJ}=1$, ② $a_{IJ}=1/a_{IJ}$ ③ $aIJ=a_{IK}/a_{JK}$, exactly ω is the weights factors.

3.7 Intelligent generating test steps based on improved Particle Swarm algorithm

(1) According to reorder the questions to test bank questions, questions can be divided into different subsets, and then renumber all the questions
(2) According to the questions, calculate the number of questions to get total number of test papers
(3) Computational Intelligence group volume weight of the index

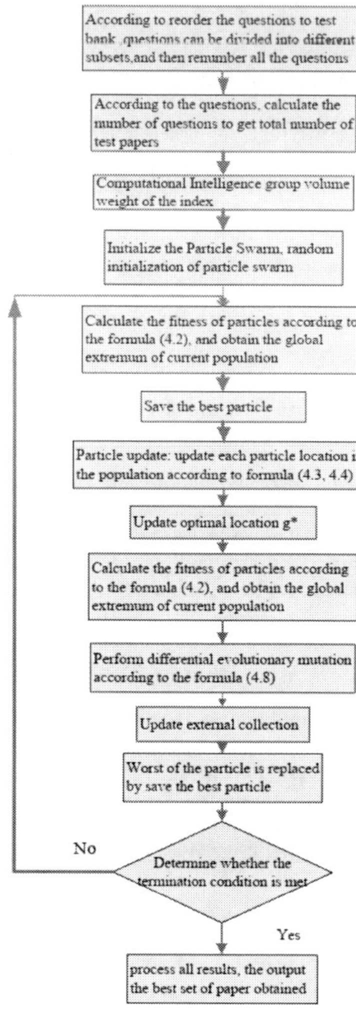

Fig 1 program flow chart

(4) Initialize the Particle Swarm, random initialization of particle swarm, each particle XI in particle swarm represents a possible group policy. In a population, the status of each particle adopt an m-dimensional vector $X = (X_1, X_2, ..., X_M)$; the dimension of the vector X depend on the total number of questions; Xi is the components of the vector X and X_I (i=1, 2, ..., m) is question number, question number cannot be repeated, according to the different type questions, number of examination questions should be arranged according to different segments
(5) Calculate the fitness of particles according to the formula (4.2), and obtain the global extreme of current population

(6)Save the best particle

(7) Particle update: update each particle location in the population according to formula (4.3, 4.4)

(8) Update optimal location g*

(9)Perform differential evolutionary mutation according to the formula (4.8)

(10) Update external collection

(11) Worst of the particle is replaced by save the best particle

(12) Determine whether the termination condition is met, if not satisfied, return the 5th step

(13) Process all results, the output the best set of paper obtained

the program flow chart is shown in Fig 1.

4 Introduction to occupational skill testing question management system

Occupational skill testing question bank management system functions should be developed as follows

4.1 Account management module

Contain users of information, a variety of roles and system permissions management.

4.2 System settings module

Administrator can set the examination subject, test type questions and the user's permissions

4.3 Online examination module

Normal users (students) can make exam online. Candidates can login system using account and password, in order to reflect the fairness of the exam, the kinds of questions randomly disrupted secondary reorganization distributed to candidates. This makes the content of the examination had the same to all candidates but with different order of items.

4.4 Manual test modules

Administrator can make basic property settings for test paper according to the different test objects and add paper strategy papers

4.5 Test management module

Mainly according to the assessment program, set the task type, test quantity, difficulty factor, automatically and quickly making test paper, and perform related operations such as editing on paper. Examination paper consists of exam papers and homework exercises paper

4.6 Item information management module

To evaluate history papers and analysis the exam, and then to modify the test information for the next exam. Administrators import questions and management database, see question distribution statistics.

4.7 Marking management modules

Mainly making judgment of the papers submitted by the candidates, there are two parts that is to say automatic marking and artificial modification.

4.8 Query management module

Administrators can view, analyze, export and other operations, and can also provide with score search and other functions for students. Students will be free to check your record after participating in the examination.

4.9 News management

Administrators can update, delete and publish news.

5 System implementation

In using Microsoft Visual Studio 2015 and SQL Server 2014 back-end databases, occupational skill testing question bank management system would be implemented for mechanical fitter, machine lathe operator, vehicle maintenance, automobile assembly workers and other categories of professional skills identification papers. Test Bank included alternative questions, choice, fill in the questions and comprehensive question and integrated application 100 questions each, through specific experiments and found that ,within a certain time, advanced particle swarm optimization algorithm can generate optimal solution to automatically generate moderate difficulty, paper covering range of knowledge, student test scores is good using this papers.

Acknowledgments

This work has been supported by Project of the Program of National Natural Science Foundation of China under the Grant Number 11772228

This work was supported by innovative projects in Tianjin University of Technology and Education of china under the Grant Number 201610066023

References

1. Han Yinghui. Improved Genetic Algorithm in Intelligent Test Paper System [J]. Changchun College of Education 2015:33-35
2. Zhang Shunmiao. intelligent test paper composition studies based on improved multi-objective Particle Swarm algorithm[J] Journal of Fujian University of Technology 2011:45-48
3. Lu Ping. Intelligent test paper composition grading in weighted knowledge point policies [J] Computer Apllication and Software 2014.

4. Duan Hong. Intelligent test paper composition studies using multi-objective Particle Swarm algorithm[D] Northeast Normal University 2013
5. Yinghui Peng. Ant Colony Optimization Algorithm In Occupational Skill Testing Management System Advanced Information Management Communicates, Electronic and Automation Control Conference, IMCEC 2016:

Table 1. Hierarchy of Difficulty

Hierarchy of difficulty	Easy	little Easy	Middle	little Difficult	Difficult
Range	0.0-0.2	0.2-0.4	0.4-0.6	0.6-0.8	0.8-1.0
Degree	0.1	0.3	0.5	0.7	0.9

Table 2. Degree evaluation criteria

Distinction	0.4 above	0.3-0.39	0.2-0.29	0.2 below
Evaluation	good	qualified	Barely changes	Worse, should be eliminated

AUTHOR INDEX

AUTHOR INDEX

AUTHOR INDEX

AUTHOR INDEX

AUTHOR INDEX

AUTHOR INDEX

AUTHOR INDEX

EDP Sciences – Web of Conferences
17, Avenue du Hoggar
Parc d'Activité de Courtabouf
BP 112
F-91944 Les Ulis Cedex A
France

ISBN 978-1-5108-5350-8

CURRAN ASSOCIATES INC.
proceedings.com